hydro- [Gk. *hydor*, water; now, confusingly, pertaining either to water or to hydrogen]: hydrogen, dehydration, hydraulic, carbohydrate, hydrolytic

hyper- [Gk. *hyper*, over, above, more than]: hyperacidity, hyperthyroid, hypertonic

hypo- [Gk. *hypo*, under, below, beneath, less than]: hypochondria, hypocotyl, hypodermic, hypoglycemia, hypothalamus, hypothesis, hypotonic

in-, im- [L. *in*, in]: imprinting, inbreeding, instinct, insulin

inter- [L. *inter*, between, among, together, during]: interaction, interbreed, intercellular, intercostal

intra-, intro- [L. *intra*, within]: intracellular, intrauterine, intravenous, introduced

-itis [L., Gk. *-itis*, inflammation of]: arthritis, bronchitis, dermatitis

leuko-, leuco- [Gk. *leukos*, white]: leukocyte, leukemia, leukoplast, leukocytosis

-logue, -logy [Gk. *-logos*, word, language, type of speech]: analogue (analogy), homologue (homology), dialogue

-logy [Gk. *-logia*, the study of, from *logos*, word]: anthropology, biology, cytology, embryology

-lysis, lys-, lyso-, -lyze, -lyte [Gk. *lysis*, a loosening, dissolution]: lyse, lysogeny, paralysis, hydrolysis, analysis, catalysis, catalytic

macro- [Gk. *makro-*, now "great," "large"]: macrocyte, macromolecule, macronucleus, macrophage

mega-, megalo-, -megaly [Gk. *megas*, large, great, powerful]: megaspore, megaton, acromegaly

-mere, -mer, mero- [Gk. *meros*, part]: blastomere, centromere, chromomere, polymer

meso-, mes- [Gk. *mesos*, middle, in the middle]: mesentery, mesoderm, mesophyll, Mesozoic

meta-, met- [Gk. *meta*, after, beyond; now often denoting change]: metabolism, metastasis, Metazoa

micro- [Gk. *mikros*, small]: microbe, micrometer, micronucleus, micropyle, microscope

myo- [Gk. *mys*, mouse, muscle]: myocardial, myocardium, myoglobin, myosin

neuro- [Gk. *neuron*, nerve sinew, tendon]: neurasthenia, neuroanatomy, neuroblast, neuron, neurotransmitter,

oligo- [Gk. *oligos*, few, little]: Oligocene, oligochaete, oligotrophic

-oma [Gk. *-oma*, tumor, swelling]: carcinoma, glaucoma, hematoma, lipoma, sarcoma

oo- [GK, *oion*, egg]: oogenesis, oogonium, oophyte

-osis [Gk. *-osis*, a state of being, condition]: arteriosclerosis, cirrhosis, halitosis

osteo-, oss- [Gk. *osteon*, bone; L. *os, ossa*, bone]: ossification, ossified, Osteichthys, osteoblast, periosteum

para- [Gk. *para-*, alongside of, beside, beyond]: paradigm, paralysis, paremedic, parasite

patho-, -pathy, -path [Gk. *pathos*, suffering; now often disease or the treatment of disease]: pathogen, pathology, pathological

peri- [Gk. *peri*, around]: pericardial, pericarp, pericycle, periosteum, photoperiod, peritoneum

phago-, -phage [Gk. *phagein*, to eat]: phagocyte, phagocytosis, bacteriophage

plasm-, -plasm, -plast, -plasty [Gk. *plasm*, something molded or formed; Gk. *plassein*, to form or mold]: plasma, plasma membrane, plasmid, plasmolysis, cytoplasm, nucleoplasm, chloroplast, chromoplast, protoplast, dermoplasty

-pod [Gk. *pod*, foot]: cephalopod, gastropod, pseudopod

poly- [Gk, *poly-, polys*, many]: polychaete, polydactyly, polygenic, polymer, polypeptide

-rrhea [Gr. *rhoia*, flow]: amenorrhea, diarrhea, gonorrhea

septi-, -sepsis, -septic [Gk. *septicos*, rotten, infected]: septic, septicemia, aseptic, antiseptic

-some, somat- [Gk. *soma*, body; Gk. *somat-*, of the body]: chromosome, ribosome

-stat,-stasis, stato- [Gk. *stasis*, stand]: metastasis, thermostat, electrostatic, hydrostatic

stoma-, stomato-, -stome [Gk. *stoma*, mouth, opening]: stoma, cyclostome, deuterostome, protostome

sym-, syn- [Gk. *syn*, with, together]: symbiont, symbiosis, symmetry, synapsis, synchrony

taxo-, -taxis [Gk. *taxis*, to arrange, put in order; now often referring to ordered movement]: taxonomy, chemotaxis, geotaxis, phototaxis

tomo-, -tome, -tomy [Gk. *tome*, a cutting; Gk. *tomos*, slice]: atom [you can't cut it], anatomy, lobotomy, appendectomy

tropho-, -troph, -trophy [Gk. *trophe*, nutrition]: trophic level, trophoblast, atrophy, autotroph, heterotroph

trop-, tropo-, -tropy, -tropism [Gk. *tropos*, to turn, to turn toward]: tropism, tropical, entropy, geotropism, phototropism

ur-, -uria [Gk. *ouron*, urine]: uracil, urea, ureter, phenylketonuria

uro-, -uran [Gk. *oura*, tail]: urochordate, anuran

Biology

THE SCIENCE OF LIFE *Fourth Edition*

Biology

THE SCIENCE OF LIFE *Fourth Edition*

Robert A. Wallace
University of Florida

Gerald P. Sanders
Grossmont College

Robert J. Ferl
University of Florida

with contributions by

Lisa Baird
San Diego University

Kenneth M. Klemow
Wilkes University

HarperCollinsCollegePublishers

Sponsoring Editor: Liz Covello
Developmental Editor: Karen Trost
Project Coordination: Electronic Publishing Services, Inc.
Design Administrator: Jess Schaal
Text and Cover Designer: Jeanne Calabrese Design
Front and Back Cover Photographs: Tony Stone Images/Art Wolfe
Illustrations and Illustration Concepting: J/B Woolsey Associates, Inc.
Photo Researcher: Leslie Coopersmith
Production Administrator: Randee Wire

Printer and Binder: R.R. Donnelley & Sons Company
Cover Printer: Phoenix Color Corp.

Library of Congress Cataloging-in-Publication Data
Wallace, Robert A.
    Biology, the science of life/Robert A. Wallace, Gerald P. Sanders,
Robert J. Ferl.—4th ed.
        p. cm.
    Include bibliographical references (p.   ) and index.
    ISBN 0-673-46774-0
    1. Biology. I. Sanders, Gerald P. II. Ferl, Robert J. III. Title.
QH308.2.W34 1996
574—dc20   95-49010
                                                       CIP

96 97 98 9 8 7 6 5 4 3 2

About the Authors

Bob Wallace was raised in Arkansas. He received a B.A. in Fine Art and Biology from Harding College, an M.A. dealing with muscle histochemistry from Vanderbilt University, and a Ph.D focusing on the behavioral ecology of island birds from the University of Texas at Austin. He has taught at a number of colleges and universities in the United States and Europe, and is the author of several textbooks in general biology and animal behavior, as well as nonfiction books on science.

Bob has worked as a nickel prospector in Alaska, a longshoreman, a martial arts instructor, a scuba diver, and an art instructor. (He specializes in 17th-century painting techniques.) His interests include marathoning, fly fishing, surfcasting, boating, shooting, skiing, and cooking.

He is a fellow of the Explorers Club based in New York, and a Contributing Editor of the *Explorers Journal*. He is a fellow of the Royal Geographical Society, and has explored in the Arctic, the Mediterranean, the Indian Ocean, the West Indies, and the Amazon. Working with the shamans and healers of Ecuador's often-feared Waorani tribe, the nearby Sacha Runas and the Shuar headhunters, he has collected medicinal plants to be tested against AIDS, cancer, malaria, and leishmaniasis. He has guided expeditions into these areas, lecturing on his work extensively both here and abroad. He has been active in promoting the conservation of the world's rain forests, and was recently awarded a medal by the government of Ecuador for his work.

He and his wife, Jayne, live with Stormy the cat in Steamboat Springs, Colorado, and Amelia Island, Florida.

Jerry Sanders spent an idyllic "Huck Finn" boyhood in the woods and waterways of rural Long Island, where his training as a biologist really began. After a hitch in the Navy, he attended San Diego State University, earning a degree in biology and a teaching credential. After a few years of teaching in high school, he went on to do graduate work at San Diego State and earned an advanced degree in biology from the University of Pennsylvania. His areas of concentration were radiation biology, cell physiology, and developmental biology.

Returning to San Diego State to join the faculty, Jerry divided his professional time for many years between teaching, developing science curriculum, directing the large general biology program, and teaching laboratory and lecture methods to students entering the secondary education field. He has also worked closely over the years with local secondary teachers, organizing refresher seminars in biology. He has been an active participant in both the National Science Teachers Association and the National Association of Biology Teachers.

In the late 1970s, Jerry began writing introductory biology materials. He has coauthored two successful textbooks, authored a third, and written several laboratory manuals. From time to time he returns to teaching, a first love, just to revisit the "real world," try out his texts and manuals, and get to know young people again.

Jerry and his wife, Mary, live in Southern California, in the foothills overlooking the Colorado Desert, where they are frequently visited by their children and grandchildren. In Jerry's spare time, he and Sam, his Australian shepherd, hike deep into the desert, enjoying its rugged beauty and solitude (when Sam isn't treeing bobcats).

Rob Ferl was born and raised in the small town of Conneaut, Ohio. He received a B.A. in Biology from Hiram College, and a Ph.D in Genetics from Indiana University, before moving to the University of Florida, where he continues his work on gene structure and function.

At the University of Florida he is a Professor in the Institute of Food and Agricultural Sciences and Assistant Director of the Biotechnology Program. He has taught introductory biology for both majors and non-majors, received the Distinguished Teaching Award, and was named Teacher of the Year. He has also served as director of the Biological Sciences Teaching Program.

Rob's research program is centered around the structure of genes and the mechanisms of gene regulation, and his research results are widely published. He is funded by grants from NSF, USDA, NIH, and NASA, supporting studies dedicated toward an understanding of the way genes work and applying that knowledge to real problems. His studies on DNA are heavily oriented toward laboratory technologies, making him a real "lab rat" in the happiest sense of the term. Application of those studies has led to such places as the beaches of Costa Rica, where his DNA sequencing to identify and tag nesting marine turtle populations, and to the mid-dock of the space shuttle, where his genetically engineered plants are used to study gene responses to microgravity environments.

Rob and his wife, Mary-B, and their son, Evan, live in Gainesville, Florida.

Preface

It is a pleasure to introduce the fourth edition of *Biology: The Science of Life.* As you may know, it follows three very successful editions, and there was, we must admit, a temptation to fine-tune the book, to improve it in small ways and let it stand largely intact. (That would also, we were keenly aware, be easier.) However, after long consideration and consultations with numerous reviewers and teachers, we decided that a major overhaul was in order. And so it began.

The first thing we decided we needed was a gunslinger—someone new in town who would take a cold, hard look at the book, compare it with what was needed in today's classroom, and make some tough decisions regarding serious change. That person turned out to be James Funston. James worked very hard (the word relentlessly comes to mind) and gave us, we believe, just what we needed. His suggestions and criticisms were far-ranging and precise. Because of his fresh ideas, we were able to envision a different kind of book, a bit shorter, with more specific definitions and more concise explanations. We also wanted the most current, even if controversial, views presented. We were helped here by our reviewers, a solid bunch of scientists and teachers with powerful notions of how modern biology should be presented. It was, to say the least, a lively process.

A Look at the Art

The second major change, you will quickly see, is in the appearance of the pages. Thumb through the book, look at the art, and you will see the input of one of the most remarkable art houses in the business. J/B Woolsey Associates is the best group of artists we've ever worked with. Incidentally, we were, in the early days, often surprised (shocked?) at how much biology the artists knew. They delved deeply into the material and often came up with new ideas on visual presentation. Because of their knowledge and willingness to research the material, we were able to work together to carefully craft each figure. We are indeed proud of their accomplishments.

Two New Friends

The third change can be viewed on the title page. We decided to bring some new talent into the text, so look for the acknowledgment of two contributors. Lisa Baird of the Uni-

versity of San Diego reorganized and punched up the botany, bringing an expert's touch to the pages. Her work is greatly appreciated. The ecology chapters have also been heavily revised and updated, thanks to the efforts of Ken Klemow of Wilkes University. Look for a very solid presentation in this dynamic area of biology.

The Twin Themes

Although the material has changed, partly because biology itself has changed, our themes remain the same: *evolution* and *adaptation.* As we describe each biological principle, we try to view it in terms of its history and its role in the pageant of life. We believe the continuing attention to these themes will provide the student with a consistent framework in which to place new concepts.

The Organization and Goal

The organization of the text has not changed very much for a very simple reason: it worked. The changes have been within the existing format in the form of adding new material, deleting and shortening other topics, updating and revising explanations. The specific changes that have been made to improve the already successful organization of the text follow:

Part opening outlines and chapter opening outlines show students the topics to be studied.

A new design and page format allows more illustrations to appear with the text discussion.

A new Chapter 26 streamlines the material covered in two chapters (30 and 31) of the previous edition.

The invertebrates and chordates chapters now appear in Part 4 with the other diversity coverage. Plant Evolution and Diversity (Chapter 25) also appears in Part 4 for a complete treatment of diversity.

Specialty reviewers examined the coverage of cell and molecular biology, evolution, and physiology to update and improve this material.

We seek to tell the story of biology in a lively manner. Biology, after all, is the study of life, and we believe the excitement of it all should be a part of telling the tale.

The Supplement Package to Accompany Biology: The Science of Life

The text is supported by a variety of supplements for both instructors and students.

For the Instructor

The *Instructor's Manual* offers instructors valuable support in preparing lectures for their introductory biology course. For each chapter there are suggested references for print and media.

The printed *Test Bank* consists of X000 multiple-choice, true or false, matching, short-answer, and essay questions that are new to the fourth edition.

TestMaster software in IBM and Macintosh formats is available to adopters who prefer a computerized testing system. *TestMaster* enables instructors to select problems for any chapter, scramble them as desired, or create new questions. New with this edition, *QuizMaster* coordinates with the *TestMaster* program. *QuizMaster* allows students to take tests at the computer. Upon completion, a student can evaluate his or her test score and view or print a diagnostic report that lists the topics or objectives that need further study. With network access to *QuizMaster,* student scores are saved on disk and instructors can use the utility program to view records and print reports for individual students, class sections, and entire courses.

Overhead Transparencies of 300 four-color figures cover key topics in the introductory biology course. These full-color illustrations are taken from the art program of the text as well as from other sources; the transparencies are available to adopters.

The *HarperCollins Encyclopedia of Biology* videodisk is an exciting way to integrate multimedia presentations into your classroom. Loaded with still images, motion footage, and animations of key biological concepts, the disk allows students to see complex biological topics in ways that are impossible to present with two-dimensional illustrations. A bar code manual accompanies the disk along with software drivers for both IBM and Macintosh computers. A guide for integrating the videodisk with the text is also available to adopters.

For the Student

Student Study Guide   *Essays on Wellness*, by Barbara A. Brehm of Smith College, is a collection of 29 essays that cover a variety of human health issues. To order use ISBN 0-06-501549-5.

The *Harper Dictionary of Biology* is a valuable reference tool. By W. G. Hale and J.P. Margham, both professors of biology at the Liverpool Polytechnic Institute, it covers all the major subjects—anatomy, biochemistry, ecology, evolutionary theories—plus has biographies of important biologists. It contains 5600 entries that provide in-depth explanations and examples. Diagrams illustrate such concepts as genetic organization, plant structure, and human physiology. To order use ISBN 0-06-461015-2.

Writing About Biology, by Jan A. Pechenik of Tufts University, is a brief guide that prepares students to meet the demands of writing at all levels of biology. Every aspect is covered: laboratory reports, research proposals, research papers, essay exams, oral presentations, and applications for jobs or graduate schools. To order use ISBN 0-673-52128-1.

Studying for Biology, by Anton E. Lawson of Arizona State University, details ways to improve basic study skills. Several chapters cover the thinking patterns that biologists use to answer questions and formulate hypotheses and theories. The book also introduces the basic postulates of the major theories in biology. To order use ISBN 0-06-50065-X.

The Biology Coloring Book is part of a new approach to learning biology. Each page shows a biological process or concept. You create the color key and color in each part of the plate. As you color, you can read an explanation about each element in the accompanying text. When completed, the colored plates are an excellent tool for reviewing what you've learned. To order use ISBN 0-06-460307-5. Other books in the series include the *Anatomy Coloring Book* (0-06-455016-8), *Physiology Coloring Book* (0-06-043479-1), *Botany Coloring Book* (0-06-460302-4), and *Zoology Coloring Book* (0-06-460301-6).

The Illustrated Five Kingdoms: A Guide to the Diversity of Life on Earth is by Lynn Margulis, University of Massachusetts, Amherst; Karlene V. Schwartz, University of Massachusetts, Boston, and Michael Dolan, University of Massachusetts, Amherst. It contains full-page, unlabeled drawings of examples of each of the major groups (phyla) illustrated in their natural habitats. Each full-page drawing is shown in a reduced size with labels for reference. Accompanying text introduces and describes both familiar and unfamiliar plants, animals, and microorganisms. To order use ISBN 0-06-500843-X.

Acknowledgments

Books have authors, but it takes a team to get them to their readers. Our team, we believe, is one of the most remarkable we've ever worked with. We would like to mention a few of the players to whom we owe an immense debt of gratitude.

The project was launched by Glyn Davies, editor-in-chief, whose unflagging faith and enthusiasm in the project, and great hospitality during our many visits to New York, were immensely appreciated.

Ed Moura, Vice President and Science Editor, has been at the helm through the critical stages of this book, and he is, indeed, a breath of fresh air. His quick intelligence, balance, and determination have steered us past many a sticky wicket. We are pleased and proud to be working with him.

Susan Katz, President and Publisher, has been supportive and decisive in forming and directing the team, and we thank her.

Susan Driscoll, editor-in-chief, has again quietly and unobtrusively been there when we needed her, watching over our efforts and helping us at every turn.

Our developmental editor, Karen Trost, is among the best in the business. Her organizational skills and attention to detail have been indispensable to us. We look forward to working with her in the future.

Our sincere thanks also to the team at EPS, Electronic Publishing Services Inc., in New York, a young and vigorous company who took on the immense task of turning manuscript pages and art into a product that pleases us enormously. Special acknowledgment goes to Ruth Randall, who headed up production, and to our highly talented copy editor with the classy name, Christina Della Bartolomea.

We owe a great debt to our reviewers, a stellar group of scientists and teachers who have helped us plan the book and steer our course. Each is an expert in some way, and we are grateful for that expertise.

R.W. Atherton, *University of Wyoming*
R. Howard Berg, *The University of Memphis*
Charles Biggers, *Memphis State University*
John Brink, *Clark University*
John Campbell, *Northwest College*
Donald Defler, *Portland Community College*
Wayne Fagerberg, *University of New Hampshire*
Piotr Fajer, *Florida State University*
Gregory Florant, *Temple University*
Susan Foster, *Mt. Hood Community College*
Robert George, *University of Wyoming*
Edward Hallman, *Daytona Beach Community College*

Marcia Harrison, *Marshall University*
Thomas Hemmerly, *Middle Tennessee State University*
Paul Hertz, *Barnard College*
Cindy Hoorn, *Western Michigan University*
Jane Huffman, *East Stroudsburg University*
Valerie Kish, *University of Richmond*
Rudy Koch, *University of Wisconsin/La Crosse*
Charles Krebs, *University of British Columbia*
Jay Langdon, *University of South Alabama*
Ann Lumsden, *Florida State University*
John Mallett, *University of Massachusetts/Lowell*
Charles Mims, *University of Georgia*
Eli Minkoff, *Bates College*
Lorraine Moran, *Iona College*
Darrel Murray, *University of Illinois/Chicago*
Steven Murray, *California State University, Fullerton*
Ken Nuss, *University of Northern Iowa*
Frances Pick, *University of Ottawa*
Robert Platt, *Ohio State University*
James Robinson, *University of Texas at Arlington*
James Rooney, *Lincoln University*
Fred Sack, *Ohio State University*
Mark Sanders, *University of California/Davis*
Ted Sargent, *University of Massachusetts/Amherst*
Gary Sarinsky, *Kingsborough Community College*
Dan Skean, *Albion College*
Stig Skreslet, *Nordland College*
Robert Leo Smith, *West Virginia University*
Kemet Spence, *Washington State University*
Turner Spencer, *Thomas Nelson Community College*
Lori Stevens, *University of Vermont*
Janet Sullivan, *University of New Hampshire*
Bob Swanson, *North Hennepin Community College*
W.H. Tam, *The University of Western Ontario*
Chris Tarp, *Contra Costa College*
F. William Vockell, *Florida Community College/Jacksonville*
Cherie Wetzel, *City College of San Francisco*
Joe Whitesell, *University of Arkansas/Little Rock*
Anne Zayaitz, *Kutztown University*

Brief Contents

PART ONE

Molecules to Cells 3

1 Mr. Darwin and the Meaning of Life 4
2 Chemistry of Small Molecules 20
3 The Molecules of Life 44
4 Cell Structure 68
5 The Plasma Membrane
   and Cell Transport 102
6 Energy and Chemical Activity in the Cell 126
7 Photosynthesis 148
8 Cell Respiration 172

PART TWO

Molecular Biology and Heredity 199

9 Eukaryotic Cell Reproduction 200
10 Mendelian Genetics 220
11 Going Beyond Mendel 236
12 DNA as the Genetic Material 258
13 Genes in Action 276
14 Gene Regulation 292
15 Genetics in Viruses and Bacteria 306
16 Genetic Engineering: The Frontier 318
17 When DNA Changes 334

PART THREE

Evolution 347

18 Natural Selection and Adaptation 348
19 Microevolution and Changing Alleles 366
20 The Origin of Species 378
21 Origin of Life 402

PART FOUR

Diversity 413

22 Viruses and Bacteria 414
23 Protists 444

24 Fungi 472
25 Plant Evolution and Diversity 490
26 Invertebrates 526
27 Chordates 564

PART FIVE

Plant Functions 591

28 Flowering Plant Reproduction 592
29 Plant Growth and Structure 610
30 Plant Transport and Nutrition 632
31 Plant Regulation and Response 650

PART SIX

Animal Functions 669

32 Support and Movement 670
33 Neural Control I: The Neuron 692
34 Neural Control II: Nervous and
   Sensory Systems 708
35 Thermoregulation, Osmoregulation,
   and Excretion 740
36 Hormonal Control 764
37 Digestion and Nutrition 788
38 Circulation 812
39 Respiration: The Exchange of Gases 832
40 The Immune System 856
41 Reproduction 884
42 Animal Development 910

PART SEVEN

Animal Behavior and Ecology 943

43 The Development and Structure of Animal
   Behavior 944
44 Adaptiveness of Behavior 958
45 Individuals and Populations 980
46 Communities, Ecosystems,
   and Landscapes 1006
47 Biosphere and Biomes 1032
48 The Human Impact 1062

Contents

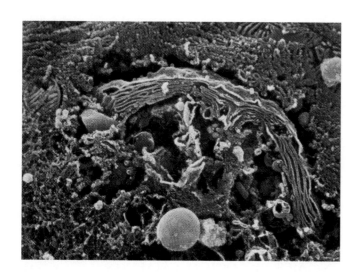

PART ONE

Molecules to Cells 3

CHAPTER 1
Mr. Darwin and the Meaning of Life 4

WHERE ARE THE RABBITS? 5
An Explanation about Rabbits and Oceans 6
THE WORKINGS OF SCIENCE 9
Inductive Reasoning 9
Deductive Reasoning 10
Scientific Method 10
Hypothesis, Theory, and Law 10
EVOLUTION: THE THEORY DEVELOPS 12
The Theory of Natural Selection 12
Publication of *On the Origin of Species* 13
Testing Evolutionary Hypotheses 14
The Impact of Darwin 14
CHARACTERISTICS OF LIFE: THE ULTIMATE
   REDUCTION 15
SIGNS OF LIFE 16
BIOLOGY, THE REALM OF LIFE 17
KEY IDEAS 18
APPLICATION OF IDEAS 19
REVIEW QUESTIONS 19

Essay 1.1 The Enchanted Isles 8

CHAPTER 2
Chemistry of Small Molecules 20

ELEMENTS, ATOMS, AND MOLECULES 21
Atoms and Molecules 21
Atomic Structure 22
Isotopes, Radioactivity, and Biology 23
ELECTRONS AND THE CHEMICAL
   PROPERTIES OF ELEMENTS 24
Electron Energy Levels and Shells 24
THE INTERACTION OF ATOMS AND
   ELECTRONS: CHEMICAL BONDS 25
Covalent Bonds 26
Ionic Bonds 28
Hydrogen Bonds 29
Why Chemical Reactions Occur 30
CARBON BONDS AND
   THE SHAPE OF THINGS 30
The Versatility of Carbon 30
THE OTHER SPONCH ELEMENTS 32
Nitrogen 32
Phosphorus 32
Sulfur 32
Functional Groups 33
THE WATER MOLECULE AND HYDROGEN
   BONDING 33
Water and the Hydrogen Bond 34
Water: Ionization, pH, and Acids and Bases 38
KEY IDEAS 41
APPLICATION OF IDEAS 42
REVIEW QUESTIONS 42

CHAPTER 3
The Molecules of Life 44

THE CARBOHYDRATES 45
Monosaccharides and Disaccharides 45
Polysaccharides 47
THE LIPIDS 49
Triglycerides 51
Saturated, Unsaturated, and Polyunsaturated Fats 52
Phospholipids 53
Other Lipids 55

THE PROTEINS   **56**
What Is a Protein   56
Amino Acids   57
Polypeptides and Proteins   59
Structural Proteins   61
Conjugated Proteins   61
THE NUCLEIC ACIDS   **62**
Nucleic Acid Structure   64
KEY IDEAS   **65**
APPLICATION OF IDEAS   **66**
REVIEW QUESTIONS   **67**

Essay 3.1 Reading Structural Formulas   *46*

CHAPTER 4
Cell Structure   **68**

CELL THEORY   **69**
WHAT IS A CELL?   **70**
The Sizes of Cells   70
THE PROKARYOTIC CELL   **74**
THE EUKARYOTIC CELL   **77**
Surface Structures   77
Internal Support: The Cytoskeleton   81
Organelles of Movement   84
Control and Cell Reproduction: The Nucleus   86
Organelles of Synthesis, Processing, and Storage   88
Endomembranal System   93
Energy-Generating Organelles   94
KEY IDEAS   **99**
APPLICATION OF IDEAS   **100**
REVIEW QUESTIONS   **101**

Essay 4.1 Looking at Cells   *71*
Essay 4.2 Evolution of the Eukaryotic Cell: Autogeny or
*   Endosymbiosis?*   *98*

CHAPTER 5
The Plasma Membrane and Cell Transport   **102**

THE PLASMA MEMBRANE   **102**
Lipid Bilayers   103
The Membrane Proteins   103
The Glycocalyx: Glycoproteins and Glycolipids   105
Cholesterol   106
The Fluid Mosaic Model   107
MECHANISMS OF TRANSPORT   **108**
Passive Transport   109

Active Transport   114
CELL JUNCTIONS   **121**
Cell-to-Cell Transport   121
Cell-to-Cell Barriers and Support   123
KEY IDEAS   **123**
APPLICATION OF IDEAS   **124**
REVIEW QUESTIONS   **124**

CHAPTER 6
Energy and Chemical Activity in the Cell   **126**

ENERGY   **127**
Forms of Energy   127
The Laws of Thermodynamics   127
APPLYING ENERGY PRINCIPLES TO LIFE   **128**
Life and the Laws of Thermodynamics   128
Free Energy   129
Exergonic and Endergonic Reactions   129
Getting Reactions Started   129
ENZYMES: BIOLOGICAL CATALYSTS   **132**
Enzyme Specificity   132
The Enzymatic Reaction   132
The Enzyme–Substrate Complex: Events at the
    Active Site   133
Influences on the Rates of Enzyme Action   134
Cofactors (Vitamins and Minerals)   135
Teams of Enzymes: Metabolic Pathways   135
Mechanisms of Enzyme Control   135
Summarizing the Activity of Enzymes   137
ATP: THE ENERGY CURRENCY OF
**   THE CELL**   **138**
ATP Structure and the Release of Energy   139
The Work of ATP   139
The Cycling of ATP   140
THE COENZYMES AND OTHER ELECTRON
**   CARRIERS**   **140**
Coenzymes   140
Redox Reactions: Oxidation and Reduction   140
Coenzymes in Action   141
Stationary Electron Carriers of Mitochondria and
    Chloroplasts   143
THE FORMATION OF ATP IN CELLS   **143**
Chemiosmotic Phosphorylation   143
KEY IDEAS   **146**
APPLICATION OF IDEAS   **147**
REVIEW QUESTIONS   **147**

CHAPTER 7
Photosynthesis   **148**

PHOTOSYNTHESIS: AN OVERVIEW   **149**
The Two Parts of Photosynthesis   150
Light: The Energy Source   150
CHLOROPLASTS   **151**
The Thylakoids   151
**THE LIGHT-DEPENDENT REACTIONS OF
    PHOTOSYNTHESIS**   **156**
The Noncyclic Events   157
The Cyclic Events   159
Chemiosmotic Phosphorylation   159
THE LIGHT-INDEPENDENT REACTIONS   **160**
The Calvin Cycle   161
**PHOTORESPIRATION: TROUBLE IN THE
    CALVIN CYCLE**   **162**
C4 Carbon Fixation: Solving the Photorespiration
    Problem   164
C4 Plants and Evolution   168
KEY IDEAS   **169**
APPLICATION OF IDEAS   **170**
REVIEW QUESTIONS   **170**

*Essay 7.1 From Willows, Mice and Candles to Electron
    Microscopes*   *166*

CHAPTER 8
Cell Respiration   **172**

TWO WAYS OF UTILIZING GLUCOSE   **174**
GLYCOLYSIS   **175**
The Highlights of Glycolysis   175
Control of Glycolysis   176
**AEROBIC CELL RESPIRATION: THE
    MITOCHONDRION**   **177**
Pyruvate to Acetyl–CoA   177
The Citric Acid Cycle   177
**ELECTRON TRANSPORT AND CHEMIOSMOTIC
    PHOSPHORYLATION**   **181**
Mitochondrial Structure   181
NADH and $FADH_2$ Electrons Build the Chemiosmotic
    Gradient   184
Reviewing the Proton Gradient and ATP Formation   187
Other Uses for the Proton Gradient   188
ANAEROBIC CELL RESPIRATION   **188**
FERMENTATION   **188**
Fermentation Pathways   190

ALTERNATIVE FUELS FOR THE CELL   **193**
The Metabolism of Fatty Acids   193
The Preparation of Proteins   194
Intermediary Metabolism: Biosynthetic Pathways   194
KEY IDEAS   **195**
APPLICATION OF IDEAS   **196**
REVIEW QUESTIONS   **197**

Essay 8.1 A Marathon Runner Meets "The Wall"   *195*

PART TWO

Molecular Biology and Heredity   199

CHAPTER 9
Eukaryotic Cell Reproduction   **200**

THE NUCLEUS   **201**
DNA, Chromatin, and the Eukaryotic Chromosomes   201
The Cell Cycle: Growth, Replication, and Division   202
MITOSIS   **203**
Prophase   203
Metaphase   204
Anaphase   206
Telophase   206
Cytokinesis   208
Advantages of Mitosis   210
MEIOSIS   **210**
Homologous Chromosomes   211

Overview of Meiosis   212
Meiosis I: The First Meiotic Division   212
Meiosis II: The Second Meiotic Division   213
MEIOSIS IN HUMANS   **214**
Meiosis in Females: A Special Case   214
Meiosis, Evolution, and Sex   215
KEY IDEAS   **218**
APPLICATION OF IDEAS   **219**
REVIEW QUESTIONS   **219**

Essay 9.1 The Work of the Spindle   207
Essay 9.2 Karyotyping   212
Essay 9.3 When Meiosis Goes Wrong   217

C H A P T E R   1 0
Mendelian Genetics

Mendelian Genetics   **220**
WHEN DARWIN MET MENDEL (ALMOST)   **220**
MENDEL'S CROSSES   **222**
Dominance   222
MENDEL'S FIRST LAW: SEGREGATION OF
   ALTERNATE ALLELES   **227**
Segregation and Probability   227
MENDEL'S SECOND LAW: INDEPENDENT
   ASSORTMENT   **228**
Mendel's Testcrosses   230
THE CHROMOSOMAL BASIS FOR MENDEL'S
   LAWS   **231**
PEDIGREES AND HUMAN TRAITS   **232**
THE DECLINE AND RISE OF MENDELIAN
   GENETICS   **232**
KEY IDEAS   **233**
APPLICATION OF IDEAS   **234**
REVIEW QUESTIONS   **235**

C H A P T E R   1 1
Going Beyond Mendel

Going Beyond Mendel   **236**
DOMINANCE RELATIONSHIPS AND GENE
   INTERACTIONS   **237**
Incomplete Dominance   237
Codominance   238
Multiple Alleles   238
Lethals   240
Gene Interactions   241
CONDITIONAL PHENOTYPES   **241**

Influences of Environment   241
Influences of Development   241
Influences of Sex   242
Influences of Age   242
Pleiotropy   243
Continuous Variation and Polygenic Inheritance   244
GENES ON CHROMOSOMES   **244**
Linkage   244
Crossing Over and Genetic Recombination   245
Mapping Genes   247
Chromosomes and Sex   248
KEY IDEAS   **254**
APPLICATION OF IDEAS   **255**
REVIEW QUESTIONS   **256**

Essay 11.1 Sex Chromosome Abnormalities   *251*
Essay 11.2 The Disease of Royalty   *253*

C H A P T E R   1 2
DNA as the Genetic Material

DNA as the Genetic Material   **258**
THE DISCOVERY OF DNA AS THE GENETIC
   MATERIAL   **259**
The Early Efforts   259
Transformation: A Hereditary Role for DNA   259
The Experiments of Hershey and Chase   261
THE STRUCTURE OF DNA   **261**
The Chemical Properties of DNA   263
The Watson and Crick Model of DNA   264
DNA REPLICATION   **265**
The Chemistry of DNA Synthesis   265
Origins of Replication   267
DNA Polymerase   267
DNA Replication in Eukaryotes and Prokaryotes   268
DNA AND GENETIC INFORMATION   **272**
KEY IDEAS   **274**
APPLICATION OF IDEAS   **275**
REVIEW QUESTIONS   **275**

Essay 12.1 Meselson and Stahl   *270*
Essay 12.2 Eukaryotic DNA: A Closer Look   *273*

C H A P T E R   1 3
Genes in Action

Genes in Action   **276**
RNA STRUCTURE AND TRANSCRIPTION   **277**
Comparing RNA to DNA   277

Transcription: RNA Synthesis   277
VARIETIES OF RNA   **278**
Ribosomal RNA and the Ribosome   279
Messenger RNA   280
Transfer RNA   283
TRANSLATION: HOW PROTEINS ARE
   ASSEMBLED   **284**
Initiation   285
Elongation   285
Termination   287
Polyribosomes   287
Free and Bound Ribosomes   287
KEY IDEAS   **289**
APPLICATION OF IDEAS   **290**
REVIEW QUESTIONS   **291**

C H A P T E R   1 4
Gene Regulation   **292**

PROKARYOTIC GENE REGULATION: THE
   OPERON   **293**
Inducible Operon for Digesting Lactose   294
Repressible Operon for Synthesis of Tryptophan   296
EUKARYOTIC GENE REGULATION   **298**
Opportunities for Regulation   298
Control of Transcription: Inducible Genes   298
DNA Packaging and Chemical Modification   301
Post-transcriptional Regulation   303
KEY IDEAS   **305**
APPLICATION OF IDEAS   **305**
REVIEW QUESTIONS   **305**

Essay 14.1 Homeotic Genes and the Homeobox   *301*

C H A P T E R   1 5
Genetics in Viruses and Bacteria   **306**

VIRUSES   **307**
The Life and Times of the Bacteriophage   307
The Lytic Cycle   307
The Lysogenic Cycle   307
Variations on the Viral Theme   308
BACTERIA   **309**
Locating Mutant and Recombinant Bacteria   309
Recombination in Bacteria   310
Plasmid Genes   316

KEY IDEAS   **316**
REVIEW QUESTIONS   **317**

C H A P T E R   1 6
Genetic Engineering: The Frontier   **318**

DNA TECHNOLOGY   **319**
Restriction Enzymes and Sticky Ends   319
Splicing and Cloning DNA Molecules   319
APPLICATIONS OF GENETIC ENGINEERING   **325**
Manufacture of Biomolecules   325
Agriculture   326
Medicine   327
GENETIC ENGINEERING: A SOCIAL ISSUE   **330**
KEY IDEAS   **332**
APPLICATION OF IDEAS   **332**
REVIEW QUESTIONS   **333**

Essay 16.1 DNA Sequencing   *324*
Essay 16.2 Electrophoresis, Hybridization, and DNA
   Fingerprinting   *328*
Essay 16.3 The Polymerase Chain Reaction   *331*

C H A P T E R   1 7
When DNA Changes   **334**

THE STABILITY OF DNA   **334**
DNA DAMAGE AT THE MOLECULAR
   LEVEL   **335**
DNA REPAIR: THE CLEANUP CREW   **336**
POINT MUTATIONS   **336**
Substitutions   337
Chain-Termination Mutations   337
Additions, Deletions, and Frameshift Mutations   339
Point Mutations in Noncoding Regions of DNA   339
CHROMOSOMAL MUTATIONS   **340**
Deletions and Inversions   340
Translocations   340
TRANSPOSITIONS   **340**
Transposon Structure   342
Transposons as Mutagenic Agents   343
KEY IDEAS   **344**
APPLICATION OF IDEAS   **345**
REVIEW QUESTIONS   **345**

Essay 17.1 Mutations, Oncogenes, and Cancer   *342*
Essay 17.2 Mutation and the Evolution of Gene Structure *344*

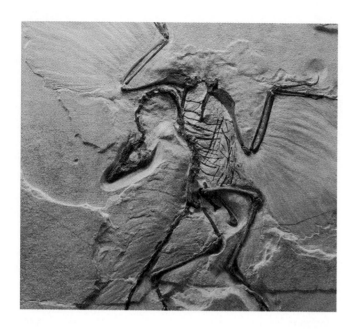

PART THREE

Evolution   *347*

CHAPTER 18
Natural Selection and Adaptation   **348**

THE HISTORY OF AN IDEA   **348**
Enter Charles Darwin   350
DARWIN'S THEORY   **350**
The Origin of Species   352
The Modern Synthesis   353
VARIATION   **353**
Sources of Variation   354
Maintaining Genetic Variation   354
NATURAL SELECTION   **355**
The Peppered Moth: Natural Selection in a Natural
  Population   355
How the Leopard Got Its Spots   357
Fitness   357
EVIDENCE OF EVOLUTION   **358**
Fossils and the Fossil Record   358
Biogeography   359
Comparative Anatomy   361
Molecular Biology   362

Artificial Selection   363
KEY IDEAS   **363**
REVIEW QUESTIONS   **365**
Essay 18.1 The Evolution of the Horse   *360*

CHAPTER 19
Microevolution and Changing Alleles   **366**

**ALLELE REPRESENTATION WITHIN
  POPULATIONS**   **367**
The Population Genetics of G. H. Hardy   367
An Implication of the Hardy–Weinberg Principle 367
Hardy–Weinberg: An Algebraic Discussion   368
The Restrictions of the Hardy–Weinberg Principle 369
**MUTATION: THE INPUT OF NEW
  INFORMATION**   **370**
Balancing Mutation and Selection   370
**THE OUTCOMES OF NATURAL
  SELECTION**   **370**
Stabilizing Selection   370
Directional Selection   371
Disruptive Selection   372
RANDOM CHANGES IN FREQUENCIES   **372**
Neutralists Versus Selectionists   374
**QUESTIONS ABOUT THE GENETIC FUTURE
  OF *HOMO SAPIENS***   **375**
KEY IDEAS   **376**
APPLICATION OF IDEAS   **377**
REVIEW QUESTIONS   **377**
Essay 19.1 The "Tail" of Two Genotypes   *371*

CHAPTER 20
The Origin of Species   **378**

WHAT IS A SPECIES?   **379**
Problems with the Species Concept   379
TAXONOMY AND SYSTEMATICS   **382**
How Species Are Named   382
Taxonomic Organization: Species to Kingdom   382
The Five Kingdoms   382
**DETERMINING RELATIONSHIPS AMONG
  ORGANISMS**   **384**

The Classical Approach   385
The Phenetic Approach   386
Cladistics   386
THE MECHANISMS OF SPECIATION   **387**
Allopatric Speciation   387
Sympatric Speciation   388
REPRODUCTIVE ISOLATING
MECHANISMS   **392**
Prezygotic Barriers   393
Postzygotic Barriers   393
PATTERNS OF EVOLUTION   **394**
Divergent Evolution   394
Convergent Evolution   395
Coevolution   395
SMALL STEPS OR GREAT LEAPS?   **396**
EXTINCTION AND EVOLUTION   **397**
Extinction and Us   397
KEY IDEAS   **399**
APPLICATION OF IDEAS   **400**
REVIEW QUESTIONS   **401**

Essay 20.1 The Galápagos Islands, Home of
Darwin's Finches   *390*
Essay 20.2 Continents Adrift   *391*

CHAPTER 21
Origin of Life   **402**
THE EARLY EARTH   **403**
OVERVIEW OF A HYPOTHESIS ON THE
ORIGIN OF LIFE   **403**
THE HYPOTHESIS TODAY   **404**
The Monomers of Life   405
The Formation of Polymers   405
Self-replicating Systems   407
THE EARLIEST CELLS   **408**
RNA: Gene and Enzyme?   408
The Progenote, Ancestor of All Cells   409
The Early Metabolic Cell   409
The Early Autotrophic Cells   409
KEY IDEAS   **410**
APPLICATION OF IDEAS   **411**
REVIEW QUESTIONS   **411**

PART FOUR

Diversity   *413*

CHAPTER 22
Viruses and Bacteria   **414**
VIRUSES   **414**
Introducing Viruses   414
Origin of Viruses   415
The Discovery of Viruses   416
Characteristics of Viruses   417
Bacterial Viruses: The Bacteriophages   417
Animal Viruses   418
Plant Viruses   421
HIV: Study of a Retroviral Killer   422
Viroids and Prions   423
KINGDOM MONERA: THE PROKARYOTES   **425**
Evolutionary Roots of Monerans   425
Eubacteria, Archaebacteria, and a New Phylogeny   425
Archaebacteria: A Brief Look   426
Eubacteria   428
Bacterial Reproduction   432
Metabolic Diversity in Bacteria   433
KEY IDEAS   **441**

APPLICATION OF IDEAS   442
REVIEW QUESTIONS   443

Essay 22.1 The Nitrogen Fixation Process   *440*

CHAPTER 23
Protists   444
CHARACTERISTICS OF PROTISTS   445
Organization   445
Nutrition   446
Movement   446
Reproduction   447
Life Cycles in Protists and Other Eukaryotes   447
Evolutionary Origins of Protists   447
Phylogeny of Protists   448
ANIMAL-LIKE PROTISTS: PROTOZOANS   448
Flagellates   448
Amebas   449
Sporozoans   451
Ciliates   451
FUNGUS-LIKE PROTISTS   454
Slime Molds   455
Oomycetes: Downy Mildews and Water Molds   457
PLANT-LIKE PROTISTS: THE ALGAE   458
Pyrrhophytes: "Fire Algae"   458
Euglenophytes   460
Chrysophytes: Yellow-Green Algae, Golden Brown
  Algae, and Diatoms   460
Rhodophytes: Red Algae   462
Phaeophytes: Brown Algae   464
Chlorophytes: Green Algae   465
KEY IDEAS   469
APPLICATION OF IDEAS   470
REVIEW QUESTIONS   471

CHAPTER 24
Fungi   472
WHAT ARE FUNGI?   472
The Fungal Body   472
Feeding in Fungi   474
Ecological Importance   474
Medical and Commercial Uses   475
Reproduction in Fungi   475
Fungal Relationships   476
KINGDOM FUNGI   476

Zygomycota: Conjugating Molds   476
Ascomycota: Sac Fungi   477
Basidiomycota: Club Fungi   484
EVOLUTION OF FUNGI   487
KEY IDEAS   488
REVIEW QUESTIONS   489

Essay 24.1 A Fungus that Spits   *487*

CHAPTER 25
Plant Evolution and Diversity   490
WHAT ARE PLANTS?   491
Alternation of Generations in Plants   491
ORIGIN OF PLANTS AND EARLY
  ADAPTATIONS TO LAND   492
BRYOPHYTES: THE NONVASCULAR PLANTS   494
Bryophyte Characteristics   494
Division Hepatophyta: Liverworts   494
Division Anthocerophyta: Hornworts   495
Division Bryophyta: Mosses   497
Evolutionary Origin of Bryophytes   500
THE VASCULAR PLANTS   501
History of Vascular Plants   501
Taxonomic Organization of Vascular Plants   502
Division Psilotophyta: Whisk Ferns   504
Division Lycophyta: Club Mosses   506
Division Sphenophyta: Horsetails   507
Division Pterophyta: Ferns   509
THE SEED PLANTS   510
Reproductive Adaptations in Emerging Seed Plants   512
The Gymnosperms   513
THE END OF AN ERA   516
Angiosperms: The Rise of the Flowering Plants   516
The Angiosperms Today   519
THE PLANT KINGDOM: A SUMMARY   522
KEY IDEAS   522
APPLICATION OF IDEAS   524
REVIEW QUESTIONS   524

Essay 25.1 An Evolutionary Lesson From Plants   *518*

CHAPTER 26
Invertebrates   526
WHAT IS AN ANIMAL?   527
ANIMAL ORIGINS   527
The Early Fossil Record   527

MAJOR MILESTONES IN ANIMAL
  EVOLUTION **531**
Landmark Evolutionary Changes 532
PHYLUM PORIFERA: AN EVOLUTIONARY
  DEAD END **532**
The Biology of Sponges 535
PHYLA CNIDARIA AND CTENOPHORA:
  TISSUES AND THE RADIAL PLAN **535**
Phylum Cnidaria 536
Phylum Ctenophora: Comb Jellies 537
PHYLUM PLATYHELMINTHES: MESODERM
  AND THE BILATERAL PLAN **537**
Phylum Platyhelminthes: The Flatworms 538
PHYLA NEMATODA AND ROTIFERA: BODY
  CAVITIES AND A ONE-WAY GUT **541**
Phylum Nematoda 541
Phylum Rotifera 543
PHYLA ECTOPROCTA, BRACHIOPODA AND
  PHORONIDA: THE LOPHOPHORATES **545**
PHYLUM MOLLUSCA: UNSEGMENTED
  COELOMATES **546**
PHYLUM ANNELIDA: SEGMENTATION
  ALONG THE LENGTH **549**
PHYLUM ARTHROPODA: SPECIALIZATION
  OF SEGMENTS AND JOINTED LEGS **551**
PHYLUM ONYCHOPHORA: LINKING
  ANNELIDS AND ARTHROPODS **558**
PHYLUM ECHINODERMATA:
  DEUTEROSTOMES WITH RADIAL BODIES **558**
KEY IDEAS **560**
APPLICATION OF IDEAS **562**
REVIEW QUESTIONS **563**

CHAPTER 27
Chordates **564**

PHYLUM HEMICHORDATA **565**
PHYLUM CHORDATA **565**
Subphylum Urochordata: Tunicates 565
Subphylum Cephalochordata: Lancelets 566
SUBPHYLUM VERTEBRATA: ANIMALS
  WITH BACKBONES **567**
Class Agnatha: Jawless Fishes 567
Class Placodermi: Extinct Jawed Fishes 568
Class Chrondrichthyes: Sharks and Rays 569
Class Osteichthyes: The Bony Fishes 570

Class Amphibia: The Amphibians 572
Class Reptilia: The Reptiles 573
Class Aves: The Birds 575
Class Mammalia: The Mammals 576
HUMAN EVOLUTIONARY HISTORY **584**
The Australopithecine Line 584
The Human Line 585
KEY IDEAS **588**
APPLICATION OF IDEAS **589**
REVIEW QUESTIONS **589**

Essay 27.1 Exit the Great Reptiles *577*

PART FIVE

Plant Functions *591*

CHAPTER 28
Flowering Plant Reproduction **592**
SEXUAL REPRODUCTION **593**
The Flower 593
Variation in Flowers 593
Flower Structure and Pollination 595
Sexual Activity in Flowers 597
SEED DEVELOPMENT **600**
Development of the Embryo and Seed 600
Seed Dormancy 604
Seed Dispersal 604
ASEXUAL REPRODUCTION **605**
Vegetative Propagation at Work 606
Apomixis: Seeds Without Sex 606

KEY IDEAS **608**
APPLICATION OF IDEAS **609**
REVIEW QUESTIONS **609**

Essay 28.1 Diversity in Flowers *594*
Essay 28.2 Flowers to Fruits, Or a Quince is a Pome *602*

CHAPTER 29
Plant Growth and Structure **610**
GERMINATION AND GROWTH OF
  THE SEEDLING **611**
Germination Conditions 611
The Seedling 611
Primary and Secondary Growth 611
TISSUE ORGANIZATION IN THE PLANT **613**
Protoderm and the Epidermis 614
Ground Meristem and Its Three Tissues 615
Procambium and the Vascular Tissues 615
PRIMARY GROWTH IN THE ROOT **617**
The Root Tip 617
Root Systems 620
PRIMARY GROWTH IN THE STEM **620**
LEAVES **623**
Anatomy of the Leaf 623
SECONDARY GROWTH IN THE STEM **625**
Transition to Secondary Growth 625
Growth of the Periderm 626
The Older Woody Stem 627
KEY IDEAS **630**
APPLICATION OF IDEAS **631**
REVIEW QUESTIONS **631**

Essay 29.1 The Problem of Differentiation *629*

CHAPTER 30
Plant Transport and Nutrition **632**
THE MOVEMENT OF WATER AND MINERALS **633**
Transpiration 633
Water Potential and the Vascular Plant 633
Root Pressure: Is the Root a Pump? 636
Water Movement in the Leaf: Transpiration and the
  Pulling of Water 636

Water Movement in the Stem 637
Water Movement in the Root 639
Guard Cells and Water Transport 640
FOOD TRANSPORT IN PLANTS **643**
Phloem Transport 643
GAS TRANSPORT IN PLANTS **644**
PLANT NUTRITION: MINERALS AND
  THEIR TRANSPORT **645**
Mycorrhiza: Fungal Association with Roots 645
Plant Mineral Nutrition 646
Insect-Eating Plants 647
KEY IDEAS **648**
APPLICATION OF IDEAS **649**
REVIEW QUESTIONS **649**

Essay 30.1 Testing the Forces that Move Water *638*

CHAPTER 31
Plant Regulation and Response **650**
LIGHT AND THE GROWTH RESPONSE **651**
PLANT HORMONES **653**
Auxin: Its Structure and Roles 653
Gibberellins 654
Cytokinins 655
Abscisic Acid 656
Ethylene 656
Leaf Abscission 657
Apical Dominance 658
Applications of Plant Hormones 658
DIFFERENTIAL GROWTH RESPONSES:
  TROPISMS **658**
Phototropism 658
Roots and Gravitropism 659
Thigmotropism 659
Solar Tracking 659
NASTIC MOVEMENT **661**
LIGHT AND FLOWERING **662**
Photoperiodicity 662
Flower Initiation and Phytochrome 664
KEY IDEAS **666**
APPLICATION OF IDEAS **667**
REVIEW QUESTIONS **667**

PART SIX

Animal Functions 669

CHAPTER 32
Support and Movement   **670**

INVERTEBRATE SUPPORT AND MOVEMENT   **671**
Hydrostatic Skeletons and Their Movement   671
Invertebrate Exoskeletons   671
Invertebrate Endoskeletons   671
VERTEBRATE TISSUES   **675**
Epithelial Tissue   675
Connective Tissue   675
VERTEBRATE SKELETONS   **675**
The Structure of Bone   677
Organization of the Vertebrate Skeleton   678
VERTEBRATE MUSCLE: ITS ORGANIZATION
  AND MOVEMENT   **681**
Smooth Muscle   682
Cardiac Muscle   682
Skeletal Muscle   682
Summing Up Contraction   687
KEY IDEAS   **689**
APPLICATION OF IDEAS   **690**
REVIEW QUESTIONS   **691**

CHAPTER 33
Neural Control I: The Neuron   **692**

CELLS OF THE NERVOUS SYSTEM   **693**
The Structure of Neurons   693
Nerves   695
THE ORIGIN AND TRANSMISSION OF
  NEURAL IMPULSES   **695**
The Action Potential   696
Ion Exchange Pumps and Gradients   696
Generating the Action Potential   697
Reestablishing the Resting Potential   698
Ion Channels and Gates   698
Myelin and Impulse Velocity   700
COMMUNICATION AMONG NEURONS AND
  BETWEEN NERVES AND MUSCLES   **701**
Electrical Synapses   701
Chemical Synapses   702
The Reflex Arc: The Simplest Model of Neural Activity   705
KEY IDEAS   **706**
APPLICATION OF IDEAS   **707**
REVIEW QUESTIONS   **707**

CHAPTER 34
**Neural Control II: Nervous and Sensory
Systems**   **708**

INVERTEBRATE NERVOUS SYSTEMS   **709**
Simple Nerve Nets, Ladders, and Rings   709
Echinoderms and Nerve Rings   709
From Ganglia to Organized Brains   709
VERTEBRATE NERVOUS SYSTEMS   **710**
THE HUMAN CENTRAL NERVOUS SYSTEM   **712**
Overview of the Human Brain   712
The Forebrain   714
The Midbrain and Hindbrain   719
The Spinal Cord   720
Chemicals in the Brain   720
Electrical Activity in the Brain   720
THE PERIPHERAL NERVOUS SYSTEM   **723**
The Autonomic Nervous System   723
THE SENSES   **723**
Sensory Receptors   725

Neural Codes 725
Tactile Reception 725
Thermoreception 727
Chemoreception 728
Proprioception 729
Auditory Reception 729
Gravity and Movement Reception 732
Visual Reception 732
KEY IDEAS **736**
APPLICATION OF IDEAS **738**
REVIEW QUESTIONS **739**
Essay 34.1 The Great Split-Brain Experiments *717*

CHAPTER 35
Thermoregulation, Osmoregulation, and Excretion **740**
HOMEOSTASIS **741**
Negative and Positive Feedback Loops 741
THERMOREGULATION **742**
Why Thermoregulate? 742
Categories of Thermoregulation 742
Endothermic Regulation 743
Thermoregulation in Birds and Mammals 744
Ectothermic Regulation 748
Physiological Adaptations to Freezing 748
OSMOREGULATION AND EXCRETION **751**
Producing Nitrogen Wastes 751
The Osmotic Environment 751
The Marine Environment 751
The Freshwater Environment 752
The Terrestrial Environment 753
THE HUMAN EXCRETORY SYSTEM **755**
Anatomy of the Human Excretory System 755
Microanatomy of the Nephron 755
The Work of the Nephron 755
Control of the Nephron Function 759
KEY IDEAS **761**
APPLICATION OF IDEAS **762**
REVIEW QUESTIONS **763**

CHAPTER 36
Hormonal Control **764**
THE CHEMICAL MESSENGERS **765**
Neural and Hormonal Control Compared 765
Molecular Structure of Hormonal Messengers 765

Characteristics of Hormonal Control 766
CHEMICAL MESSENGERS AND THE TARGET CELL **766**
Peptide Hormones and Second Messengers 767
Steroid Hormones and Gene Control 769
INVERTEBRATE HORMONES **769**
Hormonal Activity in Arthropods 770
THE HUMAN ENDOCRINE SYSTEM **771**
The Hypothalamus and Pituitary 771
The Thyroid 776
The Parathyroid Glands 777
The Pancreas 778
The Adrenal Glands 779
The Gonads: Ovaries and Testes 781
The Thymus 782
The Pineal Body 782
Prostaglandins 783
KEY IDEAS **785**
APPLICATION OF IDEAS **786**
REVIEW QUESTIONS **787**
Essay 36.1 Diabetes Mellitus and Hypoglycemia *780*
Essay 36.2 The Heart as an Endocrine Structure *781*

CHAPTER 37
Digestion and Nutrition **788**
DIGESTIVE SYSTEMS **788**
Intracellular Digestion 789
Sac-like Digestive Systems 789
Tube-within-a-Tube Digestive Systems 790
Vertebrate Digestive Systems 791
THE DIGESTIVE SYSTEM OF HUMANS **795**
The Oral Cavity and Esophagus 795
The Stomach 797
The Small Intestine 797
Accessory Organs: The Liver and Pancreas 800
The Large Intestine 801
THE CHEMISTRY OF DIGESTION **801**
Carbohydrate Digestion 802
Fat Digestion 803
Protein Digestion 803
Nucleic Acid Digestion 805
Integration and Control of the Digestive Process 805
SOME ESSENTIALS OF NUTRITION **806**
Carbohydrates 806
Fats 806
Protein 807

Vitamins 807
Mineral Requirements of Humans 807
KEY IDEAS 809
APPLICATION OF IDEAS 811
REVIEW QUESTIONS 800

Essay 37.1 The Heimlich Maneuver 800
Essay 37.2 Cholesterol and Controversy 808

CHAPTER 38
Circulation 812

**ADAPTATIONS IN ANIMALS WITHOUT
    CIRCULATORY SYSTEMS** 813
**CIRCULATORY SYSTEMS IN
    INVERTEBRATES** 813
Open and Closed Circulatory Systems 813
CIRCULATION IN THE VERTEBRATES 815
The Blood 815
Blood Circulation in the Vertebrates 817
THE HUMAN CIRCULATORY SYSTEM 819
Circulation Through the Heart 819
Control of the Heart's Contractions 821
The Working Heart 822
Blood Pressure 823
Circuits in the Human Circulatory System 824
The Role of the Capillaries 825
The Veins 827
THE LYMPHATIC SYSTEM 828
KEY IDEAS 829
APPLICATION OF IDEAS 831
REVIEW QUESTIONS 831

Essay 38.1 Cardiovascular Disease 828

CHAPTER 39
Respiration: The Exchange of Gases 832

GAS EXCHANGE SURFACES 833
The Simple Body Interface 833
Internalizing the Interface: Tracheae 836
More Complex Interfaces: The Aquatic Gill 836
More Complex Interfaces: The Vertebrate Lung 838
THE HUMAN RESPIRATORY SYSTEM 843
The Flow of Air 843
The Breathing Movements 846
The Exchange of Gases 847
Oxygen Transport 849
Carbon Dioxide Transport 851

The Control of Respiration 851
KEY IDEAS 854
APPLICATION OF IDEAS 855
REVIEW QUESTIONS 855

Essay 39.1 Lung Cancer: The "Time Bomb" Within 844

CHAPTER 40
The Immune System 856

NONSPECIFIC DEFENSES 857
Nonspecific Primary Defenses: The Body Coverings 857
Nonspecific Chemical Defenses 858
Nonspecific Cellular Defenses 861
Resistance in the Invaders 862
SPECIFIC DEFENSES 863
Lymphoid Tissues and the B- and T-Cell Lymphocytes 863
Antibodies and the Humoral Response 865
T-Cells and Cell Recognition 868
The Mature B-Cell and Its Receptor 869
**CLONAL SELECTION AND THE PRIMARY
    IMMUNE RESPONSE** 870
Clonal Selection: Antigen-Presenting Macrophages 870
Spreading the Alarm: Aroused T-Cells 871
B-Cells and the Antibody Attack 873
Suppressor T-Cells: The Battle Is Won 874
Summing Up the Primary Immune Response 874
**VIGILANT MEMORY CELLS AND THE
    SECONDARY IMMUNE RESPONSE** 874
Active and Passive Immunity 876
LYMPHOCYTE DIVERSITY 876
MONOCLONAL ANTIBODIES 877
WHEN THE IMMUNE SYSTEM GOES WRONG 878
Autoimmunity: Attack Against Self 878
Allergies 879
AIDS: The Crippled Immune System 879
KEY IDEAS 882
REVIEW QUESTIONS 883

CHAPTER 41
Reproduction 884

**REPRODUCTION AND THE SURVIVAL
    PRINCIPLE** 884
ASEXUAL REPRODUCTION 885
New Individuals from Old 885
Development from the Egg: Parthenogenesis 886
SEXUAL REPRODUCTION IN ANIMALS 886

SEXUAL REPRODUCTION IN
 INVERTEBRATES 887
External Fertilization in Invertebrates 887
Internal Fertilization in Invertebrates 887
SEXUAL REPRODUCTION IN VERTEBRATES 889
External Fertilization in Vertebrates 889
Internal Fertilization in Vertebrates 890
HUMAN REPRODUCTION AND SEXUALITY 894
The Male Reproductive System 894
The Female Reproductive System 896
Hormonal Control of Human Reproduction 898
CONCEPTION CONTROL 903
Hormonal Control 903
Implantation Barriers 903
Sperm Barriers 905
Chemical Agents 905
Unaugmented Methods 905
Rhythm 905
Surgical Intervention 906
KEY IDEAS 908
APPLICATION OF IDEAS 909
REVIEW QUESTIONS 909

Essay 41.1 Human Sexuality: The Sexual Response 906

CHAPTER 42
Animal Development 910

GAMETES 911
The Sperm 911
The Egg 911
FERTILIZATION 912
Fertilization in Echinoderms 913
Fertilization in Mammals 914
EARLY DEVELOPMENT EVENTS 915
Becoming Multicellular: The First Cleavages 916
The Insect: A Very Different Pattern 916
The Blastula 916
Gastrulation: Organizing the Germ Tissues 917
THE EMBRYO TAKES FORM:
 ORGANOGENESIS 920
Neurulation 921
Further Vertebrate Development 921
VERTEBRATE SUPPORTING STRUCTURES 921
The Support Systems of Reptiles and Birds 922
The Support System of Placental Mammals 923
DEVELOPMENT IN THE HUMAN 925
The First Trimester 925
The Second and Third Trimesters 928

Birth 928
AN ANALYSIS OF DEVELOPMENT 929
Determination in the Egg and Zygote 930
Caenorhabditis elegans: A Model of Early
 Determination 931
Tissue Interaction in Vertebrates 931
The Role of Cell Migration 933
A Developmental Program in the Insect 934
A Developmental Program in the Vertebrate Wing 935
Summing Up 938
KEY IDEAS 939
REVIEW QUESTIONS 941

Essay 42.1 The Emergence of Human Life 926

PART SEVEN

Animal Behavior and Ecology 943

CHAPTER 43
The Development and Structure of
Animal Behavior 944

THE DEVELOPMENT OF BEHAVIOR 945
Genes and Behavior 945
Hormones and Behavior 947
The Three Ways Hormones Can Influence Behavior 948
THE ETHOLOGIST'S CONCEPT OF INSTINCT ... 949
Sign Stimuli 949
Innate Release Mechanisms 950
Fixed Action Patterns 950
LEARNING 950
Types of Learning 950

HOW INSTINCT AND LEARNING CAN INTERACT 952
LEARNING AND MEMORY 953
Theories on Information Storage 954
KEY IDEAS 956
APPLICATION OF IDEAS 957
REVIEW QUESTIONS 957

Essay 43.1 Lest Ye Forget *954*

CHAPTER 44
Adaptiveness of Behavior 958
PROXIMATE AND ULTIMATE CAUSATION 959
BEHAVIORAL ECOLOGY 959
Habitat Selection 960
Foraging Behavior 961
BIOLOGICAL CLOCKS 963
The Adaptiveness of Rhythms 963
The Range of Rhythms 963
ORIENTATION AND NAVIGATION 964
Homing Pigeons 965
Migration 966
THE SOCIAL BASIS OF COMMUNICATION 967
Visual Communication 968
Sound Communication 968
Chemical Communication 969
Tactile Communication 970
SOCIAL BEHAVIOR 971
Agonistic Behavior 971
Fighting, a Form of Aggression 971
Cooperation 973
Symbiosis 974
ALTRUISM 975
Humans and Reciprocal Altruism 976
SOCIOBIOLOGY 977
KEY IDEAS 978
APPLICATION OF IDEAS 979
REVIEW QUESTIONS 979

CHAPTER 45
Individuals and Populations 980
THE ECOLOGICAL HIERARCHY 981
Basic Versus Applied Ecology 981
THE ECOLOGY OF INDIVIDUALS 982
Environmental Factors 982
Resources and Regulators 982
Limiting Resources 983

Tolerance Ranges and Optima 984
Environment and Scale 984
Habitat and the Ecological Niche 985
The Principle of Competitive Exclusion 985
Indicator Species 986
Habitat Types 986
IMPORTANT ENVIRONMENTAL FACTORS 987
Temperature 987
Water 987
Light 988
Soil 989
THE ECOLOGY OF POPULATIONS 989
Dispersion Patterns 990
Population Changes in Time 991
The Difference Equation 991
Intrinsic Rate of Population Growth 992
Population Growth Models 992
Exponential Growth 993
Environmental Resistance and Logistic Growth 993
Demography 995
THE EVOLUTION OF REPRODUCTIVE STRATEGIES 998
The Theory of r and K Selection 1000
POPULATION-REGULATING MECHANISMS 1002
Density-Dependent Factors 1002
Density-Independent Factors 1003
KEY IDEAS 1004
APPLICATION OF IDEAS 1005
REVIEW QUESTIONS 1005

CHAPTER 46
Communities, Ecosystems, and Landscapes 1006
ATTRIBUTES OF COMMUNITIES 1007
Species Composition 1008
Strata 1008
Frequency and Distribution 1009
Diversity 1009
Stability 1010
IMPORTANT COMMUNITY-LEVEL PROCESSES 1010
The Roles of Competition and Niche in Community Structure 1010
The Role of Territoriality in Community Structure 1011
The Roles of Disease and Parasitism in Community Structure 1012
The Role of Predation in Community Structure 1013

The Role of Disturbance in Community Structure   1014
COMMUNITY DEVELOPMENT OVER TIME:
  ECOLOGICAL SUCCESSION   **1014**
Primary Succession   1015
Secondary Succession   1015
Succession in Aquatic Communities   1016
Three Alternative Models of Succession   1016
Mature Communities   1016
COMMUNITIES TO ECOSYSTEMS   **1017**
Components of Ecosystem: Trophic Levels   1018
Food Chains and Food Webs   1018
Ecological Pyramids   1019
Humans and Trophic Levels   1020
Energy and Productivity   1022
NUTRIENT CYCLING IN ECOSYSTEMS   **1024**
The Nitrogen Cycle   1025
From Producer to Consumer to Decomposer   1025
The Nitrogen-Fixers   1025
The Phosphorus and Calcium Cycles   1027
The Carbon Cycle   1028
KEY IDEAS   **1029**
APPLICATIONS OF IDEAS   **1031**
REVIEW QUESTIONS   **1031**

CHAPTER 47
Biosphere and Biomes   **1032**
THE BIOSPHERE   **1033**
Physical Characteristics of the Biosphere   1033
THE DISTRIBUTION OF LIFE; TERRESTRIAL
  ENVIRONMENT   **1037**
THE BIOMES   **1039**
Deserts   1039
Grasslands   1042
Tropical Savannas   1043
Tundra   1043
Tropical Rain Forests   1045
Chaparral   1046
Temperate Deciduous Forests   1047
Taiga   1049
WATER COMMUNITIES   **1051**
The Freshwater Province   1052
Wetlands   1054
The Marine Province   1055
KEY IDEAS   **1059**

APPLICATION OF IDEAS   **1061**
REVIEW QUESTIONS   **1061**

Essay 47.1 The Changing Carbon Cycle and the Greenhouse
  Effect: A Destabilized Equilibrium   *1035*
Essay 47.2 The Destruction of Tropical Rain Forests   *1047*
Essay 47.3 An Unusual Community: The Galápagos Rift   *1057*

CHAPTER 48
The Human Impact   **1062**
THE HISTORY OF THE HUMAN
  POPULATION   **1064**
The First Population Surge   1064
The Second Population Surge   1065
The Third Population Surge   1065
THE HUMAN POPULATION TODAY   **1066**
Growth in the Developing Regions   1066
Demographic Transition   1067
THE FUTURE OF THE HUMAN POPULATION   **1067**
Population Structure   1069
Growth Predictions and the Earth's Carrying
  Capacity   1069
KEY IDEAS   **1072**
APPLICATION OF IDEAS   **1072**
REVIEW QUESTIONS   **1073**

Essay 48.1 What Have They Done to the Rain?   *1070*
Essay 48.2 Holes in the Sky   *1071*

APPENDIX A
Classification of Organisms   **A-1**

APPENDIX B
Answers to Selected Genetics Problems   **A-7**

SUGGESTED READINGS   **R-1**

GLOSSARY   **G-1**

ACKNOWLEDGMENTS   **C-1**

INDEX   **I-1**

Biology

THE SCIENCE OF LIFE   *Fourth Edition*

CHAPTER 1

Mr. Darwin and the Meaning of Life

CHAPTER 2

Chemistry of Small Molecules

CHAPTER 3

The Molecules of Life

CHAPTER 4

Cell Structure

CHAPTER 5

The Plasma Membrane and Cell Transport

CHAPTER 6

Energy and Chemical Activity in the Cell

CHAPTER 7

Photosynthesis

CHAPTER 8

Cell Respiration

PART ONE

Molecules to Cells

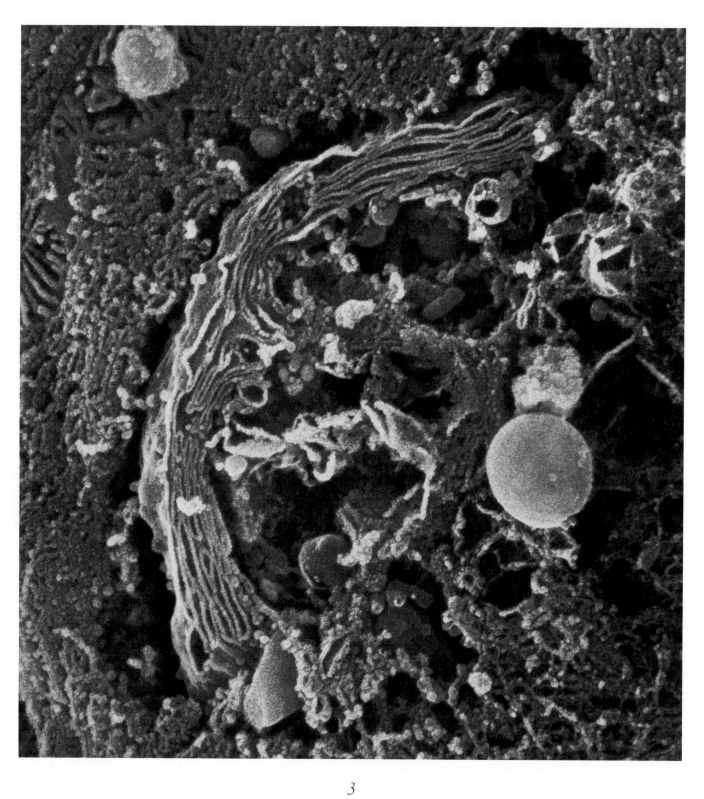

CHAPTER 1

Mr. Darwin and the Meaning of Life

CHAPTER OUTLINE

WHERE ARE THE RABBITS?

An Explanation About Rabbits and Oceans

Essay 1.1 The Enchanted Isles

THE WORKINGS OF SCIENCE

Inductive Reasoning

Deductive Reasoning

Scientific Method

Hypothesis, Theory, and Law

EVOLUTION: THE THEORY DEVELOPS

The Theory of Natural Selection

Publication of *On the Origin of Species*

Testing Evolutionary Hypotheses

The Impact of Darwin

CHARACTERISTICS OF LIFE: THE ULTIMATE REDUCTION

SIGNS OF LIFE

BIOLOGY, THE REALM OF LIFE

KEY IDEAS

APPLICATION OF IDEAS

REVIEW QUESTIONS

Something was wrong. He couldn't put his finger on it, but he knew it just the same. The feeling had nagged at him before, but he had dismissed it. After all, it was hard to be troubled while standing on those grassy hills with the fresh winds of an exotic land brushing across his face.

The good ship *Beagle* lay at anchor off the coast of Argentina, and the young naturalist was 200 miles inland, glad to be ashore, crossing the Argentine grasslands on horseback. Life had not always been a joyous adventure for Charles Darwin (Figure 1.1). Robert, his father, and Erasmus, his grandfather, had been among the wealthiest and most famous physicians in England. Charles had been expected to follow in their footsteps and, at the age of 16, was sent off to medical school. But he found that he became ill at the sight of blood, and he nearly fainted upon witnessing his first operation. He saved himself that embarrassment only by rushing from the room. At one point Charles was mortified to be told by his father that "you care for nothing but shooting, dogs, and ratcatching, and you will be a disgrace to yourself and all your family."

So Charles tried again—this time law. But he had no aptitude for it and was soon shuttled into training for the clergy. He was duly enrolled in divinity school at Cambridge, where he promptly showed almost as little aptitude for divinity as for medicine. His curriculum included classics, which he loathed, and mathematics, which he couldn't understand. He once wrote a friend about his trouble with mathematics and said, "I stick fast in the mud at the bottom and there I shall remain." Still, his college experience had its pleasant aspects; he could keep up with his insect and rock collecting. At Cambridge he found a friend in one of his teachers, the Reverend John Henslow, who was a botanist as well as a clergyman. Often the two would take long walks in the countryside around Cambridge and discuss the natural history of the area. Darwin was even known by some as "the one who walks with Henslow."

When at last Charles surprised himself and his family by passing his final examinations at Christ's College, he came home shouting, "I'm through! I'm through!" No more school. He was ecstatic. He was now expected to enter the clergy, but he spent his first postgraduate summer happily "geologizing" around the English countryside, away from difficult decisions and away from his family. When Darwin returned home in the fall he found a letter from Henslow waiting. Henslow had been offered an appointment as naturalist on a British naval survey ship that was to sail around the world. (Actually, the captain desired a companion of his own social standing.) Mrs. Henslow, however, had become so disconsolate at the idea of her husband being gone so long that he had reluctantly refused the offer. He recommended that young Darwin go in his place. Darwin thought it was a great idea.

Unfortunately, Charles' father would have none of it. Darwin was disheartened but continued to press for permission. Finally his father told him, "If you can find any man of common sense who would advise you to go, I will give my consent." Young Darwin didn't think anyone would give such advice and was prepared to decline the offer once and for all when his uncle, Josiah Wedgwood (of pottery fame), said that he thought it would be a splendid thing for his favorite nephew. The two of them together persuaded Dr. Darwin to hold to his word.

On September 5, 1831, Charles was summoned to London to be interviewed by the captain of the *Beagle,* James Fitzroy. Darwin was only a little younger than the 26-year-old captain, who nevertheless had already distinguished himself as a seaman of remarkable abilities. There was an initial awkwardness between them. (Fitzroy thought that the shape of Darwin's nose indicated a weak character.) But soon Fitzroy's doubts about Darwin's nose evaporated, and Darwin was accepted as the *Beagle*'s naturalist. In fact, Charles shared quarters with the captain himself. There was no salary, and Darwin had to pay for his own room and board throughout the voyage—a voyage wracked with seasickness for the unfortunate Darwin.

Captain Fitzroy's public mission (Figure 1.2) was to chart the waters off South America, but his private mission—his personal passion—was to find evidence that would establish once and for all the literal truth of the biblical account of the creation of the world. For that he needed a naturalist, and this amiable young divinity student who was so hungry for adventure turned out to be just right for the job.

FIGURE 1.1
Charles Darwin (1809–1882)
Here Darwin is shown at about age 27, shortly after returning from the voyage of the *Beagle.*

Where Are the Rabbits?

Now, as Darwin gazed across the lush grassland of South America, he felt that there was something unusual about the place, something not quite right. What was it? Why was he troubled? There was certainly no apparent reason to feel uneasy. The warm breeze gently smoothed the unkept grass, the sky was clear and blue, and his confidence and energy were high. But something was nagging him. What was it? Then it came.

There were no rabbits.

No rabbits. The phrase could be engraved in the consciousness of Western civilization along with $E = mc^2$ and *E Pluribus Unum.* No rabbits; such an innocuous phrase, but in a sense, it signaled the beginnings of a revolution. Darwin was not thinking about revolution, though. He was thinking about rabbits. Where were they? Darwin knew rabbit country when he saw it, and this place was a rabbit heaven. There was grass for rabbits to eat and bushes to hide in and dirt to dig in—still, there were no rabbits.

Where Darwin expected to find rabbits, there were other, strange little animals eating grass, digging holes, and hiding in bushes. They had long legs like rabbits, and large ears, and did many things that rabbits did, such as making burrows, eating succulent vegetation, and serving as prey for a variety of

FIGURE 1.2
HMS *Beagle*
The *Beagle* was one of many British vessels whose primary mission was to chart the oceans and collect oceanographic and biological information. The vessel, just 238 tons in draft, was guided through a five-year voyage around the world by James Fitzroy, her young captain. Much of the time was spent in South America, where Charles Darwin, the ship's naturalist, made extensive observations and collections.

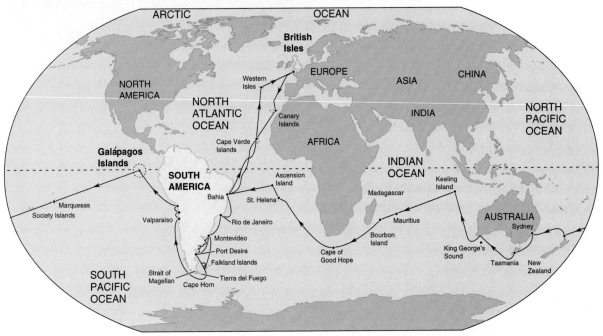

predators, yet they were clearly not rabbits. They looked more like guinea pigs but, in reality, they were Patagonian hares or mara (Figure 1.3). But where were the rabbits?

On one level Darwin knew perfectly well why there were no rabbits. There were no rabbits on the Argentine pampas because he was in South America, and rabbits don't live there. That kind of answer usually satisfies almost everyone; however, it is really no answer at all. If the question *Why aren't there any rabbits in South America?* had been pressed, another naturalist might have answered that South America was really not the proper place for rabbits, that the land couldn't support them. But Darwin thought the land *could* support them. He continued to mull over the question and at last a partial answer formed in his mind.

Perhaps there are no rabbits in South America, he thought, *because rabbits can't swim across the Atlantic Ocean.*

That question and its apparently simple answer were to change our perception of the world forever.

An Explanation About Rabbits and Oceans

If Darwin's tentative answer doesn't seem earthshaking, the first thing to consider is how the mara got to the Argentine grasslands. If it didn't swim there, its ancestors might well have originated there, or near there.

Darwin also noticed that the rabbitlike rodents of the pampas were very similar to other South American rodents, such as guinea pigs. Why? The conventional wisdom was that

South American rodents were similar because that general form of life was suited to (or designed for) life in South America. In Darwin's mind, a newer, different answer began to take shape: perhaps, he mused, South American rodents were similar to each other because they were related.

The question of relatedness continued to vex Darwin, especially any relatedness between living species and extinct forms. Darwin had dug up and reconstructed the bones of several extinct mammals, including a gigantic armadillo and even giant ground sloths. The extinct giants were clearly similar to—in fact, Darwin had to say, were apparently *related* to—the small burrowing armadillos and the tree-dwelling sloths that were around when he was there. So he had bones of animals that no longer existed and living animals that were so similar that they seemed to be related. Related? Did these creatures share some distant ancestor? Darwin focused mightily on the question of relatedness, but he did it alone, with only his books.

The Time–Life Connection. Darwin had taken with him on the *Beagle* a copy of the newly published first volume of his friend Charles Lyell's revolutionary book, *Principles of Geology.* By the time he had crossed the Atlantic, Darwin was a convert to the new geology. While in South America he received the second volume. Lyell had some rather startling things to say about the physical evolution of the earth. Based on his observations, he concluded that the world was much older than anyone had imagined; that over long periods of time, continents and mountains rose slowly out of the sea; and that they just as slowly subsided again or were washed away. Most importantly, Lyell claimed that the very forces that had so changed the earth in the past were still at work and that the world was still changing.

Darwin's own observations of South American geology seemed to confirm Lyell's position at every hand. In his adventurous climbing of the Andes, he had found fossil clam shells at 10,000 feet. Below them, near an ancient seashore at 8000 feet, he found a petrified pine forest that had clearly once lain beneath the sea because it, too, was interspersed with seashells. In fact, the *Beagle* had arrived in Peru just after a strong local earthquake had destroyed several cities, in some places *raising the ground level by 2 feet.* The earth had changed and clearly was still changing.

Darwin was excited by his developing idea, but he kept the most revolutionary of his thoughts to himself because he was sometimes uneasy with his ideas and often full of doubts. After all, he had studied for the ministry and had believed in the literal truth of the Bible, but the evidence seemed to contradict the creation account in every detail.

Captain Fitzroy had no such doubts. To him the bones of extinct mammals merely proved the account of the Flood, if one simply allowed that perhaps Noah hadn't been able to round up all the animals. If there were no rabbits in South America, it was because rabbits did not belong in South America. There was a very good reason for everything. Fitzroy believed in laws and rules, and furthermore he knew

FIGURE 1.3
The Patagonian Hare
Although rabbits are not native to South America, their ecological parallel is the Patagonian hare or mara (*Dolichotis patagonum*). Although the two animals are similar, the mara is a rodent and not at all related to rabbits.

what the laws and rules were. Darwin, on the other hand, was blessed (or cursed) with an ever-inquiring mind and was always ready to consider an alternative hypothesis. The hypothesis that was forming in his mind now was a dangerous one, and Darwin knew it could lead to trouble.

The Enchanted Isles. Darwin's ideas were to be buttressed by his observations during one momentous part of his five-year trip. The *Beagle,* on a dead run from the west coast of South America, had reached a peculiar little group of islands, and its anchor clattered into the quiet lee waters of an apparently insignificant island the English called Chatham. Chatham was one of the Galápagos Islands, a recently formed group of volcanic islands that lie astride the equator some 600 miles off the coast of Ecuador. His first view of the islands revealed a scene quite unlike anything Darwin had seen on the South American mainland: black, bleak, dry, and hot. And isolated—there were relatively few species of plants and animals. In fact, there were no mammals other than those brought by European ships, and only a few species of birds other than sea birds (Essay 1.1).

The Galápagos archipelago (group of islands) was, as Darwin would write later, a little world unto itself. Yet Darwin noticed that there was something familiar about the plants and

Essay 1.1 ● THE ENCHANTED ISLES

The Galápagos archipelago includes habitats ranging from dry, lowland deserts to wet, species-rich highlands. The islands are home to a bizarre collection of life, from grazing lizards and giant tortoises to a variety of bird life and shore dwellers. Darwin, who despised the place, was to make it famous because he saw it as an experiment of nature in progress. These are some of the places he went and some of the species he saw as he pieced together what he came to call the Great Puzzle.

(right) Opuntia Cactus. The plant is a favorite of the Galápagos tortoise

(above) Evidence of volcanic activity and the volcanic origin of the Galápagos is seen everywhere in the island

(above left) Galápagos tortoise. The species vary according to the island on which they are found.

(above right) The masked booby, largest of its kind, and a relative of the odd looking red and blue footed boobies, ranges worldwide through tropical regions.

(left) Marine iguana. They graze beneath the sea.

animals. Though most of the species were unique to the islands, Darwin had seen species like them only recently. They were similar to South American species and, as he was to learn, unlike those of Europe, Asia, North America, or Australia. Why should this be?

The usual explanation just wouldn't do. There was nothing in the physical environment to suggest that the islands were somehow appropriate for South American creatures. Indeed the environment was almost identical to that of the Cape Verde Islands, volcanic islands where the *Beagle* had tied up for nearly a month early in its voyage. The Cape Verde Islands, however, lay off the coast of Africa, and the species there were typically African, different from the species he found on the Galápagos archipelago.

Darwin reasoned that it would be difficult, if not impossible, for Asian, Australian, and North American species to cross the Pacific Ocean to settle on these volcanic islands. But perhaps from time to time, a few drifting seeds, a few reptiles on floating logs, and a pair of birds blown off course might have traveled across the 600 miles from the South American mainland. Darwin wondered if, finding a hospitable island free from competition, they could have survived and increased in number.

Darwin's attention focused on a variety of little finches fluttering about there. The finches were dark and drab and notably unspectacular. Nevertheless, Darwin collected a number of them to be examined later by specialists in England. Even while on the Galápagos, Darwin had already noticed two things about them: they were all rather similar to species on the South American mainland, and they differed from each other in many critical ways, such as bill size and foraging behavior (Figure 1.4). As Darwin worked out the details of his theory after arriving home, he wondered if the island species could have been somehow modified from an ancestral mainland stock. This idea did not fit with the prevailing idea of how species arose, but it was a cornerstone for Darwin's developing theory. We will come back to the Galápagos and the importance of the finches in Chapter 18.

The Rumblings of Revolution. By now Darwin and Fitzroy were engaging in lively, if not heated, discussions about the nature and origins of life. Fitzroy would probably have been chagrined to think he was *helping* Darwin form his "heretical" ideas by providing a sounding board. Like most of his contemporaries, Fitzroy was convinced that species did not change over time. In his mind, species were as they had always been, ever since the Creator placed them here. Darwin probably was hesitant to reveal his true thoughts, for he was beginning to think that life does, in fact, change—that living species are modifications of earlier and different species. Other people, including Darwin's own grandfather, had said the same thing. But where their musings had been unsupported by hard information, Darwin was slowly gathering evidence. (*Species,* as we will see later, refers to any of the millions of unique kinds of organisms, such as humans, red

FIGURE 1.4
The Galápagos Finches
When Darwin collected his finch specimens, he had no idea of their importance. These species, it was discovered later, all sprang from the same ancestral stock. Their bill sizes and shapes reflect their distinct modes of gathering food.

maples, and rainbow trout. Members of a species are similar enough to interbreed successfully and produce healthy young—see Chapter 20.)

Darwin saw evidence of species changing over time, that is, of *evolution,* everywhere. Could his observations prove that evolution is a fact of life on the planet? How are such facts established? For the answer we turn to the inner workings of science.

The Workings of Science

In its broadest sense science might be defined as the way one gets at the truth. Even in our freewheeling world, there are rules for getting at the truth, and some methods are held in greater esteem than others. If we witness a volcano rising from the seabed, we can learn something about how volcanoes are formed. This then is a part of how science is done. But simple observation is not enough. Today the rules for getting ideas accepted by the scientific community are rather formalized. Usually the process involves one or both of two methods, called *inductive reasoning* and *deductive reasoning.*

Inductive Reasoning

Inductive reasoning involves reaching a conclusion based on a number of observations. It moves from the specific to the general. In other words, it begins with observations of specific events, and these eventually lead to the formulation of a general statement about what the observations mean.

One of the questions of Darwin's time was: How are coral atolls formed? They often appear as broken circles of islands, formed by the hardened bodies of countless coral animals that make up immense colonies. No one knew just how they came to be, and how the broken circles were formed. The two approaches to the problem involved inductive and deductive reasoning. The inductive approach involved a group of statements (observations) that could best be explained by a larger, general statement. Table 1.1 describes the two approaches to this intriguing question.

Deductive Reasoning

Unlike the inductive process, **deductive reasoning** involves drawing specific conclusions from some larger assumption. It moves from the general to the specific, that is, from a broad idea to one or more specific statements. Deductive reasoning leads logically to predictions, which are often described as a form of "if . . . then" reasoning. In Table 1.1 we see that the deductive process begins with a general statement about atolls and, from this, more specific statements are drawn. As you can see, the process is essentially the opposite of inductive reasoning. It is important to understand that neither of these

processes, in reality, must exclude the other. The human mind works in complex ways, and in problem-solving, no matter how hard you work to embrace one philosophy, the result is likely to be a product of both.

Scientific Method

The scientific method has been explained in many ways, yet it always eludes precise definition. Essentially, it is the process of establishing new facts and of understanding mechanisms. Although there is no set of directions for accomplishing these things, the process usually begins and ends with observations about the real world.

There are the initial observations, which are presumably not understood. Darwin, for example, observed that the land birds of the Galápagos are unique species with strong physical similarities to those of mainland South America. But what did it mean?

What happens next is the least documented aspect of science, and perhaps the most important. It begins with mulling over observations and wondering. Almost no one writes down very much at this stage, which may be why it is so little understood. Perhaps most scientists are wondering and mulling over facts most of the time, and most of the time nothing comes of it. Sometimes, though, wondering leads to speculation, and speculation at its best can take the form of something called a hypothesis.

Hypothesis, Theory, and Law

At some stage a scientist with a question must formulate a **hypothesis** to explain the facts. A hypothesis is a possible explanation of some observation. It may have very little evidence to support it, but it usually suggests ways to get at that evidence. The hypothesis represents but one level of scientific activity. At a somewhat loftier level we find the **theory**, which is usually considered to be a stronger explanation of some observation—one that is supported by a considerable amount of evidence.

As more and more evidence comes in that is consistent with a given theory, the explanation may become virtually (but not quite) irrefutable and is called a **law**. Laws enjoy more confidence than theories and, in fact, seem to withstand any kind of testing we can invent. Examples are the law of gravity and the laws of thermodynamics, both universal statements from physics. Biology is notoriously short on laws because life is by nature shifty, elusive, and hard to define. Although biology has its "laws," such as Mendel's laws (Chapter 10) and the Hardy–Weinberg law (Chapter 19), biological statements wear their titles provisionally and uneasily. Most biological laws are based on mathematical descriptions rather than simple observation.

Hypothesis and Experiment. A hypothesis can be used to make **predictions** (statements regarding expected events),

---

TABLE 1.1
An Example Of Inductive And Deductive Reasoning

Inductive Reasoning

Observation:	Coral atolls usually consist of a circle of islands.
Observation:	Coral atolls form from the deposits of living animals.
Observation:	Coral animals without direct access to fresh seawater tend to die.
Observation:	The interior of an atoll seems to consist of sunken coral.
Generality:	Coral atolls are formed as coral animals secrete deposits. The animals in the center, lacking nutrient-laden water, die and sink. This leaves a ring that, in turn, breaks apart to form a circle of islands.

Deductive Reasoning

Generality:	Coral atolls form as coral animals secrete deposits. The animals in the center, lacking nutrient-laden water, die and sink. This leaves a ring that, in turn, breaks apart to form a circle of islands.
Deduction:	Coral atolls will have a sunken center.
Deduction:	Coral animals need contact with fresh seawater.
Deduction:	Seawater contains something that coral animals need.
Deduction:	Coral atolls comprised of more nearly complete rings of land are probably more recently formed than those comprised of more broken rings.

which can then be tested. The hypothesis can arrive fully formed, or in very rough outline so that modifications and refinements can be made to fit newer observations.

Testing is usually done through additional observations; such observations may take place as organized *experiments,* procedures designed to test hypotheses. As an example, assume that an investigator has observed certain abnormalities in the growth of bird embryos in an area. He or she suspects that a certain herbicide (plant-killing chemical) known to have been used to control plant growth in the area is responsible. It can never be proved that the herbicide was responsible for past events, but if the idea is tested, the results may support the hypothesis strongly enough to convince people to stop using the herbicide. At any rate, the hypothesis becomes: "Herbicide X can cause abnormalities in bird embryos."

The hypothesis immediately suggests the prediction: "If I administer herbicide X to bird embryos, I should be able to observe abnormalities as they grow." The prediction, as you can see, determines the next observation, which in this case will be an experiment. In the experiment, the investigator deliberately sets up carefully specified conditions under which certain observations or results can be expected, according to the prediction being tested. In this case, herbicide X will be administered to bird embryos and the effects observed.

Controls. The experiment must be set up in such a way that there can be only one explanation for the observations to come; frequently this involves using controls. A **control** is a replica of the experiment in which the special factor being studied is omitted. In **controlled experiments**, the experimental subjects are arranged in two groups, often labeled "control" and "experimental." Both groups are treated identically in every way except that the special condition under consideration is *not* applied to the control group. The special condition (herbicide X in our example) is often referred to as the **variable**, since it represents a variation from the usual, or normal, condition. The control group may be given an inactive substitute for the variable, called a **placebo**. Where a chemical is being administered, the placebo may be any inert substance. Where drugs are being tested on humans, the placebo often turns out to be an innocuous sugar pill. Incidentally, the number of subjects comprising each group is quite important. The greater the number of subjects, the more confidence the experimenter may feel in the results.

The control group, then, represents a standard to which the treated group is compared. Since they are treated identically in every way except for the single variable, any difference in the results can, with confidence, be attributed to that variable. Figure 1.5 illustrates some of the potential problems

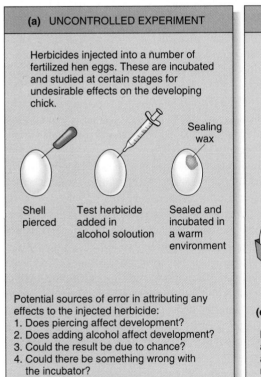

(a) UNCONTROLLED EXPERIMENT

Herbicides injected into a number of fertilized hen eggs. These are incubated and studied at certain stages for undesirable effects on the developing chick.

Shell pierced

Test herbicide added in alcohol soloution

Sealing wax

Sealed and incubated in a warm environment

Potential sources of error in attributing any effects to the injected herbicide:
1. Does piercing affect development?
2. Does adding alcohol affect development?
3. Could the result be due to chance?
4. Could there be something wrong with the incubator?

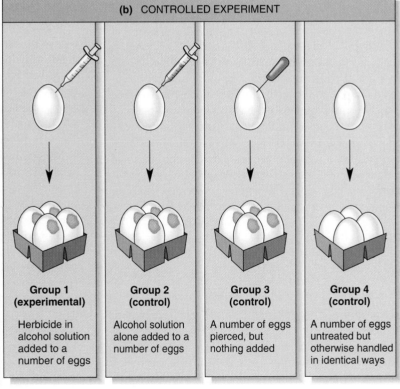

(b) CONTROLLED EXPERIMENT

Group 1 (experimental)

Herbicide in alcohol solution added to a number of eggs

Group 2 (control)

Alcohol solution alone added to a number of eggs

Group 3 (control)

A number of eggs pierced, but nothing added

Group 4 (control)

A number of eggs untreated but otherwise handled in identical ways

FIGURE 1.5
A Controlled Experiment
Both experiments are intended to determine the effect, if any, of a certain herbicide on bird development. The first (**a**), an uncontrolled experiment, involves a number of eggs pierced, treated with the herbicide, and sealed as shown. Note the list of potential errors. The second experiment (**b**) utilizes three groups that are treated differently and one that is untreated. Suggest reasons for the use of each group and explain how each helps to answer the criticisms listed in (**a**).

in our example and how they are addressed through the use of a controlled experiment.

Conclusions and Beyond. The results of the experiment may or may not support the investigator's predictions. The next step will simply be to accept or reject the hypothesis accordingly. This step, in more formal language, is the investigator's **conclusion**, a statement arrived at through the experiment. The conclusion can be tentative or firm, depending on the investigator's confidence in the strength of the evidence.

It is important to note that not all experiments are performed in laboratories. There have been many excellent field experiments—that is, those done under natural conditions. An example is a mark-and-recapture experiment, in which animals are captured, marked, released, and later recaptured, thus providing information about such things as their movement, growth, or longevity. Hypotheses can also be tested without experiments at all. An ecologist might predict that a certain relationship will be found in nature and then test the prediction by going out into the field to look for it. Whether in laboratory or field, whichever method is selected, it is necessary to first make the prediction and then test it, either by experimentation or observations.

Finally, sharing of results through publication is an important part of the scientific method. Keeping a discovery to yourself is not in keeping with the goals of modern science, which include sharing information.

Research is rarely as precise and straightforward as it would appear from reading a published article. For example, we know little about the phase that involves mulling, dreaming, or inspiration. We also know very little about the scientist's own motivation, his or her need to know and tell. The very best scientists are terribly curious, of course, but they also tend to be driving, egocentric, proud enthusiasts who may covet fame, or titles and prizes, or the adulation of others. Some are also of a romantic nature, searching for order and beauty in the natural world. There is also the driving force of the individual's very human need for the acceptance, approval, and admiration of peers. And Darwin was no exception.

Evolution: The Theory Develops

Darwin too strove for acceptance, approval, and recognition. Near the conclusion of its voyage, when the *Beagle* arrived at Ascension Island in the South Atlantic Ocean, Darwin found a letter from his sisters waiting for him. In it they mentioned that Adam Sedgwick, an English scientist, had told Darwin's father that on the strength of the letters and specimens that Henslow had already received from the *Beagle,* his son Charles would take a place among the leading scientists. Darwin wrote, "After reading this letter I clambered over the mountains of Ascension with a bounding step and made the volcanic rocks resound under my geological hammer!"

After the exciting discoveries in Galápagos, the rest of the voyage was somewhat anticlimactic. Fitzroy's official task was done, so the *Beagle* made for home as quickly as possible, which in this case meant continuing westward around the world. It called at Tahiti, New Zealand, Tasmania, Australia, Mauritius, South Africa, St. Helena, Ascension, and Brazil.

By the time the *Beagle* returned to English waters, it was late in 1836 and Darwin was 27 years old. When the boat tied up, a grateful Darwin leaped ashore and immediately took a carriage home. Since he arrived at a late hour, he decided not to rouse his family and took a room at an inn nearby. The next morning he received a joyous greeting from his family, but he was especially delighted when his dog greeted him and immediately set off down the trail on which they had last enjoyed their morning walk five years before.

Darwin did indeed take his place among the leading scientists of his day. His observations on geology and zoology were published, and he was revealed as a keen observer of nature, undoubtedly one of the greatest natural historians of all time. He was a good storyteller as well. His journal, *The Voyage of the Beagle,* became successful popular reading in England and remains a classic today. Of course, he is better known for other writings, particularly those that would seek to explain just why life changes on a changing planet.

The Theory of Natural Selection

Beginning in 1837, Darwin started to keep a journal entitled *Transmutation of Species* and was soon making entries referring to "my theory." Reading that journal today, and the journal of the voyage of the *Beagle,* is like reading a detective story after you already know whodunit—the suspense is in watching the detective sift through clues for the right answer. In his journal entries Darwin began to toy with the idea of *natural selection.* It is fascinating to watch the idea develop in his writings.

Keep in mind that Darwin was a country boy. He knew a lot about agricultural practices, and he knew about livestock breeding. Any good farmer knows that a breed can be improved by *selection,* or, in biological terms, **artificial selection**—selecting the best individuals of each generation for breeding (Figure 1.6). By longstanding folk tradition, the best of the breed (whether cattle, fowl, dog, or cucumber) were honored annually at country fairs and chosen for propagation. The reasoning was simple: like begets like, offspring tend to resemble their parents, and the "best" parents produce the "best" offspring.

But Darwin wondered, could selection operate *without* human intervention? And, if so, how? Agricultural selection involved the conscious choice of the breeder. If selection occurred in nature, who was the selector? This line of thought seemed at first to lead right back to a supernatural factor. Without a conscious selector, it seemed, the inferior individuals were as free to breed as were the most superior, in which case no improvement or change or adaptation would occur.

Did selection have to be conscious then? Perhaps not. Darwin was impressed with an essay on population that had

FIGURE 1.6
Artificial Selection
The results of selective breeding indicate that artificial selection is a powerful mechanism in changing living things, as we see here.

been written by Reverend Thomas Malthus three decades earlier. Malthus stressed that all species had enormous reproductive capabilities and that their numbers tended to expand rapidly geometrically (2, 4, 8, 16, 32, and so on) unless held in check by starvation or disease. Natural populations, he argued, reached a balance in which all but a few of the young of each generation were forced to perish. *All but a few. The environment,* Darwin thought, *could select the individuals that were allowed to breed, and these would be the hardiest. And it all happened through natural means.*

Darwin's journal shows that by 1838 he had solved the major riddle of evolution. The mechanism of evolution, he said, was **natural selection**, the process by which nature determines which individuals survive and reproduce. Briefly, natural selection includes (1) overproduction of offspring, (2) natural variation within a population, (3) limited resources and the struggle for survival, and (4) selection by the environment for those with traits that enable the individual to survive and reproduce.

In other words, Darwin noted that living things do not leave as many offspring as they otherwise might because many are killed or their reproduction is curtailed by natural forces. The environment, both the living and nonliving surroundings, determines (selects) those individuals that are to survive and reproduce. The traits that permit survival will then increase in the population as other, less advantageous, traits are weeded out.

Publication of On the Origin of Species

Recall that Darwin had seen fossil remains of extinct species of ground sloths and armadillos on the east coast of South America, and had wondered if they could be related to living species. He also had noted the similarities of South American rheas and African ostriches (Figure 1.7), and he wondered if they could be related. As Darwin was mulling these ideas over in his mind, he read Malthus' essay, and the grand idea began to come together for him. His theory of evolution by natural selection was developed slowly and methodically, even as he routinely tested each premise in rigorous conversations with a few close scientific allies. He continued gathering his evidence for some 20 years after his return from the voyage of the *Beagle,* yet he made no formal statement, publishing nothing on the topic.

Finally, in 1856, Charles Lyell and the botanist Joseph Hooker persuaded Darwin to set out his argument. Darwin began, almost reluctantly. Keep in mind that Darwin was not shy on matters relating to his science. He was already acknowledged as one of the leading scientists of his day. But for some reason, he was dragging his feet in publishing this theory. In fact, two years later, in 1858, Darwin had written only ten chapters of a work that was to include several volumes.

A half a world away, though, things were moving at a different pace. A young biologist named Alfred Russel Wallace was on an expedition to investigate life in the Malay archipelago. He too was wondering about the relationship between natural forces and the changing panorama of life. He too had read Malthus' essay and wondered how it all fit together. Then, while in bed with fever one night, it all came to him in a burst of insight. He saw that the "fittest would continue the race" after nature had weeded out the rest from among a variable population.

Wallace had corresponded with Darwin before on other matters. So he sent his notes to Darwin, and a shocked Darwin became almost despondent. He was ready to let Wallace accept all the credit for the idea until Lyell and Hooker took it upon themselves to present both papers at a scientific meeting just a few weeks later. Darwin's paper was presented first, in keeping with his much more substantial evidence, but the whole topic received very little attention that night. A year later, however, in 1859, the world was alerted to the idea, and when Darwin's *On the Origin of Species* was published, all 1250 copies sold out the first day.

Testing Evolutionary Hypotheses

Even the brilliant scientist and debater Thomas Huxley, Darwin's most pugnacious defender in the 19th century, felt that the idea of evolution would remain untested and unproven until someone directly observed the experimental creation of a species. However, Darwin did not believe such a test was possible. He thought that such events simply take too much time. One of the key ingredients in his theory is time—a great deal of time.

Keep in mind that evolution and natural selection are not the same thing. Evolution—or change over time—is the process; natural selection is one means by which it can occur. Evolution can be observed, for example, in fossils, but natural selection is exceedingly difficult to observe or test experimentally. An experimental test of natural selection in moths, under more natural conditions, is described in Chapter 19.

A difficulty with the concept of natural selection is not that it predicts too little, but that it explains too much. As soon as it became apparent that natural selection explained the adaptations of organisms to their environment, natural selection began to be used in a lazy way to explain all kinds of biological phenomena. To explain any phenomenon, it began to seem that the speculating biologist need not prove anything

new, but needed merely to dream up some halfway plausible way in which the phenomenon might benefit the organism. And if the imagination failed in this, it seemed adequate to state that even if the benefit of the phenomenon to the organism was not obvious, surely there must be one. For instance, if the Indian rhinoceros has one horn and the African rhinoceros has two (Figure 1.8), it might be argued that there must be something about the two environments that makes this arrangement the best one for all concerned.

This extravagant faith in the power of natural selection leads to the benign view that everything is always for the best and there is a reason for everything. Such a viewpoint elevates natural selection to the status of a new, all-powerful deity, and such faith is contradictory to science. It is certainly a contrast to Darwin's own conviction of the role of chance and historicity in evolution. *Historicity* is the notion that things are as they are because of events that occurred in the past. Natural selection is a powerful phenomenon, but it is limited. The adaptations of organisms are marvelous, but they are never perfect, just as evolutionary change is always opportunistic and never predictable.

The Impact of Darwin

Of course, Darwin's ideas were immediately and bitterly controversial. And, unlike the once-controversial ideas of Newton and Einstein, his writings continue to resist resolution. Almost all biologists believe in the central notions of natural selection and evolution, but they still differ greatly among themselves on such questions as what constitutes a species, how species really change, why different species can't mate, and whether most evolutionary change comes in small, continuous incre-

FIGURE 1.7
Rhea and Ostrich Compared
The rhea (**a**), a South American bird, closely resembles Africa's ostrich (**b**). Such similarities led Darwin to wonder about relatedness.

FIGURE 1.8
Natural Selection—The Answer?
Why does the African rhinoceros have two horns and the Indian rhinoceros have only one? Remember that any explanation would have to be testable. What explanations come to mind? Can they be tested? Is it possible that there is no selective advantage of one horn over two? Or perhaps there is no *longer* a selective advantage. You won't find many biologists willing to say much about questions like this.

ments or in larger and less regular leaps. Through his works, Darwin remains an active participant in the debate.

Darwin went on to publish other major theoretical works and continued his simple but first-rate experimentation. For example, he discovered plant hormones, as we'll see in Chapter 31. *The Expression of the Emotions in Man and Animals* (1872) was the foundation of the modern sciences of ethology and comparative animal behavior (the subject of Chapter 43). *The Formation of Vegetable Mould Through the Action of Worms* (1881) established the importance of earthworms in soil development. Because of his work on orchids, climbing plants, and insectivorous plants, modern botanists claim Darwin as one of their own.

But it is *On the Origin of Species* and a related work, *The Descent of Man,* that Darwin's reputation is based, and it is the idea of natural selection that has become the central concept of the science of biology.

Characteristics of Life: The Ultimate Reduction

While Darwin was able to focus on such detailed processes as earthworm diggings and snail longevity, he was also a master at seeing the grand scale, the overall picture. His theory of evolution, in fact, could not have been crafted without this ability to generalize and deduce encompassing principles. Most of the scientific progress in biology in this century has not been achieved through such grand conceptual breakthroughs as the theory of evolution by means of natural selec-

tion, but through what can be called reductionism. **Reductionism** involves the assumption that the properties of biological systems can be understood in terms of physical laws, such as those concerning physics and chemistry. In reductionist science, the questions asked are small ones that can be stated and answered in specific, precisely defined terms. Cause and effect are determined, whenever possible, by eliminating all competing explanations until one is left. Finally, the reductionist seeks to find mechanisms, not reasons, for observed phenomena.

Those dealing with overview and the big picture are called **synthesists**. They seek underlying order in other ways. In general, they seek to show that seemingly unrelated observations can be related after all. Synthesists are the intellectual and emotional descendants of Darwin, clearly interested in forming grand rules and sweeping generalities. Of course, for science to work well, the reductionist and synthesist approaches should be integrated. The synthesist, after all, is able to generalize because he or she has available so much detailed data produced by the reductionist. Furthermore, the synthesist's vision can be validated only through precise experimentation by the reductionist.

As the reductionists take us spiraling inward to ask the most fundamental questions about life, they are finally confronted with the most basic question of all: What is life? How can we know the living from the nonliving? At many different levels, the latter is a simple question. It is easy to tell living things from nonliving—until the question falls into the hands of the reductionist. Then it can become most difficult, indeed.

In fact, there are disagreements over the characteristics that mark life, but there is general agreement about some rather constant features, as we will see next.

FIGURE 1.9
Living Things Need Energy
This carnivore is seeking to avail itself of the energy in its prey's body. That energy will then be used, in part, to keep the carnivore's body organized.

FIGURE 1.10
Living Things Respond
Some living things respond to environmental changes instantly, if in a limited manner, as do these tide pool animals. Other organisms, such as plants, may respond more slowly, but all living things react to changes in their environment.

Signs of Life

Life is notoriously difficult to define, but we find that it has certain properties that, taken as a whole, characterize life in general. Let's review some of those traits here.

First, *living things use energy to stay organized.* Life exists in a disruptive world where things tend to run down, decompose, and become disorganized without an input of energy. So living things have ways of availing themselves of energy. An example would be plants that convert the energy of sunlight into molecules that contain energy. The animals avail themselves of that energy by, for example, eating the plants or other animals who eat the plants (Figure 1.9).

Second, *living things respond* (Figure 1.10). The world is a variable and changeable place, so living things must have ways of detecting certain situations and responding to them. Touch a sea anemone and it will suddenly snap shut, drawing its delicate tentacles into its protective body.

Third, *living things adapt* (Figure 1.11). Adaptation involves responding so that some benefit is gained. Adaptation thus involves becoming better suited to the environment as time passes. Individuals can adapt, for example, through learning and species can adapt, for example, through natural selection.

Fourth, *living things reproduce* (Figure 1.12). Reproduction is the means by which individuals leave their genes in the next generation and each generation is largely composed of the descendants of the best reproducers of previous generations, so reproduction is a strong tendency among living things.

FIGURE 1.11
Living Things Adapt
Living things tend to adapt, to behave in a manner beneficial to themselves when new opportunities arise. These bald eagles, our national bird, are scrounging in a dump while their colleagues, not far away, are catching fish.

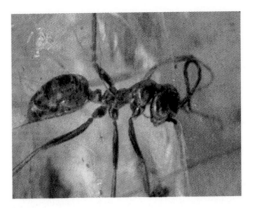

FIGURE 1.13
Living Things Evolve
The evolution of one form of life on the planet is encapsulated in this insect entrapped in amber, a fossil sap. This extinct form has traits found in some living species, indicating relatedness.

FIGURE 1.12
Living Things Reproduce
Living things reproduce in a variety of ways. Some care for their young and some don't, but if they are not successful, their genes will not be represented in the next population.

Fifth, *living things evolve* (Figure 1.13). Populations of living things must change over time to become better adapted to the world around them. Natural selection is one means of evolution.

We should not rely too strongly on any such list since there are always exceptions. Furthermore, many nonliving things possess at least one of these traits. We see, for example, that mineral crystals and oil droplets can grow as they merge, and that they can reproduce by spontaneously breaking up, that blowing sand is responsive to the wind, that rusting iron takes energy from the environment, and that many metals are highly organized. Thus our definitions of life are only inadequate guidelines as we simply admit that life exists and forge ahead. But keep in mind that the inadequacy is not just with our definition. The problem is fundamental. We really don't know

much about the basic nature of life (although you may think that seems a bit improbable, considering the size of this book).

Biology, the Realm of Life

We have seen the beginnings of modern biology through the life of the one who started it all, Charles Darwin. We have also seen some of the fundamental problems with the study of a science so evasive, changing, and influenced by philosophy. In a sense, because we are living things (if our readership is as we hope), we are confronted with the dilemma of an eye trying to see itself.

Yet we believe there are ways to proceed, ways to organize our thoughts and to take one step at a time through an examination of this thing called life. Essentially, we will first look at the basic stuff living things are made of, and how that material behaves in inheritance. Then we will take a close look at the principles of evolution before examining the vast array of species on the planet and the internal mechanisms that keep them alive. We end with a look at behavior and how all of this fits into the big picture of life on earth.

One great advantage you have as a human is your wealth of simple curiosity, a heritage that has historically sent us scattering in countless directions over the planet, probing, peering, counting, measuring, watching, and gathering as we learn more about this place called home. Sometimes we learn, it seems, just by being there. All the time, in one way or another, it seems our learning is about life itself. Our quest has been a grand celebration of inquisitiveness and, in our search, we have left our marks everywhere.

Key Ideas

WHERE ARE THE RABBITS?

1. Many of Darwin's ideas about evolution arose from observations made during his voyage in the *Beagle*.

2. Darwin's observations of life in South America started him wondering about the absence of some species (rabbits) where one would expect them (grasslands). He pondered the meaning of fossils that were different from, yet vaguely similar to, living species.

3. Charles Lyell's theories provided both a substantial time frame and far-reaching geological changes essential to Darwin's evolutionary scenario.

4. The isolated Galápagos Islands offered Darwin a veritable laboratory of animals that had originated elsewhere and had undergone change from the original stock.

THE WORKINGS OF SCIENCE

1. Scientific reasoning may take two approaches: inductive and deductive reasoning. Inductive reasoning proceeds from the specific to the general. Thus a number of observations lead to a larger, more encompassing statement. Deductive reasoning proceeds from the general to the specific. Thus a more encompassing statement leads to a number of deductions or predictions.

2. While the scientific method can be elusive and difficult to describe, certain elements are standard. It usually proceeds through observations, deduction, or induction.

   a. A hypothesis is a provisional or tentative explanation of some observed phenomenon. A theory is a larger generalization, often based on tested hypotheses but always on a larger body of evidence. A law carries even more weight; for most practical purposes, it is considered irrefutable.

   b. To be useful, a hypothesis must lead to predictions and be testable. The predictions suggest a test, which is often an experiment, or may simply call for more observations.

   c. Controlled experiments include both an experimental, or variable, group and a controlled group. The variable is the condition being tested for. The control becomes the standard of comparison and is used to prove that any difference between the two is due to the variable being tested and not to chance. The control group must be identical to the experimental group to avoid introducing confusing variables.

   d. The conclusion is the investigator's decision about the hypothesis. Based on the experimental or observed results alone, the hypothesis will be accepted or rejected.

EVOLUTION: THE THEORY DEVELOPS

1. Upon the completion of his voyage, Darwin was convinced of the idea of evolutionary change but lacked an explanatory mechanism.

2. A knowledge of livestock breeding and artificial selection suggested an evolutionary mechanism to Darwin. Malthus suggested that the environment was the agent of selection, and only the fittest individuals lived to reproduce. Darwin called selection by the environment *natural selection.*

3. Natural selection includes (a) overproduction of offspring, (b) natural variation within a population, (c) limited resources and the struggle for survival, and (d) selection by the environment for those with traits that enable the individual to survive and reproduce.

4. Evolutionary hypotheses can be tested by first making predictions and then, from the experiments or observations they suggest, accepting or rejecting the hypothesis. Natural selection is far more difficult to test.

5. Darwin's major achievement was his theory of evolution through natural selection, but he also developed many other important ideas in biology.

CHARACTERISTICS OF LIFE: THE ULTIMATE REDUCTION

1. Science includes two aspects, reductionism and synthesism. Reductionism involves breaking problems down and investigating the smaller elements. Synthesism consists of putting many smaller ideas together into new grand schemes.

SIGNS OF LIFE

1. Living things (a) use energy to stay organized, (b) respond to environmental conditions, (c) adapt to a changing environment, (d) reproduce, and (e) evolve.

Application of Ideas

1. In explaining scientific and other intellectual achievements, someone once said, "Chance favors the prepared mind." What does this mean to you, and how do you define "prepared"? As you answer, think of the background and experiences of Charles Darwin, and consider how both chance and preparation fit in with what he accomplished.

2. Set up a hypothetical problem and develop an organizational diagram that illustrates how scientific investigation might proceed. Decide whether to proceed inductively or deductively and use the word "control."

3. One important aspect of scientific progress is the rise of new technology. With new inventions and techniques, it is possible to test hypotheses that were heretofore untestable. Yet no technological innovation has replaced the human intellect. Consider the relationship between technology and intellect. Which advanced technology (say, microscopy, statistics, or protein analysis) might have been more important to Darwin?

4. An antievolutionist derides Darwinian evolution, saying that the notion of new species arising is "unscientific" since such an event has never been observed. What evidence would counter such an assertion? Are there other, perfectly acceptable, scientific mechanisms that have never been observed? Discuss one.

5. Are humans in modern society subject to natural selection or have we managed to thwart the process as it occurs in other organisms? How might we interfere with the process?

Review Questions

1. How did people in the 19th century perceive the species on the earth? What was their feeling about fossils that represented extinct life? (page 9)

2. In what way is the mara like the rabbit? How do biologists explain the similarities? (pages 5–6)

3. List three specific observations that might have been instrumental in Darwin's early thoughts about evolutionary descent. (pages 5–7)

4. In what ways were Lyell's revolutionary ideas on geology important to the emerging theory of evolution? (page 7)

5. How does a hypothesis differ from a theory? (pages 10–11)

6. What are the two most important characteristics of a hypothesis? (pages 10–11)

7. Comment on the validity of the statement, "The experimenter then set out to prove the hypothesis." (pages 11–12)

8. Suggest how a controlled experiment might be organized to test a new medicine. (pages 11–12)

9. What is the purpose of an experimental control? (page 11)

10. Upon what, specifically, must a scientist base his or her conclusions? How are such conclusions usually reached? (page 12)

11. List four important elements of the mechanism of natural selection. (pages 12–13)

12. Why is the explanatory power of natural selection a problem? (page 13)

13. How can evolution, a gradual process, ever be tested? Be specific. (page 14)

14. Distinguish between reductionism and synthesism. How is one dependent on the other? (page 15)

15. List five characteristics of life. Do any of these apply to nonliving entities? Explain. (pages 16–17)

CHAPTER 2

Chemistry of Small Molecules

CHAPTER OUTLINE

ELEMENTS, ATOMS, AND MOLECULES

Atoms and Molecules

Atomic Structure

Isotopes, Radioactivity, and Biology

ELECTRONS AND THE CHEMICAL PROPERTIES OF ELEMENTS

Electron Energy Levels and Shells

THE INTERACTION OF ATOMS AND ELECTRONS: CHEMICAL BONDS

Covalent Bonds

Ionic Bonds

Hydrogen Bonds

Why Chemical Reactions Occur

CARBON BONDS AND THE SHAPE OF THINGS

The Versatility of Carbon

THE OTHER SPONCH ELEMENTS

Nitrogen

Phosphorus

Sulfur

Functional Groups

THE WATER MOLECULE AND HYDROGEN BONDING

Water and the Hydrogen Bond

Water: Ionization, pH, and Acids and Bases

KEY IDEAS

APPLICATION OF IDEAS

REVIEW QUESTIONS

In my hunt for the secret of life, I started research in histology. Unsatisfied by the information that cellular morphology could give me about life, I turned to physiology. Finding physiology too complex I took up pharmacology. Still finding the situation too complicated I turned to bacteriology. But bacteria were even too complex, so I descended to the molecular level, studying chemistry and physical chemistry. After twenty years' work, I was led to conclude that to understand life we have to descend to the electronic level, and to the world of wave mechanics. But electrons are just electrons, and have no life at all. Evidently on the way I lost life; it had run out between my fingers.

Albert Szent-Györgi, *Personal Reminiscences*

Szent-Györgi's wry comment on his search for the secret of life simply reminds us that the "meaning of life" cannot be reduced, it seems, to basic understandable processes. It seems that the nature of life will never be grasped as a pure, crystalline gem of truth. If the secret is ever unveiled, it will probably be found in the very complexity that Szent-Györgi tried to avoid. Nonetheless, the reductionist approach is valid; the processes of life depend ultimately on the behavior of lifeless molecules, atoms, and electrons moving mindlessly in space. In order to make sense of what we do know about life, we must know of its components. It is for this reason that we find ourselves immersed, from time to time, in the precise and measured world of chemistry.

Of course, biology can be "done" without chemistry. Darwin, for example, had very little knowledge of the subject, and, even now, the woods are full of biologists with a love of the outdoors but a hatred of beakers, bases, and balanced equations. Would this love increase if they knew more about life at the molecular level? Certainly, it cannot detract from the beauty of a delicately veined leaf to know that it can make food from carbon dioxide and water. It may be true that some chemists must periodically be convinced of the existence of the platypus, but only a chemist can tell us how a fat little hummingbird is able to fly nonstop across the Gulf of Mexico.

In this chapter, we will briefly enter the chemist's world. We will begin with some basic information about atoms and small molecules. At some point we'll cross the line between the lifeless world of chemicals and the vibrant world of life— but we're not sure where that line is.

Elements, Atoms, and Molecules

Every discipline has its general terms, those that are handy to use when getting the subject off the ground. Biologists use the term "organism" a lot when they don't need to be specific. In chemistry, an equally handy term is **matter**, which includes anything that has mass and takes up space—just about everything around us. If all matter could be reduced to its pure states, we would find that there are 92 naturally occurring kinds of matter (and a number that are synthetic). These kinds of matter are the chemical elements.

A **chemical element** is a substance that cannot be separated into simpler substances through ordinary chemical means. Each element has specific properties that make it different from other elements. The properties of elements include the most common physical state (solid, liquid, or gas), color, odor, texture, boiling and freezing points, chemical reactivity, and others. Elements can actually be broken down further, such as by bombarding them with high-energy particles, but then their properties would change and the products would no longer represent that element. This is just another way of saying that the properties define the element.

Familiar elements include sulfur, phosphorus, oxygen, nitrogen, carbon, and hydrogen. These six elements, often referred to by the acronym **SPONCH**, are important because they make up about 99% of living matter. The SPONCH elements are not the only ones important to life. Table 2.1 lists a number of others. The remaining elements are rare in organisms and of less interest to biologists.

Atoms and Molecules

Elements are made up of **atoms**, the individual units of matter (Figure 2.1). As we will see, the atoms of each element have common features: each is made up of subatomic particles called protons, neutrons, and electrons. The number of subatomic particles is different in each element, which is what gives each element its unique properties. We will return to the subatomic particles shortly.

FIGURE 2.1
A Look at Atoms
Atomic structure is made visible through the scanning tunneling microscope. This is a crystal of the element silicon as viewed from the surface. The atoms are the brighter spots. The entire array is interconnected by attracting forces known as chemical bonds. Magnification is about 100 million times actual size.

TABLE 2.1
Elements Essential to the Processes of Life

Element	Percentage of SPONCH Atoms In Humans	Symbol	Atomic Number	Atomic Mass	Example of Role in Life
Calcium		Ca	20	40.1	In bone, muscle, shells; nerve and muscle function
Carbon	10. 50%	C	6	12.0	Forms structural "backbone" for molecules of life
Chlorine		Cl	17	35.5	HCl: stomach acid
Cobalt		Co	27	58.9	In vitamin B_{12}
Copper		Cu	29	63.5	O_2 carrier in mollusk blood
Fluorine		F	9	19.0	Needed for tooth enamel
Hydrogen	60.90%	H	1	1.0	Part of H_2O and all molecules of life
Iodine		I	53	126.9	In thyroid hormone
Iron		Fe	26	55.8	O_2 carrier in hemoglobin
Magnesium		Mg	12	24.3	Part of photosynthetic pigments; aid to chemical activity
Manganese		Mn	25	54.9	Aid to chemical activity
Molybdenum		Mo	42	95.9	Aid to chemical activity
Nitrogen	2.47%	N	7	14	In all proteins and nucleic acids
Oxygen	25.60%	0	8	16	In all molecules of life, in water; needed in cell respiration
Phosphorus	0.16%	P	15	31.0	In ATP: energy transfer molecules of cell
Potassium		K	19	39.1	Nerve activity
Selenium		Se	34	79.0	Aid to chemical activity
Silicon		Si	14	28.1	In some algae, diatom shells, arteries
Sodium		Na	11	23.0	Nerve activity
Sulfur	0.06%	S	16	32.1	In most proteins
Vanadium		V	23	50.9	Oxygen transport in sea squirts
Zinc		Zn	30	65.4	Aid to chemical activity

A **molecule** is a unit formed by two or more atoms joined together. The atoms of a molecule can be of the same element or of different elements. For example, a molecule of oxygen consists of two oxygen atoms bound together. The symbol for oxygen is simply O, but the molecule is symbolized by the chemical formula O_2. Similarly, hydrogen gas consists of molecular hydrogen, or H_2. And you are undoubtedly aware that two atoms of hydrogen and one atom of oxygen combine to form one molecule of water, H_2O. Water is a compound. A **compound** is any pure molecular substance in which each molecule contains atoms of two or more different elements in specific proportions.

Let's look into the use of chemical formulas for a moment. The number appearing *before* a chemical symbol in a molecule shows how many units there are of whatever follows that number. The small subscript *after* a chemical symbol indicates the number of atoms of the element directly preceding. Thus, in H_2O, we find two atoms of hydrogen and one of oxygen, combined to form one molecule of water. The symbol $12 \ H_2O$, thus refers to 12 molecules of water. If a subscript number follows a molecular symbol that is in parentheses, $(CH_2O)_3$, for instance, it indicates the number of molecules involved and assumes they are part of an even larger molecule. In this example, there are three CH_2O subunits.

Atomic Structure

Subatomic Particles. The principal subatomic particles that make up atoms are **neutrons**, **protons**, and **electrons** (Figure

2.2). Protons and neutrons cluster together to form the **atomic nucleus**. Electrons occur outside the nucleus.

Charges. Neutrons are "neutral" (0); that is, they have no electrical charge. However, protons have a positive (+) electrical charge, and electrons have an equally strong negative (–) electrical charge. The electrical charges have important implications for atomic structure. For instance, unlike charges (+ and –) attract each other, whereas like charges (+ and +, or – and –) repel each other. It is the attraction of positive protons and negative electrons that helps keep the atom together. Where atoms are in their elemental state, that is, where they exist singly, the numbers of electrons and protons are equal. Thus the charges cancel each other and individual atoms are electrically neutral.

Mass. Neutrons and protons are about equal to each other in mass and are much heavier than electrons (some 1836 times heavier). Accordingly, we can say that protons and neutrons make up most of the mass of the atom (and indeed most of the mass of the universe). **Mass**, you should know, is the quantity of matter in an object. **Weight**, a more familiar term but not a precise synonym, refers to the gravitational pull on an object. Thus your mass is the same throughout the universe, but your weight depends on which planet you are on (or whether you are in space). In other words, you can be weightless but you will never be without mass.

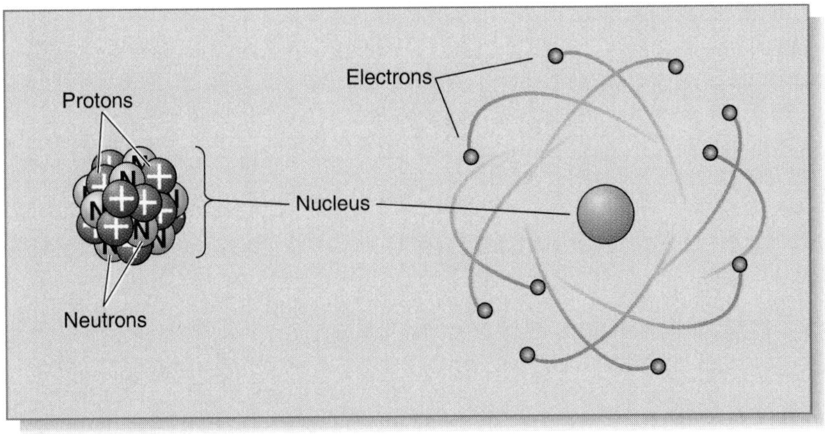

FIGURE 2.2
The Atom: A Planetary Model
(**a**) In an older but useful way of viewing the atom, the nucleus is seen as consisting of two kinds of particles: protons and neutrons. (**b**) Outside are the electrons, minute, negatively charged bodies that travel at great speed about the nucleus.

Atomic mass, the number of protons and neutrons in an atom, is measured in **daltons**[1]. An atom of hydrogen, the lightest element, has a mass of about 1 dalton, while an atom of one of the heaviest elements, uranium-238, weighs just about 238 times as much and so has a mass of about 238 daltons (Figure 2.3).

Molecular mass is the sum of the atomic masses of the atoms that make up a molecule. For example, consider water (H_2O):

$$\begin{aligned} \text{Two H} &= 2 \times 1 = 2 \\ \text{One O} &= 1 \times 16 = 16 \\ \text{Mass} &= 18 \text{ daltons} \end{aligned}$$

Atomic Number. Each element has its own **atomic number**, which equals the number of protons in the atom. For instance, the atomic numbers of the six SPONCH elements are 16, 15, 8, 7, 6, and 1, respectively. (Note that the acronym SPONCH lists the six elements in order of decreasing atomic number.) The number of electrons in the atom also equals the number of protons, but only as long as it is in its elemental form. Atomic mass and atomic numbers are listed with the elements in Table 2.1.

Isotopes, Radioactivity, and Biology

From what we have just seen, determining the atomic mass of an element seems straightforward—we just add protons and neutrons. But, as it turns out, things aren't quite as neat as we might hope. Atomic mass varies among the atoms within most elements simply because the number of neutrons varies. Such variants are called **isotopes**. Thus we can say that isotopes are particular forms of an element, differentiated by the number of neutrons in their atomic nuclei.

The nuclei of some isotopes are unstable, or **radioactive**. "Unstable" refers to the tendency of the atomic nucleus to spontaneously *decay,* or break down: during decay, radiation is released. Such radiation takes the form of minute particles and energy. As you may know, such radiation can be injurious or deadly to life. In the process of decay, some of the radioactive isotopes, or **radioisotopes** as they are also known, change from one element to another as they lose mass. The new element may or may not be radioactive.

The time required for half of the atoms of any radioactive material to decay is called the isotope's **half-life**. Half-lives can vary considerably. Some laboratory-produced radioisotopes have a fleeting half-life of seconds or minutes. On the other hand, most naturally occurring radioisotopes are extremely durable; some have half-lives of billions of years. Uranium-238, for instance, has a half-life of about one billion years, after which half of its atoms would have formed an isotope called lead-206.

Isotopes are important to scientists in a number of ways. The longer-lived ones are often used in dating fossil-bearing

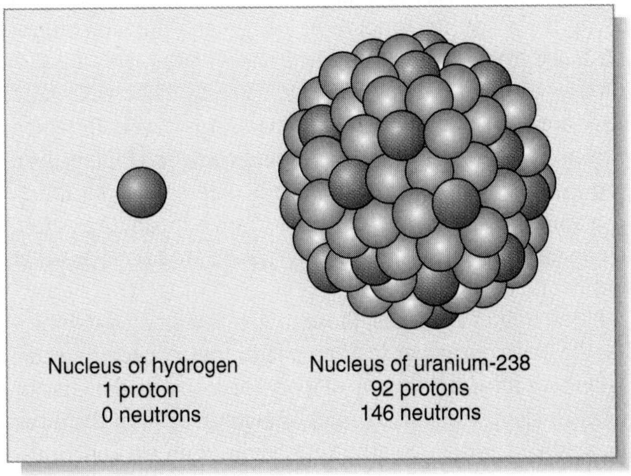

FIGURE 2.3
Atomic Nuclei
The comparatively massive nucleus of uranium-238 has 92 protons and 146 neutrons. Hydrogen's nucleus (in the most common form) consists of a single proton.

[1]By definition, a **dalton** *is a unit of mass equal to one-twelfth of the mass of an atom of carbon-12, which has six protons and six neutrons in its nucleus.* The mass of every other known element is established relative to the mass of this atom.

samples from the earth's crust or, indeed, the earth itself. Such dating yields important clues to ancient geological and evolutionary events. In medicine, powerful radioisotopes are used in radiation treatment of cancer-ridden tissues. And, of course, scientists are vitally interested in the destructive effects of radiation on all life.

Research biologists also use short-lived, low-energy radioisotopes as **tracers** to determine where certain chemicals go and how they behave in living cells. For example, the use of radioactive carbon, phosphorus, sulfur, and hydrogen was vital in determining the structure and function of DNA, the gigantic molecule that bears the hereditary information of each species.

Electrons and the Chemical Properties of Elements

Our interest here is primarily in the chemical activity of elements, and this brings us back to the electrons. As we will soon see, all chemical reactions depend on the arrangement and behavior of these tiny, fast-moving particles. **Chemical reactions** are interactions between atoms or molecules in which starting substances, the **reactants**, are changed into new substances, the **products**. The products of reactions may have very different properties from the reactants. During chemical reactions, **chemical bonds**, forces that hold molecules together, may be formed or broken. Both the formation and the breaking of chemical bonds involve electrons, so we will begin with these fast-moving, negatively charged particles.

Electron Energy Levels and Shells

Electrons occur in definite spatial arrangements around the atomic nucleus where they are in constant movement. Each electron has a specific amount of energy and this determines its distance from the nucleus. Electrons with the least amount of energy are closest to the nucleus, whereas those with more energy are further away. The various energy levels in which electrons occur are often called **energy shells** (Figure 2.4). Electrons remain in their energy shells unless they can somehow gain or lose enough energy to move to another shell. Such a shift is an all-or-none matter; there are no electrons at intermediate levels.

In the lighter elements, those of greatest interest to biologists, the electrons occur in from one to four concentric energy shells, with a maximum of two electrons in the first or innermost shell, and a maximum of eight in each of the other three shells (2-8-8-8). Shell filling begins with the innermost and continues outward. Thus leftovers form an outer shell. Chemists refer to the eight electron maximum as the "octet rule," but keep in mind that this refers to the *lighter elements only*; the rules change with the heavier elements. Table 2.2 lists the arrangement of shells in selected light elements.

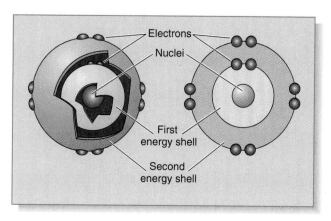

FIGURE 2.4
Energy Shells
Electrons of varying energy form concentric energy levels or shells. Their energy increases with the distance from the nucleus. At the left is a cross section through the atom, showing the electrons in concentric circles. At the right is an old but useful model known as the Bohr atom. Everything is shown in one plane, with the nucleus at the center and the electrons, arranged in pairs, at each energy level.

Electron Orbitals. Electrons in each energy shell are precisely arranged. First, most electrons occur in pairs and each pair moves in a pathway known as an **orbital**.

The innermost and lowest energy shell contains one orbital, the **1s orbital**, which holds a maximum of two electrons. (Hydrogen, with its lone electron, is the only element with an unfilled 1s orbital.) The two electrons follow a circular path about the nucleus, but since the plane of the circle is constantly changing, their total movement actually describes a small sphere. The element helium (atomic number 2) has only one shell, the innermost one, and its 1s orbital is filled to capacity with two electrons.

The arrangement of electron orbitals in the second shell also follows strict rules. Its eight electrons occur in four orbitals, each holding no more than one pair. One pair moves in a spherical **2s orbital** (like the 1s orbital, but larger), while the other three pairs occur in three dumbbell-shaped **2p orbitals**. Figure 2.5 illustrates the orbitals of neon, an element that happens to have a full outer shell. Note in Figure 2.5d that

TABLE 2.2					
Electron Shells in Selected Elements, Listed by Atomic Number					
Element	**Atomic Number**	**First Shell**	**Second Shell**	**Third Shell**	**Fourth Shell**
Hydrogen	1	1 e			
Carbon	6	2 e	4 e		
Phosphorus	15	2 e	8 e	5 e	
Calcium	20	2 e	8 e	8 e	2 e

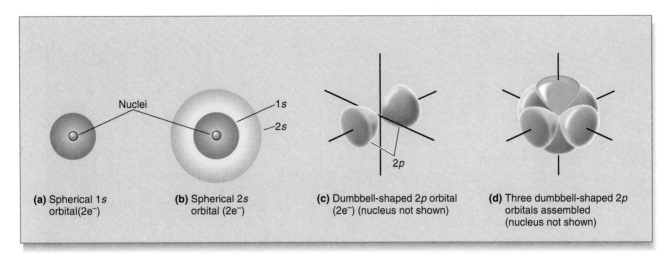

FIGURE 2.5
Electron Orbitals of Neon
Neon has a total of ten electrons. The density of shading indicates where electrons would most likely occur at any point in time. In neon, they occur in two energy shells distributed into five orbitals. (a) Two electrons occupy the innermost, spherical 1s orbit. (b) Two more electrons occur in an outer, spherical 2s orbital. (c) The remaining six electrons occur in three 2p orbitals, like the one shown here. As in the 1s and 2s orbitals, each 2p orbital of neon contains a pair of electrons. (d) When the 2p orbitals are assembled, we see that they are equidistant, taking the form of three dumbbells. Neon has even numbers of electrons, so the are always paired. Many elements have odd numbers, which means that one orbital in the atom will have a single electron.

the axes of the 2p orbitals, termed *x, y,* and *z,* are *equidistant,* meaning that they are placed as far apart as possible.

When an element has more than ten electrons, a third shell must exist, its orbitals resembling those of the second but designated **3s** and **3p**. If this is the outer shell, then it too may contain no more than eight electrons. Additional electrons will fall into a fourth shell, but this is as far as our discussion will go.

We see then that atoms have only two electrons in the first shell and, in the lighter elements, a maximum of eight in each of the next three. Figure 2.6 is a chart showing the first 20 elements. The energy shells are illustrated in the form of concentric circles containing the electrons. This is an older convention still used as a convenience. (No one wants to draw so many different kinds of orbitals.)

Valence Electrons. We've said that the chemical properties of the elements are determined by the numbers of subatomic particles. Let's refine that statement substantially by noting that it is the outer shell electrons that are important. Outer shell electrons are known as **valence electrons** and the outer shell itself is the **valence shell**. It is the valence electrons that determine much of the chemical behavior of an element.

A return to Figure 2.6 will help us make the point about valence electrons and chemical activity. The active elements are those with unfilled outermost shells. A quick look at the figure should tell you that this includes most of the 20 shown, but if you look closely at the last column on the right, you will note three exceptions: helium, neon, and argon. Each has a

complete outer shell. It turns out that the three are not chemically active at all. In fact their chemical "aloofness" led early chemists to call them the "noble elements." Today they are simply described as **inert**. Another look at the column organization shows that all the elements in a given vertical column have similar valence shells. They also have similar properties and behave similarly. For example, H, Li, Na, and K all have one valence electron. At the other side of the chart are Fl, Cl, and Br, elements that have seven valence electrons.

The Interaction of Atoms and Electrons: Chemical Bonds

Now that you have an idea of the structure of atoms, how electrons are arranged, and that outer shell electrons affect chemical activity, let's look into the way reactions occur.

A basic tendency in atoms with incomplete shells is to interact with other atoms in ways that result in the outer shell being filled. This is sometimes accomplished by shedding electrons, and at other times by taking them in. In other instances, electrons are shared between atoms, each then behaving as though its outer shell was filled. Until the outer shell requirement has been met, atoms remain unstable and reactive.

When atoms share or transfer electrons they form **chemical bonds**. These are electrostatic forces that hold molecules together. There are three kinds of chemical bonds of interest

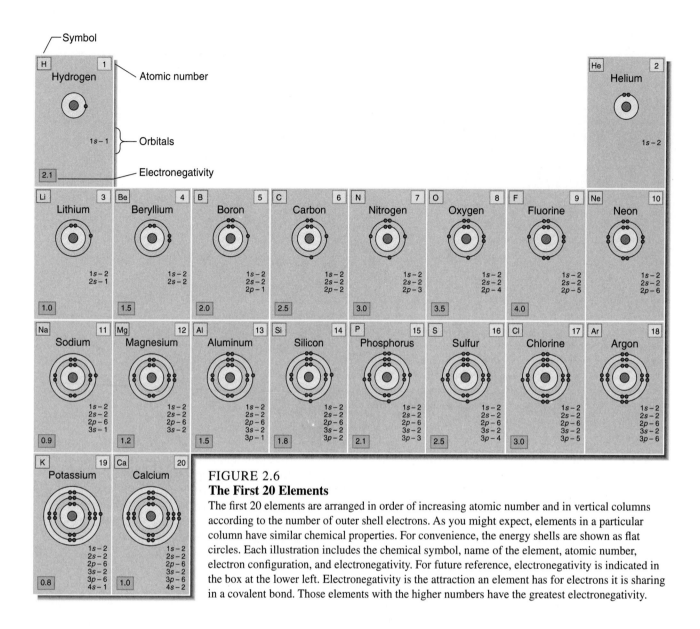

FIGURE 2.6
The First 20 Elements
The first 20 elements are arranged in order of increasing atomic number and in vertical columns according to the number of outer shell electrons. As you might expect, elements in a particular column have similar chemical properties. For convenience, the energy shells are shown as flat circles. Each illustration includes the chemical symbol, name of the element, atomic number, electron configuration, and electronegativity. For future reference, electronegativity is indicated in the box at the lower left. Electronegativity is the attraction an element has for electrons it is sharing in a covalent bond. Those elements with the higher numbers have the greatest electronegativity.

to us in this chapter: **covalent bonds**, **ionic bonds**, and **hydrogen bonds**.

Covalent Bonds

Covalent bonds are those that form through a sharing of valence electrons between atoms. The simplest molecule is formed when two hydrogen atoms come together, forming a hydrogen molecule (H_2). Recall that each hydrogen atom consists of one proton and one electron. Each requires just one electron to fill its $1s$ orbital, which they do by pooling their two electrons (Figure 2.7a). This fills their outer shells, or at least they behave as though they were filled, and they reach a new, more stable state. Chemists often represent molecules with **structural formulas**. Thus hydrogen, or H_2, becomes H—H. The dash between hydrogens represents one pair of shared electrons.

Other elements that form molecules this way include chlorine, oxygen, and nitrogen. Two atoms of chlorine share one pair of electrons, thus forming a single covalent bond (Cl—Cl) as they become a molecule of chlorine gas (Cl_2). Oxygen has six electrons in its outer shell, leaving it two short. Two oxygen atoms share four of their outer shell electrons, forming an oxygen molecule (O_2)(Figure 2.7b). The structural formula would then be O=O. The two dashes, known as a **double covalent bond**, represent two pairs of shared valence electrons. Nitrogen atoms form a **triple covalent bond** by sharing three pairs of electrons. A triple bond, designated N≡N, shows six shared electrons in the nitrogen molecule (N_2).

As we have seen, when different elements react together, compounds are formed. In carbon dioxide (CO_2), for example, one atom of carbon reacts with two of oxygen. Oxygen has six valence electrons and carbon has four. Carbon will fill

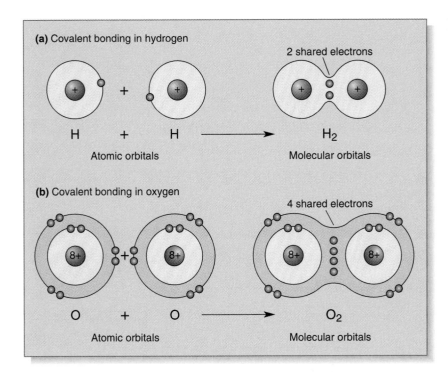

(a) Covalent bonding in hydrogen

2 shared electrons

H + H ⟶ H₂

Atomic orbitals   Molecular orbitals

(b) Covalent bonding in oxygen

4 shared electrons

O + O ⟶ O₂

Atomic orbitals   Molecular orbitals

FIGURE 2.7
Covalent Bonding
In covalent bonding, two or more atoms share electrons. As new molecular orbitals form, a molecule emerges. **(a)** The simplest example of covalent bonding is seen in the hydrogen molecule (H_2), which can be shown by simply overlapping the $1s$ orbitals. **(b)** Oxygen is more complex since more orbitals are involved. As the oxygen molecule forms, new molecular orbitals emerge wherein the shared electrons locate.

its valence shell if it can borrow two electrons from each of two oxygens and the two oxygens will fill their valence shells if they each borrow two carbon electrons. The sharing produces carbon dioxide (CO_2). The structural formula is O=C=O (two double covalent bonds). In some instances, carbon will share its four electrons with one oxygen, forming carbon monoxide, or C≡O. Carbon and oxygen, it seems, were made for each other.

One of the more universal compounds on earth is water (H_2O), which, as we will see shortly, also happens to be critical to life. Water forms when two hydrogen atoms share their lone electrons with two from oxygen.

Molecular Orbitals. As atoms react together to form covalent bonds, atomic orbitals become molecular orbitals. The way molecular orbitals take shape ultimately affects the shape of the molecule.

Hydrogen and oxygen molecules have linear forms, the only forms possible since there are only two atoms in each. Carbon dioxide also forms a linear molecule. You might expect water to have a linear shape, but it turns out to be asymmetrical, with the hydrogen groups off to one side of the molecule. This occurs for a special reason. In the reaction between oxygen and hydrogen, the new molecular orbitals take the form of four teardrops that are arranged equidistantly. The two hydrogens can interact with only two of these orbitals, which puts them off to one side of the molecule. (Figure 2.8). As we will soon see, carbon forms similarly shaped molecular orbitals.

These aspects of covalent bonding are important because the shapes that molecules take are important to their function in life. The shapes of the water molecule and the many carbon compounds are perhaps the most influential in life.

Polar and Nonpolar Covalent Bonds. In the hydrogen and oxygen molecules, and other molecules made up of just one element, both atoms are equal in the struggle to fill outer shells. Thus the sharing of electrons is fully equitable. Another way of saying this is that **electronegativity** in the two is equal. Electronegativity is the attraction an atom has for the electrons it is sharing in a covalent bond. Figure 2.6 lists the electronegativity value for the first 20 elements.

But when atoms of a strongly electronegative element form covalent bonds with those of weakly electronegative elements, the stronger atoms tend to draw the electrons away from the weaker and the sharing can become highly inequitable. When electrons frequent one area of the molecule, the result is an unequal charge distribution around the molecule. The chemical bond in such cases is described as a **polar covalent bond** (as in the + and − "poles" of a magnet). In water, the oxygen, a strongly electronegative element, attracts electrons away from weakly electronegative hydrogen. As a result, water forms slightly negative and positive regions, although the molecule as a whole has no net charge. In contrast, hydrogen and oxygen molecules have **nonpolar covalent bonds**. Actually, since elements vary in electronegativity, the polarity of molecules forms a continuum. Hydrogen is at one end and oxygen is nearer the middle. At the extreme

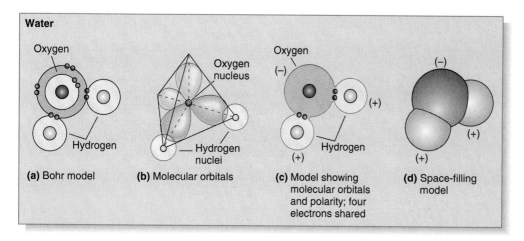

FIGURE 2.8
Water
(a) In the formation of a water molecule, two hydrogen atoms share electrons with one oxygen atom, and two covalent bonds are formed. (b) In the reaction, four teardrop-shaped, equidistant molecular orbitals emerge. The dashed lines connecting the four apices of the orbitals trace the outline of a pyramid. The two hydrogens of water are located at two apices of the orbitals, producing an asymmetrical V shape. (c) Although the oxygen and hydrogen of the water molecule are covalently bonded, the electrons are not equally shared. Oxygen's greater electronegativity draws the electrons away from hydrogen and a polar molecule emerges. (d) Each water molecule has slightly positive and negative regions.

end are chemical bonds in which sharing ceases and electrons are captured by strongly electronegative elements from weakly electronegative elements. This brings us to ionic bonds.

Ionic Bonds

In ionic bonding, electrons are not shared at all. Elements forming **ionic bonds** differ so greatly in their ability to attract electrons that those with the greater electronegativity simply capture them, adding them to their valence shell. But how do the losers fill their valence shells? A look at a common reaction between sodium and chlorine will provide the answer.

Sodium and chlorine are well known for forming ionic bonds. As Figure 2.6 shows, sodium (atomic number 11) has just one valence electron (2-8-1), whereas chlorine (atomic number 17) has seven valence electrons, a nearly complete outer shell (2-8-7). (The electronegativity values, respectively, are 0.9 and 3.0.) When they come together, chlorine quickly captures sodium's lone outer shell electron, completing its own outer shell. Then, since sodium's outer, or third, shell is gone, its filled second shell (8 e$^-$), becomes the new outer shell. But there's more.

When chlorine captures a negative electron, it loses the electrically neutral state of the single atom and becomes *negative*. Likewise, sodium loses its neutrality; but having lost a negative charge, it becomes *positive*. Such electrically charged atoms are called **ions.** The sodium ion and chloride ion, as they are known, are written as Na$^+$ and Cl$^-$, respectively. Then, as oppositely charged bodies do, the ions attract each other. The attractive force between ions is the ionic bond (Figure

2.9). (Note the spelling of "chloride." The "ide" suffix is conventional for negative ions.) We can represent the reactions so far as follows:

$$Na \quad + \quad Cl \quad \longrightarrow \quad Na^+ \quad + \quad Cl^-$$
sodium + chlorine   sodium ion + chloride ion

$$Na^+ \quad + \quad Cl^- \quad \longrightarrow \quad NaCl$$
sodium ion + chlorine ion   sodium chloride

As you see, the product of the reaction just described is sodium chloride, which is no more than common table salt.

Sodium chloride, like many other ionic compounds, does not form discrete molecules. The ions tend to gather in large numbers forming crystalline arrays (Figure 2.10a). This occurs because, oppositely charged ions tend to attract each other. The array is quite orderly, as row upon row of similarly oriented ions gather. Salt crystals vary enormously in size.

One of the characteristics of ionic compounds is that they readily **dissociate** (separate from each other) when placed in water, distributing themselves throughout (Figure 2.10b) the medium. That is, they go into solution. They do this because water molecules have slight positive and negative regions that attract the ions from the crystal. We can represent the dissociation of sodium chloride in the following way:

$$NaCl \quad \rightleftharpoons \quad Na^+ + Cl^-$$
sodium chloride   sodium and chloride ions

The double arrows indicate that the reaction can go in either direction, depending on the presence or absence of water. Should the water in a salt solution evaporate, the ions will again cluster into crystals.

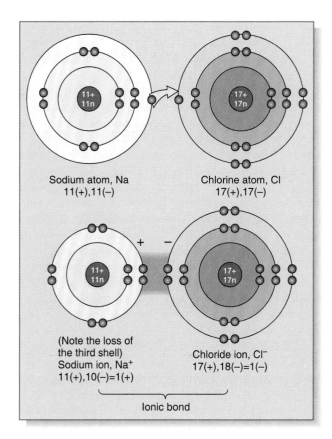

FIGURE 2.9
Ionic Bonding
When sodium and chlorine atoms meet, sodium's lone outer shell electron is attracted to chlorine. After electron transfer, both atoms have filled outer shells, but since each now has a net charge, they are called ions (Na^+ and Cl^-). Electrostatic attraction between the oppositely charged ions forms an ionic bond, holding the two together.

Importantly, some ions have multiple charges. Ions with multiple charges include the calcium ion (Ca^{2+}) and iron ion (Fe^{3+}). Furthermore, some essentially covalent molecules have groups that release ions. Molecular ions, also called **complex ions**, include the ammonium ion (NH_4^+) and the phosphate ion (PO_3^{4-}). [Note that the superscript numbers indicate the number of plus (+) or minus (−) electrical charges an ion has.] In life, the dissociation of a molecule of sodium bicarbonate ($NaHCO_3$) in the bloodstream would yield one sodium ion (Na^+) and one bicarbonate ion (HCO^{3-}). As we will see shortly, this particular reaction helps maintain constant chemical conditions in the blood and tissue fluids.

We are emphasizing dissociation because living organisms have a very high water content in which there are a great number of ions present. (You are about 70% water. A jellyfish, on the other hand, is almost all water.) Our own tissue fluids contain all the ions discussed so far, and many more. As we will see later in the text, ions are an integral part of such processes as nerve conduction and energy transfer.

(a) Salt crystal

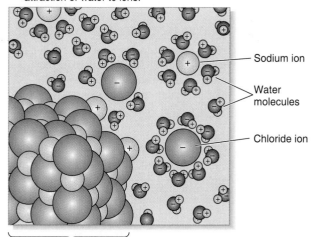

(b) Dissociation of salt and attraction of water to ions.

Sodium ion

Water molecules

Chloride ion

Sodium chloride crystal

FIGURE 2.10
Salt Crystals and Dissociation
(a) Oppositely charged sodium and chloride ions attract each other in mass, forming a crystalline array of alternating ions. Although the crystals have a very specific geometric form, the potential size range is enormous. (b) In the presence of water, the ions of sodium and chlorine dissociate from the crystal, each interacting with individual water molecules as a solution forms. Note how the positive sides of the polar water molecules are attracted to the negative chloride ions and the negative side to the positive sodium ions.

Hydrogen Bonds

Hydrogen bonds do not involve the transfer or sharing of electrons but instead are electrostatic attractions between polar molecules or polar regions in the same molecule. They are called "hydrogen bonds" because they involve hydrogen. Specifically, hydrogen bonded to a strongly electronegative atom, such as oxygen or nitrogen, in one polar molecule is attracted to a strongly electronegative atom, such as oxygen or

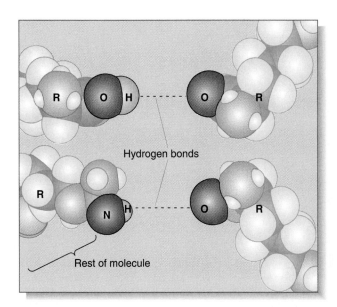

FIGURE 2.11
Hydrogen Bonding
Hydrogen bonding occurs between polar molecules with strong electrostatic elements such as a hydrogen group on oxygen or nitrogen and a neighboring nitrogen or oxygen (respectively). The hydrogen's slight positive charge attracts the slightly negative oxygen or nitrogen from adjacent (or the same) molecules.

nitrogen, in a second polar molecule, (or to such groups within the same molecule) (Figure 2.11). We can illustrate this as follows (the letters represent molecules and the dotted lines, hydrogen bonds):

$$R—OH \cdots N—R \text{ and } R—OH \cdots O—R$$
$$\text{or}$$
$$R—NH \cdots O—R \text{ and } R—NH \cdots N—R$$

Compared to covalent and ionic bonding, hydrogen bonds are very weak. However, since they often occur in great numbers, their combined effect is great. As we will learn in the next chapter, hydrogen bonds are extremely important in life because they determine the shape of essential molecules such as proteins and nucleic acids (DNA and RNA).

Hydrogen bonding is most common in water. The attraction occurs between the hydrogen end (positive region) of one water molecule and the oxygen end (negative region) of another. Such attractions extend throughout a given volume of water, from a single drop to an entire lake. As we will see shortly, hydrogen bonding helps explain many of the special qualities of water.

Why Chemical Reactions Occur

Most of what you have been reading is pretty descriptive. Electrons move here and there, outer shells fill, and bonds form. Things tend to get anecdotal when we try to cover a lot of ground in a short time. In all of this you may now be ready to ask some fundamental questions. Why do elements react? Why do chemical bonds form? The answers can be bewilderingly complex, but let's look into a few generalizations and save the complexities for Chapter 6.

In the real world, chemical reactions occur because of the inexorable tendency of matter to go from high-energy states to low-energy states or, alternatively, from unstable states to stable states. We've hinted at this from time to time, when we said that unfilled energy shells represented a chemically unstable state. Thus a system containing a lot of atoms, say, of elemental hydrogen, would tend to react, forming a lot of hydrogen molecules. For this reason, there is not much elemental hydrogen around. Actually, there isn't much molecular hydrogen either; it too reacts quite readily. Nearly all hydrogen is tied up in the earth's compounds. This is true of many elements.

Accordingly, most chemical reactions in life and nonlife occur between molecules or ions. Chemical bonds are broken, molecules are rearranged, and new chemical bonds are formed. In some instances, these reactions are spontaneous, where two substances will react on their own. A spontaneous reaction may be instantaneous, or it may occur slowly. Most spontaneous reactions in nature give off heat energy, which makes them **exergonic reactions** ("energy out"). As they give off heat, bonds are rearranged, and the products reach a lower energy state and a more stable configuration. Mixing baking soda and vinegar in a glass will bring on an instant spontaneous reaction, one in which salt, water, and carbon dioxide are formed, all at lower energy levels.

Commonly, a small quantity of energy must be provided for an exergonic reaction to begin. This is because each existing chemical bond represents a specific quantity of energy, and if that bond is to be broken, an identical quantity of energy must be applied. In reactions outside life, that energy is usually heat. Once heat energy gets an exergonic reaction going, the chances are it will continue on its own. This is because the breaking of chemical bonds and the forming of newer bonds at a lower energy level produce excess energy. For this reason, a carelessly discarded match can create the massive chemical reactions of a forest fire. Later, we will discuss other kinds of reactions, including **endergonic reactions** ("energy in"), those that actually take in energy. We will also find out how reactions, under the right conditions, can reverse themselves, and how living organisms can get exothermic reactions to occur without adding much heat.

Carbon Bonds and the Shape of Things

The Versatility of Carbon

The chemistry of life is essentially the chemistry of carbon. Carbon atoms literally form the backbone for the major molecules of life. Carbon owes its versatility to the presence of four valence electrons, which it can share with other elements in a

FIGURE 2.12
Geometry of Methane
(a) In methane (CH_4), the carbon atom's four outer shell electrons form covalent bonds with four hydrogen atoms. As in many hydrocarbons, the electrons are equally shared. Thus methane is nonpolar.
(b) When carbon interacts to form covalent bonds, its 2*s* and 2*p* orbitals undergo a transition into four teardrop-shaped orbitals.

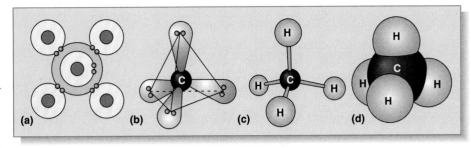

The superimposed line shows that the equidistant orbitals trace the form of a pyramid. (c) Four hydrogens will join one carbon at the corners of the pyramid. (d) The space-filling model is perhaps the most accurate, showing both the close clustering of the four atoms and the molecular geometry.

variety of ways. In addition, carbon atoms can covalently bond together, as they do in gigantic crystals of diamond and underground seams of coal—dramatically different forms of pure carbon. In life, carbon bonds with the SPONCH elements in enormously varied ways as it forms compounds. We can learn a great deal about the chemistry of carbon compounds through a brief look at **methane**, one of the simplest examples.

Methane. Methane (CH_4) is a component of marsh gas and the principal by-product of bacteria that decompose organic matter in places where there is little free oxygen. For example, methane-generating bacteria abound in the bowels of dairy cattle, which is one reason barns have to be well ventilated. (Not for aesthetics—it's a matter of fire hazard!) Methane is so common that it makes up most of the natural gas that helps fuel the world. It is just one of many **hydrocarbons**—molecules that consist essentially of carbon and hydrogen.

In the formation of the methane molecule, a carbon atom shares its outer shell electrons with four hydrogen atoms. Since there is little difference in electronegativity between the two elements (see Figure 2.6), sharing is equitable and methane is nonpolar. Its structural formula is

$$\begin{array}{c} H \\ | \\ H-C-H \\ | \\ H \end{array}$$

(Recall that each single line indicates one covalent bond, one formed by the sharing of two electrons.)

Geometry of Methane. Figure 2.12 offers several ways of looking at methane, beginning with an electron shell diagram (Figure 2.12a). Like other covalent compounds, its formation is accompanied by changes in the atomic orbitals. As carbon interacts with hydrogen, its former 2*s* and 2*p* atomic orbitals are changed into four teardrop-shaped molecular orbitals (Figure 2.12b). As we saw in water, if we connect the outer ends of the four molecular orbitals with imaginary lines, a pyramidal form takes shape. Again, this shape occurs because the orbitals are equidistant. The hydrogen nuclei themselves reside at each of the four points of the pyramid (Figure 2.12c). A final look at

methane, using a realistic space-filling model, gives a good view of what the molecule is probably like (Figure 2.12d).

In more complex carbon compounds (Figure 2.13), the basic geometry of methane is retained. Its four molecular orbitals remain equidistant. Note in the figure, for example, the angular arrangement of the carbon bonds as it forms a chain in butane. Carbon can also form branched chains and even rings. Actually, the geometry described can vary. The single covalent bonds between carbons operate like little swivels, in which the carbon atoms can rotate freely about the axis of the chain. Carbon can form double and triple bonds

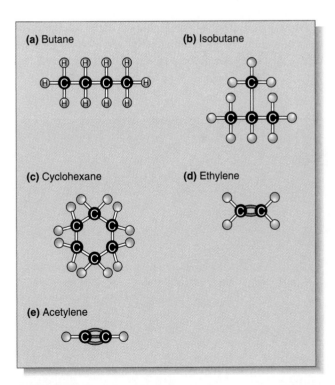

FIGURE 2.13
Carbon Versatility
Not only does carbon form chains of varied lengths (**a**), but the chains can branch (**b**). Furthermore, they can close into rings as they reach their more stable configuration (**c**). Bonds can also vary, many molecules forming double bonds (**d**) and a few triple (**e**).

with other carbons, although the latter are highly unstable because of the strained orbitals formed. Double bonds and triple bonds are inflexible and hold the molecule in a rigid shape. Finally, there seems to be no limit to how large a carbon-based molecule can be. A single giant DNA molecule, as we will see later on, can contain up to 50 billion atoms. These giants are aptly called **macromolecules** (*macro-,* "large").

The Other SPONCH Elements

We now reconsider some of the other SPONCH elements and see how they form covalent and ionic bonds.

Nitrogen

Nitrogen is critically essential to life because it is a constituent of proteins, nucleic acids (DNA and RNA), and many other of life's molecules. It is also one of the most common elements, making up some 78% of the atmospheric gases as molecular nitrogen (N_2). But in spite of its abundance, nitrogen is generally in a limited supply as far as most life is concerned. The reason is that most organisms cannot convert molecular nitrogen to useful compounds. The nitrogen-to-nitrogen linkage is a triple covalent bond ($N \equiv N$), involving the mutual sharing of six electrons. It takes a lot of energy, such as lightning or some special capability of cells, to break these bonds and get the molecule to react in useful ways.

Most organisms obtain their nitrogen in the form of complex ammonium (NH_4^+) or nitrate (NO_3^-) ions. Both ions are made available by the decomposition of dead organisms and

their wastes, so the availability is limited. Fortunately for life, a few bacteria, the **nitrogen-fixing bacteria**, can "fix" (convert) atmospheric nitrogen into a form useful to other organisms. Some of these bacteria live freely in water, whereas others live in the roots of plants such as peas, alfalfa, locust, beans, and peanuts (Figure 2.14). (A discussion of the nitrogen cycle is given in Chapter 22.)

Phosphorus

Phosphorus is another of the essential SPONCH elements. In biological systems it is always combined with oxygen as a **phosphate**. A phosphate can be a free complex ion, or it can be combined with a larger organic molecule as a **phosphate group**. In the cellular and extracellular fluids of organisms, the phosphate group ionizes, yielding $R—PO_3^{2-}$. As a free ion (PO_4^{3-}) it is given a symbol of its own, P_i, which represents **inorganic phosphate** (Figure 2.15).

ATP (**adenosine triphosphate**), is a molecule you'll get to know very well. It is an energy carrier that all organisms use. Cells make use of the special phosphate-to-phosphate bonds of ATP as a means of shuffling energy around to where it is needed. We'll have much more to say about ATP in future chapters.

Sulfur

Sulfur, our last SPONCH element, has several roles in the chemistry of life. It appears in some of the **amino acids**, small molecules that are the building blocks of proteins. The sulfur occurs in the amino acid as a **sulfhydryl group**, a sulfur atom covalently bonded to a hydrogen atom ($—S—H$). When the

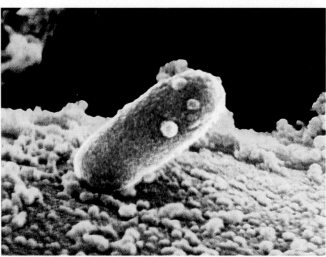

FIGURE 2.14
Nitrogen-Fixing Bacteria
The swellings throughout this root are called nodules. They harbor nitrogen-fixing bacteria that are capable of converting atmospheric nitrogen (N_2) into nitrogen compounds that are used by both the bacteria and the plant host.

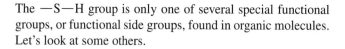

FIGURE 2.15
Phosphoric Acid
The element phosphorus commonly occurs in living organisms as phosphoric acid *(left)*, which in its ionized form is called phosphate or the phosphate ion. Phosphorus makes up the center of the compound and is covalently bonded to four oxygen atoms, three of which have hydrogens attached. When the molecule ionizes in water *(center)*, three of the hydrogens give off protons (their nuclei), leaving their electrons behind in the orbitals of oxygen. The ion *(right)*, as you can see, has three negative charges (3⁻).

amino acids are assembled into protein, pairs of sulfhydryl groups serve as a kind of interlocking "snap" or "hook" that can bond the long strands of protein together, aiding them in forming their special shapes. Two sulfhydryl groups give up their hydrogen and covalently bond together to form a reversible **disulfide linkage**, or **bridge**:

$$\text{Amino acid}-\text{S}-\text{H} + \text{H}-\text{S}-\text{Amino acid}$$
$$\rightleftharpoons$$
$$\text{Amino acid}-\text{S}-\text{S}-\text{Amino acid}$$

The —S—H group is only one of several special functional groups, or functional side groups, found in organic molecules. Let's look at some others.

Functional Groups

A **functional group** is a specific chemical group that appears rather frequently in the many kinds of organic molecules. They are important because their chemical behavior is pretty much the same, regardless of the kind of molecule to which they are attached. Some important functional groups are the carboxyl and amino groups, found in amino acids; phosphate and sulfhydryl groups (already discussed); and methyl groups, hydroxyls, hydroxides, aldehydes, and ketones. Table 2.3 discusses the structure and characteristics of the most familiar functional groups.

The Water Molecule and Hydrogen Bonding

If you were to walk across a barren reach of desert, you might see few obvious signs of life for miles. Then, in the distance, you may see green. Coming closer, you find a few cottonwood trees and scattered mesquites. You know why the plants are there. Water! You might not be able to see it, but you know that it's there, beneath the surface at least. Where there's life, there's water, and just about any place where there is water, there is life (Figure 2.16). Not only does it provide a habitat for much of life, but organisms themselves are mainly water.

FIGURE 2.16
Water and the Desert
Water has a dramatic influence in the desert. Long periods of drought and dry, searing winds leave only the best-adapted plants to dot the landscape. When water is again available, the desert bursts into a short-lived riot of color and new growth.

TABLE 2.3
Common Functional Groups and Their Chemical Properties

Name	Structure[1]	Ionic Form	Properties
Hydroxyl group $R-OH$	$R-O-H$		Common in alcohols, slightly polar, tends to form hydrogen bonds
Amino group $R-NH_2$	$R-N \begin{smallmatrix} H \\ \\ H \end{smallmatrix}$	$R-N \begin{smallmatrix} H \\ \\ H \end{smallmatrix} H^+$	Common in nitrogen bases; part of each amino acid; weak base in water—accepts a proton
Carboxyl group $R-COOH$	$R-C \begin{smallmatrix} O \\ \\ OH \end{smallmatrix}$	$R-C \begin{smallmatrix} O \\ \\ O^- \end{smallmatrix} +H^+$	Common in amino and other organic acids; weak acid in water—releases a proton
Methyl group $R-CH_3$	$R-C \begin{smallmatrix} H \\ | \\ H \end{smallmatrix} -H$		Common in organic molecules; nonpolar and rejected by polar groups; insoluble in water
Aldehyde group $R-COH$	$R-C \begin{smallmatrix} O \\ \\ H \end{smallmatrix}$		Slightly polar; soluble in water; common in sugars
Sulfhydryl group $R-SH$	$R-S-H$		Common in protein, where it forms important covalent linkages within and between polypeptides
Ketone group $R-CO-R$	$\begin{smallmatrix} R \\ \\ R \end{smallmatrix} C-O$		Slightly polar; soluble in water; common in sugars
Phosphate group $R-H_2PO_4$	$R-O-\overset{O}{\underset{O}{\overset{\|}{P}}}-O-H$	$R-O-\overset{O}{\underset{O^-}{\overset{\|}{P}}}-O^- +2H^+$	Polar, weak acid—releases two protons in water; participates in energy transfers in cells when in ATP

[1]R, undesignated molecule to which the functional group is attached

Within cells and tissues, water is the medium in which the chemical reactions of life must occur.

On an even grander scale, most biologists accept the idea that life began in water; and certainly, the association between water and life endures. Hope of discovering life on Mars faded when we found that water was virtually absent on our celestial neighbor. After all, most of life's chemical reactions occur in this medium; in fact, most living organisms are between 50% and 90% water (Figure 2.17). (You are about 70% water.)

Thus water is essential to life as we know it, and on our unique planet, water is plentiful. Three-fourths of the earth's surface is water, and an enormous volume of water constantly shifts between the earth and its atmosphere in what is called the **hydrologic cycle** (Figure 2.18). The cycle is a process that involves all organisms of the earth. Plants, animals, and the many other forms of life use water, incorporating it for a time in their cells as they go about their chemical activities. Eventually, all this water is returned to the cycle of the atmosphere and oceans as the organisms respire or die. In later chapters, we will see the importance of water to life in many different ways, but for now, let's consider some of the molecular peculiarities of this vital, life-sustaining substance, and why its characteristics encourage chemical reactions.

Water and the Hydrogen Bond

Recall from our early discussion that the water molecule is quite polar: that is, it has positive and negatively charged sides. The presence of negative and positive ends will produce weak intermolecular forces in the form of hydrogen bonds. In water, the attraction occurs between a hydrogen in one water molecule and the oxygen of another. This attraction draws water molecules together (Figure 2.19).

Water as a Solvent. For the record, a **solvent** is a fluid capable of dissolving another substance. Such dissolving substances are called **solutes**. Typically, solvents more readily

FIGURE 2.17
Animals of the Water
Aquatic animals and their embryos are particularly high in water content. The vulnerability of early growth stages to drying emphasizes the critical role of water in life. Developing fish *(left)*; aquatic egg mass *(right)*.

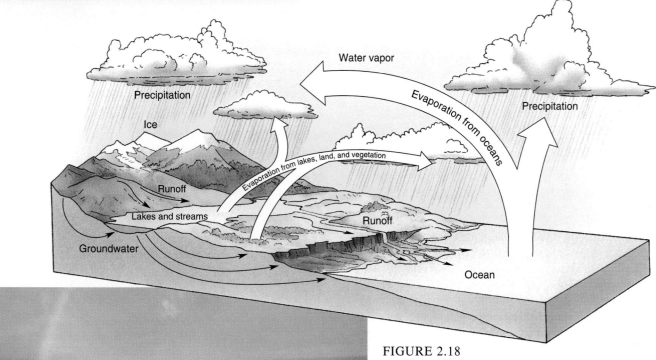

FIGURE 2.18
Cycling of Water
The hydrologic cycle is an ongoing exchange of water through evaporation and precipitation. Living organisms also participate, taking in and releasing sizable quantities of water.

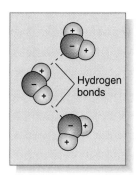

FIGURE 2.19
Interaction Of Water Molecules
Water molecules are polar, each with slightly positive and negative regions. For this reason, they tend to attract each other, forming hydrogen bonds.

dissolve substances with similar properties. Since water is polar, this means that it will readily dissolve other polar substances. Nonpolar substances such as fats and hydrocarbon chains will not easily dissolve in water, but they will dissolve in nonpolar solvents such as acetone and ether.

We've already mentioned that salts such as sodium chloride dissociate into ions when exposed to water. During this process, water forms **hydration shells**, which are layers of water molecules that come to surround the ions. (Don't confuse hydration shells with electron shells.) Water molecules orient their positive (hydrogen) ends toward negative ions, such as chloride, and their negative (oxygen) ends toward positive ions, such as sodium, as shown in Figure 2.20a, b. One of our more imaginative colleagues compared hydration shells to "groupies" clustering around a "highly charged" rock star.

Water will also form hydration shells around polar molecules. For instance, sugars contain protruding, slightly charged, hydroxyl groups (—OH; see Table 2.3) with which water can build hydrogen bonds. This keeps the polar sugar molecules from clumping together; in other words, sugar stays dissolved because hydration shells are formed (Figure 2.20c).

Water and Nonpolar Molecules. If you mix a teaspoonful of water in a jar of salad oil, you might notice that the water will soon form droplets that will coalesce and isolate themselves from the oil. The reason for this is the strong mutual attraction of water molecules. Water has very little attraction for salad oil, which is a nonpolar compound. (Salad oil generally contains vegetable oils, which are primarily composed of long carbon–hydrogen chains. This means they are nonpolar and will not form hydrogen bonds. (See Figure 3.8.)

Nonpolar molecules are also described as **hydrophobic** ("water-fearing"; repelled by water), whereas polar molecules are called **hydrophilic** ("water-loving"; attracted to water). Significantly, there are some molecules that are *both*. **Detergents** are long molecules that have both hydrophilic and hydrophobic regions. Thus detergents can interact with both polar and nonpolar molecules. Dishwashing detergents, for example, are useful cleaning agents because they form bridges between the grease left on your dishes and the surrounding water molecules, effectively lifting the grease from the dishes and dispersing it throughout the dishwater. In the

next chapter, we will come across molecules called phospholipids, which perform similar linking actions within the membranes of cells.

Water Is Wet. Water tends to get things wet. This isn't exactly news, but what does this really mean? It means that water forms hydrogen bonds with polar surface molecules of solid objects (not with oily or waxy objects, those containing nonpolar molecules). This wetting ability is the result of **adhesion**—an attraction between two dissimilar substances. **Cohesion** is the attraction between similar substances; the hydrogen bonds between water molecules give water a considerable cohesive property.

Acting together, adhesion and cohesion give water another of its special properties: **capillarity**. If a thin glass tube is lowered into a beaker of water, the water will rise in the tube. If glass tubes of different diameters are put into the same beaker, water will rise highest in the tube with the smallest bore. The rise of water, called **capillary action**, is due in part to the adhesion of water to glass and in part to the cohesion of water to itself. The two forces also explain the peculiar concave bend (meniscus) seen at the top of the water in a graduated cylinder (Figure 2.21).

Imbibition is similar to capillary action, but on a finer scale. It is the movement of water into porous substances such

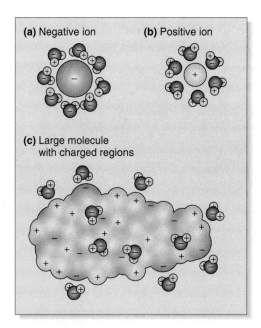

FIGURE 2.20
Hydration Shells

In its interaction with a negative ion **(a)** and a positive ion **(b)**, water forms hydration shells. Note the opposite orientation of water to the two ions. **(c)** Sugar goes into solution because it bears slight positive and negative charges, which attract water molecules. The shell of water molecules separates one sugar molecule from the next as the solution forms.

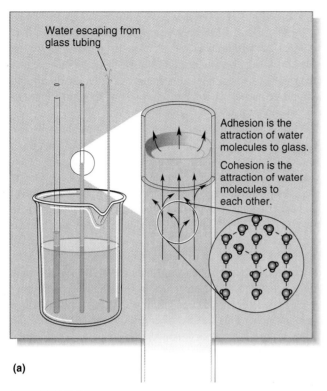

(a)

FIGURE 2.21
Capillarity
The combined effects of cohesion and adhesion in water molecules, known as capillarity, can be seen when glass tubes of different diameter are placed in water. The height to which water rises is inversely proportional to the diameter of the tubing.

as wood or gelatin through **adsorption**, or adhesion to surfaces. The substances swell as the water moves in, and, in fact, the swelling can generate a startlingly powerful force. Seeds can split their tough coats by the force of imbibition. And it has even been suggested that the great stones used in the construction of the Egyptian pyramids were quarried by driving wooden pegs into holes in the rock face and then soaking the pegs with water.

Water Has High Surface Tension. An old party trick involves carefully "floating" a needle in a glass of water. How is this possible? Surely a needle cannot float. It works because of surface tension, another aspect of cohesion. Where air and water meet, the water molecules at the interface have a much greater attraction to other water molecules than to the air above them. (Hydrogen bonding cannot occur between water and air.) The surface thus forms a tough, elastic film of hydrogen-bonded water molecules. The familiar whirligig beetle and water strider are able to walk on this film, and the needle can lay upon it, without breaking it (Figure 2.22). Surface tension is also the force that shapes raindrops. Thus spring showers are delicate and pleasurable, rather than terrifying, as they would be if the same amount of water fell in ponderous masses.

Water, Heat, and Life. One of the most important qualities of water as far as life is concerned is its temperature stability. It takes a large amount of heat energy to raise the temperature of water significantly, but once heat has been absorbed by water, its elevated temperature is just as stubbornly retained. Thus the presence of water on the earth's surface and in its atmosphere is important to its retaining heat, which makes the earth more habitable.

FIGURE 2.22
Surface Tension
Insects such as the whirligig beetle and the water strider are able to literally walk on water because of surface tension. A surprisingly tough water film occurs at the air–water interface.

Chemists refer to the amount of heat needed to raise the temperature of a given amount of a substance as its **specific heat**. Specific heat can be stated in calories; in fact, the calorie (cal) itself is defined as the heat needed to raise the temperature of 1 gram of water 1°C. (The Calorie, with a capital C, or kcal, the one dieters count, is 1000 cal.) Compared to most other substances, water has a high specific heat. For example, to raise the temperature of lead, 1°C requires only 0.03 cal, while a 1°C rise in the temperature of table sugar requires 0.30 cal, and ethyl alcohol, 0.60 cal. But liquid ammonia (NH_3) has a greater specific heat than water, requiring 1.23 cal of heat to raise its temperature 1°C (for this reason, it has been used as a refrigerant in refrigerators.) Interestingly, molecules of ammonia (NH_3), like water, tend to join by forming hydrogen bonds. (Can you tell why from the formula? Is ammonia soluble in water?)

The high specific heat of both water and ammonia is closely related to the presence of hydrogen bonds. Chemists explain that temperature is a product of the rapid movement of molecules in a substance—more specifically, the average **kinetic energy** (energy of motion) of molecules. It is important to realize that under ordinary conditions molecules are always in very rapid random motion. But such motion in water and other substances that are subject to hydrogen bonding is greatly retarded. When hydrogen bonds are present, much of the heat input is required simply to rupture or counteract these numerous attractive forces. Once this requirement is met, the heat can bring about increased molecular movement, and a thermometer will record this activity as a temperature rise. So, in a real sense, water simply captures and holds much of the heat to which it is exposed.

Because of hydrogen bonding, water resists evaporation better than most liquids. Chemists (never at a loss for terminology—like biologists) say that water has a "high heat of vaporization." The heat energy required to convert 1 gram of liquid water to water vapor (a gas) is 540 cal. This is more than twice that required for ethyl alcohol, and nearly twice that required to vaporize liquid ammonia. The vaporization (or evaporation) of water is retarded by hydrogen bonds for the same reasons stated above. Water must absorb a considerable amount of heat before the hydrogen bonds break and the molecules escape as a gas.

Water's high resistance to vaporization is also significant to animals, such as ourselves, who rely on the evaporation of water from our body surfaces as a cooling device. It works well simply because evaporating water molecules remove a considerable amount of heat as they escape. On a much grander scale, in the cycling of water into the atmosphere and back, an enormous amount of heat is also carried aloft and redistributed over the earth's surface (see Chapter 46). This is a key factor in making most of the earth habitable. Furthermore, the waters of the earth—particularly the large bodies—are quite hospitable to life since their temperatures vary only slightly compared to the wide variation in land temperatures.

The cooling of water reveals even more of its peculiarities. Water freezes slowly because of the great amount of heat that must be withdrawn. As it cools, its molecules at first move closer together, increasing in their density and reaching their greatest low-temperature density at 4°C. At this point, the hydrogen bonds become more rigid and the molecular latticework closes up.

This is what happens in a temperate zone lake in late autumn. As the surface temperatures reach 4°C, layers of cooler water sink into the lake depths, displacing warmer waters, which rise to the surface. The seasonal rotation is called **thermal overturn**. Thermal overturn is of biological importance because it carries life-supporting oxygen from surface layers downward, and nutrients from bottom sediments upward.

As the temperature of water approaches 0°C, however, the molecular lattice opens up and the molecules become widely separated, reaching their lowest density (becoming lightest) as ice crystals form. (Figure 2.23 portrays the three states of water.) This means, of course, that ice forms on the surface of the lake, rather than the bottom. It also explains why ice cubes float in your drink and why ice skating can be done without scuba gear. A biologist would be more likely to consider what would happen to lake dwellers if the lake froze from the bottom up.

Water: Ionization, pH, and Acids and Bases

Although water usually exists as a covalently bonded molecule, a very tiny fraction of the molecules in a drop of water will briefly and reversibly dissociate into a **hydrogen ion** (H^+) and a **hydroxide ion** (OH^-):

$$H_2O \rightleftharpoons H^+ + OH^-$$

That is, water molecules are continually breaking apart into ions, and the ions are continually rejoining to form the neutral molecule again. In pure water, at any one instant, something like 1 molecule in 550 million will be dissociated into a pair of positively and negatively charged ions. An instant later, they will be back together again, but another approximately 1 in 550 million water molecules will dissociate in the meantime.

The molar concentration of hydrogen ions in pure water is 0.0000001 *mole per liter*. A **mole** of any substance is the weight *in grams* that equals the molecular mass in daltons of one molecule. Thus 1 mole of hydrogen ions weighs 1 gram, and the concentration of hydrogen ions by weight in pure water is 0.0000001 *gram per liter*. The number 0.0000001, or the digit "1" seven places to the right of the decimal point, can also be written as 10^{-7}, which, in scientific notation, is the equivalent of 1 divided by 10 million. The molar concentration of hydrogen ions in pure water is then 10^{-7} mole per liter, and the same concentration of hydroxide ions is also 10^{-7} mole per liter.

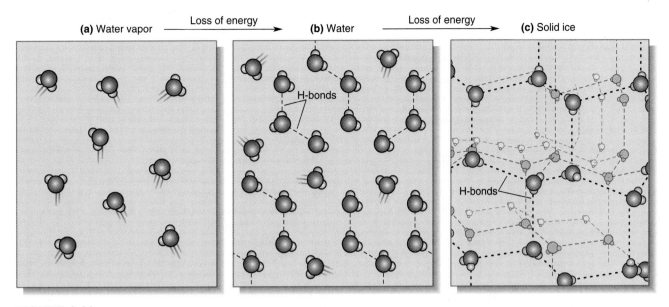

(a) Water vapor —— Loss of energy ——▶ **(b) Water** —— Loss of energy ——▶ **(c) Solid ice**

FIGURE 2.23
Hydrogen Bonding in Water
Weak hydrogen bonds attract water molecules to one another. The degree of attraction depends on heat energy, which disrupts the hydrogen bond. In the gaseous state (**a**) (above 100°C), molecular movement is rapid and random. In the liquid phase (**b**) (0–100°C), molecular movement lessens, and the association between molecules, though firmer, is still quite loose, resulting in sliding rows or lattices. Below water's freezing point (**c**) (0°C), hydrogen bonds hold the water molecules rigidly in place in the expanded, crystalline lattice called ice.

Actually, the hydrogen ion is a naked proton, and while such things exist, they do not float around very long. In reality, the hydrogen nucleus from a dissociating water molecule becomes bound up with another water molecule to make a **hydronium ion** (H_3O^+). We will keep calling it the hydrogen ion (H^+) for convenience, although we mean hydronium ion (H_3O^+). The dissociation and reassociation reaction may be symbolized as follows:

$$2\,H_2O \rightleftharpoons OH^- + H_3O^+$$

where the longer arrow pointing left indicates that most molecules are in the H_2O form at any one time.

An **acid** is a substance that releases hydrogen ions when dissolved in water. Stomach acid, for instance, is dissolved hydrochloric acid (HCl). Pure HCl is a gas, but when it is dissolved in water, HCl ionizes to become paired H^+ and Cl^- (hydrogen ions and chloride ions).

Just how strong, or *acidic,* an acid solution is depends on the concentration of the hydrogen ions. The molecular weight of hydrochloric acid is 36 (1 for the hydrogen, 35 for the chlorine). Thus 1 mole of pure HCl gas weighs 36 grams, and 36 grams of HCl dissolved in 1 liter of water would create a concentration of 1 mole of HCl per liter, a decidedly strong acid.

Actually, since the HCl ionizes almost completely, the concentration of *chloride ions* would be 1 mole per liter, and the concentration of *hydrogen ions* would also be 1 mole per liter—a concentration of hydrogen ions that is 10 million times greater than that of pure water. That's very acidic.

A shorthand notation for the strength of acid solutions, or for acidity in general, is the **pH scale** This scale uses scientific notation to express the H^+ concentration in a solution. Thus if the acidity of some rather tart orange juice is 0.01 mole of H^+ ions per liter, which is the same thing as 10^{-2} mole of H^+ ions per liter, that orange juice has a pH of 2. We just leave out the "ten to the minus" part of the number and write the exponent. On the pH scale, then, pure water has a pH of 7. The pH is considered neutral: neither acidic nor basic. (Figure 2.24 gives the pH values of various substances. Note that the smaller the pH value, the more acid the solution.)

A **base**, or **alkali**, is a substance that accepts protons or releases hydroxide ions (OH^-) when dissolved in water. Lye, or sodium hydroxide (NaOH), a familiar example of a strong base, releases hydroxide ions (OH^-) in water. Ammonia (NH_3) is another base; although it contains no hydroxide ion itself, it accepts a proton from water, leaving behind a hydroxide ion:

$$\underset{\text{ammonia}}{NH_3} + \underset{\text{water}}{H_2O} \rightleftharpoons \underset{\text{ammonium ion}}{NH_4^+} + \underset{\text{hydroxide ion}}{OH^-}$$

A solution can be acidic, neutral, or basic (alkaline), but it cannot be acidic and basic at the same time. This is because hydroxide ions and hydrogen ions join spontaneously to form water. In pure water, the concentrations of hydrogen ions and

More acidic

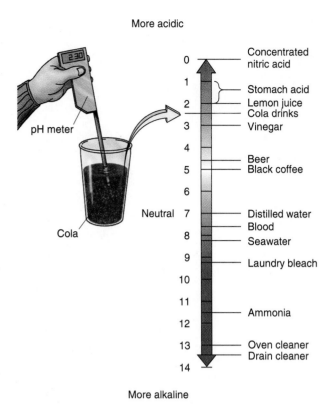

pH	
0	Concentrated nitric acid
1	
	Stomach acid
2	Lemon juice
	Cola drinks
3	Vinegar
4	
	Beer
5	Black coffee
6	
7 (Neutral)	Distilled water
	Blood
8	Seawater
9	Laundry bleach
10	
11	
	Ammonia
12	
13	Oven cleaner
	Drain cleaner
14	

More alkaline

FIGURE 2.24
pH Table
The pH range is from 0 to 14, with the low numbers representing acids and the high numbers bases. The neutral point is 7.

hydroxide ions are both 10^{-7} mole per liter. But when the concentration of hydrogen ions rises to 10^{-1} mole per liter, as in gastric juice, the molar concentration of hydroxide ions falls to 10^{-13}, which is a very small number. In general, the negative exponents of the H^+ and OH^- ion concentrations in a solution always add up to 14. Thus seawater, which is slightly basic, has a hydroxide ion concentration of 10^{-6} and a hydrogen ion concentration of 10^{-8}: $(-6) + (-8) = 14$. The seawater thus has a pH of 8.0.

Most of the acids and bases mentioned so far are "inorganic": they form outside life. Organisms also form acids and bases. "Organic acids" occur as carboxyl functional groups on molecules and are designated as —COOH (see Table 2.3). When they release a hydrogen ion, the designation becomes $-COO^- + H^+$. Organic acids include citric acid (in citrus fruit), acetic acid (in vinegar), and lactic acid (in milk products). There are also organic bases. The amino functional group, designated —C—NH_2, for example, will take in a proton, becoming C—NH_3^+. Many animals form sodium bicarbonate ("baking soda"), a base that releases hydroxide ions (OH^-) when in solution.

Most living things thrive in conditions that are around neutral (pH 7), although there are exceptions (some bacteria and some algae do very well in very acidic hot springs). Our own body fluids tend to remain slightly basic. The normal range for blood, for example, is pH 7.3–7.4. In contrast, the digestive juices of the stomach are quite acidic, ranging from pH 1.2 to pH 3.0 (see Figure 2.24). Most aquatic organisms are especially susceptible to pH changes in their surroundings, particularly the delicate larval (developing) stages. (For a look at the outcome of recent pH decreases due to "acid rain," see Chapter 48.)

Buffers. Maintaining a constant blood and tissue fluid pH is critical to humans and many other animals; should your blood pH fall to 7, for instance, nerve activity would be suppressed, possibly leading to coma and death. We are protected from such changes by many safeguards, one of which is the presence of **buffers.** Buffers are chemical substances, generally weak acids and weak bases (such as carbonic acid and sodium bicarbonate), that neutralize strong bases and acids as fast as they form. Typical reactions are the following:

$$HCl + NaHCO_3 \longrightarrow NaCl + H_2CO_3$$

| strong acid | + | weak base (sodium bicarbonate) | | salt | + | weak acid (carbonic acid) |

or

$$NaOH + H_2CO_3 \longrightarrow H_2O + NaHCO_3$$

| strong base | + | weak acid (carbonic) | | water | + | weak base (sodium bicarbonate) |

In both instances the products at the right are relatively harmless.

Key Ideas

ELEMENTS, ATOMS, AND MOLECULES

1. An **element** is a substance that cannot be separated into simpler substances. Examples common to life include the SPONCH elements.

2. An **atom** is the smallest indivisible unit of an element.

3. A **molecule** is two or more atoms chemically combined. A **compound** is a molecule made up of two or more different elements in specific proportions.

4. In the chemical structure 12 H_2O_2, which is hydrogen peroxide, the 12 indicates twelve molecules. H_2 is two molecules of hydrogen; O_2 is, of course, two atoms of oxygen.

5. Atoms consist of positively charged **protons**, uncharged **neutrons**, and negatively charged **electrons**.

6. Protons and neutrons make up the **atomic nucleus**. Electrons are in motion about the nucleus.

7. The combined masses of protons and neutrons make up the **atomic mass**. **Molecular mass** is the combined atomic masses of the atoms in a molecule.

8. The **atomic number** of an element is the number of protons in its atoms.

9. The number of neutrons in atoms of an element may vary, the variants known as **isotopes**.

10. **Radioactive** isotopes (**radioisotopes**) decay, giving off radiation.

ELECTRONS AND THE CHEMICAL PROPERTIES OF ELEMENTS

1. In **chemical reactions**, **reactants** undergo changes, forming **products**. Electrons become involved in the formation of **chemical bonds**.

2. Electrons occur at specific **energy levels** or **shells.**

3. Lighter elements follow the **octet rule**, with a maximum of 2 first shell electrons, 8 second shell electrons, and 8 third shell electrons.

4. Electrons travel in pathways called **orbitals**, each holding up to two electrons.

5. Outer shell electrons, called **valence electrons**, determine chemical activity. Elements with filled valence shells are nonreactive.

THE INTERACTION OF ATOMS AND ELECTRONS: CHEMICAL BONDS

1. Atoms react in ways that fill the outer shell: taking in electrons, losing electrons, and sharing electrons. Each results in new chemical bonds.

2. Hydrogen, oxygen, and nitrogen share valence electrons, forming **covalent bonds**.

3. One atom of carbon and two of oxygen form the compound carbon dioxide by sharing four electrons.

4. In covalent bonding, molecular orbitals replace the atomic orbitals. The form taken by the new orbitals determines the shape of the molecule.

5. Atoms of the same element share electrons equally, forming **nonpolar covalent bonds**. **Polar covalent bonds** form when elements with greater electronegativity draw electrons away from other atoms in the molecule.

6. In **ionic bonding**, an atom of one element draws electrons completely away from an atom of another.

7. Chlorine draws an electron from sodium, forming a positive **sodium ion** and the negative **chloride ion**. The ions are attracted, forming an **ionic bond**.

8. A **hydrogen bond** is an electrical attraction between a weakly electronegative hydrogen group in one molecule and a strongly electronegative atom, such as oxygen or nitrogen, in another.

9. Chemical reactions generally bring reactants to a lower, more stable energy state. The tendency to reach a lower energy state is a fundamental behavior of matter.

10. Most chemical reactions are exothermic—that is, they give off heat energy. A small amount of starting energy will initiate a reaction, with the energy given off sustaining it.

CARBON BONDS AND THE SHAPE OF THINGS

1. Carbon forms the backbone for many major molecules of life. Carbon bonds readily to itself and to other elements such as hydrogen and oxygen. Molecules of carbon and hydrogen are called **hydrocarbons**.

2. In **methane**, one carbon atom shares its four valence electrons with four hydrogens.

THE OTHER SPONCH ELEMENTS

1. Nitrogen gas (N_2), a triple-bonded molecule must be converted to ammonia and nitrate by nitrogen-fixing bacteria, if it is to be used by plants.

2. Inorganic phosphorus, or P_i, occurs chiefly as the **phosphate ion**. In the molecules of life, it occurs as a **phosphate group**.

3. Cells convert the energy of their fuels, or the energy of sunlight, into compact molecules of **ATP (adenosine triphosphate)**, which then provides small, controllable amounts of energy for the many reactions of life.

4. When incorporated into protein, amino acids containing **sulfhydryl groups (X—SH)** form **disulfide linkages (X—S—S—X)**, that help determine molecular shape.

5. **Functional groups** have special chemical characteristics. Amino, carboxyl, and phosphate groups ionize in water, forming acids and bases. Hydroxyls, aldehydes, and ketones are slightly charged, or polar, whereas methyl groups are nonpolar.

THE WATER MOLECULE AND HYDROGEN BONDING

1. Water is a medium in which many organisms live, makes up much of living material, and is a medium for the chemical reactions of life.

2. Evaporation and precipitation occur in the **hydrologic cycle**.

3. Water is polar, its slightly charged regions giving it special chemical properties.

4. Water molecules are drawn together by weak **hydrogen bonds**.

5. Acting as a solvent, water molecules form **hydration shells** around ions and polar molecules.

6. Nonpolar molecules, such as natural fats and oils and petroleum products, are repelled by water.

7. Detergents form bridges between water and oils because they have both **hydrophilic** (water-loving) and **hydrophobic** (water-fearing) regions.

8. Water's **high specific heat** (resistance to temperature change) is attributed to the stability of numerous hydrogen bonds.

9. As water cools its density increases, reaching a low-temperature maximum at 4°C. Below this, its density decreases, reaching a low-temperature minimum at 0°C (ice).

10. Pure water ionizes slightly, forming **hydrogen ions** (H^+) (actually **hydronium ions**: H_3O^+) and **hydroxide ions** (OH^-) in equal numbers.

11. Equal numbers of H^+ and OH^- represent the neutral condition. Acids form when H^+ ions outnumber OH^- ions, and bases or alkalis form when OH^- ions outnumber H^+.

12. The **pH scale** ranges from pH 0 to pH 14, with 7 representing the neutral point. Below pH 7 is acidic, with the strongest acids at pH 0. Above pH 7 is alkaline, with the strongest bases at pH 14.

13. Buffers are chemical substances that neutralize acids and bases, thus maintaining constant pH conditions.

Application of Ideas

1. Reread Szent-Györgi's lament that opens this chapter. Actually, he is an extremely successful scientist, a Nobel Prize winner, and the discoverer of many of life's secrets (for example, the structure of vitamin C and much of the biochemistry of cell respiration). In what sense then did "life" run out between his fingers?

2. Elements like hydrogen, sodium, and chlorine are rarely found in their elemental form. Explain why this is true. From a periodic table of the elements, list other elements with the same chemical characteristics and try to determine whether these are ever found in their elemental form. (The organization of the periodic table will tell you what the others are.)

3. It is interesting to compare silicon and carbon, since they both have four valence electrons. Silicon can form long chains called silicones, which are like the long chains of carbons called hydrocarbons. In addition, silicon combines with oxygen (SiO_2) to form crystals of silica (or quartz) and readily combines with fluorine to form highly soluble SiF_4. This similarity has not escaped the attention of science fiction writers, who have described life based on silicon rather than carbon. What might such life forms be like? What similarities and differences would you expect between silicon- and carbon-based life forms? What might such alien scientists have to say about the "special properties" of silicon?

Review Questions

1. Write the names, symbols, and atomic numbers of the chemical elements in SPONCH. (page 21)

2. Write a brief definition for *atom, element, molecule,* and *compound.* (pages 21–22)

3. Explain the meaning of the letters and numbers in the chemical formula 12 $C_6H_{12}O_6$. (page 22)

4. What are isotopes? Cite an example of a radioactive isotope with an extremely long half-life. (page 23)

5. State the shell-filling (octet) rule. Following the rule, draw the electron shells of the element potassium (atomic number 19). (page 24)

6. Describe the shapes of the $1s$, $2s$, and $2p$ orbitals. What is the maximum number of electrons each holds? (pages 24–25)

7. Which electrons are valence electrons? Why are they important? How many valence electrons do the following have: sodium, fluorine, neon, and oxygen? (page 25)

8. What is a chemical bond? Using two atoms of oxygen (atomic number 8), illustrate how covalent bonding occurs. (pages 25–27)

9. Using potassium and fluorine (atomic numbers 19 and 10) as examples, show the formation of an ionic bond. Name the compound. Name the ions. (page 28)

10. What would happen if the ionic compound magnesium chloride were to be placed in water? Why would this happen? (page 28)

11. State two important differences between hydrogen bonds and other kinds of chemical bonds. (pages 29–30)

12. What fundamental tendency of matter and energy explains why chemical reactions occur? (page 30)

13. What geometric form does the methane molecule take? What accounts for this particular shape? (page 31)

14. Write the structural formulas for ethane (C_2H_6), propane (C_3H_8), and pentane (C_5H_{12}), all of which have only single C—C bonds (page 31).

15. What are some of the uses of nitrogen in living things? Why is N_2 of little use to most organisms? What forms of nitrogen can they use? (page 32)

16. What commonplace use do organisms make of the element phosphorus? (page 33)

17. Which of life's molecules contains sulfur? How does it help link molecules together? (pages 32–33)

18. Draw three or four water molecules, showing how they are attracted by hydrogen bonds. (pages 34, 36)

19. Define the terms *hydrophilic* and *hydrophobic*. What is the peculiar nature of soap molecules that enables them to interact with grease and water? (page 36)

20. Explain the meaning of specific heat, and discuss two ways in which the specific heat of water is important to life. (pages 37–38)

21. What do hydrogen bonds have to do with water's resistance to temperature change? (page 38)

22. Briefly describe the pH scale and cite examples of strong and weak acids and bases. (pages 39–40)

23. What is a buffer? Using the reaction of hydrochloric acid and sodium bicarbonate as an example, explain how a buffer works. (page 40)

CHAPTER 3

The Molecules of Life

CHAPTER OUTLINE

THE CARBOHYDRATES

Monosaccharides and Disaccharides

Essay 3.1 Reading Structural Formulas

Polysaccharides

THE LIPIDS

Triglycerides

Saturated, Unsaturated, and Polyunsaturated Fats

Phospholipids

Other Lipids

THE PROTEINS

What Is a Protein?

Amino Acids

Polypeptides and Proteins

Structural Proteins

Conjugated Proteins

THE NUCLEIC ACIDS

Nucleic Acid Structure

KEY IDEAS

APPLICATION OF IDEAS

REVIEW QUESTIONS

Life is not simple. How many times we've heard that, usually from weary or worried people trying to live out their lives and hoping their problems are not indicative of some larger, more unpleasant phenomenon still to come. But even those of us who don't focus on life's problems would have to agree that life is complex and often mysterious. In fact, it is complex in a number of ways, including its origins, mechanisms, and structure. Here, we'll prepare for a look at the molecular structure of living things by focusing on their building blocks. Specifically, we will concentrate on the large molecules that comprise life. As you will see, "large" is an understatement. Whereas the molecular weight of methane is just 16, that of the hormone insulin, among the smallest of proteins, is about 5700, and the largest proteins, truly giants, have molecular weights ranging in the millions!

Although large molecules can appear hopelessly complex, they are usually just **polymers**: that is, they are composed of many identical or similar molecular subunits, or **monomers**, joined together by covalent bonding into polymers. The large polymers are also known as **macromolecules** ("large molecules").

In this chapter, we will look at structure and function in the four major classes of molecules: the carbohydrates, lipids, proteins, and nucleic acids.

The Carbohydrates

Carbohydrates are familiar to us as dietary sugars and starches. But, as we will see, they are important in other ways as well. The term "carbohydrate" literally means "hydrate of carbon" or "carbon plus water." This is made clear by the empirical formula $(CH_2O)n$. (Empirical formulas are those whose numbers are reduced to their simplest terms. Here the *n* means that there can be any multiple of CH_2O.) Actually, the simplest carbohydrates are 3-carbon compounds, but here we are primarily interested in 6-carbon carbohydrates and the way they are linked together by covalent bonds to form the large polymers. Let's begin by looking at the organization of the "single" and "double" sugars.

Monosaccharides and Disaccharides

Monosaccharides. The simplest sugars are the **monosaccharides**, or "single sugars" (*mono-*, "1"; *saccharide-*, "sugar"). There are many simple sugars, but the most familiar is the 6-carbon sugar, **glucose** ($C_6H_{12}O_6$). A monosaccharide consists of either a short chain or a ring of carbon atoms (usually five or six), with nearly every carbon having a hydroxyl (—OH) and a hydrogen (—H) side group (Figure 3.1). When present in disaccharides or polysaccharides, monosaccharides are always in their ring form.

In some sugars, those of the **aldose family**, one carbon at the *end* of each simple sugar forms a double bond with its oxygen, producing an **aldehyde** group:

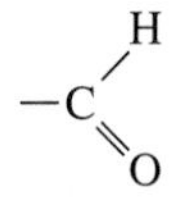

In the **ketose family** of sugars, a carbon *within* the chain forms a double bond with an oxygen group, yielding what is called a **ketone** group:

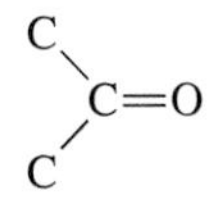

Glucose is the most common and most important monosaccharide. It is intensely involved in energy metabolism and is the building block from which most polysaccharides are built. Glucose is called blood sugar, corn sugar, or grape sugar, depending on its source. In the medical field it is called **dextrose**. There are a variety of ways of representing glucose that we will be using, so it would be a good idea to study Essay 3.1 before proceeding.

Disaccharides. The next most complex type of carbohydrate is the **disaccharide**, or "double sugar" (*di-*, "two"), which contains two monosaccharides. The most familiar disaccharide is **sucrose**, or table sugar ($C_{12}H_{22}O_{11}$). Sucrose is made up of one glucose subunit and one fructose subunit (Figure 3.2). The two 6-carbon monosaccharides are covalently bonded together through the **dehydration reaction** (Figure 3.2), a water-yielding process that we will view again and again since it is the common reaction through which all molecules of life are assembled.

Dehydration Reactions. In a dehydration reaction, —OH side groups in adjacent monomers are brought together, water is removed, and an oxygen linkage is formed:

A—OH + H—B → H—O—H + A—O—B
(2 monosaccharides yield water plus 1 disaccharide)

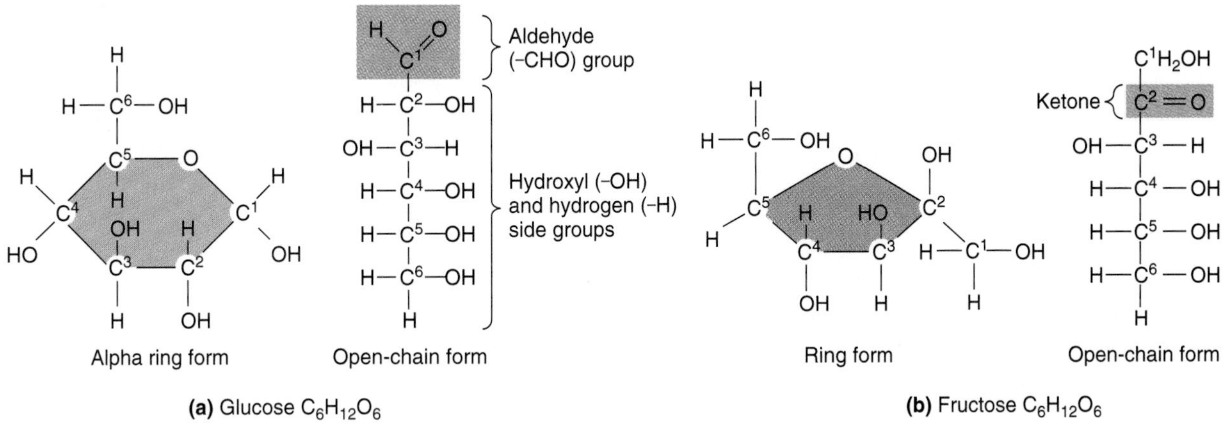

FIGURE 3.1
Glucose and Fructose
(a) Glucose occurs in both the straight chain and ring forms. (Follow the numbering of the carbons to see how one form is converted to the other.) The side groups branching from the carbon backbone include one aldehyde group and a number of hydrogens and hydroxyls. (b) Fructose has the same chemical formula as glucose, but its geometry differs. Note further that a ketone group is present, rather than an aldehyde.

Essay 3.1 ● READING STRUCTURAL FORMULAS

Chemists have several ways of representing organic molecules. We'll use glucose as an example. The chemical formula $C_6H_{12}O_6$ provides information about the elements and their proportions but conveys little about the molecule's three-dimensional form.

(a) Glucose is most often represented by two-dimensional structural formulas that take the form of a ring. They are useful in seeing all of the side groups of the ring and the position of oxygen. (b) A third

dimension is hinted at by the use of shading to raise the structural formula off the page slightly. (c) A still better look at the geometry of glucose is seen in the ball-and-stick model. Actual balls and sticks (dowels) are used to construct these for the artist. Each carbon ball has been drilled with equidistant holes, so the four bonds protrude out at representative angles. This positions the side groups at the proper angle. (d) Here, in a more symbolic representation, the three-

dimensional look is retained, but most of the side groups have been removed for simplicity. This "bare-bones" representation is often used where chains of glucose are illustrated or where the structural formula has just been reviewed. (e) Finally, like other molecules, glucose can be represented by a space-filling model, one that presents both the geometry of the molecule and a more realistic view of the position taken by its atoms.

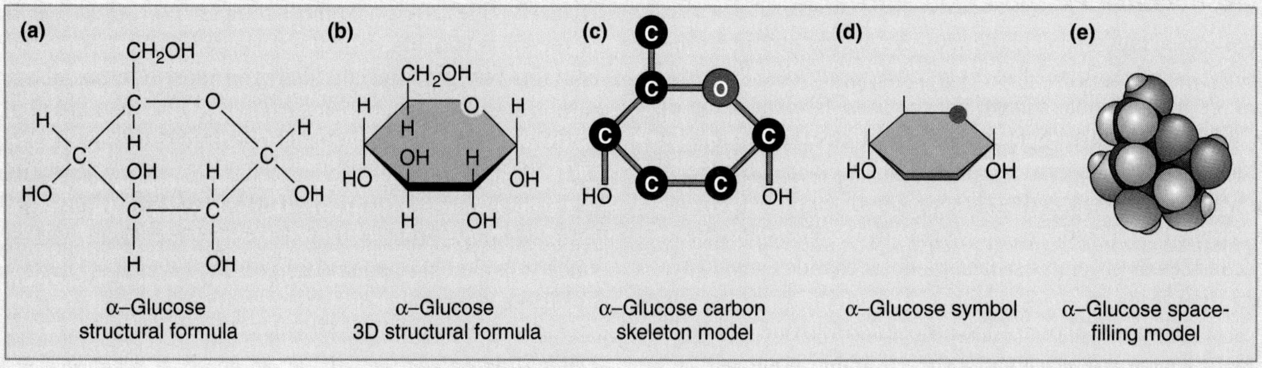

(a) α–Glucose structural formula   (b) α–Glucose 3D structural formula   (c) α–Glucose carbon skeleton model   (d) α–Glucose symbol   (e) α–Glucose space-filling model

True to its name, the dehydration reaction yields one molecule of water (H_2O). In carbohydrates, the newly formed oxygen bridge is called a **glycosidic linkage** (glyco, "sugar"). Since the number 1 carbon of glucose is linked by the oxygen bridge to the number 2 carbon of fructose, we can add that this is a 1–2 glycosidic linkage. We will be mentioning linkages such as 1–2, 1–4, and 1–6 as we proceed with the carbohydrates.

Dehydration reactions, like most chemical reactions in life, require the presence of biologically active proteins called

enzymes. Enzymes abound in the cell and are the agents that speed up the chemical reactions of life. (Enzymes are discussed in detail in Chapter 6.) Furthermore, like many reactions, this one requires a source of energy. Typically, the chemical reactions of life are driven by ATP (adenosine triphosphate), a universal energy storage molecule in organisms.

The dehydration reaction can be reversed: the sucrose is broken back down into glucose and fructose. Of course, different enzymes are involved. In the opposite reaction, called

FIGURE 3.2
The Making of Sucrose: Table Sugar

The monosaccharides glucose and fructose, shown in their "bare bones" structure (with most atoms omitted—see Essay 3.1), become linked through dehydration to form the disaccharide sucrose (table sugar) found in sugar beets and sugar cane and in fruits, vegetables, and honey. In the enzyme-mediated reaction, a hydrogen (—H) group from one sugar and a hydroxyl (—OH) group from the other combine to form water. The oxygen and carbon to which they were bonded now share electrons, forming a covalent 1–2 linkage.

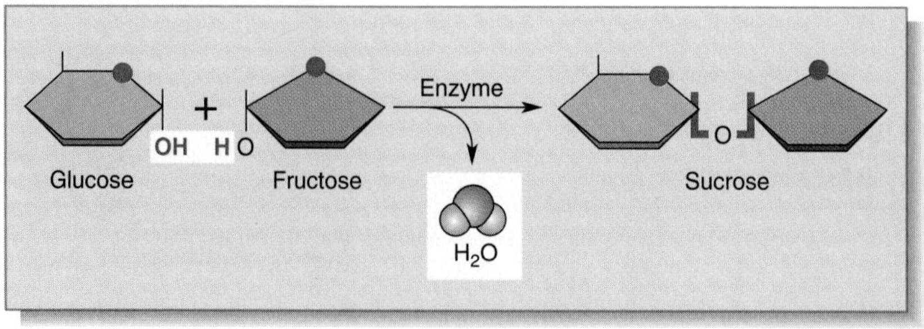

Glucose + Fructose →(Enzyme)→ Sucrose

H_2O

hydrolysis, (*hydro-,* "water"; *lysis,* "rupture"), a molecule of water is added as the glycosidic linkage is broken:

$$A—O—B \quad + \quad H—O—H \quad \rightarrow \quad A—OH + HO—B$$
(disaccharide plus water yields 2 monosaccharides)

This is what happens to foods you eat, but you probably call the process **digestion** instead of hydrolysis. Digestion is necessary because your gut can't absorb sucrose very efficiently, and whereas the cells of your body can't use the disaccharide, they can use glucose and fructose. Hydrolysis is a common reaction in the digestion of carbohydrates, lipids, proteins, and nucleic acids.

Another disaccharide of interest is **lactose,** which consists of glucose and another six-carbon sugar, **galactaose** (Figure 3.3). Since it is found only in milk it is called a "milk-sugar." Lactose is not as sweet as sucrose, and whereas babies thrive on it, many adults (including nearly all non-Caucasian adults) who consume milk products suffer from **lactose intolerance.** They lack the enzyme that breaks lactose into its two subunits—glucose and galactose—and therefore can't digest it. Instead, lactose passes into the colon (bowel), where it is attacked by gas-forming bacteria, often causing a painful accumulation of gas.

Polysaccharides

The more complex carbohydrates, the **polysaccharides** (*poly,* "many"), are longer chains of simple sugars (Figure 3.4). They may contain hundreds to thousands of monosaccharide subunits, covalently linked into polymers. The polysaccharides serve principally as food storage and structural molecules. **Glycogen,** a common polysaccharide, is formed by animals as a means of storing glucose. Foods such as potatoes, wheat and corn flour, whole grains, seed vegetables, and fruits contain polysaccharides in the form of **plant starches.** The two most common starches are amylose and amylopectin—important food reserves for higher plants. Then there is cellulose and chitin, two polysaccharides used not as storage carbohydrates but as structural material. **Cellulose** is a structural polysaccharide in plant cell walls. **Chitin** is used by insects to form their exoskeletons (external skeletons) and by fungi to form the walls surrounding their cells.

Glycogen, the storage polysaccharide of animals, is usually a much larger molecule than either amylose or amy-

lopectin and is highly branched (Figure 3.4a). Glucose units in its many chains are joined through 1–4 dehydration linkages, but the branches join the chains with 1–6 linkages. (The numbers refer to carbons in the glucose molecule—see Figures 3.1–3.3). Glycogen is a temporary, short-term storage unit in animal cells, particularly prevalent in the liver and muscles of vertebrate animals. It is broken down into glucose to meet the energy demands of the body.

Amylose is the simplest starch (Figure 3.4b). Its molecules consist of unbranched chains of hundreds of glucose subunits, joined together by 1–4 dehydration linkages. Potato starch is about 20% amylose. The unbranched chain represents the primary structure of amylose. In its secondary structure, the chain spontaneously coils into a helical, or coiled, shape and is held that way by numerous hydrogen bonds along its length.

Amylopectin (see Figure 3.4b) makes up the other 80% of potato starch. It is a large, highly branched polymer. Its main chain of 1–4 linked glucose subunits gives rise to side branches that form via 1–6 glucose linkages. Such branches, in turn, produce their own branches, again through 1–6 glucose linkages.

Starches form a large part of our diet, perhaps too large a part. Most snack foods are high-calorie foods largely because of their high starch content (along with substantial amounts of sugar and fat). Nevertheless, the starches are vitally important storage carbohydrates, which are readily metabolized for energy.

Cellulose, a structural polymer (Figure 3.4c), is the most abundant polysaccharide on earth, forming much of the cell wall structure in plants and algae. Cellulose accounts for the great structural strength of wood and plant fibers.

Cellulose is a linear polymer of glucose subunits put together with 1–4 glycosidic linkages. If that sounds familiar, it's because the storage polysaccharides just discussed also consist of 1–4 linkages. Nonetheless, the other polysaccharides and cellulose are very different. Storage polysaccharides are fairly soluble and have little structural strength, whereas cellulose is insoluble and has great tensile (stretch) strength. So what makes cellulose so different?

Our discussion so far has centered about **alpha glucose,** but there is also a **beta** form of the molecule. The difference between the two is in the orientation of hydrogen and hydroxyl (—H and —OH) groups on the number 1 carbon, as seen

FIGURE 3.3
The Making of Lactose: Milk Sugar
The monosaccharides galactose and glucose are identical except for the orientation of —H and —OH groups on carbons 1 and 4. Together, they form the disaccharide lactose (milk sugar) through the usual dehydration reaction, but note that one of the subunits (here we've made it glucose) turns over when the 1–4 linkage forms.

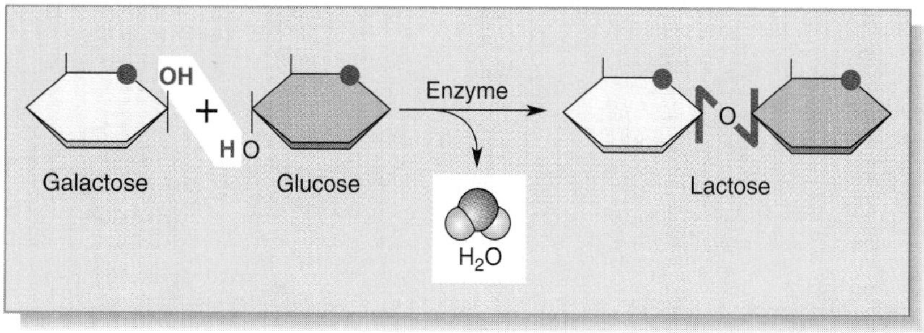

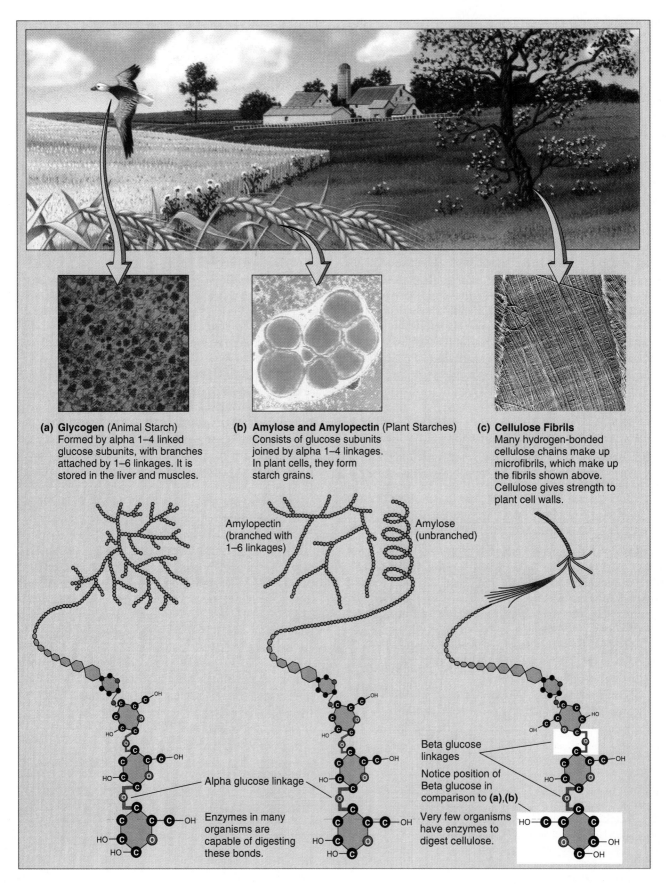

(a) Glycogen (Animal Starch)
Formed by alpha 1–4 linked glucose subunits, with branches attached by 1–6 linkages. It is stored in the liver and muscles.

(b) Amylose and Amylopectin (Plant Starches)
Consists of glucose subunits joined by alpha 1–4 linkages. In plant cells, they form starch grains.

(c) Cellulose Fibrils
Many hydrogen-bonded cellulose chains make up microfibrils, which make up the fibrils shown above. Cellulose gives strength to plant cell walls.

Amylopectin (branched with 1–6 linkages)

Amylose (unbranched)

Alpha glucose linkage

Enzymes in many organisms are capable of digesting these bonds.

Beta glucose linkages

Notice position of Beta glucose in comparison to (a),(b)

Very few organisms have enzymes to digest cellulose.

FIGURE 3.4

in Figure 3.4c. Because of this orientation in beta glucose, every other glucose in the chain is inverted (turned "upside down"). It is this difference that gives cellulose its unique properties among polysaccharides.

For example, relatively few organisms produce the enzyme **cellulase**, necessary to hydrolyze the beta glycosidic linkages. Included in the list of cellulose digesters are certain bacteria, protists, and fungi. Among the few animals with this capability are silverfish and garden snails. But what about the many kinds of grazing mammals and the wood-eating insects such as termites? Interestingly, they can't digest cellulose themselves. Instead, they harbor helpful intestinal bacteria and protists that do have the right enzymes. The animals simply absorb the digested products.

What about people? People can eat all the cellulose they want and it will just pass right through. In fact, it may go through fairly fast because fiber, which is what we are really describing, stimulates the intestinal lining, causing lubricating fluids to be released and the gut to more actively move its contents along. (Whole-bran cereals are rich in fiber.) Nutritional researchers are convinced that enhancing the expediency of nature's call through increased fiber intake reduces the time that potential carcinogens (cancer-causing chemicals) are in contact with the bowel, thereby reducing the chance for cancers to develop there.

Digestive considerations aside, the glycosidic linkages of cellulose produce other unusual characteristics. Cellulose polymers do not form coils. They remain linear, and each polymer is attracted to others by hydrogen bonds, until some 60–70 are drawn together into dense, cable-like strands called

FIGURE 3.4
Polysaccharides
(a) Glycogen is a large, highly branched polymer of glucose. Its main chain and branches contain numerous glucose subunits joined through 1–4 linkages. The branches are also formed through 1–4 linkages, but they are joined to the main chain through 1–6 linkages. (b) The plant starch amylose, a polysaccharide, consists of hundreds of glucose subunits joined into a simple chain by 1–4 linkages. When in water, the chain spontaneously forms a helix, and in this form is insoluble. In plant cells, such as those of the potato tuber, amylose chains (in combination with amylopectin chains) form readily visible starch grains (stained blue in photo). Amylopectin, like glycogen, is a branched polysaccharide. Like glycogen, its chains are glucose units joined by 1–4 linkages, with the branches linked to the main chain by 1–6. Compare the branching of amylopectin with the coiling of amylose. (c) Like other polysaccharides, cellulose consists of 1–4 linked glucose subunits, but because beta glucose is the subunit, the structure of cellulose and some of its chemical and physical characteristics are quite different from those of amylose. Note the orientation of — H and — OH side groups on the number 1 carbon of beta glucose and how this affects the positioning of each beta glucose subunit in the chain. Note also that there is an absence of branches.

microfibrils (Figure 3.5). Whereas the individual hydrogen bonds are quite weak, the great number in each microfibril provides considerable strength. Cellulose microfibrils, when organized into larger strands called **fibrils**, are used in the construction of plant cell walls. As the fibrils are laid down, they take on a laminated and crisscross arrangement. Then, as growth of the cell wall is completed, cement-like hardening and strengthening agents are added. As a result, plant cell walls are among the strongest of biological structures. If you don't believe this, go throw your best punch at a tree, and report back. The walls of countless previously living cells in a tree trunk, when impregnated with another chemical, **lignin**, become the tough, useful wood of woody plants. But in spite of its strength, cellulose is also flexible. Watching trees sway in strong winds is adequate evidence of this.

In the young, tough walls that surround plant cells, cellulose fibrils are imbedded in a gluey matrix of **hemicelluloses** and **pectins**. The hemicelluloses, despite the name, are not structurally related to cellulose. Instead, they include polymers of some of the less common 5-carbon sugars. Pectin is composed of 6-carbon sugars derived from the monosaccharide galactose. It is used by plants to cement newly formed cell walls together. Pectin extracts are used commercially and in the home to give preserves their "jelly" consistency.

Chitin is a principal constituent of the exoskeletons (external skeletons) of insects, lobsters, crabs, and other arthropods (Figure 3.6), and occurs in the cell walls of fungi. Chitin in the exoskeletons of grasshoppers and cockroaches is rather flexible and leathery, but it can become very hard when impregnated with calcium, as in lobsters and crabs. Chitin makes use of a form of beta glucose that contains an amino group (NH_2) on its number 2 carbon. Attached to the amino group is an acetyl group ($CH_3C = O —$). In the logic of biochemistry, it is given the name *N*-**acetylglucosamine**. Chitin, like cellulose, is indigestible to most animals. (Table 3.1 reviews the major carbohydrates.)

The Lipids

The **lipids** are a diverse group of molecules, defined by their solubility rather than by their structure. Lipids tend to be greasy or oily, and fat-soluble rather than water-soluble. They are generally nonpolar and therefore dissolve in nonpolar solvents, such as chloroform and ether, but not in water, which is a polar solvent. Where carbohydrates and proteins tend to be hydrophilic ("water-loving"), lipids tend to be hydrophobic ("water-fearing"; see Chapter 2).

Lipids may be small molecules, large molecules, monomers, polymers, energy storage molecules, structural molecules, hormones, lubricants, or parts of proteins and carbohydrates. Important groups within the lipid category are **fats**, **oils**, **sterols**, **waxes**, and **phospholipids**. We will begin with the fats and oils, lipids that are in a category called triglycerides.

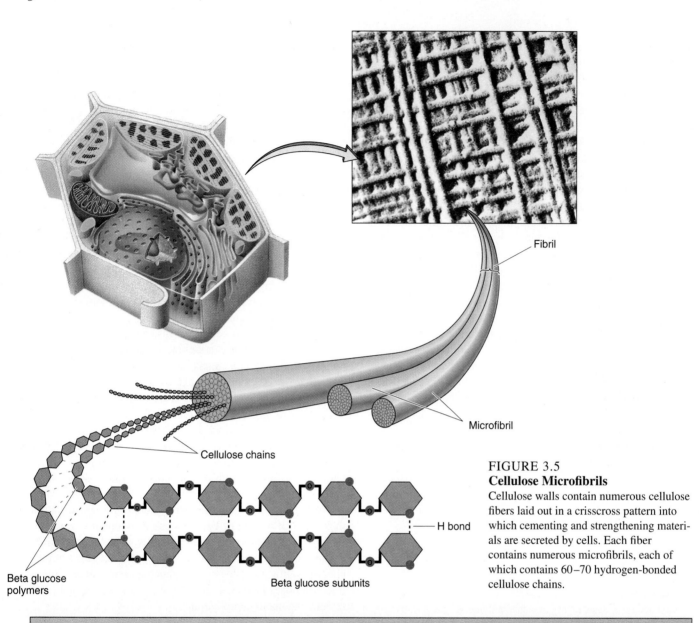

Fibril

Microfibril

Cellulose chains

H bond

Beta glucose polymers

Beta glucose subunits

FIGURE 3.5
Cellulose Microfibrils
Cellulose walls contain numerous cellulose fibers laid out in a crisscross pattern into which cementing and strengthening materials are secreted by cells. Each fiber contains numerous microfibrils, each of which contains 60–70 hydrogen-bonded cellulose chains.

TABLE 3.1
The Major Carbohydrates

Name	Structural Information	Functions
Glucose	Monosaccharide, six carbon: $C_6H_{12}O_6$	Molecular subunit or building block for most carbohydrates
Sucrose	Disaccharide, 12 carbon: $C_{12}H_{22}O_{11}$	Storage carbohydrate, used for energy
Amylose	Polysaccharide (starch); single chain with 1–4 linkages; forms alpha helix	Storage starch in plants, common in potatoes and grains
Amylopectin	Polysaccharide (starch); branched chain with 1–4 and 1–6 linkages	Storage starch in plants, common in potatoes and grains
Glycogen	Polysaccharide; heavyweight; highly branched chain with 1–4 and 1–6 linkages	Glucose storage in animals, ready supply of energy
Cellulose	Polysaccharide; heavyweight, unbranched chain of beta glucose subunits, 1–4 linkages with every other glucose inverted. H-bonds gather polymers into dense strands	Structural role; forms plant and algal cell walls
Chitin	Polysaccharide; polymer of N–acetylglucosamine subunits	Structural role; forms arthropod exoskeleton and fungal cell wall

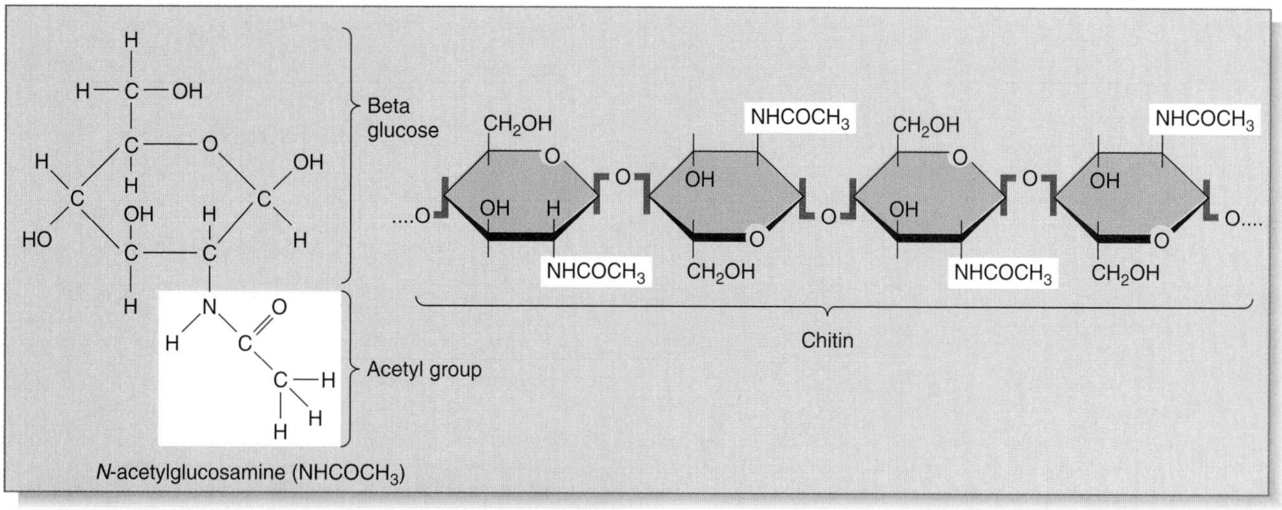

N-acetylglucosamine (NHCOCH₃)

Beta glucose

Acetyl group

Chitin

FIGURE 3.6
Chitin and the Exoskeleton
Lobsters and other arthropods have an exoskeleton containing the complex carbohydrate chitin, a polymer of *N*-acetylglucosamine subunits. In many aquatic arthropods (in both fresh and salt water), the exoskeleton is hardened by calcium.

Triglycerides

Everyone is familiar with animal fats (such as lard) and vegetable oils (such as corn oil, peanut oil, and safflower oil). The difference between a fat and an oil is that a fat has a higher melting point; fat is solid at room temperature but oil, with its lower melting point, is liquid at room temperature. Fats and oils make up the **triglycerides**—compounds consisting of three fatty acid chains attached to a molecule of glycerol. **Glycerol**, chemically an alcohol ($C-OH$), is a small molecule that has three carbons and three hydroxyl side groups. A **fatty acid** consists of a hydrocarbon chain (a chain of carbon atoms with hydrogen side groups) with a carboxyl, or acid, group at one end (Figure 3.7).

Fatty acids can be of many different lengths, occurring most commonly in chains of 14, 16, 18, or 20 carbon atoms. Free fatty acids are not very soluble in water, but if the water is alkaline (has a high pH), the fatty acid carboxyl group becomes ionized and strongly polar, whereupon it does go into solution. Sodium and potassium salts of fatty acids are known as "soap" and are soluble in water. If you can get along without the added perfumes and deodorants, and do not boast an active social life, you can make your own soap from waste kitchen grease and potassium or sodium hydroxide (our forefathers got their potassium hydroxide, or "lye," from wood ashes).

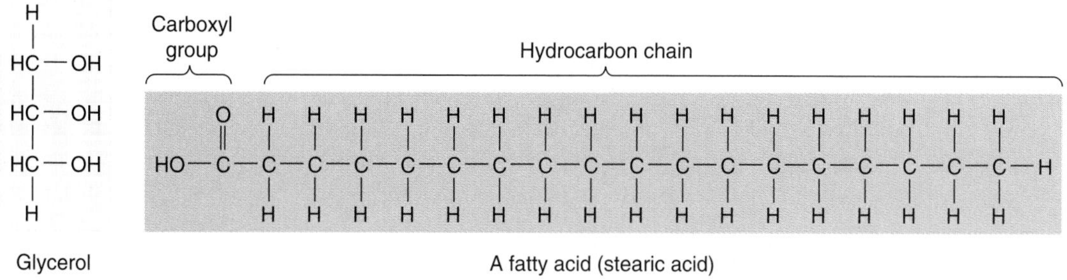

Glycerol

Carboxyl group

Hydrocarbon chain

A fatty acid (stearic acid)

FIGURE 3.7
Triglyceride Subunits
Triglycerides consist of glycerol and fatty acids. The fatty acid seen here is stearic acid, which is a constituent of animal fat. There are many different kinds of fatty acids.

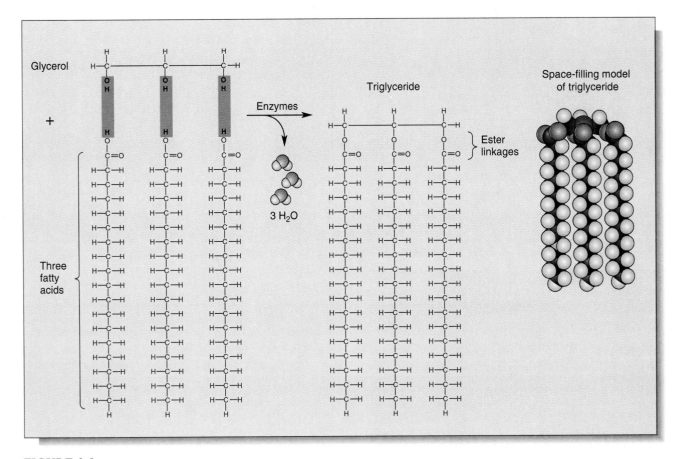

FIGURE 3.8
Formation of a Triglyceride
In the synthesis of a triglyceride, dehydration linkages form between the three hydroxyl (— OH) groups
of glycerol and the hydroxyl (— OH) groups of three fatty acids. The products are one molecule of
triglyceride and three of water.

The synthesis of a triglyceride involves the familiar dehydration reaction. In the presence of an enzyme, three fatty acids become covalently bonded to the glycerol. The reaction also yields three molecules of water (Figure 3.8). Whereas the free fatty acids are polar, triglycerides are completely nonpolar. This is because the charged carboxyl groups of the fatty acids are lost when the fatty acid becomes linked to glycerol (see Figure 3.8). A dehydration linkage between a carboxyl group and an — OH side group is called an **ester linkage**:

$$X - \overset{\displaystyle O}{\overset{\|}{C}} - O - C - X$$

Saturated, Unsaturated, and Polyunsaturated Fats

Why all the diet talk about saturated fats? The obvious question here is, saturated with what? Saturated with hydrogen, of course. A **saturated fat** is a triglyceride that contains saturated fatty acids, those containing all the hydrogen atoms possible. In saturated fatty acids, the carbon backbone is composed of carbon-to-carbon single bonds (— C — C — C — C — C — C — etc.). However, unsaturated fatty acids have double bonds between one or more pairs of carbons in the chain: (— C — C = C — C — C = C — C — etc.). Therefore the latter have fewer hydrogens (Figure 3.9). Fatty acids with more than one carbon-to-carbon double bond are described as **polyunsaturated**. Such fatty acids include linoleic acid (cottonseed oil) and linolenic acid (linseed oil). Note in Figure 3.9 that the presence of a double bond produces a rigid point in the fatty acid chain, which is represented by an angle in the hydrocarbon tail.

Highly saturated fats tend to have rather high melting points (tend to remain solid, like lard, at room temperature), while polyunsaturated fats tend to have low melting points (tend to remain liquid or "oily"). Simple vegetable oils, such as peanut oil, are largely unsaturated, which explains why the oils in old-fashioned (nonhydrogenated) peanut butter separate, forming a layer at the surface.

In spite of constant warnings about our eating too much saturated animal fat (see Chapter 37), unsaturated fatty acids are essential to good nutrition. They are precursors to impor-

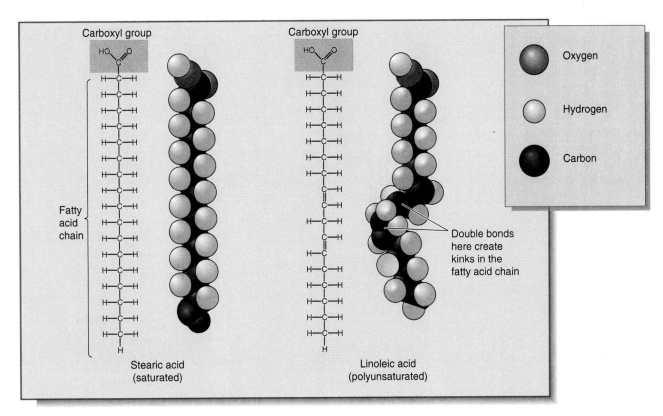

FIGURE 3.9
Saturated and Unsaturated Fatty Acids
Stearic acid *(left)* is a principal fatty acid of animal fat. As in other saturated fats, its carbon chain contains
the maximum number of hydrogens. Linoleic acid *(right)*, abundant in cotton seeds, is an unsaturated fatty
acid (technically *poly*unsaturated because there is more than one double bond). Note the kinks where the
double bonds occur in the lengthy hydrocarbon chain.

tant molecules we must produce, and the membranes of our cells are rich in unsaturated fatty acids. Furthermore, because chemical processes in humans and other mammals cannot produce double bonds beyond the ninth carbon in a given fatty acid chain, it is necessary to include a small amount of unsaturated fats in the diet. That is why linoleic and linolenic acids, two such unsaturated fatty acids, are known as *essential fatty acids.* Actually, the essential fatty acids are so common in foods that lipid nutritional deficiencies rarely arise.

Because of their molecular properties, fats make ideal energy storage molecules (gram for gram they contain about twice the calories of carbohydrates). Plants store triglycerides in their seeds, and many animals store body fat as an adaptation to lean seasons, hibernation, or migration. Humans also store fat under the skin and around internal organs, a perfectly normal situation, but one that can easily get out of hand.

Phospholipids

Phospholipids are major components of the delicate membranes of cells (see Chapters 4 and 5). The **plasma mem-**

brane, as the surrounding membrane is known, consists primarily of two layers of phospholipids, along with **glycolipids** ("sugar lipids"), **glycoproteins** ("sugar proteins"), and several kinds of proteins. As you can see, glycolipids and glycoproteins are lipids and proteins with sugar groups, actually sugar chains, attached.

While the phospholipids are structurally related to the triglycerides, there are obvious differences between them. Phospholipids have just *two* fatty acids linked to the glycerol backbone. In place of the third fatty acid of the triglyceride, they have a phosphate-bearing complex. The phosphate is commonly linked to still other molecular groups such as **choline**—as shown in Figure 3.10a,b. We refer to the whole complex as a "phosphate head." Of the two fatty acids, one is commonly saturated, while the other is unsaturated. As mentioned earlier, the carbon-to-carbon double bond produces a peculiar bend in the tail. This is important. In the last chapter we emphasized how the shape of molecules influences their function in cells. In a later chapter we will see how the presence of the angles in the phospholipid tails affects the properties of the plasma membrane (see Chapter 5).

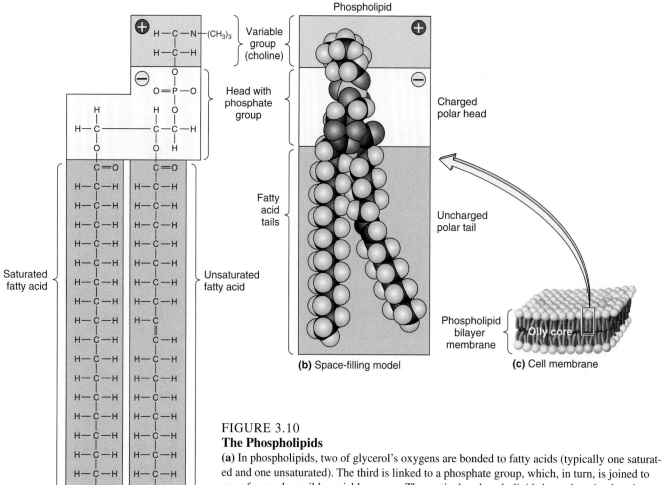

Phospholipid

Variable group (choline)

Head with phosphate group

Fatty acid tails

Saturated fatty acid

Unsaturated fatty acid

(a) Structural formula of phosphatidylcholine

Charged polar head

Uncharged polar tail

(b) Space-filling model

Phospholipid bilayer membrane

Oily core

(c) Cell membrane

FIGURE 3.10

The Phospholipids

(a) In phospholipids, two of glycerol's oxygens are bonded to fatty acids (typically one saturated and one unsaturated). The third is linked to a phosphate group, which, in turn, is joined to one of several possible variable groups. The particular phospholipid shown here is *phosphatidylcholine*. Phospholipids are common constituents of cell membranes. **(b)** The space-filling model reveals the peculiar bent form of the unsaturated fatty acid. **(c)** When assembled into cell membranes, phospholipids form a bilayer, the charged, polar heads facing outward and the uncharged, nonpolar tails inward. The tails intermingle, producing a distinctly oily core.

When assembled into membranes (Figure 3.10c), the phospholipids are arranged in bilayers, tail to tail, with the phosphate heads facing outward. The cluster of fatty acid tails gives the core of the membrane an oily characteristic, which is also important to its functions.

The phosphate group bears negative charges, which makes it hydrophilic; thus the charged heads mingle freely with water and other polar molecules. The tails themselves are uncharged and thus hydrophobic, avoiding water and mingling only with nonpolar substances. The presence of both hydrophobic and hydrophilic groups makes the phospholipid an **amphipathic** molecule. As we will see, many other components of cell membranes, including glycoproteins and glycolipids, have this property.

Left on their own in water, phospholipids spontaneously assemble themselves into lipid bilayers that take the form of spheres known as **micelles**. Under certain conditions, the bilayer forms flattened sheets similar to cell membranes (Fig-

ure 3.11). The hydrophilic heads face outward and interact with water and their hydrophobic tails face inward, forming the oily core. (Like the soap film that forms in the plastic ring you used to blow bubbles with. Remember?)

The interaction of the polar heads with water involves both electrostatic attractions and hydrogen bonding. The intermingling and packing of the phospholipid tails involves **hydrophobic interaction**. Literally, "grouping together through fear of water." Actually, the intermingling and isolation of the tails are chiefly brought about by the surrounding water molecules, which have a much higher affinity for each other than for the polar fatty acid tails. (As an analogy, think of how boys and girls tend to form separate clusters at the first junior high dance: the boys drawn together by mutual shyness and girls by mutual disdain!)

Hydrophobic interaction and the packing of the hydrocarbon tails are reinforced by weak attractions called **van der Waals forces**. The van der Waals forces occur when large

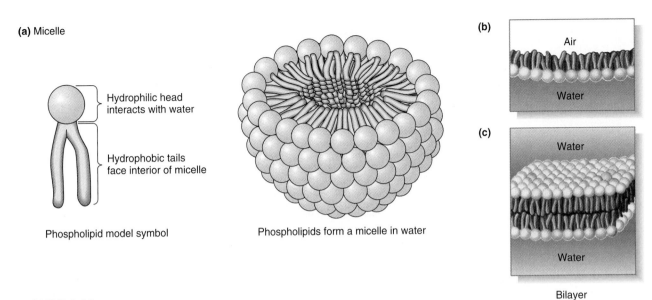

(a) Micelle

Hydrophilic head interacts with water

Hydrophobic tails face interior of micelle

Phospholipid model symbol

Phospholipids form a micelle in water

(b) Air / Water

(c) Water / Water

Bilayer

FIGURE 3.11
The Behavior of Phospholipids in Water
When introduced to water, individual phospholipids cluster together in a characteristic manner. The polar (charged) heads interact readily with water, whereas the nonpolar (uncharged) hydrocarbon tails cluster together. **(a)** Minute droplets called micelles form spontaneously. **(b)** When exposed to air, the nonpolar tails form the air–water interface; the heads interacting with water below. **(c)** A membrane-like lipid bilayer will form under certain conditions.

numbers of molecules draw close together. The attractions occur because the ongoing movement of electrons brings about transient shifts in charges that result in slight electrostatic attractions. There is a minimal distance they keep, however, because as electrons overlap they begin repelling each other.

The resulting phospholipid bilayer has three important characteristics: (1) the bilayers tend to self-perpetuate and become extensive; (2) they tend to close in on themselves, leaving none of the hydrophobic tails exposed to water; and (3) they automatically repair any holes or tears that might appear. We can conclude that the lipid bilayer, as described, is probably the most energetically favorable arrangement possible for phospholipids in water. We've dwelt on this aspect of lipid behavior for a good reason. It has important implications concerning the evolution of the earliest cells, as we will learn in future chapters.

Other Lipids

Waxes consist of one fatty acid covalently joined by an ester (C—O—C) linkage to a long-chain alcohol; there is no glycerol. Beeswax consists largely of an ester of palmitic acid and an alcohol. Because of its hydrophobic, water-repelling properties, wax is useful to a variety of living things, especially those that need to retain body water. Insects generally have waxy body surfaces. So do plants; the waxes of leaves, fruit skins, and flower petals are examples. We manufacture **ceru-men**, more commonly called "ear wax," that, along with our ear hairs, helps keep dust and small insects out of the ear canal. Waxes are generally much harder and even more hydrophobic than fats—two reasons why commercial wax makes a good protectant for surfaces.

Steroids are structurally different from other lipids; since they contain so many carbon–hydrogen groups, they are essentially hydrophobic and nonpolar. All steroids contain four interlocking rings of carbon and hydrogen (Figure 3.12) but differ in the side groups protruding here and there. In some instances the side groups, such as the —OH seen in cholesterol, are polar, which makes this region of the steroid slightly hydrophilic.

Cholesterol, a common steroid, is a substance we've all heard of because various food manufacturers have assaulted us with the message that their products don't contain any. Why should we care? We might be interested because cholesterol (see Figure 3.12) is known to contribute to arterial plaque (thickening) in the lining of these blood vessels. This reduces the vessel's internal diameter and increases blood pressure (the way one might increase water pressure by crimping a garden hose).

In spite of these facts, cholesterol is an important nutrient with several vital functions. Like phospholipids, cholesterol is a major constituent of plasma membranes in animals. There, its four polar carbon rings are submerged into the phospholipid core with only its polar —OH group exposed. Cholesterol is also used to form bile acids, which are necessary for fat digestion. When exposed to sunlight in the skin, a deriva-

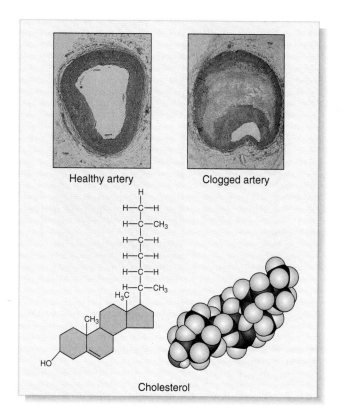

Healthy artery   Clogged artery

Cholesterol

FIGURE 3.12
Cholesterol and Arterial Disease
Cholesterol plays many roles in humans and other animals. It is modified into vitamin D and steroid hormones, and it is used as a component of cell membranes. However, cholesterol is also implicated in atherosclerosis, a vascular disease in which thickenings and rough spots called plaques form in arterial linings. In addition to restricting blood flow, the roughened linings may trigger blood-clotting, leading to stroke. Compare the healthy artery with the occluded one.

tive of cholesterol is converted to vitamin D (the "sunshine vitamin"), which is necessary for normal bone growth and maintenance. Cholesterol is also used by the body in the synthesis of steroid hormones, including sex hormones. We will come back to the pros and cons of excess cholesterol in the diet in Chapter 37. (Table 3.2 reviews the major lipids.)

The Proteins

Proteins, heavyweights among the molelcules of life, have many functions. Included are support, structure, movement, transport, communication, and disease defense. Protein-containing structures in vertebrate animals include hair, nails, hooves, cartilage, muscle, hormones, antibodies, blood proteins, and enzymes.

Enzymes are perhaps the most interesting of all proteins. We mentioned the basic role of enzymes where dehydration reactions were introduced, but here let's stress that enzymes are biologically active proteins that work in cells as catalysts—agents that speed up chemical activities, making them occur at a fast enough rate to be useful to the organism. (Many such reactions occur without enzymes, but too slowly to serve a useful purpose.) As we proceed into the chemical reactions of cells, you will soon develop an abiding respect, perhaps even an awe, for these amazing mediators of cellular activity, since they seem to be involved in nearly every conceivable aspect of life.

What Is a Protein?

A **protein** is a macromolecule composed of one or more **polypeptides**. A polypeptide in turn, is, a linear chain of covalently linked **amino acids**. The covalent linkage in this case is often called a **peptide bond**. A polypeptide can be siz-

TABLE 3.2		
The Major Lipids		
Name	**Structural Information**	**Functions**
Triglycerides	One glycerol and three fatty acids	Energy and material storage in plants and animals
Saturated	Single carbon to carbon bonds; only maximum amount of hydrogen; nonpolar	Chiefly animal fats; storage and thermal insulation
Unsaturated	Some double carbon to carbon bonds, less than maximum amount of hydrogen; nonpolar	Chiefly plant oils; storage in seeds
Phospholipids	One glycerol, with variable charged group, two fatty acids, one of which may be unsaturated polar	Chiefly structural; basic structure of plasma membrane and other cell membranes
Steroids	Four interlocking rings with side groups that are polar	Cholesterol: present in animal cell membranes; precursor to vitamin D and steroid hormones
Waxes	Chiefly palmitic fatty acid and an alcohol; highly nonpolar	Waterproofing in insects and plants; ear wax

able, often containing hundreds of amino acids. On a considerably smaller scale is the **peptide**, a short chain of amino acids joined together by peptide bonds. Then there is the **dipeptide**, two amino acids joined by a single peptide bond. Such bonds are formed essentially through the usual enzyme-mediated dehydration reaction.

Proteins fall into two general categories related to shape: **globular proteins** and **fibrous proteins**. Globular proteins, so-called because they fold spontaneously into compact globs, include enzymes, hormones, and antibodies. Fibrous protein tends to remain linear, although, as we will see, fibrous proteins take on a "zigzag" or "fan-folded" configuration along their length. Fibrous protein occurs in muscle, ligaments, and tendons. The protein **collagen**, which makes up most of the structure of ligaments and tendons, is the most common protein in the vertebrate body.

The range in molecular weight of the various proteins is enormous—from about 5700 for insulin (which has 55 amino acids) to 7–8 million in some of the largest enzymes (those with 70,000–80,000 amino acids). Let's begin with the protein subunits, the amino acids.

Amino Acids

Amino acids are the molecular subunits, the "building blocks," of proteins. There are just 20 different primary amino acids found in the proteins of living cells. Twenty may seem to be a small number considering how diverse and complex proteins are, but we might bear in mind that this book was written with only 26 letters (some may have been used more than once). Actually, there are more than 20 total amino acids but the additional ones are secondary; that is, they are derived from the primary amino acids.

All amino acids have the same core structure. All contain an **alpha carbon** to which an **amino group** (NH₂), a **carboxyl group** (COOH), a hydrogen atom, and an **R group** (variable group) are bonded:

$$\begin{array}{ccc} H & R & O \\ \diagdown & | & \diagup\!\!\!\!/ \\ N\!-\!C\!-\!C\!-\!OH \\ \diagup & | \\ H & H \end{array}$$

Figure 3.13 shows a generalized amino acid in its molecular and ionized states and reveals some of its geometry. Amino acids readily ionize under standard physiological conditions, which means within cells and in their surrounding fluids. During ionization, the carboxyl group (acid) gives off a proton, and the amino group (base) takes in a proton. Because of this, an ionized amino acid has both a negative and a positive charge.

It is in their R groups that the 20 amino acids differ. Knowing about the R groups is important because their individual properties and position in a protein strongly affect that protein's geometric shape. Shape, in turn, is critical to a protein's function. We will come back to this point shortly after a look at the R groups themselves.

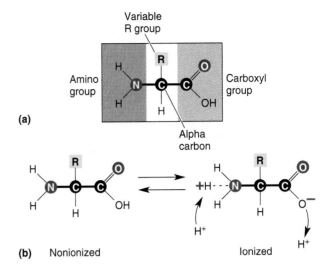

(a)

(b) Nonionized   Ionized

FIGURE 3.13
Amino Acid Structure
(a) Each amino acid contains an alpha carbon to which are attached an amino group (—NH₂), a carboxyl group (—COOH), and one of the 20 different R groups. **(b)** Amino acids commonly ionize, with a proton leaving a carboxyl group and another proton joining an amino group.

The R Groups. Figure 3.14 shows the amino acids arranged according to R-group properties, with particular emphasis on R-group polarity, lack of polarity, the tendency to ionize, and the ability to form disulfide bridges (covalent bonds).

Note in Figure 3.14 that there are eight nonpolar, uncharged amino acids; their R groups are mainly hydrocarbons, so there is no tendency for these groups to form hydrogen bonds or to ionize. Nonpolar amino acid R groups are hydrophobic, and, under the usual watery conditions in the cell, nearby nonpolar amino acids are clustered together through *hydrophobic interaction*.

The second group of seven amino acids has polar R groups. The polar R groups readily form hydrogen bonds with water, so the corresponding amino acids are hydrophilic. Polar R groups also form *hydrogen bonds* with other polar R groups in the protein, should they be close by.

The R groups of the remaining five amino acids ionize in water and thus bear positive or negative charges. These charges are in addition to those formed when the amino and carboxyl groups of the alpha carbon ionize, as explained earlier. For example, the R group of the amino acid glutamic acid has its own carboxyl group. When ionized, glutamic acid gives off two protons, one from the alpha carboxyl group and one from the R-linked carboxyl group, and thus takes on a net negative charge. Complementing these amino acids are those whose R groups have their own amino groups. When ionized, these amino acids take in two protons, one joining the alpha amino group, the other the R-linked amino group. Accordingly, these amino acids take on a net positive charge. So some amino acids have

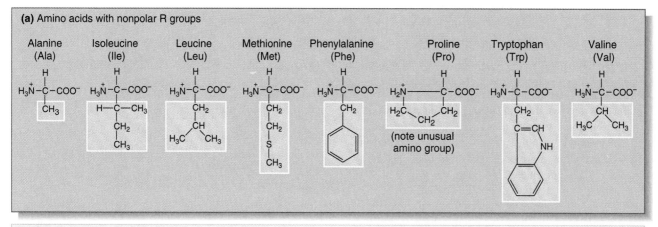

(a) Amino acids with nonpolar R groups

Alanine (Ala) · Isoleucine (Ile) · Leucine (Leu) · Methionine (Met) · Phenylalanine (Phe) · Proline (Pro) *(note unusual amino group)* · Tryptophan (Trp) · Valine (Val)

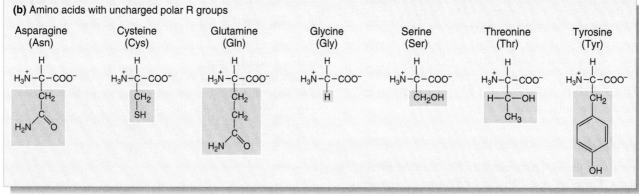

(b) Amino acids with uncharged polar R groups

Asparagine (Asn) · Cysteine (Cys) · Glutamine (Gln) · Glycine (Gly) · Serine (Ser) · Threonine (Thr) · Tyrosine (Tyr)

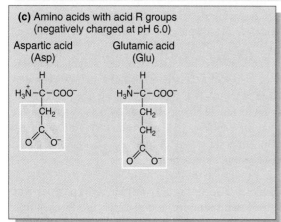

(c) Amino acids with acid R groups (negatively charged at pH 6.0)

Aspartic acid (Asp) · Glutamic acid (Glu)

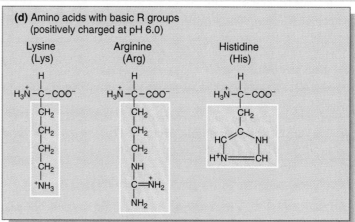

(d) Amino acids with basic R groups (positively charged at pH 6.0)

Lysine (Lys) · Arginine (Arg) · Histidine (His)

FIGURE 3.14
Amino Acids and Their R Groups
Proteins are made up of 20 different amino acids. Whereas each (except for proline) has identical amino and carboxyl groups, the R groups differ. The R groups fall into four general types: **(a)** nonpolar (green), **(b)** uncharged polar (yellow), **(c)** charged acidic (red), and **(d)** charged basic (blue). The R groups are highly significant to the final shape taken by the protein, and shape is critical to the specific function.

net charges. Importantly, oppositely charged amino acids within a protein will interact with each other, forming *ionic bonds.*

One of the polar, but uncharged, amino acids, cysteine (pronounced SIS-tuh-een), has a very special property. It contains a sulfhydryl side group ($-$SH). Under certain conditions, the sulfhydryl groups of two adjacent cysteines will react, forming a strong *covalent bond* known as a **disulfide linkage**:

$$R-SH + HS-R \longrightarrow R-S-S-R$$
(two sulfhydryl groups)   (disulfide linkage)

We see then that amino acids have four significant properties: (1) some are nonpolar and hydrophobic and these tend to cluster together in what is called hydrophobic interaction; (2) others are polar and tend to form hydrogen bonds with

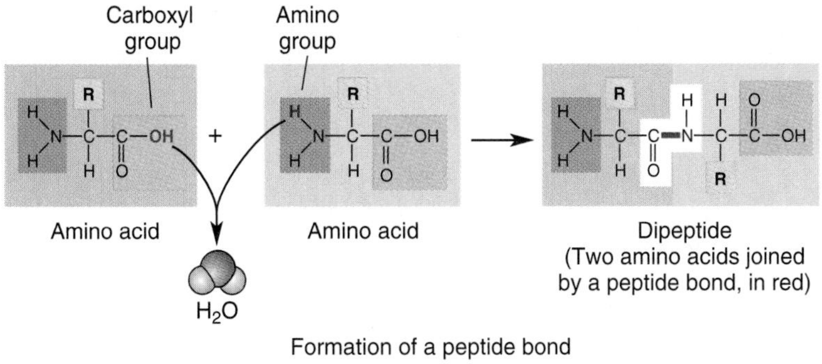

FIGURE 3.15
Dipeptides
Peptide bonds form through the dehydration process. They are linkages that join two amino acids by their carboxyl and amino groups as shown here. The product is a dipeptide.

water and with each other; (3) some ionize in water, take on charges, and form ionic bonds with each other; and (4) some tend to form covalent disulfide linkages when opportunities appear. Why is all this important?

As we have been saying, the shape taken by a protein is determined by its amino acid content. It is the interaction of amino acids—as they enter into hydrophobic interactions, form hydrogen bonds, and undergo ionic and covalent bonding—that draws the protein into its final shape. Each finished protein takes on a highly specific three-dimensional **conformation**, as its molecular shape is known, and conformation is critical to each protein's specific function. For example, many proteins do their work through two processes: recognition and binding. An enzyme, for instance, must first recognize a specific molecule, called its **substrate**, among all of the many kinds of molecules in a cell. Recognition occurs because the enzyme has a specific shape that matches only its own particular substrate. Once recognition occurs, binding of the substrate to the enzyme follows. Only then can the chemical work of the enzyme begin. We will turn to the assembly of proteins now and return to these important ideas shortly.

Polypeptides and Proteins

In polypeptides, the carboxyl group of each amino acid is linked by a **peptide bond** (specifically, the covalent bond between amino acids) to the amino group of the next amino acid (Figure 3.15); the carboxyl group of the second amino acid is linked to the amino group of the third; and so on along the polypeptide. By convention, the "beginning" of a polypeptide is its **N terminal**, or free amino end (the first amino acid in the chain), and the "end" is the **C terminal**, or free carboxyl end (the last amino acid).

Levels of Protein Structure. Proteins may have as many as four levels of structure, and most have at least three. We will refer to these as primary, secondary, tertiary, and quaternary (Figure 3.16).

The **primary structure** is determined by the *number, kind,* and *order* of amino acids in the polypeptide strand (Figure 3.16 [1]). The amino acid sequence, in turn, is genetically

determined. This means that each of the many different proteins in a cell is determined by a specific gene. More specifically, each gene contains a coding that, when translated, reads "place this amino acid here, and that amino acid there." This information is known as the **genetic code**.

The **secondary structure** of a protein (Figure 3.16 [2]) occurs spontaneously, as the polypeptide is synthesized. The secondary structure can be a right-handed coil or a zigzag pleated sheet, or both. The coil, technically an **alpha helix**, forms as a result of hydrogen bonding along the strand as seen in Figure 3.16 [2]. Hydrogen bonds form between double-bonded oxygens ($=O$) of carboxyl groups and the hydrogen of amino groups ($-NH$). The individual hydrogen bonds in an alpha helix are quite weak, but there are so many of them that the coil becomes a stable configuration. Alpha helices occur in globular proteins.

The zigzag, back-and-forth folding of a polypeptide is called a **beta sheet**. The zigzag form is maintained by hydrogen bonds between adjacent polypeptide strands (see Figure 3.16 [2]) or in the same strand where it makes a U-turn and runs parallel. As in the alpha helix, the hydrogen bonds form between double-bonded oxygens in one polypeptide and the hydrogen of $-NH$ groups in the other. The beta sheet is the principal conformation in many fibrous proteins and also occurs here and there in large, globular proteins.

The **tertiary** (third) **level** of protein structure involves highly specific looping and folding of the polypeptide (Figure 3.16 [3]), brought about by interactions between various R groups along the length of the polypeptide. Different polypeptides bend and loop in their own unique manner because each has its own sequence of amino acids. Thus the tertiary structure of the protein is rigidly set by its primary organization.

Four forces produce tertiary folding. (1) Wherever nonpolar amino acids occur in large numbers, *hydrophobic interactions* occur with the amino acids, forming a dense, hydrophobic cluster. Such interactions generally occur deep within the globular protein. (2) Next, adjacent polar R groups form *hydrogen bonds* with one another, these involving $-NH$ or $-OH$ groups and opposing $-N$ and $-O$ groups (see Chapter 2). (3) *Ionic attractions* occur between oppositely charged R groups. Each of the three forces is individually

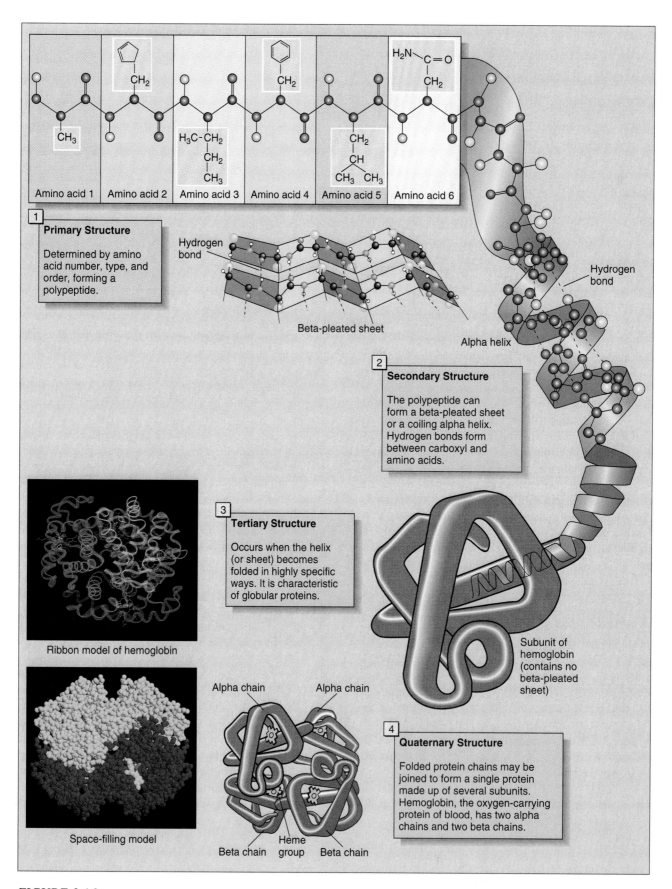

Amino acid 1 **Amino acid 2** **Amino acid 3** **Amino acid 4** **Amino acid 5** **Amino acid 6**

1

Primary Structure

Determined by amino acid number, type, and order, forming a polypeptide.

Hydrogen bond

Beta-pleated sheet

Hydrogen bond

Alpha helix

2

Secondary Structure

The polypeptide can form a beta-pleated sheet or a coiling alpha helix. Hydrogen bonds form between carboxyl and amino acids.

3

Tertiary Structure

Occurs when the helix (or sheet) becomes folded in highly specific ways. It is characteristic of globular proteins.

Ribbon model of hemoglobin

Subunit of hemoglobin (contains no beta-pleated sheet)

Space-filling model

Alpha chain   Alpha chain

4

Quaternary Structure

Folded protein chains may be joined to form a single protein made up of several subunits. Hemoglobin, the oxygen-carrying protein of blood, has two alpha chains and two beta chains.

Beta chain   Heme group   Beta chain

FIGURE 3.16

weak, but when repeated throughout the polypeptide, the combined effect is large. (4) The fourth interaction, *covalent bonding,* a much stronger force, occurs when nearby cysteine amino acids form disulfide linkages (— S — S —). Obviously those proteins lacking cysteine must rely on the other forces alone for maintaining their tertiary structure. (A few polypeptides contain cross linkages involving lysine amino acid.)

Incidentally, the permanent wave industry is based primarily on the disulfide bonds of keratin, the protein constituent of hair. Permanent wave solution contains an agent that breaks disulfide bonds in keratin, allowing the molecules to literally uncoil, becoming quite limp. Then a second agent is used to bring about the formation of new disulfide bonds, hopefully in places that will produce an aesthetically pleasing appearance.

There is another aspect to polypeptide folding and the tertiary level of structure. The actual bends or foldings generally occur at flexible "linker regions," flexible places where the alpha helix fails to form. Such regions often begin with the amino acid proline, which as you will recall, has an unusual ring-like amino group (see Figure 3.14) and thus does not support the formation of a helix. Linker regions are essential because they permit parts of the polypeptide to come together in precise ways. The tertiary level is the final level of organization for proteins containing only a single polypeptide chain.

Some proteins reach a fourth or **quaternary structural level**. Here, two or more polypeptides join to form the finished protein. Some globular proteins are virtual giants. The oxygen-carrying blood protein **hemoglobin** (Figure 3.16 4) has four polypeptides, bound to each other by forces similar to those of the tertiary level. The four polypeptides, actually two pairs, are designated alpha and beta (not to be confused

with the alpha helix and beta sheet). Also present are four very special iron-containing, nitrogen rings called **heme groups**, the actual sites of oxygen transport. The four forces involved in protein structure are reviewed in Figure 3.17.

Proteins can readily be **denatured**, their conformation or shape altered by any chemical or physical agent that affects the forces that determine shape. When this happens, a protein will generally lose its biological function. Heat is an important denaturing agent, as is excess acidity or alkalinity (low or high pH). They cause old bonds to break and new bonds to form randomly in the molecule. The results are generally irreversible. Consider the effect of heat on egg white, which is chiefly water and the protein albumen. (It's hard to "unfry" an egg!) Interestingly, intense cold can also denature protein, but its effects are generally reversible.

Structural Proteins

Structural proteins are important in maintaining the physical form of organisms. They may be *intracellular* (inside cells) or *extracellular* (outside cells). In vertebrates (animals with backbones), common extracellular proteins include **collagen** and **elastin**. These are long, fibrous molecules that wind together to make larger fibers (Figure 3.18a). Collagen, the principal component of connective tissue such as tendons, ligaments, and muscle coverings, makes up about 25% of the protein in humans.

Elastin gets its name from its remarkable ability to stretch. Elastin fibers give elasticity to connective tissue, including that of your ears and skin. Ears are very rich in elastin, so if you reach over and tug on your neighbor's ear, it will snap right back to its original form (if all goes well). You can also use pinching as an age test. If you pinch the skin on the back of your hand and it snaps back, you're young. If a ridge stands there, you're not. This is because with aging, elastin loses its elastic properties. The same phenomenon is responsible for the bagginess under the eyes and about the face and neck that awaits all of us.

Keratin is an intracellular protein (Figure 3.18b). It fills the growing cells of hair, skin, feathers, claws, nails, horns, antlers, and scales, eventually replacing the living material. As these important coverings grow, the dead cell layers are pushed outward by the growing layers below—which is why many of us periodically shave here and there and trim our nails.

Conjugated Proteins

Hemoglobin, with its four heme groups, is an example of a **conjugated protein**: one that contains chemical components other than amino acids (see Figure 3.16 4). Conjugated proteins are generally named for the added chemical components; thus hemoglobin falls in a class of conjugated proteins called **hemoproteins**. Others include **lipoproteins**, which have lipid components; **glycoproteins**, with carbohydrate

FIGURE 3.16
Protein Structural Organization
1 Once the primary structure of the protein is set (number, kind, and order of amino acids), hydrogen bonding acting within the chain produces the secondary structure. Included are two forms: 2 the alpha helix and the beta pleated sheet. The alpha helix *(right)* is produced by hydrogen bonding between amino acids four positions apart along a single polypeptide. The beta pleated sheet *(left)* is maintained by hydrogen bonding between adjacent polypeptide strands. 3 The tertiary structure of the protein is brought about by forces that include hydrogen bonding, ionic attractions, hydrophobic interactions, and covalent bonding (disulfide linkages) within the chain. They produce a highly specific folding in the finished protein. The bending occurs in places where the alpha helix is absent. 4 In the quaternary or fourth level, two or more polypeptide chains are joined through covalent bonding, thus forming a functional protein. The protein red blood cell hemoglobin, for example, includes four polypeptides, designated as alphas and betas (see the schematic). Each has a heme group (iron-containing), where oxygen binds for transport in the blood.

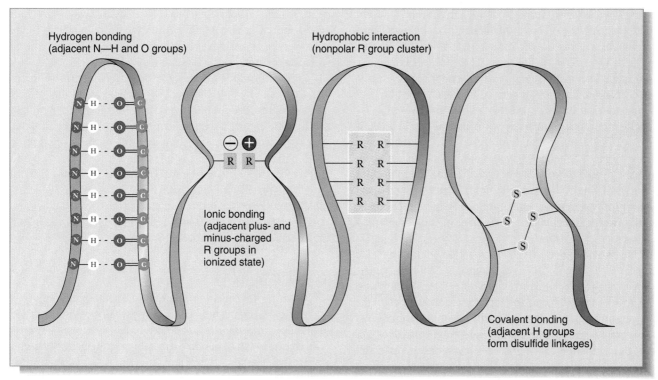

FIGURE 3.17
Summary of Tertiary and Quaternary Forces
The tertiary and quaternary levels of protein structure are brought about by four forces: hydrogen bonding
(among polar R groups), ionic interaction (among charged R groups), hydrophobic interaction (among
nonpolar R groups), and covalent disulfide linkages (between cysteines).

components; and **phosphoproteins**, with phosphate. Lipoproteins and glycoproteins are common in the cell's plasma membrane, where they function as recognition molecules. We will return to this important group of proteins in Chapter 5. (Table 3.3 reviews the major proteins.)

The Nucleic Acids

Probably by now almost everyone has heard of the nucleic acids, particularly DNA, simply because some of the recent discoveries and technical breakthroughs in DNA research have been relentlessly and breathlessly touted by the popular press. Understanding the two nucleic acids, DNA and RNA, is so fundamental to biology that they will be our focus in several chapters (Chapters 9 and 12–17). Here we will be brief, introducing the basic structure and making a few comments on how the nucleic acids do their work.

The acronyms **DNA** and **RNA** stand for **deoxyribonucleic acid** and **ribonucleic acid**, respectively. Both molecules are common to all life. DNA is often referred to as the molecular core of life, since it contains the units of heredity we call **genes**. As we mentioned earlier, genes are chemical codes that

specify the kinds of protein an organism can produce. Proteins are quite significant since, directly or indirectly, each organism's proteins determine both its visible structure and its biochemical capabilities. These are referred to as an organism's hereditary traits. In indirect ways, protein structure helps determine traits such as body build, facial structure, and hair texture. Of even greater fundamental importance, enzymes are proteins, and enzymes are responsible for promoting the countless chemical reactions that characterize life.

While DNA contains the chemical coding required for the assembly of amino acids into protein, it does not participate directly in the assembly process. Instead, the chemical coding is transcribed into RNA, which assembles the amino acids. So for any of our many thousands of genes to carry out their task, they must first bring about the construction of the RNA molecules that actually do the work. RNA, while consisting of linked nucleotides, differs from DNA in a number of ways. It is single-stranded, much shorter in length, and is short-lived. It also makes use of a slightly different sugar and replaces the nitrogen base thymine with one called uracil. We will get into all of this later in the text.

DNA has another important capability. It is self-replicating. This means it can make copies of itself. This is critical for growth, since organisms grow through cell division. When

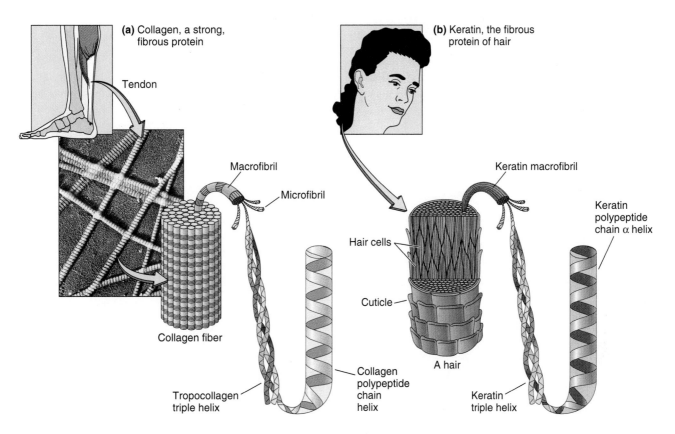

FIGURE 3.18
Structural Protein

(a) Collagen is the principal protein of connective tissues such as the tendon illustrated here. The electron micrograph reveals the rod-shaped macrofibrils that, in turn, are composed of row after row of macrofibrils, collagen molecules laid out in staggered form. (b) Keratin is the principal component of hair. Each hair consists of cells that die after becoming packed with the tough protein. Keratin is organized very much like steel cable, with several levels of twisted strands. The final level is the keratin molecule, which includes three twisted polypeptide strands. The organization provides considerable tensile strength to hair, but you can still get split ends if you don't treat yours right.

TABLE 3.3
The Major Proteins

Name	Structural Information	Functions
Enzymes	Globular; generally quaternary level of structure—two or more polypeptides; heavyweights	Cellular catalysts—speed chemical reactions in cells
Antibodies	Globular; complex quaternary proteins with four polypeptides	Defense against invading organisms and foreign matter in body
Hormones	Globular; tertiary or quaternary level of structure; one or two folded polypeptides; lightweight	Chemical messengers—stimulate responses in target cells
Fibrous protein	Polypeptides generally form side by side array in secondary level of structure, held by hydrogen bonds	Structural protein: keratin in hair, nails, hooves, and claws; collagen in tendons and ligaments
Conjugated protein	Usual polypeptides plus nonprotein side groups—hemes, sugars, lipids, and phosphates	Heme—oxygen transport; others as cell recognition molecules on cell surfaces; important in defense and communication

cells are preparing for division, each DNA molecule goes through replication, and then, at the moment of division, the replicas are properly separated and delivered to the two newly emerging cells. Thus each generation of cells has a complete set of genes. The same holds true for reproduction; DNA must be replicated and copies delivered to developing sperm and egg cells. When egg and sperm join, they bring together the genes of two different individuals.

Nucleic Acid Structure

DNA consists of structural subunits known as **nucleotides** (Figure 3.19). Each nucleotide consists of three parts: a phosphate group, a simple 5-carbon sugar known as **deoxyribose**, and a **nitrogen base**. Nitrogen bases are single- or double-ringed molecules made up of nitrogen and carbon, along with various side groups. Although the phosphates and sugars are the same

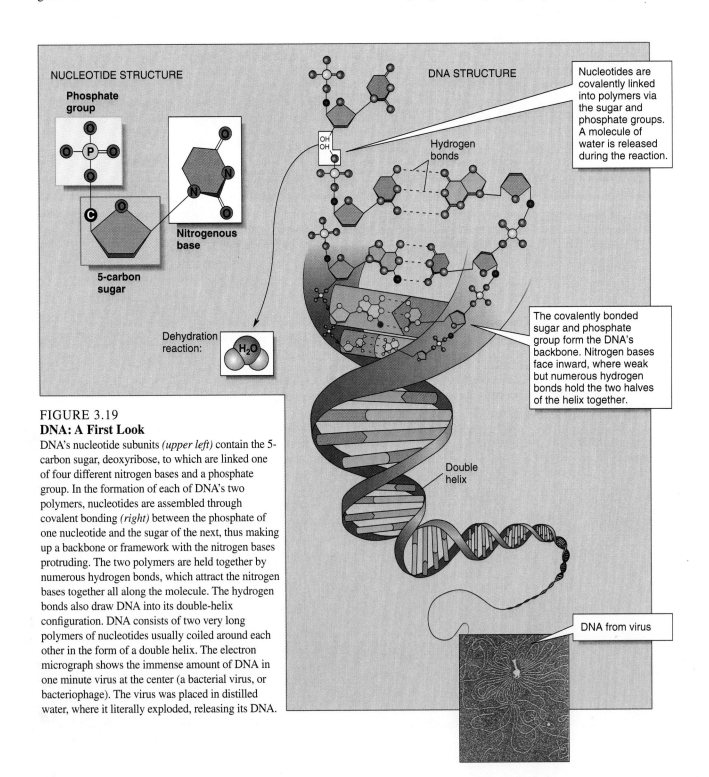

FIGURE 3.19
DNA: A First Look
DNA's nucleotide subunits *(upper left)* contain the 5-carbon sugar, deoxyribose, to which are linked one of four different nitrogen bases and a phosphate group. In the formation of each of DNA's two polymers, nucleotides are assembled through covalent bonding *(right)* between the phosphate of one nucleotide and the sugar of the next, thus making up a backbone or framework with the nitrogen bases protruding. The two polymers are held together by numerous hydrogen bonds, which attract the nitrogen bases together all along the molecule. The hydrogen bonds also draw DNA into its double-helix configuration. DNA consists of two very long polymers of nucleotides usually coiled around each other in the form of a double helix. The electron micrograph shows the immense amount of DNA in one minute virus at the center (a bacterial virus, or bacteriophage). The virus was placed in distilled water, where it literally exploded, releasing its DNA.

throughout DNA, there are four different nitrogen bases. These are called **adenine**, **thymine**, **guanine**, and **cytosine**.

The nucleotides form polymers through covalent bonding between the deoxyribose sugar of one nucleotide and the phosphate of the next. Each completed polymer may have thousands to millions of nucleotides. The polymers, in turn, are drawn together in pairs by hydrogen bonding between nitrogen bases (Figure 3.19), forming a double strand. The sugars and phosphates make up the outside of the double-stranded molecule and the paired nitrogen bases make up the core. As the strands join, they spontaneously twist into a helical form, known as a **double helix**.

We will look further into the molecular details of DNA in later chapters, but let's note here that the specific, linear arrangement of the nitrogen bases along the DNA polymer, like signal flags along a rope, makes up the genetic code mentioned earlier. When read in groups of three, the nucleotides become signals that translate into the placement of one or another of the 20 amino acids at this or that location along the polypeptide.

So we see that the genes we all like to talk about at family gatherings are basically an arrangement of DNA nucleotides that simply determines what our proteins will be like.

Key Ideas

THE CARBOHYDRATES

1. The empirical formula for **carbohydrates** is CH_2O. All carbohydrates are (roughly) multiples of this basic unit.

2. The **monosaccharide glucose** is the structural subunit of carbohydrates and is an important energy source for cells.

3. The disaccharide **sucrose** is made up of glucose and fructose.

4. All the molecules of life consist of structural subunits joined through **dehydration reactions** (enzymatic removal of water) and broken down through **hydrolysis** (enzymatic addition of water).

5. **Glycolytic linkages** (covalent bonds) join monosaccharides into disaccharides and polysaccharides.

6. Plant starches include **amylose** and **amylopectin.** The first contains 1–4 linkages; the second, 1–4 and 1–6 linkages. The animal polysaccharide **glycogen** contains both 1–4 and 1–6 linkages.

7. **Cellulose**, composed of **beta-glucose** subunits, is strong, insoluble, and digestible by a few organisms.

8. Cellulose polymers become hydrogen bonded into **microfibrils**, which form **fibrils**; the latter are strengthened by added hardening agents.

9. The structural subunit of **chitin**, a structural polysaccharide of insect exoskeletons and fungal cell walls, is *N*-**acetylglucosamine.**

THE LIPIDS

1. Lipids are used for energy reserves, membrane components, hormones, and waterproofing.

2. **Triglycerides** consist of one **glycerol** and three **fatty acids**. Free fatty acids have polar carboxylic acid groups and nonpolar hydrocarbon tails.

3. **Saturated** fats (animal fats) have no carbon-to-carbon double bonds and hold the maximum number of hydrogen atoms. **Unsaturated** fats (plant oils) contain at least one carbon-to-carbon double bond and fewer than the maximum number of hydrogens. **Polyunsaturated** fats have two or more double bonds.

4. **Phospholipids** are made up of one glycerol, two fatty acids, and a complex, electrically charged **phosphate head**.

5. Phospholipids in cell membranes are arranged in a lipid bilayer—the polar heads facing outward and the nonpolar tails inward. In water, free phospholipids spontaneously form lipid bilayers in the shape of hollow spheres or **micelles**.

6. **Waxes** are highly hydrophobic, making them suitable as waterproofing material.

7. **Steroids** have four nonpolar carbon rings with slightly polar side groups. Included are **lanolin** and **cholesterol**. Cholesterol is a constituent of animal cell membranes and liver bile and is a precursor (raw material to be used in synthesis) to steroid hormones and vitamin D.

THE PROTEINS

1. **Proteins** form cell structures, enzymes, energy sources, antibodies, and hormones and function in movement. **Enzymes** are biologically active proteins that speed up chemical activities.

2. **Amino acids** become linked via covalent **peptide bonds** into **polypeptides**. Polypeptides are further modified to form proteins.

3. Each amino acid contains an **amino group**, a **carboxyl group**, and a variable **R group**.

4. The special R-group properties bring about interactions within and between polypeptides that determine the protein's final shape.

5. **Primary structure** is the gene-specified sequence of amino acids in the polypeptide.

6. **Secondary structure** is either an **alpha helix**, produced by hydrogen bonding between amino acids in the polypeptide, or the zigzag **beta sheet**, produced by hydrogen bonding between amino acids in adjacent polypeptides.

7. **Tertiary structure** is a folding, produced by interactions among amino acid R groups [hydrophobic interactions, hydrogen bonding, ionic bonding, and covalent bonding (disulfide linkages)].

8. **Quaternary structure** is the joining of two or more polypeptides by disulfide linkages and the other tertiary forces. **Hemoglobin** contains four polypeptides and four iron-containing **heme groups**.

9. **Structural proteins** in vertebrates include **collagen** (in connective tissue), **elastin** (in flexible tissue), and **keratin** (hooves, nails, and hair).

10. **Conjugated proteins** are proteins with special groups added. They include hemoglobin, glycoproteins, lipoproteins, and phosphoproteins.

THE NUCLEIC ACIDS

1. Nucleic acids include **DNA (deoxyribonucleic acid)** and **RNA (ribonucleic acid)**.

2. DNA comprises the genes—chemical codings that determine the primary structure of protein.

3. DNA is double-stranded, each strand consisting of structural subunits called **nucleotides**.

4. DNA's chemical code is transcribed into RNA, which carries out the actual synthesis of protein. DNA is self-replicating.

5. DNA nucleotides consist of phosphate, ribose sugar, and one of four nitrogen bases: **adenine**, **thymine**, **guanine**, and **cytosine**.

6. In forming DNA, nucleotides are linked one above the other. Polymers are attracted together into a **double helix** by numerous hydrogen bonds between bases.

Application of Ideas

1. Nucleic acids and proteins are considered to be "informational molecules," but carbohydrates, lipids, and proteins are not, even though they may be extremely large and complex. Why do carbohydrates fail to qualify? Could lipids be "informational molecules"? How about proteins?

2. Animals use enzymes to transform glucose and oxygen into water, carbon dioxide, and energy. Since all enzymatic reactions are reversible, can they transform water, carbon dioxide, and energy into oxygen and glucose? Explain your answer and plan to come back to this question after having read Chapters 6, 7, and 8.

3. Discuss the nutritional consequences of various fad diets, such as the all-banana or all-egg diet, a fat-free diet, or a low-carbohydrate, low-fat, all-meat diet. How might such diets affect your health?

Review Questions

1. To what do the terms *monosaccharide, disaccharide,* and *polysaccharide* refer? What do all three have in common? (pages 45–47)

2. What is a dehydration reaction? How are these reactions important to the molecules of life? (pages 45–46)

3. What are two major uses for glucose? (page 45)

4. Name four major kinds of polysaccharides. Where would each be found? (pages 47–49)

5. How do the alpha and beta forms of glucose differ? What special properties does the presence of beta glucose bring to cellulose? (pages 47–49)

6. Describe three levels of organization in cellulose. What use do plants make of the final level? (page 50)

7. List the essential parts of a triglyceride. How do the various triglycerides differ from each other? (pages 51–52)

8. Distinguish between oils and fats; and among saturated fats, unsaturated fats, and polyunsaturated fats. (page 55)

9. Describe the structure of phospholipids. How are they arranged in cell membranes? (pages 53–54)

10. List the special uses of steroids. How do steroids differ structurally from other lipids? (pages 55–56)

11. List six different functions of protein. (pages 56)

12. What do all amino acids have in common? What makes one different from the next? (pages 57–58)

13. Describe the four organizational levels in protein and explain what forces are involved at each level. (pages 59–61)

14. In what way is DNA important to protein synthesis? How does RNA help in this? (pages 62, 64)

15. What special capability of DNA enables it to pass its genetic information along to new cells and individuals? Describe this. (pages 62, 64)

CHAPTER 4

Cell Structure

CHAPTER OUTLINE

CELL THEORY

WHAT IS A CELL?

The Sizes of Cells

Essay 4.1 Looking at Cells

THE PROKARYOTIC CELL

THE EUKARYOTIC CELL

Surface Structures

Internal Support: The Cytoskeleton

Organelles of Movement

Control and Cell Reproduction: The Nucleus

Organelles of Synthesis, Processing, and Storage

Endomembranal System

Energy-Generating Organelles

Essay 4.2 Evolution of the Eukaryotic Cell: Autogeny of Endosymbiosis?

KEY IDEAS

APPLICATION OF IDEAS

REVIEW QUESTIONS

Robert Hooke had just been appointed "curator of instruments" for the prestigious Royal Society of London, and he knew he had to come up with something good for the next weekly meeting. He was only too aware that the elite of British science would be there, and he wanted to present a demonstration that would enlighten, entertain, and impress them. It would not be an easy task.

Hooke had an idea. Perhaps he would use an exciting new technology of the 17th century—the casting and grinding of glass lenses. The world was buzzing with talk of lenses. With a pair of lenses in a frame the nearly blind were able to see—a miracle come true. Old men who had not been able to read for years had their books and letters restored to them. Earlier in the century, Galileo had pointed a lens to the sky and had started an intellectual revolution. The human eye has a voracious appetite, and Hooke knew it. But Hooke himself was interested in a new use of the lens—to look at very small things. In fact, he had built his own compound microscope, one of the first in the world.

So for a scientific demonstration, Hooke thought of using the microscope to try to see why cork floats. Cork was a mystery. It appears to be solid, yet it floats. Perhaps it is not so solid after all, Hooke thought. He decided it was worth a look.

Hooke trained his microscope on a cork, and what do you think he saw? Nothing. It turns out that microscopes do not work very well on reflected light. So out of curiosity he took out his pen knife and cut a very thin sliver of cork and shined a bright light *through* it. Then, when he peered into the microscope, he did see something. And it puzzled him. Hooke wrote at the time that the cork seemed to be composed of "little boxes." The little boxes, he surmised, were full of air, and that's why cork floats. (What Hooke was viewing were the dead, empty cells from the cork oak.) He called the little boxes "cells" because they reminded him of the rows of monks' cells in a monastery, and a new scientific field, **cell biology**, was born.

Cell Theory

The group that week was pleased by Hooke's demonstration, but a full century would pass before the scientific world would grasp the meaning and importance of Hooke's cells. One of the first people to move on the idea was the German naturalist Lorenz Oken, who wrote in 1805 what was to become known as the **cell theory**: "All organic beings originate from and consist of vesicles or cells." But in spite of this rather clear and encompassing statement, the formulation of the cell theory is usually credited to two other Germans, the botanist Matthias Jakob Schleiden and the zoologist Theodor Schwann. In 1839, they published a conclusion that they had reached more or less simultaneously; they said that all living things, from oak trees to violets, and from worms to tigers, were composed of cells. Either they were better at public relations than Oken, or perhaps the world was simply more ready to listen in 1839, but Schleiden and Schwann got the credit. Then, about 20 years after Schleiden and Schwann revealed their ideas, a fourth German, Rudolf Virchow, elaborated on the proposal to add

omnia cellula e cellula: "all cells from cells." Under the conditions on earth today, cells only arise from preexisting cells. This addition to the cell theory remains a thoroughly respectable idea today. The provision, "under today's conditions," leaves room for a prominent theory proposing that life (cells) arose spontaneously on the early earth, but under much different environmental conditions (see Chapter 21).

We know now that not only must cells come from cells, but virtually every living thing is composed of cells. They are indeed the basic units of life. Once Hooke had described his little boxes, the art of microscopy blossomed, and nothing—literally nothing—proved sacred. People were peering through their handmade microscopes at everything. Of course, they went over the human body with avid interest, as they did other animal life and plant life as well—looking at and into everything. What they found were not just the dead cell walls that Hooke had described, but variable, changing cells of astounding diversity (Figure 4.1). They could see some of these easily, but others needed to be stained before they were clearly visible. (Much of the cell is virtually transparent and must be stained with dyes for viewing. See Essay 4.1.) Living cells, they found, were full of a very active, viscous fluid—now called **cytoplasm**—a puzzling finding indeed. In time, cell researchers would also find that the cells of every creature are unique, differing not only from one species to the next but from place to place within the same individual. Yet, in spite of such diversity, what we find is that cells are remarkably similar in many fundamental ways.

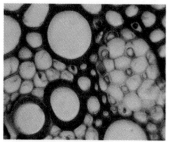
Plant root (transport cells)

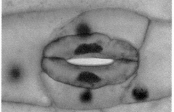
Plant leaf (guard cells around pore)

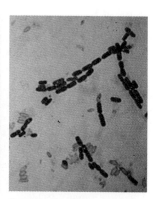
Animal respiratory liners (ciliated cells)

Bacteria (bacilli)

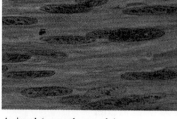
Animal (smooth muscle)

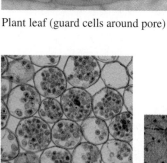
Plant root (storage cells)

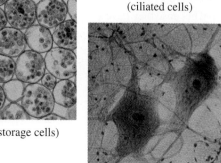
Animal (nerve cells)

Protist (ciliate)

FIGURE 4.1
Cell Diversity
Cells vary enormously in size, shape, and structure, each adapted to its own specialized task.

What Is a Cell?

We have said that the cell is the fundamental unit of life, but like most definitions, this one requires some explanation. What it really means is that the cell contains all the structures and molecular constituents needed for life. With this organization in place, the cell can (1) *take in raw materials;* (2) *make use of the raw materials to extract useful energy and to synthesize its own molecules;* (3) *grow in an organized manner;* (4) *respond to stimuli from its surroundings;* (5) *maintain a homeostatic state;* (6) *reproduce itself;* and (7) *adapt over time through evolution* (see Chapter 1).

We will return to some of these ideas shortly, and all of them from time to time in chapters to come. Now, let's consider the size range of cells.

The Sizes of Cells

How big are cells? Not very. Most cells are far too small to be seen with the naked eye. But, of course, there are exceptions: a chicken egg is technically a cell, and a frog egg is much smaller but still readily visible to the unaided eye. Neurons (nerve cells) may be over a meter long (such as those that run down a giraffe's leg), but they are still too thin to be seen. However, apart from such specialized cells, most cells fall within a surprisingly small size range. Very few are smaller than 10 micrometers (μm) in diameter or larger than 100 micrometers. (Figure 4.2 introduces the metric units.) Plant cells tend to be somewhat larger than animal cells but perhaps only because they contain large, water-filled internal cavities (or vacuoles); the average amount of cytoplasm is about the same in plant and animal cells. Bacteria, as we shall see, are much smaller than either, seldom exceeding a few micrometers in diameter or length (Figure 4.3).

Why Are Cells So Small?   This all brings up an interesting question. Why are most cells small, and why hasn't cell size increased as organisms became larger? For that matter, why aren't there gigantic single-cell creatures? We have yet to hear of an 800-pound ameba lurking in the old swimming hole, and no one has reported sighting single-cell whales and elephants. The evolutionary trend has been to keep cell size small and simply increase the number of cells to accommodate increases in size. Thus large organisms are large because they have so many more cells than small organisms. Why so many?

FIGURE 4.2
The Range of Cell Structure
The size range in life is enormous. The upper region includes the largest organisms down to those cells that can be seen with the unaided eye. The middle region of the scale includes those structures visible with the light microscope—from waving cilia to the human egg. Below this is the realm of the transmission electron microscope and the tunneling scanning electron microscope, the latter of which can discern atomic structure.

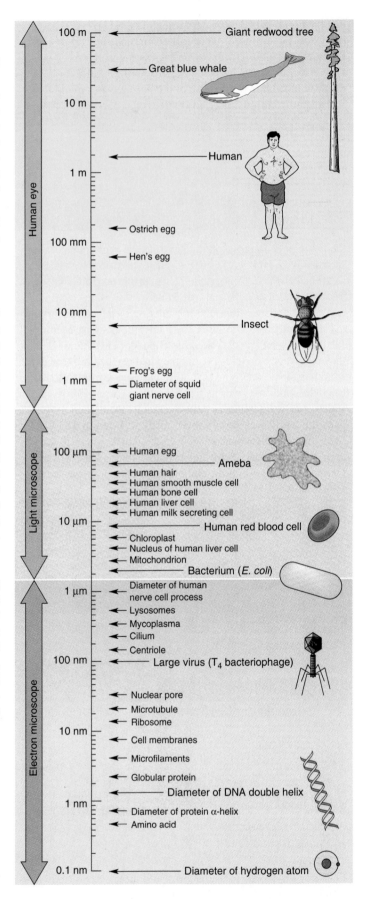

Essay 4.1 ● LOOKING AT CELLS

The Light Microscope

The light microscope was fully perfected by the start of this century. As a result, almost all cell structures that can be seen with a light microscope had been identified and cataloged before your grandparents were born.

The light microscope, or compound light microscope, consists basically of a tube with lenses at its ends. The lenses, of course, magnify the object, making its details visible. The principal magnifying lens, the objective lens, is brought very close to the object being viewed. It focuses the image on the ocular lens or eyepiece, which magnifies the object a second time. (Total magnification is calculated by multiplying the powers of the two lenses together.) An additional lens, called a condenser, lies just above the light source, where it focuses light into a cone over the object. The condenser also contains an iris diaphragm, which (like the iris of the eye) can be opened or closed to adjust brightness.

Preparation and Staining of Specimens for the Light Microscope

Cells and tissues tend to be transparent, so special stains (dyes) must be used to distinguish cellular structures. Stains may be of a general nature, simply darkening the cytoplasm or nuclei of cells, or (depending on their chemistry) stains may be highly selective for certain structures in the cell. Selective staining has been invaluable to the fields of cell biology and histology (the study of tissues), and many of the original staining methods are still in use. Generally, microscope specimens have to be quite small or very thin in order for light to pass through. Small cells such as red blood cells and bacteria can be viewed whole, but most plant and animal tissues must be cut into very thin slices, or sections, before staining and viewing.

Resolving Power: The Practical Limit

The limits of useful magnification in any microscope are determined by the instrument's resolving power. Resolving power is a measure of how close two objects can be and still be distinguished as separate. (The resolving power of the normal, unaided human eye at close range is about one-tenth of a millimeter.) There is a definite limit to the resolving power of even the finest light microscopes, and many important cellular structures are well below this limit. Optical physicists have shown that no system can resolve points that

are closer together than half the wavelength of the light used to view them, so the limitation is really light itself. The resolving power of the best light microscopes is approximately 220 nm, about 500 times the resolving power of the eye. The bottom line is that the limit of useful magnification with the light microscope is about 1400× (1400 times actual size). While this magnification can be exceeded, the resolution will not be improved. The result is called empty magnification.

The Transmission Electron Microscope

The transmission electron microscope, or TEM, has enormous powers of resolution. The object to be viewed is flooded with electrons rather than light waves, and the wavelength of electrons is much smaller than that of the shortest light wave. The TEM's useful magnification can reach 250,000×. Theoretically, its resolving power is about 0.025 nm, but technically the limits are about 0.2–0.3 nm (1000 times that of the light microscope). The principle on which the TEM works is relatively simple. It makes use of electrons shot from an electron gun similar to the electron gun in the picture tube of a television set. The term "transmission" refers to the formation of an image by those electrons that actually pass through and around the object to be viewed.

Focusing the electron microscope begins when electromagnetic condensers direct the electron beam onto the specimen. Focusing is possible because electrons have negative charges that respond to the surrounding electromagnetic field. The focusing of electrons somewhat resembles the bending of light rays by glass lenses. As the electrons pass into the specimen, some are absorbed by the nuclei of heavier atoms, and others pass through unhindered. Those that get through pass between additional electromagnetic lenses that spread them out. Magnification can be varied by changing the strength of the electromagnetic lenses. The electrons are finally focused onto a screen, where they form an image for direct viewing, or onto photographic emulsions for preparing electron micrographs (such as those seen in this text).

Preparation of Specimens for the Electron Microscope

We begin by noting that living material cannot readily be viewed in the electron micro-

scope: among other reasons, the chamber holding the object is a partial vacuum. In the preparation of tissue specimens for the electron microscope, they are first killed and fixed (preserved) by an application of electron-dense metallic salts. The heavy metal helps solidify the cytoplasmic proteins and combines with them in specific ways. Thus the metal selectively stains the specimen, its atoms later serving to absorb some electrons and permitting others to pass through. The tissue is embedded in a plastic matrix and then sliced into exceedingly thin sections. Sectioning is necessary because electrons do not pass through thick material very well.

Small objects, such as viruses, chromosomes, or the bacterial flagellum, can be viewed without slicing. They are first spread onto a film of protein that is supported by a wire screen. Then the mounted specimen is placed in a vacuum chamber, and atoms of platinum or gold are splattered onto the specimen at an angle—a procedure called shadow casting. The gold absorbs electrons, and the resulting image is a surprisingly life-like photo of three-dimensional (3-D) bodies.

One of the most specialized methods of specimen preparation is used in the study of cellular membranes. This is a process known as freeze fracturing. Tissue is frozen hard at −100°C and then sectioned. The blade follows natural weaknesses in the hardened tissue, such as between phospholipid bilayers making up the membrane. Shearing the membrane in this manner splits it between layers, thereby exposing its inner structures. Such methods have been vital in learning about the protein component of cellular membranes.

A 3-D View:
The Scanning Electron Microscope

The scanning electron microscope, or SEM, has the distinct advantage of producing three-dimensional images with unusually great depth of field. Reflected electrons from a moving beam are captured and used to generate an image electronically on a screen.

The High-Voltage Electron Microscope

The high-voltage electron microscope is a gigantic, three-story version of the TEM. With its enormous energy output (1 million electron volts), tissues can be viewed without slicing, and the image produced is three-dimensional.

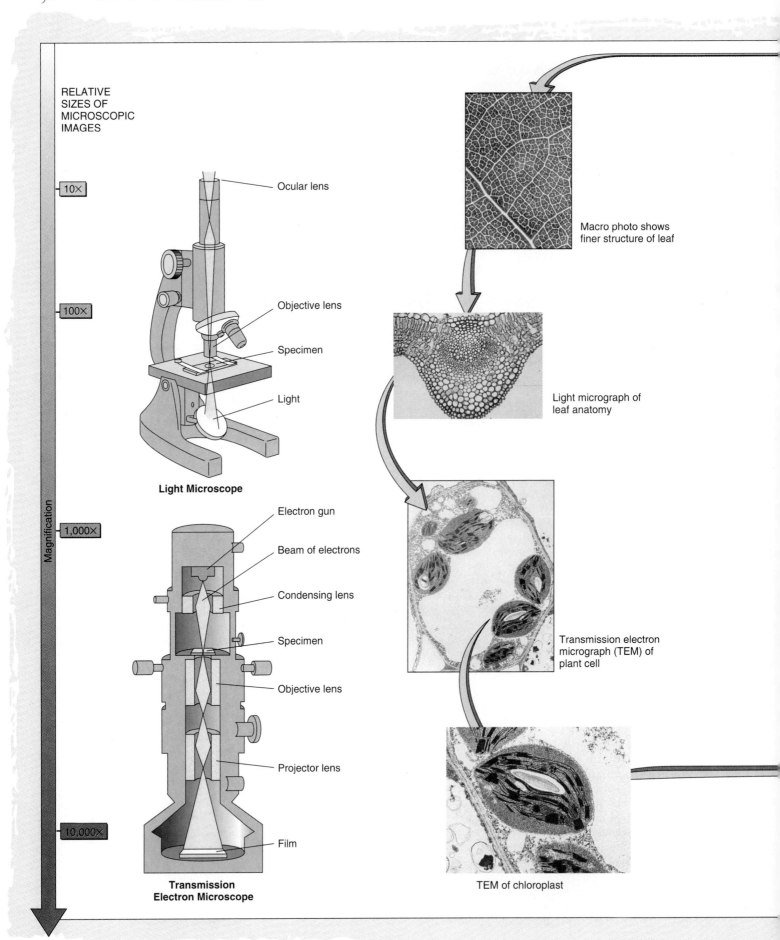

RELATIVE
SIZES OF
MICROSCOPIC
IMAGES

10×

100×

Magnification

1,000×

10,000×

Ocular lens

Objective lens

Specimen

Light

Light Microscope

Electron gun

Beam of electrons

Condensing lens

Specimen

Objective lens

Projector lens

Film

**Transmission
Electron Microscope**

Macro photo shows
finer structure of leaf

Light micrograph of
leaf anatomy

Transmission electron
micrograph (TEM) of
plant cell

TEM of chloroplast

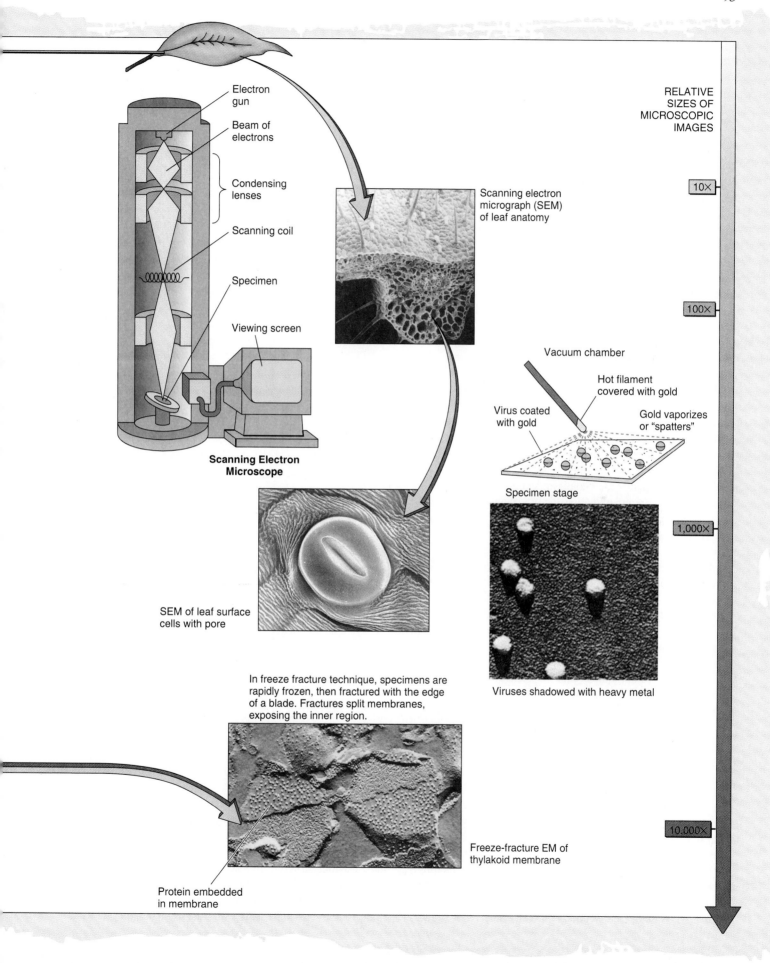

Electron gun

Beam of electrons

Condensing lenses

Scanning coil

Specimen

Viewing screen

Scanning Electron Microscope

Scanning electron micrograph (SEM) of leaf anatomy

RELATIVE SIZES OF MICROSCOPIC IMAGES

10×

100×

Vacuum chamber

Hot filament covered with gold

Virus coated with gold

Gold vaporizes or "spatters"

Specimen stage

SEM of leaf surface cells with pore

1,000×

Viruses shadowed with heavy metal

In freeze fracture technique, specimens are rapidly frozen, then fractured with the edge of a blade. Fractures split membranes, exposing the inner region.

10,000×

Freeze-fracture EM of thylakoid membrane

Protein embedded in membrane

FIGURE 4.3
Comparisons of Cell Dimensions
Plant cells are among the larger cells in multicellular organisms.
The red blood cells of the mammal are among the smallest animal
cells, not much larger than many kinds of bacteria. Some
protists are easily seen with the unaided eye.

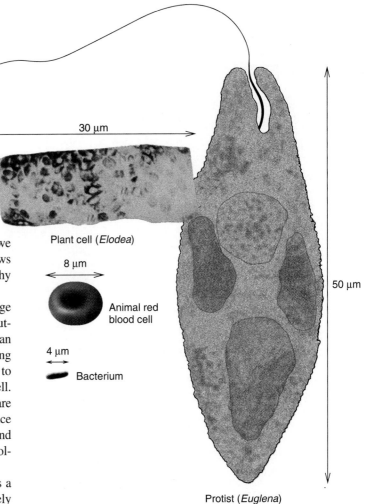

Plant cell (*Elodea*)

Animal red
blood cell

Bacterium

Protist (*Euglena*)

The Surface–Volume Theory. The generally
accepted explanation of the minute size of cells
has to do with problems related to surface area
and volume and is called the **surface–volume theory**. The surface–volume theory is based on the observation that as the volume of a body is doubled, its surface
area also increases, but not proportionately. If the surface-
to-volume ratio at the start is compared to that at the end, we
will find that relative surface lags behind. Figure 4.4 reviews
such a situation in a cuboid cell. But, given such a change, why
is this important to the maximum size a cell may attain?

Cells, especially growing cells, require an active exchange
of substances (chiefly an uptake of food and oxygen and output of wastes and carbon dioxide). Carrying out such an
exchange is a function of the **plasma membrane** surrounding
the cell. More precisely, the role of the plasma membrane is to
provide for the transport of materials into and out of the cell.
Some materials pass freely across the membrane, others are
rejected, and still others are actively helped in and out. Since
the plasma membrane represents the surface area of a cell, and
since the cell contents (the cytoplasm and nucleus) fill its volume, the surface–volume relationship is a critical one.

So the factors that restrict cell size can be restated: as a
cell increases in size, its volume increases disproportionately
to its surface area. Thus, as growth continues, a point is
reached at which the membrane can no longer service the
needs of the active cytoplasm and nucleus. Typically, before
such a point is reached, growth stops, and the cell divides in
two: consequently, a more ideal relationship between surface
and volume is established in the new cells. There are undoubtedly some powerful cues triggering cell division in growing
cells, but at the moment they are unknown.

Cells that are not simple cubes or spheres may have different surface–volume relationships. These are often highly
specialized cells or those in which special circumstances exist.
For example, some plant cells grow quite large because much
of their volume is taken up by the large water-filled **vacuole** (a
sac-like compartment). Furthermore, some plant cells make
use of **cytoplasmic streaming**, a circulating movement in
their cytoplasm, to distribute materials. Nerve cells can grow
extremely long extensions without large increases in volume.
Their massive surface area thus puts them in touch with other
cells with which they must function in concert. Cells in our
own small intestine and kidney specialize in the transport of
materials across their membranes. Such cells have numerous
microvilli, folds in their plasma membranes that create a large
surface area for absorbing digested foods (Figure 4.5). Bird
eggs, as we've said, are essentially one cell with a lot of stored

food. They are not limited by the problem of surface and volume because most of their volume consists of chemically
inactive stored food. Most metabolic activity occurs in a small
region on the yolk surface, representing the future embryo.

The Prokaryotic Cell

Prokaryotes are a diverse group of minute, single-cell or
colonial organisms, more commonly known as **bacteria**.
Their origins can be traced to the most ancient forms of life
known. All other forms of life (including human) are **eukaryotes** ("true nucleus"). You are already quite aware of prokaryotes, for among them are the familiar "germs" that linger in
all sorts of places and inspire fascinating TV commercials.

Prokaryotes tend to be structurally simple (Figure 4.6).
They generally range in size from 1 to 10 micrometers, only
one-tenth the diameter of representative eukaryotic cells. The
most obvious structure in most is a dense **cell wall**, a tough
barrier composed of either **peptidoglycan** or protein (peptidoglycan is composed of polysaccharides and short peptides).
Many eukaryotes have cell walls as well, but they consist

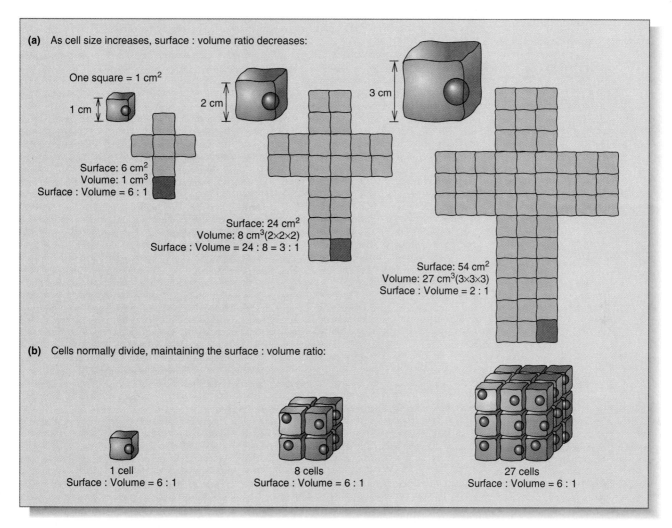

FIGURE 4.4
The Surface–Volume Relationship
(a) The relationship between surface area and volume is readily seen by studying cubes of increasing size. A cube measuring 1 unit per side has a surface-to-volume ratio of 6:1, whereas a cube twice that size has a surface-to-volume ratio of 3:1. A cube three times the original size has a surface-to-volume ratio of 2:1. The surface is obviously losing ground at a rapid rate as the "cell" grows. The problem is that the diminishing surface area, the plasma membrane, will no longer be able to provide sufficient transport to meet the cell's needs. (b) As we see, cell division continually restores the favorable surface–volume relationship.

chiefly of cellulose or chitin. Some bacteria have a surrounding sheath, and others have a slimy protective capsule. Within the cell wall lies the plasma membrane, which surrounds the cytoplasm. Many bacteria have a second membrane outside the cell wall. The cytoplasm, the living material of the cell, is featureless compared to that of the eukaryotic cell, but, nevertheless, it is a veritable cauldron of metabolic activity (much of it quite similar to what goes on in eukaryotic cytoplasm).

The genetic material of bacteria, their DNA, also referred to as a **chromosome**, lies free in the cell in an area called the **nucleoid**. (In prokaryotes there is no organized cell nucleus, hence their name, which means "before the nucleus.") The bacterial chromosome occurs as one continuous double-strand DNA molecule, which, although full of twists and

turns, forms a closed loop and can thus be described as circular. Some bacteria have small amounts of DNA incorporated into circular **plasmids** that are separate from the main chromosome. In some instances the plasmids contain genes that confer resistance against commonly used antibiotics, whereas others enable the bacterium to reproduce sexually (see Chapter 15). Slender extensions called **pili** are used to attach the cell to surfaces and, during sexual reproduction, to each other.

When prokaryotes approach the time of cell division, the DNA is replicated (copied), and the two copies are separated. Cell division occurs through a simple process called **fission**. During fission, some species form a **mesosome**, an inward extension of the plasma membrane that is believed to aid in the separation of DNA replicas. Eukaryotes, as we will see,

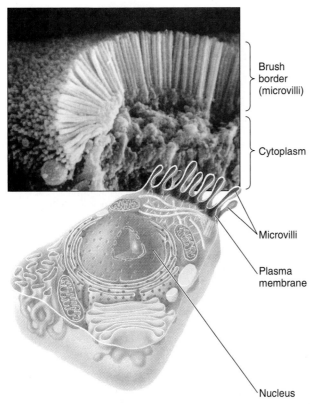

FIGURE 4.5
Increasing Membrane Area
The epithelial cells lining the small intestine carry on transport across a surface that is greatly expanded by numerous microvilli, brush-like extensions of the plasma membrane.

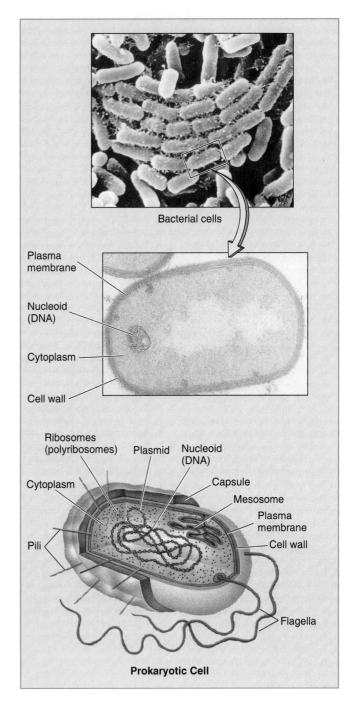

FIGURE 4.6
The Prokaryotic Cell
The electron micrograph reveals little internal detail in the prokaryotic cell, although a number of flagella are seen. Prokaryotes are structurally much simpler than eukaryotic cells (see Figure 4.7), yet they carry out many of the same chemical processes.

have special protein structures called microtubules, that function in cell division.

Many bacteria have structures that provide movement. One kind of movement occurs through the action of a **flagellum**, which in prokaryotes is a long S-shaped structure (see Figure 4.6). The prokaryotic flagellum spins on its axis like a propeller and is the only spinning structure seen in cells. Unlike the eukaryotic flagellum, which consists of protein assembled into microtubules, the prokaryotic flagellum is a solid filament of the bacterial protein **flagellin**.

In terms of their energy sources, prokaryotes (like other organisms) fall into two major categories—**heterotrophs** ("other-feeding") and **autotrophs** ("self-feeding"). The heterotrophs require complex food sources (carbohydrates, lipids, proteins), those produced by other organisms. Autotrophs produce their own foods from simple inorganic materials.

Heterotrophs include **parasites**, clinically known as **pathogens** (disease causers). Pathogens feed directly on the tissues and body fluids of a living host. But most heterotrophic bacteria are **decomposers** (decay bacteria), those that make use of dead organisms and their wastes as an energy source. Decomposers are often the cause of those pungent odors (some would say "stenches") associated with body wastes and decay, and while they thrive in our foods and other goods, they generally do not invade our bodies. In fact, decomposers are vital to life because

they recycle essential elements such as carbon and nitrogen. Without their efforts, we're told, the earth would be strewn with well-preserved, albeit weather-beaten, corpses of everything that ever died! (Makes you want to send a thank-you note.)

Autotrophic prokaryotes use simple inorganic compounds such as carbon dioxide, water, and certain minerals to form their own complex molecules. This of course requires a considerable input of energy. Some are **photoautotrophic**. Like green plants they use light as a source of energy to do their molecule building. The process itself is called **photosynthesis**. Other bacteria are **chemoautotrophs**, prokaryotes that live deeper in the earth and its waters, where they derive their energy from chemical reactions involving elements such as sulfur and iron.

The Eukaryotic Cell

Eukaryotes include plants, animals, fungi, and protists, which, along with bacteria, make up the five kingdoms of life (see Chapter 20). The cells of eukaryotes have some features in common with prokaryotes—chiefly a surrounding plasma membrane, cytoplasm, and chromosomes of DNA—but there are marked differences. Eukaryotic cells have an extensive array of membrane-surrounded **organelles** ("little organs"), each with its own specific function (Figure 4.7). Organelles are functional compartments, each specializing in a specific task for the cell. The most visible organelle is likely to be the **nucleus**, whose membranes (it has two) maintain some separation of the genetic material and the surrounding cytoplasm. The genetic material, the DNA, is organized into a number of separate chromosomes that are linear rather than circular. Furthermore, whereas prokaryotic DNA is relatively free of protein, the DNA of eukaryotes is intimately bound to proteins along its entire length. The DNA and its associated protein is called **chromatin**. As we proceed into our study of eukaryot-

FIGURE 4.7
Representative Cells
Few cells of either plants (**a**) or animals (**b**) have all the structures seen in these composite drawings. Both plants and animals are eukaryotes, which means their cells have a prominent, membrane-surrounded nucleus and several kinds of organelles that are also surrounded by membranes. Plant and animal cells have many features in common, but they also have important differences. Note the large central water vacuole of the plant cell, the dense cell wall, and the presence of chloroplasts. The latter contain chlorophyll, which takes in sunlight and begins the process of carbohydrate synthesis. Animal cells lack these structures but do contain centrioles and sometimes cilia or flagella. The three are involved in cell movement, which is notably absent in higher plants.

ic cell structures, other distinct differences will become obvious. Table 4.1 makes further comparisons between prokaryotic and eukaryotic cells.

Surface Structures

A good place to begin is with those structures that support the cell, form the necessary boundaries that maintain the cell's integrity, and provide for transport of materials into and out of the cell. From there we will progress to structures devoted to movement, control, reproduction, synthesis, storage, and, finally, energy transfer.

Cell Walls. The cells of plants, algae, and fungi are surrounded by dense cell walls (Figure 4.8). Cell walls protect the cell

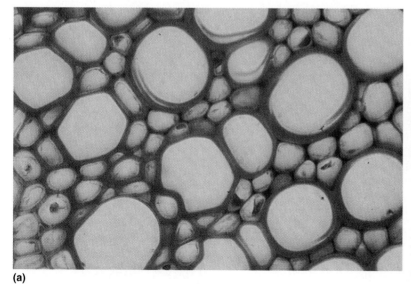
(a)

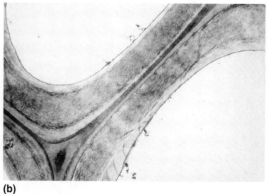
(b)

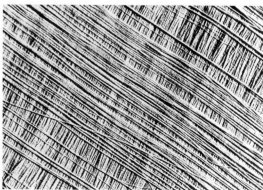
(c)

FIGURE 4.8
The Plant Cell Wall
(**a**) Plant cell walls such as those of the water-conducting xylem tissue have extremely thick walls, which helps resist some of the pressures involved in their functions.
(**b**) In the enlarged view, the density of cell walls is apparent. (**c**) In a higher power magnification, the cellulose fibrils are readily seen. They are embedded in a cement-like matrix that gives the wall its great strength and resiliency.

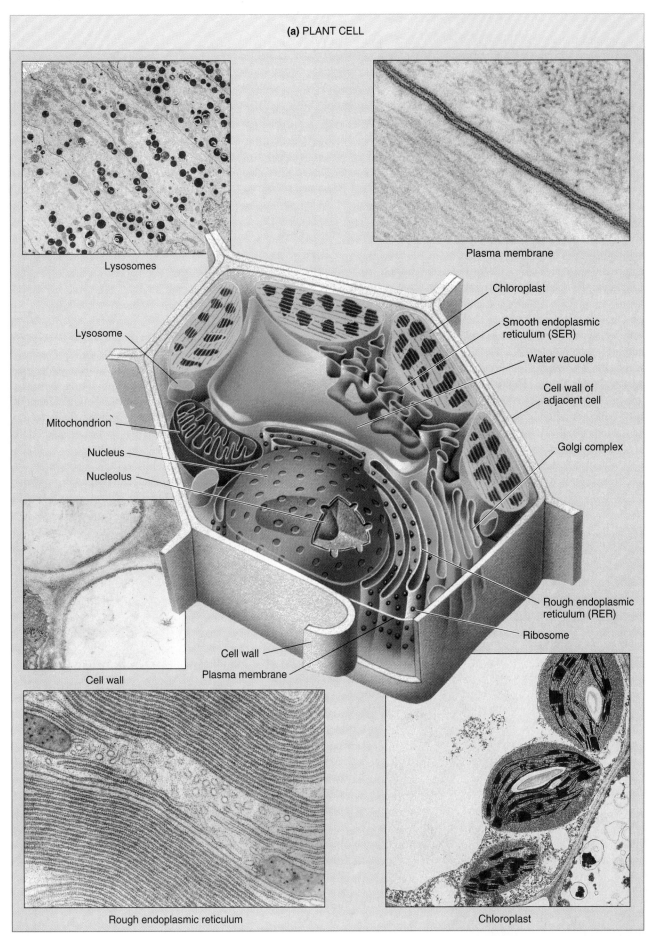

(a) PLANT CELL

Lysosomes

Plasma membrane

Lysosome

Mitochondrion

Nucleus

Nucleolus

Chloroplast

Smooth endoplasmic reticulum (SER)

Water vacuole

Cell wall of adjacent cell

Golgi complex

Rough endoplasmic reticulum (RER)

Ribosome

Cell wall

Plasma membrane

Cell wall

Rough endoplasmic reticulum

Chloroplast

FIGURE 4.7

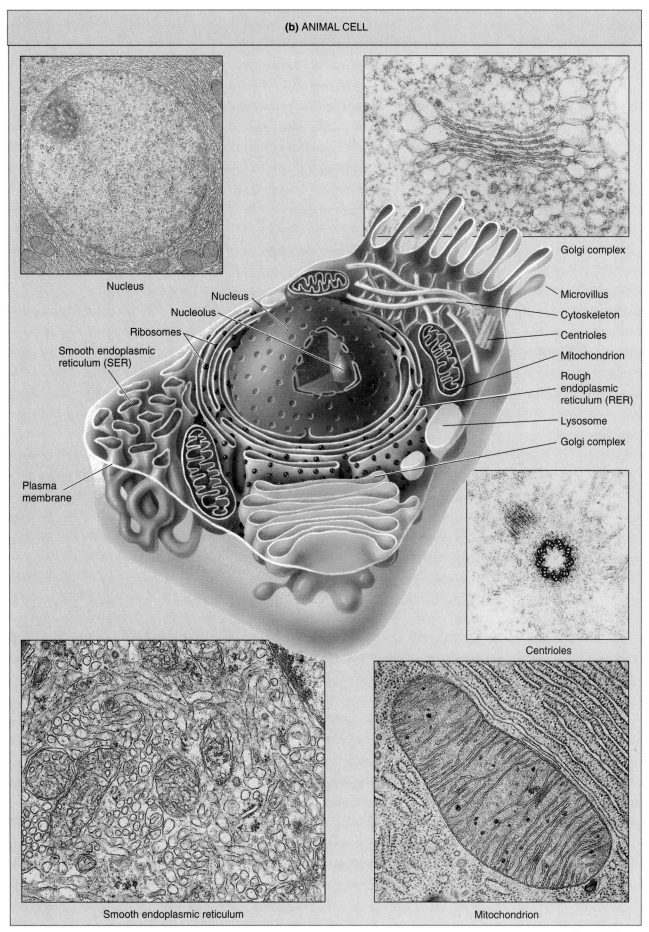

(b) ANIMAL CELL

Nucleus

Golgi complex

Microvillus

Cytoskeleton

Centrioles

Mitochondrion

Rough endoplasmic reticulum (RER)

Lysosome

Golgi complex

Nucleus

Nucleolus

Ribosomes

Smooth endoplasmic reticulum (SER)

Plasma membrane

Centrioles

Smooth endoplasmic reticulum

Mitochondrion

FIGURE 4.7

TABLE 4.1 Comparing Prokaryotic and Selected Eukaryotic Cells

Feature	Prokaryotic cell	Eukaryotic cell: Plant	Eukaryotic cell: Animal
Plasma membrane	Present	Present	Present
Cytoplasmic support	None seen	Protein cytoskeleton	Protein cytoskeleton
Nuclear envelope	Absent	Present	Present
Membrane-bounded organelles	Absent	Present	Present
Endoplasmic reticulum	Absent	Present	Present
Ribosomes	Present: smaller, free	Present: larger; both free and bound to membranes	Present: larger; both free and bound to membranes
Cell wall	Present: peptidoglycan or protein	Present: cellulose	Absent
Flagella or cilia	Present: flagellin protein, tubular, rotating	Absent[1]	Present: tubulin protein, microtubular organization, undulating
Phagocytic mechanisms	Absent	Absent	Present
Centrioles	Absent	Absent[1]	Present
Mitochondrion	Absent	Present	Present
Chloroplast	Absent	Present	Present

[1] Cilia, flagella, and centrioles are absent in most seed plants, but present in seedless plants such as mosses and ferns.

from mechanical injury and from the potentially damaging effects of water entry, a problem we will discuss in the next chapter. They also provide a supporting framework for the cell. Here, we will focus our attention on the cell walls of plants.

Plant Cell Walls. In most plants the cell walls are rigid and multilayered, consisting principally of a framework of cellulose into which a number of other substances are deposited. As we found in the last chapter, the lengthy cellulose polymers are organized into microfibrils, which are collected into fibrils. The fibrils are laid down in crisscross pattern, with each new layer at an angle to the last (see Figure 4.8c). As the cell wall forms, cement-like hemicelluloses and pectins are added, binding the fibrils together into a tough but porous covering over the cell. As you can imagine, the crisscross pattern and cement-like materials produce an extremely strong product, not unlike resin-impregnated fiberglass in a boat hull

or auto fender. Lignin, a hardening material, is also secreted into cell walls and provides much of the great strength of tree trunks and limbs. Other materials include **suberin** and **cutin**, used primarily for waterproofing. (Where would you expect to find the latter two? Right!)

The Plasma Membrane. We have already mentioned this vital cell boundary, and vital is the right word because if anything is important to the integrity of the cell, the membrane is. Basically, it keeps the inside of the cell in and the outside out, yet it lets through those materials that the cell must exchange with its environment. The membrane's complex structure permits it to effortlessly accept the passage of some substances, utterly reject others, and, at times, expend energy to actively assist the transport of still others. Thus the plasma membrane is not simply a passive covering but is highly active in cell transport.

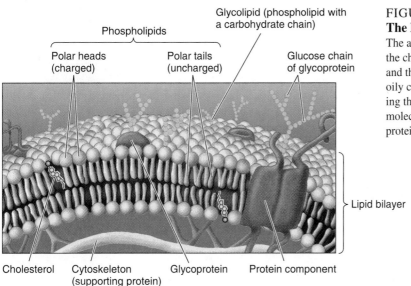

Phospholipids

Glycolipid (phospholipid with a carbohydrate chain)

Polar heads (charged)

Polar tails (uncharged)

Glucose chain of glycoprotein

Lipid bilayer

Cholesterol   Cytoskeleton (supporting protein)   Glycoprotein   Protein component

FIGURE 4.9

The Plasma Membrane

The artist's reconstruction shows the phospholipid bilayer, the charged or polar heads forming the hydrophilic surface and the uncharged, nonpolar tails forming a hydrophobic, oily core. There are numerous protein components penetrating the phospholipid bilayer and (in animal cells) occasional molecules of cholesterol. Note the network of supporting protein fibers below.

The plasma membrane is also important to cellular communication, with special surface molecules recognizing and responding to hormones and other messengers. In multicellular organisms (those with many cells), cells identify each other through membrane surface molecules, which is important during embryonic development and when the immune system tracks down invading disease agents.

These important functions, and others, are made possible by the plasma membrane's intricate structure. Plasma membranes and, in fact, all cell membranes are essentially a double layer of phospholipids, oriented in such a manner that the polar heads face outward and the fatty acid tails face inward (Figure 4.9; see also Chapter 3). This arrangement gives the membrane a hydrophilic surface and a hydrophobic, oily core. Animal membranes also have a cholesterol component. Cholesterol's four, nonpolar interlocking rings penetrate the phospholipid core while its polar hydroxyl group (— OH) interacts with the phospholipid heads. Interspersed in the lipid bilayer are a number of different kinds of molecules that carry out the various membrane functions. Included are enzymes, transport proteins, supporting proteins, glycolipids (sugar lipids), and glycoproteins (sugar proteins). We will consider their roles in Chapter 5.

Internal Support: The Cytoskeleton

Until a few years ago, little was known about structure in the less organized regions of cytoplasm, those regions between the well-defined organelles. It appeared that such places were simply viscous regions and were vaguely designated as the "ground substance." But with the advent of fluorescence microscopy and of the high-voltage transmission electron microscope, cell biologists have discovered a vast, maze-like network of supporting fibers, which they have named the **cytoskeleton** (Figure 4.10).

The cytoskeleton is made up of three types of fibrous elements: very fine **actin filaments**, chiefly composed of the protein actin (6 nm in diameter), **intermediate filaments** (7–11 nm in diameter), and **microtubules** (22 nm in diameter). Their lengths vary widely. The cytoskeletal elements provide a supporting framework for cell organelles, aid in cell movement, anchor the plasma membrane, and provide surfaces on which chemical activity and intracellular transport can occur in an organized manner.

Actin Filaments. Actin filaments are composed of two slender chains of actin wound about each other. Each chain is made up of numerous individual globular protein subunits (Figure 4.11a,b). Cells can assemble and disassemble the chains with remarkable speed and efficiency. Actin filaments have a variety of roles. For instance, along with other elements of the cytoskeleton, they provide structural support to the plasma membrane. On its own, the membrane has little strength, but when bonded to a strong fiber network on its inner side, the membrane is well supported. Such support is particularly apparent in the **microvilli**, those cytoplasmic projections of the intestinal lining (Figure 4.11c; review Figure 4.5).

In still other protein associations, actin filaments become involved in certain types of cell movement, namely, muscle contraction, cytoplasmic streaming, ameboid movement, and cytoplasmic division. Actin is joined by myosin, another filamentous protein, to form the contractile units that make up the bulk of muscle tissue. As the actin filaments interact with myosin, they slide along, forcefully shortening the entire muscle. For a closer look at this amazing process, see Chapter 32.

In cytoplasmic streaming, an active, circular movement of cytoplasm within certain plant cells, the actin filaments form a fixed boundary along which the freer cytoplasm moves (Figure 4.11d). Researchers believe that certain proteins, linked to the moving organelles, may interact with actin in a

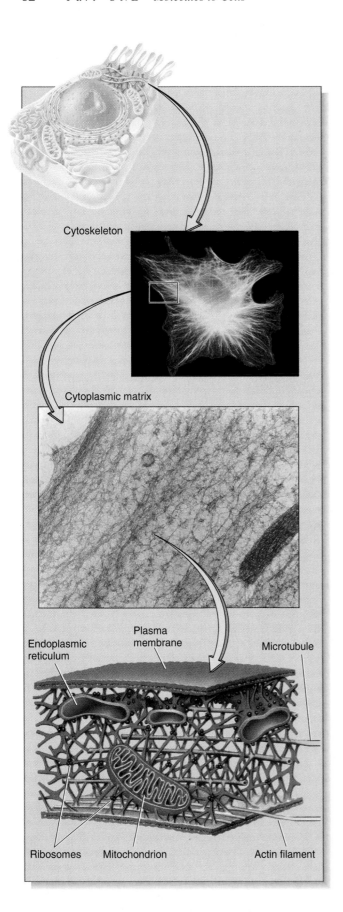

Cytoskeleton

Cytoplasmic matrix

Endoplasmic reticulum

Plasma membrane

Microtubule

Ribosomes   Mitochondrion   Actin filament

FIGURE 4.10
The Cytoskeleton
The cytoskeleton stands out clearly when special fluorescent dyes are used on the cell *(top photo)*. Details of the cytoskeleton are clearly visible when seen through the high-power electron microscope, an innovation that gives the object viewed a three-dimensional quality *(center)*. A reconstruction *(bottom)* shows that a network of microscopic fibers crisscrosses the cell, supporting some of the organelles and anchoring the plasma membrane in place. The elaborate framework also helps the cell maintain its shape.

way that promotes the ongoing streaming action, but they're just not sure how.

Ameboid cells have no definite shape and, in fact, change their shape in order to move. They do this by organizing actin filaments to form firm tubelike extensions called **pseudopodia** ("false feet"). The more liquid inner cytoplasm then flows into the pseudopods. Simultaneously, other actin filaments in the trailing region are disassembled and their protein units recycled to the advancing front (Figure 4.11e). Since the firm and fluid states are interconvertible, the cell seems to "pick itself up and follow its advancing front."

Finally, actin filaments are involved in animal cell division, specifically the division of the cytoplasm. [Cell division includes two events, the division of the nucleus and the division of the cytoplasm. In the latter event, microfilaments, acting like a drawstring, pinch the cell cytoplasm in two (see Chapter 9).]

Intermediate Filaments. Intermediate filaments are the least understood of the three cytoskeletal elements, but they are common in epithelial cells (cells that form inner and outer linings in the animal body). There, the intermediate filaments are composed of the protein keratin (also found in hair and nails), which lends strength to the important linings they form (Figures 4.12 and 3.18). In some instances, these filaments literally bind sheets of epithelial cells together. Intermediate filaments also make up part of the supporting network of microvilli (see Figure 4.11c) and appear in such diverse places as the brain, spinal cord, muscle fibers, and blood cells.

Microtubules. Microtubules, the largest of the cytoskeletal elements, are hollow and tubelike, their walls formed from numerous molecular subunits of **tubulin**, a globular protein. Each tubulin subunit is composed of two separate spherical polypeptides, designated as alpha and beta tubulin. As microtubules form, these doublets are assembled into spiraling rows (Figure 4.13). As with actin filaments, cells can rapidly assemble microtubules when needed and just as quickly disassemble them, conserving the tubulin units for future use as the cell's needs change.

Microtubules, in conjunction with actin and intermediate filaments, provide structural strength to the cell, helping to anchor the plasma membrane and reinforce the cytoplasm. Microtubules, along with actin filaments, are believed to influ-

(a) Actin of cytoskeleton (fluorescence micrograph)

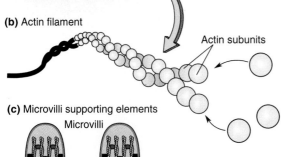

FIGURE 4.11
Actin Filaments
(a) The photograph reveals the vast number of actin filaments a cell may have.
(b) Actin filaments are assembled from spherical protein subunits that form intertwining strands. **(c)** They form a supporting framework for the microvilli of the intestine, where they are bound together by cross-linking proteins and anchored at their bases by still other proteins. **(d)** Actin filaments interact with the cytoplasm in cytoplasmic streaming and **(e)** account for the fascinating phenomenon of ameboid movement. The usefulness of actin in movement is attributed to its rapid assembly and disassembly.

(b) Actin filament

Actin subunits

(c) Microvilli supporting elements

Microvilli

Cross-linking proteins

Actin

Anchoring proteins

(d) Cytoplasmic streaming

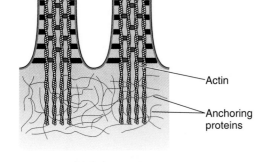

Vacuole

Moving cytoplasm with ER and other organelles

Actin filaments

Chloroplasts

Nonmoving cytoplasm with chloroplasts

(e) Ameboid movement

Actin disassembly

Actin assembly

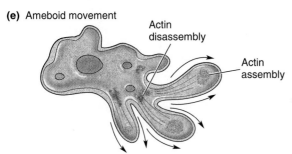

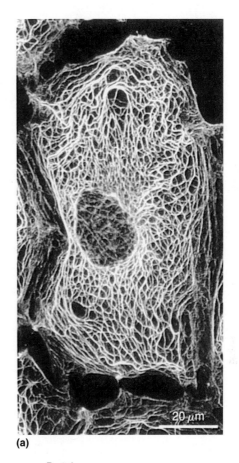

20 μm

(a)

Protein subunits

10 nm

(b) Alpha keratin

FIGURE 4.12
Intermediate Filaments
Intermediate filaments of the cytoskeleton are primarily structural, lending strength to the cell. **(a)** The filaments standing out so clearly in the photograph of epithelial cells are composed of keratin. **(b)** Intermediate filaments consist of three strands of protein spheres braided into a strand.

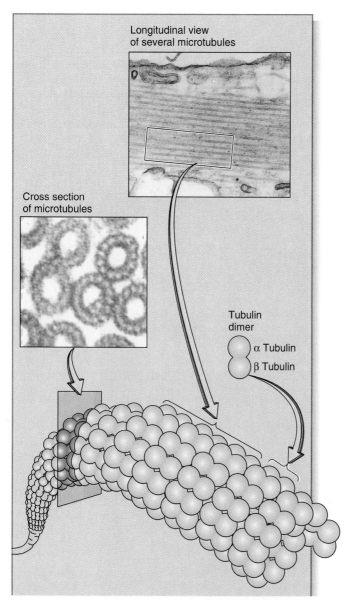

FIGURE 4.13
Microtubule Structure
Part of the cytoskeleton is made up of microtubules. Dimeric tubulin molecules, doublets of spherical alpha and beta tubulin, form a spiraling pattern as they are assembled into the microtubule.

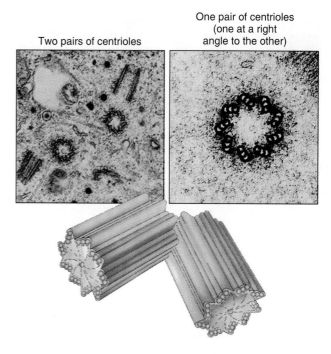

FIGURE 4.14
Centrioles
Centrioles are cylindrical bodies. The cylinders *(left photo)* generally occur in pairs, with one at a right angle to the other. In cross section *(right photo)*, the nine sets of triplet microtubules show clearly, taking on a pinwheel look. The three-dimensional illustration reveals the spatial arrangement of microtubules in each triplet.

In addition to these important roles, microtubules form **centrioles**, **basal bodies**, **cilia**, and **flagella**. These structures are directly or indirectly involved in cell movement, our next topic.

Organelles of Movement

Centrioles and Basal Bodies. Centrioles occur only in those eukaryotes whose cells develop cilia or flagella at some stage in their lives, even if it's just sperm cells. This includes animals, protists, and the more primitive seedless plants. Centrioles are not found in fungi and the more reproductively advanced seed plants, possibly having been lost through evolution.

A centriole is a short cylinder of microtubules, usually seen in pairs with one lying at a right angle to the other. Each cylinder contains nine groups of triplet microtubules, arranged in a circle, referred to as a "9 + 0" arrangement. (Compare to the 9 + 2 arrangement in cilia and flagella, discussed just ahead.) In cross section, an individual centriole looks a bit like a child's toy pinwheel (Figure 4.14).

Two functions of centrioles are known with some certainty. First, they give rise to **basal bodies**. Like the centriole, the basal body has nine groups of triplet microtubules. Basal bodies assemble the microtubules that form the basic structure of the cilium and flagellum. The other role of the centriole is a little less certain. It may determine in which plane cytoplasmic division occurs during cell division in animal cells.

ence, if not control, the shape of many types of cells. When these cytoskeletal elements are chemically disrupted, the cells lose their normal shape, often forming simple spheres.

Nowhere is the role of microtubules more dramatically visible than in the dividing cell, where many of them cross the cell, taking on the form of an old-fashioned spindle (sort of cylindrical, but wide in the middle and narrow at the ends; see Figure 4.13). These **spindle fibers**, as they are known, attach to and move the chromosomes. Thus they play a critical role in cell division by ensuring the precise delivery of DNA replicas to the two daughter cells (see Chapter 9).

FIGURE 4.15
A Study of the Cilium
The SEM *(top)* is of a protist, showing its cytoskeleton and cilia. In the artist's reconstruction *(center)*, we see the protruding cilia covered by plasma membrane. The cutaway view *(right)* reveals that cilia are composed of paired microtubules. As is evident in the electron micrographs nearby, nine pairs of microtubules occur in a circle with an additional pair at the center. The basal body, which lies at the base of the cilium, is different. It consists of nine triplets of microtubules with nothing in the center. It strongly resembles the centriole, which is not astonishing when we realize that centrioles give rise to basal bodies, which then form cilia.

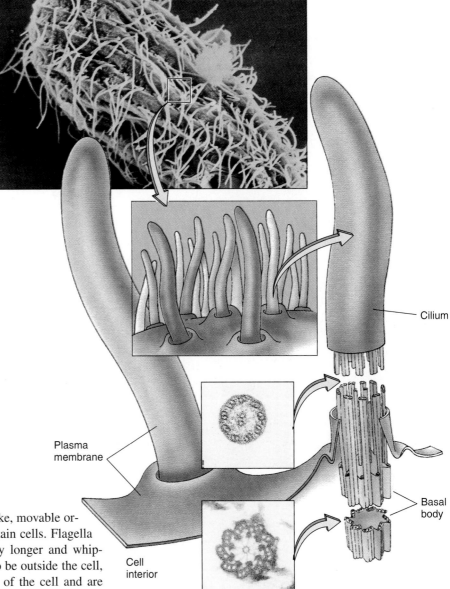

Cilium

Plasma membrane

Cell interior

Basal body

Cilia and Flagella. Cilia are fine, hair-like, movable organelles found on the cell surface of certain cells. Flagella are similar, except they are considerably longer and whip-like. Although cilia and flagella appear to be outside the cell, they really are not; they are outgrowths of the cell and are covered by the plasma membrane.

Both cilia and flagella may either propel motile cells through surrounding fluid or move surrounding fluid past the surface of stationary cells (Figure 4.15). For example, sperm cells swim by undulations (wave-like movements) of the flagellum, and ciliated protists swim along through the coordinated rowing action of rows of surface cilia. The cilia of stationary cells that line passages, such as the human trachea (windpipe) and oviduct, move materials along their surfaces. Cilia of the trachea move a dust-laden film of mucus upward to your pharynx, thus periodically clearing your respiratory passages (unless you have permanently paralyzed them with tobacco smoke). In mammals, cilia of the oviduct draw the newly released ovum inside the duct, where it encounters swimming sperm moving in the opposite direction. (Cell movement can be a big part of sexual reproduction.)

Cilia and flagella are structurally almost identical. They differ only in length. Most cilia are numerous and short, between 10 and 20 micrometers long. Also, flagella are usual-

ly fewer in number and comparatively long—in exceptional cases, several thousand micrometers (several millimeters) long. The flagellum of one fruit fly's sperm may be longer than the fly itself! Functionally, cilia have a highly coordinated, rowing motion (like oars, but since these "oars" remain under water, they must bend for the return stroke). Flagella are few in number and move by wave-like undulation.

Fine Structure and Movement in Cilia and Flagella. Under the electron microscope, cilia and flagella are shown to have an intricate structure (see Figure 4.15). The microtubular arrangement includes nine pairs of microtubules arranged in a circle with a single pair of microtubules at the center: known as the "9 + 2" arrangement. (Compare this to the "9 + 0" arrangement of microtubules in centrioles and basal bodies—see Figure 4.14.)

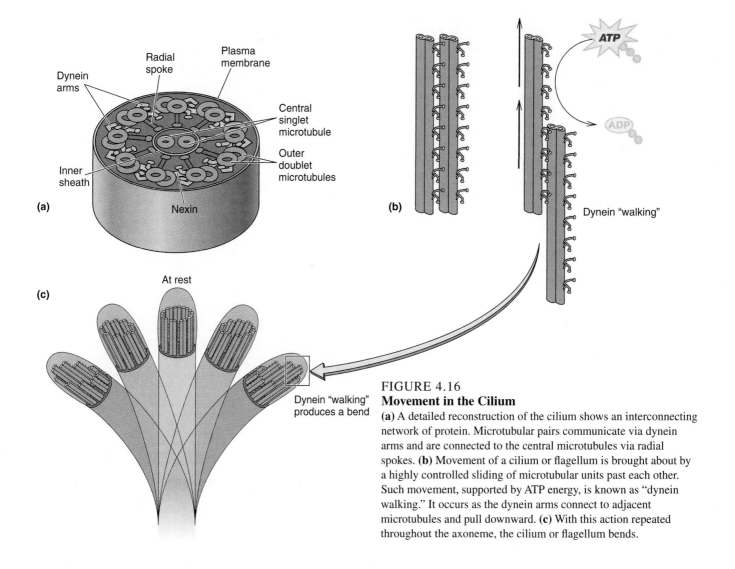

FIGURE 4.16
Movement in the Cilium

(a) A detailed reconstruction of the cilium shows an interconnecting network of protein. Microtubular pairs communicate via dynein arms and are connected to the central microtubules via radial spokes. (b) Movement of a cilium or flagellum is brought about by a highly controlled sliding of microtubular units past each other. Such movement, supported by ATP energy, is known as "dynein walking." It occurs as the dynein arms connect to adjacent microtubules and pull downward. (c) With this action repeated throughout the axoneme, the cilium or flagellum bends.

An even closer look at the cilium (Figure 4.16a) reveals many movable, paired protein extensions, the **dynein arms**, which occur along each pair of microtubules. When extended outward, the arms contact the next pair of microtubules. **Nexin**, another protein, forms supporting connections between the doublets. Finally, extending inward from each of the doublets are the **radial spokes**, which contact an **inner sheath** of protein surrounding the two central microtubules. The entire active core of the organelle—the microtubules, dynein arms, nexin, radial spokes, and inner sheath—is known as the **axoneme**. The parts interact to produce a special kind of bending movement called **dynein walking** (Figure 4.16b).

Using ATP as an energy source, the dynein arms of one doublet microtubule attach to and pull against the nearest doublet microtubule. As the pulling action goes on, the arms detach and reattach further along, repeating this "hand-over-hand" action. Apparently the central microtubules act as a stationary base, and, because of this restraint, the force exerted by the dynein arms produces a local bending of the whole structure (Figure 4.16c). This bending action, moving smoothly along the length of the structure, first on one side and then on the other, produces the familiar rowing and undulating movement of cilia and flagella, respectively.

The idea of one filamentous structure forcefully sliding against another is not unique to cilia and flagella. Although the actual mechanics differ, this is precisely the principle involved in muscle contraction (see Chapter 32).

Cilia and flagella have some sort of primitive cellular sensory function as well. Many of our own sensory receptors are believed to have evolved from cilia. Amazingly, the 9 + 2 arrangement of microtubules can still be seen in (1) the rods and cones of the retina, (2) the olfactory fibers of the nasal epithelium, and (3) some sensory hairs of the internal ear. It seems then that we see, smell, hear, and balance ourselves with highly modified cilia-bearing cells.

Control and Cell Reproduction: The Nucleus

The nucleus, the most prominent organelle of the cell, carries out the all-important functions of control and cell reproduc-

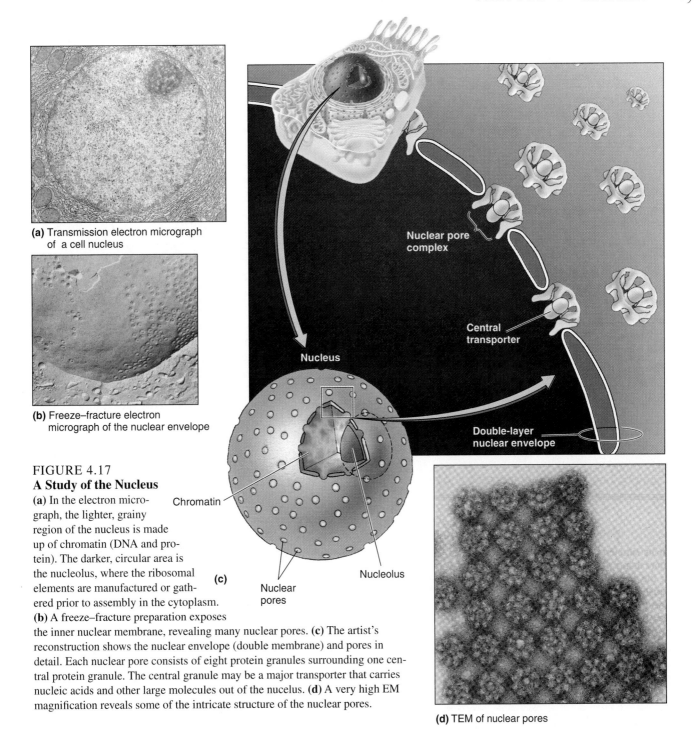

(a) Transmission electron micrograph of a cell nucleus

(b) Freeze–fracture electron micrograph of the nuclear envelope

Nuclear pore complex

Central transporter

Nucleus

Chromatin

Nucleolus

Nuclear pores

(c)

Double-layer nuclear envelope

(d) TEM of nuclear pores

FIGURE 4.17
A Study of the Nucleus
(a) In the electron micrograph, the lighter, grainy region of the nucleus is made up of chromatin (DNA and protein). The darker, circular area is the nucleolus, where the ribosomal elements are manufactured or gathered prior to assembly in the cytoplasm. **(b)** A freeze–fracture preparation exposes the inner nuclear membrane, revealing many nuclear pores. **(c)** The artist's reconstruction shows the nuclear envelope (double membrane) and pores in detail. Each nuclear pore consists of eight protein granules surrounding one central protein granule. The central granule may be a major transporter that carries nucleic acids and other large molecules out of the nucelus. **(d)** A very high EM magnification reveals some of the intricate structure of the nuclear pores.

tion. At first glance, there doesn't seem to be much structure in the nucleus, but a closer look reveals some interesting features (Figure 4.17).

Whereas most organelles have one surrounding membrane, the nucleus has two, aptly called the **nuclear envelope**. Each membrane consists of the usual phospholipid bilayer along with associated proteins. Unlike other cellular membranes, the nuclear envelope contains a great number of pores scattered over its surface (see Figure 4.17). The **nuclear pores** are not simple holes; they are a network of proteins that control the movement of materials into and out of the nucleus.

They consist of an outer ring of smaller proteins and a larger central protein, thought to be a major transporter. The transporter may be responsible for the movement of RNA, a fairly large molecule, out of the nucleus to the cytoplasm, where it engages in protein synthesis.

Another prominent nuclear structure is the **nucleolus** ("little nucleus"; plural, **nucleoli**), a dark-staining body made up of genes that are intensely involved in the synthesis of a nucleic acid called ribosomal RNA. This nucleic acid, along with certain proteins, is assembled in the nucleoli into protein-synthesizing bodies called **ribosomes**.

Because the nucleus takes up stains well, making it easy to view with the light microscope, it was described early in the history of cell study. In fact, the word "nucleus," which means "kernel," was first used in 1831 by British botanist Robert Brown, about the time the cell theory emerged (the same investigator who discovered "Brownian movement"; see Chapter 5).

The nucleus contains most of the cell's hereditary material—its DNA—which, as we've seen, is assembled into a number of linear molecules. Each DNA molecule with its associated proteins is referred to as a chromosome (*chrome,* "color"; *soma,* "body"; chromosomes are easily stained). The chromosomes are readily visible during cell division, but at other times their DNA is dispersed in the nucleus and difficult to discern.

The two functions of the nucleus—control and cell reproduction—are carried out by DNA. DNA is a repository of chemical specifications for producing proteins, and among the proteins are the all-important enzymes. Enzymes speed up chemical reactions in the cell, which means that their timely production permits the metabolic activity of life to take place. So we see that DNA's control over the cell is actually indirect.

Cells make use of DNA replication to prepare the cell for reproduction. Here two copies of every chromosome are made. In this way, both **daughter cells**, as the products of cell division are called, will have all the chemical instructions needed to carry on the life processes. We will expand considerably on the two roles of DNA in chapters to come.

Organelles of Synthesis, Processing, and Storage

There are extensive internal membranes in the eukaryotic cell. The arrangement of these membranes into organelles establishes a number of functional compartments within the cytoplasm, where very specific chemical activities can be carried out in isolation, a situation that increases efficiency of such reactions. Imagine how cumbersome and inefficient the vast array of chemical activity might be if enzymes and reactants were scattered throughout the vast eukaryotic cytoplasm (like trying to find a friend at a concert when you arrive a half-hour late). This is more or less the situation in prokaryotes, but the minute volume of cytoplasm makes compartmentalization unnecessary. The first of these compartments to be considered is the endoplasmic reticulum.

The Endoplasmic Reticulum. The **endoplasmic reticulum** (**ER**) is an extensive membrane system that takes up a large part of the cytoplasm of eukaryotic cells, especially those cells that are engaged in the synthesis of substances destined for export from the cell.

Painstaking electron microscope studies of the ER have led researchers to realize that it is one continuous membrane folded back and forth within the cytoplasm. It thus forms a closed sac whose contents are kept isolated from the general cytoplasm. The ER is a dynamic organelle, forming first in one place, then in another. In some cells, its lumen (internal space) may account for up to 10% of the entire cytoplasmic volume, and the ER membrane itself can account for more

than half the membrane area of an entire cell. It has clearly been demonstrated that the ER is continuous with the outer of the two membranes of the nuclear envelope, so actually the nuclear material is separated from the ER lumen by a single membrane only. The endoplasmic reticulum appears in two main forms: rough and smooth.

Rough Endoplasmic Reticulum. The **rough endoplasmic reticulum** (**RER**) receives its name from the presence of numerous ribosomes, dense granular bodies that tightly adhere to its outer side and make it look a little like coarse sandpaper (Figure 4.18). The primary role of the RER is the processing, transport, and temporary storage of proteins.

Rough endoplasmic reticulum occurs in most cells, but it is extremely dense in cells that manufacture proteins destined for release outside the cell (such as enzymes, hormones, and antibodies). The RER also contains proteins that are later incorporated into the plasma membrane. As the polypeptides that form these proteins are synthesized, they enter the lumen. It is within the RER that many newly synthesized polypeptides assume the higher levels of protein organization, folding and linking themselves into their final tertiary configuration (see Chapter 3): in other words, becoming full-fledged, active proteins.

Smooth Endoplasmic Reticulum. The **smooth endoplasmic reticulum** (**SER**) appears smooth because it lacks ribosomes (see Figure 4.18). It is an intricate network of tubes and sacs that is very extensive in cells that synthesize, secrete, and store carbohydrates, steroids, lipids, and other nonprotein products. Examples include newly divided plant cells, cells of the testis, oil gland cells of the skin, steroid gland cells, and cells lining the small intestine. The SER forms when sections of the RER undergo a transition from protein synthesis to the synthesis of the other materials (see Figure 4.18).

SER in liver cells is known to have still another function: it contains oxidizing enzymes that detoxify certain chemicals. The administration of large doses of alcohol or the sedative phenobarbital into experimental animals, for example, is followed by a substantial increase in both liver cell smooth endoplasmic reticulum and detoxifying enzymes. (Can you imagine what the liver cells of an alcoholic must look like?) The storage and breakdown of the polysaccharide glycogen are also associated with the SER of liver cells.

Ribosomes: Sites of Polypeptide Synthesis. We've mentioned ribosomes in passing, so before going on we will consider the structure and role of these interesting bodies. **Ribosomes** contain sites where amino acids are assembled into polypeptides. The assembly is made according to chemical instructions provided by **messenger RNA (mRNA)** that was copied earlier from DNA (see Chapter 13).

Ribosomes (Figure 4.19) are large molecular complexes consisting chiefly of **ribosomal RNA (rRNA)** and many kinds of proteins. Each ribosome is made up of two subunits that come together with mRNA just before polypeptide syn-

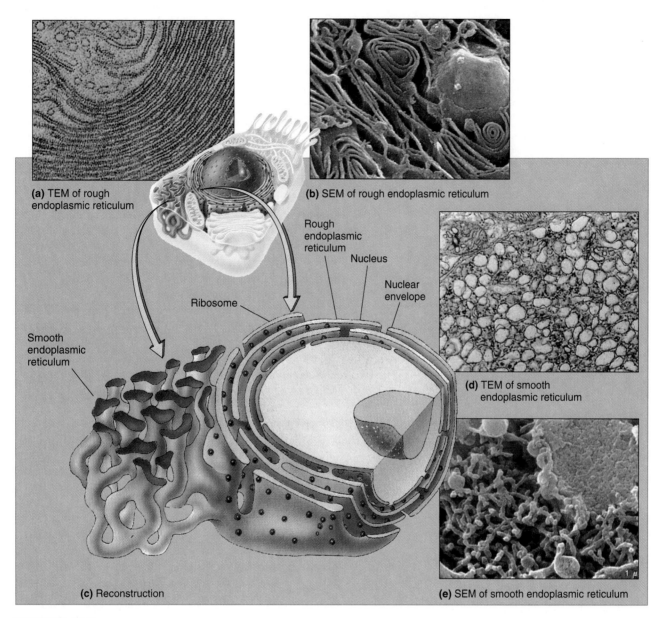

(a) TEM of rough endoplasmic reticulum

(b) SEM of rough endoplasmic reticulum

Smooth endoplasmic reticulum

Ribosome

Rough endoplasmic reticulum

Nucleus

Nuclear envelope

(d) TEM of smooth endoplasmic reticulum

(c) Reconstruction

(e) SEM of smooth endoplasmic reticulum

FIGURE 4.18

Endoplasmic Reticulum

(a) The transmission electron micrograph of a thin slice through a cell reveals a number of dotted, double lines. The lines are the membranes of the RER and the dots are ribosomes. (b) The scanning electron microscope produces a three-dimensional view of the RER, this time revealing the many enclosed channels. (c) A reconstruction helps complete the view, showing the RER to be a number of flattened channels. (d) A transmission electron microscope view of the cell reveals the SER as a number of vesicles. (e) A three-dimensional view via the scanning electron microscope clarifies the first picture, and the SER turns out to be a number of tubes. When reconstructed, the SER is quite unlike the RER, yet EM studies clearly indicate that the RER undergoes a transition into SER. As we will see, many of the membranous organelles are interchangeable.

thesis begins. The ribosomal subunits are assembled separately in the nucleolus and then transported through the nuclear envelope to the cytoplasm. These subunits remain separate until they begin functioning.

The ribosomal subunits in prokaryotes are somewhat smaller and differ chemically from those of eukaryotes. In fact, it is this difference that makes antibiotics such as tetracycline and streptomycin safe but effective against bacterial infection. The two antibiotics block protein synthesis in the bacterial ribosomes but have no effect on our own ribosomes.

Some ribosomes are free in the cytoplasm, occurring either singly or in bead-like strands called **polyribosomes** (or just **polysomes**). Others, called **bound ribosomes**, appear to be attached to the rough endoplasmic reticulum. Actually, bound ribosomes are only temporarily held to the RER while

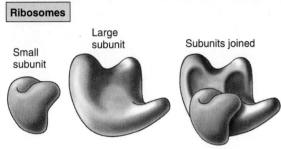

Ribosomes

Small
subunit

Large
subunit

Subunits joined

FIGURE 4.19
The Ribosome
Ribosomes are molecular assemblies of RNA and protein. Each consists of interlocking large and small subunits that join when the ribosome is active in protein synthesis. Because of their small size, ribosomes do not photograph well, but this electron micrograph reveals some of the structure.

they assemble the polypeptides that enter the endoplasmic reticulum for processing (see Chapter 13).

The Golgi Complex. In 1898, Camillo Golgi, an Italian scientist, tried a tissue-staining technique involving metallic salts. Through its use, Golgi discovered axons and dendrites—fine, lengthy, communicating extensions of nerve cells, heretofore invisible. He also observed a peculiar structure that today bears his name, the **Golgi complex**. It was one of the first cytoplasmic structures to be seen; but despite its early discovery, some 70–80 years passed before the function of the Golgi complex was fully appreciated.

Modern electron microscope studies of the Golgi complex (Figure 4.20a) show it to be a stack of flattened sacs or **cisternae**, often found near the nucleus. Some biologists refer to similar structures in plant cells as **dictyosomes**.

In Figure 4.20b, we see that individual Golgi cisternae continually form from membranous vesicles that pinch off from the nearby RER. The vesicles, laden with products treated in the RER, fuse with the nearest cisterna of the Golgi complex—at what cell biologists call the *cis* face. The *cis* face, in turn, buds its own vesicles that fuse with the next cisterna in line, and the transfer repeats progressively through the stack. This is how the Golgi complex originates and

grows. The final cisterna is the ***trans* face** (maturing face). Vesicles formed from the *trans* face leave the Golgi complex altogether, headed for various destinations.

So what is all the budding about? Biochemical studies have provided some of the answers. The Golgi stack, as it turns out, is actually an assembly line (indicated by changing colors in Figure 4.20), involved in the step-by-step modification of a variety of proteins arriving from the RER. A battery of enzymes, each specific to its own cisterna in the stack, is responsible for the many reactions of this complex assembly line. Final treatment occurs in the *trans* face, where the different products are sorted out, and each is packaged in its own vesicle. For example, certain powerful digestive enzymes receive their final modification in the *trans* face and enter vesicles that later coalesce to form dark-staining lysosomes (our next organelle). Other products are stored temporarily in storage granules, while still others go into secretory vesicles that move directly to the cell membrane, where their contents are secreted (expelled) from the cell (see Figure 4.20b). In plant cells, newly formed cell wall materials are delivered via such vesicles to the developing primary cell wall. Cell biologists know a good deal about the Golgi complex, but there are still mysteries. Their latest unsolved puzzle: How does the Golgi complex manage the complex sorting process?

Lysosomes. **Lysosomes** have been observed in nearly all types of animal cells and recently in plant cells, where they are known as **spherosomes** and are considerably smaller than most animal lysosomes. Lysosomes are membrane-bound sacs that are roughly spherical in shape, varying greatly in size and shape (Figure 4.21a).

The simple appearance of the lysosome belies the rather startling role it plays in the life of a cell. Lysosomes are bags of powerful hydrolytic enzymes, synthesized in the RER and packaged primarily by the Golgi complex (Figure 4.21b). In all, about 40 different enzymes have been detected in lysosomes. Imagine what might happen to a cell should a lysosome rupture! As a safeguard, the enzymes require an acidic condition to work efficiently. The near neutral pH of the cell cytoplasm prevents enzyme activation should the lysosome leak. To activate its enzymes, the lysosome creates its own acidic condition by taking in protons (H^+) from the cytoplasm.

Cell biologist Christian de Duve, who discovered lysosomes, described them as little "suicide bags." His poetic fancy was not entirely unwarranted, since lysosomes are known to engage in **autophagy** (self-eating), which is a specialized version of a general process known as **endocytosis**. (In endocytosis, cells literally engulf objects, taking them into the cytoplasm in vacuoles.) Autophagy is a tidying-up process wherein damaged or aged organelles within the cell are taken into a digestive vacuole and hydrolyzed by lysosomal enzymes. Actually, autophagy is not as destructive as it may seem. Sometimes, in fact, the destruction of cells is a normal part of cell activity. In some instances, lysosomes might destroy a cell that is not functioning well or a cell in a part of

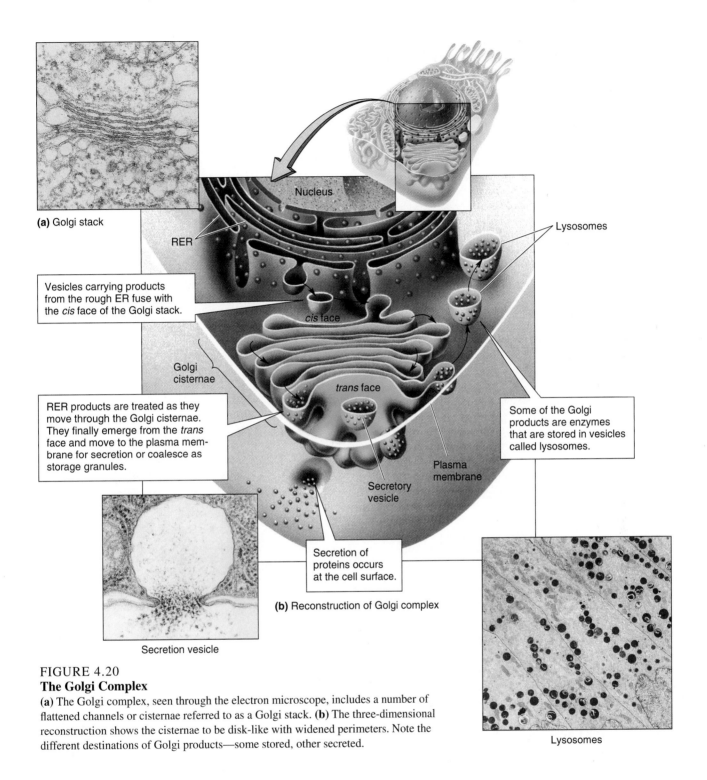

(a) Golgi stack

Nucleus

RER

Vesicles carrying products from the rough ER fuse with the *cis* face of the Golgi stack.

cis face

Golgi cisternae

trans face

RER products are treated as they move through the Golgi cisternae. They finally emerge from the *trans* face and move to the plasma membrane for secretion or coalesce as storage granules.

Lysosomes

Some of the Golgi products are enzymes that are stored in vesicles called lysosomes.

Plasma membrane

Secretory vesicle

Secretion of proteins occurs at the cell surface.

(b) Reconstruction of Golgi complex

Secretion vesicle

Lysosomes

FIGURE 4.20
The Golgi Complex
(a) The Golgi complex, seen through the electron microscope, includes a number of flattened channels or cisternae referred to as a Golgi stack. **(b)** The three-dimensional reconstruction shows the cisternae to be disk-like with widened perimeters. Note the different destinations of Golgi products—some stored, other secreted.

the body that is undergoing reduction as part of a developmental process. This is exactly what happens to tissue originally between the fingers in the developing human hand. The shrinking of the tail as the tadpole develops is another example.

Lysosomes also have specific functions in certain healthy, active cells. Such cells take in solid substances and form **digestive vacuoles**. Afterward, a lysosome fuses with the

digestive vacuole, its enzymes digesting whatever has been captured. This is the principal means of feeding in free-living amebas and many other single-cell protists, and it is the way that some of our own white blood cells destroy invading bacteria and clean up cell debris at infection sites.

There are a number of human genetic disorders called storage diseases, in which some essential lysosomal enzyme

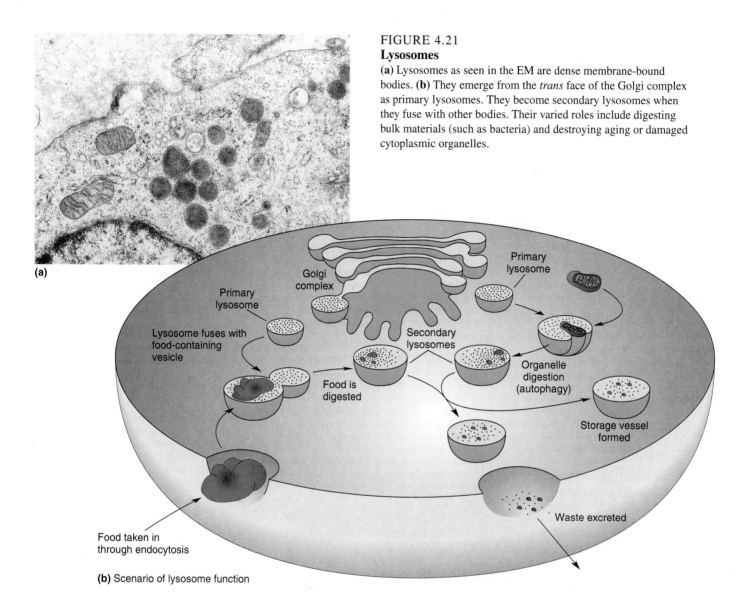

(a)

FIGURE 4.21
Lysosomes
(a) Lysosomes as seen in the EM are dense membrane-bound bodies. **(b)** They emerge from the *trans* face of the Golgi complex as primary lysosomes. They become secondary lysosomes when they fuse with other bodies. Their varied roles include digesting bulk materials (such as bacteria) and destroying aging or damaged cytoplasmic organelles.

(b) Scenario of lysosome function

is altered or absent. As a result, substances that would normally be processed and eliminated accumulate to toxic levels in the cell. Most storage diseases are fatal in the first five years of life. A well-known example, Tay–Sachs disease, is relatively common among Jews of East European descent. The absence of the critical lysosomal enzyme (*N*-acetylhexosaminidase) in the infant results in the buildup of lipids called gangliosides in brain cells, resulting in retardation, blindness, and early death.

Peroxisomes. **Peroxisomes** are found in a great variety of organisms, including plants, animals, and protists. In animals, they are most common in liver and kidney cells, where they are believed to originate from outpocketings of the SER. Some peroxisomes appear as very dense bodies with a unique crystalline array that causes the organelle to stand out unmistakably in electron micrographs (Figure 4.22). The crystals within peroxisomes are stored enzymes.

Catalase is a principal enzyme of liver and kidney peroxisomes. It catalyzes the breakdown of hydrogen peroxide into water and oxygen:

$$2\ H_2O_2 \longrightarrow 2\ H_2O + O_2$$

This is a safeguard, since hydrogen peroxide is a very destructive chemical that could easily damage the cell.

For a time it was suspected that catalysis of hydrogen peroxide was the only role of the peroxisome, but further studies indicate that peroxisomes may be involved in far more. In fact, they may regulate oxygen concentrations in tissues. Whereas oxygen is essential, excesses are toxic to the cell. In addition, peroxisomal enzymes are known to help in the regeneration of a coenzyme known as NAD. This coenzyme is important in energy-transferring reactions associated with the manufacture of ATP. Still other peroxisomal enzymes are involved in the breakdown and recycling of the nitrogen bases of DNA and RNA.

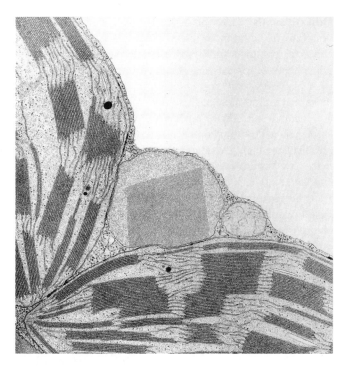

FIGURE 4.22
Peroxisomes
Peroxisomes are readily identified by the crystalline bodies inside them. This peroxisome is from a plant cell, where it shares borders with two membranous chloroplasts.

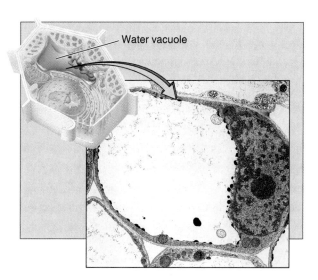

Water vacuole

FIGURE 4.23
Vacuoles
Plant cells often have large central vacuoles filled with watery sap. In leaf cells, numerous water-filled vacuoles exert pressure on their cell walls, keeping the leaf rigid and erect.

Glyoxysomes. Glyoxysomes are bodies that commonly occur in the lipid-storing regions of seeds. When seeds sprout, enzymes of the glyoxysomes convert the stored lipids to sugars, which help maintain the young seedling just long enough for it to begin making its own food through photosynthesis.

Vacuoles. The term **vacuole** is rather general. It refers merely to any membrane-bound body with little or no inner structure. Vacuoles generally hold something, but their contents vary widely, depending on the cell and the organism.

Plant cells generally have larger vacuoles than do animal cells. In many types of plant cells, a vacuole dominates the central part of the cell, crowding all other elements against the cell wall (Figure 4.23; see also Figure 4.7a). Such vacuoles fill with water that contains various substances in solution or suspension. Water-filled vacuoles in the cells of leaves and in the stems of young seedlings play important structural roles. They aid in holding the leaves and stems erect. This is possible because, as mentioned in the cell wall discussion, water exerts a considerable pressure against the surrounding cell walls. When deprived of water, leaves and young stems simply wilt.

The solutes in plant vacuoles may include atmospheric gases, inorganic salts, organic acids, sugars, pigments, or other materials. Sometimes the vacuoles are filled with water-soluble blue, red, or purple pigments (the *anthocyanins*),

which are responsible for many flower colors. Plants commonly store toxic, metabolic wastes as crystals in the central vacuoles, which in some instances serve as a protective device against herbivores. If a herbivore should eat the plant, it will get not only food but stored poisons.

Specialized vacuoles in single-cell protists include **food vacuoles** or **phagosomes**, bubble-like organelles that hold and digest food taken in through endocytosis (see Chapter 5). Many protists have **contractile vacuoles**, those that collect water from the surrounding cytoplasm and pump it out of the organism. This is a critical process for those protists that cannot resist the inward movement of water. The pumping action involves the contraction of a ring of microfilaments, their action much like squeezing a water-filled balloon.

Endomembranal System

The sharing and transfer of membranes among organelles are quite common in the eukaryotic cell. You've seen that the nuclear membrane is continuous in places with the endoplasmic reticulum, and you now know that membranes of the RER form those of the SER and contribute to the Golgi complex. The Golgi complex uses its own membranes to form specialized vesicles, some of which fuse with the plasma membrane, thus contributing their membranes to that structure.

At one time the membrane-bound organelles were believed to be more or less separate and autonomous, but, as you can see, this view is no longer held, at least for a number of organelles. The ongoing and dynamic formation of one organelle from another has led cell biologists to consider the components as part of an **endomembranal system**. The endomembranal system involves the synthesis of cell products, their modification and storage, and eventually their final disposition. In the coordination of such sequential activities, reactants and products must be isolated from other such activities, and it is the endomembranal system organelles that make this possible.

Energy-Generating Organelles

Chloroplasts and **mitochondria** are the energy-generating organelles. Their basic function is to convert available sources of energy to a form useful to the cell, and each does this in its own way. Their unique roles make the chloroplast and mitochondrion interesting on their own, but when one considers how these energetic little organelles may have gotten into cells in the first place, they can only be described as fascinating. We'll come back to their origins shortly, after we get to know them better.

Chloroplasts. Chloroplasts (Figure 4.24) are among the largest organelles and are readily visible with the lower magnifying powers of a light microscope. These colorful, green bodies occur only in the cells of plants and algae, never in animals and other eukaryotes.

Chloroplasts have a surrounding envelope, a double membrane, just like that of the cell nucleus. The clear, watery region within the chloroplast is the **stroma**. Within the stroma lies an extensive *third* membrane system that forms flattened, sac-like vesicles called **thylakoids**. Some thylakoids occur in stacks called **grana** (singular, **granum**), with membranous interconnections known as **lamellae**. The thylakoid membranes contain numerous spherical bodies, the **ATP synthases** (see Figure 4.24), that contain enzymes specializing in the synthesis of ATP.

Chloroplasts carry out **photosynthesis**, the process in which the energy of light is used to convert carbon dioxide and water to carbohydrates and other useful molecules. Central to this process are active, light-absorbing pigments known as **chlorophylls** and **carotenoids**, which are distributed in the thylakoid membranes. It is chlorophyll that imparts the familiar green color to plants and algae. Actually, all pigments absorb light, but chlorophyll manages to convert light energy into chemical energy that is used in the synthesis of carbohydrates.

Surprisingly, chloroplasts have their own DNA and ribosomes—both of which are like those of prokaryotes and unlike those of the surrounding cell. With DNA and ribosomes present, chloroplasts are able to reproduce themselves and to carry out the synthesis of some proteins, although many of their other proteins are synthesized by the cell's nuclear DNA. Chloroplasts reproduce by undergoing DNA replication and simple fission (pinching in two)—another striking similarity to prokaryotes. As we will see, these characteristics are true of mitochondria as well.

Other Plastids. Chloroplasts are but one of a group of plant **plastids**. In addition to green chloroplasts there are **leucoplasts** ("white plastids") and **chromoplasts** ("colored plastids"). The leucoplasts are starch-storing bodies, present in most plant cells, but most common in storage tissues such as those making up most of the onion and apple. Chromoplasts are named for their pigments, which include orange carotenes, yellow xanthophylls, and several pigments that are red. They are highly visible as the bright colors of flowers and fruits. All plastids are alike in that they are surrounded by a double membrane and have their own DNA and ribosomes.

Plastids are derived from a line of undifferentiated bodies called **proplastids**. Proplastids, present in the seed and plant embryo (which is how they get from generation to generation), differentiate into the various types of plastids as the young plant grows. Apparently, the kind of plastid emerging depends on cues from the cell's surroundings. In the presence of light, for instance, proplastids routinely mature into chloroplasts.

Mitochondria. Mitochondria (singular, mitochondrion) are complex, ATP-generating organelles found in virtually every eukaryotic cell. To some extent, chloroplasts and mitochondria do exactly opposite things. Put simply, chloroplasts use energy and raw materials to produce carbon compounds and oxygen, while mitochondria use carbon compounds and oxygen to obtain energy. We will deal with both these important biochemical processes in Chapters 7 and 8.

Typically, mitochondria are long and slender, more or less sausage shaped, although in electron micrographs, mitochondria often appear as oval and capsule-shaped structures (Figure 4.25). The abbreviated view is actually the result of the angle through which the organelle was sliced when prepared for the electron microscope. Like chloroplasts, mitochondria have double membranes, here designated as the **outer membrane** and the **inner membrane**. The inner membrane has numerous narrow folds known as **cristae**. The region between the two membranes is called the **outer compartment** (or **intermembrane space**), whereas the region within the inner membrane is called the **inner compartment** (or **matrix**).

The biochemical work of the mitochondrion, known as **cell respiration** or **oxidative respiration**, occurs in both compartments and along the inner membrane. Cell or oxidative respiration makes use of oxygen in the chemical processes in which ATP is produced. Supporting this process are numerous enzymes and other molecular complexes that work to transfer the energy of foods to the chemical bonds of ATP. Special electron microscope techniques reveal that the cristae of the inner membrane are dotted with spherical ATP synthases, very similar to those in the chloroplast (see Figure 4.25). Like those of the chloroplast, the mitochondrial ATP synthases are sites of ATP synthesis. The number of mitochondria in a cell varies in proportion to the cell's energy requirements. For example, a single hardworking liver cell may contain as many as 1000 mitochondria, but few can be found in most fat storage cells. As you might expect, mitochondria are abundant in muscle cells.

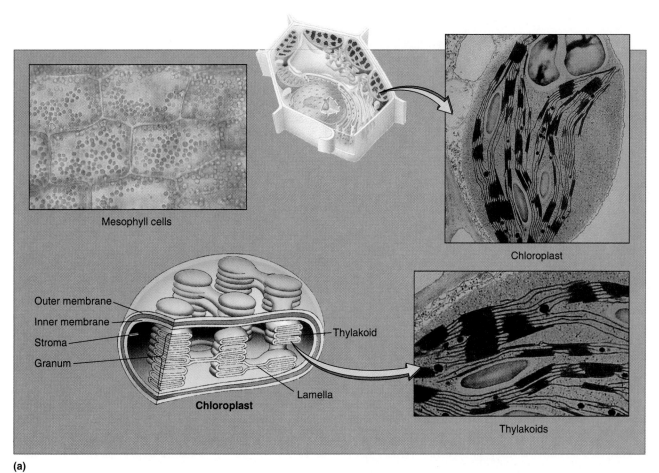

Mesophyll cells

Chloroplast

Outer membrane
Inner membrane
Stroma
Granum

Thylakoid

Lamella

Chloroplast

Thylakoids

(a)

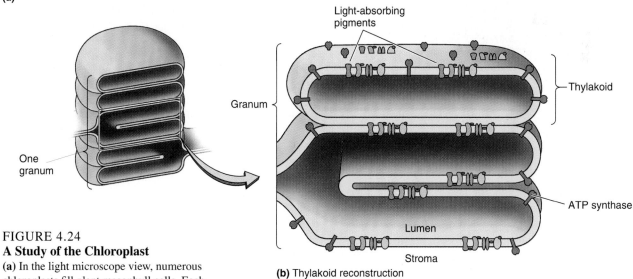

One granum

Granum

Light-absorbing pigments

Thylakoid

ATP synthase

Lumen

Stroma

(b) Thylakoid reconstruction

FIGURE 4.24
A Study of the Chloroplast

(a) In the light microscope view, numerous chloroplasts fill plant mesophyll cells. Each chloroplast is an oval, membrane-surrounded body containing a highly folded, inner membrane system and a lighter stroma. A close-up view reveals individual thylakoids, some of which occur as stacked granum thylakoids and others as connecting lamellae. **(b)** individually, each thylakoid is disk-shaped, its membranes enclosing an inner chamber, the lumen. Light-absorbing pigments are located within the membranes and ATP is synthesized within bodies called ATP synthases.

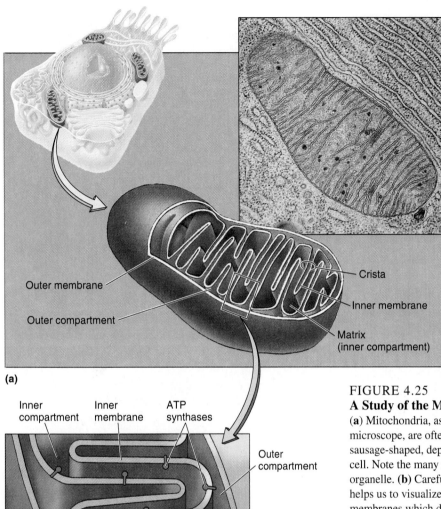

(a)

(b) Compartmental organization

Inner compartment · Inner membrane · ATP synthases · Outer compartment · Outer membrane

Outer membrane · Outer compartment · Crista · Inner membrane · Matrix (inner compartment)

FIGURE 4.25
A Study of the Mitochondrion

(**a**) Mitochondria, as viewed through the transmission electron microscope, are often oval, as seen here, although they may also be sausage-shaped, depending on how the slice was made through the cell. Note the many folded membranes, or cristae, crossing the organelle. (**b**) Careful reconstruction from photographic studies helps us to visualize the organization. Mitochondria have two membranes which divide it into an outer and an inner compartment (the latter is also called the matrix). The inner membrane is highly folded, forming numerous cristae that extend nearly across the organelle. The closer view shows the orientation of the spherical ATP synthases, sites of ATP synthesis.

Like chloroplasts, mitochondria have their own circular DNA and their own ribosomes and synthesize some of their own proteins. They also arise only from preexisting mitochondria through DNA replication and fission. All this is fascinating in that it makes the chloroplast and mitochondrion seem autonomous! They multiply and develop in an independent manner, as though they were "just visiting" the cell. As it turns out, this may not be far from the mark.

Chloroplasts and Mitochondria as Endosymbionts. Many biologists believe that both mitochondria and chloroplasts are the descendants of once-independent prokaryotic cells.

Such cells were purportedly captured by other very ancient lines of prokaryotes, in which they survived as **endosymbionts** (*endo,* "within"; *symbiosis,* "living together"). The idea is embodied in what is called the **serial endosymbiosis hypothesis**. It was through such invasions, the hypothesis maintains, that some membrane-bound organelles of the eukaryotic cell arose. Although the evidence for serial endosymbiosis is fascinating and compelling, not everyone is convinced. The **autogenous hypothesis** explains the origin of membrane-bound organelles in a very different manner. Both hypotheses are presented in Essay 4.2. The cell organelles are reviewed in Table 4.2.

TABLE 4.2 Eukaryotic Cell Structures and Their Functions

	Name	Feature	Function
	Cell wall	Cellulose fibrils embedded in matrix	Structural support of plant cells
	Centriole	Paired cylinders composed of a complex arrangement of microtubules (9:0)	Give rise to basal bodies that form cilia and flagella
	Cytoskeleton	Mictotubules, intermediate fibers, and actin filaments	Physical support of cytoplasm, cell movement
	Plasma membrane	Phospholipid bilayer: proteins, cholesterol, glycolipids, glycoprotein	Transport, boundary, cell recognition
	Cilia and flagella	Slender extensions of plasma membrane, containing a basal body and a complex assembly of microtubules (9:2)	Movement through sliding action of adjacent microtubules
	Nucleus	Double membrane containing chromosomes (DNA–protein), nucleoli	Controls cell synthesis, repository of genes
	Nucleolus	Dense body of DNA/RNA and proteins in nucleus	Synthesis of ribosomal RNA
	Endoplasmic reticulum	Rough: membranes form flattened passages, ribosomes attached	Rough: protein modification, transport
		Smooth: membranes form tubelike passages	Smooth: carbohydrate, lipid synthesis, transport
	Ribosome	Two units formed from protein and RNA, interlocked when active	Assembly of amino acid into polypeptides
	Lysosome	Membrane forms simple but tough sacs	Storage of proteolytic enzymes for intracellular digestion
	Golgi complex	Stacked membranous sacs, form from rough ER	Modification, packaging, storage, and secretion of protein
	Chloroplast	Double outer membrane, highly folded inner membrane forms saclike thylakoids that contain chlorophyll and ATP synthases	Uses sunlight energy to form ATP, which is used to produce glucose and other molecules
	Vacuole	Membrane forms simple sacs	Storage, water disposal, structural support when turgid, digestion
	Mitochondrion	Simple outer membrane, folded inner membrane with ATP synthases	Transfer energy of fuels to ATP for use by organism

Essay 4.2 ● EVOLUTION OF THE EUKARYOTIC CELL: AUTOGENY OR ENDOSYMBIOSIS?

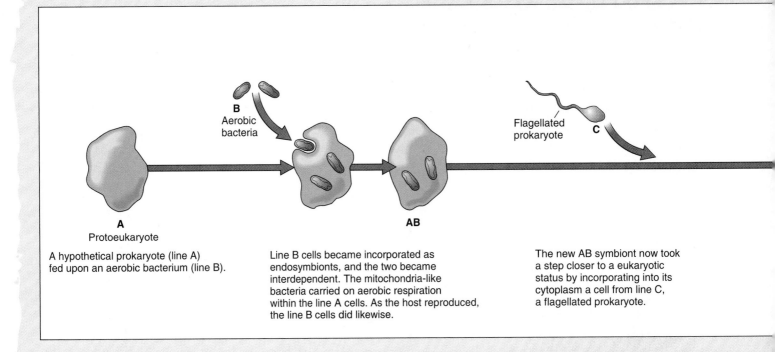

A
Protoeukaryote

A hypothetical prokaryote (line A) fed upon an aerobic bacterium (line B).

Line B cells became incorporated as endosymbionts, and the two became interdependent. The mitochondria-like bacteria carried on aerobic respiration within the line A cells. As the host reproduced, the line B cells did likewise.

The new AB symbiont now took a step closer to a eukaryotic status by incorporating into its cytoplasm a cell from line C, a flagellated prokaryote.

The evolutionary history of the eukaryotic cell, with its complex membrane-bound organelles and organized nucleus, is a subject of considerable controversy. Eukaryotes are thought to have been around as long as 2.5 billion years ago, an idea based on fossil evidence that is still very tentative. It is certain, however, that eukaryotes were well established over one billion years ago, which is still quite ancient. Ideas on the evolution of eukaryotes have changed considerably over the years, and today, two fundamentally different proposals have been brought forward: the autogenous hypothesis and the serial endosymbiosis hypothesis.

The autogenous hypothesis (*auto-*, "self"; *gen-*, "beginning") is based on the premise that the many membrane-bound structures within the eukaryotic cell were derived from the plasma membrane. The membrane underwent what was probably a gradual and lengthy evolution that began with events in simple and primitive prokaryotes. The idea is that inpocketings in the plasma membrane pinched off, floating free, as is common during feeding in ameboid cells. In time some of these structures took on specific functions. They gave rise to specialized organelles, particularly those of the endomembranal system: the nuclear enve-

lope, the endoplasmic reticulum, the Golgi complex, lysosomes, and vacuoles.

The development of mitochondria and chloroplasts, according to the autogenous hypothesis, was a bit more complex. Their double membranes would have necessitated more extensive inpocketings of the plasma membrane, or perhaps secondary inpocketings of certain organelles derived from the plasma membrane.

In light of what we know about the endomembranal system and the ongoing generation of organelles, the autogenous hypothesis is quite reasonable, but perhaps hard to test.

Difficulties arise as we try to explain the origin and peculiar characteristics of mitochondria and chloroplasts. As we see, these difficulties are solved through the next hypothesis, but we might also consider that the two ideas are not necessarily mutually exclusive.

The serial endosymbiosis hypothesis is an ingenious and well-received alternative. It proposes that the eukaryotes arose principally through a series of events wherein certain prokaryotic cells were engulfed by other larger phagocytic prokaryotes, those whose membranes evolved the ability to take in food through phagocytosis (also called endocytosis—see Chapter 5). Perhaps

because the digestive machinery was not yet particularly efficient, or the prey had developed defenses against the predator's enzymes, some of the ingested organisms survived and continued living within the predator (who then became the host).

The invaders, the hypothesis goes on, lived somewhat independently at first, but soon an interdependence was established. It was through these successive events that the newly emerging eukaryote came into possession of mitochondria, chloroplasts, and perhaps the cilium or flagellum.

Other eukaryotic structures, those of the endomembranal system, the ER and the Golgi complex, and so on, could have originated autogenously, since the two hypotheses are not mutually exclusive.

What had the host gained so far? Well, in addition to a new, highly efficient form of energy metabolism, oxidative respiration, it gained the ability to capture sunlight energy and carry on photosynthesis. Furthermore, it gained a greater degree of mobility through the action of cilia or flagella.

How strong is the evidence for serial endosymbiosis? The case for mitochondria and chloroplasts is quite convincing. Even today mitochondria retain a circular chromosome, certain types of RNA, and ribosomes,

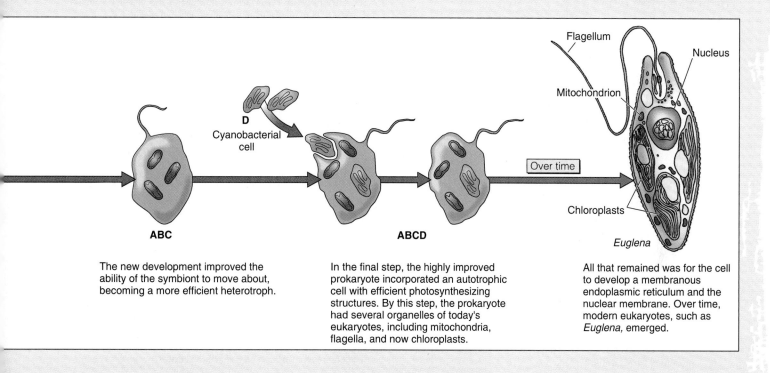

D
Cyanobacterial
cell

Flagellum

Nucleus

Mitochondrion

Chloroplasts

Euglena

ABC

ABCD

Over time

The new development improved the ability of the symbiont to move about, becoming a more efficient heterotroph.

In the final step, the highly improved prokaryote incorporated an autotrophic cell with efficient photosynthesizing structures. By this step, the prokaryote had several organelles of today's eukaryotes, including mitochondria, flagella, and now chloroplasts.

All that remained was for the cell to develop a membranous endoplasmic reticulum and the nuclear membrane. Over time, modern eukaryotes, such as *Euglena,* emerged.

which are all remarkably like those of bacteria. The fact that mitochondria arise only from preexisting mitochondria is also convincing evidence of independent origins. Furthermore, at least one mitochondrial protein, *cytochrome c,* is recognizably similar in shape and amino acid sequence to the cytochromes of certain bacteria. Biologists often use such information to establish evolutionary relationships.

What about the chloroplasts? Photosynthetic endosymbionts are found in many of the simpler marine animals and in many protists, so this mutual relationship is not in the least unusual. Also, there are many similarities between chloroplasts and modern photosynthetic bacteria, including their structure and biochemistry. In addition, chloroplasts contain DNA and ribosomes that are quite

similar to those found in prokaryotes. Both chloroplast DNA and mitochondrial DNA are susceptible to antibiotics that do not affect eukaryotic DNA—further evidence of similarity between the organelle and its prokaryote counterpart. Finally, like mitochondria, chloroplasts arise independently of other organelles from a continuing line of proplastids.

Key Ideas

CELL THEORY

1. The **cell theory** maintains that living organisms are composed of cells and that all cells come from preexisting cells.

2. Cells are extremely diverse, their differences reflecting specialized functions.

WHAT IS A CELL?

1. Cells are the basic units of life. They are highly organized, take in materials, extract energy, synthesize molecules, grow, respond to stimuli, maintain homeostasis, adapt, and reproduce.

2. As cells increase in size, the volume increases proportionally faster than the surface area. The **surface–volume hypothesis** maintains that cells must remain small in volume in order for

the **plasma membrane** to provide sufficiently for the transport of materials into and out of the cell. The solution in most cells is division, although the surface area can be increased by folding the borders into **microvilli**.

THE PROKARYOTIC CELL

1. **Prokaryotes**, or bacteria, are simple, single-cell organisms. Most bacteria have dense **cell walls** of peptidoglycan or protein. DNA occurs in a continuous, circular **chromosome**. Most reproduction is asexual. Movement in some occurs through an S-shaped rotating **flagellum**.

2. Some bacteria are **heterotrophs**, living as parasites or as **decomposers** and requiring complex foods. Others are **autotrophic**, requiring only inorganic materials. **Photoautotrophs**

make use of light energy and **chemoautotrophs** derive energy from chemical reactions.

THE EUKARYOTIC CELL

1. **Eukaryotes** include plants, animals, fungi, and protists, all of which have membranous **organelles**, including an organized **nucleus**.

2. Plants, fungi, and algal protists have dense **cell walls** that maintain shape, provide strength, and resist excessive water entry. Plant cell walls are crisscross layers of cellulose fibrils impregnated with strengthening agents.

3. The **plasma membrane**, a bilayer of phospholipids, in which proteins, glycoproteins, glycolipids, and cholesterol are interspersed, controls the passage of materials into and out of the cell and has recognition functions.

4. The cytoplasm of the cell is supported by the **cytoskeleton**, a network of protein components:

   a. **Actin filaments** provide structural support to the cytoplasm and plasma membrane and function in cell movement and cell division.

   b. **Intermediate filaments** strengthen microvilli and nerve cells and bind epithelial cells together.

   c. **Microtubules** consist of **tubulin** subunits. They support the cytoplasm and form **centrioles**, **basal bodies**, **cilia** and **flagella**, and **spindles**. The latter separate replicated chromosomes during cell division.

5. **Centrioles** are short rod-like bodies whose microtubules are in a 9 + 0 arrangement. Centrioles give rise to **basal bodies**, which produce cilia and flagella.

6. Cilia and flagella consist of microtubules in a 9 + 2 arrangement. Some move the cell or organism and others move materials past stationary cells. The two differ in length and movement pattern. Movement occurs when **dynein arms** from one set of microtubules attach to and produce a sliding action in another. The whole structure bends in response.

7. The nucleus is a prominent, spherical body, containing DNA. DNA controls cell activity through the production of enzymes, stores genetic information, and replicates for cell division. **Nucleoli** produce ribosomal RNA, which is assembled there into **ribosome** subunits. The **nuclear envelope** is a double membrane containing protein-lined pores through which materials enter and leave the nucleus.

8. The **endoplasmic reticulum**, a number of membranous channels through the cytoplasm, separates the cell into two compartments. **Rough endoplasmic reticulum** receives and processes newly synthesized polypeptides from ribosomes on its outer surface, particularly those destined for export from the cell. **Smooth endoplasmic reticulum** occurs in cells where carbohydrates, lipids, and other nonprotein products are formed.

9. **Ribosomes** consist of ribosomal RNA and protein organized into subunits that join when polypeptides are synthesized. Some are bound to the RER while others are free, the latter often grouped as **polyribosomes**.

10. The **Golgi complex**, a stack of flattened cisternae, forms continuously from protein-filled vesicles that move from the RER to the *cis* face. Proteins are modified and packaged into storage or secretion vesicles that bud from the *trans* face.

11. **Lysosomes** digest damaged or aging cell components and materials taken into the cell.

12. **Peroxisomes** contain enzymes such as **catalase**, which converts harmful hydrogen peroxide to water and oxygen. **Glyoxysomes** in sprouting seeds convert plant lipids to sugars.

13. Large water-filled vacuoles in plants lend supporting firmness to leaves and sometimes store pigments and toxic wastes. Many protists carry on digestion in **food vacuoles** and rid themselves of excess water through **contractile vacuoles**.

14. **Chloroplasts** are complex **plastids** that carry on **photosynthesis**, during which sunlight energy is used in forming carbohydrates.

15. Other plastids include starch-storing **leucoplasts** and pigment-filled **chromoplasts**. All plastids develop from **proplastids**.

16. Mitochondria carry out cell respiration, where chemical bond energy in cellular fuels is transferred to ATP. Membranes separate the mitochondria into **outer** and **inner compartments**, which are involved in **ATP synthases**.

17. The autonomous nature of chloroplasts and mitochondria prompted the **serial endosymbiosis hypothesis**. It maintains that the two organelles originated as free-living prokaryotes that were taken in by an ancient phagocytic line, probably as food, with some eventually becoming permanent residents.

Application of Ideas

1. Many exceptions to the cell theory have been noted. In view of this, how can the theory continue to be important and instructive? If you think it is not, offer reasons for this conclusion.

2. A biological principle called complementarity of structure and function holds that the two are very closely related. Cite five or six examples of how complementarity works at the cellular level.

Review Questions

1. List the fundamental activities of cells.

2. What happens to the relationship between surface area and volume when a cell doubles its size? Illustrate this with a drawing.

3. Summarize the surface–volume theory. How does periodic cell division resolve the problem? (pages 71, 75)

4. Prepare a drawing of a prokaryotic cell, labeling the following: cell wall, plasma membrane, pili, capsule, mesosome, and chromosome. (pages 75–77)

5. Summarize two important functions of the plant cell wall. (page 80)

6. Describe the structure of the plasma membrane. What are its functions? (pages 80–81)

7. What is the general role of actin filaments and intermediate filaments? (pages 81–82)

8. Describe the structure of a microtubule. List three of its uses.

9. With the aid of a drawing, explain the 9 + 2 microtubular structure of a cilium. (pages 82–84)

10. List the components of the axoneme and explain their roles in the bending action of a cilium or flagellum. (pages 85–86)

11. Summarize two important roles of DNA. Be specific. (pages 88)

12. Make a simple drawing of the nucleus and label the following: envelope, pores, chromatin, and nucleolus. (pages 86–88)

13. How do the two types of endoplasmic reticulum differ in appearance and function? (page 88)

14. Describe the appearance, formation, and function of the Golgi complex. (page 90)

15. Discuss two important roles of the large water vacuoles of plants. (page 93)

16. Using a simple drawing, illustrate the structure of a chloroplast. Label the envelope, grana, stroma, thylakoids, and ATP synthases. What is the general role of the chloroplast? (pages 94–95)

17. Make a simple drawing of a mitochondrion. Label the outer membrane, inner membrane, outer compartment, inner compartment, cristae, and ATP synthases. What is its role? (pages 94–99)

18. According to the serial endosymbiosis hypothesis, what was the origin of chloroplasts and mitochondria? List the four characteristics of mitochondria and chloroplasts that support this idea. (pages 96–97)

CHAPTER 5

The Plasma Membrane and Cell Transport

CHAPTER OUTLINE

THE PLASMA MEMBRANE

Lipid Bilayers

The Membrane Proteins

The Glycocalyx: Glycoproteins and Glycolipids

Cholesterol

The Fluid Mosaic Model

MECHANISMS OF TRANSPORT

Passive Transport

Active Transport

CELL JUNCTIONS

Cell-to-Cell Transport

Cell-to-Cell Barriers and Support

KEY IDEAS

APPLICATION OF IDEAS

REVIEW QUESTIONS

One problem that has presented an enduring challenge to biologists is determining how cells move critical materials around so that the essential processes of life can go on. After all, for cells to remain active they require a constant uptake of vital raw materials and a constant output of metabolic wastes and, in many instances, special substances they produce. Materials moving into or out of cells must pass through that delicate boundary structure called the plasma membrane, so it is here that we can begin our search for the mechanisms of such movement.

The Plasma Membrane

The plasma membrane is a critical structure and rather taken for granted in its role as a living barrier around the cell. For a long time some people refused to believe it existed, although intuition should have convinced otherwise. This is understandable; the plasma membrane is so thin that its existence, as well as its structure, was postulated entirely on circumstantial evidence, at least until the advent of the transmission electron microscope (TEM). The circumstantial evidence was gathered from observations that something was regulating the passage of materials into and out of living cells. Such evidence even suggested the composition of the membrane. For example, it had long been postulated that the membrane has a lipid core simply because lipid-soluble materials and lipid solvents readily pass through. Since water also passes through the membrane, though at a slower rate, biologists believed that the membrane also had water-admitting pores, perhaps surrounded by proteins. As we will see, these ideas were generally on target, although water-admitting pores have never been observed.

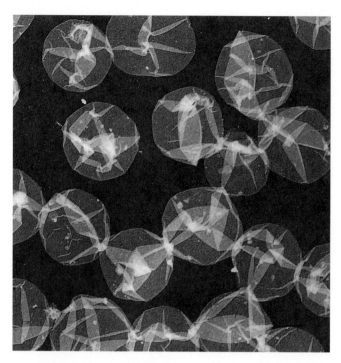

FIGURE 5.1
Electron Microscope View of the Plasma Membrane
The two dark lines seen in the electron micrograph *(top)* are the phospholipid heads of the bilayer that makes up much of the plasma membrane. The clear area between represents the hydrocarbon tails that form the core of the membrane. Red blood cell "ghosts" (empty cells, *bottom*), seen at a considerably lower magnification, reveal the incredibly thin, ethereal nature of the plasma membrane.

With the transmission electron microscope, researchers could obtain for the first time visual evidence verifying much of what the membrane theorists had proposed. Sections cut across the plasma membrane, stained, and viewed through the TEM reveal two darkly stained lines separated by a lighter, less-stained region. The total thickness is about 7–8 nm (nanometers) (Figure 5.1). The clear space, it turns out, is about 4.5 nm wide, which happens to be twice the average length of the hydrocarbon tails of membrane phospholipids. This suggested a two-layer structure, or lipid bilayer as it is called. The dark lines apparently represented the polar heads of the phospholipids, along with proteins and other molecules associated with the membrane surface. But what is the membrane really like? We will get into the technical explanations soon enough, but for a quick answer, get some bubble soap and start blowing "membranes." The analogy is a good one.

Lipid Bilayers

Phospholipids, as we saw in Chapter 3, are composed of one molecule of glycerol to which are linked two nonpolar fatty acid tails and a polar phosphate head, the heads differing according to the variable group contained. Quite often, but not always, the fatty acid tails include one that is saturated and one that is unsaturated, the latter taking on a bent configuration.

Also in Chapter 3, we described how phospholipids, when in water, spontaneously form spheres called micelles, flattened, sheet-like bilayers, resembling cell membranes, and compartment-like **liposomes**. The molecules of the bilayer align themselves with polar (charged) heads facing outward and the nonpolar (uncharged) hydrocarbon tails making up the inner portion of the bilayer. The tails are drawn together by hydrophobic interactions, thus producing an oily, nonpolar core in the structure. The polar heads form strong interactions with water, chiefly hydrogen bonds, that reinforce their orientation in the membrane and create an energetically stable organization (Figure 5.2).

Such lipid bilayers represent the fundamental organization of all biological membranes. The outer hydrophilic heads readily interact with water and other polar or charged substances, whereas the inner hydrophobic region reacts only with other nonpolar, uncharged substances.

There was a time when the phospholipid content of all cellular membranes was considered to be identical. But this idea has been discarded as more sophisticated methods of biochemical analysis appeared. The specific phospholipid content differs in various kinds of cells and organelles and among organisms. This is not unexpected, since cells and organelles commonly have different transport functions. Thus the phospholipid components of membranes in mitochondria and lysosomes differ substantially from those of the plasma membrane.

Actually, variation goes a bit further since, in the membranes of chloroplasts in plants and of nerves in animals, much of the usual phospholipid component has been replaced by lipoprotein (lipid protein). We will look into the structure of lipoproteins shortly.

As noted earlier, the cell membranes contain other molecular constituents, chiefly proteins. Whereas the phospholipids make up the framework of membranes, it is the protein constituent that carries out most of the important functions.

The Membrane Proteins

The protein component of the membrane is highly varied. Cell biologists recognize three general categories: integral proteins, peripheral proteins, and lipid-anchored proteins. **Integral proteins** are amphipathic; that is, they have both hydrophilic and hydrophobic regions. The first interacts with the hydrophilic phospholipid heads, whereas the second forms strong interactions with the hydrophobic core of the phospholipid bilayer. Integral proteins are "transmembranal," which means they extend completely through the membrane (Figure 5.3).

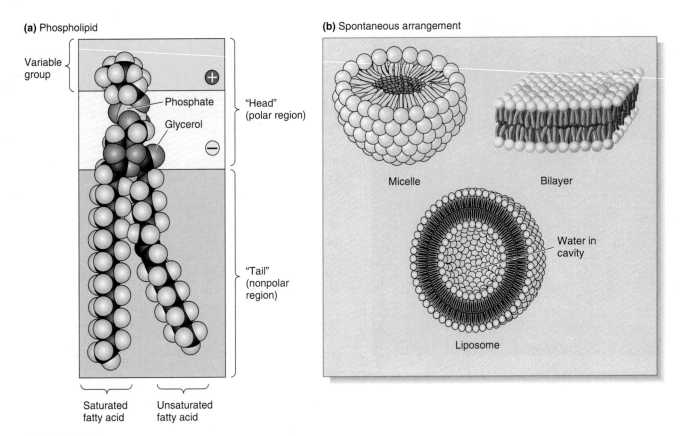

(a) Phospholipid

Variable group

Phosphate

Glycerol

"Head" (polar region)

"Tail" (nonpolar region)

Saturated fatty acid   Unsaturated fatty acid

(b) Spontaneous arrangement

Micelle   Bilayer

Water in cavity

Liposome

FIGURE 5.2
Behavior of Phospholipids in Water
(a) Phospholipids contain a glycerol molecule, to which two fatty acids and one phosphate/variable group are linked. The fatty acid tails are nonpolar hydrocarbons. Note the bent tail on the unsaturated fatty acid. The variable group and phosphate group contain net charges and are thus polar. **(b)** When placed in water, free phospholipids spontaneously arrange themselves into spherical micelles. Under certain conditions, they will also form bilayers. The bilayers can take the form of membranes or liposomes (closed bodies). In these arrangements the charged, hydrophilic heads are free to interact with water but the uncharged, hydrophobic tails interact together and are shielded from water. This is the basic arrangement seen in all cellular membranes.

Integral proteins have several important functions. Some act as specialized ports, freely admitting selected ions such as those of sodium and potassium. Others, working as **transport proteins**, facilitate the movement of selected substances into and out of the cell. Some utilize the cell's energy reserves to do this while others do not. Some of the larger, globular membrane proteins are enzymes, carrying out their reactions right in the membrane. Finally, some integral proteins are involved in support. Those on the membrane's inner side may form bridges with elements of the cytoskeleton, which we described in Chapter 4 (see Figure 5.3).

Because of their amphipathic property, integral proteins are well fixed into the membrane and are difficult for the experimenter to dislodge. Evidence of their presence is provided by freeze–fracture preparations of plasma membrane

(see Essay 4.1). When the membrane is sheared during this procedure, many of the proteins remain behind and are clearly visible in the sheared surface. The raised portion, as seen in Figure 5.4, includes matching indentations where the proteins were located.

Peripheral proteins lie on the surfaces of the membrane, some held in place by ionic interactions formed with the charged phospholipid heads, and others held by interactions with integral proteins. Surface proteins are readily extracted from the plasma membrane by treatment with salt solutions, whose ions attract the proteins away.

Lipid-anchored proteins are actually rare. They extend part way through the membrane, but well into the lipid core. Anchorage in the core is provided by side chains of lipids, which intermingle with those of the membrane.

FIGURE 5.3
Membrane Proteins
Integral proteins extend through the membrane, while peripheral proteins are located entirely on the inner and outer surfaces. Integral proteins have hydrophilic and hydrophobic regions that match those of the phospholipid bilayer. Some of the trans-membranal integral proteins also interact with the cytoskeleton, forming junctions that anchor the membrane to the cytoplasm.

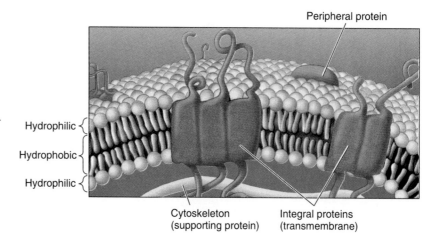

Peripheral protein

Hydrophilic
Hydrophobic
Hydrophilic

Cytoskeleton
(supporting protein)

Integral proteins
(transmembrane)

The Glycocalyx: Glycoproteins and Glycolipids

When viewed in the electron microscope, the plasma membrane of many animal cells reveals a fuzzy, indistinct outermost region that cell biologists call the **glycocalyx** (Figure 5.5). As its name implies, the glycocalyx is literally a "sugar coating" over the membrane surface. Actually, the sugars are short chains of glucose attached to the surface proteins and lipids. Glycoproteins are integral proteins with just their sugar chains protruding above the surface. Glycolipids have their lipid tails anchored in the phospholipid bilayer and their sugar chains protruding (see Figure 5.7).

Glycolipids and glycoproteins play important roles. Glycolipids are known to function in the binding of bacterial cells

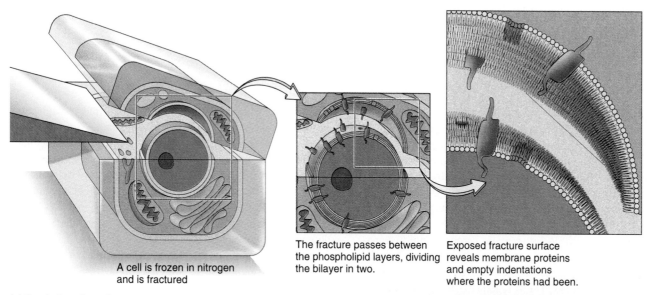

The fracture passes between the phospholipid layers, dividing the bilayer in two.

Exposed fracture surface reveals membrane proteins and empty indentations where the proteins had been.

A cell is frozen in nitrogen and is fractured

(a) Fracturing of a cell

(b)

FIGURE 5.4
Freeze Fracture Study
(a) In the freeze–fracture technique, tissues are frozen in liquid nitrogen and freon and struck randomly with a cold cutting knife. The effect is to shear the cells at points of natural weakness, which is often through the fatty acid tails of cell membranes. This opens the membrane up for close study of its internal components. The sheared regions are treated with a metallic spray, applied at an angle, to bring out more detail. **(b)** The electron micrograph shows sections of a membrane that have been sheared in this manner. The mounds seen in the sheared layer are integral membrane proteins that have been partially exposed when the upper half of the enclosing phospholipid bilayer was split away.

0.2 μm

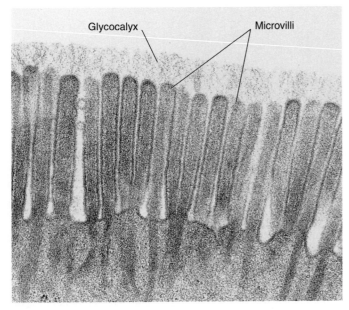

(a)

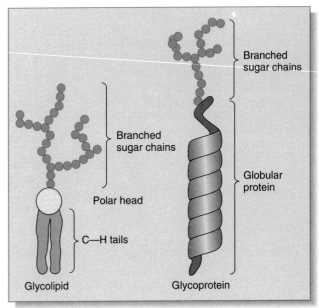

(b) Models of glycolipids and glycoproteins

FIGURE 5.5
The Glycocalyx
(**a**) Glycolipids and glycoproteins form the glycocalyx. The glycocalyx appears on the surfaces of animal cells that line internal spaces; in this instance the glycocalyx is on the surface of intestinal cells. The projections below the glycocalyx are the intestinal microvilli (see Chapter 4). (**b**) A closer view of model glycolipids and glycoproteins.

to each other and, in the case of disease-causing bacteria, to the cell surfaces of the organisms they invade. Glycolipids and glycoproteins residing on our red blood cell surfaces act as the antigens that are used clinically to determine blood types such as those in the A, B, O group. Antigens are molecules that stimulate body cells to respond defensively by forming antibodies, proteins that help neutralize the "invader." Should a person of, say, B blood type, receive type A blood (blood with the A antigen) in a transfusion, that person's immune system would react to the A antigen as it would to an invading organism. A flood of anti-A antibodies would soon clump the transfused cells together (see Chapter 11).

Some glycoproteins in animals act as receptors for chemical messengers and others function in cell recognition. Many kinds of hormones begin their work by binding to a specific, matching cell surface receptor. Once binding occurs, a series of reactions begins that brings about the cell's response to that particular hormone. Hormones and their cell surface receptors are very specific, with each receptor capable of binding to just one kind of hormone (see Chapter 36).

Cell recognition is the ability of cells to identify and interact with other cells. In our own bodies, cell recognition is vitally important to the work of our **immune system**. The immune system provides the body's response to invading disease organisms such as bacteria (Figure 5.6). Each person has

unique cell surface glycoproteins and glycolipids, which cells of the immune system recognize. For example, immune system cells known as **lymphocytes** have cell surface molecules that enable them to recognize and avoid normal body cells as they seek out invaders. However, they also have cell surface molecules that enable them to recognize cells that are cancerous or those that have been invaded by viruses. (Viruses cause detectable changes in cell surface chemistry that tip off defending lymphocytes of the immune system.) Upon such recognition, binding occurs between respective surface molecules and destruction of the afflicted cell begins.

Ironically, glycoproteins on viral surfaces help the virus bind to proteins on our own cell surfaces. This is the way many animal viruses begin their invasion. In fact, some viruses get help from the host's plasma membrane. As binding occurs, the cell is stimulated to take the virus in, suspending it within a vacuole. Next, a lysosome fuses with the vacuole and the viral protein is digested. Unfortunately, the viral genes are left intact, free in the cell, which is all the virus needs to do its deadly work.

Cholesterol

Cholesterol is a common constituent of animal cell membranes. Recall that cholesterol has four interlocking carbon

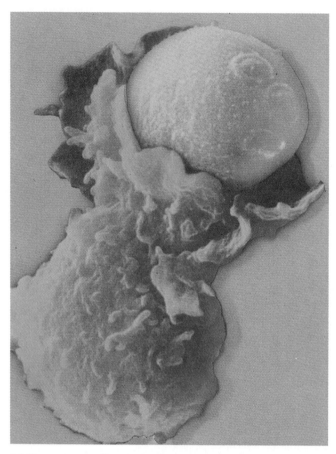

(a) Phagocyte engulfing a spherical bacterium.

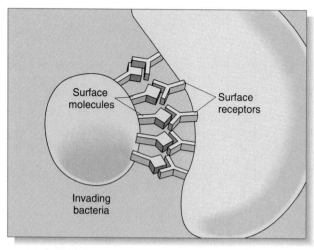

(b) Model of interaction

FIGURE 5.6
Cell Recognition
(a) The larger cell *(left)*, a phagocytic cell (macrophage) of the human immune system, is making use of its cell recognition molecules to bind to and engulf the bacterium *(above)*. This is one way the body keeps infection from spreading. (b) Phagocytes bind through the action of proteins on their surfaces, some of which match those on the cell surfaces of invading bacteria.

rings with a hydroxyl (— OH) side group (see Chapter 3). Like the membrane proteins, cholesterol is amphipathic: it has polar and nonpolar regions. The interlocking carbon–hydrogen rings are nonpolar whereas the — OH group is polar. How might cholesterol be oriented in the phospholipid bilayer? If you said the rings were buried in the lipid core and the — OH group poked out among the phospholipid heads, you have a good grasp on membrane characteristics.

The Fluid Mosaic Model

Figure 5.7 is a comprehensive diagram of what biologists believe the plasma membrane structure to be. This is the **fluid mosaic model**. The terms "fluid mosaic" refer to the fluid-like properties of the phospholipid bilayer and the patchwork arrangement of its other components. The word "model" tells us that this particular explanation is still tentative. This model, like others in science, brings various kinds of information together into a working prototype, one that can be tested further through continued observation and experimentation. The term fluid may sound unusual for a structure, so let's explore this now.

The bilayer, as previously noted, is in a stable configuration for phospholipids; nevertheless, the individual molecules move quite freely in the membrane. Such freedom is possible because the phospholipids are held in place by hydrophobic interactions, and these are very weak forces. Phospholipids can and do trade places and generally slide around in the membrane. Most movement, however, is restricted to the plane of the membrane (think of a pool full of swimmers bobbing about and intermingling as they tread water).

The degree of fluidity depends on the specific kinds of fatty acids that predominate. The greatest fluidity is seen where unsaturated fatty acid tails predominate. This is because the unsaturated fatty acids, those with bent tails (see Chapter 3), do not interact much. Conversely, membranes composed chiefly of phospholipids with saturated hydrocarbon tails, where greater interaction occurs, tend to be more rigid. Another constituent that increases fluidity is cholesterol, which you will recall is present in animal cell membranes (Figure 5.8).

Temperature also affects membrane fluidity. As you might expect, membranes become more rigid at lower temperatures. However, membranes with a preponderance of unsaturated fatty acids retain their fluidity at lower temperatures. Interestingly, some cells respond to lowered temperatures by adjusting the phospholipid content of membranes. This is necessary because an increasing rigidity can readily alter membrane function. Some plants maintain fluidity by utilizing more unsaturated fatty acids in newly synthesized membranes. Ani-

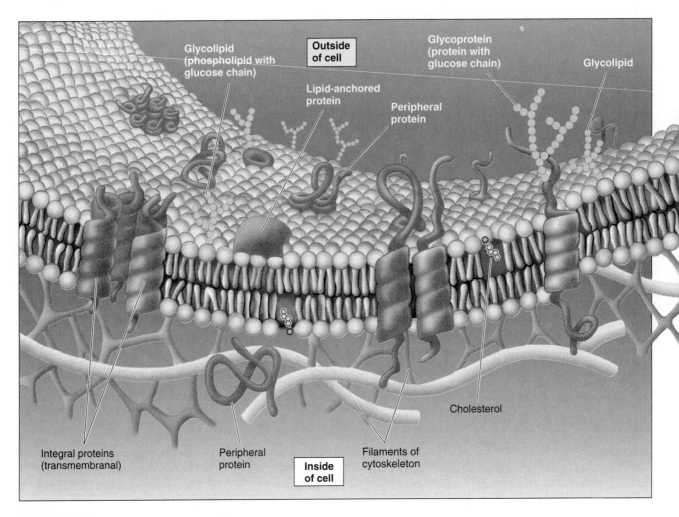

FIGURE 5.7
Fluid Mosaic Model of the Plasma Membrane
The plasma membrane is basically a bilayer of phospholipids with a varied protein component. Note how the phospholipid bilayer is similar to that spontaneously formed by phospholipids in water. Several cholesterol molecules are dispersed among the phospholipid heads, as is the case with animal membranes. The protein component includes peripheral proteins residing on the surface, integral proteins extending through the membrane, and lipid-anchored proteins partly submerged in the phospholipid bilayer. Some of the branched sugar chains are part of glycoproteins, while others are parts of glycolipids. Glycoproteins are usually transmembranal. The lipid portion of the glycolipid is nonpolar and anchors into the core of the bilayer. The plasma membrane has a very dynamic characteristic with most of the components free to move about. In spite of this movement, however, the basic construction, with the hydrophilic outer layers and hydrophobic core, is a stable arrangement.

mals retain membrane fluidity by increasing the amount of cholesterol during membrane synthesis. This adaptive response is especially important to hibernating mammals, since their body temperatures drop nearly to that of their surroundings.

Most of the protein component is also mobile, although slower moving than the phospholipids (in our swimming pool analogy, add swimmers in large inner tubes). Exceptions include those integral proteins that are anchored to the cytoskeleton; they are immobile.

We have seen that the plasma membrane is a delicate but exquisitely complex and dynamic structure. Its lipid and pro-

tein components have precise roles in the membrane's structure and function. With a clearer view of this in mind, we now turn our attention to the way materials move across the membrane.

Mechanisms of Transport

As noted earlier, the plasma membrane is highly selective in its transport role. Some substances pass readily through with little help or interference, while others can neither enter nor

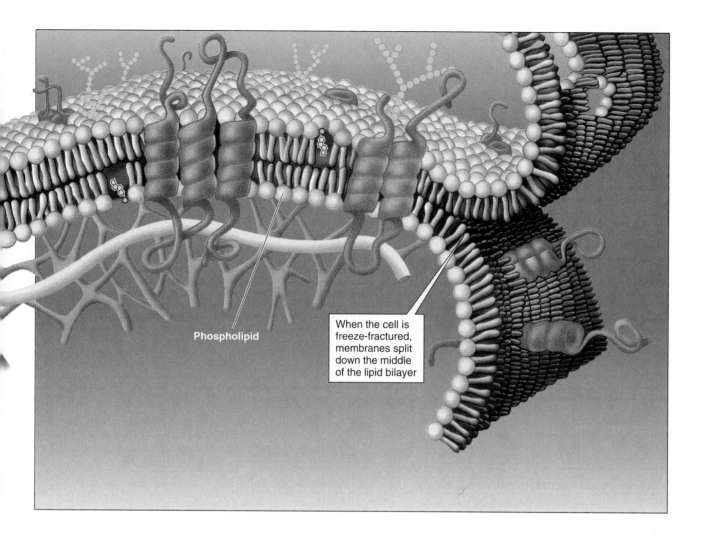

Phospholipid

When the cell is freeze-fractured, membranes split down the middle of the lipid bilayer

leave the cell without assistance. The degree to which substances can pass through a membrane determines the membrane's **permeability**. Because the plasma membrane is more permeable to some substances than others, it is said to be **selectively permeable**. The admission or rejection of substances by the plasma membrane depends on a number of factors, but the most significant are size, polarity, electrical charge, and lipid solubility. As a rule, small, nonpolar molecules (such as oxygen and nitrogen) and small, polar molecules (such as water, glycerol, and urea) pass rapidly across membranes. Somewhat larger polar molecules such as glucose and sucrose have great difficulty passing through the membrane on their own, and charged particles, such as sodium, potassium, and calcium ions, are quite often rejected unless aided across in some way. Such ions often pass through special ion channels or cross with the aid of transport proteins (which we will come to shortly). We also point out that although gases cross membranes, they must first go into solution. This requires that the membrane surfaces be moist, which gas exchange surfaces generally are. (Think of lungs, gills, and body surfaces in skin-breathers.)

Cell transport occurs in two general categories: passive transport and active transport. They differ primarily in the source of energy.

Passive Transport

Passive transport is cell transport that does not require the cell's own energy supplies. It includes diffusion, osmosis, and facilitated diffusion.

Diffusion. **Diffusion** is the net movement of molecules from a region of their higher concentration to a region of their lower concentration. By net, we mean moving mainly in one direction. Substances may be atoms or molecules; the process applies to gases, liquids, and solids. As we will see shortly, ions represent a special case.

Behind all this is the fact that matter is in constant motion (except when the theoretical temperature of absolute zero is reached, and all molecular motion ceases). That movement is random, with particles bumping each other and rebounding in all directions. However, when a substance is concentrated in

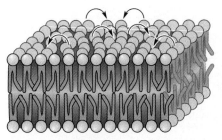
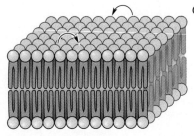
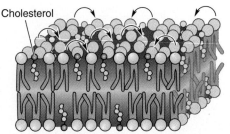

Cholesterol

(a) Unsaturated fatty acids (more fluid state)

(b) Saturated fatty acids (more rigid state)

(c) Cholesterol and saturated fatty acids (more fluid state)

FIGURE 5.8
Membrane Fluidity
The mobility of phospholipids and other membrane components depends on several factors: fatty acid composition, cholesterol content, and temperature. **(a)** Where unsaturated fatty acids predominate, the phospholipid bilayer is quite fluid and the phospholipids move readily about. Note how the bent tails break up the dense fatty acid packing. **(b)** Where the bilayer contains mainly saturated fatty acids, the membrane becomes fairly rigid and movement is restricted. Note how the straight tailed fatty acids pack solidly, increasing the strength of the hydrophobic interactions. Cold temperatures can produce the same effect, although less so in the membranes described in **(a)** and **(c)**. **(c)** Animal membranes rich in cholesterol are also quite fluid and phospholipid movement is freer than it would otherwise be. The presence of steroid carbon–hydrogen rings breaks up the fatty acid packing that might otherwise occur.

one region, it undergoes a *net* movement; that is, most of the movement is away from its own concentration. Net movement continues until the substance is randomly distributed. Molecular movement continues after that, but it is random; that is, there is *no net movement* (movement in one direction). Diffusion, by definition, will have stopped at that point (Figure 5.9).

Molecular motion is brought about by heat energy in individual particles of a substance. Thus the more heat applied to a system, the faster its particles move. Molecular movement can be observed indirectly. Tiny granules of carbon suspended in water and observed under the light microscope are seen to jiggle and jar each other. They are being buffeted to and fro by moving water molecules, which, of course, we can't see. The jiggling movement of the carbon particles is called **Brownian movement**.

We can also look at diffusion in terms of the energy present in a system. Concentrations of molecules represent systems of greater energy than systems whose components are randomly distributed (Figure 5.10). We will look further into

this idea in the next chapter, but for now let's just note that when molecules are randomly distributed, they will have reached their lowest energy state as far as their distribution in space is concerned. Let's look at some examples of diffusion.

The audible sigh coming from you as you read this fascinating account has produced a carbon dioxide **concentration gradient** between your face and your surroundings. This means that there is more CO_2 around your face than there is 2 feet away, and more 2 feet away than 4 feet distant. The tendency will be for the CO_2 you just expelled to diffuse *down* that concentration gradient. (It may be helped along by air movement as well.) The CO_2 molecules will continue to move down the gradient until they are equally dispersed throughout the surroundings, but you will have to stop breathing for that to happen (don't try this at home).

Diffusion is easy to demonstrate (Figure 5.11). If a chunk of brightly colored, soluble material (like the blue tablets used to disinfect toilet water) is placed in a glass container filled with water, the material will soon start to dissolve and spread

FIGURE 5.9
Simple Diffusion
Diffusion is the net movement of a substance from an area of greater concentration to an area of lesser concentration. Net movement means that most substances are moving in the same direction. The particles seen here are following their natural tendency, moving *down* their concentration gradient. Diffusion continues until the distribution of the substance is random, whereupon net movement, and therefore diffusion, ceases.

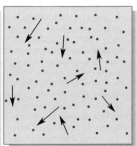

Net movement   Net movement   Random movement

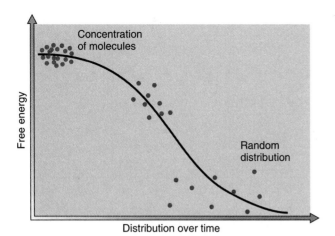

FIGURE 5.10
Diffusion and Free Energy Levels
Looking at diffusion from the point of view of energy, a heavy concentration of molecules has considerably more energy than does any lesser concentration. Molecules moving down their concentration gradient therefore lose energy as they become randomly distributed. As we will see in later chapters, the cell has important uses for the energy of concentration gradients.

FIGURE 5.11
Diffusion of a Dye
As the brightly colored, soluble material breaks down, its individual molecules diffuse throughout the water-filled container. The color gradient makes the concentration gradient visible. Eventually the molecules will reach a state of dynamic equilibrium, in which there is no net movement.

through the water. The dissolving crystal is a highly concentrated substance that forms a steep concentration gradient in its watery surroundings. As diffusion continues, the color will spread further and further from the source, and, given time, all of the substance will enter solution and the color will become uniform throughout the container. What has happened is that the individual molecules of the substance have diffused down their gradient and become randomly distributed.

The Diffusion of Ions: A Different Story. The diffusion of ions into and out of cells is complicated by the fact that they bear electrical charges. This means that their net movement is not determined by the concentration gradient alone, but by the charges already present on the other side of the membrane. If the cytoplasm just inside the membrane is strongly negative, this will hasten the movement of positive ions in that direction. Remember the basic law of electrical charges: unlike charges attract; like charges repel. If the reverse is true, and the cytoplasm just inside the membrane is positively charged, then the passive movement of positive ions will be much slower and may even cease. Because of these complications, it is more accurate to define the diffusion of ions as *their movement down their electrochemical gradient*. This takes into consideration both the concentration and the electrical charges on both sides of the membrane. We will come back to this problem when active transport is discussed.

Diffusion and Life. Living cells frequently make use of diffusion, especially in their ongoing exchange of gases. As an air-breathing organism, you undoubtedly know that your cells

require a continuous supply of oxygen, and that the oxygen is rapidly consumed during cell, or oxidative, respiration. Also during cell respiration, your cells continually produce carbon dioxide, a waste product. Thus the greatest concentrations of carbon dioxide are found within cells, and the greatest concentrations of oxygen occur outside cells. Therefore the net movement of the two gases is in opposite directions, that is, down their respective concentration gradients. This is convenient since cells need to take in oxygen and get rid of carbon dioxide.

We see then that the exchange of gases in life is a simple process; the molecules move into and out of cells by merely following their own concentration gradients. With molecules such as water, the situation is similar, but it can get complicated.

Osmosis. Because water is so crucial to life, there has always been a keen interest in its passage across membranes as it enters and leaves the organism and moves within it from place to place. Biologists refer to the diffusion of water molecules across membranes as osmosis. More succinctly, **osmosis** is the net movement of water across a selectively permeable membrane from a region of higher water concentration to a region of lower water concentration.

Osmosis can readily be demonstrated through the use of models such as the simple laboratory demonstration shown in Figure 5.12. Here, a thistle tube is filled with a 3.0% sugar solution (3% sugar, 97% water) and its large end, which is covered by a selectively permeable membrane, is immersed into a beaker of distilled (100%) water. The membrane is permeable to water but not to sugar. Let's pose some questions: On which side of the membrane is the *concentration* of water molecules greater? (On

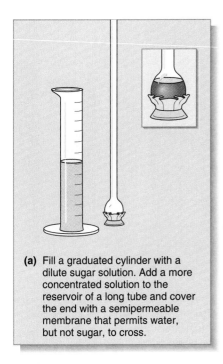

(a) Fill a graduated cylinder with a dilute sugar solution. Add a more concentrated solution to the reservoir of a long tube and cover the end with a semipermeable membrane that permits water, but not sugar, to cross.

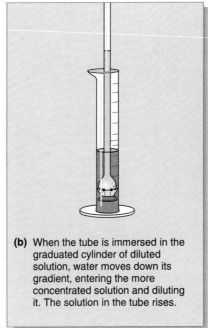

(b) When the tube is immersed in the graduated cylinder of diluted solution, water moves down its gradient, entering the more concentrated solution and diluting it. The solution in the tube rises.

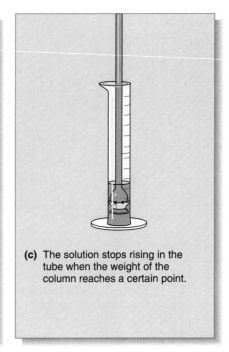

(c) The solution stops rising in the tube when the weight of the column reaches a certain point.

FIGURE 5.12
Model of Osmosis
Osmosis, the diffusion of water across a selectively permeable membrane, can readily be demonstrated. The membrane must be permeable to water and impermeable to the solute (sugar).

which side is the number of water molecules per cubic millimeter greater?) If your answer was, "in the beaker," you win. The beaker is pure water; thus it is as concentrated as it can be.

But which way then would you predict the water will move—into the thistle tube or out of it? Remember, the sugar molecules can't get out—only water can cross the membrane. Also keep in mind the definition of diffusion. You're right! The net movement of water will be into the thistle tube, and the level of fluid in the thistle tube will rise.

How long will the net movement of water continue? If we assume that the tube is tall enough, we can predict that at some point, the sheer weight of the water column in the thistle tube will force water molecules back across the membrane until, finally, the movement of water in the whole system reaches equilibrium. This means the net movement will be zero. The force (pressure) that is just enough to impede osmotic movement is known technically as osmotic pressure.

More specifically, **osmotic pressure** is defined as a measure of the tendency for water to cross a selectively permeable membrane from a less concentrated to a more concentrated solution. What determines the osmotic pressure of a solution? Osmotic pressure is directly proportional to the difference in the concentrations of the solutions: the greater the amount of solute, the greater the osmotic pressure. (A solute is any substance that will go into solution.) Another way of saying this

is that the power of osmosis—the force itself—depends on the difference in the solute concentrations on either side of a membrane that is permeable to water.

In summary, osmotic systems greatly influence the movement of water in living organisms by virtue of the selectively permeable nature of membranes and the solutes commonly found in cells. The ability of water in osmotic systems to move from one place to another is influenced by water and solute concentrations. These factors will be understood more readily as we see them at work in cells; so let's turn to osmosis in plant and animal cells.

Osmosis and the Cell. Cells are always subject to varying osmotic conditions and cope well with such changes as long as they are not too extreme. The two problems confronting both plant and animal cells are the possibilities of excess water loss and excess water gain. Plant cells can cope well with water gain by virtue of their tough, resilient cell walls, and, in fact, they make use of steep water gradients to maintain shape and rigidity in their leaves and young shoots, as we will see shortly. Animal cells, however, lack cell walls and should too much water enter, they will burst or lyse, killing the cell.

Because of their susceptibility to water intake, animal cells under study in the laboratory must be maintained in what is called an **isotonic** environment. An environment is isotonic if the surrounding solution contains the same concentration of water and solutes as does the cell. Red blood cells, for instance, will remain stable and maintain their shape for a time if they are placed in an isotonic 0.9% salt solution. Should these same blood cells be placed in a **hypotonic** environment—in a solution containing relatively *more* water (less solute) than is found within their cytoplasm—water gain through osmosis will be excessive, and the cells will literally burst. At the other extreme, in a **hypertonic** environment—one in which the water concentration is lower and the solute concentration higher than that of the cell—water is rapidly lost through osmosis. (The three conditions are illustrated in Figure 5.13.) Such a cell will shrink—a process physiologists call **plasmolysis**. The effects of isotonic, hypertonic, and hypotonic surroundings on animal and plant cells are illustrated in Figure 5.14.

We mentioned earlier that plants make use of water-filled vacuoles to support the young stem and give leaves their shape (see Chapter 4). Plants often encourage the osmotic uptake of water by retaining high solute concentrations in their large cell vacuoles. This, of course, creates a hypertonic condition there. The subsequent entrance of water and filling of the vacuole create great **hydrostatic pressure** (water pressure) against the plant cell walls. Such pressure in an animal cell would surely lyse the cell, but plant cells resist lysing by virtue of their strong, flexible cell walls. Botanists refer to the water-swelled condition as a state of **turgor** and often refer to hydrostatic pressure as **turgor pressure**. The stretching cell wall exerts an opposing force against the entering water, the **wall pressure**. When the two forces, turgor pressure and wall pressure, become equal, the net movement of water into the cell ceases and a state of dynamic equilibrium is reached. This will be maintained despite differing water concentrations inside and outside the cell.

Turgor in plants is maintained only when the water concentration in the soil around the roots exceeds the water concentration in the plant cells. If you forget to water your young tomato plants you may soon see the effects of a *shifting* water concentration. When too much soil water is lost, the water concentration in the root cells begins to exceed that of the soil, and osmosis is reversed. Water passes out of the root and the plant begins to wilt (Figure 5.15). Fortunately for the plant, the effects of wilting are reversible, up to a point, and with timely watering your tomato plants will perk up once again.

Facilitated Diffusion. Facilitated diffusion, our last example of passive transport, refers to the movement of molecules across a membrane with the assistance of special *transport proteins*—transmembranal proteins embedded in the membrane. It is similar to simple diffusion in that the energy involved is thermal energy, and the net movement of molecules is always from regions of higher concentration to

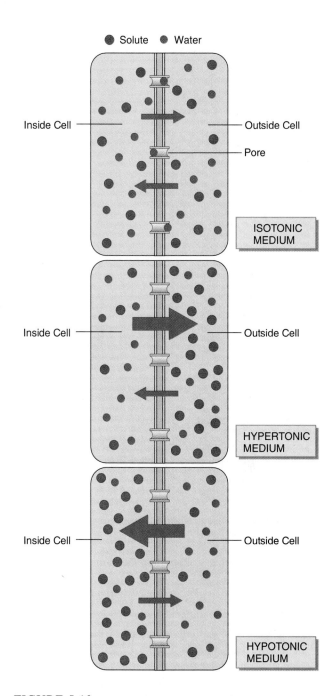

FIGURE 5.13
The Osmotic Environment
The three kinds of osmotic environments—isotonic, hypertonic, and hypotonic—are based on comparative solute concentrations or comparative water concentrations, depending on how you view them. In any case, the concentrations affect the net movement of water molecules, which is indicated by the size of the arrows. (The large arrows indicate greater movement.)

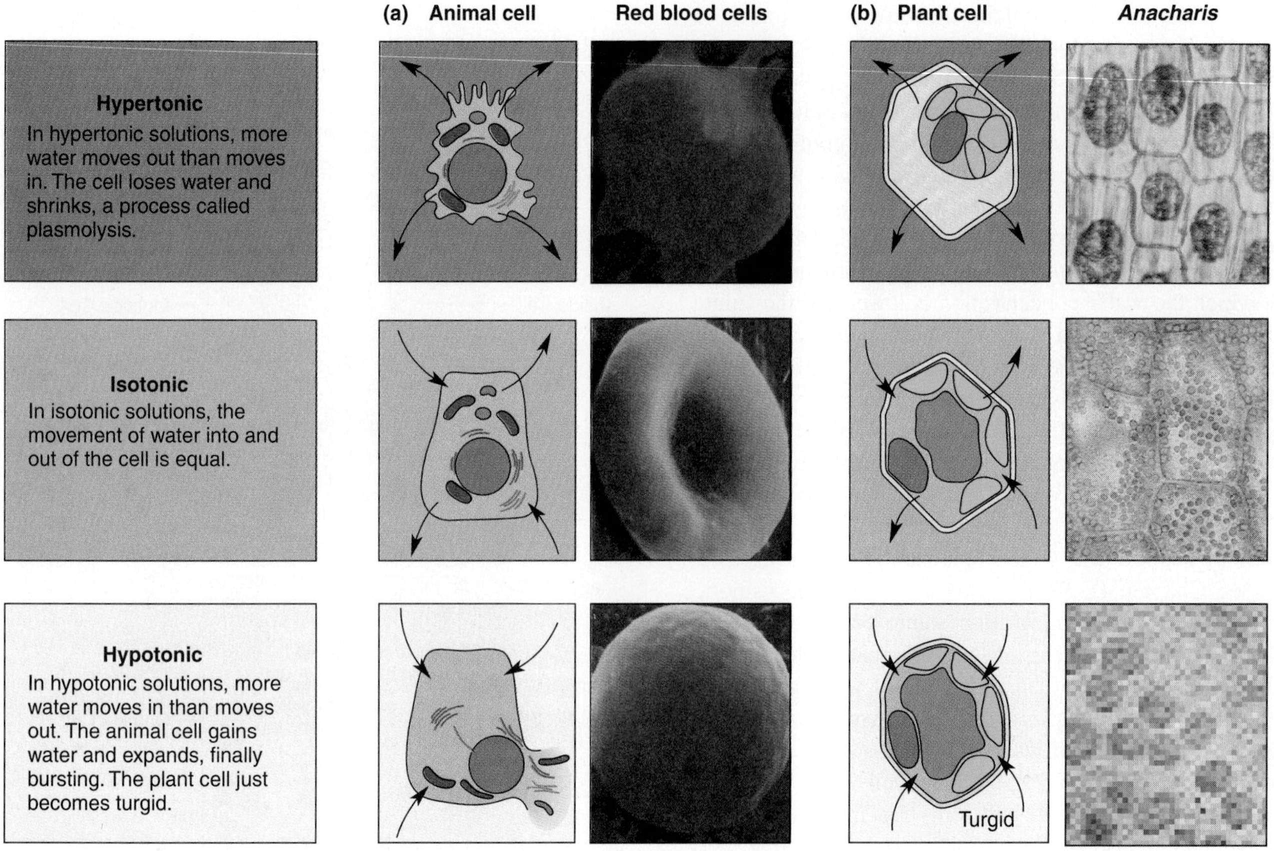

FIGURE 5.14
Osmotic Responses
(a) Animal and (b) plant cells respond differently to osmotic changes in their surroundings. Because of their cell walls, plant cells are better able to cope with osmotic changes.

regions of lower concentration. It differs from simple diffusion in several ways:

1. The rate of diffusion across the membrane is far greater than usual.

2. Facilitated diffusion is specific: each transport protein carries one kind of molecule only.

3. As in diffusion, the rate increases as the concentration increases; but, unlike diffusion, facilitated diffusion reaches a maximum rate after which the concentration has no effect. The latter indicates that each transport protein in the membrane is working at capacity.

But what are the transport proteins, and how do they do their work? Many models have been suggested, but, at the moment, the one of choice works as follows. The transport protein is a transmembranal protein that has the ability to change its shape without actually changing its location and without rotating in the membrane.

A diffusing molecule enters a specific site on the transporter, whereupon the transporter makes a shape change that can only be described as a "swallowing movement," and the molecule is released on the other side of the membrane (Figure 5.16). If this sounds vague to you, you are perceptive. There is still much to learn about facilitated diffusion.

Active Transport

Active transport is movement that definitely requires an expenditure of the cell's chemical energy, usually in the form of ATP. Active transport may involve individual atoms, ions, or molecules, each moved by transport proteins, or it may involve the uptake or expulsion of materials in bulk, so that a large segment of the membrane becomes involved. In either case, the process is characterized by (1) the movement of materials *against the concentration gradient* or, in the case of ions, *against the electrochemical gradient;* and (2) *an expenditure of the cell's chemical energy.*

The Work of Active Transport. As with facilitated diffusion, it isn't entirely clear how transport proteins do their work, but they all share four characteristics:

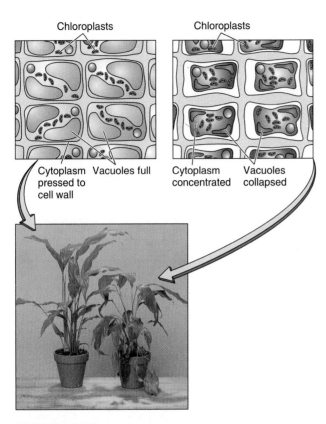

Chloroplasts Chloroplasts

Cytoplasm Vacuoles full   Cytoplasm   Vacuoles
pressed to   concentrated   collapsed
cell wall

FIGURE 5.15
Water and Structural Support in Plants
Leaves and young shoots are held erect partly by water pressure within the cells. Should the plant's loss of water exceed its uptake, the cells will lose water. The leaves and shoots will then lose their water-filled, turgid state and wilt.

1. Transport proteins are transmembranal; that is, they extend all the way across the membrane.

2. Transport proteins are specific; each type has a site or sites whose shape matches the molecule it transports.

3. Transport proteins work fastest when a large number of molecules are present.

4. Transport proteins do their work through conformational changes, that is, changes in shape.

The simplest transport proteins are **uniports**, those that move a single substance—a molecule or an ion—across a membrane. **Coports** transport two different substances simultaneously. **Symports** transport the two different materials in the same direction and **antiports** transport different materials in opposite directions (Figure 5.17). Examples of symports are found in the cells lining the digestive tract. Here the transport of glucose and amino acids, products of digestion, is *sodium dependent*. Either must be accompanied by the transport of sodium ions. One of the best known examples of the antiport involves the ions of sodium and potassium.

Sodium/Potassium Ion Exchange Pumps. The **sodium/potassium ion exchange pump**, or **Na$^+$/K$^+$ pump** (the term "pump" is not entirely figurative) actively transports potassium ions (K$^+$) into the cell and sodium ions (Na$^+$) out, both against the natural electrochemical gradient. As an antiport, the pump moves three sodium ions out of the cell for every two potassium ions brought in. This unequal transport means that the fluids just outside the membrane—with more positive charges—may take on a net positive charge, compared to the region just inside the cell, where there are fewer positive charges.

Differences in charges, or voltage differences, across a membrane are referred to as **membrane potential**. The differences can be considerable. The membrane potential in a resting nerve cell, for example, is on the order of −70 mV (millivolts: $\frac{1}{1000}$ volt), although this can reach −200 mV in some cells. The negative sign represents the negative condition inside the cell, with respect to the outside (which is typical). The presence of a membrane potential likens the cell to a storage battery—a system of considerable energy and one that can accomplish work when allowed to discharge. As we will see in future chapters, cells make good use of such energy storage systems.

The Na$^+$/K$^+$ pump is so important to cells that an estimated one-third of an animal's ATP energy reserve is expended in its operation. In nerve cells, where these exchange pumps are most active, the figure exceeds two-thirds. This is because neural impulses involve the sudden passive movement of sodium ions into the nerve cell, an event that shifts the membrane potential from −70 mV to +30 mV. This is followed by the passive movement of potassium ions out of the cell, which

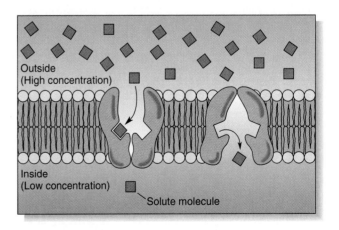

Outside
(High concentration)

Inside
(Low concentration)   Solute molecule

FIGURE 5.16
Facilitated Diffusion
In facilitated diffusion, membrane transport proteins make use of conformational changes to move substances into or out of the cell. This model suggests that the entry of a molecule triggers one event (the shape change), and its release triggers another (the change back to the original shape).

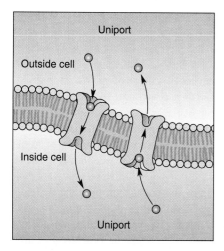
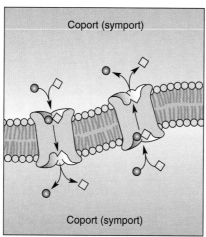
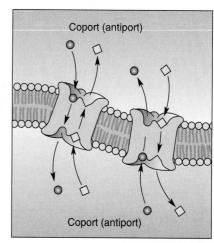

FIGURE 5.17

Uniport/Coport Transport

Membrane carriers work in various ways. Uniport carriers are the simplest, actively transporting a single type of material into or out of the cell. Coport carriers may be symport carriers (transporting two different materials in the same direction) or antiport carriers (simultaneously transporting two different materials in opposite directions).

temporarily restores the membrane potential. But to permanently restore the ion balance, the sodium/potassium ion exchange pumps in nerve cells must go to work, using ATP energy to move both sodium out and potassium back in. Considering how much neural activity goes on in an animal, the great energy expenditure is understandable.

Model of the Na⁺/K⁺ Pump. A model of the sodium/potassium ion exchange pump is shown in Figure 5.18. In the sequence of events, sodium is taken into its own entry site, and an ATP molecule transfers its outer phosphate to the transporter, providing the required energy. During this period the potassium entry site is closed. The transporter then under-

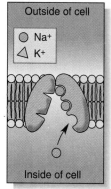
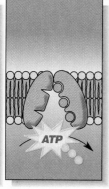
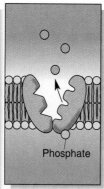
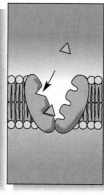
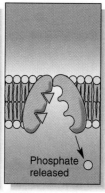
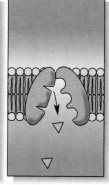

1 Three sodium ions (Na⁺) are taken up from cell cytoplasm, filling corner sites.

2 Energy is required to change shape of carrier so that it opens outside the cell.

3 Due to shape change, sodium ions are released outside the cell.

4 Two potassium ions (K⁺) from outside bind carrier sites.

5 Carrier returns to its original shape.

6 Potassium is released and cycle begins again.

FIGURE 5.18

Sodium/Potassium Ion Exchange Pump

The cotransport of sodium and potassium ions occurs through an ATP-driven sodium/potassium ion exchange pump. Two kinds of conformational changes occur in the protein as the pump works. One involves the sodium and potassium sites, the other the entire carrier.

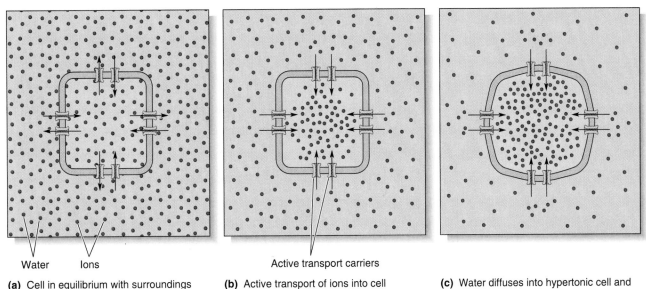

Water   Ions

(a) Cell in equilibrium with surroundings

Active transport carriers

(b) Active transport of ions into cell in great numbers

(c) Water diffuses into hypertonic cell and cell bows outward

FIGURE 5.19
Encouraging Osmotic Action
There is no known mechanism for actively transporting water across plasma membranes, so plants commonly make use of the active transport of ions to encourage its water's osmotic movement. **(a)** Active transport has not begun. Water and ions in cells with isotonic surroundings are generally in equilibrium with that of the surroundings: the amount of water and ions entering the cell roughly equals the amount leaving the cell. **(b)** Active transport begins. Ion pumps begin transporting ions into the cell, and soon the equilibrium is disturbed and the surroundings become hypotonic to the cell. **(c)** Water enters the cell through osmosis, and the cell becomes turgid.

goes a conformational change, releasing the sodium *outside the cell,* closing that sodium entry site and simultaneously opening the potassium entry site. Removal of the phosphate reverses the conformational change and the potassium escapes *into the cell cytoplasm.* The transporter then prepares for the next pumping cycle.

Maintaining Osmotic Conditions. Na^+/K^+ pumps aid cells in maintaining specific osmotic conditions. Through their continued work as coports for the two ions, the cell can be maintained in a state that is isotonic to its surroundings. There will then be no net movement of water molecules. In other instances, an ion distribution is established for the sole purpose of encouraging the diffusion of water into or out of the cell or organism. For instance, plants often actively transport ions from the soil into their root tissues, thus encouraging the entry of water. Water moves across the root cells to conducting tissues, which will direct the water up to the leaves for use in photosynthesis. The use of ion transport to influence water movement into the plant root is illustrated in Figure 5.19.

The transport of ions from one region to the next in the animal kidney sets up osmotic conditions that rid the body of excess water. Here the active transport of sodium ions, fol-

lowed passively by the movement of chloride ions, creates salty, hypertonic conditions that cause water to move into the blood for redistribution in the body.

Calcium Ion Pumps. Another important pump in animal cells is the **calcium ion pump**, which transports calcium ions (Ca^{2+}) across cell membranes and functions in nerve and muscle cells. In nerve cells, for example, the uptake of calcium ions taken in by one nerve cell aids in the relay of its neural messages to the next nerve cell. In muscle cells, calcium is essential to contraction. But between contractions, when muscle activity might be counterproductive, calcium ion pumps actively transport the ions into reservoirs, where they remain until the next contraction (see Chapter 33).

Proton Pumps. Hydrogen ion (H^+) pumps, also called **proton pumps**, play a vital role in cellular energetics. As we will see in the next three chapters, it is through the action of proton pumps that mitochondria and chloroplasts build high concentrations of protons (H^+) in some of their chambers. Such concentrations represent a steep electrochemical gradient, and therefore a substantial amount of stored energy. This energy is used to generate ATP, the universal energy transfer molecule.

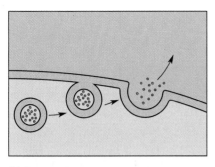

(a) Exocytosis

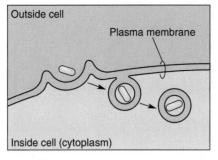

(b) Phagocytosis

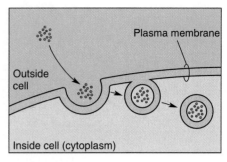

(c) Pinocytosis

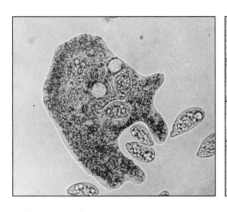

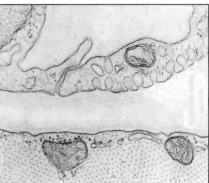

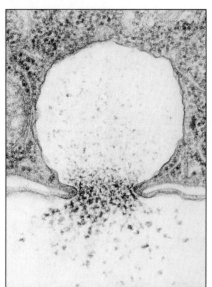

FIGURE 5.20
Endocytosis and Exocytosis
(a) Here a cell is secreting materials outside. Exocytosis involves membrane fusion, in this case the fusion of a vesicle with the plasma membrane. The precise details are not all known, but a linking protein may be involved. As the two membranes interact, the fused regions are partially disassembled and the two membranes joined. **(b)** The common pond ameba takes captured prey into food vacuoles formed through phagocytosis. Here the cytoplasm of the predator rises up to surround the prey. **(c)** Pinocytotic vesicles are seen in the cells of a blood capillary. In pinocytosis, molecules and other smaller particles of matter are taken into invaginations in the membrane. The behavior of the membrane is similar in both of these two endocytotic processes.

Exocytosis and Endocytosis. Exocytosis ("out of cell") and endocytosis ("into cell") refer, respectively, to the expulsion and uptake of substances by cells. Both involve bulk materials and a very active participation by the plasma membrane. We begin with exocytosis, a fairly simple process.

Exocytosis. **Exocytosis** is the process through which materials are expelled from the cell (Figure 5.20a). It is used by protists to rid the cell of digestive residues and by multicellular organisms for secretion. Secretion is the release of special products such as digestive enzymes, hormones, or other chemicals from the cell. In exocytosis, vesicles containing the substance to be released are drawn to the plasma membrane by contractile microfilaments. Upon contact, a vesicle fuses with the membrane, whereupon the site of contact opens up, and the two membranes become one. The substance is simply dumped outside the cell, and the inverted vesicle becomes part of the membrane.

Endocytosis. **Endocytosis**, the engulfing of a bulk material, occurs in three ways: phagocytosis, pinocytosis, and receptor-mediated endocytosis. In each, a portion of the plasma membrane becomes filled with materials and pinches off as a vesicle, bringing the material into the cell.

Phagocytosis. **Phagocytosis** (Figure 5.20b), or "cell-eating," is the uptake of solid material from outside the cell. It occurs in just a few kinds of cells, including protists, such as amebas and ciliates, and the phagocytic white blood cells of animals. The protists use phagocytosis for feeding while the white blood cells use it as a means of capturing bacteria and cleaning up cell debris from infection sites. In phagocytosis, the plasma membrane actively surrounds the solid object to be taken in, eventually pinching off and forming a **phagocytic vesicle**, or **phagosome**. This is generally followed by fusion of the vesicle with a lysosome and the digestion of its contents. Later, the protist will expel undigested residues through exocytosis.

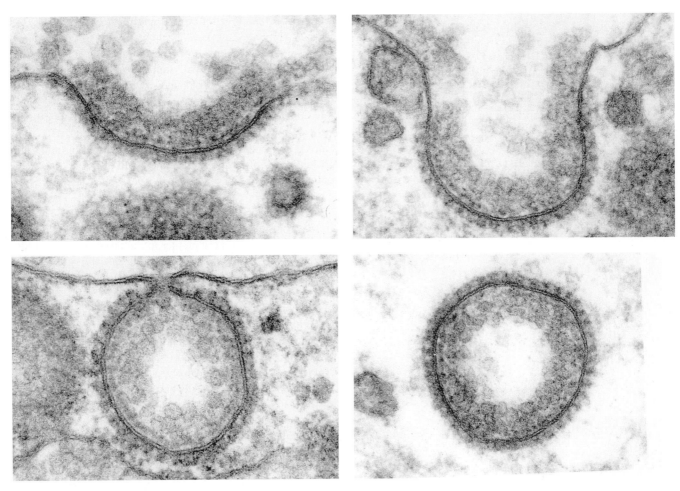

FIGURE 5.21
Receptor-Mediated Endocytosis
In receptor-mediated endocytosis, receptor sites on the plasma membrane fill with metabolites (substances
to be used by the cell). With the filling of the sites, a coated pit develops below, followed by the pinching
off of a coated vesicle. Such vesicles lose their coats, which recycle to the surface, and may next be joined
by others like them (not shown).

Pinocytosis. **Pinocytosis**, or "cell-drinking," is the uptake of fluids from outside the cell (see Figure 5.20c). Here the plasma membrane tends to fold inward, forming small vesicles containing the material. As in phagocytosis, the vesicle generally fuses with a lysosome and digestion of the contents takes place.

Cell biologists first demonstrated pinocytosis by exposing amebas to dense suspensions of protein. The amebas responded by forming many lengthy channels around the cell and taking the protein into vacuoles. The process also occurs in our own capillaries, where substances are transported from the blood to the surrounding tissue.

Recycling the Membrane. If you think carefully about the effects of endocytosis, you may recognize a problem: as cells form vacuoles they constantly lose bits of plasma membrane. How much do they lose? Some of the larger phagocytic white blood cells are so heavily involved in endocytosis that they can literally use up the equivalent of a plasma membrane in about one hour. Such cells, however, make up the loss when the opposite process, exocytosis, takes place. Here new segments of membrane are added. This is another example of the dynamic nature of the endomembranal system (see Chapter 4).

Receptor-Mediated Endocytosis. Cell biologists have identified another form of endocytosis, one that is **receptor mediated**. Here receptor molecules in the plasma membrane bind specifically to substances and transport them into the cell (Figure 5.21). Any substance capable of binding to such a receptor is called a **ligand**.

Generally, a number of ligands bind to protein receptor sites located above slight membrane indentations known as **coated pits**. The name refers to the inner surface of the indented region, which is coated with a supporting layer of protein known as **clathrin**. As the ligand–receptor complex fills the pit, the membrane sinks in, capturing the ligand. The

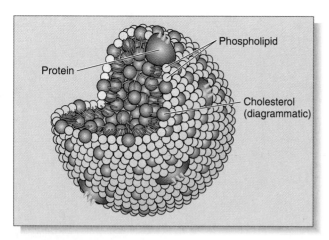

(a) LDL (low-density lipoprotein)

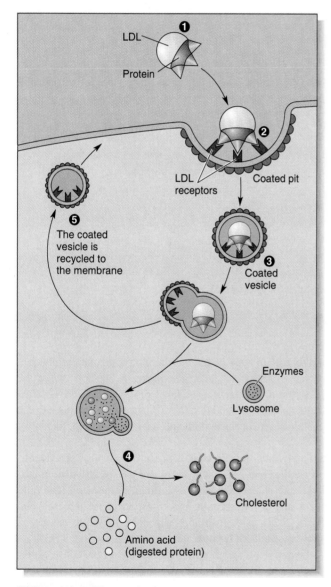

(b) Processing of lipoprotein

membrane then pinches off forming what is called a **coated vesicle** (Figure 5.21). The vesicle then loses its clathrin coat, which is recycled to the plasma membrane. Some vesicles transported in this way will simply store materials for later use. Others will interact with a lysosome and the contents will be hydrolyzed.

Receptor-mediated endocytosis was first reported in the early 1960s by Thomas Roth and Keith Porter. They determined that it was being used to transport storage proteins into the egg cells of birds and mosquitoes. In the latter, they identified 300,000 pitted sites on the oocyte surface, each of which was capable of forming a protein-laden vesicle. Other substances entering the cell through this process include cholesterol, iron, hormones, and those mammalian antibodies that are transported from the mother to the fetus late in fetal life.

One important class of membrane receptors on our own cells is the **LDL (low-density lipoprotein) receptor**. Its task is to remove cholesterol from the blood. Cholesterol circulates as part of a ligand complex containing other lipids and a large protein. A number of such complexes, when bound to LDL receptors and localized in clathrin-coated pits, form a coated vesicle. Once released inside, the vesicles fuse with lysosomes, whose enzymes hydrolyze the protein and some of the cholesterol into amino acids and simple fatty acids, respectively. The remaining cholesterol is made available for membrane synthesis, or, if in excess, it too can be converted to fatty acids. (The LDL receptors are recycled to the membrane, as usual; see Figure 5.22.)

While this is what normally happens, there are people with a genetic disease that leaves too few LDL receptors on their cholesterol-metabolizing cells. This condition leads to **hypercholesterolemia** ("high cholesterol"), in which chronically elevated blood cholesterol levels increase the chances of the arterial disease called atherosclerosis. In atherosclerosis, the plaque-roughened surfaces of obstructed arteries cause the spontaneous formation of blood clots, leading to stroke (see Chapter 3).

FIGURE 5.22
LDL Receptors at Work
(a) Low density lipoproteins (LDLs) are large complexes consisting of cholesterol and protein. **(b)** In this scheme, taken from the activity in cultured human cells, an LDL ① becomes bound to LDL receptors ② on the plasma membrane. A coated vesicle ③ forms, passing through the cytoplasm, where it fuses with a lysosome. The protein is digested ④, but the cholesterol is released. Some of the cholesterol will be used at once in assembling new cell membranes and as a precursor to other steroids. The remainder will be stored for future use. ⑤ The special coat cycles back to the plasma membrane.

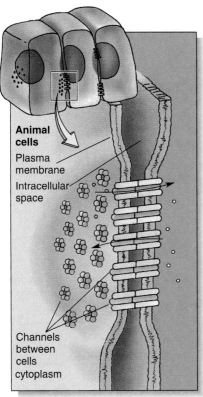

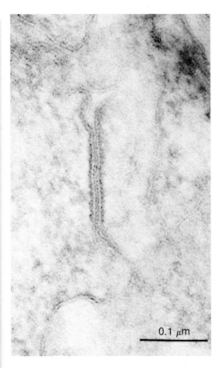

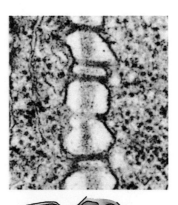

(a) Gap junctions

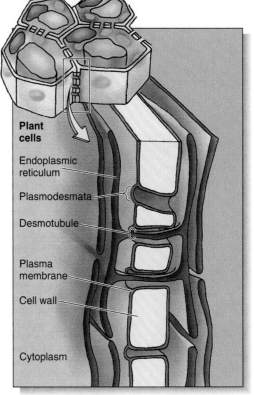

(b) Plasmodesmata

FIGURE 5.23
Passages Between Cells

(a) Materials move between certain animal cells through pores called gap junctions. Each pore is formed through the special assembly of six transmembranal proteins. The assemblies in each membrane are arranged back to back, creating a minute pipeline between cells. **(b)** The electron micrograph views plasmodesmata in a longitudinal section. The presence of pore-like plasmodesmata between plant cells permits the ready passage of materials from one cell to the next. The cytoplasmic connection often includes desmotubules, tube-like segments of endoplasmic reticulum that also pass between the cells.

Cell Junctions

Cell-to-Cell Transport

As one studies the cell for the first time, it is not unusual to form the impression that cells are isolated entities. Let's correct that now. Many types of multicellular tissues reveal communicating channels between their closely packed cells. Included are the intricate gap junctions of animal cells and the slender plasmodesmata of plant cells, both of which are very common.

Gap Junctions. **Gap junctions** are formed by rosettes of protein that firmly hold adjacent membranes together but at the

same time form minute pores between the fused cells (Figure 5.23a). The passages are 1.5–2 nm in diameter and will freely admit substances up to a molecular mass of 1000 daltons. This would include most ions, water, amino acids, sugars, vitamins, hormones, and even nucleotides. Gap junctions are believed to be important to heart muscle, where they facilitate the spread of electrical waves, and in the embryo, where they help with the passage of materials through layers of cells.

The protein-lined channels of gap junctions are not simply passive openings but are known to separate from the neighboring cells if an injury occurs. This effectively isolates the damaged cell, preventing the loss of essential materials from the continuous cytoplasm. The researchers who first observed this phenomenon also determined that it was proba-

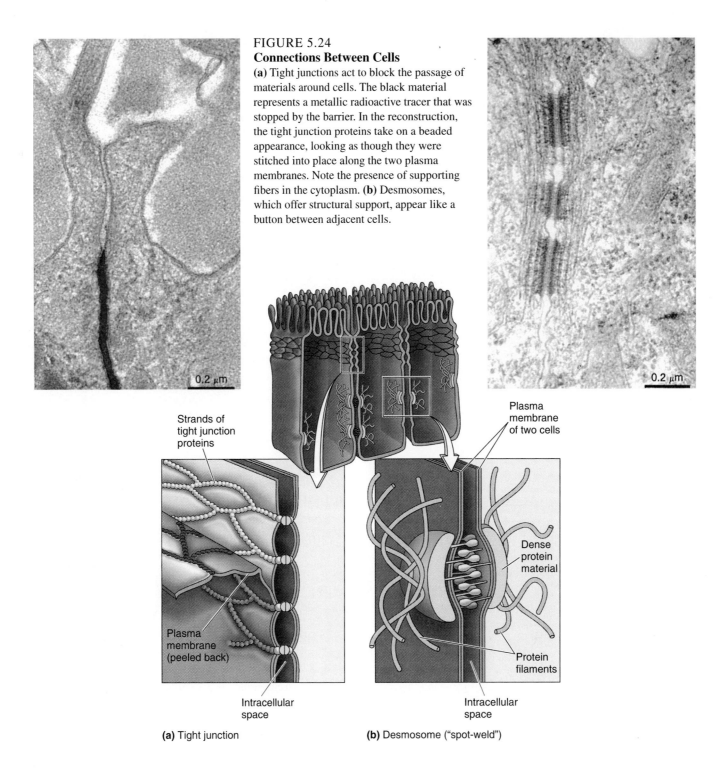

FIGURE 5.24
Connections Between Cells
(a) Tight junctions act to block the passage of materials around cells. The black material represents a metallic radioactive tracer that was stopped by the barrier. In the reconstruction, the tight junction proteins take on a beaded appearance, looking as though they were stitched into place along the two plasma membranes. Note the presence of supporting fibers in the cytoplasm. (b) Desmosomes, which offer structural support, appear like a button between adjacent cells.

Strands of tight junction proteins

Plasma membrane of two cells

Plasma membrane (peeled back)

Dense protein material

Protein filaments

Intracellular space

Intracellular space

(a) Tight junction

(b) Desmosome ("spot-weld")

bly an influx of calcium ions from the outside that caused the uncoupling. Free calcium ions are generally in short supply in the cytoplasm of cells, but when these ions are experimentally injected into the cell, the junctions are rapidly uncoupled. Eventually, a cell treated in this manner will recover, isolating the calcium ions in its mitochondria, whereupon the gap junctions will be reestablished.

Plasmodesmata. Plasmodesmata (singular, **plasmodesma**) are tube-like cytoplasmic extensions that pass from cell to cell through the cell walls of plants (Figure 5.23b). They also provide for cell-to-cell transport. The diameter of plasmodesmata ranges between 20 and 40 nm, considerably larger than the openings in gap junctions. Extending through the pore-like openings are elements of the endoplasmic reticulum (ER) of the two cells. They form **desmotubules**—slender channels within the pores—that make the lumen of the ER in the two cells continuous. Researchers have discovered thickenings at either end of the plasmodesmata and have suggested some valve-like role, permitting some control over the passage of materials. Plasmodesmata may be especially significant to the passage of water and other small molecules through dense tissues and over great distances.

Cell-to-Cell Barriers and Support

Other types of junctions in animal cells act as barriers to transport and as structurally supporting elements. These include tight junctions and desmosomes (Figure 5.24a,b).

Tight Junctions. Tight junctions are proteins that clamp cell borders together in a manner that *prevents* material from passing around them. The urinary bladder, for example, is a sac composed of flattened cells whose borders are held together by tight junctions. With this arrangement, urine cannot leak out *between* bladder cells into the surrounding body cavity. Tight junctions are also seen in cells lining the intestine. Because of their presence, digested food cannot pass

between lining cells but is directed into the cytoplasm, where its movement is controlled.

Desmosomes. Desmosomes bind cells together but do not form transport barriers. Their purpose is supportive. Each desmosome consists of dense plaques just below the plasma membranes of adjoining cells. The plaques are penetrated by tough keratin filaments that anchor the desmosome to the cell cytoplasm. These are the "intermediate filaments" mentioned in the last chapter. Desmosomes are most frequently seen in tissues that undergo physical stress, such as those of the skin and intestinal wall.

We have seen that the incredibly thin plasma membrane is a delicate but exquisitely complex structure. Its lipid, protein, and carbohydrate components have precise roles in the membrane's essential barrier, transport, and recognition functions. We've also had some clues on the role of energy in the various transport processes. As we proceed, the importance of energy will be increasingly evident. In the next chapter we will begin with some fundamentals on the nature and behavior of energy, all in preparation for the topics of photosynthesis and respiration, just ahead.

Key Ideas

THE PLASMA MEMBRANE

1. Models of the plasma membrane were derived from careful comparisons of electron micrographs, known molecular dimensions, and permeability studies.

2. The plasma membrane is a phospholipid bilayer with its polar phospholipid heads facing outward and the nonpolar fatty acid tails facing inward and intermingling.

3. **Integral proteins** pass entirely through the membrane. Some are enzymes, and others are transport proteins. **Peripheral proteins** lie at the surfaces, and **lipid-anchored proteins** extend part way through the membrane.

4. The glycocalyx consists of glucose chains emerging from glycoproteins and glycolipids. These molecules bind cells, acting as cell-surface antigens, as receptors for chemical messengers, and as cell recognition molecules.

5. Cholesterol, a steroid, is an animal membrane component. Its nonpolar rings lie submerged, with its polar —OH group at the surface.

6. Because of the membrane's fluid-like properties, individual molecules move freely. The degree of movement depends on the predominating fatty acids—the more saturated fatty acids present, the more rigid the membrane. Heat and increased cholesterol content also increase fluidity.

MECHANISMS OF TRANSPORT

1. Membranes that admit some substances and not others are **selectively permeable**. **Permeability** depends on particle size, electrical charges, and membrane composition.

2. The mechanisms of transport include **passive transport**, which does not require the cell's energy, and **active transport**, which does.

3. **Diffusion** is the net movement of atoms and molecules from areas of their greater concentration to areas of their lesser concentration. Diffusion ends when the distribution of molecules becomes random. The movement of ions is influenced by charges inside the membrane.

4. **Osmosis** is the diffusion of water across a selectively permeable membrane—from an area of greater concentration to one of lesser. **Osmotic pressure** is the force sufficient to stop the inward movement. The net movement of water into and out of cells through osmosis is determined by the comparative amounts of water and solutes on either side of the plasma membrane.

    a. When the solute concentration is equal on both sides, the system is **isotonic**, and the movement of water will be equal in both directions.

    b. When the solute concentration is less outside than inside, the surroundings are **hypotonic**, and water will move into the cell.

c. When the solute concentration is greater outside than inside, the surroundings are **hypertonic**, and water will move out of the cell.

5. When plant cells fill with water, **hydrostatic pressure** and a state of firmness called **turgor** are produced. When the opposing **wall pressure** equals turgor pressure, the net movement of water ceases. A loss of hydrostatic pressure results in wilting.

6. In facilitated diffusion, no ATP energy is used, but special membrane **transporter proteins** speed the passage of specific molecules or ions through the membrane. Each substance has a specific transporter.

7. In active transport, transport proteins move materials against a concentration or electrochemical gradient and ATP is consumed. **Uniports** move a single substance; **coports** move two substances. **Symports** move substances in the same direction; **antiports** move substances in opposite directions.

8. In the **sodium/potassium ion exchange pump**, an antiport transport protein undergoes ATP-driven conformational (shape) changes, carrying sodium ions into the cell and potassium ions out.

9. Ion pumps also set up osmotic gradients to direct the movement of water, participate in muscle contraction, and produce ATP.

10. In **exocytosis**, a vesicle fuses with the plasma membrane, which releases materials from the cell.

11. In **endocytosis** substances are taken in. (a) Phagocytosis is the active engulfing of solid material. (b) Pinocytosis is the active engulfing of extracellular fluids.

12. In **receptor-mediated endocytosis**, receptor molecules in the plasma membrane are activated by ligands. The membrane sinks inward, forming a vesicle and bringing the ligands into the cell.

CELL JUNCTIONS

1. **Gap junctions** are protein-lined pores between adjacent animal cells that permit the intercellular movement of materials.

2. **Plasmodesmata** are slender strands of cytoplasm that extend through pores in adjacent plant cell walls.

3. **Tight junctions** help seal cell borders, preventing the movement of water and other materials around cells.

4. **Desmosomes** support tissues by binding their cells together.

Applications of Ideas

1. It was once thought that frogs could not survive in distilled water—that is, until someone actually kept some in a pan of distilled water for months without ill effect. Theoretically, what should have happened to the frogs? Why do you think it did not?

2. Early methods of preserving food included drying, pickling, smoking, salting, and sugar-curing. Offer a physiological explanation, based on osmosis, for the failure of spoilage bacteria and molds to grow readily on food treated in these manners.

3. In an experiment, the leaves of the water plant *Anacharis* are treated in several ways and observed under the light microscope. The cells are normally rectangular in shape, with a large central vacuole and a thin layer of chloroplast-containing cyto-

plasm between the vacuole and plasma membrane. Explain each of the following microscopic observations and compare it with the others:

a. Leaves in pond water show no change.

b. Leaves in distilled water show no change.

c. Leaves in 1% and 2% glucose solutions show no change, but in 3% glucose, the chloroplasts gather into a tight sphere near the cell center.

d. Leaves in 1% NaCl solution show no change, but those in a 1.5% NaCl solution resemble those in the 3% glucose solution.

Review Questions

1. Prepare a simple drawing of the plasma membrane. Label phospholipids, cholesterol, and integral and peripheral proteins. (page 103)

2. What are the two positions taken by membranal proteins? How does being amphipathic help retain a transmembranal protein in its position? (page 104)

3. What molecules occur in the glycocalyx? What are their functions? (pages 105–106)

4. Discuss the characteristics of the plasma membrane as suggested by the fluid mosaic model. (pages 107–108)

5. Explain how the following terms differ: impermeable, semipermeable, and permeable. (page 109)

6. What are two fundamental differences between passive and active transport? (page 109)

7. Write a precise definition of the term *diffusion* and use an example to show how it is useful to organisms. (pages 109–111)

8. The term *osmosis* was coined to cover a special case of diffusion. Explain this case. (pages 111–112)

9. Explain what would happen if cells whose cytoplasm averaged about 1% solute concentration were placed in a 3% salt solution. What term describes the osmotic conditions in the cell? Outside the cell? (pages 111–113)

10. Explain why microorganisms might fail to grow on food that has been preserved with salt or sugar. (pages 112–114)

11. What special use do plants make of the turgor pressure in their leaf cells? (page 113)

12. List three ways in which facilitated diffusion differs from ordinary diffusion. (page 114)

13. Explain how the sodium/potassium ion exchange pump works. (pages 116–117)

14. In what ways are phagocytosis and pinocytosis similar? Different? What generally happens to materials taken in through phagocytosis? (pages 118–119)

15. What is meant by the term *receptor-mediated?* Review the process of receptor-mediated endocytosis and provide two examples. (pages119–120)

16. What are gap junctions? In what organisms do they occur? (page 121)

17. Describe the structure of plasmodesmata and state their importance to cells of the root tip and leaf. (page 122)

18. Compare the role of the tight junction with that of the desmosome. (page 123)

CHAPTER 6

Energy and Chemical Activity in the Cell

CHAPTER OUTLINE

ENERGY

Forms of Energy

The Laws of Thermodynamics

APPLYING ENERGY PRINCIPLES TO LIFE

Life and the Laws of Thermodynamics

Free Energy

Exergonic and Endergonic Reactions

Getting Reactions Started

ENZYMES: BIOLOGICAL CATALYSTS

Enzyme Specificity

The Enzymatic Reaction

The Enzyme–Substrate Complex: Events at the Active Site

Influences on the Rates of Enzyme Action

Cofactors (Vitamins and Minerals)

Teams of Enzymes: Metabolic Pathways

Mechanisms of Enzyme Control

Summarizing the Activity of Enzymes

ATP: THE ENERGY CURRENCY OF THE CELL

ATP Structure and the Release of Energy

The Work of ATP

The Cycling of ATP

THE COENZYMES AND OTHER ELECTRON CARRIERS

Coenzymes

Redox Reactions: Oxidation and Reduction

Coenzymes in Action

Stationary Electron Carriers of Mitochondria and Chloroplasts

THE FORMATION OF ATP IN CELLS

Chemiosmotic Phosphorylation

KEY IDEAS

APPLICATION OF IDEAS

REVIEW QUESTIONS

*P*lants growing peacefully on a flower-strewn hillside are the very essence of beauty and tranquillity. However, this picture is deceptive. In their own way, the plants are engaged in a battle for life or death. They must quietly fight for survival as they compete with other plants for sunlight, water, and soil minerals. The fact that they're there at all indicates a certain success in the struggle. But success itself can spell new problems. Once a plant has managed to store some energy in its own molecules, it may suddenly have to yield its hard-won gains to some casual grazing animal, who then rearranges the plant's molecules and energy to suit its own needs. The plant eater, in turn, often falls to some sharp-toothed predator, bent on fulfilling its own requirements (Figure 6.1). Some predators serve as prey for yet others, and some do not. But every living thing eventually meets its fate, and then the microbes of decay have their way. In this, the final episode, the simplest of life forms extract the last remnants of energy as they rearrange the molecules of lifeless corpses. (Enough sentiment.)

Life, it seems, is really all about energy. Energy, as we view it here, is shuffled about in molecules, where it is boosted, drained, and eventually lost. As we consider the energy of molecules, however, we must not lose sight of the "big picture." After all, these various shifts of molecules and energy all give rise to the more visible and familiar panorama we associate with life.

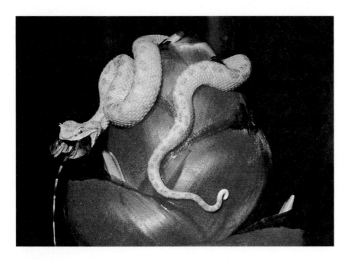

FIGURE 6.1
Transfers of Energy
Energy is transferred from one organism to another in a variety of ways.

Energy

We can begin with the obvious questions. What is energy? What are its characteristics, and how does it behave? Like life itself, energy is an elusive concept; in fact, we can only perceive the existence of energy through its effect on matter. Nonetheless, let's begin with a functional, if somewhat old and creaky, definition. **Energy** is the ability to do work. Work, presented here in its broadest context, means the application of force, for example, to move something. Work applies to every aspect of life, from the chemical work of molecule-building to the physical work in movement and transporting molecules into and out of the cell.

Energy—the ability to do work—occurs in two states, potential energy and kinetic energy. **Potential energy** is energy "stored," and thus "not doing anything." For example, the 5-ton boulder on the hill above your house represents a considerable store of potential energy. It was raised to that level in some forgotten time when geological forces carried out a great deal of work to get it there. Should the soil holding the rock give way, you may quickly come to grasp the difference between potential and kinetic energy.

Kinetic energy is energy of motion. We know this because of its effect on matter. Not only does the boulder leave a trail through the brush as its potential energy becomes kinetic, but it will undoubtedly do some "work" on your house, should it lie in the boulder's path.

Forms of Energy

Energy can take various forms. Thus we speak of chemical energy, electrical energy, magnetic energy, mechanical energy, and radiant energy. The latter includes the familiar electro-magnetic radiations, such as infrared, visible light, ultraviolet light, and so on (see Chapter 7). We will soon see that all these forms of energy are interchangeable; that is, under certain conditions, one form of energy can become another (Figure 6.2).

The Laws of Thermodynamics

The behavior of energy is incorporated into time-honored principles known as the **laws of thermodynamics.** These laws are based on certain observations about the behavior of matter and energy that are remarkably invariable from one time to the next. Such consistency leads to predictions. What happens, for example, to objects raised above the ground and then released? What happens to an object that is heated and then set aside, away from the heat? Everyone can predict that the first object will fall to the ground and that the second will

FIGURE 6.2
Energy Transformations and Human Activity
Energy can change from one form to another in spectacular ways. Here chemical energy in the rocket fuel is transformed into heat, light, and a considerable amount of thrust.

cool. Such observations, made time and again, have eventually led to the formulation of *laws* (or laws of nature). We like to think that scientific laws cannot be broken under any circumstances. Because of the almost infinite variety of life, biology is actually rather short of laws. Nevertheless, living systems are made of matter that interacts chemically and physically, and the laws of physics and chemistry are as inviolable in life as anywhere else. So like the rest of the known universe, living systems obey the laws of thermodynamics.

The First Law. The **first law of thermodynamics** states that *energy can neither be created nor destroyed, but it can be converted from one form to another.* Understandably, the first law is also called the **law of conservation of energy**. What the first law means is that the total amount of energy present in a system remains constant. The concept refers to idealized conditions that exist only in what is called an "isolated system"—one in which matter and energy cannot enter or leave. Such systems do not really exist (except perhaps as the universe itself) but are contrived as models by scientists who wish to test their ideas under hypothetical conditions that can be limited and controlled.

Energy Transitions: A Case History. Energy changes occur in a variety of ways. When you start your lawnmower engine you can begin to appreciate the idea of energy conversion. Consider that the gasoline in the lawnmower's tank is a veritable storehouse of chemical energy, locked away in the chemical bonds that hold the carbon and hydrogen of the gasoline molecules together. As you pull the cord, a mix of gasoline vapor and air encounters an electrical discharge from the spark plug, and the engine starts. Chemical energy in the fuel molecules becomes heat energy, and heat then expands the gases in the engine cylinder. The next energy transformation is to mechanical energy, or energy of motion, which comes about when the expanding gases push against the piston. The piston moves up and down, its connecting rod rotating the crankshaft, which spins the lawnmower blade. At certain points in the piston's movement, valves open, and the expanding gases escape into the surroundings, their energy dissipating as heat. The escaping gases—carbon dioxide and water—are at considerably lower energy level than gasoline, and the difference between the two energy levels is to be found in exhausted heat and the energy of motion. The latter largely becomes heat as well, as the moving parts of the lawnmower encounter friction.

What we see here is characteristic of energy as it changes form. Energy transformations are accompanied by the formation of heat, and when such heat has dissipated, it is no longer capable of work, at least as far as that system (in this case, the lawnmower) is concerned. The transition of systems with great energy to systems with low energy extends beyond lawnmowers; *it is a tendency of the universe at large.* Physicists express this general observation in the *second law of thermodynamics.*

The Second Law. The **second law of thermodynamics** tells us that energy transitions are imperfect—that some energy is always lost, usually as heat, in each transition. Again, in the transitions from chemical bond energy in gasoline to the mechanical energy in the spinning lawnmower blade, much of the original energy is lost as heat. The energy is not lost from the system; it is just that such heat energy is normally unavailable to do useful work.

The second law is also referred to as the **law of entropy.** In this context the law asserts that *energy transitions lead to an increase in disorder.* The spontaneous increase in disorder (or decrease in order) is measured as **entropy.** As energy shifts through the universe, moving from higher levels to lower, entropy increases, and the universe becomes more random. This again is an inexorable universal trend.

From this statement, it follows that there is more energy in a more organized system, and that as a system becomes less organized—or more random—its energy decreases (and its entropy increases). Carbon dioxide and water molecules represent a more random state than do molecules of gasoline, and so the lawnmower system has become a system of greater entropy. As we proceed into cell energetics, you will see that one of the fascinating things about living organisms is that they use energy to build highly ordered systems, and then, as such systems are allowed to run down, they harness the escaping energy to do useful work.

As a larger issue, we have said that the laws of thermodynamics apply to the living as well as to the physical world. Since all living organisms are highly ordered, the following question arises: How do living systems resist the inexorable trend toward increased entropy?

Applying Energy Principles to Life

Life and the Laws of Thermodynamics

Life, in accordance with the second law, requires a constant input of energy. It is this energy that enables living things to remain ordered. If anything interferes with the uptake of energy, or if anything sufficiently alters the ordered state (such as disease or injury), order gives way to the randomness of death (Figure 6.3).

Also according to the second law, transfers of energy in living things are not entirely efficient; energy is lost at each stage. Living cells are particularly efficient at extracting the energy from fuels such as glucose. Oxygen-using organisms can transfer about 40% of the chemical bond energy of glucose to chemical energy in ATP, the most common source of energy in the cell. You might derive some satisfaction from knowing that your own cells are far more fuel-efficient than is your new car, which, right after a tune-up, may have a fuel efficiency of about 25%.

Considering the energy loss in each transition, how can life sustain itself? Life goes on because a virtually endless

FIGURE 6.3
Death: Maximum Entropy
Organisms maintain their complex molecular state only by taking in energy. When energy input ceases, the molecular state is rapidly simplified and randomized, and the energy is dissipated.

supply of energy bathes the earth—energy from the sun. In fact, there is so much light energy available that all life sustains itself on less than 1% of the total reaching the earth. Life should continue as long as this energy source is available. In fact, for us the second law will eventually make its final impact when the sun becomes a "red giant" and incinerates all life on earth. At that point the earth and everything on it will be reduced to simple molecules and heat and that heat will dissipate into space. But that's some five billion years in the future so you should probably go on studying for the midterm.

Free Energy

We have noted that the complex molecules of life—the carbohydrates, lipids, proteins, and nucleic acids—have great potential energy because of their highly organized state. They remain organized because they are held together in precise configurations by bond energy. When such bonds are broken

during chemical reactions, some of their energy becomes available for work.

Such available energy is referred to as **free energy**. The term "free" has nothing to do with the "cost" of energy but refers instead to its *availability* for work (as in, "Are you free for work today?"). The free energy of reactants usually changes after a reaction, and such changes are symbolized as ΔG [Δ (Greek letter delta) meaning "change," G indicating "Gibbs," the chemist who clarified these ideas].

Glucose, or $C_6H_{12}O_6$, when metabolized by cells in the presence of oxygen (such as occurs in the mitochondrion), is broken down into CO_2 and H_2O. The difference in free energy between the reactant and its two products turns out to be 686 kcal/mole (kcal = kilocalories or 1000 calories). The reaction is stated as follows:

$$C_6H_{12}O_6 + 6\ O_2 \longrightarrow 6\ CO_2 + 6\ H_2O + {}^*Energy$$
$$\text{(reactants)} \qquad \text{(products)}$$
$${}^*\Delta G = -686 \text{ kcal/mole}$$

The negative sign of ΔG indicates that the products have less free energy than the reactants.

Exergonic and Endergonic Reactions

The reaction of glucose with oxygen is representative of an **exergonic reaction** ("energy out," Figure 6.4), one in which the products have less free energy than the reactants. Reactions of a second type, called **endergonic reactions** ("energy in"), are those in which the products have more free energy than the reactants. The formation of glucose in plants illustrates an endergonic reaction from nature.

During photosynthesis, plants utilize light energy to make glucose from carbon dioxide and water. This is the reverse of what occurs in the breakdown of glucose, so ΔG is a positive number. A chemist might write this as follows:

$$6\ CO_2 + 6\ H_2O + \text{Light energy} \longrightarrow C_6H_{12}O_6 + 6\ O_2$$
$${}^*\Delta G = +686 \text{ kcal/mole}$$

It may seem at first glance that endergonic reactions "generate energy," thus violating the first law; but let's be careful with this. Both the breaking of chemical bonds and the formation of chemical bonds require energy—in both cases, an amount equal to the energy of the chemical bond in question. The energy required to form glucose from reactants of a lower free energy level is provided by light, an outside source.

Getting Reactions Started

We sometimes find dramatic examples of exergonic reactions. Consider the airship *Hindenburg*. Back in the 1930s, Germany's Third Reich intended to impress America with this hydrogen-bloated airship, the most famous of the giant zeppelins. It indeed impressed America; while landing at Lakehurst, New Jersey, it blew up! The *Hindenburg*'s hydrogen gas

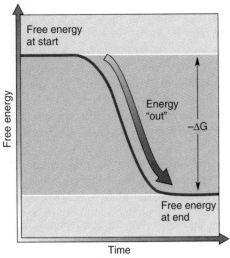

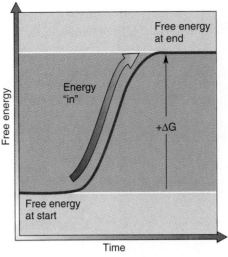

Free energy at start

Free energy

Energy "out"

$-\Delta G$

Free energy at end

Time

(a) Exergonic reaction

Free energy at end

Free energy

Energy "in"

$+\Delta G$

Free energy at start

Time

(b) Endergonic reaction

FIGURE 6.4
Exergonic Versus Endergonic Reactions
(**a**) The products of an exergonic reaction have less free energy than the reactants. The reduction in free energy is expressed as $-\Delta G$. (**b**) The reverse is true of endergonic reactions, where the products have more free energy than the reactants, and the difference is expressed as $+\Delta G$.

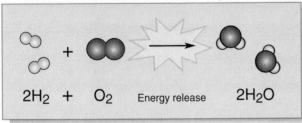

$2H_2$ + O_2 Energy release $2H_2O$

FIGURE 6.5
Chemistry of a Disaster
The hydrogen-filled zeppelin *Hindenburg* exploded in 1937. Hydrogen is rarely encountered in its pure molecular form (H_2) because it reacts readily with many other chemicals and is nearly always found combined with something. Hydrogen production is always risky because of this reactivity. The smallest spark or flame can provide the impetus needed for a rapid chain reaction such as the one shown here. We will never know what triggered the reaction behind the *Hindenburg* disaster, but today, the few remaining lighter-than-air craft (e.g., the *Goodyear* blimp) are filled with heavier, but unreactive helium.

combined violently with molecular oxygen of the atmosphere to produce water. The rest is history (Figure 6.5).

But why was the water formed so explosively? It was a matter of energy states and quantities. Water (H_2O) is in a lower energy state than an equivalent amount of H_2 and O_2. We know this because a mixture of hydrogen gas and oxygen gas confined in a space and ignited by a spark will release great heat. More specifically, if 1 mole (see Chapter 2) of H_2 and $\frac{1}{2}$ mole of O_2 (the quantities needed to produce 1 mole of water) are placed in a device called a calorimeter, and a spark is added, one mole of water (actually steam) will be produced, and a heat rise of just about 58 kcal will occur. The 17 million cubic feet of H_2 in the exploding *Hindenburg* illustrated this same principle on a gargantuan scale. When the *Hindenburg* went up, the chemical mix went quickly to a lower free energy state, producing low-energy water and releasing the excess energy as an enormous fireball of heat and light.

The molecular reaction of the tragedy, taken one mole at a time, may be written as follows:

$$2\,H_2 \ + \ O_2 \longrightarrow 2\,H_2O \ + \ \text{Energy}$$
$$^*\Delta G = -58 \text{ kcal/mole}$$

Energy of Activation. Actually, you can mix oxygen and hydrogen together all day and nothing dramatic will happen. The molecules will just sit there, their outer shell electrons nicely filled. But even a tiny spark will make things happen. Why doesn't the mixture react without the spark? The reason molecular oxygen and molecular hydrogen don't react with each other at usual temperatures is that, before they can, the atoms of each molecule must first be forced apart. The existing covalent bonds between hydrogens and between oxygens must first be broken. This requires an outside source of energy, at least initially. Each chemical bond represents a specific amount of energy and to break that bond requires the input of at least an equal amount of energy. This energy is referred to as the **energy of activation** (Figure 6.6).

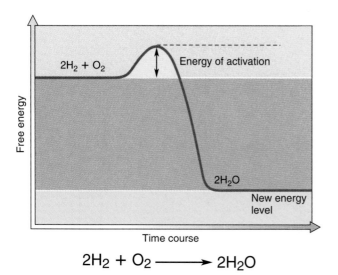

$$2H_2 + O_2 \longrightarrow 2H_2O$$

FIGURE 6.6
Activation Energy
Although molecular hydrogen and molecular oxygen are at a high initial free energy level, they will not react under ordinary circumstances without an input of energy. When this happens, the atoms within the molecules react, forming water that is at a considerably lower free energy level.

The point at which the atoms of molecules in a reaction start coming apart is called the **transition state**. During this very brief point in time, some of the hydrogen molecules (H_2) become hydrogen atoms ($H + H$) and some of the oxygen molecules (O_2) become oxygen atoms ($O + O$). This frees the electrons that were tied up in the covalent bonds of the hydrogen and oxygen molecules; and as you already know, atoms in their free elemental state are highly reactive. The individual hydrogen and oxygen atoms may then join, forming water. Since water is at a lower free energy level than the reactants, there will be excess energy (ΔG will be negative), and this energy escapes as heat and light. As we've seen, the heat and light energy released in the *Hindenburg* disaster was considerable.

Why don't the hydrogen and oxygen atoms simply form O_2 and H_2 molecules again? Actually, some do, but since much of the old bond energy is lost, a return to the original state becomes energetically unfavorable. Only an input of a similar amount of energy can bring this about. (You can generate oxygen and hydrogen from water through the use of an electrical current.)

Importantly, the initial spark—the energy of activation—need only be enough to get things started. Once the first molecules are disrupted, the liberated energy sets off a chain reaction among the other hydrogen and oxygen molecules, and more and more water forms. Again, the water molecules represent a lower free energy state with far greater stability than the original mix of $2H_2 + O_2$.

Reversible Chemical Reactions. What we've described here is a strong exergonic reaction, one in which most of the reac-

tants formed products and a great decrease in free energy occurred. In chemical terms, the reaction *went to completion.* Many chemical reactions, both inside and outside life, are far less drastic and many do not actually go *to completion.* In fact, many reactions can and do reverse themselves—some of the product molecules interacting to form the original reactants. As we proceed to discuss various metabolic pathways, you will notice the presence in equations of double arrows (pointing in opposite directions), indicating such reversibility. In fact, the reversal of reactions is one way in which the cell meets its many metabolic requirements and controls the use of its raw materials and energy.

In reversible reactions, the interacting molecules eventually reach a state of equilibrium, one in which there is no further net change. Of course if one of the products is given off as a gas, such as oxygen or carbon dioxide, or is used in a second process, then the reaction can never reach equilibrium.

In the absence of such eventualities, there are two factors that determine which way a reaction will go: (1) the comparative free energy of the reactants and products and (2) the relative concentrations of reactants and products. Where a substantial difference in free energy exists, say, more free energy in the reactants, the reaction will go to the right (toward product), but when the products contain greater free energy, then the reverse can occur (toward reactant). This is because it is free energy that drives the chemical reaction.

Should there be little or no free energy difference between the reactants and products, then the determining factor will be concentration. Should the reactants be in greater concentration, the reaction will go toward the right. Should products dominate, the reaction will go to the left. In other words, where free energy is not a factor, chemical reactions end up minimizing the difference between reactant and product.

The idea of chemical equilibrium is the antithesis of life. Should the myriad chemical activities in a cell or organism reach such a state of equilibrium, life would no longer exist. Living things are in a constant state of *disequilibrium* mainly because the products of reactions are drawn away from the reaction. Some are used for other purposes and others are released from the body as wastes. For example, the metabolic pathway in which glucose is metabolized leads to the formation of ATP and the release of water and carbon dioxide. ATP is used in other reactions and much of the water and carbon dioxide are released as wastes. Once these products have been formed, there is no way for the pathway to reverse itself. Furthermore, the free energy drop in the complete metabolism of glucose is substantial. Both factors keep the metabolic reactions going to the right. Nonetheless, as we will see, certain reactions *within* the metabolic pathway are reversible. Thus if sufficient ATP is available, or if product accumulates somewhere in the pathway, the reactions may temporarily reverse themselves.

Catalysts: Quieter Changes in Energy States There is a way to get the hydrogen and oxygen in our earlier example to

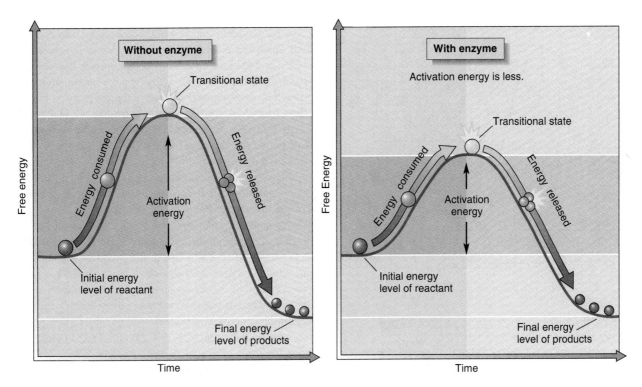

FIGURE 6.7
Enzymes and Activation Energy
Reactions that do not utilize enzymes need far more energy of activation than do reactions with enzymes.

react and form water without the spark. You can use a catalyst. A **catalyst** is a substance that accelerates chemical reactions without itself being consumed. In practical terms, catalysts greatly lower the energy of activation required for chemical reactions. For instance, hydrogen and oxygen will combine readily at room temperature in the presence of powdered platinum. The hydrogen first combines with the platinum and then with the oxygen, leaving the catalyst in its original state. As we've said before, living cells make use of a special class of catalysts, known as *enzymes.* You can make a good case for the idea that enzymes are the most important biological molecules.

Enzymes: Biological Catalysts

Enzymes are proteins that act as biological catalysts, agents that greatly accelerate chemical reactions in the organism. There are many different kinds of enzymes in any cell's biochemical arsenal, each of which functions in only one kind of reaction.

Enzymes have several important characteristics: (1) Enzymes are proteins. (2) Enzymes lower the energy of activation requirement (Figure 6.7). (3) Enzymes speed up reactions to a rate useful to the cell. (4) Each enzyme reacts with a specific molecule, known as its **substrate.** (5) Enzymes have no effect on the free energy changes (ΔG) of a given reaction. (6) Enzymes emerge unaltered from reactions, ready to act again and again. (7) Finally, enzymes can be either **catabolic**

or **anabolic**, meaning that they can break molecules down or build them up, respectively.

Enzyme Specificity

Enzyme specificity is critical to the cell. Without it, enzymatic reactions in the cell would be completely random and virtually useless. Specificity enables the cell to manage its chemical activity, generally through the timely synthesis of enzymes that are needed and the cessation of this synthesis when they are not. In addition, cells can be selective in what enzymes they produce and therefore become chemical specialists. Cells that synthesize digestive enzymes for secretion into the digestive tract are examples of such specialization.

An enzyme's specificity is due in large part to the shape of its active site. The **active site** is a specific location on the enzyme where reactions occur (Figure 6.8). It is usually a pocket or groove whose shape matches that of the intended substrate. In addition to matching the shape of the substrate, the active site is arranged in such a way as to bring reactive amino acid R groups close to reactive side groups in the substrate. Let's look at the ensuing reactions.

The Enzymatic Reaction

The enzyme-catalyzed event begins when a substrate or substrates enter the active site. How do substrate and enzyme meet? The union of substrate and enzyme occurs only

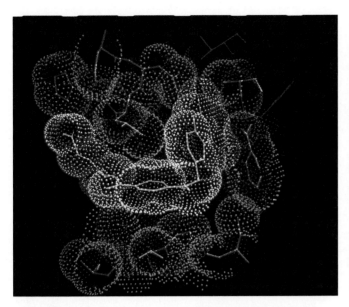

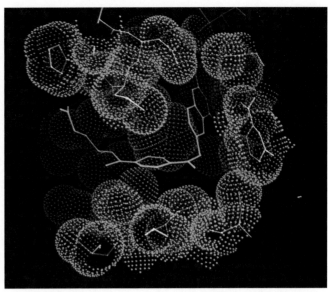

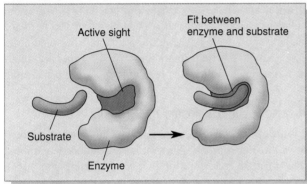

FIGURE 6.8
Active Site of an Enzyme
The computer-generated, three-dimensional view of an enzyme *(above)* reveals its globular shape and a cleft that forms the active site. On the right, the view reveals a substrate within the active site. The active site is both topographically and chemically matched for the specific substrate (substance) on which it reacts. The illustrations *(left)* provide a different view of the active site and its substrate.

because of the rapid, random motion of molecules, so the union is really just a matter of chance. Yet under the right conditions, such as suitable temperatures and concentrations, reactions occur with dazzling speed. Be mindful that in spite of our frequent references to "the enzyme," these remarkable catalysts are usually present in great numbers; thus reactions represent molecular activity on an immense scale. Let's move in for a closer look, this time focusing on the activity in a single active site. We'll also consider how activity there helps circumvent the usual energy of activation requirement.

The Enzyme-Substrate Complex: Events at the Active Site

The initial event is *binding* of substrate and active site (Figure 6.9). Here molecular groups on the substrate form weak ionic attractions and hydrogen bonds with certain groups in the active site, forming the **enzyme–substrate complex**

Next, the enzyme undergoes a slight change in shape, fitting the contours of the substrate more snugly. This change, known as an **induced fit**, brings potentially reactive groups in the substrate and active site still closer (Figure 6.9 2).

In addition, the induced fit places the substrate under physical stress, which aids in the disruption of its chemical bonds as the reaction proceeds. Here is one way that activation energy is reduced. Remember that the breaking and forming of chemical bonds require energy. The physical stress exerted on covalent bonds at critical locations during the induced fit helps decrease that energy requirement.

In some instances, the active site contains acidic or basic R groups, those that release protons or take them in, respectively. The presence of these groups alters the pH locally within the active site, which can also facilitate the breaking and making of chemical bonds at low cellular temperatures.

We used to hear in biology that "enzymes catalyze reactions without entering into them." You can already see that this early idea was wrong. In addition to induced fit and localized pH changes, enzymes go so far as to form new covalent bonds with substrate. In the hydrolytic enzyme shown in Figure 6.9, the formation of a temporary covalent bond frees one of the products (P_1), which drifts away (Figure 6.9 3). For the second product (P_2) to be released, the new covalent bond must be broken and the freed electrons taken up by — H and — OH groups from water. This done, the second product (P_2)

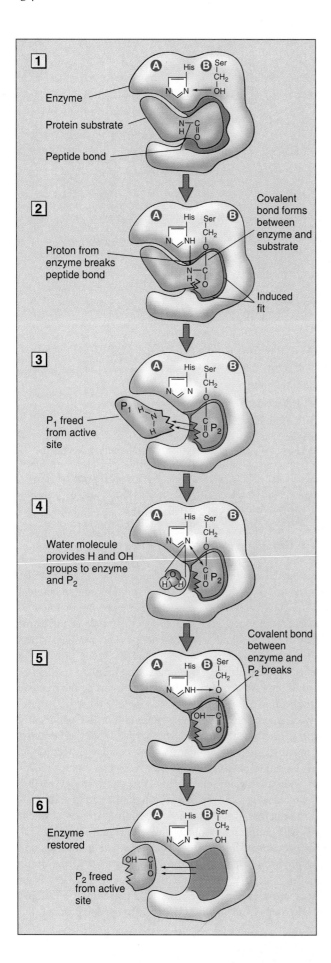

FIGURE 6.9
Action of a Hydrolytic Enzyme

⬜1 Here a hydrolytic enzyme encounters a dipeptide, which binds to the active site, held there by weak attractive forces. Note the inexact fit between cleft and substrate. ⬜2 Then a slight conformational change produces a tighter fit and places stress on the substrate. The shape change also brings reactive groups in the enzyme closer to those in the substrate. As a result, hydrogen is attracted from amino acid A in the active site to an amino group in the substrate, and a covalent bond forms between the enzyme and the substrate's second amino acid. ⬜3 This breaks the substrate's peptide bond, cleaving it into segments P1 and P2 (products 1 and 2). The action also frees P1, which drifts from the active site. ⬜4 For P2 to be released, the new covalent bond must be cleaved. This is where water comes in, and this explains what a hydrolysis reaction is about. ⬜5 Hydrogen from the water joins amino acid B and the remaining hydroxyl group of water joins P2, breaking its covalent bond with the enzyme. ⬜6 This releases P2 from the active site and fully restores the enzyme, making it ready to act again.

then drifts out of the active site (Figure 6.9 ⬜4–⬜6). These roundabout chemical steps also occur without the need for much activation energy. The entire reaction, from binding to release, happens with remarkable speed—generally thousands, sometimes millions, of reactions occurring per second.

Influences on the Rates of Enzyme Action

The rates of enzyme-catalyzed reactions are strongly dependent on three conditions. The first is substrate concentration. As the substrate concentration increases (starting with zero), the probability of an enzyme molecule colliding with a substrate molecule increases, and so the overall rate of reaction increases. At some concentration, though, a point called **enzyme saturation** is reached. In enzyme saturation, each

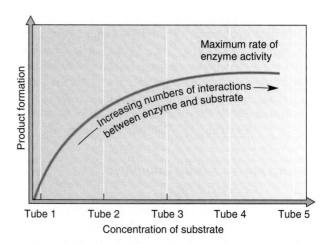

FIGURE 6.10
Increasing the Substrate Concentration

In an experiment, five tubes are prepared, each with the same amount of enzyme but with increasing substrate. After a set period of time, the amount of product in each tube is determined. The increasing rate of product formation levels off after a point (tube 5)—a point at which all the enzyme molecules are working at capacity.

active site is engaged, and the reaction rate is then determined by "turnover time." This is the time that it takes for an enzyme to bind to a substrate, react, and release the products. At enzyme saturation, any increase in substrate, without an accompanying increase in enzyme, will not affect the overall reaction rate (Figure 6.10).

Second, the pH of an enzyme's surroundings also influences enzyme action. At its optimum pH, critical amino acid R groups in the enzyme's active site will ionize, causing the enzyme and substrate to interact more readily. However, excessive acidity or alkalinity can cause trouble. It may disrupt important bonds that determine an enzyme's shape. When this happens, the enzyme may be rendered inoperative. The optimal pH depends on the specific enzyme (Figure 6.11). When enzymes are used in experiments, the investigator typically introduces the enzymes (often powdered) into *buffered solutions* set at an optimum pH. Within limits, such solutions will retain the optimal pH in spite of any hydrogen or hydroxide ions that might be released during the ensuing reactions. (Buffers, recall, are agents that maintain a given pH by neutralizing hydrogen or hydroxide ions that form. See Chapter 2.)

Temperature is a third factor. As a rule of thumb, *every increase of 10°C doubles the rate of most chemical reactions*, including enzyme activity. This is because heat in a system is expressed as molecular motion; thus an increase in the movement of the substrate and enzyme molecules favors the chances for just the right collision, that of substrate with active site. Of course, there is a limit. Recall that heat in excess can denature protein (Figure 6.12). In denaturation, the bonds that hold the enzyme into its tertiary and quaternary structure—the precise foldings—are broken, and it undergoes what can be an irreversible conformational change. Once the

original folding and coiling are disrupted, new folding and coiling occur randomly. Since shape is critical to an enzyme's function, any significant change renders it inoperative. We might note that denaturation is often reversible, sometimes just by gradual cooling.

Cofactors (Vitamins and Minerals)

Enzymes may require the assistance of **cofactors** to carry on their activities. Cofactors include atoms of metals such as zinc, iron, and copper. They may be permanently bound to the enzyme or, alternatively, may become loosely bound to the substrate. In any case, they are essential for proper enzyme function. Interestingly, excesses of these very metals may be highly poisonous.

Cofactors also include organic molecules called *coenzymes*. Most vitamins are coenzymes, or the precursors of coenzymes (raw materials from which the vitamin, or other molecules, are made). Vitamins, like metallic cofactors, can be quite harmful when in excess (see Chapter 37). We will return to coenzymes shortly.

Teams of Enzymes: Metabolic Pathways

Much of the chemical work of the cell occurs in **metabolic pathways.** A metabolic pathway is a sequence of enzyme-catalyzed reactions in which molecules may be built up or broken down. They are very much like factory assembly lines in which each worker makes some change to the product as it passes his or her station. In effect, the product of one enzyme becomes the substrate of the next, and so on along the pathway (Figure 6.13). Some teamed enzymes are bound to membrane surfaces, whereas others occur within the membrane itself. When enzymes are so fixed in place, the analogy of the production line fits even better.

An example of a metabolic pathway would be the Calvin cycle of photosynthesis, which comes up in the next chapter. Here the precursors of glucose are formed in a cycling pathway. In the initial step, an enzyme reacts with carbon dioxide, adding it to another substrate. Then in several additional steps, ATP energy is used to add hydrogen to the growing substrate. Eventually, finished 3-carbon molecules emerge, and these can then be assembled into glucose (see Chapter 7, Figure 7.15). As the term cycle implies, the sequence of reactions eventually leads back to a starting point. A considerable number of metabolic pathways occur as cycles.

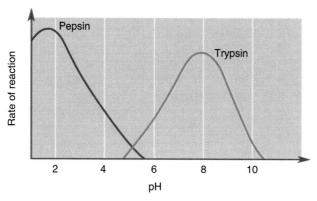

FIGURE 6.11
pH and Enzyme Activity
Enzymes have optimal pH levels at which they perform. The stomach enzyme pepsin is most active at a pH range of 1–2 (strongly acid) and begins to taper off its activity at higher (more basic) pH levels. Trypsin, a small intestinal enzyme, cannot function at the pH of the stomach. Its optimal range is between pH 7 and 9 (a neutral to slightly alkaline condition). Acidic material entering the small intestine is immediately neutralized by sodium bicarbonate secreted by the pancreas, which permits digestion to proceed.

Mechanisms of Enzyme Control

Although the activities of enzymes can be modified by general conditions such as temperature, pH, and substrate concentrations, far more specific controls exist. Such controls, working through activating and inhibiting mechanisms, regulate the accessibility of the enzyme. At the gene level, enzyme synthesis itself may be controlled by conditions in the cell such as the availability of substrate. Typically, enzyme-specifying genes

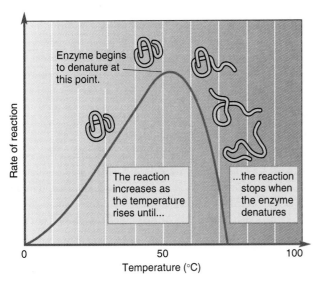

(a) Effect of heat

FIGURE 6.12
Heat and Enzyme Activity
(a) Heat accelerates enzyme action—up to a certain point, which differs among enzymes in different environments. The sudden reduction of activity at 50–60 °C indicates that the enzymes have probably been denatured, rendering them inactive. **(b)** Interestingly, organisms (bacteria and algae) in these hot springs can tolerate conditions near boiling. Apparently their enzymes are heat-stabilized in some way.

(b) Hot springs

are inactivated unless something happens to cause their activation. We will consider gene control and enzymes in detail in a later chapter, but for a fascinating example in bacteria, use the index to find the "*lac* operon."

Enzyme-Activating Mechanisms. In some instances the enzymes are available but remain in an inactive form until certain chemical signals prompt them into activity. We find a good example of such control in the vertebrate liver cell, which contains teams of enzymes that help regulate blood sugar levels. The enzymes remain in an inactive state until a regulating hormone reaches the cell surface. There it sets in motion a chain of events that activates those enzymes that hydrolyze glycogen. Glycogen is then broken down into glucose subunits, which are released into the blood (see also Chapter 36).

Enzyme-Inhibiting Mechanisms. At the opposite end of the spectrum is enzyme inhibition, where an agent interferes with the functioning of otherwise active enzymes. Two common ways this negative kind of regulation can occur are called competitive inhibition and noncompetitive inhibition.

In **competitive inhibition**, a molecule that is structurally similar to the usual substrate binds to the active site of the enzyme. Since it cannot act on the competing molecule, the enzyme simply becomes inactive. Penicillin, one competitive inhibitor, binds to the active site of enzymes that synthesize the cell wall materials in certain types of bacteria. As a result, an abnormal wall is constructed. Although the bacterial cell is normal enough in other ways, without a proper cell wall it cannot cope with the osmotic pressures to which it is subjected. Water continues to enter until the cell ruptures.

In **noncompetitive inhibition**, an agent will bind to the enzyme, but at a locale other than the active site. Such binding alters the enzyme's shape enough so that it no longer fits the usual substrate. Let's look at a few examples.

Allosteric Enzyme Systems. Allosteric enzymes are those with two kinds of binding sites—the usual active site, where substrate binds, and a secondary or **allosteric site** to which either an inhibitor or activator may bind. (*Allo-,* "other"; *stereo,* "space") (Figure 6.14). When an inhibitor binds to the allosteric site, it effects a shape change on the site. This change is transmitted over the entire protein, reaching and altering the active site so that it can no longer bind to substrate. Alternatively, an activator binding to the allosteric site stabilizes the enzyme shape, assuring its continued action. In the molecular realm of enzymes, the presence of inhibitors

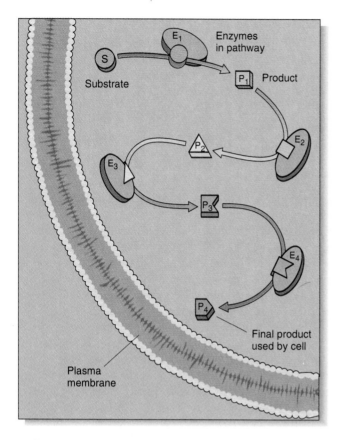

FIGURE 6.13
Enzymes in Metabolic Pathways
Enzymes commonly do their work in sequential reactions that occur in metabolic pathways. Thus the initial reactant will be acted on by several enzymes, each making a certain change before the final product emerges.

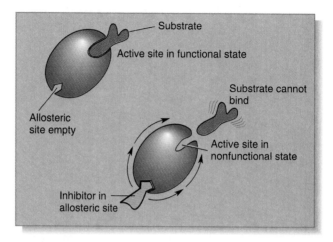

FIGURE 6.14
Allosteric Enzymes
Allosteric enzymes have, in addition to the active site, a secondary or allosteric site that will form weak bonds with an inhibitor of some kind. When the inhibiting molecule binds to the allosteric site, a conformational change occurs in the enzyme, and its active site is altered enough to stop its activity.

and activators in ever-changing concentrations acts as a sensitive switch that shifts the enzyme "on" to "off" and back. Thus the enzyme becomes fine-tuned to the cell's changing requirements.

One of the most exquisitely sensitive control systems is seen in a particular metabolic pathway in bacteria. The pathway consists of five enzymes, each of which plays a part in converting the amino acid threonine into the amino acid isoleucine (Figure 6.15). The interesting point about the pathway is that it is automated or self-regulating. Whenever the concentration of isoleucine increases beyond a certain point—indicating that it is being produced in excess of the needs of the cell—the first enzyme in the pathway slows its activity. Then, as the isoleucine concentration falls, the first enzyme speeds up again. So how does the bacterium "know" when it's producing too much isoleucine, or when it should resume full-speed production? As far as we are aware, bacteria don't really know anything; the pathway is automated. It's a matter of control by an allosteric enzyme.

In the threonine-to-isoleucine pathway, the inhibitor turns out to be isoleucine, the final product. The allosteric enzyme, or **control enzyme** as it is called, is the first enzyme of the pathway. Its allosteric site fits isoleucine. When isoleucine's concentration rises, some of the molecules find their way to the allosteric site, where they bind, thus altering the enzyme and slowing activity in the pathway. Why doesn't activity cease altogether? Binding in the allosteric site involves weak forces only and is readily reversed. But because of the high concentration of isoleucine, the molecules pop in and out of the allosteric site, turning the enzyme rapidly on and off but not stopping it altogether. Then, as the cell resumes its use of isoleucine, the concentration falls and the control enzyme is inhibited less and less until the pathway works at full-speed once again.

The inhibition of allosteric enzymes by the last product of the pathway is an example of what biologists call **negative feedback inhibition**, or sometimes a **negative feedback loop.** In negative feedback, the product of an action inhibits that action. In this case, the product of the action is isoleucine and its accumulation produces a negative feedback loop to the control enzyme. Thus control in the pathway, like so many activities at the cell level, is automated. The adaptiveness of such a system is that it tends to conserve material and energy. It wouldn't do for the cell to invest its energy and resources in the production of products that are no longer needed.

Summarizing the Activity of Enzymes

At this point let's review some of the major characteristics of enzymes and their functions:

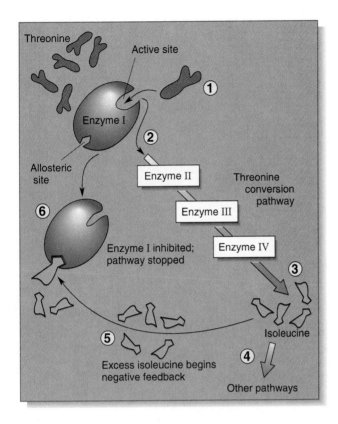

FIGURE 6.15
Allosteric Control of a Metabolic Pathway
In the metabolic pathway where threonine is converted to isoleucine, the isoleucine molecule itself is an inhibitor. When isoleucine is in excess, indicating that the cell has as much as it needs for now, it will concentrate around the pathway, some of it binding to the allosteric sites on enzyme I, at the start of the pathway. Control of a pathway by a product is an example of negative feedback. As isoleucine is again used by the cell, its concentration in the pathway falls and it leaves the allosteric site of enzyme I. Enzyme I will speed up its activity and the pathway will proceed.

1. Enzymes lower the energy of activation requirement and speed up reactions that might otherwise be too slow to be useful and require too much energy.

2. Enzymatic reactions occur at active sites that are highly specific for the particular substrate of that enzyme.

3. Amino acid R groups in the active site react chemically with substrate, breaking old bonds and forming new ones.

4. Enzymes emerge unaltered from their reactions and can react repeatedly.

5. The rates at which enzymes function are influenced by substrate and product concentration, by temperature, and by pH.

6. Enzymes may occur in metabolic pathways, where each carries out a specific step in the formation of product.

7. Enzymes in some metabolic pathways are controlled by competitive and noncompetitive inhibition.

8. Allosteric enzymes are those with secondary sites, where the binding of an inhibitor or activator can slow or speed up the action of that enzyme.

ATP: The Energy Currency of the Cell

You're probably convinced, by now, that living things must remain chemically active or perish. Even the most sedentary creature on earth is a virtual hotbed of activity, although much of its activity may take place quietly, within its cells. No matter what this creature is, all its activities are intimately tied to one kind of molecule—**adenosine triphosphate**, or **ATP**.

All living cells use ATP. It is a universal molecule of energy transfer. If energy is made available by some exergonic cellular process, such as cell respiration, it is first transferred to ATP. When that energy is needed, it is released by ATP. Recall from Chapter 3 that carbohydrates, lipids, and proteins contain energy reserves the cell can use. But energy cannot be taken directly from these molecules. First it must be transferred to the special bonds of ATP. We should note here that there are three other energy transfer molecules, very similar to ATP: thymine triphosphate (TTP), guanine triphosphate (GTP), and cytosine triphosphate (CTP), each making use of one of the other three nitrogen bases. They are reserved for special reactions, such as the synthesis of nucleic acids (see Chapters 12 and 13). ATP is by far the most broadly used.

The work of ATP can be organized into three broad categories: synthesis, movement, and active transport. *Synthesis* includes endergonic reactions ranging from the assembly of simple sugars into polysaccharides to the assembly of nucleotides into gigantic molecules of DNA and RNA. It also includes the breakdown of molecules for their chemical bond energy. Movement includes that of the organism or cell through the action of cilia or flagella or the contraction of muscle proteins. It also includes such intracellular movement as cytoplasmic streaming, ameboid action, and chromosome separation during cell division. Active transport, as we have seen, involves the movement of individual ions and molecules by membranal pumps and the bulk movement of materials through endocytosis and exocytosis.

ATP can accurately be called the "coin" of the cell's energy transactions. Like the penny, our smallest monetary unit, an ATP molecule represents a basic energy unit. When larger amounts of energy are needed, more ATP units can be used. The fact that small amounts of energy are available within each ATP molecule is very important to the processes of life. Such amounts permit the expenditure of free energy to match the task at hand. If nothing else, life is parsimonious, and wastefulness is soon dealt with by natural selection.

ATP Structure and the Release of Energy

As Figure 6.16 reveals, each ATP molecule consists of three parts: an **adenine base**, a 5-carbon sugar named **ribose**, and three **phosphates.** Adenine is a double ring of carbon and nitrogen atoms. Ribose, a carbon ring, forms a connecting link between adenine and the three phosphates that make up the "tail" of the molecule. (This arrangement of a base, sugar, and phosphate tail is also common in the nucleic acids, DNA and RNA.)

Typically, energy in ATP is made available when its outermost phosphate group is cleaved away from the molecule, forming **ADP (adenosine diphosphate)** and **P_i (inorganic phosphate).** In much rarer instances, the second phosphate is cleaved away, which makes still more energy available and yields **AMP (adenosine monophosphate).** Instrumental to this cleaving process is the enzyme **adenosine triphosphatase,** or **ATPase** for short. (Enzyme names commonly end in *ase.*)

Note in our ATP molecule that the two outermost phosphates are shown connected by wavy lines. The two outermost phosphates are special in the sense that, when hydrolyzed (broken by the enzymatic addition of water), they yield a sizable amount of free energy. In this reaction the products are ADP and P_i:

ATPase

ATP + H_2O $\longrightarrow$ ADP + P_i

(OH and H from water join ADP and P_i)

*$\Delta G = -7.3$ kcal/mole

When the terminal phosphate group has been removed, the free energy change (ΔG) between ATP and ADP + P_i, is –7.3 kcal/mole. We could also say that 7.3 kcal of energy were freed for work.

This is not an extraordinary yield of energy. Actually, several other sorts of cellular molecules yield more energy upon hydrolysis than does ATP. The difference is that the phosphate bonds in ATP, indicated by the squiggly lines, are weak. For that reason, the *net* energy yield, upon hydrolysis, is greater. It might help to think of the wavy lines in the ATP structure as tiny coiled springs, ready to pop out at the slightest provocation. The analogy is not as farfetched as it may seem. The ease with which the outer two phosphates cleave has to do with the repulsion among phosphate groups themselves, all due to the strong, closely adjacent, negative charges. Again, *the relative ease with which this occurs helps explain the high free energy yield* (a large $-\Delta G$).

Looking at it another way, the electron repulsion does not exist between the innermost phosphate and the ribose of ATP. Should that phosphate bond be hydrolyzed, the free energy change would be far less:

AMP + H_2O $\longrightarrow$ Adenosine + P_i

(OH and H from water join adenosine and P_i)

*$\Delta G = -3.4$ kcal/mole

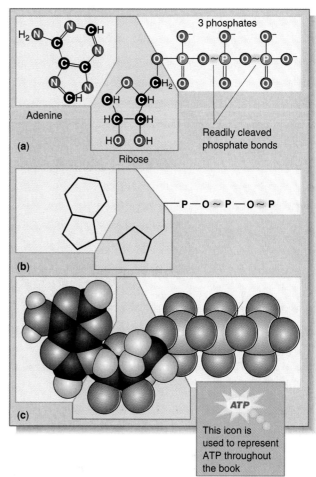

FIGURE 6.16
The ATP Molecule
ATP consists of the nitrogen base adenine, the pentose sugar ribose, and three phosphates. The phosphate bonds of the two outer phosphates (curly lines) are readily cleaved, making the energy-rich phosphates available for other reactions.

Work of ATP

The simple hydrolysis of an ATP molecule, as described so far, would do little for the cell except provide some heat and use up some water. In life, these strongly exergonic reactions are closely coupled with endergonic reactions in which some of the free energy given up by ATP is put to work. Such reactions often begin with **phosphorylation**—the addition of the terminal phosphate to another molecule. When another molecule accepts one of the phosphates, it gains new free energy of its own. The newly energized molecule is then more likely to enter into yet other reactions (Figure 6.17). Therefore much of ATP's value to the cell is the readiness with which its phosphates can be transferred to other molecules, what biochemists call its "high group transfer potential."

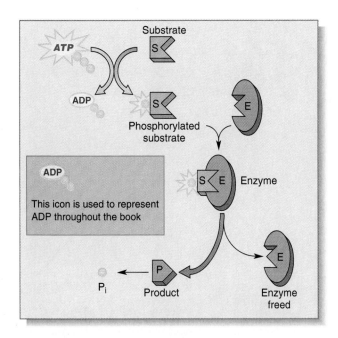

FIGURE 6.17
A Useful Phosphorylation Reaction
In this scenario, ATP reacts with substrate, phosphorylating it and preparing it for reaction with an enzyme. With the substrate's free energy increased, the enzyme is able to effect a change, yielding a new product.

Frequently, the phosphorylated molecule enters into a **substitution reaction,** one in which the newly acquired phosphate is traded for some other chemical group or perhaps for just a hydrogen atom. It is precisely through a number of such phosphorylations and substitutions that sugars are formed during the Calvin cycle of photosynthesis (see Chapter 7).

The Cycling of ATP

Since ATP supplies the major share of the cell's energy requirements, one might think that living things would be bulging with the stuff. One would be wrong. Humans, in fact, have only about 2 ounces of ATP in the entire body and most ATP is expended within one minute of its synthesis! Quite simply, cells don't need much ATP because they recycle ADP so efficiently. While we're at it, why is it that organisms don't simply convert all energy sources into ATP and store it around the body. One reason is that energy-rich molecules such as glucose can readily be converted to storage polysaccharides, such as starch and glycogen, and stored away in compact masses until needed. For another, constructing ATP is metabolically expensive and there are limits as to how much the cell can afford (review ATP structure in Figure 6.16). It is much more efficient to simply recycle the limited amount.

When ATP is hydrolyzed in the cell, the newly formed ADP is recycled—sent through energy-yielding processes in which the terminal phosphate is restored (Figure 6.18). The most common means of ATP regeneration are glycolysis, cell

respiration, and photosynthesis, processes we will look into in the next two chapters. An overall summary of ATP cycling would be the following:

$$ATP \longrightarrow ADP + P_i + Energy$$
$$ADP + P_i + Energy \longrightarrow ATP$$

We will come back to the synthesis of ATP shortly, but first we set the stage by introducing a few additional molecules with related roles.

The Coenzymes and Other Electron Carriers

Coenzymes

Coenzymes, you'll recall, work together with enzymes. But unlike enzymes, coenzymes are not proteins. They are much smaller than proteins, and, as a matter of fact, the components of some are similar to those of ATP. Coenzymes are involved in many kinds of cellular reactions, but one of their major roles is to participate in reactions that lead eventually to the synthesis of ATP.

The two most common coenzymes are **NAD** and **NADP** (nicotinamide adenine dinucleotide and nicotinamide adenine dinucleotide phosphate, respectively). We will also consider a third one, **FAD** (flavin adenine dinucleotide). These are commonly pronounced as "nad," "nad-phosphate," and "fad." In our discussions of cell energetics, we will focus on NADP's role in photosynthesis and NAD's role in cell respiration. FAD, as we will see, functions in both processes.

Chemically, NAD and NADP are quite similar, but NADP contains one more phosphate (Figure 6.19a). Like ATP, both contain the nitrogen base adenine, along with ribose sugar and phosphate. But, in addition, NAD and NADP contain a compound known as **nicotinamide,** a component of vitamin B. Nicotinamide is the active group in both coenzymes. FAD is similar to the other two coenzymes, but its active group is isoalloxazine, a subunit of riboflavin (also from the B vitamins) (see Figure 6.19a).

The three coenzymes carry out similar functions. They team up with enzymes that are involved in the *oxidation* and *reduction* of substrates. Let's see what the two new terms mean.

Redox Reactions: Oxidation and Reduction

It may seem strange that so much of life depends on simply passing electrons along (giving it almost an *electrical* quality), but that is precisely what we're about to see. Electron shifting takes place in what are called **redox reactions,** coupled reactions involving both oxidation and reduction. **Oxidation** is the *removal of electrons* (e⁻) from a substance, and **reduction** is the *addition of electrons* (e⁻) to a substance. (These are two of the most troublesome terms in chemistry; but take heart,

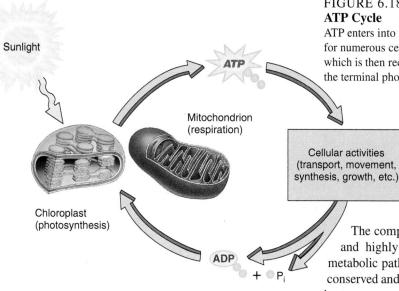

FIGURE 6.18
ATP Cycle
ATP enters into many exergonic reactions and provides the energy for numerous cellular processes. In so doing, it is converted to ADP, which is then recycled through endergonic reactions, during which the terminal phosphate is restored.

similarities between the two "fuels," cell respiration differs substantially from combustion, a difference for which we can all be thankful. The main difference is in the speed with which the redox reactions proceed. Combustion is almost instantaneous. The complete oxidation of glucose in the cell is gradual and highly controlled, utilizing many steps in lengthy metabolic pathways. In this way, much of the energy can be conserved and transferred to ATP, instead of being released as heat, as occurs in combustion.

generations of students have survived them.) When something is **oxidized** it loses electrons, but those electrons do not just wander about. They quickly (almost instantaneously) join with another molecule, which becomes **reduced.** (Because oxidation and reduction are usually paired this way, the two are referred to as redox reactions.) Thus when something is oxidized, something else is reduced. We have already looked into reactions where electrons are lost or gained, such as those involving ionic bonding (see Chapter 2). So now we can review one of our examples using the new terminology.

In the formation of common table salt, atoms of sodium lose an electron to atoms of chlorine. In "redox language," sodium is *oxidized* and chlorine is *reduced.* Furthermore, because sodium is an electron donor, it can be thought of as a **reducing agent**—here, one that gives electrons to chlorine. It follows then that chlorine is an **oxidizing agent**, one that accepts electrons, in this case from sodium. Chlorine, as you may be aware, is an excellent oxidizing agent, which is why it is effective as a laundry whitener and water purifier, where it removes electrons from stains and bacteria, respectively. In fact, chlorine gas is such a strong oxidizing agent that when mixed with iron filings or steel wool, they will readily burn.

In some cases, particularly during chemical activity in cells, when electrons are removed during oxidation, protons (H^+) tag along. Thus oxidation involves the *removal of electrons and protons* (hydrogen equivalent), and reduction involves the *addition of electrons and protons* (hydrogen equivalent). (Remember, one proton plus one electron equals a hydrogen atom equivalent.) Let's turn to the role of oxidation and reduction in the use of fuels.

Redox reactions in cells often start with an energy source, or fuel, such as glucose. You may not have thought much about this, but gasoline and glucose have a lot in common. They are both fuels and both are loaded with carbon–hydrogen bonds. In fact, people often speak about "burning" foods in their bodies. But in spite of the obvious

Coenzymes in Action

Coenzymes have a critical role in the transfer of energy. When they team up with oxidizing enzymes, they become reduced, that is, they take in the electrons (and protons) removed by the enzyme. Eventually, they pass them along to other molecules. Because of their role, coenzymes are often referred to as **electron carriers**.

Prior to teaming up with an oxidizing enzyme, NAD and NADP are in their oxidized form. Since they bear positive charges at this time, they are designated as NAD^+ and $NADP^+$. In this state, both NAD^+ and $NADP^+$ can attract two electrons and two protons (Figure 6.19b). The electrons and one of the protons join the nicotinamide group. Thus we commonly show the reduced forms as NADH and NADPH. The second proton cannot join nicotinamide but will be found in the surrounding medium. To indicate this, the reduced forms of the two coenzymes can be written as $NADH + H^+$ and $NADPH + H^+$. We will generally use the simpler form.

What about FAD, you anxiously mutter? You will be glad to hear that FAD doesn't give us the same problems. First, in its oxidized state, it is written *FAD;* and second, it actually takes in two electrons and two protons. Since this is the equivalent of two hydrogen atoms, the reduced state of FAD is written $FADH_2$ (Figure 6.19b).

To return to the combined action of an enzyme, a coenzyme, and a substrate, we can symbolize a reaction in the following manner:

$$\text{Substrate} + NAD^+ \xrightarrow{\text{Oxidizing enzyme}} \text{Oxidized substrate} + NADH + H^+$$

(The enzyme isn't changed in the oxidation reaction, so by convention it is listed on the arrow and not among the reactants or products.)

It is important to note that when a coenzyme is reduced, its free energy state is increased. Predictably then, the reduced

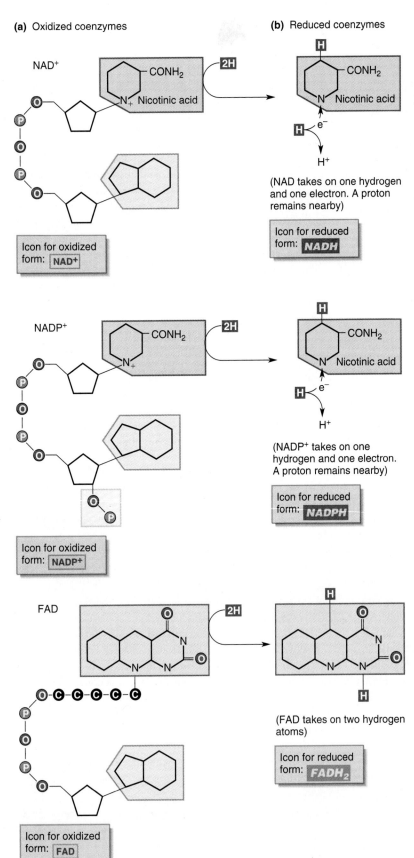

(a) Oxidized coenzymes

NAD+

Icon for oxidized form: NAD+

NADP+

Icon for oxidized form: NADP+

FAD

Icon for oxidized form: FAD

(b) Reduced coenzymes

(NAD takes on one hydrogen and one electron. A proton remains nearby)

Icon for reduced form: NADH

(NADP+ takes on one hydrogen and one electron. A proton remains nearby)

Icon for reduced form: NADPH

(FAD takes on two hydrogen atoms)

Icon for reduced form: FADH$_2$

FIGURE 6.19
Oxidized and Reduced Coenzymes
(a) The oxidized coenzymes. The active group in coenzymes NAD+ and NADP+ is the nicotinic acid ring (upper right). The only difference between NAD and NADP is the number of phosphate groups. The active group in FAD is the triple ring structure known as isoalloxazine. The active groups in the three coenzymes gain electrons and protons when a substrate is oxidized and lose them when a substrate is reduced. **(b)** When NAD+ and NADP+ are reduced, the nicotinic acid group accepts the equivalent of two hydrogens: two electrons and one proton join the nicotinic acid ring, and an additional proton enters the surrounding medium. The two coenzymes are then designated as NADH + H+ and NADPH + H+. FAD, however, receives two electrons and two protons (or two hydrogens), thus becoming NADH$_2$. As the arrows indicate, the reactions are fully reversible.

coenzyme will pass its electrons and hydrogens off to some molecule that is in a lower free energy state. The ability to reduce other substances is often called **reducing power.**

Whereas NAD, NADP, and FAD are small, mobile carriers, we find electrons are also passed along by large, stationary carriers that are also important in the ATP-generating processes.

Stationary Electron Carriers of Mitochondria and Chloroplasts

Stationary electron carriers are found in strategic locations in the membranes within mitochondria and chloroplasts. Actually, we find them in the *inner membrane* of the mitochondrion and in the *thylakoid membranes* of the chloroplast, structures that were discussed in Chapter 4. Most membrane-bound electron carriers are proteins with active prosthetic groups attached. Many fall into a family of proteins called **cytochromes.** As in the case of the giant oxygen-carrying protein hemoglobin, the active group of the cytochrome is a heme group, that is, a ring of nitrogen with an iron atom at its center. The heme group is specifically involved with redox reactions. As the iron of the heme group receives and passes electrons, it switches back and forth between two ionic states: Fe^{2+}, the reduced state, and Fe^{3+}, the oxidized state. (When it gains an electron its net positive charge is changed from +3 to +2, and when it loses that electron it reverts back.)

A series of such carriers within a membrane, arranged according to their reducing power, is aptly called an **electron transport system** (**ETS**)(Figure 6.20). Thus carrier 1 in an ETS can reduce carrier number 2, and number 2 can reduce carrier number 3, and so on through the system. Generally, the first fixed carrier to be reduced receives its electrons from a mobile electron carrier such as NADH or NADPH, whereupon the members pass the electrons along, like buckets of water in a bucket brigade, down the system. And "down" is the right word since the sequence occurs in an energetically "downhill" direction. In electron transport systems, the free energy of moving electrons decreases as they pass from carrier to carrier. Such drops in free energy, as we have seen, provide an opportunity for work to be done should that energy be harnessed. In fact, it is by harnessing the free energy in electron transport systems that ATP is produced.

The Formation of ATP in Cells

We have been leading up to just how ATP is made for some time, but the background information just given is important to understanding this critical process. Here we can tie together the action of oxidizing and reducing enzymes, the mobile coenzymes, and the membrane-bound carriers of electron transport systems. We will bring these entities together in what is called the theory of chemiosmotic phosphorylation.

As we consider the production of ATP, let's first remind ourselves that ATP is a cycling molecule; that is, the phosphory-

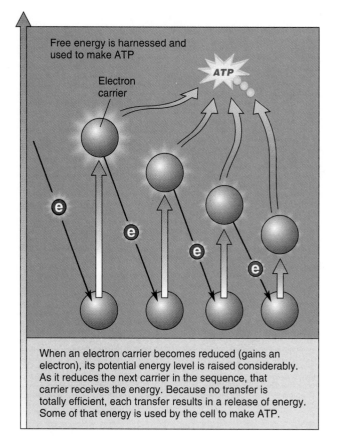

When an electron carrier becomes reduced (gains an electron), its potential energy level is raised considerably. As it reduces the next carrier in the sequence, that carrier receives the energy. Because no transfer is totally efficient, each transfer results in a release of energy. Some of that energy is used by the cell to make ATP.

FIGURE 6.20
Electron Transport System
Electron carriers (cytochromes) in an electron transport system form a sequence within membranes according to their reducing power (those with the greatest reducing power are shown at the left).

lation of ADP forms ATP, which, when used by the cell, is broken back down to ADP (see Figure 6.18). Now we will see that ADP can be "recharged" to ATP through two distinct processes: substrate-level phosphorylation and chemiosmotic phosphorylation. **Substrate-level phosphorylation** involves the transfer of phosphates from phosphorylated compounds to ADP. We will look further into this more direct process in Chapter 8.

Chemiosmotic Phosphorylation

The theory of **chemiosmotic phosphorylation,** or **chemiosmosis** as it is also known, was proposed by Peter Mitchell in 1961. Mitchell, it seems, had his own peculiar way of looking at things. While other biochemists were trying to understand ATP synthesis by analyzing each specific component in its manufacture, Mitchell took a broader view and looked at ATP production in terms of the intact chloroplasts and mitochondria. His first clue was a basic and well-substantiated observation: neither chloroplasts nor mitochondria can make ATP unless their membranous compartments are physically intact. From this and other lines of evidence, Mitchell devised the scheme that became known as chemiosmotic phosphorylation.

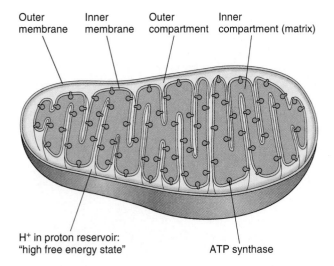

Outer membrane   Inner membrane   Outer compartment   Inner compartment (matrix)

H+ in proton reservoir: "high free energy state"   ATP synthase

FIGURE 6.21
The Mitochondrion
The mitochondrion contains two compartments separated by a membrane—the "sac-within-a-sac" arrangement of all eukaryotic chemiosmotic systems. As chemiosmosis proceeds, the outer compartment fills with protons, forming a system of great free energy. The proton-filled compartment takes on a strong positive charge, whereas the inner compartment becomes negative through a buildup of hydroxide ions (OH^-). The free energy of the system is used to phosphorylate ADP, yielding ATP.

Mitchell proposed that ATP synthesis during cell respiration in the mitochondrion, and during photosynthesis in the chloroplast, is dependent on the creation of steep concentration gradients created by isolating positively charged hydrogen ions (H^+), or protons, on one side of a membrane and negatively charged hydroxide ions (OH^-) on the other. A quick review of the mitochondrion, this time emphasizing its compartments, will help here (Figure 6.21). Note that the folded inner membrane forms a barrier between an outer and an inner compartment. The protons are pumped from the inner compartment, to the outer compartment, where their concentration represents a system of great free energy. Should the protons be permitted to escape in a controlled manner, their free energy could be used to phosphorylate ADP. Let's look at this process through the use of a model. (For a closer look at the chloroplast and mitochondrion see Figures 4.24 and 4.25.)

Charging the System. Our model of chemiosmotic phosphorylation is seen in Figure 6.22. The proton (H^+) gradient, or chemiosmotic gradient as it is also known, is produced through the work of electron transport systems located in the membrane separating the two compartments. Activity begins when mobile carriers transport electrons to the first carrier in the electron transport system. As electrons pass through the carriers, their free energy is used to power proton pumps. The pumps actively transport protons across the membrane into the proton reservoir on the other side (the outer compartment

in Figure 6.21). Proton pumping, which we briefly considered in Chapter 5, is a critical step in chemiosmosis. Also of importance is the fact that as protons are selectively removed from a compartment, they leave behind a number of hydroxide ions (OH^-).

The Proton Gradient: A System of Great Free Energy. Since an accumulation of positively and negatively charged ions (H^+ and OH^-) is involved, the chemiosmotic gradient can be characterized as an **electrochemical gradient**. It is an electrical gradient (separated positive and negative charges—a potential or voltage difference); and it is a chemical gradient—actually a pH gradient (the protons constituting an acid and the hydroxide ions a base). It is also a simple concentration gradient (so the ions have a strong tendency to diffuse down their gradient). Conceptually, they all represent a system of great free energy, and free energy can do work. The free energy of such proton gradients isn't imaginary—it has been measured. For example, the electrical potential across such membranes in the mitochondrion has been measured at 0.14 volts, quite sufficient to power the phosphorylation of ADP.

Needless to say, if the free energy were released in an uncontrolled manner, the result could be disastrous to the cell, perhaps even deadly. But suppose the protons were allowed to trickle down their gradient, their energy being released gradually. Then suppose that such an exergonic event were coupled to an endergonic one, a reaction in which phosphate joins ADP to form ATP. That's essentially what's going on.

The sac-within-a-sac arrangement in nature is represented by membranous compartments in chloroplasts and mitochondria. As we've emphasized, both have an inner and an outer compartment, thus providing for a proton reservoir. The free energy of the proton gradient of these bodies is tapped by letting the protons escape down the steep concentration gradient and move out of the reservoir.

Tapping the Free Energy of the Gradient. The free energy of the system is released as the protons move down their gradient (see Figure 6.22). They do this by passing through special protein-lined channels that lead into spherical enzyme complexes known as ATP synthases (mentioned in Chapter 4). ATP synthases use the free energy of the escaping protons to bring phosphate and ADP together, forming ATP.

A Quick Review. In summary, chemiosmotic phosphorylation involves the following:

1. *Electron transport*—the passage of energy-rich electrons down their energy gradient, from carrier to carrier, in electron transport systems.

2. *Proton pumping*—the use of free energy from the electrons to power proton pumps that concentrate protons inside a reservoir.

3. *Free energy system*—the subsequent establishment of a system of great free energy.

FIGURE 6.22
Model Chemiosmotic System
The simplest model of a chemiosmotic system makes use of two membrane-surrounded compartments (sac within a sac). Protons, pumped across the inner membrane by energy from electrons passing along an electron transport system, accumulate within the outer compartment, charging the system. The concentration of protons there creates an electrochemical gradient of great free energy. For work to be done, the protons must be permitted to escape down their concentration gradient to the inner compartment. They escape through the protein-lined channel into the bulbous protein complex known as an ATP synthase. There the free energy of the protons is used to phosphorylate ADP, yielding ATP.

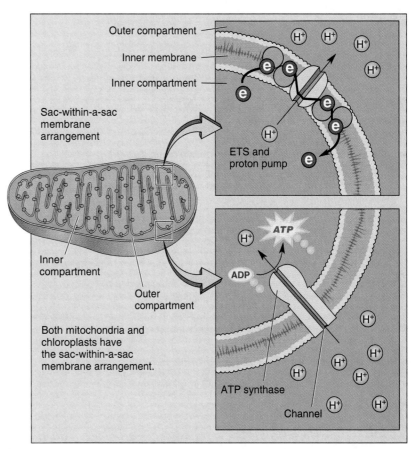

4. *Formation of ATP*—the protons escape from the reservoir through ATP synthases, where their free energy is used to phosphorylate ATP.

This last topic probably leaves you with a lot of questions. For example, where do the electrons used to charge the chemiosmotic system come from and where do they go? And how does charging ATP in the mitochondrion differ from that in the chloroplast? To answer in brief, the electrons used in the mitochondrion come from oxidized foods and end up in water molecules. Electrons flowing in the chloroplast, however, come ultimately from water, and they end up in the food molecules the plant makes. The flow of electrons in the chloroplast begins as sunlight energy is captured by molecules of chlorophyll. As you see, there are some interesting but opposite parallels in the functioning of the two organelles. But, this gets us into the next chapter, the story of the remarkable process called photosynthesis.

Key Ideas

ENERGY

1. **Energy** is the ability to do work, and work is the application of force. **Potential energy** is energy stored. **Kinetic energy** is energy in motion.

2. Forms of energy include chemical, mechanical, magnetic, and electromagnetic—all of which are interchangeable.

3. The *first law* (the law of conservation of energy) states that energy can neither be created nor destroyed, but it can be transformed from one form to another.

4. The *second law* (law of entropy) states that the free energy of a system is always decreasing. Complex, organized systems tend to reach disorder and randomness, and a measure of the disorder is called **entropy**.

5. Transfers of energy from one form to another, in both living and nonliving systems, are always accompanied by losses.

6. The organization necessary for life is possible only because of sunlight energy. Without it entropy would soon increase.

APPLYING ENERGY PRINCIPLES TO LIFE

1. Energy available within the chemical bonds of the molecules of life is released as free energy when the bonds are broken. **Free energy** is energy available for work.

2. **Exergonic reactions** are those in which the free energy of the product(s) is less than the free energy of the reactants: ΔG is negative. In **endergonic reactions** the free energy of the product(s) is greater than the free energy of the reactants: ΔG is positive.

3. The *Hindenburg* disaster was a massive exergonic reaction. Hydrogen gas, ignited by a spark, united with oxygen to form water, both reaching a lower, more stable energy state. The energy difference was represented by light and heat in the massive fireball.

4. **Energy of activation** is the energy required to get stable molecules to react. Once started, the energy generated is sufficient to keep the reaction going.

5. Many reactions are reversible. When the concentration or free energy of the reactants equals that of the products, equilibrium is reached. When the two quantities are greater in the products, the reaction reverses.

6. **Catalysts** are agents that lower the energy of activation and speed up reactions without themselves being used up.

ENZYMES: BIOLOGICAL CATALYSTS

1. Enzymes are biological catalysts. Enzymes (a) are proteins, (b) lower the **energy of activation** requirement, (c) speed up reactions, (d) are highly specific for their **substrate**, (e) do not affect free energy changes, and (f) emerge unaltered from reactions.

2. Enzyme specificity is due largely to the **active site,** a groove whose shape fits only one kind of substrate.

3. Hydrolytic reactions include (a) binding, (b) induced fit, (c) weakening of bonds by acidic or basic R groups in the site, (d) bonds breaking and reforming, (e) release of first product, (f) reaction with water, and (g) release of second product. The enzyme is then freed to react again.

4. Increased substrate concentration increases the rate of enzyme activity until saturation is reached, whereupon the rate is governed by turnover time.

5. With few exceptions, enzymatic reactions occur at a near-neutral pH.

6. Heat increases the chances of collision between substrate and active sites, thus speeding up enzymatic reaction. Greater heat denatures the enzyme, and the reaction ceases.

7. **Cofactors,** such as zinc, iron, and copper, facilitate enzyme action. Organic cofactors are called **coenzymes**.

8. **Metabolic pathways** are teams of enzymes that perform sequential reactions.

9. In **competitive inhibition** substrate look-alikes block active sites.

10. **Allosteric enzymes** have **allosteric sites** that, when filled, alter the active site, either inhibiting or enhancing its action. Where allosteric enzymes work in metabolic pathways, a product, when in high concentration, may fill the allosteric site and act as an inhibitor. When the product's use resumes, the allosteric site is cleared and the pathway resumes. This is an example of **negative feedback inhibition**.

ATP: THE ENERGY CURRENCY OF THE CELL

1. ATP provides energy for cellular activity such as synthesis, movement, and transport.

2. ATP consists of the nitrogen base adenine, ribose sugar, and three phosphate groups. The phosphate groups are readily transferred, making free energy available for many uses.

3. The hydrolysis of ATP yields free energy, ADP (adenosine diphosphate), and P_i.

4. The products ADP and P_i are recycled—sent through reactions that provide energy to restore ATP.

THE COENZYMES AND OTHER ELECTRON CARRIERS

1. **Coenzymes** work in concert with enzymes. **NAD, NADP,** and **FAD** operate in oxidation–reduction reactions.

2. Electrons are passed from molecule to molecule through redox reactions (coupled reduction and oxidation reactions). **Reduction** is the gain of electrons, and **oxidation** is their loss. Oxidation and reduction can involve protons (H^+) as well as electrons.

3. The coenzymes in their oxidized and reduced forms, respectively, are:

    a. $NAD^+ \longrightarrow NADH + H^+$,

    b. $NADP^+ \longrightarrow NADPH + H^+$.

    c. $FAD \longrightarrow FADH_2$.

4. Many electron carriers are fixed in place, primarily in the membranes of the chloroplast and mitochondrion, where they form **electron transport systems.** Stationary carriers include iron-protein complexes called **cytochromes**.

5. In their passage through electron transport systems, electrons give up free energy, which is used in the synthesis of ATP.

THE FORMATION OF ATP IN CELLS

1. ATP is formed through **substrate-level phosphorylation**, the generation of ATP directly from substrate, and through **chemiosmotic phosphorylation**.

2. For chemiosmotic phosphorylation to occur, chloroplasts and mitochondria must have high concentrations of protons (H^+).

The proton gradient, an electrochemical gradient, is produced by proton pumps. The proton gradient has great free energy, which can be used in converting ADP to ATP.

3. The free energy of the mitochondrion and chloroplast is harnessed as protons pass down their electrochemical gradients through ATP synthases, giving off free energy as they go. Phosphorylating enzymes conserve some of that free energy system, using it to form ATP.

Application of Ideas

1. Many inventors have attempted to design "perpetual motion machines." All failed. The reasons for such consistent failure were not understood until the laws of thermodynamics were formulated. Explain how they apply to the problem.

2. When asked in a essay exam to define life, a thoughtful student once wrote an answer totaling six words. "Life is an interruption in entropy." Is this true? Using the laws of thermodynamics for background, comment on this statement.

3. Reviewing what has been presented so far on chemiosmosis in mitochondria and chloroplasts and the electrical differential generated across their membranes, develop an analogy using a rechargeable battery. How are they similar, and where does the analogy fail?

Review Questions

1. Distinguish between the terms potential energy and kinetic energy. (page 127)

2. Summarize the main ideas in the first and second laws of thermodynamics. (pages 127–128)

3. Carefully define *free energy*. To what does ΔG refer? (page 129)

4. Distinguish between *exergonic* and *endergonic reactions* and cite an example of each. (page 129)

5. What does *energy of activation* have to do with chemical reactions? (page 130)

6. What two factors determine whether a reaction will reach equilibrium? What happens when it does this? (page 131)

7. List five important characteristics of enzymes. (page 132)

8. Summarize the events in an enzymatic reaction, beginning with substrate entering the active site. (pages 133–134)

9. Describe the effect of heat on enzymatic reactions. What rule applies, and what are the limits? (page 135)

10. Using the threonine-to-isoleucine pathway as an example, explain how allosteric sites and negative feedback operates. Why is such an automated process valuable and necessary? (pages 136–137)

11. Describe the structure of ATP and explain why the presence of phosphates makes ATP so useful. (pages 138–139)

12. Describe a phosphorylation reaction. Why is this type of reaction used by the cell? (pages 139–140)

13. Draw an ATP–ADP cycle, and explain what is happening. (pages 140–141)

14. Carefully define oxidation and reduction and explain the role of a coenzyme in these processes. (pages 141–142)

15. List the three common coenzymes of cells and write their shorthand formulas in the oxidized and reduced states. (page 141)

16. What is an electron transport system? What function do such systems serve in the mitochondrion and chloroplast? (page 143)

17. What is a proton gradient? Explain how a proton gradient can provide the energy needed to resynthesize ATP. (pages 143–144)

CHAPTER 7

Photosynthesis

CHAPTER OUTLINE

PHOTOSYNTHESIS: AN OVERVIEW

The Two Parts of Photosynthesis

Light: The Energy Source

CHLOROPLASTS

The Thylakoids

THE LIGHT-DEPENDENT REACTIONS OF PHOTOSYNTHESIS

The Noncyclic Events

The Cyclic Events

Chemiosmotic Phosphorylation

THE LIGHT-INDEPENDENT REACTIONS

The Calvin Cycle

PHOTORESPIRATION: TROUBLE IN THE CALVIN CYCLE

C4 Carbon Fixation: Solving the Photorespiration Problem

Essay 7.1 From Willows, Mice, and Candles to Electron Microscopes

C4 Plants and Evolution

KEY IDEAS

APPLICATION OF IDEAS

REVIEW QUESTIONS

*O*ur frequent bantering about the merits of solar energy has an almost touching naiveté about it. We seem to treat solar energy as a recent idea, a new concept. We seem to forget that the sun is an ageless source of energy. The fossil fuels—oil, gas, and coal—are simply releasing solar energy stored away in hydrocarbons formed from the bodies of long-dead plants and algae. So while we continue to wrestle with the maze of engineering problems associated with harnessing the sun's energy to produce electricity, perhaps we should turn to the real experts for some ideas. And the real experts are likely to be green (Figure 7.1). [We will look into photosynthesis in the prokaryotes (bacteria) in Chapter 22.]

Plants get their energy from sunlight. But it's less well known that plants long ago evolved the ability to convert **photons** (energy units of radiant energy), first, into a flow of electrons and, ultimately, into chemical energy. Light energy is captured in those tiny, highly organized bodies called chloroplasts. **Chloroplasts** convert sunlight energy into chemical bond energy by first passing this energy to electrons, which, in a new high-energy state, move along pathways known as **electron transport systems** (see Chapter 6). This flow of electrons has many of the characteristics of a current of electricity moving along a conductor. We will return to electrons shortly, but first let's step back for a broader look at the process of photosynthesis.

FIGURE 7.1
Capturing Sunlight
Plants are specialists in the capture and utilization of solar energy.

Photosynthesis: An Overview

Photosynthesis is the process by which the energy of sunlight is used to bond certain molecules together to produce glucose and other substances (Figure 7.2). The raw materials of photosynthesis are carbon dioxide (CO_2) and water (H_2O). The products are glucose ($C_6H_{12}O_6$), water (H_2O), and molecular oxygen (O_2). Since glucose is at a considerably greater free energy level than carbon dioxide and water, the reactions are highly endergonic. The process can be expressed by the following general equation:

$$6\ CO_2\ +\ 12\ H_2O\ \xrightarrow[\text{light energy}]{\text{chlorophyll}}\ C_6H_{12}O_6\ +\ 6\ H_2O\ +\ 6\ CO_2$$

$$^*\Delta G = +686 \text{ kcal/mole}$$

When translated, the reaction simply tells us that water and carbon dioxide, in the presence of light and chlorophyll, form glucose, water, and oxygen, with a substantial gain in free energy. The gain comes from light energy, which is absorbed by the pigment chlorophyll. This venerable old equation, memorized by generations of biology students, hides a vast amount of detail. For example, the equation doesn't indicate how something as ethereal as sunlight could possibly power the formation of chemical bonds, nor does it suggest that certain wavelengths of light are more effective than others. There is no suggestion that the water at the right side of the equation is not the same water as that at the left (which it isn't), or that the oxygen at the right comes from the water at the left (which it does). Furthermore, as we will see, the final product of photosynthesis is actually the 3-carbon carbohydrate **glyceraldehyde-3-phosphate** ($C_3H_5O_3$), which is later converted to glucose or other useful substances in the cell cytoplasm. Finally, the equation hides the fact that before glucose can be synthesized, the energy of light is used to form

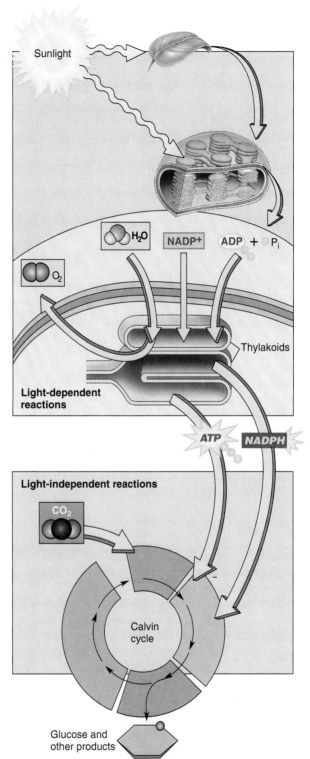

FIGURE 7.2
Overview of Photosynthesis
The scenario shows the reactants of the light-dependent reactions and light-independent reactions on the background of the chloroplast. The light reactions occur in the thylakoid, where light provides the energy to convert ADP and $NADP^+$ to ATP and NADPH. Water is broken down and oxygen is released. ADP and NADH, along with carbon dioxide, are used in the production of glucose in the light-independent reactions.

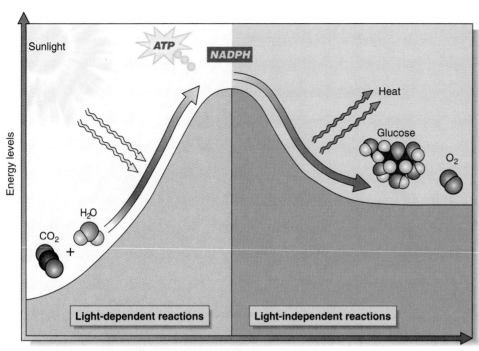

FIGURE 7.3
Charging the System and Tapping the Energy
The plot shows the various free energy levels of the participants in photosynthesis. Compare the free energy of carbon dioxide and water with that of glucose. Light energy, captured in the light reactions of photosynthesis, is used to generate ATP and NADPH, products with considerable free energy. Oxygen is a byproduct. The ATP and NADPH provide the energy and hydrogen to the light-independent reactions in which carbon dioxide is used in the synthesis of glucose.

such energetic substances as ATP and NADPH—which are not even mentioned in the formula. So let's tie up some of these loose ends. Much of what we will consider here represents hard-won information from discoveries made mostly over the past 60 years, and a few from earlier periods. For the historical highlights, see Essay 7.1, pages 166–167.

The Two Parts of Photosynthesis

The events of photosynthesis can be divided into two parts—the **light-dependent reactions** (those that require light) and the **light-independent reactions** (those that do not). In a sense, the first "charges" the system, and the second lets it run back down (see Figure 7.2). As we emphasized in the last chapter, a system of free energy, when running down, can accomplish a great deal of useful work.

In the light reactions (Figure 7.3), sunlight energy is used in two ways: to convert ADP to ATP and to reduce $NADP^+$ to NADPH. Both substances will be needed in the light-independent reactions—ATP for its energy and NADPH for its protons and electrons. Considering the last two participants and water alone, the light reaction becomes:

$$H_2O + ADP + P_i + NADP^+ \xrightarrow[\text{chlorophyll}]{\text{light energy}} ATP + NADPH + H^+ + \tfrac{1}{2}O_2$$

The ATP and NADPH formed in the light reactions become reactants in the light-independent reaction. After ATP and NADPH make their contributions to carbohydrate-building, they are represented by ADP and $NADP^+$. The two are recycled to the light reactions to act again.

Light: The Energy Source

Photosynthesis makes use of **visible light**, by coincidence the portion of the electromagnetic spectrum that we and most animals detect with our eyes. Visible light, as we see in Figure 7.4, is but a small part of all the radiant energy reaching the earth. The **electromagnetic spectrum**, as radiant energy is known, is a continuum that ranges from the highest energies, which are gamma radiations, all the way to the lowest energies, which are radio waves.

Visible light, like other electromagnetic radiations, is hard to describe in nontechnical terms. One reason is that it can be considered as two entities: as light waves and as photons. According to the wave theory, light travels in a wave-like path, not unlike the waves created by a stone that is dropped into a pond. Wavelength refers to the distance from the crest of one wave ("pond ripple") to the crest of the next. A photon can be considered as a discrete packet or unit of light energy. Its value is the inverse of its wavelength. Thus photons with short wavelengths contain more energy than those with long wavelengths.

The visible spectrum contains wavelengths ranging between 400 and 760 nm, which, when translated into color, means between violet and red, respectively. Just outside the violet range is ultraviolet (UV), which we cannot see (though bees can) but which is dangerous to the skin and eyes. Fortunately for life, most of the UV region is absorbed by the ozone layer of the atmosphere—at least it was until recently (see Chapter 48). Just outside the red light range is infrared, which we detect as heat.

Between the violet and red regions of the spectrum are blue, green, yellow, and orange. Relating this back to the photon concept, we see that the violet end of the range is the most energetic and the red is the least energetic. Plants make use primarily of the two ends of the spectrum in photosynthesis. Much of the central region, the yellow and green, is simply reflected—which is why leaves appear in various shades of green to us (other, unrelated pigments may also be present).

As we proceed into the light reactions, we will make use of these fundamental ideas about light energy. We will mainly be interested in the effects of visible light energy on matter—in particular, how it is captured by chlorophyll and how light energy is converted to chemical energy that eventually ends up in the chemical bonds of carbohydrate. To get ready for this, let's set the stage with a look at the intricate structure in which these transitions take place.

Chloroplasts

Figure 7.5 takes you through several levels of organization, from the leaf into the chloroplast. Chloroplasts are relatively large organelles, easily seen through the light microscope. They have double membranes, or envelopes, surrounding them. The outer membrane is relatively permeable to small molecules and solutes but the inner membrane acts more as a barrier. The inner membrane contains a large protein complement, much of which is involved in active transport. Included are coports that specialize in the import of ADP and the export of ATP. Another transport protein brings in inorganic phosphate.

Within the second membrane is a watery **stroma**, which is fairly structureless but often contains starch grains, ribosomes, and lipid bodies. (Recall that chloroplasts and mitochondria contain DNA and ribosomes and thus synthesize some of their own proteins.) Of more direct importance to photosynthesis, the stroma contains those enzymes required for the synthesis of the simple three-carbon carbohydrates that are used in the manufacture of glucose and other materials. Also within the stroma lies an extensive third membrane system whose foldings form the thylakoids.

The Thylakoids

The thylakoids occur in stacks called **grana**; each stack is referred to as a **granum** (Figure 7.6). Here and there, thylakoid membranes called **lamellae** extend from one granum to another. Both the thylakoids and the lamellae contain an extensive, and apparently continuous, channel-like interior known as the **lumen**. Since the lumen and the stroma are separated by a membrane, the two are, in effect, separate watery compartments. They form the all-important "sac-within-a-sac" arrangement necessary for chemiosmotic phosphorylation. Recall that both chloroplasts and mitochondria utilize this mechanism of ATP formation (see Chapter 6).

The thylakoid membrane contains an exceptionally large protein content, with proteins making up about 60% of the total mass. The proteins and other special elements are organized into photosystems, electron transport systems, and ATP

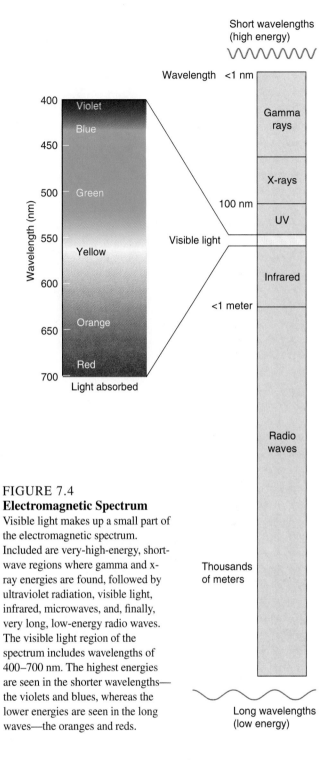

FIGURE 7.4
Electromagnetic Spectrum
Visible light makes up a small part of the electromagnetic spectrum. Included are very-high-energy, short-wave regions where gamma and x-ray energies are found, followed by ultraviolet radiation, visible light, infrared, microwaves, and, finally, very long, low-energy radio waves. The visible light region of the spectrum includes wavelengths of 400–700 nm. The highest energies are seen in the shorter wavelengths—the violets and blues, whereas the lower energies are seen in the long waves—the oranges and reds.

synthases, each playing a critical role in the light reactions of photosynthesis. Let's look closer.

The Two Photosystems. The active units in the thylakoid membranes contain two kinds of photosynthetic centers, designated **photosystem I (PSI)** and **photosystem II (PS II)**. Each photosystem includes a **light-harvesting complex** with a **reaction center** (Figure 7.7), and each is closely associated

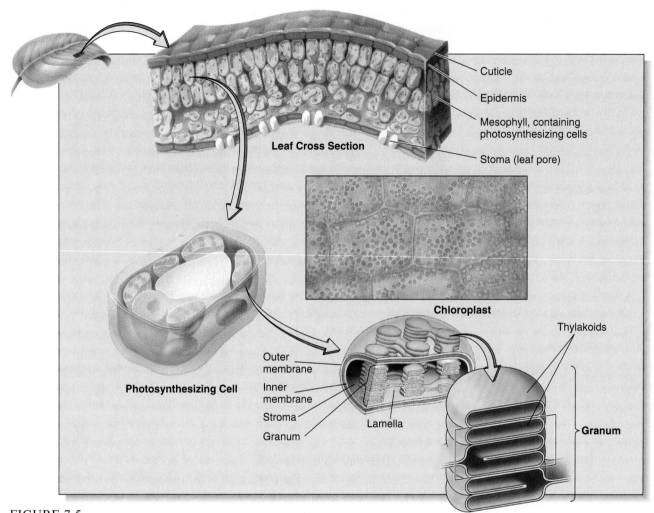

FIGURE 7.5
Organization of the Chloroplast
By moving down through several levels of organization, we can begin to understand the chloroplast and its intricate internal organization. Chloroplasts are commonplace in the mesophyll cells of the leaf. Most of the chloroplast's structure is clearly visible when viewed through the electron microscope. Note the double surrounding membrane, or envelope. Beyond the envelope is a clearer, watery stroma, which houses the complex, highly folded, inner membrane system that makes up the thylakoids. The denser membranous regions are stacks of thylakoids called grana. The membranes of some thylakoids, known as lamellae, extend from one granum to the next.

with an electron transport system. The photosystems contain chlorophyll and other pigments whose function is to absorb light and begin the flow of electrons.

Chlorophyll and the Light-Harvesting Complex. Light-harvesting complexes absorb and concentrate light energy. They contain many molecules of **chlorophyll *a***, **chlorophyll *b***, and **beta-carotene** (Figure 7.8a). The long, nonpolar, hydrocarbon tails on the chlorophylls are their "anchors." They lie deeply submerged in the lipid bilayer of the thylakoid membrane, while their active, polar heads protrude above the surface. The two chlorophylls are quite similar in their chemical structure, but they absorb different light wavelengths. Together they

manage to cover the two ends of the visible light spectrum, orange–red (600–700 nm) and violet–blue (400–500 nm). The major carotenoids of plants, called accessory pigments, absorb strongly in the blue wavelengths and slightly in the green. Most of the green is not absorbed, however, and it is this reflected color that is seen in the plants around us. We should note that the relative amounts and kinds of light-absorbing pigments differ among the photosynthetic organisms, providing a wide range of colors in plants and especially in algae.

The absorption of light by the chlorophylls and carotenoids is described in Figure 7.8b, where an **absorption spectrum** is illustrated. An absorption spectrum reveals those wavelengths (colors) that are actually absorbed by a pigment.

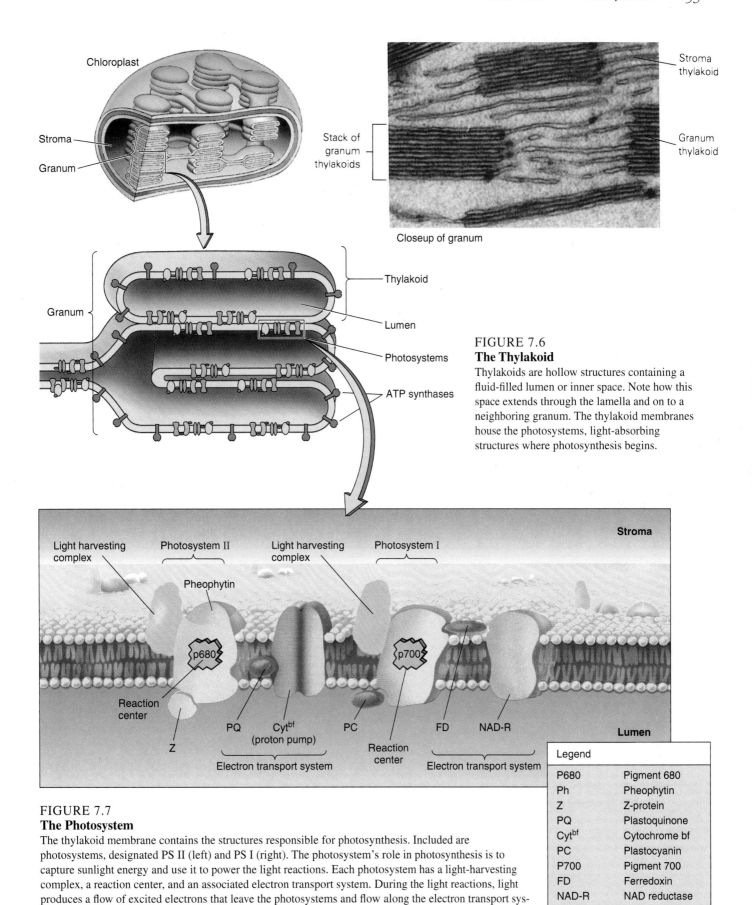

FIGURE 7.6
The Thylakoid

Thylakoids are hollow structures containing a fluid-filled lumen or inner space. Note how this space extends through the lamella and on to a neighboring granum. The thylakoid membranes house the photosystems, light-absorbing structures where photosynthesis begins.

Legend	
P680	Pigment 680
Ph	Pheophytin
Z	Z-protein
PQ	Plastoquinone
Cyt^bf	Cytochrome bf
PC	Plastocyanin
P700	Pigment 700
FD	Ferredoxin
NAD-R	NAD reductase

FIGURE 7.7
The Photosystem

The thylakoid membrane contains the structures responsible for photosynthesis. Included are photosystems, designated PS II (left) and PS I (right). The photosystem's role in photosynthesis is to capture sunlight energy and use it to power the light reactions. Each photosystem has a light-harvesting complex, a reaction center, and an associated electron transport system. During the light reactions, light produces a flow of excited electrons that leave the photosystems and flow along the electron transport systems. Their energy will be tapped for use in generating NADPH and ATP.

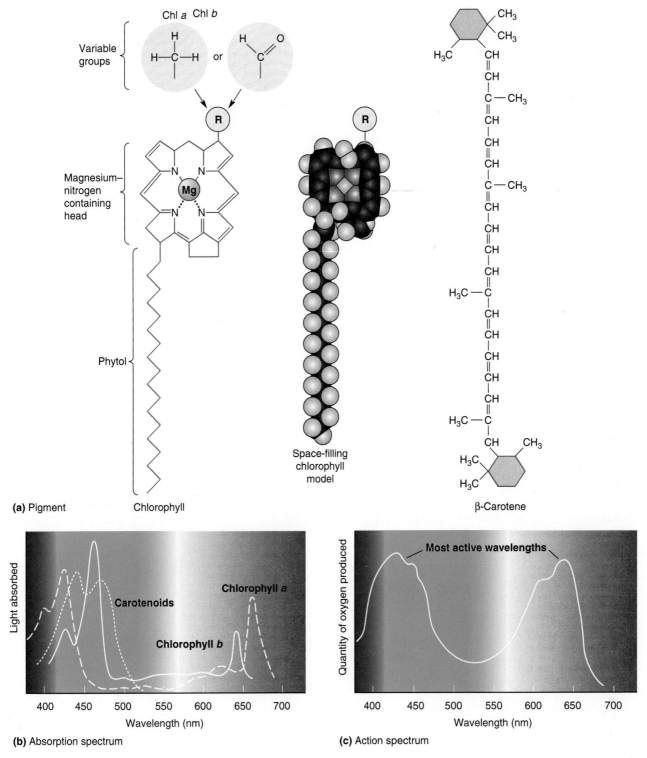

(a) Pigment   Chlorophyll   β-Carotene

(b) Absorption spectrum   **(c)** Action spectrum

FIGURE 7.8
Photosynthetic Pigments and the Absorption of Light

(a) The chlorophylls are similar in structure, differing only in one side group on the porphyrin head (color). Note the single magnesium atom surrounded by four nitrogen atoms. They make up the photoactive core of the molecule. Beta-carotene is one of the carotenoids. **(b)** The absorption spectrum is obtained by using an instrument that determines how much of each wavelength of visible light is absorbed by a chlorophyll extract. As you can see, the two ends of the spectrum are intensely absorbed by chlorophyll *a* and *b,* whereas the carotenoids absorb strongly in the blue range and some of the green wavelengths. **(c)** An action spectrum is the quantity of oxygen produced under varying colors of light. Oxygen is produced as a by-product of the light reactions. If you compare the action and absorption spectra, you can see that there is a good fit. This indicates that light absorbed is actually light used in photosynthesis.

The accompanying **action spectrum** in Figure 7.8c makes the connection between light absorbed and the photosynthetic process itself. As we saw in the general formula, each molecule of glucose produced in photosynthesis is accompanied by the liberation of six oxygen molecules. This makes oxygen a convenient experimental "handle" on the process; that is, we can determine the rate of photosynthesis by collecting and measuring the oxygen being given off. In developing the action spectrum, oxygen output was measured a number of times, each trial making use of a different color of light. When these data are graphed, we see that the greatest output of oxygen occurs when violet and red lighting is used. These are also the colors that show the strongest absorption in the absorption spectrum. Thus these data support the conclusion that light absorbed represents light used.

There are hundreds of chlorophyll and carotenoid molecules in each light-harvesting complex. They are bound into a special pattern by a framework of protein (Figure 7.9a). Most of them channel energy to the reaction center.

The Reaction Center. Each reaction center consists of one molecule of chlorophyll *a* and an associated protein. The protein is closely linked to the nearby electron transport system. The chlorophylls and carotenoids absorb light and shunt the energy into the reaction centers. The reaction centers thus act as "energy sinks" (regions where energy concentrates). Although the reaction centers of PS I and PS II each contain a molecule of chlorophyll *a,* they absorb light most strongly at wavelengths 700 nm and 680 nm, respectively. For this reason they are often referred to as **P700** and **P680**.

It is in the reaction center that light energy is transformed into chemical energy. Many of these reactions are not completely understood, but we're beginning to grasp some of the goings-on. When chlorophyll *a* of the reaction center absorbs sufficient energy from the surrounding light-harvesting complex, an electron from this molecule becomes excited and escapes from its orbital (Figure 7.9b). This is where the transformation of sunlight energy to chemical energy begins.

In the everyday world of the molecule, the ejection of a light-energized electron is not an unusual event (Albert Einstein won a Nobel Prize for the discovery in 1920); it happens routinely as the free energy of a molecule changes. Usually, though, an ejected electron simply gives off its absorbed energy as light and heat and falls back to its lower energy "ground state" (Figure 7.10). When an excited electron of chlorophyll *a* leaves its orbit, however, it does not at once release its newfound energy, nor does it fall back to its ground state—not yet. *Instead, the activated electron is immediately captured by an electron acceptor in the closely associated electron transport system,* where its newly gained free energy can be put to use doing chemical work. Here is the crucial event that provides the energy link between the physical and biological world. This is the specialness of photosynthesis.

Actually, one photoactivated electron cannot accomplish much photosynthetic work. But considering the lightning speed of such an event, the many photosystems at work in each thylakoid, the great number of thylakoids in a chloroplast, and the number of chloroplasts in a leaf cell, we begin to develop a great respect for what's going on in those unapplauded green leaves.

Electron Transport System. Electron transport systems consist of a number of electron carriers precisely arranged in the thylakoid membrane (see Figure 7.7). These assemblages have two roles. First, they use the energy of light-activated electrons to concentrate protons in the thylakoid lumen, where their free energy can be used to generate ATP. Second, the electrons (along with protons from the surroundings) are used to reduce $NADP^+$ to $NADPH + H^+$. Both ATP and NADPH, you will recall, are essential to the light-independent reactions.

Briefly, the electron carriers in the thylakoid electron transport system, in the order of their actions, are:

1. **Pheophytin.** Pheophytin is a modified form of chlorophyll *a* that forms a link between the reaction center and **plastiquinone (PQ)**. PQ is not a protein at all, but is a small, very mobile lipid.

FIGURE 7.9
Model of the Light-Harvesting Complex
(a) Each light-harvesting complex contains a number of molecules of chlorophyll *a,* chlorophyll *b,* and carotenoids (mainly beta-carotene and xanthophyll). At its base is the reaction center, which contains one molecule of chlorophyll *a* and an associated protein. (b) Light energy absorbed by molecules of the complex is shunted into the reaction center, an energy sink. There it is absorbed by chlorophyll *a,* from which a photoactivated electron is ejected. The first electron acceptor is known as pheophytin.

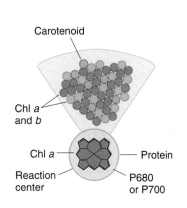

(a) Light-harvesting complex

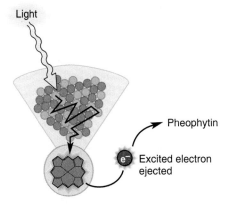

(b) Light energy transferred to reaction center

(a) Usual situation

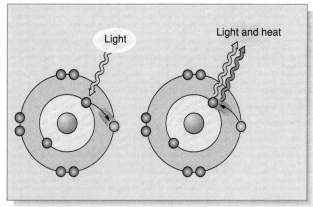

(b) In the chloroplast

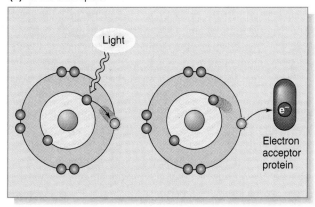

FIGURE 7.10
Energy and Electrons
Under nonphotosynthetic conditions, **(a)** an electron absorbing excess energy will be ejected from its orbit, whereupon it will release the excess energy as heat and/or light and return to its orbit. Luminescence would be an example. **(b)** Within a photosystem, however, a photoactivated electron ejected from a reaction center will have a different fate. It will be captured by a nearby element of the electron transport system and begin its productive journey through the light reactions.

2. **Cytochromes (Cyt).** Electron carriers known as **cytochromes** are proteins with iron-containing **heme** groups (also present in hemoglobin). It is the iron within the heme groups that accepts and transfers electrons.

3. **Plastocyanin (PC).** PC is a small, highly mobile, protein carrier whose active, electron-accepting group is copper.

4. **Ferredoxin (Fd).** Iron is also the active group in an iron–sulfur (Fe–S) protein carrier called ferredoxin.

5. **NADP reductase (FAD).** The last carrier in the chain is NADP reductase, which is from a molecular group known as flavoproteins. It contains bound coenzyme FAD, which contains the active, electron-accepting "flavin" group.

As we will see in the next chapter, these electron carriers are quite similar to those in the mitochondrion.

The electron carriers are arranged so that each one along the route has more reducing power than the next one in line. Thus as electrons are received by one carrier, they are immediately passed to the next—at an energetically lower level—in an ongoing sequence of reductions and oxidations.

ATP Synthases. In addition to the photosystems, the thylakoid membrane contains a large number of **ATP synthases** (see Figure 7.6 and look ahead to Figure 7.14). Recall that these protein assemblies play a key role in producing ATP. The base of each ATP synthase anchors the structure into the thylakoid membrane. It also forms a channel, originating in the lumen and leading across the thylakoid membrane into the enlarged head of the synthase. As we will see shortly, its arrangement permits protons to move down their concentration gradient, through the ATP synthase, and into the stroma. The head contains the important enzyme **ATP synthase**, which utilizes the free energy of the escaping protons to bond phosphates to ADPs, forming ATP. Now let us see how the various elements of the photosystems—the light-harvesting complexes, the electron transport systems, and the ATP synthases—cooperate to capture light energy and convert it into chemical energy.

The Light-Dependent Reactions of Photosynthesis

Light falling on a green leaf passes into its photosynthetic cells and enters thousands upon thousands of green chloroplasts. Within the numerous thylakoids, countless light-harvesting complexes and reaction centers respond. As we take a close look at just one set of photosystems, the events of the light reactions begin to unfold.

There are two ways in which photosystems I and II can act: (1) they can work cooperatively to build the chemiosmotic gradient and reduce $NADP^+$ to NADPH, an effort known as **noncyclic photophosphorylation**; or (2) photosystem I can act alone, simply building the chemiosmotic gradient for ATP synthesis in what is called **cyclic photophosphorylation**. We will refer to these simply as the noncyclic and cyclic events. In both, light-activated electrons, ejected from the reaction centers, pass through electron transport systems, where their free energy is used to accomplish work.

The term *photophosphorylation* was developed before the chemiosmotic hypothesis appeared, when ATP synthesis was believed to be more directly connected to the absorption of light energy. Biologists are now convinced that the generation of ATP in the chloroplast occurs through *chemiosmotic phosphorylation,* a separate event.

The Noncyclic Events

The noncyclic events begin as light is absorbed by the light-harvesting complex of photosystem II. A look at Figure 7.11a will help as an overview of the entire process. (Why do we start with PS II? It's a quirk of discovery: the photosystems were named before the order of their activity was fully understood. We're stuck with it.)

Photosystem II. As light energy is shunted to chlorophyll *a* of the P680 reaction center, an electron is ejected from its orbit. It quickly passes to nearby electron acceptor phenophytin, the first member of the associated electron transport system (Figure 7.11b). We will come back to the electron in a moment, after tending to a problem in P680.

Water and the Restoration (Reduction) of P680. The loss of an electron from chlorophyll *a* leaves the molecule in an oxidized ("electron hungry") state. Its lost electron will be replaced, preparing it to react again. The actual chemistry is poorly understood, but we know that the newly oxidized chlorophyll *a* is reduced by electrons from a nearby manganese-containing protein, referred to only as **protein Z**. Protein Z regains its own lost electron by oxidizing a water molecule (see Figure 7.11b).

Although we are considering one electron at a time, in the complete disruption of each water molecule (as seen below) *two* electrons are released, as are two protons and one atom of oxygen:

$$H_2O \longrightarrow 2\,e^- + 2\,H^+ + 1\;oxygen$$

The oxygen atom (designated as $\frac{1}{2}O_2$) joins another from a similar disruption of water, and oxygen gas (O_2) is given off (which contributes to the oxygen we breathe). Importantly, *for each electron* taken from water, *a proton is released* directly into the thylakoid lumen, where it helps build the proton gradient.

Back to the P680 Electron. The electron ejected from P680 and accepted by pheophytin is immediately transferred to a small, mobile carrier called **plastoquinone (PQ)**. As a lipid, PQ moves quite freely within the phospholipid bilayer of the thylakoid membrane. PQ moves to the large cytochrome complex, which it reduces, then cycles back to P680 to receive another electron (Figure 7.11c). The free energy of the electron is used by the cytochrome complex to pump a proton from the stroma to the lumen.

The P680 electron, now at a considerably lower free energy state, will move on to photosystem I. This transfer is made by another mobile carrier, **plastocyanin (PC)**, a peripheral membrane protein that moves freely in the lumen (Figure 7.11d).

Photosystem I: The Reduction of NADP⁺. P680 electrons can only enter PS I when it is ready to receive them, which means that chlorophyll *a* in the P700 reaction center must be oxidized. This occurs as light is absorbed and an excited P700 electron is captured by the first acceptor in the photosystem I electron transport system. The P680 electron now reduces the P700 reaction center, restoring it for the next light-activated event.

In the *noncyclic* reactions, the energy of light-activated P700 electrons is used for one purpose only—to reduce NADP⁺ to NADPH. There is no proton pumping at this time. Thus electrons from P700 pass directly through the electron transport system to NADP⁺ (Figure 7.11e). The complete reduction of NADP⁺ to NADPH actually requires two electrons from the reaction centers, along with two protons from the watery stroma. The formation of NADPH is the final step in the noncyclic events.

In summary, we have seen that for each electron ejected from P680, the chemiosmotic gradient is increased by two protons—one pumped in by the photosystem II electron transport system and one released during the disruption of water. We have also seen how electrons from P700 of photosystem I are used to reduce NADP⁺ to NADPH. The tit-for-tat accounting of electrons and protons is important, but, in reality, the noncyclic events can be viewed as a continuous process, with a steady stream of electrons flowing from water to a long line of NADP⁺s, and a similar stream of protons entering the thylakoid lumen.

Traditionally, the shifting energy levels of electrons and carriers and the current-like flow through the photosystems have been shown somewhat differently, in what is called the **Z-scheme** (Figure 7.12). It may help to look this over and gain a somewhat different perspective on the process. Note that the vertical scale suggests the reducing power of the participants. And, by the way, don't confuse the name Z-scheme with the Z protein that reduces P680.

Finally, we will note that the noncyclic events of the light reactions must occur many times to provide enough NADPH and ATP for the synthesis of one glucose molecule in the light-independent reactions.

Summing Up the Noncyclic Events. Now let's briefly recap what goes on in the noncyclic light reactions. In the order discussed, the events are as follows:

1. Absorption of photons by P680 (PS II).
2. Movement of excited P680 electrons into the electron transport system.
3. The reduction of P680, through the oxidation of water, adding protons to the lumen and releasing oxygen gas.
4. Use of the free energy of P680 electrons to power proton transport from the stroma to the lumen.
5. Absorption of photons by P700 (PS I).
6. Replacement of P700 electrons by those of P680.
7. Movement of excited P700 electrons into the electron transport system.
8. Use of P700 electrons and free protons from the stroma to reduce NADP⁺ to NADPH.

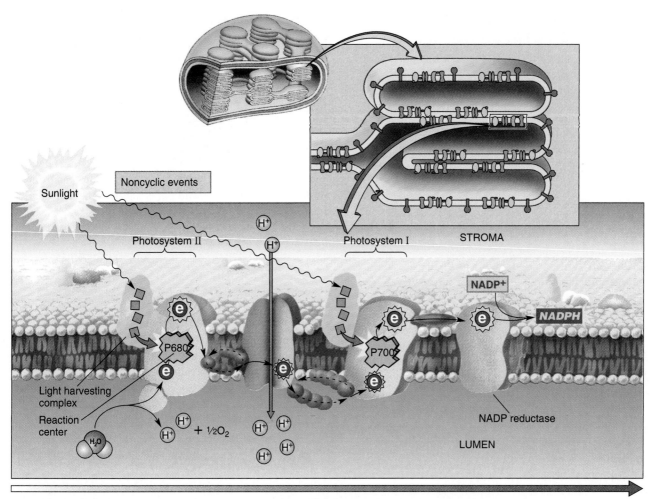

(a) Scenario of noncyclic reactions. Light-activated electrons from P680 provide the energy to pump protons into the lumen before passing to P700. Electrons ejected from P700 pass to the end of the system, where they reduce NADP$^+$ to NADPH. During the process, electrons from water continually replace those lost from P680. The steps and details follow in **(b)–(e).**

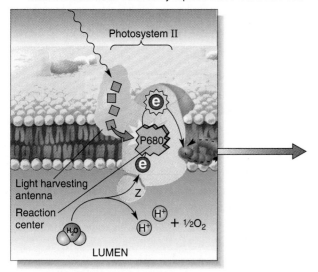

(b) Light activated electron is ejected from P680. The electron is replaced by one from Z-protein which receives its electrons from water. Protons (H+) from water are released into the lumen.

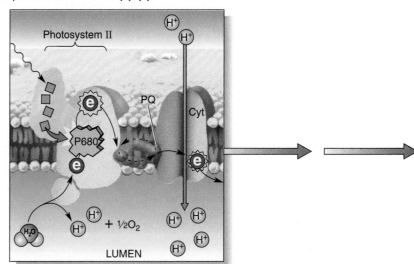

(c) Excited electron passes to PQ, which ferries it to the cyt complex. The cyt complex makes use of the electron's free energy to pump a proton (H+) into the lumen.

The most important point to keep in mind is that the light reactions provide ATP and NADPH for the light-independent reactions.

The Cyclic Events

As we indicated earlier, the photosystems act in a cooperative manner during the noncyclic events. But in the cyclic events, *photosystem I acts on its own.* The term *cyclic* refers to the circular pathway of P700 electrons. When activated by light, each electron leaves P700, passes through the electron transport system, and returns directly to P700. Along the way, the electron's energy is used to pump one proton into the lumen. At present, we can only make informed guesses about the exact path of cycling electrons, so the membrane scenario in Figure 7.13 is only a rough idea.

As you can see, there is no NADPH produced during the cyclic events. Instead, the energy of P700 electrons is used *strictly for pumping protons* from the stroma to the lumen. Thus the cyclic events help only in building the proton gradient.

No one is quite sure how the cyclic events came to be, but there are some prevailing ideas. One of these is that the cyclic events occur when $NADP^+$ is in short supply. In other words,

photosystem II goes into "idle," and photosystem I switches to proton pumping, continuing until the light-independent reactions can cycle oxidized NADP ($NADP^+$) back to the light reactions. This effort is certainly not wasted, since any increase in the chemiosmotic gradient means a potential gain in ATP.

Chemiosmotic Phosphorylation

Now that we have seen how the chemiosmotic gradient is built up during the noncyclic and cyclic events, let's turn to chemiosmotic phosphorylation and the production of ATP.

As positively charged protons accumulate in the lumen, their concentration there builds a steep electrochemical gradient. As mentioned in Chapter 6, the gradient is named for its electrical and chemical characteristics. Its electrical characteristic results from an accumulation of positive charges (protons, H^+) in the lumen and negative charges (hydroxide ions, OH^-) left behind in the stroma. Chemically, it is a pH gradient with an acidic environment within the lumen (about pH 4 to pH 5) and an alkaline environment in the stroma (about pH 8). And lest we forget, it is also a diffusion gradient.

Whatever way we view it, such a system has great free energy that can be used to accomplish useful chemical work.

FIGURE 7.11
The Noncyclic Light Reactions
(a) In the noncyclic light reactions, sunlight energy produces an electron flow from photosystem II, through photosystem I, and on to $NADP^+$, which is reduced to NADPH + H^+. In each event a proton is pumped across the membrane and water is broken down, liberating still more protons into the lumen. The purpose of the cyclic reactions is thus to generate NADPH and to increase the proton gradient in the lumen. The key events are described stepwise in (b) through (e).

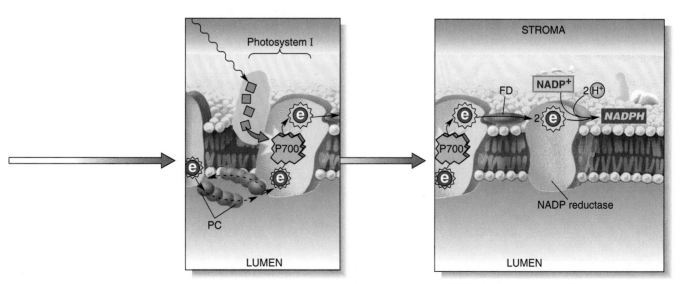

(d) Cyt complex passes the electron to a surface carrier PC, which transfers it to PSI. It will replace a light-activated electron that has been ejected from P700.

(e) P700 electron passes to FD and on to the NADP-reductase complex, where two electrons from P700 and two protons from the stroma reduce NADP to NADPH.

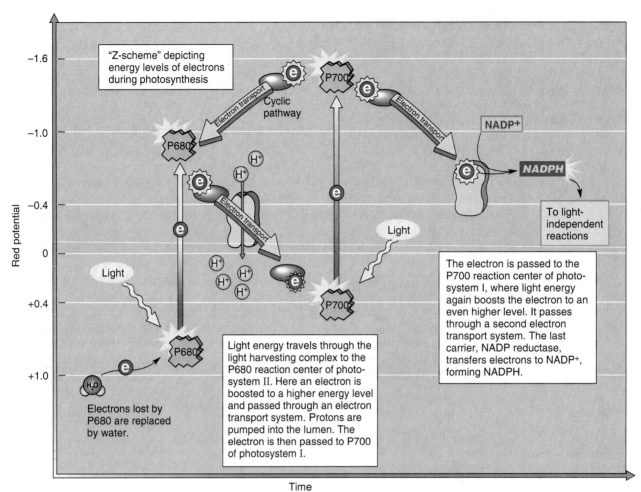

FIGURE 7.12
The Z-Scheme
The traditional Z-scheme is arranged to show the reduction potential, or reducing power, of the major participants in the light reactions. It is not meant to convey position in the thylakoid membrane. The vertical axis
indicates the reduction potential, which increases with negative charge. Light activates P680, raising its
reducing power substantially. It sends electrons down the electron transport system, where their energy is
used to pump protons into the lumen. A second light event raises the P700 reduction potential to the greatest
level in the system. Electrons emitted from P700 pass down their electron transport system to reduce $NADP^+$
to NADPH. A close look at the scheme will reveal an arrow extending from P700 back toward P680. This is
the pathway of electrons from P700 in the cyclic light reactions, which comes up next.

For this to happen, the protons must be permitted to escape
down their gradient to a lower free energy level. The chief
avenue of escape is through channels leading into the ATP
synthases, where the ATP-synthesizing enzymes are located.
As they pass through the ATP synthases, their free energy is
used to join ADP and P_i, forming ATP (Figure 7.14). The consensus is that two protons passing through an ATP synthase
provide enough free energy to account for one ATP.

It takes considerable time to relate these events, a fact
that might conjure up the wrong image of the light reactions
as plodding, methodical processes. In reality, they occur with
dazzling speed and in countless repetitions in innumerable
chloroplasts. Once again, the flow of electrons, driven by
beaming photons, can best be portrayed as a continuous

stream, like an electrical current, from water to $NADP^+$. In
conjunction, a steady flow of protons enters the thylakoid
lumen, building the electrochemical gradient there. Finally,
protons escape two-by-two, through the ATP synthases, generating a steady output of ATP.

The Light-Independent Reactions

Throughout the light reactions, the chloroplast builds up a
store of potential energy and reducing power in the form of
ATP and NADPH. Thus, with an input of carbon dioxide, the
events that lead to glucose production can begin. Glucose is

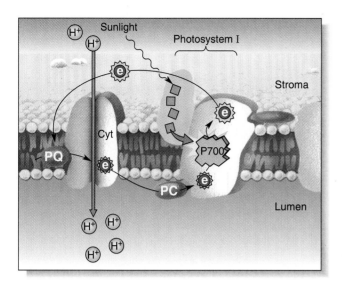

FIGURE 7.13
The Cyclic Light Reactions
The actual events aren't well understood, but in this tentative view of the cyclic reactions, an excited P700 electron is ejected from its reaction center. From carrier PQ, it reaches the large cytochrome complex—the proton pump—which responds by transporting a proton to the lumen. The electron then joins PC and cycles back to the P700 reaction center.

actually synthesized outside the chloroplast in the cytoplasmic fluids, but the reactions that provide the raw materials occur in the stroma of the chloroplast. The essentials of the light-independent reactions, ignoring this geographical technicality, can be summarized as follows:

$$6\ CO_2 + 12\ NADPH + 12\ H^+ + 18\ ATP \xrightarrow{\text{enzymes}}$$
$$Glucose + 12\ NADP^+ + 18\ ADP + 18\ P_i + 6\ H_2O$$

(The reactants and products have now been numerically balanced.)

The process seems simple enough, just a matter of adding hydrogen to carbon dioxide until the 6-carbon sugar forms. Not quite! Carbon dioxide is at a far lower free energy level than glucose and therefore would be quite reluctant to form glucose spontaneously. Only the free energy of ATP and NADPH from the light reactions and a battery of synthesizing enzymes make such an event possible. All this happens in a circular pathway called the **Calvin cycle**.

The cycle begins as CO_2 molecules enter the stroma, where, with the help of a key enzyme, each is joined to a 5-carbon "resident molecule." This crucial step begins the synthesis of glucose. Let's look at the details.

The Calvin Cycle

The Calvin cycle is named in honor of Melvin Calvin, who was one of its principal discoverers. He described the nature

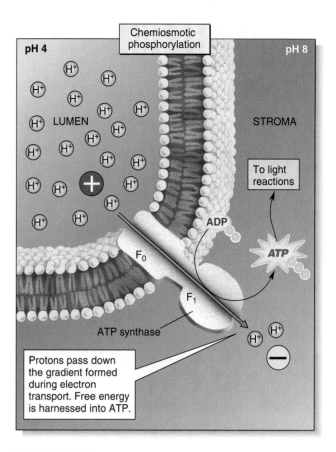

FIGURE 7.14
Chemiosmotic Phosphorylation
Protons, concentrated in the acidic lumen, pass down their gradient through the F_0 channels into the F_1 portion of an ATP synthase. There the free energy of the proton gradient is used in generating ATP from ADP and P_1. Note that the charged thylakoid has both electrical (+ and −) and chemical (pH) characteristics.

of carbon dioxide incorporation in the late 1940s, an accomplishment that earned him a Nobel Prize (see Essay 7.1).

The Calvin cycle is a biochemical pathway in which each enzyme in the team takes its turn at altering the substrate. With each reaction, the substrate takes on a new form and identity until the work is complete. The cycle has five key events, which we label "carboxylation," "phosphorylation," "reduction," "yield," and "regeneration," to help you keep track of the important events. The discussion to follow is illustrated in Figure 7.15.

Carboxylation. Carboxylation is the addition of carbon dioxide (also called **carbon fixation**). One molecule of carbon dioxide reacts in each complete turn of the Calvin cycle. The enzyme responsible for carboxylation is called **ribulose-1,5-bisphosphate carboxylase**, named for its job: adding carbon dioxide to the 5-carbon molecule, **ribulose-1,5-bisphosphate (RuBP)**. The product is an unstable 6-carbon intermediate that spontaneously cleaves into two 3-carbon molecules known as **3-phosphoglycerate (3PG)**.

We won't be discussing each individual enzyme, but the carbon dioxide adding enzyme "rubisco" as biochemists like to call it, is one of the most significant enzymes on earth (and perhaps the most abundant protein). For most living things, it represents the chemical link between the physical and biological worlds. This may sound bold, but the vast majority of the earth's organisms rely, directly or indirectly, on carbon compounds first formed by plants and algae in the Calvin cycle. And rubisco does the first step.

Phosphorylation. The second key event is phosphorylation, the transfer of a high-energy phosphate group from ATP to the two 3PGs from the first reaction. Recall that phosphorylation (see Chapter 6) is a common preliminary reaction, one that increases the free energy of the substrate, thus setting it up for further reactions. The new products are two 3-carbon molecules of **bisphosphoglycerate (BPG)**.

Reduction. The next key reaction involves reduction, in this instance, the addition of both electrons and protons. Two NADPHs reduce the two BPGs, forming two 3-carbon molecules of **glyceraldehyde-3-phosphate (G3P)**. In the reaction, one of the phosphates in each molecule is replaced by hydrogen. The newly freed phosphates (P_i), the two $NADP^+$s, and the two ADPs from the previous reaction are recycled back to the light reactions.

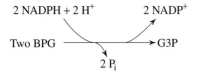

Regeneration or Yield. At this point a major division in the pathway occurs. G3P, the last product, can proceed in either of two directions. The **yield pathway** leads out of the chloroplast to glucose production, whereas the other, the **regeneration pathway**, makes the continuation of the Calvin cycle possible. We will see how both pathways can occur in a moment, but first let's summarize the light-independent reactions so far:

1. *Carboxylation* occurs when carbon dioxide is added into the Calvin cycle by the enzyme rubisco, forming 3PG.

2. *Phosphorylation* by ATP forms BPG, setting the substrate up for reduction by NADPH.

3. *Reduction* by NADPH yields G3P, which can be drawn away to form glucose or recycled to keep the Calvin cycle going.

4. *Recycling* occurs when ADP, P_1, and $NADP^+$ are recycled to the light reactions.

We'll try once more to incorporate all the reactants and products into a summarizing equation:

(a) Calvin cycle:
$$6\ CO_2 + 6\ RuBP + 12\ NADPH + 12\ H^+ + 18\ ATP \longrightarrow$$
$$12\ G3P + 12\ NADP^+ + 18\ ADP + 18\ P_i + 6\ H_2O$$
(b) Yield:
$$2\ G3P \longrightarrow C_6H_{12}O_6 + 2\ P_i \text{ (in the cytoplasm outside the chloroplast)}$$
(c) Regeneration:
$$10\ G3P + 6\ ATP \longrightarrow 6\ RuBP + 6\ ADP$$

The Yield Pathway: Glucose and Other Products. Two G3Ps are used in the synthesis of each glucose. In the yield pathway, two G3Ps leave the chloroplast and enter the cytoplasm. There they are first assembled into **fructose-1,6-bisphosphate (F-1,6-BP)**, which is then converted to **glucose-1-phosphate (G1P)**. Next, the phosphate is cleaved away (and recycled), and free glucose is formed. Alternatively, glucose-1-phosphate units can enter dehydration reactions and be converted to starches (see Chapter 3). With additional changes, glucose can be linked into polymers of cellulose. G3P can follow still other pathways, where it is modified and used in the synthesis of fatty acids, glycerol, or amino acids. It can be modified further and sent into the mitochondrion, where its free energy will be used to produce more ATP. The various pathways available for G3P are a portion of an even greater metabolic network included in what is called **intermediary metabolism**. We will look further into this in the next chapter.

Why a Regeneration Pathway? It may have occurred to you that drawing the two G3Ps from the cycle to form glucose would, for all practical purposes, end the cycle. This doesn't happen because glucose is only formed when the Calvin cycle reactions have occurred six times. This is because in five out of six turns of the cycle, the ten G3Ps formed simply enter the regeneration pathway, where they are used *to keep the cycle going*. Only in the sixth turn can two G3Ps be drawn off to form glucose. Figure 7.16 reviews the cycle once again, this time focusing on the number of carbons. (Actually, there are so many Calvin cycles going on simultaneously in the chloroplast that things aren't quite this neat. However, the net result is as we have described it.)

Photorespiration: Trouble in the Calvin Cycle

If everything went along in the Calvin cycle as we have described it, the chemical efficiency of photosynthesis (cap-

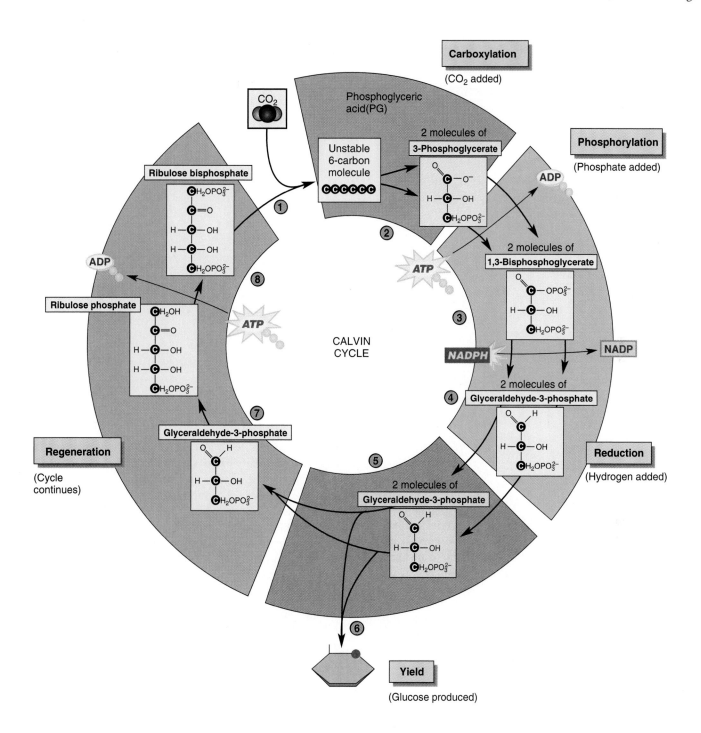

FIGURE 7.15
The Calvin Cycle
The cycle begins ① when carbon dioxide enters and is combined with the 5-carbon resident, ribulose-1,5-bisphosphate, forming an unstable 6-carbon molecule which simultaneously splits, ② forming two 3-phosphoglycerates (3PG). The two are phosphorylated by ATP, ③ forming two bisphosphoglycerates (BPG). These are each reduced ④ by NADPH, forming two glyceraldehyde-3-phosphates (G3P) ⑤. This last product is at a crossroads, some leaving the chloroplast for the cytoplasm ⑥ to form glucose and some ⑦ continuing in the Calvin cycle, to be phosphorylated ⑧, which restores the original ribulose-1,5-bisphosphate, the 5-carbon resident. As we will see, it requires six turns of the cycle to yield enough G3P to form one glucose molecule.

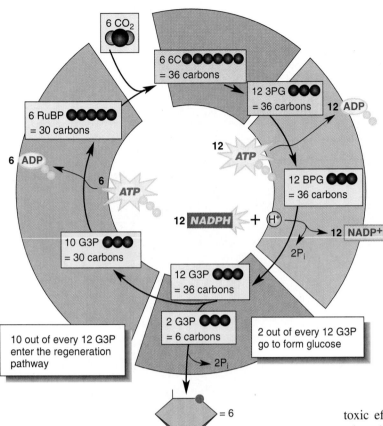

FIGURE 7.16
Balancing the Calvin Cycle
In this trip through the Calvin cycle, the cycle is balanced so that a yield of one glucose molecule and continuation of the cycle are both possible. The numbers represent six complete turns of the cycle at one time. (The RuP step is omitted.)

tured light energy versus free energy in glucose) would be a respectable 38%. But in the real world things do not always go this way. It turns out that the ability of rubisco to bind with carbon dioxide depends to a large extent on the amount of carbon dioxide present. When carbon dioxide concentrations are low, as would be the case where it is used faster than it can be replaced from the atmosphere, the enzyme begins to bind with oxygen instead of carbon dioxide, and a process known as **photorespiration** begins. [Atmospheric oxygen is in a far greater concentration (20%) than carbon dioxide (0.04%).] But why oxygen?

Rubisco, it turns out, works with either gas, depending on concentrations. But what is photorespiration? Its name comes from the fact that the process is light activated, uses oxygen, and releases carbon dioxide: except for the light activation, this is similar to cell respiration.

In photorespiration, oxygen joins ribulose-1,5-bisphosphate and the product goes through several pathways involving peroxisomes and mitochondria. Along the way, carbon dioxide is lost and ATP is consumed, so in terms of photosynthetic yield, photorespiration is considered a waste of resources. Biochemists estimate that photorespiration can waste almost 50% of the carbon taken in at the Calvin cycle. Some biologists take a different view, pointing out that by tying oxygen up in harmless compounds, photorespiration protects plants from the

toxic effects of its high concentrations. Others suggest that where low concentrations of carbon dioxide exist, the use of NADPH slows. Then, as it becomes concentrated, it enters into destructive reactions. Photorespiration may be one way of safely oxidizing NADPH. Both ideas will need more testing, but meanwhile photorespiration remains a scientific enigma.

Interestingly, natural selection has been at work over the ages, and, consequently, some plants have a way of avoiding the problem altogether.

C4 Carbon Fixation: Solving the Photorespiration Problem

Photorespiration is more likely to be a problem where light is intense, and CO_2 fixation is accelerated. Not surprisingly, it is a major problem in tropical and subtropical regions. In response, many tropical and subtropical plants have evolved pathways that help assure that carbon dioxide remains in high enough concentrations so as to avoid photorespiration altogether. Plants such as maize, sugar cane, sorghum, and Bermuda grass (Figure 7.17) are examples. They are known as **C4 plants**. All others are called **C3 plants**. The C4 and C3 designations refer to 4-carbon or 3-carbon compounds formed when carbon dioxide is first taken in. Accordingly, the light-independent reactions discussed earlier pertain only to C3 plants.

There are distinct anatomical and functional differences in the two kinds of plants (Figure 7.18). Where C3 plants carry on the light reactions and the Calvin cycle throughout the **leaf mesophyll** (most of the internal cells of the leaf), C4

Sorghum

Sugarcane

Corn

FIGURE 7.17
C4 Plants
C4 plants include over 100 genera distributed in at least ten families, and the list is growing. C4 plants have made structural and biochemical adaptations to the intense sunlight of the tropics and subtropics.

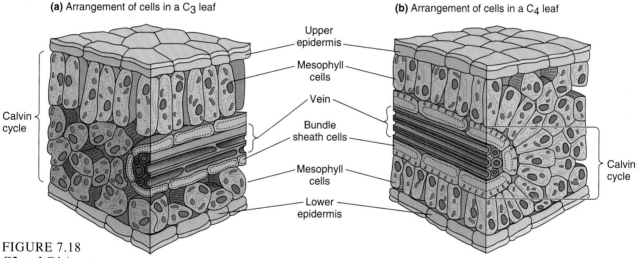

(a) Arrangement of cells in a C₃ leaf   **(b)** Arrangement of cells in a C₄ leaf

Upper epidermis
Mesophyll cells
Vein
Bundle sheath cells
Mesophyll cells
Lower epidermis
Calvin cycle
Calvin cycle

FIGURE 7.18
C3 and C4 Anatomy
Both illustrations show the arrangement of the leaf mesophyll, the bundle sheath cells, and a leaf vein. The organization of these tissues in the C3 and C4 plants are obviously very different. **(a)** Note the layered arrangement of mesophyll in the C3 plant, and the far less obvious bundle sheath cells. The light reactions *and* the Calvin cycle of the C3 plant occur throughout the mesophyll (the neatly stacked and loosely arranged photosynthetic cells). The bundle sheath cells surrounding the veins are nonphotosynthetic. **(b)** In contrast, the bundle sheath cells are much more prominent in the C4 plant, and they are surrounded by a dense layer of mesophyll. *Only* the light reactions occur in the mesophyll of the C4 plants. The main light-independent role of the mesophyll is collecting CO_2 and concentrating it in the bundle sheath cells, where the Calvin cycle occurs. **(c)** Compare the chloroplasts in the two types of cells. Note the absence of grana in the bundle sheath chloroplasts, whose main task is the light-independent reactions.

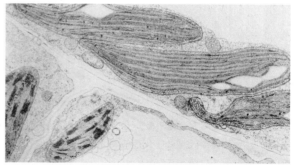

Mesophyll chloroplasts   Bundle sheath chloroplasts

Essay 7.1 ● FROM WILLOWS, MICE, AND CANDLES TO ELECTRON MICROSCOPES

It's the Water!

Prior to the time of Jan Baptiste van Helmont, a 17th century Belgian physician, it was commonly accepted that plants derived their matter from materials in the soil. (Probably many people who haven't studied photosynthesis would go along with this today!) We aren't sure why, but van Helmont decided to test the idea. He carefully stripped a young willow sapling of all surrounding soil, weighed it, and planted it in a tub of soil that had also been carefully weighed. After five years of diligent watering (with rainwater), van Helmont removed the greatly enlarged willow and again stripped away the soil and weighed it. The young tree had gained 164 pounds. Upon weighing the soil, van Helmont was amazed to learn that it had lost only 2 ounces. van Helmont's conclusion was inescapable: "It's the water!" he exclaimed, but van Helmont was only half right. (Carbon dioxide hadn't been discovered yet.)

Oxygen from Plants

The next scientific breakthrough occurred late in the 18th century, around 1771, when British clergyman Joseph Priestley made an astute observation. In a simple but classic series of experiments, Priestley learned that plants growing in a sealed container would support the burning of a candle and breathing of a mouse longer than either would go on without the plant's presence. What Priestley had found, in addition to the reciprocity of gases in plants and animals, was that plants produce oxygen.

Light Is Also Essential

Priestley's discovery was followed almost immediately by the ambitious efforts of a Dutch engineer named Jan Ingenhousz.

Ingenhousz carried out some 500 experiments, which supported the conclusion that light was essential for photosynthesis. Ingenhousz determined that plants give off oxygen only in the light and that, under darkened conditions, they even produced some carbon dioxide. By the mid-1800s, the basic equation for photosynthesis was well established.

Photosynthesis Occurs in Chloroplasts

Toward the end of the 1800s, Theodore Engleman determined that the chloroplasts were the sites of the generation of oxygen, known to be a part of photosynthesis. This represented the first time anyone had associated a cellular process with a specific organelle. In an experiment elegant in its simplicity, Engleman made use of a strain of bacteria that were attracted to oxygen and a filamentous green alga, *Spirogyra*. The alga has a large, ribbon-like chloroplast that spirals the length of the cell. When shining a fine point of light on different parts of the chloroplast, Engleman was able to observe the oxygen-seeking bacteria congregating at those locations.

Oxygen Generated in the Light Reactions

Many of the mysteries in the biochemistry of photosynthesis were resolved in the 1930s, particularly through the efforts of C. B. van Niel. In van Niel's time, it was held that, during photosynthesis, light energy broke down carbon dioxide, with oxygen escaping as a gas and carbon joining water to form CH_2O—the empirical carbohydrate. Van Niel knew from his studies of sulfur bacteria that photosynthesis could occur without the production of oxygen. In sulfur bacteria, the source of hydrogen is hydrogen sulfide, and, when oxidized, the by-product is elemental

sulfur rather than oxygen. This fact suggested to van Niel that the source of oxygen released during plant photosynthesis was water rather than carbon dioxide. He proposed a general equation for all photosynthesis:

$$CO_2 + H_2A \longrightarrow (CH_2O) + H_2O + 2\,A$$

In this equation, the H_2A represents the source of electrons, which could be either hydrogen sulfide, water, or some other similar substance, but not carbon dioxide.

While van Niel's arguments were certainly persuasive and logical, persuasion and logic are not proofs. There was at the time no way of knowing whether bacterial photosynthesis was similar enough to photosynthesis in algae and plants for van Niel to be certain of his hypothesis. In fact, the final confirmation had to await the development of new research techniques, which in this case turned out to be the use of radioisotopes of oxygen and carbon and ways of detecting their presence.

Radioisotopes on the Trail of Hydrogen.

In the early 1940s, Samuel Ruben and Martin Kamen made use of a heavier isotope of oxygen, ^{18}O, whose presence in molecules produced during photosynthesis could be detected by a device called a mass spectrophotometer. In one experiment, the two researchers raised algae in ordinary carbon dioxide but used water containing the heavy isotope of oxygen ($H_2\,^{18}O$). They found that the oxygen gas given off contained the heavy isotope of oxygen, whereas the carbohydrate formed did not. In a second experiment, ordinary water was used, but the carbon dioxide gas contained the heavy isotope of oxygen ($C\,^{18}O_2$). In this case, the oxygen gas given off was free of the heavy isotope of oxygen, but it was subsequently found to be present in the carbohydrate product of photosynthesis. Van Niel's hypothesis was strongly supported.

The Calvin Cycle.

The carbon-fixing pathway was discovered in the 1940s by Melvin Calvin and his associates. They made use of radioactive isotopes of elements involved in photosynthesis. Radioactive carbon ($^{14}CO_2$) was administered to green algae, which were then exposed to bright light for a few seconds. The algae were quickly killed (to stop any biochemical reactions), and the cells were disrupted and searched for new, radioactive, molecular intermediates. Calvin found the radioactive carbon incorporated into 3-phosphoglycerate (3PG), which turned out to be the first stable intermediate in the pathway. Continued research eventually determined the other intermediates of the carbon-fixing pathway, including the key carboxylating enzyme rubisco. Calvin and his associates went on to propose a cyclic pathway leading to the formation of glucose, soon dubbed the "Calvin cycle."

The Two Parts of Photosynthesis.

In the late 1930s, British biochemist Robert Hill provided strong experimental evidence that also supported van Niel's theorizing. Hill was among the first to isolate successfully and experiment with chloroplasts. From his experiments, it was determined that oxygen production in chloroplasts could occur independently of carbon dioxide fixation. In fact, it occurred in the absence of carbon dioxide. All that was required was the presence of an electron acceptor. From these observations, it became clear that photosynthesis occurs in two separate series of reactions: the light reactions involving water and the light-independent reactions involving carbon dioxide fixation. From Hill's discoveries, the experimental focus turned to the role of light in creating a flow of electrons and the pathways taken by the electrons. By 1951, the role of $NADP^+$ (first called TPN^+) as an electron acceptor was known. In 1954, Daniel Arnon was able to demonstrate the entire photosynthetic process in isolated chloroplasts. His work was obviously a milestone in photosynthesis research.

The Details.

Arnon and his associates at the University of California at Berkeley, taking cues from Hill's brilliant work, used intact chloroplasts as their "experimental organism." Much of the essential material in this chapter has come down from Arnon's initial discoveries. For example, in an early experiment, he determined that if carbon dioxide was withheld, chloroplasts could carry out all the known photosynthetic reactions except carbohydrate synthesis. In other words, the process yielded only ATP, $NADPH + H^+$, and oxygen—but no glucose. From this observation, Arnon was able to propose that photosynthesis occurred in two distinct phases, the "light" and "dark" reactions.

Arnon then determined that if he withheld both carbon dioxide and NADP, providing his chloroplasts with ADP, P_i, water, and light only, then the only photosynthetic product would be ATP. There was, as expected, no glucose, no $NADPH + H^+$, and, perhaps surprisingly, *no oxygen*. The lack of oxygen was significant in that it indicated that a simpler cyclic process might be going on in the absence of NADP, one that could still yield ATP. We now call this independent process the cyclic light reactions.

In a final experiment, elegant in its conceptual simplicity (although technically very difficult), Arnon placed his chloroplast suspension in the dark for a time, washed away any late-forming products, and waited for chemical activity to stop. He then supplied the chloroplasts with CO_2 along with the products normally produced in the light-dependent reactions: ATP and $NADPH + H^+$. Once again, the amazing little organelles became active and went on to produce carbohydrate. As you can see, we owe much of what we understand about photosynthesis to the efforts of Robert Hill and Daniel Arnon and his associates.

Chemiosmosis.

Research in photosynthesis has been consistently intense over the years, but recently the interest has moved from the specific chemical reactions to the structures in which they occur. Motivating the switch has been the chemiosmotic theory of Peter Mitchell, the British biochemist. The chemiosmotic theory has become well entrenched in both photosynthetic and respiratory biochemistry today.

Recent studies of the thylakoid membrane, using freeze–fracture techniques, reveal details that help reinforce the chemiosmotic hypothesis. As Mitchell's hypothesis predicted, the ATP synthases are structurally independent of all other membrane proteins. Mitchell had predicted that phosphorylation was a separate process from the photoactivation of electrons and their subsequent transport through electron transport systems. Recent freeze–fracture studies have revealed an elaborate protein array in the thylakoid membranes, and biologists are certain that they have identified specific proteins and other elements of photosystem I and photosystem II.

plants use these cells for the light reactions and CO_2 capture only. They carry on the Calvin cycle reactions in their **bundle sheath cells** (cells that form a sheath around the leaf veins) (Figure 7.18b). While bundle sheath cells occur in both C3 and C4 plants, those in C4 plants have dense, relatively air-tight walls. The electron micrograph in Figure 7.18c shows ultrastructural differences as well. Note the absence of grana and the presence of starch grains in the C4 bundle sheath, both characteristics indicating that the chief role of these chloroplasts is carbohydrate synthesis.

Figure 7.19 traces the events of C4 photosynthesis, beginning with a brief review of the Calvin cycle in C3 plants (Figure 7.19a). There we see carbon dioxide joining rubisco, the carboxylating enzyme, and the Calvin cycle commencing. In the C4 plants (Figure 7.19b), carbon dioxide entering the C4 leaf mesophyll is met there by **PEP carboxylase** (phosphoenolpyruvate carboxylase), an enzyme that works far more efficiently than rubisco in low CO_2 concentrations. The enzyme links carbon dioxide to a 3-carbon resident molecule, **phosphoenolpyruvate** (**PEP**), forming a 4-carbon acid known as **oxaloacetate**. As the C4 pathway continues, a second enzyme converts the oxaloacetate to **malate**, another 4-carbon acid. Malate then passes into the bundle sheath cells, where a **decarboxylating enzyme** (one that removes CO_2) breaks it down into 3-carbon **pyruvate** and carbon dioxide. Pyruvate recycles back into the mesophyll cells, where it is converted to PEP for reuse. More importantly, CO_2 is delivered in quantity to rubisco, and the regular Calvin cycle begins. This completes the C4 pathway. In summary:

1. $\overset{\text{PEP carboxylase}}{3C\ PEP\ +\ CO_2 \longrightarrow 4C\ Oxaloacetate}$

2. $\overset{\text{enzyme}}{4C\ Oxaloacetate \longrightarrow 4C\ Malate}$

3. $\overset{\text{decarboxylase (enzyme)}}{4C\ Malate \longrightarrow 3C\ Pyruvate\ +\ CO_2}$

4. $3C\ Pyruvate\ +\ ATP \longrightarrow 3C\ PEP\ +\ ADP$

So, through the ongoing reactions of the C4 pathway—repeated many times and at great speed—the Calvin cycle enzyme, rubisco, is assured of an adequate concentration of carbon dioxide. In this way, C4 plants avoid the problem of photorespiration. What's the catch, you ask? Well, all this occurs at no small cost. The C4 pathway requires an additional investment of ATP, beyond the usual amount required in the Calvin cycle. But, apparently, the plants generate sufficient ATP under their brightly lit conditions to afford the splurge.

C4 Plants and Evolution

It would be tidy to say that C4 plants dominate the sunny tropics and deserts, and C3 plants are banished to those regions of

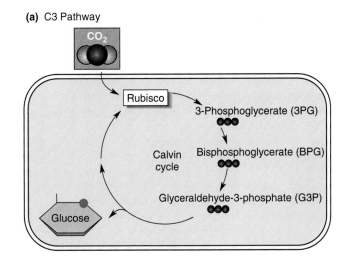

(a) C3 Pathway

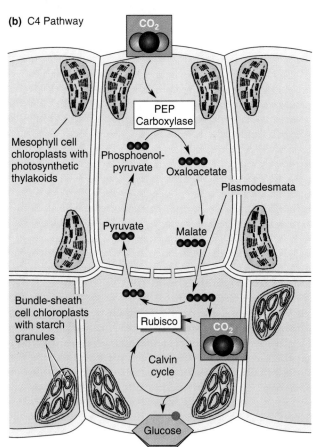

(b) C4 Pathway

FIGURE 7.19
Leaf Cells and the C4 Pathway
(a) In C3 plants, the uptake of CO_2 and its use in the Calvin cycle occur in chloroplasts of the leaf mesophyll cells. The C3 pathway is named for the three-carbon compounds that are formed. **(b)** In C4 plants, the mesophyll cells incorporate CO_2 into 4-carbon acids. The acids then enter the bundle sheath cells, where carbon dioxide is extracted and used in the Calvin cycle.

the earth where light is less intense. However, no such simple distribution exists. The vast majority of plants, well over 99%, are C3 plants, many of which are alive and well in the tropics. Thus C3 and C4 plants grow side by side in sunny climates. Furthermore, there are no clear-cut evolutionary divisions between the two. C3 and C4 plants exist in the same plant families, and even within the same plant genera. Evidently, C4 plants represent a rare but interesting evolutionary venture. Whether they are on the increase or not is unknown. What we do know is that some C4 species represent the most productive domestic plants known. The average growth rate of sugar cane and corn can be twice that of tobacco and hay. Geneticists have shown an interest in the possibility of breeding C4 capabilities into valuable C3 crop plants.

Evolutionary theorists disagree about the origin of the C4 pathway, but most agree that it arose far more recently than the C3. In fact, the C3 pathway is believed to be ancient.

Rubisco, as a primeval enzyme, may have been present in the earliest photosynthetic cells, those of primitive bacteria that thrived at a time when the atmosphere was far richer in carbon dioxide and oxygen was absent or nearly so (see Chapter 21). This may explain rubisco's inability to discriminate between CO_2 and O_2. Later, as water-utilizing plants and algae evolved, the oxygen they released accumulated in greater and greater abundance. Since the enzyme simply wasn't able to discriminate between the new gas and the old, photorespiration has plagued plants and algae ever since.

In Chapter 28 we'll look into another special photosynthetic adaptation in which certain desert plants, the CAM plants, use their C4 pathway to take in and store carbon dioxide *during the nighttime hours*. This strategy enables the plants to keep their leaf pores closed tightly during daylight hours, thus avoiding excessive water loss to the superheated, parching, desert winds, while at the same time heading off photorespiration.

Key Ideas

PHOTOSYNTHESIS: AN OVERVIEW

1. The general equation for **photosynthesis** is:

$$6\ CO_2 + 12\ H_2O \xrightarrow[\text{chlorophyll}]{\text{light}} C_6H_{12}O_6 + 6\ O_2 + 6\ H_2O$$

2. Free energy in the chloroplast is increased in the **light reactions** and decreased as the synthesis of carbohydrate occurs in the **light-independent reactions**.

3. In the light reactions, H_2O, ADP, P_i, and $NADP^+$ are the reactants, and O_2, ATP, and $NADPH + H^+$ are the products.

4. In the light-independent reactions, ATP, NADPH, H^+, and CO_2 are the reactants, while carbohydrate, ADP, P_i, $NADP^+$, and water are the products.

5. Most of the light energy used in photosynthesis is in the violet-blue and orange-red ends of the spectrum. The greens and yellows are reflected or transmitted.

CHLOROPLASTS

1. Within the envelope of the **chloroplast** is a clearer area, the **stroma**, and an extensive inner membrane whose folds form **thylakoids**, the latter arranged in stacks called **grana**. The **lumen** (inner space) is a proton reservoir.

2. **Photosystems** include a **light-harvesting complex**, a **reaction center**, and **electron transport systems**. A light-harvesting complex contains **chlorophylls *a* and *b*** and **carotenoids.** Each pigment absorbs light of slightly different wavelengths. Reaction centers, made up of chlorophyll *a* and protein, are designated **P700** and **P680**.

3. Light-activated electrons from reaction centers pass through electron transport systems, where their free energy is used for reducing $NADP^+$ to NADPH, and converting ADP to ATP.

THE LIGHT-DEPENDENT REACTIONS OF PHOTOSYNTHESIS

1. In the noncyclic events, excited electrons from P680 pass to pheophytin and then into the ETS. They are replaced by electrons from water, which breaks down into oxygen, protons, and electrons. The protons are released into the lumen, adding directly to the proton gradient, and oxygen is given off as O_2.

2. Light-excited electrons in P700 leave PS I and pass to NADPH, reducing it to $NADPH + H^+$.

3. The noncyclic light reactions enrich the chemiosmotic gradient and reduce $NADP^+$ to NADPH, thus providing ATP and hydrogen for carbohydrate synthesis.

4. Photosystem I acts independently in the cyclic events. Energy from light-activated P700 electrons is used for pumping protons before cycling back to P700.

5. The proton concentration, an electrochemical gradient, is both a concentration gradient and a pH gradient.

6. As protons escape the lumen through ATP synthases, their free energy is used in the phosphorylation of ADP, yielding ATP.

THE LIGHT-INDEPENDENT REACTIONS

1. ATP and NADPH are used in the reduction, or *fixing,* of carbon dioxide into 3-carbon carbohydrates from which glucose can be made. The carbon fixation pathway, known as the **Calvin cycle,** occurs in the unstructured stroma. Glucose is synthesized in the cell cytoplasm.

2. The Calvin cycle includes the following chemical steps:

    a. **Carboxylation:**
       $RuBP + CO_2 \longrightarrow$ two 3PG

    b. **Phosphorylation:**
       two $3PG$ + two $ATP \longrightarrow$ two 1,3-BPG + 2 ADP

    c. **Reduction:**
       two $1,3\text{-BPG} + 2\ NADPH + H^+ \longrightarrow$ two $G3P + 2\ NADP^+$

    d. **Yield:** There is no yield until six CO_2 molecules react (six turns of the cycle). From the 12 G3P molecules thus formed, two go to form glucose.

    e. **Regeneration:** Ten of the twelve G3P molecules yielded in six turns are used to regenerate the six RuBP molecules. The other two are used to form glucose and other products.

    f. Summary of six turns of the Calvin cycle:
       $$6\ CO_2 + 6\ RuBP + 18\ ATP + 12\ NADPH + 12\ H^+ \longrightarrow$$
       $$12\ G3P + 18\ ADP$$

       $$2\ G3P \longrightarrow C_6H_{12}O_6 + 2\ P_i$$
       $$10\ G3P + 6\ ATP \longrightarrow 6\ RuBP + 6\ ADP$$

PHOTORESPIRATION: TROUBLE IN THE CALVIN CYCLE

1. Photosynthesis in **C3 plants** becomes inefficient when carbon dioxide gas is in low concentration, a time when the enzyme *rubisco* cannot readily incorporate carbon dioxide.

2. At such times, **photorespiration** ensues. RuBP is changed to **glycolate,** which is then converted to glyoxylate, which next forms 3PG and carbon dioxide.

3. **C4 plants** use an alternative pathway, where CO_2 enters **leaf mesophyll cells** and is incorporated into 4-carbon organic acids by the more efficient enzyme **PEP carboxylase.**

4. The acids enter the **bundle sheath cells**, where enzymes liberate the CO_2 for the regular Calvin cycle.

5. Despite the greater efficiency of C4 plants in intense light, C3 plants still dominate in tropical and desert environments. C4 photosynthesis appears to be a newer adaptive response. The C3 pathway is apparently much older; rubisco was probably present in the early photosynthetic bacteria.

Application of Ideas

1. Earlier in this century, plant physiologists determined the precise role of water and carbon dioxide in photosynthesis by using the radioactive isotopes carbon-14, oxygen-18, and tritium (hydrogen-3). Using your new knowledge of photosynthesis, suggest how these isotopes might be used in such determinations, and what you might expect in the results.

2. Sucking fluids from the moist inner pulp of the desert barrel cacti, a C4 plant, has quenched the thirst of many a desperate desert wanderer. When would you expect the fluids to taste best—during daytime or during night? Why?

Review Questions

1. Write the general formula for photosynthesis and list several factors missing from this simple representation. (page 149)

2. Photosynthesis can occur in isolated chloroplasts, but when their membranes are disrupted, the process stops. What does this indicate? (general)

3. Write a detailed equation for the light reactions that includes $NADP^+$, ATP, ADP, P_i, and $NADPH + H^+$. (page 150)

4. Write a detailed equation for the light-independent reactions that includes glucose, $NADP^+$, ADP, P_i, CO_2, and NADPH. (page 161)

5. Prepare a drawing of a chloroplast, labeling envelope, granum, lamella, lumen, and stroma. (pages 151–153)

6. Prepare a simplified drawing of paired photosystems I and II, carefully labeling the following: lumen, stroma, light-harvesting complex, reaction centers, and ETS. (page 153)

7. Summarize the events of the noncyclic light reactions, beginning with the absorption of light by P680: (page 157)

    a. Path of electrons.
    b. Role of water.
    c. Protons gained in (a) and (b).
    d. Final electron acceptor.

8. What is gained by the noncyclic events? (page 157)

9. How are the ATP synthases oriented in relation to the lumen and stroma? What does this tell you about the path of escaping protons? (pages 159–161)

10. Using a simple diagram of an ATP synthase, show how ADP becomes ATP. (page 161)

11. Write a short overview of the Calvin cycle, just hitting the highlights. (page 161)

12. What are the roles of carbon dioxide, NADPH, ATP, and enzymes in the Calvin cycle? (page 162)

13. Briefly explain why it requires six turns of the Calvin cycle to generate one molecule of glucose. (page 162)

14. From a plant's "point of view," what is wrong with photorespiration? Under what conditions does it occur? (page 164)

15. Write a summary of events in the C4 cycle. Mention the two kinds of cells involved and what generally happens in each. (page 168)

CHAPTER 8

Cell Respiration

CHAPTER OUTLINE

TWO WAYS OF UTILIZING GLUCOSE

GLYCOLYSIS

The Highlights of Glycolysis

Control of Glycolysis

AEROBIC CELL RESPIRATION: THE MITOCHONDRION

Pyruvate to Acetyl–CoA

The Citric Acid Cycle

ELECTRON TRANSPORT AND CHEMIOSMOTIC PHOSPHORYLATION

Mitochondrial Structure

NADH and $FADH_2$ Electrons Build the Chemiosmotic Gradient

Reviewing the Proton Gradient and ATP Formation

Other Uses for the Proton Gradient

ANAEROBIC CELL RESPIRATION

FERMENTATION

Fermentation Pathways

ALTERNATIVE FUELS FOR THE CELL

The Metabolism of Fatty Acids

The Preparation of Proteins

Intermediary Metabolism: Biosynthetic Pathways

Essay 8.1 A Marathon Runner Meets "The Wall"

KEY IDEAS

APPLICATION OF IDEAS

REVIEW QUESTIONS

Life is a celebration of captured energy, a notable interruption in an otherwise inexorable, and seemingly unpleasant, trend toward disorder and entropy. We have seen that life can only exist through intense efforts to keep its molecules organized. As soon as this organization ceases, so does life. A corpse is a once-organized entity, gradually becoming disorganized until finally its molecules have no more to do with each other than they do with any other molecules. Before the final episode, though, the organism lives. And it does so by staying organized. Here then we will look into another way in which living things obtain energy to maintain that organization.

We should begin by reminding ourselves that the energy available to living things is used in a variety of ways. Energy is required for all cellular work: movement, active transport, communication, and molecular synthesis. A key participant in these activities is ATP, the direct source of energy for most of the cell's activities.

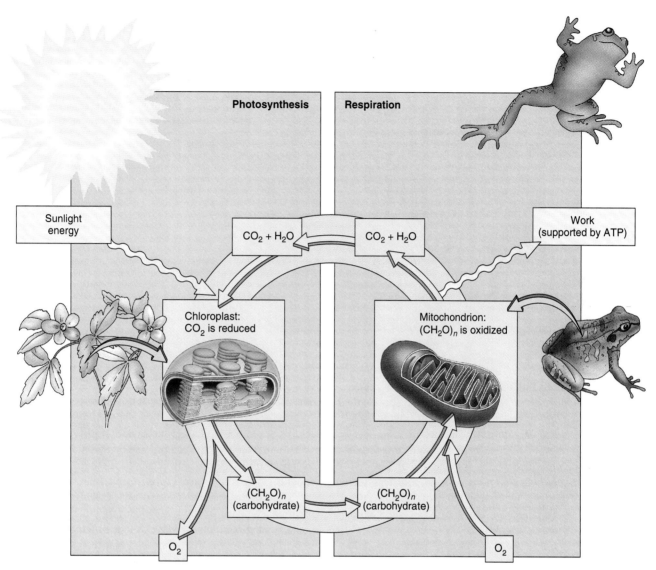

FIGURE 8.1
Photosynthesis Versus Carbohydrate Metabolism
In an overall view, the reactions of photosynthesis and carbohydrate metabolism are the reverse of each other.
Energy, carbon dioxide, and water (left) are taken into the chloroplast and used in the formation of carbohydrate.
Carbohydrate is used as a fuel in the mitochondrion (right), yielding energy, carbon dioxide, and water.

With the exception of ATP produced directly through photosynthesis and chemosynthesis, most of the earth's organisms produce their ATP by using chemical bond energy in organic molecules, chiefly carbohydrates, fats, and proteins. We will focus primarily on the carbohydrate called glucose, the most familiar of these "cellular fuels."

Before going on, we should also remind ourselves of just where glucose comes from—there is only one source. It was manufactured during photosynthesis. Interestingly, the breakdown of glucose for its energy is roughly the opposite of its manufacture. Compare the two overall reactions and then study Figure 8.1 for a time:

Photosynthesis:
$$6\ CO_2 + 12\ H_2O + \text{Light energy} \longrightarrow C_6H_{12}O_6 + 6\ O_2 + 6\ H_2O$$
$$^*\Delta G = +686\ \text{kcal/mole}$$

Respiration:
$$C_6H_{12}O_6 + 6\ O_2 \longrightarrow 6\ CO_2 + 6\ H_2O + \text{Chemical energy}$$
$$^*\Delta G = -686\ \text{kcal/mole}$$

The reactions are not simply reversed, though. For example, the energy for photosynthesis originates as photons of light, whereas the energy extracted during respiration is primarily from the chemical bonds linking hydrogen to carbon ($C - H$).

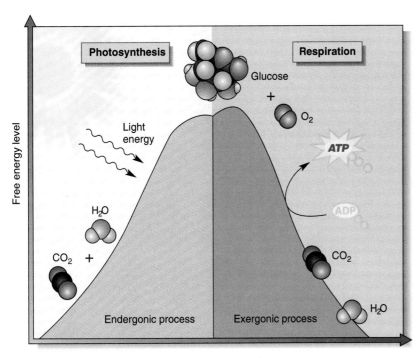

FIGURE 8.2
Uphills and Downhills in Cell Energetics
Photosynthesis is portrayed as an uphill (endergonic) process in which sunlight energy is utilized to power the synthesis of glucose from carbon dioxide and water. Respiration is a downhill (exergonic) process in which glucose is broken down to carbon dioxide and water, with some of its free energy conserved in ATP.

The free energy changes (ΔG) show another important difference. The overall photosynthetic reaction is an endergonic, or energetically *uphill,* process, with the product, glucose, having much more free energy than the reactants—carbon dioxide and water (Figure 8.2). The overall respiratory process, the breakdown of glucose, is an exergonic or energetically *downhill* process; that is, the free energy of the products—carbon dioxide and water—is considerably less than the free energy of glucose. Of course, much of that free energy is conserved in the phosphate bonds of ATP—the goal of respiration. The rest dissipates as heat.

It is popular to use terms like "burning" to describe how the body releases energy from fuels such as glucose, but let's correct this right now. In the metabolic breakdown of glucose, it is important for the cell to break the chemical bonds of glucose in such a way as to avoid any sudden release of energy, since this would result in the wasteful (even dangerous) release of heat, an unsatisfactory situation for living cells. The transfer of free energy from glucose to ATP proceeds through a gentler, more gradual series of coupled reactions, some losing, others gaining energy. And that, in a word or two, is what the chapter is about.

Two Ways of Utilizing Glucose

The fundamental process in energy metabolism is the conversion of the large amount of free energy in glucose and other cellular fuels to the small, more easily utilized, energy units of ATP. (It's like changing a $100 bill into $5 bills—smaller, more spendable units.) There are two general pathways that organisms use in the transfer of glucose energy to ATP: *anaerobic cell respiration* and *aerobic cell respiration*. Our main focus in this chapter will be on aerobic cell respiration, the oxygen-using process. We will, however, look briefly into anaerobic respiration as it occurs in bacteria and into a very short anaerobic pathway called **fermentation**. The most familiar example is alcohol fermentation, the yeast-mediated process that yields ethyl alcohol and carbon dioxide. It is used by bakers for its CO_2, which causes bread to rise, and by brewers and vintners, producers of beer and wine, for its alcohol.

Each of the pathways begins with **glycolysis**, a universal preliminary phase in the metabolism of glucose. Glycolysis yields a small amount of ATP through **substrate-level phosphorylation**, the transfer of phosphates directly from substrate to ADP. The final products of glycolysis, still energy-rich, are then metabolized in cell respiration or fermentation.

Most organisms, including many bacteria, most fungi, protists, plants, and animals (us humans included), utilize **aerobic cell respiration**, the metabolic pathway that requires oxygen. Our in-depth discussion of glucose metabolism will be divided into three parts, as summarized in Figure 8.3. The first is glycolysis, which occurs in the cytoplasm. The other two, both parts of aerobic cell respiration, occur in the mitochondrion. In brief:

Part I, glycolysis: the breakdown of glucose and substrate-level phosphorylation and the formation of energy-rich products.

Part II, citric acid cycle: the complete oxidation of the glycolytic products to CO_2 and H_2O, yielding protons and electrons.

Part III, electron transport and chemiosmotic phosphorylation: use of the free energy of electrons to produce a proton gradient, and use of the proton gradient to form ATP.

Glycolysis

Since glycolysis is common to most of the earth's organisms, and since it does not require oxygen, it has been suggested that the glycolytic pathway arose early in the evolution of life. This notion agrees with the conviction among earth scientists that there was little oxygen available in the early atmosphere (see Chapter 21). Life today has universally retained the simple glycolytic pathway, but, as we've noted, the process is preliminary to cell respiration.

As an overview, we can note that (1) the glycolytic pathway begins with a glucose molecule and ends with two pyruvates, and (2) along the way two coenzyme NAD^+s are reduced to NADH, and four ADPs are converted to four ATPs. Thus some of the free energy of the original glucose is conserved in NADH and ATP (Figure 8.4).

The Highlights of Glycolysis

A study of Figure 8.5, the complete glycolytic pathway, will reveal that there are nine principal reactions in glycolysis, each catalyzed by a specific enzyme. The reactions can conveniently be divided into "preparatory phases" and "yielding phases." In the preparatory phases, the free energy of the substrate (or fuel) is increased, which causes it to become more reactive. In the yielding phases, the free energy of the fuel decreases as the substrate-level phosphorylation of ADP occurs. Here phosphates are transferred from the activated substrate to ADP, forming ATP. We'll confine our discussion of glycolysis to just the key reactions in each phase. (The numbers in the following discussion refer to reaction numbers in Figure 8.5.)

The Preparatory Phase: Priming the Fuel. The first five reactions are preparatory. In reactions 1 and 3, the substrate is phosphorylated, using phosphates from ATP. While costly, this investment of ATP serves to increase the free energy of the glucose substrate, preparing it for reactions to come.

Following this, the doubly phosphorylated fuel, now in the form of **fructose-1,6-bisphosphate**, is cleaved (reaction 4), leading to the formation of two **glyceraldehyde-3-phosphates (G3P)**. They enter into a two-part reaction (reaction 5) in which they are first oxidized and then phosphorylated once more (hydrogen removed and replaced by inorganic phosphate—P_i). The products, two **1, 3-bisphosphoglycerate (1, 3-BPG)** have highly elevated free energy levels. The newly added phosphate group is in a high-energy form, a state that primes it for transfer to another substrate. Importantly, the oxidations are coupled with the reduction of two NAD^+s to NADH.

The Yielding Phase: Forming ATP. The doubly phosphorylated 1, 3-BGPs are now fully primed for substrate-level phosphorylation (reaction 6). In substrate-level phosphorylation, phosphates are transferred directly from the substrate to ADP. Here an energy-rich phosphate group from each 1, 3-BPG is

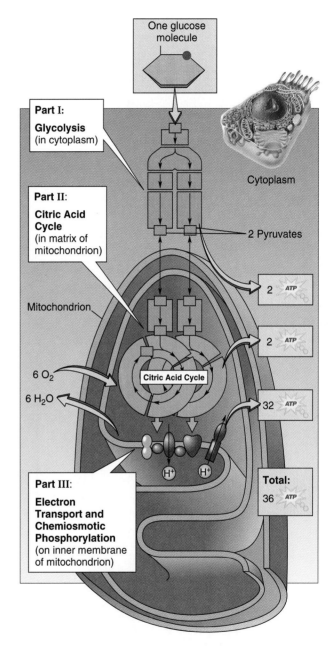

FIGURE 8.3
Three Parts of Glucose Metabolism
The three parts of glucose metabolism are incorporated into a cell scenario. **Part I**: Glycolysis occurs in the cytoplasmic fluids, where a small amount of ATP is generated and the glucose is broken down into two molecules of the organic acid, pyruvate. **Part II**: Pyruvate, the product of glycolysis, enters the mitochondrion, where it is oxidized in the citric acid cycle. **Part III**: Electrons and protons from the citric acid cycle are sent through electron transport systems, where their free energy is used to bring new protons into the chemiosmotic proton gradient. Finally, the free energy of the gradient is used to phosphorylate ADP in ATP synthases, where most of the cell's ATP is produced.

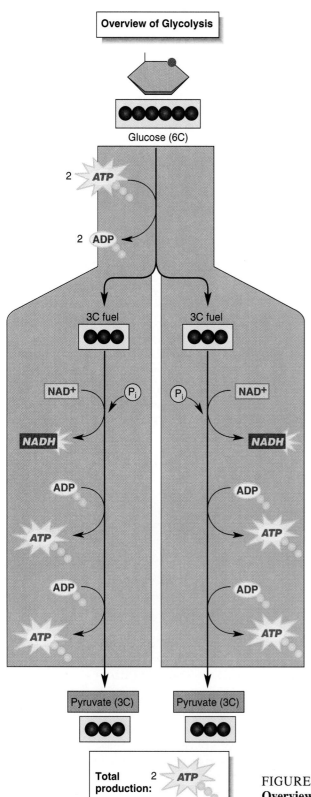

Overview of Glycolysis

Glucose (6C)

2 ATP

2 ADP

3C fuel   3C fuel

NAD+   Pᵢ   Pᵢ   NAD+

NADH   NADH

ADP   ADP

ATP   ATP

ADP   ADP

ATP   ATP

Pyruvate (3C)   Pyruvate (3C)

Total production: 2 ATP

2 NADH

2 Pyruvate

transferred to two ADPs, yielding two ATPs. The substrates, now at a substantially lower free energy level, are two **3-phosphoglycerates** (**3PG**).

In the overall glycolytic scheme, the yield of two ATPs so far is really just a "payback," since they only replace the two ATPs invested earlier (reactions 1 and 3). But a second yield is coming up.

More Preparatory and Yielding Activity. The two 3PGs enter reactions that prepare them for a second yield (reactions 7 and 8). In the second of these, a dehydration reaction, the substrates' energy distribution is altered, centering much of it in the remaining phosphate group. When this phosphate bond is cleaved, ΔG will be large (−14.8 kcal/mole), sufficiently so to drive two more substrate-level phosphorylations. The remaining phosphates join two ADPs, forming two more ATPs (reaction 9). The second two represent the *net yield* from glycolysis. The other products of this important step are two pyruvates, which brings glycolysis to an end.

The Net Energy Yield. Using some fairly simple calculations, we can determine the caloric value of the energy yield. Such calculations are made on a mole-to-mole basis, as was described in Chapter 6. Recall that when ATP is hydrolyzed to ADP under standard laboratory conditions, the free energy change is −7.3 kcal/mole. Replacing the terminal phosphate requires the same amount of energy. So the two ATPs formed in glycolysis represent a net free energy gain of 14.6 kcal (Table 8.1).

How much of the original free energy in glucose does this represent? In other words, how energy efficient is glycolysis? The free energy of glucose, when the sugar is broken down all the way to CO_2 and H_2O, is 686 kcal/mole. The energy conserved in ATP during glycolysis then is only about 2.1% of the total ($14.6/686 \times 100 = 2.13\%$). As we will find, this is a low figure compared to the energy yield from the aerobic process to come. The free energy remaining at the end of glycolysis is now present in the two pyruvates and two NADHs.

Control of Glycolysis

Like so many biochemical processes, glycolysis is under a number of delicate regulatory mechanisms. As a general observation, when the cell is using its ATP at a rapid rate, glycolysis speeds up, and conversely, if its ATP use is curtailed, glycolysis slows down. But what adjusts such rates? Obvious-

FIGURE 8.4
Overview of Glycolysis

An overview of glycolysis shows glucose entering a pathway, where it is twice phosphorylated by ATP and then split into two 3-carbon fuels. The fuels are oxidized by enzymes and phosphorylated, which prepares them for two substrate-level phosphorylations in which 4 ATP are produced. The net end products of glycolysis are 2 *ATP,* 2 *NADH,* and 2 molecules of 3-carbon *pyruvate.*

ly, the availability of glucose is one controlling factor, but there are far more subtle controls.

A major regulatory factor is the responsiveness of a key participant, **phosphofructokinase**, an allosteric enzyme responsible for the conversion of fructose-6-phosphate to fructose-1,6-bisphosphate (reaction 3 in glycolysis). As an allosteric enzyme (see Chapter 6), phosphofructokinase is affected by both activators and inhibitors. Logically, it is inhibited by high cellular concentrations of ATP and is stimulated by high concentrations of ADP (either can fit the allosteric site; Figure 8.6). This makes sense, since an accumulation of ATP would indicate that metabolic activity had slowed and there wasn't much of a need. On the other hand, an accumulation of ADP would suggest the opposite.

Rates of reaction in this versatile key enzyme can change radically. When animal muscle is working at a maximum, for example, the activity of phosphofructokinase increases several hundred times. This is important because glycolysis provides much of the ATP utilized by hard-working muscles.

Aerobic Cell Respiration: The Mitochondrion

The two energy-rich products of glycolysis are pyruvate and NADH. Pyruvate will enter the mitochondrion where the remainder of its free energy will be extracted. NADH cannot enter the mitochondria, but there is a mechanism, a shuttle transport protein, to transport its energetic electrons to carriers on the inside. We will get back to this later. Pyruvate is helped into the mitochondrion by a transport protein. Once inside, pyruvate is altered so that it can take part in the citric acid cycle.

Pyruvate to Acetyl–CoA

Pyruvate is first transformed into an energy-rich substrate called **acetyl–coenzyme A** (**acetyl–CoA**) (Figure 8.7). This transition is a complex process involving a gigantic three-part enzyme–coenzyme complex that is bound to the inner membrane. Like the other coenzymes we have introduced, **coenzyme A** includes a vitamin group known as pantothenic acid. (So those caring people who pestered us about eating our vegetables all those years were right after all. One wouldn't want to run out of acetyl–CoA.) With so many large interacting molecules, the total CoA–enzyme complex turns out to be almost as large as a ribosome. The intricate details are beyond our discussion, but the overall reaction for each pyruvate entering the complex is as follows:

$$\text{Pyruvate} + \text{NAD}^+ + \text{CoA} \longrightarrow \text{Acetyl-CoA} + \text{NADH} + \text{H}^+ + \text{CO}_2$$

As you see, pyruvate loses a carbon, which is given off as carbon dioxide (the CO_2 we exhale). The product, a 2 carbon acetyl group (CH_3C —), is covalently bonded to coenzyme A, forming acetyl–CoA (CH_3C — CoA).

TABLE 8.1
Energy Yield in Glycolysis (per mole of glucose)

Starting free energy in glucose: 686 kcal.mole	Loss in kcal	Gain in kcal
Reactions 1–3: −2 ATP	−14.6	
Reaction 6: +2 ATP		+14.6
Reaction 9: +2 ATP		+14.6
Net gain		+14.6
Overall efficiency: 14.6 kcal ÷ 686 kcal × 100 = 2.1%		

The Citric Acid Cycle

The **citric acid cycle**, also called the **Krebs cycle**, was first described by Hans A. Krebs in 1937. Ironically, although his paper has become a classic in biochemistry, it was rejected by the disbelieving editors of *Nature*, the prestigious British journal to which it was first submitted. In spite of the early setback, Professor Krebs was later awarded a Nobel Prize for his efforts.

One of the key functions of the citric acid cycle is the oxidation of fuel molecules. The oxidation reactions are coupled to the reduction of coenzymes NAD^+ and FAD, which then transfer captured energetic electrons and protons to the electron transport system (Figure 8.8). The free energy of the electrons powers proton pumps that produce the all-important proton gradient used in chemiosmotic phosphorylation.

Principal Reactions of the Citric Acid Cycle. Figure 8.9 reviews each of the nine citric acid cycle reactions, so here we will mention only those steps where key reactions occur. The numbers in the discussion to follow refer to reaction numbers in Figure 8.9.

Enter Acetyl CoA. Acetyl–CoA enters the citric acid cycle (reaction 1) by reacting with 4-carbon **oxaloacetate**, what might be called the "resident molecule," because it must be present for each new turn of the cycle. The enzyme that prompts this reaction is called citrate synthetase, and the products are 6-carbon **citrate**, or **citric acid**, and CoA. Citrate is the same acid found in citrus fruits and the one for which the cycle is named. Because of the high free energy level of acetyl–CoA, the reaction occurs quite readily. When released from the acetyl group, the CoA simply recycles to react with another pyruvate and generate more acetyl–CoA.

Carbon Dioxide Release. Since the pathway is cyclic, oxaloacetate will later form again, bringing the series of reactions back where it began. To proceed from 6-carbon citrate back to 4-carbon oxaloacetate, however, there will have to be two decarboxylations (CO_2 removals; see reactions 4 and 5). Although these reactions do not bring about an energy yield, they are interesting because this is the CO_2 we exhale with each breath—the CO_2 all organisms release as they carry on cell respiration.

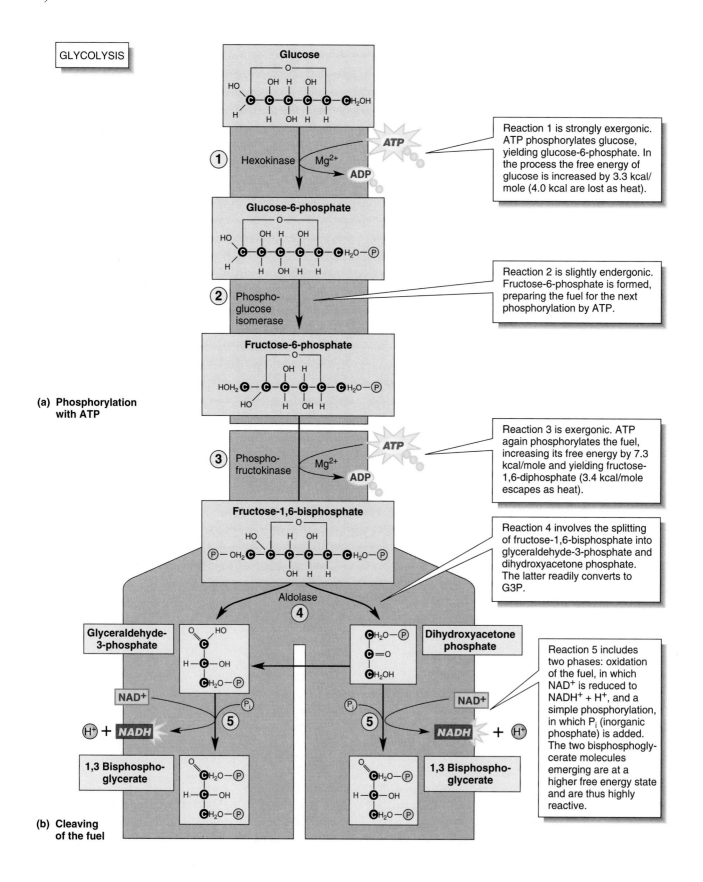

GLYCOLYSIS

Glucose

Reaction 1 is strongly exergonic. ATP phosphorylates glucose, yielding glucose-6-phosphate. In the process the free energy of glucose is increased by 3.3 kcal/mole (4.0 kcal are lost as heat).

(1) Hexokinase Mg^{2+} ATP → ADP

Glucose-6-phosphate

Reaction 2 is slightly endergonic. Fructose-6-phosphate is formed, preparing the fuel for the next phosphorylation by ATP.

(2) Phospho-glucose isomerase

Fructose-6-phosphate

(a) Phosphorylation with ATP

(3) Phospho-fructokinase Mg^{2+} ATP → ADP

Reaction 3 is exergonic. ATP again phosphorylates the fuel, increasing its free energy by 7.3 kcal/mole and yielding fructose-1,6-diphosphate (3.4 kcal/mole escapes as heat).

Fructose-1,6-bisphosphate

Reaction 4 involves the splitting of fructose-1,6-bisphosphate into glyceraldehyde-3-phosphate and dihydroxyacetone phosphate. The latter readily converts to G3P.

Aldolase
(4)

Glyceraldehyde-3-phosphate

Dihydroxyacetone phosphate

NAD^+

(P) P_i

(5) P_i (P)   (5)

H^+ + NADH   NADH + H^+

Reaction 5 includes two phases: oxidation of the fuel, in which NAD^+ is reduced to $NADH + H^+$, and a simple phosphorylation, in which P_i (inorganic phosphate) is added. The two bisphosphoglycerate molecules emerging are at a higher free energy state and are thus highly reactive.

1,3 Bisphospho-glycerate

1,3 Bisphospho-glycerate

(b) Cleaving of the fuel

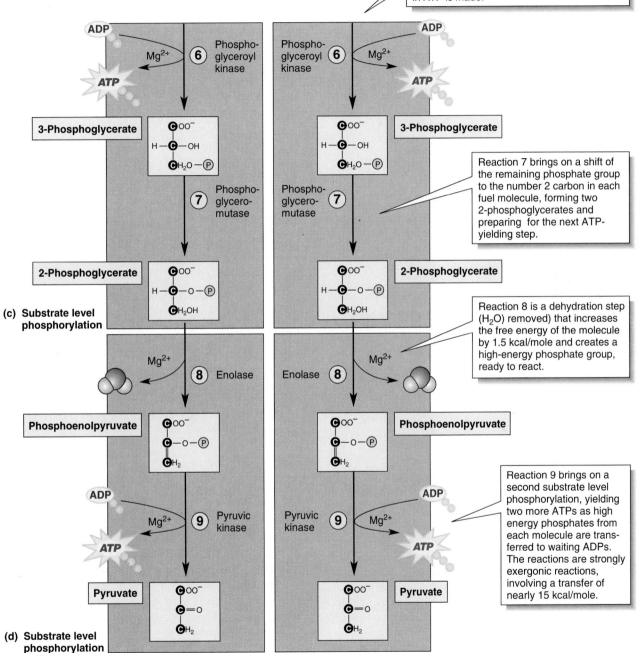

Reaction 6 brings on the first substrate phosphorylation as a high energy phosphate group from each fuel molecule is transferred to ADP. The reaction is highly exergonic with a ΔG of −9.0 kcal/mole. Actually, the gain of 2 ATP pays off the initial investment, so no net gain in ATP is made.

Reaction 7 brings on a shift of the remaining phosphate group to the number 2 carbon in each fuel molecule, forming two 2-phosphoglycerates and preparing for the next ATP-yielding step.

Reaction 8 is a dehydration step (H₂O removed) that increases the free energy of the molecule by 1.5 kcal/mole and creates a high-energy phosphate group, ready to react.

Reaction 9 brings on a second substrate level phosphorylation, yielding two more ATPs as high energy phosphates from each molecule are transferred to waiting ADPs. The reactions are strongly exergonic reactions, involving a transfer of nearly 15 kcal/mole.

FIGURE 8.5
Glycolysis

Glycolysis occurs in nine principal reactions, each involving a specific enzyme. The first five are preparatory. **(a)** They include two phosphorylations by ATP, **(b)** cleaving of the fuel into 3-carbon products, and oxidation and phosphorylation. **(c)** These are followed by two substrate-level phosphorylations and the first ATP yield. A second preparatory phase includes two reactions that lead to **(d)** more substrate-level phosphorylations (reaction 9), and thus the second ATP yield. The net yield, after the payback of 2 ATP is only 2 ATP. The final products are 2 ATP, 2 NADH, and 2 pyruvate.

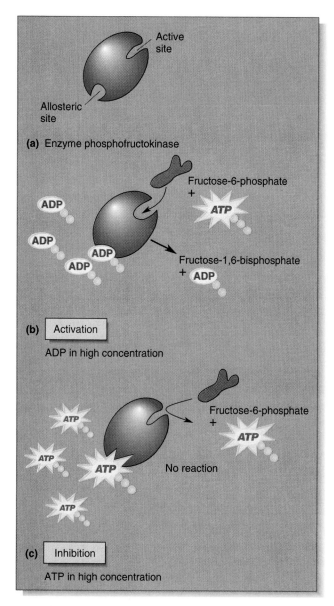

FIGURE 8.6
Control of Glycolysis
(a) The allosteric enzyme phosphofructokinase catalyzes the conversion of fructose-6-phosphate to the product fructose-1,6-bisphosphate. As an allosteric enzyme it has both an active site and a secondary, or allosteric, site, the latter of which will accept and bind loosely with either ATP or ADP, depending on which is in the greatest concentration. (b) When ADP binds to the allosteric site, the enzyme increases its reaction with fructose-6-phosphate and ATP, thus speeding up glycolysis. (c) When ATP is more highly concentrated and bound to the allosteric site, the active site is altered in such a way that it is no longer able to bind with the substrate fructose-6-phosphate, the reactions diminish, and glycolysis slows.

Oxidations in the Cycle. Some of the reactions result in subtle molecular rearrangements and free energy shifts, each of which prepares the molecule for major reactions to follow. However, we will focus on the oxidation reactions (reactions

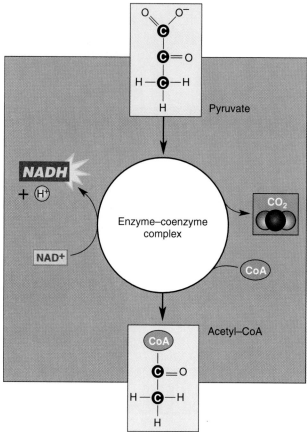

FIGURE 8.7
The Acetyl–CoA Step
The conversion of pyruvate to acetyl–CoA requires the activity of a large enzyme–coenzyme complex. The products are carbon dioxide, NADH, and a 2-carbon acetyl group bonded to a molecule of coenzyme A.

4, 5, 7, and 9). In each, the oxidation of the fuel is coupled with the transfer of electrons and protons to coenzyme NAD^+ or coenzyme FAD, reducing them to NADH and $FADH_2$, respectively. The free energy change in these redox reactions is great, with ΔG sometimes in the neighborhood of –7 kcal/mole. But most significantly, the newly reduced coenzymes themselves have the reducing power needed to pass their electrons and protons to the mitochondrial electron transport systems.

Substrate-Level Phosphorylation. Another event of special interest is found in reaction 6. Here we see a substrate-level phosphorylation, much the same as occurs in glycolysis. In this reaction, guanosine diphosphate (GDP) is phosphorylated to form guanosine triphosphate (GTP), which in turn phosphorylates ADP to form ATP. GTP is chemically similar to ATP, but with the nitrogen base guanine replacing adenine. The free energy exchange in the reaction generating GTP is high (ΔG = –8.0 kcal/mole).

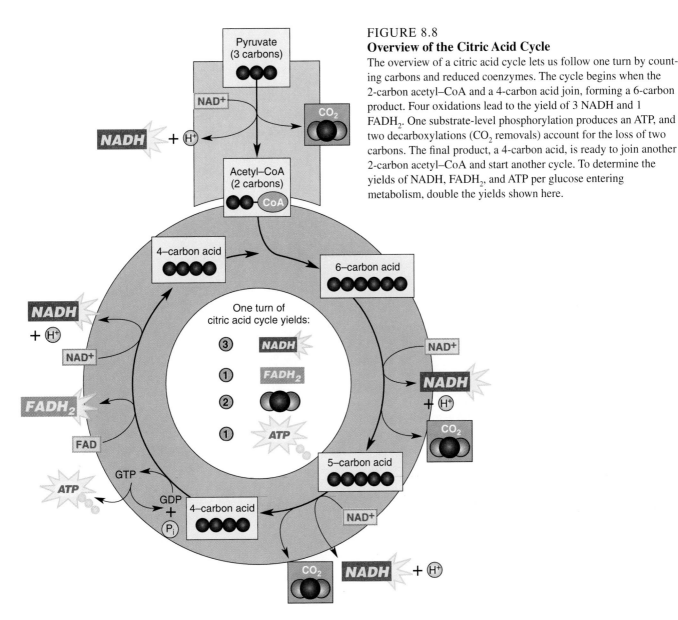

FIGURE 8.8
Overview of the Citric Acid Cycle

The overview of a citric acid cycle lets us follow one turn by counting carbons and reduced coenzymes. The cycle begins when the 2-carbon acetyl–CoA and a 4-carbon acid join, forming a 6-carbon product. Four oxidations lead to the yield of 3 NADH and 1 $FADH_2$. One substrate-level phosphorylation produces an ATP, and two decarboxylations (CO_2 removals) account for the loss of two carbons. The final product, a 4-carbon acid, is ready to join another 2-carbon acetyl–CoA and start another cycle. To determine the yields of NADH, $FADH_2$, and ATP per glucose entering metabolism, double the yields shown here.

In summary, the key reactions of the citric acid cycle are decarboxylations, oxidations/reductions, and substrate-level phosphorylations.

Glucose Metabolism in the Mitochondrion So Far. Let's stand back now and look at the mitochondrial reactions so far. The two important kinds of yields are the reduced coenzymes NADH and $FADH_2$, and the ATP. For each pyruvate entering the mitochondrion, there is the following yield: (1) one NADH from the acetyl–CoA step, and (2) three NADHs, one $FADH_2$, and one ATP from the cycle.

However, since we started respiration with glucose, and each glucose yields two pyruvates, let's double the yields: (1) eight NADH, (2) two $FADH_2$, and (3) two ATPs. In the final accounting to come—for all glucose metabolism—we will consider the two NADH molecules formed in glycolysis.

Keep these yields in mind as we turn to the next events: electron transport and proton pumping.

Electron Transport and Chemiosmotic Phosphorylation

You may recall from Chapter 7 that the light reactions of photosynthesis were dependent on the internal structure of the chloroplast. Similarly, cell respiration depends on the intact internal structure of the mitochondrion. In fact, it may occur to you as we proceed that there are many structural parallels between the two organelles.

Mitochondrial Structure

We saw in Chapters 4 and 6 that the mitochondrion contains an outer and an inner membrane—the "sac-within-a-sac" arrangement typical of chemiosmotic systems (Figure 8.10). The outer membrane is readily permeable to many small molecules and ions and it contains transport proteins that make

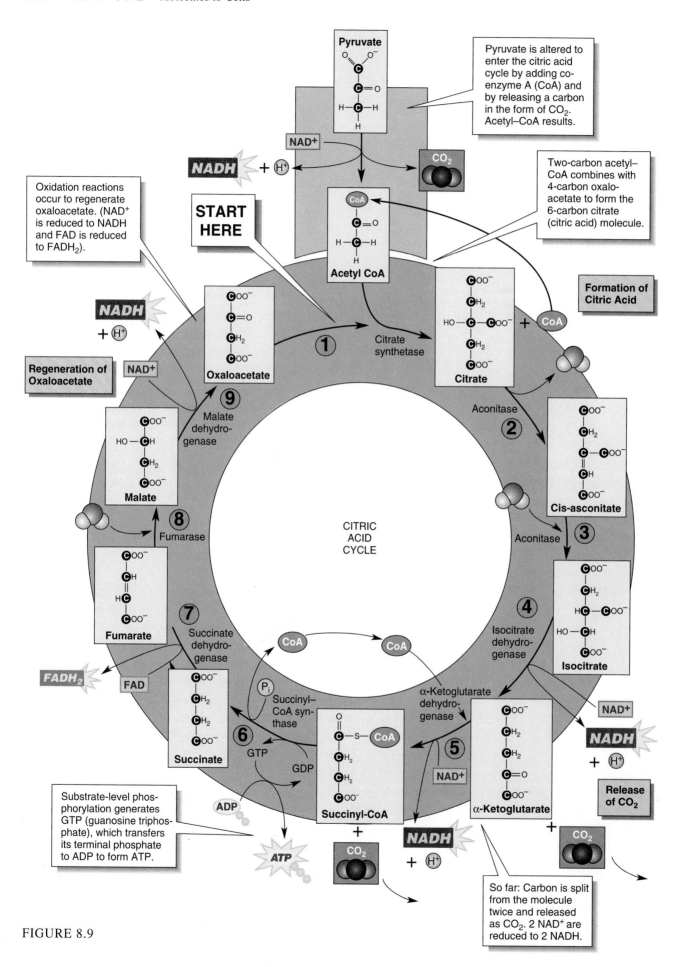

FIGURE 8.9

use of active transport to shuttle materials in and out. The **outer compartment** (also called the **intermembranal compartment**) lies between the two membranes.

The inner membrane surrounds the **inner compartment**, or **matrix**. This membrane forms extensive folds or **cristae** that vastly increase its surface area, thus providing an enormous reaction surface. The inner membrane is selectively permeable and most materials that move across must do so through transport proteins that abound there. Transporters include those that ferry P_i (inorganic phosphate), ADP, calcium ions, and sodium ions into the inner compartment. Others transport metabolites, or substrates, of the citric acid cycle itself. We will come back to these later in the chapter. Then, of course, there are the electron transport systems, proton pumps, and ATP synthases (see Figure 8.10). Thus we see that the inner mitochondrial membrane, like the thylakoid membrane, is one of the most metabolically active membranes in the cell.

It is important that you understand that the specific events to follow, and even the roles of the electron and proton transporters, are provisional. There is still much to learn about this aspect of cell respiration and there are different hypothetical schemes. As researchers continue to probe this important process, many of the details will probably change.

Electron Transport Systems and ATP Synthases. The inner mitochondrial membrane, like the thylakoid membranes of the chloroplast, contains numerous electron transport systems (Figure 8.11). The electron carriers pass electrons along from one carrier to the next, with some carriers serving as proton pumps. The pumps transport protons from the inner to the outer compartment. Included in the carriers are large, transmembranal protein complexes containing active prosthetic groups (special non-amino acid groups). The prosthetic groups are directly involved in the sequential oxidation–reduction reactions that characterize the systems. The carriers of the mitochondrial ETS are the following:

1. **Flavoproteins.** In protein carriers called *flavoproteins,* the active electron acceptors are either FAD or **FMN (flavin mononucleotide)**, which is bound to protein.

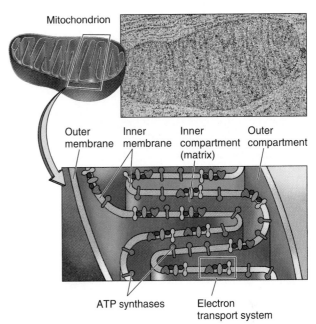

FIGURE 8.10
The Mitochondrion
A three-dimensional view of the mitochondrion provides clues to its functions. Note the two compartments, separated by the inner membrane. Along the inner membrane are numerous aggregations of electron carriers that form highly organized electron transport systems. Also embedded in the inner membrane are ATP synthases, their channels leading from the outer compartment into the bulbous protein complexes where ATP is synthesized.

2. **Coenzyme Q.** A small carrier called coenzyme Q, or **CoQ**, is not a protein at all, but a lipid. If this sounds familiar, it should. CoQ is very similar to PQ of the thylakoid (see Chapter 7).

3. **Cytochromes.** Protein carriers known as cytochromes have iron-containing *heme* groups (like hemoglobin; see Chapter 4). It is the iron within the heme groups that accepts and transfers electrons. Cytochrome c, like PC of the thylakoid, is a small, mobile carrier, located on the surface of the membrane.

4. **Iron–Sulfur Proteins.** Iron is also the active group in carriers known as iron–sulfur proteins.

The electron carriers are arranged in sequence, so that each carrier has greater reducing power than the next one in line. Thus as electrons are received by the first carrier, they are immediately passed to the next, at an energetically lower level, in an ongoing sequence of reductions and oxidations.

Other prominent elements of the inner membrane are the ATP synthases, bodies that contain the ATP-synthesizing enzymes. Like those of the thylakoid, each is bound to the membrane by proteins that form a channel into the opposite compartment.

Most of the carriers are integrated into four large protein complexes in the membrane (see Figure 8.11). Three of the

FIGURE 8.9
Citric Acid Cycle
The citric acid cycle has nine major steps, each catalyzed by its own enzyme. Each turn of the cycle begins and ends with oxaloacetate. In the first reaction, oxaloacetate joins with acetyl–CoA. Water enters, and coenzyme A is released. The product formed is citrate (citric acid). As each enzyme does its job in the succeeding steps, the fuel changes, CO_2 is released, oxidations occur with NAD^+ and FAD receiving electrons and protons, H_2O enters and leaves, and one ATP is formed. The main purpose of the citric acid cycle is oxidation, the removal of electrons and protons that are then made available to the electron transport systems in the inner membrane of the mitochondrion. Their free energy will be used there to enrich the proton gradient.

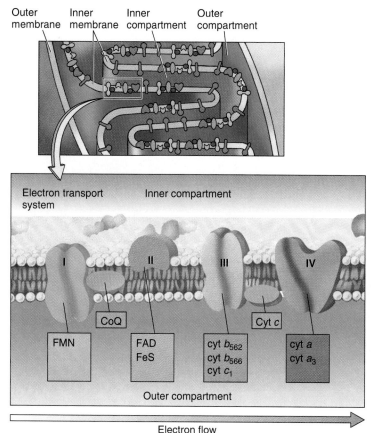

Legend: Electron carriers

Complex I	FMN	Flavin mononucleotide
Complex II	FAD	Bound coenzyme FAD
	FeS	Iron-sulfide protein
Complex III	Cyt	Cytochromes:
		cyt b_{562}
		cyt b_{566}
		cyt c_1
Complex IV	Cyt	Cytochromes:
		cyt a
		cyt a_3
Mobile carriers	CoQ	Lipid carrier moves within membrane
	Cyt c	Cytochrome carrier moves in outer compartment

FIGURE 8.11
Electron Transport System
The inner membrane of the mitochondrion has numerous electron transport systems. Each consists of four protein complexes, each containing active prosthetic groups. Iron-containing groups called cytochromes are particularly abundant. Complexes I, III, and IV act as proton pumps as well as electron carriers. Complex II, which includes bound coenzyme FAD, receives electrons and protons directly from the citric acid cycle. In addition, there are two mobile carriers, coenzyme Q and cytochrome c. The first, a lipid, moves freely in the oily core of the phospholipid bilayer, while the second moves in the watery fluids, outside the membrane. The large arrow indicates the direction of electron flow and also diminishing reducing power.

complexes, denoted I, III, and IV, double as proton pumps. Complex II contains coenzyme FAD that is bound to an iron–sulfur protein. This is the FAD that is reduced in the citric acid cycle. CoQ and Cyt c, as we've seen, are small, mobile carriers. CoQ, a lipid, moves within the fatty acid tails of membrane phospholipids, whereas Cyt c is free to move along the inner membrane surface.

NADH and FADH$_2$ Electrons Build the Chemiosmotic Gradient

We're now ready to see just how electrons and protons carried by NADH and FADH$_2$ build the chemiosmotic gradient. We'll begin with the arrival of NADH at the first carrier. The numbers in the discussion that follows refer to those in Figure 8.12.

The flow of electrons begins when NADH passes two electrons and two protons ① to **flavin mononucleotide** or

(FMN), which is part of a large protein complex. The two protons from NADH pass to the far side of the carrier and are released into the outer compartment ②, leaving the two electrons behind. Thus we see that the FMN complex is the first proton pump.

The electrons next reduce the mobile lipid carrier, coenzyme Q ③, which passes freely through the lipid membrane, bypassing complex II and on to complex III ④, where it reduces cytochrome b. Complex III is also a proton pump, so more of the electron energy is put to work, pumping a second pair of protons ⑤ from the inner to the outer compartment.

From complex III, the electrons pass to mobile carrier Cyt c ⑥, which moves over to complex IV, reducing it. Complex IV, which contains several cytochromes, is the final proton pumping site and two more protons ⑦ are pumped across the inner membrane. A *total of six protons have now been shuttled from the inner to the outer compartment.*

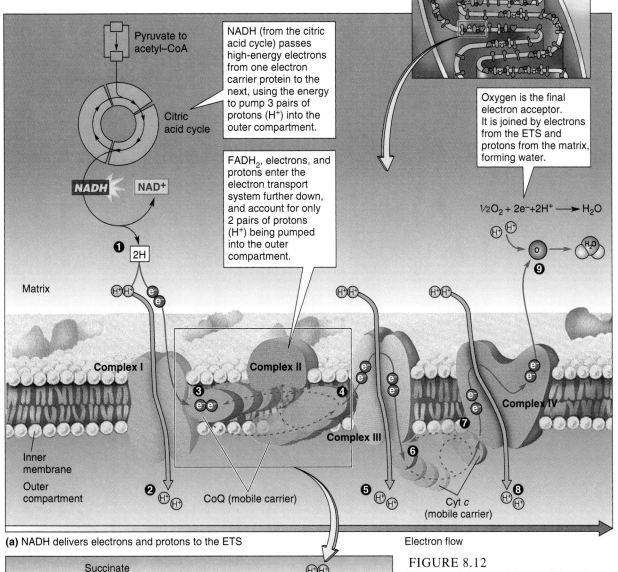

Pyruvate to acetyl–CoA

Citric acid cycle

NADH (from the citric acid cycle) passes high-energy electrons from one electron carrier protein to the next, using the energy to pump 3 pairs of protons (H^+) into the outer compartment.

$FADH_2$, electrons, and protons enter the electron transport system further down, and account for only 2 pairs of protons (H^+) being pumped into the outer compartment.

Oxygen is the final electron acceptor. It is joined by electrons from the ETS and protons from the matrix, forming water.

$$\tfrac{1}{2}O_2 + 2e^- + 2H^+ \longrightarrow H_2O$$

NADH → **NAD+**

❶ 2H

Matrix

Complex I

Complex II

❸

❹

Complex III

❺

❻

❼

Complex IV

❽

❾

Inner membrane

Outer compartment

❷ CoQ (mobile carrier)

Cyt *c* (mobile carrier)

(a) NADH delivers electrons and protons to the ETS

Electron flow

Succinate

2H

FAD **FADH₂**

Complex II

CoQ (mobile carrier)

Complex III

(b) $FADH_2$ passes electrons and protons to mobile carrier CoQ, which transports them to complex III. There the protons are pumped to the outer compartment and the electrons passed to Cyt*c*.

FIGURE 8.12
Electron Flow and Proton Pumping
(a) The flow of electrons through the mitochondrial electron transport system begins ① when NADH reduces FMN in complex I. It receives two electrons and two protons. ② The two protons are transported into the outer compartment, and the two electrons ③ reduce coenzyme Q. CoQ then reduces cytochrome *b* ④ in protein complex III, which makes use of the free energy ⑤ to transport two more protons across to the outer compartment. From complex III, the electrons pass to Cyt *c* ⑥, which moves to complex IV, which it reduces ⑦. The remaining free energy of the electrons is used to power the transport ⑧ of two more protons. The electrons then react with oxygen ⑨ and protons, forming water. In all, there are three proton-pumping events with a total of six protons added to the proton gradient in the outer compartment. **(b)** $FADH_2$ electrons and protons are passed directly to CoQ, which then reduces complex III. Thus $FADH_2$ electrons power the transport of just four protons.

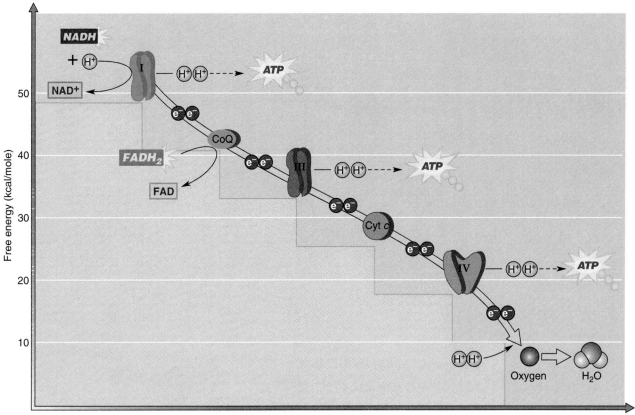

FIGURE 8.13
Energetics of the Electron Transport System
Free energy changes in electrons traversing the mitochondrial electron transport system take the form of a descending stairway. Note that there are three steps in which the free energy change is substantial. These changes represent points along the system where protons are pumped and thus can be equated to energy available for the generation of ATP.

Before moving on, let's consider the FADH$_2$ that was produced in the citric acid cycle. FADH$_2$ is bound to protein complex II, to which it passes its two electrons and two protons ⑧. Complex II is not a protein pump, so it must make use of versatile CoQ to pass its cargo along. Coenzyme Q passes the electrons and protons to protein complex III, from which the protons cross to the outer compartment. The electrons follow the usual route, powering the transfer of two more protons at complex IV, before becoming energy depleted. So, whereas NADH electrons power the transport of six protons, *those of FADH$_2$ account for just four.*

Oxygen: The Final Electron Acceptor. We are close to the finish of aerobic cell respiration, but if you think about it, we have yet to mention oxygen. Now's the time. Oxygen enters cell respiration near its very end, acting as the final electron acceptor ⑨ by collecting the electrons after their free energy has been depleted. As the oxygen and electrons join protons from the inner compartment, water is formed. If our arithmetic is right, the formation of a water molecule requires two electrons, two protons, and one oxygen atom ($\frac{1}{2}O_2$):

$$\tfrac{1}{2}O_2 + 2\,e^- + 2\,H^+ \longrightarrow H_2O$$

What becomes of all that water, or metabolic water, as it is known? In our own bodies, much of it goes directly into the blood for a fast trip around the body, and ultimately to the kidneys. Then, depending on how much water you've been drinking, some metabolic water will be recycled to the blood, and the rest forms urine. Things are simpler in plants. A small amount is used in photosynthesis, but most metabolic water moves upward through the plant and, upon evaporation, escapes through pores in the leaves.

As a final note on electron transport, we should emphasize that the overall process, beginning with NADH and ending with oxygen, is highly exergonic. The free energy change from one end to the other is –53 kcal/mole. Some idea of the nature of this energy change is suggested by Figure 8.13. Let's keep in mind, however, that the –53 kcal/mole is the difference between the initial free energy state and the final free energy state. It doesn't tell us where the energy went. Of course, you know that a considerable amount ends up as the free energy of the proton gradient.

Summing up the Proton Yields. Two pyruvates entering the mitochondrion (representing one glucose), generate 8 NADH and 2 FADH$_2$. (See Figure 8.8.) We can now express these products in terms of protons pumped into the outer compartment.

FIGURE 8.14
Chemiosmotic Phosphorylation
As protons escape the steep proton gradient of the outer compartment, their free energy is used within the ATP synthases to form ATP from ADP and P_i. Each head contains the phosphorylating enzyme of ATP.

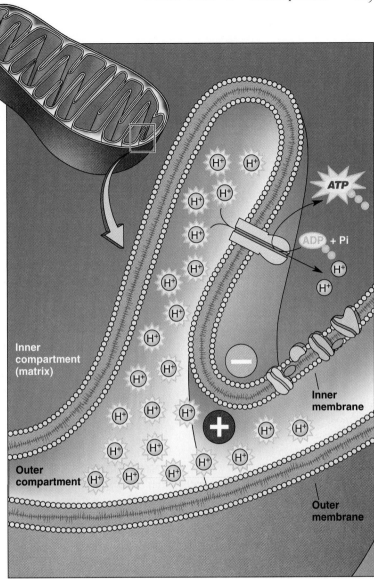

Electrons from each of the 8 NADH pump six protons, and those from each $FADH_2$ pump four, for a total of 56. Thus the chemiosmotic equivalent of two pyruvates turns out to be 56 protons added to the proton gradient, $(8 \times 6) + (2 \times 4)$. But now let's bring in the two NADHs from glycolysis (see Figure 8.4). Of course, we left them in the cytoplasm, and NADH cannot enter the mitochondrion so there's more to tell.

Cytoplasmic NADH reduces a mitochondrial coenzyme, either FAD or NAD^+, via a shuttle system in the outer mitochondrial membrane. The coenzyme reduced depends on the kind of cell involved, but we'll consider it to be FAD, the coenzyme reduced in the animal muscle mitochondrion.

So we find that two cytoplasmic NADHs generate two mitochondrial $FADH_2$s (and, in so doing, free NAD^+ to recycle back to glycolysis, where it is needed). The two $FADH_2$s account for eight more protons pumped, *bringing the grand total to 64* $(56 + 8)$. Keep this number in mind as we turn once again to the chemiosmotic gradient and its use in the phosphorylation of ADP.

Reviewing the Proton Gradient and ATP Formation

The proton concentration in the mitochondrion, as in the thylakoid lumen, represents an electrochemical gradient. The electrical component involves positively charged protons (H^+) of the outer compartment, opposed on the other side of the inner membrane by a preponderance of negatively charged hydroxide ions (OH^-) in the inner compartment. This also translates into a voltage potential (140 mV in the fully charged mitochondrion). The chemical component is a pH gradient, with the outer compartment at a lower pH than the inner compartment. What is important is that an electrochemical gradient has a great deal of free energy, and this free energy can be put to work.

As we've seen, the free energy of the chemiosmotic gradient can be harnessed as its protons move to a lower free energy state—which means across the inner membrane to the matrix. They do this through channels leading to ATP synthases. It is there that the free energy of the proton gradient is used to phosphorylate ADP, thus generating new ATP (Figure 8.14).

Although there is some question about the numerical relationship between escaping protons and ATP produced, it appears that each pair of protons passing into an ATP synthase generates one ATP. This fits well with older calculations, wherein the complete oxidation of glucose was said to lead to the synthesis of 36 ATP. Actually, this number takes into account both chemiosmotic phosphorylation and substrate-level phosphorylation, and the arithmetic is fairly straightforward (although open to debate). It is generally thought that the total number of ATP molecules per glucose is determined in the following manner:

- Substrate-level phosphorylation
  Glycolysis   2 ATP
  Citric acid cycle   2 ATP

- Chemiosmosis
  64 protons gained   32 ATP

- Total per glucose   36 ATP

The yields in terms of protons, NADH, $FADH_2$, and ATP are summarized in Table 8.2.

How efficient is the entire process of glucose metabolism? In other words, how much of the free energy of glucose ends

TABLE 8.2
ATP Yields from Glucose Metabolism

For each Glucose entering metabolism:

Phases	NADH	FADH2	ATP
Glycolysis[1]	(2 NADH)	2 $FADH_2$	2 ATP
Citric acid cycle (including acetyl-Co-A step)	8 NADH	2 $FADH_2$	2 ATP
	8 NADH	**4 $FADH_2$**	**4 ATP**
Proton equivalents	48	16	
ATP subtotals	24 ATP	8 ATP	4 ATP

ATP Total   36 per Glucose

[1] NADH formed in glycolysis converts to $FADH_2$ when transported into the mitochondrion.

up in the phosphate groups of ATP? The free energy increase in ADP as it is phosphorylated to ATP is, again, +7.3 kcal/mole. Since there are 36 ATP produced for each glucose, the total kcal/mole value is about 263. As we've seen, the free energy value of glucose is 686 kcal/mole. Thus the efficiency of glucose metabolism in the cell turns out to be about 38% ($263 \div 686 \times 100 = 38.3\%$). (In cells where cytoplasmic NADH shuttles its electrons and protons to mitochondrial NAD^+, a total of 38 ATP is possible and the percentage comes out to about 40%.) By most standards, 38–40% represents an efficient energy utilization. The automobile, when tuned to perfection and driven with utmost care, is less than 25% fuel-efficient! Just in case you've missed anything along the way, have a look at Figure 8.15, where glucose metabolism is summarized.

Other Uses for the Proton Gradient

We referred earlier to a number of transport proteins in the inner mitochondrial membrane. They also make use of proton gradient energy. One coport carrier exchanges hydroxide ions (OH^-) for P_i, thus bringing phosphate to the inner compartment. Another coport exchanges newly made ATP for ADP, which also keeps ATP synthesis going. A uniport moves calcium ions into the inner compartment, which acts as a reservoir for the valuable ion. The transport of these materials is a routine use of the proton gradient, but there is another that is quite unusual, one involving "brown fat" and body heat.

The Mitochondrion as a Heat Machine. The mitochondria in brown fatty tissue are used almost exclusively to generate body heat. This rather remarkable use of the proton gradient is found in mammals that are born relatively hairless—humans being good examples. This special type of fatty tissue, called **brown fat**, accumulates in the neck and upper back regions of the young of these species. Its dark color is attributed to a

high concentration of cytochromes, indicating a great number of mitochondria. Fatty tissue normally doesn't contain a lot of mitochondria, but in brown fat their presence is an adaptation toward generating body heat. Heat is generated as protons from the chemiosmotic gradient escape to the inner compartment, but not through the ATP synthases. They escape through special ("uncoupling") proteins that lack the phosphorylating enzymes. Without these enzymes to make use of the energy, it is simply released as heat, and this heat is used in keeping the young mammals warm. Incidentally, if you have trouble accepting the notion of a proton-filled compartment having great free energy, the observed and recorded use of such a gradient to generate body heat should dispel any doubts.

Anaerobic Cell Respiration

Anaerobic cell respiration is respiration carried out in the absence of molecular oxygen. Most organisms utilizing the anaerobic process are bacteria. They are specifically adapted to anaerobic conditions that exist in such places as the muddy bottom sediments of anaerobic lakes or in airless pockets far down in the soil. Much of the respiratory process is similar to aerobic respiration, including electron transport and chemiosmotic phosphorylation. However, bacteria, as you may recall, do not have mitochondria. It turns out that this doesn't make much difference since *the whole cell,* in both aerobic and anaerobic bacteria, *acts just like a mitochondrion* (Figure 8.16). Both glycolysis and the citric acid cycle occur within the cytoplasm, and coenzymes pass the electrons and protons to electron transport systems situated in the plasma membrane. Proton pumps in the membrane pump protons outside the cell, where they accumulate. They pass down their gradient, reentering the cell through the ATP synthase bodies, also integrated into the plasma membrane. As usual, the free energy of these protons is used to produce ATP (see Chapter 22).

You're probably asking, "If anaerobes don't use oxygen, what happens to the electrons?" We won't keep you waiting. Whereas these bacteria cannot make use of free oxygen (O_2), they can make use of other substances that contain oxygen. Some anaerobic bacteria use sulfate ions (SO_4^{2-}), others use carbon dioxide (CO_2), and still others use nitrate ions (NO_3^-). The end products, respectively, are water and sulfur, water and methane gas (CH_4), and water and nitrogen (N_2). We'll have one more look at bacterial metabolism in Chapter 22.

Fermentation

Organisms that utilize fermentation as a means of generating ATP are metabolically simple. Fermentation is identical to glycolysis except for the addition of a few final reactions. Thus fermenters gain all their ATP through the use of the glycolytic

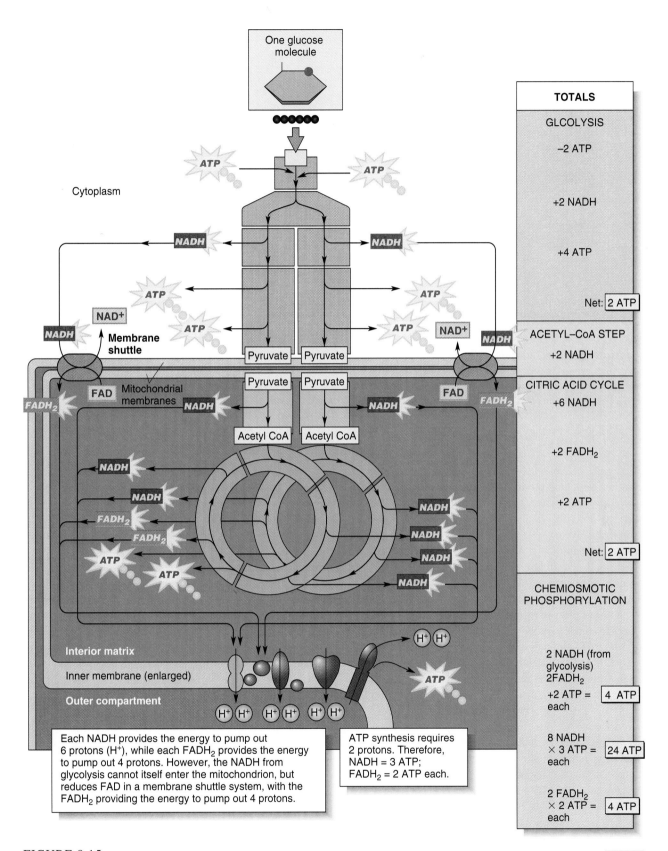

FIGURE 8.15
Summing Up Glucose Metabolism
The entire process of glucose metabolism, from glycolysis to chemiosmotic phosphorylation, is summed. ATPs are totaled at the right.

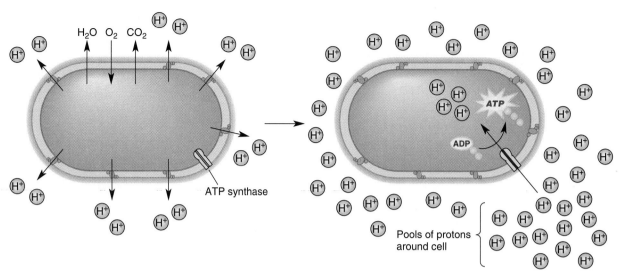

Cellular respiration(results in protons
being pumped out across the membrane)

Protons move back in through ATP synthase,
their energy being used to make ATP

FIGURE 8.16
Chemiosmotic Systems in Bacteria
Aerobic bacteria carry out the full range of glucose metabolism seen in eukaryotes. In so doing, they produce
a steep proton gradient *in their immediate surroundings;* thus they actually live within a pool of protons.
They generate ATP by admitting the protons back into the cell through their own versions of ATP synthases.

pathway. The yield, a small one, occurs through substrate-level phosphorylation. The final products of glycolysis, rather than entering the ATP-generating pathways of cell respiration, are simply converted to energy-rich organic wastes. With this, the fermentation pathway ends. It's important to note that the whole purpose of the fermentation pathway, as far as the organisms are concerned, is to oxidize NADH back to NAD^+. NAD^+ is urgently needed to keep the ATP-generating process going.

It's tempting to conclude that fermenters are simply "inefficient" in their use of glucose metabolism, the proof being energy-rich wastes from which other "more efficient" organisms can extract still more ATP through cell respiration. But if one considers the cost of cell respiration in terms of added structures and materials (mitochondria, enzymes, and electron transport systems), we may come to a different conclusion. The sheer simplicity of substrate-level phosphorylation has an obvious efficiency of its own. It's like comparing the energy efficiency of bicycles and automobiles, and you already know about that.

By the way, industrialists don't consider fermentation wasteful at all. In fact, fermentation wastes from microorganisms are often commercially valuable. Included are several common alcohols (isopropyl, butyl, and ethyl) and organic acids (acetic, lactic, propionic, and formic).

Fermentation Pathways

There are several fermentation pathways, but here we will look into just two: **alcoholic fermentation** and **lactate fermenta-**

tion. The first occurs in yeasts and certain bacteria, whereas the second occurs in bacteria and animal muscle tissue.

Alcoholic Fermentation. The transformation of pyruvate into ethyl alcohol occurs in two steps, each mediated by a specific enzyme (Figure 8.17). The pyruvate is first acted on by a decarboxylase enzyme, which releases carbon dioxide. The product, called **acetaldehyde**, is reduced by NADH, forming ethyl alcohol and NAD^+. The regenerated NAD^+ is recycled back to the glycolytic pathway.

Baker's and brewer's yeasts are **facultative anaerobes**. Depending on conditions, they can switch the metabolism of pyruvate between anaerobic alcohol fermentation and fully aerobic cell respiration. When they go aerobic, carbon dioxide and water are the end products and the total energy yield is 36 ATP per mole of glucose. Aerobic cell respiration in yeast is biochemically identical to human aerobic respiration.

We've been saying that the fermentation pathway ends with energy-rich wastes, and ethyl alcohol (ethanol) certainly qualifies. Ethyl alcohol is familiar as the alcohol of beverages, but it also has medical and industrial importance. You should know that alcoholic beverages are loaded with calories, a fact familiar to dieters. Because it has so much free energy, ethyl alcohol can be combined with gasoline, forming "gasohol." (Gasohol is one alternative to pure fossil fuels.) Fermentation is also important to the baking industry, but bakers are primarily interested in the carbon dioxide, which causes bread to "rise." The ethyl alcohol simply evaporates in the oven.

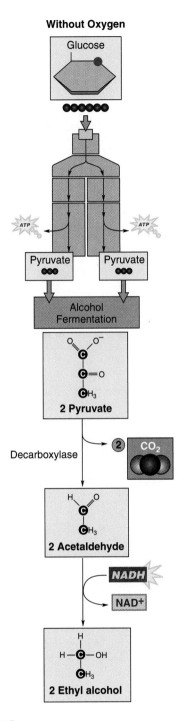

FIGURE 8.17
Alcoholic Fermentation

Alcoholic fermentation, common to yeasts, begins with pyruvate. Carbon dioxide is removed, yielding acetaldehyde, which is reduced by NADH, yielding ethyl alcohol. The reduction step converts NADH to NAD⁺, which is then free to recycle back to the glycolytic pathway. The regeneration of NAD⁺ is important since it permits glycolysis to continue.

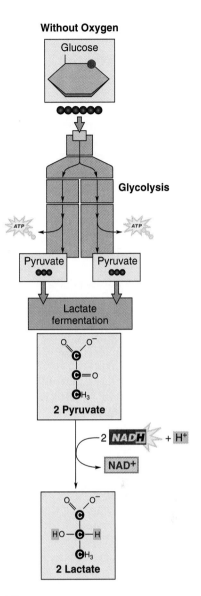

FIGURE 8.18
Lactate Fermentation

In lactate fermentation, pyruvate is reduced by NADH in the presence of the enzyme lactate dehydrogenase, yielding lactate and NAD⁺. The latter recycles back to the glycolytic pathway.

Lactate Fermentation. In the lactate fermentation pathway (Figure 8.18), pyruvate is reduced by NADH, forming lactate and NAD⁺ As usual, the NAD⁺ recycles back to the glycolytic pathway, which again is critical to keeping the pathway going. Among the lactate fermenters are certain bacteria of genera *Lactobacillus* and *Streptococcus,* whose lactate secretions are used to create the unique flavors of rye breads, yogurt, and some cheeses. As we have mentioned already, lactate is also produced in the active muscles of vertebrates and invertebrates. The story of lactate fermentation in our own muscles is interesting, since it explains a lot about exercise, fatigue, and physical conditioning.

Lactate Fermentation in Muscle Tissue. During heavy exertion, a great deal of ATP is used up quickly. In smaller animals, a continuously adequate supply of ATP is provided by aerobic cell respiration, but in larger animals, the circulatory system cannot supply oxygen fast enough to meet the demands. Fortunately, muscle cells have two backup systems. The first backup system is a store of high-energy phosphate in **creatine phosphate**, a molecule that is abundant in muscles (also known as **phosphocreatine**). Creatine phosphate is a much more compact molecule than ATP, and the cell can store a great deal of readily available energy in this form. Creatine phosphate doesn't provide muscle contraction energy directly, but it can transfer its phosphate to ADP through substrate-level phosphorylation:

Creatine phosphate + ADP $\longrightarrow$ ATP + Creatine

As the creatine phosphate is gradually used up, the muscle tissue falls back on its second quick energy source—glycolysis. The ATP generated can then be used to phosphorylate the creatine, restoring the creatine phosphate. But this leads to a buildup of pyruvate and NADH. Muscles have mitochondria so some of this can enter the aerobic cell respiration pathway for the generation of more ATP, but, as we have noted, this is limited by the body's inability to provide sufficient oxygen.

To keep NAD$^+$ recycling, muscle cells immediately activate the lactate fermentation pathway. There NADH reduces pyruvate to lactate, which frees up the NAD$^+$. The lactate, however, accumulates rapidly during intense muscular activity and contributes to muscle fatigue. To counter this, much of the lactate is carried out of the muscles by the bloodstream and delivered to liver cells. There lactate is converted back to pyruvate. The pyruvate is then sent through a special biochemical process known as **gluconeogenesis**, a pathway that is roughly the reverse of glycolysis. There, through an expenditure of ATP, the pyruvate is converted to glucose (Figure 8.19).

A Look at Fatigue and Oxygen Debt. After a period of heavy exertion, the muscle tissues in humans and other vertebrates will be depleted of creatine phosphate, and both the liver and muscles will be loaded with lactate. It takes a period of time and a large amount of oxygen and ATP for the lactate to be metabolized and for the creatine to be regenerated as creatine phosphate. During this time a person will continue to breathe hard, taking in as much oxygen as the lungs can handle. The state of oxygen and creatine phosphate depletion creates what is known as **oxygen debt**, which can be expressed as the amount of extra oxygen needed to restore the system to its pre-exertion equilibrium. (Also see Essay 8.1, page 195.) How long this takes in humans ultimately depends on physical conditioning, which determines heart and lung capacities. Whereas the seasoned athlete can recover from vigorous activity in a few minutes, some of us might need the rest of the afternoon.

Interestingly, many smaller animals generate all or most of their muscle ATP aerobically and have no problem with lactate accumulation and oxygen debt. Examples include

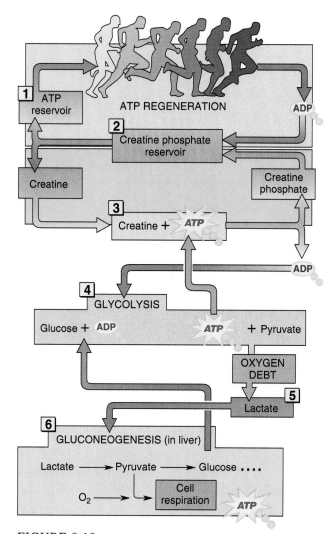

FIGURE 8.19
Energy for Muscular Activity
[1] The energy available for prolonged physical activity comes from limited ATP stores. These stores are replenished [2] through the use of creatine phosphate, which regenerates ATP from ADP. [3] Leftover creatine is then phosphorylated by ATP, which was produced through glycolysis [4]. The pyruvate formed during glycolysis is converted to lactate and sent to the liver [5]. There, during gluconeogenesis [6], some lactate is changed to pyruvate, and converted back to glucose and the rest is used in cell respiration to provide energy needed to make the uphill conversion.

seemingly tireless migratory birds and small, fast-running mammals. But the great body mass of larger animals does not allow their circulatory systems to keep up with the increased oxygen demands. Nile crocodiles, for instance, are generally sluggish creatures. But when threatened or when stalking prey on land, they can make astonishingly fast charges and can lash their tails about with results that are legendary. Following such outbursts though, the giant reptiles must remain still, requiring many hours to repay the oxygen debt.

Alternative Fuels for the Cell

Biologists speak of glucose so much that it might appear that this is the only cellular fuel. While glucose is a common cellular fuel, it is not the only one, and sometimes not even the most important one. For instance, we all rely heavily on stored fats as another primary energy supply. We maintain a supply of glucose in the form of glycogen, but only enough to keep us going for about a day of normal activity. Unless the glycogen is replenished, our body fats soon become mobilized, broken down into fatty acids and made ready for use by cells. Unless you've found a way to eat in your sleep, for instance, nighttime is a time of fasting. By morning your body will have switched to fats, so that most of the ATP energy produced in the mitochondria comes from the oxidation of fatty acids. (If you are counting calories, this might be a good excuse for sleeping in!) It should be good news to dieters that body fats can be sent through the respiratory mill. In fact, fats are excellent energy sources, yielding about twice the amount of energy as carbohydrates on a gram per gram basis. Interestingly, fatty acids are the preferred chemical fuel of heart muscle, but they are utterly rejected by the brain, whose chief energy source is glucose. Proteins can also be used as energy sources, but their use is complicated by their much more varied composition.

We've seen how the carbohydrate glucose is metabolized, so let's consider the two other fuels, the fats and proteins. Both can be used as fuels in the mitochondrion, but first they must be converted to a form that the mitochondrion can handle (Figure 8.20). The pathways through which such conversions occur are part of what is called **intermediary metabolism**.

The Metabolism of Fatty Acids

To enter the mitochondrion, fatty acids must first be activated by coenzyme A (CoA). Once inside, they are prepared for the citric acid cycle. Activation, which takes place on the outer mitochondrial membrane, consists of joining the fatty acid to coenzyme A. The activated fatty acid is then ferried across the membrane by special transport proteins, ending up in the matrix. Once there, a 2-carbon acetyl–CoA is split away to enter the citric acid cycle. Then another coenzyme A joins the remaining fatty acid fragment, and the reaction is repeated, freeing a second 2-carbon acetyl–CoA. This continues until the fatty acid chain has been fragmented completely, two carbons at a time, into acetyl–CoAs.

Palmitic acid, a 16-carbon fatty acid, goes through seven reactions, yielding eight acetyl–CoAs. This might seem like a lot of bother, but there is a special bonus. Each reaction includes an oxidation and a reduction, so along with the acetyl–CoAs, seven NADHs and seven FADH$_2$s also form. The reduced coenzymes, as you might expect, greatly enrich the chemiosmotic gradient of the mitochondrion, leading to the production of ATP. This explains why fats are such a high-energy food.

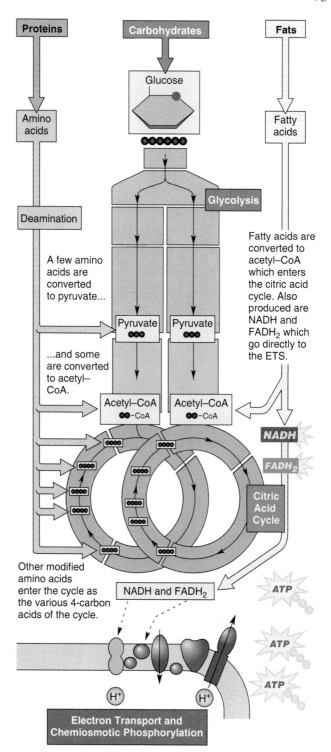

FIGURE 8.20
Alternative Fuels
The use of alternative fuels by the mitochondrion is commonplace. Modified amino acids (left) enter the mitochondrial processes as any of several metabolites. Some are converted to pyruvate, others to acetyl–CoA, and still others to acids of the cycle itself. The least productive would be fumarate, which yields just one NADH. Fatty acids (right) are converted to acetyl–CoA. The large gain in NADH is made available to the electron transport system. The acetyl–CoA itself joins oxaloacetate, entering the citric acid cycle at its beginning.

FIGURE 8.21
Biosynthesis
The versatility of the citric acid cycle is impressive. Its metabolites can be used to form fatty acids, steroids, amino acids, purines, heme groups, chlorophyll, pyrimidines, and even glucose itself. So the next time a friend says "you are what you eat," tell him or her about biosynthesis. (You can always find a new friend.)

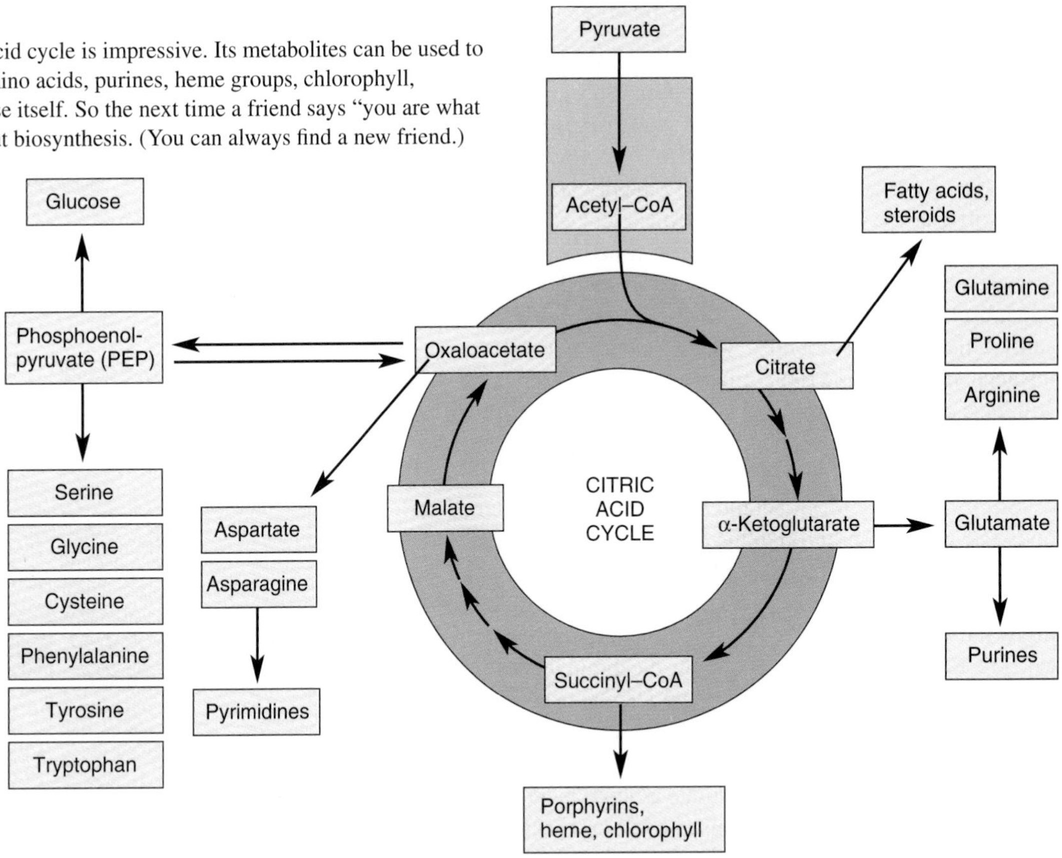

The Preparation of Proteins

In vertebrates, proteins are hydrolyzed by digestive enzymes of the stomach and small intestine into the usual 20 amino acids. They are then transported to the liver for processing and distribution. For use in cell respiration, they must next have their amino groups (NH_3^+) removed, a process called **deamination**, and these groups must be rendered harmless. Unattended, the amino groups would accumulate as dangerous ammonium ions. In mammals, the ammonia is converted primarily to urea, which is excreted by the kidneys.

The valuable amino acid fragments, or metabolites as we have called them, can then enter the mitochondria. Five of the 20 different amino acid metabolites are simply converted to pyruvate, which is then converted to acetyl–CoA. Six more skip the pyruvate step and enter as acetyl–CoA itself. The other amino acids are converted into specific acids, or metabolites, of the citric acid cycle, joining it directly (see Figure 8.19).

The amount of energy available from each type of amino acid depends, as you might suspect, on where its conversion product joins the citric acid cycle. This makes sense because the point of entry also determines how much NADH is generated. The five amino acids converted to pyruvate have the most to offer, since they yield the maximum number of NADH possible. Those entering as acetyl–CoA are next in yield, while those entering as citric acid cycle substrates are, of course, the least productive.

Intermediary Metabolism: Biosynthetic Pathways

The citric acid cycle, while central to respiration, serves another purpose. Its enzymes provide raw materials for a variety of uses. Several of its acids can be diverted from the cycle for use in the synthesis of new molecules, a part of intermediary metabolism called **biosynthesis**. This is possible because the reactions described so far are reversible, and because there are many pathways leading into and out of the citric acid cycle. For example, we see that although fatty acids may enter the cycle as acetyl–CoA, under certain conditions (with energy provided), the reactions can be reversed and the acetyl groups can be converted to fatty acids.

Perhaps the greatest versatility is in the biosynthetic pathways involving amino acids (Figure 8.21). Many of the amino acids can actually be manufactured from citric acid cycle metabolites. For example, alpha-ketoglutarate (see Figure 8.20) can be drawn off and used in the synthesis of glutamic acid, proline, or arginine. Alternatively, alpha-ketoglutarate can be used to form purines such as thymine and cytosine, components of DNA nucleotides (see Chapter 3). Oxaloacetate, the resident molecule, can be converted to no fewer than eight different amino acids. Conversely, amino acids entering the reactions as acetyl–CoA can be turned around and converted directly into fatty acids. (Such information might be important to dieters starting on some exclusive protein or car-

Essay 8.1 ● A MARATHON RUNNER MEETS "THE WALL"

There is a point in the marathon when many runners hit the dreaded "wall." The wall usually looms before them at about mile 18 or 20, and it is here that many flounder—falling prostrate, staggering, or simply walking. It is here that other, better-trained runners begin to pass those whose dust they had been eating, on their way to the conclusion of the 26-mile, 385-yard race. (Some runners say that a marathon is divided into two equal parts: the first twenty miles and the last six.)

There has been a lot of discussion over the years about what the wall really is, but everyone who has encountered it can attest to its reality. Physiologically, it seems to correspond to the period when muscle glyco-gen, the most readily available reserve of glucose, is depleted. Glucose is normally depleted after two to three hours of slow running, at about an eight-minute-per-mile pace. The runner's body at this point must switch to metabolizing fat molecules. The blood pH lowers, and the transition is, for some reason, exceedingly stressful. It has also been suggested that women switch to fat metabolism more easily than men, a suggestion supported by their increased relative numbers in 50- and 100-mile "ultra-marathons."

Interestingly, when you encounter the wall, sheer willpower has little effect. You simply can't go on. The muscles refuse. Any effort is extremely painful and fatiguing. In addition, blood sugar levels drop, and you may suffer the psychological depression associated with hypoglycemia.

Marathoners have been able to beat the wall by training so hard that it no longer exists for them. This training includes long runs of 20 miles or more, during which muscle glyco-gen is repeatedly depleted and then restored in even greater abundance during the following week. Some runners believe they are able to move the wall back by "carbo loading," that is, eating primarily foods high in carbohy-drates for the week preceding the race. The idea is that the excess carbohydrates will be stored as glycogen that can then be called on at mile 20 to get them through to mile 26.

bohydrate diets—it seems that *practically anything we eat can end up as fat.*) In a real sense, because of such versatility, glucose can be considered as a starting molecule for many important molecular subunits. Once it has entered respiration, its products can be used in a great variety of ways. But, more to the point, the importance of glycolysis and the citric acid cycle to the cell goes much farther than providing energy.

This completes our focus on cell structure, function, and energetics, although certainly our interest in these important topics is not over. Here, we've laid a foundation on which to build. As we venture into higher levels of organization in biology, from organisms to ecology, we'll return time after time to the cell and the fundamental considerations of energy.

Key Ideas

1. Photosynthesis is basically an uphill, endergonic process, whereas glucose metabolism is downhill and exergonic.

TWO WAYS OF UTILIZING GLUCOSE

1. Cells metabolize glucose in **aerobic cell respiration** and **anaerobic cell respiration**. **Fermentation** is a brief anaerobic process. All are preceded by **glycolysis**, a preliminary process with a small ATP yield. The three parts of glucose metabolism are glycolysis, citric acid cycle, and electron transport/chemios-motic phosphorylation.

GLYCOLYSIS

Since it is a preliminary metabolic phase in all organisms, glycolysis may be an ancient process. The pathway includes nine principal reactions.

1. The **preparatory phase** includes the phosphorylation of the glucose by ATP, cleaving of the molecule into two 3-carbon fuels, oxidation, and phosphorylation by P_i. During oxidation, NAD^+ is reduced to NADH.

2. The **yielding phase** occurs in two reactions, when high energy phosphates in the fuel are transferred directly to ADP. The net gain is two ATPs, which represents about 2.1% of the available energy. The end products are two ATPs, two NADHs and two pyruvates.

3. The allosteric enzyme **phosphofructokinase** is inhibited by ATP and excess citrate, and stimulated by ADP.

AEROBIC CELL RESPIRATION: THE MITOCHONDRION

1. Pyruvate is acted on by coenzyme A, yielding CO_2, NADH, and **acetyl–CoA.**

2. The important events of the cycle are the oxidation reactions. Coenzyme NAD^+ and FAD are reduced by electrons and protons from the substrate. A small amount of ATP is also generated.

3. In more detail, acetyl–CoA joins 4-carbon **oxaloacetate**, forming 6-carbon citrate with coenzyme A recycling.

4. Four oxidations in the cycle bring about the reduction of three NAD^+s to $NADH + H^+$, and one FAD to $FADH_2$.

5. A substrate-level phosphorylation generates one ATP.

6. For each two pyruvates entering the mitochondrion, the yield is 8 NADH, 2 FADH$_2$, 2 ATP, and 6 CO$_2$.

ELECTRON TRANSPORT AND CHEMIOSMOTIC PHOSPHORYLATION

1. The double membrane provides for an **outer** and **inner compartment**. The outer compartment is the proton reservoir. The **inner membrane** is highly folded and contains many electron transport systems and ATP synthases.

2. A number of electron carriers are iron proteins called cytochromes. The iron groups take in electrons and pass them along.

3. Electrons and protons from NADH join the electron transport system at **protein complex I**. At this site and at **complexes III** and **IV**, further along, protons are pumped into the outer compartment, for a total of six protons per NADH.

4. Electrons and protons from FADH$_2$ enter the electron transport system at **protein complex II**, powering the transport of four protons across the inner membrane at complexes III and IV.

5. Spent electrons join oxygen, and, with protons from the matrix, they form water.

6. NADH and FADH$_2$ from two pyruvates account for the pumping of 56 protons. Eight more are provided by the two NADHs from glycolysis, which are shuttled into the mitochondrion as FADH$_2$. Grand total: 64 protons per glucose.

7. Every two protons passing through an ATP synthase provide the free energy for generating one ATP.

8. The ATP yield from all of respiration is 36 per glucose—2 from glycolysis, 2 from the citric acid cycle, and 32 (64 protons ÷ 2 protons per ATP) from chemiosmotic phosphorylation. Thirty-six ATPs represent an efficiency of 38%.

9. In brown fatty tissue there are no ATP synthases and the proton gradient is used to generate body heat.

ANAEROBIC CELL RESPIRATION

1. Anaerobic bacteria carry out cell respiration in a similar manner but without molecular oxygen. The whole cell acts as a mito-chondrion, with electron transport and proton pumping going on in the plasma membrane, and a proton concentration forming just outside the cell. ATP is produced when protons move back into the cell through ATP synthases.

2. Anaerobic bacteria use sulfate ions (SO$_4^{2-}$), carbon dioxide (CO$_2$), or nitrate ions (NO$_3^-$) as oxygen acceptors, yielding sulfur, methane gas, or nitrogen, respectively.

FERMENTATION

1. In **alcoholic fermentation**, pyruvate loses CO$_2$, and the product, acetaldehyde, is reduced to ethyl alcohol by NADH. NAD$^+$ recycles.

2. In **lactate fermentation**, pyruvate is reduced by NADH to form **lactate**. NAD$^+$ recycles. Lactate fermentation occurs in lactic acid bacteria and in animal muscle.

3. During muscle contraction, ATP is provided as follows:
   a. Small reserves of ATP get the muscle going, generating ADP.
   b. ADP is phosphorylated by **creatine phosphate**, regenerating ATP.
   c. Creatine phosphate is replenished by ATP from glycolysis.

4. During glycolysis, lactate accumulates in the muscle tissue. Some is converted back to glucose through **gluconeogenesis** and some is oxidized. The oxygen needed to restore muscle is known as **oxygen debt**.

ALTERNATIVE FUELS FOR THE CELL

1. Cells convert fats into fatty acids and glycerol with a large NADH yield. They enter respiration as pyruvate and acetyl–CoA. Proteins are broken down into amino acids, which are further processed for cell respiration. Some enter as pyruvate, others as acetyl–CoA, and still others as acids of the citric acid cycle.

2. The respiratory pathways also act as biosynthetic pathways, wherein the one type of molecular subunit can be used in the synthesis of another. Examples include the use of acetyl-CoA formed from amino acids or pyruvate for the synthesis of new fatty acids. Several citric acid cycle acids can be taken from the mitochondrion and converted to amino acids.

Application of Ideas

1. What does it mean for a runner to "go anaerobic"? Who depends more on anaerobic respiration—a sprinter or a marathon runner? Explain.

2. Is anaerobic respiration really inefficient? Consider the question in terms of the free energy actually available to dwellers of the anaerobic environment. Also consider the answer in terms of whether or not anaerobic organisms are successful, well-adapted organisms.

3. A study of quail, notably permanent residents and primarily ground birds, reveals that their flight muscles (breast) are nearly white—like those of chickens. Similar observations in migratory ducks reveal that the flight muscles are dark colored. The dark color, it turns out, is due to heavy concentrations of cytochromes and myoglobin (an oxygen-carrying pigment). What can you conclude about the biochemistry of flight in the two birds?

Review Questions

1. What is the primary difference between anaerobic and aerobic cellular respiration? (page 174)

2. In what way is fermentation like anaerobic cellular respiration? What is the biggest difference? (pages 188–190)

3. Describe five things that happen to substrate molecules in the preparatory phase of glycolysis. Name the product formed just before substrate-level phosphorylation. Describe its energy state. (pages 175–176)

4. What percentage of the free energy of glucose is represented in the phosphate bonds of ATP at the end of glycolysis? (page 176)

5. Write a word equation that summarizes the acetyl–CoA step of respiration. What happens to each product? (page 177)

6. What is the overall purpose of the citric acid cycle? What molecules transfer the fuel energy to the electron transport system? (page 177)

7. Determine the total number of NADHs, $FADH_2$s, and ATPs produced when two pyruvates complete mitochondrial respiration. (page 181)

8. Use a simple line drawing to illustrate the arrangement of compartments in the mitochondrion. What is the significance of the two compartments to chemiosmosis? (page 183)

9. For what is the free energy of electrons in the ETS used? (page 183)

10. Compare the proton yield from NADH with that of $FADH_2$. (page 184)

11. What becomes of spent electrons leaving the electron transport systems? Where in our own bodies does the final product end up? (page 186–187)

12. Explain how the free energy of the proton gradient of the mitochondrion is used. (pages 187–188)

13. Why is it so important for fermenters to oxidize their NADH? (page 190)

14. List four sequential steps used in replenishing ATP during muscle contraction. (pages 191–192)

15. Explain how an "oxygen debt" arises and how it is paid off. (pages 191–192)

16. Explain why fats have such a high energy value. (page 193)

17. What is biosynthesis? Explain how the mitochondrion is used in biosynthesis. (page 194)

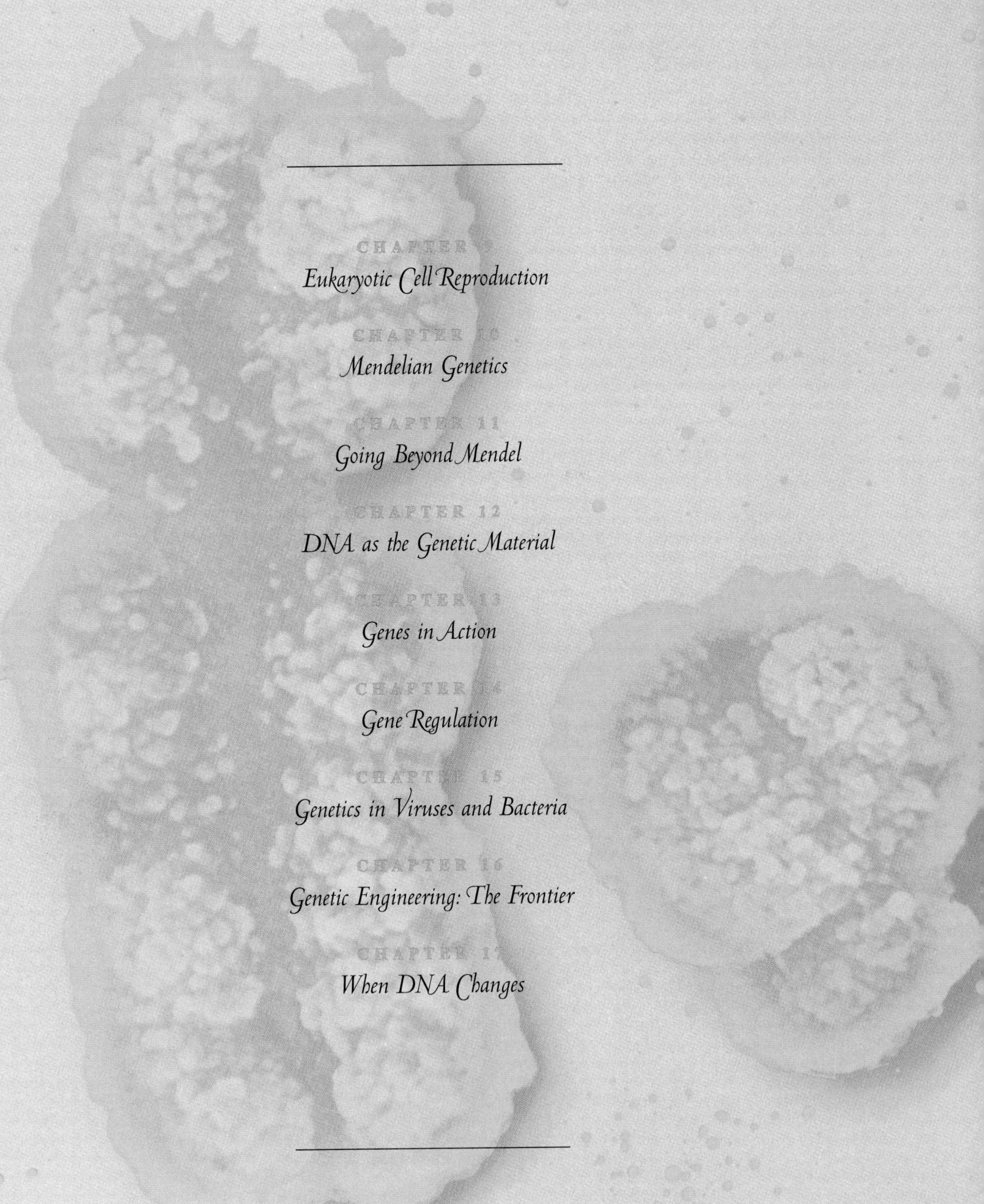

CHAPTER 9

Eukaryotic Cell Reproduction

CHAPTER 10

Mendelian Genetics

CHAPTER 11

Going Beyond Mendel

CHAPTER 12

DNA as the Genetic Material

CHAPTER 13

Genes in Action

CHAPTER 14

Gene Regulation

CHAPTER 15

Genetics in Viruses and Bacteria

CHAPTER 16

Genetic Engineering: The Frontier

CHAPTER 17

When DNA Changes

PART TWO

Molecular Biology and Heredity

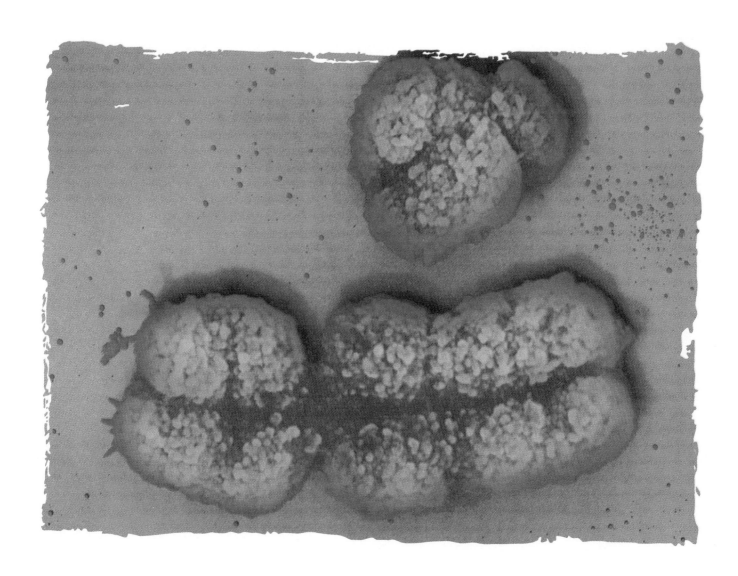

CHAPTER 9

Eukaryotic Cell Reproduction

CHAPTER OUTLINE

THE NUCLEUS

DNA, Chromatin, and the Eukaryotic Chromosomes

The Cell Cycle: Growth, Replication, and Division

MITOSIS

Prophase

Metaphase

Anaphase

Telophase

Essay 9.1 The Work of the Spindle

Cytokinesis

Advantages of Mitosis

MEIOSIS

Homologous Chromosomes

Essay 9.2 Karyotyping

Overview of Meiosis

Meiosis I: The First Meiotic Division

Meiosis II: The Second Meiotic Division

MEIOSIS IN HUMANS

Meiosis in Females: A Special Case

Meiosis, Evolution, and Sex

Essay 9.3 When Meiosis Goes Wrong

KEY IDEAS

APPLICATION OF IDEAS

REVIEW QUESTIONS

*Y*ou are not the same person you were a few years ago. In fact, you aren't the same person you were a few seconds ago. This is not to say that these sentences are so profound that you will never be the same for having read them. It's just that even as you sit here reading, your cells are growing, dying, and reproducing. Your body is coping with a rigorous world, and it must change or perish. Some cells, of course, reproduce more rapidly than others. For example, the parts of your body that receive a great deal of friction, such as the palms of your hands or the lining of your small intestine, must constantly replace cells that have died or been worn away. Other parts of your body have very low cell turnover. And some, such as the skeletal muscles and nervous system, are never replaced in adults. Nonetheless, at this very moment, your cells are quietly accomplishing the processes of life, many of them slowly and methodically reorganizing themselves and dividing again and again, duplicating themselves ever so precisely in their unending pageant until the day it all stops.

While this precision rules most of the time to produce the harmonic symphony of life, a dissonant note in this orchestrated performance, just a single miscue, can spell disaster for the resulting cells. Thus it is important in our understanding of life to be aware of how cells reproduce so unerringly. Here then we will see just how closely regulated and precise the process of eukaryotic cell reproduction really is, beginning with the events in the nucleus. After discussing the structure of the nucleus and the chromosomes within the nucleus, we will examine in detail the mechanisms that guide the duplication and distribution of those chromosomes.

The Nucleus

The eukaryotic cell nucleus, its structure, and its contents are central to our considerations of cell reproduction, so we begin with a general review of this prominent organelle shown again in Figure 9.1. Recall that the nucleus contains the cell's primary hereditary information (DNA), assembled into chromosomes. In addition to chromosomes, the nucleus generally contains one or more prominent nucleoli, dense bodies that produce a special class of RNA called ribosomal RNA. The **nuclear envelope** is a double membrane that is continuous with the endoplasmic reticulum (see Chapter 4) and surrounds the nucleus through most of the life of a cell.

The nuclear envelope is a highly selective barrier, readily admitting some substances and rejecting others. Transport of molecules between the nucleus and the cytoplasm is accomplished via **nuclear pores**. Nuclear pores are not simply openings in the nuclear envelope, nor are they similar to the gap junctions or plasmodesmata of the plasma membrane. Instead, nuclear pores are areas with complex protein mechanisms that facilitate the passage of certain macromolecules. Because of this selectivity, materials within the nucleus can be quite different from those of the cytoplasm. During cell reproduction, the nuclear envelope disappears, only to be reestablished after the completion of cell division.

DNA, Chromatin, and the Eukaryotic Chromosomes

DNA, you recall, is a double helix composed of two extremely long strands of nucleotides. In fact, these molecules can be so long that if the DNA molecule of the largest human chromosome were stretched out, it would be approximately 12 cm (4.7 in.) in length. But fortunately for our physical integrity, this never happens. Instead, the DNA is condensed and packaged into manageable bundles. This packaging is accomplished by the formation of **chromatin**, which is DNA that is associated with an array of proteins. These proteins are of two general types—the histones and others known (logically enough) as nonhistone chromosomal proteins.

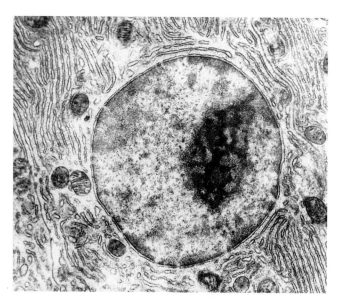

FIGURE 9.1
The Eukaryotic Cell Nucleus
The most prominent features of the nucleus are the nucleolus (darkest area) and the chromatin net, both of which stain darkly in typical electron microscope preparations.

Chromatin is formed in several steps (Figure 9.2), each increasing the compactness of the chromosome. First, the negatively charged DNA molecule is wound two and a half times around globules consisting of two copies each of four different kinds of histone protein (called H2a, H2b, H3, and H4). Histones carry a strong positive charge, and this aids in their association with the negatively charged DNA. The units formed in this way are called **nucleosomes**. The millions of nucleosomes along the length of a chromosome are connected by short stretches of histone-free "linker" DNA to form a "beads-on-a-string"-like structure. This nucleosome-linker-nucleosome structure is the basic unit of DNA packing.

Higher levels of chromosome condensation are accomplished by further packaging of the beads-on-a-string (see Figure 9.2). First, an additional histone molecule, called H1, assists in holding the nucleosomes closer together. Then the nucleosomes are stacked into a coil with a width of 30 nm, logically called the 30-nm fiber. Higher levels of condensation are accomplished with the aid of the nonhistone chromosomal proteins by further coiling of the 30-nm fiber, until the familiar eukaryotic chromosome appears, which is condensed (potentially) to $\frac{1}{10,000}$ of its original length!

The mass of chromatin in nondividing cells, which resembles a bundle of yarn subjected to the attentions of a demented kitten, is sometimes referred to as the chromatin net because of its net-like appearance under the electron microscope (see Figure 9.1). Parts of the chromatin net are thin and diffuse, while other parts are thicker and more tangled. This varied appearance is due to the fact that the chromatin of nondividing cells is found in many different stages of condensation.

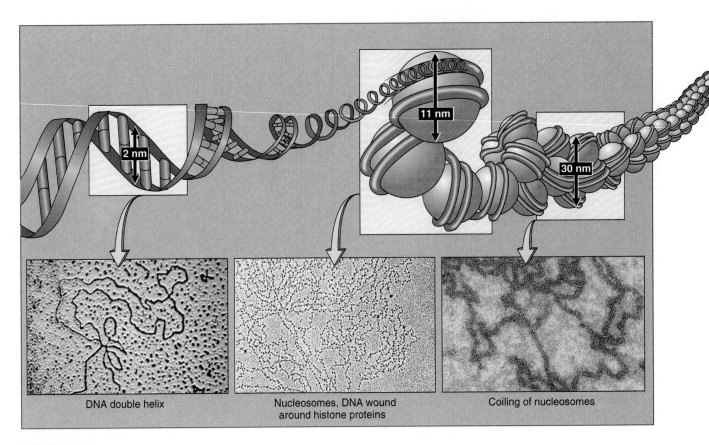

FIGURE 9.2
Chromosomal Organization
The familiar chromosome is the ultimate result of several levels of organization and condensation. Naked DNA associated with histone proteins forms nucleosomes that give a "bead-on-a-string" impression. Coiled packing of nucleosomes forms the 30-nm fiber. Further twisting of the 30-nm fiber forms increasingly more dense coils, which eventually intertwine to form the chromosomal strands that become the arms of the chromosome.

The Cell Cycle: Growth, Replication, and Division

The life of a healthy growing cell begins with its formation from a parent cell and ends when the cell divides to produce progeny cells. This is called the **cell cycle**, which is an endless repetition of growth, reproduction, growth, reproduction, and so on (Figure 9.3). The cell cycle is traditionally divided into two phases, **interphase** and **mitosis** (also called M phase). Each of these phases, in turn, is made up of several different parts. The cell actually spends most of its time in interphase, the period of metabolic activity and growth. After briefly discussing interphase, we will concentrate on mitosis because it is the process that results in the division of one cell into two progeny or daughter cells.

Interphase may be defined as the stage of the cell cycle between successive mitoses. It includes three different parts: the G_1 phase, the S phase, and the G_2 phase (see Figure 9.3). When cells cycle, G_1 follows immediately on the heels of mitosis. This is a very active period, a time when the cell synthesizes its vast array of proteins, including the enzymes and structural proteins it will need for growth. In G_1 each chromosome consists of a single molecule of DNA and its proteins, but this soon changes.

During the S phase (or synthesis phase), all of the DNA within the nucleus will be replicated. This means that each chromosome will faithfully be copied so that by the end of the S phase there will be two DNA molecules for each one present in G_1. The importance of this step in the cell cycle is obvious. It ensures that each emerging progeny cell will have the same genetic content as the parent cell.

G_2 is the time after DNA replication and before mitosis. It is an important preparatory period, as the apparatus for nuclear division is assembled in preparation for mitosis. Specifically, this is a time of tubulin synthesis. The two-part protein (see Chapter 4) will be used in assembling microtubules, as we will see shortly.

The amount of time the cell spends in interphase varies greatly according to the cell's type and stage of development. The shortest periods generally are found in the rapidly growing embryo. The large, newly fertilized frog egg, for example, will complete a cell cycle every 30 minutes, thereby producing about 8000 cells in 12 hours.

Some cells break out of their cycle, as shown by option 2 of Figure 9.3. As an example, fingernail cells become filled with keratin (a tough protein) and then die to form the actual

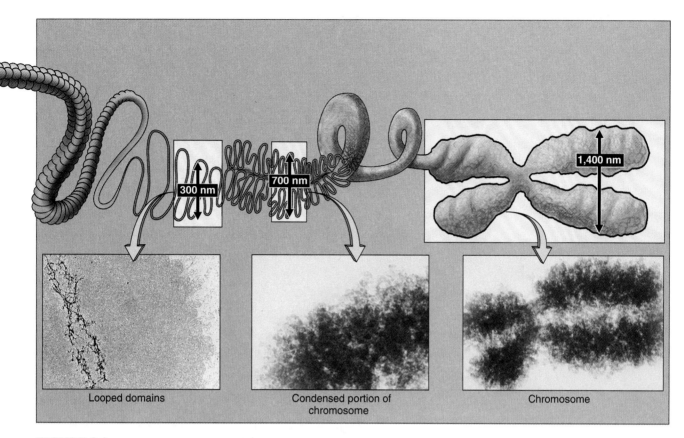

Looped domains   Condensed portion of   Chromosome
                              chromosome

FIGURE 9.2

fingernail, only to be pushed along by dividing cells further back. The red blood cells of mammals, which become specialized to the point that they have no nuclei, will eventually die without reproducing. Still other cells, most notably some types found in the liver and all nerve cells, essentially cease dividing when we are quite young and usually cannot be replaced. (It is worth noting, however, that recent research has made great strides toward getting nerve cells to divide to regenerate nerves.) Although such cells are said to "last a lifetime," it may well be that the gradual cumulative effects of the deaths of such cells contribute to aging.

Mitosis

During interphase, the nucleus is truly a busy place, but all the activity is invisible when viewed with the light microscope. However, in mitosis, the activity within the nucleus becomes dramatically visible, as the nuclear material shifts, reorganizes, and moves around.

Mitosis is simply the process of cell division. It is a continuous process with one event leading gradually to the next, yet biologists have described the process in phases: prophase, metaphase, anaphase, telophase, and cytokinesis. (Cytokinesis is sometimes viewed as a stage of its own, with the term mitosis limited to division of the nucleus. However, we will keep with the traditional definition of mitosis.) As we move

through our discussion of the phases of mitosis (Figure 9.4), note that each phase ends with a smooth transition to the next.

Prophase

The onset of mitosis, **prophase**, is marked by the condensation of the dispersed chromatin net into the visible structures of chromosomes (Figure 9.4). Remember that the chromosomes were just replicated in S phase, so that there are now two copies of each chromosome. The two copies are called **chromatids** (or sister chromatids) and are joined together at a region known as a **centromere** (Figure 9.5). Generally, the individual chromatids, connected by their centromeres, are not visible until late in prophase. As condensation of the chromosomes proceeds, the spindle begins to form. The **spindle** is an elaborate structure, formed of protein fibers, that is involved in chromosome movement during mitosis. It is constructed anew for each cell cycle, then dismantled after it has been used. You may recall from Chapter 4 that the mitotic spindle fibers are microtubules, which are long polymers of tubulin subunits.

As prophase continues, the chromosomes begin to move in an agitated manner, presumably being tugged about by the spindle. Following the breakdown of the nuclear envelope, each pair of sister chromatids is attached to **centromeric spindle fibers** from opposite directions. The spindle is attached to the centromere region of the chromatids by a special structure known as the **kinetochore** (see Figure 9.5). One

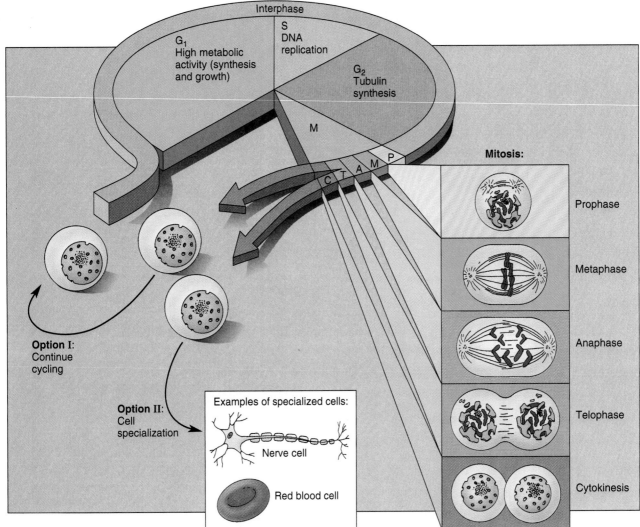

FIGURE 9.3
The Cell Cycle
There are two generally recognized phases to the cell cycle, interphase and mitosis. Interphase is divided into G_1, S, and G_2. Mitosis is subdivided into prophase, metaphase, anaphase, telophase, and cytokinesis. After mitosis, a cell may cease to cycle and may permanently become a particular cell type (option II); or it may continue to go through the cell cycle (option I), generating more daughter cells.

of the sister chromatids is attached to a spindle fiber emanating from the "north pole" of the dividing cell; the other becomes attached similarly to the "south pole." The centrioles (where present, as in animals, many protists, and a few plants—see Chapter 4) migrate to opposite sides of the cell. The centrioles then become surrounded by a cluster of microtubules that radiate outward, taking on a "starburst" form called an **aster**. Other microtubules form **polar spindle fibers**, fibers that are not attached to chromosomes but span the entire distance between the two poles.

Lastly, in late prophase, movement becomes more organized, as each chromosome moves to the cell's center (see Figure 9.4), finally coming to rest in the middle of the cell at the equatorial plane. There the paired chromatids line up across the center of the cell, thus marking the arrival of the next phase, metaphase. The nucleoli degrade, and in the final stages of prophase the nuclear envelope disappears.

In summary, prophase is an incredibly busy time in which the chromatin acheives its most condensed state, the spindle apparatus forms, the nuclear envelope breaks down, and the many individual chromosomes become arranged across the cell. The key feature is the precise and opposite attachment of centromeric spindle fibers to sister chromatids. This sets the stage for equitable division of the replicated DNA.

Metaphase

Metaphase is simply that stage of the process that is marked by the arrival of the chromosomes at the equatorial plane, which is also referred to as the **metaphase plate** (see Figure 9.4). Here the chromosomes are momentarily held in place and appear to be at rest. Their static position, however, is maintained by strong opposing forces that keep the centromeres on the metaphase plate.

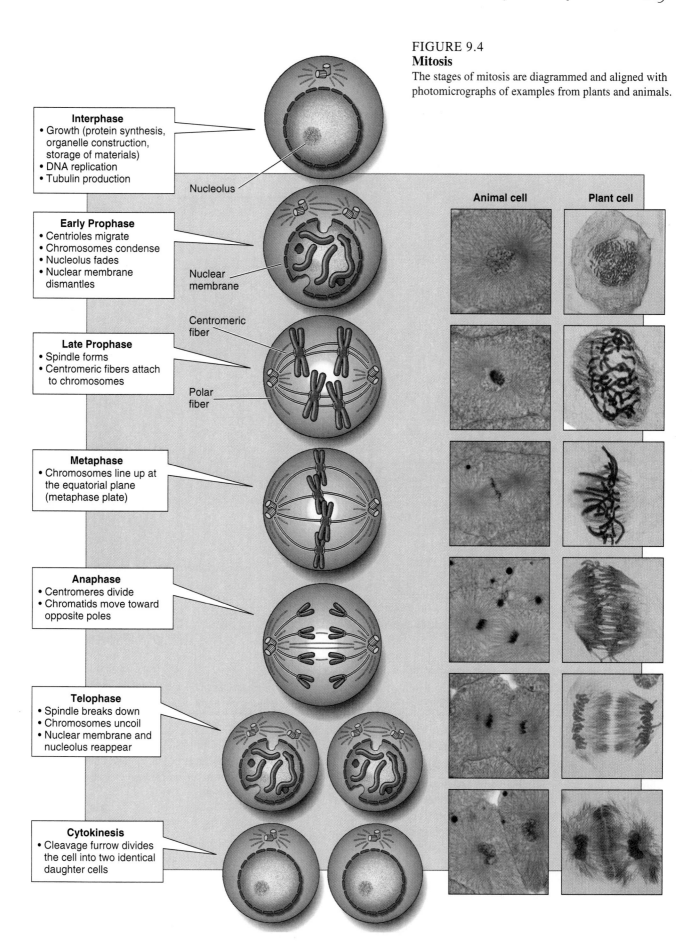

FIGURE 9.4
Mitosis
The stages of mitosis are diagrammed and aligned with photomicrographs of examples from plants and animals.

Interphase
• Growth (protein synthesis, organelle construction, storage of materials)
• DNA replication
• Tubulin production

Nucleolus

Early Prophase
• Centrioles migrate
• Chromosomes condense
• Nucleolus fades
• Nuclear membrane dismantles

Nuclear membrane

Late Prophase
• Spindle forms
• Centromeric fibers attach to chromosomes

Centromeric fiber

Polar fiber

Metaphase
• Chromosomes line up at the equatorial plane (metaphase plate)

Anaphase
• Centromeres divide
• Chromatids move toward opposite poles

Telophase
• Spindle breaks down
• Chromosomes uncoil
• Nuclear membrane and nucleolus reappear

Cytokinesis
• Cleavage furrow divides the cell into two identical daughter cells

Animal cell

Plant cell

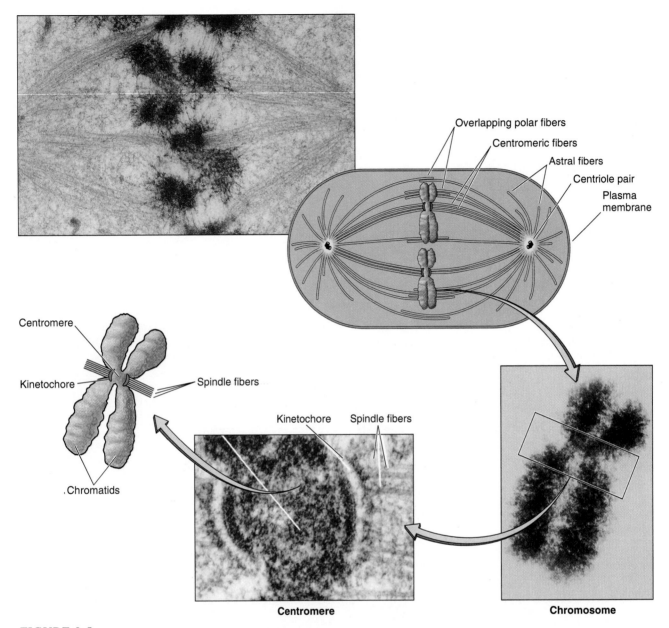

FIGURE 9.5
The Mitotic Spindle Apparatus and Spindle Fiber Attachment.
Note the relationship of the spindle fibers, asters, centrioles, and chromosomes. The microtubules of the spindle fibers
are seen attached to the curved kinetochore, which is physically associated with the centromere of the chromosome.

Anaphase

Anaphase ends abruptly as the centromeres holding the sister chromatids together split apart. The chromatids can now be referred to as chromosomes. They can take on rough V or J shapes (depending on the position of the centromere) as they are drawn to their respective poles (see Figure 9.4), centromere first, by the spindle fibers. Chromosome movement requires several minutes. The centromeric fibers seem to pull the chromosomes along while the polar fibers elongate, thereby pushing the poles apart. Essay 9.1 provides a closer look.

It is in anaphase that the precise and equal separation and distribution of chromosomes take place, even though the cell has yet to divide. One sister chromatid (one copy of each chromosome) has been brought to each end of the cell. This must happen with utterly faithful precision if the two emerging cells are to survive.

Telophase

In **telophase** (see Figure 9.4), many of the events of prophase are essentially reversed. The chromosomes begin to decondense, their sausage-like form fading into the diffuse chromatin seen at interphase. The spindle is dismantled, its proteins to be recycled in forming the new cytoskeleton. A new nuclear

Essay 9.1 ● THE WORK OF THE SPINDLE

We've seen that the spindle apparatus is architecturally at the very core of mitotic movement. It consists primarily of microtubules composed of the protein tubulin, which is also a principal component of the cytoskeleton and cilia and flagella. In fact, in the late G_2 phase of the cell cycle, tubulin accounts for as much as 10% of the cell's protein. The assembly of tubulin into the microtubules comprising the spindle occurs in early prophase. The site of microtubule assembly is a region at each spindle pole. Curiously, these regions, aptly called microtubular organizing centers (MTOCs), lie directly around the centrioles in animal cells. Yet a functional relationship between the MTOC and centriole has yet to be established.

For years, biologists have made two heretofore unexplained observations about mitosis: (1) as the chromatids separate and the newly liberated chromosomes move to opposite poles, the kinetochore spindle fibers get shorter, and (2) as this happens, the entire spindle gets longer. Now there seems to be some some solid understanding of the related processes, as well as an explanation of how chromosomes move during anaphase of mitosis and meiosis.

The kinetochore spindle fibers shorten through tubulin disassembly, which is now believed to occur at the kinetochore itself. Further, as the disassembly process occurs, dynein arms extend from anchoring points in the kinetochore, attaching to the microtubular fibers just past the disassembly point. Then, through an ongoing tugging, detaching and reattaching process, the dynein arms creep along the fiber, pulling the chromosomes toward the pole. A reasonable analogy would be standing in the bow of a rowboat and pulling it ashore by a rope attached there, except that the rope would have to fall apart instead of gathering at your feet. You may recall a similar tugging action of dynein that causes cilia and flagella to bend.

The polar fibers or microtubules operate quite differently. First, let's note that the microtubules do not actually extend from pole to pole, but rather, they extend out from each pole, meeting and overlapping each other at mid-cell. There, dynein arms, or some similarly acting motor proteins, reach out from one microtubule to the next, attach, tug and reattach further along, thereby pulling the spindle fibers past each other and lengthening the spindle. A second process at work in the overlapping region is the continuous lengthening of the polar fibers through the assembly of tubulin subunits. This process keeps the polar fibers from "derailing" as they slide past each other.

So we see that the kinetochore fibers shorten through the disassembly of tubulin subunits, and the polar fibers lengthen by assembly of tubulin subunits. The action of motor proteins account for both the movement of chromosomes and the lengthening of the spindle.

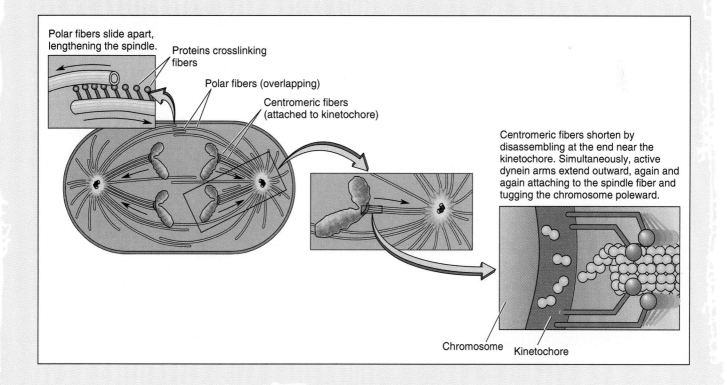

Polar fibers slide apart, lengthening the spindle.

Proteins crosslinking fibers

Polar fibers (overlapping)

Centromeric fibers (attached to kinetochore)

Centromeric fibers shorten by disassembling at the end near the kinetochore. Simultaneously, active dynein arms extend outward, again and again attaching to the spindle fiber and tugging the chromosome poleward.

Chromosome   Kinetochore

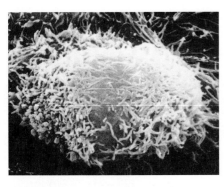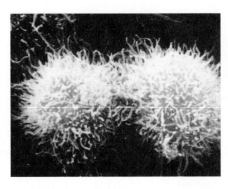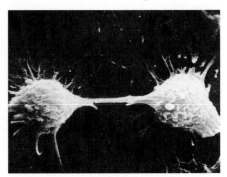

FIGURE 9.6
Cytokinesis in Animal Cells
Following the division of chromosomes and their migration to opposite ends of the cell (**a**), cytokinesis is accomplished in animal cells by a contraction of microfilaments beneath the cell membrane (**b**) around the center of the cell. This results in the sequestering of the divided chromosomes into separate progeny cells (**c**).

envelope is formed from remnants of the old, and soon the nucleoli are reestablished.

Cytokinesis

Cytokinesis is the division of the cytoplasm after the chromosomes have been separated to the opposite poles of the parent cell. In other words, this is the point at which two new, separate progeny cells are formed (see Figure 9.4). The process of cytokinesis is fundamentally different in plants and animals because of the presence of the cell wall in plants.

Animals.   In animals, cytokinesis generally begins with an equatorial furrow—an indentation in the surface of the cell, usually at about the plane of the metaphase plate. It looks as though someone had looped an invisible thread around the cell and were pulling it tight, pinching the cell in two. For this reason, animal cytokinesis is often referred to as cleavage. The furrow is created by the contraction of a ring of microfilaments (made up of a protein called actin) in the cytoplasm just beneath the plasma membrane. The remains of the spin-

dle may be caught by the tightening ring of contracting microfilaments, but regardless of what is in the way, the constriction continues until eventually the cell divides in two (Figure 9.6). Mitosis in animals is reviewed in Table 9.1.

Plants.   The problem of cytokinesis in plants is complicated by the presence of tough, rigid cell walls. Cell walls cannot bend and flex very much, so cleaving is out of the question. In plants then cytokinesis proceeds quite differently. At about the time of telophase, a new cell wall begins to be assembled in the plane of the metaphase plate. Small membranous vesicles containing cell wall precursors are formed from the Golgi complex and accumulate along a plane in the center of the cell. There they fuse and form a virtually continuous double membrane, but one that contains materials that will be used in cell wall construction. This dense, double-membrane structure is called the **cell plate**.

The cell plate begins forming in the middle of the cytoplasm and then, as more Golgi vesicles join, it grows outward to fuse with the plasma membrane at the periphery of the cell (Figure 9.7), unlike the animal cell division furrow, which begins at the periphery and extends toward the middle.

TABLE 9.1 **Mitosis**		
Prophase	**Metaphase**	**Anaphase**
Chromosomes condense	Spindle is fully formed	Centromeres divide
Separate chromatids may become visible	Chromosomes are aligned with their centromeres on the metaphase plate	Sister chromaids separate to become chromosomes
Asters and mitotic spindles begin to form	Centromeres begin to divide	Daughter chromosomes go to opposite poles
Nuclear membrane breaks down		Centromeric spindle fibers shorten; the spindle as a whole elongates
Nucleolus disperses		
Centromeres become attached to the centromeric spindle fibers		
Chromosomes migrate toward the metaphase plate		

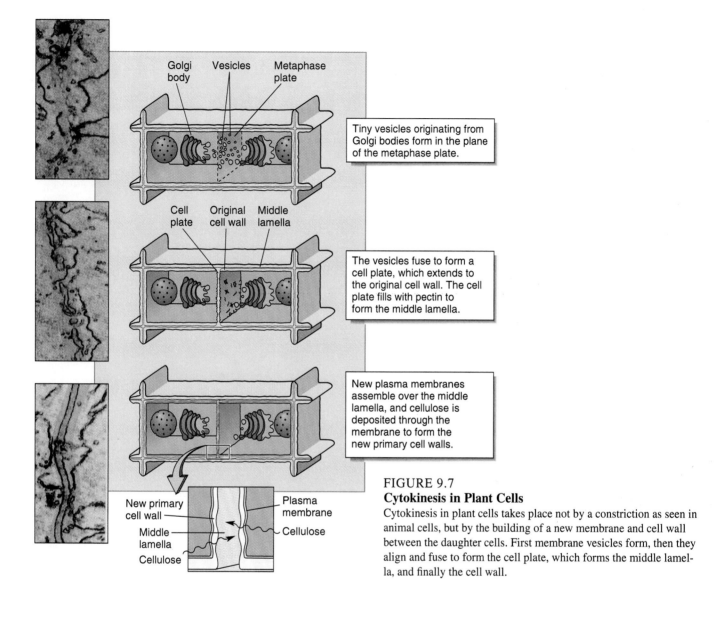

Golgi body   Vesicles   Metaphase plate

Tiny vesicles originating from Golgi bodies form in the plane of the metaphase plate.

Cell plate   Original cell wall   Middle lamella

The vesicles fuse to form a cell plate, which extends to the original cell wall. The cell plate fills with pectin to form the middle lamella.

New plasma membranes assemble over the middle lamella, and cellulose is deposited through the membrane to form the new primary cell walls.

New primary cell wall   Plasma membrane
Middle lamella   Cellulose
Cellulose

FIGURE 9.7
Cytokinesis in Plant Cells
Cytokinesis in plant cells takes place not by a constriction as seen in animal cells, but by the building of a new membrane and cell wall between the daughter cells. First membrane vesicles form, then they align and fuse to form the cell plate, which forms the middle lamella, and finally the cell wall.

Telophase	Cytokinesis in Animals	Cytokinesis in Plants
Chromosomes begin to decondense	Microfilaments associated with the cell membrane form a circular band around the cell	Small membrane vesicles fuse in the plane of the previous metaphase plate to form the cell plate
Nuclear membranes reform		Cell plate grows to separate daughter cells
Nucleoli reappear		Cell walls are laid down
Spindle disappears	Microfilaments contact, pinching apart daughter cells	
Cytokinesis (cytoplasmic cell division) usually begins		

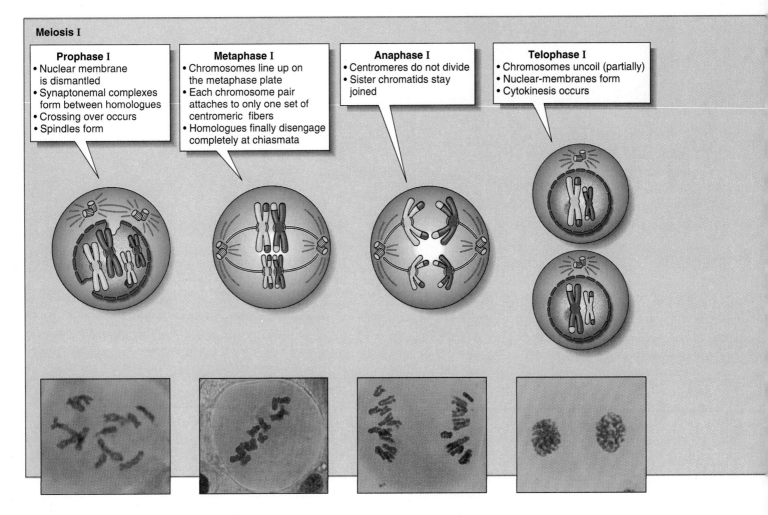

Meiosis I

Prophase I
• Nuclear membrane is dismantled
• Synaptonemal complexes form between homologues
• Crossing over occurs
• Spindles form

Metaphase I
• Chromosomes line up on the metaphase plate
• Each chromosome pair attaches to only one set of centromeric fibers
• Homologues finally disengage completely at chiasmata

Anaphase I
• Centromeres do not divide
• Sister chromatids stay joined

Telophase I
• Chromosomes uncoil (partially)
• Nuclear-membranes form
• Cytokinesis occurs

FIGURE 9.8
Meiosis
The stages of meiosis are diagrammed. Representative stages from the lily are shown in the photomicrographs.

Advantages of Mitosis

Mitosis is the critical process that allows cells to divide, to produce two progeny cells that are identical to the parent cell. Mitosis is a fundamental property of the many different kinds of eukaryotic life, but its mechanisms are basically the same wherever it occurs. But what are the advantages of mitosis?

In many single-cell eukaryotes, cell division by mitosis is a way of reproducing asexually (without combining genetic material from two sources). Since mating is not involved, the organisms can make completely identical copies of themselves.

In multicellular organisms, mitosis permits growth through the formation of new cells. In some cases these replace older, worn cells. In humans, for instance, billions of new red blood cells must be produced daily to keep up with the rapid loss of these cells. New cells can also provide material for cell specialization. Here the new cells take different developmental pathways and become biochemically and structurally specialized to perform different tasks.

In some species, mitosis permits very precise wound healing, as damaged cells are replaced; in others, regeneration of lost body parts is possible through complex processes involving mitosis. Plants can reproduce asexually, regenerating whole organisms from small pieces that grow through mitosis. Obviously, mitosis is a critical and tightly controlled process in the lives of cells. Now we will consider another process, seemingly similar and also tightly controlled, but with an entirely different function.

Meiosis

In most familiar plants and animals, individuals are **diploid**. This means that every cell in the organism has two sets of chromosomes, one set of hereditary information from each parent. In sexual reproduction, the two partners produce **gametes** (sex cells, usually eggs or sperm), which join

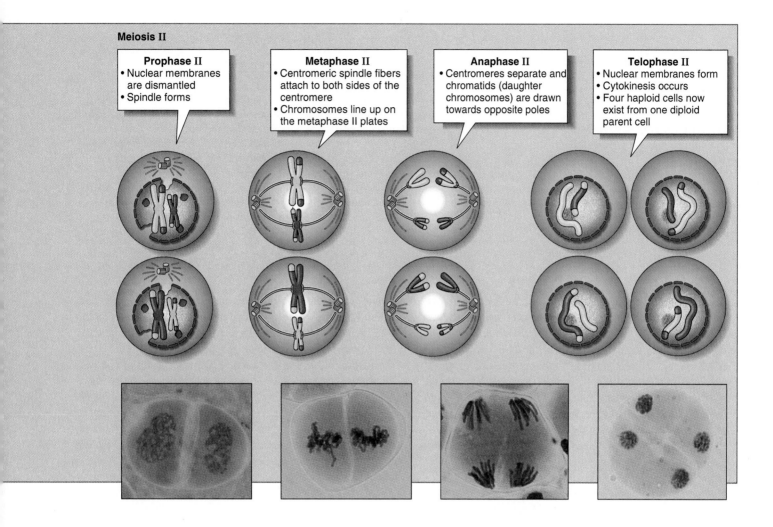

Meiosis II

Prophase II
- Nuclear membranes are dismantled
- Spindle forms

Metaphase II
- Centromeric spindle fibers attach to both sides of the centromere
- Chromosomes line up on the metaphase II plates

Anaphase II
- Centromeres separate and chromatids (daughter chromosomes) are drawn towards opposite poles

Telophase II
- Nuclear membranes form
- Cytokinesis occurs
- Four haploid cells now exist from one diploid parent cell

through fertilization to produce an offspring. Thus sexual reproduction ensures that the offspring will receive chromosomes from each of the two parents.

Two related questions come to mind. First, if fertilization joins the genetic material of two parents, why is it that the amount of genetic material is not doubled in every generation? Second, if each parent has two sets of genes and chromosomes, how is it that the offspring receives only one set from each parent?

It turns out that the answer to the second question also answers the first. Unlike most of the cells in a diploid organism, gametes are **haploid**—that is, they have only one set of chromosomes. Thus each gamete (sperm or egg) enters into fertilization with only one set of chromosomes instead of two. Because of this, the number of chromosomes does not double each generation. How is the number of chromosome halved during gamete formation? In other words, how do gametes with one set of chromosomes per cell come from cells with two sets of chromosomes? The answer involves **meiosis**, a specialized form of cell division that reduces the diploid chromosome contents to a haploid state within the gametes. A highly simplified scheme of meiosis is seen in Figure 9.8.

In diploid organisms, meiosis guarantees (1) that the chromosome number will remain constant from generation to generation and (2) that upon fertilization each sexually reproduced offspring will receive two complete sets of genetic instructions, one from each parent. We will soon see that the meiotic process also virtually guarantees that each gamete will be different from every other gamete. In humans there are 46 chromosomes in nearly every cell of the body—23 originating from the father's sperm (the paternal chromosomes) and 23 from the mother's egg (the maternal chromosomes). It is in meiosis that the 46 chromosomes per cell are reduced to 23 chromosomes per gamete.

It is important to note that most of what we will say about meiosis applies to animals, most plants, and a few protists—specifically to those organisms that spend most of their life as diploids. The fungi and most protists differ in many aspects of sexual reproduction, as we will see in Chapters 23 and 24.

Homologous Chromosomes

As we said, the 46 chromosomes in each human diploid cell come in 23 pairs; one member of each pair comes from the

Essay 9.2 ● KARYOTYPING

A karyotype is a detailed and accurate graphic representation of the chromosomes of any organism. In a karyotype, images of individual chromosomes are systematically arranged according to size and shape. Each species has its particular karyotype, and so we know the number and kinds of chromosomes found in carrots, fruit flies, and people. Karyotyping in humans is done by a simple and straightforward method.

A blood sample is drawn and the white blood cells are separated and transferred to a culture medium. Along with nutrients, the medium contains chemical agents that first induce mitosis and then stop the process when the chromosomes are at their maximal condensation at metaphase, the stage at which the chromatids are seen most easily. The cells are then put on microscope slides and stained.

In the (not so) old days, cells showing all the chromosomes were photographed through a microscope. Then a large print was made and the chromosomes were cut out with scissors, sorted by size and shape, then mounted with rubber cement. Computer-aided image analysis has modernized the process. Now technicians can view the chromosomes, identify them, and sort them all electronically to produce the karyotype.

father's sperm and one comes from the mother's egg. The two members of each pair are called **homologues**, or homologous chromosomes. The two homologues are *functionally equivalent;* they contain the information for the same features, but they may carry different versions of the information for those features. For instance, the homologue from the father might carry the information that tends to make eyes blue, and the homologue from the mother might carry information that tends to make eyes green.

Each chromosome within a cell of an individual is distinctive and can be identified through a process called **karyotyping** (Essay 9.2). In the human karyotype, each of the 23 different chromosomes is identified by its size, the position of its centromere, and its unique pattern of dark-staining bands. In Figure 9.8, as an example of the simplified approach we take in this text, homologous chromosomes are identified by their same size and color.

Overview of Meiosis

At this point we can review the meiotic process as detailed in Figure 9.8. We begin by noting that, at the start of meiosis, the chromosomes of cells entering meiosis will have doubled, and each chromosome now consists of two identical chromatids. Thus, in the human meiotic cell, there are actually 92 chromatids at this stage, the same as in mitosis. The cell will then go through two divisions, called meiosis I and meiosis II, in order to produce gametes with 23 chromatids apiece. (In the arithmetic of meiosis, one human cell with 92 chromatids produces two with 46 each, which, in turn, produce four cells with 23 chromatids each.)

Meiosis I: The First Meiotic Division

This first division in meiosis sets the stage for all that follows. Remember that the main function of meiosis is to reduce the genetic material in half, so that fertilization can restore the diploid state. In the first meiotic division, meiosis I, the homologous chromosomes undergo **synapsis**, a process

whereby the homologues are physically paired with each other. Then they align at the metaphase plate (as a pair) so that one member (in this case, two chromatids attached together by the centromere) of the pair is sent to each pole. This is distinct from mitosis, where nothing like the pairing of chromosomes occurs (see Figure 9.4). Thus each pole receives one of the homologues—one pole receives the paternal homologue, while the other pole gets the maternal one. It is important to understand that the alignment of homologous pairs is random, so that the homologues arriving at each pole are a random assortment of maternal and paternal chromosomes. Understanding this paragraph is the key to understanding meiosis and its impact on inheritance. The two concepts to remember are: (1) homologous chromosomes pair and (2) relative alignment of chromosomes is random.

Prophase I.   Prophase of meiosis I is longer and more complex than mitotic prophase. The process of chromosome condensation occurs early in prophase I and is quite similar to that of mitosis. The main difference is that the homologues pair in synapsis. Synapsis can be seen to occur when homologous chromosomes touch each other in the right places, but first there seems to be a lot of random groping. However, over a period of time, the two homologues seem to come together like the two halves of a zipper. In fact, in synapsis there is a specialized structure, the **synaptonemal complex** (made up of protein and RNA), that bridges the two homologues in meiotic prophase, and it even looks like a zipper (Figure 9.9).

Because of the pairing up of chromosomes, each grouping (joined homologues) actually consists of four chromatids called a **tetrad**. The meiotic cell of a human has 23 such tetrads. Each tetrad, remember, has two centromeres, one from each of the two members of the homologous pair.

While the four chromatids are held in close contact by the synaptonemal complex, the stage is set for a very important process. While they are so closely bound, they may exchange segments of their chromosomes through a process called **crossing over** (also referred to as crossover) (Figure 9.9). At any point along the DNA backbone of a chromatid, special

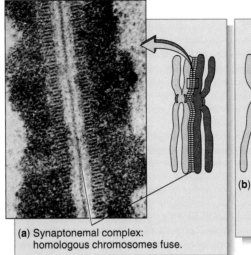

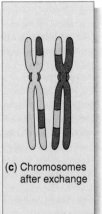

FIGURE 9.9
Pairing, Crossing Over, and Chiasma Formation
In early prophase I, the maternal and paternal homologous chromosome pairs are intimately associated and bound together by the synaptonemal complex, shown in the electron micrograph and diagrammatically in (**a**). During this association, regions of chromosomes are exchanged between the homologues in the process of crossing over (**b**). The homologues separate from each other

(**a**) Synaptonemal complex: homologous chromosomes fuse.

(**b**) Crossing over and repulsion: genetic material is exchanged on the molecular level.

(**c**) Chromosomes after exchange

but are held together at chiasmata, which are the places where crossing over has occurred. If the four chromatids could be unwound and separated, and if the regions of maternal and paternal origin could be indicated separately, they would appear as shown in (**c**). Note that exchanges involve only two of the four chromatids.

multienzyme complexes called recombination nodules can cause the chromatids to undergo crossing over. The region of crossover becomes known as a **chiasma** (plural, **chiasmata**), a joined area of the chromosomes that is visible under the light microscope.

In human cells, there is an average of about ten crossovers for every meiotic tetrad; in other species, there may be fewer. But one thing is clear: after meiotic prophase, one cannot properly refer to individual chromatids as maternal or paternal. The crossovers produce new kinds of chromatids, and it should be mentioned that this is an important source of genetic variation among individuals born to the same parents. This constantly renewed variation is crucial to the evolutionary process because it is from such variation that nature will select those individuals whose genes will pass into the next generation.

As the synaptonemal complex "unzips," the homologues begin parting. But at chiasmata the homologues tend to cling together. So for a brief period, the chiasmata reveal where crossing over has occurred.

Metaphase I. In the first metaphase of meiosis (see Figure 9.8), the tetrads are brought to the metaphase plate just as in mitosis, except that the homologues arrive as a pair. In metaphase I meiosis, centromeric spindle fibers from just one pole attach to each centromere, such that anaphase will result in separation of the homologues. (In mitosis, you'll recall from Figure 9.4, centromeric spindle fibers from *both* poles attached to each centromere.)

Anaphase I. Anaphase I follows quickly (see Figure 9.8). Because of the way in which the centromeric spindle fibers are attached to the chromosomes, the results are quite different from those of mitotic anaphase. In mitosis, the pull on each chromo-

some came from two directions, and as the centromeres divided, sister chromatids moved to opposite poles. In metaphase I of meiosis, the pull is from one direction only, and the centromere does not divide. Thus, when migration begins, homologue separates from homologue, and each chromosome, still composed of sister chromatids joined by a centromere, is drawn to a pole. In humans, this means that 23 chromosomes go to each pole and that the homologues—the chromosome pairs—are forever separated. So following the events of anaphase I, we can see that the number of chromosomes arriving in each progeny cell has been reduced by half. For this reason, the first meiotic division is also known as the **reduction division**.

Telophase I and Meiotic Interphase. Anaphase I is followed by telophase I (see Figure 9.8), including the reorganization of nuclei, decondensation of the chromosomes, and duplication of centrioles, followed, in turn, by cytoplasmic division. The cell then enters a meiotic interphase.

Meiotic interphase is unlike mitotic interphase in one very important detail: there is no S phase. In other words, there is no DNA replication. Depending on the species, meiotic interphase may be extremely brief, or it may be exceedingly long.

Meiosis II: The Second Meiotic Division

Meiosis II is much easier to follow than meiosis I because it is essentially the same as mitosis (see Figure 9.8).

Prophase II. In prophase II, we immediately see similarities to mitotic prophase. The chromosomes, each containing two chromatids still attached by their centromeres, condense. This time the centromeric microtubules become attached in standard mitotic fashion: that is, the kinetochores of each chromosome attach to fibers from opposite poles. The chromosomes then move to the metaphase plate.

Metaphase II and Anaphase II.   In metaphase II the chromosomes line up on the metaphase II plate in preparation for separation. Anaphase II is marked by the division of the centromeres and the separation of the chromatids into individual chromosomes. These chromosomes are then pulled apart as they are in mitotic anaphase, with the same forces and attachments. The chromosomes that have separated in anaphase II very likely have been rearranged by crossing over.

Telophase II.   As anaphase II proceeds into telophase II, nuclear envelopes once again form around the four groups of decondensing chromosomes. The four resulting cells then enter interphase, where they are destined to remain in the G_1 phase. Although cellular activities continue as the egg and sperm develop, DNA replication is over until fertilization triggers new cell cycles.

To sum up then, we see that each of the four chromatids of the prophase I tetrad (that have potentially been rearranged by crossing over) ends up in one of the four cells that are the products of meiosis. As in mitosis, the amount of genetic material was doubled before the process started (during the S phase), but in two divisions the products were distributed to four cells. Each cell now has one-fourth of the normal G_2 amount of DNA. Each of these is now haploid; that is, each cell has only one of each of the different types of chromosomes of the species (for example, 23 chromosomes in human gametes). Meiosis and mitosis are directly compared in Figure 9.10.

There are two major take-home lessons from the story of meiosis. First, meiosis halves the chromosome number of eggs and sperm, such that fertilization restores the original number of chromosomes. Second, meiosis provides a means of shuffling and reorganizing chromosomes, thus increasing genetic variation in offspring. This shuffling and reorganization takes place in two major ways: (1) the prophase I exchange of chromatid parts by crossing over and (2) in the random lining up of homologous chromosomes in metaphase I, so that maternal and paternal centromeres are randomly distributed to the two poles. This is not to mention the somewhat random way two parents can get together in the first place.

Meiosis in Humans

Before we get into the details of meiosis in humans, we should make a few preliminary points about the production of gametes as it occurs in most animals. First, meiosis and gamete production occur in tissues called **germinal epithelium** (*germ,* "seed"). All other tissue is called **somatic tissue**. The germinal epithelium begins with a small group of germinal cells in the embryo. These cells will become part of the **gonads**, the organs (either ovaries or testes) responsible for producing gametes. The germinal epithelium of the gonads, like any other

tissue, undergoes repeated mitosis during development until its cells become meiotic cells and finally gametes. Understand that not all the tissues of the gonads are germinal. Some gonad tissue plays a supporting role but does not undergo meiosis.

Now on to meiosis in humans. During fetal development in human males, the germinal epithelium of the testes forms the lining of long, convoluted tubes—the seminiferous tubules (see Chapter 41)—but both mitosis and meiosis in this tissue are suppressed until puberty. Then the germinal epithelium undergoes a type of asymmetrical mitotic division. With each mitotic division, one of the progeny cells remains a germinal epithelium cell, capable of further mitosis, but the other progeny cell becomes differentiated as a **primary spermatocyte**, a specialized diploid cell ready to begin meiosis. The primary spermatocyte will proceed through two divisions. The first meiotic division yields two **secondary spermatocytes**, and the second meiotic division yields four sperm cells (Figure 9.11). Billions of sperm will be produced in the germinal epithelium daily for the rest of the man's life.

Meiosis in Females: A Special Case

Meiosis in human females is quite unlike that in males. During fetal development in human females, germinal epithelium forms in the ovaries. The cells of the germinal epithelium then begin to undergo meiosis (see Figure 9.11). They develop more or less simultaneously during the final months of fetal development, completing most of the first prophase of meiosis. Then meiosis suddenly stops. Thus when a baby girl is born, her germinal cells, now referred to as **primary oocytes**, are well along in the process of gamete formation, but frozen in their development. And there they will remain, suspended in this state for 10–12 years. Then, when the menstrual cycle begins, they will be released, one or sometimes two at a time, during each monthly cycle for the next 35 years or so. Interestingly, meiosis will not be completed unless the oocyte is fertilized. Although a woman may have several thousand oocytes, only about 400–500 will become mature, and, with any luck, most of these will not be fertilized.

Since all oocytes are suspended in meiotic prophase in human females, meiotic prophase can last 45 or perhaps 50 years. (Actually, the long prophase I in females is typical of many animals, including all vertebrates.)

In spite of such differences in the beginning stages of meiosis in males and females, once it is underway it proceeds in much the same fashion in both sexes: that is, at least, until meiotic cytokinesis. Here sexual differences again appear.

Whereas each germinal cell in males produces four sperm, each germinal cell in females produces only one viable egg. This happens as the cell constituents are divided unevenly among the products of meiosis, with most of the cytoplasm ending up in one cell (see Figure 9.11). Only that one cell gets to be the egg, and it is indeed fat and opulent and swollen with nutrients that will sustain it until it can be fertilized and begin to draw nutrients from the mother's tissues. The unequal division begins about the time of ovulation, where the oocyte's

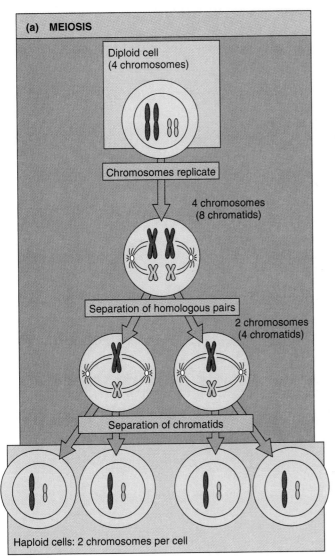

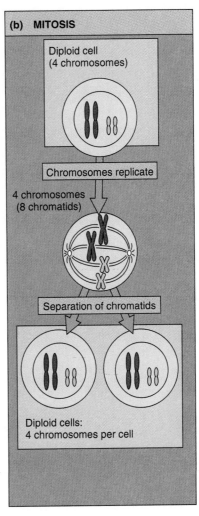

FIGURE 9.10
Meiosis and Mitosis Compared
Simplified schemes of meiosis (**a**) and mitosis (**b**) are compared side by side to highlight similarities and differences. It should be clear why the second divison of meiosis is often called a "mitotic" division.

meiotic spindle forms off to one side of the cell, actually just under the surface. A normal first meiotic chromosome separation and cell division occur, but one of the two progeny cells is very small and is called a **polar body**. The other cell is known as the **secondary oocyte**. After fertilization, the second unequal meiotic division occurs, pinching off another tiny cell as a secondary polar body, which emerges just under the first.

The first polar body normally does not bother to complete its own meiosis II division. Generally, the polar bodies can be regarded as no more than convenient little garbage cans into which an unwanted three-quarters of the genetic material of the primary oocyte can be dumped, leaving one large oocyte with its haploid complement of chromosomes and the lion's share of its predecessor's cytoplasm and nutrients.

Meiosis, Evolution, and Sex

Meiosis, we see, is a complicated process. Furthermore, its complexity means that all too frequently something goes wrong (Essay 9.3). So why do organisms bother? Maybe because of sex. Meiosis, after all, is part of sex. So the next question is, why have sex?

In the long run, of course, we will all die. But our genes live on. When we leave offspring ("we" includes all us sexual beings), we pass our genetic information on to our descendants. The problem is that our descendants may not find the world to be the same as the one we lived in; the weather changes, new diseases show up, and unexpected competitors eat our food. Logically, no one particular combination of genes, including

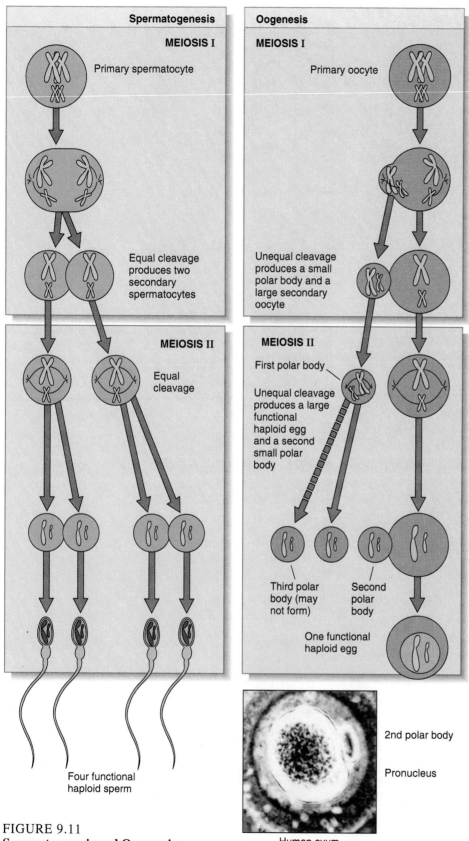

FIGURE 9.11
Spermatogenesis and Oogenesis
Spermatogenesis occurs as expected, with one cell yielding four haploid products that will become sperm.
However, oogenesis is different. In each of the two divisions, the metaphase plate forms to one side of the
oocyte, and unequal cleavage occurs. Thus only one large ovum results from each meiotic event.

Essay 9.3 ● WHEN MEIOSIS GOES WRONG

Meiosis is a complicated process. When you consider the process of pairing in prophase and the accuracy needed for two cell divisions, you might not be surprised to find that often things do go wrong. In humans, for example, about one-third of all pregnancies spontaneously abort in the first two or three months. When these naturally aborted embryos are karyotyped, it turns out that most of them have the wrong number of chromosomes. Failure of the chromosomes to separate correctly is called nondisjunction.

Not all failures of meiosis result in early miscarriage. There are late miscarriages and stillbirths of severely malformed fetuses with wrong chromosome numbers. In addition, about one live-born human baby in 200 has the wrong number of chromosomes, often accompanied by severe physical and/or mental abnormalities.

About one baby in 600 has three copies of chromosome 21. Such persons may grow into adulthood, but many suffer abnormalities. The syndrome is known as trisomy 21 or Down syndrome, the latter after the 19th century physician who first described it. Characteristics of the syndrome are general pudginess, rounded features, an enlarged tongue that often protrudes, and various internal disorders. Often a peculiar fold in the eyelid is seen, and in the past this was erroneously equated with the characteristic eye fold of Asians (thus the earlier name "mongoloid").

Trisomy 21 individuals also have a characteristic bark-like voice and usually happy, friendly dispositions.

Trysomy 21 occurs more frequently among babies born to women over 35 years old. At that age, the incidence is about 2 per 1000 births. By age 40, this climbs to about 6, and by age 45, 16 children with trisomy 21 are born for each 1000 births. The age of the father has little if any effect, though the nondisjunction that eventually leads to trisomy 21 can occur in males.

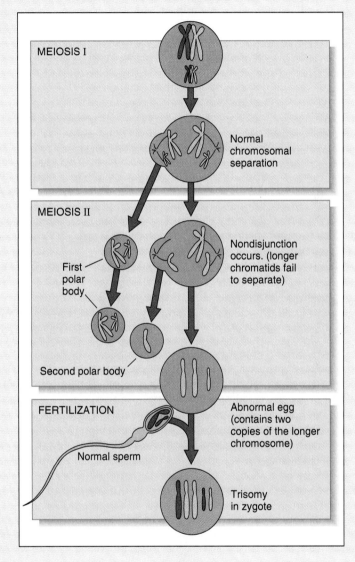

ours, will be adequate to meet all new situations. So, according to evolutionary theory, sexual reproduction comes to the rescue.

Meiosis and sexual unions keep reshuffling the genes of all successful individuals in the population, producing variation such that virtually infinite possible combinations of genes are produced. The reshuffling occurs when crossing over and random assortment of the homologous pairs occur during meiosis I. Most of the new combinations will be less successful than the old ones because, whereas the old ones are tried and true, the new ones are genetic experiments. In the long run this is the cost of success. The winning evolutionary strategy is to cover as many bets as possible by having highly variable offspring. Then, when the weather changes or the Creeping Purple Flu shows up, perhaps not all of our descendants will be wiped out. Some of them are likely to have the right combinations of genes to be able to live under these conditions.

We can see this principle at work in certain organisms that can reproduce either sexually or asexually. For instance, *Daphnia,* the little water flea, reproduces asexually for generation after generation, females producing only females. In fact, sometimes one can see an asexual female embryo within an asexual female embryo within an asexual adult! The little animals keep up this kind of reproduction as long as conditions are stable. But when their pond begins to dry up, or things otherwise get dangerous and unpredictable, they change strategies. They begin to produce both males and females. These go on to mate and reproduce sexually in the usual way. Thus, in hard times, a few of their highly variable offspring may have that particular combination of genes that will allow them to cope with their new environment.

The silent dances of mitosis and meiosis are indeed intricate pageants in the drama of life, and critical to the evolutionary direction on the planet. An understanding of these mechanisms sets the stage for our next story, the story of how such mechanisms are used by cells to parcel out genetic information.

Key Ideas

THE NUCLEUS

1. The nuclear structures include a double membrane called the **nuclear envelope** and the **nuclear pores** that selectively admit the passage of materials.

2. The nuclear envelope disappears from sight during much of cell division.

3. The nucleus contains the chromosomes.

4. Chromosomes consist of **chromatin**, which includes DNA and nuclear proteins.

5. The DNA of chromosomes is wound about numerous bead-like **nucleosomes**, which consist of histones.

6. In cell division, chromosomes are highly condensed. When not in division, the chromosomes are uncoiled and spread out, forming the chromatin net.

7. The cell cycle includes growth, replication, and **mitosis**.

8. Mitosis includes formation of the mitotic **spindle**, separation of chromosome replicas, and division of the cell into two. The time between successive mitoses is **interphase**, consisting of G_1, S, and G_2.

9. G_1 is a period of protein synthesis, while the **S** phase involves DNA replication, and the G_2 phase includes synthesis of the mitotic spindle protein.

MITOSIS

1. Mitosis includes five phases: **prophase**, **metaphase**, **anaphase**, **telophase**, and **cytokinesis**.

2. Chromosome condensation facilitates the movement of chromosomes during mitosis.

3. The term *chromatid* applies only to DNA replicas held together at the centromere. Upon centromeric division, chromatids become chromosomes.

4. The mitotic spindle consists of microtubules made of the protein tubulin. Spindle fibers attach to centromeres at a region called the **kinetochore**.

5. At metaphase, the chromosomes pause after aligning on the **metaphase plate**. Centromeric microtubules are seen on each side of each centromere.

6. At anaphase, the centromeres divide, and the newly separated chromosomes migrate to opposite poles.

7. Telophase is the reverse of prophase and includes chromosome uncoiling, reestablishment of the nuclear envelope, and the formation of nucleoli.

8. In animals, cytokinesis, or cleavage, involves the contraction of microfilaments below the cell membrane, forming a cleavage furrow that tightens, dividing the cytoplasm. In plants, cytoplasmic division requires that a new wall form between progeny nuclei.

9. Mitosis and cell division are the primary means of asexual reproduction in single-cell organisms. In multicellular organisms, mitosis accommodates growth, cell replacement, repair, and cell specialization.

MEIOSIS

1. **Diploid** organisms must undergo **meiosis** to form **haploid** sex cells. Meiosis reduces the number of chromosomes in half (by the separation of pairs) and permits new gene combinations to arise.

2. Chromosomes occur in pairs (in the original fertilized cell, one from each parent. Members of a chromosome pair are called **homologues**, or homologous chromosomes, each of which has the same gene complement.

3. Meiosis requires two divisions, meiosis I and meiosis II, which in humans reduces the chromosomes in half—from diploid to haploid.

4. Meiotic prophase includes the time of chromosome condensation, synapsis, and crossing over.

5. During synapsis, homologues are bridged by the **synaptonemal complex**, which provides for crossing over. In crossing over regions along chromatids break and are exchanged, creating new gene associations in the chromosomes.

6. In metaphase I the highly condensed chromosomes are aligned on the metaphase plate. Centromeric spindle microtubules from one pole attach to *only one side of each centromere*.

7. In anaphase I, each set of homologues separates, going to opposite poles.

8. During telophase, the reorganization of the nucleus occurs, followed by cytoplasmic division. There is no DNA replication during meiotic interphase.

9. In anaphase II of meiosis II, centromeres divide and single chromosomes (former chromatids) are drawn to opposite poles.

MEIOSIS IN HUMANS

1. Meiosis occurs in germinal epithelium, which, in animals, gives rise to gametes (sperm and egg).

2. In the human male embryo, meiosis in the germinal epithelium does not begin until puberty.

3. In the human female embryo, germinal epithelium develops in the ovaries and by the time of birth meiosis has begun in all germinal cells or **oocytes**. Meiosis ceases during prophase I and remains suspended in each oocyte until it actually undergoes ovulation.

4. While each meiosis in males produces four functional sperm, meiosis in females results in only one oocyte forming. The other cells are lost as **polar bodies**.

5. Meiosis provides variation through (a) random genetic exchange during crossing over, (b) random chromosome lineup at the metaphase plate, (c) random discarding of chromosomes into polar bodies, (d) random selection of oocytes for development and ovulation, and (e) random success of genetically different sperm.

Application of Ideas

1. Single-cell organisms that reproduce asexually through mitosis die in enormous numbers. But potentially, at least in some sense, such organisms are immortal. Multicellular organisms that can reproduce by fission (splitting) or budding (asexually producing miniatures of themselves) are also potentially immortal in the same sense. What do these statements mean? What kind of a trade-off might there have been in the evolution of sex? Is death more of a reality to sexual beings than to asexual beings?

Review Questions

1. Using the terms *DNA, chromatin, histone,* and *nucleosome,* describe the organization of a chromosome. (page 201)

2. List the stages of a cell cycle, and briefly describe the events of each. (pages 202–203)

3. What is the relationship between the chromosome and the chromatid? (page 204)

4. In a few sentences, clearly explain the function of mitosis. (page 208)

5. List the phases of mitosis, and summarize the events of each. (page 204)

6. Describe the centromere. To which part of a centromere do the spindle microtubules attach? (pages 204–205)

7. Describe the arrangement of chromosomes and the centromeric spindle fibers at mitotic metaphase, and also explain the significance of the arrangement. (pages 204–207)

8. Summarize the events of mitotic anaphase. What do these events ensure about the chromosomes of progeny cells? (page 204)

9. Contrast cytokinesis in animal cells and plant cells. (pages 204, 207–208)

10. Discuss the importance of mitosis to single-cell organisms such as algae and protozoans. (page 208)

11. Humans have 46 chromosomes. How many chromosomes would one find in a cell in G_1, a cell in mitotic metaphase, a sperm cell, a cell in meiotic prophase I, a cell in meiotic telophase I, and a cell in meiotic telophase II? (page 210)

12. Using humans with their 46 chromosomes as an example, explain the "number problem" solved through meiosis. (page 210)

13. In a few words, define the terms synapsis, crossing over, and chiasma. (pages 210–211)

14. Compare the attachment of centromeric microtubules in metaphase I of meiosis with that of mitosis. What specific difference will this make in anaphase? (page 211)

15. Summarize the events of meiosis II, comparing each phase to those of mitosis. (Figure 9.10)

16. Briefly summarize the development of germinal epithelium in the human female. How does this development differ in the male? (page 214)

17. Using a diagram, suggest how meiotic cytokinesis in the female vertebrate differs from cytokinesis in the male. Offer a logical reason for the difference. (Figure 9.11)

18. List four ways in which genetic variability is increased through sexual reproduction. (pages 214–215)

CHAPTER 10

Mendelian Genetics

CHAPTER OUTLINE

WHEN DARWIN MET MENDEL (ALMOST)

MENDEL'S CROSSES

Dominance

MENDEL'S FIRST LAW: SEGREGATION OF ALTERNATE ALLELES

Segregation and Probability

MENDEL'S SECOND LAW: INDEPENDENT ASSORTMENT

Mendel's Testcrosses

THE CHROMOSOMAL BASIS FOR MENDEL'S LAWS

PEDIGREES AND HUMAN TRAITS

THE DECLINE AND RISE OF MENDELIAN GENETICS

KEY IDEAS

APPLICATION OF IDEAS

REVIEW QUESTIONS

The silent dance of chromosomes at meiosis is precise and intricate, but the implications of this delicate shuffling are profound. Since chromosomes carry the genetic material of an individual, their segregation and combinations directly influence the traits of the offspring. Indeed, the relationship between chromosomes and inherited traits was deduced long before the discovery of the function of DNA. As you wend your way through these next few chapters, you have the benefit of knowing beforehand the relationship of DNA and chromosomes. Keep in mind though that scientists of the 19th century could only go by what they could see.

As the preeminent biologist of his time, Charles Darwin was led to wonder about just how traits are inherited. Like other biologists of those days, he was particularly struck by the observation that offspring tended to look like a combination of the two parents.

When Darwin Met Mendel (Almost)

Charles Darwin, as we will see, developed evolutionary biology as one of the cornerstones of modern science. It has been suggested that his basic knowledge of certain principles of inheritance helped set the tone of his thinking as he worked out his great theory of evolution. As the son of a gentleman farmer, Darwin was well aware of the simple fact that offspring tend to resemble their parents. The offspring of heavier animals, he knew, tended toward heaviness; the daughters of higher milk-yielding cows tended to produce more milk. Even though Darwin was able to formulate some of the most powerful ideas in modern evolutionary biology, most of his ideas about hereditary mechanisms were wrong.

Darwin's chief mistake was to accept the only intellectually respectable theory of his day, which was "blending inheritance." It was believed that the "blood," or hereditary traits, of both parents blended in the offspring, just as two colors of ink blend when they are mixed. The blending hypothesis appeared to work reasonably well for some traits, such as height or weight, that are continuously graded, but it couldn't account for other observations.

It is always surprising, in hindsight, to recognize the degree to which a strongly held belief will blind its proponents to obvious contradictions. The blending theory would predict that if a yellow labrador retriever were crossed with a black one, the offspring would usually be some kind of yellow black, not black or yellow (Figure 10.1). Something obviously was wrong with the theory, but no one seemed to notice, or if they did, they tried to ignore it.

When Darwin published his theory of natural selection, some of his sharper critics used the weakness of his genetic arguments against him. They pointed out, quite correctly, that natural selection will not work with blending inheritance because any variation will be blended away. Try as he might, Darwin could not come up with an answer for them, and the tired old man, so ill in his later years, began to backtrack. In later editions of his famous book, he began to give more and more discussion to issues involving the mechanisms of inheritance of traits without ever solving the puzzle.

The remarkable part of this story is that Darwin's thinking was on the right track and that he came very close indeed to solving some of his most vexing problems with explaining the mechanisms of inheritance. In his experiments, Darwin crossed two strains of snapdragons. The two strains differed in only one respect: one strain had abnormal, radially symmetrical flowers (which botanists called *peloric* flowers), and the other had the normal, bilaterally symmetrical snapdragon flowers (Figure 10.2). Of the progeny of this first cross, Darwin wrote: "I thus raised two great beds of seedlings, and not

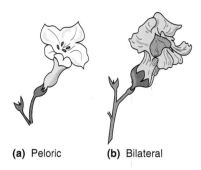

(a) Peloric   **(b)** Bilateral

FIGURE 10.2
Darwin's Snapdragons
One of Charles Darwin's experiments included breeding snapdragons. One of his observations was that when true-breeding peloric-flowered snapdragons **(a)** were crossed to true-breeding bilateral flowers **(b)**, the peloric flower trait was lost in the progeny. He termed the dominance of one trait over another prepotence.

one was peloric." Darwin called this tendency of one trait in a cross to suppress another *prepotence*. We now call it dominance, and we will consider it in depth shortly.

While Darwin was working on his snapdragon experiment, scientific history was being made across the English Channel. The self-effacing but incredibly bright and dedicated Gregor Mendel was crossing strains of garden peas. Darwin and Mendel performed almost exactly the same experiments. The two great scientists, unknown to each other, obtained almost exactly the same results. But only Mendel was able to understand what had happened.

The Abbot Gregor Johann Mendel (1822–1884) was a member of an Augustinian order in Brunn, Moravia (now part of the Czech Republic). Mendel (Figure 10.3) sometimes seems like a 20th-century biologist somehow displaced into the 19th century, probably because his controlled, numerically precise, and analytical approach contrasts sharply with the descriptive observational science usually associated with his historical period.

Early in his life Mendel began training himself, and he became a rather competent naturalist. To support himself during those years, he worked as a substitute high school science teacher. The professors at the school, noting his unusual abilities, suggested that he take the rigorous qualifying examination and become a regular member of the high school faculty. Mendel took the test and did reasonably well, but he failed to qualify. He joined a monastic order instead.

His superiors, confident of his abilities, sent him in 1851 to the University of Vienna for two years of concentrated study in science and mathematics. There he learned about the developing science of probability and statistics.

When Mendel returned to the monastery he began his plant-breeding studies in earnest. He developed new varieties of fruits and vegetables, kept abreast of the latest developments in his field, joined the local science club, and became active in community affairs.

FIGURE 10.1
Traits Are Not Blended
For many traits, it is clear that the offspring are not simply a blending of the characteristics of the parent; offspring can also resemble one parent or the other.

FIGURE 10.3
Gregor Mendel (1822–1884)
Mendel discovered the underlying fundamental principles that
govern the inheritance of observed traits. His work, though
published in 1866, was essentially unnoticed and unappreciated
until the turn of the century.

Mendel's Crosses

Mendel began to experiment by observing the effects of
crossing different strains of the common garden pea. To begin
with, he based his research on a very carefully planned series
of experiments *and* on a careful numerical analysis of the
results. The use of mathematics to describe biological phenomena was a new concept. Clearly, Mendel's two years at
the University of Vienna had not been wasted.

The careful planning that went into Mendel's work is
reflected in his selection of the common garden pea as his
experimental subject. There were several advantages in this
choice. Since others had successfully used the garden pea in
experimental crosses, he wasn't proceeding entirely on guess-
work. Furthermore, pea plants were readily available and fairly
easy to grow, and Mendel was able to purchase true-breeding
strains. **True-breeding** refers to the fact that the offspring are
the same from generation to generation. The strains differed

from each other in very pronounced ways, so that there could
be no problem in identifying the results of a given experiment.
Mendel chose seven different characters (a name he used for
genetic traits) to work with, each of which has two variations or
conditions, as shown in Table 10.1.

Keep in mind that each pea in a pod is a seed—essentially a new plant, an individual with its own *genotype* and *phenotype*. An organism's **genotype** refers to an organism's actual
genes—its genetic constitution, whereas **phenotype** refers to
observable traits. (Often "genotype" and "phenotype" refer to
the limited set of genes and traits under investigation.) Thus
the phenotype of a plant's offspring—traits such as seed
shape and coat color—can easily be determined by simply
examining the peas.

The pea, however, is not without its problems as an experimental organism. Mendel's artificial crosses relied on tedious
manipulation. Left to themselves, most pea plants will simply
self-fertilize. To get a cross between two pea plants, it was necessary to open the pea flower, remove the pollen-producing
anthers (to prevent self-pollination), and apply the foreign
pollen with a small paintbrush (Figure 10.4). This way, Mendel
could control his crosses, prevent accidental contamination,
and allow self-pollination only when it suited his needs.

Mendel's analytic approach involved observation of the
inheritance of one character at a time, such as would occur in
a **monohybrid cross,** a cross in which the parents differ by a
single trait. We now call the original parent generation P_1 and
designate the first generation offspring as F_1 (F_1 = first "filius"
or "filial," for first son/daughter). When the F_1 plants are
crossed with each other or are allowed to self-pollinate, the
resulting offspring are called the F_2 generation, and so on.

Dominance

Mendel was aware that, for some traits, crosses between pea
plants could produce offspring whose characters were not a
blend of the two parents. Mendel chose to study those traits
that did not blend and found, for example, that when plants
grown from yellow seeds were crossed with plants grown
from green seeds, the F_1 offspring were not intermediate in
color (Figure 10.5). Instead, all of them were yellow seeds.
Mendel termed the trait that appears in the F_1 generation the
dominant trait, and the one that had failed to appear the
recessive trait. But he was now left with a vexing question.
What had happened to the recessive trait? It had been passed
along through countless generations in his true-breeding
lines, so it couldn't have just disappeared.

Mendel then allowed his F_1 pea plants to pollinate themselves. In the offspring of the F_1 generation (which, remember, we call the F_2 generation), Mendel found that roughly
one-fourth of the peas were green and that about three-fourths
were yellow (see Figure 10.5). The recessive trait had reappeared! He repeated the experiment with other pea strains and
obtained similar results with other traits. He crossed a round
pea strain with a wrinkled pea strain. All the F_1 peas were

TABLE 10.1 Mendel's F₂ Generations

The dominant and recessive traits analyzed by Mendel are shown, along with results of the F₂ generations. Note the large numbers with which he worked. How does a large sample size (large numbers) improve the validity of the conclusions? How well do his numbers in the last two columns agree with what we would expect in the crosses. The proportion of the F₂ generation showing recessive forms is in the far-right column.

Dominant Form of Trait in One Parent Plant	× Recessive Form of Trait in One Parent Plant	Number of Dominant Plants in F₂	Number of Recessive Plants in F₂	Total Examined	Ratio of Dominant to Recessive (avg. is 3:1)	Proportion of F₂ That Is Recessive (avg. is 25%)
Round seeds	Wrinkled seeds	5474	1850	7324	2.96:1	25.3
Yellow seeds	Green seeds	6022	2001	8023	3.01:1	24.9
Gray seed coats	White seed coats	705	224	929	3.15:1	24.1
Green pods	Yellow pods	428	152	580	2.82:1	26.2
Inflated pods	Constricted pods	882	299	1181	2.95:1	25.3
Long stems (tall)	Short stems (dwarf)	787	277	1064	2.84:1	26.0
Axial flowers and fruit	Terminal flowers and fruit	651	207	858	3.14:1	24.1

round, but in the F₂ generation about one-fourth of the peas were wrinkled again. Instead of merely noting this recurring ratio of ¾ to ¼ (or 3:1), Mendel was convinced that this ratio provided fundamental insights into the mechanism of inheritance and that any proposed mechanisms must explain the occurrence of this ratio of traits.

One of Mendel's early observations was critical to the development of his genetic model. Mendel found that there were, in fact, two kinds of yellow peas: the true-breeding kind, like the original parent stock, which would grow into plants that would bear only yellow peas; and another type, which when grown and self-pollinated would produce pods

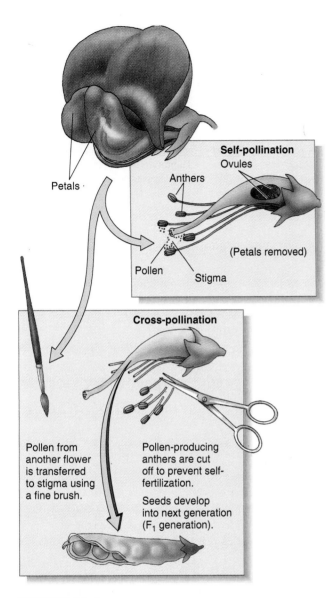

FIGURE 10.4
The Pea Flower and Pollination

The petals of the garden pea generally ensure self-pollination. To cross-pollinate a plant, Mendel had to open the young flower and remove the pollen-bearing anthers. Then he transferred pollen from another flower to the stigma (female, pollen-receptive structure) to accomplish the cross he wanted.

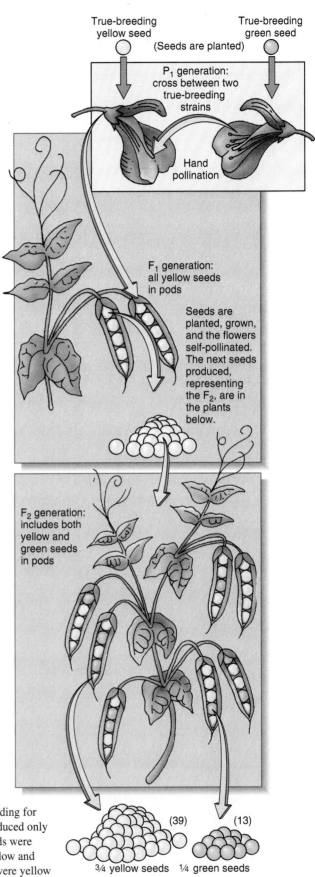

FIGURE 10.5
Single-Factor Cross: Yellow × Green Peas

In one of his early experiments, Mendel crossed pea plants that were true-breeding for yellow peas with plants that were true-breeding for green peas. This cross produced only yellow seeds in the pods of the F_1 generation progeny. When those yellow seeds were grown and allowed to self-pollinate, they produced F_2 plants carrying both yellow and green seeds within their pods. When those seeds were counted, three-fourths were yellow and one-fourth green. Thus there was a 3:1 ratio of yellow to green seeds.

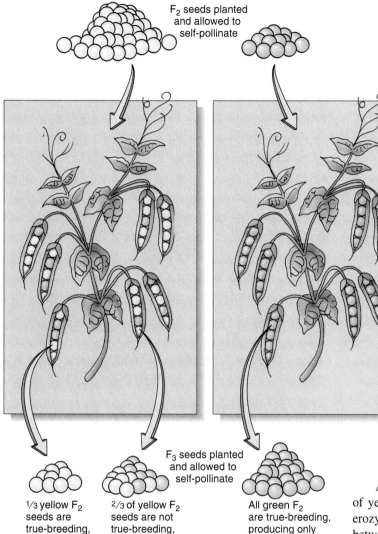

F₂ seeds planted and allowed to self-pollinate

F₃ seeds planted and allowed to self-pollinate

¹/₃ yellow F₂ seeds are true-breeding, producing only yellow

²/₃ of yellow F₂ seeds are not true-breeding, producing yellow and green

All green F₂ are true-breeding, producing only green

FIGURE 10.6
The F_3 Generation

Mendel took the piles of green and yellow seeds from the F_2 generation of the single-factor cross of Figure 10.3, grew them up, and allowed them to self-pollinate and produce an F_3 generation. He found that all the green seeds (which were one-fourth of the F_2) were true-breeding, in that they produced only green progeny. However, only one-third of the yellow seeds (again, one-fourth of the F_2) were true-breeding. The remaining two-thirds of the yellow seeds (which amounts to one-half of the F_2) continued to produce both yellow and green seeds.

containing both yellow and green peas. At this point we can almost hear Mendel musing: "two kinds of yellow peas—one true-breeding, one not." Were there also two kinds of green peas? He found that there were not. When green peas were cultivated and allowed to self-pollinate, they always bore only green peas (Figure 10.6).

Mendel concluded that there were two hereditary "factors" for each character in an individual, one contributed by each of the parents, and that factors might exist in different forms or variants. Thus seed color is determined by the "seed color factor"; there is one variant for green seeds and another for yellow seeds. Mendel's factors have since been renamed **genes**, and the variants or forms have been termed **alleles**. Thus each gene in a diploid organism is represented by two alleles. If the two alleles are identical, the individual is said to be **homozygous** for that gene. If the individual has two different alleles for the gene, the individual is referred to as **heterozygous** for that gene.

As we've seen, Mendel realized that there were two kinds of yellow peas in the F_2 generation—homozygous and heterozygous—and that it was impossible to tell the difference between the two kinds of yellow peas unless they were allowed to mature and reproduce. He used a procedure called **progeny testing** (see Figure 10.6) to determine which was which. This test would allow two seemingly identical yellow seeds to show just how different they really were. The kind we call homozygous would be true-breeding: if allowed to self-pollinate, they would produce one type of pea (yellow in this case). However, those that were heterozygous would produce two types of peas. He planted the yellow peas of the F_2 generation, waited another year for them to grow and bear pods, and then opened the pods to determine which plants were true-breeding. He found that one-third of the F_2 yellow peas had been true-breeding (were homozygous) and that two-thirds had not (and thus were heterozygous).

Mendel was intrigued again by the ratios he observed. He had already determined that three-fourths of the F_2 generation had been yellow and now one-third of those yellow F_2 had been true-breeding. One-third of three-fourths is one-fourth. So one-fourth of the F_2 generation were homozygous and yellow. Furthermore, when he considered all the F_2 peas together, he found the following: three-fourths were yellow and one-fourth green, which is a 3:1 ratio of yellow to green.

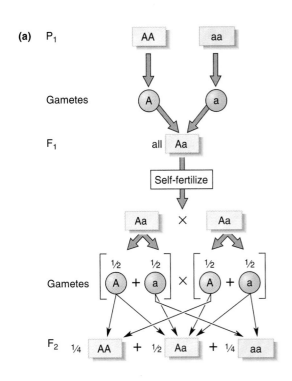

(a) P₁

Gametes

F₁   all Aa

Self-fertilize

Aa × Aa

Gametes

F₂ ¼ AA + ½ Aa + ¼ aa

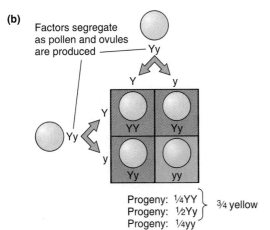

(b)

Factors segregate as pollen and ovules are produced — Yy

Progeny: ¼YY } ¾ yellow
Progeny: ½Yy }
Progeny: ¼yy

FIGURE 10.7
Following Alleles in a Cross

(a) In a generalized scheme for following alleles through a series of crosses, we can see that the outcome of any cross of a homozygous dominant plant (**AA**) with a homozygous recessive plant (**aa**) can only be a generation of heterozygous plants (**Aa**). When the F_1 heterozygous plants are bred, we see that the final outcome of the F_2 generation is strictly dependent on the possible combinations of gametes produced by meiosis. **(b)** Punnett squares can be used as a means of keeping track of alleles in simple crosses. In this example, two heterozygous yellow-seeded individuals are shown. Each produces gametes that are arranged on the axes of the square. In each quadrant of the square, the intersection of the gametes of the parents determines the genotype (and hence the phenotype) of the offspring. In this example, each square represents one-fourth of the offspring. Compare this picture of an F_1 cross with part **(a)** to be sure that you see that these are all just different ways to view the same thing.

$\frac{1}{4}$ of the total F_2 were yellow and true-breeding (homozygous)

$\frac{1}{2}$ of the total F_2 were yellow and not true-breeding (heterozygous) } $\frac{3}{4}$ yellow

$\frac{1}{4}$ of the total F_2 were green and true-breeding (homozygous) } $\frac{1}{4}$ green

The same thing worked for round and wrinkled peas. There were two kinds of round and one kind of wrinkled. Mendel looked at all seven characters (see Table 10.1), and the same principle worked in each case; roughly the same ratios were generated. In every case, one form, bearing what Mendel called the dominant trait, was present in all the F_1 peas or pea plants. In every case, three-fourths of the F_2 seeds or pea plants showed the dominant form of the trait (we would say the dominant phenotype), and one-fourth of the F_2 seeds or pea plants showed the contrasting, recessive trait, for a ratio of 3:1.

In every case, a progeny test revealed that, in actuality, one-fourth of the F_2 were the dominant form and true-breeding (homozygous), one-half of the F_2 were the dominant form but not true-breeding (heterozygous), and one-fourth of the F_2 were the recessive form and true-breeding (homozygous). Clearly, something determining the recessive form was passed through the hybrid F_1 to the recessive F_2.

Mendel used algebraic symbols to help in his analysis (Figure 10.7). He let **A** represent the allele that determines the dominant form, and he let **a** represent the allele that determines the recessive form. He hypothesized that the F_1 hybrid, and in fact all the non-true-breeding plants, must have both alleles present, as represented by the symbol **Aa**. Since there are two parents, Mendel figured that, in the hybrid, **A** comes from one parent and **a** from the other. Mendel determined experimentally that it didn't matter which parental strain bore the peas and which provided the pollen. He let the capital letters signify dominance and the lowercase letters signify recessivity, which merely meant that the **Aa** individuals would look exactly the same as **AA** individuals.

Early in the 20th century, these relationships were put into a graphic form by a fan of Mendel named Reginald Crandall Punnett. Figure 10.7b shows a **Punnett square**, a method that can be used to keep track of a cross between two F_1 heterozygous individuals. Each little square represents the simultaneous occurrence of the event directly above it and to its left. Or we might simply say that gametes from one parent are written at the left and the gametes from the other parent at the top (as shown). The symbols within the square represent possible fertilizations. Note that there are two **Aa** squares, which must be added together to account for all the **Aa** individuals in the F_2 progeny.

If the heterozygous plants get **A** from one parent and **a** from the other parent, and are symbolized by **Aa**, it makes sense that the true-breeding dominant forms get two **A** alleles—one from each parent—and can be symbolized by **AA**. In the same way, the true-breeding recessive forms get **a** alleles from both parents and can be symbolized by **aa**. We can use

AA to symbolize the dominant homozygote, **aa** to represent the recessive homozygote, **Aa** to represent the heterozygote, and **A** to symbolize those plants with the dominant phenotype whose complete genotype is not known.

In summary then:

A = dominant allele

a = recessive allele

AA = homozygous dominant

Aa = heterozygous

aa = homozygous recessive

A_ = dominant phenotype; genotype unknown, **AA** or **Aa**

Now Mendel was ready for some conclusions. In time they were codified into two laws, cleverly called the First Law and the Second Law. Let's see what these laws stated.

Mendel's First Law: Segregation of Alternate Alleles

In the first law, also called the **law of segregation**, Mendel's most fundamental principle proposes that each genetic character is produced by a pair of alleles, and that alleles segregate (became separated) during meiotic formation of pollen and ovule (or sperm and egg). Members of a pair of alleles may be the same or they may differ, with one dominant over the other.

The combination of alleles in the next generation is a matter of chance. Let's look further into these principles through a brief foray into probability itself.

Segregation and Probability

Mendel was the first to realize that when a heterozygote reproduces, its gametes will be of two types in equal proportions:

Now, how did he come up with that? He did it by playing with the numbers that his experiments were generating. He figured that when the heterozygous plant was allowed to pollinate itself, it would produce half **A** ovules and half **a** ovules. It would also produce half **A** pollen and half **a** pollen. When it self-pollinated, what proportion of all the fertilized ovules would be **aa**?

The probability of any given ovule being **a** is $\frac{1}{2}$. The probability of this ovule being fertilized by an **a** pollen is also $\frac{1}{2}$. One of the basic laws of probability, the **multiplicative law**, states that the probability of two *independent* events (that is, those that have no influence on one another) both occurring is equal to the product of their individual probabilities (Figure

10.8). Since the probability of an ovule being **a** and the pollen being **a** have no influence on each other, the probability that an **a** pollen will meet an **a** ovule during self-pollination of an **Aa** parent is $\frac{1}{2} \times \frac{1}{2} = \frac{1}{4}$. (If the probability of your passing your next exam is $\frac{1}{6}$ and the probability of rain on that day is $\frac{1}{5}$, then the probability that both these things will happen is $\frac{1}{30}$, unless your test performance influences the weather, or vice versa.)

The same reasoning applies to the one-quarter of the F_2 progeny that are **AA**: a $\frac{1}{2}$ chance of **A** pollen times a $\frac{1}{2}$ chance of **A** ovule gives a $\frac{1}{4}$ chance of a pea seed (a newly fertilized ovule) being **AA**.

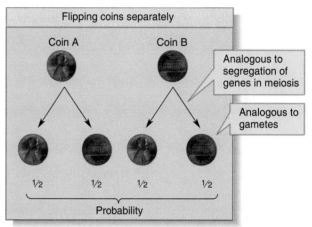

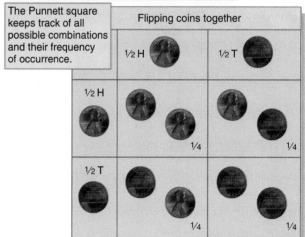

FIGURE 10.8
Probability and Genetics
It is easy to see why Mendel's ratios happen to be such simple fractions. In his crosses there were only two different alleles for any one gene, and for nearly all cases meiosis produces an equal number of gametes of, for example, **A** and **a**. The probability of an individual receiving either an **A** or an **a** from any parent is one-half, the same as the probability of betting heads (or tails) from the flip of a penny. When two pennies are flipped together, we can, in effect, see a simulation of a cross of two **Aa** individuals.

But half of Mendel's F_2 peas were **Aa**. How do we account for that? We find that there are two different ways that **Aa** zygotes are formed in an **Aa** × **Aa** cross. Either an **a** pollen fertilizes an **A** ovule, or an **A** pollen fertilizes an **a** ovule. The probability of the first combination of events is $\frac{1}{2} \times \frac{1}{2} = \frac{1}{4}$, and the probability of the second combination of events is also $\frac{1}{4}$.

There is no other way of getting an **Aa** zygote from such a cross, and the two possibilities are mutually exclusive (they can't both happen to the same zygote). Furthermore, Mendel had shown that the **Aa** heterozygotes were identical regardless of which parent contributed which allele.

This leads to another basic law of probability, the **additive law**: the probability of either one or another of two mutually exclusive events occurring is equal to the sum of their individual probabilities. [If there is a $\frac{1}{2}$ probability of a coin coming up heads, and a $\frac{1}{2}$ probability of it coming up tails, then you can be absolutely sure ($\frac{1}{2} + \frac{1}{2} = 1$) that it will come up either heads or tails.]

But back to the question: Why were half the F_2 peas **Aa**? We know that the two mutually exclusive possibilities are (1) **a** pollen and **A** ovule and (2) **A** pollen and **a** ovule. If either one or the other of these events occurs, the zygote will be an **Aa** heterozygote. The probability of a zygote from such a cross being an **Aa** heterozygote is $\frac{1}{4} + \frac{1}{4} = \frac{1}{2}$.

We have already seen another example of the additive law in probability: the events "zygote is **AA**" and "zygote is **Aa**" are mutually exclusive. The probability of the first in an **Aa** × **Aa** cross is $\frac{1}{4}$, and the probability of the second is $\frac{1}{2}$. Thus the probability that an individual will have the dominant phenotype **A** is $\frac{1}{4} + \frac{1}{2} = \frac{3}{4}$, which, of course, is what Mendel observed.

It is important to realize that, in science, models and hypotheses cannot be *proved* with experimental data. We can only say that the data are consistent with the model. Mendel did not prove his model of the segregation of alternate traits, but the simplicity of the model and the exellence of the fit enabled him to make new predictions and test them. His success came very close to a proof, at least as far as he was concerned. However, others were unconvinced until after the discovery of chromosomes and meiosis. Have you noticed how well Mendel's findings fit with what you already know about meiosis, that homologous chromosomes separate from each other? Imagine how elated Mendel would have been if meiosis had been discovered in his own lifetime.

Mendel's Second Law: Independent Assortment

In his second law, the **law of independent assortment**, Mendel states that the inheritance of a pair of alleles affecting one character occurs independently of the simultaneous inheritance of alleles affecting any other character. Let's see how he arrived at this.

P_1:   **YYRR** × **yyrr**
F_1:   All **YyRr**

where **R** and **r** are symbols for the two alleles of the round-wrinkled gene and **Y** and **y** are symbols for the two alleles of the yellow-green gene. (The letter chosen for the gene is usually the first letter of the dominant phenotype, hence **R** for round, and **r** for wrinkled.)

Now let's see what happened in the F_2 generation when Mendel conducted a **dihybrid cross**—that is, crossed plants that were different in two ways (Figure 10.9). First, as we have said, the F_1 peas were round and yellow. In the F_2, the offspring of $F_1 \times F_1$ (**RrYy** × **RrYy**), Mendel classified 556 peas into four groups (recall that **R** and **Y** indicate that the individual displays the dominant phenotype, but the second allele isn't known).

315 round and yellow	**R_Y_**
101 wrinkled and yellow	**rrY_**
108 round and green	**R_yy**
32 wrinkled and green	**rryy**

Note that 133 peas altogether were wrinkled, and 140 peas all together were green. Both numbers are close to 139, which is one-fourth of 556. So the results supported Mendel's

TABLE 10.2   Mendel's Predictions and Results for F_2 Phenotype (556 seeds observed)

Phenotype of F_2		Fraction Predicted	Number Predicted Out of 556	Number Actually Observed
Round and yellow	○	9/16	312.75	315
Wrinkled and yellow	⬡	3/16	104.25	101
Round and green	○	3/16	104.25	108
Wrinkled and green	⬡	1/16	34.75	32

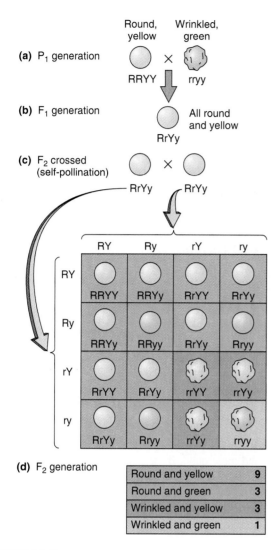

(a) P₁ generation — Round, yellow (RRYY) × Wrinkled, green (rryy)

(b) F₁ generation — All round and yellow (RrYy)

(c) F₂ crossed (self-pollination) — RrYy × RrYy

(d) F₂ generation

Round and yellow	9
Round and green	3
Wrinkled and yellow	3
Wrinkled and green	1

FIGURE 10.9

Independent Assortment of Two Pairs of Alleles

Mendel developed his second law by observing the behavior of two traits simultaneously. The traits being considered here are color (yellow or green) and shape (round or wrinkled) of seeds. **(a)** When true-breeding round-yellows (**RRYY**) are crossed with true-breeding wrinkled-greens (**rryy**), they produce an F_1 that is round and yellow **(b)**, heterozygous for both traits (**RrYy**). The F_2 is diagrammed in the Punnett square **(c)**. There are four distinct phenotypes, which include all possible color and shape combinations. These phenotypes occur in a 9:3:3:1 ratio, but the important thing to remember is that any individual trait (if you just look at yellow versus green, for example) still is represented in a 3:1 ratio.

first law since about one-fourth of the F_2 peas were wrinkled and one-fourth were green, while three-fourths were round and three-fourths were yellow.

But since round wasn't always inherited with yellow, and wrinkled wasn't always inherited with green, Mendel had also shown that the color and shape characteristic were inherited independently. We have already mentioned the multiplicative

law of probability, which gives the probability of two independent events both occurring. If the probability of being round is $\frac{3}{4}$ and the probability of being yellow is also $\frac{3}{4}$, the probability of being both round and yellow is $\frac{3}{4} \times \frac{3}{4} = \frac{9}{16}$ *if* the two events (seed shape and seed color) are independent. Nine-sixteenths of 556 peas is 312.75 peas. When Mendel counted he found 315 round, yellow peas—remarkably close (Table 10.2)!

Mendel thought the fit between expected and observed numbers was impressive. He went on to determine which of these peas were true-breeding and which were not. Here again, the fit between expected and observed was good. He expected one-fourth to be **RR**, one-half to be **Rr**, one-fourth to be **rr**, and one-fourth to be **YY**, one-half to be **Yy**, and one-fourth to be **yy**, so that:

$$\frac{1}{4} \times \frac{1}{4} = \frac{1}{16} \text{ should be } \mathbf{RRYY}$$
$$\frac{1}{2} \times \frac{1}{4} = \frac{1}{8} \text{ should be } \mathbf{RrYY}$$
$$\frac{1}{2} \times \frac{1}{2} = \frac{1}{4} \text{ should be } \mathbf{RrYy}$$

and so on. Mendel was able to get 529 of his 556 F_2 peas to bear F_3 progeny. The breakdown of their genotypes is listed in Table 10.3.

Again, Mendel felt that the numbers fit his model quite well. He tried combinations of other traits; he even tried three traits together. In each case, the different pairs of alternative traits behaved independently. We know why, of course: maternal and paternal chromosome pairs line up and separate independently during meiosis. (See Figure 10.10, and refer back to

TABLE 10.3 Mendel's Predictions and Results for F₂ Genotypes (529 peas classified by progeny test)

Genotype of F₂	Fraction Expected	Number According to the Hypothesis	Number Actually Observed by Progeny Testing
RRYY	1/16	33	38
RRYy	2/16	66	65
RRyy	1/16	33	35
RrYY	2/16	66	60
RrYy	4/16	132	138
Rryy	2/16	66	67
rrYY	1/16	33	28
rrYy	2/16	66	68
rryy	1/16	33	30

Note: Keep in mind that Mendel had to perform a progeny test for each of the 529 F₂ peas in order to learn their genotype. There is no other way to prove the genotype of a heterozygote.

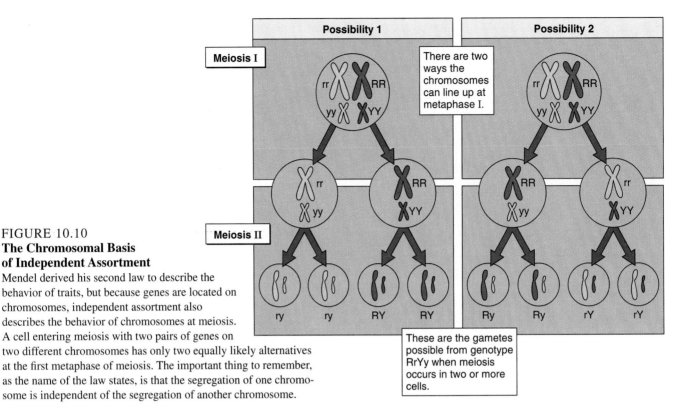

Possibility 1

Possibility 2

Meiosis I

There are two ways the chromosomes can line up at metaphase I.

Meiosis II

These are the gametes possible from genotype RrYy when meiosis occurs in two or more cells.

FIGURE 10.10
**The Chromosomal Basis
of Independent Assortment**
Mendel derived his second law to describe the
behavior of traits, but because genes are located on
chromosomes, independent assortment also
describes the behavior of chromosomes at meiosis.
A cell entering meiosis with two pairs of genes on
two different chromosomes has only two equally likely alternatives
at the first metaphase of meiosis. The important thing to remember,
as the name of the law states, is that the segregation of one chromosome is independent of the segregation of another chromosome.

Figure 9.9.) But Mendel had never seen DNA or the movement
of chromosomes. In Mendel's words: "The behavior of each
pair of differing traits in a hybrid association is independent of
all other differences between the two parental plants..." This
then became known as Mendel's second law, or the law of
independent assortment. Now we can restate it quite simply:
each gene will segregate independently of other genes.

What Mendel didn't know was that the law of independent assortment works only for genes that are not on the same
chromosome. Simply put, if **R** and **Y** are on the same chromosome, they are said to be **linked genes**, and Mendel's law of
independent assortment simply doesn't hold for linked genes.
We will further explore linked genes in the next chapter.

Mendel's Testcrosses

We have seen that Mendel was able to determine the genotypes
of members of his F_2 generations that were showing the dominant phenotype by permitting the individuals to self-pollinate
and then observing the phenotypes of the F_3. Mendel also
devised a simpler procedure to determine the genotype of any
dominant individual. In what is called a **testcross**, such individuals are crossed with subjects that are true-breeding (homozygous) recessives for the trait or traits under consideration.
(Remember, the genotype of an individual with the recessive
trait is unambiguous, such as **aa**.) For instance, if the phenotype of a seed is "round," its genotype could be either **RR** or
Rr. Crossing it with a homozygous recessive individual, wrinkled **rr**, would readily reveal the actual genotype. If the geno-

type is **RR**, the cross would be **RR** × **rr**, and all offspring must
be **Rr** and thus round. If the genotype is **Rr**, then the cross is
Rr × **rr**, and the phenotypic ratio in the offspring is one-half
round and one-half wrinkled. The principles of the testcross
are demonstrated in Figure 10.11.

Testcrosses can also be used to determine the genotype of
dominant individuals where two traits are involved. Such testcrosses, as a matter of fact, greatly aided Mendel in further
testing the idea of independent assortment. The individual
used as the recessive test parent would, of course, be doubly
recessive; thus in our last example, the individual used would
be wrinkled and green, or, **rryy**. Any dominant F_2 individual
from Mendel's dihybrid cross could have its genotype determined in this manner. For instance, a round, yellow seeded
individual, **R_ Y_**, could be any of the following: **RRYY,
RrYY, RrYy,** or **RRYy**

What testcross results would you predict in each case?
You may see the answers intuitively, or, like most of us, you
may have to draw Punnett squares and carry out each cross (if
all else fails, see the example in Table 10.4). Incidentally,
Mendel, never one to be easily satisfied, made it his practice
to carry out additional crosses—actually testcrosses on testcross progeny. Such endless probing and checking could
explain why, in spite of seemingly endless challenges, his
laws are as valid today as they were well over 100 years ago.

The results of Mendel's experiments were published in
1866. However, the work remained obscure for some time and
appreciation for the tremendous importance of his findings
did not begin until the turn of the century.

Round

Q: Is the genotype RR or Rr?

A: Testcross:

R_   rr

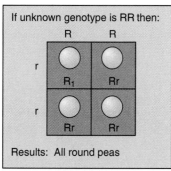

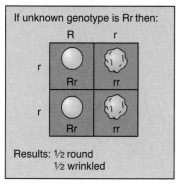

FIGURE 10.11
The Testcross

The two types of round seeds (heterozygous and homozygous) can be distinguished in the results of a cross with homozygous recessive **rr**. The homozygous **RR** produces all round offspring. The heterozygous **Rr** produces 1:1 round:wrinkled.

The Chromosomal Basis for Mendel's Laws

While Mendel was completing his work, others were making advances in new directions. One of them was Theodore Boveri, a German who, working with sea urchin eggs and sperm, determined that the chromosomes were an essential part of fertilization and development. Another important advance came from a bright young graduate student at Columbia University, Walter Sutton. In 1902 he published a paper in which he reported a relationship betwen Mendelian inheritance and meiosis. Sutton also reasoned that each chromosome must carry many genes, since there are only a small

number of chromosomes and there are many hereditary factors. But keep in mind that Mendel didn't know about meiosis and that those studying chromosomes and their movement weren't sure how they fit into the big picture.

The realization that there is a relationship between the physical behavior of chromosomes in meiosis and Mendel's mathematically based laws is rather straightforward for us today (see Figure 10.10). However, the initial finding required that investigators observe the actual random lineup of homologous chromosomes on the metaphase plate and their subsequent random separation during anaphase. At first, this may seem easy because, in our figures here, we often show the chromosomes as clearly distinguishable from each other. However, this is a necessary fiction. In the real world under the microscope, homologous chromosomes are not clearly distinguishable from each other. All the same, the researchers needed to actually "see" chromosomes engaged in segregation and in independent assortment.

This is exactly what they did. Investigators found chromosome pairs in which the two homologous chromosomes have some visible difference, such as an extra knob of chromatin on one end, or a dark or faint band here or there. When two sets of visibly different chromosome pairs were present in the same organism, they were observed in the microscope to segregate independently in different meiotic cells, giving visible evidence of Mendel's second law. Thus, while Mendel drew his inferences from analysis of numbers, the cytologists could actually view alternate segregation and independent assortment in the meiotic process through their light microscopes. These experiments took place some 40 years after Mendel's experiments and, together with a renewed interest in genetics, contributed to a celebrated "rebirth" of Mendel's laws.

TABLE 10.4	The Dihybrid Testcross on Plants with Round and Yellow Seeds of Unknown Genotype (R_Y_)		
Possible Genotype	**Testcross**		**Results Expected**
RRYY	RRYY × rryy		All offspring round and yellow
RrYY	RrYY × rryy		1/2 round and yellow
			1/2 wrinkled and yellow
RrYy	RrYy × rryy		1/4 round and yellow
			1/4 round and green
			1/4 wrinkled and yellow
			1/4 wrinkled and green
RRYy	RRYy × rryy		1/2 round and yellow
			1/2 round and green

Pedigrees and Human Traits

Mendel's principles are so basic that they can be applied to the analysis of hereditary patterns in virtually any plant or animal. But it is important to know that his analytical methods apply best where large numbers of progeny are produced. Individual pea plants (at least in the strains Mendel used) produce 30 or so seeds per cross, and even with this seemingly large number, Mendel's records revealed that one of his plants produced 30 yellow seeds and only one green! To obtain his famous 3:1 and 9:3:3:1 ratios, Mendel had to combine the results from many identical crosses.

Statisticians today refer to Mendel's problem with the aberrant (30:1) plant "random statistical variation" or, more commonly, **sampling error**. Because of such problems, they recommend sample sizes of at least 100. Imagine the impact of this when it comes to determining hereditary patterns in humans, where family sizes are commonly only two or three offspring! As you see, it is virtually impossible to reach valid conclusions about ratios of offspring on the basis of a single human mating.

To overcome this problem in the study of inheritance in humans, scientists use what is called **pedigree analysis**, charting the passage of a trait through many generations. In pedigree charts, squares and circles (□ = male and ○ = female) represent individuals and darkening them indicates the occurrence of the trait being analyzed. The pedigree shown in Figure 10.12 examines the inheritance of a hypothetical genetic disease.

The first step in any pedigree analysis is to apply logic in deciding whether the trait is dominant or recessive. Specifically, an individual must be homozygous to express a recessive trait but need only be heterozygous to express a dominant trait. More importantly, individuals expressing a dominant trait must have at least one parent who also expresses that trait. The disease in this example is clearly not caused by a dominant allele, since the parents of the affected children do not have the trait. It must be recessive.

If you are wondering why this trait is so common in this family, consider that the parents of the affected individuals were related, so that both could have received the allele from a common ancestor. (The double lines signify marriages between closely related individuals, called consanguineous or "of common blood" marriages.) In this pedigree the disease allele must have occurred originally in one of the two original parents and passed with a probability of $\frac{1}{2}$ to a child in each subsequent generation, hidden or "carried" in the heterozygous condition, until a consanguineous marriage between two heterozygotes resulted in a $\frac{1}{4}$ probability of children being homozygous.

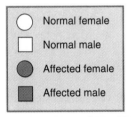

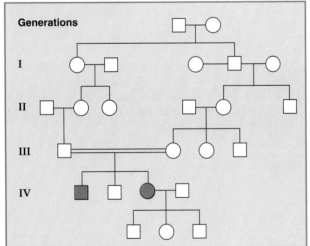

FIGURE 10.12
Pedigree of a Human Family
This hypothetical family has members that suffered a rare disease. It is a recessive condition and only occurred in this family when close relatives were married. Marriages of closely related individuals are termed consanguineous and are indicated in pedigrees by a double horizontal line connecting the couple. The convention is that □ = male and ○ = female, and that affected individuals are indicated by a darkening of the circle or square.

The Decline and Rise of Mendelian Genetics

In 1865, after seven years of experimentation (at the very time Darwin was beating his brains out over the enigma of heredity), Mendel presented his results at a meeting of the Brunn Natural Science Society. His audience of local science buffs probably did not understand a word of what they were hearing. After a polite applause, they burst into a vigorous discussion of the hot new idea of the day: Darwin's ideas of natural selection. Mendel's single paper was published in the society's proceedings the following year and was actually distributed rather widely. However, the learned scientists of the day were just as baffled and just as uninterested as Mendel's original audience. Eventually, a German botanist included an abstract of Mendel's work in an enormous encyclopedia of plant breeding. Again, the world responded with silence. Apparently, no one had the foggiest notion of the importance of the Austrian monk's experiments and analysis; the minds of 1865 were just not ready for 20th century mathematical biology.

In Darwin's huge library, historians found a one-page account of Mendel's pea work that appears in a German encyclopedia of plant breeding. Some relatively obscure work on the facing page was extensively annotated in Darwin's handwriting. The page describing Mendel's work, however, contains no notes to indicate that Darwin paid it any attention. Had he read and understood Mendel's work, it would have explained his own snapdragon data and, more important, could have clarified his theory of natural selection, saving him years of agony and uncertainty. But even Darwin was not ready for mathematical biology, and he too failed to grasp Mendel's simple but profound ideas.

Darwin apparently had a very different kind of mind than Mendel. Mendel was not only one of the first mathematical biologists familiar with statistics, but he also knew how to isolate small parts of great problems. Darwin's genius, on the other hand, was in the enormous breadth and scope of his ideas and the ability to fit seemingly unrelated details into a grand scheme.

Mendel's work continued to be ignored until 1900. In that year, three biologists in three different countries, all trying to work out the laws of inheritance, recognized the importance of Mendel's paper. Science had changed in 35 years. The obscure monk became one of the most famous scientists of all time. But he had been dead for 16 years.

Later in the 20th century, Mendelian genetics was applied to the theory of natural selection, and Darwin's reputation, which had faded considerably, ascended to new heights. Darwin had been right all along, if only he had gotten his genetics straight. It is often too easy for those of us who have just had something explained to us to say, "but of course." However, breaking new conceptual ground, even a little, can be stultifyingly difficult. Mendel's work and his reasoning, like Darwin's, are so obvious to us now that we often forget how formidable the monk's task was as he placed those first pea seeds in the carefully cultivated soil of that monastery garden.

Key Ideas

WHEN DARWIN MET MENDEL (ALMOST)

1. Darwin subscribed to the idea of blending inheritance, which held that traits from both parents are "blended" in the offspring.

2. Critics of natural selection pointed out how blending would destroy variations—a key part of Darwinian evolution.

3. Darwin crossed true-breeding strains of snapdragons, in which he observed **dominant** traits, which he then called prepotent.

MENDEL'S CROSSES

1. Mendel reported on experiments with seven different traits in true-breeding garden peas.

2. In peas, each pea in a pod has its own **genotype** and **phenotype**. (*Genotype* is total combination of an organism's genes; *phenotype* is the combination of observed or measured traits, generally what is readily visible.)

3. The symbols P_1, F_1, and F_2 are used to designate first and subsequent generations in crosses.

4. Mendel perceived that the **recessive** trait disappeared in the F_1 generation and reappeared in the F_2. He noted that the reappearance of a *recessive* trait occurred with a definite frequency of one-fourth of the F_2.

5. He also determined that dominant individuals could be **heterozygous** or **homozygous**. To determine this required **progeny testing**—breeding an F_3 generation by self-fertilizing F_2 round peas. He determined that one-third of the round F_2 peas were true-breeding (homozygous), and two-thirds were heterozygous.

6. Mendel concluded that characteristics were controlled by factors in two forms—dominant and recessive. In modern terms, factor is replaced by gene. The alternate forms of a gene are called **alleles**.

MENDEL'S FIRST LAW: SEGREGATION OF ALTERNATE ALLELES

1. Heterozygous (**Aa**) individuals produce two kinds of gametes (sex cells) in equal proportions:

$$\mathbf{Aa}$$

$$\tfrac{1}{2}\mathbf{A} \qquad \tfrac{1}{2}\mathbf{a}$$

2. Following this, the **multiplicative law** from the laws of probability can be applied to crosses: "the probability of two independent events both occurring is equal to the *product* of their individual probabilities."

3. Therefore the F_1 cross **Aa** × **Aa** can be stated as $(\tfrac{1}{2}\mathbf{A} + \tfrac{1}{2}\mathbf{a}) \times (\tfrac{1}{2}\mathbf{A} + \tfrac{1}{2}\mathbf{a})$. Multiplying produces $\tfrac{1}{4}\mathbf{AA} + \tfrac{1}{2}\mathbf{Aa} + \tfrac{1}{4}\mathbf{aa}$. This $\tfrac{1}{2}\mathbf{Aa}$ is determined algebraically but can be explained genetically. The combination **A** + **a** can occur two ways: **A** + **a** or **a** + **A**.

4. The **additive law** is stated as follows: "the probability of either one or another of two mutually exclusive events occurring is equal to the *sum* of their individual probabilities."

5. Therefore the probability of an **A** pollen and an **a** ovule combining is one-fourth, as is the probability of **a** + **A**. Since they are mutually exclusive events, the probability of a heterozygous F_2 individual is $\tfrac{1}{4} + \tfrac{1}{4} = \tfrac{1}{2}$.

6. Punnett illustrated Mendel's principles using squares:

	A	**a**
A	AA	Aa
a	Aa	aa

When summed up the results are:

Genotype: $\frac{1}{4}\mathbf{AA} + \frac{1}{2}\mathbf{Aa} + \frac{1}{4}\mathbf{aa}$

Phenotype: $\frac{3}{4}$ dominant $+ \frac{1}{4}$ recessive

MENDEL'S SECOND LAW: INDEPENDENT ASSORTMENT

1. Mendel crossed two alleles of the round locus, **R** and **r**, and two alleles of the yellow locus, **Y** and **y**. **Locus** refers to a specific gene location on a chromosome.

    RRYY × **rryy**   (P_1 cross)

    all **RRYY**   (F_1 offspring)

    RrYy × **RrYy**   (F_2 dihybrid cross)

    To predict the results, consider the following:

    a. You know that **Rr** × **Rr** produces $\frac{1}{4}\mathbf{RR}, \frac{1}{2}\mathbf{Rr},$ and $\frac{1}{4}\mathbf{rr}$. Likewise, **Yy** × **Yy** produces $\frac{1}{4}\mathbf{YY}, \frac{1}{2}\mathbf{Yy},$ and $\frac{1}{4}\mathbf{yy}$.

    b. To predict the results when both are considered simultaneously, follow the multiplicative law for all possible **R** and **Y** combinations.

2. In addition to being consistent with his expectations, Mendel's findings indicate that two traits, pea shape and color, are inherited independently. If they were not, the multiplicative law would not have worked.

3. Today we know that all the characters that Mendel studied in this way were located on different chromosomes. Because of the random way in which pairs of chromosomes align at metaphase and separate at anaphase, genes separate independently of each other.

4. Mendel used testcrosses to determine whether a dominant type was homozygous or heterozygous, and whether assortment was independent. Suspected heterozygotes are crossed with homozygous recessive individuals.

THE CHROMOSOMAL BASIS FOR MENDEL'S LAW

1. Sutton proposed that the hereditary factors were contained on the chromosomes and that each chromosome carried many factors.

2. Establishing the relationship between Mendel's first and second laws and meiosis required following two heteromorphic pairs of chromosomes through meiosis.

PEDIGREES AND HUMAN TRAITS

1. Pedigree analysis allows insight into genetic effects in situations (as in human genetics) where controlled crosses are impractical.

THE DECLINE AND RISE OF MENDELIAN GENETICS

1. Mendel's findings, which were not understood in his time, were rediscovered about the turn of the century.

Application of Ideas

GENETICS PROBLEMS

To the dismay of many biology students, most educators agree that solving genetics problems is a tremendously instructive means toward truly understanding the ins and outs of Mendelian genetics. It is widely held that ability to do genetics problems is directly related to one's understanding of the principles, AND that a thorough understanding of the principles requires doing genetics problems. Below you will find a series of genetics problems designed around the main points of this chapter. They begin with the obvious and straightforward type, then move onto the more subtle kind that require some sleuthing. If you work them through from beginning to end, you will see that there are a limited number of possible answers to any particular part of a problem, and that the more difficult problems represent a collection of simple parts. The key is seeing through to those simple parts. The key is based on Mendel's laws.

1. Answer the following questions regarding a cross of two heterozygous (**Aa**) individuals:

    a. If they have a child, what is the probability that it is homozygous recessive?

    b. If they have two children, what is the probability that both will be heterozygous?

    c. If they have two children, what is the probability that the first one will be homozygous recessive and the second one will be homozygous dominant?

    d. f they have two children, what is the probability that one will be homozygous recessive and one will be homozygous dominant?

    e. What is the difference between parts (c) and (d)?

2. Answer the following questions regarding a cross of a homozygous dominant (**AA**) with a homozygous recessive (**aa**):

    a. If they have a child, what is the probability that the child will be heterozygous?

    b. What is the probability that the child will be homozygous recessive?

    c. If they have two children, what is the probability of the first one being homozygous dominant and the second one being homozygous recessive?

3. Answer the following questions regarding a cross of two homozygous dominant individuals (**AA**):

    a. If they have a child, what is the probability that the child will be heterozygous?

    b. If they have a child, what is the probability that the child will be homozygous dominant?

4. In humans, albinism is caused by a single gene. The recessive allele (**a**), when homozygous, results in no pigment deposition in the skin and therefore a very pale skin color. Ann Smith is albino. She married Mike Jones, and they had two children, Sara (who had skin pigment) and Martha (who was albino).

   a. Write out the genotypes of Ann, Mike, and their children, Sara and Martha.

   b. What is the probability that their next child would be albino?

   c. What is the probability that, if they have two more children, both children will be normally pigmented?

   d. If they have two more children, what is the probability that one will be albino and one will be normally pigmented?

5. Ann and Mike divorce. Ann remarries, this time to Ralph Johnson. They have six normally pigmented children.

   a. What would you assume was Ralph's genotype?

   b. Assuming, however, that Ralph was heterozygous, what is the probability of their having those six normally pigmented children?

6. Given two separate, independently assorting genes, **A** and **B**, with **A** dominant to **a** and **B** dominant to **b**, answer the following questions:

   a. What is the probability of getting an **AB** gamete from an individual who is heterozygous at both loci?

   b. What is the probability of getting an **Ab** gamete from an individual who is heterozygous at both loci?

   c. What is the probability of getting an **AB** gamete from an **AABb** individual?

   d. What is the probability of getting an **AABB** child from a cross of two people heterozygous at both loci?

   e. From a cross of two people heterozygous at both loci, what is the probability of having a child that expresses the dominant phenotype for both loci?

Review Questions

1. What kind of preparation did Mendel have for mathematical biology? What other characteristics led to his success? (page 221)

2. Describe Mendel's general procedure from P_1 to F_2. What kinds of crosses did he make? (page 222)

3. What important question confronted Mendel as he observed the F_1 generation in each cross? (page 222)

4. Distinguish between the terms *genotype* and *phenotype*. (page 222)

5. Write out Mendel's yellow-versus-green crosses from F_1 to F_2 and verify his ratios in the F_2. State the phenotypic ratio of the F_2. (Figure 10.7)

6. Distinguish between the terms *heterozygous* and *homozygous*. (page 225)

7. Explain how Mendel used progeny testing to determine the *genotype* of his F_2 peas. What did he learn? (pages 225–226)

8. Using the cross **Aa** × **Aa** and applying the multiplicative law, answer the following: (pages 227–228)

   a. What is the probability of an **A** sperm fertilizing an **A** egg? Why?

   b. What is the probability of an offspring carrying the **Aa** combination? Explain carefully.

   c. What is the probability of an **a** sperm fertilizing an **A** egg? Why?

   d. How does the *additive law* apply to predicting the **Aa** offspring?

9. Carry out Mendel's dihybrid cross from P_1 through F_2. Write the *phenotypic* ratio of the F_2. (pages 228–229)

10. State Mendel's second law and explain what it has to do with the results of the above cross. (page 228)

11. Using diagrams of chromosomes to represent the cross **AaBb** × **AaBb**, show how independent assortment works. (Review meiosis in Chapter 11.) *Hint:* There are two ways the homologous pairs of chromosomes can align on the metaphase plate. (pages 228–229)

12. Show the two types of testcrosses or backcrosses Mendel made for the double heterozygote **RrYy**. What is the purpose of a *testcross?* (page 230)

Going Beyond Mendel

CHAPTER OUTLINE

DOMINANCE RELATIONSHIPS AND GENE INTERACTIONS

Incomplete Dominance

Codominance

Multiple Alleles

Lethals

Gene Interactions

CONDITIONAL PHENOTYPES

Influences of Environment

Influences of Development

Influences of Sex

Influences of Age

Pleiotropy

Continuous Variation and Polygenic Inheritance

GENES ON CHROMOSOMES

Linkage

Crossing Over and Genetic Recombination

Mapping Genes

Chromosomes and Sex

Essay 11.1 Sex Chromosome Abnormalities

Essay 11.2 The Disease of Royalty

KEY IDEAS

APPLICATION OF IDEAS

REVIEW QUESTIONS

Mendel frankly attributed his success to his deliberate decision to work only with factors that always produced large, dramatic effects with clear and distinct phenotypes. He examined such traits, one or two at a time, in true-breeding strains. Only in such simple systems could he have worked out his famous ratios. Mendel's discoveries were valuable because nearly all genes, in fact, are transmitted according to his principles, at least on a genotypic level. But most genetic variation is not so simple at the level of the phenotype. While there are clearly situations where a gene simply controls the presence or absence of a trait, it is also true that a genetic trait can be governed by a number of genes that interact with each other and the environment in some fashion to produce the phenotype. Most genetic issues involve complexities of development and gene expression that Mendel had so fortuitously avoided.

In this chapter we will consider those circumstances that influence many phenotypic ratios, causing them to be something other than a perfect 3:1 in a cross of heterozygous individuals. The complications do not violate or change the laws of heredity, but they do influence our ability to interpret the number and classes of offspring in a cross. There are two main issues that influence phenotypic ratios resulting from a cross. The first is the complexity of the production of a phenotype from the genotype. Thus we begin our discussion with a consideration of phenomena necessary for an understanding of the production of a genetic trait. The second issue is linkage, a concept that we initially defined in the last chapter. Clearly, when genes are physically close together on the same chromosome, they cannot behave as the completely independent entities envisioned by Mendel. In addition to complicating our picture of inheritance, linkage offers its own unique insights into the mechanisms that govern inheritance.

As you consider this chapter, you will begin to see that these situations have a serious impact on the ideal ratios generated by Mendel's laws. However, this does not invalidate the laws themselves. Because after all is said and done, essentially all genes in the nucleus behave according to the laws of random segregation and independent assortment.

To illustrate how genes, following such clear-cut laws, can produce unexpected results, consider the following. Most of the genetic differences we see in our friends—differences in height, weight, body build, skin color, temperament, facial features, athletic ability, and hairiness—appear to be due to normal variation in the genes that govern these traits (along with the modifying effects of the environment). Even blue and brown eye color, a popular example of Mendelian inheritance in humans, turns out to be quite complex and seemingly unpredictable. People do not merely have blue or brown eyes; they may have gray, light blue, deep blue, hazel, flecked, or green eyes. In this chapter, we will look beyond Mendel to learn more about how genes behave in non-Mendelian ways.

Dominance Relationships and Gene Interactions

The various ways in which two alleles of a gene can affect the phenotype are called **dominance relationships**. You are aware, of course, that one allele comes from each parent. Thus each parent contributes to the phenotype controlled by a gene.

Dominant means that the phenotype of the heterozygote (an individual with two different alleles) will be exactly like that of one of the homozygotes. The allele that is expressed is called the dominant allele, and the allele that is not expressed in the heterozygote is called the recessive allele.

There are several different ways in which dominance can occur. In most cases the recessive allele isn't expressed because it simply isn't doing anything; its instructions aren't

being read. This is clearest in rare medical disorders, where the absence of an enzyme can have a severe and often lethal effect. In a relatively benign example shown in Figure 11.1, albinos lack an enzyme that is necessary to make melanin pigments. They didn't get a functioning gene from their father, and they didn't get one from their mother, so their cells cannot make the pigment. Such individuals are homozygous for recessive alleles, which in this case are alleles that aren't functioning. Heterozygotes for albinism or other enzyme deficiencies, on the other hand, have one working allele and one that doesn't work, and so they produce only half the usual amount of enzyme; but in most cases half the normal amount of enzyme is still enough to metabolize all the enzyme's substrate, and the phenotype of the heterozygote will be perfectly normal.

In contrast to the seemingly clear-cut nonfunctioning of the recessive allele, there are often situations where a recessive allele may seem to function a little or interact with the dominant allele in some other way.

Incomplete Dominance

Whenever the phenotyope of the heterozygote is intermediate between the phenotypes of the two homozygotes, and not exactly like either one of them, we call the relationship **incomplete dominance** (or sometimes *partial* dominance). A classic example of incomplete dominance is found in snapdragons, in crosses between strains with red flowers and strains with white flowers (Figure 11.2). Here the two alleles can be symbolized by c^R and c^W. (Historically, incompletely dominant alleles are designated by lowercase letters. Hence we use c^R and c^W here, instead of **R** and **r**, which would imply complete dominance.) When the homozygous $c^R c^R$ red-flowered snapdragons are crossed with homozygous $c^W c^W$ white-flowered snapdragons, the $c^R c^W$ heterozygotes in the F_1 generation all have pink flowers. If the pink-flowered F_1 plants are self-pollinated to pro-

FIGURE 11.1
Albinism in a Human Family
Albinism, the absence of pigmentation in hair and a light skin color, is a striking, simple recessive genetic trait.

FIGURE 11.2
Incomplete Dominance

Flower color in snapdragons presents a good example of incomplete dominance. The hallmark of incomplete dominance is that the heterozygotes appear to have a phenotype that is somewhere in between the homozygous dominant and the homozygous recessive. Another distinguishing feature is that the phenotypic ratio of the progeny is always the same as the genotypic ratio. Can you see why?

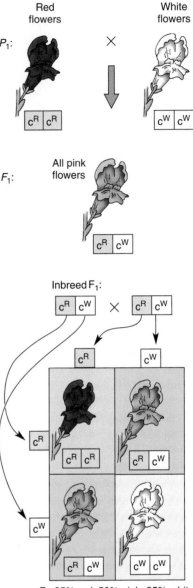

duce an F_2 generation, we have an interesting Mendelian ratio. Instead of a 3:1 or 75% red to 25% white, the F_2 snapdragons are 25% red, 50% pink, and 25% white. Since all three genotypes are easily distinguished, the F_2 phenotypic ratio is exactly the same as the F_2 genotypic ratio (see Figure 11.2).

Sometimes the phenotype of the heterozygote is not simply a compromise between the two homozygous phenotypes but has some unique characteristics of its own. Consider Roy Rogers' horse, Trigger, for instance (Figure 11.3). Trigger was a beautiful palomino horse, with a golden coat and blonde mane and tail. (You can see him for yourself, stuffed and mounted, in the Roy Rogers and Dale Evans Museum in southern California.) All palomino horses are heterozygotes. Crosses between palomino horses yield brown, palomino, and white foals in an approximate 1:2:1 ratio.

Codominance

Codominance is the equal expression of both alleles that results in a mixed phenotype. This kind of coexpression is seen most easily in some biochemical phenotypes like the blood types that we will soon discuss. But, for now, consider roan cattle. The roan phenotype results from a cross of red cattle and white cattle. The phenotype of the heterozygote is a kind of pink color called roan (Figure 11.4). But if you look closely at a roan-colored cow you will find, not pink hairs, but

a mix of red and white hairs. So here we have an equal expression of the two alleles.

Perhaps now you see that dominance is a question of degree. At one end of the spectrum we have completely dominant and completely recessive alleles. In between we have alleles that interact as incomplete dominant alleles or as codominant alleles.

Multiple Alleles

Fortunately for Mendel, he had to deal with only two alleles at any gene locus; otherwise he might have become hopelessly entangled in complex genetic systems. This is because a gene can have many alleles (although any one individual can have only two alleles—one from each parent—for each gene).

FIGURE 11.3
Trigger, the Palomino Horse
The beautiful golden palomino color, here shown in the most famous of palomino horses (who is stuffed and on display), is a result of heterozygosity of alleles for both brown and white coat color in horses. Happy trails.

FIGURE 11.4
Codominance in Cattle
The roan-colored cow with its white and red hairs is an example of codominance, the equal expression of two alleles.

Let's look at a simple example of multiple alleles. Coat color in rabbits is determined by a single gene (Figure 11.5). The agouti coloration is the brown mottled color that you find in wild rabbits and is determined by a dominant allele, **C**. Albino rabbits are white due to a lack of pigment and are homozygous for the recessive **c** allele. Agouti (**C**) is completely dominant to albino (**c**). There is a third type of coat color that you may be familiar with, called himalayan. Himalayan rabbits are mostly white but have black fur on their noses, paws, and ears. The allele that causes the himalayan phenotype is **cʰ**. The **cʰ** allele is dominant to **c**. Therefore a cross of two heterozygous individuals (**cʰc × cʰc**) will produce himalayans and albinos in a 3:1 ratio. But himalayan is recessive to agouti! **Ccʰ** individuals are agouti. A cross of individuals heterozygous for the agouti and himalayan alleles (**Ccʰ × Ccʰ**) will produce agoutis and himalayans in a 3:1 ratio.

Thus we can see, once again, that the concept of dominance is often a question of point of view. Even trying to be more precise by using uppercase letters to show dominance can be troublesome. More importantly, however, we can see that while a population may have many alleles for any given gene, any individual can have only two alleles for that gene.

Another example of multiple alleles is the familiar ABO blood group system in humans. Human blood type is determined by polysaccharides on the outer surfaces of the red blood cells. These cell surface polysaccharides are called antigens because they can produce an immune system response if introduced into another body. Part of this immune response is

(a)   (b)   (c)

FIGURE 11.5
Multiple Alleles and Rabbit Coat Color
There are three alleles that can interact to produce coat colors in rabbits, **C**, **c**, and **cʰ**. **C** rabbits are agouti (**a**), **cc** rabbits are albino (**b**), while **cʰcʰ** and **cʰc** rabbits are himalayan (**c**). Therefore **C** is dominant to **cʰ** and **c**, **cʰ** is recessive to **C** but dominant to **c**, and **c** is recessive to both **C** and **cʰ**.

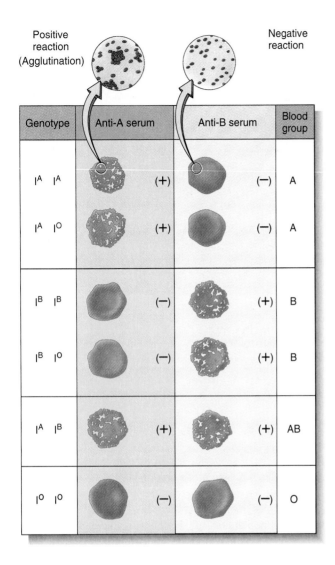

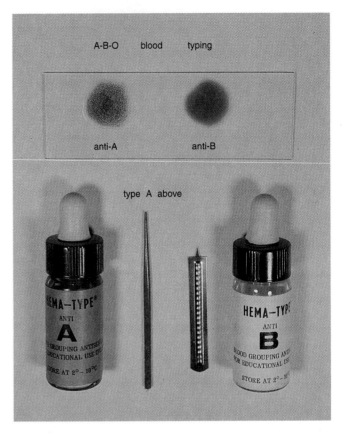

FIGURE 11.6
ABO Blood Typing
The presence of A and B antigens in human blood can be determined in a simple test using anti-A and anti-B test reagents. A positive reaction (agglutination) is characterized by a distinct graininess occurring in the mixture.

the production of proteins called antibodies, which bind to the introduced antigens, clumping them together for easier destruction by certain white blood cells. The red blood cells of a type A person, for instance, have A antigens on their cell surface, and the blood cells of a type B person have B cell surface antigens. The **I** gene (**I** stands for isoagglutinogen, which is another term for the cell surface polysaccharide) is responsible for the production of these polysaccharides. Type AB people have both A and B antigens, so the two alleles for A and B are codominant. The two alleles are usually written I^A and I^B to distinguish the allele that makes the antigen from the antigen itself. Individuals with the fourth blood type, type O, have neither the A antigen nor the B antigen on the surfaces of their red blood cells. A third allele then is I^O (or **i**), which is recessive to both I^A and I^B. The possible genotypes and their corresponding phenotypes are shown in Figure 11.6.

The three alleles can combine into six different genotypes, but since $I^A I^A$ and $I^A I^O$ are both type A, and $I^B I^B$ and $I^B I^O$ are both type B, there are only four ABO blood types or phenotypes: A, B, AB, and O. Tests of blood types are carried

out with drops of blood and antibodies to the A and B antigens, on glass microscope slides (see Figure 11.6).

The blood types of donors and recipients have to be matched very carefully. For example, type A blood transfused into a type B person will be agglutinated by the antibodies of the recipient, forming possibly fatal clumps. In emergency cases, type O blood can be transfused into persons of other blood types, since the introduced blood cells have neither A nor B antigens.

Lethals

There are alleles that when homozygous cause the death of the individual, yet have no effect when carried in a heterozygote. These are called **recessive lethal** alleles. A good example is the white seedling allele of maize. Homozygous **ww** individuals fail to produce green chloroplasts; hence the seedlings don't survive. They grow just fine as long as they are small and can live off the stored reserves in the seed (Figure 11.7), but they die when those reserves are spent. Het-

FIGURE 11.7
White Seedlings, a Recessive Lethal Trait

This white seedling trait in corn is recessive, here segregating in a 3:1 ratio. White seedling is also lethal, since the seedlings fail to produce chloroplasts and die when the energy reserves in the seeds are exhausted.

erozygous **Ww** plants produce normal chloroplasts and are indistinguishable from the homozygous **WW** plants.

If an allele is both lethal and *dominant,* all individuals carrying the allele (heterozygous as well as homozygous) would die . . . so how would the allele ever be transmitted? The answer is that the individuals carrying the allele must survive at least long enough to reproduce. A famous example is Huntington's disease, a debilitating and lethal genetic disease in humans. Death is the ultimate result, but the disease does not begin to have its effects until age 40 or so, leaving plenty of time for the allele to be passed to offspring.

Gene Interactions

You will recall that Mendel found that the F_2 of the round-yellow and wrinkled-green dihybrid cross yielded a 9:3:3:1 phenotypic ratio (see Table 10.2). This worked out very nicely for Mendel because the round, wrinkled characteristics do not influence the inheritance of the yellow, green characteristics. But this is not true of all pairs of genes. Consider the coat color in mice. In this case, we find that, at one gene, **B** is dominant to **b**, such that **BB** and **Bb** mice are black and **bb** mice are brown. A cross between a homozygous black (**BB**) mouse and a homozygous brown (**bb**) mouse will produce nothing but black heterozygotes (**Bb**). A cross between two heterozygous black mice produces an F_2 generation with three-fourths black mice and one-fourth brown mice. So far so good, since this is a simple Mendelian pattern.

But at a separate gene located on a different chromosome, **C** is dominant to **c**, such that **CC** and **Cc** mice can make pigment (either black or brown) but the **cc** mice cannot

and are thus albinos. The **C** allele, in effect, allows for coat color. Hence the two separate genes at **B** and **C** control two different steps in the biochemical pathway that produces the pigment normally present in mouse fur. Thus the **C** locus is said to be epistatic to the **B** locus. **Epistasis** is the interference, by one gene, of the expression of another gene.

Now consider a mating between a true-breeding white mouse and a true-breeding brown mouse. What would you expect? You might not expect the entire litter to be black. But you shouldn't be too surprised either, because of the possibility of this cross:

P_1:   **CCbb** (brown) × **ccBB** (white)
F_1:   **CcBb** (black)

The black F_1 are heterozygous at two different genes. The real surprise comes at the next cross. If you mate two double heterozygotes, such as those shown in Figure 11.8, you will find that in the F_2 generation one-fourth are **cc** (white), regardless of what's happening at the **B** locus. Of the remaining colored mice, three-fourths are black and one-fourth are brown. The phenotypic classes of the F_2 are 9/16 black, 3/16 brown, and 4/16 white (see Figure 11.8). This is just the old 9:3:3:1 ratio with the last two terms combined (9:3:4), because in mice that can't make pigment, you can't tell whether it might have been brown or black. Perhaps we should put this another way: you cannot distinguish between **BBcc**, **Bbcc**, and **bbcc** without doing a progeny test.

Conditional Phenotypes

For many traits, a single genotype hardly ever results in exactly the same phenotype in any two individuals, or for that matter the same way in different tissues of a single individual. We will see now that just how a gene is expressed is conditional—it depends on the conditions under which it exists.

Influences of Environment

Environmental influences on gene expression may be very obvious or very subtle. Let's consider a straightforward example: the Siamese cat. One of the enzymes in its pigmentation pathway is temperature-sensitive; it won't function when it is warm. As a result, pigmentation of the fur occurs primarily in the colder extremities of the cat—the ears, the tail, the feet, and the nose. While these parts are black or dark brown, the rest of the cat is tan or almost white, which is why Siamese cats look like Siamese cats (Figure 11.9). If you keep your Siamese cat in the warm house, it may grow to be almost white, but if you put it out at night, it will get to be quite dark.

Influences of Development

In some cases an individual may have a dominant genotype without showing it, a condition known as **incomplete pene-**

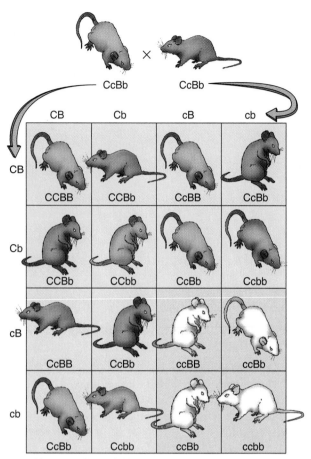

cc = white, regardless of other gene
CC or Cc = black or brown, depending on other gene as follows:
BB or Bb = black
bb = brown
Resulting ratio = 9 : 3 : 4

FIGURE 11.8
Epistasis
In this scheme, we follow the path of the two pairs of genes that influence coat color in mice. The **B** (black) and **b** (brown) alleles occur at one locus, and the **C** (color) and **c** (albino) alleles occur at another locus. The alleles of the two genes assort independently but produce what seem to be strange ratios when two heterozygous mice are inbred. The result is a 9:3:4 ratio of black/brown/white because of the epistatic interaction between the two loci.

trance. For instance, the gene that gives one the ability to roll his or her tongue is dominant, yet not all persons with the dominant allele will be able to roll their tongues (Figure 11.10). But for any one person it is all or nothing: there are no people who can roll their tongue only part way. In fact, in a room full of people with the dominant allele, only about 8/10 will actually be able to roll their tongue. Therefore we say that the tongue-rolling allele has 80% penetrance.

There is a related situation called **variable expressivity**, where an allele will be expressed to varying degrees. For instance, a rare dominant trait in human genetics is polydactyly, the tendency to have extra fingers or toes (Figure 11.11).

FIGURE 11.9
Conditional Phenotypes
Environmental temperature can have a drastic effect on the expression of coat color genes in the Siamese cat. Siamese cats have their distinctive color pattern because the enzyme that produces color functions well only in the cooler extremities of the cat.

Persons carrying this dominant allele show variable expressivity in that all four extremities may be affected, or only one hand or one foot. Both hands and both feet of a given carrier have the same genes and the same general environment but may have either normal or abnormal numbers of digits.

The basis of both incomplete penetrance and variable expressivity lies in subtle variation in the complex process of producing a mature body from a single fertilized egg. There are many, many situations where certain tissues must grow to certain sizes and where certain nerves must reach certain muscles. If, by chance, a certain nerve did not grow all the way to your tongue, you would not be able to roll your tongue even if you had the right genotype.

Influences of Sex
A dominant gene is known to be responsible for a rare type of cancer of the uterus. Since, needless to say, the gene affects women only, it controls a **sex-limited trait**. A sex-limited trait then is a trait that shows up in *only* one sex or the other. **Sex-influenced** traits are another matter. Sex-influenced traits can affect both sexes, but the effect is different. The most common kind of middle-aged male baldness, for example, is caused by a dominant allele that produces only thinning of the hair in women.

Influences of Age
Baldness, muscular dystrophy, and Huntington's disease are examples of genetic traits that can have variable ages of onset. Muscular dystrophy is a degenerative condition affecting

muscles and can begin at very different ages even in affected brothers, who would have received exactly the same abnormal allele. We have already seen that Huntington's disease is lethal after a certain age. However, the exact age of onset is quite variable.

Pleiotropy

Of the thousands of children who show up in hospital emergency rooms with fractured bones, a few become frequent visitors. Some are accident prone, and others may be the victims of physical abuse, but occasionally a sharp-eyed physician, one who paid close attention in genetics courses, will spot a very rare genetic abnormality called blue sclera–brittle bone disease. Blue sclera refers to a bluish tint in "the white of the eyes"; the brittle bones that accompany the blue sclera are the product of defective calcium metabolism. Both are examples of **pleiotropy** ("many turnings"), the tendency for an allele to be expressed in different ways in different tissues. While this example is rare, pleiotropy is itself commonplace.

Perhaps the best example of pleiotropy in humans is a disease of newborns called phenylketonuria, or PKU (see Chapter 12). Because of an enzyme deficiency, PKU victims cannot metabolize the amino acid phenylalanine, so it accumulates in the blood. Its effects are widespread, depending on the tissues. For example, the accumulation causes brain damage leading to mental retardation, and the head fails to grow to normal size. Furthermore, a shortage of pigments normally produced at the end of the phenylalanine pathway can result in light hair and skin color.

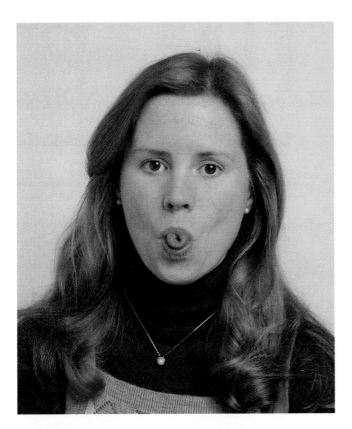

FIGURE 11.10
Incomplete Penetrance: Tongue Rolling
The ability to roll your tongue, while genetically controlled, is not always expressed even in persons with the proper genotype.

FIGURE 11.11
Polydactyly and Variable Expressivity
Polydactyly, the inheritance of extra fingers or toes, is a dominant trait. However, it shows variable expressivity in that a person with the allele may show polydactyly on all four extremities, on hands only, on feet only, or in any combination, or they may not show the trait at all. The numbers below the individuals of this pedigree indicate the number of digits on the hands and feet.

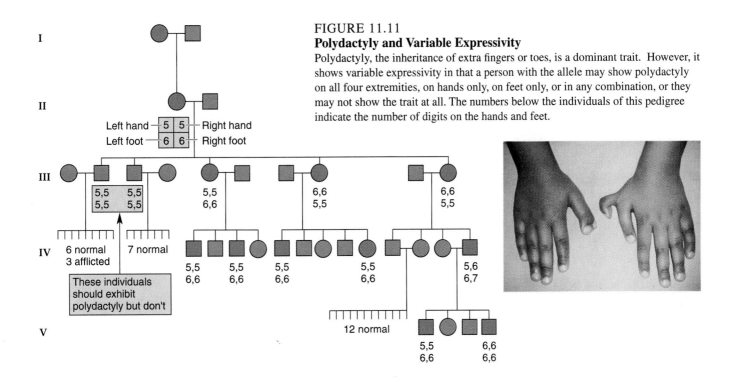

Continuous Variation and Polygenic Inheritance

Many of the phenotypic traits that are most important to biologists, and especially to plant and animal breeders, do not fit into "either–or" categories such as round or wrinkled peas. Instead, these traits occur in a gradient, a situation called **continuous variation**. Of course, all alleles for any condition still occur in pairs, and they still segregate and assort according to Mendelian law. The continuous variation often exists because alleles at more than one gene are involved. When alleles at more than one gene contribute to the same trait, this is called **polygenic inheritance** and the trait is called a **polygenic trait**. (Remember, polygenic means more than one gene controlling a single trait, while multiple alleles means that any one gene may have more than two alleles.) Examples in humans include skin color, foot size, nose length, birth weight, height, and intelligence. Let's look more closely at one example.

A great many genes determine human height, but to simplify things we'll assume that height is determined by only three different genes (three genes on three separate chromosomes). Also, in our example, we'll assume only two alternatives are available: "short" alleles and "tall" alleles. Furthermore, we will assume that the presence of a "tall" allele rather than a "short" allele increases adult height by 5 centimeters (about 2 inches). People with only the "short" alleles (six in all) grow to be about 160 cm (5′3″), while those with only "tall" alleles (again six) grow to 190 cm (6′3″). In the middle, with three "short" alleles and three "tall" alleles, are the average individuals about 175 cm tall (5′9″).

Figure 11.12 summarizes the seven height categories possible with this model. We have added the relative frequencies of the seven height categories that would be predicted in the offspring from a large number of heterozygous couples (**AABbCc × AaBbCc**). Note that the distribution approximates a "bell-shaped curve" or normal distribution.

Genes on Chromosomes

In science, it is often said, the answer to one question gives rise inevitably to new questions. And this was certainly the case with early 20 century genetics. For example, not long after Sutton and others concluded that chromosomes bore the genes and many genes were contained in each chromosome, they began to notice that Mendel's principle of independent assortment didn't work out with all genes.

Linkage

The question was: Why? They concluded that if traits didn't randomly assort, they must be kept together, somehow *linked.* They further concluded that such traits were inherited together because the genes are part of the same chromosome, and so

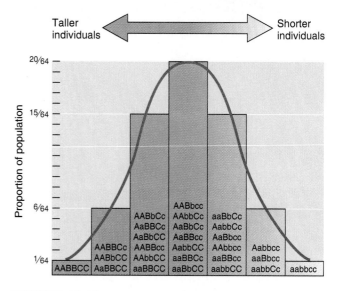

FIGURE 11.12
Continuous Variation and Polygenic Inheritance
Traits that can be measured quantitatively, such as height in humans, can vary over a considerable and continuous range, as demonstrated by a bell-shaped curve of human height variation. The chart at the bottom presents a model for the inheritance of height in humans, with three loci that each contribute equal increments to the final height.

these genes moved together as part of a **linkage group**. A linkage group came to be defined as any group of genes that tended to be inherited together because they were on the same chromosome (Figure 11.13).

William Bateson and Reginald Punnett (of Punnett square fame), while trying to confirm Mendel's findings, got some puzzling results that turned out to be due to linkage. They started with two true-breeding strains of sweet peas: one with blue flowers (**BB**) and long pollen grains (**LL**), the other with red flowers (**bb**) and round pollen grains (**ll**). The F_1 offspring of this cross had blue flowers and long pollen grains, hence "blue" and "long" were known to be dominants. Their P_1 cross was:

$$P_1: \quad \textbf{BBLL} \times \textbf{bbll}$$
$$F_1: \quad \textbf{BbLl}$$

So far, there were no surprises. Then, as shown in Figure 11.13, they sought to reconfirm the law of independent assort-

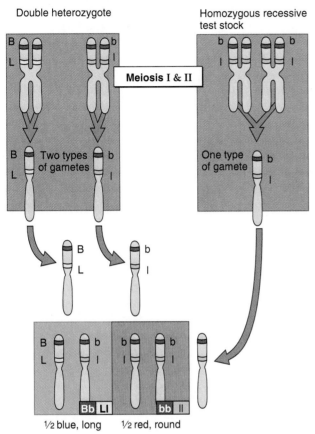

Assumption:
Two pairs of alleles linked on the same chromosome pair will not segregate in a Mendelian fashion

Double heterozygote

Homozygous recessive test stock

Meiosis I & II

Two types of gametes

One type of gamete

½ blue, long

½ red, round

Results: 1:1 ratio with no new combinations appearing

FIGURE 11.13
Linked Genes
If the allele pairs for flower color and pollen shape are on the same chromosome, they cannot assort independently. With this being the case, the phenotypic ratio of the offspring of this testcross should be 1:1, blue-long/red-round. In reality, few genes are linked so very tightly.

ment by crossing the doubly heterozygous F_1 back to the doubly recessive parental stock in a standard testcross:

BbLl × bbll

Mendel's second law predicted that they should get equal numbers of all four possible phenotypes—a 1:1:1:1 ratio of blue-long, blue-round, red-long, and red-round (you may want to confirm this for yourself), but this is not what Bateson and Punnett observed. Instead they found that the alleles from the two genes tended to be inherited together because the genes for flower color and pollen grain length are on the same

chromosome. If the genes were indeed linked, with the alleles for blue and long on one chromosome, and the alleles for red and round on its homologue, the results should have been 50% blue-long, 50% red-round (see Figure 11.13). Blue flowers would always appear with long pollen grains and red flowers with round pollen grains.

Crossing Over and Genetic Recombination

Recall that the process of crossing over brings about the exchange of parts of chromosomes. This requires the breaking of chemical bonds in the DNA and actual physical exchange of chromatid regions (see Chapter 9). Now considering the fact that genes occupy certain locations on those chromatid arms, the physical exchange of chromatid regions can have a profound effect on the arrangement of alleles present on a chromosome. As shown in Figure 11.14, crossing over between the genes can result in a change in which alleles are linked together. This change in linkage relationship between alleles is called **genetic recombination**. Recombination is the genetic result of crossing over.

In the Bateson–Punnett example, 12% of the testcross progeny were **recombinants**; that is, they showed a different linkage of alleles than did the parents. The percent recombination is determined by dividing the number of recombinant offspring by the total number of offspring. In Figure 11.15, the linkages shown by the parents were **B** with **L** and **b** with **l**. The recombinant offspring are those that have **B** with **l** or **b** with **L**.

Thus crossing over allows the alleles of linked gene loci to occasionally change their linkage relationship with each other. **Crossing over** is the physical, cellular phenomenon that is responsible for the genetic phenomenon of recombination.

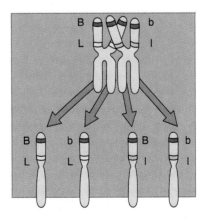

FIGURE 11.14
Genetic Recombination
Continuing from Figure 11.13, we look at the same testcross, but with crossing over occurring in the double heterozygote. Note that, with crossing over, the types of gametes double. The result is two new classes of offspring, but not necessarily or even generally in a frequency equal to the expected phenotypes. (See Figure 11.15.)

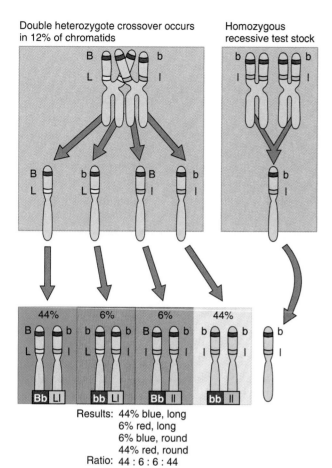

FIGURE 11.15
Twelve Percent Crossing Over
In this example of crossing over, the frequency of occurrence is
12%, which means the non-crossing over frequency is 88%. We still
end up with the four types of gametes that we saw in Figure 11.14,
but now we must consider the percentage at which each gamete is
produced. The end result is near a 7:1:1:7 ratio, which is a far cry
from the 1:1:1:1 that would be expected of independent assortment,
or the 1:1 ratio of very tight linkage.

FIGURE 11.16
T. H. Morgan and the Fruit Fly
Morgan's discoveries brought the lowly fruit fly to the forefront of
genetic studies.

By the way, crossing over can occur without recombination (if
the genes of interest on the homologous chromosomes are
homozygous, genetically you cannot tell if there is recombination), but recombination cannot occur without crossing over.

Genes located far apart on the same long chromosome
may seem to obey Mendel's second law. Although they are
physically part of the same molecule at the beginning of
meiosis, they are so far apart that the probability of the genes
ending up in the same gamete is just about the same as if they
had been on separate chromosomes to begin with, 50%.
Genes that lie closer together on a chromosome have a corre-
spondingly smaller probability of a crossover event happening between them. They tend to be shunted around as a unit
during meiosis and thus tend to be inherited as a linkage

group in testcrosses. For example, the alleles of two **B** and **L**
are so close together on a chromosome that there is only a
12% chance that they will be separated by crossing over and a
88% chance that they won't. That is, there is a 12% chance of
recombination in the region between the two genes. The total
recombination between two genes is expressed as the percent
recombination, which is often referred to as the **recombination frequency**, or **R**. Gene pairs that had very low percentages of recombination have come to be known as "tightly
linked genes"; those with higher percentages are termed
"loosely linked."

R is related to the probability that crossing over will
occur, but it is not exactly equal to this probability. Also, the
farther apart two genes are, the more **R** becomes an underesti-

mate of the distance. This is because of double crossing over. If two crossing over events occur between **B** and **L**, the first would separate **B** and **L**, and the second would put them back together again. The result would be no genetic recombination. Thus **R** is equal to the probability of an *odd number* of crossing over events.

Mapping Genes

The laboratory of Thomas Morgan was responsible for bringing the fruit fly, *Drosophila melanogaster,* to its pedestal as the premier eukaryotic organism for genetic study (Figure 11.16). Morgan and his associates were quite successful in their ongoing search for new *Drosophila* traits. Fruit flies were easily reared in the laboratory, in great numbers and with little care, and thus served the geneticists well as an experimental organism. As these new traits were uncovered, crosses were carried out in a variety of ways, often involving pairs of genes. In some, the results conformed nicely to Mendel's law of independent assortment, while in others, gene linkage was clearly established; for some gene pairs the recombination frequency was low (they were tightly linked), while in other cases the frequency was high.

Morgan's group found that each pair of linked genes had its own characteristic recombination frequency; no two linked gene pairs gave the same results. Morgan himself was the first to propose an explanation. Intuitively, he reasoned that the rate of crossing over, the recombination frequency, should be proportional to the distance separating two genes on a chromosome. Then Alfred Sturtevant, a gifted young undergraduate at the time, made a highly logical extension to Morgan's assertion. In Sturtevant's own words: "It would seem . . . that the proportion of crossovers could be used as an index of the distance between any two factors. Then, by determining the distances between A and B, and B and C, one should be able to predict AC. For if proportion of crossovers really represents distance, AC must be approximately either AB plus BC, or AB minus BC."

With this idea in mind, we can follow the lead in Figure 11.17 and construct a map of several genes in *Drosophila.* To begin, we see that the mutant alleles for yellow body (**y**) and white eyes (**w**) have a very low recombination (1%), and so they must be very close together. On the other hand, the recombination between the yellow body and vermillion eye (**v**) alleles is 32.2%, and the recombination between white eye and vermillion eye is 30.0%. When mapping, recombination percentages are called **map units**. The three recombinations fit together reasonably well if we begin mapping with vermillion, placing white 30 units away, and yellow 1 unit further still.

y—w————————————————————v
1   30

You might ask yourself why yellow is placed 1 map unit to the left of white, instead of one unit to the right. A moment's

Genes		Recombination frequency (% crossover)
Yellow body	white eyes	1.0%
Yellow body	vermillion eyes	32.2%
Yellow body	miniature wings	35.5%
Vermillion eyes	miniature wings	3.0%
White eyes	vermillion eyes	30.0%
White eyes	miniature wings	32.7%

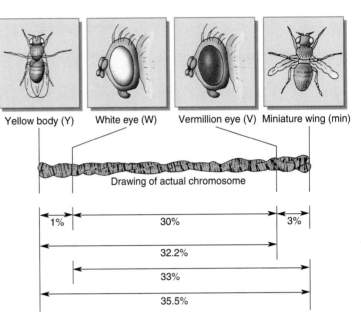

Yellow body (Y) White eye (W) Vermillion eye (V) Miniature wing (min)

Drawing of actual chromosome

1%   30%   3%

32.2%

33%

35.5%

FIGURE 11.17
Distance Between Genes and Gene Mapping
By studying the recombination frequency or percent crossing over between genes, their relative locations on the chromosome can be determined. Here we list the percent crossing over observed for crosses involving yellow body color, white eyes, vermillion eyes, and miniature wings in *Drosophila.*

reflection would reveal that the distance between **y** and **v** has to be greater than that between **v** and **w**.

It follows that miniature wing must be three units to the other side of vermillion: its position is apparently farther from white and yellow than is vermillion's (by 3 map units). So with the addition of the fourth gene, the map becomes:

y—w————————————————————v——min
1   30   3

You'll notice that the individual recombination fractions are nearly additive, as they should be if crossing over and genetic recombination are random events that depend only on the distance between genes. Geneticists since Morgan and

Sturtevant have used this basic method to determine the relative positions of thousands of genes on the chromosomes of *Drosophila.*

As we see then, a genetic map yields information regarding not only the order in which genes occur on the chromosome, but the distances between the genes as well (Figure 11.18). Mapping enables the discoverer of a new mutant allele to determine whether the discovery is a variant of a known gene. Finally, mapping gives an identifiable location to a gene. In genetic terms, this location is called the gene's **locus** (plural, loci). In fact, once a gene has been assigned a position on a chromosome, the terms gene and locus become synonymous, and one could refer to the vermillion *locus* just as easily and correctly as referring to the vermillion *gene.*

Chromosomes and Sex

When Morgan first began his search for variation in the traits of *Drosophila,* all the flies looked alike, except that males were visibly different from females. However, as Morgan carefully scrutinized each new generation, he eventually turned up one variant. Among a group of flies with normal brick-red eyes, Morgan found a single male with white eyes. He carefully nurtured his little white-eyed specimen and crossed it with several of its red-eyed sisters (see Figure 11.19, *left.*) From these matings, all the F_1 were red-eyed, to the surprise of none of the "new Mendelians." Obviously, white eyes were the expression of a recessive trait.

P_1: white male ($\male$) $\times$ red female ($\female$)
F_1: all red, $\male$ and $\female$ alike

The experiments continued, and when the F_1 flies were mated with one another to produce an F_2 generation, sure enough, about one-fourth of the F_2 flies were white-eyed and about three-fourths were red-eyed (see Figure 11.19, *left side*). The actual numbers were not as close to this expected ratio as Morgan had hoped because, as it turned out, the white-eyed flies have a somewhat lower rate of survival than the normal flies. But there was something more peculiar about the F_2 flies. Every single white-eyed fly was a male! In fact, the F_2 ratio approximated the following:

$\frac{1}{4}$ red-eyed males : $\frac{1}{2}$ red-eyed females : $\frac{1}{4}$ white-eyed males

At this point, you may have decided that only males can be white-eyed. You would be wrong. Morgan discovered this when he first did a testcross, mating his original, now-geriatric, white-eyed male to its own red-eyed F_1 daughters (Figure 11.19, *right side*). A simple testcross should have provided a 1:1 ratio of dominant to recessive and this one did. In fact, the testcross offspring consisted of approximately equal numbers of red-eyed males, red-eyed females, white-eyed males, and white-eyed females. So females could have white eyes. But even more surprises were in store. When Morgan mated white-eyed females to red-eyed F_1 males, in what is called a reciprocal testcross, again, half of the offspring were red-eyed and

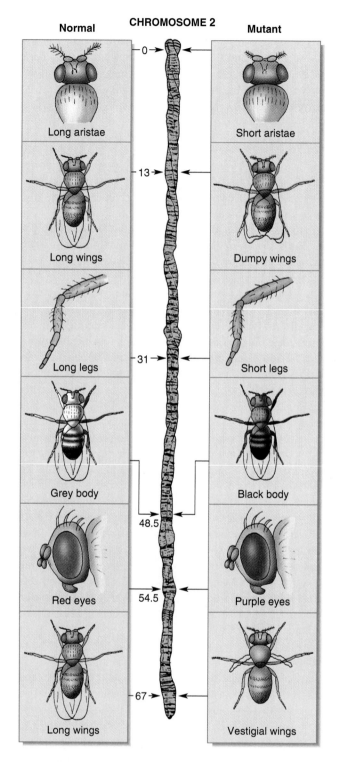

FIGURE 11.18
Chromosome Maps

Maps of a *Drosophila* chromosome can take several forms, two of which are shown here. First, the positions of the loci can be arranged as a genetic map, where the positions of the loci are listed and positioned according to their recombination relative to each other, and numbered from one end of the chromosome. Second, the positions of the genes can be described on the cytological map, a map of the visible bands on the polytene chromosome.

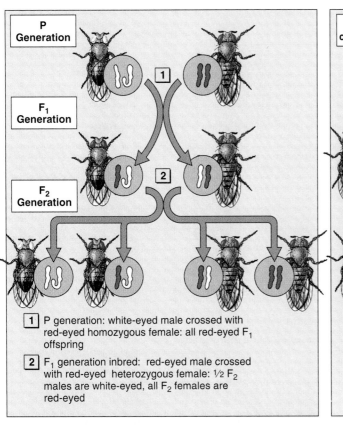

P Generation

F₁ Generation

F₂ Generation

1 P generation: white-eyed male crossed with red-eyed homozygous female: all red-eyed F₁ offspring

2 F₁ generation inbred: red-eyed male crossed with red-eyed heterozygous female: ½ F₂ males are white-eyed, all F₂ females are red-eyed

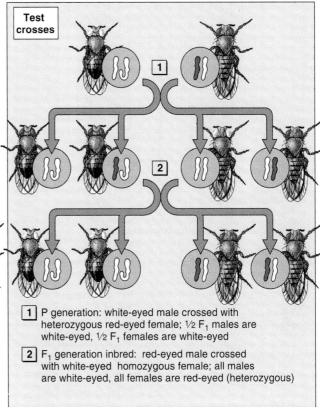

Test crosses

1 P generation: white-eyed male crossed with heterozygous red-eyed female; ½ F₁ males are white-eyed, ½ F₁ females are white-eyed

2 F₁ generation inbred: red-eyed male crossed with white-eyed homozygous female; all males are white-eyed, all females are red-eyed (heterozygous)

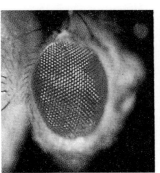

Wild eye (normal allele)

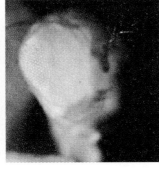

White eye

FIGURE 11.19
White Eyes and Sex Linkage

In his first cross, Morgan mated his newly discovered white-eyed male with a normal, red-eyed female. All the F_1 offspring were red-eyed. Inbreeding the F_1 produced an F_2 that was three-fourths red-eyed and one-fourth white-eyed. But all the white-eyed flies were males. A testcross between the original white-eyed male and a red-eyed female from the F_1 showed that the female was heterozygous. Another testcross was done, with a red-eyed male and a white-eyed female. All the male offspring were white-eyed.

half were white-eyed. But now every single one of the males was white-eyed, and every female had red eyes.

Morgan was aware of Sutton's suggestion that a single chromosome may carry a number of hereditary factors. Morgan surmised that the sex-determining factor and the eye-color factor are linked together, since these traits did not follow the law of independent assortment in the F_2 or in the testcrosses.

Sex Chromosomes in *Drosophila*. By this time, Morgan knew something about the chromosomes that determine gender. He knew that male and female *Drosophila* were different with regard to one of their four chromosome pairs, and he guessed that this chromosome difference was the cause and

not the result of sex differences. Female flies have two X chromosomes, while the males have one X and one Y (similar to humans). At metaphase in *Drosophila*, the X appears as a long, rod-shaped chromosome and the Y appears as a shorter, J-shaped chromosome (Figure 11.20).

We have been saying that you have two of every gene—one from your father and one from your mother. But now we see that this is only partly true. While females are truly diploid, males are only partly diploid. Actually, at about 10% of their gene loci, men have one gene from their mother—period. Although the X and Y chromosomes behave like homologues in meiosis, lining up together, they do not carry the same genes. While the X chromosome has many genes—

those determining growth patterns, enzymes, and so on—the little Y chromosome bears little more than a few genes relating to male sexual development.

A gene on an X chromosome has no homologous gene on the Y chromosome to interact with or yield to. Thus the Y chromosome behaves as if it has a recessive allele for virtually all the X chromosome loci. For a recessive allele to be expressed in a female, it must be present on both of her two X chromosomes. It follows that since a male has only one copy of any X-linked gene, it will express itself, whether it is recessive or dominant in females. For X-linked genes in males, the term **hemizygous** (*hemi-,* "half") is used instead of homozygous or heterozygous.

If you now consult Figure 11.19 for a fresh look at Morgan's crosses, you'll probably have an easier time following the hereditary patterns produced by sex-linked genes. First, note that the F_1 offspring clearly established that red eye color is dominant over white. Next, if the eye color gene is carried on the X chromosome, Morgan's original female must have been $X^R X^R$, and his white-eyed mutant male would have been $X^r Y$. (Note the use of the superscript on the **X** to denote alleles located on the X chromosome, and remember that females must be XX, males XY). It follows that each F_1 female would have received an X^R from her mother and an X^r from her father, so the F_1 females were all red-eyed heterozygotes. Each F_1 male would have received a Y chromosome from his father and X^R from his mother, so all F_1 males were red-eyed ($X^R Y$). In the F_2, however, each male would have received a Y from his father and *either* an X^R or X^r from his mother, the chances of the latter being 50:50. Thus half the F_2 males were red-eyed, and half were white-eyed. Finally, the two test crosses show us that females do express sex-linked traits, but *only if each parent carries the allele in question.* Interestingly, as sex-linked traits pass down through generations, they seem to show up in every other one, following a "crisscross" pattern. That is, the traits are most commonly hidden in females, expressed in half their sons, hidden again in the granddaughters, only to be expressed once more in half the great-grandsons. The primary rules regarding these X-linked traits are: (1) *males expressing an X-linked recessive trait received the allele from their mother,* and (2) *daughters expressing an X-linked recessive trait have a father with that trait.*

Sex Determination. Although normal females are XX and normal males are XY in both *Drosophila* and humans, the physiological mechanisms determining sex are somewhat different in the two species. In *Drosophila,* sex is determined by the number of Xs: two Xs, female; one X, male. Abnormal XXY flies are fully functional females, and XO flies (one X, no Y) are sterile males that look normal. In humans and other mammals, the presence of a Y determines the development of testes, which, in turn, determines male development.

At about the sixth week of development, a gene on the human Y chromosome becomes active. It is the testis determination factor (TDF gene), and its product stimulates testis formation. The testes subsequently secrete male sex hormone,

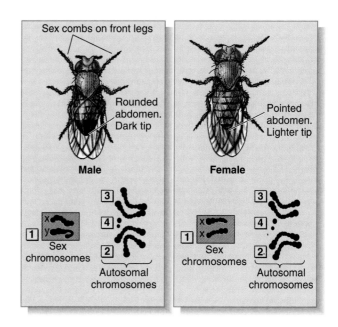

FIGURE 11.20
Drosophila **Chromosomes**
Drosophila has four pairs of chromosomes. In males only three of the four pairs are homologous. The fourth pair consists of an X chromosome (which is identical to that of the female) and a J-shaped Y chromosome (which is not). Other sexual differences are shown.

which prompts further male development. Humans lacking a Y chromosome and having only an X chromosome are designated XO and are sterile, abnormal females, whereas XXY humans are sterile, abnormal males. You will recall from Essay 9.3 that such abnormalities are the result of nondisjunction during meiosis. For more on X and Y chromosome abnormalities, see Essay 11.1.

X Chromosomes and the Lyon Hypothesis. The "saliva test" is sometimes used in women's athletic competitions as a test of whether a competitor is, in fact, a woman. Actually, the test has nothing to do with saliva and a great deal to do with X chromosomes. A swab of the inner surface of a person's cheek will pick up a few cells from the mucous membrane lining. When stained, the nuclei in cells from females have a dark-staining **Barr body,** which is a condensed X chromosome. The condensed X chromosome can also be viewed microscopically in certain white blood cells, where it forms a characteristic projection from the cell nucleus called a drumstick (Figure 11.21). But what are condensed X chromosomes and why do they occur only in females?

Since every female has two X chromosomes and males get along fine with just one, we can presume that there would be a physiological imbalance if both X chromosomes were functional in females. Females inactivate one of the two X chromosomes in each cell during embryonic development. Which of the two X chromosomes is inactivated—the one from the father or the one from the mother—seems to be largely a mat-

Essay 11.1 ● SEX CHROMOSOME ABNORMALITIES

As with trisomy 21, or Down syndrome, abnormal numbers of sex chromosomes are brought about by nondisjunction—the failure of chromosomes to assort properly during meiosis (Chapter 9). A leading factor causing this problem is believed to be aging in the oocytes. While most nondisjunctions in autosomal chromosomes are fatal to the embryo, wrong numbers of sex chromosomes in humans result in live babies and sterile adults. There are many varieties of sex chromosome conditions. The designations of normal individuals and the most common abnormalities are as follows:

XX Normal female (two X chromosomes)
XY Normal male (one X, one Y)
XO Turner's syndrome female (one X, no homologue)
XXY Klinefelter's syndrome male (two Xs, one Y)
XYY Extra Y, or XYY syndrome male
XXX Trisomy X, or XXX female

In addition to the above, there are many more extreme situations, such as XXYY, XXXY, XXXYY, XXXX, XXXXX, XXXXYY, and so on, each syndrome having its own distinguishing phenotypic characteristics. However, we can make four generalizations. First, one must have at least one X chromosome to live. Second, the presence of a Y chromosome causes the individual to develop male sexual characteristics, and the absence of a Y causes the individual to develop female sexual characteristics. Third (and this is probably why these syndromes are not fatal), all but one of the X chromosomes will condense into a visible Barr body, so the XO females lack a Barr body, XXY males have one, XXX females have two, XXXXYY males have four, and so on. Fourth, the more sex chromosomes a person has, the taller he or she will be, so that XO females are shorter than average, while XXX, XXY, and XYY individuals are usually much taller than chromosomally

normal men and women, and XXYY men are huge. XO (Turner's) individuals are phenotypically female but do not develop ovaries. They remain sexually immature as adults unless given hormones. XXY (Klinefelter's) males are tall and have small, imperfect testes and low levels of male hormones. They may have female-like breast development and somewhat feminine body contours. XXX females are tall and frequently sterile but otherwise appear normal. XYY males appear normal except for their extreme height and for a tendency toward severe acne. They are also generally sterile. On the average, they have somewhat reduced IQ levels and, in common with other low-IQ groups, they average significantly increased criminal arrest records. At one time there was speculation the XYY males had "genetic criminal tendencies," but other analysis suggested that an XYY male is no more likely to be arrested than an XY or XXY male of the same IQ.

ter of chance. Every tissue in the adult female is a mosaic of cell lines in which one or the other X has been inactivated. The highly condensed X is replicated normally and is passed from cell to daughter cell, but it is inactive with regard to genetic function. The peculiar behavior of the X chromosome was first established by the English geneticist, Mary Lyon.

Incidentally, inactivation of extra X chromosomes probably accounts for the fact that extra numbers of sex chromosomes do not have the disastrous effects that autosome imbalances have.

XXX females, for instance, have two Barr bodies but are otherwise normal. Humans with sex-chromosome number abnormalities survive rather well, although they are usually sterile (see Essay 11.1).

Colorblindness. Colorblindness in humans is usually caused by a recessive allele at either of two closely linked gene loci on the X chromosome. Human color vision depends on the differential sensitivity of three groups of light receptors in the retina,

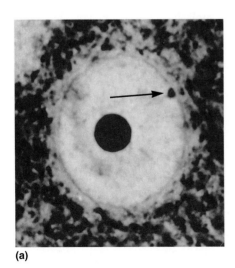

(a)

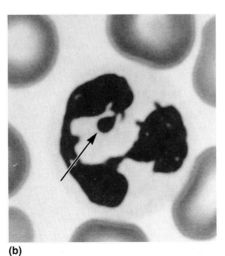

(b)

FIGURE 11.21
The Condensed X
One of the X chromosomes in each human female somatic cell is represented by permanently condensed heterochromatin. The condensed X chromosome can appear as a simple blob at the side of the nucleus (Barr body) (**a**) or in the form of a drumstick (**b**).

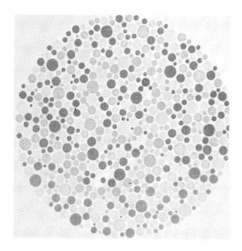

FIGURE 11.22
Colorblindness Test
Color vision is tested using colored plates such as the one shown here. Actually, several plates are required for the complete test. If you are having trouble seeing the number 9 (in red and orange) you may want to take the complete test.

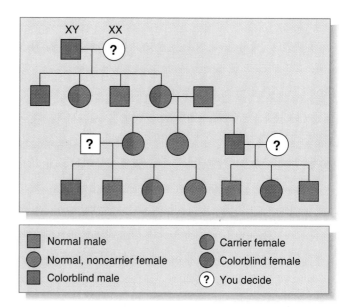

FIGURE 11.23
Colorblindness in a Family
In this hypothetical family tree, colorblindness can be traced back to the great-grandfather and great-grandmother, although the problem was intensified by the marriage of a colorblind man and a carrier woman, as shown at the right. We did not complete the great-grandmother's circle and two of the parents' spaces because we want you to decide.

called cones. One group of cones is maximally sensitive to blue light, one to red light, and one to green light. Perception of other colors and of subtle hues depends on the relative stimulation of these three types of cones. Persons homozygous (or hemizygous) for a recessive allele at one of the two X-linked loci lack the cones that are more sensitive to green, while homozygotes for recessive alleles at the other X-linked locus lack the cones that are maximally sensitive to red. By the way, the locus controlling the blue-sensitive group is autosomal, and blue-insensitive colorblindness is very rare. Both X-linked defects are called red–green colorblindness because both types of colorblind people have trouble distinguishing many shades of red and green (Figure 11.22). The defect was first described in a little boy who couldn't learn how to pick ripe cherries. He always brought home a mix of red and green fruit.

Somebody in the British Navy was aware of this situation well over a century ago and set about developing "running lights" on the boats that even colorblind men could tell apart. The green (starboard) side had a touch of blue, and the red (port) side had a touch of orange. When traffic lights were introduced on railroads and city streets, these readily recognizable hues were the logical choice, and we now see them daily.

About 8% of American men have one form or another of X-linked colorblindness: about 6% are hemizygous for a recessive allele at the locus coding for green-sensitive cones, and about 2% are hemizygous for a recessive allele at the locus coding for red-sensitive cones. Women are affected far less often: only about 0.4% of American women are red–green colorblind. The reason for the sex difference is that, to be affected, a man need only receive one recessive allele from his mother. But an affected woman must receive recessive alleles

from both her mother and her father. (Recall that every daughter expressing a sex-linked recessive trait has a father with that trait.) The chance that both parents will carry the rare allele is much smaller than the chance that just one of them will.

A woman who is heterozygous for colorblindness shows no symptoms of the condition. However, she can expect half her sons to be colorblind and half her daughters to be carriers like herself—assuming that she marries a man with normal vision. Figure 11.23 shows the appearance of colorblindness in one family.

Other Sex-Linked Conditions. Many sex-linked genetic conditions have had great impact on people's lives. Two of the most common, but also dramatic, of all genetic disorders are hemophilia—the bleeder's disease—in which the blood doesn't clot normally, and muscular dystrophy, in which muscle tissue breaks down in late childhood. The usual forms of both of these are sex-linked. There are actually two common forms of X-linked hemophilia, governed by different X-linked loci. Most hemophiliac males formerly bled to death in their youth. But in recent years, modern medicine has allowed affected hemophiliacs to survive and reproduce, thanks to blood transfusions and to infusions of a blood-derived substance known as antihemophilic factor, which supplies the critical substance missing in hemophiliacs. Hemophilia has had interesting implications in European history, as we find in Essay 11.2.

Essay 11.2 ● THE DISEASE OF ROYALTY

It was the practice of ruling monarchs to consolidate their empires through marriage alliances. As a result, a highly restricted "royal mating population" was created, and hemophilia was common throughout the royal families of Europe. Hemophilia is a sex-linked recessive condition in which the blood does not clot properly, so that any small injury can result in severe bleeding and, if the bleeding cannot be stopped, in death. Hence it has sometimes been called the *bleeder's disease.*

The hemophilia of European royalty has been traced back as far as Queen Victoria, who was born in 1819. One of her sons, Leopold, Duke of Albany, died of the disease at the age of 31. Apparently, at least two of Victoria's daughters were carriers, since several of their descendants were hemophilic.

Hemophilia also played an important historical role in Russia during the reign of Nikolas II, the last czar. Alexis, the only son of Nikolas II, was hemophilic, and his mother, the czarina, was convinced that the only one who could save her son's life was the monk Rasputin, known as the "mad monk." Through this hold over the reigning family, Rasputin became the real power behind the disintegrating throne.

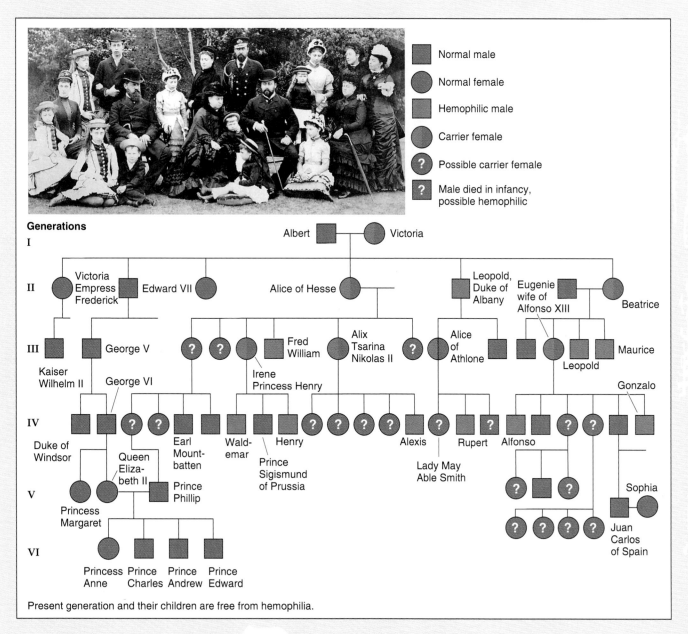

Legend:
- ■ Normal male
- ● Normal female
- ▨ Hemophilic male
- ◑ Carrier female
- ? Possible carrier female
- ? Male died in infancy, possible hemophilic

Present generation and their children are free from hemophilia.

A few sex-linked mutant alleles are dominant. Dominant brown spotting of the teeth, for instance, is passed from affected men to all their daughters and none of their sons; affected women pass the condition on to half their daughters and half their sons. (Work that one out.) Dominant brown spotting affects about twice as many women as men, since they have two chances of receiving the X chromosome with the dominant brown-spotting allele. Another curious X-linked genetic disease is known as the oral–facial–digital syndrome, which involves irregularities of the mouth, face, fingers, and toes. This condition affects only women, who get it only from their mothers. Such women pass the condition on to half their daughters but to none of their sons. It turns out that, as a group, affected women have twice as many daughters as sons. We can therefore surmise that the allele in the hemizygous state is lethal.

This chapter has covered a lot of ground. Be sure to keep track of the underlying themes: Mendel's laws. They are the basis for all of these phenomena. In Chapter 10 we covered the laws of segregation and independent assortment in their pure form. In this chapter we covered the ways in which those laws *seem* to be violated or altered. But what we found, in fact, is that those laws hold up perfectly well, and we merely have to be careful with our interpretation of what we see.

Key Ideas

DOMINANCE RELATIONSHIPS AND GENE INTERACTIONS

1. Recessivity may mean that a gene is simply not functioning (as in albinism). Alternatively, recessive alleles may be functioning, but to a lesser degree than dominant genes.

2. Specific **dominance relationships** are **incomplete dominance** and **codominance**. Incomplete dominance is seen when red and white snapdragons are crossed, producing pink offspring. Codominance occurs when one homozygote expresses a trait differently from the other homozygote, but the heterozygote shows both traits.

3. The term *multiple alleles* means that even though each individual in a population gets two alleles for a trait, more than two different alleles are present in the population's genes.

4. The blood types of the ABO system are determined by cell surface antigens. Antigens are large molecules that are capable of reacting with specific antibodies. In terms of dominance, antigens I^A and I^B are codominant, and both are dominant over I^O (the absence of an antigen). I^A, I^B, and I^O are multiple alleles.

5. From the alleles I^A, I^B, and I^O there are six possible genotypes. From these six genotypes there are only four blood groups or phenotypes.

6. The immune systems of individuals produce either anti-A or anti-B antibodies. The type A person produces anti-B, type B produces anti-A, type AB produces neither, and type O produces both.

7. Sometimes pairs of alleles control different steps in producing the same trait. These are known as **epistatic** genes or alleles. An example is mouse hair color. One pair of alleles produces brown or black, but another pair (**cc**) permits or prevents the presence of color. In the epistatic cross **BbCc** × **BbCc**, the phenotypic ratio in the offspring turns out to be 9:3:4 instead of the classic 9:3:3:1 because every offspring with a **cc** is white.

CONDITIONAL PHENOTYPES

1. Many alleles may be subject to environmental influence, sexual influence, or influences not yet understood.

2. Dark hair pigment in Siamese cats increases in response to cold.

3. The dominant allele for polydactyly (extra digits), an example of **incomplete penetrance**, may not be expressed at all, or the degree of expression (number of affected digits) may vary widely.

4. **Sex-limited** traits can only affect one sex, while **sex-influenced** traits (such as baldness) tend to be expressed in one sex and not in the other.

5. Some traits have variable ages of onset, appearing most often near middle age.

6. In many instances a trait is produced by the cumulative effect of genes from different loci. This is called **polygenic** inheritance, and its effect is to produce continuous variation in a trait.

GENES ON CHROMOSOMES

1. Linkage and crossing over were discovered when expected Mendelian ratios failed to appear.

2. In the testcross **AaBb** × **aabb**, the predictable Mendelian ratio in the progeny would be 1:1:1:1. If the genes are on different chromosomes, no other combinations are possible.

3. If, however, the genes in the above testcross were linked on the same chromosome, then the ratio in the progeny would be 1:1. If they are fully linked, no other combinations are possible.

4. Frequently, even when linkage is established, testcrosses can reveal "impossible" new combinations and strange ratios. For example, the progeny might be 9:9:1:1. Apparently, an exchange of alleles between homologous chromosomes has occurred. Here crossing over in meiosis occurred often enough between the **A** and **B** loci for 10% of the gametes to bear the results of recombination (and did not occur the other 90% of the time).

5. **Recombination frequencies** are used to construct genetic maps. Map distances are measured in recombination percentages between loci.

6. In 1910 Morgan's discovery of a mutant white-eyed male led to the discovery of sex linkage. In his experimental crosses he observed the following:

   a. P_1: White-eyed males × red-eyed females. F_1: all red-eyed.

   b. F_1: Red-eyed males × red-eyed females. F_2: $\frac{3}{4}$ red, $\frac{1}{4}$ white, but *all white-eyed flies were males.*

   c. Testcross: P_1 white-eyed male × F_1 red-eyed female: $\frac{1}{2}$ red, $\frac{1}{2}$ white-eyed offspring evenly distributed between males and females.

   d. **Reciprocal testcross:** Red-eyed male × white-eyed female: $\frac{1}{2}$ white $\frac{1}{2}$ red, but *all females were red-eyed and all males were white-eyed.*

   e. The solution to the above was the linkage between the white eye allele and the alleles that determine sex: both were on the X chromosome.

7. In *Drosophila* the X is rod-shaped, and the Y is J-shaped. Females have two X chromosomes, while males have an X and a Y.

8. In *Drosophila,* sex is determined by the number of Xs; thus the abnormal XXYs are functional females in the fly, but sterile males in humans (*Klinefelter syndrome*). So the presence of a Y chromosome in humans produces maleness. Furthermore, the absence of a second sex chromosome (XO) in fruit flies produces a sterile male, while this combination in humans results in a sterile female (*Turner's syndrome*).

9. In each cell of the human female, one X chromosome remains condensed. It is detected as a Barr body in cheek cells and a drumstick figure in the nucleus of certain white blood cells. The selection of such Xs is random, and a mosaic of maternal and paternal Xs occurs in the tissues.

10. Red–green colorblindness in humans, an abnormality in the cones of the retina, is caused by a recessive allele at either of two closely linked loci on the X chromosome.

11. Women heterozygous for a sex-linked trait do not express the trait, although half their sons do.

Application of Ideas

1. In one breed of chickens, *pea* comb (**PP**, **Pp**) is dominant to *single* comb (**pp**). In another breed of chickens, *rose* comb (**RR**, **Rr**) is dominant to *single* comb (**rr**). When a homozygous *pea* comb chicken (**PPrr**) is mated with a homozygous rose comb chicken (**ppRR**), the F_1 progeny all have yet a fourth phenotype, *walnut* comb (**PpRr**). What are the expected phenotypic ratios among the F_2 progeny of two F_1 walnut comb chickens? What about the progeny of an F_1 walnut comb chicken (**PpRr**) cross to a single comb chicken (**pprr**)?

2. The Japanese geneticist, Hagiwara, crossed two true-breeding strains of Japanese morning glories, both of which had blue flowers. The plants of the F_1 generation all had purple flowers. When Hagiwara crossed the purple-flowered F_1 back to one of the parental strains, the progeny of the backcross were approximately 50% blue-flowered and 50% purple-flowered. When the purple-flowered F_1 progeny were crossed to the other parental strain, the results were the same. But when the F_1 purple-flowered progeny were crossed to each other, the F_2 offspring included purple-flowered, blue-flowered, and scarlet-flowered plants. Hagiwara concluded that two gene loci were involved, an **Aa** locus and a **Bb** locus. The scarlet-flowered plants must then have been **aabb**.

   a. What were the genotypes of the parental strains, the F_1 hybrid, and the progeny of the two backcrosses?

   b. List all phenotypes and all genotypes that appear in the F_2.

3. Eleanor Perkins is phenotypically normal, but her family has its share of sex-linked abnormalities. Her husband, Garvey, has the X-linked dominant allele for brown-spotted teeth but is otherwise normal. Her brother, Arthur, and her son, Little Chester, both suffer from hemophilia A, the most common type of sex-linked bleeder's disease; her father, Grandpa, is not hemophilic, but he is colorblind.

   a. Draw and label a pedigree diagram of this family.

   b. From this information alone, Eleanor knows that she herself is a carrier. Which X-linked mutant gene or genes does Eleanor carry?

   c. Once Grandpa came to breakfast wearing one red sock and one green sock and couldn't understand why Eleanor, Garvey, and Arthur were all laughing at him. But when Little Chester didn't get the joke, Eleanor knew that her son was both hemophilic and colorblind, and she suddenly realized what must have happened in one of her own oocytes shortly before her birth. Explain.

   d. Eleanor and Garvey are expecting another child. If it is a daughter, what is the probability that she will be hemophilic? Colorblind? Have brown teeth?

   e. If their child is a son, what is the probability that he will be hemophilic? Colorblind? Have brown teeth?

   f. What is the probability that such a son will be normal—that is, he will have none of the traits mentioned? (Note: This last question requires more information. Haldane and Smith calculated that the recombination frequency between the red–green colorblindness and hemophilia A locus is about 10%. List the expected frequencies of all possibilities for Eleanor and Garvey's sons, and be sure they add up to one.)

4. In many plant species, male and female floral parts are borne on separate plants. Bauer and Shull took pollen from a narrow-leaved male *Lynchnis alba* and dusted it on the flowers of a broad-leaved female plant. In the F_1 both males and females were all broad-leaved. The F_1 male and female plants were crossed to each other, and in the F_2 the females were broad-leaved, but the males were both broad- and narrow-leaved.

   a. Explain these results.

   b. Is breadth of leaf sex-linked or sex-limited?

   c. What do you expect would happen in the F_1 and F_2 of a cross between a male of a true-breeding broad-leaved strain and a female of a true-breeding narrow-leaved strain?

Review Questions

1. Explain in terms of pigmentation how dominance works in albinisim. (page 237)

2. Define and give an example of each of the following: (a) incomplete dominance (b) recessive–lethal, and (c) codominance. (pages 237–238)

3. Explain what is meant by multiple alleles. (pages 238–240)

4. List both the blood phenotypes (types) and the genotypes possible in the ABO blood system. (pages 239–240)

5. Carry out the cross between two **BbCc** mice (as described in the text). From your results, derive the 9:3:4 ratio and explain why the familiar 9:3:3:1 phenotypic ratio did not show up. (page 241)

6. List several ways in which the *Drosophila* species is well suited for the study of heredity. (page 247)

7. Carry out the calculations that explain the behavior of the white-eye trait in fruit flies. Cross the white-eyed male with a homozygous red-eyed female. Inbreed the F_1 and explain the results in the F_2. (pages 248–249)

8. Can females express sex-linked traits? Explain your answer. (pages 248–250)

9. Using a Punnett square and the symbols XX (female) and XY (male), determine the probability of male or female offspring. (page 250)

10. Explain why males always inherit X-linked traits from their mothers and never from their fathers. (page 250)

11. Explain the statement: Males are hemizygous for X-linked traits. (page 250)

12. What is the probability of colorblindness in the children of a colorblind carrier woman and her normal-vision mate? (pages 251–252)

13. Briefly describe two examples of how physical conditions in the surroundings determine whether or not a trait will be expressed. (page 241)

14. How do sex-limited traits differ from sex-influenced traits? (page 242)

15. How does polygenic inheritance differ from that of multiple alleles? From that of epistasis? (page 241)

16. Using the testcross **GgOo × ggoo**, what genotypes in what ratios might one expect if (a) the genes were unlinked, (b) the genes were linked but could not cross over, and (c) the genes were linked but crossed over 20% of the time? (pages 245–246)

17. Using a diagram, carefully explain why only odd numbers of crossovers are revealed in recombination frequencies. (page 247)

18. What is the basis for map distances in genetic or recombination mapping? Explain the logic. (pages 247–248)

19. Construct a simple recombination map from the following alleles and recombination frequencies: genes **P** and **L**, 30 map units; genes **L** and **X**, 5 map units; and genes **X** and **P**, 35 map units. (pages 247–248)

CHAPTER 12

DNA as the Genetic Material

CHAPTER OUTLINE

THE DISCOVERY OF DNA AS THE GENETIC MATERIAL

The Early Efforts

Transformation: A Hereditary Role for DNA

The Experiments of Hershey and Chase

THE STRUCTURE OF DNA

The Chemical Properties of DNA

The Watson and Crick Model of DNA

DNA REPLICATION

The Chemistry of DNA Synthesis

Origins of Replication

DNA Polymerase

DNA Replication in Eukaryotes and Prokaryotes

Essay 12.1 Meselson and Stahl

DNA AND GENETIC INFORMATION

Essay 12.2 Eukaryotic DNA: A Closer Look

KEY IDEAS

APPLICATION OF IDEAS

REVIEW QUESTIONS

Not so very long ago, as recently as the early 1950s, the chemical basis of heredity was unknown. Scientists had generally agreed that the gene had a chemical composition, but they did not know what genes were, how they were put together, or how they worked. As early as the 1940s there was some good evidence that the genetic material was deoxyribonucleic acid (DNA), but at midcentury, most biologists would have said that genes were probably made of protein. After all, they knew that enzymes were proteins, and they believed that genes must be much more complex even than enzymes. And enzymes seemed to be so enormous and complex that their own structure would be difficult to unravel. How could we ever hope to understand genes? It had been pointed out that genes not only had to direct all the life processes of the cell and the development of the organism, but they also somehow had to make exact copies of themselves every cell generation.

James Watson and Francis Crick showed that the structure of genes and the mechanism of gene function and replication were essentially not so hard to understand after all. In fact, the secrets lay in the simplicity of DNA, a molecule whose presence in the cell had been known for about 80 years. The gene itself turned out to be a surprisingly simple, four-element chemical system for the coding of protein synthesis. The order of the four elements of DNA determines the order in which the 20 amino acids are arranged in any protein.

One might ask, what does protein synthesis have to do with genes and heredity? The answer is, everything. Remember that proteins form much of the structure of living things, as well as the

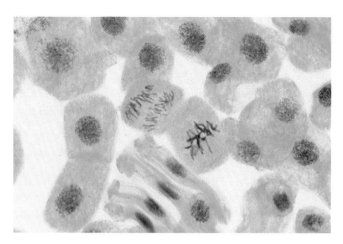

FIGURE 12.1
Feulgen Staining of DNA in Nuclei
DNA within the nuclei of cells becomes deeply colored by the stain.

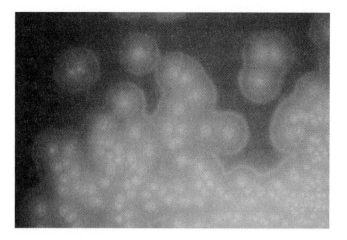

FIGURE 12.2
Bacterial Colonies
Bacterial colonies are just that—large groups of bacterial cells piled on top of one another. Shown here are *Pneumococcus,* organisms like those used in Griffith's experiments.

enzymes that regulate the processes of life. Phenotypic traits are then produced by proteins that form structures of life and regulate the processes of life.

As we will see, the story of Watson and Crick, their friends, their personal lives, and their luck and successes has turned out to be one of the classical, even legendary, stories of biology. In this chapter, we will review how they first assembled their ideas on DNA structure and function. What they reveal to us is pleasingly logical and understandable when it is told one step at a time.

The Discovery of DNA as the Genetic Material

A fundamental question in biology in the first half of this century was what, precisely, was the material that constituted the factors described by Mendel? Although there were strong indications of ties between chromosomes and inheritance, what were these chromosomes made of, and how did they hold the genetic information?

The Early Efforts

Surprisingly, DNA itself was discovered while both Darwin and Mendel were working. In 1869, Friedrich Miescher, a Swiss chemist, isolated a strange, phosphorus-containing material from the nuclei of white blood cells taken from pus of open wounds. In a private letter, Miescher actually speculated that this material might be the stuff of heredity. Later, in 1914, a German chemist named Robert Feulgen invented a still widely used staining procedure, Feulgen staining, that is specific for DNA. The Feulgen procedure stains cell nuclei

more or less strongly according to how much DNA is present (Figure 12.1). Thus the amount of DNA present can be calculated by measuring the strength of the color. A few key experiments revealed that virtually every cell nucleus in a given plant or animal has the same amount of DNA, except for gametes (eggs and sperm), which have half the amount.

Most biologists still couldn't bring themselves to consider seriously the possibility that DNA was the hereditary material, for a few very convincing reasons. In the first place, the composition of DNA is very simple: just four different nucleotides are present (refer to Chapter 3). How could anything so simple be the physical basis of anything so wonderful as the gene? How could only four nucleotides produce the complex variations of life, or contain the information for the arrangement of the 20 different amino acids known to be in proteins? In fact, the chemist P. A. Levene, who initially described the chemical composition of DNA and first described the nucleotide, mistakenly assumed that because the four nucleotides occurred in approximately equal proportions, DNA was a simple, repeating polymer of the four nucleotides. In the second place, DNA didn't seem to do anything. It just sat there, some scientists said, probably holding the chromosome together, or making it acidic, or doing something even less significant.

Transformation: A Hereditary Role for DNA

In 1928, bacteriologist Fred Griffith conducted an experiment that proved to be a classic. He was studying the virulence (disease-producing capability) of two strains of *Pneumococcus,* a bacterium that causes pneumonia. One strain was dangerous and one harmless. The cells of the virulent (disease-producing) strain synthesized a smooth, gummy polysaccharide coat that seemed to protect them from the host's defenses; the harmless strain did not. When grown in the laboratory (Figure 12.2), the

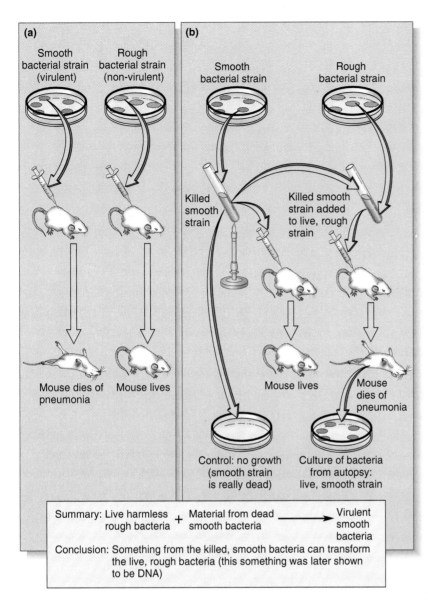

(a)

Smooth bacterial strain (virulent)

Rough bacterial strain (non-virulent)

Mouse dies of pneumonia

Mouse lives

(b)

Smooth bacterial strain

Rough bacterial strain

Killed smooth strain

Killed smooth strain added to live, rough strain

Mouse lives

Mouse dies of pneumonia

Control: no growth (smooth strain is really dead)

Culture of bacteria from autopsy: live, smooth strain

Summary: Live harmless rough bacteria + Material from dead smooth bacteria → Virulent smooth bacteria

Conclusion: Something from the killed, smooth bacteria can transform the live, rough bacteria (this something was later shown to be DNA)

FIGURE 12.3
Transformation
Griffith's experiments in 1928 clearly laid the groundwork for the discovery of the genetic material. **(a)** His smooth strain of *Pneumococcus* was virulent and caused pneumonia in mice—that is, unless heat was used to kill the bacteria. Griffith's rough strain was harmless. **(b)** But the rough strain could be "transformed" into the virulent, smooth strain by the addition of some substance or "factor" from the heat-killed smooth strain. Work by Avery and his co-workers in 1944 indicated that DNA might be the factor responsible for transformation.

Griffith thought that perhaps he had erred in his experimental technique, so he repeated the experiment with great care, again and again. The results were clearly not due to faulty techniques, nor were they due to accidental contamination; the dead smooth-strain pneumococci were indeed dead. As Sherlock Holmes said, when you have eliminated the impossible, whatever is left must be the truth, no matter how improbable. Apparently the living rough-strain *Pneumococcus* had somehow been *transformed*. That is, they had incorporated hereditary material from an outside source and, in so doing, expressed a new (smooth) trait. This process came to be called **transformation**.

Others improved on the experiment, trying to discover what was behind these results. They found that transformation could occur in test tubes, as well as in living mouse hosts, and further tests for transformation monitored colony phenotypes on plates rather than mice autopsies. Various materials from the smooth bacteria were isolated and purified in attempts to identify the mysterious transforming substance. Not until 1944, through the work of O. T. Avery, C. McLeod, and M. McCarty, was it demonstrated that pure DNA extracted from smooth-strain *Pneumococcus* could transform rough-strain bacteria, giving them the ability to synthesize the necessary enzymes for making the smooth polysaccharide coat. Avery and his co-workers further showed that purified proteins would not cause transformation, nor would carbohydrates or lipids. After years of research we now know that the harmless rough cells actually take in pieces of smooth-cell DNA. With a low but measurable frequency, the DNA-repairing enzymes of the rough cell insert the deadly smooth-strain DNA fragments into the cell's own chromosome, replacing the rough strain's

virulent strain produced "smooth"-looking colonies; the harmless strain lacked the proper enzymes to coat themselves and produced colonies that appeared "rough."

When Griffith injected smooth-strain bacteria into mice, the mice died. When he injected rough-strain bacteria into mice, the mice did not die. He then killed some smooth-strain bacteria by heating them and injected their bacterial corpses into more mice. The mice did not die (Figure 12.3a). But then Griffith killed some smooth-strain bacteria and mixed them with live rough-strain bacteria (both of which, in separate experiments, had proved to be harmless) and injected the mixture into still more mice. These mice came down with severe pneumonia and died (Figure 12.3b). Did the chemical remains of the smooth-strain bacteria help the rough strain do its deadly work? To further confuse things, autopsies showed that the dead mice were full of virulent, living, smooth bacteria! Where did they come from?

gene. The few rough bacteria in a mixture that are transformed in this way are then able to synthesize new enzymes to produce the phenotype of smooth colonies and virulence in mice.

By 1944 it was becoming clear that DNA was the genetic material—clear, that is, to a few of the more visionary biologists. Many researchers were still far from convinced, even after some 30 different instances of bacterial transformation by purified DNA had been reported. But some caution and a critical attitude are good traits in scientists.

The Experiments of Hershey and Chase

Alfred Hershey and Martha Chase performed another classic experiment that, in retrospect at least, firmly established that DNA was the genetic material. They were pioneers in genetic research of the bacteriophage (called phage for short; see also Chapters 15 and 22). Most phages consist of a DNA chromosome contained in a coat made of protein. Protein and DNA then are the only components of a phage. This little fact is important, because the way that phages reproduce is to "infect" bacteria. When phages infect a bacterium, they cause the bacterium to produce hundreds of progeny phages (Figure 12.4).

The phage that Hershey and Chase worked with infects *Escherichia coli*, the common, rod-shaped bacterium that lives harmlessly in your intestine. Hershey and Chase set out to see which of the two phage components (protein or DNA) was passed into the bacterium in order to produce the next generation of phage particles.

The important point discovered by Hershey and Chase is that only the DNA of the phage enters the host cell; the protein portion of the phage stays outside. How did they discover this without the help of pictures such as the one in Figure 12.4? They labeled either the protein or the DNA of phages by infecting bacteria grown in two types of media, one containing radioactive sulfur (^{35}S), the other radioactive phosphorus (^{32}P). Since proteins contain sulfur but no phosphorus, and nucleic acids contain phosphorus but no sulfur, the protein and nucleic acids conveniently labeled themselves by incorporating the radioactive sulfur and phosphorus, respectively. They then had two kinds of phages; one with ^{32}P-labeled DNA and the other with ^{35}S-labeled protein.

Hershey and Chase then infected "cold" (normal, nonradioactive) bacteria with their "hot" (radioactive) bacteriophages and allowed enough time for the phages to attach themselves and inject their DNA. Then they put the phages and bacteria into a kitchen blender. The empty "ghosts" of the bacteriophages were dislodged from the bacterial surface and could be separated by centrifugation. It turned out that nearly all the radioactive sulfur was found in the empty bacteriophage ghosts; nearly all the radioactive phosphorus was found inside the infected bacteria. Thus Hershey and Chase proved that only the DNA entered the cell, and that DNA would then be responsible for the production of the next generation of phages. More important, they later showed that the purified phage DNA has all the genetic information necessary to

enable the host to make new bacteriophage DNA, and to specify the synthesis of new bacteriophage protein as well.

The experiments of Griffith, Avery, and Hershey and Chase stand out as the pivotal set of experiments that began to establish firmly the growing idea that DNA was indeed the genetic material.

Still, no one had much of an idea of how DNA worked. How did its chemical composition and structure relate to its role as the hereditary material? Genes were known to produce enzymes, which are complex proteins. The big question was, how could a molecule with only four different subunits determine the specificity of proteins, which are composed of 20 different amino acids?

The Structure of DNA

We saw in Chapters 3 and 9 that DNA molecules are incredibly long polymers, each containing millions of the structural units called nucleotides. DNA nucleotides, you recall, consist of three parts: phosphate groups, the 5-carbon sugar deoxyribose, and one of four different nitrogen bases. The sugar links the phosphate and the nitrogen base:

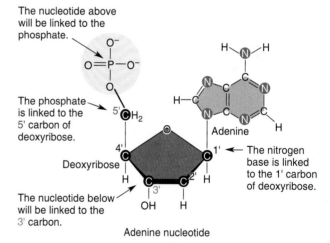

Adenine nucleotide

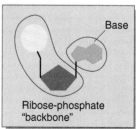

The phosphate groups are the familiar acidic groups we saw in ATP (see Chapter 6). When incorporated into the DNA polymer, only the single phosphate group attached directly to the sugar ring will remain. The 5-carbon sugar, deoxyribose, is quite similar to ribose, the 5-carbon sugar also familiar from our discussion of ATP. As you see, deoxyribose has a simple —H (hydrogen) on its 2′ carbon, instead of the

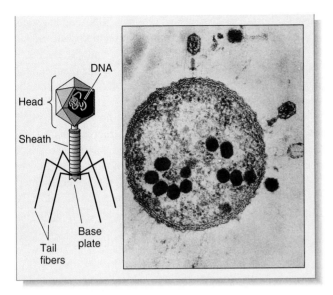

(a) The bacteriophage

—OH (hydroxyl group) seen in ribose. Note that the five carbons of deoxyribose are traditionally numbered 1′ through 5′ (read "one prime through five prime").

The four nitrogen bases of DNA are known as **adenine**, **guanine**, **thymine**, and **cytosine**. Thymine and cytosine, called **pyrimidines**, consist of a single six-cornered ring of nitrogen and carbon. Adenine and guanine are double-ringed molecules, called **purines**. (It may help to keep the purines and pyrimidines straight if you recall that the smaller molecules have the longer names.) Adenine, guanine, thymine, and cytosine are also known simply by the letters A, G, T, and C, respectively. Interestingly, both adenine and thymine are named for the thymus, a gland from which they were first isolated; *adeno* means "gland." Cytosine uses the prefix *cyto-*, which, of course, means "cell." The least romantic is guanine, first isolated from *guano,* a name that refers to the fecal droppings of bats or seabirds.

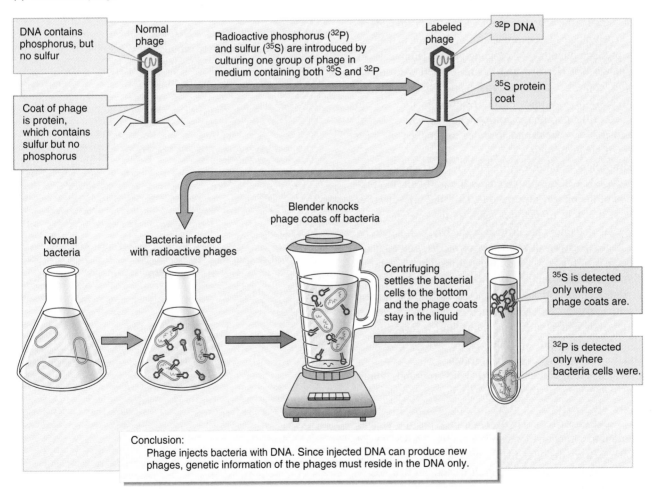

Conclusion:
Phage injects bacteria with DNA. Since injected DNA can produce new phages, genetic information of the phages must reside in the DNA only.

FIGURE 12.4
The Bacteriophage: Hershey and Chase (1950)
Phage viruses are composed of only a protein coat surrounding a DNA core. **(a)** The virus attacks a bacterial cell, causing the cell to mass-produce a new generation of viruses. **(b)** Hershey and Chase utilized the fact that proteins could be uniquely labeled with ^{35}S and DNA could be labeled with ^{32}P to answer the question: "Is it protein or DNA that is passed into the bacterium to direct the formation of the next generation?" They found that it was DNA.

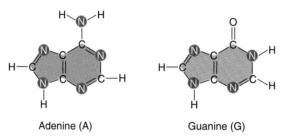

Thymine (T)   Cytosine (C)

Pyrimidines (single ring)

Adenine (A)   Guanine (G)

Purines (double ring)

The Chemical Properties of DNA

Remember that one of the arguments against DNA as the hereditary molecule was that nothing so simple and *repetitive* could carry the information needed to genetically define an organism. But in 1950, Erwin Chargaff showed that DNA from different sources had different base ratios, that is, different frequencies of the four subunits. DNA from *Escherichia coli,* for instance, has about 25% A, 25% T, 25% C, and 25% G, fitting Levene's original theory that DNA was a simple repeating polymer; but the DNA of humans and other mammals is about 21% C, 21% G, 29% A, and 29% T. As Chargaff examined the DNA of more and more organisms, he found different ratios, but he also discovered one general rule: the amounts of A and T in his samples were always equal, and the amounts of G and C were always equal (Table 12.1). In shorthand, A = T and G = C. (This finding became known as Chargaff's rule.) Regardless of the source of DNA, exactly half of the nucleotide bases are purines (adenine and guanine) and exactly half are pyrimidines (thymine and cytosine): A + G = T + C = 50%.

We now know that the reason for these equalities is the specific pairing of nitrogen bases, but to Chargaff they were only intriguing quantitative statements without any clear explanation for the genetic capabilities of DNA. Nevertheless, they were key observations that, as we will see, enabled Watson and Crick to work out the structure and function of DNA.

Additional information on the physical structure of DNA was to come from x-ray crystallography. Essentially, the technique involves aiming x-rays at a crystal and noting how the rays are bent (diffracted) by the regular, repeating molecular structure within the crystal. The closer together the regularly repeated atomic structures are within the crystal, the greater the angle through which the x-rays are bent or diffracted. The pattern produced on a photographic plate consists of whorls and

	Base Composition (mole %)			
Organism	**A**	**T**	**G**	**C**
Animals				
Human	30.9	29.4	19.9	19.8
Sheep	29.3	28.3	21.4	21.0
Hen	28.8	29.2	20.5	21.5
Turtle	29.7	27.9	22.0	21.3
Salmon	29.7	29.1	20.8	20.4
Sea urchin	32.8	32.1	17.7	17.3
Locust	29.3	29.3	20.5	20.7
Plants				
Wheat germ	27.3	27.1	22.7	22.8
Yeast	31.3	32.9	18.7	17.1
Aspergillus niger (mold)	25.0	24.9	25.1	25.0
Bacteria				
Esherichia coli	24.7	23.6	26.0	25.7
Staphylococcus aureus	30.8	29.2	21.0	19.0
Clostridium perfringens	36.9	36.3	14.0	12.8
Brucella abortus	21.0	21.1	29.0	28.9
Sarcina lutea	13.4	12.4	37.1	37.1
Bacteriophages				
T7	26.0	26.0	24.0	24.0
λ	21.3	22.9	28.6	27.2
φX174, single strand DNA[2]	24.6	32.7	24.1	18.5
φX174, replicative form	26.3	26.4	22.3	22.3

TABLE 12.1
Chargaff's Rule[1]

[1] By determining the composition of nitrogen bases in the DNA of a variety of organisms, Chargaff and his contemporaries were able to provide vital information. Pay close attention to the relative quantities of A and T, and G and C (but note that the values are not exactly equal due to experimental error.)

[2] This virus has single strand DNA, which does not follow Chargaff's rule. Why not?

Source: Adapted from A.L. Lehninger, *Biochemistry,* 2d ed. New York: Worth, 1975.

dots, with those farthest from the center of the plate representing the most closely spaced repeating structure (Figure 12.5).

In Watson and Crick's time, people were just starting to use x-ray crystallography to look at DNA. Chief among them were Maurice Wilkins (who received a Nobel Prize with Watson and Crick) and Rosalind Franklin, who died before the Nobel Prize was awarded. Their studies revealed a few repeated intramolecular distances, namely, 2.0 nm, 0.34 nm, and 3.4 nm, numbers that showed up again and again in the x-ray image. Wilkins and Franklin recognized that the patterns were caused by a helical molecule with the phosphates on the outside. Franklin even argued that there were probably two strands, not one or three, as had been suggested by others. Although Wilkins and Franklin had a general idea of what the

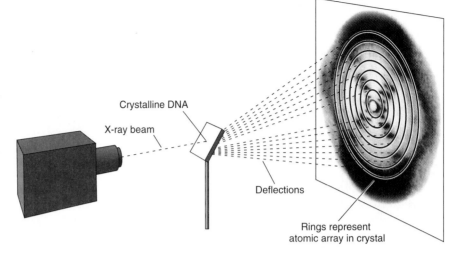

FIGURE 12.5
X-Ray Diffraction of DNA
The actual image of an x-ray diffraction pattern is a subtle pattern of light and dark spots, caused by the manner in which the crystal lattice diffracts the incoming x-ray beam. The pattern of spots gives critical information about the structure and dimensions of the crystal lattice that causes the diffraction. In the case of DNA, the **X**-shaped crossing pattern indicated a helix, and the spacing of the spots in the dots provided clues to the dimensions of the helix and its repeating substructure.

molecule was like, they still didn't know the specifics. It was at this point that Watson and Crick stepped in.

The Watson and Crick Model of DNA

Watson and Crick knew several things when they tackled the DNA problem in the early 1950s. They knew that DNA was a polymer consisting of four different nucleotides, they understood the chemical structure of the nucleotides, and they realized that since DNA was acidic, the phosphate groups must be exposed. From Chargaff's data, they knew that somehow the number of adenines had to equal the number of thymines and that the number of guanines had to equal the number of cytosines. They also knew of Wilkins and Franklin's ideas about DNA and the results of their x-ray diffraction studies—the intramolecular measurements 2.0 nm, 0.34 nm, and 3.4 nm, as well as the idea of a helix.

Watson and Crick's experimentation was based on physical models made of wire, sheet metal, and nuts and bolts (Figure 12.6). Their models provided a graphic, "hands-on" representation of the emerging molecule. (Sometimes the fingers can grasp what the mind cannot.) Biological intuition, plus biophysical and chemical calculations, along with trying model pieces this way and that, seemed to indicate that there might be two strands wrapped around one another, with the phosphate–sugar backbone on the outside and the nucleotide bases inside, facing one another. But how were the bases arranged inside?

It was Watson who insightfully grasped the true meaning of Chargaff's strange data about the ratios of A, T, G, and C. In the sheet-metal and wire model, two purines would not fit opposite one another within the 2.0 nm confines of the double helix, and two pyrimidines would leave a gap. But one purine and one pyrimidine could fit opposite one another. Thus if a purine were always found across the helix from a pyrimidine, the distance between the sugar–phosphate backbones would remain constant. Watson also saw that adenine and thymine would form two hydrogen bonds, while guanine and cytosine would form three hydrogen bonds (Figure 12.7). This was the

only way the components of the DNA molecule could fit together. Thus the specifics of base pairing represented one of Watson and Crick's major findings. Whatever nucleotides were in one strand, they rigidly fixed the sequence of nucleotides in the other strand. We say that the two strands are **complementary**; they aren't identical, but they fit together like a hand in a glove.

It was about this time that Wilkins and Franklin's numbers began to make sense. The 2.0-nm measurement represented the total width of the double helix. The 0.34 nm represented distance from one base to the next; that is, if the bases were stacked one on top of the other, like pennies in a roll, each

FIGURE 12.6
Watson, Crick, and Their Model of DNA (1953)
Using wire, bits of metal, and intuition, Watson (left) and Crick (right) put all available information together to deduce the structure of DNA. Here they are shown with one of their original models of DNA.

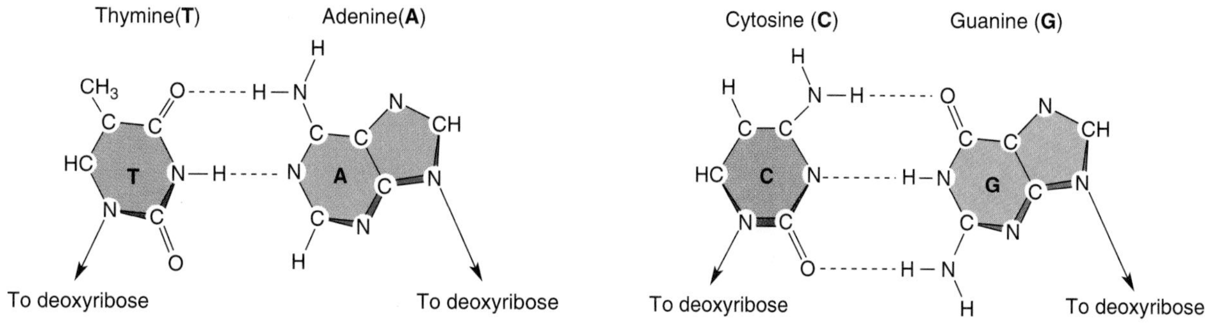

FIGURE 12.7
Concept of Base Pairing
Simply put, base pairing in DNA always occurs between A and T, and G and C. The pairing of the bases is a result of hydrogen bonding (indicated by the dashed lines) between the bases that span the double helix. The arrangement of atoms to produce the hydrogen bonds is a specific requirement for pairing, and mismatches will not form stable hydrogen bonds.

layer would be 0.34 nm thick. The 3.4-nm measurement was a tough one, but it turned out that if each base was set slightly off from the one above like steps in a circular staircase, the double backbone would make one complete twist every 3.4 nm along the axis of the molecule. Furthermore, with a little arithmetic (3.4 nm/0.34 nm = 10), we find that there would be exactly ten nucleotide pairs in each helical turn (Figure 12.8).

The configuration of the two strands of DNA wound around each other is that of the famous double helix. Within the double helix, the two strands of DNA are **antiparallel**; that is, they run in opposite directions. The end of a strand with the free 5′ phosphate is known as the **5′ end**, and the end with the free 3′ — OH group is called the **3′ end**. So if you pictured a DNA molecule vertically on a page, one of the chains would run from top to bottom in its 5′-to-3′ direction, and the other would run from bottom to top in its 5′-to-3′ direction (Figure 12.9). The two antiparallel strands are held together by the hydrogen bonds formed between base pairs across the helix.

When you look closely at the space-filling model of DNA (see Figure 12.8, *bottom*), you might note that the double helix produces two grooves, called the **major groove** and the **minor groove**. As we will discuss later, the order of bases in the major groove is thought to be very important in the control of gene action (Chapter 14).

DNA Replication

Watson and Crick immediately recognized that the structure of DNA, particularly the complementary nature of the paired nucleotides, suggested the core of an elegant mechanism that would ensure the correct replication of the genetic material. "It has not escaped our notice that the specific pairing we have postulated immediately suggests a possible copying mechanism for the genetic material," they said at the close of their initial paper on the double helix. They conceptualized a simple replication model in which the two strands of the double helix separated, allowing a new complementary strand to be synthesized using each of the old strands for templates (Figure 12.10). In essence, they were correct. The details, though, are only now becoming clear.

Synthesis means making something, and **replication** means making an exact copy of something. With very few exceptions, DNA synthesis and DNA replication are the same thing, since DNA molecules are made only by copying other DNA molecules. Historically, it was realized that one of the properties of the unknown genetic material is the ability to replicate when a cell divides. A complete set of genes (within the chromosomes) must be passed down to each new cell, and any reproducing cell or organism must pass along a full set of genes to its descendants.

The Chemistry of DNA Synthesis

As Figure 12.10 shows, the concept of DNA replication is quite straightforward. The two complementary strands of DNA are separated from each other, and then those strands are used as templates from which new strands are constructed. The details of all this, however, are a bit more involved. It all begins with a large number of free nucleoside triphosphates; dATP, dGTP, dTTP, and dCTP (the "d" signifying deoxyribose).

In the synthesis of DNA chains, the high energy of the triphosphates (recall ATP) is used to form a covalent bond between nucleotides. In the process, the two outermost phosphate groups are liberated as pyrophosphate, leaving the innermost phosphate still attached. This remaining phosphate forms a link between two deoxyribose subunits as follows. The free 3′ — OH group of deoxyribose in the first nucleotide reacts with the first 5′ phosphate of the second nucleotide, forming a sugar–phosphate–sugar linkage. The 3′ — OH group of the second nucleotide is free to interact with the 5′ phosphate group of the third nucleotide, and so on (Figure 12.10). Perhaps the most important thing to remember about this chemistry is that the 5′ phosphate of the incoming nucleotide reacts with the 3′ — OH of the previous nucleotide. The identity of the incoming nucleotide (A, G, T, or C) depends, of course, on the information in the complementary template strand.

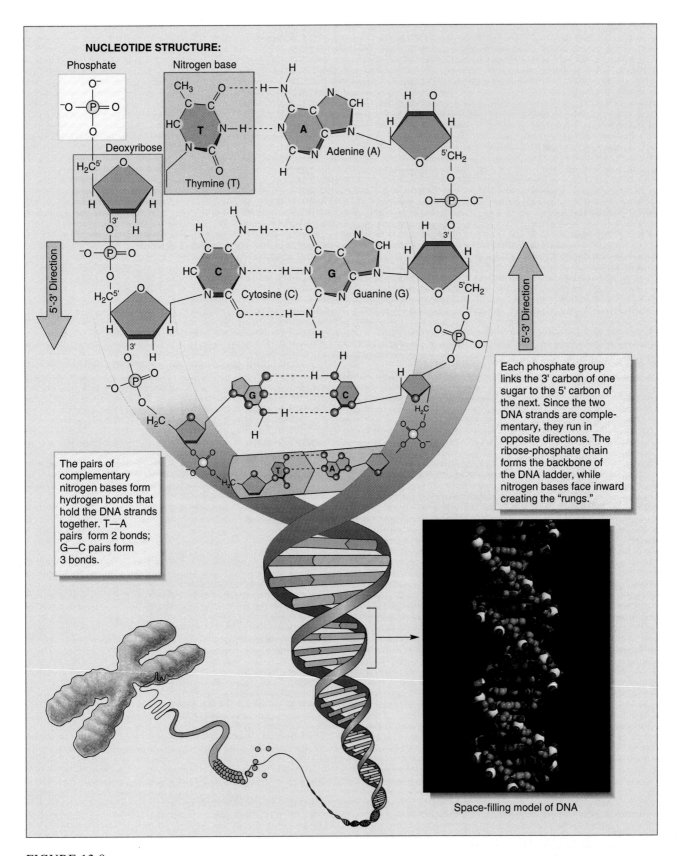

The pairs of complementary nitrogen bases form hydrogen bonds that hold the DNA strands together. T—A pairs form 2 bonds; G—C pairs form 3 bonds.

Each phosphate group links the 3' carbon of one sugar to the 5' carbon of the next. Since the two DNA strands are complementary, they run in opposite directions. The ribose-phosphate chain forms the backbone of the DNA ladder, while nitrogen bases face inward creating the "rungs."

Space-filling model of DNA

FIGURE 12.8
Three-Dimensional Structure of DNA
Images that we use to describe DNA are all based on the actual structure of DNA; depending on the depiction, more or less detail is visible.

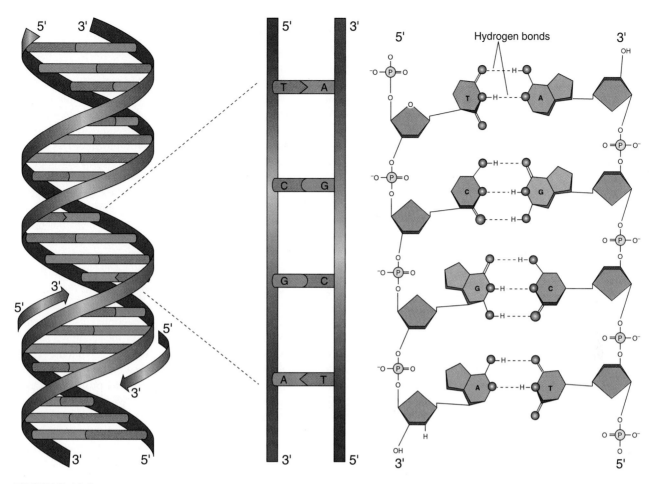

FIGURE 12.9
Diagrammatic Representations of DNA
These are but a few of the many ways that the chemical complexity of the structure of DNA can be
reduced to shorthand notations. The most important features required of any shorthand representation are
two antiparallel strands separated by nitrogen bases that have specific hydrogen bonds.

Eventually, a long polymer of nucleotides is produced. The covalently bonded backbone of the chain consists of alternating sugars and phosphates linked by shared oxygen atoms. The nitrogen bases are side groups hanging off the sugars and are not part of the backbone.

Origins of Replication

The replication process begins within the DNA molecule, rather than at one end, at sites known as **origins of replication** (Figure 12.11). There, hydrogen bonds are broken, and the paired bases are separated. The helix begins to pull apart—much as two pieces of string wound about each other would separate at the middle. The unwinding is facilitated by an unwinding enzyme called **helicase**. Helicase is a part of the **replication complex** (also called the **replisome**), a group of enzymes and other proteins that take care of the replication process. Two such replication complexes are present at each replication origin, and as unwinding continues they move off in opposite directions, creating two Y-shaped replication forks (Figure 12.11; see also Essay 12.1, pages 270–271).

As the strands of the double helix are unwound from each other, an enzyme called gyrase prevents accumulation of twists by making temporary single-strand nicks in the DNA. Where the DNA is nicked by the enzyme gyrase, the helix can unwind around the connected strand. After the tension of unwinding is relieved (for a time), the nick is sealed by the enzyme **ligase**. Replication complexes also contain single-strand binding proteins to help keep the template strands apart, and, of course, DNA polymerase—the most prominent enzyme in DNA synthesis.

DNA Polymerase

The main function of the primary enzyme responsible for replication, **DNA polymerase**, is to add the 5′ phosphate of a new nucleotide to an existing 3′ — OH group. Indeed, for our purposes here, this polymerizing activity is the most important of several enzymatic activites possessed by DNA polymerase. Yet this single polymerizing activity imposes critical limitations on the replication process as a whole.

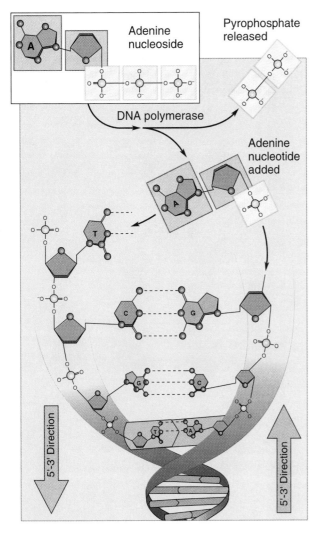

FIGURE 12.10
DNA Synthesis
Each strand of the double helix can serve as a template on which to synthesize a new complementary strand. Nucleotides are added to a growing DNA chain by the enzyme DNA polymerase. Although not completely shown here, there is always a preexisting strand that serves as a template or guide to decide which of the four nucleotides is to be added. Newly added nucleotides are attached at the 3′ carbon of the deoxyribose, forming 3′–5′ linkages and free pyrophosphates as the strand grows.

First, DNA polymerase cannot begin a new daughter strand all by itself; it requires an existing 3′ — OH to add the next nucleotide. So replication actually begins when a different enzyme called **primase** begins to copy the template strand by forming a primer, a short sequence of a few bases of RNA. These few bases of RNA provide the free 3′ — OH for DNA polymerase to begin the polymerization process, but they will later be removed and replaced with DNA.

The second limitation is that DNA polymerase can only add new bases to the 3′ — OH end of a growing strand; that is, new DNA strands always grow from their 5′ to their 3′

ends. Yet, as we have seen, the two strands of the double helix are antiparallel. So as a replication complex moves off from an origin, it frees up two template strands, and they have opposite polarity. This means that while replication can occur in the 5′-to-3′ direction on one of them, it seemingly must occur in the 3′-to-5′ direction on the other. So how is this problem solved? This question plagued molecular biologists in the early years, and they searched for a form of DNA polymerase that could somehow add nucleotides in the opposite direction. The search was in vain; no such enzyme exists. Eventually, researchers determined that the replication events are quite different on the two template strands.

As the replication complex moves out from the origin along an unwound segment of DNA, synthesis on one strand can proceed continuously from the 5′ to the 3′ direction, to create a *leading* daughter strand (the upper strand in Figure 12.11). For this reason, synthesis on the part of the strand is referred to as *continuous* replication. But the other parent strand (the lower strand in the figure) is copied *discontinuously* and is called the *lagging strand*. As the leading strand is being synthesized continuously, unpaired template bases accumulate on the lagging strand until a long enough segment of template has been exposed to allow first primase and then polymerase to begin copying it in its own 5′-to-3′ direction. (You'll note that this copying is in a direction opposite to that of the movement of the replication fork.) These short segments of newly assembled DNA are called **Okazaki fragments** (after their discoverer). As replication proceeds and nucleotides are added to the 3′ ends of the Okazaki fragments, they come to meet each other. When DNA polymerase meets the RNA primer from the previous segment, it excises the primers, fills in the gap with DNA, and then leaves the scene. Finally, the ends of the adjacent fragments are joined together through the action of the enzyme ligase.

DNA Replication in Eukaryotes and Prokaryotes

In eukaryotes, each chromosome contains one very long, linear DNA molecule prior to DNA replication. At the initiation of DNA replication, hundreds of origins are formed along the length of each DNA molecule (Figure 12.12). As the bubbles enlarge, the replication forks of different bubbles run into each other, and the bubbles merge. Eventually, two long, linear DNA molecules are formed.

Prokaryotes are different in that most require only one origin of replication. Multiple replication sites are essential to the eukaryote because its replication complexes work more slowly than do those of prokaryotes. While bacterial replication can proceed at a rate of 1 million base pairs added per minute, eukaryotic rates only range from about 500 to 5000 base pairs per minute. At that rate it would require about one month for a eukaryotic cell to complete replication, but because of multiple origins of replication, each growing bidirectionally, replication in eukaryotes averages just a few hours. *Drosophila*, the fruit fly, may be a record holder, since the newly fertilized egg

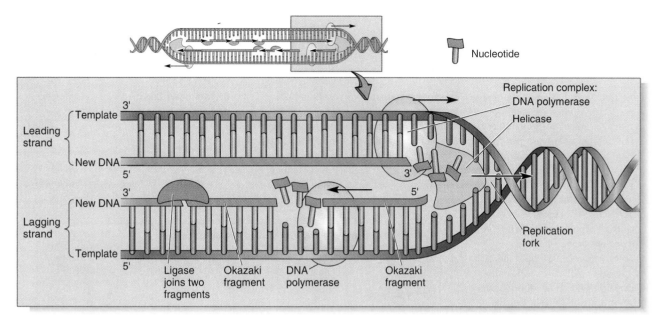

Nucleotide

FIGURE 12.11
DNA Replication
Localized separation of the two DNA strands at an origin of replication allows the beginning of DNA synthesis. As DNA synthesis proceeds in the 5′-to-3′ direction of the new strand, the "bubble" grows through the movement of two replication forks. However, the limitations of DNA polymerase dictate that only one of the new strands at a fork is continuously synthesized. The other strand can only begin synthesis after the template strand becomes single-stranded. A detailed view of a replication fork shows that the discontinuous strand (the bottom one in this case) is synthesized as small fragments. Ligase then joins the fragments together. The replication complex (including gyrase, helicase, and single-strand binding proteins) leads DNA polymerase through for continuous 5′-to-3′ synthesis of the leading (upper) strand.

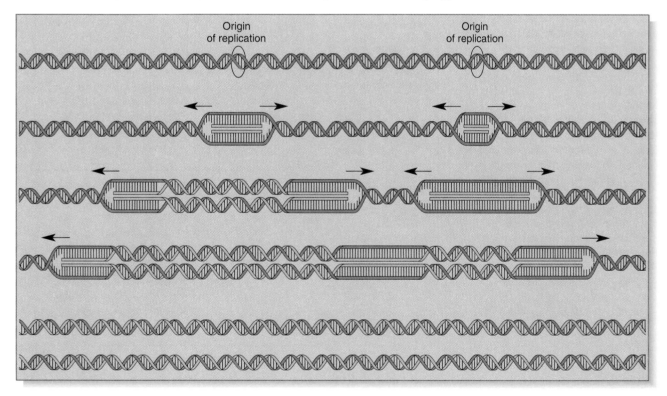

FIGURE 12.12
Multiple Origins of Replication
The task of replicating all of the very long DNA molecules in a eukaryotic nucleus is immense. Part of the problem is solved by having replication begin in many places, as these several replication forks on one *Drosophila* chromosome demonstrate.

Essay 12.1 ● MESELSON AND STAHL

Although the structure of DNA was clarified by Watson and Crick in 1953, the details of replication remained something of a mystery. Watson and Crick's information strongly suggested a *semiconservative* mechanism, in which the two strands of the double helix separate during replication but each remains intact, acting as a template for the assembly of a new partner. It was not inconceivable, however, that DNA replication might be *conservative,* or perhaps even *dispersive.* According to the conservative replication hypothesis, the entire molecule would remain intact and another double strand of new material would be replicated alongside. The dispersive replication hypothesis stated that the strand would be dismantled piece by piece and replication would occur along the pieces.

In 1957, Matthew Meselson and Franklin Stahl set out to determine how replication was accomplished. Their procedure used isotope-labeled biological chemicals and the ultracentrifuge. Using special techniques for isolating DNA from bacteria, they were able to centrifuge the DNA in a cesium chloride (CsCl) solution, which bands each molecule according to its specific gravity (density). Since bacteria are normally exposed only to nutrients containing the common, nonradioactive nitrogen-14 isotope, a ^{14}N reference line was available. In other words, DNA containing ^{14}N settled in the centrifuge tubes at a certain position, forming what was to be the first reference line. Next, a second reference line was established, this time for bacterial DNA extracted from cells that had been grown for many generations in culture media containing ^{15}N-labeled nucleotides. ^{15}N is not radioactive, but since it is heavier than ^{14}N, the ^{15}N DNA settled at a somewhat lower position in the centrifuge tube. With these reference points established, the experiment could begin.

Bacteria were grown in a culture medium containing ^{15}N nucleotides for several generations so that DNA containing ^{15}N would be present in the vast majority of bacterial chromosomes. Bacteria from this culture were then removed, washed, and resuspended in a culture medium containing only nucleotides with ^{14}N. One round of replication was permitted. Cells from this stage were then removed, and the DNA was extracted and centrifuged in the CsCl gradient.

Prediction:

1. If DNA replication is *conservative,* then two regions of DNA will be seen in the centrifuge tube. One will contain only

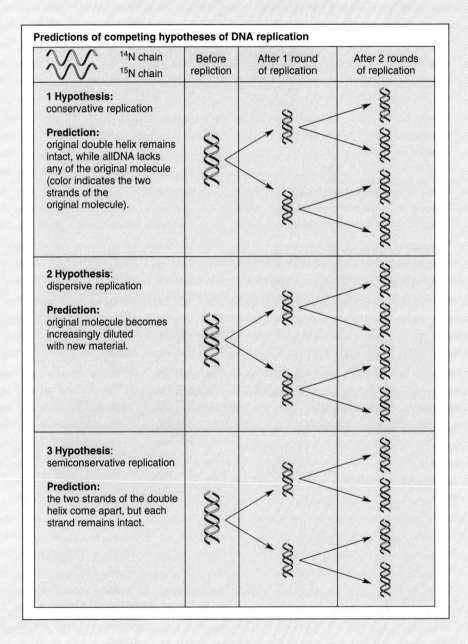

Predictions of competing hypotheses of DNA replication

		Before repliction	After 1 round of replication	After 2 rounds of replication
	^{14}N chain ^{15}N chain			
1 Hypothesis: conservative replication **Prediction:** original double helix remains intact, while allDNA lacks any of the original molecule (color indicates the two strands of the original molecule).				
2 Hypothesis: dispersive replication **Prediction:** original molecule becomes increasingly diluted with new material.				
3 Hypothesis: semiconservative replication **Prediction:** the two strands of the double helix come apart, but each strand remains intact.				

¹⁴N DNA, while the other will contain only ¹⁵N DNA.

2. If replication is *dispersive,* all DNA after one round of replication will be halfway between the reference densities of ¹⁴N and ¹⁵N DNA.

3. If DNA replication is *semiconservative,* then only one region of DNA sediment will be found. This region will contain *hybrid* DNA, each molecule containing a ¹⁴N strand and a ¹⁵N strand. The position will be halfway between the two reference lines established earlier.

As you can see from the predictions, one round of replication could not distin-guish between the dispersive and semicon-servative hypotheses. But in another experi-ment, bacteria were allowed to continue into a second round of replication. The cells were then removed, and the DNA was isolated and centrifuged in the CsCl gradient.

Prediction:

1. If DNA replication is *conservative,* then the same two bands will appear as they did in prediction 1, but the ¹⁴N band will contain three times as much DNA as the ¹⁵N band.

2. If DNA replication is *dispersive,* then all DNA after two rounds of replication will be uniform; namely, 25% will be ¹⁵N and 75% will be ¹⁴N. These should appear as a single band appropriately spaced between the two reference points.

3. If DNA replication is *semiconservative,* then two equal sedimentation bands will appear. One will contain *hybrid* DNA and will form a band at the same location as did hybrid DNA in prediction 2. The other will contain ¹⁴N DNA only and will settle out at the ¹⁴N reference line established earlier.

Results: The results, as diagrammed here, were clear. What Meselson and Stahl saw were the bands that supported only the semi-conservative hypothesis.

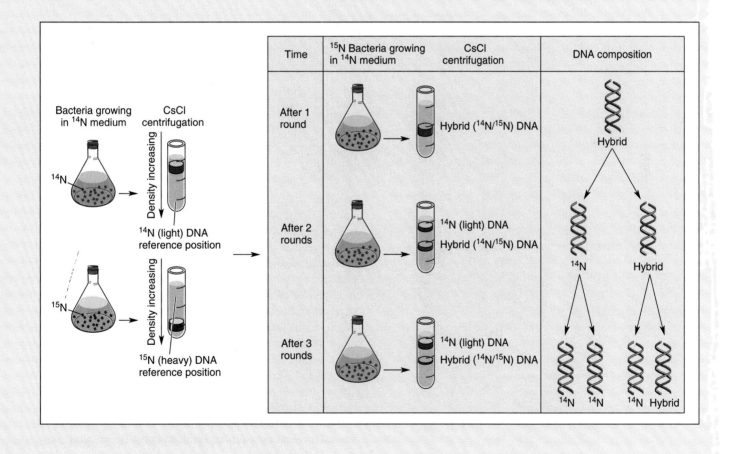

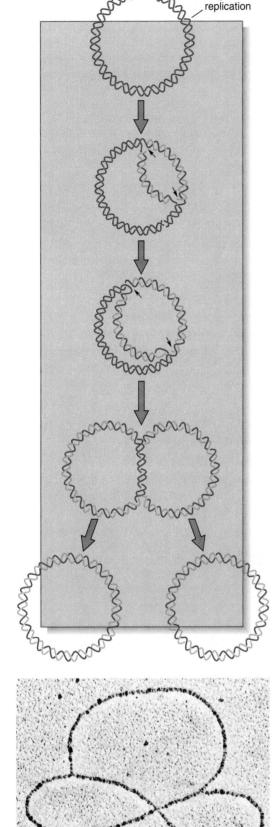

Origin of replication

forms some 50,000 origins of replication, and its four pairs of chromosomes can be replicated in about three minutes.

In prokaryotes, most or all of the DNA is in a single circular molecule that is not nearly as long as a eukaryotic DNA molecule (you have about 1400 times more DNA per cell than an average bacterium; see Essay 12.2). There is no end and no beginning of a circle, of course, but replication in the bacterium *Escherichia coli* always begins at a single origin of replication (Figure 12.13). Again, unwinding proteins form a bubble of single-stranded DNA, and the unpaired strands are replicated to form new helices. The two replication forks from the single origin of replication travel in opposite directions around the circular DNA molecule, eventually meeting each other at a specific termination point about halfway around the circle.

DNA and Genetic Information

We have seen now how DNA meets two of the requirements of the genetic material. It has the capacity to replicate itself precisely. It can contain information by having specific sequences of bases. These properties of DNA are a direct result of its base composition and double helical structure. Yet as the story of DNA as the genetic material began to unfold, it was realized that somehow the information held within DNA must be expressed as proteins. The specificity of the gene must determine the specificity of the protein.

In fact, the idea that genetic information was expressed in protein had been around since 1908. In that year, A. E. Garrod, a physician influenced by Mendel's work, published a book called *Inborn Errors of Metabolism.* His subject was *Homo sapiens,* which, at the time, was an unusual experimental organism for genetics research. Garrod was interested in metabolic defects, breakdowns in the complicated biochemical processes of life. He searched for the abnormal products of such defects in the urine, where many metabolic products are excreted from the body. Of special interest to Garrod was alkaptonuria, a disease in which the urine contains metabolites (products of metabolism) called alkaptones—substances that happen to turn black upon oxidation and so are easily revealed. Infants with alkaptonuria are usually detected as soon as their diapers start turning black. As the child grows older, black pigments begin to settle in cartilage and other tissues, blackening the ears and even the whites of the eyes. Another more serious effect is a form of arthritis, caused by the accumulation of the metabolite in the cartilage of the joints.

Garrod observed that the disease tended to be found in several brothers or sisters in a single family. By studying fam-

FIGURE 12.13
Replicating Circular DNA
Many prokaryotic chromosomes are circular. As replication proceeds, the two replication forks travel in opposite directions, eventually meeting at the opposite side of the circle to produce the two replicas.

Essay 12.2 • EUKARYOTIC DNA: A CLOSER LOOK

Earlier, we noted that eukaryotes have about 1400 times more DNA than bacteria. You may not be surprised at such a difference, considering how much more complicated we are than bacteria, but molecular biologists took a different, more searching view of this seemingly simple observation. They wanted to know why.

At least part of the answer comes easily; we have about 60 times more genes than bacteria. But why then do we have 1400 times more DNA and not 60 times more? There must be something essentially different about the eukaryotic and prokaryotic chromosome structure. And there is. Bacterial chromosomes consist mainly of long stretches of DNA that are unique; eukaryotic chromosomes contain thousands of repetitive base sequences that occur tens, hundreds, thousands, or even millions of times. Let's take a closer look at their distribution, variety, and possible function.

Satellite DNA

Highly repetitive or satellite DNA makes up about 10% of the total DNA in mouse chromosomes (but its relative concentration varies from species to species). It consists of a relatively small number of simple nucleotide sequences like ATATATATAT, CGCGCGCGCG, and so on—but each may be present from 100,000 to over 1,000,000 times in the DNA of a single cell. Researchers have discovered that this form of DNA is highly localized in centromeric regions and at the tips of chromosomes. Although no one knows its function for certain, experiments suggest that it may be necessary for structural stability of chromosomes.

Middle Repetitive DNA

Another form of repetitive DNA found in eukaryotic chromosomes is called middle repetitive DNA, because each individual base sequence is present from 20 to 100,000 times in a single cell's genome.

Middle repetitive DNA, like highly repetitive DNA, is common in centromeric regions, where it is thought to help maintain their highly coiled structures. But it is also found interspersed between the regions of single-copy DNA that represent functional genes. Some middle repetitive DNA may contribute to the control of gene activity. Yet another fraction of a cell's middle repetitive DNA belongs to a class of mobile genetic elements called transposons. Transposon biology is discussed more fully in Chapter 17. But for now you should note that these so-called "jumping genes" are able to insert copies of themselves throughout the genome and are an important source of mutation. As such, they may have served an important role in the evolutionary process. Geneticists have been theorizing for years about their function. But whatever the answer, they make up a substantial portion of a cell's genome. One, called Alu, is repeated 30,000 times and makes up 3% percent of the human genome!

A final example of middle repetitive DNA represents genes that may be repeated from a few to hundreds of times per cell. Such genes usually code for molecules that the cell needs to produce in vast quantities within a short period of time. For instance, in certain active oocytes there are thousands of copies of the genes that produce ribosomal RNA, an essential component of the apparatus that translates genetic messages into polypeptides.

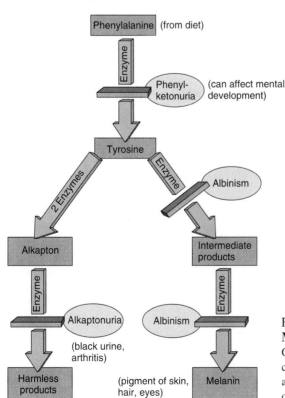

ily histories, he correctly inferred that alkaptonuria was an inherited genetic trait. The problem, he deduced, was caused by the absence of a specific enzyme that is necessary for the biochemical reactions to break down alkaptones. If the enzyme for the reaction is absent, no reactions can take place past the point where the enzyme is normally a part of the pathway, and the substance that the enzyme acts on builds up.

Other inborn errors of metabolism create albinism (a complete or partial lack of melanin pigment in the hair, skin, iris, and sometimes the retina) and phenylketonuria (which also affects hair and skin pigmentation but has a much more severe effect on mental development because of the accumulation of toxic metabolites in the nervous system).

As it turns out, albinism, phenylketonuria, and alkaptonuria are all caused by defects in enzymes that act in the metabolism of the amino acids phenylalanine and tyrosine (Figure 12.14). These sorts of genetic defects are called

FIGURE 12.14
Metabolic Pathways of Tyrosine and Phenylalanine
Garrod's work established the relationship between genes and enzymes. Three conditions caused by absent or abnormal enzymes that block metabolic pathways are albinism, alkaptonuria, and phenylketonuria. The blocked pathways normally operate on intermediate breakdown products of the amino acid phenylalanine.

mutations. While the term may call to mind more sinister images, all mutations are nothing more than changes in DNA that cause some sort of a change in the protein encoded by the gene. Once changed by mutation, a gene is simply inherited in the normal fashion, passed on to succeeding generations as an allele of the normal gene.

Significantly, Garrod discovered that heredity played a role in enzyme activity and that there was a definite connection between heredity and the presence of normal or abnormal enzymes—quite an accomplishment for his day. But he was ignored. Biochemistry was still an infant science, and geneticists of the time were more interested in how the genes influenced visible traits, not metabolism.

So even after the discovery of DNA as the genetic material, the remaining issue was: "How does the information stored in the DNA molecules get expressed as proteins?" That issue would lead researchers finally to the core of gene structure and function.

Key Ideas

THE DISCOVERY OF DNA AS THE GENETIC MATERIAL

1. Mendel helped establish that heredity was controlled by "factors," and chromosomes were soon suspected of carrying the factors (genes).

2. Miescher identified DNA in 1869, and in 1914 Feulgen perfected a specific DNA stain (Feulgen stain).

3. In 1928, Griffith, experimenting with virulence in *Pneumococcus,* determined that nonvirulent strains could be transformed (genetically changed) to virulent strains if the remains of dead virulent bacteria were made available. In 1944, Avery concluded that the transforming material was DNA.

4. Using bacteriophages and bacteria, Hershey and Chase were able to show that only viral DNA entered the host; thus it was DNA that directed the production of new viral particles. This strongly suggested that DNA was the genetic material.

THE STRUCTURE OF DNA

1. Four different deoxynucleotides, or nucleotides, the structural units of DNA, are assembled into long polymers of DNA strands, or nucleic acids. Prior to assembly, they are in the form of *nucleotide triphosphates.*

2. Each nucleotide contains three parts: phosphate, deoxyribose, and one of four nitrogen bases: **adenine** (dATP), **guanine** (dGTP), **thymine** (dTTP), or **cytosine** (dCTP).

3. In 1950, Chargaff developed the principle of base pairing. He determined the relative amounts of A, T, G, and C in a variety of cells, proving that A = T and G = C and that there is exactly as much **purine** in the nucleus as there is **pyrimidine**.

4. Through the use of x-ray crystallography, Wilkins and Franklin determined that DNA was double stranded, probably formed a helix, and had intramolecular measurements of 2.0 nm, 0.34 nm, and 3.4 nm.

5. In 1953, by using critical information from the work of others and by constructing models of their own, Watson and Crick determined the structure of DNA, including its phosphate-sugar backbone, specific (A-T, G-C) base pairing of purines and pyrimidines.

DNA REPLICATION

1. Replication is the preparation of DNA copies. Because of base pairing, each strand, upon separation, can reproduce the other.

2. Nucleotides are joined by their phosphates and sugars, which form the backbone of the polymer with the nitrogen bases projecting off the side.

3. Synthesis of DNA polymers proceeds from the **5′ end** to the **3′ end**. In its finished form, DNA is a double strand of nucleotides wound into a double helix.

4. Replication is carried out by **replication complexes** (**replisomes**), which include the unwinding enzyme **helicase** and the nucleotide-adding enzyme **DNA polymerase**, as well as other enzymes and proteins.

5. The helix is unwound, the separated strands form replication forks, and new nucleotides are added according to base-pairing rules.

6. The addition of bases to the leading end of a polymer occurs smoothly, one base at a time, but at the lagging end, **Okazaki fragments** must first be assembled in the 5′-to-3′ direction, and then, utilizing the enzyme **ligase**, they are added in.

7. In eukaryotic replication, multiple replication forks work simultaneously. In prokaryotes, only two replication forks form along the circular chromosome, but replication in prokaryotes is much faster.

DNA AND GENETIC INFORMATION

1. Garrod identified metabolic disorders such as alkaptonuria by the presence of abnormal metabolites such as alkaptones in the urine. He determined that such conditions were inherited and surmised that they involved abnormalities in the enzymes of metabolic pathways. He correctly associated the abnormal metabolites with abnormal enzymes and such enzymes with abnormal genes.

Application of Ideas

1. In the prehistory of life, nucleic acids might have come before proteins, or proteins before nucleic acids (this is still being argued). How could purely protein organisms store information or reproduce such information for the next generation? Is there some other way around the problem?

Review Questions

1. Briefly summarize Griffith's observations. What was his conclusion, and what would we add to this conclusion today? (pages 259–260)

2. Describe how the use of radioactive tracers by Hershey and Chase led to the identification of DNA as the hereditary agent of the phage virus. (page 261)

3. If Chargaff had worked with the bacterium *Escherichia coli* only, how would this have affected the discovery of base pairing? Describe what he actually found. (page 263)

4. What did the measurements 2.0 nm, 0.34 nm, and 3.4 nm mean to Wilkins and Franklin? What do they actually represent? (page 263)

5. List three aspects of DNA structure, previously discovered by others, that led Watson and Crick to construct an accurate model of DNA. Which of these led to the prediction of how replication would work? (pages 264–265)

6. Using a simplified drawing, identify each component of a nucleotide and explain where they are bound together. (page 265)

7. List the four nucleotides in DNA, and identify whether they are purines or pyrimidines. How do new nucleotides attach and in what direction is the chain synthesized? (page 265)

8. Explain how a nucleoside triphosphate is added to a growing polymer of DNA. (page 265)

9. Describe the final structure of DNA. Include its geometric form, what holds it together, and some idea of its length. (page 266)

10. Using the terms *hydrogen bonding, pyrimidine, purine,* and the letters G, A, C, and T, explain the manner in which nitrogen bases fit together in the complete DNA molecule. (page 267)

11. Explain why base pairing is so essential to the process of replication. (page 265)

12. List the three components found in a replication complex, and briefly explain what each does. (pages 267–268)

13. In what direction *must* replication proceed? Since replication on each DNA strand proceeds in both directions, what problem does this introduce? How is the problem solved? (pages 267–268)

14. Compare prokaryote and eukaryote replication in terms of replication forks and bubbles. Suggest reasons why the *rate* of replication in prokaryotes may be so much faster than it is in eukaryotes. (page 268)

CHAPTER 13

Genes in Action

CHAPTER OUTLINE

RNA STRUCTURE AND TRANSCRIPTION

Comparing RNA to DNA

Transcription: RNA Synthesis

VARIETIES OF RNA

Ribosomal RNA and the Ribosome

Messenger RNA

Transfer RNA

TRANSLATION: HOW PROTEINS ARE ASSEMBLED

Initiation

Elongation

Termination

Polyribosomes

Free and Bound Ribosomes

KEY IDEAS

APPLICATION OF IDEAS

REVIEW QUESTIONS

The story of the search for the "stuff of heredity" and the unveiling of its molecular structure indeed makes a good tale. But there is more of the story to tell—the part about how genes actually work to produce a phenotype. Here we will tell the rest of the story. We will see how the information stored in genes dictates the formation of proteins, which by their structure or by their enzymatic function determine the phenotype of the organism.

A full description of DNA as the genetic material requires two major points. The first point is that DNA contains all the information needed to produce new cells and even entire individuals, and this information is faithfully replicated prior to cell division. The second major point involves the production of the phenotype based on the genetic information in DNA (Figure 13.1). This is accomplished by the processes of transcription and translation, as we will see.

The discovery that genes are composed of DNA and that their information lies in the linear arrangement of DNA nucleotides was indeed an enormous breakthrough, but it also left many questions unanswered. For example, how exactly are the genetic instructions stated? And how do they determine the order of amino acids in proteins? It didn't take researchers long to determine that DNA is not *directly* involved in making proteins. It couldn't be, because, in eukaryotes, DNA remains in the nucleus while protein assembly occurs in the cytoplasm. So DNA directs protein synthesis by long distance. But how? The answer is by encoding its instructions in molecules of RNA.

FIGURE 13.1
DNA to Proteins
This scheme of information flow is the cornerstone of our understanding of the way in which the hereditary data are expressed as protein. The processes involved are transcription and translation. Transcription is the process by which the information stored in DNA is converted to RNA, while translation (which quite logically involves changing languages—from nucleotides to amino acids) is the process that converts the information from RNA into protein.

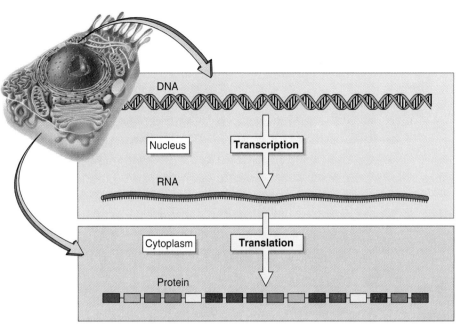

RNA Structure and Transcription

RNA and DNA are very similar molecules. However, their roles in protein synthesis are distinct and dependent on the differences that *do* exist between them.

Comparing RNA to DNA

Let us begin our discussion of RNA with a brief review of its structure (Figure 13.2).

1. The 5-carbon sugar in RNA is *ribose* instead of *deoxyribose*. All this really means is that ribose has a hydroxyl group ($-$OH) attached to the $2'$ carbon instead of a hydrogen ($-$H). (Thus DNA lacks an oxygen there, hence the term *deoxy*.)

2. While both RNA and DNA contain adenine (A), guanine (G), and cytosine (C), the fourth nucleotide base differs in the two molecules. DNA contains thymine (T), whereas RNA contains **uracil** (U), a closely related but slightly different nitrogen base that base pairs with A, the same as thymine.

3. DNA almost always occurs as a double-stranded helix. RNA almost always occurs as a single-stranded molecule, which can have complex, twisted, and folded secondary and tertiary structures.

4. DNA molecules are almost always much longer than RNA molecules—typically.

5. DNA is generally more stable than RNA; it is more resistant to spontaneous and enzymatic breakdown, and damage can be repaired because the opposite strand contains the complementary information. RNA is more reactive partly because of the additional reactive $-$OH side group of ribose, and direct repairs are not possible if it is single-stranded.

6. There are several classes of RNA, each with its own function.

Transcription: RNA Synthesis

The process by which the chemical information encoded in DNA is copied into RNA is called **transcription**. Generally, only one strand of a gene's DNA is transcribed. Therefore the segment of the DNA molecule from which an RNA molecule is transcribed is in a very real sense *the* gene. In the past few chapters, we've been discussing genes as units of information. Now we can begin to tighten our definition and say that a gene is a segment of DNA that (through RNA) specifies the primary structure (sequence of amino acids) in a protein. Most genes are composed of two general parts: a **coding region** that is transcribed into RNA and a **regulatory region** to regulate transcription of the coding region.

The enzymes involved in transcription are **RNA polymerases**, which are among the largest known enzymes. Eukaryotes contain three different types of RNA polymerases, each of which helps form a specific type of RNA.

RNA polymerase can begin assembling chains of new RNA bases only after it identifies and binds to a specific DNA sequence in a regulatory region known as a promoter. A **promoter** is defined as the DNA within the regulatory region that directs the binding of RNA polymerase and the subsequent

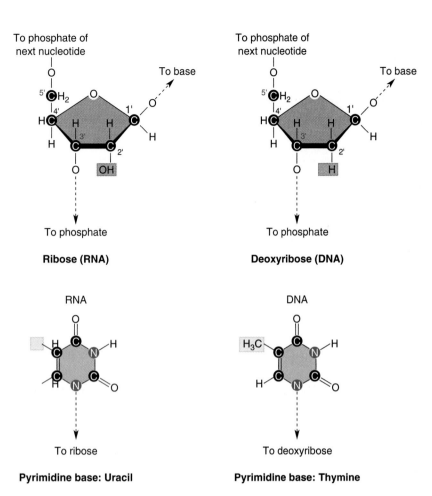

To phosphate of next nucleotide

To base

Ribose (RNA)

To phosphate of next nucleotide

To base

Deoxyribose (DNA)

RNA

To ribose

Pyrimidine base: Uracil

DNA

To deoxyribose

Pyrimidine base: Thymine

FIGURE 13.2
RNA Versus DNA
RNA nucleotides differ from those of DNA in two chemical ways. First, RNA contains the sugar ribose, which contains an — OH group at the 2′ carbon rather than the — H at the 2′ carbon of deoxyribose. Second, in RNA uracil replaces thymine. Uracil and thymine are structurally similar (thymine has a methyl group where uracil has a hydrogen) and both base pair with adenine.

transcription of a coding region. (We will return to promoters in the next chapter.)

Following the binding of RNA polymerase to the promoter, the giant enzyme apparently begins to break the hydrogen bonds of the complementary bases, unwinding the DNA. It unwinds about one full turn of the DNA helix, exposing that segment of unpaired DNA bases. Nucleosides (ATP, UTP, GTP, CTP) can base pair with the exposed DNA strand, and RNA polymerase catalyzes the formation of the RNA polymer. As a segment is transcribed, the opened helix rewinds while a new segment ahead unwinds (Figure 13.3). Interestingly, as the new RNA bases are brought in and base paired with the template DNA strand, they temporarily form a DNA:RNA double polymer, as though a kind of replication were occurring. However, the original DNA helix soon reforms. The RNA, which is now referred to as the **primary transcript**, is displaced by the nontranscribed DNA strand.

Except for the substitution of uracil for thymine, the base-pairing rules and chemical aspects of RNA synthesis are the same as those of DNA replication, and the behavior of RNA polymerase is similar to that of DNA polymerase. The RNA strand is synthesized in the 5′ ⟶ 3′ direction, using ribonucleoside triphosphates (CTP, GTP, ATP, and UTP). By way of summary then, for every C in the template DNA strand, RNA polymerase puts in a G ribonucleotide; for every G, a C; for every T, an A; and for every A, a U. When the process is completed, the new RNA has the same order of bases as the appropriate stretch of the nontranscribed strand of DNA, with U substituting for T (see Figure 13.3).

Transcription of a DNA strand continues until the moving polymerase encounters what is called a **transcription termination signal**. Like the promoter, this consists of a special sequence of DNA bases, but it acts to dislodge the growing RNA strand and to release the RNA polymerase from the DNA and from the RNA.

Many RNA molecules can be transcribed from the same gene simultaneously, for as soon as the first few bases of a sequence have been copied and the RNA polymerase has physically moved out of the way, another RNA polymerase can bind to the promoter and initiate transcription (Figure 13.4).

Varieties of RNA

We have stated, in general terms, that RNA plays a key role in the process of producing proteins. Let's now examine the different classes of RNA to see what specific role each of them

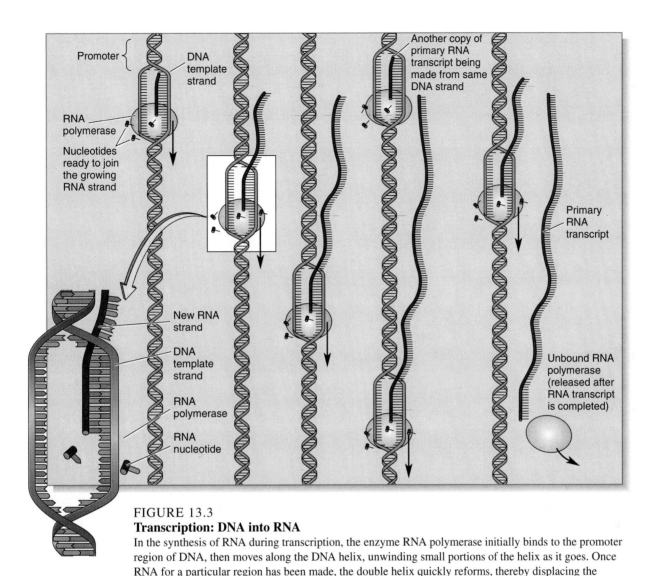

Promoter

DNA template strand

RNA polymerase

Nucleotides ready to join the growing RNA strand

Another copy of primary RNA transcript being made from same DNA strand

Primary RNA transcript

New RNA strand

DNA template strand

RNA polymerase

RNA nucleotide

Unbound RNA polymerase (released after RNA transcript is completed)

FIGURE 13.3
Transcription: DNA into RNA
In the synthesis of RNA during transcription, the enzyme RNA polymerase initially binds to the promoter region of DNA, then moves along the DNA helix, unwinding small portions of the helix as it goes. Once RNA for a particular region has been made, the double helix quickly reforms, thereby displacing the growing single strand of RNA.

plays. Once we have described these players, we will put them all on stage to describe the process itself.

There are three kinds of RNA in most cells: ribosomal RNA (rRNA), messenger RNA (mRNA), and transfer RNA (tRNA). Each is transcribed from DNA, as described previously, by one of the three classes of RNA polymerase. Ribosomal RNA and transfer RNA may be thought of as "bit players," but in their supporting roles they function in the expression of virtually every gene. **Ribosomal RNA** contributes significantly to the structure of ribosomes (the site of protein synthesis, as we will see). **Transfer RNA** is the key intermediary, responsible for bringing the proper amino acid to be put into the protein. **Messenger RNA** is the star. It is in the direct line of information flow between DNA and protein. As its name implies, mRNA carries the coded message that will determine the polypeptide to be produced. The messenger RNA for each gene is unique to that gene, so there are many thousands of different mRNAs.

Ribosomal RNA and the Ribosome

Ribosomal RNA is found in ribosomes, the molecular complexes that translate mRNA into protein. In eukaryotes, ribosomal RNA, unlike other types of RNA, is transcribed exclusively within the nucleolus, whereas the other types of RNA are synthesized throughout the nucleus. In addition, there are multiple copies of the genes for rRNA. Why are so many identical genes needed? It seems that at certain times the demand for rRNA is much greater than at other times. Thus when large amounts of rRNA are needed, a large number of RNA polymerases can travel along copies of the transcribing genes, spinning off strand after strand of rRNA.

What will become three major pieces of rRNA are all transcribed as one very long primary transcript of the rRNA gene. Following its production, the long primary rRNA transcripts are immediately processed by an enzyme to yield the specific shorter strands of ribosomal RNA needed for ribo-

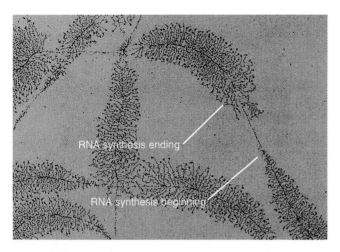

FIGURE 13.4
Simultaneous Transcription
Once an RNA polymerase molecule has begun transcription, there is no reason why a second one cannot begin transcription before the first is finished. In fact (as seen in this electron micrograph), if much RNA is needed in a short time, a gene can become loaded with a large number of polymerase molecules, all transcribing at the same time.

some assembly. In eukaryotes the three forms are called **18S**, **5.8S**, and **28S rRNAs**. A fourth member of the group, **5S rRNA**, is transcribed from a separate gene and prepared outside the nucleolus. ("S" is a sedimentation or density unit used in describing the results of ultracentrifugation and reflects the size and shape of a molecule or particle. Basically, the larger the S value, the larger the particle.) While the various rRNAs will form the skeleton of the ribosome, the remainder will consist of special ribosomal proteins assembled in the cytoplasm. Such proteins enter the nucleus and find their way to the nucleolus, where they join the rRNA. But while ribosomal assembly begins in the nucleolus, it must be completed in the cytoplasm.

Completed ribosomes are made up of two different-sized subunits (Figure 13.5). The smaller **40S subunit** fits into the larger in an elaborate interlocking manner, and when assembled will clamp itself over part of a messenger RNA molecule. The larger, **60S subunit** contains the 28S, 5.8S, and 5S rRNAs, while the smaller, 40S subunit contains the 18S rRNA. Proteins make up about half the ribosomal mass, with the smaller subunit containing some 30 different proteins and the larger containing 45–50, all tightly bound to the rRNAs. The complete ribosome has a "size" of 80S. (Because the S value is an operationally defined unit rather than a reflection of the intrinsic molecular weight, the combination of the 60S and 40S subunits does not give a simple additive S value for the complete ribosome.) Prokaryotic ribosomes, by the way, are somewhat smaller than those of eukaryotes but still consist of a large and small subunit. (In fact, many of the antibiotics used to fight bacterial infections take advantage of the differences between bacterial and eukaryotic ribosomes in order to kill bacteria without harming the eukaryotic host.)

Ribosomes are the only places where proteins are synthesized. The ribosome is where mRNA, tRNA, and the growing polypeptide chain of a protein all work together to synthesize proteins. The ribosome itself is not passive in this process. Its surface has specific attachment sites (specially shaped cavities) that will allow tRNAs and mRNA to be in the proper close contact. There is also a site where the enzyme peptidyl transferase works to form peptide bonds between adjacent amino acids (see Figure 13.5). (You may want to briefly review protein structure in Chapter 3.)

Messenger RNA

As its name suggests, messenger RNA carries the message—in this case, information regarding the sequence of amino acids of the polypeptide to be produced.

mRNA and the Genetic Code. The linear amino acid sequence of every protein that a cell produces is encoded in the DNA of a specific gene. But DNA, as we have noted, does not make proteins directly—it can only direct the synthesis of RNA. In eukaryotes, this RNA then moves to the cytoplasm to direct protein synthesis, while DNA remains in the nucleus. Messenger RNA is the physical link between the gene and the protein. The mRNA molecule is synthesized on DNA and incorporates the information necessary to specify a protein.

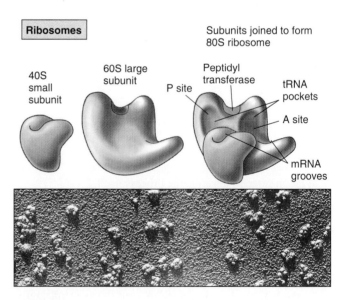

FIGURE 13.5
Ribosome
Ribosomes are composed of two different subunits that are physically joined only during the process of translation. The eukaryotic ribosomes consist of a larger 60S subunit and a smaller 40S subunit. The assembled 80S ribosome has two sites or pockets (identified as A and P) for the binding of tRNAs and a groove between the subunits for the binding of mRNA. Within the ribosome is also the enzyme peptidyl transferase, which actually joins amino acids together.

The information contained in mRNA is written in the genetic code, which is the same for all organisms. Each code word, or **codon**, is made up of three adjacent nucleotides. The sequence of three nucleotides specifies one of the 20 common amino acids (Table 13.1). The sequence GAG in mRNA, for instance, specifies the amino acid glutamic acid. Its nucleotide *complement* in DNA would be CTC.

The information in DNA, mRNA, and proteins is referred to as colinear; that is, there is a linear relationship between the order of nucleotides in DNA and the order of amino acids in the protein (Figure 13.6). The beginning of the mRNA is always its 5′ end, and the 5′ end of the mRNA always corresponds to the beginning or amino terminus of the resulting polypeptide. However, while there is a one-for-one correlation between the DNA and RNA nucleotides, it takes three nucleotides to code for one amino acid.

Since there are four different RNA nucleotides that can occur in any of the three positions of a codon, there are $4 \times 4 \times 4 = 64$ different codons. Three of these, **UAA**, **UAG**, and **UGA**, are **stop codons** that specify the end of a protein, like the period at the end of a sentence. The remaining 61 codons specify the 20 amino acids. Obviously, there are more types of amino acid-specifying codons than there are types of amino acids, so most amino acids are coded by more than one codon. For instance, GGU, GGC, GGA, and GGG all code for one amino acid, glycine. These are called **synonymous codons**.

One codon, **AUG**, is quite special. It can either specify the amino acid methionine in the middle of a protein or serve

TABLE 13.1
The Genetic Code

FIRST BASE	SECOND BASE				THIRD BASE
	U	C	A	G	
U	UUU Phe UUC	UCU UCC Ser UCA UCG	UAU Tyr UAC	UGU Cys UGC	U C
	UUA Leu UUG		UAA Stop UAG Stop	UGA Stop UGG Trp	A G
C	CUU CUC Leu CUA CUG	CCU CCC Pro CCA CCG	CAU His CAC	CGU CGC Arg CGA CGG	U C A G
			CAA Gln CAG		
A	AUU AUC Ile AUA	ACU ACC Thr ACA ACG	AAU Asn AAC	AGU Ser AGC	U C
	AUG Met Start		AAA Lys AAG	AGA Arg AGG	A G
G	GUU GUC Val GUA GUG	GCU GCC Ala GCA GCG	GAU Asp GAC	GGU GGC Gly GGA GGG	U C
			GAA Glu GAG		A G

Amino acid abbreviations: alanine, Ala; arginine, Arg; asparagine, Asn; cysteine, Cys; glutamic acid, Glu; glutamine, Gln; glycine, Gly; histidine, His; isoleucine, Ile; leucine, Leu; lysine, Lys; methionine, Met; phenylalanine, Phe; proline, Pro; serine, Ser; threonine, Thr; tryptophan, Trp; tyrosine, Tyr; valine, Val.

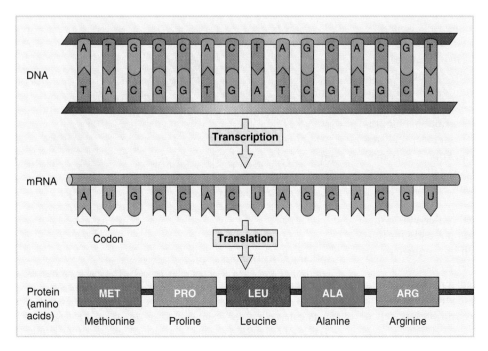

FIGURE 13.6
Colinearity Between DNA, RNA, and Proteins
There is a clear, fundamental relationship among the molecules involved in transcription and translation. They are colinear, in that the nucleotides of DNA and of RNA and the amino acids of the resulting protein are found in the same order, from beginning to end. As you can see, the resulting chain of amino acids is colinear, though not in a one-for-one ratio with nucleotides. We will see that it takes three nucleotides to specify an amino acid.

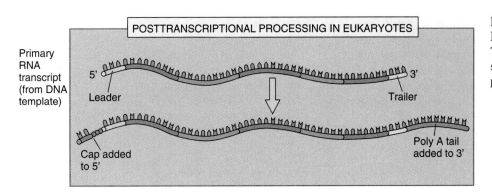

Primary
RNA
transcript
(from DNA
template)

Leader

Trailer

Cap added
to 5'

Poly A tail
added to 3'

FIGURE 13.7
Eukaryotic mRNA
The mRNA of eukaryotes undergoes
several stages of processing in order to
prepare for translation.

as an **initiation** or **start signal**. Ribosomes begin the translation of any mRNA at the first AUG they find in the sequence. Hence all proteins at least start out with a methionine as their first amino acid. However, for some proteins this first methionine is enzymatically removed after the protein is synthesized.

mRNA Structure.   There is more to mRNA structure than the sequence of bases that will specify a protein. In both prokaryotic and eukaryotic mRNAs, there is a portion of the 5' end of the mRNA, before the initiation codon, that is not translated into protein. This is called the **5' leader** (or simply the **leader**). These bases offer the physical space to which ribosomes will bind so that they then can move down the mRNA to the initiation codon and the part of the mRNA that actually encodes the protein. That part of the mRNA that actually codes for the protein can be called the **coding region**. After the termination codons of the coding region, toward the 3' end of the mRNA, comes another series of bases called the **3' trailer** (or simply the **trailer**). The entire mRNA, from the beginning of the 5' leader to the end of the 3' trailer, is transcribed from the coding region of a gene.

In eukaryotes, the primary mRNA transcripts cannot be translated without post-transcriptional modification or RNA processing (Figure 13.7). First, the mRNA must undergo **capping**, a process in which a special methylated version of triphosphate guanine nucleoside is added to the 5' end. Capping apparently aids the mRNA in later binding and positioning on the ribosome. The second post-transcriptional modification consists of adding a series of 50–150 adenines (called a **poly-A tail**) to the 3' end of the molecule. Poly-A tails may help, in some way, to transport the mRNA out of the nucleus and to determine the number of times an mRNA can be translated before it is degraded.

In addition to the 5' leader and the 3' trailer, eukaryotic mRNAs also contain other regions that do not code for protein. These regions are called **intervening sequences** or **introns**. After the primary mRNA transcript has been produced, the introns must be identified and removed, leaving only expressed regions, or **exons**, in the mature mRNA. The removal of introns to produce the mature mRNA is called **splicing** (Figure 13.8).

Splicing is accomplished in the nucleus with the aid of large RNA–protein complexes called **spliceosomes**. Spliceosomes apparently contain many different enzymes and several different kinds of RNA that help with the splicing process, but one particular component caught the attention of molecular

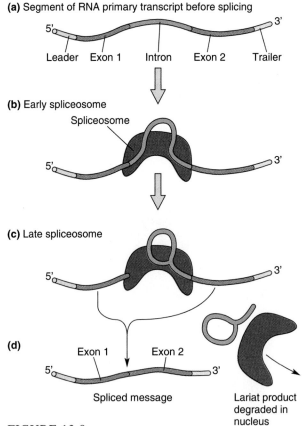

(a) Segment of RNA primary transcript before splicing

Leader   Exon 1   Intron   Exon 2   Trailer

(b) Early spliceosome

Spliceosome

(c) Late spliceosome

(d)

Exon 1   Exon 2

Spliced message

Lariat product
degraded in
nucleus

FIGURE 13.8
Splicing Out Introns
Removal of introns (**a**) is a complex process. The major points are that an mRNA becomes associated with a multiprotein–RNA complex called a spliceosome (**b**) that clips out the intron while holding the ends of the exons in close proximity (**c**). The end result (**d**) is a spliced mRNA and the lariat-shaped remains of the intron.

biologists. This is a short piece of RNA that contains a core base sequence that complements short base sequences found at the two ends of nearly every intron. Earlier researchers decided that splicing must involve pairing between the spliceosome RNA and the RNA at the ends of an intron. This would bring the two ends of adjacent exons together so that they could be covalently bonded together end-to-end. The looped intron could, at the same time, be discarded. We now know that splicing is a two-step process and that it involves several different spliceosome RNAs (see Figure 13.8).

Prokaryotic mRNA is quite different from its eukaryotic counterpart. For one thing, it lacks the methylated cap and the poly-A tail. Also, there are no introns in prokaryotic mRNA. Thus there is no post-transcriptional processing before the message is ready for translation, and the mRNA is ready to be translated as soon as it is made. In fact, translation can even begin on the 5′ end of the mRNA before the 3′ end is transcribed. In addition, it is common for prokaryotic mRNA to be **polycistronic**—to contain coding regions for producing more than one polypeptide. Each of these coding regions contains its own initiation and stop codons, so that separate proteins are produced. We will see in Chapter 14 that grouping these protein coding regions together is a way to ensure equal production of several proteins that are needed at the same time.

Transfer RNA

As we will see, the mRNA carries the coded genetic message to the ribosome, where it is decoded into protein. But the ribosome itself cannot tell one codon from the next. Deciphering the codons, one at a time, is the job of our last type of RNA: transfer RNA, or tRNA. This small RNA molecule is a critical contributor to the success of translation, as it is responsible for translating the language of nucleotides into the language of amino acids. Let's examine just how the molecule accomplishes this feat.

Each tRNA molecule is a relatively short length of RNA, consisting of about 90 nucleotides. In its primary form, when it is first transcribed from DNA, the tRNA is somewhat longer, but before it becomes active it undergoes some post-transcriptional modification. Many primary tRNAs contain introns, which must be excised. In addition, the molecule is "tailored," as special enzymes remove segments from each end. Other enzymes make chemical modifications to some of the bases in special places on the different tRNAs so that the completed molecule contains "exotic" RNA bases in addition to the usual four. Yet another enzyme adds three more nucleotides to the 3′ end of every tRNA, so that all completed tRNAs end with the sequence -CCA (Figure 13.9).

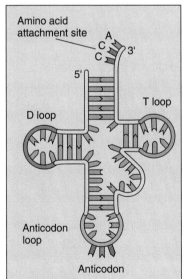
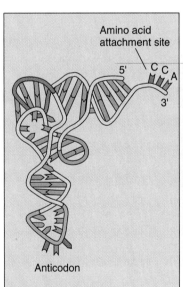
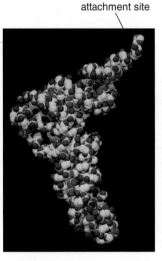
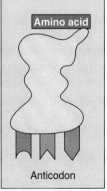

Cloverleaf model showing basic structure

Model showing tertiary structure and folding

Space-filling model

Generalized icon used in this chapter

FIGURE 13.9
tRNA Structure
Transfer RNA molecules are transcribed in the nucleus and are generally about 90 bases long. Some of the bases are enzymatically modified. In its final form, the tRNA molecule has a folded secondary structure caused by extensive base pairing with itself. This folding creates several loops, or unpaired regions, the most important of which is the anticodon loop, which pairs with the codon of the mRNA. The 3′ end of the tRNA is unpaired and always contains the sequence CCA for the attachment of the amino acid. In its tertiary form, we see that the amino acid attachment site and the anticodon loop are at opposite ends of a molecule that is shaped like a three-dimensional upside down capital L, and that the other loops help to form the structure.

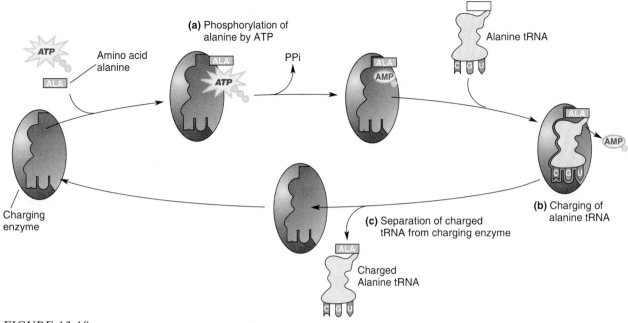

FIGURE 13.10
Charging
Charging is the process by which an amino acid is attached to the proper tRNA. (**a**) In the first step, the energy-rich AMP from an ATP is transferred to the amino acid and a pyrophosphate is released. (**b**) Then the energy in that phosphate bonding is used to form a covalent bond between amino acid and its tRNA and the AMP is freed. (**c**) The charged tRNA is then released from the enzyme.

The mature tRNA molecule is precisely coiled and loops back on itself in a characteristic conformation. The folded tRNA has three loops and a stem, and finally the whole molecule is held in a twisted, upside down L-shaped configuration by hydrogen bonds (see Figure 13.9).

One of the two key features of a tRNA molecule is that it can be linked covalently to an amino acid. **Charging enzymes** (Figure 13.10) attach amino acids to the -CCA 3′ ends of specific tRNAs. As would be expected, there are tRNAs for each of the 20 amino acids. It follows that there are 20 different enzymes and at least 20 different tRNAs (actually more, because of synonymous codons). The charging enzymes recognize each particular type of tRNA and link it to its appropriate amino acid with a high-energy covalent bond. For example, an alanine tRNA charging enzyme binds an alanine tRNA in one of its receptive sites, an alanine amino acid in another site, and (with the aid of ATP) joins alanine to its specific tRNA (see Figure 13.10). The tRNA with its bound amino acid is called a charged amino tRNA.

The second key feature of the tRNA is the **anticodon**. The anticodon is a series of three bases, physically located on the "opposite" side of the mature tRNA from the amino acid attachment site, that can base pair with the codons found in mRNA. Each tRNA has its own special anticodon, which recognizes a specific codon found in the genetic code. Do you see how important tRNA is to the process of translation? One end of the molecule recognizes a codon (written in the language of nucleic acids), while the other end carries the specific amino acids (to write in the language of proteins) that correlates with that codon.

The "recognition" between codon and anticodon is yet another example of the elegance and power of specific base pairing by hydrogen bonds. In this case, U in the codon pairs with A in the anticodon, C pairs with G, and so on. Thus in the example shown in Figure 13.11, the tRNA anticodon 5′-UGC-3′ pairs with the mRNA codon 5′-GCA-3′, which codes for the amino acid alanine. Remember that in all cases (DNA with DNA, mRNA with DNA, or tRNA with mRNA) the strands run in the opposite directions for base pairing to occur.

Translation: How Proteins Are Assembled

Having described all the players in the show, let's now look at how they interact within the cytoplasm to accomplish the process of **translation**: converting the information in mRNA into a protein. Translation is divided into three parts: initiation, elongation, and termination.

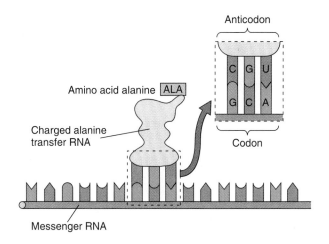

Anticodon

Amino acid alanine ALA

Charged alanine transfer RNA

Codon

Messenger RNA

FIGURE 13.11
Codon and Anticodon
The codon is the series of three bases on the mRNA that encode a specific amino acid (see Table 13.1). The anticodon of the tRNA is a complementary series of three bases that form specific base pairs with the codon. When the correct tRNA is present, complementary hydrogen bonds form between the codon and anticodon, with the end result being the placement of an amino acid (in this case alanine) in its proper position in a protein.

Initiation

Initiation, the start of protein formation, begins when the leading end of the mRNA associates with the small ribosomal subunit. This step is quite precise, and, if all goes well, the initiator codon AUG will have aligned itself in the left-hand pocket, or **P site** as it is called. The other pocket is called the **A site.** Thus the first amino acid to be incorporated into a polypeptide is always methionine. (Or, in prokaryotes, it is a modified methionine called *N*-formyl methionine.) A special

initiator tRNA with a 3′-UAC-5′ anticodon recognizes and pairs with the initiation codon 5′-AUG-3′. It also binds with the small ribosomal subunit in a reaction involving at least three specific initiation proteins and ATP. Only after this **initiation complex** (the three initiation proteins, the smaller ribosomal subunit, a charged methionine tRNA, and the mRNA initiator codon) is formed can the large subunit join the complex to form a functional, intact ribosome (Figure 13.12).

The methionine will form the **amino-terminal** (or **N-terminal**) end of the growing polypeptide. This simply means that the amino group (NH_2) of the first amino acid in a polypeptide will be exposed, while the carboxyl group ($-COOH$) of the first amino acid will react with the amino group of the second amino acid to form a peptide bond. Thus the final amino acid of any polypeptide will have its carboxyl group ($-COOH$) exposed, making up the **carboxy-terminal** (or **C-terminal**) end of the polypeptide. By convention the N terminus is written at the left and the C terminus at the right.

The initiation step is most critical, as it sets the beginning point from which every subsequent codon will be read during the translation process. The genetic code has no commas or any other punctuation to tell the ribosomes which group of three letters to use as codons. The only thing that marks the beginning of the coding region is the initiation codon. After that first AUG, *all* subsequent codons are read in register, three bases at a time.

Elongation

With the formation of the complete initiation complex, protein synthesis has begun. The next question is, how does it grow from this point? How are additional amino acids added to elongate the protein?

Elongation is the process by which all the rest of the amino acids in the protein are joined, in order, during transla-

INITIATION

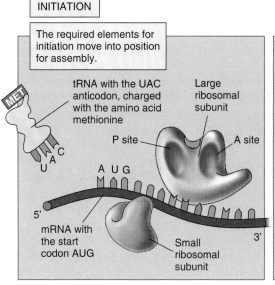

The required elements for initiation move into position for assembly.

tRNA with the UAC anticodon, charged with the amino acid methionine

Large ribosomal subunit

P site

A site

A U G

5′

mRNA with the start codon AUG

Small ribosomal subunit

3′

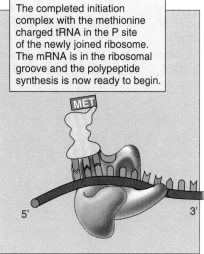

The completed initiation complex with the methionine charged tRNA in the P site of the newly joined ribosome. The mRNA is in the ribosomal groove and the polypeptide synthesis is now ready to begin.

MET

5′

3′

FIGURE 13.12
Initiation
The required elements are mRNA, the two ribosomal subunits, and charged methionine tRNA. The initial event is base pairing between the mRNA initiator, AUG, and the anticodon, UAC, which is found only on methionine tRNA. As the base pairing occurs, the smaller ribosomal subunit joins the RNAs. Only then can the larger subunit move in and complete the polypeptide assembling complex.

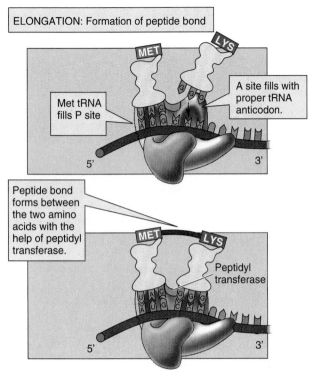

ELONGATION: Formation of peptide bond

Met tRNA fills P site

A site fills with proper tRNA anticodon.

Peptide bond forms between the two amino acids with the help of peptidyl transferase.

Peptidyl transferase

FIGURE 13.13
Elongation: Formation of Peptide Bond
The initial part of elongation involves the juxtaposition of the two charged tRNAs, then the formation of a peptide bond between the two amino acids. The key here is that after the first amino acid is joined to the second, they remain attached to the second tRNA. The first tRNA is ready to be released.

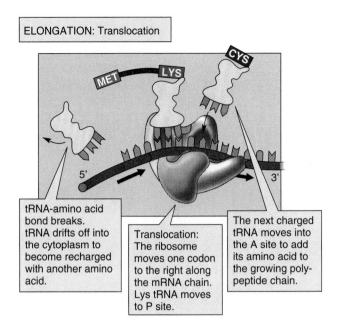

ELONGATION: Translocation

tRNA-amino acid bond breaks. tRNA drifts off into the cytoplasm to become recharged with another amino acid.

Translocation: The ribosome moves one codon to the right along the mRNA chain. Lys tRNA moves to P site.

The next charged tRNA moves into the A site to add its amino acid to the growing poly-peptide chain.

FIGURE 13.14
Elongation: Translocation
After releasing the first tRNA, the ribosome moves one codon along the mRNA, so that the second tRNA (with its growing chain of amino acids attached) now occupies the pocket of the P site. The A site is now open to receive the charged tRNA for the next codon. The process then cycles from peptide bond formation through translocation as the ribosome moves further down the mRNA.

tion. It begins immediately with the charged tRNA whose anticodon complements the next codon of the mRNA.

In Figure 13.13 observe that CCC is the next codon, and the tRNA with the GGG anticodon bearing the amino acid proline joins the complex in the ribosome's A site. During the time that the A site was empty, all sorts of small molecules randomly bumped into and out of the empty site in the ribosome, including any tRNA molecules that might have happened to be in the neighborhood. Sooner or later a charged tRNA wandered in, and it fit so well that it stuck. The good fit resulted from a combination of the shape of the site, which fits all charged tRNAs, and the matching of the anticodon with the codon.

Figure 13.13 shows the two tRNA sites in the ribosome occupied by the methionine and lysine tRNAs. The methionine has been transferred from the stem of its tRNA to the amino group of the proline, as a peptide bond is formed between the carboxyl group of the methionine and the amino group of the lysine. The energy for peptide bond formation was provided by the bond that initially held the amino acid to its tRNA (and thus, indirectly, by ATP in the tRNA charging reaction).

Methionine tRNA, now no longer "charged" with its amino acid, loses its affinity for the ribosome, which only binds to charged tRNAs. The methionine tRNA will soon drift

out of the pocket, eventually to be recharged with another methionine so that it can participate in translation once again.

In Figure 13.14, the lysine tRNA has moved from the pocket of the A site to the pocket of the P site, bringing along with it both the amino acids and the mRNA, which is still bound to its anticodon. This step is called **translocation**. Translocation, you will notice, moves the ribosome three nucleotides to the right along the mRNA. As a consequence, the next codon now lies along the floor of the newly empty pocket of the A site. Soon a charged tRNA will randomly bump into place, and the elongation process will proceed by repeatedly moving down one codon and using a new charged tRNA to bring the next amino acid to the growing protein. This then is why we can compare the ribosome with the head of a tape recorder: the ribosome not only "reads" the mRNA but moves it along. (It is arbitrary whether we visualize the mRNA moving past the ribosome, or the ribosome traveling down the length of the mRNA.)

A question that students often ask about translation is: How does everything know where to go? The scheme just described is indeed straightforward and elegant, but it prompts questions. For example, the movement of the molecules appears to be totally random, and there is every reason to believe that it is. Does this mean then that the base pairing of codon and anticodon is just a matter of chance contact? Of course, no combination but the correct one will work, and all others will be rejected. One way of improving the odds is for a great deal of charged tRNA to be around and in motion. Actually, it seems that most chemical events in cells depend on random motion and accidental but predictable collision.

Termination

Termination is the point at which a translating ribosome reaches a stop codon, which signals the end of the protein coding information. There are three such codons: UAA, UAG, and UGA. Sometimes there are double stops (for example, UAA-UAG), apparently just to be sure that the ribosome gets the idea.

None of the tRNAs have anticodons that are complementary to the three stop codons. Instead, there are specific proteins that occupy the A site once a stop codon has been reached. Without a tRNA in the A site, elongation stops. Then yet another enzyme frees the C-terminal carboxyl group from the last tRNA. Following this, the completed polypeptide is released, and the ribosome falls apart into its two components (Figure 13.15). The entire translation process is reviewed in Figure 13.16.

Polyribosomes

Polyribosomes (also called polysomes) consist of several ribosomes, usually 5–10 (up to 40 in some cells), bound together on an mRNA molecule. They look like little strings of beads (Figure 13.17). One might wonder why they are bound together in groups. It seems that different ribosomes are reading the same mRNA molecule and are spaced along it at appropriate intervals—a minimum of 25 nucleotides apart. Each ribosome will travel the length of the mRNA, from the initiation codon to the termination codon; then each will fall apart and drop off. Meanwhile, other ribosomes will assemble themselves at the initiation codon and begin moving along the mRNA.

Free and Bound Ribosomes

In Chapter 4, we noted that ribosomes occur either floating free in the cytoplasm or bound to membranes. In higher organisms, bound ribosomes are attached to one side of the membranes of the rough endoplasmic reticulum; in fact, their pebbly appearance gives the rough endoplasmic reticulum its name. You may recall that the endoplasmic reticulum, or ER, is part of the membrane system that specializes in storing, transporting, and modifying newly synthesized substances within and outside the cell. Bound ribosomes are also seen in bacteria, but prokaryotes lack an ER; the ribosomes are bound to the inner surface of the plasma membrane itself.

The polypeptides that are produced by ribosomes bound to the ER (or bound to the membrane of a bacterium) experience a different fate from those produced by the free ribosomes. Whereas polypeptides produced by free ribosomes are simply released into the cytoplasm, those produced on the bound ribosomes are moved across the membrane into the

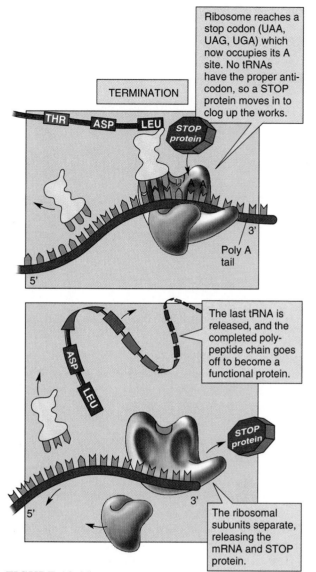

Ribosome reaches a stop codon (UAA, UAG, UGA) which now occupies its A site. No tRNAs have the proper anticodon, so a STOP protein moves in to clog up the works.

TERMINATION

THR ASP LEU STOP protein

3'
Poly A tail
5'

The last tRNA is released, and the completed polypeptide chain goes off to become a functional protein.

ASP LEU

STOP protein

5' 3'

The ribosomal subunits separate, releasing the mRNA and STOP protein.

FIGURE 13.15
Termination
The ribosome has reached a terminator codon (UAA). There is no opposing anticodon, since no AUU-bearing tRNA exists. The amino acid just added will be the last. The "derailing" of the ribosome occurs when special proteins then occupy the A site.

lumen of the rough endoplasmic reticulum (Figure 13.18) (or, in the case of bacteria, into the external environment).

It is thought that bound ribosomes are no different from free ribosomes, and that their presence on the ER is a function of the polypeptide they happen to be producing, rather than any specialization of their own. In other words, if they are producing polypeptides destined to enter the ER, then they become bound to the ER.

How does this happen and how does the polypeptide get from the bound ribosome into the lumen of the ER?

Polypeptides destined for the ER contain at their N terminus a special short segment of amino acids, called a **signal**

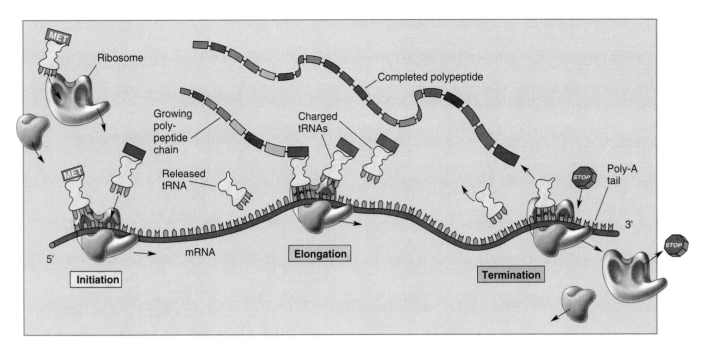

FIGURE 13.16
Scenario of Polypeptide Synthesis
The entire translation process is summarized here, with initiation at the left, elongation toward the center, and termination at the right. While the events seen here appear to be highly ordered, everything happening is thought to be completely random, and the proper interaction of all molecules is really just a chancy affair. The odds are improved by large numbers of charged tRNAs and the amazingly high speed at which events occur.

sequence. Most of the amino acids of the signal sequence are hydrophobic (water-fearing; see Chapter 3), so they will have a natural affinity for the lipid-rich membrane of the ER. The signal sequence acts to identify a specific receptor site on the ER. Once contact is made, the ribosome is bound to the membrane and the polypeptide is actively transported, bit by bit while it is translated, through the membrane into the lumen. There the signal sequence, which is of no further use, is

clipped off by an enzyme called signal peptidase. The growing polypeptide strand continues to move into the ER as fast as it is synthesized.

The evidence for the role of signal peptides and receptor sites is compelling. For example, proteins or polypeptides that are customarily synthesized and used in the free cytoplasm never contain a signal sequence. Furthermore, when polypeptides known to enter the ER for packaging are produced in a test tube rather than in a cell, the proteins always contain the signal sequence. These same proteins normally synthesized in the cell and found in the ER lack the signal peptide, which, as we mentioned, has been enzymatically removed. Geneticists, always eager to apply their own tools toward the solution of such problems, have isolated mutant bacteria and provided some answers. Certain mutant strains of *Escherichia coli* produce polypeptides with faulty signal peptides, and these polypeptides remain in the cell cytoplasm instead of becoming integrated into the plasma membrane as they normally would. In addition, geneticists have succeeded in hybridizing membranal proteins containing the signal peptide with cytoplasmic proteins, and these hybrids find themselves inserted into the membrane.

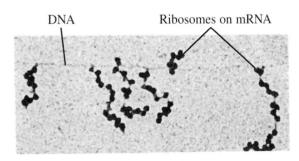

FIGURE 13.17
Polysomes in a Prokaryote
Polyribosomes (or polysomes) are shown in this electron micrograph. The amount of protein synthesis is greatly increased when several ribosomes read the mRNA code at the same time. In *Escherichia coli*, polyribosomal translation is so rapid that ribosomes often move along mRNA that is still being transcribed along the chromosome.

We can now see how transcription and translation, processes by which the information stored as genes in DNA becomes expressed as proteins, are accomplished. By producing proteins that are responsible for phenotypes, translation finishes a story that was begun when Mendel first gathered data on the inheritance of traits.

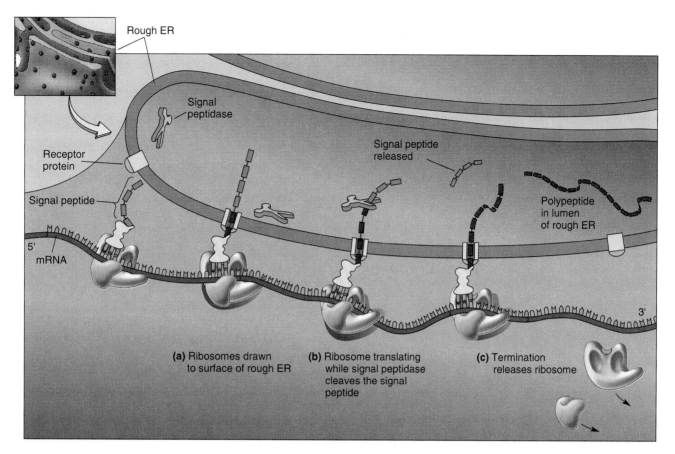

FIGURE 13.18
Bound Ribosomes
Recent discoveries presented in this scenario strongly suggest that the ribosomes studding the rough endoplasmic reticulum are bound there in a complex process that at least partially involves the specific polypeptide that they are synthesizing. A free-floating ribosome has begun the synthesis of a polypeptide. (**a**) At its leading (N-terminal) end, the polypeptide contains a signal peptide that participates in the binding to a specific surface receptor protein (**b**) on the membrane of the rough ER. Once bound, the signal peptide is drawn into the rough ER. A bound ribosome continues translating its polypeptide, which is continually drawn into the lumen of the rough ER. (**c**) Within the lumen, an enzyme called *signal peptidase* cleaves the signal peptide from the growing polypeptide. Completed polypeptides accumulate within the rough ER, which will later form vesicles of the Golgi body where final protein-forming modifications will occur. Once a polypeptide nears completion and mRNA reaches its stop codons, termination occurs and the ribosomal subunits separate and are freed from the rough ER.

Key Ideas

RNA STRUCTURE AND TRANSCRIPTION

1. RNA differs from DNA in that it contains the sugar ribose, substitutes uracil for thymine, is usually single-stranded, breaks down more easily, and exists in several forms.

2. In **transcription**, **RNA polymerase**, the principal enzyme, binds to a DNA **promoter** region, the helix unwinds, and, proceeding in the 5′-to-3′ direction, RNA triphosphate nucleosides are paired to the exposed DNA bases. U is always substituted for T.

3. Transcription ends when RNA polymerase encounters a **termination signal**. Multiple transcription along a DNA strand is common.

VARIETIES OF RNA

1. The three kinds of RNA are **ribosomal (rRNA)**, **messenger (mRNA)**, and **transfer (tRNA)**. All must undergo modification from the precursors to the final mature form.

2. Ribosomal RNA makes up most of the ribosome, a two-part, interlocking organelle that is the site of protein synthesis. Each ribosome has attachment sites for mRNA and tRNA.

3. In eukaryotes, the long primary mRNA transcript undergoes post-transcriptional modification, where noncoding segments called **intervening sequences**, or **introns**, are removed, leaving only coding segments called **exons.** Introns are absent in prokaryotes.

4. Eukaryotic mRNA contains a **capped** or methylated 5′ end, a **leader**, a **trailer**, and a **poly-A tail**. Prokaryotic mRNA lacks the cap and tail and can be **polycistronic** (contains several cistrons, polypeptide coding regions).

5. The linear arrangement of bases in mRNA constitutes the genetic code. The amino acids are specified by **codons—** nucleotides in groups of three. There are 64 possible codons, so most amino acids have more than one (**synonymous**) specifying codon. One codon, the initiation codon, specifies both *start* and an amino acid, while three codons specify *stop.*

6. **Charging enzymes** bond tRNAs to their specific amino acids. Each tRNA has a specific amino acid binding site, a ribosomal binding site, and an **anticodon** loop. The anticodon matches a codon on mRNA, assuring that the amino acid carried by a tRNA will be inserted correctly in the polypeptide.

7. The primary tRNA transcripts are first tailored, then folded into a cloverleaf secondary shape, and finally folded again into a tertiary L shape.

TRANSLATION: HOW PROTEINS ARE ASSEMBLED

1. Translation includes **initiation, elongation,** and **termination**.

2. Initiation requires the smaller ribosomal subunit, an initiation tRNA (with a 5′-CAU-3′ anticodon), and the mRNA initiator codon (5′-AUG-3′), all of which form the ribosomal **initiation complex**.

3. When each component is in place, the larger ribosomal subunit joins the complex, and a second amino acid can be inserted.

4. Polypeptides in eukaryotes all begin with methionine, and each has an **N-terminal** (NH_2) end and a **C-terminal** carboxyl (COO^-) end.

5. The elongation of polypeptides occurs through **translocation—** the formation of the peptide bond and the movement of a tRNA from the A to the P site.

6. As a polypeptide grows, a charged tRNA whose anticodon matches the mRNA codon in the A pocket becomes attached. A peptide bond forms between its amino acid and the last one in the polypeptide above, and translocation occurs again.

7. Following translocation, the tRNA in the P site is released and drifts away to be recycled.

8. When the ribosome reaches a chain termination (**stop**) codon, proteins block the pockets, and the final tRNA is released along with the completed polypeptide.

9. **Polyribosomes**, or **polysomes**, occur in clusters on one mRNA molecule, where they carry on simultaneous translation.

10. Bound ribosomes are located along the rough endoplasmic reticulum (and on the plasma membrane of prokaryotes). Their polypeptides enter the ER.

11. Polypeptides destined to enter the ER are tipped by a signal sequence. The growing polypeptide enters the lumen, where its signal sequence is removed. As the polypeptide grows, it binds the ribosome to the ER.

Application of Ideas

1. It has recently been shown that UGA specifies tryptophan and CUA codes for threonine in the mitochondrial translation systems of yeasts and hamsters. What do these codons usually specify? Comment on the significance, if any, of these minor departures from the code.

Review Questions

1. List four ways in which RNA differs from DNA. (page 277)

2. If the sequence of nucleotides in a transcribed strand of DNA was 5'-A-A-G-C-C-T-T-A-G-G-C-A-3', what would be the sequence of nucleotides in the RNA transcript? (pages 277–278)

3. Briefly describe the process of transcription. Include mention of the primary enzyme and the role of the promotor and termination signal. (pages 277–278)

4. Describe the organization of the ribosome as we now see it. (pages 279–280)

5. Briefly discuss what happens to eukaryotic mRNA during its post-transcriptional modification. What are introns and exons? (pages 282–283)

6. Describe the three regions of the mature eukaryotic mRNA transcript. In what three ways does prokaryotic mRNA differ? (pages 281–282)

7. Describe the organization of the genetic code. What is a codon? How many codons are there? Since only 20 codons are required, what happens to the extras? In what molecules do we find the code written? (pages 281–282)

8. List the steps and the molecules involved in the charging of a tRNA. (pages 283–284)

9. Briefly describe the form taken by tRNA and identify key regions. (page 284)

10. List the different elements of the initiation complex. Explain how they get together to initiate translation. (page 285)

11. Using a simple drawing, show your understanding of chain elongation by illustrating the key steps involved. (page 286)

12. Briefly explain how chain termination occurs. (page 287)

13. What are polysomes? How does their presence affect the efficiency of protein synthesis? (page 287–288)

14. Are bound ribosomes really part of the ER, as they seem to be? Explain fully. (pages 287–289)

15. Briefly describe the role of signal sequences. (pages 288–289)

CHAPTER 14

Gene Regulation

CHAPTER OUTLINE

PROKARYOTIC GENE REGULATION: THE OPERON

Inducible Operon for Digesting Lactose

Repressible Operon for Synthesis of Tryptophan

EUKARYOTIC GENE REGULATION

Opportunities for Regulation

Control of Transcription: Inducible Genes

DNA Packaging and Chemical Modification

Essay 14.1 Homeotic Genes and the Homeobox

Post-transcriptional Regulation

KEY IDEAS

APPLICATION OF IDEAS

REVIEW QUESTIONS

*M*ost organisms contain thousands of genes: even the lowly bacterium *Escherichia coli* has some 2500 different genes. But in all species, most of the genes, most of the time, are shut down. Only a small portion are busy transcribing. This, after all, makes sense. Since most cells in an individual contain the same genes, if all those genes were transcribing at once, every cell would make the same products; and all cells would be the same. How then could a muscle cell be different from a nerve cell? Indeed, although any cell *can* make just about any proteins that any others *can* make, it makes only certain ones, and so cells, tissues, and organs can become specialized for their different roles in the life of an organism.

The genetic adjustments to specific cellular roles, as we will see, are even further refined by the various rates at which genes are transcribed. Add to this increasingly complex picture the fact that cells must be able to respond to changes in their environment by adjusting the expression of their genes, and the stage becomes set for us to examine the processes by which genes are selectively and precisely regulated. Indeed, without the precise coordination of genes, cellular metabolism would be more like biochemical chaos than ordered enzymatic pathways.

We have seen in earlier chapters that enzyme products can be regulated through feedback inhibition. (As a product builds up, the process that produced it is slowed.) This kind of pathway regulation is virtually instantaneous and very effective. However, feedback inhibition is somewhat wasteful as a long-term strategy. If the enzymes in a pathway are needed only at certain times, it is much more efficient to synthesize those enzymes only when they are needed, saving the resources and energy that would otherwise be required for constant transcription and translation.

In this chapter we will explore some of the ways cells regulate the expression of their genes. We will begin with discussions of the ways in which prokaryotic cells regulate genes, then move on to discussions of eukaryotic gene regulation. Before we get into it, though, you should be aware of some of the important terminology. As shown in Figure 14.1, we note that the "gene" is actually composed of two basic parts: a **coding region** that codes for the mRNA and protein product, and a **regulatory region** that controls transcription of the coding region. We will use

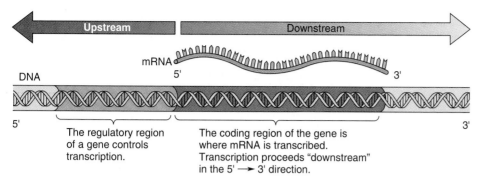

FIGURE 14.1
General Gene Structure and Terminology
Most genes consist of two separate parts: a coding region that actually is transcribed into mRNA and a regulatory region that controls the process of transcription. The position of the 5′ end of the mRNA marks the boundary between the regions. Movement toward the 3′ end of the mRNA is in the "downstream" direction, while movement in the opposite direction is "upstream."

the convention that orients the coding region such that the 5′ end of the resulting mRNA is to the left, and the 3′ (naturally) to the right. It is this orientation of the mRNA that provides directional references for terminology used to refer to parts of genes. In most cases, and in all the cases that we will examine, the regulatory region of the gene is found at the 5′ end of the coding region. The term **promoter** has become accepted as another way to refer to the regulatory region, though it is sometimes used to refer only to that portion of the regulatory region that actually binds the RNA polymerase to start transcription (see Chapter 13). Also, because RNA polymerase transcribes genes such that the mRNA grows from 5′ to 3′, the direction toward the 3′ end of the mRNA is called "downstream" and toward the 5′ end of the mRNA is "upstream." Always keep this orientation of the mRNA in mind, and you will avoid confusion regarding directionality within genes.

We should also be aware that not all genes are highly regulated. After all, some enzymes and proteins are needed in all cells at all times, such as those of the glycolytic pathway. So the genes for these enzymes are continuously transcribed, and are referred to as **constitutive**. It is important to note also that not all constitutive genes are transcribed at the same rate. For any gene, the *rate* of transcription is determined by the relative ability of the promoter to bind RNA polymerase (Figure 14.2).

Prokaryotic Gene Regulation: The Operon

Bacterial gene systems provided our only early insights into the regulation of transcription. Even though we now have some ideas as to how eukaryotic genes are regulated (as we'll see in the next sections), we still better understand many of these processes in bacteria.

There are two recurring themes that characterize coordinated gene regulation in bacteria. The first is that the coding regions that produce the enzymes of the various steps in a single metabolic pathway can be grouped together, occupying adjacent segments of DNA and all under the control of a single regulatory region. Since all the coding regions for the pathway are under the control of that one regulatory region, the transcription of all the enzymes in the pathway is automatically synchronized. The second underlying theme is that transcription of genes can be prevented by the binding of a *repressor* to a certain area of the regulatory region. A **repressor** is a protein that binds to DNA to inhibit transcription. If a repressor is bound to the regulatory region, no transcription occurs. If the repressor is not bound to the regulatory region, transcription may occur, and the enzymes encoded by the grouped coding regions are all transcribed at the same time. In fact, all the coding regions are transcribed into one long mRNA that is referred to as *polycistronic* (see Chapter 13).

In 1961, two scientists at the Pasteur Institute in Paris, Francois Jacob and Jacques Monod, did a series of experi-

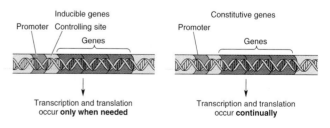

FIGURE 14.2
Gene Regulation
In inducible genes, the protein products are produced only when needed. Constitutive genes, on the other hand, produce their products at a controlled but unchanging rate.

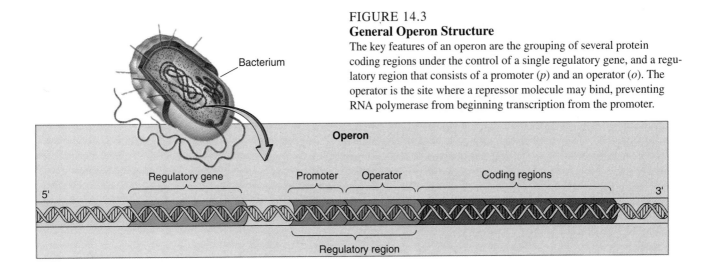

FIGURE 14.3
General Operon Structure
The key features of an operon are the grouping of several protein coding regions under the control of a single regulatory gene, and a regulatory region that consists of a promoter (*p*) and an operator (*o*). The operator is the site where a repressor molecule may bind, preventing RNA polymerase from beginning transcription from the promoter.

ments that resulted in the combination of these two themes into a model for gene regulation called the *operon*. An **operon** is a set of coding regions (usually encoding enzymes involved in a single biochemical pathway) clustered together under the control of a single regulatory region, and transcription is regulated by the presence or absence of a repressor protein bound to a segment of the regulatory region called the **operator** (hence the name "operon"). All these basic themes are presented in Figure 14.3.

Let's examine two operons and see just why they are so effective in gene regulation.

Inducible Operon for Digesting Lactose

Escherichia coli can grow well with glucose as the only source of carbon, which is a good thing because glucose is so common in the intestine where *E. coli* normally makes its home. However, the bacterium can also utilize another sugar, lactose (milk sugar). The enzymes that act on lactose are usually not made by *E. coli,* because the high energy cost of protein synthesis would only be wasted if the enzymes were not used. Only when lactose becomes the most abundant sugar in the environment does the bacterium even initiate the production of the enzymes that will digest lactose.

It takes three enzymes to allow *E. coli* to digest lactose: a beta-galactosidase to cleave the lactose, a permease to bring the lactose into the cell more efficiently, and a trans-acetylase whose function is not clearly defined. The coding regions for these three enzymes are called *z, y,* and *a* and are linked together under the control of a single regulatory region. This complex is called the **lac operon** (Figure 14.4). The *lac* operon is referred to as an **inducible** operon because it is normally not transcribed but can be *induced* (turned on) when an **inducer** molecule is present. In this case, the inducer molecule is lactose.

Linked quite closely to the *lac* operon is a gene called the *i* **gene**, which makes the *lac* repressor protein. The *i* gene con-

stitutively transcribes mRNA for the repressor protein at a low rate, resulting in a constant level of about ten repressor protein molecules per cell. The key characteristic of the repressor molecule that allows it to act as a regulatory protein is that it can bind to either the operator DNA or lactose, but not both at the same time. The binding of a repressor protein to either DNA or lactose is reversible, so that at any given moment the repressor is bound most often to whichever of the two it finds in highest concentration.

By itself, the *lac* repressor protein has a strong affinity for the operator sequence of the *lac* promoter. As mentioned before, when the repressor is bound to the operator, no transcription occurs. You might think of this as a simple blocking mechanism, an immovably large repressor boulder sitting on the track of the RNA polymerase train.

If lactose levels rise to high concentrations, the lactose molecules will quickly bind to the free repressor proteins in the cell. Also, when the repressor that is bound to the operator releases (because of the reversibility of the binding), chances are that the repressor will bind to one of the many lactose molecules rather than to the one operator. The binding of repressor to lactose is also reversible, but when lactose levels are high the chances are that the repressor will bind another lactose rather than the operator. The net effect is that the operator is no longer occupied by the repressor. Thus transcription can proceed from the promoter through to the coding regions—at precisely the time when it is needed most.

The coding region of the *lac* operon is transcribed as a single polycistronic mRNA, which is translated into the three lactose metabolizing enzymes described above. As the enzymes become available, they begin the process of digesting the lactose for carbon and energy. They will continue to be produced as long as there is enough lactose present in the environment to keep the repressor molecules from binding to the operator.

But by their very nature, the *lac* operon enzymes remove lactose from the environment. Thus if the external supply of lactose diminishes or disappears, these enzymes will soon

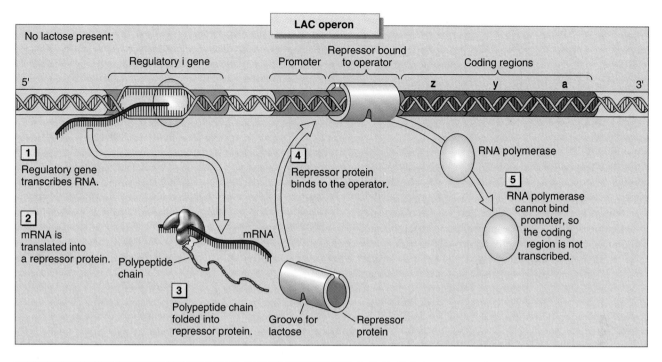

LAC operon

No lactose present:

Regulatory i gene — Promoter — Repressor bound to operator — Coding regions

5′ — z — y — a — 3′

1 Regulatory gene transcribes RNA.

2 mRNA is translated into a repressor protein.
Polypeptide chain — mRNA

3 Polypeptide chain folded into repressor protein.
Groove for lactose — Repressor protein

4 Repressor protein binds to the operator.

RNA polymerase

5 RNA polymerase cannot bind promoter, so the coding region is not transcribed.

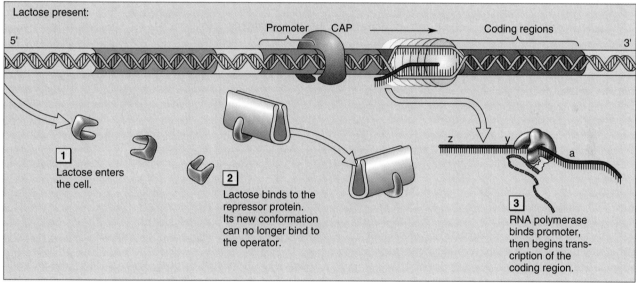

Lactose present:

Promoter — CAP — Coding regions

5′ — 3′

1 Lactose enters the cell.

2 Lactose binds to the repressor protein. Its new conformation can no longer bind to the operator.

z — y — a

3 RNA polymerase binds promoter, then begins transcription of the coding region.

FIGURE 14.4

The *lac* Operon: The Jacob–Monod Operon Model

The production of the three inducible enzymes responsible for the metabolism of lactose is under control of a system known as the *lac* operon. The operon consists of three principal parts, as shown in the diagram: the regulator gene (*i*), the promoter/operator region, and the coding regions for *z, y,* and *a.* The repression of the structural genes is accomplished by an interaction between regulator gene *i* and the operator. The regulator gene transcribes messenger RNA that is translated into a repressor protein, which binds to the operator, blocking the action of RNA polymerase. Thus the transcription of mRNA in the coding region is inhibited. When lactose enters the cell it acts as an inducer, tying up the repressor protein. Transcription is now allowed, but very little takes place, unless the catabolite activating protein (CAP) is also present.

digest all the lactose within the cell. Because of the reversibility of the binding between lactose and the repressor, as the concentration of lactose gets low, even the lactose molecules that were bound to repressor molecules can be degraded. This action frees up the repressor molecules to once again bind to the operator and shut down transcription. Thus. when the enzymes are no longer needed (because the lactose is depleted), the operon ceases transcription.

This level of regulation by the repressor is an elegant response to the need for transcription of the *lac* operon only during the times that lactose is available to be metabolized. However, the situation is not quite as simple as it seems, and the *E. coli* regulatory system is even more subtle and elegant than we have described so far. Basically, the presence of lactose itself is not reason enough to switch over to lactose metabolism, nor, as it turns out, is it sufficient to induce the *lac* operon. Lactose is a complex sugar and it makes energetic sense to utilize glucose, a simpler sugar, when there is a choice. Thus the bacterium should not waste energy making lactose-metabolizing enzymes when there is plenty of glucose present. The *lac* operon then should be induced only when there is lactose present *and* no glucose around.

This additional adaptive level of control is accomplished by yet another protein–DNA interaction in the promoter sequence. The *lac* promoter by itself binds RNA polymerase quite poorly. Thus, even if the repressor is not bound to the operator, transcription does not occur readily. Therefore the mere presence of lactose does not induce much transcription. There is, however, an accessory protein that can help the promoter bind RNA polymerase efficiently. This accessory protein is called **CAP, catabolite activator protein**. As opposed to the repressor (whose binding has a negative effect on transcription), CAP is a positive regulatory protein. When it binds to the promoter, it greatly stimulates the interaction of RNA polymerase and thus transcription. But CAP binds to the promoter only when there is no glucose present in the cell. How does CAP know whether there is glucose around?

When the level of glucose in the cell falls to a low level, there is an increase in the level of a molecule called **cyclic AMP (cAMP)**, a metabolic derivative of ATP. cAMP will bind to CAP, and only then will CAP bind to the *lac* promoter and stimulate transcription. The bottom line is that the *lac* operon will be transcribed only when lactose is present *and* there is no glucose.

Thus we have seen that the *lac* operon is inducible only under conditions that make outstanding energetic sense, and that the induction is accomplished by the interaction of regulatory proteins that have the dual capacity to respond to the cellular environment (in this case, by binding to lactose or cAMP) and to bind to DNA. We have already seen examples of both positive and negative control exerted by proteins. The repressor exerts negative control by blocking RNA polymerase, while CAP exerts a positive influence by stimulating the binding of RNA polymerase.

Repressible Operon for Synthesis of Tryptophan

This same kind of metabolic logic that applies to the *lac* operon can work in an opposite manner. We might imagine a situation when the products of an operon are normally needed, and only under unusual circumstances would the operon need to be shut down. A repressor protein can accomplish this kind of regulation as well.

Escherichia coli normally must manufacture its own tryptophan (one of the amino acids) because tryptophan is rarely found in its environment. The biosynthetic pathway for the production of tryptophan consists of five enzymatic steps. The coding regions for each of the enzymes in the pathway are

FIGURE 14.5
The *trp* Operon
The repressible tryptophan (*trp*) operon works in the opposite way of the inducible lactose operon. In this system, cells grown in the absence of the amino acid tryptophan continually produce the enzyme tryptophan synthase, which is essential to synthesize the amino acid tryptophan from precursor molecules. Although the regulator gene produces the repressor protein, it cannot bind the operator. Therefore the system remains turned on with transcription normally occurring. If the cells are fed the amino acid tryptophan, the system immediately shuts down by ceasing transcription. Tryptophan joins the repressor protein to form a tryptophan–repressor complex. This complex, in turn, binds to the operator, blocking the action of RNA polymerase along the promoter. This shuts down transcription. As you can see, this is a conserving process. Why produce the enzyme when its product is abundant?

located within the polycistronic ***trp* operon** (Figure 14.5). The promoter of the *trp* operon has a relatively high affinity for RNA polymerase and needs no additional activating protein (like CAP) to efficiently transcribe the operon. The operon is normally actively transcribed in order for the cell to produce enough tryptophan for its use. But if tryptophan appears in the environment, it makes sense for the bacterium to utilize the free tryptophan rather than invest in the enzymatic work to produce it internally. The trick then is to inactivate the operon when tryptophan is present.

This is accomplished by a twist of the repressor theme. Much like the *lac* repressor, there is a constitutive gene near the *trp* operon that produces a constant but small amount of *trp* repressor protein. But unlike the *lac* repressor, the *trp* repressor is, by itself, incapable of binding to the *trp* operator. But when there is excess free tryptophan around, the tryptophan acts as a corepressor and binds to the repressor protein. The tryptophan–repressor complex acquires the ability to bind to the operator and shut down transcription.

Thus when tryptophan is present in the surrounding environment, the bacterium makes use of that free source of the amino acid. If the tryptophan levels decline, the repressor is less likely to be bound to a tryptophan molecule, the operator site is freed, and transcription can begin again. Once again, a repressor molecule offers the opportunity for finely tuned regulation—the more tryptophan available, the less the operon is transcribed; the less tryptophan available, the more the operon is transcribed. All of this carefully balanced biochemical regulation is accomplished by the properties of the *trp* repressor protein.

The fundamental characteristics of the repressor molecules are the distinguishing features of the two basic types of operons. These features are summarized in Figure 14.6.

We have now seen that bacteria have evolved fairly complex and subtle ways to regulate the expression of certain genes. The most obvious is the grouping together of coding regions for pathway enzymes and their coordinated regulation by way of operons. This type of physical arrangement is

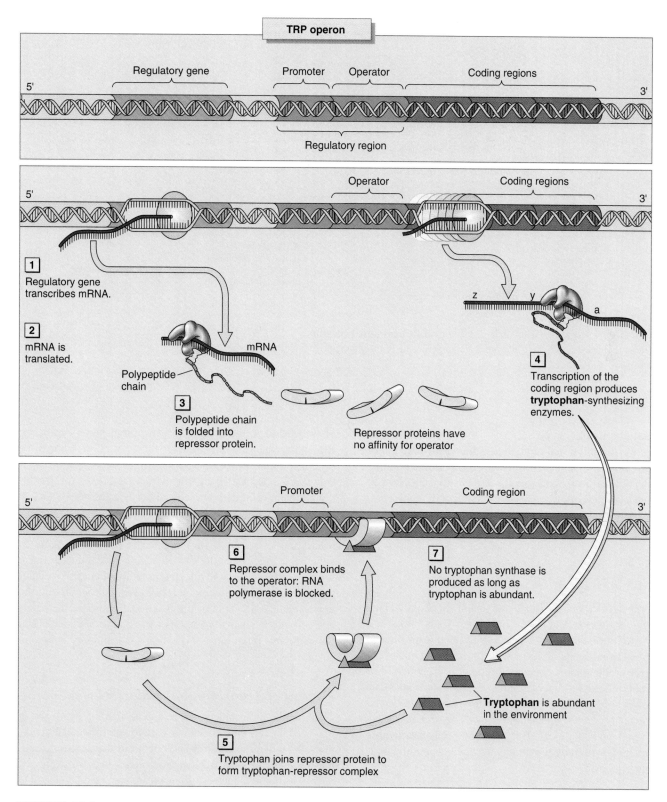

TRP operon

Regulatory gene

Promoter   Operator

Coding regions

5'   3'

Regulatory region

Operator   Coding regions

5'   3'

1 Regulatory gene transcribes mRNA.

2 mRNA is translated.

z   y   a

mRNA

Polypeptide chain

3 Polypeptide chain is folded into repressor protein.

Repressor proteins have no affinity for operator

4 Transcription of the coding region produces **tryptophan**-synthesizing enzymes.

Promoter   Coding region

5'   3'

6 Repressor complex binds to the operator: RNA polymerase is blocked.

7 No tryptophan synthase is produced as long as tryptophan is abundant.

Tryptophan is abundant in the environment

5 Tryptophan joins repressor protein to form tryptophan-repressor complex

FIGURE 14.5

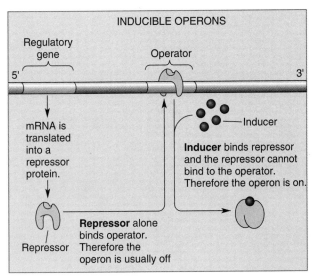

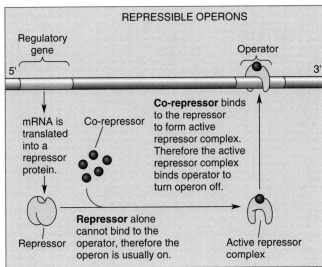

FIGURE 14.6
Distinguishing Characteristics of Inducible and Repressible Operons
The differences between inducible and repressible operons lie in the characteristics of the repressor and
the things to which it binds.

apparently limited to the prokaryotes. However, the principal features of transcriptional regulation—the negative action of repressor molecules and the positive action of stimulating factors such as CAP—are more generalized phenomena. We will see some examples of the eukaryotic version of these processes in the following section.

Eukaryotic Gene Regulation

Evolution had a relatively simple problem in the design of bacterial gene regulation. But multicellular eukaryotes consist of a variety of differentiated cell types with different capacities to transcribe different sets of genes. This immediately makes the regulatory problem more complex. In addition, each individual eukaryotic cell can be under the same demands and constraints that we've discussed for prokaryotic cells. There are the metabolic demands of energy acquisition as well as other responses to environmental conditions. But eukaryotic cells face a more complex task, especially if the individual cell is part of a multicellular organism.

Opportunities for Regulation

In prokaryotes, the mechanisms for gene control are elegant but fairly direct; that is, when the messenger RNA is made, the protein is translated immediately. In fact, the protein translation can begin even before the mRNA is completely assembled.

The presence of the nuclear membrane in eukaryotes necessitates the completion of transcription and the transport of the mRNA to the cytoplasm before translation can occur. But rather than an impediment, these additional processes offer a complete set of new opportunities to control the eventual expression of a gene. After all, we don't see the phenotypic effects of a gene just because it is transcribed; we see them only after an active protein is formed.

Figure 14.7 outlines the steps in the expression of a eukaryotic gene. Each step represents a possible point at which the process could be selectively blocked—in a real sense, regulated. We will take a look at these possible regulatory levels in the following sections. We will begin with the direct regulation of transcription, a situation where the parallels with the prokaryotic operons will be most apparent. Then we will move on to those levels that are seemingly unique to eukaryotes.

Control of Transcription: Inducible Genes

There are many instances where a gene need be expressed in only certain cell types or under certain environmental conditions. We will discuss two eukaryotic genes that serve as examples of selective gene transcription: genes that are expressed only in response to increases in environmental temperature and genes that are expressed only in certain cells in highly specialized organs or tissues.

Heat Shock Genes.   One of the well-studied eukaryotic gene regulatory systems involves a general cellular response to the environmental temperature. Most cells induce, or express, a specific set of genes, the **heat shock protein (*Hsp*)** genes, when they experience a sudden shift to elevated temperatures.

FIGURE 14.7
Opportunities For Gene Regulation in Eukaryotes
This chart points out the several steps in the production of a eukaryotic protein where regulation may occur.

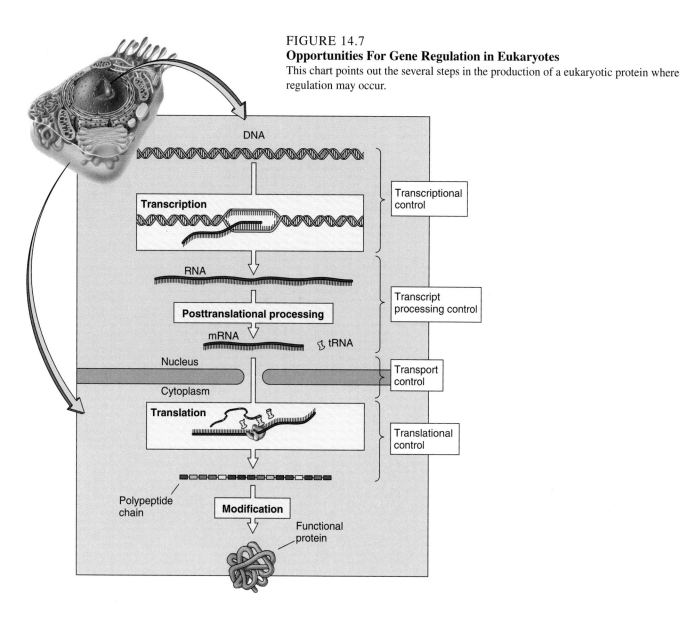

While the precise mechanisms are not well understood, the products of these genes provide important protective features that allow the cells to continue living at temperatures that would potentially destroy many biochemical processes. Let's examine the details of the thermal induction of a particular *Hsp* gene from *Drosophila* as an example.

As we might expect from their name, *Hsp* genes are not usually transcribed but are very rapidly induced within minutes of a rise in cellular temperature of a few degrees. This induction of gene activity involves protein interactions with the DNA of the promoter, which (like the prokaryotic situation) lies just upstream of the coding region of the gene (Figure 14.8). There is an area of the promoter, located about 35 bases upstream from the beginning of the mRNA, called the **TATA box**, which serves as the binding site for RNA polymerase. It gets its name from the fact that its sequence consists of the bases TATAAA. The TATA box appears to be a pervasive feature of many, if not all, genes. The TATA box

apparently serves a fairly general function in all eukaryotic genes, to bind specific proteins called **TATA binding proteins** (**TAB proteins**) that, in turn, help direct the binding of RNA polymerase. While the presence of these TAB proteins is a necessary prerequisite for transcription, TAB proteins cannot work alone to stimulate RNA polymerase.

In the *Drosophila Hsp* genes, the TAB proteins appear to be bound to the TATA box at all temperatures. This reinforces the idea that their presence alone is not sufficient to accomplish transcription but that they do act to set the stage for transcription. When the cells experience elevated temperatures, a second protein, the **heat shock factor**, or **HSF**, binds to a specific sequence (called the **heat shock element**) of the promoter DNA, right next to the TAB protein (see Figure 14.8). The binding of both the HSF and the TAB proteins stimulates RNA polymerase to transcribe the gene. If the cells return to a lower temperature, the HSF will no longer bind to the promoter, and transcription of the *Hsp* gene ceases.

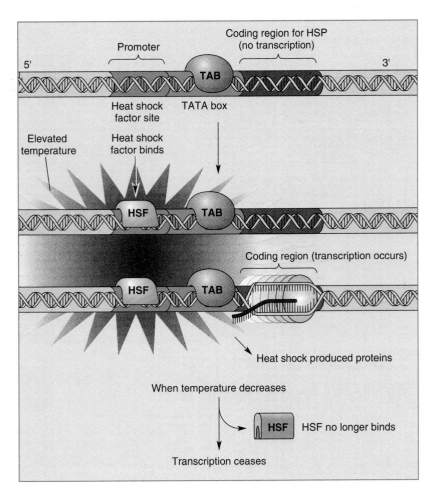

FIGURE 14.8
Heat Shock Protein (*Hsp*) Gene Structure
The regulatory region of a heat shock protein (*Hsp*) gene consists of two parts: a TATA box and the *heat shock element* where the heat shock factor binds to stimulate transcription by RNA polymerase. Transcription occurs only when the heat shock factor is bound to the HSF site *and* when TAB proteins are bound to the TATA box. The binding of these regulatory proteins stimulates transcription of the coding region by RNA polymerase.

The exact mechanism(s) by which the HSF is selectively able to bind to the promoter is not yet known. It appears that protein modifications, such as phosphorylation, may play a role in specifying the particular binding characteristics of the HSF. In addition, there is more than one *Hsp* gene in *Drosophila,* and each one is activated by the binding of HSF to its promoter. Thus a whole bank of genes that need to be expressed during heat shock is activated by the HSF. This provides one eukaryotic answer as to how to coordinately regulate a number of separate genes.

The action of the HSF serves as an appropriate example of eukaryotic versions of regulatory proteins that respond to environmental signals. There is also a type of specific gene regulation unique to eukaryotes, the type of gene regulation that dictates gene expression in response to developmental signals.

Steroid-Hormone-Induced Genes. The steroid hormones, such as estrogen, are produced by specific endocrine glands (see Chapter 37), released to the bloodstream, and carried throughout the body of all animals. But each hormone can influence the gene activities of only a few specific cell types called **target cells**. For example, one of the main estrogen tar-

get cells in the chicken is a part of the lining of the oviducts. When the oviduct cells are stimulated by estrogen, they begin to make and secrete egg albumin (the main protein in the whites of eggs). The presence of estrogen in the environment of an oviduct cell stimulates albumin gene transcription. But how? This fairly complicated system of control is shown in Figure 14.9. Estrogen target cells have a special estrogen receptor protein. These are the only cell types that have this protein, which identifies or "tags" them as estrogen-responsive cells.

When an estrogen molecule contacts such a receptor, they form a complex. Then the estrogen–receptor complex apparently migrates to the albumin gene regulatory region. Once there, it binds to the albumin gene promoter to stimulate transcription by RNA polymerase.

In summary, the presence of the proper hormone receptor protein gives a cell the ability to be stimulated by the hormone; this stimulation takes the form of transcription activated by the hormone–receptor complex.

Other examples of eukaryotic gene regulation exist, and more are coming to light. In particular, some of the more complex questions of regulated gene expression, such as the details of how regulatory proteins recognize and bind to specific sequences of DNA and how the intricate modifications of

Essay 14.1 ● HOMEOTIC GENES AND THE HOMEOBOX

As we will see in later chapters, there are many questions that can be asked as an organism develops from the fertilized egg. After just a few cell divisions, we could ask: Which end will be the head and which the tail? How many segments will be made, and which will carry what appendages?

It has been known since the early work in *Drosophila* genetics that there are several genes that can drastically alter the developmental fate of parts of the embryo. In fact, we now know that there are some eight different genes that control critical developmental steps in fruit fly maturation. These genes are generally referred to as *homeotic* genes (*homeo-*, "same"), since they can result in the replacement of organs by structures similar to those found elsewhere on the

fly. For example, the *antennapedia* gene can result in the production of legs on the head, where antennae should be. Other homeotic genes control the number of segments in the body, the number of wings, and the head to tail orientation of body segments. In short, at least eight different homeotic genes control key developmental processes that require the coordinated action of many different genes. Thus the homeotic genes are considered to be a class of regulatory genes, genes that control the expression of other genes.

Researchers are now beginning to understand the way in which these regulatory genes exert their controlling effects. While the protein products of the various homeotic genes are distinctly different, they all share one feature. They all contain a section of

approximately 60 amino acids called the *homeobox*. Computer analysis of the three-dimensional structure of the proteins reveals the structure of the homeobox amino acids, and why it may be important in their role as regulatory proteins. The homeobox region of the proteins is structurally similar to other proteins known to bind to DNA. In fact, it is the same basic design that the CAP protein uses to bind to the *lac* promoter in *E. coli*.

So part of the mystery of how homeotic genes act is solved. The homeotic genes apparently exert their regulatory functions through protein products that bind to DNA, probably the promoter DNA of the many genes required to accomplish the elaborate tasks of building organs and body parts.

---

development and differentiation occur, are being unraveled (Essay 14.1).

DNA Packaging and Chemical Modification

We saw in Chapter 9 that eukaryotic DNA is packaged and organized within the nucleus by specific associations with histone proteins. Now we will see that the characteristics of the packaging have a direct impact on gene regulation.

Chromatin. Chromosomes, we have seen, consist of chromatin that is condensed into large, highly visible bodies. At least, this is the familiar form they take during cell division. During interphase, the chromosomes relax into the diffused and moderately condensed tangle of the chromatin net. We now know that it is only the DNA in the diffuse portions of the chromatin net that is actually transcribing RNA. The condensed portions are inactive. After all, imagine molecules of RNA polymerase trying to reach specific DNA sequences that are buried within the highly condensed chromatin network. They just can't get in. It is important to note, however, that chromatin can be condensed to different degrees and that such differences can act as a form of gene regulation.

Recall from Chapter 9 that the primary organizational unit of chromatin is the nucleosome (see Figure 9.2). We might expect that DNA wrapped around a nucleosome would have some difficulty interacting with specific proteins such as the HSF. Indeed it appears that areas of DNA that are important for regulation (such as promoters) are often kept free of nucleosomes, or that removal of nucleosomes from promoters is necessary for transcription.

Chromatin in its most highly condensed state becomes quite prominently visible in the light microscope and is called **heterochromatin**. Since it is condensed well beyond the 30-nm stage (see Figure 9.2), all heterochromatin is considered to be transcriptionally inert, with no genes being actively transcribed. There are two basic types of heterochromatin. One, called **constitutive heterochromatin**, is found near the centromeres of all chromosomes and, as its name implies, never decondenses and probably has no active genes. The other is called **facultative heterochromatin** because its state of condensation can change. Presumably, facultative heterochromatin can contain genes. **Euchromatin** is the much less condensed chromatin that contains potentially transcribable genes. But even in euchromatin the majority of the chromatin is condensed beyond the 30-nm stage. Condensation must be relieved before regulatory proteins and RNA polymerase can bind to promoters and initiate transcription.

There are two striking examples of the relationship between the degree of condensation and gene activity. The first example is the inactivation of one of the X chromosomes in mammalian females to form the transcriptionally inert, highly condensed Barr body (see Chapter 11). The other conspicuous example is gene-specific decondensation during the larval stages of *Drosophila* development. In fruit flies, and many other insects, the larval stage is a period of intense gene activity. The chromosomes of the *Drosophila* salivary gland, for example, offer a rather unique view. The DNA has been replicated many times, but without segregation into daughter cells. These **polytene** chromosomes consist of thousands of copies of the DNA strands lying next to each other and are therefore much more visible under the microscope.

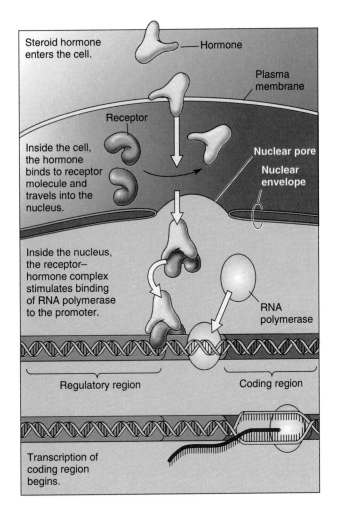

FIGURE 14.9
Steroid Hormone Regulation of Gene Expression

In this model, steroid hormones, such as estrogen, bind to cellular receptor proteins. The protein–hormone complex can stimulate transcription by binding to the regulatory region of a gene.

In Figure 14.10 we see that gene activity in larval salivary glands is signaled by visible changes in chromatin structure. The figure shows the regions of transcribing genes as decondensed "puffs" along a *Drosophila* polytene chromosome. Puffs actually consist of loops of DNA that are much less condensed than the DNA in neighboring areas; this means the DNA is accessible to RNA polymerase and regulatory protein.

FIGURE 14.10
Decondensation of Chromatin in the Puffs of *Drosophila* Salivary Chromosomes

Puffs are regions of the polytene salivary chromosomes where transcription is known to occur. **(a)** An actual photomicrograph of a puff. **(b–d)** Interpretations of the structure of a puff, showing that several stages of decondensation accompany transcription of this area of the chromosome.

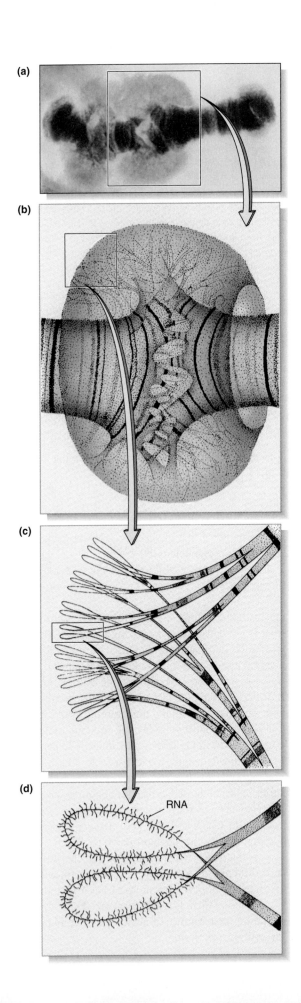

Unmethylated
cytosine

FIGURE 14.11
5-Methyl Cytosine
The presence of this modified base can reduce gene expression.

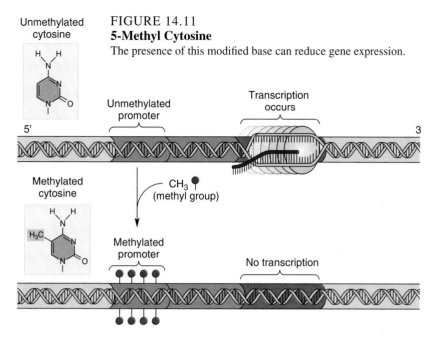

Methylated
cytosine

Chromatin that is able to decondense seems to have less H1 histone, the histone protein that appears to bind adjacent nucleosomes together. Consequently, the nucleosomes should be less stable and less tightly bound together than in chromatin stabilized by H1. It may even be that in this less tightly bound state, nucleosomes can be displaced more easily from a promoter by regulatory proteins.

Chemical Modification of DNA: Methylation. Eukaryotes have a second generalized method for inactivating DNA. This involves **methylation**, the addition of a methyl group ($-CH_3$) to certain bases. The primary example is methylation of cytosine nucleotides. Depending on the source of the DNA, 5% or more of the cytosine nucleotides within the genome may be modified by the addition of a methyl group to the pyrimidine ring (Figure 14.11). This additional chemical group does not influence the hydrogen bonding characteristics of the cytosine, but it seems to have a rather drastic effect on gene transcription. Genes that contain a high number of methylated cytosines are simply not transcribed.

The underlying mechanism relating gene activity and methylation is as yet unknown. One possibility is that methylation plays some role in the packaging of DNA into higher-order, nontranscribable chromatin. Another possibility is that the extra methyl group on a cytosine might well interfere with the binding of specific regulatory proteins.

Post-transcriptional Regulation

As we discussed earlier, the presence of the nuclear membrane offers eukaryotes the ability to separate the two processes of transcription and translation in time and space (see Figure 14.7). The many steps that accompany this separation of tran-

scription and translation offer several opportunities for regulation.

Nuclear Storage of Primary Transcripts. The first of these opportunities involves the processing of introns from the initial RNA transcript (see also Chapter 13). Since mRNA is not transported to the cytoplasm until the introns are removed, some transcripts could be held in the unprocessed state as a sort of ready reserve. When the gene product is needed, the only thing to be done is to process the message and transport it to the cytoplasm.

Alternative Processing of Primary Transcripts. It is even possible to enlist these intron processing steps in determining the eventual structure of the protein. Alternative splicing schemes can be used to generate two different proteins from the same transcript. For example, using alternative splicing, different troponin proteins can be derived from the same gene (Figure 14.12). By including different exons in the final mRNA, the activities of the resulting proteins can be modified.

The alternative selection of exons can be even further expanded—alternative promoters can be used to drive transcription of a single gene. Take, for example, the mouse alpha-amylase gene shown in Figure 14.12. It has two promoters, one that is used in the salivary gland and one that is used in the liver. The two different promoters individually control the transcription of the gene, depending on the organ. The promoters are separated by some distance at the beginning of the gene, with the salivary promoter located upstream of the liver promoter. When transcription occurs from the salivary promoter, the liver promoter is transcribed as part of the mRNA. However, in the salivary gland the RNA of the liver promoter region is removed from the primary transcript as an intron. Therefore the coding region of the mRNA is constant, whichever promoter is utilized for transcription. The final liver and salivary mRNAs differ only in the sequence of their first exon, which consists of $5'$ leader sequences. The use of alternative promoters is one way of accomplishing organ-specific gene regulation; by using two different promoters for the same gene, the need for a single promoter that is capable of responding to multiple and different biological signals is avoided.

Stored mRNA. Another form of post-transcriptional regulation involves the storage of mature mRNAs in the cytoplasm. When the mature mRNA is stored, it can be activated very quickly and with no nuclear involvement. Consider the following example. The rapid, initial DNA replications and cell cleavages that occur shortly after the fertilization of eggs preclude very much transcriptional activity. Therefore many mRNAs that are translated at these early times are activated from stores of maternal mRNAs that were deposited in the

(a) Alternative splicing patterns in the rat troponin T gene.

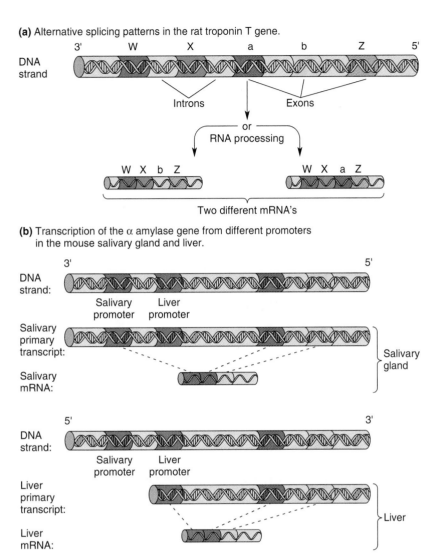

(b) Transcription of the α amylase gene from different promoters in the mouse salivary gland and liver.

FIGURE 14.12

Alternative Splicing and the Use of Alternate Regulatory Regions

(a) Alternative splicing patterns can result in two different proteins being translated from the same initial gene transcript. In this example of the rat troponin T gene, variable processing includes either exon alpha or exon beta as part of the final mRNA. **(b)** The mouse alpha-amylase gene has two different regulatory regions. The one furthest upstream is active in the salivary gland, while the one closer to the coding region is used in the liver. In the salivary gland, the primary transcript includes the liver regulatory region. But that region is now processed out as part of the first intron. The portions of the mRNA that code for the alpha-amylase protein are identical. Only the 5′ leader is changed.

egg. These maternal mRNAs are sequestered as stable protein–RNA complexes, stored until certain conditions, brought on by fertilization, make the mRNA available for translation.

Different mRNAs and Proteins and Their Different Lifetimes. Our discussion of post-transcriptional control must finally consider the metabolic lifetime of initial transcripts, mature mRNAs, and the proteins they encode. Different mRNAs have different lifetimes in the cell. All else being equal, an mRNA that is around longer will result in more translation and a higher concentration of the protein product. Two genes that produce mRNA at the same rate may differ widely in their apparent gene expression simply because of the

different lifetimes of their respective mRNAs. It is also known that different mRNAs can be translated at different rates, and that a given mRNA may be translated at different rates in different cells. So, too, the relative lifetime of the protein product will also play a role. Given equal mRNA levels and translation rates, proteins that degrade faster will appear at a lower concentration in the cell relative to more stable proteins.

Even with these brief discussions, it becomes clear that gene expression is not just a simple matter of regulating transcription. Indeed there is a whole host of regulatory possibilities that don't directly involve transcription at all, and it is very likely that the ultimate expression of all eukaryotic genes is subjected to some degree of management at each of these levels.

Key Ideas

1. Of the thousands of protein coding genes in humans and bacteria, many are active only part of the time, suggesting they are controlled by some mechanism.

2. While prokaryote gene-controlling mechanisms are well understood, the details of eukaryotic transcriptional regulation are now being elucidated.

PROKARYOTIC GENE REGULATION: THE OPERON

1. The **operon** model of prokaryotic gene control was reported by Jacob and Monod in 1961.

2. The *lac* operon in *Escherichia coli* is inducible, its three lactose-metabolizing enzymes (symbolized *z, y,* and *a*) are only produced when lactose is present in the cell.

3. Elements of the operon include the coding regions for producing the enzymes, the **operator** (*o*), the **promoter** (*p*), and the **repressor** gene (*i*).

4. The lactose operon functions as follows:

    a. Enzyme synthesis is repressed. The regulator produces a **repressor** protein that binds with the operator, blocking RNA polymerase so that it cannot interact with the promotor and begin transcription.

    b. Enzyme synthesis is induced. Lactose enters the cell and binds the repressor protein, freeing the operator. RNA polymerase interacts with the promoter, transcription begins, and mRNA is translated into the three enzymes. Transcrip-

tion is efficient, however, only if **CAP** is also bound to the regulatory region.

    c. Enzyme synthesis ends. The lactose-metabolizing enzymes hydrolyze lactose, breaking it down, thus releasing the repressor protein, which once more binds the operator and stops synthesis.

5. The *trp* operon controls the production of enzymes that synthesize tryptophan. In this system, the enzymes are synthesized continuously, even though a repressor protein is produced. Here the repressor protein cannot bind the operator unless it first joins a tryptophan molecule.

EUKARYOTIC GENE REGULATION

1. Comparatively little is known about eukaryotic gene control, but it is becoming evident that control systems utilizing regulatory proteins that bind to DNA, much like the operon repressors, also exist in eukaryotes.

2. Eukaryotic heat shock protein (*Hsp*) genes appear to require the binding of a TATA box binding protein and a heat shock factor. Steroid-induced genes require the binding of a steroid to a receptor, and subsequent binding of this complex to DNA.

3. Eukaryotic genes can be regulated by the state of packaging into chromatin, as well as chemical modification of DNA.

4. Post-transcriptional regulatory phenomena such as stored mRNA, alternative splicing of introns, and use of alternative promoters allow increased diversity in eukaryotic gene expression.

Application of Ideas

1. We tend to think of gene regulation and gene expression only as the result of selective transcription. At what other "stages" or "levels" might the biology of living cells regulate the phenotype of an individual?

Review Questions

1. How does a repressible operon differ from one that is inducible? Be specific. (pages 293–294)

2. Explain the role of the following in the lactose operon of *E. coli:* (a) promoter, (b) repressor, (c) *i* gene, (d) coding region, and (e) operator. (page 294)

3. Describe the proteins bound to a heat shock protein gene promoter under normal temperature conditions and after heat shock. (pages 298–299)

4. Describe how the presence of a steroid hormone in the bloodstream activates a gene that responds to that hormone. (page 300)

5. What is chromatin and what are its levels of organization? How does chromatin packaging relate to gene activity? (page 301)

6. What is the chemical modification referred to as methylation? What is the effect of methylation on gene activity? (page 303)

7. Describe how the processing of introns can profoundly influence apparent gene activity. (page 303)

CHAPTER 15

Genetics in Viruses and Bacteria

CHAPTER OUTLINE

VIRUSES

The Life and Times of the Bacteriophage

The Lytic Cycle

The Lysogenic Cycle

Variations on the Viral Theme

BACTERIA

Locating Mutant and Recombinant Bacteria

Recombination in Bacteria

Plasmid Genes

KEY IDEAS

REVIEW QUESTIONS

Over the years, different organisms have tended to dominate the study of genetics. First we had Mendel's true-breeding pea plants, then the hardy, prolific, and amazingly versatile fruit fly, *Drosophila melanogaster.* Later, the cutting edge of genetic research focused on the corn plant, *Zea mays,* after which the pink mold, *Neurospora crassa,* took over the spotlight. Then, in the late 1940s, a somewhat surprising organism gained the attention of geneticists. They began to focus on *Escherichia coli,* the common colon bacterium, and the viruses that infect it.

Perhaps the emergence of *E. coli* should not have been so surprising. After all, the thrust of molecular biology was to reduce problems to their simplest terms, and bacteria, when all is said and done, are structurally much simpler than fruit flies, corn, or even mold. The viruses that infect bacteria are simpler still. It was thought that the wisest move would be to try first to understand these simpler organisms and then to progress to more complex ones.

Not only are *E. coli* bacteria genetically simpler (they are essentially haploid, with only one chromosome), but their cells have little internal structure and are easily broken open so that their cellular machinery can be isolated and analyzed biochemically. More important, bacteria and viruses can be grown in enormous numbers in very short periods of time—*E. coli* populations can double in number in 20 minutes—so experiments can be done quickly. Billions of such organisms can be grown in a test tube, so statistical sampling is never a problem, and even very rare events (such as the occurrence of changes in DNA sequence that lead to new genes or alleles) will occur at a dependable frequency. Furthermore, *E. coli* can get sick—it can be experimentally infected with a virus.

In this chapter we will examine bacteria and viruses from a genetic perspective, in order to develop an appreciation for these powerful tools that scientists have used to unravel the complexities of hereditary information and DNA function.

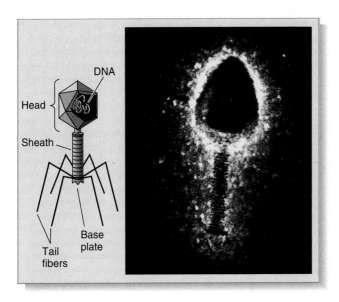

FIGURE 15.1
Bacteriophage
The T$_2$ bacteriophage is an excellent example of one type of viral structure. The polyhedral head contains the DNA, and the rest of the bacteriophage is an intricate structure and system designed to deliver the DNA into the host cell.

Viruses

Viruses, as we will see later, fall into that nether world between life and nonlife. They have many of the traits of living things, yet they can form apparently inanimate crystals, with all the charm of table salt. Part of the reason that viruses can seem to be nonliving is that they are exceedingly simple in structure. They usually contain only a protein coat, some RNA or DNA, and maybe a few enzymes, and their diameters are about one-tenth that of a bacterium. The reason that viruses can be considered living is because they become "active" within the host that they infect. Viruses rely on the biological mechanisms of the host cell to provide all the energy and molecular building blocks to construct new viruses. Once a large number of progeny viruses have been constructed within the host, the viruses usually burst out of the host cell, killing it in the process.

In the conduct of their parasitic lives, viruses attack all kinds of living cells. However, each kind of virus can attack only a certain kind of host. The viruses that attack eukaryotic cells are different from those that attack prokaryotes. Nonetheless, certain kinds—specifically bacteriophages—do infect bacteria, and experiments using them have not only told us a great deal about prokaryotic genetics but have provided tremendous insights into the basic structure and function of all viruses.

The Life and Times of the Bacteriophage

The **bacteriophage**, or phage virus, a parasite of bacteria, has been mentioned with respect to the classic experiment of Hershey and Chase, which helped establish DNA as the hereditary material (see Chapter 12). Let's look more closely at the bacteriophage and see what happens when it finds its host. We will consider a typical bacteriophage called T$_2$, one that commonly invades *E. coli.*

As viruses go, the T$_2$ phage consists of a DNA core surrounded by a polyhedral protein head (made of coat proteins), which ends in a cylindrical tail (Figure 15.1). Attached to the tail are a peculiar set of leg-like, protein tail fibers. Upon "touching down" on its host, the tail firmly attaches itself, and the viral DNA is injected into the cell (Figure 15.2).

The Lytic Cycle

If the viral DNA successfully enters the *E. coli* cell, it may well encounter the host's RNA polymerase. RNA polymerase isn't able to distinguish between its own DNA and foreign DNA. It reacts to the viral DNA as if it were its own and transcribes mRNA from it, producing viral mRNA. The viral mRNA is then blindly translated by the host's ribosomal machinery into viral proteins.

Among the first proteins produced from the viral mRNA are enzymes that can readily distinguish between host DNA and viral DNA. These enzymes chop the host DNA into fragments, presumably to destroy host control and release the nucleotides to be recycled into viral DNA. Other viral enzymes make copies of the viral DNA; some even make more viral mRNA to speed up the process. The viral genes also make coat proteins to form the heads and the tails with the leg-like landing gear. The final act of the viral proteins is to lyse (rupture) the envelope of the host cell, now hardly more than a bag of virus particles. Lysis of an *E. coli* cell may release hundreds of new, infectious viral particles (see Figure 15.2) that can then move on to infect more host cells, in a continuing cycle called the **lytic cycle**.

Viruses are completely parasitic and can only grow in living cells; they cannot grow by themselves in laboratory medium. However, they can be grown in the laboratory if they are diluted and then spread on agar plates that have been specially prepared with a "lawn"—a continuous surface layer—of susceptible bacteria. Each active virus particle infects just one bacterium, but after lysis its progeny spread to the neighboring bacteria until there is a clear hole in the lawn where a virus and its progeny have lysed the bacteria. These holes, called **plaques**, are then counted to determine how many viruses there were (Figure 15.3).

The Lysogenic Cycle

Although the life histories of most bacteriophages are based on the lytic cycle, many, such as bacteriophage lambda (λ), can go through a different, perhaps more insidious, sequence called the **lysogenic cycle**. Instead of destroying the host chromosome, the phage DNA cuts into and joins the host DNA in a process called **lysogeny** (Figure 15.4). The viral DNA then undergoes replication with the rest of the bacterial chromosome and is passed on to all the bacterium's descendants. While in the chromosome, it does very little except to transcribe and translate a repressor protein that protects the

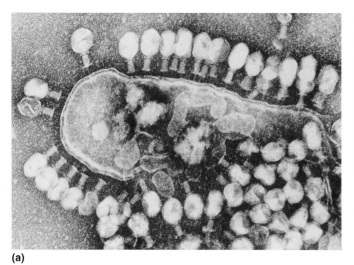

(a)

FIGURE 15.2
Bacteriophage in Action
(a) This electron micrograph shows a bacteriophage attacking a bacterial cell. The end result is a bacterial cell full of newly formed bacteriophage particles, and eventually cell lysis. **(b)** This releases a "new generation" of viral particles that can then go on to infect other host cells, starting the lytic cycle again.

cell from further attack by other viruses. Then, in some future cellular generation, the viral DNA may be induced to a lytic phase, breaking away from the host DNA and making copies of itself. The cell will then be lysed, releasing hundreds of new virus particles.

Variations on the Viral Theme

There are many different kinds of viruses, and virtually every kind of cell is subject to viral infection. In spite of such an array of these parasites, most utilize either the lytic or the lysogenic life cycle. However, there are variations on this theme. The most significant sort of variation involves whether the virus uses DNA or RNA as its primary genetic material, and if it uses RNA, how it handles the replication of that RNA genome.

Some viruses indeed use RNA as the nucleic acid of the mature viral particle. The RNA remains single-stranded, but even so, it is perfectly capable of carrying the genetic information for the virus. RNA viruses known as **retroviruses** first convert that information into DNA, which is then transcribed by the host cell to produce viral protein as well as more copies of the RNA. So RNA makes DNA, which is used to make more RNA (as well as proteins). This process utilizes a very special enzyme, encoded by the viral RNA, known as **reverse transcriptase** (Figure 15.5). The name of this enzyme tells you what it does. It makes the DNA copy of RNA. Once the first strand complement of the viral RNA is made, the host cell's DNA polymerase can make the second strand. This double-stranded DNA copy of the viral RNA can then be used to direct transcription of more viral RNA. Also, the double-stranded DNA copy of the viral RNA can be incorporated into the host's genome in a process similar to lysogeny.

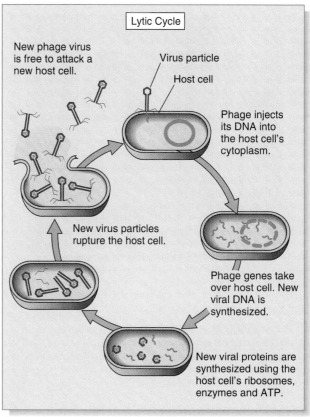

(b)

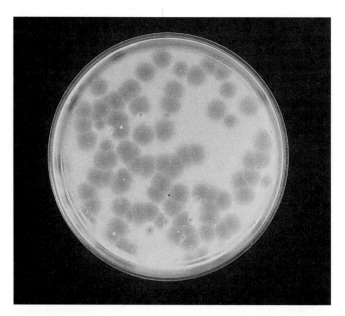

FIGURE 15.3
Growing Bacteriophage
Unlike bacteria, bacteriophages and other viruses will not grow on organic media—they require a living host. In this figure, a mixture of bacteriophages has been spread out over a layer (or "lawn") of bacterial cells on the surface of nutrient agar in a Petri dish. The clear circles are places where the bacteria are now absent, an indication that they have been lysed by the bacteriophages, which continue to invade the bacterial cells in an ever-widening circle.

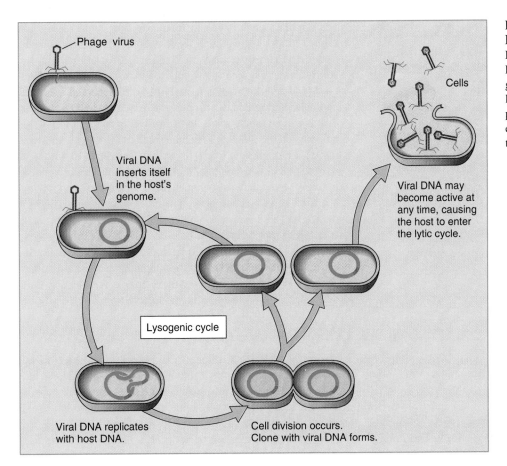

FIGURE 15.4
Lysogeny
During the process of lysogeny, viral DNA is incorporated into the host's genome, where it can remain essentially inactive and can be replicated as part of the host's genome, until it enters a stage of rapid division and then lyses the cell.

Figure labels:
Phage virus
Cells
Viral DNA inserts itself in the host's genome.
Viral DNA may become active at any time, causing the host to enter the lytic cycle.
Lysogenic cycle
Viral DNA replicates with host DNA.
Cell division occurs. Clone with viral DNA forms.

There are also RNA viruses that take things a step further and encode an enzyme called **RNA replicase**. This enzyme is capable of making RNA copies from RNA templates, thus eliminating entirely the need for DNA.

Bacteria

In order to work with genes and genetics, we must be able to detect variation. After all, Mendel had yellow and green peas which enabled him to follow the inheritance of color (Chapter 10); and Garrod correctly deduced that genes act through proteins by observing the mutations that cause certain metabolic enzymes to lose function (Chapter 12). Thus most genetic work depends on genetic variation in traits. What traits do bacteria have? How can bacteria be manipulated in order to conduct genetic experiments such as gene mapping?

Locating Mutant and Recombinant Bacteria

Genetic experiments with bacteria require the ability to identify bacterial cells with special traits. Among the most easily located "abnormal" bacteria are nutritional mutants—those unable to grow on minimal media that are readily used by wild-type, or unmutated, bacteria. Minimal media contain only sources of the very basic nutrients, such as sugars and salts. If, for example, a certain strain of bacteria will not grow on a min-

imal medium, various supplemental nutrients can be added, one at a time. When the nutritional mutant strain shows signs of growth, the researcher knows that its metabolic machinery was unable to provide that nutrient, and it is labeled accordingly. For example, a strain unable to grow unless the medium is supplemented with tryptophan would have a mutation in the pathway that produces tryptophan and would be labeled trp-.

But how do you get nutritional mutants in the first place? One technique for isolating nutritional mutants involves adding penicillin to the minimal medium. Penicillin is a powerful antibiotic that kills many kinds of actively growing bacteria by interfering with cell wall synthesis (see Chapter 22). In this procedure, the experimenter puts a large, known quantity of wild-type bacteria into a minimal medium containing penicillin (Figure 15.6). Among the bacteria are a small but unknown quantity of nutritional mutants. The normal cells metabolize the medium, grow, try to divide, and die. But the mutants survive. They remain inactive because they are unable to metabolize and cannot grow and enter cell division. They therefore don't need new cell wall material. So the penicillin doesn't kill the mutant cells; they just sit there, inactive but alive. In this way, the experimenter can essentially select nutritional mutants from the subject population; these mutants can then be grown on supplemented media and tested on minimal medium.

When a dilute sample of bacteria is spread on a Petri plate containing agar and the proper nutrients, individual bacteria will begin to grow and divide. On the semisolid agar,

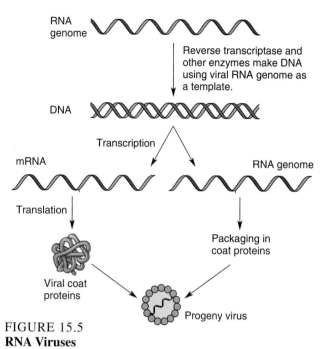

RNA
genome

Reverse transcriptase and
other enzymes make DNA
using viral RNA genome as
a template.

DNA

Transcription

mRNA   RNA genome

Translation

Packaging in
coat proteins

Viral coat
proteins

Progeny virus

FIGURE 15.5
RNA Viruses
These viruses constitute a diverse group, but all have only RNA in
their genomes.

each invisible bacterium essentially grows in place, forming a
visible mound of bacteria that is called a **colony**. This colony
is a small population of bacteria that are all direct progeny and
identical to that first, individual bacterium that was placed on
the Petri plate. For every bacterium present in the original
sample, a colony will form on the plate if the bacterium can
grow in the medium present in the agar.

Another ingenious way of recovering mutations is by
replica plating, a technique developed by Esther and Joshua
Lederberg. As you can see in Figure 15.7, replica plating is a
sort of "now you see it, now you don't" process. Here a population of bacteria are plated on a fully supplemented medium,
and a large number of colonies will form. A velvet-covered
disk is placed on the plate, and some bacteria from each of the
colonies on the plate will become stuck to the velvet while the
rest of the colony will be left behind. The velvet disk is then
used to innoculate a plate of minimal medium. The velvet disk
preserves the pattern of colonies on the original plate, creating
a replica on the second plate. The nutritional mutants are recognized as the bacterial colonies that do not grow after being
transferred. But because the colonies have been replica plated,
the mutants can be recovered from the original plate and tested to see what nutrients must be supplemented. Among other
things, the replica plate experiment proved for the first time
that nutritional mutants occur spontaneously.

Recombination in Bacteria

Sex in bacteria is so different from sex in even the lowliest
eukaryotes that you may not think they really have sex at all.
If sex is defined as any mechanism that combines genetic
material from two cells into one cell, then bacteria have sex.
In fact, they have a fairly varied sexual repertoire.

Different strains of bacteria are able to exchange and
recombine their DNA in various ways: transformation, transduction, and conjugation. (Here we broaden our definition of
the term *recombination* to include events that do not occur
during meiotic prophase.) However, recombination doesn't
involve sex as we know it in eukaryotes. What passes for sex

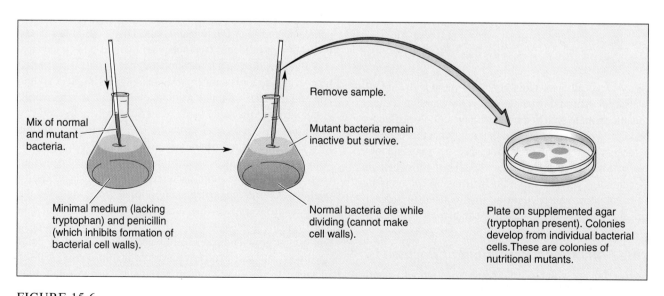

Mix of normal
and mutant
bacteria.

Remove sample.

Mutant bacteria remain
inactive but survive.

Minimal medium (lacking
tryptophan) and penicillin
(which inhibits formation of
bacterial cell walls).

Normal bacteria die while
dividing (cannot make
cell walls).

Plate on supplemented agar
(tryptophan present). Colonies
develop from individual bacterial
cells. These are colonies of
nutritional mutants.

FIGURE 15.6
Recovering Bacterial Mutants
Bacterial mutants can be recovered through a technique involving penicillin. In a mixture of normal and
nutritionally mutant cells, the normal cells attempt to grow but die. Because the nutritional mutants don't
grow, they survive. When the surviving cells are placed in correctly supplemented media, they grow and
divide to produce a colony of the mutant for further study.

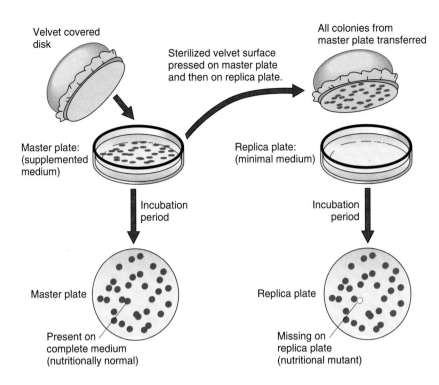

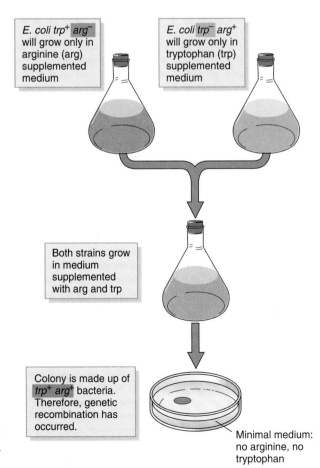

FIGURE 15.7
Replica Plating
Replica plating is a method of locating mutant bacteria. A *master plate (left)* containing completely supplemented medium that can support any growth, is coated with a mix of mutant and normal bacteria. The resulting bacterial colonies form a random pattern of dots on the agar. After growth has occurred, a velvet-covered disk is used to transfer the pattern of colonies to a minimal medium *replica plate (right)*. A comparison of the plates soon reveals the mutants, which can be recovered from the master plate.

in bacteria is varied, bizarre, and infrequent—so infrequent that the odds of a bacterium having sex are about one in a million. In order to observe such infrequent behavior, it's obviously necessary to take some shortcuts.

Suppose you have two nutritionally mutant strains of *E. coli,* each of which requires a different amino acid in its nutrient medium. One strain needs tryptophan (trp), and another needs arginine (arg). The genotype of the first strain is symbolized *trp⁻, arg⁺*, because it requires trp but doesn't require arg. (The minus means the the strain is deficient for the synthesis of the nutrient and hence the nutrient is required in the medium for growth to occur.) Then it follows that the second strain is *trp⁺, arg⁻*. Neither can grow on minimal medium. To find that one-in-a-million recombinant, you would mix 50 million or so bacteria of each type, let them mate (or undergo whatever other kind of genetic recombination event they can manage), and transfer them to minimal medium. The *trp⁻, arg⁺* bacteria and the *trp⁺, arg⁻* bacteria will both just sit there, unable to grow. However, if the two strains have managed to mate and recombine their DNAs, some colonies of the *trp⁺, arg⁺* genotype will appear and grow on the minimal medium (Figure 15.8). If such colonies are formed, this indicates that a genetic exchange has taken place between the two strains of bacteria.

Now let's look at some of the ways that bacteria manage to exchange DNA.

FIGURE 15.8
Bacterial Sex and Nutritional Mutants
Biochemical and genetic evidence for sexual reproduction in bacteria was first gathered with the techniques shown in this scheme. Two strains, *trp⁺, arg⁻* and *trp⁻, arg⁺*, which were selected through replica plating, were grown together in a supplemented medium. Then samples were plated out on minimal medium that lacked tryptophan and arginine. If colonies grew, they represented genetic recombinants with the *trp⁺, arg⁺* genotype.

Transformation. The first kind of recombination observed in bacteria was transformation (see Griffith's experiment, Chapter 12), which involves bacteria taking up fragments of DNA from the environment into their cells and incorporating them into their own chromosome. Such recombination takes place very rarely. Interestingly, the fragment is incorporated into the bacterial chromosome by what amounts to double crossing over (Figure 15.9). The fragment aligns with the homologous section of the host's chromosome, and the two sections are exchanged. The leftover host segment is then simply dismantled by enzymes. The new addition is passed along to the cell's descendants when the cell divides.

Transduction. **Transduction** involves the transfer of genetic material from one bacterium to another using a bacteriophage as the carrier. What happens is this. Sometimes, when the phage coat protein envelope has already formed and is beginning to fill with DNA, there may be fragments of the host's DNA floating about. The developing viruses may take up this bacterial DNA in place of or in addition to viral DNA. The developing virus coat doesn't distinguish between them. After lysis, the viral particle with the bacterial DNA attacks a new host, injecting the piece of foreign bacterial DNA. When the new DNA encounters the DNA of the host, recombination can take place. Essentially, the protein body of the virus has brought bacterial DNA from one strain into another (Figure 15.10).

Conjugation. The types of genetic exchange that we have discussed so far, transformation and transduction, involve the fairly random uptake of DNA fragments. These are quite rare processes that require DNA to be released into the medium or carried by a phage. However, there is a more direct means for bacteria to exchange DNA. **Conjugation** is the direct transfer of DNA from one bacteria to another that involves a specialized bacterial sexual apparatus and replication of bacterial DNA before transfer.

Some bacteria contain, in addition to their large circular chromosome, one or more very small circular chromosomes called **plasmids**. Plasmids are hundreds of times smaller than the main bacterial chromosome. However, they contain genes, generally replicate in synchrony with the main chromosome, and are passed on to progeny of the cell at the time of cell division.

Some plasmids have genes that code for the proteins that make up special structures called **pili** (singular, *pilus*). While most pili are simple proteinaceous spines, some are larger and tube-like. The simpler pili aid the cell in maintaining contact with surfaces. The larger pili are believed to be instrumental in sexual reproduction, drawing cells together and forming a hollow **conjugation tube**, through which DNA can pass. For this reason, the larger pili are often called **sex pili** (Figure 15.11).

If a plasmid-containing (plasmid-plus) bacterium and a plasmid-free (plasmid-minus) bacterium come close to one another, a sex pilus is formed, and the plasmid-plus bacterium snuggles up against the plasmid-minus bacterium. The plas-

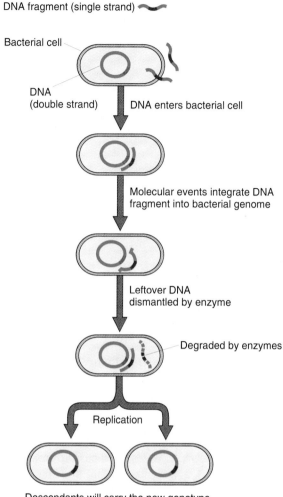

DNA fragment (single strand)

Bacterial cell

DNA
(double strand)   DNA enters bacterial cell

Molecular events integrate DNA
fragment into bacterial genome

Leftover DNA
dismantled by enzyme

Degraded by enzymes

Replication

Descendants will carry the new genotype

FIGURE 15.9
Bacterial Transformation
In the process of transformation, a bacterium takes in a fragment of DNA. Once inside, the fragment aligns itself along the homologous segment of the circular DNA and becomes integrated through double crossing over. The original part of the chromosomal DNA is degraded. All of the bacterium's descendants will then carry the new recombinant chromosome.

mid DNA undergoes an extra round of replication, and as it does so, one of the new copies opens up and is transferred (as a linear DNA molecule) through the sex pilus into the plasmid-minus bacterium (see Figure 15.11). Once inside its new host, the plasmid DNA becomes circular again and is replicated as a normal plasmid in the new cell. Other plasmid genes then produce a cell surface substance that prevents further plasmid invasions. Both cells are now plasmid-plus. Thus conjugation represents a direct mechanism for transferring DNA between bacteria.

Some plasmids are transferred quite easily from one species of bacteria to another. This is unfortunate for us humans, because some plasmids, as we will learn, carry genes

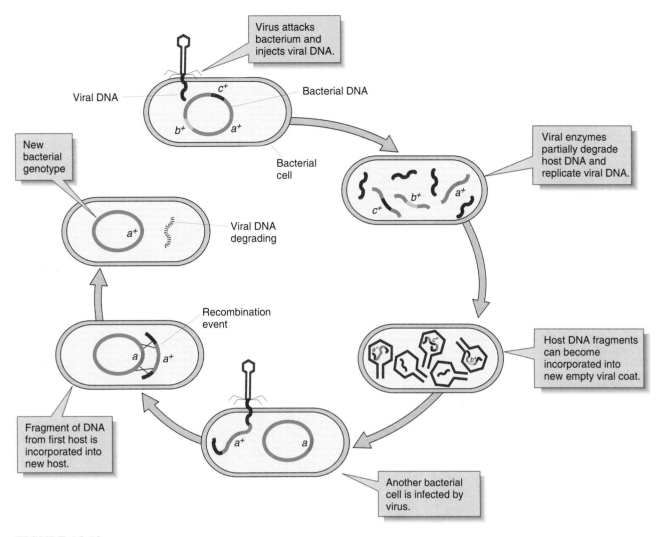

FIGURE 15.10
Bacterial Transduction
Transduction is a sort of DNA hitchhiking transfer. During phage replication, a segment of the host's
fragmented DNA is packaged in the phage head together with the phage DNA. When that virus infects a new
host, the segment of bacterial DNA that it carries can be added to the new host's DNA through recombination.

that make bacteria simultaneously resistant to many different antibiotics.

***F* Plasmids as Agents of Gene Transfer.** Our understanding of another kind of genetic recombination is based directly on the work of Joshua Lederberg and Edward Tatum. They worked out the behavior of a certain kind of plasmid now known as the *F* plasmid (*F,* for *fertility,* since presence of the plasmid apparently conferred the ability to mate). As you might expect, there was also an opposite strain, the "fertility minus" strain. These bacteria were particularly interesting because genetic material from the positive strain always moved across the sex pilus into the negative strain. Furthermore, after such a union, the negative strain changed, itself becoming positive. Originally, bacterial strains were divided into two groups, which were promptly (perhaps too promptly) dubbed "male" and "female." The names seemed appropriate because males or females mated only with bacteria of the opposite sex, and mixtures of two male strains or two female strains never produced recombinant progeny. These days the "male" strain is simply called F^+, and the "female" strain is called F^-.

What makes the *F* plasmid so interesting? It turns out that, while simple plasmid transfer is relatively common, in rare cases the *F* plasmid can become inserted into the main chromosome of its host. The plasmid has no preferred site of insertion and can show up anywhere in the host's circular chromosome. After insertion, the *F* DNA is replicated right along with the host DNA and can be passed on to all the bacterial offspring. So far, this sounds like phage lysogeny. But later, when the plasmid forms a conjugation tube and attempts the transfer to an F^- bacterium, strange things happen. First, the *F* plasmid

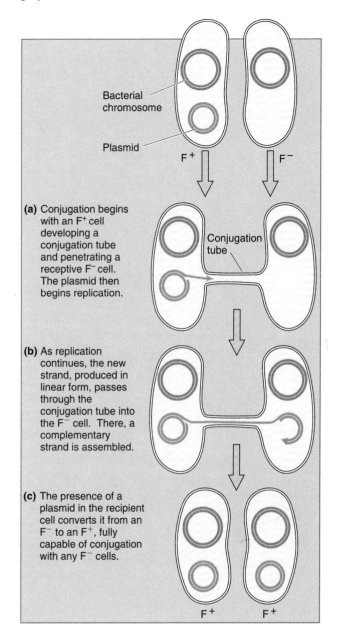

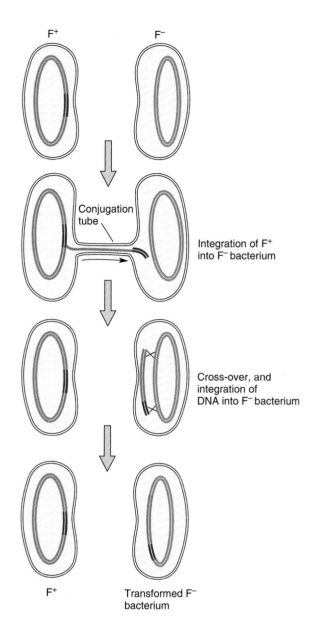

FIGURE 15.11
The Plasmid and Conjugation
(a) The slender tube between the two bacteria of different strains is known as a sex pilus or conjugation tube. (b) Once the tube forms, the plasmid undergoes a round of replication. The new plasmid passes through the tube into the plasmid-minus bacterium. (c) The arrival of the plasmid changes the recipient into a plasmid-plus bacterium.

FIGURE 15.12
Transferring the Integrated Plasmid
In rare cases, the *F* plasmid joins the entire bacterial chromosome. When this type of *F*⁺ bacterium mates with an *F*⁻ type, the entire chromosome replicates, beginning with the integrated plasmid. The replicated plasmid starts through the pilus, dragging the lengthy bacterial chromosome with it. Any portions of the bacterial chromosome that make it through the pilus before the DNA breaks may be incorporated into the recipient by recombination.

begins to replicate, as usual. But since it is now part of a larger circle, the entire structure—host chromosome and all—is replicated into a linear DNA molecule. Then, when one end of the *F* plasmid goes through the conjugation tube, it starts to pull the entire host chromosome in after it. Because the chromosome is very long, it takes about 89 minutes for the whole thing to get through. In fact, the tube usually breaks apart before transfer is completed. When the tube breaks and the cells separate, the

recipient cell usually gets only a piece of the *F*⁺ sequence and it isn't transformed into an *F*⁺ bacterium. However, the bacterial DNA that is brought in can recombine, by crossing over, with the DNA of the recipient bacterium (Figure 15.12).

Bacterial strains with integrated *F* plasmids are called ***Hfr***, for *h*igh *f*requency of *r*ecombination, because of the ability of the *F* plasmid to drag along genes from the main bacterial chromosome.

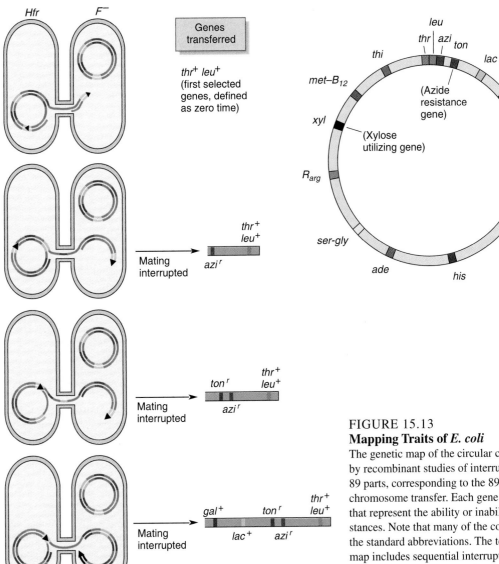

FIGURE 15.13
Mapping Traits of *E. coli*
The genetic map of the circular chromosome of *E. coli,* produced by recombinant studies of interrupted *Hfr* matings, is divided into 89 parts, corresponding to the 89 minutes required for a complete chromosome transfer. Each gene locus is identified by code letters that represent the ability or inability to synthesize various substances. Note that many of the codes represent amino acids using the standard abbreviations. The technique used in developing the map includes sequential interruption of gene transfer and, following this, identifying the genes that made it through by plating on different media.

Mapping the *E. coli* Chromosome: The Great Kitchen Blender Experiment. Progress in molecular biology often follows the development of new technology. In this case, the technology was provided by Fred Waring, a popular band leader of the 1950s. When he was not leading his group, The Pennsylvanians, in song, he was busy inventing the Waring blender. The blender, of course, beats food into a mush, and in so doing it disrupts cells. Molecular biologists needed a good way to separate mating cells, so they tiptoed away with the blenders from their kitchens.

As we have mentioned, it takes about 89 minutes for the *Hfr* chromosomes to transfer to the F^- bacterium. Thus, if the transfer is interrupted by the blender after just a few minutes, only part of the chromosome would be transferred and only a few of the known gene loci would make it across, while most of the others would remain behind (Figure 15.13). As the peri-

ods of mating were extended, more and more loci would be transferred and could be recovered in the progeny, although at lower efficiencies because of spontaneous breakups of the bacterial couples. The last genetic loci come through, but with extremely low efficiency, toward the end of the 89 minutes.

Using matings that are interrupted by the blender, it was possible to map genes on *Hfr* bacterial chromosomes. On a standard *Hfr* chromosome the first locus passed through in about two minutes, and each subsequent marker was given a position appropriate to the time it entered the recipient cell. Figure 15.13 also shows the genetic map of *E. coli* derived from such experiments.

Because the F plasmid can integrate into the main chromosome at essentially any position, different *Hfr* strains can be developed, each with the F plasmid located at a different position and oriented in either direction. Mapping can be done with each *Hfr* strain, and the order of genes transferred can be combined into a complete map. In fact, it was genetic mapping data such as these that first showed that the *E. coli* chromosome was circular, because the overlapping maps from different *Hfr* strains demonstrated a repeating sequence of genes that could be explained by making repeated turns around a circle.

Plasmid Genes

The *F* plasmid is only one of many kinds of bacterial plasmids that have been discovered. The others do not transfer the host's chromosomal genes to other bacteria, but they do play roles in bacterial genetics. These plasmids have genes of their own, often including genes for making pili and for coating the cell surface to protect the host (and the plasmid) from infection by other plasmids. In addition, some have genes that protect their hosts (and thus themselves) in other ways. For example, plasmids are responsible for some of the deadly toxins produced by disease-causing bacteria and some plasmids even carry genes that make their hosts resistant to antibiotics.

One plasmid, known as *R6*, endows its host with resistance to six important antibiotics (Figure 15.14). *R6* can also be transferred between different species of bacteria. Time and again physicians have successfully treated bacterial infection with antibiotics, only to find that the same disease turns up again but is no longer cured by the same antibiotic. What has happened is that selection has been at work, and our very success at battling microbes has paved the way for the rise of new, more potent strains. Fortunately, the drug industry has more or less kept up with the challenge by continually developing new or altered antibiotics.

By now it seems clear that "bacterial sex" is, in fact, an aberration of viral infection or picking up DNA from the environment, and plasmid transfer has little to do with sex as we eukaryotes know it. But it remains a very useful tool in genetics and molecular biology. It may also prove to play a signifi-

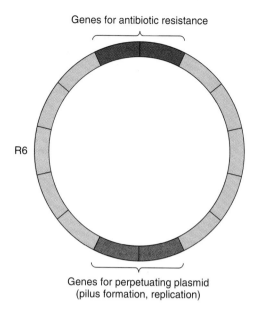

Genes for antibiotic resistance

R6

Genes for perpetuating plasmid
(pilus formation, replication)

FIGURE 15.14
The *R6* Plasmid
The *R6* plasmid is a circle of DNA that contains genes that endow the bacterial cell with resistance to certain antibiotics.

cant role in long-term bacterial evolution. Particularly, it has not escaped the attention of biologists that genetic transfer by such omnipresent agents as viruses could have important evolutionary significance in the transfer of genetic information between organisms.

Viruses, plasmids, and bacterial sex have taken on additional significance in molecular biology. As we will see in Chapter 16, the small genomes of viruses and plasmids are readily isolated and manipulated in the laboratory, yet they can easily be reintroduced into bacteria by reconstructing viruses or by simple transformation.

Key Ideas

VIRUSES

1. **Bacteriophages** are viruses that inject their DNA into a bacterial host. It then uses the host's RNA polymerase to transcribe the viral DNA and synthesize viral enzymes.

2. Bacteriophage enzymes disrupt host DNA and recycle its nucleotides for use in replicating many copies of viral DNA. In addition, viral DNA is transcribed into RNA in order to produce viral coat proteins. When viral reproduction is complete, the host cell will be lysed, and many infectious bacteriophages released.

3. Phage activity in a bacterial plate can be recognized as **plaques**—clear areas in an agar plate that otherwise has heavy bacterial growth.

4. An alternative to lysis is **lysogeny**. The viral DNA is incorporated into the host chromosome, and the viral genes are replicated along with the host genes as reproduction occurs. In some future generation, the virus may again enter its lytic cycle.

5. There are viruses whose genome is made of RNA. Some of these viruses replicate their RNA directly using **replicase** while others make a DNA copy of their RNA by using **reverse transcriptase**.

BACTERIA

1. Mutations that affect nutritional requirements of bacteria are readily identified. In one method, **replica plating**, colonies are transferred from a supplemented to a minimal medium, and the

experimenter looks for colonies that grow on the supplemented medium but not the minimal medium.

2. Although bacterial sex is infrequent, true sex—the recombination of DNA between two individuals—can occur through the following mechanisms:

   a. **Transformation**—the uptake and incorporation of fragments of DNA from the environment into the bacterial chromosome.

   b. **Transduction**—the carrying of DNA fragments into the bacterial cell by a phage virus, with incorporation into the bacterial chromosome.

3. **Plasmids** are small, supplemental DNA molecules in bacteria. One kind of plasmid codes for the synthesis of the sex pili that act as conjugation tubes.

4. Bacteria with the sex pilus coding plasmid are labeled plasmid-plus; those without are labeled plasmid-minus. Such plasmids can be transferred to plasmid-minus bacteria, changing them to the plus form.

5. The *F* plasmid represents the first thoroughly investigated sexual activity in bacteria. It codes for sex pili, conjugation tubes through which copies of the plasmid can be transferred.

6. In rare instances, the *F* DNA will insert into the host's main chromosome and, as in phage lysogeny, can be replicated and passed along to progeny. Once inserted, the *F* plasmid can also transfer all or part of host's main chromosome to another bacterium during conjugation. Bacteria that did this were isolated and grown in pure cultures and termed **Hfr** (high frequency of recombination).

7. The chromosome transferring ability of *Hfr* strains made genetic mapping of the circular *E. coli* chromosome possible.

8. Plasmids such as *R6* contain genes, among which are antibiotic resistance genes. There is some concern that overuse of antibiotics is leading to more resistant strains of bacterial diseases because of the ability of plasmids such as R6 to be transferred to different bacterial strains.

Review Questions

1. Suggest several advantages for using the bacterium *E. coli* in studies of genetics. (pages 307–308)

2. Briefly outline the general procedure for isolating nutritional mutants of bacteria. (page 309)

3. Suggest a procedure that could be followed to determine whether genetic recombination had occurred in bacteria. (pages 309–311)

4. Suggest several ways in which sexual reproduction in bacteria is different from sexual reproduction in the eukaryote. (pages 310–311)

5. Explain how a bacteriophage reproduces. (page 312)

6. Distinguish between lytic and lysogenic cycles in the bacteriophage. (page 308)

7. Outline a procedure for detecting the presence of bacteriophages. (page 309)

8. In what way is transformation similar to transduction? How are they different? (page 312)

9. What is a plasmid? How might a plasmid get from one bacterium to another? (page 312)

10. Using the terms F^+ and F^-, explain how conjugation occurs in *E. coli*. (page 313)

11. List two things that happen to a F^- bacterium that has undergone recombination with an F^+ bacterium. (pages 313–314)

12. Briefly explain how the F^+ plasmid is able to transfer genes between cells. (pages 313–314)

13. What special capabilities are there in *Hfr* strains? (pages 314–315)

14. Explain the kitchen blender experiment, including the purpose, the timing, and the reasoning behind it. (page 315)

15. What is an *R6* plasmid? In what way can such plasmids be threatening to humans? (page 316)

Genetic Engineering: The Frontier

C H A P T E R   O U T L I N E

DNA TECHNOLOGY

Restriction Enzymes and Sticky Ends

Splicing and Cloning DNA Molecules

Essay 16.1 DNA Sequencing

APPLICATIONS OF GENETIC ENGINEERING

Manufacture of Biomolecules

Agriculture

Medicine

Essay 16.2 Electrophoresis, Hybridization, and DNA Fingerprinting

Essay 16.3 The Polymerase Chain Reaction

GENETIC ENGINEERING: A SOCIAL ISSUE

KEY IDEAS

APPLICATION OF IDEAS

REVIEW QUESTIONS

Genetic engineering burst upon the public consciousness with an impact unusual for such technical matters. In the past, the tremendous attention given by the public to some scientific issues had proved to be transient, finally leaving scientists working quietly in their labs as public interest diminished. Genetic engineering, however, has proved to be a bit different. The public has continued to be fascinated by the idea of working directly on the genetic makeup of living things.

Indeed the promise of genetic engineering is a great one, and the potential benefits of the technology will continually be revised upward as new findings open other avenues for development. We no longer have to squeeze infinitesimal amounts of growth hormone from the pituitaries of cadavers. Thanks to genetic engineering, we can now make it by the vatload. We have the potential to supply normal genes to individuals crippled by inherited maladies, or to diagnose other diseases long before they begin to take their toll. We can alter the amino acid balance in fruits and vegetables and grow crops that will be less susceptible to chilling frosts. Truly, we are in the process of a genetic revolution of tremendous magnitude. No topic in biological science is more likely to impact our personal lives and is therefore more worthy of our attention and understanding.

Before going any further, we might ask just what genetic engineering is. **Genetic engineering** is the process of altering biological systems by the purposeful manipulation of DNA. It is a subset of the more general kind of endeavor known as biotechnology, the use of biological systems for the production of materials. The focus of this chapter is the technology of genetic engineering. We will take a look at the methods by which scientists can manipulate the genes of any organism, then we will address some of the applications and implications of this capability.

Ordering Information 7
Explanation of Quality Controls 8
Setting Up a Restriction Digest 9

Restriction Endonucleases

pages 12 to 49

Aat II	Bsm I	EcoR V	Pac I
Acc I	BsmA I	Esp3 I	PaeR7 I
Acc65 I	BsmF I	Fnu4H I	PflM I
Aci I	Bsp120 I	Fok I	Ple I
Afl II	Bsp1286 I	Fsp I	Pme I
Afl III	BspD I	Hae II	Pml I
Age I	BspE I	Hae III	Ppu10 I
Alu I	BspH I	Hga I	PpuM I
Alw I	BspM I	Hha I	Psp1406 I
AlwN I	Bsr I	Hinc II	Pst I
Apa I	BsrB I	Hind III	Pvu I
ApaL I	BsrD I	Hinf I	Pvu II
Apo I	BsrF I	HinP1 I	Rsa I
Asc I	BsrG I	Hpa I	Rsr II
Ase I	BssH II	Hpa II	Sac I
Ava I	Bst1107 I	Hph I	Sac II
Ava II	BstB I	Kas I	Sal I
Avr II	BstE II	Kpn I	Sap I
BamH I	BstN I	Mbo I	Sau3A I
Ban I	BstU I	Mbo II	Sau96 I
Ban II	BstX I	Mlu I	Sca I
Bbs I	BstY I	Mnl I	ScrF I
Bbv I	Bsu36 I	Msc I	SfaN I
Bcg I	Cla I	Mse I	Sfc I
Bcl I	Csp6 I	Msl I	Sfi I
Bfa I	Dde I	Msp I	Sma I
Bgl I	Dpn I	MspA1 I	SnaB I
Bgl II	Dpn II	Mun I	Spe I
Bpm I	Dra I	Mwo I	Sph I
Bpu1102 I	Dra III	Nae I	Ssp I
Bsa I	Drd I	Nar I	Stu I
BsaA I	Eae I	Nci I	Sty I
BsaB I	Eag I	Nco I	Taq I
BsaH I	Eam1105 I	Nde I	Tfi I
BsaJ I	Ear I	NgoM I	Tsp509 I
Bsg I	Ecl136 II	Nhe I	Tth111 I
BsiE I	Eco47 III	Nla III	Xba I
BsiW I	Eco57 I	Nla IV	Xcm I
Bsl I	EcoN I		

Nucleases
Nuclease BAL-31 71
Exonuclease III 72
Mung Bean Nuclease 73

Other
T4 Polynucleotide Kinase 74
recA 75
Alkaline Phosphatase (CIP) 75

Nucleic Acids, Linkers, Primers

Cloning Vectors
pBR322 DNA 79
pUC19 DNA 79
pNEB193 DNA 79
M13mp18 and 19 DNA 79

Phage DNAs
φX174 RF I, RF II, and Virion DNA 80
Lambda DNA 81
Lambda DNA (N6-methyl-adenine-free) 81

DNA Size Standards
Lambda DNA–Mono Cut Mix 81
Lambda DNA–Hind III Digest 82
Lambda DNA–BstE II Digest 82
φX174 RF DNA–Hae III Digest 83
pBR322 DNA–BstN I Digest 83
pBR322 DNA–Msp I Digest 84
Mid Range PFG Markers I & II 84
Yeast Chromosome PFG Marker 85
Lambda Ladder PFG Marker 85
Low Range PFG Marker 85

Primers, Linkers, Custom Oligonucleotides
M13 Primers 86
pBR322 Primers 88
Lambda gt10, gt11 Primers 90
Transcription Promotor Primers 91
Random Primer 91
M13 Hybridization Probe Primer 91
malE Primer 91
Linkers 92
Adaptors
Reagents for mRNA Isolation
Custom Oligonucleotide

FIGURE 16.1
Mail Order Molecular Biology

This is a reproduction of the table of contents of one of many companies in the business of selling reagents for molecular biology research. Note the large number of different restriction enzymes available. You will also find many of the other enzymes mentioned in this and previous chapters, such as DNA and RNA polymerases and DNA ligase.

DNA Technology

Basically, what genetic engineers do is introduce specific pieces of DNA into cells in such a way that the introduced or "foreign" DNA is replicated in the cell. As the cell containing the newly introduced DNA divides, a **clone** or exact copy is produced. In a fashion similar to the way in which a bacterium grows into a colony of identical bacteria, all the cells in a clone colony are identical and, in this case, contain copies of the introduced DNA. Cells containing that new DNA can be grown in any quantity. The foreign DNA is then available in large amounts for study or manipulation. So too the protein product of the genes of introduced DNA can also be made in large quantities as each cell containing the introduced gene can now make the protein encoded by the gene.

The trick then is to attach the DNA segment of interest to a proper carrier DNA molecule, called a **vector**, so that the host cell will treat the combination of vector and foreign DNA (the **recombinant DNA**) as a natural part of its genetic constitution.

In the discussion that follows, we'll examine the enzymes used in recombinant DNA research and then show just how these enzymes are used to clone genes. All the enzymatic tools required to perform the process are natural parts of biological systems, and all the vectors and DNA transfer technologies are extensions of normal genetic components and processes.

Restriction Enzymes and Sticky Ends

While the basic biology of DNA had been studied for decades, it wasn't until the discovery of restriction enzymes that recombinant DNA became possible. **Restriction enzymes** (also called restriction endonucleases) are enzymes capable of cutting the phosphate backbones of DNA molecules at specific base sequences called the **restriction site**.

Genetic engineers did not invent restriction enzymes. These enzymes exist naturally in bacteria, where they degrade invading viral DNA. Restriction enzymes are of great use to genetic engineers because, although their function in nature is to cut viral DNA, the enzymes will also cut those same sequences of nucleotides along any strand of DNA, from any source.

Each species of bacterium produces at least one restriction enzyme. Hundreds of different restriction enzymes that recognize and cut hundreds of different restriction sites have been isolated. Many restriction enzymes (as well as other enzymes used in cloning and molecular biology) are commercially available from molecular biology supply companies (Figure 16.1). As an example of how one restriction enzyme works, let's take a look at *Eco* R1. Most restriction enzymes are given names that are derived from the species from which they were isolated. *Eco* R1, for example, was the first (thus "1") restriction enzyme isolated from *E. coli*.

In the laboratory, when *Eco* R1 is added to isolated DNA, it will cut the DNA everywhere it finds its restriction site. In the case of *Eco* R1, that restriction site is six bases long:

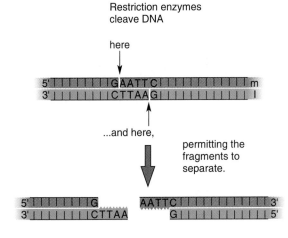

The enzyme cleaves this sequence between the G and A in the staggered manner shown, so that the ends of each resulting **restriction fragment** (cut piece) of DNA contain short, sin-

TABLE 16.1 RESTRICTION ENZYMES AND THEIR SITES OF ACTION

Enzyme	Restriction Site
BamHI	G⌐G A T C C C C T A G⌐G
EcoRI	G⌐A A T T C C T T A A⌐G
HindIII	A⌐A G C T T T C G A⌐A
HpaII	C⌐C G G G G C⌐C
MboI	⌐G A T C C T A G⌐

recognize a four-base sequence instead of the six-base sequence recognized by *Eco* R1. In fact, there are also restriction enzymes that recognize five- and eight-base restriction sites, and some that recognize the same site as others.

Splicing and Cloning DNA Molecules

Any restriction fragment of DNA can be "cloned" by inserting the fragment into a compatible restriction site in a vector DNA molecule. Usually the restriction fragment will contain the gene that we are interested in studying or engineering. The most commonly used types of vector molecules are bacterial plasmids and DNA viruses that have themselves been engineered to facilitate the production of recombinant DNA. In either case, cloning a fragment is fairly straightforward and is outlined in Figure 16.2.

First, the vector molecule is cleaved by the same restriction enzyme used to generate the fragment of foreign DNA (see Figure 16.2). That way, the sticky ends of the vector will be the same as those of the DNA fragment to be inserted. Then the cleaved vector molecules are mixed with the foreign DNA under conditions that allow the complementary bases of the sticky ends to hydrogen bond, or anneal. At this point, the enzyme ligase is added (see Figure 16.2). Ligase covalently joins the sugar–phosphate backbones, making the foreign DNA an integral part of the plasmid.

The newly ligated DNA plasmid is added to bacterial cells that have been treated with calcium to make their cell walls per-

gle-stranded tails that are called "sticky ends." The term "sticky" comes from the fact that the tails can easily realign with tails from other fragments with the same sticky end because the sequence of the tail is complementary. Fragments can be held together, temporarily, by the hydrogen bonding between complementary bases of the sticky ends.

The restriction sites of some other restriction enzymes are presented in Table 16.1. You will note that some enzymes

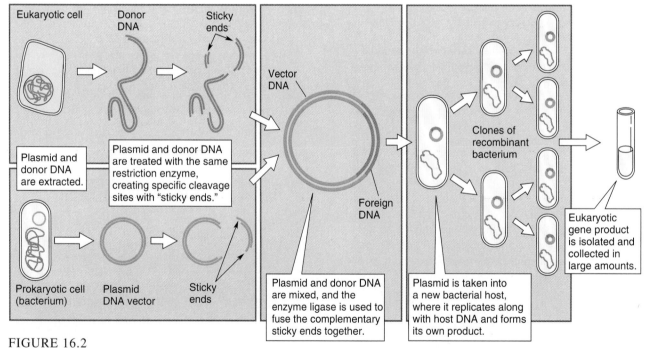

FIGURE 16.2
Recombinant DNA
Donor DNA from nearly any source can be inserted into a plasmid where it will be replicated and transcribed as part of the bacterial genome. Bacterial plasmids and segments of donor DNA are obtained by using the same restriction enzyme. The plasmid and donor DNAs are mixed, and the enzyme ligase is then used to fuse the matching sticky ends of the donor DNA to the sticky ends of the plasmid. Next, the plasmid is transformed into a bacterial host, where it will be replicated along with the bacterial DNA.

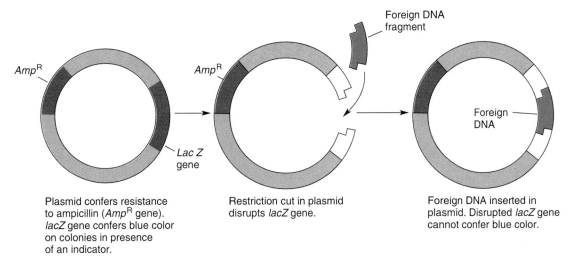

Plasmid confers resistance
to ampicillin (*Amp*^R gene).
lacZ gene confers blue color
on colonies in presence
of an indicator.

Restriction cut in plasmid
disrupts *lacZ* gene.

Foreign DNA inserted in
plasmid. Disrupted *lacZ* gene
cannot confer blue color.

FIGURE 16.3
pUC19: A Widely Used Plasmid For Recombinant DNA
The ampicillin resistance gene in plasmids such as pUC19 (developed by J. Messing and his associates) allows only those bacteria harboring this plasmid to grow on media containing the antibiotic ampicillin. The *lacZ* gene produces an enzyme capable of turning the compound X-GAL from colorless to deep blue. Thus bacteria containing this plasmid will become blue. When foreign DNA is inserted into the cloning restriction sites, the *lacZ* gene becomes nonfunctional, and the bacterial colony cannot turn blue. This enables the experimenter to identify those plasmids containing bacterial colonies that are carrying the newly inserted foreign DNA.

meable. Thus some of the treated bacteria will absorb the DNA. (This is essentially an intentional version of the process of transformation described in Chapters 12 and 15.) The introduced DNA will then be replicated each time the bacteria divide.

Since transformation is not a very efficient process, there is a problem at this stage in distinguishing those cells that have taken up plasmid from those that haven't. This distinction is accomplished by using plasmids that contain genes that confer antibiotic resistance to their host cells (Figure 16.3). So the mixture of cells in a transformation experiment are simply plated on media containing an antibiotic that kills those cells that have not taken up the plasmid. The antibiotic ampicillin is commonly used for this purpose. A bacterial cell that does take up the plasmid will be able to grow in the presence of ampicillin to form a colony, and that colony can be used to inoculate culture flasks to produce essentially unlimited numbers of identical, cloned cell—each one carrying the introduced gene.

It may have occurred to you that the sticky ends of the vector could simply rejoin without incorporating the new DNA at all. And you are absolutely right; this happens. So how do we separate those bacteria harboring the reconstituted plasmid from those that have the plasmid with the newly inserted DNA?

The solution is quite simple in concept. The vector plasmids are engineered so that the restriction site into which the DNA is to be inserted is in the middle of a gene. When DNA is inserted into the middle of a gene, that gene becomes nonfunctional. Therefore plasmids that have simply rejoined without inserting the new DNA will retain the activity of that gene, while those plasmids that have the extra DNA will now have a nonfunctional gene. We only have to look for loss of that gene activity to know those bacteria that have the recombinant DNA.

In practice, one method of identifying bacteria having plasmids with inserted DNA is to use a cloning restriction site that

lies within a gene whose protein product is an enzyme capable of turning a colorless indicator chemical dark blue. In the plasmid pUC19 of Figure 16.3, the intact *lacZ* gene produces such an enzyme. Those bacterial colonies that turn blue contain only the reconstituted original plasmids. As shown in Figure 16.4, those colonies that have the recombinant DNA plasmids are white because the *lacZ* gene is interrupted by the inserted DNA. A white colony can be picked off the indicator plate and grown in test tubes, liter bottles, or 20-gallon vats, all the while producing millions of copies of the engineered plasmid DNA.

Libraries. We can see then that it is actually quite simple to produce molecular clones of any piece of DNA: that is, if you already have that piece of DNA. The hard part is getting precisely the segment of DNA you want.

So how do genetic engineers find that molecular needle in the genetic haystack? They may check it out of a library. This obviously demands a bit of explanation.

A **library** is a collection of various clones. Each clone bears a specific piece of DNA. Creating a library is simply an expanded version of the process we outlined in Figure 16.2. First, the source DNA (for example, from a eukaryotic cell) is cleaved with a restriction enzyme. Then, in a mass ligation reaction, all the fragments are ligated separately into a vector molecule that has been cleaved with that same restriction enzyme. Next, all the resulting recombinant molecules in the ligation reaction are transformed into numerous host bacterial cells, and all the bacterial colonies that have recombinant DNA are collected (for example, all the white colonies that grow on ampicillin). This collection of clones, each one potentially representing a different fragment of DNA inserted into the vector, constitutes our library. It can simply be stored in a small test tube in the refrigerator. If the source DNA were, say, human nuclear DNA, we would call

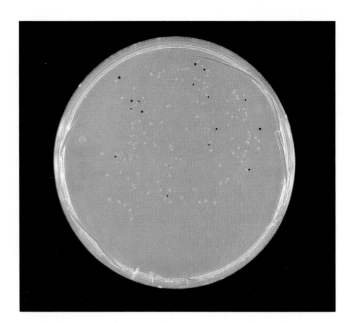

FIGURE 16.4
Identifying Colonies With Recombinant Plasmids
Because this Petri plate contains ampicillin, only bacterial colonies
containing the pUC19 plasmid are able to grow. Those with the
nonrecombinant plasmid have a functional *lacZ* gene and are
therefore blue. The white colonies contain recombinant plasmids
with donor DNA inserted within the *lacZ* gene.

it a human genomic library, because it would potentially contain
clones of an entire human genome.

It is important to realize that the number of clones necessary to represent a complete library depends on the size of the
genome and the carrying capacity of the vector. In developing
a library of the human genome, for example, producing plasmid clones averaging about 10,000 base pairs each, well over
a hundred thousand individual clones would have to be in our
library in order to contain all the sequences of the human
genome. If we were making a library of the much smaller
genome of *E. coli*, a few thousand clones would suffice.

In some cases we may not want to recover clones of the
intact gene (complete with introns), or we may want a library
that consists not of the entire genome but only of those genes
that are expressed, say, in the liver. Such specificity can be
accomplished by constructing what is called a library based
on complementary DNA (cDNA), a **cDNA library**. The
cDNA is a complementary copy of mRNA. The mRNA can
easily be isolated from a particular tissue or organ, so the
cDNA library will contain only clones of those genes that are
expressed as mRNA in that tissue. Also, since mRNA is the
starting material, no intron or promoter regions will be represented in the library.

To make cDNA for cloning into a library (Figure 16.5),
mRNA is first treated with the enzyme reverse transcriptase.
As we saw in the last chapter, this enzyme does transcription
"backward"—it makes a DNA copy of the mRNA. Once the
copy of the message—the cDNA—is made, the mRNA is

removed, and the DNA is made double-stranded by incuba-
tion with DNA polymerase and nucleotides. The double-
stranded cDNA is then ligated into a plasmid vector to create
the cDNA library.

Remember that the term *library* implies a collection of
many different things. So once we have any library, we are
faced with the problem of sorting it out. We must be able to go
through it to find that one clone of the DNA we are interested
in from among the thousands in the library.

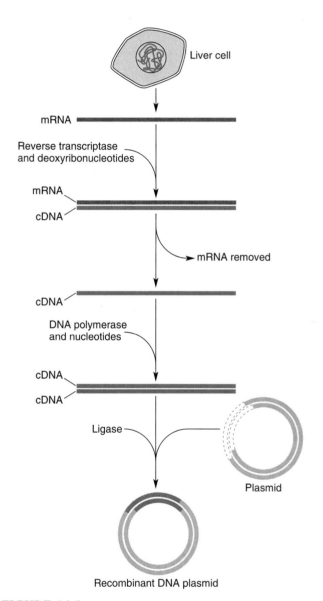

FIGURE 16.5
cDNA
To make cDNA, a complementary copy of mRNA, one needs a
preparation of mRNA. The addition of the enzyme reverse transcriptase and deoxyribonucleotides results in the copying of the mRNA
strand into its DNA complement. The mRNA is removed, and the
second strand of the DNA is made by adding the enzyme DNA
polymerase and more deoxyribonucleotides. The result is a double-
stranded DNA molecule, with the base sequence complementary to
the original mRNA, that can be cloned into a plasmid.

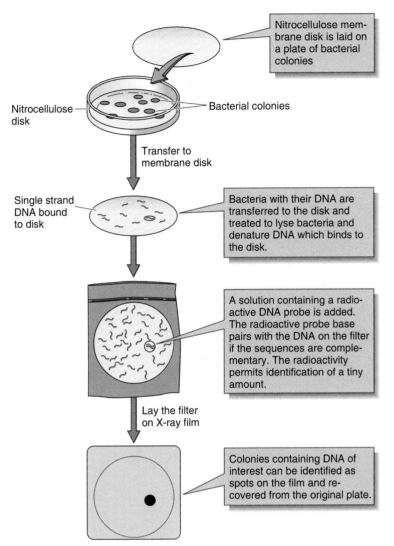

Nitrocellulose membrane disk is laid on a plate of bacterial colonies

Nitrocellulose disk

Bacterial colonies

Transfer to membrane disk

Single strand DNA bound to disk

Bacteria with their DNA are transferred to the disk and treated to lyse bacteria and denature DNA which binds to the disk.

A solution containing a radioactive DNA probe is added. The radioactive probe base pairs with the DNA on the filter if the sequences are complementary. The radioactivity permits identification of a tiny amount.

Lay the filter on X-ray film

Colonies containing DNA of interest can be identified as spots on the film and recovered from the original plate.

FIGURE 16.6
Screening a Recombinant DNA Library
When a disk of nitrocellulose membrane is placed on the surface of the Petri plate, most of the bacterial colony is transferred to the membrane disk, but some of the colony is left behind. Treatment of the membrane with alkaline detergents releases the DNA from the bacteria, denatures the DNA to single-stranded form, and bonds the DNA to the membrane. A radioactive DNA probe solution is added to the membrane. The radioactive probe will bind to DNA from those colonies harboring plasmids that contain sequences complementary to the probe.

Screening a Library. Much like the libraries that you are used to sorting through, it is possible to search a recombinant DNA library for the fragment of interest. If you were looking for a volume in a book library, you could look at each book one by one, until you came upon the one you wanted. This would be very tedious and time consuming. A better approach is to search the whole library at once, using a computer or card catalog. In a similar fashion, it is possible to screen the whole recombinant DNA library at once.

Figure 16.6 outlines the procedure. The first step is to grow the bacterial colonies containing the fragments of the library on Petri dishes. Then a sterile membrane disk is placed down on top of the colonies. A portion of each colony sticks to the disk; the rest remains on the plate. The cells on the disk are then lysed by treatment with alkaline solutions. During this process the DNA is released from the cells, is denatured (becomes single-stranded), and binds directly to the membrane disk. The disk now has a spot of DNA at every place where there was a colony on the plate.

The actual search of the library is accomplished by placing the membrane disk in a solution that contains a single-stranded piece of radioactive DNA that is complementary to the DNA we are after. This is called a **DNA hybridization probe**. This solution allows the DNA probe to come into contact with the DNA from the bacterial colonies. For most of the colonies, the DNA probe will not base pair (that is, hybridize) with the DNA from the clone. However, in those few cases where the DNA sequences are complementary, the radioactive DNA will hybridize with the cloned DNA and thereby be attached to the filter. After hybridization, the filters are washed to remove the excess, nonhybridized radioactive probe. The filter is then put next to x-ray film, which will get exposed at the areas of radioactivity by a process called **autoradiography**. When developed, the film will have spots where the radioactive probe hybridized to the DNA on the filter, indicating the position of bacterial clones with plasmid DNA homologous to the probe. The x-ray film can be placed under the original Petri dish, and the bacterial clone over the spot can be picked up and grown in quantity.

But where do you get the hybridization probe? There are several possibilities, and the choice of probes depends on the experiment and the amount of information available.

One possibility is simply to build a probe in a **gene machine** as described in Figure 16.7. A gene machine is an automated chemical synthesis apparatus that will build a sequence with A, T, G, and C in any order specified by the engineer (Essay 16.1). Therefore, if you know the amino acid sequence of the protein corresponding to the gene you want, you simply use the genetic code to translate the amino acid sequence into the nucleotide sequence, enter the sequence into the keyboard, and have the machine build it. (At present, the gene machines can produce only a hundred bases or so at a time. Therefore only part of the necessary sequence for most proteins can be built during any one run of the machine.) Once the DNA is made and recovered from the machine, an enzyme named **kinase** will add a radioactive phosphate group to the 5′ end, and your radioactive probe is ready for use.

You can still recover a bacterial clone containing a particular DNA fragment even if you don't know the sequence of the amino acids. If purified mRNA for the specific gene is available, it can be labeled by kinase and used as a probe. Yet another option is to use the gene from one organism to search a library for clones of that gene in another organism. This is possible in those cases where sequence similarities have been conserved through evolution. There are other more intricate procedures for identifying clones of DNA fragments, and a major amount of time in a cloning laboratory can be spent on the process of determining new and different ways to identify clones.

Essay 16.1 ● DNA SEQUENCING

Without a doubt, the ability to determine the exact sequence of bases in a gene or restriction fragment has substantially contributed to our knowledge of how molecular genetics works. The developer of the technique described here, Dr. F. Sanger, received the Nobel Prize for his contribution. A cursory scan of the preceding chapters shows that DNA sequence information forms one of the cornerstones of our comprehension of gene structure and regulation. Let's briefly examine one method by which it is possible to determine the sequence of bases in a DNA fragment.

Sequencing begins by dividing the DNA into four test tubes, one for each of the reactions that will identify the positions of the A, T, G, and C residues in the fragment. In each tube a short, radioactive piece of DNA called a primer is annealed to the end of the DNA. Then the DNA is replicated by the enzyme DNA polymerase. This replication process is essentially the same as happens in nature, except that an additional component is added to each reaction tube. The additional component is a dideoxynucleotide triphosphate. Dideoxynucleotides lack two oxygens and, when incorporated into a growing DNA chain, cause the reaction to terminate because there is no 3′ OH group for the next nucleotide to join. Thus, when a dideoxy adenosine triphosphate (ddATP) gets incorporated, a DNA fragment is created that begins at the primer and ends where there is an A in the sequence. Each of the reactions is analyzed by electrophoresis to determine the order of the bases in the fragment.

The process of DNA sequencing is now automated to a large extent. There are computer-driven x-ray scanners to interpret the results of the electrophoresis. There are also machines that use robots to run the sequencing reactions, lasers to scan the gels, and computers to interpret the results!

(a)
DNA replication experiments are set up at the template, dNTP's, a radioactive primer and a dideoxy NTP. In this example, we will follow the tube with dideoxy CTP (ddCTP), but parallel reactions will be done with ddATP, ddGTP and ddTTP.

(b)
Replication proceeds, but the new strands will terminate when a ddCTP is incorporated instead of a dCTP. Some strands will be shorter, some will be longer, but all will end with ddCTP.

(c)
When each of the four reactions are separated side-by-side using gel electrophoresis, the sequence can be deduced from size and position of the radioactive bands.

5′ GGCTATTAGCCGTACTCGCA 3′

ddATP   ddTTP   ddGTP   ddCTP

5′ GGCTATTAGCCGTACTCGCA 3′
                    3′ GCGT 5′ — RNA primer

5′ GGCTATTAGCCGTACTCGCA 3′
3′ CCGATAATCGGCATGAGCGT 5′

5′ GGCTATTAGCCGTACTCGCA 3′
3′ CGATAATCGGCATGAGCGT 5′

5′ GGCTATTAGCCGTACTCGCA 3′
              3′ CGGCATGAGCGT 5′

5′ GGCTATTAGCCGTACTCGCA 3′
               3′ CATGAGCGT 5′

ddCTP ddGTP ddATP ddTTP

Longer DNA fragments

C C G A T A A T C G G C A T G A

Shorter DNA fragments

Autoradiogram

Sequence as shown on autoradiogram   Sequence of primer

3′ CCGATAATCGGCATGAGCGT 5′
5′ GGCTATTAGCCGTACTCGCA 3′

The complementary sequence is the sequence of the DNA at the top

His	Ser	Arg	Lys	Leu	Amino acids
CAC	AGC	CGC	AAG	UUA	Codons
GTG	TCG	GCG	TTC	AAT	Complementary DNA

FIGURE 16.7
Constructing a Probe with a Gene Machine
In some cases, knowledge of the amino acid sequence of a protein can be used to deduce the sequence of bases necessary to construct a hybridization probe for detecting clones of the protein. Ambiguities must sometimes be built into such probes, where, for example, the amino acid serine can be coded for by more than one possible codon. Gene machines are capable of producing short segments of DNA (up to about 100 bases long) of any base sequence. The technician has only to enter the sequence of bases into the control keyboard.

So we see how DNA libraries are produced and how genes or parts of genes are recovered. We've covered a number of complex ideas to this point, so now let's see how genetic engineers put their skills and techniques to useful work.

Applications of Genetic Engineering

Now that we have some insights into the process, let's look at some examples of the application of genetic engineering to practical problems.

Manufacture of Biomolecules

Early in the development of DNA technology, it was realized that the major strength of genetic engineering is the capacity to produce large amounts of what was originally a rare piece of DNA. We have seen that any one of the clones from our human DNA library represents a very minor part of the human genome; yet as part of a plasmid we can recover large quantities of that piece of DNA. We can do this simply by growing the bacteria that harbor the plasmid, then isolating the plasmid DNA. Because DNA can code for proteins, that

idea can be extended to the production of proteins that are otherwise rare or hard to obtain.

Take, for example, insulin. Insulin is a protein hormone, naturally produced in human bodies. However, people with diabetes must take regular injections of insulin. Real human insulin was once available only by harvesting human pancreatic tissue from cadavers. Since this source was insufficient to provide our medical needs, insulin was also isolated from pigs and cows and then prepared for use by diabetics. Insulin from other animals is not identical in amino acid sequence to human insulin, but it works in the same way. However, because it *is* subtly different, some people develop allergic reactions to injection of animal insulin. Since the supply of authentic human insulin was very small indeed, people allergic to the pig or cow insulin were in trouble.

Genetic engineers saw this as a good opportunity to apply their technology. A complete cDNA clone for human insulin was recovered from a library, providing all the coding information for true human insulin. In order to have that cDNA information expressed as insulin protein, the cDNA was placed in a plasmid next to a promoter sequence. In fact, the now familiar *lac* operon promoter (see Chapter 14) can serve to transcribe the mRNA for insulin. Now instead of producing the enzymes to digest lactose, that promoter initiates the production of human insulin within the bacteria. This insulin is human in every detail, each amino acid is the right one, in the right order. Since these bacterial clones can be grown in any quantity desired (Figure 16.8), the amount of authentic human insulin is now essentially unlimited.

FIGURE 16.8
Large-Scale Growth of Bacteria
Industrial fermenters allow the production of essentially unlimited amounts of bacteria containing recombinant DNA.

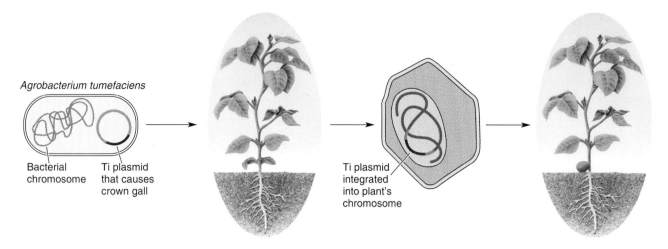

FIGURE 16.9
Plant Genetic Engineering
Agrobacterium tumefaciens contains a plasmid that the bacterium transfers to plant chromosomes in the process of infection. Normally, the presence of this plasmid in the plant chromosome would cause crown gall disease, but the plasmid can be engineered to remove the genes responsible for the gall.

In fact, the process of producing insulin, or any other protein, can be streamlined even more by further genetic engineering. The recombinant DNA plasmid can be modified further by adding a short DNA segment between the promoter and the cDNA, so that the resulting protein has something called a signal peptide at its N-terminal end. This signal peptide turns the insulin into a secretory protein (see Chapter 13). As the insulin is synthesized, it is secreted across the bacterial membrane into the surrounding medium. In the process, the signal peptide is removed, thus leaving the medium rich in authentic human insulin. So now we don't even have to harvest the bacteria to recover the insulin. Large growth vats of the engineered bacteria can be supplied with fresh medium at one end, while the medium containing the secreted insulin can be drawn off at the other end.

We have seen one immediate promise of genetic engineering—the production of insulin—and this technology can be applied to a large number of other useful substances. The immediate benefit of such technology to people is pretty clear. But the technology holds other promises that, in the long run, may have other implications. In particular, it may improve human life through its influences on additional areas of medicine and agriculture.

Agriculture

For centuries, farmers have been waging war with plant pests and diseases, infertile soils, and the ongoing need to produce more to feed the world's burgeoning population. Genetic engineering offers new hope as it raises the possibility of making plants resistant to insects and bacterial and fungal disease and able to grow in previously unsuitable conditions. Productivity may also be improved by selectively increasing the size of the edible plant parts, such as storage roots, seeds, or fruits. Fur-

thermore, those foods can be made more nutritious through alterations in their essential amino acid content. (Most plant foods lack one or more of the "essential amino acids," those our bodies cannot manufacture.) So we are on the threshold of a genetically engineered agricultural revolution.

In one sense, plant genetic engineers have a real advantage: several important species (carrots, cabbage, citrus, and potatoes) can be grown from single somatic cells. Potentially, after a plant cell has been modified by the introduction of a novel gene, the researcher can grow a clone of altered cells, and from that clone of cells an entire plant can be generated.

The first step in producing plants with engineered genomes is to select a vector capable of carrying the foreign gene into the plant cell. So far, the most reliable vector of plant genes is a plasmid called **Ti**, found in the bacterium *Agrobacterium tumefaciens* (Figure 16.9). This bacterium infects plant cells and causes a disease called **crown gall**. During infection of a plant, the Ti plasmid is transferred to the plant's chromosomes. From then on it is a stable part of the plant's genome. In nature, the transfer of the Ti plasmid brings on the symptoms of crown gall disease. However, genetic engineers have now developed Ti plasmids that have had the disease-causing genes removed. So we now have a harmless natural gene vector.

Genes can be spliced into the Ti plasmid, by the methods described above, and reintroduced into the *A. tumefaciens* which then infects plant cells and transfers the DNA to the plant genome. Using this technology, many new varieties of plants have been produced and are on their way to being used in farmers' fields (Figure 16.10).

The problem is that only certain plants can be transformed with Ti plasmids. These do not include the important grains such as corn, wheat, rye, and oats, by far the world's most important crops. However, research into extending the range of plants infected by *A. tumefaciens* continues with

FIGURE 16.10
Engineered Plants
Practical applications of genetic engineering in plants are now available, such as these plants that have been engineered to survive treatment with an otherwise potent herbicide.

some promise, and advances have recently been made using miniature needles to microinject DNA directly into plant and animal cells, as well as microprojectile "guns" that shoot DNA-coated metal particles into cells.

Plant genetic engineering is difficult for another reason: most characteristics that need improvement, such as growth rate, relative size of edible parts, and a balance of essential amino acids, are polygenic—that is controlled by many genes. Furthermore, the identities of such genes are generally unknown. But even if the genes are identified, you can imagine how difficult the replacement of 5 or 10 or even 100 genes might be.

Still, there have been some impressive examples of successful genetic engineering in plants. For example, plants resistant to certain herbicides possess special enzymes that protect those plants from the killing action of the herbicide. Genetic engineers have cloned the enzyme responsible for the resistance and are now in the process of transferring that gene into valuable crops in hopes of solving major weed control problems by spraying infested fields with herbicides that will kill the weed but not the engineered crops. The list of such examples continues to grow, and it probably won't be long until genetically engineered plants are quite common (see Figure 16.10).

Medicine

Equally astounding progress has been made in the application of genetic engineering technology to medically related fields. This list of accomplishments continues to grow and can be summarized under two types of use: diagnosis or treatment.

Diagnosis: Genetic Engineering on the Trail of Killers. There are well over 2000 metabolic genetic errors in the human

population. Some, such as Tay–Sachs disease and sickle cell anemia, occur with high frequency in certain races or ethnic groups. Others show no racial or ethnic bias, plaguing all groups about equally. Some disorders appear in the young and mark them for life, while others, such as Huntington's disease, have late ages of onset. In the latter, the individual may have reproduced and passed on the genes for the disease long before being aware of his or her own fate (see Chapter 11). In the past, we could predict the probability of a disease by pedigree analysis and then, often, just wait for it to occur.

Things are changing—rapidly. Genetic diseases are the result of changes in the DNA sequence of specific genes, in the manner of the mutations in the enzymes of metabolism studied by Garrod (see Chapter 12). Many of the mutations that lead to genetic disease are detectable by modern DNA methods. For example, researchers are now able to detect sickle cell carriers (heterozygotes) by examination of DNA samples, because the mutation leading to sickle cell disease occurs within a restriction site in the beta-globin gene. When the changes alter a restriction site, the corresponding restriction enzyme will no longer cleave the DNA there. Thus, in the case of the sickle cell allele, shown in Figure 16.11, there will be one large fragment of DNA containing the beta-globin gene, instead of the normal situation where the gene is carried on two smaller ones. This change in fragment size is called **restriction fragment polymorphism**, or **RFLP**. In other cases, mutation may create a restriction site, thereby increasing the number of fragments. All this analysis can be done with DNA collected from blood samples, using electrophoresis and specific DNA probes recovered from gene libraries. (See Essay 16.2 on electrophoresis and hybridization.)

Restriction fragment analysis is a powerful diagnostic tool. People who are heterozygous for sickle cell disease or who have Huntington's disease can be identified and counseled, perhaps in time to prevent the tragedy of afflicted children—all this because of the advent of restriction enzymes and the ability to clone DNA.

This ability to determine the molecular genotype of an individual has two more applications that extend the utilization of DNA technology. First, these DNA analyses can be performed on cells recovered during amniocentesis. Thus fetuses at risk of having an inborn error can be identified unambiguously at a very early stage of development. Second, forensic scientists have applied the DNA technologies to identify criminals—rapists are the prime example. Using DNA that can be isolated from minute quantities of blood or semen, molecular biologists can use restriction enzymes and a variety of DNA probes to create a molecular "fingerprint" of the criminal (see Essay 16.3). These DNA fingerprints are extraordinarily accurate: it is essentially impossible for two people on the earth (except identical twins) to have identical DNA fingerprints. DNA fingerprints have already helped put rapists in jail. It is equally important that DNA fingerprint analysis also has resulted in acquittals of men mistakenly imprisoned for rape.

We can see then that DNA analysis can be a powerful diagnostic implement, potentially able to detect genetic disor-

Essay 16.2 ● ELECTROPHORESIS, HYBRIDIZATION, AND DNA FINGERPRINTING

In many situations it may be impossible, or simply not necessary, to clone the gene that we would like to examine. For example, we may wish to characterize a gene from many individuals, without cloning the gene from each person. Much information can be derived from simple restriction digests and an application of the hybridization technology that we described for screening libraries. In this application, we use radioactive probes to tell us the size of the restriction fragment(s) carrying the gene. The applications of these techniques has produced advances in many areas of genetic science, but the most celebrated application is in the process of DNA fingerprinting.

Electrophoresis

After digesting the DNA sample with the restriction enzyme, the myriad of fragments produced can easily be sorted by size, using the process called *electrophoresis*. Electrophoresis, as its name implies, uses electrical current to move DNA molecules. (The DNA, because of its backbone containing negatively charged phosphate groups, tends to migrate to the positive electrode.) The electrical field provides the motive force, but a gel matrix is required to achieve separation of DNA fragments. The gel acts like a three-

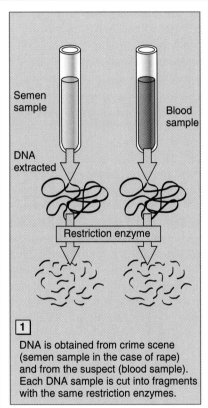
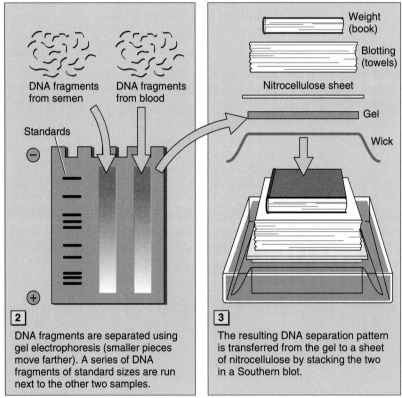

1 DNA is obtained from crime scene (semen sample in the case of rape) and from the suspect (blood sample). Each DNA sample is cut into fragments with the same restriction enzymes.

2 DNA fragments are separated using gel electrophoresis (smaller pieces move farther). A series of DNA fragments of standard sizes are run next to the other two samples.

3 The resulting DNA separation pattern is transferred from the gel to a sheet of nitrocellulose by stacking the two in a Southern blot.

ders such as phenylketonuria, sickle cell anemia, and Huntington's disease. But diagnosis is not treatment. Is gene treatment possible? We are on the threshold of a time when defective genes will be replaced with normally functioning genes.

Gene Replacement Therapy Hypogonadism is a relatively common recessive condition in mice and humans. Homozygotes for the allele responsible, the *hpg* allele, have underdeveloped gonads and (in mice anyway) seem to have no idea of how to go about mating. Such mice lack the ability to produce a chemical messenger called **gonadotropin-releasing hormone**. Its task is to stimulate the release of pituitary hormones that, in turn, prompt the formation of gametes and sex hormones by the gonads. Hypogonadism has been "cured" in mice by injecting eggs with copies of the normal gene—a procedure called **gene replacement therapy**. Because eukaryotic cells often fail to correctly process and translate messages that lack promoters and introns, the starting point for gene replacement therapy is the genomic library, where the gene is still complete. When eggs were taken from

dimensional sieve; in order to move through the gel, the DNA fragments must work their way through, over, and around the gel matrix material. Naturally, the smaller a DNA fragment is, the less its progress is impeded by the gel. The larger a fragment is, the harder it is for it to make it through the matrix. The end result is that the DNA fragments are nicely separated in the body of the gel according to their length.

Hybridization

When genomic DNA is subjected to this analysis, the resulting distribution of fragments is incredibly complex. There are so many fragments produced that it is impossible to distinguish individual bands on the gels. So to detect the size of the DNA fragment bearing the gene or genes of interest, we must use a hybridization probe. The hybridization probe can identify bands on the gel corresponding to the genes (or parts of genes) in the same way that it can point to clones in a library.

To locate the fragment with the hybridization probe, the DNA is separated on the gel, and then the gel is treated with alkaline solutions to denature the DNA into single strands. Then the DNA is transferred to a filter membrane by blotting. The filter membrane then contains a molecular image of the separation achieved by the gel, and the image consists of the denatured, single-stranded DNA fragments. Single-stranded, radioactively labeled probe DNA is then added to the membrane. After a few hours, the probe DNA will have hybridized to the DNA fragments on the membrane that have the complementary base sequence. After washing, the membrane can be placed next to x-ray film to detect the positions of DNA fragments that have hybridized. The result is called a *Southern blot*, named for Dr. E. M. Southern, the person who first described the technique.

Thus it is possible to detect and characterize the pattern of restriction sites in or near a gene from any source without going through the hassle of cloning the gene each time. If tissue samples are available from a crime scene and a suspect, the patterns of DNA fragment fingerprints can be compared.

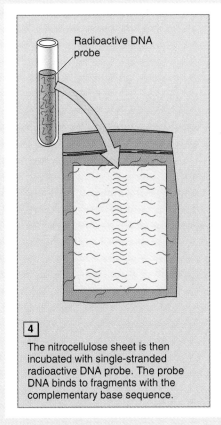

Radioactive DNA probe

4
The nitrocellulose sheet is then incubated with single-stranded radioactive DNA probe. The probe DNA binds to fragments with the complementary base sequence.

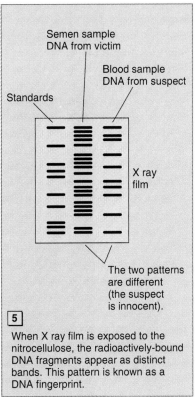

Semen sample DNA from victim

Blood sample DNA from suspect

Standards

X ray film

The two patterns are different (the suspect is innocent).

5
When X ray film is exposed to the nitrocellulose, the radioactively-bound DNA fragments appear as distinct bands. This pattern is known as a DNA fingerprint.

females and microinjected with copies of the normal allele (Figure 16.12), about 20% of them incorporated at least one copy of the gene in one of their chromosomes. After fertilization and reimplantation in a surrogate mother, the altered embryos grew into normal fertile mice. Several other gene replacements have been tried with less success.

Gene replacement therapy is the newest application of genetic engineering, so there are many unknowns. In the first place, treatment of an existing individual might cure them, but the defective gene would still be present in their sperm or eggs, and thus the treatment might have to be reperformed on offspring. While replacing defective genes in cells of an existing body is a more acceptable idea, no one is seriously considering using human eggs for genetic engineering (think of the possibilities). One instance is the treatment of immune deficiencies. In such cases, the cells will be obtained from bone marrow and cultured in the laboratory. (Bone marrow houses cells that give rise to cells of the immune system.) A viral vector will then be used to introduce the normal gene into such cells, and once this succeeds, the cells will be rein-

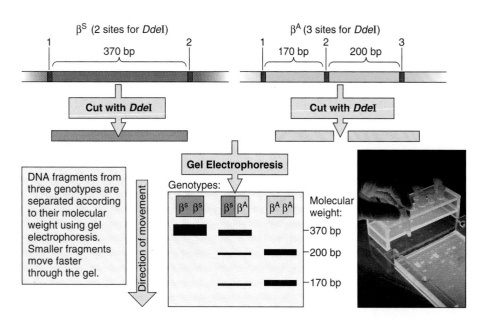

β^S (2 sites for *Dde*I)

1   2
370 bp

Cut with *Dde*I

β^A (3 sites for *Dde*I)

1   2   3
170 bp   200 bp

Cut with *Dde*I

Gel Electrophoresis

DNA fragments from three genotypes are separated according to their molecular weight using gel electrophoresis. Smaller fragments move faster through the gel.

Direction of movement

Genotypes:

$\beta^S \beta^S$ $\beta^S \beta^A$ $\beta^A \beta^A$

Molecular weight:

—370 bp

—200 bp

—170 bp

FIGURE 16.11
Restriction Fragment Detection of Sickle Cell Alleles
Electrophoresis of DNA fragments separates them by size, with the smaller fragments moving further through the gel. A map of the β-globin gene from humans shows that the normal β^A-globin allele region contains three sites where the restriction enzyme DdeI will cut the gene. In the β^S allele that is responsible for the sickle cell trait, the base sequence does not contain the central DdeI cutting site. DNA isolated from persons homozygous ($\beta^S\beta^S$, $\beta^A\beta^A$) and heterozygous ($\beta^S\beta^A$) for the sickle cell allele, as it would appear after digestion with DdeI (see also Essay 16.2).("bp" means base pair.)

troduced into the patient's bone marrow. The results of preliminary experiments with mice indicate that the gene should function properly, correcting the deficiency. Any cells descended from that line will have the corrected genome, so the cure should be permanent for that individual. In this case, since his (or her) reproductive cells weren't treated, the defective gene could still be passed on to children.

Genetic Engineering: A Social Issue

When the use of plasmids for genetic engineering was introduced by Paul Berg and others in the late 1970s, the possibilities were so dramatic and so bizarre that the new technology immediately spawned a raging controversy. Trouble started when the very people who invented the techniques called for a research moratorium so that possible dangers of the new technology could be evaluated.

Supporters of the new gene-splicing technology claimed that the principal benefit is that scientists like themselves will be able to do more experiments and to learn things faster—and that's nothing to scoff at. There are also more practical benefits, some of which have been realized and some of which are on the horizon. For example, there are thousands of growth-hormone-deficient people who, with the help of human growth hormone, can attain normal height. In the past, growth hormone had to be extracted from the pituitary glands of human cadavers and was extremely expensive. Even though the glands of 50 cadavers were needed for only one dose, a few thousand seriously undersized adolescents were treated with some beneficial results, as well as one man who grew to normal height after having been only 4 feet tall until the age of 35. Perhaps, with genetically engineered growth

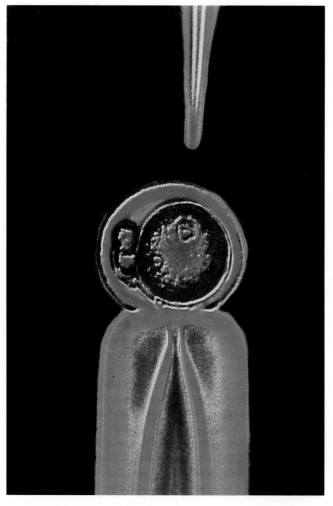

FIGURE 16.12
Microinjection of DNA into a Fertilized Mouse Egg
The injecting needle is at the bottom. The egg itself is held in place by a suction pipette from the top.

Essay 16.3 ● THE POLYMERASE CHAIN REACTION

Much of the whole business of DNA cloning and recombinant DNA technology revolves around producing usable quantities of a rare piece of DNA. Gene cloning is designed to put a gene into a plasmid, then have the plasmid replicated in a bacterium to produce thousands of copies. While this is a very effective process, it is time consuming and labor intensive for the experimenter.

However, recent advances in DNA technology include a remarkable process called the polymerase chain reaction, or PCR for short. PCR is a method for making large numbers of copies of a gene or any other piece of DNA in just a few hours and without the process of cloning, and it is therefore very useful for minute samples (such as those left at a crime scene). PCR involves successive rounds of DNA replication.

First, the target DNA fragment is heated to denature the DNA strands, then cooled in the presence of short DNA primers that are complementary to the ends of the target DNA. The primer–target DNA hybrids are then subjected to DNA replication, using DNA polymerase. At this point, there are now two copies of the original fragment. This process of heating, cooling in the presence of primers, and DNA replication is repeated over many cycles (20–50) until 2^{20}–2^{50} copies of the target DNA are produced.

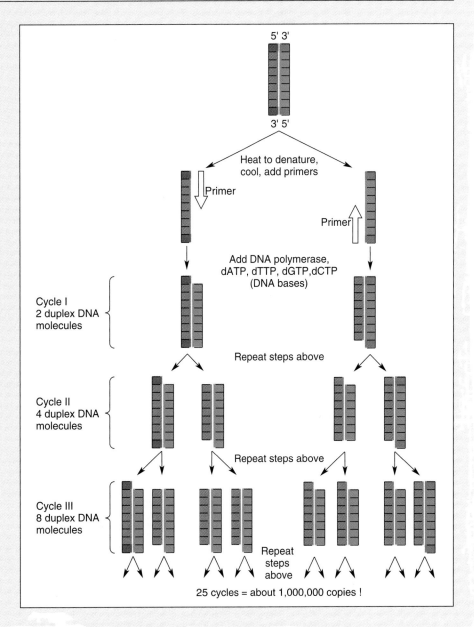

Heat to denature, cool, add primers

Primer

Primer

Add DNA polymerase, dATP, dTTP, dGTP, dCTP (DNA bases)

Cycle I
2 duplex DNA molecules

Repeat steps above

Cycle II
4 duplex DNA molecules

Repeat steps above

Cycle III
8 duplex DNA molecules

Repeat steps above

25 cycles = about 1,000,000 copies !

hormones now available, no one will have to be any shorter than he or she wants to be. Does that seem to you like a blessing, or like cavalier tampering with nature?

There were, and still are, critics of the new techniques. Their fears range from the possible release of newly created disease organisms—genetic monsters capable of creating uncontrollable plagues—to new kinds of cancer. Some critics accused the new biologists of "playing God." However, in response to the potential fears of the public and the requests of the scientists involved, the National Institutes of Health set up rigid guidelines as to what kind of research was to be allowed and what precautions were to be taken.

Can you see why some people were alarmed? It's possible to put the gene for botulism toxin into *E. coli*, a bacterium that is already adapted for thriving in your gut. It's also possible to clone the entire genome of a cancer-causing virus, or the specific cancer-causing genes. Unlikely, you think? This too has been done. But now that it has actually happened, the isolation of cancer-causing genes has been hailed as the research breakthrough that may lead science to a final understanding of cancer and possibly to a dependable cure. The controversy has not completely died out, however, and scientists continue vigorously to monitor themselves and their colleagues for any dangers arising from genetic engineering.

We see then that the potential of genetic engineering is immense. There probably isn't a genetic defect that can't be reversed in some way, a protein that cannot be made in bacterial cultures, or a gene that cannot be replaced. Of course, most genes will not be altered, but it is important to know that we have the technology to do it. The only question is whether technology will outpace our abilities to use such powerful tools wisely and well.

Key Ideas

DNA TECHNOLOGY

1. DNA technology is the altering of biological systems through the manipulation of DNA (genes). Specific DNA segments are introduced into cells, where they are integrated into the genome and begin functioning as any gene would.

2. **Restriction enzymes** cleave DNA at specific DNA sequences known as **restriction sites**.

3. A major gene-splicing technique involves obtaining plasmids from bacteria; using restriction enzymes to cleave plasmid DNA, introducing foreign DNA strands to be cloned, and using the enzyme *ligase* to reattach the sticky ends; preparing recipient cells and permitting the new plasmids to be taken in; and producing large clones and harvesting the gene product.

4. Gene libraries consist of collections of many individual clones, each one representing a different DNA fragment. A specific gene can be recovered from a library through the use of a **hybridization probe**, a radioactive piece of DNA complementary to the desired gene.

5. The technology is available for determining the amino acid sequence of any protein, then synthetically producing a DNA strand that will code for that sequence. The DNA can then be spliced into a plasmid and the protein produced in quantity.

APPLICATIONS OF GENETIC ENGINEERING

1. The production of protein products from genetically engineered microorganisms is already a billion dollar industry.

2. It is possible to genetically engineer plants using cloned genes and the natural plant vector *Agrobacterium tumefaciens*. *Agrobacterium* inserts part of its plasmid genome directly into the chromosomes of some plants.

3. Medical science now employs cloned genes and gene segments to diagnose carriers of certain genetic diseases. Gene replacement therapy in mammals has been partially successful, in that it is possible to microinject genes into fertilized eggs and have them be incorporated into the genome. However, all the ramifications of proper gene regulation are not well understood, and it will be some time before gene replacement therapy will be a viable part of human medicine.

GENETIC ENGINEERING: A SOCIAL ISSUE

1. Genetic engineering remains controversial, although industry now employs many of the technical procedures. There is an element of medical and genetic risk, but much of the controversy centers around the moral and philosophical issues.

Application of Ideas

1. Consider the abilities of genetic engineers to produce proteins in vast quantities, and also consider the fact that proteins can be altered simply by introducing changes in the DNA sequence. Now, what applications can you see for such products?

2. As science develops DNA probes that are useful for detecting changes in gene sequences, it is becoming possible to define a person's genetic predisposition to certain diseases and conditions. What if a person was denied a job in (for example) the chemical solvent industry because it was determined that he or she was potentially at medical risk by working there? Just how far should genetic screening go?

Review Questions

1. Outline the general steps involved in cloning any particular piece of DNA. (pages 320–321)

2. Define the enzymatic activity and application in genetic engineering of restriction enzymes and ligase. Be specific. (pages 319–320)

3. What is a gene library? How is a gene library screened? (pages 321–322)

4. What is a gene machine? What are its uses in genetic engineering? (pages 323–324)

5. Describe the process of cloning the gene of a particular protein, and then the engineering of that gene for industrial-level expression of the protein. (pages 325–326)

6. Describe the way in which *Agrobacterium tumefaciens* acts as a natural genetic engineer. (pages 326–327)

7. What is the basis for the use of gene probes as diagnostic tools in medicine? (pages 327–328)

8. What is gene replacement therapy, and how might it be accomplished? (pages 328–330)

CHAPTER 17

When DNA Changes

CHAPTER OUTLINE

THE STABILITY OF DNA

DNA DAMAGE AT THE MOLECULAR LEVEL

DNA REPAIR: THE CLEANUP CREW

POINT MUTATIONS

Substitutions

Chain-Termination Mutations

Additions, Deletions, and Frameshift Mutations

Point Mutations in Noncoding Regions of DNA

CHROMOSOMAL MUTATIONS

Deletions and Inversions

Translocations

TRANSPOSITIONS

Essay 17.1 Mutations, Oncogenes, and Cancer

Transposon Structure

Transposons as Mutagenic Agents

*Essay 17.2 Mutation and the Evolution
of Gene Structure*

APPLICATION OF IDEAS

KEY IDEAS

REVIEW QUESTIONS

*I*t is hard to imagine the genetic diversity of life on this planet. Not only are all the species different, but within the species—even within populations and families—each individual varies genetically from the next. It is just this variation that is so important to the process of natural selection, as some genetic combinations prove reproductively more advantageous than others and so come to predominate in future generations. From where does all this genetic variation come? What causes such diversity? Fundamentally, it begins with the genetic changes called mutations, gene sequence alterations that may result in the altered phenotypes that we have seen over the last chapters. All different phenotypes have their origins in mutation, from the white eyes of Morgan's fruit fly to the alkaptonuria of the patients studied by Garrod (Figure 17.1).

We should first note that mutations can have quite different effects, depending on where and when they appear. Mutations in the cell that develops into sperm or eggs may result in seriously ill, malformed (or dead) offspring. If the mutations do not prevent reproduction and growth, the changes may be expressed in the phenotype of the organism and passed on to the next generation. Mutations in other cells of the body, say, those of the skin or liver, are called **somatic** mutations (Figure 17.2) and will not be passed on to the next generation. Somatic mutations, however, are passed on through cell lineages and can cause cell death or even cancer. Fortunately, many mutations are benign; that is, they neither help nor harm the individual.

In general, some mutations must occur in order to generate variability. Mutations provided the raw material for the very evolution that has given rise to the earth's various life forms. In addition, some mutations may be beneficial to individuals. But these positive aspects of mutations must be balanced against the deleterious effects of certain mutations.

The Stability of DNA

The stability of DNA is largely due to the way its nucleotides are arranged. On the outside lies only the sugar–phosphate backbone, with all the potentially reactive side groups of the sugar already covalently bonded. The nucleotide bases lie protected inside this backbone, their reac-

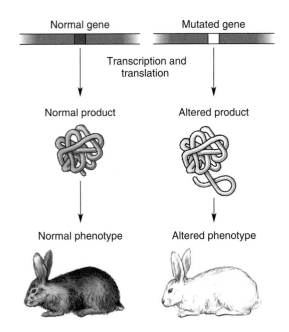

Normal gene   Mutated gene

Transcription and translation

Normal product   Altered product

Normal phenotype   Altered phenotype

FIGURE 17.1
Mutation and Phenotype
Mutations arise as changes in DNA. Their ultimate expression depends on what effect (if any) the change has on the protein product encoded by the gene.

FIGURE 17.2
Somatic Mutations
Mutations that occur in nongerminal cell lineages are called somatic mutations and will not be passed on to the next generation. Here we see an example of a somatic mutation in a human being. Cells in part of this person's scalp are descendants from a cell that acquired a mutation in the genes that produce hair color. Those cells form a patch of skin that has hair with no color.

tive side groups usually immobilized by hydrogen bonds. The stability of the structure is further enhanced by the geometric tightness of the double helix.

DNA in higher organisms has an additional means of protecting itself from chemical damage. It is tightly wound in nucleosomes (see Chapter 9). The bound histones and the specific configuration of the higher-order chromosome package offer few opportunities for direct access to the DNA.

In all organisms, DNA is also protected from the permanent effects of damage by complementary base pairing (see Chapter 12). Any change in the bases or the base sequence in one strand has the potential for being corrected as long as the other strand is unaltered. It's like having photographic prints and negatives. If either is damaged, the other can be used to replace it.

In spite of these seemingly formidable protective barriers, changes in DNA occur with an ongoing regularity. Here we will look into several categories of changes, the agents responsible, and the specific way in which they alter DNA. Then we'll see that much of the initial damage is quickly and efficiently reversed. Although chromosomes have ways of protecting themselves against mutations, mutations do occur.

Basically there are three levels of mutational change: (1) **point mutation**, which involves changes in one or a few nucleotides in DNA, and (2) **chromosomal mutation**, which includes rearrangement, loss, or duplication of whole sections of chromosomes. Point mutations and chromosomal mutations are the result of unrepaired or improperly repaired damage to DNA. Finally, there are (3) **transpositions**, or insertions of copies of DNA into new positions in the genome. Before we consider each of these three classes, we'll look into some causes of DNA damage and their molecular effects.

DNA Damage at the Molecular Level

Any agent that causes a mutation is called a **mutagen**. The chief mutagens are radiation and chemicals. Mutagenic radiations are forms of energy that are capable of penetrating cells to do damage to DNA and may include electromagnetic waves or subatomic particles. High-energy electromagnetic radiations such as gamma rays and x-rays are extremely powerful mutagens, as are alpha and beta particles. Ultraviolet radiation, such as that emanating from the sun, although far less "powerful" than x-rays or gamma rays, is a highly significant source of mutation.

Chemical mutagens are highly reactive agents that damage DNA through chemical reactions. The problem of potential chemical mutagens is not isolated to murky, foul-smelling puddles in chemical dumps. For example, a number of common food additives are mutagenic under certain circumstances. Sodium nitrite, a common additive that prevents bacterial spoilage of bacon, hot dogs, and lunch meats, has the potential for conversion within the body into a chemical mutagen.

Whatever the mutagen, its effect is to generate chemical changes in DNA. Such changes are known as **primary lesions**. Fortunately, for reasons we will get into shortly, most primary lesions are temporary and are quickly eliminated from the gene by highly efficient DNA repair processes. Those that are not repaired, or are improperly repaired, become mutations.

One of the more common forms of primary lesions is a break in a single strand of the double helix, which, like a break in a zipper, interferes physically with its functions (Figure 17.3). Single-strand breaks are particularly prevalent in DNA that has been exposed to x-rays and other high-energy radiations. They interfere with the polymerase enzymes, compromising transcription and replication. Such high-energy radiations may also cleave both strands of the DNA molecule, leaving the way open to chromosome-level mutations of the sort we will discuss shortly.

When DNA is exposed to ultraviolet (UV) light, thymines (and sometimes cytosines) that are adjacent on the same strand can become linked together, side-by-side to form **pyrimidine dimers** (see Figure 17.3). Pyrimidine dimers interfere with replication and transcription, in much the same way that sewing together two teeth in a zipper would interfere with its operation.

Chemical mutagens may also cause random single- and double-strand breaks in the DNA, but most have more specific biochemical effects. We can distinguish three basic types of chemical mutagens. **Intercalating agents** are usually flat molecules that can squeeze between (intercalate) adjacent base pairs of the DNA helix. For instance, molecules of the dye acridine orange lodge between bases along the length of the double helix (see Figure 17.3). This distorts the molecule, potentially interfering with replication and transcription. During replication, the distortion can result in addition or deletion of single bases in the daughter strands. **Base analogs** are compounds that are quite similar to the normal nucleotides and are therefore readily incorporated into growing DNA chains. However, many base analogs can exist in two forms, forms that differ in their hydrogen-bonding character and thus may cause the wrong bases to be inserted in the next round of replication. **DNA-modifying agents** chemically react with the DNA, adding chemical groups to the nitrogen bases. Nitrous acid, for example, changes cytosine to uracil. Some DNA-modifying agents change the hydrogen-bonding character of bases and cause polymerases to misread bases, while others interfere with transcription and replication in other ways.

We have seen that gamma rays and x-rays can directly affect DNA, but they can also cause damage indirectly. These forms of radiation are deeply penetrating and energetic enough to dislodge electrons from many of the cell's molecules. This ionizing effect has led to their being labeled **ionizing radiation**. Ionizing radiation creates a variety of highly reactive, electrically charged molecular groups called free radicals. The free radicals, in turn, create genetic damage by reacting with and altering the nitrogen bases of DNA.

DNA Repair: The Cleanup Crew

Primary lesions in DNA occur at a fairly regular pace and are not at all unusual. In fact, the sources of primary lesions are a normal part of our daily existence. Every day we encounter various levels of mutagenic chemicals and radiation. For that matter, many of the processes of life produce potentially harmful compounds. There are many ways in which primary lesions are repaired, but we'll consider two of the best known, each involving the repair of pyrimidine dimers.

Photoreactivation repair is a repair mechanism effective against UV-induced pyrimidine dimers in a wide range of organisms from bacteria to humans. Blue wavelengths of visible light activate a repair enzyme that cleaves the abnormal bonds that form dimers, restoring the proper arrangement of adjacent pyrimidines. If photoreactivation doesn't repair the dimers directly, the primary lesion can still be fixed by a process

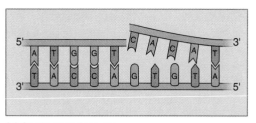

(a) Single-strand breaks

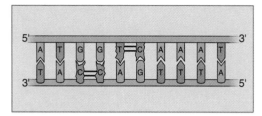

(b) Pyrimidine dimer

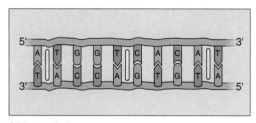

(c) Intercalation

FIGURE 17.3
Primary Lesions in DNA
These modifications of DNA, if left unattended, will result in mutations. The major structural modifications are **(a)** breaks in the sugar–phosphate backbone, **(b)** pyrimidine dimers, and **(c)** intercalated molecules.

known as **excision repair** (Figure 17.4) in which an enzyme moving along the DNA excises the dimer (or, in fact, some other anomalies) along with a few bases on either side. Then DNA polymerase and ligase (Chapter 12) fill in the gap with the proper nucleotides, using the opposite strand for a template.

People with a rare recessive disorder called **xeroderma pigmentosum** ("dry pigmented skin") have an extremely high rate of skin cancer and must constantly protect themselves from sunlight—even to the extent of becoming nocturnal. The cause of xeroderma pigmentosum is a failure of excision repair enzymes to repair primary lesions, particularly UV-induced pyrimidine dimers.

Point Mutations

Now that we have some idea of how DNA is altered by physical and chemical means, and how at least some of these damages can be repaired, we can look into the ramifications of mutations. We will begin with point mutations, those small changes that may involve only a single base pair. We will see that even these small alterations can have sweeping effects on the phenotype.

Substitutions

Base substitution mutations are just what they sound like, substitutions or replacements of a single base pair. There are three outcomes of base substitutions, each of which can be illustrated using the first six codons of the human beta-globin gene (Figure 17.5), which produces one of the protein subunits that comprise hemoglobin.

One outcome would have absolutely no effect on the structure of the protein and thus would be called a **silent mutation**. For example, consider amino acid number 3, leucine, that is encoded by the beta-globin sequence presented in Figure 17.5b. The codon specifying leucine is UUG. If a substitution occurs in the third letter of that codon, leucine can still be put in its proper place in the protein. One need only consult the genetic code table (see Table 13.1) to see that the redundant, synonymous codons of the genetic code will allow the substitution of an A for the third letter for leucine, and that, in fact, most amino acids are specified by more than one codon.

Another outcome would change the structure of the protein but would have no effect on its performance and thus would be called a **neutral mutation**. An example is the Toguchi allele of beta-globin (see Figure 17.5c). Here a substitution occurs in the first letter of the second codon such that the amino acid tyrosine is encoded instead of the normal histidine. (Consult Table 13.1 to be sure you see why.) However, as chance would have it, the substituted amino acid has no discernible effect on the performance of the hemoglobin protein.

While many point mutations are silent or neutral, many others can have drastic effects, as illustrated by the sickle cell allele of beta-globin (Figure 17.5d). In the sickle cell allele, the codon for amino acid number 6 is mutated from GAA (glutamic acid) to GUA (valine). It seems a small change, but it has serious effects on the future structure and performance of hemoglobin. In fact, the hemoglobin of humans bearing this substitution in both of their beta-globin alleles (homozygous persons) periodically undergoes changes that distort the red blood cells into peculiar crescent (sickle) shapes. These altered blood cells have a drastic effect on the health of the individual and cause the symptoms of sickle cell anemia. The irregular cells become trapped in the smaller blood vessels, forming blockages that seriously damage the organs being served. Sickling generally occurs any time the red blood cells have a low oxygen content, as they would in metabolically active regions of the body where they must give up oxygen. Unfortunately, blocking the flow to such organs by the sickled cells further decreases the oxygen content of the surroundings, amplifying the sickling effect. In addition, the body is further deprived of oxygen due to anemia, which results because the affected red cells have shorter than normal life spans.

Chain-Termination Mutations

Some of the most interesting base substitutions involve termination codons. If the new codon produced by a mutation is one of the three termination codons, the base substitution mutation is called, not surprisingly, a **chain-termination mutation**.

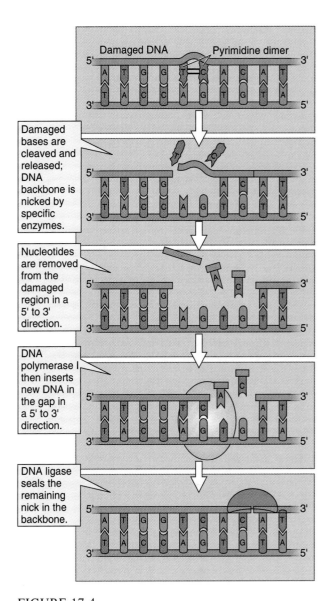

FIGURE 17.4
Excision Repair of a Pyrimidine Dimer
The repair enzyme complex detects a pyrimidine dimer. The damaged section of DNA, plus about 10 bases on either side of the dimer (about 20 bases in all), is removed. A repair polymerase resynthesizes the excised region, using the bases present on the remaining strand as a template. The newly synthesized segment is joined to the original strand by the enzyme ligase.

Unless the terminating codon is very close to the normal end of the coding region, premature chain terminations usually result in totally nonfunctional polypeptides. The majority of lethal mutations in microorganisms are of the chain-termination type.

Sometimes, the normal termination codon mutates to some other form, say, UAA to GAA. Now, instead of terminating, a glutamic acid is added to the chain, and the ribosome simply continues to add amino acids coded by the sequence of bases on the trailing end of the mRNA. Following the normal stop codon is (probably) a random assemblage of bases, fol-

(a) Normal configuration

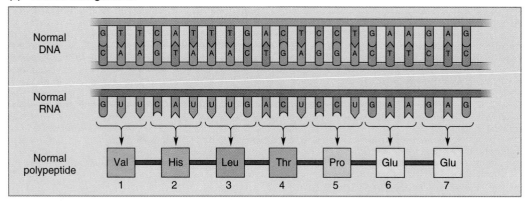

(b) Silent mutation

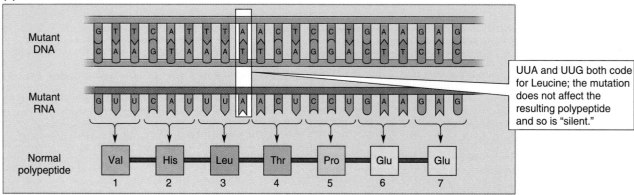

UUA and UUG both code for Leucine; the mutation does not affect the resulting polypeptide and so is "silent."

(c) Neutral mutation (Toguchi)

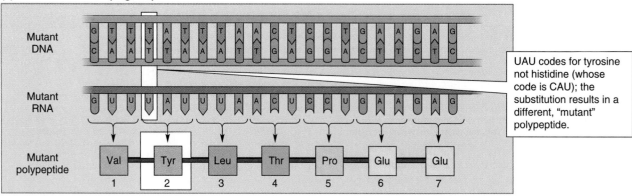

UAU codes for tyrosine not histidine (whose code is CAU); the substitution results in a different, "mutant" polypeptide.

(d) Sickle-cell mutation

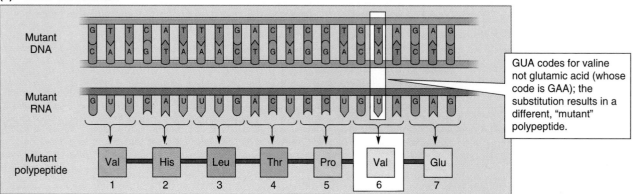

GUA codes for valine not glutamic acid (whose code is GAA); the substitution results in a different, "mutant" polypeptide.

FIGURE 17.5

FIGURE 17.5
Base Substitutions and Mutations in Beta-Globin

The beta-globin gene offers excellent examples of the results of base substitutions. These examples are taken from real life—they are actual sequences of alleles found in human beings. **(a)** The first 21 bases of the coding region of the normal beta-globin gene, and the resulting first 7 amino acids. **(b)** A silent mutation, where the change of a base pair has no effect whatsoever on the resulting amino acids. **(c)** A neutral mutation, where the change of a base does alter the sequence of the amino acids by substituting a tyrosine for a histidine. However, this substitution is called "neutral" because it has no physiological effect on the performance of the hemoglobin molecule. **(d)** Unfortunately, the base change that results in the substitution of a valine for the glutamic acid at amino acid position 6 produces sickle cell disease, which has a drastic effect on the ability of hemoglobin to carry oxygen.

lowed by the poly-A tail mentioned in Chapter 13. The ribosomes would continue translation until, by chance, they encountered a stop codon or the end of the mRNA.

Additions, Deletions, and Frameshift Mutations

It will not surprise you to learn that an **addition** involves the insertion of a base into a nucleotide sequence, and a **deletion** involves the subtraction of a nucleotide. The profound changes caused by these simple alterations may be surprising, however. In fact, additions or deletions of base pairs have more serious effects than do most substitutions because they change the reading frame of the message (see Chapter 13). In Figure 17.6, for example, the insertion of a single base pair changes the sense of the message from there on and creates a completely novel peptide chain. Because additions and deletions cause shifts in the reading frame of messages (Chapter 14), they are called **frameshift mutations**. Frameshift mutations are also likely to cause misreading of translation stop words and to create new stop words in

the middle of the message. They nearly always result in formation of a grossly abnormal polypeptide.

Point Mutations in Noncoding Regions of DNA

What about point mutations that occur in introns (intervening sequences)? Can they possibly affect gene function? One might expect that most mutations in introns would be silent, having no effect on the resulting protein. Indeed most have no effect. But, surprisingly enough, it has been shown that mutations within introns may cause dramatic changes in gene function. For example, consider the genetic condition known as thalassemia, a disorder that leads to serious anemia due to very low amounts of hemoglobin. Thalassemia can be caused by any of 30 known point mutations due to substitutions in the globin gene. Several are located within introns, and apparently these affect the efficiency of intron splicing. Reduced splicing efficiency results in reduced hemoglobin formation and hence the disease symptoms.

We should also mention the effects of substitution mutations in promoter regions. Five of the thirty substitutions that lead to thalassemia are located in the promoter of the globin gene. Apparently these mutations impair promoter function and result in very reduced levels of hemoglobin. All in all, less than half of the substitutions that cause thalassemia are in exons (expressed sequences).

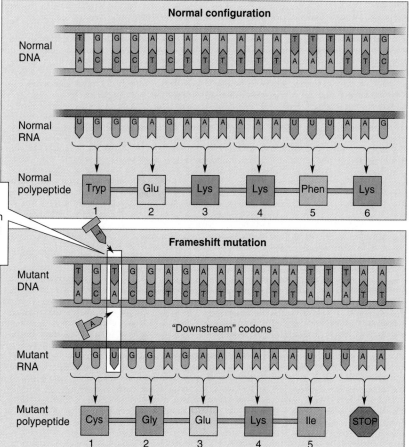

When an additional base pair gets inserted, all subsequent bases shift and the downstream triplet codons change, resulting in a completely different polypeptide.

FIGURE 17.6
The Frameshift Mutation

Addition or deletion of bases usually has a very drastic effect on the resulting protein. For example, the addition of a base in this sequence creates a totally different protein, since each codon is "shifted" in the reading frame. The resulting protein bears no resemblance to the original protein.

Chromosomal Mutations

Chromosomal mutations involve gross changes in chromosome structure, rather than the simple DNA sequence changes we have discussed so far. They involve the complete breaking of the double helix in one or more chromosomes. Usually, the two broken ends simply come back together and are rejoined by DNA repair systems typical of those that correct other kinds of damage. In such breaks in *Drosophila* sperm chromosomes, it has been shown that more than 99.9% are corrected by repair mechanisms.

Deletions and Inversions

Problems arise when chromosome fragments are not properly rejoined. One problem arises when one of the two fragments has a centromere (the **centric** fragment) while the other has none (the **acentric** fragment). This means that during anaphase of mitosis or meiosis, when chromosome separation occurs, the centric fragment moves normally; because it has no centromere to attach to the spindle fibers, the acentric fragment tends to lag behind, moving too slowly to end up in the daughter cell where it belongs. The daughter cell lacking the lost fragment suffers what is referred to as a **chromosomal deletion**.

While some deletions have no apparent effect, others can cause serious problems. For example, the cri-du-chat ("cat cry") syndrome occurs in individuals that lack a small terminal portion of their number 5 chromosomes (Figure 17.7).

Consider what can happen when two breaks occur in the same chromosome. This means that there will be three fragments, one of them with two unhealed ends. The two end fragments may rejoin and leave out the middle fragment, resulting in a deletion of genetic material. A small deletion affecting only one or a few genes may be passed on to the next generation, but most larger deletions are immediately lethal, and there is no next generation. Also, the middle fragment with two broken ends can form into a ring. If this fragment is large and contains a centromere, it may become an abnormal structure called a **ring chromosome** (Figure 17.8). There are reasonably healthy people walking around today with ring chromosomes in every one of their cells. Another possible outcome is that the middle fragment may flip over before being rejoined to the two end fragments, resulting in a chromosome with an **inversion**, or inverted segment (see Figure 17.8).

If the two breakpoints occur in separate but homologous chromosomes, and in different positions on the two homologues, abnormal fusion repair can result in one chromosome being too short and the other being too long. The short chromosome will have a deletion, and the long homologues will have a duplication of genetic material.

Translocations

A chromosome mutation called a **translocation** results when two nonhomologous chromosomes break, and an incorrect fusion repair results in the attachment of part of one chromosome to another. If both chromosomes break more or less in the

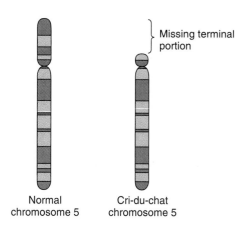

FIGURE 17.7
Cri-du-Chat Deletion
The cri-du-chat condition is a result of deletion of part of chromosome 5. The plaintive voices of those afflicted with the condition remind one of the cry of a cat. While the loss of this part of the chromosome is not fatal, such individuals suffer severe mental and physical defects. In most of the cases studied, the parents' chromosomes were perfectly normal, indicating that most incidents of cri-du-chat emerge from newly mutated gametes.

middle, such misrepair can result in two reciprocal translocation chromosomes. When the breaks happen to be close to the ends in both chromosomes, the tiny end fragments may become lost, and the only surviving chromosome will be a chromosome containing most of the genetic material of both chromosomes. There are several medical problems associated with particular translocations. For example, there is a type of leukemia called chronic myelogenic leukemia that has been shown to be the result of a translocation involving chromosomes 9 and 22 in humans. Over 90% of patients with this leukemia have what is called the Philadelphia chromosome. The Philadelphia chromosome is an altered chromosome 22, which carries a small portion of chromosome 9 attached to its long arm (Figure 17.9). It turns out that the breakpoints of the translocation involve an oncogene, a gene whose action can lead to cancer. (See Essay 17.1.)

Transpositions

The third kind of mutation that we will consider is transposition. Transpositions are insertions of copies of DNA segments into new positions in the genome. They are fundamentally different from the kinds of mutations we've already seen. Some of the *results* may be similar, but the cause is not chemical or radiation damage. The cause is the actual movement of pieces of DNA from one place in the chromosome to another.

"Jumping genes" is a fanciful reference to **transposons**, those pieces of DNA capable of moving from place to place in the genome. The story of how these jumping genes were discovered is one of the more compelling tales in modern biology.

It all started some 50 years ago when Barbara McClintock (a shy but tenacious researcher whose demands for experimental equipment included "a decent pair of eyeglass-

(a) Ring chromosome

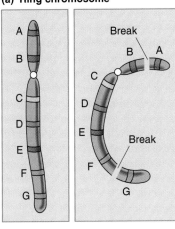
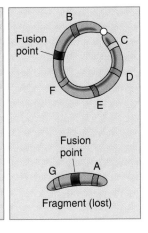

FIGURE 17.8
Ring Chromosomes and Inversions
(a) If both chromosome arms break, the two arms may join to form a ring chromosome and a small fragment that lacks a centromere. The small fragment will be lost, but the ring chromosome could be replicated and carried in a fairly normal fashion. If one chromosome arm breaks in two places **(b)**, the internal piece may rejoin in a way that creates a simple inversion. An inversion is just what it sounds like: the piece joins back in the inverted orientation. Since no genetic material is lost, most inversions are not harmful, although they may produce difficulties at synapsis. Prophase I of meiosis and crossing over can result in weird chromosomes that will be broken or lost.

(b) Inversion chromosome

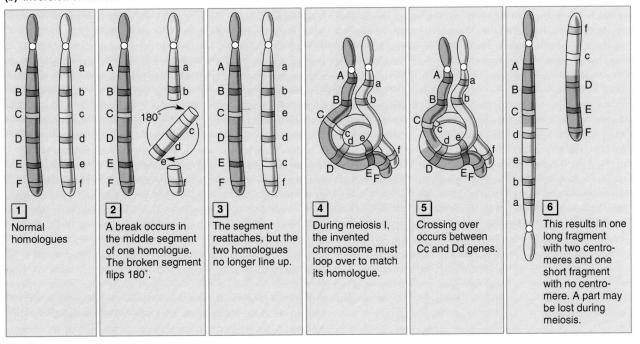

es") asked questions about the mutations associated with complex inheritance of color patterns of Indian corn, the corn with multicolored kernels sometimes used as autumn decoration. She concluded that this could only be accounted for if genes moved around from place to place on the chromosomes of corn plants, a radical conclusion at the time.

McClintock found that some of these mutations were unstable. To illustrate, some strains, that by her calculations should have had all red kernels, had many yellow ones instead. At first she thought the results may have been due to some kind of mutation back to the original yellow condition, but such mutations are relatively rare, and her observations were quite common. Eventually, she conceived of the existence of a mobile genetic element. She called it a "controlling element." Such an element, she proposed, moved around and upon settling next to a color gene promptly inactivated it. She

also found that the element could move during the development of the kernel, resulting in part of the kernel being one color and another part of the kernel being another color (Figure 17.10), a condition referred to as **variegated coloration**.

McClintock published her first paper on jumping genes in the 1940s. But even though some biologists were interested in her results, most felt that they were supremely unimportant in the overall scheme of things. But she doggedly followed her interests, without benefit of funding and without help, for most of her professional life. Unlike Mendel, who suffered the same kind of neglect throughout his lifetime, McClintock was finally vindicated. In the late 1960s and early 1970s, other researchers began to report similar mobile elements in other organisms. In time, jumping genes became a major focal point for genetic research. In 1983, Barbara McClintock received the Nobel Prize, 40 years after her first research on jumping genes.

Essay 17.1 • MUTATIONS, ONCOGENES, AND CANCER

For most of this century, cancer research has focused on viruses as possible cancer-causing agents. Although they have now been ruled out as a major cause of human cancer, virus research has helped identify certain human genes that are implicated in at least some cancers.

Researchers discovered that some cancerous cells contain genes that can be incorporated into viruses and then transmitted to normal cells. Once there, the activity of these genes can be associated with cancer. It has also been discovered that virtually those same genes are actually a normal part of the cell's genome; that is, they occur in normal cells. In normal cells, however, the *proto-oncogenes,* as the cancer-related genes are known, are under tight control. So far, more than 30 proto-oncogenes have been discovered.

Since the hallmark of cancer is uncontrolled cell proliferation, it is not surprising that most proto-oncogenes function in normal cell division. Apparently, these genes can be mutated or activated inappropriately, converting them to functional, cancer-causing *oncogenes* and causing the rapid and unrestrained growth of cancer cells, which subsequently invade and displace normal tissue.

How does the virus-transported oncogene affect the normal cells? One possibility is that the viruses may also pick up and transmit the DNA sequence that can activate uncontrolled cell proliferation. Or it may be that the viruses transmit the proto-oncogenes into healthy cells without also passing along the regulatory components of the genes, those that restrain the gene and keep it from acting inappropriately as an oncogene.

The mechanism of triggering cancer suggested by viral research could also help explain how some chemical carcinogens and ultraviolet rays operate to cause cancer. In fact, we now suspect that several different cancers (particularly leukemias) begin when x-rays, UV light, or certain chemical carcinogens cause chromosome breaks or mutations near or within certain proto-oncogenes.

Chromosome-level mutations such as inversions, translocations, or deletions may convert proto-oncogenes to oncogenes by severely affecting their expression. Point mutations can also convert proto-oncogenes into oncogenes. For example, researchers think that genes called *ras* code for regulatory proteins that normally switch back and forth between active and inactive states

depending on conditions. But one of two different substitution mutations (in either the 12th or the 61st codon) is all that is needed to create the oncogene whose product is locked in the active state. Mutation of *ras* is likely to lead to common forms of lung, pancreatic, and bladder cancer. *Ras* is particularly susceptible to mutation by chemicals found in tobacco smoke.

Such findings are promising, but final answers are not around the corner. Before we can say that we truly understand cancer, some hard questions have to be answered. For instance, the notion that cancer begins when a single oncogene is moved from one place to another or mutated is incompatible with the long lapse between exposure to a carcinogen and development of the disease. (Many researchers feel that cancer begins only after two or more oncogenes have been created. This helps explain the time lapse, but it must be proved.) It might also be possible that the stage is set for cancer when a single oncogene is created, but that cancer initiation occurs only when normal changes in our bodies, such as those associated with aging, provide the right conditions for oncogene expression.

Transposon Structure

But why have these mobile control elements become so important to our understanding of genetics? What are they? What do they do? Transposons (there are hundreds of kinds) consist, basically, of two parts as shown in Figure 17.11. The central part is the gene for **transposase**, an enzyme that enables a transposon to insert a copy of itself almost anywhere in the genome. On either side of the transposase gene are sequence elements called **inverted repeats** (because the nucleotide sequence of these regions is identical, but in opposite orientations). Simple transposons may consist of these inverted repeats and the transposase gene only. Others, though, are more complex, containing up to 40,000 nucleotide pairs. At the transposon's core may be one to several complete, functional genes in addition to the transposase.

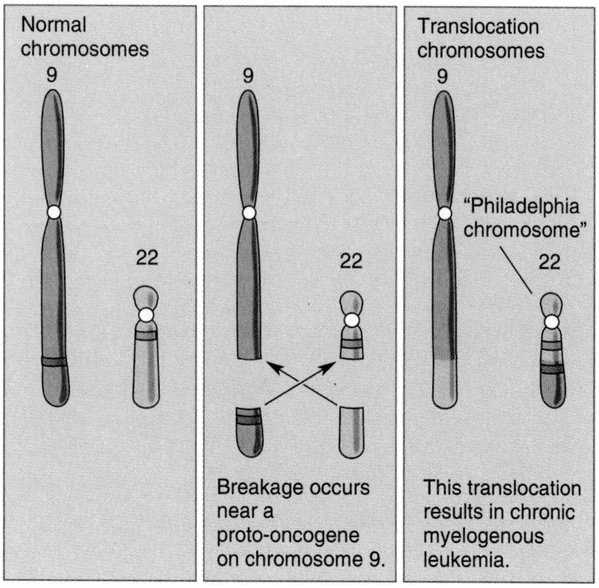

Normal chromosomes 9   22

Breakage occurs near a proto-oncogene on chromosome 9.

Translocation chromosomes 9   "Philadelphia chromosome"   22

This translocation results in chronic myelogenous leukemia.

FIGURE 17.9
Chromosome Translocations
In a documented medical genetic example, a translocation between chromosomes 9 and 22 has resulted in what is called the Philadelphia chromosome. This chromosome is mainly chromosome 22, but carries a small part of the long arm of chromosome 9. Unfortunately, the break point of the translocation on the Philadelphia chromosome often causes leukemia, because the area of translocation involves an oncogene.

FIGURE 17.10
Barbara McClintock and Transposable Elements
Transposable elements can disrupt gene function. Here, what should have been colored kernels lack pigmentation because a transposable element lies in or near the gene for color production. During development of the kernels, sometimes the element leaves the gene, restoring color production in that cell and all its descendants. This results in the colored stripes.

In addition, the process of transposition itself can be mutagenic. Often the excision of a transposable element is not clean, and small insertions of several base pairs may be left behind. Also, some elements leave behind broken chromosomes, so the action of transposons may occasionally cause chromosomal mutations as well.

Much of the time, transposon movement appears to be random. However, McClintock suggested in her Nobel lecture that, since the transposition rate increases manyfold when cells are placed under life-threatening stress, transposition may be an organism's last-ditch attempt to generate new variability. Thus, under some circumstances, jumping genes may be vital to the continuity of life.

So you see that mutation can occur in a wide variety of ways, with an even wider array of possible effects. Mutations can indeed be harmful, leading to disease, physical abnormalities, and a good deal of suffering. (There are over 3000 human genetic abnormalities, and prevention of increasing the occurrence of these is the major reason for the concern over the rate at which we are adding mutagens to our environment.) But mutations can also be beneficial, in that they provide the variation on which natural selection can act, and so they have been the basis for the evolution of life as we see it today (see Essay 17.2).

Transposons as Mutagenic Agents

How do transposons act as mutagens? Obviously, it is because they move around. It is easy to see how insertion of DNA elements within a chromosome can disrupt gene function (Figure 17.12). First, insertion into promoter sequences could completely inactivate transcription. Second, insertion of large elements could physically separate the promoter from its coding region by thousands of bases. And finally, insertion into exons could cause frameshifts or insertions of completely new amino acids.

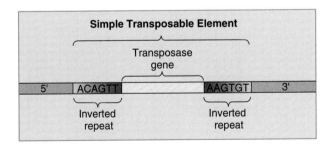

FIGURE 17.11
A Simple Transposable Element
The simplest form of a transposable element is composed of a pair of inverted repeat sequences (same DNA sequence, but reversed in one of the pairs) that flank a transposase gene. The transposase gene encodes a protein that is capable of recognizing the inverted repeat sequences of the element, cutting the element out of the surrounding DNA, and reinserting the element at a new location. Larger transposable elements can have additional genes located between the inverted repeats.

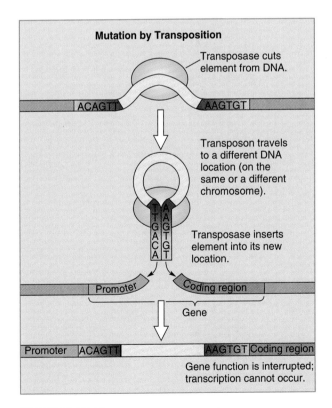

FIGURE 17.12
Mutation by Transposable Element
Movement of a transposable element into an existing gene may drastically alter the structure of the gene and thereby destroy or alter its function.

Essay 17.2 ● MUTATION AND THE EVOLUTION OF GENE STRUCTURE

We have estimated the number of genes in the genome of various kinds of organisms, such as *Escherichia coli* (about 2000), *Drosophila* (about 5000), and humans (about 120,000). Obviously, gene numbers vary tremendously from species to species.

But at some time in the distant past there were fewer forms of life with fewer genes each. What is responsible for the changes from this primordial condition, and by what means have they occurred? Intuitively, we might deduce that increases in organism complexity were associated with more enzymes, structural proteins, and regulatory proteins, and therefore more genes. So how did the number of genes in evolving populations increase? Most of the answers to this question are based on what we know of (1) transposition and (2) gene duplication. Consider the human hemoglobin gene complex as an example.

The hemoglobin molecules of adult humans are made up of the products of separate genes that code for the alpha- and the beta-globin polypeptide chains. You may recall that in its quaternary level of organization, each hemoglobin molecule consists of two alpha and two beta subunits. But at some time in the past, an ancient version of animal hemoglobin is thought to have consisted only of a pair of alpha-like subunits. So the initial step in the evolution of hemoglobin beta subunits must have been gene duplication of a sort that resulted in an extra copy of the alpha gene. This could occur by the action of translocations or transposons.

Following the creation of two separate alpha genes (by whatever means), point mutations must have accumulated in the separate genes to change the structures of the two polypeptides slightly, finally creating a pair of molecules that could combine and work together efficiently to carry oxygen in some primordial creature's body fluids.

But the story of the development of hemoglobin doesn't end there. Mammals have at least four different kinds of hemoglobin; each is used during a different period of development and life. Embryonic hemoglobin appears soon after the embryo is formed. This molecule is well suited to pick up oxygen from the surrounding fluids. Then, somewhat later, embryonic hemoglobin is no longer transcribed, and the gene for fetal hemoglo-

bin is activated. Fetal hemoglobin is well suited to exchange gases with the mother's circulation. Finally, a little before birth, the genes for adult hemoglobin begin to be transcribed. Adult hemoglobin is especially well suited for picking up oxygen from the lungs and transporting it to the tissues. Just how these important changes were engineered was a good mystery until it was discovered that each of the mammalian globin genes is actually a tightly linked gene complex. There are four beta-globin genes on one chromosome and three alpha-globin genes on another.

Thus gene duplication, transposition, and evolution have worked together to allow an elegant solution to the human body's changing oxygen-carrying demands during development.

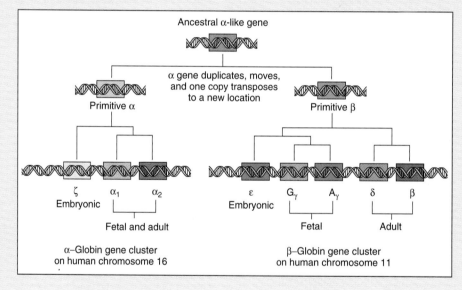

Key Ideas

THE STABILITY OF DNA

1. Because it lacks reactive side groups and its bases are tucked into the double helix, DNA is not very reactive. However changes in DNA do occur. The three levels of mutational change are **point mutation**, **chromosomal mutation**, and **transposition**.

2. The presence of histones in eukaryote chromosomes and the redundancy of base pairing reduce the incidence of mutation.

DNA DAMAGE AT THE MOLECULAR LEVEL

1. Changes that do occur in DNA are called **primary lesions**, and unrepaired lesions become mutations. Agents that cause mutations are called **mutagens**.

2. Single-strand breaks are often caused by high-energy radiations. UV light causes **pyrimidine dimers** to form.

3. Chemical mutagens cause double-strand breaks in DNA. Examples include **intercalating agents** (that lodge within DNA),

base analogs (copycat molecules that cause faulty base pairing), and **DNA-modifying agents** (that cause replication and transcription errors).

DNA REPAIR: THE CLEANUP CREW

1. Most lesions are repaired by DNA repair systems. Two mechanisms exist to repair UV damage, **photoreactivation repair** and **excision repair**.

POINT MUTATIONS

1. Point mutations include base substitutions, additions, and deletions.

2. **Base substitutions** change a codon, and unless the change produces a synonymous codon (**silent** substitution), an amino acid substitution will occur in the polypeptide. The result may be negligible (**neutral**), or it may produce a serious condition such as sickle cell anemia. In this disorder, a single base substitution brings about the collapse of red blood cells.

3. Base substitutions producing an unwanted *stop* codon, or **chain-termination mutations**, stop polypeptide synthesis too early. If it is the *stop* codon that is altered, the mRNA poly-A tail will add amino acids to the polypeptide, and it may "lock up" the ribosome as termination fails.

4. Base insertions and deletions can produce highly abnormal proteins with lethal effects on the cell. They cause frameshift mutations, in which all codons from the insertion or deletion onward will be altered and misread.

CHROMOSOMAL MUTATIONS

1. Chromosome breakage and fusion repair are very common, and most repairs occur without problems.

2. When the wrong sticky ends are rejoined, **deletions**, **ring chromosomes**, **inversions**, and **translocations** can result.

3. In translocations, the greater part of a broken chromosome is fused to a normal chromosome. Generally, minor portions are lost.

TRANSPOSITIONS

1. There are naturally occurring genetic elements, that can physically move around the genome. These transposable elements, or **transposons**, were first described by Barbara McClintock, who observed the effects of transposon insertion into the genes of corn.

2. Transposons act as mutagenic agents. When they insert into DNA they can disrupt gene activity, and when they leave a gene they can leave behind base changes in the DNA.

Application of Ideas

1. Recall for a minute the various kinds of mutations and mutagens that have been discussed in this chapter. Then consider the environment that we live in, both the "natural" environment and the one we humans have created. How do your thoughts impact on the idea that there are two kinds of mutation, "naturally occurring variation" and "induced mutations"?

2. One of the most frightening things about life in today's world is the tremendous number (70,000-plus) and density of potential mutagens our society is producing. Write a letter to your congressional representative explaining why you are worried about this problem. (You might even want to send it!)

3. From the very first, mutagens have been a fact of life. Solar and other forms of radiation have always been with us, as have chemical mutagens (in the form of chemicals that plants produce to protect themselves against being eaten, and even in the form of oxygen, which, as it crosses over cell membranes, generates DNA-damaging "free radicals"). Yet, evolution has manifestly failed to derive foolproof means of avoiding DNA damage or repairing it. Why should this be? (Imagine that it most certainly must have been possible and consider the fact that fruit flies are less well-protected against mutation than humans.)

Review Questions

1. In what ways does the double helix itself help prevent primary lesions or mutations from occurring? How do nucleosomes help? (page 335)

2. Describe DNA repair systems. What characteristic of DNA makes such systems feasible? (pages 336–337)

3. How does a point mutation differ from a chromosomal mutation? Which tends to be more serious? (pages 337–339)

4. Describe an instance in which a base substitution would have no effect on a polypeptide. What, if any, adaptive significance does this suggest? (pages 337–339)

5. Using an example, explain how a single base substitution could be disastrous. (pages 337–339)

6. Why would a base addition or deletion tend to be more serious than a base substitution? (page 339)

7. What effect on a polypeptide would a mutation in the *start* or initiator codon have? Could this be detected? Explain. (page 339)

8. Describe the events that lead to inversions and translocations. (page 339)

9. What is the structure of a simple transposon? A complex transposon? (pages 340–341)

10. In what ways may a transposon act as a mutagen? (page 343)

CHAPTER 18
Natural Selection and Adaptation

CHAPTER 19
Microevolution and Changing Alleles

CHAPTER 20
The Origin of Species

CHAPTER 21
Origin of Life

PART THREE

Evolution

CHAPTER 18

Natural Selection and Adaptation

CHAPTER OUTLINE

THE HISTORY OF AN IDEA

Enter Charles Darwin

DARWIN'S THEORY

The Origin of Species

The Modern Synthesis

VARIATION

Sources of Variation

Maintaining Genetic Variation

NATURAL SELECTION

The Peppered Moth: Natural Selection in a Natural Population

How the Leopard Got Its Spots

Fitness

EVIDENCE OF EVOLUTION

Fossils and the Fossil Record

Biogeography

Essay 18.1 The Evolution of the Horse

Comparative Anatomy

Molecular Biology

Artificial Selection

KEY IDEAS

REVIEW QUESTIONS

Among the oldest questions concerning biology are those that seek explanations for the abundant diversity of life forms on earth while recognizing that all life forms have characteristics in common. Are the shared characteristics a result of a common heritage? Do species change over time? Is the diversity among species the result of accumulated change within individual species? These questions are about as old as philosophy itself. In fact, both Plato and his student, Aristotle, had considered the question long before Darwin and had resolutely declared that species do not change.

The History of an Idea

Plato believed in ideals, perfect forms that were stable. Variation from ideal form was imperfection. Aristotle (Figure 18.1), Plato's best known protégé, developed a "natural scale" in which organisms are arranged in an ascending order from simple to complex, with humans at the very

FIGURE 18.1
Aristotle's View of Life
The Greek philosopher and naturalist, Aristotle, shown here in a painting, believed that humans were at the pinnacle of a simple ladder of increasing complexity.

top, just above the Indian elephant. Both Aristotle and Plato believed, though, in the immutability of species. How, they asked, could you pass on to your offspring something different from what you are?

Biological thinking continued to be firmly embedded in Aristotelian philosophy for some 2000 years. The notion of immutability was bolstered further by Judeo-Christian theology, which described all life as having been created in a few days. The dogma of immutability stemmed largely from the logical extension of the alternative: if life changed through time, and if one species could arise from another, then unless humans were somehow exempt from natural law, they *could* have arisen from some other life form. This, of course, flew in the face of the biblical teaching that humans were created in the likeness of God.

Immutability was certainly the view of the Swedish botanist Carl von Linne (1707–1778), who devised a system of classification for all living organisms and, in his fondness for Latin, even called himself Carolus Linnaeus. Some argued that Linnaeus' designations, which involved lumping the species together on the basis of their similarities, were artificial and based on personal whim. Few, however, considered criticizing

his assumption of special creation or his concept that each species was fixed in its original form, never to change.

One small departure was suggested by a French contemporary, George-Louis Leclerc De Buffon (1707–1788). In 1753, Buffon proposed that, in addition to those animals that had originated in the Creation, there were also lesser families "conceived by Nature and produced by Time."

Other scientists were also beginning to toy with the notion of the heritability of change. In France, Jean Baptiste de Lamarck (1744–1829), a protégé of Buffon, boldly suggested that not only had one species given rise to another, but humans themselves had arisen from other species. In his view there was a "force of life" that caused an organism to generate new structures and organs to meet its biological needs. The transmission of such structures to subsequent generations was referred to as the *inheritance of acquired characteristics.* Lamarck cited as an example the long neck of the giraffe (Figure 18.2), which he maintained had evolved as each generation of giraffes had stretched their necks in an effort to reach the topmost branches of trees and then transmitted genes for this longer neck to their offspring. Lamarck's notions, called the "law of use and disuse," have largely been discarded.

FIGURE 18.2
Acquired Characteristics and Inheritance
A narrow interpretation of one of Lamarck's ideas on evolution suggests that organisms inherit acquired traits, an idea that became known as the "law of use and disuse." It maintains, for example, that the giraffe's long neck evolved through generations of "neckstretching." It's unfortunate that we only remember Lamarck for his error, since he clearly challenged the predominant view of immutability of species and stimulated a lively interest in evolution.

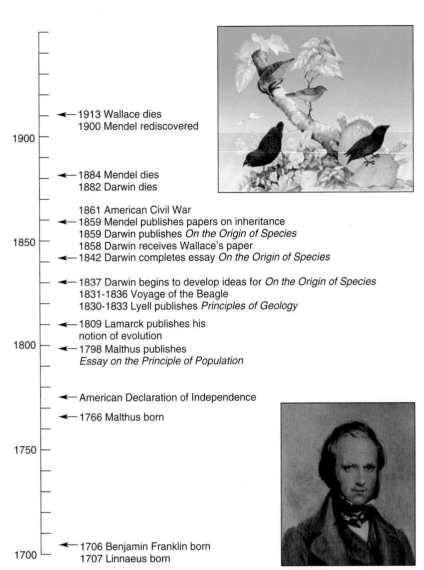

1913 Wallace dies
1900 Mendel rediscovered

1900

1884 Mendel dies
1882 Darwin dies

1861 American Civil War
1859 Mendel publishes papers on inheritance
1859 Darwin publishes *On the Origin of Species*
1858 Darwin receives Wallace's paper
1850
1842 Darwin completes essay *On the Origin of Species*

1837 Darwin begins to develop ideas for *On the Origin of Species*
1831-1836 Voyage of the Beagle
1830-1833 Lyell publishes *Principles of Geology*

1809 Lamarck publishes his
notion of evolution
1800
1798 Malthus publishes
Essay on the Principle of Population

American Declaration of Independence

1766 Malthus born

1750

1706 Benjamin Franklin born
1700
1707 Linnaeus born

FIGURE 18.3
Darwin's Galápagos Visit and His Place in History
A key insight for Darwin was the realization that the finches that inhabited the various islands of the Galápagos were, in fact, very similar but had changed in order to make use of the specialized food sources of their island. Species do change! A time line depicts events that mark Darwin's contributions within the advances in evolutionary thought.

At the time, Lamarck's arguments did little to persuade his lecture audiences. With the foment of the French Revolution, there was some lively discussion in intellectual circles, but in society at large the firm conviction that each form of life had arisen through special creation ruled out serious consideration of any other concept.

The intellectual climate into which Darwin was born (1809) was far more conservative than that in France. The English had been horrified by the French Revolution, and ideas such as those held by the "French atheists" were either dismissed out of hand or viewed with extreme suspicion. Partly for this reason, the church continued to hold strong

sway over the sciences, and biologists in England continued to adhere rigidly to the traditional tenets. This was the philosophical milieu into which Charles Darwin's ideas were thrust.

Enter Charles Darwin

Charles Darwin, as one of the most influential figures in modern history, has by now probably generated as much research about himself as about his work (Figure 18.3). Historians and biologists alike are intrigued with such questions as: How did he come to his ideas? When was he struck with the concept of natural selection? What triggered his thinking along such lines? Obviously, his visit to the Galápagos Islands was pivotal in his thinking. The usual myth has the whole thing turning on his observations of those Galápagos finches. This can hardly be a simple case, however, since he was clearly influenced by his observations in South America and he didn't even learn that the Galápagos birds *were* finches until some years after his return to England. He wasn't even particularly taken with the little birds while he was in the Galápagos Islands. In fact, much of what he reported about them was admitted hearsay by members of the crew. Many of them had taken the opportunity to roam the strange little islands and had offhandedly witnessed things that Darwin hadn't seen. He may have been inspired by his visit to the islands, but many of the details of his theory were pounded out by sheer intellectual labor after his return to England.

It has been suggested from time to time that the subject was "in the air," that the stage was set by circumstance, and that the notion of species changing over time by "descent with modification" (Darwin didn't use the word "evolution" at first) through natural selection was inevitable. However, Darwin vigorously disagreed that his concept of evolution was an inevitable extension of the thinking at the time. He noted that he had often tried to explain his idea to able people but had "signally failed" to make them understand.

Darwin's Theory

Beginning in 1837, Darwin started to keep a journal entitled *Transmutation of Species* and was soon making entries refer-

FIGURE 18.4
Artificial Selection
Generations of selective breeding have produced some amazing variations in *Canis familiaris,* the dog species.

ring to "my theory." It is fascinating to watch his ideas develop in his writings. Keep in mind that Darwin knew a lot about agricultural practices and livestock breeding. Any good farmer knows that a lineage of animals can be improved by selecting only certain individuals of each generation for breeding (Figure 18.4). The approach is called artificial selection. By longstanding folk tradition, the best animals, whether cattle, fowl, or dog, were honored annually at country fairs and chosen for further breeding. The reasoning was simple: like begets like, offspring tend to resemble their parents, and the "best" parents produce the "best" offspring.

But, Darwin wondered, would selection operate without human intervention? Did selection have to be conscious? Perhaps not. Darwin was impressed with an essay on populations that had been written by the Reverend Thomas Malthus three decades earlier (1798). Malthus stressed that all species had enormous reproductive capabilities and that their numbers tended to expand geometrically unless their numbers were held in check by processes that kill a certain portion of the population. Natural populations, he argued, reached a balance in which all but a few of the young of each generation were forced to perish. So, mused Darwin, inferior individuals in a population were not free to reproduce after all, precisely because they would be removed from the population before breeding. The environment then could indeed select the individuals for breeding, and these would be the hardiest because they were selected on the basis of survival. And it all happened through natural means (Figure 18.5).

Darwin's journal shows that by 1838 he had solved the major riddle of evolution. The mechanism of evolution, he said, was **natural selection**. His idea of natural selection was based on four points:

1. There is variation among the individuals of most natural populations.
2. Some of that variation is inherited.
3. Populations tend to produce more offspring than the environment can support.
4. Those individuals whose traits best adapt them to the environment will survive better and leave more offspring than those with less adaptive traits.

A penciled manuscript in 1842 laid out the entire theory of the origin of species through natural selection. Darwin did not publish it, however. He knew that the idea would arouse fierce resistance and that he would have to back up every part of his idea with evidence. He set about preparing to defend his argument.

Darwin's health began to fail during the years of compiling information on his theory. In those times when he was able to work, Darwin suggested further tests. Darwin found that a variety of species can survive extended periods of time in seawater and that airborne organisms can easily move from one place to another, perhaps by being transported by storms. Such evidence was important because much of the foundation of his developing theory depended on the ability of species to disperse.

Working only a few hours a day, Darwin continued to develop his theory of natural selection. He showed his early manuscript on natural selection to only one close friend, the botanist Joseph Hooker, who remained doubtful. Twenty years

Natural Selection

Darwin suggested that from among any population of giraffes, some are born with longer necks than others. These are able to reach the leaves.

Better-fed animals would then most likely be able to survive and rear their young.

Longer necks would come to predominate in subsequent generations.

FIGURE 18.5
Natural Selection
Darwin maintained that the environment selected the longer-necked giraffes from a population of variants. Thus those with longer necks survived to pass their "long-neck" genes along.

passed, and finally in 1856 Darwin began writing his major work, to be called *Natural Selection*. It was planned as an enormous, six-volume monograph. But in 1858, his work on the manuscript was suddenly interrupted. Unexpectedly, another manuscript, a very brief one, arrived in the mail from Alfred Russel Wallace, a young naturalist working in Malaya. Wallace asked politely whether Darwin would care to make any comments on the manuscript. The article was a short but well-written statement about evolution and natural selection—Darwin's ideas had been quite independently deduced by someone else! Darwin was mortified. In a letter to his friend Hooker, he lamented: "So all my originality, whatever it may amount to, will be smashed. . . . Do you not think that this sketch ties my hands? . . . I would far rather burn my whole book, than that he or any other man should think that I have behaved in a paltry spirit." Darwin was about to be scooped; it is not an uncommon experience in science. Would his own work now be thought to have been based on the ideas of another?

Darwin's friends persuaded him to allow his previously unpublished 1842 summary and the text of a long letter describing his ideas to be presented together with Wallace's paper before the Linnean Society of London. Because of Darwin's much more substantial evidence, his paper was given first. The two were presented in July and were published in August 1858. Now Darwin went furiously to work and he quickly finished the famous *On the Origin of Species*, published in 1859. The first edition sold out on the first day.

The Origin of Species

In the first chapter of *On the Origin of Species*, Darwin sought to establish that animals and plants under domestication are extremely variable, that the variation is heritable, and that the many domestic varieties have arisen, under artificial selection, from wild ancestors. In the second chapter, he drew together what evidence he could find to show that plants and animals in nature were variable too. In the third chapter, he discussed Malthus' idea of the struggle for existence, the idea that the natural reproductive capacities of living things greatly outreach the ability of the environment to support them (Figure 18.6), so that many organisms perish by starvation or disease without reproducing. In the fourth chapter, Darwin introduced natural selection:

> How will the struggle for existence . . . act in regard to variation? Can the principle of selection, which we have seen is so potent in the hands of man, apply in nature? I think we shall see that it can act most effectively. . . . If such [variations] do occur, can we doubt [remembering that many more individuals are born than can possibly survive] that individuals having any advantage, however slight, over others, would have the best chance of surviving and of procreating their own kind? On the other hand, we may feel sure that any variation in the last degree injurious would be rigidly destroyed. This preservation of favourable variations and the rejection of injurious variations, I call Natural Selection.

One hundred beetles weighing 10 mg each can produce 6.1×10^{28} offspring in only 82 weeks. Their weight would be 6.1×10^{21} metric tons at that time, equal to the weight of the entire earth.

FIGURE 18.6
Reproductive Output Is Set by Natural Selection
The number of offspring produced by most organisms commonly exceeds the environment's ability to support them. This accounts for the fact that many will die or fail to reproduce.

The Modern Synthesis

Charles Darwin developed a cornerstone of modern biology in his description of evolution through natural selection. But he had a great deal of trouble convincing his contemporaries of this relationship. Most scientists of his time were quick to accept the fact that evolution occurs, but many of them failed to see just how natural selection was involved. Darwin was unable to convince them, primarily because he had no *mechanism* whereby natural selection could proceed. The problem stemmed from the fact that Darwin could not explain the mechanism that produced variation in a species. He didn't even know genes *existed.* So as he set about trying to explain how natural selection might work, he began grasping at straws. He continued to believe that inherited factors were somehow blended in the offspring. Furthermore, with each revision of *On the Origin of Species,* he lapsed deeper into Lamarckian explanations that even his contemporaries knew didn't hold water.

We now know that, not far away, Gregor Mendel was generating precisely the information that would clear up Darwin's great dilemma. But since Mendel was the only modern geneticist in the world, he had trouble making people see the importance of his work—including Darwin. In fact, as you may recall from Chapter 10, a copy of Mendel's paper was found in Darwin's library, completely unmarked.

As we saw in Chapter 10, Mendel's work was rediscovered in the early part of this century, long after the hard-working monk had gone to his reward. The scientific world, newly

fascinated with genes and how they are carried along from generation to generation, began to wonder just what genes have to do with natural selection. By the 1930s, the broad principles of the relationships between genes and natural selection had been worked out. Then in the 1940s, Ernst Mayr and others finally produced what they called the *modern synthesis,* the *integration* of the principles of genetics and evolution.

The modern synthesis explained a great deal and laid the framework for further investigation. That investigation, not unexpectedly, has resulted in the alteration of some of the original statements and has, in fact, produced arguments over rates of evolution and just how evolution proceeds (see Chapter 19). Many scientists see such disagreements as no attack at all, but simply as fine-tuning an established principle that will remain, perhaps altered, but alive and well and a linchpin of modern scientific thinking. With this critical integration of ideas in mind, let's see just how evolution does, in fact, proceed by natural selection working on variation, favoring some forms while weeding out others. We can begin by noting some of the many ways that variation can exist in populations.

Variation

In any population, individuals are likely to vary from each other as each bears novel traits or combinations of traits. We are familiar with some of the ways in which our own species can vary (Figure 18.7), but we must keep in mind that variations in other species may not be so obvious.

There are two basic points here. The first is that variation can occur in a wide array of traits. The second is that we, with our own special sensibilities, may not be readily aware of the variation. Whether the variation is apparent to us, though, doesn't matter. If one type confers any survival or reproductive advantage, natural selection can act on it.

For example, we humans are particularly keyed to physical variations of the face. We can often distinguish identical twins whom we know well. Dogs, on the other hand, are less sensitive to our facial features, but very attentive to differences in our scent. Apparently our scents are quite distinctive, but those differences go largely unnoticed by us, at least in better social circles.

Animals can vary not only in their appearance but in their behavior. Some dogs, for example, may bite you even if they are fully aware of who you are, while others are simply not disposed to bite at all. If you approached a herd of fighting bulls, some will walk toward you while others trot away. (Of course, this difference can be due to mood as well as genetics, but the breeders prefer those that routinely challenge.) There is even behavioral variation among ants of the same caste. Some work efficiently while others run back and forth along the line, accomplishing nothing. Other variations are even less apparent. Researchers have found that we vary biochemically in any number of ways, from our blood proteins to digestive enzymes.

FIGURE 18.7
Variation Within Populations
Penguins, like humans, form large populations of sexually reproducing individuals.
No two penguins are exactly alike—at least to another penguin. Every king
penguin here knows its mate, its offspring, and all its nesting-ground neighbors.
Human beings, like all large populations of sexually reproducing organisms, are
extremely variable. No two individuals are genetically alike, with perhaps the
exception of monozygotic twins.

It is important to note that natural selection operates on
those genetic variations that are related to the ability to survive and reproduce. In this way, certain genes or combinations of genes can be favored and come to increase in succeeding populations.

Sources of Variation

Variation in a population arises from two primary sources,
genetic recombination (Chapter 11) and mutation (Chapter
17). You will recall that genetic recombination is the shuffling
of the genes of an individual as the cells enter meiosis. This
allows unique combinations of the genes from that individual's
parents to reach the next generation. Mutations are heritable
changes in the genetic material that provide a source of new
variation within a population. Thus genetic recombination
continually produces variation within a population by shuffling existing genes and alleles into new combinations, while
mutation provides a steady input of new genes and alleles into
the population.

Maintaining Genetic Variation

Successful populations must have ways of maintaining their
genetic variation in the face of natural selection. There are several mechanisms that help populations to maintain variation.

Part of the basis for variability lies in the tendency of
recessive traits to be maintained in the population because the
alleles can be carried by heterozygous individuals without
expressing the trait and thus without natural selection operating on them. There are cases in which an allele continues in a
population even though it produces a trait that is selected
against when homozygous. One process that inhibits the loss
of variant alleles (even when they are detrimental) is called
balancing selection, and the phenotypic result is called a **balanced polymorphism** (*poly-*, "many"; *morph,* "form"). Balanced polymorphism is marked by populations with variant
individuals where the ratios of those variants do not change
noticeably from one generation to the next. Each distinct form
or variant is called a **morph**. In some cases, each morph may
have some particular advantage under certain conditions (Figure 18.8). For example, balanced polymorphisms can occur in
"patchy" environments, where a population extends over a
number of distinct habitat types. The habitat of the British land
snail may vary greatly, some areas being shaded, some lighted,
some brown, some green, some mottled, and so forth. Since the
snails' shells differ in color and marking, some individuals will
be difficult for predatory birds to find in each of these habitat
types; thus the great variation in these snails is maintained.

Genetic diversity can also be maintained by **heterozygote
advantage**, in which the heterozygous genotype has greater
reproductive success than either homozygote. One example of
heterozygote advantage lies within the sickle and normal alleles of beta-hemoglobin in populations where malaria is relatively common. It turns out that persons carrying one allele for
sickle cell hemoglobin are not as susceptible to malaria. Thus

FIGURE 18.8
Genetic Variation in Snails
The land snail, *Capaea nemoralis,* varies markedly in two ways: shell color and banding patterns. The two traits are inherited independently, so each color can occur with each banding pattern. The result is a dramatic variability in appearance.

there is an advantage to having a sickle cell allele, and persons with two normal alleles would tend to be selected against. However, carrying two sickle cell alleles presents the symptoms of sickle cell disease, and such persons would also tend to be selected against. Thus the superior genotype is heterozygous, having a single sickle cell allele to protect against malaria and a normal allele to prevent sickle cell disease. Since each of the two homozygous types is inferior to the heterozygote, natural selection keeps both alleles in the population.

Frequency-dependent selection occurs when the reproductive success of a genotype depends on its frequency in the population. The term *frequency* simply refers to the ratio or proportion of an allele in a population. If the success of a genotype is dependent on how frequently it appears in a population, the results may be a stable polymorphism. This happens if a genotype has a net advantage when it is rare and a net disadvantage when it is more common. The problem is that net advantage, when rare, will result in an increase in allele frequency; when this occurs, the genotype becomes less rare, and its advantage decreases.

For example, some tropical freshwater fish populations are polymorphic for a common gray morph and a relatively rare red morph. The red fish, by their color alone, intimidate the other fish and almost always win out in fish-to-fish competitions. On the other hand, the red morph is easier to see and is more subject to predation by birds. When the red fish are rare, they have a net benefit because of their powers of intimidation. However, if the red forms were to become common, their gray competitors would encounter them frequently and become aware that those red fish aren't so tough after all. The gray fish that are harder to intimidate would produce more offspring than the more timid gray fish, so a bolder gray morph would evolve. Thus each morph has its advantages, but their relative numbers are important. The red forms are kept at a frequency at which the benefits of being red just balance the disadvantages (Figure 18.9).

Now that we have seen how variation can arise and how it can be maintained in a population, we can discuss how natural selection actually operates on that variation.

Natural Selection

Natural selection has come to be defined as the process by which those genotypes that are best adapted to the prevailing conditions are more likely to survive and reproduce, thereby making a larger relative contribution to the next generation. Thus, from among the various genes and combinations of genes in a population, some will confer certain advantages to the individual that bears them. If those advantages lead to greater reproductive success (more offspring), then those kinds of genes will come to increase proportionately in the population. We now consider two ways in which natural selection can result in evolution.

The Peppered Moth: Natural Selection in a Natural Population

Biston betularia is a British moth, commonly called the peppered moth. It occurs in two morphs, as shown in Figure 18.10. One morph is light and mottled (or peppered), and the other morph is black. The British have a long tradition of butterfly and moth collecting, and records on the peppered moth go back two centuries. The black morph, whose color is controlled by a single dominant allele, originally showed up in 18th century collections as a rare, highly prized variant. In the early stages of the industrial revolution (in the 1840s), the black form began to show up in greater frequencies in collections, especially near cities. The black morph continued to become more and more common in industrialized areas, until it greatly outnumbered the light peppered morph. In Manchester, England's industrial center, the dominant black morph achieved a frequency of 98%. Meanwhile, the light peppered morph remained the predominant form in rural areas.

The environment had changed, and the species, through differential mortality and change in allele frequencies, adapted to it. The environmental factor was soot from burning coal. Industrial England, as the 19th century proceeded, quietly submitted to its dark cloak of carbon. Bird predation is probably the principal cause of death in these moths. Over the long course of evolutionary time, the moth had achieved a camouflaging coloration that blended well with the light, peppered appearance of lichen-covered tree trunks. But pollution killed the lichens and blackened the trees, making the light peppered morph highly visible and extremely vulnerable to predation.

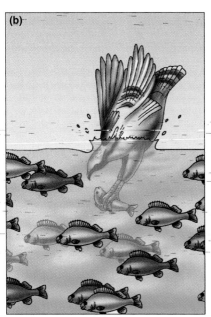

By virtue of their color, the red variety of fish are more reproductively successful.

However, their bright colors make them easier for predators to find.

Thus, the numbers of red fish never overwhelm the gray.

FIGURE 18.9
Frequency-Dependent Selection
By virtue of their bright color, the red variety of fish **(a)** is more reproductively successful. **(b)** However, as their numbers increase, their bright colors also make them easier for predators to find; thus the numbers of red fish never overwhelm the gray **(c)**. The red color also helps in competitive encounters with the gray fish, as long as the red morph does not become too common.

FIGURE 18.10
The Peppered Moth
In this "bird's-eye view," the black and peppered morphs of Biston betularia are seen on the natural lichen-covered and unnatural soot-covered backgrounds. Their frequencies are drastically affected by predatory birds and can change quickly, over a few generations.

In industrialized areas, the black morph achieved a significant selective advantage because it was less easily spotted by birds.

A British naturalist, H. B. D. Kettlewell, performed the crucial experiment. He released known numbers of marked black and light peppered moths in unpolluted woodlands and in polluted, soot-blackened woodlands. In each habitat, after a period of time had elapsed, he recaptured a portion of the released moths. Kettlewell's mark-and-recapture data are seen in Table 18.1. For the first data set, released in the unpolluted woodland, almost exactly twice as many light forms survived as black forms. That's equivalent to a 100% advantage of the light type in a brief exposure to predation. In the second data set, selection was against the light peppered morph. Almost exactly twice as great a percentage of the favored black type survived.

Incidentally, England has been doing pretty well of late in its battle against air pollution. The woodlands near the cities are once again becoming covered with lichens, and the soot is disappearing. As one might predict, the black morphs of *Biston betularia* are now declining in frequency.

With only one allele for natural selection to work on, evolution can proceed quickly when selection is intense, as with the peppered moth. But usually it proceeds much more slowly as natural selection operates on traits governed by large numbers of alleles, and selection is not always as obvious as in this example. Let's consider then how such a process might produce evolutionary changes.

How the Leopard Got Its Spots

Actually, we don't know how the leopard got its spots, but we can consider an imaginary scenario that shows how natural selection might operate on a trait controlled by many alleles and thereby mottle the cat. Assume that somewhere in the leopard's past, its ancestors were a nondescript tawny color, like lions. Then suppose one group of this ancestral cat moved into an area where the ability to climb trees was important. Climbing might be important, say, to an animal that was fond of eggs, or to one that needed to climb to escape larger predators. It might also need to climb trees to cache its kills there, safe from other carnivores that couldn't climb. In any case, climbing became important, so the cats began to spend more of their time in trees. It is easy to see that once the pattern of tree climbing was established, it would be to the cat's advantage to be inconspicuous in the leafy limbs (Figure 18.11).

Among the offspring of each generation, some kittens might tend to be a bit more mottled than the others, and these individuals would be less conspicuous among the splotchy, sun-bathed leaves. Since they would be camouflaged from both prey and predators, they would tend to survive more frequently than their tawny siblings and to leave more spotted offspring in the population. In each generation, the more mottled individuals would survive to reproduce more frequently and so, in time, the entire population would be strongly spotted and hard to see. It is important to understand that an adaptive trait does not have to appear full-blown and completely developed before it confers some advantage to its bearer. A slight change can bestow slight benefits and then become magnified in the population through time.

Fitness

It is sometimes said that natural selection favors the fittest individuals. Actually, natural selection *determines* the fittest individuals. The subtle distinction arises because of the technical definition of the word *fit*. **Fitness** is defined as a measure of an organism's relative contribution to the gene pool of the next generation and is functionally equivalent to the relative

TABLE 18.1 **Kettlewell's Mark-and-Recapture Experiment with Moths**	Peppered Morph	Black Morph
Dorset, England **Unpolluted Woodland**		
Marked and released	496	473
Recaptured after predation	62	30
Percentage recaptured	12.5%	6.3%
Relative survival	1.00	0.507
Birmingham, England **Soot-Blackened Woodland**		
Marked and released	137	447
Recaptured later	18	123
Percentage recaptured	13.1%	27.5%
Relative survival	0.477	1.00

FIGURE 18.11
A Leopard and Its Spots
The mottled fur of the leopard may have developed gradually but at any stage probably confers some advantage in camouflage, as evident here.

survival listed in the data of Table 18.1. Thus it relates directly to reproductive success.

Incidentally, the term "fitness" itself has evolved since
Darwin's time. When Darwin spoke of the "fitness" of an
individual, or of selection favoring the most "fit," he meant
that evolutionary processes favored the organism whose phenotype and behavior were most appropriate to its role and
habitat. The idea of the "survival of the fittest" has worked its
way into our language, and we now may speak of "feeling fit
as a fiddle" or of "physical fitness," equating *fitness* with
health and vigor. The *fittest* may be closest to the average,
when the average phenotype is one that is already well adapted to the environment.

In those cases in which a trait is controlled by two alleles,
different genotypes can be assigned different numerical fitness values and therefore relative fitness can be compared. In
such a case, the reproductive output of organisms with either
of two traits is measured. If one is reproductively more successful, that genotype is given the number 1.00. The reproductive output then of the organisms with the *other* genotypes
is expressed as a percentage of that figure. For example, if the
less successful reproducer leaves only 50% as many offspring
as its more successful colleague, its relative fitness is 0.5.

Evidence of Evolution

How good is the evidence supporting the theory of evolution?
How well has Darwin's idea held up? Pretty well, it turns out.
Let's consider some of the most powerful supporting evidence
for the theory, drawing specifically from the fossil record, biogeography, comparative anatomy, molecular biology, and artificial selection.

Fossils and the Fossil Record

Fossils are the impressions or petrified remains of organisms,
or other evidence of past life, found within or as part of rock
formations. Fossils are most often found within the layers of
sedimentary rock that were produced by the settling of mud
and sand at the bottom of ancient seas and lakes. Because the
layers of sedimentary rock must form on top of one another, it
is logical to assume that the lower layers of sedimentary rock
contain fossils that are older than those found in the newer,
upper layers (Figure 18.12).

Fossils generally form from hardened body parts, such as
bone, shells and exoskeletons, and woody plant parts, so softer-
bodied organisms and softer body parts are largely missing
from the record. Paleontologists suggest that two-thirds of the
forms that ever lived did not form fossils. Even those that did
fossilize have often left us scant evidence. In many cases, the
fossils were simply destroyed by erosion or pressure, while
others were scattered about or altered by moving waters or
shifting earth. (The best fossils come from quiet waters in geologically stable areas.) The fossil record has indeed told us a
great deal about the life that preceded us.

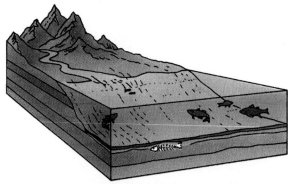

A fish swimming in an ancient sea dies. Hard parts of the
dead fish are preserved by quick burial in layers of sand or
mud. Soft parts may be replaced by minerals (fossilization).

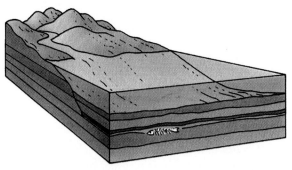

Additional layers of sediment may bury the fossil remains
deeper.

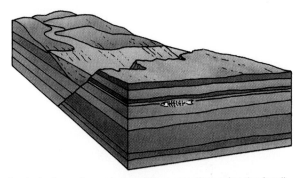

Geological events, such as uplifting, may then raise the fossil-
bearing layer.

Erosion, mining or road construction expose the fossil.

FIGURE 18.12
Fossils and the Fossil Record
The steps in fossilization can include the deposition of organism
remains in the soft mud of a lake bottom and the fossilization of
those remains as additional layers of mud are deposited over time. A
later geological event may raise the level of the original lake bottom.

Modern Armadillo

The glyptodon is thought to be the
extinct ancestor to the modern armadillo.

FIGURE 18.13
The Extinct Giant Armadillo
Fossils such as that of *Glyptotherium,* the giant armadillo,
suggested to Darwin that the usual 19th-century explanations of life
were not enough.

The fossil record has provided fascinating indirect evidence of just what kinds of selective pressures might have been operating on ancient populations. For example, in some reptile species and in some island situations, natural selection appears to be operating on species, just as it does on individuals. The idea is that, as new species arise from existing strains, those that survive the longest and generate the most new species will have the greatest impact on evolutionary trends. This process is called **species selection**. Note in Figure 18.14 that each species branches, giving rise to new species even as each continues its own existence.

Biogeography

The study of the distribution of living things over the earth—**biogeography**—offers another line of supporting evidence for evolution. Why does each geographically isolated region have its own peculiar assemblage of plants and animals? Recall from Chapter 1 that Darwin noticed that the mara, or Patagonian hare, exists where one would expect to find rabbits, and that it has, in fact, many rabbit-like characteristics. South America, Darwin deduced, doesn't have rabbits because rabbits can't get there. Instead, another kind of animal has evolved to occupy what would be the rabbit's unique position in the ecosystem.

Darwin also noticed that remote oceanic islands have few mammals except those that travel well over water, such as bats. With the exception of bats and introduced mammals, Australia, an isolated island continent, is populated only by marsupial mammals (pouched mammals that lack a true pla-

Darwin is best known as a biologist, but he was also one of the foremost geologists of his time. His interest in rocks and strata and his enthusiasm for digging fossils contributed powerfully to his development of evolutionary theory. In his "geologizing" in South America, he uncovered the fossil remains of an extinct giant armadillo (Figure 18.13). The remains were found deep in ancient rocks, deposited long ago. Above were smaller armadillos of a different sort, alive and scurrying over the stony graves of the giants. The extinct species, Darwin realized, must have been related, perhaps even ancestral to the living species.

Many fossils have been discovered since the time of Charles Darwin, and the known record is now immense and largely consistent with a Darwinian view of natural selection. Fortunately for an aging Darwin, one of the most amazing records was uncovered during his lifetime. In 1879, Yale University paleontologist O. C. Marsh published a comprehensive study on the evolution of the modern horse, *Equus* (Essay 18.1). These findings lent considerable credibility to Darwin's theory. Thomas Huxley, master debater and Darwin's greatest defender, put these findings to good use in arguing the case for evolution through natural selection.

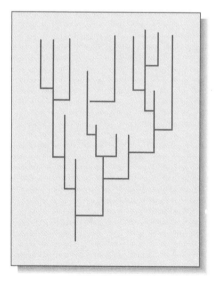

FIGURE 18.14
Species Selection
New species arise as branches from older stems. Those that survive longest and give rise to the most branches have the greatest impact on later generations.

Essay 18.1 ● THE EVOLUTION OF THE HORSE

The evolution of the horse is an often cited example of the potential paths of evolutionary change for a species. While it is now recognized that the horse did not necessarily evolve in a linear progression involving the species shown here, these forms do demonstrate the type of adaptations that the horse line had to progress through. Tracing the origin of the horse has taken paleontologists back to the early Eocene epoch of the Cenozoic era. In that distant time lived *Hyracotherium,* a timid, dog-sized woodland creature that browsed on the soft parts of plants and literally tiptoed around the forest.

Like the dog, it had multiple toes and footpads, but in contrast to the carnivorous dog, its teeth were clearly developed for grinding plants. From this decidedly unhorse-like creature arose a succession of forms that gradually took on the appearance of a respectable horse. The changes represented a continuing adaptation to a new source of energy, grass. Grasses first appeared in the Eocene epoch and began a gradual spread that culminated in today's vast grasslands.

Adapting to the grasslands required many changes in the horse's anatomy, not the least of which was a continuing change in

teeth structure to an ever more efficient grinding surface. The grinding surface of the molars enlarged to become broad and hardened for better grinding, while the front incisors lengthened for effective nipping. Of course, as the primitive horses left the forests, predators followed. Because concealment was more difficult in the open spaces, natural selection favored "early warning systems" in the form of longer necks and better vision and hearing. It also favored the ability to respond with great speed; thus we see major changes in the lower legs and feet, particularly the single toe that would become the hoof.

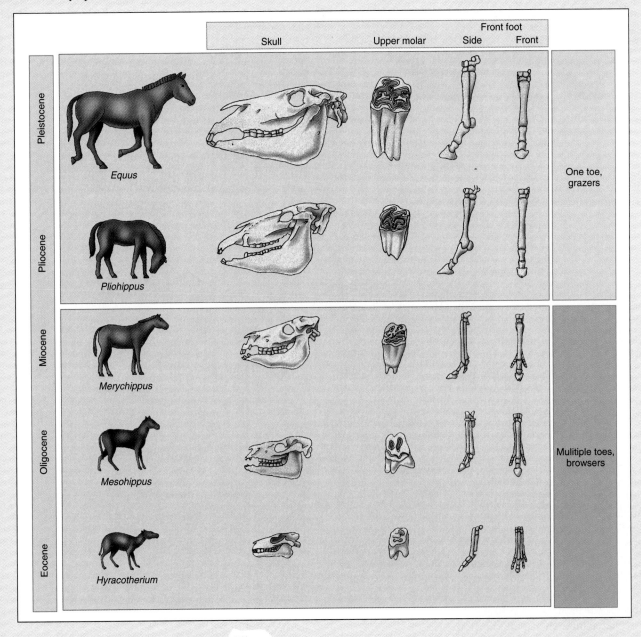

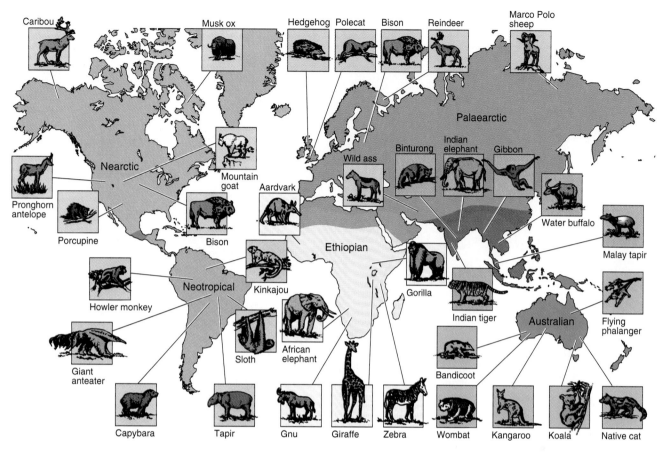

FIGURE 18.15
Biogeography and Evolution
Distribution of species in relation to geography lends insights into the processes by which species evolve.

centa), such as the kangaroo, and monotremes, such as the duck-billed platypus and spiny anteater (Figure 18.15). Marsupials are rare over the rest of the earth, with the exception being the common opossum. Since studies of today's marsupials show that they are indeed all related to one another, their distribution lends support to the idea that animals are where they are because of their evolutionary history. Related animals are likely to be found in the same area.

Comparative Anatomy

An important source of information about evolutionary relationships comes from the study of **comparative anatomy**, in which the structures of modern species are compared. Such studies are particularly useful where the fossil record is too poor to help, but often they reinforce our interpretation of that fossil record. For instance, studies of the skulls of various vertebrates (animals with backbones) have provided clues to the evolutionary trends that led to species living today. The relatedness of mammals, for example, is supported by the fact that almost all of them, from bats to whales, have seven neck (cervical) vertebrae. Also, all vertebrate embryos (including humans) have gill arches. Studies of development reveal that

although gill arches give rise to gills in fishes, they become highly modified into wholly different structures in other vertebrates (ours become part of the lower jaw, middle ear, larynx, and tongue). The study of embryos provides many other evolutionary clues as well (see Chapter 43).

In the comparative study of anatomy it is necessary to distinguish structures that are similar. **Homologous** structures are those with common evolutionary origin. Actually, though, they may look quite different and may even have taken on different functions. **Analogous** structures are those that have similar functions and often appear similar but are presumed to have different evolutionary origins. The wings of insects and birds are often cited as examples of analogous structures. What inferences can you make from a consideration of the forelimbs of the cat, whale, bat, bird, and horse and the upper arm of a human (Figure 18.16)? A study of the embryos of these vertebrates reveals that, in each case, the limbs emerge from similarly formed limb buds.

A closer look at the whale's anatomy and that of the python presents another slant to the idea of homology. Both lack hindlimbs, but tucked away in the body are useless vestigial bones that are clearly remnants of what was once a functional pelvic (hip) girdle and even vestigial hindleg bones.

Human	Whale	Bat	Bird	Horse
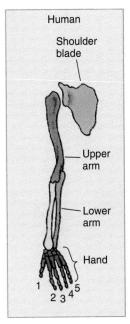	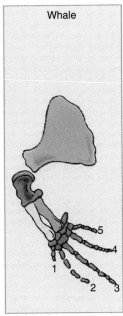	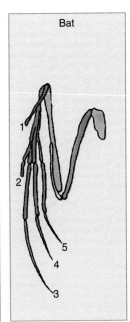		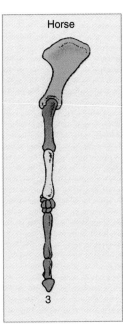

FIGURE 18.16
Homologous Structures
The forelimbs of several representative vertebrates are compared here with those of the suggested ancestral type.
Since they all have the same embryological origin, they are said to be homologous. The most dramatic changes
can be seen in the horse, bird, and whale. In these animals, many individual bones have become smaller, and
some have even fused. Another interesting modification is seen in the greatly extended finger bones of the bat.

Basilosaurus, an extinct whale

Python

Baleen whale

FIGURE 18.17
**Vestigial Limbs in Modern
Animals and in the Fossil Record**
Within the bodies of both the whale and the   python are bones
that demonstrate the evolutionary presence of the pelvic girdle and
hindlimbs of their four-legged ancestors. Fossils from an extinct
whale show an even more obvious presence of vestigial hind legs.

Both animals descended from four-legged land dwellers.
Vestigial organs are looked on as rare but welcome
records of evolutionary events and evidence of such
organs in the fossil record is cited in support of the idea
of evolutionary transitions in form (see Figure 18.17).

Molecular Biology

The study of molecular biology has shown that the
idea of homologous structures actually exists
within the genes of all life forms. Indeed there are
proteins that have virtually the same structure and
function in every life form on earth. Each species may have
acquired mutations within the genes for these proteins, some
later becoming favorable adaptations, but there is no question that versions of the same genes occur in every life
form. The acquisition of mutations within a gene is a
random event. Therefore the longer two species
have been separated from their common ancestor,
the greater the chance that different mutations will
occur in the two species. In fact, the number of
different mutations between two species can be
used as a measure of their evolutionary diver-
gence. Thus it is possible to catalogue the mutations that have
accumulated in various evolutionary lineages, and it is even
possible to use DNA sequence data to deduce evolutionary
relationships (Figure 18.18).

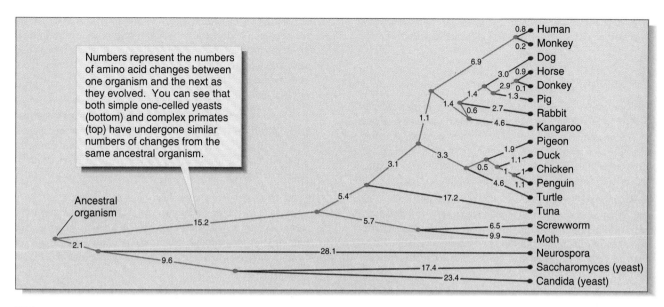

Numbers represent the numbers of amino acid changes between one organism and the next as they evolved. You can see that both simple one-celled yeasts (bottom) and complex primates (top) have undergone similar numbers of changes from the same ancestral organism.

FIGURE 18.18
An Evolutionary Tree Based on Mutations in DNA
This computer-generated phylogenetic tree represents amino acid differences in the polypeptide sequence of respiratory pigment cytochrome *c,* which is found in all aerobic organisms. Each amino acid difference represents at least one DNA mutation that has occurred in evolution. The number of DNA changes (numbers on the tree) determines the evolutionary distance from one organism to another. In general, the tree derived from this gene is consistent with other trees based on morphology or the fossil record, although the computer seems to think that the turtle is a bird.

Molecular biology also lends considerable support to the most basic tenet of Darwinian thought. The fact that all life forms utilize virtually the same genetic code (Chapter 12) and share most of the basic cellular and molecular structures supports Darwin's thoughts that all life is related, probably through a very ancient, but common, ancestor.

Artificial Selection

Finally, the principles of artificial selection have been as valuable to us in understanding natural selection as they were to Darwin when he developed the idea. We have relentlessly molded the genes of almost every species we have had an interest in, sometimes for practical reasons (as when we created stronger and faster horses), sometimes whimsically (as when we developed hairless, mouse-like dogs or wrinkled rats). Nonetheless, we have proved time and again that by selecting individuals with certain traits for breeding, those traits can become magnified in the population. The primary difference between our choices and nature's is that nature selects only those traits that lead toward successful reproduction, while we are often more capricious.

In this chapter we have followed the concept of evolution from the time the idea was developed to the theory's acceptance and integration into biological thought. Finally, we considered the modern evidence that supports the theory. Next, we will consider the development of the earth's far-flung species in terms of the accumulation of small changes at the molecular level.

Key Ideas

THE HISTORY OF AN IDEA

1. Aristotle placed animals in a "natural scale," with humans (the most complex) at the top. He and Plato believed species did not change (immutability), an idea that held until the time of Lamarck and Darwin.

2. Linnaeus, the Swedish naturalist who devised a system for classifying plants and animals, also believed that species were immutable.

3. Buffon held with immutability but proposed that some lesser organisms arose independently of Creation and changed over time.

4. Lamarck, a strong proponent of evolution, cited the fossil record as evidence of change, including change in humans. As a mechanism of change, he proposed that constant use and improvement of a structure led to that structure's improvement, and that such improvements were then inherited. This notion became known as "the law of use and disuse."

5. By Darwin's time the intellectual climate was right for serious consideration of the idea of "descent with modification," or evolution.

DARWIN'S THEORY

1. Darwin's thinking on evolution was partly a product of his knowledge of animal breeding, wherein the traits to be retained are determined by the breeder. He began to consider whether such selection could occur in other ways. The writings of Thomas Malthus, who maintained that populations tended to increase exponentially, provided some insight.

2. Darwin proposed **natural selection** as the mechanism of evolution, citing as evidence that (a) individuals vary, (b) variations are inherited, (c) more offspring are produced than can survive, and (d) those offspring with the most adaptive variations survive and reproduce better than others, thus leaving more offspring of their own type.

3. Darwin experimented with various aspects of his theory, particularly on island colonization and the dispersal of organisms. He published several works on natural history and geology.

4. Darwin presented his ideas in his book, *On the Origin of Species.* Darwin built his case by first discussing genetic variability in domestic plants and animals, then going on to variability in wild species, and then to the subjects of overproduction of offspring and natural selection.

5. The weakest part of Darwin's theory was his inability to explain how genetic variation arose and how it was maintained in populations, both of which were poorly understood in his time.

6. The "modern synthesis" incorporates the principles of genetics into the theory of evolution.

VARIATION

1. Variation includes behavioral and biochemical traits as well as the common visible traits. Such variation must be genetically based to be significant to evolution. The two principal sources of variation are mutation and genetic recombination.

2. Genetic variation is maintained in several ways.

   a. Recessive alleles are retained because their expression is masked by the dominant alleles.

   b. Genetic variation is also maintained by **balancing selection**, which occurs in spotty, variable environments where first one allele and then its alternatives are favored. The resulting phenotype is called a **balanced polymorphism**.

   c. The frequency of alternate alleles is also maintained through **heterozygote advantage**, wherein individuals heterozygous for a certain gene survive better than either homozygote.

   d. Diversity is also maintained by **frequency-dependent selection**, in which a particular combination of alleles offers a definite advantage, but when the numbers increase the combination of alleles is selected against by other factors.

NATURAL SELECTION

1. Natural selection is defined as selection by the environment for genotypes that best adapt an organism to present conditions. The test of any genotype is its effect on reproductive output.

2. A classic case of natural selection is seen in *Biston betularia,* the peppered moth. A black variant, or morph, is rare in rural areas but predominant in industrial regions, where it blends in with soot-covered trees. The peppered morph is much more common in the clean forests, where it blends in with lichen-covered trees.

3. Release-and-recapture studies of both moths in both environments clearly revealed that background coloration and predation were the factors affecting survival. In more recent times, air pollution control has meant a decrease in the black moth in industrial regions.

4. Spots in the leopard's coat, like other such traits, might have originated as minor variations that offer modest advantages when accompanying other changes, such as in behavior (tree climbing). Then, through continued selection, such traits are favored more and more, until they become a prominent aspect of the phenotype.

5. Natural selection determines **fitness**—an individual's contribution to the gene pool relative to that of other individuals. In determining relative fitness, the fittest individual (the one leaving the most offspring) is assigned a value of 1.00, and values less than 1.00 are assigned to those less fit.

EVIDENCE OF EVOLUTION

1. Natural selection is a respectable theory today.

2. The fossil record, such as that of the modern horse, provides important information about evolutionary change. Much of the fossil record is far less complete than that of the horse. Soft-bodied creatures do not fossilize well, and geological conditions must be just right for the formation and maintenance of fossils.

3. The record shows evidence of **species selection**, in which species that exist longer give rise to more lines. Such species are seen as side branches in the trees.

4. The distribution of plants and animals provides evidence of past evolutionary activity. Where a range is continuous, although extending over great distances, there are strong similarities among the related inhabitants. Where geographical isolation has occurred, we sometimes find greater differences among related species, the degree of difference depending on the time in isolation.

5. Evidence of evolution is seen in anatomical similarities among adult animals, particularly in the vertebrates. **Homologous** organs indicate evolutionary relatedness. Those that appear similar or function similarly, but are not of common origin are **analogous** and are not evolutionarily related.

6. Differences in the sequence of amino acids in proteins support evolutionary schemes. Analysis further indicates that mutations occur at a regular rate over time.

7. Artificial selection also provides evidence for natural selection. Great differences from ancestral stock are seen in animals produced in deliberate breeding programs where certain traits are selected for by the breeder.

Review Questions

1. Discuss immutability of species. Where did this idea originate? Why did it persist for so long? (page 349)

2. What was Lamarck's view on species? In what important way did his ideas differ from those of Charles Darwin? (pages 349–350)

3. Discuss three aspects of Darwin's background that influenced his ideas on evolution. (page 350)

4. List the four principal observations Darwin used to support natural selection. (page 351)

5. Darwin was able to experiment with certain aspects of evolutionary theory. Describe one such experiment. (page 351)

6. What prompted Darwin to publish his ideas long before he had originally intended? What did he have to offer that Wallace did not have? (pages 352–353)

7. What was the greatest weakness in Darwin's thesis? Why was this important, and what was he able to do about it? (page 353)

8. When did the "modern synthesis" begin? What is it about? (page 353)

9. Why must variation be genetically based to be significant? Relate this to a major fault in the evolutionary mechanism proposed by Lamarck. (pages 354–355)

10. What is the source of most genetic diversity in sexual organisms? (pages 354–355)

11. How does a highly varied environment affect genetic diversity? (pages 354–355)

12. Explain heterozygote advantage and provide an example. (pages 354–355)

13. Using the text example, explain how frequency-dependent selection maintains genetic diversity. (page 355)

14. Define evolution in terms of allele frequencies. What does this mean? (pages 355–357)

15. Briefly summarize the "peppered moth" episode. What would you expect if the British had not reduced the output of industrial smoke? (pages 355–357)

16. Explain the meaning of "fitness." (pages 357–358)

17. In what way is the fossil record the most significant evidence of evolution? Why would you expect the record to be incomplete? (pages 358–359)

18. According to the idea of species selection, what should an evolutionary tree look like? (page 363)

19. With the uniqueness of Australian animals in mind, predict what you might find if you were to compare hooved mammals from northwestern North America with those from Northeastern Asia. Explain your reasoning. (pages 359–361)

20. What is the obvious difference between artificial and natural selection? In what way can the products of artificial selection be used as evidence for natural selection? (page 363)

CHAPTER 19

Microevolution and Changing Alleles

CHAPTER OUTLINE

ALLELE REPRESENTATION WITHIN POPULATIONS

The Population Genetics of G. H. Hardy

An Implication of the Hardy–Weinberg Principle

Hardy–Weinberg: An Algebraic Discussion

The Restrictions of the Hardy–Weinberg Principle

Mutation: The Input of New Information

Balancing Mutation and Selection

THE OUTCOMES OF NATURAL SELECTION

Stabilizing Selection

Essay 19.1 The "Tail" of Two Genotypes

Directional Selection

Disruptive Selection

RANDOM CHANGES IN FREQUENCIES

Neutralists Versus Selectionists

QUESTIONS ABOUT THE GENETIC FUTURE OF *HOMO SAPIENS*

KEY IDEAS

APPLICATION OF IDEAS

REVIEW QUESTIONS

*E*volution is one of the most pervasive and explanatory themes in modern biology. As an intellectual lever, it has been used to pry loose countless gems from the complex matrix of life. In fact, it is so useful that one wonders how biology could even have been done without a clear understanding of its principles. Of course, as we know, much biology was done without it, but, as we also know, much of it was wrong. So as we continue to look at the venerable old idea, we find that some parts of it have weathered, aged, hardened, and cured, while other parts have been changed, and new parts—parts that Darwin could never have imagined—have been added as the concept of evolution has itself evolved.

We have defined evolution as descent with modification. It is important to remember, however, that individuals do not evolve—at least not in the sense that we will consider evolution here. Instead, evolution occurs within groups of individuals called *populations*. A **population** is a group of interbreeding individuals of a single species. All the alleles found in such a population make up the **gene pool** of that population, and evolution occurs as the relative abundance of certain alleles in a population increases or decreases over time. These are the types of changes that we will consider here.

We begin with a few fundamental ideas in **population genetics**, the study of changes in the representation of alleles within populations.

Allele Representation Within Populations

Allele frequency refers to the proportion of a given allele in a population. As allele frequencies change, evolution occurs. As stated earlier, evolution involves changes in allele frequencies. The term *frequency,* when used in genetics, has little to do with how frequently something happens, or how often something occurs in time, such as when we refer, say, to the frequency of tornadoes in the spring. Instead, frequency is a *proportion,* namely, a proportion of items of a particular kind in a more general class. For example, if there are 10,000 registered voters in town, and 4000 are Republicans, the *frequency* of Republicans in this town is 4000/10,000 = 0.40. Frequencies are always numbers between 0 and 1 because they are always a fraction consisting of a part divided by the whole. Both the numerator and the denominator are counts of individual items, and any individual that appears in the numerator must also appear in the denominator. For instance, the 10,000 total registered voters include the 4000 registered Republicans.

The situation is the same when counting the alleles present in a population. For example, when considering the **A** gene (which has only two alleles, **A** and **a**) in a population of 5000 people, there is a total of 10,000 alleles at the **A** locus. The **A** alleles can be found in two genotypes (**AA** and **Aa**), as can the **a** alleles (**Aa** and **aa**). The distribution of genotypes might look like this:

1000 people are **AA**

2000 people are **Aa**

2000 people are **aa**

Total number of people = 5000

Total number of alleles = 10,000

The distribution of alleles is 4000 **A** alleles (2000 from the **AA** genotype and 2000 from the **Aa** genotype) and 6000 **a** alleles (2000 from the **Aa** genotype and 4000 from the **aa** genotype). The allele frequency of **A** is then 4000/10,000 or 0.40, and the frequency of the **a** allele is 0.60. The sum of the frequencies of the alleles within a population must always equal 1. When there are two alleles for a locus, as the frequency of one allele increases (**A** rises to 0.55), the frequency of the alternate allele decreases (**a** falls to 0.45), so as to always total 1 (0.55 + 0.45 = 1.0).

The situation is also the same when counting the genotypes present in a population. For example, when considering the **A** gene in a population of 5000 people, there may be a certain number of people with **AA**, a certain number with **Aa**, and a certain number with **aa**. The frequency of the **AA** genotype would simply be the number of individuals with the **AA** genotype divided by the total number of individuals within the population (1000/5000 = 0.20 in the original example above). Again by definition, the frequencies of **AA**, **Aa**, and **aa** must total 1.

The Population Genetics of G. H. Hardy

G. H. Hardy, an eminent mathematician, had few professional interests in common with R. C. Punnett, the young Mendelian geneticist, but they frequently met for lunch or tea at the faculty club of Cambridge University. One day in 1908 Punnett was telling his colleague about a small problem in genetics. Someone had noted that there was a dominant but rare gene for abnormally short fingers, while the allele for normal fingers was recessive. In view of the dominance of short fingers and the famous 3:1 Mendelian ratio, shouldn't short fingers become more and more common with each generation, until no one in Britain had normal fingers at all? Punnett didn't think this argument was correct, but he couldn't explain why.

Hardy thought the problem was simple enough and wrote a few equations on his napkin. He showed that the relative numbers of people with normal fingers and people with short fingers ought to stay the same for generation after generation, *as long as there were no forces—such as natural selection—to change them.*

Punnett was excited. He was amazed that his friend had solved so casually what seemed to him to be such a complex puzzle and he wanted to have the idea published (on something besides a napkin) as soon as possible. But Hardy was reluctant. He felt that the idea was so simple and obvious that he didn't want to have his name associated with it and risk his reputation as one of the great mathematical minds of the day. But Punnett prevailed, and the relationship between genotypes and phenotypes in populations quickly became known as *Hardy's law.* As fate would have it, Hardy, who was indeed one of the great mathematical minds of the day, is now best known for his rather reluctant contribution to biology.

Within weeks of the publication of Hardy's short paper, the same principle was described by the German physician Wilhelm Weinberg. Eventually, the formula became known as the *Hardy–Weinberg principle.* (Later, some chose to call it the Castle–Hardy–Weinberg principle, in recognition of the belated discovery that an American, William Castle, had published a neglected exposition of the same observation in 1903.) The principle, restated, is that *both allele and genotype frequencies will remain unchanged—in equilibrium—unless outside forces change those frequencies.*

An Implication of the Hardy–Weinberg Principle

To rephrase the problem faced by Punnett and the other Mendelians, let's consider human eye color. Although the genetics of eye color aren't completely understood (several gene loci are thought to be involved), we will ignore the complications and state that the allele for brown eyes is dominant over the allele for blue. So why are blue eyes still around?

From basic Mendelian genetics, we know what happens in a cross between a homozygous blue-eyed person and a homozygous brown-eyed person. But in population genetics, we must consider not just one cross, but large numbers of matings and all their outcomes taken together. Imagine a population consisting of an equal number of blue-eyed homozygotes and brown-eyed homozygotes. The frequency of the **B** allele is 0.5, and the frequency of the **b** allele is also 0.5 (Figure 19.1). If the blue-eyed homozygotes (**bb**) mate only with brown-eyed homozygotes (**BB**), the resulting children will all be brown-eyed, **Bb** heterozygotes. Blue eyes have disappeared, just as feared by the early Mendelians. However, since each of the individuals of this generation are heterozygous, the allele frequency of **b** has remained at 0.5.

Now imagine that all these genetically identical F_1 heterozygotes grow up, randomly choose mates, and manage to reproduce. Now it becomes apparent that the blue-eyed alleles didn't disappear after all. In fact, about one-fourth of their children will have blue eyes (see Figure 19.1). Now if we let the F_2 generation grow up and choose mates, things seem to get complicated if you try to pay attention to all possible mating combinations. However, if you use a modified Punnet square to keep track of allele frequencies instead of individuals, the problem remains quite simple. When we look over the next generation, we see that once again one-fourth of the children will be blue-eyed (**bb**) and three-fourths will be brown-eyed (**Bb** or **BB**) (see Figure 19.1). And so it will continue for generation after generation.

We see then that the gene for blue eyes didn't disappear at all. What did happen? Since the heterozygotes harbored hidden blue-eye genes, the actual *frequency* of the blue-eyed allele remained at 50%, just as at the start. The idea is that, whether genes are expressed in homozygotes or hidden in heterozygotes, they are distinct physical entities that don't just go away without some reason. In the language of population genetics, eye color (and the alleles for finger length) are in **genetic equilibrium**; that is, their allele frequencies do not change over time. The crux of the Hardy–Weinberg principle then is that, barring selection or other biasing factors to be considered shortly, in randomly mating populations, allele frequencies remain in equilibrium; they do not change.

Hardy–Weinberg: An Algebraic Discussion

In the example above, the two alleles start out at the same frequency—half **B** and half **b**. To illustrate this algebraically, let p be the frequency of allele **B**, while q is the frequency of allele **b**. If there are only two alleles the following *must* be true:

$$p + q = 1$$

In our eye color example, 0.5 (frequency of **B**) plus 0.5 (frequency of **b**) equals 1. The Hardy–Weinberg law also works in populations in which the alleles have different frequencies, such as 0.60 for the frequency of **B** and 0.40 for the frequency of **b**. Let's consider all possible ways in which the **B** and **b** alleles can combine in a population. Males have two kinds of alleles, **B** and **b**, and females have the same two kinds, **B** and **b**, so all possible matings can be represented as a kind of Punnet square and an algebraic equivalent as shown in Figure 19.2.

The expression $(p + q)^2 = p^2 + 2pq + q^2$ is often called the Hardy–Weinberg formula, and it is actually quite useful. It

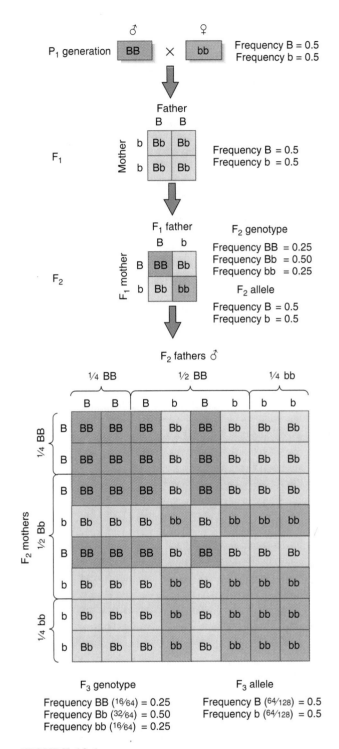

FIGURE 19.1

Tracking Allele Frequencies and the Elusive Recessive Allele

In an F_2 generation, a count of alleles proves that the frequency is as it was in the P and F_1 populations. Carrying out all possible crosses among F_2 individuals leads to an F_3. A simple summing of the dominant and recessive alleles in the table proves, once again, that the recessive allele does not simply diminish and that the allele frequencies remain the same.

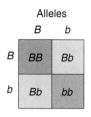

Alleles

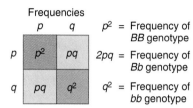

Frequencies

p^2 = Frequency of *BB* genotype

$2pq$ = Frequency of *Bb* genotype

q^2 = Frequency of *bb* genotype

(a) $p^2 + 2pq + q^2$ = total population

Allele *B* frequency ($p = 0.6$) Allele *b* frequency ($q = 0.4$)

Allele *B* frequency = p (0.6)

| Genotype *BB* frequency = p^2 $0.6 \times 0.6 = 0.36$ | Genotype *Bb* frequency = pq $0.6 \times 0.4 = 0.24$ |

Allele *b* frequency = q (0.4)

| Genotype *Bb* frequency = pq $0.6 \times 0.4 = 0.24$ | Genotype *bb* frequency = q^2 $0.4 \times 0.4 = 0.16$ |

(b) An example where *B* frequency = 0.6, *b* = 0.4

FIGURE 19.2
The Hardy–Weinberg Equilibrium
(a) If there are two alleles, **B** and **b** occurring in relative frequencies p and q (the values here are 0.60 and 0.40), respectively, and mating is random, a proportion p of the sperm will carry the **B** allele, and p of the eggs will also carry the **B** allele. **(b)** For each zygote formed, the probability that the egg and sperm will both carry **B** alleles is $p \times p$ or p^2. Similarly, the probability that both of the uniting gametes will carry **b** alleles is q^2. There are two ways that a **Bb** zygote can be formed: **B** sperm uniting with **b** egg, or **b** sperm uniting with **B** egg. The total probability of one of these two events occurring is $pq + pq$, or $2pq$. The relative frequencies of the three kinds of genotypes in the population will be equal to the individual probabilities of each kind of event: p^2, $2pq$, and q^2 will be the frequencies of genotypes **BB**, **Bb**, and **bb**, respectively. (The dimensions of the blocks [colors] represent a situation where the frequency of p is 0.60 and that of q is 0.40.)

tells us that we can find the frequency of the homozygous dominant (**BB**) individuals in a population by squaring the frequency of the dominant allele ($p \times p = p^2$). Similarly, we can find the frequency of the recessive (**bb**) individuals by squaring the frequency of the recessive allele ($q \times q = q^2$). Finding the frequency or proportion of heterozygotes is just a little different. To do this, we multiply the frequency of the dominant and recessive alleles and then multiply that answer by 2 ($2pq$).

The Hardy–Weinberg formula has very specific implications. For example, if we know the prevalence of a recessive trait such as albinism (the absence of normal melanin pigment) in the population, we can predict, within limits, the probability that any couple in that population will have an albino baby. Here's how it works. Normal skin and eye pigment in humans, *A*, is dominant over the albino condition *a*. Albinism, the **aa** genotype, occurs in about one of every 20,000 people. According to the Hardy–Weinberg formula, the frequency of genotype **aa** is represented by q^2 so that

$$q^2 = 1/20{,}000$$

and the frequency of a single allele (**a**) for this trait is thus

$$q = \sqrt{1/20{,}000} = 1/141 = 0.007$$

The frequency of the dominant allele **A** would then be

$$p = 1 - q \quad \text{or} \quad p = 1 - 1/141 = 140/141 = 0.993$$

The heterozygous condition **Aa** would therefore occur in the population with a frequency of

$$2pq = 2 \times 0.993 \times 0.007 = 0.0142$$

Since $0.0142 \times 20{,}000$ is 284, about 284 people in every 20,000 will be carrying a recessive allele for albinism, while, as we have seen, only one will be affected. Hence, in the absence of a family history of this characteristic in either parent, the chance that any couple will have an albino child is very slim.

The Restrictions of the Hardy–Weinberg Principle

It may have occurred to you that such purely mathematical constructions may be a little short on realism, or that they may operate only under very limited conditions. You are correct. In fact, the Hardy–Weinberg principle has a number of stringent restrictions:

1. Mating must be completely random.
2. There can be no mutation.
3. There can be no immigration or emigration.
4. The alleles must segregate according to Mendel's first law.
5. The expectations are *exact* only if the population and the sample are infinitely large (and this is never the case).
6. There can be no selection operating on the population.

In a sense then, the Hardy–Weinberg predictions are most useful when they don't come true, because when they don't, something is happening to the allele frequencies: that is, *evolution may be occurring*. As we see, two of the factors that can disrupt the Hardy–Weinberg equilibrium are mutation and natural selection. As we found in the last chapter, evolution is fundamentally based on these two events. So we can now consider them in terms of population genetics.

Mutation: The Input of New Information

Natural selection is a primary force determining the direction of evolution, but it must have something with which to work. That raw material is genetic variation. Of course, the genetic recombination that occurs in meiosis, crossing over, and sexual reproduction continually reshuffles genes to ensure new variation in combinations of existing alleles, but the source of *new* variations in species is *mutation*. In Chapter 17 we saw how mutations arise through rare, unrepaired changes in base sequences and through chromosome breakage and rejoining. Now we are interested in the fate of mutations as they enter the gene pool.

Balancing Mutation and Selection

In some cases a mutation can be beneficial, but we must keep in mind that this is an extremely rare event. Experiments have shown that few mutations will make a gene work better than it did before; the great majority make genes work less well or not at all. Still, randomly occurring changes provide the variability on which natural selection will ultimately act.

It is also important to realize that, evolutionarily, mutations have the same effect whether they kill an individual or just prevent reproduction. In either case, that individual's genes are not passed along. Their representation within the population dies when the individual does. Thus, while lethal alleles are constantly fed into the gene pool by mutation, they remain rare because they are not passed along. However, if the mutant allele is only partially limiting in its effect, afflicted individuals may reproduce but at a reduced rate. That is why minor genetic abnormalities are relatively common: a "mutant" allele for, say, buck teeth is passed from generation to generation, producing whole families with malocclusion but with otherwise good health. (Essay 19.1, a story about Manx cats, may help in understanding how aberrant genes reach equilibrium.)

Without regard to the severity of the impact on an individual or population, a given gene in any population will mutate at one time or another. In fact, each gene undergoes mutation in a statistically regular and predictable manner. Thus there is a constant and measurable input of new genetic information into the gene pool. For example, one type of human dwarfism, achondroplasia, occurs in one out of about 12,000 births to normal parents. Since the condition is readily visible and also dominant, each dwarf born to normal parents represents a newly mutated gene. Some of the most severe mutations go unnoticed, however, because no offspring are produced. (It has been estimated that *one-third* to *one-half* of all human zygotes fail to develop because of dominant lethal mutations. The potential parents are usually quite unaware that anything untoward has happened, since the lethal gene destroys the embryo very early in development.)

Once a mutation enters the population, the genetic equilibrium will be altered by natural selection according to whether the change is beneficial or detrimental. Over time the change will diminish in the gene pool if it interferes with successful reproduction, or increase if it contributes to reproductive success.

The Outcomes of Natural Selection

When we analyze the structure of a population for a single trait, say, height in humans, we generally find that most individuals are intermediate for the trait (so most humans are of about average height). There are increasingly fewer individuals farther from the intermediate (or average) condition. This observation then gives us a standard against which we can determine change. For example, if we plot graphically a single trait for all the individuals of a population, we find that the resulting graph will often form a bell-shaped curve approximating the statistician's **normal distribution**. In considering the effects of natural selection, we ask the following question: How do the individuals in the middle of the distribution thrive compared with those on either extreme? Depending on how they fare, we can find three trends.

Stabilizing Selection

Stabilizing selection (Figure 19.3a) is a situation where the extremes of the range of distribution have the lowest fitness and is often associated with a population that has become well adapted to its particular surroundings. Although genetic variability still exists, selection tends to favor the mean, or average, individual. Because many populations are well adapted to their environments most of the time, stabilizing selection is a commonly observed form of natural selection.

Perhaps the best-studied example of stabilizing selection is that of birth weight in human babies. (The data are readily available from hospital obstetric wards.) If we plot survival rate against birth weight, we find that, not surprisingly, abnormally small babies have relatively low rates of survival. But abnormally large babies also have lower survival rates (Figure 19.4). The highest survival rate is for babies around 3.4 kg (7.4 lb). In this case, the optimal birth weight (as determined by survival rate) is almost exactly the average birth weight. Selection then is working against genes for both high and low birth weight. In essence, the average tends to be the best, and "survival of the fittest" becomes "survival of the most average."

Essay 19.1 ● THE "TAIL" OF TWO GENOTYPES

To see how natural selection can act in rather unexpected ways, consider what might happen in the relatively simple case of lethal alleles. To illustrate, consider Manx cats. Manx cats are peculiar genetic anomalies. They have rather large hind legs and no tails (or very short tails). No one has ever been able to develop a strain of true-breeding Manx cats for the simple reason that the tailless Manx cats are all heterozygotes. Normal cats, with tails, are **TT** homozygotes; Manx cats, without tails, are **Tt** heterozygotes. The homozygous **tt** genotype is an embryonic lethal; that is, it kills the embryo. So already we see a strange thing: the **t** allele is dominant for one trait and recessive for another. It is dominant for the absence of a tail, and it is recessive for the absence of a kitten. Thus two Manx (**Tt**) cats, mated, produce $\frac{1}{4}$ normal cats, $\frac{1}{2}$ Manx cats, and $\frac{1}{4}$ dead (lost or "absorbed" as early embryos). Actually, the only litter you would see from such a cross would be $\frac{2}{3}$ **Tt** Manx and $\frac{1}{3}$ **TT** alley cat.

Now suppose that someone should populate a remote island with a whole shipload of Manx cats, which should then run wild, yowling and scratching and mating randomly, as cats are wont to do. What would happen? The frequency of the Manx allele, **t**, starts out at $q = 0.5$. In one generation the frequency is reduced to $q = 0.33$ ($\frac{1}{3}$). What happens then? With random mating, the third generation of *zygotes* will be $p^2 = \frac{4}{9}$ **TT**, $2pq = \frac{4}{9}$ **Tt** (Manx), and $q^2 = \frac{1}{9}$ **tt** homozygous lethal. But when the recessive homozygotes are removed by natural selection, the remaining cats are now half Manx and half alley cats. The frequency of the recessive lethal allele **t** has gone from $\frac{1}{2}$ to $\frac{1}{3}$ to $\frac{1}{4}$ in three generations, and in succeeding generations it will fall further to $\frac{1}{5}$, $\frac{1}{6}$, and so on. Meanwhile, the proportion of homozygous lethal **tt** zygotes will decrease accordingly (applying the Hardy–Weinberg

formulas): $\frac{1}{4}, \frac{1}{9}, \frac{1}{16}, \frac{1}{25}$, and so on. Eventually, Manx cats will be fairly rare on our hypothetical island. But, significantly, the severe selection against the **t** allele will diminish. The accompanying graph plots the course of the genotypes over 70 generations. Note that the allele frequency of **t** has fallen to 0.014 ($\frac{1}{70}$) by the end of 70 generations, and that about 0.028 of the cats will be Manx. It would take another 70 generations to bring the recessive allele frequency down to 0.007.

Selection does not have to be so severe, of course. As a general rule, the speed of frequency change is proportional to the amount of selection against the unfit genotype. For instance, suppose that the recessive genotype **tt** was not lethal but merely reduced the individual cat's reproductive ability by one-tenth. Then it would take 700 generations, rather than 70, to go from $q = 0.5$ to $q = 0.014$.

You might wonder why there are any Manx cats at all. The truth is that people are impressed by anything bizarre in cats and preferentially keep and breed the Manx kittens, while disposing of the alley kittens. So, in the final analysis, it is the human intervention—what can be called artificial selection—that keeps the Manx gene going.

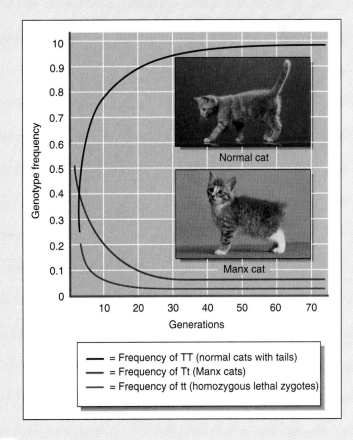

= Frequency of TT (normal cats with tails)
= Frequency of Tt (Manx cats)
= Frequency of tt (homozygous lethal zygotes)

Directional Selection

Directional selection (Figure 19.3b) removes one extreme of the phenotypic range, one end of the curve, and thus shifts the distribution of the population. This is the kind of selection practiced by dairy breeders who want only the offspring of the cows that give the most milk. In nature, directional selection

may be a response to a change in the environment that begins to favor individuals at one extreme. For example, a population may find itself in an unfamiliar territory, a new place offering new challenges. Or the species may suddenly lose a competitor and have new food sources open to it. In such cases, the formerly aberrant individuals at one end of a curve may be

(a) Stabilizing selections

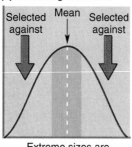

Extreme sizes are selected against.

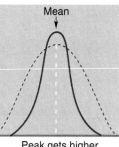

Peak gets higher and narrower.

(b) Directional selection

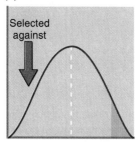

Larger individuals are favored, smaller ones are selected against.

Strong selection shifts peak to the right.

(c) Disruptive selection

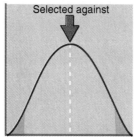

Large and small individuals are favored, medium ones are selected against.

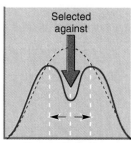

Two peaks begin to form.

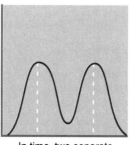

In time, two separate peaks occur.

FIGURE 19.3
Three Results of Selection
Given a starting population with a normal distribution, the distribution can change according to the selection placed on the population. **(a)** Stabilizing selection for a trait maintains a cluster about the mean. **(b)** In directional selection, the mean shifts to the right or left, as some alleles become scarce and others more common. **(c)** Disruptive selection can result in the two extremes in phenotypic expression being emphasized in a population.

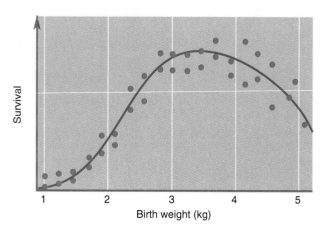

FIGURE 19.4
Selection for Birth Weight
Plotting the survival rate of babies of different birth weights indicates that there is an optimal weight of about 2.7–3.8 kg (6–8 lb).

Disruptive Selection

Disruptive (or diversifying) **selection** (see Figure 19.3c) selects against the intermediate types, and those at the extremes are favored. In this way, disruptive selection produces a *bimodal* (or two-hump) distribution curve.

One of the best known examples of disruptive selection is that of the bent-grass plants of Wales. In one area where copper is mined, copper-laden soil is piled up into mounds of waste called spoils. Botanists noticed that the mounds were bare; apparently, they thought, the copper is so toxic that nothing can grow in such places. Then they found spoils covered with bent-grass plants thriving on the copper-laden soil. Further investigation revealed that the copper-tolerant plants were freely exchanging pollen with nearby nontolerant plants. The result was that plants containing genes from both groups were sprouting all over the area, but those with the tolerance trait survived only on the spoils and those without the trait grew everywhere else. Disruptive selection then was favoring the two extremes.

better adapted to the new conditions than are those at the center of the curve. Previously unfavored traits may then become the new optimum. Subsequent evolution can be rapid, as the population quickly adapts to its new environmental demands. The peppered moth story, discussed in Chapter 18, is an excellent example of directional selection at work. Also, the early evolution of giraffes represents a classic case often attributed to directional selection (Figure 19.5).

Random Changes in Frequencies

So far, we have been considering the effects of mutation in large populations that were then subjected to the effects of natural selection. Now we will consider different scenarios by which evolution might proceed.

You will recall that one of the conditions by which the Hardy–Weinberg equilibrium would be maintained is a large population (infinitely large). This is because equilibrium

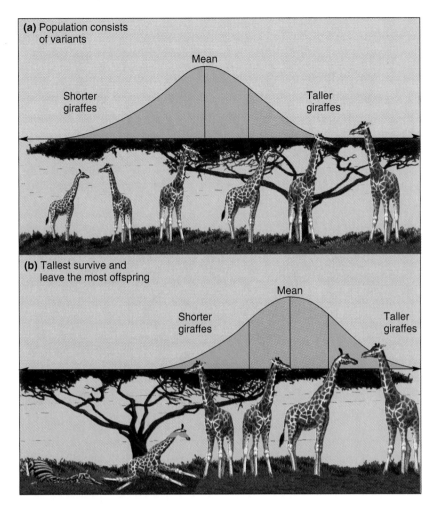

(a) Population consists of variants

Mean

Shorter giraffes

Taller giraffes

(b) Tallest survive and leave the most offspring

Mean

Shorter giraffes

Taller giraffes

FIGURE 19.5
Selection History of Giraffes
In the evolutionary history of the giraffe, an animal that browses on tall trees, height has been the critical factor. Among the ancestors **(a)**, height was variable, but as competition for the foliage increased **(b)**, natural selection began to favor taller individuals that could reach the still-untapped resources of the trees. (Natural selection was probably favoring taller trees as well.) As we see, the mean shifted. The taller individual survived, leaving the most offspring. The average height increases over the course of one generation (exaggerated here). Over many generations, giraffes become taller and taller. (And so, incidentally, do the trees, as only the tallest trees escape defoliation by giraffes.)

depends on the laws of chance. (Recall that if you flip a penny ten times, you *may* get heads nine times—90% of the time—but if you flip it 10,000 times, you are more likely to get heads closer to 50% of the time, as the laws of probability work their way.) This is why a large population size can hamper the speed of evolution. As some random mutations arise, moving an allele frequency in one direction, they are likely to be canceled out by other mutations moving the frequency in the other direction. In some cases, though, they are not canceled out, and the allele frequency of the population shifts.

A change in allele frequency in a population due simply to random chance is called **genetic drift**. This phenomenon is observed most easily in smaller populations. In very small populations, in fact, small shifts in allele frequency can have momentous effects on the population. To take an extreme example, suppose in a population of 10,000 individuals, 5%, or 500 individuals, bear a certain allele. If some disaster wipes out 100 of them, the allele is still present in 400 individuals. Now suppose the population size is 100 individuals and 5%, or five individuals, carry the allele. If all five happen to be standing on the wrong side of the mountain some day, a landslide would wipe them out all at once, and the allele would be lost entirely from the population. The larger the population

then, the greater the probability that at least some individuals carrying the allele will survive any catastrophic event.

Random changes in allele frequency can also occur in populations that experience catastrophes, such as flood or famine, that can wipe out most of the individuals in an entire population. As the greatly reduced population recovers, its allele frequencies might be altered through the chance loss of certain alleles. Thus the allele frequency in the renewed population would depend on just which alleles happened to have been carried by the few survivors (Figure 19.6). A **population bottleneck** is the result of a population becoming reduced in a brief period of time and produces a random change in gene frequencies.

Population bottlenecks can result in sudden increases in the frequency of deleterious alleles. For example, mammologist Lloyd Ingles noted in the 1950s that a species of California deer, the dwarf or tule elk, suffered a drastic reduction in population size. Whereas the herd once ranged throughout California's vast central valleys and mountains, it became restricted to one 1100-acre park in Kern County. The tule elk has recovered but now suffers an increased frequency of shortened lower jaw (Figure 19.7), a condition that produces grazing difficulties. It is quite possible that the change can be attributed to the effects of a population bottleneck. The poten-

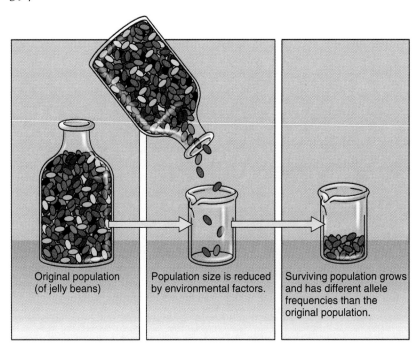

FIGURE 19.6
Population Bottleneck
In any large, diploid population, most individuals will be heterozygous carriers for rare recessive alleles at several different gene loci. At the time of a population bottleneck, only a relatively few individuals survive. The survivors will carry a random sample of the rare alleles that were present in the formerly large population. After the bottleneck, the few survivors will produce large numbers of descendants. Many of these progeny will carry some of the same recessive alleles, which will no longer be rare. Numerous individuals will become homozygous for these alleles. On the other hand, many rare recessive alleles from the original population will not occur at all in the new population.

Original population (of jelly beans)

Population size is reduced by environmental factors.

Surviving population grows and has different allele frequencies than the original population.

tial of such negative effects is now a major concern for many species whose numbers are becoming restricted or largely limited to parks and zoos.

Something like bottlenecks can also occur when a few individuals stray out of their normal species range and establish a successful colony in a new habitat. This situation can produce what is called the **founder effect**, the establishment of new allele frequencies depending on the sample present in the founding individuals (Figure 19.8).

There are many examples of population bottlenecks and founder effects in human history. The Afrikaaners of South Africa all descend from some thirty 17th century European families. As you know, all humans carry their share of recessive mutations, and most of these are harmful, but they are very rare and are seldom expressed in homozygotes. However, the Afrikaaners' genes have gone through a bottleneck of only 30 families, so present-day Afrikaaners suffer from a unique set of recurrent recessive genetic diseases that are seldom seen in other populations. For example, a normally rare condition called porphyria variegata is common among the South African settlers of Dutch ancestry. It is a metabolic disorder that is characterized by excess iron porphyrins in the blood (the heme group of blood hemoglobin), red urine, acute sensitivity to light, and eventual liver damage. On the other hand, Afrikaaners are almost completely free of other recessive genetic diseases; the 30 families obviously did not carry those genes to Africa.

Neutralists Versus Selectionists

In the 1960s and 1970s there was some controversy over how much genetic variation in natural populations is due to neutral mutations, that is, mutations that have no effect on the function of the protein encoded by the gene. The argument has produced two camps. On the one hand, there are the **selectionists**,

those who attribute most, if not all, evolutionary change to natural selection. The **neutralists** say that much variation on the molecular level is due to selectively neutral mutations that accumulate through genetic drift (see Chapter 17). For example, there are frequently several slightly different allelic forms of any enzyme in a population, differing from each other by only one or two amino acids. Neutralists hold that most of this structural variation has no effect on the function of the enzyme. They think that functionally equivalent alleles just

FIGURE 19.7
Effects of a Population Bottleneck
California tule elk suffered a drastic population decline in the 1950s. Today's survivors have an increased incidence of shortened lower jaw.

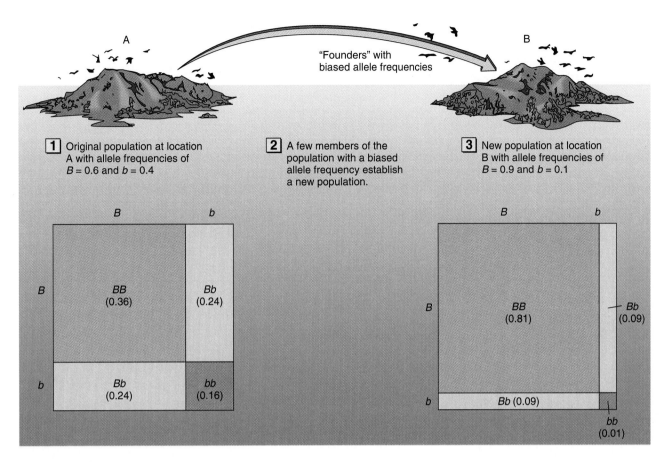

FIGURE 19.8
The Founder Effect
The founder effect is a type of genetic bottleneck, one that does not require much of the population to die off. Instead, a few members of a population, representing a small sample and a potentially biased frequency distribution, establish an entirely new population. The new population will have a new frequency distribution that depends entirely on the frequencies present in the founders, not on the frequencies of the original population.

happen by chance mutation and that the allele frequencies drift around meaninglessly. Such random changes, they say, help explain the long periods of rather constant rates of evolution. Take note that the disagreement centers over normally invisible traits, not over obvious differences in phenotypes. Neither side maintains that any *visible* phenotypic variation is likely to be meaningless. The neutralist and selectionist hypotheses are restricted to the question of whether normally *invisible* details of molecular structure are subject to natural selection.

It can be shown mathematically that a completely neutral mutation would usually disappear by chance, but that it does have a finite probability of spreading through the population and thus creating a molecular polymorphism that would be revealed as changes in protein or DNA sequences. Let's look at some possible examples of genetic drift at the molecular level.

Horse beta-globin and human beta-globin are exactly the same at 129 out of 146 amino acid positions but are different at the 17 remaining positions. Since the time of the last common ancestor of the two species, which was a little insectivore that lived about 80 million years ago, the beta-globin gene has

undergone *at least* 17 evolutionary changes—17 occasions on which a new mutant has replaced an older allele, with some changes occurring in the horse line and some occurring in the human line. Has all this change been adaptive? Are these changes examples of the replacement of an inferior allele by a superior new one through natural selection? Or have some or most of the changes been the result of meaningless neutral mutations and chance alone? We don't know, but the question has prompted vigorous academic arguments.

Questions About the Genetic Future of Homo Sapiens

Now we come to a philosophical and moral question that inevitably arises about genetic variation in future human populations. If better medical and social care saves the lives of persons with adverse genetic conditions, thus allowing them to reproduce, what is to become of us?

It's not hard to find examples that illustrate the problem. The genetic condition **pyloric stenosis**, an abnormal overgrowth of a stomach valve muscle, was once invariably fatal in infancy. Since the 1920s, a simple surgical procedure has saved the lives of nearly all affected infants (in developed countries). But about half the offspring of the saved people are also affected and also need surgery. In this way then, the alleles for the condition have actually increased in frequency. What has happened, in effect, is that a *severe* genetic condition for "certain death in infancy by intestinal obstruction" has been transformed to a relatively *mild* genetic condition for "simple abdominal surgery needed in infancy." Is this good or bad?

With selection dampened, what *is* to become of us? The question simply cannot be answered at present—we will have to wait and see. But we should keep in mind that selection may well be operating on our population in new ways that are not yet apparent. Perhaps, for example, there is an advantage in being able to function well under crowded or regimented conditions, or being less sensitive to carcinogens. Just what other kinds of traits will be important in this new kind of world we are preparing for ourselves? We don't know, but we can be sure that selection is operating on us, even if we aren't sure how.

Key Ideas

ALLELE REPRESENTATION WITHIN POPULATIONS

1. A **population** is defined as a group of interbreeding individuals, and **frequency** refers to a fraction or proportion of one type of item to all the items in a general class.

2. Evolution occurs in populations, not in individuals. It begins with changes in the **gene pool** (all the population's genes).

3. Changes in genes occur through mutation, while changes in **allele frequencies** occur through **natural selection** and random events.

4. The **Hardy–Weinberg (HW) principle** predicts the behavior of alleles and allele frequencies in a model population. It explains why in the absence of selective forces, allele frequencies do not change. Such populations are in genetic equilibrium for those alleles and remain so unless acted on by forces of change.

5. In the absence of outside influences, allele frequencies remain constant. This can be proved by using an individual cross of **Aa** × **Aa**, and carrying the cross through any number of generations. The frequencies of **A** and **a** remain 0.5 or $\frac{1}{2}$.

6. The algebraic equivalent of a population in terms of the alleles for any gene can be stated as $p^2 + 2pq + q^2 = 1$, with p and q being the allele frequencies. In the formula, p^2 equals the frequency of the homozygous dominant genotype, $2pq$ is the frequency of the heterozygote genotype, and q^2 is that of the recessive genotype.

7. From the HW principle, it is possible to predict, within its limits, the probability and outcome of matings in a population.

    a. This requires, first, that the allele frequencies of the gene in question be determined.

    b. Such frequencies can be determined through application of the formula $p + q = 1$. Here p is the frequency of the dominant allele.

    c. Together their frequencies make up all those alleles in the population, or 1.

    d. Allele with frequency q is usually detectable simply because it is recessive. First, all the recessives in a population are identified and counted. Since the number equals qq or q^2, q is the square root of that number.

    e. When q is known, p is readily determined, since $p = 1 - q$.

    f. When both p and q are known, the genotype frequencies can be determined by substituting the numbers for factors in the general HW formula: $p^2 + 2pq + q^2 = 1$.

8. The HW formulas can be used for predicting the probability of any pair in a population producing certain genotypes in their offspring. Albinism is an example:

    a. Recessive genotype frequency: $q^2 = 1/20,000$.

    b. Recessive allele frequency: $q = 1/141$.

    c. Dominant allele frequency: $p = 1 - 1/141 = 140/141$. Since knowing p and q permits the genotype frequencies to be determined, applying the multiplicative law enables predictions of certain matings to occur.

9. The HW principle operates under stringent restrictions in the model population: random mating, no mutation, no immigration or emigration, simple Mendelian segregation of alleles, infinite population size, and no selection.

10. Any change in allele frequencies indicates that evolution is occurring.

MUTATION: THE INPUT OF NEW INFORMATION

1. Genetic variation is maintained by recombination, but new variations arise through mutation.

2. Most mutations are harmful. The more harmful the effect, the stronger the selection. Less harmful alleles may be retained in the gene pool for long periods.

THE OUTCOMES OF NATURAL SELECTION

1. When height in a human population is analyzed and graphed, the data form a bell-shaped curve.

2. **Stabilizing selection** occurs in populations that are well adapted to their surroundings, favoring the intermediate phenotype and selecting against the extremes. Human birth weight is an example.

3. When changes occur in the environment or perhaps in a predator or a major food item, selection may begin to favor less common phenotypes, those at one end of the scale. Such **directional selection** probably occurred in the evolution of the giraffe.

4. In **disruptive selection**, when the environment becomes unstable, selection may shift from the intermediate phenotype to both extremes.

RANDOM CHANGES IN FREQUENCIES

1. **Selectionists** attribute most evolutionary change to natural selection. **Neutralists** believe a considerable amount of change at the *molecular level* occurs randomly, particularly where such changes do not affect the performance of an enzyme or other active molecule. Studies of hemoglobin from horses reveal that many such innocuous changes have occurred.

QUESTIONS ABOUT THE GENETIC FUTURE OF *HOMO SAPIENS*

1. Medical intervention in human genetic disorders, such as pyloric stenosis, circumvents the usual negative effects of natural selection, thus promoting increases in the frequency of such alleles.

Application of Ideas

1. In a herd of wild mustangs, Greg Meddlesome counted 10 palominos, 29 dark (brown or black), and one white.

   a. If the genotypes are **Aa**, **AA**, and **aa**, respectively, what are the respective genotype frequencies?

   b. What is the allele frequency of **a**?

   c. Is the group in approximate HW equilibrium? Calculate the HW expectations.

   d. Greg further noticed that there was only one stallion, which happened by pure chance to be the white (**aa**) horse. The rest constituted a harem of mares—not an unusual situation with groups of wild horses, but certainly a nonrandom situation. What will be the approximate allele frequency of **a** in the next generation, assuming that the group remains isolated?

   e. Would you expect the next generation to be in HW equilibrium? Explain.

2. Among Americans of European descent, about 70% find weak solutions of the chemical phenylthiourea (also called phenylthi-ocarbamide, or PTC) bitter and distasteful. The remaining 30% are unable to taste the chemical unless it is extremely concentrated. The difference in ability to taste the chemical is genetically determined by a single gene locus with two alleles. Tasting the chemical is dominant over not tasting, so tasters are designated as **TT** and **Tt**, while nontasters are **tt**. What is the frequency of each allele (**T** and **t**) in the above population? What are the frequencies of the three genotypes (**TT, Tt, tt**)? What is the probability that any taster in the population is homozygous?

3. It is always fascinating to speculate on the future course in the evolution of life, particularly of human life. Using what you have learned so far, prepare a scenario depicting humans (or what humans will have become) a few million years into the future. Try to base your assumptions on what you know about the directions of selection today. Which of the human attributes (strength, muscular coordination, intelligence, craftiness, and so on) would you expect to persist and perhaps be further emphasized? Which will be lost? Will the pace of evolution have increased over what it was in the past few million years? Why?

Review Questions

1. Carefully define the term *population*. (page 366)

2. What exactly constitutes a *gene pool*? (page 366)

3. What is population genetics? (page 366)

4. Explain what the term *frequency* refers to in population genetics. (page 367)

5. Using the logic of Hardy, explain why, in the absence of selection, the frequency of a recessive gene remains constant. (page 367)

6. Using Punnett squares and the alleles **B** and **b**, prove that the recessive **b** allele remains constant through three generations. (pages 367–368)

7. Define each of the terms of the following expression:
   $p^2 + 2pq + q^2 = 1$. (pages 368–369)

8. A certain recessive genotype appears in 16% of the individuals in a population. Determine the frequency of its two alleles, **T** and **t**. (pages 368–369)

9. What is the probability of two heterozygotes mating in the population in question 8? Of two heterozygotes mating and producing an offspring that is also heterozygous? (pages 368–369)

10. The HW formula applies to model populations. List the six assumptions that must be made to make the application valid. (page 369)

11. Considering your answer to the last question, how can the HW principle be of any use to geneticists? (page 370)

12. List three ways the HW principle can be applied. (page 370)

13. Explain the relationship among evolution, variation, mutation, and natural selection. (pages 370–371)

14. List several human maladies that can be traced to mutations. (page 370)

15. Why is it that nearly all mutations are harmful? (page 370)

16. Draw three graphs, showing the frequency of a certain allele undergoing directional, stabilizing, and disruptive selection. (pages 370–372)

17. Suggest a situation that might tend to encourage disruptive selection. (page 372)

18. What kinds of environmental changes might encourage directional selection? What other influences might favor this? (pages 371–372)

19. Specifically, how does genetic drift differ from selection? (page 373)

20. Mention ways that populations might go through a "bottleneck." How might such occasions affect the allele frequencies of the parent population? (page 373)

21. What do the neutralists maintain about variation? The selectionists? (pages 374–375)

CHAPTER 20

The Origin of Species

WHAT IS A SPECIES?

Problems with the Species Concept

TAXONOMY AND SYSTEMATICS

How Species Are Named

Taxonomic Organization: Species to Kingdom

The Five Kingdoms

DETERMINING RELATIONSHIPS AMONG ORGANISMS

The Classical Approach

The Phenetic Approach

Cladistics

THE MECHANISMS OF SPECIATION

Allopatric Speciation

Sympatric Speciation

Essay 20.1 The Galápagos Islands, Home of Darwin's Finches

Essay 20.2 Continents Adrift

REPRODUCTIVE ISOLATING MECHANISMS

Prezygotic Barriers

Postzygotic Barriers

PATTERNS OF EVOLUTION

Divergent Evolution

Convergent Evolution

Coevolution

SMALL STEPS OR GREAT LEAPS?

EXTINCTION AND EVOLUTION

Extinction and Us

KEY IDEAS

APPLICATION OF IDEAS

REVIEW QUESTIONS

*I*t is, in a sense, unfortunate that Charles Darwin removed himself from most social interaction after he married his cousin and retired to work in the countryside. It can be argued that his chronic ill health simply did not allow him the luxury of much socializing. But if his writings are any indication, he would have been a marvelous storyteller. Consider this line he wrote in his diary after seeing the Galápagos Islands:

> Both in space and time, we seem to be brought somewhat nearer to that great fact—that mystery of mysteries—the first appearance of new beings on this Earth.

What had elicited such prose from this most observant of men was the life forms he saw on those islands. Even then, he suspected that he was witnessing the *beginnings* of certain forms—the origin of species, as it were. And he knew that in such origins arose the diversity of life.

Here we will consider the ways in which the different life forms could have appeared. We will also examine **macroevolution**, the grand changes in living things that generally take many generations and result in the formation of such groups as genera and families. Our discussion includes the processes by which new species are formed and processes that lead to extinction. We will also consider how species are named and catalogued relative to higher classifications of living things, after we take a look at the sticky old question of just what *is* a species.

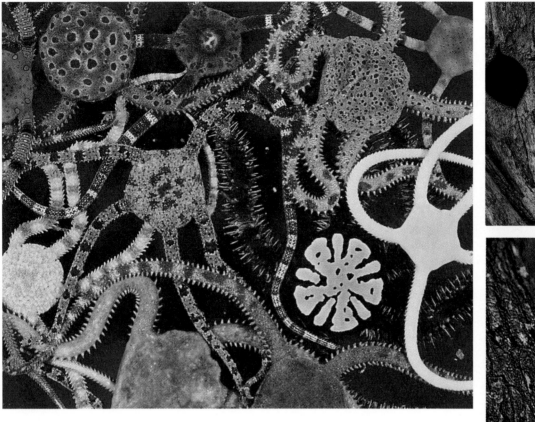

FIGURE 20.1
Variations
The outward appearance of the organism is not always reliable in identifying its species. Whereas the brittle stars are of the same species, the golden fronted and red-bellied woodpeckers are not.

What Is a Species?

One of the simplest questions in biology is also one of the most confounding: What is a species? You would think that the question would have been decided and dismissed ages ago, but it almost seems that the more we learn about life, the more complex the question becomes. We will first consider a working definition of a species, and then we will see some of the problems associated with the definition. Later in the chapter, we will see how new species arise and, finally, how they remain distinct.

Species is Latin for "appearance," and indeed organisms are put into this species or that species based on what they look like. In most cases, in fact, this works quite well, but it doesn't solve all the problems. For example, a single species of brittle star (a spindly starfish) comes in a wide range of appearances (perhaps as a way to confuse predators) yet some clearly distinct animal species have remarkably similar appearances (Figure 20.1).

The problem of distinguishing different species became a bit easier with the development of population genetics in the early 1900s. Population genetics introduced the notion of interbreeding. A shared gene pool became the fundamental criterion for defining a species. Then in 1942, the great biologist Ernst Mayr proposed that a species be defined as *a reproductively isolated group of actually or potentially interbreeding natural populations that produces fertile offspring.*

"Actually" refers to populations that do indeed interbreed. "Potentially" means they *could* interbreed if they were able to reach each other, even though they may never do so. Texas grackles (large, black, raucous birds) are larger than the grackles of Puerto Rico, but their general appearance and behavior suggest that they might well be able to interbreed if their populations were somehow joined. They are placed in separate species for now, based on size differences, but if breeding experiments proved that they could produce generations of healthy offspring, they might be considered the same species.

Problems with the Species Concept

Even Mayr's carefully crafted definition of the species can run into trouble. For example, one might ask: Is the golden jackal a different species from the timber wolf? That seems like an easy enough question. In the first place, jackals don't

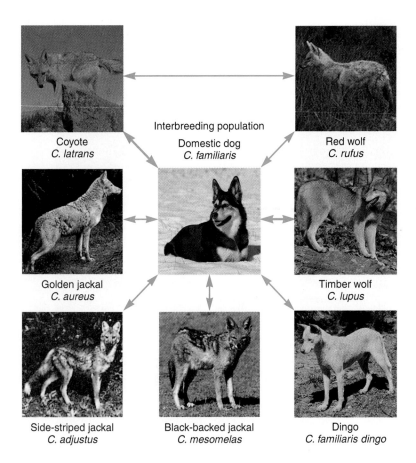

Interbreeding population

Coyote
C. latrans

Domestic dog
C. familiaris

Red wolf
C. rufus

Golden jackal
C. aureus

Timber wolf
C. lupus

Side-striped jackal
C. adjustus

Black-backed jackal
C. mesomelas

Dingo
C. familiaris dingo

FIGURE 20.2
The Genus *Canis*
Certain problems with the species concept are shown here. Each of the *Canis* species here is considered separate, but each is known to mate readily with the domestic dog. However, they generally do not mate with the others in the group, even if they are given the opportunity. So are they all members of one highly variable species? The arrows between recognized groups indicate well-documented interbreeding.

look much like wolves. Furthermore, jackals are adapted to hot, dry, open country, and timber wolves are adapted to cold, damp forests. Jackals live in Africa, and timber wolves live in northern Europe, Siberia, and North America. They are apparently two different species.

But it turns out that jackals can mate with domestic dogs, and the offspring are fertile. Furthermore, domestic dogs can and do successfully interbreed with wolves. They can also interbreed with coyotes and dingos. Thus we see that, according to the criterion of "potential interbreeding," timber wolves, jackals, coyotes, and dingos could reasonably be considered one large, highly variable species—because they can all breed with domestic dogs (Figure 20.2).

But the story grows even more complicated. It turns out that there are three quite distinct species of jackals: the golden jackal (*Canis aureus*), the side-striped jackal (*Canis adjustus*), and the black-backed jackal (*Canis mesomelas*). And while they will all breed with dogs, they apparently don't interbreed with each other. In the Serengeti of eastern Africa, the ranges of all three jackals overlap, and there they simply ignore each other. A highly territorial jackal of one species won't even bother one of a different species that wanders through its personal domain. On the other hand, the range of the coyote overlaps slightly with that of the now-rare red wolf (*Canis rufus*) in Texas, and there,

wolf–coyote hybrids have been found (Figure 20.3). Still, wolves remain wolves and coyotes remain coyotes, and the two are physically quite distinct animals. Aside from this and the promiscuity of "man's best friend," the lack of interbreeding among the eight species of the genus *Canis* appears to be more one of unwillingness rather than inability. Is unwillingness then as important a barrier as geographic separation?

To illustrate another kind of complexity with naming species, consider the case of a group of salamanders in California. Their range extends over a long, circular route (Figure 20.4). They interbreed continuously along their range, except at the southerly end, where two variants overlap. There the variants are dissimilar in appearance, and they do not interbreed. However, according to our definition, they are not different species. The entire group is considered one species consisting of a number of distinct geographical units called **subspecies**.

In many cases then, the definition of species is rather loosely held. Particularly troublesome are cases of asexually reproducing species. (How can interbreeding be a factor if they don't breed?) Here then we must fall back on physical appearance or even to the relatedness of DNA sequences. Especially when the functional test of interbreeding is removed from the definition of species, the question of whether two organisms are of the same or separate species can be difficult.

(a) Red wolf — Coyote hybrid

(b) Timber wolf — Dog hybrid

FIGURE 20.3
Hybrids in the *Canis* Genus
(a) Hybrids sometimes occur between species in nature. In Texas, where the range of the red wolf (*Canis rufus*) overlaps with that of the coyote (*Canis latrans*), natural hybrids are common, such as this 16-month-old male. **(b)** Interbreeding between wolves and domestic dogs is fairly common, even though they are recognized as separate species. The parents of this superb animal were a dog (*Canis familiaris*) and a timber wolf (*Canis lupus*).

FIGURE 20.4
***Ensatina* and the Species Problem**
Populations of the salamander *Ensatina eschscholtzi* are found in California along the coast and in the inland mountains. Skin color and size vary more or less continuously along the range, which forms a circular shape, as shown on the map. In the southernmost end of the range, two variants overlap, sharing the same territory. While individuals of each type may live just a few yards from each other (see *E. eschscholtzi eschscholtzi* and *E. eschscholtzi klauberi*), there is no evidence of interbreeding between types. At the same time, studies made along the range of the salamander indicate that other neighboring variants can and do interbreed. If variants along a range are of the same species but individuals in the extremes of the range do not interbreed, are they two of the same species? Many scientists disagree on this issue; others believe the question to be meaningless, arguing that the species concept itself is invalid.

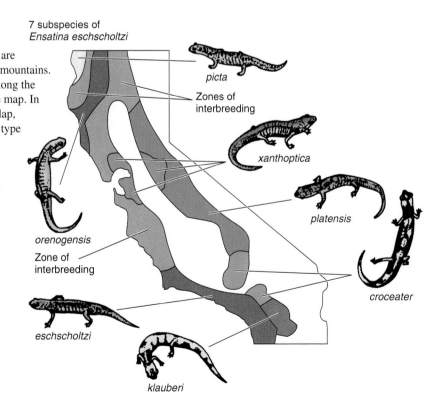

7 subspecies of
Ensatina eschscholtzi

picta

Zones of
interbreeding

xanthoptica

platensis

orenogensis

Zone of
interbreeding

croceater

eschscholtzi

klauberi

Taxonomy and Systematics

Humans come in two kinds: those who like asparagus and those who don't. Or those who like country music and those who don't. Or those who like biology and those who don't. Furthermore, those who fall into one group or another are usually given a label that identifies them with their group.

Pigeonholing living things has proved difficult, however, because people disagree on how large any "hole" is and what should go into it. We have already mentioned part of the problem: living things vary. How much can a life form vary from the others before it goes into a different hole (that is, before it gets a different name)? Another problem is that in each pigeonhole are yet smaller holes, and these have holes that are smaller yet. It's not too difficult to place most things in the largest holes. For example, if it has leaves and roots, we can safely place it in the great pigeonhole labeled "plants." If it has ears and it bites, it's definitely an "animal." But not all plants have leaves, and not all animals have ears, so the story already grows complicated. At some point, though, decisions are made about just what collection of traits defines this species or that. The next question is how to name it.

How Species Are Named

Each species is given what is called a scientific name. Scientific names follow a **binomial** ("two-name") **system**, wherein the designation includes two terms: **genus** (usually a noun) and **species** (also called the *specific epithet,* often an adjective but sometimes a person's name). For example, living humans are called *Homo sapiens.* Our genus, or generic name, is *Homo.* Our specific epithet is sapiens (wise). Our species, or specific name, is *Homo sapiens.* The scientific name assigned to each species is used only for that species. By convention, the names are italicized. Generic names are often abbreviated, especially after having been written in full; thus we see our name written as *H. sapiens.* If a subspecies is named, the subspecies name follows the species name and is also italicized.

The two-name system of classification was developed by Carl von Linne (1707–1778), who latinized his own name to Carolus Linnaeus. Linnaeus is responsible both for establishing categories within categories and for naming things with two Latin names. Latin was chosen because it is a dead language, not commonly spoken anywhere outside a few academic halls or in religious ceremonies. Thus it isn't likely to change much. Also, because Latin is the root of a number of present-day languages, latinized names can transcend many language barriers.

In a monumental undertaking, Linnaeus classified and named many of the earth's creatures. Today, however, the rules for assigning names are rigidly enforced by international commissions. In fact, Charles Darwin sat on the very first such commission. There are now international nomenclature commissions for zoology, botany, bacteriology, and virology (not to mention enzymology, organic chemistry, and just about any other branch of science in which names are important).

Taxonomic Organization: Species to Kingdom

First, we should say that the science of identifying, naming, and classifying organisms is called **taxonomy**. **Systematics** is the science of determining evolutionary relationships among organisms. Systematics and taxonomy are not (necessarily) the same thing. While it would seem to make sense to classify organisms as similar when they share an evolutionary heritage, it is not the only way to determine groupings. Systematists and taxonomists deal with determining which species group together into which genus and which genera group together into which **families** (the next highest category), and so on. The categories form a hierarchy, which, beginning at the lowest (most exclusive), becomes *species, genus, family, order, class, phylum* (or *division,* in plants), and *kingdom* (Figure 20.5).

The **kingdom** then is the largest, most inclusive, category of life. Mostly out of historical convenience and preference we divide life into five great kingdoms. The increasingly finer divisions, along with the divisions and subdivisions, can then be likened to boxes within boxes or as branches of a tree (Figure 20.6).

We must keep in mind that, in essence, these are just names applied to groups of organisms that are assumed to be related, to one degree or another. Each taxonomic group, from subspecies to kingdom, is called a taxon (plural, *taxa* from *taxis,* "to put in order"). Naming, of course, is necessary, but we must keep in mind that names are human contrivances, created for human purposes, and have no other importance and no separate reality in themselves. At the same time, if a name is going to be useful, it should not be totally arbitrary but should reflect *some* kind of reality. When all is going well then, taxonomy should reflect systematics; that is, names and groupings into higher categories should reflect evolutionary relationships. Thus two species within the same class should share more ancestry (are more closely related) than two species in different classes. The hierarchy of names, the increasingly larger boxes or branches of the tree, define the **phylogeny** or evolutionary history of an organism. Therefore the evolutionary or phylogenetic relatedness among different organisms should also be apparent in the names and the trees.

The Five Kingdoms

We must be aware that, because it is a human concept, the definition of life's kingdoms may undergo revision. One prevailing notion is that there are five kingdoms and that the five kingdoms are related to each other as seen in Figure 20.7. The kingdoms are as follows:

1. **Monera**. The prokaryotes or bacteria. As we will see in Chapter 22, biologists are now in the process of reorga-

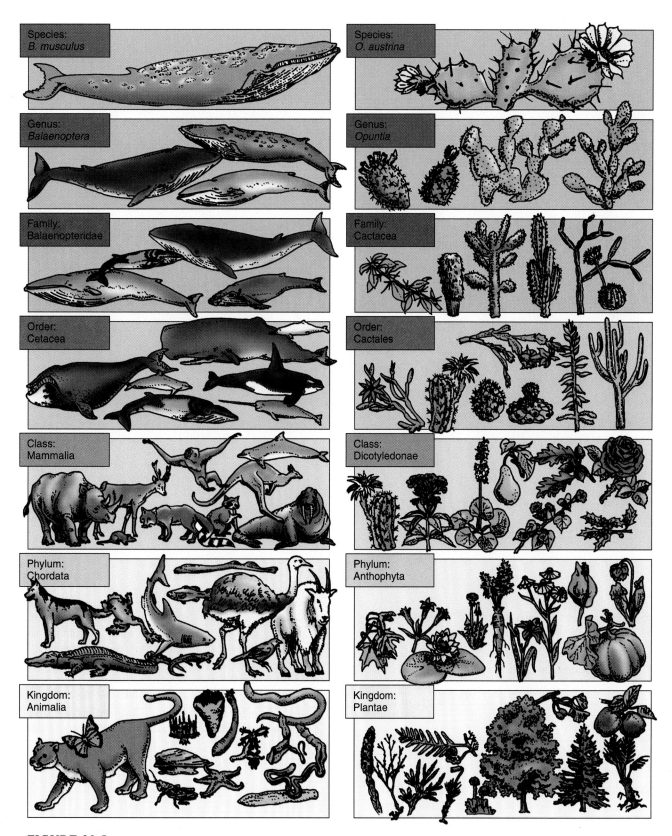

FIGURE 20.5
Taxonomic Designations
The hierarchical organization of taxa, from kingdom to species can be accurately likened to the boxes within boxes seen here. The largest being the kingdoms, the smallest, the species.

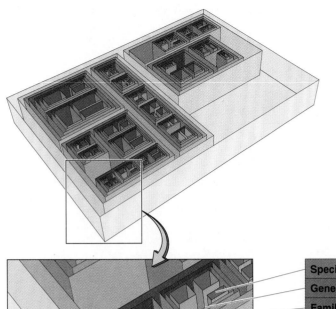

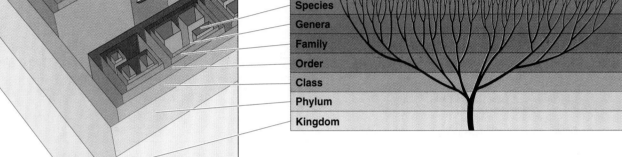

FIGURE 20.6
Boxes Within Boxes and Branches on a Tree
The hierarchy of taxonomic groups comprising a taxon may be thought of as "nested boxes." Inside each box are additional boxes, and inside those, more still. The smallest groupings within any box might represent members of the same genus. What might the largest boxes represent? The boxes can also be viewed as the top view of an evolutionary tree. Two small branches are close to each other and related if they emerge from the same larger branch.

nizing the bacteria into two kingdoms, Archaebacteria and Eubacteria.

2. **Protista**. A grouping of various, mostly single-cell eukaryotes, including both *protozoa* (nonphotosynthetic protists) and *algae* (photosynthetic protists). Protists may include multicellular forms, but these do not have pronounced tissue differentiation.

3. **Fungi**. Mostly multicellular, parasitic, and scavenging organisms (the molds and mildews, mushrooms and toadstools, yeasts, and one element of the lichens).

4. **Plantae**. Multicellular, photosynthetic organisms ranging from evolutionarily simple mosses to the evolutionarily advanced flowering plants.

5. **Animalia**. Multicellular animals, including sponges (parazoa) and other animals (metazoa).

One of the problems with this organization, you may have noted, is that there is no mention of the viruses. It has been argued that they should have their own kingdom since they are unlike anything else (see Chapter 22). There are additional problems, particularly but not exclusively in the protists. All members of kingdoms should be traceable to a common evolutionary ancestor. Yet the protozoan and algal protists may well have evolved from different ancestors. However, they are still lumped together in the five kingdom system (see Figure 20.7), even though new insights are now reshaping our perceptions of the kingdom concept.

Determining Relationships Among Organisms

As we think about taxonomy and systematics, we should carefully avoid the assumption that primitive species are giving way to advanced species. There is no such thing as a primitive

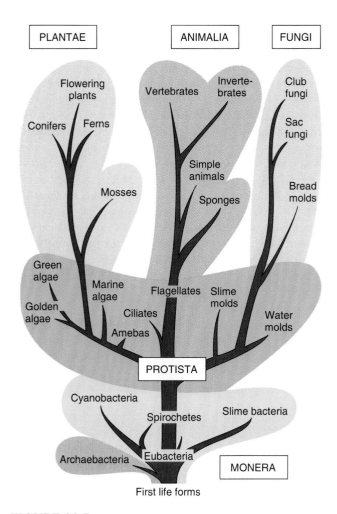

PLANTAE

ANIMALIA

FUNGI

Flowering plants

Vertebrates

Invertebrates

Club fungi

Conifers   Ferns

Sac fungi

Simple animals

Mosses

Bread molds

Sponges

Green algae

Marine algae

Flagellates   Slime molds

Golden algae

Ciliates

Water molds

Amebas

PROTISTA

Cyanobacteria

Spirochetes   Slime bacteria

Archaebacteria   Eubacteria

MONERA

First life forms

FIGURE 20.7
Kingdoms of Life
The five kingdoms are arranged in a manner that suggests their phylogenetic relationship to each other. The underlying kingdom, known as Monera, is now being reorganized into two kingdoms, Archaebacteria and Eubacteria, containing two quite diverse groups of prokaryotes. Above this is the large assemblage provisionally called Protista. Some of its members are descended from the same lines that gave rise to plants, fungi, and animals.

species. After all, a slime-mold's ancestry is just as long as our own. Thus we do not speak of primitive and advanced species. However, we may speak of *primitive* and *derived traits*. **Primitive traits** are those that arose relatively early in evolutionary history and have been inherited with little change from remote ancestors. **Derived traits** are those that have undergone recent change. Most animals are "mosaics" of primitive and derived characters. For example, the platypus is heavily endowed with primitive traits, such as egg-laying and the presence of a cloaca. But its mouth is very recently derived and highly specialized. Humans, on the other hand, have a number of derived traits, such as bipedalism and a sim-

ple aortic arch, but our jaw is remarkably primitive. Such differences in the rates of evolution of various structures have presented evolutionists with great problems of naming and classifying the various life forms.

Another problem in developing a unified theory of evolution is that scientists must make judgment calls. Such decisions often reflect a very personal philosophy, since they are simply a matter of how you choose to look at the problem. Some scientists, for example, will look at two groups and be impressed by their similarities. Others will be impressed by their differences. The former (the "lumpers") will tend to place the two groups in the same category; the latter (the "splitters") will place them in different categories. For example, some put tigers and house cats in the genus *Felis,* while others, impressed with the size differences, segregate the big cats into the genus *Panthera.*

These then are some of the problems confronting taxonomists and systematists. Nevertheless, the challenges have been accepted, and these researchers have forged ahead, trying to make some sense of the vast array of life, past and present. We will now consider three approaches to the question— the *classical, phenetic,* and *cladistic.*

The Classical Approach

Historically, the **classical** approach to classifying organisms has been based on homologies (similarity due to common descent of a structure, that is, descent from a common ancestor). The first steps in discovering homologies generally involve comparative anatomy and are based simply on what the organism looks like. If it looks like a bear, it's going to be treated by classical taxonomists as a bear—at least at first. If, however, the fossil ancestors were not bear-like, the animal's classification will be reconsidered. For example, it was once thought that bears were distinct carnivores, not closely related to any other living carnivore. But, more recently, fossil evidence indicates that bears are, in fact, rather closely related to dogs. Thus bears have been repositioned on the evolutionary tree in light of the newer findings.

Classical taxonomists also consider the embryological development of the animal in question. Its embryology will be compared to that of other species. Similarities in embryological development are considered evidence of homology and indicate a fundamental relatedness. Some of the results of embryological comparisons can be unexpected. You may be surprised to find, for example, that the echinoderms (such as sea stars) are placed near our own phylum (chordata) in evolutionary trees. Obviously, this is not based on any outward resemblance between us and them, but rather on very fundamental embryological considerations (see Chapters 26 and 27).

Classical taxonomists thus rely on a "constellation of characters," an approach that involves comparative anatomy, paleontology, and embryology (always with an eye toward discovering homology). However, proponents of the next two approaches reject the classical approach as being entirely too subjective.

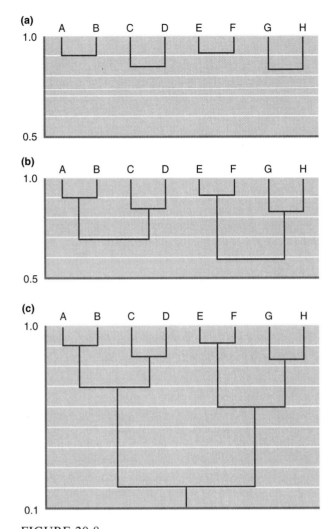

FIGURE 20.8
Species Designation Through Phenetics

Problem: to determine the relationships among eight species—A, B, C, D, E, F, G, and H. **(a)** Each species is paired by computer with the species with which it has the most in common, according to a large number of phenotypic traits. The number 1.0 indicates complete agreement, 0.5 indicates agreement in half of the characters. The level of agreement is shown by the horizontal lines joining the groups. **(b)** Here each pair is matched with the pair it most closely resembles in the range of characters being considered. **(c)** Each grouping of pairs is then compared to other such groupings and placed with the groups with which they have the most phenotypic traits in common. In this way, descendancy is determined by the sheer number of observable traits. Any analogous structures that are included will, according to the theory, be swamped by the greater number of homologous structures.

The Phenetic Approach

Phenetics is an approach based entirely on measurable differences and similarities. It is not based on assumptions about homology. Furthermore, the approach does not presume anything about relatedness. Since any statement regarding relatedness is regarded as conjectural, pheneticists let numbers speak for themselves. Essentially, this process examines a great many phenotypic characters (say, about 100), using a numerical scoring system as shown in Figure 20.8. Furthermore, no assumption is made about the evolutionary significance or derivation of any individual character, or about any other factor regarding its development. The idea is that homologous structures will be distinguished from analogous structures by sheer numerical weight of the first over the second. Thus, when enough data are available, the numerical pheneticists argue that computers can score the organisms according to the number of characters they share, thereby revealing the relationships of organisms.

Critics reply that phenetics looks for relatedness based on phenotype alone and that such outward appearances can mask a host of internal or genetic differences. There are probably few strict pheneticists today, but their quantitative techniques have had a strong impact on other approaches to taxonomy.

Cladistics

Cladistics (*clados,* "branch") classifies organisms according to *when* they branched from common ancestors. It does not take into account *how much* they diverged from the ancestral group.

The **cladogram**, the evolutionary tree produced by cladistics, is based on particular kinds of homologous traits called *shared derived characters.* These derived characters must have appeared *after* a branch diverged from the ancestral stock. Those traits that are common to all species on a cladogram are called shared primitive characters. For example, all the groups shown in Figure 20.9 have a nerve cord. This trait (a shared primitive character) was present in their last common ancestor, and so it is useless in helping to differentiate one group from another. The branch points in the cladogram are based on shared derived characters. These are new, or novel, structures that are homologous in all the groups that appeared after the branch.

The problem is that the relationships indicated in some cladograms are in conflict with some of our usual evolutionary assumptions. For example, according to the classical and phenetic views, crocodiles are more closely related to lizards than they are to birds. This view would seem logical given what we know about the life of lizards and crocodiles. However, the cladists believe that crocodiles are more closely related to birds than to lizards, that birds and crocodiles share a more recent evolutionary ancestor than crocodiles and lizards (Figure 20.10). Their reasoning is that the same line that diverged from ancestral lizards gave rise to both crocodiles and birds, such that crocodiles and birds share many

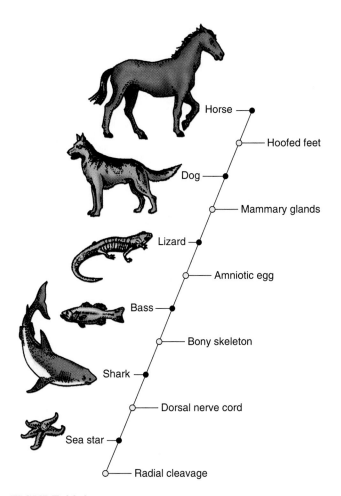

FIGURE 20.9
The Cladogram: Species Designation Through Cladistics
Each branch, represented by the dark circles, indicates the presence of an evolutionary novelty that sets it apart from what went before. All the animals started with a nerve cord. Thus they are all set apart from organisms with diffuse, net-like nervous systems, and they all appear on the branch determined by the shared derived character (open circle) that is dorsal nerve cords. In a similar manner, all mammals are found on the same branch of the tree, the branch that is defined by the shared derived character of mammary glands.

advanced characters that are absent from lizards. The confusion arises because birds evolved far more rapidly and added so many additional novel traits along the way that they now appear vastly different from their close relatives, the crocodilians. Cladistic arguments have been so persuasive that the closer relationship of birds and crocodiles is a view now held by most systematists.

We see then that the business of grouping living things with their nearest relatives is not an easy matter. Furthermore, tracing lines of descendancy is more difficult still. Nonetheless, such efforts continue, since our understanding of evolutionary processes is strongly based on just such information.

The Mechanisms of Speciation

One of the most interesting and challenging questions in biology is: How does **speciation** (the formation of new species) come about? A critical factor, it turns out, involves geography, particularly when a geological process creates separate populations of a single species. We will first see how species can arise in geographically separated populations and then in overlapping populations.

Allopatric Speciation

Speciation can occur when a geographical process splits or fragments a population such that the groups evolve separately to the point that they can no longer interbreed. The formation of new species after the geographic separation of once continuous populations is known as **allopatric speciation**. The idea is that, when a population is split, each subgroup takes its own distinct evolutionary route until finally the two subgroups have diverged so much that interbreeding is no longer possible, even if they should rejoin. It is at this point that they are regarded as separate species.

There are two primary ways that populations of a species might become geographically isolated. One way is that a group of individuals (or a seed, or even one inseminated female) might find itself in a new but hospitable place. Indi-

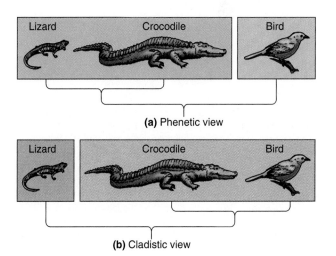

FIGURE 20.10
Two Views of Classification
(a) In the phenetic view, lizards and crocodiles share more similarities with each other than either does with birds. It follows that they are more closely related to each other than either is to birds. (b) The cladistic view considers shared derived characters. This tree is based on the belief that the last common ancestor of crocodiles and lizards predated the last common ancestor of crocodiles and birds.

FIGURE 20.11
Grand Canyon Squirrels
These two species of squirrels were once a single species and population, that is, until the Colorado River cut an impassable gorge between them. Now each species occupies one side of the geological barrier known as the Grand Canyon.

viduals can reach such places in a number of ways. Ocean islands, for instance, are occasionally populated by the descendants of unwilling, drenched, and bedraggled passengers on driftwood logs. Species may also reach islands by air, as when birds and flying insects are blown to some island by powerful winds. The finches that Darwin observed on the Galápagos islands may have arrived there in just such a way (see Essay 20.1, page 390).

The second primary way that populations can be separated is by the slow process of geological change dividing a previously continuous group. This can happen on a local scale, as when the Grand Canyon gradually divided a population of squirrels (Figure 20.11), or it can occur on an even larger scale. Historically, the most dramatic separation of populations has been due to **continental drift**, the separation and moving apart of immense land masses (see Essay 20.2, page 391). It is believed that there have been at least two such episodes: the first in Precambrian times and the second during the Mesozoic era, about 230 million years ago. The final stage of the latter episode began some 180 million years ago when Africa and South America parted company. The unique array of species on the various continents, and particularly those on Australia and South America, has been the result of millions of years of allopatric speciation following continental separation.

Sympatric Speciation

Speciation can also occur in populations occupying the same geographical area. Whereas allopatric is by far the most common type of speciation, **sympatric speciation**—speciation within a population occupying a single habitat—does indeed occur. The best examples are found in plants.

Sympatric Speciation in Plants. Sympatric speciation is quite common in some plants. In particular, it occurs through two main processes—**hybridization** and **polyploidy**. Among the flowering plants, in particular, there are many examples of new species arising by hybridization (interbreeding between two species).

We have mentioned that in animals hybrid offspring, such as mules, are usually sterile. The mule's sterility can be traced to gametogenesis, where, during prophase I of meiosis, the horse and donkey chromosomes are so different that they fail to synapse (see Chapter 9). (Pairing up is essential for both crossing over and for the proper alignment and separation of homologous chromosomes in the first division of meiosis.) Such a failure results in abnormal, random chromosome separations in anaphase and, eventually, in abnormal gametes that are unable to form viable embryos (Figure 20.12).

New species can result from **polyploidy**, an increase in the number of chromosomes, usually caused by the doubling of chromosome sets. This happens spontaneously from time to time in the mitotic divisions of plants. The chromosomes double normally, in preparation for cell division, but for some reason the cell fails to divide. Thus the abnormal cell and all its progeny will have double sets of chromosomes. Self-fertilization of a polyploid flower results in **autopolyploid** ("auto," self) offspring, plants that are normal except that they now have twice as many chromosomes as the original plant and can produce fertile crosses only with other polyploids.

A more common form of polyploidy occurs in the formation of **allopolyploid** ("allo," other) plants, plants that result from the doubling of chromosomes in otherwise infertile hybridizations. One process that can result in allopolyploid plants is the spontaneous doubling of the chromosomes in an infertile hybrid. The doubling produces flowers with an even number of chromosomes, each able to pair during meiosis (Figure 20.12).

Interestingly, polyploidy can occur repeatedly. Thus we find tetraploids (four genomes), hexaploids (six genomes, or three doubled copies of three different parental genomes), octaploids, and so on (the term polyploid covers any level of chromosome duplication higher than diploid). All this may seem freakish, but, in fact, it is not even a rare process in flowering plants, and many plant species, both wild and domestic, are polyploids. Wheat, for example, is actually an allohexaploid, a genetic combination of the chromosome complements of three entirely different Middle Eastern species of wild grasses. *Triticale* (Figure 20.13) is a human-engineered hybrid of wheat and rye grasses, which have a natural tendency to form allotetraploids. In the final analysis, it turns out that polyploidy is an excellent mechanism for increasing plant diversity through rapid speciation.

For reasons not fully understood, polyploid species may have much greater tolerance for harsh climatic conditions. In one study, for instance, only 25% of the plants sampled in lush tropical regions were polyploids, while 85% of all the flowering plant species in the raw environs of northern Greenland were polyploid.

Thus polyploidization creates instant, sympatric species of flowering plants, ready to be tested by natural selection. This may help to explain how flowering plants arose rather abruptly in evolution and very quickly radiated out over the landscape to create the incredible diversity of plant species that dominate our world.

Sympatric Speciation in Animals. Sympatric speciation in animals is probably much rarer than it is in plants. In fact, for years theorists have argued over whether sympatric speciation occurs at all in animals. A few rather clear-cut examples have been found; however, the processes in animals are quite unlike those in plants. For example, in animals, polyploidy is not likely to be involved. One reason is that many animals have sex chromosomes, the number and types of which are crucial to normal development.

One of the best known cases of sympatric speciation in animals involves flightless grasshoppers. Across their range,

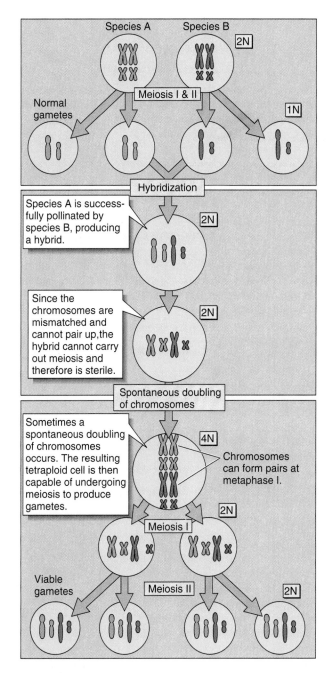

FIGURE 20.12
Hybrid Sterility and Polyploidy
While a plant hybrid may readily be produced in nature, it is often sterile. Sterility may arise when, during meiosis, unmatched chromosomes fail to find their homologues. Here species A is successfully pollinated by species B, producing a hybrid. Since there is no match in the chromosome complement, the hybrid cannot carry out meiosis and is therefore sterile. However, should a spontaneous doubling of chromosomes occur, as it sometimes does, normal pairing-up and meiosis occur, and normal gametes are produced. Following self-fertilization, a new viable species is produced.

Essay 20.1 ● THE GALÁPAGOS ISLANDS, HOME OF DARWIN'S FINCHES

The variety of finches of the Galápagos islands (see Chapter 1) is a result of allopatric speciation. Compared to the giant tortoises, strange flightless cormorants, and impish sea iguanas living there, Darwin's 13 species of finches are not particularly interesting—that is, not until the saga of their evolution is revealed.

The finches are all 10–20 cm long, and both sexes are drab-colored browns and grays. Six are ground species, each feeding on different, appropriately sized seeds or cacti, while seven are tree finches. In each species the beak has become modified for its specific diet. One of the strangest tree-dwellers is the woodpecker finch. Lacking the long, piercing tongue of the woodpecker, it uses barbed cactus spines to pry insect grubs out of cracks and crevices in the trees.

On the *Beagle's* historic visit to the Galápagos islands, Darwin collected everything he could find or catch, including the ordinary little brown finches. He took no special interest in them, but a London bird taxonomist later examined the specimens and noted that the inhabitants of different islands, though very similar to each other, were clearly different species. Ordinarily, birds living in one locality tend to be very different from those of another locality. Whereas these birds were all clearly finches, many were seen doing things that finches don't ordinarily do. Why have these little birds become so important to our understanding of speciation? Darwin concluded (and his conclusions are backed by years of careful research by others since) that the different species of birds were all descended from the same stock. Long ago, about 10,000 years ago by current estimates, the volcanic islands were colonized by South American finches that probably were blown out to sea by a storm. Apparently, conditions on the islands were favorable, and the "castaways" flourished. Their descendants eventually populated all the islands by occasional island hopping. However, the island hopping was rare enough to ensure the virtual isolation of each population. What followed then is referred to as adaptive radiation—the branching of an evolutionary line through the invasion

of new environment, accompanied by adaptive evolutionary changes.

According to our scenario, the little birds had the islands to themselves, as far as they were concerned. There was a great variety of food. They were already well adapted for foraging for small seeds on the ground, but there were other plentiful untapped food resources—food not ordinarily eaten by finches.

Soon the expanding populations were depleting the available supply of small seeds. Thus natural selection began to favor birds that could also cope with larger seeds and with other food resources. In time, the bird's bill sizes began to change as each population began to adjust itself more closely to the different kinds of food found on each island. Natural selection was at work. Eventually, the isolated birds of the different islands differed genetically to the extent that any island hoppers would find themselves to be reproductively incompatible with the residents of other islands. Clearly, speciation, the formation of new species, had occurred.

As differences in lifestyles and specialization became magnified among speciating populations, competition would have been reduced. If the genetic differences led to differences in utilization of resources, competition might become so low that two emerging species eventually could coexist on the same island. Their coexistence would also set the stage for further change. For example, with two species of finches trying to survive on one island, natural selection would favor the individuals in each population that were as different as possible from those in the other population, thereby further reducing competition between them. The tendency for differences in competing species to become exaggerated as each specializes in different directions is called *character displacement*.

After thousands of years of finches occupying the Galápagos islands and separating, changing, specializing, and rejoining, the different populations today are totally unable to interbreed. This means that several of the species can exist side-by-side on every island.

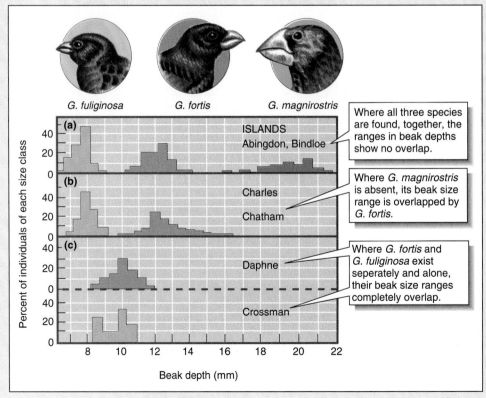

Essay 20.2 ● CONTINENTS ADRIFT

Have you ever noticed that the bulge of eastern South America would fit rather nicely against the western coast of Africa? So has almost every fourth grader who ever played with a globe. But after the initial observation no one usually pursues the matter further.

No one, that is, except people like geologist Alfred Wegener. Wegener was captivated by the shape of the continents surrounding the Atlantic Ocean, and he too noted the particularly good fit between South America and Africa. In a work first published in 1912, he was able to coordinate this jigsaw puzzle analysis with other geological and climatological data to propose the idea of continental drift. He proposed that about 230 million years ago, all the continents were joined together into one enormous land mass, which he called *Pangea*. In the ensuing millennia, according to Wegener's hypothesis, Pangea broke apart, and the fragments began to drift northward (by today's compass orientation) to their present locations.

Wegener was not treated well by his colleagues. His contemporaries attacked his naiveté as well as his supporting data, and the theory was pretty much discarded until about 1960. About that time, a new generation of geologists revived the idea and found new data to support it. The most useful data have been the determination of magnetism in ancient lava flows. When a lava flow cools, it permanently fossilizes a small sample of the earth's magnetic field, recording for future geologists both its north–south orientation and its latitude. Detailed maps of the positions of the continents through the ages can then be made. It now seems that Wegener's (and all those fourth-graders) insight was absolutely right. Not only did continental drift occur as Wegener hypothesized, but it continues to occur today.

Geologists have long maintained that the earth's surface is a restless crust, con-

stantly changing—sinking and rising through incredible unrelenting forces below. These constant changes are now known to involve large distinct segments of the crust known as *plates*. At the edges of these immense masses, new ridges are constantly built up and, in response, some edges are ground down. Where continents or pieces of continents are slammed together, mountain chains have formed. When ridges are built up in the ocean floor, the oceans expand. For example, astoundingly precise satellite studies reveal that the Atlantic Ocean is growing 5 cm wider each year. In addition to its fascinating geological implications, the theory of continental drift is vital to our understanding of the distribution of life on the planet today. It helps explain the presence of fossil tropical species on Antarctica, for example, and the uniqueness of animal life on the Australian continent and South America.

As the composite maps indicate, the disruption of Pangea began some 230 million years ago in the Paleozoic era. By the Mesozoic era, the Eurasian land mass, now named Laurasia, had moved away to form the northernmost continent. Gondwanaland, the mass that included India and the southern continents, had just begun to divide. Finally, during the late Mesozoic, South America and Africa completed their separation. For a time Australia and South America remained connected through Antarctica, which was more temperate then. During this time the marsupials spread over all three continents but were driven to near extinction by placental mammals in all but Australia. Both the North and South Atlantic Oceans would continue to widen considerably up into the Cenozoic, a trend that is continuing today. So we see that whereas the bumper sticker "Reunite Gondwanaland" has a trendy ring to it, it's an unlikely proposition.

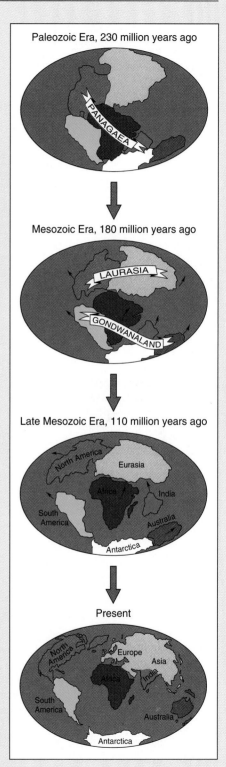

Paleozoic Era, 230 million years ago

PANAGAEA

Mesozoic Era, 180 million years ago

LAURASIA

GONDWANALAND

Late Mesozoic Era, 110 million years ago

North America   Eurasia   India   Africa   South America   Australia   Antarctica

Present

North America   Europe   Asia   India   Africa   South America   Australia   Antarctica

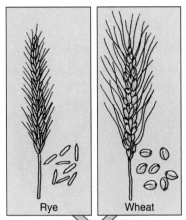

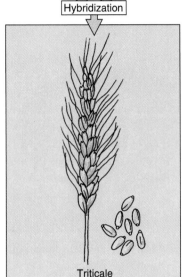

Rye

Wheat

Hybridization

Triticale

FIGURE 20.13
Miracle Grain

Triticale is a hybrid grain produced by crossing wheat (*Triticum*) and rye (*Secale*). Actually, the hybrid formed is sterile because of the usual chromosome incompatibility. But agricultural scientists solve the problem by treating the hybrid with *colchicine,* a mitosis inhibitor that causes a doubling of chromosomes—synthetic tetraploidy. Following this treatment, the treated plant will carry out a normal meiosis and produce viable gametes and viable F_2, F_3, and so on. *Triticale* combines the vigor of rye with the high grain yield of wheat. Furthermore, the genes of the rye add the amino acid lysine to the protein composition of the hybrid, making it more nutritious than wheat.

these grasshoppers are morphologically homogeneous—they all look alike—but chromosomal analysis revealed that the group is comprised of two different species that had come to occupy different parts of the range. Apparently, a random change had occurred in the parent population that enabled those grasshoppers bearing the change to adapt more precisely to a part of the parent population's former range. The change was great enough, however, to preclude further mixing of the genes from the old population and the new. Thus a new species apparently arose in the midst of an existing one.

A case of apparent sympatric speciation now in progress involves the maggot fly, a pest of the North American hawthorn tree. When apple trees were introduced to North America in the 19th century, the maggot fly began to infect them too. Now, it seems, the flies have become specialized, one line infecting hawthorn trees and the other preferring

apple trees. The speciation is not complete because the two lines can still interbreed, but they have come to differ in a number of ways. For example, not only do they show pronounced differences in fruit preference, but the apple flies mature in the laboratory in 40 days while the hawthorn flies mature in 54–61 days.

Reproductive Isolating Mechanisms

There are a number of means of enabling species to maintain their integrity and thereby continue as separate species without geographical isolation. These means are called **reproductive isolating mechanisms** and either prevent the formation of hybrid offspring or ensure that hybrids fail to be successful.

Prezygotic Barriers

Prezygotic means "before the zygote," and so **prezygotic barriers** are those mechanisms that ensure reproductive isolation by preventing the formation of a zygote. Here we will consider four such barriers.

1. **Ecological barriers**. When two similar species share the same territory, they may occupy different parts of it and therefore rarely come into contact for breeding, even though breeding might be successful. For example, when lions and tigers mingle in zoos, they do interbreed, producing fertile *tiglons* (when the tiger is the father) (Figure 20.14). Until the last century, the ranges of tigers and lions did overlap in India, yet no hybrids were discovered. One reason may have been that lions tended to live in open grassland (just as they do in Africa today), whereas tigers hunted in the deep shadows of the Asian forests.

2. **Behavioral barriers**. Animals that might otherwise attempt to interbreed may not be drawn to each other because of differences in their behavior. In some cases, those critical differences may not be related to reproduction. For example, when the lions and tigers of Asia overlapped, the species may have rarely encountered each other because not only were they attracted to different habitats but they had quite different social tendencies. The lions mingled easily in family groups known as prides, while tigers tend carefully to avoid others of their kind, except when breeding. In other cases, the behavioral barriers may be directly related to reproduction. Bird song is a reproductive advertising device that attracts potential mates. However, it does not attract those species not cued to respond to the message. Such reproductive advertising thus brings individuals of a species together without attracting other species. The ranges of the strikingly similar eastern and western meadowlark (Figure 20.15), for example, overlap, yet they are not attracted to each other, apparently in part because of quite distinctive songs. Behavioral barriers may also involve temporal (time) isolation. Brown and rainbow trout may live in the same waters in the Rockies, but they cannot normally interbreed because the browns mate in the spring and the rainbows in the fall.

3. **Mechanical barriers**. Some species cannot interbreed simply because their genitalia don't fit. Notably, insects have quite distinct reproductive parts. The male structure may be a tortuous device with hooks, barbs, and protuberances that would seem to discourage any but the most precise of fits. Interestingly, some flowers bar hybridization through marked specialization in floral parts; they are compatible only with specific pollinators that do not visit other species of plants (Figure 20.16).

4. **Gametic barriers**. Even if individuals of two species should mate successfully, fertilization might not take place because of incompatibility between eggs and sperm. In animals with internal fertilization, the reproductive tract of the female is often a hostile environment for the sperm of another species. Hybridization may also be blocked by the presence of species-specific receptors on the surfaces of gametes.

FIGURE 20.14
A Tiglon
Tigers and lions have been crossed successfully in captivity, although the offspring are probably sterile. One product, the tiglon, had a Siberian tiger father and lion mother. (The reciprocal would be a liger.) Differences in geography and other considerations preclude such hybridization from occurring in the wild.

If all these prezygotic barriers should be crossed, a hybrid zygote will form. Then an array of postzygotic barriers will come into play. It seems that the world does not welcome those of confused lineage.

Postzygotic Barriers

Postzygotic means "after the zygote," and so **postzygotic barriers** are those mechanisms that ensure reproductive isolation by acting after the zygote is formed. Here we will consider three such barriers.

1. **Hybrid inviability**. Even if the chromosomes of the two species pair properly after fertilization, genetic differences can lead to such severe problems for growth and development that, after a number of mitotic divisions, the zygote dies.

2. **Hybrid breakdown**. In hybrid breakdown, the F_1 hybrids are vigorous and fertile and reproduce easily. The problem comes with a weak and defective F_2 generation that cannot produce viable offspring.

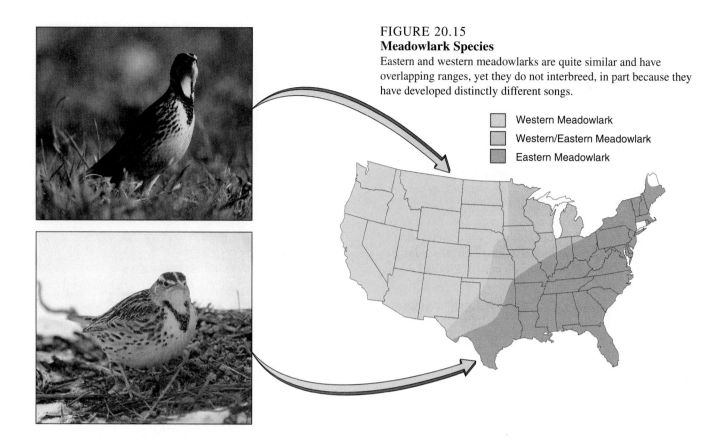

FIGURE 20.15
Meadowlark Species
Eastern and western meadowlarks are quite similar and have overlapping ranges, yet they do not interbreed, in part because they have developed distinctly different songs.

Western Meadowlark
Western/Eastern Meadowlark
Eastern Meadowlark

FIGURE 20.16
Specialized Pollination
Some flowers, such as this orchid with its specialized colors, forms, and odors, specifically attract only a single species of insect as pollinators.

3. **Hybrid sterility**. In some cases, a vigorous healthy hybrid organism will develop; however, it is unable to reproduce successfully because abnormal meiosis in the hybrid produces abnormal gametes. Mules are hybrids between horses and donkeys (Figure 20.17). They are indeed robust and powerful beasts but because of problems within their gametes, they usually cannot reproduce.

Patterns of Evolution

Now that we have seen how the far-flung species on the planet can appear and how they can be maintained as distinct entities, let's step back and take a broader look at patterns of change created by natural selection. We will see first how species come to be increasingly different from each other, then how they become more similar, and, finally, how they sometimes determine each other's course of evolution.

Divergent Evolution

Divergent evolution is the process whereby species tend to become increasingly different from each other over time. The idea of divergence has strong ecological overtones, since it often occurs when a species finds some new and different way

FIGURE 20.17
The Mule
Some modern-day philosopher described the mule (foreground) as a sad creature, one that has no past and no future—both genetically accurate observations. Although prized for their stamina and strength, mules are evolutionarily dead-ended because of sterility. (Compare the mule to the horse behind it.)

of using the environmental resources, thus establishing a new territory or, in Darwin's phrase, a new "place in the polity of nature." The opening of new habitats is greatly encouraged by natural selection, since it relieves (for a time) the unrelenting competition for resources.

Adaptive radiation is a form of divergent evolution in which a species spreads out into new territories and adapts through evolution so that a new species arises in the new territories. Such speciation appears to have occurred in the Galápagos finches first scientifically observed by Darwin (see Essay 20.1, page 390). The finches spread into new habitats on different islands, and, in the process of adapting to their new environments, they became more different from one another.

Convergent Evolution

While one may be struck with trends toward divergence in evolution, in some cases different species may grow more alike. The process, called **convergent evolution**, is due to different species undergoing similar adaptations to the same kind of environment. Such convergence occurs on all levels, from biochemical to morphological. Thus organisms may

grow more alike in enzyme systems or photosensitive pigments, as well as in coloration or body shape.

Darwin was struck by the evidence of convergent evolution as he traveled. As we saw earlier, he recorded in his journal that the South American mara, or Patagonian hare, *Dolichotis patagonum,* was quite similar in both appearance and behavior to the European rabbit (see Chapter 1). It turns out, however, that the mara is evolutionarily quite distinct from the rabbit.

Convergence is dramatically illustrated in comparisons between placental mammals and the distantly related, geographically isolated marsupial (pouched) mammals of Australia. Marsupials occupy habitats and have developed adaptations that are very similar to those of many placental mammals of other continents. Through long periods of isolation, selection, and adaptation, unrelated species—placental and marsupial—have often taken on a striking resemblance to one another. Thus in Australia we have the rabbit bandicoot, the marsupial mouse, the marsupial mole, the flying phalanger, the Tasmanian wolf, and the banded anteater (Figure 20.18). All have placental counterparts on other continents.

Coevolution

Quite often the direction of evolution in one species is strongly influenced by what is happening in the evolution of another species, particularly if the two species are somehow dependent on each other. The reciprocal influence of two species on each other's evolutionary direction is called **coevolution**. The most obvious examples of coevolution are to be found in predator–prey relationships. As natural selection improves the predator's skill, selection also favors greater evasiveness in the prey. As the prey develops better defenses, the predator must develop better means of detection or pursuit. The prey population responds once again, and the "evolutionary chase" goes on.

In some cases, coevolution has been responsible for great specificity in reproductive modes. For example, flowering plants and their insect pollinators may come to be intricately adapted to each other through coevolution. In fact, some flowering plant species are pollinated by only one insect species. The yucca is entirely dependent on the yucca moth for pollination, and the yucca moth, in turn, is entirely dependent on the yucca flower, where it lays its eggs and where its larvae grow on a diet of yucca seeds (Figure 20.19). The yucca moth must ensure pollination, since only properly pollinated flowers will produce seeds that the moth larvae need.

We see then that natural selection molds populations in a variety of ways: its influences can stem from a variety of sources, and it can proceed in a number of directions. One of the precepts of Darwinian evolution is that these changes occur gradually, that they are the result of a relentless accumulation of small changes. However, this idea has recently been challenged.

Placental hare

Placental wolverine

Placental flying squirrel

Marsupial rabbit bandicoot

Tasmanian devil

Marsupial sugar glider

FIGURE 20.18
Convergent Evolution
Convergent evolution is well supported by selected comparisons of certain North American placental mammals with marsupial counterparts in Australia. Each pair has made similar adaptations to the environment and shows striking similarities in particular body structures.

Small Steps or Great Leaps?

Charles Darwin's assumption that species change slowly through time because of the accumulation of small changes is now called **gradualism** (Figure 20.20a). Darwin was fully aware that there were certain problems with this concept, particularly as evidenced in the fossil record. He noticed that many species seemed to go unchanged for great periods of time while other forms seemed to change drastically over geologically brief periods. He was convinced that the problem lay in the "incompleteness" of the fossil record. After all, he reasoned, we are likely to find only pitifully small remnants of the vast array of life that once existed. Furthermore, the remnants themselves are often incomplete and so severely treated by time that it can be hard to judge what the living thing might have looked like or how rapidly its kind had changed.

As research continued into the 20th century, there was increasing evidence in the fossil record that species go unchanged, sometimes for millions of years, and then suddenly and dramatically undergo marked changes—so marked, in fact, that they produce new species. Niles Eldridge of the American Museum of Natural History and Stephen Jay Gould of Harvard University proposed that these abrupt and great changes in the fossil record were due not to its incompleteness, but because of the fact that evolution is not a slow, gradual process. They suggested, quite simply, that evolution generally proceeds in fits and starts, that species, once formed, go unchanged for great periods, and that when they do change, they are likely to change markedly. This concept is now called **punctuated equilibrium** (Figure 20.20b). The "suddenness," they stress, must be viewed in geological time, perhaps tens of thousands of years. This may seem like a long time, but keep in mind that species last, on the average, several million years on the planet, so even tens of thousands of years is but a small part of that time.

Extinction and Evolution

Just as speciation marks the birth of species, extinction heralds the death of species. Extinctions have always been an ongoing part of the evolving drama of life on the planet, from the first forms of life to some of the more publicized modern species that have dwindled into oblivion.

The history of life has also been marked by dramatic levels of extinction, those involving a great number and variety of species. Specifically, some scientists estimate that there have been five great extinctions since life appeared. The most sweeping extinction marked the demise of a great many species at the close of the Permian period (some 225 million years ago). It is estimated that more than 90% of marine life became extinct at that time. A major extinction occurred at the end of the Mesozoic era (some 65 million years ago) and marked the end of the dinosaurs.

However frequently and for whatever reasons the great extinctions occurred, they certainly took a great toll on existing life and opened new and unchallenged evolutionary directions for the survivors. Now, however, the question of extinction has taken on a new and more personal significance.

Extinction and Us

There is disconcerting evidence that we are apparently in the midst of another, truly devastating mass extinction, and this time there is little question that humans are the cause. Indeed the extinction rate is presently estimated by a number of experts to be between a thousand and several thousand species per year. If we continue on our present course of environmental destruction, by the year 2000 the extinction rate could be about 100 species per day. Within the next several decades, we could lose one-quarter to one-third of all species now alive. This rate of loss is unprecedented on the planet. Furthermore, the problem is qualitative as well as quantitative. The earlier mass extinctions involved only certain groups of species, such as the cycads and the dinosaurs. The other species were left more or less intact. The current extinction affects all major groups of species. Of particular importance is the fact that in this contemporary extinction the terrestrial plants are involved. In the past, such plants provided resources that the surviving animals could use to launch their comeback. With the plants also devastated, any comeback (marked by a period of rapid expansion and speciation) by animals will be greatly slowed.

The present great extinction will also prove to be devastating to the resurgence of species for another reason. This time we are killing the systems that are particularly rich in life, including tropical forests, coral reefs, saltwater marshes, river systems, and estuaries. In the past, these systems have provided genetic reservoirs from which new species could spring and replenish the diversity of life on the planet. In effect, we are drying up the wellspring of future speciation.

FIGURE 20.19
Coevolution
The yucca flower is pollinated by only one insect species, the yucca moth. In turn, the yucca flower is the only home for the yucca moth larvae. Therefore each is dependent on the other for survival, and any evolutionary change in one species must be matched by a compensating change in the other if the intricate interdependence is to continue successfully.

The loss of such richness and diversity could alter some of the prevailing major trends in evolution. One of these trends has been a net accumulation of species. As one species died out, it would leave an opening for one or more species to take its place. Now, however, we are losing so many species, so quickly, that they are not likely to be balanced by speciation.

In the past, extinctions have been due to two major processes—environmental forces (such as climatic change) and competition. For the first time, a single species (our own) has had the opportunity to cause mass extinctions at both levels. Much of the habitat destruction has been due to humans needing the land where other species lived, forcing them into extinction as they failed in their competition with us. And now we find ourselves in the remarkable position of being able to alter environmental forces. As the Amazon basin is destroyed, the trees are no longer available to cycle water back to the atmosphere, causing many experts to predict sweeping changes in the weather.

We are continuing to interact in new ways with the environment. As we continue to release chlorofluorocarbons into the atmosphere, the ozone holes grow larger and the delicate veil of life on earth becomes increasingly bathed in destructive radiation. And we are learning that the oceans have become perilous places for many forms of life because we continue to use the great waters as dumps for dangerous or unknown chemicals.

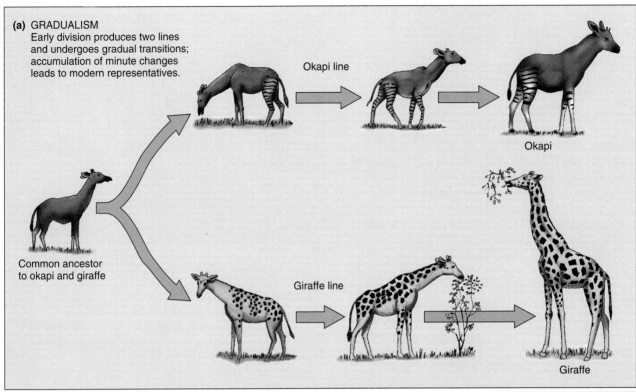

(a) GRADUALISM
Early division produces two lines
and undergoes gradual transitions;
accumulation of minute changes
leads to modern representatives.

Okapi line

Okapi

Common ancestor
to okapi and giraffe

Giraffe line

Giraffe

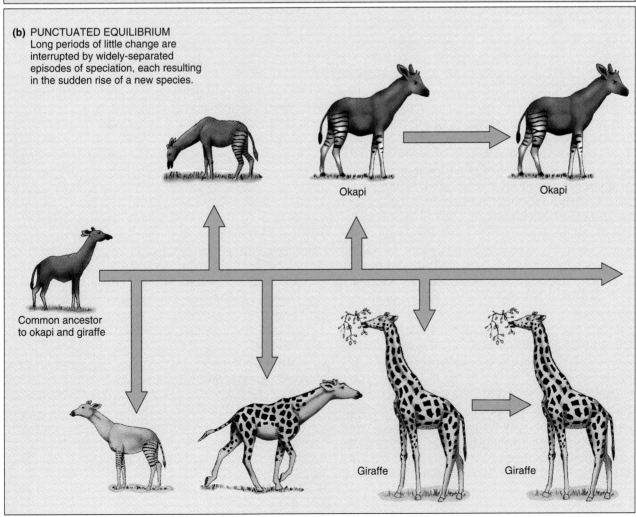

(b) PUNCTUATED EQUILIBRIUM
Long periods of little change are
interrupted by widely-separated
episodes of speciation, each resulting
in the sudden rise of a new species.

Okapi

Okapi

Common ancestor
to okapi and giraffe

Giraffe

Giraffe

FIGURE 20.20

FIGURE 20.20
Gradualism Versus Punctuated Equilibrium
The okapi and giraffe are believed to have descended from a common ancestor. **(a)** According to the traditional, gradualistic model, an early division produced two lines, each of which underwent gradual evolutionary transitions. The accumulation of countless minute changes leads to the modern representatives. **(b)** The punctuated model proposes that the giraffe line witnessed long periods with little change, interrupted by widely separated episodes of speciation, each resulting in the sudden rise of new species. The most recent episode of speciation produced the modern giraffe. The okapi line branched early from the ancestral line and, following its initial changes, remained relatively unchanged over time.

There are those who say that our course, by now, is irreversible, that the damage is done and now we must sit back and prepare to reap what we have sown. Others, though, argue that there is still time to alter our course, to salvage much of what is left and to protect it from further damage. The danger is in accepting the first alternative if the second is really the case.

The concept of evolution encompasses a range of complex and interactive ideas. Since its inception, it has also provided us with a host of questions, enduring puzzles that continue to resist our intellectual attacks. Someday we may know the answers to all our vexing questions about how and why life changes and how the various forms are related. Yet, perhaps not. It is possible that some of the events of our biological past will never be unraveled, and that these same questions will remain forever as unweathered monuments to human limitation. Whenever our species passes into extinction, we may carry with us many of the same questions that we ask today.

Key Ideas

WHAT IS A SPECIES?

1. A **species** is a group of actually or potentially interbreeding natural populations, each group reproductively isolated from the others.

2. The criterion of interbreeding in defining species is not always sufficient. Several species in the dog family successfully interbreed. Populations of a salamander species successfully interbreed all along their very long range, but when individuals occupying the two ends of the range are brought together, interbreeding fails.

TAXONOMY AND SYSTEMATICS

1. The **binomial system** is followed when assigning scientific names. Each species receives a **genus** and **species** designation (for example, *Homo sapiens*) with both terms written in Latin. The binomial system was developed by the Swedish naturalist Linnaeus (Carl von Linne).

2. **Taxonomy** involves identifying, naming, and classifying organisms; **systematics** involves establishing evolutionary relationships.

3. The hierarchy of taxa into which organisms are assigned is **species**, **genus**, **family**, **order**, **class**, **phylum** (**division** in plants), and **kingdom**.

4. The five kingdoms include **Monera** (prokaryotes, including Archaebacteria and Eubacteria), **Protista** (primarily single-cell and colonial eukaryotes, some multicellular), **Fungi** (mostly multicellular, parasitic, and scavenging eukaryotes), **Plantae** (multicellular, photosynthetic eukaryotes), and **Animalia** (multicellular, heterotrophic eukaryotes).

DETERMINING RELATIONSHIPS AMONG ORGANISMS

1. Systematics involves establishing relationships based on common ancestry. The terms "primitive" and "advanced" pertain to traits rather than to species. In establishing taxa, some systematists ("lumpers") tend to organize many organisms into large groups, while other taxonomists ("splitters") favor the establishment of many smaller groups.

2. In establishing evolutionary relationships through traditional methods, systematists concern themselves with homologies (structures of similar origin), paleontology (fossil record), and embryology (developmental patterns).

3. Pheneticists examine many characters in a group of organisms and use a numerical scoring system to draw comparisons. **Phenetics** usually stresses visible phenotype, which raises problems where convergent evolution is encountered.

4. Cladists organize their taxonomic trees, or **cladograms**, according to primitive and derived characters. All organisms in the cladogram exhibit shared primitive characters. (All chordates have a dorsal nerve cord and a notochord.) Branches are then developed by grouping organisms according to the shared derived characters.

THE MECHANISMS OF SPECIATION

1. The formation of new species from old is called **speciation**. It occurs in several ways.

2. **Allopatric speciation** is the rise of new species following geographical isolation and geological events such as erosion, earthquake, volcanism, and **continental drift**.

3. **Sympatric speciation** occurs within a continuous population.

4. In plants, sympatric speciation occurs through hybridization and **polyploidy**.

5. Whereas hybrids may be sterile because of chromosome incompatibility at meiosis, this can be overcome through polyploidy—a spontaneous chromosome increase by doubling, tripling, and so on of chromosomes. Whereas **autopolyploidy** (doubling of chromosomes in nonhybrids) produces abnormalities in meiosis, **allopolyploidy** in hybrids results in compatible meiotic chromosomes and fertile offspring. New species thus arise. *Triticale* is an allohexaploid. Polyploids seem to tolerate harsh climates well, and 85% of Greenland's flowering plants are polyploids.

6. Sympatric speciation in animals is rare, but examples include the Australian flightless grasshopper and the maggot fly. One line of maggot fly infects apple trees, while another infects hawthorn trees, and although they can still interbreed, several significant differences have come about.

REPRODUCTIVE ISOLATING MECHANISMS

1. Several barriers prevent diverging species from rejoining.

2. **Prezygotic barriers** (before the zygote) include **ecological** (different habitat preferences), **behavioral** (different social tendencies such as mating behaviors), **mechanical** (incompatible genitalia and pollination structures), and **gametic** (sperm barriers or incompatible pollination mechanisms).

3. **Postzygotic barriers** include **hybrid inviability** (problems with development), **hybrid sterility** (failure in meiosis and thus abnormal gametes), and **hybrid breakdown** (segregation in F_2 yields weakened or nonviable offspring).

PATTERNS OF EVOLUTION

1. **Divergent evolution** involves newly formed species becoming progressively different.

2. In **convergent evolution,** different species living in similar environments take on chemical and physical similarities. There are many examples between placental and marsupial mammals.

3. **Coevolution** involves reciprocal influences between two species. It is commonly seen in close relationships between species.

SMALL STEPS OR GREAT LEAPS?

1. The notion of evolution occurring through the gradual accumulation of small changes is called **gradualism**. An opposing view, called **punctuated equilibrium**, maintains that evolution occurs in intermittent spurts of activity following long periods of inactivity. Some gaps in the incomplete fossil record may be attributed to punctuated equilibrium.

EXTINCTION AND EVOLUTION

1. Evolution is marked by ongoing low-level extinctions and by dramatic mass extinctions. The fossil record reveals one that occurred 500 million years ago and another, 225 million years ago. The last great extinction occurred some 65 million years ago.

2. Evidence today suggests that we are in another period of mass extinction, this one related to environmental destruction by humans.

Application of Ideas

1. The fossil record for nearly all forms of multicellular life on earth begins rather abruptly shortly before the Cambrian period of the Paleozoic era. This raises numerous questions and, in the minds of some, adds fuel to the creationist's arguments. Offer two different hypothetical explanations for the seemingly abrupt appearance of such life.

2. Until recently, phylogenetic trees were hypotheses that could only be tested in terms of the characteristics contrived by the tree's originator. This is no longer true. Describe two important discoveries of this century that offer other ways of testing these hypotheses and explain how such testing is done.

Review Questions

1. Describe two situations where the interbreeding provision of the species definition fails. (pages 379–380)

2. Distinguish between the work of the taxonomist and systematist. (page 382)

3. Explain how the binomial system is used in naming species. Why is the Latin language used? (page 382)

4. Name the five kingdoms and list examples of organisms in each. (pages 382, 384)

5. Why is it erroneous to characterize a species as primitive or advanced? (pages 384–385)

6. List three main sources of information used by the traditional systematist when establishing taxonomic relationships. (pages 384–385)

7. Explain how pheneticists establish phylogenetic relationships. What problems does convergent evolution bring to them? (pages 385–386)

8. How does the cladist make use of shared primitive and shared derived characters in establishing phylogenetic trees? (pages 386–387)

9. What did the cladists find that was surprising about the evolutionary relationship between birds and crocodiles? (page 386)

10. Using the Galápagos finches as an example, explain how allopatric speciation occurs. (pages 386–387)

11. What is sympatric speciation? What organisms would one best look to for examples? (page 388)

12. Compare autopolyploidy and allopolyploidy in plants. (page 389)

13. In what kinds of environments do polyploids seem to survive best? (page 389)

14. List examples of sympatric speciation in animals. (pages 389, 392)

15. What are reproductive isolating mechanisms? How are they significant to speciation? (pages 392–393)

16. Briefly discuss three prezygotic barriers. (pages 392–393)

17. Briefly discuss three postzygotic barriers. (pages 393–394)

18. Using the Galápagos finches as an example, explain divergent evolution. (page 395)

19. List three kinds of relationships in which coevolution might occur. (page 395)

20. Characterize the way in which evolution occurs according to both gradualism and punctuated equilibrium. (page 396)

21. Summarize the history of mass extinctions. Which was the most recent? (pages 396–399)

CHAPTER 21

Origin of Life

CHAPTER OUTLINE

THE EARLY EARTH

OVERVIEW OF A HYPOTHESIS ON THE ORIGIN OF LIFE

THE HYPOTHESIS TODAY

The Monomers of Life

The Formation of Polymers

Self-replicating Systems

THE EARLIEST CELLS

RNA: Gene and Enzyme?

The Progenote, Ancestor of All Cells

The Early Metabolic Cell

The Early Autotrophic Cells

KEY IDEAS

APPLICATION OF IDEAS

REVIEW QUESTIONS

We have now had a look at some of the principles governing evolution, and we have explored its classical triad of forces: mutation, variation, and natural selection. There is indeed a substantial body of information on such processes, yet many of the most fundamental questions remain, including the ultimate question: How did life begin? The question is as old as humanity itself. The ancients throughout the world were absorbed by this mystery, and their conclusions have lingered in our consciousness, often forming the basic premises of many philosophies and religions.

Undoubtedly, many of the ancient cultures had great faith in their ideas. But faith in an idea, while it may serve as some sort of motivating force, is only a starting point. The faith must stand ready to be tested, and the burden of proof falls on the faithful.

But how can we investigate an improbable event that may have happened only once, several billion years ago? It is clear that we can never *prove* how life came to be on this planet. We can, however, examine any number of seemingly plausible notions of how life *might* have arisen. All speculations about the origin of life must involve assumptions. These assumptions can lead to predictions, and it is the predictions that can be tested. Many such speculations have been trotted out, and many of their predictions have been tested. Most of the proposed schemes of how life might have originated couldn't possibly have worked; they have depended on assumptions about nature that have proved to be untrue. But through the process of elimination, we have narrowed the range of possibilities and have highlighted the most promising avenues for further examination. It is likely that *life successfully arose only once* in the earth's history and that the conditions at that time were unique, probably never to occur again.

The Early Earth

The best estimates suggest that the earth took form about 4.6 billion years ago, along with the sun and the other planets of our solar system. The precursor of the solar system was a vast, flattened cloud of gases, dust, and other debris (Figure 21.1). Recent theories maintain that the cloud was cold, but that as the sun and planets coalesced, a great deal of heat was generated from the press of gravitational forces, supplemented with heat from radioactivity. When it took form, the earth's crust was a molten semiliquid. Some 600–800 million years were to pass before it solidified. (Figure 21.2 is a timetable of earth history.)

When the crust finally cooled to below the boiling point of water, torrential rains began to fall, initiating the formation of the oceans. Based in part on data collected from space probes of the solar system, the constituents of the early earth atmosphere are thought to have been water vapor (H_2O), carbon dioxide (CO_2), carbon monoxide (CO), methane (CH_4), ammonia (NH_4), molecular nitrogen (N_2), hydrogen sulfide (H_2S), and possibly some hydrogen (H_2).

All the elements necessary to create life are contained within these simple molecular constituents of the atmosphere. In fact, the only difference between these simple molecules and life as we know it is in the organization of these elements into the macromolecules that serve as the structures of living cells, and a continuous metabolism that uses energy to maintain that organization.

As we know from earlier chapters, today essentially all the energy to maintain life comes from the sun, and visible light is used directly in photosynthesis. On the primitive earth, other important energy sources included lightning, heat, and what geologists call shock energy (from earthquakes and tremors). So all the components for life existed on the primitive earth. The simple molecules were there, and there was plenty of energy. But how was that energy converted into the molecular organization that can be thought of as life?

FIGURE 21.1
The Young Solar System
The hypothetical events in the formation of the solar system from a flattened cloud of cold gases: (*top*) the flattened dust cloud, (*center*) sun and planets forming, and (*bottom*) the completed solar system.

Overview of a Hypothesis on the Origin of Life

Although Darwin speculated that life might have arisen in a warm, phosphate-rich pond, the first serious proposals concerning the origin of life began to appear some 60 years ago. Such schemes were presented by a Scottish scientist, J. B. S. Haldane, and by A. P. Oparin, his Russian colleague. They proposed that soon after the earth's formation, a period of spontaneous chemical synthesis began in the warm ancient seas. During this era, amino acids, sugars, and nucleotide bases—the structural subunits of some of life's macromolecules—formed spontaneously from the hydrogen-rich molecules of ammonia, methane, and water. Such spontaneous synthesis, explained Haldane, was only possible because there was little oxygen in the atmosphere, an important condition since molecular oxy-

gen will react with and destroy many organic molecules. The energy for this spontaneous synthesis was, at this time, readily available in the form of lightning, ultraviolet light, heat, and higher energy radiations. Since there were no organisms and no molecular oxygen to degrade the spontaneously formed organic molecules, they accumulated until the sea became a "hot, thin soup" (of, perhaps, about the consistency and nutritive quality of chicken bouillon) (Figure 21.3).

Both Haldane and Oparin suggested that polymerization—the joining of the structural subunits into macromolecules—was speeded by the ever-increasing concentrations of precursor molecules. Amino acids joined to form the first polypeptides and eventually the first complete proteins. Certain versions of these proteins became the earliest enzymes. The rate of chemical interaction must have been very slight at

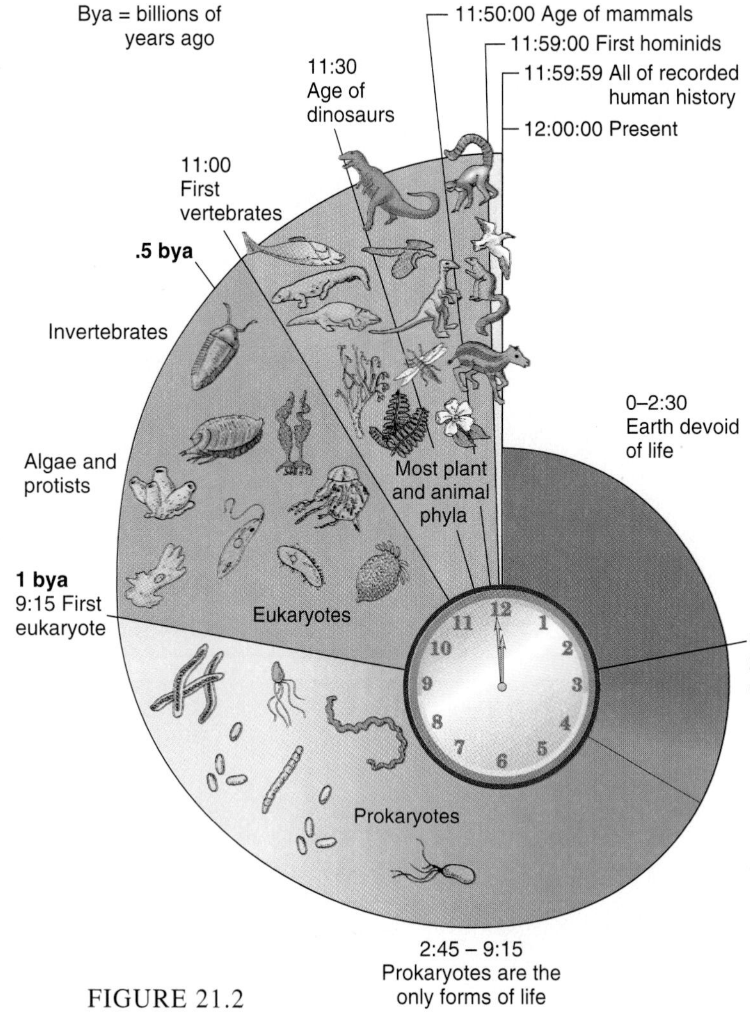

FIGURE 21.2
An Earth Calendar
Here the earth's history is shrunk into a period of 12 hours—from midnight to noon. The events are in chronological order, and the spacing indicates elapsed time.

Bya = billions of years ago

11:50:00 Age of mammals
11:59:00 First hominids
11:59:59 All of recorded human history
12:00:00 Present

11:30 Age of dinosaurs

11:00 First vertebrates

.5 bya

Invertebrates

Algae and protists

1 bya
9:15 First eukaryote

Eukaryotes

Most plant and animal phyla

0–2:30 Earth devoid of life

3.5 bya
2:45 First cellular life

Prokaryotes

2:45 – 9:15
Prokaryotes are the only forms of life

first, but there was plenty of time. The early enzymes may not have been particularly effective, but in time they were able to catalyze a number of kinds of activity, including that which would have produced more proteins like themselves.

A major advance took place when collections of these new catalysts then became enclosed by simple membrane-like, water-resistant protein or lipid shells. Oparin referred to these early cell-like, membrane-bound containers as coacervates (Figure 21.4). **Coacervates** are tiny spheroid bodies that form when certain macromolecules are introduced to water under very specific ionic and pH conditions. Each droplet includes a cluster of the large molecules and a surrounding membrane-like shell of water molecules. An interesting property of coacervates is that they "grow." A coacervate continually takes in selected substances from its surroundings, thus increasing in size. Then, at some critical mass, the coacervate divides (shears in two). Then its growth resumes, only to divide again. In this way, its numbers increase. In some instances, the droplets become even more selective through the formation of a more complex membrane beneath the watery shell. So we see that coacervates have some of the properties of life discussed in Chapter 1—the selective intake of materials, growth, division, and increased numbers.

The reason that this is such a major step is that, for the first time, there would be an *inside* and an *outside*. This created an opportunity to concentrate molecules within the confined and protected environment of the coacervates and increase the rate and number of chemical reactions necessary to continue the formation of life.

Eventually the coacervates developed series of chemical reactions that became metabolism, and increasingly complex mechanisms to regulate their growth and division. Winnowed by the forces of natural selection, the tiny, membranous droplets eventually produced the first simple cellular life.

The Haldane–Oparin hypothetical scheme remained a neglected intellectual curiosity for more than a quarter of a century. But in 1952, Nobel laureate Harold Urey (the discoverer of deuterium, a heavy isotope of hydrogen) and Stanley Miller, a graduate student, began testing some of the assumptions that form the basis of the hypothesis.

The crucial proposition of the Haldane–Oparin hypothesis was that the precursors of the molecules of life would form spontaneously on the primitive earth. Miller and Urey created a laboratory apparatus at the University of Chicago that attempted to simulate the conditions of the primitive earth (Figure 21.5). They introduced a small amount of water and a mixture of gases including methane, ammonia, water vapor, and hydrogen—but no free oxygen—into the apparatus in concentrations thought to be present in the primitive atmosphere. As an energy source, they produced repeated electrical discharges to simulate lightning through the atmosphere of the upper flask as a source of energy. After a week, they analyzed the sediments that collected in the lower flask. Among the various molecules they found aldehydes, carboxylic acids, and, most interestingly, amino acids, all of which are commonly found in living cells.

Although these small molecules were a far cry from living things, the fact that they were produced at all provoked a lively revival of interest in the Haldane–Oparin hypothesis.

The Hypothesis Today

Scientists continue to test hypotheses regarding the origins of life. Efforts to understand the conditions of the primitive earth continue, and thoughts focus on three key aspects of the hypothesis that the molecular monomers of life could arise from the constituents of the primitive earth, that the monomers could be joined together to form the macromolecular polymers that characterize living things, and that the ability to direct replication could be developed.

FIGURE 21.3
Scenario of Earliest Earth Conditions
Torrential rains formed the young seas, while volcanoes released gases into the developing atmosphere. Heat and electrical discharge provided energy for chemical synthesis in which many precursors to life's molecules arose (inset).

The Monomers of Life

The early work of Miller and Urey gave rise to a host of similar experiments. The list of monomers synthesized in the laboratory under primitive earth conditions now includes all the nucleotide bases of DNA and RNA, along with the essential sugars, all the amino acids, and most essential vitamins. Adenine, the most widely occurring nucleotide base in today's life forms, is a common and abundant product in these experiments. Thus research supports the idea that many of the monomers associated with life were produced through spontaneous generation on the early earth.

The Formation of Polymers

It is one thing to produce such monomers and quite another to induce them to form polymers, that is, without the assistance of preexisting enzymes. All biological polymerizations involve dehydration linkages between the monomers—that is, removal of water to produce the linkage. Researchers agree that such reactions would *not* have been energetically favorable in the primitive sea. In water, biological polymers slowly dissociate back into monomers, and heat just accelerates the process. Without the catalyzing effects of enzymes, spontaneous polymerization is possible only when the concentration of monomers in water is high.

With such problems then, how could polymerization have been achieved? In one scenario, the inside of coacervates would allow the accumulation of high concentrations of monomers. Some researchers have turned to clay particles as an answer to some of the problems of polymerization. Why clay? Under experimental conditions, organic monomers adhere to clay particles, forming high concentrations on the particle surfaces. Because clay particles carry electrical charges, they interact with monomers and in some instances

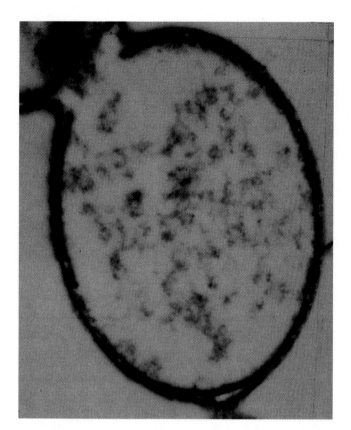

FIGURE 21.4
Coacervates, the Earliest Cellular Organization?
When proteins or other polymers are introduced into water, they tend to cluster together into distinct droplets called coacervates. The coacervate surrounds itself with a boundary layer that is selective in admitting certain kinds of molecules. When coacervates reach a critical size and mass, they divide spontaneously, a process characteristic of cells. As one theory has it, when coacervate droplets formed, the world made its first step toward biological organization. There was an inside and an outside to the soup of the primordial sea, and on the inside there could be concentration and coordination of the chemical reactions that characterize life.

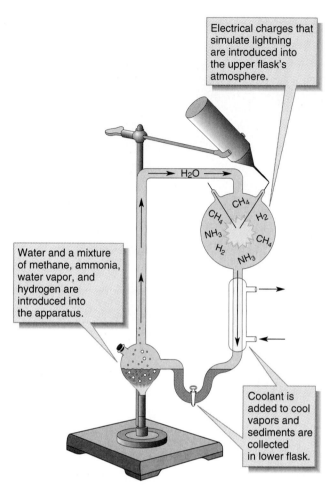

FIGURE 21.5
The Classic Miller–Urey Experiment
Heated gases of the theoretical primitive atmosphere were subjected to electrical discharges in a sealed, sterile environment. Residues were periodically collected in the lower chamber and analyzed. Results indicated that some of the simple monomers of life could be produced spontaneously under the test conditions.

catalyze their polymerization. Amino acids have polymerized to form polymers on the surface of clay particles. Interestingly, the amino acids favored by clay particles (and, as a matter of fact, the sugars that also bind to clay) are of the same chemical structure as those found in life.

A different view of the formation of proteins is proposed by Sidney Fox, who has demonstrated that polymerization of amino acids occurs readily under hot, drying conditions such as might be found along the edges of volcanoes or even on the hot beaches of ancient seas. Pools of organic precursors, rich in amino acids, could have been concentrated by evaporation and heated to allow the formation of polypeptides. Fox has succeeded in producing polymers of 200 or more amino acids, he called **thermal proteinoids**, under hot, drying conditions such as might have been present on the primitive earth.

Sidney Fox's thermal proteinoids have also given us alternative indications of how cells may have arisen. When placed in water, the proteinoids cluster into bodies similar to coacervates. Fox calls these bodies **proteinoid microspheres** (Figure 21.6). Such spheres automatically form two-layer membranes, isolating themselves from their water surroundings, again in the manner of coacervates. Furthermore, the spheres take up molecules from the surrounding environment, behaving as though their surrounding membrane were selective and also demonstrating some of the properties of photosynthesis. They grow, fuse with other microspheres, form tiny buds on their surface, and under certain conditions produce energy and divide their mass.

One of the greatest problems in reconstructing the development of life lies in explaining how the nucleic acids, DNA

FIGURE 21.6
Proteinoid Organization
Proteinoids are polypeptides that polymerize spontaneously from evaporating concentrations of amino acids. Proteinoid microspheres could have formed in regions of intense volcanic activity.

and RNA, might have formed. While polymers of amino acids can form, no such polymerization has been observed for nucleic acids under the primitive earth conditions. Investigators have boiled and dried concentrated energy-primed nucleotide triphosphates in the presence of single-stranded templates of DNA, producing an energetically favorable direction of reaction. However, without the appropriate enzymes, no second strand of DNA is produced. The units do not link to form polymers. Linkages can be forced, but they occur in the wrong places. To date, there is no evidence that DNA polymers can be produced under primitive earth conditions. Hence some scientists are now theorizing that the first synthesis of DNA might have come long after the actual origin of life.

Self-replicating Systems

Oparin's work and Fox's work suggest how a wide variety of organized droplets might have arisen, perhaps differing in content and chemical activity. Extending such observations, we can surmise that, in the primitive earth, there may have been competition among the droplets for the energy-rich monomers in the thin, hot soup. As less efficient droplets failed, their constituents would be used by their more stable competitors. Of course, stability through generations of droplets could only be achieved by the droplets having some means of faithfully duplicating themselves.

Imagine a system in which the molecules inside a droplet might act as a template (mold) that could organize smaller molecules to form a replica of the template molecule prior to division (Figure 21.7). In this way the droplet could remain functionally stable through generations of its kind (roughly the way genetic mechanisms work today). Of course, this is a

giant step in our scenario of how cells came to be. There is little indication that simple molecules can behave in such a way.

Of course, not everyone is convinced that such active protein droplets, or protocells as they became known, developed first. Some biologists are convinced that self-perpetuating genetic systems arose first, probably in the form of DNA or RNA. Such nucleic acids have the self-duplicating or "replicating" quality needed. At some time, such molecules would have then become surrounded by an active droplet and membrane. The intriguing thing about this notion is that there are such models around today in the form of viruses. The simplest of these consist only of a nucleic acid core (the genes) and a protective protein coat. But there are also difficulties with this hypothesis: every virus known today is parasitic, unable to carry on vital metabolic and reproductive activity outside a living host.

So one of the questions that bothers some investigators is which macromolecules formed first: nucleic acids or proteins? The problem is that contemporary organisms use nucleic acids—DNA, mRNA, tRNA, and rRNA—to synthesize the polypeptides incorporated into enzymes; yet at the same time, enzymes appear to be necessary to synthesize nucleic acids. It's the ultimate chicken-and-egg problem. Those who favor proteins as primitive macromolecules point out the relative ease with which peptide bonds are produced and the great generality of protein's enzymatic activity. Those who favor nucleic acids dwell on the information-carrying feature of the genetic material and the fact that some enzyme-like characteristics are associated with certain RNA molecules. They suggest that rRNA and tRNA are the remnants of a once-large class of nucleic acid "enzymes" or enzyme-like molecules. It is possible that the two systems originated together.

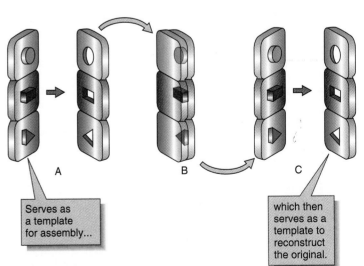

Serves as a template for assembly...

which then serves as a template to reconstruct the original.

FIGURE 21.7
Self-Replicating Mechanisms
The figure at the left serves as a template from which the figure at the right can be formed. Such self-perpetuating mechanisms were a necessary step in the evolution of early life.

The Earliest Cells

By taking the giant step from metabolically active coacervates and proteinoids to self-reproducing **protocells**, we can apply some informed speculation to many questions about early cellular life. What were the earliest cells like? What were their energy sources? How did they replicate? How do we get from the earliest protocells to cyanobacteria, complex photosynthetic prokaryotes that may have been around in various forms for 3.5 billion years?

RNA: Gene and Enzyme?

A growing number of researchers are convinced that the first genes were encoded in RNA rather than DNA. RNA, like DNA, has information-storing capabilities, yet it is chemically simpler and exists as a single strand. Whereas DNA nucleotides do not polymerize under simulated primitive conditions (Figure 21.8a), researcher Leslie E. Orgel has succeeded in producing RNA polymers by concentrating RNA nucleotides in a saline environment. Such polymers can also replicate, forming copies of themselves through RNA base pairing. The process, admittedly slow and inefficient, nonetheless happens without the intervention of enzymes. In other words, we now know that RNA can make RNA (Figure 21.8b). So a major obstacle in the understanding of molecular evolution may have been overcome. In addition, this replication process is much simpler than the complex, multienzyme process of DNA replication, which probably arose much later in the history of life.

Researchers have also noted examples of catalytic properties in RNA. In the processing of transfer RNA by *Escherichia coli*, an enzyme containing a small molecule of RNA is used. When the RNA is removed and used alone, it succeeds in cutting and splicing the raw tRNA transcripts. In other examples, RNA molecules "enzymatically" cleave transcripts by forming hydrogen-bonded secondary structures that break the backbone of the transcript and release the catalytic RNA. Furthermore, recent studies of intron processing reveal that the same transcripts are capable of removing their own introns, also without the aid of an enzyme. Speculating on the significance of this, James E. Darnell, Jr. suggests that the early RNA genes, more or less randomly organized at first, must have contained various introns (intervening sequences) that had to be clipped out and exons (expressed sequences) that had to be spliced together (Figure 21.8c). Darnell is thus suggesting that introns arose very early in the history of life.

The first DNA, suggest the researchers, may well have been produced through reverse transcription (Figure 21.8d), a biochemical trick well known in single-stranded RNA viruses. For a time, biologists thought that reverse transcriptase, the enzyme responsible for making DNA from an RNA template, was unique to certain viruses. But now similar enzymes are known to be present in all eukaryotic cells, a finding that suggests that the enzyme is indeed ancient. But why did DNA arise at all? DNA, it turns out, offers advantages over RNA. Included are greater stability and certainly a built-in system of redundancy (its double strand) that aids in the correction of replication errors and in the correction of mutations.

The conversion of RNA to DNA early in the history of life potentially explains even the presence of introns in eukaryotic cells today. Scientists suggest that introns have been present since the first genes formed and that prokaryotes,

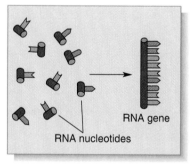

(a) Polymerization of RNA nucleotides

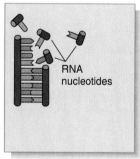

(b) Replication of RNA gene

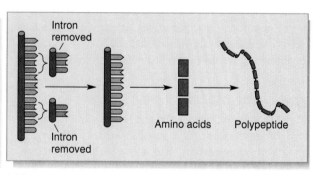

(c) Introns removed

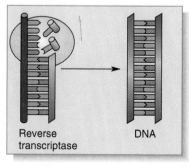

(d) Reverse transcription

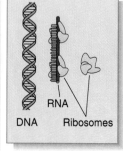

(e) Modern DNA

FIGURE 21.8
The First Genetic System?
(a) The first genes may have formed from the spontaneous polymerization of RNA nucleotides. **(b)** An early capability of the crude RNA gene would have been replication, which can occur in RNA without the presence of protein enzymes. **(c)** Such RNA would most likely have contained introns, which were clipped out before translation of the coding into a polypeptide. **(d)** The presence of reverse transcriptase, which would have been one of the first enzymes in this system, made the formation of DNA possible. **(e)** The earliest cell may well have contained DNA, still containing the introns collected randomly in earlier events.

which are intron-free, have since streamlined their protein-synthesizing machinery by eliminating the introns and thus becoming far more efficient protein synthesizers. This viewpoint implies that present day bacteria are not as unchanged and primitive as their simpler structure might initially suggest.

The Progenote, Ancestor of All Cells

Along this line of informed speculation, researchers are suggesting that the very first cell, which biologist Carl Woese calls the **progenote**, had undergone the change to using DNA, complete with its introns. It also had ribosomes for translating the genetic message into polypeptides (Figure 21.8e). The progenote may have given rise to three separate lines of life, two of which produced the prokaryotes—the archaebacteria and eubacteria—and a third that gave rise to the eukaryotes (Figure 21.9). This new thinking is a clear departure from early concepts wherein eukaryotes are thought to have arisen from prokaryotes.

The Early Metabolic Cell

There is considerable disagreement as to the metabolic characteristics of the earliest cells. Some hold that the first life forms were **autotrophic**, deriving their energy from light (phototrophic) or inorganic chemicals (chemotrophic) and living independently of other organisms. Others propose that the earliest cells were **heterotrophic**, deriving their energy by consuming other living material. These life forms probably relied heavily on the comparatively simple process of anaerobic respiration. (Remember that there was no available oxygen for oxidative respiration.)

The original energy supply for the early heterotrophic cells may well have come from the organic molecules that were still available in the ocean. But we can surmise that expanding populations of the new, living cells soon began to use up the available resources, and the resulting competition led natural selection to favor those cells able to exploit new energy sources or to exploit old ones more efficiently. Cells could prey on one another, but this simply redistributed the limited and dwindling supply of organic nutrients. The next major advance was to break out of this circle altogether. The cells that could begin to do this, even somewhat inefficiently, would begin a line that would replace others. Thus we have the rise of autotrophic cells.

The Early Autotrophic Cells

Some primitive autotrophic cells today are photosynthetic: they obtain the energy to make macromolecules directly from sunlight. It's a good bet that the earliest successful autotrophic cells also used light energy. But it is unlikely that early photosynthesis involved the splitting of water to supply the hydrogen atoms that power the synthetic reactions. Instead, like a few present-day forms of bacteria, early autotrophic cells probably used hydrogen sulfide as a source of hydrogen.

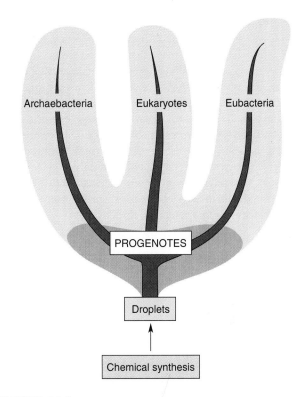

FIGURE 21.9
Progenote: Ancestor to All Life
The hypothetical progenote, the first organism with an organized DNA-based genome, is presumed to be the ancestor to all other life.

But the number of places in the world that provide both hydrogen sulfide and abundant sunlight is severely limited. At some point, some ancestral cyanobacterium began to obtain its photosynthetic hydrogen supply from an energetically less favorable but far more abundant source: water. This step required a complex photosystem, as we discussed in detail in Chapter 7. But the accomplishment was a success, and it was to change the earth forever.

Where water is the source of hydrogen in photosynthesis, the waste product is molecular oxygen. Although many oxygen-sensitive organisms undoubtedly became extinct, being literally driven into the mud, new forms less sensitive to oxygen were to emerge through mutation and natural selection. Eventually, the power of oxygen was actually utilized, put to work in extracting the energy from organic foodstuffs. Thus oxidative aerobic respiration came into being.

It took about 2 billion years, but eventually photosynthesis changed the strongly reducing atmosphere of the early earth to the oxygen-rich, strongly oxidizing atmosphere of today. The early earth contained many chemical forms that absorbed oxygen as fast as it could be produced. For example, the elemental iron, elemental sulfur, and abundant iron sulfide of the early earth's crust took in enormous amounts of oxygen as they were transformed into iron oxide and various sulfates. Only when these reactive materials were finally saturated could free oxygen exist in the air. By the time oxygen became a significant atmospheric gas, organisms of a new kind—the eukaryotes—had already made their presence felt.

New modes of nutrition arose too. The burgeoning cyanobacteria themselves represented an abundant new source of food for any heterotroph able to engulf and assimilate their stored energy. Thus the world saw the emergence of the first herbivores. Those anaerobes unable to cope with oxygen and eat other organisms were soon relegated to the backwaters of life's great progression. They could exist only if hidden away from poisonous oxygen in pockets of the earth's crust, in nutrient-rich muds and in deep recesses of stagnant waters. And it is in such places that they remain to this day.

In our imaginative scenario, we have seen two new metabolic forms of life emerging: the photosynthetic, oxygen-producing phototroph and the aerobically respiring, oxygen-utilizing heterotroph (Figure 21.10). Now that we have arrived at the time of the first fossil evidence of life on earth, we can leave this hypothetical world and begin to consider the known world.

We will move on now to learn more about the prokaryotes, the organisms believed to be the most direct descendants of the earliest forms of life on earth. We have visited this remarkable group before, specifically, in our comparisons of cells and our discussion of prokaryotes as experimental subjects. In the next chapter, we will examine the prokaryotes as our first subjects in our consideration of the diversity of life.

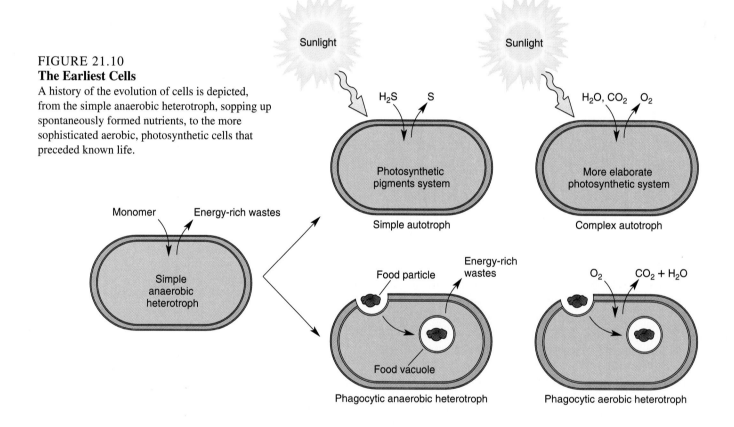

FIGURE 21.10
The Earliest Cells
A history of the evolution of cells is depicted, from the simple anaerobic heterotroph, sopping up spontaneously formed nutrients, to the more sophisticated aerobic, photosynthetic cells that preceded known life.

Key Ideas

1. Compelling evidence suggests that life arose spontaneously on the earth, at a time when unique conditions were present.

THE EARLY EARTH

1. The atmosphere in which life arose was likely to have consisted of water, carbon dioxide, carbon monoxide, molecular nitrogen, and some free hydrogen.

2. Energy sources in the primitive atmosphere may have included ultraviolet light, lightning, heat, and geological shock energy.

3. Estimates place the origin of life some 3.8 billion years ago; fossils of cyanobacteria are believed to be 3.5 billion years old.

OVERVIEW OF A HYPOTHESIS ON THE ORIGIN OF LIFE

1. Haldane and Oparin were the first to speculate on the spontaneous origin of life. The **Haldane–Oparin hypothesis** proposed that the precursors of life's molecules formed from inorganic sources, and these underwent polymerization to form macromolecules. Such macromolecules included primitive versions of

enzymes. Droplets, or **coacervates**, formed and were enclosed by protein or lipid shells.

2. Miller and Urey tested the Haldane–Oparin hypothesis with a device that simulated the primitive environment. Using a mix of gases and applying an electrical discharge, they succeeded in synthesizing amino acids, aldehydes, and carboxylic acids, all known to be monomers of cells.

THE HYPOTHESIS TODAY

1. Additional experiments have shown that many of the monomers of life are able to be produced under simulated primitive earth conditions, and that under specialized conditions many monomers will self-polymerize to form macromolecules similar to those that exist today.

2. Sidney Fox demonstrated that simple polymers could form **proteinoids**, that could further form **proteinoid microspheres**, which had some of the characteristics of coacervates.

3. One of the major problems to be overcome in our understanding of early life is the question of whether genes or enzymes came first, and how the first self-replicating systems worked.

THE EARLIEST CELLS

1. Because RNA has both information-storing and catalytic capabilities and polymerizes and replicates spontaneously under the early earth conditions, it is a good candidate for the first genetic system.

2. The earliest life form, called the **progenote**, is believed to have had a DNA genome.

3. The progenote gave rise to Archaebacteria, Eubacteria, and the eukaryote kingdoms.

4. There is disagreement on the metabolic characteristics of the early *protocells*. Some suggest they were chemotrophic or phototropic, while others claim they were simple anaerobic heterotrophs. The continued autocatalytic synthesis of monomers may have provided carbon compounds for early heterotrophs.

5. Keen competition for energy-rich molecules may have encouraged variants that could extract hydrogen from inorganic sources, using light energy. The utilization of water as a hydrogen source probably came later, since it requires complex photosystems.

6. When the use of water in photosynthesis occurred, oxygen joined the gases of the atmosphere for the first time.

7. The buildup of significant amounts of oxygen in the atmosphere required an enormous time period because of the presence of elemental substances such as sulfur and iron that readily combined with the gas.

8. As cell populations grew, new modes of heterotrophic nutrition arose.

Application of Ideas

1. Considering the fact that clearly nonliving entities can have traits associated with life, such as growth, reproduction, oxidation, and responsiveness, at what point do you think it is justifiable to say that life appeared on the planet?

2. It can be argued that the formation of cell-like structures, such as coacervate droplets, were critical to the development of life. Can you justify the position, considering the concentration of cellular materials today and the nature of a semipermeable membrane?

Review Questions

1. Briefly describe the Haldane–Oparin hypothesis of the origin of life. (pages 403–404)

2. According to most biologists, the spontaneous origin of life occurred only once in the earth's history. What is the basis for this restricted statement? Why not several times? (pages 402–403)

3. What steps are involved in going from inorganic matter to the coacervate level? What, according to Haldane and Oparin, were the energy sources? (page 404)

4. Describe the experimental apparatus and materials used by Miller. What did the experiments actually establish? (page 404)

5. What problems, if any, do researchers find with attempts at spontaneously synthesizing nucleic acids? Did nucleic acids necessarily precede enzymes? Explain. (pages 406–407)

6. Briefly describe the polymerization experiments of Sidney Fox. Why didn't he use a water solution such as might have been found in the early oceans? (page 406)

7. Describe Oparin's experimental work with coacervates. (page 404)

8. List several reasons why RNA is a logical choice as the first genome. (pages 407–408)

9. What is a progenote? To what cell lines did it give rise? (page 409)

10. Why might the earliest phototrophs have used hydrogen sulfide as a hydrogen source rather than water? In what way did the later shift to water "change the earth forever"? (page 409)

11. According to the latest thinking, it took 2 billion years for significant oxygen to accumulate in the atmosphere. Why was such a long period of time required? (pages 409–410)

CHAPTER 22
Viruses and Bacteria

CHAPTER 23
The Protists

CHAPTER 24
The Fungi

CHAPTER 25
Plant Evolution and Diversity

CHAPTER 26
Invertebrates

CHAPTER 25
Chordates

PART FOUR

Diversity

CHAPTER 22

Viruses and Bacteria

CHAPTER OUTLINE

VIRUSES

Introducing Viruses

Origin of Viruses

The Discovery of Viruses

Characteristics of Viruses

Bacterial Viruses: The Bacteriophages

Animal Viruses

Plant Viruses

HIV: Study of a Retroviral Killer

Viroids and Prions

KINGDOM MONERA: THE PROKARYOTES

Evolutionary Roots of Monerans

Eubacteria, Archaebacteria, and a New Phylogeny

Archaebacteria: A Brief Look

Eubacteria

Bacterial Reproduction

Metabolic Diversity in Bacteria

Essay 22.1 The Nitrogen Fixation Process

KEY IDEAS

APPLICATION OF IDEAS

REVIEW QUESTIONS

*I*n our survey of the kingdoms of life we will start small—with the *microorganisms*. The term is general, referring to viruses and small, simple organisms such as bacteria and protists. The bacteria represent the simplest forms of cellular life and, as we will see, the most ancient. Viruses are included for convenience and because they are incredibly small and simple, not because of any evolutionary relatedness. We'll explain their peculiar status in the world of life by asking the obvious question. What is a virus?

Viruses

Introducing Viruses

Viruses are minute, biologically active particles, made up of a nucleic acid core, a protein covering, and often an enzyme or two. Some also have a surrounding membrane, or **envelope** as it is known. But as it turns out the envelope doesn't really belong to them.

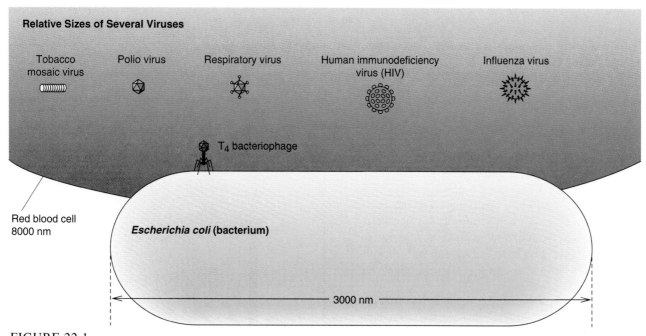

Relative Sizes of Several Viruses

Tobacco mosaic virus

Polio virus

Respiratory virus

Human immunodeficiency virus (HIV)

Influenza virus

T$_4$ bacteriophage

Red blood cell 8000 nm

Escherichia coli (bacterium)

3000 nm

FIGURE 22.1
Viral Dimensions
Comparing several viruses with the colon bacterium *Escherichia coli* and a human red blood cell gives us some idea of size ranges in the viral realm.

The use of the term "active particle" is deliberate, since viruses cannot be characterized as cells. Other than their nucleic acid core, the viral equivalent of a nucleus, they lack almost everything cells have. Because of this, viruses cannot carry out metabolic activity or reproduce outside a living host. By definition, viruses are completely parasitic.

The fact is that biologists aren't sure viruses are even living. By the criteria we have been using (see Chapter 1), they are not. Nevertheless, they can indeed be active, as anyone with an annoying head cold can tell you. Yet some can be crystallized like common table salt and stored in vials for prolonged periods. Then, when introduced into living cells, viruses become active again.

So how should we classify viruses? They are definitely not members of the five recognized kingdoms of life. Some biologists suggest organizing another kingdom, but a survey of viral characteristics shows the futility of this. It turns out that they are so genetically diverse that, by all taxonomic criteria, there is no common evolutionary thread, no known common ancestor. For now, viruses remain a taxonomic enigma.

The only characteristics common to viruses are their simple internal structure, parasitic lifestyle, and minuscule size. Yet even their size range is considerable: from 30 to 300 nm, a range that extends from large molecules to the smallest bacterial cells. Viral dimensions are compared with bacteria and red blood cells in Figure 22.1.

Our interest in viruses is not entirely academic. Each of us is susceptible to several hundred viral diseases, not to mention those that infect our crops and domestic animals. In addi-

tion to the common cold, the list of human maladies caused by viruses includes smallpox, polio, German measles, chickenpox, mumps, and several kinds of influenza. Viruses known as **oncogenic viruses** have been implicated in certain human cancers ("onco" refers to tumor or cancer), and now humans are confronted by HIV, the agent of AIDS, a most confounding viral invader.

What do viruses do to cells? This is chiefly what the coming discussion is about; but let's have a brief preview. A few have no apparent effect, but they are the exceptions. For example, researchers estimate that 95% of the American population harbors the Epstein–Barr virus, which appears to be harmless to most people. In people with defective immune systems, however, there is evidence that Epstein–Barr causes cancer.

In their infective state, viruses take over the cell's metabolism, depriving it of all it needs to live. If this doesn't kill the cell, it may later be lysed, as great numbers of newly formed viral particles rupture the plasma membrane and escape. If enough cells are killed, the host will probably die. Viruses may also bring about random changes in the host cell's genome, sometimes causing a wild proliferation of cells that is a form of cancer.

Origin of Viruses

The evolutionary origin of viruses is an ongoing puzzle to biologists. Over the years, three competing hypotheses have come forward. One favors the idea that viruses are the ultimate parasite, former cells that have become so specialized that the only structures retained are those giving it the ability to invade cells and reproduce. There is precedent for this notion. Among animal parasites, such as the tapeworms, many organs and organ systems have degenerated to the point where they are virtually without function. All that remains are systems specialized for feeding and reproduction. However, the cells of all such parasites are fully functional, so they are still far more complex than viruses.

A second hypothesis is that viruses represent the oldest forms of life on earth, a precellular state. As we saw in the last chapter, theorists propose that the earliest precursors to life, the proteinoids and coacervates (see Chapter 21), were simple membrane-surrounded spheres, containing only the barest essentials. Their sources of energy were the spontaneously forming molecules in the surrounding primordial waters. Yet each of these entities must have carried out the metabolic functions needed to make use of these sources. Metabolism is completely lacking in viruses.

A more favored hypothesis takes another direction, that viruses originated as bits of genetic material from living organisms. Such fragments, as we know from bacterial transformation (see Chapter 15), are frequently taken in by cells. Over time, the host cells may have provided the genetic wanderers with their protein coverings, perhaps needed to isolate them from host cytoplasm. From there they evolved the ability to reproduce themselves (with the help of the host's metabolic machinery), escape, and penetrate other cells.

Support for the third hypothesis comes from the fact that viruses are highly specific to the organisms they invade. Studies of DNA sequences in the viruses and their hosts reveal strong similarities in portions of their DNA. And, as we have pointed out, the various types of viruses are genetically more similar to their hosts than they are to each other. The third hypothesis may be testable. If viruses originate from fragments of cellular DNA, then it may be possible to duplicate the conditions under which this happens and observe their origin! As you read about *viroids* and *prions* later in the chapter, you'll find additional evidence supporting the third hypothesis.

The Discovery of Viruses

Most of what we know about viruses has been learned in this century, primarily in the past 50 years. But the story begins even before that. The discovery of viruses came at a time when the newly emerging field of bacteriology was making its first inroads into the conquest of disease. By the late 1800s, Louis Pasteur, Robert Koch, and others had proposed the **germ theory**, the proposition that bacteria are the agents of contagious disease. However, it was soon discovered that some contagious diseases are not caused by bacteria at all. In these instances, methods that proved so successful in growing and identifying bacterial disease agents simply didn't work. In frustration, the early researchers named the mysterious agents "viruses," a term derived from a Latin word meaning "poison." Despite this difficulty, there was success in develop-

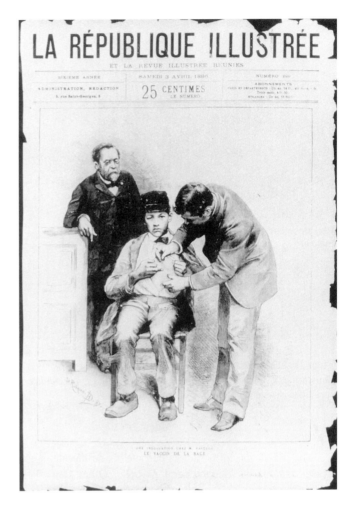

FIGURE 22.2
Early Vaccination
Although Louis Pasteur failed to grow the rabies virus, he was able to prepare a vaccine that could be used to fight off the deadly disease. Here the great scientist supervises the administration of the vaccine to a dog-bite victim.

ing vaccines against devastating viral diseases such as smallpox, rabies, and hoof-and-mouth disease (Figure 22.2).

The Tobacco Mosaic Virus. By 1892, studies of a contagious disease in tobacco revealed several important clues to the nature of viruses. In one experiment, Russian biologist Dimitri Iwanowski extracted juices from infected plants and strained the liquid through an extremely fine filter (the type used to remove bacteria from growth media). Then, applying the filtrate to healthy plants, he succeeded in spreading the disease. Iwanowski did this by infecting one plant and then transferring fluids from that plant to another, and so on, through a series. He reasoned that if the agent was a simple poison it would lose its potency through dilution in the plant fluids as the series progressed. He soon found that the agent did not lose its potency—it was as strong at the end of the sequence as it was at the beginning. From his work, Iwanowski was able to

(a) TMV-infected leaf

(b) Crystallized TMV particles

FIGURE 22.3
Tobacco Mosaic Virus
(a) The leaves of an infected tobacco plant are mottled in color and wrinkled in texture. **(b)** In this electron microscope view, the crystallized virus takes on a cylindrical, rod-shaped appearance.

reach three conclusions about the agent: (1) it was smaller than a bacterium, (2) it was contagious, and (3) it somehow maintained its strength as it spread in host plants. The agent later became known as **tobacco mosaic virus** (Figure 22.3a).

By the turn of the century, the filtration technique had been broadly applied, and the list of **filterable viruses**, as they became known, grew substantially. Such progress was not without its problems, however, for despite persistent efforts, the early bacteriologists were unable to grow the filterable agents in the usual bacterial culture media. This meant that well-established techniques could not be applied to the tiny viral agents.

In 1935, American microbiologist Wendell Stanley trained his light microscope on a drop of filtrate from infected tobacco plants and became the first person to see a virus. Actually, what he saw was the crystalline form—the long, slender crystals of the dormant tobacco mosaic viruses. In that same year, the large, rod-shaped virus particle itself was first observed through the newly invented electron microscope (Figure 22.3b). Stanley also proved that the tobacco mosaic virus could reproduce if it was introduced into healthy tobacco plants. Stanley's procedure was straightforward. In a variation of Iwanowski's procedure, he introduced a small drop of dilute viral filtrate into a healthy plant, waited a time, and recovered a much greater quantity of the viral substance. The search was on to learn more about the biology of these peculiar little entities.

Characteristics of Viruses

As we've seen, the individual viral particle, known today as a **virion**, consists of a core of nucleic acid (either DNA or RNA), a protein coat called a **capsid**, and, in some viruses, an enzyme or two. The herpes, influenza, AIDS viruses, and other viruses also have a surrounding envelope closely resembling a plasma membrane. The resemblance to a cell membrane is not a coincidence, since such viruses capture a portion of the host cell's plasma membrane as they make their exit from the infected cell. Those lacking an envelope are often called "naked viruses."

Viral Shapes. Viruses are usually helical or polyhedral (many-sided) in shape, but a few are cuboidal or rectangular. The shape of a virion depends ultimately on the arrangement of protein subunits, or **capsomeres**, in the capsid. Several typical examples are seen in Figure 22.4.

Genetic Differences. Viruses also differ in the kind of nucleic acid making up their genomes. In some, the genetic material is standard *double-stranded DNA*, just as it is in true cells. But other viruses may have *single-stranded DNA double-stranded RNA,* or *single-stranded RNA*. Some of the differences in genetic composition are used in classifying viruses, as seen in Table 22.1.

Bacterial Viruses: The Bacteriophages

We introduced the **bacteriophages** and their infection cycles in Chapter 15, so we can be brief here. Recall that the phage is quite complex. Typically, it has a polyhedral head from which a lengthy tail extends, giving rise to tail fibers that look like spider legs. The tail fibers bind to the bacterial host cell and enzymes of the tail penetrate the cell (Figure 22.5a). The phage then introduces its double-stranded DNA into the host.

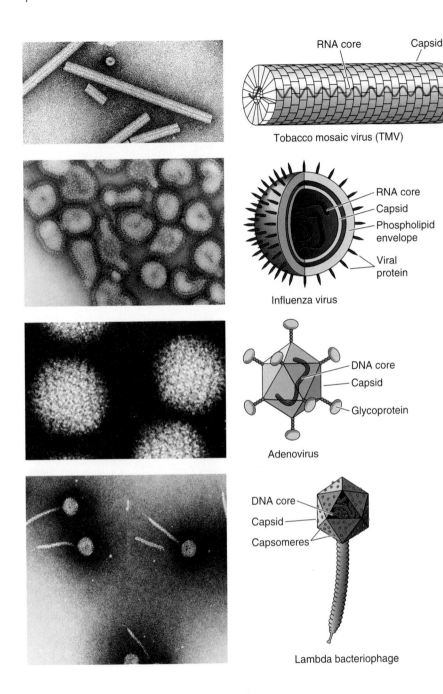

RNA core   Capsid
Capsomeres
Tobacco mosaic virus (TMV)

RNA core
Capsid
Phospholipid envelope
Viral protein
Influenza virus

DNA core
Capsid
Glycoprotein
Adenovirus

DNA core
Capsid
Capsomeres
Lambda bacteriophage

FIGURE 22.4
Viral Diversity

The tobacco mosaic virus is a lengthy naked particle with an RNA core and a dense capsid of bean-shaped capsomeres. The influenza virus is an enveloped particle. Its RNA core consists of eight segments of double-stranded RNA. The protein capsid is surrounded by a phospholipid bilayer membrane into which spike-shaped viral glycoproteins are inserted. The adenoviruses, from a family of infectious respiratory viruses, is a naked particle. The arrangement of its capsomeres gives the capsid a polyhedral (many-sided) appearance. The core is double-stranded DNA. The lambda bacteriophage, a parasite of bacteria, is naked and complex. In addition to its polyhedral head, it contains a tail with a single tail fiber for binding to its host cell. The core consists of double-stranded DNA.

Upon entering a host cell, the T_4 bacteriophage proceeds immediately into the destructive **lytic cycle** (see Chapter 15). (*Lytic* comes from the term lyse, which literally means "dissolve" or "destroy.") The lytic cycle begins as the invader disrupts the host cell's DNA, replacing it with its own. As transcription and translation occur, the bacterial cell obligingly produces new coat proteins, tail assemblies, and, following many rounds of replication, new viral DNA. Soon the phage virions are assembled, their burgeoning numbers rupturing the cell. The bacteriophages escape to infect other cells (Figure 22.5b).

Temperate bacteriophages are those that are capable of entering a quiescent, **lysogenic state** (see Chapter 15). Here the phage integrates its DNA into the DNA of the host, where it remains as a **prophage**. As the bacterial host reproduces, it unwittingly replicates prophage DNA along with its own, and a large clone of infected cells soon forms. Upon activation, prophages enter the usual lytic cycle and the clone is destroyed.

Animal Viruses

Animal viruses have all the genetic variations, from double-stranded DNA viruses to single-stranded RNA viruses. Most animal viruses begin their attack on host cells in the same general manner, following seven steps: *binding, penetration, uncoating, transcription, replication, assembly,* and *release*. The events

TABLE 22.1
Classification of Viruses

Viral Group	Host	Nucleic Acid Strands	Examples	Thousands Of Bases	Coverings
DNA Viruses					
Coliphage	Bacteria	Double	Bacteriophages, T-even (2,4,6)	Variable	Naked
Papovavirus	Primates	Double	SV40, human warts	5.8	Naked
Adenovirus	Vertebrates	Double	Respiratory disease	35–40	Naked
Herpesvirus	Vertebrates	Double	Herpes simplex I and II	120–200	Enveloped
Poxvirus	Vertebrates	Double	Vaccinia, smallpox	120–300	Enveloped
Coliphages	Bacteria	Single	Phange lambda φ×174	47	Naked
Parovirus	Vertebrates	Single	Dog parvo	1–5	Naked
RNA Viruses					
Reovirus	Vertebrates	Double	Diarrhea, Colorado tick fever	18–30	Naked
Rhabdovirus	Vertebrates	Single (−)	Rabies, Newcastle disease	12–15	Enveloped
Picornavirus	Primates	Single (+)	Poliomyelitis, hepatitis A common cold	7	Naked
Orthomyxovirus	Mammals	Single (−)	Influenza	14	Enveloped
Paramyxovirus	Mammals	Single (−)	Measles, mumps	15	Enveloped
Plant virus	Plants	Single and Double	Tobacco mosaic virus, potato yellow dwarf, wound tumor	6.4	Naked
Retrovirus	Vertebrates	Single (+)	HIV, Rous sarcoma	5–10	Enveloped
Togavirus	Vertebrates	Single (+)	Encephalitis, rubella, yellow fever		Enveloped

[1] Plus (+) single RNA strands act directly as mRNA, whereas minus (−) single RNA strands must first be transcribed into a plus strand.

are summarized in Figures 22.6 and 22.7, where reproduction in a naked *plus strand RNA virus* and an *enveloped DNA virus* are illustrated.

Binding. Molecular complexes in the capsid recognize and bind to specific sites on the host plasma membrane. For example, the AIDS virus has a glycoprotein complex that binds to a human cell surface protein designated CD4. In an adenovirus known to cause respiratory infections, binding proteins are at the corners of the polyhedral-shaped capsid. In the German measles virus, binding proteins are on "spikes" on the envelope. Such binding is highly specific, which is why people don't usually contract viral diseases of horses and chickens. The absence of proper binding sites on a would-be host membrane indicates that an organism has natural immunity against the virus.

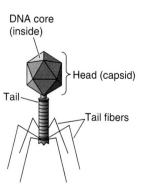

(a) T-even bacteriophage (a "moon lander")

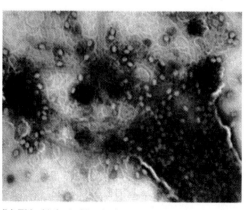

(b) EM of infected bacterium

FIGURE 22.5
A T₄ Bacteriophage
(a) Some people will swear that NASA was inspired by the T-even bacteriophage when the "moonlander" was first conceived. The head contains the double-stranded DNA bacteriophage genome. The tail fibers bind with surface molecules on the bacterium and, with the aid of a cell-penetrating enzyme, the genome is literally injected into a host cell. **(b)** The EM caught an infected bacterium with many T₄-even phages bound to its surface. Then, as the lytic cycle ends, the cell becomes filled with viral particles and literally bursts.

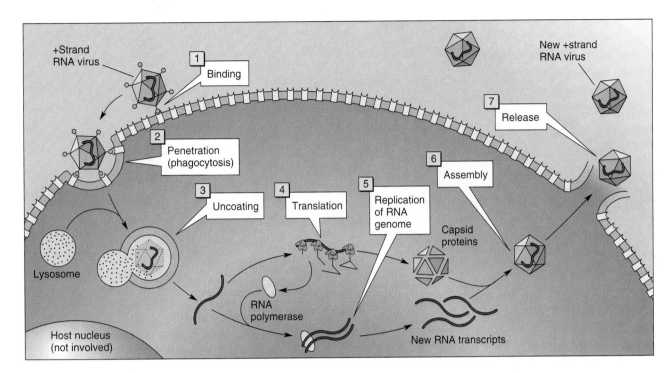

FIGURE 22.6

Attack by a Naked Animal Virus

A naked, plus strand RNA virus binds to a host cell ☐1 and enters through phagocytosis ☐2. Uncoating ☐3 occurs through the action of a lysosome that digests the protein coat, releasing the RNA genome. In this instance, the RNA acts as mRNA and ☐4 begins the process of viral protein synthesis. The RNA will then synthesize its own RNA polymerase and replication ☐5 will begin. The plus strand makes many new minus complementary strands, which then act as a template for forming many new plus strands for the next generation of virions. Assembly ☐6 of viral capsids and RNA cores follows, and the new virions escape ☐7 through exocytosis.

Penetration. When a naked animal virus binds to its host cell, the cell is stimulated into taking the virus in through phagocytosis. Enveloped viruses use a different method. They fuse with the plasma membrane, which opens the envelope and releases the capsid and core inside.

Uncoating. Uncoating is the removal of the capsid and release of the infective nucleic acid core. Ironically, the cell itself is responsible for this crucial step. A newly entering capsid attracts one of the host's lysosomes, which digests the capsid, inadvertently freeing the core. It turns out this way because lysosomal enzymes can digest protein but cannot digest nucleic acids. The invasion is then complete.

Transcription and Replication. The next event is viral reproduction. This requires transcription of the viral DNA into viral messenger RNA, which, during translation, makes capsid proteins and viral enzymes. To accomplish this, the viral mRNA makes use of the host cell's ribosomes, amino acids, and energy. Eventually, the viral genome undergoes many rounds of replication, after which the assembly of new virions begins.

The actual events surrounding transcription will, of course, be very different, depending on the type of genome. In double-stranded DNA viruses, such as herpes simplex and smallpox, transcription and replication are much like that of cellular life. In what are called **plus strand** RNA viruses, the entering RNA acts as messenger RNA, coding directly for the viral protein. **Minus strand** RNA viruses, however, must first transcribe their RNA into a complementary strand, which then becomes their mRNA. For this reason, minus strand RNA viruses carry along their own RNA polymerase, which does the job.

When it comes time for replication, plus strand RNA viruses produce the RNA polymerase enzyme, which then forms many minus strands. The minus strands are then used as templates for making new plus strands, needed for reproduction. In the minus strand RNA viruses, the newly formed mRNAs can act as templates for this purpose.

A most unusual process is seen in the retrovirus, which, as we have seen in earlier chapters, uses the enzyme reverse transcriptase to convert its single-stranded RNA into DNA (see Chapter 16). This is then inserted into a host chromosome, where it will remain in an inactive state for a time. We will look at the details of reverse transcription in the AIDS virus shortly.

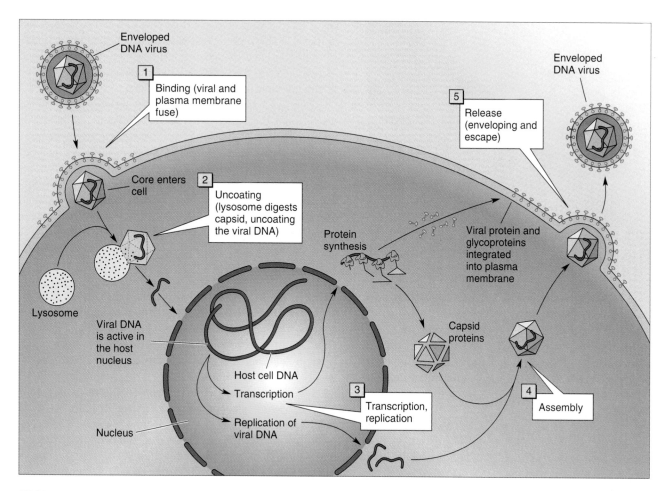

FIGURE 22.7
Attack by an Enveloped Animal Virus

A more complex virus, one bearing an envelope, fuses to its receptor molecules on the host cell's plasma membrane ☐1, whereupon the envelope opens up and the capsid enters the cell. There, with the help of a lysosome ☐2, uncoating occurs. Once in the host's nucleus, ☐3 the viral genome, double-stranded DNA in this case, enters a frenzy of transcription and protein synthesis, forming new capsid and membrane proteins. Then, after many rounds of DNA replication, the virions are assembled ☐4. They escape by capturing a bit of the altered plasma membrane ☐5 and budding off the surface.

Assembly and Release. When sufficient capsid proteins and viral nucleic acids have been synthesized, they are spontaneously assembled into new virions. Naked viruses, those with the capsid covering only, are released through either cell lysis or exocytosis. Enveloped viruses are released by budding, which is how they get their envelopes. The newly assembled capsid fuses with the plasma membrane at special sites where viral proteins have already been inserted. The virion, now surrounded by a bit of modified host plasma membrane, finds itself outside—ready to infect another cell. Some of the viral proteins adorning the newly appropriated envelope will enable the escaping virus to bind to the next host cell.

By now, you are probably wondering how hosts like ourselves ever survive such insidious invasions. We, like many other animals, have a variety of defense measures including physical barriers, chemical agents, and armies of highly effective defense cells. And, fortunately, ongoing random changes in the viral genome can result in a reduced virulence, so some viral diseases seem to just "go away." We'll explain it all in Chapter 40.

Plant Viruses

Viruses also invade plants, many becoming serious agricultural and forestry pests (Figure 22.8). Unlike animal viruses, plant viruses do not seem to have special surface molecules for binding to the plant cell. Their transmission from plant to plant is carried out chiefly by plant-eating and pollinating insects, although any injury to the plant's surface tissues represents an avenue for infection. Researchers simply scratch the leaf or stem surface with sand and apply a virus suspension with a cotton swab or atomizing spray. Typically, the

FIGURE 22.8
Plant Viruses
Viruses can affect plants in several ways. Many cause a mosaic effect—a mottled coloring in the leaves resulting from the destruction of chloroplasts.

viruses are carried from cell to cell through plasmodesmata (cytoplasmic connections between cells).

Plant viruses differ from animal viruses in a number of other ways. In many RNA viruses, the genome occurs in several separate portions, each packaged into its own capsid. This means that all of the capsids must be present if the virus is to reproduce successfully. Curiously, the maize streak virus of corn, a double-stranded DNA virus, forms paired capsids, each containing only a single strand of the replicated DNA!

Viral infections are particularly destructive to crops, especially potatoes, sugar beets, wheat, soybeans, and tobacco. Infected plants are easily recognized by the presence of mottled and wrinkled leaves, loss of color, tumors, dying tissue, and stunted growth. The tobacco mosaic virus (see Figure 22.3) fares so well in the domestic tobacco plant that the minute viral particles can account for *one-tenth* of an infected plant's dry weight. That's a lot of viruses!

HIV: Study of a Retroviral Killer

Our list of human retroviruses includes three recently discovered groups: human T-cell leukemia virus I and II (HTLV-I and HTLV-II) and HIV. The first two may be responsible for two rare forms of leukemia, but the third is by far the best known. It causes AIDS.

HIV (human immunodeficiency virus) shows a special preference for certain cells of the immune system. In particu-lar, it attacks white blood cells known as **macrophages** and **helper T-cell lymphocytes**, both of which play pivotal roles in initiating and organizing the immune system's usual response. When these cells fail, the immune response is feeble at best, and other invaders soon have their way.

This is all spelled out in the name **AIDS**, which, by the way, is not the name of a disease; it is the acronym for **acquired immunodeficiency syndrome**, a set of symptoms that occur because of the crippled immune system's inability to perform. AIDS victims suffer from many routine diseases, which become increasingly difficult to control. They also suffer from truly rare diseases, such as pneumocystic pneumonia and Kaposi's sarcoma. Before the AIDS epidemic, physicians often knew about these exotic afflictions only from medical school textbooks. Such afflictions were rare because they are stopped cold by the immune system.

Biology of HIV. Like other human retroviruses, HIV is spherical in shape (Figure 22.9). Its surrounding envelope contains many glycoprotein units, each penetrating the bilayer of the lipid membrane below. Within the envelope is the usual protein capsid, which surrounds a core of single-stranded RNA and reverse transcriptase.

The history of an HIV invasion is illustrated in Figure 22.10. Envelope glycoproteins, designated gp120/gp41, bind to CD4, the human cell surface protein (see Figure 22.10, ⬚1). Once the match is made, the two membranes fuse, and the naked protein capsid containing the RNA viral genome is released into the cytoplasm. Following uncoating, the retrovirus begins its deadly work.

We should mention that the highly specific virus–host match represents a possible weak link in the life cycle of HIV. One means of treatment could be a vaccine containing antibodies that would coat the viral glycoprotein, thus preventing binding. A second would be to flood the body with look-alike CD4 chemicals, which would tie up the virus' binding sites. Unfortunately, neither idea has proved practical. To date, the treatment of choice is **AZT** (azido-dideoxy-thymidine), which inhibits HIV's reverse transcriptase by competing with the normal thymidine nucleotide. Unfortunately, it is not completely effective and can only extend the life of AIDS sufferers. Furthermore, it is toxic to bone marrow cells and can cause anemia.

Following a successful invasion, the single-stranded RNA genome of the retrovirus undergoes reverse transcription (see Figure 22.10, ⬚3). As in the temperate bacteriophage, the viral DNA is incorporated into the host cell's genome, somewhere in its chromosomes. There the HIV genome remains, quite inactive, but replicating right along with the host chromosome and passing from one generation of host cells to the next.

Fortunately, dormancy is the most common state of HIV. No one knows how many humans carry the virus this way, but medical researchers are desperately seeking the answer so that they may better understand the severity of the present epidemic (see Chapter 40). It is known that years may pass (some say

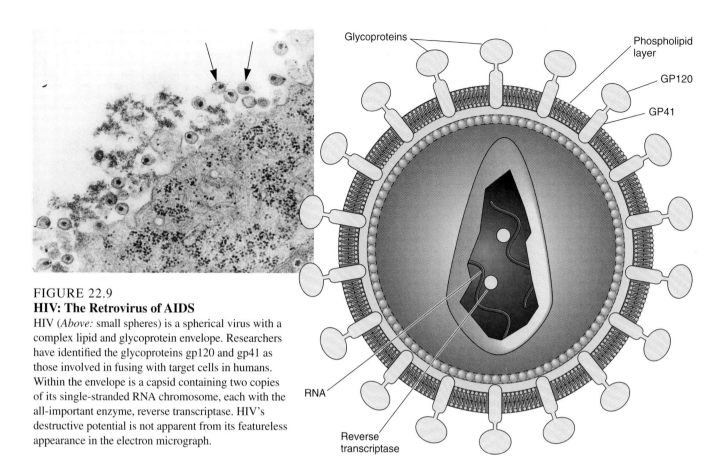

FIGURE 22.9
HIV: The Retrovirus of AIDS

HIV (*Above:* small spheres) is a spherical virus with a complex lipid and glycoprotein envelope. Researchers have identified the glycoproteins gp120 and gp41 as those involved in fusing with target cells in humans. Within the envelope is a capsid containing two copies of its single-stranded RNA chromosome, each with the all-important enzyme, reverse transcriptase. HIV's destructive potential is not apparent from its featureless appearance in the electron micrograph.

up to ten) before HIV becomes fully active. When active, implanted viral DNA in each infected cell enters into a frenzy of transcriptional activity (see Figure 22.10, 4), producing copy after copy of the single-stranded RNA that makes up the HIV genome. This RNA codes directly for viral envelope and capsid proteins, and more reverse transcriptase. As usual, these activities are carried out at the expense of the host cell's supply of amino acids, ribosomes, energy, and whatever else is needed.

The assembly of new virions begins when newly transcribed viral RNA and raw proteins gather at the host cell's plasma membrane. Some of the proteins, modified into glycoproteins, are integrated into the membrane while the remainder are assembled into new capsids and reverse transcriptase. The assembled virions then bud from the cell surface, breaking away as new, infective particles. They are fully ready to invade more lymphocytes, or new hosts, should the opportunity arise.

The formation and budding of viral particles are so intense that the plasma membrane is virtually riddled with holes. As you might expect, this kills the cell, effectively eliminating a key participant in the immune reaction and thus leaving the victim open to the infections that will soon end his or her life.

Like any successful parasite, HIV must have a means of infecting a second host; otherwise the host's death will bring about its own demise. There are two primary avenues of transmission. Like gonorrhea and syphilis, HIV is a sexually transmitted disease. The virus can pass from person to person via semen or vaginal secretions. It is also transmitted through contaminated blood, chiefly in hypodermic syringes used to inject street drugs. On rare occasions, it is introduced during blood transfusion and other medical procedures. HIV can also be passed to the fetus across the placenta and to infants in mother's milk (see Chapter 40).

This brief account of one of modern society's most urgent unsolved medical problems has undoubtedly left you with many questions. Where did HIV come from? How does the virus single out specific kinds of cells? How do lymphocytes normally perform their functions? What are some of the difficulties in overcoming AIDS? Chapter 40 takes up some of these questions.

Viroids and Prions

We must leave you with two more puzzles: viroids and prions. These are infectious agents of plants and animals, respectively, but they are neither viruses nor cells.

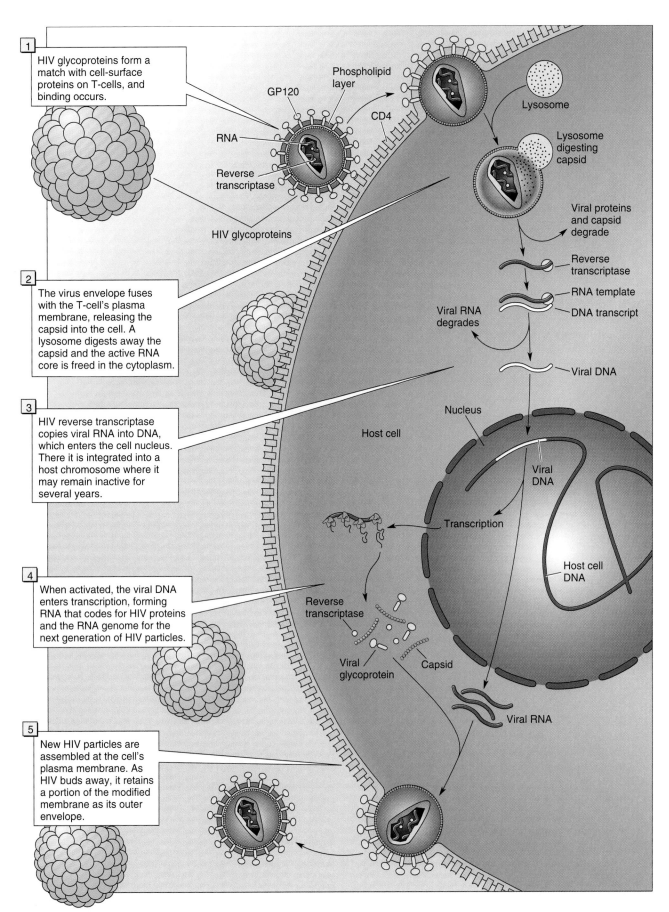

1 HIV glycoproteins form a match with cell-surface proteins on T-cells, and binding occurs.

2 The virus envelope fuses with the T-cell's plasma membrane, releasing the capsid into the cell. A lysosome digests away the capsid and the active RNA core is freed in the cytoplasm.

3 HIV reverse transcriptase copies viral RNA into DNA, which enters the cell nucleus. There it is integrated into a host chromosome where it may remain inactive for several years.

4 When activated, the viral DNA enters transcription, forming RNA that codes for HIV proteins and the RNA genome for the next generation of HIV particles.

5 New HIV particles are assembled at the cell's plasma membrane. As HIV buds away, it retains a portion of the modified membrane as its outer envelope.

GP120

Phospholipid layer

RNA

CD4

Reverse transcriptase

HIV glycoproteins

Lysosome

Lysosome digesting capsid

Viral proteins and capsid degrade

Reverse transcriptase

RNA template

DNA transcript

Viral RNA degrades

Viral DNA

Nucleus

Host cell

Viral DNA

Transcription

Host cell DNA

Reverse transcriptase

Viral glycoprotein

Capsid

Viral RNA

FIGURE 22.10

Viroids. In 1951, the chrysanthemum industry in the United States was virtually destroyed by a mystery disease, and some years later, some 12 million coconut palms died unexpectedly on Philippine plantations. Other mystery plant diseases included potato spindle tuber (small, cracked, spindle-like potatoes) and citrus exocortis (stunting in citrus). Concentrated sleuthing at several universities revealed that the agents responsible were short, naked fragments of RNA, *period.* They were dubbed **viroids**.

The potato spindle viroid is simply a fragment of RNA some 359 nucleotides in length, about $\frac{1}{20}$ the length of the smallest virus. Apparently that short fragment can wreak havoc with the metabolic machinery of the cell since its effects are often fatal to the plant. The viroids are transmitted from plant to plant much the way viruses are, by insects and through damaged surface tissues. Viroids may also be responsible for some of the mysterious, slow-forming diseases of animals.

Prions. Prions are even stranger than viroids. **Prions** are empty protein shells, yet they can penetrate cells, reproduce themselves, and exit, killing the cell. They have been well-established as the agents of several neurological diseases in humans and other animals, including kuru and Creutzfeldt-Jakob disease in humans and scrappie in sheep. All destroy brain tissue and are eventually fatal. More recently, researchers have been investigating the possibility of prions causing Alzheimer's disease.

If you're a little baffled by the idea of a protein shell reproducing, you are in good company. But, as you might expect, the stymied scientists have some ideas. Some suggest that, once in the cell, portions of the protein agent undergo "reverse *translation*": the use of the amino acid sequence to form RNA (as is done by gene machines). Accordingly, RNA is then converted to DNA, which then directs the synthesis of more protein agents. However, "reverse translation," or whatever such a process might be called, has never been observed in life. Others suspect that the prion is a gene activator: that is, once it makes its way into the host cell nucleus, it activates—

directly or indirectly—regions of the genome that code for the desired proteins. It has not escaped the notice of some biologists that viroids and prions may also represent steps in the evolution of a virus. (Had this occurred to you?)

Kingdom Monera: The Prokaryotes

Kingdom Monera includes the prokaryotes, more traditionally known as bacteria. (Some taxonomists suggest the name Prokaryota for this kingdom.) The term prokaryote ("before a nucleus") refers to the fact that bacteria have no organized nucleus. We have also noted in previous chapters that prokaryotes lack most of the organized, membrane-bound organelles of the eukaryotic cell.

Evolutionary Roots of Monerans

Bacteria are indeed ancient, their fossils having been found in deposits 3.5 billion years old (between 1 and 1.5 billion years before the first eukaryotes). The best evidence of prokaryote antiquity is seen in laminated, column-like deposits of sedimentary rock known as stromatolites (Figure 22.11a). Stromatolites formed in the shallows of ancient Precambrian seas, where layer after layer was produced by dense populations of bacteria resembling today's cyanobacteria. (We'll discuss this group later in the chapter.) The cell layers were fossilized by calcium carbonate and other minerals from the surrounding water.

Some authorities doubted that these strange columns actually represented fossilized life forms until it was found that there are still active stromatolite-forming cyanobacteria at work along the shores of Shark's Bay on Australia's west coast (Figure 22.11b). Similar mineral-incorporating formations are also seen in Yellowstone National Park in the United States.

Eubacteria, Archaebacteria, and a New Phylogeny

As recently as 1975, biologists were quite comfortable with the organization of prokaryotes into the Kingdom Monera, as offered above. The kingdom contained two major groupings of prokaryotes—bacteria and cyanobacteria. But as you are aware, taxonomic groupings continually change in biology as new information becomes available. Researcher Carl R. Woese has applied modern molecular investigative techniques such as nucleotide sequencing to the problem, thus changing the way we view these microorganisms. As a result, the vast realm of prokaryotes is now being divided into two new kingdoms: **Archaebacteria** ("first, or ancient, bacteria") and **Eubacteria** ("true bacteria"; Figure 22.12). While both are definitely prokaryotic, they are different in enough ways to suggest separate progenotic origins. Recall that progenotes are believed to represent the earliest forms of life and may also have produced the first eukaryotes (see Chapter 21). The two bacterial groups look much alike when viewed through

FIGURE 22.10
Life Cycle of a Killer

[1] HIV fuses with the host cell membrane, releasing its capsid inside. Fusion occurs between glycoprotein gp120 and the human membrane protein CD4. Following uncoating [2], the single-stranded RNA is released. The viral RNA then uses reverse transcriptase to make a complementary strand of DNA [3]. The viral DNA is then inserted into a host chromosome. The insertion may remain intact through many rounds of cell division by the host cell, or, alternatively, it may become activated [4], entering a period of intense transcription and replication [5] and producing many single-stranded RNA transcripts. The necessary viral proteins are then produced. Assembly of the new virions occurs at the plasma membrane, which has been modified by the addition of viral proteins and glycoproteins. Upon assembly, each new, infectious viral particle buds from the surface of the doomed cell.

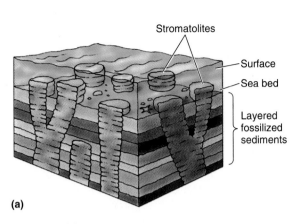

(a)

(b) Stromatolites at Shark's Bay

FIGURE 22.11
Early Prokaryote Fossils
(a) Stromatolites arose on rocky ledges in the tidepools of shallow Precambrian seas. As new populations arose at the surface, their predecessors underwent fossilization below, leaving a permanent record of past struggles. **(b)** Fossilized stromatolites are thought to be the most ancient evidence of life on earth.

the light microscope, but when the electron microscope and molecular analysis are used, basic physical and chemical differences are seen. Table 22.2 lists some of the distinctions between archaebacteria and eubacteria. Most of what we know about bacteria pertains to the eubacteria, so our survey of archaebacteria will be brief.

Archaebacteria: A Brief Look

Methanogens. The largest archaebacterial group, the **methanogens** (methane generators), are found in habitats where carbon dioxide and hydrogen are readily available but where there is little oxygen (Figure 22.13). Such places include anaerobic marshes, mucky anaerobic sea and lake bottoms (such as in the Black Sea), and the sediments in sewage treatment plants. Methanogens living in such environments are usually **obligate anaerobes**, organisms that cannot

tolerate oxygen. Some are more tolerant of oxygen, and one group lives side by side with the familiar colon bacterium, *Escherichia coli,* in the animal bowel. (Yes, we are included and ours do generate methane gas.)

Methanogens use hydrogen gas to reduce carbon dioxide to methane gas (CH_4, also called "marsh gas") and water. The overall reaction, which requires a battery of enzymes, is

$$4\ H_2\ +\ CO_2\ \longrightarrow\ CH_4\ +\ 2\ H_2O$$

Methane, incidentally, is one of the greenhouse gases, those that contribute to global warming. According to environmentalist G. Tyler Miller, Jr., methane produced by human activity accounts for some 8% of the global warming trend. The primary man-made sources include rice growing, landfills, cow and sheep flatulence, and leaks that occur as methane is used as a fuel. (Dairy cattle in barns pass enough methane gas to create a fire hazard!)

TABLE 22.2
Comparison of Archaebacteria and Eubacteria

Structure	Archaebacteria	Eubacteria
Cell wall	Variety of substances often proteinaceous	Peptidoglycan, teichoic acid
Plasma membrane	Modified branched fatty acids	Straight chain fatty acids
Ribosomes	30S, 50S subunits, more similar to eukaryotic than eubacterial	30S, 50S subunits, unlike archaebacterial or eukaryotic
Flagella	Unknown	Solid, rotating, composed of flagellin
Photosynthetic pigments	Bacteriorhodopsin	Bacteriochlorophyll and chlorophyll a

FIGURE 22.12
Phylogeny of Prokaryotes
Recent phylogenetic studies recognize two prokaryotic kingdoms, Archaebacteria and Eubacteria. The distinction is based on striking biochemical differences. The ancestors of both, and of the eukaryotes, were the progenotes, the first life forms to emerge from a long period of chemical synthesis. The arrows indicate theoretical steps in serial endosymbiosis in which bacterial symbionts in early eukaryotes gave rise to mitochondria and chloroplasts (see Essay 4.2).

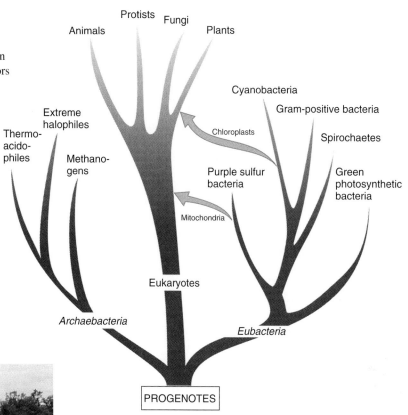

FIGURE 22.13
An Anaerobic Environment
Many archaebacteria are right at home in anaerobic bogs such as this one.

Halophiles. Other archaebacteria include the **halophiles** ("salt lovers"). Photosynthetic halophiles thrive in Great Salt Lake, Utah, the Dead Sea, and in sea-salt evaporation facilities. There, using simple photosystems containing the purple archaebacterial pigment **bacteriorhodopsin**, they make use of the sun's energy to pump protons out of the cell. This creates a steep chemiosmotic gradient between the cell and its environment, just as happens within mitochondria and thylakoids, and the bacterium uses the gradient to generate ATP (Figure 22.14). Although the end result is the same (producing ATP and NADPH), photosystems in archaebacteria are unlike those of any other organism.

Thermophiles and Thermoacidophiles. By all standards, thermophiles and thermoacidophiles live under the harshest of conditions. **Thermophiles** ("heat lovers") are known to thrive at 90°C or higher (near boiling). They can be found in hot mineral springs and in manure and compost heaps, where temperatures can be surprisingly high. The **thermoacidophiles** ("heat" and "acid lovers") live under even harsher conditions, some in acid solutions registering pH 2 at temperatures near boiling—about the same conditions used in boiling out engine blocks in preparation for rebuilding! Surprisingly, the internal pH of such archaebacteria remains close to neutral. Other microorganisms exposed to such conditions would perish in seconds.

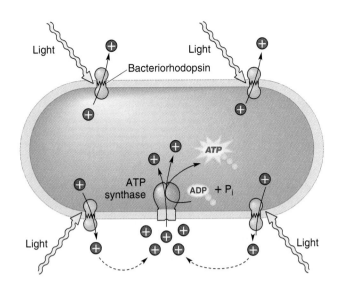

FIGURE 22.14
Photosynthesis in Archaebacteria
The elaborate photosystems of the photosynthetic eubacteria are absent in archaebacteria. Instead, protons are pumped by light-activated pigments known as bacteriorhodopsin. The chemiosmotic gradient builds up *just outside the cell* and as the protons pass back in—through ATP synthases—ATP is generated.

Eubacteria

Everyone knows a little about eubacteria. They are the organisms most people call "germs," those generally thought of as harmful. They're the ones people go to so much trouble and expense to avoid. Yet, in spite of endless gargling, spraying, and bathing, the fact remains that we are literally awash in a sea of the tiny creatures. It may help to know that most of the inhabitants of our own personal crevices and crannies are harmless, merely partaking of our oily body secretions and digestive wastes. In addition, many kinds of bacteria are ecologically important and some are directly helpful to us. We'll get to all of this shortly, but first let's focus on the bacterial cell, beginning with appearances.

Cell Form and Arrangement. The most common shapes of bacterial cells are rod-like, spherical, spiral, comma-shaped, and filamentous. Arrangements vary from single cells to pairs, octets, chains, and clusters (Figure 22.15).

1. The rod-like cells, called **bacilli** (singular, *bacillus*), occur as single cells and chains, with some chains enclosed in sheaths.

2. The spherical cells, or **cocci** (singular, *coccus*), have the most varied arrangements, occurring singly (**monococcus**) or paired (**diplococcus**), in bead-like chains (**streptococcus**), in grape-like clusters (**staphylococcus**), and sometimes in cuboid groups of eight (**sarcina**).

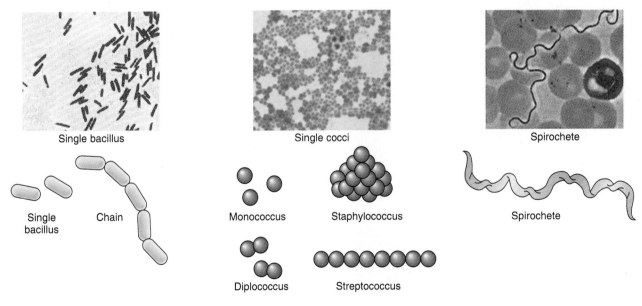

FIGURE 22.15
Eubacterial Form and Arrangement
The three principal cell shapes in bacteria are rod-shaped, spherical, and spiral. Whereas rod-shaped cells, or bacilli, occur singly and in chains, the spherical or coccoid form may occur as singles, doubles, chains, and clusters.

3. Spiral-shaped cells fall into three categories according to motility. **Helical bacteria** occur as filaments that move through a gliding action. **Spirochetes**, a second group, have flexible walls, moving through the action of axial filaments beneath an outer membrane. The third, spiral-shaped group, the **spirilla**, have rigid cell walls and move through the action of flagella.

4. The **vibrio**, or comma-shaped, bacteria are a curved version of the rod shape, a distinction that is important to clinical microbiologists. A particularly nasty vibrio produces an exotoxin (released poison) that is the cause of cholera.

5. The **filamentous forms**, including a group of soil bacteria called *Actinomycetes*, grow in lengthy filaments. Some are important sources of antibiotics.

Cell Structure. One of the most striking characteristics of all prokaryotic cells (Figure 22.16) is their small size. Many are only 1 μm (micrometer) in length or diameter. You could line up about 2000 of these across the head of a straight pin. As we mentioned earlier, there are no membrane-surrounded organelles, so the eubacterial cytoplasm has little discernible structure. (People are rarely impressed by their first look at bacteria through a light microscope.)

Surrounding the cytoplasm is a plasma membrane that is similar in many respects to the eukaryotic version. It contains the usual bilayer of straight-chain phospholipids with numerous proteins interspersed (see Chapter 5). Cholesterol is notably absent in nearly all eubacterial membranes, and some bacteria, as we will see shortly, have a second "outer membrane."

The plasma membrane of many bacteria has important metabolic functions. In aerobic bacteria it contains assemblages of respiratory and phosphorylating enzymes, analogous to those of the eukaryotic mitochondrion. In some photosynthetic bacteria the plasma membrane becomes greatly extended and folded internally, forming thylakoids, complete with photosynthetic pigments.

The **mesosome** is another kind of membranous infolding. Its functions haven't been resolved, but they may help chromosome replicas to separate and new cross-walls to form during cell division. Others may be involved in the secretion and metabolism of hydrocarbons. Some evidence suggests that they may really be artifacts; that is, they form as a result of the treatment used in preparing the cells for viewing. We will await the outcome of this controversial development.

Bacterial cells often contain a number of **pili**, slender surface extensions composed of the protein *pilin*. Pili aid the parasitic bacterium in adhering to the host cell (the agent of gonorrhea uses its pili in this manner). Other pili, called **sex pili**, aid in the transfer of genes during rare episodes of sexual reproduction (see Chapter 15). Other visible features include bacterial ribosomes, which often occur in chains as polyribosomes, and aggregations of stored materials, including glycogen and lipids. Bacteria are also known to store phosphorus, iron, and sulfur compounds.

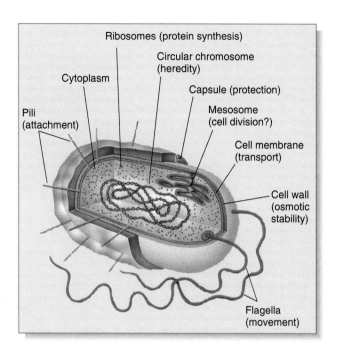

FIGURE 22.16
Eubacterial Cell Structure
The generalized eubacterium, a prokaryote, lacks the membrane-surrounded organelles of the eukaryote, including the organized nucleus. The cell wall is often surrounded by a slimy sheath or capsule. Free ribosomes and polyribosomes are common, as are fibrous protein projections known as pili. Flagella are corkscrew-shaped, tubular, rotating structures. Inclusions are often bodies of stored materials, such as sulfur, that are being isolated from the cytoplasm.

Bacteria do not form a nuclear envelope, nor do they have anything comparable to the highly organized eukaryotic nucleus. The single chromosome is a continuous DNA molecule (effectively circular), lying in the cytoplasm. It is described as "naked" because it lacks the chromosomal histone proteins seen in eukaryotic chromosomes. Finally, some bacteria have smaller, circular DNA molecules called **plasmids** (discussed earlier in Chapters 15 and 16).

Cell Walls and Antibiotics. Eubacterial cell walls contain carbohydrate polymers called **peptidoglycans**. In some eubacteria, the peptidoglycan wall is quite thin, but overlying it is an outer membrane. The outer membrane consists of a single layer of phospholipids, covered by a layer of lipopolysaccharides and proteins (Figure 22.17b). In other bacteria, however, the peptidoglycan layer is quite thick and contains a polysaccharide known as **teichoic acid** (Figure 22.17a). Then to remind us of the inevitable exceptions in biology, one group of bacteria, the mycoplasmas, lack the cell wall completely.

Cell wall composition is interesting for a special reason. Many of those bacteria with the thicker walls are susceptible

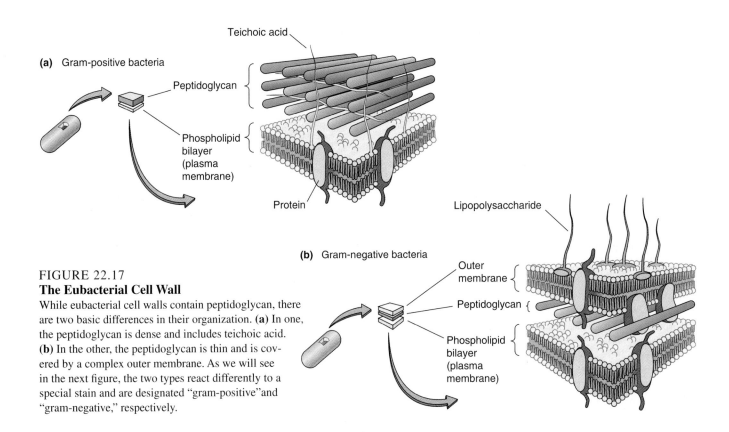

(a) Gram-positive bacteria

Teichoic acid

Peptidoglycan

Phospholipid bilayer (plasma membrane)

Protein

Lipopolysaccharide

(b) Gram-negative bacteria

Outer membrane

Peptidoglycan

Phospholipid bilayer (plasma membrane)

FIGURE 22.17
The Eubacterial Cell Wall
While eubacterial cell walls contain peptidoglycan, there are two basic differences in their organization. **(a)** In one, the peptidoglycan is dense and includes teichoic acid. **(b)** In the other, the peptidoglycan is thin and is covered by a complex outer membrane. As we will see in the next figure, the two types react differently to a special stain and are designated "gram-positive"and "gram-negative," respectively.

to penicillin and related antibiotics. In the presence of penicillin, peptidoglycan-synthesizing enzymes in newly divided cells are inhibited and cell walls cannot be formed. Without the tough, resistant wall, the bacterial cell, rich in solutes, cannot resist the natural osmotic entrance of water, and the unsupported membrane soon bursts. Thus the bacteria's fate, in the presence of the antibiotic, is the inverse of our own—their misfortune means that we survive a troubling or perhaps life-threatening infection.

Gram-Positive and Gram-Negative Bacteria. Bacteria with the thicker cell wall readily absorb and retain crystal violet, a deep purple cellular dye used in what microbiologists call the **Gram stain**. Such bacteria are designated **gram-positive** (Figure 22.18). Those bacteria with the thinner walls do not retain the crystal violet of the Gram stain and are designated **gram-negative**.

The Gram reaction has an important medical application. Many pathogenic bacteria identified as gram-positive can quickly be controlled with penicillin or penicillin-like antibiotics. But if the physician knows that the invading bacterium is gram-negative, some time is saved because alternative antibiotics, such as tetracycline and streptomycin, are prescribed. They kill bacteria by inhibiting protein synthesis. The Gram reaction has also been used by microbiologists for years as an important criterion in bacterial classification.

Capsules. In addition to the cell wall, many bacteria are surrounded by a **capsule** consisting of complex carbohydrates or polypeptides. Such coats tend to be sticky, which aids bacteria in adhering to surfaces that provide nutrients, an important ability for both parasitic and free-living types. Some parasitic types have slippery capsules that help the bacterial cells avoid being engulfed by the host's phagocytic white blood cells. On the other hand, capsule carbohydrates are often antigenic, which means they bind with host antibodies. This clumps invading cells together, making phagocytosis by defensive white blood cells much easier (see Chapter 40).

Endospores. When unfavorable conditions such as food shortages or excessive dryness arise, many bacteria become transformed into highly resistant, thick-walled **endospores** (Figure 22.19). Endospores are dehydrated cells in which all metabolic activity has ceased; the cellular components are held in a state of dormancy. Some endospores are so resistant that they can survive brief periods of boiling. One study revealed that spores 150,000 years old were able to germinate. **Germination** in bacteria is the return to the metabolically active state. The ability or inability to form endospores is important in the classification of bacteria.

Bacterial Movement. Many bacteria are motile, including all spiral types and about half of the bacilli. Coccoid bacteria

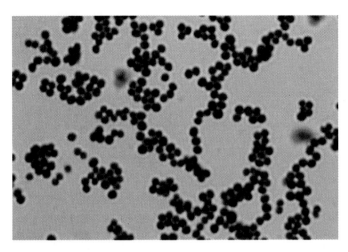 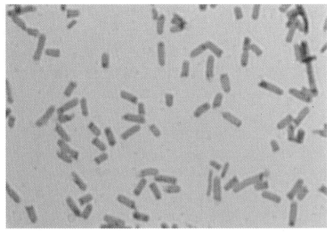

FIGURE 22.18
Gram Reactions
The bacteria on the left are gram-positive (purple), while those on the right are gram-negative (red). The reaction to Gram staining is an important diagnostic tool, particularly in clinical use. The red coloring within gram-negative bacteria is a "counterstain" that is masked by the purple stain in gram-positive bacteria.

are nonmotile. The most common propulsive device is the flagellum; the number varies from one to many, depending on the species.

Rotating Flagellum. The bacterial flagellum is structurally and functionally quite different from its eukaryotic counterpart. Recall that the eukaryotic flagellum moves with whip-like undulations, based on sliding microtubules. In contrast, the bacterial flagellum is a stiff, tubular, rod-like structure, composed of the protein **flagellin** (Figure 22.20).

The bacterial flagellum consists of three regions—an outer **filament**, a short **hook**, and a **basal body** (unlike the eukaryotic basal body). The basal body anchors the flagellum in the plasma membrane. It contains a series of rings and a central rod. When the flagellum moves it does so in a rotating manner that originates in the inner rings, those embedded in the plasma membrane. This spins the entire structure, much like a ship's propeller. Interestingly, studies reveal that the rotational force is not provided by ATP. Instead, it is an influx of protons, moving down the chemiosmotic gradient established by the cell, that provides the power.

Chemotaxis. Flagellar motion helps explain a long-observed bacterial phenomenon known as **chemotaxis**: the ability of bacteria to move toward or away from different stimuli (Figure 22.21). Researchers have discovered that the flagellum is bidirectional. When it moves in a counterclockwise direction, the helical coil is tight and the bacterium moves in a straight line. When the direction of rotation is reversed, the helical shape flies apart and the bacterium tumbles.

Studies show that counterclockwise, straight-path motion occurs when known chemical attractants are present. Recep-

tors on the cell surface apparently send chemical signals to the basal body, affecting the rotation, and thus drawing the cell to the stimulus. When chemical retardants bind to the cell, the reverse spin occurs, and the bacterium tumbles away.

Gliding Bacteria. A less familiar form of movement in bacteria is gliding. The so-called **gliding bacteria** are thought to

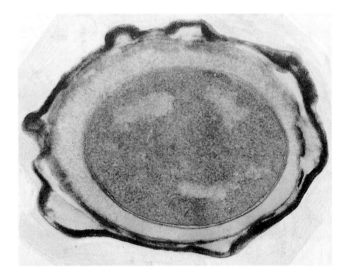

FIGURE 22.19
The Endospore
Many bacteria form tough-shelled resistant endospores. Within each endospore is a chromosome and the dehydrated cytoplasm. Under more ideal conditions the endospore will take in water, and the cell will resume activity.

FIGURE 22.20
The Bacterial Flagellum
(EM: *left*) Flagella are arranged in many ways in bacteria. This one has them alongside. A close-up EM of a dislodged flagellum (EM: *lower right*) clearly reveals its parts. In the illustrated view, two sets of rings are revealed: one embedded in the plasma membrane and the other in the outer membrane. The rings are stationary, but the rod passing through rotates, its power provided by an influx of protons from the chemiosmotic gradient. A hook connects the rod with the filament. (This is a gram-negative bacterium. Can you see why?)

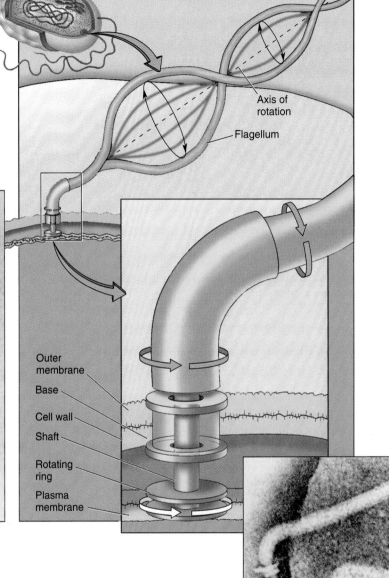
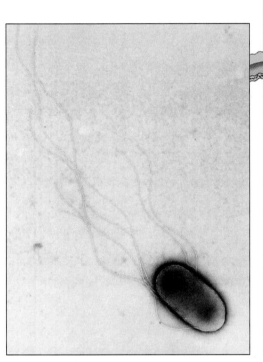

move by exuding a slippery mucilage and then moving over it. Alternatively, some researchers believe the mucus is secreted in a concentrated form, and, as it takes up water, it expands, pushing the organism along.

One group of gliding bacteria, the **myxobacteria**, are unusual in other ways as well. Under drying conditions some types mass together and form an upright stalk that bears a spore-forming body at its tip. (In this case, the term "spore" refers to an asexual reproductive body.) Under favorable circumstances, the **myxospores**, as they are known, will germinate and form new, active cells. In this respect, myxobacteria are very much like some of the protists and fungi.

Bacterial Reproduction

By far, the most common mode of reproduction in bacteria is asexual. It occurs through a primitive process called **fission**: simple cell division without mitosis (Figure 22.22). Fission is preceded by DNA replication, and if all goes well during the division process, each daughter cell will receive a replica of the original bacterial chromosome. How is an equitable division of replicated DNA ensured without mitosis? Some of the details are unknown but the observations seem clear enough. Following chromosome replication, the cell takes in water and elongates. The replicas, which attach separately to the cell

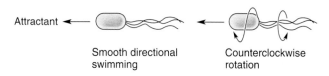
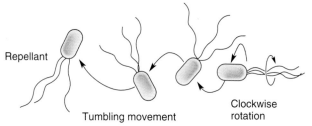

Attractant ← Smooth directional swimming ← Counterclockwise rotation

(a) Counterclockwise

Repellant — Tumbling movement — Clockwise rotation

(b) Clockwise

FIGURE 22.21
Chemotaxis
Many motile bacteria demonstrate a chemotactic ability: the ability to move toward or away from a chemical stimulus. **(a)** When attractants bind to the cell surface, the flagella respond by spinning counterclockwise, moving the cell in a straight line down the chemical gradient. **(b)** When retardants bind, the flagella reverse, setting the bacterium into a tumbling action that eventually gets it away from the irritant.

membrane, are thus moved apart. Next, the membrane extends inward, and a new cell wall is laid down between the two daughter cells (see Figure 22.22). Again, the mesosome may or may not play a role.

The rate of growth and cell division in bacteria can be phenomenal. Under ideal conditions, our own colon bacterium, *Escherichia coli,* can double its numbers every 20 minutes. At this rate, 72 generations could form in just one day! (That's 4.7×10^{21} or 4.7 sextrillion bacteria, literally millions of pounds.) You may be relieved to know this never happens. After a few hours of such unrestricted growth, the colony runs out of food, and its wastes may poison its environment. In the case of a pathogen, such unrestricted growth is usually prevented by the host's immune defenses.

On rare occasions, prokaryotes undergo a simple kind of sexual process called **conjugation**, during which a one-way transfer of genes may occur (see Chapter 15). This requires the aid of a sex pilus, which draws the cells together and permits the transfer to occur. While conjugation is commonly observed in highly selected laboratory strains, not much is known about this sexual exchange under natural conditions. (Other mechanisms of gene transfer among bacteria were discussed in Chapter 15.)

Metabolic Diversity in Bacteria

Many bacteria are heterotrophs ("other feeding"); thus their nutrients must come from other organisms. Among the heterotrophs are the **decomposers** and **parasites**. The first group

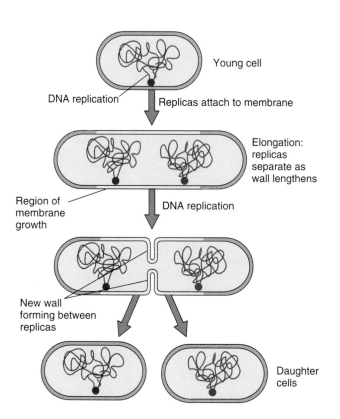

Young cell — DNA replication — Replicas attach to membrane — Elongation: replicas separate as wall lengthens — Region of membrane growth — DNA replication — New wall forming between replicas — Daughter cells

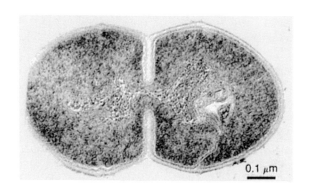

0.1 μm

FIGURE 22.22
Fission: Asexual Reproduction in Bacteria
Cell division in bacteria occurs through fission. Note the attachment of chromosome replicas to the cell membrane, the membrane's inward growth, and the synthesis of a new wall. Some bacteria form mesosomes after replication. These membranous complexes attach to the newly replicated chromosomes and aid with separation.

brings about the decay of dead organisms and their wastes. The second group, often referred to as **pathogens**, causes many of the diseases of plants and animals, including humans. Several groups of bacteria are autotrophs ("self-feeding"), living either as *photoautotrophs* or *chemoautotrophs*. The **photoautotrophs** synthesize their essential molecules from simple inorganic molecules, using the energy of light to drive the reactions. The **chemoautotrophs** make use of simple inorganic molecules from the earth's crust, deriving their energy from the exergonic chemical reactions they bring about.

Bacteria also vary in other aspects of their metabolism, such as the use of oxygen. As we've seen, some live as aerobes, while others are anaerobes. Those that must have oxygen are called **obligate aerobes** (they're "obligated" to use oxygen), whereas those that cannot live in the presence of oxygen are aptly called **obligate anaerobes**. Anaerobes that function with or without oxygen are called **facultative anaerobes**. (For an in-depth review, see Chapter 8.)

A Word About Bacterial Energetics. Aerobic (oxygen-using), heterotrophic bacteria carry on cell respiration in a manner surprisingly similar to what we see happening in the mitochondrion (see Chapter 8). In fact, according to the serial endosymbiosis hypothesis, mitochondria are derived from aerobic bacteria that invaded or were phagocytized by primordial cells (see Essay 4.2).

Bacterial oxidative respiration utilizes high-energy electrons from foods oxidized in the cytoplasm. The electrons are sent through electron transport systems in the plasma membrane, powering proton pumps that eject the protons from the cell. As a result, the protons establish a steep chemiosmotic gradient between the immediate surroundings and the cytoplasm, representing substantial potential energy for the cell. The membrane also contains the familiar ATP synthases (Chapters 7 and 8), where ATP is synthesized as the protons pass through (Figure 22.23).

Bacterial Heterotrophs. Bacterial heterotrophs are so diverse that there is hardly an organic molecule that some bacterium cannot use. We'll consider the decomposers as an example of their diversity.

Most heterotrophic bacteria gain their energy through decomposition or decay. They secrete enzymes that cause the breakdown of organic matter in dead organisms and their wastes, a practice that has earned them an alternative name, *saprobe* (*sapros* from Greek, meaning "rotten" or "putrid"). However offensive it might seem to humans, this activity is ecologically crucial and is undoubtedly the most important of all bacterial activities. Decomposition releases such key complex ions as nitrates, phosphates, and sulfates, which then cycle back to plants and algae. In turn, plants and algae bring solar energy into the living realm, forming the energy base for all life. (Now *that's* important!) Without decomposers, life would be seriously curtailed. Sometimes its hard to maintain our objectivity and

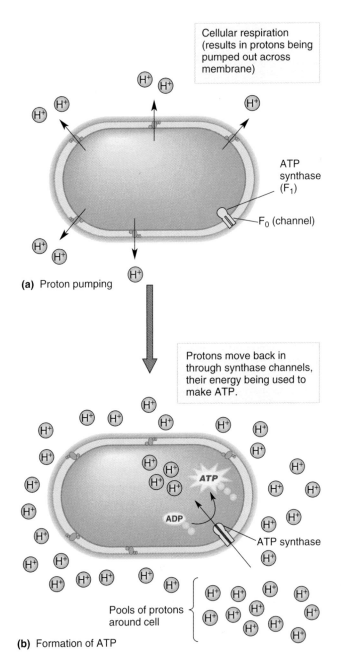

(a) Proton pumping

(b) Formation of ATP

Pools of protons around cell

FIGURE 22.23
Cell Respiration and Chemiosmosis in Eubacteria
Aerobic eubacteria make use of a steep chemiosmotic gradient to form ATP. **(a)** Protons gained through the oxidation of fuels are pumped from the cell, forming a higher concentration outside. **(b)** The protons are admitted back into the cell via ATP synthases, where the energy of the gradient is used to make ATP.

keep this "big picture" in mind—especially when decomposers get to tomorrow's roast beef dinner before we do.

Inhibiting Decomposers. We go to a lot of trouble to keep food-spoiling decomposers from getting to our food and other

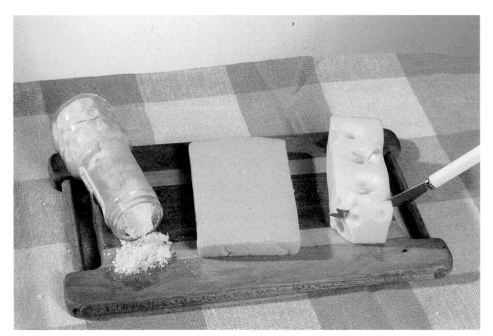

FIGURE 22.24
Microbial Delights
Parmesan, Cheddar, and Swiss cheeses owe their distinctive flavors and other characteristics to specific bacterial agents. The holes in Swiss cheese are naturally produced by trapped gases from the same propionic acid bacteria that are responsible for the tangy flavor.

goods. In general, the optimal growth conditions for decomposers include an appropriate *food* supply (carbohydrates, lipids, and protein—our foods do nicely), *warmth, moisture, and adequate* oxygen *for the aerobes or its absence for the anaerobes. Importantly, an abundance of oxygen will retard the growth of anaerobes, some of which are serious pathogens as well as food-spoilers.*

This information naturally suggests ways to inhibit the growth of the decomposers. Among the most common means are cooling, freezing, drying, salting, and sugaring. Whereas cooling slows all metabolic activity, the other methods remove water from the food or set up osmotic conditions that dehydrate bacterial cells. Of course, we can begin by killing the bacterial cells that already exist on the foodstuffs. This is done through sterilization, which involves the use of intense heat, gases such as ethylene oxide, and even irradiation. In canning, food is heat-sterilized and sealed in airtight containers. Pasteurization, a far gentler process involving limited heating, is usually reserved for milk and other fragile foods (no one wants boiled milk). It is intended to kill pathogens. Many decomposers, especially spore-formers, survive pasteurization, and even an unopened container of milk soon spoils. Chemical preservation, common in the food industry (just read the labels), includes the use of small amounts of sorbic acid, sodium nitrate, calcium propionate, or other chemicals to retard bacterial growth.

Economically Useful Decomposers. Before we continue our long litany of accusations against bacteria, we are compelled to note that they can be quite useful. For example, we owe the tartness of pickles and sauerkraut to the metabolic wastes of lactic acid bacteria. Still other bacteria provide flavor to yogurt, buttermilk, sour cream, and cheeses such as Parmesan, Cheddar, and Swiss (Figure 22.24).

Bacteria are also used commercially to generate enzymes such as amylase, the starch-splitter used in brewing, and those used in laundry detergents and pharmaceuticals. Additional harvested bacterial products include amino acids, hormones, vitamins, and antibiotics.

Let's not forget that bacteria are important to genetic engineering (see Chapter 16). On a more personal level, the vast populations of *E. coli* wriggling in your bowel this very minute help make vitamin K (a blood-clotting aid) available. This is an example of *mutualism*, where both organisms—ourselves and our colon bacteria—benefit by the relationship. Before someone asks—yes, this is the same species that causes occasional outbreaks of deadly food poisoning; at least, certain strains are known to do so. Actually, *E. coli* has always been potentially infectious, especially when introduced into the wrong places, but new strains finding their way into our foods are gaining notoriety. They also give us one more reason for thoroughly cooking our meat.

Finally, by their very presence, the bacterial flora living harmlessly on and in our bodies compete with would-be pathogens, helping to restrict their success. Some even produce an antibiotic effect, chemically retarding pathogen growth.

Pathogens of Animals. Many bacteria cause disease in animals, ourselves included. Pathogens occur in each of the three bacterial forms. Among the bacilli, for example, are agents of

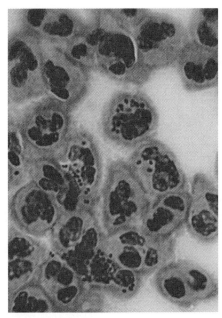

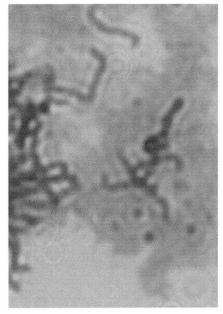

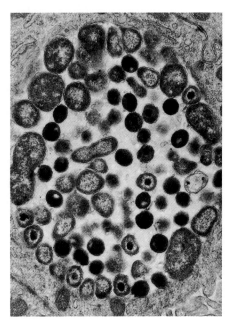

(a) Agent of gonorrhea   (b) Agent of syphilis   (c) Agent of chlamydia

FIGURE 22.25
Agents of Sexually Transmitted Disease
Two common agents of sexually transmitted disease are (a) *Neisseria gonorrhoeae* and (b) *Treponema pallidum,* the agents of gonorrhea and syphilis, respectively. Gonorrheal bacteria are diplococci, occurring in pairs with a capsule. The agent of syphilis is a spirochete. (c) The agent of chlamydia, *Chlamydia trachomatis,* is unlike other bacteria although theorists believe it evolved from gram-negative bacteria. Chlamydia is an obligate parasite because it cannot generate its own ATP.

such dread diseases as leprosy, typhus, black plague, diphtheria, and tuberculosis.

Over the years, medicine has responded to the threat of pathogens through intensive campaigns, and the response has been notably successful. In spite of such progress, though, we are still threatened by some bacteria. You may have read recently that tuberculosis is now on the rise. And, even as you read this, thousands of infants are now dying from a disease called *salmonellosis*. You won't see much about this on the evening news because it's not a big problem in economically advanced nations. But in others, a lack of public sanitation and clean water condemn millions of infants to death from the disease each year.

Other problem bacilli include the agents of *botulism* and *tetanus*. The first is an extremely dangerous form of food poisoning and the second is an infection that occurs in deep puncture wounds. The two organisms responsible, *Clostridium botulinum* and *Clostridium tetani,* are similar in that they require anaerobic conditions for growth, their spores are highly resistant to heat and other standard food preparation treatments, and they secrete the most potent toxins known.

Coccoid pathogens include certain staphylococci, clusters of spherical bacteria that cause minor skin infections, including boils and pimples. Under certain conditions, these relative-

ly harmless bacteria can bloom into enormously dangerous mass infections. Streptococci (chains) cause scarlet fever, rheumatic heart disease, and rheumatic nephritis, which are still significant causes of death.

Sexually Transmitted Diseases (STDs). In a class of their own, these days, are the bacterial **STDs**—the sexually transmitted diseases. (STDs also include viral diseases such as AIDS and herpes.) The major bacterial STDs are gonorrhea, syphilis, and chlamydia. The first two ran rampant in the World War II era but, thanks to the development of penicillin, were brought well under control. Today they are again reaching epidemic proportions, mostly because of antibiotic-resistant strains, greater sexual promiscuity, and ignorance of the problem.

The agent of gonorrhea is *Neisseria gonorrhoeae,* a diplococcoid organism (Figure 22.25a). Signs of its presence are readily apparent in men (pus discharge from the penis and painful urination) but are not readily apparent in women. Because of the latter fact, gonorrhea in women can quickly bloom into **pelvic inflammatory disease** (**PID**), a serious problem that can lead to sterility. The gonorrhea organism can be passed along to infants during the birth process and if undetected can cause blindness and other problems.

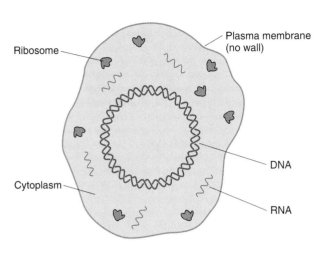

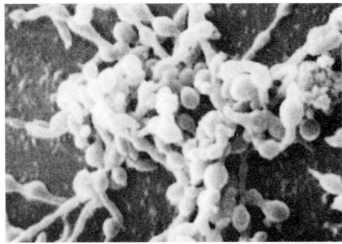

Pneumonia organism

FIGURE 22.26
Mycoplasmas
Mycoplasmas are the smallest known cells. Although they are prokaryotes, they lack a cell wall. The mycoplasma in the photograph causes a form of human pneumonia.

The agent of syphilis is a spiral-shaped organism known as *Treponema pallidum* (Figure 22.25b). The symptoms of syphilis, occurring in three distinct stages, are far more complex than those of gonorrhea. In primary syphilis, the chief symptom is the appearance of chancres—small ulcerated sores on the genitals or mouth. They soon subside, to the relief of the concerned host, but the respite is temporary. In secondary syphilis, widespread rashes with pimple-like eruptions are accompanied by flu-like symptoms, kidney damage, and hair loss. They too subside after a time, but may reoccur at intervals. The tertiary form of syphilis is devastating. Included are widespread tumors, pain in the legs and trunk, and, as the organisms invade the central nervous system, severe mental deterioration. Often, the weakened walls of the great arteries rupture and the sufferer dies from massive hemorrhage. Sadly, the syphilis organism can cross the placenta and infect the fetus, producing serious birth defects.

Chlamydia, a newly important, rapidly spreading STD, is caused by the bacterium *Chlamydia trachomatis* (Figure 22.25c). The symptoms in men are painful urination, penile discharge, and painfully swollen testicles, this last symptom distinguishing it from gonorrhea. Women experience vaginal discharge, cervical irritation, and pelvic pain. The pain is most intense during menstruation and intercourse. If untreated in women, chlamydia also leads to PID and all its accompanying complications.

Mycoplasmas: The Smallest Cells. Before we move along, let's consider what are perhaps the oddest pathogens of all, the **mycoplasmas**. Mycoplasmas are among the smallest living things, with the cells of some species being less than 0.16 µm in diameter—about one-sixth the length of the smallest bacteria (Figure 22.26). In fact, they are smaller than some viruses. Oddly, they lack a rigid cell wall and therefore have no definite shape.

Mycoplasmas are parasites of plants and animals, including humans. One form is responsible for a relatively mild form of human pneumonia, another is implicated in urinary tract infections, and a third may cause a form of arthritis. Many plant diseases formerly attributed to viruses are now known to be caused by mycoplasmas.

Reproduction in mycoplasmas is also unusual. The cells join into filaments in which spherical bodies form. They are released when the filaments fragment, forming the next generation of cells.

Pathogens of Plants. Most plant diseases that make the evening news are viral or fungal in origin, but bacterial diseases also occur. Included are such earthy-sounding names as blights, wilts, galls, and rots. The effects range from simple discoloration of fruit and leaves to plant death. Virtually all bacterial plant pathogens are bacilli, and most are flagellated.

Blight is heralded by necrotic (dead or dying) tissue on flowers, leaves, and stems. Fire blight quickly kills young pear and apple trees. The drooping of leaves and stems suggests what is happening—**wilt**. Wilt bacteria invade the vascular system, hindering the movement of water, minerals, and food in the plant. Wilts may threaten such significant crops as alfalfa, beans, squash, and watermelons, each of which is invaded by a different species. *Gall* is a form of plant tumor.

(a) Crown gall in tree

FIGURE 22.27
Bacterial Disease In Plants
(a) Crown gall is a tumorous growth brought on by the insertion of DNA from the bacillus *into* some of the plant's chromosomes. The plant responds through rapid cell division, forming the tumor.
(b) The agent of crown gall, *Agrobacterium tumefaciens,* is useful in genetic engineering experiments as a plasmid vector.

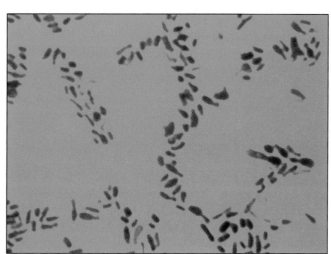

(a) Bacterial agent of crown gall

Crown gall, among the most common, is initiated by *Agrobacterium tumefaciens.* While not generally fatal, gall is quite disfiguring (Figure 22.27). The tumors have an interesting origin. Plant pathologists have known for some time that the afflicted cells become genetically transformed when bits of bacterial DNA are incorporated into their chromosomes. The cells then divide out of control, a typical cancerous response. You may recall our mentioning this organism as an important gene vector in genetic engineering (Chapter 16).

Bacterial Autotrophs. Photoautotrophs, those autotrophs that utilize sunlight energy in photosynthesis, are ecologically important organisms. As part of the plankton (floating aquatic organisms), bacterial photoautotrophs, along with algae, make up the energy base of ocean and freshwater food chains. They also give off oxygen, which supports aerobic life.

Cyanobacteria. The best known of the photoautotrophic or photosynthetic bacteria are the **cyanobacteria**, a widely distributed group containing some 2500 species. They live in aquatic environments including oceans, ponds, lakes, tidal flats, moist soil, fish tanks, swimming pools (where they are called "algae"), and even around leaky faucets. Some, like archaebacteria, flourish in hot springs where temperatures reach 75°C (167°F). Some live symbiotically with fungi as lichens. Many cyanobacteria have blue-green color and were formerly called "blue-green algae," a doubly erroneous name since they aren't algae and they can be red, green, and black as well as blue-green. Cyanobacteria abound in the marine environment, where they release significant amounts of oxygen and are used for food by many small marine organisms.

Cyanobacteria occur as single cells, in long chains or filaments (Figure 22.28), and in gelatinous masses. Some

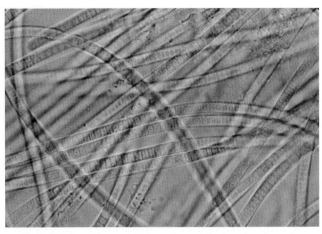

(b) *Oscillatoria*

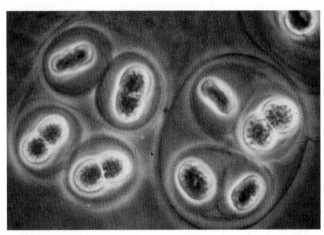

(c) *Gloeocapsa*

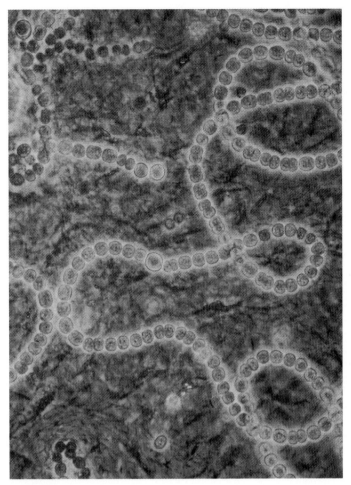

(a) *Nostoc*

FIGURE 22.28
Three Cyanobacteria
A few of the diverse cyanobacteria are **(a)** the filamentous, bead-like *Nostoc,* **(b)** the slowly undulating, filamentous *Oscillatoria,* and **(c)** the spherical *Gloeocapsa,* enclosed in its gelatinous sheath. All are inhabitants of fresh water. The larger, thick-walled cells of *Nostoc* shown in **(a)** are known as heterocysts, cells that specialize in nitrogen fixation.

cyanobacteria can move: filamentous forms such as *Oscillatoria* rotate in a screw-like manner, while the gelatinous forms glide along in a mucus-like slime they produce. Reproduction is asexual, through binary fission.

The cells of cyanobacteria can be surprisingly complex (Figure 22.29). Their chlorophyll is integrated into thylakoids, making the whole cell comparable to a eukaryotic chloroplast. Biochemically, photosynthesis in the cyanobacteria is nearly identical to that of algae and plants and very different from that in archaebacteria. Such characteristics make these complex organisms a likely predecessor to modern plants, or at least to modern chloroplasts.

Some filamentous cyanobacteria produce specialized, nitrogen-fixing cells called **heterocysts** (see Figure 22.28a). Their role is to incorporate atmospheric nitrogen (N_2) into

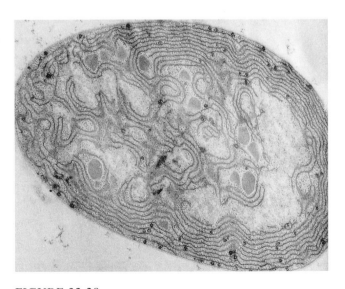

FIGURE 22.29
Cyanobacterial structure
Cyanobacteria are unusual prokaryotes in that they have internal membranous organelles. Taking up most of the cell in this EM view are the thylakoids, chlorophyll-containing extensions of the plasma membrane.

Essay 22.1 ● THE NITROGEN FIXATION PROCESS

The chemical process of nitrogen fixation can briefly be summarized by the simple formula:

$$N_2 + 3\,H_2 \longrightarrow 2\,NH_3\ \text{(ammonia molecule)}$$
$$NH_3 + H_2O \longrightarrow NH_4^+ + OH^-$$
(ammonium and hydroxyl ions)

But such simplicity is truly deceptive. The atoms of nitrogen gas (N_2) are held together by three extremely stable covalent bonds, and so molecular nitrogen does not readily enter into chemical reactions. For instance, consider the widely used Haber industrial process of nitrogen reduction, in which synthetic ammonia fertilizer is produced. This process uses *iron as a catalyst, heat exceeding 500°C, and pressure exceeding 300 atmospheres.* All this is needed just to pry the nitrogen molecule apart so that its fragments can be reduced by hydrogen. So how do the tiny nitrogen-fixing bacteria overcome these seemingly impossible requirements?

Bacteria fix nitrogen through the use of a two-part enzyme complex, one of which is the iron-containing protein reductase, and the other an enzyme known simply as nitrogenase. The reactions converting nitrogen gas to ammonium ions occur in two main stages, summarized in the illustration. Here we see that electrons from the respiratory chain, along with energy from ATP, provide the reducing power. Reductase then passes its electrons to nitrogenase, and this potent enzyme then reduces atmospheric nitrogen, forming the desired ammonium ions. For the record, the source of electrons in photosyn-

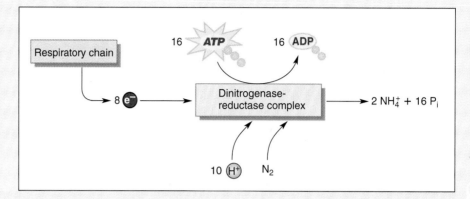

thetic nitrogen fixers (cyanobacteria) is the light-activated electron transport system. In chemoautotrophs (soil bacteria), it is the electron transport system of cell respiration.

How expensive is the process in terms of ATP? The nitrogen-fixing bacteria in the roots of the common garden pea use fully *one-fifth of the ATP produced by the plant.* Each ammonia molecule (or ammonium ion) produced requires 9 ATP molecules (some authorities estimate 12).

There is an interesting epilogue to this story, having to do with genetic engineering and the nitrogenase enzyme complex. It is clear that the industrial reduction of nitrogen for chemical fertilizer is an energy-costly process. But what are the alternatives in a hungry world? Why not turn crop plants into nitrogen fixers? That is, why not use recombinant DNA technology to provide corn, wheat, and rice with their own nitrogenase

coding genes? Result: no more chemical fertilizers needed for these major crops!

Geneticists have been trying to do just that. The nitrogenase genes, present in the nitrogen-fixing bacterium *Klebsiella pneumoniae,* have been excised and spliced into an *E. coli* plasmid. The engineered plasmid was then reintroduced into colonies of the common colon bacterium, and some nitrogen-fixing activity was detected. The next step, a formidable one and one that has not been accomplished, is to get the plasmid to work in plant cells. There are serious problems to solve. One is that the enzyme is readily inactivated by oxygen, so its surroundings must be kept anaerobic. Another drawback is the heavy ATP requirement mentioned earlier. Such a demand might not be easy for corn or wheat to fulfill. Work today is stalled around these problems.

ammonium ions (NH_4^+), a form plants can use for producing amino acids and other nitrogen-containing molecules. Interestingly, the formation of nitrogen-fixing heterocysts is inhibited in many species when alternate sources of nitrogen—ammonia or nitrates—are available. Nitrogen fixation, by the way, also occurs in heterotrophic bacteria, those that live mutualistically in the roots of leguminous plants (peas, beans, alfalfa, and others). The relationship is considered mutualistic because the plant makes use of surplus ammonium ions and the bacterium uses sugars made by the plant. Essay 22.1 takes us further into the nitrogen-fixing process.

Chemoautotrophs. Chemoautotrophs obtain their energy from the oxidation (removal of electrons or hydrogen) of simple

inorganic compounds in the earth's crust. Much of this energy is then used in molecule building, and as with many photoautotrophs, the usual starting molecule is carbon dioxide.

One group of chemoautotrophs obtains its energy through the oxidation of sulfur or sulfur compounds. Among these are the myxobacteria—gliding bacteria that oxidize hydrogen sulfide (H_2S), using the hydrogen to form carbohydrates and storing the sulfur in granules in their cells. One sulfur-using group produces sulfuric acid, thus creating a highly acidic soil.

Other chemoautotrophs get their energy by oxidizing iron and manganese, where they occur in the reduced state, while others make use of hydrogen gas. (For a look at how one hydrogen-sulfide-utilizing bacterium provides a source of energy for the strange organisms of the great oceanic rifts, see Essay 47.3).

Key Ideas

VIRUSES

1. Viruses are minute parasitic particles consisting of a nucleic acid **core**, a surrounding protein **capsid**, and, in some, a phospholipid **envelope**. They range in size from that of large molecules to the smallest bacteria and are responsible for many human, animal, and plant diseases.

2. A favored hypothesis is that viruses originated as bits of nucleic acid from cells. When taken in by cells, they evolved protein coverings and envelopes. Genetically, viruses are more closely related to their own hosts than to each other.

3. Pasteur and Koch had developed vaccines against viral diseases long before anyone knew what viruses were. Iwanowski experimented with the **tobacco mosaic virus** in the late 1800s, transferring a virulent filtrate from plant to plant. Since the virulent material increased while in the plant it was thought to be alive. Wendell Stanley identified tobacco mosaic virus crystals in 1935.

4. Each virus particle or **virion** contains a nucleic acid core and a protein coat or **capsid**. They may be helical, polyhedral, or cubic and brick-shaped (smallpox). Shape depends on the arrangement of coat proteins or **capsomeres**. Some have surrounding envelopes. Viruses may have double- or single-stranded DNA or RNA.

5. **Bacteriophages** bind to the host cell via tail fibers, dissolve an opening, and inject the DNA genome. **Lytic** phages disrupt the host chromosome, generate many new virions, and lyse the cell. **Temperate bacteriophages** may enter a **lysogenic state**, where the viral DNA inserts into a host chromosome as a **prophage**. The prophage replicates along with the host DNA, infecting a large clone population. When activated, it enters the lytic state.

6. Animal viruses follow seven steps in their invasion:

   a. *Binding.* Capsid or envelope molecules bind to host surface molecules.

   b. *Penetration.* Naked viruses are taken in by phagocytosis, but enveloped viruses fuse with membrane, releasing the capsid inside.

   c. *Uncoating.* Lysosomes digest the protein capsid, releasing the nucleic acid core.

   d. *Transcription and replication.* DNA viruses follow the usual transcription process. **Plus strand** RNA viruses use their genome as mRNA. To reproduce they make RNA polymerase in order to replicate new plus strands. **Minus strand** RNA viruses use their RNA polymerase to form mRNA. Their mRNA is used to replicate new minus strand RNA for reproduction. Retroviruses use reverse transcriptase to convert their single-stranded RNA genome to double-stranded DNA, which is then inserted into the host chromosome.

   e. *Assembly and release.* Naked viruses assemble new genomes into capsids and escape by lysis or exocytosis. Enveloped viruses modify the host membrane and, after assembling their genomes into capsids, bud from the cell. Portions of the membrane become their new envelopes.

7. Plant viruses are transmitted by insects and across damaged surfaces. Many crops are susceptible. Their genomes may consist of fragmented DNA or RNA.

8. **HIV** (**human immunodeficiency virus**) infects **helper T-cells**, a type of lymphocyte, thus crippling the immune system and bringing on the fatal condition called **AIDS** (**acquired immunodeficiency syndrome**). HIV is spherical and enveloped, and its genome is single-stranded RNA. It enters the host cell at receptor sites matching those of its glycoprotein envelope. Following a transformation to double-stranded DNA, it inserts into a host chromosome, where it may remain dormant for as long as ten years. When activated, it transcribes the viral genome, the host cell is lysed, and infectious virions are released. Transmission is through blood, semen, vaginal secretions, and mother's milk.

9. **Viroids** are short fragments of RNA that, when in cells, cause metabolic disorders characteristic of some plant and animal diseases. They spread in the same way as plant viruses.

10. **Prions** are protein shells that cause several diseases in humans. They somehow penetrate the host's genome, which responds by synthesizing new prions.

KINGDOM MONERA: THE PROKARYOTES

1. Prokaryotes make up the kingdom Monera. They lack the membranous organelles and histone protein of the eukaryotes.

2. Recent revisions divide the older kingdom Monera into the kingdoms **Archaebacteria** and **Eubacteria**. Differences between the two include cell wall content, membrane lipids, ribosomal RNA, and photosynthetic pigments.

3. Archaebacteria include, (a) **methanogens**—anaerobic methane generators that live in marshes, in lake and sea beds, and in the animal colon; (b) **halophiles**, **thermophiles**, and **thermoacidophiles**—bacteria that live in dense salt concentrations, very hot water, and hot, acidic conditions, respectively; and (c) photosynthetic archaebacteria—those that make use of **bacteriorhodopsin** to harness light energy and pump protons.

4. Bacteria occur as rods (**bacillus**), spheres (**coccus**), spirals (**spirochaete** and **spirillum**), comma-shapes (**vibrio**), and strands (**filamentous**). Cells are arranged in singles (**mono-**), doubles (**diplo-**), chains (**strepto-**), clusters (**staphylo-**), and groups of eight (**sarcinae**).

5. Some eubacteria have thylakoids and **mesosomes**. Others form cellular extensions called **pili** that aid in attachment and DNA transfer.

6. The eubacteria genome is a single, circular, protein-free chromosome. Some bacteria have additional circles of DNA called **plasmids**.

7. Cell walls are composed of **peptidoglycan** in most eubacteria and protein in archaebacteria. In gram-positive cells, the peptidoglycan layer is thick and contains **teichoic acid**, but in gram-negative types, it is thin and has a complex, overlying **outer membrane**. **Gram-positive** bacteria are often susceptible to penicillin, whereas **gram-negative** bacteria usually are not. Penicillin inhibits cell wall synthesis.

8. A few bacteria form carbohydrate or polypeptide **capsules** that help the cell resist phagocytosis.

9. Soil and water bacteria form **endospores**, highly resistant, thick-walled dehydrated survival structures.

10. The bacterial flagellum is a coiled S shape, is solid and rotating, and is made up of **flagellin**. **Gliding bacteria** use a slippery mucilage in movement. One group, the myxobacteria, develop upright spore-forming bodies.

11. Asexual reproduction occurs through **fission**, which includes DNA replication, replica separation, inward extension of the membrane, and daughter cell separation. A life cycle may occur in as few as 20 minutes. Sexual reproduction is rare. One form, **conjugation**, includes a one-way transfer of DNA, aided by the **sex pilus**.

12. Heterotrophic eubacteria include **decomposers** and **parasites** or **pathogens**. **Photoautotrophs** and **chemoautotrophs** synthesize food from inorganic molecules and many are important ecologically since they bring energy into the aquatic environment. **Obligate aerobes** require oxygen; **obligate anaerobes** are poisoned by oxygen, while **facultative anaerobes** tolerate it.

13. Aerobic bacteria utilize electron transport systems to build a chemiosmotic proton gradient, which is then utilized in generating ATP. The gradient also powers flagellar movement and active transport.

14. Because they break down dead organisms, **decomposers** (**saprobes**) are ecologically important as recyclers of essential elements such as sulfur and nitrogen. Decomposition can be inhibited by altering growth requirements. Methods include cooling, drying, heating, and using chemical agents. Some decomposers are used in the production of foods and other products. Bowel species of *E. coli* secrete vitamin K, but some *E. coli* strains are serious pathogens.

15. Animal pathogens of importance today include the agents of the intestinal disease salmonellosis and agents of botulism, tetanus, gas gangrene, and strep throat. Important sexually transmitted bacterial diseases include gonorrhea, syphilis, and chlamydia. Mycoplasmas, the smallest of cells, cause several diseases in humans.

16. Plant pathogens cause discoloring blights, deadly wilts, and disfiguring galls.

17. Cyanobacteria occur as single cells, filaments, and masses. Nitrogen fixers reduce atmospheric nitrogen, forming useful nitrogen compounds. Nitrogen fixation is carried out by free-living water bacteria and by soil bacteria that live as mutualists in the roots of leguminous plants.

18. Chemoautotrophs obtain energy through the oxidation of simple inorganic compounds such as those of sulfur or iron.

Application of Ideas

1. There are strong arguments against the liberal use of antibiotics. Numerous kinds of bacteria live on and inside the body, often in an innocuous manner. Many are heterotrophs while others are marginal parasites, and still others are a constant threat as virulent parasites. What, if anything, does the frequent use of antibiotics have to do with these normal populations? Why not kill all the bacteria possible?

2. Despite the great care taken in the commercial canning of foods, occasional cans of spoiled food appear in supermarkets. A common form of spoilage can be identified by a general bulging of the can. What causes the bulging? Suggest ways in which bacteria might survive the canning process. What kind of bacteria might live in a sealed environment?

3. You are in remote backcountry in your spanking new 46-foot self-contained motor home when you become mired in the mud. Adding to your problems is a hole punched in the fuel tank—you've lost your gas, and your electrical generator is useless. It will be at least a week before help arrives. You've plenty of fresh fruits, vegetables, milk, and meats, but they are quickly warming up in the fridge. Fortunately, you read about bacterial growth in your introductory biology course. Describe the steps you will take to keep your food from spoiling. Include at least three methods.

Review Questions

1. What, precisely, are viruses? How does your answer explain why viruses require a living host? (pages 414–415)

2. List the four general shapes taken by viruses. What determines the geometry? (pages 417–418)

3. List four variations in the nucleic acids occurring in viruses. (page 417)

4. Briefly summarize the events in the lytic and lysogenic states of the bacteriophage. (page 418)

5. List the seven steps followed by most animal viruses in the infection of a host cell. What are two ways in which animal viruses enter the host cell? (pages 418–421)

6. Compare the way plus strand RNA viruses and minus strand RNA viruses begin translation in a host cell. (page 420)

7. Compare the manner in which naked and enveloped viruses exit from the host cell. (page 421)

8. How do plant viruses enter the plant tissues and how do they get from plant cell to plant cell? (pages 421–422)

9. Why are rare and exotic diseases a sign of HIV infection? (page 422)

10. Prepare a drawing of an HIV virion. Label the envelope, glycoprotein, protein capsid, single-stranded RNA, and reverse transcriptase. (page 423)

11. What specific cells does HIV tend to attack? What, in its active phase, does it do to cells? (page 423)

12. What is the difference between a prion and a viroid? What is their medical importance? (page 425)

13. Using simple drawings, illustrate the four shapes of bacterial cells. Give their technical names. (pages 428–429)

14. Prepare a simple line drawing of a bacterium, adding and labeling the following: cell wall, outer membrane, capsule, plasma membrane, flagellum, chromosome, thylakoid, and mesosome. (page 429)

15. Describe an endospore. Of what adaptive value is the endospore to the bacterium? (page 430)

16. Distinguish between anaerobe and aerobe, between obligate anaerobe and obligate aerobe. Which of the latter two would most likely employ an electron transport system in its respiratory activity? Explain. (page 434)

17. Using a diagram, show how bacteria create and make use of a proton gradient to produce ATP. (page 434)

18. Why are gonorrhea and syphilis increasing today when they were well under control 40 years ago? (pages 436–437)

19. What is the basic difference between the photoautotroph and the chemoautotroph? (pages 438–440)

CHAPTER 23

Protists

CHAPTER OUTLINE

CHARACTERISTICS OF PROTISTS

Organization

Nutrition

Movement

Reproduction

Life Cycles in Protists and Other Eukaryotes

Evolutionary Origins of Protists

Phylogeny of Protists

ANIMAL-LIKE PROTISTS: PROTOZOANS

Flagellates

Amebas

Sporozoans

Ciliates

FUNGUS-LIKE PROTISTS

Slime Molds

Oomycetes: Downy Mildews and Water Molds

PLANT-LIKE PROTISTS: THE ALGAE

Pyrrhophytes: "Fire Algae"

Euglenophytes

Chrysophytes: Yellow-Green Algae, Golden Brown Algae, and Diatoms

Rhodophytes: Red Algae

Phaeophytes: Brown Algae

Chlorophytes: Green Algae

KEY IDEAS

APPLICATION OF IDEAS

REVIEW QUESTIONS

*I*n dimensions that fall between those of bacteria and the smallest of animals and plants, there exists a fascinating realm of life so diverse in its makeup that we are at a loss to find enough common ground for a satisfying description. A few drops of pond water, especially one including nutrient-rich bottom sediments, seen under the light microscope, can quickly confirm the existence of the microscopic eukaryotes that make up much of **Kingdom Protista**. While protists are chiefly creatures of the earth's waters, places where their numbers may become astronomical, some have found ways to exist in the soil and on the trunks of trees, while others hide out in the bodies of plants and animals. But not all protists are microscopic. In fact, as we'll see, Kingdom Protista includes several groups that are primarily multicellular, and one group spawns giants, albeit simple ones.

FIGURE 23.1
Antony van Leeuwenhoek
(**a**) The elegantly coiffured gentleman is Antony van Leeuwenhoek (about 1670). (**b**) The device is a microscope of his own design. The specimen was placed on the apparatus at ①, brought into position vertically by turning the lower screw ②, and moved toward or away from the lens ③ by turning the shorter screw ④. (**c**) The magnification and resolving power of van Leeuwenhoek's crude microscope are surprisingly good.

(a) Antony van Leeuwenhoek

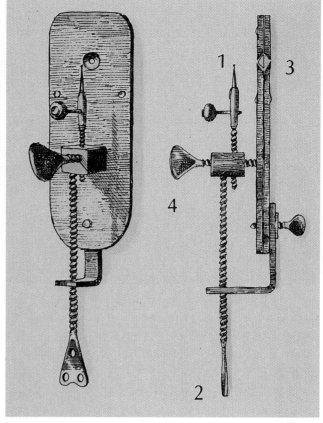

(b) Leeuwenhoek's microscope

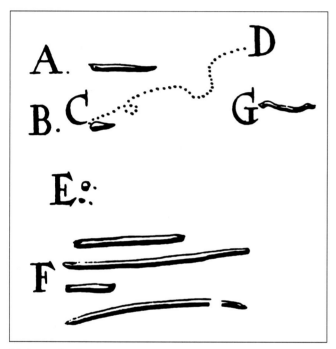

(c) A sample of van Leeuwenhoek's many drawings

that the latter occurred without mating—a form of asexual reproduction called *parthenogenesis.*)

As you will see, our familiarity with protists has increased enormously since Leeuwenhoek's day. In fact, today we know of more than 100,000 living species, and of 35,000 more by their fossils. We'll begin with a look at the diverse characteristics of protists.

Characteristics of Protists

Organization

Although most protists are clearly **unicellular** (single-cell) organisms, others form colonies of cells, and still others are multicellular (Figure 23.2). A **colony** is a grouping of cells in which members become specialized for tasks such as locomotion, feeding, and reproduction. Such specialization may improve the colony's efficiency, but the actual degree of interdependence is limited. In fact, each cell can generally get along on its own if circumstances require it. In **multicellular** organisms, the cells are interdependent. If they are experimentally separated, they generally do not survive. In some situations, separated cells, if given the opportunity, will reaggregate, resuming their former organization. Multicellular protists such

It is downright peculiar that the first person to report having seen protists was a 17th century Dutch cloth merchant, but this person had an insatiable curiosity and a penchant for grinding lenses. His name was Antony van Leeuwenhoek. Leeuwenhoek built his own microscope, and although it contained only one lens and was very unrefined by today's standards (Figure 23.1), it enabled him to make some amazing discoveries. He described certain minute organisms that could only have been bacteria. Their actual existence was not verified by professionals for many years. (If you've had the opportunity to focus a modern light microscope on living bacteria you can better appreciate Leeuwenhoek's effort.) In addition to his many writings on protists, which he called "cavorting beasties," this skilled observer made original observations of mites, red blood cells, human sperm, muscle structure, and aphid reproduction. (He accurately concluded

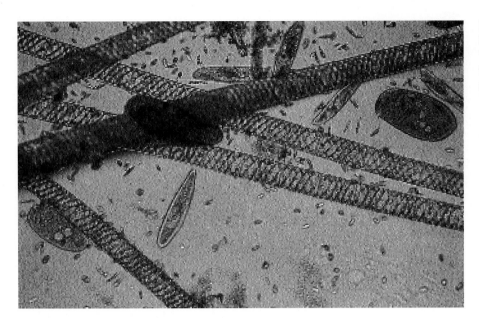

FIGURE 23.2
Protists
This mix of pond protists, viewed through the light microscope, gives you a small sample of this highly diverse kingdom.

as the seaweeds even exhibit a tissue level of organization (tissues are groups of similar cells sharing a function).

Nutrition

Protists practice all common modes of feeding and nutrition. Many are heterotrophic, requiring complex organic substances such as amino acids, carbohydrates, and fats. Some heterotrophs simply absorb nutrients from their surroundings, whereas others, as predators, attack and devour living prey (Figure 23.2). In addition, there are a number of parasites that obtain their nourishment to the detriment of their larger eukaryotic hosts. Some of the parasites are devastating to humans. Other protists are photoautotrophs; like plants, they use the sun's energy to drive photosynthetic reactions (Figure 23.3b).

Movement

Movement, where it occurs, is through the action of cilia or flagella or through ameboid movement. The cilia and flagella are typically eukaryotic, their action based on sliding microtubules. Ameboid movement occurs through the flow of cytoplasmic extensions called pseudopods ("false feet"). Both types of movement are described in Chapter 4.

FIGURE 23.3
Protistan Nutrition
(a) The protist sporting two rings of cilia is *Didinium*, a predatory species *(below)*. It is engulfing *Paramecium*, another ciliate *(above)*. (b) *Spirogyra* is readily identified by its spiraling chloroplast. Like most algae, *Spirogyra* is a photoautotroph, carrying on photosynthesis much the way it is done by plants.

(a) *Didinium*

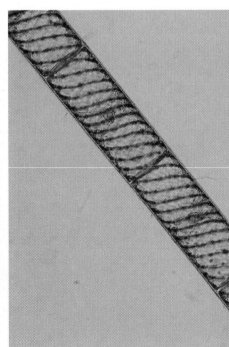

(b) *Spirogyra* with spiral choloroplast

Reproduction

Asexual reproduction in unicellular protists is a means of increasing population size and dispersing the species. In the simpler protists it occurs through mitosis and binary fission—an efficient method for single-cell types. A variation of this process, called **fragmentation**, is observed in some parasitic protists. Here mitosis occurs, but cytoplasmic division is delayed. Later the cytoplasm fragments into tiny individual cells.

Sexual reproduction has not been observed in a large number of protists. Where it happens, the process varies greatly. In some, entire cells fuse, whereas in others, the cells fuse temporarily and exchange haploid nuclei. A few follow the familiar pattern of meiosis, gamete formation, and fertilization. But even here there is variation (Figure 23.4). Some form flagellated **isogametes**, those in which there is no morphological difference among the gametes. Others form **heterogametes**, with large stationary eggs and small flagellated sperm. In between, in an evolutionary sense, are the **anisogametes**. Here we find large and small flagellated cells, but they are otherwise identical. The larger is considered to be the egg equivalent; the smaller, the sperm. In an unusual variation of sexual reproduction called **autogamy**, meiosis is completed, but the daughter cells simply pair up and fuse back together. Autogamy is a form of self-fertilization, which precludes new genetic input. But because of crossing over in meiosis and the random union of daughter cells, there is some shuffling of the parental genes.

Despite the sexual variation described so far, biologists have organized the life cycles into a few basic types, some familiar, others new to our discussions.

Life Cycles in Protists and Other Eukaryotes

There are three fundamental kinds of life cycles in eukaryotic organisms: zygotic, gametic, and sporic. They are represented in diagrammatic form in Figure 23.5. We will begin with the most familiar.

The Gametic Cycle. The **gametic cycle** (Figure 23.5a) occurs in some of the animal-like protists, a few algae, and in all animals including humans. It is direct and uncomplicated. Here the cells are always diploid except for the gametes, which are always haploid. In the life cycle of animals, for example, diploid cells in the gonad undergo meiosis to form gametes. The gametes join in fertilization, restoring the diploid state.

With this conceptually comfortable cycle in mind, consider one that is just the opposite.

The Zygotic Cycle. The **zygotic cycle** (Figure 23.5b) is common in fungi and some algae. Nearly all of the organism's life is spent in the haploid state, often with new haploid individuals produced by mitosis. When sexual reproduction occurs, the haploid cells, or just their nuclei, fuse, bringing on the diploid state. Typically, newly diploid zygotes immediately enter meiosis, restoring the haploid state.

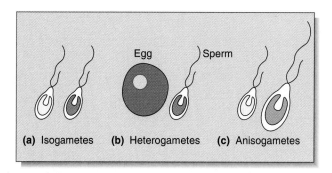

FIGURE 23.4
Gametes
(a) Isogametes have little if any structural distinction. There is no sperm and egg as such. Those shown are flagellated, but many are ameboid or nonmotile. (b) With heterogametes, the egg is large and stationary and the sperm is small and flagellated. (c) Anisogametes are of two different sizes, but are otherwise identical.

The Sporic Cycle. The **sporic cycle** (Figure 23.5c) is seen in many of the multicellular algae and in all plants. The novel aspect of the sporic cycle is that, in many species, it includes two distinct generations. A **sporophyte generation** ("spore plant") produces spores, and a **gametophyte generation** ("gamete plant") produces gametes. In its more primitive form, in algae and a few plants, the generations occur in *two completely separate individuals*. Cells in the sporophyte generation, which is always diploid, undergo meiosis, giving rise to haploid cells. But instead of forming gametes, they become spores. The haploid spores then give rise to the multicellular haploid gametophytes. Eventually, cells in the gametophytes form gametes (through mitosis). When the gametes fuse in fertilization, a new sporophyte generation begins. Since sporic organisms alternate between the diploid and haploid state, the life cycle is often called an **alternation of generations**.

There's one final note. The sporic cycle varies in different species. In some instances, the haploid gametophyte is by far the larger and more conspicuous of the two. In others, the diploid sporophyte state is the larger and dominates the cycle. In still others, the gametophytes and sporophytes are identical in size and appearance and neither dominates.

If all of this has left you somewhat befuddled, the examples to follow in this and the next two chapters should help.

Evolutionary Origins of Protists

When did the protists arise? The oldest fossils indicate that the first protists, which means the first eukaryotes, made their appearance about 1.5 billion years ago. The evidence is scant and somewhat indirect since it consists of quartets of fossil cells whose positioning suggests that they had just completed meiosis, a definitely eukaryotic process. Other early eukaryote fossils yield little information about origins, since such cells have well-defined nuclei and the usual array of organelles.

There is no fossil evidence substantiating the transition from prokaryotic to eukaryotic life, but as you may recall that

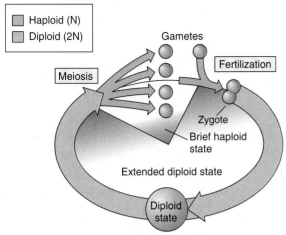

(a) Gametic life cycle (animals, some protists)

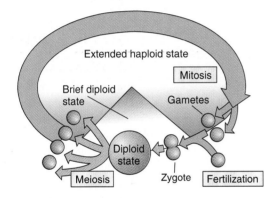

(b) Zygotic life cycle (fungi, some protists)

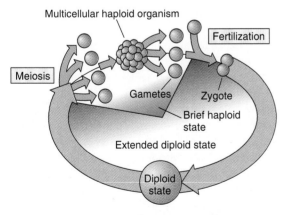

(c) Sporic life cycle (most plants, some protists)

FIGURE 23.5
Three Life Cycles
Most of the earth's eukaryotes fit into one of the three principal kinds of life cycles: **(a)** gametic, **(b)** zygotic, and **(c)** sporic. Here the haploid and diploid states are distinguished by different colors. The diploid phase always ends with meiosis, which ushers in the haploid phase. The haploid phase ends with fertilization (or its equivalent) and the resumption of the diploid phase.

there is a well-accepted theory on how this came about. The serial endosymbiosis theory (see Essay 4.2) maintains that at least some eukaryotic organelles arose through a series of events in which phagocytic prokaryotes fed on other prokaryotes, including aerobic cells, photosynthetic cells, and ciliated cells. Whereas the object of such phagocytic activity was most likely to obtain food, some of the captured cells avoided this fate and established a mutual relationship with the host cell. The establishment of an organized, membrane-surrounded nucleus and the intricate intermembranal system of eukaryotic cells is explained through the theory of autogenesis (again, see Essay 4.2).

Phylogeny of Protists

We will begin by dividing the protists into three groups. By tradition, those with animal-like characteristics are called **protozoans** and those with plant-like characteristics are called **algae**. The few that share characteristics with fungi comprise the **slime molds** and **water molds**. Subdividing the kingdom in this manner is quite useful but it doesn't begin to cover all cases. For example, the puzzling little protist *Euglena gracilis* and many others like it staunchly resist categorizing as alga or protozoan. They clearly have some characteristics of each.

It may have occurred to you that the protists are a disparate group, perhaps difficult to conceive of as a single kingdom. If so, you are in good company. Biologists have contrived the kingdom Protista as a temporary assemblage of species that will be sorted out in time. There is always room for debate on phylogenetic relationships, and protists are hotly debated.

It is clear, however, that in their relationship to other eukaryotes, protists gave rise to the three other kingdoms. Thus fungi, plants, and animals each have their roots in protist stock. Plants, it is believed, evolved from green algal stock, whereas animals and fungi arose from flagellated protozoans.

Animal-like Protists: Protozoans

A favored taxonomic scheme divides the protozoan protists into seven phyla, three that are significant to us (Table 23.1). Included are the phylum **Sarcomastigophora**, which includes two groups—the flagellates and amebas; the phylum **Apicomplexa**, the nonmotile spore formers; and the phylum **Ciliophora**, mostly ciliated species. As you can see, locomotion plays an important role in protozoan classification.

Flagellates

Flagellated protozoans are readily recognized by the presence of one or more flagella; but beware—some of the algae also have flagella. The protistan flagellum has the 9 + 2 microtubular arrangement typical of eukaryotes and moves through an undulating motion. In some flagellates, the propulsive force of the single flagellum is increased by its attachment to

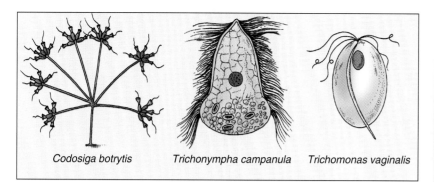

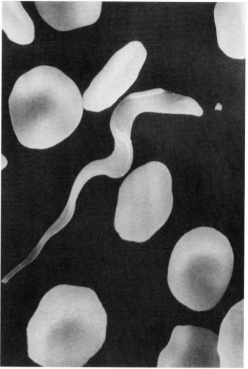

Codosiga botrytis   *Trichonympha campanula*   *Trichomonas vaginalis*

Trypanosoma gambiense and red blood cells

FIGURE 23.6
Flagellates
Some of the flagellates of the subphylum Mastigophora include *Codosiga,* a colonial, with collar-like structures used in feeding. The multiflagellated *Trichonympha campanula* lives as a symbiont in the termite gut, trading its cellulose-digesting powers for a safe living place and a steady food supply. *Trichomonas vaginalis* causes irritating vaginal infections. *Trypanosoma gambiense (*photo) is a notorious flagellate that causes African sleeping sickness. The flagellum of *T. gambiense,* attached along the body by a thin, membranous flap, produces an undulating motion as the parasite makes its way through the host's blood cells. When humans are infected, the trypanosome reproduces in the blood and lymph glands and eventually enters the cerebrospinal fluid, where it brings on the lethargy of African sleeping sickness (*trypanosomiasis*). As the parasites invade the central nervous system, tremors, apathy, and convulsions commence, leading to coma and death.

an **undulating membrane**, a flap that extends along the body. In others, the flagella are surrounded by a funnel-like collar, a structure that filters food from the surrounding water.

When feeding, some flagellates absorb nutrients through the body wall while others are active predators. Several are seen in Figure 23.6. Reproduction is primarily asexual, with mitosis followed by longitudinal division of the body. Little is known about sexual reproduction.

One species of particular importance to humans is the parasite *Trypanosoma gambiense,* the agent of **African sleep-**ing sickness. *Trypanosoma gambiense* is carried by the infamous tsetse fly, whose bite injects the parasite into the victim. Controlling the disease in East Africa, where it is endemic, is difficult because nearly all hoofed mammals, including the domestic cattle of wandering, nomadic peoples, are infected.

Amebas

The best-known amebas occur in four groups: the *sarcodines, heliozoans, radiolarians,* and *foraminiferans.* **Sarcodines**

TABLE 23.1 The Animal-Like Protists	
Phylum	**Typical Characteristics**
Protozoans	Protists with animal-like movement and feeding
Sarcomastigophora (flagellates and amebas)	Flagellates: movement by flagella; feed by absorption and phagocytosis; asexual reproduction by binary fission; sexual reproduction rarely observed
	Amebas: ameboid movement; feeding by phagocytosis; asexual reproduction through binary fission; sexual reproduction through meiosis and fusion of cells; some cause amebic dysentery; some with glassy or calcareous skeleton
Apicomplexa (sporozoans)	Nonmotile as adults; feed by absorption; many parasitic *Plasmoditum vivax,* the agent of malaria, reproduces asexually through binary fission and fragmentation in the bird or mammalian host, sexual reproduction is through fusion of gametes in the mosquito
Ciliophora (ciliates)	Movement and feeding involve highly coordinated cilia; asexual reproduction by fission; sexual reproduction through conjugation and exchange of haploid nuclei, some are parasites

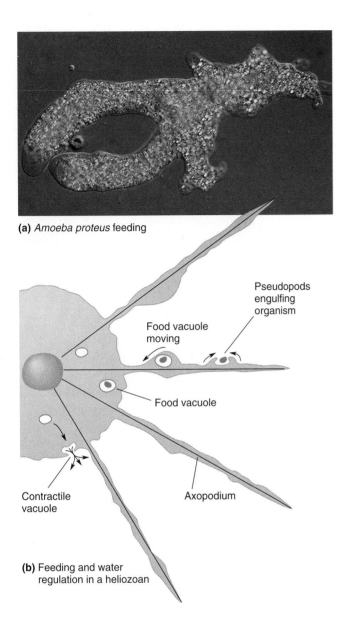

(a) *Amoeba proteus* feeding

Pseudopods
engulfing
organism

Food vacuole
moving

Food vacuole

Contractile
vacuole

Axopodium

(b) Feeding and water
regulation in a heliozoan

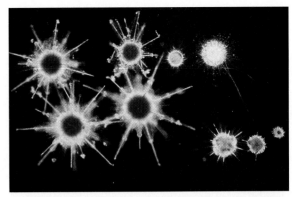

(c) Radiolarians

(d) Foraminiferans

FIGURE 23.7
Amebas

(a) *Amoeba proteus* is a "naked ameba," one without a hardened outer covering. Here we see it using two pseudopods to surround and engulf prey. **(b)** *Actinosphaerium* has a spherical body covered with long, fine axopodia. They are made up of numerous microtubules and surrounded by streaming cytoplasm that is used in feeding. **(c)** Radiolarians are typically spherical, with highly sculptured glassy skeletons. **(d)** The calcareous shells of foraminiferans take on spiral shapes reminiscent of tiny snails.

include the familiar pond and laboratory amebas. They are sometimes called "naked amebas" because they lack a hardened covering. They have no definite shape, constantly changing as they form the pseudopods that are used in movement and prey capture (Figure 23.7a). Ameboid movement, as this is known, was discussed at length in Chapter 4. Ameboid movement is also utilized by phagocytic white blood cells in the animal body.

Sarcodines feed through phagocytosis, where food particles such as algae, bacteria, or other protozoans are surrounded by converging pseudopods and taken in. The prey end up in phagosomes, or food vacuoles, that fuse with lysosomes containing digestive enzymes (see Chapter 4). The products of digestion enter the cytoplasm, and undigested residues are expelled through exocytosis.

Sarcodines increase their numbers asexually through mitosis and cell division. Sexual reproduction is rare but, when observed, includes meiosis and simple cell fusion.

One of the naked amebas, *Entamoeba histolytica,* is the agent of ameboid dysentery, a debilitating disease that remains important on a worldwide scale. *Entamoeba histolytica* is passed from person to person chiefly through water or food that has been contaminated by feces. If you think this mode of transmission is unlikely here in the United States, consider the recent, widely publicized outbreak of *E. coli* food poisoning in fast food restaurants. *Escherichia coli,* you should know, gets from place to place primarily in animal feces. Careless handling by meat packers permits the meat to become contaminated.

Since **heliozoans** have coverings, they could be called "clothed amebas," but we won't call them that. They produce a hardened capsule, or **test,** of silicon dioxide (glass) from which extend long, slender **axopodia** (Figure 23.7b). Axopodia are slender pseudopods with a central axis of microtubules. The microtubules are assembled and disassembled to extend and withdraw the structure. As the cytoplasm creeps back and forth along the axopod, food is taken in through phagocytosis. Sexual reproduction includes withdrawal into cyst-like states where meiosis occurs, followed by cell fusion.

Radiolarians and **foraminiferans** (Figure 23.7c,d) are marine organisms, a part of the drifting plankton that makes up the base of many marine food chains. They form tests of silicon dioxide and calcium carbonate, respectively. The empty tests, settling to the bottom over countless years, make up a substantial part of the ooze covering much of the ocean floor. "Forams" were once so numerous that their remains formed the immense, white sedimentary deposits that make up the famous White Cliffs of Dover, England.

Sporozoans

Sporozoans are protists with three major characteristics: (1) most form spores or have spore-like stages, (2) they commonly live as parasites, and (3) they are nonmotile, at least in the mature stages. "Spores" are defined here as cells enclosed in a protective casing and capable of infecting a host. Sporozoans may have incredibly complex life cycles, with highly prolific asexual and sexual phases that may even occur in different hosts. The best-known sporozoan is *Plasmodium vivax,* the agent of malaria. There are several species of *Plasmodium,* each infecting a different mammal or bird host.

Malaria is of great historic significance and remains one of the most important diseases of humans today. The malarial parasite is spread by the female *Anopheles* mosquito that, in spite of vast eradication efforts, is alive and well throughout much of the world. Today we are confronted by insecticide-resistant strains of *Anopheles*. In its life cycle (Figure 23.8), *P. vivax* reproduces asexually in its human host and sexually in the mosquito. The cycle begins when a mosquito pierces the skin with its sucking mouth parts, releasing **sporozoites**. The sporozoites, ameboid cells from the mosquito's salivary glands, are carried by the blood to the liver, where they enter liver cells and quietly undergo mitosis. Each ameba produces a multinucleate mass that undergoes fragmentation, forming thousands of tiny cells called **merozoites**. The merozoites emerge from the liver in huge numbers and invade the red blood cells. There, another round of asexual reproduction disrupts the red cells, releasing toxins and new infectious merozoites throughout the body. This brings on the familiar fever and chills of malaria. As the merozoites reinfect other red blood cells, the chills and fever are repeated in cycles.

The sexual phase of *Plasmodium* begins when the merozoites undergo a different transformation, this time into **gamonts**, the gamete-forming cells. Before gametes actually form, the gamonts must reenter the mosquito. This is accomplished in the next feeding cycle. (Can you see how keeping infectious malaria victims in isolation breaks this link in the parasite's life cycle?) Fertilization occurs in the mosquito gut, whereupon the zygotes transform into sporozoites. They prepare for the next host by migrating to the salivary glands.

Ciliates

The ciliates are a highly diverse group of protozoans, some species representing the most complex single cells on earth. The size range is enormous—from 10 to 3000 μm: on a different but comparable scale, the range between a shrew and blue whale.

The cilia that characterize this group (Figure 23.9) are used in movement and feeding. They occur in rows, each performing a precise rowing motion. The activity is not helter-skelter but is highly coordinated through an elaborate cytoskeletal network interconnecting the basal bodies. Cilia are quite efficient: one study clocked the swimming rate of a ciliate at 1 mm/sec, which is about 10,000 body lengths per hour. The equivalent in humans would be a person swimming 10 miles/hr!

In the slipper-shaped *Paramecium caudatum,* a laboratory favorite, some cilia are fused into a membrane-like arrangement that lines the **cytostome**, a funnel-shaped feeding structure. We will use *P. caudatum* (Figure 23.10) as our representative ciliate, referring to it simply as paramecium.

Paramecium. As in many protozoans, the body covering of paramecium is a tough elastic **pellicle** consisting of secreted material outside the plasma membrane. Its elasticity permits the ciliate to bend and wriggle as it passes around obstructions in its aquatic environment.

Alternating with the ciliary basal bodies are the **trichocysts**, slender, thread-like bodies with barbed heads that can be forcefully discharged *en masse* when the protist is disturbed. Some ciliates use the trichocysts in prey capture; others use them for anchoring themselves in place while feeding.

Like many freshwater protozoans, paramecium has **contractile vacuoles**. These are versatile pumping structures that periodically fill and contract, expelling excess water from the body. This is essential since there is no way for the cell to resist the inward osmotic movement of water. Both filling and emptying cycles appear to operate by ATP-powered contractions of cytoplasmic actin filaments.

Many ciliates feed by taking bulk food, such as bacteria, into the body. In paramecium, cilia sweep bacteria along its **oral groove**, a depression that leads into the cytostome. Digestion occurs in phagosomes that form at a thin region of the pellicle. Residues leave the cell through exocytosis.

Paramecium has two types of nuclei—a larger **macronucleus** and a smaller **micronucleus** (some species have several micronuclei). The macronucleus contains multiple copies of chromosomes whose genes are actively involved in transcription for protein synthesis. (The macronucleus is similar in ways to the nucleolus, where multiple copies of certain genes continually transcribe large amounts of ribosomal RNA. See Chapter 4.) The small nucleus is inactive between cell divisions, acting as a repository for the organism's genome, which is held in reserve for reproduction.

Both asexual and sexual reproduction occur in ciliates. In asexual reproduction (Figure 23.11a), mitosis occurs in the micronucleus, but the macronucleus is simply divided in half. Mitosis is followed by transverse division of the cell near the

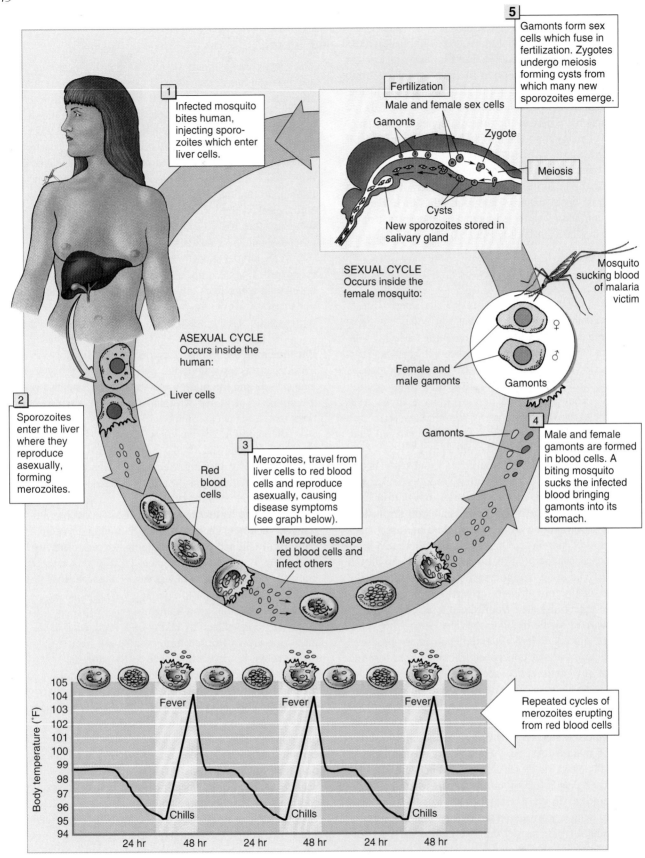

FIGURE 23.8
***Plasmodium* Life Cycle**
The *Plasmodium* life cycle includes episodes of asexual reproduction, in which the number of parasites vastly increases (producing the symptoms of malaria), and of sexual reproduction, which occurs inside the mosquito. The graph below tracks the intermittent episodes of fever and chills so characteristic of this dreaded disease.

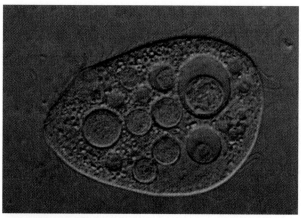

(a) *Tetrahymera*

(b) *Euplotes*

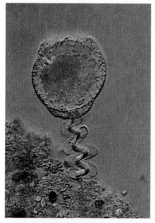

(c) *Vorticella*

FIGURE 23.9
A Sample of Ciliate Diversity
The ciliates are enormously varied in both size and structure. One of the smallest is **(a)** *Tetrahymena,* about 20 μm in length. **(b)** *Euplotes* is considerably larger, measuring about 90 μm. **(c)** *Vorticella* is of intermediate size but has a long, contractile stalk. When included, the stalk brings its length to 150 μm. The stalk is springlike, permitting the protist to retract its cell when disturbed.

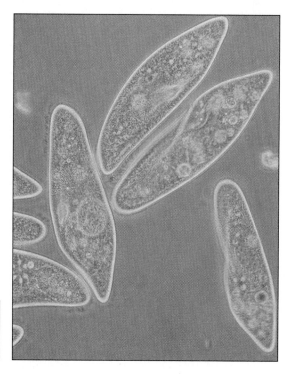

Contractile vacuole

H_2O H_2O H_2O
H_2O H_2O
Filling

H_2O H_2O
H_2O H_2O
Filling

H_2O H_2O
H_2O H_2O
Emptying

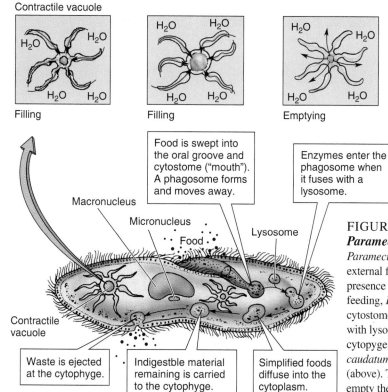

Food is swept into the oral groove and cytostome ("mouth"). A phagosome forms and moves away.

Enzymes enter the phagosome when it fuses with a lysosome.

Macronucleus

Micronucleus

Food

Lysosome

Contractile vacuole

Waste is ejected at the cytophyge.

Indigestble material remaining is carried to the cytophyge.

Simplified foods diffuse into the cytoplasm.

FIGURE 23.10
Paramecium caudatum

Paramecium caudatum is a slipper-shaped ciliate with two prominent external features: a dense covering of cilia and an oral groove. Note the presence of two types of nuclei, a macronucleus and micronucleus. In feeding, *P. caudatum* sweeps minute particles into its oral groove and cytostome to a region where phagosomes form. The phagosomes fuse with lysosomes and digestion occurs. Residues are extruded through the cytopyge, a special region in the surface. Like many other protists, *P. caudatum* uses contractile vacuoles for maintaining its cell water balance (above). They balloon out when filling and contract suddenly as they empty their contents to the outside.

region of the oral groove. Missing organelles are produced anew in daughter cells.

Sexual reproduction in *Paramecium* occurs through an interesting type of conjugation (Figure 23.11b). Cells of different mating types (there are no sexes) fuse along their oral grooves, and meiosis begins in the micronuclei. Three of the four haploid daughter micronuclei disintegrate and the fourth divides by mitosis, forming two new micronuclei that are still haploid. Next, an exchange begins. Each cell sends one of its new micronuclei to the other. Fertilization is completed as each entering micronucleus fuses with the one left behind. This restores the diploid condition, but, more significantly, it produces a new combination of genetic material in each cell. The macronuclei, incidentally, are completely dismantled to

be later reconstructed. Despite its complexities, the life cycle of *Paramecium is gametic and animal-like.*

Fungus-like Protists

Those protists with fungus-like characteristics include the slime molds of the phyla **Myxomycota** (MIX oh my coat uh) and **Acrasiomycota** (a CRAZY oh my coat uh), and the water molds and downy mildews of the phylum **Oomycota** (OH OH my coat ah) (Table 23.2). Each includes stages in which spore-forming structures emerge and spores are formed. It is this characteristic that is shared with the fungi.

FIGURE 23.11
Reproduction in Paramecium
(**a**) Asexual reproduction occurs through mitosis and cytoplasmic division. The macronucleus simply pinches in half. (**b**) Sexual reproduction requires conjugation. During conjugation, micronuclei undergo meiosis, and an exchange of the haploid products occurs between cells. At about this time, the macronucleus is broken down to be later reconstituted from the genetically recombined micronuclei.

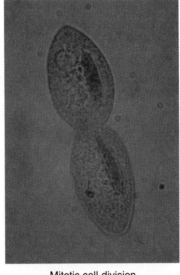

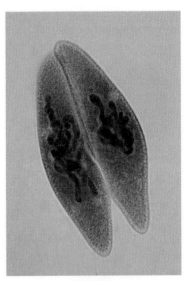

Mitotic cell division

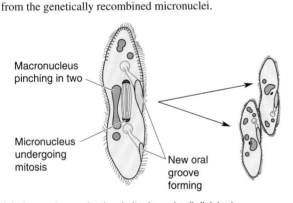

Macronucleus pinching in two

Micronucleus undergoing mitosis

New oral groove forming

(**a**) Asexual reproduction (mitosis and cell division)

Conjugation

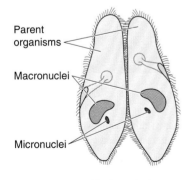

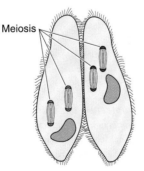

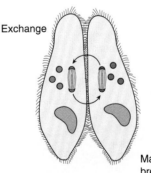

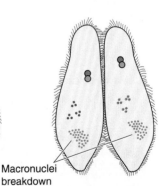

Parent organisms

Macronuclei

Micronuclei

Conjugants fuse together.

Meiosis

Meiosis occurs in the micronuclei producing four haploid products.

Exchange

After meiosis, one micronucleus divides again and is exchanged with the partner.

Macronuclei breakdown

Micronuclei fuse, restoring the diploid state. The macronucleus and extra micronuclei disintegrate.

(**b**) Sexual reproduction (conjugation)

TABLE 23.2
Slime and Water Molds: The Fungal-like Protists

Protists with fungus-like characteristics but with motile cells: spores formed in sporangia and feeding by absorption

Phylum	Typical Characteristics
Acrasiomycota (cellular slime molds)	Small ameboid cells feed by phagocytosis; asexual reproduction begins with swarming and the formation of a large cellular mass, which produces spores, sexual reproduction via fusion of amebas
Myxomycota (plasmodial slime molds)	Multinucleate ameboid mass feeds by phagocytosis; asexual reproduction includes spore formation; sexual reproduction by fusion of amebas or flagellated cells
Oomycota (water molds)	Fungus-like filamentous growth but with cellulose cell walls and flagellated stages; asexual reproduction by formation of flagellated cells; sexual reproduction includes formation of large eggs and small, nonmotile sperm; parasites of plants and animals; cause late blight in crops

Slime Molds

Some biologists include the slime molds with the fungi, but we have opted for classification schemes where motility is the deciding factor. In our chosen scheme, none of the true fungi are motile and all lack centrioles.

There are two distinct phyla of slime molds: **Myxomycota**, the "plasmodial slime molds," and **Acrasiomycota**, the "cellular slime molds." Slime molds join with decomposer bacteria in the ecologically vital activity of recycling mineral nutrients. A few are plant parasites.

Plasmodial Slime Molds. One of the best-known **plasmodial slime molds**, *Physarum polycephalum* (Figure 23.12), has all the characteristics of a huge ameba except that it is multinucleate, its many nuclei the product of numerous mitotic events without cytoplasmic division. This is the **plasmodial stage**, the term for which the group is named. The plasmodium of *P. polycephalum* is usually found flowing over the moist underside of rotting tree trunks. As it moves along, it feeds by phagocytizing bacteria and bits of organic matter.

FIGURE 23.12
Life Cycle of a Plasmodial Slime Mold

(a) Life cycle. The diploid phase of the plasmodial slime mold cycle includes ⬜1 a plasmodium—a multinucleate, ameboid feeding stage during which mitosis occurs but the cytoplasm does not divide. This is followed by the growth of sporangia ⬜2 in which meiosis and spore formation occur. ⬜3 The haploid spores form either myxameba or flagellated swarm cells that, upon fusion ⬜4, bring on the next diploid, plasmodial phase of the cycle. (b) Both the plasmodial and spore-forming stages *(photos)* can be quite colorful.

Sporangia

Plasmodium

(a) Life cycle of plasmodial slime mold

FIGURE 23.13
Life Cycle of a Cellular Slime Mold

1 Cellular slime molds include an ameboid phase in which the cells remain separate. 2 They mass together forming an aggregate that later 3 takes the shape of a slug. 4 This is followed by the formation of a spore-forming structure, which produces spores that 5 later give rise to new amebas. Since meiosis does not occur, this is an asexual cycle and is used here only to form resistant spores. A sexual phase 6 has also been reported. Here amebas fuse, forming a diploid individual that engulfs other amebas and finally forms a large macrocyst. Meiosis occurs and the new haploid amebas are released.

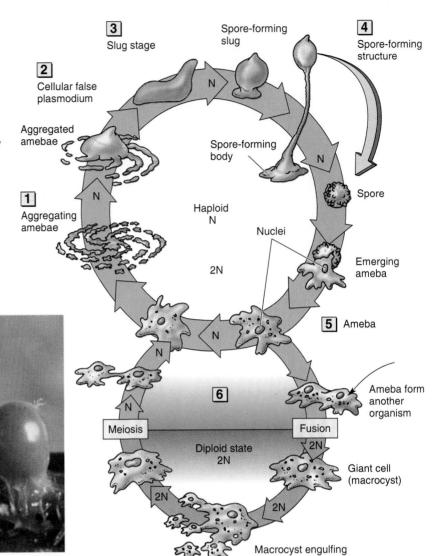

Spore-forming structures emerging

The plasmodium shows some responsiveness as it avoids obstacles, bright light, and dry areas. Should food run out or drying conditions prevail, the plasmodium will change drastically, its fungal characteristics emerging.

The drying mass produces slender vertical props called **sporangiophores**, each of which forms a **sporangium** at its tip. Mitosis within the sporangia yields large numbers of spores. Each spore then undergoes meiosis, producing a larger number of haploid spores. The haploid spores germinate to produce either an ameboid cell called a **myxameba** or a flagellated cell called a **swarm cell**. If conditions aren't quite suitable, either the myxameba or swarm cell can suspend activity, forming a dormant cyst from which it will emerge when things improve. Otherwise the cells pair off and fuse, restoring the diploid state. The diploid cells then form a new multinucleate ameboid stage and the life cycle repeats.

Cellular Slime Molds. The cellular slime molds (Figure 23.13) differ fundamentally from the plasmodial group in that the amebas live as individuals. The individual stage is main-

tained even when **swarming**, an act for which the cellular slime molds are famous. Swarming, which can be brought on by starvation, involves the massing of numerous individuals, forming a large, many-celled **false plasmodium**. Studies reveal that the release of **acrasin**, a common chemical messenger, known as cyclic AMP in animal cells (see Chapter 36), is responsible for drawing the individuals together. Once merged, the aggregation becomes surrounded by a cellulose sheath, forming what is called a **slug**. The entire slug migrates for a time, whereupon some of the amebas form erect processes while others move into these processes and form spores. Upon dispersal, the spores germinate into new amebas. The formation and release of spores in this manner help the species disperse itself and find more favorable habitats.

The cellular slime mold's sexual cycle is not well understood, but according to some authorities, it begins when two of the haploid amebas fuse together, forming a diploid phase. The diploid cell then proceeds to engulf a number of other amebas, eventually forming a **macrocyst** ("large" cyst—but actually quite small compared to the slug) that becomes sur-

FIGURE 23.14
Downy Mildews
Downy mildews infect a large number of crop plants such as this.

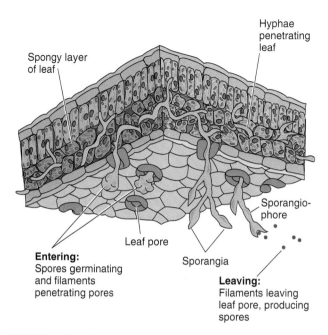

FIGURE 23.15
Late Blight
When a downy mildew or late blight infection occurs in a plant, the fuzzy filamentous growth can be seen on the underside of the leaves. *Phytophthora infestans,* the agent of late blight, sends its filaments throughout the spongy tissue of the leaf, where they absorb the plant's nutrients. Reproductive filaments then emerge through the leaf pores, form sporangia at their tips, and release spores that can then carry the infection to other plants.

rounded by a cellulose wall. Within the macrocyst, the diploid zygote undergoes meiosis, the haploid products divide mitotically, and a number of new, genetically recombined amebas escape and resume the feeding stage.

Oomycetes: Downy Mildews and Water Molds

Members of the phylum **Oomycota** are named partly for the large eggs that some species produce (*oo* means "egg") and partly for their filamentous body (*-mycetes* means "thread-like"). Because of the thread-like or filamentous growth form, water molds are sometimes classified as fungi, but in our chosen scheme their motility prohibits membership in that phylum.

One group of oomycetes helped change the course of history, at least for the Irish. If you are an American of Irish ancestry, you just might owe your U.S. citizenship to the activity of the highly destructive **downy mildew**, *Phytophthora infestans* or "late blight." This parasite causes a disease that kills potato plants. Late blight was the cause of the famous

Irish potato famine, which lasted from 1843 through 1847. Many Irish starved, and many others left their homeland, emigrating to the United States.

Downy mildews can be found on the undersides of infected leaves. You can expect them on your house or garden plants if you tend to overwater or undercultivate. Crops such as potatoes, beans, melons, sugar beets, and grains are especially susceptible to infection (Figure 23.14). One species of *Phytophthora* causes over 1000 root rot diseases including that of citrus and avocado.

Phytophthora grows on moist leaves by penetrating the numerous stomata (leaf pores) with its filaments. These natural openings on leaves permit the mold to penetrate the photosynthesizing cells, absorbing the nutrients as they are produced. Eventually, an asexual reproductive stage begins, and the filaments grow back out through the stomata, producing sporangia at their tips (Figure 23.15).

Some of the **water molds** are saprobes, feeding on dead organisms, whereas others are parasites. One parasite infects salmon. In their asexual cycle, the water molds produce sporangia from which flagellated **zoospores** (motile spores) emerge and begin to swim about. If they encounter a food source, growth occurs and a new colony develops (Figure 23.16a).

Sexual reproduction in water molds (Figure 23.16b) begins with the emergence of unusually thick filaments that produce spherical **oogonia**. Cells in the oogonium undergo meiosis,

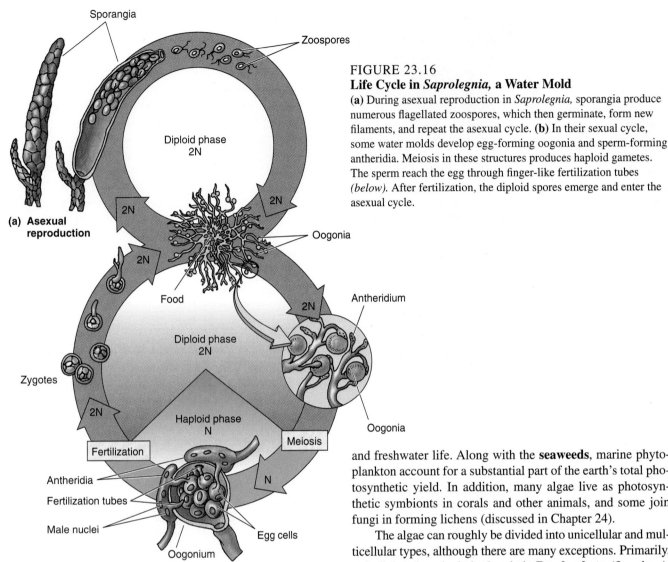

FIGURE 23.16
Life Cycle in *Saprolegnia,* a Water Mold
(a) During asexual reproduction in *Saprolegnia,* sporangia produce numerous flagellated zoospores, which then germinate, form new filaments, and repeat the asexual cycle. **(b)** In their sexual cycle, some water molds develop egg-forming oogonia and sperm-forming antheridia. Meiosis in these structures produces haploid gametes. The sperm reach the egg through finger-like fertilization tubes *(below).* After fertilization, the diploid spores emerge and enter the asexual cycle.

and freshwater life. Along with the **seaweeds,** marine phytoplankton account for a substantial part of the earth's total photosynthetic yield. In addition, many algae live as photosynthetic symbionts in corals and other animals, and some join fungi in forming lichens (discussed in Chapter 24).

The algae can roughly be divided into unicellular and multicellular types, although there are many exceptions. Primarily, unicellular algae include the phyla **Pyrrhophyta** (fire algae), **Euglenophyta** (euglenoids), **Chrysophyta** (golden, yellow-green algae and diatoms), and some of the **Chlorophyta** (green algae). Multicellular algae include the phyla **Rhodophyta** (red algae), **Phaeophyta** (brown algae), and some of the Chlorophyta. Table 23.3 summarizes the main features.

forming haploid egg cells. Fertilization is accomplished in an unusual way. Filaments growing near an oogonium send finger-like branches over the spherical body. Within these branches are the **antheridia,** wherein sperm nuclei arise through meiosis. The branches form fertilization tubes that penetrate the oogonium, permitting the sperm nuclei to reach the eggs.

Plant-like Protists: The Algae

Many of the algae are plant-like in that they are photosynthetic, have similar chlorophylls, and generally have cellulose walls. Algae are extremely widespread and diverse, living in all aquatic habitats and even a few terrestrial ones. Significantly, algal protists make up much of the **phytoplankton** (small, floating, aquatic photosynthetic organisms). Again, phytoplankton are of immense ecological importance because they form the energy base that supports heterotrophic marine

Pyrrhophytes: "Fire Algae"

Visitors to tropical waters (and some northern waters as well) may be surprised and delighted as they are rowed back to their anchored cruise ship after a rousing night ashore. Each time the oar slides beneath the surface, the water seems to explode with tiny iridescent lights. It seems like magic, but it is the magic of the pyrrhophytes, more commonly called **dinoflagellates** ("whirling flagella"). The name "Pyrrhophyta" means "fire plants."

Biochemists tell us that the mysterious light, known technically as **bioluminescence,** forms in a cycle involving a substance called **luciferin** (as in *Lucifer*), ATP, and an enzyme called **luciferase.** In the cycle, luciferin is phosphorylated by ATP, whereupon the product, reacting with oxygen in the

TABLE 23.3
Algea: Plant-like Protists

Plant-like protists are photosynthetic, containing chloroplasts and most having cell walls.

Phylum	Typical Characteristics
Pyrrhophyta ("fire algae," includes dinoflagellates)	Unicellular; many move by flagella; photosynthetic pigments include chlorophyll *a* and *c* and carotenoids; cellulose cell walls; carbohydrate is starch. Many heterotrophs. Asexual reproduction through binary fission; sexual reproduction by cell fusion. Some dinoflagellates cause red tide
Euglenophyta (euglenas)	Unicelluar; flagellated; prominent eyespot; photosynthetic pigments include Chl *a* and *c* and carotenolds; carbohydrate is paramylum; asexual reproduction is through binary fission; sexual reproduction unknown
Chrysophyta (yellow-green, golden brown algae, and diatoms)	Mostly unicellular; photosynthetic pigments include Chl *a* and *c* and carotenoids; cell walls of cellulose or glass; carbohydrate is chrysolaminarin; asexual reproduction by binary fission; sexual reproduction through fusion of gametes in diatoms
Rhodophyta (red algae: seaweeds)	Mostly multicellular; nonmotile; photosynthetic pigments include Chl *a* and *c*, phycocyanin, and phycoerythrin; cellulose cell walls; carbohydrate is floridean starch; asexual reproduction through spores; sexual reproduction through fusion of gametes. Clear alternation of generations
Phaeophyta (brown algae: seaweeds)	Multicellular, motile gametes; Chl *a* and *c* fucoxanthin; cellulose cell walls; carbohydrates are laminarin and mannitol; asexual reproduction through spores; sexual reproduction through fusion of gametes. Clear alternation of generations
Chlorophyta (green algae)	Unicellular colonial, or multicellular; some with flagella; photosynthetic pigments are Chl *a* and *b* and carotenoids; cellulose cell walls; carbohydrate is starch; asexual reproduction through flagellated spores; sexual reproduction through fusion of gametes. Clear alternation of generations

presence of luciferase, gives off light energy and becomes luciferin once again (we've skipped a few steps).

Some dinoflagellates are important as photosynthetic symbionts of larger organisms. Through photosynthesis, for example, they provide coral animals with much of the energy expended in the formation of the great coral reefs. A great many dinoflagellates are heterotrophs, some feeding on other organisms and others living as parasites. Several examples are seen in Figure 23.17.

Many dinoflagellates have two flagella, which lie in prominent grooves, one acting as a rudder, the other supplying propulsion. Their activity gives the cell a spinning motion.

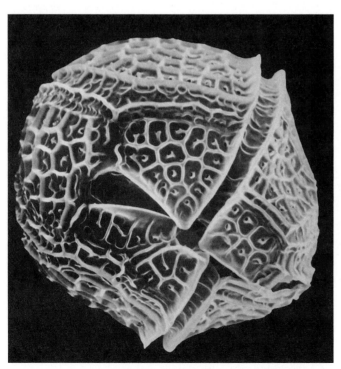

Gonyaulax catanella

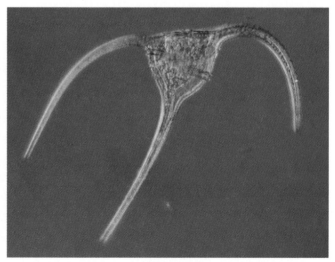

Ceratium tripos

FIGURE 23.17
Common Pyrrhophytes
Pyrrhophytes, also called dinoflagellates, are characterized by cell walls made up of interlocking plates that contain cellulose and pectin. *Gonyaulax catanella* is a red tide species.

Fish kill

FIGURE 23.18
Red Tide
Dinoflagellates can multiply to incredible densities with seasonal increases in mineral nutrients. These densities, called blooms, are known as "red tide." Certain dinoflagellates, when responsible for red tides, produce powerful nerve toxins that make clams, mussels, and other bivalve mollusks unsafe to eat. The enormous metabolic activity in the bloom may produce oxygen deficits that can kill great numbers of fish.

The photosynthetic species have chloroplasts containing chlorophyll *a* and *c* along with carotenoids. Their cell walls are stiff cellulose plates whose arrangements give some cells the appearance of armored helmets. Like *Paramecium,* some dinoflagellates have trichocysts.

The dinoflagellate nucleus is unique, with large, permanently condensed chromosomes that contain little protein and are attached to the nuclear envelope. Furthermore, the nuclear envelope is not dismantled during mitosis and appears to play a role in chromosome separation. Mitosis occurs without the elegant interplay of spindle elements within the nucleoplasm. Instead, bundles of microtubules simply form a cytoplasmic channel between grouped chromosomes.

Asexual reproduction in the pyrrhophytes occurs through mitosis and cell division. Sexual reproduction commonly occurs through the fusion of cells, producing a brief diploid state (the dinoflagellate life cycle is primarily haploid). After this, meiosis occurs and the haploid state resumes. As you may have realized from this, the life cycle is zygotic.

Some of the dinoflagellates are responsible for the dramatic and dangerous phenomenon of **red tides**. When conditions are right, such as with sudden increases in water temperature and mineral nutrients, certain dinoflagellates multiply to incredible densities, described as "blooms." Species with reddish carotenoid pigments may dominate such blooms, which

explains the name "red tide." Some species contain a potent nerve poison, which is taken in by bivalve (two-shelled) mollusks such as clams, oysters, and mussels. The bivalves are unaffected by the organisms, but they often accumulate the poisons, making their flesh very toxic to humans. The illness, called paralytic shellfish poisoning, can be quite serious and even fatal. Other results of red tide blooms are extensive fish kills (Figure 23.18), which are caused by nighttime depletion of dissolved oxygen by the combined cellular respiration of countless algal cells.

Euglenophytes

Members of the phylum **Euglenophyta** are unicellular and flagellated. They are named for the prominent red "eyespot" (*Euglena,* "true eye″) found in many species. It is really not a true eye at all, but a shield of red pigment lying next to a light-absorbing **photoreceptor**. Because of the shield's arrangement, euglenophytes can detect light and move toward it—an adaptation for photosynthesis.

Actually, only about a third of euglenophyte species are photosynthetic. The remainder lack chloroplasts and are heterotrophs that strongly resemble flagellate protozoans. In fact, some theorists maintain that the euglenophytes are actually protozoans that have acquired chloroplasts through endosymbiosis. They propose that the original ancestors were colorless, as are some euglenophytes today.

A well-known euglenophyte, *Euglena gracilis* (Figure 23.19), mentioned earlier as having both protozoan and algal characteristics, can serve as our example. The flask-shaped body of *Euglena* is covered by a flexible pellicle rather than the cellulose cell wall of many algae. A single, long flagellum arises from a depression. A second very short, rudimentary flagellum of unknown function is also present. Curiously, the nuclear envelope of *Euglena* is a *triple* membrane.

Chloroplasts in *Euglena* contain chlorophyll *a* and *b* and carotenoids—all quite similar to what we find in plant cells. It stores its carbohydrate in a type of starch known as *paramylon.*

So far, no one has observed sexual activity in *Euglena,* but asexual reproduction, as you might expect, occurs by mitosis and cell division. They reproduce so rapidly that they may impart a green color to their pond habitat—except for one ruddy species that colors it red.

Chrysophytes: Yellow-Green Algae, Golden Brown Algae, and Diatoms

Both the **yellow-green algae** and the **golden brown algae** make up part of the phytoplankton of marine and fresh waters, also forming a vital part of food chains. The names are derived from the color of their light-absorbing pigments. All members of the phylum **Chrysophyta** ("golden plants") contain chlorophyll *a* and *c* and high concentrations of carotenoids, the most abundant of which is called *fucoxanthin.* Fucoxanthin absorbs light strongly in the blue range. **Chrysolaminarin** is the storage carbohydrate.

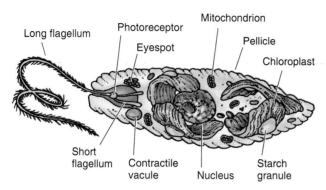

Long flagellum
Photoreceptor
Eyespot
Mitochondrion
Pellicle
Chloroplast
Short flagellum
Contractile vacule
Nucleus
Starch granule

FIGURE 23.19
***Euglena gracilis,* a Well-Known Euglenophyte**
Note the lengthy, tinsillated flagellum. A second flagellum is present but does not protrude. Surrounding the flagellar base is a space, the reservoir, where excess water is directed for expulsion. There is a prominent red eyespot, a shield that covers a photoreceptor. Photosynthesis involves large, well-organized chloroplasts. The products are stored as starch in special paramylon bodies.

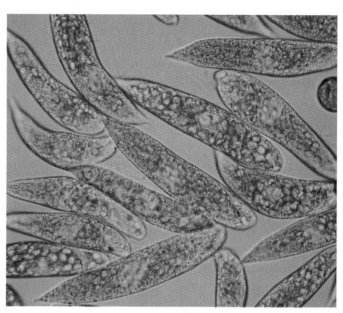

Euglena gracilis

Some chrysophytes surround themselves with cell walls that are essentially glass boxes (silicon dioxide). Some are flagellated, others are ameboid, and some are nonmotile. Most are solitary, but a few species are colonial. Although they are photosynthetic, under certain conditions some species of golden brown algae feed on bacteria, which they take in through phagocytosis.

Diatoms are common photosynthetic marine and freshwater chrysophytes that make up much of the phytoplankton of the sea (Figure 23.20). Their glassy walls consist of an inner box and an outer lid. The walls of marine diatoms are highly ornamented and beautiful to the human eye. The intricate designs are produced by the arrangements of tiny holes

through which gas and water are exchanged. The sculpturing is so fine that diatoms are used to test the quality of light microscope lenses.

When they reproduce asexually, diatoms undergo mitosis and divide within their shells. The two old halves of the shell become the outer lids of the daughter cells and new inner lids are secreted. A little thought on this process tells us that generations of diatoms reproducing this way get *smaller and smaller.* At some critical size, however, sexual reproduction may solve the problem.

Sexual reproduction (Figure 23.21) begins with meiosis and, in some species, the formation of ameboid isogametes. In others, a small flagellated sperm and a larger egg are formed.

FIGURE 23.20
Diatoms: Sculptured Glass
Diatoms are well known for their diverse, beautifully sculptured, glassy skeletons.

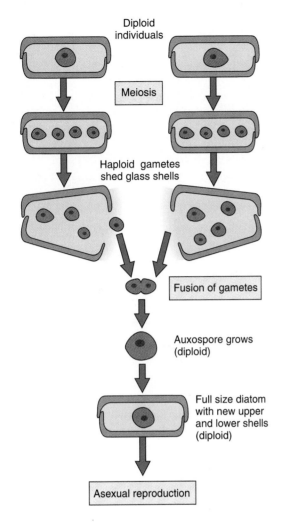

FIGURE 23.21
Sex in Diatoms
When continued cell division results in cells that become critically small, meiosis occurs and gametes form. In this species of diatom, ameboid isogametes from different diatoms fuse, forming diploid auxospores. When the spores form the next generation of diatoms, they produce new full-size shells that are large enough to go through many asexual divisions before the sexual process must be repeated.

Both types of gametes shed their glassy boxes and become free, naked sexual forms. Gametes meet and fuse to produce diploid zygotes, which form auxospores. As new individuals mature, they expand in size and secrete new, full-sized glass coverings. Thus the cycle continues. The diatoms, we see, follow a modified gametic life cycle.

Ocean floor sediments consist largely of glassy diatom coverings. Deposits of ancient diatom coverings are so vast they form **diatomaceous earth**, which is quarried for use in toothpaste, swimming pool filters, and insulating material.

Rhodophytes: Red Algae

A true multicellular condition is common among the red, brown, and green algae, although the reds and greens include many unicellular species. As we'll see, the size range in these algae is enormous, from minute, single cells as small as 2 μm to ocean kelps as long as 50 m.

The phylum **Rhodophyta**, or **red algae** (*rhodo-,* "red"), includes over 4000 species. Despite the name, they all contain blue-green and green pigments (chlorophyll *a*), as well as red. The red pigments happen to mask the others. The pigment combination is particularly well suited for absorbing blue light that penetrates deeper into the sea than other wavelengths. Some of these pigments also occur in cyanobacteria, and this, along with structural similarities in the chloroplasts, has convinced some theorists that red algae emerged from ancient cyanobacteria.

All but a few red algae are aquatic, and most are marine (Figure 23.22), the latter comprising a large portion of what we commonly call "seaweed." Marine red algae grow primar-

FIGURE 23.22
Red Algae
Red algae range in length from a few centimeters to perhaps a meter in length. Most are marine organisms and are found in warm waters, usually attached to rocks by their holdfasts. Some of the branched bodies form widened, flat blades, but most are frilly and delicate.

ily on rocky coasts, where they attach firmly to the seabed by specialized structures known as **holdfasts**. Holdfasts are simply anchors. Unlike true roots, they do not conduct materials.

Red algae store their food in the form of a starch called **floridean starch** and produce a number of other polysaccharides. One of these, **agar-agar**, is used by Indonesians to thicken soup and by biologists as a jelly-like medium on which to grow bacteria. Although red algal polysaccharides cannot be metabolized by most microorganisms (nutrients must be added to agar-agar) or by anything else, we end up eating them anyway. Why? It seems that red algae are also harvested in enormous quantities for the polysaccharide **carrageenan**, which, among other uses, gives a fake richness to chocolate-flavored dairy drinks and fast-food milkshakes.

The sporic life cycle, with a pronounced alternation of generations, is quite common in the larger algae. In some, the sporophytes and gametophytes are identical in appearance, although the first is diploid and the second haploid. This condition is called **isomorphic**. Others are **heteromorphic**, where the two generations are structurally different. The dominating phase is usually larger and longer lived. The other phase may be greatly reduced, often to just a few cells.

The common red algae *Polysiphonia* follows an isomorphic, alternating life cycle (Figure 23.23), but one that actually has *three generations* instead of the usual two. Both sporophytes and gametophytes are multicellular, with identical branching growth forms that can only be distinguished by a trained observer using a microscope.

In its life cycle, cells within the sporophyte of *Polysiphonia* undergo meiosis, producing haploid spores that are released into the surrounding sea. Successful spores then develop into male or female gametophytes that, in turn, produce gametes. Sperm arise in **spermatangia** while eggs form in **carpogonia**. All red algae are nonmotile right down to their sperm, or **spermatia**, as they are called. Lacking flagella, the spermatia float passively to receptive female cells. A lengthy, hair-like extension protrudes from the carpogonium and provides a surface to which sperm adhere. Fertilization occurs within the confines of the female gametophyte.

The third generation in the red algae life cycle is an intermediate stage, a **carposporophyte**, that is actually a second sporophyte. The carposporophyte produces numerous spores that, when released, develop into the larger sporophyte, and a new cycle begins.

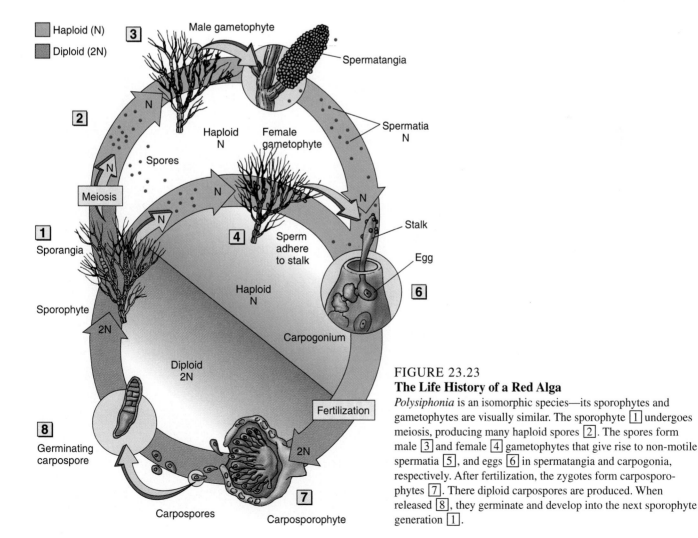

FIGURE 23.23
The Life History of a Red Alga
Polysiphonia is an isomorphic species—its sporophytes and gametophytes are visually similar. The sporophyte 1 undergoes meiosis, producing many haploid spores 2 . The spores form male 3 and female 4 gametophytes that give rise to non-motile spermatia 5 , and eggs 6 in spermatangia and carpogonia, respectively. After fertilization, the zygotes form carposporophytes 7 . There diploid carpospores are produced. When released 8 , they germinate and develop into the next sporophyte generation 1 .

Sargassum

Phaeophytes: Brown Algae

The phylum **Phaeophyta** includes about 1500 known species. The **brown algae** are distinguished by their characteristic brown pigment, fucoxanthin, along with the usual chlorophyll *a* and *c*. They store carbohydrates in the form of *laminarin* and *mannitol*, and they have characteristic structural polysaccharides as well. One of these, *algin*, is a filler and smoothing agent, used in commercial ice cream and frozen custards.

Most brown algae live in the ocean, particularly in cold coastal waters. An exception of sorts is the genus *Sargassum,* which drifts in great masses in the warm Gulf Stream waters that flow through the middle of the Atlantic Ocean, an area called the Sargasso Sea (Figure 23.24).

Brown algae come in all sizes, from microscopic deep-water filaments to the famous giant kelps living in dense marine forests in shallow offshore waters (Figure 23.25).

FIGURE 23.24
The Brown Alga *Sargassum*
Sargassum, a floating seaweed, originates along the coastline, but much of it breaks away and is carried out to sea by currents. As it grows, *Sargassum* forms branching stipes and flat, leaf-like blades. The spherical floats keep the seaweed near the surface where light is plentiful.

(a) *Macrocystis*

FIGURE 23.25
The Giant Kelps
The giant kelp *Macrocystis* thrives in colder ocean regions.
(a) Seen here are its highly branching stem-like stipes and gas-filled bladders at the base of each blade. **(b)** Some kelps reach enormous sizes, forming undersea forests that provide habitats for many kinds of sea life (and great places for scuba diving).

(b) Undersea kelp forest

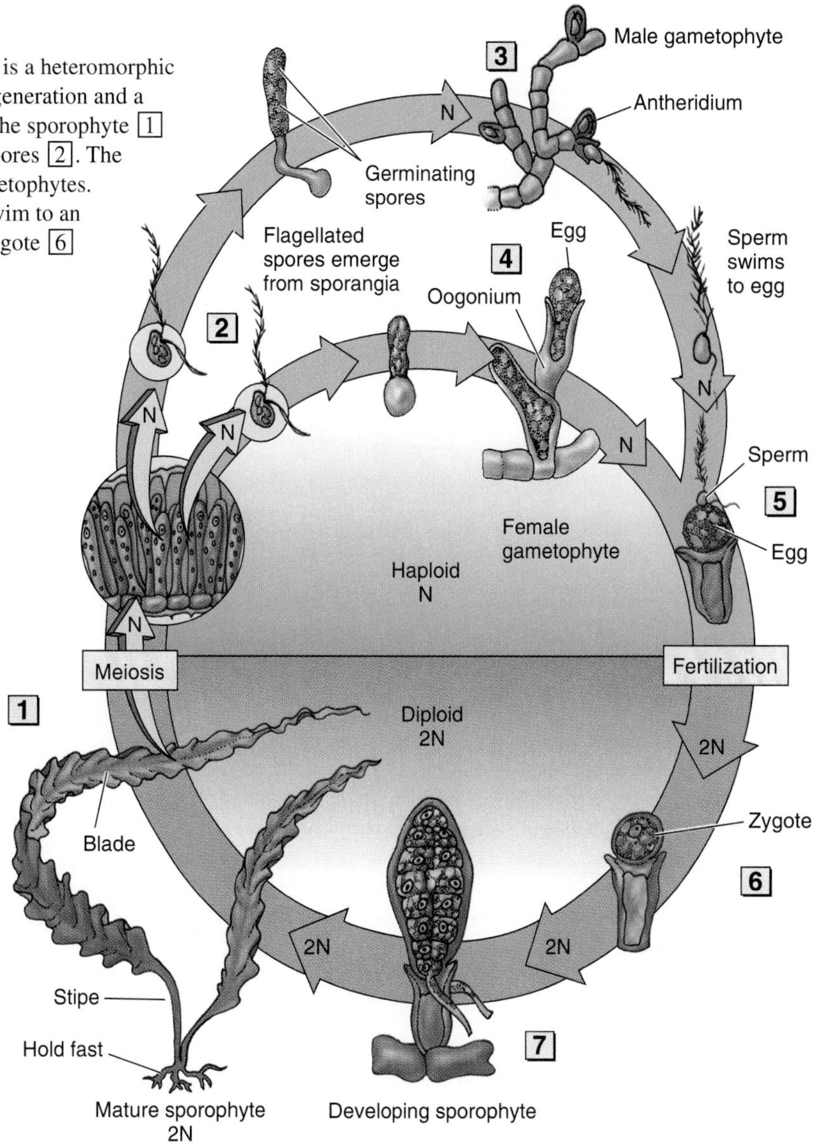

FIGURE 23.26
Life Cycle of *Laminaria*
The life cyclic of *Laminaria,* a brown alga, is sporic. It is a heteromorphic species, characterized by a very dominant sporophyte generation and a brief, much smaller, gametophyte generation. Cells in the sporophyte ☐1 undergoe meiosis, producing numerous biflagellated spores ☐2. The spores then form separate male ☐3 and female ☐4 gametophytes. Flagellated sperm cells are released from antheridia, swim to an oogonium, and fertilize the egg cell ☐5. The diploid zygote ☐6 develops into the next generation of sporophytes ☐7.

Some of the kelps are enormous; *Macrocystis* and *Nereocystis* are record holders, exceeding 60 and 40 meters in length, respectively. *Macrocystis* is regularly harvested for its carbohydrates and minerals by special kelp-cutting boats that travel along the southern California coast. Kelp regrowth is rapid, thanks to actively dividing tissues just below the blades.

The life histories of brown algae also vary considerably, but most are sporic, with a clear alternation of generations. The kelp *Laminaria* (Figure 23.26) has a strongly heteromorphic alternation of generations, with dominating sporophytes and separate microscopic gametophytes. The presence of large stationary eggs and small motile sperm indicate that *Laminaria* is a reproductively advanced species.

The rockweed *Fucus* is different from most brown algae. There is *no* alternation of generations, and its life cycle is *gametic* (as is our own). That is, the products of meiosis *are gametes, not spores.* The diploid individuals form hollow **conceptacles** at the tips of their highly branched blades (Fig-

ure 23.27). Cells in the conceptacles give rise to antheridia and oogonia, and it is within these that meiosis occurs and sperm and eggs mature.

The kelps appear to be the most structurally complex protists. For example, *Macrocystis,* the giant kelp, has highly specialized tissues and organs. These include the anchoring **holdfasts**, stem-like **stipes** that support leaf-like **blades**, and spherical **floats** that keep the photosynthetic cells of the kelp near the surface. Some brown algae even have specialized internal tissue for transporting food from place to place, a capability that closely resembles that seen in advanced land plants.

Chlorophytes: Green Algae

The phylum **Chlorophyta** contains about 7000 named species. Although many of the chlorophytes, or **green algae**, are freshwater forms, there are about 900 marine species as well. A few terrestrial species appear on the surface of melt-

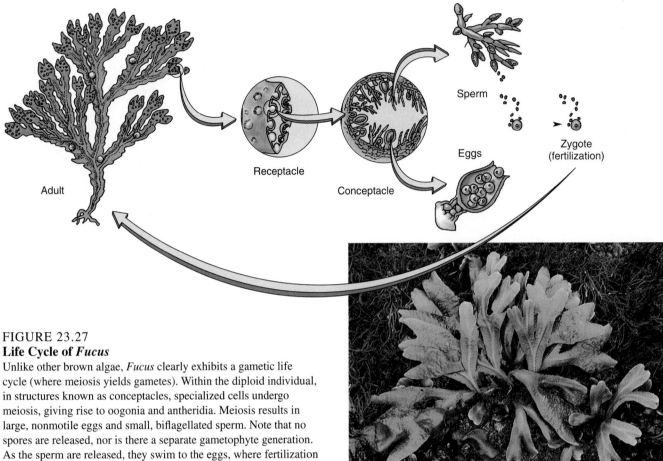

FIGURE 23.27
Life Cycle of *Fucus*

Unlike other brown algae, *Fucus* clearly exhibits a gametic life cycle (where meiosis yields gametes). Within the diploid individual, in structures known as conceptacles, specialized cells undergo meiosis, giving rise to oogonia and antheridia. Meiosis results in large, nonmotile eggs and small, biflagellated sperm. Note that no spores are released, nor is there a separate gametophyte generation. As the sperm are released, they swim to the eggs, where fertilization occurs. The diploid zygotes form new adult individuals. Thus *Fucus* is almost animal-like in its reproductive habits.

Fucus

ing snow, on the moist sides of trees, or in the soil. Chlorophytes include unicellular, colonial, and multicellular species. Many unicellular green algae have become photosynthetic symbionts in lichens, ciliates, and invertebrates.

Biochemically, the green algae closely resemble plants. They contain chlorophyll *a* and *b* and carotenoids. A chief storage carbohydrate is starch, and their cell walls contain cellulose impregnated with hemicellulose and pectin.

The green algae occur as unicellular flagellated or nonflagellated forms, as chains or filaments, and as delicate flattened blades. The group is so diverse that it is difficult to find a representative species, so we'll arbitrarily choose a few examples.

Unicellular Green Algae. *Chlamydomonas* (Figure 23.28) is a favorite organism of many biologists. It is easily grown and its genetics and physiology have been studied in detail. The cells have two flagella, a single light-sensitive **red eyespot**, and one cup-shaped chloroplast. Like *Euglena, Chlamydomonas* is positively phototactic—it moves toward light. *Chlamydomonas* usually exists in a haploid state, wherein it reproduces asexually by mitosis and, under ideal conditions, rapidly builds huge clone populations. It will not enter into a sexual cycle unless conditions are right.

The right conditions include the presence of opposite mating types [called plus (+) and minus (−) strains] and some form of environmental stress, such as the absence of nitrogen compounds. Then the haploid *Chlamydomonas* produces isogametes (identical gametes) mitotically. Isogametes pair up and fuse "head-to-head," entering a wildly spinning "nuptial dance" as their nuclei slowly fuse. Following the union, the diploid zygote forms a tough resistant wall, becoming a **zygospore**. The zygospore may remain dormant for a considerable time, or, if conditions are right, it may immediately enter into meiosis, producing four flagellated, haploid cells. As we see then, *Chlamydomonas* follows the more primitive zygotic life cycle, in which the haploid state dominates and the diploid state is but a brief interlude.

Colonial Green Algae. *Volvox* is another remarkable green alga (Figure 23.29). Although it is not related to any truly multicellular form, its history tantalizingly suggests what the earliest beginnings of multicellularity *might* have been like. *Volvox* is a spherical colony of virtually identical cells, each of which is very similar to an individual *Chlamydomonas* organism. So, in a sense, *Volvox* is a group of individuals in a sphere behaving as a single organism.

FIGURE 23.28
Zygotic Life Cycle of *Chlamydomonas*
The diploid zygospore is a brief interlude in an otherwise lengthy haploid life cycle. Zygospores $\boxed{1}$ enter meiosis and produce flagellated, haploid zoospores $\boxed{2}$. They may then enter an asexual phase $\boxed{3}$, producing enormous cloned populations through repeated mitotic divisions, or a sexual phase $\boxed{4}$, producing many isogametes. Sexual reproduction occurs when different (+ and –) mating strains meet. Flagellated isogametes fuse head-to-head $\boxed{5}$, producing diploid zygospores once again.

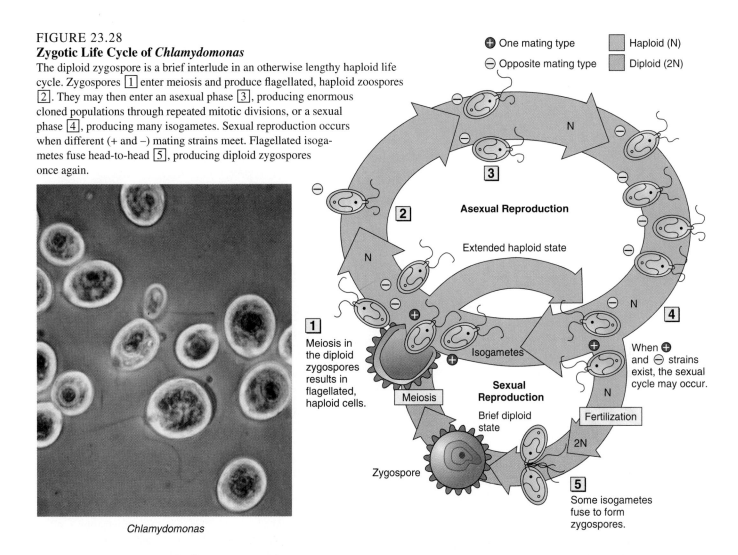

⊕ One mating type   ▨ Haploid (N)
⊖ Opposite mating type   ▨ Diploid (2N)

Asexual Reproduction

Extended haploid state

$\boxed{3}$
$\boxed{2}$
$\boxed{4}$

$\boxed{1}$
Meiosis in the diploid zygospores results in flagellated, haploid cells.

Meiosis

Zygospore

Isogametes

Sexual Reproduction

Brief diploid state

When ⊕ and ⊖ strains exist, the sexual cycle may occur.

Fertilization

2N

$\boxed{5}$
Some isogametes fuse to form zygospores.

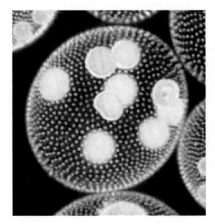

Chlamydomonas

FIGURE 23.29
***Volvox,* a Colonial Alga**
Volvox, a green alga, is a spherical colony of tiny, interconnected, flagellated cells. The dense spherical bodies seen within the colony are concentrations of cells that were produced asexually. The result is a miniature of the original colony, soon to escape and live on its own. In sexual reproduction, flagellated cells of the sphere differentiate into sperm- and egg-producing structures. With fertilization, a zygospore is produced. *Volvox* colonies are zygotic, so the zygospores enter meiosis, and the long haploid state is restored.

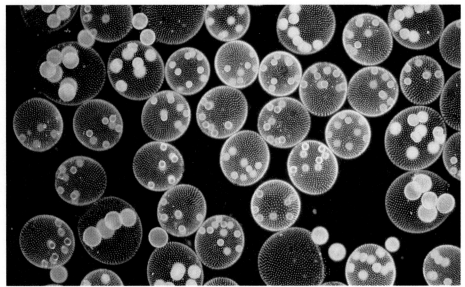

Volvox

FIGURE 23.30
Ulva **Life Cycle**

The cycle begins ☐1 when cells in the leafy, diploid sporophyte undergo meiosis, producing motile, haploid spores ☐2. *Ulva* is isomorphic, so the haploid spores form gametophytes ☐3. that are visually identical to the sporophyte. The gametophytes give rise to flagellated isogametes ☐4. When separate plus and minus mating types are present, the gametes fuse ☐5, forming zygotes ☐6. Fertilization represents the start of the diploid phase once again.

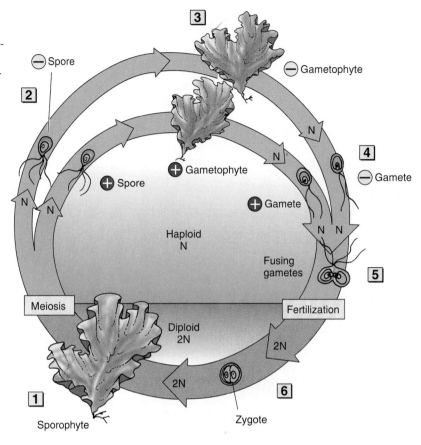

Volvox is interesting not only because it suggests a way in which multicellularity could have arisen, but also because it shows a kind of rudimentary differentiation within the colony (see Figure 23.29). The sphere swims in an organized manner, with the flagellated cells in front pulling and those in the rear pushing. In the sexual phase, some cells become specialized for reproduction, producing a few large eggs or many smaller sperm. In asexual reproduction, pockets of the sphere depress inward, and new spheres of cells pinch off and come to lie inside the parental sphere. The new, young spheres are inverted, however, with their flagella directed inward. The daughter *Volvox* colony has to turn itself inside-out within the parent structure. Eventually it is released through a hole in the parent colony.

Multicellular Green Algae. *Ulva* and *Ulothrix* are among the more structurally complex green algae. *Ulva,* familiar to tidepool enthusiasts as "sea lettuce," is broad and leaf-like, although only two cells in thickness (Figure 23. 30). Its leafy form is the product of cell division in two dimensions. On the other hand, cell division in *Ulothrix* and other filamentous algae occurs in one dimension only, forming hair-like filaments. The life cycle of the two species is also quite different. *Ulva* exhibits a clear-cut alternation of generations and, like the red alga *Polysiphonia,* is isomorphic. Both the spores and the gametes are flagellated and almost identical.

The themes of alternation of generations and of the tendency of one phase to become dominant continue in the evolution of the land plants. Although there is no strong fossil evidence strongly supporting this thesis, most authorities concur that the ancestors of the land plants were closely related to the ancestors of multicellular green algae. The green algae were probably the dominant form of aquatic life throughout the early Paleozoic era, giving rise to terrestrial plants some 400 million years ago. The brown and red algae remained in the sea, their evolutionary adaptations leading them in other directions.

Key Ideas

CHARACTERISTICS OF PROTISTS

1. The earliest reports of protists were made by amateur scientist Antony van Leeuwenhoek. Today there are 100,000 known living species.

2. Protists exist as individuals, in colonies, and as multicellular organisms. Nutritional types include photoautotrophs and heterotrophs, the latter including saprobes, parasites, and predators. Protists make use chiefly of cilia, flagella, and ameboid movement.

3. Asexual reproduction in protists includes the usual mitotic cell division and a variation called **fragmentation**. Sexual reproduction occurs through conjugation with (a) complete cell fusion and (b) exchange of haploid nuclei. In **autogamy**, self-fertilization occurs through the fusion of meiotic daughter cells. Gametes include **isogametes** (identical in form) or **heterogametes** (large stationary egg and small motile sperm).

4. The **gametic cycle** of animals and protists includes a lengthy diploid phase, followed by meiosis and gamete formation, fertilization, and resumption of the diploid state.

5. The **zygotic cycle** of protists and fungi includes a lengthy haploid phase, followed by sexual union, with or without gametes, meiosis, and resumption of the haploid state.

6. The **sporic cycle** of plants and the more complex algal protists includes a diploid, or **sporophyte**, phase, ending with meiosis and the formation of haploid spores. Spores form a haploid, multicellular **gametophyte** phase, where gametes are formed. Fertilization leads to resumption of the diploid phase.

7. Kingdom Protista includes **protozoans** (animal-like), **algae** (plant-like), and **slime molds** and **water molds** (fungus-like).

ANIMAL-LIKE PROTISTS: PROTOZOANS

1. Most flagellates have the 9 + 2 flagellum. They may capture food or simply absorb it through the body wall. *Trypanosoma gambiense,* the agent of **African sleeping sickness**, is one of several important flagellate parasites.

2. *Entamoeba histolytica,* an ameboid protist, is the water-borne agent of amebic dysentery. Most ameboid protists move about by the formation of pseudopods (false feet) and feed mostly through phagocytosis. **Heliozoans** have retractable microtubular **axopodia**, spine-like extensions covered by a moving ameboid cytoplasm used in feeding. **Radiolarians** and **foraminiferans** have hardened skeletons of silicon dioxide and calcium carbonate, respectively. Asexual reproduction in **sarcodines** is through mitosis and cell division. Sexual reproduction is rarely seen but usually involves cell fusion or the union of gametes.

3. **Sporozoans** are mainly spore-forming, nonmotile parasites. *Plasmodium vivax,* the agent of malaria, is transmitted from victim to victim by the female *Anopheles* mosquito. The mosquito injects **sporozoites** of *P. vivax* into the victim when it feeds. Each form thousands of **merozoites** in the liver. They invade the red blood cells, divide again, and release toxins that bring on the alternating fever and chills of malaria. Some enter a sexual stage, forming **gamonts** that undergo transformation into sperm and eggs in the mosquito.

4. In ciliates, cilia are used in movement and in feeding. Some ciliates release thread-like **trichocysts** for capturing prey and for mooring. Water balance is maintained by **contractile vacuoles**. Food drawn through the oral groove to the **cytostome** is taken into phagosomes for digestion. Wastes are expelled through exocytosis.

5. In the ciliate *Paramecium,* germ-line DNA (for reproduction) is maintained in one or more **micronuclei**, while synthetically active DNA is maintained in the **macronucleus**. Asexual reproduction includes mitosis and cell division. Sexual reproduction occurs through the exchange of haploid micronuclei during conjugation.

FUNGUS-LIKE PROTISTS

1. One **plasmodial slime mold**, *Physarum polycephalum,* has a giant multinucleate, ameboid feeding stage. Unfavorable conditions bring on the formation of sporangia, where meiosis gives rise to haploid spores. The spores germinate into ameboid **myxamebas** or flagellated **swarm cells** that fuse with others, becoming diploid and forming another feeding stage.

2. **Cellular slime molds** consist of small separate amebas that, under stressful conditions, swarm together into a spore-forming **slug** stage. Sexual reproduction may occur via fusion of haploid amebas, which then form **macrocysts**. The latter undergo meiosis and resume the ameboid feeding stage.

3. Many aquatic water molds are parasitic. Terrestrial species of oomycetes include the parasitic **downy mildews** and blights. The downy mildew *Phytophthora infestans* caused the Irish potato famine by infecting the potato plant. *Phytophthora* invades the leaf through the stomata and returns through these pores to produce spores.

4. Water molds have large egg cells and filamentous bodies. Although they resemble fungi, they produce flagellated spores and have cellulose cell walls. In asexual reproduction, they produce flagellated spores. Sexual reproduction include the fusion of sperm and egg.

PLANT-LIKE PROTISTS: ALGAE

1. Algae, the photosynthetic protists, range from unicellular to complex multicellular forms. Most are aquatic, living in all of the earth's waters. A few live on land. Many are part of the **phytoplankton**.

2. **Dinoflagellates** (phylum **Pyrrhophyta**), produce the famous **red tides**, where immense blooms deplete oxygen in the water, bringing about the death of fish and other marine life. Their toxins accumulate in shellfish and can be fatal to humans. **Bioluminescence** occurs through a chemical cycle involving **luciferin**, ATP, and **luciferase**. Dinoflagellates are flagellated and unicellular. There is a primitive form of cell division in which the single nuclear membrane does not break down. Reproduction is chiefly asexual, through binary fission.

3. *Euglena,* a unicellular representative euglenophyte, is a flagellated swimmer that orients to light through a light-sensitive **photoreceptor**. Its chloroplast contains chlorophyll *a* and *b* and carotenoids. Its storage carbohydrate is **paramylon**. It reproduces asexually by mitosis and transverse cell division: no sexual reproduction has been observed.

4. Chrysophytes inhabit fresh and salt water. The carotenoid **fucoxanthin** provides their yellow color. **Chrysolaminarin** is the storage carbohydrate. Some chrysophytes have glassy cell walls. Some are flagellated individuals; others live in colonies or are multicellular. As **diatoms** divide, their glass coverings get progressively smaller. The problem is solved during sexual reproduction, where motile gametes leave the wall, fuse in fertilization, and synthesize a full-size wall. Over time, vast accumulations of glassy skeletons formed **diatomaceous earth**.

5. Most **red algae** (**Rhodophyta**) are marine seaweeds that grow on rocky coasts, using **holdfasts** to fasten themselves in place. The storage polysaccharide is **floridean starch**. Other polysaccharides include **agar-agar** and **carrageenan**. During alternation of generations in *Polysiphonia,* meiosis in the sporophyte results in the production of haploid spores and the start of the gametophyte phase. The gametophyte produces nonmotile haploid gametes that fuse in fertilization, yielding a **carposporophyte**. Its spores later develop into the main sporophyte.

6. **Brown algae** (**Phaeophyta**) contain the pigment fucoxanthin and store their carbohydrates as **laminarin** and **mannitol**. *Fucus* has no alternation of generations and the life cycle is gametic, like our own. Brown algae include the giant kelps. Their specialized bodies have anchoring **holdfasts**, stem-like **stipes**, and large, leaf-like **blades**.

7. **Green algae** (**Chlorophyta**) occur as single cells, filaments of cells, and flattened multicellular blades. They contain chlorophyll *a* and *b* and produce cell walls of cellulose and hemicellulose—all plant characteristics.

8. Unicellular algae: *Chlamydomonas* is motile with two flagella, a **red eyespot**, and a zygotic life cycle. In its predominating haploid state, it reproduces asexually through mitosis. Sexual reproduction occurs when opposite mating types form isogametes that fuse in fertilization and form diploid **zygospores**. They may immediately undergo meiosis resuming the haploid phase.

9. Colonial algae: *Volvox* forms a spherical colony of small flagellated cells. Some specialize in movement, some as gametes, and others form small asexual spheres.

10. Multicellular green algae: *Ulva,* a multicellular marine form, has a distinct isomorphic alternation of generations. Filamentous green algae are important, since they are believed to represent the ancestral group from which plants arose.

Application of Ideas

1. The use of DDT has been greatly curtailed in the United States for important ecological reasons, yet it has proved in the past to be the greatest malaria deterrent known in parts of the world where other measures have been ineffective. Many organizations would like to see this form of mosquito control resumed. Discuss the issue of human health versus preservation of the environment. (Read *The Silent Spring,* by Rachel Carson, for some insight into the DDT controversy.)

2. Euglenophytes—in particular, *Euglena gracilis*—have been classified as plants by botanists, animals by zoologists, and protists by others. Review the characteristics of *Euglena* and discuss the basis for the taxonomic disagreement. In what ways does the protist designation help? What does such a problem reveal about the categories contrived by taxonomists?

3. One of the problems of controlling African sleeping sickness has been the movement of nomadic tribes and their cattle in and out of endemic areas. Your job as an official attempting to control the spread of the disease is to explain to the tribes why they can no longer move about freely and why some of their cattle have to be disposed of for health reasons. How would you explain the life cycle of the trypanosome and the spread of sleeping sickness to these unsophisticated people?

4. Some protists, particularly very large ciliates, seem to represent exceptions to the cell theory (see Chapter 4), and the term "unicellular" or "single-celled" seems inappropriate. Using one of the larger ciliates as an example, make a case for using the term "acellular." What might substitute for the term organelles in an acellular organism?

Review Questions

1. List the three organizational states seen in protists. From an evolutionary viewpoint, which seems to be most advanced? (page 445)

2. Write brief definitions for the following: fragmentation, isogamete, heterogamete, and anisogamete. (page 447)

3. Using a circular diagram, characterize the zygotic life cycle. List two groups of organisms in which this cycle occurs. (page 447)

4. Briefly discuss the one overriding characteristic of sporic life cycles. To which groups of organisms does this type of cycle pertain? (page 447)

5. What causes African sleeping sickness? Why has it been so hard to control its spread? (page 449)

6. Explain what axopodia are, where they are found, and how they are used. (pages 450–451)

7. List three characteristics of the sporozoans. (page 451)

8. Briefly summarize the life cycle of *Plasmodium vivax* and relate the events to the symptoms of malaria. (pages 451–452)

9. Discuss the peculiar organization of DNA in *Paramecium*. (page 451)

10. Describe the manner in which late blight infects potato plants. (page 457)

11. List several animal-like and several plant-like characteristics of *Euglena*. (page 460)

12. Describe the problem created by asexual reproduction in diatoms and explain how it is solved. (pages 461–462)

13. List the characteristic pigments and polysaccharides of the brown algae. (page 464)

14. Briefly characterize the life history of the brown alga *Fucus*. Why is its sexual reproduction considered advanced? (page 465)

15. Summarize the life history of *Chlamydomonas*. Is this protist primitive or advanced? Explain your answer? (pages 466–468)

16. Using *Volvox* as an example, explain the colonial form of life. How does this differ from multicellularity? (page 468)

CHAPTER 24

Fungi

CHAPTER OUTLINE

WHAT ARE FUNGI?

The Fungal Body

Feeding in Fungi

Ecological Importance

Medical and Commercial Uses

Reproduction in Fungi

Fungal Relationships

KINGDOM FUNGI

Zygomycota: Conjugating Molds

Ascomycota: Sac Fungi

Basidiomycota: Club Fungi

Essay 24.1 A Fungus that Spits

EVOLUTION OF FUNGI

KEY IDEAS

REVIEW QUESTIONS

*H*ave you ever been walking in a damp woodland and suddenly come upon a cluster of delicate, colorful mushrooms on the forest floor (Figure 24.1)? It can be an intriguing experience, especially if you don't know much about mushrooms. They have a sense of mystery about them, possibly because they are so often associated with dark, wet forests—primal places. If you have experienced a moment like this, you may be surprised to hear that the mushroom is a fungus. A fungus! The word *fungus,* for most people, does not conjure up pleasant images. Fungi are fuzzy things that grow on our food and lurk on shower floors. Some even invade our bodies where they may cause embarrassing infections.

But let's regain some objectivity and have a closer look. Maybe we can convince you that fungi are important and useful. Like bacteria, fungi are among nature's recyclers—those organisms whose activities help keep essential mineral nutrients available to plant life. Some, as you will learn, are quite useful to us in more direct ways. Imagine life without sautéed mushrooms, antibiotics, bread, and some of our favorite cheeses—not earth-shattering losses in every case, but the thought does give you an idea of how important some fungi are to us.

What Are Fungi?

The Fungal Body

Most fungi are multicellular, but with relatively simple bodies. The typical fungal body, or **mycelium** as it is known, often consists of extensive, spreading, thread-like filaments called **hyphae** (singular **hypha**) (Figure 24.2). A hypha may be a number of individual cells joined end to end, or it may be tube-like without cell crosswalls. In such hyphae the cytoplasm is multinucleate, or **coenocytic**; mitosis occurs without cytoplasmic division (as in plasmodial slime molds; see Chapter 23). In many fungi, the hyphae mass together at times, forming large, dense reproductive structures such as the familiar mushroom cap or the "shelf" of the shelf fungus.

The extensive growth of the mycelium must be supported by an efficient nutrient supply system. In fungi, cytoplasmic streaming distributes food and other materials throughout the mycelium. Fungal growth is so extensive in fertile soil that it has an important effect on soil texture itself, greatly increasing its water-holding characteristics. Fungi tolerate a variety of growth conditions, but the optimal conditions include the presence of moisture, food, and warmth. They also grow best in darker places, as they are inhibited by bright sunlight, which can parch the delicate mycelium.

Mycelial growth rates can be phenomenal; a single mushroom can produce up to a kilometer of new hyphae in only one day. Most of the metabolic activity during growth occurs at the very tips of the hyphae (Figure 24.3). Studies of this region reveal that most of the nuclei, mitochondria, and synthetic organelles occur just behind the fast growing tip. The tip, or **dome** as it is known, is crowded with newly forming vesicles carrying materials and enzymes needed for plasma membrane and cell wall synthesis. The membrane expands as it is joined by vesicles that secrete cell wall material outside. As the dome proceeds forward, cell wall hardening occurs just behind. Fungal cell walls are composed of chitin, coincidentally, the material of the arthropod exoskeleton.

FIGURE 24.1
The Fungal World
The cluster of mushrooms probably originated when spores from some distant mushroom were carried by air currents to this ideal location. Here dead vegetation, moisture, and warmth made growth possible. Unseen beneath the surface is a vast, spreading maze of fungal growth.

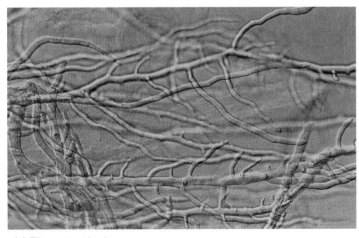

(a) Filamentous hyphae

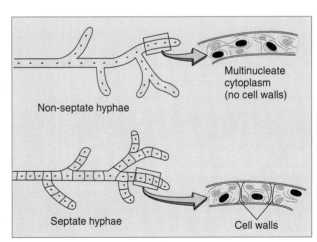

(b) Types of hyphae

FIGURE 24.2
Fungal Body
(a) Most fungi grow by producing branching, filamentous hyphae. These are fine tubes of cytoplasm with a surrounding chitinous wall that penetrate the substrate to obtain food. The entire mass of hyphae is known as the mycelium. **(b)** There are two types of hyphae, one that is multinucleate and nonseptate (without crosswalls), and one that is septate (with crosswalls).

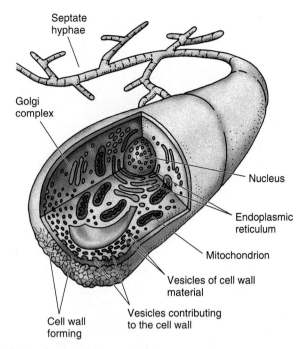

FIGURE 24.3
Metabolic Activity in the Hypha
The most metabolically active parts of the mycelium are the growing tips. Located there are most of the cellular organelles including nuclei, mitochondria, Golgi complexes, and rough endoplasmic reticulum. Vesicles budding from the Golgi complexes carry materials to the tips, where they are deposited into the membrane and wall to accommodate growth.

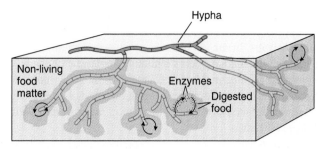

(a) Feeding in saprobic fungi

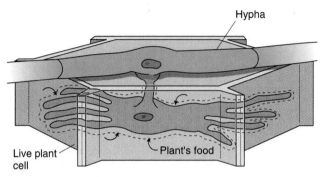

(b) Feeding in parasitic fingi

FIGURE 24.4
Feeding Hyphae
(a) Saprobic bacteria feed by secreting enzymes from the hyphae. The enzymes break down organic matter into simpler components that subsequently enter the hyphae through diffusion and active transport. **(b)** Many parasites produce specialized hyphae called haustoria. They penetrate the cell and branch out, absorbing nutrients directly from the host cytoplasm.

Feeding in Fungi

All fungi are heterotrophs, which means they must have organic substances—carbohydrates, amino acids, fatty acids, vitamins, and minerals, although the exact requirements differ from group to group. These essentials must be obtained from other organisms. There are two principal ways this happens. **Saprobes**, or decomposers, feed off the dead remains and wastes of other organisms, whereas **parasites** feed from the cells and tissues of a living host. A small number can accurately be classified as **predators**, since they capture and feed on live prey—a surprising state for a fungus.

Like bacteria, fungi feed by **absorption** (Figure 24.4a). The hyphae secrete digestive enzymes into their surroundings, which happen to be their foods, and the products of digestion diffuse or are actively transported back into the cells. Many parasitic fungi produce special hyphae called **haustoria**, which penetrate cells, making it possible for those fungi to absorb nutrients directly from the host's cytoplasm (Figure 24.4b).

Ecological Importance

Decomposition. As we've indicated, fungi share with bacteria the role of decomposers, breaking down the wastes and remains of organisms. Their metabolic activities release carbon compounds and ions of phosphorus, calcium, potassium, and sulfur into the soil and water, making them available first to plants and then to organisms that feed on plants. Eventually, things come full cycle as the organisms die and their nutrient-rich bodies are returned to the decomposers. How widespread are the decomposers? People who know about such things estimate that there is about 1 ton of bacterial and fungal life present in the upper 8 inches of an acre of fertile soil. We will consider the cycling of essential nutrients in more detail when we come to ecology.

Of course, decomposition has another side as well. Fungal heterotrophs cannot distinguish among fallen trees, dead squirrels, and valuable refrigerator leftovers. In fact, they attack anything that is organic, so add our clothing, stored goods, books, and houses to the list.

Parasitism in Fungi. The list of parasitic activities among the fungi is long and infamous, ranging from rose mildew to athlete's foot. The fungal parasites of plants can be particularly devastating, so much so that our well-being, on a worldwide scale, depends on our ability to control the parasites of food crops, and especially those of wheat, rye, corn, rice, and other grains. To the list, we must add that parasitic fungi continue to threaten the very existence of certain trees, including the graceful American elm and chestnut tree that were once so common in our eastern forests.

Mutualism in Fungi. Most land plants (estimates reach 95%) depend on a mutualistic (mutually beneficial) association between their roots and the mycelia of certain fungi. The fungus–root associations are called **mycorrhizae** (singular, *mycorrhiza*; "fungus roots"). Here the fungal hyphae surround or penetrate the root, vastly increasing the total absorbing area. The fungi expend energy in the absorption of minerals, which they deposit around the root hairs or within the root cells, thereby placing the minerals at the plant's disposal. In turn, they help themselves to sugars and other nutrients the plant produces. We will look into the details of this amazing relationship in Chapter 30.

Medical and Commercial Uses

We have learned to put heterotrophic fungi to work commercially. Various species are used in the manufacture of cheeses, antibiotics, linen, bread, wine, and beer. **Cyclosporine** is one of a long list of pharmaceutical products derived from soil fungi. It is widely used today in organ transplantation since it can suppress the body's natural immune response with relative safety, thus preventing transplant rejection. Earlier immunosuppressant drugs caused a variety of unacceptable side effects.

Reproduction in Fungi

Asexual reproduction in fungi usually involves the formation of **spores**, a fast way of increasing numbers and disseminating the species. Fungi are prolific spore formers, with each individual capable of forming great numbers of the minute bodies (Figure 24.5) (a single puffball, for instance, can produce trillions of spores). Spores are formed by mitosis and cell division throughout most of the life cycle, and through meiosis in sexual phases. In fungi, asexual spores usually consist of a haploid nucleus, a cytoplasm greatly reduced by dehydration, and a protective covering or spore case. The minute spores are readily transported by air currents and water. Since the fungal mycelium is fixed in place, such dispersal provides a means for the organism to establish itself in new food sources, often at some distance away. Spores can also survive long periods where growth conditions are not favorable (for example, dryness, heat, and intense sunlight), so they also represent a survival mechanism.

Sexual reproduction also occurs in fungi. They follow what is clearly a zygotic life cycle (Figure 24.6a). Recall from Chapter 23 that zygotic organisms remain haploid through most of their life cycle, reaching a diploid state only when fertilization occurs. We also noted that the diploid state is notoriously brief, followed immediately by meiosis (which in fungi usually leads to more spores) and the return to the haploid

Penicillium

FIGURE 24.5
Asexual Fungal Spores
Asexual reproduction in fungi is generally through spore formation. Asexual spores are resistant cells, produced through mitosis, that are released from the organism in vast numbers.

state. Such an extensive haploid state is, as we've seen, common in the protists as well (see Chapter 23). While it is the rule in fungi, there is an important, peculiar, variation on this theme: the *dikaryotic state.*

The Dikaryotic State. Sex in fungi occurs through simple conjugation (the fusion of cells), which occurs between different mating strains, often referred to as (+) and (–) (there are no males and females). The result is a new cell receiving a haploid nucleus from each mating strain. In some fungi the haploid nuclei simply fuse and fertilization is complete. But other fungi undergo a peculiar, *delayed fertilization.* Following the coming together of cells in conjugation, the actual fusion of the two haploid nuclei is delayed while an extensive new growth occurs and reproductive structures emerge. Cells bearing such paired nuclei are said to be in a **dikaryotic** ("double-nucleate") state, what we might call the "N + N" condition. The two nuclei, chastely residing side by side, go through round after round of mitosis as the mycelium grows. When nuclear fusion (fertilization) finally does occur, the zygotes, in typical fungal fashion, enter meiosis and resume the haploid state (Figure 24.6b).

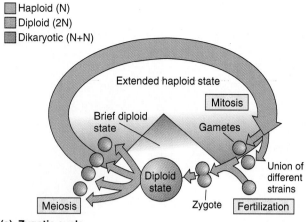

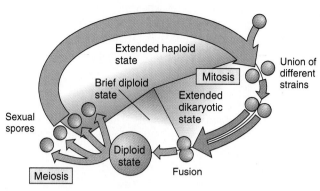

(a) Zygotic cycle

(b) Zygotic cycle with extended dikaryotic state

FIGURE 24.6
Zygotic Life Cycle
The fungal zygotic life cycle occurs in two ways. **(a)** In the more direct cycle, most of the fungal life cycle is spent in a haploid state, with only the briefest diploid pause following fertilization. Asexual spores are produced mitotically throughout the extended haploid state, and sexual spores are produced following meiosis in diploid individuals. **(b)** In a number of fungi, the fusion of hyphae between different strains does not lead directly to fertilization and the true diploid state. Instead, nuclei brought together remain separate in what is called a dikaryotic state. A considerable time may go by before the nuclei fuse and fertilization is completed.

Fungal Relationships

Fungi were once included in the plant kingdom, mainly because of their lack of motility, the presence of cell walls, and the production of spores in special structures. But as researchers unveiled more and more differences between fungi and plants, the two were finally placed in different kingdoms. The differences are indeed numerous. Plants are almost all phototrophs, whereas fungi are heterotrophs. And whereas plant cell walls are essentially cellulose, fungal cell walls are composed of chitin. Unlike plants, many fungi carry on a primitive form of mitosis and meiosis seen elsewhere only in protists such as dinoflagellates (see Chapter 23). Here the nuclear envelope is not dismantled during mitosis but remains intact, constricting between the clusters of daughter chromosomes. In some species, the nuclear envelope is only partially broken down or is dismantled late in the process. Curiously, the mitotic spindles are formed *within* the nucleus. Furthermore, it was recently discovered that, compared to those of plants and other eukaryotes, the fungal chromosomes have very little histone protein bound to them (reminiscent of prokaryotes). Like most plants, fungi lack centrioles. This is not all that surprising, however, since centrioles are characteristic of organisms that have cilia and flagella in at least some stage, and these are absent in the fungi. In the past, some of the flagellated protists were assigned to the fungi, but in the scheme chosen for this text, all flagellated fungal-like species have been assigned to the protists. We will look further into the evolution of fungi later in the chapter.

Kingdom Fungi

The 100,000 named species of fungi occupy three divisions[1] and a provisional group. They are as follows:

1. **Zygomycota** (conjugating molds, including bread molds).
2. **Ascomycota** (the sac fungi, including truffles and baker's yeast).
3. **Basidiomycota** (the club fungi, including mushrooms).

There is a fourth grouping, the fungi imperfecti, listed as division **Deuteromycota** in some schemes. This is only a provisional grouping of fungi whose sexual cycles are unknown. Typically, as species become better known, they are classified as sac fungi (Table 24.1).

Zygomycota: Conjugating Molds

The zygomycota are characterized by hyphae that do not form complete crosswalls, or septa, and the cytoplasm is continuous, or coenocytic. The name "zygomycota" itself is used because newly formed zygotes enter a spore stage called a **zygospore**. Zygospores are also common among algae such as *Chlamydomonas,* where we find the zygotic life cycle as well. One of the most unusual zygomycota, *Pilobolus,* has a bizarre way of disseminating its spores. You'll want to read about it in Essay 24.1, page 487.

[1]The designation "division" is traditionally used by botanists, where zoologists use the term "phylum."

TABLE 24.1
The Fungi

Division	Characteristics
Zygomycota (bread molds)	The familiar bread molds, common on all starchy foods. Simple, tubular hyphae without crosswalls; asexual reproduction by spores; sexual reproduction by conjugation and formation of zygospores
Ascomycota (sac fungi)	Includes powdery mildews, blue and green molds, and others. Hyphae with crosswalls; asexual reproduction through spores; sexual reproduction by conjugation and establishment of a dikaryotic state; often elaborate sexual structures; fertilization occurs in sac-like asci, followed by meiosis and spore formation
Basidiomycota (club fungi)	Includes familiar mushrooms, bracket and shelf fungi, along with corn smuts and wheat rusts. Hyphae with crosswalls; asexual reproduction by spores; sexual reproduction by conjugation and establishment of a dikaryotic state; often elaborate, sexual structures; fertilization occurs in club-like basidia, followed by meiosis and spore formation
Deuteromycota	Mixed group whose sexual stages are incompletely known. Many have been added to the sac fungi, includes penicillin mold, parasite of athlete's foot, thrush , and parasitic yeast. Asexual reproduction through spore formation

Rhizopus: **Black Bread Mold.** The common bread mold, *Rhizopus stolonifer,* turns up in everyone's refrigerator at one time or another (Figure 24.7). But don't look only on bread for *Rhizopus;* it grows vigorously on many foods. During its growth, the bread mold develops long horizontal hyphae known as **stolons**, which venture along the surface of the food, sending root-like growths called **rhizoids** deep into the food (Figure 24.7a).

Asexual Reproduction. Much of the bread mold growth is invisible to the unaided eye, but another part of the mold's growth is quite prominent. You may have noticed the fuzzy growth and tiny black dots along the edge of your sandwich just after you took your first bite (a feeling not unlike finding half a worm in your apple). The fuzzy growths are aerial hyphae called **sporangiophores**, whereas the black dots atop them are **sporangia** (see Figures 24.7 and 24.8). In addition to suggesting a late night trip to the market, their presence is a sure sign that *Rhizopus* has begun asexual reproduction. The black spores arise through mitosis in the cells of the sporangium.

When released, the spores are lofted by air currents, and the rest is up to chance. If favorable growth conditions are encountered, the spores germinate. They begin by taking in water and swelling (see Figure 24.7). Then metabolic activity, including protein synthesis, begins. Next, a region on the spore wall softens and a cytoplasmic growth protrudes—the start of a new hypha. Since many of the spores fall near the sporangium that bore them, the food may become covered with new growth almost overnight.

Sexual Reproduction. Rhizopus has a fairly simple sexual phase involving a special form of conjugation (Figure 24.8).

When plus (+) and minus (−) mating types are grown in the same medium, they release chemical attractants that stimulate the growth of sexual structures. The two strains respond by producing club-shaped outgrowths that form closed chambers called **gametangia**, each containing many haploid nuclei. Next, the plus and minus gametangia merge, permitting haploid plus and minus nuclei to fuse, completing fertilization. The common chamber, now containing a number of diploid nuclei, then forms a tough-coated, resistant **zygospore**.

The diploid period may be short, or the zygospore may remain dormant for a considerable period. But before anything else happens, meiosis will occur within the zygospore and many haploid nuclei will form. This ends the diploid state. Shortly afterward, the zygospore will germinate and a hypha will emerge and immediately produce a sporangium. The haploid nuclei produced earlier are then incorporated into spores and released. The new spores are quite varied. They represent the union of two genetically different strains and are the product of the usual gene shuffling that goes on during crossing over in meiosis.

Ascomycota: Sac Fungi

The divisions Ascomycota and Basidiomycota are often referred to as *higher fungi.* This elevation in status is primarily due to their **septate** condition; that is, cell divisions are routinely accompanied by the formation of **septa** (crosswalls). Such septa may not be complete, however, and both cytoplasm and nuclei may still pass from cell to cell.

The **Ascomycota**, or **sac fungi** as they are also known, include some familiar members such as the edible morels and truffles, the powdery mildews (not to be confused with the "downy mildews"; see Chapter 23), and the blue and green

(a) *Rhizopus* on strawberries

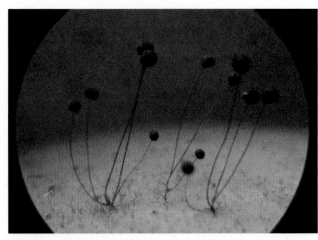

(b) *Rhizopus sporangia*

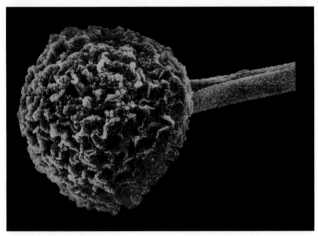

(c) *Rhizopus* spores

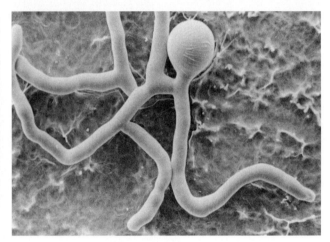

(d) Germination in *Rhizopus* spores

FIGURE 24.7
Rhizopus, A Black Bread Mold
(a) The haploid mycelium of *Rhizopus* grows in a tangled mat throughout the medium and is the typical thread-like growth. **(b)** Black asexual spores are produced on rounded sporangia that form at the tips of slender, upright sporangiophores. **(c)** Each sporangium produces a great many spores. **(d)** When a spore finds its way into the right environment, germination begins. The spore takes in water and swells, whereupon the first hypha emerges.

molds of citrus fruit. Some yeasts are also placed in this group. The sac fungi also include well-known parasites such as the infamous chestnut blight, *Endothia parasitica,* an Asian mildew that brought about the destruction of vast forests of the American chestnut. Dutch elm disease, an equally disastrous plague of American forests, is produced by the sac fungus *Ceratocystis ulmi.* In the southwestern United States, a human sac fungus disease called **coccidiomycosis,** or "valley fever," affects the lungs in much the same way as tuberculosis. Fortunately, most people are resistant. *Claviceps purpurea,* a parasite of rye and other grasses, is a sac fungus with rather ghastly implications for our species. Let's consider this one carefully.

Claviceps produces a plant disease called **ergot,** which is characterized by the formation of a compact, black mycelium in rye plants. Bread that has been made from "ergoted" flour contains certain alkaloids with the peculiar ability to constrict blood vessels in the body extremities. People who continue to eat bread made from such flour suffer from **ergotism.** Behavioral symptoms include hallucinations, psychotic delusions, and convulsion. Some historians believe that the peculiar behavior of the "witches of Salem" and the "voices" heard by Joan of Arc were actually products of ergot poisoning. Ergotism may also be accompanied by burning sensations of the hands and feet and, in extreme cases, gangrene. The burning sensation led to the name "Saint Anthony's fire." Although

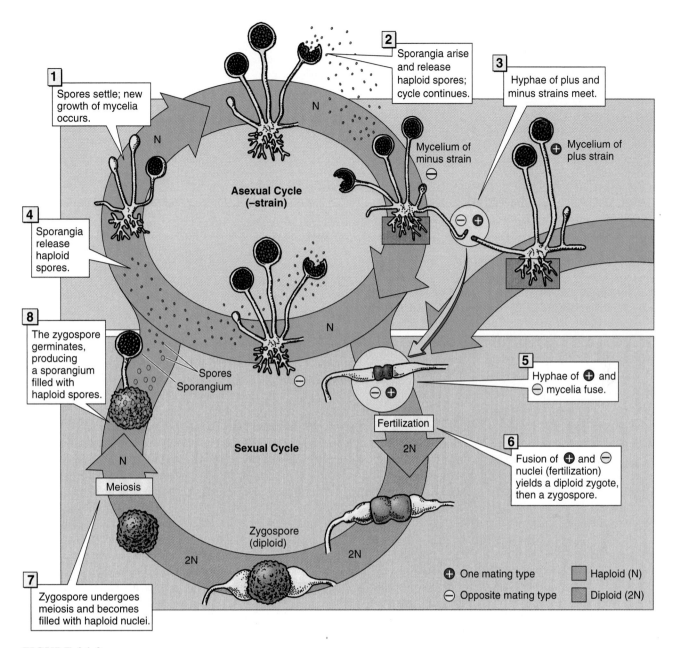

FIGURE 24.8

Asexual and Sexual Reproduction in *Rhizopus*

In its asexual phase, ⌐1⌐–⌐3⌐), *Rhizopus* goes through round after round of spore formation, germination, and growth of new haploid mycelia. This is the stage we see when the bread mold attacks our stored foods. The sexual phase begins ⌐4⌐ when mycelia from opposite mating strains, plus (+) and minus (−), meet. Then, special, club-shaped sexual structures called gametangia form ⌐5⌐. Fusion of the gametangia ⌐6⌐ permits plus and minus nuclei to join, bringing on a brief diploid state. The surrounding structure then forms a zygospore. Meiosis then occurs in the diploid nuclei ⌐7⌐. Upon germination of the zygospore ⌐8⌐, a slender hypha will emerge, produce a sporangium, and begin forming haploid spores which are released into the air. The spores then enter the asexual phase.

ergotism is rare, the ongoing possibility requires vigilance on the part of agricultural authorities.

Can there be anything good to say about *C. purpurea?* The destructive fungus is the source of some useful drugs, including tranquilizers and migraine headache remedies. In cases of uncontrolled bleeding after childbirth, ergot deriva-tives are used to promote contraction of the uterine muscles, which helps close the bleeding vessels.

Asexual Reproduction. During asexual reproduction, sac fungi produce spores known as **conidia** (singular, *conidium*). Conidia are produced mitotically in long chains at the ends of

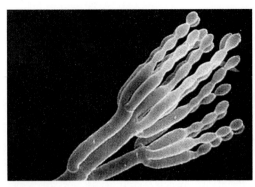
Conidiophore and spores

FIGURE 24.9
Asexual Spore Formation in Sac Fungi
The spore-forming bodies seen here are called conidiophores. On each conidiophore, the terminal cell rounds out and matures into a spore called a conidium, followed by the one below and so on, until a long chain of such haploid spores forms. They then will break away and become airborne, some landing on new food sources.

numerous specialized hyphae known as **conidiophores** (Figure 24.9).

Sexual Reproduction. Sexual reproduction among the sac fungi is quite varied, but there are some underlying themes. The name "sac fungus" itself comes from the sac-like **ascus** (plural, **asci**) in which **ascospores** are produced during sexual reproduction. Asci are commonly found in a structure known as an **ascocarp**, which occurs in a variety of shapes. To view the life cycle we have chosen the sac fungus *Peziza* as our representative example (Figure 24.10).

In the asexual cycle, intermittent periods of asexual spore formation occur, with spores giving rise to new haploid hyphae. But when the hyphae of opposite mating types of *Peziza* come into contact, each produces large multinucleate swellings (see Figure 24.10). One, the **ascogonium**, the equivalent of a female sexual structure, contains many haploid nuclei that remain in place. The other, the **antheridium**, a male structure, also contains numerous haploid nuclei, but they do not remain in place. A bridge-like conjugation tube forms between the two bodies, and haploid nuclei from the antheridium cross to enter the ascogonium. But unlike the situation in *Rhizopus,* the haploid nuclei do not fuse. They remain separated, side by side in each cell, as they enter into the dikaryotic (N + N) state mentioned earlier. They remain separate until the prominent reproductive structure, the ascocarp, is completed (like an engaged couple, putting off the nuptials until their dream house is finished).

The dikaryotic cells, along with hyphae of the two original mating types, cooperatively produce the ascocarp. In *Peziza,* the ascocarp is a hollow, cup-like chamber containing the young dikaryotic asci (sacs) (see Figure 24.10). It is within the asci that haploid nuclei finally fuse, ending the dikaryotic state. The

long-awaited union may immediately be followed by meiosis, or it may be delayed. When meiosis does occur, the four haploid daughter cells undergo one round of mitosis each, and the eight emerging cells then begin the formation of ascospores.

When the ascospores are fully mature, the ascus breaks open, and the spores disperse. Unlike the asexual spores, which are genetically identical, each ascospore contains a unique genome, the product of genetic recombination that began with fertilization and continued during meiosis. Each is capable of producing a new mycelium, which will remain haploid until another sexual encounter ushers in the brief diploid phase.

Yeasts. Sac fungi also include certain **yeasts**. "Yeast" is a general category, referring to any unicellular fungus that reproduces asexually by budding. Actually, yeast-like species are seen in all fungal groups. Of interest here are those yeasts that carry out the fermentation process so important to bakers and brewers.

Budding includes mitosis and cytokinesis, but the cytoplasmic division, incomplete at first, is highly unequal. The buds (the parts receiving the lesser amount of cytoplasm) remain attached, usually growing to full size before separating from the parent cell. Common baker's yeast can be diploid or haploid, and either form can undergo asexual reproduction by budding (Figure 24.11).

When yeasts reproduce sexually, the diploid cells undergo meiosis (see Figure 24.11) and four haploid ascospores appear inside the original cell wall. Thus the cell itself becomes an ascus (which is why these yeast's are classed as sac fungi).

Two of the four ascospores will be the alpha mating type, and two will be the *a* mating type. The alpha and *a* can undergo immediate fusion to regain the diploid state, or fusion can occur much later, as long as the two types are present. Generally, diploid strains do not undergo meiosis when conditions are good; meiosis usually is induced by food shortages and drying. The formation of many genetically varied spores is the yeast's way of coping with an uncertain future.

Like other eukaryotes, yeasts normally have mitochondria. When oxygen is available, they go aerobic, oxidizing sugars completely to carbon dioxide and water. But mitochondria cannot function under anaerobic conditions, and in the absence of oxygen, yeasts derive their energy principally from simple alcoholic fermentation (see Chapter 8). The end products of alcoholic fermentation are ethyl alcohol and carbon dioxide, along with a small yield of ATP. The first two products are immensely important to humans and have been throughout history (the ATP is important to the yeast). The carbon dioxide gas given off is what makes bread rise and gives some beverages their bubbles. The ethyl alcohol is what brewers and vintners are after when they make beer and wine. Yeasts don't know any of this so they just continue to ferment until their own alcohol waste inhibits further growth. In genetically selected vintner's yeast, growth stops when the concentration of alcohol reaches about 13%, which is the percentage of alcohol in unfortified ("light") wines. (Wild yeast strains can only tolerate 4–5% alcohol before their metabolism stops.)

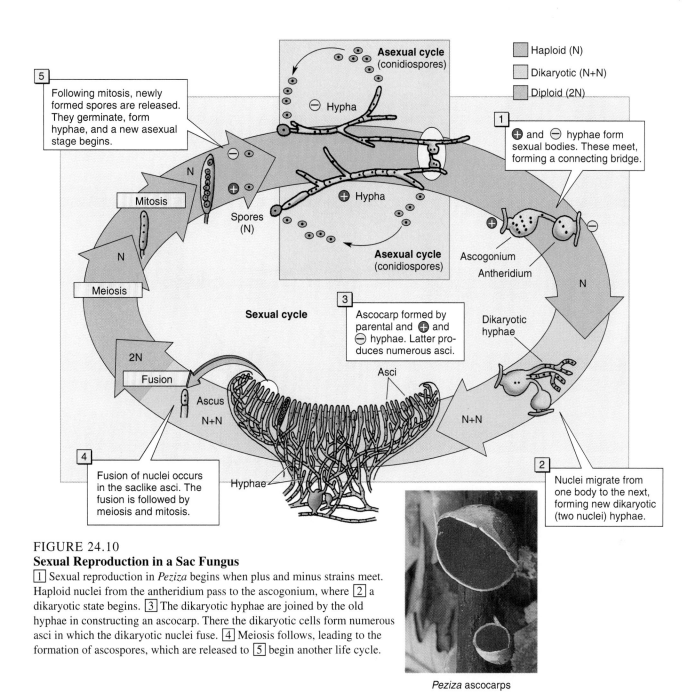

Haploid (N)
Dikaryotic (N+N)
Diploid (2N)

Asexual cycle (conidiospores)

⊖ Hypha

5 Following mitosis, newly formed spores are released. They germinate, form hyphae, and a new asexual stage begins.

1 ⊕ and ⊖ hyphae form sexual bodies. These meet, forming a connecting bridge.

N

Mitosis

Spores (N)

⊕ Hypha

Asexual cycle (conidiospores)

⊕

Ascogonium
Antheridium

⊖

N

N

Meiosis

Sexual cycle

3 Ascocarp formed by parental and ⊕ and ⊖ hyphae. Latter produces numerous asci.

Dikaryotic hyphae

2N

Fusion

Ascus

N+N

Asci

N+N

4 Fusion of nuclei occurs in the saclike asci. The fusion is followed by meiosis and mitosis.

Hyphae

2 Nuclei migrate from one body to the next, forming new dikaryotic (two nuclei) hyphae.

FIGURE 24.10
Sexual Reproduction in a Sac Fungus

1 Sexual reproduction in *Peziza* begins when plus and minus strains meet. Haploid nuclei from the antheridium pass to the ascogonium, where 2 a dikaryotic state begins. 3 The dikaryotic hyphae are joined by the old hyphae in constructing an ascocarp. There the dikaryotic cells form numerous asci in which the dikaryotic nuclei fuse. 4 Meiosis follows, leading to the formation of ascospores, which are released to 5 begin another life cycle.

Peziza ascocarps

Deuteromycota: Fungi Imperfecti. A fungal group with the name "fungi imperfecti" will need more explaining. (*Imperfecti* is Latin for "imperfect," a term mycologists and botanists use when referring to plants with sexual structures missing.) At one time, some 24,000 named species of fungi with no known sexual phase were placed in this provisional group. Eventually, as their sexual stages became known, most were placed with the sac fungi.

Fungi imperfecti include a number of well-known human parasites, including *Trichophyton mentagrophytes,* the agent of athlete's foot, and *Candida albicans,* the yeast that causes troublesome vaginal infections and a mouth infection called

"thrush." Yeast infections often follow prolonged episodes of antibiotic therapy. During such treatment bacteria and fungi that are harmless symbiotic residents of the body are destroyed. Without their presence, there is little competition, and the antibiotic-resistant *C. albicans* becomes a successful invader.

On a more positive note, there are also some useful species. The genus *Penicillium* is a source of the important antibiotic penicillin. From the same genus, consider *Penicillium roquefortii* and *Penicillium camembertii,* species whose names give them away as agents used in the flavoring of two famous cheeses. Then there is *Aspergillus oryzae,* which, along with lactic acid bacteria, is responsible for the special flavor of soy

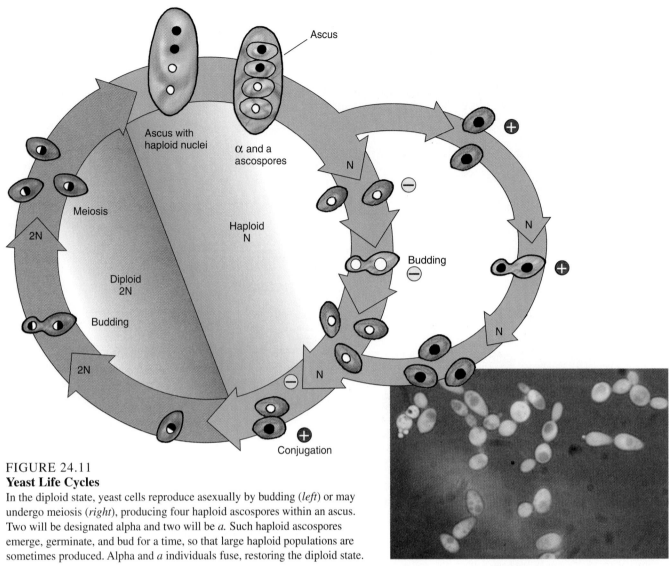

FIGURE 24.11
Yeast Life Cycles
In the diploid state, yeast cells reproduce asexually by budding (*left*) or may undergo meiosis (*right*), producing four haploid ascospores within an ascus. Two will be designated alpha and two will be *a*. Such haploid ascospores emerge, germinate, and bud for a time, so that large haploid populations are sometimes produced. Alpha and *a* individuals fuse, restoring the diploid state.

Budding yeasts

sauce and, in part, for the fermenting of saki. *Aspergillus oryzae* is also used to improve the nutritional value of livestock feeds.

Some fungi imperfecti are predators; they actually catch prey! As you read this, certain tiny roundworms that live in the soil are being captured by a strange fungus called *Arthrobotrys* (Figure 24.12). The fungus develops ring-like hyphae that constrict like an inflated noose when a hapless roundworm inadvertently passes through. Then fungal haustoria enter the prey and digestion begins.

Lichens Many sac fungi, and some Basidiomycota, live in intimate associations with certain algae or cyanobacteria, forming the unique growth referred to as **lichens**. The example in Figure 24.13 shows the organization of one lichen and how it reproduces. Note the layered arrangement of algal and fungal components and the formation of ascospores. Lichens also exhibit a vegetative form of asexual reproduction. **Soredia**, minute bodies consisting of algal cells surrounded by

fungal hyphae (Figure 25.13c,d), are released into the surroundings. Under suitable conditions, soredia will begin the growth of a new lichen complex. Lichens also reproduce asexually through fragmentation, where less organized particles form new growths.

The algal component of the lichen provides photosynthetic products (sugars) to the fungus in exchange for moisture, housing, and anchorage. This symbiotic association is usually considered to be mutualistic—beneficial to both parties. But it has been demonstrated that the algal or cyanobacterial partner can get along without the fungus, so, in reality, one benefits more than the other. The fungus, however, thrives only if supplied with the complex nutrients it normally gets from its partner. For this reason, some authorities consider this often-touted example of pure mutualism to be actually parasitism. (If they are correct, we have just lost another long-cherished notion.)

About 17,000 lichen combinations are known. They are distributed worldwide, some living on bare rock, others on

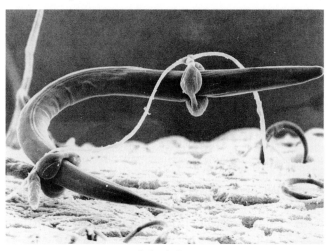

Arthrobotrys

FIGURE 24.12
A Predatory Fungus

Arthrobotrys, a microscopic fungus, bears many snares (open loops). These swell rapidly upon touch, trapping anything inside. Gotcha! The nematode worm becomes ensnared by a rapidly closing noose. Following the capture, haustoria from the fungus will penetrate the worm's body and digestion and absorption will begin.

trees, and still others on bare soil. A few do well under water. Lichens are arranged into three groups, mainly according to their habitat and structure (Figure 24.14).

Crustose species form dry, but colorful, colonies on sun-scorched rocky surfaces, where they act as hardy pioneers, hastening the weathering process and beginning the long soil-building process. Where cyanobacteria make up the photosynthetic component, their nitrogen-fixing activities further enrich the soil. **Foliose** ("leafy") lichens lead a more aerial existence, growing in trees where they bear some resemblance to Spanish moss (a flowering plant). A few grow on rock surfaces. **Fruticose** ("shrubby") lichens live mainly on the soil surfaces, often forming tangled, mat-like growths. A few grow on trees. One fruticose lichen, misnamed *reindeer moss*, forms great expanses of ground cover in Arctic tundra. As the name suggests, it is a source of food for reindeer.

Lichens are known to take up substances rapidly from the surrounding air and from rainfall. This is generally a very useful adaptation, except that in recent times they have been taking in industrial contaminants such as sulfur dioxide, heavy metals, and radioactive materials. As such substances are accumulated, these highly sensitive organisms are quickly destroyed (the photosynthetic partner dying first). For this reason, lichens have become "indicators" of air pollution problems. Observing

FIGURE 24.13
The Lichen Complex

(a) The lichen shown here is a ground or fruticose type. **(b)** The enlarged drawing reveals the general body plan—a multilayered arrangement of algae and fungi. At the surface are a number of asci containing ascospores. **(c)** A section through the lichen wall shows asexual bodies called soredia. When the wall erupts, the soredia escape. Should they find a suitable area, they will produce a new growth. **(d)** A close-up view through the scanning EM shows the details of a soredium.

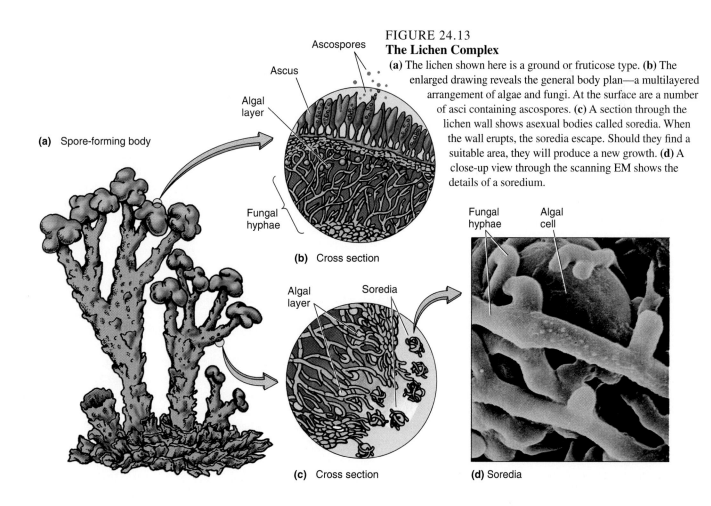

Ascospores

Ascus

Algal layer

(a) Spore-forming body

Fungal hyphae

(b) Cross section

Algal layer

Soredia

(c) Cross section

Fungal hyphae   Algal cell

(d) Soredia

(a) Crustose lichen on rock

(b) Foliose lichen on a tree

(c) Fruticose lichen on soil

FIGURE 24.14
Lichens
(a) A crustose lichen forms a crusty colorful mat over bare rock surfaces. (b) Foliose lichens, such as *Parmotrema tinctorus,* often appear on tree surfaces, whereas (c) *Cladonia cristotelli,* a fruticose lichen, lives on the soil surface.

their general condition and determining their chemical content are now routine parts of environmental monitoring.

Basidiomycota: Club Fungi

The **Basidiomycota**, or **club fungi**, are significant to us in a number of ways. Although the group includes such familiar varieties as mushrooms and shelf fungi, it also includes some devastating parasites (Figure 24.15). Among these are wheat rust and corn smut. Considering the worldwide dependence on wheat and corn, it is easy to appreciate the concern over these parasites. In fact, the development of rust-resistant strains of wheat is one of the great success stories of applied genetics. Saprobic club fungi are also partly responsible for the decomposition of dead trees and litter on forest floors.

Like the sac fungi, the club fungi are septate (or partially so). Club fungi reproduce sexually, and, like many sac fungi,

some produce large, elaborate sexual structures. We will use the mushroom as our representative Basidiomycota.

Growth and Sexual Reproduction in the Mushroom. The familiar "mushroom" with its thick **stalk** and dome-like **cap** is really just the reproductive part of the mushroom's life cycle. Long before the cap appears, a dense, deeply penetrating subterranean growth occurs. Its hyphal mass actively digests and absorbs nutrients from organic matter in the soil. In some club fungi, the mycelium may also form a mutualistic, mycorrhizal relationship with plants mentioned earlier. This is also true of the zygomycota.

For sexual reproduction to begin, hyphae from different mating strains must meet (Figure 24.16). As in some sac fungi, conjugation brings on the dikaryotic state, which may persist in the soil for some time. In the meantime, the dikaryotic hyphae feed, grow, and divide. Mitosis here introduces a logistics problem, which most club fungi have solved in an interesting manner.

The Clamp Connection. If the dikaryotic state is to be retained, then each of the new daughter cells formed must have both types of nuclei following division. But how can mitosis accurately sort out four nuclei? It wouldn't do for new daughter cells to end up with two plus nuclei or two minus nuclei, or worse. In the club fungi, the **clamp connection** ensures that a proper distribution happens. A look at Figure 24.17 will reveal how the fungal cell isolates the two nuclei for mitosis—first, by making use of branching growth, and later, through the strategic formation of cell walls that separate the products. As a result, two cells emerge, each with a plus and minus nucleus, and the dikaryotic state is retained.

The Basidiocarp. At some point, the dikaryotic hyphae go on to form the dense, fleshy reproductive structure, the **basidiocarp** (stalk and cap), which appears aboveground (see Figures 24.15 and 24.16). Thus the familiar mushroom, the one some of us eat, is a sexual structure, one in which the nuclear union will finally occur. Fertilization occurs in minute **basidia** (singular, basidium, club-like structures arranged in rows along thin **gills** on the underside of the cap. [The basidium ("little club") is, of course, the source of the names "Basidiomycota" and "club fungus."] Fertilization is followed by meiosis and spore formation. Each haploid nucleus becomes incorporated into a **basidiospore**, and enormous numbers of such basidiospores may be released from each mushroom. Each spore can begin another mushroom life cycle.

Picking Your Own. We cannot leave the subject without a word about picking your own wild mushrooms for the table. A few years ago, the news carried a story about two people who used an illustrated mushroom picker's guide to gather wild mushrooms. One of the two was a veteran picker. Their carefully gathered harvest was taken home, sautéed in butter and garlic, and eaten. It turned out that while their selection

(a) Toadstool (poisonous mushroom)

(b) Coral fungus

(c) Shelf fungus

FIGURE 24.15
Several Club Fungi
Club fungi can be large and colorful, as is **(a)** the poisonous mushroom, *Amanita muscaria.*
Poisonous mushrooms are often referred to as toadstools. **(b)** The cap of the coral fungus is
unlike that of other club fungi in that it has a lacy appearance. **(c)** Shelf fungi are often seen
emerging from the trunks of living trees, where they cause wood rot.

closely resembled those labeled "edible" in the guide, they
had actually collected a highly toxic species, aptly named "the
death cap." The two gourmets became critically ill, hovering
near death for a time. (About half of such cases end in death.)
Both survived, their lives spared when suitable donors were
found and their badly damaged livers replaced. The best bet, it
seems, is to pick mushrooms only from the source expert
mycologists use—the corner supermarket.

Wheat Rusts We'll briefly consider the wheat rust, *Puccinia
graminis,* which annually is responsible for the loss of millions of metric tons of wheat in North America alone. There
are many different strains of *P. graminis,* some of which infect
wheat and some others that attack barley, oats, rye, and wild
grasses. Wheat infected by *P. graminis* develops a blackened
stem and is invariably weakened or killed.

Puccinia graminis is a species with a truly complex life
cycle, most of which is spent in the dikaryotic state and which
involves two different hosts and several kinds of spores. The
two hosts are wheat and the common barberry plant, *Berberis
vulgaris.*

While in the wheat, the rust fungus produces dikaryotic
spores of two types. One—a red spore, or **uredospore**—simply infects other wheat plants, while the other—a black spore,

or **teliospore**—is the rust's investment in future generations.
Teliospores survive the winter season and germinate in moist
soil in the spring. Until then these black spores remain in a
dormant state. As such they are not active and cannot infect
the wheat or the barberry. When the teliospores become
active, they undergo meiosis in the usual fashion, producing
basidiospores. Upon their release, the basidiospores infect the
wild barberry.

It is only in the barberry leaf that the sexual phase
occurs. When plus and minus strains meet, the dikaryotic
state resumes. Then another round of spore formation
occurs, this time with dikaryotic **aeciospores** emerging. It
is the aeciospore that infects the next wheat crop. Their
growth leads to the formation of **uredinia**, structures in
which the red urediniospores take form. The cycle then
repeats.

After the wheat rust's intricate life history was discovered, it became common practice to break the infection cycle

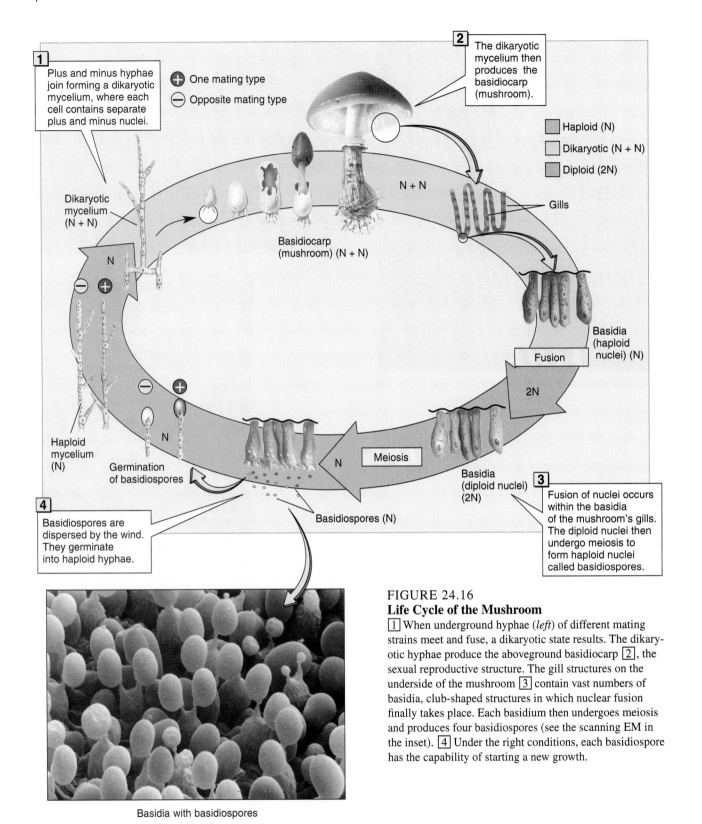

1 Plus and minus hyphae join forming a dikaryotic mycelium, where each cell contains separate plus and minus nuclei.

⊕ One mating type

⊖ Opposite mating type

2 The dikaryotic mycelium then produces the basidiocarp (mushroom).

Haploid (N)

Dikaryotic (N + N)

Diploid (2N)

Dikaryotic mycelium (N + N)

N + N

Gills

Basidiocarp (mushroom) (N + N)

Basidia (haploid nuclei) (N)

Fusion

2N

Haploid mycelium (N)

Germination of basidiospores

Meiosis

N

Basidia (diploid nuclei) (2N)

3 Fusion of nuclei occurs within the basidia of the mushroom's gills. The diploid nuclei then undergo meiosis to form haploid nuclei called basidiospores.

Basidiospores (N)

4 Basidiospores are dispersed by the wind. They germinate into haploid hyphae.

Basidia with basidiospores

FIGURE 24.16
Life Cycle of the Mushroom

☐1 When underground hyphae (*left*) of different mating strains meet and fuse, a dikaryotic state results. The dikaryotic hyphae produce the aboveground basidiocarp ☐2, the sexual reproductive structure. The gill structures on the underside of the mushroom ☐3 contain vast numbers of basidia, club-shaped structures in which nuclear fusion finally takes place. Each basidium then undergoes meiosis and produces four basidiospores (see the scanning EM in the inset). ☐4 Under the right conditions, each basidiospore has the capability of starting a new growth.

by burning contaminated fields and by systematically destroying barberry bushes, the intermediate host. This fails when wheat is grown year round over large geographic regions, such as from Mexico to Canada. Since wheat plants are grown year round, all that is required for the continued growth of the fungus are for the seasonally shifting winds to carry urediniospores north and south.

There has been success with rust-resistant strains of wheat, but, unfortunately, the parasite mutates rapidly enough to make these agricultural advances short-lived. So the battle goes on.

Essay 24.1 ● **A FUNGUS THAT SPITS**

Talents needed for survival on this planet take many forms but rarely (except for small boys and professional baseball players) does it involve spitting. However, *Pilobolus*, a zygomycota, depends on the ability to expectorate in an accurate and timely manner. It turns out that shooting out liquid streams is its primary way of ensuring that the "spit"—actually its spores—will find their way into the same fortunate circumstance that previously led to the success of the spitter. In this case, the fortunate circumstance involves a manure pile. Let's see what's going on here.

After gorging on delicacies in the dung of a herbivore, *Pilobolus* shifts from feeding to asexual reproduction. It produces a typical upright sporangiophore, tipped by a sporangium filled with hardy, resistant spores. But this fungus doesn't wait for wind or water to carry its spores away to shift for themselves. Below each sporangium lies a swollen, bulbous structure, the *subsporangial swelling,* not seen in other zygomycota (see photo). Under the right conditions, it fills to the bursting point with water. Then it literally explodes, ejecting the sporangium for a distance of 2 meters or more. If fortune smiles on the sporangium, it will land in and adhere to a blade of grass. If fortune smiles twice, the grass will be eaten by a herbivore, and the spores will survive the rigors of the digestive tract to emerge with the animal's feces, thus guaranteed a food supply and a good prospect for future success.

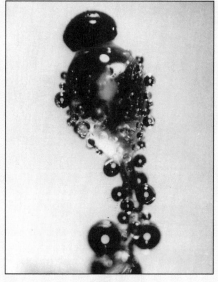
Pilobolus

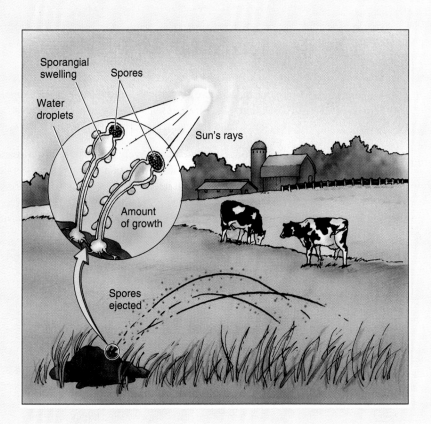

It turns out that sporangium ejection is not a random process but is closely oriented to the direction from which it receives light. In aligning itself, the stalk-like sporangiophore bends first in one direction and then in another, but eventually the sporangium is aimed into the light. At this point, an explosion in the subsporangial swelling is triggered and off goes the sporangium toward the light. The adaptive value of this aiming system seems obvious. It is far more advantageous to send the spores upward and outward than it would be to shoot them down into the exhausted food reserves.

Precisely what triggers the rupturing of the subsporangial swelling? Researchers have suggested a lens effect, whereby light is focused directly on a photoreceptor that controls the direction of growth in the sporangiophore. When the sporangium is aimed directly at the light, the photoreceptor receives maximum stimulation. However, this is all speculation, and researchers have yet to put their finger on the precise mechanisms involved.

Evolution of Fungi

Tracing the origin of fungi back to the ancestral stock is at best a conjectural exercise. There are many possibilities, but none, so far, is strongly supported by the fossil record or by comparative biochemical analysis of amino acids, nucleic acids, or other compounds. The principal guesses have included the following:

1. Fungi arose from colorless, heterotrophic, flagellated protists. The connection is represented by the slime molds, which, evolving from amebas, developed aerial spore formation as an adaptation to land. Fungi then emerged from

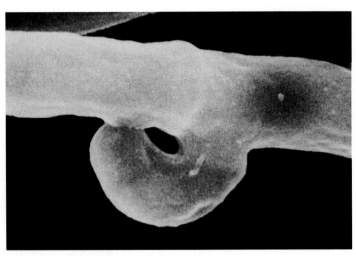

Clamp connection

FIGURE 24.17
Dikaryotic Cells and the Clamp Connection
During cell division in a dikaryotic cell, a cytoplasmic bridge called a clamp connection ensures the proper distribution of newly formed plus and minus nuclei into different daughter cells. **(a)** The cell before mitosis. **(b)** mitosis begins in both nuclei along with the start of a clamp connection. One of the nuclei enters the growing protrusion. **(c)** The migrating nucleus is isolated by a new cell wall. **(d)** The migrating nucleus is released into the second cell. Now each cell has both nuclei of the dikaryon.

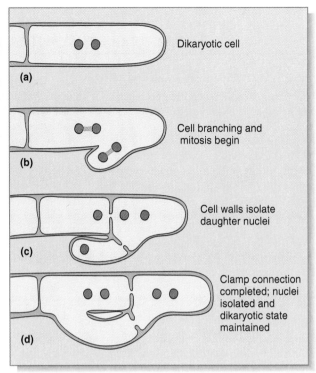

Dikaryotic cell

(a)

Cell branching and mitosis begin

(b)

Cell walls isolate daughter nuclei

(c)

Clamp connection completed; nuclei isolated and dikaryotic state maintained

(d)

an early slime mold line. (This is currently the favored hypothesis.)

2. Fungi are derived from green algal stock. The derivation was accompanied by a loss of the photosynthetic pigments of algal forerunners and the retention of cell walls.

3. Fungi arose from ancient red algae. Like the red algae, the fungi lack centrioles and motility.

4. Fungi have several origins, some having been derived from algal ancestry and others from flagellated protists. Accord-ingly, aerial spore formation is an example of convergent evolution, since it represents a very successful adaptation to life on land for many organisms, fungal and otherwise.

Fossil fungi are dated tentatively back 900 million years, to the Precambrian era, but the evidence isn't firm. Fossils from the Ordovician period, between 450 and 500 million years ago, have clearly been identified as fungi, and mycorrhizal associations have been preserved in Silurian deposits dating back over 400 million years.

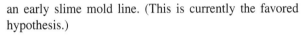

Key Ideas

WHAT ARE FUNGI?

1. Fungi are mostly multicellular organisms, often consisting of a **mycelium** made up of thread-like tubular cells called **hyphae**. Hyphae may be cellular or **coenocytic**. Hyphae mass to form large reproductive structures. Optimal growth conditions include food, warmth, moisture, and shade or darkness.

2. All fungi are heterotrophs, some living as decomposers, others as parasites. Most feed by **absorption**—secreting enzymes outside and absorbing the digested food. Parasitic fungi feed via penetrating hyphae called **haustoria**.

3. Decomposers join soil bacteria in breaking down nutrients and recycling essential mineral ions. Their mycelia contribute to soil fertility and water-holding ability. Many fungal parasites are important agricultural pests, threatening grain and other crops. Some fungi form a mutualistic association with plant roots called **mycorrhizae**, an arrangement that helps roots absorb minerals and water.

4. We use fungi as food, in food production, and for manufacturing pharmaceutical products.

5. Asexual reproduction often occurs through mitotic **spore** production. Spores are easily dispersed and highly resistant.

6. Fungi are haploid except for brief diploid states after fertilization. In the **dikaryotic** state, which in many fungi follows conjugation, the + and – haploid nuclei remain separated during the growth of elaborate sexual structures. Eventually they fuse, only to enter meiosis and form haploid spores.

7. Fungal characteristics such as chitinous cell walls, primitive mitosis and meiosis (within the nuclear envelope), an absence of centrioles and thus of motility, and chromosomes with little protein, make the kingdom unique and separate from others.

KINGDOM FUNGI

1. Kingdom Fungi contains divisions **Zygomycota**, **Ascomycota**, **Basidiomycota**, and a nontaxonomic grouping of convenience: **Fungi Imperfecti**, formerly the division **Deuteromycota**.

2. Zygomycota acquire their name from the **zygospore**. *Rhizopus stolonifer* (black bread mold), a common contaminant of foods, forms surface **stolons** (horizontal hyphae) that send **rhizoids** (penetrating hyphae) into food. Asexual reproduction occurs when **sporangiophores** (upright hyphae) produce **sporangia**, in which mitotic spore formation occurs. Sexual reproduction occurs through a form of **conjugation** involving plus and minus strains. Haploid nuclei gather in **gametangia**, which then join, permitting the nuclei to fuse. The diploid nuclei form a thick-walled **zygospore**. The emerging hypha produces a sporangium where meiosis and haploid spore formation occur.

3. The Ascomycota, or sac fungi, include morels, powdery mildews, citrus molds, and certain species of yeasts. In asexual reproduction, **sac fungi** form **conidiophores** on which haploid spores called **conidia** emerge through mitosis.

4. The sac fungus *Claviceps purpurea* produces **ergot**—a compact, black mycelium—in the rye plant. When ergoted rye flour is eaten, the symptoms of **ergotism**—restricted blood flow to extremities, burning sensations, hallucinations, and gangrene—may develop. Other destructive Ascomycota cause chestnut blight and Dutch elm disease.

5. Sexual reproduction in the sac fungus *Peziza* requires plus and minus strains. One produces an **ascogonium**, the other an **antheridium**. These are joined by a tube through which nuclei migrate. The plus and minus nuclei enter a dikaryotic state, and an **ascocarp** forms. Fusion of plus and minus nuclei in sac-like **asci** is followed by meiosis and mitosis, resulting in the formation of haploid **ascospores**. Germination yields a new mycelium.

6. **Yeasts** are unicellular sac fungi that increase through **budding** (mitosis and highly unequal division). A population can be diploid or haploid, and the diploids may undergo meiosis, producing ascospores within the old cell. Some yeasts are impor-

tant in producing bread, wine, beer, and liquor, while others are parasites, causing serious infections.

7. **Fungi imperfecti** are mainly sac fungi. Members include the athlete's foot fungus and several *Penicillium* species, the latter used in producing penicillin and in flavoring foods. Another species captures and feeds on soil roundworms.

8. The **lichen** is a symbiotic relationship of an alga or cyanobacterium and a fungus. The photosynthesizer supplies nutrients, while the fungus provides moisture and anchorage. (Recent studies suggest that it is only the fungus that benefits.) All are important soil builders.

9. **Club fungi** include saprobic mushrooms, shelf fungi, puffballs, and parasites such as wheat rust and corn smut. Most of the mushroom mycelium is below ground, while above is the **basidiocarp**, a fruiting body that produces **basidiospores**. Following the subterranean union of plus and minus mushroom hyphae, a dikaryotic mycelium forms. The dikaryotic mycelium emerges from the ground to produce the basidiocarp. Its cap contains **gills** lined with **basidia** in which fusion of plus and minus nuclei occurs, followed by meiosis and haploid basidiospore formation.

10. The parasite *Puccinia graminis* requires two hosts, the wheat plant and barberry, to complete its life cycle. In wheat it produces red **uredospores**, which infect other wheat, and black **teliospores**, which winter over and in the spring produce basidiospores that infect the barberry. A third spore, the **aeciospore**, emerges from the barberry and infects the next wheat crop.

EVOLUTION OF FUNGI

1. Fungi probably arose from colorless, heterotrophic, flagellated protists, emerging from a slime mold line. The fossil record of fungi is tentatively traced back 900 million years to the Precambrian era. A solid record begins 500 million years ago in the Ordovician period.

Review Questions

1. List four characteristics shared by fungi. (pages 472–473)

2. Describe the typical fungal mycelium, its organization, and its growth. (page 473)

3. List two nutritional modes of fungi, and describe the absorptive method of feeding. (page 474)

4. What is unusual about the cell walls of fungi? How has this been used as an argument against including fungi in the plant kingdom? (page 473)

5. Explain how a mycorrhizal association works. Why is it considered mutualistic? (page 475)

6. List four reason why biologists have stopped grouping the fungi with the plants. (general)

7. From what reproductive characteristic is the name zygomycota derived? (page 476)

8. Describe asexual reproduction in the common bread mold, *Rhizopus stolonifer.* (page 477)

9. What is the "sac" of the sac fungi? (page 480)

10. What is ergoted wheat? Ergotism? (pages 478–479)

11. Trace events in sexual reproduction in the sac fungus, using such terms as *ascogonium, dikaryon, plus* and *minus strains, antheridium, asci,* and *ascospore.* (page 480)

12. List two ecologically significant activities of lichens. Why are they useful pollution indicators? (pages 482–483)

13. Describe the life cycle of the common yeast. (page 480)

14. What is the basis for assigning fungi to the Fungi Imperfecti? List three familiar examples. (pages 481–482)

15. Trace the events leading to the production of the mushroom basidiocarp. What reproductive event happens in the basidiocarp? (page 484)

CHAPTER 25

Plant Evolution and Diversity

CHAPTER OUTLINE

WHAT ARE PLANTS?

Alternation of Generations in Plants

ORIGIN OF PLANTS AND EARLY ADAPTATIONS TO LAND

Bryophytes: The NonVascular Plants

Bryophyte Characteristics

Division Hepatophyta: Liverworts

Division Anthocerophyta: Hornworts

Division Bryophyta: Mosses

Evolutionary Origin of Bryophytes

THE VASCULAR PLANTS

History of Vascular Plants

Taxonomic Organization of Vascular Plants

Division Psilotophyta: Whisk Ferns

Division Lycophyta: Club Mosses

Division Sphenophyta: Horsetails

Division Pterophyta: Ferns

THE SEED PLANTS

Reproductive Adaptations in Emerging Seed Plants

The Gymnosperms

THE END OF AN ERA

Angiosperms: The Rise of the Flowering Plants

Essay 25.1 An Evolutionary Lesson From Plants

The Angiosperms Today

THE PLANT KINGDOM: A SUMMARY

KEY IDEAS

APPLICATION OF IDEAS

REVIEW QUESTIONS

*W*ith our arrival at the plant kingdom we come to those organisms that are familiar to everyone. They are the life forms that lend soft hues to our surroundings, grace our lives, provide our shelter, and warm our hearths (Figure 25.1). But perhaps most important, plants (along with algae) are our ultimate source of food and oxygen.

As we introduce you to the plant kingdom, our emphasis here will be on the origin of plants, major trends in their evolutionary history, and adaptations that made them successful in today's environment. Later, we will look closely at plant reproduction, growth, physiology, and responses (see Part Five).

FIGURE 25.1
The Plant World
Humans have a strong bond with the plant world, finding comfort in cool green foliage and colorful flowers.

What are Plants?

Plants are multicellular autotrophic organisms with specialized cells and tissues. They are best described as nonmotile, or stationary, for with the exception of those plants with swimming sperm, there are no moving stages. Plants lead quiescent lives, responding to outside stimuli and internal chemical signals mainly through growth processes that are inherently slow. There are mechanisms for rapid response in a few plants, but such responses occur through changes in turgor (water pressure) in highly specialized tissues (see Chapter 31). In other words, there is nothing comparable to animal responses, which are made possible by nervous, muscular, and skeletal systems.

As you know, plants are photosynthetic. They make use of light-absorbing pigments, which include chlorophyll *a* and *b* and carotenoids. The principal storage polysaccharide is starch, and the most common structural polysaccharide is cellulose. Cellulose is used in cell wall construction and is often joined there by strengthening agents such as pectins and lignins.

As you might have noticed, some of these characteristics are shared with certain algae, but unlike almost all algae, plants produce *multicellular* embryos, which are housed within *multicellular* gametophyte tissues. This is a fairly clear departure since the typical algal zygote, with the exception of certain red algae, is on its own in the watery environment. Finally, all plants exhibit a sporic life cycle, one involving an alternation between multicellular diploid sporophyte and multicellular haploid gametophyte phases. We introduced you to alternating generations and other life cycles in Chapter 23, but it will be useful to begin with a brief review of the sporic cycle.

Alternation of Generations in Plants

The plant life cycle alternates between diploid sporophyte and haploid gametophyte states in what has traditionally been called an **alternation of generations** (Figure 25.2). In most of the plants we will be discussing, the sporophyte is the larger, more prominent individual whereas the gametophyte is commonly small and inconspicuous, sometimes consisting of just a few cells. The sporophytes of the pine, maple, date palm, and tulip, for example, are the familiar plant itself. The gametophytes are tucked away in the cones or flowers. As we will see shortly, there is one important exception among plants, where the gametophyte is large and prominent and the sporophyte is reduced.

You may recall that organisms displaying an alternation of generations have a complex life cycle (see Chapter 22). The diploid sporophyte, for example, contains tissues that undergo meiosis, but no gametes are produced. Instead, meiosis leads to the formation of haploid spores. The spores then divide mitotically to form the multicellular haploid gametophyte generation, and it is within the gametophyte that tissues give rise to sperm and egg. Since the gametophyte is already haploid, the gametes form through mitosis. When sperm and egg meet in fertilization, a new diploid sporophyte generation begins. In summary:

1. Sporophyte (2N) — meiosis $\longrightarrow$ Spores (N)
2. Gametophyte (N) — mitosis $\longrightarrow$ Gametes (N) $\longrightarrow$ Fertilization $\longrightarrow$ Sporophyte (2N)

This is a very general overview of alternation of generations in plants, and as we will soon see, there are many variations on this theme.

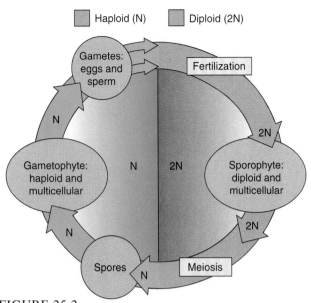

FIGURE 25.2
Alternation of Generations

Plants exhibit a sporic life cycle with a definite alternation of generations. In all but the bryophytes, the diploid sporophyte generation dominates. The sporophyte stage ends with meiosis and the emergence of haploid spores. The spores then give rise to a multicellular gametophyte generation. The gametophyte produces gametes through mitosis. When the gametes join in fertilization, the diploid sporophyte generation begins again. Sporophytes are the familiar plant life we see. The gametophyte is usually inside the flower or cone, although in reproductively primitive plants, it may be separate but very small.

Origin of Plants and Early Adaptations to Land

The evolutionary origin of plants can be traced to ancient aquatic green algae, perhaps similar to what we see in today's complex green alga *Coleochaete* (Figure 25.3). *Coleochaete* and plants have certain features in common, including the plant-like manner in which mitosis and meiosis occur in the alga. As in the plants, the nuclear envelope is dismantled during cell division. In addition, a barrel-shaped, plant-like spindle forms, and following nuclear division, formation of the cell plate between daughter cells occurs at right angles to the spindle and through the coalescing of Golgi vesicles (see Chapter 9). Furthermore, the embryo of some species of *Coleochaete,* like that of plants, is surrounded by a protective layer of cells.

Paleobotanists trace the fossil record of plants back 430 million years ago, to the Silurian period of the Paleozoic era. Changes from the simpler life form of aquatic algae were most likely spurred on by the transition from aquatic to terrestrial life. This transition must have been fraught with problems, since the primeval seas provided many of the requirements needed by life in ways not possible on land. For example, most of the individual cells of aquatic plants take their minerals and water directly from their surroundings. Aquatic plants also make use of the water to disperse their spores and gametes. Furthermore, the surrounding water supports the

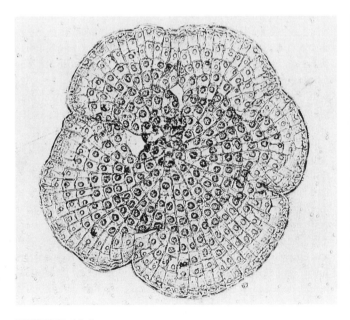

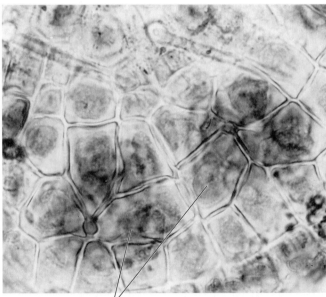

Protected embryos

FIGURE 25.3
Algal Ancestors

The direct ancestor of plants is now believed to have been a green alga that was similar to today's *Coleochaete.* The individuals form a disk of cells, one cell layer thick. They retain their zygotes, protecting them in surrounding cells as seen here.

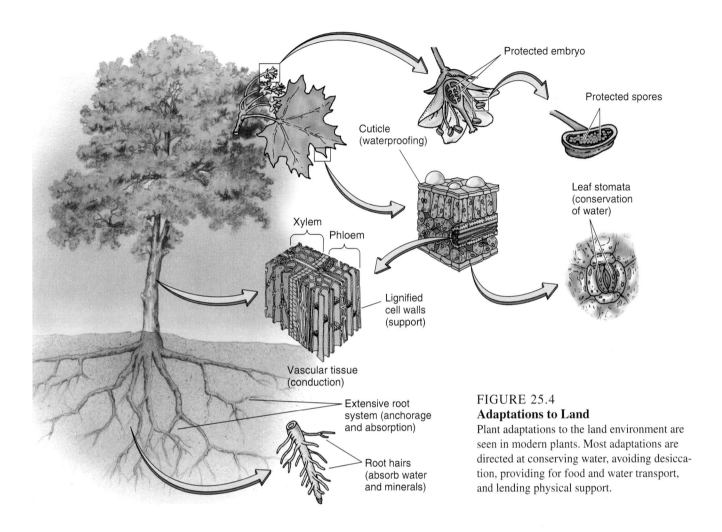

FIGURE 25.4
Adaptations to Land
Plant adaptations to the land environment are seen in modern plants. Most adaptations are directed at conserving water, avoiding desiccation, providing for food and water transport, and lending physical support.

weight of the aquatic organism. Remember that kelps (brown algae) grow quite large without much of an investment in supporting tissues (see Chapter 23). Without the amenities of the watery environment, the early plants had to make many adaptive changes (Figure 25.4). Let's look at some of these, beginning with the problem of avoiding water loss.

All terrestrial life, whether plant or animal, is subject to desiccation. One adaptive response in plants is the production of a **cuticle**, a waterproof, protective coating secreted by the **epidermis** (outermost layer of cells). Such a coating protects the plant from excessive water loss through evaporation and, in Paleozoic times, may also have protected it against harmful ultraviolet radiation.

The spores and pollen also received a protective coating of **sporopollenin**, a sealing material that permits their dispersal by air currents without the danger of drying. Such a coating may also have permitted spores to pass safely through the digestive system of an animal that had eaten the sporophyte.

As plants became larger, they needed ways of insuring a sufficient supply of water and carbon dioxide for photosynthesis. The first problem was solved through the evolution of vascular tissue. Most plants produce two types of vascular tissue: xylem for transporting water and phloem for transporting sugars and other nutrients. The evolution of specialized pores known as **stomata** (singular, stoma) assured an adequate supply of carbon dioxide without undue water loss. Stomata do this by opening during photosynthetic activity and closing at other times.

The support problem, so readily solved through buoyancy in the aquatic environment, required far-reaching structural adaptations as plants grew larger. Part of the solution was to thicken and strengthen cell walls through the addition of lignin. Lignin, a secreted, hardening substance, is the chief strengthening material of plants; its presence makes the growth of great trees possible.

Reproductive adaptations to land life were dramatic. No longer could the plant simply release its gametes into the watery surroundings. Some plants, as we will see next, developed structures that simply make use of raindrops to splash the sperm to the egg, but evolution has provided most with more elaborate mechanisms to accomplish this critical task. We will look into these as we proceed into the major plant divisions. (For a preview of the divisions, see Table 25.1, page 503.)

Bryophytes: The Nonvascular Plants

Bryophyte Characteristics

The **bryophytes**, which include 15,000 species of **mosses**, **liverworts**, and **hornworts**, are plants with simple tissue organization. They are called nonvascular plants because they lack the xylem and phloem tissues that occur in other plants. They do, however, have simpler types of conducting tissue that some theorists believe evolved from the same source as xylem and phloem. Bryophytes lack true roots, stems, and leaves, which by definition must have vascular tissue; but they have analogous structures that carry out similar functions. For example, simple **leaf scales** carry out photosynthesis, and thread-like **rhizoids**, just one cell in thickness, anchor the plants to the soil. Although most bryophytes are terrestrial, they retain a reproductive feature of the aquatic algae: a swimming sperm that must have water to reach the egg. However, that water need only be a few drops of dew or splashing raindrops, so bryophytes are among those plants with the simplest solution for getting sperm to eggs in the terrestrial environment.

The bryophyte life cycle (Figure 25.5) is unique among the plants. Whereas a clear alternation of generations exists, it is the *gametophyte that is prominent.* The gametophyte in mosses, for example, is the dark green carpet-like growth seen in damp places such as the water-splashed rocky surfaces along streams and in the deep shade of trees. The gametophyte is easy to see, but observing the sporophyte requires closer scrutiny. It is usually far less conspicuous and shorter-lived.

Bryophytes are generally small plants, some ground-hugging and others growing on the surfaces of trees. They lack the strong, lignin-filled supporting tissues found in most other plants. As we've mentioned, lignified cells lend great strength to most vascular plants, permitting them to grow quite tall. But in spite of their structural simplicity, bryophytes have been quite successful. They apparently fill ecological niches that have not been exploited by vascular plants. While evolutionary trends in many vascular plants brought about large size and adaptations to drier conditions, most bryophytes became specialists in inhabiting more limited, moist places, perhaps less suitable for many vascular plants. A few have successfully adapted to hot, dry desert conditions, windswept mountain outcrwoppings, and the frigid polar regions. It seems the only conditions that bryophytes cannot tolerate are those produced by humans—they are notably absent in areas of severe air pollution.

There are about 16,500 named species of bryophytes, two-thirds of them mosses. The others, primarily liverworts, are less common. Chances are you haven't heard much about liverworts, but we are about to change this.

Division Hepatophyta: Liverworts

There are some 6000 species of liverworts. The division name, **Hepatophyta**, refers to the liver (*hepato,* "liver"; *phyta,*

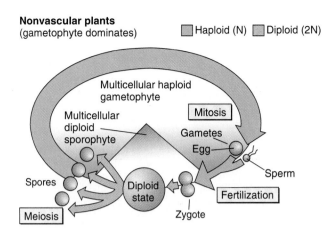

Nonvascular plants (gametophyte dominates)   ■ Haploid (N) ■ Diploid (2N)

FIGURE 25.5
The Bryophyte Life Cycle
The bryophyte life cycle is opposite to that of vascular plants. The haploid gametophyte dominates the life cycle. It gives rise to gametes through mitosis. When gametes join in fertilization, a new diploid sporophyte emerges. The sporophyte ends with meiosis, in which a number of haploid spores are formed. The spores will give rise to new gametophytes, and the cycle repeats.

"plant"), as does the common name, "liverwort" ("wort" means herb). This peculiar nomenclature dates from the 9th century, when, following the quaint logic of the times, the liver-shaped plant was popularly used for the treatment of liver ailments.

Just about everyone's favorite liverwort is *Marchantia* (Figure 25.6). Its dark green gametophyte is a branching, ribbon-like growth that lies flattened against the soil. (Someone likened it to green corn flakes.) The gametophyte contains simple pore-like openings that admit carbon dioxide, so essential to the photosynthetic cells within. Although they are not true stomata, the pores help the plant avoid excessive water loss by narrowing when the humidity is low and dilating when it is high. Simple thread-like rhizoids anchor the plant to the soil.

Asexual reproduction in *Marchantia* occurs in two ways. The first is **fragmentation**, where portions of the plant body that have come loose may grow into a complete gametophyte. Gardeners commonly do this with cuttings. The second is through the formation of **gemmae** (singular, *gemma*). These are small multicellular bodies that form in concave structures called **gemmae cups** (Figure 25.6). Like a gametophyte fragment, a dislodged gemma is capable of forming a new liverwort. Since this is an asexual process, the new individual will be genetically identical to the old.

The Liverwort Life Cycle. The life cycle of *Marchantia* (Figure 25.7) begins when the gametophyte produces egg-forming **archegonia** and sperm-forming **antheridia**. They occur within peculiar stalked structures that resemble miniature palms and parasols, respectively (Figure 25.8). Archegonia and antheridia may occur on the same individual, or they

FIGURE 25.6
Marchantia Gametophyte

The gametophyte of *Marchantia* consists of flattened, leafy blades that are anchored to the soil by rhizoids. A cross section through one of the blades *(drawing)* reveals a dense layer of cells bearing chloroplasts on the upper surface. Pores in the leafy surface admit carbon dioxide for photosynthesis. The pores consist of tiered cells, some of which change shape in response to dryness, closing the pore. The circular structures on the leaf surface *(lower right)* are gemmae cups. Within are gemmae, vegetative bodies produced through mitosis, that, when released, can produce new gametophytes.

Pores

Rhizoids

may occur on separate individuals, depending on the species. In the plant world, the two conditions are termed **monoecious** ("one house"), with male and female parts on the *same individual;* and **dioecious** ("two houses"), with male and female parts on *separate individuals.*

When the sexual structures mature, swimming sperm are dislodged by raindrops, and as the drops splatter, the sperm are literally splashed against the archegonia. They swim along moist surfaces and enter an archegonium, where a sperm will fertilize the egg within. With this union, a new diploid sporophyte phase in the life cycle begins (see Figure 25.7).

The tiny liverwort sporophytes remain small and attached to the gametophyte. When mature they develop **sporangia**, structures in which meiosis occurs and haploid spores are formed. The sporophyte may be photosynthetic, but it is not independent. It depends on the gametophyte for water and minerals. Liverworts, like other bryophytes, are **homosporous**, which means that the spores are all similar. We will soon come to **heterosporous** plants, where two kinds of spores are produced. With meiosis and spore formation, the brief diploid sporophyte phase of the life cycle ends and the gametophyte phase begins again. Because they have just completed meiosis, spores contain a great deal of genetic variability.

Spore dispersal is assisted by the presence of **elaters**, cells with spiral thickenings (see Figure 25.7). The elaters respond to minute changes in humidity by coiling and uncoiling, the twisting movement dislodging the spores. Each spore, under suitable conditions, can produce a new gametophyte.

Division Anthocerophyta: Hornworts

The hornworts are a relatively minor division, with only 100 or so named species. Although the gametophyte somewhat resembles certain liverworts, its lobed growth forms a rosette rather than a ribbon (Figure 25.9). Furthermore, in *Anthoceros,* a representative genus, each cell contains one large

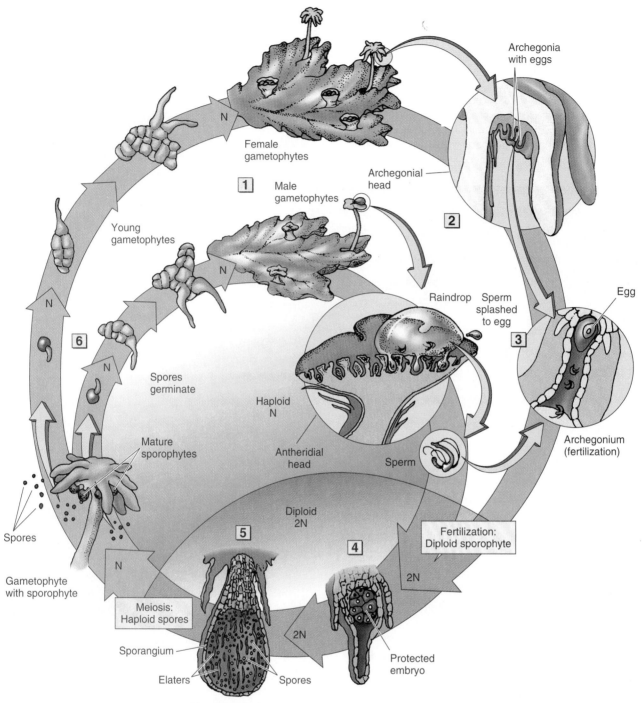

FIGURE 25.7
***Marchantia* Life Cycle**
In its life cycle, the haploid *Marchantia* gametophyte produces reproductive structures on its surfaces. ☐1 Miniature palm-like (female) and parasol-like (male) structures produce archegonia and antheridia, respectively ☐2. The mature archegonium has a tube-like appearance with the egg located at its base ☐3. Antheridia produce numerous sperm that swim to the egg and fertilize it when sufficient water is present. The diploid sporophyte ☐4 then grows out of the archegonium. When mature, cells of the sporophyte undergo meiosis, giving rise to numerous haploid spores ☐5. Elaters assist in spore dispersal. When the spores germinate ☐6, they produce a new leafy gametophyte and the cycle repeats.

chloroplast along with a starch-storing pyrenoid. These are very primitive traits, characteristic of some algae. *Anthoceros* is also known to harbor symbiotic colonies of nitrogen-fixing cyanobacteria that provide the gametophyte with an ongoing source of valuable nitrogen compounds.

The Hornwort Life Cycle. Sexual reproduction in hornworts includes the formation of antheridia and archegonia. Sperm swim to the eggs when rainwater or dew is present. Following fertilization, a number of long, slender, horn-like sporophytes grow from each gametophyte (which explains the

FIGURE 25.8
Marchantia **Sexual Structures**
The miniature "palm trees" emerging from the gametophyte are female reproductive structures. They will form archegonia on the underside. The nearby "parasols," male reproductive structures, produce antheridia on the upper side.

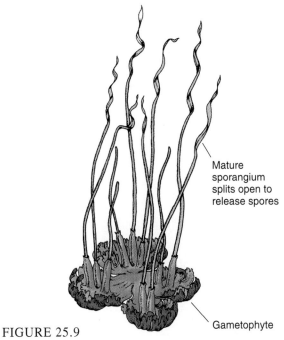

Mature sporangium splits open to release spores

Gametophyte

FIGURE 25.9
Hornworts
The leafy hornwort gametophyte produces tall, upright sporophytes. Numerous spores, produced through meiosis, are released as the sporophyte dries and splits open. The sporophyte is far less dependent on the gametophyte than in other bryophytes and can exist independently if something happens to the gametophyte.

name "hornwort") (see Figure 25.9). Spores are produced in long, slender sporangia and, as the sporophyte dries and splits open, are released to the air currents.

The hornwort sporophyte is unusual in that it can grow continuously, producing spores over a long period of time. Furthermore, it is photosynthetic and less dependent on the gametophyte than are other bryophyte sporophytes. Under exceptional conditions it can survive on its own should the supporting gametophyte die. The hornwort sporophyte has a water-resistant coating or cuticle and exchanges gases with the air through true **stomata**. These pores open and close through changes in water pressure.

Division Bryophyta: Mosses

The mosses are the most numerous and common of the bryophytes, with about 9500 named species (second in species number only to the flowering plants). They grow best in moist conditions but are found in a variety of habitats. Mosses thrive even in deserts, some having made a simple adaptation to the drying conditions. Between rains, desert mosses simply dry out, a time when their metabolic activity becomes almost nonexistent. With the next rainfall, water diffuses into the dry tissues, and the moss literally "comes to life," resuming where it left off.

The moss gametophyte (Figure 25.10) is anchored in the soil or to a rocky substrate by rhizoids, elongated threads of cells that emerge from the base of the leafy plant. The moss

equivalent of a leaf is a flattened, scale-like growth, only one cell in thickness and lacking in supporting and conducting elements. In some species, a waterproofing layer of waxy *cutin* covers the leafy scales.

The upright, stem-like growth to which the leaves attach is several layers thick. Included is an outer layer of photosynthetic epidermal cells, a middle layer of slender storage cells, and an inner region known as the **central cylinder**. The central cylinder contains thick-walled **hydroids** and **leptoids**, cells that conduct water and foods, respectively. They are analogous to the xylem and phloem of vascular plants, which also transport water and foods.

Mosses are of considerable ecological importance, particularly in Arctic regions where, along with lichens, they provide much of the ground cover. As with lichens, they are also well known as **pioneer plants**, species responsible for colonizing previously uninhabited regions and producing the first soils. Thus we often find mosses and lichens growing side by side on bare rocky outcroppings. Gardeners commonly use mosses for decorative purposes, and *Sphagnum* (peat moss) is used as a mulching or bedding material for lawns and gardens. In Ireland and a few other places, *Sphagnum* in the form of peat is dried and used as heating and cooking fuel. It is also burned to provide the smoky (some suggest "wet newspaper") flavoring of Scotch whiskey.

(a)

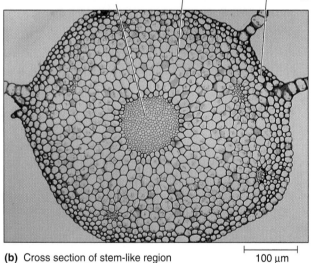

Conducting tissue   Cortex   Epidermis

(b) Cross section of stem-like region   100 µm

FIGURE 25.10
Gametophyte of the Moss *Polytrichum*

(a) The minute leafy stalks are the gametophyte of the moss *Polytrichum*. The stalked, capsule-like growths are sporophytes. Cross-sectional views of the stem-like region reveal a rather intricate organization. (b) In the scanning EM, a ring of small cells around the perimeter marks the epidermis. Within this is a cylinder of larger, thin-walled cells that make up the cortex. The dense central cylinder of cells is the simple conducting system of the moss, including an outer region of food-conducting leptoids and an inner region of larger, water-conducting hydroids.

The Moss Life Cycle. The moss gametophyte begins life as a haploid spore, which produces a thread-like growth, known as a **protonema** ("first thread"). The protonema will bud in several places, each bud producing a new, leafy gametophyte. At maturity, the moss gametophyte enters into sexual reproduction, with antheridia and archegonia forming on each plant (Figure 25.11). Both monoecious and dioecious species occur. As in other bryophytes, and all plants, the archegonium is surrounded by a protective layer of nonreproductive cells. These cells later protect the fragile embryo.

If water is available, sperm emerge from the mature antheridium, enter a neighboring archegonium, and fertilize the egg. As in the liverworts, it is usually just splashing raindrops that carry the sperm from the antheridia to the archegonia. Evidence suggests that the sperm are directed down into the neck of the archegonium by chemical attractants.

Fertilization ushers in the diploid sporophyte generation. The typical moss sporophyte (Figure 25.12a) consists of a broad base, firmly anchored in the archegonium, and a slender stalk that supports a sporangium at its tip. Within the sporangium, cells undergo meiosis, producing vast numbers of haploid spores. Since they are haploid, they represent the new gametophyte generation.

When the spore capsule in some species is mature and spore-filled (Figure 25.12b), an **operculum** (cap) falls away, and the spores escape. The likelihood of any one spore landing in a suitable place is not very good, but the low odds are

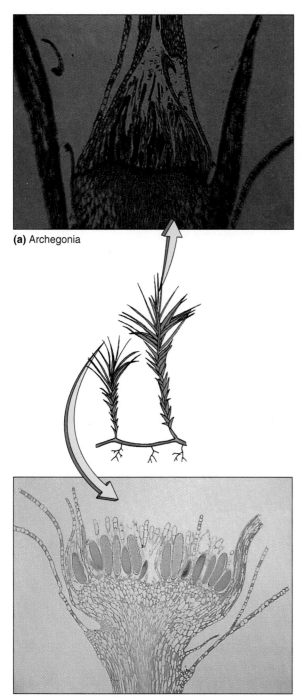

(a) Archegonia

(b) Antheridia

FIGURE 25.11
Moss Sexual Structures
Archegonia and antheridia in this species emerge from the gametophyte tips. **(a)** Archegonia are vase-shaped structures, each with an egg cell at the base of a narrow opening called the neck. **(b)** The club-shaped structures are antheridia. The dots within are huge numbers of sperm cells. Splashing raindrops serve to transport the sperm to the archegonium.

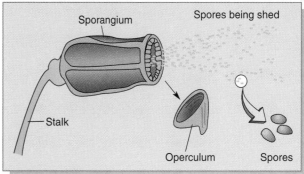

FIGURE 25.12
Moss Sporophyte
The moss sporophyte emerges from the archegonium as a lengthy stalk with a capsule-like sporophyte at its tip. Cells in the sporangium undergo meiosis, forming millions of haploid spores. As the sporophyte dries, the operculum falls away from the capsule, dislodging the spores.

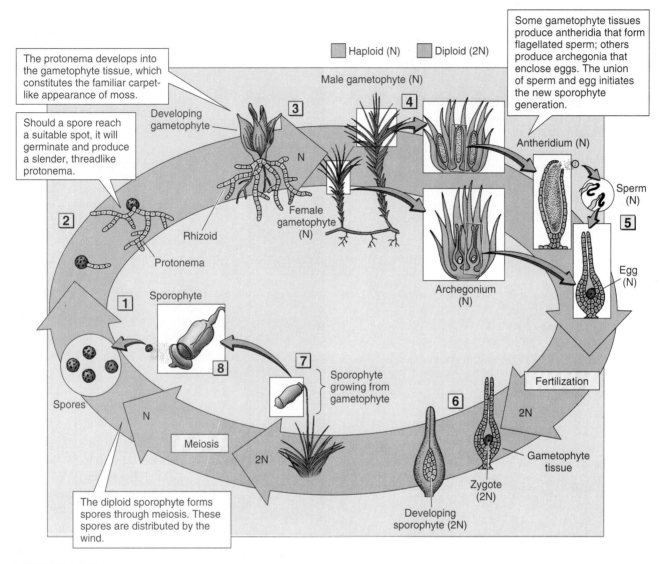

The protonema develops into the gametophyte tissue, which constitutes the familiar carpet-like appearance of moss.

Should a spore reach a suitable spot, it will germinate and produce a slender, threadlike protonema.

Some gametophyte tissues produce antheridia that form flagellated sperm; others produce archegonia that enclose eggs. The union of sperm and egg initiates the new sporophyte generation.

The diploid sporophyte forms spores through meiosis. These spores are distributed by the wind.

Haploid (N)   Diploid (2N)

Male gametophyte (N)

Developing gametophyte

Rhizoid

Protonema

Female gametophyte (N)

Antheridium (N)

Sperm (N)

Archegonium (N)

Egg (N)

Sporophyte

Spores

Meiosis

Sporophyte growing from gametophyte

Fertilization

Gametophyte tissue

Zygote (2N)

Developing sporophyte (2N)

FIGURE 25.13
Moss Life Cycle

1 The gametophyte generation begins at the end of meiosis, when vast numbers of haploid spores are formed. 2 Upon germination, the moss spore gives rise to a protonema, a slender chain of cells. 3 Buds emerging from the protonema form new leafy gametophytes. 4 In some species, archegonia and antheridia form at the tips of separate gametophytes. 5 Sperm from antheridia, splashed to the vase-like archegonia, make their way down the slender neck to fertilize the eggs. 6 The embryo develops in the archegonium, 7 emerging as the stalked sporophyte. 8 Meiosis in the sporangium leads to the formation of spores and the start of a new gametophyte generation.

countered by the great number of spores each capsule produces—up to 50 million! Should a haploid spore become established in a suitable place, it will absorb water and sprout, and a new gametophyte will arise. The moss life cycle is summarized in Figure 25.13.

In summary, bryophytes are nonvascular plants, lacking xylem and phloem tissues. Thus bryophytes do not have true roots, stems, and leaves. The absence of lignin limits upright growth. The gametophyte lives independently and dominates

the life cycle, incorporating the sporophyte within its tissues. All bryophytes release single-cell aerial spores that produce new gametophytes.

Evolutionary Origins of Bryophytes

How do the bryophytes fit into the evolution of plants? Some botanists maintain that the mosses, liverworts, and hornworts evolved independently of the vascular plants—each from its

FIGURE 25.14
Vascular Tissues
Vascular tissues include food-conducting
phloem and water-conducting xylem. The
phloem, a living tissue, conducts sugars and
other foods from cell to cell in its dense
cytoplasm. Phloem conduction occurs in all
directions as sugars and other foods are
distributed to regions where they are needed.
In its mature, functional state, the xylem
consists of lignified cell walls only and is thus
in a nonliving state. In addition to conducting
water and minerals, xylem provides a great
deal of the supporting material of trees, where
it is referred to as "wood." The compact layer
of cells lying between phloem and xylem is
called "vascular cambium." It produces new
xylem and phloem.

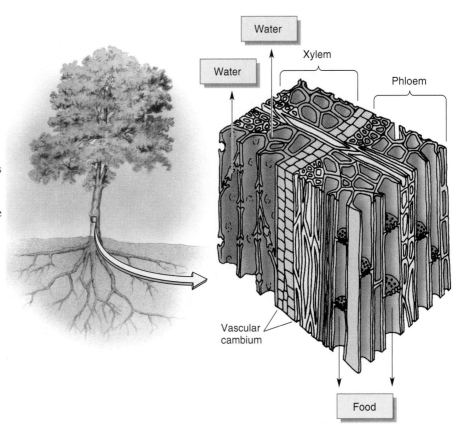

own green algal ancestor. They cite a number of similarities
between bryophytes and green algae, especially the moss protonema, photosynthetic pigments, and the presence of
pyrenoids (in hornworts).

Other botanists maintain that the bryophytes and vascular
plants share a common green algal ancestor. They propose
that the bryophytes diverged from the early vascular plant line
some 350 million years ago, about 80 million years after plant
evolution began. The presence of simple vascular tissue in
bryophytes, they maintain, is an example of evolutionary simplification—a loss of whatever vascular development had
occurred in their ancestors.

The Vascular Plants

Bryophytes are interesting, but if someone were to ask you to
name a plant, "hornwort" would probably not come to mind.
More likely, you would think of an oak, rose, or pine—a vascular plant. The vascular plants are sometimes referred to by
the more technical term **tracheophyte** (*tracheo-*, "tubes"; -
phyton, "plants"). This term refers to the tube-like vascular
elements of the **xylem** and **phloem**, which transport water and
foods, respectively (Figure 25.14). With conducting tissue
present, vascular plants, by definition, produce *true* roots,
stems, and leaves.

The most important point to be made here is that vascular-
ity was a smashing evolutionary success, providing answers to
some of the difficult problems of plants adapting to the terrestrial environment. In addition to its efficient transport of water
and minerals, the xylem tissues, when impregnated with
lignin, provide woody plants with exceptional strength. Large
size provided a novel way of adapting to the terrestrial environment—one not available to the bryophytes.

History of Vascular Plants

The actual origin of vascular plants is not well understood
right now, but well-preserved fossils have been found in 400-
million-year-old rocks of the Silurian period of the Paleozoic
era. Among the earliest vascular plants is an extinct group
known as the **rhyniophytes**. They made their debut in the
same period as the first vertebrates—those clumsy, unimpres-
sive armored fishes with jawless, sucking mouths.

Rhynia, a representative rhyniophyte, grew in stagnant
marshes, probing the air with simple leafless stems arising
from horizontal stems known as **rhizomes** (Figure 25.15).
The primitive stems contained a system of tiny tube-like, nonliving, water-filled elements called **tracheids**, a part of the
water-conducting xylem common in today's vascular plants.
With the tracheids came the promise of new horizons for, as
we have seen, xylem not only conducts water, but its tough,
lignified elements provide vital support that permits vascular

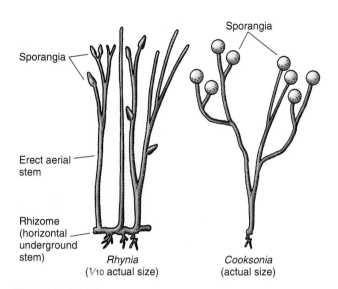

FIGURE 25.15
First Vascular Plants
Rhynia, a marsh plant, grew to about 50 cm (20 in.) in height and
produced slender, upright growths from an underground stem. The
stems were photosynthetic, bore stomata, and had a waterproofing
cuticle. *Cooksonia* is believed to be the oldest vascular plant. Like
Rhynia, it produced slender, branching stems, but it only grew to
about 6–7 cm (1.6–2.4 in.) in height. Although these species lacked
leaves, the presence of xylem and phloem conducting tissues
signaled the start of an enduring trend. Typically, the first vascular
plants produced sporangia at the tips of their slender stems. These
plants became extinct some 390 million years ago.

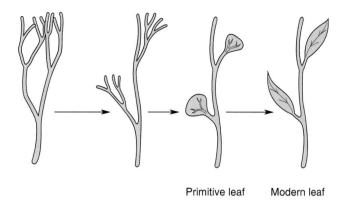

FIGURE 25.16
Evolution of Leaves
The highly branched stems of primitive vascular plants are believed
to have been the forerunner to leaf structure. The growth of
photosynthetic tissue between the branches formed the leaf blade.
The former branches themselves became leaf veins.

plants to grow tall. The peculiar growth of the finger-like
stems was also important for the future. The branching and
rebranching provided a vascular framework for the later evolution of the flattened leaf blade (Figure 25.16).

Today vascular plants inhabit nearly every part of the
land environment. As they radiated out over the landscape,
they adapted to their environments in different ways and thus
produced a wide variety of types. Nearly all developed true
leaves, stems, and roots. A finishing touch was the development of mechanisms for fertilization that did not require the
flagellated sperm or an external supply of water for sperm
travel. As we will see, the more recently evolved plants produce only nonmotile sperm. In fact, their cells lack centrioles
entirely, and centrioles, recall, give rise to basal bodies that
form cilia and flagella.

Taxonomic Organization of Vascular Plants

Most plant biologists subscribe to a taxonomic scheme in
which there are nine divisions of vascular plants. (Botanists
prefer the term "Division" over the term "Phylum.") Four of
these are **seedless plants**, those that do not produce seeds, and
the remaining five are **seed plants**. All have very prominent
sporophytes and small, inconspicuous gametophytes (the

opposite of bryophytes). There are important reproductive
differences between seedless and seed plants. The spores of
seedless plants develop into *separate and independent* gametophytes, so, like some of the algae, the life cycle actually
includes *two separate individuals.* The gametophytes of seed
plants, however, are *integrated into the sporophyte tissues*—
in the cones and flowers, as mentioned earlier.

Of the nine divisions, six probably will be unfamiliar to
all but botanists and a few knowledgeable gardeners (see
Table 25.1). The nine divisions are:

I. **Seedless plants** (primitive spores; minute, separate
    gametophytes)
    Division Psilotophyta ("naked plants"): whisk ferns
    Division Lycophyta ("spider-like plants"): club mosses
    Division Sphenophyta ("wedge plants"): horsetails
    Division Pterophyta ("winged plants"): ferns

II. **Seed plants** (advanced spores; gametophyte
    incorporated into sporophyte)

    A. **Gymnosperms** (naked seeds)
       Division Cycadophyta: cycads, sago palms
       Division Ginkgophyta: maidenhair tree
       Division Gnetophyta: *Ephedra*
       Division Coniferophyta: pines and other conifers
       ("cone-bearers")

    B. **Angiosperms** (seeds with fruit)
       Division Anthophyta: flowering plants

During the Paleozoic era, the seedless plants of the first
three divisions were an important part of the earth's greenery,
but their heyday is well over, and they are far less common
today. We have included them because they are interesting and
because some of their characteristics may represent significant
stages in the evolution of the seed plants. Of an estimated

TABLE 25.1
Summary of the Plant Kindgom

Divisions	Examples	Characteristics	Life Cycle
NON-VASCULAR PLANTS: BRYOPHYTES	Liverworts, hornworts, mosses	Little or no supporting or vascular tissue (xylem and phloem)	Very dominating gametophyte, dependent sporophyte, swimming sperm
Hepatophyta (6,000 species)	Liverworts (*Marchantia*)	Some with flattened, ribbonlike growth, gemmae cups, simple rhizoids	Sperm formed in parasol-shaped antheridia, splash to palm-shaped archegonia
Bryophyta (9,500 species)	Mosses (*Polytrichum, Riccia*)	Upright or hanging body, numerous leaf-like scales, simple rhizoids	Sperm formed in antheridia, splash to eggs in archegonia, stalked sporophyte emerges
SEEDLESS VASCULAR PLANTS:		Organized vascular tissue conducts water, minerals and foods. True roots, stems and leaves	Separate generations; very dominant sporophyte, swimming sperm
Psilophyta (4 species)	Whisk ferns (*Psilotum*)	Primitive vascular tissue, no true leaves, tropical	
Lycophyta (4 species)	Ground or club pine (*Lycopodium, Selaginella*)	Well developed vascular tissue, widespread, true leaves	
Sphenophyta (15 species)	Horsetails (*Equisetum*)	Well developed vascular tissue, widespread, hollow aerial stems, rhizomes	
Pterophyta (11,000 species)	Ferns (*Platycerium*)	Well developed vascular tissue, large complex leaves common, rhizomes, widespread	
SEEDED VASCULAR PLANTS:		Pollen and seeds, includes cone-bearing and flowering plants	Dominant sporophyte, retains gametophyte which is reduced to a few cells, mostly nonmotile sperm
Gymnosperms	All cone-bearing plants	Cone bearers, naked seeds (no fruit)	
Ginkgophyta (1 species)	Maidenhair tree (*Ginkgo*)	Large tree, fan shaped leaves, highly cultivated, but none known in wild	Swimming sperm (plant fluids)
Cycadophyta (100 species)	Sago palms or cycads (*Zamia, Cycas*)	Palm-like foliage, thick, partially subterranean stem, very large cones, mainly tropical and subtropical	Swimming sperm (plant fluids)
Gnetophyta (71 species)	(*Gnetum, Welwitschia, Ephedra*)	Angiosperm features, including xylem vessels, double fertilization, flower-like cones, primarily desert	Non-swimming sperm

(continued)

TABLE 25.1 (Continued)
Summary of the Plant Kindgom

	Divisions	Examples	Characteristics	Life Cycle
	Coniferophyta (550 species)	Conifers (pines, redwoods, firs, larch cypress, hemlock)	Many are large trees, needle and scale leaves, drought resistant, widespread in subarctic, alpine regions	Non-swimming sperm, wind pollinated
	Angiosperms Anthophyta	All flowering plants	Flowers, fruit surrounding seeds	All non-swimming sperm, wind, water, insect pollinated
	Class Dicotyledonae (170,000 species)	Dicots (magnolia, cabbage tobacco, cotton, willow, apple, oak, maple, bean)	Two cotyledons, net-veined broad leaves, floral parts in 4s or 5s or their multiples; many have secondary growth, worldwide distribution	
	Class Monocotyledonae (65,000 species)	Monocots (grasses, orchids, iris, yucca)	Single cotyledon, parallel-veined leaves, floral parts in 3s or their multiples, little secondary growth, worldwide distribution	

258,000 named species of vascular plants, only about 1000 species comprise the first three divisions, and of these, all but 19 are found in Division Lycophyta. Thus we are looking at a very few survivors from another time, before the seed plants became the most widespread forms of plant life on earth. The fourth seedless division, the ferns, are widely distributed today.

Of the seed plants, the first four divisions are more familiarly known as **gymnosperms**, whereas the fifth group, the flowering plants, are called **angiosperms**. Both terms are in common use but are no longer true taxonomic designations. Figure 25.17 summarizes the evolutionary history of the vascular plants.

Division Psilotophyta: Whisk Ferns

Today's **psilotophytes**, members of Division Psilotophyta, are remarkably similar to those found in the fossil record of 375 million years ago, making them a very persistent group of plants. The psilophytes, along with the lycophytes and sphenophytes, formed the vast, swampy Paleozoic forests (Figure 25.18).

The division is represented today by two genera and only four species, all found mainly in the tropics. Sporophytes of *Psilotum nudum,* the "whisk fern" (Figure 25.19), do not produce true roots, only hairy rhizoids similar to those of the bryophytes. The stems are fairly well developed, with simple, primitive xylem and phloem. Although leafless, the stems are photosynthetic. They bear sporangia above tiny scale-like appendages. The gametophyte, a tiny, separate, underground plant, produces both antheridia and archegonia. The swimming sperm make use of rainwater to reach the egg. Following fertilization, the sporophyte emerges from the gametophyte in a manner reminiscent of the mosses. However, it soon becomes independent, and the tiny gametophyte withers and dies.

Psilotum lives where subtropical and tropical conditions prevail. In the United States, it ranges along the southernmost states and occurs in Hawaii and Puerto Rico. Its sister genus, *Tmesipteris,* occurs only in Australia, New Caledonia, New Zealand, and some southern Pacific islands. It lives chiefly as an **epiphyte** (a plant that grows on the surface of larger plants, but not as a parasite).

FIGURE 25.17
Evolutionary History of Vascular Plants
The history of vascular plants is superimposed on a geological timetable. The ancestral line is at the bottom of the diagram, emerging from the Silurian period of the Paleozoic era. The divisions of seedless plants are seen at the left and the seed plants at the center and right. The varying width of each plant line indicates changes in the number of species over time. As you can readily see, the number of species of seedless plants and gymnosperms was considerably greater in past time than it is today. Also note that the flowering plants are relative newcomers to the plant kingdom, emerging only 135 million years ago. Their numbers are greater today than ever before in their short evolutionary history. The absence of connections between flowering plants and older groups is intentional since there are no fossils that establish these links.

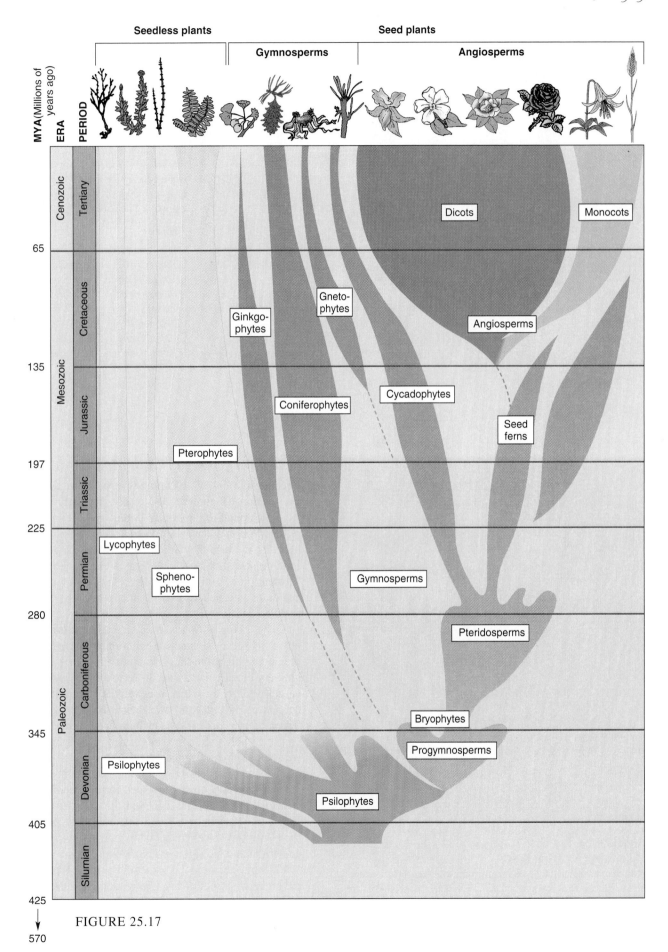

FIGURE 25.17

FIGURE 25.18
Paleozoic Coal Age Forest
The reconstruction of a coal age forest was made from a rich source of fossil evidence. The marshy lowland was typical of the Paleozoic landscape. Many of the forest giants seen here are represented today by comparative dwarfs. Some represent lines that are extinct today. Such forests disappeared during the great Permian extinction, brought on by worldwide drought and extensive glaciation.

Animals	Plants
①*Eusthenopteron*	Ⓐ *Eospermatopteris*
②*Eogyrinus*	Ⓑ *Calamites*
③*Diplovertebron*	Ⓒ *Lepidodendron*
④*Meganeuron*	Ⓓ *Sigillaria*
⑤*Eryops*	Ⓔ *Cordaites*
⑥*Seymouria*	
⑦*Limnoscelis*	
⑧*Varanosaurus*	

FIGURE 25.19
Psilophyte: A Living Fossil
Psilotum, a vascular plant with many primitive traits, produces an erect, branching sporophyte with simple vascular tissue. Spores are produced in clustered sporangia. Its branching form is clearly reminiscent of the rhyniophytes, the first vascular plants (see Figure 25.15).

Division Lycophyta: Club Mosses

The most familiar **lycophytes** are those in the genus *Lycopodium* (Figure 25.20), whose members range widely from the Arctic to the tropics. Many tropical species are epiphytes. You may have seen *Lycopodium* sporophytes without knowing it, since they are often mistaken for pine seedlings. In some parts of the United States they are known as both "ground pine" and "club moss." They are green year-round and quite conspicuous in the winter when other ground cover has died back.

The leaves of *Lycopodium* are called **microphylls** (literally, "small leaves") and are believed to have evolved as simple outgrowths from the stem epidermis. They contain a single vein of vascular tissue. In contrast, the leaves of higher vascular plants are called **megaphylls** ("large leaves"). They contain many veins and owe their evolutionary origin to the fusion of branching stems (such as those seen in *Rhynia*).

In lycophytes, spores are formed in sporangia associated with **sporophylls** ("spore leaves," specialized microphylls). The sporophylls are arranged along the stem in some species, but in others they cluster together, forming **strobili** (singular, **strobilus**; also called a cone) at the ends of the branches. When the haploid spores are released, they germinate and grow into tiny, independent gametophytes that produce both antheridia and archegonia. Fertilization occurs when there is sufficient water over the gametophyte for sperm to swim to the archegonia. Following fertilization and development, the diploid sporophyte emerges from the gametophyte, becomes independent, and begins a new life cycle.

Selaginella: **A Reproductively Advanced Lycophyte.** The life cycle of *Selaginella* is seen in Figure 25.21. The genus is of considerable evolutionary interest because it is **heterosporous**: it produces two types of spores: **microspores** and **megaspores**. The plants we have considered so far, including *Lycopodium,* are **homosporous**; only one kind of spore is produced.

We've already seen that plants produce spores in structures called sporangia. It is logical then for the structures in which microspores and megaspores are produced to be called **microsporangia** and **megasporangia**. The microsporangia and megasporangia of *Selaginella* are located within its strobili. Taking the terminology one step further, let's note that microspores form **microgametophytes** and megaspores form **megagametophytes**. The two structures produce antheridia and archegonia, respectively. (Table 25.2 organizes the terms and compares *Selaginella* with *Lycopodium.*)

When sufficient water is available, the antheridia burst, releasing motile sperm that swim into the neck-like opening of each archegonium and fertilize the egg cell within. The young embryo sporophyte is supported by nutrients from cells in the surrounding megagametophyte, as in seed plants, but, unlike seed plants, *seed coats do not form* and there is *no seed dormancy.*

We've invested some time on *Selaginella* because, with the exceptions noted, its life cycle is surprisingly like that of seed plants. We'll come to these shortly, so you will be hearing more about microspores and megaspores, and microgametophytes and megagametophytes.

The lycophytes were not always the humble, dwarfed plants we see today. One group produced great trees over 50 meters tall and 2 meters in diameter. In fact, the lycophytes were the most conspicuous plants of the Devonian and Carboniferous forests that persisted 300–400 million years ago (see Figure 25.17). But change is inexorable, and with the drying climate of the Permian period, the primitive giants died, possibly replaced by newly evolving gymnosperms. Today they are represented by a few remnants of a once dom-

FIGURE 25.20
A Lycophyte
Lycopodium is common in eastern woodlands. It is generally inconspicuous, but because it is evergreen, it stands out against the snowy, winter landscape. In this species, the sporangia are clustered in terminal strobili (lighter green regions).

inating life form. Yet their ghosts still haunt us, for the corpses of these great plants helped form vast coal and oil reserves. The ancient lycophytes influence modern global politics as humans squabble over their remains.

Division Sphenophyta: Horsetails

The **sphenophytes**, like the lycophytes, had their day in the late Paleozoic, when they too contributed to the lush Carboniferous forests. The few surviving species, known familiarly as "horsetails," are distributed worldwide, often favoring

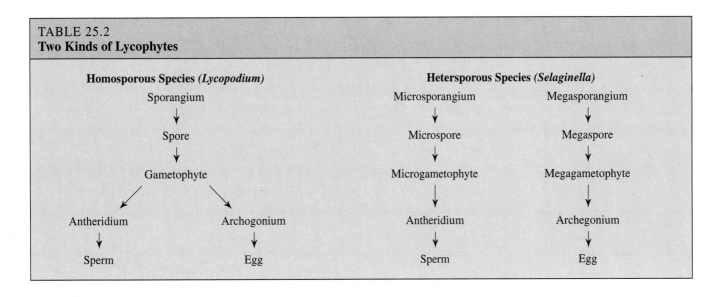

TABLE 25.2
Two Kinds of Lycophytes

Homosporous Species (*Lycopodium*)	Heterosporous Species (*Selaginella*)	
Sporangium	Microsporangium	Megasporangium
↓	↓	↓
Spore	Microspore	Megaspore
↓	↓	↓
Gametophyte	Microgametophyte	Megagametophyte
Antheridium / Archogonium	↓	↓
Antheridium — Archogonium	Antheridium	Archegonium
↓ — ↓	↓	↓
Sperm — Egg	Sperm	Egg

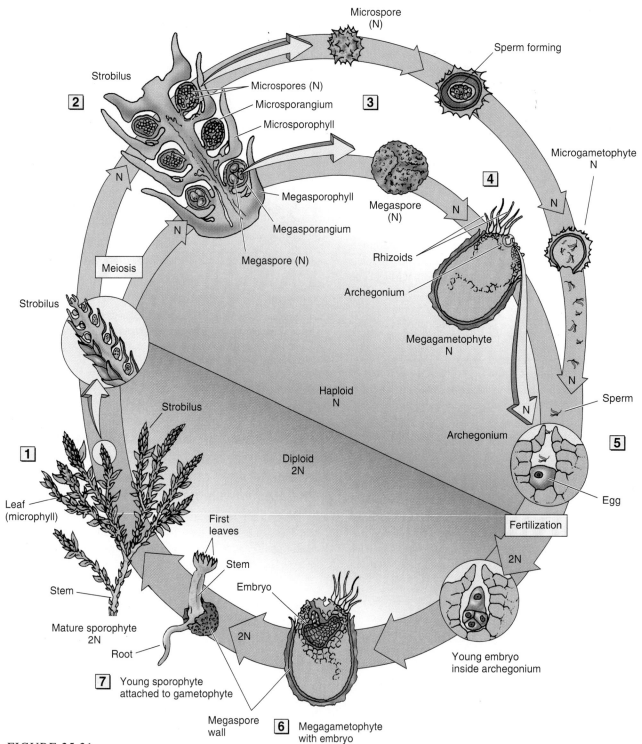

FIGURE 25.21

Life Cycle of *Selaginella*

Selaginella has advanced reproductive characteristics, similar in many ways to seed plants. ⓵ The sporophyte produces megasporangia and microsporangia in the strobili ⓶. The megaspores and microspores ⓷ develop into megagametophytes (female) and microgametophytes (male), respectively ⓸. Eggs and sperm are formed within archegonia and antheridia, and fertilization occurs in the archegonia ⓹. The embryo ⓺ is housed and nourished within the megagametophyte. It emerges to begin developing into a new sporophyte ⓻. The events are very much like those in seed plants; however, the developing embryo does not enter a dormant period, nor does it receive protective seed coats.

moist locales along the banks of rivers and streams. The only living genus, *Equisetum* (*equi-*, "horse"; *setum*, "bristle"), contains 15 or so species.

Sphenophytes have been called "scouring rushes," a name reflecting their use as pot cleaners in the days before steel wool soap pads. They are well suited for this because their cell walls contain glassy silica, a good abrasive.

Most sphenophytes grow to less than a meter in height. Some species produce two kinds of shoots. Vegetative shoots bear scale-like microphylls and whorls of short, lateral branches. Unbranched reproductive shoots produce spores in prominent strobili (Figure 25.22).

Sphenophytes are homosporous. The spores develop into pinhead-sized, independent gametophytes that bear both antheridia and archegonia. Once again, flagellated sperm make use of dew or rainfall to swim from antheridia to archegonia, and upon fertilization a new sporophyte generation begins.

Division Pterophyta: Ferns

Pterophytes (*ptero,* "wing"), or ferns (Figure 25.23), are undoubtedly among the most enchanting of plants. The spring forest, dripping with cool rain, is accented by delicate ferns originating as tiny, coiled **fiddleheads** from the damp floor. Later, the plants will lend an exotic touch to the woods as they stand full grown, their leaves splayed, as if placed there for decoration.

Ferns have survived in great numbers since Paleozoic times, apparently having adapted more readily to changing environments than have most other seedless plants. Today there are some 11,000 named species of ferns. They are widely distributed, growing mainly in tropical and temperate regions.

Fern Life Cycle. The fern life cycle is seen in Figure 25.24. Note that, as is typical of seedless plants, the sporophyte and gametophyte stages occur in separate generations.

The fern sporophyte usually consists of a thick rhizome that produces a number of leaves and tiny roots. The leaves are true megaphylls, having evolved from branch systems. The leaves often emerge from the rhizomes as tightly coiled fiddleheads**,** each of which uncoils into a large leaf that is sometimes frilly and highly subdivided into leaflets. A few, such as the tree ferns, have tall, upright stems supporting a leafy rosette (see Figure 25.23).

Fern sporangia (Figure 25.25) are commonly clustered into structures called **sori** (singular, **sorus**). In their early stages, the sori are sometimes hidden under a scale-like cover called an **indusium**. You have undoubtedly seen sori on the underside of the leaves of decorative fern leaves, where they often occur as rows of brown dots. Sori are often mistaken for insect or fungal invasions, a conclusion that has sent many an alarmed gardener scurrying to the nearest plant nursery for pesticides.

Sporangia in some species contain elaborate mechanisms for spore release. The spores are forcefully ejected when changes in the thick-walled cells of the **annulus** cause thin-walled **lip cells** to split open (see Figure 25.24 3 and 4).

FIGURE 25.22
The Horsetails
Some species of *Equisetum*, the horsetail, produce two types of shoots. The frilly shoot with numerous leaves is vegetative and does not enter into reproduction. The naked shoot with tiny, scaly leaves and a large strobilus at its tip is the reproductive shoot.

Following their release, the spores are distributed by air currents. Those spores that settle into moist, well-protected, surfaces will germinate, forming delicate gametophytes. The fern gametophyte, typically flattened and heart-shaped, is known as a **prothallus** (Figure 25.26). The prothallus is photosynthetic and independent, anchored to the soil by minute rhizoids. Typically, prothallia produce both antheridia and archegonia. When water is available, the flagellated sperm escape from the antheridia and swim to the archegonia, where fertilization occurs. The diploid embryo, housed and protected for a brief time by the prothallus, develops into the familiar leafy sporophyte.

Self-fertilization is fairly common, but cross-fertilization is considered more adaptive since it increases genetic variability. Accordingly, many ferns have evolved mechanisms for inhibiting self-fertilization. For example, in some species, the first prothallus to mature produces a hormone known as

A tree fern

Madienhair fern

Staghorn fern

FIGURE 25.23
Fern Diversity
This small sample of fern diversity includes sporophytes of the gracious tree fern *(top right)*, the prolific maidenhair fern *(left)*, and one epiphytic species, the staghorn fern *(right)*. Many ferns emerge from underground rhizomes in the form of "fiddleheads," a descriptive name because the coiled leaf resembles the head of a violin.

antheridogen. It *suppresses* male (antheridial) development in the gametophyte in which it is produced but *stimulates* male development in neighboring gametophytes. This process, in effect, results in prothallia becoming male or female, which promotes cross-fertilization.

Summarizing the seedless vascular plants, let's note that the sporophytes, which are mainly what the discussion is about, have true roots, stems, and leaves, and their cell walls are lignified. The latter permits plants to grow to great size. Unlike the bryophyte, the vascular plant has a large, dominating sporophyte and highly reduced gametophyte. Like the bryophyte, the seedless vascular plant releases single-cell aerial spores that give rise to tiny, but completely independent, gametophytes. Much of this changes in the seed plants.

The Seed Plants

The primitive vascular plants and their less conspicuous contemporaries, the earliest seed plants, persisted through the Devonian, Carboniferous, and Permian periods (see Figure 25.17). Toward the close of the Permian period, which was the end of the Paleozoic era (about 225 million years ago), the earth experienced dramatic environmental changes. Powerful geological movements produced a different landscape and climate. The earth's surface in the Paleozoic had been relatively smooth, permitting countless warm and shallow lowland seas to exist. Then soggy lowlands were uplifted to form vast mountain ranges. The long-lasting warm climate gave way to a general cooling, even as the new uplands dried. The primitive, marsh-loving giants of

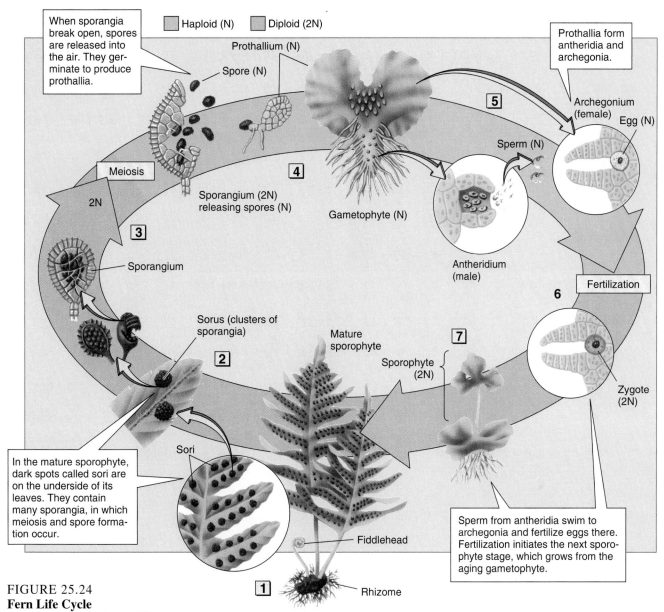

Haploid (N)   Diploid (2N)

When sporangia break open, spores are released into the air. They germinate to produce prothallia.

Prothallia form antheridia and archegonia.

Prothallium (N)

Spore (N)

Archegonium (female)

Egg (N)

5

Sperm (N)

Meiosis

4

2N

Sporangium (2N) releasing spores (N)

Gametophyte (N)

3

Antheridium (male)

Fertilization

Sporangium

6

Sorus (clusters of sporangia)

Mature sporophyte

7

Sporophyte (2N)

2

Zygote (2N)

Sori

In the mature sporophyte, dark spots called sori are on the underside of its leaves. They contain many sporangia, in which meiosis and spore formation occur.

Sperm from antheridia swim to archegonia and fertilize eggs there. Fertilization initiates the next sporophyte stage, which grows from the aging gametophyte.

Fiddlehead

1   Rhizome

FIGURE 25.24
Fern Life Cycle
The large, leafy sporophyte $\boxed{1}$ often produces its sporangia on the undersides or margins of the leaflets $\boxed{2}$ in structures called sori. Cells within the sporangia $\boxed{3}$ undergo meiosis, generating many haploid spores. The spores germinate, forming heart-shaped, green prothallia $\boxed{4}$. Prothallia represent the brief, ground-hugging gametophyte generation. In many species, the prothallus forms both archegonia and antheridia $\boxed{5}$. The eggs are fertilized within the archegonia $\boxed{6}$, where the zygote gives rise to a new sporophyte $\boxed{7}$. The sporophyte then repeats the life cycle.

the late Paleozoic era perished, leaving only a few scattered remnants of those once-prominent groups. One direct reason for their demise may relate to the life cycle. The small, ground-hugging gametophyte, so successful in a marshy world, may have been unsuited to the harsh, drying Mesozoic landscape.

With the demise of the ancient forests, the competitive edge in the plant world passed to the previously inconspicuous gymnosperms, seed plants that were better prepared to survive on this new kind of earth. They continued to evolve,

becoming larger and better adapted to the land, and soon appeared everywhere over the Mesozoic landscape. One secret of their newfound success was their ability to draw water from deep below the dry surface of the earth, transport it in efficient vascular systems, and conserve it in water-resistant stems and leaves.

The changing earth also presented certain problems for the emerging seed plants, such as how to get the sperm from one tall plant to the egg in another.

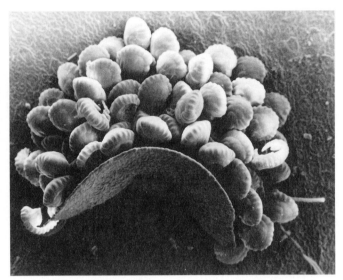

(a) Sporangia on leaf undersides

(b) Sorus with many sporangia

FIGURE 25.25
Fern Sporangia
(a) The fern sporophyte produces numerous sporangia, often appearing on the underside of the leaflets.
(b) Sporangia often occur in clusters called sori. In some species, the sori are covered by a scale-like indusium. As the sporangia dry, changes in shape cause the them to split open, releasing the spores.

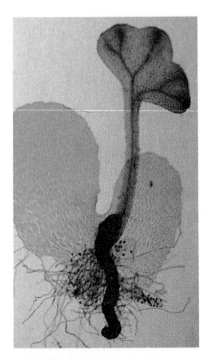

FIGURE 25.26
Fern Gametophyte
The heart-shaped fern gametophyte is a brief interlude in the otherwise dominating sporophyte generation. The emerging sporophyte remains attached to the gametophyte for a time but soon produces its own roots as the gametophyte withers away.

Reproductive Adaptations in Emerging Seed Plants

Pollen. In large part, seed plants succeeded because they are heterosporous. In the seed plant, microspores are formed through meiosis, as usual, but each meiotic daughter cell then undergoes mitosis and develops into a minute, resistant body known as **pollen** (Figure 25.27a). Pollen contains the male gametophyte, which produces the sperm. As hay fever sufferers can attest, the pollen of some species can easily be carried in the air. Pollen is also transported by water currents or by animals, especially insects. The deposition of pollen on the receptive cone or flower, known as **pollination**, solves the problem of sperm transport. Let's look closer at the male gametophyte, or microgametophyte, as we have called it.

Mitosis in the microspore produces two cells that differentiate into a **generative cell** and a **tube cell**. (The presence here of more than one cell helps us distinguish pollen grains from the simple spores in the seedless plants.) The generative cell forms two sperm cells, and the tube cell directs the growth of a **pollen tube**. The pollen tube, another device unique to seed plants, is a fine tubular growth that emerges from a pollen grain and grows into the female gametophyte. Its growth forms a safe, internal pathway for sperm to travel to the egg. The pollen tube and the cells within make up the seed plant's mature microgametophyte (see Figure 25.27a).

The newly evolved pollen system not only worked in earlier times but is carried on efficiently in seed plants today. With this system established, motility in the sperm was no longer important. In fact, swimming sperm are absent in near-

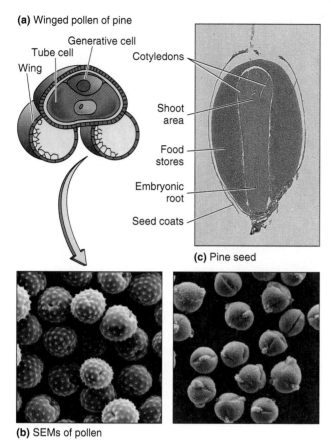

(a) Winged pollen of pine

Tube cell
Generative cell
Wing
Cotyledons
Shoot area
Food stores
Embryonic root
Seed coats

(c) Pine seed

(b) SEMs of pollen

FIGURE 25.27
Pollen and Seeds
Unique features of the seed plants include pollen and seeds. **(a)** Pollen represents the multicellular gametophyte. In the winged pine pollen, we see several cells, but the critical ones are the generative cell (future sperm) and the tube cell. **(b)** Scanning electron micrographs are of pollen from two different seed plants. Pollen forms a protective surrounding wall, which in each species has a distinctive shape. Palynologists (people who study pollen types) can identify most species from their pollen, including fossil pollens. Their techniques help anthropologists determine the vegetation types present in ancient civilizations, giving useful clues to climatic changes. They have also been of great importance to determining the origin of flowering plants. **(c)** The seed includes a dormant embryo, a supply of food, and tough, protective seed coats. The pine seed includes multiple cotyledons (seed leaves) above the embryo. The mound between represents tissue that will produce the shoot, whereas the region below will produce the root.

ly all seed plants. (We'll look at exceptions in a moment.) Thus we see that the last reproductive link to the ancient watery plant environment was finally broken.

Seeds. Another novel reproductive adaptation in the seed plants was the **seed** itself (Figure 25.27c). A seed consists of an embryo, a region of stored food, and hardened seed coats—quite a change from anything we've described before. Following fertilization, the embryo develops along with a quantity of food that will later support its early growth. The embryo then becomes dehydrated, entering a dormant state, and the surrounding sporophyte tissues form the hardened, protective seed coats. The seeds of gymnosperms form within the cones, whereas those in flowering plants (angiosperms) form within the ovary of the flower. In flowering plants, the ovary later forms the fruit that surrounds the seeds.

The Gymnosperms

The name *gymnosperm* literally means "naked seed," which refers to plants that have seeds without fruit. The gymnosperms are divided into four divisions—**Ginkgophyta, Cycadophyta, Gnetophyta,** and **Coniferophyta** (see Table 25.1). The first three are not very widespread today, but the

fourth division includes the familiar conifers. **Conifers** are the common "evergreens" that cover our mountains, adorn our homes at Christmas, and form a vast belt across many northerly regions of the world. We will return to the conifers after a brief look at the other gymnosperms.

Ginkgophytes. In the early Mesozoic era, great parts of the earth were covered by ginkgophytes. Today, one species remains. Long believed extinct, *Ginkgo biloba,* the maidenhair tree (Figure 25.28a), was subsequently found on the grounds of Oriental temples and, finally, in 1946, growing wild in China. It's no wonder then that the single surviving species, like the rare seedless plant, *Psilotum,* is referred to as a "living fossil." *Ginkgo* is now commonly cultivated as decorative plants throughout the world.

From the appearance of its smooth bark and fan-shaped leaves, you might think that *G. biloba* is just another flowering plant—it even sheds its leaves in autumn after they turn from green to lovely hues of yellow. However, its fruitless seeds reveal its membership in the "naked seed" club. Male trees (the species is dioecious: separate male and female plants) produce airborne pollen grains that generate swimming sperm. *Ginkgo* is one of the few seed plants retaining this primitive reproductive trait. Fluids for the very short distance the sperm must travel are produced by the female sporophyte itself. Following fertilization, the female *Ginkgo* produces seeds in the manner of a typical gymnosperm.

Cycads. Plants resembling modern **cycads** had their heyday during the late Triassic period of the Mesozoic, some 200 million years ago. We know from fossils that they were among the most common plants in those ancient forests. They may even have been an important part of the diet of giant reptiles

(a) The *Ginkgo*

(b) Male cycad

(c) *Welwitschia*, a gnetophyte

FIGURE 25.28
The Lesser Known Gymnosperms
Three of the four gymnosperm divisions are represented here.
(a) *Ginkgo biloba* is the sole survivor of a once important division. Its flat-bladed leaves are deceptive, resembling those of angiosperms. Likewise, the fleshy seeds resemble the fruit of a flowering plant. It completes the deception by shedding its leaves in winter. **(b)** The large pollen cone in the center of the male cycad reminds us that this plant is definitely not a palm. Palms are flowering plants. **(c)** The gnetophyte *Welwitschia*, a native of the Namibian deserts of southwest Africa, produces just two leaves. They grow continuously, however, splitting again and again and forming the tangled growth seen here. Like other gymnosperms it produces separate pollen and seed cones.

of that era. Today the 100 or so remaining species are found mainly in tropical regions (Figure 25.28b). You may have seen cycads in museums and parks, and you may have mistaken them for palms or ferns. In fact, many are sold in plant nurseries as "sago palms."

In some cycads, pollen is produced in gigantic cones. Sexes are separate, and pollen is carried by the wind from male to female cones. The windborne pollen produces sperm cells that, like those of the *Ginkgo,* are flagellated. This is our last look at swimming sperm in plants.

Gnetophytes. The 70 or so species of gnetophytes alive today are probably more closely related to flowering plants than are any of the other gymnosperms. There are several reasons for this belief. As indicated earlier, this group, like angiosperms, has nonmotile sperm. Within the stems of gnetophytes are water-conducting **xylem vessels**, an advanced type of water conductor that otherwise is seen only in the angiosperms.

Other gymnosperms produce only the more primitive, water-conducting **tracheids**. There's more.

The gnetophyte *Gnetum* reproduces as a true gymnosperm but has fleshy seeds, and its blade-like leaves resemble those of the cherry. *Welwitschia,* of the southwestern African deserts, produces pollen cones that resemble flowers and has a bizarre spreading leaf growth (Figure 25.28c). *Ephedra* undergoes **double fertilization**, another hallmark of angiosperms. Double fertilization involves both of the sperm cells formed by the generative cell. In other gymnosperms, just one of the sperm cells is active.

If you live in the North American desert, you've probably seen *Ephedra* many times (perhaps you call it "Mormon tea" or "joint fir"). A finely branched shrub with tiny scale-like leaves, *Ephedra* was used by American Indians, wandering prospectors, and homesteaders to produce a hot, bracing drink. An Asian variety is the source of the drug ephedrine used in the treatment of asthma.

Conifers. The **conifers** (Division Coniferophyta) include nine families containing 550 or so species, not a large number in spite of their great northerly and mountainous populations (Figure 25.29). The most common and best known are in the pine family, Pinaceae, which includes pines, firs, spruces, hemlocks, and larches.

Vast populations of conifers live in the cold northerly climates of the earth, where they form a worldwide belt known as the **boreal** (northern) **forest**. Cold weather species are also found in higher altitudes further south. But conifers are not restricted to the colder climates. For example, there are great pine forests over much of the southeastern United States, and one of the tallest plants known, the coast redwood (*Sequoia sempervirens*), thrives in the milder, foggy climate of coastal California and Oregon.

The needle-like or scaly leaves of conifers (Figure 25.30) are adaptations to arid conditions brought on by both limited precipitation and extreme cold (where water is tied up as ice and snow). Needles and scales present minimal surface area from which water can be lost. As a further safeguard, the leaves have thick, waterproof cuticles and deeply recessed stomata. Typically, conifers produce their needle-like or scale-like leaves seasonally, but unlike other seasonal plants, the shedding of old dead leaves is gradual and continuous so that the trees remain green ("evergreen") all year round.

Reproduction in a Conifer. Conifers typically bear both **pollen cones** (pollen-bearing) and **ovulate cones** (ovule-bearing). So while the tree is monoecious, a cone is either male or female. The **ovule**, we should note, is a structure containing the female gametophyte and egg cell. When mature it becomes the seed. Each species has distinctive cones. Here we will focus on the complex reproductive events in the pine.

As you can see in the pine life cycle (Figure 25.31), events begin in sporophyte cells located in the scales of the cone. **Microsporocytes** in pollen cones undergo meiosis, giving rise to numerous microspores. Within the hardening walls of each microspore, the haploid cell undergoes mitosis to form a generative cell and a tube cell. The resulting pollen grain is the immature male gametophyte. It is at this time that pollen is carried by wind from the pollen cone to the ovulate cone, and pollination occurs. Meanwhile, things have been happening in the female counterpart.

The scales of the female counterpart, the ovulate cone, bear two ovules each. Each ovule consists of sporophyte tissue containing a **megasporocyte**. Upon meiosis, the megasporocyte gives rise to four haploid megaspores. Three of the four megaspores disintegrate, and the remaining one produces the female gametophyte. The gametophyte houses two or three archegonia in which egg cells are produced.

Curiously, pollination and pollen tube growth commence long before meiosis in the megaspore mother cell. Pines take their time with all of this, and it will be some 15 months from the time pollen arrives at the ovulate cone to the moment of fertilization. Prior to pollination, the scales of young ovulate

FIGURE 25.29
Coniferous Forest
The northern coniferous forest, or boreal forest, is a vast intercontinental expanse of conifers. Unlike the highly variable flowering plant forests, coniferous forests often have great populations of just a few species, extending mile after mile.

cones are obligingly parted, and drops of sticky fluid are exuded. The fluids trap the windblown pollen and, as dehydration occurs, draw the trapped grains through the **micropyle**, an opening near the female gametophyte. The scales then close, and pollination is complete. As you can see, with these added preparations, wind pollination can be quite efficient.

Next, the pollen grains form pollen tubes, which slowly penetrate, carrying the generative cell along. When the generative cell divides, it yields two sperm cells, which are deposited in the egg. One sperm fertilizes the egg (the other disintegrates). The zygote gives rise to an embryo, which, following a period of growth and differentiation, becomes dormant, and seed formation begins. Each pine seed consists of an embryo, some stored food, and hardened seed coats. By the time seed

(a) Pine leaves and seed cone

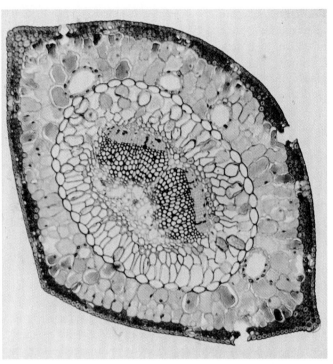

(b) Cross section through pine leaf

FIGURE 25.30
Adaptations for Dryness
The conifers are well adapted to dry conditions. This adapts them to the northerly and high-altitude climates where liquid water is unavailable throughout most of the winter months, and to southerly regions where summers are hot and the sandy soil does not hold the water very well. **(a)** The leaves of pines are needle-like. **(b)** A cross-sectional view reveals a thick cuticle, recessed stomata, and restricted surface area.

development has been completed, a second year will have passed—as we said, pines take their time.

In summarizing the gymnosperms, let's remember that the sporophytes are vascular and thus have true roots, stems, and leaves. The sporophyte is large and dominating, and, unlike the seedless plant, the gametophyte is incorporated into sporophyte tissues. In gymnosperms, gametophytes occur in the cone. There are no single-cell aerial spores formed. The male gametophyte, or pollen grain, consists of a generative and tube cell. Some of the gymnosperms have lost the swimming sperm and the centrioles that make cilia and flagella formation possible. Gymnosperm seeds are considered "naked"; that is, they do not develop within fruit.

The End of an Era

The lives of the gymnosperms and the great dinosaurs were once strangely intertwined. For eons, these great beasts foraged through immense Mesozoic forests of gymnosperms, but then, for reasons that are not clear, they both began to die off. The dinosaurs had marched to extinction by the end of the Mesozoic, but some gymnosperms managed to survive. With the arrival of the Cenozoic era about 65 million years ago, the angiosperms came to prominence. With an apparently greater adaptability to shifting climatic and geological conditions, the angiosperms fanned out over much of the planet in those distant days. Although the cold northerly and mountainous regions remained the domain of the hardy gymnosperms, the angiosperms abounded throughout the tropical and temperate regions. Yet today, the gymnosperms seem to be on the comeback trail. In the last few million years, extensive northern spruce and pine forests have displaced vast tracts of angiosperms.

Angiosperms: The Rise of the Flowering Plants

The astounding diversity of flowering plants today—some 235,000 species—stands in stark contrast to the sameness of the conifers. Although biologists are certain that angiosperms

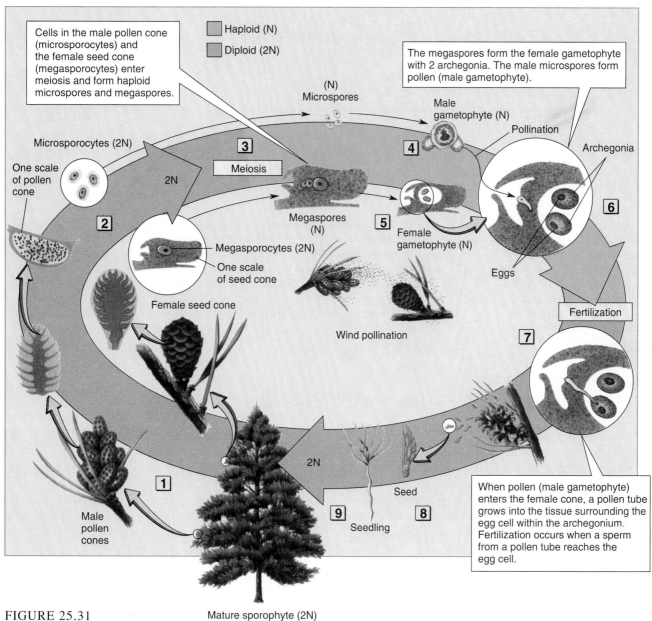

FIGURE 25.31
Pine Life Cycle
The pine tree, the sporophyte ⬜1, produces male pollen cones and female seed cones ⬜2 that contain diploid microsporocytes and megasporocytes, respectively. Meiosis gives rise to microspores and megaspores ⬜3. The microspores divide to form numerous pollen grains (male gametophytes) ⬜4 containing sperm. The megaspores enter meiosis with the once haploid product (cell) forming the female gametophyte ⬜5 in which archegonia develop ⬜6. Eggs form in the archegonia. ⬜7 Fertilization occurs when pollen tubes reach the archegonium and a sperm joins an egg. Following fertilization ⬜8, the zygote develops into an embryo that is surrounded by stored food and hardening seed coats. Upon germination ⬜9, the dormant embryo grows to form the new sporophyte.

evolved from one of several extinct gymnosperm ancestors, there are no clear fossil links between the two.

The apparent rarity of early angiosperm fossils suggests that the first flowering plants populated drier upland regions of the earth where fossils could not readily form. The lowlands were ideal locations for formation of fossils, and many fossilized gymnosperms are found in the sedimentary rocks

originating in such regions. Fortunately, pollen is readily fossilized in most regions, so in spite of the extremely spotty record of other plant structures, we know from their pollen that flowering plants diverged from their gymnosperm ancestors some 125 million years ago in the Cretaceous period of the Mesozoic era. This knowledge provides us with a reasonable time frame to explain the enormous diversification

Essay 25.1 ● AN EVOLUTIONARY LESSON FROM PLANTS

As you read about each group of plants in the chapter, it may have occurred to you that evolution is a spotty process and not as orderly as we often perceive. For example, bryophytes seem to be primitive in most ways, yet some have well-formed stomata-like pores, potentially independent sporophytes, and water-resistant cuticles—all attributes of advanced plants. Furthermore, we found, among the so-called primitive plants, some more-primitive leaves and other, well-advanced leaves. In addition, one group had undergone a degree of sexual specialization, producing separate male and female gametophytes. This is also an advanced condition. In some, the formation of simple aerial spores persists, but the mechanisms for spore release had become quite complex, again representing an advanced state though not one shared with flowering plants and conifers. Finally, going back to Chapter 23, consider the brown alga *Fucus*, a species with a decidedly nonalgal mode of reproduction. Recall that the sperm and egg form within the blades of the large, dominating sporophyte. Sound familiar? (But then, in typical algal fashion, it releases its eggs and sperm into the surrounding water.) There is a lesson to be learned from all of this, one that might give us a fresh perspective on the evolutionary process and, in particular, a more objective way of looking at the terms "primitive" and "advanced."

The concepts *primitive* and *advanced* are always troublesome. We must keep in mind that every living organism has exactly as long an evolutionary history as every other living organism—some 3–3.5 billion years—even though some groups may have changed rapidly while others appear to have been marking time. So perhaps it is best not to label any living species or other taxonomic group as either primitive or advanced. Instead, we may consider *individual features* within a group to be primitive (present in the original founders of the group) or advanced (evolved subsequently by only some members of the group and different from the earlier condition). If only specific features are considered as primitive or advanced, many semantic and philosophical problems disappear. We might then keep in mind that every living organism has a mix of primitive and advanced features (including, as we will see, present company).

experienced by angiosperms at the close of the Mesozoic (see Figure 25.17).

What prompted the rise of angiosperms? The emergence of flowering plants has been linked to vast geological changes at the end of the Mesozoic era and drastic climate changes during the early Cenozoic era that followed. Consider for a moment, the potential effects of a drop in the *average* temperature of the earth of 20°C! (Scientists today are deeply concerned with the possibility of just a few degrees change due to greenhouse gases. See Chapter 47.)

The changes in climate were brought about by three events. The first was a period of mountain building similar to that experienced at the end of the Paleozoic. The second has to do with the most recent episode of continental drift. According to geological evidence, the major land mass of the earth began breaking up and changing position at about the start of the Mesozoic era, some 220 million years ago (see Essay 20.2). Most of the great fragments drifted northward, so that by the Cenozoic era the continents had assumed roughly their present positions. Apparently, the angiosperms were versatile enough to have survived both of these changes.

In the third event, the end of the Mesozoic era may have been marked by a catastrophe of major proportions. According to the **Alvarez theory** (after paleontologist Walter Alvarez and his physicist father, Luis Alvarez), a gigantic asteroid collided with the earth some 65 million years ago. The dust it raised blocked the sun for several months, causing massive weather changes that led to large-scale extinctions. The extinction of so many species would have provided countless opportunities for surviving organisms to expand and diversify. As fantastic as the asteroid theory may seem, it has received a great deal of attention and is now provisionally accepted by many scientists. The Alvarez theory is discussed further in Chapter 27.

Were climatic and geological events the only factors influencing the rise of angiosperms? Perhaps not. Perhaps they only provided the opportunity. The great diversity seen in flowering plants can be attributed to other factors as well. It has been suggested that flowering plants can undergo faster speciation than most other living things. In a time of mass extinction, this capability would be critical, both from a survival and an opportunistic view. As we found in Chapter 20, angiosperms have the ability to form fertile hybrids, something that doesn't happen frequently in other organisms. Hybrids are usually sterile because their parental chromosomes are incompatible and meiosis fails. But flowering plants frequently experience a spontaneous doubling of chromosomes, producing a polyploid individual. When this happens in a hybrid, each parental chromosome gains a new homologue, and meiosis can proceed normally. Since plants can self-fertilize, it isn't even necessary for two polyploids to exchange pollen. The offspring of polyploids are quite different from their parental lines, representing what we have called "instant speciation."

Another contributing factor to the rise of flowering plants was an evolving partnership with the insects. Insects, drawn to flowers by the lure of sugary nectar, were probably the first animal pollinating agents. The ongoing mutual adaptation of insects and flowers has presented biologists with many classic cases of coevolution (see Chapter 20). The advent of insect pollination has an ecological twist to it as well. Wind pollination is quite effective in grasslands and in coniferous forests, where there are great numbers of the same species, but wind

FIGURE 25.32
Angiosperm Diversity
Angiosperms make up the most diverse plant kingdom. Flowering plants are unique in that their seeds are surrounded by fruit, the ripened ovary.

is certainly an inefficient vehicle in a mixed forest, where the closest individual of one's own species may be quite distant. Not only is insect pollination more efficient than wind pollination under these conditions, but it is much more selective. Once a bee finds nectar in a flower, it will return again and again to that kind of flower. Such specificity on the part of some insects makes it easier for new species to reproduce and for many different species to coexist. We will return to these points in Chapter 28.

The Angiosperms Today

In a sense, the angiosperms have inherited the earth—for the time being. Today they certainly constitute the vast majority of plant species. In spite of such diversity, Division Anthophyta, the flowering plants, contains only two living classes— **Dicotyledonae** (the **dicots**) and **Monocotyledonae** (the **monocots** (Figure 25.32). The terms refer to the presence of either one or two cotyledons, food storage areas in the seed.

The dicots are clearly in the majority, with 170,000 species. Briefly, the dicots form about 23 orders (botanists differ on the number) whose membership is typified by such familiar examples as magnolias, oaks, maples, buttercups, cabbages, hibiscus, potatoes, willows, mints, roses, apples, melons, beans, locusts, carrots, and dandelions. The 65,000 species of monocots make up 7 orders, represented by such familiar plants as lilies, date palms, rye, wheat, corn, sugar cane, irises, and orchids. The two classes represent distinct evolutionary lines, the monocots presumably having emerged from the dicots. The two differ in many ways; some of the more obvious of which are seen in Figure 25.33.

Life Cycle of a Representative Angiosperm. What follows can be considered typical or representative of flowering plants. With 235,000 species, you can expect to see a lot of variation on the basic plan. We begin our brief survey of the angiosperm life cycle (Figure 25.34) with the sporophyte, which, as in all vascular plants, dominates the life cycle. The sporophyte of a typical flowering plant includes a complex root and shoot system. The root anchors the plant in the soil, takes in and transports water and minerals, and stores varying amounts of foods. The shoot is also involved in storage and transport, but, in addition, it produces leaves and flowers.

At some point in the life cycle, angiosperms prepare for sexual reproduction by producing flowers, which are actually highly modified leaf structures. It is within the flowers that meiosis occurs and the gametophyte generation arises. We will be brief with this discussion, since reproduction in flowering plants will be treated in detail in Chapter 28.

Most flowers consist essentially of two parts (see Figure 25.34): accessory structures that include **sepals** and **petals**, and sexual structures that include the **pistil**, or *carpel*, and the

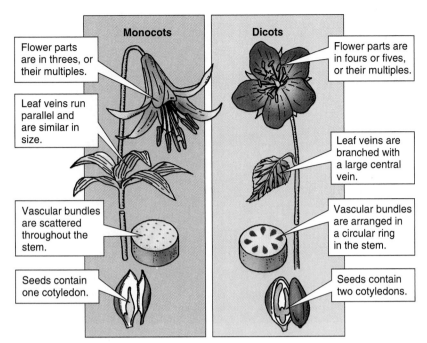

Monocots

Flower parts are in threes, or their multiples.

Leaf veins run parallel and are similar in size.

Vascular bundles are scattered throughout the stem.

Seeds contain one cotyledon.

Dicots

Flower parts are in fours or fives, or their multiples.

Leaf veins are branched with a large central vein.

Vascular bundles are arranged in a circular ring in the stem.

Seeds contain two cotyledons.

FIGURE 25.33
Monocot Versus Dicot
Several distinguishing characteristics of monocots and dicots are seen. Their major differences include leaf venation, seed anatomy, number of floral parts, and vascular arrangements.

stamens. Sepals protect the young flower bud, whereas the petals, which are often colorful, attract animal pollinators. The base of the petals may include glandular nectaries that secrete scents and nectar, which also attract pollinators.

The pistil, which may be subdivided into two or more parts called carpels, includes a sticky or fuzzy region, the **stigma**, at its tip and an **ovary** at its base, the two connected by the slender **style**. Within the ovary lie the ovules. Each ovule contains a megasporocyte. The megasporocyte undergoes meiosis, which yields four haploid megaspores, cells that represent the start of the gametophyte generation. Three of the megaspores disintegrate and the fourth enters several rounds of mitosis, producing the megagametophyte. Typically it will consist of seven cells, but the essential ones include an **egg cell** and a **central cell**. The central cell will have two nuclei. Both of these cells will be fertilized by sperm cells. At this time the seven-cell (eight nucleate) megagametophyte is referred to as an **embryo sac** (see Figure 25.34).

Each of the male counterparts, the stamens, consists of a lengthy, narrow **filament** tipped by a capsule-shaped **anther**. The anther contains many microsporocyte. During meiosis each microsporocyte gives rise to four haploid microspores. Each, in turn, undergoes mitosis, giving rise to pollen—the microgametophyte (see Figure 25.34). As in gymnosperms, the microgametophytes consist at first of just two cells, a generative cell and a tube cell. Later, the generative cell will divide to produce two sperm cells. This usually happens within the pollen tube.

Pollination in flowering plants involves deposition of pollen on the stigma of the pistil. A pollen tube penetrates the style and ovary, and when it reaches an ovule, its two sperm cells are released. One sperm fertilizes the egg cell and the other, the binucleate central cell. This is the **double fertilization**, described earlier as unique to flowering plants (and to the gnetophyte *Ephedra*). Note also that, upon fertilization, the central cell will contain three haploid nuclei. The three haploid nuclei join up to produce the *triploid* **endosperm cell** (see Figure 25.34).

The fertilized egg cell or zygote goes on to form the plant embryo. The endosperm becomes a food storage organ. Both are surrounded by seed coats. Finally, in true angiosperm fashion, the ovary, which still surrounds the seeds, forms the fruit. You'll need to partially revise your concept of fruit at this time, since in botanical terms, the fruit is not always sweet and fleshy as it is in a peach or cherry. In fact, string beans, pea pods, corn grains, coconuts, poppy capsules, acorns, and tomatoes qualify as fruits—all derived from the ovary.

From an evolutionary viewpoint, fruit provides a means of seed dispersal—away from the parent plant—where competition might be less intense. Some fruits, for example, are attractive to animals as food. The seeds usually pass through the animal gut, to be released somewhere else with the feces. Other fruits, such as maple and dandelion, are well adapted for "air travel"; and still others are literally shot from the seed pod as it dries and changes shape. The coconut is well known for its ability to travel great distances in the sea without damage.

Typically, seeds enter a period of dormancy during which they become dehydrated and hardened. Then, when sufficient water is absorbed, the cells are reactivated in what

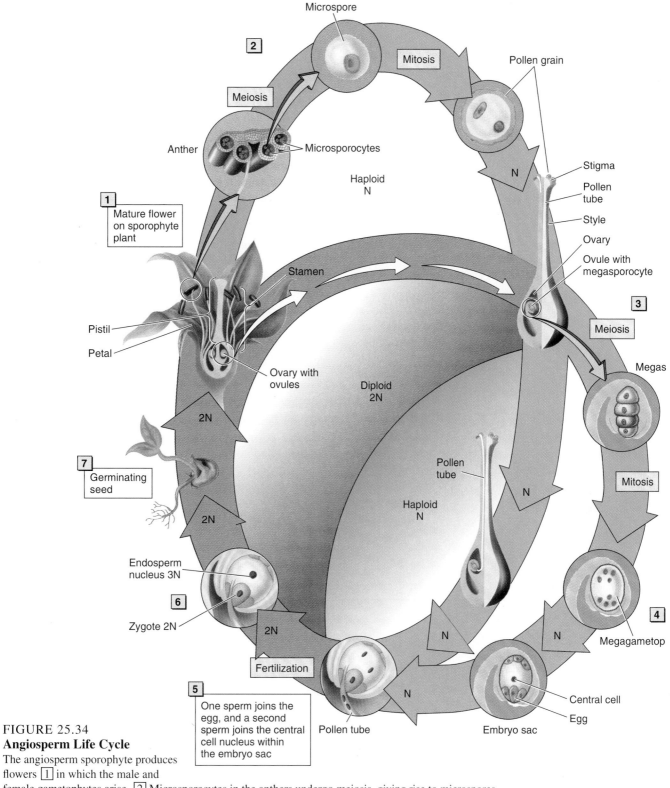

Microspore

2

Mitosis

Meiosis

Pollen grain

Anther

Microsporocytes

Haploid
N

Stigma

Pollen
tube

Style

1

Mature flower
on sporophyte
plant

N

Ovary

Ovule with
megasporocyte

Stamen

3

Meiosis

Pistil

Petal

Ovary with
ovules

Diploid
2N

Megas

7

Germinating
seed

2N

Pollen
tube

Haploid
N

N

Mitosis

2N

4

Endosperm
nucleus 3N

N

N

Megagametop

6

Zygote 2N

2N

Central cell

Egg

Fertilization

N

Embryo sac

5

One sperm joins the
egg, and a second
sperm joins the central
cell nucleus within
the embryo sac

Pollen tube

FIGURE 25.34
Angiosperm Life Cycle
The angiosperm sporophyte produces
flowers 1 in which the male and
female gametophytes arise. 2 Microsporocytes in the anthers undergo meiosis, giving rise to microspores
that form the pollen. Mature pollen represents the male gametophyte. 3 Megasporocytes in the ovules
also undergo meiosis, giving rise to megaspores. 4 One megaspore develops into the female gametophyte or embryo sac. 5 The two sperm from the male gametophyte fertilize the egg cell and the binucleate central cell. 6 The fertilized egg forms the diploid embryo, and the fertilized central cell forms the
triploid endosperm. The dormant embryo, cotyledons, and seed coats make up the seed. 7 Upon germination, the seed forms the adult sporophyte.

is called **germination**. When the seed sprouts it forms roots, stems, and leaves. When flowers are produced, the life cycle will be complete.

In summarizing the flowering plants we will note that (1) they are, of course, vascular, with xylem and phloem tissues. The xylem includes a second type of element, the vessel, which is present almost exclusively in flowering plants. (2) The sporophyte, like that of other vascular plants, is large and very prominent. (3) Gametophyte tissue is incorporated within the flower and (4) there are no single-cell aerial spores. (5) The pollen grain contains two nonmotile sperm cells that participate in a double fertilization, another flowering plant characteristic. Lastly, following fertilization, the ovary surrounding the developing seed enlarges to form the fruit. This means that angiosperm seeds are not naked.

The Plant Kingdom: A Summary

Plants arose from algal ancestors some 430 million years ago. Their early evolution included many changes that made their emergence onto the dry terrestrial environment possible. Among the critical adaptations was a water-resistant cuticle, structures to house and protect the embryo, specialized tissues for anchoring the plant body and transporting water and minerals from the soil to the photosynthetic tissues, and a means of transporting the sperm to the egg.

Bryophytes—liverworts, hornworts, and mosses—probably represent an early offshoot from the main line of plant evolution. Bryophytes evolved a simple conducting system for water and food but otherwise retained many of the ancestral algal characteristics. Included were an uncomplicated structural organization and swimming sperm with a dependence on external water for sperm travel. Perhaps their most primitive characteristic is the dominating gametophyte and reduced sporophyte.

In vascular plants the trend is clearly opposite—toward a highly reduced gametophyte generation and a very prominent sporophyte. This is clearly seen in the modern seedless plants, where the gametophyte occurs as a minute, separate individual. But seedless plants are reproductively primitive. Surviving members of this once prominent group still release single-cell, haploid spores in the manner of the algal ancestors. Their gametophytes still require an external supply of water so that the sperm can swim to the eggs. Spore dispersal, by the way, is actually an efficient way of disseminating the species. Spores can travel great distances in currents of air or water.

As the trend toward sporophyte dominance continued into more recent times, the gametophyte became fully incorporated into the sporophyte body, most likely an adaptation to drastic environmental change. In angiosperms, the most recently evolved plants, the gametophyte consists of just a few cells hidden away in the flower. Of great importance to the success of this venture was the evolution of a pollen system for transporting the sperm to the egg. A second innovation in the seed plant was the seed itself. Not only does the modern plant embryo develop safely within the large sporophyte, but upon completion of development, it enters dormancy and becomes surrounded by protective seed coats. In this state, the embryo can remain viable for many years—or until optimal growth conditions prevail. Furthermore, the evolution of the fruit provided a means of dispersal—away from the parent plant where competition might be less intense (making up for the sacrifice of the primitive but efficient spore dispersal).

Key Ideas

WHAT ARE PLANTS?

1. Key plant characteristics include (a) multicellularity with specialized cells and tissues, (b) chlorophyll *a* and *b* and carotenoids, (c) amylose starch and cellulose, (d) sporic life cycle—alternating haploid and diploid phases, and (e) multicellular embryos housed in gametophyte tissues.

2. In the sporic life cycle, sporophyte cells undergo meiosis, producing spores that form multicellular haploid gametophytes in which gametes arise. Gametophytes dominate in the **nonvascular plants**, sporophytes in the **vascular plants**.

ORIGIN OF PLANTS AND EARLY ADAPTATIONS TO LAND

1. Plants originated 430 million years ago in the Silurian period from green algal ancestors, perhaps similar to *Coleochaete.* In the transition to land, plants evolved ways of avoiding desiccation, reproducing on land, supporting the body, and transporting water and other materials. Adaptations included a waterproof cuticle, spore coverings, vascular tissue, leaf stomata, strengthening lignin, and mechanisms for sperm travel.

BRYOPHYTES: THE NONVASCULAR PLANTS

1. Nonvascular plants include **liverworts**, **hornworts**, and **mosses**, collectively called **bryophytes**. They lack true roots, stems, and leaves. Although terrestrial, they require external water for reproduction. An absence of lignified supporting tissue limits the size of bryophytes. Typically, they thrive in moist habitats, but many have adapted to rocky places, deserts, and tundra.

2. *Marchantia,* a liverwort, produces a ribbon-like, branching gametophyte. Pore-like epidermal openings permit gas exchange and water conservation, and **rhizoids** and **leaf scales** absorb water. Asexual reproduction is through **fragmentation** and **gem-**

mae production. In sexual reproduction, gametophytes form **antheridia** and **archegonia**, which produce sperm and eggs. Fertilization leads to a small, short-lived sporophyte that produces spores through meiosis. The haploid spores form gametophytes.

3. The gametophyte of the hornwort *Anthoceros* is a low rosette form. The sporophyte is horn-like and undergoes continuous growth and spore formation. Hornwort sporophytes form a water-resistant cuticle with pore-like stomata.

4. Mosses make up the greatest number of nonvascular plants. Some have primitive transport tissues, **hydroids** and **leptoids**, that conduct water and food.

5. The moss gametophyte produces antheridia and archegonia. Rain splashes sperm to the archegonia, where fertilization occurs. The diploid sporophyte forms a sporangium at its tip, where meiosis and spore formation occur. Spores germinate, producing thread-like growths called **protonema**, that develop into new gametophytes.

6. A recent theory is that bryophytes branched off the vascular line and through evolutionary simplification lost their vascular tissue and took on other bryophyte characteristics.

THE VASCULAR PLANTS

1. Vascular plants, often called **tracheophytes,** have tube-like vascular tissues (**xylem** and **phloem**), a highly dominant sporophyte, and a greatly reduced gametophyte. Water-conducting vascular tissue impregnated with strengthening lignin makes it possible for vascular plants to grow very large and to draw water from deep in the soil.

2. The **rhyniophytes** of the Silurian period, among the earliest vascular plants, had leafless, upright shoots arising from **rhizomes** (horizontal stems). These simple stems contained tough, supporting xylem tissue with water-conducting **tracheids**.

3. Vascular plant divisions include the seedless plants—**Psilotophyta, Lycophyta, Sphenophyta,** and **Pterophyta** (ferns)—and the seed plants. **Gymnosperms** include **Coniferophyta** (conifers), **Cycadophyta, Ginkgophyta,** and **Gnetophyta. Angiosperms,** or **Anthophyta,** are the flowering plants.

4. The dominant sporophyte of **psilotophytes**, or whisk ferns, has anchoring rhizoids (but no true roots), rhizomes, photosynthetic stems, and sporangia. They produce tiny separate gametophytes, which, in turn, produces sperm in antheridia and eggs in archegonia.

5. The dominant sporophyte of **lycophytes** (ground pines or club mosses) has true stems and true leaves. Spores give rise to separate gametophytes that produce the gametes. Spore formation in some is advanced and heterosporous (spores of different types) with **microspores** (small) and **megaspores** (large). Microspores form **microgametophytes,** which give rise to antheridia. **Megagametophytes** give rise to archegonia, each forming an egg.

6. One surviving genus of Sphenophyta, *Equisetum,* forms slender photosynthetic sporophytes. Homosporous spore formation occurs in a strobilus. The gametophyte is tiny and independent.

7. Ferns are widespread and numerous. The typical dominant sporophyte consists of thick rhizomes (underground stems), which give rise to fiddleheads that uncoil to form leaves. Homosporous spores are produced in sporangia with elaborate spore-ejecting mechanisms. When a haploid spore germinates, it develops into a photosynthetic gametophyte—the **prothallus**, which forms archegonia and antheridia. Following fertilization, the young independent sporophyte emerges.

THE SEED PLANTS

1. The demise of the ancient forests and the rise of the seed plants may have been brought about by vast geological changes—upheavals that raised the land and subjected it to drying and cooling trends.

2. Pollen provides the means for the sperm to reach the egg in the cones and flowers of seed plants. Seed plants produce microspores, which develop into **pollen grains** (the male gametophyte) consisting of **generative** and **tube cells. Pollen tubes** provide a passage into the female gametophyte for the sperm to follow. This eliminated the need for swimming sperm and outside water. The seed includes an embryo, stored food, and surrounding seed coats.

3. Gymnosperms, now an informal term, include **ginkgos, cycads, gnetophytes,** and **conifers. Ginkgophytes** include just one species, *Ginkgo biloba,* the maidenhair tree. It has broad leaves, produces seeds, and retains the swimming sperm. **Cycads** are palm-like plants with separate sexes. Pollen is produced in large strobili. They have motile sperm. **Gnetophytes** have certain angiosperm characteristics. One species has broad leaves, one produces flower-like cones, and another has advanced water-conducting elements called **xylem vessels.** A third has double fertilization.

4. **Conifers** form great forests in the northern hemisphere and in mountain regions. The leaves, typically needle-like or scale-like, are replaced continually. The leaf is adapted to dryness. After meiosis and microspore production in the male or **pollen cone**, mitosis yields a microgametophyte consisting of a generative cell and a tube cell. The microgametophyte becomes enclosed in a tough covering, forming a pollen grain. Before fertilization, the generative cell divides to form two sperm nuclei. The tube cell guides the development of a pollen tube, which penetrates the female gametophyte, where one sperm fertilizes the egg. In the female pine cone, the megagametophyte develops in the **ovule**—a megasporangium. After meiosis, a single surviving megaspore produces the megagametophyte, which develops archegonia in which eggs form.

THE END OF AN ERA

1. The gymnosperms and the dinosaurs fell from prominence together at the close of the Mesozoic. The period is marked by the rise and divergence of the angiosperms.

2. Flowering plants are far more diverse than conifers. Flowering plants probably originated in the Mesozoic, but until the Cenozoic they were apparently restricted to drier uplands, where fossils rarely formed. The sudden rise of flowering plants is attributed to a rapidly changing climate accompanied by mountain building and continental drift. According to the **Alvarez theory**, the collision of a gigantic asteroid with the earth darkened

the atmosphere with dust, created massive weather changes, and brought about many extinctions. The replacement of gymnosperms by angiosperms may have also been due to the angiosperms' greater variability, efficient insect pollination, and the coevolution of pollinators and plants, both of which made pollination more efficient.

3. The 235,000 angiosperm species form two major classes, **Monocotyledonae** (**monocots**) and **Dicotyledonae** (**dicots**). They differ in seed structure (number of cotyledons), flower anatomy, and vascular tissue distribution in stem and leaf.

4. The flower consists of **sepals**, **petals**, **pistils**, and **stamen.** Microspores form through meiosis in the **anthers.** They next enter mitosis to produce pollen, the male gametophyte. Megaspores form in the ovules. Following meiosis, one haploid megaspore goes through mitotic divisions, forming a seven-cell **embryo sac.**

5. Upon pollination, the generative cell of the male gametophyte passes along the pollen tube toward the ovary and divides, forming two sperm. One fertilizes the egg, the other the central cell. The central cell forms the triploid **endosperm cell.**

6. The seed contains an embryo, stored food, and seed coats. During seed development, the ovary usually enlarges, forming the fruit. The fruit is important to the dispersal of seeds. After a period of dormancy, the seed germinates, the embryo resuming metabolic activity and growth. The seedling then emerges.

Application of Ideas

1. Present an argument for dividing the plant kingdom into two kingdoms. Why would one consider this in the first place? What organisms would the kingdoms contain, and from what ancestral types would they have emerged?

2. Describe the conditions (climatic, topographic, and so on) that would have to prevail for the bryophytes to return to the prominence they once held. Which animal groups might persist under these conditions?

3. The fossil record clearly demonstrates the presence and even prominence of fern-like but seed-producing plants in the Devonian forests. The significance of these plants to the evolution of today's seed producers is unknown. Present two competing hypotheses that might account for seed evolution having occurred twice. Support the more plausible of the two.

4. The involvement of animals in plant reproduction may have been a key factor in the rise and spread of angiosperms. Discuss this proposition and explore two aspects of angiosperm reproduction that involve animals today. How might these have influenced animal evolution?

Review Questions

1. List five significant characteristics of plants. Which, if any, is (are) unique to plants? (page 491)

2. Why is *Coleochaete* considered representative of the earliest plants? (page 492)

3. List four problems in the transition of plant life to land, and explain how they were solved. (pages 492–493)

4. What is the significance of lignin to the success of land plants? How is its absence in bryophytes evident? (page 492)

5. What are "true" roots, stems, and leaves? What characteristic of bryophytes permits them to get along fine without these? (page 494)

6. How does the hornwort sporophyte differ nutritionally from the liverwort sporophyte? How does its growth differ from that of other bryophyte sporophytes? (pages 494–497)

7. Explain how the presence of vascular tissue has influenced the distribution and size of vascular plants. (page 501)

8. List four divisions of vascular plants that do not produce seeds. In what two ways is reproduction in these groups primitive? (pages 502–504)

9. Why is spore formation in *Selaginella* considered advanced over spore formation in *Lycopodium?* Which of the two species appears to be the most similar to seed plants? (page 507)

10. Characterize the geographic distribution of ferns. What does this indicate about their evolutionary success? (page 509)

11. Summarize the geological changes that may have brought an end to the ancient Paleozoic forests. How might the presence of small independent gametophytes have added to the survival problems of the early seedless plants? (pages 510–511)

12. What constitutes a seed? How might the evolution of seeds have given the gymnosperm an advantage over earlier vascular plants? (page 513)

13. In what ways are the ginkgo and cycad reproductively primitive? (pages 513–514)

14. List three advanced characteristics of gnetophytes. What does this suggest about their relationship to angiosperms? (page 514)

15. What specific adaptation do the needle-like and scale-like leaves of the conifer represent? In what other ways are conifer leaves adapted to this condition? (pages 515–516)

16. When did the angiosperms rise to prominence? Where were their forebears living during the time gymnosperms were prominent? (page 515)

17. What were the geological conditions at the close of the Mesozoic period and how did they affect animal life? (pages 516–518)

18. How might hybridization have affected the success of angiosperms? (page 518)

19. What role might insect pollinators have played in the success of angiosperms? (pages 518–519)

20. Outline the events in the formation of the microgametophyte generation in a flowering plant. What is another name for a male gametophyte? (page 520)

21. Outline events in the ovary leading to the megagametophyte generation. What is another name for a female gametophyte in a flowering plant? (page 520)

CHAPTER 26

Invertebrates

CHAPTER OUTLINE

WHAT IS AN ANIMAL?

ANIMAL ORIGINS

The Early Fossil Record

MAJOR MILESTONES IN ANIMAL EVOLUTION

Landmark Evolutionary Changes

PHYLUM PORIFERA: AN EVOLUTIONARY DEAD END

The Biology of Sponges

PHYLA CNIDARIA AND CTENOPHORA: TISSUES AND THE RADIAL PLAN

Phylum Cnidaria

Phylum Ctenophora: Comb Jellies

PHYLUM PLATYHELMINTHES: MESODERM AND THE BILATERAL PLAN

Phylum Platyhelminthes: The Flatworms

PHYLA NEMATODA AND ROTIFERA: BODY CAVITIES AND A ONE-WAY GUT

Phylum Nematoda

Phylum Rotifera

PHYLA ECTOPROCTA, BRACHIOPODA AND PHORONIDA: THE LOPHOPHORATES

PHYLUM MOLLUSCA: UNSEGMENTED COELOMATES

PHYLUM ANNELIDA: SEGMENTATION ALONG THE LENGTH

PHYLUM ARTHROPODA: SPECIALIZATION OF SEGMENTS AND JOINTED LEGS

PHYLUM ONYCHOPHORA: LINKING ANNELIDS AND ARTHROPODS

PHYLUM ECHINODERMATA: DEUTEROSTOMES WITH RADIAL BODIES

KEY IDEAS

APPLICATION OF IDEAS

REVIEW QUESTIONS

"Worms don't feel pain," is the self-absolving explanation of why many fishermen feel no qualms about skewing an earthworm on a number 6 hook. The recoil and writhing of the worm seem not to weigh heavily in forming such pronouncements. Texans are told not to put their faces too close to female tarantulas to get a better look, because the females tend to jump on intruders' faces, causing them to spill their beer. In the jungle, some ants tend to bite, causing moderate pain, while others sting, causing incredible pain. An intrigued traveler might want to know which is which. Actually, we may have a lot of questions about those animals that do not share with us those supportive structures called backbones. Here then, we will take a closer look at these creatures and, as we move from one group to the next, we will focus on important evolutionary advents that led to the remarkable diversity we see among the invertebrates today. To set the stage for this chapter and the next (where we discuss the vertebrates), we should first see just what animals are, anyway.

What Is an Animal?

If someone asked you what an animal is, you could probably come up with a pretty fair definition. But you would probably be wrong just the same. The reason is that animals are so varied that exceptions can be found to almost every rule. So what kinds of rules can we make? Some might begin by simply saying that animals move around and have mouths and eat things. That's not a bad start. If we add "and are multicellular," we've pretty much covered the animal kingdom. But even then some animals would shout from the back of the room, "Not me!" So let's tighten our definition and say that animals have the following traits:

1. *Multicellularity.* Animals are made up of numerous eukaryotic cells.

2. *Organization.* Most animals have well-organized organ systems that carry out the various life functions (such as digestion, reproduction, and gas exchange).

3. *Movement.* Animals are able to move by the contraction of organized contractile elements, usually muscle.

4. *Support.* Cells and tissues are bound together by the complex animal protein collagen, which, in vertebrates, is the most common protein.

5. *Nutrition.* Animals are heterotrophic and require complex nutrients, including carbohydrates, fats, proteins, and vitamins.

6. *Reproduction.* Animals are essentially diploid, with gametic life cycles (discussed later). Meiosis leads directly to gamete formation, characterized by large non-motile eggs and minute, motile sperm. Many species reproduce asexually.

As we said, these are general statements. There are some very conspicuous exceptions to these rules.

Most of us are probably familiar with at least a few representatives of the most common phyla. Interestingly, even members of the same phylum may be quite different, so although we may be somewhat familiar with two animals, we may not be aware that they are related. For example, humans are chordates (in the phylum **Chordata**), yet we share the phylum with creatures with odd names like "sea squirts" and "salps." (When you're looking at a salp it can take all your concentration to convince yourself it's a phyletic relative.) Other commonly encountered phyla are the insects (phylum **Arthropoda**), earthworms (**Annelida**), and snails (**Mollusca**). You probably know that your dog may have roundworms (**Nematoda**), and if you frequently swim in the sea, you may have felt the sting of a jellyfish (**Cnidaria**). Coastal dwellers also know that tidepools abound with sea stars (**Echinodermata**), sponges (**Porifera**), and flatworms (**Platyhelminthes**).

We have just named the major animal phyla. Each phylum has certain characteristics that set it apart from others, and each has a considerable range of diversity within it. All are quite fascinating, and some may include species that might be regarded as startling. In this chapter we will deal with the phyla that lack backbones, the invertebrates. Then we will consider the phylum that includes the vertebrates (the chordata). Let's begin, though, with a brief review of some ideas about where animals came from.

Animal Origins

The questions "Who are we?" and "Where did we come from?" have long interested scientists, as well as philosophers, mystics, theologians, drunks, and others of bad habit. From the scientific view, it seems certain that today's plants, animals, and fungi sprang from ancient protists. The consensus among biologists is that the animal kingdom has its roots in the ancestors of two different protist lines, both flagellate protozoans (see Chapter 23). One produced the subkingdom that includes only the sponges. The other gave rise to the rest of the earth's animals.

Unfortunately, there is no fossil evidence that bears on the issue at all. In fact, the oldest eukaryotic fossils are similar to some of today's simpler animals. These animals, which lived some 680 million years ago, were not very different from modern jellyfish. This leads us to the following question: "How did the first multicellular animals come into being?" Let's see if we can marshal our informed ignorance to develop a scenario that might put together some of these ideas.

The basic question is how organisms go from the unicellular state to the multicellular one. According to one theory, this change occurred through the aggregation of unicellular forms. As unicellular organisms joined, they formed groups that would have become increasingly coordinated and interdependent. The interdependence, according to the scenario, grew stronger as the various cells began to specialize. Some, for example, would have specialized in movement, others in feeding, and still others in reproduction. With such specialization, the coordination of the cells became ever more diverse and efficient. Then, as the cells reached a state of total interdependence, they achieved true multicellularity.

Most biologists agree that the earliest developments toward multicellularity in animals must have occurred in the ancient seas, among marine protozoans. Perhaps the simplest crawling or stationary forms were followed by floating or swimming versions, or maybe animals anchored by long stalks—any adaptation that could enable the organism to begin to exploit new food resources. Figure 26.1 shows an existing animal that is thought to be similar in some ways to the earliest animal life.

The Early Fossil Record

Most authorities agree that animal life extends back no further than one billion years. The earliest indisputable animal fossils, the **Ediacara fauna**, are found in late Precambrian deposits from southern Australia. The deposits are 600–700 million years old and marked by scattered remains of numerous fossil jellyfish, soft-bodied corals, and worms (Figure 26.2). (See Table

FIGURE 26.1
Marine Flatworm
Thin-bodied marine flatworms are seen gliding over rocky surfaces of the seabed. Their simple organization suggests what some of the first metazoan animals might have looked like.

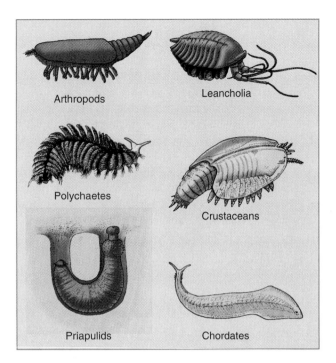

FIGURE 26.3
Burgess Shale
Burgess Shale fossils include familiar animals and some that are completely unfamiliar to zoologists. Early arthropods and polychaete worms abound, as do large worm-like priapulids in U-shaped burrows and complex crustaceans. The first chordates, ancestors to today's vertebrates are also seen in these rich fossil beds.

26.1 for a review of geological time and the history of life.) Still older rocks in the area bear what seem to be fossilized tracks and burrows that might have been made by marine worms.

An even richer fossil deposit, estimated to be about 530–550 million years old, occurs in the **Burgess Shale Formation** of western Canada (Figure 26.3). Its remarkable fossils, many of which are of soft-bodied animals, include representatives from *all* of today's major animal phyla. Many of the animals were apparently more complex than those of the Ediacara fauna.

The great diversity of animal life seen in the late Precambrian appears to have arisen very suddenly, at least in geological terms. The most compelling explanation for this sudden burst in speciation and evolutionary divergence is a widespread and massive animal extinction known to have occurred some 650 million years ago. Such extinctions have occurred again and again in the history of animal life, and each is followed by new bursts of speciation among the survivors. Recall that such sudden episodes of intense evolutionary activity are explained through the theory of punctuated equilibrium (see Chapter 20).

FIGURE 26.2
Ediacara Fauna
The Ediacara fossils from the late Precambrian seas are dominated by the fossilized remains of thin-bodied cnidarians. **(a)** Stalked, feather-like corals are nearly identical to today's "sea pens." **(b)** The bottom dwellers include several species of annelids, segmented worms that are identifiable by the lines crossing the body. A shelled mollusk **(c)** and a fossil from a phylum that is now extinct **(d)** are also seen.

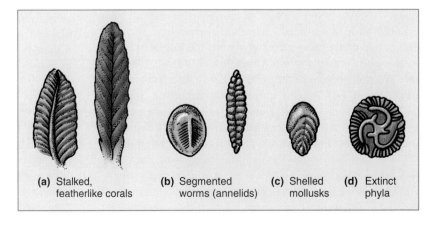

(a) Stalked, featherlike corals   **(b)** Segmented worms (annelids)   **(c)** Shelled mollusks   **(d)** Extinct phyla

TABLE 26.1
A Brief History of Life

ERA	Period (and epoch)	Years Before Present	Life Forms
Cenozoic	(Recent)	10,000	Increase in human population; rise of human civilization / Angiosperms dominant; herbaceous (soft-bodied) plants increase
Cenozoic	Quaternary (Pleistocene: ice age)	3 Million	Human societies; extinction of many great mammals such as wooly mammoth / Many plant species become extinct
Cenozoic	Tertiary	63 Million	First humans; mammalian divergence; land dominated by insects, birds, mammals / Dominance of angiosperms pronounced
Mesozoic	Cretaceous	135 Million	Primates appear / Dinosaurs reign, then become extinct; spread of insects; / Angiosperms dominate; gymnosperms decline
Mesozoic	Jurassic	181 Million	First birds and mammals; dinosaurs common; cartilaginous fish decrease temporarily / Gymnosperms dominate
Mesozoic	Triassic	230 Million	Mammallike reptiles; first dinosaurs / Burgeoning of gymnosperms and ferns; decline of club mosses and horsetails
Paleozoic	Permian	280 Million	Amphibians decline; reptiles expand / Gymnosperms and possibly angiosperms evolve
Paleozoic	Carboniferous	345 Million	Amphibians reign; first reptiles appear; insects diverge rapidly / Forests that will produce coal; ferns, mosses, and horsetails continue
Paleozoic	Devonian	405 Million	Fishes reign (jawed); amphibians appear / Radiation of land plants; forests of ferns, club mosses, and horsetails
Paleozoic	Silurian	425 Million	First air-breathing terrestrial animals; Jawed fishes common / Vascular plants, Fungi appear
Paleozoic	Ordovician	500 Million	Many marine invertebrates; first vertebrates (jawless fish) / Possible invasion of land by water plants
Paleozoic	Cambrian	600 Million	Marine invertebrates reign; first skeletal creatures; trilobites common / Algae appear
Proterozoic	Precambrian	1.5 Billion	Multicellular animals / Eukaryotes and protists appear
Proterozoic	Precambrian	2.5 Billion	Prokaryotes diversify / Anaerobic and photosynthetic bacteria evolve
Archeo-zoic		4.5 Billion	Formation of solar system

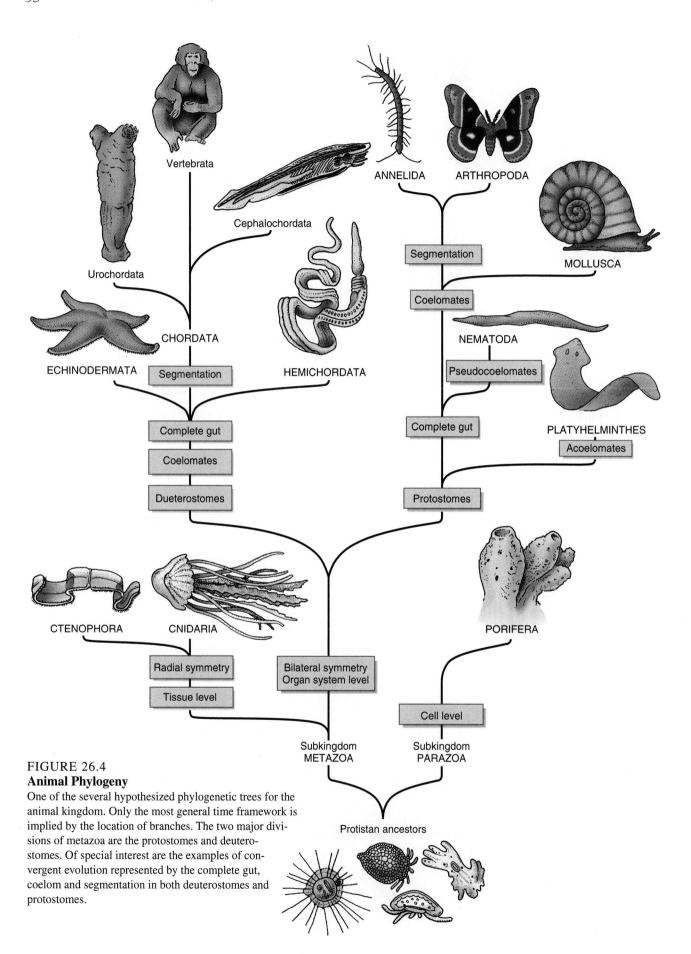

FIGURE 26.4
Animal Phylogeny
One of the several hypothesized phylogenetic trees for the
animal kingdom. Only the most general time framework is
implied by the location of branches. The two major divisions of metazoa are the protostomes and deuterostomes. Of special interest are the examples of convergent evolution represented by the complete gut,
coelom and segmentation in both deuterostomes and
protostomes.

Major Milestones in Animal Evolution

The evolutionary history and relationships among animal phyla can be represented by a phylogenetic tree. The tree shown in Figure 26.4 is a generally accepted scheme that indicates what are believed to be some of the major milestones of animal evolution. As with all such trees, it uses several sources, including the fossil record, comparative biochemistry and anatomy, physiology, and embryology. You may want to refer back to this figure as we proceed through the animal kingdom, focusing on the major evolutionary developments from one group to the next.

Note the tree's major branchings. Each branch marks significant evolutionary changes that gave rise to the various major animal phyla. In particular, you can see that the largest, most fundamental divergence is between what are called *deuterostomes* and *protostomes,* two major groups that include most of the animal phyla. Our own phylum, Chordata, is located on the left branch (the deuterostomes) along with echinoderms and hemichordates. The right branch (the protostomes) includes the annelids, arthropods, and mollusks. Strange as it seems, the profound differences between protostomes and deuterostomes exist largely because of the timing of the appearance of the mouth in the developing embryo and its origin, as well as the way the early embryo undergoes cleavage and other, even more subtle, distinctions.

The terms "protostome" and "deuterostome" literally mean "mouth first" and "mouth second," respectively. In **protostomes** the mouth opening forms earlier in development than it does in deuterostomes. The anus forms later. In **deuterostomes** the anus, or digestive outlet, forms first, the mouth later (Figure 26.5, bottom). Furthermore, in protostomes, by about the third round of cell division in the embryo (see Chapter 42) the new cells come to lie at an angle above the dividing line of the cells below, rather than directly over them, the product of **spiral cleavage** (Figure 26.5, top). In deuterostomes, on the other hand, the newly formed cells of the early embryo come to lie directly above the cells underneath, resulting in a **radial cleavage** (Figure 26.5b).

The third difference between protostomes and deuterostomes is seen in the way the body cavity (the space between the gut and the body wall) forms. We will see shortly that when this cavity is fully lined with tissue derived from an embryonic layer called mesoderm, it is called a **coelom**, and both protostomes and deuterostomes have a coelom. They differ, howev-

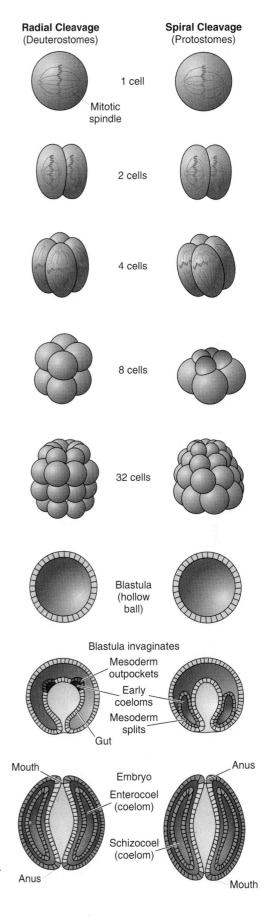

FIGURE 26.5
Differences in the Development of Protostomes and Deuterostomes
Note the spiral cleavage of protostomes and the radial cleavage of deuterostomes. Also note the differences in the formation of the coelom, mouth, and anus in the two groups.

er, in the way this cavity forms. In protostomes, the coelom forms as a split in the mesoderm. Hence this cavity is called a **schizocoel** (*schizo,* "split"). In deuterostomes the mesoderm forms two pouches along the hollow center of the embryo (see Figure 26.5), which spreads into the developing body cavity until it is fully lined with mesoderm, forming an **enterocoel**.

The two groups also differ in the origin of the skeleton. In the protostomes, the exoskeletons of the arthropods, the shells of the mollusks, and the bristles of the annelids consist of non-living, noncellular materials secreted by the ectoderm-derived epithelium. In contrast, most deuterostomes produce **endo-skeletons**—stiff, internal skeletons that are formed by meso-dermal cells. In most vertebrates the endoskeleton is bone, a hardened matrix containing numerous living bone cells.

With relatively few exceptions, the highly diverse proto-stome line (most invertebrates) emphasizes small bodies, extremely rapid reproductive cycles, and great population size, all characteristics that more than compensate for limited learn-ing ability.

The early deuterostomes (the group that now includes the echinoderms and us chordates) weren't much different—if any-thing, they were less intellectually gifted and less adventurous. They were also small, simple creatures with extremely limited nervous systems. Yet somehow there was amazing potential in this kind of animal. However, hundreds of millions of years separated the early deuterostomes from the vertebrates.

The important question for evolutionists is: How could both protostomes and deuterostomes have arisen on the earth? One would think that the first animal species to have devel-oped such a major advantage as the coelom and tube-within-a-tube body plan would have taken over the world. Yet these two kinds of coelomates—protostomes and deuterostomes—coexisted. Both these evolutionary lines underwent massive adaptive radiation in the late Precambrian, and the phyla that descended from each are today in competition with one another. However, ecological and evolutionary theory sug-gests that the earliest protostomes and deuterostomes must somehow have failed to fall into competition with one anoth-er at the time they first became established. There are at least two ways this might have happened.

The first hypothesis is that the early protostomes and early deuterostomes were protected from direct competition because they were *ecologically* isolated. Whereas the earliest protostomes were burrowing worms, the primitive deutero-stomes were upright suspension-feeders. So while the annelid–arthropod–mollusk ancestors were crawling about in the muck, the echinoderm–chordate ancestors—our own fore-bears—were sitting on stalked rear ends straining seawater. Who could have predicted which of the second line would eventually produce eagles and astronauts?

The second hypothesis is that the earliest protostomes and deuterostomes were protected from direct competition because they were *geographically* isolated. You will recall from our dis-cussions of continental drift (Essay 20.2) that the present-day land masses were produced by the breakup and drift of the supercontinent, Pangaea, which began about 230 million years

ago. Studies of magnetic lines and forces in ancient lava beds indicate that continental drift also occurred at an even earlier time. Current thinking is that Pangaea itself was formed through the union of two great land masses or supercontinents. The northern supercontinent, Proto-Laurasia, consisted of what is now North America, Siberia, and China; the southern supercon-tinent, Proto-Gondwanaland, included everything else. But what has this to do with protostomes and deuterostomes? Simply that some early fossil evidence suggests that the protostomes began in the shallow shelf waters of the southern supercontinent and that the deuterostomes radiated out from the shallow waters of the northern supercontinent. Thus they were firmly established, diverse, and occupying different niches before the shifting land masses brought them the pleasure of each other's company.

Landmark Evolutionary Changes

In noting the landmark evolutionary changes as we move from one group to the next, we will be looking at six conditions:

1. Level of organization.
2. Body symmetry.
3. Cleavage patterns.
4. Digestive traits.
5. Segmentation and specialization of segments.
6. Sequence of mouth and anus formation (this may seem odd but, as previously stated, it's critical).

To begin, let's note that, anatomically, most animals have several levels of organization: cell, tissue, organ, and organ system (Figure 26.6). The cell constitutes the fundamental level that still has the properties of living systems. The second level, the **tissue**, consists of a group of specialized cells with a common function. The lining of the intestine, for instance, is a tissue that specializes in food absorption. Tissues, in turn, are organized into **organs**, such as the intestine itself, which per-forms all or part of a major function. The intestine contains absorbing tissue along with muscle, nerve, and secretory tis-sues; hence it consists of many different cell types. Organs generally interact with other organs to form the **organ system** or, simply, **system**. The digestive system, including the mouth, esophagus, stomach, intestine, and so on, is an organ system, as are the respiratory and reproductive systems. We will consider the other evolutionary landmarks as they appear in the various phyla we will consider. You may wish to refer to Table 26.2 as we proceed.

Phylum Porifera: An Evolutionary Dead End

Sponges (phylum **Porifera**) exhibit only the cellular level of organization, making them the simplest animals. Essentially, a sponge consists of only a few specialized but loosely organized

TABLE 26.2
Characteristics of the Major Animal Phyla

Phylum	Level of Organization	Body Symmetry	Cleavage Pattern	Larval Type	Digestive Tract	Body Cavity	Segmentation
SUBKINGDOM PARAZOA							
Porifera	Cellular	Radial	NA	Flagellated	None	Primitive spongocoel	No
SUBKINGDOM METAZOA							
Cnidaria	Tissue, some organs	Radial	NA	Ciliated planula	Gastrovascular cavity	None	No
PROTOSTOMES							
Platyhelminthes	Organ system	Bilateral	NA	Similar to trochophore	Gastrovascular cavity	None (acoelomate)	No
Nematoda			Spiral cleavage	None	Complete gut	Pseudocoelom	
Mollusca				Trochophore		Coelom (schizocoel) (Greatly reduced in mollusks)	Yes (greatly reduced or absent in mollusks)
Annelida							
Arthropoda				Nauplius in some crustaceans			
DEUTERSTOMES							
Echinodermata	Organ system	Pentaradial	Radial cleavage	Dipleurula or similar	Complete gut	Coelom (enterocoel)	No
Hemichordata							
Chordata		Bilateral		Unique where occurring			Yes

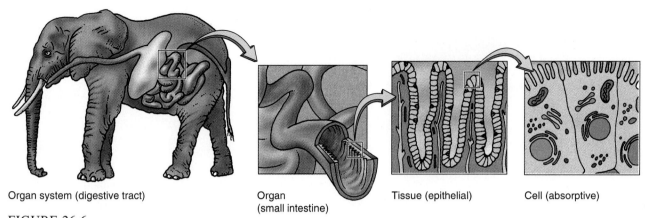

Organ system (digestive tract)   Organ (small intestine)   Tissue (epithelial)   Cell (absorptive)

FIGURE 26.6
Levels of Organization
Most animals have several levels of organization: cell, tissue, organ, and organ system.

types of cells that display little coordination. In fact, sponges can be separated completely into individual cells in a laboratory dish, and their cells will move back together and form new sponges. Restriction to the cellular level of organization means that functions are carried out by individual cells or aggregates of cells, without the coordination seen in animals with true tissues. As adults, the sponges' ability to respond to the environment is extremely limited. There are no nerve cells although a

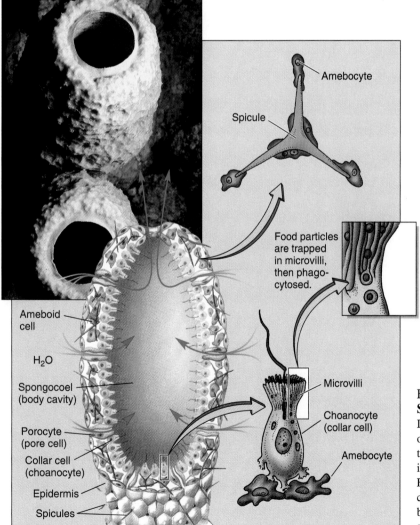

FIGURE 26.7
Sponge Anatomy
In the simple, vase-like sponges, the body wall consists of two organized cell layers surrounding a central cavity, the spongocoel. Between lies the gelatinous region in which are embedded supporting calcareous spicules. Flagellated choanocytes (see inset) sweep in water that contains microorganisms and bits of organic debris, both of which are used as food. Amebocytes produce spicules that form skeleton elements (inset).

rudimentary means of cell-to-cell communication may exist. Because the ancestral sponges have not given rise to any other major group, they are considered an evolutionary dead end.

The Biology of Sponges

Most sponges live in clear, shallow ocean waters around the world, although a few live at great depths and some live in fresh water. The adults are *sessile* (fixed in place), living quietly attached to the bottom, where they exhibit no apparent movement. While sponges are often small, some are more than 2 meters across.

In the simpler sponges, the body is vase-like, with a large central cavity or **spongocoel**, and an opening, the **osculum**, at the top (Figure 26.7). The body wall is composed of two cell layers, the outer *epidermis* and the inner *gastrodermis,* between which lies a noncellular, jelly-like region. The epidermis is riddled with pore-like cells called **porocytes**, which admit water into the central cavity. Motile cells called **amebocytes** wander through the mesohyl. The inner wall is made up of numerous flagellated collar cells, the **choanocytes**, which line the central cavity and there create water currents that draw seawater in through the porocytes and move it out through the osculum. The incoming water brings with it fresh supplies of oxygen and food. The outgoing water carries CO_2 and waste. As filter-feeders, sponges feed on microscopic organisms and bits of organic matter that become trapped in mucous secretions of the choanocytes. These trapped particles pass down the collar of the choanocytes to the base of the cell below, where they are phagocytized and digested in food vacuoles.

Sponges maintain their body shape by **spicules** or fibers that are scattered through the body wall and that act as skeletal elements. The spicules, produced by amebocytes, differ in each of the three classes of sponges. In **calcareous sponges**, the spicules contain calcium carbonate, while those of the beautiful **glassy sponges** consist of silicon dioxide (glass). In a third group, the **proteinaceous sponges** or "bath sponges," the skeletal material is **spongin**, a fibrous protein. (*Note:* Most cleaning "sponges" sold today are synthetic.)

Sponges are usually **hermaphroditic**, meaning that each individual produces both sperm and eggs. Sperm release (often in clouds) is aided by the currents from the collar cells. The egg cells are often located just beneath the collar cells, where the sperm can easily penetrate. A fertilized egg escapes as a highly flagellated, swimming larva, which after a brief period settles to the bottom and begins to sit out its adult life. Some sponges can also reproduce asexually, releasing **gemmules**, clusters of cells surrounded by a resistant wall. The gemmules of freshwater sponges may remain dormant through hard times, such as cold or dry spells, before becoming active and producing a new sponge body.

Earlier, we noted that sponges probably evolved from a line of flagellated protists, but the time in which this occurred is not known. Fossil sponges, as we've seen, were present in the Ediacara fauna from Precambrian times. By the Paleozoic era,

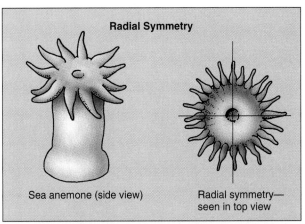

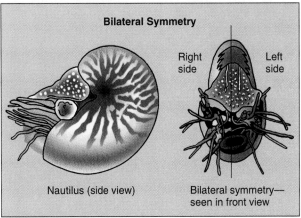

FIGURE 26.8
Body Symmetry
In radial symmetry, the body is essentially spherical, disk-shaped, or cylindrical. Any plane drawn through the center produces roughly equal left and right halves. Although two such planes are shown, there are infinitely more. Bilateral symmetry, on the other hand, means that the body can be divided equally by one plane only. This division produces mirror-image left and right halves that contain similar structures.

calcareous (chalky) sponges were common, and the siliceous (glassy) sponges became widespread by the mid-Paleozoic.

Now we move to animals with tissues, organs, and symmetry.

Phyla Cnidaria and Ctenophora: Tissues and the Radial Plan

The metazoans, the animals that are not parazoans, boast a variety of body types and life histories. The body plan of metazoan animals reveal two basic kinds of symmetry, **bilateral** and **radial** (Figure 26.8). Both are conceptually simple, but the differences have far-reaching evolutionary implications. The bilateral body plan has only *one* plane of symmetry,

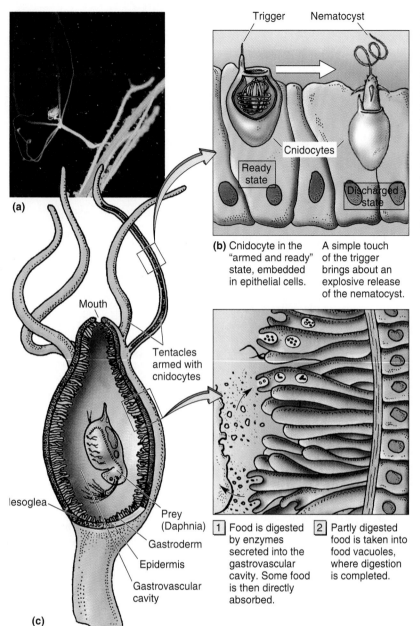

Trigger   Nematocyst

Cnidocytes

Ready state

Discharged state

(b) Cnidocyte in the "armed and ready" state, embedded in epithelial cells.

A simple touch of the trigger brings about an explosive release of the nematocyst.

Mouth

Tentacles armed with cnidocytes

Mesoglea

Prey (Daphnia)

Gastroderm

Epidermis

Gastrovascular cavity

(a)

(c)

1 Food is digested by enzymes secreted into the gastrovascular cavity. Some food is then directly absorbed.

2 Partly digested food is taken into food vacuoles, where digestion is completed.

FIGURE 26.9
Hydra, A Representative Cnidarian

(a) The hydra stuns its prey with stinging cells located on the tentacles, using tentacles to draw the prey into its gastrovascular cavity. **(b)** The stinging cells have barbed, harpoon-like, toxic threads that are released with great force and number when the triggers are brushed. **(c)** Note the thin-walled, hollow body, characteristic of cnidarians. It consists of an outer epidermis (with stinging cells), an inner digestive lining, and a middle layer, the mesoglea, that contains nerve and muscle fibers. The digestive lining of the gastrovascular cavity has specialized cells for the secretion of enzymes and phagocytosis of food particles.

which basically divides the body into right and left sides. We humans provide a good example of a bilateral animal. The radially symmetrical form is quite different. The body is essentially disk-like or cylindrical (sometimes spherical), perhaps with radiating arms. Theoretically, there are unlimited planes of symmetry. In other words, any plane that passes through the center body from top to bottom will divide the body into roughly equal halves.

Radial symmetry in today's animals exists in only a few phyla, most markedly in the cnidarians and ctenophores. There is no agreement about how the radial body form arose, but many zoologists theorize that the direct ancestor of all metazoans, including the radial ones, was probably a bilateral inhabitant of the muddy ocean bottom, where bilateral sym-

metry works well. (Can you see why this would be the case?) It may have resembled the *planula larva,* an early developmental stage of cnidarians that we will look at shortly. The theorists go on to suggest that a radial stage of these simple animals did well as a swimming form, and that this gave rise to the radiates.

Phylum Cnidaria

Phylum Cnidaria (pronounced "nidaria") includes hydrozoans, jellyfish, corals, and anemones. Most cnidarians are organized along the tissue level, but some have what could be called organs. For example, some have tentacles, and tentacles are composed of several tissues (including skin, nerve cells, and contractile fibers), which seems to qualify them as organs. In most species of cnidarians the tentacles are armed with stinging cells called **cnidocytes** ("nettle cells"). When triggered, the cnidocytes release harpoon-like stinging structures called **nematocysts** ("thread pouch") Figure 26.9 illustrates the general cnidarian features using *Hydra*, a freshwater hydrozoan.

Cnidarians are thin-walled animals; their bodies are composed essentially of an outer layer of cells (the epidermis) and an inner layer (the gastrodermis). Sandwiched between these layers is cellular, jelly-like matter—the **mesoglea**. Wandering ameboid cells, nerve processes, and contractile fibers are found in the mesoglea. The outer cells (or *epidermis*) are mainly protective, while the inner cells (or *gastrodermis*) form a sac-like gut called the **gastrovascular cavity** because it serves both digestive and circulatory functions. A simple opening serves as both mouth and anus—or, to put it another way, cnidarians have a mouth but no anus and must spit out the undigested remnants of whatever they swallow (see Figure 26.9).

As with other radial animals, there is little centralization of the nervous system. In fact, it takes the form of a **nerve net**, a diffuse network of nerve cells and their extensions, covering the entire animal. In addition, the neurons or nerve cells of cnidarians are quite unusual in that nerve impulses can travel in either direction from one cell to the next. In bilateral animals, most neural conduction is one way. The cnidarian nerve cells are organized to receive impulses from sensory cells (cells that receive external stimuli) and bring about responses in the contractile fibers and cnidocytes. As you might expect, any contraction of the cnidarian is a general response and not delicately coordinated between different parts of the body.

Cnidarians may take either of two forms: the **polyp** or the **medusa**. Polyps are usually sedentary, attached to rocks, wharf pilings, and other objects, whereas medusae can swim. They include the familiar jellyfishes. In some cnidarian species, both polyp and medusa stages exist in an unusual alternation of generations. Unlike alternation of generations in plants, however, both polyp and medusa forms are diploid.

Class Hydrozoa. In **hydrozoans** the polyp state usually dominates, and although some solitary species occur, most form dense colonies. They reproduce asexually by **budding**, that is, by outgrowths in the form of miniature polyps. Some polyps are specialized for capturing and digesting food, which is then moved by ciliary action throughout the continuous hollow body of the colony. Others specialize in sexual reproduction, many producing swimming **hydromedusae** that form sperm and eggs that are released to join in the open water.

The life cycle of the hydrozoan *Obelia* is shown in Figure 26.10. Note the stage called a **planula**, a ciliated, swimming larva. In a sessile animal, the swimming larval stage makes it possible for the species to disperse through the environment.

Class Scyphozoa. In the **scyphozoans**, or jellyfish, the dominant stage is the swimming medusa, which in this class is called a **scyphomedusa** (Figure 26.11a). Polyps are present in most species, but they are highly reduced. In effect, the jellyfish life cycle is opposite that of the hydrozoan.

Most jellyfish have a bell-shaped body with the mouth centrally located on the underside, surrounded by tentacles. For the most part, they are drifters, but they can swim by contracting muscle fibers around the bell. While most are modest in size, some, such as *Cyanea,* attain large proportions—up to 2 meters in diameter with tentacles 70 meters long. These enormous jellyfish are indeed dramatic sights when drifting in clear, blue waters, far out at sea. Sexes are separate in nearly all jellyfish, and fertilization is usually external.

Class Anthozoa. Anthozoans—the corals and anemones—exist only as polyps (Figure 26.12). The most familiar anthozoans in temperate waters are the heavy-bodied sea anemones. Anemones abound in rocky tide pools and in shallower coastal waters. Corals are colonial anthozoans whose polyps secrete surrounding walls of limestone (calcium carbonate). The lovely coral formations sold in curio shops are actually their dead skeletons. The cumulative effects of coral secretions (along with significant contributions by certain mollusks, tube worms, and coralline algae) over millions of years have produced coral atolls (islands), as well as the barrier reefs that surround many islands in the Pacific. Although all corals can feed with tentacles, most tropical corals play host to photosynthetic, symbiotic algae that provide most of their nutrients.

It was mentioned earlier that polyps can reproduce asexually by simply budding new polyps. In their sexual reproduction, female and male polyps produce eggs or sperm separately, and fertilization occurs in the water. The zygotes develop into swimming planula larvae that form more anthozoans.

Phylum Ctenophora: Comb Jellies

Ctenophorans (Figure 26.13)—the "C" is silent—commonly called "comb jellies," appear to be close relatives of cnidarians, but their similarities may be a case of convergent evolution. Like the cnidarians, comb jellies are radially symmetrical marine animals. Also, the body consists of two cell layers separated by a jelly-like mesoglea, with the innermost layer surrounding a gastrovascular cavity. Unlike cnidarians, comb jellies have eight rows of **combs** consisting of fused cilia. The coordinated beating of these combs moves the animals along. And whereas cnidarians have tentacles armed with stinging cells, the tentacles of comb jellies are armed with "glue cells." They snare the prey rather than sting it. Their chief food is plankton, minute drifting marine organisms. Ctenophorans have another distinguishing, even enchanting, trait. They are bioluminescent; that is, they glow.

Phylum Platyhelminthes: Mesoderm and the Bilateral Plan

In the animal kingdom, most animals are bilaterally symmetrical (see Figure 26.8). With the evolution of two sides, the stage was set for **cephalization**. Cephalization refers to developing a head or leading end and a tail or trailing end. At first

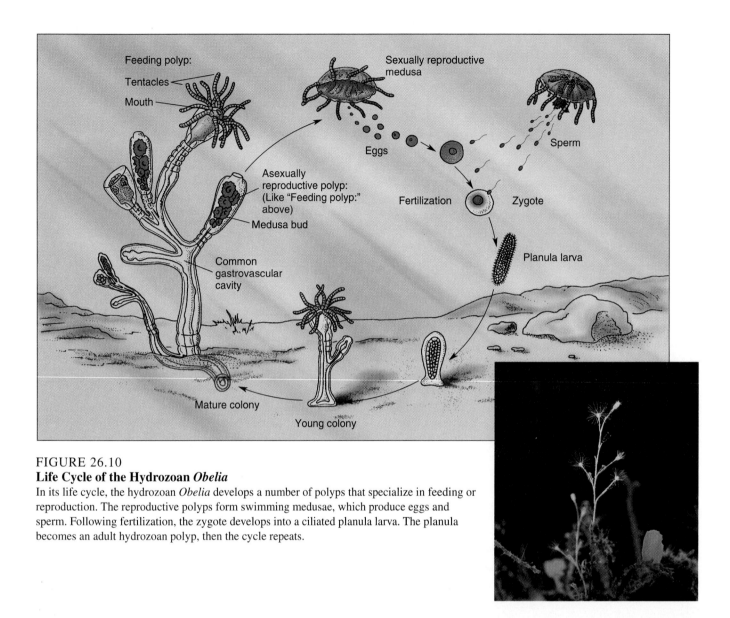

FIGURE 26.10
Life Cycle of the Hydrozoan *Obelia*
In its life cycle, the hydrozoan *Obelia* develops a number of polyps that specialize in feeding or reproduction. The reproductive polyps form swimming medusae, which produce eggs and sperm. Following fertilization, the zygote develops into a ciliated planula larva. The planula becomes an adult hydrozoan polyp, then the cycle repeats.

the head may have been simply a concentration of muscles, an adaptation for burrowing in the soft sea beds of ancient oceans, which is where the oldest fossils of bilateral animals are found. But leading ends soon became equipped with sensory structures for the detection of food, light, sound, and other stimuli. With senses concentrated in the leading end, an animal could more quickly perceive what sort of environment it was moving into. It wouldn't do to go around backing into new environments. Such sensory structures would have required neural support and integration; thus clusters of nerve cells were concentrated at the head end, and the evolution of the brain was underway.

The cnidarians and ctenophorans we have just seen are much simpler than the other metazoans. Their simplicity is related to their having only two fundamental types of embryonic tissues, or **germ layers**, as they are called. These are the **ectoderm** ("outer skin") and **endoderm** ("inner skin"). All other metazoans have a third germ layer called the **mesoderm**

("middle skin"). Each of the germ layers contributes to specific portions of the developing body, laying down the basic framework from which that region develops. For example, whereas the nervous system is essentially ectodermal in its origin, most of the inner linings such as those of secretory glands, the gut, the blood vessels, and those of the respiratory and reproductive systems are endodermal in origin. The mesoderm, on the other hand, largely contributes to muscle and, in vertebrates, to the internal bony skeleton and the blood. Actually, once the basic framework is laid down, each organ eventually receives tissues derived from the other germ layers as well.

Phylum Platyhelminthes: The Flatworms

The phylum **Platyhelminthes** contains the flatworms, including one class that is free-living and two that are parasitic. Flatworms have bilateral body symmetry, so they have an anterior (head) end and a posterior (tail) end. They have also clearly

FIGURE 26.11
Life Cycle of a Scyphozoan
Sexes are separate, and after fertilization a ciliated planula larva forms. It attaches as a polyp to the seabed. The polyp undergoes an asexual budding process that yields many young medusae. Each will mature into an adult jellyfish.

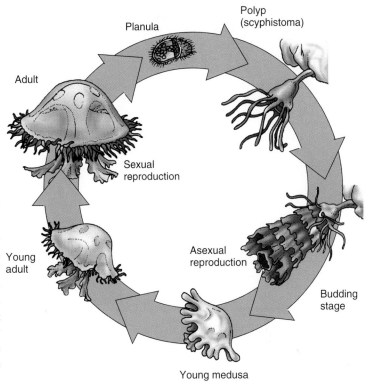

arrived at the organ-system level of development, although some of the organ systems are quite simple.

Flatworms lack special respiratory and circulatory structures. Because of their flattened bodies, most cells are close to the body surface, making simple cell-to-cell diffusion of gases sufficient to accommodate metabolic needs. In the parasitic species, the nervous and digestive systems are greatly reduced or virtually absent. (Do you see why this might be the case? Remember, tapeworms usually live in the gut of a host animal.)

Unlike most other animals with mesoderm, flatworms have solid bodies. Although a gastrovascular cavity is present in many, there is no internal cavity between the gut and body wall. For this reason flatworms are called **acoelomate** ("without a cavity"). As we will see later in the chapter, the presence of a body cavity, or coelom, is a major evolutionary feature.

Class Turbellaria: Free-Living Flatworms. Although many of the **turbellarians**—the free-living flatworms—are marine, the most familiar of the group are the very common

FIGURE 26.12
Representative Anthozoans
Corals live in vast colonies, inside a secreted calcarious covering. Anemones are often solitary anthozoans and are usually attached by their bases to the substrate.

FIGURE 26.13
Ctenophorans
Delicate comb jellies are propelled through the water by rows of cilia making up the combs. The combs are in rows on vertical ridges around the body wall, and the central pouch-like object is the gastrovascular cavity.

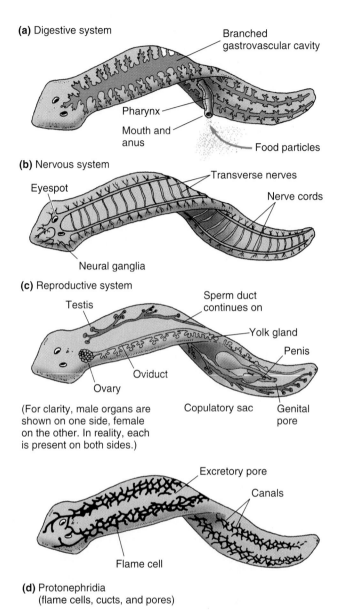

(a) Digestive system

Branched gastrovascular cavity

Pharynx

Mouth and anus

Food particles

(b) Nervous system

Transverse nerves

Eyespot

Nerve cords

Neural ganglia

(c) Reproductive system

Testis

Sperm duct continues on

Yolk gland

Penis

Oviduct

Ovary

Copulatory sac

Genital pore

(For clarity, male organs are shown on one side, female on the other. In reality, each is present on both sides.)

Excretory pore

Canals

Flame cell

(d) Protonephridia
(flame cells, cucts, and pores)

FIGURE 26.14
Planarian Anatomy
Dugesia a free-living planarian flatworm, is only a few millimeters long, yet it has several organ systems that permit the animal to feed, reproduce, and respond to external stimuli. *Dugesia* is about 15 mm long. **(a)** Digestion occurs in a highly branched gastrovascular cavity, which begins with a complex, muscular pharynx. **(b)** Light-sensitive eyespots help the flatworm respond to light, and a ladder-like nerve network coordinates movement of its muscles. **(c)** The reproductive system includes both male and female parts. **(d)** Osmoregulation is carried out in a system of tubules containing many flame cells.

freshwater planarians, including the species *Dugesia* (Figure 26.14). While many species are quite small—a few millimeters in length—their well-developed organ systems make them attractive as laboratory subjects. Thus planarians have been the subject of anatomical, developmental, and behavioral studies for many years.

The nervous system is concentrated at the anterior end, where large **ganglia** (clustered nerve cell bodies) and eyespots (light receptors) are located. Two ventral nerves run the length of the body, sending out branches from a number of smaller ganglia in a ladder-like arrangement (Figure 26.14b).

Planarians are predators and scavengers, feeding through a protrusible, muscular **pharynx**, while pinning the unlucky victim beneath the flat body in a slimy mucous layer secreted from its undersurface. Food enters the highly branched gastrovascular cavity, where digestion begins (Figure 26.14a), but bits of food are also phagocytized by cells lining the cavity, and digestion is completed in food vacuoles within the cells. Thus digestion is both extracellular and intracellular. Waste exits through the same opening through which food enters (thus they have an incomplete digestive tract).

In freshwater flatworms, the osmotic uptake of water over the body's surface is balanced by specialized cells called **protonephridia** that get rid of excess water. In many species, the protonephridium consists of two highly branched systems of tubules with a number of blind sacs containing the **flame bulbs** (Figure 26.14d). Each bulb is made up of a group of ciliated **flame cells** (the waving or "flickering" of the cilia, as

seen under the microscope, inspired their name). Their primary function is getting rid of excess water that continually enters the body because of a natural osmotic gradient.

Planarians swim in a curious sort of crawl. The smaller species secrete a layer of mucus underneath them and then beat their way through it with the cilia that abound on their ventral body surface. Larger species rely on an undulating movement of the flat body.

Planarians reproduce both asexually and sexually. In asexual reproduction, the individual undergoes fission—simply pinching in half across the body. Each half then regenerates the missing parts. The regenerative power of planarians is legendary; almost any portion of the body can be restored following its removal. Planarians are hermaphroditic; each individual has both male and female sexual structures. They include well-defined ovaries, testes, and copulatory organs, the latter permitting internal fertilization (Figure 26.14c). Yet planarians do not usually self-fertilize. During sexual reproduction, two individuals copulate with the simultaneous exchange of sperm. Although planarians develop directly into miniature adults, many marine turbellarians produce a ciliated larval stage.

Class Trematoda: The Flukes. The **trematodes**, or **flukes**, are parasites. The flukes are equipped with large suckers, an anterior "oral" sucker and a ventral sucker, both of which are used to attach to their host's tissues. Flukes invade many kinds of animals and make their home in various parts of the host, including the gills, lungs, liver, and intestines. One species has even found its niche in the inhospitable environment of the urinary bladder of frogs. One human liver fluke (*Opisthorchis sinensis*), common to Asia, has an enormously complex life cycle involving three very different hosts—human, snail, and fish. Since the sexual reproduction portion of the liver fluke life cycle occurs in the human, we are the **primary host**. The snail and fish are **intermediate hosts**. However, each host is essential to the parasite's life cycle.

Class Cestoda: The Tapeworms. Members of class **Cestoda**, the **tapeworms**, are quite different from other flatworms. These parasites have no digestive cavity and must absorb digested food of the host directly across the body wall. To this end, the outermost external membrane is highly folded, greatly increasing the surface area. Virtually all vertebrates harbor tapeworms, generally with little harm to the host animal, although the tapeworms do absorb nutrients from the host's digestive tract. Humans are host to seven different species of tapeworm. Someone has pointed out that in any classroom, several students are likely to be wormy. (Can you pick them out?)

Typically, the sexually mature tapeworm (Figure 26.15) consists of an extremely small, rounded, anterior **scolex** that is equipped with suckers and hooks for attachment to the host's intestinal wall (Figure 26.15a). Extending from the scolex is a lengthy chain of flattened blocks called **proglottids**, produced continually through budding. The proglottid is a complete reproductive unit, each containing ovaries, testes, and male and female copulatory structures (Figure 26.15c). Whereas cross-fertilization with other individuals is the general rule, self-fertilization among proglottids and even within a single proglottid is known to happen. Fertilized proglottids enlarge, and finally the mature proglottids, each containing thousands

of fertilized eggs, break away from the chain and pass out of the host's body in the feces. The fertilized eggs must be ingested by the next host in order for the life cycle to continue.

In addition to the primary host, where sexual development and reproduction occur, most tapeworms have an intermediate host. It is usually a different species, typically an animal that is eaten by the primary host. For example, pigs, cattle, and fish serve as intermediate hosts to some human tapeworms. In the pork tapeworm the fertilized eggs, passing out of the human host, are ingested by the pig as it feeds. Upon entering the pig's intestine, the egg develops into a tunneling creature that bores its way through the intermediate host's tissues, generally ending up in the muscles. There it forms an encapsulating, protective cyst around itself and develops into a stage known as a **bladderworm**. The bladderworm becomes dormant, remaining so until the pork is eaten uncooked or undercooked, whereupon it emerges as a young tapeworm, migrates, attaches to the host intestine, and repeats its life cycle. For the trendy folks, sushi (uncooked fish) is now implicated in the transmission of such parasites.

Phyla Nematoda and Rotifera: Body Cavities and a One-Way Gut

As animal life on our planet continued to evolve and specialize, ever adapting to particular habitats and lifestyles, new landmark evolutionary changes appeared. We have already mentioned differences in symmetry and the importance of a third germ layer, the mesoderm. Now we will see how the development of a body cavity was associated with the formation of a more efficient one-way gut and how this arrangement permitted yet other novel developments on the animal scene. With the development of a one-way gut, we find a digestive tract with a mouth, intestine, and anus (Figure 26.16). The resulting body plan resembles a "tube-within-a-tube," since the gut lies within the body wall. (Keep in mind that "gut" is not an indelicate term in biology.)

The body cavity is the space between the gut and body wall. As we mentioned earlier, such a body cavity completely lined with mesoderm is called a **coelom.** Not all hollow spaces within the body qualify as coeloms. In the *nematodes* and *rotifers,* our next two phyla, we encounter a body cavity that is lined partly by mesodermally derived tissue and partly by tissue derived from ectoderm. A body cavity incompletely lined with mesoderm is called a **pseudocoelom** ("false coelom"). Recall that an animal that lacks a body cavity is called an **acoelomate.** Figure 26.17 compares the three conditions, acoelomate, pseudocoelomate, and coelomate.

Phylum Nematoda

The **nematodes** or roundworms are best known for their parasitic members, some of which are truly horrendous. Simply

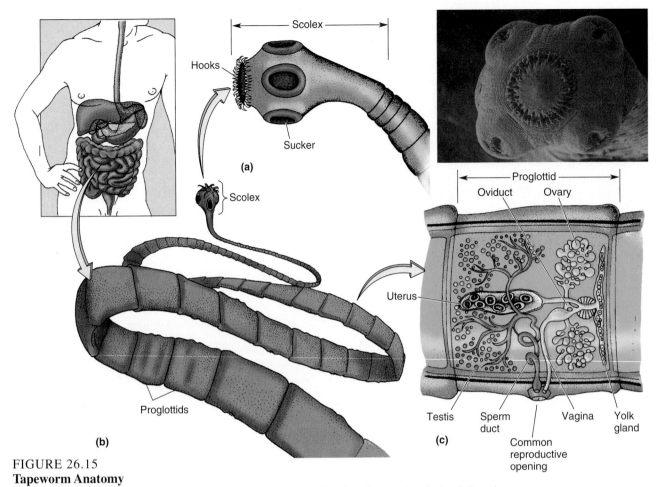

FIGURE 26.15
Tapeworm Anatomy
(a) The tapeworm scolex is equipped with a ring of hooks and several suckers for grasping the host's intestinal wall. (b) The remainder of the body, consisting of multiple reproductive units known as proglottids, buds continuously from the scolex. (c) Each proglottid contains well-defined ovaries and testes, along with related ducts and glands.

reciting the life cycle of some of them can cause nightmares. But let's stay calm and just note that roundworms are slender, cylindrical, and usually tapered at both ends. They come in a range of sizes, but they are usually easy to recognize since they all look like smooth, glistening worms (Figure 26.18). They are bilaterally symmetrical, although from outside appearances they lack a distinctive head. Nematodes move by flexing longitudinal muscles first on one side and then on the other, producing a wriggling or undulating motion. This wouldn't get them very far in the open water, but many nematodes live in the soil, and the wriggling action helps them thread their way between the solid particles.

Nematodes thrive in nearly every conceivable moist and aquatic habitat, from the soil of flower gardens to the world's oceans. Most of the estimated half-million species of nematodes are free-living, feeding on protozoans, rotifers, small earthworms, and each other. Some capture their prey with a paralyzing saliva. Others make use of a tiny blade-like device that impales the hapless prey while the sucking lips and pumping pharynx drain its juices. Because of their astounding

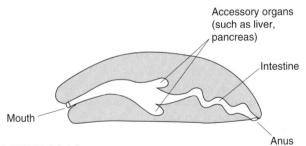

FIGURE 26.16
Generalized Digestive Tract with Mouth, Intestine, and Anus
The generalized tube-within-a-tube body structure is typical of most complex animals.

number, free-living nematodes are an essential link in the earth's ecological organization.

Many nematodes are very important parasites of plants. Plant nematodes may have a devastating effect on agriculture

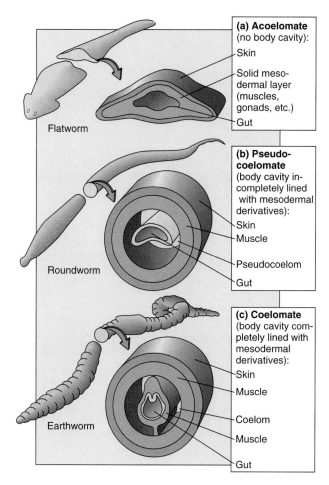

FIGURE 26.17
Body Cavities
The coelom is a mesodermally derived and lined body cavity between the gut and the body wall. **(a)** Flatworms lack a body cavity and are designated acoelomate. **(b)** Roundworms are pseudo-coelomates. Although their body cavities are extensive, they are not fully lined by tissue of mesodermal origin. **(c)** In coelomates, the coelom is fully lined by mesodermally derived tissue.

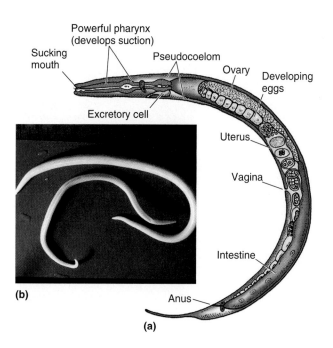

FIGURE 26.18
Soil Nematodes
Most nematodes are scavengers or predators.

and require constant control at great expense. In fact, sometimes farmers must shift to more resistant crops in order to make contaminated land usable again.

Parasitic roundworms probably infest all vertebrate species. In fact, we harbor at least 50 rather harmless species and 10 more dangerous ones. One species that is particularly dangerous to humans is *Trichinella spiralis.*

The roundworm *Trichinella spiralis* is a common inhabitant of pigs, rats, and humans. It is passed from one animal to the next when flesh that contains larvae is eaten. The larvae bore into the intestinal lining, where they mature and reproduce. Each female can produce about 1500 young. The young then begin to migrate through the lymph channels, eventually finding their way into the voluntary muscles, tissue rich in blood and oxygen. There the worms mature and surround themselves with a protective cyst (Figure 26.19). The cycle ends there unless the hog is eaten by another hog or some other

carnivore, such as ourselves. If we eat infected pork or pork products without first killing the worms, they may travel through our own lymph channels, to encyst in our own muscles. It has been estimated that an ounce of heavily infected sausage can contain 100,000 or more encysted larvae, which could produce 100 million venturesome offspring in their new host. Pork or pork products can be made safe by thorough cooking (58°C, 137°F) or deep freezing (–23°C, –10°F) for 20 days. (Well-cooked pork is always gray—never pink.) The symptoms of this infestation, called **trichinosis**, include muscle pain, fever, blood disorders, edema (swelling), and gastrointestinal disturbances. There is no way of killing the parasites, but most people survive the invasion unless other medical complications set in.

Phylum Rotifera

The rotifers, we can report, are not parasitic. In fact, their elaborate feeding system is unlike that of all other pseudocoelomates. They feed by sweeping microscopic organisms into their mouths through the action of double rings of cilia around their head area. Food is swept into a grinding gullet, or **gizzard**, which prepares it for digestion. The digestive system contains other specialized organs, including a stomach, two digestive glands, intestine, and anus (Figure 26.20).

Female rotifers, though microscopic, are much larger than the males. Males are also scarce, and, in some species, they have never been found. One reason may be that the male digestive system is so rudimentary and inadequate that, apparently, they die soon after mating. The short interlude of the

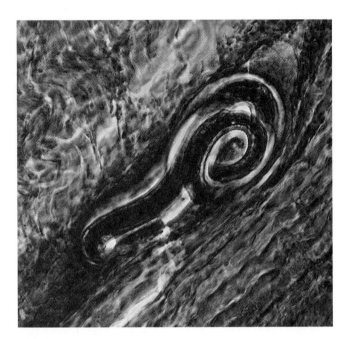

FIGURE 26.19
Source of Trichinosis

Trichinella spiralis is known to encyst in pork. If such meat is eaten without sufficient cooking, trichinosis may result. The worms will leave their cysts and venture through the body, forming new cysts, generally in the diaphragm, ribs, tongue, eye muscles, and larynx.

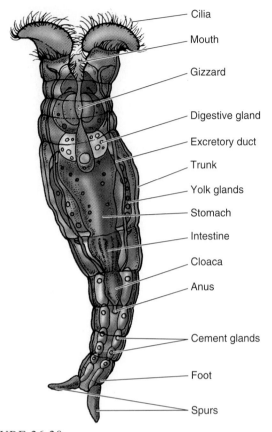

FIGURE 26.20
Rotifer Anatomy

Rotifers, extremely common freshwater animals, are about the same size as many protists (most less than 1 mm in length) but are much more complex. Note the well-defined digestive and reproductive systems. Cement glands at the base of the foot permit the animal to temporarily stop its whirling motion and anchor itself to the bottom.

male's existence would seem to threaten the survival of these animals were it not for a surprising ability of the females to compensate. The females, it turns out, are of two types: those that mate with males and those that don't.

In rotifers that reproduce sexually, females may first go through several generations of reproduction without fertilization. Asexual reproduction by females without the eggs being fertilized is called **parthenogenesis**. The eggs do not enter meiosis, so they remain diploid. These begin to divide, rapidly developing into more females. Rotifer populations increase quickly in this fashion, some doubling every two days. Then, when some (unknown) environmental cue is received, the females switch to meiosis, and haploid eggs are formed. If unfertilized, these eggs develop into males. Upon maturity, the haploid males produce sperm that fertilize any haploid eggs they encounter. Such zygotes, interestingly, become surrounded by thickened, resistant walls and can remain dormant for prolonged periods. Some of these winter over and develop only when warm weather returns. These produce females. So, in the world of rotifers, males are reduced to the simplest of roles. Their contribution is an occasional input of genes, adding some genetic variability to an otherwise all-female population.

Here then we have reviewed a number of different types of animals with a variety of evolutionary advances. In particular, we have seen the development of bilaterality and a one-way gut. Next, we will see species in which this gut passes through the mesoderm-lined cavity called a coelom. Because the rest of the species we will consider are coelomates, let's

briefly consider this important cavity that we find in so many diverse kinds of animals. We should first say that the invertebrate coelomates are sometimes referred to as "advanced invertebrates," because they generally have a number of other advanced features, traits that evolved later in the history of life.

The coelom itself is lined with a tough, glistening, mesodermally derived **peritoneum** from which arise broad, membranous folds, the **mesenteries** (Figure 26.21). The mesenteries hold the organs within the coelom in place while allowing a certain freedom of movement. In species with circulatory systems, blood vessels run through the mesenteries, carrying blood to and from the intestine.

In many species, the coelom is fluid-filled and lined with beating cilia that keep the fluids in motion. Such movement can be an efficient means of transporting digested foods and metabolic wastes. The oldest function of the coelom, one involving both primitive and modern-day worms, was probably locomotion. In such soft-bodied animals, the coelomic fluids form a hydrostatic skeleton, as muscular pressure against the fluids gives the body a certain rigidity and provides a resistant base for the action used in burrowing. Fos-

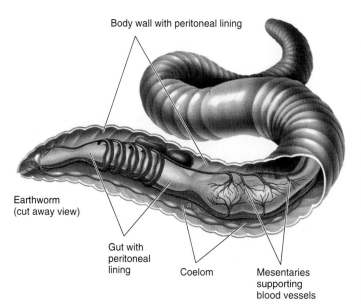

Body wall with peritoneal lining

Earthworm
(cut away view)

Gut with
peritoneal
lining

Coelom

Mesenteries
supporting
blood vessels

FIGURE 26.21
Peritoneum and Mesenteries
The coelom is lined with the peritoneum, from which arises the membranous mesenteries. The mesenteries cover many of the internal organs and are permeated by blood vessels.

silized worm burrows, you may recall, are the oldest evidence of animal life on earth.

The principal phyla of invertebrate coelomates that we will consider here are the mollusks, annelids, arthropods, and echinoderms. The first three phyla are protostomates (as are the flatworms, nematodes, and rotifers), whereas the echinoderms are deuterostomes (along with hemichordates and chordates). We will also look at the **lophophorates**, a peculiar group that is considered protostomate by some zoologists and deuterostomate by others.

While both protostomes and deuterostomes may form a coelom, the coelom is believed to have evolved independently in the two groups and, in fact, the coelom actually forms differently in the embryos of the two groups (see Figure 26.5). (The presence of a coelom in both animal groups is thus a major example of convergent evolution.) Let's now take a look at the coelomate protostomes.

Phyla Ectoprocta, Brachiopoda, and Phoronida: The Lophophorates

We will begin our look at protostome coelomates with the lophophorates. What comes to mind when you think of a lophophorate? Probably not much. The lophophorates are comprised of three rather obscure phyla. Some of these phyla, though, include some rather familiar forms of life. They are all distinguished, however, by a **lophophore**, a curved ridge to which ciliated tentacles are attached.

FIGURE 26.22
A Lophophorate
One of the phyla, Ectoprocta, forms either crusty colonies (on rocks) or seaweed-like fronds.

The three lophophorate phyla are **Ectoprocta, Brachiopoda,** and **Phoronida** (Figure 26.22). Ectoprocts, a group containing some 4000 species, are known more familiarly as bryozoans. These animals are often seen along rocky coasts at low tide, where they form branched, crusty structures that look something like seaweed but are actually colonies of individuals formed by asexual budding. Some feathery ectoprocts are dried and sprayed with green paint, and then sold as "living air ferns—requiring no care."

Brachiopods are shelled lophophorates that superficially resemble a clam or scallop. However, the two valves or shells hinge so as to form a top and bottom, rather than two sides (dorsoventral rather than lateral). The shell encloses a coiled ridge of ciliated tentacles that comprise the lophophore.

The phoronids are small worm-like creatures that live in tubes buried in the oxygen-starved sediments of estuaries and bays. They also bear ciliated lophophores. Phoronids, brachiopods, and ectoprocts all produce free-swimming larvae, typical of marine invertebrates that are fixed in place as adults.

Such larvae make up part of the marine plankton, tiny, drifting organisms that make up a basic element of aquatic food chains.

While today's lophophorates are only minor phyla in terms of their species numbers, they were once more important. In fact, they are well represented as the oldest fossils. Both brachiopods and ectoprocts are found in Cambrian deposits over 500 million years old. Some rock strata are composed almost entirely of their fossilized shells. The brachiopods have dwindled from about 3000 named species in Ordovician fossil beds to about 200 named species alive today.

Phylum Mollusca: Unsegmented Coelomates

The earth is burgeoning with mollusks. In fact, we already know of about 50,000 species, which, in terms of sheer numbers, puts them behind only the nematodes and arthropods. Some mollusks are minute, living in tiny inconspicuous shells, while the giant North Atlantic squid that swims confidently through the cold ocean waters reaches over 18 meters (59 feet) in length. Of the seven classes, we will consider only the four larger classes as representatives of the phylum (Figure 26.23): gastropoda (snails and slugs), polyplacophora (chitons), bivalvia (bivalves, such as clams), and cephalopoda (octopuses, squids, and nautiluses). There is some discussion about the relationship of mollusks to other coelomate protostomes. Whereas their ciliated larvae are similar to other protostomes, they lack one critical feature—segmentation. Some zoologists suggest that the mollusks are derived from a segmented ancestor, but most believe that they appeared well before the segmented phyla arose. (Segmentation will be introduced shortly.)

The term *mollusk,* derived from Latin, means "soft-bodied," though for an invertebrate, this isn't very descriptive. Aside from being soft-bodied, all mollusks are distinguished by a muscular **foot** that contains sensory and motor systems and may be used for swimming, creeping, digging, holding on, or capturing prey. Some have external **shells** produced by secretions of a fleshy covering known as the **mantle**, which is also present in nonshelled members of the phylum. The **mantle cavity** (the space enclosed by the mantle) houses feathery respiratory **gills** in aquatic mollusks. In terrestrial mollusks, the lining of the mantle cavity is highly vascular and serves as a respiratory membrane across which oxygen and carbon dioxide can pass (Figure 26.24).

In mollusks, the coelom is conspicuous in embryos but is very much reduced in adults, often present only as an open region surrounding the heart. Except for the cephalopods (squids and octopuses), mollusks have an **open circulatory system** in which the blood leaves the vessels, in some cases to percolate freely through the tissues. Unlike **closed circulatory systems**, those in which the blood remains within vessels, the

Tree snail

FIGURE 26.23
Representative Mollusks
A sample of molluskan diversity—some 100,000 species are known.

vessels in open systems end some distance from the heart, the blood then flowing into sponge-like sinuses and cavities before returning to the heart. The distinction between the two kinds of systems is an important evolutionary difference, so keep your eye peeled for the first appearance of a closed system.

The digestive system is, of course, tube-like, with both a mouth and an anus. Mollusks (except for bivalves and tuskshells) have a rasping tongue-like structure called the **radula** with which they file their food into small particles (Figure 26.25). Mollusks make use of a specialized organ called the **nephridium** to carry on osmoregulation and excretion. We will look more closely at the nephridium in the earthworms.

The evolutionary relationship of mollusks to other protostome coelomates (annelids and arthropods) is uncertain, mainly due to their apparent lack of segmentation and their highly reduced coelom. Yet some mollusks produce a ciliated, pear-shaped swimming larva, like that of the marine annelids. In addition, studies of certain fossil mollusks indicate that the coelom may at one time have been much more prominent than it is now. It is because there is no convincing evidence of segmentation in the fossils that some scientists argue that mollusks diverged from coelomate ancestors before segmentation became established.

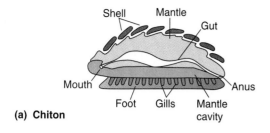

(a) Chiton

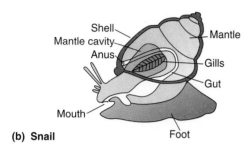

(b) Snail

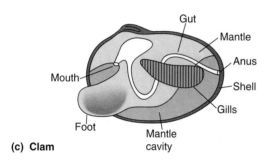

(c) Clam

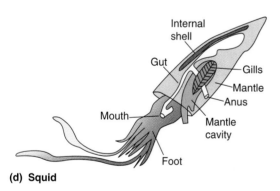

(d) Squid

FIGURE 26.24
Mollusk Body Plan
All mollusks share a basic body plan, which includes a fleshy foot and mantle and, commonly, a shell or its remnants. Members of the phylum are classified according to variations in the basic structures, shown with color coding.

The **chiton**, class **Polyplacophora**, is a mollusk with a flattened body, an elongated foot, and a shell composed of eight simple plates (see Figure 26.24a). The mantle forms a skirt-like arrangement just above the foot and houses numerous feathery gills.

Chitons are slow-moving creatures that creep along over the rocky bottoms of tide pools through wave-like contrac-

tions of muscles in the foot. They graze as they go, using their sharp radula to scrape the algae from the surfaces of the rocks. The gut, beginning with a radula-equipped mouth, extends the length of the body. When disturbed, their defense is to put powerful foot muscles to work gripping the rock, making them very difficult to dislodge. Alternatively, they can curl up, presenting only their tough plates to a predator. Chitons very similar to those of today appear in the fossil record in 500-million-year-old Ordovician deposits.

Gastropods, class **Gastropoda**, include the snails and their relatives. They glide along on a large, extendible, muscular foot much as the chitons do. The foot lies just below most of the digestive organs, thus giving rise to the poetic name "gastropod" or "stomach foot" (see Figure 26.24b). Some, such as snails and slugs, have successfully adapted to land life, while others glide over the seabed, their beautifully sculptured shells accenting the ocean floor. Whereas marine gastropods generally possess well-formed feathery gills, gas exchange in terrestrial and freshwater snails and slugs is provided by a lung consisting of a heavily vascularized region within the mantle cavity. Interestingly, freshwater snails, and a few marine species, have lungs instead of gills and must come to the surface to breathe. It is theorized that, like the marine reptiles and mammals, such species arose from terrestrial lines, returning to the water after lungs had evolved.

The basic organization of the snail is unusual in that, during development, its body undergoes a 180° twist—a process called **torsion** (twisting or rotation) brought about by uneven muscle growth. This brings the mantle cavity forward and the anus around to a position just above the head (certainly not the most esthetic arrangement). The overall configuration provides considerable space for retraction of the entire head-foot into the shell, which may be the chief advantage—protection of the delicate gastropod head. Attached to the foot is a tough lid, or **operculum**, composed of a horny material. Once the operculum is in place across the opening into the shell, the snail is safe from many predators. Interestingly, the coiled shell itself is believed to have evolved independently of torsion, perhaps even preceding it.

There are a lot of gastropods around, somewhere in the neighborhood of 35,000 species. They vary from those with elaborate twisted shells to shell-less creatures such as the nudibranch, as well as the cone snail, and the keyhole limpet (Figure 26.26).

The bivalve (two-valve) mollusks, class **Bivalvia** (**Pelycypoda**) (see Figure 26.24c), differ from gastropods primarily in that the shell is organized into two hinged **valves**. The head and sensory appendages are greatly reduced, although the foot is fleshy and highly extensible—used by bivalves such as clams for burrowing. The hinged shell is drawn closed by two powerful muscle groups.

Bivalves lack the radula and are chiefly filter-feeders, using mucus to trap small particles of food from the water. Currents of water bearing plankton are drawn into the mantle cavity through an **incurrent siphon** by the action of cilia that

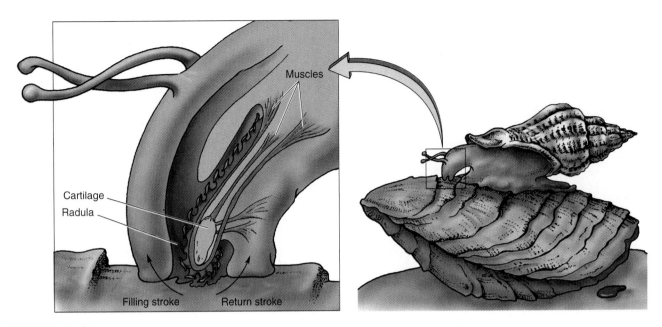

FIGURE 26.25
The Radula
The radula is a ribbon-like membrane containing a row of teeth, all pointing backward. It lies over a bed of cartilage and slides back and forth as it is pulled first by one set of muscles, then by another. The rasping action tears and shreds food or, in this case, bores through an oyster shell. Photograph shows a surface view of the radula.

line the large gills (Figure 26.27). The food particles are then trapped in heavy mucous secretions, which are moved by cilia along food grooves to the **labial palps**. The labial palps then move the food into the mouth. The gill surface also serves in the exchange of carbon dioxide for oxygen, with water passing on its way to the **excurrent siphon**. The heart is located in the coelomic cavity. As you can see in Figure 26.27, the intestine passes through the cavity; in fact, the heart is wrapped around the gut.

FIGURE 26.26
Representative Gastropods
The typical gastropod has a coiled shell, gliding foot and stalked eyes.

The **cephalopods**, class **Cephalopoda**, "head-foot" mollusks, include the squid, octopuses, cuttlefish, and nautiluses (Figure 26.28 and see 26.24d). All but the nautiluses lack the external shell. Squid and cuttlefish have greatly reduced internalized shells.

In many ways, cephalopods are highly specialized creatures. They are active and voracious predators, feeding on fast-moving invertebrates and vertebrates. Octopuses and squid move by forcefully ejecting water through their siphons in a jet-like action. The siphon can be turned in any direction. They may also quickly change colors or release a cloudy "ink" to foil predators. Octopuses often hide in burrows and can squeeze their large bodies through and into quite narrow cracks and crevices. Like other cephalopods they have highly developed sensory receptors, including eyes that are very similar to those of vertebrates even though they evolved independently (Figure 26.29). The octopuses also have a large, well-developed brain and are surprisingly intelligent invertebrates. Their circulatory system is closed, an advanced condition that provides the rapid circulatory efficiency needed by large fast-moving animals. They have a centralized, pumping heart but in addition there are two **branchial hearts**, chambers that provide an additional boost to deoxygenated blood as it moves into the gills. Water is moved over the gills by the pumping of the muscular mantle, in contrast to the ciliary gill currents found in other marine mollusks. In cephalopods, the head is greatly enlarged and is covered by the mantle itself. The thick muscular molluscan

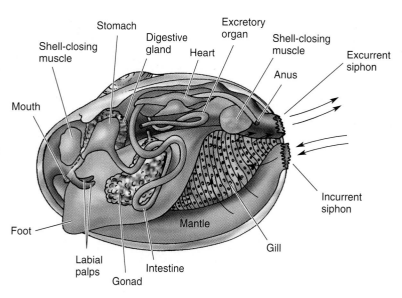

FIGURE 26.27
Anatomy of a Clam
In this diagram, one valve, part of the gills, and part of the body wall have been cut away. The foot makes up much of the body. Clams bring in a steady flow of water through the incurrent siphon, pass it through pores in the gills, and then send it back out through the excurrent siphon. Rows of cilia sweep any particles forward, and labial palps detect whether an object is edible. The mouth opens into a short gullet that leads into a stomach pouch. Food then moves into a lengthy intestine, which actually passes through the heart and on to the anus.

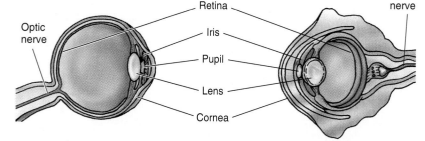

FIGURE 26.28
Cephalopod
The octopus is among the most versatile of all cephalopods.

foot is highly modified into tentacles, some complete with suckers. The mouth is armed with a horny beak used in ripping and tearing prey.

Phylum Annelida: Segmentation Along the Length

Along with the coelom and a newly specialized digestive system, a third evolutionary milestone appeared in most of the coelomates: **metamerism**, the segmented body plan. Here both the body and the coelom are divided by transverse septa (crosswalls) into a sequence of units, or **metameres**. In the simplest forms of segmentation, the units are more or less repetitious. Such structural repetition is most obvious in certain annelids and in arthropods such as millipedes and centipedes. A more complex version of the theme is also found in other coelomate animals. In vertebrates, for example, we find a form of segmentation in the repeated vertebrae, ribs, spinal nerves, and trunk muscles. However, vertebrate segmentation is believed to have evolved independently of that of invertebrates. Nonetheless, the theme prevails in all coelomate animals, thereby attesting to its evolutionary importance. As we will see, many adaptive variations have sprung from this basic body plan.

The extreme segmentation in the annelids is believed to be an adaptation for burrowing. The trait developed hand in

FIGURE 26.29
Cephalopod Vision
The eyes of cephalopods (such as the octopus and squid) are remarkably similar to those of vertebrates.

hand with the fluid-filled coelom, which produces the hydrostatic pressure found in both earthworms and roundworms. The hydrostatic skeleton was an early alternative to the hardened skeleton that arose later.

The annelids include three classes: **Oligochaeta** (earthworms), **Hirudinea** (leeches), and **Polychaeta** (marine worms).

Oligochaetes, class Oligochaeta, include the terrestrial earthworms, freshwater worms, and a few marine forms. Oligochaetes have little differentiation in the head region, lacking eyes and other elaborate sensory structures. Of course, when you use your head for digging, such elaborations might create a few problems. The most obvious characteristic of the earthworm body, whether viewed externally or internally, is segmentation. Internally, each segment contains elements of the circulatory, digestive, excretory, and nervous systems (Figure 26.30). Their circulatory system, like our own, is closed. The earthworm's circulatory system includes five pairs of **aortic arches** (hearts), which are essentially pulsating vessels, along with arteries, veins, and capillaries. The blood of many annelids contains hemoglobin, an oxygen-carrying protein, but it is not bound within blood cells, as is our own hemoglobin. Instead, it is dissolved in the circulating fluid.

The excretory system is also highly developed in the terrestrial worms. Nearly all segments have paired **nephridia**. A nephridium not only takes up fluids, ions, and nitrogen wastes, but it recovers most of the water and valuable ions,

sending them back into the blood. (As a terrestrial animal, the earthworm must be a water conserver.)

The earthworm is hermaphroditic, with complex male and female reproductive organs present in each individual, but is not self-fertilizing. When earthworms copulate, they exchange sperm reciprocally, and each stores the sperm in its **sperm receptacles** until the proper time for fertilization. The **clitellum**, the smooth, whitish cylinder of external tissue on earthworms, secretes a mucous cocoon that will slide forward along the body, receiving eggs and sperm from special reproductive pores. Fertilization occurs within the cocoon, which is then shrugged off. It will house the embryos while they develop. There is no larval stage, and the young hatchlings resemble the adults.

The digestive system is complete, with separate mouth and anus. The digestive tube is subdivided in the first 20 or so segments into swallowing, storing, and grinding regions. The intestine, a digestive and absorptive structure, continues to the anus. A prominent fold in the intestinal wall, the *typhlosole,* increases the digestive surface (see Figure 26.30, lower right). Earthworms feed on organic matter in the soil.

The earthworm's nervous system includes a brain and a fused pair of nerves forming a **ventral nerve cords**. The brain consists of a ring of nerve tissue around the pharynx, just behind the mouth. This nerve ring includes enlarged paired ganglia above and below the pharynx, with those below giving rise to the ventral nerve cords. The nerve cords, arranged closely parallel to

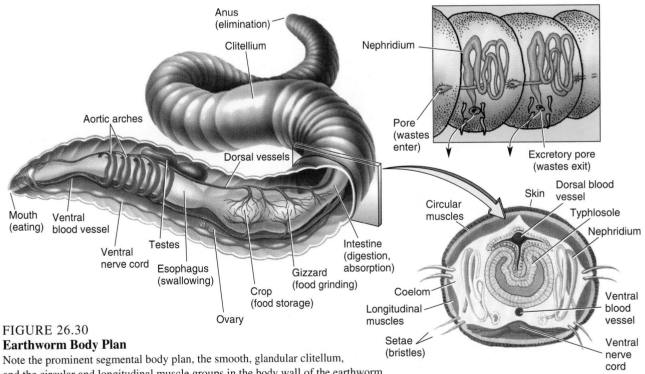

FIGURE 26.30
Earthworm Body Plan
Note the prominent segmental body plan, the smooth, glandular clitellum, and the circular and longitudinal muscle groups in the body wall of the earthworm. Transport is carried out by a closed circulatory system that includes five paired "hearts" (aortic arches) and an extensive system of blood vessels. The nervous system consists of a pair of enlarged ganglia above the pharynx and a lengthy ventral nerve that gives rise to ganglia in each segment. The gut, suspended in the coelom, contains several specialized regions (*lower right*). Nearly all segments contain paired nephridia, complex tubular structures that eliminate body water (*upper right*).

each other, extend along the body to the most posterior segment. Paired *ganglia* (clumps of nerve cell bodies) form in each segment, with branches that innervate the surrounding tissues.

Burrowing and other movements are accomplished by two layers of body wall muscles—circular and longitudinal—which extend and shorten the body, respectively. Acting as a hydrostatic skeleton, the turgid, fluid-filled body provides a flexible but resistant base for muscle action while maintaining the earthworm's shape. During burrowing, anchorage is provided by the **setae** (singular, **seta**), chitinous bristles found on most segments, which can be inserted into the burrow to hold some segments fast while other parts of the body are extended or contracted.

Leeches, class Hirudinea, live primarily in fresh water, although some species are found in marine and moist terrestrial habitats. Although commonly thought of as parasitic, there are many predatory and scavenging species. The parasites are **ectoparasites**—that is, they attach themselves to the skin of their hosts, which include humans, and draw blood.

The segmented body of the parasitic leech is usually flattened, with suckers at the anterior and posterior ends (Figure 26.31). Suction in the anterior sucker is applied by a muscular pharynx, and some species have horny teeth that cut through the skin of the host organism. When they have made an incision, they secrete an anticoagulant called **hirudin** into the wound. Leeches were once sold in pharmacies as a popular remedy for the swelling and discoloration of "black eyes," and in earlier times to "bleed" patients as a common treatment for illness.

The body of the leech is formed of modified segments. The segmentation is apparent in the nervous, reproductive, and excretory systems. Like the earthworm, leeches are hermaphroditic but not self-fertilizing. They are also similar to earthworms in their copulation, egg laying, and development.

Polychaete worms, class Polychaeta, are segmented marine worms that generally have well-formed head regions with eyes, specialized sensory structures, and numerous fleshy extensions called **parapodia**. Although clearly segmented, some polychaete worms have specialized body regions, unlike earthworms and leeches. Some are active swimmers and burrowers, while others are sedentary, remaining in burrows or tubes and exposing only their anterior parts for feeding and gas exchange. A number are particle-feeders, trapping minute marine plankton and bits of detritus (loose material) in cilia or mucus covering numerous tentacles.

Among the polychaetes are a number of tube-dwelling species. Some construct tubes of sand particles, cemented with mucus or calcium secretions. They are easy to see in their burrows because their feathery plumes (which are actually feeding and respiratory devices) constantly pop in and out. These colorful ciliated tentacles have inspired common names such as "fan worm," "feather worm," and "peacock worm."

Recall that burrowing worms may have a truly ancient evolutionary history. Fossilized burrows from Precambrian strata have been dated at about 700 million years, representing the oldest signs of animal life. The earliest tracks are random and dispersed, but later ones appear aggregated, possibly signifying changes in the social behavior of the worms.

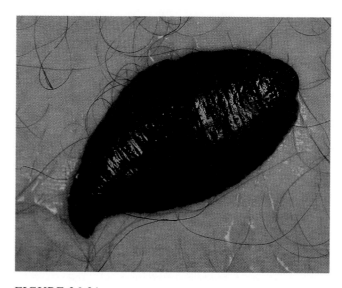

FIGURE 26.31
The Leech
Leeches are armed with sharp, piercing mouthparts and suckers for holding onto their hosts and drawing blood.

Phylum Arthropoda: Specialization of Segments and Jointed Legs

There are a lot of arthropods. You can get the idea by imagining Noah trying to load his ark with animals. If, by chasing, swatting, and kicking, he could have gotten one pair aboard per minute, he would have spent 18 months, night and day, simply loading his quota of crickets, crabs, lice, flies, centipedes, aphids, wasps, weevils, dragonflies, lobsters, ticks, and the like into the vessel.

To date, some 900,000 species of arthropods have been described, and some authorities estimate that about 1 million more await identification. We must conclude that, as a phylum, **Arthropoda** is the most diverse on earth.

So what are these pervasive creatures? The name *arthropod* tells us that they have "jointed feet," actually jointed legs. They are also segmented, but these segments are not simply repeating units as in earthworms; they may instead be highly specialized for different tasks such as walking, feeding, reproduction, and sensing. Arthropods have an **exoskeleton** (*exo*, "outside") made of chitin, often hardened with calcium salts. It serves as a protective covering and for the attachment of muscles used in movement. Joints are formed at thin, flexible regions of the exoskeleton. As arthropods grow they must **molt**, that is, periodically shed their protective coverings and form new exoskeletons. After all, with a rigid external skeleton, there is limited room for growth. Most arthropods shed, or molt, several times during development, and some continue the process throughout their adult life.

Arthropods have successfully invaded most of the earth, often developing amazingly narrow specializations. (Consider the adult mayfly, a delicate creature that emerges without mouthparts with which to feed, and which must mate and leave offspring in the few precious hours of life allowed it.)

Arthropods have open circulatory systems in which the *hemolymph* (the arthropod version of blood) is pumped by a heart into spaces called sinuses, from which it reenters the heart through pores.

Arthropods have varying diets, living as omnivores, herbivores, carnivores, scavengers, ectoparasites, and endoparasites. They live on, under, and above the surface of land and water, from deep ocean trenches to the highest mountain peaks. In many cases, because of their highly specialized needs, numerous species can exist side-by-side without seriously competing with each other.

Arthropod evolutionary success is due to other important factors as well. Their reproductive capacity can be phenomenal. In addition, many produce larvae (immature stages) that have an entirely different diet from the adults, thus avoiding competition between parent and offspring. Because of the

large brood size, particularly in insects, the opportunity for genetic diversity is great. With more offspring, there is more opportunity for variation. This, along with a very short generation time, helps the insect overcome all sorts of environmental deterrents. (For example, DDT-resistant mosquitoes have evolved in only a few generations.) The phylum Arthropoda is divided into four subphyla: *Trilobita, Chelicerata, Crustacea,* and *Uniramia* (Figure 26.32).

There are no living **trilobites**. Subphylum Trilobita originated in Precambrian times and extended into the Cambrian and Ordovician, where they peaked, undergoing extinction some 300 million years ago during the Carboniferous period. In their peak they dominated ancient Paleozoic seabeds, burrowing or shoveling their way through bottom mud and sand. To date, over 4000 fossil species have been described, most of which are rather small, 5–7.5 cm (2–3 in.), but some species got to be nearly 70 cm (about 28 in.) in length. The taxonomic position of trilobites is unsettled. Some zoologists maintain

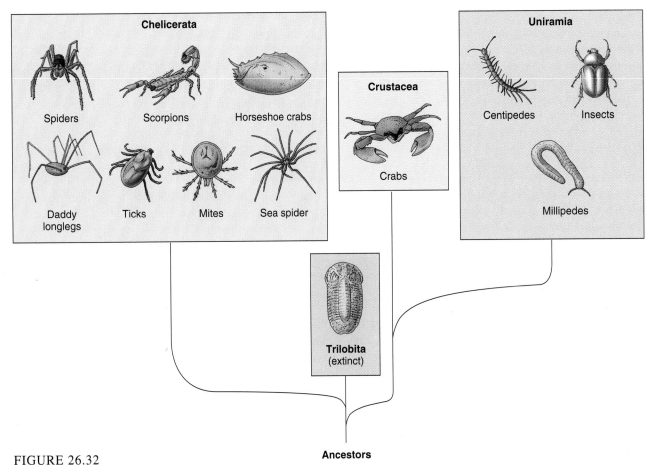

FIGURE 26.32
Phylogeny of Arthropods
The phylum Arthropoda can be organized into three contemporary subphyla on the basis of mouthparts and appendages. The chelicerates, shown in the left branch, lack jaws and antennae but produce claw-like chelicerae (which in the spiders are venomous fangs). In addition, they have four pairs of walking legs. The crustacea and uniramia, branching off to the right, have jaws and three or more pairs of walking legs. Most have antennae. Trilobites (below) form an extinct subphylum.

that they were ancestral to the crustaceans, but others consider them a separate line.

Trilobites have a characteristic three-lobed body that is divided by two grooves running its length. They are easily recognized by their shield-like shape and highly segmented body with its numerous similar paired appendages (see Figure 26.33), which were apparently used in movement and respiration.

Subphylum Chelicerata includes horseshoe crabs (Figure 26.32), daddy longlegs, spiders, ticks, mites, and scorpions. They all lack jaws and have six pairs of appendages, including four pairs of legs. The first two pairs of appendages are sensory palps and **chelicerae** (fangs). None of the chelicerates have antennae.

The **arachnids**, class **Arachnida**, don't win much affection with their habit of sucking the juices of other organisms. Spiders, equipped with hollow fangs and venom, are all carnivores. Typically, when a spider bites its prey, it injects venom, which weakens the unfortunate animal. Next, it injects digestive enzymes into its host's body that liquefy the tissues there. Then it sucks the fluid out.

Spiders show little evidence of external segmentation. They follow a two-part body plan, which includes a cephalothorax (fused head and thorax) and abdomen. Spiders also differ from other arthropods in that they respire through a structure known as the **book lung**. Book lungs are internal sacs that open to the outside through slits. The sacs are lined with leafy folds, similar to the pages of a book. Blood passing through the thin "pages" exchanges gases with air in the sac.

Most spiders are equipped with silk-producing glands connected by ducts to external devices called **spinnerets** (Figure 26.34). Both the **web** and the **cocoon** of the spider are produced by this system. The silk of spiders (and insects) is a protein known as **fibroin**, composed of the amino acids glycine, alanine, and tyrosine. Spiders as a group are known to produce seven different kinds of silk, each with its own function. Webs have both sticky parts with which to ensnare prey and safe, dry parts along which the spider can run.

Members of the subphyla Crustacea and Uniramia have **mandibles** (jaws), as well as antennae and various numbers of paired appendages, including three or more pairs of walking legs. They also have sensory antennae and usually a pair of **compound eyes** (multifaceted with separate imaging elements). Crustacea include marine and freshwater crabs, lobsters, copepods, barnacles, and sowbugs. Uniramia include centipedes, millipedes, and insects. Figure 26.35 illustrates the same variation within these subphyla.

Crustaceans live in both marine and fresh water. Their size ranges from microscopic ostracods and copepods to giant crabs from the western North Pacific. Species in this diverse group, such as the familiar pillbugs (or sowbugs), clearly demonstrate their relationship to the primitive annelids, while others (like the crabs) have become much more advanced, departing dramatically from the ancient plan. The segments of the most primitive crustaceans bear paired appendages, each

FIGURE 26.33
Trilobites
Although now extinct, trilobites were commonplace through the early Paleozoic era.

somewhat like the next. In fact, they are reminiscent of the polychaete worms in this respect. They are the only arthropods with two pairs of antennae. Three or more pairs of appendages form mouthparts. They have walking legs on the thorax and, unlike the insects, they also have appendages on the abdomen.

Small crustaceans exchange gases across thin areas of the exoskeleton, but larger crustaceans are gill breathers, with gill cavities partly covered by folds of the body wall. The active movement of certain appendages creates water currents that pass over the feathery gills (see Chapter 39). Because they are encased in a semirigid, secreted exoskeleton, to allow for growth crustaceans frequently molt, shedding their exoskeleton and growing a new one in a hormonally regulated manner (see Chapter 36).

The compound eye is often borne on a movable or retractable stalk. The eye, similar to that of insects, is composed of a large number of visual units called **ommatidia**. Each has a corneal lens and crystalline cone lens below. A light-sensitive translucent cylinder below each set of lenses responds to incoming light (Figure 26.36). The compound construction and stalked position enable the animal to obtain a 180° view of its surroundings.

Many of the marine forms of the diverse crustaceans are of great ecological importance. For example, much of the ocean's minute floating life, the plankton, consists of microscopic crustaceans. They are so numerous that one order, Euphausiacea, commonly known as "krill," is the main diet of the largest whales, as well as an important part of the ocean's longer food chains.

Among the Uniramia, **centipedes** (class Chilopoda) are sometimes confused with **millipedes** (class Diplopoda). But centipedes have flattened bodies, while millipede bodies tend

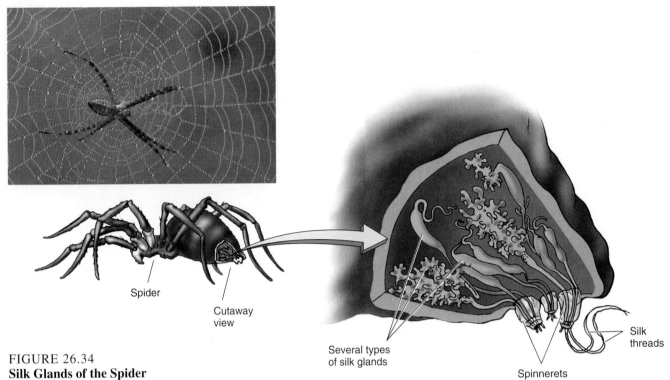

FIGURE 26.34
Silk Glands of the Spider

The orb spider, a master web-maker, can produce several kinds of silk. The silk glands, as shown here, are located in the abdomen. They open into the spinnerets, which emit threads of different diameters. The web in the photograph belongs to an orb spider. The spider can move about quickly on the web, but it must know where the sticky parts are or it will become ensnared in its own trap.

to be cylindrical; and centipedes have only one pair of appendages per segment (except for the first), while most segments of millipedes have two pairs. If it stings you, it's a centipede. Millipedes are harmless herbivores, while centipedes are predators and scavengers. The centipede's first pair of legs are modified into perforated claws with which they inject poison into prey. The sting of common temperate zone centipedes is painful but probably no more life-threatening to humans

than a wasp sting. However, the sting of larger tropical species can put an adult in bed with a fever for several days, detracting from what might have been a wonderful walk in the rain forest.

The majority of arthropods, and indeed the majority of animals, are **insects**, class Insecta. Except in the sea where the crustaceans hold sway, insects rule the earth in terms of numbers and kinds. We are continuously engaged in conflict with them and, even today, we often lose. Insects are of such importance

(a) Long-arm crab

(b) Javanese leaf insect

(c) Hercules beetle

FIGURE 26.35
Crustaceans and Uniramians

These species can give you some idea about diversity in the two subphyla. One crustacean is **(a)** the long-arm crab. **(b)** The exotic Javanese leaf insect contrasts with **(c)** the Hercules beetle, whose long jaws serve to attract females.

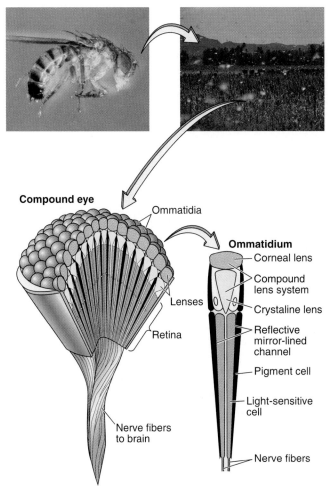

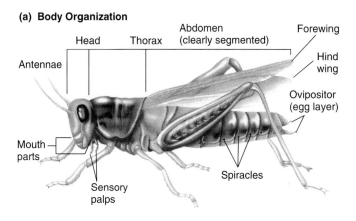

(a) Body Organization

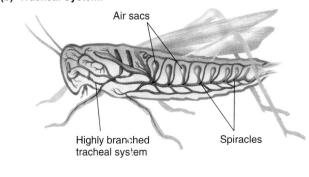

(b) Tracheal System:

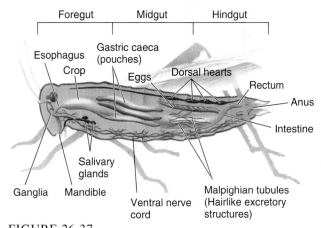

(c) Other Systems:

Compound eye

Ommatidium

FIGURE 26.36
The Compound Eye

Many light-receiving ommatidia make up the compound eye. Their arrangement provides a very wide visual field, over 180° in the stalked crayfish eye. Each ommatidium has two lenses, an outer corneal lens and a crystalline cone lens below. The two focus light into a translucent cylinder that contains light-sensitive pigments. Slender pigment cells surrounding the unit keep light from passing into adjacent ommatidia—as long as the light is bright. Each active ommatidium contributes to the final image. In dim light, the pigment cells withdraw, and light from several lens systems finds its way to one crystalline cone lens. Whereas the image is not as acute this way, it does provide enough light stimulus for the eye to respond.

ecologically and economically that we literally could not have reached our present population size and prominence among the world's creatures without understanding something about them.

The segmented body of insects has undergone considerable modification from the ancestral form, primarily through the fusion and alteration of segments. The modification has been so sweeping that the ancestral segmented form is only seen in the abdominal region and in the embryo. The exoskeleton, like that of many other arthropods, consists basically of chitin, but it is more flexible than that of the crustaceans.

The bodies of insects have three major regions: the **head**, the **thorax**, and the **abdomen** (Figure 26.37). The thorax gives

FIGURE 26.37
Insect Body Plan

(a) The insect body consists of three main parts: head, thorax, and abdomen. Segmentation is most apparent in the abdomen. The thorax and head are fused. The grasshopper has a pair of compound eyes, two or three simple eyes, and a pair of sensory antennae. Its mouthparts include the jaws, lips, and sensory palps. The thorax gives rise to three pairs of walking legs and two pairs of wings, while the abdomen contains 12 hinged segments. **(b)** Paired spiracles open into each segment, admitting air into a branched tracheal system. **(c)** Digestion occurs in a three-part gut, aided by secretions from salivary glands. Excretion is carried out by hair-like Malpighian tubules. The last segments bear the reproductive structures and the anal opening.

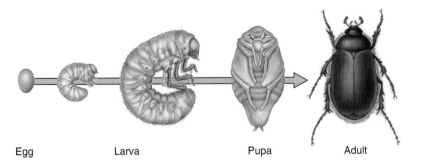

FIGURE 26.38
Insect Metamorphosis
In the beetle's complete metamorphosis, an egg develops into a larva, a feeding stage. Following a period of intense growth and development, the larva is transformed into a pupa, and development is completed. The complete adult emerges from the pupa. Growing juveniles resemble adults.

Egg   Larva   Pupa   Adult

rise to three pairs of legs and, in many species, wings. The legs are modified in different species for running, jumping, catching, holding, or simply resting. The appendages on the abdomen have been modified to form the genitalia and anal structures. In many species, females have an **ovipositor**, a hollow appendage that is used to dig into the soil, or bore holes in plants, or puncture insect prey, in which they lay their eggs.

During development, most insect species (about 88%) pass through a **complete metamorphosis**. That is, their life cycles include four stages: **egg**, **larva**, **pupa**, and **adult** (Figure 26.38). Insect larvae, known variously as caterpillars, grubs, and maggots, are worm-like and often have chewing mouthparts. The larvae are usually voracious eaters, and it is in this stage when those identified as agricultural pests do most of their damage. As the larval stage continues, the wings are formed within the body. Eventually, the larvae form a surrounding case or cocoon and enter the pupal stage. The pupa may become dormant, spending the winter in this state, but it

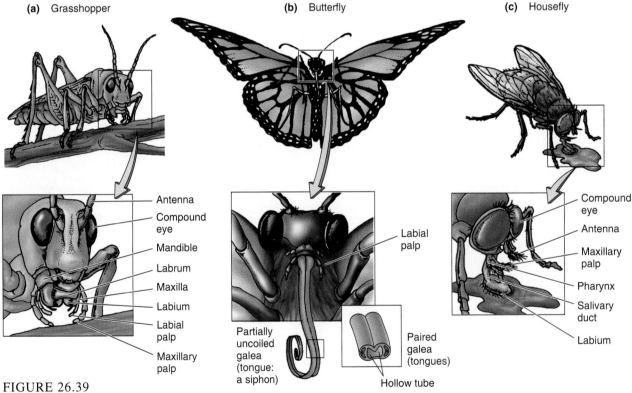

FIGURE 26.39
Insect Mouthparts
(a) The grasshopper's chewing mouth is nicely adapted for eating foliage. Two sets of palps contain taste receptors, while biting is done with the paired mandibles and maxillae. The labium and labrum help by holding the food. (b) Butterflies use their lengthy mouthparts in siphoning up plant juices. The long, extendible paired maxillae are held together like the sides of Ziploc bags, forming a long drinking tube. When not in use, the tube is kept tightly coiled by flexible rods. (c) The housefly has its own feeding strategy. It salivates on its food, stirring it with the large sponge-like labium. Its labrum, part of a tube-like device, sucks up the partially digested mess.

is in this stage that development is completed. A full-grown adult emerges from the pupal case.

Other insects, such as grasshoppers, mayflies, and mantids, undergo an **incomplete metamorphosis**, going from the egg to a **nymph** to adult. Nymphs tend to resemble the adult but lack wings. The wings develop externally as the nymphs gradually increase in size, eventually reaching the adult stage.

In a relatively small number of insects, chiefly the wingless types such as silverfish and springtails, development is direct. They emerge from the egg as juveniles—miniatures of the adult—and begin their growth to full adult size.

Gas exchange is accomplished through a complex system of tubules. The air passes through external openings known as **spiracles** into highly branched **tracheae**, which subdivide into smaller **tracheoles**, finally reaching individual cells where gas exchange occurs.

The major sense organs and mouthparts are in the head, where you would expect them to be. Consider, for example, the head of the grasshopper. It has one pair of sensory antennae, a pair of compound eyes, and three simple eyes, or **ocelli**. Its mouthparts consist of a single **labrum** (upper lip) and a **labium** (lower lip) bearing **sensory palps**, and there are two laterally movable **mandibles** (jaws) that do most of the chewing. These are assisted by a pair of **maxillae** also bearing sensory palps (Figure 26.39a). The entire apparatus is magnificently adapted for biting and chewing leaves. The sensory palps detect texture and flavor, the jaws rip off and chew bits of vegetation, and the lips hold the food in place during chewing.

Mouthparts in other insects are often highly specialized to fit their specific feeding needs. For example, the maxillae of butterflies are modified into a long, flexible siphoning device (neatly coiled when not in use—Figure 26.39b). On the other hand, the cicada has a piercing and sucking mouth, while the housefly has a large, complex, tongue-like apparatus used for licking (Figure 26.39c).

Returning to our grasshopper, let's examine its locomotive structures. First, although some grasshoppers are great jumpers, they are also flying insects. They have two pairs of wings; one pair actually propels the insect, and the other serves as a protective wing cover. The desert locust, a close relative of the grasshopper, can fly over 200 miles in a day.

The legs of insects are also highly specialized for specific modes of life. The grasshopper, for example, has two pairs of legs of similar size for walking and grasping, while its third pair is greatly enlarged. The combination of extremely powerful muscles in the third pair and the light exoskeleton produces a great mechanical advantage, enabling the grasshopper to jump great distances from a standing start. The praying mantis has very strong, spiked forelegs that snatch and hold prey, which is then ripped apart by powerful jaws (Figure 26.40). Honeybees have walking legs modified for pollen collection. Pollen is a major source of bee food. And then, consider the sensitive legs of roaches, which detect delicate air currents set up by anything that might be after them, such as an irate apartment dweller.

FIGURE 26.40
Praying Mantis
The praying mantis is a voracious predator. This mantis has torn its insect prey open with its powerful barbed forelegs and is feeding with its strong jaws.

Of course, these are only a few examples of insect diversity. It is estimated that one of four animals roaming the earth is a beetle and most of the rest of the animals fall into one of the other insect orders. We will discuss additional aspects of their biology elsewhere.

Arthropods have been around a very long time. Fossils of marine forms are found in Cambrian strata and possibly in the Precambrian. By the Ordovician period, some members had left the sea and had begun to explore the terrestrial environment. This escape from their marine enemies was only temporary, though; by the Devonian, the first amphibians had struggled ashore and have been eating insects ever since. But the arthropods did not fall easily before the hungry vertebrates. They ran and crawled away, multiplying in great numbers, invading every part of the globe and establishing all sorts of new niches.

Phylum Onychophora: Linking Annelids and Arthropods

Phylum **Onychophora**, which includes only about 65–70 named species, is so minor that it could well be ignored if it weren't for the fact that some researchers consider it an evolutionary link between Annelida and Arthropoda, or perhaps representative ancestors of both. The onychophorans are believed to represent an ancient animal that arose from the primitive arthropod line very shortly after the arthropods diverged from the early annelids. Onychophorans are seen among the fossils of the famed Burgess Shale fauna (see Figure 26.3). After leaving the arthropod line, they made the difficult transition to land, where they assumed the rather restricted existence we see today (Figure 26.41).

The best-known onychophoran is the peculiar *Peripatus,* which lives among the damp leaves of the tropics in Africa, Southeast Asia, New Zealand, Australia, and the West Indies. *Peripatus* has a thin, soft cuticle covering a muscular body wall. It has two muscle layers that run in circular and longitudinal directions. The animal is highly segmented, revealing its annelid heritage, but the segmentation is largely confined to its appendages and internal structure; its body as seen from the outside is fairly smooth. However, it also has several features characteristic of arthropods. Tiny claws or small pincers arm its soft appendages, and its head even has antennae. It breathes through a series of spiracles along its body, which open into tiny pits connected to a number of tracheae. This arrangement permits air to reach the internal organs easily, but there is no way of controlling the size of the opening, so *Peripatus* avoids water loss by living in moist habitats.

We have reviewed the major groups in the prostostome line, those animals in which the mouth appears before the anus. Now we will review the invertebrate deuterostomes. As you will see, we have considered only one invertebrate phylum in this group, the echinoderms. The deuterostome line is composed principally of the echinoderms, such as sea stars, and the chordates. We will consider the chordates in the next chapter.

Phylum Echinodermata: Deuterostomes with Radial Bodies

Your first look at **echinoderms** ("spiny-skinned" animals) may lead you to believe that you are on the wrong road to the vertebrates. Sea stars, brittle stars, sea urchins, sea cucumbers, and sand dollars (Figure 26.42) appear unlikely to be relatives of amphibians, fishes, mammals, and birds, not to mention humans.

Annelid (marine worm)

Arthropod (centipede)

Onycophoran (*Peripitus*)

FIGURE 26.41
An Evolutionary Link
The phylum Onychophora has many characteristics of both the annelids and the arthropods. The thin-walled, segmented onychophoran body (*bottom*) and the serial duplication of internal structures are reminiscent of the annelids (*top*). On the other hand, onychophorans have jointed legs tipped with claws and an open circulatory system, both of which are arthropod characteristics (*center*).

The phylum consists of five classes: **Crinoidea** (sea lilies), **Holothuroidea** (sea cucumbers), **Echinoidea** (sea urchins and sand dollars), **Asteroidea** (sea stars), and **Ophiuroidea** (serpent stars and brittle stars). Each class differs considerably from the others, with perhaps the greatest departure being the soft-bodied, worm-like sea cucumbers. A comparison of the anatomy of echinoderms and vertebrates also won't instill much confidence that they are, in fact, related. To see the relationship between the echinoderms and the vertebrates, we must look at their earliest embryonic development, when the first cleavages occur, and later at the developing mouth and coelom. It is there that the echinoderms are unveiled as deuterostomes. They show radial cleavage, their blastopore develops into an anus instead of a mouth, and their coelom develops as an enterocoel.

Echinoderms are exclusively marine and brackish water animals. Many species are found in the shallow waters of the continental shelf, particularly just below the reaches of the lowest tides. However, one group, the brittle stars, abound in the deepest trenches of the ocean.

Echinoderms are unusual in many respects. Their spiny, crusted covering is actually an endoskeleton, even though it seems to be on the outside. Echinoderms are covered by an epidermis, beneath which lie the calcareous plates of mesodermal origin that are secreted by the dermis. (Recall that the shells of mollusks and the exoskeletons of arthropods are both secreted by epidermal cells that stem from ectoderm.)

The body of adult echinoderms is essentially of a pentaradial (five-part radial) construction, a scheme found in no other phylum. The larvae, however, are bilateral. Echinoderms show no obvious segmentation, indicating that they are not closely related to the annelid line.

One of the most unusual features of the echinoderm is the **water vascular system**, which is composed of a series of water-filled canals, which by hydraulic pressure help extend their numerous muscular **tube feet**. In the sea star each tube foot is equipped with a terminal sucker that is used in grasping. The remarkable tube feet are located in double rows in the grooved underside (the oral side, since the sea star's mouth is on the bottom) of each arm (Figure 26.43). Each tube foot is attached to a short **lateral canal**, which, in turn, connects with a pipe-like **radial canal** that extends along the arm. The radial canals from each arm all connect to the hollow **ring canal** in the central disk. Changing pressures in the water vascular system may be brought to equilibrium by a vertical tube, the **stone canal**, which has a sieved opening called a **madreporite** (mother pore) on the upper side of the sea star.

Each tube foot contains longitudinal muscles that can contract, shortening the delicate structure. Above each foot is a rounded sac, an **ampulla**, which is similar to a squeezebulb. When muscles contract a water-filled ampulla, water is forced into the foot, extending it hydraulically. The sucker end of the foot attaches to a surface, and the longitudinal mus-

Brittle star

Feather star

FIGURE 26.42
Representative Echinoderms
One of the most striking features of the echinoderms is their fivefold radial symmetry, clearly seen in the brittle stars but not apparent in the feather star.

cles then contract, shortening the tube and exerting a pull. Thus the tube feet, contracting in a highly coordinated fashion, can attach to an oyster or draw the sea star over the rocky bottom of its habitat.

The sea star, like all echinoderms, lacks a centralized nervous system. The greatest concentration of nerves is in a ring around the mouth, with branches extending into the arms. Sensory receptors are limited in the sea star to one or more sensory tentacles and a light-sensitive region at the tip of each arm.

Echinoderms are either male or female, but it is hard to tell which is which. Sperm or eggs, produced in the well-defined testes or ovaries found in each arm, are shed into the water through tiny openings in the arms. Fertilization takes place in the water.

With the echinoderms then, we have set the stage for our entrance into a more familiar world, that of backboned animals and their considerably less familiar chordate relatives.

FIGURE 26.43
Water Vascular System

The water vascular system of echinoderms extends from the central disk into each arm. It begins with a sieve-like madreporite and includes a number of canals and numerous tube feet. Water is used to fill the ampullae, which, in turn, are used to extend the tube feet. Each tube foot ends in a sucker tip. When several of the tube feet contact a surface, they can contract by muscular action. Working in a series, the tube feet pull the sea star along the sea bottom. They also assist in holding prey.

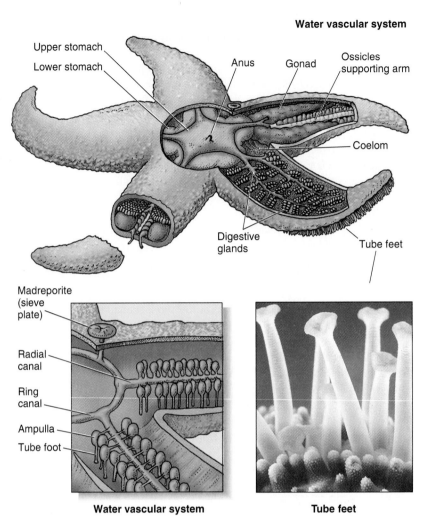

Water vascular system

Tube feet

Key Ideas

WHAT IS AN ANIMAL?

1. Animals are multicellular, eukaryotic heterotrophs, usually organized at the organ-system level with specialized organs, tissues, and cells. Most exhibit active movement through muscle action and are heterotrophic. Animals produce a cellular matrix of collagen. They reproduce sexually, producing large, stationary eggs and small, flagellated sperm.

ANIMAL ORIGINS

1. Reliable animal fossils date back to the Precambrian era some 580–680 million years ago, although fossil worm tubes and tracks may be as old as 700 million years. The **Ediacara fauna** fossils included jellyfish and worms: the slightly newer **Burgess Shale formation** includes fossils from most of today's phyla.

MAJOR MILESTONES IN ANIMAL EVOLUTION

1. The phylogenetic tree representing animal evolution reveals an early major split that produced the **protostomes** ("mouth first") and **deuterostomes** ("mouth second"), the latter of which include the vertebrates.

2. Two hypotheses seek to explain how such divergence occurred:

   a. In one, *ecological* isolation occurred as one group specialized in burrowing for food and the other formed stalks and filtered food from the sea.

   b. A second hypothesis proposes that isolation was *geographical.* Prior to the formation of Pangaea, there existed Proto-Laurasia and Proto-Gondwanaland, two supercontinents in which, respectively, the deuterostomes and protostomes arose.

3. Within organisms, the cell represents the fundamental **level of organization**, followed by the **tissue level**, the **organ level**, and the **organ system level**.

PHYLUM PORIFERA: AN EVOLUTIONARY DEAD END

1. The sponges of phylum **Porifera** represent the cellular level. Sponges are all aquatic and mostly marine. Their body consists of a few kinds of cells, including **porocytes**, **choanocytes**, and **amebocytes**. The amebocytes secrete skeletal elements—**spicules**—of calcium carbonate, silicon dioxide, or protein.

2. Sponges are filter-feeders that draw water in through the body wall and phagocytize tiny food particles.

3. As **hermaphrodites**, sponges produce sperm and eggs, and, upon fertilization, a ciliated larva forms.

PHYLA CNIDARIA AND CTENOPHORA: TISSUES AND THE RADIAL PLAN

1. Most animals are **bilateral**; but a few groups are **radial**.

2. Phylum **Cnidaria** is organized primarily on the tissue level with some organ development. They are thin-walled, sac-like animals with tentacles and stinging cells—**cnidocytes**—that release **nematocysts**. The central **gastrovascular cavity** carries on digestion and respiration. Responses are facilitated by a widespread **nerve net** of nerve cells and processes.

3. Cnidarians include two basic forms, the stationary, attached **polyp** and the swimming **medusa** (jellyfish). **Hydrozoans** are mostly marine, forming a stationary feeding and asexually reproducing colonial polyp. **Scyphozoans** (jellyfish) drift in the ocean, capturing prey with stinging tentacles. **Anthozoans** (corals and anemones) are strictly polyps. Most coral polyps secrete limestone coverings, forming coral beds and reefs.

4. **Ctenophorans**, the comb jellies, are organizationally similar to cnidarians but form the fused cilia or **combs** and tentacles bearing glue cells.

PHYLUM PLATYHELMINTHES: MESODERM AND THE BILATERAL PLAN

1. Part of the bilateral trend included **cephalization**, an emphasis on a head or leading end, bearing feeding and sensory structures.

2. Metazoans have two embryonic **germ layers**, **endoderm** and **ectoderm**, from which all body tissues are derived. A third layer, the **mesoderm**, first seen in the flatworms, makes many new tissues possible, including organized muscle, blood, and internal skeleton (endoskeleton).

3. Phylum **Platyhelminthes**, the flatworms, includes free-living **turbellarians** such as *Dugesia* (**planaria**), a complex species with an extensive, branched gastrovascular cavity; a ladder-like nervous system, feeding through a complex protrusible **pharynx**, with digestion beginning in the gastrovascular cavity and being completed within the lining cells; osmoregulation (water management) through ciliated **flame cells** within **flame bulbs**; move-

ment by cilia; and asexual reproduction through fission and sexual reproduction involving copulation and complex sexual organs.

4. **Flukes** (class **Trematoda**) are all parasitic, using a sucker mouth to feed on the host. **Tapeworms** (class **Cestoda**) attach to the host intestine via a small hooked and suckered **scolex**. They produce numerous **proglottids** in which the sex organs and eggs and sperm develop.

PHYLA NEMATODA AND ROTIFERA: BODY CAVITIES AND A ONE-WAY GUT

1. **Coelomate** animals have a mesodermally lined body cavity—the **coelom**—and a tube-like, one-way muscular gut, complete with mouth and anus. In roundworms, the cavity is a **pseudocoelom** or false coelom. It is not mesodermally lined, and the gut lacks muscle. **Acoelomates** lack such body cavities.

2. The **nematodes**, or roundworms, are slender, cylindrical pseudocoelomates with little cephalization. Most are free-living predators, but others are important parasites of plants and animals.

3. Coelomates have a muscular gut and a fluid-filled coelom that forms a hydrostatic skeleton.

PHYLA ECTOPROCTA, BRACHIOPODA, AND PHORONIDA: THE LOPHOPHORATES

1. Lophophorates have characteristics of both protostomes and deuterostomes. The three phyla include **Ectoprocta** (bryozoans), **Brachiopoda**, and **Phoronida**. The **lophophore** is a group of ciliated tentacles attached to a curved ridge. Brachiopod and ectoproct fossils occur in Cambrian strata.

PHYLUM MOLLUSCA: UNSEGMENTED COELOMATES

1. **Mollusks** have a variation of a **foot**, **mantle**, **mantle cavity**, **shell**, and, in most aquatic forms, a **gill**. The coelom is highly reduced, an **open circulatory system** is present in most, and many feed with a rasping **radula**.

2. Most **gastropods** ("stomach feet") are snails or snail-like, although some lack shells. Many undergo **torsion** (a 180° twist) as they develop. Bivalves, clam-like mollusks with paired, hinged shells, often a digging foot, and are filter-feeders. Gases are exchanged in gills. **Cephalopods** have reduced and internalized shells, well-developed brains, image-forming eyes, closed circulatory systems, muscular mantles, and grasping tentacles.

PHYLUM ANNELIDA: SEGMENTATION ALONG THE LENGTH

1. In annelids, extreme segmentation and a hydrostatic skeleton may have evolved as adaptations to burrowing. In **metamerism**, the segmented body plan, the body is composed of repeated segments or **metameres**. Segmentation is also seen in vertebrates.

2. Earthworms have a **closed circulatory system** with five pairs of pumping **aortic arches** and blood that contains hemoglobin.

Nephridia carry on excretion and osmoregulation. Earthworms are hermaphroditic. The gut includes specialized areas for swallowing, storing, grinding, digesting, and absorbing. The nervous system includes a brain (enlarged ganglia) and a **ventral nerve cord** with branches in each segment.

3. **Leeches** attach themselves via toothed suckers to warm-blooded hosts for feeding.

4. The **Polychaete** worms are annelids of the marine environment.

PHYLUM ARTHROPODA: SPECIALIZATION OF SEGMENTS AND JOINTED LEGS

1. Common **arthropod** characteristics include segmented body, jointed legs, ectodermally secreted **exoskeleton**, open circulatory system, and ventral nerve cord.

2. The **trilobites**, an extinct arthropod group had a segmented, three-lobed, shield-like body, with numerous paired appendages.

3. Chelicerates lack antennae and jaws and have four pairs of appendages, and venom-injecting fangs known as **chelicerae**. **Arachnids**, the spiders, are venomous carnivores that exchange gases in **book lungs** and produce silk for **webs** and **cocoons**.

4. **Crustaceans** are mainly aquatic gill breathers, with a chitinous exoskeleton made rigid with calcium salts. They have compound eyes, consisting of multiple **ommatidia**.

5. Chilopods and diplopods are represented, respectively, by the venomous, predatory **centipede** and the herbivorous **millipede**.

6. **Insects** characteristics include a three-part, segmented body (head, thorax, and abdomen), a flexible chitinous exoskeleton, three pairs of legs, wings (in many), one pair of antennae, and both simple and compound eyes. Gas exchange is through a tracheal system with branched tracheae and air sacs. Mouthparts are highly varied. All reproduce sexually and have complex genitalia. During development, some insects follow complete metamorphosis (egg–larva–pupa–adult), whereas others have incomplete metamorphosis (egg–nymph–adult).

PHYLUM ONYCHOPHORA: LINKING ANNELIDS AND ARTHROPODS

1. The onychophorans have both annelid and arthropod characteristics and may have diverged from ancient arthropods shortly after that group diverged from annelids.

PHYLUM ECHINODERMATA: DEUTEROSTOMES WITH RADIAL BODIES

1. The phylogenetic link between **echinoderms** and vertebrates is seen in the embryo, where similarities in cleavage planes and mouth development are seen.

2. The spiny skin of the echinoderm is secreted by mesodermally derived tissue and is covered by an epidermis. Although adult echinoderms are pentaradial, their larval forms are bilateral. The **water vascular system** with its **tube feet** is an exclusive feature.

Application of Ideas

1. Interestingly, one group of animals and one group of plants represent an aside from the main line of animal and plant evolution. Both emerged from protist ancestors, although very different ones, and each is doing well today but in restricted environments. Describe these two groups and how they are related to the main lines. In what way does each differ from main line organisms?

2. Radially symmetrical animals are all aquatic. Suggest reasons why this type of symmetry is not found in animals of the terrestrial environment. If it were, how might such an animal function?

3. Using the mollusks as examples, explain the evolutionary terms *primitive, specialized, generalized,* and *advanced.* Keep in mind that "advanced" refers to more recently evolved *traits,* and "primitive" to more ancient traits. Also, be aware that "specialized" refers to the ability to avail oneself of a narrow range of offerings, while "generalized" applies to more opportunistic creatures that can take advantage of a wide range of resources. Why are such terms often inappropriate when applied to entire organisms?

Review Questions

1. List several important characteristics of animals, two of which are exclusive to animals. (page 527)

2. What evidence suggests that the animals evolved from flagellate ancestors? (page 527)

3. Approximately when does the *undisputed* animal fossil record begin? What animal groups were present then? Were these the first multicellular animals? Explain. (pages 527–528)

4. What do the Burgess Shale fossils tell us about the state of animal evolution at the end of the Precambrian? (page 528)

5. What is the difference between a polyp and a medusa? In which cnidarian class is each of these prominent? (page 537)

6. In what two significant ways does the anthozoan life cycle differ from that of most hydrozoans and scyphozoans? (page 537)

7. Briefly review the life cycle of the human liver fluke, *Opisthorchis sinensis.* How might such a complex cycle both help and hinder our control of the parasite? (page 541)

8. Using simple drawings, carefully distinguish between the terms *coelomate, acoelomate,* and *pseudocoelomate.* (page 541)

9. Describe the usual method of reproduction in a rotifer population. Under what conditions do males become important? (page 544)

10. Briefly summarize an ecological hypothesis and a geographical hypothesis that explain how the protostome and deuterostome evolutionary lines became established without conflict on the early earth. (page 532)

11. Briefly explain metamerism and give an example of a highly metameric animal. (page 549)

12. What evidence of cephalization is found in the earthworm? (page 550)

13. List the four molluskan structures that vary greatly from class to class. (page 546)

14. Suggest how the following may be significant to the success of arthropods: rate of reproduction, potential variability, life cycle with distinctly different stages, and small size. (page 551)

15. List three characteristics of the chelicerates that are not present in crustaceans and uniramians. (page 553)

16. Describe the structure of the compound eye and explain what it actually enables arthropods to see. (page 553)

17. List both the arthropod and annelid characteristics of the onychophorans. Where do these strange animals seem to fit in invertebrate phylogeny? (page 558)

18. How does the spiny covering of the echinoderm qualify as an endoskeleton? (page 559)

19. List the elements of the water vascular system of sea stars and briefly explain how the tube feet extend, fasten on, and shorten. What are two roles of this unique system? (pages 559–560)

Chordates

CHAPTER OUTLINE

PHYLUM HEMICHORDATA

PHYLUM CHORDATA

Subphylum Urochordata: Tunicates

Subphylum Cephalochordata: Lancelets

SUBPHYLUM VERTEBRATA: ANIMALS WITH BACKBONES

Class Agnatha: Jawless Fishes

Class Placodermi: Extinct Jawed Fishes

Class Chrondrichthyes: Sharks and Rays

Class Osteichthyes: The Bony Fishes

Class Amphibia: The Amphibians

Class Reptilia: The Reptiles

Essay 27.1 Exit the Great Reptiles

Class Aves: The Birds

Class Mammalia: The Mammals

HUMAN EVOLUTIONARY HISTORY

The Australopithecine Line

The Human Line

KEY IDEAS

APPLICATION OF IDEAS

REVIEW QUESTIONS

The hairy-nosed wombat cannot run very fast. Neither can its cousin, the naked-nosed wombat. But coyotes can, and cheetahs can outrun coyotes. None of these animals jump from the tops of trees, but flying squirrels and birds do. And a bird may go up as well as down. Both birds and chimpanzees build nests in trees, but only one gives milk, and neither can hold its breath as long as a turtle. Most fish don't have to come up for air at all, but fish and turtles may share their habitat. Surprisingly, all these animals have enough traits in common that they are placed in the same phylum: **Chordata**.

The chordates are indeed a fascinating and highly diverse group of animals that, while seemingly quite different from each other, have certain fundamental traits in common. It may be a little hard to see, at first, how a species as grand and noble as our own could be related to something that sits like a leathery little pouch on a wharf piling or one that wriggles through the sandy ocean bottom. But, as we will see, the backboned animals do share a number of characteristics with such creatures. Furthermore, all chordates share certain traits with another group, even more distantly related—the hemichordates. The hemichordates were once considered a subphylum of Chordata, but that was before scientists discovered that the rod-like supporting structure running along the length of the animal was neither analogous nor homologous to a true notochord. Before we get into the chordates, let's briefly consider the hemichordates.

Phylum Hemichordata

The **hemichordates** ("half-chordates") are represented by the acorn worm, a burrowing marine animal with a conical (acorn-like) proboscis or nose, and just behind it a collar (Figure 27.1). Note the rows of pharyngeal gill slits in the wall of the pharynx, behind the collar. These slits reveal the worm's affinity to the chordates, since chordates have pharyngeal gill slits at some time during their development. In the acorn worm, the apparatus acts as a kind of gill and helps them strain food particles from the water.

While the gill slits link the hemichordates with the chordates, another seemingly insignificant trait links the acorn worms to the echinoderms—specifically, the close resemblance of their larvae (see Figure 27.1). In addition, the embryos of echinoderms, hemichordates, and chordates undergo radial cleavage. It seems that, in the distant past, the ancestors of echinoderms, chordates, and hemichordates were rather closely related. The theoretical descendancy of these groups, all deuterostomes (see Chapter 26), is shown in Figure 27.2.

Phylum Chordata

Now we turn to the chordates, a group that shares three traits that, taken together, distinguish them from all other animals.

1. **Chordates** all have **pharyngeal gill slits** at some stage in their lives. These slits characteristically appear in the embryo as a series of slits through the body wall in the pharyngeal (throat) region. Some researchers believe that in the earliest chordates, the gill slits were used as a sieve for filtering out food particles. In terrestrial species the pharyngeal gill slits form other structures, such as the ear canal.

2. Chordates have a **dorsal, tubular nerve cord**. Our own spinal cord is a good example of such a structure. (Recall from the previous chapter that some invertebrates had *paired, ventral, solid* nerve cords.) We will learn in Chapter 42 how the dorsal nerve cord is formed during embryological development.

3. Chordates have a **notochord** at some stage in their lives. The notochord is a flexible, turgid rod running along the back that serves as a kind of skeletal support. It consists of large cells, apparently under considerable hydrostatic pressure, encased in a tight covering of connective tissue. In nearly all living chordates the notochord exists only in the embryo or larva. In vertebrates it begins to dissipate early in embryonic development as the backbone appears.

Most chordates also have a **postanal tail**, a muscular projection that extends posteriorly past the anus, an endoskeleton, and a ventral heart that pumps blood through a closed circulatory system. Let's now consider three subphyla of the phylum Chordata: *Urochordata*, *Cephalochordata*, and *Vertebrata*.

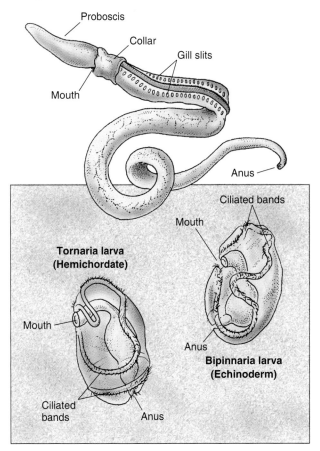

Acorn Worm

FIGURE 27.1
A Hemichordate
The acorn worm has a large number of pharyngeal gill slits, a chordate trait. It burrows in mud flats, using its proboscis somewhat in the manner of an earthworm. Water and organic debris enter the mouth. The nutrients in the debris are sorted and digested by the gut, while the water passes out through the gill slits. Evidence of the hemichordates' affiliation with echinoderms is seen in the similarity of their larvae

Subphylum Urochordata: Tunicates

Urochordates include the tunicates and their relatives. **Tunicates**, also called *ascidians* or simply "sea squirts," are chordates with a thick covering called a tunic. Oddly, the tunic is composed largely of cellulose, a material found mainly in plants. Tunicates are revealed as chordates by their tadpole-like larvae, which bear a notochord, a dorsal, tubular nerve cord (which enlarges anteriorly), gill slits, and postanal tail (Figure 27.3). In its transition to the adult form, the tunicate loses its obvious chordate characteristics. Its tail, notochord, and nerve cord are absorbed into the body, leaving only the enlarging *gill sac* (or gill basket) as a clue to its chordate relationship. Tunicates are filter-feeders, using their gill clefts as strainers. The adult circulatory system is open, consisting of little more than a bizarre heart that pumps blood first one way, then the other.

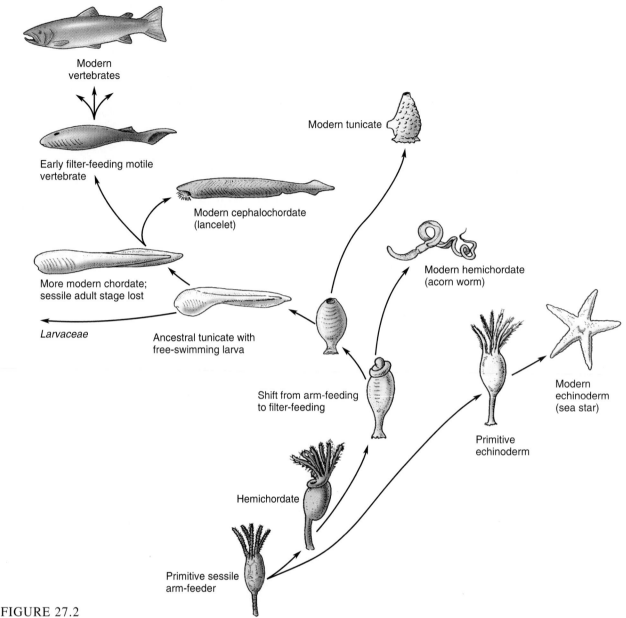

FIGURE 27.2
Deuterostome Evolution
According to one theory, echinoderms, hemichordates, primitive chordates, and vertebrates arose from a sessile, deuterostome, arm-feeder. These simple animals produced a swimming larva. From these early creatures arose the echinoderms and hemichordates, both of which still retain the free-swimming larva. Following these divergences, the deuterostome line divided once more. One branch produced the urochordate line leading to modern tunicates (right). The other arose from certain free-swimming larvae that had somehow retained their juvenile body form yet reached sexual maturity. From this evolutionary experiment arose the cephalochordates (right) and the vertebrates (left).

Subphylum Cephalochordata: Lancelets

Cephalochordates include the **lancelets** (or amphioxus), small fish-shaped chordates. Whereas they look like fish (Figure 27.4), unlike fish, the notochord is prominent in adults. In fact, the notochord protrudes beyond the brain. (This is how the creature got its name—*cephalo,* "head".) The relationship between cephalochordates and vertebrates is best seen by comparing the lancelet to the larval ammocoete (lamprey), a primitive vertebrate (Figure 27.5). The adult cephalochordates retain the pharyngeal gill slits and develop a dorsal, tubular nerve cord. Lancelets are filter-feeders. Beating cilia draw water into a complex mouth whose tentacles separate out food particles to be taken in for digestion. Water taken into the mouth passes through the gills where oxygen and carbon dioxide are exchanged, enters an outer chamber, the *atrium,* and exits through an opening near the tail.

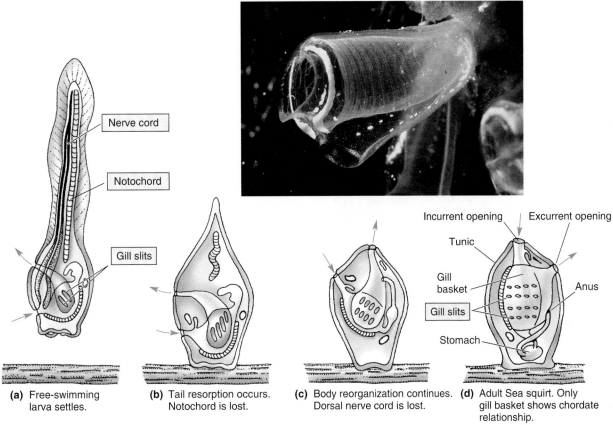

(a) Free-swimming larva settles.

(b) Tail resorption occurs. Notochord is lost.

(c) Body reorganization continues. Dorsal nerve cord is lost.

(d) Adult Sea squirt. Only gill basket shows chordate relationship.

FIGURE 27.3
The Tunicate

(a) As a larva, the tunicate has the pharyngeal gill slits, a lengthy notochord, and a dorsal, hollow nerve cord—all hallmarks of the chordates. **(b, c)** Eventually the larva enters a transitional period in which it approaches the adult state. **(d)** The body form of the adult is rather simple and has no specialized sensory devices, although it maintains the pharyngeal gill slits throughout adult life. Currents of water are produced by cilia in the gill structure, drawing food particles into the digestive tract. Respiration occurs as water exits through the body wall.

Subphylum Vertebrata: Animals with Backbones

The **vertebrates** are animals with backbones or **vertebral columns** that enclose the spinal cord. At the anterior end of the spinal column, the neural tissue elaborates into the *brain,* enclosed in a cranium. The *ventral heart* pumps blood through a system of blood vessels and is oxygenated by *gills* or *lungs.* In addition, vertebrates have no more than *two pairs of limbs, two eyes, two kidneys,* and *two separate sexes.*

There are seven classes of living vertebrates (Figure 27.6). The first three are fishes; the others, amphibians, reptiles, birds, and mammals. The classes are: *Agnatha,* jawless fishes; *Chondrichthyes,* cartilaginous fishes; *Osteichthyes,* bony fishes; *Amphibia,* amphibians; *Reptilia,* reptiles; *Aves,* birds; and *Mammalia,* mammals.

Class Agnatha: Jawless Fishes

The oldest well-defined vertebrate fossils on Earth are a group of **agnathans** ("without jaws") called the **ostracoderms**. These were heavy, jawless fishes with their bodies covered by heavy plates of scale-like "armor" (see Figure 27.8). Their fossils date back half a billion years, occurring in the sediments of the Ordovician, Silurian, and Devonian periods (see Appendix A). Their fin structure, which does not include the two pairs of fins seen in later fishes, tells us that they were sluggish swimmers. They probably depended on their armor and bottom-dwelling habits for protection against the large invertebrate predators of the day. Without jaws, the ostracoderms were probably not predacious but more likely strained their food from the water. The armored fishes died out at the end of the Devonian period.

Today's agnathans, boneless descendants of the armored fishes, remain jawless. Instead of bones, their skeletons are

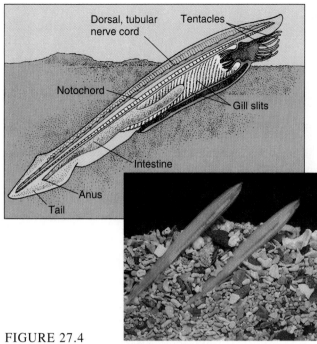

FIGURE 27.4
The Lancelet
In its body plan, *Branchiostoma* is clearly similar to the basic vertebrate pattern. The notochord runs the length of the body and functions as an anchor for the period muscle segments (myostomes). There are also a dorsal, hollow nerve cord and pharyngeal gill slits.

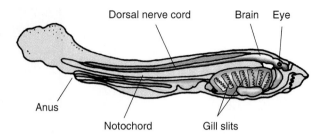

FIGURE 27.5
An Ammocoete Larva
The ammocoete larva of the lamprey bears many similarities to the lancelet and to the tunicate larva.

Class Placodermi: Extinct Jawed Fishes

The **placoderms** are extinct jawed fishes. They are believed to have descended from one line of ostracoderms, but the evidence is scanty. They originated back in the Silurian period, their numbers swelling during the Devonian and dwindling out some 150 million years later. Clearly, they must be considered a successful group in terms of their tenure on Earth.

Placoderms were also armored, but unlike their agnathan ancestors, they had hinged jaws. There is little doubt that the

composed of cartilage. The lampreys and hagfish are members of the subclass **Cyclostomata** ("rounded mouths"). Their bodies are long and cylindrical, with simple median fins adapted for wriggling along the ocean bottom. These cyclostomes lack the paired fins of other fishes, their skeleton is cartilaginous, and, oddly, their notochord persists throughout life. Parasitic species of lamprey feed with the aid of a rounded sucker mouth armed with horny spikes and a rasping tongue (Figure 27.7). The hagfishes, or slime hags as some species are called, lack the sucker mouth and feed by a rasping device, boring into the bodies of dead or dying animals. By most standards they aren't a particularly admirable group.

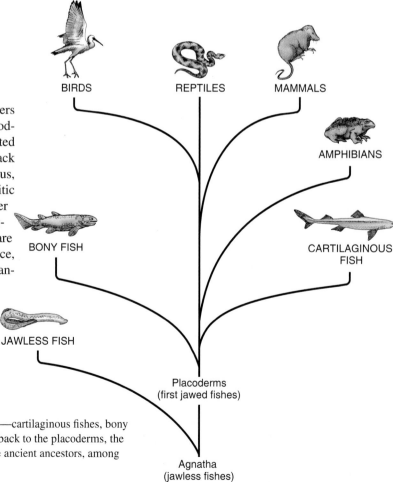

FIGURE 27.6
Vertebrate Relationships
Because of a rich fossil record, six of the seven vertebrate classes—cartilaginous fishes, bony fishes, amphibians, reptiles, birds, and mammals—can be traced back to the placoderms, the first jawed fishes. The seventh, the jawless fishes, have even more ancient ancestors, among the early agnathans.

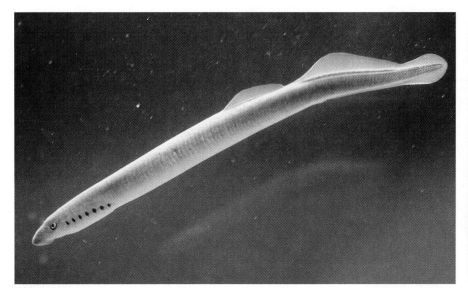

(a) Lamprey

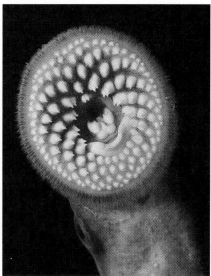

Rasping mouth of lamprey

FIGURE 27.7
The Clyclostomes
Today's jawless fishes, lampreys and hagfish, lack a bony skeleton. A notochord persists in the adult along with the rudimentary cartilaginous skeleton. **(a)** The parasitic lamprey uses rasping and sucking mouthparts (*top right*) in feeding. **(b)** The hagfish is a bottom scavenger known to attack weakened or injured fish.

(b) Hagfish

evolution of movable, biting jaws was one of the most significant events in vertebrate history, quickly permitting the establishment of many new patterns of predation (and generally enabling vertebrates to make life miserable for each other).

The placoderm jaw is known to have derived from the gill arches (the bony structure that supports the gills) (Figure 27.8). This, of course, required a considerable modification in both the position and strength of the gill arches, as well as the pharynx. Today, the upper and lower jaws of most fish form a hinge-like structure that is only loosely attached to the cranium. The hinge improved jaw mobility and enabled the primitive vertebrates to become predators, a role formerly dominated by invertebrates.

From the placoderm line evolved the two large classes of jawed fishes we see today—the cartilaginous fishes (class Chondrichthyes) and the bony fishes (class Osteichthyes).

Class Chondrichthyes: Sharks and Rays

Class **Chondrichthyes** includes the fishes with skeletons made up of cartilage (see Chapter 32). The chondrichthyes is a fascinating group of predators and scavengers made up mainly of sharks and rays. Many of these modern predators have replaced the protective armor and the heavy skeleton of their ancestors with a tough skin, light frame, and great speed. Their precise origin remains a puzzle, but their fossils first appear in the early Devonian period. By the Mesozoic era,

such fossils dwindled as the numbers of bony fishes burgeoned. But beginning in the Jurassic period, they again increased gradually, and since then they have held their own.

The cartilaginous skeleton of sharks and rays was once taken as evidence of their primitive nature. Evolutionists more recently have established that the cartilaginous skeleton of sharks is a *derived* condition—that is, the ancestors of cartilaginous fishes had bony skeletons. (Recently, scientists have found exciting new evidence of residual bony tissue in the spinal column of sharks.) Some biologists believe the evolution of a lightweight cartilaginous skeleton is an adaptation to deep-water life. In support, they note that sharks and rays lack the hollow, buoyant swim bladder found in bony fishes, although some buoyancy is provided by a large store of lightweight lipids in the liver. In addition to their cartilaginous skeleton, sharks and rays are unique in several other respects. For example, their body and tail shape is unlike that of most bony fishes (Figure 27.9). The shark's tail is asymmetrical, more like that of the extinct placoderms than those of modern bony fishes. The body, particularly at the anterior, is dorsoventrally flattened, rather than laterally as in typical bony fishes.

The skin of sharks and rays is very rough, embedded with miniature "teeth" known as *placoid scales* (see Figure 27.9c). Each is anchored into the dermis by a basal plate. Embryolog-

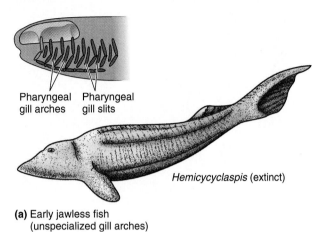

(a) Early jawless fish
(unspecialized gill arches)

Hemicycyclaspis (extinct)

Pharyngeal gill arches   Pharyngeal gill slits

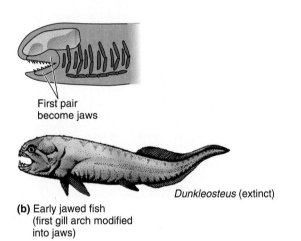

First pair become jaws

Dunkleosteus (extinct)

(b) Early jawed fish
(first gill arch modified into jaws)

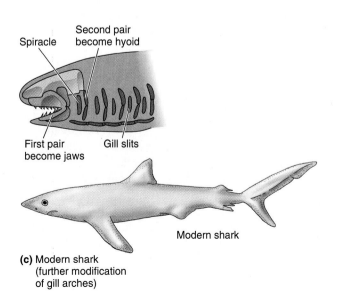

Spiracle   Second pair become hyoid

First pair become jaws   Gill slits

Modern shark

(c) Modern shark
(further modification of gill arches)

ically, these scales form the same way as teeth, and, in fact, the shark's teeth are simply larger versions of the scales. The teeth themselves develop continuously throughout life, with newly formed teeth migrating forward from rows inside the mouth to replace older or broken teeth. The older teeth fall out and sink to the bottom, often washing up on beaches, where artistically inclined beachcombers may collect them for jewelry-making. The rays and several kinds of sharks have heavy, flattened "paved" teeth, useful for feeding on shellfish. The largest sharks, the whale sharks and basking sharks, are filter-feeders that eat small zooplankton they strain from the sea with baleen—feathery strainers in place of hard teeth.

Sharks have keen senses and locate prey by smell, by sight, and by certain patterns of vibrations (distress movements) that are picked up by the **lateral line organ** (see Figure 34.18). The lateral line organ, characteristic of all fish, is formed of canals running the length of the body, containing groups of sensory cells that respond to water movements. Other sensory canals on the head are sensitive to bioelectrical disturbances or weak electrical fields created by nearby animals, although little is known about the receptor mechanism.

Contrary to the concern of some scuba divers, the toothsome, gaping grin on the mouth of an approaching shark is not necessarily anticipatory. It is generally accepted that its open mouth ensures a continuous flow of oxygen-laden water over the gills and out through the gill slits (Keep in mind the "not necessarily!").

Fertilization in sharks and rays is internal, but embryo development varies from simple primitive egg-laying in some species to a more advanced regime of protecting the independent egg and embryo in the uterus until hatching in others. A few shark species are quite reproductively advanced, retaining the egg and embryo in the uterus and providing nourishment, removing wastes, and exchanging gases, just as mammals do. In some shark species, embryos are produced in a steady assembly line, but only the oldest survive. They feed by devouring their younger brothers and sisters in the uterus as fast as they appear.

Class Osteichthyes: The Bony Fishes

The name **Osteichthyes** means "bony fishes" (*os,* "bone"; *ichthyes,* "fishes"), the fish with a skeleton formed of bone. Like many modern animal groups, the bony fishes have expanded into just about every conceivable aquatic niche.

FIGURE 27.8
Evolution of Jaws
(a) The earliest fishes, such as *Hemiclapsis,* were jawless, but their seven pairs of pharyngeal gill arches held promise for the future. **(b)** In the early jawed fishes, such as *Dunkleosteus,* the first pair of gill arches had undergone a transition into the first jaw elements, and a major evolutionary trend began. **(c)** In the modern shark, more of the arches have undergone similar transitions, producing more supporting elements in the jaw region.

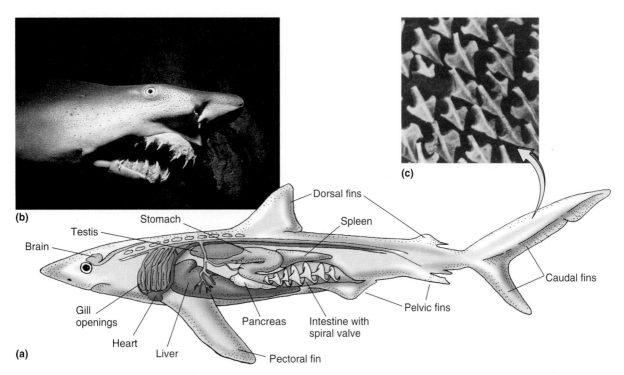

FIGURE 27.9
Shark Characteristics
(a) The shark body shape is an excellent lesson in streamlining. (b) The teeth occur in rows with those further in moving outward to replace lost and damaged teeth from the front row. (c) The skin is tough and flexible, consisting of placoid scales, each a miniature tooth-like structure.

They are a very diverse group (Figure 27.10) with well over 20,000 named species. It should be made clear that bony fishes are not so closely related to sharks as a first glance might indicate. The constraints of an aquatic life on a swimming vertebrate tend to limit body and fin shape, so there are general similarities, such as low-friction bodies and flattened fins.

Bony fish have a *swim bladder,* a gas-filled bubble-like organ that improves balance and permits the animal to remain

Angler fish

Lionfish

FIGURE 27.10
Class Osteichthyes
Bony fish exhibit a great diversity of shapes. Most of the structural differences are in fin and body shape. Some species depart drastically from other forms, such as the lionfish (*right*), and the creatures of the abysses such as the angler fish (*left*).

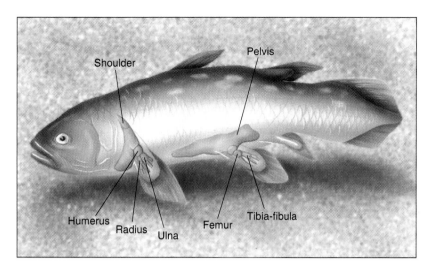

Shoulder

Pelvis

Humerus

Radius

Ulna

Femur

Tibia-fibula

FIGURE 27.11
A Living "Fossil"
The coelacanth *Latimeria* was found off the East
African coast. Since their discovery in 1939, a
number of *Latimeria* have been caught and subjected
to intense study. Note the fleshy lobed fins.

stationary at varying depths. Most oxygenate their blood by drawing water into the mouth and pumping it over the gills. The gill structures include five rows of gills on each side, located in a common *gill chamber* and each protected by a bony *operculum,* a movable external flap. Each gill consists of a supporting gill arch with toothed *gill rakers* (that keep food out of the gills) and rows of *gill filaments.* Gas exchange occurs in the gill filaments, which are extremely thin-walled with a rich supply of fine blood vessels just beneath the surface (see Figure 39.6). Respiratory structures include a closed circulatory system, and a two-chamber heart. In its circuit through the body, blood is pumped from the heart by the single, muscular ventricle (a large pumping chamber) directly to the gills, where it exchanges carbon dioxide for oxygen before being circulated to the body and returned to the heart (see Chapters 38 and 39).

Lobe-finned fishes, those with heavy, fleshy fins, had strong, fleshy bases on their pectoral and pelvic fins. These species were always few in number, but they have had enormous evolutionary influence. They gave rise to the amphibians—the first animal land invaders—and from these arose the reptiles, and from these, the birds and mammals. A remarkable example exists today—the rare, deep-sea **coelacanth** (*Latimeria chalumnae*), the survivor of a once rather large and diverse group that was long believed to be extinct (Figure 27.11). Latimerians themselves are not ancestral to anything alive today and do not even belong to the specific (crossopterygian) group from which land vertebrates evolved. *Latimeria* is a distant relative indeed, but it is nonetheless more closely related to us than is any other living fish.

Class Amphibia: The Amphibians

The Devonian world was an unpredictable place indeed. Torrential rains pounded the Earth's surface only to be followed by long periods of parching drought, and the struggling life below was forced to adapt to the changes. One group, the lobe-finned fishes with lungs or lung-like structures, may have been among the most successful groups. Their sturdy fins would have enabled them to move about on exposed mud, looking for better conditions, and their lungs would have enabled them to gulp air to get the oxygen they needed. This group led to the **amphibians**, the group that would spend part of its life on land, part in water.

The first successful vertebrates to be able to live on land were the stubby, four-legged amphibians called **Labyrinthodonts** (Figure 27.12); they ranged in size from small salamander-like creatures to animals the size of crocodiles. These, it is generally believed, gave rise not only to modern frogs and salamanders but to the earliest reptiles as well.

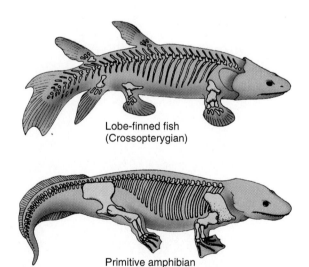

Lobe-finned fish
(Crossopterygian)

Primitive amphibian
(Labyrinthodont)

FIGURE 27.12
Transition to Land
The labyrinthodont may have been the primitive amphibian from which both reptiles and amphibians emerged. Note its resemblance to the lobe-finned fishes.

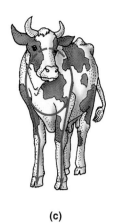

(a)   (b)   (c)

FIGURE 27.13
Vertebrate Limbs and Posture
The adaptation to terrestrial life is seen as a transition in the positioning of limbs in the tetrapods (four-legged vertebrates). **(a)** Many amphibians have their thin, lightly muscled legs angled or splayed out to the side. Thus the body weight is borne on flexed joints. **(b)** Reptile legs are alongside the body, but they still angle out to the side. **(c)** Mammals have their bodies raised above the ground with the limbs alongside or below the body.

Amphibians today are represented by three orders: **Urodela** (the salamanders), **Anura** (frogs and toads), and **Apoda** (the worm-like caecilians). Nearly all amphibians reproduce and develop in aquatic habitats, though some are well adapted to drier environments and a few even live in deserts.

Note the differences in the skeletons of a lobe-finned fish and a fossil amphibian in Figure 27.12. Among the skeletal modifications necessary for terrestrial life is increased development and size in the vertebral columns and the shoulder and hip girdles. In the evolution of other terrestrial vertebrates, particularly those to whom speed was important, the legs gradually came to be located alongside or under the body and projected downward rather than sideways. Locomotion came to depend on the muscles of the limbs and limb girdles, rather than on lateral undulations of the entire body (Figure 27.13).

Amphibians have three-chamber hearts and go the fish one better by having a blood circuit through the lungs that is separated from the rest of the body. This arrangement permits oxygenated blood to be returned to the heart for a second pump before entering the body circuit. The separation is imperfect, however, and some mixing of oxygenated and deoxygenated blood occurs in the single ventricle (see Chapter 38).

Amphibians have moist, highly vascularized skin, which, along with the lungs, is an important organ of respiratory exchange. (Amphibian lungs have little surface area since they are essentially simple, vascularized sacs and not spongy like those of other vertebrates.) A moist skin is essential for gas exchange, but if gases can cross the skin, so can water. Thus the amphibian is always at risk of dehydration, and, accordingly, many live a semiaquatic life. There are important exceptions, though. The desert spadefoot toad, for example, survives its desert habitat by spending long dry spells in a fully dormant state, buried in the soil. The amphibian egg must be kept moist; thus most amphibians must return to the water to reproduce.

Class Reptilia: The Reptiles

The **reptiles** are the vertebrates with scales that live an essentially terrestrial existence. (Some, however, return to the water to feed.) Modern reptiles probably descended from four sepa-

rate branches of the earliest reptiles and today form four orders (Figure 27.14). Order **Chelonia** includes the turtles and side-necked turtles. Order **Crocodilia** contains alligators and crocodiles. Order **Sphenodonia** contains only one species, *Sphenodon punctatum* or tuatara. The fourth order, **Squamata**, includes the lizards and snakes. Snakes are believed to have evolved away from one branch of the lizard line about the start of the Jurassic, midway through the reptile era. The four lines survived the rigors of the Cenozoic era and are represented today by 6000–7000 species. The new land dwellers continually adapted to the dry environment and, in so doing, created a wide variety of specializations. Changes in their sexual reproduction and development were especially significant. Reptiles retained the cloaca, but males developed a penis for efficient copulation and internal fertilization, a prerequisite for full-time land dwelling.

A highly significant change was the evolution of the amniotic egg, in which the embryo is surrounded by a fluid-filled, membranous sac known as an amnion (characteristic of reptiles, birds, and mammals). This change marked an important adaptation to terrestrial life in that the egg carried its own water. The amniotic egg of the reptile is a porous, leathery case, containing a food supply, the **yolk**, and a number of **extraembryonic membranes**, produced by the embryo, that carry on life-supporting activities (Figure 27.15). A **yolk sac**, growing over the yolk, absorbs nutrients and transports them to the embryo via a vast network of blood vessels. The embryo itself is surrounded and cushioned in the fluid-filled **amniotic cavity**, surrounded by another membrane, the **amnion**. Waste removal is provided by the **allantois** and gas exchange (carbon dioxide for oxygen) is taken care of by the **chorion**. The amniotic egg system works so well that it has been retained by the birds and, in a modified form, by the mammals.

The Earth doesn't harbor as many reptiles now as it once did. In fact, the Mesozoic era, which lasted over 160 million years, is known as the "Age of the Reptiles," largely because of the great numbers of dinosaurs in those days. However, the reptiles actually originated much earlier, branching off from the amphibian line in the Carboniferous period of the Paleozoic era (see Table 26.1). They peaked during the Jurassic and dwindled down to the comparatively few species of today.

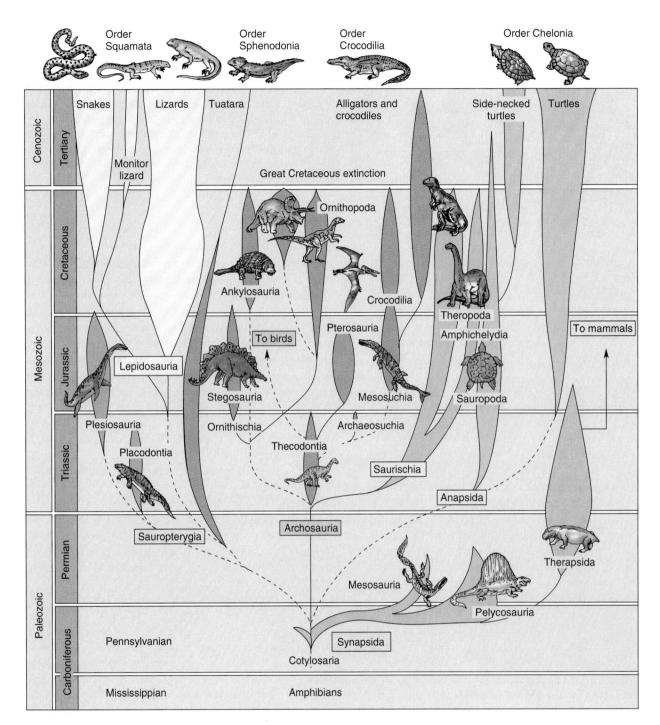

FIGURE 27.14

Survivors of the Reptile Age

Today's reptiles are believed to have evolved through four separate lines. The snakes and lizards descended from a line of smaller reptiles. The lizard-like tuatara of New Zealand is the lone remnant of another group. Right out of the central line of dinosaurs emerged the crocodilians, forming their own line. The side-necked turtles have a long separate history of their own. Both kinds of turtles, like the snakes and lizards, sidestepped the giants and emerged from ancient stock early in the Mesozoic. Many extinct groups are also shown here. The widths of the colored lines roughly indicate the relative numbers of species at any point in time; dashed lines indicate our best guesses in the absence of fossil data.

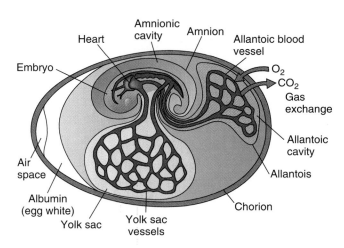

FIGURE 27.15
The Amniotic Egg
One of the most significant land adaptations achieved by vertebrates was the amniotic egg, which fully equipped the reptile embryo for safe and successful development on land. In addition to its stored food and water, the amniotic egg contains four extraembryonic membranes, which provide the necessary exchanges for sustaining life. The four membranes, produced by the embryo, are the amnion, yolk sac, allantois, and chorion.

If the reptiles were so successful (Figure 27.16), why did they die off? Their demise came with the closing events of the Cretaceous period, and no one is sure what brought it about. We do know, however, that the delicate webs of life are highly vulnerable to change, and the loss of one member affects all. Perhaps a change in the climate started the process of their extinction. We know, for example, that their habitat was cooling at about this time. Some scientists blame the emergence of the early mammals, which might have preyed on the eggs of the giant reptiles. An exciting and well-substantiated theory has to do with the vast effects brought about by a giant asteroid that collided with the Earth at the close of the Mesozoic. This idea, the Alvarez hypothesis, is discussed in Essay 27.1. Another recent theory suggests that the sun has a companion star, named *Nemesis,* that, in cycles of about 26 million years, causes a rain of comets that destroys most life on Earth. The geologic record roughly suggests such a cycle but the star has not been found.

Class Aves: The Birds

The fossil ancestors of modern birds can be traced back to newly discovered *Protoavis,* from the late Triassic period, about 225 million years ago. It had feathers and presumably could fly. The best known fossils are of *Archaeopteryx* (Figure 27.17), whose skeleton is, interestingly, almost indistinguishable from that of the thecodonts, the reptilian stock that

produced crocodiles. Modifications for flight are found throughout the modern bird's body (Figure 27.18). The skeleton is light and strong. Many bones are hollow, containing extensive air cavities crisscrossed with bracings for strength. The reptilian jaw has drastically been lightened, and teeth have been replaced by a light horny beak. The neck is long and flexible, and the bones of the trunk (pelvis, backbone, and rib cage) have become fused into a semirigid unit. The breastbone (sternum) is greatly enlarged and possesses a large keel to which the large flight (breast) muscles attach in flying species. The tail is greatly reduced and, with the exception of a few individual bones, has become fused. The feet are often specialized, for example, for perching or grasping.

Feathers, another important modification for flight, are believed to have evolved from reptilian scales. Feathers are hollow and rod-like, which makes them unusually strong for their weight. Part of their strength results from their interlocking barbs (branches). Softer, noninterlocking down feathers provide insulation in all young and many adult birds. The feathers act as insulation, helping to retain the heat produced by the bird's metabolism (see Chapter 35).

There are about 8600 named species of birds, divided into about 27 orders. In spite of such extensive speciation, however, all birds have rather similar general structures. Apparently, the demands of flight have superimposed a specific set of design requirements. Birds differ most strikingly in the development of their beaks and feet, both of which are often related to their feeding behavior (Figure 27.19).

Class Mammalia: The Mammals

Mammals are hairy animals with glands that give milk. In fact, it is the milk-producing **mammary gland** for which the class is named. The hair or fur of the mammal is produced in follicles and is lubricated by numerous oil glands. Hair provides an insulating covering that helps the mammal maintain a relatively constant body temperature. Mammals generate body heat, thus maintaining a constant body temperature. The young of most mammals develop in the uterus, where they are nourished and sustained by a *placenta* (see Chapter 42). Mammals have a muscular diaphragm that helps move air in and out of the lungs.

While the limbs of amphibians and reptiles are commonly directed outward and then down, the mammal's (and the bird's) limbs are directed generally downward, raising the body clear of the ground. Variations on this basic theme have produced mammals that swim, run, climb, burrow, leap, and even fly. In some species, such a limb arrangement also permits greater efficiency in catching, holding, and killing.

The evolution of the mammalian jaws and teeth has been equally dramatic. Except for the crocodiles, only the mammals have socketed teeth and most have differentiated teeth (incisors, canines, premolars, and molars). And only the mammals have a fully mobile jaw. The teeth, of course, vary with the animal's diet. Herbivores generally have massive, grinding molars, while carnivores have large canine teeth and high-ridged, bone-crushing molars. The chisel-like incisors of rodents are specialized for gnawing, while insectivorous mammals have sharp, pointed teeth for crunching bugs. The massive tusks of walruses (modified incisors) are used both defensively and for raking up clams. Elephants use their tusks (also modified incisors) in defense. The great peg-like teeth of the killer whale are specialized for grasping and crushing large prey such as other whales and seals. Human teeth are among the least specialized, so we are capable of eating about anything we come across, from lettuce to spider crabs.

FIGURE 27.16
Ruling Reptiles
The so-called "ruling reptiles" of the Mesozoic include some of the largest animals ever to rove the land. Several plant species are also identified.

1. Araucarites
2. Ramphorhyncus
3. Allosaurus
4. Schizoneura
5. Matonidium
6. Archaeopteryx
7. Brontosaurus
8. Ginkgo
9. Pteranodon
10. Quercus
11. Cornuss
12. Pandanus
13. Anatosaurus
14. Sassafras
15. Ankylosaurus
16. Palmetto
17. Sabalites
18. Salix
19. Struthiomimus
20. Magnolia
21. Triceratops
22. Stegosaurus
23. Tyrannosaurus
24. Neocalamites

Jurassic Period

Essay 27.1 ● EXIT THE GREAT REPTILES

The demise of the great reptiles, or dinosaurs, and the many species that died out with them has long been a puzzle. Hundreds of hypotheses have been considered. Some people once believed that dinosaurs were simply outcompeted by brainier mammals, or that the little rascals ate the dinosaur eggs. Others say that drastic climate changes marked the close of the Mesozoic or that a companion star to the sun, Nemesis, causes comets to rain down every 26 million years. (Cartoonist Gary Larsen suggests it's because they took up smoking.) The problem is that none of these ideas can account for the truly dramatic suddenness of the mass extinction in which the dinosaurs disappeared.

Perhaps the answers can be found in the rocks. There are places on the Earth where the 65-million-year-old rock stratum that marks the boundary between the Mesozoic and the Cenozoic can be found. In the waters off Gubbio, Italy, the marine deposits are particularly clear. What can they tell us? Below the boundary, layer after layer of Mesozoic carbonate rocks show a clear assemblage of Mesozoic plankton skeletons, dominated by very large foraminiferans (protists; see Chapter 23). But above the Mesozoic rocks, in the strata that were deposited in the Cenozoic, the fossil life forms are dramatically different. In layer after layer of Cenozoic rocks, we find fewer, much smaller species. Even more interesting, at the boundary between the two layers of plankton deposits is a single layer of clay about a centimeter thick.

In 1980 the paleontologist Walter Alvarez and his Nobel Prize-winning physicist father, Luis Alvarez, reported that this boundary layer has about 50 times more of the elements iridium and platinum than would be expected. Iridium is extremely rare in the earth's crust but is a common metal in meteorites and asteroids. The Alvarez hypothesis, as it has come to be known, states that an asteroid collided with the Earth 65 million years ago, and that's what ended the Mesozoic. Astronomers calculate that enough asteroids pass through the Earth's orbit that we can expect one to hit our planet every 100 million years or so. From the amount of iridium they found in that thin layer, they calculated that the asteroid was about 10 kilometers in diameter. Computer simulations tell us that the dust thrown up by the impact would bring the world into total darkness—much darker than the darkest night—for about three months, accompanied by intense cold. The effect of such darkness would be enormous. Since the sun's rays could not reach the Earth, photosynthesis would stop, and plants would begin to wither and die as the weather changed drastically. Tropical species would have experienced subfreezing temperatures for the first time. Further research has corroborated the Alvarez hypothesis, and the iridium-rich Mesozoic–Cenozoic boundary layer has been found in many places. (The data are confounded, however, by Alaskan deposits that tend to refute the hypothesis. Still, the weight of the evidence supports the Alvarez's idea. New evidence suggests that the collision may have triggered the worldwide eruption of volcanoes, adding to the atmospheric debris.)

Few organisms were equipped to survive such a catastrophe. Seed plants and marine plankton that formed resistant cysts were seemingly unaffected. But animal species changed significantly; for example, no animal species weighing more than 26 kg (57 lb) survived into the Cenozoic. Most of the Earth's great dinosaurs then were killed by the Earth's great darkness.

The demise of the reptiles was obviously not quite complete, since we have a considerable number with us today. As you watch the small lizard on your porch going about its business, keep in mind that it is a descendant of survivors of the great dinosaur extinction.

Cretaceous Period

FIGURE 27.16

FIGURE 27.17
Archaeopteryx
Although bird-like in some ways, *Archaeopteryx* is not ancestral to today's birds. Note the clawed fingers in the reconstruction. Were it not for the feathers, *Archaeopteryx* would probably have been identified as just one more small dinosaur.

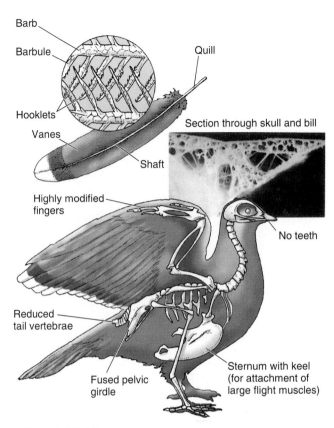

FIGURE 27.18
Design for Flight
Modifications for flight are seen in nearly every aspect of the bird's anatomy and physiology, from the streamlined form to the elevated metabolic rate. In spite of the demands placed on it, the skeleton is extremely light. The frigate bird, for instance, has a wingspan of just over 2 m (7 ft), yet its skeleton weighs an average of just 113 g (4 oz). In general, the slender, hollow bones of birds have a deceivingly delicate appearance; in fact, they are strong and flexible, containing numerous triangular bracings (see photograph). Part of the skeletal strength is due to fusion, as is seen in the hip girdle, tail vertebrae, and, most spectacularly, in the long finger (wing) bones. Flight feathers, which may weigh more than the skeleton, owe their extreme strength and flexibility to numerous vanes. These have an interlocking arrangement of hook-like barbules.

Mammalian Origins. The mammals had their reptilian origin somewhere in the Permian period, branching off very early from a large group that preceded the ruling reptiles, known as the **therapsids**, sometimes known as the "mammal-like reptiles" (Figure 27.20). The transition into the full mammalian condition was gradual, with no sharp dividing line seen in the fossil record. Some early mammals are said to have been "dog-like" in appearance. Whereas this image may not be entirely accurate, you could probably clear out a bar if you walked in with one on a leash.

The therapsids were flesh eaters that developed the forelegs and musculature necessary for running. The angle of their elbows was quite different from that of the first terrestrial reptiles, whose elbows angled out to the side. Other indications of their mammalian association include tooth structure as well as skull and jaw features.

Survival in the developing mammals may have been enhanced by the ability to carry embryos within the body and to bear live young, rather than deserting the eggs to predators or having to defend a stationary nest. Another advancement was the production of milk, a very convenient food that provided the young with high-protein, high-calorie nutrition dur-

ing the critical stages of growth. In turn, the young would have tended to remain with at least one parent in order to be fed, and this continued association would have afforded the opportunity for the offspring to learn from the parent. This opportunity would have reinforced selection toward braininess.

The dawning Cenozoic era ushered in the "Age of Mammals." By the end of the Mesozoic era, three types of mammals had diverged from the original line: **monotremes** (order Monotremata: the egg-laying mammals), **marsupials** (order Marsupalia: the pouched mammals), and **placental mammals** (mammals that form a true placenta). Figure 27.21 shows examples of each.

(a) Hawk talon

(b) Lobed toes of the coot

FIGURE 27.19
Diversity in Bird Feet
(a) The hawk uses its sharply curved talons to capture prey. **(b)** The lobed toes of the coot are well suited to their aquatic habitats.

The monotremes never became widespread, and today there are just two species: the duck-billed platypus and the spiny anteater (also called the echidna). The marsupials are much more similar to the placental mammals, differing primarily in some aspects of development. Whereas in the placental mammals, the placenta provides for the needs of the embryo throughout its development in the uterus, marsupials produce only a "pseudoplacenta," a modification of the yolk sac. It supports the young for a brief time in the uterus before a very premature birth occurs. The newborn marsupial then migrates to the pouch, where development is completed.

The three modern mammalian subclasses radiated from an explosion of species in the early Cenozoic era. By the Eocene period, which began 54 million years ago, the main lines of mammalian evolution had already become established.

The Cenozoic era was a disruptive time indeed, marked by drastically fluctuating climatic patterns and intense selection. Many Cenozoic mammals didn't survive the severe stress; however, those that did underwent rapid speciation and divergence, filling many new and highly specialized niches. Today's 4500 species are organized into 19 orders. Each represents a major line that appeared in the early Cenozoic (Figure 27.22).

The Primates. Modern primates, of the order **Primates**, are divided into two major suborders, the first consisting of the more primitive tree shrews, tarsiers, lemurs, and lorises (**Prosimii**), and the second, the monkeys, apes, and humans (**Anthropoidea**). The latter is further subdivided into the New World monkeys (Ceboidea), Old World monkeys (Cercopithecoidea), and apes plus humans (Hominoidea). New

FIGURE 27.20
The Earliest Mammals
The mammals arose from the therapsids, an early reptile line. The earliest therapsids retained many reptilian traits, including scales, but by the Triassic period (197–225 million years ago), features such as the limbs and stance had already changed. Raising the entire body from the ground was an adaptation for running and leaping, but it introduced new problems in balance and coordination—problems that required simultaneous adaptive changes in the brain.

Duckbill platypus

Kangaroo

Mare with foal

FIGURE 27.21
Three Types of Mammals
Three lines of mammals evolved from the ancestral therapsids. Monotremes (duck-billed platypus, *top left*) are hairy egg layers. Their young feed by simply lapping milk off the mother's fur. The marsupials (kangaroo, *right*) give birth to immature young, which climb to the pouch, where they fasten to a nipple and remain there as they complete their development. The placental mammal (horse, *left*) completes its embryonic development in the uterus, supported by membranes that form the placenta, and nurses for a time after birth.

World, western hemisphere monkeys are those with the prehensile (grasping) tails and nostrils set apart by a wide nasal septum, typified by the familiar "organ-grinder monkey," or spider monkey. The Old World, Eurasian and African monkeys are more likely to have short tails and dog-like snouts with the nostrils close together, as in the baboons. The great apes (gibbons, orangutans, chimpanzees, and gorillas) and humans, members of superfamily **Hominoidea**, are called *hominoids*. The great apes belong to family **Pongidae**, whereas humans are in the family **Hominidae**. All members of our family, whether extinct or extant, are thus called (note the spelling) **hominids**.

Primates can be traced through the fossil record as far back as the Paleocene epoch, some 65 million years ago. The earliest primates were unlike modern types in many ways. For example, they had long snouts and claws and actually looked more like rodents. In fact, it is believed that they lived their lives much as modern rodents, some in trees, some scurrying about on the ground, and some in burrows. The divergence of the various primate groups is represented in Figure 27.23. Note how recently humans have diverged from common ancestry with the gorillas and chimpanzees. How do primates differ from other mammals? This would seem to be an easy question, and it is unless you attempt to become too specific. Primates are often described as generalized mammals, meaning that they have many unspecialized features, some even quite primitive.

The limbs and body of *most* primates are well adapted for arboreal (tree-dwelling) life. For example, all have prehensile (grasping) hands divided into clawless (nailed) fingers. Most have long arms and some have prehensile tails. Primates have also developed binocular, stereoscopic vision with eyes locat-

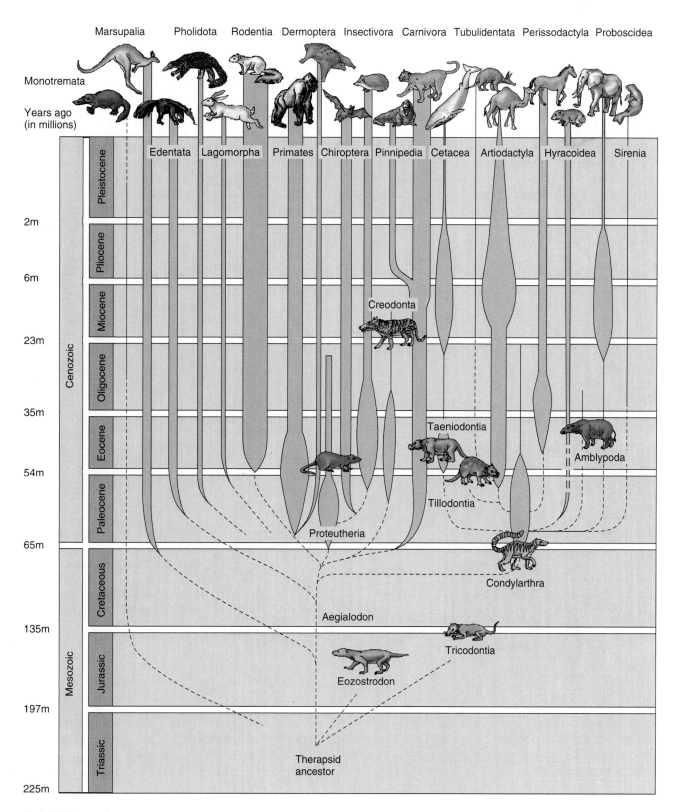

FIGURE 27.22
The Divergence of Mammals

Today's mammals occupy 19 orders and are dispersed into innumerable niches over the earth. Mammalian roots reach back to the early Mesozoic era but do not appear to branch out significantly until the Cenozoic, an era of great climatic changes that saw the end of the ruling reptiles. Numbers of species are indicated by line thickness. Dashed lines represent hypothetical relationships.

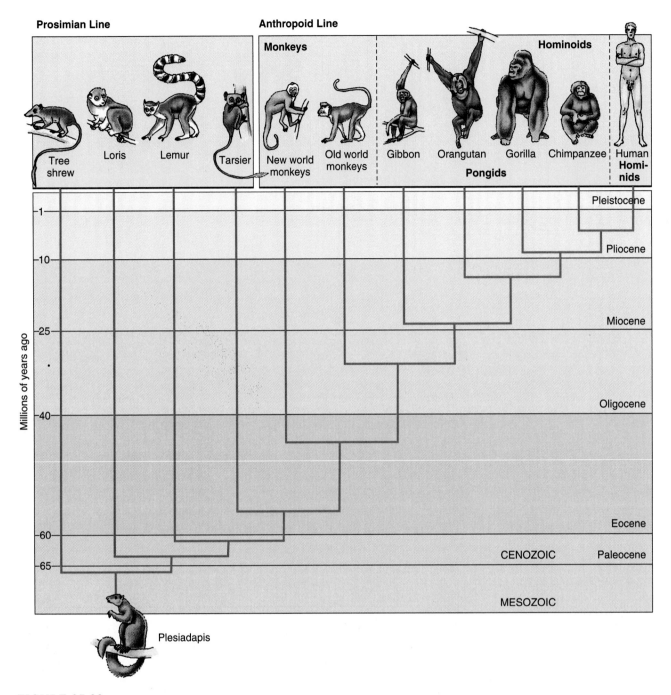

FIGURE 27.23
Primate History
This phylogeny of the primates places *Plesiadapis* at the base. The earliest branch produced the prosimian
line, followed by the New World monkeys. Old World monkeys diverged some 10 million years later, subsequently followed by the great apes. The most recent divergence occurred between humans and chimpanzees.

ed in a forward position. If you are going to swing on limbs,
you need good eye–hand coordination—and the ability to
judge distances (Figure 27.24).

Of course, all these anatomical specializations require a
substantial degree of brain development, with strong emphasis
on centers dealing with coordination and vision. But more sig-

nificantly, the primate brain is highly adapted for learning.
This ability is especially developed in chimpanzees and gorillas (and the capuchin, a South American monkey), but humans
are without parallel in intelligence and relative brain size.

Primates tend to be omnivores, eating all sorts of food, so
their teeth are relatively unspecialized. However, with the

FIGURE 27.24
Brachiation
Good brachiation (swinging from limb to limb) requires prehensile hands, long arms, and keen hand–eye coordination. The gibbon is one of the most talented brachiators. Note the lengthy arms and the four-fingered grasp.

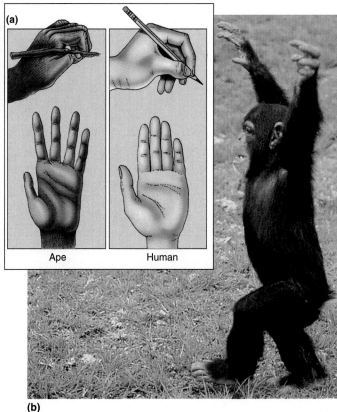

FIGURE 27.25
Small but Important Differences
(a) The hands of humans and chimpanzees appear generally similar, but important differences include the length and musculature of the thumb. The thumb on a chimp's hand has an "out of the way" location. This is important to a brachiator but makes the handling of tools far less precise. (b) While chimpanzees can assume a bipedal posture, neither hip nor leg structure supports this stance very well.

exception of humans, primates, especially males, have rather large canine teeth, used primarily in fighting, threat, and defense. Humans have reduced canines and have substituted tools such as rocks, baseball bats, and nuclear missiles. So today, an officer and a gentleman would never bite the enemy; his canines are too short.

Human Specialization. In some cases the specializations that are unique to humans are really only refinements of traits that also exist in other primates. Several primates, for example, have an opposable thumb that can touch the fingers, but ours has a slightly different musculature and is much more dexterous. With this manual ability we can make precise tools, and although other animals make and use very primitive tools, none can build a watch (Figure 27.25a).

And then there is our upright posture, which is quite unlike that of any other modern primate. Savanna chimpanzees stand to see over tall grass, and many chimpanzees walk bipedally (on two legs), a short way, but normally they move on all fours. It is amusing to see a chimpanzee or other primate

running on its hind legs because it is so human-like yet so ungainly (Figure 27.25b). Humans, however, are beautifully adapted to bipedal walking and running, surpassing any other vertebrate in this accomplishment. We owe this ability partly to our highly specialized foot, with its arched construction, and partly to our enlarged *gluteus maximus* (buttock) muscles.

We pride ourselves on our intelligence, but its evolution has not been without cost. The large skull that houses our brain produces all sorts of birth difficulties. And, even more than most other mammals, we are born in a helpless state. A baby hare will lunge and hiss at an intruder, and some baby birds will crouch and "freeze" at the mother's warning call. But our newborns are much more defenseless and seem totally witless. (New evidence, however, indicates that human babies begin learning immediately after birth and within months are behaving in such a way as to alter the behavior of their parents, as any mom can tell you.)

To some extent, humans have sacrificed a "prewired" brain, one that is programmed to react in certain ways under given situations, for one with greater potential. That is, we rely less on genetically based instincts and more on raw intelligence than do other species. Certainly this makes it possible for us to cope with a more variable and unpredictable environment. But we must undergo long periods of learning and therefore have evolved extended parental care. The fact that we have adapted to our world through intelligence, learning, and nurturing has undoubtedly altered our social patterns to an enormous degree.

Aside from our magnificently overgrown brain, the principal and unique specialization of the human species is our capacity for language. Language is not easily separated from intelligence because much of our brain is directly related to this function. Despite highly publicized and controversial studies, there is no evidence whatsoever that dolphins or apes can tell each other about their experiences in any but the most basic terms. Language, the capacity for creating abstract symbols to communicate concrete ideas, appears to be present only in humans.

With language, evolution entered a new dimension. Humans are capable of *cultural* evolution in addition to genetic evolution. We can tell our children the truths (and lies) that we learned from our parents. Factual information can be transmitted via word of mouth over tens of generations; the historically accurate account of the siege of Troy, handed down by illiterate singing bards for hundreds of years before being committed to paper, is but one example. Language also allows for politics: even unlettered tribes manage social organization far above the level that could ever be achieved by gibbering apes. Written language, of course, has extended the scope and power of cultural evolution by yet another order of magnitude.

Human Evolutionary History

As is usual when humans try to assess themselves, their genes, or their heritage, the air becomes full of indignation and accusations. The issue has been a volatile one since Darwin's day, when the very idea that humans had an evolutionary ancestry at all was offensive to many. Even Alfred Wallace, the codiscoverer of the principle of natural selection, couldn't bring himself to believe that human beings had originated in the same way other species had. Darwin himself firmly believed that humans and apes had evolved from a common ancestor, but there was no fossil evidence for this contention. Opponents of the idea of human evolution made much of the absence of a "missing link" and continued to make much of it long after many hominid fossils had been discovered.

In the early 1960s, biochemists began comparing primate proteins and DNA sequences. Their data clearly indicated that humans were more closely related to the chimpanzees and gorillas of Africa than to the orangutans of Asia. They were aware that molecules in these substances change due to mutation at a steady and dependable rate and thus provide us with a kind of "molecular clock." This clock suggested that humans and African apes diverged from a common ancestor not more than 6 million years ago—a startling notion at the time. Anthropologists had routinely assumed a much more ancient divergence and were mostly hostile to the new ideas. Another furious scientific feud was underway. However, other data, both biochemical and paleontological, backed up the biochemists rather than the traditional anthropologists, causing us to adjust our notions of human lineage. Let's review the current theory.

The Australopithecine Line

About 2–3 million years ago, a number of human-like forms lived in relatively dry open grasslands of Africa. Their fossils have been given a great number of contradictory names, but when the dust settles, several early species can be recognized, five of which are *Australopithecus robustus, Australopithecus africanus, Australopithecus boisei, Australopithecus afarensis,* and *Homo habilis.* It is not conclusively established that any of these forms was directly ancestral to *Homo sapiens,* but it is clear that the first three *Australopithecus* species were widespread and successful and persisted almost unchanged until at least 1 million years ago, long after the genus *Homo* was well established. Still, our own ancestors could have evolved from early offshoots of the *Australopithecus* line.

The **australopithecines** (members of the *Australopithecus* line) were rather small-boned, light-bodied creatures, about 1–1.5 m (3.5–5 ft) tall. They had human-like teeth and jaws with small incisors and canines, and they walked upright (Figure 27.26). They apparently hunted baboons, gazelles, hares, birds, and giraffes. Their weapons were clubs made from the long bones of their prey. (Near some *Australopithecus* bones, anthropologist Raymond Dart found a fossil baboon skull with a fossil antelope leg bone jammed into it.) The australopithecine **cranial capacity** (a measure of brain size) ranged from 450 to 650 cc (cubic centimeters), as compared with 1200 to 1500 cc for modern adult humans.

The first specimen of *A. afarensis,* the second oldest known hominid after the recently discovered and little-known *A. amanensis,* was found in the northern Ethiopian desert region in 1974 by Donald Johanson. The fossil remains, estimated to be over 3 million years old, consisted of little more than half of the skeleton of an upright-walking female. The following year, Johanson's luck blossomed. He unearthed the remains of 13 more "Lucy" types, all in one area. After several years of puzzling over these slightly built creatures, Johanson reached a decidedly unorthodox yet insightful conclusion. He proposed that *A. afarensis* was indeed more primitive than other australopithecines, placing it at the base of the hominid tree. The idea, hotly contested at first, is now well accepted by many researchers.

A number of physical anthropologists go along with the scheme seen in Figure 27.27. According to this scheme, *A. afarensis* produced two or three branches, one of which led to the genus *Homo,* which includes modern humans. The other line or lines produced *A. africanus, A. robustus,* and *A. boisei.*

Australopithecus boisei and *A. robustus,* as the latter's name implies, were larger boned—though not taller—than *A.*

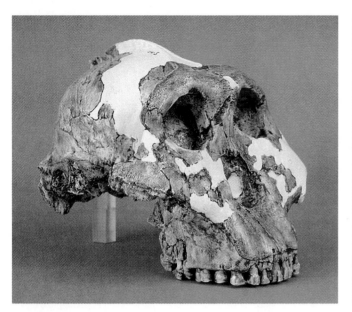 

FIGURE 27.26
Early Hominids
The autralopithecines were small but muscular creatures. Their heads were more ape-like than human in appearance, with a low cranial profile and little or no chin. The jaws were large and forward-thrusting.

africanus, the oldest of the three. Most strikingly, the two newer arrivals had much larger jaws and teeth with greatly expanded cheekbones to accommodate the massive jaw muscles. From a traditional view, this progression may seem to be backward—toward an ape-like condition—but other physical evidence and dating are to the contrary. The shape of the teeth and the curve in the jaws are decidedly human, not ape-like. The tooth wear, incidentally, suggests that much of the diet of these australopithecines consisted of plant food.

The Human Line

Homo habilis is the name given by famed anthropologist Louis Leakey to certain fossils from the Olduvai Gorge of East Africa. As the genus name indicates, these fossils seem to be close to the modern human form, but with an average cranial capacity of only about 656 cc. Critics have argued that the new find is just a variant of *Australopithecus* and that Leakey was unjustified in trying to place his fossils within the genus *Homo.*

Somewhat later, Leakey's widow, Mary, and his son, Richard, found a hominid skull of unusual interest at Kenya's Koobi Fora. At 1.6 million years old, it is older than some *Australopithecus* fossils, but it appears to be much closer to the human line than is *Australopithecus.* For one thing its cranial capacity—over 800 cc—is greater than those of either *Australopithecus* or the earlier disputed *Homo habilis* finds. The Leakeys, apparently fed up with the sterile arguments that have raged over the naming of hominid fossils, simply identified their unique find by its arbitrary field identification: fossil skull 1470.

Homo erectus. The fossil beds of the eastern shore of Lake Turkana have more recently yielded fossils that are closer to the modern human. The new species has been called *Homo erectus* and is considered to be an extinct member of our own genus. They are most interesting in that they appear to be of the same species as some of the first hominid fossils ever found—those once called "Java Ape Man" and "Peking Man" and designated *Pithecanthropus.* The earliest fossil *Homo erectus* skulls are known to be more than 1.5 million years old. But the Lake Turkana fossils were only about half a million years old, and some *Homo erectus* fossils may be even less than 200,000 years old. In other words, *Homo erectus* flourished, relatively unchanged, for well over a million years. During its first 300,000 years it coexisted with other, more primitive hominid species, including the australopithecines and possibly *Homo habilis.* In at least the last 100,000–200,000 years of its existence, *Homo erectus* overlapped with yet another upstart species, *Homo sapiens.* Fossils of *Homo erectus* have been found in China, Europe, southern Africa, and East Africa (Figure 27.28).

Homo erectus had a small brain, heavy brows, and strong jaws and teeth. But from the neck down, they apparently were very much like us. In addition to their fossilized bones, *Homo erectus* left behind crude stone tools. The tools, consisting of hand axes and scrapers, are designated *Lower Paleolithic* ("Old Stone Age"). There is evidence that these Lower Old Stone Age people built shelters, hunted small game, and gathered plant foods. The era of *Homo erectus* apparently ended just after the first glaciers of the Pleistocene receded.

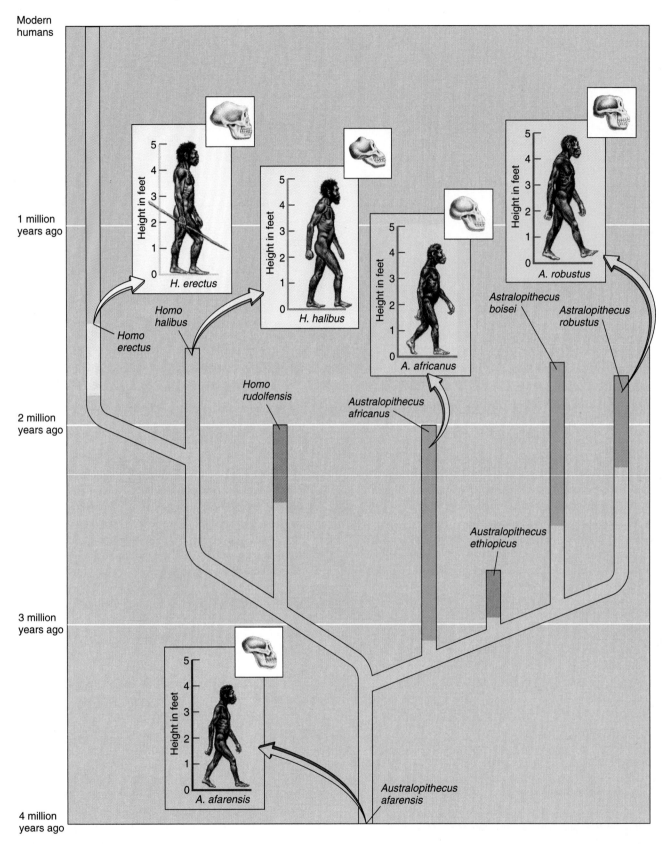

FIGURE 27.27
Hominid History
The known hominid history spans a period of nearly 4 million years. In a scheme based partly on the conclusions of Donald Johanson, a line represented by *Australopithecus afarensis* produced one branch containing other australopithecines (right) and another branch leading to *Homo (left)*.

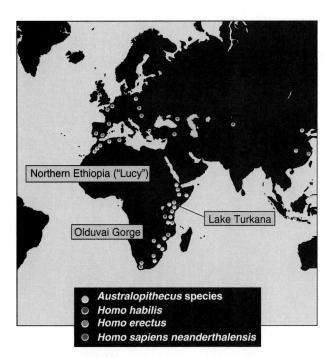

FIGURE 27.28
Hominid Distribution
The richest australopithecine finds are located in Ethiopia, Tanzania, and Kenya. *Homo erectus* was far-ranging, with fossils recovered in eastern and southern Africa, Europe, China, and Indonesia. Neanderthal fossils have been unearthed throughout much of Europe, the Mideast, North Africa, and China.

Recent findings at a well-known site at Zhoukoudian, China, reveal that one large cave was continuously inhabited by *H. erectus* for over 200,000 years and finally abandoned 230,000 years ago. During this incredibly long period of continuous residence, both the anatomy and the technology of *H. erectus* changed significantly. Skulls unearthed in the oldest debris of the cave revealed a cranial capacity of about 915 cc, while those in the most recent layers had reached 1140 cc. The stone tools had progressed from large, crude, hastily fashioned choppers and scrapers in the oldest, deepest layers, to smaller, much more refined tools in the newer surface layers. The cave inhabitants used fire from the start, hunted both large and small game, and ate a variety of nuts, fruits, seeds, and other plant matter.

Neanderthals and Us. Even while Darwin was studying and writing at his country estate, workers in a steep gorge in the Valley of Neander (in German, *Neanderthal,* or sometimes Neandertal) were pounding at something that turned out to be a skeleton, inadvertently smashing it to bits but leaving enough for researchers to see evidence of a new and different kind of human. (A similar skull had been unearthed at Gibraltar a few years earlier but had not created much of a stir.)

Scientists are rarely at a loss for words, and an explanation was immediately forthcoming. A professor Mayer of Bonn examined the heavy-browed skull and proclaimed that the skull and bone fragments belonged to a Mongolian cossack chasing Napoleon's retreating troops through Prussia in 1814; an advanced case of rickets had caused him great pain and his furrowed brow had produced the great ridges; he was so distraught he had crawled into a cave to rest but, alas, had died there. This fanciful scenario has been rejected in favor of the idea that the bones were those of a member of an early form of human that became extinct—the **Neanderthal**.

Members of *Homo sapiens neanderthalensis* thrived, or survived, until around 40,000 years ago. They left a rich record of their fossil remains and some indication of their tools and culture. Their geographic range was large, including Europe, Asia, and Africa (see Figure 27.28). Neanderthals had large brow ridges and sloping foreheads but were not as radically different from modern humans as was once believed. Their chins did not protrude as far as those of *Homo sapiens*, but their necks and bodies—when they didn't have rickets—were like ours and indeed like those of *Homo erectus* (Figure 27.29). There is no way to know whether they had a spoken language. However, we do know, from fossil pollens found in their remains, that Neanderthals sometimes covered their dead with flowers before burial. Whether or not that's sophisticated, it certainly is human—touchingly so.

Neanderthal-type fossils are not found in rocks younger than about 40,000 years. Why did the Neanderthals disappear? One school has proposed that the classic Neanderthal line and the line of modern humans diverged as long ago as 250,000 years, perhaps when early *Homo sapiens* was separated geographically by the great glaciers of the Pleistocene. About 40,000 years ago, the scarce *Homo sapiens sapiens*—the modern human—suddenly burgeoned, and Neanderthal became extinct, possibly wiped out by *Homo sapiens sapiens*, a subspecies famous for its eagerness to go to war.

A second theory minimizes the differences and states that *Homo sapiens* in the Neanderthal age was one highly variable species that included both the classic Neanderthal types and individuals much like ourselves. This theory places the Neanderthals squarely in the mainstream of human evolution. Modern humans, the theory maintains, simply evolved from somewhat archaic Neanderthal ancestors. Therefore the Neanderthals didn't die out; they merely changed. Those who believe this interpretation suggest that earlier anthropologists simply put too much emphasis on some extremes of normal variation. Much of the remainder of human evolution is documented and is called *history*.

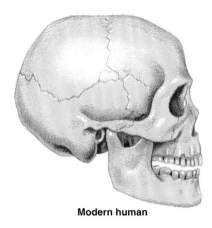

Modern human

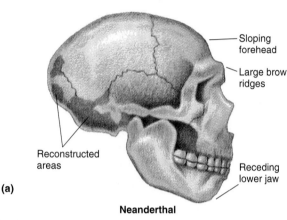

Sloping
forehead

Large brow
ridges

Reconstructed
areas

Receding
lower jaw

(a)

Neanderthal

(b)

FIGURE 27.29
Face-to-Face
A modern human and Neanderthal reveal many general similarities and some striking differences. **(a)**
Notable in Neanderthal is the slope of the cranial vault (forehead), the large brow ridges, and receding
lower jaw. The jawbone itself has less of an angle, where it curves upward to form its articulation with the
skull. **(b)** The facial reconstruction of the Neanderthal produces a decidedly modern human image.

Key Ideas

PHYLUM HEMICHORDATA

1. **Hemichordates** include the burrowing acorn worm, named for
   its conical proboscis protruding from a collar. While its gill slits
   suggest its relatedness to chordates, its larvae are quite similar
   to those of echinoderms.

PHYLUM CHORDATA

1. **Chordates** have pharyngeal gill slits; dorsal, hollow nerve cords;
   a notochord; and many have a postanal tail. In primitive chordates,
   the **pharyngeal gill slits** are used as sieves, while in fishes, the gill
   arches between them support the gills. The **dorsal, tubular nerve
   cord** in chordates contrasts to the ventral, solid nerve cord of
   many invertebrates. The **notochord**, a firm supporting rod, is seen
   mainly in the chordate embryo. The **postanal tail** is present in
   most adult chordates and in the embryos of the rest.

2. **Tunicates** are the best-known **urochordates**. While the swimming, tadpole-like, larval tunicate has all of the chordate traits,
   the only one remaining in the stationary adult is the gill sac.

3. **Cephalochordates** (lancelets) are fish-like in shape and exhibit
   all of the chordate characteristics. **Lancelets** use cilia for filter-
   feeding and have a complex mouth surrounded by tentacles.

SUBPHYLUM VERTEBRATA:
ANIMALS WITH BACKBONES

1. **Vertebrates** have a **vertebral column** that encloses a spinal
   cord, a cranial brain, a ventral heart that pumps blood entirely
   through blood vessels, gills or lungs, no more than two pairs of
   limbs, two eyes, two kidneys, and two separate sexes.

2. The oldest vertebrate fossils are **agnathans**. Their descendents are
   the jawless and boneless **cyclostomes**—the lampreys and hagfish.

3. The extinct **placoderms**, armored fishes with hinged jaws, arose
   during the Silurian, giving rise to the cartilaginous and bony fishes.

4. **Chondrichthyes** have cartilaginous skeletons, an asymmetrical
   tail, a dorsoventrally flattened body, and unique, placoid scales.
   Sharks have a highly developed **lateral line organ**, which senses
   water movement. Sharks utilize internal fertilization; some are
   egg-layers, others retain the embryos throughout development.

5. Most bony fishes of class **Osteichthyes** have an adjustable
   swim bladder for maintaining buoyancy at various depths. Gas
   exchange occurs in thin-walled gills. Blood flowing through
   fine vessels in the gills receives oxygen, which is then distributed to the body. Circulation involves a two-chamber heart. The
   lobe-finned fish line is ancestral to the terrestrial vertebrates.

6. **Amphibians** today occur in three orders: **Urodela** (salamanders), **Anura** (frogs and toads), and **Apoda** (ceacilians). Most amphibians have lungs but to some extent also use their skin for gas exchange. A circuit through the lungs is made possible by a third heart chamber. Reproduction in amphibians usually involves external fertilization and development in water although a few use internal fertilization.

7. Adaptations of reptiles for dry terrestrial life include internal fertilization and an amniotic egg named for the amnion, a membrane surrounding the embryo. The orders of reptiles are **Chelonia** (turtles and side-necked turtles), **Crocodilia** (crocodiles and alligators), **Sphenodonia** (the *Sphenodon* or tuatara), and **Squamata** (lizards and snakes).

8. The earliest known bird, *Archaeopteryx,* appeared in the early Jurassic. Reptilian traits include scales on the legs and the general reptilian skeletal framework. Flight modifications in modern birds include forelimbs modified into wing bones, a light, strong skeleton with fused units and reduced tail, a light beak (no teeth), enlarged breastbone with a keel, perching and grasping feet, and scales modified into feathers used for flight surfaces and insulation.

9. **Mammals** produce hair and give milk. Nearly all young develop in the uterus supported by the placenta. Breathing is assisted by a muscular diaphragm. The jaws are quite mobile and contain socketed teeth that vary widely with feeding and dietary habits.

10. Three groups of mammals include **monotremes** (egg-layers), **marsupials** (pouched mammals), and **placental mammals**. The first is minor today with only a few species, while the second, the marsupials, are concentrated in Australia. Most mammals are placentals.

11. Order **Primates** includes New World monkeys, Old World monkeys, and apes, plus humans. Primates are generalized mammals whose characteristics include long forelimbs and prehensile hands with fingernails. Some have prehensile tails. Vision is binocular and stereoscopic, and the brain is highly adapted to learning. Most are omnivores.

12. Human specializations include an opposable thumb, a naturally upright posture, an arched foot, and bipedal gait. Most striking is human intelligence, the capacity for abstract thought, learning, and a facility for language.

13. The known fossil hominid line begins with the australopithecines, *Australopithecus afarensis, A. robustus, A. africanus,* and *A. boisei.* Donald Johanson maintains that *A. afarensis* is ancestral to both the other australopithecenes and humans. *Homo habilis,* is tentatively placed near the start of the human line. Fossils of *Homo erectus* are found throughout much of the Old World, persisting for as long as 1.3 million years and overlapping *H. sapiens.* The earliest *H. sapiens,* the **Neanderthals**, ranged across the Old World as recently as 40,000 years ago. All fossils since the Neanderthal's demise are *Homo sapiens sapiens* (modern humans).

Application of Ideas

1. Compare a bony fish (a vertebrate) with the lancelet (a cephalochordate). In the comparison try to determine which characteristics in each represent primitive or advanced states.

2. The Cenozoic era is often referred to as the Age of Mammals. Did the mammals evolve in the Cenozoic? If not, why the label? When did mammals first appear, and what did some of the earliest representatives look like?

3. Suggest three anatomical differences that clearly distinguish humans from other primates. Are the human characteristics abrupt departures from general primate characteristics, or are they differences in degree? Cite several examples. Comment on how some of these differences may have been responsible for the uniqueness of human accomplishments.

Review Questions

1. List and briefly describe the three main chordate characteristics. (pages 565)

2. Why are the placoderms so significant in the history of vertebrates? Briefly explain how their major evolutionary innovation came about. (pages 568–569)

3. Compare the gill structure of bony fishes and sharks and explain how gills are ventilated in each. (pages 570–572)

4. In what way were the lobe-finned fishes significant to vertebrate evolution? (page 572)

5. Name the three groups of living amphibians and list common examples of each. (pages 572–573)

6. What separate circuit does the three-chamber amphibian heart provide that was not present in fishes? Would you expect pulmonary branches of the circulatory system to enter the skin? Why? (page 573)

7. List two examples of adaptations that permit amphibians to reproduce out of water. (page 573)

8. Trace the history of reptiles. When did they originate and in what geological period were they most prominent? Briefly, how is the demise of the ruling reptiles explained? (pages 573–575)

9. List five specific ways in which the bird's body is modified for flight. (pages 575, 578)

10. List the three subclasses of mammals and describe reproductive differences among them. (pages 578–579)

11. Trace the evolutionary history of the primates, mentioning the original type, and listing the main branches as they occurred. (pages 578–579)

12. Name the australopithecene species, briefly describe them, and suggest how they fit into the hominid evolutionary line. Where does *Homo habilis* fit in? In your opinion, is it properly named? (pages 584–585)

13. Compare the physical features of Neanderthal to modern humans. (pages 587–588)

Flowering Plant Reproduction

Plant Growth and Structure

Plant Transport Mechanism

Plant Regulation and Response

PART FIVE

Plant Functions

Flowering Plant Reproduction

C H A P T E R   O U T L I N E

SEXUAL REPRODUCTION

The Flower

Variation in Flowers

Essay 28.1 Diversity in Flowers

Flower Structure and Pollination

Sexual Activity in Flowers

SEED DEVELOPMENT

Development of the Embryo and Seed

Essay 28.2 Flowers to Fruits, or a Quince is a Pome

Seed Dormancy

Seed Dispersal

ASEXUAL REPRODUCTION

Vegetative Propagation at Work

Apomixis: Seeds Without Sex

KEY IDEAS

APPLICATION OF IDEAS

REVIEW QUESTIONS

$\mathcal{A}$s plants began to expand over the earth's often parching surface, they were met by a host of unexplored opportunities. At the same time, they encountered perhaps as many real threats to their very existence. Among those threats was the problem of reproducing in the dry air (see Chapter 25). As we will see here, that problem was largely solved by flowers.

In the flowering plants—the angiosperms—the role of sexual reproduction falls on the flower itself. The flower is, to many of us, the most notable part of an angiosperm, and we extol its beauty and fragrance. We must keep in mind, however, that they are not simply decorations; they have a job to do. Furthermore, we should keep in mind that, after fertilization, the flower's petals drop off, and part of the flower enlarges to become the seed-bearing fruit. The seed, we know, is the flowering plant's investment in the future. So let's begin our consideration of sexual reproduction in angiosperms with the flower.

Sexual Reproduction

The Flower

A typical flower is composed of as many as four **whorls** of parts around the **receptacle** or **base** (Figure 28.1). The term whorl means simply that the floral parts are repeated in concentric circles. Actually, each member of a whorl is a modified leaf, but some are much more modified than others. The outermost whorl consists of the **sepals**, which are usually green and leaf-like. They surround and protect the flower bud before it opens and are commonly photosynthetic. The whorl of sepals is collectively known as the **calyx** (meaning "cup"). Just within the calyx lie the **petals**; these are the second whorl of floral parts. They are often large and colorful, but most are still somewhat leaf-like. A whorl of petals is known as the **corolla** (meaning "garland").

The third whorl of floral parts are the **stamens**, which collectively make up the **androecium** ("house of man"). The stamens, highly modified leaves, can be considered the male floral parts. Each stamen usually includes a slender stalk, the **filament**, which leads to an enlarged bilobe structure, the **anther**. The two lobes contain four microsporangia, or **pollen sacs**, as they are also known. As we've seen, cells in the microsporangia produce **microspores**, which, in turn, develop into **pollen grains** (see Chapter 25).

The fourth region of the flower is the **gynoecium** ("house of woman"), which includes the **pistil**. The pistil, which often takes the shape of a bowling pin, can be considered the female floral part. Like stamens, pistils are also highly modified leaves. There are many variations in number and arrangement, depending on the species. In many species, the pistils are subdivided into two or more units called **carpels** that may be arranged in a whorl of separate individuals. In others, the carpels may be partially or completely fused into a single unit. The pistil consists of three parts: the **ovary**, the **style**, and the **stigma**. The ovary produces the ovules, which contain the megasporangia. Extending from the ovary is the style, a stalk that supports and elevates the stigma, an enlarged structure at its tip. The stigma is often hairy or sticky and specializes in receiving pollen.

In summary, the flower consists of a calyx (whorl of sepals), corolla (whorl of petals), androecium (whorl of stamens), and gynoecium (pistil). The last two whorls are the most intimately involved in reproduction. Sepals and petals are called **accessory floral parts**, since they play no direct role in sexual reproduction. However, they can be quite important to the process, particularly in insect-pollinated species.

Variation in Flowers

It seems that natural selection has molded, changed, hidden, and amplified flower parts in endless ways (Essay 28.1). For example, there tend to be clear differences between monocots and dicots (classes Monocotyledonae and Dicotyledonae; see Chapter 25). Most dicot floral parts are arranged in fours and

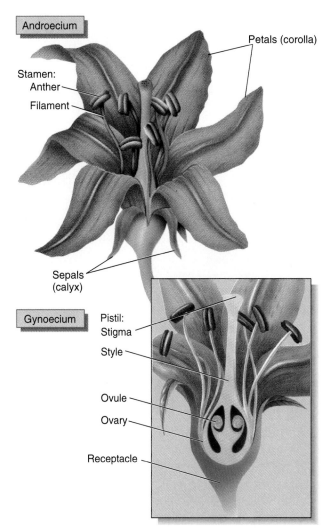

FIGURE 28.1
A Generalized Flower
Flowers generally consist of four major parts occurring in whorls and supported on a receptacle. They are the calyx (sepals), corolla (petals), androecium, which includes the stamens (filaments and anthers), and gynoecium, which is the pistil (ovary, style, and stigma).

fives or multiples of these numbers, whereas monocot flowers often have parts arranged in threes and multiples of three. The most primitive floral types, in evolutionary terms, have well-defined parts; the sepals, petals, pistils, and stamens occur in a **radially symmetrical** pattern—that is, circular and disk-like. A line drawn through the center of the disk produces equal halves. Such flowers act as platforms on which a pollinator may land.

Highly advanced flowers, in contrast, are often **bilaterally symmetrical**; they have right and left halves that are essentially mirror images. A few advanced flowers seem to have lost their symmetry altogether. Bilateral flowers sometimes fit the pollinator's body like a lock-and-key mechanism, and they tend to attract very specific animal pollinators—an example of coevolution.

Essay 28.1 ● **DIVERSITY IN FLOWERS**

Daffodils (*left*) and lilies (*right*) are monocots. Note the number of floral parts (six). Both are radially symmetrical and simple in design.

The San Diego hibiscus (*left*) and the red pimpernel (*right*) are both dicots, have floral parts in fives, are radially symmetrical, and follow simple designs.

Another dicot, the sunflower, a composite flower (actually a whole bouquet), occurs as a broad, disk-shaped head. Whereas most of its members are tiny disk flowers, large ray flowers form the periphery.

The snapdragon (*left*), a dicot, and lady slipper (*right*), a monocot, are bilaterally symmetrical (have right and left sides) and have very complex flowers. Both represent a clear departure from the more primitive, simple radial flowers.

Other variations include flowers with missing parts. A flower with one or more basic parts, that is, sepals, petals, stamens, or pistils, absent is called **incomplete**, as opposed to the **complete** flower, which has all of the usual parts. If either (or both) of the sexual parts, the pistil or stamens, is absent, the flower is referred to as **imperfect**. If they are both present, the flower is **perfect**. Obviously, flowers that are imperfect must also be incomplete, and flowers that are complete must be perfect. Cornflowers are imperfect and are either **pistillate flowers** (female: lacking stamens) or **staminate flowers** (male: lacking pistils), but both occur on the same plant, which tells us the corn plant is actually monoecious ("one house," both sexes in one plant). Date palms are dioecious ("two houses," male and female plants), so pistillate and staminate flowers grow on separate trees. To provide for pollination a date-grower usually plants one staminate tree per ten or so pistillate trees. (To avoid confusion, think of the "perfect" and "imperfect" as referring to flowers and "monoecious" and "dioecious" as referring to the entire plant.)

In some instances, flowers are grouped together into clusters, or **inflorescences**, which take various forms such as the heads of daisies, marigolds, and sunflowers. Thus a single sunflower is really a bouquet of flowers. **Composite flowers** are inflorescences composed of tiny, individual flowers, each of which produces a single seed-bearing fruit. In the daisy, marigold, and sunflower heads, the outer rows of **ray flowers** are usually pistillate, or sometimes sexless flowers, their large petals serving to attract pollinators to the central **disk flowers.**

Flower Structure and Pollination

We all know that flowers have distinctive shapes, coloring, and fragrance. Some also produce a sugary fluid called **nectar**. Color, shape, fragrance, and nectar all have a role in flowers that depend on animals such as insects, bats, and birds for transferring pollen from one flower to another. The color, shape, and fragrance attract pollinators to the flowers with the promise of sweet, energy-rich nectar (usually secreted from *nectaries* near the base of the floral parts). It is interesting that we share the insect's appreciation for flowers, because insects and people sense quite different things according to their different sensory abilities. For example, some insects see ultraviolet light, so they may find purple patterns in flowers that to us appear white. And since most insects can't see red at all, a beautiful red rose appears to them as black. As certain flowers evolved, it must have become adaptive for them to increasingly specialize according to the sensory abilities of various kinds of pollinators. Flower structure, appearance, and scent were modified to attract and assist specific kinds of pollinators, whether bees, beetles, flies, moths, butterflies, bats, or others (Table 28.1).

Animal Pollinators and Flower Specializations. From an evolutionary viewpoint then, insects and other animal pollinators may "specialize"; that is, they may be attracted to only one kind of flower. A pollinator that has been rewarded with nectar even once may become "keyed" to that particular type of flower, ignoring all others, and will transfer pollen only between flowers of the same species, or even of the same strain. Natural selection may thus tend to favor plants with the most distinctive and easily identified flowers. A new strain with a slightly modified flower can therefore become genetically isolated if it is not recognized by the insect that pollinates the parental strain. In this way, the flower might be directed along a new evolutionary pathway without ever having undergone the usual geographical isolation from its parents as speciation occurred (see Chapter 20). Such evolutionary separation would lead to specialization, of course, as each strain came to adapt more precisely to the most efficient pollinator available.

Actually, there are many examples of insect specialization that have led to angiosperm diversity. You may recall from Chapter 20 our discussion of coevolution between the *Yucca,* a desert plant, and a specific insect, the yucca moth, to accomplish pollination; but the interaction goes far beyond that (see Figure 20.19). When in bloom, the *Yucca* bears large clusters of white flowers. As the yucca moth pays its nightly visits to yucca flowers, it gathers pollen and rolls it into small balls. At succes-

TABLE 28.1
Animal Pollinators and Floral Adaptations

Animal vector	Visual cues (human perception)	Chemical cues (human perception)
Bees	Bright colors, UV color guides. Highly divided floral parts with uneven outline.	Sweet odors
Beetles	Not significant, flowers dull colored or white	Strong odors: fruity, spicy, or foul
Flies	Large flowers, dull or flesh colored	Musk to rotting odors
Butterflies	Bright colors: red, orange, yellow, blue	Strong, sweet odors
Birds	Bright flowers: red and yellows	Sugary nectar, odors not significant
Bats	Large, dull colored flowers, color not significant (night flyers)	Sugary nectar, fruity and fermenting odors

Skunk cabbage

FIGURE 28.2
A Foul-Smelling Flower
Lysichiton americanum (western skunk cabbage) gives off a
pungent "skunk" odor. It is pollinated by flying beetles.

sive stops, it pierces the ovary using its sharp ovipositor (egg-
burying structure) and injects a number of eggs among the
young ovules. It then places a ball of sticky pollen into the
opening (perhaps a form of "sealing behavior," since many
insects entomb their eggs). Upon hatching, the larvae eat their
way through the pistils and reproductive tissue, but they emerge
before they destroy too much. The remaining undamaged
ovules form seeds and drop to the ground about the same time
the moth larvae become pupae. Both moth and plant have
evolved to a state where they are entirely dependent on each
other. The moth cannot survive without the plant, and the plant
is not naturally pollinated by any other means.

Another example of specialization is seen in western
skunk cabbage (Figure 28.2). This marsh plant produces a
flesh-colored structure, which bears flowers within. It has a
pungent, unpleasant odor, like that of rotting flesh. Honeybees
avoid this flower, but flies swarm over it. Flies, in fact, lay

their eggs in the flower and in so doing pollinate it. Some
large, drab, nectar-rich, night-opening flowers are pollinated
by bats, which are attracted to the strong fruit-like or fermenting odor. The coevolving relationship has gone even further.
In one species, the pollen has a much higher than usual protein content, which, as it turns out, is an important dietary
supplement to the bat pollinator of that species.

Two examples of adaptations to insect pollinators involve
nectar guides and mimicry. Nectar guides are lines of color
that guide the insect into the flower. Some nectar guides are
invisible to humans but clearly visible to bees who have
broader-ranged vision (Figure 28.3). In mimicry, which some
might consider the "ultimate effort," the flower has taken the
form of a female insect. When the male attempts to mate with
the flower, it doesn't accomplish much, but the plant gets
pollen transferred to its pistil (Figure 28.4).

Not all flowering plants are such specialists. Plants that
are generalists take a different adaptive route and depend on a
variety of animal species. They thus develop traits that are
attractive to a number of pollinators. For example, many
insect pollinators are attracted to flowers with distinctly separated petals as well as aromatic secretions. The adaptations
for attracting a variety of pollinators are, of course, tempered
by other factors such as environmental influences.

Wind Pollination. There is a very large group of flowering
plants that are wind-pollinated. Among the wind-pollinated
monocots are grasses such as bluegrass, wheat, rye, and corn.
Wind-pollinated dicots include most species of temperate-zone
trees, for example, oak, birch, poplar, and alder. Of course, the
conifers described in Chapter 25 are largely wind-pollinated.

It may seem strange that grasses produce flowers, but
remember, they are angiosperms. Grasses have stamens and
pistils, but they are plain, small, and inconspicuous (Figure
28.5). There would be little adaptive value in such energy-
costly amenities as colorful petals and sweet nectaries in
wind-pollinated flowers. Instead, these flowers produce copious amounts of very light, easily dispersed pollen and elaborate stigmas for trapping it. Despite this, wind pollination

(a) What we see

(b) What bees see

FIGURE 28.3
Insect Pollination Devices
Some plants evolved elaborate devices for
ensuring pollination. *Wedelia* has nectar
guides, pigment lines, that are easily visible to
insects and guide them into the nectaries. In
some cases, the guides are not visible to us
(a) but show up under ultraviolet light, which
is visible to bees **(b)**.

FIGURE 28.4
Insect Imitator
(**a**) One orchid, *Oncidium,* which has specialized anthers with very adhesive surfaces, mimics male bees, and transfer of the pollinium (pollen sac) occurs when an aggressive, territorial male bee is deceived into picking a fight with it. (**b**) Flowers of the orchid *Ophrys speculum* mimic female wasps. Male wasps try to copulate with them but manage only to pollinate the orchid.

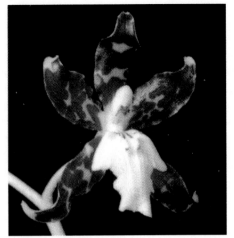

(**a**) *Oncidium* mimics male bee

(**b**) *Ophrys speculum* mimics female wasp

seems intuitively to be inefficient. It's true that much of the pollen is lost; however, recent studies involving the use of wind tunnels with gymnosperm cones and grass flowers indicate that wind pollination may be more efficient than we think. Pine cones and grass inflorescences tend to slow and trap wind-borne pollen, just like snow fences stop snow from blowing across highways. Some trees improve the odds by producing their flowers and releasing their pollen early, before their leaves and those of other trees grow and block the wind. As a final point, wind pollination is most efficient where members of a species occur in dense clusters, as is common in grasses and conifers.

Sexual Activity in Flowers

Earlier we noted that a few of the primitive vascular plants (for example, *Selaginella;* see Chapter 25) and all seed plants are **heterosporous**; that is, they produce two kinds of spores—megaspores and microspores. We also saw that as plants evolved, the gametophyte phase of the life cycle became more and more reduced and dependent on the sporophyte. In the flowering plants, the female gametophyte and at least the younger phase of the male gametophyte are tucked away in the ovaries and anthers of the flower. With these points in mind, let's look into the process of sporogenesis, the formation of spores, and their development into the gametophyte in flowering plants.

Events in the Ovary. You will want to study Figure 28.6 carefully as we proceed with development of the female gametophyte. We'll begin in the soft tissues of the flower's ovary, where the young **ovules** have formed. The ovules originate from a surrounding tissue known as the **placenta** and for a time remain attached to it by a stalk-like **funiculus.** Each ovule consists of a **megasporangium**, or **nucellus**, as it is known in flowering plants, and two **integuments**—skin-like protective coverings that will much later form the seed coat.

At the exposed end of the integument is a small opening called the **micropyle**. But most importantly, each ovule contains one large cell called the **megasporocyte**, and it is this cell that undergoes **megasporogenesis**, which we will now describe.

The megasporocyte, like its male counterpart in the stamen, the **microsporocyte**, will undergo meiosis, producing haploid spores. Typical of plants, these meiotic products are not gametes, but instead they will enter mitosis to produce the

FIGURE 28.5
Wind-Pollinated Flowers
Wind-pollinated flowers lack color and fragrance. Also absent are sepals and petals. They often have feathery, plume-like, pollen-catching stigmas, stamens usually in threes, and pollen that does not tend to stick together as it does in insect-pollinated species. Wind pollination is efficient where many plants of the same species are clustered.

gametophyte, which, in turn, will produce the gametes (see Figure 28.6, left).

Meiosis in the megasporocyte involves the usual two divisions, so at the end of the process there are four haploid spores—the megaspores. Usually, three of the megaspores simply disintegrate. The surviving megaspore enlarges considerably before its haploid nucleus undergoes three successive mitotic divisions, giving rise to the gametophyte or, more accurately, the **megagametophyte** (the female gametophyte generation of the flowering plant). After three mitotic divisions, the megagametophyte will contain eight identical haploid nuclei, all in one greatly enlarged cell (one divides into two, two into four, and four into eight). With the mitotic events completed, the megagametophyte reaches an eight-nucleate stage. The events we are describing here may vary according to the species, so this is only a generalized account.

In most cases, the eight nuclei are equally distributed to the two ends of the developing megagametophyte, after which two of the nuclei, one from each end, migrate to the center of the cell. Finally, new cell walls are laid down, and the cytoplasm is divided into seven separate, unequal cells. The result of all this is that one cell, often the largest one, ends up in the center with two nuclei (see Figure 28.6, left). The binucleate cell will later give rise to the **endosperm** (a food storage tissue in the seed), but for now it is known as the **central cell**. Its nuclei are called **polar nuclei**. Of the remaining six cells, the one nearest the micropyle (site of sperm entry) will probably become the **egg cell**. The two cells that flank it are called **synergids**. Synergids may be the last remnant of the ancient archegonium, so prominent in the seedless plants and gymnosperms. The three cells at the opposite end are called **antipodal cells**. The completed megagametophyte is now called an **embryo sac**. (We should point out once again that the flowering plants are diverse, and whereas the eight-nucleate embryo sac is commonplace, it is not universal.) There will be no more nuclear divisions in the embryo sac until after fertilization.

Events in the Anther. While the embryo sac has been undergoing its development, similar but less complex events have been occurring in the anthers, the sites of **microsporogenesis** (see Figure 28.6, right). Typically, each anther contains four pollen sacs—chambers that contain numerous diploid microsporocytes, each of which will enter meiosis. By the end of meiosis, each microsporocyte will have produced four microspores. The microspores will then begin their development into male gametophytes (microgametophytes), which will then produce sperm cells.

The haploid nucleus within each microspore will undergo mitosis, and when the two resulting cells differentiate, they become a **generative cell** and a **tube cell**, both of which remain within the microspore. With mitosis, the male gametophyte's (microgametophyte's) formation is underway. Meanwhile, each emerging microgametophyte develops a resistant coating and becomes a pollen grain. In some species, the generative cell will soon undergo mitosis again, yielding two sperm cells, thus completing the male gametophyte's formation while in the pollen grain. In others, this event occurs later, after pollination.

Pollination and Double Fertilization. Technically, **pollination** occurs when pollen is deposited on the stigma. Fertilization occurs when sperm and egg fuse. The source of the pollen may be the male parts of the same flower, other flowers on the same plant, or flowers of another plant. We are aware of the dangers of inbreeding, and since self-fertilization is obviously inbreeding of the severest sort, it is not surprising to learn that some plants have ways of avoiding self-pollination, such as the position of their anthers, the timing of microspore and megaspore production, or even physiological incompatibilities. Nonetheless, strangely enough, self-pollination is a regular event in some species.

Following pollination, the pollen grains germinate (Figure 28.7), forming **pollen tubes**. As a pollen tube lengthens, it progresses through the style toward the ovary. The tube cell makes this growth possible through the production of proteolytic enzymes that literally digest the tissue ahead of the lengthening tube. It is about this time in many species that mitosis occurs in the generative cell and two sperm cells emerge. This completes development of the male gametophyte.

Finally, the tube penetrates the ovule at the tiny, pore-like micropyle. The two sperm cells are then released into the embryo sac, where that uniquely angiosperm event called **double fertilization** occurs (see Figure 28.7). One sperm fertilizes the egg cell, forming the diploid zygote. The zygote will develop into the plant embryo. The second sperm penetrates the central cell, where its nucleus joins the two polar nuclei already present, producing a triploid **primary endosperm nucleus**. (While a triploid primary endosperm nucleus is common in many species, it may in others be diploid or even pentaploid, depending on the number of polar nuclei.) The endosperm provides food for early development after seed germination occurs.

After fertilization, the flower begins to change, its petals and other nonessential parts withering and falling away. Then the ovary itself will swell, sometimes to enormous propor-

FIGURE 28.6
Megasporogenesis and Microsporogenesis

Both processes are shown beginning at the top of the diagram. Megasporogenesis proceeds to the left; microsporogenesis to the right. [1a] Megasporogenesis begins in the immature ovule (see inset) in which resides the megasporocyte, surrounded by the nucellus. Meiosis in the megasporocyte gives rise to four haploid cells, one of which enters three rounds of mitosis to form the eight-nucleate megagametophyte [2a]. As cell walls are laid down, a seven-cell embryo sac emerges. One cell will be the egg, another will be the large central cell with two polar nuclei. [1b] Microsporogenesis proceeds as microsporocytes enter meiosis. Each of the four haploid products undergoes a round of mitosis, forming a generative cell and a tube cell [2b]. Coats form around the cells, and a pollen grain emerges.

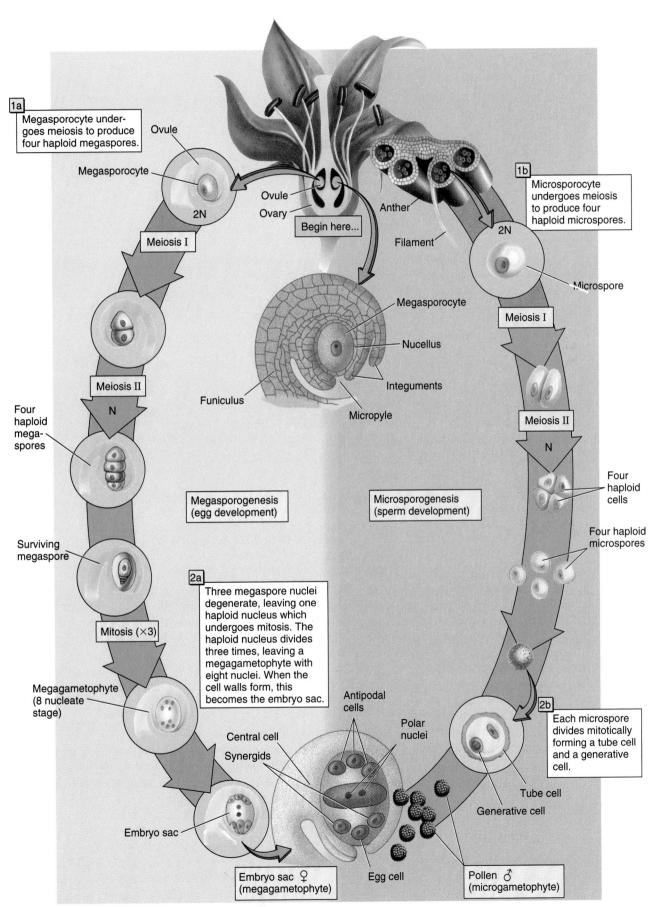

1a Megasporocyte undergoes meiosis to produce four haploid megaspores.

Ovule

Megasporocyte

2N

Meiosis I

Meiosis II

N

Four haploid megaspores

Surviving megaspore

Mitosis (×3)

Megagametophyte (8 nucleate stage)

Embryo sac

Megasporogenesis (egg development)

2a Three megaspore nuclei degenerate, leaving one haploid nucleus which undergoes mitosis. The haploid nucleus divides three times, leaving a megagametophyte with eight nuclei. When the cell walls form, this becomes the embryo sac.

Ovule
Ovary

Begin here...

Anther

Filament

Megasporocyte

Nucellus

Integuments

Funiculus

Micropyle

1b Microsporocyte undergoes meiosis to produce four haploid microspores.

2N

Microspore

Meiosis I

Meiosis II

N

Four haploid cells

Four haploid microspores

Microsporogenesis (sperm development)

2b Each microspore divides mitotically forming a tube cell and a generative cell.

Tube cell

Generative cell

Antipodal cells

Polar nuclei

Central cell

Synergids

Embryo sac ♀ (megagametophyte)

Egg cell

Pollen ♂ (microgametophyte)

FIGURE 28.6

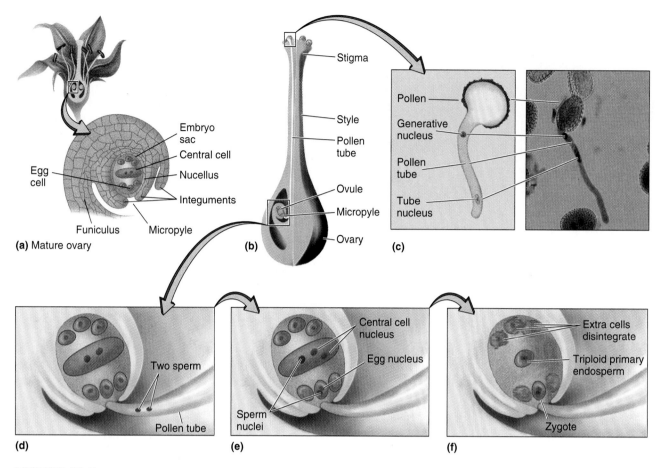

FIGURE 28.7
Pollination and Fertilization
(a) Among the cells of the mature ovule is the egg cell and the central cell. (b) Pollination is followed by pollen tube growth. (c) The pollen tube contains the generative nucleus and tube nucleus, the latter controlling growth. (d) As the pollen tube penetrates the style and ovary, the generative cell undergoes mitosis, producing two sperm cells. (e) The sperm enter the ovule, where (f) one fertilizes the egg cell and the other the central cell. Note the triploid nucleus in the newly formed primary endosperm.

tions. This begins the development of **fruit**, another structure characteristic of angiosperms. In everyday usage, "fruit" usually refers to a sweet and juicy structure such as the familiar apple, grape, or banana. Botanically, however, the fruit is the ripened or mature ovary. Whereas it does commonly consist of a sweet fleshy structure, it may manifest itself as the familiar string bean, cereal grain, pumpkin, or tomato. It even includes certain hardened parts of coconut, walnut, and what we commonly call a sunflower "seed." For a more complete look at fruits and their development, see Essay 28.2.

Seed Development

Development of the Embryo and Seed

After fertilization, the new diploid zygote, the primary endosperm nucleus, and all the associated maternal tissues begin mitosis and the differentiation that will produce the

plant **embryo** and its life support systems. The finished product—the seed—will consist of (1) an embryo, (2) some kind of food supply, and (3) hardened, protective **seed coats** or their equivalent. Seed development varies widely among flowering plants, so we will follow the process in selected representatives from the dicots and monocots.

Development in the Bean: A Dicot. The most familiar dicot seeds are probably foods such as peas, beans, peanuts, and sunflower seeds—those eaten without much alteration (except, perhaps, for getting rid of the thin, papery peanut seed coat). The embryo, a miniature sporophyte with two large cotyledons, takes up most of the dicot seed. Consider the peanut again. The shell in this case is the dry fruit wall you peel away. Inside are two seeds (usually), each of which is covered by a papery thin, reddish seed coat. If you peel off the seed coat, you can easily separate the two fleshy cotyledons. Look closely and you can see the embryo, a tiny, cylindrical object still attached to one of its cotyledons. With this ele-

FIGURE 28.8
Seed and Embryo Development in a Dicot
After fertilization, the embryo sac contains a diploid cell, the zygote, and a triploid primary endosperm nucleus. As seed development nears completion, the two cotyledons take up much of the embryo. Below is the embryo proper, with an upper epicotyl and lower hypocotyl. The first will produce the shoot and leaves, whereas the second produces the root.

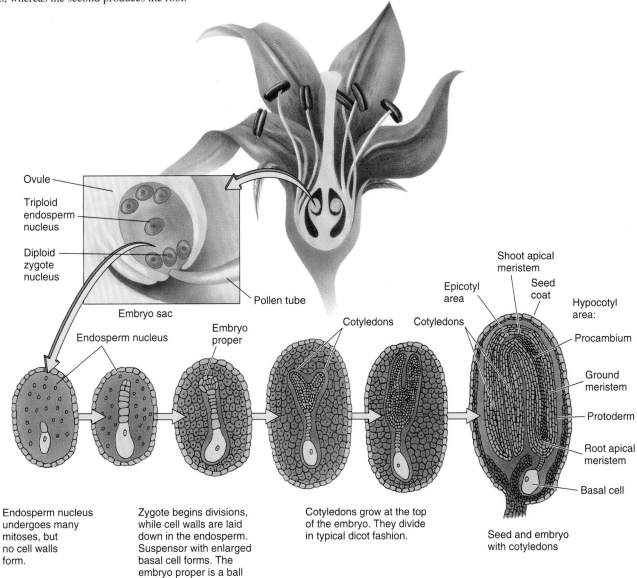

Endosperm nucleus undergoes many mitoses, but no cell walls form.

Zygote begins divisions, while cell walls are laid down in the endosperm. Suspensor with enlarged basal cell forms. The embryo proper is a ball of cells at the top.

Cotyledons grow at the top of the embryo. They divide in typical dicot fashion.

Seed and embryo with cotyledons

mentary lesson in mind, we'll look into the events that lead up to seed formation in the bean. You will want to refer to Figure 28.8 as we proceed.

After fertilization, the ovule contains an embryo sac with a diploid zygote, a triploid primary endosperm nucleus, and several layers of surrounding and supporting diploid cells of the maternal tissue. For convenience, what follows can be divided into three events: (1) developments in the endosperm, (2) growth of the cotyledons, and (3) growth of the rest of the embryo, or embryo proper.

Within the primary endosperm, the nuclei divide continuously until a multinucleate mass surrounds the zygote (see Fig-

ure 28.8). (This stage is carried to an extreme in the coconut since "coconut milk" is nothing but a fluid endosperm cytoplasm with free nuclei floating in it.) Ultimately, in most dicot seeds, cell walls are laid down around the endosperm nuclei.

The Cotyledons. At this time, we see a fundamental difference between the monocot and dicot embryos. In the rapidly growing dicots, the cells of the embryo form a two-pronged, heart-shaped structure. In monocots, cell division produces a simple, elongated structure. The two lobes of the heart will produce the two cotyledons of dicot plants; the single growth will form the single cotyledon of monocots.

Essay 28.2 • **FLOWERS TO FRUITS, OR A QUINCE IS A POME**

A fruit is essentially a ripened ovary (sometimes including other floral parts: see the apple). There are three basic types of fruits: simple, aggregate, and multiple, depending on the number of ovaries in the flower or the number of flowers in the fruiting structure.

Simple fruits may be derived from a single ovary or, more commonly, from the compound ovary of a single flower. They can be divided into two groups according to their consistency at maturity: simple fleshy fruits and simple dry fruits.

Simple fleshy fruits include the **berry**, **pome**, and **drupe**. The berry has one or several united fleshy carpels, each with many seeds. Thus the tomato (a) is a berry, and each of the seed-filled cavities is derived from a carpel. Watermelons, cucumbers, and grapefruits are also berries (but, technically, blackberries, raspberries, and strawberries are *not* berries—Whew!). Pome (b) means "apple," and the group includes apples, pears, and quinces. In the pome, only the inner chambers (roughly, the "core") are derived from the ovary, and most of the flesh comes from accessory tissue (most of the flesh comes from the receptacle, which grows up over the ovary). A drupe—what a wonderful word—is also derived from one to several carpels, but only a single seed in each develops to maturity. The ripened ovary consists of an outer fleshy part and a hard, inner stone, containing the single seed. Peaches and cherries are drupes, as are coconuts.

There are many kinds of simple dry fruits, but they are neatly categorized as follows: (1) those with many seeds, which split open and release their seeds, and (2) those with few seeds, which do not split open or release seeds. The first group is called **dehiscent** (c), from the verb dehisce, "to split" or "to open," and includes poppies, peas, beans, milkweed, snapdragons, and mustard. Dehiscent fruits split at maturity. The second group is called **indehiscent** (d) (nonsplit-

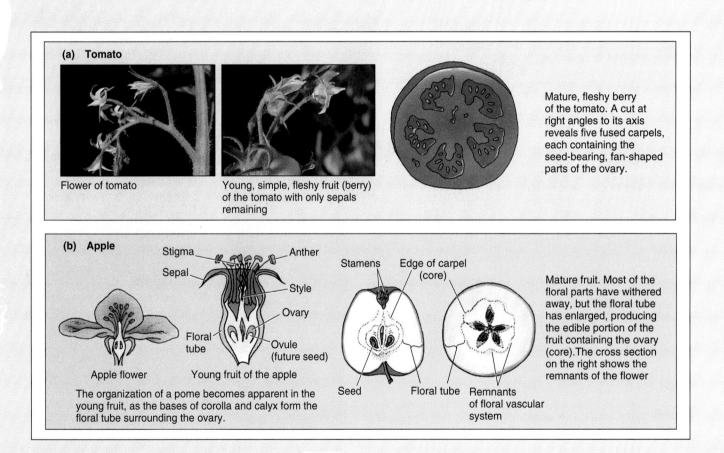

(a) Tomato

Flower of tomato

Young, simple, fleshy fruit (berry) of the tomato with only sepals remaining

Mature, fleshy berry of the tomato. A cut at right angles to its axis reveals five fused carpels, each containing the seed-bearing, fan-shaped parts of the ovary.

(b) Apple

Stigma
Sepal
Anther
Style
Ovary
Floral tube
Ovule (future seed)

Apple flower

Young fruit of the apple

The organization of a pome becomes apparent in the young fruit, as the bases of corolla and calyx form the floral tube surrounding the ovary.

Stamens
Edge of carpel (core)
Seed
Floral tube
Remnants of floral vascular system

Mature fruit. Most of the floral parts have withered away, but the floral tube has enlarged, producing the edible portion of the fruit containing the ovary (core). The cross section on the right shows the remnants of the flower

The bean cotyledons continue to grow, absorbing all or nearly all of the endosperm and filling most of the space of the growing embryo sac. As they grow, the cotyledons form a U-turn, growing back toward the base of the seed (see Figure 28.8). We should add that not all dicot cotyledons are so fleshy. Those of the castor bean, for instance, are thin and more leaf-like, surrounded by a large amount of endosperm that is not incorporated into cells.

The Embryo Proper. As the cotyledons form, the remainder of the embryo begins to take shape. The zygote beings a series of cell divisions that, at first, produces a simple row of cells.

ting). Its members include sunflowers, dandelions, maples, ash, corn, and wheat. Indehiscent fruits do not split at maturity.

Aggregate fruits (e) are derived from numerous separate carpels of a single flower. In other words, they are many simple fruits clumped together on a common base. Blackberries, raspberries, and strawberries are aggregate fruits.

Multiple fruits (f) are formed when the single ovaries of many flowers join together, as seen in the mulberry, fig, and pineapple. The pineapple starts out as a cluster of separate flowers on a single stalk, but as the ovaries enlarge, they coalesce to form the giant multiple fruit (the commercial variety, the kind we most commonly see, is a seedless hybrid).

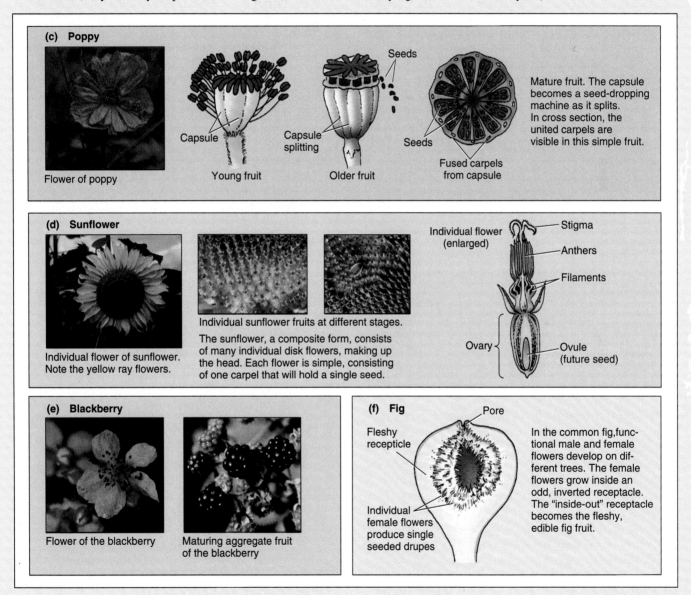

(c) Poppy

Flower of poppy

Young fruit — Capsule

Older fruit — Capsule splitting — Seeds

Mature fruit. — Seeds — Fused carpels from capsule

Mature fruit. The capsule becomes a seed-dropping machine as it splits. In cross section, the united carpels are visible in this simple fruit.

(d) Sunflower

Individual flower of sunflower. Note the yellow ray flowers.

Individual sunflower fruits at different stages.

The sunflower, a composite form, consists of many individual disk flowers, making up the head. Each flower is simple, consisting of one carpel that will hold a single seed.

Individual flower (enlarged) — Stigma — Anthers — Filaments

Ovary — Ovule (future seed)

(e) Blackberry

Flower of the blackberry

Maturing aggregate fruit of the blackberry

(f) Fig

Fleshy recepticle — Pore

Individual female flowers produce single seeded drupes

In the common fig, functional male and female flowers develop on different trees. The female flowers grow inside an odd, inverted receptacle. The "inside-out" receptacle becomes the fleshy, edible fig fruit.

Then, at the inner end of the row, cell divisions speed up and, more important, begin to occur in many planes, producing a three-dimensional ball of cells that constitutes the embryo proper (see Figure 28.8). Unlike the cells in animal embryos, which lack walls and grow between, around, and over each other as the embryo takes form, plant embryo cells must do this chiefly through changes in division planes. Cells on the underside of the ball divide more slowly, forming a **suspensor**, a single file of cells that extend down to a much larger **basal cell** below. The suspensor and basal cell anchor the embryo in place and aid in the transfer of food to its rapidly growing cells

The Finished Embryo. Soon the embryo comes to fill the surrounding space, its cotyledons to one side and the rest on the other (see Figure 28.8). At its upper portion, where the cotyledons divide, a naked dome of tissue, the **shoot apical meristem**, forms. It will give rise to the future shoot and leaves. **Meristematic tissue** in plants is special in that it is made up of simple, undifferentiated cells that are generally heavily involved in cell division and can be recognized in microscope sections as the cells where different stages of mitosis are seen. Meristems are the source of cells for the ongoing formation of new tissues in the embryo and in the adult plant. The shoot apical meristem, and anything it produces in the embryo, is called the **epicotyl** ["above the cotyledon(s)"]. In some embryos the epicotyl is no more than simple meristem, but in others, the bean embryo included, the epicotyl includes a **plumule**, tiny embryonic leaves. In the emerging seedling, the plumule is regarded as the first leaf. That region of the embryo axis below the cotyledons is logically called the **hypocotyl** ["below the cotyledon(s)"]. At the lower portion of the hypocotyl is another region of meristematic tissue, the **root apical meristem**, which may be covered by a cap of cells. This meristem will provide an ongoing supply of cells as new root tissue is formed in the seedling and adult plant. In some embryos, root development begins early and is clearly visible. The embryonic root is called a **radicle**.

At this period in the development of the plant embryo, cells arising from the apical meristems begin their differentiation into the plant's three primary tissue systems: **protoderm**, **ground meristem**, and **procambium**. The protoderm will provide surface tissues, while the procambium provides vascular tissue. The remaining tissues arise from ground meristem. We will return to the primary tissues, and the epicotyl and hypocotyl, in the next chapter, where germination and early growth are introduced.

Comparing Development in the Bean Seed and Corn Grain. Corn and bean embryos are compared in Figure 28.9. It's important to note that the bean seed and corn grain are not equivalent structures. The corn grain is actually a fruit, so the comparative bean structure would be the bean pod with its seeds inside. Another pronounced difference is in the cotyledons, which in the bean seed, having greatly enlarged late in development, take up most of the space. The bean embryo, now comparatively tiny, is located toward one end of the seed, between the cotyledons. Furthermore the single cotyledon of corn and other monocots takes on a cylindrical shape, rather than a heart shape, and becomes quite large. The cotyledon here is called the **scutellum** (see Figure 28.9). Unlike the bean cotyledons, which function in food storage, digestion, and absorption, the scutellum of corn carries out digestion and absorption only. The food lies adjacent to, but outside, the scutellum. Food storage occurs primarily as undifferentiated starchy endosperm, which takes up much of the space in the kernel.

The corn embryo undergoes a considerable degree of differentiation during its formation, with the epicotyl soon taking the form of a well-developed multilayered plumule, and the hypocotyl giving rise to a dominating, well-formed radicle. It seems that, in corn and other grains, leaf and root development get a head start (see Figure 28.9). Another striking feature is the presence of two protective sheaths, the **coleoptile**, which surrounds the plumule, and the **coleorhiza**, which surrounds the young root. You've seen the fine, green, coleoptile-covered shoots if you have watched the emergence of corn seedlings in your garden or grass seedlings in a newly seeded lawn. The plumule or young leaf soon breaks through the coleoptile and begins to unfurl into the sunlight. The corn root is very fast growing and likewise soon breaks through the coleorhiza.

The large corn endosperm is surrounded by an outer, protein-filled layer of cells, called the **aleurone**. The covering of the corn grain or kernel itself is the **pericarp**, which is actually part of the ovary. A similar organization is seen in wheat. What you have heard referred to as "wheat germ" is the wheat embryo, including its scutellum. The wheat germ, along with the pericarp and aleurone, is commonly removed before milling, and only the starchy endosperm becomes flour. Wheat germ is highly nutritious, but it also spoils easily, which is one reason why it is removed. The pericarp and aleurone become what are called "bran." Bran is primarily cellulose, which cannot be digested, but as we are now aware, it is a good source of dietary fiber. Because people have recently become more aware of the need for good nutrition, both wheat germ and bran are now frequently added back into foods. They are also used in making animal feed.

Seed Dormancy

Following development, most seeds and grains enter a dormant period. They lose most of their water and enter a state of greatly reduced metabolism that lasts until germination occurs. They do not lie unprotected, however, because their outer coverings (seed coats and pericarps) have hardened. While in their dehydrated, hardened state, seeds and grains resist the growth of bacteria and fungi, and, as we will see in the next section, they travel well.

The dormant period varies greatly, and although the seeds of the majority of species are viable for just a few years (some much less), those of some species remain viable for much longer. Lotus seeds found in peat deposits near Tokyo germinated after 2000 years of dormancy. The record, though, is held by a delicate flower of the Yukon, *Lupinus arcticus.* Some of its seeds grew into fine seedlings after having lain dormant in frozen soil for over 10,000 years.

Seed Dispersal

In order for seeds to succeed, they must begin their growth under suitable conditions—generally away from the competition of parents and other seeds. Thus seed dispersal is a particularly critical factor in the life cycles of plants, and they have intricate means for accomplishing this end.

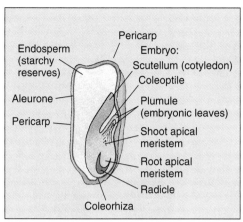

(a) Corn kernel

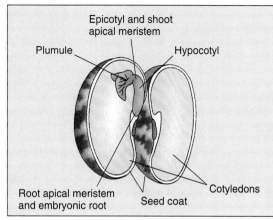

(b) Bean seed

FIGURE 28.9
Monocot and Dicot Seeds
(a) In the corn kernel (a fruit), food reserves occur in the form of an unorganized starchy endosperm. During germination and early growth, the lengthy cotyledon, or scutellum, will act as a food-absorbing structure, taking in digested nutrients from the endosperm. **(b)** In the bean seed, food is stored in cells of the paired cotyledons, which occupy much of the seed. Upon germination, the bean cotyledons, which emerge from the soil, develop chlorophyll and for a time carry on photosynthesis. More detail is seen in Figure 28.8.

Annual plants, those that live for one growing season, have a brief time to produce and disperse their seeds. Perennials, plants that live on for many seasons, particularly the large trees, have a different kind of problem. Undispersed seeds that germinate beneath their parent's boughs must compete with it for water, mineral resources, and, especially, sunlight.

But, back to the questions of dispersal. How do seeds bearing the young embryos get around? The primary natural seed carriers are water, wind, and animals (Figure 28.10). Perhaps the best-known example of a water-borne seed is the coconut. Coconut palms have become established on even the most remote and tiny islands of the Pacific and are found throughout the tropical and subtropical regions of the earth. Their seeds simply fall from the trees along beaches. When captured by the tides, they float about, and some eventually wash up on hospitable shores. There they germinate and establish themselves. Wind dissemination is common in the maples and elms, which produce winged seed-bearing fruits. As they fall from trees they gain speed until at a certain rate of descent they begin to fly sideways, spiraling to new frontiers away from their parents. Dandelion fruits are carried in the wind on fine plumes derived from the calyx. An odd adaptation to wind dispersal is the tumbleweed: the whole plant forms a large and very light ball that dries up, breaks off at its base, and rolls across the prairie, scattering seeds and giving rise to country-western songs.

Seeds are carried by animals in two ways—on the outside and on the inside. Those that are carried on the outside often have spiny, barbed fruit (the geranium and bur clover are familiar examples) or barbed or sticky seed coats. Animals transport seeds on the inside, usually, by eating the digestible

fruit that surrounds indigestible seeds. When eaten, the seeds are carried for a while in the animal's digestive tract to be deposited, intact and ready to germinate, in the animal's excrement. This is what bright-colored, fleshy, sweet-tasting fruits and berries are all about—at least from a plant's point of view. Many such fruits also contain powerful laxatives to help the process along.

In some cases, the seed itself is attractive and nutritious to the dispersing animal. (Birds at feeders are attracted to the seeds.) Since these seeds are small, a plant can produce many of them, and, although they can be digested, a number will escape the animal's digestive processes to germinate later in a nitrate- and phosphate-rich dropping. Some animals also store seeds for later use, and many an oak has grown from an acorn that some absentminded squirrel has buried and forgotten.

Asexual Reproduction

Asexual reproduction in flowering plants is commonplace. The most common form it takes is **vegetative propagation**, the development of new individuals from a fragment of the parent plant. Of course, vegetative propagation results in individuals that are genetically identical to the parent plant. Agriculturists have long realized certain advantages in the use of vegetative propagation over seeds, where practical. Vegetative propagation ensures that the desirable qualities of certain crop plants will be maintained from one crop to the next with little effort on the grower's part. Retaining those qualities through seed production can be complex and costly, often requiring

(a) Wind-dispersed

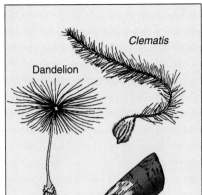

(b) Water-dispersed

(c) Animal-dispersed

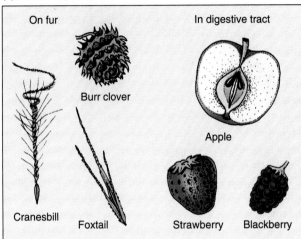

FIGURE 28.10
Diversity in Seed Dispersal
Seeds and seed-bearing fruits are dispersed by wind, water, and animals. Plumes and wings help in lofting the seed-bearing structures of *Clematis* and dandelion into breezes that will carry the offspring well away from their parents. The coconut seed, well protected by its tough fruit (husk), is uniquely adapted for drifting in the sea to the next island. The cranesbill, foxtail, and bur clover have adaptations for clinging to the fur of animals. The seeds within apple, blackberry and strawberry fruits may be carried in an animal's digestive tract for a time but are eventually deposited in the feces.

carefully managed genetic breeding programs for each generation of seeds to be used. We'll consider some of the specifics.

Vegetative Propagation at Work

It is generally known that many plants can be propagated from **cuttings**, pieces, usually stem tips, of the desired plant. When people attempt to propagate plants from cuttings, they usually keep them in water until roots appear and then plant them. Sometimes, they add the plant hormone **auxin** to the water in which the plant is kept. Root growth is generally preceded by the appearance of **callus** tissue, which is an undifferentiated mass of thin-walled cells produced at the site of injury.

New plant growth can also emerge from what are known as **adventitious buds**. These commonly formed on severely damaged stems or on cut off stems. They are also found on the roots of some plants, such as the silver leaf poplar and black locust. In these cases, hidden buds along underground roots begin to sprout, sending young stems upward and suffusing the ground beneath with tiny roots. They are often referred to as *suckers*.

In a few instances, shoots can emerge vegetatively by **leaf generation**, such as in *Kalanchoe* (Figure 28.11a). A single leaf may produce a number of new plants along its margin. Home gardeners often propagate *Bryophyllum* by simply laying the leaf on moist soil until the young shoots appear. You can obtain about 12 new plants in this way.

Some plants send **runners** or **stolons** along the ground. These are wispy stems that snake away from the parent plant, send down roots, and eventually develop into independent plants (Figure 28.11b). The runner stems have intermittent nodes, where adventitious roots and shoots are produced. Hardly anyone grows strawberries from seeds, since it is such a simple matter to pinch off a runner and transplant the shoots arising from it.

Tubers are also stems, but they are typically thick and are found underground. The domestic white, or Irish, potato is our best example (Figure 28.11c). The potato doesn't look much like a stem, but careful examination reveals typical stem features. The most significant is the eye, which is actually a bud, fully capable of producing stem and adventitious root growth. In fact, a new potato crop can be produced by planting pieces of potato that include at least one eye.

Crab grass and Bermuda grass can reproduce vegetatively through underground stems known as **rhizomes**. These stems produce nodes at various distances from each other. Each node sends roots down and stems up. In addition to their insidious method of spreading, these grasses are C4 plants with improved photosynthetic efficiency, so that propagating Bermuda grass is easy. A cylindrical core cutter is used to obtain "plugs" of the grass, which can then be set into new ground.

Apomixis: Seeds Without Sex

Sexual reproduction in higher plants is always through seeds, but interestingly, not all seeds are produced sexually. In some instances they are produced asexually—that is, without the union of sperm and egg nuclei. This form of asexual reproduction in plants is known as **apomixis**, and the phenomenon can

(a) Leaf with plantlets along margin

(b) Strawberry plants with runners

(c) New shoots and roots sprouting from potato

FIGURE 28.11
Vegetative Propagation
Vegetative propagation, an asexual process, occurs in many ways. **(a)** *Kalanchoe* produces plantlets along the margins. **(b)** Strawberry plants produce new plants by the development of runners. **(c)** A potato tuber produces new plants from its "eyes," which are lateral buds.

occur in several ways. Usually, the megasporocyte fails to carry out meiosis; thus the diploid condition is retained, and the embryo begins development without the intervention of the male gamete. Of course, the new embryo has exactly the same chromosomes and genetic characteristics as the single parent sporophyte on which it is borne. Thus all descendants of an *apomict* line are essentially clones. Examples of apomict species include Kentucky bluegrass, hawthorn, and blackberry.

It is interesting that while completely asexual plant species are very common, none of these species appears to have been around very long in evolutionary terms. Unless the environment remains exceptionally stable, none of them can be expected to hold out in the evolutionary long run. Most apomicts, however, can be expected to survive because they can also reproduce sexually. It's as though these species are "hedging their bets," going the asexual route when it is advantageous but always maintaining their main evolutionary sexual line.

Natural Selection and Apomixis. Evolutionary theorists have an explanation for the appearance of this peculiar phe-

nomenon. Natural selection seems to proceed on the profound and unbiased principle that whatever works, works. And sometimes asexual reproduction is extremely adaptive to a plant. For example, a particularly successful genotype can be saved from the vagaries of genetic recombination that occur during meiosis and when unrelated haploid cells are brought together in fertilization. One hundred percent of an apomict's genes are passed along to all its offspring, and this favored genotype can quickly spread through a habitat. But, one might ask, under what conditions could a lack of variability be adaptive? Isn't variability in a population the key to success?

Plants living under drastic conditions, such as those in Arctic regions, cling to existence through the most precarious and fragile adaptations. Any disruption of its precise genotypes could spell disaster for a population. Furthermore in such rigorous environments, plants may be so far apart as to make pollination difficult, so asexual reproduction through apomixis is highly adaptive.

Apomixis may also speed up the speciation process by increasing the chances of successful hybridization. Recall from our earlier discussion that hybrids are frequently sterile because of the failure of chromosomes to synapse at prophase I of meiosis. There are no homologues available, and gametogenesis fails. In our earlier example, we pointed out how spontaneous chromosome doubling provides one solution. Apomixis is an even simpler solution, because meiosis is not necessary. Thus apomict hybrids undergo seed development without a hitch. This has happened time and again with the highly variable Kentucky bluegrass.

The problem with apomixis should be apparent. When the environment changes, as it inevitably will, the apomictic population may not contain the variability needed to ensure even minimal survival. Those plants that can shift to a sexual phase, however, can shuffle genes, producing new combinations and leaving more variable offspring. These kinds of plants will have an improved chance of success when the environment changes.

Since sexual organisms have more variability built into their genetic systems, they have a long-term advantage and will once again take over the habitat. Then, as the environment stabilizes, the newly succeeding variants may become apomictic. We can see then how apomixis could have arisen, but judging from the fact that strict apomicts don't have long evolutionary histories, it seems that sex is the safer bet in the long run.

Key Ideas

SEXUAL REPRODUCTION

1. The flower consists of modified leaves arranged into **whorls**, which include (a) **accessory parts**—the **receptacle**, the **calyx** (**sepals**), and **corolla** (**petals**)—and (b) **sexual parts**—the **androecium** (**stamens** or microsporophylls, consisting of **filaments** topped by **anthers**) and the **gynoecium** or **pistil**(s), each of which contains an **ovary**, a neck-like **style**, and a **stigma**. Divided ovaries are known as **carpels**.

2. Floral structure varies with the pollinating agent. Attractants include odor, **nectar**, pollen, color, and structural pattern.

3. Floral parts in monocots occur in threes and its multiples, while dicot floral parts occur in fours or fives or their multiples. More primitive flowers are **radially symmetrical**, while advanced types are often **bilaterally symmetrical**.

4. **Complete flowers** have all four basic parts, but **incomplete flowers** may have one or more missing. **Perfect flowers** have both pistils and stamens, but **imperfect flowers** are **pistillate** (female—lacking stamens) or **staminate** (male—lacking pistils).

5. Some flowers are clusters, or **inflorescences**, and if they contain individual flowers, each with a single fruit, they are **composite flowers**. Composite flowers have many central **disk flowers** and modified, surrounding **ray flowers**.

6. Through coevolution, many flowers and insects reach a specialized relationship that assures the plant of pollination and the insect of a food supply. *Yucca* requires the yucca moth, which lays its eggs in the flower's ovary. Skunk cabbage attracts the fly as a pollinator.

7. Wind-pollinated flowers are inconspicuous, lacking colors and odors. Floral parts are modified for capturing pollen.

8. Ovules originate in **placental** tissue to which they are connected by a **funiculus**. An ovule is a megasporangium or **nucellus**, containing a **megasporocyte** and accompanied by skin-like **integuments**. An opening, the **micropyle**, occurs at one end.

9. In the ovary, megasporocytes undergo meiosis, producing four haploid products. Three disintegrate, but the fourth divides mitotically three times, producing an eight-nucleate **megagametophyte**.

10. Upon cell division, seven, single-nucleate cells emerge. Included is the **egg cell** and the large **central cell**. The latter, which contains two **polar nuclei**, will later give rise to the **endosperm**. The completed megagametophyte is called an **embryo sac**.

11. **Microsporogenesis** begins in the **anthers**, where each **microsporocyte** undergoes meiosis, producing four haploid **microspores**. The nucleus of each divides mitotically, producing a **generative cell** and a **tube cell**. They become the **pollen grain**.

12. **Pollination** is the transfer of pollen onto a stigma. Germinating pollen produces a **pollen tube**, which penetrates the ovule, releasing two sperm.

13. In **double fertilization**, one event produces the diploid zygote, and the second, the triploid **primary endosperm nucleus**.

14. Following fertilization, the ovary matures into the seed-bearing **fruit**.

SEED DEVELOPMENT

1. Mitosis and differentiation in the primary endosperm and zygote will produce the **embryo**, which along with a food supply and **seed coat** make up the seed.

2. After repeated mitosis, the early embryo becomes a ball of cells anchored by the large **basal cell** of the **suspensor**.

3. The bean embryo takes on a heart shape as two cotyledons take form. In corn, the cotyledon is single.

4. The embryo produces a **shoot apical meristem** at its tip and a **root apical meristem** at its base. Growth above the origin of the cotyledons is called **epicotyl**, which in some species gives rise to a **plumule**. Most of the embryo axis in the bean is **hypocotyl**, which may contain a radicle at its base. The meristems give rise to the three primary tissues: **protoderm**, **procambium**, and **ground meristem**.

5. In corn, each kernel develops from an ovary, so it is technically a fruit. The fruit includes a starchy endosperm, an embryo, and a digestive structure, the **scutellum** (cotyledon). The endosperm forms an **aleurone**, which is surrounded by the **pericarp**. The corn radicle and plumule are protected by a **coleorhiza** and **coleoptile**, respectively.

6. Many seeds enter a dormant period in which they harden through dehydration and are metabolically inactive.

7. Seeds are dispersed by water, wind, and animals. Wind-dispersed seeds may have wing-like structures or plumes. Some seeds are eaten and distributed in the animals' wastes.

ASEXUAL REPRODUCTION

1. **Vegetative propagation** is the development of a new individual from some fragment or portion of another plant. It is agriculturally useful, since the absence of genetic recombination ensures the precise continuation of desirable qualities.

2. New plants can be raised by **cuttings**, **adventitious buds**, and **leaf generation**. **Runners** or **stolons** put down roots, establishing new growths. Potatoes and other **tubers** are underground stems that produce growth at buds ("eyes"). Underground stems called **rhizomes** produce new growths along their lengths.

3. Apomicts are plants that can produce embryos and seeds without fertilization. Theorists suggest that **apomixis** is an adaptation to very harsh conditions, where genetic variation might be disadvantageous to the plant. It may also be adaptive to sparsely distributed populations. Apomixis may speed up speciation, since otherwise infertile hybrids avoid meiosis altogether, thereby avoiding the problem caused by unmatched homologues. In the long run, sexual systems, those that produce variability, are probably more adaptive.

Application of Ideas

1. Characteristically, the great diversity in angiosperm floral parts can partly be explained through the dependence of plants on animals for pollen transfer. However, one large group has become adapted for pollination by wind. Which group is this? Would you expect much diversity in flowers of this group? In what other ways might wind-pollinated flowers differ from those pollinated by insects? Compare the specific structures in a typical wind-pollinated angiosperm flower with that of an insect-pollinated flower.

Review Questions

1. List the four major regions of the flower. Which are accessory, and which are the "male" and the "female" parts? (page 593)

2. What colors and odors would you expect in flowers that are pollinated by the following: bats, carrion flies, and bees? (page 595)

3. What is an imperfect flower? Can an imperfect flower be complete? Explain. (page 595)

4. From the viewpoint of natural selection, explain why grasses lack colorful, fragrant flowers. (pages 596–597)

5. Using flower symmetry or numbers of floral parts as a basis, distinguish between primitive and advanced plants and between monocotyledons and dicotyledons. (pages 593–595)

6. Using the example of *Yucca*, explain how plants and insects specialize through coevolution. (pages 595–596)

7. Describe events that lead to the production of the eight-nucleate, seven-cell embryo sac. (pages 597–598)

8. List the parts of an embryo sac. Which participate in fertilization? (page 598)

9. Trace the events in the formation of a pollen grain. What cells does it contain? (page 598)

10. Beginning with pollination, describe fertilization in the flowering plant. (page 598)

11. Give a technical definition of the term *fruit*. Describe the fruit of corn and string beans. (page 598, Essay 28.2)

12. Trace the development of the dicot embryo. Include a description of the suspensor, the heart-shaped embryo, and the cotyledons. What are the three primary tissues? (pages 600–604)

13. What are two basic differences between the monocot and dicot embryo and seed? (page 604)

14. List three agents of seed dispersal. Why is dispersal important? (pages 604–605)

15. Provide three examples of vegetative propagation. Why is it favored in agriculture? (page 606)

CHAPTER 29

Plant Growth and Structure

CHAPTER OUTLINE

GERMINATION AND GROWTH OF THE SEEDLING

Germination Conditions

The Seedling

Primary and Secondary Growth

TISSUE ORGANIZATION IN THE PLANT

Protoderm and the Epidermis

Ground Meristem and Three Tissues

Procambium and the Vascular Tissues

PRIMARY GROWTH IN THE ROOT

The Root Tip

Root Systems

PRIMARY GROWTH IN THE STEM

LEAVES

Anatomy of the Leaf

SECONDARY GROWTH IN THE STEM

Transition to Secondary Growth

Growth of the Periderm

The Older Woody Stem

Essay 29.1 The Problem of Differentiation

KEY IDEAS

APPLICATION OF IDEAS

REVIEW QUESTIONS

*I*t is hard to believe that the giant redwoods that tower 80–100 meters above our heads started out as seeds only 5–6 mm in length. There is a fascinating story to be told about how all this comes about. It involves profound and highly organized changes as the seed is stirred from dormancy and begins to draw sustenance and energy from its stored reserves. It builds tissue, constantly growing and reorganizing until it takes its place among the other forest stalwarts. Soon it will produce its own seeds, which will then play their own role in the ongoing saga of life. Considering the importance of plants to all of us in our everyday life, the story of growth and development is well worth knowing.

Germination and Growth of the Seedling

Germination marks the beginning, when the seed, essentially a dormant stage, is activated, beginning a period of intense metabolic activity. This period is triggered by a rather specific set of conditions.

Germination Conditions

Nearly all plants have similar germination requirements: an adequate supply of water, favorable temperatures, and the availability of oxygen. Metabolism resumes when water is absorbed into the dry interior of the seed and its enzymes are mobilized. While these conditions are common, a few plants have additional requirements for germination. The seed coats of some species are especially tough and resistant. Some require a period of freezing temperatures to split their coats; some need the heat of fire; and others might need the abrasive action of sand particles carried by flowing water. A fair number of seeds require the hydrolyzing action of animal digestive enzymes. On the island of Mauritius there are 11 huge Tambalacoque trees, a very rare species. All of the trees are about 300 years old; there are no younger trees. Every year they put out a crop of huge seeds with very thick seed coats. But the seeds never germinate. Normal germination requires that the seed be passed through the crop of another native of the island, a bird called *Raphus cucullatus,* better known as the dodo. But the dodo has been extinct for about 300 years, so the seeds just lie around waiting for the bird that will never come. (Recently, the naturalist who worked out this curious relationship managed to get several of the seeds to germinate by passing them through a turkey.)

The Seedling

In a number of monocots such as the grains, those with large unorganized endosperm, the uptake of water is followed by the mobilization of the starchy reserves stored therein. This is an indirect process involving hormones known as gibberellins produced by the embryo (see Chapter 31). The target of the gibberellins is the aleurone cell layer of the endosperm, just below the seed coats. The aleurone cells respond to the hormone by producing alpha-amylase, a starch-digesting enzyme that hydrolyzes the starch to mobile, soluble sugars that the embryo can put to use for growth.

Corn. As germination occurs in corn, the embryonic root or radicle of this monocot emerges first, breaking through the coleorhiza and pushing down into the soil (Figure 29.1a). Next, the shoot, surrounded by the protective coleoptile, elongates rapidly, pushing upward through the soil. Soon after, the first leaf breaks through the coleoptile and emerges into the sunlight. The rapidly growing primary root below is soon accompanied by more roots. Oddly, in corn and some other monocots, the newcomers, which are called adventitious roots, will emerge not from the primary root but from the stem region above.

Onion. The emergence of the onion seedling, another monocot, is quite different. Its lengthy cotyledon elongates rapidly, emerging from the seed in a U-shape. At its free lower end, a fast-growing root develops. The U-shaped cotyledon, acting as a "bumper," grows toward the surface. After the cotyledon emerges from the soil it straightens, drawing what is left of the seed out of the soil and holding it aloft. The first leaf emerges far below the seed remnant, from the base of the cotyledon. Eventually the cotyledon, an embryonic structure, withers, and the adult plant takes form (Figure 29.1b).

Bean. As germination in the bean proceeds, the young root emerges and quickly penetrates the soil, forming many lateral roots and anchoring itself securely. The hypocotyl then elongates rapidly (Figure 29.1c). (The intake of water into cells is an important factor in elongation, since it is the resulting force exerted against the young cell wall that lengthens the cell.) As the hypocotyl grows it forms a sharp curve—the **hypocotyl hook**—which acts as a protective bumper for the delicate parts below. In its upward growth, the hypocotyl draws the cotyledons along behind it. Upon breaking through the surface the hypocotyl straightens, the cotyledons open up, and the young plumule is exposed to the light. At this point the epicotyl begins its elongation, lifting the expanding leaves up to the light.

When the bean seedling breaks the surface, its chloroplasts mature, turning the plant a deep green. The embryonic leaves are exposed to energy-laden light as their growth continues. At this time, the food reserves in the cotyledons diminish, and they wither and fall away, leaving telltale scars on the young shoot. The young bean seedling is then fully on its own.

Primary and Secondary Growth

Seed germination and the early growth of the shoot are examples of **primary growth**. Primary growth originates in apical meristems and goes on throughout the life of a plant. It is responsible for continued growth in the length of shoots and roots, and for the production of leaves, flowers, and new branches. In contrast, **secondary growth** occurs in tissues derived from primary meristems and accounts for growth in girth or thickness. Secondary growth in flowering plants is seen mainly in **perennial** dicots, those plants that live on year after year. **Annuals**, in contrast, germinate, grow, and reproduce in a single season. **Biennials**, we should note, are those plants that live for two seasons, generally reproducing in the second. Secondary growth is virtually universal in conifers and other gymnosperms.

Perennial plants are also unique among living things in that they exhibit what is called **open growth**, or indeterminate life span. This means that theoretically, barring fire, injury, disease, or human intervention, they may live forever. There are the usual arguments over which plant holds the record, and for

FIGURE 29.1
Germination and Growth

(a) As the corn germinates, the radicle breaks through the coleorhiza to form the primary root. Next, the coleoptile, with the plumule enclosed, pushes up through the soil, whereupon the plumule breaks through its sheath. The primary root is soon obliterated by numerous adventitious roots, which emerge directly from the stem. (b) In the onion seedling, the hypocotyl curves downward with the young root emerging. The cotyledon breaks through the soil, raising the seed coat aloft. Note the position of the first leaf. (c) In the bean, the rapidly elongating hypocotyl forms a curved "bumper," grows upward, and draws the cotyledons behind it. Upon emerging from the soil, the hypocotyl straightens, raising the cotyledons and plumule. Just below the plumule, the epicotyl raises the young foliage even higher. Meanwhile, the young primary root has produced many lateral roots as the root system matures.

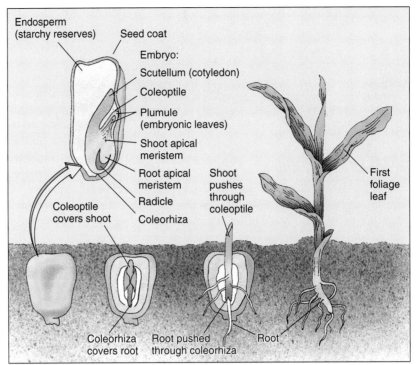

(a) Corn

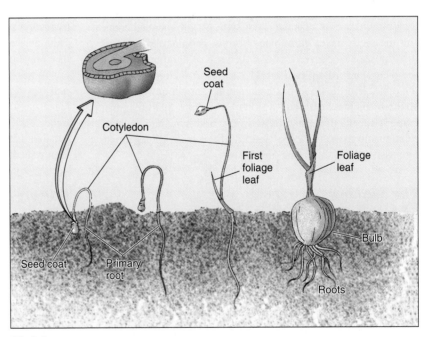

(b) Onion

years the documented champion was a 4900-year-old bristle-cone pine from the White Mountains of California (Figure 29.2). Today, we're not so sure. Some cottonwoods may have reached an age of 8000 years, and the age of a creosote bush in the Mojave Desert, the so-called King Clone, is estimated to be about 11,700 years! In any case, we can be sure that somewhere on earth grows a plant that is the world's oldest living thing.

In a sense, plants with open growth never completely mature. They always have a reserve of undifferentiated meristematic tissue. In addition, many cells that have differentiated and begun to function in a specific role are capable, in a manner of speaking, of having their clocks set back. They *dedifferentiate*. Here tissues return to an immature state from which they have the potential to mature or differentiate all over again,

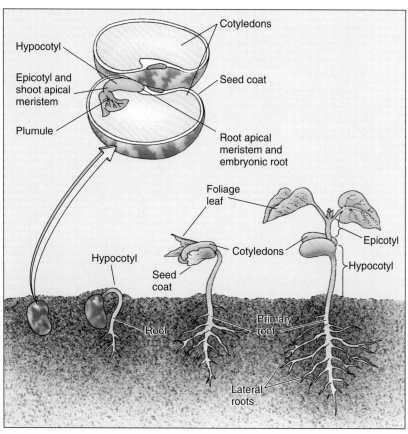

(c) Bean

often along new routes. Essay 29.1 (page 629) explains this further. We'll prepare for a closer look at primary and secondary growth by learning something about plant tissues.

Tissue Organization in the Plant

The mature plant body can be broken down into the shoot system, consisting of the stems, leaves, and flowers, and the root system. Stems support the foliage, carry on transport, and store foods. Young green stems carry on photosynthesis, as do the leaves. As we've seen, flowers carry on sexual reproduction. Root systems anchor the plant, take in water and minerals, and store foods. The two systems are, of course, interdependent.

The shoot and root systems are composed of many types of cells organized into seemingly countless cell layers. Yet the entire plant body is composed of just a few basic tissues. As we've seen, all tissues arise from apical meristems in the root and shoot tips. As the meristems divide, they contribute cells to three primary tissues: **protoderm**, **ground meristem**, and **procambium**. (We noted earlier that these are the first specific tissues formed in the embryo.) Each of these, in turn, differentiates into cells and tissues that carry out the many specialized functions that characterize plant life. This straightforward organization is illustrated in Figure 29.3.

FIGURE 29.2
One of the Oldest Plants
One of the world's oldest trees, the bristlecone pine, grows at the timberline of California's White Mountains. One specimen reached 4900 years of age.

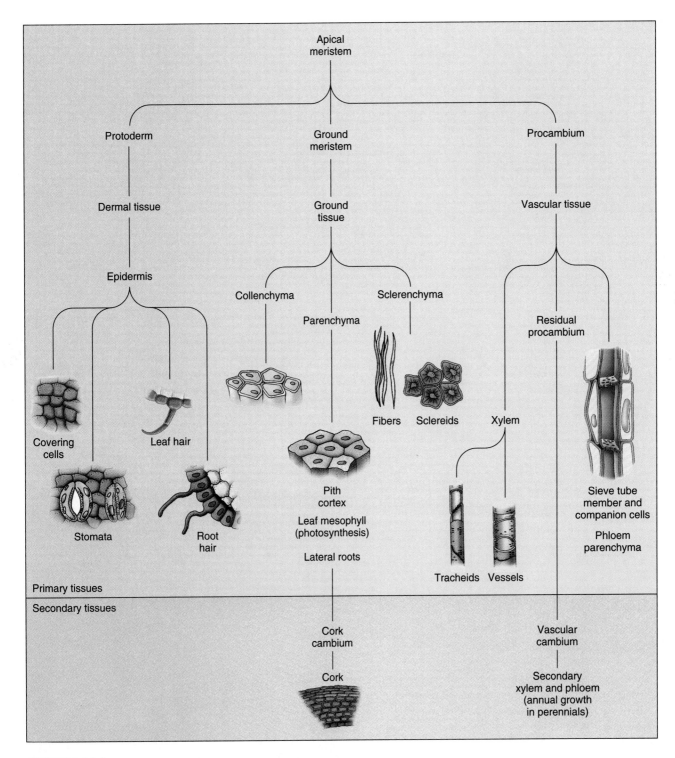

FIGURE 29.3
Primary Meristems and Their Tissue Derivatives
During primary growth, the apical meristem gives rise to the three primary tissues: protoderm, ground meristem, and procambium. Derivatives of the primary tissues differentiate into all the cells of the plant body.

Protoderm and the Epidermis

The **protoderm** contributes to **dermal tissues**. These are the outermost layers of cells, which, in the young root and shoot, form the **epidermis**. In the leaf and younger stem, epidermal cells are commonly flattened and irregular in shape (see Figure 29.3). They are often covered by a waxy, waterproof cuticle of **cutin**, which prevents water loss from the more delicate tis-

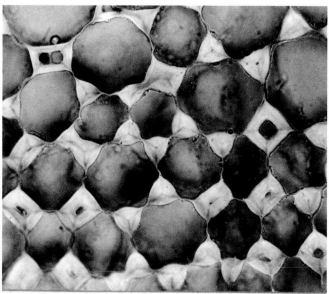

(a) Collenchyma in leaf petiole

FIGURE 29.4
Ground Meristem Derivatives
(a) Supporting collenchyma tissue in the leaf petiole is recognized by its unevenly thickened walls. **(b)** One form of sclerenchyma, the branched sclereid (at *center*), occurs in leaves and fruit.

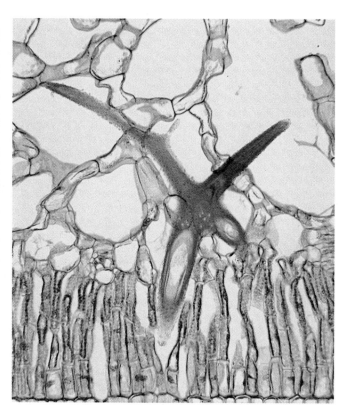

(b) Branched sclereid (dark structure) in a leaf

sues below. The epidermis also contains paired **guard cells** that make up **stomata** (singular, *stoma*), pore-like structures that permit the exchange of gases between photosynthetic cells and the atmosphere. Epidermal cells just above the root tip have the vital job of absorbing water. Many produce lengthy **root hairs** that greatly increase the absorbing surface. Where secondary growth occurs, the epidermis is replaced by a multi-layered tissue called **periderm**. The periderm is made water-proof by the addition of **suberin**, another waxy substance.

Ground Meristem and Its Three Tissues

Ground meristem differentiates into three important tissues: *parenchyma, collenchyma, and sclerenchyma* (Figure 29.4). **Parenchyma** is widely distributed throughout the plant. It is an often loosely arranged tissue consisting of large, thin-walled, irregularly shaped cells. Parenchyma takes in a consid-erable amount of water, producing the turgid condition impor-tant in holding the leaves and shoots of young plants erect. It is also a food storage tissue, often packed with starches. Leaf parenchyma (also called chlorenchyma) contains most of the chloroplasts and is the most important site of photosynthesis. Parenchyma tissue is capable of cell division and differentia-tion into other tissue during secondary growth and is responsi-ble for the growth of new tissues in wound healing.

Collenchyma and **sclerenchyma** tissues have some roles in common, but structurally they are quite different. Col-lenchyma cells have very thick primary walls of cellulose that lack lignin. Unlike sclerenchyma, they remain alive at maturi-ty. Collenchyma forms layers of supporting tissue just within the epidermis of young stems, and cylinders of supporting tis-sues in the petioles and midribs of leaves. Those "strings" in celery are collenchyma. The thickness and flexibility of col-lenchyma make it useful as supporting material in tissues whose cells are still increasing in length.

Sclerenchyma tissue is often nonliving, its cells having lost their nuclei and cytoplasm. Its cellulose cell walls are usually thick and hardened by lignin, making them especially useful as supporting elements in tissues that have finished their growth. Sclerenchyma occurs in two forms, **fibers** and **sclereids**. Because of their great strength, the fibers make up an important structural component in vascular plants. The strong sclerenchyma fibers of jute and hemp are used for manufacturing string, rope, and sacking. Sclereids are short branching cells. They are common in seed coats and nut shells, and you've probably come across sclereids called **stone cells** in the gritty parts of pear flesh.

Procambium and the Vascular Tissues

Procambium, the third primary tissue, has two major roles. During primary growth, it forms the primary xylem and phloem. Then, in plants with secondary growth, it forms **vascu-lar cambium**, a tissue that produces all new xylem and phloem.

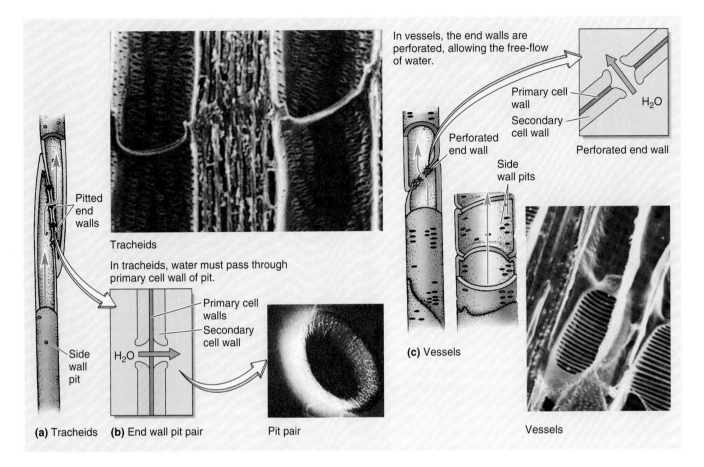

FIGURE 29.5
Xylem Tissue
Xylem includes conducting elements known as tracheids and vessel elements. **(a)** Tracheids are long, slender elements, usually with angular end walls and pitted side and end walls. **(b)** Pit pairs are thin-walled regions consisting of only the two adjacent primary cell walls. **(c)** Vessel elements have pit pairs in their side walls, but their end walls are completely perforated. Vessel elements are usually much wider and less tapered than tracheids.

Xylem. Water and minerals move upward from the root to the stem and leaves through the xylem. Xylem in angiosperms includes **tracheids** and **vessel elements** along with nonconducting fibers and parenchyma. The xylem fibers, which can be long and tough, have a supporting role. Both tracheids and vessels are dead at functional maturity. Prior to their cytoplasm degenerating, intricately patterned walls of lignin are deposited, giving the mature cell great support and rigidity.

Tracheids are long, slender, tube-like cells (Figure 29.5a). Since their end walls are tapered, they form angular connections where they join. You might expect the end walls to be absent or made up of open pores, but this isn't the case. They contain very thin, porous regions called pits or pit pairs, which consist of primary cell wall material only (Figure 29.5b). The pits are aligned so that water readily passes from one tracheid to the next. The side walls of tracheids are commonly pitted as well; thus fluid can move laterally.

Vessel elements differ considerably from tracheids. They tend to be shorter and much wider (see Figure 29.5c), and endwall tapering, if it occurs at all, is much less noticeable. Fur-thermore, the end walls of vessel elements are heavily perforated or even entirely absent. As a result, vessel elements lying end to end, known as **vessels**, form uninterrupted pipelines through which water readily moves. There is a small evolutionary lesson here. You may recall that vessels are absent in nearly all gymnosperms, those vascular plants of more ancient lineage (see Chapter 25). Thus vessels are considered to be advanced evolutionary structures. Like tracheids, the side walls of vessel elements are heavily pitted and readily admit water.

Phloem. While xylem conducts water and minerals, phloem is responsible for the movement of photosynthetic products, primarily sucrose. Unlike xylem, phloem must remain alive to carry out its function. This is a complex tissue that is composed of **sieve elements**, **companion cells**, and **phloem parenchyma** (Figure 29.6). Also included are **phloem fibers**, which, like xylem fibers, provide structural support. Sieve elements occur as **sieve cells** in the gymnosperms and as **sieve tube members** in the flowering plants. Sieve tube members,

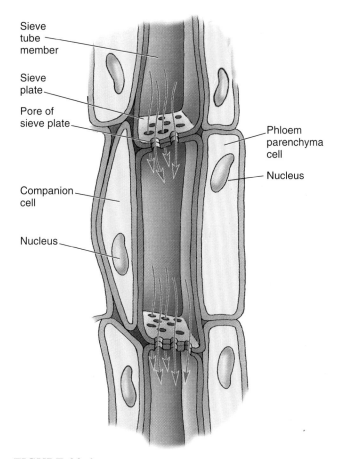

FIGURE 29.6
Phloem Tissue
In flowering plants, phloem includes sieve tube members bordered by companion cells and phloem parenchyma cells. Sieve tube members conduct the phloem stream, while the other two provide physiological support and lateral transport. The end walls of sieve tube members contain sieve plates whose pores are penetrated by strands of cytoplasm from the adjacent cells.

Sieve tube member

Sieve plate

Pore of sieve plate

Companion cell

Nucleus

Phloem parenchyma cell

Nucleus

like the vessel elements of xylem, represent an evolutionary advancement, so we will confine discussion to these.

Sieve tube members (individual cells), when arranged end to end, form the **sieve tubes**. (Just as vessel elements form vessels.) It is in the sieve tubes that foods are conducted from one location in the plant to another. The term "sieve" is appropriate since sieve tube members have pores in their side and end walls. The largest pores are seen in **sieve plates**, which form the end walls between sieve tube members. Thin streams of cytoplasm extending through the enlarged pores of sieve plates are believed to conduct substances from one member to the next. Sieve plates, we should note, are absent in the more primitive sieve cells, which have pores evenly distributed along the side and end walls.

Companion cells are aptly named, since they lie adjacent to the sieve tube members and also function in transport. The exact relationship between the two isn't entirely clear, but we know that while companion cells contain nuclei, sieve tube members do not. Presumably, the companion cell takes care

of any needs related to protein synthesis for both types of cells. Companion cells function importantly in phloem transport by actively transporting sugars into and out of the sieve tube members. Phloem parenchyma cells also transport foods into and out of the sieve tube members but are involved chiefly in storage. We will look further into xylem and phloem transport in the next chapter.

Primary Growth in the Root

The Root Tip

Primary growth in the root takes place in the **root tips**. A look at Figure 29.7 will reveal the organization of a typical root tip. The very end of the root tip is the **root cap**. The root cap aids in penetration of the soil by the root, is believed to function in perception of gravity, and protects the actively dividing cells of the apical meristem of the root. The **root apical meristem**, behind the cap, produces the three primary meristems—the protoderm, procambium, and ground meristem—and also replaces cells of the root cap that are sloughed away during growth. The root apical meristem also contains a region of cells that divide much more slowly than meristematic cells. This area is called the **quiescent center**. In experiments where a portion of the apical meristem was surgically removed, cells of the quiescent center begin to divide more rapidly, replace the damaged meristematic cells, and become quiescent again. We believe that the quiescent center is a reservoir of cells that can respond if the root apex is damaged.

The root tip is commonly divided into three functional zones, the first of which includes the meristem and forms the **region of cell division**. Above this lies the **region of elongation** and, further back, the **region of maturation**. The boundaries of these regions are not distinct, with one blending into the next. This is because root tip growth is continuous, with new regions being left behind as the root moves through the soil. Thus today's region of elongation is tomorrow's region of maturation.

In the apical meristem, the region of cell division provides a constant supply of cells for the root cap below and for the zone of elongation above. The region of elongation is usually only a few millimeters long, but the elongation of cells in this zone results in the "downward push" we associate with root growth. The cells above the region of elongation do not increase in length. The mechanics of cell elongation are seen in Figure 29.8.

Differentiation in the Young Root. Once elongation has been completed, the older cells, which now form the region of maturation, are left behind the growing root tip to mature (Figure 29.9). The outer tissues, those made up of protoderm, will form the epidermis, and many of these young covering cells will produce lengthy root hairs. Their primary task is the uptake of water and minerals. Root hairs greatly increase the active area of absorption. Ground meristem just within the protoderm forms the **cortex**, a region of thin-walled parenchyma.

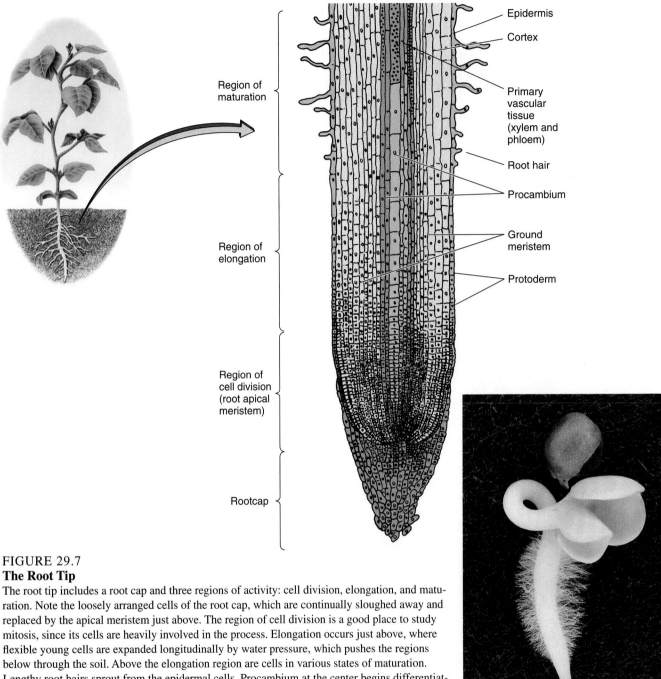

Region of
maturation

Region of
elongation

Region of
cell division
(root apical
meristem)

Rootcap

Epidermis

Cortex

Primary
vascular
tissue
(xylem and
phloem)

Root hair

Procambium

Ground
meristem

Protoderm

FIGURE 29.7
The Root Tip
The root tip includes a root cap and three regions of activity: cell division, elongation, and maturation. Note the loosely arranged cells of the root cap, which are continually sloughed away and
replaced by the apical meristem just above. The region of cell division is a good place to study
mitosis, since its cells are heavily involved in the process. Elongation occurs just above, where
flexible young cells are expanded longitudinally by water pressure, which pushes the regions
below through the soil. Above the elongation region are cells in various states of maturation.
Lengthy root hairs sprout from the epidermal cells. Procambium at the center begins differentiating into xylem and phloem. Between the two regions, ground meristem forms parenchymal cells
that make up the cortex.

Radish root with root hair

Toward the central region of the root, some of the ground
meristem will mature into a cylindrical ring of cells known as
the **endodermis** (inner skin). The endodermis performs an
active role in the uptake of water and minerals by the plant.
Layers of impregnated waterproof material, strategically located in radial and tangential walls of the endodermal cells, form
the Casparian strip (look ahead to Figure 30.6). Because of their
presence, all incoming water is detoured from the usual cell
wall route into the endodermal cell, where it must pass through
the cytoplasm before moving on. We will return to this special

arrangement in the next chapter. Just within the endodermis lies
the stele. It comprises the developing vascular system.

The Stele. The **stele** is the central part of the root, and includes
the xylem and phloem. It is surrounded by a cylinder of cells
called the **pericycle**, which lies just inside the endodermis (see
Figure 29.9). The pericycle conducts water and minerals
inward to the vascular tissue, but in dicots it has another function as well. It has the capacity to produce lateral roots (also
called branch roots). The young lateral roots emerging from the

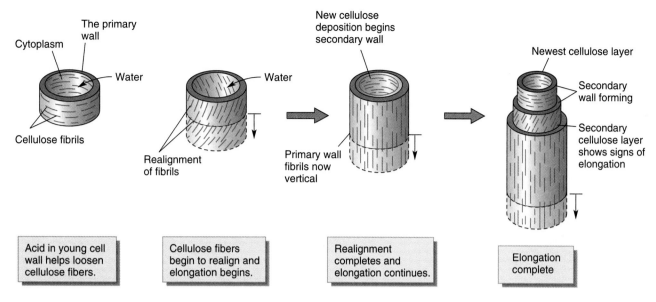

Acid in young cell wall helps loosen cellulose fibers.

Cellulose fibers begin to realign and elongation begins.

Realignment completes and elongation continues.

Elongation complete

FIGURE 29.8
Mechanics of Cell Elongation
Cell elongation is a highly precise process. Acidification of the wall loosens the cellulose fibers, and the uptake of water creates sufficient turgor pressure to cause cell lengthening. As elongation continues and new secondary walls are laid down, cellulose microfibril patterns change.

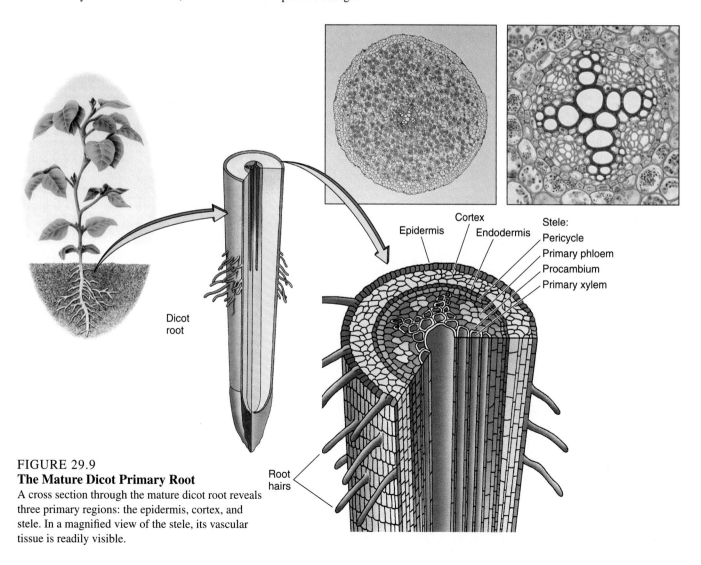

FIGURE 29.9
The Mature Dicot Primary Root
A cross section through the mature dicot root reveals three primary regions: the epidermis, cortex, and stele. In a magnified view of the stele, its vascular tissue is readily visible.

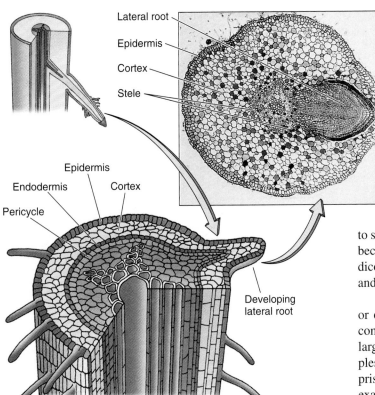

Lateral root
Epidermis
Cortex
Stele

Epidermis
Endodermis
Cortex
Pericycle

Developing
lateral root

FIGURE 29.10
Growth of Lateral Roots
Lateral roots originate in the pericycle of mature roots. Cells there undergo repeated divisions, producing what at first appears to be a formless mass. As the emerging root takes form it closely resembles the original root tip. Soon it elongates, making its way through the cortex and breaking through the epidermis.

pericycle closely resemble primary root tips in their organization. Their growth has a seemingly brutal aspect, since they must digest and push their way through the endodermis and cortex, crushing many cells and finally bursting through the epidermis of the primary root (Figure 29.10). Eventually, each branch root will develop its own vascular tissue, which will become joined to that of the primary root.

In some dicots, the organization of the vascular tissue within the stele looks (in cross section) like a four-armed star (see Figure 29.9). The broad arms contain the primary xylem, which is made up of large, thick-walled, water-conducting cells. The primary phloem, with its food-conducting sieve tubes, is found between the arms of the star. In plants capable of secondary growth, undifferentiated procambium produces a strip of vascular cambium between the xylem arms and phloem regions. Vascular cambium will later produce the secondary xylem and secondary phloem.

Root Systems

Root systems differ between dicots and monocots. While both begin by producing a fast-growing main or primary root, the primary root in monocots does not usually persist as it does in dicots. For instance, in corn, the primary root is replaced early in life by **adventitious roots**, those that emerge from the stem. These, in turn, produce root branches that originate from the root pericycle in the usual manner of lateral roots. You may have seen the adventitious roots in corn, since they are often exposed. From their origin in the stem's base, they curve downward into the soil. Adventitious roots arising from above ground are called **aerial roots**. Where their function is

to support the heavy stem and leaves (as they do in corn), they become **prop roots**. Aerial adventitious roots are also seen in dicots, for example, in English ivy, tropical red mangroves, and banyan trees.

Roots systems can also be categorized as either **tap roots** or **diffuse roots** (Figure 29.11), although some plants have combinations of the two. Tap roots are characterized by a large, main root from which lesser branches emerge. Examples are the carrot and sweet potato. Diffuse roots are comprised of a number of roots of roughly the same size. An example is the fibrous, matted roots of grasses.

The extent to which roots grow and the power of their growth are legendary. Everyone has seen sidewalks and streets literally lifted up by a meandering tree root. The direction and extent of root growth depend not only on individual species but on physical factors as well. Among these are soil moisture and composition and certainly soil temperature. The most active water absorption goes on in what are called "feeder roots," usually concentrated in the top meter of soil.

Whereas the roots of some trees, such as beeches, spruces, and poplars, are relatively shallow, those of others, such as oaks and pines, grow quite deep. Even the adventitious roots of corn penetrate the soil to a surprising degree, venturing down $1\frac{1}{2}$ meters. Alfalfa roots reach even further— 6 meters into the soil. For now, the record holders are desert species: an Egyptian acacia tree's roots reached over 30 meters (almost 100 ft), and an Arizona mesquite (a low-growing, thorny tree), uncovered in an open-pit mine, sent its roots down an astonishing 53 meters (174 ft)!

Primary Growth in the Stem

The organization of the apical meristem of the shoot is somewhat more complex than that of the root (Figure 29.12). As we mentioned earlier, a small mound of tiny, actively dividing cells, the apical meristem, is found at the very tip of the shoot. As cell division is occurring, cells at the lower side of the meristem undergo extensive elongation, carrying the meristem upward or outward. Primary tissues—protoderm, ground meristem, and procambium—form and go through phases

FIGURE 29.11
Two Kinds of Root Systems
(a) Diffuse root systems are common in monocots such as the bluegrass seen here. This vast, spreading system consists of adventitious and lateral roots. (b) In tap root systems, the primary root maintains dominance, leading the way and producing a profusion of lateral roots.

(a)

(b)

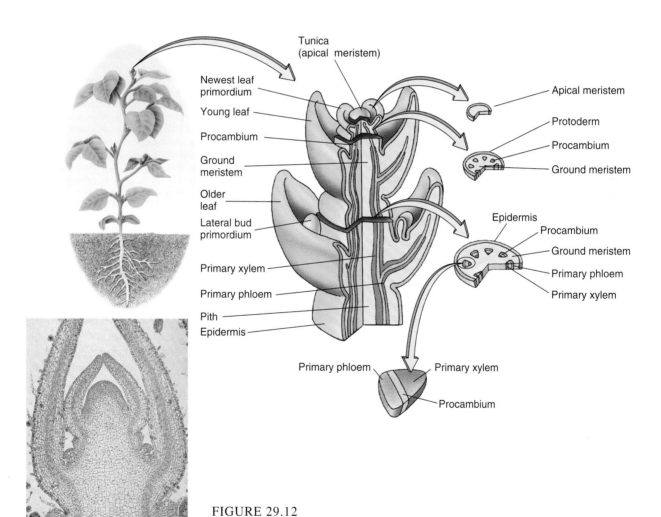

Tunica (apical meristem)

Newest leaf primordium

Young leaf

Procambium

Ground meristem

Older leaf

Lateral bud primordium

Primary xylem

Primary phloem

Pith

Epidermis

Apical meristem

Protoderm

Procambium

Ground meristem

Epidermis

Procambium

Ground meristem

Primary phloem

Primary xylem

Primary phloem

Primary xylem

Procambium

Light microscope view (longitudinal section)

FIGURE 29.12
The Dicot Shoot Tip
A dicot shoot tip includes the shoot apical meristem, a number of leaf primordia, and the three primary tissues. The primary tissues—protoderm, procambium, and ground meristem, respectively—form the epidermis, vascular system, and cortex of the maturing stem.

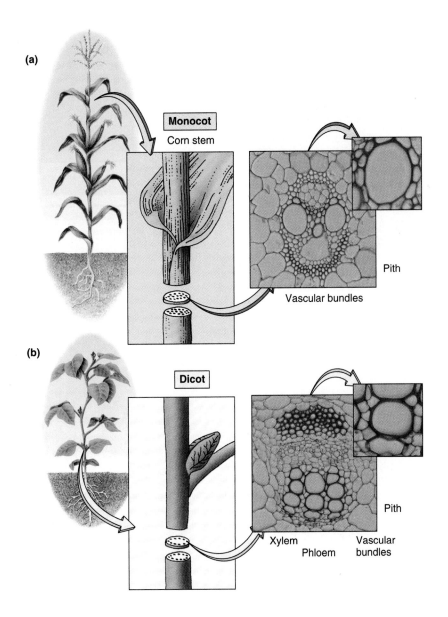

(a)

Monocot
Corn stem

Pith

Vascular bundles

(b)

Dicot

Pith

Xylem   Vascular
Phloem   bundles

FIGURE 29.13
Monocot and Dicot Stems
(a) A cross section of the monocot stem reveals a scattered distribution of vascular bundles. Each bundle (see detail) contains xylem and phloem surrounded by tough sclerenchyma cells (fibers). **(b)** The vascular bundles of the dicot, occur as a cylinder organized around the stem's perimeter. The primary phloem and primary xylem are separated by a region of procambium. Most of the stem's interior is soft pith.

similar to those in roots, but the regions of cell division, elongation, and maturation are not as clearly demarcated. Newly elongated cells undergo differentiation as usual, but here the cortex, a region of thin-walled parenchyma cells, fills the center of the shoot, while, just outside, strands of procambium differentiate into the primary xylem and phloem that is gathered into distinct bundles. At the close of primary growth in dicots, the bundles of vascular tissue will have become arranged in a ring around the perimeter of the cortex. Each bundle (actually a cylinder) contains a strand of procambium, with xylem on its inner side and phloem on its outer side. The vascular bundles are surrounded by cortex, which extends to the young differentiating epidermis. In many young, green stems, the cortical cells just below the epidermis contain chloroplasts and carry on photosynthesis.

A central region called **pith** is seen in older portions of the stem. Pith is chiefly parenchymal in origin and can be dis-

tinguished from the surrounding cortex by its loose organization, including many intercellular spaces.

The distribution of vascular tissue in the monocot shoot, as you may recall from Chapter 25, is fundamentally different from that in the dicot shoot. While the vascular bundles commonly form a rather neat ring within the dicot stem, in monocots they typically are spread throughout the stem. Figure 29.13 compares the arrangement in corn and sunflower plants.

As the shoot grows, it leaves behind differentiating tissue as we described above, but it also leaves behind patches of meristem that give rise to **leaf primordia** and **branch primordia**. The primordia are embryonic cells that give rise to new branches, leaves, and flowers. Such regions of potential leaf production or leaf attachment are known as **nodes**. The stem region between successive nodes is called an **internode**.

Nodes and internodes are most easily seen in the growing tips of deciduous woody plants—seasonally growing plants

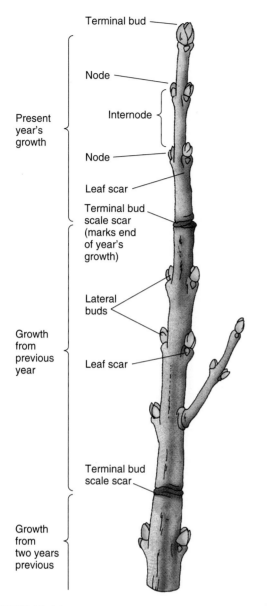

Terminal bud

Node

Internode

Node

Leaf scar

Terminal bud scale scar (marks end of year's growth)

Lateral buds

Leaf scar

Terminal bud scale scar

Present year's growth

Growth from previous year

Growth from two years previous

FIGURE 29.14
Shoot Anatomy
This dormant deciduous woody stem tip reveals two years of growth history. The dormant terminal bud will emerge as the next season's growing tip. The last season's growth can be determined by measuring the distance between the terminal bud and the first group of terminal bud scale scars.

that shed their leaves in the autumn, such as oak, maple, or hickory trees (Figure 29.14). When a leaf falls from such a plant, it leaves a scar that remains visible for a time on the young branch. Above each **leaf scar** is a tiny **lateral** or **axillary bud**, which in the younger region may produce leafy stems in the spring. The age of a young woody stem can be determined by counting the number of groupings of **terminal bud scale scars** below the **terminal bud** at the stem tip.

Lateral buds may be inhibited from sprouting into lateral branches by the presence of the terminal bud. This phenome-

non is called **apical dominance**. Under such inhibition, the uppermost growing tip grows fastest, with a decreasing gradient of inhibition forming below. Thus a plant may take on a triangular shape, or at least it will have a somewhat stepped arrangement in its branches. This, of course, provides for greater exposure to the light. But, under certain conditions, depending on the position of the bud relative to the apex and other factors, the lateral branch bud may be released from its dormancy and sprout into a lateral branch. Or, if a principal branch is lost, a nearby dormant apical meristem, suddenly released from its inhibited state, will begin a growth surge, thus replacing the branch. The hormonal basis for apical dominance is discussed in Chapter 31.

Leaves

In its development, each leaf requires contributions from the three primary meristems. Protoderm contributes to the highly specialized epidermis, which, you will recall, has the conflicting tasks of slowing water loss yet admitting air. The light-trapping photosynthetic cells within the leaf are essentially parenchyma, arising from ground meristem. Also originating from ground meristem are the many collenchyma and sclerenchyma cells that form important supporting elements. And finally, because of the activity of the procambium, the leaf contains an extensive vascular system, with xylem and phloem elements, which brings in water and minerals and takes away newly produced sugars, respectively.

Anatomy of the Leaf
The typical dicot leaf (Figure 29.15) is attached to the stem by a stalk-like **petiole**, which extends to the flattened **leaf blade**. (Are there people who wonder if plants are dangerous because they have pistils and blades?) The vascular system of the stem passes into the leaf through the petiole and into the blade along a large central vein called a **midrib**. The central midrib has a number of major branching veins that penetrate the blade on either side, branching and rebranching so that no cell in the blade is very far from the vascular system. The smaller veins are surrounded by specialized cells that form **bundle sheaths**, and all materials passing in or out of the veins must pass through the bundle sheath cells. (You may recall the special role of the bundle sheath cells in C4 plants discussed in Chapter 7.)

Looking at the remaining tissue organization of the dicot leaf (Figure 29.16), we see that the **upper epidermis** (that side most exposed to the sun) consists of large, simple, flattened cells whose outer walls are typically shiny—coated with a waxy layer of cutin. These cells lack chloroplasts and are often translucent. Within the leaf are layers of photosynthetic parenchyma, or **leaf mesophyll**, including the tightly packed, vertically arranged, **palisade parenchyma** and, below this, the

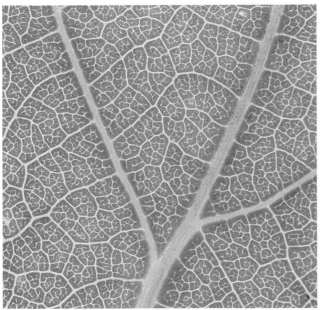

Branching veins (light microscope)

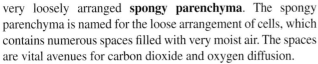

FIGURE 29.15
Dicot Leaves
Dicot leaves (*top*) are typically net-veined, with a major, central vein giving rise to smaller veins at either side. The smaller veins subdivide to form finer and finer veins. (*Bottom*) A variety of dicot leaves.

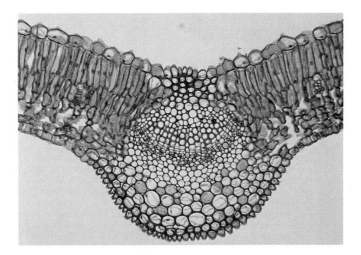

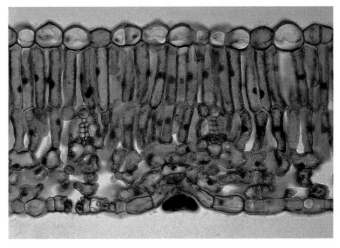

FIGURE 29.16
Dicot Leaf Anatomy
A cross section through the midrib of the dicot leaf (*top*) reveals the central vein of xylem and phloem and supporting cells, and a dense area of strengthening collenchyma below. In a cross section through a leaf blade (*bottom*), the tightly packed layer of vertical cells is the palisade parenchyma. Below this is the loosely arranged spongy parenchyma, with its many spaces containing water and moist air. The lower epidermis is interrupted by a number of pore-like stomata, formed by paired guard cells.

very loosely arranged **spongy parenchyma**. The spongy parenchyma is named for the loose arrangement of cells, which contains numerous spaces filled with very moist air. The spaces are vital avenues for carbon dioxide and oxygen diffusion.

Carbon dioxide enters and water vapor exits the leaf through numerous pore-like stomata. Each stoma is a tiny pore formed by two curved guard cells, which, through changes in shape, can open and close the pores (see Figure 30.8). Technically, each pair of guard cells and the pore they form are called a **stomatal apparatus**, but we will usually just call it a stoma.

In many plants, the stomata are found primarily in the **lower epidermis**—the underside of the leaf. Lower (and often upper) epidermal cells also often produce lengthy **leaf hairs**, which give the surface a fuzzy appearance. Leaf hairs tend to impede the movement of air over the pore-marked surface, cutting down on evaporation in windy weather. In addition, some hairs are sharp and hooked and can repel voracious insect larvae. Some hairs are glandular, containing chemical toxins or sticky, trapping substances. Some even contain "appetite suppressants" that inhibit insect feeding behavior.

The monocot leaf (Figure 29.17) is quite different from the dicot in external appearance. Grasses such as corn have no petioles. The leaf consists of two parts: the **leaf sheath**, which surrounds the stem, and the blade, which extends outward. The sheath is extremely tough, as you can attest if you have ever tried to tear off a corn or bamboo leaf. In addition, the veins do not branch from a central midrib. Monocots have parallel veins, interconnected by smaller veins.

Secondary Growth in the Stem

While primary growth provides the growing plant with its basic tissues and accounts for growth in length, secondary growth produces growth in thickness. As you know, secondary growth among angiosperms is characteristic of dicots and is most notable in woody shrubs and trees. It arises chiefly from regions of undifferentiated procambium that become active, forming vascular cambium. Some monocots (for example, palms) also have secondary growth, but it is not derived from vascular cambium. Since secondary growth in roots and stems is similar in many ways, we will confine our discussion to the stem.

Transition to Secondary Growth

As we have seen, primary growth in the dicot stem ends with the cylindrical vascular bundles becoming arranged into a ring. A tiny region of procambium remains between patches of xylem and phloem elements within the bundles. Between the bundles are areas of cortex consisting of versatile parenchyma cells. We'll focus our attention on theses two regions. A close look at Figure 29.18, the highlights of secondary growth, will help as you proceed.

As the procambium between the xylem and phloem becomes active, rapid cell division ensues, and the expanding tissue links up with the parenchyma cells *between the vascular bundles*. This forms a complete cylinder of active tissue, the **vascular cambium**. As the cells in the ring of newly formed vascular cambium divide, those inside will differentiate into new xylem, and those outside into new phloem (just as the procambium did in primary growth). It is more accurate now to refer to these tissues as **secondary xylem** and **secondary phloem**. As growth continues, the vascular cambium forms a ring of rapidly expanding tissue (Figure 29.19), with secondary xylem forming inside the ring and secondary phloem forming outside. As you might expect, pressure resulting from the rapid production and growth of new cells in internal regions of the stem can cause mechanical problems. As new phloem and xylem emerge, their outwardly expanding front reaches delicate older phloem and parenchyma tissues, crushing them and eventually obliterating this primary organization. Soon only the tougher primary xylem remains in the pithy center to mark the earlier stages. The newly emerging xylem enlarges and matures, and as it grows it pushes the vascular cambium outward. Likewise, the new phloem pushes outward, crushing the fragile parenchyma tissue of the cortex. Eventually, the ring of vascular cambium and phloem comes to rest near the perimeter of the stem.

How does the plant handle these pushing and crushing forces? Wouldn't such force split the stem? As it turns out, vas-

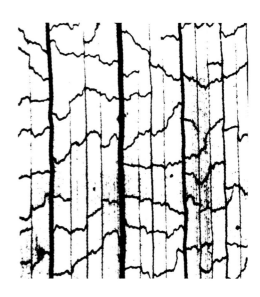

FIGURE 29.17
Monocot Leaf Venation
In corn and other grasses, the leaf is connected to the stem by a tough sheath. Monocot leaves are parallel-veined, with a number of similar sized veins running alongside each other. The highest magnification reveals that the parallel veins are connected by a fine network of smaller veins.

FIGURE 29.18
From Primary to Secondary Growth
Primary growth begins at the stem tip and is
completed over time. **(a)** In the youngest area, the
vascular system comprises a number of bundles,
forming a ring around the stem. **(b)** In the
transition, procambium in the bundles expands,
linking up with parenchyma outside and forming a
ring of vascular cambium. **(c)** Here and in older
regions, vascular cambium produces secondary
xylem on its inner side and secondary phloem on
its outer. The remnants of the primary xylem and
phloem are still visible, but as the years go by they
will be obliterated by new growth.

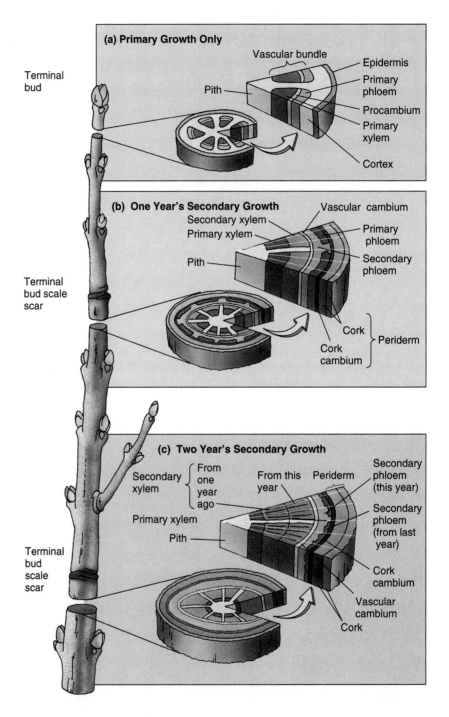

cular cambium also contributes spoke-like lines of parenchyma
cells called **vascular rays**, or simply *rays*. Vascular rays store
materials and provide a means of lateral transport, but, more to
the point, their expansion helps accommodate the stress created
by expanding cylinders of xylem. A woody stem showing the
results of two seasons of growth is seen in Figure 29.20.

Growth of the Periderm

Secondary growth in the vascular cambium is accompanied by
changes in the shoot epidermis, which is replaced by the **peri-**
derm. Ever versatile parenchymal cells in the cortex below
the epidermis form **cork cambium** and take on a new role
(Figure 29.21). The cork cambium, like the vascular cambium,
undergoes continuous cell division, contributing layer after
layer of **cork**. Cork consists of large, thin-walled parenchyma
cells. As they form, they expand outward, rupturing and
replacing the old epidermis. This new tissue forms the shoot's
periderm, whose function is to provide a tough, water-resistant
covering over the maturing stem. The outer cells of the periderm, what people mistakenly call bark, will be impregnated
with waterproofing suberin. (Technically, bark refers to all

stem tissue outside the secondary xylem, including the vascular cambium, phloem, and periderm.) As cork cambium divides in some plants, it produces inner layers of cells as well, forming a tissue called phelloderm (see Figure 29.21).

The cork in wine bottles is true cork. But it doesn't occur naturally, not even on the cork oak from which it is obtained. Cork growers must remove the natural periderm of the oak, causing the tree to respond to the injury by forming a new, smoother regrowth. This is then stripped away periodically, cut into cylinders, and placed into wine bottles.

In nature, the cork layer is highly water-resistant, but it is not entirely sealed. The living tissues below require a constant supply of oxygen, and they must release carbon dioxide, just as we would expect from metabolically active cells. It turns out that this vital gas exchange occurs through minute openings in the periderm known as **lenticels** (see Figure 29.20). Lenticels are most numerous over metabolically active stem regions, and they are also found in the outer skin of pears, apples, and some other fruit.

The Older Woody Stem

Older regions of the woody stem, those that have gone through several seasons of secondary growth, are mostly composed of nonliving xylem tissue—the "wood." Some of this is **heartwood**, a region, commonly darker in color, that no longer conducts water and minerals. (Heartwood is responsible for the beauty of natural wood products such as fine furniture and paneling.) The most recent layers of xylem, the **sapwood**, are those that conduct water and minerals.

The vascular cambium, phloem, and periderm—all lying outside the xylem—form a rather thin cylinder of living material around the perimeter of the stem. In fact, while many large trees can survive a burned out interior or woody region, removing a strip of bark all the way around the trunk (called girdling) will kill even the most robust forest giant. Why is girdling so effective? Consider what the loss of phloem might mean to the roots of a tree.

Annual Rings. Cross sections cut through the stems of temperate zone trees, trees that undergo seasonal growth, reveal very prominent **annual rings** (see Figure 29.20). Annual rings form because temperate zone trees grow rapidly in spring and summer, tapering off in autumn and stopping growth in winter. The earliest layers of xylem cells are very large in diameter, while those produced later in the season tend to be smaller. There is normally less water available and less time for cell enlargement as summer passes and autumn draws near. The tree's annual rings can be counted to determine the age of the tree. Annual rings can be compared over the years, telling us something about climatic change, particularly in very old trees, such as the sequoia and bristlecone pine, or in fossilized tree trunks. The poor growing seasons result in smaller cells and thus thinner annual rings. More recently, researchers have begun to correlate ring thickness with another factor, increasing levels of air pollutants such as ozone and acid rain.

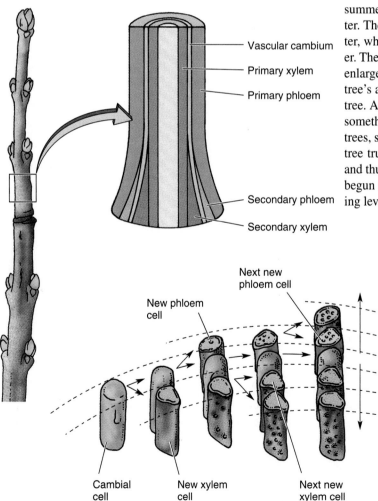

Vascular cambium
Primary xylem
Primary phloem

Secondary phloem
Secondary xylem

Next new
phloem cell

New phloem
cell

Cambial
cell

New xylem
cell

Next new
xylem cell

FIGURE 29.19
Mechanics of Secondary Growth
The activity of the vascular cambium is seen in a continuing scenario. In one division, a new xylem cell is added to the inner tissues, and in the next, new phloem is added to the outer tissues. As this continues, the woody plant becomes thicker and thicker.

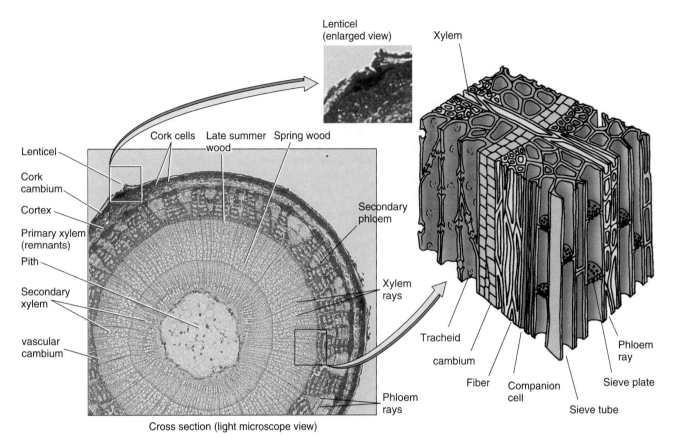

Cross section (light microscope view)

FIGURE 29.20
Older Woody Stem
Several different tissues are visible in the stained cross section of a two-year-old woody stem. The outer
ring of cork and cork cambium is followed within by a region of cortex, then by a ring of triangularly
arranged phloem and phloem rays. Just within this is a thin line of dark cells, the vascular cambium, and
on its inner side, wide, distinct bands of xylem, the annual rings. The lines radiating outward in the xylem
are the xylem rays. An enlarged view of a lenticel shows its cellular structure. The lenticel is an avenue of
gas exchange between the active tissues of the stem and the outside.

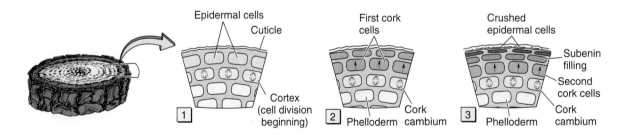

FIGURE 29.21
Formation of the Periderm
Secondary growth also includes replacement of the epidermis by the periderm. This happens when
parenchyma just outside the phloem is transformed into cork cambium. Continued cell division in the cork
cambium contributes cells to the cork layer outside and to the phelloderm within. The outermost cork
layer is impregnated with waterproofing materials and constantly cracks and sheds as growth goes on.

Essay 29.1 ● THE PROBLEM OF DIFFERENTIATION

Scientists have indeed uncovered detail after detail about plant development until it seems that we surely must have the big problems solved. But many questions remain: "How do plant cells differentiate?" This, of course, is a most fundamental question, so we might wonder what we really do know. The answer is essentially the same for plants and animals: we are just beginning to find answers.

Botanists, in particular, have provided valuable basic information. This may well be because in some ways plants make ideal subjects for studying cell growth and proliferation under controlled conditions. For one thing, many of them can regenerate from bits and pieces. Many plants can regenerate roots from stem cuttings, stems from bits of root, and even entire plants from leaves. This knowledge has been invaluable in agriculture through the ages, but for biologists it suggests clues to the puzzle of differentiation. It supports the idea, for example, that plant tissues are *totipotent;* that is, their cell nuclei retain the capability needed to produce the entire organism from which they come.

Plant researchers turned to tissue culture methods in the 1930s and 1940s. A pioneer in these efforts, Johannes van Overbeek, discovered that a suitable medium was coconut milk. Besides important nutrients, coconut milk contains some critical hormones needed for plant growth. Van Overbeek (Experiment 1) succeeded in growing individual cells that he had separated out from young carrot embryos. He cultured them in his coconut milk and later planted them in soil. They produced normal adult carrot plants. The results indicated that cells in the embryo, at least, were totipotent.

It wasn't until the late 1950s, however, that mature tissue was first cultured in a similar manner by F. C. Steward at Cornell University. Mature carrot cells were successfully removed and grown in a nutrient medium. The mature cells grew into root-like structures that, when planted, produced entire carrot plants. More recently, there has been excellent progress in culturing redwood trees, orchids, and many other plants. Plant physiologists have also succeeded in producing clones of potato varieties from cells in the leaf. In so doing, they have provided agriculture with a potentially economical way of growing potatoes that are free of viruses. The potato clone also promises to lend itself to investigations into the unyielding mysteries of development.

Extending the earlier work, researchers at Kansas State University obtained experimental material from leaf cells that were first converted to protoplasts. Protoplasts (Experiment 2) are plant cells with no cell walls. Enzymes were used to remove cell walls and intercellular materials, and the cells are left behind. The new protoplasts then massed together, entered cell division, and produced new cell walls. The resulting tissue was then transferred sequentially through several types of culture media. Each medium contained certain mixes of hormones, each of which had its own effect on differentiation. The first mix produced a large undifferentiated mass of cells called "callus." The second encouraged shoot growth and differentiation, and in response to a third mix, the tissue produced roots. Newly differentiated plants were then transferred to regular soil beds, where they were grown to maturity. This procedure shows once again

that mature plant cells lose none of their original genetic potency.

The benefits of such work are clear. Through cloning, agriculturists can produce highly selected, unvarying crop plants (such as virus-free strains). In keeping with the new era of genetics and molecular biology, protoplasts can be used in gene splicing and recombinant DNA studies. Researchers have succeeded in fusing nuclei in potato and tomato protoplasts, producing a hybrid that would not readily occur in nature or through sexual, genetic crosses. Such readily manipulated experimental organisms hold great potential for the ongoing study of development.

A more immediate goal in such efforts is to confer the disease resistance of one species onto another, less resistant one. For example, the tomato plant used in the fusion described above carries with it genes that will enable the potato to resist the water mold that causes *late blight* (see Chapter 23). Not only could this eliminate a constant and serious threat to an important food supply, but it also could eliminate the widespread use of certain dangerous chemicals—the fungicides now used to control the blight.

Plant geneticists can also use their new techniques to retain the original genetic variability of older potato strains. Who needs variable potatoes? The answer is, we do. In fact, scientists have begun developing "gene banks" to preserve the variation in the original phenotypes. Some biologists are alarmed at what might happen if our highly selected food strains should be decimated by some new fungal or bacterial mutant or per-

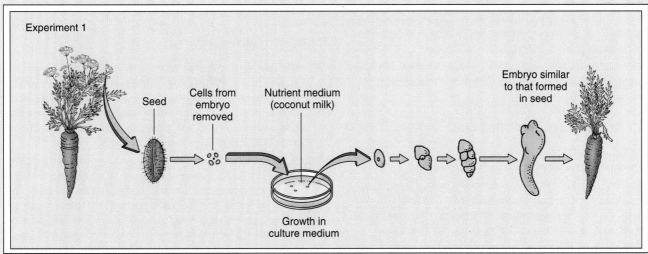

Experiment 1

Seed — Cells from embryo removed — Nutrient medium (coconut milk) — Growth in culture medium — Embryo similar to that formed in seed

continued

Essay 29.1 ● **THE PROBLEM OF DIFFERENTIATION** *continued*

haps some environmental variable that we can't control.

We see then that many cells retain the capability necessary to produce the entire plant. Second, when removed from their cell associations, cells can regress to an undifferentiated state from which they can then proceed through normal embryonic development.

How does this answer the question about differentiation? It tells us both that genetic information is not lost as development proceeds, and that it is not irreversibly repressed.

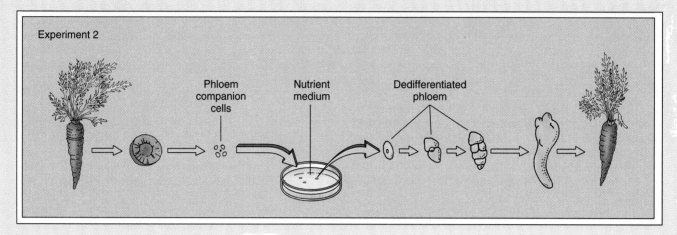

Experiment 2

Phloem companion cells

Nutrient medium

Dedifferentiated phloem

Key Ideas

GERMINATION AND GROWTH OF THE SEEDLING

1. For many seeds, **germination** simply requires the uptake of water, suitable temperatures, and the availability of oxygen; but others require special conditions such as exposure to freezing temperatures, fire, abrasion, or exposure to animal digestive enzymes.

2. In the corn seedling, the radicle emerges from the coleorhiza and grows downward, absorbing water and anchoring the seedling. The coleoptile-enclosed shoot grows upward, breaks free, and unfurls the first leaf. Seedling growth in the bean, a dicot, begins with the emergence of the hypocotyl, which forms a bumper-like **hypocotyl hook**. Its growth raises the seed out of the soil, whereupon the epicotyl rapidly elongates and spreads the plumule out to face the sunlight.

3. **Primary growth** includes growth in length in the shoot and root, along with the production of leaves, flowers, branches, and branch roots. **Secondary growth**, seen primarily in dicots (and in conifers), provides for growth in thickness.

4. Perennial plants have **open growth**, or **indeterminate life span** (unlimited growth). Continuous growth depends on undifferentiated meristematic tissue.

TISSUE ORGANIZATION IN THE PLANT

1. Apical meristem produces three kinds of primary meristems: **protoderm**, **ground meristem**, and **procambium**.

2. Protoderm produces the leaf **epidermis**; **guard cells**, which surround the pores, forming the **stomata**; root epidermis with **root hairs**; and the complex **periderm**. Waterproofing materials include **suberin** (in the periderm) and **cutin** of the leaf or stem cuticle.

3. Ground meristem differentiates into the widely distributed, thin-walled **parenchyma** of leaves, stems, and roots; the thick-walled,

supporting **collenchyma fibers** of stems; and the **sclerenchyma**, including **sclereids**, which occur as **stone cells** in pears.

4. Xylem includes water-conducting **tracheids** and **vessel elements** and surrounding xylem fibers. The tracheids are long and slender and have pitted end and side walls. Vessel elements—the cellular units of **vessels**—are wider and shorter, with open-ended walls and pitted side walls.

5. Phloem consists of food-conducting **sieve elements**, **phloem parenchyma**, **phloem fibers**, and **companion cells**. Sieve elements include **sieve tube members**, the latter of which have **sieve plates**. When assembled, sieve tube members form **sieve tubes** with a continuous cytoplasm. Nuclear functions are carried out by companion cells, which are also involved in the active transport of nutrients into sieve tubes. Phloem parenchyma specializes in storage and active transport.

PRIMARY GROWTH IN THE ROOT

1. The **root tip** contains the root apical meristem, which gives rise through active cell division to replacement cells for the **root cap** region ahead and primary tissues behind. The **root apical meristem** also contains an area of less rapidly dividing cells called the **quiescent center**. Behind the meristem, cells stimulated by the hormone auxin elongate, pushing the root tip through the soil.

2. In the primary tissues, protoderm produces the young epidermis, which forms numerous root hairs, while ground meristem begins the formation of the root **cortex** and **endodermis**. Procambium forms the xylem and phloem of the **stele**, a cylinder of conducting and supporting tissue. The stele includes an outer **pericycle** and regions of primary xylem and primary phloem. In plants capable of secondary growth, procambium forms regions of **vascular cambium**, which later gives rise to **secondary**

xylem and **secondary phloem**. The pericycle gives rise to branch roots.

3. Monocot root systems often include **adventitious roots** that emerge from the stem. **Tap root systems** have a prominent main root and many branching or secondary roots, while all of the roots in **diffuse root systems** are similar in size.

PRIMARY GROWTH IN THE STEM

1. Apical meristem produces new shoots. When the young primary tissues differentiate, they form a thin epidermis along the outside and islands of vascular tissue that surround an inner region of thin-walled parenchyma called the **pith**.

2. Meristem also gives rise to leaves, **leaf primordia**, and **branch primordia** at **nodes**. The expanses of stem between nodes are called **internodes**. **Lateral branch buds** or **axillary buds** give rise to branches unless suppressed by hormones. **Terminal buds** form **terminal bud scale scars** at the end of each growth season.

LEAVES

1. Protoderm contributes to the complex leaf epidermis, ground meristem differentiates into photosynthetic parenchyma (**leaf mesophyll**), and procambium gives rise to the leaf vascular tissues.

2. The structures of leaves include a supporting **petiole**, the flattened **leaf blade**, and a central vein called a **midrib**. Smaller

veins are surrounded by **bundle sheath cells**, through which materials must pass.

3. A cross section through the leaf reveals a simple **upper epidermis**, then **palisade parenchyma**, followed by **spongy parenchyma** with air spaces, and then the more complex **lower epidermis**. The lower epidermis includes cells with extensive **leaf hairs** and the pore-like stomata.

4. While veins in dicot leaves tend to be branching (net-veined), those in monocots generally run parallel (parallel-veined).

SECONDARY GROWTH IN THE STEM

1. During secondary growth, secondary xylem and phloem arise from **vascular cambium** (**secondary cambium**). The growth of newly produced **secondary xylem** and **secondary phloem** expands outward, forming a continuous ring and crushing older growth in its path.

2. In secondary growth, the epidermis is replaced by a **periderm**, which includes the **cork cambium** and **cork**. The cork cells are suberized, but gases can be exchanged through the **lenticels**.

3. Older woody stems are composed mainly of xylem or "wood" (nonconducting **heartwood** and conducting **sapwood**). Living tissues are restricted to a thin outer layer called bark.

4. **Annual rings** in the wood represent seasonal growth patterns. Spoke-like patterns—the **vascular rays**, or **rays**—aid in transport and help the stem expand.

Application of Ideas

1. An important characteristic of many plants is indeterminate growth. How does the organization of plants provide for such growth? Do any animals share this characteristic? Is there anything comparable in the other kingdoms? (Consider a clone of protists.)

2. A young boy carved his initials about eye-level in a small tree. Returning as a man, he noted that both he and the tree had grown and that the initials were still there. In what position

were the man's eyes when he read the initials—looking up, down, or straight ahead? Explain your answer with a brief but technical discussion of growth in the woody stem.

3. Dendrochronology involves the study of growth rings in wood stems—both intact and fossilized. Of what do growth rings consist, and what kinds of information might a dendrochronologist gain by their study in plant fossils? How might this information be useful?

Review Questions

1. Compare the early growth of the corn seedling with that of the bean. (page 611)

2. Distinguish between primary and secondary growth. List two groups of plants in which secondary growth is possible. (pages 611–612)

3. List three tissue derivatives of protoderm and ground meristem. (pages 613–615)

4. Compare the structure of tracheids and vessel elements. (page 616)

5. List the four kinds of cells in phloem tissue and state a function for each. (pagea 616–617)

6. What is the function of the sieve tube? List two provisions that make transport possible from one sieve tube member to the next. (pages 616–617)

7. Prepare a diagrammatic drawing of a root tip, labeling the following: *root cap, root hairs, epidermal cells, apical meristem,*

quiescent center, stele, cortex, regions of elongation and maturation, procambium ground meristem, and *protoderm.* (page 618)

8. Draw a cross section of a mature region in a root labeling the *stele, epidermis, endodermis, pericycle, cortex, primary xylem, vascular procambium,* and *primary phloem.* (page 619)

9. Contrast the tap root system with the diffuse root system. In what plants would one most likely find the latter? (page 620)

10. Draw simple representations of a dicot leaf and a monocot leaf, labeling midrib, blade, petiole, leaf sheath, net venation, and parallel venation, as appropriate. (pages 623–625)

11. Explain how secondary growth changes the arrangement of vascular tissue in the stem from bundles to rings. (pages 625–626)

12. Construct a "pie-slice" drawing of a three-year-old stem, and label the important tissue areas. Include the *cork* and *cork cambium, cortex, secondary phloem, phloem rays, vascular cambium, secondary xylem,* and *annual rings.* (pages 626–627)

CHAPTER 30

Plant Transport and Nutrition

CHAPTER OUTLINE

THE MOVEMENT OF WATER AND MINERALS

Transpiration

Water Potential and the Vascular Plant

Root Pressure: Is the Root a Pump?

Water Movement in the Leaf: Transpiration and the Pulling of Water

Water Movement in the Stem

Essay 30.1 Testing the Forces that Move Water

Water Movement in the Root

Guard Cells and Water Transport

FOOD TRANSPORT IN PLANTS

Phloem Transport

GAS TRANSPORT IN PLANTS

PLANT NUTRITION: MINERALS AND THEIR TRANSPORT

Mycorrhiza: Fungal Association with Roots

Plant Mineral Nutrition

Insect-Eating Plants

KEY IDEAS

APPLICATION OF IDEAS

REVIEW QUESTIONS

Humans are awed and fascinated by large organisms. For example, our interest in dinosaurs, their structure and life-style, is almost unquenchable. Among the plants, our attention turns to giants such as the redwoods. Why have so many of the plants become so large? One reason is that there are decided advantages to being tall, especially if one is a forest plant, since here there is intense competition for sunlight. The lofty giants are able to expose their leaves to unobstructed light rays. Also, the vast root system typically produced by large plants allows them good access to water and minerals and provides firm anchorage. Furthermore, tall trees can disperse their seeds and fruit farther than can their shorter contemporaries. Thus there are a number of advantages to being a forest giant.

Great size, however, has its price. The scaly leaves of a giant redwood constantly lose water to the atmosphere. That water must be replaced by a root system in the soil, which is about 100 meters below. On the other hand, those roots must be fed by carbohydrates manufactured in the distant leaves. The plant must bring water and minerals from the soil up to its leafy canopy and food from the leaves down to the roots. Thus large plants have had to pay for their large size by developing extensive vascular systems for the transport of water and nutrients.

The cost of large size does not end there. The high forest canopy must be supported, sometimes against strong winds and heavy rains or snow. Therefore much of the plant tissue must be devoted to physical support. The forest giants, and other higher plants as well, provide both support and water transport with the same tissue—the woody **xylem**. Xylem is well adapted for its functions. It is marvelously strong supporting material, and through it water can be raised great distances above the earth. In past years, this ability to lift water mystified engineers since it is done with little apparent direct expenditure of plant energy. It is important to point out that there are no known active transport mechanisms that directly move individual water molecules.

The other part of the vascular (transport) system is the **phloem**, which specializes in carrying food molecules from their site of synthesis in the leaves to other parts of the plant where they are needed (for example, developing leaves, seeds, fruits, and roots). Food molecules can be moved by known processes that involve energy expenditures, but there are many aspects to this movement that also remain a mystery. For example, food molecules do not haphazardly pass from the leaves but are somehow routed according to the specific requirements of the different tissues in the plant. Furthermore, the distribution can shift as conditions change. Food can even be moved upward from storage regions in the root; in spring, trees and bushes that have gone through the winter with bare, leafless branches routinely move food molecules up from their roots to the newly forming leaves and twigs. Indeed, there are a number of questions regarding the movement of nutrients through the phloem. We will come back to these later, but first let's look at the role of xylem.

The Movement of Water and Minerals

The upward movement of water through a plant (Figure 30.1) begins when water enters the plant through the root epidermis, where there are countless thin-walled root hairs that vastly increase the absorbing surface (see Figure 29.7). Water then crosses the root cortex mainly by moving freely in the very hydrated matrix provided by the porous cellulose cell walls—what is called the apoplastic route. (The area outside the plasma membrane is called the "apoplast.") The water then passes through the cytoplasm of cells of the endodermis and moves into the *stele* (the root vascular cylinder that contains both xylem and phloem; see Figure 29.9). Upon entering the stele, water passes into the xylem, the vessels and tracheids of the root vascular system. It then continues through the xylem of the stem and on to the foliage. There it passes out of the xylem, moving through numerous cells to eventually enter the air spaces in the leaf's spongy mesophyll. It is in the moist air spaces that evaporation occurs, and water, in its gaseous state, diffuses out of the plant through the leaf *stomata*, except for a relatively meager but vital amount that is used in photosynthesis.

Transpiration

Water loss from the leaf through evaporation and diffusion is called **transpiration**. A great deal of water is shifted from the soil to the atmosphere through transpiration. A single potato plant, for instance, can lose about 95 liters of water in one growing season, and in the same amount of time 206 liters of water can evaporate from the leaves of a single corn plant (Table 30.1). Later, in our study of ecology, we will find that the transpiration from a forest has immense ecological and meteorological significance, affecting both organisms and climate. In fact, recent estimates are that a full 50% of the rainfall in the Amazon basin represents water transpired by the vast rain forest.

How is so much water moved through a plant? And more to the point, why is so much water moved? Answering the second question is relatively simple. Plants require water to carry on photosynthesis and to provide the turgor that holds leaves and young stems erect. (Actually, only about 1% of the water passing through a plant is used in photosynthesis.) The first question is a little tougher. The problem of water moving up through a tall tree can be addressed by two hypotheses: (1) something is *pushing* the water up, or (2) something is *pulling* it up. Let's begin to consider water potential and the vascular plant.

Water Potential and the Vascular Plant

Forces that bring about the movement of water in plants, regardless of their origin, do so by affecting **water potential**. Water potential is the potential energy of water in a particular system compared to that of pure water at the same atmospheric pressure and the same temperature. Water potential is abbreviated by the Greek letter psi (ψ). The concept of water potential is based on the free energy of water. Free energy, as you recall, is the thermodynamic property that predicts the direction in which chemical or physical changes will occur, as well as the ability of a system to do work. In plant cells, water potential is

TABLE 30.1 **Transpiration Rates per Day for Selected Plants (in Liters)**	
Cactus	0.02
Tomato	1
Sunflower	5
Ragweed	6
Apple	19
Coconut palm	75
Date palm	450

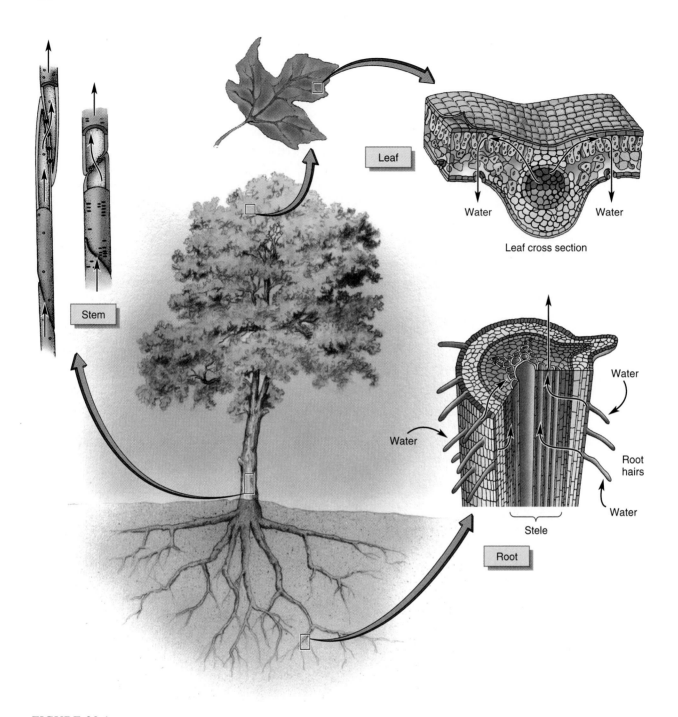

FIGURE 30.1
The Passage of Water Through the Plant

The movement of water through a plant begins at the root, where epidermal cells in the root tip, aided by slender root hairs, take in water from the surrounding soil. From there the water passes across the root to the stele, where it enters the vascular system for the trip upward through the stem to the leaves, and out of the plant through its many stomata.

largely determined by two factors: (1) the concentration of dissolved solutes in a cell, which reduces water potential, and (2) the hydrostatic pressure, which increases water potential. Because water potential is related to the free energy of water, it is measured in units of pressure called megapascals (MPa): 1 MPa is equal to about 10 atmospheres of pressure (an atmosphere is the pressure exerted at sea level by an imaginary col-

umn of air). As the potential energy of water is increased, the water potential increases. Decreasing the potential energy of water will similarly decrease the water potential. The movement of water in plants is mediated by water potential differences, so plant physiologists use water potential measurements to predict which way water will move in a plant and to assess whether a plant is under stress from lack of water.

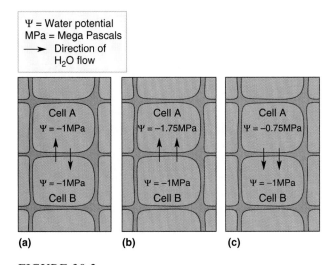

Ψ = Water potential
MPa = Mega Pascals
→ Direction of H₂O flow

(a) Cell A Ψ = −1MPa / Ψ = −1MPa Cell B

(b) Cell A Ψ = −1.75MPa / Ψ = −1MPa Cell B

(c) Cell A Ψ = −0.75MPa / Ψ = −1MPa Cell B

FIGURE 30.2
Water Potential in Cells
Water potential (Pa) is measured in megapascals (MPa), represented by the Greek letter psi (ψ). Note that all values here are negative. This is because the highest value of ψ, that of distilled water, is 1. In each pair of cells water moves from areas of higher to lower water potential, which in the cell is inversely related to the amount of solute present. The greater the solute the less the water potential, and the higher the negative number.

This whole matter is a lot easier to understand if you keep in mind that water will always move from a region of higher water potential to a region of lower water potential. The words "pressure," "free energy," and "potential" should give you the idea that water has the ability to do work as it moves from a region of higher water potential to a region of lower water potential. That work includes expanding cells during growth, opening stomata, and affecting some leaf movements.

Of the several factors that interact to determine the water potential in a cell, perhaps the most familiar way in which water potential may be altered is through changes in its solute concentration. Recall that solutes are ions or molecules in solution. Consider two cells with the same water potential (Figure 30.2). Addition of solutes to pure water (where by definition ψ = 0) has the effect of lowering the water potential. So the water potential of any solution at atmospheric pressure will be a negative number. If the first cell has its solute concentration increased (perhaps by active transport), then that cell's water potential (for example, ψ = −1.75 MPa) would be lower than the unchanged cell alongside. Water from the second cell would pass into the first. On the other hand, if the first cell had its solute concentration decreased, its water potential would be greater than that of the second cell, and the reverse movement of water would occur. Again, water moves from regions of higher water potential to regions of lower potential.

In relating water potential to the functioning plant, we can point to a well-substantiated observation: water moves from the soil into the root and from the root through the stem to the leaves, and finally out of the leaves to the surrounding air. Thus we can infer that the water potential is high in the

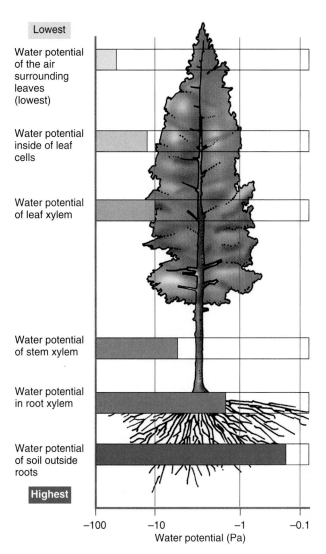

FIGURE 30.3
Water Potential and the Total Plant
The movement of water and minerals through a plant follows an established water potential gradient from soil water to air surrounding the leaf. As the measurements indicate, the highest water potential is in the soil water and the lowest is in the air surrounding the plant. In between is found a steadily decreasing gradient.

soil, somewhat lower in the plant, and lowest in the surrounding atmosphere (Figure 30.3). When, for any reason, this gradient is disrupted, the plant is in trouble. This can happen, for example, when solutes in the soil water (such as nitrates and phosphates in fertilizers) are too concentrated around the plant. This would decrease the water potential in the soil water outside the root. Thus water would tend to move out of the root. In response, the leaves would wilt, and with the subsequent loss of leaf cell turgor, they would collapse. Fortunately, the loss of turgor also occurs in guard cells, which would thereby change shape, closing the stomata and slowing transpiration. If the normal water potential gradient is not disrupted too long, the leaf changes would be reversed.

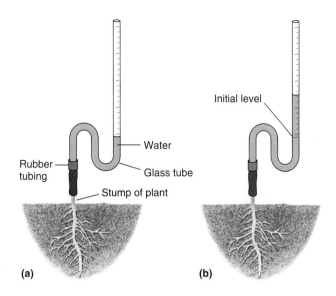

FIGURE 30.4
Detecting Root Pressure
(**a**) Root pressure can be measured by substituting a manometer for the stem and foliage of a potted plant. (**b**) Water passing into the root and up through the stem rises in the tube according to the pressure exerted.

Root Pressure: Is the Root a Pump?

Have you ever observed moisture on the lawn first thing in the morning? At the time, you may have thought the moisture was caused by condensation of water from the air. In some cases it is, but often the presence of water drops on the grass is the result of its having been pushed up by the roots through the plant's vascular system. The blades of grass have special openings near their tips, where excess water can escape. Such loss of water in its *liquid* phase is known as **guttation**, not to be confused with transpiration, which involves the evaporation of water. Guttation occurs mainly at night, when evaporation is reduced and water excesses occur. Furthermore, water droplets forced out through guttation occur in a definite pattern, not seen in dewdrops. The water has been forced up to the tips of the grass blades by *root pressure* from below.

Root pressure, a water-moving force developed in the root, occurs when ions are actively secreted into the stele from nearby living cells. The presence of ions in the xylem fluids lowers the water potential, and water moves into the stele. The endodermis, we know, has the ability to actively transport ions in one direction, resisting their diffusion back into the cortex. If the root is a pump, then the pressure the pump generates can be traced back to an energy-requiring mechanism, the active transport of ions. It creates a condition favoring the passive inward movement of water.

Root pressure can be demonstrated by removing the stem from a small plant and attaching a mercury manometer (a device for measuring fluid or air pressure) in its place (Figure 30.4). A rise in the manometer's column of mercury is used as a measure of pressure. With this apparatus it has been possible

to demonstrate that roots can generate pressures of about 3–5 atmospheres (that is, 3–5 times greater than that of the atmosphere). So the immediate question is: Is this sufficient to push water to the top of a tall tree?

The answer is no. The weight of a 100-meter column of water is at least two times too great to be supported by a root pressure of even 5 atmospheres. That is, over 10 atmospheres is the minimal requirement, twice the pressure demonstrated experimentally. The weight of such a column ought to push water right out of the vascular cylinder, through the root cortex, and back into the earth. Yet trees continue to stand there with water constantly moving upward through their enormously tall stems and out the lofty foliage.

Most plant physiologists now believe that root pressure, when it occurs, is only an indirect effect of the active transport of inorganic nutrients, such as nitrates, potassium, and phosphates, into the roots. So we find that roots can provide only a modest push, certainly not enough to explain water transport through tall plants. Therefore we must consider the alternate explanation, the possibility of a pulling force. Is it possible to pull water?

Water Movement in the Leaf: Transpiration and the Pulling of Water

Plants have to be porous; they cannot be waterproof. If photosynthesis is to occur, carbon dioxide must be able to diffuse from the surrounding air into the leaves. Any such avenue permitting the inward diffusion of gases must likewise permit the escape of gases, and such gases include water vapor. So the loss of water by photosynthesizing plants is unavoidable. Let's review this common avenue of gas exchange.

The tissue organization of the typical dicot leaf, as you will recall (see Figure 29.16), includes an upper epidermis, a dense palisade parenchyma just within, and a very loosely arranged spongy parenchyma below this. Both groups of parenchymal cells are photosynthetic. Finally, there is the lower epidermis. Numerous stomata occur in the lower epidermis and often the upper epidermis as well (Figure 30.5).

The open moist air spaces between parenchymal cells are avenues of diffusion, both for incoming carbon dioxide and for outgoing water vapor. The rate of water loss depends primarily on two factors: (1) temperature at the leaf surface and air spaces and (2) relative humidity in the air outside the leaf. Each of these factors profoundly affects the rate of transpiration from the leaf. As many a dismayed gardener has found, the rate of water loss in plants on hot, dry days far exceeds such losses on humid days. Evaporation from the leaf surface immediately lowers water potential in tissues just within the leaf, setting in motion the movement of water throughout the plant. Let's look at the details.

On a "nice" day, 20°C (about 68°F.) and 50% relative humidity, the water potential of the air is approximately –93.5 MPa. That of a typical leaf epidermal cell is about –2.0 MPa. Water will move from an area of higher water potential to an area of lower water potential, so there is a very strong gradient that pulls water out of the leaf through the stomata. Water evaporating from the leaf spaces is replaced by water from the cells surrounding the spaces (see Figure 30.5). This leaves these cells with a reduced water potential. The result is that water will move into these cells from others even closer to the source. The source, of course, is the water-filled xylem of the leaf veins. Thus the continued evaporation of water establishes a steep water potential gradient between the vascular system and the spongy region of the leaf, and water tends to move rapidly down its gradient. In addition to its movement within the hydrated cell walls, water movement through the leaf cells is facilitated by the presence of numerous **plasmodesmata**—minute cytoplasmic connections passing through pores in adjacent cell walls and connecting the cytoplasm of one cell to that of another (see Chapter 5).

Transpiration at the leaf surface translates into a powerful water potential gradient within the leaf, which produces a major pulling force. Measurements reveal that this force, in water-depleted leaf cells, creates pressures up to 12 atmospheres—enough to lift a xylem-sized column of water 130 m high! This force is therefore quite sufficient to raise water from the soil to the tops of the tallest trees.

We have seen that the movement of water from cell to cell in the leaf is always along a gradient from higher water potential to lower water potential, and that it is a *passive* process, not a direct result of biochemical work done by the plant. The process is powered by the water potential gradient set up by transpiration, the evaporation of water from the air spaces of the leaf.

Water Movement in the Stem

Four important forces influence the movement of water in the xylem elements: *transpiration*, *adhesion*, *cohesion*, and *tension*. We have already discussed transpiration as a major force in creating the water potential gradient, so let's briefly look at the roles of the others. The second force, **adhesion**, refers to the attraction of water molecules to materials such as glass or the cellulose in plant cell walls. **Cohesion** is the attraction of certain molecules to each other. (We considered both in detail in our discussion of water in Chapter 2. You may recall that it is the polarity of water that attracts the molecules to each other.) **Tension**, the fourth force, refers to the stress produced by exerting a pull on

a narrow column of water—namely, the pull on water in the vessels and tracheids of xylem during transpiration.

The adhesion of water to cellulose cell walls may be a major water-moving force throughout the plant since it produces water's "creeping" movement from cell to cell and through the cellulose-lined vessels and tracheids of the xylem.

Cohesion—the attractive force between the molecules of a substance—is particularly evident in water. The positive (hydrogen) ends of one molecule are attracted to the negative (oxygen) ends of the next, forming countless, shifting hydrogen bonds. It is cohesion that gives water its physical viscosity ("liquid-stickiness"). When water is present in a cylinder of narrow diameter such as that of the vessel or tracheid, its cohesive forces are greatly magnified. Such a column has significant tensile strength—the ability to be pulled without breaking.

Tension, as we mentioned, is the stress placed on an object by a pulling force. As you might expect, when a plant is heavily transpiring, the tension on the water columns within the many xylem vessels and tracheids can be enormous. In fact, the pull has a measurable squeezing effect on the trunks of trees; their trunk diameters actually decrease during intense transpiration! (Some observers report a faint sucking sound when the xylem is punctured.) Essay 30.1 explains how such delicate trunk measurements are made. The important point is that as tension increases, the cohesive strength of water in the xylem elements and their adhesion to the xylem walls prevent breakage of the column of water molecules. The pull on the column is transferred downward, creating a decrease in water

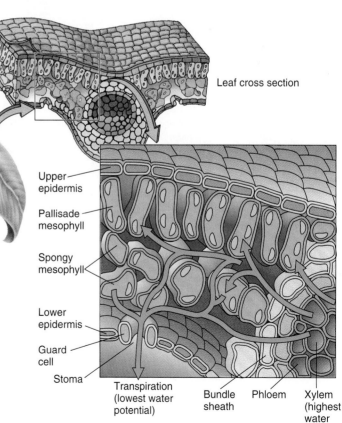

Leaf cross section

Upper epidermis

Pallisade mesophyll

Spongy mesophyll

Lower epidermis

Guard cell

Stoma

Transpiration (lowest water potential)

Bundle sheath

Phloem

Xylem (highest water potential)

FIGURE 30.5
Water Potential in the Leaf
Evaporation from the leaf sets up a water potential gradient between the outside air and the leaf's air spaces. The gradient is transmitted into the photosynthetic cells and on to the water-filled xylem in the leaf vein.

Essay 30.1 ● TESTING THE FORCES THAT MOVE WATER

Elements of the four forces involved in water movement in plants can readily be tested. To demonstrate cohesion and tension, simple techniques similar to those used to demonstrate root pressure can be applied. (a) The stem of a plant is cut (under water so as not to break the columns of bonded water molecules) and connected to a water-filled glass tube. (If dye is added to the water, it will soon appear in the leaves and flowers; florists sometimes use this trick to dye flowers odd and unnatural colors.) If the lower end of the water-filled glass tube is put into a dish of mercury, the column of mercury rises as the water evaporates above, indicating a strong pull. But how can it be demonstrated that it is evaporation and not some other force that is involved in transpiration? A clever mechanical model seems to support the evaporation principle. (b) The setup is the same as before, but this time the top of the glass tube is attached to a porous clay *potometer* instead of the crown of a plant. (Potometer—from *potare,* "to drink"—is a device for permitting water to escape by evaporation.) There is no air in the potometer, only water. The wetted microscopic pores of the clay potometer permit water to literally creep out to the surface of the clay cylinder through the forces of adhesion and cohesion (much like what is believed to go on in the spongy tissue and air spaces of the leaf). Moisture evaporates from the damp outer surface, and the column of water rises, creating a partial vacuum in the tubing that is filled by the rising column of mercury. The rise is much faster if a fan is used to blow air around the potometer, just as transpiration would be more rapid on a dry, windy day.

D. T. MacDougal, a plant physiologist, wondered about the enormous tension that must be placed on water columns in the immense xylem of a tree if the transpiration pull idea were correct. He reasoned that such a tension might even affect the stem diameter, perhaps causing an actual decrease during peak activity. (c) MacDougal tested this idea by devising an instrument that could measure minute changes in the stem. The instrument, a *dendrometer* (*dendro,* "tree"; a device for measuring tree girth) did, in fact, detect the expected changes. Not only did the stem decrease in diameter, but it did so in regular day–night cycles. It is common knowledge that plants transpire much less at night—in the absence of sunlight—when evaporation rates are low, and MacDougal's measurements coincided with this very nicely. How might increased temperature accompanied by decreased humidity have affected MacDougal's dendrograph tracings?

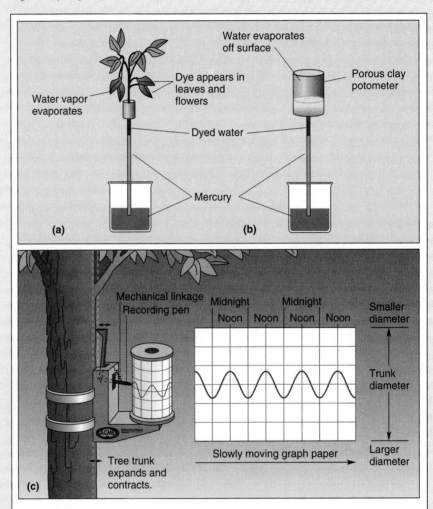

(**a**) The pulling force of transpiration can be demonstrated. As the foliage transpires, dye appears in the leaves and water is drawn into the tube with a force that lifts the mercury up the tube. (**b**) Evaporation from a clay potometer is analogous to evaporation from the leaf surface. (**c**) The dendograph tracing shows changes in a tree's diameter over several days. The peaks and valleys occurring in 12-hour intervals correspond to periods of maximum and minimum transpiration. Plants transpire most at midday and least at night.

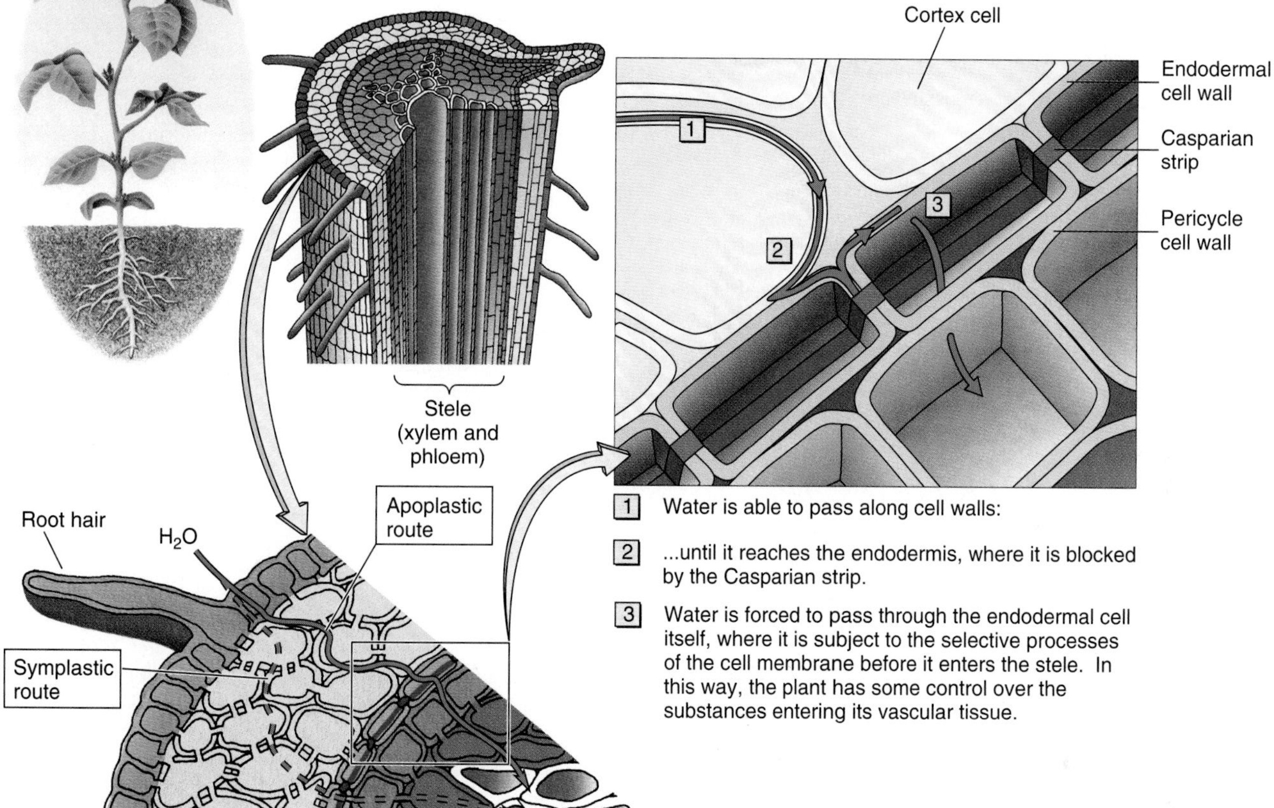

FIGURE 30.6
Movement of Water in the Root
Most water in the root moves via the apoplastic pathway, along cell walls, until it reaches the Casparian strip of the endodermis. The Casparian strip forces water and minerals to pass through the endodermal cell cytoplasm, a symplastic route. The strip also prevents backflow from the stele into the cortex.

1 Water is able to pass along cell walls:

2 ...until it reaches the endodermis, where it is blocked by the Casparian strip.

3 Water is forced to pass through the endodermal cell itself, where it is subject to the selective processes of the cell membrane before it enters the stele. In this way, the plant has some control over the substances entering its vascular tissue.

potential in the root stele below. In response, water moves into the stele, permitting the upward movement to continue.

We see then that water movement in the plant is caused by the combined effect of four cooperating forces. In review, we've seen that *transpiration* produces a water potential gradient, which along with *adhesive forces* creates a powerful pull or *tension* on minute water columns in the leaf xylem. Because of incredibly strong *cohesive forces* characteristic of water in minute columns, the columns remain intact, and water is literally lifted through the entire vascular system. This all works very well, but as you might expect, it can only continue if water intake by the root can keep up with the movement of water through the stem and out of the leaf.

Water Movement in the Root

Within the root, the ongoing pull occurring in the xylem creates a water potential gradient in the other cells of the stele that is transmitted across the cortex to the extensive root hairs of

the epidermal cells (Figure 30.6). As long as soil water is plentiful, water will passively enter the epidermal cells and pass across the root to the xylem. In its transit through the cortex, water passes mainly along the highly porous parenchyma cell walls, along the **apoplastic route**. Some water passes along the **symplastic route**, through the cytoplasm. (The "symplast" is within the plasma membrane.) In the cortical cells this movement is facilitated by the presence of many plasmodesmata, cytoplasmic connections between cells. Upon reaching the endodermis, however, all water bound for the stele must follow the symplastic route—through the endodermal cytoplasm. Let's see why.

The endodermal cell walls contain a waxy region—the **Casparian strip**—arranged in such a way as to direct the water flow across the plasma membrane and through the cell cytoplasm rather than along its cell walls (see Figure 30.6). The cytoplasm, of course, is a living and highly regulated substance, so the passage of water and solutes may not be a simple process there. Because of the Casparian strip, the endodermis regulates the movement of nutrients into the vascular system.

It is possible that one function of the endodermis is to help maintain an inward water potential gradient by actively transporting mineral solutes into the stele. This is important in preventing water loss when transpiration slows or when water

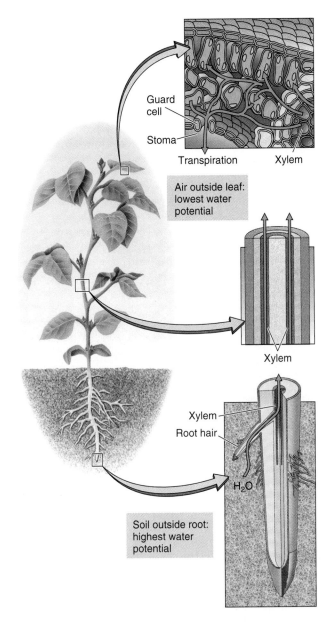

FIGURE 30.7
Summary of Water Transport
Water movement begins with transpiration from the leaves.
Decreased water potential gradient in the leaf is translated into a
pull on water columns in the xylem. Adhesion also plays a role in
the movement of water along cell walls in the leaf. Cohesive forces,
in the form of hydrogen bonding, in water hold the column together,
creating tension throughout the xylem. The pull is transferred to the
root tips, where water enters via the epidermis and its root hairs.

potential in the soil drops. In a real sense, the Casparian strip
waterproofs the stele. Of course, the endodermis also transports ions into the stele, where they can be carried to the
leaves for metabolic use.

  Before leaving the subject, we should note that the water
gradient potential responsible for water's inward movement
around the root is frequently lost. This usually happens toward
the end of an active day—when the root may literally "run

dry" and water can no longer be obtained from the surrounding soil. But roots are not passive organs; they are capable of
rapid growth, particularly at night. This growth increases
when new water resources are encountered and slows where
dry soil is encountered. So root growth itself is also an important factor in the transport of water. Water transport is summarized in Figure 30.7.

Guard Cells and Water Transport

When plants made the evolutionary transition to land, one
problem of immediate importance was the danger of drying.
The gradual evolution of an epidermis with a cuticle was a key
adaptive solution to this problem, but in view of the terrestrial
plant's dependence on atmospheric carbon dioxide, it was just
half a solution. The second half involved a means of permitting
carbon dioxide to enter the leaf, but in a controlled manner,
since, as we've seen, the inward diffusion of carbon dioxide is
unavoidably accompanied by the outward diffusion of water.
The adaptive solution in this case was the versatile *stoma* (plural, *stomata*) (see Chapter 25). The stoma is a pore, formed by
the pairing of **guard cells** that are distributed in the epidermis
(Figure 30.8). The whole assemblage makes up the **stomatal
apparatus**. Unlike other cells of the epidermis, guard cells contain chloroplasts. The amount of starch in these chloroplasts
decreases during the day and increases during the night.

  Guard cells can increase and decrease the size of the
stoma, freely admitting an exchange of gases at one time but
restricting such an exchange at others. In this way the stomatal
apparatus helps curtail the excessive loss of water from the
plant interior. Stomata are located in the leaf epidermis, usually
concentrated in the lower epidermis, but not restricted to that
location. They also occur on the photosynthetic, green stems of
herbaceous plants such as cacti and on some green fruit.
Whereas the average density is about $10,000/cm^2$, some leaves
contain as many as 100,000 in the same area.

Turgor and the Stoma. The stomatal opening is regulated
by turgor in the guard cells (Figure 30.9). Put simply, a pair of
guard cells pressed together reduce the stoma size, and the
same pair, when bent apart, increase it. At first glance, it
might appear that an increase in turgor should press the guard
cells together, closing the stomatal opening. But this doesn't
happen, for a good reason. The inner or facing walls of the
paired guard cells are fused at both ends. Therefore, in spite of
turgor increases, the inner facing walls must remain the same
overall length. Furthermore, the orientation of cellulose
microfibrils in the walls of the guard cells is radial (see Figure
30.9). As turgor increases, this orientation permits the outer
margin of the turgid guard cells to lengthen but keeps them
from bulging out or getting thicker. What happens if such
cells get longer along their outer margins but remain the same
length on their inner margins? They bend. It is this bending
that increases the stomatal opening. When turgor is lost, the
cells straighten, resuming their previous shape, and the stomatal opening decreases. So when guard cells are flaccid, short,

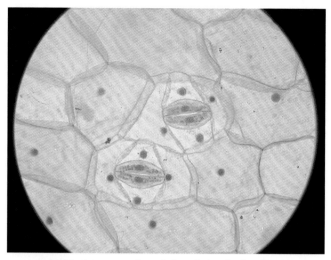

Light microscope view of stomata in a leaf surface

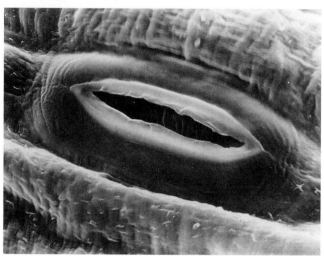

SEM view of a stoma

FIGURE 30.8
Stomata

Stomata, pores formed by paired guard cells, occur in the leaf epidermis, although they also appear on green, photosynthetic stems. They are often recessed, as seen in the scanning EM view.

and straight, the stomata are closed; but when they are turgid, long, and bent, the stomata are open. In general, the turgor increases during daylight and decreases at night. In this manner, gases are allowed to pass through the stomata when the plant is engaged in photosynthetic activity, and water loss is restricted when it is not.

Potassium Ions and the Stoma. How does turgor change in guard cells? As you might expect, water enters and leaves guard cells in response to changes in solute concentrations, which affect water potential. When solutes increase in guard cells, water potential lessens, and water enters from surrounding cells. When solutes decrease, water potential increases, and water leaves the guard cells. What then causes the changes in solute concentration?

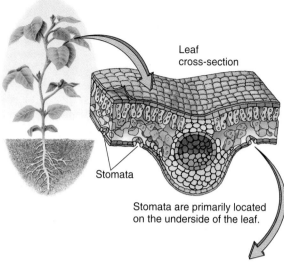

Leaf cross-section

Stomata

Stomata are primarily located on the underside of the leaf.

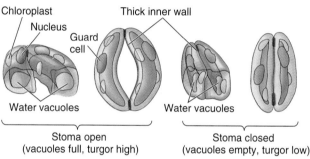

Chloroplast
Nucleus
Guard cell
Thick inner wall
Water vacuoles
Water vacuoles

Stoma open (vacuoles full, turgor high)
Stoma closed (vacuoles empty, turgor low)

(a) Opening and closing

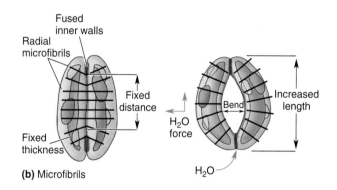

Fused inner walls
Radial microfibrils
Fixed distance
Fixed thickness
H_2O force
Bend
Increased length
H_2O

(b) Microfibrils

FIGURE 30.9
The Stomatal Mechanism

(a) The stoma opens and closes in response to changing turgor pressure within the guard cells. This coincides with the presence of light and darkness. Under lighted conditions, water enters the guard cells, creating a turgid condition that bends the cells, opening the pore. In darkness, water leaves the guard cells, and they straighten, closing the pore. **(b)** The peculiar bending of turgid guard cells is attributed to the orientation of microfibrils in the cell walls. Their binding effect permits expansion in one direction only, producing the bending action.

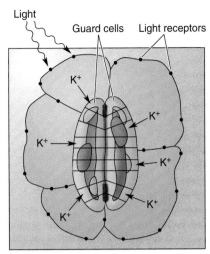

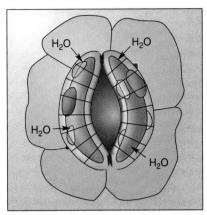

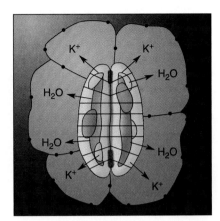

Light is absorbed by photoreceptors which initiates K^+ transport into guard cells. Water potential in the guard cells decreases.

H_2O follows, moving down its concentration gradient. Increased hydrostatic pressure causes the stoma to open.

As light fades, K^+ diffuses out of guard cells and H_2O follows. Hydrostatic pressure decreases and the stoma closes.

FIGURE 30.10
Role of Potassium in Stomatal Opening and Closing
Light, when absorbed by receptors on the surface of epidermal cells, activates potassium ion (K^+) pumps that transport the ions into the guard cells. Their growing presence there decreases the water potential gradient, and water enters from neighboring cells. This bends the cells, opening the stoma and permitting the critical exchange of gases to go on. As the light fades, the pumps cease, and potassium diffuses down its gradient out of the guard cells and into the surrounding epidermal cells. As their water potential decreases, water reenters from the turgid guard cells, which then straighten, closing the stoma.

In the late 1960s, physiologists found that plants in direct light show an increase in the active transport of potassium ions into the guard cells. It turns out that only the blue part of the light spectrum is effective in stimulating potassium transport because the process is triggered by a blue-sensitive pigment. The entrance of potassium increases the solute concentration of the cell, speeding the uptake of water and resulting in stomatal opening (Figure 30.10). Loss of potassium from the guard cells reverses the process, causing the stomata to close. Severe water loss by the plant apparently affects the potassium transport mechanism. As plants become "water stressed" through developing lower water potential in their leaves, a hormone known as abscisic acid (or ABA) accumulates in the guard cells. In the presence of ABA, potassium transport is reversed, and the ion leaves the guard cells, which rapidly lose turgor, and the stomata close.

Carbon dioxide has a positive, reinforcing effect on the light-mediated transport of potassium ions, but in an indirect way. Experiments with corn reveal that the stomata open widely when internal carbon dioxide levels are experimentally *decreased,* even when this is done in the dark. This indicates that as CO_2 in the leaf cells diminishes, K^+ transport into the guard cell increases, although the relationship between the two hasn't been clarified. The increase of this solute, as we've seen, results in the further entrance of water into the guard cell and, subsequently, an increase in the stomatal opening. Although unexplained, the response to CO_2 is logical, since a

diminishing level of the gas is associated with intense photosynthetic activity, and it is at this time that having wide open stomata is adaptive, since it makes more CO_2 available.

Considering the variables involved, physiologists have come to the conclusion that the stomatal apparatus has multiple controls. Chief among the factors influencing turgor are light, K^+, CO_2, H_2O, and ABA. Finally, there are the **CAM plants**, a group that have developed another strategy.

CAM Photosynthesis, or How to Hold Your Breath All Day. In some plants the opening and closing cycle of the stomata is reversed. The stomata are closed all day and open all night. Included are many cacti and other succulents (fleshy-bodied plants common to the desert) and members of the family Crassulaceae—the "stonecrops" (like the Jade plant). This strange adaptation helps prevent intolerable water loss in the excessively arid desert environment in which the plants live. It also means that access to atmospheric carbon dioxide is cut off during the day when most plants need to obtain CO_2 for photosynthesis. Yet the plants carry out the light-independent or sugar-synthesizing reactions of photosynthesis in a manner similar to other plants. How do they handle their carbon dioxide requirements?

The plants use what is called **crassulacean acid metabolism**, or **CAM**. Carbon dioxide is admitted at night and temporarily incorporated into certain organic acids. Then, in the daylight, the chemical reactions are reversed, and the carbon dioxide is released within the cell. From there it is used in the production of glucose according to the usual biochemical pathway—the Calvin cycle of the light-independent reactions (see Chapter 7). The energy and hydrogen sources, ATP and NADPH, are provided, as usual, from the light reactions.

It turns out that, in terms of water used to produce living material, CAM plants are far more efficient that C3 and C4 plants. This begs the question: "Why aren't all plants CAM plants?" The answer: because where there is plenty of water, the CAM plants cannot compete well with the others. This is the price of narrow adaptation.

Food Transport in Plants

In 1671, just six years after he described the cells in cork, Robert Hooke and a colleague, Robert Brotherton, "girdled" a tree. That is, they removed a ring of bark around the trunk of the tree, including the soft, moist layer beneath the dead outer cork. The tree eventually died, but they observed that it continued to grow for some time, putting out new leaves and branches. They also noted something odd. The trunk of the tree increased in diameter above the ring *but not below it*. The roots did not grow and, in fact, began to shrivel. Hooke and Brotherton concluded that the material a plant receives from the leaves is transported downward in the bark, while the material it receives from the roots is transported upward in the wood.

Actually, sugar and other nutrients move through the phloem from source to sink. A **source** is a plant organ that is actively producing sugars by photosynthesis. Leaves are usually sources in a plant. A **sink** is any plant organ that consumes or stores sugars from photosynthesis. Sinks in the plant may be roots, growing leaves, fruits, or seeds. It is now clear that some form of active transport is involved in the movement of sugar and other nutrients through the phloem. We will look at the role of active transport after making a few general observations about food transport in the phloem.

Phloem Transport

We learned earlier that, in the movement of water from roots to foliage through the xylem, plants make use of the energy of evaporation, water potential gradients, and those molecular properties called adhesion and cohesion. These forces provide what is essentially a pulling force that creates great tension within the tracheids and vessels. Only the thickness and strength of the lignified xylem walls prevent these conducting elements from collapsing under this great force. The situation is quite different in the phloem, though. The content of the sieve tubes, the **phloem sap**, is instead pushed along. The pushing force, as we will see, is provided by hydrostatic pressure, which can be considerable in the sieve tubes.

A Little Help from the Aphids. In 1953, insect physiologists J. S. Kennedy and T. E. Miller observed that aphids inserted their needle-like mouth part, the stylet, into individual sieve tubes during feeding. The tiny insects then relax and let the nutrient-laden sap, which is under pressure, to flow into their bodies. In fact, the sap often flows right through their bodies and accumulates in drops at the other end, where it is euphemistically called "honeydew." (If you have parked your car under an aphid-infested tree, you may already know about honeydew.) Plant physiologists sometimes take advantage of the aphids' drilling technique to study the chemical composition of phloem sap and its transport to various parts of the plant. The procedure involves letting aphids drill into the phloem and then anesthetizing them with a gentle stream of carbon dioxide. Next, a sharp razor blade is used to cut the imbedded mouthparts away from the head, and the resulting liquid flow becomes a source of pure sap for analysis (Figure 30.11). The amount of sap that can be captured in this manner is considerable, up to a milliliter of fluid per hour, the flow lasting up to about four days. Of particular interest to the plant physiologist, the aphid stylets do little if any harm to the plant, which helps maintain experimental integrity. Although Kennedy and Miller were really interested in studying aphid nutrition, they immediately became aware of the applications possible to their plant physiology colleagues. Such methods have been continually in use since that time.

It turns out that sap is actually a rather dense fluid, especially when the plant is photosynthetically active. Nonetheless, it moves rather rapidly through the phloem stream, up to 1 meter per hour. Phloem sap is mainly water and dissolved solutes, with carbohydrates making up about 90% of the latter. In most plants, the carbohydrates move largely as sucrose, a nonreactive low molecular weight disaccharide. Other sugars, along with nutrients, hormones, and amino acids, are also

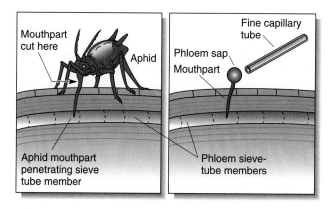

FIGURE 30.11
Aphids and Phloem Analysis
Aphids use their fine, tube-like mouthparts to pierce the phloem and intercept the sugar-laden phloem stream, which is under pressure. Taking advantage of this veritable "artesian well," plant physiologists simply anesthetize the busy aphid and cut it away from its mouthparts. Fine capillary tubes are then touched to the oozing sap, and a sample is available for analysis.

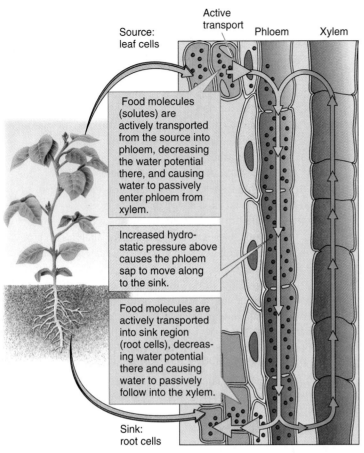

Source: leaf cells

Active transport

Phloem

Xylem

Food molecules (solutes) are actively transported from the source into phloem, decreasing the water potential there, and causing water to passively enter phloem from xylem.

Increased hydro-static pressure above causes the phloem sap to move along to the sink.

Food molecules are actively transported into sink region (root cells), decreasing water potential there and causing water to passively follow into the xylem.

Sink: root cells

FIGURE 30.12
Pressure Flow in the Phloem

The movement of sugars in the phloem begins at the source (*upper left*), where sugars are loaded (actively transported) into the phloem sieve tube. Loading of the phloem sets up a water potential gradient that facilitates the movement of water into the dense phloem sap from the neighboring xylem (*upper right*). As hydrostatic pressure in the phloem sieve tube increases, bulk flow begins, and the sap moves through the phloem. Meanwhile, at the sink (*lower left*), incoming sugars are actively transported out of the phloem. The loss of solute produces a high water potential in the phloem, and water passes out (*lower right*), returning to the xylem.

transported in the phloem. The movement of such substances in the plant is referred to as **translocation**.

Flow from Source to Sink. As we mentioned previously, phloem sap moves from source to sink. At both ends of this pathway active transport is essential. Sugars in the leaf cells, for instance, are often at a much lower concentration than they are in the nearby phloem elements into which the sugars are being loaded. In turn, such sugars may be unloaded from the phloem into food storage cells with even higher concentrations of sugar. Moving materials against their concentration gradient takes energy, and this is provided by ATP. When the supply of sucrose in some sinks is abundant, the sucrose may be converted through a multistep process into starch for stor-

age. Because of their size, the large starch molecules are more apt to stay put than are the smaller, more mobile sucrose molecules. If the plant has need of additional carbohydrate, starch reserves may be hydrolyzed for use by the plant.

Mechanism Of Transport in the Phloem. We have seen that active transport is used in loading nutrients into the sieve tubes at the source and unloading them at the sink. But what accounts for the flow of such materials in between? How do they pass through the sieve tubes? Many hypotheses have been proposed to account for phloem sap flow. Interestingly, the most favored one—the **pressure flow (bulk flow) hypothesis**—is far from the newest. Proposed back in 1927 by Ernst Munch, a German physiologist, it is based on differences in water potential between the phloem and xylem—differences that are created by the active transport of nutrients (Figure 30.12).

The idea is straightforward: the active transport of sugars into the phloem stream greatly decreases the water potential there in comparison to the high water potential in the nearby water-filled xylem elements. Water responds by moving out of the xylem into the phloem. The inward movement of water raises the hydrostatic pressure within the phloem sieve tubes. This pressure forces along the phloem sap with its load of nutrients and water. Then, at the sink, sugars are moved out of the stream by active transport. This loss of solutes increases the water potential within the phloem, and the water escapes from the phloem stream, eventually finding its way back into the adjacent xylem.

As we can see, water apparently circulates in plants, moving from xylem to phloem at the source, and moving from phloem back to xylem at the sink. The sieve tube elements themselves may play a passive role in transport, with the companion cells and phloem parenchyma providing the energy for active transport and determining whether nutrients are to be loaded or unloaded.

A half-century-old idea that explains such a vital aspect of plant biology as food transport reminds us that our own generation has no monopoly on creativity. But, venerable though it is, the pressure flow hypothesis is not without its problems and critics. Scientists would still like to know, for instance, how the phloem sap is isolated from the phloem cytoplasm. How does it move through the sieve pores without disturbing the cytoplasmic extensions there? And what determines whether nutrients are to be loaded or unloaded? In other words, what controls translocation? Obviously, science still needs bright and inquiring minds to address such questions.

Gas Transport in Plants

Plants continually exchange gases with the environment while carrying out both photosynthesis and respiration. Few plants have developed ways to move gases around in their own tissues. In monocots, the vascular bundles commonly contain an air channel along with the xylem and phloem and supporting fibers. So at least some plants have made provision for the passage of air. Otherwise, gases simply diffuse across thin-walled cells, such as those of the root tip epidermis, those just within the lenticels (pores along the woody stem), and the spongy parenchyma bordering the air spaces within the leaves. Varying amounts of oxygen and carbon dioxide are also dissolved in water being transported in the xylem stream.

Root hairs are good gas exchangers when the soil is well aerated. Since root hairs arise near areas of intense metabolic activity, including growth, maturation, and active transport, their oxygen requirement is great. The exchange of gases takes place across the cells of the root tip and the root hair surfaces. The exchange is easy here because the thin-walled cells have no resistant cuticle, and the tremendous surface area of root hairs provides a large interface for gas exchange.

Where roots are submerged in water, gas exchange and oxygen availability are adversely affected. For this reason, the roots of plants that are grown in tanks of water must be aerated just as one would do for aquarium fishes. This is also why you can kill your potted plants by overwatering them: without gaseous oxygen in the soil, they simply drown. Some plants, however, have adapted to flooded, oxygen-poor soils. These include marsh and swamp plants, such as red maple and swamp alder, and domesticated plants, such as rice. Many of these plants have large, hollow air channels in their stems (not to be confused with the much smaller channels associated with the vascular bundles of monocots). The *pneumatophores* ("air roots") of the mangrove tree extend above the surface of the water, permitting air to diffuse into the loosely arranged tissue and then down into the roots (Figure 30.13).

Lenticels, common in woody stems, provide an avenue of gas exchange for the very active tissues in the bark (see Figure 29.20). Recall that both the vascular cambium and the cork cambium are intensely active during growth. Furthermore, as we have just seen, the phloem elements and their associated cells, also part of the bark, are metabolically active, using ATP for transporting food materials into and out of the phloem stream. Thus an adequate supply of oxygen is essential.

FIGURE 30.13
Gas Exchange in the Mangrove
The mangrove often grows in water that is low in oxygen content. It has adapted to its environment through the evolution of pneumatophores, porous root extensions that reach above the surface, where air penetrates their spongy tissues.

Plant Nutrition: Minerals and their Transport

Minerals, dissolved in water, enter the root via the root hairs and epidermal cells. Their route takes them along the apoplastic and symplastic pathways across the cortex to the stele (Figure 30.14). Apparently most of their movement is passive until they reach the endodermis, whereupon active transport and a considerable investment of ATP is required. Active transport in the endodermal cells passes the ions to the pericycle, which then secretes them into the water columns for transport upward.

Mycorrhiza: Fungal Association with Roots

The uptake of mineral ions is not always straightforward. In most plant families, the plant gets a little help from a fungus that is directly associated with the root. An association of plant root and fungal mycelium known as a *mycorrhiza* ("fungus root") is one of those fascinating instances of true mutualism: both members benefit. Plants provide the fungus with complex carbon compounds such as sugars and amino acids. But what does the fungus offer the plant?

It turns out that the mycorrhizae efficiently absorb and concentrate certain ions, notably phosphate. In some instances the fungal mycelium with its mineral load actually penetrates the root cortex and deposits ions there. In other cases, the

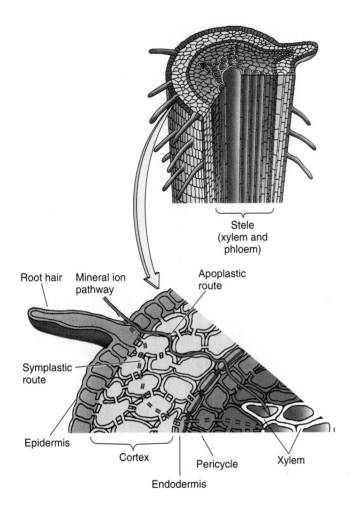

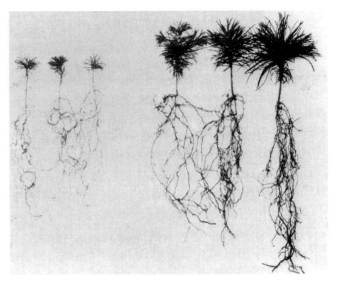

Stele
(xylem and phloem)

Root hair   Mineral ion
pathway

Apoplastic
route

Symplastic
route

Epidermis

Cortex

Pericycle

Xylem

Endodermis

FIGURE 30.14
Passage of Ions Through the Root
Ions enter the root along with water, some swept in by the flow and others pumped across by active transport. The water and ions then move through the root cortex via both the apoplastic (cell wall) and symplastic (cytoplasmic) pathways. Movement along the symplastic route is facilitated by numerous plasmodesmata in the cortical parenchyma cells. Apoplastic movement cannot occur at the endodermis because of the Casparian strip, which directs all materials to the endodermal cell membranes and cytoplasm. Thus only ions that can move passively or be transported into the endodermal cytoplasm can enter the xylem. This gives the plant some selectivity over what enters the xylem, what does not, and how much.

FIGURE 30.15
A Study of the Fungus–Root Association
In this experiment, all the seedlings were first grown in a nutrient solution with all essential materials provided. The group on the left were then transplanted directly into prairie soil. The group on the right were planted first in forest soil, where mycorrhizal associations are common, and then, after such associations had formed, transplanted to prairie.

mycelium simply surrounds the root epidermis and brings the plant into close association with the ions. In either case, the fungus extends its many mycelial "fingers" out into the soil, acting like a second root system for the plant. In transporting phosphates directly into the plant or concentrating them near the epidermis, the fungus makes this valuable ion more avail-

able to the plant. So the trade-off, phosphates for food, is a good one! Of the many plant species studied for mycorrhizal relationships, this fungus–plant-root association has turned up in about 90%. Figure 30.15 reveals the critical role of the mycorrhizal association.

Plant Mineral Nutrition

Agricultural scientists have long been interested in the mineral requirements of plants. While it is technically easy for researchers to establish a long list of chemical constituents of plants, the challenge is to determine which are truly essential nutrients and which are coincidental, with no role in plant nutrition. One approach to the question has involved deprivation experiments. Plants are grown in aerated distilled water of rigidly controlled purity. The only mineral ions present are those added by the experimenter. Once the minimal nutrients needed for growth are known, researchers can determine the particular effect of any mineral simply by withholding it from the solution (Figure 30.16).

From such experiments botanists have accumulated a list of important mineral nutrients essential to plant growth. Note in Table 30.2, that the list is divided into **macronutrients** and

FIGURE 30.16
Iron Deficiency
The iron-deprived tomato plants (*right*) clearly reveal problems of iron deficiency through stunted growth and foliage lacking the deep green color of the control plants. The plants were grown as a student experiment in introductory biology.

TABLE 30.2			
Minerals Essential to Plants			
Element	**Source**	**ppm (Dry Tissue)**	**Percentage (Dry Tissue)**
Macronutrients			
Carbon (C)	CO_2	450,000	45
Oxygen (O)	H_2O	450,000	45
Hydrogen (H)	H_2O	60,000	6
Nitrogen (N)	NH_4^-	15,000	1.5
Potassium (K)	K^+	10,000	1.0
Calcium (Ca)	Ca^{2+}	5,000	0.5
Magnesium (Mg)	Mg^{2+}	2,000	0.2
Phosphorous (P)	$H_2PO_4^-$ HPO_4^{2-}	2,000	0.2
Sulfur (S)	SO_4^{2-}	1,000	0.1
Micronutrients			
Chlorine (Cl)	Cl^-	100	0.010
Iron (Fe)	Fe^{3+} Fe^{2+}	100	0.010
Manganese (Mn)	Mn^{2+}	50	0.005
Boron (B)	H_3BO_3	20	0.002
Zinc (Zn)	Zn^{2+}	20	0.002
Copper (Cu)	Cu^{2+} Cu^+	6	0.0006
Molybdenum (Mo)	MoO_4^{2-}	0.1	0.00001

Note: Where two ions are shown, the first is more common.

micronutrients. In general, the first group is involved in structure and the second participates in chemical reactions such as enzyme activation. (Since there are exceptions, some plant physiologists limit the term micronutrient to substances needed in quantities less than 100 mg/kg of dry plant matter, and macronutrients to those needed in excess of 1000 mg/kg.) Micronutrients are sometimes called *trace elements*, and for good reasons. Ions such as chloride (Cl^-) are required in such minute quantities that they are very difficult to detect, yet photosynthesis cannot proceed without them. Trace elements often cause problems for plant scientists in trying to control experiments—if you can't detect it, you can't always keep it out.

Many of the mineral nutrients of plants are made available through cycles that are often quite intricate. These are usually referred to as *biogeochemical cycles*, some of which are described in Chapter 46. Key elements such as nitrogen, sulfur, phosphorus, and calcium, as well as carbon, oxygen, and hydrogen, are constantly recycled between the living and nonliving realms of the earth. When these cycles fail, the result can be infertile soil and limited plant growth.

Insect Eating Plants

One of the more intriguing ways for a plant to fulfill its nutrient requirements is that seen in **insectivorous plants**, those that feed on insects. Actually, these plants aren't after the animal for food in the usual sense of the word, they are after the nitrogen compounds they extract from the food. It turns out that insect capture by plants is an adaptation to life in nitrogen-poor soils such as those of swamps and bogs where the insectivorous plants live. There, nitrogen compounds are leached away by water or consumed by anaerobic bacteria. (Oxygen is in short supply as well.)

Examples of insectivorous plants include the bladderwort (*Utricularia*), the Venus flytrap (*Dionaea*), the pitcher plant (*Sarracenia*), and sundew (*Drosera*), some of which are seen in Figure 30.17. Both the bladderwort and the Venus flytrap actively capture prey. Interestingly, if essential nitrogen compounds are provided to these plants, they go right on trapping insects. (But then, just to further prove that variation and adaptation are endless in life, there are some insectivorous plants that stop growing traps when sufficient nitrogen is available.)

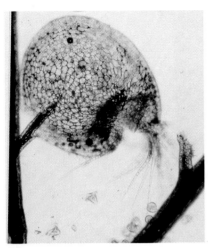

(a) *Utricularia* (bladder wort)   **(b)** *Sarracenia* (pitcher plant)   **(c)** *Drosera* (sundew)

FIGURE 30.17
Insectivorous Plants
(a) The bladderwort (*Utricularia*) is a water plant that forms many bladder-like chambers, each equipped with a trigger-activated trapdoor. When the trap is tripped, the door opens inward, the prey is sucked in by rushing water, and the door shuts behind it. **(b)** The pitcher plant (*Sarracenia*) has evolved a one-way passage into its trap. Insects venturing into the vase-like structure find their return blocked by downward-pointing hairs. **(c)** The sundew (*Drosera*) makes use of a thick, sticky fluid that traps the insect.

Key Ideas

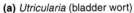

1. The vascular system of plants has the dual function of transporting water through its **xylem** and foods through its **phloem**. The first is "free," but the second requires considerable ATP energy.

THE MOVEMENT OF WATER AND MINERALS

1. Water travels from the soil through the plant to the air around the leaf via root hairs, root cortex, **stele**, root xylem, stem, leaf, leaf mesophyll, air spaces, and **stomata**.

2. The evaporation and subsequent diffusion of water from the leaf is called **transpiration**.

3. **Water potential** is based on the free energy of water. Pure water has the greatest water potential, stated as 1, and all solutions have negative values—less than 1.

4. Water potential is influenced by (a) solute concentration (the addition of solutes to cells lowers water potential) and (b) hydrostatic pressure.

5. Water always moves from regions of greater water potential to regions of lesser water potential. The normal water potential gradient of a plant is from high potential in soil water to low potential in air surrounding the leaf.

6. Root pressure is produced when ions are actively transported into the stele, setting up a water potential gradient, but it is not a major force in the movement of water to the tops of tall plants.

7. Water loss is unavoidable if CO_2 is to be available for photosynthesis. Plants make use of the continued water loss to initiate the mechanism that raises water from the roots.

8. Transpiration creates a steep water potential gradient through the cells of the leaf, and water enters from the leaf xylem. The cell-to-cell movement is facilitated by numerous **plasmodesmata**.

9. The water potential gradient in the leaf produces a pulling force that creates pressures of up to 12 atmospheres, enough to raise water through the tallest trees.

10. The movement of water through the stem and leaves is provided by the free energy of evaporation. Indirectly, this energy comes from the sun.

11. **Transpiration** and **adhesion** set up the pulling force that raises water through the xylem. Transpiration involves the evaporation of water from the leaf spaces. Adhesion, the attraction of one type of molecule to another (such as water to those of surrounding surfaces), is a major force, since the adhesion of water to the cellulose walls of leaf cells assists in a pulling action.

12. **Cohesion**, the mutual attraction of similar molecules, is critical in the tiny water columns of the xylem. It provides the tensile strength needed to hold them together when under tension. **Tension** is the stress placed on the water columns by the pulling forces above.

13. Water entering the root mainly follows the **apoplastic** (cell wall) **route**. At the **Casparian strip**, water is directed through the endodermal cytoplasm—a **symplastic route**.

14. Paired **guard cells** form stomata in the leaf epidermis. Because of the radial arrangement of cellulose microfibrils in their cell

walls, guard cells bend when turgid, increasing the stomatal opening. A loss of turgor closes the **stoma**. Stomata are generally open in daylight and closed in darkness.

15. Turgor changes in guard cells are brought about by shifting solute concentrations. In the potassium ion transport mechanism, light activates transport of the ions into the guard cell. This decreases their water potential, water enters, and the turgid cells bend apart. A decreased CO_2 concentration also increases K^+ transport. This usually occurs during intense photosynthetic activity—a time when more CO_2 is needed. Water stress causes the release of ABA, which causes guard cells to lose K^+ and then water. This brings about closing of the stomata, and the plant is protected.

FOOD TRANSPORT IN PLANTS

1. Phloem transport makes use of **pressure flow.** The active transport of sugars into the phloem stream at the source, followed by water, increases hydrostatic pressure, which forces the dense **phloem sap** along to the sink. There, the solutes are actively transported out, followed again by water. The movement of foods from **source** to **sink** is called **translocation**.

GAS TRANSPORT IN PLANTS

1. Plants must carry on gas exchange in their metabolically active tissues. Some oxygen and carbon dioxide are exchanged through leaves, lenticels, open air channels, and root epidermis. Some marsh plants have hollow, air-conducting channels in their stems. Black mangrove trees have **pneumatophores** that extend above the water surface, permitting gases to diffuse into the roots.

PLANTS NUTRITION: MINERALS AND THEIR TRANSPORT

1. The transport of mineral ions through root cells is primarily active.

2. Most plants are assisted in mineral uptake by a fungus–root association called a mycorrhiza. The fungus absorbs and concentrates ions near the root hairs or within the root and, in turn, absorbs useful compounds from the plant.

3. Nutrients are divided into **macronutrients** and **micronutrients**. Mineral nutrients are commonly provided through **biogeochemical cycles**, where the ions cycle between organisms and the physical environment.

4. **Insectivorous plants** obtain their nitrogen from the digestion of insects.

Application of Ideas

1. The statement is often made that vascular plants have literally "evolved around the characteristics of water." Discuss the full meaning of this statement. Begin by explaining exactly how vascular plants have taken advantage of water's peculiarities. Then rephrase the statement in such a way as to introduce the terms *natural selection* and *adaptation*. Complete your discussion by considering how this evolutionary direction made great increases in size possible, and how this result has been advantageous to evolving plants.

Review Questions

1. Starting with the epidermis, trace the movement of water through the plant, naming the important tissues through which it passes. (page 633)

2. What is transpiration? How could transpiration possibly affect the weather? (page 633)

3. Explain why root pressure cannot be responsible for the rise of water to the tops of tall trees. What function does root pressure more likely fulfill? (page 636)

4. What is the rule about water potential and the direction in which water will flow? What effect do solutes and hydrostatic pressure have on water potential? (page 635)

5. Describe the water potential gradient in various parts of the plant, beginning with the root. (page 635)

6. Assuming that transpiration and adhesion create the necessary pulling force, what is the role of cohesion in water transport? How is the diameter of xylem vessels and tracheids related? (pages 637–639)

7. What two kinds of evidence do physiologists find that suggest the presence of tension in the transpiring tree? (page 638)

8. What is the usual cycle of stomatal opening and closing in the plant? (page 640)

9. Describe the shape of a pair of guard cells in the turgid condition. What factor apparently influences turgor? (pages 640–641)

10. Explain the potassium transport mechanism of guard cell function. (pages 641–642)

11. Briefly describe how events at the source produce a flow in the phloem. Is this a push or a pull? (page 644)

12. Explain what happens to the nutrients and to the hydrostatic pressure when the phloem stream reaches the sink. (page 644)

13. What function do lenticels, pneumatophores, and root hairs have in common? (pages 644–645)

14. Explain how the formation of a mycorrhizal association is beneficial to the plant. How is it beneficial to the fungus? (pages 645–646)

CHAPTER 31

Plant Regulation and Response

CHAPTER OUTLINE

LIGHT AND THE GROWTH RESPONSE

PLANT HORMONES

Auxin: Its Structure and Roles

Gibberellins

Cytokinins

Abscisic Acid

Ethylene

Leaf Abscission

Apical Dominance

Applications of Plant Hormones

DIFFERENTIAL GROWTH RESPONSES: TROPISMS

Phototropism

Roots and Gravitropism

Thigmotropism

Solar Tracking

NASTIC MOVEMENTS

LIGHT AND FLOWERING

Photoperiodicity

Flower Initiation and Phytochrome

KEY IDEAS

APPLICATION OF IDEAS

REVIEW QUESTIONS

Charles Darwin's investigations into evolution led him into a wide array of scientific fields, from shellfish taxonomy to plant behavior. Actually the "behavior" that Darwin studied was, in fact, plant responses to stimuli. Darwin began his studies with one of the most interesting of all plant behaviors, that of the Venus flytrap, one of the rare insect eating plants. In his 1875 publication *Insectivorous Plants,* he described in exquisite detail the triggering of the plant and its response, but he was at a loss to explain just *how* such rapid movement was accomplished. So he changed the focus of his research; he looked into the slower and seemingly simpler behavior of climbing plants.

The tips of climbing plants, he found, move about in circles while they grow, "as if seeking something to twine around." If no object is encountered, the tip eventually grows straight upward. But if a twig or a pole is encountered, the slowly growing tip begins to spiral around it and to tighten up. In this way, vines and climbing plants can reach great heights without investing energy in thick woody trunks (Figure 31.1).

FIGURE 31.1
Climbing Plants
Climbing plants are adapted to rapid upward growth using other plants or objects for support.

Darwin described many variations of this basic pattern and (not surprisingly) interpreted them in terms of natural selection. He showed that the underlying mechanism was the elongation of first one and then another part of the stem below the growing shoot, causing it to bend first one way and then another (*Climbing Plants*, September 1875). But what brought about this selective cell elongation?

Light and the Growth Response

At age 71, Darwin worked on the problem with his son, Francis. Their experimental organism was canary grass. Like other grasses, in its early stages of growth, it produces a tubular sheath, the **coleoptile**. As the shoot emerges from the soil, the primary (first) leaf remains within the protective coleoptile for a brief time. It is this young stage that researchers have found so interesting. Darwin and his son found that when the coleoptile is grown in the dark and then is illuminated from one side, bending occurs toward the light (Figure 31.2a). The bending, they noted, does not actually occur at the tip, but rather in the elongating part of the coleoptile, well below the tip. Was light acting directly on the bending part? Apparently not. Darwin illuminated only the growing tip of the coleoptile, and it bent as before. With a small piece of foil, he covered the tip and exposed the rest of the plant to light. Nothing happened; it didn't curve. Darwin then knew that something was happening in the growing tip that caused certain areas of the shoot beneath to elongate. In *The Power of Movement in Plants* (1880), Darwin ascribed this fundamental discovery to "some matter in the upper part which is acted upon by light, and which transmits its effect to the lower part. When seedlings are freely exposed to a lateral light some influence is transmitted from the upper to the lower part, causing the latter to bend." Thus, in an indirect way, Charles Darwin and his son were the first to propose the existence of a plant hormone.

Early in this century, a Dane, Peter Boysen-Jensen, and a Hungarian, Arpad Paal, extended Darwin's observations. Boysen-Jensen (Figure 31.2b) decapitated a coleoptile of an oat seedling a few millimeters from the tip, put a tiny block of gelatin (a porous material) on the stump, and replaced the tip, which he then illuminated from one side. It bent as before, showing that Darwin's "influence" was something that could move through gelatin. To verify this, Boysen-Jensen placed tiny slivers of impermeable mica between the coleoptiles and the stumps. There was no bending. When he inserted mica into slits cut part way through coleoptiles and then illuminated them from different sides, he found that the active substance did not move down the illuminated side, but only down the side away from the light. Thus the stimulus that caused the normal cell elongation and bending was present on the darkened side of the coleoptile only.

In his efforts (Figure 31.2c), Paal decapitated oat coleoptiles and then, under darkened conditions, restored the tips on the growing stubs, *but off to one side or the other.* Elongation was always greater just below the offset tip. In other words, the shoot bent in a direction opposite the side containing the replaced tip. This clearly suggested that there was a substance moving from the tip to the tissues just below, which stimulated the elongation of cells there. The next logical step was to isolate this material and see precisely what it could do.

By 1926, Fritz Went, a Dutch scientist, succeeded in the isolation of the growth substance in coleoptiles. Went's technique consisted of decapitating oat seedlings and placing the tips on agar blocks (Figure 31.2d). (Agar is also porous.) He then took tiny squares of the agar and placed them on the side of the decapitated seedlings. Went kept his subjects in a darkened environment throughout the experiment, so that any bending response could be attributed to substances in the agar block alone. Typically, bending occurred within an hour after the blocks were applied, proving that an active, collectible agent was the cause of bending. Although he did not chemically characterize the active substance, Went named it **auxin** (from the Greek word *auxein,* "to increase"). Later, when more of its chemistry became known, it would also be called **indoleacetic acid**, or **IAA**.

Any biochemical isolation requires an **assay system**, a way of measuring whether a substance is present and in what concentrations. Such an **assay** may be a **chemical assay** if the substance causing the effect is known and can be chemically measured. If the substance is unknown or if measurement via chemical means is difficult, a **bioassay** is used. A bioassay is a procedure in which a biologically active compound is quantified by measuring the amount of an effect it elicits in a biological test system. A bioassay system for the coleoptile growth substance was soon worked out with oat seedlings, using Went's basic methods. Hormone concentrations are estimated by the

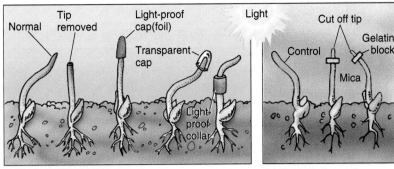

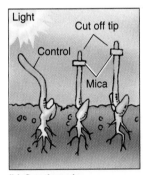

FIGURE 31.2

Experimenting with Light and Seedling Growth

In a series of experiments over some 60 years, experimenters from Darwin to Went tried to understand the reaction of plants to light. Each experiment uncovered some new bit of information that culminated in an understanding of the relationship of auxin and the bending of plants toward a light source.

(a) Darwin's Experiment

Charles Darwin and his son, Francis, determined that a light-dependent agent in the tip of canary grass seedlings caused them to bend toward a light source. The collar helped prove that the light-activated substance was formed above, at the tip.

(b) Boysen-Jensen's Experiment

Boysen-Jensen elaborated upon Darwin's experiments, using gelatin slabs and slivers of mica. (1) Cut off growing tips were placed on tiny blocks of gelatin balanced on the stem. The plants grew towards the light as usual, suggesting that a diffusible substance had passed through the gelatin.

(b) Continued

(2) When impermeable mica was used instead of gelatin, no response to light was seen.

(b) Continued

(3) When mica slivers were inserted partway into the stems on opposite sides of plants, the response depended upon which side was lighted. This indicated that the agent passed down the side *away from the light* (or was perhaps inactivated by light).

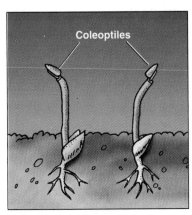

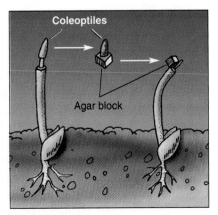

(c) Paal's Experiment

Paal carried out his key experiment in the dark. He used severed coleoptile tips to direct the bending of oat seedlings, showing that the coleoptile was the source of the active agent.

(d) Went's Experiment

Fritz Went, also working in the dark, removed coleoptiles and placed them on agar blocks for varying times. He then used the blocks to produce curvature, and related the amount of curvature to the amount of auxin present.

degree of bending seen in the oat seedlings: the greater the angle, the more hormone is present in the block (Figure 31.3).

Plant Hormones

We've mentioned hormones, so let's look more closely at these amazing chemicals. Plant hormones are organic compounds that play important roles in regulating many plant activities. Some are synthesized in one part of a plant and transported to another, where they elicit a response. Others act within the tissues in which they are produced. One special characteristic of plant hormones is that they work in incredibly small amounts, often in concentrations measured in micrograms (millionths of a gram).

Hormones can either stimulate or inhibit responses and sometimes it is the concentration that determines which of these will occur. In fact, the particular response may depend on the specific target tissue for which the hormone is destined. We will look into the workings of five hormone groups: auxin, gibberellins, cytokinins, abscisic acid, and ethylene (Table 31.1).

Auxin: Its Structure and Roles

Unknown to Boysen-Jensen, Paal, or Went, the active substance of auxin had been isolated in 1885 by a pair of biochemists, E. and H. Salkowski. However, the Salkowskis didn't know about its biological significance. The substance was not successfully isolated again until 1934. We now know that the mysterious coleoptile growth substance is actually a widespread and fundamental plant growth hormone. The chemical structure of auxin is very similar to the amino acid tryptophan; in fact, tryptophan is its primary building block. IAA is unique among plant hor-

TABLE 31.1 **Plant Hormones**	
Name of Hormone	**Action**
Auxins (a major one is IAA: indoleacetic acid)	Stimulate cell elongation in response to light (phototropic), and gravity (gravitropic); leaf abscission (with ethylene); and apical dominance; prompt differentiation (in combination with cytokinins)
Gibberellins	Promote stem elongation (with auxin); germination of grains (mobilize digestive enzymes)
Cytokinins	Stimulate cell division and differentiation (in combination with auxin); inhibit aging
Abscisic acid (ABA)	Inhibits K+ transport during water stress; promotes bud dormancy
Ethylene	Speeds fruit ripening, maintains hook shape in emerging hypocotyl of seedling

mones in that there is a distinct polarity to its movement and the movement involves active transport. In stems, auxin is transported toward the base of the stem regardless of whether the stem is right side up or upside down. In roots IAA is transported toward the root tip. This very specific movement involves protein carriers located in the plant cell membrane. We will have a closer look at this precise movement later in the chapter.

Effects of Auxin. Auxin doesn't cause cells to proliferate by mitosis; rather, it promotes cell enlargement through elongation (Figure 31.4). Cell elongation, a primary plant growth mecha-

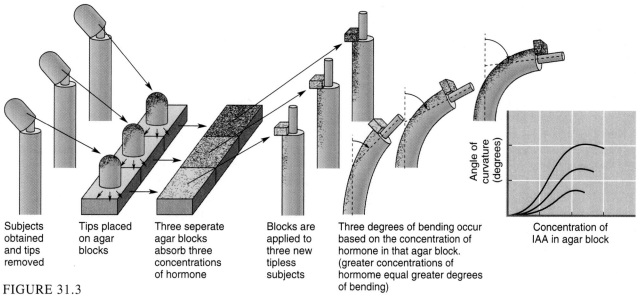

Subjects obtained and tips removed

Tips placed on agar blocks

Three seperate agar blocks absorb three concentrations of hormone

Blocks are applied to three new tipless subjects

Three degrees of bending occur based on the concentration of hormone in that agar block. (greater concentrations of hormome equal greater degrees of bending)

Angle of curvature (degrees)

Concentration of IAA in agar block

FIGURE 31.3
Went's Auxin Assay
In his assay procedure, Fritz Went used varying concentrations of auxin in agar blocks. The bending of growing tips was then carefully measured and recorded, forming a standard to which bending growth in other plants could be compared.

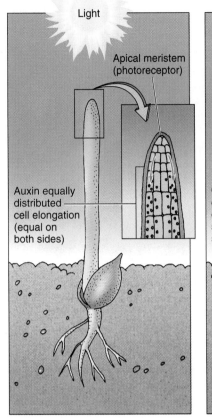

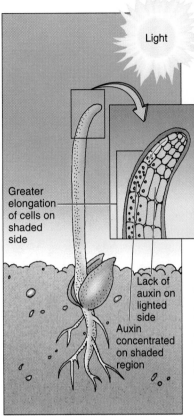

(a) Overhead light   **(b) Light from one side**

FIGURE 31.4
The Action of Auxin
Auxin is known to promote cell elongation in the growing shoot. **(a)** When light is multidirectional, auxin
is evenly distributed and elongation proceeds equally about the stem. **(b)** When light is directed to one
side, auxin concentrates on the opposite, unlighted side, where the greatest elongation occurs.

nism, occurs in cells produced by the apical meristems, right after mitosis (see Chapter 29). In auxin's presence, the plasticity of the cell wall increases. This loosening effect permits the cell to take up additional water by osmosis and elongation occurs (see Figure 29.8). How does auxin increase the plasticity of plant cell walls? According to the **acid growth hypothesis**, IAA stimulates the pumping of hydrogen ions (H^+) from the plasma membrane into the area of the cell wall. The acidic condition there activates enzymes that loosen bonds between the microfibrils of cellulose.

Auxin has other functions. It promotes the formation of adventitious roots at the cut base of a stem. Accordingly, horticulturists may dip cuttings in rooting media (like Rootone®, a commercial product), which contain synthetic auxins. Auxins from developing seeds stimulate fruit development in many plants. Synthetic auxins like 2,4-D disrupt the normal balance in plant growth and can function as herbicides.

Gibberellins

The **gibberellins**, a large family of molecules, received their name from the source of their discovery, *Gibberella fujikuroi*, a fungus. The fungus produces a condition the Japanese call "foolish seedling disease." The symptoms in rice plants include abnormal flowers and strangely elongated, weak stems that tend to break as the grains form. At one time the fungus threatened rice harvests in Japan.

Between 1926 and 1935, Japanese botanists had already isolated and purified the active substances from the fungus. This chemical was given the name gibberellin. It wasn't until the 1950s that other nations took an interest in these strange molecules. Today, we are continuing to study the natural production of gibberellins to learn more about how they control growth in plants, and some of the knowledge is being applied to agriculture. For example, gibberellins can be used to promote stem growth in sugar cane and celery, to increase starch digestion during the malting process in brewing, and to bring about a loosening effect on grape clusters, which makes the spraying of fungicides more effective.

Gibberellins are formed in young leaves around the growing tip, and possibly in the roots of some plants. (However, we don't know what role they play in root activity.) The power of gibberellins in stem elongation was dramatically illustrated in experiments in 1951, as the photos in Figure 31.5 attests. Their growth-promoting effects can clearly be

FIGURE 31.5
Gibberellins and Stem Elongation
The dramatic effects of gibberellins on stem elongation are seen in kidney bean seedlings. On the left are two untreated controls. The two plants on the right were treated with gibberellins (500 ppm solution) two days earlier.

| Hormone released | Hormone reaches the aleurone layer | Aleurone releases starch-splitting enzyme alpha amylase, which makes glucose available | Glucose, taken in by embryo supports its rapid growth |

FIGURE 31.6
Gibberellins and Seed Germination
Gibberellins bring about the conversion of starch to sugar, which is then taken in by the embryo.

demonstrated with genetic dwarfs. Dwarf corn, for instance, can be induced to grow to normal height after the application of gibberellins. This indicates that a hormonal failure can cause the plants' shortness and, except for this abnormality, some dwarf corn has the potential to grow tall. Incidentally, the degree of growth in a dwarf depends on the quantity of gibberellins applied. Thus the botanist has an excellent method of gibberellin bioassay.

Gibberellins also have a role in the germination of some seeds. They act as chemical messengers to stimulate the synthesis of an enzyme called **alpha-amylase** and other hydrolytic enzymes in grains such as barley and corn (see Chapter 29). As these grains germinate, the embryo secretes gibberellins. They move to the **aleurone**, a cell layer surrounding the starchy endosperm. Their presence stimulates the aleurone to produce alpha-amylase, which then breaks down starch and makes sugars available to the growing plant (Figure 31.6). There may be many steps or only a few between the arrival of the hormone messenger and the transcription activity required for producing the enzyme. As you may recall from past discussions (see Chapter 14), the specific hormonal gene-activating mechanisms are still under investigation.

Cytokinins

Much of what we know about **cytokinins** springs from work begun in the 1950s when research scientists found that in

order to get the cells in cultures of plant tissues to divide, they had to add coconut milk to the medium. What was so special about coconut milk? Coconut milk is the liquid endosperm of coconut seeds and contains many nucleic acids. So they began to examine chemicals related to nucleic acid to see if any would stimulate cell division. The mysterious substance turned out to be a group of hormones they called cytokinins.

The first thing biologists learned about cytokinins was that they stimulated cell division, although to do this they had to be combined with auxin. Researchers mixed a cytokinin called **kinetin** with auxin in different ratios and then applied them to callus tissue removed from tobacco plants (a callus is a mass of undifferentiated tissue). The interaction of the two hormones is quite complex. Auxin encouraged root growth, whereas kinetin encouraged shoot growth—but in each case, only when the opposing hormone was in low concentration. Furthermore, when very low concentrations of both hormones were present, little callus growth and no differentiation occurred. At high concentrations of both, the callus grew but no differentiation occurred (Figure 31.7). Thus cytokinins play a role in plant differentiation.

The studies indicated that the undifferentiated and ordinary cells of the pith, the parenchyma cells, contain all the genetic information necessary to develop into other kinds of plant cells, producing a variety of tissues and organs. This capability in cells is referred to as **totipotency**, which means the cells had retained all of the growth capabilities of cells in the original zygote. That is, no matter how specialized the cell becomes, it never loses its original genetic capability. The retracing of original developmental events by mature cells usually begins with **dedifferentiation**, the return to a simple undifferentiated state. This is followed by differentiation, but this time along a pathway that differs from the original. More on this subject is found in Essay 29.1.

Another role of the cytokinins is associated with **senescence**, or aging, in plants, which is a sequence of irreversible

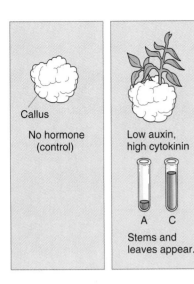
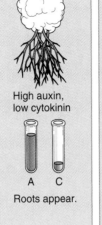
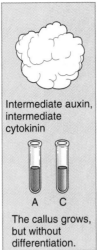

Callus

No hormone (control)

Low auxin, high cytokinin

A   C

Stems and leaves appear.

High auxin, low cytokinin

A   C

Roots appear.

Intermediate auxin, intermediate cytokinin

A   C

The callus grows, but without differentiation.

FIGURE 31.7
Cytokinins and Differentiation
In this experiment, the pith is treated with varying combinations of auxin and cytokinin, each of which produces some variation in growth and differentiation.

changes that eventually result in death. In leaves, this process involves chlorophyll breakdown and other leaf pigments filling the leaf cells. This happens quickly when leaves are removed from the plant, but if they are treated with cytokinins, senescence is retarded. As long as they are supplied with water, the chlorophyll does not break down (so the leaves stay green), protein synthesis continues, and carbohydrates do not break down. Synthetic cytokinins have been applied to harvested vegetable crops such as celery, broccoli, and other leafy foods in order to extend their storage life.

Abscisic Acid

All the hormones discussed so far are somehow involved in the promotion or stimulation of plant growth. In the 1940s and 1950s some botanists had evidence that plants contained inhibitors that blocked such effects. Since these compounds were studied in dormant buds, they were named dormins. About the same time that dormins were being discovered, F. Addicott discovered a compound that promoted abscission in cotton. He called the compound abscisin. When it was discovered that dormin and abscisin were the same compound, it was agreed that the compound should be named **abscisic acid** or **ABA**. The molecular structure of ABA consists of a carbon ring with a short carbon chain containing a carboxyl (acid) group.

As physiologists learned more about the effects of ABA, they discovered that it actually doesn't control abscission at all, and even its proposed role in controlling bud dormancy has been questioned. Probably the best documented role of ABA involves the stoma. Recall that under periods of water stress (see Chapter 30), ABA enters guard cells, where it brings about the outward transport of potassium ions. The outward movement of water follows, and the flaccid guard cells shrink, closing the leaf's stomatal openings. The plant is thus protected from further water loss. A plant's ability to tolerate drought may relate to its ability to synthesize ABA. ABA is also known to regulate germination of seeds in some plant species and will inhibit the stimulatory effects of other plant hormones.

Ethylene

Ethylene is a relatively simple compound when compared to other plant hormones: ethylene is a gas—one you can smell around ripening fruit. In fact, it controls the ripening process. It is synthesized from the amino acid methionine, commonly found in cells. Since it is a gas, it readily diffuses out of plants; thus the concentration of ethylene in plant tissues depends on its rate of production and escape.

In addition to its role in initiating the ripening process, ethylene is believed to play an important part in the emergence of seedlings from the soil. As some seeds sprout, the upper portion forms a sharp curve, or hook, which shelters the fragile, developing leaves underneath as the shoot plows its way up through the soil (Figure 31.8) . Somehow, the presence of ethylene inhibits the plant from straightening and keeps the fragile leaves from unfolding to the sky until the shoot is free of the ground. Once this happens, the light ethylene gas readily escapes into the air, permitting the stem to straighten and the leaves to expand.

Ethylene and Senescence. Like cytokinins, ethylene plays a role in senescence, but instead of inhibiting the process, it speeds it up. It may be involved in the withering of leaves and flower petals and cell death. Most research, however, has focused on ethylene's role in fruit ripening and leaf abscission.

In its fruit-ripening role, ethylene works in the breakdown of cell walls, which causes the fruit to soften, and the loss of chlorophyll, which causes the fruit to turn from green to whatever color is associated with ripeness. Ethylene is now known to affect its own rate of synthesis, so a little ethylene from one rotten (over-ripe) apple will stimulate its release from other nearby fruit. So, you see, one rotten apple *can* spoil the barrel.

Ethylene gas has long been used to initiate the ripening of fruits in transit to the world's markets. The banana industry in particular owes much of its success to this simple chemical. It is now possible to ship unripened fruit to world markets in prime condition without worrying about untimely ripening.

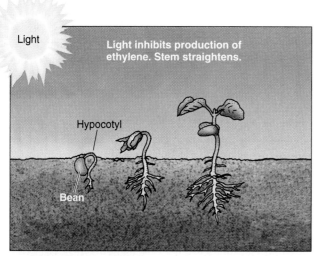

(a) Hypocotyl hook in bean

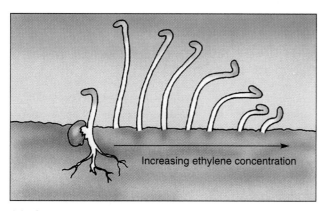

(b) Curvature in pea experiment

FIGURE 31.8
Testing Ethylene's Role in the Hypocotyl Hook
(a) Many dicot seedlings form a curved, protective "bumper"—the hypocotyl hook—as they push their way through the soil. The curve is maintained by a surrounding concentration of ethylene. When the seedling breaks through the soil, the curve straightens. **(b)** The effects of ethylene on curvature in pea seedlings are demonstrated by this experiment, in which seedlings on the right were exposed to increasing ethylene concentrations.

The release of ethylene over green bananas in warehouses or in ship cargo holds will produce ripe, yellow bananas at just the time they reach the consumer.

Leaf Abscission

All leaves have a limited life span and will at some point undergo senescence and be shed from the plant by a process called leaf abscission. The processes of senescence and abscission are probably the best known and most spectacular in the deciduous trees, which turn color and lose their leaves in the autumn.

Before leaves fall, a number of changes occur in the plant. Chlorophyll is broken down and existing or newly synthesized yellow, red, and orange pigments are revealed. Valuable materi-als, such as proteins and sugars, are transported out of the leaf to tissues in the stem and root for recycling. At the base of the petiole, in a region of cells called the abscission zone (Figure 31.9), ethylene triggers the synthesis of enzymes that break down the polysaccharides of the cell walls. As this happens, a layer of cells on the stem side accumulates a high content of a waxy substance called suberin, effectively sealing off the plant's vascular system, preventing invasion by bacteria, fungi, and insects. Then, thin-walled parenchyma cells along the abscission zone enlarge, weakening the remaining connection. Finally, the leaves break away (and the raking begins). Abscission is controlled by several factors including environmental stimuli such as day length and temperature, a change in the balance of auxin and ethylene, and the age and nutritional status of the plant.

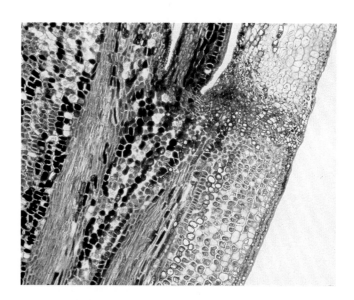

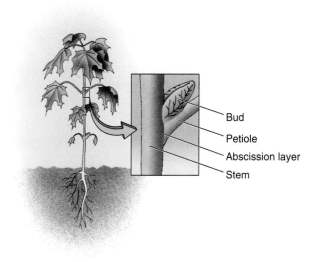

FIGURE 31.9
The Leaf Abscission Layer
Before a leaf falls, the cells of the abscission zone (arrow) die and harden. Eventually, the leaf petiole will break away and fall. A suberized leaf scar is left on the stem.

Apical Dominance

The familiar triangular profile of conifers (Figure 31.10) is the result of a phenomenon known as **apical dominance**, the ability of a terminal bud to suppress the outgrowth of lateral buds. Flowering plants, particularly trees, often lack a dominant growing tip and therefore reveal a spreading almost spherical growth form. If a plant is pruned and its dominant tip removed, the lateral branches will begin to grow. Constant pruning will result in a dense, heavily branched plant (Figure 31.11).

The inhibitory effect of the terminal bud was long believed to be due to the activity of auxin. Experiments with decapitated pea plants show that apical dominance can be restored by applying auxin to the shoot. However, the situation in most plants appears to be more complex and to involve the interaction of auxin and cytokinin, and perhaps ethylene, in regulating the growth of lateral buds.

Applications of Plant Hormones

Since the discovery of auxins, chemists have developed an entire family of analogous compounds, some of which have even greater growth-promoting effects than the original. One such compound is 2,4-dichlorophenoxyacetic acid, known as 2,4-D. While 2,4-D promotes growth at very low concentra-

FIGURE 31.10
Apical Dominance in Conifers
The presence of the terminal branch inhibits growth in the lateral branches below. The older branches, however, have had longer to grow, even at a reduced rate, so the result is the familiar triangular appearance in plants of this group.

tions, it kills plants at higher concentrations. This compound is a very common and potent agent of weed control. Dicots, including the broad-leaved plants we often think of as "weeds," are much more sensitive to the herbicide than monocots (wheat, oats, and so on). This means that you can kill the clover in your lawn with 2,4-D, while sparing the bluegrass and Bermuda grass. Of major ecological importance, 2,4-D offers the advantage of rapid **biodegradability**. Biodegradability means that a substance ceases biological activity when exposed to the elements and organisms of the soil.

Weed control agents are not uncontroversial, however. For example, there have been serious questions about the effects of a chemical defoliant called 2,4,5-T (2,4,5-trichlorophenoxyacetic acid), closely related to 2,4-D and used during the Vietnam War to defoliate trees. (The trees hid the activities of the North Vietnamese soldiers.) Actually, the real risk, as we now see it, did not arise so much from the control agents themselves as from a lack of quality control in the manufacturing process. In the past, 2,4,5-T is known to have been contaminated by dioxin, the infamous and frightening environmental pollutant common to illegal (and legal) chemical dumps. We now know that the 2,4,5-T used in the defoliant *agent orange* of Vietnam fame was heavily contaminated with dioxin. The potential consequence of the frequent exposure of American military personnel and Vietnamese citizens to agent orange during the Vietnam conflict is still a volatile issue. Veterans' groups have claimed that such exposure has caused an increase in leukemia and birth defects in their children. The use of 2,4,5-T was banned in the United States by the Environmental Protection Agency in 1979.

Differential Growth Responses: Tropisms

As we now know, auxin and other plant hormones are responsible for many of the specialized growth responses resulting in the movements that so intrigued Darwin. Some of the visible responses are called **tropisms** ("to turn"). As the derivation implies, tropisms involve bending toward or away from a directional stimulus. For example, stems or leaves orienting themselves with respect to directional light, such as the auxin response discussed earlier, is **phototropism**. There are others. Unequal growth in response to gravity is called **gravitropism** (formerly "geotropism"). And some plants exhibit uneven growth in response to touch, a response called **thigmotropism** (*thigma*, "touch").

Phototropism

Most researchers believe that during the phototropic response in stems, light stimulates something in the growing tip and causes the auxin to move laterally in the apical meristem. The

As with many hormones, auxin is short-lived. Instead of accumulating in the stem, it is inactivated by specific enzymes in the lower parts of the plant. The inactivation process tightly regulates the level of auxin within the plant.

Roots and Gravitropism

When a seedling is placed on its side, the shoot tip eventually will bend upward, whereas the root tip will bend downward (Figure 31.12a). The root is said to be positively gravitropic (growing in the direction of the stimulus, gravity) and the stem negatively gravitropic. What accounts for this differential response, and why do the two plant regions respond differently?

Plants are believed to perceive gravity through the settling of large starch grains called **statoliths** (Figure 31.12b). In roots these statoliths are found in cells located in the root cap. Researchers believe that when a root is placed on its side, statoliths reaggregate to the new low point of the cells. This is believed to cause a change in calcium distribution in the cells, which triggers transport of auxin to the lower side of the root (Figure 31.12c). Roots are very sensitive to auxin concentration, and when it begins to increase above a certain level, cell elongation there is inhibited. The concentration in the upper side of the root, however, is just right for causing elongation in the cells there. As a result, the root bends downward.

Gravitropism is less well studied in stems, but statoliths are believed to be present in cells outside vascular bundles and to reorient when a stem is placed on its side. Calcium and auxin are believed to be redistributed. Stems, however, respond differently to auxin and the high concentration of auxin on the lower side *promotes* elongation until the stem is again upright.

The phenomenon of gravitropism is an active area of plant research and the hypotheses presented here will undoubtedly change and become refined as data from experiments continue to come in. Some of these data will be collected from observations of gravitropism aboard the Space Shuttle, or perhaps the planned Space Station. How might weightlessness be useful in the search for answers?

Thigmotropism

The slender, coiling growths at the tips of young grape and pea plants are known as **tendrils**. Tendrils, which are modified leaves and stems, are highly sensitive to touch. This response is a **thigmotropism**. When tendrils encounter a solid object, their tips respond by growing toward it, perhaps coiling around the object, thereby providing a firm anchor that permits further growth (Figure 31.13). Compared to most plant responses, the coiling can be quite rapid, with a complete turn forming in less than an hour (so the coiling action is faster than the minute hand of a clock).

As you might expect, coiling is produced by unequal cell elongation, with the greatest cell lengthening occurring oppo-

(a)

(b)

FIGURE 31.11
Apical Dominance in Broad-Leaved Plants
(a) Broad-leaved plants often lack a single dominant tip and produce a spreading growth. (b) Very dense decorative hedges are produced through constant pruning. With repeated loss of growing tips, each stem continues to produce new lateral branches.

process appears to be controlled by a photoreceptor, probably a yellow pigment. Although this initial photoreceptor has not yet been identified, it is known to be most responsive to blue light. The entire growing tip produces auxin in quantity, but as it diffuses down the shoot it crosses over to the unlighted side, where cells respond by elongating and thereby bending the tip toward the light (see Figure 31.4).

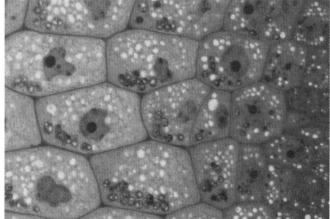

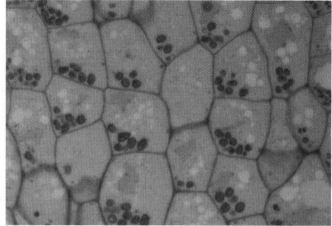

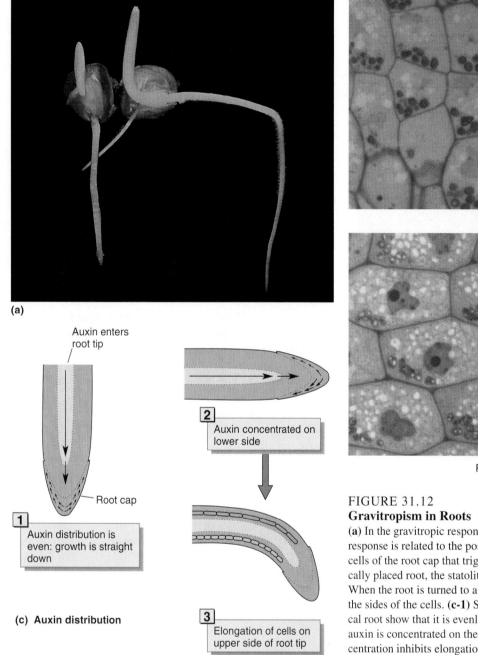

(a)

Auxin enters root tip

Root cap

1 Auxin distribution is even: growth is straight down

(c) Auxin distribution

2 Auxin concentrated on lower side

3 Elongation of cells on upper side of root tip

Root vertical

Root horizontal

(b)

FIGURE 31.12

Gravitropism in Roots

(a) In the gravitropic response, roots curve toward gravity. **(b)** The response is related to the position of statoliths, large starch grains in cells of the root cap that trigger the movement of auxin. In the vertically placed root, the statoliths are located at the base of the cells. When the root is turned to a horizontal position, the statoliths sink to the sides of the cells. **(c-1)** Studies of auxin concentration in the vertical root show that it is evenly distributed. **(c-2)** In the horizontal root, auxin is concentrated on the lower, "gravity" side. **(c-3)** The new concentration inhibits elongation on the lower side, and normal elongation on the upper side bends the root downward, "toward gravity."

site the touching surface. While the actual mechanism hasn't been satisfactorily explained, it is believed to be influenced by IAA and ethylene. Their application to tendrils, in the absence of any touching object, induces the usual curvature.

Solar Tracking

Consider the familiar sunflower. Observe a field of sunflowers on a sunny morning. The flowers are strikingly beautiful, and they are all facing in the same direction—toward the morning

sun (Figure 31.14). Return the same afternoon, though, and you will see that the flowers all face the opposite direction, toward the afternoon sun! This can be disconcerting until you understand what's going on. What you have observed is called **solar tracking**. How does this differ from the usual phototropic response? Phototropism, as we've seen, involves growth—cell division and elongation—which is a generally slow process. Solar tracking in sunflowers (along with cowpeas, soybeans, cotton, and others) is much faster and involves changing turgor in the tissues of the stem. It is also readily reversible, which is why the flowers may face you on your way to work and, again, on your way home. Yet solar tracking is a true tropism because the orientation of the flowers or leaves depends on the direction of the light.

Negative solar tracking responses are also known. The leaves of some drought-resistant plants regularly fold or expose only their edges when struck by bright sunlight. This keeps the surface temperature down, an important water-saving adaptation.

Nastic Movement

Nastic movements are movements in response to a stimulus, but they are not directed toward or away from that stimulus. Nastic movements (*nast,* "to press down") usually involve the movement of leaves and, like solar tracking, are produced by changes in turgor pressure in certain cells in the base of the leaf petiole.

One example is the **nyctinastic movement** (*nyct,* "night"), the so-called "sleep movement," wherein leaves or leaflets that are splayed out to the sunlight in the daytime, fold or turn into a vertical position at night (Figure 31.15). Like other nastic responses, this one involves groups of cells at the base of the leaf petiole or the base of leaflets that form a special structure called a **pulvinus** (plural, *pulvini*). Its components, known as **motor cells**, specialize in the pumping of potassium ions (K^+) from one place to another, thereby alter-

FIGURE 31.13
Thigmotropism
Upon touching a surface, grape tendrils begin a coiling growth, which soon fastens the plant securely to the object.

FIGURE 31.14
Solar Tracking
Sunflowers turn to follow the sun throughout the day, a response to light that involves the shifting of water from one tissue to another, rather than a slower, auxin-directed growth response.

Morning glory "awake" and "asleep."

FIGURE 31.15
Sleep Movements
Morning glory flowers exhibit sleep movements. In daylight hours, the petals are fully exposed to sunlight, but at night they turn or fold into a vertical position.

ing the water potential in such cells. As we've seen again and again: where ions go, water quickly follows, and the resultant changes in turgor bring about the raising and lowering of petioles that characterize sleep movement. (Recall that a K$^+$ transport mechanism also regulates the turgor in guard cells.)

The abrupt response of plants to touch, called a **thigmonastic response**, has fascinated people for ages. The mechanism, again involving changes in turgor in pulvini, is similar in principle to sleep movement.

Touching or pressing any of the many fine leaflets of *Mimosa pudica* sends its leaves into an almost spasm-like reaction. The leaflets fold, and the leaf petioles suddenly droop (Figure 31.16a). Some theorists suggest that this response may discourage browsing animals, while others maintain that it simply helps the plant avoid excessive water loss when hot, dry winds blow. The second hypothesis is supported by the finding that *M. pudica* is heat-sensitive, responding to heat as it does to touch. Like the usual nastic

response, the thigmonastic response in *M. pudica* involves potassium pumping that brings about turgor changes in motor cells of its pulvini. Cells losing turgor collapse, permitting the petiole to bend, whereas cells gaining turgor restore the petiole to its upright position.

Dionaea muscipula, the Venus flytrap (Figure 31.16b), responds to touch by rapidly closing its toothed leaves (see Chapter 30). Two hypothetical mechanisms have been proposed. One involves acid-induced expansion, where rapid acidification occurs in the walls on the outer surface of the flytrap leaves. These cells then elongate, while the inner cells do not, and the trap closes. The trap reopens only after growth of the inner surface "catches up" with the elongation of the outer.

Another hypothesis maintains that trap closure results from a release of tension built up in the mesophyll cells of the leaf while the trap is open. Further research will be needed to test these two hypotheses and determine the mechanism.

Light and Flowering

The response of organisms to changing lengths of day and night is known as **photoperiodism**, a phenomenon important to most forms of life. Throughout much of the Northern Hemisphere, spring is heralded by dormant buds breaking and spring flowers blooming. As the days lengthen, species after species bursts into bloom in a rather predictable sequence. As summer gives way to fall, and days again begin to grow shorter, still other species will unfurl their colors. How do these plants perceive changing seasons?

Photoperiodicity

As to how a plant knows what the season is, we have some of the answers. The critical factor for many species is the *length of night,* rather than the length of day, as was once believed. Of course, the length of night varies with location and with the season, but plants somehow are able to "count the hours." When the period of darkness reaches a certain length, the plant responds in a predictable manner, as though it "knows" April, June, or January has arrived.

The response of plants to changes in the length of night can roughly be organized into three categories. Plants that begin the flowering process before the summer solstice, June 21, are traditionally called **long-day plants** (*short night*) (Figure 31.17a) because they flower only when the lengthening days reach some critical length (for that species). Examples include henbane, radish, lettuce, and some varieties of wheat, potatoes, and spinach. Plants that do not flower until after the days shorten to some critical length, generally after the summer solstice, are called **short-day plants** (*long night*) (Figure 31.17b,c). Included are strawberries, poinsettias, chrysanthemums, and primroses. The terminology is a little misleading

(a) *Mimosa pudica* (sensitive plant)   *M. pudica* disturbed

(b) *Dionaca muscipula* (venus flytrap)   *D. muscipula* with trap "sprung"

FIGURE 31.16
Touch Responses in Plants
(a) *Mimosa pudica*, the touch-sensitive plant, visibly "cringes" when touched along its leaflets. **(b)** *Dionaea muscipula*, the Venus flytrap, springs its trap on a hapless insect if its triggers are touched in a certain way.

because, as we said, day length is not really the critical factor. It is actually the length of night that matters. We'll show you the evidence shortly.

The third category of flowering plants are the **day-neutrals**, which appear to be indifferent to cycles of light and darkness. They may flower continuously or respond to other stimuli. Examples include cucumber, sunflower, and dandelions.

The Dark Clock and Flowering. How was it determined that it was the duration of *darkness* that triggers flowering, and not the duration of daylight? Actually, the demonstration is rather simple and can be done in laboratories where the light and dark periods can be regulated. Under such conditions it was found

that the period of light can be changed without causing any flowering changes, but if the length of darkness is tampered with, the plant responds. Quite dramatically, a single, relatively brief exposure to intense light in the middle of the night can "trick" a long-day (short-night) plant into responding as if the short nights of spring had arrived. Such a plant can then be made to bloom in any season. The same treatment, if done every night, can prevent a short-day (long-night) plant from flowering at its normal late summer or autumn time, since it will react as if the nights were still too short for such activity.

This bit of academic tinkering was immediately put to practical use by chrysanthemum growers. For many years they had extended the chrysanthemum season into early win-

(a) Henbane

(b) Chrysanthemum

(c) Poinsettia

FIGURE 31.17
Photoperiodicity
(a) For its flowering, the henbane requires a long day (short night), usually in excess of 12 hours. **(b)** The chrysanthemum is a short-day plant that produces flowers in autumn. **(c)** Poinsettias are also short-day plants and will bloom when the night is longer than 13 hours.

ter, artificially lengthening the days by keeping bright lights turned on for several hours after sunset. But they later found they needed only to turn on the lights for a few minutes each night in order to get the same results. In another application, sugar cane growers stall flowering in their fields by turning floodlights on at night. Sugar-laden sap that might otherwise be used to produce commercially useless flowers can then continue to accumulate in the stems.

Flower Initiation and Phytochrome

The cocklebur has proved to be an ideal organism for studying the initiation of flowering. It is so hardy that researchers have been able to prune, trim, and graft it in drastic ways and still count on its survival. The cocklebur is a short-day (long-night) plant and will put out homely little green flowers if exposed just once to a period of darkness longer than $8\frac{1}{2}$ hours.

Armed with that knowledge, plant physiologists began to toy with the plant. They found that varying the wavelengths of the light affected their results (Figure 31.18). The most effective light for establishing the photoperiod was found to be in the red region of visible light. Far-red light, with its longer wavelength, had the opposite effect; that is, a flash of far-red light actually reversed the effect of either white or red light if it immediately followed either of them. However, if the far-red flash *preceded* the red flash, or was *delayed* for more than 35 minutes after the red flash, the red flash had its full effect.

The flash experiments gave rise to the hypothesis that there was some receptive pigment involved. The receptive pigment could not be chlorophyll (which is abundant in the photosensitive leaves) because it responds to different wavelengths from those produced by the floodlights.

The investigators hypothesized that the pigment (P) occurred in two forms: Pr, the red-absorbing form, and Pfr, the far-red-absorbing form. (Pr absorbs maximally at 660 nm, whereas Pfr absorbs maximally at 730 nm.) Absorption of light of the appropriate wavelength would change this hypothetical pigment from one form to another. During the day, the predominance of red light over far-red would convert all the pigment from the Pr form into the Pfr form, but at night there would be a spontaneous reversion of the pigment back to the Pr form. A midnight flash of red light would immediately convert the pigment once again into the Pfr form, but in this state it could again be reversed to Pr by absorbing far-red light.

$$Pr \text{ (in red light)} \longrightarrow Pfr$$

and

$$Pfr \text{ (in far-red light)} \longrightarrow Pr$$

The hypothetical pigment was eventually isolated and was named **phytochrome**. It turned out to be a large membrane-bound protein complex. It now seems that phytochrome is involved in a number of other light-induced phenomena, such as the turning of leaves toward light, germination of light-sensitive seeds, and the rapid orientation of chloroplasts.

It is important to remember that phytochrome is a light receptor, not a plant hormone or other effector. Plant physiologists are currently working to understand the link between phytochrome and the various responses that it mediates. It is hypothesized that gene expression or activity may be altered and the movement of hormones may also be affected.

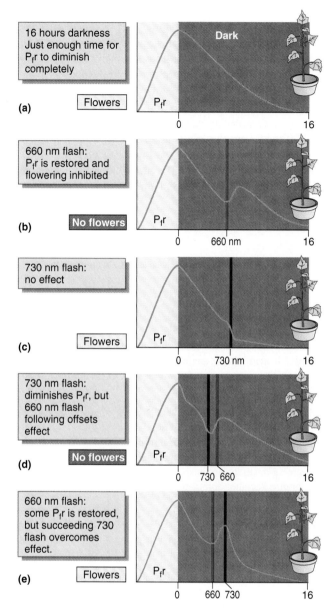

(a) 16 hours darkness Just enough time for P$_f$r to diminish completely — Flowers

(b) 660 nm flash: P$_f$r is restored and flowering inhibited — No flowers

(c) 730 nm flash: no effect — Flowers

(d) 730 nm flash: diminishes P$_f$r, but 660 nm flash following offsets effect — No flowers

(e) 660 nm flash: some P$_f$r is restored, but succeeding 730 flash overcomes effect. — Flowers

FIGURE 31.18
The Dark Period Controls Flowering
Experiments show that the dark period is critical and that phytochrome far-red (*Pfr*) has an inhibiting effect on flowering in short-day plants such as the cocklebur. The following graphs hypothesize the effects of varying levels of the far-red form of phytochrome. (**a**) Under natural conditions sixteen hours or more of darkness is required for phytochrome *Pfr* to diminish sufficiently to permit flowering. (**b**) A flash of red light (660nm) during the dark period increases the *Pfr* enough to inhibit flowering. (**c**) A similar flash of far-red light (730nm) has no inhibitory effect, and even accelerates the shift of residual *Pfr* to *Pr*, permitting flowering. (**d**) However, if the flash of far-red is followed immediately by flash of red, enough new *Pfr* is generated to cancel the effect of far-red light and no flowering occurs. (**e**) Support for the last statement comes from a reversal of the procedure: a flash of red followed immediately by a flash of far-red produces the same result as a flash of far-red alone, and flowering occurs.

Transmitting the Stimulus. The dark clock and its photoreceptor, phytochrome, are located in the leaves. The signal that actually induces flower development has to be delivered from the leaves to other parts of the plant. In the resilient cocklebur, an isolated, amputated leaf can be subjected to an appropriate flower-inducing photoperiod regimen and subsequently grafted back onto a plant, causing the plant to flower. The signal is evidently hormonal, but no flower-inducing hormone has been isolated. In some experiments using plants other than the cocklebur, it appears that there are *repressing* hormones produced by the leaves at all times *except* when the dark clock indicates that the appropriate time for flowering has arrived.

The hormone-like transmission of the photoperiod effect has been demonstrated most convincingly by an experiment in which six cocklebur plants were grafted together in a row (Figure 31.19). One leaf of the plant at one end of the line was enclosed in a box and given the appropriate photoperiod treatment, after which *all six plants* flowered, one after the other, right down the line. In other grafting experiments, the cambia of the two plants were separated by a sheet of paper, but the message got through anyway, just as we would expect if the signal were hormonal.

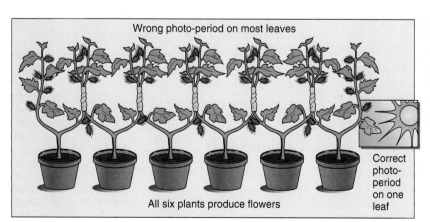

Wrong photo-period on most leaves
All six plants produce flowers
Correct photo-period on one leaf

FIGURE 31.19
The Leaf Contains the Photoreceptor
This dramatic experiment lends strong support to the idea that the photoreceptor is in the leaf and that a transmitted agent is involved. Only one leaf in this series of six grafted cocklebur plants is exposed to the proper photoperiod, yet all six plants produce flowers.

Because the hypothetical flower-inducing hormone, **florigen**, has proved to be so elusive, most plant physiologists believe that there is no special flower hormone. Instead, the message for flowering is conveyed by particular levels and combinations of auxin, gibberellins, and other known plant growth hormones.

We are aware that, paradoxically, auxin has an inhibitory effect on flowering. Since growing tips of shoots produce auxin in quantity, flowering can sometimes be induced or increased by simply cutting off the shoot tips. Rose growers have discovered this, and the clicking of pruning shears can be heard throughout the countryside at the beginning of the floral season.

Pineapples, Mangos, and Ethylene. The hormone ethylene promotes flowering in pineapples and mangos. As tropical plants, they live where there is little seasonal change in day length. Not surprisingly, they are day-neutral, blooming and bearing fruit all year long on irregular schedules of their own. In the past, growers used smoke from bonfires to synchronize flowering so that all the fruit could be harvested at the same time. We know now that ethylene in the smoke is the active agent, so growers synchronize the flowering of whole fields by the application of ethylene. The plants flower and ripen all at once and can be harvested efficiently.

Key Ideas

LIGHT AND THE GROWTH RESPONSE

1. Working with canary grass, Darwin found that the young **coleoptile**-covered shoot would respond to light by bending toward it. By experimenting, he found that the agent responsible was formed in the tip and passed downward.

2. Boysen-Jensen decapitated oat seedlings, mounted tiny blocks of gelatin on the stump, and placed the tips on the gelatin. He also used chips of mica to separate the tip from the stump. The shoots bent toward light only in the group where gelatin was used, indicating something from the tip had to pass downward for the bending response to occur.

3. Paal decapitated oat seedlings and used the tip alone for experiments. He placed the tip to one side of the stump, and the bending response always occurred on that side. He concluded that some influence from the tip caused cell elongation.

4. Fritz Went placed oat seedling tips on agar for a time and then used agar blocks in his experiments. Arranging the agar blocks on the seedling stumps in various ways, and keeping his seedlings in darkness, he observed the usual bending, indicating that the active substance had diffused into the agar.

5. Went named the active substance **auxin** (later identified as **indoleacetic acid** or **IAA**). He also worked out a bioassay system where curvature in subjects could be compared to a standard and the auxin content of the subject determined.

6. Auxin's primary action is to stimulate cell elongation. The mechanism may involve lowering the cell wall pH and thus activating microfibril-loosening enzymes.

7. Auxin is involved in **abscission** (leaf fall). Prior to leaf fall, changes in cells at the base of the petiole produce an **abscission zone**. Then, as auxin diminishes, materials in the leaf are reclaimed by the plant and cell walls in the abscission zone break down.

PLANT HORMONES

1. **Gibberellins** cause stem elongation. When released by the grain embryo, they stimulate the aleurone to produce starch-digesting **alpha-amylase**.

2. Some **cytokinins** stimulate cell division and retard **senescence**, or aging, in plants. Together, auxin and **kinetin** influence differentiation in the callus formed from tobacco pith cells. The relative proportions of the two determine whether the growth will form shoots, roots, both, or neither.

3. **Abscisic acid** inhibits auxin and gibberellins, bringing on plant dormancy in preparation for winter. It also participates in the stomatal water-stress response.

4. **Ethylene** hastens fruit ripening and helps retain the hypocotyl hook in some emerging seedlings.

5. Auxin plays an inhibitory role in **apical dominance**, whereby one growing shoot elongates and the rest are inhibited to some degree. Loss of the dominant tip results in inhibition, and another branch becomes dominant.

DIFFERENTIAL GROWTH RESPONSES: TROPISMS

1. Plant growth responses, called **tropisms**, include **phototropism** and **gravitropism**, which are directional.

2. In phototropism, a blue-sensitive photoreceptor detects light direction, and, through processes still unknown, the migration of auxin is directed to the unlighted side of the shoot, where elongation then occurs.

3. **Gravitropism** occurs when unequal, gravity-driven distributions of auxin occur in the root tip. When a seedling root is placed horizontally, auxin concentrates on the lowermost side, inhibiting elongation, whereas the lighter concentration on the upper side permits elongation, and thus the root bends toward gravity.

4. Touch-induced growth responses occur in **tendrils**, which respond by uneven cell elongation opposite the touching part. This produces a coiling growth about solid objects.

5. The continuous daily turning of leaves or flowers toward or away from the sun is called **solar tracking**.

NASTIC MOVEMENTS

1. **Nastic responses**, movements based on reversible turgor changes, are generally nondirectional. Nastic responses include the folding and turning of leaves and flowers. They are brought

about by turgor changes in **motor cells** of **pulvini**, caused by the transport of potassium ions from one part of the petiole base to another. Included are **nyctinastic movements** (or sleep movements), the daily unfolding and folding or turning of leaves.

2. Touch-initiated nastic responses, as seen in *Mimosa pudica* and *Dionaea muscipula,* are rapid. In the first, touch or heat causes a current flow that activates potassium pumps in the pulvini. Subsequent turgor changes bring about the collapse of leaf petioles and leaflets. In *D. muscipula,* the trap—a modified leaf—is triggered by the movement of hairs, whereupon standard cell elongation in the outermost cells causes the trap to close.

LIGHT AND FLOWERING

1. **Photoperiodism** is any response to changing length of night or day. Flowering in plants is often photoperiodic.

2. In their response to photoperiods, plants fall into three categories: **long-day, short-day**, and **day-neutral**.

3. Plants respond to the length of night rather than the length of day. A brief exposure to light during the night will induce flowering in long-day plants and stall flowering in short-day plants.

4. The most effective light for interrupting flowering is in the red region of the spectrum, while far-red reverses the red effect if it follows within 35 minutes. Two incontrovertible variations of a light receptor, *Pr* and *Pfr,* have been proposed. The receptor has been identified as **phytochrome**, a large, membrane-bound protein, but its operation as a "dark clock" isn't understood.

5. Experiments with the cocklebur strongly suggest a hormonal mechanism that may work through the repression of flowering at all times except during the proper photoperiod.

Application of Ideas

1. Formerly, plant physiologists attributed the success of plant roots in reaching water to hydrotropism, a water-initiated growth response. While the idea has now been discarded, root growth is known to be most dense in moist areas of soil. Suggest an alternative hypothesis to explain this observation. How do roots happen to find water? How might your hypothesis be tested?

2. Many angiosperms respond to photoperiods in their flowering activities. What might be some advantages to complex flowering systems that relate to day or night length? If photoperiodism is adaptive to plants, how can the success of the day-neutral plants be explained?

Review Questions

1. Describe Charles Darwin's experiments with the Canary grass. What conclusion did he reach? (page 651)

2. What information did Boysen-Jensen and Paal add to what Darwin had already determined? (page 651)

3. Describe the technique used by Went, and explain how his assay procedure worked. (page 651)

4. Describe the specific action of auxin on plant cell elongation. (pages 652–653)

5. List the main steps involved in leaf abscission. What role does auxin play? (pages 653, 657)

6. Describe the role of gibberellins on stem growth and in the germination and growth of grain embryos. (pages 653–654)

7. What effect do cytokinins have on senescence? How might cytokinins interact with auxins in this role? (pages 654–656)

8. How might abscisic acid help temperate zone plants survive? (page 656)

9. Using the idea of a photoreceptor and the pattern of auxin diffusion, present an explanation of phototropism. (pages 658–659)

10. Suggest two basic differences between tropic and nastic responses and provide an example of each. (page 658)

11. Explain how the distribution of auxin in vertically and horizontally arranged seedling roots differs, and how this affects root tip growth. (page 659)

12. Review an example of the nastic response and explain the role of pulvini and motor cells. (pages 661–662)

13. Briefly explain how plant physiologists determined that the period of darkness was the significant factor in the photoperiodic flowering responses. (pages 662–664)

14. Describe the experimental observations that led to the conclusion that the flowering photoreceptor had two forms, *Pr* and *Pfr.* Write a simple equation showing the effects of red and far-red light. (pages 664–665)

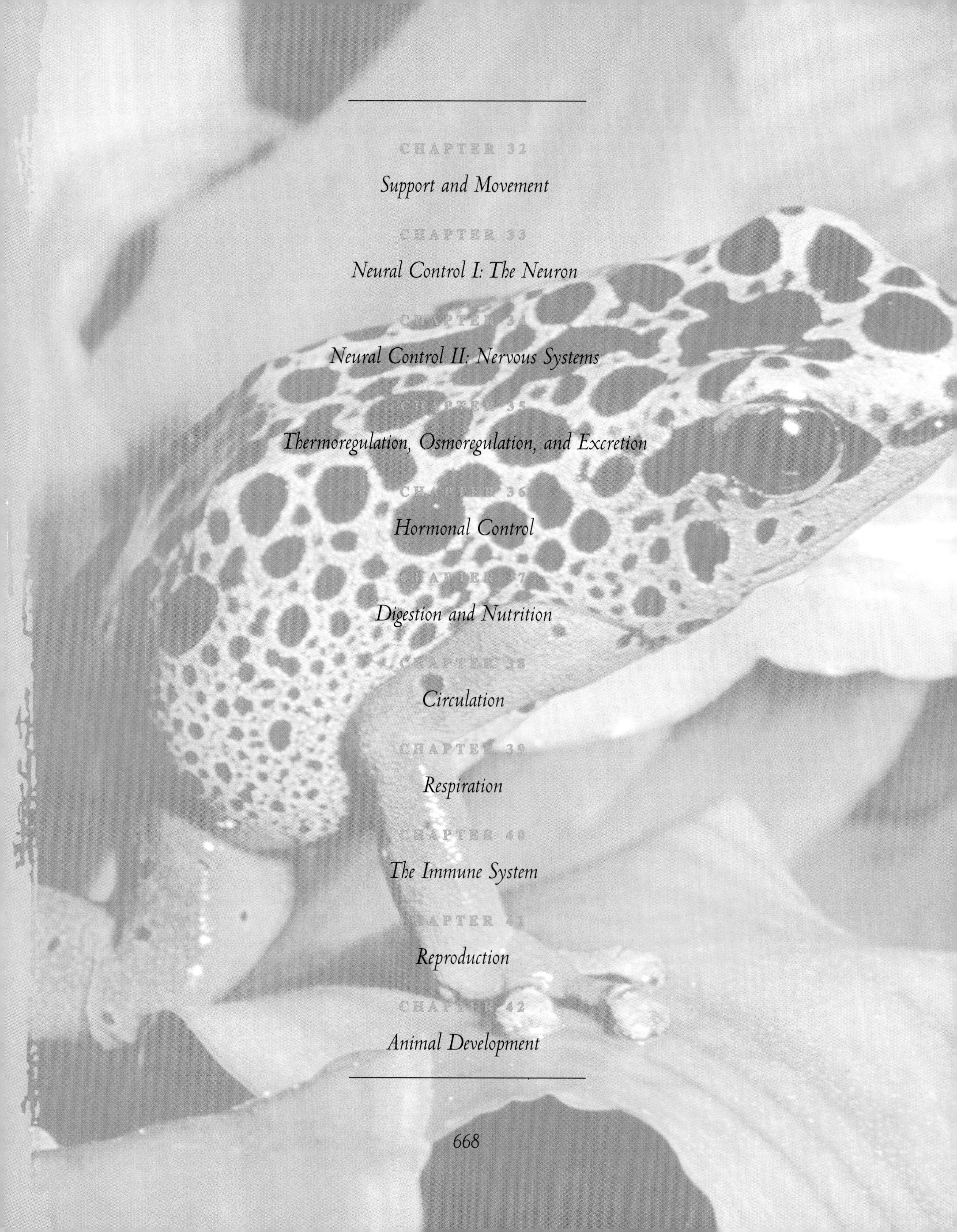

CHAPTER 32

Support and Movement

CHAPTER 33

Neural Control I: The Neuron

CHAPTER 34

Neural Control II: Nervous Systems

CHAPTER 35

Thermoregulation, Osmoregulation, and Excretion

CHAPTER 36

Hormonal Control

CHAPTER 37

Digestion and Nutrition

CHAPTER 38

Circulation

CHAPTER 39

Respiration

CHAPTER 40

The Immune System

CHAPTER 41

Reproduction

CHAPTER 42

Animal Development

PART SIX

Animal Functions

CHAPTER 32

Support and Movement

CHAPTER OUTLINE

INVERTEBRATE SUPPORT AND MOVEMENT

Hydrostatic Skeletons and Their Movement

Invertebrate Exoskeletons

Invertebrate Endoskeletons

VERTEBRATE TISSUES

Epithelial Tissue

Connective Tissue

VERTEBRATE SKELETONS

The Structure of Bone

Organization of the Vertebrate Skeleton

VERTEBRATE MUSCLE: ITS ORGANIZATION AND MOVEMENT

Smooth Muscle

Cardiac Muscle

Skeletal Muscle

Summing Up Contraction

KEY IDEAS

APPLICATION OF IDEAS

REVIEW QUESTIONS

Compared to the atmosphere of other celestial bodies, the earth is a rather unspectacular place. Even our temperatures are moderate compared to those of other planets. And then consider how tranquil the earth's surface is compared to the great storms of Jupiter and the scorching surface temperatures of Venus. Perhaps, in fact, it is precisely because the earth is so moderate and unspectacular that life has formed here. That life, though, has had to face some problems. After all, the earth is not entirely a benign place.

One challenge of life arises because the earth is not only large, but it differs greatly from one place to another. It also changes with time. This means that some areas are more suitable for life than others, and areas that are quite hospitable at one time may be unsuitable (even deadly) at another time. Thus the success of many animals depends partly on their ability to move from one place to another. Some travel easily, such as the plover, a bird whose migratory distance between its wintering grounds and breeding grounds would take it around the world each year. Other animals, such as marine snails, creep slowly over wave-washed rocks, scraping off morsels of slimy food as they go. Nonetheless, movement is as important to one as it is to the other.

Hand in hand with the usefulness of movement goes the need for support. This problem arises because massive celestial bodies tend to pull objects toward their surfaces. So because of earth's gravity, not only do apples fall, but animals find it hard to jump. Gravity exerts a continuous tug on our bodies, seeking to flatten us against our planet. But we resist with muscles that contract and defy that pull, and with skeletons that support our bodies and give our muscles something to pull against. Among the earth's inhabitants, natural selection has produced in animals a vast array of systems that serve in support and movement. We will begin our survey of these systems with the animals without backbones.

Invertebrate Support and Movement

We begin with the invertebrates, where we find support systems ranging from the pressure of fluids to crusty plates—some just under the skin, some on the outside. Then we will move to the many species with internal skeletons of cartilage and bone. The first groups we will see are protostomes (see Chapter 20), then we go on to the deuterostomes beginning with the echinoderms.

Hydrostatic Skeletons and Their Movement

The **hydrostatic skeleton**, as its name implies, works through pressure generated by fluids (*hydro*, "water"; *stasis*, "maintenance"). Many soft-bodied animals make use of the pressure within fluid-filled spaces for maintaining body form and for support and movement. We find hydrostatic skeletons in cnidarians, nematodes, mollusks, and annelids. How can fluids provide these important functions? Let's consider the simplest example, the hydrostatic skeleton at work in a free-living soil nematode.

Nematodes. Recall from Chapter 26 (see Figure 26.18) that nematodes are slender, hair-like worms that move through moist soil particles with a whip-like, thrashing motion, the body bending first one way and then the other. In water, where solid particles are absent, their movements are not very effective, but in soil they are able to move along efficiently. Recall also that nematodes have an extensive, fluid-filled body cavity, the pseudocoelom. The body wall surrounding the pseudocoelom contains muscle fibers that run longitudinally (along the length), and these are encircled by a tough, elastic, cuticle (Figure 32.1). The worm's firmness, and thus its body shape, is maintained through pressure exerted by the cuticle against the noncompressible fluids within. The hydrostatic skeleton is reinforced in these worms by a network of collagen fibers that crisscross the pseudocoelom.

Movement in nematodes occurs as muscle fibers along one side of the turgid body contract, bending the body toward that side and stretching the muscles of the opposite side. Upon relaxation of the muscle, the cuticle rebounds, and the displaced fluids return the body to its original shape. Then opposing muscles contract, and the body bends the other way. So we see that the trapped fluids and the muscles work in opposition along the body to complete this simple motion. The fluids resist the bending and tend to return the body to its original shape.

Mollusks. A variety of other animals support their bodies by exerting pressure on fluids. The clam, for example, pumps blood into its fleshy foot, extending and forcing the foot into the soft seabed. There the foot swells, acting as an anchor. Next, powerful retractor muscles draw the heavy body toward the firmly anchored foot. The clam thus is able to slowly and powerfully dig its way through the seabed.

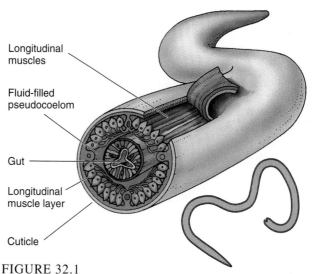

Longitudinal muscles

Fluid-filled pseudocoelom

Gut

Longitudinal muscle layer

Cuticle

FIGURE 32.1
The Hydrostatic Skeleton
A turgid, fluid-filled pseudocoelom provides the resistant base against which the nematode's longitudinally arranged muscles work. Contraction flexes the body, and the rebounding of the stretched cuticle on the extended side returns it to its former shape. Through coordinated contractions, the undulating movement of the roundworm is produced.

Annelids. The earthworm has a fluid-filled coelom and a muscular body wall, so its hydrostatic skeleton is fundamentally similar to that of the nematode, but much more complex. Earthworms are segmented internally as well as externally, so fluids in each segment are isolated or partitioned off from those in the next. Whereas nematodes have only longitudinal muscles, earthworms have two layers: an inner longitudinal layer and an outer circular layer (Figure 32.2). Contraction of the longitudinal layer brings about a shortening and thickening of the body, while contraction of the circular layer extends and narrows the body as coelomic fluids are compressed from different directions.

The use of the hydrostatic skeleton for burrowing, although very efficient, is quite ancient. Recall that burrowing worm tracks represent some of the earliest evidence of animal life. From these findings it seems that the hydrostatic skeleton was most likely the first known to animals.

Invertebrate Exoskeletons

We now leave the concept of watery skeletons to consider those that are more familiar: hard skeletons. These can occur as *exoskeletons* and as *endoskeletons*. **Exoskeletons** are hardened support and protective structures that lie outside soft parts. **Endoskeletons** are hardened support and protective structures that lie within soft parts. Exoskeletons and endoskeletons are further distinguished by their embryological development. Exoskeletons are formed by secretions of epidermal cells, derived in the embryo from *ectoderm.* Endoskeletons in vertebrates are formed essentially by cartilage and bone cells. Such cells originate for the most part in embryonic *mesoderm.*

(a)

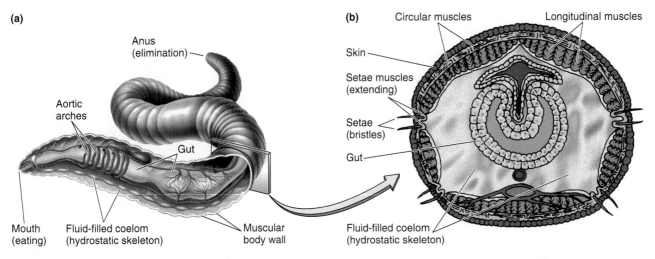

Anus (elimination)

Aortic arches

Gut

Mouth (eating)

Fluid-filled coelom (hydrostatic skeleton)

Muscular body wall

(b)

Circular muscles   Longitudinal muscles

Skin

Setae muscles (extending)

Setae (bristles)

Gut

Fluid-filled coelom (hydrostatic skeleton)

(c)

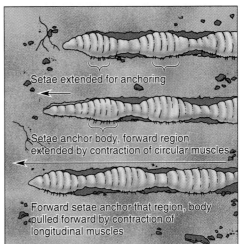

Setae extended for anchoring

Setae anchor body, forward region extended by contraction of circular muscles

Forward setae anchor that region, body pulled forward by contraction of longitudinal muscles

FIGURE 32.2

Movement in an Annelid

(**a**) The soft-bodied earthworm. (**b**) Each of the earthworm's segments contains longitudinal and circular muscle layers and paired setae. Its fluid-filled coelom acts as a resistant base for muscle action. (**c**) The inching movement is far more complex than shown and can be thought of as an ongoing series of actions: extension, anchoring, and inching up. This requires highly coordinated alternating contractions of the longitudinal and circular muscles, and anchoring by the setae. Where the longitudinal muscles contract, the circular muscles must relax and vice versa.

Hard skeletons function in movement by providing a firm base for the attachment of muscles that move limbs and other parts. Such muscles are arranged in opposing pairs referred to as **antagonists**. Since antagonists create opposing movements, one member of the pair extends a limb or part, and the other draws it back. Opposing muscle groups are essential, since animal muscles do their work through contraction only. (Muscles pull; they do not push.)

The protective function of an exoskeleton is quite obvious. It acts as a hardened shield around the body. Consider, for example, the shell of a clam or the hard, tough exterior of

an insect. Let's look at the exoskeleton and associated musculature of an insect.

Hardened, jointed exoskeletons are characteristic of the phylum Arthropoda, the gigantic phylum that includes insects, crustaceans, spiders, and other joint-legged types. In addition to providing a firm muscle base and protection for the animal, the arthropod exoskeleton, or cuticle as it is more specifically known, helps retard water loss in terrestrial species. You may recall (Chapter 26) that its principal constituent is chitin, a complex carbohydrate. The cuticle at the joints remains thin and flexible, but tough like leather.

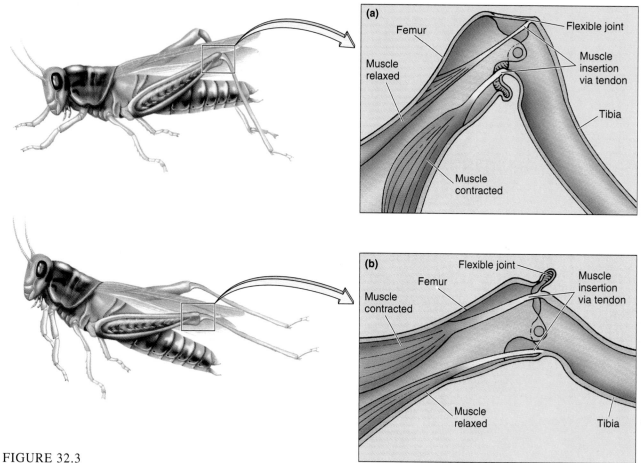

FIGURE 32.3
How a Grasshopper Jumps
The jumping legs of the grasshopper contain powerful muscle groups responsible for its impressive, sudden leaps. The muscles originate within the femur and insert, via tendons, on the tibia below. **(a)** As the muscle on the underside of the femur contracts, it brings the tibia close to the femur, and the insect assumes a crouched position. **(b)** Rapid contraction by the upper muscle extends the tibia with startling suddenness, raising the insect's body as a jump begins.

As we see in the insect body in Figure 32.3a, each muscle has an **origin** and an **insertion**. The origin is the end of the muscle that is attached to a more stationary base, whereas its insertion is toward the opposite end (the end attached to the more movable part of the skeleton). In the arthropod limb, muscles originate in an inner segment of the exoskeleton, which may house all of the muscle. The insertion is in the next segment out—the segment to be moved. The connection between the end of the muscle and its point of insertion occurs via a long tendon that passes through the joint and makes a firm attachment to the exoskeleton.

It might seem that exoskeletons would be cumbersome, limiting the movements of the animal (like the armor used by knights of the Middle Ages). However, just the opposite is true. Insects, we know, are quite agile and, in fact, are capable of feats of great strength. We've all been impressed by ants dragging burdens that are heavier than they are. And you may have discovered that a flea can jump from the floor all the way up into your pants leg (usually making its presence known just as you drive into heavy traffic). So imagine how high a horse-sized grasshopper could jump! Actually, it couldn't—in fact, it would hardly be able to move. The reason is that arthropod exoskeletons are essentially hollow, thin-walled, jointed tubes with complex muscle attachments inside. Such tubes are quite strong and light—as long as they remain small. For a substantial increase in size to occur without a loss in strength, the exoskeleton would have to undergo a corresponding increase in thickness. Thin tubes of large diameter would simply buckle when placed under stress. So at some critical size, the sheer weight of an exoskeleton would curtail most movement, and the "armored horse" would just lie there, feebly kicking at the air.

Although like arthropods, the mollusk exoskeletons are secreted by epidermal tissue, unlike arthropods, mollusks don't shed their exoskeletons as their bodies grow. Instead, their shells grow by ever-widening increments at the edges. Such

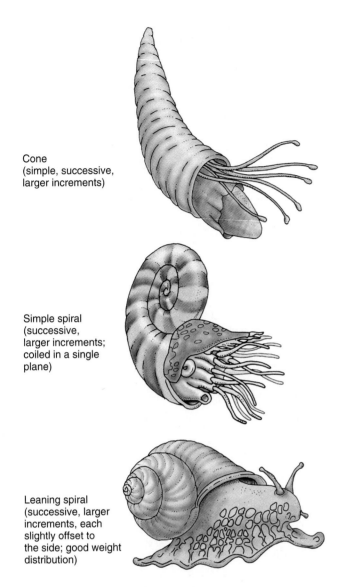

Cone
(simple, successive,
larger increments)

Simple spiral
(successive,
larger increments;
coiled in a single
plane)

Leaning spiral
(successive, larger
increments, each
slightly offset to
the side; good weight
distribution)

FIGURE 32.4
Shell Growth
Mollusk shells grow continuously through the addition of shell
material at the margins. The most primitive shells are simple cones
and cones spiraled in a single plane. The shell of the land snail
forms a far more complex leaning spiral, which provides a more
even distribution of shell weight.

growth follows some interesting patterns. In the most primitive
mollusks, the shells are simple, straight "dunce cap" cones
with new shell added evenly around the edge. In others the conical shape remains but is modified by simple coiling, that is,
coiling in a single plane. In the chambered *Nautilus,* for example, new shell growth occurs fastest on one side, so the simple
spiraling coil forms. This works fine in a swimming aquatic
animal where the weight is supported by water, but a chambered *Nautilus* living on land, or remaining on the seabed,
would tend to fall over on its side. Most snails and other gastropods have evolved shells with a complex "leaning coil,"

where the angle of the coil as well as the size of the shell opening undergoes constant change. Such growth ensures a more
equalized distribution of shell and body weight (Figure 32.4).

Invertebrate Endoskeletons

Endoskeletons lie within the body and are initially deposited
by embryonic mesoderm. We are most familiar with this sort
of skeleton because ours is an endoskeleton. The spiny-
skinned echinoderms, such as sea stars, also have endoskeletons. This may not be too surprising if you remember that
both groups are deuterostomes, quite distant from the protostomes with their exoskeletons. Under the echinoderm epidermis, the thick dermis (of mesodermal origin) secretes a series
of plates that fit closely against each other. The plates themselves are composed of many tightly interlocked little spined
bodies, formed chiefly of calcium carbonate, a common material in marine animals. In the sea star, as the skeletal plates
form, they erupt over the entire upper surface, creating short,
blunt, immovable spines (Figure 32.5).

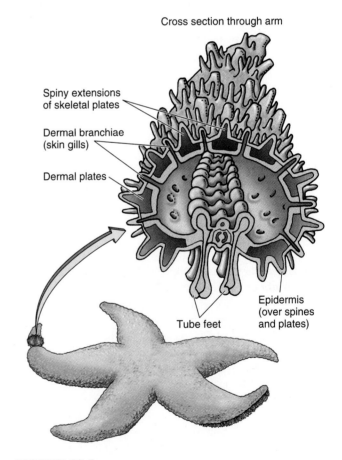

Cross section through arm

Spiny extensions
of skeletal plates

Dermal branchiae
(skin gills)

Dermal plates

Epidermis
(over spines
and plates)

Tube feet

FIGURE 32.5
Sea Star Endoskeleton
In the sea star, the skeleton consists of numerous plates of calcium
carbonate. Some plates produce blunt spines. Note the thin
epidermis overlying the endoskeleton.

Vertebrate Tissues

Before we get into vertebrate supporting structures, we will look briefly at some kinds of tissues. **Tissue**, you will recall, is an aggregation of similarly specialized cells. There are just four basic tissue types that contribute to the many organs and organ systems of all metazoans: **epithelium**, **connective tissue**, **muscle**, and **nerve** (Figure 32.6). Whereas the animals with exoskeletons are covered by a tough layer of chitin, vertebrates are covered by an epithelial layer that extends to all body surfaces, inside and out. And the vertebrate endoskeleton, as we shall see, is composed of connective tissue. So let's take a closer look at epithelial and connective tissue, then focus on tissues that can contract, and, later, tissues that can carry signals.

Epithelial Tissue

Epithelial tissue, or epithelium, forms most surface linings or coverings of the organism, both interior and exterior. Epithelial cells arise from a *basement layer* underneath, while above they lie in contact with either air or fluids. Thus the **epidermis**—the outermost layer of the skin—is epithelial. Epithelial tissue lines the mouth, the nasal cavities the respiratory system, the coelom, the tubes of the reproductive system, the gut, and the body cavities, and forms the inside lining of blood vessels. The tissues lining glands contain a thickened epithelium called *glandular epithelium.* The cells of the glandular epithelium both line the ducts (openings) and secrete whatever it is that gland secretes. We will look more closely at glandular epithelium in chapters to come.

Epithelium may be classified according to the number of cell layers and the shape of cells that lie on the surface. The cell layers may be described according to whether they are arranged in a single layer, called **simple epithelium**, or in more than one layer, called **stratified epithelium**. There is even a pseudostratified epithelium that looks like it's stratified because the bulk of some cells lie near the surface and that of others near the basement membrane. The cells at the surface may be classified further into **squamous** (flattened), **cuboidal** (cube-like), and **columnar** (elongated) (see Figure 32.6).

Connective Tissue

Connective tissue is found throughout the vertebrate body. This is a good thing since its role is essentially to hold the body together. Connective tissue consists of cells that produce a **connective tissue matrix** noncellular material (Figure 32.7). Examples of familiar connective tissues include bone, cartilage, ligaments, tendons, adipose (fat) tissues, and even blood.

The principal and most abundant substance in most connective tissue matrices is **collagen**, the fibrous protein that accounts for at least a third of the total body protein in the larger vertebrates. Collagen is the "glue" of the animal body and forms extremely tough structures, such as the cornea of the eye, the tendons and ligaments, and the all-important disks that cushion the spine. Collagen is also the principal nonmineral component of bone.

Collagen is produced by cells known as **fibroblasts**. As the protein is secreted, it takes the form of tough, lengthy fibers that form the connective tissue matrix (Figure 32.8 and see Figure 3.18). The fibroblasts are vitally involved with wound healing and the mending of broken bones. In wound healing, a number of different kinds of connective tissue cells cooperate, reforming normal structures and, when necessary, forming tough collagenous scar tissue.

In the mending of bones, fibroblasts form a thick sheet of connective tissue around bones and temporarily fill the fracture site with tough fibers of collagen. Eventually, the new bone forms and hardens, permanently fusing the break.

Another important protein of connective tissue is **elastin**, a major component of elastic fibers. Such fibers have "memory"—they can stretch to several times their length and snap back to their original size. Elastin is particularly important to arteries, notably the larger ones whose flexibility helps maintain blood pressure. Interestingly, elastin is common in the necks of grazing mammals—those that spend so much time with their heads near the ground. Can you see the role of elastin here?

With this overview of selected vertebrate tissues, we can now take a closer look at vertebrate skeletons.

Vertebrate Skeletons

As vertebrates, when we think of supporting structures, we normally don't think of cuticles, shells, or spiny plates. We think of bone. Bones are the light, hardened structures that hold us erect (keeping us from lying about, flattened against the ground) and shield our more delicate parts from damaging environmental objects. So, we might well ask, what is bone, and what gives it its toughness?

Except for the cartilaginous fishes (see Chapter 27), vertebrates have skeletons composed primarily of bone (although some cartilage usually persists, as we shall see). The skeletons of the jawless fishes (lampreys and hagfishes) and the sharks and rays are composed of **cartilage**. It has a clear, almost glassy consistency and is made up of **chondrocytes** (cartilage cells) embedded in a surrounding gel-like protein matrix that is heavily imbued with collagen fibers. In the vertebrate embryo, cartilage is the forerunner of bone, but in the mature vertebrate it persists only in a few places such as the ears, nose, and joints.

The bony skeleton is important in at least five ways: (1) it supports the body; (2) it serves as an attachment for muscles; (3) certain parts of the skeleton, in particular, the skull and rib cage, are protective; (4) some bones produce red blood cells in their interior spaces; and (5) all bones act as a natural reservoir for the body's calcium.

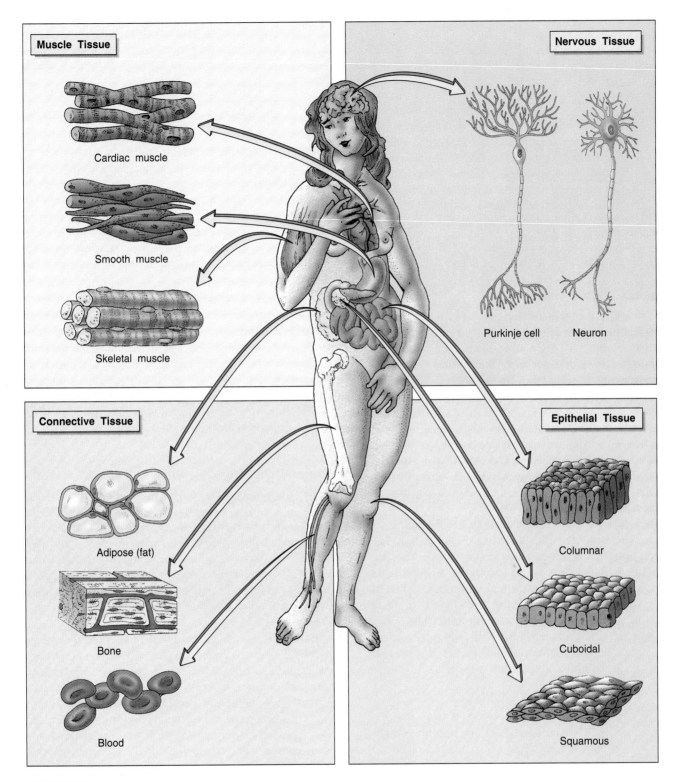

Muscle Tissue

Cardiac muscle

Smooth muscle

Skeletal muscle

Nervous Tissue

Purkinje cell   Neuron

Connective Tissue

Adipose (fat)

Bone

Blood

Epithelial Tissue

Columnar

Cuboidal

Squamous

FIGURE 32.6
Vertebrate Tissue
The complex vertebrate body, which includes our own, is composed of only four types of tissue. The sampling seen represents some of the many specialized sybtypes.

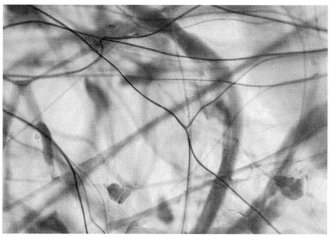

(a) Loose connective tissue of the skin

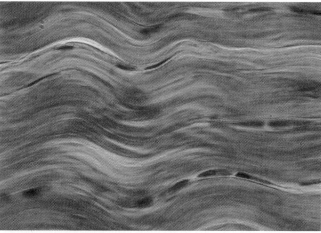

(b) Dense connective tissue in a tendon

FIGURE 32.7
Connective Tissue
(a) In loose connective tissue, the cellular elements lie scattered in a loosely arranged matrix of dense collagen and fine elastic fibers. The blue objects are nuclei. **(b)** Dense connective tissue contains much more of the fibrous matrix arranged in parallel rows, with the living cells crowded among individual collagen fibers. The blue objects are nuclei.

(a) SEM of collagen fibers

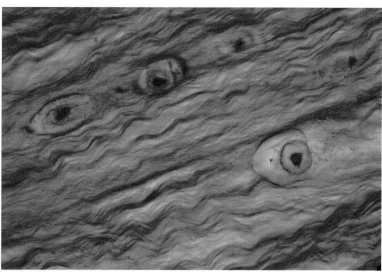

(b) Collagen fibers in a tendon

FIGURE 32.8
Collagen
(a) Collagen fibers are the primary material of tendons—strong cord-like structures attached to bones. **(b)** Individual collagen fibers in tendons form a wavy pattern, with distinct cross-banding of dense protein. Fibroblasts lie embedded in the collagen.

The Structure of Bone

Typically, the long bones of the limbs are surrounded by a thin membrane of connective tissue called the **periosteum**. Within, we find two kinds of bone, **spongy** and **compact** (Figure 32.9a). Spongy bone is well named since it is porous and sponge-like, made up of a web-like structure of hard bone (laid down along lines of stress so that this seemingly fragile bone is very strong).

The spaces in the web are filled with soft tissue. Spongy bone is found at the enlargements at each end of long bones. Some bones, such as the ribs, sternum, vertebrae, and hip bones, contain **red marrow**, the site of red blood cell production.

Compact bone makes up the shaft—the cylindrical, lengthy portion of the long bone—and is thick and dense. Its central cavity contains **yellow marrow**, a dense fatty material. Much of the red marrow in young vertebrates is replaced by yellow marrow in the adult. Despite its stone-like appearance, compact bone is definitely a living tissue. It contains numerous metabolically active bone cells, as well as nerve cells and blood vessels. Most of the bony mass, however, consists of hardened calcium phosphate in a collagen matrix. The source of compact bone's great strength is seen best in its microscopic organization.

Under the microscope, thin sections of compact bone reveal intricate, repeated units of structure called **Haversian systems**. Each Haversian system consists of a **central canal**, containing blood vessels and nerves, and a surrounding region of concentric rings or **lamellae**, consisting of calcified bone (Figure 32.9b). The rings are, in effect, tubes within tubes—a laminated tubular construction that imparts great strength and resiliency to the bone.

Within tiny cavities or **lacunae** (singular, *lacuna,* "little lake") we find the bone cells, or **osteocytes**. Each lacuna is interconnected with others via minute canals, aptly named **canaliculi** ("little canals"), which pass throughout the hardened bone. The osteocytes touch each other via long projections through the canaliculi. Some of these extensions reach the central canal, permitting the cells to carry on an exchange of materials with the circulatory system. The osteocytes, which cannot divide in their chambers, are formed from cells called **osteoblasts**, which lay down the bony matrix in a constantly rebuilding and renewing process. Other cells, called **osteoclasts**, which are formed from white blood cells called monocytes (Chapter 38), constantly destroy and reabsorb the bone as part of the normal development and maintenance of the skeleton and the redistribution of body calcium.

Organization of the Vertebrate Skeleton

Anatomists divide the vertebrate skeleton into two regions: the *axial skeleton* and the *appendicular skeleton* (Figure 32.10a). The **axial skeleton** includes the **cranium** (skull), the **vertebral column** (backbone), the **rib cage**, and **sternum** (breastbone). The appendicular component includes the limbs (always two or fewer pairs) and the two **girdles—pelvic** (hip) and **pectoral** (shoulder).

The Backbone. The backbone consists of a series of bones that form a flexible axis for the body. Each bone is individually known as a **vertebra**, and their number varies greatly from one class of vertebrate to another. Reptiles have the record, with as many as 400 (most of them very similar to each other). Birds, on the other hand, have comparatively few, partly because the posterior ones are fused. All mammals, from giraffes to shrews to whales, have exactly seven vertebrae in the neck, although the number in the remainder of the column varies.

Humans have 33 vertebrae (Figure 32.10b,c), although there may be one more or one less, depending on how many make up the **coccyx** or "tailbone." They are organized into five regions. All vertebrae are separated by **intervertebral disks** of cartilage. Each disk has a softer, compressible center that provides cushioning and helps make lateral movement possible. Intervertebral disks take a lot of wear, and painful ruptures ("slipped disks") sometimes occur.

In addition to its supporting role, the vertebral column protects the spinal cord. The cord passes through what is called the **spinal canal**, which is formed by the central openings in the individual vertebrae. Vertebrae in the thoracic or chest region also provide attachments for ribs.

The **appendicular skeleton** includes the pectoral and pelvic girdles (rings of bones) and the bones of the limbs (arm and leg bones), as shown in Figure 32.11. Humans (and birds) are unusual among vertebrates in that they are bipedal (walk on two feet). Therefore our pectoral girdles, greatly relieved of their weight-supporting burden, are considerably reduced. The human pelvic girdles are also different even from the other primates, who are not well adapted to an upright gait. As you can see in Figure 32.11, our upright posture is associated with a distribution of weight that is unusual among primates. The upper torso rests entirely on the **sacroiliac joint** (the joint between the **sacrum** and the **ilium**), with some help from certain back muscles that distribute the weight to the ilium.

The pectoral girdle consists of two **scapulae** (shoulder blades) and two **clavicles** (collar bones). The shoulder blade is a triangular, flattened bone whose outer margin is thicker and forms part of the shoulder joint, a ball-and-socket arrangement (see Figure 32.10).

Joints. The various bones of the body meet at **joints** (Figure 32.12a). Each joint allows only a certain kind or degree of movement, and some, such as the **sutures** between the bones of the skull, allow no movement at all (try flexing your skull). The pubic symphysis is slightly movable, which accommodates childbirth. The more movable joints—for example, the knee, elbow, or hip—are both complex and elegant. The articulating (contacting) surfaces are covered with a thin layer of exceedingly smooth cartilage. The bones are held together by short bands of white fibers of collagen called **ligaments** (familiar because they often suffer sprains—or tears). Where the ligaments form a capsule enclosing the joint they produce what is called a **synovial joint** (Figure 32.12b). The synovial joint is constantly lubricated by fluid, secreted inside the capsule. The presence of smooth cartilage and lubricating fluid is highly advantageous. It helps keep the joint from wearing out by reducing friction and allows for quieter movement. (As you can imagine, a hungry lion creaking and squeaking around in the grass is likely to come up short.)

The movable joints are generally named according to the type and degree of their movement. Thus there are **gliding** (wrist bones; slight back and forth movement), **pivotal** (head and neck; rotation in one plane), **ball-and-socket** (shoulder; free rotation), and **hinge** (knee; hinge-like extension and flexion in one place).

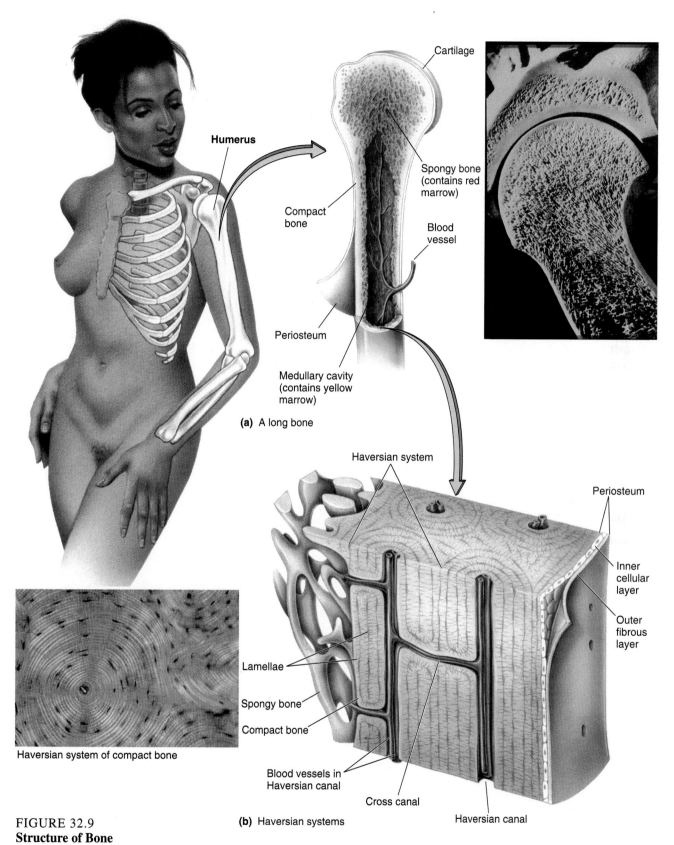

Humerus

Cartilage

Spongy bone (contains red marrow)

Compact bone

Blood vessel

Periosteum

Medullary cavity (contains yellow marrow)

(a) A long bone

Haversian system

Periosteum

Inner cellular layer

Outer fibrous layer

Lamellae

Spongy bone

Compact bone

Blood vessels in Haversian canal

Cross canal

Haversian canal

Haversian system of compact bone

(b) Haversian systems

FIGURE 32.9
Structure of Bone
(a) The sectioned long bone reveals a hard shaft of compact bone and a central cavity of yellow marrow.
Spongy bone is seen near the joints. **(b)** Numerous central canals, responsible for carrying blood vessels and
nerves, are surrounded by hardened concentric rings, or lamellae, containing bone cells (osteocytes). Such cells
are entombed in lacunae, tiny cavities that communicate with each other via minute crevices called canaliculi.

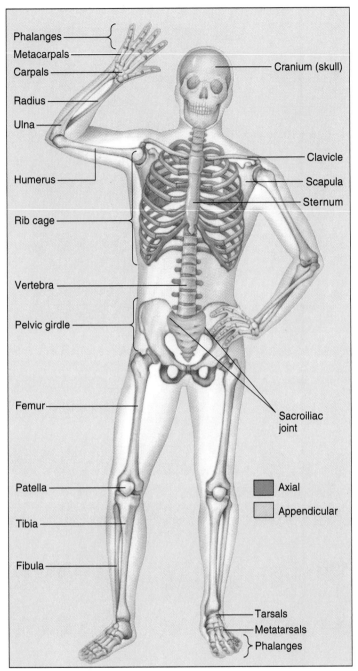

Phalanges
Metacarpals
Carpals
Radius
Ulna
Humerus
Rib cage
Vertebra
Pelvic girdle
Femur
Patella
Tibia
Fibula

Cranium (skull)
Clavicle
Scapula
Sternum
Sacroiliac joint

Axial
Appendicular

Tarsals
Metatarsals
Phalanges

(a) The human skeleton

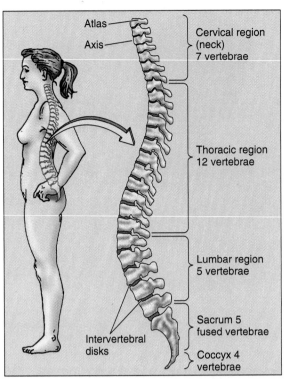

Atlas
Axis
Cervical region (neck) 7 vertebrae
Thoracic region 12 vertebrae
Lumbar region 5 vertebrae
Sacrum 5 fused vertebrae
Intervertebral disks
Coccyx 4 vertebrae

(b) Vertebral column

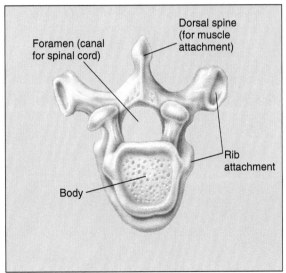

Foramen (canal for spinal cord)
Dorsal spine (for muscle attachment)
Rib attachment
Body

(c) Thoracic vertebra

FIGURE 32.10

The Human Skeleton

(**a**) The axial skeleton includes the skull, vertebral column, rib cage, and sternum. The appendicular skeleton (shown in yellow) includes the pectoral and pelvic girdles and the appendages. (**b**) The vertebral column, in profile, takes on a gradual S-shape. It is divided into the cervical, thoracic, and lumbar regions, plus the fused sacrum and coccyx. (**c**) Vertebrae in the thoracic region contain special indentations (facets) for the attachment of the ribs. The protruding dorsal spine is an attachment point for back muscles.

FIGURE 32.11
Comparison of Gorilla and Human Skeletons
Of all the primates, only humans are truly bipedal. Compare the pectoral and pelvic girdles, position of skull, relative length of arms and legs, and the shape of the foot in the human and the gorilla. Note the more horizontal position of the pelvis in the gorilla compared to its vertical position in the human. The pectoral girdles are more generally similar, with greater mass in the scapula of the gorilla, as might be expected in a heavy, brachiating primate.

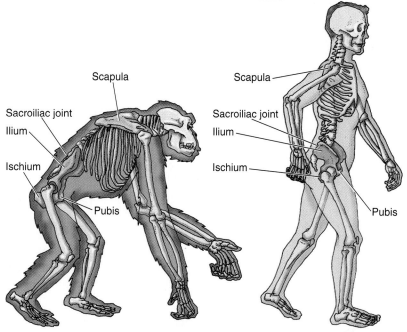

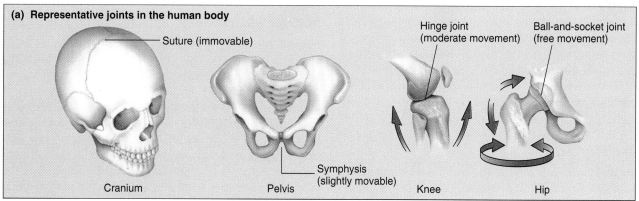

(a) Representative joints in the human body

Suture (immovable)

Cranium

Symphysis (slightly movable)

Pelvis

Hinge joint (moderate movement)

Knee

Ball-and-socket joint (free movement)

Hip

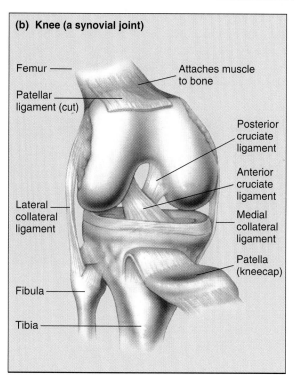

(b) Knee (a synovial joint)

Femur

Patellar ligament (cut)

Attaches muscle to bone

Posterior cruciate ligament

Anterior cruciate ligament

Medial collateral ligament

Lateral collateral ligament

Patella (kneecap)

Fibula

Tibia

FIGURE 32.12
Joints in the Human Body
(a) Representative joints include the immovable skull sutures, slightly movable pelvic symphysis, moderately movable hinges, and free-moving ball-and-socket joints such as the hip and shoulder. (b) Synovial joints, such as the knee, have a surrounding capsule (not seen here) and moist, lubricating synovial fluid within.

Vertebrate Muscle: Its Organization and Movement

Vertebrate muscles can be divided into three types according to their microscopic structure, characteristics of contraction, and means of control. The three types are *smooth muscle, cardiac muscle,* and *skeletal muscle.* Whereas each is composed of muscle cells, such cells have become so specialized that they are generally referred to as muscle fibers. Here we will focus on human muscles as representative of vertebrates.

Smooth Muscle

Smooth muscle is involuntary muscle that is found in the digestive tract, in blood vessels, at the base of each hair, in the iris of the eye, and in the uterus and other reproductive structures. Involuntary here means that we have little conscious control over it. You may have noticed that the hair on the back of your neck rises when you look a politician in the eye, or that you can't stop your stomach from growling on a date, or that you can't call back a blush once it has begun. These are all due to involuntary actions of smooth muscle. Smooth muscle even influences sex, since it is the muscular control of blood entering and leaving the genitals that accounts for changes occurring during sexual arousal. Smooth muscle is under the control of the *autonomic nervous system,* a branch of the nervous system that controls many activities below the conscious level (see Chapter 34). The autonomic nervous system can both inhibit and stimulate smooth muscle contraction.

Under the microscope, smooth muscle tissue is, logically enough, smooth in appearance, at least in comparison to the other types of muscle (Figure 32.13a). The individual cells or fibers are spindle-shaped and tapered at their ends, and each has a single nucleus. Smooth muscle tends to occur in sheets rather than in the dense bundles seen in skeletal and cardiac muscle. Smooth muscle contains the same contractile proteins as other muscle types, but its contractile proteins do not have the same highly ordered organization.

Cardiac Muscle

Cardiac muscle is heart muscle. Its remarkable strength and endurance are legendary, and rightfully so. Whereas it is unique in some ways, cardiac muscle has some things in common with the other muscle types. Like smooth muscle, cardiac muscle fibers contain a single, centralized nucleus. Also like smooth muscle, cardiac muscle is largely under autonomic control. But like fibers of skeletal muscle, cardiac muscle fibers are cylindrical and *striated* (striped) as seen in Figure 32.13b. The stripes reflect the highly ordered arrangement of contractile proteins.

Unlike both smooth and skeletal muscle, cardiac muscle fibers are branched and present a woven appearance. For this reason, heart muscles contract in a twisting motion that, when magnified throughout the heart, results in a "wringing" or "squeezing" action in the chambers with each beat. Where individual cardiac muscle fibers join, their membranes are highly folded, forming interlocking junctions named **intercalated disks**. These disks are extremely "leaky"; that is, they permit an easy flow of electrical currents between cells and thus throughout the tissue. The "leaks" actually occur at gap junctions, pore-like connections between the cells (see Chapter 4). Finally, cardiac muscle will contract without outside stimulation; contraction is intrinsic—it comes from within. Contraction will even occur in isolated cells, those that have been teased apart and kept alive in a tissue culture medium. We will look further into the unique properties of cardiac tissue in Chapter 38. Again, the most impressive feature of cardiac muscle is its tireless, rhythmic beating.

Skeletal Muscle

Skeletal muscle is voluntary muscle. That is, it is largely under conscious control. (Because of its striped appearance under the microscope, it is sometimes called *striated* muscle.) It is controlled by *motor neurons* that are part of the *somatic* (voluntary) *nervous system.* As for its three names, it is *skeletal* because these are the muscles that move the skeleton; it is *striated* because of its molecular organization; and it is *voluntary* because it is *possible* to move the muscles at will. Skeletal muscle can contract much more rapidly and forcefully than the other types, but it cannot sustain such contraction without tiring.

Skeletal muscle cells are striated, unbranching, cylindrical, and multinucleate, with the nuclei lying just beneath the cell surface (see Figure 32.13c). The fibers may also be quite long, sometimes several centimeters in length. Each fiber is packed with precisely arranged contractile proteins. We will return to the details of this arrangement shortly.

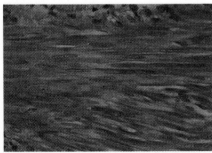 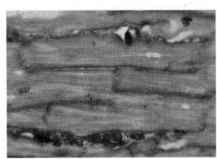 

(a) Smooth muscle   **(b)** Cardiac muscle   **(c)** Skeletal muscle

FIGURE 32.13
Three Types of Muscle Tissue
(a) Smooth muscle consists of long, spindly cells, each containing a single nucleus. **(b)** Cardiac muscle branches and rebranches and is interrupted by intercalated disks. Like skeletal muscle, it is striated. **(c)** Skeletal muscle is heavily striated and multinucleate.

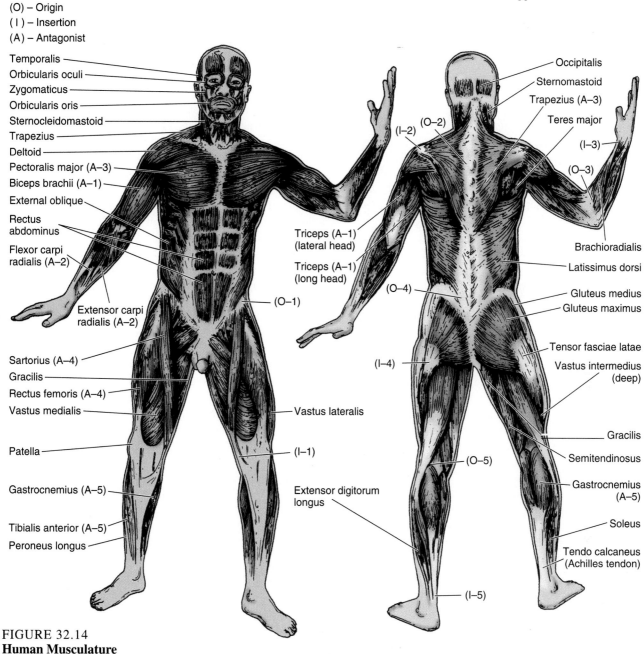

(O) – Origin
(I) – Insertion
(A) – Antagonist

Temporalis
Orbicularis oculi
Zygomaticus
Orbicularis oris
Sternocleidomastoid
Trapezius
Deltoid
Pectoralis major (A–3)
Biceps brachii (A–1)
External oblique
Rectus abdominus
Flexor carpi radialis (A–2)
Extensor carpi radialis (A–2)
Sartorius (A–4)
Gracilis
Rectus femoris (A–4)
Vastus medialis
Patella
Gastrocnemius (A–5)
Tibialis anterior (A–5)
Peroneus longus

Triceps (A–1) (lateral head)
Triceps (A–1) (long head)
(O–1)
Vastus lateralis
(I–1)
Extensor digitorum longus

Occipitalis
Sternomastoid
Trapezius (A–3)
Teres major
(I–2) (O–2)
(I–3)
(O–3)
Brachioradialis
Latissimus dorsi
(O–4)
Gluteus medius
Gluteus maximus
Tensor fasciae latae
Vastus intermedius (deep)
(I–4)
Gracilis
Semitendinosus
(O–5)
Gastrocnemius (A–5)
Soleus
Tendo calcaneus (Achilles tendon)
(I–5)

FIGURE 32.14
Human Musculature
The externally visible muscles of the human illustrate the various types and arrangements and give us some idea about the origins and insertions (as on the biceps brachii, for example). Only some of the major muscles have been named. By finding tendons of origin and insertion, and the general orientation of a muscle, you can determine just what it does. Keep in mind that although muscles can contract forcefully, they cannot extend with force. For this reason, they work in opposing units known as antagonists.

Gross Anatomy of Skeletal Muscle. Skeletal muscle fibers are bound into bundles known as **fascicles**, which are surrounded by a tough sheath of connective tissue. The fascicles in turn are bound together to form the muscle, and it is the fascicles that give meat its stringy appearance. Blood vessels run throughout the fascicular bundle, supplying oxygen and nutrients and carrying off wastes. Nerves also penetrate the bundle, their branches dividing ever more finely until they reach each fiber. These nerves will carry the messages that cause the fibers to contract.

Skeletal muscles are enclosed in a tough casing of connective tissue, the **fascia**, which is continuous with the inner connective tissue sheaths. (*Fascia* is Latin for a "band" or "bandage.") At the ends of the muscle, the fascia coalesces into denser collagenous tissue, forming the cord-like **tendons**, which may be broad and short or thin and long. Tendons, in turn, integrate with the periosteum covering the bones, thus ensuring the continuity of force from muscle to bone. Most muscles have an *origin* (attached to a fixed base) and an *insertion* (attached to the movable end). Figure 32.14 shows the arrangement of the major muscles.

Not all skeletal muscles move bones. For instance, there are the facial muscles, which in humans are quite significant since we rely so heavily on facial expressions in communicating. (A simple raised eyebrow has ruined more than one career, it is said.) Tongue muscles also have complex origins and insertions. Thus the tongue has considerable dexterity, which is essential for eating, speaking, and cleaning the teeth in expensive restaurants. Some muscles form **sphincters**, rings that surround passages within the gut and at the anus and mouth. Other muscles may form into flattened sheets that join with broad, thin tendons called **aponeuroses**. A familiar example is the sheet of abdominal muscles that hold your paunch in as you walk on the beach.

The Ultrastructure of Skeletal Muscle. Now let's take a look at the details of muscle ultrastructure and contraction. We will look into the finer structural levels of muscle and find out how it contracts and how ATP actually works at the contractile site.

Muscle structure is best understood by mentally dissecting it down through its levels of organization, beginning with the gross tissue itself (Figure 32.15), and moving down to the molecular level. We will see that the muscle is subdivided into bundles, each containing numerous fibers. Each muscle fiber is composed of a single cell. Describing the individual muscle fibers in detail (see Figure 32.15) brings us into the ultrastructural level of organization, where most of our knowledge comes from electron microscope studies. We can review a few aspects about muscle fiber first and then look at its contractile structure.

Skeletal muscle fibers range from a few millimeters in length to several centimeters—enormously long as cells go. Each muscle fiber is surrounded by its equivalent of a plasma membrane, the **sarcolemma**. Motor neurons relay impulses from the central nervous system (the brain and spinal cord) to the muscles. The sarcolemma receives the endings of motor neurons at **neuromuscular junctions**, the nerve–muscle interfaces where motor impulses are received. These impulses stimulate muscle contraction, as we will see. Distinct neuromuscular junctions, by the way, are absent in smooth muscle; there the junctions are marked only by slight swellings in the motor neurons.

Within the muscle fiber, just below the sarcolemma, are found a number of nuclei and mitochondria and many glycogen granules—just what you would expect for such an active tissue. Lying just beneath the sarcolemma is the **sarcoplasmic reticulum**, which, like the endoplasmic reticulum of other cells, is a membranous, hollow structure. Extending from the sarcolemma or plasma membrane are numbers of rod-shaped tubules, also hollow and membranous but somewhat larger than the tubes making up the sarcoplasmic reticulum. These are called *transverse tubules,* or **T-tubules**. (They are poorly developed in smooth muscle.) So far, we have named only supporting cellular structures. The active contractile elements of the muscle fiber are actually cylindrical **myofibrils** (or simply *fibrils*). The myofibrils form the visibly striped pattern seen in Figure 32.15 (bottom) that suggested the name *striated muscle*. To understand the pattern, we must first realize that each myofibril contains numerous rod-shaped filaments, or

myofilaments, of the protein **myosin**, and an even greater number of thinner filaments of the protein **actin**. When viewed in cross section, it is common to observe each thick filament surrounded by six thinner filaments.

We can now look at the arrangement of myofilaments from the surface of the myofibril, in a relaxed muscle, beginning with the very prominent **Z disks**. The region between the Z disks is the contractile unit, also known as the **sarcomere**. Moving inward from either Z disk, we come to a very light zone called the **I band**. It consists of actin only, with no overlapping of myosin. Further inward is a broad, dark **A band**, with a smaller and lighter central strip, the **H zone**. The two darker parts of the A band consist of overlapping actin and myosin myofilaments. The lighter central H zone represents myosin alone unless the muscle is contracted, whereupon the H zone disappears.

The Contraction of Skeletal Muscle. When the muscle fiber is stimulated, the Z disks move closer together and the sarcomere is shortened, although the filaments themselves don't change in length. This shortening changes the banding, as you can see in Figure 32.16, darkening the region where the H zone was. What actually happens is that actin myofilaments slide inward past the myosin. When the muscle is fully contracted, the light I bands are greatly reduced because the actin and myosin are then almost completely overlapping. Notice that the width of the A band remains constant—it is exactly equal to the length of a myosin filament.

All this leaves us with a host of questions. What causes actin filaments to slide? What is actin, anyway? What is the relationship between its movement and ATP? With the development of electron microscopes researchers have found numerous tiny bridges extending from the myosin filaments to the surrounding actin filaments. The bridges are globular

FIGURE 32.15
Skeletal Muscle Organization
Whole muscles (*top*) are surrounded by a fascia, which is continuous with the tendons of origin (the anchor) and insertion (the part to be moved; not shown). Within the muscle are fascicles (bundles), each fascicle consisting of many fibers bound together in a connective tissue matrix. Each muscle fiber (*middle*) is surrounded by a plasma membrane called a sarcolemma. Below are numerous T-shaped extensions called transverse tubules. Within the sarcolemma are nuclei, the membranous sarcoplasmic reticulum, and mitochondria. Each muscle fiber contains many myofibrils, composed of actin and myosin myofilaments, which form the contractile units. The myofibrils reveal the banding typical of skeletal muscle (*bottom*). The Z disks define the limits of each sarcomere. Each sarcomere in this myofibril is bordered by two Z disks. Just inside each Z disk are the lighter I bands, containing actin only. Between the I bands is the darker A band, containing both actin and myosin. Its lighter center is the H zone, with myosin only. The end view (EM to the right), taken through the darkest part of an A band, shows the arrangement of actin myofilaments around each myosin.

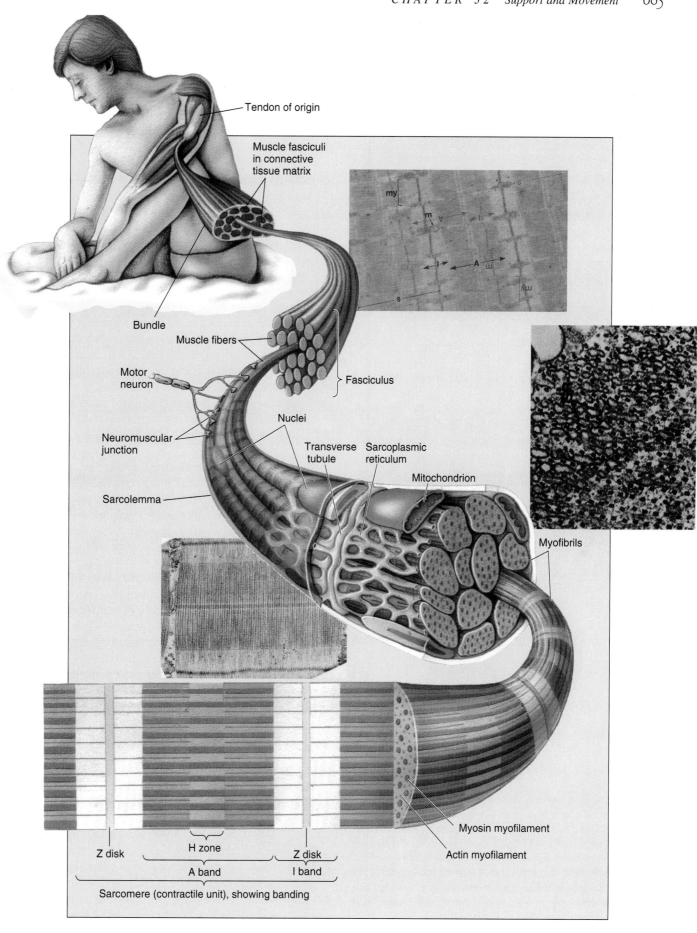

Tendon of origin

Muscle fasciculi
in connective
tissue matrix

Bundle

Muscle fibers

Fasciculus

Motor
neuron

Neuromuscular
junction

Sarcolemma

Nuclei

Transverse
tubule

Sarcoplasmic
reticulum

Mitochondrion

Myofibrils

Myosin myofilament

Actin myofilament

Z disk

H zone

Z disk

A band

I band

Sarcomere (contractile unit), showing banding

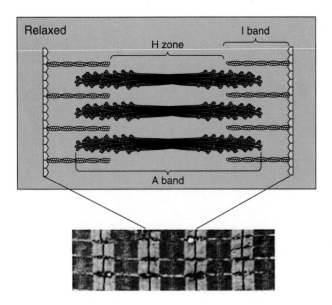

(a) Relaxed

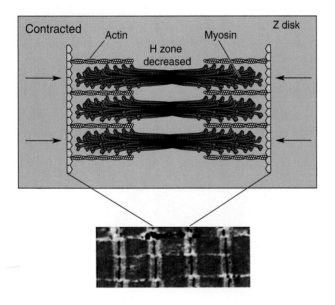

(b) Contracted

FIGURE 32.16
Relaxed and Contracted Muscle
Electron micrographs of muscle show what happens at contraction.
(a) The muscle in a relaxed state: note the distance between Z disks,
the width of the I bands, and the density of the A band. **(b)** Now
observe the muscle in a contracted state: the lines have been included as guides to reveal changes in Z disk distance and I band width.
Note these changes and the increase in density in the A band. Has
the A band width changed? What has happened in the H zone? How
would these changes affect an entire muscle?

extensions of myosin, called **myosin heads** (Figure 32.17).
During muscle contraction the myosin heads apparently reach
out at an angle, attach to the nearest actin filament, forming
bridges, and then they straighten, pulling on the actin and
causing it to slide along. The sequence is repeated again and
again—the myosin head leaning out along the length of the
actin, attaching, straightening, releasing, and then leaning out
to get a new grip further along. The pulling action is not
unlike the hand-over-hand movements a line of sailors might
use to raise the mainsail.

The myosin heads contain ATPases, enzymes that can
hydrolyze ATP, yielding ADP, P_i, and some free energy. The
energy is used in the formation and movement of myosin
bridges. The pulling action occurs on each side of the H zone,
which explains why the Z disks draw together as each contractile unit shortens. When the action is repeated throughout
a muscle, the entire muscle shortens.

We've seen how the action of numerous myosin heads
within the contractile units can bring on muscle contraction.
This bit of information, we might add, represents a major
finding by biochemists and cell biologists. But the myosin
bridges are only half the story. The rest has to do with control.
What causes the bridges to form, and what prevents them
from forming when they are not needed? The answers are
found at the molecular level of muscle action.

Molecular Interactions During Contraction. In their search
for controlling mechanisms, muscle physiologists turned their
attention to isolated actin and myosin. Applying chemical
analysis and electron microscopy (this time using magnifica-
tions of up to half a million diameters), they found the molecular detail needed to develop the necessary models of what
occurs in living organisms.

What they found was that each myosin filament consists
of numerous spirally wound, rod-like proteins. Where individual protein molecules end, they turn outward from the filament,
forming the club-shaped myosin heads (Figure 32.18a). The
heads occur in clusters at specific distances along the filament.

The thin actin filaments are more complex, each consisting of three protein components: **actin**, **tropomyosin**, and
troponin (Figure 32.18b). The largest component is the actin
itself, which consists of two long strands arranged in a gradual helical twist (Figure 32.18b). Each strand contains numerous globular polypeptide subunits. The protein tropomyosin is
also long, double-stranded, and helical but is much thinner
than actin. It lies along the grooves formed by the actin helix.
The third component, the globular protein troponin, occurs in
regularly spaced clusters along the tropomyosin strands,
which are bound to the actin strands.

The Role of Calcium. As we've seen, actin does the actual
moving during contraction. It provides binding sites for the
myosin bridges, and it is this binding that triggers ATPase
activity in the myosin heads. Tropomyosin and troponin control muscle contraction because they block the actin–myosin

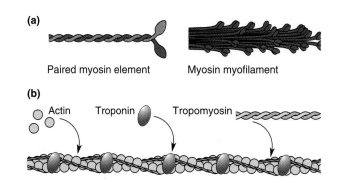

FIGURE 32.18
Actin and Myosin
(a) The thicker myosin myofilament consists of many paired, proteins spirally wound, rod-like proteins. At the end of each rod there emerges a thick, club-shaped tip—the myosin head. (b) The thinner actin myofilaments contain three kinds of protein. The major protein is globular actin. The individual actin spheres are arranged in a long, slender, helical filament. Tropomyosin, a short filamentous protein, follows the curve of the helix for a distance, ending at a molecule of troponin. The tropomyosin–troponin–actin complex of the actin filament controls the formation of myosin bridges.

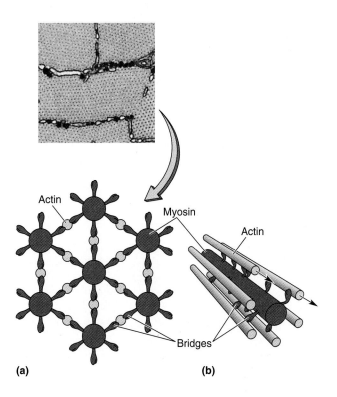

FIGURE 32.17
Myosin Bridges
(a) The electron micrograph of a cross section through the myofibril and the accompanying drawing show the bridges between myosin and the surrounding actin. (b) Note the relationships of the myofilaments and their bridges in the three-dimensional drawing. During contraction, these bridges actively pull the actin fibers inward, thus shortening the contractile unit.

binding sites—thus inhibiting contraction (Figure 32.19). The sites will stay blocked, in fact, until the appearance of the calcium ion (Ca^{2+}). When Ca^{2+} is present in the contractile unit, it binds to troponin, which somehow alters tropomyosin so that it can no longer block the binding sites. Actin's myosin-binding sites are then exposed, and the myosin heads attach and become active. Upon the release of energy from ATP, the bridges draw the actin filaments inward, and the muscle contracts.

The next question is: What controls the calcium? In the resting muscle, calcium ions are stored in the sarcoplasmic reticulum (see Figure 32.15), continually transported there by ATP-powered membranal pumps. The pumps thus keep the contractile units clear of calcium until muscle contraction is needed.

We already know that neural impulses bring on muscle contraction, so all that remains is to fill in one more blank. When a neural impulse from a motor neuron reaches the neuromuscular junction on the sarcolemma, it sets in motion an electrical disturbance. The disturbance spreads over the sar-

colemma and then along the T-tubules to the membranous sarcoplasmic reticulum—within which the calcium ions are stored. The disturbance causes calcium ion channels in the sarcoplasmic reticulum to open, and as the ions suddenly leak out, they diffuse rapidly down their gradient into the surrounding contractile units. The muscle then contracts. All this occurs so rapidly that all the contractile units in each muscle fiber can contract almost simultaneously.

The muscle will contract as long as calcium ions are present, and calcium remains as long as the neural impulses continue. When the impulses cease, the events of contraction reverse themselves. The calcium ion pumps quickly clear the ions from the contractile units, the tropomyosin–troponin complexes block the myosin-binding sites of actin, and the muscle relaxes.

Summing Up Contraction

We've covered a lot of ground here, so let's briefly review the major points, starting with the neural impulse.

1. Neural impulses create electrical disturbances that travel down the sarcoplasm and down the T-tubules, causing the release of calcium ions from the sarcoplasmic reticulum.
2. Calcium ions enter the contractile units, where they alter the troponin–tropomyosin inhibitors, exposing actin's myosin-binding sites.
3. Myosin heads bind to actin, energy from ATP is released, the heads change shape, and the actin myofilaments are drawn along.

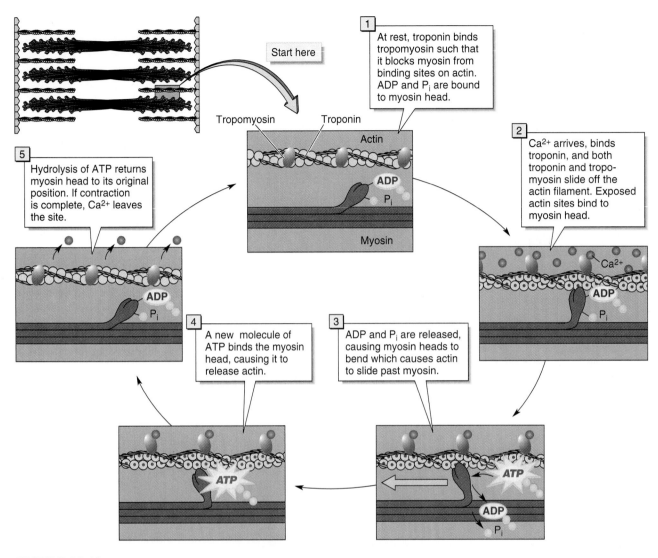

Start here

At rest, troponin binds tropomyosin such that it blocks myosin from binding sites on actin. ADP and P_i are bound to myosin head.

1

Ca^{2+} arrives, binds troponin, and both troponin and tropomyosin slide off the actin filament. Exposed actin sites bind to myosin head.

2

Tropomyosin Troponin

Actin

ADP

P_i

Myosin

Ca^{2+}

ADP

P_i

5

Hydrolysis of ATP returns myosin head to its original position. If contraction is complete, Ca^{2+} leaves the site.

ADP

P_i

4

A new molecule of ATP binds the myosin head, causing it to release actin.

3

ADP and P_i are released, causing myosin heads to bend which causes actin to slide past myosin.

ATP

ATP

ADP

P_i

FIGURE 32.19
Calcium and Control

Calcium is the local controlling agent in skeletal muscle contraction. In its absence, myosin cannot bind to actin and begin the contraction process. When a muscle is stimulated, all of this changes: calcium ions flood the contractile units, myosin bridges form, and the actin myofilaments slide together. When the muscle is no longer stimulated, the calcium pumps move the ion back into its reservoir in the sarcoplasmic reticulum, and further contraction is again inhibited.

4. ATP provides the energy for restoring the myosin heads so that they may reattach and continue pulling.

5. When the neural impulses cease, the calcium ions are pumped back into the sarcoplasmic reticulum, and the inhibited (relaxed) state is restored.

Graded Contractions of Muscles. A single muscle fiber will contract completely when stimulated by a motor neuron. But we know that entire muscles can contract completely or incompletely—the contraction can be strong or weak. To illustrate, we can flex our arm part way or entirely and we can

flex with greater or lesser power. So how do we produce these graded contractions?

One way is by changing the frequency of the action potentials in the muscle. For example, if an action potential, lasting less than 100 milliseconds, is generated in a motor fiber, and a second one is generated before the muscle entirely relaxes, the two will be summed to produce a stronger muscle contraction. If a rapid-fire barrage of impulses arrives, the muscle will finally reach full contraction (called *tetanus*).

A graded muscle response can also be achieved because any motor neuron may split repeatedly to innervate many muscle cells. However, some may innervate only a few muscle cells while others may innervate many, even hundreds. If the nervous system sends a signal over the former kind of neuron, the response will be weaker than if the signal is sent over the latter. So the graded response is generated in the nervous system as it chooses which motor neuron to activate.

Fast and Slow Muscle Contractions.   Once an action potential triggers the all-out contraction of a muscle fiber, that fiber may contract slowly or rapidly, depending on its type. The slower, so-called "slow-twitch," fibers tend to remain contracted longer—in fact, up to five times longer. One reason for the difference is the calcium level within the muscle cell. The calcium stays in the contractile units longer in those muscle cells with less sarcoplasmic reticulum, thereby producing slow muscle contractions. The steady contraction produced in these cells is supported by a rich blood supply, great numbers of mitochondria, and an abundance of myoglobin, a reddish pigment that binds oxygen more efficiently than does hemoglobin. This support provides slow-twitch muscle fibers with more oxygen. As you can see, slow-twitch muscle fibers are highly aerobic (see Chapter 8).

Slow-contracting muscle fibers are important in those muscles that must maintain a more constant level of contraction, such as those that keep the jaw from hanging slack and ruining first impressions, and those that keep us sitting straight, winning praise from our instructors. Animals that make use of slow-acting, aerobic muscle fibers include migratory birds such as the plover, a long-distance champion.

Fast-acting muscle fibers, the other type, are used in quick, strong, but time-limited movements, such as sprinting. These fibers have more sarcoplasmic reticulum and more glycogen. Glycogen is quickly converted to glucose and run through glycolysis for a quick energy supply. Fast-acting fibers, then, tend to be anaerobic.

In this chapter, we have considered skeletal systems and the muscles that move them. Now let's take a closer look at the neurons, such as those that carry signals to the muscles, keeping in mind that just as the answers are often fascinating, so in many cases are the questions themselves.

Key Ideas

INVERTEBRATE SUPPORT AND MOVEMENT

1. The **hydrostatic skeleton** firms the animal body through fluid pressure against a resistant wall. It also provides a base for muscle action.

2. Roundworms are fluid-filled, with muscles directed longitudinally, permitting lashing movements only.

3. Earthworms have complex muscle tissue arranged in two directions around their fluid-filled coelom. The body is extended by the squeezing action of circular muscles and shortened by contraction of longitudinal muscles.

4. Certain mollusks make use of the hydrostatic property for making digging movements with the muscular foot.

5. The arthropod **exoskeleton**—the cuticle—provides a muscle base and lever-like parts, prevents water loss, and protects soft body parts. The cuticle is leathery and flexible at the joints. Muscles that move the body have fixed **origins** and movable **insertions**. Muscles occur as **antagonists**, groups that have opposing movements.

6. In mollusks, external shells are continuously secreted by the mantle and take several forms.

7. The dermis of echinoderms secretes calcareous plates and spines that originate below the epidermis.

VERTEBRATE TISSUES

1. The vertebrate body consists of four **tissue** types: **epithelial**, **connective**, **nerve**, and **muscle**.

2. Epithelial tissue covers the body as **epidermis** and lines internal organs as well as blood vessels and glands.

3. Connective tissue contains cells in a noncellular **connective tissue matrix** as seen in bone, cartilage, ligaments, tendons, and blood. **Collagen** is a common binding or tough matrix material. Cells called **fibroblasts** secrete collagen and are active in bone mending. **Elastin** fibers provide elasticity to artery walls.

VERTEBRATE SKELETONS

1. The bony endoskeleton provides support, a muscle base, and protection, and acts as a source of blood cells and a calcium reservoir.

2. Bone is surrounded by the sheath-like **periosteum**. Within are regions of web-like **spongy bone**, containing **red marrow** where red cells form, and regions of **compact bone**, containing fatty **yellow marrow**. Bone consists of hard calcium phosphate in a collagen matrix containing numerous bone cells, blood vessels, and nerve.

3. Compact bone is organized into **Haversian systems**, each with a **central canal** and hard, concentric, laminated **lamellae**. **Osteocytes** (bone cells) reside in hollow **lacunae**, interconnected by web-like **canaliculi**.

4. Vertebrate skeletons consist of two divisions. The **axial** division contains the **cranium** (skull), **vertebral column** (backbones), and **ribs**. The **appendicular** division contains the two **girdles**—**pelvic** (hip bones) and **pectoral** (shoulder bones and limbs).

5. The number of **vertebrae** in the **vertebral column** varies among vertebrate classes. The 33 human vertebrae include four in the **coccyx**. The free vertebrae are separated by compressible, soft-centered **intervertebral disks**. The vertebral column houses and protects the spinal cord, which passes through the **neural canal**.

6. Each half of the human pelvic girdle contains three bones. On each side, the three bones form the hip socket. The two **iliums** and the **sacrum** form the **sacroiliac joint** in the back. The pectoral girdle includes the two **scapulae** and **clavicles**, a loose assemblage that forms the shoulder joint.

7. Immovable joints include the fused **sutures** in the cranium. Movable joints have articulating surfaces of cartilage, and the bones are held in place by **ligaments** that often form the capsules of **synovial joints**. Lubrication is provided by fluid secretions.

8. Movable **joints** include **gliding**, **pivotal**, **ball-and-socket**, and **hinge** (knee; hinge-like extension and flexion in one place).

VERTEBRATE MUSCLE: ITS ORGANIZATION AND MOVEMENT

1. **Smooth muscle** is involuntary, under the unconscious control of the autonomic nervous system. Individual cells occur in sheets, are long and tapered, and have a single nucleus.

2. **Cardiac muscle** (heart) is involuntary. Its fibers are cylindrical and branching, with interlocking borders called **intercalated disks**. It has a rhythmic and tireless contractile characteristic.

3. **Skeletal muscle** is voluntary, striated, and controlled at the conscious level through the somatic nervous system. Cells, called fibers, are striated and multinucleated.

4. Muscle fibers form bundles called **fascicles**, and a number of fascicles bound by connective tissue form muscle bundles. A sheath-like **fascia** surrounds the muscle, forming the **tendons** at its ends. Tendons connect muscle to bone, forming insertions on the part to be moved and origins on the stationary part.

5. Muscles also move the face and the tongue, and others form ring-like **sphincters** in the gut. Sheet-like muscles in the abdomen originate and insert via flattened tendons called **aponeuroses**.

6. Muscle fibers are surrounded by the **sarcolemma**, which receives motor neurons at **neuromuscular junctions**. Each fiber contains several nuclei, many mitochondria, and an extensive **sarcoplasmic reticulum**. Hollow **transverse tubules**, or **T-tubules**, form boundaries for each contractile unit. The patterns of contractile **myofibrils**, or fibrils, produces the striations.

7. Myofibrils contain **myofilaments**, or **filaments**, made up of two kinds of protein: **myosin** and **actin**. Each myosin filament is surrounded by six actins. The striations are seen in the longitudinal view. Prominent **Z disks** mark the **sarcomere**, or contractile unit. Within the Z lines lies a light **I band** (actin alone), and further in lies a darker **A band** (actin and myosin), ending in a lighter central **H zone** (myosin alone).

8. Upon contraction, the Z disks move closer together, the I bands are reduced, and the H zone darkens. However, the A band remains constant, indicating that it is the actin that moves, while the myosin is stationary. The patterns change because minute **myosin heads** form bridges with the actin filament, pulling it inward.

9. The energy for bridge movement is provided by ATP; ATP is hydrolyzed within the **myosin heads**, which contain ATPases.

10. Myosin heads emerge at regular intervals from spirally wound protein filaments. Actin filaments contain the proteins **actin**, **tropomyosin**, and **troponin**. Tropomyosin and troponin control contraction by inhibiting myosin attachment.

11. In contraction, a neural impulse reaches the muscle fibers and travels along the **T-tubules** to the sarcoplasmic reticulum, which responds by suddenly releasing calcium ions into the contractile unit. The ions alter the inhibition state, and the bridges connect and straighten sequentially along the actin filaments, pulling them inward as contraction continues.

12. Restoration of the resting state begins when neural stimulation stops and the calcium ions are actively transported back into the **sarcoplasmic reticulum**.

Application of Ideas

1. The evolution of skeletons has occurred in two distinctly different directions in the protostomes and deuterostomes. How do skeletons in the two groups differ? What causes the drastic terrestrial size restriction in one of these trends? Is there a restriction in the other? Are such restrictions present in aquatic species? Explain.

2. Describe how muscle studies have involved the merging of different scientific disciplines: for example, how the observed aspects of contraction have now been interpreted on the biochemical level.

3. Levers are classified according to the position of the fulcrum, the force applied, and the weight moved. *First, second,* and *third class levers* are described respectively as follows: weight and downward force at opposite ends, fulcrum near weight (pry bar lifting a stone); fulcrum at very end, weight near fulcrum, upward force at other end (wheelbarrow); fulcrum and weight at very ends, lifting force near the center (hand shovel). Illustrate the levers with simple drawings, and then find examples of each in the human skeleton and muscles. Determine which, if any, offers the best mechanical advantage (the least amount of force needed to move a part the greatest distance).

Review Questions

1. Explain how fluids in the nematode body provide the resistant base needed for muscle movement. Describe the arrangement of muscle fibers and the movement they provide. (page 671)

2. Compare the location, manner of formation, embryological origin, and functions of exoskeletons and endoskeletons. (pages 671, 674)

3. Using an example, explain the antagonistic arrangement of muscles in the arthropod. Why are antagonists essential? (page 672)

4. Describe the endoskeleton of the echinoderm. Considering that much of it protrudes through the epidermis, what qualified it as an endoskeleton? (page 674)

5. What are two roles of collagen? How is it produced? (page 675)

6. List five important functions of the vertebrate skeleton. (page 675)

7. Compare spongy and compact bone. In which are the red and yellow marrows found? (page 677)

8. With the aid of a simple diagram, illustrate a Haversian canal, labeling *lamellae, canaliculi, lacunae, osteocytes,* and the *central canal.* Which is (are) nonliving (noncellular)? (page 678)

9. Describe the organization of the human vertebral column, naming the main regions and listing characteristics of vertebrae in each. Also describe the structure of the disks separating the vertebrae. (pages 677–678)

10. List one unique characteristic each for skeletal, cardiac, and smooth muscle tissue. (pages 681–682)

11. List two ways in which cardiac and smooth muscles are similar, and one way in which cardiac and skeletal muscles are similar. (pages 681–682)

12. How do the roles of ligaments and tendons differ? (pages 678, 683)

13. Starting with the complete muscle, define the following terms: complete muscle, myofilament, troponin, fiber, actin, myofibril, tropomyosin, sarcomere, myosin, and fascicle. (pages 683–686)

14. Using a simple drawing, explain the organization of a contractile unit. Begin with two Z disks, and fill in between. Explain the composition of each part of the pattern. (page 684)

15. Describe the arrangement of actin and myosin in a contractile unit, and in so doing explain how the actin moves during contraction. (page 686)

16. What keeps the myosin bridges from spontaneously attaching to actin? What happens to ATP when attachment does occur? (pages 686–687)

17. How does the absence of Ca^{2+} affect the contractile unit? Where are these ions when the muscle is at rest? (pages 686–687)

CHAPTER 33

Neural Control I: The Neuron

CHAPTER OUTLINE

CELLS OF THE NERVOUS SYSTEM

The Structure of Neurons

Nerves

THE ORIGIN AND TRANSMISSION OF NEURAL IMPULSES

The Action Potential

Ion Exchange Pumps and Gradients

Generating the Action Potential

Reestablishing the Resting Potential

Ion Channels and Gates

Myelin and Impulse Velocity

COMMUNICATION AMONG NEURONS AND BETWEEN NERVES AND MUSCLES

Electrical Synapses

Chemical Synapses

The Reflex Arc: The Simplest Model of Neural Activity

KEY IDEAS

APPLICATION OF IDEAS

REVIEW QUESTIONS

*O*ur planet is a lively place, and one of mixed blessings. It provides us not only with opportunities but with dangers. In fact, that is what much of life is about: taking advantage of the opportunities and avoiding the dangers. This means, of course, that living things must be able to assess the nature of the environment accurately and to respond to it appropriately. Deer, rabbits, and grasshoppers are able to sense the presence of young, tender plants. As the plants quietly submit to the ravages of the plant eater, they may be avenged by some sharp-tooth predator peering from behind a rock. The predator is able to detect and evaluate signals emanating from the prey as surely as the plant eaters can find grass. Both kinds of animals share the same environment, but their sensory abilities are specialized to react to different aspects of it. In this chapter and the next, we will see not only how signals are detected but also how they move through the body, activating a pattern of responses that help animals succeed in a complex and changing world.

What we will see is a network of highly specialized cells that are devoted to the task of transmitting information. These are the nerve cells, or **neurons**. Their function is to conduct signals from one part of the body to another. The signals are brought about by the movement of ions across the neural membrane. The complexity we will see here is due to the elaborate and intricate arrangements and interconnections of billions of neurons.

Functioning nervous systems accomplish three things: they detect, they integrate, and they respond. They *detect* events in their surroundings and send a signal to coordinating centers. These centers, which consist of other neurons, *integrate* the information and initiate a reaction. Addi-

FIGURE 33.1
Animal Responses
The ability to respond rapidly to incoming environmental cues is an outstanding characteristic of animals. Specialized receptors detect events in their surroundings, and this "raw information" is directed to the animal's brain for sensing, integration, and response.

tional neurons carry this decisive message to some body part, where a *response* is initiated. As you can see, the three key terms are "detection," "integration," and "response." The environmental cue that started all this is referred to as a **stimulus**, the event that initiates a reaction in the nervous system. Figure 33.1 illustrates the three aspects at work in a real-life situation.

Cells of the Nervous System

Neurons come in many sizes and shapes but they are characterized by having a *cell body* that contains the nucleus, from which a number of lengthy structures extend. We will describe these shortly, but you can see these processes can be rather long. Some, in fact, must carry messages between a distant part of the body (such as the foot) and the spinal cord.

The nervous system consists of two kinds of specialized cells, the neurons we have just mentioned and **glial cells** (also called *neuroglia*), which comprise some 90% of the cellular component of the vertebrate brain. Glial cells are believed to perform supporting functions and, in the brain and spinal cord, to produce a fatty covering called the myelin sheath of some cells. In addition, glial cells provide a structural framework for the fragile neurons and carry out helpful metabolic functions.

Neurons fall into three categories according to their general functions in the three fundamental phases of neural activity: *sensory neurons, interneurons,* and *motor neurons*.

As their name implies, **sensory neurons** (also called *afferent neurons*) (Figure 33.2a) receive such information from **sensory receptors**, highly specialized structures that are responsive to gravity, touch, light, chemicals, heat, and other environmental stimuli. Sensory neurons carry information concerning both the external and internal environment; thus we are also able to receive signals from our internal organs. Sensory neurons have lengthy processes called dendrites that originate at the sensory receptor and conduct signals to the cell body, from where they are relayed to an **axon**.

Interneurons (Figure 33.2b) are nerve cells that communicate only with other nerve cells, eventually directing impulses to motor neurons. Initially, interneurons receive input from sensory neurons, usually relaying their signals to other interneurons. The complex circuitry of interneurons in the vertebrate spinal cord and brain is largely responsible for the integration of stimuli and coordination of responses.

Motor neurons (also called *efferent neurons*) (Figure 33.2c) receive signals from interneurons and transmit signals called **motor impulses** to effectors, such as muscles and glands, which then respond. In the last chapter we learned about the complex response of skeletal muscle to motor impulses arriving at neuromuscular junctions on the surface of muscle fibers.

The Structure of Neurons

The **cell body** contains the nucleus and, generally, most of the cell's cytoplasm (Figure 33.3a). The cytoplasm contains the typical organelles, such as ribosomes, endoplasmic reticulum, and numerous secretory bodies. The cell body can produce **neurotransmitters**–chemical messengers that must be present for a **neural impulse** to be relayed from one neuron to another.

The processes that extend from the cell body are of two principal types: **dendrites** and **axons**. The highly branched dendrite ("little tree") is the receiving end of the neuron. It receives stimuli from its surroundings, often other neurons, and conducts this information in the form of an electrical signal toward the cell body.

Axons are usually lengthy processes that are specialized for transmitting signals in the animal body. Axons function by converting signals from the cell body into a special kind of signal, or *neural impulse,* produced by an *action potential.* An axon may communicate with other neurons, relaying its signal along to them, or it may directly stimulate an **effector**, such as a muscle or gland.

A neuron often has many dendrites, but it usually has only one axon. The single axon, however, may branch at any point along its length. An axon commonly divides and redivides at its tip, forming a terminal **axon tree**, each branch of which forms a knob-like ending, called the **synaptic end bulbs**. It is from these endings that the axon releases a neurotransmitter that either triggers an impulse in the next neuron or activates an effector. Where the tips of the axons innervate a muscle fiber, they spread to form **neuromuscular junctions** (see Chapter 32).

The axon seen in Figure 33.3a is surrounded (jelly-roll fashion) by a **myelin sheath**, flattened layers of fatty material found in many types of vertebrate neurons. Like any lipid,

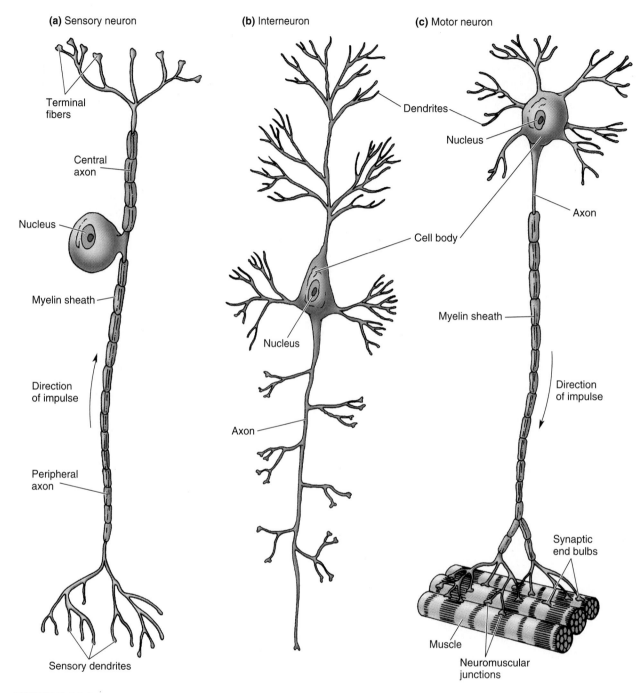

(a) Sensory neuron

Terminal fibers

Central axon

Nucleus

Myelin sheath

Direction of impulse

Peripheral axon

Sensory dendrites

(b) Interneuron

Dendrites

Cell body

Nucleus

Axon

(c) Motor neuron

Dendrites

Nucleus

Axon

Myelin sheath

Direction of impulse

Synaptic end bulbs

Muscle

Neuromuscular junctions

FIGURE 33.2
Variation in Neurons
Three neurons found in human beings show the diversity of these cells. Note the differences in the cell bodies, axons, and dendrites. **(a)** A motor neuron with axons that run from the nervous system to the effector (in this case, a muscle). A single nerve cell may be 9 feet long, such as those that run from the base of a giraffe's spine down its hind leg. **(b)** A pyramidal cell is an interneuron of the brain. **(c)** A sensory neuron is one that runs from the receptor to the spine.

myelin has great electrical resistance and acts as an insulator. But the sheath is interrupted at frequent intervals by the **nodes** (also called nodes of Ranvier), where the axon is in direct contact with the surrounding intercellular fluid. The nodes are small spaces between the end of one wrapping cell, called a

Schwann cell, or **internode** as the wrapped regions are called, and the beginning of the next. The fatty myelin sheath is composed of 50–100 layers of the flattened and rolled plasma membrane of a specialized axon-encasing cell called the Schwann cell, as depicted in Figure 33.3b.

FIGURE 33.3
Structure of a Neuron

(a) Basically, the neuron is a cell body containing slender processes or extensions called dendrites and axons, the first carrying impulses to the cell body, the second away. The cell body has the usual cell organelles. Note that dendrites have simple endings, but axons have elaborate synaptic end bulbs, where they communicate with other neurons or with muscles and other effectors. Many vertebrate neurons are myelinated; that is, they are covered by wrappings known as myelin sheaths. The sheath is not continuous but is interrupted by naked spaces called nodes. The myelinated region between is the internode. (b) Electron micrograph of cross-section. The sheaths are actually the membranes of special supporting glial cells; Schwann cells outside the brain, and oligodendrites inside. Note how the wrappings are produced through a tunneling process by the Schwann cell.

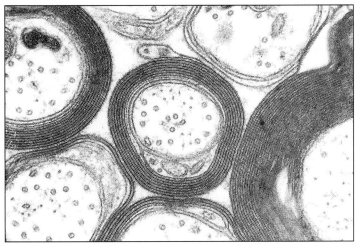

Neurons with myelin sheaths (cross section)

Nerves

The axons of many neurons often travel together throughout the body in parallel arrangements forming larger structures called **nerves** (Figure 33.4). A nerve can be likened to a telephone cable carrying many individual lines (neurons), each insulated from the others. The large glistening nerves are surrounded by their own coverings of tough connective tissues. As you are probably aware, damaging or severing a major nerve can produce devastating effects such as paralysis, loss of sensitivity, or both.

The Origin and Transmission of Neural Impulses

Neural impulses are generated by what is called an action potential. As was mentioned earlier, an **action potential** is created when a polarized neuron—with an unequal distribution of positive and negative charges inside and outside the neuron—is suddenly depolarized in a wave along its length. In order for this to happen, ions bearing positive and negative

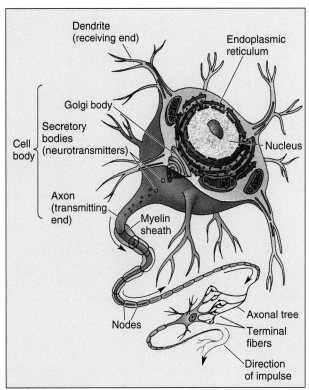

(a) The neuron

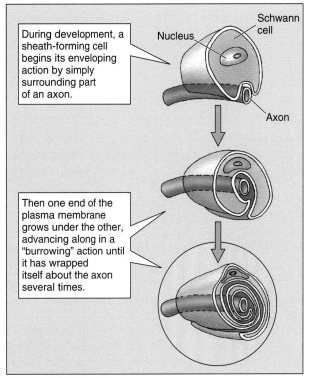

During development, a sheath-forming cell begins its enveloping action by simply surrounding part of an axon.

Then one end of the plasma membrane grows under the other, advancing along in a "burrowing" action until it has wrapped itself about the axon several times.

(b) The myelin sheath

charges must cross the neural membrane. Such a depolarization is rapidly followed by repolarization as the ions are moved back to their original positions, as we shall see in our closer look at the details.

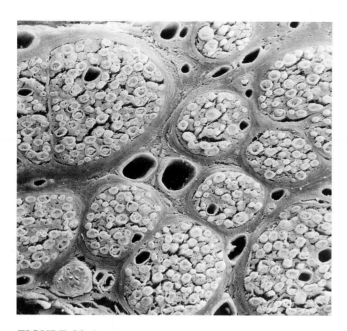

FIGURE 33.4
A Group of Nerves
Several nerves, as seen in cross section through the scanning electron microscope, contain numerous neural processes, possibly both axons and dendrites. Each contains its own insulating sheath, and the entire ensemble is surrounded by dense connective tissue that binds it into its cable-like structure. Blood vessels penetrate the nerve, providing the exchanges needed to maintain the neurons.

The Action Potential

An action potential, while electrical in nature, does not behave as an electrical current moving along a wire. While an electrical current diminishes with distance, an action potential moves without loss or gain in strength, seemingly regenerating itself as it travels. For some idea of the distance action potentials may travel, consider what happens when you wiggle your big toe. Impulses originating in your brain are relayed to a motor neuron in the spinal cord. The motor impulses follow the axon out of the cord, through the leg, and finally to neuromuscular junctions in the big toe muscles. There they trigger contraction; quite a feat for a single cell.

The action potential involves highly selective and controlled ion movement, and such movement requires versatile membrane channels. Both active transport and diffusion are involved. To understand precisely how an axon transmits a signal so efficiently along its length, we must look more closely at the special nature of its membrane as well as the ions involved. We can divide the events surrounding the action potential into three stages:

1. Establishing a polarized, resting state in the neuron.
2. Generating an action potential and propagating it along the axon.
3. Restoring the polarized, resting state.

Ion Exchange Pumps and Gradients

Actually, the term "resting state" is not very appropriate. One thing that living neurons never do is "rest." We shall understand that a "resting" neuron is in a polarized state, that is, a condition in which the electrical charges outside the axon's plasma membrane are different from those inside. Specifically, resting axons are more negative inside than outside. The maintenance of this charge difference involves the positioning of certain ions, with their electrical charges, on one side of the membrane or the other. When the axon is polarized in this way, it is in a precarious state indeed. If the properties of the neuron are disturbed, the ions will rush to equilibrium. This tendency is the basis for the action potential. But before we get to how the impulse travels, let's look at the conditions behind the polarized resting state.

The ions most important to the neural impulse are sodium ($Na+$) and potassium (K^+), along with much larger, negatively charged proteins. Although the small sodium and potassium ions are quite mobile, the proteins, because of their large size, cannot cross the plasma membrane and so must remain within the axon. The potassium ions are also much more abundant within the resting neuron than outside it. However, in the neuron's resting state, the sodium ions just outside the membrane outnumber those inside by about ten to one, resulting in a steep sodium ion gradient.

This distribution is largely maintained by many **sodium/potassium ion exchange pumps** found in the axon's plasma membrane. These pumps are powered by the energy of ATP; they move three sodium ions out of the axon and two potassium ions into the axon in each pump cycle (Figure 33.5).

Other factors that maintain the two ion gradients have to do with membrane permeability and electrical charges. When at rest, the axon's plasma membrane is not very permeable to Na^+, so as these ions are pumped out, most remain outside the neuron. However, the axon's membrane is much more permeable (leaky) to K^+, and as these ions are pumped in, many diffuse right back out across the membrane to the cell exterior. Because the membrane is so leaky to K^+, one might expect K^+ to reach equilibrium (to equalize in concentrations on both sides of the membrane). This doesn't happen, though, because of the fundamental behavior of charged particles: like charges repel and unlike charges attract. Thus the many positively charged sodium ions already outside the axon repel the positively charged potassium ions; similarly, the many negatively charged proteins on the inside of the axon attract the potassium ions. Thus potassium remains in a greater concentration inside the axon than outside (Figure 33.5).

The Resting Potential. Now that we've considered how the ion distribution produces the polarized state in an axon, let's look at the axon's electrical characteristics. Again, the net electrical charge inside a resting neuron, despite the presence of K^+, is negative relative to the outside, where Na^+ ions abound. The polarized state of the resting neuron can be verified experimentally by placing a tiny electrode on the surface

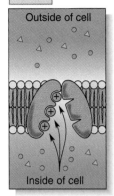
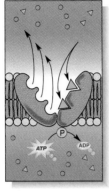

1 Three sodium ions bind carrier sites.

2 Phosphorylation of the Na⁺–K⁺ pump by ATP causes a shape change to the pump. Two K⁺ from the outside bind to carrier sites, Na⁺ is released outside the cell.

3 Phosphate release causes return to original state and K⁺ enters cell.

FIGURE 33.5
Axonal Ion Exchange Pumps and Ion Distribution
ATP-powered Na⁺/K⁺ ion exchange pumps transport Na⁺ out of the neuron and K⁺ into the neuron in a 3:2 ratio. Both ions form steep concentration gradients across the membrane.

of the axon and inserting another electrode into the cytoplasm within (Figure 33.6). Typically, the charge difference across the membrane results in an electrical voltage of about −70 mV (millivolts: $\frac{1}{1000}$ of a volt). This voltage across a neuron is called the **resting potential**. The minus sign indicates the negatively charged state of the axon's interior relative to the outside surface. The electrical and chemical characteristics of the polarized resting neuron, as you may have noticed, are not unlike those of the fully charged chemiosmotic system in chloroplasts and mitochondria (see Chapter 6). At that time we described the conditions as an "electrochemical gradient," a fitting term since it takes into account both the chemical and the charge gradient. You can see then that the term "resting neuron" is indeed misleading. (Think of how a set mouse trap is also "at rest.")

The electrochemical gradient of the polarized neuron represents a significant amount of energy. To tap this energy, some of the system must be allowed to "run down," to move toward equilibrium. And this brings us back again to the action potential.

Generating the Action Potential

The axon is able to generate and conduct an action potential, which, as we've seen, is a wave of *depolarization* that moves

along the axon to its end. When a neuron is stimulated, say, on a dendrite or on the cell body, a passive electrical signal spreads to the beginning of the axon. There the action potential arises and sweeps along the axon to its terminal branches. The terminal branches then can activate another neuron or an effector.

An action potential begins with a sudden increase in the permeability of the axon's plasma membrane to sodium; the sodium ions, no longer held in check, rush inward. At that point, the balance of charges is suddenly changed, and the membrane potential (the measure of its polarity) shifts rapidly from −70 mV to about +30 mV. Thus that area of the membrane is depolarized.

Action potentials at first affect only a small part of the membrane. But once they begin, they form an ongoing depolarizing wave that sweeps along the axon. The wave is quite dramatic when seen on an oscilloscope (Figure 33.7). Following the passage of an action potential, the neuron is repolarized and the resting potential is restored. An action potential

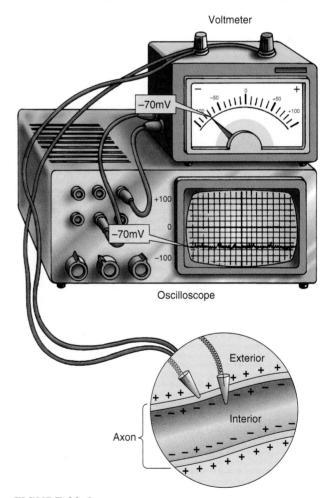

FIGURE 33.6
Resting Potential
In its resting state a neuron is polarized. Its inside surface, just within the plasma membrane, is negative relative to the outside. The resting potential for many neurons is about −70 mV. The voltage on either side of the neural membrane is measured by a voltmeter and registered on an oscilloscope.

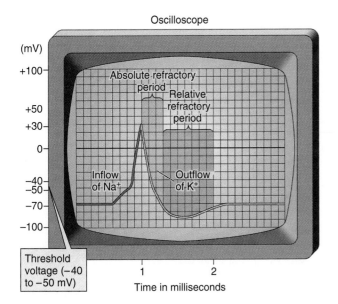

Oscilloscope

FIGURE 33.7
The Electrical Characteristics of an Action Potential
The oscilloscope tracing reveals the electrical characteristics of an action potential. Because of a sodium ion influx, the membrane becomes depolarized, the membrane potential rising from −70 mV to +30 mV. Repolarization, involving an outpouring of potassium ions, occurs rapidly. Until the resting potential is attained, the neuron is in a refractory period. The small dip below the resting potential, a slight overshoot, is the relative refractory period.

at any given point along the axon is extremely short-lived, the period from depolarization to repolarization requiring about 2 milliseconds ($\frac{2}{1000}$ of a second).

Threshold for Generating an Action Potential. An important characteristic of the action potential is called the "all-or-none" principle. Axons have a **threshold voltage** requirement that must be attained if action potentials are to begin. For many neurons, the threshold requirement is a depolarization to about −40 mV to −50 mV. Anything less, and the action potential will not be sustained; it will simply fade out. Once the threshold requirement is reached, however, the action potential will proceed unfalteringly along the entire length of the axon.

Reestablishing the Resting Potential

So the neuron has fired. An impulse has passed. This is not a one-time event, though. It may have to fire again. But it can't do so until it reestablishes its resting potential. So the rapid influx of Na$^+$ in an action potential is cut short as the plasma membrane once again becomes impermeable to that ion. (We'll see how shortly.) This renewed impermeability marks the start of repolarization—restoring the resting potential. However, the essential ion of recovery is not sodium, but potassium. In repolarization, potassium ions rapidly diffuse out of the neuron, until their numbers roughly balance the

number of sodium ions that entered. Why potassium? Recall that in the resting neuron, outward potassium diffusion was curtailed by the attractive force of negative proteins within the neuron. However, with sodium now competing for these negative charges, potassium is more free to diffuse to the outside. Thus it is the *loss* of positively charged potassium that restores the membrane potential to −70 mV. So now there is excess K$^+$ outside the neuron and excess Na$^+$ inside, but the Na$^+$/K$^+$ ion exchange pumps quickly restore the former ion distribution, using the energy of ATP.

The time of repolarization, when potassium ions are leaving the axon and the membrane potential is again moving toward −70 mV, is called the **refractory period**. Figure 33.7 shows two parts of the refractory period. In the **absolute refractory period**, new action potentials cannot be generated at any point along the axon where recovery is occurring, regardless of the intensity of a stimulus. In the **relative refractory period**, which is characterized by a dip or "overshoot" below the resting potential, new action potentials can be generated, but only by a very strong stimulus. As you see, neurons have a maximum rate of firing that is limited by the period required for recovery.

Ion Channels and Gates

You may wonder, at this point, just how a membrane can suddenly change its permeability to certain ions. The neuron, it turns out, contains large numbers of special protein-lined **ion channels** all along its axon, each capable of admitting a specific kind of ion. For example, there are channels that admit ions of sodium, potassium, calcium, or chlorine. The channels are not simple pores. They have movable regions in their protein constituents, the *ion gates,* that can open and close the channels in response to potential (voltage) changes in the cell. Keep in mind, also, that these gates open and close using the energy of ATP.

The resting potential of the axon, its depolarization, and its repolarization all involve the operation of **sodium ion gates** and **potassium ion gates**, both of which are voltage-sensitive. Each sodium ion channel actually has two kinds of gates, an outer *activation gate* and an inner *inactivation gate.* Although the Na$^+$ activation and inactivation gates are both voltage-sensitive, they respond to different voltages. Figure 33.8 shows a model of the Na$^+$ and K$^+$ gates and how they work.

The Sodium Gates. Recall that the resting axon's plasma membrane is impermeable to the sodium ion. This means that at least one of the two sodium gates must be closed at this time. It turns out that it is the sodium activation gates that are closed; the inactivation gates remain open during the resting state (Figure 33.8, [1]). When an electrical signal of sufficient strength reaches the axon, the nearest sodium activation gates respond by opening. Sodium ions thus rush into the axon, starting the depolarization that characterizes the action potential (Figure 33.8, [2]). This depolarization generates still more electrical signals, and sodium activation gates a little further along the

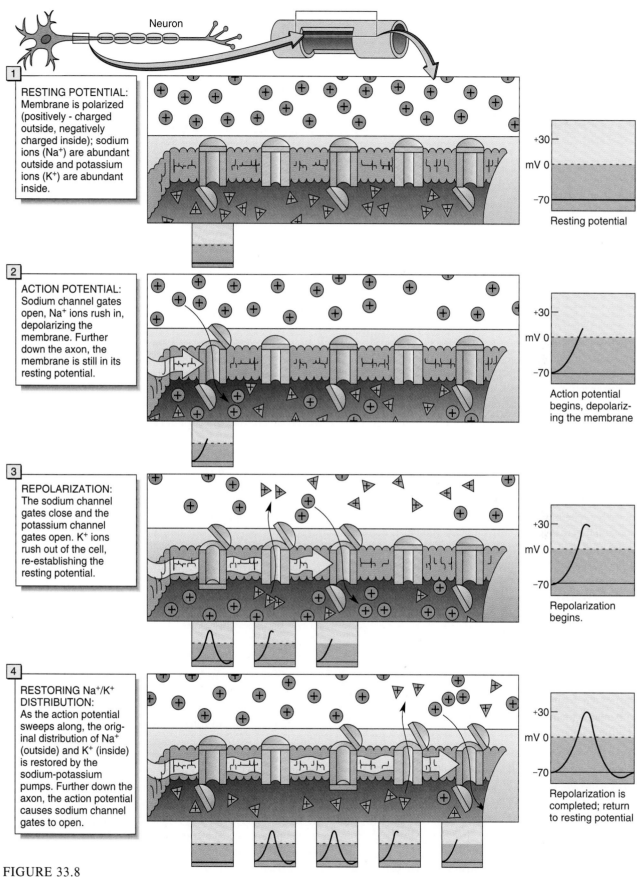

FIGURE 33.8
Gated Sodium and Potassium Channels
Gated, voltage-sensitive sodium and potassium channels bring about changes in plasma membrane permeability to these ions that characterize the action potential.

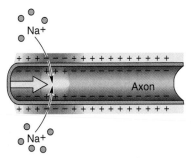

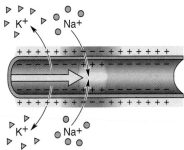

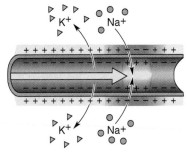

Action potential begins with
Na⁺ entering the axon

AP moves along axon as K⁺ leaves,
restoring the region behind

The two events continue in the
same sequence along the neuron

FIGURE 33.9
Summary of an Action Potential
The ongoing scenario illustrates how the displacement of sodium and potassium ions leads to a moving
action potential.

axon open in response. Thus the sweeping movement of an action potential is due to a chain reaction of gate openings.

As the membrane potential reaches +30 mV, the sodium inactivation gates close, blocking the further entry of Na^+ (Figure 33.8, ③). The inactivation gates will remain closed until the surrounding region of the neuron is repolarized. The repolarization, you will recall, is brought on by shifts in potassium, so let's look at the gates through which it passes.

The Potassium Gates. Potassium gates also respond to the initial electrical disturbance, but the K^+ gates respond much more slowly—opening fully only when the +30 mV membrane potential is reached, or just about the time that the sodium inactivation gates close. With the sudden exit of positively charged potassium ions, that region of the axon is repolarized, and the resting potential of −70 mV is restored (Figure 33.8, ③, ④). When that voltage is reached, the sodium activation gates and potassium gates close (the latter barring the further escape of K^+), and the sodium inactivation gates open (Figure 33.8, ④). Thus the resting state is fully restored.

Let's summarize what we know about the action potential so far. (Take a look at Figure 33.9.)

1. During the resting state, the axon is polarized (−70 mV), with many more negative charges inside than outside. The ion balance is such that Na^+ is in greater concentration outside, while K^+ is in greater concentration inside.

2. When a neuron is stimulated, the resulting electrical signal reaches the axon and causes sodium activation gates to open. Sodium ions enter, depolarizing the region (to +30 mV), and this new disturbance opens Na^+ activation gates further along. Thus the action potential sweeps along the neuron (see Figure 33.9).

3. Repolarization begins at +30 mV, with the closing of Na^+ inactivation gates and the complete opening of the K^+ gates. This transition marks the beginning of the refractory period.

4. The outpouring of K^+ restores the resting potential (−70 mV). The K^+ gates and Na^+ activation gates close, and the Na^+ inactivation gates open.

Speed and Magnitude of an Action Potential. An important characteristic of action potentials is that, once generated, their speed and their magnitude cannot be altered by increases or decreases in the stimulus (the event that triggered the sudden increase in Na^+ permeability). Action potentials remain constant from one end of the axon to the other. Thus, in a given neuron, all action potentials look about the same on the oscilloscope, and they all move at the same speed.

One factor that is affected by stimulus intensity is the *rate* at which action potentials are formed. The brain interprets a greater frequency of action potentials as being due to a stronger stimulus. Stepping on a thumb tack generates more action potentials than stepping on a pea (your verbal response to the two stimuli may also reflect the difference).

Myelin and Impulse Velocity

As we've seen, axons in vertebrates are commonly surrounded by myelin sheaths—the membranes of Schwann cells. These act both as insulators and as a mechanism for speeding up impulses. In myelinated human neurons, for example, impulses are conducted at a speed of up to 100 meters per second, while the speed of conduction in nonmyelinated neurons is much slower. But how do myelin sheaths function? We can see how the fatty sheaths would insulate neurons, keeping their activities separate from nearby neurons, but how do they increase impulse velocity? And for that matter, how can an action potential be generated at all in an axon that is insulated from the surrounding, sodium-rich extracellular fluid?

The answer to both questions lies in the arrangement of the myelin sheath. At this point we must tell you that the smooth-flowing, wave-like nerve impulse that we just described only pertains to nonmyelinated fibers. In the myelinated neurons,

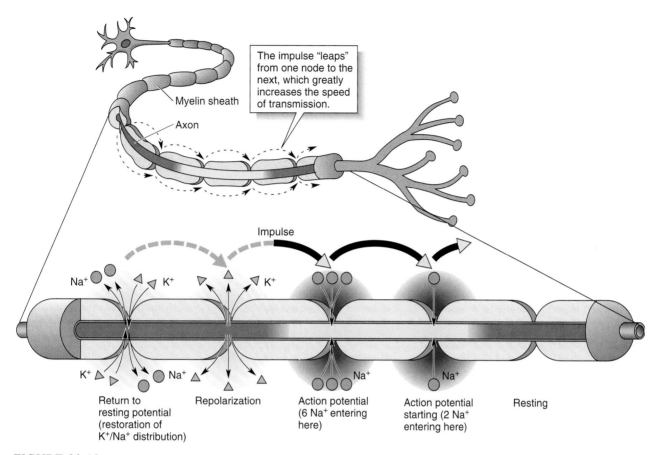

The impulse "leaps" from one node to the next, which greatly increases the speed of transmission.

Myelin sheath

Axon

Impulse

Na⁺   K⁺   K⁺

K⁺   Na⁺   Na⁺   Na⁺

Return to resting potential (restoration of K⁺/Na⁺ distribution)

Repolarization

Action potential (6 Na⁺ entering here)

Action potential starting (2 Na⁺ entering here)

Resting

FIGURE 33.10
Saltatory Propagation
Action potentials occur only at the nodes of myelinated axons—jumping from one node to the next. The internodal "leaps," called saltatory propagation, are possible because electrical currents generated in one node during depolarization reach voltage-sensitive sodium activation gates in the next node, starting a new action potential there. Thus the impulse jumps from node to node.

the action potentials occur only at the nodes, those gaps in the myelin found at regular intervals along the axon.

The increased speed of the impulse in myelinated fibers occurs because the neural impulse virtually "jumps" from one node to the next all along the axon. Neurobiologists call this **saltatory propagation** (*salto,* "jump") (Figure 33.10). The generation of an action potential in one node produces a minute current that spreads quickly throughout the internode. When this current reaches voltage-sensitive sodium activation gates in the next node, those gates open, a new action potential is generated, and the impulse continues down the axon.

Like a skillfully thrown rock skipping along a lake surface, saltatory propagation permits a greatly increased velocity. Also, this method of propagation is more energy-efficient, because there are so few ions to be pumped in and out later, relatively few having passed across the neuromembrane at the nodes. In one comparison between myelinated and nonmyelinated axons conducting at the same speed, it was estimated that the non-myelinated axon required 5000 times as much ATP to restore the distribution of sodium and potassium ions.

Communication Among Neurons and Between Nerves and Muscles

Neurons communicate with each other at special sites called **synapses**. At a synapse, one neuron triggers an impulse in another neuron (or in an effector, such as a muscle). In neural transmission, typically, the axon of one neuron will form a synapse with a dendrite or cell body of a second neuron. Although a neuron typically has but one axon, that axon's highly branched ending may provide hundreds to thousands of synaptic connections with other neurons. There are two distinct kinds of synapses: electrical and chemical. Chemical synapses are far more common, but both have unique advantages. Let's consider the electrical synapses first.

Electrical Synapses

Electrical synapses occur at gap junctions between neurons (Figure 33.11a). You may recall from Chapter 4 that gap junctions are places where adjacent plasma membranes press together and minute, protein-lined openings form. The gap junctions make the cytoplasm of one cell continuous with that of another. Such an intimate relationship between neurons permits current to pass readily from an active neuron to a resting neuron, thus simply continuing the action potential in the second neuron.

The main advantage of the electrical synapse is speed. Chemical synapses require several steps for action potentials to be generated in a second neuron, and this requires more time than the current flow associated with electrical synapses. Electrical synapses have been found in several invertebrates and are associated with escape movements. In the crayfish, for instance, electrical synapses innervate the abdominal muscles that produce a powerful thrust in the paddle-like tail that propels the crayfish backward, away from danger. Electrical synapses also occur in vertebrates. In fishes, for example, electrical synapses activate the sudden flip of the tail and make possible the quick starts and turns used in the feeding chase by both predator and prey.

In humans, electrical synapses are found in the brain and spinal cord and at neuromuscular junctions (between nerves and muscles) in those smooth muscles and cardiac muscles in which cells are joined by gap junctions. Gap junctions, you will recall, occur frequently in the intercalated disks that join heart muscle fibers together end to end (see Figure 32.13).

Chemical Synapses

Unlike electrical synapses, the neurons in **chemical synapses** do not actually make physical contact, and action potentials do not simply pass from one neuron to the next. Action potentials must be generated anew in receiving neurons. In chemical synapses, neurons are separated by a minute, 20-nm space known as the **synaptic cleft** (Figure 33.11b). Communication occurs through the action of chemicals called neurotransmitters. Neurotransmitters released by the first or **presynaptic** neuron diffuse across the synaptic cleft, where they activate specialized receptors on the second or **postsynaptic** neuron. Receptor activation triggers a new action potential.

While slower than electrical synapses, chemical synapses offer much more versatility in their responses. This is made possible by a number of variables, such as a variety of neurotransmitters, differences in the number of synapses reaching a single neuron, and a range of potential receptors. Let's focus now on what actually goes on at a chemical synapse.

Action at the Synapse. Within each axonal knob are a large number of membrane-bound vesicles that store the chemical neurotransmitter, such as **acetylcholine**, one of about 50 neurotransmitters. Each neurotransmitter is released by a specific type of neuron. Its release involves calcium ions (Ca^{2+}) that are present in extracellular fluids outside the knob. When an action potential reaches the axonal knob, voltage-sensitive gates of calcium ion channels in its membrane open, and calcium ions

diffuse inward. In their presence, the transmitter-laden vesicles fuse with the **presynaptic membrane** (the membrane of the first, or *presynaptic, neuron*), where they rupture, releasing acetylcholine molecules into the synaptic cleft. The molecules then diffuse across the cleft and bind to specialized receptors on the **postsynaptic membrane** (the membrane of the second, or *postsynaptic, neuron*) (see Figure 33.11b).

The acetylcholine receptor sites of the postsynaptic membrane open their gates in response to the presence of the neurotransmitter, thereby admitting sodium ions (or, in some cases, potassium ions). When a sufficient number of such channels has opened in the postsynaptic membrane, an action potential begins. Whereas one such event would probably have little effect, the presence of many synapses on a single postsynaptic neuron helps ensure that enough channels can be opened.

By its very organization, the synapse acts as a one-way valve between most neurons. The transmission is one way because the postsynaptic membrane has no neurotransmitters to release, and the presynaptic membrane has no receptor sites. One-way transmission is an important means of coordinating the work of the nervous system. Coordination is accomplished in other ways as well, such as through inhibitory synapses.

Inhibitory Synapses. So far we have been discussing **excitatory synapses**, the synapses that cause action potentials in the postsynaptic neuron. Another kind of synapse does not trigger action potentials, but *inhibits* them, making them less likely to occur. These are called **inhibitory synapses** (Figure 33.12). One way such inhibition occurs is through *hyperpolarization*, in which the net negative charge inside the neuron becomes considerably increased. This may happen when specialized receptors respond to a neurotransmitter by opening chemically gated chloride ion channels. Once opened, they admit negative chloride ions (Cl^-) into the neuron's interior, which increases its polarized state. Consequently, the hyperpolarized neuron requires considerably more than the usual threshold voltage for action potentials to start.

Inhibition can also occur if the neurotransmitter causes gated potassium channels to open—channels that permit the escape of potassium ions. Their escape counteracts the uptake of sodium ions, brought about by nearby excitatory synapses, effectively canceling out depolarization.

You might wonder why a cell would go to all that trouble to keep from firing. Why would a system exist in which there

FIGURE 33.11
The Synapse
(a) In electrical synapses, gap junctions between neurons result in a continuous cytoplasm that permits neural impulses to pass smoothly from one neuron to the next. (b) In chemical synapses, minute spaces called synaptic clefts exist. Neurotransmitters from presynaptic neurons must cross the cleft and open ion gates on the postsynaptic neuron. This permits an inflow of sodium ions, prompting a neural impulse in the second neuron.

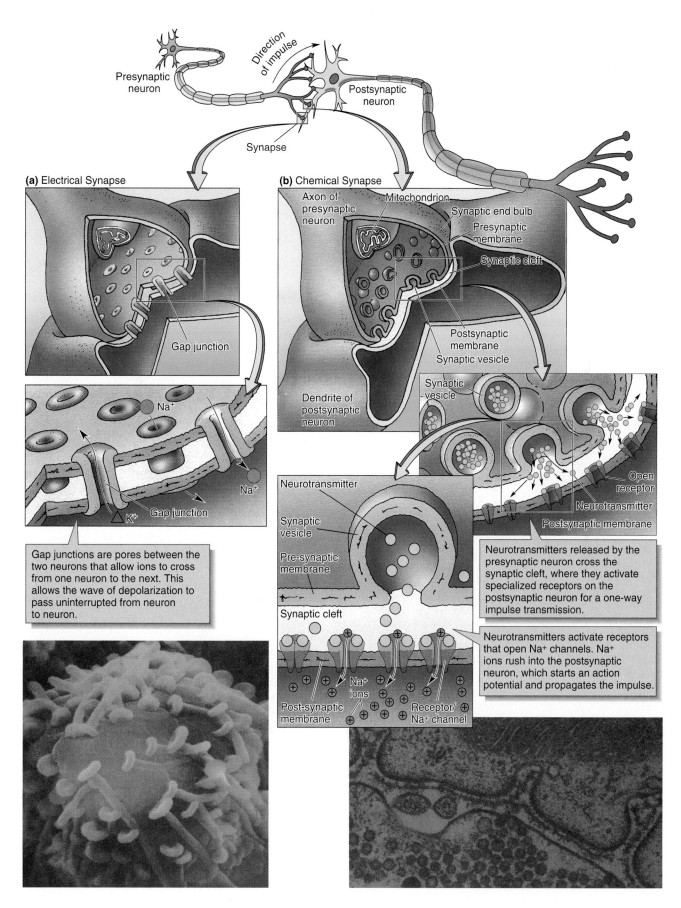

Direction of impulse

Presynaptic neuron

Postsynaptic neuron

Synapse

(a) Electrical Synapse

Gap junction

Na+

Na+

K+

Gap junction

Gap junctions are pores between the two neurons that allow ions to cross from one neuron to the next. This allows the wave of depolarization to pass uninterrupted from neuron to neuron.

(b) Chemical Synapse

Axon of presynaptic neuron

Mitochondrion

Synaptic end bulb

Presynaptic membrane

Synaptic cleft

Postsynaptic membrane

Synaptic vesicle

Dendrite of postsynaptic neuron

Synaptic vesicle

Open receptor

Neurotransmitter

Postsynaptic membrane

Neurotransmitter

Synaptic vesicle

Pre-synaptic membrane

Synaptic cleft

Post-synaptic membrane

Na+ ions

Receptor/ Na+ channel

Neurotransmitters released by the presynaptic neuron cross the synaptic cleft, where they activate specialized receptors on the postsynaptic neuron for a one-way impulse transmission.

Neurotransmitters activate receptors that open Na+ channels. Na+ ions rush into the postsynaptic neuron, which starts an action potential and propagates the impulse.

FIGURE 33.11

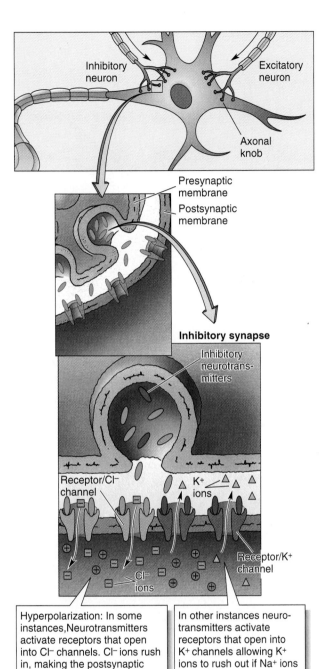

Presynaptic membrane
Postsynaptic membrane

Inhibitory synapse

Inhibitory neurotrans- mitters

Receptor/Cl⁻ channel

K^+ ions

Receptor/K⁺ channel

Cl^- ions

Hyperpolarization: In some instances,Neurotransmitters activate receptors that open into Cl⁻ channels. Cl⁻ ions rush in, making the postsynaptic neuron more negative internally. Much higher levels of Na⁺ ions are required to start an action potential.

In other instances neuro- transmitters activate receptors that open into K⁺ channels allowing K⁺ ions to rush out if Na⁺ ions rush in, cancelling out depolarization.

FIGURE 33.12
The Inhibitory Synapse

Inhibitory synapses work by increasing the negativity of the resting potential in postsynaptic neurons and thus raising the response threshold. This makes them less likely to fire when stimulated by synapsing excitatory neurons. On the left, an inhibitory neurotrans- mitter (lighter colored ovals) prompts the opening of chloride ion gates, and the introduction of the negative ions hyperpolarizes the neuron. In the alternative just to the right, a different inhibitory neu- rotransmitter (darker colored ovals) prompts the opening of potassi- um ion channels, which permits the positive ions to rush out, also hyperpolarizing the neuron. (Both types of inhibiting synapses would probably not occur in the same activating neuron.)

are both inhibitory and excitatory synapses coming from other neurons? In part, the advantage of such a system lies in fine-tuning the control. The inhibitory synapses can have a dampening effect, reducing the frequency with which a post- synaptic neuron may fire—quietening its activity when con- stant stimulation may not be necessary. In such a system, whether the second neuron fires or the effector reacts depends on the net effect of both types of synapses. At times, such inhibition can be vital. It is inhibition, for example, that per- mits sleep as certain inhibitory neurons can screen out routine incoming stimuli—background noises or random unsettling thought patterns—whatever might disturb your sleep. On the other hand, a moan from the closet might stimulate enough excitatory neurons to overwhelm the inhibitory neurons. In the next chapter we'll see that the cerebellum, the brain struc- ture responsible for coordinating voluntary movement, does much of its work through inhibition.

Recovery at the Synapse. It is essential that a neurotrans- mitter not be allowed to linger in the synaptic cleft. Should this happen, the receiving neuron would continue to be excit- ed or inhibited, and the coordination of neural activity would be lost. Some neurotransmitters are simply recycled back to the presynaptic membrane for uptake and storage. Others are enzymatically broken down. For example, acetylcholine is dismantled into choline and acetyl groups by **acetyl- cholinesterase**, an enzyme that is often bound to collagen fibers within the synaptic cleft. The choline group is then taken up by the presynaptic membrane and recycled. The removal of a neurotransmitter from the synaptic cleft is so rapid that its presence is quite fleeting, lasting less than a mil- lisecond. Interestingly, nerve poisons such as pesticides of the

FIGURE 33.13
A Reflex Arc
In the knee-jerk reflex, striking the tendon causes the muscle above to extend slightly. Stretch receptors sense this change and carry impulses to the spinal cord, where they synapse with motor neurons. Impulses from the motor neurons bring about contraction in the muscle without the brain's intervention.

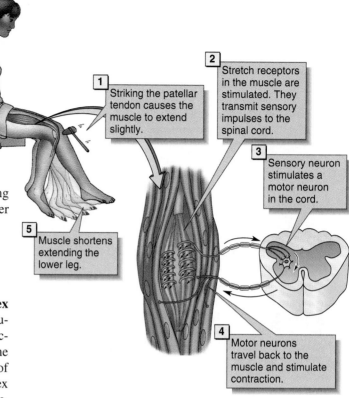

1 Striking the patellar tendon causes the muscle to extend slightly.

2 Stretch receptors in the muscle are stimulated. They transmit sensory impulses to the spinal cord.

3 Sensory neuron stimulates a motor neuron in the cord.

4 Motor neurons travel back to the muscle and stimulate contraction.

5 Muscle shortens extending the lower leg.

organophosphate class do their deadly work by blocking acetylcholinesterase activity. How might this affect other functions in the animal?

The Reflex Arc: The Simplest Model of Neural Activity

The simplest behavioral response occurs through the **reflex arc**, a neuronal circuit in which impulses from a sensory neuron may be relayed directly to a motor neuron. Such a reaction involves as few as two neurons, but, in most cases, the impulse is simultaneously transferred to a number of interneurons. Because of this simplicity and speed, the reflex arc can be quite important to adaptive behavior. For example, should you step on a hot coal at the beach, your reaction may involve a sensory neuron that perceives the stimulus, and a motor neuron that brings about the graceless "high-stepping" move that entertains your friends and gets you off the coal. By the time your brain is aware of the problem, having been alerted via additional impulses shunted there by interneurons, your reflexive behavior has prevented a more serious burn.

The *patellar response,* known also as the "knee-jerk reflex," is also simple, directly involving just two neurons (Figure 33.13). Your physician taps you just below the knee cap with a small rubber hammer, and in response, your lower leg kicks out slightly. This may seem to be for the physician's amusement, but the good doctor is actually trying to rule out certain neurological disorders. What happens is that the blow causes the tendon below the knee to stretch slightly. This activates stretch receptors (neurons that detect slight changes in skeletal muscle) in the large quadriceps muscle above the knee, which flash a message to the spinal cord. There the sensory neuron synapses directly with a motor neuron that sends an impulse back to the quadriceps, causing it to contract and making you kick. The reflexive response is sudden and invol-

untary, and the brain is only indirectly involved (as if it were being notified out of courtesy). The reflex can be important at other times, as when the knee inadvertently buckles, activating the stretch receptor that causes the leg to suddenly straighten, keeping you from falling.

We have seen then the fascinating events involved in neural behavior. By the time we get through analyzing the processes involved in such events, we may have the feeling that the nervous system is so complex and ponderous that it sets the stage for error and inefficiency. We should keep in mind, though, that all these things happen routinely and with dazzling swiftness and accuracy. The evolution of neural mechanisms, such as those we've just seen, indeed underscores the remarkable efficacy of natural selection.

We'll move now to other aspects of neural function, looking particularly at the organization of neurons and neuroglia into complex animal nervous systems, including that of the human. Then we will explore the fascinating sensory organs, those structures with which organisms perceive their surroundings.

Key Ideas

CELLS OF THE NERVOUS SYSTEM

1. **Neurons** consist of a **cell body** with the usual cellular organelles and receiving and sending processes called **dendrites** and **axons**, respectively. One neuron activates a second, or an **effector**, through electrical impulses or through chemical substances called **neurotransmitters** that are produced in the cell body and released at knobby ends of the **axon tree**. Neurotransmitters that activate muscles are released at **neuromuscular junctions**. Dendrites are receiving processes that conduct signals to the cell body. Axons conduct **neural impulses** over long distances.

2. Many vertebrate neurons are wrapped in **myelin sheaths**, produced by Schwann cells outside the brain and spinal cord and oligodendrocytes within. The sheaths, or **internode** regions, contain minute gaps called nodes.

3. Neurons include **sensory neurons** (also *afferent neurons*), **interneurons**, and **motor neurons** (also *efferent neurons*). Sensory neurons receive stimuli such as light and heat and communicate with interneurons, which integrate the response and transmit it to the motor neurons, which produce the response. **Glial cells** are extremely common in the brain. They supply structural support, regulate potassium, and act as a selective barrier to materials moving into and out of the brain's neurons.

4. **Nerves**, composed of numerous axons and dendrites, form cable-like structures surrounded by connective tissue.

THE ORIGIN AND TRANSMISSION OF NEURAL IMPULSES

1. Neural impulses are generated by action potentials. **Action potentials** involve a wave of depolarization sweeping along a polarized neuron.

2. Action potentials do not diminish with distance. They occur in axons where controlled ion movement, active transport and diffusion, are involved. Their three aspects are (a) polarization (resting state), (b) propagation, and (c) repolarization.

3. In the resting state, the axon is polarized; the interior, just within the membrane, is more negative than the exterior, just outside. The charges involved come from negatively charged proteins and from positively charged sodium and potassium ions. **Sodium/potassium ion exchange pumps** transport Na^+ out of the axon and K^+ in. Although the membrane is more permeable to K^+ (because of "leak channels"), its attraction to the negatively charged proteins within keeps potassium from reaching equilibrium.

4. The charge difference across the membrane of the resting neuron is -70 mV.

5. Neural impulses begin, where a sudden change in sodium permeability brings on an inrush of the ion and depolarization. The membrane potential changes from -70 mV to $+30$ mV.

6. The action potential proceeds as a fast-moving wave, followed immediately by repolarization.

7. Action potentials require a specific **threshold voltage** for their generation. Once generated their magnitude and speed are constant, although the rate at which they are generated depends on the strength of the stimulus.

8. Repolarization is brought on by an efflux of K^+, which continues until the -70 mV, the **resting potential**, is attained.

9. The period of recovery, between $+30$ mV and -70 mV, is the **absolute refractory period**, a time when a second action potential cannot occur.

10. The movement of sodium and potassium ions during a neural impulse is controlled by the behavior of voltage-sensitive **ion gates** that control **ion channels**. The gates operate in response to small current flows in the neuron.

11. Action potentials at the node of myelinated neurons create enough current flow to activate the sodium gates in the next node, so the impulse jumps from node to node in what is called **saltatory propagation**.

COMMUNICATION AMONG NEURONS AND BETWEEN NERVES AND MUSCLES

1. Neurons communicate with each other at **synapses**, the gap between the **presynaptic** and **postsynaptic** neurons.

2. **Electrical synapses** occur at gap junctions between neurons. A moving action potential simply continues in the second cell.

3. **Chemical synapses** involve the secretion of chemicals called neurotransmitters from one neuron that brings on action potentials in a second neuron.

4. Impulses reaching the **axonal knobs** trigger the opening of calcium ion channels. Calcium entering the knob brings on the fusion of vesicles with the presynaptic membrane and the release of the neurotransmitter.

5. In many neurons the neurotransmitter **acetylcholine** brings on an influx of ions, creating the current necessary to produce an action potential.

6. **Inhibitory synapses** hyperpolarize the second neuron, thus reducing the likelihood of action potentials.

7. **Reflex arcs** may involve as few as two neurons—a sensory neuron and a motor neuron. In the patellar response, a sensory neuron detects muscle lengthening and flashes an impulse to the spinal cord, where it synapses with a motor neuron whose impulses cause the muscle to shorten.

Application of Ideas

1. There are a number of types of neural cells in the vertebrate body. Some of those are myelinated, using rapid saltatory propagation, and some are unmyelinated and conduct impulses more slowly. Using what you know of the principle of natural selection and evolution, how might myelinated fibers have evolved? Remember, there are many kinds of cells in the body (some with high fat content), many of them lying in close association but with unrelated roles.

2. A recent finding has been the discovery of the roles of ion channels and gates. Explain the roles of these structures in terms of depolarization and repolarization and graphically relate their behavior to the waves of an action potential shown on an oscilloscope.

Review Questions

1. Name the three functional parts of a neuron and explain what each does. (pages 692–693)

2. List the three types of neurons and state their general functions. (page 693)

3. List the probable functions of neuroglia. (page 693)

4. Describe the structure of a nerve. (pages 693–695)

5. Describe the arrangement of sodium and potassium ions in the resting neuron. What two factors account for this distribution? (page 696)

6. What happens to sodium at the start of an action potential? How does this affect the membrane potential? (page 696)

7. Explain the "all-or-none" principle. How might this be adaptive to an animal? (pages 696–700)

8. Describe the events that bring about repolarization. What is the refractory period? (pages 698–700)

9. What happens to the sodium and potassium gates during an action potential? How about during repolarization? (pages 698–700)

10. Explain how a neural impulse proceeds through a myelinated axon. How does this affect speed and energy cost? (pages 700–701)

11. Describe the electrical synapse. What are its advantages over the chemical synapse? (pages 701–702)

12. Using a simple drawing, describe the organization of a chemical synapse, labeling the *axonal knob, postsynaptic membrane, synaptic cleft, neurotransmitter, vesicle, receptor sites, calcium gates,* and *presynaptic membrane.* (pages 700–704)

13. Briefly describe the activity at the synapse, starting with the arrival of a neural impulse at a synaptic knob and ending with the opening of gates in the next neuron. (pages 700–704)

14. Suggest two ways in which a neurotransmitter can inhibit the generation of an action potential in the next neuron. (pages 700–704)

15. Explain the organization of a reflex arc. Of what adaptive value are such simple arrangements? (page 705)

CHAPTER 34

Neural Control II: Nervous and Sensory Systems

CHAPTER OUTLINE

INVERTEBRATE NERVOUS SYSTEMS

Simple Nerve Nets, Ladders, and Rings

Echinoderms and Nerve Rings

From Ganglia to Organized Brains

VERTEBRATE NERVOUS SYSTEMS

THE HUMAN CENTRAL NERVOUS SYSTEM

Overview of the Human Brain

The Forebrain

Essay 34.1 The Great Split-Brain Experiments

The Midbrain and Hindbrain

The Spinal Cord

Chemicals in the Brain

Electrical Activity in the Brain

THE PERIPHERAL NERVOUS SYSTEM

The Autonomic Nervous System

THE SENSES

Sensory Receptors

Neural Codes

Tactile Reception

Thermoreception

Chemoreception

Proprioception

Auditory Reception

Gravity and Movement Reception

Visual Reception

KEY IDEAS

APPLICATION OF IDEAS

REVIEW QUESTIONS

We humans pride ourselves, rightly or not, on our intelligence, and we are aware that the seat of that intelligence resides in that great gray structure we call the brain. We are also aware that the brain and the long stem that descends from it, the spinal cord, are very complex associations of the sorts of neural cells that we have just discussed. So here then we will focus on the structure and activities of both the brain and the spinal cord.

Before we launch into a discussion of the human brain, we want to emphasize that many kinds of animals seem to do just fine without either. So let's begin with them and see if we can place all this in some kind of evolutionary perspective. The brain, after all, is the result of an extreme development of one end of the nervous system. Note that in talking about ends, we must be talking about animals with bilateral body plans. Sea stars don't have ends. Neither do sea anemones. Being radially symmetrical, these animals are as likely to move off in one direction as another. Worms, on the other hand, have ends and they tend to move along in the direction of their body axis. In the course of evolutionary history, it seems that it would have become advantageous for animals with bilateral bodies, as they move about, to lead with the same end, and that this leading end would be the site of a concentration of sensory receptors that could tell the animal the nature of the environment into which it was moving. After all, it wouldn't do to back into an unfamiliar place. So *cephalization,* the development of an anterior end, set the stage for the formation of the brain by localizing sensory neurons in the leading end of a bilaterally symmetrical body.

Let's now consider a few of the specializations in a selected sequence of animals, moving from those that don't have centralized brains to those that do. We will arrange these species from simple to complex; however, keep in mind that each representative animal has had its own long evolutionary history and does not represent some sort of evolutionary stepping stone leading toward humans. We are only looking for principles that might help lead us through the history of the brain's development.

Invertebrate Nervous Systems

Simple Nerve Nets, Ladders, and Rings

The simplest metazoan neural organization, the **nerve net**, appears in the radially symmetrical animals of phylum Cnidaria, such as the hydra (Figure 34.1a). Theirs is a diffuse nervous system in which there are no clumps of neurons, although the neurons are more concentrated about the tentacles and mouth. The arrangement of axons in the cnidarian gives the appearance of a net, but often the crisscrossing neurons do not actually touch. Presumably, such separation would be more versatile than a true net since it would permit pathways to become established that could innervate certain tissues or organs.

Echinoderms and Nerve Rings

Echinoderms, represented here by sea stars, are among the oddest of all animals because of their unique pentaradial (*penta,* "five") body symmetry. Furthermore, they have a central nerve ring circling the mouth that gives rise to branches that enter each arm (Figure 34.1b). Their pentaradial body symmetry has not led to the development of the type of brain common to bilateral animals, but they are capable of coordinated, if slow, movements.

The most primitive flatworms have a net-like nervous system similar to that of the cnidarian hydra, but in others the system is somewhat modified. For example, some flatworms may have up to five nerve cords, tracts, or bundles of neurons. In others, such as many planarians, the nervous system resembles a ladder, consisting of only two parallel nerve cords with ring-like cross connections and a concentration of **ganglia**, clusters of cell bodies, in the head region (Figure 34.1c). Flatworms also boast cellular specializations in their nerve cells, having motor neurons, sensory neurons, and interneurons.

From Ganglia to Organized Brains

Annelids, Mollusks, and their Ganglia. The annelids are segmented worms with aggregations of neural tissue at the anterior end. In the earthworm, for example, large ganglia surround the pharynx in the head of the worm (a common arrangement in annelids, arthropods, and other complex invertebrates), and each segment along the length of the worm has a pair of fused ganglia emanating from the *ventral nerve cord* (Figure 34.1d). (The position of the nerve cord, we will see, has marked evolutionary significance.)

What about mollusks? If you have ever owned a pet clam, you may have noticed that it really wasn't much company. Clams and other bivalves are notoriously short on personality, perhaps because their nervous systems are quite simple. The nervous system of sedentary mollusks, such as the chiton, for example, takes on a ladder form roughly similar to that of a flatworm, but much more complex. The greatest concentration of nerves forms a nerve ring around the mouth and a dense ladder-like arrangement along the foot.

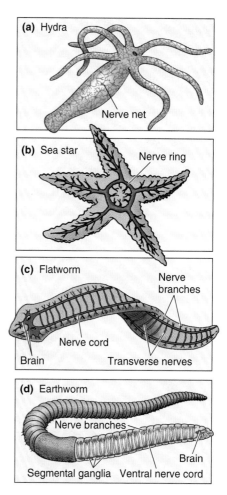

FIGURE 34.1

Nerve Nets, Ladders, and Ventral Nerve Cords

(a) In the hydra's net-like arrangement of nerve fibers, neurons are scattered over its body, with the greatest number near its mouth and tentacles. **(b)** In the sea star, the nervous system consists of a central ring with major nerves radiating into each arm. There is little cephalization. **(c)** In planarians, many neurons accumulate in the head region, forming ganglia. Two nerves emerge from the ganglia, forming a ladder-like arrangement. **(d)** The earthworm has a complex, segmented nervous system with large paired ganglia above and below the esophagus. A ventral, solid nerve cord produces paired ganglia in each segment.

Cephalopod Mollusks and the Appearance of a Brain. Surprisingly enough, another mollusk has been called "smarter" than some vertebrates. We're referring to the octopus, a remarkably active predator with a life-style that would demand a complex nervous system. It, like many invertebrates, has a ganglionic arrangement of neural tissue around its esophagus that may rightfully be called a brain, a neural mass with differentiated areas, each with a specific function. For example, a distinct and identifiable part of that complex serves as a respiratory center; another part controls the animal's rapid color changes; other parts deal with spatial associations, and other parts control eye movement (Figure 34.2). In fact, because of its mental abilities, the octopus has been an important research animal in the study of learning and memory.

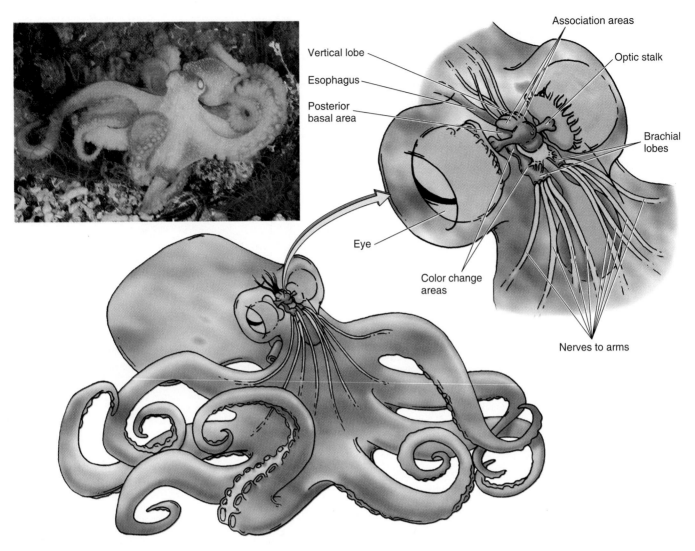

FIGURE 34.2
The Cephalopod Brain
The octopus has a well-developed brain with specialized regions related to vision, rapid movement, and
intricate muscle control.

Arthropods and a Segmented Nervous System. Like annelids, arthropods' bodies are segmented, but unlike annelids, the
segments are more likely to be distinct, with special functions,
not repeating units. In many species the anterior neural ganglia
are prominent, qualifying as a brain, with segmental ganglia
reduced.

The insect brain is composed of a large ganglion above
the esophagus and a smaller one below (Figure 34.3). Each of
these brain parts has very specific roles, with much of the
larger mass devoted to integrating information from the eyes,
antennae, and other sensory structures so abundant on the
insect body. The ganglion below the esophagus controls the
mouthparts and salivary glands. Interestingly, in arthropods
that have a simple eye that crudely registers only light intensi-

ties, the visual centers take up about 3% of the brain. But in
highly visual species, such as the housefly, the visual centers
may occupy as much as 80% of the brain.

Vertebrate Nervous Systems

So far, the nerve cords we have seen have been solid structures running along the ventral part of the body. With the vertebrates, we find a dorsal hollow nerve cord and, at its bulbous
anterior end, a large, complex, brain.

We can begin our discussion of the vertebrate nervous
systems with a comparison of the various classes. The living
classes of vertebrates include the jawless lampreys and hag-

fish, the sharks and rays, bony fishes, amphibians, reptiles, birds, and mammals.

Among the vertebrates, there is a pronounced tendency toward increased cephalization. That is, the more recently evolved classes, the birds and mammals, generally have larger brains in relation to body size.

As you can see in Figure 34.4, vertebrate brains have three parts: the *forebrain, midbrain,* and *hindbrain.* The **fore-**

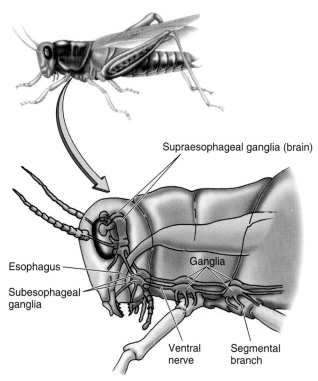

FIGURE 34.3
An Insect Brain
The insect nervous system includes ganglia (brain) above and below the pharynx, a ventral, solid nerve cord, and ganglia serving most segments.

FIGURE 34.4
The Vertebrate Brain
Comparing the anatomical structures of the brain in five classes of vertebrates (fish, amphibian, reptile, bird, and mammal) reveals general evolutionary trends and specific trends in specialization. **(a)** Note the relatively large olfactory lobes in the fish's brain. It clearly reflects the importance of chemical detection to this predator. **(b)** The amphibian feeds by visual means, as does the reptile shown in part **(c).** Note the relative size of their optic lobes. The trend toward increasing dominance of the cerebrum in vertebrate evolution is also apparent, beginning with the bird **(d)** and becoming more pronounced in the mammal **(e).** The trend toward specialization is greatest in the mammal, with increased convolutions in the cortex. Convolutions or foldings are a way of increasing cerebral size without greatly increasing cranial size.

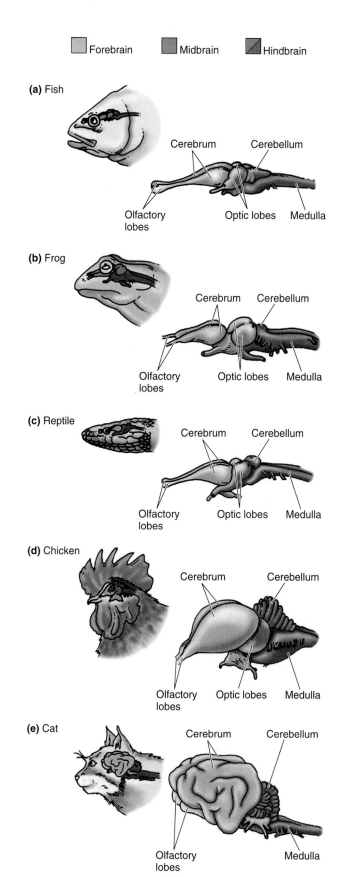

brain, the anterior part of the brain, includes the prominent **cerebrum** and certain regions below. In some vertebrates the *forebrain,* the anterior part of the brain, is disproportionately smaller than the other two brain regions. The fish brain, for example, is dominated by the midbrain, but they also have a relatively large hindbrain. In many vertebrates, the *midbrain* is important in analyzing visual and chemical (olfactory) stimuli, but in humans it serves primarily as a communicating bridge between the forebrain and *hindbrain,* the posterior part that joins the spinal cord. Compare the brains of the vertebrates in Figure 34.4. You can see that, in mammals, the midbrain and hindbrain have been overgrown by the forebrain's cerebrum. So it is in the mammals that we find the most complex cerebral development. In addition, the wrinkled outer region of the cerebrum—the *cerebral cortex*—is far more prominent in the mammals. As representative of the mammalian nervous system, we will turn to that brainiest of all creatures, the human.

The Human Central Nervous System

The human nervous system, like that of other vertebrates, is divided anatomically into two major parts: the **central nervous system** (CNS) and the **peripheral nervous system** (PNS) as shown in Figure 34.5. The CNS is composed of the brain and spinal cord, while the PNS (discussed later) includes all neural structures outside these two.

Overview of the Human Brain

The human brain is an absolutely fascinating structure given to self-congratulation. (In these words, in fact, we find one complimenting itself.) Actually, an enlarged brain and an enlarged gluteus maximus are two of the most prominent traits of our species. One might wonder what the planet would look like had the human brain not motivated us to action and had we spent more time resting dolefully on our glutei maximi. With all the world's problems to solve, the human brain has devoted a great deal of time to thinking about itself, but to this day many of its processes remain unknown. However, some of our discoveries are hard to believe. For example, there is evidence that every word you have ever uttered or heard in your entire life may be filed away in your brain, even though you will go to your grave having recalled hardly any of that information. (How nice it would be to be able to retrieve the exact words of our last conversation with a loved one.)

The human brain (Figure 34.6) weighs about 1.4 kilograms (3 pounds), has a volume of about 1300–1500 cubic centimeters (cc), and contains some 100 billion or more neurons and at least 10 times that many supporting glial cells. Each neuron may form hundreds of synapses, producing a veritable harmonic neural symphony.

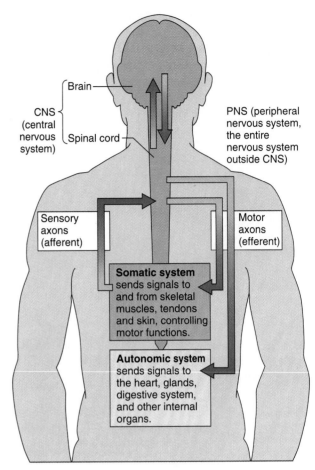

FIGURE 34.5
Organization of the Human Nervous System
The human nervous system is divided into the central and peripheral nervous system. The latter is further subdivided into the somatic and autonomic systems. The somatic system includes both motor and sensory functions, but the autonomic system is strictly motor.

The delicate brain with its gelatin-like consistency requires an effective means of support and protection. In addition to the surrounding bony skull, it is enclosed by three tough, protective connective tissue membranes called **meninges**. The spaces within the meninges, and any open spaces within the brain itself, are filled with pressurized, shock-absorbing **cerebrospinal fluid**. Finally, the billions of fragile neurons themselves are embedded in an even more vast matrix of glial cells, which make up much of the mass of the brain.

Metabolically, the brain is highly active and makes great demands on the circulatory system. For example, it normally receives about 15% of the blood output from the heart. It has a substantial oxygen demand and quickly malfunctions when oxygen-deprived. Glucose supplies most of the brain's energy, and, in fact, the brain utilizes about 75% of the body's intake of that sugar.

The arteries carrying blood into the brain have a special arrangement that helps ensure a sufficient supply of blood to this vital organ. They form a circle, the *circle of Willis,* at the

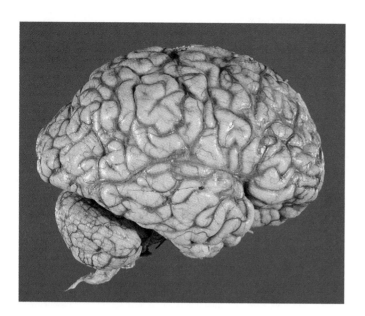

FIGURE 34.6
The Human Brain
The cerebrum, with its two hemispheres and highly convoluted surface, dominates the human brain. The equally wrinkled cerebellum lies on either side of the medulla oblongata, which itself gives way to the spinal cord.

base of the brain, there giving rise to smaller arteries that serve various parts of the brain. Each branch serves more than one part, however. Thus, should an artery branching from the circle become blocked, the brain region served by that artery will receive blood from another branch of the circle, thus minimizing disruption and possibly averting a stroke (brain damage due to a blocked vessel).

The brain is also largely protected against toxic substances normally carried in the blood. To enter the brain tissue, substances carried by the blood must first cross the **blood–brain barrier**, a group of mechanisms that restrict the passage of materials from the blood into the brain. Part of this barrier is provided by the capillaries themselves, since the endothelial cells that make up the brain's capillaries have tightly joined and overlapping plasma membranes. In addition, materials that do leave the blood must first pass into one kind of highly selective cells called **astrocytes**, glial cells that attach to the capillaries via foot-like processes (Figure 34.7). The presence of the effective blood–brain barrier, incidentally, makes it difficult for researchers to carry out experiments testing the effects of new drugs on the brain. Most such chemicals simply do not reach the brain tissue.

You may be surprised to learn that our 3-pound brain isn't the largest on earth; whales and porpoises have larger brains. But, in comparing species, intelligence (since we invented the term) seems to be related to the *relative size* of the brain and the body, and here we humans lead the pack.

We have mentioned that, in all vertebrates, the brain consists of three regions, clearly named the forebrain, midbrain, and hindbrain (Figure 34.8). We will take a close look at these now, beginning with the forebrain.

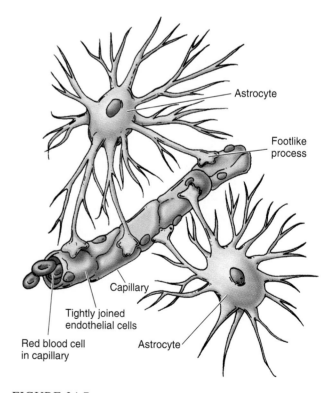

FIGURE 34.7
The Blood–Brain Barrier
Materials entering and leaving the brain tissue must pass across what is called the blood–brain barrier. Note the tightly joined cells of the capillary wall and the foot-like attachments of the astrocyte. Most materials pass through the astrocytes prior to entering the brain tissue.

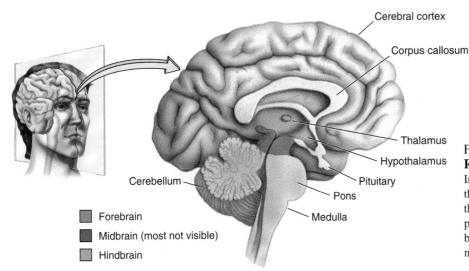

Cerebral cortex

Corpus callosum

Thalamus

Hypothalamus

Pituitary

Pons

Medulla

Cerebellum

Forebrain

Midbrain (most not visible)

Hindbrain

FIGURE 34.8
Regions of the Brain
In humans, the forebrain—the cerebrum and the structures it encloses—make up most of the brain. The midbrain contains connecting pathways between the forebrain and hindbrain. The hindbrain includes the pons, medulla oblongata, and cerebellum.

The Forebrain

The forebrain, the largest and most dominant part of the human brain, is responsible for conscious thought, reasoning, memory, language, sensory reception and decoding, and conscious movement of the skeletal muscles. In addition to the **cerebrum**, the forebrain includes the **thalamus**, **reticular system**, **hypothalamus**, and the **limbic system**. As we'll see, some of the so-called systems are overlapping. We will focus first on the cerebrum, the human brain's "thinking center."

The Cerebrum. To most people, the word *brain* conjures up an image of two large, deeply convoluted gray hemispheres. Those hemispheres actually make up only part of the brain, the cerebrum—the largest and most prominent part of the human brain.

The left and right halves of the cerebrum are the **cerebral hemispheres**, and the outer layer of gray cells is the **cerebral cortex**. The convoluted cortex is divided into distinct parts by grooves and ridges as we see in Figures 34.6 and 34.8. The cerebral cortex consists of a thin but extremely dense gray layer of nerve cell bodies (about 15 billion in all) and their dendrites. Thus the brain is sometimes erroneously referred to as gray matter, usually in the context of people cheerfully accusing each other of not having any. Underneath is the white matter, consisting of myelinated nerve fibers, reaching in all directions throughout the brain. These fibers bring information to the cortex for processing and carry the integrated messages from the cortex to other parts of the brain. All vertebrates have a cerebrum, but it differs markedly from class to class, especially in the degree of development of the cortex (see Figure 34.4). In some kinds of vertebrates the cerebrum may serve essentially to refine behavior that could be performed to some degree without it. These kinds of animals are not generally known for their learning abilities but rely more on genetically programmed behavior emanating from the "old brain"—that is, the noncerebral brain, the part that evolved first. In more recently evolved animals, the cerebrum takes on greater importance and, as in the case of the visual and auditory centers, often takes over functions that were once the responsibility of other, older parts of the brain.

For example, if the cerebrum of a frog is removed, the frog will show relatively little change in behavior. If it is turned upside down, it will right itself; if it is touched with an irritant, it will scratch; it will even catch a fly. Also, in frogs, sexual behavior can occur without the use of the brain—but it is probably best not to dwell on that. A rat is more dependent on its cerebrum. A rat that is surgically deprived of its cerebrum can visually distinguish only light and dark, although it seems to move normally. A cat with its cerebrum removed can meow and purr, swallow, and move to avoid pain, but its movements are sluggish and robot-like. Dogs treated this way are more helpless and just stand around, eventually starving unless food is thrust into their mouths. A monkey whose cerebrum has been removed is severely paralyzed and can barely distinguish light and dark. The result of massive cerebral damage in humans may be total blindness and almost complete paralysis. Although such persons can breathe and swallow, they soon die.

It seems that, from an evolutionary standpoint, more and more of the functions of the lower brain are transferred to the cerebrum in the more intelligent species. Generally, the degree to which the cerebrum has taken over neural control is reflected in its size. Thus more intelligent animals tend to have relatively larger cerebrums (see Figure 34.4). However, size is not the only indicator of the cerebrum's complexity. Convolutions, for example, increase the surface area of the cortex without enlarging the braincase. The deep convolutions seen in the human brain are lacking in the brain of the rat (but note that rats, with their smooth and tiny brains, are notoriously hard to foil when they come into conflict with humans). The highly touted and undoubtedly intelligent dolphin has a highly convoluted cerebrum, but with fewer layers than in the human brain.

The Lobes of the Cerebrum. In humans, each cerebral hemisphere is divided into four lobes, occipital, temporal, frontal, and parietal (Figure 34.9). At the posterior is the **occipital lobe**. It contains a region that receives raw, visual sensory input from the optic nerve and begins the analysis of that input. If the occipital lobe is injured, black "holes" appear in the part of the visual field served by the injured part.

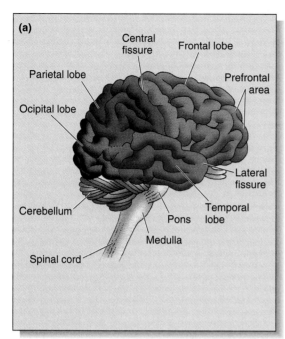
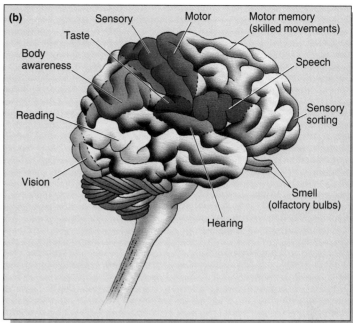

FIGURE 34.9
Cerebral Lobes
(a) The human cerebrum is divided into four prominent lobes: frontal, temporal, parietal, and occipital.
(b) Each is somewhat specialized in its functions, and all represent the so-called higher centers of the brain.

The **temporal lobes** are at either side of the brain, under the temples. Each lobe roughly resembles the thumb of a boxing glove and is bordered anteriorly by a deep groove, the **lateral fissure**. The temporal lobe helps to process input from senses relating to hearing and smell. This lobe also helps with the processing of visual information.

The **frontal lobe** is right where you would expect to find it—at the front of the cerebrum. It underlies that part of the skull that people hit with the palm of their hand when they suddenly remember what they forgot. One part of the frontal lobe regulates precise voluntary (motor) movement. Another part controls the bodily movements that produce speech and is considered to be part of the speech center.

The area at the very front of the frontal lobe is called the **prefrontal area**. Whereas it was once believed that this area was the seat of the intellect, it is now apparent that its principal function is sorting out sensory information. In other words, it places information and stimuli into their proper context. The gentle touch of a mate and the sight of a hand protruding from the bathtub drain might both serve as stimuli, but they would be sorted differently by the prefrontal area.

The **parietal lobe** lies directly behind the frontal lobe; the two are separated by a deep cleft called the **central fissure**. The parietal lobe contains the sensory areas for the skin receptors and the cortical areas that detect body position. Even if you can't see your feet right now, you probably have some idea of where they are, thanks to receptors in your mus-

cles and tendons that innervate centers in the parietal lobe (and the cerebellum as well). Damage to the parietal lobe can cause numbness and a sense that one's body is wildly distorted. In addition, the victim may be unable to perceive the spatial relationships of surrounding objects.

Sensory and Motor Regions of the Central Cortex. By stimulating various parts of the cortex with electrodes, investigators have located two large regions that are specialized for certain motor or sensory activities. The sensory and motor areas are located in specific **gyri** (singular, *gyrus*) prominent raised folds between fissures (see Figure 34.9). The motor area lies just anterior to the prominent central fissure. It generates the motor impulses that bring about voluntary muscle movement. Just posterior to the same fissure lies a second gyrus, one that receives sensory input from the skin (touch, pressure, and so on) and taste receptors.

Mapping the two gyri reveals that each is subdivided according to the region of the body served. The map is reconstructed in Figure 34.10. The distortion of the organs has not been done out of an appreciation for the macabre, but to indicate the relative amount of motor and sensory cortical regions devoted to each body part shown. For example, much of the sensory gyrus area is organized for the reception of touch input from just the tongue, face, fingers, and genitals. Whereas much of the motor gyrus is devoted to movement of the tongue, face, and fingers, little is involved with the genitals.

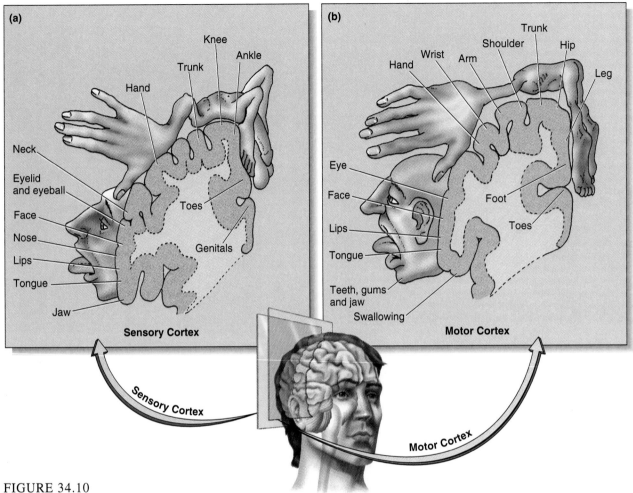

FIGURE 34.10
Sensory and Motor Cortex
The sensory (**a**) and motor (**b**) areas of the brain are located on opposite sides of the central fissure, the prominent groove between the frontal and parietal lobes. In the caricature, the size of the body parts reflects the relative number of sensory or motor neurons serving that part.

We can conclude from this that even though the genitals are very sensitive to touch, we have little voluntary control over them, a fact you may have already discovered.

Two Brains. Though the two cerebral hemispheres are roughly equal in size and potential, they are not completely identical in form and are far from identical in function. For example, the control of certain learned patterns takes place primarily in only one hemisphere. Other evidence of a difference between the hemispheres is seen in handedness.

It is interesting that other species, such as rats and parrots, also show right- and left-handedness, although not in the proportions seen in humans. About 89% of humans are right-handed, but about half of rats and parrots are right-handed. Nerve tracts (bundles of axons or dendrites) cross from one side of the body to the other side of the brain, so the right side of the body is controlled by the left side of the brain (and vice versa). Thus in right-handed people, the left half of the brain is dominant. It is also slightly larger than the right half. There

are many mysteries to handedness. For example, no one can account for the unusual prevalence of southpaws among professional tennis players and artists. Still, they are quick to tell us that only left-handed people are in their right minds.

Other functions, such as speech, perception, and certain aspects of IQ test performance, are also more likely to be located in one hemisphere than the other, and in some cases, the distribution of centers is related to handedness. About 95% of right-handers and 70% of left-handers have the speech center in the left hemisphere. (About 15% of people have bilateral speech centers.) The left hemisphere also seems to be the seat of analytical thinking, while spatial perception is right-hemisphere function. Left-hemisphere brain damage can result in *aphasia*—the inability to speak or understand language—while right-hemisphere brain damage may result in the inability to draw the simplest picture or diagram (but does not affect the ability to write letters and numbers, which is a left-hemisphere function). Essay 34.1 describes experiments with the "split" brain.

Essay 34.1 ● THE GREAT SPLIT-BRAIN EXPERIMENTS

Some years ago, Roger W. Sperry and Robert E. Meyers experimentally separated the two hemispheres of the brains of cats by severing the corpus callosum. Intensive testing then showed that the animals behaved as if they had two separate brains. A cat could be trained to perform a task by using one eye (the other covered) and when presented with the same task viewed by the other eye, the animal would respond as if no learning had occurred.

Later, Sperry and Michael Gazzaniga severed the corpus callosum in human epilepsy patients as treatment for their condition. The epileptic seizures diminished and the patients seemed to behave normally, but the researchers decided some testing was in order. The patients were then tested in an apparatus that allowed a particular image to fall entirely on only one-half of the visual field (see fig-ure). This meant that only half of the image could be transmitted to the brain from either eye. The results were intriguing. Sperry projected a word such as "kitten" onto a screen so that only half the word would fall onto half of each retina, "kit" on the left visual field "ten" on the right (see figure). He then asked the patient to read the word. The left visual field controls language, and the patient responded "ten." Sperry then asked the patient to write the word with the left hand, and "kit" was the response. The right hemisphere knew the other half of the word but couldn't communicate it to the left hemisphere.

It was quickly discovered that the results were not due to visual phenomena alone. The patients could only tell researchers what fell on the left halves of their brains because that is where the centers of speech are located. On the other hand, if they were allowed to express themselves nonverbally, they could immediately describe what the right half of the brain "saw." They also learned that because the right side of the cortex controls the left hand, and vice versa, a person can first feel an object, say, with the left hand, and then pick out that object with the same hand from a collection hidden behind a screen, simply by feel. However, because the information is traveling to the right half of the brain and that half cannot communicate with the left half, the person cannot name the object. Interestingly, the person will be able to pick out the object faster by using the left hand than the right because the right cortex is far superior at dealing with spatial relationships. Further experimentation has revealed that the human brain is, indeed, effectively divided into two halves—two brains, as it were.

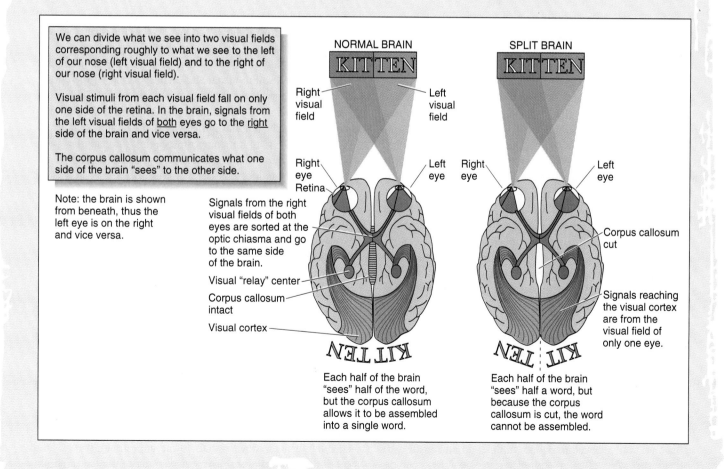

We can divide what we see into two visual fields corresponding roughly to what we see to the left of our nose (left visual field) and to the right of our nose (right visual field).

Visual stimuli from each visual field fall on only one side of the retina. In the brain, signals from the left visual fields of <u>both</u> eyes go to the <u>right</u> side of the brain and vice versa.

The corpus callosum communicates what one side of the brain "sees" to the other side.

Note: the brain is shown from beneath, thus the left eye is on the right and vice versa.

NORMAL BRAIN

KITTEN

Right visual field

Left visual field

Right eye
Retina

Left eye

Signals from the right visual fields of both eyes are sorted at the optic chiasma and go to the same side of the brain.

Visual "relay" center

Corpus callosum intact

Visual cortex

Each half of the brain "sees" half of the word, but the corpus callosum allows it to be assembled into a single word.

SPLIT BRAIN

KITTEN

Right eye

Left eye

Corpus callosum cut

Signals reaching the visual cortex are from the visual field of only one eye.

Each half of the brain "sees" half a word, but because the corpus callosum is cut, the word cannot be assembled.

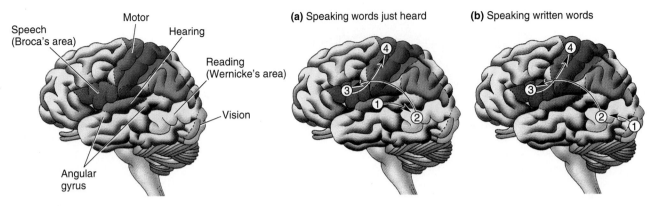

FIGURE 34.11
Language Processing
Using language requires the cooperation of neural centers in two cortical regions, known as Broca's area
and Wernicke's area, along with visual, hearing, and motor centers. The principal pathways you use in
repeating what you have just heard **(a)** and in reading aloud **(b)** are indicated by the arrows. The numbers
indicate the sequence of activity.

In spite of these specializations, the right and left hemispheres operate as an integrated functional unit. The primary route of communication between the right and left cerebral hemispheres is the **corpus callosum** (see Figure 34.8). It seems that if one side of the brain learns something—for instance, by feeling an object with just one hand—the information will be transferred to the other hemisphere. However, if the corpus callosum has been severed in an accident, the left brain literally doesn't know what the left hand is doing (Essay 34.1).

It should be pointed out that a certain degree of compensation is possible between the two halves of the brain. For example, if the dominant hand is injured, it is possible to learn to use the other hand with almost equal facility. If one hemisphere of a young child's brain is severely injured, the other side will eventually take over some or all of its functions. This ability to shift, like so many other abilities, declines with age.

Language Centers in the Cortex. The human brain has highly developed areas devoted to the processing of language. We are just now beginning to understand how the centers interact, and much of what we are learning comes from studies of stroke victims. When a stroke occurs, part of the brain suffers an oxygen and nutrient loss, causing the death of brain tissue. Among the more typical results of stroke damage is aphasia, the loss or impairment of speech, especially when the damaged area is in the left side of the cortex. From stroke studies, neurophysiologists conclude that three regions of the left hemisphere are important to speech—*Broca's area, Wernicke's area,* and the *angular gyrus* (Figure 34.11). Of course, other regions of the cortex are also important to language, since linguistic articulation also involves hearing, seeing, even writing—not to mention thinking (although some people seem to have no trouble speaking without thinking much at all).

Spoken words, like all sounds, are received by the auditory receptors and the impulses are sent on to the brain, but unless the incoming neural information is processed in **Wernicke's area**, the sound will only be heard as noise; its meaning will not be deciphered. Spoken words also originate in Wernicke's area; from there they are transmitted to **Broca's area** along a special pathway. Broca's area then activates the motor cortex, properly activating the vocal cords, mouth, lips, tongue, and jaw, so that the proper sounds may be uttered.

Reading aloud involves the third region, the **angular gyrus**, as well as the visual and auditory areas of the cortex. Neural messages from the eyes reach the visual field of the cortex for processing, and this processed visual information is transferred to the angular gyrus, where the symbols are translated into words. In a manner of speaking, the angular gyrus "hears" the words we have just seen, by associating the visual message with auditory patterns in Wernicke's area. (Can you "hear" the words you see?) Speaking the words then requires the transfer of neural patterns to Broca's area and on to the motor cortex as usual. Poor readers sometimes mouth the words they read, but most of us learn to turn off our motor cortex unless we are deliberately trying to read aloud.

The Thalamus. Leaving the cerebrum, we can now consider other parts of the forebrain, beginning with the thalamus. The thalamus is located at the base of the forebrain. It is a paired structure, with one-half on either side of a central fluid-filled cavity (see Figure 34.8). The thalamus has been rather poetically called the brain's "great relay station." It consists of densely packed clusters of cell bodies, through which most sensory input to the cerebrum must pass.

The thalamus integrates the sensory information that constantly bombards the body, subconsciously sorting it out and channeling the various signals to the appropriate parts of the cerebrum. The thalamus also receives signals from the cerebrum and sends the appropriate information to other parts of the brain.

In addition, the thalamus has the special task of relaying impulses to the cortex that help maintain consciousness. In this last function, the thalamus relies on the **reticular system**, a region composed of clusters of neurons that run throughout the thalamus as far as the lower part of the forebrain. The reticular system is still somewhat of a mystery, but several interesting facts are known about it. For example, it "bugs" your brain, tapping virtually all incoming and outgoing communications. It seems that the reticular network is something of an arousal system that serves to activate the appropriate parts of the brain upon receiving a stimulus. The more messages it intercepts, the more a specific part of the brain is aroused.

The reticular system also seems to function importantly in sleep. You may have noticed that it is usually easier to fall asleep when you are lying on a soft bed in a dark, quiet room than on the floor of a noisy bar. Under the quieter conditions, there are fewer incoming stimuli; as a result, the reticular system receives fewer messages, and the brain is allowed to relax and initiate the processes associated with sleep.

The Hypothalamus. Just below the thalamus lies the hypothalamus (see Figure 34.8). It is densely packed with cells that help regulate the body's internal environment and certain general aspects of behavior. The hypothalamus receives sensory input from many internal organs via the thalamus and makes use of the input in coordinating such internal conditions as heart rate, appetite, water balance, blood pressure, and body temperature. It also influences such basic drives as hunger, thirst, sex, and rage. Electrical stimulation of various centers in the hypothalamus can cause a cat to act hungry, angry, cold, hot, benign, or sexy. In humans it is known that a tumor pressing against a "rage center" can cause the person to behave violently and even murderously.

Another major function of the hypothalamus is the coordination of the nervous system with the hormonal (endocrine) system (see Chapter 36). In fact, the hypothalamus prompts much of the hormone secretion that goes on in the pituitary, a major endocrine gland.

The Limbic System. The hypothalamus and the thalamus, along with certain pathways in the cerebral cortex, are functionally part of what is called the **limbic system** (Figure 34.12). The limbic system links the forebrain and midbrain and controls certain aspects of muscle tone, such as the positioning of an arm so that the fingers are in the best position to type or play the piano. Since it includes the hypothalamus, the limbic system encompasses hypothalamic functions and so influences emotions such as fear, rage, sexual arousal, aggressiveness, and motivation. For example, the **amygdala**, a structure within the limbic system, can produce rage if it is stimulated and docility if it is removed. Finally, the amygdala, along a region of the temporal lobe called the **hippocampus**, figures importantly in memory storage and recall. Memory processing will be discussed in Chapter 44.

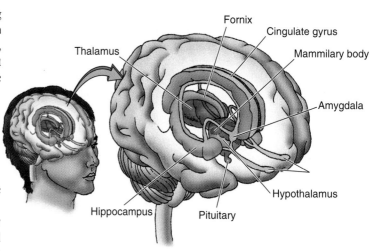

FIGURE 34.12
The Limbic System
The major components of the limbic system include the amygdala, hippocampus, thalamus, hypothalamus, and cingulate gyrus.

The Midbrain and Hindbrain

Now that we've learned something about the forebrain, let's consider the other two major areas, the midbrain and hindbrain. Essentially, the **midbrain** joins the hindbrain and forebrain, by numerous connecting nerve tracts between the two (see Figures 34.4 and 34.8). Certain parts of the midbrain receive sensory input from the eyes and ears. All auditory input of vertebrates is processed here before being sent to the forebrain. In most vertebrates, visual input is first processed in the midbrain, but in humans and other mammals the visual information is sent directly to the forebrain. And while the midbrain is involved in complex behavior in fishes and amphibians, many of these same functions are assumed by the forebrain in reptiles, birds, and mammals.

The hindbrain consists of the *medulla oblongata,* the *cerebellum,* and the *pons* and is continuous with the spinal cord. It contains the lower, more posterior brain centers (see Figures 34.6 and 34.8). As a rough generality, the unconscious, involuntary, and mechanical body processes are directed by these lower regions.

The lowest part of the brain, the **medulla oblongata** (or simply the medulla), contains centers that control breathing rate, heart rate, and blood pressure. In addition, all communication between the spinal cord and the brain must pass through the medulla.

The **cerebellum** is a small, bulbous, paired structure with about the general appearance of the two halves of an enlarged walnut. It lies above the medulla and toward the back of the head. The function of the cerebellum is the coordination of movement.

The cerebellum coordinates all voluntary movement in the limbs and body and aids in the maintenance of posture and balance. It does this chiefly through inhibition, limiting the force with which a muscle contracts and the distance a limb travels, and generally dealing with several muscle groups at the same time. To coordinate this activity, the cerebellum requires constant sensory input informing it about the degree of muscle contraction and the position of the limbs and body. Such input comes from the eyes, balance organs, and from the muscles themselves. A particularly skilled gymnast can leave us quite impressed with his or her performance. Interestingly, those smoothly executed movements are not entirely voluntary. Much of what we see is made possible by the cerebellum, which operates below the level of consciousness.

The **pons** (Latin for "bridge") is the part of the hindbrain just above the medulla oblongata. It contains ascending and descending neural tracts that run between the brain and spinal cord and between the cerebrum and cerebellum, linking the functions of the cerebellum with the more conscious centers of the forebrain. The pons also aids the medulla in regulating breathing.

The Spinal Cord

The spinal cord serves as a major link between the brain and the peripheral nervous system. The spinal cord begins as a narrow continuation of the brain, emerging from the **foramen magnum** ("big hole") at the base of the skull (Figure 34.13a). It lies sheltered within the **spinal canal**, a continuous channel formed by the neural arches of the vertebrae (see Figure 32.10c). The spinal cord, like the brain, is surrounded by a tough, three-layered sheath, the meninges, which contains the cushioning cerebrospinal fluid.

The spinal cord contains two regions that, in cross section, appear white and gray to the eye (Figure 34.13c). The outer region of the cord is white because it contains immense numbers of myelinated axons. They run in very specific pathways or tracts to and from the brain, some extending far from their neural cell bodies. The butterfly-shaped inner region of the cord is gray because it is made up chiefly of the neural cell bodies of countless interneurons and motor neurons, along with glial cells.

Paired spinal nerves emerge from the great cord through the spaces that lie between adjacent vertebral arches. Each of these spinal nerves is formed from two roots in the cord, a **dorsal root** (toward the back—the sensory root) and a **ventral root** (toward the belly—the motor root; see Figure 34.13b,c). The cell bodies of the motor neurons, like the cell bodies of interneurons, lie in the gray matter of the cord, with their lengthy axons passing out through the spinal nerves and onward to the voluntary muscle. On the other hand, the cell bodies of sensory neurons are clumped in large **dorsal root ganglia** just outside the cord.

You may recall from Chapter 33 that the spinal cord has a certain degree of autonomy. Through a reflex arc, it can carry out synaptic reflexes between sensory and motor neurons without consulting the brain.

Chemicals in the Brain

The brain is a staggeringly complex structure, containing billions of neurons playing innumerable roles. Yet these neurons interact in a delicate and coordinated manner, integrating and shunting information from one place to another in a harmonious symphony of interaction. This interaction includes both excitation and inhibition of neurons through very specific action in trillions of synapses. All this is accomplished by each neuron releasing its own specific neurotransmitters, of which there are at least 50, with more being discovered all the time. Perhaps the best known is acetylcholine, a neurotransmitter released in synapses within both the central nervous system and the peripheral nervous system.

The known neurotransmitters can be placed into any of several chemical groups. For example, they may be simple **monoamines**, such as norepinephrine, dopamine, histamine, and serotonin, each of which is a modified amino acid. (In some instances, unmodified amino acids, such as glycine and glutamate, act as neurotransmitters.)

The **neuropeptides** are a more recently discovered class of neurotransmitters. As their name suggests, they are chains of amino acids (rather short, from two to about 40). Among the more interesting brain neuropeptides are the *enkephalins* and *endorphins*. These modify our perception of pain and also have an elevating effect on mood. Anything acutely painful, such as running a marathon race or shooting oneself in the foot, will stimulate the release of enkephalins. The "runner's high," a kind of euphoria that may appear after about a 10-mile run, is also attributed to the release of enkephalins (although some runners simply may be ecstatic at covering that distance without dying). These mood-elevating peptides are also called the *opioid neurotransmitters* because morphine and other opiates will also bind to the neural enkephalin receptors and mimic their action.

Our rapidly expanding knowledge of the neurotransmitters is also shedding some light on the action of certain other drugs that affect the central nervous system (and vice versa). Amphetamines, for example, are believed to imitate norepinephrine, a neurotransmitter that stimulates the brain and accelerates blood pressure and heart rate. Cocaine has a different, more drastic effect. It amplifies the action of neurotransmitters such as serotonin, norepinephrine, and dopamine by blocking the enzymes and reuptake mechanisms that normally clear them from the synaptic cleft after they have acted. Drugs such as LSD, mescaline, and psilocybin mimic the role of other natural neurotransmitters. They produce their hallucinogenic effects by artificially stimulating action at the synapses.

Electrical Activity in the Brain

Each active neuron in the brain creates electrical impulses, although about a million neurons must fire simultaneously in order for this energy to be detected from outside the body. The instrument used in such detection is called an *electroencephalograph,* and the record obtained is an *electroen-*

FIGURE 34.13
The Spinal Cord
(a) The spinal cord extends from the base of the brain into the lumbar region of the spine, where it branches into many descending nerves. The spinal nerves emerge from between vertebrae. (b, c) Detailed views of the spinal nerves reveal that they emerge in pairs as dorsal and ventral roots. Nerves forming the ventral root contain the axons of motor neurons, emerging from cell bodies within the gray matter of the cord. They carry impulses to effectors such as muscles and glands. Nerves forming the dorsal root are sensory. The cell bodies of the individual sensory neurons are located within the spinal ganglia. From there, the neural processes continue into the white matter of the cord, where they join tracts traveling to the brain. Some sensory neurons synapse with motor neurons within the cord, forming very short pathways associated with spinal reflexes. We see then (c) that the white matter of the cord consists of myelinated neurons traveling to and from the brain, whereas the gray matter consists of cell bodies of motor neurons.

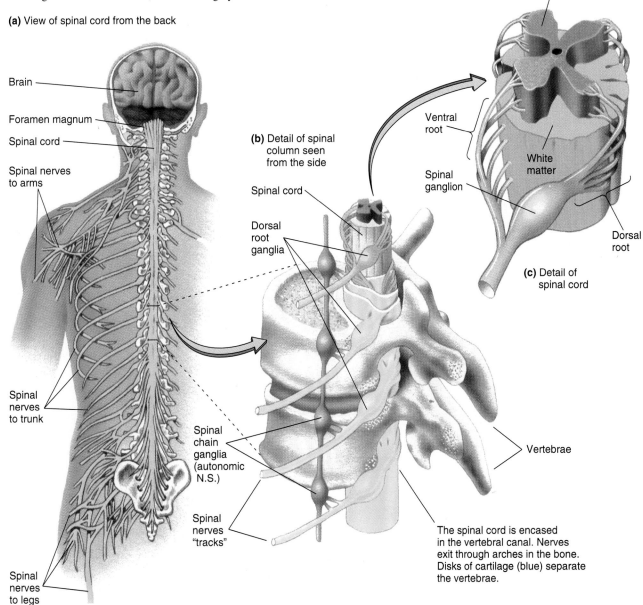

(a) View of spinal cord from the back

Brain

Foramen magnum

Spinal cord

Spinal nerves to arms

Spinal nerves to trunk

Spinal nerves to legs

(b) Detail of spinal column seen from the side

Spinal cord

Dorsal root ganglia

Spinal chain ganglia (autonomic N.S.)

Spinal nerves "tracks"

Gray matter

Ventral root

Spinal ganglion

White matter

Dorsal root

(c) Detail of spinal cord

Vertebrae

The spinal cord is encased in the vertebral canal. Nerves exit through arches in the bone. Disks of cartilage (blue) separate the vertebrae.

cephalogram (EEG). Electrodes leading to the device are fastened at various places on the head, and the tiny currents they pick up are amplified and recorded.

The electroencephalograph is not very useful in determining what, specifically, is going on in the brain, but it is useful in detecting abnormal electrical discharges and other gross changes in brain activity, such as the differences between normal activity and epilepsy, and even between wakefulness and sleep. Figure 34.14 compares the EEG of a normal person with that of a person with epilepsy. As the

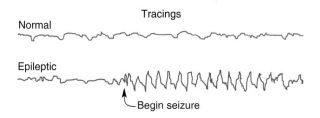

Tracings

Normal

Epileptic

Begin seizure

FIGURE 34.14
The EEG
The normal electrical activity of the brain is seen, where several EEG tracings from different electrode locations have been made. In the epileptic, normal electrical activity in the brain may occur between seizures, but when the episodes do occur, the electrical disturbances are quite obvious.

recording shows, epileptics are subject to sudden, random bursts of electrical activity. In a few instances, this abnormality can be traced to apparent physical defects, such as scars or lesions in the brain tissue, but most forms of epilepsy are simply not understood.

Sleep. Electrical activity in the brain does not cease with sleep. EEGs reveal that sleep is accompanied by a considerable amount of brain activity. Sleep has several distinct phases (Figure 34.15). By far the most interesting of these is called **REM sleep** (rapid eye movement sleep). During REM sleep the other skeletal muscles are in a highly relaxed state, but the eyes dart about beneath the eyelids. The EEG recording at this time is more similar to that produced by wakefulness. If a person is awakened at this time, he or she will report vivid dreams (that will ordinarily be forgotten if the person is allowed to awaken normally). REM sleep appears to be essential, although no one really knows why, just as no one really knows why dreaming might be useful. Experimental studies reveal that people deprived of REM sleep wake up tired, and in subsequent sleep periods they will extend the REM period (contrary to myth, you *can* make up for lost sleep). If the deprivation is continued, anxiety sets in and concentration becomes difficult. Eventually, personality disorders arise.

Psychologists and biologists generally agree that, for whatever reason, sleep appears to be highly essential and is somehow restoring. Theorists suspect that the sleep period,

FIGURE 34.15
The EEG in Sleep
By studying the electrical activity of the brain and episodic eye movement, researchers have identified five sleep stages: (1) drowsiness, (2) light sleep, (3) intermediate sleep, (4) deep sleep, and (5) REM sleep. Each has its own pattern of electrical activity as seen in the EEG. Typically, sleep is interrupted by a burst of electrical activity whose patterns resemble wakefulness. Since they are associated with rapid eye movement, they are called REM periods or REM sleep. The graph is derived from a number of night-long recordings. It shows that several periods of REM sleep occur in a typical night, each followed by resumption of other stages.

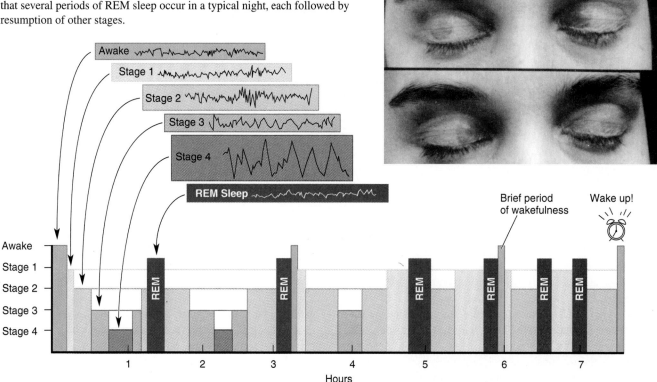

and REM sleep in particular, is a period of information sorting and storage and may be essential to long-term memory.

The Peripheral Nervous System

The peripheral nervous system is a vast network of nerves that spreads from the central nervous system to all parts of the body. It contains motor neurons and sensory neurons, often running side by side within the same nerves. The peripheral nervous system is further divided into the somatic nervous system and the autonomic nervous system (see Figure 34.5). There are motor neurons in both systems, but only the somatic system has sensory neurons. The **somatic nervous system** carries the impulses of which our conscious minds are most aware: sensations that inform us of events in the world around us and the commands to our voluntary muscles to act. The **autonomic nervous system** is more concerned with our internal workings and with our less conscious, less voluntary reactions.

The Autonomic Nervous System

The autonomic nervous system (ANS) is essentially a *motor* system, carrying impulses from the brain and spinal cord to the organs it serves. These include not only the prominent contents of the thoracic and abdominal cavities (the heart, lungs, digestive organs, kidneys, and bladder), but also the smooth muscle of the blood vessels, the iris, nasal lining, sweat glands, salivary glands, and even the tiny muscles that cause our hairs to stand on end when we see who's been elected.

The ANS is critical to homeostasis. Working under the direction of the central nervous system, especially the medulla and hypothalamus, the ANS coordinates the adjustments needed to maintain an overall stability or constancy in the face of changing conditions. The efficient function of such a system requires a considerable amount of sensory feedback from the organs served. This is carried out by sensory nerves of the somatic nervous system, which, by keeping the brain informed, also play an important role in the process of homeostasis.

The two divisions of the autonomic nervous system, the sympathetic and parasympathetic, have opposite effects and operate in a highly coordinated manner to produce an overall adaptive effect (Figure 34.16). One familiar example is seen in heart rate. The human heart, without outside influence, contracts about 70–80 times per minute but will speed up when stimulated by a sympathetic nerve and slow down on a signal from a parasympathetic nerve. How can neural impulses from two different nerves have opposing effects on heart rate? The sympathetic neurons secrete the neurotransmitter norepinephrine at their target organs, while neurons of the parasympathetic system secrete acetylcholine. Norepinephrine accelerates heart rate, while acetylcholine slows it down.

An interesting example of how the two ANS divisions effect such changes in the viscera and muscles is seen in the "fight or flight" response we will discuss later (see Chapter 36).

This is a series of responses to an emergency, instigated by the sympathetic nervous system and the endocrine system. After the emergency has passed, the parasympathetic division takes over, and conditions return to normal. The autonomic nervous system has a variety of effects on other organs as well (Table 34.1).

As you might expect, the divisions of the autonomic nervous system differ anatomically as well as physiologically. As seen in Figure 34.16, most of the nerves that make up the parasympathetic division originate in the brain and at the lower region of the spinal cord. Some go directly to their target organs; others end in external ganglia, where they relay impulses to nerves continuing toward the target organ. The sympathetic neurons all originate in the gray matter of the spinal cord. Furthermore, all the sympathetic nerves leaving the cord pass through or synapse within rows of **sympathetic ganglia** just outside the cord. Some sympathetic nerves synapse once again in ganglia at various locations in the body. One of these, the *solar plexus,* is often the target of a blow from a pugilistic opponent, bringing about a temporary paralysis of the diaphragm and a momentary adjustment of attitude.

The Senses

Here's a question: What is our world like? Of course you have an answer. We can indeed describe the world around us. Another question is: How do we know? We know because our senses tell us. Notice, though, that we're talking about "our" world—the world of the human experience—and our senses. Other species sharing the planet with us might describe an entirely different world based on what *their* senses tell *them.* A fly might describe a delicious array of swirling eddies on each surface, and a kaleidoscope of images. A dog might tell

TABLE 34.1 **Responses to Sympathetic and Parasympathetic Nerves**		
Organ	**Sympathetic Effect**	**Parasympathetic Effect**
Pupil of eye	Dilates	Constricts
Heart	Accelerates	Slows
Intestine	Decreases movement	Increases movement
Salivary glands	Decreases secretion	Increases secretion
Stomach glands	Decreases secretion	Increases secretion
Lungs	Dilates air passages	Increases secretion
Blood vessels		
Respiratory system	Dilates	Constricts
Heart	Dilates	Constricts
Skin	Constricts	Dilates
Stomach	Constricts	Dilates
Metabolism	Increases	Decreases

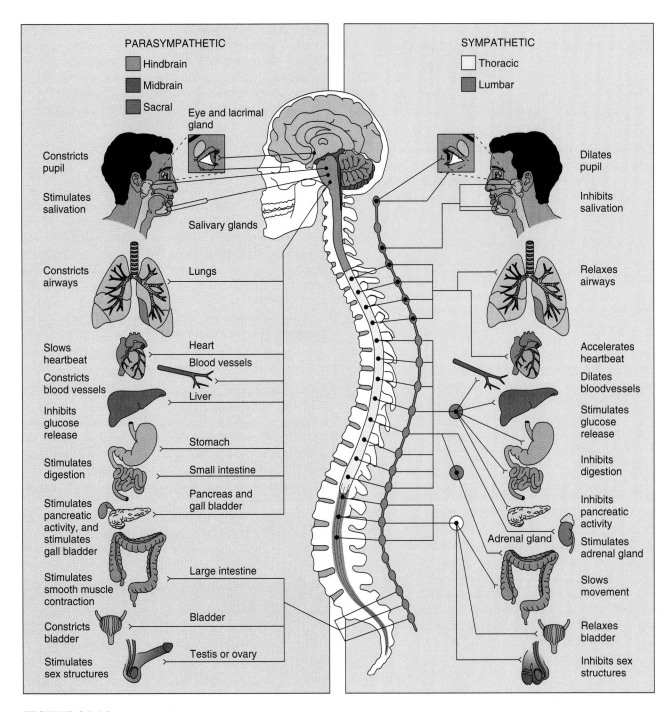

FIGURE 34.16
Autonomic Nervous System

The organization of the autonomic nervous system, showing the various organs it serves. The autonomic system is essentially a motor system, so its nerves transmit impulses from the brain and spinal cord to the organs served. Its principal role is regulatory—keeping the internal organs operating in a carefully coordinated manner in response to constantly shifting conditions and needs.

about a world of powerful odors and high-pitched sounds, while never once mentioning color. An earthworm might describe a gritty, dark planet with pockets of comforting dampness. And all these species might live in your colorful, quiet yard. The point is, species can sense what they need to

sense according to the life they live, us included. Our senses tell us only what we need to know about the world around us. So keep in mind that any species can sample only a part of what's around it and that this planet is a far richer place than we can ever know.

Sensory Receptors

Stimuli from the environment are detected by **sensory receptors**. These can be quite simple, with little more than a specialized dendritic region, as with free neural endings that lie scattered beneath the surface of the skin. Other sensory receptors, such as the vertebrate eye, can be incredibly complex, where a great many individual receptors along with other structures are organized into **sensory organs**.

Sensory receptors are *transducers*—that is, they convert the varied stimuli they receive into electrical signals. Receptors, like neurons, undergo depolarization and generate what are called **receptor potentials**. Receptor potentials, in turn, synapse with sensory neurons in which they stimulate action potentials. Unlike the "all-or-none" ("go" or "don't go") action potential, receptor potentials vary in intensity. Increases in receptor potential intensity become translated into a higher frequency of action potentials in the sensory neuron, where the rate of firing has information value.

It is through such "graded" responses that we are able to make very fine distinctions in light intensity and sound volume, and to discern between delicate aromas and *eau de goat*. We will briefly consider how such distinctions are made. It's an interesting problem because, as we've seen, neurons only transmit action potentials along their axons, and all action potentials are the same. So how do we make the distinctions? The answer lies in neural coding.

Neural Codes

Action potentials, instigated by sensory receptors and transmitted in sensory neurons, must be deciphered if the animal is to get a clear reading of its surroundings. The sorting, identifying, interpreting, and integrating are done primarily in the brain. As we've seen, one way the brain can decode incoming impulses is through their *frequency* (the number of impulses per second). A mild stimulus, such as light touch, might produce just a few impulses per second and might even be ignored by the brain. A much stronger stimulus, such as hitting your finger with a hammer, produces a virtual barrage of impulses in a very short time and registers as pain. The intensity of a stimulus can also be perceived according to the *number* of neurons being stimulated. The hammer stimulated many sensory neurons at once, making the stimulus difficult to ignore.

Decoding, however, involves more than just distinguishing between degrees. It involves making distinctions in "kind" as well. Distinguishing specific kinds of incoming information—touch, light, taste, sound, temperature, pain—depends on how the sensory neurons and the integrating interneurons are connected or "wired." Partly through learning and experience, regions of the brain specialize in interpreting sensory information, and each sensory neuron connects via specific tracts to its corresponding brain region. As we've seen, some centers of the brain deal only with sound and when they are stimulated, "hearing" results. Interestingly, stimulating such brain centers directly with electrodes will also be interpreted as sound. Finer levels of distinction and integration also exist; thus most of us can distinguish a reasonably accomplished tenor from an alley cat hitting the same note.

Tactile Reception

Tactile reception, or the sense of touch, is due to extremely sensitive, fast-firing neurons that respond to any force that deforms or alters the shape of their plasma membrane. In some cases, nonliving bristles or hair-like structures protruding from the surface of the plasma membrane touch objects in the environment before the organism's body actually reaches the objects. Such structures act as "early warning systems" that, when activated, permit the animal to make appropriate, timely responses such as in avoiding danger or finding food.

Touch in Invertebrates. A number of arthropods, from caterpillars to spiders, are fuzzy. One adaptive advantage of their covering is related to an increased sensitivity to touch, since the bristles extending from the body are connected to tactile receptors.

Tactile sensitivity may be important for a variety of reasons. For example, web-building spiders have hairy legs that respond to vibrations set up by prey caught in the web (Figure 34.17). Cockroaches have tiny hairs protruding from their abdomen that, when stimulated by very light air currents, send them scurrying for safety. Many aquatic insects have bristles on their heads that are sensitive to water currents. The famed biologist R. S. Stimson Wilcox has accumulated a fascinating body of evidence that indicates how certain aquatic

FIGURE 34.17
Touch Receptors in an Invertebrate
Touch receptors in spiders are often associated with hairs. When a hair is bent or moved, it activates a sensory neuron that transmits its impulse to an associated ganglion for integration. Some sensory hairs are fine enough to be moved by light air currents.

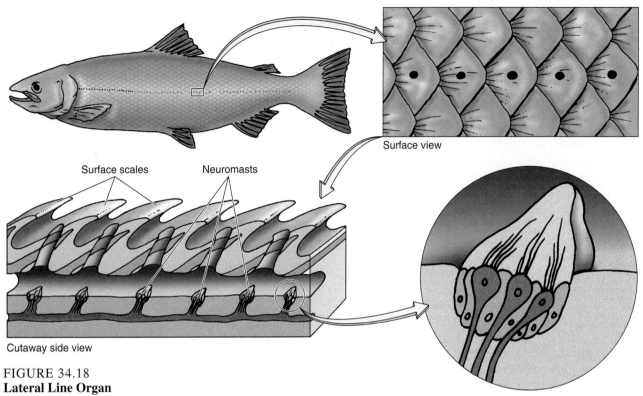

Surface view

Surface scales   Neuromasts

Cutaway side view

FIGURE 34.18
Lateral Line Organ
The canal-like lateral line organs of bony fishes and sharks can detect water movement. Such disturbances are picked up by receptors known as neuromasts, clusters of sensory hairs embedded in a gelatinous mass. Movement of water in the canal bends the neuromast, moving the hairs, which stimulates associated neurons to fire and sends impulses to the brain.

insects use water vibration in communicating with each other. Touch sensitivity, in general, is common among a wide range of invertebrates and is usually employed in finding food, mating, and avoiding predators.

Touch in Vertebrates. Vertebrates have two kinds of tactile receptors. *Distance receptors* detect signals from a distance, as when a movement sets up water motion, which is then detected some distance away. An example of distance receptors is the lateral line organs of fishes (Figure 34.18), which are sensitive to water motion. *Contact receptors,* on the other hand, respond to direct touch by the source of the stimulus. Humans and other mammals have tactile receptors that respond to both pressure and touch. The pressure receptors are located deep in the skin and consist of encapsulated nerve endings called **Pacinian corpuscles**. Light touch is detected by **Meissner's corpuscles**, which lie near the surface of the skin (Figure 34.19). Among primates, sensitivity to touch is greatest around the lips, nipples, eyes, and fingertips. In addition to touch receptors, the skin abounds with simple, unspecialized, free nerve endings that register pain. A special kind of tactile receptor found in the arteries, the **baroreceptor**, detects changes in blood pressure within the vessels.

In humans, as in most mammals, touch sensitivity is particularly great around hairy areas. As you may have noted, this includes the hairline around the face, and the genitals. The "whiskers" of many animals are especially sensitive to touch. While the scant body hairs of humans are of limited value in keeping us warm, they are good sense organs: try moving a body hair without feeling anything.

Until 1982 it was generally believed that hairs, in their sensory function, were simply dead, mechanical levers that when touched would jostle the sensory nerves surrounding their roots. Then another theory was developed based on information that the hair protein keratin is so highly structured as to be essentially crystalline. In fact, it was found that each hair comprises a single *piezoelectric crystal.* A piezoelectric crystal discharges electricity when it is deformed. Researchers found that, like other transducers, a perfectly dead hair generates a small electrical discharge when it is bent, and that nerve endings respond to this electricity.

Incidentally, the outer layer of skin is also made primarily of keratin, and it too has a piezoelectric effect. Bending or depressing the epidermis generates detectable electric discharges that are picked up by the touch-sensitive free nerve endings.

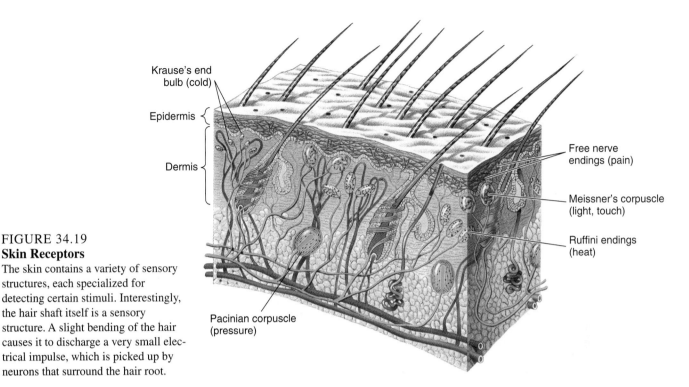

Krause's end
bulb (cold)

Epidermis

Dermis

Free nerve
endings (pain)

Meissner's corpuscle
(light, touch)

Ruffini endings
(heat)

Pacinian corpuscle
(pressure)

FIGURE 34.19
Skin Receptors
The skin contains a variety of sensory structures, each specialized for detecting certain stimuli. Interestingly, the hair shaft itself is a sensory structure. A slight bending of the hair causes it to discharge a very small electrical impulse, which is picked up by neurons that surround the hair root.

Thermoreception

Thermoreception, the detection of temperature, is important in a wide range of animals. Many of the processes of life are possible only within a specific and often narrow range of temperatures, and animals must be able to sense the temperature of their environment in order to place themselves under the most nearly optimal conditions.

Thermoreception in Invertebrates. While thermoreception may be important to many invertebrates, little is known about its mechanisms. Heat detection is particularly important to ectoparasites of birds and mammals. Thus leeches, fleas, mosquitoes, ticks, and lice must find a warm-blooded host. Their thermoreceptors are generally located on the antennae, legs, or mouthparts.

Thermoreception in Vertebrates. Many vertebrates have specific temperature receptors. As an example, the pits of a pit viper, such as a rattlesnake, are a pair of indentations between the eyes and nostrils. These are loaded with heat (infrared) receptors that tell the snake when it is facing a living thing that is generating metabolic heat (Figure 34.20).

There is some disagreement over whether humans and other mammals have specific receptors for detecting heat and cold, or whether free nerve endings register these stimuli. It has even been suggested that a single neuron can register both heat and cold, simply by firing in different patterns. However, some physiologists believe heat and cold receptors are distinct and that heat receptors are located deeper in the skin than cold receptors. Furthermore, whether the brain registers "warm" or "cold" (within limits) depends on the immediate previous

experience of the receptors. If the skin has been warm and touches something less warm, the object will seem cool, but if the skin has been cold and touches something cool, the second sensation will be exaggerated warmth. You can demonstrate this perception for yourself by switching your hands from ice water to moderately cold water and vice versa.

FIGURE 34.20
Heat Sensors in the Pit Viper
Heat sensors, used for detecting prey by a pit viper, are located in depressions near the eyes. Each pit consists of an outer chamber that ends in a thin, highly innervated membrane covering an inner chamber below. By moving its head back and forth, the pit viper is able to use incoming thermal cues from the two pits to zero in on its warm-bodied prey.

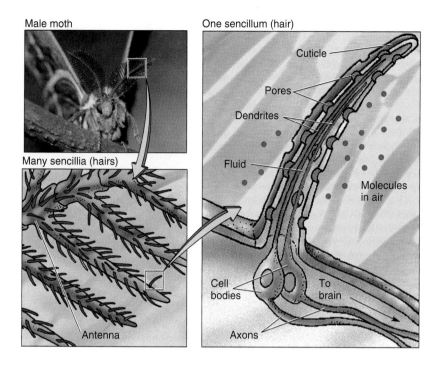

Male moth

Many sencillia (hairs)

Antenna

One sencillum (hair)

Cuticle

Pores

Dendrites

Fluid

Molecules in air

Cell bodies

To brain

Axons

FIGURE 34.21
Chemoreceptors in the Moth
The antennae of the male moth are remarkably large, complex, and sensitive. Each large bristle of an antenna has numerous hair-like sensilla extending outward. Within each sensillum is a fluid-filled cavity containing sensory neurons. Airborne molecules stimulate dendritic endings, producing impulses that are sent to the brain.

Other physiologists ascribe thermoreception to the free nerve endings and to two specialized skin receptors. The first, called **Krause's corpuscles**, consist of a branched nerve fiber in a capsule of connective tissue. They are sensitive to cold. The second, called **Ruffini's corpuscles**, lie deeper in the skin and are more sparsely distributed. These are also formed from free nerve endings but are embedded in a granular material. They are sensitive to warmth (see Figure 34.19).

Chemoreception

Chemoreception is the ability to perceive specific molecules. These molecules are often important cues to the presence of specific entities in the environment. Nearly all animals and a great number of protists exhibit chemoreception. It is essential to many animals in finding food, locating a mate, and avoiding danger.

Chemoreception in Invertebrates. Planarians locate food by following chemical gradients in their aquatic surroundings. Their simple **chemoreceptors** are located in pits on their bodies, over which they move water with their beating cilia. Certain insects have taste and smell receptors of astounding sensitivity. Among insects, chemoreceptors may be found in the body surface, mouthparts, antennae, forelegs, and, in some cases, the ovipositor (since many insects lay their eggs only on certain plants).

Among the incredible stories of insect olfactory (sense of smell) abilities we find the tale of certain moths. The moth smells with its antennae, and adult males have enormous antennae with thousands of sensory hairs called sensilla (singular, sensillum) (Figure 34.21). About 70% of the adult male receptors respond to only one complex molecule called *bom-*

bykol, a sex attractant released by females. As the molecules drift downwind finally to touch the antennae of a wandering male, they enter the tiny pores of a "hair," which houses his olfactory receptors. The molecules then dissolve in the fluid that moistens the receptor and interact with its membrane. Upon perception of the molecule, the male becomes excited, reorients his flight, and heads upwind, a behavior that sooner or later should lead him to the waiting siren.

Chemoreception in Vertebrates. Vertebrates detect chemicals in a number of ways, employing three major means: general receptors and two types of specialized receptors, **gustatory** (taste) and **olfactory** (smell) **receptors.** For example, many aquatic vertebrates have generalized chemical receptors scattered over the body surface, while in the terrestrial vertebrates, olfaction and gustation are usually accomplished by moving chemical-laden water or air into a special canal or sac that contains the chemoreceptors.

In land vertebrates, the keenest olfactory senses are found among the mammals, especially the carnivores and rodents. Good olfaction is essential to carnivores; wolves and lions, for example, often locate their prey by smell. It is less clear why rodents have evolved such strong olfactory ability, since their plant food sources are not usually widely dispersed and certainly do not run away. However, their sense of smell may be associated less with food-finding than with their complex social organization. Rats, for instance, may use their sense of smell to discriminate among families and the individuals within them.

Except for primates, vertebrates have olfactory receptors called *Jacobson's organs* in the roofs of their mouths. In amphibians, Jacobson's organ simply gives the animal a better notion of what it's eating, but snakes and lizards actually smell

FIGURE 34.22
Olfactory Receptors in Humans
The olfactory receptors in the human nose synapse with neurons in the olfactory bulb of the brian. These receptors are able to distinguish a wider variety of stimuli than are the taste buds. The olfactory receptors are part of the nasal epithelium dispersed with other cells. Each receptor has numerous olfactory "hairs," actually modified cilia, that protrude from the epithelium.

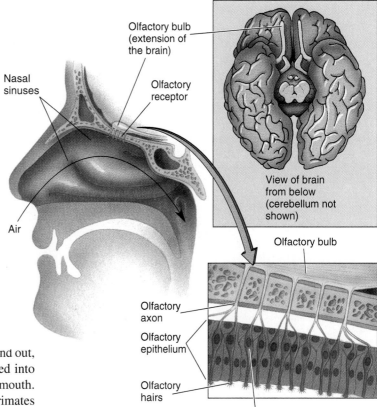

with the structure. As they flick their forked tongues in and out, the moist tips capture scent molecules that are delivered into the twin openings of their Jacobson's organ inside the mouth.

Compared to many other mammals, olfaction in primates is rather poor. For example, chimpanzees smell about like you do (no offense). The olfactory neurons of primates, like those of other mammals, are located in **olfactory epithelium**, where molecules in the air encounter a moist surface (Figure 34.22). Scattered among the epithelial cells are specialized olfactory receptors that synapse with neurons of the **olfactory bulb**, an extension of the brain located just above the bony roof of the nasal cavity.

In all land-dwelling vertebrates, taste receptors, or **taste buds**, are confined to the mouth area (Figure 34.23). Humans experience four basic tastes: sweet, sour, salty, and bitter. Humans are omnivorous; that is, we'll eat almost anything, plant or animal. So it has undoubtedly been important that through our long evolutionary history, we have become able to distinguish among a variety of tastes and smells. To illustrate, many alkaloid poisons taste bitter, and unless we deliberately train ourselves otherwise, we tend instinctively to avoid them. Sweet, on the other hand, is the taste of carbohydrates. Many kinds of fruit taste sweet, especially when they are ripe and have their highest nutritional value. Sour is not one of our favorite tastes, being the taste of acid and of unripened fruit that will benefit us more if we wait until it ripens. Gustation then is not only a matter of taste but of survival.

Proprioception

Proprioception is the ability to determine the position of the body, or the position of one part of the body relative to another. It is made possible by sensory receptors in the muscles, joints, and tendons.

Proprioceptors in Invertebrates. It is particularly important for animals with many body parts, such as arthropods, to be able to coordinate those parts—to know what each is doing. Sensors in arthropods may be located peripherally, for example, at the base of "hairs," or deep within the muscles. Cockroaches, which are running animals, have proprioceptive hairs on the side of their legs. These respond to flexion of the knees, so they are aware of whether a leg is extended or flexed. Other arthropods—the crabs and lobsters—have proprioceptors located within the muscles themselves.

Proprioceptors in Vertebrates. Some proprioceptors in vertebrates, as well as in some invertebrates, take the form of stretch receptors (mentioned earlier in our discussion of the reflex arc). These respond to the stretching of skeletal muscle and tendons that is produced as limbs are extended or flexed. You may not be able to see your foot under the table, but you could probably point to it because your proprioceptors tell your brain where it is. Proprioception is well developed in mammals such as the tree-dwelling primates, who must move with incredible agility through their forest canopy.

Auditory Reception

Hearing, or **auditory reception**, occurs when a distant stimulus is transmitted to the receptors via a medium (air or water). In fact, evolutionary theory holds that the structure of balance

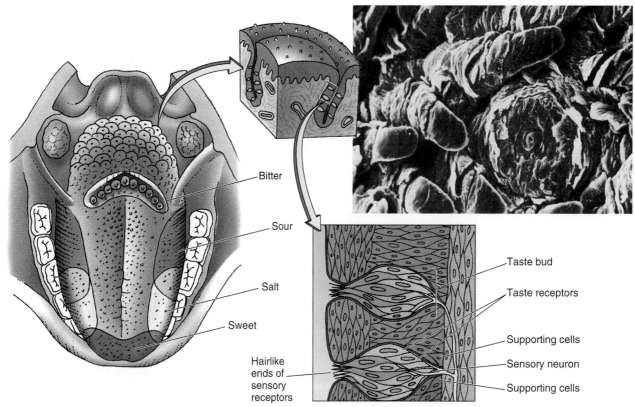

FIGURE 34.23
Taste Buds
Taste buds on the tongue respond most strongly to one of four flavors: bitter, salty, sweet, and sour. The taste buds are located in recessed pits. The photograph, taken with a scanning electron microscope, reveals the flattened columns that contain the taste buds.

and hearing in the vertebrates evolved from increasingly complex lateral line organs.

Hearing in Invertebrates. Most invertebrates don't have specialized sound receptors, but many are sensitive to vibrations in the air, water, or soil in which they live. A notable exception among the invertebrates is the insect, which boasts several types of sound receptors. For example, most species have sensory hairs that respond to low-frequency vibrations of air, but others have a specialized organ, located in a leg, that is sensitive to movements of whatever the insect is standing on. Yet others, including grasshoppers and crickets, have **tympanal organs** on the thorax that respond to high-frequency vibrations, much like the human eardrum.

Hearing in Vertebrates. The sound receptors in most species of vertebrates are located in the **inner ear**; however, vibrations reach the inner ear in a variety of ways. In many fishes, sound vibrations in water are conducted directly to the inner ear by vibrating water, but in others (such as the minnows, catfishes, and suckers), a set of tiny bones connect the swim bladder to the inner ear. Sound vibrates the air-filled bladder, which moves the bones and stimulates the receptors in the

inner ear. The fishes that have such an apparatus can detect a much wider frequency of sound than those that lack it.

In humans and other mammals, the auditory apparatus consists of the **outer, middle,** and **inner ear** (Figure 34.24a). The outer ear includes the **pinna** ("the ear"), the **auditory canal,** and the **tympanum,** or eardrum. Most mammals are able to move the pinna so as to maximize sound input and locate its source. (Humans have largely lost this ability, and those who can wiggle their ears are often in great demand for social events.)

The three bones of the middle ear are the **malleus, incus,** and **stapes** (which translate to *hammer, anvil,* and *stirrup,* respectively). Acting as a jointed lever, they transfer vibrations of the eardrum to a thin membrane in the cochlea of the inner ear.

The inner ear of mammals consists of the **cochlea** and the **vestibular apparatus.** The cochlea is a lengthy fluid-filled tube, doubled back on itself and then coiled in the manner of a snail shell. It is hard to visualize how the structure works, but if we can imagine it to be straight—as it is, in fact, in birds—we see what is actually a U-shaped tube containing the sensory neurons involved in hearing. One end of the tube holds the **oval window,** to which the stapes is affixed, while the other end contains the highly flexible **round window.**

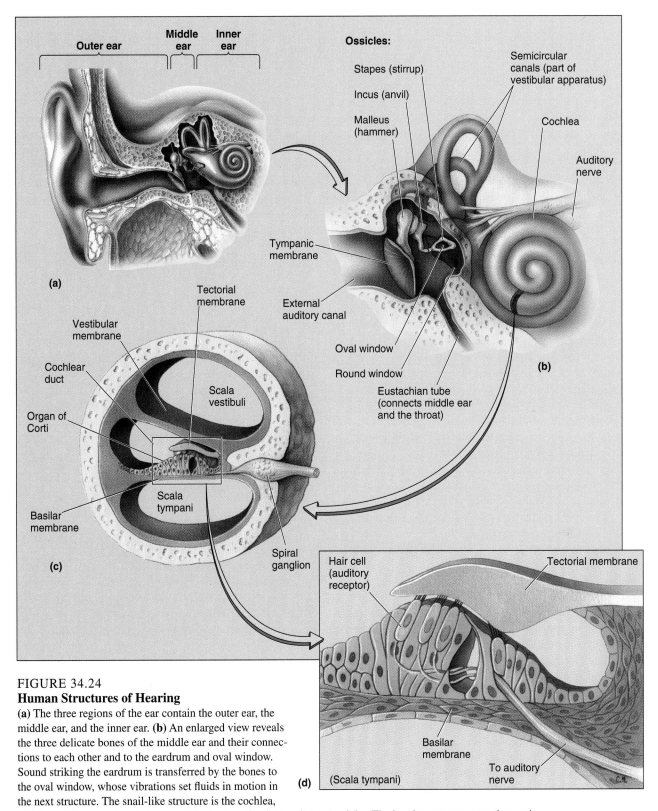

FIGURE 34.24
Human Structures of Hearing
(a) The three regions of the ear contain the outer ear, the
middle ear, and the inner ear. (b) An enlarged view reveals
the three delicate bones of the middle ear and their connections to each other and to the eardrum and oval window.
Sound striking the eardrum is transferred by the bones to
the oval window, whose vibrations set fluids in motion in
the next structure. The snail-like structure is the cochlea,
which contains receptors that convert the fluid motion into action potentials. (The looping structures are the semicircular canals, which are part of the vestibular apparatus.) (c) A cross section through the cochlea reveals three
compartments, the central one of which contains the organ of Corti, where the auditory receptors are located.
(d) One more enlargement shows how numerous hair cells in the organ of Corti contact the tectorial membrane.
It is the fluid-driven movement of the tectorial membrane that stimulates the hair cells. They synapse with sensory neurons that carry action potentials along nerve pathways to the brain, which decides what has been heard.

As sound waves strike the eardrum, the resulting vibrations are transferred by the three middle ear bones to the oval window, which, in response, vibrates rapidly. This creates a wave movement in the fluid of the outer tube and a compensating bulging of the round window at the far end of the tube. (So the round window serves to dissipate the sound energy.) As fluid pulsates within the tube, it activates what is called the **organ of Corti**. As you see in Figure 34.24c,d the organ of Corti contains a **basilar membrane**, which bears sensory neurons called **hair cells** (modified cilia). The hair-like tips of these cells are embedded in the gelatinous **tectorial membrane**. As the basilar membrane moves, the sensory hairs are bent, creating receptor potentials, which, in turn, activate neurons leading to the cochlear nerve. This nerve leads to the midbrain, where the neurons synapse with other neurons that carry impulses to be processed in the hearing centers of the cortex.

Sounds, we know, vary in intensity and pitch. The perception of intensity, or loudness, seems to depend on the *number* of auditory neurons that fire, as well as the *frequency* of their firing. Differences in pitch (highness and lowness) depend on which auditory neurons are stimulated. The basilar membrane is narrow at the broad base of the cochlea and wide at its apex end. Hair cells at its thin, more rigid beginning respond better to higher frequencies, which are interpreted by the brain as higher pitched sounds. Hair cells in the wider, more flexible apex of the snail-shaped chamber respond better to lower frequencies, those that translate into lower-pitched sounds. As we have described, sound waves travel up one side of the U-shaped tube and down the other. Each particular sound frequency traveling up one side of the basilar membrane will be exactly in phase with the same frequency traveling down the other side in only one region of the membrane. The in-phase resonance (or reinforced vibration) at that region vibrates the basilar membrane, producing the sensation of pitch.

Gravity and Movement Reception

Gravity receptors reveal the direction of gravity. (Some might say they tell us which way is "down.") They are important in determining body position and maintaining balance, two obviously related functions. Most invertebrates detect such stimuli through organs called **statocysts**. In some mollusks and in the crayfish, the statocyst is a chamber lined with sensory hairs on which lie a number of fine sand grains. As the body changes position, the sand grains move against the hairs, setting off impulses. The interpretation of these impulses enables the animal to determine its position with respect to gravity.

In humans and other mammals, body movements, position, and balance are detected in the inner ear by the **vestibular apparatus** (Figure 34.25a). It is composed of the **semicircular canals**, their **ampullae**, the **saccule**, and the **utricle**. These three structures are closely associated with those of hearing. Each semicircular canal lies in a different plane, at right angles to each of the other two. This arrangement permits the sensing of movement—acceleration or deceleration—in any direction.

Each canal is filled with fluid, and its movement jostles sensory hairs located in the three ampullae (Figure 34.25b). As the fluids move, they bend the hairs, creating receptor potentials, which, in turn, activate sensory neurons that send impulses to the brain.

The saccule and utricle contain sensory hairs coated with fine granules of calcium carbonate (Figure 34.25c). As in the crayfish statocyst, shifts in these granules pressing on the sensory hairs change the rate of neural impulses, providing information about the position of the head with respect to gravity. Some impulses travel to the spinal cord, where body position can be adjusted by reflex action; others are sent to the cerebellum, where other reflexive muscular coordination is orchestrated; and yet others move on to higher centers involved with the control of eye movement. Input from the eyes is important in maintaining balance. (Close your eyes and stand on one leg as you count slowly to 30. Do you see why some people object to certain field "sobriety" tests by law officers?)

Visual Reception

Visual reception involves sensitivity to light waves. However, light receptors are sensitive only to a particular part of the spectrum of electromagnetic energy—the part we call visible light, although some other animals can detect ultraviolet light. Visible light wavelengths range from about 430 to 750 nm. As far as is known, considerably shorter wavelengths, such as x-rays and gamma rays, can't be detected by any animal. Neither can the very long ones, such as radio waves, although many animals can detect infrared (heat).

Visual Reception in Invertebrates. The planarian flatworm (Chapter 26) has two eyespots that are shaded on opposite sides. It can therefore tell which direction light is coming from according to which eye is being stimulated. Such ability is, of course, important to these bottom-dwelling and dark-seeing animals.

Among invertebrates, the cephalopod mollusks—the octopus, squid, and others—are unique in that they have image-forming eyes, quite like those of vertebrates (see Figure 26.29). Arthropods, such as spiders, crayfish, and insects, have exceptionally good vision, especially in detecting movement in their visual fields. Spiders have eight eyes (two rows of four each), but in most species the eyes lack enough photoreceptors to form clear images. However, the jumping spiders (Salticidae) have a relatively large number of photoreceptors. These are undoubtedly beneficial; the spiders leap from a distance onto their prey; it wouldn't do for a weak-eyed spider to leap onto a hungry bird.

Crustaceans, such as crayfish, have two kinds of eyes: **simple eyes** and **compound eyes**. The simple eyes, characteristic of larval stages, lack lenses and detect only the presence of light. The compound eyes have lenses with which to gather light and focus images. In many species, the adult's sensitive compound eye rests on movable stalks. The term "compound" refers to the numerous visual units known as **ommatidia** (see

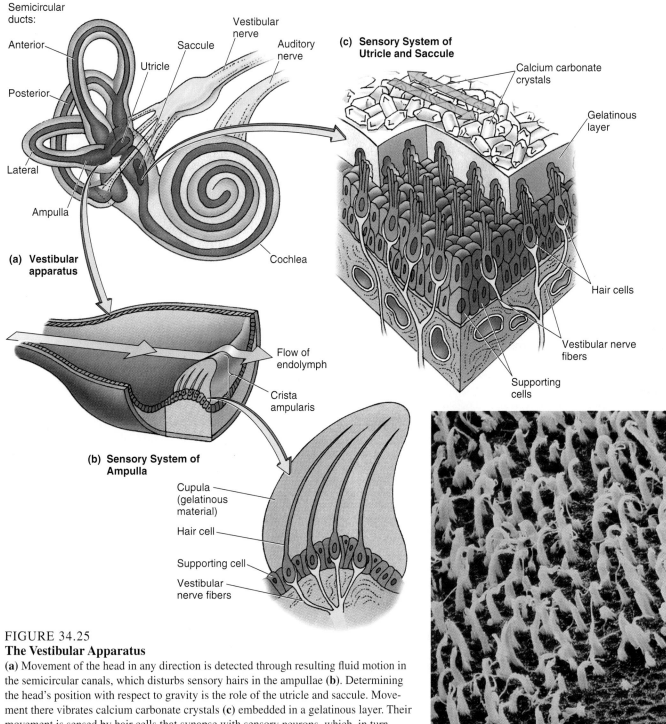

FIGURE 34.25
The Vestibular Apparatus
(a) Movement of the head in any direction is detected through resulting fluid motion in the semicircular canals, which disturbs sensory hairs in the ampullae (b). Determining the head's position with respect to gravity is the role of the utricle and saccule. Movement there vibrates calcium carbonate crystals (c) embedded in a gelatinous layer. Their movement is sensed by hair cells that synapse with sensory neurons, which, in turn, send action potentials to the brain for interpretation.

Figure 25.36). The ommatidia can't move to follow an image, so each stares blankly in its own direction until something moves into its visual field. Then it fires a signal to the central nervous system. Any movement across the animal's field of vision stimulates a series of ommatidia in turn, so that even very slight movements are detected. The convex surface of such eyes gives a visual field of about 180°.

Visual Reception in Vertebrates. We can consider vision in humans as representative of vertebrates. We are indeed a highly visual species. Humans, after all, rely on vision to help them avoid danger, coordinate movement, maintain equilibrium, read the faces of other humans, avoid stepping in unpleasant things, and enjoy a range of pleasure and creativity. While much is known about human eyes and their operation, how the brain handles all this information is an enormously complex process, and many questions remain.

We will begin with a brief discussion of the anatomical aspects of vision, where things tend to be straightforward. The human eyeball (Figure 34.26a) is roughly spherical, but protruding somewhat in front. It is surrounded by three layers. The outermost is a tough, white layer, the **sclera**, which gives way at the anterior bulge to the transparent **cornea**. The sclera maintains the shape of the eyeball and serves as an attachment for the voluntary muscles that move the eyes. Because of its curvature, the cornea aids in focusing light. The cornea has no blood supply and must receive oxygen by diffusion from outside the eye. This is why contact lenses must be porous if they are to be left in place over a long time. Incidentally, the absence of a blood supply means that immune cells are also absent, and thus corneal transplants are not as easily rejected by the body's immune defenses as are organs with a blood supply such as the heart or liver.

Just within the sclera is the **choroid layer**, containing numerous blood vessels. At its anterior portion the choroid supports the **iris** of the eye, a thin diaphragm of smooth muscle. At the center of the iris lies the **pupil**, an opening whose size the iris regulates. The ring-like iris contains pigments we recognize as eye color. Behind the iris is the transparent **lens**, held in place by another ring of muscle, and then a large space filled with transparent gel. Finally, there is an innermost layer of tissue at the back of the eye—the **retina**, a complex region of light-sensitive and supporting neurons. The neural fibers form the **optic nerve**, a common pathway out of the eye. Let's now see how each of these structures works.

The Path of Light. Light entering the eye passes through the transparent cornea, and then through the **anterior chamber**, filled with a fluid called the **aqueous humor**, and through the pupil of the iris. The iris adjusts the pupil size according to the brightness of entering light—like the light-controlling diaphragm of a microscope or camera, but far more delicate and responsive. It does this through two very fine muscle layers, one circular and the other longitudinal. The first constricts the pupil while the second dilates it; each is under the control of a different part of the autonomic nervous system. From the pupil, light traverses the lens and then passes through the **posterior chamber**, which is filled with a clear gel called the **vitreous humor**. At the back of this chamber the light falls on the retina.

The eye is often compared to a camera—the lens to the camera lens and the retina to camera film. However, the camera lens adjusts to close or distant objects by moving back and forth (interestingly, sharks focus this same way), but our lenses accommodate for distance by changing shape. This is done by a ring of precisely coordinated **ciliary muscles** and their delicate supporting ligaments, whose contraction forces the lens to assume a more rounded shape. As we grow older, the flexibility of the lens decreases, and many of us must wear reading glasses. Actually, the flexibility of the lens peaks early in life, at about age 10.

The Retina. The retina consists of four layers of cells (see Figure 34.26b). The deepest layer, attached to the inner surface of the choroid coat, is pigmented. It absorbs light that might otherwise reflect within the eye and create confusing neural signals. Overlying the pigmented tissue are the **rods** and **cones**, the actual light receptors. You might expect them to be in the direct path of incoming light, but they are covered by two more layers of rather transparent neurons. When the rods and cones are stimulated by light, they activate the overlying **bipolar cells**, which, in turn, synapse with the layer of **ganglion cells** just above. Ganglion cells gather from all parts of the retina to form the **optic nerve tract**, which carries impulses to the brain.

The slender rod cells (Figure 34.26c) far outnumber the cones, by about 18 to 1 (125 million rods to 7 million cones in each eye). Impulses from the rods are decoded by the brain into black, white, and gray images; colors are not distinguished. However, the rods make up for this limitation by being highly responsive to dim light, far more so than cones. In fact, the rods provide most of our night vision.

It may come as no surprise to learn that cones are cone-shaped (see Figure 34.26c). These cells are scattered over the retina but concentrated in a depression in the retina called the **fovea**. When concentrating on some detail we generally focus the light in such a way that the image falls on the fovea. We are able to see better this way because cones are capable of much finer discrimination of detail in bright light than are rods, although they hardly respond at all to dim light (which is why we can see light but not much color at night).

According to the prevailing theory of color vision, there are three types of cones: red-sensitive, green-sensitive, and blue-sensitive. However, we can also see many variations of color because the sensitivity ranges of these receptors overlap. The color sensitivities of the cones depend on what sorts of proteins are attached to the visual pigments (light absorbers).

The actual light-detecting regions in rods are the stacked disks that form the upper part of the cells. (Each rod may have up to a thousand such disks.) The disks contain **rhodopsin**, a photoactive pigment that breaks down in the presence of light into two colorless products: the protein **opsin** and a derivative of vitamin A called **retinal**. The chemical reaction itself sets events in motion that create neural signals in the bipolar and ganglionic cells above (see Figure 34.26).

Signal Transduction. Rods are only active in the absence of light. At that time their gated sodium channels are open, and they are in an ongoing state of depolarization. In this state they secrete an inhibitory neurotransmitter where they synapse with the bipolar cells above, thus inhibiting the bipolar cells from firing. When light is absorbed and rhodopsin is broken down, the sodium gates close, and the rods become hyperpolarized. The secretion of neurotransmitter slows considerably, and passive or graded receptor potentials begin in the bipolar cells. The bipolar cells synapse with ganglion cells, and any resulting action potentials pass through the optic nerve to the brain.

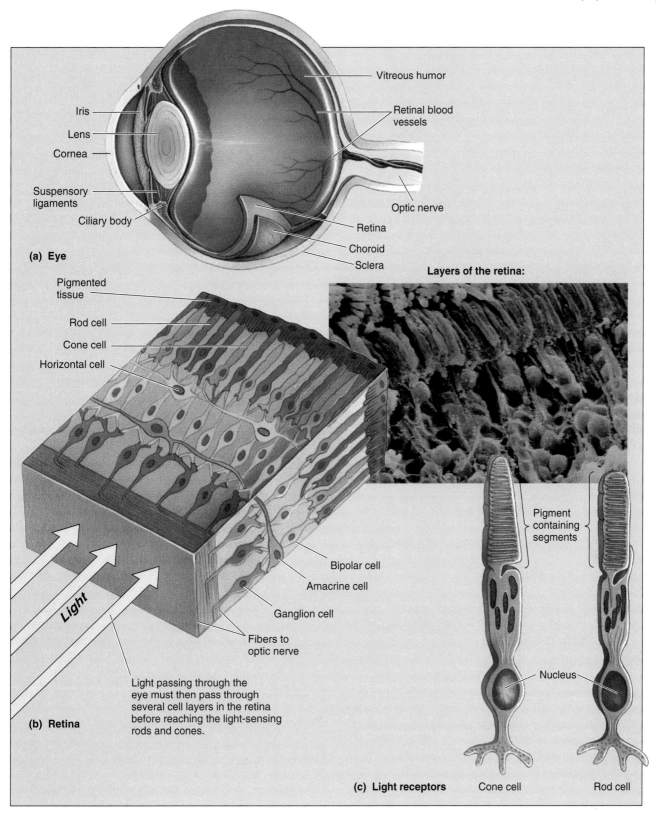

(a) Eye

Layers of the retina:

(b) Retina

Light passing through the eye must then pass through several cell layers in the retina before reaching the light-sensing rods and cones.

(c) Light receptors   Cone cell   Rod cell

FIGURE 34.26
The Human Eye
(a) The human eye is multilayered, with two major fluid-filled chambers separated by the lens. The amount of light entering the eye through the pupil is regulated by the iris, which is able to change the diameter of the pupil. The light is focused on the retina by changes in the shape of the pliable lens. **(b, c)** The retina contains two types of photoreceptors: rod cells and cone cells. Overlying the retina are the bipolar, amacrine, and ganglion cells, the latter forming the optic nerve.

Neurobiologists have yet to determine how such a negative system of operation might be adaptive, but they marvel over the acute sensitivity of rods. It is now believed that the rods are so sensitive because they amplify the signal. In fact, a single photon of light can produce a detectable electrical signal in the retina, and the human brain will actually respond to a cluster of five photons.

Opsin and retinal eventually enter into a chemical pathway in which they are recombined into rhodopsin. Apparently, dietary vitamin A must be supplied continually to keep the pathway going in the right direction. If the body is deprived of vitamin A, severe night blindness can result. If you go from a very brightly lit room into near total darkness, you may experience a temporary form of night blindness. Under intense light, most of the rhodopsin in the rods bleaches to the opsin/retinal form, and a certain period of time is required for the reconversion to rhodopsin to catch up. While you are waiting for the rods to function, the cones—which aren't very useful at night—at least help you peer dimly into the darkness.

We see then that it is the chemical, mechanical, and electrical interplay of specialized neural cells that enables us to sense some small part of the world around us and to respond appropriately to our situation. The success of the animal kingdom on this planet has been due, in large part, to the very precise sensitivity and reactivity of its members. We have seen that this highly organized and adaptive responsiveness is based on the peculiar irritability of a constellation of cells called neurons. We know something about the kinds of neurons that exist, their supporting structures, and even certain details about how neurons work. Yet neural biology remains one of the most intriguing challenges in all of science. Nonetheless, researchers continue to tell us more about just how animals are able to respond so precisely to both their internal and external environments.

Key Ideas

INVERTEBRATE NERVOUS SYSTEMS

1. Cnidarians are radially symmetrical and contain a **nerve net** whose cells concentrate at the mouth and tentacles.

2. The flatworm nervous system resembles a ladder with two **ganglia** (clusters of nerve cell bodies) at the anterior end and the ladder arrangement proceeding posteriorly.

3. Echinoderm nerves are clustered into a ring with branches extending into each arm.

4. The annelids have large ganglia above and below the esophagus and a ventral nerve cord with segmental ganglia.

5. The cephalopod has a functionally differentiated brain that permits complex behavior.

6. The insect brain includes an enlarged ganglion above the esophagus, which, in those with **compound eyes**, includes greatly enlarged visual integrating centers.

VERTEBRATE NERVOUS SYSTEMS

1. Vertebrate nervous systems include a pronounced trend toward cephalization and a dorsal, hollow nerve cord.

2. The vertebrate brain follows a three-part plan of **forebrain**, **midbrain**, and **hindbrain**. The **cerebrum** reaches its greatest development in mammals.

3. The human nervous system is divided into the **central nervous system** (CNS)—the brain and **spinal cord**—and the **peripheral nervous system** (PNS)—the **somatic** and **autonomic systems**. The somatic system is largely associated with voluntary motor nerves that move skeletal muscle, while the autonomic system is motor and involuntary.

4. Cell bodies form its gray outer regions, while myelinated fibers make up its white inner mass. **Cerebrospinal fluid** and the **meninges** cushion and protect the brain, respectively.

5. Toxic substances carried by the blood cannot enter brain cells because of the **blood–brain barrier**—tightly junctured capillary cells and selective **astrocytes** that communicate with them.

6. The forebrain is involved in conscious thought, sensory reception, voluntary movement, and other voluntary acts.

7. The cerebrum, the largest region of the forebrain, is divided into left and right **cerebral hemispheres**, which are covered by the gray, wrinkled **cerebral cortex**. The white matter below includes the myelinated fibers.

8. The importance of the cerebrum to common voluntary acts is minimal in amphibians such as frogs but increasingly important in mammals such as rats, cats, dogs, monkeys, and humans.

9. The cerebrum is made up of a number of lobes. The **occipital lobe** receives and analyzes visual information, while the **temporal lobe** processes auditory input and some visual information. The **frontal lobe** regulates voluntary movement and speech. The **prefrontal area** of the frontal lobe sorts sensory input, putting it into proper context. The **parietal lobe** processes sensory information, including body position.

10. The left cerebral hemisphere predominates in right-handed persons and in the more common type of left-handedness. Speech is primarily a left-hemisphere function, as is analytical thought. Damage to this hemisphere can cause a loss of language abilities.

11. Connections between hemispheres occur through the **corpus callosum**.

12. Three left-hemisphere regions, **Broca's area**, **Wernicke's area**, and the **angular gyrus**, are involved in language. When heard, words are processed by Wernicke's area. Words to be spoken begin in this area but are transmitted to Broca's area, which activates the motor regions involved in voice. Wernicke's area produces the basic sentence structure and coordinates vocalization.

13. The **thalamus**, which is a relay structure, connects various parts of the brain and includes the **reticular system**, which taps incoming and outgoing communications. It also acts as an alarm system and suppresses irrelevant stimuli, thus permitting sleep.

14. The **hypothalamus**, which monitors many functions, acts as a homeostatic regulator (heart rate, blood pressure, body temperature, thirst, hunger, sex drive) and stimulates hormonal activity in the pituitary.

15. The **limbic system** links the forebrain and midbrain and is involved in emotion (for example, when areas of the **amygdala** are stimulated, we may experience rage).

16. The midbrain forms connections between the hindbrain and forebrain and receives sensory input from auditory and visual receptors.

17. The hindbrain consists of the **medulla oblongata**, **cerebellum**, and **pons**. The first controls breathing and heart rate and contains many pathways, including some from **cranial nerves**.

18. The **cerebellum** coordinates and refines voluntary muscle movement, chiefly through delicately applied neural inhibition.

19. The **pons** contains nerves traveling between the forebrain and cord and to and from the cerebellum.

20. The spinal cord includes pathways between the brain and much of the peripheral nervous system. It is bathed in the cerebrospinal fluid and surrounded by the meninges, three layers of supporting connective tissue. The white regions are myelinated neurons, while the gray are nerve cell bodies. Many reflexive acts occur at the cord level.

21. Motor neurons arise from cell bodies in the cord to pass out through the ventral **motor root** of the spinal nerves. Sensory neurons enter the cord via the dorsal **sensory root** from rows of ganglia outside, where their cell bodies cluster.

22. The brain has at least 50 different neurotransmitters. The **monoamines**, modified amino acids, include norepinephrine, dopamine, histamine, and serotonin. **Neuropeptides**, short chains of amino acids, include enkephalins and endorphins, important in pain and mood. They are called opioids because their effects can be mimicked by opiates.

23. EEGs reveal that a considerable amount of electrical activity accompanies sleep. Electrical activity during **REM** (rapid eye movement) **sleep**, a time of dreaming, is similar to that of wakefulness.

THE PERIPHERAL NERVOUS SYSTEM

1. The peripheral system includes the somatic and autonomic systems. The autonomic system is made up of the sympathetic and parasympathetic divisions.

2. The **autonomic nervous system** (ANS) is essentially homeostatic in function. Its motor neurons carry impulses from the brain and spinal cord to the viscera, blood vessels, irises, secretory glands, and other involuntary structures, bringing about fine adjustments as necessary.

3. The sympathetic and parasympathetic divisions have opposing functions, with the former generally increasing an action and the latter decreasing it. In "fight or flight" events, the sympathetic division increases heart and breathing rate and blood pressure and sends more blood to the muscles and brain. Afterward, these actions are reversed by the parasympathetic division.

THE SENSES

1. **Sensory receptors** act as transducers, converting external stimuli first into **receptor potentials** and then into action potentials. Receptor potentials are graded, increasing from threshold level to an intense level.

2. The number and frequency of action potentials can act as a code in the central nervous system. Interpretation of sensory information depends on neural organization in the brain, along with experience, memory, and learning.

3. **Tactile** (touch) or mechanoreceptors respond to deforming force. In invertebrates, they are often associated with sensory hairs. They detect moving solid objects and air and water currents.

4. In vertebrates mechanoreceptors include the distance receptors and contact receptors that specialize in touch or pressure. In humans, **Pacinian corpuscles** are activated by pressure, while **Meissner's corpuscles** respond to light touch. **Baroreceptors** in the arteries respond to changes in blood pressure. Body hairs and the outer dead skin layer act as piezoelectric crystals, discharging current when deformed.

5. **Thermoreception** is the detection of changes in temperature. In vertebrates, heat sensors include the pit of pit vipers, and **Ruffini's corpuscles** and **Krause's corpuscles** in mammals.

6. **Chemoreception** is widespread in animals. Insects have chemoreceptors on many body parts, but the antennae of certain moths is the most acute known. Vertebrate chemoreceptors include general receptors and **gustatory** and **olfactory receptors** (taste and smell). In mammals, an **olfactory epithelium** contains sensory neurons that respond to chemicals, synapsing with neurons in the **olfactory bulb** that extends from the brain. In humans, taste receptors are located in **taste buds** on the tongue, where there is some specialization for sweet, sour, salty, and bitter.

7. **Auditory reception** in insects includes the specialized **tympanal organ**, a drum-like device. In land vertebrates, the hearing organs include an **auditory canal** and one to three **middle ear bones** that transmit sound from the **tympanum** to the inner ear.

8. The human hearing organ includes the **outer ear** (**pinna**, auditory canal, and tympanum). The middle ear bones include the **malleus**, **incus**, and **stapes**, which connect the tympanum with the **cochlea**. The inner ear includes the snail-shaped cochlea and the **vestibular apparatus**.

9. The cochlea is a fluid-filled, U-shaped tube with the **oval window** (stapes attachment) on one end and the **round window** at the other. Sound sets up motion in the fluids, which activates the **organ of Corti**—a **basilar membrane** containing **hair cells** embedded in the **tectorial membrane**. Bending of the hair cells creates receptor potentials, which, in turn, start action potentials in neurons from the cochlear nerve. Differences in intensity and pitch are produced by activity at various parts of the basilar membrane.

10. Some invertebrates have **statocysts**, chambers containing grains that move against sensory hairs when changes in body position occur.

11. In humans and other mammals, movement, position, and balance are detected by the vestibular apparatus. Three fluid-filled **semicircular canals**, lying in three planes, detect acceleration and deceleration as the fluids move against sensory hairs. Fine granules in the **saccule** and **utricle** shift with movement that activates sensory hairs, which inform the brain about position.

12. Visual receptors in arthropods include **simple eyes** and **compound eyes**, the first of which detect only light and lack lenses. Compound eyes contain immovable units called **ommatidia** that register images.

13. The vertebrate eyeball consists of a tough, outer **sclera**, with a forward, transparent **cornea** and more internal **choroid**. The choroid supports the **iris**, which surrounds the **pupil**. The **lens** separates two fluid-filled chambers. The light-sensitive **retina** within the eyeball communicates with the **optic nerve**.

14. Light entering the eye passes through the **anterior chamber** (filled with **aqueous humor**), lens, and **posterior chamber** (**vitreous humor**), to the retina. Focusing is done through changes in lens shape brought about by the **ciliary muscles**. The amount of light entering is altered by the iris, which determines the size of the pupil.

15. The retina consists of four cell layers, including an innermost pigmented layer, a layer of **rods** and **cones** (the sensory receptors), a layer of **bipolar cells**, and a final layer of **ganglion cells**. The bipolar cells and ganglion cells are neurons.

16. Rods respond to all wavelengths of the visible spectrum and function well at night, while each cone responds to either red, green, or blue light wavelengths and has little night function. Cones are most concentrated in the **fovea**.

17. Upon exposure to light, **rhodopsin** in rods breaks down to **opsin** and **retinal**, and the chemical change activates the synapses with the neurons above, where action potentials begin. The two products recycle, using vitamin A to produce rhodopsin.

Application of Ideas

1. Why would parts of the human brain be called primitive and others advanced? What traits would merit such descriptions?

2. The central nervous system, particularly the conscious center, exercises what might be called "noise control," as it receives, integrates, and responds to a barrage of sensory messages. The autonomic system, on the other hand, can be referred to as a "quiet" system, since its activity is generally below the conscious level. Elaborate on these ideas by considering how the two systems respond to problems in osmoregulation and in thermoregulation (in the latter include problems of both heating and cooling).

3. Compare the functioning of the image-producing human eye with that of the compound eye of arthropods. We know what *we* see, but what might the arthropod actually see? What are some of the advantages of each kind of eye?

4. Compare the degrees of development in special senses in various vertebrates. Explain how an emphasis on certain of these can adapt the animal to its specific environment.

Review Questions

1. Name the three parts of the vertebrate brain and describe the general evolutionary trends in each part. (page 711)

2. What is the function of the cerebrospinal fluid? Where is it found in the brain? (page 712)

3. What is the blood–brain barrier? How are materials actually transferred from capillaries to the brain cells? (page 713)

4. Compare the importance of the cerebrum in the rat, frog, and primate. What general trend in the localizing of functions does your answer suggest? (page 714)

5. State a function that is localized in each of the following cerebral lobes: prefrontal, occipital, temporal, and frontal. (page 715)

6. Describe the organization of sensory and motor areas around the central sulcus. Compare the relative amounts of sensory and motor control area devoted to the hand, genitals, leg, tongue, and toes. (page 715)

7. Explain the manner in which the three speech areas of the cortex interact when you read aloud. In general, what are the roles of the Wernicke and Broca areas? (page 718)

8. What are the main functions of the thalamus and its reticular system? (pages 718–719)

9. List five processes that are regulated by the hypothalamus. With what other system does it closely interact? (page 719)

10. What structures, in addition to the hypothalamus and thalamus, make up the limbic system? Name some of their activities. (page 719)

11. List the three regions of the hindbrain and state their general functions. (pages 719–720)

12. Describe the organization of the spinal cord, including its covering membranes, its gray and white regions, and the manner in which spinal nerves enter and leave. (page 720)

13. What is the role of enkephalins and endorphins? Why are they called opioid neurotransmitters? (page 720)

14. During which sleep episode is brain activity close to that of wakefulness? Why is this period of sleep so important? (page 722)

15. Describe the organization of the autonomic nervous system, naming its divisions. (page 723)

16. Compare the role of the autonomic divisions in the "fight or flight" response. (page 723)

17. In what major way does a receptor potential differ from an action potential? How is this difference adaptive? (page 725)

18. In what specific instance might sensitive thermoreceptors be highly adaptive to invertebrates? (page 727)

19. Describe the mammalian olfactory epithelium and its connections to the brain. (page 728)

20. Describe the structures and events of hearing, beginning with air waves moving the tympanum and ending with an action potential in the auditory nerve. (pages 729–732)

21. Explain how a crayfish senses its body position. In what ways is this similar to the operation of the saccule and utricle of humans? (page 732)

22. List the structures or regions of the eye through which light passes on its way to the retina. Where appropriate, explain how each structure functions. (pages 733–735)

23. Describe the organization of the retina. In which region is color vision most acute? Why is this? (pages 733–735)

CHAPTER 35

Thermoregulation, Osmoregulation, and Excretion

CHAPTER OUTLINE

HOMEOSTASIS

Negative and Positive Feedback Loops

THERMOREGULATION

Why Thermoregulate?

Categories of Thermoregulation

Endothermic Regulation

Thermoregulation in Birds and Mammals

Ectothermic Regulation

Physiological Adaptations to Freezing

OSMOREGULATION AND EXCRETION

Producing Nitrogen Wastes

The Osmotic Environment

The Marine Environment

The Freshwater Environment

The Terrestrial Environment

THE HUMAN EXCRETORY SYSTEM

Anatomy of the Human Excretory System

Microanatomy of the Nephron

The Work of the Nephron

Control of the Nephron Function

KEY IDEAS

APPLICATION OF IDEAS

REVIEW QUESTIONS

About 200 years ago, Charles Blagden, then the Secretary of the Royal Society of London, proved that he was one of the most persuasive people on earth. He managed to talk two friends into joining him in a small room in which the temperature had been raised to 126°C (260°F). Being aware that water boils at 100°C, they naturally were a bit reluctant. But Blagden prevailed, and the men, taking along a small dog and a steak, entered the room. They emerged 45 minutes later and were surprised to find that they were not only alive, but in good shape. The dog was fine, too. But the steak was cooked!

And now, a young woman who lives in the city had thought she was in good shape, but she is finding it pretty tough going while backpacking high in the mountains. The level parts of the terrain are easy enough, but when climbing, she has to stop repeatedly just to catch her breath. Her heart races, she feels light-headed and sick to her stomach, and now she has a sharp pain in her left side. After each rest, her heavy breathing subsides, she feels more comfortable, and she is able to resume climbing. But that night she weakly crawls into her tent without dinner. Her companion, who has been in the mountains all summer and is having no problem, tells her that she'll get used to the altitude in a week or two. In fact, by the next morning she is already feeling a bit better, and, sure enough, in a couple of weeks she hardly notices the altitude at all.

In these two examples, we see the remarkable ability of the human body to adjust to different, even extreme, conditions. The Blagden group was forced to adapt quickly while the climber more gradually met the new challenge. Yet in each case the bodies made the changes necessary to survive. In this chapter we will see how the body makes the adjustments necessary to continue under a variety of changing conditions dealing with temperature and osmosis.

Homeostasis

Both Blagden and the climber demonstrated what is called homeostasis. **Homeostasis** ("same status") is the maintenance of relatively constant internal conditions. Homeostatic mechanisms keep physiological conditions within a certain range that is conducive to life.

Our definition is necessarily broad since homeostasis pervades all aspects of life and functions in any number of ways on a variety of organ systems. In fact, every organ system functions, in some way, through homeostatic mechanisms, including the nervous system, respiratory system, circulatory system, and immune system. Even behavior has homeostatic significance. Homeostasis then is one of the major themes in biology.

We should also keep in mind that different homeostatic responses can operate at the same time. For example, in Blagden's experiment, the dog probably sought to leave the room—a behavioral response. Then he probably gave up and lay there panting, as physiological changes began to occur. Regarding the novice mountain climber, her pause by the trailside to "catch her breath" was a response to falling levels of oxygen in her blood. Her behavior was changing even as certain physiological responses were kicking in (Figure 35.1). The sharp pain in her side was an indication of another homeostatic mechanism. Oxygen levels are low at high altitudes, so with exertion increasing the oxygen demand, the spleen undergoes spontaneous and painful muscular contractions that release red blood cells, enabling the blood to carry more oxygen. The nausea is due to the shunting of blood from her gut to her oxygen-demanding brain and muscle tissue. Over time, continued exposure to high altitudes stimulates the climber to produce more red blood cells. How is the body able to respond sensitively to changes in the environment? The responses are usually automatic, brought on through what is called feedback, which can be either negative or positive. *Feedback,* as we saw in Chapter 34, occurs when the result of some process affects the process itself, negatively or positively.

FIGURE 35.1
Homeostatic Mechanisms at Work
As the backpackers climb higher onto the mountain, their bodies undergo many physiological changes that make it possible for them to meet new demands.

Negative and Positive Feedback Loops

The **negative feedback loop** refers to a situation in which a process produces a reaction that ultimately inhibits the process. Thus the negative feedback loop is inherently stabilizing, maintaining the organism in a steady state. Negative feedback loops tend to be delicate in nature, their operation surging and fading intermittently as an optimal condition is reached (Figure 35.2, left). The furnace and air-conditioning thermostats in a house also work on the negative feedback principle.

Let's consider an analogy. A cruise control on an automobile keeps the vehicle moving at a constant speed. Once the speed is set, when the car slows, the carburetor feeds the engine more fuel, causing the car to accelerate. The increased speed then acts to inhibit the flow of fuel from the carburetor

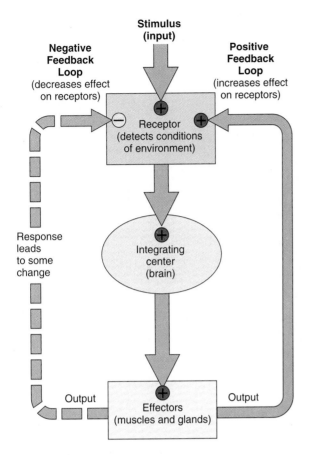

FIGURE 35.2
Negative and Positive Feedback Mechanisms
In negative feedback the product of an action inhibits the very mechanisms that led to its formation. In positive feedback, the opposite is true. The product of an action intensifies the mechanism that brought about its formation

and the car slows. In this way, the speed is kept relatively constant. You can see how a living system could utilize such a system to regulate body function.

Another kind of feedback can produce an entirely different effect. In a **positive feedback loop**, the process produces a reaction that further accelerates the process (see Figure 35.2, right). Positive feedback loops are uncommon in nature and are usually associated with illness. High blood pressure, for instance, can damage arteries and arterioles, and the damaged vessel walls can become infused with lipid materials, scar tissue, and cellular growth. This, in turn, restricts the size of the vessel opening, further increasing blood pressure. So high blood pressure can actually raise the blood pressure further. As another example, we all know of people who are depressed because they are overweight, and so they eat because they are depressed.

In this chapter, we will consider two important examples of homeostasis. One is *thermoregulation,* the regulation of body temperature. The other is *osmoregulation,* the regulation of internal osmotic conditions.

Thermoregulation

Thermoregulation is the ability of an organism to maintain its body temperature either at a constant level or within an acceptable range. Some animals have very little of this ability, yet others are highly specialized for it.

Why Thermoregulate?

Animals must thermoregulate if they are to remain active beyond certain environmental temperatures. Many physiological processes can optimally proceed only within certain temperature limits, and frigid temperatures stop certain processes. Furthermore, if cells should freeze, their water would be bound up as ice, dangerously concentrating other cellular material and also forming ice crystals that could disrupt delicate membranes and destroy tissue.

Overheating, on the other hand, often causes more problems than does overcooling. For example, high temperatures can accelerate rates of biochemical reactions to unacceptable levels. Very high temperatures can denature enzymes and other critical proteins, bringing metabolic activity to a halt and severely altering cell structure. Such heat damage is often irreversible, whereas damage done by cold, within certain limits, is more likely to be temporary.

As we will see, there are many adaptive solutions to the problem of maintaining optimal body temperatures. Some of these solutions are metabolic, and others are behavioral.

Categories of Thermoregulation

Traditionally, those animal species that thermoregulate most (those retaining a rather constant internal temperature) are called **homeotherms**. Those that do not actively thermoregulate (those whose body temperatures fluctuate to near that of their surroundings) are called **poikilotherms**. (The lay terms for the two conditions are "warm-blooded" and "cold-blooded," respectively.) At one time only birds and mammals were thought to be homeothermic. All other animals were believed to be poikilotherms. But as this idea was tested, it soon became clear that things weren't quite so simple. Although birds and mammals are certainly warm-blooded, ongoing studies have resulted in a growing list of other animals, such as some reptiles, that retain surprisingly constant body temperatures. Such studies have also unveiled a third category of thermoregulation, called *heterothermy.* Those in this category, the **heterotherms**, thermoregulate part of the time and allow their temperatures to fluctuate with that of the environment the rest of the time.

Animals warm their bodies in two major ways: *endothermy* and *ectothermy.* **Endothermy** involves the generation of heat through internal metabolic processes. You may recall that metabolic energy transfers are not 100% efficient and that, at each step, some energy is lost as heat. That heat, when generated in sufficient amounts, can then be used in thermoregulation.

In contrast, **ectothermy** involves the utilization of external sources of heat. Ectothermic animals absorb heat from the environment. A frequent method used by ectothermic animals to gain heat is to expose themselves deliberately to heat. The process can be as simple as a dog walking over to lie in the sun. Heat can be gained in other ways as well, as when a fish living in a hot spring moves closer to the vent or on a cold day when a flea migrates to a dog's warmer body parts. We can say then that in endothermy heat comes from the inside, whereas in ectothermy it comes from the outside.

Endothermic Regulation

If an animal maintains a constant temperature, the chemical reactions associated with metabolism can go on at a more constant rate, not being slowed drastically by cold or greatly accelerated by heat. Thus the animal can remain active at a wide range of environmental temperatures. The advent of endothermy has been critical in evolutionary history, since it provided a new way for animals to cope with new environments, from blazing deserts to the frigid poles.

The generation of heat through endothermy is important in maintaining temperature constancy within an animal. Keep in mind that when we talk of temperature constancy, we're referring to a *range* of internal temperatures. After all, even among the strictest homeotherms, body temperatures can vary—both from time to time and place to place within the body. Our temperatures, for example, vary throughout the day and our skin and extremities are cooler than our internal organs.

The temperature deep within the body, called the *core temperature,* is most constant. But even core temperature varies widely among endotherms. Consider hibernators (such as dormice and bats) in which the core temperature may drop drastically (Figure 35.3). Such inactivity coupled with low temperature is adaptive in that it reduces the need for energy when food may not be abundant.

In fact, a principal disadvantage of endothermy is that is requires a great deal of energy. Since the source of energy is food, endotherms must feed frequently. But there are other disadvantages to endothermy. With few exceptions, birds and mammals cannot tolerate much change in core body temperature without dying. (Variations of two or three degrees in our own internal temperatures, for instance, send us running to our physicians.) Endothermy also places body size restrictions on warm-blooded animals, primarily limiting how small they can be. Below a certain body size, heat production cannot keep up with heat loss (because of the proportionately greater surface area of smaller animals).

Countercurrent Heat Exchangers. Many animals conserve heat through a circulatory arrangement called a **countercurrent heat exchanger**. In such arrangements, warm blood leaving the core region of the animal body passes near cooler blood returning from the outer regions. Because the fluids are moving in opposite directions, generally in closely adjacent vessels, heat is shunted from the warmer to the cooler blood

FIGURE 35.3
Hibernating Dormouse
During hibernation the usual body thermoregulatory controls are overridden, body core temperature falls as much as 30°C, and metabolism slows drastically.

along the entire flow. By the time the core blood reaches the outer regions it has cooled down close to environmental temperature, its heat having been recycled to incoming blood. Such heat exchanges are possible anytime fluids of different temperatures pass close by in opposite directions.

Most fish of the sea have little thermoregulatory ability. Those that do thermoregulate rely almost exclusively on endothermy, the metabolic generation of body heat. (Obviously, sunbathing would be a problem, and there are few concentrated heat sources beneath the surface.) The best known endothermic fish is the bluefin tuna, a fast-moving, cold-water predator that is among the fastest swimmers in the sea. Furthermore, its swimming muscles are able to function actively even in the coldest waters of its range. It is able to function so well in cold water because it is a true endothermic homeotherm.

Compare in Figure 35.4 the circulatory systems of nonregulating and thermoregulating species of fish found in similar habitats. Note that in the first, the major blood vessels give rise to branches that serve the propulsion muscles and other portions of the body. Such an arrangement can do little to retain body heat in a central place. On the other hand, the major arteries and veins of the homeothermic bluefin tuna tend to run parallel to each other with blood flowing in opposite directions, thus forming a countercurrent heat exchanger.

Much of the bluefin's body heat is generated by its large, dark, swimming muscles, which make up most of the body core. The body core is some 10 degrees warmer than the skin, which remains at about the temperature of surrounding seawater. As intense muscle activity generates heat, this heat is

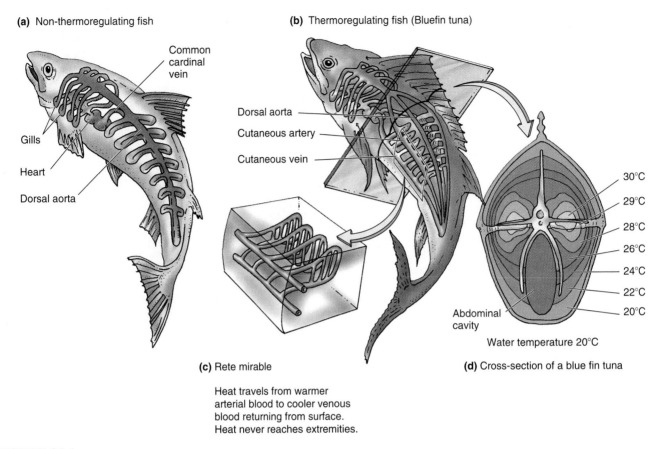

(a) Non-thermoregulating fish

Common cardinal vein

Gills

Heart

Dorsal aorta

(b) Thermoregulating fish (Bluefin tuna)

Dorsal aorta

Cutaneous artery

Cutaneous vein

30°C
29°C
28°C
26°C
24°C
22°C
20°C

Abdominal cavity

Water temperature 20°C

(c) Rete mirable

Heat travels from warmer arterial blood to cooler venous blood returning from surface. Heat never reaches extremities.

(d) Cross-section of a blue fin tuna

FIGURE 35.4

Circulation in a Thermoregulating Fish

(a) In most fishes the arrangement of blood vessels is such that core heat is quickly distributed to the rest of the body. **(b)** In the bluefin tuna, the retention of heat in the core is aided by a countercurrent heat exchange among major vessels, provided by the rete mirable **(c)**. Note the parallel arrangement of vessels carrying warmed blood out of the core region with those bringing cooler blood in from the extremities. **(d)** The tuna retains heat in the core region around the dark swimming muscles, so its body core is much warmer than its extremities and its cold water surroundings.

efficiently retained in the core region by use of the countercurrent heat exchanger.

Note in Figure 35.4d the presence of a second, much more complex heat exchanger in the *rete mirabile* ("wonderful net"). The dark muscle making up much of the tuna's body core, and the warmest part of its body, is served by a vast network of arteries and veins, all arranged in a parallel fashion. They make up a dense, highly efficient countercurrent heat exchanger that keeps the dark swimming muscles quite warm.

Thermoregulation in Birds and Mammals

Mammals and birds have developed endothermy into a fine art. They have evolved very effective ways of generating and conserving heat. In fact, they can develop and retain so much heat that they must have efficient cooling mechanisms as well. We will first see how heat can be generated, then look at how body size is important to heat generation, how heat can be conserved, and how excess heat can be given off.

Generating Heat. Birds and mammals generate heat in a variety of ways, all involving the conversion of chemical energy into heat energy. There are three principal mechanisms involved. The first is oxidative respiration, a process that produces heat during ATP synthesis (discussed in Chapter 6). The second is by the generation of heat through shivering, which involves the rapid alternate contraction of opposing muscle groups. The third is the utilization of brown fat. Let's focus briefly on this last mechanism.

Brown fat is brown because of the presence of numerous mitochondria (actually it is the cytochrome oxidase they contain that causes the brown color). In such mitochondria, the free energy of the chemiosmotic gradient isn't used to generate ATP but is simply converted to heat. Unlike most other fat storage tissue, brown fat has a rich blood supply, so the heat generated in the fat is rapidly dispersed throughout the body.

Brown fat heat production occurs in a number of animals. We find it, for example, in human infants, and especially in hibernating mammals. Hibernating mammals eat prodigiously before hibernating, laying in great reserves of fat, including brown fat. When they enter that deep sleep, their body temperatures drop by as much as 30°C. Regaining normal body temperature would require a considerable amount of time were it not for brown fat. The metabolism of brown fat, which is neurally induced by the autonomic nervous system in the torpid animal, readily provides the body heat needed for a quick arousal.

Body Size and Heat Generation. In birds and mammals, metabolic rate, the total body energy produced per unit time, is generally related to body size. Specifically, the smaller the body size, the greater the metabolic rate (Figure 35.5). And the greater the metabolic rate, the more heat produced. The hummingbird, for example, supports its intense metabolic activity through a large sugar (nectar) intake, so high in fact that at night when it can no longer feed, it survives only by falling into a metabolic slump. Thus it may be regarded as heterothermic. Among the mammals, the small shrews (about the size of the large cockroaches on their menu) have the highest metabolic rates, whereas elephants have the lowest. Although such shrews have about one millionth the mass of the elephant, they consume about 100 times as much oxygen per gram of body weight. For a carnivore, their food consumption is also monumental, their daily intake being about 75% of their body weight.

Physiologists are now beginning to challenge the traditional explanation of the link between metabolic rate and heat loss. It has long been held that small animals, with their proportionately larger surface areas, must maintain high metabolic rates just to keep up with heat loss. If this were so, then one would expect small *ectotherms* to have cooler bodies than

large ectotherms. This isn't the case; both tend to be in equilibrium with their surroundings. So more is involved in heat loss than simply body mass and surface area. We will continue to watch these developments with interest.

Slowing Heat Loss. Once heat is produced, it must be retained, and birds and mammals have developed a number of ways to slow heat loss. One way they do this is by countercurrent heat exchangers that help retain core heat. Another way involves shunting warm blood away from the extremities.

Humans, for example, have very effective countercurrent heat exchangers in the arms. These involve veins (which carry blood toward the heart) and arteries (which carry blood away from the heart). When the body is warm, blood returns from the hands through veins near the skin (the ones you can see in your arm), thus permitting heat to escape. As the body becomes chilled, though, blood is shunted into deeper veins that run parallel to the arteries, thus enabling the body to conserve heat through a countercurrent heat exchange between the opposing vessels (Figure 35.6). Shunting blood deeper into the body to conserve heat is a costly mechanism. Cooler limb muscles do not work as efficiently, and the extremities, deprived of warming blood, can become numb and eventually frostbitten. Through frostbite a person can actually lose fingers, toes, and ears before the core temperature reaches a danger level. The appendages are sacrificed as the internal organs receive first priority in the battle for survival.

Birds also employ countercurrent heat exchangers. Watching wading birds in the dead of winter can send a shiver through the most stalwart of souls. The birds casually wade in water that is near freezing. However, they still maintain their normal body temperature, due to countercurrent heat exchangers in the feet. A high level of unsaturated fats in cell membranes of tissues in the feet is an additional adaptation to cold. Such fats stay liquid even at very low temperatures.

Mammals also utilize fat for protection from extreme cold. Marine mammals such as whales, seals, and walruses are protected from heat loss by thick layers of *blubber,* a specific type of fatty tissue (Figure 35.7). Blubber, an excellent insulator, is not simply a passive insulator; it plays a versatile role in thermoregulation. Unlike most fatty tissue (but like brown fat), blubber has a rich blood supply. In frigid polar waters the vessels close down, and the blubber is virtually blood-free. In that state, it resists nearly all heat loss. But when a whale encounters warmer waters, or when a sea lion basks on a sunny shore, reduction of the core temperature becomes essential. Blood vessels in the blubber then open and blood enters, heat escapes, and overheating is avoided.

Most mammals rely on body hair, and all birds rely on feathers for insulation. Both can be "fluffed," or made to stand on end, thereby producing tiny, heat-trapping air spaces. Perhaps you've seen chickadees perched at your bird feeder on a cold winter day, their feathers fluffed, looking like tennis balls with beaks (Figure 35.8).

Many mammals have dense, soft *underhair* that acts as insulation, and long, coarser *guard hair* over it that protects

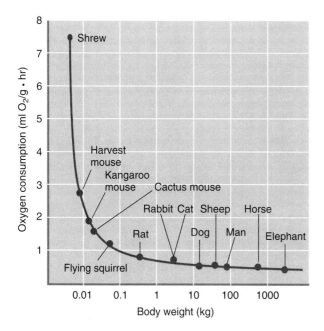

FIGURE 35.5
Body Size and Metabolic Rates
The metabolic rate in selected mammals, as determined through oxygen consumption per gram of body weight, indicates that this rate is greater in the smaller animals than in the larger.

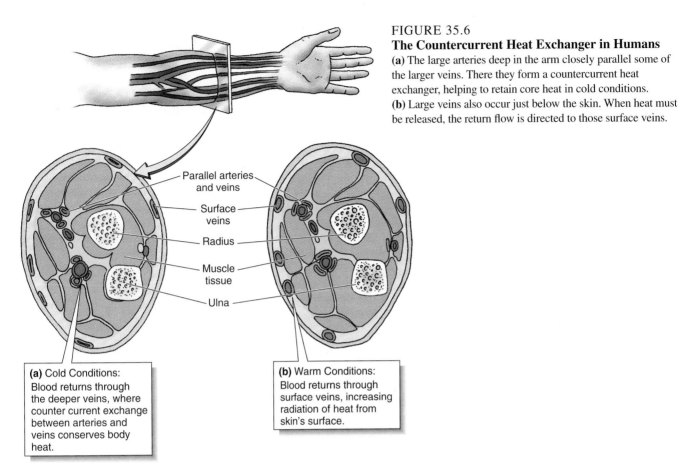

FIGURE 35.6
The Countercurrent Heat Exchanger in Humans
(a) The large arteries deep in the arm closely parallel some of the larger veins. There they form a countercurrent heat exchanger, helping to retain core heat in cold conditions.
(b) Large veins also occur just below the skin. When heat must be released, the return flow is directed to those surface veins.

Parallel arteries and veins
Surface veins
Radius
Muscle tissue
Ulna

(a) Cold Conditions: Blood returns through the deeper veins, where counter current exchange between arteries and veins conserves body heat.

(b) Warm Conditions: Blood returns through surface veins, increasing radiation of heat from skin's surface.

against moisture and abrasion and provides for coloration. In aquatic mammals, the layer of guard hair is virtually waterproof, and thus they remain almost dry and conserve heat this way (Figure 35.9). The combination of hair and blubber insulating the Arctic harp seal holds in virtually *all* of the animal's body heat. Canadian researchers were recently thwarted in their attempts to use airborne infrared tracking devices to follow the seals' movements because, surprisingly, the animals simply do not give off enough body heat to be detected by infrared imaging. Furthermore, when the researchers switched to ultraviolet detectors, they could find the seals all right, but they appeared as negative images only. This was a startling situation. It turned out that, whereas the surrounding snow reflected ultraviolet light, the seals absorbed it. Further studies revealed that the individual hairs on seals, polar bears, and many arctic mammals act as excellent solar collectors. Polar bear hairs, they learned, have fibrous cores that, upon absorbing ultraviolet light, act as tiny fiber-optic cables, conducting the energy to the black, light-absorbing skin below. Engineers are now trying to apply the same principle to manufactured solar collectors.

When it comes to using body hair for insulation, humans are exceptions. Your hirsute cousin aside, humans don't have enough hair on their bodies for effective insulation. Anthropologists tell us that humans evolved in hot climates where the furry coat of our ancestors was traded for the ability to sweat profusely, thereby cooling the body—something few other mammals can do as well. Thus we owe much of our success in colder climates to behavioral, not physiological, strategies.

We'll turn now to another aspect of thermoregulation in endotherms, the problem of overheating.

Increasing Heat Loss. Just as animals must have ways to gain heat, they must also have ways to lose it, to cool their bodies. The most important avenues for heat loss in birds and mammals are the skin and the respiratory passages. Heat escapes from such surfaces in three major ways: through *radiation, conduction,* and *convection.*

Walruses basking

FIGURE 35.7
Blubber
A dense layer of fat forms in polar mammals.

Robin(*Turdus migratorius*)

FIGURE 35.8
Feather Fluffing
Birds increase the insulating effects of feathers by fluffing them, thereby creating minute heat-retaining air spaces around the skin. Fluffing increases with drops in temperature.

Heat radiates into cooler surroundings. **Radiation** involves heat moving from a warm surface into the air and is important in dissipating heat from the peripheral circulation, such as through the skin. In addition, heat escapes through **conduction** (in which heat passes by contact to a cooler solid, as when you place your hand on a cool tile wall). In **convection**, heat is carried away by the movement of air molecules. Convection figures prominently in cooling through evaporation, as when a cool breeze passes over your face.

Evaporative cooling is a vital part of thermoregulation in both birds and mammals. The evaporation of water from the surface of an animal is accompanied by a transfer of heat to the surroundings and a subsequent cooling of the body surface. To encourage such cooling, animals may wet themselves, some making use of standing water, others spreading saliva or even urine on their warm bodies. Most mammals have sweat glands in the skin that become active as the body warms, wetting and cooling the skin. Sweat glands are particularly important in humans, horses, and certain apes. (A human in a hot desert may lose as much as 1.5 liters of water per hour through sweating. At this rate, and without replacement, you would be near death by a single day's end.)

Mammals and birds also carry out evaporative cooling through the respiratory passages, generally exhausting most of the heat through the open mouth. Birds actively pant, but, in addition, they vibrate the thin-walled floor of the mouth, accelerating evaporative cooling there. As dogs pant, they inhale through the nose (you didn't know that, did you?) and exhale through the open mouth, their wet tongues extended to increase the area of evaporative cooling. The use of evaporative cooling, of course, depends on the availability of body fluids, which must be maintained at certain levels in all animals. An excessive loss of water can quickly bring an end to evaporative cooling and result in overheating.

Behavioral Adaptations in Endotherms. Endotherms not only regulate their temperatures physiologically, but behaviorally as well. Behavioral regulation simply involves doing those things that result in cooling or heating the body. For example, at midday, birds may stop singing and rest quietly in the trees, while some may bathe in standing water. Dogs like to dig holes in newly planted lawns, using the cool, moist, freshly dug soil to take up heat from their warm bellies. Desert rodents, unable to sacrifice much body water for cool-

Polar bear

FIGURE 35.9
Cold-Adapted Mammals
Mammals of the Arctic are well adapted for normal activity in common subzero air temperatures. Their well-insulated bodies keep in most of the heat that is generated metabolically.

ing their bodies, retreat to cooler burrows until nightfall. One species of mammal simply turns on the air conditioner.

Certain activities can overheat the body, and cessation of those activities often can reverse the effect. Some time back, one of your authors (Wallace) suffered heat stroke in a marathon. He reports:

> The temperature was 75°F, and the humidity was 80%. Hardly ideal, but all seemed well until the 24th mile. My time was about 6 ½ minutes per mile when suddenly I felt strange—remote, dizzy, and spacy. I slowed down, thinking I would quickly recoup. But I didn't recoup. My goal for the 26 miles was anything under 2 hours and 50 minutes, and with 2 miles to go my time was 2 hours and 36 minutes. I tried to mentally subtract 36 from 50 to see how fast I would have to run the final 2 miles, but the mental calculations were impossible. I began to realize I was in trouble when my skin felt dry and chilled. A wave of dismay swept over me because I knew continued stress could cause kidney or brain damage, but I wanted to finish, so I kept running, more slowly with each minute. Then I began to fall asleep on my feet. It happened again and again while I kept stumbling along. I managed to stay awake and made it in 2:55. As soon as I sat down and began to take water and mineral-laden drinks the recovery began. Strangely enough, a blanket was needed to fend off what felt like a chill wind. Some physiologists said later that I had interfered with my thermoregulatory mechanisms so that I had lost too much heat, and that the sleepiness was the same sort described by people who are freezing. The consensus was I should have stopped sooner.

The point is, bad judgment can lead to behavior that can overheat the body and disrupt homeostasis, while other behavior can restore the balance necessary for life—if it is done in time.

The Mechanisms of Human Thermoregulation. An organism can change in many ways to meet the challenge of variable environmental temperatures. The question then is: How are these responses regulated and integrated? In particular, how do humans maintain an average internal temperature of about 37°C (98.6°F) within an amazingly narrow range? Surely some form of sophisticated internal control is required since we have so many independent ways to regulate temperature.

To begin with, the brain constantly monitors the temperature of blood flowing through it. The responses initiated by the brain override all other controls; hence the brain can be thought of as the real thermostat of the body.

The heat-monitoring part of the brain is the **hypothalamus**. It indirectly receives temperature information originating in sensory receptors in the skin and, more directly, through its own sensors, which detect the temperature of blood in the vessels that pass through it.

The hypothalamus, in its constant monitoring of blood temperature, can elicit a variety of responses to increase the rate of heat production or heat loss (Figure 35.10). For example, through its influence of the autonomic (sympathetic and parasympathetic) nervous system, it can increase the rate of heat production by stimulating the adrenal gland (a hormone-secreting body), causing it to release the hormone epinephrine into the bloodstream. Epinephrine speeds up the conversion of glycogen to glucose in the liver. The release of glucose into general circulation and its subsequently increased availability to cells enables cells to increase their respiratory activity, which creates heat. Epinephrine also stimulates thermogenesis in brown fat. In addition, the hypothalamus can prompt the autonomic nervous system to reduce blood flow in the skin.

The hypothalamus also stimulates the pituitary gland to release thyroid-stimulating hormone, which, in turn, causes the thyroid gland to increase the output of its hormone, thyroxin (see Chapter 36). The effect of thyroxin is to increase cellular respiratory activity and thereby produce body heat. Increases in body heat are subsequently detected by the hypothalamus, which then eases off on the pituitary, slowing heat output—another example of a negative feedback loop.

Thus information arriving from peripheral and internal temperature sensors results in a considerable variety of hypothalamic actions. As a homeostatic control structure, the hypothalamus keeps the organism in a finely tuned and responsive state, maintaining body temperature at an optimal level.

No one is quite sure how the hypothalamus works in thermoregulation. In fact, after all is said and done, perhaps the hypothalamus doesn't actually measure heat. Maybe it measures heat-related chemical changes in the blood instead. A leading hypothesis suggests that it monitors calcium ions.

It is known that the thermostat can be artificially altered by suffusing the brain with calcium solutions. Whereas it is not known whether calcium directly affects the thermostat, the effect of the presence of calcium cannot be argued. In fact, calcium suffusion works so well that, with it, body temperature can be chemically lowered enough to permit **cryogenic** (deep-cooled) **surgery**, duplicating the effect of submerging the body in ice water. The use of calcium removes the need to lower body temperature by removing body heat; the thermostat is merely set to a lower level.

Ectothermic Regulation

The chief advantage of ectothermy is its low metabolic cost. It's like living in a house that's not heated or air-conditioned. If it's cold outside, the house is cold; if it's hot, the house is hot. But your energy bills are low. Because they do not require extra food just for keeping body temperature up, ectothermic animals need far less food than endotherms, and many can fast for long periods. There are, however, marked disadvantages to ectothermy. Since ectotherms rely mainly on outside sources of heat, their activities may be curtailed at certain times, such as at night and in the colder seasons. We will begin our look at ectotherms with problems related to low temperatures.

Physiological Adaptations to Freezing

Animals can adapt to cold in a variety of ways. But interestingly, animals are not irreversibly affected by extreme cold unless freezing occurs—that is, unless ice crystals form within cells. There, ice crystals can disrupt membranes and cellular organelles and cause the death of the tissue. The formation of

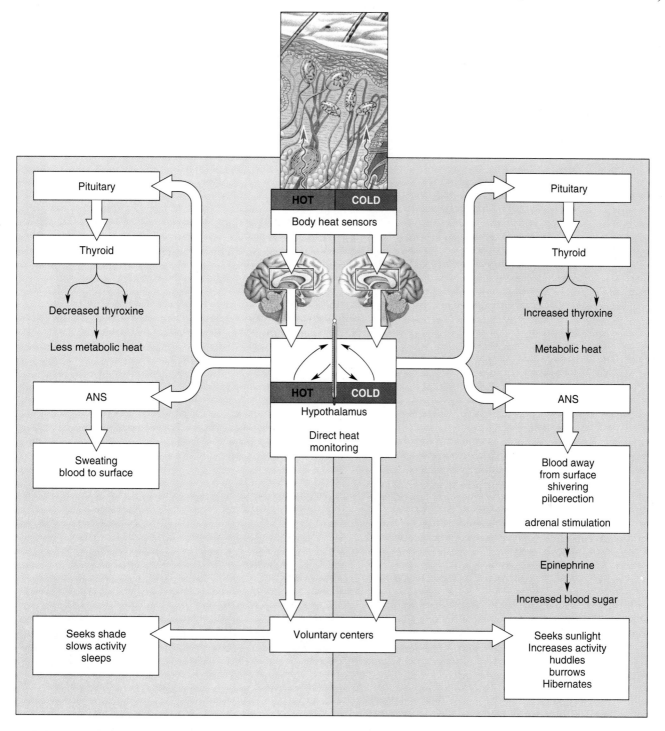

FIGURE 35.10
The Center for Thermoregulation
The hypothalamus receives thermal information from thermoreceptors in the skin and by direct sensing of
blood arriving from the core of the body. Its response to changing temperatures is carried out in two ways:
through activation of the autonomic nervous system and of the endocrine system. Negative feedback
occurs through the continued monitoring of blood temperature.

ice crystals outside the cells, in the intercellular spaces, is not
as serious. In fact, such freezing can be part of a protective
mechanism in cold-adapted species. Researchers report that, in
certain beetles and midge larvae, the freezing of extracellular
fluids is encouraged by the presence of particles that act as ice

nuclei. How does this protect the cells? The cells themselves
are protected because as water is locked up as ice in the extracellular spaces, the solutes left behind produce an osmotic
effect. That is, their presence encourages the osmotic movement of water out of nearby cells, which then become dehy-

FIGURE 35.11
Fish with Antifreeze
Concentrations of glycoproteins in the body of Antarctic ice fish prevent the formation of ice crystals. Ocean temperatures below the ice hover between −1° and −2°C throughout much of the year.

drated. So even though ice forms in the extracellular spaces, the dehydrated cells avoid the problem, surviving nicely until warmer temperatures return.

A number of ectotherms resist freezing through the presence of specific antifreeze substances in their body fluids. The parasitic wasp, *Brachon cephi,* for example, increases its body's glycerol content, which lowers the freezing point of its body fluids to about −17°C. Certain insect larvae do even better, resisting ice formation at temperatures as low as −47°C. Another form of antifreeze, a glycoprotein, protects Antarctic ice fish from freezing (Figure 35.11).

Behavioral Adaptations in Ectotherms. Since the body temperature of ectotherms depends on the temperature of their surroundings, behavioral thermoregulation is very important to this group. They must go where the ambient temperature suits their needs, and once there, they must do the things that cause the loss or gain of appropriate amounts of heat. Among insects, for example, the desert locust crawls into the sunlight at dawn and orients its body sideways to the sunlight so that the greatest possible surface area is presented to the warming rays. Honeybees concentrate heat by huddling together in cold weather and, in hot weather, they promote evaporative cooling of the hive by bringing in water and fanning it with their wings.

Most deep-water fishes have little opportunity for thermoregulation, because, in their habitat, environmental temperatures vary little from place to place. Fishes in shallower fresh waters, on the other hand, can move from warmer, sunbathed shallows to cooler, deeper waters and back, as circumstances demand.

It turns out that the degree to which reptiles can regulate their body temperature depends on where they live. Reptiles from the tropics, for example, can't regulate their body temperature nearly as well as those from temperate regions, where weather changes more drastically. Historically, the success of reptiles in invading temperate zones from the tropics has depended heavily on their ability to thermoregulate behaviorally. The strategies of temperate zone lizards are apparently very successful, since these reptiles maintain remarkably constant body temperatures as they move about.

Reptiles often absorb heat by basking in the sun, and the temperate desert lizards have developed the technique into an art. They avoid excessive heat gain by avoiding the heat of midday and by simply facing the sun's direction, thus exposing as little surface as possible. On the other hand, they increase heat absorption by presenting more of the body to the sun (Figure 35.12).

Questions about the extent of thermoregulation in reptiles also extend to species that are no longer around. There is an ongoing controversy over whether dinosaurs were endotherms or ectotherms. The current scarcity of these creatures only adds to the intrigue. The dinosaurs may have, in any case, been efficient thermoregulators. For example, the strange, highly vascularized backbone plates of the stegosaurs were excellent heat radiators and absorbers (see Figure 27.16). Other dinosaurs (for example, *Dimetrodon*) had high dorsal fin structures that probably served both to capture solar heat and to radiate excess heat, depending on the animal's orientation to the sun and the surrounding temperatures.

FIGURE 35.12
Thermoregulation Through Basking
Early in the day, the lizard's problem is to warm its muscles in the sun but not expose its sluggish body to predators. It often begins by exposing its head, warming it first before exposing its entire body. At midday it may linger in shadows or dig into the sand to avoid overheating. Then in the afternoon, it may again expose its entire body to the waning sun.

Osmoregulation and Excretion

Now let's look at another problem of regulation in animals—**osmoregulation**, the process of maintaining the proper water and ion balance in the body. Most cells are about two-thirds water, but in some cases, "about" won't suffice. In certain cells, the amount of water is critical, as is the relative abundance of various ions in the cell fluids. This discussion necessarily leads us into the methods by which excesses of water or ions are removed from the body by **excretion**, the removal of metabolic wastes from the body.

Producing Nitrogen Wastes

Amino acids, you recall, are the constituents of proteins. Typically, proteins must be broken down into individual amino acids during digestion so that they can be absorbed across the gut wall and into the bloodstream. Those same amino acids can be used to build new proteins in the cells; they may also be converted to fatty acids or carbohydrates for storage, or used as fuel in respiration. Some of these chemical conversions produce leftover nitrogen, in the form of ammonia, urea, and uric acid. If these compounds were to accumulate in cells, they would be extremely poisonous.

During **deamination** of an amino acid (Figure 35.13), the amine group—NH_2—is removed as NH_3, or ammonia. In the presence of protons from water, it readily forms ammonium ions (NH_4^+). Ammonia is highly toxic, but many organisms can safely handle it if they live in fresh water and are small enough to exchange materials easily with their environment. However, in many terrestrial animals, a group that generally must conserve water, the volume needed to dilute ammonia to a safe concentration may not be available. In such cases, the ammonia must be converted to something more manageable.

In many animals, ammonia is converted to *uric acid* or to *urea*. Uric acid is a semisolid, insoluble waste. It is the primary nitrogen waste of insects, reptiles, and birds and is usually produced as a dense paste. Urea is highly soluble and is the primary nitrogen waste of earthworms, many fishes and amphibians, and all mammals. Urea is dissolved in water and released as urine. Frogs are interesting in that their aquatic young, the tadpoles, excrete their nitrogen as ammonia. Only when they metamorphose into the partially terrestrial, water-conserving adults do they switch to urea production. So the precise way that nitrogen wastes are handled is also dictated by the need to conserve water. Let's look into this.

The Osmotic Environment

Each kind of environment presents its own osmoregulatory problems for animals. Over the eons, as species attempted to explore new environments, they faced a multitude of water-regulation problems. The results today are varied, as we will find as we review how natural selection has resulted in widely

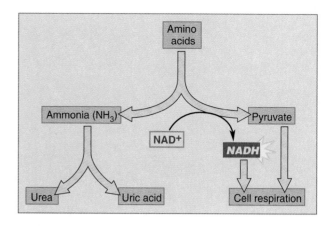

FIGURE 35.13
Nitrogen Wastes
In the metabolic breakdown of amino acids for cell respiration, the waste product, ammonia, is formed. Ammonia may be excreted directly, or it may be converted to urea and uric acid.

differing adaptations in animals that live in salt water, fresh water, and on land.

The Marine Environment

When we think of salt water we usually think of the 3.5% sodium chloride content of that solution, but we can't neglect other chemicals. Animals of the sea live immersed in a complex solution of ions, including sodium, potassium, calcium, magnesium, chloride, and others. In their totality, such solutes create osmoregulatory problems for the ocean's animal inhabitants. To survive in the marine environment, organisms must adapt to the troublesome osmotic conditions. The surrounding salt water is hypertonic to their bodies. It contains a greater concentration of sodium, chloride, and other ions than do their body fluids. Conversely, their body fluids are hypotonic to the sea, that is, they contain a higher concentration of water than do the surroundings. So there are two tendencies: the ions tend to diffuse in, and the water tends to diffuse out. If unabated, either can be life-threatening.

To survive this double threat, marine animals have evolved a variety of adaptive strategies, most of which place the animal in one of two broad categories: they are either *osmoconformers* or *osmoregulators*.

Osmoconformers. As the term **osmoconformer** implies, the solute concentration within the body conforms to that of its surroundings. Thus it matches the solute concentration of, or is *isotonic* to, the seawater. Many marine animals are osmoconformers, as this method requires little energy expenditure.

As an example, the limpet *Acmaea limatula* is a resident of the intertidal zone (Figure 35.14a). As such, it commonly experiences changes in salinity as the tides change or as runoff from the land dilutes its surroundings. Rather than

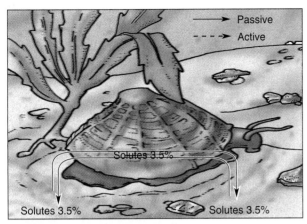

(a) The limpet, *Acmaea limatula*, is an osmoconformer. It remains isotonic to changing salinity by allowing the free diffusion of ions in and out.

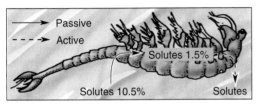

(b) The brine shrimp, *Artemia salina*, is an osmoregulator. It retains a stable solute balance in very salty water by actively transporting Na^+ across its gill membranes.

FIGURE 35.14
Solutions to the Salinity Problem
(a) Osmoconformers simply permit body salinity to vary with the surroundings, whereas osmoregulators (b) maintain a specific body salinity through active transport.

fighting to maintain any particular internal salinity, the limpet simply conforms to these salinity changes. Remarkably, its body fluids can remain isotonic in a salinity range of 1.5–5%. For some reason, these changes have little effect on the animal. A similar change in our body solutes would produce instant death.

While most vertebrates are osmoregulators, sharks and rays are osmoconformers, or at least partly so. Their adaptation to the hypertonic marine environment is to retain sufficient urea in the blood and tissue fluids to create an isotonic condition with the seawater outside. The nitrogen waste then is put to use. If this seems peculiar, keep in mind that as far as osmosis is concerned, it doesn't make any difference what the solutes are as long as the total concentrations of *water* inside and outside the organism are equal. When they are unequal, there will be a net movement of water down its gradient, in or out, until an equilibrium is reached. Incidentally, sharks and rays help avoid the problem of excessive salt intake by active-

ly transporting it back out. The site of transport is a special gland opening into the rectum.

Osmoregulators. Osmoregulators maintain a relatively constant internal solute concentration despite the salinity around them. Osmoregulation, though, involves work—active transport. One of the best-known regulators is the crustacean *Artemia salina,* which lives in salt ponds (see Figure 35.14b). It can maintain constant solute concentrations in its body when environmental salinity varies from less than 1% up to 30%! Only a highly specialized and efficient osmoregulator could survive under such varying conditions. The osmoregulatory organ of *Artemia salina* is the gill, which has salt glands that actively transport salt out of the blood. Interestingly, similar structures are used by marine vertebrates.

Most marine vertebrates are osmoregulators. The bony fishes have the same problem as the invertebrates: they lose water and gain salt. They make up for water losses by drinking seawater, but only enough to replace what water they lose. A terrestrial mammal that drank seawater would subsequently require huge quantities of fresh water to remove the salt from its body. But because of the special salt-secreting cells in its gills, the fish has no such problem. With salt actively transported from the body, seawater becomes perfectly suitable for drinking. The other ions are either excreted by the kidneys or passed through the gut unabsorbed.

Marine birds and reptiles also take in seawater with their food. However, the solute concentrations of their bodies are about the same as those of terrestrial species. So what do they do with all those salts? Unlike their terrestrial cousins, they do not rely on the kidney; instead, they actively excrete salt through special glands located near the eye that drain through a duct into the nose. Their kidneys, in fact, are not very efficient salt removers. Studies of gulls reveal that whereas the salt concentration of the excretory waste is well below 1%, the fluid exuded by their salt glands is about 5%.

Unlike marine fishes and birds, marine mammals have highly efficient kidneys and can actively transport ions into their urinary collecting ducts. For them, the kidney is the salt-excreting gland. However, most marine mammals avoid drinking seawater and rely on the relatively low-solute body fluids of the fish they eat.

The Freshwater Environment

Just as salt water presents osmoregulatory problems, so does freshwater. Freshwater is hypotonic to its inhabitants, which tend to take in the fresh water through osmosis. If the organism isn't able to keep excess water out, its cells may swell and rupture. But there is another problem. Freshwater doesn't provide its inhabitants with the ions they need. In particular, sodium, potassium, and magnesium are in short supply. Thus in freshwater environments, the ions tend to obey the laws of diffusion and leak out of the organisms into the environment. The organisms must expend energy to pump these errant mol-

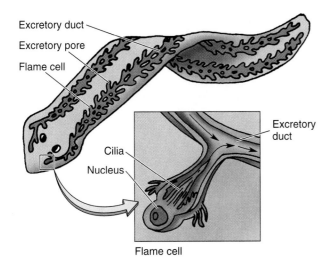

FIGURE 35.15
Flame Cells
Freshwater planarians have an extensive osmoregulatory system composed of ciliated flame cells that expel fluids into tubules leading out of the body.

ecules back in by active transport. Such transport may occur across the membranes of the skin, gills, or kidney tubules.

A freshwater existence also has certain advantages. For example, the animals there generally don't have much of a problem with handling nitrogenous wastes. In fact, many of them simply produce ammonia and flush it out with water, which dilutes it to safe concentrations. Let's now take a closer look at freshwater planaria, fishes, and amphibians.

One of the simplest organized osmoregulatory systems is the *flame cell system* of the planarians, freshwater flatworms. This system is chiefly an osmoregulatory device and possibly an excretory system as well (Figure 35.15). The flame cell system consists of numerous minute blind sacs leading into a complex network of tubules that empty to the outside via excretory pores. Water from intercellular fluids is transported into the flame cells by pinocytosis ("cell drinking"). Each flame cell contains numerous cilia facing the tubular lumen of

its *flame bulb.* The cilia set up currents that carry the fluids through the tubules to pores leading outside.

The kidneys of freshwater fishes, as you might expect, are not adapted for conserving water, and they produce copious amounts of highly dilute urine. Their kidneys are, however, extremely efficient at retaining salts, which would otherwise be quickly depleted. Despite such efficiency, the kidney cannot prevent some loss of salts. To counter this loss, the freshwater fish must transport salt into its body against its gradient, using active transport. Interestingly, once again the active transport of salt takes place in the gill, but in the opposite direction of what we saw in marine fishes.

The excretory system of amphibians is also rather highly developed. In frogs, the tadpole produces and excretes its nitrogen waste in the form of ammonia, much the way a fish does. As mentioned earlier, adult frogs shift to water-conserving urea as the principal nitrogen waste, but even so, most frogs drink and excrete huge amounts of water—up to 30% of their body weight daily. Like fishes, amphibians tend to lose body salts in the urine. Also like fishes, they actively transport salts from their water surroundings back into the body, but the site of the active transport in amphibians is the skin.

The Terrestrial Environment

In one important way, the terrestrial environment presents a problem similar to that of the oceans: how to avoid water loss. Terrestrial animals solve their water-retention problem through high water intake, highly efficient water-conserving excretory systems, watertight skin, and behavioral patterns that help deal with the problem of desiccation. We will first consider a few land-dwelling invertebrates.

Invertebrates of the Terrestrial Environment. Though the earthworm is considered terrestrial, it lives in a perpetually moist environment. Nonetheless the earthworm is a water conserver. The fluids in each segment are constantly filtered and recycled by pairs of rather complex structures known as **nephridia.** As you can see in Figure 35.16, each paired nephridium drains fluid from the segment just ahead of it. Any fluid moving through the

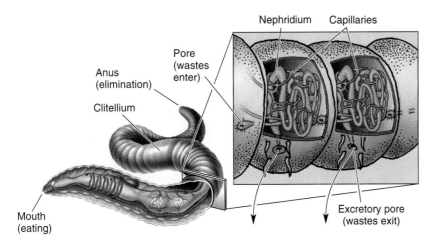

FIGURE 35.16
Nephridia
Paired nephridia occur in most segments of the earthworm. Their function is to excrete wastes and excess water. As the coelomic fluids pass through a nephridium, water, ions, and other valuable materials are returned to the blood via capillaries surrounding the nephridium.

ciliated funnel must pass a rich supply of capillaries in the tubule wall, where materials can be exchanged between the fluids and the blood. Water, minerals, and other essential materials are reabsorbed into the blood, while dissolved wastes pass to the outside through an opening known as the **excretory pore**. Here then osmoregulation and excretion are accomplished by the same organ, just as in the vertebrates. (In some ways, the nephridium is similar to the nephron, the filtering structure of the vertebrate kidney, which we will consider shortly.) But before leaving the invertebrates, let's consider some specializations in that enormous group called arthropods.

The arthropods of dry land have developed some very distinct ways of solving their osmoregulatory and excretory problems while reducing water loss. Their adaptations involve a water-resistant, waxy cuticle, a well-protected respiratory surface, and the elimination of nitrogen wastes in the form of semisolid uric acid.

The excretory system in insects consists of many blind, hollow, tubular structures known as *Malpighian tubules*. They emerge at about where the midgut and hindgut join (Figure 35.17). The tubules extend into the coelomic fluids and here absorb body wastes. Inside the tubules, nitrogen wastes from the coelomic fluids are converted to insoluble uric acid crystals, which pass through the lumen of the Malpighian tubules directly into the gut. Once in the gut, the uric acid joins the

digestive wastes, passing to the outside at defecation. This method of excretion conserves water in two ways. First, the nitrogen waste is a solid that doesn't require water for dilution. Second, fluids from the Malpighian tubules can be reabsorbed by cells in the lining of the hindgut. Thus the insects produce a dense, fairly dry digestive and excretory waste. The disadvantage of the uric acid route is the metabolic expense, requiring energy and a complex pathway.

Vertebrates of the Terrestrial Environment. Vertebrates that roam the land must constantly replace water that is lost during breathing and during the removal of nitrogen wastes. Aside from an essentially waterproof skin, their most important means of conserving water is a specialized kidney. Two vertebrate groups, the reptiles and the birds, conserve water through the production of uric acid, which can be eliminated in a semisolid state. Uric acid production is also an adaptation to development in the land egg, since it can safely be sequestered and isolated from the embryo. (Imagine the problem in the bird or reptile egg if water-soluble ammonia collected and then diffused through the embryo's tissues.) The primary nitrogen waste in mammals is urea, which requires a constant supply of water for its elimination. This is why a water-conserving kidney is so important in mammals. The human kidney serves as an example of this efficient organ.

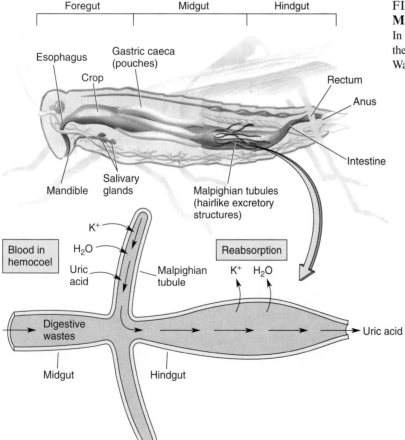

FIGURE 35.17
Malpighian Tubules
In insects, Malpighian tubules collect nitrogen wastes from the coelomic fluids (blood), before emptying into the gut. Water is reabsorbed from the fecal wastes as they form.

The Human Excretory System

Whereas humans may be very wasteful of water, their kidneys are not. (Consider that, in the home, Americans generally use about 3 gallons of highly purified water to flush away a few ounces of urine.) Like other mammals, humans have centralized, complex, and efficient kidneys, whose primary roles are the excretion of the nitrogen waste urea and the conservation and control of the body's water and mineral content. Let's first concentrate on the structure of the kidneys themselves and then discuss how they work and how their function is controlled.

Anatomy of the Human Excretory System

The human **excretory system** (Figure 35.18a) includes the **kidneys**, **ureters**, **urinary bladder**, **urethra**, and the blood vessels of the **renal circuit**. The renal circuit includes the paired **renal arteries**, leading to a great number of capillaries, and the paired **renal veins**. Blood entering the renal arteries comes directly from the aorta—the body's largest artery; thus it is under considerable pressure. High pressure is essential to the kidney's operation, and should blood pressure fall drastically, the kidney's vital functions will begin to fail.

Figure 35.18b illustrates the major features of the human kidney. Its anatomical regions include the **cortex** (the outer layers), the **medulla** (the inner area), and the **renal pelvis** (a cavity into which urine collects before it is eliminated). The dense cortex consists of the filtering units of the kidney—the **nephrons** and their related blood vessels. The medulla is largely composed of collecting ducts.

Briefly, urine formation begins when the nephrons receive materials that have been filtered out of the blood. It is from this filtrate that the urine will form. Once formed, the urine passes into the collecting ducts to empty into the cavities of the renal pelvis. From there, urine passes to the bladder via the ureter. From the bladder, urine is voided from the body through the urethra.

Microanatomy of the Nephron

The nephrons, which number about 1 million per kidney, are the functional units of the kidney (Figure 35.18c,d). Their structure reflects their function, as is so often the case in life. The nephron begins with **Bowman's capsule**, a hollow bulb or cup surrounding a ball of capillaries known as a **glomerulus**. Extending from the capsule is a lengthy tubule with a peculiar hairpin loop in its midsection. We will return to the loop, but first let's take a more detailed look at the glomerulus and the other blood vessels of the nephron.

The glomerular capillaries (tiny, thin-walled vessels through which many materials can pass) arise from the **afferent** (incoming) **vessel**—tiny branches of the renal artery. In the glomerulus, the blood passes along a lengthy, twisted, tortuous route under great pressure. As the blood emerges from the glomerulus, it enters the **efferent** (outgoing) **vessel**. The efferent arteriole branches into a second capillary network, the **per-itubular capillaries**, which form a fine network over the entire nephron. The peritubular capillaries, joined by those from other nephrons, form small veins that merge to form the renal vein.

Bowman's capsule leads into a lengthy tubule that makes up most of the nephron. The first region, the **proximal convoluted tubule**, winds a meandering route and then forms the long hairpin bend called the **loop of Henle** (see Figure 35.18c). This loop is found in all water-conserving kidneys and is extremely prominent in the kidneys of desert-dwelling mammals for reasons we will make clear. After forming the loop of Henle, the tubule again begins to twist and contort into another convoluted section, the **distal convoluted tubule**. It then joins a collecting duct, which also receives tubules from a number of other nephrons.

Note in Figure 35.18d that the afferent arteriole forms a passing connection with the distal convoluted tubule. This is the **juxtaglomerular complex**, an association that is vital to the kidney's role in sodium reabsorption and maintaining blood pressure.

In summary, the elements of the nephron proper, in the order of urine flow, are (1) Bowman's capsule, surrounding the glomerulus, (2) the proximal convoluted tubule, (3) the descending limb of the loop of Henle, (4) the ascending limb of the loop of Henle, and (5) the distal convoluted tubule, which leads to a collecting duct.

The Work of the Nephron

The critical nature of the nephron's role is underscored by its use of several transport processes. For example, we will encounter force filtration, active transport, endocytosis, exocytosis, diffusion, and osmosis. We will also see a special countercurrent exchange mechanism at work, one that is, in principle, quite similar to the countercurrent heat exchangers discussed earlier. These forces are summarized in Table 35.1 and Figure 35.19. To see how all this comes together in excretion, let's begin at Bowman's capsule.

Bowman's Capsule. Recall that afferent arterioles emerging from the renal artery enter Bowman's capsule, where they form a maze-like tuft of capillaries called the glomerulus. Arterial blood is pushed into the glomerulus with considerable force and is thus under great hydrostatic pressure. About one-fifth of its volume is forced through the capillary walls (the process being enhanced by the presence of small spaces between cells that make up the capillary walls) and into the cavity of Bowman's capsule. This filtering process is called **force filtration**, a truly descriptive term, and the fluid that passes across is called filtrate (see Figure 35.19).

Force filtration is nonselective, and only the larger molecules of the blood, and the blood cells themselves, escape its effects. Substances force-filtered from the blood include most water, ions and smaller molecules such as glucose, amino acids, and urea. Of the blood plasma, only the large proteins (molecular weight above 30,000) escape force filtration.

While the crude filtrate entering the glomerulus is considerable, not much of it ends up as urine. The adult kidneys

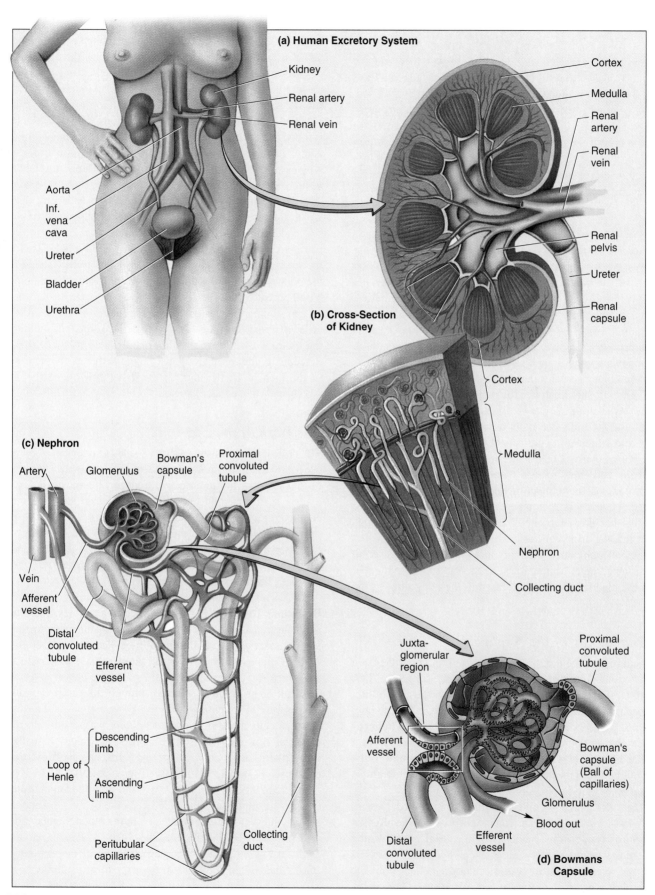

(a) Human Excretory System

Kidney

Renal artery

Renal vein

Aorta

Inf. vena cava

Ureter

Bladder

Urethra

Cortex

Medulla

Renal artery

Renal vein

Renal pelvis

Ureter

Renal capsule

(b) Cross-Section of Kidney

Cortex

Medulla

Nephron

Collecting duct

(c) Nephron

Artery

Glomerulus

Bowman's capsule

Proximal convoluted tubule

Vein

Afferent vessel

Distal convoluted tubule

Efferent vessel

Descending limb

Loop of Henle

Ascending limb

Peritubular capillaries

Collecting duct

Juxta-glomerular region

Afferent vessel

Distal convoluted tubule

Efferent vessel

Proximal convoluted tubule

Bowman's capsule (Ball of capillaries)

Glomerulus

Blood out

(d) Bowmans Capsule

FIGURE 35.18

TABLE 35.1
The Nephron at Work

Region	Process	Substances in Transit
Bowman's capsule	Force filtration from blood	Water, ions, glucose, urea, amino acids
Proximal convoluted tubule	Active transport out	Glucose, amino acids, Na^+
	Diffusion out	Cl^-, water
	Secretion into tubule	H^+, NH_3, K^+ (varies with aldosterone)
Descending loop of Henle	Diffusion out	Water
	Countercurrent concentration of salt and urea in kidney medulla	
Ascending loop of Henle	Active transport out	Cl^-
	Diffusion out	Na+, K+
Distal convoluted tubule	Active transport out	Na^+
	Diffusion out	Cl^-, water
	Secretion into tubule	H^+, NH_3, K^+
Collecting duct	Diffusion out of duct	Water (varies with ADH), urea joins salt in kidney medulla
	Secretion into duct	NH_3, H^+, K^+

on an average day (depending on fluid intake) filter about 180 liters of fluid from the blood, but usually no more than 1.2 liters of urine are formed. Thus more than 99% of the water filtered out of the blood (and much of the salt) is returned to the blood before urine leaves the collecting ducts.

Reabsorption in the Proximal Convoluted Tubule. The filtrate enters the nephron, passing into the proximal convoluted tubule (see Figure 35.19). Surrounding the proximal convoluted tubule is a network of efferent arterioles carrying blood that is now highly concentrated and therefore hypertonic. This hypertonic condition results in a steep osmotic gradient that causes the reentry of water from the dilute primary filtrate. Thus much of the water from the proximal convoluted tubule reenters the blood by osmosis. In addition, valuable substances

FIGURE 35.18
The Human Excretory System
(a) The human excretory system includes paired kidneys, their blood vessels, the ureters, urinary bladder, and urethra. (b) The kidney contains an outer cortex, an inner medulla, and a final collecting region known as the renal pelvis. Note the relationship between the cortex and medulla of the kidney and nephrons, including the loop of Henle. (c) The functional units of the kidney are the nephrons. Each includes the following: Bowman's capsule, proximal convoluted tubule, loop of Henle, and distal convoluted tubule. Each nephron is joined to a nearby collecting duct. (d) Blood from the renal artery enters the nephron at the Bowman's capsule, where an afferent vessel has branched into a mass of smaller vessels that comprise the glomerulus. Emerging from the glomerulus is an efferent vessel that immediately branches to form the extensive peritubular capillary network over the entire nephron. The juxtaglomerular region of the nephron is essential to the control of sodium reclamation.

in the filtrate, such as sodium, glucose, and amino acids, are actively transported out of the tubule to enter the surrounding cells and then cross the walls of the peritubular capillaries, returning to the blood. Chloride ions passively follow sodium ions and when they recombine, salt (NaCl) is restored to the blood. The active transport of sodium ions is carried out by the familiar sodium/potassium ion exchange pump (see Chapter 5). So in addition to sodium being pumped out of the filtrate, potassium is pumped in. But potassium, too, is a valuable ion, so much of it will be reclaimed before it passes out.

As the filtrate leaves the proximal convoluted tubule it contains urea, miscellaneous toxic substances (small amounts of ammonia, creatine, and uric acid), some salt, and still much of the original water.

Activity in the Loop of Henle. It is in the loop of Henle and in the collecting ducts that most of the remaining water and salt will be returned to the blood (see Figure 35.19). As far as we know, water cannot be actively transported by membranal pumps in biological systems, yet most of the water in the filtrate returns to the blood. This is where *countercurrent exchange* comes into play.

The **countercurrent exchange** mechanism of the nephron is a means of concentrating salt (sodium and chloride ions) and urea in the surrounding kidney medulla by an opposing flow of fluids. The hypertonic environment thus formed encourages the diffusion of water out of the tubule into the salty surroundings. From there, the water enters the nearby peritubular capillaries, where the blood is in an even greater hypertonic state than the salty surroundings. The important point is that while water cannot be actively transported, the active transport of salt through a countercurrent mechanism provides an osmotic environment that brings about the passive movement of water in the neces-

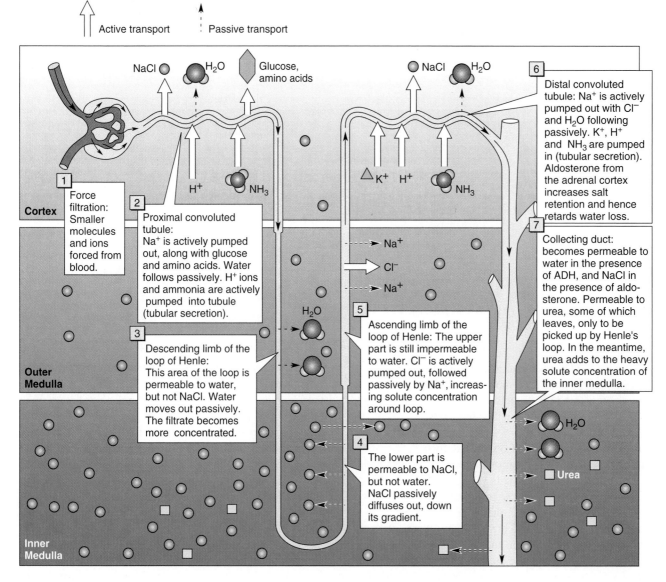

Active transport   Passive transport

NaCl   H₂O   Glucose, amino acids   NaCl   H₂O

1 Force filtration: Smaller molecules and ions forced from blood.

Cortex

H⁺   NH₃   K⁺   H⁺   NH₃

2 Proximal convoluted tubule: Na⁺ is actively pumped out, along with glucose and amino acids. Water follows passively. H⁺ ions and ammonia are actively pumped into tubule (tubular secretion).

6 Distal convoluted tubule: Na⁺ is actively pumped out with Cl⁻ and H₂O following passively. K⁺, H⁺ and NH₃ are pumped in (tubular secretion). Aldosterone from the adrenal cortex increases salt retention and hence retards water loss.

3 Descending limb of the loop of Henle: This area of the loop is permeable to water, but not NaCl. Water moves out passively. The filtrate becomes more concentrated.

Outer Medulla

H₂O

Na⁺
Cl⁻
Na⁺

5 Ascending limb of the loop of Henle: The upper part is still impermeable to water. Cl⁻ is actively pumped out, followed passively by Na⁺, increasing solute concentration around loop.

7 Collecting duct: becomes permeable to water in the presence of ADH, and NaCl in the presence of aldosterone. Permeable to urea, some of which leaves, only to be picked up by Henle's loop. In the meantime, urea adds to the heavy solute concentration of the inner medulla.

4 The lower part is permeable to NaCl, but not water. NaCl passively diffuses out, down its gradient.

H₂O

Urea

Inner Medulla

FIGURE 35.19
Transport Processes in the Nephron
Each region of the nephron makes use of its unique anatomy and various transport processes to carry out its function. As a result, wastes, excess water, and salts end up in the urine. Through selected recovery mechanisms, valuable blood constituents are recovered, and the body's water and ion balances are maintained. Note the steep solute gradient produced around the loop of Henle, a product of the countercurrent flow of salt and urea. Also note the opposing process of tubular secretion ②, ⑥, ⑦.

sary direction. How then is the countercurrent mechanism set up? Let's take a look.

Anatomically, the descending limb of the loop of Henle, the ascending limb of the loop, and the collecting tubules form a curious S-shape, with all three portions lying more or less side-by-side (see Figures 35.18c and 35.19). (Filtrate moves in opposite directions in the descending and ascending limbs, thus establishing a countercurrent.) Bowman's capsules and the convoluted tubules lie in the cortex (outer part) of the kidney, while the loops and the collecting tubules extend into the medulla (inner part). The tissue fluids of the

medulla are relatively salty (hypertonic), so as the dilute filtrate flows down the descending limb of the loop, it loses water by passive diffusion. Then, at the lower region of the ascending limb of the loop, salt moves passively out into the salty surroundings (see Figure 35.19 ④).

Next, in the upper region of the ascending limb of the loop, salt is actively transported from the nephron into the surrounding tissue (see Figure 35.19 ⑤). [Actually, only chloride (Cl⁻) is pumped out; sodium (Na⁺) ions follow the negative chloride ions.] This active transport both enables the blood to recapture the salt and maintains a high salt concentration in the kidney medulla, thus further encouraging the movement of water out of the descending loop. The water and ions then enter the nearby peritubular capillaries.

Reabsorption in the Distal Convoluted Tubule and Collecting Ducts. In the distal convoluted tubule, as in the proximal tubule, sodium ions are actively transported out, followed by chloride ions, adding still more solute to the surroundings. As before, water follows the ion gradient. Following the loss of salt and water, filtrate in the tubule tends to become isotonic with blood in the nearby peritubular capillaries, so there is no further tendency for water to move in or out. But the newly formed dilute urine next flows through the collecting ducts and has to traverse the salty medulla one more time (see Figure 35.19). As the collecting duct passes through medullary regions with an increasing outward osmotic gradient, the urine may lose yet more water. We say "may" since, as we will see, a special hormonally regulated mechanism is at work in the collecting ducts. In addition, urea itself diffuses out of the collecting duct and into the medulla, further increasing the hypertonic state of the medulla.

Tubular Secretion. While materials are being reabsorbed into the peritubular capillaries, another process, **tubular secretion**, is also occurring. Tubular secretion actively transports substances *out* of the peritubular capillaries and *into* the proximal and distal convoluted tubules and collecting duct. Not all of the blood's low molecular weight solutes pass into Bowman's capsule during filtration. Certain molecules and ions left in the blood after it leaves Bowman's capsule, such as ammonia, hydrogen ions, potassium ions, organic acids, and creatine, must be actively transported from the capillaries into the nephron. Tubular secretion then is another mechanism for getting rid of wastes from the blood. Table 35.1 reviews the structures of the nephron, the forces at work in each structure, the contents of the nephron at key points, and the contents of the peritubular capillaries.

Control of the Nephron Function

Control of Water. The movement of materials in the nephron is obviously complex, involving diffusion, osmosis, and active transport. It should not be surprising to learn that these processes can be adjusted through various control mechanisms. Let's look at some of these controlling mechanisms, beginning with the hypothalamus. If the blood registers as hypertonic (meaning the body is beginning to run short of water and osmotic pressure is high), the kidney is put on a water-rationing regime. In this case, the hypothalamus secretes **antidiuretic hormone**, or **ADH**, into capillaries of the posterior lobe of the pituitary, and the hormone is then distributed by into the blood. Its targets are the epithelial cells of the collecting ducts (Figure 35.20a).

ADH increases the permeability of the collecting duct to water. Thus more water leaves the collecting duct to reenter

the blood. When the osmotic consistency of the blood is restored to a normal range, the stimulation of the hypothalamus slows and ADH secretion falls off, producing an increase in urine volume. The stimulation of the hypothalamus also creates the sensation of thirst. As the individual drinks, the blood becomes diluted, and, again, the hypothalamus stops sending out antidiuretic hormone and the kidney increases its output of fluid. Eventually, another sensation is produced, this time from the stretch receptors of the bladder. And a new behavior is initiated.

Control of Salt and Blood Pressure. Another hormonal mechanism operates to control the retention of sodium chloride, which is closely related to blood pressure. In this case the hormone is the steroid **aldosterone**, which is released by the cortex of the adrenal gland (see Chapter 36).

Cells of the juxtaglomerular complex of the nephron's distal tubule monitor minute changes in blood pressure in the nearby afferent vessel (the one carrying blood into the glomerulus—Figure 35.20b). If blood pressure is below optimum, the complex stimulates the release of the enzyme **renin** into blood passing through the afferent vessel. Renin has no effect on the nephron, but it does activate a blood protein that, on reaching the adrenal cortex, stimulates aldosterone secretion. Aldosterone affects cells in the nephron's distal convoluted tubule and the collecting ducts. The target cells respond by speeding up their transport of sodium ions out of the tubule and back to the blood. As expected, water follows, and the additional water increases blood pressure enough to form a negative feedback loop back to the juxtaglomerular complex.

This mechanism has two important aspects. First, it ensures that the blood pressure in the kidney itself is great enough to maintain an efficient force filtration phase. Second, the aldosterone-related increase in sodium transport also means an increase in the excretion of potassium ion, mentioned earlier. The active transport mechanism involved is identical to the familiar sodium/potassium ion exchange pump of nerve cells.

Control of Acidity. We should add that the excretory system is also involved in other homeostatic mechanisms—for instance, in the control of blood acidity. The pH of our blood can change, depending on what we eat, but it is generally kept within fairly narrow limits. Since the acidity of urine can vary tremendously, from pH 4 to 9, this indicates that excess hydrogen and hydroxide ions in the blood, like toxic substances, are eliminated through the kidneys. Aldosterone, by the way, also speeds the movement of hydrogen ions into the urine.

In this chapter we have seen only a few of the many delicate, interacting, and highly coordinated mechanisms that keep the body's internal environment within the extremely precise limits critical to life. Remember, we live in what is essentially a disruptive environment. To remain organized in

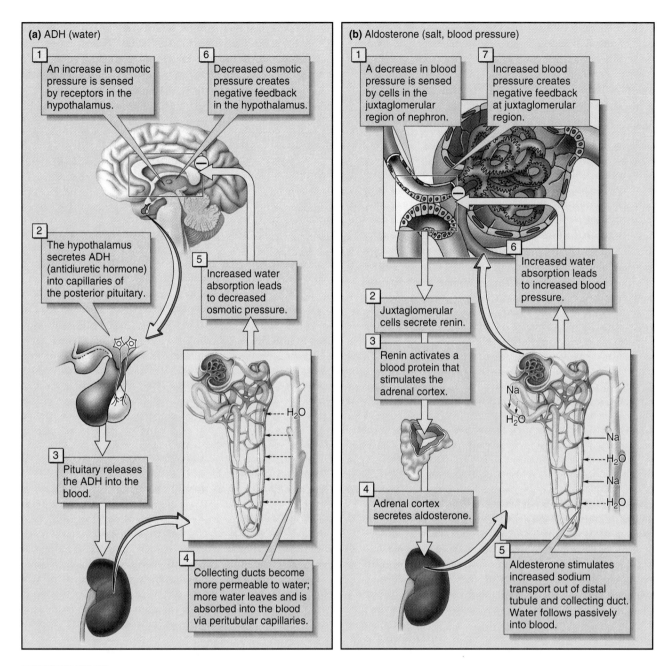

FIGURE 35.20
Hormonal Control of Osmoregulation
(a) Much of osmoregulation is under the influence of the hormone ADH, which is produced by the hypothalamus and secreted by the posterior pituitary when osmotic pressure is high. The system is regulated through negative feedback, which occurs as the water content of the blood reaches the optimal range. (b) Both osmoregulation and blood pressure (which are closely related) are influenced by the hormone aldosterone, which is secreted by the adrenal cortex. The control of blood pressure is indirect, since aldosterone promotes greater sodium reabsorption by the nephrons, and this results in more water leaving the crude filtrate and reentering the blood. Increased water means increased blood pressure.

the face of potential disruption requires an ongoing, uphill battle against entropy-increasing forces. That battle is best fought under optimal physiological conditions, and the body's homeostatic mechanisms help ensure those conditions.

Next, we will focus on a general mechanism that controls the body's interaction and coordination and promotes the stability necessary for complex systems, such as those we've seen here, to function properly.

Key Ideas

HOMEOSTASIS

1. **Homeostasis** is defined as the tendency for a physiological system to maintain internal stability through the coordinated response of its parts to anything tending to disturb such stability.

2. Many homeostatic mechanisms are regulated by **negative feedback loops**, whereby a process produces a reaction that, in turn, inhibits the process.

3. **Positive feedback loops** also occur, but they are often a sign of physiological trouble. In this case a process creates a reaction that, in turn, accelerates the process.

THERMOREGULATION

1. **Thermoregulation** is the ability of an animal to regulate its body temperature. Through thermoregulation, animals maintain metabolic activity at low temperatures, and avoid the denaturing effects of high temperatures.

2. Whereas the core body temperature of **homeotherms** is relatively constant, that of **poikilotherms** varies with external temperatures. Probably few pure poikilotherms exist. **Heterotherms** are part-time thermoregulators.

3. Ectotherms utilize external heat, employing behavioral strategies for both heating and cooling the body. Endotherms release more heat energy and thus require greater food intake.

4. **Endothermy** places limitations on small size, because small animals have proportionally larger surfaces from which heat can escape.

5. The bluefin tuna retains a warm core temperature around its swimming muscles through the use of countercurrent heat exchangers. Through its retention of heat, the tuna remains active and fast-moving in very cold water.

6. Thermogenesis in birds and mammals includes oxidative respiration, shivering, and brown fat metabolism. Brown fat thermogenesis is important to thermoregulation in human infants and in those mammals recovering from low hibernation temperatures.

7. Birds and mammals resist heat loss by shunting blood away from the skin and extremities and through the use of countercurrent heat exchangers. In our own arms and legs, blood has two alternative routes: along the skin (for heat loss) or alongside deep arteries (for heat retention through the countercurrent exchange).

8. Many mammals, particularly marine species, make use of thick layers of blubber to provide insulation from frigid surroundings.

9. Shunting blood toward the skin aids in cooling the body through radiation, conduction, and convection. Evaporative cooling, involving the use of external water or water from sweat glands, is an efficient means of speeding heat loss.

10. The body also regulates metabolically—increasing or decreasing heat output by varying the rate of cell respiration. To warm the body the hypothalamus may (a) stimulate the adrenal medulla to release the hormone epinephrine, which prompts the liver to release glucose for increased cell respiration and heat output; (b) prompt the pituitary to release thyroid-stimulating hormone, which, in turn, causes the thyroid gland to release a metabolism-elevating hormone, thyroxin.

11. **Ectothermy**, absorbing heat from the environment, has the advantage of low energy (food) costs. A clear disadvantage is the failure of behavioral strategies when it is cold.

OSMOREGULATION AND EXCRETION

1. Osmoregulation is the ability of animals to regulate ions and water in the body. It is closely related to excretion, the removal of metabolic wastes.

2. During **deamination**, the amino groups are removed from amino acids forming toxic ammonia, NH_3, which is excreted.

3. In insects, reptiles, and birds, the primary excretory waste is uric acid, while for many other invertebrates and the fishes, amphibians, and mammals, the primary excretory waste is urea.

4. The marine environment is a hyperosmotic medium in which organisms tend to lose water and gain ions. The tissues of **osmoconformers** conform to the surroundings, becoming isotonic. The tissues of **osmoregulators** maintain constant osmotic conditions in the body.

5. The freshwater environment is hypotonic, so water tends to enter the body through osmosis. Conversely, ions tend to diffuse out of the body. Since water conservation is not necessary, nitrogen wastes can readily be flushed out with water.

6. Some freshwater invertebrates use the flame cell system, where excess fluids are transported out of the body.

7. The kidney of the freshwater fish recovers some ions, but some must also be actively transported in through gill structures.

8. Osmoregulatory problems in the terrestrial environment primarily involve conserving water. All terrestrial vertebrates have specialized, water-reabsorbing kidneys

THE HUMAN EXCRETORY SYSTEM

1. The human **excretory system** consists of the **kidneys, ureters, urinary bladder, urethra,** and the **renal circuit** (**renal arteries** and **renal veins**).

2. The kidney includes an outer **cortex**, middle **medulla**, and the **nephrons**. The nephrons include a capsule and a looping tubule that joins others to form the collecting ducts. The ducts empty

into the **renal pelvis**, which empties into the ureters, and then the urinary bladder. The urethra voids the urine from the body.

3. The nephron begins with **Bowman's capsule**, which surrounds the **glomerulus**, a ball of capillaries arising from an **afferent vessel** of the renal artery. Leaving the glomerulus is an **efferent vessel**, which forms the **peritubular capillaries**, where reabsorption takes place. These spread over the nephron to later form a small vein that joins others to make up the renal vein.

4. Bowman's capsule leads to the **proximal convoluted tubule**, the **loop of Henle**, and the **distal convoluted tubule**, which joins a collecting duct. The afferent arteriole also connects with the distal convoluted tubule, forming the **juxtaglomerular complex**.

5. **Force filtration** in Bowman's capsule causes much of the water and ions and smaller molecules to leave the blood and enter the proximal convoluted tubule.

6. In the proximal convoluted tubule, the peritubular capillaries contain blood in a hypertonic state, so much of the water filtrate reenters the blood by osmosis. Active transport also returns sodium (chloride follows passively), glucose, and amino acids to the blood.

7. The ascending loop of Henle actively transports chloride ions (sodium ions follow passively) into the surrounding area, recycling salt and creating a hypertonic state in the kidney medulla.

8. In the distal convoluted tubule, the active secretion of sodium ions occurs with chloride ions and water passively following. Potassium ions enter the tubule.

9. Water leaves the collecting ducts in response to **antidiuretic hormone** (**ADH**), which is secreted by the posterior pituitary in response to osmotic conditions in the blood (actually detected by the hypothalamus).

10. **Tubular secretion** forces ammonia, hydrogen ions, potassium ions, organic acids, and creatine into the tubule.

11. Nephron control is hormonal, with water reabsorption controlled by ADH from the posterior pituitary and sodium chloride reabsorption controlled by **aldosterone** from the adrenal medulla. Sodium chloride transport and blood pressure are monitored by the juxtaglomerular complex. The arteriolar cells secrete **renin**, which stimulates the adrenal cortex to secrete aldosterone. Aldosterone increases the reabsorption of sodium chloride, which is followed by water, increasing blood pressure.

Application of Ideas

1. Many of the animal body regulatory functions are automated, occurring with little, if any, conscious intervention. What are some adaptive advantages to this kind of control? What is the alternative, and how might it affect the general efficiency of an animal's metabolic activities?

2. Homeostatic mechanisms involve far more than thermoregulation and osmoregulation. Describe examples of homeostasis in other systems, such as digestive, respiratory, and circulatory, and in the action of skeletal muscle.

3. Is thermoregulation in an endotherm more likely to fail under conditions of extreme heat or extreme cold? Explain fully.

4. Once the problem of cross-matching tissue types is solved, transplanting a kidney is a comparatively simple and straightforward procedure. Knowing what you do about kidney function, explain this statement. Exactly what is required for a kidney to function, and why are its needs so simple?

Review Questions

1. Give an example of negative and positive feedback. In biological systems, what do most instances of positive feedback indicate? (pages 741–742)

2. Distinguish between homeothermy, poikilothermy, and heterothermy. (page 742)

3. How do heat sources differ in endotherms and ectotherms? (pages 742–743)

4. State one advantage and one disadvantage of endothermy and of ectothermy. (pages 742–743)

5. Using a simple drawing of paired pipes, one carrying cold fluid and the other warm fluid, explain the principle of the countercurrent heat exchanger. (pages 743–744)

6. State the relationship between body size and metabolic rate in mammals. Name a mammal at each extreme. What is the principal disadvantage of a very high metabolic rate? (page 745)

7. Explain how heat production through brown fat metabolism works. Why is this source of heat important to hibernating mammals? (page 745)

8. List an advantage and a disadvantage in the shunting of blood away from the body extremities during extreme cold. (page 747)

9. What is the role of the thyroid gland in maintaining a constant internal body temperature? Explain how it is monitored. (page 748)

10. Define osmoregulation. What substances are involved? (page 751)

11. List the three common nitrogen wastes. Through what chemical activity are such wastes produced? (page 751)

12. What are two osmoregulatory problems of animals living in the marine environment? (pages 751–752)

13. What osmoregulatory problems confront the freshwater animal? (pages 752–753)

14. Describe the gross anatomy of the human excretory system, listing five significant parts. Trace the flow of urine through these parts. (pages 754–755)

15. List the five parts of a nephron in their functional order, and carefully explain the arrangement of related blood vessels. (page 755)

16. In general, what is the function of the loop of Henle? What is the significance of its hairpin shape? (pages 755–757)

17. Explain how the secretion of aldosterone is regulated. (page 759)

CHAPTER 36

Hormonal Control

CHAPTER OUTLINE

THE CHEMICAL MESSENGERS

Neural and Hormonal Control Compared

Molecular Structure of Hormonal Messengers

Characteristics of Hormonal Control

CHEMICAL MESSENGERS AND THE TARGET CELL

Peptide Hormones and Second Messengers

Steroid Hormones and Gene Control

INVERTEBRATE HORMONES

Hormonal Activity in Arthropods

THE HUMAN ENDOCRINE SYSTEM

The Hypothalamus and Pituitary

The Thyroid

The Parathyroid Glands

The Pancreas

The Adrenal Glands

Essay 36.1 Diabetes Mellitus and Hypoglycemia

Essay 36.2 The Heart as an Endocrine Structure

The Gonads: Ovaries and Testes

The Thymus

The Pineal Body

Prostaglandins

KEY IDEAS

APPLICATION OF IDEAS

REVIEW QUESTION

Living things respond. If they didn't, they wouldn't exist—plain and simple. Responses take many forms, some sudden and fleeting, others slower and longer lasting. Some responses are even permanent. Whatever the time frame, though, many of life's responses are quite complex and tightly coordinated. If you poke a sleeping dog with a stick and scream, he may leap straight up into the air. The response may seem simple enough, but unseen within his body, chemicals and neural changes interact in immediate and complex ways to help him adjust to this new situation. In the longer term, other neural and chemical changes may be associated with his changing attitude toward you. We have seen how neural mechanisms help us respond to stimuli from both the external and internal environment. Here we will look at chemical mechanisms, focusing primarily on hormones. We will also see how neural and chemical events interact to produce adaptive responses.

The Chemical Messengers

Many biologists have chosen their profession simply because they are fascinated by the complexities of life and want to help unravel the mysterious weave that makes up life's fabric. The tighter the weave (the more aspects to the problem), the greater the challenge. And among the greatest of biological challenges has been to try to understand how cells come to respond so precisely when quietly touched by molecules of a particular configuration (Figure 36.1). In many cases, the molecules act as "messengers," sent perhaps from somewhere else in the body.

Until recently, all chemical messengers were known simply as **hormones**. They were defined as chemical substances produced in one part of the body and transported in the blood to other parts of the body, where they produced a specific effect. While attractive in its simplicity, this definition now must be broadened. We can still agree that hormones are chemicals that regulate the actions of cells, that they are effective in small dosages, and each affects only specific target cells. But today we know that many hormones don't travel very far at all. Some affect nearby cells and tissues, and others affect only the cell in which they are produced. Now endocrinologists speak of **endocrine hormones** (chemical messengers that act in areas other than where they are produced), **paracrine hormones** (those that act in adjacent cells), and **autocrine hormones** (those that act in the cell in which they are produced). In nearly all instances in which we use the term "hormone" here, we will be referring to endocrines, and most of our attention will be directed to this group. Endocrine hormones are produced by **endocrine glands**, the *ductless* glands that secrete their hormones into the blood. In contrast, *ducted* glands such as sweat and mucous glands are termed *exocrine glands*. Later, in Chapter 43, we will discuss *pheromones,* chemical messengers produced in one animal that affect the behavior of another. With our terminology sorted out now, we will begin our look into hormonal control by contrasting it with neural control.

Neural and Hormonal Control Compared

Neural and hormonal control generally work in close concert and often have overlapping functions, particularly in homeostasis. Interestingly, they have similar chemical characteristics. In fact, some vertebrate hormones are identical to certain neurotransmitters.

The similarities between neural and hormonal control have strong evolutionary implications. According to some biologists, neural control preceded hormonal control, the latter evolving as non-neural cells took on the task of synthesizing neurotransmitters. Other biologists believe that hormonal control preceded neural control. They point out that true nervous systems require at least a tissue level of organization, but organisms below the tissue level rely heavily on purely chemical control. The evolutionary sequence has not been resolved completely.

FIGURE 36.1
Hormones at Work
Salmon swim upriver to spawn and die, all under the influence of hormones.

As if to emphasize the close relationship between neural and hormonal control, some hormones are secreted by neurons. Such secretions, fittingly called **neurosecretions** (or *neurohormones*), are produced in neurons and travel through axons to their terminals, where, like neurotransmitters, they are released. But unlike neurotransmitters, which are released into synaptic clefts, neurosecretions are released into the blood. Neurosecretions are quite common in invertebrates, and less common in vertebrates.

A primary difference between neural and hormonal control involves the time frame in which they work. Neural activity is generally instantaneous—neural impulses arise and subside with startling swiftness. Chemical control is generally much slower, and its effects are longer lasting. Whereas the action of a neurotransmitter is restricted to a particular synapse, endocrine hormones are usually released into the bloodstream or body fluids for distribution to other cells throughout the body.

Molecular Structure of Hormonal Messengers

Most hormones fall into one of two general categories: peptides or lipids (Figure 36.2). The peptide hormones contain amino acids and range in size from parathyroid hormone and insulin, which contain 84 and 51 amino acids, respectively, to lightweights such as thyroxin, with two amino acids, and epinephrine, with just one. In the lipid category are the familiar four-ring steroids such as the sex hormones and a group of relative "newcomers" to biologists, the complex, long-tailed lipids known as *prostaglandins.*

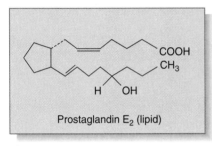

Peptide hormones

Lipid hormones

FIGURE 36.2
Chemical Structure of Hormones
Two classes of hormones are peptides and lipids. Some peptides are very simple, while others are large enough to qualify as proteins. Lipids include steroids and prostaglandins.

Characteristics of Hormonal Control

Over 50 hormones are likely to be circulating through the human body at any given time. Although present in very low concentrations, hormones are extremely powerful molecules, able to cause sweeping changes. Obviously, they must be under some form of closely coordinated control if the body is to avoid complete chaos. Much of this coordination is afforded by the nature of hormonal control itself.

Although hormones may circulate widely in the animal, they elicit responses only in specific **target cells**. Hormones encounter a variety of cells in their travels and in such a system it is up to the cell to identify the messenger. The identification is accomplished by the presence of very specific **receptor sites** on or in the target cells, sites that are able to bind only with certain hormones.

The precision of hormonal control is also made possible by the brevity of the chemical's existence. Hormones remain active in many cases for less than an hour and are usually enzymatically degraded after they have stimulated the target cell. The short hormonal lifetime is quite essential; after all, it wouldn't do to have every hormonal response repeated endlessly. Furthermore, the rapid degradation of hormones enables the body to promote both rapid, short-term effects, such as the familiar urgent responses to danger, and longer-term effects, such as those involving growth and development. The short-term effect is accomplished through a burst of hormone release, whereas the long-term effect occurs through a slow but steady release of hormones that are not so rapidly degraded.

Cells may secrete hormones in response to a number of cues. Hormone release may be stimulated by neurons, by hormones from other cells, or by conditions in the extracellular surroundings, and even in response to cues from the external environment. Commonly, hormone secretion is regulated through negative feedback (see Chapter 35). Recall that in negative feedback, the product of a process acts to slow the process so that when the product is abundant, the process slows; then with less product, the process is allowed to accelerate again, keeping the product at rather constant levels. So when a certain level of hormone concentration is reached, the gland cells releasing the hormone are inhibited, and hormone release slows. Then, when the hormone level falls sufficiently, gland inhibition lessens and hormone secretion again increases.

Chemical Messengers and the Target Cell

The next question is: Just what happens at the target cell? How do cells respond to hormones? The target cell's response

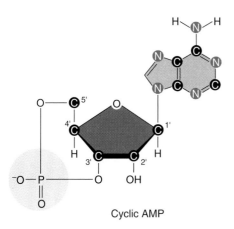

FIGURE 36.3
Cyclic AMP, A Common Second Messenger

Cyclic AMP is a modified version of AMP (adenosine monophosphate). Note the bonds linking the single phosphate back to ribose, forming a ring structure.

depends on the nature and location of the receptors that will receive the hormones. Do the hormones bind to the surface of the cell and cause changes within? Or do they enter the cell and affect the activity inside the nucleus itself?

Peptide Hormones and Second Messengers

Most peptide hormones attach to the target cell surface and cause changes within that cell. Those changes are due to a sequence of events triggered by "messengers" within the target cell. So the peptide hormones attached outside the cell are referred to as *first messengers* and the messengers within the cell as *second messengers*.

One of the most common second messengers is a molecule called **cyclic adenosine monophosphate** (Figure 36.3), usually shortened to "cyclic AMP" when spoken, and to **cAMP** when written. There are other second messengers, but let's take a look at cAMP and how it can regulate blood sugars.

cAMP, Epinephrine, and the Liver Cell. Epinephrine is produced in the adrenal medulla and released in response to stress. One of the effects of epinephrine is a sudden increase in the blood glucose level, resulting from the enzymatic breakdown of glycogen reserves in the liver. The chain of events leading to the elevated glucose level is now rather well understood.

Epinephrine is carried by the blood to its target cells, such as those of the liver (Figure 36.4). Like all target cells of peptide hormones, liver cells have very specific hormone receptor sites on their plasma membranes. Once epinephrine (the first messenger) binds to its specific membrane receptor on a liver

Secondary Messenger Systems (peptide hormones)

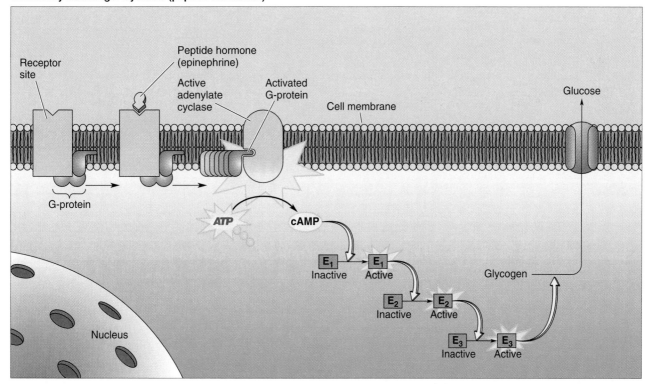

FIGURE 36.4
Action of Cyclic AMP

When epinephrine binds to its plasma membrane receptor on a liver cell, it sets in motion a sequence of reactions that lead to the breakdown of glycogen and the transport of glucose out of the cell.

cell, it activates an enzyme known as **adenylate cyclase** (the second messenger). This enzyme is part of the membrane system itself, but it remains inactive until the epinephrine binds to the receptor. But adenylate cyclase is situated on the cytoplasmic (interior) side of the plasma membrane. So how can it be activated by epinephrine? Another protein, called the **G protein**, also located on the cytoplasmic side of the plasma membrane can be stimulated by the epinephrine-activated receptors to act as an intermediary between the epinephrine and the adenylate cyclase. Actually, there are two kinds of G proteins involved in adenylate cyclase activity. One stimulates adenylate cyclase, the other inhibits it, a very effective means of hormonal control. When adenylate cyclase is activated by the epinephrine, it immediately converts cytoplasmic ATP to cAMP, which, in the case of epinephrine-stimulated liver cells, activates critical enzymes and eventually leads to the breakdown of glycogen.

Specifically, cAMP activates an enzyme called **protein kinase**. The protein kinase activates a second enzyme, phosphorylase kinase, which, in turn, activates glycogen phosphorylase *a*. Glycogen phosphorylase *a* then cleaves each glycogen molecule into many glucose units, actually glucose-1-phosphate and glucose-6-phosphate. The phosphate is then removed, and simple glucose passes through the plasma membrane of the liver cell and enters the blood.

Whereas in the liver protein kinase activates phosphorylase kinase, other kinds of cells contain other proteins capable of being activated by protein kinase. In this way, protein kinase can have a variety of effects depending on the kind of cell it activates.

An important aspect of the second messenger reaction is called a **cascade**. The initial reaction—the binding of a hormone to its receptor site—involves relatively few molecules. But with each following step, many more molecules are involved; thus a few molecules of epinephrine bound to the plasma membrane activate many cAMP molecules, which, in turn, activate a great number of enzyme molecules. Each enzyme, acting over and over and with lightning speed, as enzymes do, produces immense numbers of product molecules. One estimate is that, through the cascading effect, a single molecule of epinephrine can account for the release of 100 million molecules of glucose.

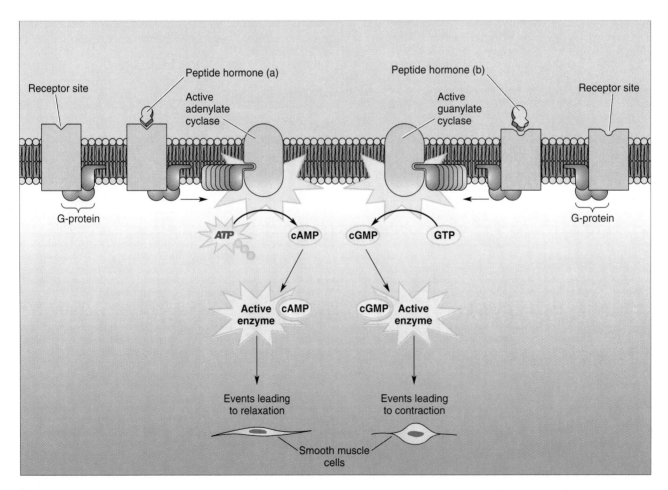

FIGURE 36.5
Multiple Second Messengers
In this model of a double second-messenger system, a smooth muscle cell relaxes or contracts in response to two different hormones. Each has its own receptor, membrane enzyme, and second messenger.

Another important characteristic of the second messenger operation is rapid degradation. The cyclic bonds of cAMP are readily cleaved, producing AMP. AMP is recycled into the usual respiratory pathways (Chapter 8), and ATP is regenerated. Furthermore, when the second messenger is deactivated, the protein kinases return to an inactive state. In this way, the entire system shuts down, remaining inactive until another barrage of epinephrine reaches the target cells.

It is now known that, except for thyroxin and insulin, all peptide hormones work through second messengers. This raises an important question. We know that cAMP is involved in the action of several hormones and that each of these hormones has a different effect. But how can one kind of agent bring on different effects? The answer lies in the ways that various kinds of cells differ in the enzyme systems awaiting activation by protein kinase. For instance, whereas the response to cAMP in liver cells may be the breakdown of glycogen and liberation of glucose, the same second messenger activated in follicle cells of the ovary will bring on the release of a sex hormone.

Dual Second-Messenger Operation. In many tissues, two different hormones can elicit opposing effects. Each has its own kind of receptor sites, and each stimulates the formation of its own second messenger. Such a dual control system works in both smooth and cardiac muscle.

In smooth muscle, that of the gut and blood vessel walls, for example, one hormone activates adenylate cyclase, which, in turn, catalyzes the conversion of ATP to cAMP, just as we described earlier. This brings on muscle relaxation. A second hormone activates the membrane enzyme guanylate cyclase, which, in turn, catalyzes the conversion of guanine triphosphate (GTP) to another second messenger, **cyclic guanine monophosphate** (*cGMP*). cGMP's effect is to cause smooth muscle to contract (Figure 36.5). The hormone in greater concentration (filling more receptor sites) has an inhibiting effect on the lesser hormone's ability to stimulate its second messenger. Thus as cAMP increases, cGMP decreases, and vice versa.

Steroid Hormones and Gene Control

The other class of chemical messengers, the lipid hormones, includes the steroids. This group does not employ second messengers, nor does it involve enzymes already present in the cytoplasm. Instead, steroid hormones penetrate the plasma membrane, enter the nucleus, and direct the action of genes there.

Upon entering the cytoplasm of a target cell, the steroid hormone becomes tightly bound to a **cytoplasmic binding protein**. The newly formed protein–hormone complex then moves into the nucleus, where it is able to bind with an acceptor protein positioned at certain locations along the chromatin. The complex of hormone, cytoplasmic binding protein, and acceptor protein activates certain genes, causing them to begin transcription (see Chapter 14). The resulting messenger RNA molecules move into the cytoplasm where they direct the formation of new proteins (Figure 36.6).

Gene Activators (steroid hormones)

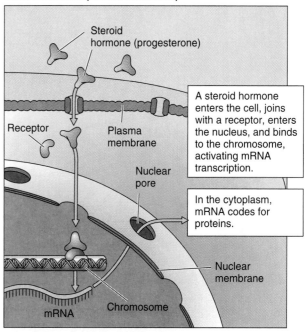

FIGURE 36.6
Steroids in the Cell
Progesterone, a lipid-soluble steroid, readily passes through the plasma membrane and joins a receptor called a cytoplasmic binding protein. Together they penetrate the nucleus, where the hormone–protein complex joins a chromosomal receptor, which then activates a DNA segment. RNA is transcribed, and a protein is produced along ribosomes.

One might wonder how a single hormone could cause different reactions in different cells, as when the male hormone testosterone promotes both muscle growth and voice changes. Apparently, although the hormones and cytoplasmic binding proteins are the same, the acceptor protein is located at different sites along the chromatin, each coding for a different protein.

Now that we've seen the primary ways that hormones function at the cellular level, we can consider, in both invertebrates and vertebrates, some specific hormones, their sources, and their effects on target cells.

Invertebrate Hormones

Probably all invertebrates rely on some form of chemical regulation. In many cases the chemical messengers are neurosecretions, those secreted by neurons rather than by specialized gland cells. In other cases they are hormones, glandular secretions. Because human fortunes are so intimately tied to those of the arthropods, particularly insects, we know more about chemical messengers in the jointed-leg creatures than we do about those in the other invertebrates.

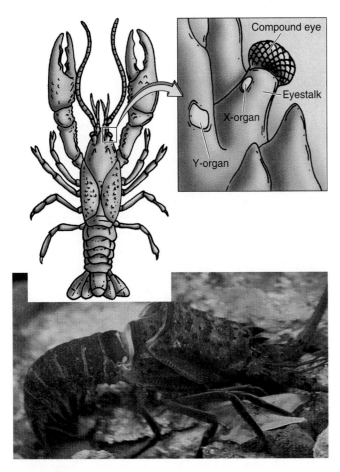

FIGURE 36.7
Molting Hormones
The lobster (*below*) has molted under the influence of molting hormones. The molting hormone, MH, is secreted by the Y-organ, whereas molt-inhibiting hormone, MIH, is secreted by the X-organ.

Hormonal Activity in Arthropods

Examples of chemical control abound in the arthropods. We know that hormones are involved in growth, reproduction, pigmentation, osmoregulation, and metabolism. In particular, researchers have worked out many of the hormonal actions involved in **ecdysis** ("to strip off") or molting—the periodic shedding of the exoskeleton seen in crustaceans and insects (Figure 36.7). Molting is necessary because the hard, confining exoskeleton does not provide space for growth. Molting occurs most frequently in the larval stage, but in some arthropods, it occurs at intervals throughout life.

In the crustacean, molting begins with preparatory steps, including a thinning and weakening of the hard, tough exoskeleton brought about by the absorption of calcium and organic substances from the exoskeleton into the body. Simultaneously, cells in the underlying epidermis divide and grow rapidly, forming a soft, new cuticle covering. It is then that actual shedding begins, usually with the old exoskeleton split-

ting down the back. Typically, the animal then arches its body and works its way out, leaving a ghostly shell of its former self. The body then absorbs water, stretching its new flexible cuticle. Finally, a "tanning" process begins. Calcium salts are added to the cuticle, thickening and hardening it as a new exoskeleton takes form.

Molting in the lobster is influenced by the interplay of two hormones, *MIH* (*molt-inhibiting hormone*) and *MH* (*molting hormone*). **MIH** is a neurosecretion produced by the **X-organ** in the eyestalk. **MH** is produced by the Y-organ, a true endocrine gland (one of the few true endocrine glands in invertebrates) located in the head (see Figure 36.8). Between molts, the ongoing secretion of MIH inhibits molting activity, but in response to some cue, generally external, its secretion slows and MH, the molting hormone, dominates.

Researchers have identified a number of external cues that influence molting. Included are light, temperature, injury, nutritional state, and reproductive activity. Some of the cues tend to be specific to certain groups of arthropods. For instance, some burrowing crabs will not molt if kept in constant light, whereas crayfish will not molt if kept in continuous darkness. Some environmental conditions influence molting in all groups. For example, low temperatures retard molting in all arthropods, as does starvation.

Hormonal Control of Development in Insects. Insect development has received a great deal of scientific attention, partly because some people would like to bring much of it to a screeching halt. It is difficult to overemphasize the importance of this research area, since our ability to control insects—probably our leading ecological competitors—depends on an exact knowledge of all aspects of their lives.

Much of what we know about their development has come from a few in-depth studies such as those on the American silkworm, *Hyalophora cecropia*. The silkworm undergoes a developmental program called a complete metamorphosis. In a complete metamorphosis, the insect passes through all developmental stages: egg, larva, pupa, and adult (see Chapter 26). The structural changes from one stage to another are strikingly dramatic. The silkworm larva goes through several molts before entering pupation. It spends a winter in the pupal stage, emerging from the pupa case in the soft light of spring. It has carefully conserved its stored energy through the winter by existing at a very low metabolic rate in **diapause**, a common developmental state in insects.

Metamorphosis in all insects is controlled through a complex interplay of two hormones *ecdysone* and *juvenile hormone* (Figure 36.8). **Ecdysone** is secreted by the **prothoracic gland**, which, in turn, is stimulated by a neurosecretion called **brain hormone** or **BH**. BH, not surprisingly, is secreted by neurons of the brain. Ecdysone has two roles: it supports molting and instigates the changes that bring on new developmental stages. Specifically, it brings the larval stage to an end by prompting the emergence of the pupal stage. Later, it triggers the appearance of the adult stage. The second hormone, **juvenile hormone** (**JH**), is produced by neurons in paired brain extensions known as **corpora allata** (singular, **corpus allatum**). JH counters the effects of ecdysone. Whereas it does not interfere with molting *per se,* it does maintain the juvenile state. That is, as long as the level of JH is high enough, it overrides the effects of BH and ecdysone, and the final transition into pupa and adult stages cannot occur.

The Human Endocrine System

As the vertebrates evolved, both the nervous system and hormonal system became increasingly important as means of maintaining homeostasis. The hormones work in a rather similar fashion in many of the vertebrates, so, for the most part, we will again draw on that representative mammal with the generalized teeth and the untalented toes.

We should begin by noting that, anatomically, the human endocrine system consists of a number of distinct glands as well as less organized hormonal tissues in various parts of the body (Figure 36.9 and Table 36.1). These glands are highly interactive, and their coordination is critical to the normal development and functioning of the animal.

The Hypothalamus and Pituitary

The first of the human endocrine structures we will consider is the **pituitary gland**. It is a bilobed structure about the size of the tip of your little finger, located at the base of brain, where it is cradled in a saddle-like pocket of bone rising from the floor of the skull (see Figure 36.10). If you point one finger directly between your eyes and stick another into your ear, you will not only gain the attention of other people on the bus, but the lines will intersect at about the location of the pituitary. Considering its small size, it may seem surprising that the pituitary is one of the busiest of the endocrine glands, releasing at least ten hormones.

In humans, the pituitary has two distinct lobes or structures, the **anterior pituitary**, a true gland, and the **posterior pituitary**, which is a neural structure. Although the first manufactures hormones, the second receives neurosecretions carried there by the axons of neurons.

In addition to their functional differences, the two regions are also embryologically distinct: the anterior pituitary develops from embryonic ectoderm from the embryonic mouth, whereas the posterior pituitary (or neurohypophysis) forms as an outgrowth of the developing forebrain, so it is indeed a neural structure.

The pituitary interacts intimately with the hypothalamus, a part of the forebrain mentioned several times in earlier chapters. It is located just above the stalk on which the pituitary is suspended (see Figure 36.10).

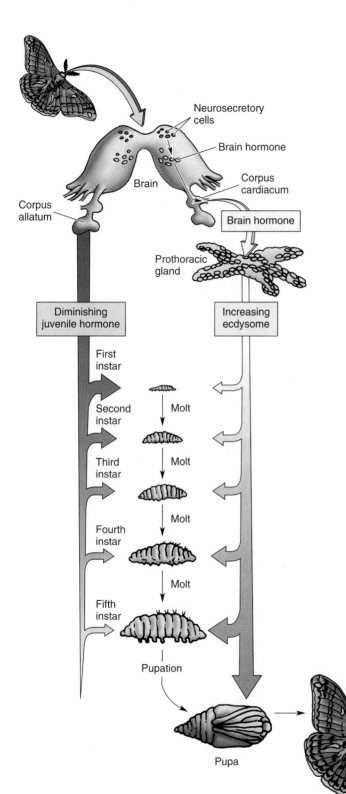

FIGURE 36.8
Hormones of Metamorphosis
When enough juvenile hormone is present, it dominates metamorphosis, permitting larval growth, but inhibiting pupal formation. As its concentration wanes, ecdysone dominates, a pupa forms, and the adult soon emerges. Pupation time varies greatly among different species.

Control by the Hypothalamus. In Chapter 35 we described how the hypothalamus plays a key role in maintaining body temperature and in osmoregulation, two critical homeostatic functions. In its relationship with the pituitary we find even more homeostatic functions. Specifically, the hypothalamus determines *what* hormones the two pituitary lobes will release and *when* they will be released. The hypothalamus actually interacts with the pituitary in two ways. Its relationship to the anterior pituitary is hormonal, but its relationship to the posterior pituitary is strictly neural. A close look at the anatomy will help with this distinction.

The Anterior Pituitary Connection. As we see in Figure 36.10 (*left*), the hypothalamus is anatomically linked to the anterior pituitary by a special circulatory arrangement. Capillaries in the hypothalamus merge to form a short vein (a *portal vessel*) that passes directly into the anterior pituitary. Neurons of the hypothalamus release neurosecretions into nearby capillaries. The secretions are then carried by the blood directly to the anterior pituitary. There they make contact with the glandular endocrine cells of the anterior pituitary, where they regulate the release of pituitary hormones.

There are nine hypothalamic secretions that control the anterior pituitary, each having a specific effect. Among these are **releasing hormones** and **inhibiting hormones**, secretions that encourage or inhibit, respectively, the release of hormones from the anterior pituitary. For example, **growth hormone releasing hormone (GHRH)** prompts the release of **growth hormone (GH)**, whereas **growth hormone release-inhibiting hormone** does just what you would expect—it inhibits the release of GH. A list of releasing and inhibiting hormones is given in Table 36.1. Note from the table that the secretion of three hormones of the anterior pituitary (GH, prolactin, and MSH) is controlled through *releasing hormones* and *release-inhibiting hormones,* whereas the others are controlled through negative feedback.

Releasing and inhibiting hormones are released into the blood in incredibly small amounts. Were such amounts to be released into the main circulation they would be too diluted to be effective. But because of the short and direct circuit, these key messengers move unerringly from the hypothalamus to their target cells in the anterior pituitary.

The Posterior Pituitary Connection. The relationship between the hypothalamus and the posterior lobe of the pituitary, as we see in Figure 36.10 (*right*), is even simpler and more direct. As we've indicated, the lobe itself is not glandular and does not synthesize hormones. Instead, it receives and then releases neurosecretions formed in the hypothalamus. Axons originating in nerve cells in the hypothalamus enter the posterior pituitary and make contact with its capillaries. The neurosecretions are then released into the capillaries for distribution to the body.

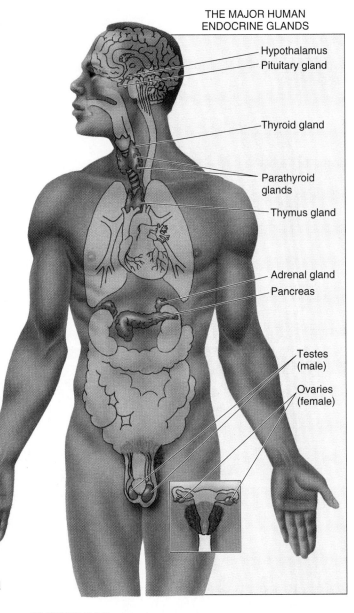

THE MAJOR HUMAN
ENDOCRINE GLANDS

- Hypothalamus
- Pituitary gland
- Thyroid gland
- Parathyroid glands
- Thymus gland
- Adrenal gland
- Pancreas
- Testes (male)
- Ovaries (female)

FIGURE 36.9
The Human Endocrine System
The hormone-secreting structures in humans consist of several distinct ductless glands and a number of less organized tissues.

Thus we see that the pituitary is subservient to the hypothalamus. Does this mean that the hypothalamus is autonomous? Not at all. The hypothalamus is controlled by an ongoing barrage of neural information from the central nervous system, as well as by a number of hormonal negative feedback loops.

Hormones of the Anterior Pituitary. We will begin our look at the anterior pituitary hormones with those that are controlled by negative feedback: ACTH, TSH, FSH, and LH (see Table 36.1). **ACTH,** or **adrenocorticotropic hormone,** is targeted for the **adrenal cortex,** the outer region of each **adrenal gland** of

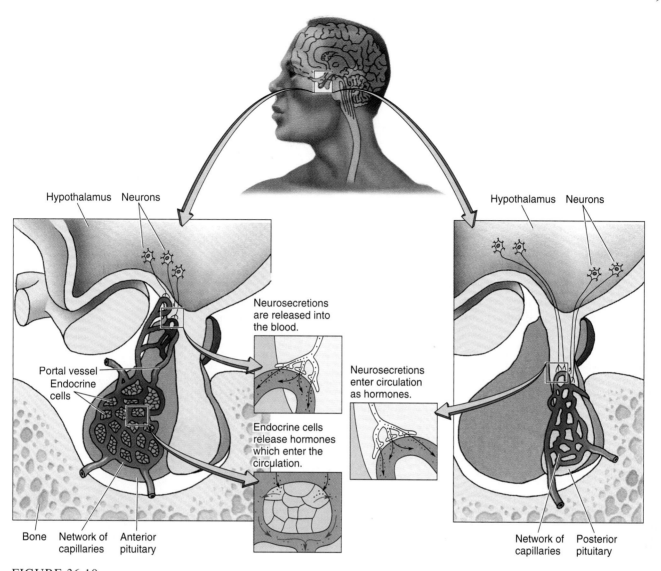

Hypothalamus Neurons

Hypothalamus Neurons

Neurosecretions are released into the blood.

Portal vessel
Endocrine cells

Neurosecretions enter circulation as hormones.

Endocrine cells release hormones which enter the circulation.

Bone | Network of capillaries | Anterior pituitary

Network of capillaries | Posterior pituitary

FIGURE 36.10
The Pituitary and Hypothalamus
The hypothalamus stimulates hormone secretion in the anterior pituitary *(left)* through its neurosecretions (releasing and inhibiting hormones). After they are released, the neurosecretions pass through the portal vessel to the anterior pituitary, entering cells there and stimulating hormone release. The posterior pituitary *(right)* receives its hormones from axons emerging from cell bodies in the hypothalamus; thus it acts only as a reservoir.

the kidney (see Figure 36.11). When stimulated by ACTH, the adrenal cortex secretes a number of steroid hormones that have various effects. As the hormone levels in the bloodstream rise, negative feedback begins. Since hormones enter the general circulation, they reach all parts of the body, including the hypothalamus and the anterior pituitary. Both are sensitive to rising levels of the hormones and at critical levels become inhibited. As a result the release of ACTH is slowed (Figure 36.11). In addition to its specific role in stimulating release of hormones from the adrenal cortex, ACTH also has another, more general, role in influencing the metabolism of fats, regulating their release from storage sites for redistribution in the body.

Thyroid-stimulating hormone, or **TSH** (also known as *thyrotropin*), is responsible for stimulating the **thyroid**, anoth-

er endocrine gland, to release the hormones **thyroxine** and **triiodothyronine**. These hormones regulate the metabolic rate of the body; we will have a closer look at their functions when we consider the thyroid gland later in the chapter. Thyroxine and triiodothyronine are widely targeted, so many tissues respond.

Follicle-stimulating hormone (**FSH**) and **luteinizing hormone** (**LH**) are both involved in stimulating the gonads to produce sperm or eggs and sex hormones. Since FSH and LH affect the gonads (testes and ovaries) they are known as **gonadotropins**. In human females, the two hormones play a regulating role in the monthly menstrual cycle. We will return to the specifics of FSH and LH action in Chapter 41.

The **melanocyte-stimulating hormone (MSH)** is controlled more directly by the hypothalamus through both releasing and inhibiting hormones. Releasing hormones, we've seen, stimulate hormone secretion, whereas inhibiting hormones slow such secretion. Another anterior pituitary hormone, **GH** or **growth hormone** (also called *somatotropin*), has very broad influences, activating many tissues. It stimulates growth by promoting the uptake of amino acids into cells, thereby accelerating protein synthesis. In addition, GH stimulates the release of fats from fatty tissues, in effect shifting the body more toward the metabolism of fats as an energy source. GH also stimulates the breakdown of liver glycogen into glucose (as does epinephrine, one of the adrenal hormones).

Proper levels of GH in the body are essential to normal growth. This becomes distressingly obvious when something goes wrong. Excessive secretion (such as might occur with a pituitary tumor) during the normal growing years can produce **pituitary giants**. Many people with this condition range from 7 to 9 feet tall. Conversely, in young people, lower than normal GH levels produce **pituitary dwarfs**, often called midgets (Figure 36.12a). There are several thousand pituitary dwarfs in the United States today. Until recently, treatment for growth deficiencies was difficult and expensive. Stimulating growth in even one pituitary dwarf required a daily injection of all the GH that could be isolated from several human bodies. But since GH is a protein, new sources are now available through recombinant DNA technology.

In an adult human, should GH levels rise suddenly, the resulting condition is known as **acromegaly**. In this abnormality, growth that has stopped on time mysteriously resumes, but it is restricted to areas where cartilage persists, such as the hands, feet, jaw, nose, and some internal organs (Figure 36.12b).

Prolactin promotes milk production in mammals, including humans. Toward the end of pregnancy, the blood levels of prolactin increase to 28 times that in nonpregnant females. But milk is still not produced, because it is inhibited by high estrogen and progesterone levels associated with pregnancy. These hormone levels fall off dramatically at birth and milk flow then begins. The actual release of milk is in part under the influence of *oxytocin* from the posterior pituitary. Thus while milk *production* is influenced by prolactin, milk *release* is brought on by oxytocin.

Because prolactin is found in so many vertebrate groups, it is believed to have appeared early in vertebrate evolution. Theorists suggest that its diverse roles in the vertebrate groups arose through evolutionary changes, not in the hormone, but in its target cells.

In humans, **melanocyte-stimulating hormone**, or **MSH**, is secreted by the anterior pituitary, but little is known about its function in our bodies. In other vertebrates, however, MSH influences the migration of pigment granules in cells called **melanocytes**. When MSH binds to cell surface receptors, melanin granules disperse throughout the melanocytes, bringing about a darkening of the skin. As the MSH concentration diminishes, the granules begin to cluster about the nucleus, and the skin lightens in color (Figure 36.13). This reversal

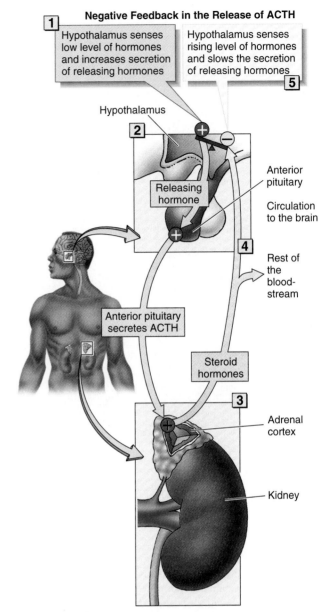

FIGURE 36.11
ACTH and the Negative Feedback Loop
In negative feedback loops, the product of an action has a suppressing effect on that action. Here the hypothalamus [1] has stimulated the pituitary [2] to release ACTH. In response, the target organ—the adrenal cortex—secretes steroid hormones [3]. The rise of steroids in the blood [4] acts as a negative feedback loop. [5] When sufficient quantities are present, they inhibit the hypothalamus, and the system slows or shuts down.

may be brought about by another hormone, melatonin from a structure called the pineal body (discussed later).

Hormones of the Posterior Pituitary. As we've seen, the "hormones" of the posterior pituitary are actually neurosecretions. They are produced by neurons of the hypothalamus and delivered to the posterior pituitary via axons of those neurons.

(a)

(b)

FIGURE 36.12
Growth Abnormalities
(a) When growth hormone levels are too low, the result can be growth retardation and the development of a pituitary dwarf. Oversecretion during adolescence can have the opposite effect, producing a pituitary giant. **(b)** Sudden increases in GH after maturity and when bone lengthening has ended produce growth in certain body parts, such as the face and hands. Here we see the result, a condition known as acromegaly.

A number of neurosecretions are distributed in this way. They form a chemical family known as **neuropeptides**, each composed of nine amino acids and differing slightly from each other in their amino acid composition.

The posterior pituitary releases two general categories of neuropeptides: *uterotonics* and *vasopressins*. The **uterotonics** promote contractions of the uterus and oviducts and prompt the release of milk. The best known of the milk releasers is **oxytocin**. In women, oxytocin is released by, among other things, the stimulation of nipples and by sexual intercourse. It causes uterine contractions during orgasm. The uterine activity, known as "tenting," may help ensure fertilization by actively drawing semen into the uterus (see Chapter 41). And we've seen that oxytocin functions with prolactin to permit milk flow.

Oxytocin also plays an important role in bringing about the contractions of the uterine muscles during labor and delivery (see Essay 41.1). Once the baby is born, suckling stimulates further oxytocin production, which helps bring the distended uterus back down to normal size. Synthetic oxytocin, known clinically as "pitocin," is usually administered to stimulate uterine contractions if labor is unusually prolonged or the contractions are

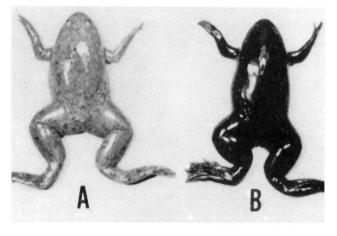

FIGURE 36.13
MSH and Skin Pigmentation
Coloration in the two frogs depends on the distribution of melanin pigments that occur in versatile melanocytes. Lighter coloration occurs when melanin granules form tight clusters. In the presence of MSH, the granules spread out, distributing melanin throughout each skin cell and bringing on the darkened appearance.

too feeble. Although the role of oxytocin in men hasn't been fully clarified, it is thought to stimulate the rhythmic smooth muscle contractions associated with ejaculation. When injected into male sheep and cattle, oxytocin increases semen output.

In humans, the posterior pituitary receives oxytocin and vasopressin from hypothalamic neurons. Human **vasopressin**, known more familiarly as **antidiuretic hormone** (**ADH**), helps adjust blood pressure levels by bringing about an increase in the recovery of water in the kidney, prior to the final formation of urine. ADH is secreted in response to impulses from hypothalamic pressure sensors that constantly monitor the osmotic condition of blood (Chapter 35).

The Thyroid

In humans, the thyroid gland is shaped somewhat like a bow tie and is located in an appropriate place for one, slightly below the larynx ("Adam's apple," see Figures 36.9 and 36.14).

The thyroid produces three hormones. Two are nearly identical, thyroxine and triiodothyronine, both of which are made up of two covalently bonded molecules of the amino acid tyrosine, with the addition of iodine. Whereas **thyroxine**, also called T_4, contains four atoms of iodine, **triiodothyronine**, or T_3, contains only three. This caused some puzzlement among endocrinologists until recently, when it was determined that T_4 is relatively inactive and stable and is really a storage form of T_3. T_3 is both active and unstable at the target cell, where it does its job and quickly breaks down, just as a hormone should. We will keep things simple by just referring to the thyroid hormones as "thyroxine." These hormones operate in a manner similar to that of the steroid hormones discussed earlier.

Thyroxine increases the metabolic rate, specifically by accelerating activity in biochemical pathways where carbohydrates are oxidized. Precisely how it does this is not known, but unlike most peptide hormones, thyroxine passes through the cell membrane, becoming active in the cytoplasm. Since it is known to enter the mitochondrion, it may express its effect there.

Overactive and underactive thyroid glands produce **hyperthyroidism** and **hypothyroidism**, respectively. The symptoms of hyperthyroidism are nervousness, hyperactivity, insomnia, and weight loss. Thin, active people are often accused of being hyperthyroid, but in fact most thin, active people are perfectly normal—they're just thin and active. In Graves' disease, a hyperactive thyroid produces, in addition to the usual symptoms, a characteristic bulging of the eyes brought about by fluid accumulation in the tissues behind them (Figure 36.15).

A **goiter** may seem to be an abnormal growth (see Figure 36.15b), but it is actually a normal response to low levels of iodine in the diet. The prominent, bulging overgrowth of the thyroid gland increases the efficiency of iodine utilization and retention so that more nearly normal levels of thyroxin can be maintained. Before the introduction of iodized salt early in this century, such goiters were common where foods grown in iodine-depleted soil could not provide the necessary supply.

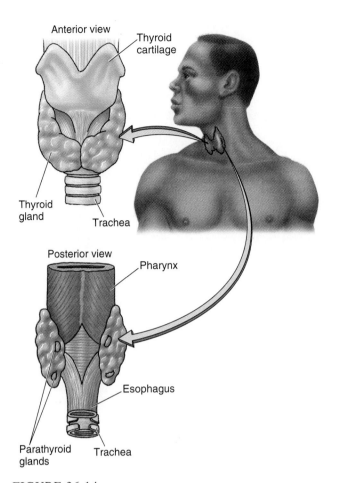

FIGURE 36.14
Human Thyroid and Parathyroids
The thyroid gland is a bilobed structure nearly surrounding the trachea at a point just below the larynx. Its hormones include thyroxine, triiodothyronine, and calcitonin. Note the position of the parathyroid glands in the posterior view.

Seafood has always been an excellent source of dietary iodine, but today an adequate supply is available in iodized salt sold in grocery stores.

Hypothyroidism (underactive thyroid) causes a general slowing of the metabolic rate, bringing on a condition in adults called **myxedema**. Such individuals are usually overweight, physically and mentally sluggish, and have a puffy, bloated appearance due to fluid and protein accumulation beneath the skin.

In some cases, genetic abnormalities can cause a lack of thyroid hormone from birth. If left untreated, **cretinism**, a condition marked by extreme mental and physical retardation, may develop. At present, if the sluggish thyroid condition is diagnosed early enough, cretinism can be treated effectively by the administration of a daily dose of thyroid hormone.

Thyroid control occurs through a negative feedback loop. When thyrotropin releasing hormone (TRH) from the hypothalamus reaches the anterior pituitary, the latter responds by

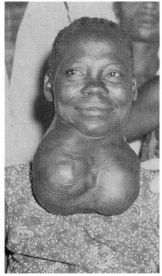

(a) Graves' disease   **(b)** Goiter (enlarged thyroid)

FIGURE 36.15
Thyroid-Related Abnormalities
A malfunctioning thyroid can cause a number of important clinical problems. A hyperactive thyroid may produce rapid metabolic rate, general nervousness, and a failure to gain weight irrespective of diet. **(a)** In Graves' disease, the symptoms also include bulging eyes. **(b)** Iodine deficiency can lead to the formation of goiter, an enlargement of the thyroid.

secreting TSH (thyroid-stimulating hormone or thyrotropin). TSH stimulates the thyroid to secrete T_3 and T_4, the latter of which elevates the rate of aerobic respiration. But rising levels of T_3 and T_4 also begin to inhibit the hypothalamus and anterior pituitary. Then TRH and TSH secretions subside, and the thyroid, no longer stimulated, slows its own hormone secretion (Figure 36.16).

The third thyroid hormone, **calcitonin**, with secretions from the parathyroid, regulates calcium ion levels in the body as part of calcium homeostasis.

The Parathyroid Glands

The **parathyroid glands** are pea-sized bodies embedded in the tissue of the thyroid, two or three in each lobe of the "bow tie" (see Figure 36.14). You will not be startled to learn that they produce **parathyroid hormone** (**PTH**). PTH is a giant among peptide hormones, with its 84 amino acids.

PTH, working in concert with calcitonin from the thyroid, regulates calcium ion levels in the blood. Abnormally low levels of blood calcium ion disrupt calcium-dependent processes such as blood clotting, membrane permeability, enzyme action, and muscle function. In the latter case, calcium ion deficiencies can bring on muscle spasms and convulsions. Abnormally high levels of calcium lead to serious deterioration of bone, which, considering the high calcium content of bone, seems paradoxical.

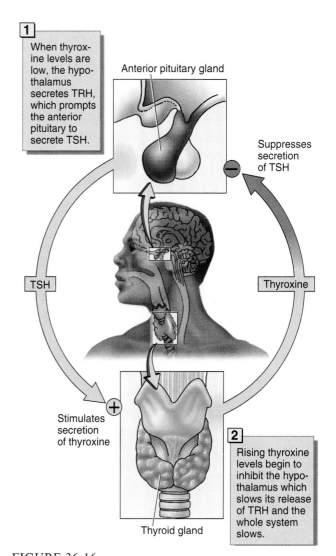

1 When thyroxine levels are low, the hypothalamus secretes TRH, which prompts the anterior pituitary to secrete TSH.

Anterior pituitary gland

Suppresses secretion of TSH

TSH

Thyroxine

Stimulates secretion of thyroxine

2 Rising thyroxine levels begin to inhibit the hypothalamus which slows its release of TRH and the whole system slows.

Thyroid gland

FIGURE 36.16
Negative Feedback and Metabolic Rate
The thyroid hormone thyroxine helps the body maintain a metabolic rate that is consistent with changing needs. Secretions are stimulated by hormones from the anterior pituitary and inhibited through a delicate feedback loop.

The two calcium-regulating hormones have opposite effects—PTH increases the blood calcium level, and calcitonin decreases it. When calcium levels fall below optimal, the parathyroid responds by releasing PTH, but when calcium levels rise above optimal, the thyroid increases its release of calcitonin (Figure 36.17).

The cells most directly involved in calcium management, and the targets of the two hormones, are those of the kidneys and bones. (In the case of PTH, we can add the lining cells of the small intestine.) Cells in the nephrons of the kidney respond to PTH by decreasing calcium excretion, and to calcitonin by increasing calcium excretion. In bone, the body's main calcium reservoir, PTH stimulates activity in bone-dissolving cells called *osteoclasts*. When they are active, calcium

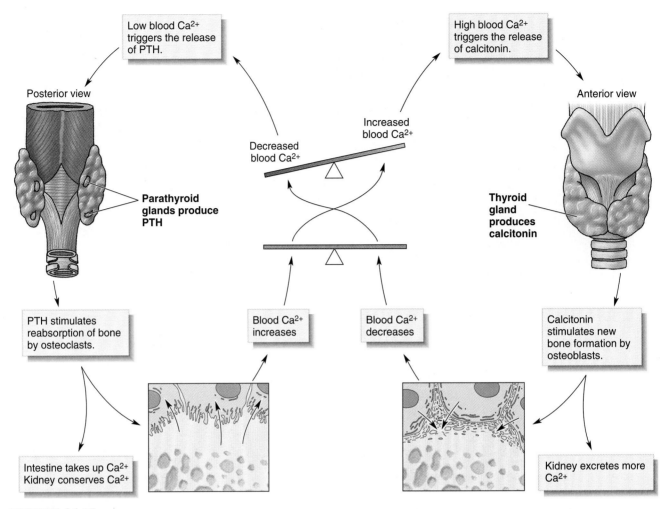

FIGURE 36.17
Calcium Regulation
Parathyroid hormone (PTH) and calcitonin cooperatively regulate body calcium. PTH secretion (*left*)
increases when blood calcium drops. It brings that level up by withdrawing calcium from the bones,
increasing intestinal uptake of calcium, and slowing the excretion of calcium by the kidneys. Calcitonin
has the opposite effect (*right*). When blood calcium is high, it decreases the rate at which calcium leaves
the bone and increases excretion of calcium through the kidneys.

is released into the blood. Calcitonin, on the other hand, stimulates *osteoblasts,* the bone-forming cells. They withdraw calcium from the blood and deposit it in bone. Calcitonin also prompts the kidney to send more calcium into the urine. Finally, cells lining the small intestine absorb calcium in response to PTH, but in its absence they permit the calcium to pass unabsorbed through the digestive tract. Vitamin D, we might add, aids PTH in the uptake of calcium from the digestive tract. This, of course, is why mothers buy vitamin D-fortified milk and pester their children to drink it; mothers know about vitamin D's influence on PTH.

The Pancreas

The pancreas is essentially an exocrine gland, responsible for the secretion of a battery of digestive enzymes. But within the pancreas lie clusters of endocrine cells known as **islet cells,** or more formally, *islets of Langerhans* (Figure 36.18). Within the islets, cells designated "alpha" cells secrete the hormone *glucagon,* and cells designated "beta" secrete *insulin.* **Glucagon** is a polypeptide consisting of a single chain of 29 amino acids. Its principal role is to stimulate target cells in the liver to break down stored glycogen into glucose. Glucagon is released when blood glucose levels fall below optimum.

Insulin, with its 51 amino acids, has the opposite effect on blood glucose—it decreases blood glucose levels. The two islet hormones complement each other in maintaining the delicate balance of glucose in the blood (Figure 36.19).

Insulin starts off as a protein that contains 81 amino acids. It is inactive in this form and must be altered by enzymes before it can do its job. The enzymes remove a number of the amino acids from the middle of the sequence, pro-

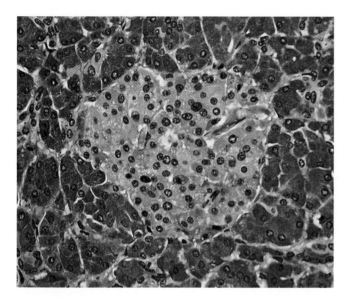

FIGURE 36.18
Islets of Langerhans
The islets of Langerhans are endocrine tissues clustered in the pancreas (an exocrine gland). Cells in the islets secrete insulin or glucagon.

ducing active insulin composed of two short chains with disulfide cross-linkages (see Figure 36.2). Insulin's best-known role is to help move glucose across plasma membranes, but it is also thought to encourage glycogen synthesis in the liver. Along with growth hormone, insulin promotes the uptake of amino acids and their incorporation into protein, while at the same time inhibiting the conversion of amino acids into glucose.

Insulin promotes glucose uptake across plasma membranes and so is broadly targeted: most cells have insulin receptor sites. The mechanism by which insulin works is not well understood. This may seem surprising considering the many years of clinical studies of this common hormone. We do know that when insulin finds its binding site in the membrane-bound receptor, the plasma membrane becomes much more permeable to glucose. Upon entry into the cell, glucose is usually converted to glucose-6-phosphate, a step that prevents it from diffusing out of the cell. Any failure in glucose regulation, whether through an under- or overabundance of insulin, or through deficiencies or defects in its cell receptors, creates havoc with the body. Such failures produce the wide-ranging symptoms of sugar diabetes, or diabetes mellitus (Essay 36.1).

The Adrenal Glands

Hormones of the Adrenal Cortex. In humans, the paired adrenal glands are situated atop each kidney (*ad*, "upon"; *renal*, "kidney"; Figure 36.20), whereas in other vertebrates they may simply be near the kidney. The **corticosteroids**, steroid hormones of the adrenal cortex, include three groups, the **mineralocorticoids**, **glucocorticoids**, and **sex steroids**. The most familiar mineralocorticoid is **aldosterone**, which, as we saw in Chapter 35, affects blood pressure by increasing the reabsorption of sodium by the kidney, which subsequently results in the recovery of water. Since ADH also promotes water recovery in the kidney, we see that the two hormones have overlapping functions. Aldosterone also decreases the reabsorption of potassium and hydrogen ions by the kidney, and slows sodium excretion. Any change in hydrogen ion concentration, of course, alters the delicately balanced blood pH. For another look at the management of blood pressure, see Essay 36.2.

A common glucocorticoid, **cortisol**, promotes the conversion of certain amino acids to glucose, increases blood glucose levels, and speeds the mobilization and utilization of fatty acids. Cortisol also has anti-inflammatory and anti-allergenic qualities. Hydrocortisone, a pharmaceutical preparation of cortisol, is available as an over-the-counter drug commonly used to reduce inflammation.

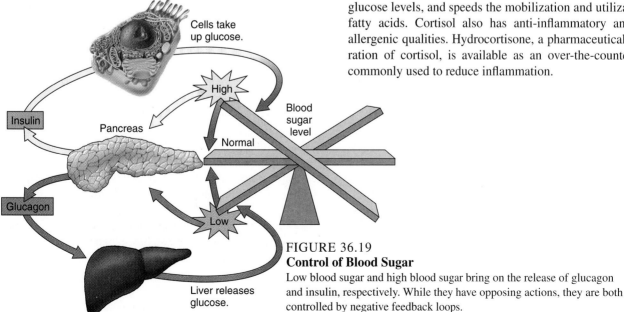

FIGURE 36.19
Control of Blood Sugar
Low blood sugar and high blood sugar bring on the release of glucagon and insulin, respectively. While they have opposing actions, they are both controlled by negative feedback loops.

Essay 36.1 ● DIABETES MELLITUS AND HYPOGLYCEMIA

Deficiencies in insulin activity produce *hyperglycemia* (high blood sugar), a condition that characterizes *sugar diabetes* or *diabetes mellitus*. In some diabetics, the ultimate cause of the disease is an actual deficiency of insulin, which can be caused by insufficient insulin production or by increased levels of *insulinase,* an enzyme that destroys insulin. However, most adult diabetics have normal levels of insulin but a deficiency of receptor sites on the membranes of the target cells. In some cases, the target cells have enough receptor sites but simply fail to respond properly to the hormone.

The symptoms of diabetes include glucose in the urine, greatly increased urine volume, dehydration, constant thirst, excessive weight loss, exhaustion, and many other problems. Some symptoms are the result of glucose starvation of the cells. Though glucose may be present "in the blood," an insulin deficiency prevents it from crossing the plasma membranes into the cells, where it is needed. In addition, the untreated diabetic has a remarkable halitosis (bad breath). This is because the exhaled breath contains ketones, especially acetone (one of the strong-smelling chemicals in some fingernail polish removers).

Mild cases of some forms of diabetes in adults can be controlled by a well-regimented diet in which carbohydrates and fats are carefully regulated. More advanced cases can often be controlled by insulin injections, as long as the problem does not involve insufficient cell surface receptors.

Diabetes in young people (those under 20), called *juvenile-onset diabetes,* accounts for about 10% of all cases of the disease. It is quite unlike *adult-onset diabetes,* in that its cause is singular, centering specifically about a degeneration of the beta cells of the pancreas. This, of course, leads to a chronically high blood sugar, since not enough insulin is produced to bring about glucose uptake by cells. With glucose unavailable, the body mobilizes its fat reserves, which brings on the accumulation of ketones and other organic metabolites in the blood. This lowers the blood pH, sometimes to a dangerous or even fatal level. The brain cells, which depend on glucose as a fuel, are the first to be affected by low blood sugar, resulting in dizziness, tremors, violent temper, and sometimes blurred vision and fainting.

Studies of identical twins show that juvenile-onset diabetes does not have a genetic cause, but adult-onset diabetes may

have a strong genetic basis. Ongoing studies suggest several nongenetic possibilities, including autoimmune disease (where the immune system attacks the beta cells) and viral infection of the beta cells.

Glucose starvation can be due to other causes than diabetes. For example, it can be due to *hypoglycemia,* or low blood glucose. Also, a pancreatic tumor may cause too much insulin to enter the blood. This results in blood glucose being rapidly transported out of the blood and sequestered as intracellular glycogen.

In the absence of pancreatic tumors, some people can have milder (but sometimes serious) types of chronic hypoglycemia. In such cases, one of our much-praised negative feedback mechanisms goes haywire. An increase of blood glucose brings about an increase in insulin secretion, as you might suspect. The insulin, in turn, facilitates the conversion of blood glucose into tissue glycogen, bringing the glucose level back down. Unfortunately, there is a time lag, and the pancreatic beta cells may overproduce insulin, causing the blood glucose level to fall sharply.

The third group of corticosteroids are the **sex steroids**. These closely resemble the estrogens and testosterone secreted by the gonads (as we will see shortly). Adrenal sex steroids influence sexual behavior, particularly the female libido, and some aspects of sexual development. Excessive adrenal androgen secretion, seen more commonly in older women, is known to produce masculinizing effects such as the growth of coarse, dark facial hair.

Hormones of the Adrenal Medulla. The adrenal medulla produces two fascinating modified amino acid hormones: **epinephrine** and **norepinephrine** (commonly called *adrenalin* and *noradrenalin,* old trade names originally coined by a pharmaceutical company). The two differ slightly in that epinephrine has one more methyl (CH_3) group than does norepinephrine. Their actions in the body are nearly identical, but epinephrine's effects are widespread, while those of norepinephrine are more limited. (See Chapter 34 for a review of their work as neurotransmitters in the autonomic nervous system.)

Epinephrine and norepinephrine cause a wide variety of dramatic changes in the body when an emergency arises. Their

release, usually a reaction to danger, fright, or anger, brings on what is called the "fight or flight" response. The two hormones affect circulation by accelerating heart rate and by shunting blood into the skeletal muscles and away from other organs such as those of digestion and reproduction. This is accomplished by selected vasoconstriction and vasodilation, which closes and opens capillary beds, respectively. Blood flow in the capillaries near the body surfaces is usually restricted (causing the skin to pale and reducing the extent of bleeding). In addition, blood pressure increases as glucose, a readily available "fuel" of the body, is suddenly released into the bloodstream from the liver. We have all heard stories of people exhibiting superhuman feats of strength under stressful conditions, and you may have had such an experience yourself.

Adrenal Hormones and Physiological Stress. Stress is a topic of increasing concern to a host of people, from physiologists to psychologists, sociologists, and apologists. If you're sitting there trying to get through a book of this size and finals are coming up Thursday, you have some idea of the concept. As you haul the book around, you may gain a different appre-

Essay 36.2 ● THE HEART AS AN ENDOCRINE STRUCTURE

The heart pumps blood. That's its job, and we've known about it since William Harvey's 1628 essay on circulation. (Actually, Harvey may have been scooped in 1553 by a Spaniard, Miguel Servet, who Latinized his name as *Servetus*. His novel ideas received little attention except from the members of the Inquisition, who condemned Servetus for heresy and had him burned at the stake. His book joined him in the fire.)

Now, though, we've learned the heart has another role: it helps regulate the excretion of both sodium and water, and it does so by secreting a hormone called atrial natriuretic factor, or ANF.

The presence in the body of a substance such as ANF has long been suspected, simply because the known regulatory process couldn't account for all known changes in water and sodium retention. In 1935, John Peters of Yale University suggested that there must be some other mechanism, in or near the heart, to monitor the volume of fluid passing through the heart and to regulate that volume. Indeed, it was found that water and sodium excretion in the kidneys were directly related to the degree to which the atria were distended when filled with blood. Seemingly, some factor in the atria was operating in conjunction with the secretion of aldosterone, whose role in regulating blood pressure and blood volume was known.

In 1956, very dense granules were discovered in the cells of guinea pig atria and were later found to be present in the atria of other mammals. In 1974, Marc Cantin and Jacques Genest of the University of Montreal found that the granules were similar to storage granules in the cells of certain endocrine glands. Key experiments then revealed the granules to be sites of hormone concentration. For example, when a rat atrium was homogenized and injected into another rat, the recipient showed a brief, but massive, surge in water and sodium excretion.

We now know that ANF acts in complex ways in various centers throughout the human body, including the hypothalamus, the pituitary, the adrenals, and the kidneys. Essentially, it regulates the volume of blood passing through the heart by influencing the excretion of water from the kidneys. ANF probably works by causing the glomeruli to become more permeable so that more water and sodium can be filtered from the blood. ANF is part of a complicated negative feedback loop that operates as the ANF inhibits the production of angiotensin II, a powerful constrictor of smooth muscle, such as that around the arterioles that pass into the kidney. Angiotensin II also prompts the adrenal gland to release aldosterone, which stimulates both the kidney and the posterior pituitary to inhibit water loss.

Interestingly, people with congestive heart failure usually have high blood pressure, high water retention, and high levels of sodium in the blood. However, their ANF release is also very high. It seems that, in some cases, the kidneys simply do not respond to ANF.

---

ciation of the term. Stress then can arise on many levels, such as psychological and physical. Furthermore, it is not an exclusively human malady; stress can occur in a variety of living things. What does stress do and what causes it?

When a human or other vertebrate is stressed, the body responds in an effort to meet the problem and counteract the effects of the stress. Both psychological and physical stress are associated with sudden increases in the levels of certain adrenal hormones. Their surge begins as higher brain centers prompt the hypothalamus to increase its secretion of corticotropin releasing hormone into the anterior pituitary. The anterior pituitary responds by increasing its ACTH output, which stimulates the adrenal cortex to increase considerably its output of cortisol, and to a lesser extent aldosterone. This results in elevated blood glucose levels, suppression of the immune and allergic responses (an effect of cortisol), and elevated blood pressure through increased water and sodium reabsorption (an effect of aldosterone). Whereas some of these reactions have straightforward survival value, others do not. What possible use would it be for the immune system to be suppressed following severe injury? It may be that the suppression merely keeps the immune response under control so that it can function over the long term, or maybe it's only a side effect. We just don't know.

Stress also increases the output of hormones from the adrenal medulla, including epinephrine and norepinephrine, which further elevate blood sugar, heart rate, and blood pressure, and in general prepare the body for some form of exertion. Vasopressin (ADH) secretion by the posterior pituitary is another result of stress, so water retention and blood pressure are further increased. Increased blood pressure, of course, would become important following an injury that resulted in massive bleeding. Falling blood pressure is a principal cause of physiological shock and kidney failure. (Of course, increased blood pressure would also temporarily increase the loss of blood in the case of wounding.)

From a biological point of view, stress responses are adaptive—*if they are transient*. A situation arises, the body responds, and the response is adaptive. But ongoing or long-term stress, such as humans might experience in a fast-paced, problem-ridden, and highly competitive life-style, is apparently different. The prolongation of such dramatic physiological responses can bring on permanent damage to the mind and body. When rats are stressed by being artificially crowded together, they develop a range of aberrant behaviors, from asexuality to cannibalism and infanticide. Monkeys forced to make decisions amid unpleasantness develop ulcers. What sort of experiments might we be currently running, unaware, on our own species?

The Gonads: Ovaries and Testes

The endocrine function of the ovaries and testes is discussed in Chapter 41, so let's just briefly note some of the major points here. First, the hormones of the ovaries—**estrogens** and **progesterone**—and the principal hormone of the testes—**testosterone**—are all steroids. Estrogen, by the way, includes a whole family of molecules, among them **estradiol**, **estriol**, and **estrone**. Estradiol is the major estrogen secreted by the ovary. For a look at the significance of sex hormones in embryonic development, see Chapter 42.

In addition to their reproductive function, sex hormones have an important role in skeletal development. Their sudden increase in the blood at the onset of puberty stimulates the lengthening of the long bones in humans and causes a marked spurt of growth. (Interestingly, it also causes a simultaneous spurt in mental growth as measured by test performance—if not by behavior.) Toward the end of puberty, the sex hormones have the opposite effect on growth. They promote the final fusion of growth regions in the long bones, after which no further increase in height is normally possible. Interestingly, the onset of puberty is variable: some of us are precocious and some are definitely late bloomers. Thus late-maturing adolescents, those in which bone fusion is delayed, may grow to be unusually tall. Low levels of estrogen secretion may delay puberty and produce tall girls. The same relationship, by the way, holds for testosterone and boys. Boys may remember that tough, well-muscled kid who matured way ahead of everyone else—and then quit growing, so that by high school he was a "short guy" and quit bullying people whose own hormones were now kicking in.

The Thymus

The **thymus**, located just behind the sternum in adult mammals (see Figure 36.9), is a rounded, spongy organ. Its only known function, a vitally important one, is providing cells for the immune system, particularly a class of cells called lymphocytes. Such cells originate in the red bone marrow and apparently migrate to the thymus before birth. Some of the lymphocytes venture out into other places, such as the spleen and lymph nodes. Soon after birth, the thymus begins to secrete the hormone **thymosin**, which targets certain members of the resident lymphocyte population. Thymosin stimulates the development of T-lymphocytes. The human thymus changes with time. It is most prominent in children, peaking in its growth at puberty, whereupon it begins to atrophy, becoming almost unrecognizable in older people.

The Pineal Body

In humans, the **pineal body** is a pea-sized gland, located deep in the brain, just above the brain stem (Figure 36.21a). Some biologists have serious reservations about whether the pineal body has an endocrine function, but others are convinced that it does. The first strong evidence of an endocrine function was the discovery that when a pineal extract from cattle is injected into frogs, a general blanching of the skin occurs. Scientists were intrigued with the notion that cattle extracts could cause

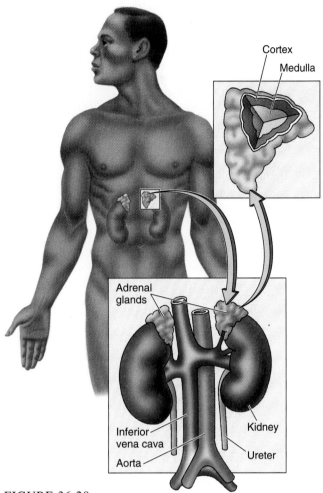

FIGURE 36.20
Adrenal Glands
The adrenal glands are roughly triangular structures situated on the top of the kidneys. The gland is divided into two secretory parts, an outer cortex and an inner medulla. Each produces its own battery of hormones.

this effect in frogs. Because of its pigment-related effects, the hormone found in the extract was called **melatonin**. Further observations revealed that such color changes were the result of a clustering together of melanin pigments—just the opposite of the MSH effects mentioned earlier. Recall that MSH brings on pigment dispersal and a darkening of the skin.

Melatonin, a modified amino acid, is also thought to play a role in **circadian rhythms** (*circa,* "about"; *dia,* "day"; see Chapter 43). Circadian rhythms are behavioral and physiological changes that take place on a daily basis. It is known that many circadian rhythms in vertebrates are regulated by melatonin. These rhythms include daily cycles of sleep and activity, and they even affect receptivity to medicines or other hormones. If the pineal body is removed in birds, the cyclic behavior ceases; if it is implanted in a noncycling bird lacking the body, cycling activity resumes. Researchers have found that

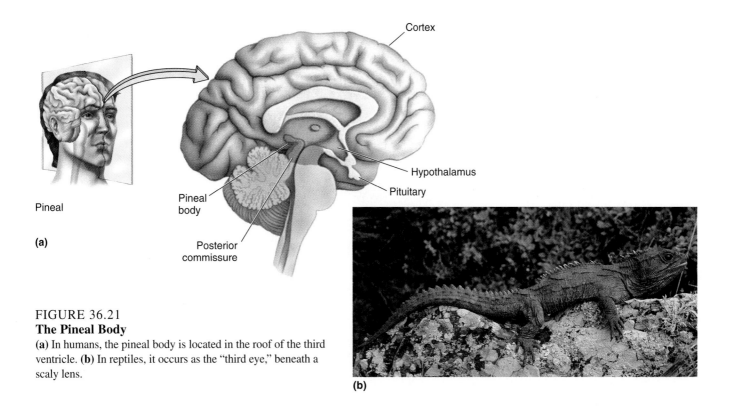

FIGURE 36.21
The Pineal Body
(a) In humans, the pineal body is located in the roof of the third ventricle. (b) In reptiles, it occurs as the "third eye," beneath a scaly lens.

when even a few pineal cells are grown in a culture dish in darkened surroundings, they secrete melatonin, roughly on a cyclic 24-hour schedule, just as they do in the living organism. Scientists, who often travel a lot, are now exploring the use of melatonin in altering human clocks to alleviate jet lag.

Melatonin may also be a gonadal inhibitor in mammals. In rats exposed to long periods of light, melatonin secretion is slowed, and the rat enters reproductive readiness. In continuous light, the rat remains in constant readiness. The connection between light and the pineal is indirect, with nerve pathways relaying their impulses between the retina of the eye and the pineal body. In birds, the light receptor is not the eye at all but some unknown region of the brain.

People have continued to speculate about the pineal, particularly about its evolutionary history. The answer is a bit startling. It turns out that it may be the remains of an ancient and vestigial second pair of eyes. Here is a case in which an understanding of evolutionary origins was a real help in working out the physiology and function of an organ. The fossil skulls of the earliest ostracoderms (jawless fishes) have holes for three eyes. The third eye is in the middle, right on top of the skull. Only a remnant of the eye remains in modern fishes, but some amphibians and reptiles have retained a rudimentary but recognizable eye. In the New Zealand reptile *Sphenodon punctatus,* called the tuatara (Figure 36.21b), the small median or third eye even has a lens and retina. Some very common American lizards also have a tiny, degenerate third eye on the top of their heads, complete with light receptors and an eye-

hole in the skull covered by a special translucent scale. This third eye can't form an image, but it is sensitive to light. It is, in fact, the light receptor by which the lizard sets its biological clock, behaving one way when days are long and another way when short winter days draw on.

Physiologists are notably vague in discussing the role of the pineal body in humans. It is known to secrete melatonin, and the quantity released follows a day–night cycle, rising in the night and falling in the daytime.

Prostaglandins

The **prostaglandins** make up the most recently discovered class of vertebrate hormones. Some of the prostaglandins are among the most potent biological materials known. Perhaps the reason for their late discovery is that they are not produced by specialized organs. In fact, prostaglandins are produced by most kinds of tissues.

The release of prostaglandins can be brought about by other hormones or by almost any irritation of the tissues, including mechanical agitation. Not surprisingly, some prostaglandins are involved in inflammatory responses and in the sensation of pain. For instance, one type causes uterine contractions.

Blood platelets, which are involved in blood clotting reactions, produce a prostaglandin called **thromboxane** that causes platelets to stick together and the walls of arteries to squeeze shut. Another prostaglandin, **prostacyclin**, has exactly the opposite effect—it prevents clots, keeping blood fluid, and

TABLE 36.1
Human Endocrine System

Structure and Secretions	Target	Action
HYPOTHALAMUS		
Releasing and inhibiting hormones		Stimulate and inhibit hormone release
ANTERIOR PITUITARY		
Adrenocorticotropic hormone (ACTH)	Adrenal cortex	Secretes steroid hormones
	Fat storage regions	Fatty acids released into blood
Growth hormone (GH)	General (no specific organs)	Stimulates growth; amino acid transport
Thyroid-stimulating hormone (TSH), also called thyrotropin	Thyroid	Secretes T_3 and T_4
Prolactin	Breasts	Promotes milk production
Follicle-stimulating hormone (FSH)	Ovary and testis	Stimulates growth of follicle and estrogen production in females, spermatogenesis in males
Luteinizing hormone (LH)	Mature ovarian follicle	Stimulates ovulation, conversion of follicle to corpus luteum, and production of progesterone
	Interstitial cells of testis	Stimulates sperm and testosterone production
Melanocyte-stimulating hormone (MSH)	Melanocytes	Pigment dispersal (skin darkening)
POSTERIOR PITUITARY (FROM HYPOTHALAMUS)		
Oxytocin	Breasts	Stimulates release of milk
	Uterus	Contraction of smooth muscle in childbirth and orgasm
	Seminal vesicles	Ejaculation
Antidiuretic hormone (ADH)	Kidney	Increases water uptake in kidney (decreasing urine volume)
THYROID		
Thyroxine (T3)	Many regions and organs	Increases oxidation of carbohydrates; stimulates (with GH) growth and brain development
Calcitonin	Kidney	Decreases blood calcium level; increases excretion of calcium by kidney, stimulates uptake by bone cells
PARATHYROID		
Parathyroid hormone (PTH)	Intestine, kidney, bone	Increases blood calcium level; decreases excretion of calcium by kidney and speeds absorption by intestine and release from bones
HEART		
Atrial natriuretic factor (ANF)	Kidneys	Increases excretion of Na^+ and H_2O
ISLETS OF LANGERHANS *(Pancreas)*		
Alpha cells: glucagon	Liver	Stimulates liver to convert glycogen to glucose; elevates glucose level in blood
Beta cells: insulin	Plasma membranes	Facilitates transport of glucose into cells; lowers glucose level in blood
ADRENAL CORTEX		
Mineralocorticoids—aldosterone	Kidneys	Increased recovery of sodium and excretion of K^+ and H^+; uptake of Cl^- and H_2O
Glucocorticoids—cortisol	General (no specific organs)	Increases glucose synthesis through protein and fat metabolism; reduces inflammation
Sex steroids	Many regions and organs	Promotes secondary sex characteristics
ADRENAL MEDULLA		
Epinephrine Norepinephrine	Many regions and organs	Increases heart rate and blood pressure; directs blood to muscles and brain; "fight or flight mechanism"
OVARIES		
Estrogen Progesterone (ovarian source replaced by placenta during pregnancy)	Many regions and organs	Development of secondary sex characteristics; bone growth; sex drive with male hormones regulates cyclic development of endometrium in menstrual cycle; maintenance of uterus during pregnancy
TESTES		
Testosterone	Many regions and organs	Differentiation of male sex organs; development of secondary sex characteristics; bone growth; sex drive
THYMUS		
Thymosin	Lymphocytes	Stimulates development of T-lymphocytes
PINEAL BODY		
Melatonin	Melanocytes, hypothalamus and pituitary	Pigment aggregation (blanching of skin); may have some influence over hypothalamus or pituitary in cyclic activity
NONSPECIFIC ORIGIN		
Prostaglandins	Many regions and organs	Aid birth through uterine contraction and cervical relaxation; affect blood clotting

prevents arteries from closing. Since prostacyclin is produced by the cells that line the blood vessels, it seems that prostaglandins may be important in maintaining normal circulation.

We see then that animals have developed highly complex and specific ways of chemically coordinating their activities and responses. We should add that our discussion has not been exhaustive. There are other chemical messengers. Some of these we discuss elsewhere, such as the digestive hormones (Chapter 37) and hormones secreted by the mammalian placenta (Chapter 41). Nevertheless, you should now be able to appreciate the important regulatory, responsive, and homeostatic roles of chemical signals in the daily lives of animals, including us.

Key Ideas

THE CHEMICAL MESSENGERS

1. Hormones form three categories: **endocrine** hormones (acting at some distance, **paracrine** hormones (acting in cells nearby), and **autocrine** hormones (acting within the cell of origin).

2. Although both neural and hormonal control involve chemicals, in neural control the chemicals are secreted into the synaptic cleft. In endocrine hormonal control, they are secreted into the blood. Neural control is rapid and transient; hormonal control is slower and longer lasting.

3. Chemically, most hormones are peptides or lipids. The lipid group includes steroids and prostaglandins.

4. Each kind of hormone has specific **target cells** that bear matching receptor sites, either on the cell surface or in the cytoplasm.

5. Hormones are short-lived. Most are chemically degraded in less than one hour. Short-term versus long-term effects are determined by the rate of secretion.

6. Physical and chemical cues prompt hormonal secretion, whereas they are controlled by negative feedback loops.

CHEMICAL MESSENGERS AND THE TARGET CELL

1. Most peptide hormones interact with membrane receptors on the target cell, thereby activating second messengers. Lipid hormones act directly at the gene level.

2. The many second messengers include **cyclic AMP**, and **cyclic GMP**.

3. When epinephrine binds to its receptor site on a liver cell, it activates G-protein, which, in turn, activates the enzyme **adenylate cyclase**, which converts ATP to cAMP. cAMP activates a cascade of reactions that ultimately lead to glycogen breakdown into glucose, which diffuses out of the cell. The action of second messengers brings on a hormonal **cascade**, with more and more molecules involved as the steps proceed.

4. Some cells have dual second-messenger systems, each activated by its own hormone. Their actions are opposite; thus in smooth muscle the formation of cAMP causes relaxation, and the formation of cGMP brings on contraction.

5. Steroids pass through the plasma membrane and join **cytoplasmic binding proteins**. The new association then activates a gene, and protein synthesis ensues. Typically, the protein is an enzyme whose appearance enables the cell to carry out the required hormonal response.

INVERTEBRATE HORMONES

1. The source of hormones in many invertebrates is the nervous system.

2. In the molting crayfish, the **Y-organ** secretes **MH,** the molting hormone, while the nearby **X-organ** secretes **molt-inhibiting hormone (MIH)**.

3. Many insects pass through a complete metamorphosis, including egg, larva, pupa, and adult states. The juvenile state is maintained by **juvenile hormone (JH)**. Final molting and the end of the larval state occur when the prothoracic gland secretes **ecdysone**, which brings on the pupal and adult stages.

THE HUMAN ENDOCRINE SYSTEM

1. The hypophysis, or **pituitary gland**, has two functionally and embryologically distinct regions, the adenohypophysis, or **anterior pituitary**, and the neurohypophysis, or **posterior pituitary**.

2. The hypothalamus controls secretion by the pituitary. It interacts with the anterior pituitary through neurosecretions that are released into blood.

3. Hypothalamic neurosecretions include a number of specific **releasing hormones**, each of which stimulates the anterior pituitary to release one of its seven corresponding hormones. The release of some anterior pituitary hormones is inhibited by negative feedback alone, but several are controlled by other neurosecretions called **inhibiting hormones**.

4. The two hormones of the posterior pituitary are not synthesized there but are received as neurosecretions from axons originating in the hypothalamus.

5. The secretion of **ACTH, TSH, FSH,** and **LH** is inhibited by negative feedback only. ACTH stimulates the adrenal cortex to release its steroid hormones and helps regulate the metabolism of fats. TSH stimulates the thyroid to release its two hormones, thyroxin and triiodothyronine. FSH and LH stimulate the gonads to produce sex hormones and gametes.

6. The secretion of **GH, prolactin,** and **MSH** is inhibited by inhibiting hormones. GH promotes growth through increased uptake of amino acids, stimulates the release of fats, and prompts the conversion of liver glycogen to glucose. **Prolactin** promotes milk production, but milk release requires oxytocin. MSH stimulates the dispersal of pigment melanocytes, bringing about darkening of the skin.

7. Vertebrate **neuropeptides** have two general functions. One group is **uterotonic**, causing contractions in the uterus and oviducts; the best-known member is **oxytocin**. The second group, the **vasopressins** (ADH in humans), helps regulate blood pressure.

8. In humans, oxytocin is involved in orgasm, ejaculation, the birth process, and milk release.

9. In humans, the thyroid hormone thyroxine influences carbohydrate metabolism. **Calcitonin**, the third thyroid hormone, works with the parathyroid hormone in regulating the distribution of body calcium.

10. The **islets cells** are clusters of insulin-secreting beta cells, and glucagon-secreting alpha cells.

11. Glucagon stimulates the liver to convert glycogen to glucose, which elevates the blood glucose level. Insulin stimulates cells to take up glucose, thus lowering blood levels. Deficient insulin levels cause high blood sugar, commonly called **diabetes mellitus**.

12. The **adrenal cortex** secretes **mineralocorticoids** (such as **aldosterone**), **glucocorticoids** (such as **cortisol**), and **sex steroids**. Aldosterone decreases potassium and hydrogen ion reabsorption by the kidney and slows sodium excretion. Cortisol increases blood levels of glucose, amino acids, and fatty acids and is an anti-inflammatory agent. The sex steroids influence libido and development.

13. Adrenal medullary hormones include **epinephrine** and **norepinephrine** (adrenalin and noradrenalin). Their general effect is to prepare the body for emergencies.

14. The ovaries produce **estrogens** and **progesterone**, while the testes produce **testosterone**. Sex hormones influence reproduction, body development, and bone growth.

15. The **thymus** secretes **thymosin**, which stimulates lymphocyte development.

16. Pineal extracts containing melatonin cause pigment withdrawal in frog melanocytes, thus blanching the skin. **Melatonin** may also influence **circadian rhythms**—affecting the alternating periods of sleep and activity. Light may influence melatonin action, via the optic nerves in some groups, the "third eye" in frogs and lizards.

17. **Prostaglandins** are produced almost universally in the body. Some are involved in inflammation, others cause uterine contractions; one (**thromboxane**) aids blood clotting, while another (**prostacyclin**) slows clotting.

Application of Ideas

1. As two 19th century physiologists attempted to determine the function of the pancreas, they observed flies swarming around the urine of dogs from which the pancreas had been removed. This observation was an initial step in our understanding of the role of insulin in sugar metabolism. What might the next steps have been in isolating insulin, and what rigorous rules should the investigators have followed in establishing its function?

2. Assuming that insect hormones could be synthesized in sufficient quantity, suggest a specific hormone that might be applied to insect control. Explain your choice, how it might work, and reasons why this method of insect control might be more ecologically suitable than the use of chemical insecticides.

3. While the anterior pituitary and thyroid glands are known to control growth, malfunctions in any of the endocrine glands can also affect growth. Considering each of the endocrine glands, explain why this is so.

4. The controlling functions of the endocrine system often overlap those of the nervous system. Cite examples of this overlap and comment on the adaptive significance, if any, of this.

Review Questions

1. List a similarity and a difference between neural and hormonal control. (page 765)

2. Outline the events that follow the binding of epinephrine to its cell membrane receptor. Which substance represents the second messenger? How does the system shut down? (pages 767–768)

3. Outline the events that follow the entry of a steroid hormone into the cell. (page 769)

4. What are the roles of juvenile hormone and ecdysone on insect development? (pages 770–771)

5. Compare the functional relationship of the anterior and posterior pituitary lobes with the hypothalamus. What does this tell you about the posterior lobe as a gland? (page 771)

6. Review the negative feedback loop established between the hypothalamus, pituitary, and adrenal cortex. (page 774)

7. List the targets of the anterior pituitary hormones ACTH, TSH, GH, and prolactin. (page 774)

8. What are the effects of too much GH in the child? In the adult? (page 774)

9. What are the roles of oxytocin and prolactin? (page 774)

10. Compare the calcium-regulating action of calcitonin with that of PTH. What are the symptoms of an overabundance of the latter? (page 777)

11. Briefly discuss the interaction of glucagon and insulin in the precise regulation of blood sugar. (pages 778–779)

12. What are some possible causes of hypoglycemia? What symptoms accompany the abnormality? (Essay 36.1)

13. Describe the action of the hormone cortisol. Which of its actions are also produced by the drug hydrocortisone? (page 779)

14. What are the two known actions of melatonin, the pineal secretion? (page 782)

15. Where are the prostaglandins produced? List two actions of prostaglandins in the circulatory system. (pages 783–785)

CHAPTER 37

Digestion and Nutrition

CHAPTER OUTLINE

DIGESTIVE SYSTEMS

Intracellular Digestion

Sac-like Digestive Systems

Tube-within-a-Tube Digestive Systems

Vertebrate Digestive Systems

THE DIGESTIVE SYSTEM OF HUMANS

The Oral Cavity and Esophagus

The Stomach

The Small Intestine

Essay 37.1 The Heimlich Maneuver

Accessory Organs: The Liver and Pancreas

The Large Intestine

THE CHEMISTRY OF DIGESTION

Carbohydrate Digestion

Fat Digestion

Protein Digestion

Nucleic Acid Digestion

Integration and Control of the Digestive Process

SOME ESSENTIALS OF NUTRITION

Carbohydrates

Fats

Protein

Vitamins

Mineral Requirements of Humans

Essay 37.2 Cholesterol and Controversy

KEY IDEAS

APPLICATION OF IDEAS

REVIEW QUESTIONS

*I*t's always a little risky to generalize about evolutionary rates in a single species because evolutionary change runs in many directions at very different rates. And in some cases, species with many advanced (more recently evolved) traits can also harbor some very primitive characteristics. For example, we humans consider ourselves advanced in many ways, but we have very primitive mouths, much like those of our ancestral species. The specialized bill of the seemingly primitive platypus, on the other hand, is considered one of the most advanced mouth structures among vertebrates. Whereas our entire digestive system may not be unique, we can at least be proud of our gut. We have a very highly developed gut—not as highly evolved as that of the cow, perhaps, with its four-chamber stomach, but still quite advanced. It is only fitting, then, to learn more about the specializations of such a nice gut, but first we'll look at those of other animals and see how they have fared.

Digestive Systems

Digestion refers to the mechanical and chemical breakdown of complex nutrients (foods) into smaller molecular components. After digestion those components undergo **absorption**, movement through the gut wall and into the bloodstream. Once absorbed, the products of digestion are used as an energy source and as raw materials for synthesizing the molecules required for life. Animals go about getting food in a great variety of ways, and, in fact, much of the pageantry of life is centered on acquiring something to digest and in avoiding being so acquired (Figure 37.1).

First we will consider the mechanics of digestion, then the chemistry of digestion, and finally the nutritive character of the major food molecules.

FIGURE 37.1
Food-Getting in Animals
Animals spend much of their time in the search for and capture of food. Countless variations occur in this vital activity.

Intracellular Digestion

Most of the more familiar species (ourselves included) utilize **extracellular digestion**, digestion outside the cell. In some cases, the food material is ingested and held within the diges-tive tract, where digestive enzymes are secreted, breaking down the food molecules so that they can pass across the lin-ing of the digestive tract and into the bloodstream. In a few cases, such as with sea stars, the food is digested where it is encountered. (In the sea star, a shellfish may be pried open, the sea star's stomach is everted to cover the food, digestive juices are secreted, and the digested material passes into the sea star's body.)

A number of species show **intracellular digestion**, di-gestion inside the cell. In these cases, the organism must have a way of getting the food material into the digestive cells where it can be broken down. This is usually accomplished by the cells engulfing the food through phagocytosis. Intracellu-lar digestion can be seen in the sponges and cnidarians.

We will focus on the sponges to introduce not only intra-cellular digestion but also the sac-like digestive system com-mon to several phyla.

Sac-like Digestive Systems

Sac-like digestive systems are called *incomplete digestive sys-tems* in that they lack separate openings for food entering and waste exiting. Sponges, with their vase-like shapes, obviously have incomplete digestive systems.

Food-getting in the sponges—animals that lack organ systems—is the responsibility of the flagellated **collar cells**, or *choanocytes* (Figure 37.2a; see also Chapter 26). These active cells line the body canals, and their beating causes sea-

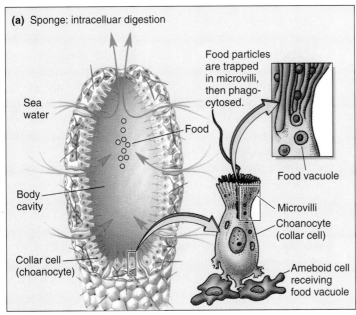

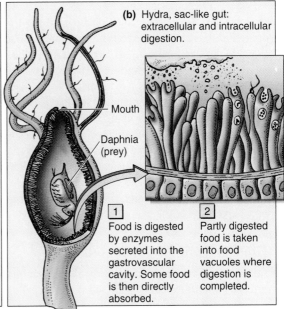

FIGURE 37.2
Simple Digestive Structures
(a) Sponges obtain their food by trapping food particles from water drawn in by choanocytes (collar cells). Food particles pass down the collar of the choanocyte, ending up in food vacuoles for digestion and dis-tribution. **(b)** In the cnidarian *Hydra,* stinging cells subdue prey, which is then drawn into the gastrovascu-lar cavity, where extracellular digestion begins. Phagocytosis and intracellular digestion follow.

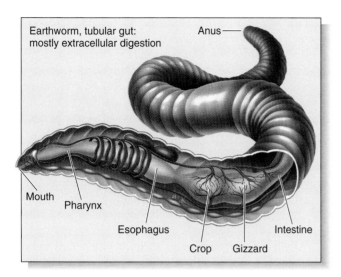

Earthworm, tubular gut: mostly extracellular digestion

Anus

Mouth
Pharynx
Esophagus
Crop   Gizzard
Intestine

FIGURE 37.3
Specializations in the Earthworm Gut
The complex earthworm digestive system is "complete," following a tube-within-a-tube plan. Food moves onward through a tube from mouth to anus. The muscular gut has several specialized regions, including the pharynx, crop, gizzard, and lengthy intestine. The absorbing surface is greatly increased by an intestinal fold known as the typhlosole.

water to swirl and eddy throughout the body of the sponge. The water carries in tiny food particles, which are trapped by a layer of mucus on each collar, so that they can be passed down to the cell below. From there the food is transported to other cells, where it becomes enclosed in digestive vacuoles. After digestion takes place, molecules of digested food are then transported throughout all the cells of the sponge. So the sponge, a simple multicellular animal, carries on intracellular digestion without a well-defined digestive system.

The cnidarians (Figure 37.2b) have somewhat more complex digestive systems, including a prominent sac-like gut or **gastrovascular cavity** with a complex lining of enzyme-secreting, phagocytic (food-engulfing), and ciliated cells. Digestion begins extracellularly when enzymes are released into the gastrovascular cavity, but after the food is partially broken down, particles are engulfed by phagocytic cells lining the gut, and digestion is completed intracellularly. (For a review of phagocytosis, see Chapter 5.)

Tube-Within-a-Tube Digestive Systems

Beyond the flatworms and cnidarians, nearly all other animals have the tube-within-a-tube body plan and use extracellular digestion. Yet some intracellular digestion persists in many invertebrates, such as mollusks and crustaceans, that have intestinal side passages lined with phagocytic cells that conduct intracellular digestion. We will consider two quite different invertebrate groups with *complete* digestive systems—having separate openings for food entering and waste exiting: earthworms and insects.

Earthworm Digestive System. As you can see in Figure 37.3, the gut in earthworms is well defined and associated with several specialized organs. As the earthworm burrows through the soil, it feeds on decaying organic matter. Its mouth is a simple opening, followed by a very muscular **pharynx** that swallows the organic matter and forces it along. A short **esophagus** directs the mass into the thin-walled **crop**, which is a temporary storage organ. The ingested material is next moved into the thick-walled, muscular **gizzard**. The gizzard is filled with bits of gravel and, like the gizzard in seed-eating birds, is used to grind the food.

From the gizzard the food moves into the **intestine**, where it is digested by enzymes secreted by cells in the intestinal lining. The digestion and uptake of food occur along the entire length of the intestine. The *typhlosole,* a large, fold-like extension of the gut wall, speeds up absorption by greatly increasing the surface area (see Figure 26.30). From the absorptive surface, the food molecules pass into the circulatory system for distribution throughout the body. Although the earthworm has nothing to compare with the vertebrate liver, groups of epithelial cells known as *chloragogen* cells perform some of the same functions, including the conversion of glucose to glycogen, the *deamination* (removal of nitrogen) of amino acids, and the formation of urea. Worms get their color partly from the accumulation of some of these products. In addition (strangely enough), roaming phagocytes pick up whole bits of soil, which pass through the earthworm's intestine and then move to the skin, with the resulting earthy coloration presenting a major visual problem to the early bird.

Insect Digestive System. It is difficult to generalize about insect feeding structures. In fact, the group serves as a lesson in digestive variation, one that illustrates the tremendous range of adaptations to the varied energy sources of the earth. It is difficult to think of anything that at least some insects don't eat. This versatility has been particularly troublesome to humans since the dawn of agriculture; the availability of food in large patches has made it possible for some insect species to increase in number dramatically.

Biting and chewing mouthparts are very common in insects, particularly in predators and foliage eaters. They can clearly be seen in the grasshopper, which has essentially a five-part mouth (see Figure 26.39). The grasshopper tastes its food with its **palps,** pairs of sensory structures attached to the mouthparts. Biting and shearing are done with the large saw-toothed **mandibles,** while the smaller **maxillae** deftly handle the food and do some finer grinding. The upper and lower lips assist by holding food so that the shearing and grinding can occur efficiently. Saliva from well-formed salivary glands in grasshoppers helps with the grinding and begins digestion.

The grasshopper's digestive system, like that of many insects, consists of three parts, the *foregut, midgut,* and *hindgut* (Figure 37.4). The **foregut** is divided into a pharynx, esophagus, and crop. The crop functions primarily in food storage. In some insects this region has a roughened surface

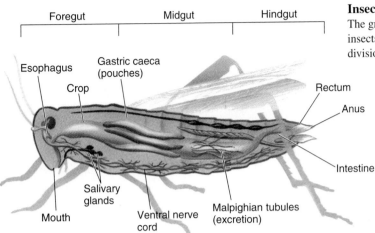

Foregut | Midgut | Hindgut

Esophagus
Gastric caeca (pouches)
Crop
Rectum
Anus
Intestine
Mouth
Salivary glands
Ventral nerve cord
Malpighian tubules (excretion)

FIGURE 37.4
Insect Digestive Tract
The grasshopper's digestive system is similar to that of many other insects. The digestive tract typically consists of three main divisions—the foregut, midgut, and hindgut.

that is used to further grind the food. The **midgut** is the digestive area where both digestion and absorption occur.

Surrounding the junction of the foregut and midgut are 12 **gastric ceca** (singular, *cecum*)—pouch-like accessory structures that secrete digestive enzymes into the midgut. The **hindgut** is mainly a water-absorbing structure that receives both digested food and cellular nitrogenous wastes. As digestive residues pass through, the water is removed and conserved, and the feces leave in a semidry condition, an important adaptation to the dry terrestrial environment.

The digestive system of the housefly is slightly different; it carries on some digestion outside the gut. You should be aware that the housefly walking across your chocolate cake is doing two things. First, the fly is tasting the icing with its feet—a place where many insects have taste receptors. Second, when it bends its head down, it is disgorging its enzymes onto the surface and stirring it all around with its hairy labium (see Figure 26.39). But don't worry—it will suck up any mess it makes with its hollow, tube-like mouthparts. The rest is yours.

Plant eaters frequently have piercing and sucking mouthparts. Recall the aphids that suck up the phloem sap (see Chapter 30). There are also fruit-eating moths that tear the fruit first, then suck up the exposed fluids. Of course, piercing and sucking mouthparts are found among the blood-sucking and predatory species as well. When the ant lion sinks its huge curved mandibles and maxillae into the ant's vulnerable abdomen, it doesn't chew. It uses channels formed between the paired mouthparts to suck out the ant's body fluids.

Vertebrate Digestive Systems

The jaws, teeth, and beaks of vertebrates provide us with clues to what they eat. For example, if you find a large skull with conical teeth, you've probably found a fish eater. If the skull is *very* large, though, the conical teeth may have once graced a killer whale who used them to crush the bodies of large prey such as seals. On the other hand, if the skull is very small and the teeth more pointed, the animal probably ate

insects. You can rest assured, however, that the creature with the pointed teeth was not a grazer. Let's find out about feeding and digestive specializations in one of the sharper-toothed types from the briny depths.

Sharks and Bony Fishes. Sharks are certainly carnivores, despite the fact that they swallow practically anything. One of your authors, while on an expedition, helped dissect a 12-foot tiger shark and found a Polaroid negative that had been thrown overboard six days earlier and 200 miles away! Finding a picture of oneself in a shark's stomach can lessen the pleasures of seafaring life.

Two aspects of the shark's food-getting and digestive apparatus are especially noteworthy. First, its teeth occur in rows, most of which are around the periphery of the mouth (Figure 37.5). As the teeth grow, the rows migrate outward so that the outermost ones fall out (and sink to the ocean floor, thereby providing tourists with necklaces). The jaw itself is a hinge joint, moving in one plane only. Instead of chewing, the shark uses the sideways head-thrashing and tearing action typical of shark attacks.

The second unusual aspect of the shark's digestive system is the **spiral valve** in its relatively short intestine. The valve (Figure 37.5) resembles a circular stairway within a tube. Apparently, its function is to slow the passage of food through the shark's short intestine to give digestive enzymes time to act and to increase the surface area for absorption.

Sharks, with their cartilagenous skeletons, are distinct in many ways from the much more numerous bony fishes. Bony fishes are extremely diverse, with over 20,000 species described, several of which will be neglected here. Like the sharks, the jaws of the bony fishes can only move in one plane and are used for biting and sucking only. The teeth are numerous, and in many species they grow from the roof of the mouth as well as from the upper and lower jaws.

The herbivorous fishes have remarkable mouthparts. For one thing, they lack tearing teeth. They have a row of fine teeth borne on a horny ridge. Some patches of teeth are on the

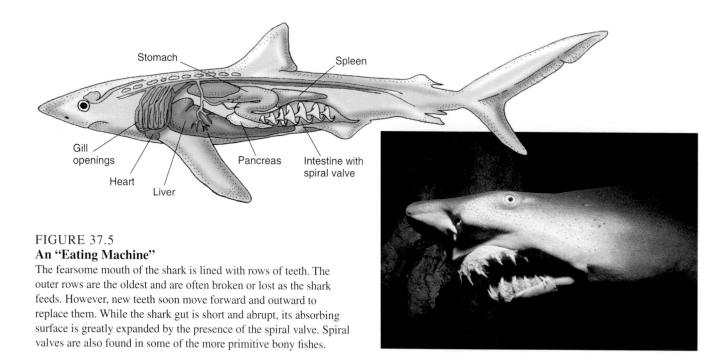

Stomach
Spleen
Gill openings
Pancreas
Intestine with spiral valve
Heart
Liver

FIGURE 37.5
An "Eating Machine"
The fearsome mouth of the shark is lined with rows of teeth. The outer rows are the oldest and are often broken or lost as the shark feeds. However, new teeth soon move forward and outward to replace them. While the shark gut is short and abrupt, its absorbing surface is greatly expanded by the presence of the spiral valve. Spiral valves are also found in some of the more primitive bony fishes.

floor of the mouth, where they grind bits of plant material against leathery patches on the roof. Other herbivorous fishes filter minute particles from the water with highly branched gill rakers through which water passes as it enters the gill chambers. Herbivorous fishes usually have very long, highly coiled intestines, typical in animals that extract nourishment from cellulose-enclosed plant cells.

Amphibians, Reptiles, and Birds. The amphibians have a digestive system similar to that of bony fishes but with some

important differences. For example, whereas fish have rather immobile tongues, the tongue in amphibians is often highly movable. In fact, frogs use their tongue to capture insects. The tongue is attached to the front of the mouth, permitting the sticky organ to be flicked far out with considerable speed and accuracy. The prey is then crushed against a peculiar patch of teeth on the roof of the mouth and swallowed whole. Like sharks and bony fishes, amphibians can't chew.

The reptiles are a rather diverse group and boast a wide variety of techniques in detecting and catching prey. For

Rattlesnake skull

FIGURE 37.6
Feeding in the Snake
Snakes cannot tear or chew their food. Captured prey, sometimes quite large, is swallowed whole. This is possible because the lower jaw is loosely attached to the skull, and the bones of the palate are movable, as are the left and right halves of the lower jaw. Stretching of the esophagus and stomach and the absence of a sternum help in moving the prey through the digestive tract.

example, some reptiles have many teeth, while, in others, the teeth are fewer in number but more specialized. Crocodiles and alligators are unusual in that their teeth are seated in sockets in the bone, much like our own. Evolution has provided some venomous snakes, including the North American pit vipers, with retractable grooved or hollow **fangs** with intricate injecting structures. Other venomous snakes—cobras, kraits, mambas, and coral snakes, for instance—have permanently erect fangs and must "chew" the venom into their prey. All snakes, by the way, are carnivorous and have undergone a unique modification of the jaw for swallowing whole prey (Figure 37.6).

Snakes and lizards find food with the aid of well-developed eyes and a specialized olfactory (odor-sensing) organ at the roof of the mouth. This structure, called **Jacobson's organ**, detects molecules (odors) from the air, which are brought to it by the moist, flicking tongue. The pit vipers orient to warm-blooded prey through the use of acutely sensitive heat-detecting receptors in the facial pits (see Chapter 34).

The dentition of birds is as scarce as hen's teeth. In fact, dentition (toothiness) in birds no longer exists. All of today's birds have bills rather than teeth. The bill or beak is a bony structure with a tough, continuously growing covering of keratin, a horny protein also found in fingernails and hair. While the basic structure of the bill is similar in all birds, there is considerable variation in its shape and size. Each type strongly reflects the feeding habits of the bird (Figure 37.7).

In some instances, feeding habits are reflected in the feet as well. Typically, birds of prey have long, curved talons for grasping prey. Ground-foraging species such as grouse and pheasants have heavy, strong feet for scratching the soil, while the feet of the ostrich and emu can be deadly weapons.

The digestive system of seed-eating birds includes a large **crop** for storing and moistening food, followed by a two-part **stomach**. The first portion—the very glandular **proventriculus**—secretes gastric juices into the coarse food. The food then passes into a very muscular **gizzard**, where it is pulverized with the aid of a hardened lining and abrasive sand grains the bird routinely swallows. From the gizzard the food passes into the intestine.

Foraging and Digestive Structures in Mammals. Mammalian teeth grow in sockets in the jaws, a trait that distinguishes them from almost all other vertebrates. Generally, the mam-

White ibis

Hummingbird

Bald eagle

FIGURE 37.7
Bird Food-Getting Specializations
The bills of birds provide clues to their feeding habits.

mal begins life with temporary teeth, or milk teeth, which are later replaced by permanent teeth. The basic structure of each permanent tooth is the same in all mammals (Figure 37.8).

While the teeth of most fishes and reptiles show little specialization, mammals have four types of teeth, each usually with its own specialized function. The front teeth, or **incisors**, highly developed in the rodents and rabbits, are chisel-shaped for gnawing and cutting. The second group, the **canines** (sometimes referred to as fangs), are used for capturing and killing prey, tearing food, and defense. They are very prominent in carnivores and completely absent in some of the herbivores. (Imagine feeding a carrot to a horse with fangs.) The last two groups, the **premolars** and **molars**, are specialized for grinding. They are large and flat in the herbivores and are used to break down the cellulose walls and fibers of plants. Some carnivores, such as dogs, have large, sharp-edged premolars specialized for shearing tissues and crushing bones.

Dentition in humans is rather generalized, lacking the prominent, specialized, exaggerated features often found in other mammals. Because of the lack of specialization, some zoologists regard human dentition as primitive (see Figure 37.8).

The Ruminants: Grazing Mammals. Plant-eating mammals, particularly foliage eaters, have solved the problems of cellulose digestion. You may recall that, although cellulose is a potentially rich source of energy, breaking the beta linkages between the glucose subunits requires enzymes that few animals possess. Cellulose-digesting enzymes are present principally in microorganisms such as fungi, protists, and bacteria. Those mammals that carry out cellulose digestion have evolved special relationships with microorganisms that can digest cellulose.

Cattle, horses, and other herbivores begin the digestive process with heavy, flat teeth that are specialized for grinding. The digestive tracts of cattle and other grazers, or **ruminants** (including deer, giraffes, antelope, and buffalo), have four-chamber "stomachs," including the **rumen** from which the name *ruminant* is derived, and the **reticulum**, **omasum**, and **abomasum**. Cattle don't produce enzymes that can digest cellulose but, instead, harbor certain protozoans and bacteria in

FIGURE 37.8
Dentition in Mammals
Nearly all mammals share the same basic tooth structure, a hardened layer of enamel surrounding a softer dentin region. Within the dentin is a pulp cavity, penetrated by blood vessels and sensory nerve endings. The arrangement and individual shapes vary considerably, however, according to the diet to which the species has adapted. Note the simple pointed teeth of the insect-capturing insectivore. Compare the gnawing and grinding teeth of a rodent with the tearing and crushing teeth of a lion. Also note the nipping lower incisors of the hooved herbivore and the absence of upper incisors—a horny covering meshes with the lower incisors for nipping foliage. Human dentition is quite uniform and relatively unspecialized.

(a) Shrew (insectivore)

(b) Beaver (rodent)

Enamel
Dentin
Pulp
Pulp canal (blood vessels, nerves)
Tooth structure

(c) Impala (herbivore)

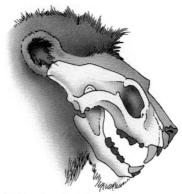
(d) Lion (carnivore)

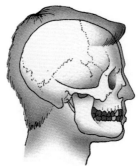
(e) Human (omnivore)

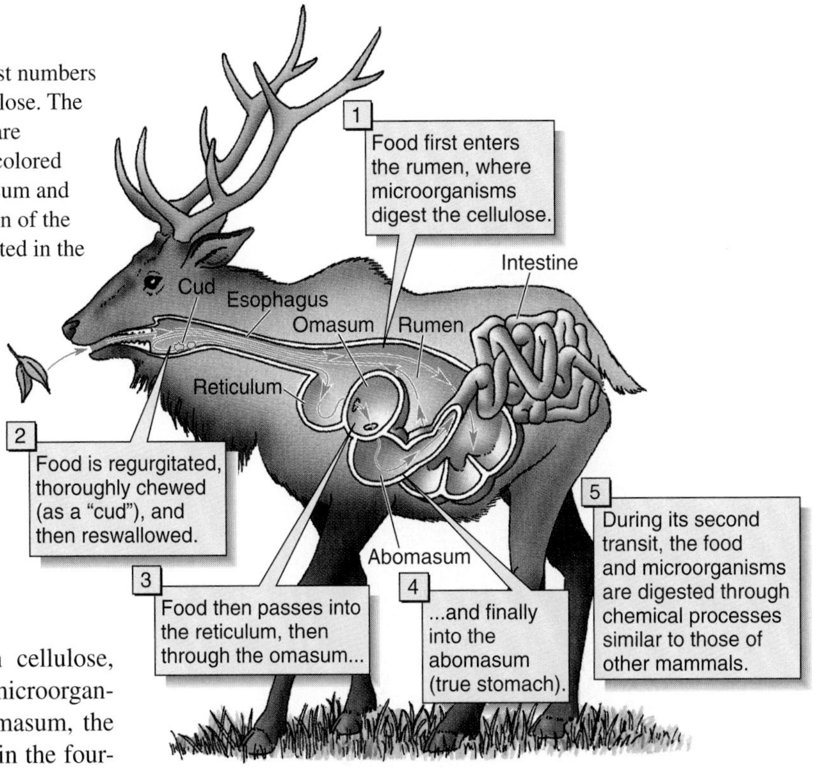

FIGURE 37.9
The Four-Part Ruminant Stomach
Digestion in the elk begins in the large rumen, where vast numbers of protozoans and bacteria begin the breakdown of cellulose. The products are regurgitated as the cud, are rechewed, and are swallowed, this time entering the reticulum (follow the colored arrow). The partially digested mass then enters the omasum and finally the abomasum (the true stomach), where digestion of the microorganisms themselves begins. Digestion is completed in the small intestine.

1 Food first enters the rumen, where microorganisms digest the cellulose.

Intestine

Cud Esophagus Omasum Rumen

Reticulum

2 Food is regurgitated, thoroughly chewed (as a "cud"), and then reswallowed.

Abomasum

5 During its second transit, the food and microorganisms are digested through chemical processes similar to those of other mammals.

3 Food then passes into the reticulum, then through the omasum...

4 ...and finally into the abomasum (true stomach).

their rumen. These organisms *can* break down cellulose, forming nutritious products that, along with the microorganisms themselves, can begin digestion in the abomasum, the actual stomach. The unusual aspects of digestion in the four-part stomach, including "regurgitation" and "cud chewing," are described in Figure 37.9.

The Digestive System of Humans

The human digestive system (Figure 37.10) is rather representative of the mammalian system in many respects. We will now follow the enchanting path of food through the human tract, discussing the specialized structures along the way. We will then take a close look at the chemistry of digestion and some aspects of nutrition.

The Oral Cavity and Esophagus

We have already discussed human dentition, so now let's concentrate on some of our other feeding structures, such as the lips and tongue. In all mammals, the lips have a rather essential role in eating. If you don't know what that role is, try eating a meal without closing your lips (but do it somewhere else). Fortunately, even those souls who chew with their mouths open must close them to swallow. So the lips hold food in and seal the mouth to permit swallowing.

In addition to its role in moving food into position for chewing, swallowing, and sorting out fish bones, the tongue constantly monitors the texture and nature of foods. This is a valuable chore, since it warns us about peculiarities in time to prevent our swallowing something we shouldn't. Furthermore, the tongue *tastes;* that is, it can distinguish certain chemicals by specialized chemoreceptors clustered in sense organs called **taste buds**. These receptors distinguish four

basic tastes: salty, sour, sweet, and bitter (but see Figure 34.23). In addition to informing us of flavor, the stimulation of taste buds enhances the flow of **saliva**.

There are three pairs of salivary glands (see Figure 37.10). Each salivary gland empties its secretions into the mouth. These secretions collectively form the saliva, which consists mainly of water, ions, lubricating mucus, and the starch-splitting enzyme **salivary amylase**. Because it helps break down starchy food particles caught between the teeth, salivary amylase is probably less important in digestion than it is in oral hygiene. Saliva itself is mainly important for moistening and lubricating food.

The pharynx, the throat region in the rear of the oral cavity, forms a common passageway with the **nasopharynx**. Just below the base of the **tongue** the pharynx divides, forming the anterior **larynx** (Adam's apple) and the posterior **laryngopharynx**. During swallowing, food is pressed by the tongue against the **soft palate**, closing the nasopharynx; the larynx is then raised, bending the **epiglottis** over the **glottis** (laryngeal opening), thus closing the air passageway to the lungs and directing food into the esophagus (Figure 37.11). Should this action fail, food will enter the larynx, producing violent spasms of coughing or even choking (Essay 37.1, page 800.).

The **esophagus** (the muscular tube between the pharynx and stomach) has no digestive function but simply moistens food and moves it to the stomach. Its upper portion contains skeletal (voluntary) muscle, while the rest has smooth (involuntary) muscle. Thus swallowing begins as a conscious act but soon becomes automatic.

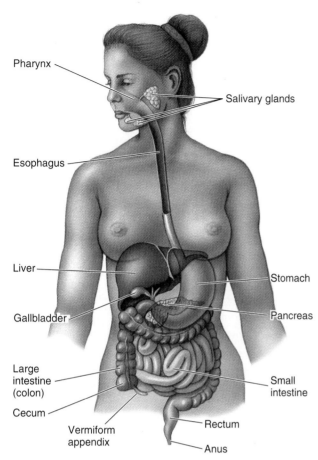

FIGURE 37.10
The Human Digestive System
The three salivary glands open into the oral or mouth cavity. The pharynx and esophagus are primarily simple, muscular passages secreting only lubricating mucus. The stomach is quite complex, secreting hydrochloric acid and protein-digesting enzymes. The small intestine is both secretory and absorptive, its villi and microvilli greatly increasing its absorbing surface. It also receives the secretions of the liver and pancreas. The colon, or large intestine, is chiefly involved in water absorption and the compacting of digestive wastes. It also provides a suitable environment for enormous populations of bacteria, some of whose waste products are useful to the host.

The lower esophagus, by the way, is very similar in its tissue organization to the rest of the digestive tract. As you can see in Figure 37.12, there are three principal layers of tissue plus a fibrous outer coat, or **serosa**. The inner lining, the **mucosa**, contains a mucus-secreting epithelium. The second tissue layer, the **submucosa**, contains connective tissue with blood and lymphatic vessels, nerves, and glands. The third layer, the **muscularis**, is a region of smooth muscle with an inner circular layer and an outer longitudinal layer.

The muscles of the esophagus move food along by wave-like contractions called **peristalsis**. Food being swallowed is pressed into a **bolus** (clump) by the tongue and pharynx. As the bolus approaches the involuntary muscles of the esopha-

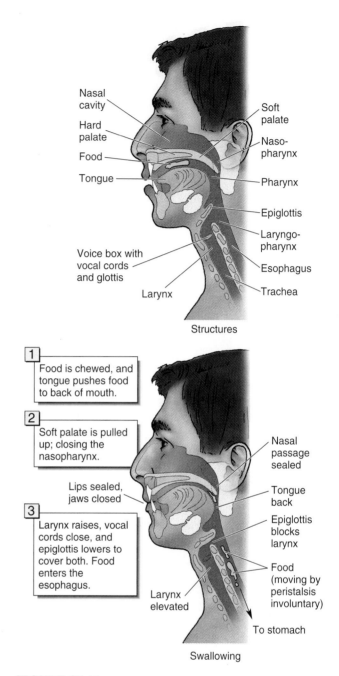

FIGURE 37.11
Swallowing
Swallowing begins as a food mass is pressed upward by the tongue against the soft palate and back toward the pharynx. As the involuntary stages begin, the soft palate elevates, closing off the nasopharynx. The larynx elevates, bending the epiglottis over the glottis and sealing off the breathing passage. Food then enters the esophagus.

gus, the circular layer just ahead of it relaxes and the muscles behind the bolus contract, pushing food along. The longitudinal muscles ahead help by contracting also. (See Figure 37.12.) The muscles are coordinated by nerve endings that are activated by the presence of food. Because peristalsis is under local (intrinsic), hormonal, and autonomic control, it is an involuntary action, and unless you have an embarrassingly noisy intestine you are probably not aware of the process.

The Stomach

The human stomach is structurally and functionally similar to that of most other vertebrates. It temporarily stores food and begins its digestion. In addition, its acids and enzymes kill many of the microorganisms we swallow. The stomach can be closed off at either end by two muscular sphincters, the *cardiac sphincter* and the *pyloric sphincter* (Figure 37.13). When contracted, these circular groups of muscles permit the stomach to churn and liquefy food without forcing it back into the esophagus or on into the intestine before it is ready. You may have noticed that the upper ring of muscle, the **cardiac sphincter**, sometimes fails when the stomach is overfull or filled with gas and allows acidic fluids to enter the esophagus, creating "heartburn." The overproduction of these acids is also related to emotional stress.

The muscle layers of the stomach are more complex than those of the esophagus. Here we encounter a third, diagonal layer of muscle just inside the circular layers. This one produces a twisting action that accompanies the usual wringing (by circular muscles) and shortening (by the longitudinal muscles). Also, peristalsis in the stomach is nondirectional and moves food back and forth in a sequence of contortions until it reaches the well-churned, liquefied state that enters the intestine.

The stomach lining (mucosa) contains many long, tubular gastric glands that secrete the gastric juices, watery solutions containing hydrochloric acid (HCl) and proteins. The glands contain **chief cells**, which secrete the protein **pepsinogen**, and **parietal cells**, which secrete HCl (see Figure 37.13, enlarged view). The acid activates the pepsinogen, forming the digestive enzyme **pepsin**, which initiates the digestion of protein. Other glands secrete water, mucus, and small quantities of **gastric lipase**, a fat-splitting enzyme. The presence of HCl produces a very low pH of 1.6–2.4 in the stomach fluids. The acid not only kills many bacteria and other microorganisms taken in with food but also opens the folds and turns of globular proteins in our food, better exposing their peptide bonds to pepsin. But the acidity of the stomach and the potency of its enzymes could (and sometimes do) endanger the lining itself. One reason we don't digest our stomachs is that a layer of **mucin**, an insoluble mucoprotein, forms a coating over the stomach lining.

The Small Intestine

Once the food reaches a liquefied state, now referred to as **chyme**, the **pyloric sphincter**, another ring of muscle, relaxes

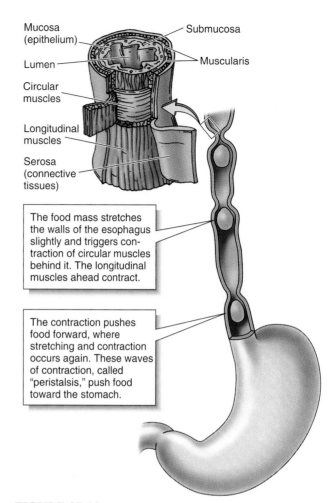

Mucosa (epithelium)
Submucosa
Lumen
Muscularis
Circular muscles
Longitudinal muscles
Serosa (connective tissues)

The food mass stretches the walls of the esophagus slightly and triggers contraction of circular muscles behind it. The longitudinal muscles ahead contract.

The contraction pushes food forward, where stretching and contraction occurs again. These waves of contraction, called "peristalsis," push food toward the stomach.

FIGURE 37.12
Organization of the Gut and Peristalsis
The basic structure of the tube includes four tissue layers—the mucosa, submucosa, and muscularis—and the outer coat or serosa. Food moves onward through peristalsis. Circular muscle layers behind the food contract.

a bit, allowing a small amount of liquefied food to move into the **small intestine**. The small intestine (see Figure 37.10) is about 6 m (20 ft) long in humans, consisting of three regions: the **duodenum**, **jejunum**, and **ileum**. The small intestine has two roles: digestion and absorption. Most of the digestion and all of the absorption of foods occur in the small intestine. Much of the absorption involves active transport, but passive transport is also involved. In fact, about 90% of the water present in foods diffuses into the small intestine.

The small intestine (Figure 37.14) is uniquely adapted for absorption, having an enormous surface area (estimated at about 700 m², or the floor area of four or five three-bedroom houses). Its inner surface has the appearance of a badly wrinkled bath towel. Closer examination of the surface reveals that there are many folds, each covered with tiny projections called **villi**. Each villus is in turn covered with columnar epithelial cells bristling with **microvilli**, finger-like projections of the plasma membrane (see Figure 37.14, SEM).

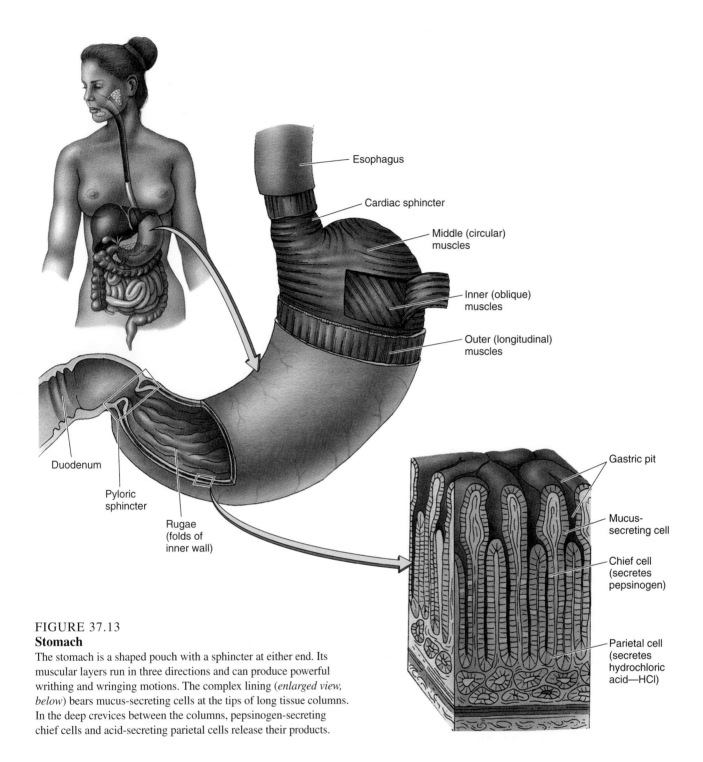

FIGURE 37.13
Stomach
The stomach is a shaped pouch with a sphincter at either end. Its muscular layers run in three directions and can produce powerful writhing and wringing motions. The complex lining (*enlarged view, below*) bears mucus-secreting cells at the tips of long tissue columns. In the deep crevices between the columns, pepsinogen-secreting chief cells and acid-secreting parietal cells release their products.

Within each villus is a **capillary bed** and a **lacteal** (a lymph vessel). The villi are capable of rather vigorous movement, like millions of wiggling fingers. They help in mixing food and enzymes in the gut and assist in absorption by creating pressures that tend to concentrate digested food in the many recesses of the lining.

Until recently, it was believed that digestive enzymes were free in the intestinal lumen. But careful studies of the epithelial cells have led physiologists to believe that most of the enzymes are actually bound to the plasma membranes, forming a frilly surface that is part of the **glycocalyx** (see Chapter 5). This precise positioning of the enzymes presumably helps prevent the lining of the small intestine from being digested.

The cellular lining of the small intestine is thought to be very temporary and always in a state of replacement. This isn't surprising if you consider the amount of frictional wear that it suffers. Abrasion scuffs away an estimated 17 million cells each day. However, new cells are continuously produced at the base of each villus, and these migrate upward in an orderly fashion, replacing those worn away at the tip.

FIGURE 37.14
The Small Intestine's Lining
The inner surface or mucosa of the small intestine contains a number of folds, each of which is covered by the finger-like intestinal villi. Each villus consists of a column of cells containing a capillary network and a lacteal (a blind ending of a lymph duct), into which digested foods move. As the scanning electron micrograph shows, the surface cells of the villus have their membranes folded into numerous microvilli. At the surface, membrane-bound enzymes form the hazy glycocalyx.

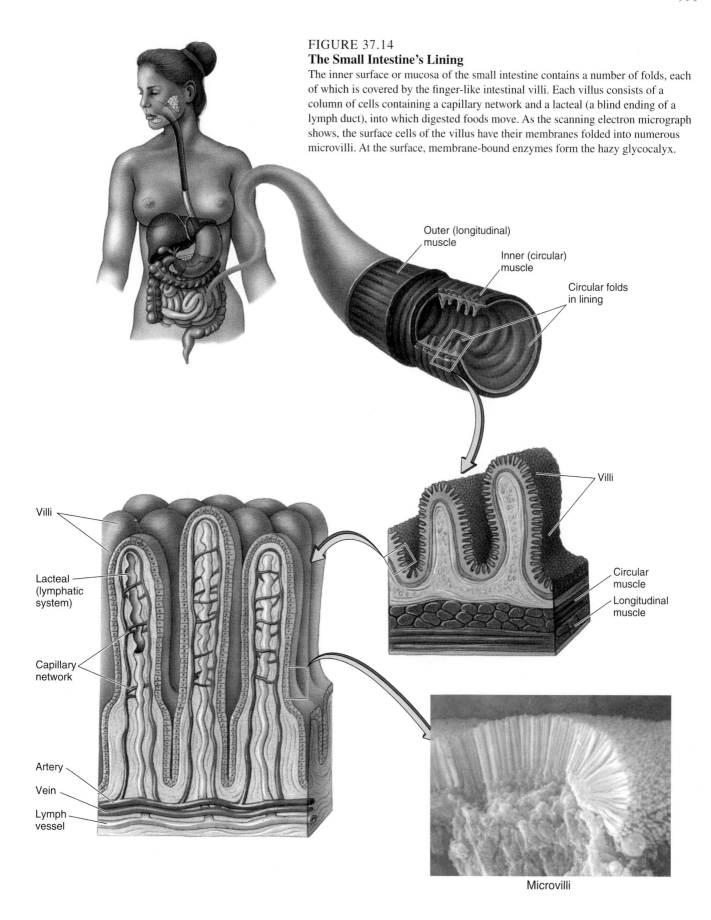

Outer (longitudinal) muscle
Inner (circular) muscle
Circular folds in lining
Villi
Circular muscle
Longitudinal muscle
Villi
Lacteal (lymphatic system)
Capillary network
Artery
Vein
Lymph vessel
Microvilli

Accessory Organs: The Liver and Pancreas

The duodenum of the small intestine receives the secretions of two accessory organs, the *liver* and the *pancreas* (Figure 37.15). The **liver**, an organ of many functions, aids in digestion by secreting a slightly alkaline fat emulsifier known as **bile** that breaks up fats into minute globules. Bile is a complex substance containing cholesterol, bile salts and pigments, water, and modified amino acids. The bile salts are actually steroids and are important in the digestion and absorption of fats. The bile pigments are products of hemoglobin destruction, since the liver (along with the spleen) is the red blood cell graveyard. These breakdown products of hemoglobin become part of the digestive wastes, providing the characteristic color to feces.

After being produced in the liver, the bile is stored in the **gallbladder** (see Figure 37.15). The release of bile is brought about by a hormone whose name defies pronunciation, *cholecystokinin-pancreozymin*, which causes the gallbladder to contract (and also stimulates the release of pancreatic enzymes). Bile reaching the duodenum via the **bile duct** is joined by highly alkaline fluids from the merging **pancreatic duct**, just before it enters the intestine.

In humans, the **pancreas** is a long glandular organ lying nestled in the first turn of the small intestine (see Figure 37.15). One of its products, **sodium bicarbonate**, neutralizes the acid accompanying partially digested food from the stomach, protecting the small intestine and raising the pH, thereby enhancing the activities of the next series of enzymes. The pancreas secretes an entire battery of digestive enzymes that are involved in the breakdown of fats, carbohydrates, proteins, and nucleic acids. The pancreatic enzymes, together with those of the small intestinal lining, carry out most of the digestive process.

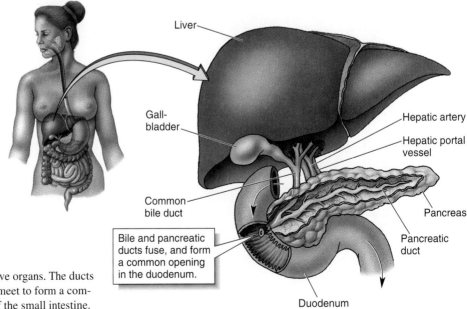

FIGURE 37.15
Accessory Digestive Organs
The liver and pancreas are accessory digestive organs. The ducts from both the gallbladder and the pancreas meet to form a common duct that empties into the duodenum of the small intestine.

Liver

Gall-bladder

Common bile duct

Bile and pancreatic ducts fuse, and form a common opening in the duodenum.

Hepatic artery

Hepatic portal vessel

Pancreas

Pancreatic duct

Duodenum

TABLE 37.1
Digestive Enzymes and Their Functions

Sources and Enzymes	Substrate	Product
Salivary glands		
Salivary amylase	Starch	Maltose and starch fragments
Stomach lining		
Pepsin	Protein	Peptides
Rennin	Casein	Insoluble curd
Gastric lipase	Triglyceride	Fatty acids + glycerol
Pancreas		
Trypsin	Peptide linkage	Shorter peptides
Chymotrypsin	Peptide linkage	Shorter peptides
Carboxypeptidase	C-terminal bond	Free amino acids
Ribonuclease	RNA	Nucleotides
Deoxyribonuclease	DNA	Deoxynucleotides
Alpha–glucosidase	1–6 linkages of amylopectin	Starch fragments
Pancreatic amylase	Starch	Maltose
Pancreatic lipase	Triglyceride	Fatty acids, glycerol, monoglycerides
Intestinal lining		
Aminopeptidase	N-terminal bond	Free amino acids
Dipeptidase	Dipeptide	Free amino acids
Nuclease	Nucleotide	5-carbon sugar + nitrogen base
Maltase	Maltose	Two glucose units
Sucrase	Sucrose	Glucose + fructose
Lactase	Lactose	Glucose + galactose

The Large Intestine

The small intestine joins the **large intestine**, also called the **colon** or **bowel**, on the right side of the abdominal cavity. The large intestine consists of the **cecum, ascending colon, transverse colon, descending colon,** and **sigmoid** (S-shaped) **colon** (see Figure 37.10). The last portions of the digestive tract are the **rectum**, the **anal canal**, and the **anus**—where the story ends.

At the union of the small and large intestines is the **ileocecal valve**, a one-way valve that prevents a backflow of the food residues into the small intestine. Below the point where the two parts of the intestine join, the large intestine forms a blind pouch, the cecum. Protruding from the cecum is the **appendix**—a hollow, finger-like extension.

The primary function of the colon is to absorb water and minerals into the blood and to prepare the feces to leave the digestive tract. The colon takes up about 10–20% of the water in the liquid digestive residues that enter. The mucosae of the colon are glandular and secrete mucus, which is used in the production of feces and lubricates the lower portions of the colon. Very little, if any, digestive activity takes place in the colon itself, although a rapid digestion of food residues is carried on by teeming populations of microorganisms. *Escherichia coli,* along with a number of methane-producing bacteria (see Chapter 22), are prominent inhabitants. Strange as it may seem, a sizable amount of fecal material, perhaps one-third by dry weight, consists of bacteria. These intestinal bacteria play an important and rather unexpected role in nutrition. A flourishing bacterial colony can help prevent vitamin deficiencies because they synthesize vitamin K, biotin, and folic acid, which leak out of the bacteria and are then available for absorption.

The final region of the digestive system is the rectum. Its structure is basically the same as that of the rest of the colon, except for the presence of folds in its walls. These are the **rectal valves**, which help support the weight of feces so that gravity doesn't become the enemy of propriety. The anus is the external opening. The movement of wastes through the anal sphincter is usually, thankfully, under voluntary control.

The Chemistry of Digestion

From our previous discussions recall that proteins, carbohydrates, and fats are often very large, complex molecules composed of repeating molecular subunits (see Chapter 3). Except for certain triglycerides, these macromolecules can't be absorbed through the lining of the gut. So in the digestive process, enzymes attack the linkages of the molecular subunits, breaking them down into their components.

During digestion, carbohydrates are hydrolyzed into simple sugars, fats into fatty acids and glycerol, proteins into their various amino acids, and nucleic acids into free nucleotides. The digestive enzymes and their functions are listed in Table 37.1.

The bonds are broken through **hydrolysis**, a process in which water is added enzymatically to the linkages, disrupting

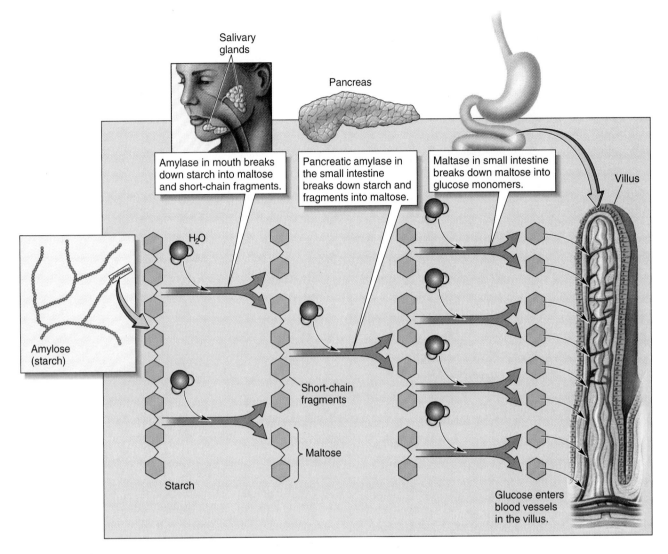

FIGURE 37.16
Starch Digestion
Amylose, a straight chain starch, is readily digested by amylase, a starch splitter that breaks 1–4 linkages.
Maltose, the product, is broken down by maltase into glucose.

them and yielding the molecular subunits. Following digestion, the nutrient molecules cross the gut lining by a process called **absorption**. While water and some ions move by osmosis and diffusion, respectively, most nutrient molecules must be actively transported through membranal carrier mechanisms into the blood or lymphatic fluids lying within the intestinal villi.

Carbohydrate Digestion

All carbohydrate digestion begins in the mouth, where the enzyme salivary amylase breaks some linkages, producing the disaccharide (two glucose units) maltose and many longer polysaccharide fragments. Starch digestion is then temporarily stalled by the acidity of the stomach but resumes in full swing in the small intestine. What happens next depends on the type of starch.

Fragments of amylose, a simple, straight chain polysaccharide, are all broken down by amylase from the pancreas (called **pancreatic amylase**) into maltose, whereupon the intestinal enzyme **maltase** completes the job by splitting the double sugars into glucose units (Figure 37.16).

Sucrose (table sugar) and lactose (milk sugar)—both disaccharides—are also split into simple sugars in the small intestine. The enzyme **sucrase** splits sucrose into glucose and fructose, whereas **lactase** splits lactose into glucose and galactose. Glucose and fructose are actively transported into the capillaries of the villi in the small intestine, and from there they are carried by the bloodstream to the liver for storage.

Fat Digestion

With some minor exceptions, fat digestion occurs in the small intestine. Since the digestive environment there is quite watery, we might expect fats and oils to cluster in masses, rejecting their surroundings and making it difficult for hydrolyzing enzymes to do their work. This would be true were it not for bile salts. When fats enter the small intestine, the gallbladder contracts, and bile passes through the bile duct and joins the fats in the small intestine (Figure 37.15). Bile is a fat emulsifier. It works like a laundry detergent, physically breaking down the large fat clusters into tiny droplets, which can be attacked by fat-splitting enzymes, dispersing the droplets in the watery surroundings.

Fats are digested by the pancreatic enzyme **lipase**. Lipase breaks down neutral fats (triglycerides) into fatty acids, glycerol and some monoglycerides, which are carried in droplets (micelles) formed by bile across the lipid-soluble core of the plasma membrane of intestinal epithelial cells. Once inside a lining cell, these products are reassembled into triglycerides that, along with absorbed cholesterol, then gather into **chylomicrons**, minute bodies enclosed by a thin protein envelope. In this form, they pass through the intestinal epithelial cells and enter the lacteals of the villi, from which they are transported via lymph vessels to the circulatory system.

Protein Digestion

Proteins are among the largest and most complex molecules known. It is not surprising then that their digestion is also complex, requiring a number of different enzymes that form a family called **proteases**.

The first step in protein digestion (Figure 37.18), as we have mentioned, occurs in the stomach, where acids open up protein molecules, exposing their subunits. Here also an enzyme precursor, pepsinogen, is activated by hydrochloric acid to form the active enzyme, pepsin. Pepsin attacks peptide

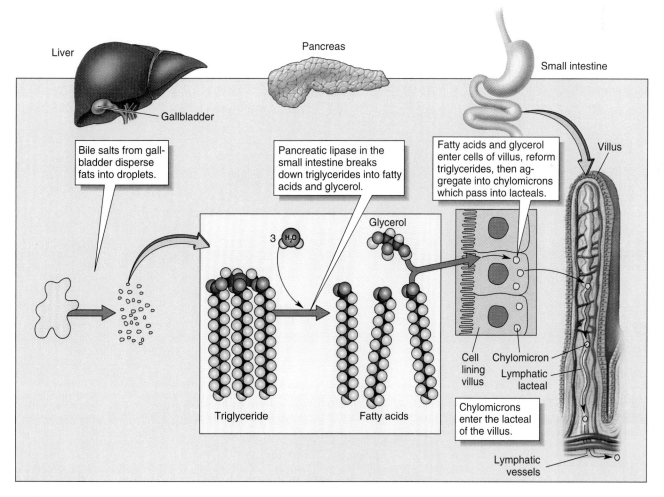

FIGURE 37.17
Fat Digestion
Fats are digested by the enzyme lipase, which attacks the bonds between each of the fatty acids and the glycerol. Three water molecules are consumed as each triglyceride is digested.

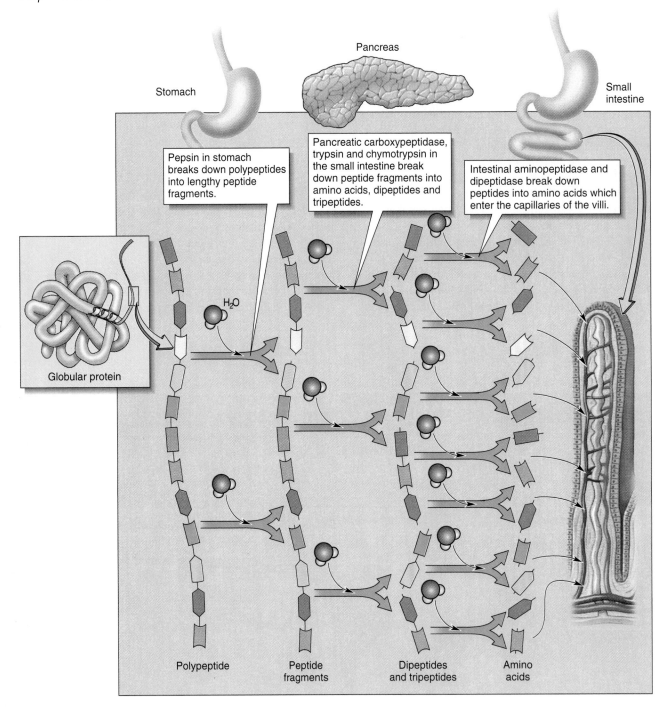

Pancreas

Stomach

Small intestine

Pepsin in stomach breaks down polypeptides into lengthy peptide fragments.

Pancreatic carboxypeptidase, trypsin and chymotrypsin in the small intestine break down peptide fragments into amino acids, dipeptides and tripeptides.

Intestinal aminopeptidase and dipeptidase break down peptides into amino acids which enter the capillaries of the villi.

H_2O

Globular protein

Polypeptide

Peptide fragments

Dipeptides and tripeptides

Amino acids

FIGURE 37.18
Protein Digestion
The digestion of protein in the stomach begins with the action of pepsin. The peptide fragments formed are then attacked in the small intestine by the pancreatic enzymes. The dipeptides and tripeptides formed are then hydrolyzed by enzymes from the small intestine itself. A residue of single amino acids emerges, and protein digestion is complete.

bonds within the protein rather than near its ends. Pepsin has a very limited action, attacking the peptide bonds of only four of the 20 amino acids. As a result, the products of pepsin's action are usually still lengthy peptides.

The peptides pass into the small intestine, where they are acted on by three pancreatic enzymes: *trypsin, chymotrypsin,* and *carboxypeptidase.* **Carboxypeptidase** cleaves peptide bonds at the carboxy-terminal end of the peptide, one after the other in a nibbling action, thus freeing individual amino acids. (Recall from Chapter 3 that one end of a protein chain contains a free carboxyl group, and the other end contains a free amino acid group.) **Trypsin** and **chymotrypsin**, however, are very selective. Each attacks the peptide bonds on the carboxyl side of just five of the 20 amino acids. The products of action by these three pancreatic enzymes include single amino acids and small peptide fragments of perhaps two to ten amino acids.

The final steps in protein digestion are carried out by *aminopeptidases* and *dipeptidases* originating in the intestinal lining. **Aminopeptidase** cleaves peptide bonds at the amino-terminal end of peptides. **Dipeptidase** attacks peptide bonds in dipeptides, yielding two amino acids (see Figure 37.18). These final steps free the remaining amino acids, which are then ready for absorption.

The absorption of amino acids, like that of glucose, involves an ATP-powered carrier. Once inside the cells of the epithelial lining, amino acids diffuse across them and out the inner walls into capillaries of the villi. After being distributed throughout the blood and utilized where they are needed, the remainder are carried to the liver, where they are removed from the blood.

In the cells of the liver, amino acids may follow various metabolic pathways, depending on the body's needs. Some are used by the active liver cells to form proteins required there, since the liver has a very rapid protein utilization rate. Other amino acids are utilized as a source of energy—see Chapter 8. Some amino acids are simply converted to other amino acids or used in the formation of the nitrogen bases of nucleotides. Still others are distributed to various cells of the body for their use in protein synthesis. A large number of amino acids are assembled into serum proteins (the proteins of blood plasma), which are released into the bloodstream. We'll look further into the serum proteins in Chapter 38.

The Activation of Proteases. It might have occurred to you that the presence of so many protein-cleaving enzymes could present a danger to cells in their surroundings. After all, cells are largely composed of protein, and proteolytic enzymes cannot really distinguish one protein from the next. (Recall that the danger posed by hydrochloric acid to the lining of the stomach and small intestine is averted by the stomach's mucin coating and the small intestine's sodium bicarbonate.)

One safeguard against damage to the digestive lining by proteases is that they are secreted in an inactive form. This is particularly important to the pancreas, where many enzymes are manufactured and stored. The pancreatic enzymes trypsin, chymotrypsin, and carboxypeptidase are secreted as inactive trypsinogen, chymotrypsinogen, and procarboxypeptidase until they enter the small intestine. There trypsinogen is activated by the intestinal enzyme **enterokinase**. (Recall that *kinases* activate enzymes.) Activated trypsin then modifies the other two enzymes, and they too become active. For this reason the proteases are usually harmless to the pancreas.

But occasionally the safeguards break down, for instance, when the pancreatic duct is blocked and the enzyme concentration builds. Then the proteases may become active while in the pancreas. This brings on a condition called acute pancreatitis, a potentially fatal disease in which the enzymes attack and destroy the pancreatic tissue itself. Thankfully, pancreatitis is rare—for a special reason. The pancreas produces a trypsin inhibitor, a protein that blocks trypsin's action. This is a particularly effective safeguard because trypsin activates the other proteases. Thus even if the proteases become active while in the pancreas, the protease actions cannot proceed.

Nucleic Acid Digestion

Almost everything we eat contains nucleic acids. So, one might wonder, how are they digested? Enzymatically, as would be expected (see Table 37.1). The enzymes that hydrolyze nucleic acids are called **nucleases**, often referred to as RNAses and DNAses. The nucleases, secreted by the pancreas, break the nucleic acids down into individual nucleotides (see Chapter 12) or into small chains of three or four nucleotides. The enzymes are divided into two classes on the basis of where they attack the giant molecules. **Exonucleases** cleave off the end nucleotides, while **endonucleases** attack bonds within the molecule.

Integration and Control of the Digestive Process

The release of digestive enzymes and related substances must be very precisely timed. Such timing is under at least three types of control—mechanical, neural, and hormonal—with some interaction among the three types. For example, saliva flow can be stimulated mechanically by chewing a tasteless substance like paraffin. Yet even the thought of eating chocolate cake can evoke salivation. Thus neural factors are clearly involved. Further along the digestive tract, gastric secretions can also be stimulated by our simply sensing the presence of food. A neural message is sent along a branch of the **vagus nerve** from the brain to the stomach lining.

The presence of food in the stomach mechanically evokes gastric secretions by stimulating sensory neurons in the stomach wall itself. Impulses reach the brain, and again the brain sends down the word via the vagus nerve to resume the flow of gastric secretions.

A third mechanism is hormonal and independent of the nervous system (Figure 37.19). For example, when food, particularly concentrated protein, is present in the stomach, the hormone **gastrin** is released into the bloodstream by cells of the stomach wall itself. Once in the blood, the hormone circulates freely but is ignored by all cells except highly specific target cells, the gastric glands (see Chapter 36). The gastrin mechanism, like many that are hormonally regulated, works through a negative feedback principle. The release of gastric juices, which include HCl, lowers the stomach pH. When it reaches pH 2, gastric secretion slows greatly. Thus the product of the hormone's action—the HCl—has a negative effect, shutting down the hormone's release. The digestive hormones secretin and cholecystokinin pancreozymin (CCK-PZ are also involved in the integration and control of digestion (see Table 37.2 and Figure 37.19).

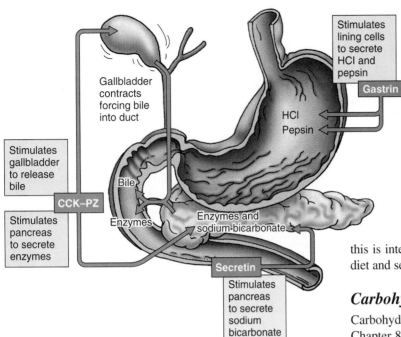

Stimulates lining cells to secrete HCl and pepsin

Gastrin

Gallbladder contracts forcing bile into duct

HCl
Pepsin

Stimulates gallbladder to release bile

CCK–PZ

Bile

Stimulates pancreas to secrete enzymes

Enzymes

Enzymes and sodium bicarbonate

Secretin

Stimulates pancreas to secrete sodium bicarbonate

FIGURE 37.19
Control of Gastric Secretions
The flow of gastric juices can be stimulated in several ways. Just the thought, the smell, or the taste of food can activate the autonomic nervous system, with the vagus nerve stimulating gastric juice release. More directly, the presence of bulk food in the stomach can stimulate release. The presence of protein in the stomach stimulates gastrin (a hormone) release into the blood. Its targets, the glandular cells of the stomach, respond by releasing gastric juices. Two other hormones, secretin and cholecystokinin pancreozymin (CCK-PZ), operate in the release of sodium bicarbonate and bile.

Some Essentials of Nutrition

The subject of nutrition is of increasing interest to the general public. Unfortunately, interest in food doesn't necessarily imply knowledge about **nutrition** (any process involved in nourishment). Many food faddists are almost ignorant of the subject. Some do not seem to realize that "organically grown" food is no better for you than food grown with chemical fertilizers; that rose hip vitamin C is no different from the commercial vitamin C produced by bacteria; that large doses of certain vitamins may not be healthful and may be harmful; that dietary protein in excess of body needs is simply deaminated and converted to glucose or fatty acids in the liver—and winds up as fat on the hips. Actually, most Americans eat far more protein than they need, and tests with long-distance runners show that athletes have more stamina on a high-carbohydrate diet than on a high-protein diet. On the other hand, some health faddists prescribe diets so low in protein as to produce virtually the only remaining cases of protein-deficiency disease in America. All

this is interesting but confusing, so let's have a closer look at diet and see if the scientific findings have any relevance for us.

Carbohydrates

Carbohydrates are common sources of energy. As we saw in Chapter 8, they are used in generating ATP, but they are used in other ways as well. For instance, once glucose makes its way from the gut into the circulatory system, it is immediately removed and stored as glycogen by the liver. From there it is meted out according to need. Thus the blood glucose level remains rather constant, with a temporary 10–20% increase immediately after a high-carbohydrate meal. Glycogen is also stored in the muscles, where it is broken down into glucose and used as an energy source. Long-distance runners build additional glycogen reserves in muscles by continually depleting their reserves in long training runs and letting them build back up at ever higher levels. For those who have no intention of becoming involved in heavy exercise, it is important to know that excess glucose is readily converted into body fat.

Fats

Fats in the diet are not all bad. The polyunsaturated fats remain an essential nutrient to humans as well as to other animals, although the quantity needed is open to debate. Fats are excellent concentrated energy sources with, gram for gram, twice the energy value of carbohydrates. In addition, fatty foods contain

TABLE 37.2 **Digestive Hormones**			
Hormone	**Source**	**Stimulation**	**Function**
Gastrin	Stomach	Protein in stomach Vagus nerve	Stimulates release of gastric juice
Enterogastrone	Small intestine	Acid state in intestine Fats in intestine	Inhibits gastric juice release and slows stomach contractions
Secretin	Small intestine	Acid state in intestine Peptides in intestine	Stimulates release of sodium bicarbonate
Cholecystokinin- pancreozymin	Small intestine	Food entering small intestine	Stimulates secretion of pancreatic enzymes and contraction of gallbladder

the essential fat-soluble vitamins A, D, E, and K. Unsaturated fats are necessary constituents of all plasma membranes, yet vertebrates and most other animals cannot synthesize all the types needed. Some, the essential fatty acids, must be provided in food. Enlightened dieters thus include at least some fats in the diet. However, fats can be synthesized from carbohydrate and protein excesses, so cutting down on these nutrients is also required if you wish to lose fat from your body. For this reason, high-carbohydrate or high-protein weight-reducing diets may defeat their own purpose if the intake is not carefully regulated.

You are probably aware of the persistent controversy over saturated fats and cholesterol and may faithfully avoid these in favor of the unsaturated fats heavily touted in advertisements for margarine and cooking oils. Essay 37.2 reveals a startling twist to some prevalent ideas on this subject.

Protein

Because our bodies don't store and release protein well, we should include it in our daily diet. The amino acids of protein are used in several ways: they are the building blocks of our own protein, some kinds are used to form the nitrogen bases of the nucleic acids DNA and RNA, they can be oxidized for energy, and, as mentioned, they can be converted to fats and carbohydrates.

It is interesting that humans (along with all other animals, protozoans as well) can convert some amino acids to others. In fact, we can synthesize 12 of the 20 amino acids we require for growth and maintenance. The remaining eight must be supplied, intact, in the diet. These are referred to as the **essential amino acids**, a phrase that delights advertising agencies, although it is something of a misnomer, since all 20 are biologically essential at some level. The essential amino acids are leucine, isoleucine, lysine, methionine, phenylalanine, threonine, tryptophan, and valine (see Chapter 3).

Dietary proteins vary in "quality." High-quality proteins contain more kinds of amino acids than do low-quality proteins. Protein synthesis in your own cells cannot proceed unless all the constituent amino acids are present, so the usefulness of a protein is limited by its scarcest essential amino acid. In general, animal proteins are of higher quality than plant proteins. Although plants do store proteins in their seeds for the development of new seedlings, these proteins are usually of comparatively low quality for human consumption. Plant storage proteins are typically deficient in tryptophan, methionine, or lysine, and lysine deficiency in particular can become a problem for vegetarians. Nevertheless, a carefully managed vegetable diet can be quite sufficient for human needs, if the right combinations of foods are consumed. A combination of peas and beans, for example, provides the essential amino acids, so these foods or other combinations with a similar balance of amino acids must become a regular part of a healthy vegetarian diet.

Vitamins

In addition to essential fatty acids and essential amino acids, a variety of substances, known as vitamins, are also essential in the diets of animals. Most of the vitamins have been clearly identified by biochemists so that we know both their molecular structure and their general function (Table 37.3). A few are only vaguely understood, however. All we can say about them is that if you don't have them, you will develop something-or-other, and it will probably be bad. Shortages of niacin, vitamin B_6, and riboflavin, for example, can lead to serious illness because the coenzymes NAD and FAD, necessary for glycolysis and cell respiration, are derived from these vitamins. Vitamins C and E have a somewhat more general function: both are **antioxidants** that remove spontaneous free radicals that would otherwise cause damage through random oxidation reactions. (An ongoing argument centers over whether antioxidants can help prevent certain kinds of cancer. Some are suggesting that the antioxidants must be given with other compounds from the plant in which they occur naturally. These compounds are called *phytochemicals*. Yet others have asked, why not just eat the plant?)

Vitamins can be categorized as fat-soluble or water-soluble. The two behave quite differently in the body. For example, water-soluble vitamins function as coenzymes; fat-soluble vitamins do not. Whereas water-soluble vitamins function in most animals, the fat-soluble vitamins function only in vertebrates. Finally, excess water-soluble vitamins are excreted, but the fat-soluble ones are stored in the body fat. This is why it is dangerous to overdose on fat-soluble vitamins—they remain in the body's fat and can be released as that fat is metabolized.

"Recommended Daily Allowances" or RDAs were established by federal agencies as a guide to how much of each vitamin we need daily. These are usually still listed on vitamin bottles, although more recent research indicates that in many cases these recommendations are far too low and a number of health problems could be alleviated by higher vitamin intake.

Mineral Requirements of Humans

In addition to the organic nutrients mentioned, animals require a variety of inorganic ions that are generally referred to as **minerals** (Table 37.4). Some of those needed in substantial amounts are calcium (a major constituent of bone and of many cellular processes), magnesium (necessary for many enzyme activities), and iron (a constituent of hemoglobin and of the cytochromes of cell respiration). Sodium, potassium, and chloride are also needed in fairly substantial quantities; they are involved in ion balance, and sodium and potassium are also involved in nerve cell conduction.

Trace Elements. Elements that are necessary for life but in only very small amounts are called **trace elements**. Iodine is perhaps the best-known trace element, since an iodine deficiency produces **goiter**, a highly visible overgrowth of the thyroid gland (see Chapter 36). Iodine is essential in producing thyroxine, the thyroid hormone, and the overgrowth is the thyroid gland's odd way of compensating for the iodine shortage. Many other minerals are constituents of coenzymes (for example, cobalt in vitamin B_{12}), are otherwise necessary for enzyme function, or are involved in synthetic processes (for example, zinc is needed for insulin synthesis). The functions of many

Essay 37.2 ● CHOLESTEROL AND CONTROVERSY

Cholesterol is a vital constituent of plasma membranes. Our own liver can make it readily. Nevertheless, the amount of cholesterol circulating in the blood is greatly influenced by dietary intake. Not only is cholesterol in food taken up directly, but the amount of cholesterol synthesized and its concentration in the blood can be decreased dramatically by reducing either the amount of cholesterol or the amount of saturated triglycerides in the diet, and especially by switching to unsaturated fats.

How did cholesterol get its unsavory reputation? The plaques that plug arteries in chronic arteriosclerosis are heavily infiltrated with cholesterol. The fact that persons with abnormally high levels of circulating cholesterol have abnormally high rates of arterial disease and heart disease suggests a cause-and-effect relationship. Thus it seems perfectly reasonable to conclude that anyone will benefit from reducing cholesterol intake

and replacing saturated (animal and hydrogenated vegetable) fats with unsaturated (nonhydrogenated vegetable) fats. Some doctors still give such advice, but new information is suggesting other routes to good health.

In a recent study, a large group of middle-aged men was divided into an experimental population and a control population. The experimental population was given a low-cholesterol, high-unsaturated diet. Sure enough, their blood cholesterol levels went down, and over the years, their rates of arterial disease and heart disease were lower than those of the control population. Case proved? Not quite. While the death rate due to heart attacks was lower in the experimental group, the overall death rate of the experimental group was significantly *higher* than that of the control group. It turns out that the experimental group had a higher cancer death rate.

What happened? One possibility is that the higher levels of cholesterol in the control

group gave them healthier plasma membranes that protected them from cancer and other disease, or perhaps gave them a better-balanced steroid metabolism.

Another, even more likely, possibility is that the higher levels of unsaturated fats in the experimental group were actively toxic. Unsaturated fats spontaneously form free radicals, highly reactive molecules that contain unpaired electrons. Free radicals tend to react with and degrade other molecules in a random manner. Interestingly, ionizing radiation damages cells in a similar manner. It creates free radicals that attack DNA and other cellular constituents. In any case, the study indicates what can happen to those who radically change their diets in response to even a seemingly well-founded scientific opinion. A generation or two of Americans have avoided cholesterol like the plague and, overall, may have gained nothing from their efforts.

TABLE 37.3
Vitamins

Vitamin	Source	Function	Daily Requirement	Result of Deficiency
A, retinol	Fruits, vegetables, liver, dairy products	Synthesis of visual pigments	1500–5000 IU[1], 3 mg	Night blindness, crustiness about eyes
B1, thiamine	Liver, peanuts, grains, yeast	Respiratory coenzyme	1–1.5 mg	Loss of appetite, beriberi, inflammation of nerves
B2, riboflavin	Dairy products, liver, eggs, spinach	Oxidative chains in cell respiration	1.3–1.7 mg	Lesions in corners of mouth, skin disorders
Niacin, nicotinic acid	Meat, fowl, yeast, liver	Part of NAD and FAD cell respiration	12–20 mg	Skin problems, diarrhea, gum disease, mental disorders
Folic acid	Vegetables, eggs, liver, grains	Synthesis of blood cells	0.4 mg	Anemia, low white blood cell count, slow growth
B6	Liver, grains, dairy products	Active transport	1.4–2.0 mg	Slow growth, skin problems, anemia
Pantothenic acid	Liver, eggs, yeast	Part of coenzyme A of cell respiration	Unknown	Reproductive problems, adrenal insufficiency
B12	Liver, meat, dairy products, eggs	Red blood cell production	5–6 µg	Pernicious anemia
Biotin	Liver, yeast, intestinal bacteria	In coenzymes	Unknown	Skin problems, loss of hair and coordination
Choline	Most foods	Fat, carbohydrate, protein metabolism	Unknown	Fatty liver, kidney failure, metabolic disorders
C, ascorbic acid	Citrus fruits, tomatoes, potatoes	Connective tissues and matrix antioxidant	40–60 mg	Scurvy, poor bone growth, slows healing (colds?)
D	Fortified milk, seafoods, fish oils, sunshine	Absorption of calcium	400 IU[a]	Rickets
E	Meat, dairy products, whole wheat	Antioxidant	Unknown	Infertility, kidney problems
K	Intestinal bacteria	Blood-clotting factors	Unknown	Blood-clotting problems

[1]IU = International units: mg = milligram (1/1000 g); µg = (1/1,000,000 g).

TABLE 37.4
Mineral Nutrients

Mineral	Food Source	Function	Daily Requirement	Result of Deficiency
Calcium	Dairy foods, eggs	Growth of bones and teeth, blood clotting, muscle contraction, nerve action	800 mg	Tetany, rickets, loss of bone minerals and muscle coordination
Cobalt	Common in foods, water	Vitamin B_{12}	1 mg	Anemia
Copper	Common in foods	Production of hemoglobin, enzyme action	2 mg	Anemia
Fluorine	Most water supplies	Prevents bacterial tooth decay	Unknown	Tooth decay, bone weakness
Iodine	Seafood, iodized salt	Thyroid hormone	0.15 mg	Hypothyroidism
Iron	Meat, eggs, nuts, raisins	Hemoglobin (oxygen transport)	10 mg (men) 18 mg (women)	Anemia, skin problems
Magnesium	Green vegetables	Enzyme function	350 mg	Dilated blood vessels, irregular heartbeat, loss of muscle coordination
Manganese	Liver, kidneys	Enzyme function	Unknown	Loss of fertility, menstrual irregularities
Phosphorus	Dairy foods, eggs, meat	Growth of bones and teeth, ATP nucleotides	800 mg	Loss of bone minerals, metabolic disorders
Potassium	Most foods	Nerve and muscle activity	2–4 g	Muscle and nerve disorders
Sodium	Most foods, salt	pH balance, nerve and muscle activity, body fluid balance	0.5 g	Weakness, muscle cramps, diarrhea, dehydration
Zinc	Common in foods	Enzyme action	15 mg	Slow sexual development, loss of appetite, retarded growth

trace elements are unknown—for instance, it is not known why small amounts of fluoride retard cavity formation in teeth.

Among the most bizarre recently discovered mineral requirements are arsenic, silicon, and selenium. Arsenic is an extremely deadly poison, yet laboratory animals have died of arsenic deficiency! Silicon is one of the most abundant elements (found in sand and rocks). Yet silicon deficiencies may also be widespread. The mineral is needed as a cross-linking agent in the elastic walls of major arteries. Finally, selenium is known to be necessary for the functioning of at least one enzyme. Human selenium deficiency disease is common in some parts of China. (In other parts of China, the population suffers from selenium poisoning.)

So what's a body to do? We've seen that low-cholesterol and high-cholesterol diets can both be harmful, that the balance between saturated and unsaturated fats mustn't be tipped too far in either direction, that too much protein is harmful but too little is worse. Vegetarianism can be dangerous, but so can meat. A tiny amount of arsenic is deadly, but an even tinier amount may help keep you alive; you can get sick from too much vitamin A or D or from not enough. Most people consume too much sodium, but everyone needs some. Too little selenium is bad, but so is too much. How can one make intelligent decisions in the face of conflicting information?

There is no simple answer. We must resort to generalities and clichés, such as "moderation in all things" or "variety is the spice of life" or "don't eat so much." No one ever suffered from eating sensible amounts of fresh fruits and vegetables. Get enough fiber. Avoid faddism. Avoid fats, but don't be a fanatic about it (unless you are overweight, in which case a little antifat fanaticism may be in order). Part of the problem is that many of us have more than enough food available. We can choose from a range of foodstuffs, some traditional, others recently and chemically contrived. With the opportunity for choice, however, comes the responsibility of being informed. And since much of what we know has just been learned recently, the responsibility remains a continuing one.

Next, we move to the circulatory system, one role of which is to distribute the food that the body has digested.

Key Ideas

DIGESTIVE SYSTEMS

1. **Digestion** is the mechanical and chemical breakdown of nutrients into absorbable parts, while **nutrition** refers to nourishment and the characteristics of essential nutrients.

2. **Extracellular digestion** refers to digestion outside cells, as seen in bacteria and fungi. **Intracellular digestion** occurs in cells, generally in **food vacuoles**.

3. Sponges phagocytize food particles for intracellular digestion. Cnidarians and flatworms begin digestion extracellularly in the **gastrovascular cavity**, with food phagocytized and digestion completed intracellularly.

4. The earthworm gut contains specialized regions. Food is swallowed by the **pharynx** and passes through the **esophagus** to the **crop** for storage. Grinding occurs in the **gizzard**, and digestion

and absorption occur in the long **intestine**. The **typhlosole**, a deep fold in the intestinal wall, increases surface area.

5. The grasshopper makes use of chewing mouthparts and salivary glands. The digestive system includes a **foregut** consisting of a pharynx, esophagus, and crop for swallowing and grinding; a **midgut** for digestion and absorption; **gastric ceca** for enzyme secretion; and a water-absorbing **hindgut**.

6. Carnivorous sharks have continuously growing rows of teeth. In the gut, a winding flap, the **spiral valve**, increases the surface area for digestion and absorption. Herbivorous fishes generally have lengthy coiled intestines suitable for the time-consuming cellulose digestion process.

7. Crocodiles and alligators have socketed teeth, while some venomous snakes have hollow or grooved retractable **fangs** for injecting venom. Snakes and lizards capture air molecules with the tongue and analyze them with the olfactory **Jacobson's organ**. Pit vipers use their heat-sensitive pits to orient themselves to their prey.

8. The beaks and feet of birds are often specialized for their source of food. Digestive specializations in seed-eaters include a storage crop and a two-part **stomach** composed of the glandular, enzyme-secreting **proventriculus** and the gravel-filled and muscular gizzard.

9. Mammalian teeth are socketed and include temporary, or milk teeth, and permanent teeth. Four types of teeth include **incisors**, **canines**, **premolars**, and **molars**. The shape and size of each group commonly relates to diet.

10. **Ruminants** are grazing mammals whose four-part stomach includes a **rumen**, where microorganisms digest cellulose. The products and microorganisms are themselves then digested.

THE DIGESTIVE SYSTEM OF HUMANS

1. **Taste buds** located on the tongue detect salt, sour, sweet, and bitter flavors. **Saliva**, a watery solution of ions, mucus, and **amylase**, a starch-digesting enzyme, are produced in the salivary glands.

2. The pharynx divides to form the **larynx** and **laryngopharynx**. Some of the structures involved in swallowing are the **tongue**, **esophagus**, **soft palate**, **epiglottis**, **glottis**, and **larynx**. The larynx is raised during swallowing, preventing food from entering the air passage.

3. The esophagus, a food-conducting tube, is made up of an innermost, secretory **mucosa**, a vascular and glandular **submucosa**, and a circular and longitudinal layer of smooth muscle, the **muscularis**, all of which are surrounded by a fibrous **serosa**. Wave-like contractions called **peristalsis**, coordinated by the autonomic nervous system, move the food **bolus** to the stomach.

4. The stomach stores and mixes food, and its acidity helps destroy microorganisms. The **cardiac** and **pyloric sphincters** close off the stomach during its churning peristalsis. A third, oblique muscle layer produces a twisting action.

5. The glandular lining contains **pepsinogen**-secreting **chief cells** and **parietal cells**, whose hydrochloric acid secretions activate pepsinogen to the active form, **pepsin**.

6. The **small intestine** consists of the **duodenum**, **jejunum**, and **ileum**. Its role is digestion and absorption. In addition to what is provided by length, the surface area of the intestine is increased by folding, by projections called **villi**, and by cellular projections called **microvilli**. Each villus contains a **capillary bed** and a **lacteal** and is capable of movement.

7. The **liver** secretes **bile**, an alkaline secretion that emulsifies fats. Bile is stored in the **gallbladder** and secreted through the **bile duct**.

8. The glandular **pancreas** secretes sodium bicarbonate and digestive enzymes.

9. The **large intestine**, or **colon**, includes the **ascending colon**, **transverse colon**, **descending colon**, and **sigmoid colon**. Further along are the **rectum**, **anal canal**, and **anus**. The colon absorbs water, concentrates the feces, and provides a suitable environment for bacteria that secrete useful vitamin K, biotin, and folic acid.

THE CHEMISTRY OF DIGESTION

1. The bonds connecting subunits of complex foods are broken through **hydrolysis**, releasing the simple subunits that can be absorbed or actively transported into cells lining the intestine.

2. Starches are hydrolyzed into maltose by salivary amylase in the mouth and by **pancreatic amylase** in the small intestine. **Maltase** hydrolyzes maltose into glucose, which is transported to the liver for storage.

3. Following emulsification by bile, fats are hydrolyzed by **lipase** into fatty acids and glycerols. These are carried in micelles into the lining cells, reform as triglycerides, and join cholesterol to form **chylomicrons**, which pass into the lacteals.

4. Protein digestion begins in the stomach, where pepsin hydrolyzes inner peptide bonds, forming shorter peptides.

5. Pancreatic **trypsin**, **chymotrypsin**, and **carboxypeptidase** produce peptides and amino acids. Intestinal **aminopeptidase** forms amino acids that are transported to the liver.

6. **Nucleases**, nucleic acid hydrolyzing enzymes, include **exonucleases** that work on the ends and **endonucleases** that work within the chain.

7. Digestive enzymes are released through mechanical processes (physical presence of food), neural processes (thoughts and detection of food), and hormonal processes. Neural messages travel to the stomach via the **vagus nerve** when the gut lining senses the chemical presence of food. The cells react by releasing a hormone such as **gastrin** into the blood. This hormone includes secretion and CCK-PZ. When hormones reach their specific target cells or tissues, they stimulate them to release enzymes into the gut.

SOME ESSENTIALS OF NUTRITION

1. Carbohydrate is a vital energy source, but in excess it is converted to body fat.

2. Fats are excellent energy-providing foods and sources of fat-soluble vitamins. Unsaturated fats are needed for plasma membranes. Essential fatty acids must come from the diet as is.

3. Protein must be provided daily since excess amino acids are deaminated and used for energy or converted to fats and carbohydrate.

4. **Essential amino acids** must be provided by the diet, while the others are interconvertible by the body. High-quality dietary protein contains sufficient amounts and kinds of amino acids to provide all the amino acids the body needs.

5. Fat-soluble vitamins are important in forming coenzymes (NAD and FAD), and water-soluble vitamins are important as **antioxidants** (removing free radicals) and other uses. The former are harmful in overdoses.

6. Major essential **minerals** (inorganic ions) include calcium, magnesium, iron, sodium, potassium, and chloride and are used in structure, enzyme action, respiration, and osmotic regulation.

7. Trace elements include iodine, cobalt, and silicon. A shortage of iodine can cause **goiter**, an overgrowth of the thyroid.

Application of Ideas

1. Select five distinct variations in mammalian dentition (including its absence) and relate these variations specifically to the food consumed by each animal. Suggest how such variation might arise.

2. The feeding habits of animals are highly varied. Omnivores commonly feed on a wide variety of foods, while more specialized species, whether herbivore or carnivore, may feed mainly on one specific item with little deviation. Discuss examples of each and suggest short- and long-term evolutionary advantages and disadvantages in either direction.

3. Careful studies of herbivorous animals reveal that they rarely produce the enzyme cellulase, yet a considerable amount of their diet consists of cellulose. Using examples, explain how mammals extract nutrients from cellulose. Considering that a large part of the carbohydrate produced on earth is in the form of cellulose, suggest reasons why most animals haven't evolved the ability to synthesize the enzyme. What advantage would such a capability offer humans?

Review Questions

1. Distinguish between extracellular and intracellular digestion and provide an example of each from the invertebrates. (page 789)

2. List the three parts of the insect gut and explain what each does. (pages 790–791)

3. Describe two anatomical feeding specializations in the shark and explain how they operate. (page 791)

4. List the four types of mammalian teeth and describe specializations in each. (page 793–794)

5. List the three salivary glands and explain how saliva functions in digestion. (page 795)

6. Explain how peristalsis works and how it is controlled. (page 796–797)

7. Describe the muscular lining of the stomach and its action in moving food. (page 797)

8. Describe the stomach's secretory lining and state the specific functions of the parietal and chief cells. Name a secretion that protects the stomach lining. (page 797)

9. Describe the four levels of structure that provide the small intestine with its enormous surface area. (page 797)

10. Draw a simple villus. Label its parts, including any specialized cell surface features. (page 798)

11. What is the digestive function of the liver? How does the release of its secretion take care of two problems at once? (page 800)

12. List two general functions of the pancreatic secretions. (page 800)

13. Briefly discuss three functions of the colon. (page 801)

14. Name and describe the common chemical mechanism of digestion. What is the opposite process called? (pages 801–802)

15. List the steps in the digestion of carbohydrate. What is the end product? (page 802)

16. List the steps in protein digestion and the places in which they occur. (pages 803–805)

17. What determines the quality of dietary protein? What vegetable foods contain high-quality proteins? (page 807)

18. List two specific ways in which vitamins are used by the body. (page 807)

19. List four major and two trace mineral requirements and state their uses. (pages 807, 809)

CHAPTER 38

Circulation

CHAPTER OUTLINE

ADAPTATIONS IN ANIMALS WITHOUT CIRCULATORY SYSTEMS

CIRCULATORY SYSTEMS IN INVERTEBRATES

Open and Closed Circulatory Systems

CIRCULATION IN THE VERTEBRATES

The Blood

Blood Circulation in the Vertebrates

THE HUMAN CIRCULATORY SYSTEM

Circulation Through the Heart

Control of the Heart's Contractions

The Working Heart

Blood Pressure

Circuits in the Human Circulatory System

The Role of the Capillaries

The Veins

Essay 38.1 Cardiovascular Disease

THE LYMPHATIC SYSTEM

KEY IDEAS

APPLICATION OF IDEAS

REVIEW QUESTIONS

Watching the early development of a vertebrate embryo is a fascinating and often reaffirming experience, even to crusty old biologists who have spent their adult lives considering the various manifestations of life. It might seem that the feeling of wonder would eventually diminish, but somehow it doesn't. The sight of the quick pulses of a tiny, unformed heart signaling a new life is perpetually intriguing. The heart is, in a sense, the very symbol of animal life.

In its earliest days, the vertebrate heart gives little indication of how complex it will become since at that time it is only a simple, twitching tube. Curiously, the beating begins hours before there is any blood to pump. In fact, the circulatory system is among the first systems to take form. Why should this be?

Actually, if we consider that the embryo needs an inflow of nutrients and oxygen and the prompt removal of metabolic wastes, we should expect the circulatory system to develop early, and to grow and elaborate as the individual matures, assuming a number of critical roles throughout the life of the animal, as we shall see. We can begin by reminding ourselves that the primary mission of the circulatory system is transport: transport of oxygen, carbon dioxide, nutrients, water, ions, hormones, antibodies, and metabolic wastes and blood cells. Essentially, this is accomplished as the heart pumps fluid out through arteries to the tiny, thin-walled capillaries, across which these materials can pass. The fluid then collects again into increasingly larger veins on its way back to the heart.

Adaptations in Animals Without Circulatory Systems

As living things evolved, they became diverse, often more complex and, in many cases, larger. With greater size and complexity came new problems. Some of these problems involved how to better service the various tissues of the body; that is, how to efficiently transport food and oxygen to them and to carry away their metabolic wastes.

The sponges (see Chapter 26) are multicellular animals that also rely on their cell surfaces for exchanging materials with the environment. Although some sponges reach considerable size, the direct exchange method still works for them because they have such a thin body wall that no cell is far from the surrounding water. Recall that the sponge's choanocytes (collar cells) constantly pump water through its body, bringing in food and oxygen and washing away metabolic wastes. The sponges, however, do make use of ameboid cells in the distribution of foods (see Chapters 26 and 37), providing an assist to cell-to-cell transport.

Cnidarians—the hydrozoans, jellyfish, and corals—also lack an organized circulatory system. As is true for sponges, cnidarians are very thin-walled and hollow; thus most of their cells carry out exchanges directly with the watery surroundings (Figure 38.1). The extensive gastrovascular cavity provides a means of food distribution. Even the flatworms, with their much denser bodies, rely heavily on cell-to-cell transport and direct exchanges with their environment, aided by an extensive gastrovascular cavity.

Circulatory Systems in Invertebrates

The body mass of the larger, more complex animal is too great for its requirements to be met by direct environmental exchanges and simple cell-to-cell transport. Such exchanges occur far too slowly to provide a sufficient supply of oxygen and food and to remove carbon dioxide and nitrogenous metabolic wastes before they become toxic. The evolution of larger animals thus included provisions for rapid transport in the form of circulating fluids contained within circulatory systems. There are essentially two types of circulatory systems: *open* (those in which the arteries empty into intercellular spaces to percolate through the tissues) and *closed* (those in which the blood remains virtually enclosed in vessels).

The circulatory system, along with the lymphatic system, which we will discuss later in the chapter, constantly shifts water, ions, and proteins about, thus helping to regulate osmotic conditions in the body. Osmotic conditions are somewhat complex since there are actually three fluid regions involved: the fluid of blood itself, the interstitial fluid (free,

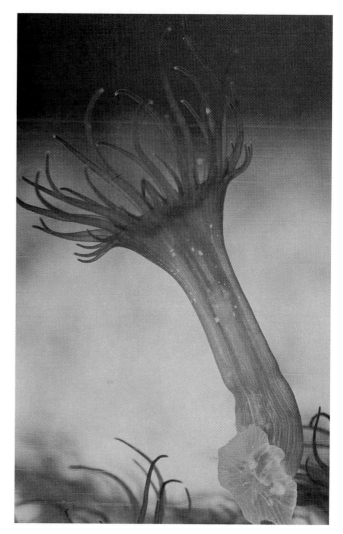

FIGURE 38.1
The Direct Exchange
Thin-bodied aquatic creatures such as the sea anemone have no need for elaborate circulatory systems. Cell-to-cell transfers of materials and a direct exchange between body cells and the environment suffice.

watery fluids between cells), and the intracellular fluid (fluid within the cell). Because capillaries are usually surrounded by interstitial fluids, all substances passing from the blood to the cells (and vice versa) must first pass through this watery matrix. We will learn more about the nature of interstitial fluids after we look a little more closely at the circulatory system.

Open and Closed Circulatory Systems

In **closed circulatory systems**, the blood remains within the confines of vessels throughout its entire circuit. In **open circulatory systems** blood does not remain in vessels. We'll use the open circulatory system of the arthropod as an example, but keep in mind that as a group arthropods vary immensely due to the sheer size of the phylum.

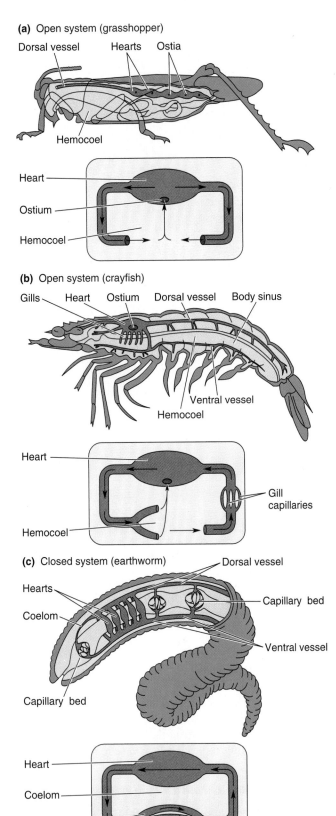

(a) Open system (grasshopper)

Dorsal vessel　Hearts　Ostia

Hemocoel

Heart

Ostium

Hemocoel

(b) Open system (crayfish)

Gills　Heart　Ostium　Dorsal vessel　Body sinus

Ventral vessel

Hemocoel

Heart

Gill capillaries

Hemocoel

(c) Closed system (earthworm)　Dorsal vessel

Hearts

Coelom

Capillary bed

Ventral vessel

Capillary bed

Heart

Coelom

Capillaries

In many arthropods, the blood is pumped by the heart (or hearts—there are often several) into an artery, a vessel that carries blood away from the heart. But the artery soon ends, and the blood enters channels leading to sinuses or cavities sometimes called **hemocoels** (blood cavities). Because these cavities are lined with a layer of flattened cells called **endothelium**, they can actually be regarded simply as expanded forms of blood vessels. As the blood percolates through such hemocoels, its fluids are forced around and through the body's various tissues. (Thus there is no clear separation of blood and interstitial fluids as in closed systems.) As the blood continues to move, it eventually reaches a cavity surrounding the heart. From there the blood is drawn into the heart through openings called **ostia** (singular, *ostium*) and pumped out again for another circuit. If you know about boats or flooded cellars, the blood-collecting work of the arthropod heart may remind you of a bilge pump, a submersible pump that draws water from its surroundings and directs it into hoses.

It is difficult to generalize about the many species of arthropods, but we can say that probably all of them have open circulatory systems. If you compare the circulatory systems of the crayfish and the grasshopper (Figure 38.2a,b), you will see that the system is more complex in the gill-breathing crayfish. Crayfish blood is involved in respiration as well as transport, whereas in the grasshopper, the circulatory and respiratory systems are independently structured. The hearts of the two arthropods are also quite different. Note that the grasshopper heart is no more than a long dorsal blood vessel with muscular thickenings along its posterior region. In the crayfish, the heart is compact, more central, and more complex.

Open circulatory systems are also found in certain mollusks as well as in other kinds of invertebrates. Some invertebrates, however, have closed circulatory systems. As you may recall (Chapter 26), the annelids have a remarkably well-

FIGURE 38.2
Invertebrate Circulatory Systems

(a) The grasshopper's open circulatory system includes a lengthy dorsal vessel subdivided into a number of hearts. The hearts draw in blood from the surrounding hemocoel and pump it back into the open spaces of the body. One-way valves called ostia assure the onward movement of blood through the dorsal vessel. Insect circulation has no respiratory function, so the blood moves directly back into the heart region once again. **(b)** The crayfish has essentially the same system as the grasshopper, but because the blood serves a respiratory function, the system is more elaborate. Blood is drawn through ostia into a centralized heart and then pumped out into the hemocoel. On the return trip, some of the blood is shunted through the gills on its way back to the heart. **(c)** In the earthworm's closed circulatory system, blood is pumped along by five pairs of aortic arches (hearts). The blood travels from the heart to the ventral blood vessel, which directs it forward and backward. Segmental branches of the ventral blood vessel send the blood into capillaries in all parts of the segments, including the skin where gas exchange occurs. The capillaries then rejoin, with the blood entering the dorsal vessel for its trip back to the heart. Although many exchanges occur with the surroundings, the blood does not leave the vessels and enter open spaces as is the case in the insect.

developed closed system with distinct vessels, tubular hearts, and hemoglobin-rich blood. In addition, they have a fluid-filled coelom that assists in the distribution of materials. Note in Figure 38.2c that the major vessels include a contractile dorsal vessel, a ventral vessel, and five pairs of hearts (actually, contractile vessels) in the earthworm. Valves in the aortic arches ensure a one-way flow of blood. Closed systems are also found in cephalopod mollusks, such as the squid and octopus.

Blood flows much more rapidly in closed systems than in open ones; thus it would seem to have a marked adaptive advantage in active species that need a ready supply of oxygen to meet their greater metabolic needs. Yet insects, among the most active of all animals (as you know if you have ever tried to catch a fly), have an open circulatory system. They are allowed this evolutionary indulgence for a very simple reason. They do not rely on their blood to carry oxygen. Instead, oxygen is transported through a system of open tubes that penetrate every part of the body, as described in Chapter 39.

Circulation in the Vertebrates

The Blood

Blood in most vertebrates is a tissue composed of plasma and several types of cells. More precisely, it is a connective tissue (see Chapter 32) with the fluid, **plasma**, as its interstitial (between cells) matrix (ground substance), and three kinds of cells: **erythrocytes** (red blood cells), **leukocytes** (white blood cells), and **platelets** (anucleate cell fragments important in clotting).

Plasma in humans, a straw-colored fluid, comprises about 45% of whole blood. About 90% of plasma is water, and the remaining 10%, the plasma solids, includes a long list of substances. By weight, about 70% of the plasma solids are proteins, and the remaining 30% includes urea, amino acids, carbohydrates (mostly glucose), organic acids, fats, hormones, and various inorganic ions.

Three of the major plasma proteins are the **albumins**, **globulins**, and **fibrinogen**. The albumins are large proteins that bind miscellaneous impurities and toxins in the blood and aid in the transport of certain hormones, fatty acids, and metal ions. They are also involved in maintaining the osmotic conditions of the blood, so important to capillary functions. The globulins include the *antibodies* (or *immunoglobulins*), which are important in the immune response (Chapter 40), and certain proteins involved in the transport of lipids and fat-soluble vitamins. Fibrinogen is important in the blood-clotting process.

Red Blood Cells. Red blood cells in human and other mammals are quite small (about 6–8 μm in diameter), biconcave, and disk-like in shape (Figure 38.3) and, when mature, lack nuclei and therefore cannot undergo further division. In all other vertebrates the red blood cells are oval, nucleated, and often much larger. Normally, humans have about 5 million red

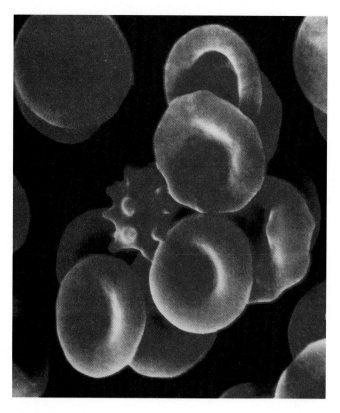

FIGURE 38.3
Human Red Blood Cells (SEM)
Mature human red blood cells take the form of biconcave disks. (The spiked cell, left center, has shrunk.)

blood cells per cubic millimeter of blood. Red and white blood cells are produced continuously in the *red bone marrow*. In adult humans, red blood cell formation occurs almost exclusively in the axial skeleton. Red blood cells have about a four-month life expectancy. When they age, they become brittle and some rupture, spilling their hemoglobin into the blood. Most aging or damaged red blood cells are phagocytized by certain white blood cells in the spleen and liver.

White Blood Cells. Unlike red blood cells, white blood cells (leukocytes) do not discard their nuclei as they mature. The white blood cells are active in the immune system, which will be discussed in detail in Chapter 40. For now, though, we can say that the *neutrophils,* whose numbers make up the majority of circulating leukocytes, are important phagocytic cells. They aggregate at infection sites, engulfing invading microorganisms. *Basophils* are involved in the inflammatory response. *Eosinophils* are also involved in inflammatory responses, and, in addition, they help destroy larger parasites. *Lymphocytes* are the backbone of the immune system; they fight off invaders and confer immunity. The *monocytes,* once activated, usually by infection, develop into *macrophages,* which phagocytize foreign cells and cellular debris. One group of cells derived from lymphocytes, the *natural killer (NK) cells,* are unspecialized leukocytes that destroy diseased cells.

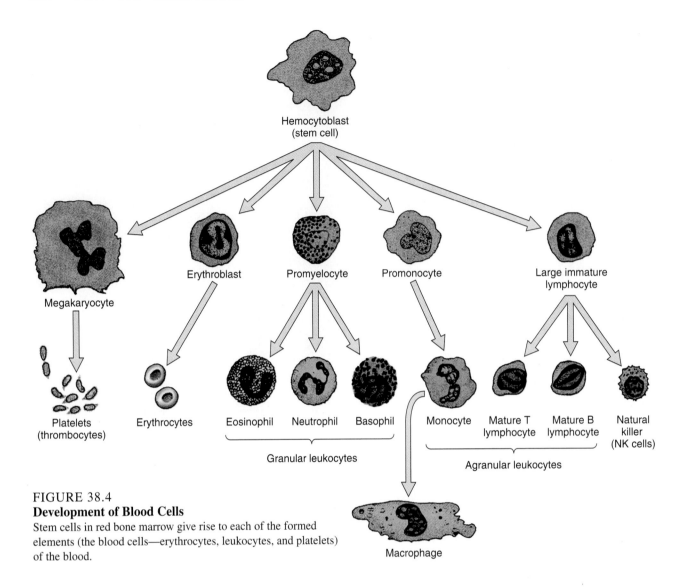

FIGURE 38.4
Development of Blood Cells
Stem cells in red bone marrow give rise to each of the formed
elements (the blood cells—erythrocytes, leukocytes, and platelets)
of the blood.

Platelets. Platelets (or *thrombocytes*) are tiny, numerous
structures. Unlike erythrocytes, which are whole cells that
have since lost their nuclei, platelets are only fragments of
cells. The small, disk-shaped vesicles are formed directly
from large *platelet mother cells (megakaryocytes),* which also
form in red bone marrow. When these large cells mature, they
fragment without mitosis—a most peculiar process. Platelets
play an important role in blood clotting.

Figure 38.4 illustrates the cellular origin of the blood cells
and platelets in humans. Curiously, each of the formed elements
of the blood—the red and white blood cells and the platelets—
has the same origin, the **stem cells** (or *hemocytoblasts*).

Clotting. Blood clotting is essential to any organism that
relies on circulating fluids. The world is full of objects that
can pierce our bodies and cause our fluids to leak out, and
clotting minimizes such leakage.

Blood clotting is an extremely complex and only partially understood process. At least 15 substances are involved,
some of which are part of an intricate arrangement that pre-

vents accidental clotting. Two proteins are basic to the
process: **prothrombin**, an inactive clotting protein present in
the plasma, and **fibrinogen**, one of the major plasma proteins.
Platelets and damaged cells also play a role. Generally, the
clotting process proceeds as follows:

1. A vessel is damaged.

2. Platelets attach at the wound site, form lengthy extensions, and adhere to collagen fibers, forming what is
   essentially a "plug."

3. The platelets rupture, releasing (a) **vasoconstrictors** that
   cause nearby vessels to constrict, thus reducing blood
   loss, and (b) **thromboplastins** (enzymes).

4. In the presence of thromboplastins and calcium ions, prothrombin in the plasma becomes **thrombin**, a specific
   endopeptidase.

5. Thrombin breaks apart the large fibrinogen molecules of
   the plasma, the smaller pieces forming a fibrous, sticky
   protein called **fibrin**.

6. Fibrin fibers, along with damaged platelets, red blood cells, and white blood cells, form a network that solidifies, becoming a clot that stops the bleeding.

7. The clot contracts, pulling the wound together, further preventing bleeding and encouraging healing.

The absence of any of the many factors participating in clotting, or their failure to perform their function, is quite serious. Earlier we discussed the genetics of one such condition, hemophilia A (see Chapter 11). In this recessive condition, factor VIII (antihemophilic factor) is absent. In hemophilia B, the absent substance is factor IX (plasma thromboplastin component). Another reason for failure of the blood-clotting mechanism is an insufficiency of calcium ions. (Citric and oxalic acids readily take up calcium and are added to blood samples to prevent clotting.) An insufficiency of vitamin K is an important cause of slow clotting. This vitamin is required by the liver for the synthesis of prothrombin.

Blood Circulation in the Vertebrates

All vertebrates have closed circulatory systems (if we ignore the fact that blood leaves the vessels to percolate through the liver) and a centrally located heart. The most complex of these is the four-chamber heart of crocodiles, mammals, and birds. But let's first consider the simpler two- and three-chamber arrangements in fish, amphibians, and reptiles.

Fishes and the Two-Chamber Heart. Fish are the most simple-hearted of all vertebrates. Their heart essentially consists of two pumping chambers, a single **atrium** (once called the *auricle*) and a single **ventricle**. The atrium is generally a thin-walled structure that receives blood and transfers it to the ventricle. The ventricle is a larger, thick-walled chamber that sends blood into arteries with considerable force.

Note in Figure 38.5a that blood leaving the fish's heart goes directly to the blood vessels and capillaries of the gills. From the gills, freshly oxygenated blood flows through the body and from there returns to the heart. Blood returning to the heart moves sluggishly, since much of it has passed through capillary beds both in the gills and in the body tissues (systemic capillaries), where frictional resistance is greater.

Some of the blood has passed through yet other capillary beds, further dissipating the energy it had when it left the ventricle, and decreasing its pressure even more. In one instance, blood that has already passed through the gills later enters capillary beds in the intestine, where it picks up the products of digestion. The capillaries then rejoin, forming the **hepatic portal vein**, which goes to the liver. Here it divides once more into capillaries. It then enters a large vessel, the hepatic vein, for its return to the heart.

In summary, the fish's blood flows from heart to gills via the ventral aorta, through the gill arches, from the gill to the tissues via the dorsal aorta, and from the tissues to the heart. Again, during each cycle, all the blood flows through the gills and continues on through the other body tissues. Oxygen is picked up in the gills and released in the tissues prior to the blood's return to the heart.

Amphibians, Reptiles, and the Three-Chamber Heart. In the evolution of animals that could survive on land, a new emphasis on lungs and a deemphasis on gills meant that the simple one-cycle circulatory system of the fish was no longer enough. As oxygen demands increased, greater blood pressure and a new way of oxygenating the blood were in order. New circuits in the system arose, with the single atrium becoming divided into two in the amphibians and reptiles (Figure 38.5b). Also, some of the blood pumped from the heart was directed through a new **pulmonary circuit** to the lungs, which became separate from the **systemic circuit** to the other body tissues. Today, we find the three-chamber heart in all amphibians and nearly all reptiles.

As an example we can consider the frog heart. We see in Figure 38.5b *two atria*, making possible the two-circuit system mentioned above. Let's follow the flow of blood in Figure 38.5b to see how the system works.

Deoxygenated blood returning from the body enters the **right atrium** and from there is pumped into the single ventricle from where it is directed out to the skin and lungs for oxygenation. Nearly simultaneously, oxygenated blood returning from the lungs and skin enters the **left atrium** and is also pumped into the ventricle. From there it is directed out to blood vessels that carry it to oxygen-needy tissues in the body. This system has produced a vexing question for biologists. Is the oxygenated blood from the lungs mixed with the deoxygenated blood from the body? The evidence indicates that it does mix, but only partially. In the amphibian heart, total mixing is prevented by flaps and partial valves that separate the oxygenated and deoxygenated blood somewhat and direct it into the proper arteries. Reptiles also have a partial **septum**, or partial partition, in the ventricle, which helps keep the two kinds of blood separated. Even a hint of such mixing would be physiologically unacceptable for endothermic birds and mammals, who have much greater oxygen demands than do amphibians and reptiles.

In summary, a major structural change is seen in the circulation of amphibians and reptiles. We now see a two-part system because of the added pulmonary circuit. As a result, the extra boost provided by the single ventricle after oxygenated blood returns to the heart from the lungs keeps the blood pressure strong as the blood is sent to the tissues.

Birds, Crocodiles, Mammals, and the Four-Chamber Heart. The evolution of the four-chamber heart was not a sudden thing. It occurred by a succession of tiny steps. We can see indications of some of these intermediate steps among living amphibians and reptiles. First, there was the amphibian three-chamber heart with its directing valves. Then a partial septum evolved in the single ventricle, further dividing oxygenated and deoxygenated blood. Then, in the crocodiles, the ventricular division became complete, and birds and mammals were to share this evolutionary innovation.

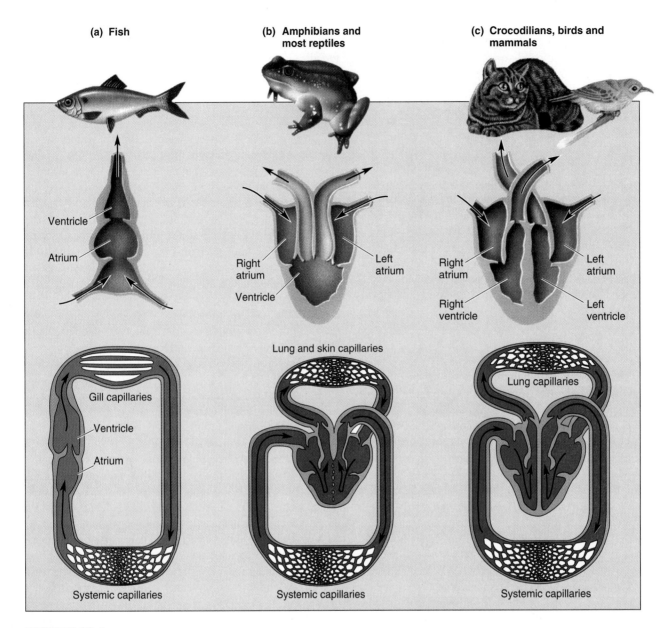

FIGURE 38.5
Circulation in Vertebrates
(a) The fish heart has two chambers, a receiving atrium and a sending ventricle. Deoxygenated blood enters the atrium, which pumps it to the ventricle. From there the blood passes to the gills for oxygenation and then through a vast network of systemic capillaries, where exchanges occur before it comes back to the heart. (b) The amphibian and reptile system adds one new feature—a third chamber. Here deoxygenated blood enters the right atrium and oxygenated blood the left atrium. They pump it into the ventricle, where the two are fairly well separated by flaps and valves. Deoxygenated blood goes to the lung and skin capillaries for oxygenation, and then back to the heart, whereas oxygenated blood goes to the systemic capillaries, where it gives up its oxygen before returning to the heart. (c) Mammals, birds and crocodilians have a fourth chamber, another ventricle, provided by a complete division of the old single ventricle. Here deoxygenated blood stays on the right side on its trip through the heart, whereas oxygenated blood stays on the left. There is no mixing.

The four-chamber heart of birds, crocodiles, and mammals includes a right atrium and ventricle and a left atrium and ventricle (Figure 38.5c). In the four-chamber heart, some of the blood is oxygenated in the lungs, and some is deoxygenated in the body tissues in each complete circuit. The cru-cial difference between this and fish circulation is that a complete circuit in birds and mammals involves two trips through the heart, once through the left side and once through the right side. In fact, it is convenient to think of the four-chamber heart as two separate pumps.

FIGURE 38.6
The Structure of Blood Vessels.
Veins and arteries are similar in structure, except that
veins have one-way valves and arteries are generally
much thicker walled. The valves help direct blood back to
the heart. The difference in thickness is attributed to a
denser region of smooth muscle and the presence of two
layers of elastic connective tissue in the artery. Arteries
must accommodate considerably greater pressure and
help to maintain that pressure. Both vessels have a very
smooth endothelium, an innermost layer of flattened cells,
and a very tough outermost connective tissue sheath. Capillaries are quite different, consisting only of an endothelium and a thin surrounding basement membrane. Since
nearly all exchanges between the blood and cells occur in
capillaries, a thin wall is essential.

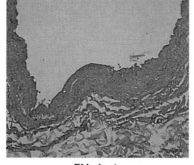

EM of vein

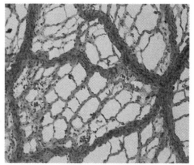

EM of capillary

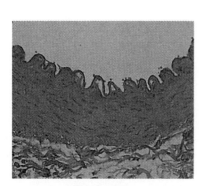

EM of artery

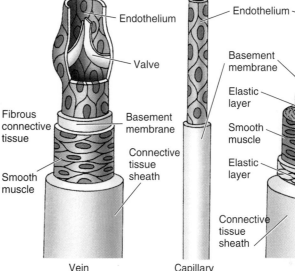

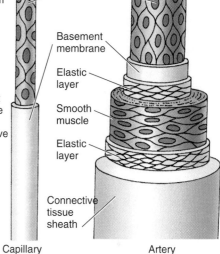

Vein   Capillary   Artery

The Human Circulatory System

Humans, like all mammals, have a four-chamber heart and a
closed circulatory system. The system consists of the heart,
arteries, capillaries, and veins (Figure 38.6). The **arteries** are
muscular vessels that carry blood away from the heart. With
the exception of the **pulmonary arteries** (which carry blood
to the lungs), they always carry oxygenated blood. Arterial
blood is usually bright red, because of the oxygenation. Arteries tend to be round and thick-walled and are invested with
heavy layers of smooth muscle and elastic connective tissue.
Constriction or dilation of arteries and, especially, arterioles
controls the blood flow in specific tissues. The arteries divide
into increasingly smaller **arterioles**, which give rise to highly
branched **capillaries**. The capillaries can form extensive *capillary beds* in the tissues. All exchanges between the blood
and the cells of the body occur through the capillaries.
Venules (small veins) receive the blood from capillaries and
flow into ever-larger **veins**, which carry blood toward the
heart. With the exception of the **pulmonary veins** (those leading from the lungs to the heart), all veins carry deoxygenated

blood. Veins, on the other hand, tend to be thin-walled and
flattened, with much larger openings in cross section. They
also generally lie nearer the surface, so most of the vessels
you see under your skin are veins.

Circulation Through the Heart

The Right Side. Circulation through the four-chamber heart
begins much like that in the three-chamber heart. Deoxygenated blood coming from the body arrives at the right atrium and, simultaneously, oxygenated blood coming from the
lungs arrives at the left atrium. Blood arriving at the right atrium is carried by two major veins, the **superior** and **inferior
venae cavae** (Figure 38.7). (They are named "superior" and
"inferior" because in humans one is higher than the other—a
result of our upright posture. In other mammals, the terms
anterior and *posterior vena cavae* are appropriate.) From the
right atrium the blood flows into a chamber below it, the muscular **right ventricle**. About 80% of the blood entering the
right ventricle gets there by flowing into the expanding chamber; the rest is pumped by contraction of the right atrium.

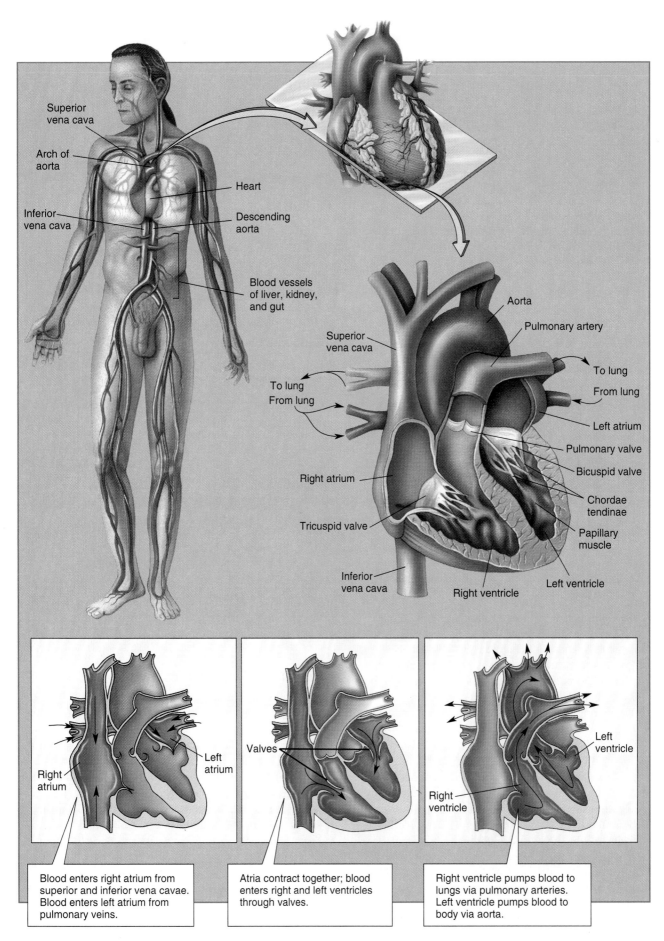

Superior
vena cava

Arch of
aorta

Inferior
vena cava

Heart

Descending
aorta

Blood vessels
of liver, kidney,
and gut

Aorta

Pulmonary artery

Superior
vena cava

To lung
From lung

To lung

From lung

Left atrium

Pulmonary valve

Bicuspid valve

Right atrium

Chordae
tendinae

Papillary
muscle

Tricuspid valve

Inferior
vena cava

Right ventricle

Left ventricle

Left
atrium

Valves

Right
atrium

Left
ventricle

Right
ventricle

Blood enters right atrium from
superior and inferior vena cavae.
Blood enters left atrium from
pulmonary veins.

Atria contract together; blood
enters right and left ventricles
through valves.

Right ventricle pumps blood to
lungs via pulmonary arteries.
Left ventricle pumps blood to
body via aorta.

FIGURE 38.7

FIGURE 38.7
The Human and Circulatory System
Major vessels emerging from the heart include the aorta, the great artery that carries blood out of the heart, and the superior and inferior venae cavae, major veins that return blood to the heart. The arrows indicate the direction of blood flow. In this view, the atria and ventricles have been cut away to reveal the valves. As in other four-chamber hearts, the right side receives deoxygenated blood from the body whereas the left side receives oxygenated blood from the lungs. The contraction sequence below shows blood entering the atria; blood moving from atria to ventricles; and blood being pumped into the pulmonary arteries and aorta.

When the right ventricle contracts, it forces blood into the pulmonary arteries and on to the lungs for gas exchange.

One-way valves between the atria and ventricles, and between the ventricles and the arteries leaving the heart (Figures 38.7 and 38.8), ensure that the blood flows in only one direction. The valve between the right atrium and ventricle, the **tricuspid valve**, checks any backward flow created by the ventricular contraction. When the tricuspid is closed, it resembles a three-part parachute with the "shroud lines" leading from the parachute to the sweaty hands of the unfortunate soul who got talked into jumping. The shroud lines are similar to the **chordae tendineae**, tendinous cords that prevent the three flaps from collapsing backward. The chordae tendineae originate in cone-shaped **papillary muscles** in the floor of the ventricle. As blood passes into the muscular pulmonary artery, its backflow is prevented by the **pulmonary semilunar valve**, a three-part curved flap that closes when the artery is full and expanded. The name *semilunar* refers to the half-moon shape of each of the three flaps.

The Left Side. Blood from the right and left branches of the pulmonary arteries finally wends its ways into the capillaries surrounding the alveoli of the lungs, where gases are exchanged (Chapter 39). The capillaries then rejoin to form the pulmonary veins, which return oxygen-rich blood to the left atrium. When the left atrium contracts, blood passes through the double-flapped **bicuspid**, or **mitral**, **valve** into the **left ventricle**. (The name *mitral valve* comes from the valve's shape, similar to a bishop's hat, or mitre.) This valve has two flaps, as opposed to the tricuspid of the right side, but otherwise has the same "parachute-like" appearance, complete with its own chordae tendinae. The contraction cycles are illustrated in Figure 38.7 (bottom).

The walls of the left ventricle are much thicker than those of the right, which probably gave rise to the myth that the heart is on the left side of the chest. (Actually, it is in the center but is tilted, so when you pledge allegiance, your hand is actually over your left lung.) The left ventricle is large and muscular enough to force the oxygenated blood throughout all the tissues of the body and into the veins that eventually return it to the right side of the heart—an enormous task. Contraction of the left ventricle sends blood through the **aortic semilunar valve** and on into that great artery, the **aorta** (see Figure 38.7). The aorta makes a U-turn to the left, giving rise to branches that serve the head, arms, and internal organs as it turns and passes down through the trunk.

Control of the Heart's Contractions

Some might say that the subject of the human heart has been overworked. After all, not only have volumes been devoted to describing its relentless and tireless activity, but we have eulogized and venerated it as the center of love and emotion. In reality, though, it needs no romanticizing, since no discussion can be cold and clinical enough to drain it of its wonder.

In Chapter 32 we saw that the heart muscle is unique because of its unusual contractile tissue. This is an understatement. There are many physiological differences between heart muscle and skeletal muscle, including the branching fibers, interlocking membranes (intercalated disks), and tireless contractions of cardiac muscle cells. One of their most interesting features is that the contraction itself and its rhythmicity are inherent in the cells themselves. The heart, in other words, is capable of beating without outside stimulation. This is dramatically illustrated in transplanted hearts, which sometimes go on beating for many years with no nerve connections whatsoever.

Extrinsic and Intrinsic Control. In no sense, however, is the normal heart functionally isolated, beating merrily away, independent of the rest of the body. The body's constantly changing demands wouldn't allow that. The heart is, in fact, greatly affected by a number of conditions, even including the emotions. We will find that the heart rate is under both **extrinsic** (from without) and **intrinsic** (from within) control.

Extrinsically, the heart can be fine-tuned by the autonomic nervous system. The nerves that serve the heart originate in the spinal cord and in the medulla of the brain. One group, the sympathetic nerves, accelerates the heart rate. The second, the parasympathetic nerves, act to slow the heartbeat. (We considered these nerves in Chapter 34.) The two kinds of nerves permit tight regulation of this vital organ so that it operates in concert with the rest of the body. The heart may be influenced further by hormones from the endocrine system. Epinephrine (adrenalin) from the adrenal gland has the same effect on heart rate as does the accelerator nerve. Also recall that the heart has blood pressure receptors within its atria and, accordingly, secretes a hormone that helps in the regulation of that pressure.

You may wonder how the heart directs itself if it is influenced by all these factors. The answer is through intrinsic (internally generated) control. The origin of the heartbeat is in the modified muscle of the right atrium, at a region known as the **sinoatrial (SA) node** (Figure 38.9). More commonly called the **pacemaker**, this is also the region influenced by the autonomic nerves. The pacemaker transmits impulses through the atrial walls (causing the two atria to contract simultaneously) and on to a second node—the **atrioventricular (AV)**

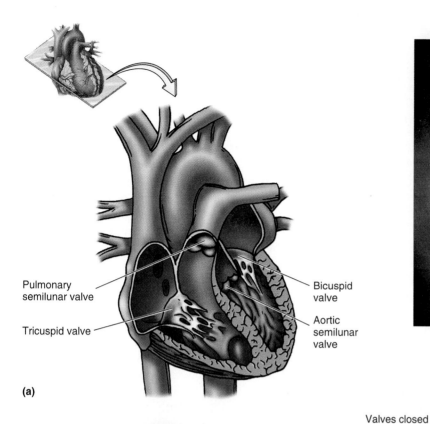

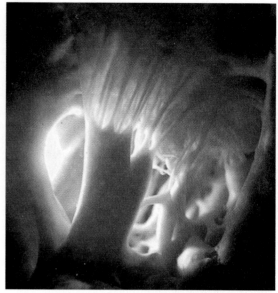

Papillary muscles and chordae tendineae
of bicuspid or tricuspid valve

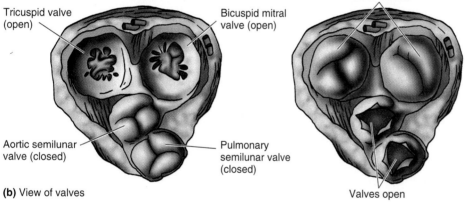

(b) View of valves

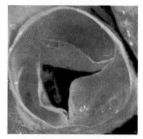

Semilunar heart valve open

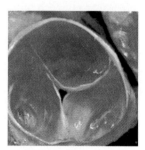

Semilunar heart valve closed

FIGURE 38.8
Valves of the Human Heart
(a) The larger tricuspid and bicuspid valves between the atria and ventricles have thin-walled flaps
restrained by strong, cord-like, chordae tendineae that extend from papillary muscles. **(b)** The aortic and
pulmonary semilunar valves have small but thick curved flaps.

node. There is enough delay in the transmission of the impulse for the atria to complete their contraction before the ventricles begin theirs. The AV node initiates contraction of the two ventricles. Impulses from the AV node pass through a strand of specialized muscle in the ventricular septum known as the **bundle of His** (pronounced *hiss*). The bundle branches into right and left halves that travel to the pointed base of the heart, where they branch into **Purkinje fibers** that initiate contraction there. The ventricular muscle fibers are arranged in a spiraling fashion so that their contraction produces a twisting, wringing motion that "squeezes" the blood out.

The Working Heart

When we listen to a beating heart through a stethoscope, we hear two major heart sounds, usually described as "lubb" and "dupp." The first is that of the sudden closing of the tricuspid and bicuspid valves as they shudder under the tremendous force of the contracting ventricles. The second sharp sound occurs as the aortic and pulmonary semilunar valves are snapped shut by arterial backflow and pressure when the ventricles relax. Then there is a brief pause. Then (if all is going well), another *lubb–dupp*. The period of ventricular contrac-

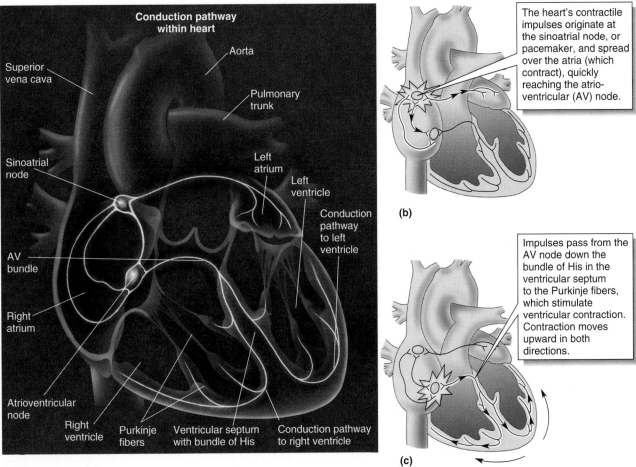

Conduction pathway within heart

Superior vena cava

Aorta

Sinoatrial node

Pulmonary trunk

Left atrium

Left ventricle

AV bundle

Conduction pathway to left ventricle

Right atrium

Atrioventricular node

Right ventricle

Purkinje fibers

Ventricular septum with bundle of His

Conduction pathway to right ventricle

(a) Control network

The heart's contractile impulses originate at the sinoatrial node, or pacemaker, and spread over the atria (which contract), quickly reaching the atrio-ventricular (AV) node.

(b)

Impulses pass from the AV node down the bundle of His in the ventricular septum to the Purkinje fibers, which stimulate ventricular contraction. Contraction moves upward in both directions.

(c)

FIGURE 38.9
Control of Heartbeat
(a) Control involves the SA and AV nodes and nerve fibers that extend into the ventricle. (b) In each heart cycle, impulses from the SA node initiate atrial contraction and then (c) activate the AV node. There, impulses are regenerated and sent down the bundle of His, through the septum to Purkinje fibers below. Ventricular contraction begins at the bottom and moves upward, producing a wringing action.

tion is known as **systole**, while the longer period during which the chambers of the heart fill up once again is called **diastole**.

It has been calculated that the average resting heart rate is 72 contractions per minute, and that each contraction forces about 80 ml (2.7 oz) of blood into the aorta. The amount of blood passing through the heart with each heartbeat is called the **stroke volume**, described as ml/min. **Cardiac output**, the amount of blood pumped each minute, is about 5–6 liters (11–13 pints). During strenuous activity, cardiac output can be increased up to 30–35 liters per minute—a remarkable reserve power! In case you have sensed an arithmetic problem here, we hasten to explain that your heart doesn't have to increase its beat rate by six to get a sixfold increase in cardiac output—*if* you are in good condition. During vigorous activity a conditioned athlete's heart rate may climb, but so does the stroke volume in a heart with enlarged chambers, due to exercise. A more sedentary person may not have this ability, and he

or she must achieve increased cardiac output primarily through increased heartbeat, with less increase in stroke volume.

Blood Pressure

Blood pressure is the force of blood against the arterial walls. It is this force that moves blood from the heart through the body. The aorta and major arteries receive the full impact of the powerful surges of blood from the left ventricle. A great many body functions depend on a rather constant pressure in the circulatory system, so the arterial walls have adapted accordingly. When the ventricles contract, blood enters the arteries faster than it can leave, so the sudden swell of blood expands the elastic walls of arteries. As the blood moves onward through the arterial tree, the expanded arterial walls contract through their elasticity, squeezing the blood volume, and raising blood pressure. That pressure is also increased by

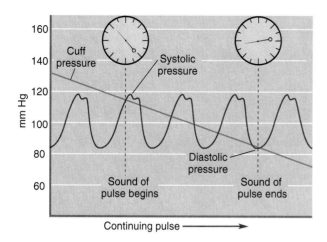

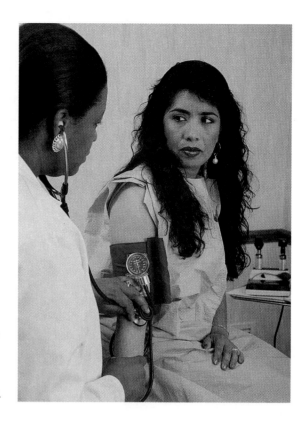

FIGURE 38.10
Measuring Blood Pressure
Blood pressure is commonly measured in the brachial artery with a device known as a sphygmomanometer *(photo)*. As the cuff is inflated, the brachial artery is collapsed, and no sound is heard. Pressure in the cuff is slowly reduced, and when a beating sound begins, the systolic pressure is noted. When the sound can no longer be detected due to reduced pressure, the diastolic pressure is noted. In the graph, the peaks represent systolic pressure and the valleys diastolic pressure. The area between the dashed lines is the pressure range at which the arterial sounds are heard. The diagonal line represents pressure decreasing in the sphygmomanometer.

muscular contractions in the walls of the arterioles. Blood pressure is recorded as two numbers, measured as millimeters of mercury (mm Hg—how far, in mm, a force can raise a column of mercury in a tube at sea level). The first number (called the *systolic pressure*) describes the pressure caused by the contraction of the ventricle (recall *systole*). The second number (the *diastolic* pressure) describes the pressure when the heart relaxes, allowing the chambers to fill with blood (recall *diastole*). A typical systolic pressure in a young, healthy person is 120 mm Hg; a typical corresponding diastolic pressure is 80 mm Hg. Your doctor would call this set of blood pressure measurements "120 over 80." You've probably had your blood pressure taken with a *sphygmomanometer* (Figure 38.10). The pressure measurement in systole, the time of ventricular contraction, reflects the sudden force exerted against the arterial walls. In diastole, the time between the heart's contractions, pressure is maintained by the return of the elastic arterial wall to its original shape. (Pay particular attention to the diastolic pressure since it is less subject to fluctuation from temporary influences; a pressure, say, of 190/120 is a severe warning.) (See Essay 38.1, page 828, for a discussion of cardiovascular diseases related to blood pressure.)

Vasoconstriction and Vasodilation. The smooth muscle lining the walls of arterioles plays a vital role in maintaining and directing the flow of blood into capillary beds. Opening of arterioles, called **vasodilation**, reduces blood pressure, and their

constriction, called **vasoconstriction**, increases blood pressure. In the arterioles, these two mechanisms function in a delicate and balanced interplay that keeps the overall blood pressure remarkably constant in spite of great surges of pressure in the arteries feeding arterioles. Vasodilation and vasoconstriction are partly under the control of the autonomic nervous system and partly under the control of hormones. In addition, oxygen, carbon dioxide, hydrogen ions, and a variety of drugs can bring about local changes in the diameter of arterioles.

Circuits in the Human Circulatory System
We are not about to discuss all the pathways of the human circulatory system here, but we would like to consider a few interesting circuits. A **circuit** can be defined as a distinct, major pathway of blood supply and return, generally contributing to the performance of some function. Figure 38.11 shows the four major circuits: pulmonary, hepatic portal, renal, and systemic (the general body circuit, which includes muscles, skin glands, and so on). The systemic circuit can be divided into the upper (arms and head) and the lower (from the trunk down, excluding the kidneys and liver). To these, we can add the vital **cardiac circuit**, the circulatory system of the heart itself. The pulmonary circuit, already described, is a good example of a relatively simple circuit.

Hepatic Portal Circuit. In the **hepatic portal circuit**, arteries branching from the abdominal aorta fan out over the intestinal membranes and enter the digestive organs. Those entering the small intestine form capillaries within the intestinal villi. Digested foods are collected in the capillaries, which then merge together and, along with vessels from the rest of the intestine, form the **hepatic portal vein** leading to the liver. (In general, portal veins lie between two capillary beds.) Upon entering the liver, the portal vein divides to form a second capillary bed. Actually, the liver contains not only capillaries but many branched, epithelium-lined cavities called **sinusoids** through which the blood passes. (As was mentioned, it could be argued that this makes the human circulatory system a partially open one.) From there, the blood, its cargo of nutrients now reduced and cleansed by the liver of toxic materials, returns directly through the inferior vena cava to the heart. (The liver, by the way, has its own fresh blood supply, which arrives via the hepatic artery.)

Renal Circuit. As it passes through the abdominal cavity, the aorta sends right and left branches, the **renal arteries**, into the kidneys. There, in one of the most complex filtering systems imaginable, the urea, excess water, miscellaneous metabolic by-products, and some salts are removed from the blood (see Chapter 35). Of equal significance, the kidneys carefully adjust water and salt levels, a vital process in the maintenance of the precise osmotic conditions and blood pressure required by the body. Blood returns from the kidneys via the **renal veins** to the inferior vena cava, and then to the heart.

Cardiac Circuit. The heart is a muscle and muscles need blood, so the **cardiac circuit** includes the arteries, capillaries, and veins of the heart muscle. In fact, you might say the heart is self-serving because this hard-working organ receives first crack at oxygenated blood leaving the left ventricle. The first branches of the aorta are the right and left **coronary arteries**, arteries servicing the heart. They emerge just beyond the aortic semilunar valve, dividing and sending branches to the right and left as they curve around the outer surface of the heart, finally joining on the dorsal side (Figure 38.12).

As the coronary arteries branch and rebranch, they rejoin to form web-like arrangements called **anastomoses**. The web-like network is quite significant, because it helps ensure that no sizable part of the heart muscle will be without a blood supply should a blockage, such as a **coronary thrombosis** (a blood clot), occur in some vessel.

Following its transit through the capillary networks of the heart, blood enters the **coronary veins**, whose routes parallel those of the arteries. The veins finally join each other to eventually open into the right atrium.

Systemic Circuit. The systemic circuit includes the rest of the body. Here we refer to capillary beds in muscles, glands, bones, the brain, and so on. As in the other circuits, CO_2 and O_2 are exchanged, required substances for cell maintenance are delivered, wastes are removed, and so on. Capillaries are so profusely distributed that no cell lies far from the blood

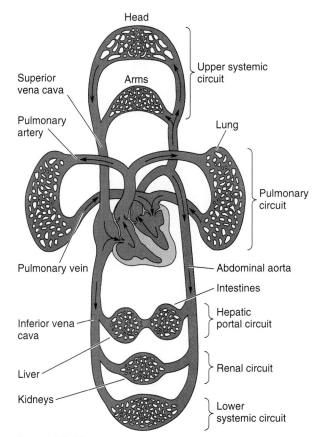

FIGURE 38.11
Major Blood Circuits
Each circuit carries out exchanges of nutrients, waste products, and oxygen between the blood and body cells, but most have other, more specialized functions.

supply. The walls of the capillaries are usually composed of only one layer of flattened endothelial cells. Let's take a closer look at these fascinating little vessels.

The Role of the Capillaries

As we learned earlier, the capillaries are, in a sense, the functional units of the circulatory system—all exchanges take place in these vessels. One might even say that the role of every other part of the system is simply to assist them. Their walls are made up of a single layer of interlocking endothelial cells, so the surroundings are never more than one flattened cell away from the blood (Figure 38.13). Furthermore, except for brain capillaries (see the "blood–brain barrier"; Chapter 34), the endothelial cells making up the capillary wall fit inexactly in places, leaving minute crevices or pores through which small molecules can pass. Finally, electron micrographs of capillaries show that their flat, thin cells engage in extensive pinocytosis. Recall from Chapter 5 that in pinocytosis ("cell drinking"), the membrane becomes actively involved in surrounding and taking in dense solutes.

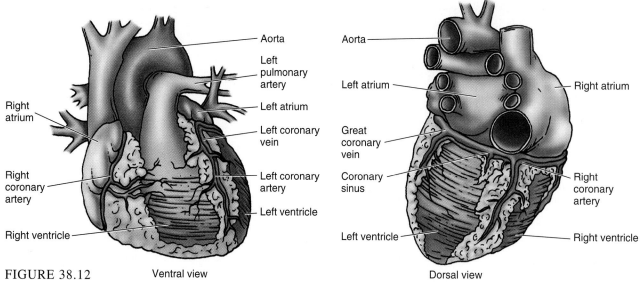

FIGURE 38.12
The Cardiac Circuit
The heart muscle has its own vessels, the coronary arteries and veins. The coronary arteries branch
directly from the aorta, sending their branches deep into the heart muscle. The coronary veins emerge
from heart muscle to form the coronary sinus, which empties into the right atrium.

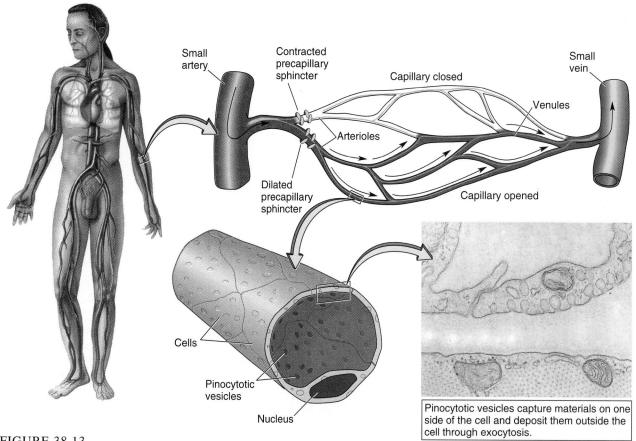

Pinocytotic vesicles capture materials on one
side of the cell and deposit them outside the
cell through exocytosis.

FIGURE 38.13
Control of Capillary Flow
Capillaries are composed principally of endothelial cells. Because their walls are only one cell thick, materials pass rapidly through them.
Capillary beds arise where arterioles branch. The beds can be partially shut down in various regions by action of precapillary sphincters. In
this instance, the lower region is open while the upper is shut down. Capillaries are far more than minute tubes. The flattened cells comprising their walls accommodate simpler diffusion, but their versatile plasma membranes carry on a considerable amount of pinocytosis—
surrounding and engulfing solutes and suspended materials.

As we pointed out in our earlier discussion of vasodilation and vasoconstriction, one of the more interesting capabilities of the circulatory system is its ability to open and close selected capillary beds. Such changes are made possible by the presence of smooth muscle **precapillary sphincters** located in the tiny arterioles at the point where they divide into the capillaries (see Figure 38.13). For example, in an emergency situation the vasoconstriction of such sphincters in the digestive area will partially close down circulation in the digestive system so that more blood can be sent to the muscles and brain.

As blood passes through a capillary bed, the movement of water and various molecules to and from the capillary depends on several opposing forces. (Measurements of the forces involved are summarized in Figure 38.14.) Blood entering a capillary bed will lose nutrients, ions, water, and oxygen. Blood leaving a capillary bed will have picked up ions, water, carbon dioxide, and other metabolic wastes. As you can see, water and ions leave the capillaries and enter the interstitial fluids at one end of the bed, only to return at the other. (The factors involved in the exchange of oxygen for carbon dioxide in capillaries are more complex; see Chapter 39.) Diffusion gradients can account for some of this peculiar behavior, but hydrostatic pressure is also important. Hydrostatic pressure is great at the arteriolar end of a capillary bed, and many substances are simply forced through the thin capillary walls in a filtration process. But things change at the venule end. Hydrostatic pressure is reduced, and since water has been forced out of the capillary, the protein-dense plasma left behind becomes a hypertonic medium, and a concentration gradient arises.

Thus most of the water simply returns to the blood through osmosis, and the ions diffuse down their gradient as well. Any excess fluid remaining outside the capillary will eventually be returned to the blood by the lymphatic system.

The Veins

Capillaries join, forming **venules** ("small veins"), which, in turn, coalesce to form **veins**, vessels that return blood to the heart. Because of frictional resistance in arterioles and capillary beds, blood pressure is lowest in the veins, where it can measure as low as 5 mm Hg. Thus, although blood volume leaving the arteries is equal to the volume entering the veins (except for water loss to the lymphatic system), the force of its movement is greatly depleted. Consider our earlier analogy: if you block the water leaving a hose with your thumb, the pressure in the hose (arteries) is considerable, but as the spray (capillaries) strikes the surface of the ground and the water trickles down the gutter (veins), the pressure is dissipated.

Because of the greatly reduced blood pressure in veins, the blood needs help in getting back to the heart. Many of the veins have one-way, flap-like valves that allow blood to move in only one direction—toward the heart. The walls of the veins, though much thinner than arterial walls, do contain some smooth muscles that can contract and help push the blood along. But venous flow receives assistance in other ways as well. In the simple movement of the limbs, for instance, the muscles squeeze and massage the veins, moving blood along. The one-way valves prevent backflow. In the

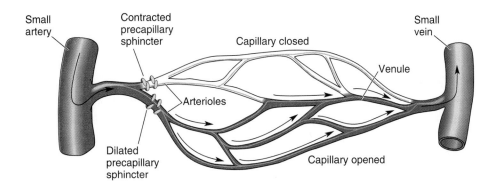

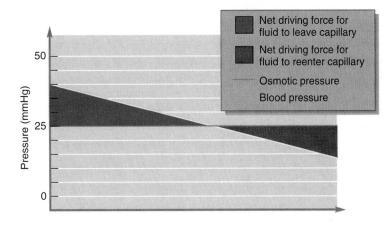

FIGURE 38.14
Forces in the Capillary Bed
The movement of water and solutes in and out of the circulatory system occurs in the capillary beds. This movement, in turn, is controlled by pressure relationships in the bed. The two forces at work are hydrostatic pressure and osmotic pressure, and both operate inside and outside the capillary. The numbers are averages.

Essay 38.1 • CARDIOVASCULAR DISEASE

Here's a little fact you may not be aware of: you will probably die of cardiovascular disease, disease of the heart and blood vessels. (The odds are better than 50:50.) In fact, the final blow is likely to be a heart attack or a stroke. Heart attacks may be brought on by a clot (embolus) moving through the bloodstream and eventually becoming lodged in a vessel of the heart. The heart tissue downstream from the clot is thus deprived of nutrients and oxygen and may die. If the damaged tissue is in an area that controls neural signals to the heart, the heart may begin to beat erratically and may stop altogether. If the clot lodges in vessels serving the brain (a notoriously delicate organ), brain tissue may be damaged, bringing on paralysis, or inability to think normally, and possibly death.

Let's focus here on problems with the heart. A number of factors can increase the likelihood of cardiovascular disease, and among the most common of these is hypertension, which simply means high blood pressure. Another problem is *atherosclerosis,*

marked by the development of *plaques,* places where lipids penetrate the walls of arteries. These plaques can become calcified, producing a kind of atherosclerosis called *arteriosclerosis* (or "hardening of the arteries"). The vessels narrow wherever plaque exists, making them likely places for a moving clot to lodge. In fact, the plaques themselves may be the sites of clot formations. Interestingly, hypertension promotes atherosclerosis, and the two conditions together cause most deaths in the United States. Unfortunately, neither hypertension nor plaque formation produce symptoms at first, so a person may not be aware that he or she is a prime candidate for cardiovascular disease.

A predisposition for hypertension and atherosclerosis may be inherited, but both can be controlled to some extent by diet (one low in animal fat), exercise, not smoking, and reducing stress (easier said than done). In particular, victims may want to lower their levels of cholesterol, which may contribute to the arterial plaques that form

inside blood vessels and can break loose with atherosclerosis. This effort can be somewhat confusing because there are two kinds of cholesterol: *low-density lipoproteins* (*LDLs*) and *high-density lipoproteins* (*HDLs*). The LDLs are believed to help form the plaques, while the HDLs may *reduce* the level of cholesterol in plaques. In fact, researchers now suggest that the ratio of LDLs to HDLs may be a better predictor of cardiovascular risk than cholesterol levels alone. Unfortunately, many people are tested for cholesterol levels at walk-right-up tables manned by technicians using tabletop analyzers that have proved to be notoriously unreliable. So if you want to get a precise reading, go where your blood will be read by larger, more dependable, instruments.

There are things that can be done if you think you are a candidate for cardiovascular disease, and they are not hard to guess. Don't smoke, get plenty of exercise, don't smoke, don't eat too much fat, and don't smoke.

---

chest area, breathing movements squeeze the walls of the vessels, sending the blood toward the heart.

The Lymphatic System

Now let's consider that "other" circulatory system, the one we don't hear much about unless something goes wrong. This is the lymphatic system and, instead of transporting blood, it transports **lymph**, the colorless interstitial fluid of the blood. The lymph, when in the blood circulation, is called plasma. It is called lymph when it seeps from the blood vessels into the lymphatic circulation, leaving the red blood cells behind.

The lymphatic system has four essential roles: (1) it helps maintain fluid and electrolyte (ion) balances in the body; (2) it transports certain fatty acids from the intestinal villi to the blood; (3) it is part of the immune system; and (4) it provides a route by which interstitial fluids can return to the circulatory system. In its first role, the lymphatic system drains tissue spaces and cavities of fluid and ions that have not been recovered by the capillaries. Such fluids are returned to the bloodstream through ducts near the heart.

The lymphatic system (Figure 38.15) consists of **lymph vessels** and clusters of **lymph nodes** (often called "lymph glands"). Lymph vessels include countless tiny blind endings called **lymph capillaries** (which include the lacteals of the

intestinal villi; see Chapter 37). Lymph capillaries collect fluids, solutes, and foreign materials from tissue spaces, emptying into **lymphatic collecting ducts**, which join to form the **lymphatics**, major vessels that carry the lymph to the bloodstream. The watery lymph is pushed along by the pressing action of nearby muscles in the limbs, and changes in pressure within the thoracic cavity brought about by breathing. Like the veins, lymph vessels have one-way check valves that help keep the lymph from backing up so that the flow continues toward the heart (Figure 38.15). Along the way, much of the lymph filters through lymph nodes.

The lymph nodes are located throughout the body, but their greatest concentrations are in the head, neck, armpits, abdomen, and groin. Each node is a compartmentalized mass of tissue that harbors multitudes of lymphocytes—primary cellular agents of the immune system. Foreign materials, bacteria, and viral particles carried into the vessels are swept into the nodes, where they are attacked by resident white blood cells. Activated lymphocytes cause the node to enlarge, so swollen lymph nodes are a telltale sign of infection.

Occasionally, the ducts and vessels in the lymphatic system are turned into deadly avenues for the spread of cancer. While cancer cells entering the lymph nodes are commonly attacked and killed by highly specialized lymphocytes, often some survive and continue their rapid cell division as they are carried by the lymphatic stream throughout the body.

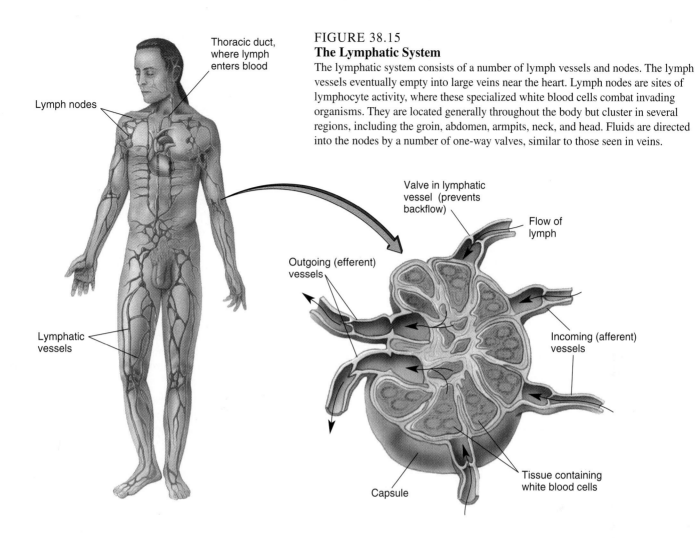

FIGURE 38.15

The Lymphatic System

The lymphatic system consists of a number of lymph vessels and nodes. The lymph vessels eventually empty into large veins near the heart. Lymph nodes are sites of lymphocyte activity, where these specialized white blood cells combat invading organisms. They are located generally throughout the body but cluster in several regions, including the groin, abdomen, armpits, neck, and head. Fluids are directed into the nodes by a number of one-way valves, similar to those seen in veins.

Key Ideas

ADAPTATIONS IN ANIMALS WITHOUT CIRCULATORY SYSTEMS

1. Simpler, thin-walled, aquatic animals lack circulatory systems, utilizing cell-by-cell transport and direct exchanges with their environment. Included are sponges, cnidarians, and flatworms.

CIRCULATORY SYSTEMS IN INVERTEBRATES

1. In **open circulatory systems**, blood is pumped into vessels but leaves them to percolate through spaces called **hemocoels**. Blood returns to the heart to be drawn up for another circuit. In **closed circulatory systems**, the blood elements remain within vessels.

2. The gill-breathing crustacean uses its circulatory system to transport oxygen and carbon dioxide, and accordingly the system is more complex than in the insect, where the circulatory system does not have a respiratory function.

3. Annelids have closed circulatory systems, distinct vessels, and five pairs of tubular aortic arches (hearts) with one-way valves.

CIRCULATION IN THE VERTEBRATES

1. **Blood** (a connective tissue) consists of cells (mainly **erythrocytes**, but also **leukocytes** and **platelets**) and **plasma** (a watery matrix with proteins, nutrients, ions, hormones, and wastes).

2. Plasma proteins include **albumins**, which aid in transport; **globulins**, which include antibodies and **fibrinogen**, which functions in blood clotting.

3. Red blood cells are constantly replaced by production in the **red bone marrow**. Aging cells are phagocytized by macrophages in the spleen and liver.

4. White blood cells include neutrophils, basophils, eosinophils, lymphocytes, monocytes, and macrophages, each of which plays a role in the immune response.

5. Platelets, or thrombocytes, are cellular fragments formed from platelet mother cells (megakaryocytes).

6. All blood cells originate from one cell type, the hemocytoblast or **stem cell**.

7. Clotting is quite complex, with some 15 steps. Following damage to a vessel, platelets gather at the wound, form a collagen plug, and release vasoconstrictors and enzymes called **thromboplastins**. The latter converts **prothrombin** to active **thrombin**. Thrombin cleaves fibrinogen, forming **fibrin**, which becomes the clot.

8. The fish heart consists of one **atrium (auricle)** and one **ventricle**. Blood leaving the fish heart enters the **ventral aorta**, which directs it to the gills, where gas exchange occurs. The blood then passes through the head and body. All blood returning to the heart has traveled through at least two capillary beds.

9. Terrestrial vertebrates have a second atrium and a **pulmonary circuit**, which carries blood to the lungs and back. In both amphibians and reptiles, some mixing of oxygenated and deoxygenated blood occurs in the single ventricle. Partial separation is provided by flaps and partial valves in the heart.

10. The four-chamber heart of crocodiles, birds, and mammals includes a right and left atrium and a right and left ventricle. The pulmonary circuit is completely separated from the systemic (body) circuit.

THE HUMAN CIRCULATORY SYSTEM

1. The human circulatory system consists of the four-chamber **heart**, **arteries**, **capillaries**, and **veins**. The muscular arteries carry blood away from the heart, and, except for the **pulmonary arteries**, this blood is oxygenated. All exchanges occur in the thin-walled capillaries, following which blood returns to the heart through the veins. While arteries tend to be thick-walled and muscular, veins tend to be larger, to have thinner walls, and to contain less smooth muscle.

2. Deoxygenated blood from the body enters the right atrium from the **superior vena cava** and **inferior vena cava**. From the right atrium, blood enters the muscular **right ventricle**, which pumps it to the lungs for gas exchange. Backflow into the right atrium is prevented by the **tricuspid valve**, a one-way valve whose thin flaps are held in place by **chordae tendineae**. Backflow from the pulmonary artery to the right ventricle is prevented by the **pulmonary semilunar valve**. Deoxygenated blood from the pulmonary artery enters the capillaries of the lung, where its gases are exchanged.

3. Oxygenated blood returns from the lungs via the **pulmonary veins**, enters the left atrium, and moves on to the thick-muscled **left ventricle**, which pumps it into the **aorta**. Backflow into the left atrium is prevented by the **bicuspid valve (mitral valve)**, while backflow into the left ventricle is prevented by the **aortic semilunar valve**.

4. While heart muscle has an inherent contractile nature, control of its rate and effort is both **extrinsic** (external) and **intrinsic** (internal). Extrinsic control is through the autonomic nervous system, which accelerates heartbeat via sympathetic nerves and slows it via parasympathetic nerves. Epinephrine from the adrenal glands also accelerates the heart.

5. Intrinsic control originates in the **sinoatrial (SA) node**, or **pacemaker**, which sends contractile impulses across the atrial walls, causing contraction there. The impulse reaches the **atrioventricular (AV) node** and is relayed to the **bundle of His** in the ventricular septum. The bundle's two branches pass to the base of the ventricles and up their outer walls, giving rise to many branched **Purkinje fibers**.

6. The period of ventricular contraction is **systole** while the period between contractions is **diastole**.

7. The amount of blood pumped per contraction is the **stroke volume**, while the **cardiac output** is the amount of blood pumped each minute.

8. Elasticity in the major arteries maintains **blood pressure** during diastole.

9. Arterioles are capable of **vasodilation** (opening) and **vasoconstriction** (closing), shunting blood where the need is greatest.

10. **Circuits** are circulatory pathways where some special function is performed by the blood. They include the **pulmonary circuit** (gas exchange); the **hepatic portal circuit** (food carried from gut to liver for storage and distribution; the **renal circuit (renal arteries** and **veins** bring blood to and from the kidneys where wastes, excess water, and other substances are removed from the blood); the **cardiac circuit** (blood supply to heart muscle, includes **coronary arteries** and **coronary veins**, and the **systemic circuit** (a catch-all for the rest of the body).

11. All transport functions are carried out by the capillaries, which are composed of a single thickness of interlocking cells. Blood is directed into or away from capillary beds by smooth muscle **precapillary sphincters** in arterioles.

12. The functioning of capillaries depends on diffusion gradients, hydrostatic pressure, and active transport (commonly, pinocytosis).

    a. Hydrostatic pressure and, to a lesser amount, diffusion gradients account for losses of water, ions, and nutrients at the arteriolar end of a capillary bed.

    b. Steep osmotic and diffusion gradients bring water and ions back into the capillaries at the venule end.

    c. The exchange of oxygen and carbon dioxide follows the diffusion gradient, and water not reclaimed is recycled by the lymphatic system.

13. Blood pressure is lowest in the veins. The onward movement of venous blood is assisted by one-way valves, muscular squeezing, and breathing movements.

THE LYMPHATIC SYSTEM

1. The lymphatic system redistributes body fluids and ions, transports lipids, and cooperates with the immune system.

2. Lymphatic structures include **lymph vessels** and **lymph nodes**. The vessels begin as **lymph capillaries**, which lead to **collecting ducts** and then the larger lymphatics. Fluids move through the squeezing action of muscles and by breathing movements.

3. Lymph nodes are the sites where disease organisms and cancer cells are destroyed by lymphocytes and other cells of the immune system. Cancer commonly spreads via the lymphatic system.

Application of Ideas

1. Tracing the embryological development of the four-chamber heart in humans reveals stages when the heart appears as a simple tube, as two-chambered, as essentially three-chambered, and finally four-chambered. What does this suggest about the genetic framework on which the human heart is organized? In what way does the fate of the pharyngeal arches in humans help support your hypothesis?

2. The gravest danger in coronary embolism (blockage) immediately follows its onset, a time when tissue death is occurring. This is true even when only a small amount of heart tissue is destroyed. Why is such loss significant to the normal heart function? What characteristic of the cardiac arteries helps minimize such damage?

Review Questions

1. Explain how open and closed circulatory systems differ and list an example of each from the invertebrates. (pages 813–814)

2. In what way is the circulatory system of a crustacean more complex than that of an insect? (page 814)

3. Draw a simplified scheme of the fish circulatory system and explain the major circuits. What is a portal circuit? (page 817)

4. Beginning with the sinus venosus, trace the flow of blood through the amphibian heart, naming each major structure through which it passes. (page 817)

5. In what groups of animals has the heart progressed to four chambers? Prepare a simple diagrammatic drawing of the four-chamber heart, naming the chambers and (with arrows) illustrating the flow of blood. (pages 817–818)

6. Compare the structure of arteries, veins, and capillaries. Which is directly involved in exchanges with the tissues? (page 819)

7. Trace the flow of blood from the venae cavae to the lungs, naming chambers, valves, and vessels along the way. (pages 819–821)

8. Starting with the SA node, describe the pathway followed by a contractile impulse as it passes through the heart. (pages 821–822)

9. Explain how blood pressure is maintained during diastole. (pages 823–824)

10. Describe the renal circuit and hepatic portal circuit and explain the special function of each. (pages 824–825)

11. Discuss the cellular construction of the capillaries and relate this to their function. (pages 825, 827)

12. What happens to blood pressure when blood reaches the veins? In view of pressure changes, explain how blood is moved back to the heart. (page 827)

13. List six components of blood plasma. Which is the most common? (page 815)

14. List five types of white blood cells and briefly describe their functions. (page 815)

15. Summarize the clotting process, mentioning the role of platelets, thromboplastins, prothrombin, and fibrin. (page 816)

16. Describe the composition of the lymphatic system. What are its three main functions? (page 828)

CHAPTER 39

Respiration: The Exchange of Gases

CHAPTER OUTLINE

GAS EXCHANGE SURFACES

The Simple Body Interface

Internalizing the Interface: Tracheae

More Complex Interfaces: The Aquatic Gill

More Complex Interfaces: The Vertebrate Lung

THE HUMAN RESPIRATORY SYSTEM

The Flow of Air

Essay 39.1 Lung Cancer: The "Time Bomb" Within

The Breathing Movements

The Exchange of Gases

Oxygen Transport

Carbon Dioxide Transport

The Control of Respiration

KEY IDEAS

APPLICATION OF IDEAS

REVIEW QUESTIONS

*T*his may not be the best of all possible worlds, but it's the only one we've got. In this world, in a very real sense, the processes of life are involved with overcoming various sorts of problems imposed by our environment and taking advantage of the array of opportunities that our world has set for us.

Oxygen, for example, is provided on our planet, and living things have devised ways to use this peculiar gas. This is not to say that oxygen has always been a blessing. In fact, much of the oxygen present today may have arisen as a form of atmospheric pollution—a toxic product of early photosynthetic life on the earth. Of course, it took many millions of years for living things to adapt to the increasing levels of atmospheric oxygen. At first it must have been a dangerous game indeed, since oxygen can play havoc with the usual chemical reactions of life unless its effect is somehow neutralized. Furthermore, at the time oxygen first appeared, there was little in the atmosphere to absorb ultraviolet radiation, so much of the atmospheric oxygen was quickly converted to ozone (O_3), a highly poisonous oxidizing agent. (It still arises spontaneously in the upper atmosphere.) The advent of oxygen then transformed the surface of an already dicey earth into an even more dangerous place.

Today, however, ozone in the atmosphere helps protect us from dangerous ultraviolet radiation from the sun. Furthermore, most organisms can actually utilize molecular oxygen (O_2). Interestingly, perhaps because of common evolutionary descent, these species tend to utilize oxygen in the same general way—as an acceptor of spent electrons during cell respiration. (Recall that energy-depleted electrons leave the respiratory electron transport systems of the mitochondrion to join oxygen and eventually form water—a metabolic waste product. See Chapter 8.)

The other respiratory gas is carbon dioxide. For each glucose molecule metabolized, cells generate six molecules of carbon dioxide. As a waste gas, it must be quickly expelled. Carbon dioxide readily dissolves in tissue fluids, rendering them acidic, a condition that is toxic to the organism, so carbon dioxide, like oxygen, must be dealt with—in this case, released into the surroundings. The uptake of oxygen and release of carbon dioxide is a one-for-one exchange in the animal—one CO_2 for one O_2, as the life-sustaining process of cell respiration goes on.

Many species have even developed very elaborate systems to distribute oxygen to the cells and pick up carbon dioxide. In many animals, including humans, the vital exchange occurs in the *respiratory system,* while the *circulatory system* transports the dissolved gases throughout the body. In more complex organisms, the respiratory process includes an **external exchange** of gases, one between the organism and the environment, and an **internal exchange**, one between the blood and the cells.

In this chapter, we'll look at the respiratory systems of a variety of animals, focusing on structures that illustrate both unity and diversity. We will find that solutions to problems of gas exchange cut across all taxonomic lines, and that similarities occur not only because of evolutionary relatedness, but because of adaptations to similar environments. Thus lobsters and fishes both utilize feathery gills for the exchange of oxygen and carbon dioxide with the surrounding seawater. In a word, we should be able to find examples of convergent as well as divergent evolution in the respiratory system.

Gas Exchange Surfaces

The exchange of gases in animals requires structures with two basic characteristics. First, the body must contain some kind of permeable surface area where an external exchange can occur. We'll call this the **respiratory interface**. The respiratory interface must be thin-walled and of sufficient surface area to meet the animal's requirements. It must be permeable to oxygen and carbon dioxide because nearly all gas exchange in animals occurs through diffusion. Second, this surface area must be moist, since gases can't normally cross dry membranes (such as our own dry, dead, outer layer of skin cells). For oxygen and carbon dioxide to cross membranes, they must first go into solution. In animals of the terrestrial environment, the problem of maintaining sufficient moistness is often solved through the secretion of mucus by cells of the respiratory interface.

The basic requirements have been met through evolution in an interesting variety of ways (Figure 39.1). As you might suppose, adaptations in aquatic animals differ substantially from those of the terrestrial environment. In small aquatic animals, for example, the respiratory interface may simply be the entire body wall, or surface. The body surface in such cases is sufficient because of the great surface-to-volume ratio in small animals (see Figure 4.4). Larger aquatic animals require additional surface area, which they meet with external or internal gills. In the terrestrial environment, where excessive

water loss across the respiratory interface is always a problem, evolution has provided internalized surfaces. Internalized respiratory surfaces occur chiefly in two forms: tracheae (tube-like systems) and lungs. We'll begin with some of the animals that use their body surfaces for respiration. Then we will consider more complex structures.

The Simple Body Interface

A number of simple animal groups make use of the body surface as the sole respiratory interface. You may recall the general structure of some of the sponges discussed earlier (see Chapter 26). Their thin-walled, vase-like bodies permit a direct exchange of gases with the water both inside and outside the body. The flagellated cells lining the body cavity help by continuously moving seawater through the animal. However, we should remind ourselves that the body of the more complex sponges can be very dense. How does oxygen reach the deeper cell layers? It turns out that body density is no problem since the animal is riddled with small canals. Flagellated collar cells lining the canals create strong currents that carry water past the cells (Figure 39.2a).

Cnidarians also make use of the body wall as the respiratory interface. The body wall exchange works equally well for tiny hydroids and large jellyfish and sea anemones. This is because the hollow cnidarian body is only two cell layers in thickness. Furthermore, the body wall surrounds an extensive gastrovascular cavity that opens to the outside. Thus both inner and outer cell layers are in contact with the watery environment (Figure 39.2b).

Though flatworms are considerably more dense than cnidarians, they are, nevertheless, flat, and this shape yields a large surface area for its mass. An additional exchange surface area occurs in the highly branched gastrovascular cavity (Figure 39.2c). These surfaces provide for a direct exchange with the water environment, and in denser regions, cell-to-cell diffusion is sufficient. Of course, as long as an organism relies on diffusion alone for gas exchange in deeper tissues, it can never grow very large—its volume would quickly outgrow its surface area. So if you ever see an 8-foot planarian gliding toward you in your favorite swimming hole, ignore it. It doesn't exist.

In the spiny echinoderms, an expansion of the respiratory interface is provided by the **dermal brancheae**. These are protrusions of the thin coelomic wall that extend through pores in the endoskeleton. Fluids are moved in and out by the action of cilia (Figure 39.2d).

More Complex Skin Breathers. A few terrestrial invertebrates and vertebrates, such as the earthworm and small lungless salamanders, make use of their moist skin as a respiratory interface (the salamander also uses its highly vascular pharynx). Whereas the risk of desiccation is too great for most terrestrial animals to use the skin for gas exchange, earthworms and salamanders represent special cases. For example, the earthworm's habitat is really only semiterrestrial since it restricts itself to moist soil. Lungless salamanders are limited to damp places, spending their days underground (or under logs), venturing out mainly at night

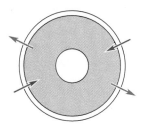

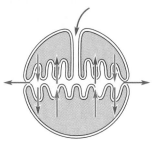

Because of its extremely thin body wall, each cell in the sea anemone can exchange gases directly with surrounding sea water.

This salamander has permanent, thin-walled external gills in which gas exchange with the surrounding water goes on (external respiration). The gases are transported to and from the cells by the circulatory system. A second exchange occurs in the cells (internal respiration).

The fish gill carries on the initial exchange of gases with the blood (external respiration). The gases are transported to and from the body cells by the circulatory system. A second exchange occurs in the cells (internal respiration).

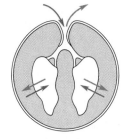

The vast tracheal system in insects reaches all of the cells. It thus carries on a more-or-less direct exchange without the involvement of the circulatory system.

The lung, like the gill, carries on an exchange of gases with the blood. Note the in-and-out movement of air. The gases are transported to and from the body cells by the circulatory system. A second exchange occurs in the cells (internal respiration).

The bird lung is unique in that air flows through rather than in-and-out. As in other types of lungs, gases are exchanged first in the lungs and second in the body tissues. The circulatory system transports the gases between the two.

FIGURE 39.1
Animal Respiratory Interfaces
Body size, complexity, and environment all play roles in determining the type of respiratory interface required to meet the needs of an organism. In simple animals, most cells are exposed to the environment, so a direct exchange of gases occurs. In more complex types, gases are exchanged across special respiratory interfaces such as gills, tracheae, and lungs.

(a) *Euspongia,* a complex sponge

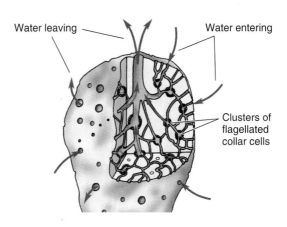

Water leaving

Water entering

Clusters of flagellated collar cells

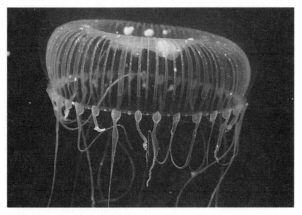

(b) Thin-bodied jellyfish

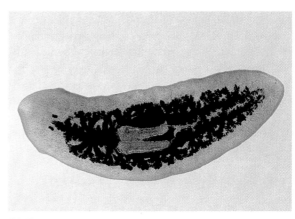

(c) Flatworm

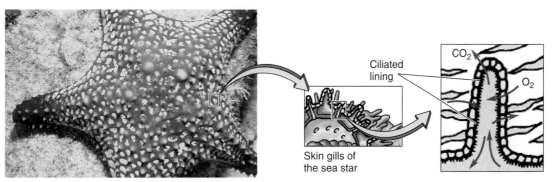

(d) Sea star

Ciliated lining

CO_2

O_2

Skin gills of the sea star

FIGURE 39.2
The Simple Body Interface
(a) The more complex sponges, such as *Euspongia,* still manage a direct exchange of gases between their cells and the environment. The cutaway view explains how this is possible. The many canals carrying water throughout the animal provide the exchange surface for the cells. Flagellated cells provide the force to keep the currents moving. **(b)** Because of their simple, two-cell-layer thick bodies, jellyfish and other cnidarians are able to exchange gases directly with water outside the body and inside the gastrovascular cavity. **(c)** The large surface area and the extensive gastrovascular cavity (shown in black) permit the thicker bodied flatworms to carry on a direct exchange of gases. **(d)** The echinoderms exchange gases with the seawater across minute projections of the coelomic membrane that penetrate the skeletal plates. A ciliated lining keeps coelomic fluids moving in and out of the so-called skin gills. The many tube feet of seastars also provide a large exchange area (see Figure 26.43).

when the risk of desiccation is lessened. Furthermore, the skin of earthworms and lungless salamanders is highly glandular and is kept moist by the secretion of a slimy layer of mucus.

Although such skin breathers have no special structures with which to exchange gas with their environments, their larger density prevents them from relying on the cell-by-cell diffusion alone. Oxygen could not reach the deeper tissues, and carbon dioxide couldn't penetrate the surface tissues, fast enough for an adequate exchange. In these animals, as in most others of greater density, the exchange of oxygen and carbon dioxide is enhanced by efficient circulatory systems that make use of the blood to transport the two gases between the respiratory interface and the cells. The exchange is diagrammed in Figure 39.3.

The blood of most animals also contains respiratory pigments, molecules that bind temporarily to oxygen and carbon dioxide, thus greatly increasing the gas-carrying capacity. In the earthworm and lungless salamander, that pigment is the protein hemoglobin. Hemoglobin in the lungless salamander, as in all vertebrates, occurs in red blood cells, but in the earthworm, the hemoglobin is free in the blood. Even in its free form, as seen in the earthworm, hemoglobin carries roughly 70 times as much oxygen as could be carried by an equal amount of water. Extending this idea, we could make a logical case for respiratory pigments being a key factor in the evolution of large forms of animal life. We will look further into the precise role of hemoglobin shortly. We'll shift our attention now to terrestrial insects with more complex solutions to gas transport.

Internalizing the Interface: Tracheae

Respiration in terrestrial insects is interesting in that they do not make use of a fluid-based transport system at all. Although

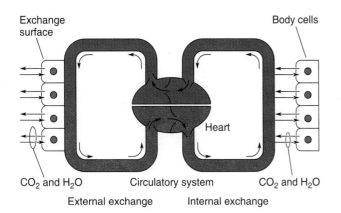

FIGURE 39.3
External and Internal Phases of Gas Exchange
This very schematic view of respiration shows the external and internal phases of gas exchange. The external phase occurs between the respiratory interface (gill, skin, or lung) and the external environment. There carbon dioxide is released and oxygen is taken in, both by diffusion. The gases are transported to and from the cells by the circulatory system, which in this instance includes a four-chamber heart and blood vessels. The internal phase is the exchange of gases between the blood and the cells, where cell respiration is going on.

insects have blood and efficient circulatory systems, the blood serves primarily in distributing food and removing metabolic wastes. For gas exchange, they make use of a hollow **tracheal system**. The insect body contains many tiny, air-filled tubes called **tracheae** (singular, *trachea*) (Figure 39.4). The tubes branch and rebranch, forming minute endings called **tracheoles**. The tracheoles pass among and into the cells, thereby assuring each of a sufficient exchange. Some tubes end in enlarged air sacs that store air, which in some aquatic insects also increases buoyancy. The tracheal system in larger insects is ventilated through a muscular bellows action of the body wall, particularly the abdomen. The external openings of the tracheae are often protected by valves known as **spiracles**. The spiracles remain closed between ventilating movements, thereby conserving body water and keeping dust out. In aquatic insects, the tracheae may terminate in external **tracheal gills**—thin-walled, hair-like extensions of the exoskeleton that contain finely branched tracheal networks. As with internal branching, the external branching increases the surface area.

More Complex Interfaces: The Aquatic Gill

Aquatic animals such as mollusks, crustaceans, and fishes make use of the gill for external respiration, carrying on gas exchange with the water. So there is no misunderstanding, the oxygen available in water is oxygen gas (O_2) *dissolved* in the water, not the oxygen from the water molecule (H_2O). The gill is intimately associated with the circulatory system where fluid containing respiratory pigments is used to promote the internal exchange, that is, transporting gases to and from the cells. The presence of similarly constructed gills in widely divergent animal groups such as crustaceans and vertebrates is an excellent example of convergent evolution. Here the problem of gas exchange in the water environment has presented unique problems whose solution in very different animals has been strikingly similar.

Gills are thin-walled, finely divided structures containing extensive capillary beds. Capillaries, you will recall, are the smallest blood vessels, with walls only one cell thick. The moving bloodstream within the gill is thus separated from the watery environment by a membrane or two. It is across this delicate interface that the one-for-one exchange of carbon dioxide and oxygen occurs.

Gills of Arthropods. Arthropod gills are varied, but those of the lobster, a crustacean, will serve as our example (Figure 39.5). Lobster gills are feathery extensions of the body wall, attached partly to the bases of the legs. Although they are external, the gills are covered by the **carapace**, a portion of the exoskeleton that is open at both ends. The carapace serves to protect the delicate gills and to help direct the flow of water. Paddle-like appendages just forward of the carapace beat rapidly, drawing water forward, under the carapace and across the gills toward the head.

Each gill contains capillaries that receive blood from a hemocoel (an open blood passage) in the thorax floor. Follow-

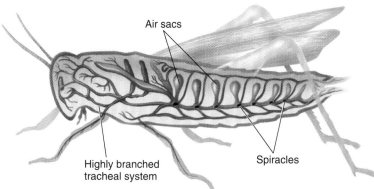

Air sacs

Spiracles

Highly branched
tracheal system

(a) Tracheal System

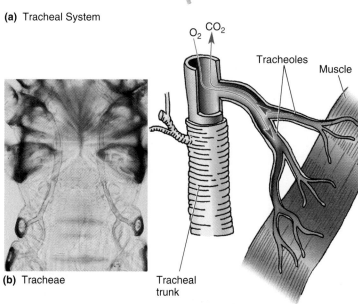

O_2 CO_2

Tracheoles

Muscle

(b) Tracheae

Tracheal
trunk

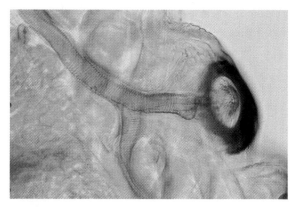

(c) Spiracles (Silkworm larva)

FIGURE 39.4
The Insect Tracheal System
Insects make use of a direct gas exchange with their environment.
(a) The respiratory interface consists of numerous tube-like tracheae,
many of which lead into balloon-like air sacs. **(b)** The tracheae
branch into thinner and thinner tracheoles that pass near or into cells.
Air moving in and out of the tracheoles is brought close to every
cell, where the direct exchange occurs. Note the reinforcing rings
along the larger tubes. They keep the tubes from collapsing as air is
drawn in and out of the system. **(c)** The tracheae begin at spiracles,
valves on the surface of the insect. The spiracles open as air is
pumped in and out and remain closed between times.

ing the exchange of carbon dioxide for oxygen, the blood
returns to the heart to be pumped through the body again. As
we saw in the last chapter, such a circulatory system—with
blood leaving the vessels and percolating through open spaces—
is called an open circulatory system.

Crustaceans and mollusks make use of a copper-containing
pigment called **hemocyanin** for transporting oxygen. In hemocyanin, which is a protein like hemoglobin, oxygen is bound to
copper rather than iron. Upon oxygenation, the pigment turns
from colorless to blue.

Gills of Fishes. The fish gill (Figure 39.6) is made up of rows
of feathery **gill filaments** that attach individually to supporting
gill arches. Each filament contains many disk-shaped **lamellae**
that provide most of the surface area of the gill. Each lamella
contains a network of capillaries in which the exchange of
gases occurs. In bony fishes the gills are protected by a covering known as an **operculum**. The flap opens and closes intermittently as water is pumped over the gills.

The exchange of carbon dioxide for oxygen
occurs as deoxygenated blood from the body passes
through the capillary network in each lamella. As in
other vertebrates, the red blood cells are literally crammed
with hemoglobin. When oxygenated, hemoglobin shifts
from a dark bluish-red color to bright red. Freshly oxygenated
blood, bright red in color, leaves the lamellae and passes out of
the gill to the body for distribution to the tissues. (For a review
of fish circulation, see Figure 38.5a.)

A special arrangement facilitates the exchange of gases
within the capillary networks of the gill: blood and water
move in opposite directions. The movement of blood inside
the lamella opposes the movement of water outside. This produces a highly efficient **countercurrent gas exchanger** (Figure 39.6c), similar in principle to the countercurrent heat
exchanger (see Chapter 35). Here, blood passing along the
capillary network is constantly confronted by a fresh supply
of water, which creates a continuous concentration gradient
favoring the inward diffusion of oxygen and the outward diffusion of carbon dioxide. Such efficiency is especially important to active animals in aquatic environments, where getting
enough oxygen can be difficult. Let's see why.

Oxygen is not really plentiful in water, at least when compared to air. Under the best of conditions, oxygen makes up
only about 1% of the volume of water, whereas the percentage
by volume in air is about 21%. The best of conditions include
low temperatures; the solubility of oxygen in water is inversely proportional to temperature (cold water can hold more oxygen than warm water). In addition, the diffusion rate of oxygen through water is vastly slower than in air ($\frac{1}{10,000}$ the rate).

The limited oxygen content and slow diffusion rate mean
that most active aquatic animals cannot tolerate stagnant water
conditions. Furthermore, they must devote considerable energy
to the movement of large volumes of water over their respiratory surfaces. The faster-swimming bony fishes and the sharks
solve their respiratory problems in a direct manner: they keep
moving. This creates a continuous flow through the mouth,

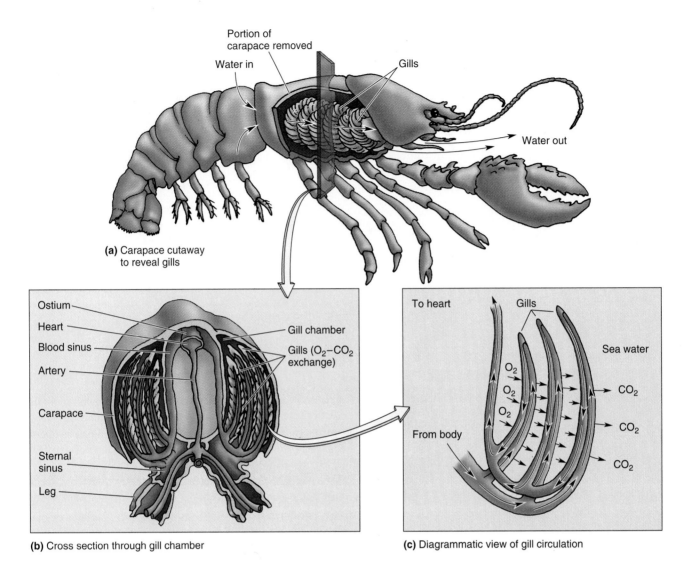

(a) Carapace cutaway to reveal gills

(b) Cross section through gill chamber

(c) Diagrammatic view of gill circulation

FIGURE 39.5
Gills in the Lobster
Like most crustaceans, lobsters make use of gills for gas exchange with the surrounding water. **(a)** A row of gills, attached to the walking legs, occurs on each side. The gills are covered by the carapace (removed here). Water is moved over the gills by the action of appendages just forward of the carapace. Movement of the legs also stirs the water around the gills. **(b)** The arrangement of blood vessels and the thoracic blood sinus are seen in a cross-sectional view of the body. **(c)** Diagrammatic view of gill circulation. By following the arrows, we can see that deoxygenated blood (blue) enters the gills from the sinus below. Following gas exchange, the oxygenated blood (red) flows through vessels back to the pericardial chamber (around the heart). The blood, drawn into the heart through minute openings, is then pumped out to the body.

over the gills, and out through the gill openings. Other fishes rely on a pumping action involving muscles of the pharyngeal cavity. Water is drawn intermittently into the mouth and forced over the gills. That's why motionless aquarium fish look as though they are gulping. If you watch carefully you can discern the opening and closing sequence of the mouth and gill flaps.

More Complex Interfaces: The Vertebrate Lung

Evolution of the Vertebrate Lung. The vertebrate lung has had a strange evolutionary history. Early in vertebrate history,

perhaps as early as the Precambrian era, the ancestor of modern bony fishes lived in a freshwater habitat, where it obtained oxygen simply by gulping air. A pocket of highly vascularized tissue in the *pharynx* of this ancestral fish became specialized as a place to hold air between gulps. Then, through millions of years of natural selection, the air pocket grew and branched, until a primitive version of the vertebrate lung emerged.

Some of the descendants of that early ancestral fish species retained the lungs as accessory organs, although they developed gills as well. Those descendants included the ancestral lungfishes and lobe-finned fishes (see Chapter 27). Lungfishes are a

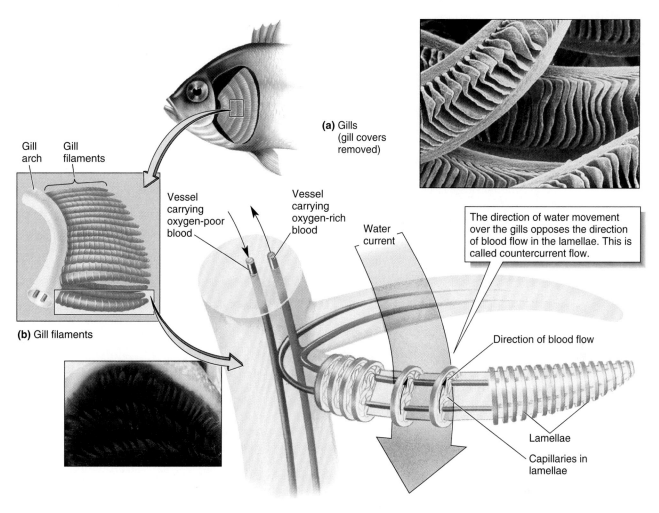

(a) Gills (gill covers removed)

Gill arch   Gill filaments

Vessel carrying oxygen-poor blood

Vessel carrying oxygen-rich blood

Water current

The direction of water movement over the gills opposes the direction of blood flow in the lamellae. This is called countercurrent flow.

(b) Gill filaments

Direction of blood flow

Lamellae

Capillaries in lamellae

(c) Direction of water and blood flow across the lamellae

FIGURE 39.6
Gills in the Bony Fish
(a) The photograph presents an impressive view of the respiratory interface of the fish gill. The bright red color indicates the presence of a great many capillary beds carrying blood that has been oxygenated by oxygen in the surroundings. In this view, the operculum (gill cover) has been eliminated to show the arrangement of the gills in their cavity. (b) The expanded view of one gill arch shows numerous gill filaments. The rings in the filaments represent lamellae. (c) An even closer view shows the arrangement of blood vessels carrying deoxygenated blood into the gill filament and oxygenated blood out. The connecting capillaries are in the gill lamellae. The large blue arrow shows the path of water across the lamellae, which opposes the flow of blood in the capillaries. The opposing flow produces a countercurrent exchange, which makes the exchange of gases highly efficient.

peculiar assemblage of freshwater fishes that are uniquely adapted for survival in ponds that periodically dry up. Some of the early lobe-finned fishes had a different fate. They emerged from the freshwater habitat to become the first terrestrial vertebrates. For this group, the possession of lungs was a vital **preadaptation**. Preadaptations are adaptations to one way of life that, by coincidence, enable an organism to survive in another.

The ancient lobe-finned fishes achieved a second evolutionary breakthrough—literally. They evolved a pair of **internal nares**—openings between the nasal cavity and the mouth. The nasal cavities in most fish are blind pouches that house chemoreceptors, used only for "smelling" the water. It's not

clear that the lobe-fins actually used their internal nares for breathing, but the development of this feature proved to be yet another preadaptation to land life. The early terrestrial vertebrate was able to take in air smoothly, ventilating its lungs without the crude gulping movements.

There was one more innovation. To this day, the internal nares of amphibians and most reptiles open directly into the mouth cavity. As mammals and crocodilians evolved from their reptilian ancestors, however, a bony ledge, the **palate**, finally divided the mouth cavity into the mouth and nasal cavities. This arrangement enables mammals to chew and breathe at the same time. If you don't think this is important

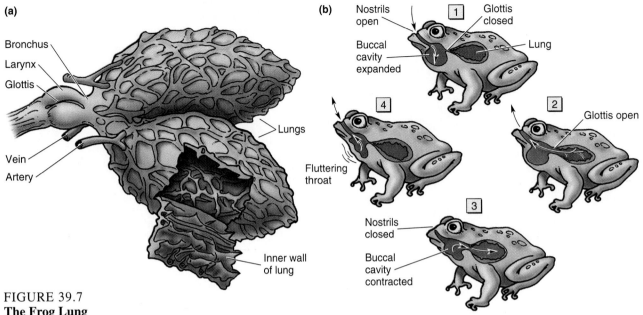

(a)

Bronchus

Larynx

Glottis

Vein

Artery

Lungs

Inner wall
of lung

(b)

Nostrils
open

Glottis
closed

Buccal
cavity
expanded

Lung

1

4

Fluttering
throat

Glottis open

2

Nostrils
closed

Buccal
cavity
contracted

3

FIGURE 39.7
The Frog Lung
(a) The frog lung is balloon-like rather than spongy. Blood vessels branch over the surface, forming capillary beds where exchanges occur. Note the slit-like glottis, a valve that admits air to the lungs. **(b)** Respiratory movements in the frog require the timely opening and closing of the glottis and valves in the nasal openings. Most of the ventilating movement involves contraction of the throat muscles. ☐1 The frog inhales by expanding its throat, which draws fresh air into the mouth cavity. During this time the nostrils are open and the glottis is closed (note how the nostrils open into the mouth). ☐2 Next, with the nostrils and glottis open, air is forced from the lungs, passing over the fresh air mass in the mouth cavity, and out through nostrils (because of the air flow characteristics, there is little mixing). ☐3 The frog then closes the nostrils, contracts the throat, and forces air into the lungs. ☐4 Finally, the frog closes the glottis and flutters the throat, forcing any residual air out of the mouth. It is then prepared for a new inhalation cycle.

try eating something really chewy with your nostrils pinched together.

Most of today's bony fish retained the primitive lung structure, but it now functions as the **swim bladder**, a flotation device (see Chapter 27). The gills are the sole gas exchange organ in most modern fishes.

The Vertebrate Lung. In most vertebrates, the lung is a highly branched, tree-like arrangement of tubes, each ending in a cluster of air sacs called **alveoli.** The air sacs resemble bunches of grapes with each alveolus the equivalent of one grape. The vast number of alveoli give the lung its spongy characteristic. The thin, moist membranes of alveoli are all that separate the bloodstream from the external environment. Ventilation is provided by various types of bellows mechanisms involving muscles of the throat, ribs, abdominal wall, and, in some instances, a muscular partition known as the diaphragm. Of course, some moisture loss is unavoidable; the air exhaled by terrestrial vertebrates is generally quite moist.

Amphibians and Reptiles. The evolutionary pathway from an aquatic to a terrestrial respiratory system is suggested by today's amphibians and reptiles. Nearly all adult amphibians have lungs, but they are hollow, balloon-like structures (Fig-

ure 39.7a), without the spongy texture seen in the lungs of most other terrestrial vertebrates. Most amphibians also exchange gases across their moist, highly vascularized skin. In fact, studies show that the escape of carbon dioxide across the skin equals that of the lungs. The ability to exchange gases through the skin enables amphibians to remain submerged for long periods when in danger and to hibernate in the mud throughout winter. The desert spadefoot toad survives through the long, hot summer by burying itself in the pond bottom until the next rains return.

The simple frog lung is ventilated by the raising and lowering of the floor of the mouth. As illustrated in Figure 39.7b, this crude pumping action requires the timely opening and closing of "check valves" in the nostrils and floor of the pharynx. The pharyngeal valve is a muscular slit called the **glottis**.

Since the dry reptilian skin is impervious to air, reptiles are strictly lung breathers. The lungs of reptiles are essentially sac-like, but with greater subdivision than is seen in the amphibian. The lungs of the large monitor lizards, crocodiles, and turtles are heavily subdivided. To ventilate the lungs, the reptile makes use of contractions in rib muscles, an action somewhat similar to that of mammals. Crocodiles also make use of a muscular shelf below the liver, somewhat like the mammalian diaphragm, to expand and compress the lungs.

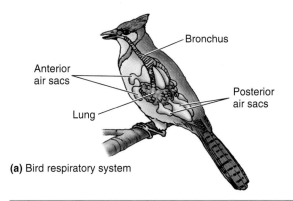

Bronchus

Anterior air sacs

Lung

Posterior air sacs

(a) Bird respiratory system

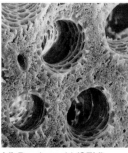

(d) Parabronchi (SEM)

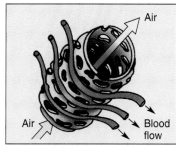

Air

Air

Blood flow

(e) Crosscurrent flow around parabronchi

FIGURE 39.8

The Bird Respiratory System

(a) The bird lung is surrounded by several posterior and anterior air sacs. A lengthy bronchus extends between the lung and the nostrils and mouth. **(b)** Inhalation begins when the chest cavity expands. Air is drawn down the bronchus into the posterior air sacs. Simultaneously, air is drawn from the lungs into the anterior sacs. **(c)** Exhalation occurs when the chest cavity contracts. This forces air from the posterior air sacs into the lung, and air from the anterior sacs into the bronchus. **(d)** A scanning EM view of the lung shows the prominent parabronchi, cylindrical passages lined with very spongy tissue. **(e)** Blood passing through the lungs is directed around the parabronchi at right angles to the air flow as seen in the diagram. This provides a crosscurrent exchange.

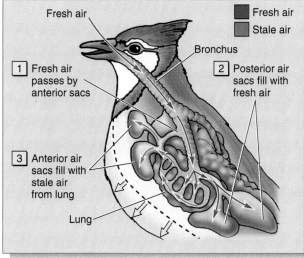

Fresh air

Fresh air

Stale air

Bronchus

1 Fresh air passes by anterior sacs

2 Posterior air sacs fill with fresh air

3 Anterior air sacs fill with stale air from lung

Lung

(b) Inhalation (air sacs fill)

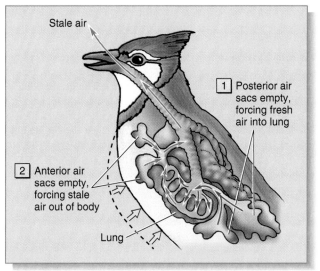

Stale air

1 Posterior air sacs empty, forcing fresh air into lung

2 Anterior air sacs empty, forcing stale air out of body

Lung

(c) Exhalation (air sacs empty)

The Unique Bird Lung. The respiratory system of the bird is unlike that of any other vertebrate. The lung, like that of the mammal, is penetrated by many air passages and is quite spongy, but most of the similarity ends there. The passages in the bird lung, known as **parabronchi**, do not end blindly in tiny clustered alveoli, as in the mammal, but form open-ended

tubes. Because of this, air flows *through* the bird lung rather than *in and out!*

Another unique feature is the presence of numerous balloon-like **air sacs** outside the lungs (Figure 39.8). The air sacs have little to do with actual gas exchange, functioning instead as air reservoirs and as bellows that force air through the lung. There are three pairs of air sacs anterior to the lung and two pairs posterior to the lung. Some air sacs form extensive branches, a few of which pass right into the larger bones. Let's follow the breathing movements.

Inhalation (Figure 39.8b) occurs when the contracting muscles in the ribs and abdomen expand the thoracic cavity. In response, the posterior air sacs expand and fill with fresh air, which is drawn into the mouth and nostrils and the lengthy **bronchus** (the bird equivalent of a trachea). The anterior sacs also expand and fill, but with stale air drawn from the lung passageways. Exhalation (Figure 39.8c) occurs when the breathing muscles relax and the thorax contracts. The collapsing posterior air sacs force fresh air into the lung and the collapsing anterior air sacs send stale air into the bronchus and out through the mouth and nostrils. Throughout all of this, the lungs show very little change in volume. In most other vertebrates, the lungs undergo large changes in volume as they inflate and deflate.

We see then that incoming air bypasses the lung at first, entering air sacs at the posterior end. It then flows through the many passages and exits into the anterior air sacs and out of the body—definitely *a one-way flow.*

Gas exchange itself occurs between the blood capillaries and the parabronchi. Because the capillaries are affixed *at*

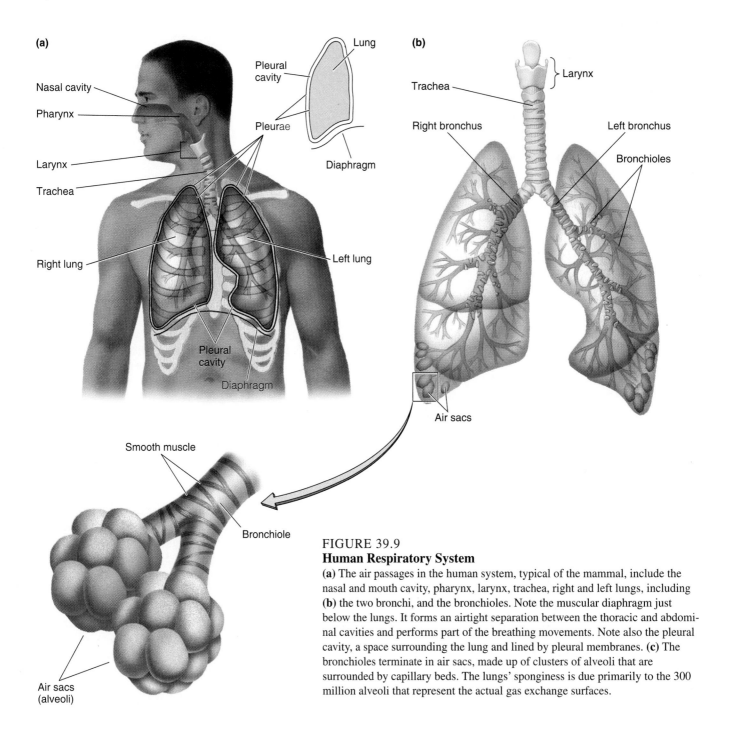

(a)

Nasal cavity
Pharynx
Larynx
Trachea
Right lung

Lung
Pleural cavity
Pleurae
Diaphragm

Pleural cavity
Left lung
Pleural cavity
Diaphragm

(b)

Trachea
Right bronchus

Larynx
Left bronchus
Bronchioles

Air sacs

Smooth muscle
Bronchiole
Air sacs (alveoli)

FIGURE 39.9
Human Respiratory System
(a) The air passages in the human system, typical of the mammal, include the nasal and mouth cavity, pharynx, larynx, trachea, right and left lungs, including **(b)** the two bronchi, and the bronchioles. Note the muscular diaphragm just below the lungs. It forms an airtight separation between the thoracic and abdominal cavities and performs part of the breathing movements. Note also the pleural cavity, a space surrounding the lung and lined by pleural membranes. **(c)** The bronchioles terminate in air sacs, made up of clusters of alveoli that are surrounded by capillary beds. The lungs' sponginess is due primarily to the 300 million alveoli that represent the actual gas exchange surfaces.

right angles to the parabronchi, the blood flow forms a **crosscurrent exchange** (Figure 39.8d,e), a third unique feature of the bird lung. While this is different from the countercurrent exchange of the fish gill, where the two currents move in opposite directions, it accomplishes the same end. In the crosscurrent exchange, the moving bloodstream is confronted by a continuous flow of fresh air, an arrangement that maintains a steep diffusion gradient for both oxygen and carbon dioxide. This provides for a highly efficient exchange of gases.

The unique features of the bird respiratory system are interesting, but how is this plan adaptive? If you're thinking

that it has to do with flight, you are correct! The efficient exchange of gases certainly supports the vigorous activity of flying. It also supports the rate of cell respiration needed to maintain a constant, high internal body temperature in cold climates. But mammals, with their traditional "in-and-out" flow of air, are also capable of vigorous movement and they too cope well with cold climates. There must be more.

Researchers believe that the primary advantage of the bird lung is that it adapts the bird to high altitudes where oxygen levels are low. Maintaining rigorous activity at higher altitudes requires the most efficient gas exchange possible, a requirement that is well met by the unique bird respiratory system. High-altitude flying is particularly important to birds that migrate, and while many birds migrate at lower altitudes, some birds have been tracked by radar at 7000 m (about 23,000 ft). A mammal attempting equally strenuous activity at this altitude without an oxygen bottle would fall into a metabolic stupor. Bats (mammals), should you be wondering, are low-altitude migrators.

We now turn to a close look at the mammalian respiratory system. Once again, our representative mammal will be our favorite—us.

The Human Respiratory System

The respiratory system in humans (Figure 39.9) is representative of that in other mammals. The major structures include the mouth, nasal cavity, pharynx, larynx, trachea, and lungs. The lungs are lobed, are roughly triangular in shape, and have broad bases. They receive the bronchi (singular, bronchus), which divide into numerous bronchioles that terminate in clustered air sacs, or alveoli. Much of the mass of the lungs can be attributed to the highly branched blood vessels making up the pulmonary circuit. Two moist, bag-like membranes, the **pleurae**, form airtight enclosures around the lungs. The inner pleura is attached firmly to the spongy surface of the lung; the outer pleura is attached to the wall of the thoracic (chest) cavity. The space between the two is the **pleural cavity**. The pleural cavity is bounded at its lower portion by that dome-shaped muscular shelf known as the muscular **diaphragm**. The muscular diaphragm, a uniquely mammalian structure, divides the body cavity into thoracic and abdominal regions and functions in ventilation.

The Flow of Air

We'll trace the path of air through the respiratory system and then focus on the muscular activity that brings it about.

Air first enters the nasal passages where it is cleansed and warmed. Cleaning is done through the filtering action of nasal hairs and the action of mucus-secreting cells and ciliated cells of the nasal epithelium (Figure 39.10). A similar lining is seen in the trachea, bronchi, and bronchioles. A film of mucus containing trapped dust, pollen, smoke, and other particulate matter is constantly swept by the waving cilia toward the throat—for swallowing. The air is warmed by blood in dense capillary beds that line the passageways. In fact, the capillary flow opposes the air flow, setting up an efficient countercurrent heat exchanger (see Chapter 35). Without this warming process the body would lose far too much heat through the lungs. Thus the process helps the mammal maintain a constant, warm internal temperature.

When you inhale, air moves from your nasal passages into your **pharynx** (the posterior mouth region) and from there through the larynx. The **larynx** contains the **voice box** and the **vocal cords** (Figure 39.11). The larynx remains open,

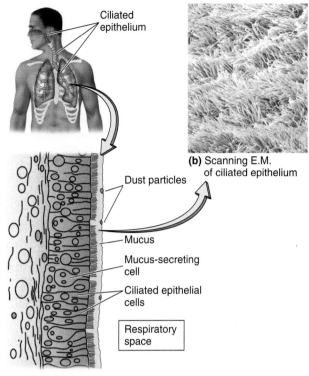

(b) Scanning E.M. of ciliated epithelium

(a) Ciliated cells and mucus-secreting cells

FIGURE 39.10
The Respiratory Lining
(a) The nasal passages and respiratory tree contain a lining made up of ciliated and mucus-secreting cells. The mucus traps particulate matter and the cilia sweep it toward the throat for swallowing. Once swallowed, the particles are disposed of by the digestive system. Among the particles are spores of bacteria and fungi. Most are destroyed by the stomach acids and enzymes, but some pathogens may cause infections. **(b)** The scanning EM provides an impressive view of the countless cilia in the respiratory linings.

unless you are swallowing, whereupon it elevates and seals its opening against the **epiglottis**, which keeps materials out of the trachea. (The role of the larynx in swallowing was discussed in Chapter 37.)

Air moves from the larynx to the **trachea**, or windpipe, a tube that contains many C-shaped rings of stiff cartilage that hold the airway open (like the rings of a flex hose). The trachea branches into right and left **primary bronchi**, which branch into **secondary** and **tertiary bronchi**, which branch, in turn, into the **bronchioles**. Because of such continuous branching, the air passages are referred to as the **respiratory tree** (see Figure 39.9b). Like the nasal passages, much of the respiratory tree contains mucus-secreting and ciliated cells. The mucus, along with trapped particles, is constantly swept upward by the cilia to the pharynx, where it is swallowed. This cleansing action is lost to heavy smokers, whose respiratory linings are subject to drastic change as the years and cigarette packs go by. (Essay 39.1 makes the connection between

Essay 39.1 ● LUNG CANCER: THE "TIME BOMB" WITHIN

More men and women now die of lung cancer than from all other forms of the disease. Each year, about 150,000 people develop lung cancer, and most die from this fast-moving killer within five years of its onset. The statistical link between lung cancer and cigarette smoking is now indisputable, and 80% of the new cases can be attributed to that practice. Historically, the rise of lung cancer in men closely paralleled the increased popularity of cigarette smoking. But what was once heralded as a disease primarily of males now strikes women with nearly equal frequency (see graph). The social acceptance of smoking by women has apparently made the difference. What's the attitude of the tobacco industry? Perhaps it's

summed up in the ad slogan: "You've come a long way, Baby!"

It's true that a significant number of this year's lung cancer victims will have been nonsmokers, a fact that the tobacco industry is ready to trot out at a moment's notice. But this doesn't weaken the statistical link between cigarette smoking and cancer. Heavy cigarette smokers are *twenty times more likely to die of lung cancer than nonsmokers*. And while our subject here is lung cancer, let's not ignore the 40,000 deaths per year from smoking-related heart disease, emphysema, chronic bronchitis, and other cancers. Then consider the known health risks to unborn babies.

Lung cancer doesn't develop overnight; if it did, cigarette smoking would quickly lose its popularity. It begins early in some, whereas others may complete a nearly normal life span before it sets in. In smoking-related lung cancer, it is the constant irritation from inhaled smoke that begins the process. The respiratory lining responds to the irritation through an excessive secretion of mucus. This renders the cilia less and less effective at moving the irritants out of the lungs, so they gather there. The accumulation of smoke chemicals begins to kill the alveolar phagocytes, so important in clearing particles from these blind passages. The ongoing mucus accumulation causes congestion, which provokes the typical, early

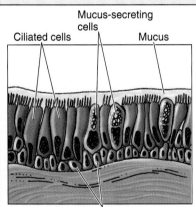

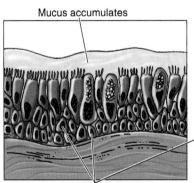

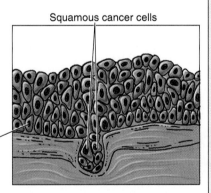

The normal lining of the respiratory passages includes mucus-secreting cells and ciliated cells. In normal function, much of the particulate matter is trapped by mucus and swept out of the lung.

Constant irritation by smoke, chemicals, and other pollutants causes the secretion of an abnormally thick mucous coating and the multiplication of basal cells at a much higher rate than normal. This precancerous condition may reverse itself if the smoker quits.

Squamous cancer cells proliferate, replacing all other cells in the respiratory linings. The cluster breaking through the basal cell layer below can cause additional trouble in distant places if the diseased cells enter the circulatory or lymphatic system.

morning (first cigarette) "smokers cough," which may be the only warning a person gets of events to come.

It is about this time that emphysema may begin. In emphysema, the accumulation of mucus ruptures the alveoli, destroying the gas exchange surfaces. This too is a very slow process, and breathing may not be outwardly affected for many years. The damage, however, is irreversible. With medical care and the proper precautions, people with advanced emphysema can live a reasonably long time; however, the quality of that extended life leaves much to be desired. Typically, emphysema victims end up bedridden, with an oxygen bottle never very far from their reach.

What about lung cancer? Cigarette smoke contains coal tars and other constituents that become carcinogens when trapped in the body. As time passes, the basal cells underlying the respiratory lining begin an abnormal response to the irritant. They divide much more rapidly than usual, some breaking through the basement membrane below, and others replacing the normal lining cells. Such cells have undergone the transition to cancer. The affected area is small at first, but lung cancer is one of the most aggressive of all cancers, and the aberrant cells spread rapidly, crowding out normal surrounding tissues (see illustration and photographs).

Medical intervention at this time may save the life, but the chances are slim. Even

the removal of a diseased lung is no guarantee, since the cancer may have already **metastasized**, a chilling term if you are the patient, since it means that the cancer has spread to other parts of the body. Once such cells find a new locale, their unchecked growth begins once more.

In its advanced state, lung cancer is known to be excruciatingly painful and the last days are generally spent in a drug-induced haze with intermittent periods of pain and semiconsciousness. Death is often the direct result of the ravaged lungs filling with fluid and the victim asphyxiating. In its final phases, the only good news is that the end comes quickly.

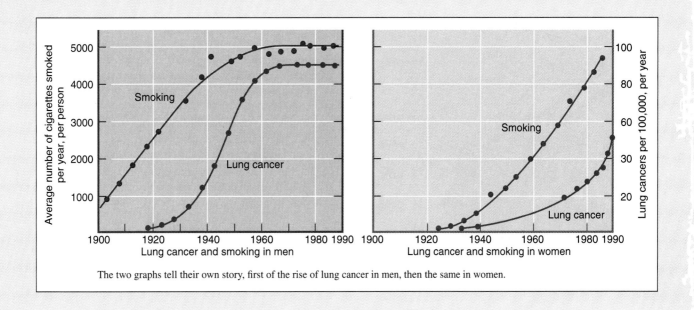

The two graphs tell their own story, first of the rise of lung cancer in men, then the same in women.

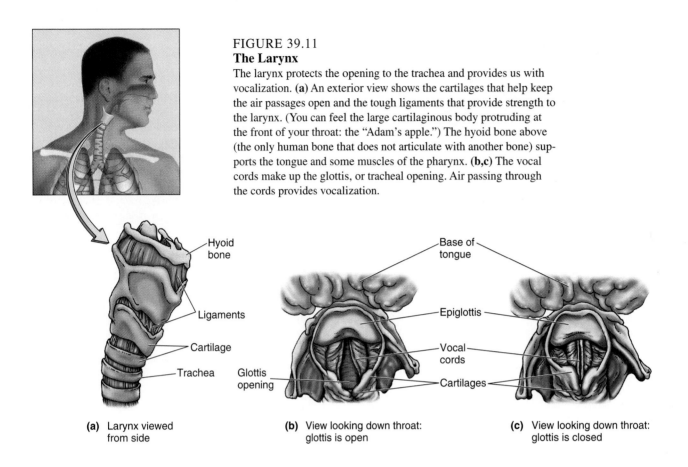

FIGURE 39.11
The Larynx
The larynx protects the opening to the trachea and provides us with vocalization. **(a)** An exterior view shows the cartilages that help keep the air passages open and the tough ligaments that provide strength to the larynx. (You can feel the large cartilaginous body protruding at the front of your throat: the "Adam's apple.") The hyoid bone above (the only human bone that does not articulate with another bone) supports the tongue and some muscles of the pharynx. **(b,c)** The vocal cords make up the glottis, or tracheal opening. Air passing through the cords provides vocalization.

(a) Larynx viewed from side

(b) View looking down throat: glottis is open

(c) View looking down throat: glottis is closed

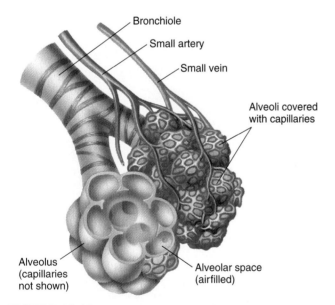

FIGURE 39.12
The Alveoli
In this view, the structure of the alveoli and the relationship of the circulatory system are clearly shown. The bronchioles branch into clustered alveoli that form the ends of the respiratory tree. Here a pulmonary arteriole (blue), a branch of the pulmonary artery, forms capillary beds that spread over the alveolar surfaces. The capillaries then merge to form pulmonary venules (red) that join others to form the pulmonary veins that carry oxygenated blood back to the heart.

smoking and cancer.) Without the cleansing action, the lungs of heavy smokers, and people subjected to heavily polluted air, are blackened and clogged by soot and dust particles. This greatly reduces the exchange of gases so vital to life.

From the bronchioles, air enters its destination, the clustered alveoli. Each alveolus is enclosed in a dense capillary bed, and here the atmosphere and blood are only two membranes apart (those of the alveolus and the capillary) (Figure 39.12). The alveoli provide an enormous respiratory interface. In fact, the total surface area of the approximatley 300 million alveoli of the human lungs has been estimated at nearly 70 m^2 (750 ft^2), or about the area of a tennis court. The alveolus has its own means of cleaning up particulate matter that might escape the mucus secretions and sweeping cilia of bronchioles. Phagocytes housed there patrol the surfaces, taking in any foreign matter. Unfortunately, lung phagocytes are killed by cigarette smoke.

The Breathing Movements

Breathing, or ventilation, in humans is brought about by activity in the muscular diaphragm and the muscles of the rib cage and abdomen (Figure 39.13). In the relaxed condition, the diaphragm rises into a dome shape and protrudes into the thoracic (chest) cavity; the rib cage is in its lowered position. During inspiration, the diaphragm is contracted, drawing it downward and flattening its otherwise dome shape. At the

FIGURE 39.13
Breathing Movements
(a) At inspiration, the external rib muscles contract, elevating the rib cage. The diaphragm muscles contract, drawing the dome downward. Both movements expand the thoracic cavity, drawing the lungs outward all the way around. The expansion decreases air pressure in the alveolar spaces, which, in turn, fill with air forced into the respiratory tree by higher atmospheric pressure outside the body. (b) At expiration, the muscles relax, and the diaphragm resumes its dome-like shape. This compresses the thoracic cavity, and the lungs respond by collapsing. The increased pressure forces air out of the lungs.

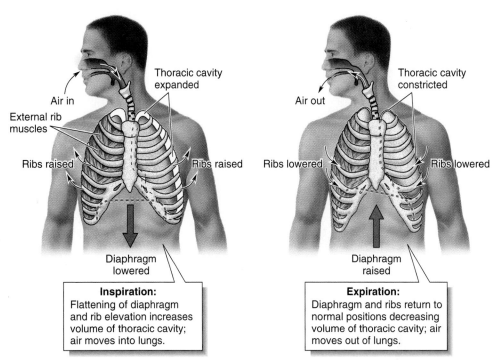

Inspiration:
Flattening of diaphragm and rib elevation increases volume of thoracic cavity; air moves into lungs.

Expiration:
Diaphragm and ribs return to normal positions decreasing volume of thoracic cavity; air moves out of lungs.

same time, the external rib muscles contract, which elevates the rib cage. The two actions expand the airtight thoracic cavity, creating a decrease in air pressure there. This, in turn, pulls on the walls of the lungs, thus expanding them and decreasing air pressure in the alveoli. Since air pressure is greater outside the body, air moves into the respiratory passages, inflating the lung. During inspiration then, it is a decrease in air pressure in the lungs that brings in air from the outside. The lungs themselves are passive in all of this; they simply respond by filling and emptying as pressures dictate.

Expiration, when passive (not forced), is produced by the relaxation of the rib cage and diaphragm, actions that decrease the volume of the thoracic cavity. The elastic lungs rebound, resuming their former volume, and air, now in a somewhat compressed state, is forced out through the respiratory passages. During forced or labored breathing, expiration is not at all passive. It involves contraction of the *internal* rib muscles and the abdominal muscles. The first forces the rib cage downward, and the second forces the diaphragm upward. The resulting movement is much greater than in passive exhalation and far more air is moved.

While we are at rest, about half a liter of air is moved into the lungs with each breath we take. But when we fill our lungs to their greatest capacity, about 4 liters are moved in. This maximal measurement, the **vital capacity**, depends mainly on body size and varies considerably among individuals and between the sexes. In a well-trained male athlete, the vital capacity can exceed 6 liters. But no matter how hard you try to expel all the air from the lungs, about 1.5 liters always remain. This **residual air** is highly significant since the carbon dioxide it contains is essential in maintaining an adequate

respiratory rate. Even in the super-efficient bird lung, the main bronchus, by virtue of its great length, helps retain a critical amount of carbon dioxide in the respiratory system. We'll return to the role of carbon dioxide in respiratory control later in the chapter.

The Exchange of Gases

Partial Pressure. To understand some of the basic aspects of the transport of oxygen and carbon dioxide, it is helpful to know some things about the behavior and characteristics of gases in general. As you know, our atmosphere is composed of a mixture of gases—primarily nitrogen and oxygen, with a much smaller amount of carbon dioxide and other gases. In total, the gases of the atmosphere exert a pressure—the total atmospheric pressure. The pressure exerted by any one gas in the mixture constitutes what is called a **partial pressure**, designated as P_g (the letter g represents an undesignated gas). At sea level, and under what chemists call **standard conditions**, the total atmospheric pressure is known to be 760 mm Hg (mercury). For instance, air is about 21% O_2; therefore the partial pressure of O_2 at sea level would be 21% of 760, or 160 mm Hg. The shorthand designation is: $P_{O_2} = 160$. Carbon dioxide gas accounts for only about 0.04% of the atmosphere; thus it exerts a partial pressure of 0.3 mm Hg (0.04% of 760), or $P_{CO_2} = 0.3$. Most of the remaining total atmospheric pressure is due to nitrogen gas.

Since atmospheric pressure decreases with altitude, so do the partial pressures of the atmospheric gases. On a mountaintop, the air is still 21% O_2 and 0.04% CO_2, but the total atmospheric pressure is much less than at sea level. (At 14,000 feet,

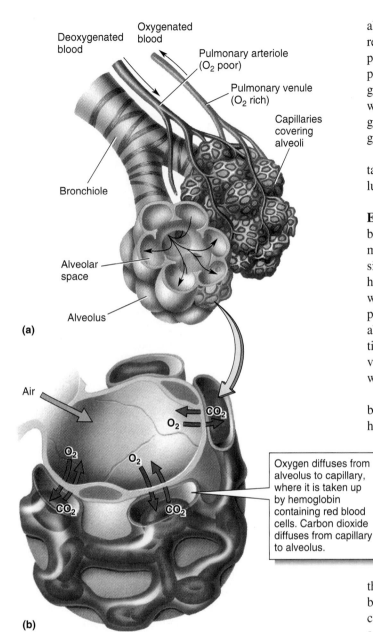

Deoxygenated blood

Oxygenated blood

Pulmonary arteriole (O$_2$ poor)

Pulmonary venule (O$_2$ rich)

Capillaries covering alveoli

Bronchiole

Alveolar space

Alveolus

(a)

Air

CO$_2$

O$_2$

O$_2$

O$_2$

CO$_2$

CO$_2$

Oxygen diffuses from alveolus to capillary, where it is taken up by hemoglobin containing red blood cells. Carbon dioxide diffuses from capillary to alveolus.

(b)

FIGURE 39.14
Gas Exchange in the Alveolus
(**a**) The intimate association of the alveolar capillaries and the alveolus is seen. Incoming arterial blood (blue, above) has a high CO$_2$ content and a low O$_2$ content. As blood passes across the alveolar surface, the two gases diffuse down their gradients, carbon dioxide (blue) escaping and oxygen (red) entering. (**b**) The exchange is shown in an enlargement depicting one alveolus. Here oxygen diffuses from the alveolar space into red blood cells and carbon dioxide diffuses out of the plasma and into the alveolar space.

the total atmospheric pressure may be only 450 mm Hg; thus PO$_2$ would be only 95 mm, and PCO$_2$ would be only 0.18 mm. For this reason, we must breathe faster and deeper up there.)

The concept of partial pressure is important to our understanding of gas exchange because diffusion of dissolved gases always occurs from regions of higher partial pressure to regions of lower partial pressure. For instance, suppose we placed an open container of a fluid that was rich in CO$_2$ and poor in O$_2$ (as measured by the partial pressures of these two gases) inside a closed container with a mixture of gases that was rich in O$_2$ and poor in CO$_2$. In due time the fluid and the gas would equilibrate, so that the partial pressure of both gases would be the same in both the fluid and the air.

We find then that partial pressure differences are important in setting up diffusion gradients, but, as we will see, evolution has added a few twists to the simple diffusion of gases.

Exchange in the Alveoli. We breathe in air rich in O$_2$ and breathe out air rich in CO$_2$. The exchange takes place on the moist surfaces of the alveoli, primarily through simple diffusion (Figure 39.14). The blood that enters the lung from the heart has previously been routed through the body tissues, where mitochondrial respiration has depleted the oxygen supply, creating a very low partial pressure of oxygen—a PO$_2$ of about 40. The same metabolic activity has increased the partial pressure of CO$_2$ in the body tissues, and carbon dioxide in venous blood leaving the tissues and entering the alveoli would have a partial pressure of about 45 mm Hg.

So compared to partial pressures in the atmosphere, blood entering the lungs has a low partial pressure of O$_2$ and a high partial pressure of CO$_2$. Conversely, in the air spaces of the alveoli, the partial pressure of oxygen is high—about 100 mm Hg—and the partial pressure of carbon dioxide is about 40 mm Hg. (The high partial pressure of carbon dioxide in the alveoli occurs because alveolar air is not completely replaced during inhalation but is a mixture of old and new air. Recall that this leftover air is called residual air. In the brief time that the blood and air are in near contact on opposite sides of the thin alveolar membrane, nearly complete equilibrium takes place. That is, the partial pressures of O$_2$ and CO$_2$ in the blood and air become almost equal as molecules of O$_2$ diffuse in and molecules of CO$_2$ diffuse out. Then the equilibrated air is exhaled, and fresh air is inhaled while the blood flows continuously through the lung. Blood leaving the lung is relatively high in oxygen, with a partial pressure of about 105 mm Hg. The corresponding partial pressure of carbon dioxide is down to about 40 mm Hg.

The partial pressure of oxygen in metabolically active tissue shifts dramatically as oxygen is consumed. There its partial pressure may fall to 25 mm Hg. The partial pressure of carbon dioxide, as you would expect, increases, perhaps reaching 46 mm Hg. Figure 39.15 illustrates this complex situation.

Thus the respiratory gases diffuse down their pressure or concentration gradient, and the organism takes advantage of this "free" means of exchange to keep its oxygen and carbon dioxide at optimal levels. For some organisms, this is about all there is to respiration, but for many, including the vertebrates, there is more. Passive diffusion is quite important, but it is only a part of the total gas exchange process. Other aspects of the process include complex biochemical mechanisms.

Oxygen Transport

Because red blood cells are literally packed with hemoglobin, they are well adapted for the transport of oxygen. So what precisely is this special molecule called hemoglobin, and how does it work? Hemoglobin consists of four polypeptide chains, each containing a heme group (Figure 39.16). Each heme group contains an iron atom that can bind with one molecule of oxygen (O_2). Thus each hemoglobin molecule can hold four oxygen molecules when fully saturated. The association and dissociation of O_2 and hemoglobin (Hb) are usually simplified as:

$$4\,Hb + 4O_2 \rightleftharpoons 4\,HbO_2$$

The left side of the formula shows hemoglobin in its deoxygenated form and the right side shows the oxygenated form, or **oxyhemoglobin**.

One key to the efficiency of our respiratory system is the behavior of hemoglobin itself. Hemoglobin is highly specialized to associate and dissociate with oxygen under certain conditions. Whereas the affinity of water for O_2 is simply dictated by partial pressure and temperature, hemoglobin has the remarkable ability to "change its mind" about how much oxygen it will accept. To put it a bit more scientifically, the affinity of hemoglobin for oxygen is affected by pH and the partial pressure of CO_2. A variable affinity for oxygen allows the hemoglobin to give up much of its oxygen in its journey through the body and to become quickly saturated with oxygen again in the lungs. Let's see how local conditions in the cells and tissues affect the role of hemoglobin.

Bohr Effect. Although the concentration of oxygen is a principal factor in determining how tightly oxygen binds to hemoglobin, there are others. One of these, the **Bohr effect**, involves acidity. Any time the red blood cells encounter more acidic conditions in tissues and tissue fluids, their hemoglobin loses its affinity for O_2. The acidity of the tissues and tissue fluid is directly proportional to metabolic activity. Two principal contributors to acidity are carbon dioxide, which forms carbonic acid in the cell and tissue fluids, and the presence of DPG (diphosphoglycerate), a 3-carbon acid produced during

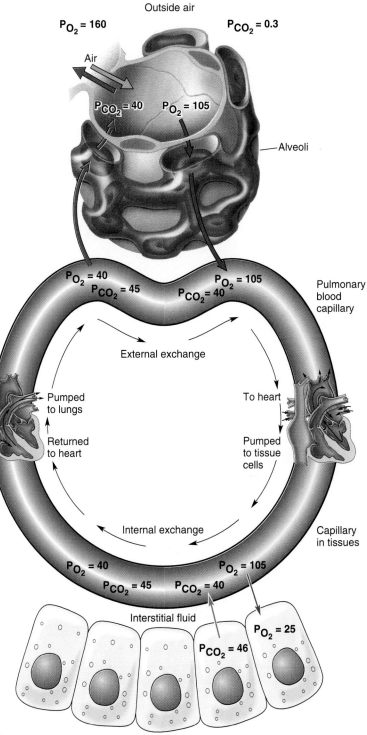

FIGURE 39.15
Partial Pressures of Oxygen and Carbon Dioxide in the Body
Partial pressures are compared in the outside air (*top*), alveoli, blood entering the lungs, blood leaving the lungs, and blood in metabolically active tissues (*bottom*). The greatest partial pressure of oxygen in the body is within the alveolar spaces (*top*). Thus oxygen diffuses across the alveolar membranes into the capillaries. Blood from the lungs contains the highest oxygen partial pressure in the bloodstream. Upon arriving at active tissues (*bottom*), where oxygen partial pressures are lowest in the body, oxygen dissociates from hemoglobin, entering those tissues. Carbon dioxide's partial pressure is highest in the active tissues, so it diffuses out of the cells and into the bloodstream. The venous bloodstream (*left*) thus contains the highest partial pressures of carbon dioxide in the body. The blood eventually reaches the alveoli of the lungs, whose spaces contain the lowest carbon dioxide partial pressures in the body. Carbon dioxide thus leaves the blood and enters the alveolar spaces for exhalation (*top*).

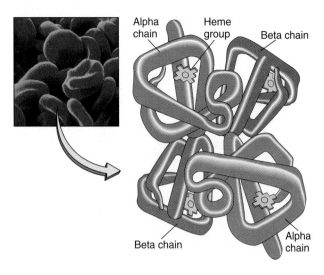

FIGURE 39.16
Hemoglobin
The hemoglobin molecule consists of two alpha- and two beta-globin chains. Each unit has its own heme group, and each can bind with one molecule of oxygen.

glycolysis (see Chapter 8). Conversely, when cells are less metabolically active, the acidity decreases and hemoglobin remains tightly bound to O_2 (Figure 39.17a).

The Bohr effect has clear adaptive significance. Metabolically active cells have great oxygen demands, and the acidic conditions they produce cause oxygen-rich blood passing by to give up more of its oxygen (Figure 39.17b). On the contrary, blood passing less metabolically active tissue will be in less acidic surroundings and will release less of its oxygen (Figure 39.17c).

We'll consider one more point about the oxygenation of blood. At sea level, blood passing over the alveoli is nearly saturated with O_2; that is, most of the hemoglobin molecules bind to their full allotment of four O_2 molecules. Increasing the O_2 concentration in the lungs, such as by breathing pure oxygen, has very little effect on the amount of oxyhemoglobin formed. There is a considerable safety margin in this relationship because it also works the other way: reducing the partial pressure of oxygen doesn't alter the amount of oxygen saturation in the blood as much as you might expect. At half the partial pressure of O_2 found at sea level (or $P_{O_2} = 80$), the

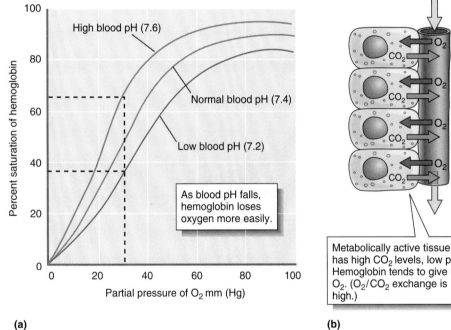

(a)

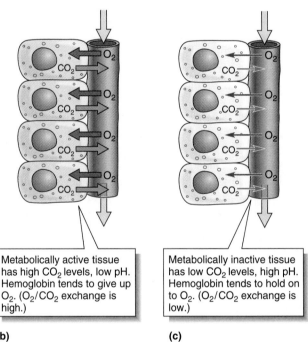

(b)   (c)

FIGURE 39.17
Bohr Effect
(a) The graph shows the effects of pH on the percentage of hemoglobin saturation (vertical axis) at different partial pressures of oxygen (horizontal axis). Where the blood is less acidic (pH is high) it retains more of its oxygen (the steeper curve means the red blood cells are more saturated). But where the blood is more acidic (pH is lower), the Bohr effect takes over and the blood gives up its oxygen (the less steep curve means the red blood cells are less saturated—have lost more oxygen). (b) The Bohr effect is evident in red blood cells passing more metabolically active cells, those that are more acidic (lower pH). The red blood cells give up more oxygen there than in the metabolically inactive tissues. (c) The Bohr effect is not in evidence in blood passing less metabolically active, less acidic, cells.

blood will still become 80% saturated with oxygen as it passes through the lungs. This is why humans can live comfortably at varying altitudes, including high mountains where O_2 partial pressure is low.

Furthermore, at high altitudes, additional red blood cells are produced—a long-term adaptation. The rate of red cell production is controlled through a feedback system originating in the kidney. Certain cells there detect the oxygen content in blood passing by, and when it is not up to a certain level, the hormone **erythropoietin** is released. Its targets are blood-forming elements in red bone marrow, which respond by producing more red blood cells. Added red blood cells mean an increase in blood oxygen and less stimulation of the oxygen-sensing cells, which then slow their erythropoietin release (another example of negative feedback). Now let's look at carbon dioxide transport.

Carbon Dioxide Transport

The transport of carbon dioxide in the blood is far more complex than that of oxygen. First, about 8% of the carbon dioxide to be transported simply goes into solution in the blood plasma. Second, the remaining CO_2 enters the red cells, where it follows one of two paths. Some enters into a loose association with the hemoglobin, but instead of combining with the heme groups as oxygen does, carbon dioxide reacts with amino side groups in other parts of the protein. The combination of carbon dioxide and hemoglobin, called **carbaminohemoglobin**, can be represented as:

$$Hb + CO_2 \rightleftharpoons HbCO_2 \text{ (carbaminohemoglobin)}$$

The remaining carbon dioxide reacts with water in the red cells, forming carbonic acid (H_2CO_3):

$$CO_2 + H_2O \rightleftharpoons H_2CO_3$$

Carbonic acid, in turn, dissociates into hydrogen ions (H^+) and bicarbonate ions (HCO_3^-):

$$H_2CO_3 \rightleftharpoons H^+ + HCO_3^-$$

This reaction can also occur in the plasma, but only very slowly. Red cells contain the enzyme **carbonic anhydrase**, which not only speeds up the formation of carbonic acid but can also rapidly convert carbonic acid back to carbon dioxide and water. This is quite important, because as blood passes through the lungs, the carbon dioxide must quickly reform if it is to diffuse out of the alveoli.

The ions formed through the dissociation of carbonic acid cannot go unattended. When carbonic acid dissociates in the red cells, the hydrogen ions remain in the red cell where they are buffered (pH changes resisted) by the protein hemoglobin itself. The bicarbonate ions diffuse out into the plasma, where they are joined by ever present sodium ions, forming sodium bicarbonate ($NaHCO_3$).

So in addition to providing a means of transporting carbon dioxide, sodium bicarbonate forms an important part of the body's **acid–base buffering system**. It helps neutralize any acids or bases that might form in the blood, keeping the pH constant—near neutral. (This sodium bicarbonate, by the way, is identical to commercial baking soda and to the main ingredient in sodium-based stomach acid neutralizers.) The reactions so far can be summarized as follows:

1. $Hb + CO_2 \rightleftharpoons \text{Carbaminohemoglobin}$

2. $CO_2 + H_2O \rightleftharpoons H^+ + HCO_3^-$
   carbonic anhydrase

3. $Na^+ + HCO_3^- \rightleftharpoons NaHCO_3$

As we see by the arrows, the reactions are all reversible. In each case, it is the local concentration of carbon dioxide that determines the direction of the reaction (typical of chemical reactions; see Chapter 6).

So in active tissues, where carbon dioxide levels are high, the direction of the reactions is toward carbaminohemoglobin and toward the formation of hydrogen and bicarbonate ions and sodium bicarbonate. But in the alveolar capillaries, any free carbon dioxide present diffuses rapidly out of the blood, so the CO_2 concentration becomes low. This prompts a speedy cascade of reversing chemical events:

1. The carbaminohemoglobin releases its carbon dioxide.

2. Sodium bicarbonate in the blood increases its dissociation into sodium and bicarbonate ions, and the bicarbonate ions reenter the red cells.

3. The bicarbonate ions next join hydrogen ions to reform carbonic acid.

4. With a boost from the enzyme carbonic anhydrase, carbonic acid is rapidly converted back to carbon dioxide and water.

5. Much of the carbon dioxide then diffuses out of the capillary and into the alveolus for expiration out of the body.

Figure 39.18 summarizes these complex events.

The Control of Respiration

The never-ending changes that characterize homeostasis, a hallmark of living things, are usually only minor adjustments—fine tuning—in response to shifting conditions within and without. For example, the need for O_2 and the production of CO_2 are always changing in every part of the body—now a little more here, then a little less there. Considering all such adjustments, however, the body's homeostatic task is enormous and its ability to respond is impressive. How can it handle a chore as complex as regulating oxygen and carbon dioxide levels in literally millions of places at once?

First, we should be aware that both the respiratory and circulatory systems must respond in a coordinated way to changing oxygen levels. After all, increasing the breathing rate is of little value without a corresponding increase in the rate of blood flow. Second, respiratory control is exceedingly com-

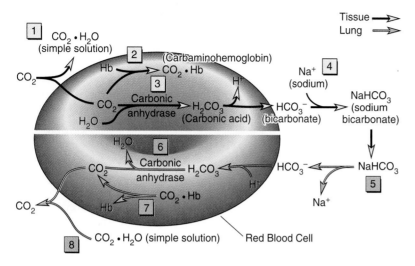

Tissue ➡
Lung ⇒

FIGURE 39.18
The Transport of Carbon Dioxide

The transport of CO_2 begins when it enters the blood as it passes active tissues. ☐1 Some CO_2 enters simple solution with water in the plasma. ☐2 More of it combines with hemoglobin, but ☐3 most is converted by the enzyme carbonic anhydrase to carbonic acid. The acid loses a hydrogen ion and enters the plasma as the bicarbonate ion, where ☐4 it is joined by sodium ion, forming sodium bicarbonate. As the blood reaches the alveoli, the processes are reversed. ☐5 The sodium ion dissociates and the bicarbonate ion enters the red blood cell where it is joined by a hydrogen ion, once more forming carbonic acid. ☐6 The same enzyme breaks the acid down into water and CO_2. At the same instant, hemoglobin gives up its CO_2 ☐7 , as does the plasma solution outside, and the waste gas diffuses ☐8 out of the blood into the alveolar space for exhalation.

FIGURE 39.19
Work of the Breathing Control Centers

(a) When the body is at rest and breathing is relaxed, the inspiratory center works alone, bringing on inspiration at two-second intervals, and remaining inactive for three-second intervals. (b) When demands increase, as during exercise, the expiratory center is activated. It activates other muscles, resulting in more forceful exhalation.

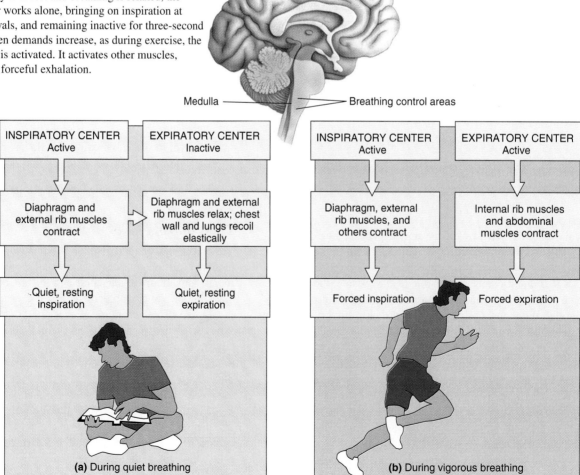

plex, involving several types of sensors and a variety of neural responses. We will look briefly at the control of breathing.

Control of Breathing. We can consciously control the rate and depth of our breathing, but only up to a point. If your little brother holds his breath to get his way, don't worry. He may begin to lose his rosy complexion, but the ruse won't work. No matter how hard he tries, the rising CO_2 level in his blood will prompt his autonomic nervous system into action, and he will be forced to breathe. To explain this, we need a look at both the neural and chemical aspects.

Breathing Centers. It is known that breathing movements are coordinated by voluntary centers in the cortex (those that permit your little brother to make the threats about holding his breath) and by involuntary centers in the pons and medulla (those that defeat his strategy). While the anatomy of the involuntary centers is still under investigation, there appear to be **inspiratory centers** and **expiratory centers**, both located in the medulla. The inspiratory center regulates breathing during quieter, relaxed periods, while the expiratory center operates in periods of strenuous breathing.

During quiet breathing, the inspiratory center is self-excited, creating impulses on its own. Its impulses pass through the spinal cord to the diaphragm and external rib muscles, which contract and bring on inspiration. After about two seconds of activity, the inspiratory center spontaneously rests for three seconds, the muscles relax, and expiration occurs. As you can see, expiration at this time is a passive process.

During periods of more strenuous activity, the breathing rate increases sharply. Whereas the inspiratory center still brings on inspiration, now at a faster rate, other factors come into play. For instance, stretch receptors in the lungs—activated by deeper inspiration—fire inhibiting signals back to the inspiratory center, thus stopping inspiration and preventing overinflation. With the inspiratory center held in check, the expiratory center can act. It sends its messages to internal rib muscles and abdominal muscles, whose contraction adds considerable force to expiration, thus increasing the expulsion of air from the lungs. Figure 39.19 summarizes the respiratory center's work.

Chemoreceptors. The breathing centers receive neural input from a number of chemoreceptors in the body. Chemoreceptors monitor the blood's carbon dioxide and oxygen levels and its pH. Two of the chemoreceptors are in major arteries and one is in the brain (Figure 39.20).

The **carotid bodies** and **aortic bodies**, located in the carotid arteries and aorta, respectively, are aroused when blood CO_2 levels rise and pH decreases (acidity increases). They also sense decreases in oxygen, but, surprisingly, to a much lesser extent. Sensitive chemoreceptors in the medulla oblongata in the brain monitor pH levels in the cerebrospinal fluid, becoming active when blood pH falls. When neurons in any of these regions become active, their impulses are relayed to the respiratory centers in the pons and medulla, and the rate

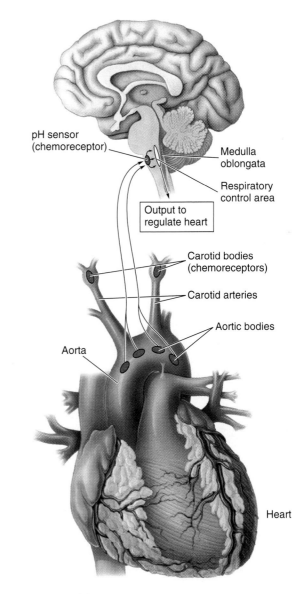

FIGURE 39.20
Respiratory Control
Chemosensors in the medulla oblongata and in the carotid and aortic bodies monitor the chemical state of the blood. They activate the respiratory control center when blood pH falls, which happens when carbon dioxide levels rise; or to a lesser extent when oxygen levels fall beyond a set point. Their activity is quelled when the chemical state of the blood returns to a resting level.

and depth of breathing increase. Such breathing, of course, decreases the levels of CO_2 and acid in the blood, and increases the blood's oxygen content. This produces a negative feedback effect, and the chemoreceptors slow their output of neural impulses to the breathing centers. Thus a balance is set that retains the blood in an optimum chemical state and assures that breathing rates will meet changing demands. Interestingly, during vigorous exercise this straightforward chemical regulatory mechanism is supplemented by new input from the cerebrum and input from other kinds of sensors, possibly proprioceptors, located in the joints and muscles.

You may have noticed that our discussion of gas exchange has often led to mention of the circulatory and nervous systems. It is quite difficult, if not impossible, to discuss one system to the exclusion of all others. In large part this is because animal systems function in an integrated and interdependent manner, and no system functions alone. As we turn to the immune system in the next chapter, we will again weave in our findings from other discussions as we continue to try to understand the basic fabric of life.

Key Ideas

1. Oxygen is essential in aerobic life as an electron acceptor in cell respiration, a process that also produces carbon dioxide as a waste. Oxygen and carbon dioxide are exchanged in the **respiratory system** and transported in the circulatory system.

GAS EXCHANGE SURFACES

1. Gas exchange in animals requires extensive, thin, moist membranes, or **respiratory interfaces**, such as the exchange surfaces of the gill and lung. Moisture is critical since gases must go into solution to cross the membrane.

2. In complex sponges, gas exchange across the body interface works well because of the extensive canal system. In cnidarians the extremely thin (two-cell layer) body wall is the exchange surface. The body surface and the highly branched gastrovascular cavity provide the exchange surfaces in flatworms. In echinoderms, the surface is increased by **dermal brancheae**.

3. Their moist environments and their closed circulatory systems containing hemoglobin make it possible for the earthworm and salamander to adequately exchange gases through the skin interface.

4. Most terrestrial animals have internalized respiratory interfaces that help resist desiccation. Insects have a **tracheal system**, which includes highly branched **tracheae** and thin-walled **air sacs**. Gas exchange occurs in the sacs and fine branches. The tracheae connect to external **tracheal gills** in aquatic insects.

5. In the **gill**, the large respiratory interface is intimately associated with a circulatory system.

6. The lobster gills lie in chambers below a protective **carapace**. The circulatory system is open, but blood vessels occur in the gills. Oxygen transport is aided by the copper-containing protein **hemocyanin**.

7. The fish gill include rows of **gill filaments**, each containing many capillary beds in plate-like **lamellae**. CO_2 is exchanged for O_2 in the lamellar capillaries. Water and blood flow in opposite directions: an efficient **countercurrent gas exchange**.

8. The earliest fishes used a vascularized **pharynx** for gas exchange. This system evolved into the lung. In mammals, the **palate** separates the pharynx into mouth and nasal cavities. In fishes, the primitive lung evolved into the **swim bladder**, a flotation device.

9. The vertebrate lung is a highly branched inpocketing that, in many species, contains numerous air sacs, intimately associated with capillaries, and inflated and deflated by a bellows action.

10. The amphibian lung is a simple paired sac-like affair, inflated and deflated by a pumping action in the muscular mouth floor. Amphibians also use the skin as an exchange surface.

11. Reptile lungs are more spongy and complex, representing the respiratory interface in most. Breathing occurs through a bellows-like action of the entire body wall.

12. In birds, air passes *through* the lung. Air first enters posterior air sacs and from there enters the lung, and then the anterior air sacs. The bird respiratory system represents a specific adaptation to flight at higher altitudes, where oxygen is limited.

THE HUMAN RESPIRATORY SYSTEM

1. Major structures of the human respiratory system include the mouth, nasal passages, pharynx, **larynx**, **trachea**, and **lungs**. The lungs include the **bronchi**, **bronchioles**, **alveoli**, and the blood vessels of the pulmonary circuit. Coverings called pleurae form an airtight chamber around the lungs.

2. The nasal passages moisten, warm, and filter air entering the lungs. Much of the respiratory system contains a mucus-secreting and ciliated epithelium that traps dust particles and sweeps them out of the system for swallowing.

3. The lungs lie in the **thoracic cavity,** enclosed by double membranes, the **pleurae**. The dome-shaped **diaphragm** forms the lower part of the cavity.

4. In inspiration, outermost rib muscles contract, elevating the rib cage, and the diaphragm flattens, reducing pressure in the thoracic cavity. The lungs passively expand in response to the inward movement of air. In exhalation, the rib cage drops, the diaphragm resumes its relaxed dome shape, and the elastic lungs resume their former shape, forcing air out.

5. While the minimal (resting) exchange of air is about 0.5 liters, the maximum, or **vital capacity**, is about 4.0–6.0 liters. Some **residual air** always remains in the lungs.

6. While all gases of the atmosphere contribute to total atmospheric pressure, each gas exerts a **partial pressure**—P_g. P_g decreases with altitude.

7. The diffusion of a gas follows its partial pressure gradient, with oxygen diffusing into the blood and carbon dioxide out of the blood.

8. Each of hemoglobin's four heme groups reversibly associates with one O_2, forming **oxyhemoglobin**. The association and dissociation of hemoglobin and oxygen are complex, depending on pH, P_{CO_2}, and temperature.

9. At sea level the blood leaving the lungs is nearly saturated with oxygen, and increasing the Po_2 has little effect. But at half the sea-level value of Po_2 (high altitude), the blood will be 80% saturated. In addition, when a consistently low Po_2 occurs in the blood, the hormone **erythropoietin** is released, stimulating new red blood cell formation, a slower, long-term adjustment.

10. In the **Bohr effect**, increased acidity in the tissues, a product of increased metabolic activity, hastens the dissociation of oxygen from oxyhemoglobin.

11. In the transport of CO_2, some 8% goes into solution in the plasma, the remainder enters the red blood cells, where some forms carbaminohemoglobin, and some forms carbonic acid, which dissociates into its ions. The enzyme carbonic anhydrase speeds the dissociation and reassociation of carbonic acid, the direction depending on the concentration of CO_2. The carbonate ions enter the plasma, where they join with Na^+, forming sodium bicarbonate. This acts as an acid–base buffer.

12. The CO_2 reactions are all reversible, going toward sodium bicarbonate in the tissues and toward CO_2 in the lungs.

13. When the body is at rest, the inspiratory center of the medulla stimulates rhythmic inhalation in an off–on manner. During vigorous exercise, the inspiratory center becomes more active. Stretch receptors in the lungs inhibit the inspiratory center, permitting expiration. The expiratory center activates inner rib muscles and abdominal muscles, causing a greater expulsion of air.

14. The carotid and aortic bodies detect changes in CO_2 and O_2 levels in the blood, whereas chemoreceptors in the medulla detect decreasing pH. Their neural signals are relayed to the breathing centers, where activity is increased.

Application of Ideas

1. All aquatic mammals, birds, and reptiles are air breathers. No matter how well they have adapted to the water in other ways, they must come to the surface to breathe. What does this clearly indicate about their ancestors? Given enough time, is it probable that these animals would evolve a gill-like device? Explain your conclusion. What is one physiological disadvantage of gills in an aquatic endotherm?

Review Questions

1. List two basic requirements of a gas exchange surface. What problems does this suggest for terrestrial animals? (page 833)

2. List three invertebrates that use the simple body interface for gas exchange, and explain how each satisfies the basic requirements listed above. (page 833)

3. Earthworms and lungless salamanders make use of the skin for gas exchange. What two additional factors make this possible in such dense animals? (page 833, 836)

4. Describe the insect tracheal system. How does the relationship of respiration and circulation differ from that of the earthworm and salamander? (page 836)

5. List the structures making up the fish gill. Explain how a countercurrent flow occurs in the fish gill. What is there about the aquatic environment that makes a countercurrent gill flow advantageous? (page 837)

6. Compare amphibian and reptilian lungs. Which provides the greater surface area? How does the amphibian supplement its limited lung surface? (page 840)

7. List several important differences between the bird respiratory system and that of reptiles and mammals. What does the bird's efficient respiratory system permit it to do that other vertebrates cannot do? (pages 841–843)

8. List all the structures through which air passes during inhalation in humans. (page 843)

9. With the aid of a drawing, show the relationship of the alveolus to the pulmonary capillaries. Add arrows to show the movement of CO_2 and O_2. (page 846)

10. Sum up the movements of the rib cage, diaphragm, and lungs during inspiration and expiration. What actually causes the lungs to fill and empty? (pages 846–847)

11. To what does the "partial" in partial pressure refer? What happens to partial pressures at higher altitudes? Does altitude affect the percentage of each gas? (pages 847–848)

12. State the partial pressures of oxygen and carbon dioxide in capillaries entering and leaving active tissues. Do the same for capillaries entering and leaving an alveolus. (Figure 39.15)

13. Write the formula representing the reaction between oxygen and hemoglobin. Do they form strong chemical bonds? Why is this important? (page 849)

14. What is the Bohr effect? How is it adaptive? (pages 849–850)

15. Describe what happens to carbon dioxide when it enters a red blood cell. What is the role of carbonic anhydrase? What determines in which direction the reactions go? (page 851)

16. List the changes that occur during carbon dioxide transport when blood reaches the alveolus. (page 851)

17. Describe the workings of the inspiratory center when the body is at rest. What two additional factors in breathing control come into play when we exercise vigorously? (pages 851–853)

18. What do the carotid and aortic bodies monitor in passing blood? What do chemical sensors in the medulla monitor? (page 853)

19. What causes the pH of the blood to decrease? How does this affect the chemical sensors, and what response does the body make? (page 853)

CHAPTER 40

The Immune System

CHAPTER OUTLINE

NONSPECIFIC DEFENSES

Nonspecific Primary Defenses:
The Body Coverings

Nonspecific Chemical Defenses

Nonspecific Cellular Defenses

Resistance in the Invaders

SPECIFIC DEFENSES

Lymphoid Tissues and the B- and T-Cell
Lymphocytes

Antibodies and the Humoral Response

T-Cells and Cell Recognition

The Mature B-Cell and Its Receptor

**CLONAL SELECTION AND THE PRIMARY
IMMUNE RESPONSE**

Clonal Selection:
Antigen-Presenting Macrophages

Spreading the Alarm: Aroused T-Cells

B-Cells and the Antibody Attack

Suppressor T-Cells: The Battle Is Won

Summing Up the Primary Immune Response

**VIGILANT MEMORY CELLS AND THE
SECONDARY IMMUNE RESPONSE**

Active and Passive Immunity

LYMPHOCYTE DIVERSITY

MONOCLONAL ANTIBODIES

WHEN THE IMMUNE SYSTEM GOES WRONG

Autoimmunity: Attack Against Self

Allergies

AIDS: The Crippled Immune System

KEY IDEAS

REVIEW QUESTIONS

*I*n every living cell there is a veritable storehouse of energy-rich nutrients. In the natural world, no nutrient source escapes the attention of other living things for long. And so we find various life forms attacking others simply to fulfill their own energy needs. The attack can occur on a larger scale, as when a tiger seal avails itself of the energy stored in a penguin, or it can be in the microcosm when a pathogen, such as a virus, attacks a human liver cell.

Here we will focus on the various ways in which living things defend themselves against the smaller invaders. Our focus will be on viruses and bacteria, but we will be mentioning pathogenic protists, fungi and, strangely enough, even roundworms. We will see that invaders can do harm in many ways. Some viruses rob organisms of vital cellular components, literally wrecking a cell when they reproduce. Others disrupt gene function by inserting their own DNA into the host's chromosomes. Some bacteria secrete enzymes that break down living tissues and produce toxins (poisons) that weaken and sometimes kill the host. At best, the pathogen leaves the weakened host open to still other invaders. Yet living things are not defenseless against pathogenic invasion. In fact, they are protected by a most remarkable defensive system.

In mammals, the **immune system** consists of an array of specialized cells, tissues, and organs, distributed throughout in the body (Figure 40.1). Its primary function is to protect the body against invading organisms and the harmful substances they produce. Much of what is known about the immune system pertains to humans and a few other mammals, although some information about immunity in other animal groups does exist. We find, for example, that certain defense processes occur across the spectrum of animal life. These defenses take two forms, grouped into what are called *nonspecific* and *specific defenses*. For a preview and a ready reference for later, see Table 40.1.

TABLE 40.1
Defenses of the Immune System

Nonspecific Defenses

First-line defense	Skin
	Mucous membranes: secrete mucus—a trapping agent
	Sweat, lysozyme, and other toxic body secretions
	Body flora: harmless bacteria compete with invaders
Chemical defenses	Histamines: inflammatory agents, aid work of other agents, induce fever
	Kinins: inflammatory agents, aid other agents
	Complement: antibacterial agent
	Interferons: antiviral agents
Cellular defenses	Phagocytes: engulf bacterial cells and debris; include eosinophils, neutrophils, and monocytes
	Basophils: secrete inflammatory agents
	Natural killer cells: nonspecific lymphocytes, antiviral and anticancer agents

Specific Defenses

Chemical (humoral) response	Antibodies secreted by plasma B-cells, aid phagocytes
	Memory B-cells: act as a ready reserve
Cellular (cell-mediated) response	Helper T-cells; activate other components of specific defense
	Cytotoxic T-cells kill infected and cancerous cells
	Suppressor T-cells: slow immune response
	Memory T-cells: act as a ready reserve

Nonspecific Defenses

Nonspecific defenses involve physical barriers and chemical and cellular responses that are general in nature—not directed at a specific invader. Such defenses are always in a fully prepared state, needing only to be triggered into action. **Specific defenses** are not in an immediate state of readiness. Furthermore, they are highly precise, requiring molecular identification of the intruder and involving a much more time-consuming arousal process. Specific defenses operate against antigens, those carried on the surfaces of invaders and those free in the body.

The nonspecific response can be likened to front-line defenses that include frontier barriers, behind which reside fast-reacting military counterstrike forces, those capable of fast but blunt counterattack. The specific response is more like a reserve force of specialists, one waiting to be directed at precise weak points in the aggressor's attack. We'll focus first on the barriers, that part of the nonspecific defense that keeps invaders out.

Nonspecific Primary Defenses: The Body Coverings

The primary defenses in our own bodies and those of many other animals include the epithelial tissues that make up the body coverings. In humans and other animals, skin covering the outer surfaces and mucous membranes lining the internal surfaces provide an effective barrier, aided in this by a host of friendly microorganisms.

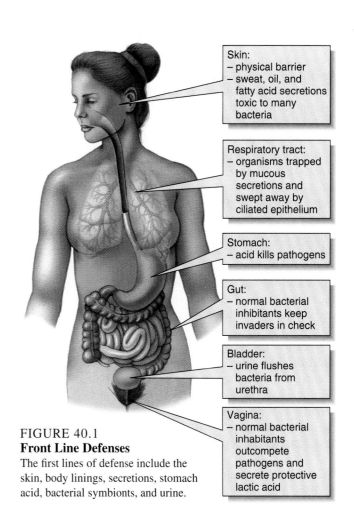

Skin:
– physical barrier
– sweat, oil, and fatty acid secretions toxic to many bacteria

Respiratory tract:
– organisms trapped by mucous secretions and swept away by ciliated epithelium

Stomach:
– acid kills pathogens

Gut:
– normal bacterial inhibitants keep invaders in check

Bladder:
– urine flushes bacteria from urethra

Vagina:
– normal bacterial inhibitants outcompete pathogens and secrete protective lactic acid

FIGURE 40.1
Front Line Defenses
The first lines of defense include the skin, body linings, secretions, stomach acid, bacterial symbionts, and urine.

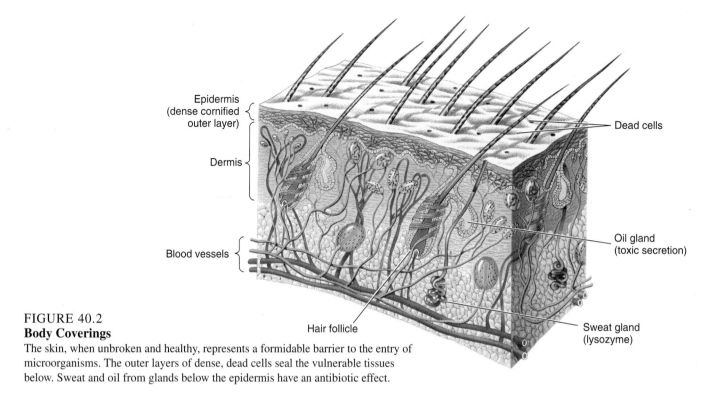

Epidermis
(dense cornified
outer layer)

Dermis

Blood vessels

Dead cells

Oil gland
(toxic secretion)

Hair follicle

Sweat gland
(lysozyme)

FIGURE 40.2
Body Coverings
The skin, when unbroken and healthy, represents a formidable barrier to the entry of microorganisms. The outer layers of dense, dead cells seal the vulnerable tissues below. Sweat and oil from glands below the epidermis have an antibiotic effect.

The Skin. The skin has two interesting defenses. In a sense we might say that it is tough and toxic. It is tough because its outer region, the epidermis, contains dense layers of dead, keratinized cells that are constantly replaced by an actively dividing cell layer below (Figure 40.2). Keratin alone resists the potent enzymes of would-be bacterial invaders. Dry, unbroken skin provides the best defense. When the skin is constantly wet or when it is damaged, this defense begins to break down. The toxic nature of the skin is to be seen in the effects of its fatty acids on certain bacteria and the secretions from sweat and oil glands.

Mucous Membranes. The internalized surfaces, those of digestive, respiratory, and reproductive passages, have effective defenses in the form of mucous membranes. You will recall that the mucous membrane lining the respiratory passages (see Figure 39.10) secretes a layer of mucus that traps microorganisms, holding them until they can be swept away by ciliary motion and swallowed. Those organisms that escape this defense and enter the alveoli of the lungs are met there by resident phagocytic cells that engulf and digest them (see Chapter 39). Organisms in the food we eat, or those trapped in respiratory mucus, find themselves in the hostile, acidic environment of the stomach, where many quickly die. The urinary passages, another potential avenue of invasion, are naturally protected from bacteria by the flushing action of urine.

Friendly Microorganisms. Finally, we get a little help from friendly microorganisms. Throughout our own evolution, a variety of bacterial, fungal, and even animal species have adapted to living harmlessly on the body surfaces. Whereas they are no threat to us, they and their secretions do present a competitive and toxic barrier to parasites. The skin contains a variety of such harmless organisms as hair follicle mites (arthropods), yeasts (fungi), bacteria of the bacillus and coccus form, and others. The large intestine contains massive colonies of friendly colon bacteria, such as *Escherichia coli,* that keep invaders in check by essentially starving them out. (Yet some strains of *E. coli* can become highly toxic to us if they are permitted to grow in our food. Even the innocuous strains can cause trouble should they penetrate the colon wall or make their way into the urethra.) Populations of harmless bacteria in the vagina feed on glycogen secretions there and secrete lactic acid that renders the vagina inhospitable to many other organisms. These symbiotic defenses, however, are often lost after prolonged antibiotic therapy in which the useful bacteria are killed. With the competition down, invaders may then establish themselves, producing stubborn infections in the mouth, urethra, and vagina. So an antibiotic may solve one problem, only to create others.

As effective as the body linings are, they can be penetrated. When microorganisms penetrate the primary defenses, secondary defenses come into play.

Nonspecific Chemical Defenses

In the **nonspecific chemical defenses**, injured cells and tissues release or activate chemicals that help destroy the intruder or impede its progress. We'll look into five of these: *lysozymes, histamines, kinins, complement,* and *interferons.*

Lysozymes. The **lysozymes** are a group of antibacterial enzymes that occur in tears and other body secretions. They quickly kill some types of bacteria by breaking down their cell walls, thus eliminating their ability to resist the osmotic uptake of water. As a result, water enters until the cell ruptures.

Histamine, Kinins, and the Inflammatory Response. The **inflammatory response** (Figure 40.3) includes the familiar redness and swelling of tissues at an infection or injury site, so it is quite visible. A primary agent in the inflammatory response is **histamine**, a modified version of the amino acid histidine. (*Antihistamines* are agents that reduce the inflammation.) Histamines are released by damaged basophils (a type of white blood cell) and mast cells (connective tissue cells). These chemicals cause inflammation by dilating arterioles, which increases blood flow at the site, and by increasing the permeability of capillaries, which permits more fluids to leave the blood. Both actions also promote the movement of phagocytic white blood cells and blood clotting agents out of the blood and into the damage site.

Fever is a more generalized part of the inflammatory response. It is induced when white cells and others release **pyrogens**, chemical agents that reset the body's thermostat in the hypothalamus (see Chapter 35). An elevated body temperature, both locally and generally, is effective because many pathogens are highly temperature sensitive and are inhibited by an increase of just a few degrees. This tells us that fever, *if not prolonged or too great,* is a natural part of the healing process.

Injured cells also release polypeptides known as **kinins**. Like histamine, kinins add to the inflammatory response by increasing local circulation and capillary permeability, and attracting phagocytic white cells to the damage site. Kinins also affect nerve endings, causing pain and tenderness—which may simply encourage you to protect the part and thus avoid further injury until healing occurs.

FIGURE 40.3
The Inflammatory Response

Inflammation is an essential part of the nonspecific immune response. Its source is primarily chemical in nature, with agents released by damaged cells, blood elements, leukocytes, and mast cells. The purpose is to increase blood and liquid flow into a damaged area, which aids in the functioning of antibodies and phagocytes. The increased blood flow elevates the local temperatures above those tolerated by some pathogens. The secretion of pyrogens brings on fever.

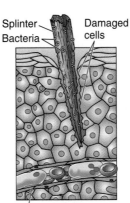

Splinter — Damaged cells
Bacteria —
Blood vessel

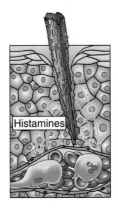

Histamines

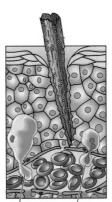

Macrophage   Neutrophil

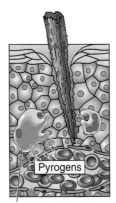

Pyrogens

Phagocyte devouring bacteria

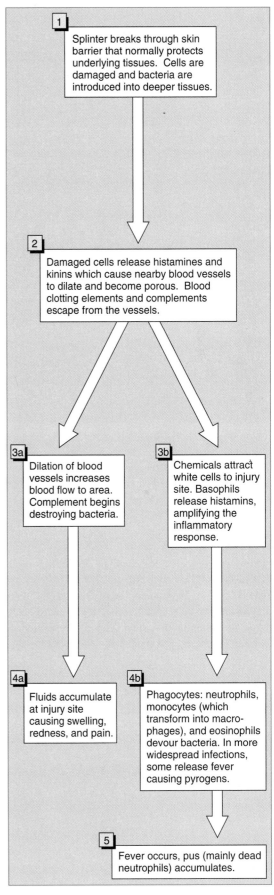

1 Splinter breaks through skin barrier that normally protects underlying tissues. Cells are damaged and bacteria are introduced into deeper tissues.

2 Damaged cells release histamines and kinins which cause nearby blood vessels to dilate and become porous. Blood clotting elements and complements escape from the vessels.

3a Dilation of blood vessels increases blood flow to area. Complement begins destroying bacteria.

3b Chemicals attract white cells to injury site. Basophils release histamins, amplifying the inflammatory response.

4a Fluids accumulate at injury site causing swelling, redness, and pain.

4b Phagocytes: neutrophils, monocytes (which transform into macrophages), and eosinophils devour bacteria. In more widespread infections, some release fever causing pyrogens.

5 Fever occurs, pus (mainly dead neutrophils) accumulates.

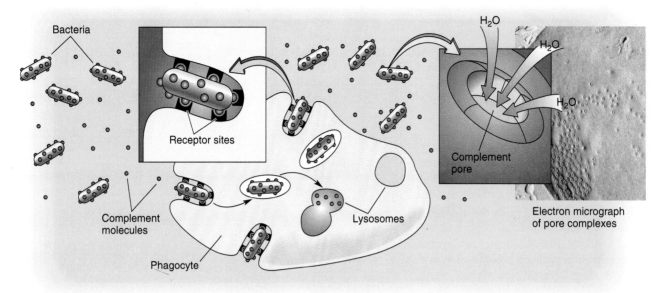

(a) Complement molecules coat bacteria.

Complement on bacteria binds to receptor sites on phagocyte. Phagocyte engulfs bacteria.

(b) Complement may also produce pore complexes in bacteria. Water rushing in bursts the bacterial cell.

FIGURE 40.4
Action of Complement
(a) The coating of bacteria by complement, called opsonization, prepares them for phagocytosis. Phagocytes contain receptors that bind to complement. The binding process makes phagocytosis far more efficient than it otherwise might be. **(b)** Another way complement works is to produce pores in the bacterial cell membrane. The subsequent inrush of water from the surroundings lyses the cell.

Complement. A battery of about 20 proteins that circulate in the blood plasma make up a chemical agent called **complement**. It kills bacteria in two ways. First, it carries out *opsonization*, a process in which the agent coats the surfaces of bacteria, making them far easier for defensive phagocytes to engulf. Phagocytes have specific receptor sites on their membranes that bind to complement, and the binding action literally locks the phagocyte and bacterial cell together (Figure 40.4a). Second, molecules of the protein adhere in ringlike formations to the bacterial cell wall and dissolve holes in the plasma membrane below. The newly formed pores allow water to freely enter the cell, which destroys it (Figure 40.4b).

Interferons. Another family of defensive proteins, the **interferons**, is released in response to viral infection. Their pres-

ence became known after researchers noticed that during the time humans are infected by one type of virus, they aren't very susceptible to most other kinds. It was eventually determined that a particular cell under attack releases substances that render other body cells resistant. Such substances, subsequently named interferons, work by blocking the initiation step in viral protein synthesis. The result is that the virus cannot synthesize its enzymes and coat proteins, which means it cannot reproduce.

Interferons can now be made in large quantities by recombinant DNA techniques. There has been some success in using engineered interferons to treat chronic hepatitis B, some herpes infections, and various papillomas (warts, including common, laryngeal, and genital types). Interferons also show promise as a treatment for rabies, a potentially fatal disease. Researchers

TABLE 40.2
The Nonspecific Chemical Defenses

Agent	Source	Effect
Lysozymes	Tears, saliva, other secretions	Break down bacterial cell walls, permits lysis
Histamine	Damaged tissue, basophils	Inflammatory response: dilates vessels resulting in swelling, faster entry of phagocytes
Kinins	Damaged tissue	Inflammatory response as above
Complement	Blood plasma	Promotes phagocytosis of bacteria, dissolves holes in bacterial plasma membranes
Interferons	Infected cells	Block viral reproduction in neighboring cells

Phagocytic Leukocytes. Some leukocytes are phagocytes, white cells that specialize in engulfing and destroying invading organisms and cell debris from infection sites. Phagocytes are commonly associated with connective tissues, particularly those that form linings within the liver, kidney, lung, capillaries, spleen, and lymph nodes. Some circulate in the blood, and others are residents of the brain and lung alveoli. As we noted earlier, phagocytes are drawn to the site of an infection by chemicals such as kinins and complement, although the attracting substance may be one produced by the invader itself. Within an hour of the release of such chemicals, great numbers of phagocytes squeeze through the capillary walls, enter the afflicted area, and begin engulfing bacterial cells, viruses, and cellular debris (Figure 40.5). The phagocytized materials are isolated within vesicles, which fuse with lysosomes. Lysosomes contain potent digestive enzymes that soon degrade the captured matter, effectively neutralizing it. The three types of phagocytes are neutrophils, monocytes, and eosinophils (Figure 40.6)

Neutrophils are generally the first to congregate at an infection site. Their great numbers make them quite effective. About 100 billion neutrophils are produced in the bone marrow each day, but these survive only a few days and suffer great casualties with every infection. In fact, the pus that accumulates at infection sites is largely dead neutrophils and debris.

Monocytes arrive after the neutrophils. Once at the site they undergo remarkable changes, growing into huge **macrophages** ("big eaters"), which take in invading cells and cellular debris (see Figure 40.5). These phagocytes are also critical to the specific defenses, particularly to the primary immune response, which we will discuss shortly. The macrophages

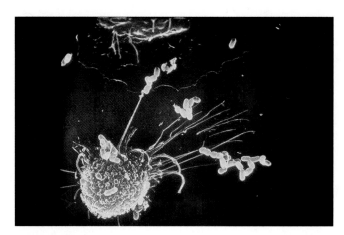

FIGURE 40.5
Action of Phagocytes
In this scanning EM view, a macrophage, one of the phagocytes, extends lengthy pseudopods that adhere to bacteria in the surroundings. When the pseudopods retract, the bacteria are taken in for destruction.

had hopes of using these versatile proteins as growth inhibitors in cancer cells, but success has been limited. Table 40.2 summarizes the nonspecific chemical responses.

Nonspecific Cellular Defenses

Nonspecific cellular defenses are carried out by leukocytes (white blood cells). Like erythrocytes (red blood cells), leukocytes are produced continually by stem cells in regions of red bone marrow (see Figure 38.4).

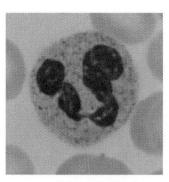

Neutrophil (phagocytosis)

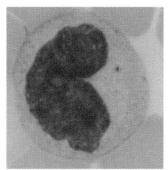

Monocyte (phagocytosis, specific defenses)

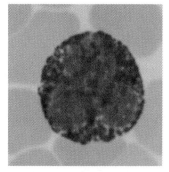

Eosinophil (phagocytosis, destroys worm larvae)

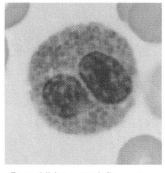

Basophil (secretes inflammatory agents)

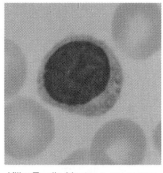

Killer T-cells (destroys cancerous cells)

FIGURE 40.6
The Leukocytes
There are five major types of leukocytes. The strongest phagocytes are the neutrophils and monocytes, the latter of which become macrophages when activated. The names of some are quite old and reflect staining properties of the cell (basophil and eosinophil) rather than conveying information about their specialized functions.

TABLE 40.3
The Nonspecific Cellular Responses

Name of Cell	Role
Neutrophil	Principal phagocyte: destroys bacteria, fungi, cellular debris at infection site
Monocyte	Phagocyte: forms large macrophage at infection site; some roving, others fixed in lymphoid tissue; key role in specific immune defenses
Eosinophil	Phagocyte; also secretes agents that kill parasitic worms
Basophil	Inflammatory response: secretes histamines
Natural killer cell (nonspecific lymphocyte)	Roving lymphocyte; detects virus-infected and cancerous cells, perforates and lyses

described here are "wandering macrophages." There are also "fixed macrophages," those that remain in place in various tissues. They occur in the liver, alveoli of the lungs (referred to earlier), brain, lymph nodes, and other places.

Eosinophils are weakly phagocytic cells that congregate at sites where chemical defense agents have clumped invaders together into masses. Once there, they devour the whole complex. Eosinophils also attack and kill parasitic worm larvae. To do this, they bind to the larval surfaces and deposit numerous toxic granules that they carry. Chemicals in the granules then kill the larvae by disrupting their soft body walls.

Nonphagocytic Leukocytes. The remaining leukocytes are nonphagocytic. They include basophils and lymphocytes (see Figure 40.6). **Basophils** are involved in both inflammatory and allergic responses. They migrate out of blood capillaries and enter damaged tissues, where they add to the inflammatory response by secreting histamines and other amplifying substances.

Most lymphocytes participate in the specific defenses, but there is an exception. **Natural killer cells** (**NK cells**) roam the body, contacting cells of all types, but ignoring most. But should a cell be cancerous or infected with viruses, it will bear telltale changes in its surface molecules. When such cells are encountered, they are immediately attacked and killed. NK cells attack only diseased body cells, ignoring invading bacterial and fungal cells. The NK cells are known to lose their effectiveness in older people, which may help explain why we tend to be more cancer prone if we live long enough. Table 40.3 summarizes the nonspecific cellular responses.

Resistance in the Invaders

With such complex and effective nonspecific lines of defense at work, one might wonder why we get sick at all. How do invading organisms slip by our defenses? How do bacteria escape deadly chemicals and efficient phagocytes? There are a few answers. For example, capsule-forming bacteria are very difficult for phagocytes to capture (like trying to pick up a slippery pumpkin seed with your fingers). In addition, some bacteria, when engulfed by a phagocyte, have the ability to inhibit mechanisms that release the lysosomal enzymes. Such parasites live inside body cells, out of reach of most defenses. Unpleasant examples include the agents of tuberculosis (a dis-

ease that is once again on the rise) and leprosy. A few bacteria secrete **leukocidins**—chemical agents that destroy leukocytes. So we see once more that natural selection is a "two-way street": hosts evolve defenses and pathogens evolve counterdefenses. We'll move on now to the specific defense mechanisms, those that often represent our last line of defense.

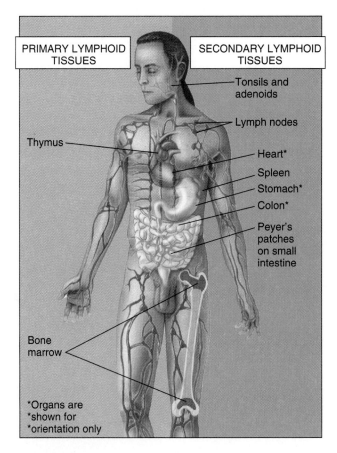

FIGURE 40.7
The Lymphoid Tissues
The highly decentralized immune system includes scattered lymphoid tissues. The primary lymphoid tissues, where lymphocytes first form, includes the thymus and red bone marrow. The secondary lymphoid tissues are regions where the lymphocytes reside after maturing and where they await a very selective activating process.

Specific Defenses

Specific defenses are carried out by lymphocytes in what is known as the **primary immune response**. Its two main aspects are *recognition* and *response*. The first requires exposure to an invading agent, while the second involves an immense searching process in which a few appropriate lymphocytes are selected. Selected cells then produce large clones, an army of similar cells, each of which can identify and react to that particular invader. The searching process, an amazingly precise one, is called **clonal selection**. As you can see, the specific defenses may require time before the lymphocytes are ready to come to the aid of the rest of the immune system.

You may not realize it, but you have experienced the primary immune response many times. As you grew up, you acquired immunity to certain childhood diseases, either because you contracted them and recovered, or because you participated in an immunization program. In both instances, the experience left you with a degree of resistance against a specific disease. We will begin our look at the primary immune response by focusing on its key participants: the lymphocytes.

Lymphoid Tissues and the B- and T-Cell Lymphocytes

Two kinds of lymphocytes are involved in the specific defenses: B-cells and T-cells. In humans and other mammals, the two originate in the red bone marrow. They then undergo differen-tiation in the **primary lymphoid tissues** (Figure 40.7), which includes the thymus (a bilobed structure under the sternum) and red bone marrow itself. Stem cells that migrate into the thymus differentiate into **T-cells** (the "T" stands for "thymus"). Those that continue their development in the red bone marrow differentiate into **B-cells**. (The "B" doesn't refer to "bone," but to the *bursa of Fabricius,* a structure in the chicken where B-cells were first recognized.) As we will see, it is during their stay in the primary lymphoid tissues that the lymphocytes develop special recognition sites on their membranes.

Once differentiation occurs, the B- and T-cells migrate into the **secondary lymphoid tissues**, including the lymph nodes, spleen, tonsils, adenoids, and Peyer's patches (regions along the intestines) (see Figure 40.7). In the inactive state, B- and T-cells are nearly impossible to distinguish visually, but, once activated, the B-cells form an extensive rough endoplasmic reticulum, an unmistakable sign that they are engaged in intense protein synthesis (Figure 40.8).

The Lymph Nodes. The confinement of great numbers of T-cells, B-cells, and macrophages to the lymph nodes aids greatly in the work of these defending cells (Figure 40.9). This is because the lymphatic system acts as an extensive filtering system, which receives fluids from the blood and from the vast tissue spaces throughout the body, eventually directing them back to the blood (see Chapter 38). Along with fluids entering the nodes are bacteria and viruses that have evaded the primary and secondary defenses. As the slow-moving

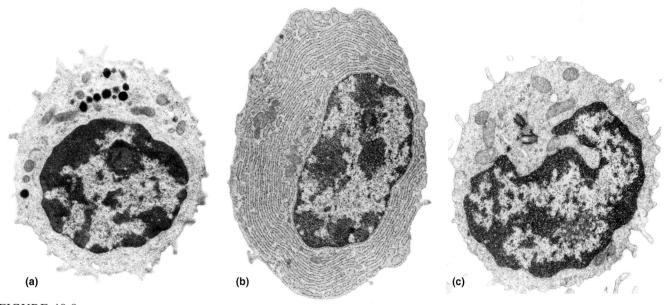

(a)   (b)   (c)

FIGURE 40.8
The Lymphocytes
(a) Immature lymphocytes, such as those located in the red marrow and thymus, are all fairly similar, but upon maturity and activation, the B-cells **(b)** become greatly enlarged, forming an extensive rough endoplasmic reticulum (ER). The ER, along with the Golgi complex, assists in the processing of great quantities of antibody that the cell synthesizes. **(c)** The T-cell lymphocyte is not nearly as large and lacks the extensive ER.

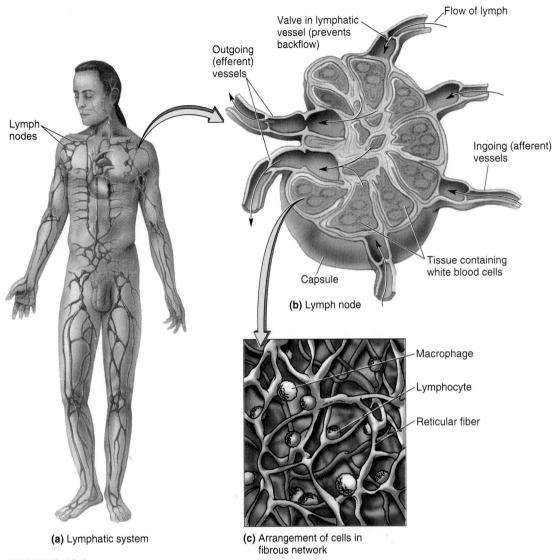

Flow of lymph

Valve in lymphatic
vessel (prevents
backflow)

Outgoing
(efferent)
vessels

Lymph
nodes

Ingoing (afferent)
vessels

Tissue containing
white blood cells

Capsule

(b) Lymph node

Macrophage

Lymphocyte

Reticular fiber

(a) Lymphatic system

(c) Arrangement of cells in
fibrous network

FIGURE 40.9
The Lymph Node
(a) The lymphatic system collects fluid from all parts of the body, sending much of it through the lymph nodes. **(b)** Lymph nodes are capsule-like structures containing numerous, maze-like passageways. Fluid enters via afferent vessels and leaves via efferent vessels. The highly divided structure forms a sieve through which the lymphatic fluid passes before reentering the general circulation. **(c)** An enlargement of one passage reveals the presence of lymphocytes and macrophages, residents in waiting. They will become active if they encounter free antigens, or antigens on the surface of invading organisms.

stream passes along, it brings the invaders into contact with the waiting defenders, thus providing opportunities for the immune response to occur.

The Work of B- and T-Cells. The B- and T-cells have quite different roles in the immune response (Table 40.4). B-cells fight invaders and their products *outside cells,* whereas T-cells carry the fight to invaders *inside cells.* More specifically, B-cells make and secrete antibodies (since these are proteins, this explains the extensive endoplasmic reticulum). The secretion of antibodies, a specific chemical defense, is called the **humoral response** (*humor* is an archaic term referring to "body fluid").

Antibodies are directed against invading viruses, bacteria, and foreign substances that are free in the blood and body fluids.

T-cells specialize in the **cell-mediated response**, which is directed against body cells that have been invaded by pathogens. For the most part, this means viruses. The cell-mediated response is also effective against cells that have become cancerous. In addition, T-cells activate B-cells, moderate the primary immune response, and suppress the process when the work is done.

Some of the B- and T-cells specialize as **memory cells**. Memory cells form a small, long-lived reserve army, capable of a very rapid response should the same pathogen strike

TABLE 40.4
Lymphocyte Activity

Cell	Response	Result
B-cells	Humoral response: synthesis of antibodies	Free antibodies bind to foreign cells, viruses, and chemicals outside cells in preparation for engulfing by phagocytes; activate complement
T-cells	Cell-mediated response: direct attack	Orchestrate the specific response; bind to and destroy virus-infected and cancerous cells
Memory T- and B-cells	Initiate very rapid secondary immune response: large clones produced in a short time	Humoral and cell-mediated response

again. When you were immunized against childhood diseases, you formed clones of memory cells against each one. This quick response is called the **secondary immune response**. Table 40.4 sums up the work of lymphocytes.

Antibodies and the Humoral Response

As we've seen, the work of B-cells is secreting antibodies. **Antibodies** (also called **immunoglobulins**) are proteins that bind to antigens. There are many kinds of antibodies, and each binds to a specific antigen.

What Is an Antigen?   For the record, an **antigen** (also called an *immunogen*) is any molecule that can stimulate the antibody response. Antigens may be any molecule that the immune system recognizes as foreign, including protein, nucleic acid, and carbohydrate, although they are generally molecules with a molecular weight greater than 5000. They can be free molecules or part of the surface of a virus or cell.

Antigens contain localized chemical groups called **antigenic determinants**; it is these groups, specifically, that trigger a response by a matching antibody. There may be several antigenic determinants on an antigen, and each of these may react with a different antibody. It only requires the presence of one antigenic determinant to elicit a response from the immune system, so the extras may just provide a little insurance that a response will be forthcoming.

Antibody Structure.   Antibodies are made up of four polypeptides, two identical long chains of amino acids called **heavy chains**, and two identical short chains called **light chains**. The chains are held together by disulfide linkages, with the finished protein assuming a Y-shaped configuration (Figure 40.10). The two branches of the Y are called **variable regions**. Variable regions contain **antigen binding sites** that bind the antibody to the antigenic determinant. The term "variable" is appropriate, since each kind of antibody differs from the next in the amino acid sequence of this region. Fur-

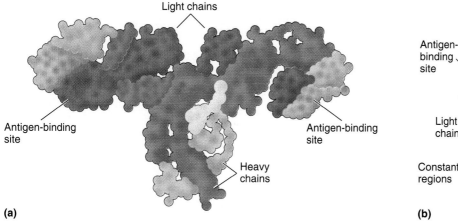

(a)

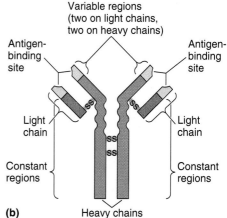

(b)

FIGURE 40.10
Antibody Structure
(a) The space-filling model clearly shows the Y-shape of the antibody. Its two antigen binding sites are at the tips of the Y. (b) Antibodies consist of two heavy and two light chain polypeptides, connected by disulfide bonds. The heavy chain is quite flexible, permitting the antibody to bend freely. At the ends of the light and heavy chains, in the variable regions, are the antigen binding sites. The remainder of the two chains form the constant regions, where differences occur only among different classes of antibody.

thermore, such variation is seemingly unlimited. Humans, for example, produce literally millions of antibodies with different variable regions. (For a preview of how this is possible, look ahead to Figure 40.23.)

Below the variable region of each chain is the **constant region**, so named because this portion is the same in all antibodies of a given class. There are five classes of antibodies in humans, each differing in the amino acid sequence of its constant region, and each with its own specialized function. The constant region proteins also have special sites known as **Fc regions**. As we will see, the presence of Fc regions makes phagocytosis more effective.

Antibody Classes. The five antibody classes are designated IgG, IgA, IgM, IgD, and IgE ("Ig" stands for immunoglobulin). Each class has a different general function in the humoral response. IgG, the most abundant class, helps lyse foreign cells, whereas IgA interferes with the ability of an invading cell to attach to a host cell. More information about the individual antibody classes is given in Figure 40.11.

As you see in Figure 40.11 antibodies occur as monomers (single "Ys"), dimers (paired "Ys"), and pentamers (five "Ys"). Those in class IgA, for example, include both monomers and dimers, whereas those of the IgM class occur as monomers and pentamers. In addition, some antibodies are

FIGURE 40.11 **Antibody classes**

	% Total immunoglobulins	Body location	Form	Target	Action
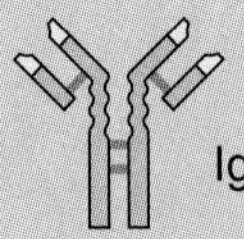 IgG	75%	Blood, lymph, intestine	Monomer	Bacteria, viruses, toxins	Facilitate phagocytosis, activates complement, neutralizes antigen
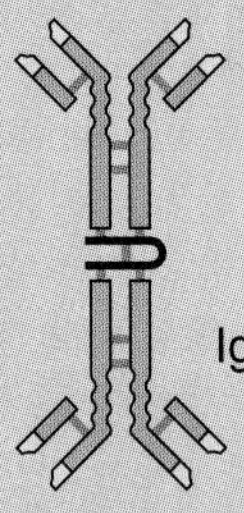 IgA	15%	Tears, saliva, mucus, milk, digestive secretions, blood, lymph	Monomers, dimers	Bacteria	Facilitate phagocytosis
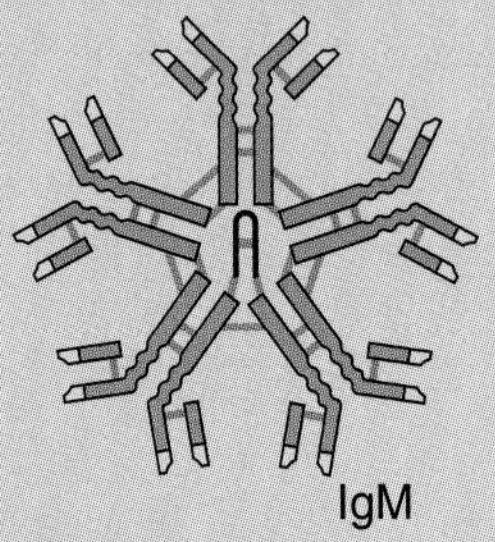 IgM	5–10%	Secreted by B-cells in blood and lymph	Pentamers, monomers-on B-cell surface	Microorganisms, ABO antigens	Agglutination of cells
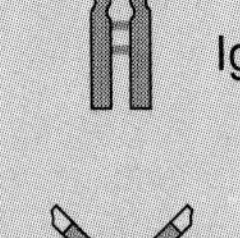 IgD	Below 1%	Blood, lymph, B-cell surface	Monomers	B-cells	Activate B-cell to secrete antibodies
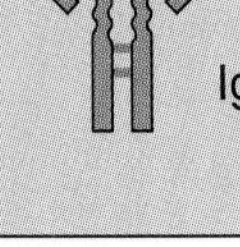 IgE	Below 0.1%	In mast cells, basophils	Monomers	General	Involved in allergic reactions, trigger histamine release

FIGURE 40.11
Antibody Classes
Although there is some functional overlapping, each class of antibody occurs in special areas and carries out specific functions.

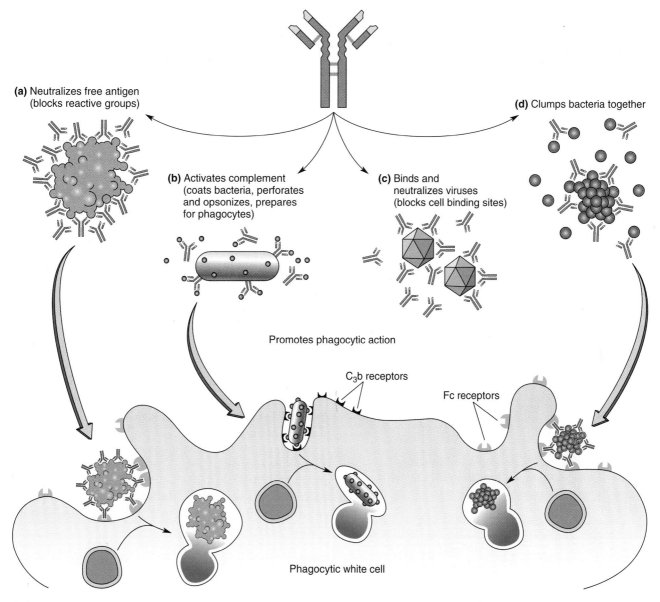

(a) Neutralizes free antigen (blocks reactive groups)

(b) Activates complement (coats bacteria, perforates and opsonizes, prepares for phagocytes)

(c) Binds and neutralizes viruses (blocks cell binding sites)

(d) Clumps bacteria together

Promotes phagocytic action

C₃b receptors

Fc receptors

Phagocytic white cell

FIGURE 40.12
Work of Antibodies
Antibodies carry out their work in a variety of ways, each involving antigens. As you can see, they operate outside cells, where they bind to various antigenic materials that occur free or on the surfaces of pathogenic viruses, bacteria, and fungi. Much of the work of antibodies prepares the antigenic material for phagocytic action.

free whereas others are bound to the surfaces of B-cells. In its pentamer form, IgM is carried along in the bloodstream, but in its single form it is bound to the surface of B-cells, where it plays an important recognition role.

Antibody–Antigen Interactions. Antibodies interact with other agents of the immune system in several ways (Figure 40.12):

1. *Neutralizing.* When antibodies encounter free antigens, such as the deadly toxins secreted by diphtheria or tetanus bacteria, many of them will bind to the antigen. This blankets the toxin's dangerous chemical side groups, effectively neutralizing the offensive molecule.

2. *Activating complement.* When antibodies such as IgM and IgG bind to a bacterial cell surface, they may also activate complement, the chemical agent that perforates and lyses cells. The presence of complement on the invading cell also sets it up for opsonization, the nonspecific chemical process described earlier (see Figure 40.4).

3. *Viral binding.* Antibodies bind to viruses, blocking their host-cell binding sites and rendering them harmless. Researchers are hopeful that they can develop an HIV binding antigen, perhaps through genetic engineering.

4. *Bacterial binding.* Antibody complexes, such as the IgM pentamer, use their multiple antigenic sites to fasten to

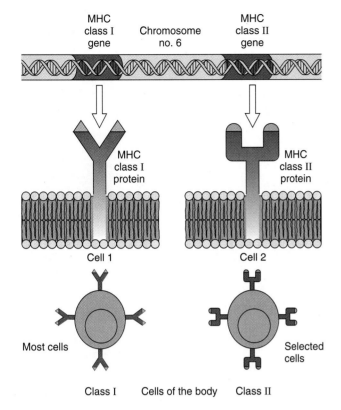

MHC class I gene   Chromosome no. 6   MHC class II gene

MHC class I protein

MHC class II protein

Cell 1

Cell 2

Most cells

Selected cells

Class I   Cells of the body   Class II

FIGURE 40.13
Models of MHC Proteins
MHC proteins identify self. Cells carry such proteins on their surfaces and it is through them that T-cells identify friendly cells. MHC proteins are coded by genes that make up the major histocompatibility complex.

several bacterial cells, making phagocytosis more efficient (one big "gulp" for the phagocyte instead of many tiny "nibbles").

5. *Opsonization.* Finally, as mentioned earlier, antibodies of the monomer (single) form may coat a bacterial cell, which exposes large numbers of the antibody Fc regions. Phagocytes with matching Fc receptors bind themselves to Fc regions on antibodies protruding from the bacterial cell, firmly gripping the bacterial cells and making phagocytosis very efficient.

We see then that antibodies are quite diverse. This great diversity makes it possible for them to recognize many different antigens and to work in many ways. Because of this, the number of B-cells carrying a particular antibody is, by necessity, very small. Yet, to be effective, the humoral response must occur on a massive scale, with vast numbers of just the right antibody molecules secreted into the blood. Meeting this requirement involves two factors. First, the right B-cell population must be selected; and second, that population must greatly increase its numbers. In some instances, the simple confrontation of a B-cell and an antigen will bring about this response, but in most instances, the interaction with a T-cell

prepares the B-cell to do its work. This brings us back to the T-cell lymphocytes.

T-Cells and Cell Recognition

The work of T-cells is entirely dependent on cell recognition—the ability of lymphocytes to recognize antigen, to recognize each other, and to distinguish healthy body cells from diseased body cells. To understand how such recognition works, we need to first see how it develops.

The MHC and Recognition of Self. Unless you have an identical twin, you are chemically unique. Each of us bears many cell surface proteins that are slightly different in their amino acid sequences from those of any other human being. This fact complicates organ transplants because the recipient's immune system treats the graft as it would any foreign material—it attacks, and the grafted tissue is rejected. Ironically, much of what we know about the immune response comes from failures in organ transplantation.

The molecular differences in our cell surfaces are genetically determined. Our cell surface proteins are determined by several hundred genes on our number six chromosome. These genes make up what is called the **major histocompatibility complex**, or **MHC**. The MHC has been found in all mammals studied so far.

Although the human MHC codes for many proteins, just two groups concern us here. They are designated as class I and class II MHC proteins (sometimes referred to as "MHC antigens"). The two protein classes have distinct configurations (Figure 40.13), and each group plays a very specific role in the complex lymphocyte interactions to come.

During fetal development each of our cells becomes "labeled" by class I and/or class II proteins coded by the MHC. Actually most body cells receive class I MHC proteins only; the MHC class II proteins are reserved for selected cells such as macrophages, B-cells, T-cells, and a few others. The logic of this will become clear when we get to the immune response itself.

The MHC proteins act as a kind of badge that identifies the cells as "self." This identification is crucial since it prevents our own T-cells from inadvertently attacking our own normal cells. This works because T-cells have **protein recognition sites** on their plasma membranes that fit the MHC proteins like a lock and key. With self-identification assured, healthy body cells are largely ignored by the T-cells. But should one of our cells harbor a virus or should it become cancerous, then the T-cell will go on the attack. How do the T-cells "know" which cell to attack? Let's see.

The T-cell can identify infected cells in two ways. First, such cells bear antigens on their surfaces. They are there because infected cells incorporate bits of antigenic material into newly forming MHC class I cell surface proteins—part of a "telltale" or "help me" warning system that enables the immune system to respond. Second, the T-cell protein recognition sites are actually dual sites, one portion recognizing

MHC protein, the other recognizing an antigen (Figure 40.14). When the T-cell dual recognition site encounters both MHC protein and antigen, it attacks, quickly killing the cell. We will elaborate more on this process shortly.

The T-cell dual recognition sites can recognize virtually any antigen that is present in the body, a fact that tells us that there are a great number of T-cells with different kinds of dual recognition sites. But how do so many recognition sites arise? Here again, we are treading on the threshold of knowledge, but immunologists think they know the answer.

During the maturation process in the thymus, a great variety of potential antigen recognition regions arise in the newly forming dual sites among vast numbers of immature T-cells. Since this is a random process, such sites must carefully be sorted out and any antigen recognition site that, by coincidence, matches normal body proteins must be eliminated. This is a crucial time in T-cell development. As the T-cells encounter the normal body cell population, any whose dual receptors both find a match are immediately killed (Figure 40.15). Those that do not find such a match will join the permanent T-cell populations.

We are looking at a very sensitive period when the immune system accomplishes what immunologists call **self-tolerance**—a euphemism for, "Don't kill normal, healthy body cells." If self-tolerance is not accomplished precisely, the result could be chaos. But cells having successfully completed the deadly sorting period will now attack any cell bearing an antigen they recognize. We will look at one more aspect of T-cell maturation and return to recognition and B-cells.

The Dual Receptor and MHC Restriction. The receptors on the mature T-cells match either a class I *or* a class II MHC protein, never both. Because of this, they are said to be **MHC restricted**. Those T-cells whose receptors fit only class I MHC proteins are now designated as **cytotoxic T-cells (T$_c$ cells**, also called *T$_8$* and *CD8 cells*), whereas those whose receptors match with class II MHC proteins become **helper T-cells (T$_h$**, also called *T$_4$* and *CD4 cells*).

The work of the two types of T-cells depends on the MHC receptor they bear. The class I MHC receptors on cytotoxic T-cells, or T$_c$ cells, match most of the body cells; thus they are the fighters, the ones that seek out and attack diseased cells. The class II MHC receptors on helper T-cells, or T$_h$ cells, only recognize macrophages, B-cells, and a few others, the few with class II MHC proteins. So T$_h$ cells interact only with selected cells of the immune system and not with diseased cells. We have already mentioned that T$_h$ cells coordinate the immune response, so being able to recognize other immune system cells is critical.

Immunologists are still struggling to explain how the great variety of original antigen recognition regions arises in the dual sites of T-cells. As with the immense variety of antibodies, the answer resides in the genes and how they code for the proteins that make up the sites. We will return to the problem later in the chapter, but you may again want to look ahead to Figure 40.23.

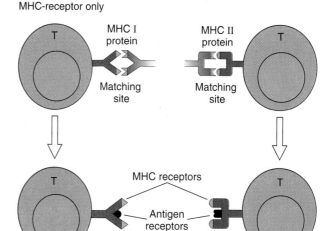

MHC-receptor only

Dual receptors MHC/antigen

FIGURE 40.14
Dual Recognition Sites
T-cells have dual recognition sites. The MHC protein receptor binds with class I or II MHC protein and the antigenic site with antigens. There is immense variation in the antigen receptors, giving the immune system the capability of interacting with just about any antigen imaginable.

Virgin T-Cells. Those T-lymphocytes that have completed the development of dual receptors are called **virgin T-cells**, a name that indicates something is yet to come. Each will retain its "chaste" state until it encounters a matching antigen. Actually, most never do, but it's comforting to know that such a large reserve army is always available should a need arise. The reserve army of virgin T-cells migrates from the thymus to the lymph nodes and other lymphoid tissues, where they are joined by mature B-cells.

The Mature B-Cell and Its Receptor

B-cells also have cell surface antigen receptors. Their receptors are antibodies, some 10,000 copies of them, bound to the plasma membrane of each cell (Figure 40.16). Each surface antibody is a copy of the specific antibody that the B-cell is capable of producing. The cell surface antibodies are monomers

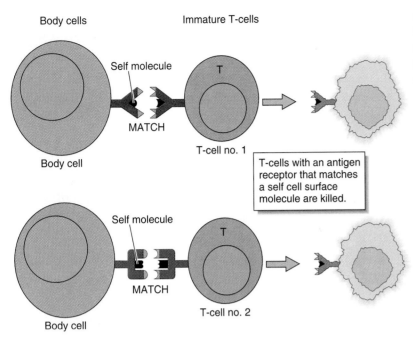

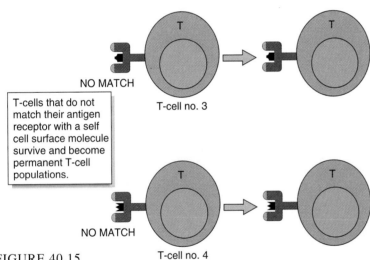

FIGURE 40.15
Learning Self
A critical phase in the maturation of T-cells is learning not to react to self. T-cells start out with antigenic receptors for just about any chemical structure possible. At first, many of these will match with and bind to many kinds of our own cell surface proteins, treating them like antigens. But at this time in development, forming a match means the destruction of that particular T-cell. In T-cells that survive this early, critical screening process, the antigenic recognition sites will recognize foreign antigens only.

(single Ys) from the IgM and IgD classes, but with an extra bit of stem used in anchoring the antibody into the plasma membrane.

Like the antigen recognition sites of T-cells, the surface antibodies of B-cells are highly diverse, each differing in its variable region. Immunologists now have a better handle on

the source of this diversity, which, as we will see later, is also genetically based. Once equipped with their surface antibodies, **virgin B-cells**, as they are known, migrate to the lymphoid tissues to join the virgin T-cells for a period of quiet waiting.

Clonal Selection and the Primary Immune Response

The primary immune response is a complex set of events that occurs when the body first encounters an invading organism. We've noted before that it requires time, but let's see why.

The virgin B- and T-cells comprise a vast army of specialists, but there are comparatively few individuals of a single type. If the body is to mount an effective response against infection, clonal selection must first occur. Clonal selection, as we have mentioned, is the proliferation of single lines (clones) selected from the vast lymphocyte army. It involves two processes: (1) cells bearing just the right antigen receptor must be located, and (2) those cells must increase in number.

Clonal Selection: Antigen-Presenting Macrophages

As versatile and efficient as T-cells are, they are incapable of responding on their own. They cannot be activated directly by antigens but require an intermediary to trigger clonal selection. This intermediary is often a macrophage.

Macrophages are the voracious eaters at infection sites, readily engulfing invading organisms, cellular debris, and even free antigens, all of which are dealt with by the macrophage's potent lysosomal enzymes— well, not quite all. The macrophage reserves portions of phagocytized material to be used for a very special purpose (Figure 40.17)—to form cell surface sites that can rouse specific T-cells to action.

The processing occurs in two ways. In one, materials captured from *extracellular* pathogens, generally bacteria, are combined with newly formed MHC class II proteins. In the other, materials from *intracellular* pathogens, usually viruses, are joined with newly formed MHC class I proteins. In both cases, the new cell surface site becomes the reciprocal of a specific T-cell dual recognition site. (Note that macrophages are among the few cell types that bear both kinds of MHC surface proteins.)

With its new cell surface sites in place, the macrophage becomes an **antigen-presenting cell** (see Figure 40.17). Antigen-presenting macrophages move through the body randomly contacting numerous virgin T-cells as they go. When they finally encounter T-cells whose dual receptors form a match,

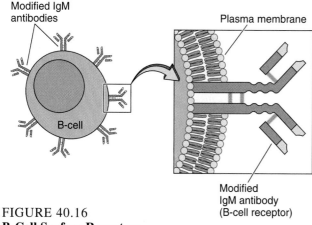

FIGURE 40.16
B-Cell Surface Receptors
Most B-cells make use of IgM class antibodies as cell surface receptors. With almost limitless diversity in the variable regions, individual B-cells have the potential to react to just about any foreign molecule. The lengthy tail (constant region) of the antibody anchors it to the plasma membrane.

the two cells bind. Since the macrophage combines antigens with both its class I and II MHC proteins, it can bind with both cytotoxic T-cells (T_c) and helper T-cells (T_h).

Spreading the Alarm: Aroused T-Cells

When T-cell and macrophage bind (Figure 40.18), the macrophage releases a chemical messenger called **interleukin-1**. The messenger activates the attached T_c cell or T_h cell, which immediately undergoes round after round of cell division. The process is greatly accelerated when aroused T_h cells secrete their own messenger, interleukin-2, which also prompts T_c cells to divide, and the activated T_c cells soon produce millions of their own kind. We see then that the identification of just a few T-cells leads to the formation of great numbers of descendants, all of which can recognize the antigen that brought about their formation. A small number is set aside as a reserve of **memory T-cells**.

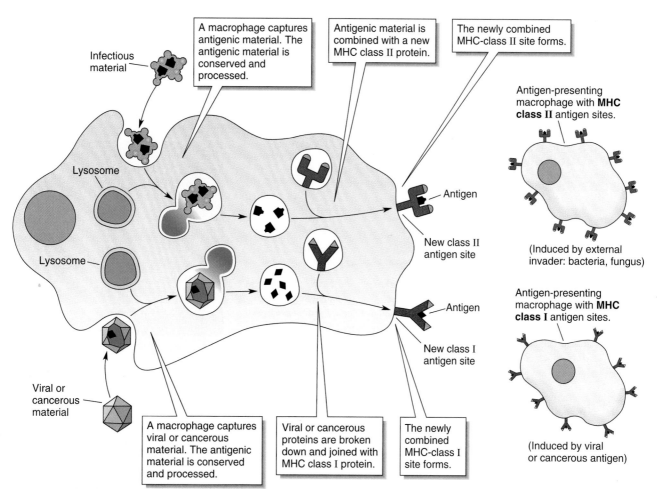

FIGURE 40.17
Antigen-Presenting Macrophage
A macrophage captures and processes antigenic material, which it uses to supplement one of its MHC class I or MHC class II sites. The antigen is fixed to the MHC protein, forming a surface complex that will match the dual surface receptors on one of the virgin T-cells. The MHC protein component of the complex determines whether it will bind to a cytotoxic T-cell or a helper T-cell.

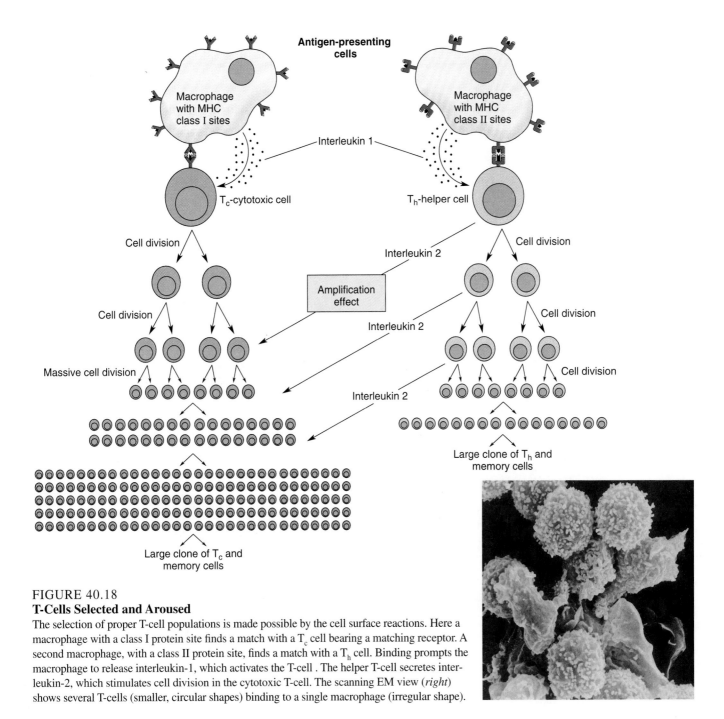

FIGURE 40.18
T-Cells Selected and Aroused
The selection of proper T-cell populations is made possible by the cell surface reactions. Here a macrophage with a class I protein site finds a match with a T_c cell bearing a matching receptor. A second macrophage, with a class II protein site, finds a match with a T_h cell. Binding prompts the macrophage to release interleukin-1, which activates the T-cell. The helper T-cell secretes interleukin-2, which stimulates cell division in the cytotoxic T-cell. The scanning EM view (*right*) shows several T-cells (smaller, circular shapes) binding to a single macrophage (irregular shape).

Cytotoxic T-Cells: Killing Infected Cells. The vast clone of activated T_c cells now goes about its deadly task—killing cells that bear matching antigens. Finding a match for its dual site involves a lot of random, undirected movement, touching body cell after body cell (we can't help but think of this as "frisking"). If a cell bears no antigen, the T_c cell goes on its way, but when an antigen is detected, the two cell surfaces bind (Figure 40.19a), and the offending cell is immediately lysed. Lysis occurs when the T_c cell secretes **perforin**, an agent that, like complement, produces holes in the afflicted cell's plasma membrane. Water enters the perforated cell and

it bursts (Figure 40.19b). In some instances, T_c cells secrete **lymphotoxin**, a chemical agent that activates DNA-fragmenting enzymes within the infected cell. Bound T_c cells are also known to secrete **gamma-interferon**, a chemical that stimulates the phagocytes into action, clearing the cellular debris. So T_c cells accomplish three results: (1) they lyse infected cells, exposing the viral invaders; (2) they prevent the virus from multiplying; and (3) they attract phagocytes that complete the destruction.

Cytotoxic T-cells also attack and destroy cancer cells. This is possible because cancer cells also produce cell surface

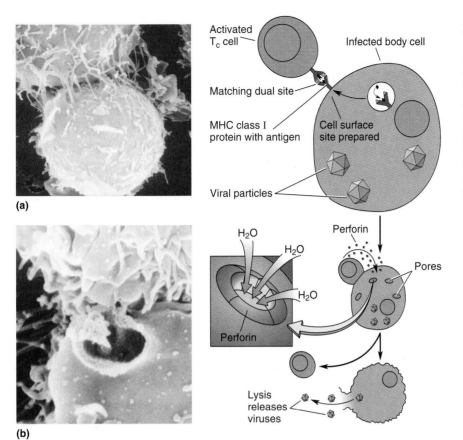

(a)

(b)

FIGURE 40.19
Attack of the Cytotoxic T-Cells
(a) The scanning EMs show how T_c cells destroy infected or cancerous cells. Note the binding in the first photograph and disruption of the cell in the second. (b) When a T_c cell binds to a virus-infected body cell, it releases perforin, a chemical substance that produces holes in the cell membrane. The holes freely admit water, which lyses the cell. The debris can then be acted on by antibodies and by phagocytic cells.

Labels in figure: Activated T_c cell; Infected body cell; Matching dual site; MHC class I protein with antigen; Cell surface site prepared; Viral particles; H_2O; Perforin; Pores; Perforin; Lysis releases viruses

antigens. But while such matches often occur, studies indicate that the T_c cells tend to fail if the cancer has metastasized, that is, spread from its original source. Researchers believe that such traveling cancer cells may undergo changes in their MHC cell surface proteins, and a match with the MHC protein receptors is no longer possible. Without the other part of the dual match, the T_c cell cannot act, and an effective cancer fighter is neutralized.

Helper T-Cells: Activating B-Cells. In addition to stimulating the growth of the T-cell clones, T_h cells activate B-cells, whose role, you will recall, is the secretion of antibodies. Antibodies are effective against any antigen not hidden away in body cells. Thus the B-cell's role is broader than that of T-cells. But clonal selection must occur before an effective B-cell attack can take place. The proper virgin B-cells must be identified and large clones formed.

The arousal of B-cells begins when the surface antibody of a virgin B-cell binds to a matching free antigen (Figure 40.20). The antigen is taken into the B-cell, where it is incorporated into newly synthesized class II MHC surface protein, which moves to the cell surface. The B-cell now has a receptor that will match its reciprocal on a T_h cell that has been aroused by the same antigen. Upon contact, the cells bind, and the helper T-cell secretes **interleukin-2**. This prompts the B-cell to enter cell division, forming its own enormous clone. As in T-cells, part of the clone is set aside as memory cells.

B-Cells and the Antibody Attack

Activated B-cells are known as **plasma cells**. Plasma cells are short-lived, lasting only four to five days, but during this brief period they secrete copious amounts of antibody—up to 2000 antibody molecules per second during a short but intensely active life. As we learned earlier, antibodies attack invaders in a variety of ways, presenting a formidable defense (see Figure 40.12). In review, antibodies coat and neutralize viruses, clump bacterial cells together for easier phagocytosis, coat bacteria to permit opsonization, neutralize free antigen, and activate complement. Before we leave the B-cells, there is one more way in which they may become active—one not involving T-cells.

Antigen Arousal of Virgin B-Cells. The direct arousal of B-cells is a random process in which antigen passes cell after cell in the lymph nodes or blood. Then, by chance, a B-cell bearing just the right receptor meets and binds to an antigen. The B-cell responds through a frenzy of cell division, during which large clones form. Each cell in the clone will secrete the precise antibody needed and the antigens are soon inundated.

Immunologists are still arguing about the significance of direct antigen arousal; some believe it is highly limited. The antigens responsible may be limited to a few that are present on bacteria. One familiar antigenic material is the protein flagellin, which comprises the rotating bacterial flagellum. Although there are many unanswered questions about this

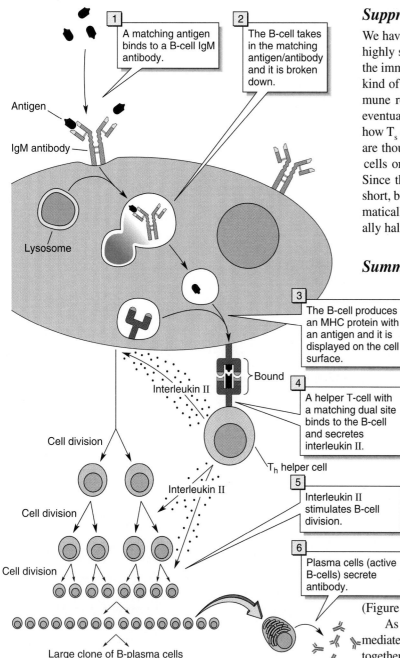

1 A matching antigen binds to a B-cell IgM antibody.

2 The B-cell takes in the matching antigen/antibody and it is broken down.

Antigen

IgM antibody

Lysosome

3 The B-cell produces an MHC protein with an antigen and it is displayed on the cell surface.

Interleukin II
Bound

4 A helper T-cell with a matching dual site binds to the B-cell and secretes interleukin II.

T_h helper cell

Cell division

Interleukin II

5 Interleukin II stimulates B-cell division.

Cell division

6 Plasma cells (active B-cells) secrete antibody.

Cell division

Large clone of B-plasma cells and memory cells

FIGURE 40.20
Activation of a B-Cell by a Helper T-Cell
The activation of B-cells begins when the B-cells capture antigen. The antigen is used along with MHC class II protein to build a cell surface recognition site. When a matching T_h cell binds to the B-cell, it secretes interleukin-2, which stimulates cell division in the B-cell. Soon a clone of plasma cells arises and begins secreting antibody. Some of the cells are reserved for memory cells.

means of B-cell arousal, such activation is an efficient means of combating some bacterial intruders. Far more often, though, the arousal of virgin B-cells requires the aid of helper T-cells.

Suppressor T-Cells: The Battle Is Won

We have seen how the immune system is activated and how a highly specific but massive response is generated. But how is the immune response stopped once the danger is past? A third kind of T-cell, the **suppressor T-cell** (T_s), modulates the immune response, keeping it from running out of control and eventually bringing it to a halt. Researchers do not yet know how T_s cells arise, nor are they certain how they work. T_s cells are thought to inhibit any remaining and unaroused virgin T-cells or B-cells that are capable of selection and activation. Since the lifetime of activated T-cells and plasma B-cells is short, blocking the continuing activation of reserves will automatically limit the number of active lymphocytes and eventually halt the immune response.

Summing Up the Primary Immune Response

As we see, the body has an immensely diverse army of virgin B- and T-cells, capable of identifying and responding to just about any antigen. When an invasion occurs, macrophages combine captured antigen with their class I and class II MHC surface proteins, thereby becoming antigen-presenting cells (APCs). APCs form matches with specific T_c and T_h cells, binding to them and secreting interleukin-1. T_h cells respond by dividing and secreting interleukin-2, which stimulates T_c cell clone formation. The T_c cells perform the cell-mediated response, attacking and killing infected body cells. T_h cells bind to selected B-cells and secrete interleukin-2, which prompts the B-cells to divide and produce vast plasma cell clones that secrete antibodies. The antibodies coat and neutralize viruses, clump bacterial cells together for easier phagocytosis, coat bacteria to permit opsonization to occur, neutralize free antigen, and activate complement. Finally, T_s cells stop the primary immune response. (Figure 40.21 reviews the response once more.)

As we've emphasized, the two-pronged humoral and cell-mediated attack is effective because B- and T-cells work well together. Antibodies cope with anything present in the blood, lymph, and tissue spaces. T-cells cope with hidden pathogens, those that find their way into cells and out of reach of antibodies.

But now that we are well, what if the same invader should again break through our defenses? Does the complex primary response have to occur all over again?

Vigilant Memory Cells and the Secondary Immune Response

The first time we fall ill to a new invader, our immune reaction may be relatively slow—clonal selection takes considerable time. But given that time, and if all goes well, the offenders

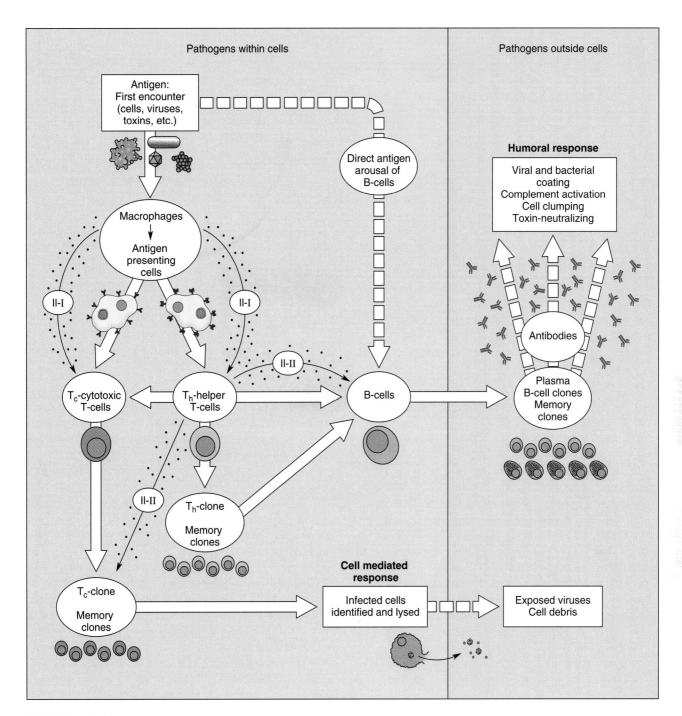

FIGURE 40.21
Summary of the Primary Immune Response
The events of the primary immune response at a glance (well, maybe a stare).

will be dealt with and we will recover. We are left temporarily weaker but "wiser." That is, our immune system is wiser. (*We may go right on doing whatever it was that got us sick!*) This new wisdom is found in the memory B- and T-cells, those responsible for the **secondary immune response**.

Memory cells have the ability to recognize the same antigen that earlier aroused their sister cells. While the other activated B- and T-cell clones live short, busy lives, memory cell

lines may go on, sometimes for decades. This is important, for should the same invader show up a second time, it will immediately be recognized. The selected memory B- and T-cells undergo round after round of cell division, and soon a massive new army of active T-cell and B-cell lymphocytes arises to repel the second invasion. We may not even be aware of the renewed struggle, for the symptoms are often quite mild or absent. So thanks to our memory cells, our suffering may

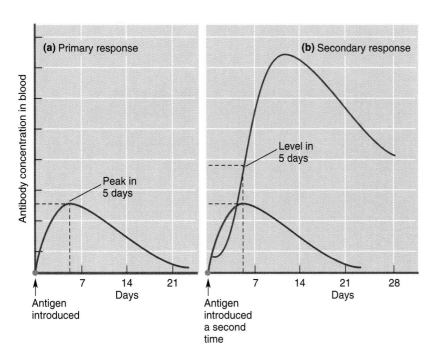

FIGURE 40.22

Comparing the Primary and Secondary Response
The secondary immune response, as indicated by the level of antibody production, is faster and more intense than the primary immune response. **(a)** With the first introduction of antigen, antibody levels peak at about day five. **(b)** The second introduction of the antigen produces a much greater quantity of antibody response in the same, five-day period. Note the final peaking of antibody and prolonged high level.

occur just once. (Figure 40.22 compares the primary and secondary responses over time.)

Active and Passive Immunity

When you received your childhood vaccinations, the vaccine contained weakened or killed disease agents that, while harmless to you, still maintained their antigenic properties. The primary immune response occurred in the usual way but at a much reduced level. You skipped most of the unpleasantness of the infection as you developed **artificial active immunity**. The response soon subsided, but the all-important memory cells remained behind. So, producing a small reserve army of selected memory T- and B-cells is the goal of the vaccination process. Had you later confronted the real disease agent, your immune system would have gone right into the streamlined secondary immune response, rousing memory clones into action, and the invader would have been destroyed quickly.

Where vaccines aren't available or where the disease has already begun, alternatives are possible. One is the introduction of an **antiserum**, a substance containing specific antibodies against the antigenic agent. Such antibodies are routinely obtained from animals (and humans) exposed to the disease agent under carefully controlled laboratory conditions. When injected, the antibodies go about the task of immobilizing the invader's cells and toxins. Antibodies are short lived, and therefore the protective effects are temporary, but they can get you through some dangerous times. Antiserum is available for the treatment of poliomyelitis, hepatitis A, tetanus, and many other serious diseases. Since the primary immune response is not activated by the antiserum, and no memory cells form, the treatment is known as **passive immunity**.

We've still left a few questions unanswered. How do so many different types of antibodies and antigen receptors arise? That is, how does the body provide for the incredible diversity seen in virgin lymphocytes?

Lymphocyte Diversity

All mammals and birds, and perhaps all vertebrates, can produce antibodies against virtually any large molecule. If you inject a rabbit with crocodile hemoglobin, the rabbit will make hundreds of different specific antibodies against crocodile hemoglobin (which tells us that hemoglobin has hundreds of antigenic determinants). How does the rabbit's immune system do it? We know that antibodies are proteins and proteins are specified by genes. But are there enough antibody-coding genes to enable rabbits to react against irrelevant molecules? After all, should a rabbit get close enough to a crocodile for exposure to its antigens, making antibodies might be the least of its concerns. More to the point, are there enough genes to code for antibodies against all the millions upon millions of potential antigens?

The answer to both questions is yes, but things are complex. First, there are not nearly enough genes initially in the human genome to provide for such diversity. To make such diversity possible, the antibody-coding genes of B-cells undergo a *vast amount of gene rearrangement*. The genes that code for the variable regions of antibodies occur in only about 300 DNA segments located throughout the chromosomes. These segments are rearranged during the immune system's development, as the cell synthesizes the variable regions of the antibody light and heavy chains (Figure 40.23). They are

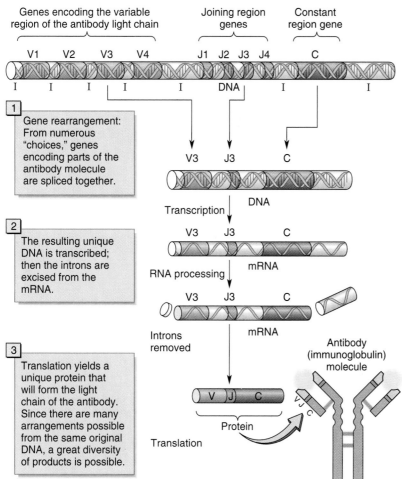

Genes encoding the variable region of the antibody light chain

Joining region genes

Constant region gene

V1 V2 V3 V4   J1 J2 J3 J4   C

I   I   I   I   I   DNA   I   I

1 Gene rearrangement: From numerous "choices," genes encoding parts of the antibody molecule are spliced together.

V3   J3   C

DNA

Transcription

2 The resulting unique DNA is transcribed; then the introns are excised from the mRNA.

V3   J3   C

mRNA

RNA processing

V3   J3   C

Introns removed   mRNA

3 Translation yields a unique protein that will form the light chain of the antibody. Since there are many arrangements possible from the same original DNA, a great diversity of products is possible.

V J C

Protein

Translation

Antibody (immunoglobulin) molecule

FIGURE 40.23
Origin of Antibody Diversity
This highly simplified illustration shows how diversity arises in the variable region of an antibody light chain. The scenario begins with a chromosome in one of the B-cells.

broken apart and recombined in a seemingly infinite number of ways (estimated at about 18 billion). You may recall that we've said that all the cells of the body have the same genetic information. Obviously, a correction is in order. You can now see that each population of differentiated B-cells had its antibody-specifying DNA rearranged so that it has certain tailor-made genes that other cells don't have. All this happens during each B-cell's maturation period, and long before it ever encounters antigen.

Antibody diversity also can be produced in other ways. One way is by rearranging the gene products, the proteins themselves. In addition, somatic mutations can produce single DNA base substitutions (see Chapter 17), causing slight variations in the antibody (known to occur in the genes coding for IgA and IgG classes).

Now what about the T-cells? How do they produce surface receptors so varied that they can recognize virtually any antigen? It turns out that much less is known about the genetics of the antigen-detecting region of the T-cell's dual recognition system. However, immunologists are generally convinced their production is similar to that of antibodies—through rearrangements of DNA segments within the genes that code for their antigen recognition molecules.

Monoclonal Antibodies

Antibodies have long been used to detect the presence of specific disease antigens and other molecules of research interest. In theory, antibodies can be used to locate just about any molecule, since once they bind to that molecule, it can readily be separated as a precipitate. The chief difficulty is in obtaining a pure sample of the specific antibody. The traditional research technique for producing antibodies was to introduce a small amount of known antigen into a laboratory animal and, after a time, to draw blood and isolate the antibodies. Unfortunately, antigens usually contain many different antigenic determinants, so a single antigen would activate many different B-cell lines, and thus a number of different antibodies would form. What was clearly needed was a way to sort out B-cells according to their specific antibody production.

In 1976, researchers found a way of doing just that—and began producing antibodies of a single type, which came to be known as **monoclonal antibodies**. The procedure is described in detail in Figure 40.24.

There are many exciting uses for monoclonal antibodies. They can be used to locate certain cells, mitochondria, ATPases, tubulin, and other cell constituents, and to locate hormone receptors. The antibody–antigen complexes are often made visible by first making them radioactive or fluorescent.

Other applications include the clinical diagnosis of viral and bacterial diseases. Antigens associated with a sexually transmitted disease organism can be detected in a matter of minutes. Previously, the diagnosis required days, during which cells were cultured and testing was done (a long time for the anxious patient to wait).

Monoclonal antibodies can be used as drug delivery systems, that is, as a means for getting medication to the right place in the body. In the treatment of cancer, for example, physicians are experimenting with the concept of physically linking potent cancer-treatment drugs to monoclonal antibodies that will bind only to the cell surface antigens of cancerous cells. By targeting the drug this way, the dosages can be greatly reduced and the harsh side effects eliminated. It seems the potential applications for monoclonal antibodies are limited only by the creativity of researchers.

When the Immune System Goes Wrong

We've marveled so much over endless antibody diversity and the incredible preciseness of the immune system that we may have left an impression of perfection at work. This is hardly the case. As with other body systems, things can go wrong. The results can be disastrous.

Autoimmunity: Attack Against Self

Autoimmunity, or *autoimmune disease*, is the reaction of lymphocytes against self, the treating of normal cells and tissues as though they are foreign. The effects are widespread, producing many different syndromes. Known or suspected autoimmune diseases are multiple sclerosis (destruction of the myelinated sheaths covering neurons in the central nervous system), rheumatoid arthritis (destruction of joint linings), myasthenia gravis (disruption of muscle function, particularly in facial and neck muscles, due to antibody reaction against acetylcholine receptors at the neuromuscular junctions), Grave's disease (excessive thyroid hormone), systemic lupus erythematosus (destruction of the kidneys and other organs), and juvenile diabetes (destruction of insulin-secreting cells).

Autoimmune disease can arise as an aftereffect of infection. When the surface molecules of disease organisms are similar enough to our own, the defensive antibodies we produce sometimes cross-react with our own tissues. The *Streptococcus* bacterium that causes strep throat is notorious for

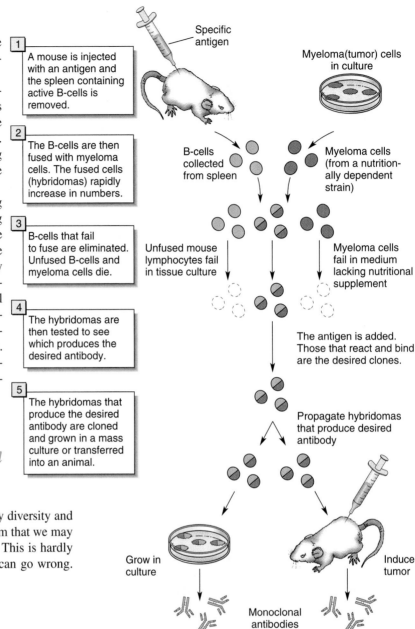

1 A mouse is injected with an antigen and the spleen containing active B-cells is removed.

2 The B-cells are then fused with myeloma cells. The fused cells (hybridomas) rapidly increase in numbers.

3 B-cells that fail to fuse are eliminated. Unfused B-cells and myeloma cells die.

4 The hybridomas are then tested to see which produces the desired antibody.

5 The hybridomas that produce the desired antibody are cloned and grown in a mass culture or transferred into an animal.

FIGURE 40.24
Producing Monoclonal Antibodies
In brief, antigen is used to arouse and proliferate B-cells, which are removed through screening. To do this, they are fused with myeloma cells, which, after more sorting, are separated for antigen testing. Those reacting with antigen are set aside to be used for cloning and producing more antibody. They may either be grown in tissue culture and their antibodies harvested, or be introduced into a reservoir animal that will produce the antibody.

this. After the usual throat infection runs its course, the antibodies we have produced may begin to attack our own tissues. Typically, the targets are the heart valves and the kidneys, both with serious and sometimes fatal results. The condition is known as rheumatic fever.

Recent advances in therapy show promise to sufferers of autoimmune disease, particularly of multiple sclerosis. In this

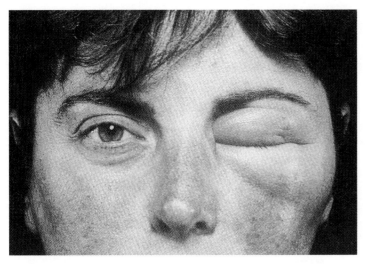

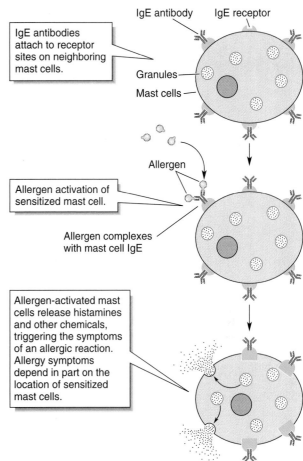

IgE antibodies attach to receptor sites on neighboring mast cells.

IgE antibody   IgE receptor

Granules

Mast cells

Allergen

Allergen activation of sensitized mast cell.

Allergen complexes with mast cell IgE

Allergen-activated mast cells release histamines and other chemicals, triggering the symptoms of an allergic reaction. Allergy symptoms depend in part on the location of sensitized mast cells.

FIGURE 40.25
A Hyperallergic Reaction
Bee stings (*photo above*) can produce sudden, severe swelling at the site of injected venom. Reactions by sensitized persons can be life-threatening. Sensitization begins when allergens such as bee venom, household dust, pollen, dust mites, and others activate virgin B-cells that produce antibodies of the IgE class (*top right*). The antibodies bind to receptors on mast cells, which prepares them for the next encounter with that allergen. When this happens (*center*), the IgE antibodies of the sensitized mast cells bind to the allergen and begin their response. The mast cells release histamines and other chemicals (*bottom*), which bring on the symptoms of the reaction. Often referred to as "hay fever," the symptoms may include runny nose and eyes, skin redness and puffiness, skin rashes, asthmatic attacks, and others.

debilitating disease, aberrant T-cells attack and destroy the myelin sheaths of nerve fibers. Some of the symptoms are general muscle weakness, impaired vision, tremor, vertigo, seizures, and loss of bladder control. The T-cells responsible for multiple sclerosis can now be neutralized through the injection of genetically engineered antibodies. The antibodies bind to specific molecules present on their surface, and that population of cells then declines markedly. Patients treated this way show some relief of outward symptoms, such as strengthening of muscles and improved walking and vision. Magnetic resonance scans of the brain reveal a definite regression in the damaged areas.

Allergies

What we call **allergies** are the body's occasional overreaction to marginal antigens, those that are basically harmless to most people. Such antigens, or **allergens** as they are known, include pollen, animal hair, household dust, milk fat, eggs, antibiotics, our own skin bacteria, and many others. Upon exposure, the allergen prompts the release of histamines, prostaglandins, and other agents that bring on a typical inflammation response. The usual results are difficulty in breathing and copious mucus secretion. Antihistamines are

often a solution. The events of an allergic response and the role of various cells are outlined in Figure 40.25.

Some allergic reactions can be quite dangerous. An example is hypersensitivity to bee sting. In susceptible persons, the inflammatory response can be so rapid and massive that the dilated blood vessels release too much plasma. This leads to an abrupt drop in blood pressure, palpitations of the heart, and often physiological shock. If untreated, the onslaught can be fatal. Treatment includes injections of epinephrine and corticosteroids, the latter reversing the inflammatory response.

AIDS: The Crippled Immune System

It's been a decade and a half since we became aware of the **AIDS (acquired immunodeficiency syndrome)** problem, and a somewhat shorter period since the agent, **HIV (human immunodeficiency virus)**, was identified (Figure 40.26). What seemed at first to be chiefly a problem of homosexual men has since become recognized as universal. Because the AIDS virus has spread into virtually every corner of the globe, and into all walks of life, it is now described as "pandemic." The *Global AIDS Policy Coalition* estimates that, at the end of 1992, some 19.5 million persons had contracted HIV. By that

FIGURE 40.26
HIV
The scanning EM is of a helper T-cell coated with new HIV particles that are budding from its cytoplasm. This gives you some idea of the reproductive potential of these deadly invaders.

same period, the total number of active AIDS cases reached three million. Most have since died. Projecting into the year 2000, the number of HIV infections is expected to range between 40 and 110 million.

After a decade and a half of AIDS, trends have changed. Most new infections, one every 15 seconds, occur in heterosexuals. In fact, heterosexual women now account for 40% of new AIDS cases and some of these women are middle-aged. Much of the sharp rise in the incidence of AIDS in heterosexuals can be attributed to the use of illicit street drugs, but there is also the natural progression of the disease into the general population itself. Of major concern is the growing number of HIV-positive infants and children who now account for 10% of all new cases. Infection in the newborn infants can be attributed to HIV crossing the placenta or directly infecting the baby during the birth process. HIV can also be transmitted to an infant through breast milk. Early detection of HIV in infants is complicated by the fact that HIV antibodies, generated by an infected mother, cross the placental barrier and can be detected in the infant for up to a year. During this period there is no way of determining whether or not the baby has been infected, since the mother's and infant's antibodies are, as yet, indistinguishable.

Before going on let's make a necessary distinction. Many people are HIV positive; that is, they harbor the virus in their bodies. These people may be quite healthy in all other respects and the symptoms of AIDS may not appear for many years. AIDS itself, or "full blown AIDS" as it is called, refers to a point in which the helper T-cell count drops below 200/mm^3 (per cubic millimeter) of blood. The biology of HIV was discussed in Chapter 22. Here we will look into the immune system aspects: the targets of HIV, problems of communicability, and the search for a cure.

Targets of HIV. HIV preferentially attacks helper T-cells and macrophages. A likely reason is the presence of protein CD4 among surface proteins of these cells. CD4 is used as a binding site by surface receptors on the HIV envelope. The CD4 protein is also present on about 5% of the B-cells, selected intestinal cells, skin cells, and the brain's glial cells.

The number of T_h (helper T-cells) lymphocytes dwindles dramatically over the course of HIV infection, from a normal state of 1200/mm^3 to less than 200, a time when the symptoms of active AIDS become evident. With the decline of T_h cells, the body's resistance to disease falls off dramatically. Not unexpectedly, this decline is also accompanied by a decrease in B-cell activity. Without the B-cell antibodies, the body is quickly made vulnerable to opportunistic infections.

No one yet knows exactly how the virus affects T_h cells. Some researchers have suggested that it renders their receptor sites inoperative. Others believe that HIV brings about an autoimmune response in which lymphocytes begin attacking each other, thus leaving the victim defenseless. Newer studies are casting some doubt on T_h cells harboring most of the HIV particles. It is now known that only about 1 in 40 T_h cells is actually invaded by HIV. The attack may be far more complex. The new studies indicate that, over the years, HIV centers its attack on the lymph nodes, literally destroying them . Lymph nodes, as we've seen, harbor large populations of virgin B-cells, T-cells, and macrophages.

Another important target may be the macrophages. These cells may act as reservoirs for the virus during the inactive period. The seclusion of HIV in macrophages would help explain why some older antibody-screening tests failed to detect the virus in donated blood. Antibodies form only after macrophages become antigen-presenting cells and the primary immune response is underway. Newer tests can detect the presence of HIV in macrophages. Whatever its mode of action, HIV is an insidious invader, destroying the very group of cells that organize the body's resistance.

The Transmission of HIV. Well over 90% of HIV transmission is via semen, vaginal fluids, and blood, although HIV has also been found in tears, sweat, and mucus. It also occurs in mother's milk. Sexual transmission of HIV almost always involves viruses from infected semen or vaginal fluids finding their way into the blood of the new victim. Direct sexual transmission is known to occur from males to males, males to females, and females to males. All that is required is a minute cut or abrasion in any of the delicate body linings—and life suddenly changes.

The risks of sexual transmission can be reduced through the use of *latex* condoms (animal membrane condoms will not block the virus), since they capture the semen and thus prevent the escape of the virus. Use of the virus-killing spermicide *nonoxynol-9* increases the protection. To be effective, the condom must be in place prior to intercourse.

It is reassuring to know that the risk of HIV transmission in ways other than those mentioned is nearly zero. Studies of

FIGURE 40.27
The Progression of HIV
The concentration of HIV particles and T-cells is charted over a ten-year period. The rapid increase in T-cells and the equally rapid decrease in the HIV count during the first year show that the immune system is very much in charge—at first. Over the next nine years the opposite occurs: HIV increases and T-cells diminish. The T-cell count drops below 200 (per cubic millimeter) at about year five (it varies) and continues to decline. At this low count, the patient experiences episodes of debilitating illness, until finally, in a very weakened condition, the patient succumbs.

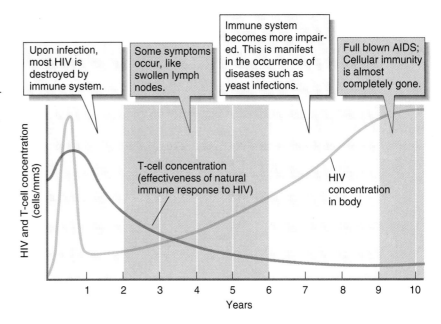

Upon infection, most HIV is destroyed by immune system.

Some symptoms occur, like swollen lymph nodes.

Immune system becomes more impaired. This is manifest in the occurrence of diseases such as yeast infections.

Full blown AIDS; Cellular immunity is almost completely gone.

T-cell concentration (effectiveness of natural immune response to HIV)

HIV concentration in body

HIV and T-cell concentration (cells/mm3)

Years

AIDS health care workers and households with infected members reveal virtually no nonsexual transmission of the disease agent to healthy individuals.

The Progress of HIV in the Body. Following the progress of the virus in the average patient over a ten-year period (Figure 40.27), we see the following. Upon the initial infection, the immune system begins a vigorous defense. The activated T_c cells (cytotoxic T-cells) seek out and destroy HIV-infected cells, and activated B-cells secrete antibodies that react with the viral proteins. However, some virus particles get by the defenses, and by the time antibody tests reveal their presence, the HIV infection is well-entrenched.

During the initial stages, many patients experience mild, flu-like symptoms: some weight loss, swollen lymph nodes, prolonged colds, fatigue, diarrhea, and persistent fever. At this time, and often before the person is even aware of the problem, the blood contains large amounts of the virus and the person is quite contagious. But the symptoms soon subside, and the HIV-positive individual resumes a relatively healthy state. The virus will have entered a quiescent period, its numbers in the blood falling off dramatically.

The quiescent period may last a few years or many—a decade or more in some cases. For most, new symptoms appear by the fifth year of infection. During the previous quiescent period, the T_h cell population steadily declined whereas the number of HIV particles in the blood increased. As the T_h cell count falls below the critical number, the opportunistic infections begin. In addition to ordinary diseases, there are very rare cancers and pneumonias, along with unusual fungal invasions, all diseases easily discouraged by the unimpaired immune system. Recently, there has been an increase in tuberculosis among AIDS patients. With the immune system effectively out of action, the victim usually dies within a few years from one disease or another.

Rx for AIDS? The CD4 surface protein was once thought to be the "Achilles heel" of HIV—the weak link in its life cycle. Medical workers believed that by administering monoclonal antibodies that bind the CD4 protein of the T-cells, they might block the entry of HIV and thus have an effective therapy. Alternatively, the administration of large amounts of free "decoy CD4" was suggested as a way to bind the viral receptors, thereby preventing them from linking up to their target cells. Such procedures were subsequently tried but the results were discouraging.

A major problem in the development of immune-based treatments is the rapid mutation rate in the virus. HIV reverse transcriptase is relatively sloppy at assembling DNA, committing errors at a rate of one per 20,000 nucleotides. Since each error potentially represents a new point mutation, the amino acid sequences in HIV proteins are constantly changing. This means that there may actually be many *different variants of HIV in each patient.* Consequently, highly focused medical attacks based on the specific chemistry of HIV produce temporary effects at best. Confronted by HIV's rapid mutation rate, immunologists believe that we are still years away from an AIDS solution, and any such solution will have to be broader based and not aimed at a single link in the HIV life cycle.

At the moment, the best direct treatment for AIDS in the United States is the antiviral drug **AZT** (azidothymidine). AZT acts by blocking the enzyme reverse transcriptase, the one responsible for converting viral RNA to DNA. During reverse transcription, the enzyme incorporates *azido*thymidine instead of the normal thymidine nucleotide into the growing DNA chain. This blocks the addition of the next DNA nucleotide, so reverse transcription stops and the viral life cycle cannot proceed. Unfortunately, AZT's effectiveness is transient.

Researchers are looking at other possibilities. They now know that T_h cells must be activated before HIV can integrate its genes into their chromosomes. Some have turned to immunosuppresants such as cyclosporine and FK 506, used in organ transplantation, to temporarily block immune system activation. Others are focusing on HIV transcription, particularly on control proteins that activate the viral genome when viral reproduction begins. One drug company has developed an agent that can inhibit such viral control proteins.

Early in 1995, a promising new treatment for HIV-positive persons was announced. The $3\frac{1}{2}$-year study involved the introduction of genetically engineered interleukin-2, the T_h chemical messenger that prompts cell division among T_h, T_c, and B cells. Among the ten persons described in the study, six showed a 50% increase in T_h cells over the 3-year course of the study. The procedure, which is extremely expensive (about $12,000 a year), consists of the introduction of IL-2 every two months. Unfortunately, the treatment is not recommended, and may even be dangerous, for persons whose T_h count has dropped below 400 cells/mm^3.

Other efforts have been directed at the opportunistic infections themselves, and at preserving the general physical and mental health of the patient. Here, there is some promise. With new drugs directed against the bacterial and viral agents of the common opportunistic infections, and with people committed to proper nutrition, remaining drug-free and to maintaining more positive attitudes, the life expectancy is increasing. But, the fact remains, there is no cure for AIDS, and no one is predicting when any breakthrough will come.

Key Ideas

NONSPECIFIC DEFENSES

1. Nonspecific defenses include the body linings, a variety of chemicals, and several kinds of white blood cells. The intact skin, its keratin layer, and sweat and oil secretions, make up first-line defenses, as do mucous secretions, stomach acids, urine flow, and populations of harmless microrganisms. Tears, saliva, and other body secretions contain bacteria-killing lysozymes. Chemicals include **histamines** and **kinins** (inflammatory response), **pyrogens** (fever inducers), **complement** (**opsonizing**, or coating, agent), and **interferons** (stop viral protein synthesis). Cells include **neutrophils**, **eosinophils**, **monocytes** (phagocytes; monocytes grow into **macrophages**, and eosinophils kill worms), and **natural killer cells** (destroy virus-infected and cancerous cells).

2. Bacterial defenses against the immune system include slimy capsules, lysosome inhibitors, and **leukocidins**.

SPECIFIC DEFENSES

1. Lymphocytes form in **primary lymphoid tissue** (red bone marrow and thymus). **T-cells** differentiate in the thymus and **B-cells** differentiate in the red bone marrow. Mature lymphocytes reside in **secondary lymphoid tissues** (lymph nodes, tonsils, adenoids, spleen, small intestine). B-cells respond to infection with the **humoral response**, the formation and secretion of antibodies. T-cells carry out the **cell-mediated response**, the activation of T-cells. Both are prompted by antigens.

2. Antibodies contain **light chains** and **heavy chains**, the latter ending in **variable regions** that bind to antigenic determinants on antigens. Antibodies differ in their variable regions. The **constant regions** are all the same within each of the five antibody classes. Antibodies work through neutralizing (masking) antigens, activating complement, binding to viruses, clumping bacteria together, and opsonization.

3. Lymphocytes form highly specific cell surface receptor sites that include class I and class II **MHC proteins**, which are essential to identifying invaders, normal and infected body cells, and each other.

4. During their development in the thymus, T-cells form dual cell surface receptors that identify class I or class II MHC proteins on body cells, and potential cell surface antigens. **Virgin T-cells** become specialists in binding to virus-infected and cancerous body cells, macrophages, and other lymphocytes.

5. During maturation in the red bone marrow, each B-cell incorporates thousands of IgM or IgD antibodies in its plasma membrane. Each B-cell then has the potential to form a match with an as yet unconfronted antigen.

CLONAL SELECTION AND THE PRIMARY IMMUNE RESPONSE

1. **Clonal selection** is the process in which highly specific lines of **virgin T-cells** and **B-cells** are selected among millions of variants for activation and cell division. Those activated have cell surface receptors that match an antigen, either free or on the surface of a pathogen that has invaded the body. A key participant is the **antigen-presenting cell**.

2. Macrophages incorporate free or partially digested antigen into the class I and II MHC proteins, forming a dual site that will match those of virgin cytotoxic and helper T-cells.

3. Antigen-presenting cells, when bound to T-cells, secrete interleukin-1, a messenger that stimulates cell proliferation. Clones of T-cells arise, including **memory cells**.

4. Cytotoxic T-cells (T_c) use their dual recognition sites to identify, attack, and kill infected cells bearing the antigen that brought about their arousal.

5. Helper T-cells (T_h) secrete **interleukin-2**, which stimulates cell division in other T-cells.

6. B-cells incorporate the antigen into one of their class II MHC proteins, forming a receptor that will match that of an aroused helper T-cell. On binding, the helper T-cell releases interleukin-2, and the aroused B-cell forms a large clone of plasma cells and memory cells.

7. **Plasma cells** secrete antibody that matches the antigen of the invader. The antibody reactions prepare invading cells and free antigen for phagocytosis and prompt complement release.

8. **Suppressor T-cells** coordinate the immune response, slowing it by blocking T-cell arousal.

VIGILANT MEMORY CELLS AND THE SECONDARY IMMUNE RESPONSE

1. In the **secondary immune response** the memory T- and B-cells instantly respond to new invasions by the pathogen that brought on their initial formation.

2. Vaccines promote **artificial active immunity** by bringing on a much reduced version of the primary immune response.

LYMPHOCYTE DIVERSITY

1. B- and T-cell diversity is genetically based, achieved by gene rearrangement involving variable region and joining genes. Versatility in the final arrangement of variable gene products and somatic mutations also result in antibody diversity.

MONOCLONAL ANTIBODIES

1. Clones of selected B-cells, those producing the desired antibodies, or **monoclonal antibodies**, are produced through cell culturing techniques. To produce large clones, subject B-cells are fused with myeloma cells to form **hybridomas**. These are separated and screened, and the desired cells are cultured and the antibodies collected. Monoclonal antibodies have many clinical applications.

WHEN THE IMMUNE SYSTEM GOES WRONG

1. **Autoimmunity**, self-destruction by one's own immune cells, arises when lymphocytes are sensitized by antigens that are chemically similar to our own MHC surface proteins. Aroused B-cells produce antibodies that interact with our own cells.

2. **Allergies** arise when normally harmless substances become **allergens**, chemicals capable of instigating the inflammatory response. Severe reactions can be life-threatening.

3. **AIDS (acquired immunodeficiency syndrome)** is a condition in which the immune system is suppressed. The agent is **HIV (human immunodeficiency virus)**, a retrovirus that destroys helper T-cells and others. HIV preferentially attacks macrophages and helper T-cells, although other cells may also be targets. It is spread chiefly through contaminated body fluids. Any surface cut or abrasion is sufficient for transmission of the virus. HIV often follows a 10-year course, during which there is a gradual decline in helper T-cells and rise in HIV particles, ending in the rise of opportunistic infections commence. There is presently no effective treatment for AIDS. AZT can slow its progress by blocking reverse transcription, but its effect is temporary and it often produces serious side effects. New treatment with interleukin-2 shows promise.

Review Questions

1. Briefly summarize the work of the skin and mucous membranes in warding off would-be pathogens. (page 858)

2. Describe the inflammatory response and list three agents that are involved. (page 859)

3. List three types of phagocytes and explain how they deal with invaders. (page 861–862)

4. What are the two specializations of natural killer (NK) cells? (page 862)

5. Distinguish between the humoral and cell-mediated specific immune responses. Which immune cells are responsible for each? (page 864)

6. Prepare a simple drawing of an antibody, labeling the following: heavy chain, light chain, disulfide bond, hinge, variable region, constant region, and Fc region. Describe four kinds of antibody–antigen interactions. (pages 865–867)

7. Briefly summarize how T-cell receptors form their dual recognition proteins and what each recognizes. (pages 868–869)

8. What comprises the B-cell's surface receptor? (pages 869–870)

9. The primary immune response occurs through the clonal selection process. What is clonal selection, and what does it accomplish? (page 870)

10. What determines whether an antigen-presenting macrophage binds to a virgin cytotoxic T-cell or a virgin helper T-cell? What is the first thing that happens when such binding occurs? (page 871)

11. Explain how cytotoxic T-cells kill their target cells. (page 872)

12. What are two functions of helper T-cells? (page 873)

13. Describe the activation of B-cells. (page 873)

14. Why does the secondary immune response occur so much more quickly than the primary immune response? (pages 875–876)

15. Explain how the enormous variability in T- and B-cells arises. (pages 876–877)

16. What are monoclonal antibodies? Why is it useful to have cultures of cells that produce only one kind of antibody? (pages 877–878)

17. Describe the steps in an allergic response. In what instances are they life-threatening? (page 879)

18. List the body fluids through which HIV is spread from person to person. In each case, what must happen for the virus to infect a second person? (page 880)

19. Which cells does the HIV preferentially attack? Name the cell surface molecule to which it binds. (page 880)

20. Outline the course of HIV infection. At what point does active AIDS usually begin? (page 881)

Reproduction

C H A P T E R   O U T L I N E

REPRODUCTION AND THE SURVIVAL PRINCIPLE

ASEXUAL REPRODUCTION

New Individuals from Old

Development from the Egg: Parthenogenesis

SEXUAL REPRODUCTION

SEXUAL REPRODUCTION IN INVERTEBRATES

External Fertilization in Invertebrates

Internal Fertilization in Invertebrates

SEXUAL REPRODUCTION IN VERTEBRATES

External Fertilization in Vertebrates

Internal Fertilization in Vertebrates

HUMAN REPRODUCTION AND SEXUALITY

The Male Reproductive System

The Female Reproductive System

Hormonal Control of Human Reproduction

CONCEPTION CONTROL

Hormonal Control

Implantation Barriers

Sperm Barriers

Chemical Agents

Unaugmented Methods

Rhythm

Essay 41.1 Human Sexuality: The Sexual Response

Surgical Intervention

KEY IDEAS

APPLICATION OF IDEAS

REVIEW QUESTIONS

We may be very proud of ourselves and think we fit quite well in our world. However, this is true only of our present world environment—we cannot assume that we would fit into some future world with equal success. If, when we reproduce, we were to make replicas of ourselves, with each of our offspring bearing 100% of our genes, they would presumably do well as long as the world remained as it is now. But should things change, our little replicas might not fit at all.

However, should we mix our genes randomly with someone else's, those genes having first been shuffled around through meiosis, we will produce a more diverse assortment of offspring. Then should new conditions arise, at least some of our offspring would be likely to survive and propel our genes into yet another generation. This is the essence of one prevailing view on why sex evolved—it promotes genetic variation. That variation is the raw material on which natural selection can act. Once sexual reproduction took hold, it was a smashing success. Over time, the sexual process itself has received enormous attention from the selective forces of nature. As a result, there are nearly countless variations in the way sperm and egg come together.

Reproduction and the Survival Principle

As we proceed, you will be introduced to diverse, often bizarre, examples of animal reproduction. We really didn't do this just to amaze and amuse you: there is a principle to be learned. We will see a great range in the numbers of offspring species produce—from just one or two, to millions. We will also note that some species, those producing millions of offspring, provide little if any maternal or paternal care to their offspring, whereas others, those producing just a few, may devote a great part of their lives to this endeavor. In addition, mating may be simple and direct,

(a)

(b)

FIGURE 41.1
Playing the Odds
The most fundamental rule of life is: *reproduce or face extinction*. There are varied ways of complying.
(a) Some species produce large numbers of offspring and leave the rest to chance, whereas **(b)** others
produce just a few and dedicate much of their time and energy to assuring their survival.

with a species simply aggregating and releasing its gametes, or it may involve elaborate precoital mating behaviors. Yet, regardless of the numbers, the care, and the mating regimen, the outcome is the same. The point is, in the long run, the average number of offspring to survive and reach sexual maturity is one for each reproducing adult. This number has proved sufficient for the perpetuation of the species. (Are humans still reaching the "long run," or are our current excesses an exception? See Chapter 48.)

So the various modes of reproduction are merely alternatives in the expenditure of reproductive energy. And natural selection seems to have favored many directions. If gametes are simply to be released into the environment and offspring left on their own, then an animal's reproductive energy goes into the production of enormous numbers, and a game of chance ensues. Thus the loss of a few offspring means very little (Figure 41.1a). On the other hand, if safe development and care of offspring are the directions that evolution has taken, then, out of necessity, relatively few offspring are produced, and the reproductive energy is spent in ways that tend to ensure a favorable outcome (Figure 41.1b). The loss of even a few offspring in this case, however, may have severe and tragic consequences.

Asexual Reproduction

If we want to start with the simplest case of reproduction, we will have to set sexual reproduction aside for the moment and

look into its alternative. We've seen it before in our perusal of the kingdoms, where reproduction involves just one individual and there is no new input of genes. This is, of course, **asexual reproduction**. As far as we know, all animals are capable of sexual reproduction, so asexual reproduction is an alternative that some species use. There are several ways in which this can be achieved by animals.

New Individuals from Old

You may remember that the reproductively primitive soft-bodied amebas, which are actually protists, undergo **fission**: they simply divide in half (Figure 41.2a). Fission in the ameba is preceded by DNA replication and mitosis, processes that assure that daughter cells will have all the genes of the species. Fission is simply a means of increasing the population and limiting cell size (see Chapter 23).

Some simpler animals reproduce asexually by **budding**. In the common *Hydra,* a freshwater cnidarian, for example, certain body cells simply produce a new body, one that usually branches from the parent animal (Figure 41.2b). Some flatworms, bryozoans, and annelids undergo **fragmentation**, a process in which the body breaks into two or more parts; each then undergoes **regeneration**, which means building an entire body from a portion. Sea stars and flatworms also undergo extensive regeneration (see Chapter 26). Freshwater sponges survive winter by producing **gemmules**, collections of cells set aside and surrounded by a new body wall. Each gemmule will later develop into a new sponge.

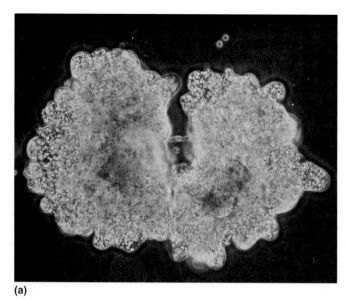

(a)

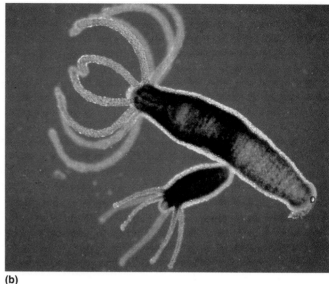

(b)

FIGURE 41.2
Asexual Reproduction
(a) The protist, *Amoeba proteus,* reproduces asexually through binary fission, which includes mitosis and cell division, preceded, of course, by DNA replication. Sexual reproduction has not been observed in this species. **(b)** *Hydra* reproduces asexually through budding. The miniature *Hydra* growing from the adult body wall will eventually break away and begin independent life.

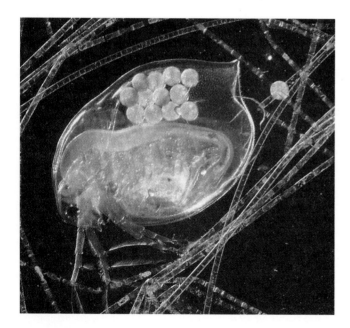

FIGURE 41.3
Parthenogenesis
Parthenogenesis in *Daphnia* (the water flea), one of the smallest crustaceans (1.6 mm), is the rule rather than the exception. During easier times, the population grows rapidly through this process. Note the presence of partly developed embryos in the female brood pouch. *Daphnia* retains its eggs, giving birth to well-developed, but miniature individuals.

Development from the Egg: Parthenogenesis

A number of animals (and many plants) reproduce asexually through **parthenogenesis**, the development of an offspring from an unfertilized egg. The process is common among insects (bees, wasps, ants, and beetles). Species such as the crustacean *Daphnia,* an important member in the food chains of northern lakes, routinely alternate between parthenogenesis and sexual reproduction. Parthenogenesis most often begins when populations are small and growth conditions optimal (as one songwriter put it, "when the livin' is easy"). At such times, *Daphnia's* unfertilized eggs begin development within the female body, with only female offspring produced (Figure 41.3). But in stressful times, as when populations are large and food supplies short, or when the water temperature falls, some eggs develop into males, and sexual reproduction resumes.

Sexual Reproduction in Animals

Sexual reproduction in animals requires an input of genes from two individuals, generally through the union of sperm and egg in fertilization. Two fundamental ways in which this is accomplished, each with a great many variations, are external fertilization and internal fertilization. They occur in both invertebrates and vertebrates.

In **external fertilization**, which almost always occurs in a watery environment, the eggs and sperm are shed into the water. Sperm swim to the eggs, sometimes drawn by chemical attractants, and fertilization occurs. In some species, there is no further association between the parents and offspring. This sounds like a very chancy process. Consider the fate of the 2.5 million eggs released in one spawning by the sea star. Most simply contribute to the marine food chain. But recall the principle: for a species to survive, individuals simply have to replace themselves. In other externally fertilizing species, protective measures may be taken by the parents.

With **internal fertilization**, the egg is retained in the female body, where it is fertilized. In most instances, internal fertilization includes some form of **copulation**, the coupling of a male and female, where sperm are introduced into the female genital tract. A penis and vagina, or their equivalents, are generally involved, but as some 8600 species of birds might attest, these specialized copulatory structures aren't absolutely essential.

Internal fertilization has at least three outcomes not found in external fertilization: (1) it reduces the number of eggs and sperm necessary; (2) it vastly increases the probability that all the eggs will be fertilized; and (3) it provides the developing embryo with some measure of protection, should it develop within the mother. As the mother looks after herself, she looks after her developing offspring.

There are other variations. Many species release their fertilized eggs to develop externally. Some species then protect the eggs; others simply abandon them.

The evolution of internal fertilization provided distinct advantages to animals making the transition from aquatic and semiaquatic life to fully terrestrial life. With the gametes shielded from the drying surroundings, and the zygote safe in the mother's body, reproduction could succeed on dry land. (In Chapter 25, we learned of a parallel step taken by plants.) The transition to land, made hundreds of millions of years ago, opened up a host of new evolutionary opportunities for animals. But a word of caution: the more complex modes of reproduction to be discussed shortly were not reserved for the terrestrial animal groups alone, many occurred in animals that remained in the water.

Sexual Reproduction in Invertebrates

External Fertilization in Invertebrates

External fertilization is commonplace in the echinoderms. In fact, generations of biology students have observed sea urchins or sand dollars shedding their gametes into laboratory beakers (Figure 41.4). External fertilization works well where populations are very dense and currents can bring the gametes togeth-

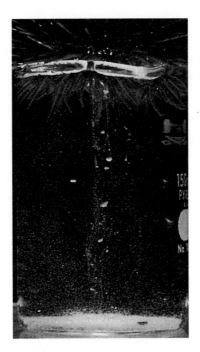

FIGURE 41.4
External Fertilization
Echinoderms usually shed their gametes into the sea, where fertilization and development take place. The presence of sperm or eggs from one individual stimulates others to release their gametes. Here a sea urchin has been prompted to release its gametes (eggs) through the introduction of a calcium chloride solution. When the process ends, a mound of brightly colored unfertilized eggs will have formed in the bottom of the beaker.

er. Success also requires the simultaneous release of gametes by many members of the population. In echinoderms, the initial release by a few individuals is prompted by cues from the surroundings; the presence of gametes released by the first individual prompting others to follow suit.

Internal Fertilization in Invertebrates

Sponges. Surprisingly, the simple sponges make use of internal fertilization. Most sponges are **monoecious** ("same house"); that is, each individual produces both sperm and eggs. Yet sponges do not self-fertilize. The eggs remain within the body wall, where they are fertilized by sperm released in clouds by other sponges and swept in by the collar cells. The zygote develops into a swimming larva that later escapes from the adult body and undergoes the transition to a new adult.

Snails and Barnacles. Internal fertilization without copulation, as seen in the sponge, is somewhat unusual. Most internally fertilizing animals copulate, making use of an intromittent organ, usually a **penis**, for transferring sperm; and a receptacle, commonly a **vagina**, for receiving them. *Helix,* the com-

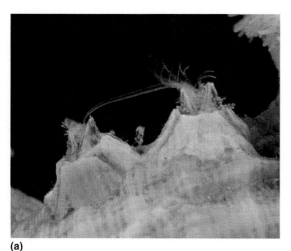
(a)

(b)

FIGURE 41.5
Internal Fertilization in Simpler Invertebrates
(a) Barnacles are monoecious. Here a barnacle has extended its lengthy penis (the light-colored, hair-like object) and inserted it into the sex opening of a neighbor. The sperm it will release will internally fertilize eggs in the recipient. **(b)** These land snails are copulating. Each of these monoecious animals has penetrated the other with its penis. Internal fertilization will result, after which the snails will part company.

mon land snail, and *Balanus,* the giant barnacle, are both admirably equipped for copulation.

Barnacles are fixed in place, living in crowded colonies on rocks and wharf pilings. To transfer sperm, they make use of a long, probing penis to reach a neighboring barnacle (Figure 41.5a). Since the barnacle is monoecious, any neighbor will do. Monoecious species are fairly common among the invertebrates, especially the sessile (stationary) ones.

In its mating activity, *Helix* has an interesting behavioral repertoire. Like *Balanus,* it is monoecious. As two reproductively ready land snails draw near, they extend their bodies, gently touching and intertwining. Then suddenly, each thrusts a tiny, sculptured dart into the body of the other. Both recoil, at first, but they are now strongly stimulated to proceed. A white spot appears near an eye stalk in each—the emerging penis. Like the barnacle, the land snail has an enormous penis—almost as long as the body when extended. Each probes the other's body for the reproductive opening, inserts its penis, and releases its sperm (Figure 41.5b). By most standards, copulation in snails is a lengthy process—on a slow day, one of your authors clocked the process at two hours. We can even find an adaptive advantage for the monoecious life: since separate sexes do not exist, the search for a mate is greatly simplified, saving a lot of time and energy.

Insects. Copulation and internal fertilization are also the rule in insect species, but most are **dioecious** ("two houses"); sperm and eggs are produced by different individuals. Male insects are particularly distinct from females in their reproductive organs (Figure 41.6). The typical male reproductive structures include a penis, for copulation; paired, internalized **testes**, within which sperm develop; paired **sperm ducts**, for conducting sperm;

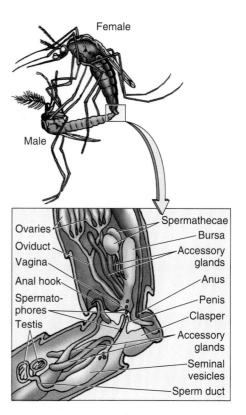

FIGURE 41.6
Internal Fertilization in Insects
The reproductive system of the male mosquito is complex. It includes claspers, hooks, and barbs, all for holding the female in position. The female mosquito's reproductive system is the reciprocal, providing places to receive the claspers, hooks, and barbs. Insects assume a tail-to-tail mating position. When the male's parts are fully engaged, he transfers his sperm into the female receptacle.

seminal vesicles for sperm storage; and **accessory glands**, to produce seminal fluid for sperm transport. The sperm are transported in packets called **spermatophores**. The female anatomy includes a vagina for receiving the penis, **spermatheca** for sperm storage, **oviducts** for conducting the eggs from the **ovary**, and **accessory glands** that secrete egg coverings. Many female insects also come equipped with **ovipositors**, digging or boring devices used to hollow out a chamber in the soil or a plant in which to lay the fertilized eggs.

During mating, which in many insects occurs in a tail-to-tail attitude (see Figure 41.6), the male extends the penis and inserts it into the female vagina. The sperm packets are then ejected through the sperm ducts. Male insects commonly use elaborate clasping organs for holding the female in position while copulating. Furthermore, the penis is usually barbed, hooked, or braced, while the vagina is fashioned to form the exact reciprocal of these features. (The overall effect is somewhat like the coupling devices of railroad cars, but more complex.) Students of evolutionary theory maintain that this lock-and-key arrangement is adaptive in that it prevents interspecific matings. This is just one means of reproductive isolation among species (see Chapter 20). (Remember that most animal hybrids, the product of interspecific mating, are sterile.) Considering the vast number of insect species, it follows that many are quite similar (at least to us), so any mechanism that helps a species avoid wasting its reproductive efforts would be adaptive.

Sexual Reproduction in Vertebrates

We've seen that invertebrates make use of a wide range of reproductive repertoires, involving external and internal fertilization, and the monoecious and dioecious condition. What can we expect in the vertebrates? If you are thinking that vertebrates have advanced reproductively well beyond invertebrates, it's time to dispel that idea. You will see many of the same reproductive modes at work, from the simple shedding of gametes into the sea, to highly elaborate copulatory behavior. What we do see is the same ecological overtones, with external fertilization restricted to the aquatic environment and internal fertilization an absolute requirement on land.

External Fertilization in Vertebrates

Fishes. Typically, female fishes release their eggs first and the males then release their sperm, or milt as it is known, over the eggs. The cod, an ocean fish, sheds its gametes into the sea and from that point on ignores the young. Fertilization is not left completely to chance, for the cod, like many other externally fertilizing animals, congregate in large groups before discharging their eggs and sperm. Nevertheless, the unreliability of this routine—the losses are astronomical—requires the production of great numbers of gametes just to assure that each individual replaces itself in the next generation. (Six million eggs were counted in one female cod.)

FIGURE 41.7
External Fertilization in Fish
Typically, the release of eggs by the female stimulates the male to release his milt right over them. Whereas some fishes release their gametes *en masse,* a practice that requires great numbers to assure success, others demonstrate elaborate behaviors. Grunion mating behavior greatly increases the chances of success, so fewer gametes are required. The female in the center has buried her tail in the moist sand of a Southern California beach. The stimulated male writhes around her, releasing his milt around the eggs. When finished, the two will flap their way back to the sea. The offspring will hatch out and follow in about one month. Biologists collecting the eggs have learned that just shaking them in seawater will hasten the hatching process.

California grunion (Figure 41.7) also engage in external fertilization, but, oddly enough, they increase the odds of reproductive success by *coming ashore to mate!* The grunion spawns seasonally, three or four days after a full moon—with the highest tides of the month. Shortly after the tides begin to recede, the females, hotly pursued by the males, ride the incoming waves, thrash their way up onto the wet beach, wiggle their tails into the sand, and release the eggs. The males writhe around the females, releasing their milt. The fertilized eggs are left to develop, buried in the sand, until the waters of the next high tide stimulate hatching and wash the young out to sea. Grunion "fishing" is permitted in California, but only hands and buckets are allowed!

Some externally fertilizing species exhibit parental care. As the breeding season of the male three-spined European stickleback (Figure 41.8) draws near, it builds an elaborate, tunnel-like nest of water plants. Then, using precise, elaborate movements—a zigzag dance, body quivering, and some anxious prodding and nudging—the male guides the female into the grassy tunnel where she releases her eggs. The excited male immediately enters the nest and releases his sperm over the eggs. The female shows no further interest, but no matter, the male guards the nest, fans fresh water over the eggs, and later protects the hatchlings. This species has thus greatly improved the efficiency of external fertilization and therefore requires far fewer eggs.

(a)   **(b)**   **(c)**

FIGURE 41.8
Improving the Odds: Elaborate Mating Behavior and Parental Care
Although fertilization and development in the European three-spined stickleback are external, complex
behavior patterns help assure survival. Phases of the reproductive behavior include nest building, brief
courtship, and mating. **(a)** The male prods the female into the nest, a grassy tunnel, where she is prompted
to release her eggs. **(b)** He follows her in and releases his milt. **(c)** Later, the male fans the eggs to provide
oxygen, and protects the nest.

Amphibians. Many amphibian species spend time on land
but must return to the water to reproduce. Frogs and toads do
this and, like many aquatic animals, are external fertilizers.
However, these amphibians increase the odds of reproductive
success by pairing off and mating. The male clasps the female
from behind (Figure 41.9), and as she releases the eggs, he
sheds sperm directly over them. Their reproductive openings
are very close to each other during this process, and it doesn't
take much imagination to see that the next evolutionary step
would be for the male to deposit the sperm directly into the
female's reproductive tract. Despite the advantages of this
careful positioning, there is no parental care, and 25,000 eggs
are required of each female bullfrog just to keep the species
going. This brings us to internal fertilization in vertebrates.

Internal Fertilization in Vertebrates

Fishes. Many fish species make use of internal fertilization.
Most male bony fishes of this category have rudimentary copulatory structures, specifically a cone-like extension of the body
opening. Males of the family Phallostethidae (*phallos* means
penis), however, have well-developed, fleshy intromittent
organs. Some species simply release the fertilized egg for external development, but others, such as the aquarium favorite
Gambusia, retain the eggs throughout development. *Gambusia,*
like sharks, are live-bearers.

Male sharks lack true penises but copulate successfully
through the use of **claspers** (Figure 41.10). These are modified
pelvic fins that are used to penetrate the female cloaca and
introduce sperm (see Chapter 27). The **cloaca** is present in
fishes, amphibians, reptiles, and birds and, in each case, repre-

sents a simple cavity that has digestive, excretory, and reproductive functions. Some sharks are egg-layers, but others are
"live-bearers," they give birth to developed young. Of the latter category, most species retain their eggs for internal devel-

FIGURE 41.9
External Fertilization in the Frog
Although frogs fertilize externally, the chances of each egg being
fertilized is improved by their mating behavior. The male stimulates
the female to release her eggs and, almost simultaneously, releases
his sperm over the egg mass. The darkly pigmented eggs are
released in a jelly-like mass where development takes place.

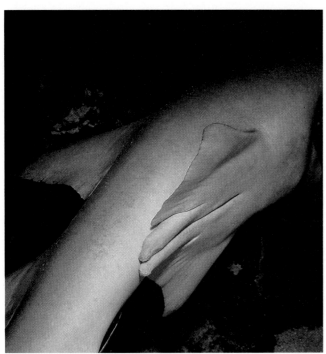

Claspers of the male shark

FIGURE 41.10
Internal Fertilization in Sharks
The male shark makes use of claspers for copulation. When these modified fins are held together, they form a grooved passage. In mating, the male inserts the claspers into the cloaca of a female and introduces sperm into the oviduct within.

opment without actually nourishing the embryos. However, a few rival the placental mammals; the female nourishes the embryos as they develop. We're not sure how to categorize sand tiger sharks. The first embryos to develop eat their younger brothers and sisters as fast as they form.

Amphibians and Reptiles. With the exception of the limbless salamander, amphibians do not have copulatory organs. Yet, a few manage internal fertilization. The tailed frog uses its tail in copulation. In most salamanders, the male releases packets of sperm, which the female retrieves with her cloaca (Figure 41.11).

Among reptiles, male turtles and crocodiles have true penises, complete with erectile tissue, basically similar to that of the mammal. In the unaroused condition, the penis appears as two spongy ridges and an enlargement called a **glans**, all within the cloaca. During sexual excitement, the spongy tissue fills with blood, and the ridges close together, forming a sperm-conducting groove. The enlarged glans, which protrudes from the body at this time, is inserted into the female cloaca, and the sperm are deposited into the oviduct.

Male snakes and lizards have two penises, or **hemipenes**, which can be everted—turned inside out like the fingers of a glove (Figure 41.12). This accommodates a "side-by-side" mating position with only one penis used at a time.

Birds. Male birds, with the exception of certain ducks, ostriches, and a few others, lack penises. Their mating

(a) Salamanders mating

FIGURE 41.11
Internal Fertilization in Amphibians
Salamanders begin their mating with a pronounced courtship procedure. **(a)** After a number of nudges by the male, the two perform a circular "waltzing" maneuver; the female then prods the male, who releases his sperm packets **(b)**. The female moves over the packets and retrieves them with her cloacal lips. Fertilization then takes place inside the female.

(b) Retrieval of sperm packet

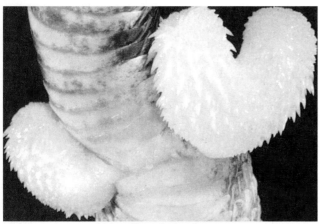

Male rattlesnake hemipenes

FIGURE 41.12
Reproductive Structures in the Snake
The copulatory organ in the snake is the hemipenis, actually a pair of hemipenes. In the inactive state, the hemipenes are tucked away in the cloaca, turned inside out. During sexual excitement, the hemipenes are everted, like the fingers of a glove, emerging from the cloacal opening. The barbed surface ensures that the penis will remain in place in the female cloaca while sperm are released. Parting may present special problems.

Great Blue Herons

FIGURE 41.13
Copulation in Birds
Copulation in birds occurs in a fleeting moment wherein the male and female cloacas are brought together and sperm transferred. This seemingly crude technique works surprisingly well.

techniques can best be described as "unlikely"—at least from a human perspective. Insemination is accomplished by the male pushing his cloaca against the female's and, in a flurry of wing-flapping, transferring his sperm (Figure 41.13). Swifts and bald eagles, which are known as great fliers, are even better fliers than we thought—they do this in midair!

Birds are well known for the care they give their offspring. Typically, only a few eggs are produced each season, and then one or both parents lavish them with warm attention. In quail, chickens, ducks, geese, and other ground-nesting birds, newly hatched young are ready at once to feed on their own (Figure 41.14a). In tree-nesting birds such as robins and finches, the young hatch in a completely helpless, semideveloped state, doddering weakly about and demanding feeding and protection for many days (Figure 41.14b). The reproductive success of these species hinges on the successful rearing of just a few well-protected offspring.

Mammals. Mammals reveal several unique reproductive features. Fertilization is of course internal, and following fertilization, the young are retained in the uterus where their metabolic needs are met by the placenta. The **placenta** is a sponge-like mat of tissue, produced cooperatively in the uterus by the embryo and the mother. It connects to the embryo via the umbilical cord (Figure 41.15). It is through the placenta that the developing placental mammal receives its oxygen and nourishment and eliminates its metabolic wastes. Following its birth, the offspring feeds on milk produced by the mother in

mammary glands. There are, of course, the inevitable exceptions, in this case the monotremes and marsupials.

Monotremes. Monotremes, represented by the spiny anteater and the duck-billed platypus (Figure 41.16), are bird-like in that they have cloacas, copulate by pressing their cloacas together, and lay large, bird-like eggs in which development occurs. The female echidna, or spiny anteater, tucks the egg away in a temporary groove in her abdomen, whereas the female platypus hides her eggs in a lengthy burrow, where they are periodically incubated. Female monotremes produce milk, but, lacking nipples, they secrete milk from simple pores.

Marsupials. Reproduction in the marsupials (kangaroos, opossums, and other pouched mammals) is more typical of mammals, except for one aspect. By mammalian standards, birth of the marsupial is quite premature, by comparison, about like a human embryo just a few weeks old. The largely unformed and glistening offspring, looking like a tiny slug with legs, leaves the security of the uterus and makes its way to the pouch (Figure 41.17). There it firmly attaches to a nipple, where it remains until its development is complete. Furthermore, the marsupial placenta is quite rudimentary and lasts just a brief time. After the birth is complete, the female may immediately become pregnant again. In fact, she may support several offspring of different ages within the pouch, making marsupial reproduction quite efficient.

(a) Precocial young fend for themselves

FIGURE 41.14
Bird Reproductive Strategies
(a) Young in ground-nesting birds are born in a precocious state,
ready to fend for themselves right after hatching. For a time a brood
will follow its mother closely, as she protects them and demonstrates
some survival techniques. They rely on protective coloration and
quickly learned behaviors to avoid predators. **(b)** In songbirds,
generally tree-nesting species, the young are born in a much more
immature state. During this helpless period, they are hidden away in
nests and fed continuously by the busy parents. As a rule, the clutch
size in tree-nesting bird is considerably smaller than ground-nesting
species, but the time required for rearing is much longer.

(b) Altricial young are helpless at hatching

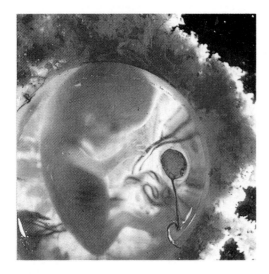

FIGURE 41.15
The Placenta
The placenta, as seen in this view of the human embryo, is the frilly,
surrounding tissue. It consists of tissues contributed by both the
embryo and the mother. Where they come together, they form
extensive blood sinuses where nutrients and oxygen can enter the
embryo's circulation and metabolic wastes can enter the mother's.

FIGURE 41.16
Duck-billed Platypus and Young
Like other mammals, monotremes have hair, fertilize internally, and
produce milk. However, like birds and reptiles, they lay eggs.

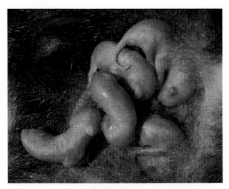

Young marsupials attached to nipples

FIGURE 41.17
Pouched Mammals
In marsupials, the young begin development in the uterus but are born early, finishing development in the pouch. The very young marsupials seen here have firmly attached themselves to nipples and will remain that way for weeks to come.

Placental Mammals. As we've seen, the placental embryo completes its development in the secure and protected environs of the uterus, its metabolic needs satisfied by the placenta. Since humans are in many ways similar to other placental mammals, we will focus our own reproductive biology.

Human Reproduction and Sexuality

We will first explore the reproductive anatomy of males and females and, along the way, focus briefly on the sperm and egg and how they form. Next, we'll turn to hormonal control—that often puzzling but always fascinating aspect of sex. With these topics behind us, we can turn an informed eye to the human sexual response. Finally, we will have a look at contraception. We have tried to anticipate some of your questions and to fill in informational gaps that are often bothersome and sometimes embarrassing to ask about.

The Male Reproductive System

External Anatomy. The **genitalia** (external organs) of the human male, typical of mammals, includes the **penis** and the **testes** (Figure 41.18).

The Penis. In humans, the penis is an erectile shaft tipped by an enlarged region, the **glans**. In those who have not been circumcised, a fold of skin known as the **foreskin** covers the glans. In the aroused state, the glans enlarges, protruding past the foreskin. The penis must be more or less erect and turgid before it can be introduced into the female vagina.

As in most mammals, the human penis contains **erectile tissue**, which makes erection possible. It occurs as three cylindrical, spongy regions (Figure 41.19a). During sexual excitement, the blood into the spongy tissue increases, and venous flow back to the body is retarded. In its turgid, blood-filled state, the penis enlarges and becomes erect (Figure 41.19b). A complete or partial loss of erection almost invariably follows ejaculation. In the flaccid, unaroused state, the entry of blood into the spongy tissue is more restricted and a steady drainage is permitted. In cattle and sheep, erection is brought about simply by extending the penis, which is already stiffened by cartilage or bone, from deep within the body where it usually rests. For a discussion of both female and male sexual arousal, see Essay 41.1.

The Testes. The testes (singular, **testis**) are paired oval bodies that produce both sperm and sex hormones. Shortly before birth, they descend from the abdomen into the sac-like **scrotum**, where development continues. The descent of the testes is important to all male mammals because the high internal body temperature is harmful to sperm. The scrotum is active in regulating the temperature of the testes. When the body is overly warm, the scrotal muscles relax and the testes become suspended further from the body. When the testes become cold, the muscles contract, drawing them closer to the body.

Much of the testis (Figure 41.20) consists of the highly coiled **seminiferous tubules**, where sperm are produced. The network of tubules and ducts leads to the **epididymis**, where the sperm are stored. Clusters of **interstitial cells**, lying outside the seminiferous tubules in a connective tissue matrix,

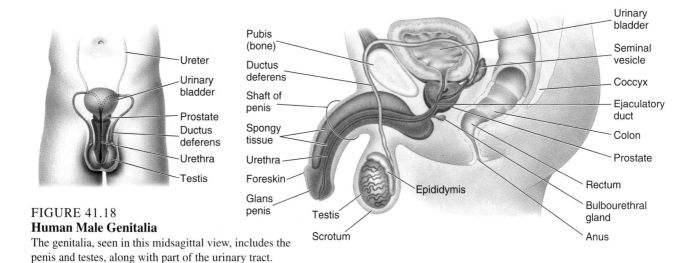

FIGURE 41.18
Human Male Genitalia
The genitalia, seen in this midsagittal view, includes the penis and testes, along with part of the urinary tract.

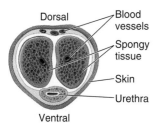

(a) Cross section of shaft

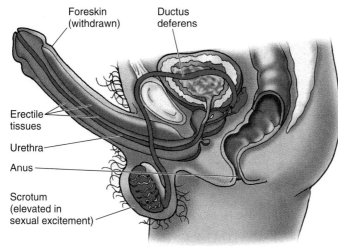

(b) Erection (the sperm pathway is highlighted)

FIGURE 41.19
Erectile Tissue of the Penis
(a) The penis has three regions of erectile tissue, as seen in the cross section. (b) The retention of blood in the erectile tissue produces the erection.

secrete the male sex hormone **testosterone**. Each testis contains a rich supply of sensory receptors that can effectively register both pleasure and, as only an experienced male can attest, excruciating pain.

Spermatogenesis. Spermatogenesis (sperm production) occurs in the **germinal epithelium** lining the seminiferous tubules (Figure 41.21). There, diploid cells known as **spermatogonia** undergo mitosis, with about half of the daughter cells committed to meiosis and sperm formation. The uncommitted cells continue mitosis, thus providing a continuous source of new cells for the meiotic process.

When a spermatogonium enters the meiotic process it becomes known as a **primary spermatocyte**. Following the first meiotic division, it becomes a **secondary spermatocyte**. Meiosis II follows, with the emergence of four **spermatids**, each with the haploid chromosome number (see Chapter 9). Each spermatid will then undergo the complete reorganization that is necessary to form mature **spermatozoa**. The spermatids lose most of their cytoplasm at this time and each receives metabolic support from surrounding "nurse" cells called **Sertoli cells**.

In the mature spermatozoan (see Figure 41.21), the chromatin will be condensed into the minute **sperm head**. The cytoplasm itself will differentiate into a midpiece and tail, and excess cytoplasm will be sloughed off. The midpiece contains a peculiar sheath of spiral-shaped mitochondria and a centriole. The tail is a very long, tapered flagellum, which propels the sperm along. At the tip of the sperm head is the **acrosome**, a cap-like structure that arises from an enzyme-laden lysosome and overlays the compact nucleus. We'll see in the next chapter how the enzymes of the acrosome function in egg penetration and fertilization. Following their development in the seminiferous tubules, the immature sperm move into the epididymis, where they are stored.

FIGURE 41.20
The Testis
The seminiferous tubules and epididymis make up most of the testis. Outside the tubules lies the interstitial cells, which consists chiefly of connective and endocrine tissues. Cells of the seminiferous tubule generate sperm, while the endocrine tissue produces testosterone, the male sex hormone.

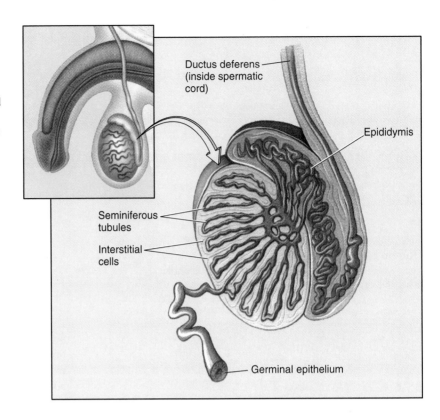

The Sperm Route. Upon **ejaculation**, the human male forcefully expels some 2–5 ml of semen from the penis, each milliliter containing between 50 and 150 million sperm. The journey begins with sperm leaving their storage area in the epididymis and entering the **ductus deferens** (or **vas deferens**), a lengthy sperm-conducting tube located within the **spermatic cord** emerging from each testis (see Figures 41.18 and 41.19). The spermatic cord itself is a sheath containing blood vessels and nerves as well as the ductus deferens. The sperm-conducting tubes pass into the body, uniting just below the bladder. It is here that the **semen** begins to gather. Ducts from the seminal vesicles add the first secretions to the sperm, making up about 60% of the semen volume. The secretions are an alkaline fluid, containing sugars, prostaglandins, and a coagulant. The alkaline fluid protects the sperm from the normally acidic state of the vagina. The sugars are taken up by the sperm for use in producing ATP (there isn't much storage room in a sperm cell). Prostaglandins activate the sperm, and the coagulant thickens the semen.

Surrounding the junction of the two ductus deferens is the fleshy tissue of the **prostate gland**. The prostate secretions, making up some 25% of semen volume, further mobilize the sperm and produce the characteristic color and odor of semen. The late activation of sperm—during ejaculation—is vital since each sperm has limited energy resources and, once released in the vagina, still has a lengthy journey to complete. (On a different scale: assuming your car is about 15 ft long, an equivalent journey would be about 200 miles.)

Following the addition of fluids from the prostate, the moving semen receives its final secretions from the **bulbourethral glands** (also called *Cowper's glands*). These secretions further increase the alkalinity of the semen and add slippery mucus that protects the delicate sperm by lubricating the urethra. The semen, now in its final state, follows the route of the **urethra** through the penis, from which it is forcefully ejected during ejaculation.

Ejaculation is brought about by peristaltic contractions of smooth muscles in the seminal vesicles and prostate, and by contractions in skeletal muscles at the base of the penis. The ejaculatory contractions and the intensely pleasurable sensations associated with them comprise the male orgasm (see Essay 41.1, page 906).

The Female Reproductive System

External Anatomy. The most prominent part of the **external genitalia** (also called the *vulva*) (Figure 41.22) of women is the **mons pubis** (also, *mons veneris*: "mountain of love"), a hair-covered mound of fatty tissue overlying the pubic arch. Below the mons pubis lie the larger outer folds, or **labia majora** ("major lips"; singular, *labium majorum*), of the vulva. Just inside the labia majora are the thinner, inner folds, or **labia minora** ("minor lips"; singular, *labium minorum*). In an unaroused state, the labial folds cover a number of rather sensitive structures. The inner lips join together at their upper portion to form a hood over a small prominence, the **clitoris**. This

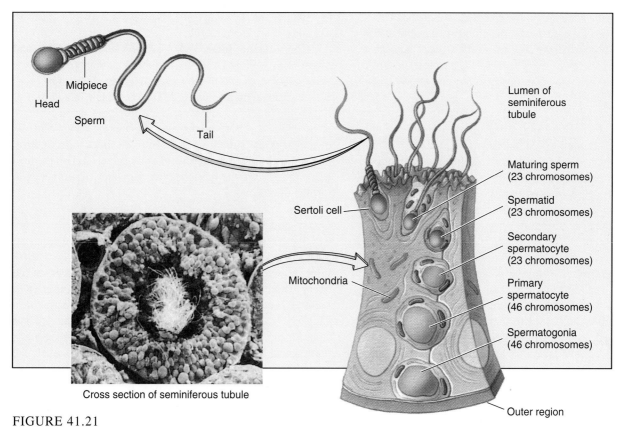

FIGURE 41.21
Spermatogenesis

In the simplified illustration, we see that meiosis proceeds from the outer region of germinal epithelium toward the lumen of the seminiferous tubule. Spermatogonia enter meiosis, forming primary spermatocytes. They undergo division, forming secondary spermatocytes, which then divide one more time to form spermatids. The spermatids mature with the aid of nurse cells. The mature spermatozoon is little more than a highly condensed nucleus (the head) and a means of propulsion (the midpiece and tail). A sheath of spiral mitochondria provides the ATP needed for the flagellum to work. The scanning EM shows numerous sperm within the lumen (note the tangle of flagella).

FIGURE 41.22
Human Female Genitalia

The female genitalia includes the mons pubis, labia majora and minora, the clitoris, and the vaginal orifice. Not visible are the openings of ducts from Bartholin's glands, near the vaginal opening. (The labia are drawn open in an exaggerated manner to better illustrate the various structures).

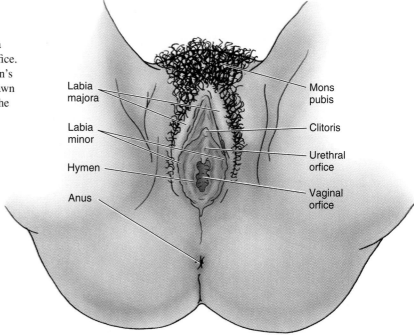

cylindrical organ contains spongy erectile tissue and numerous sensory neurons. Like the penis, the clitoris is capable of filling with blood and becoming firm during sexual arousal, a time when it is also exquisitely sensitive. The labia minora also retain blood at this time, the lips opening and exposing the vagina.

Two openings are located between the labia minora. The upper, much smaller opening is the raised, external opening of the urethra, through which the urine is voided. The second is the vaginal orifice, or opening. In virgins, the vaginal orifice may be partially blocked by membranous adhesions known collectively as the **hymen**. While there is much variation, a certain amount of discomfort and bleeding are commonly experienced during the first intercourse, when the hymen is ruptured.

Internal Anatomy. The vagina is a muscular tube about 8 cm (about 3 in.) long in its relaxed state (Figure 41.23). It serves both as the copulatory organ, receiving the erect penis during copulation, and as the birth passageway. The vagina is marvelously adapted for the two diverse functions. Its walls consist of three tissue layers: an inner mucous membrane, a middle muscular layer, and an outer layer of fibrous connective tissue. The mucus-secreting lining presents a moistened, yet firm, stimulatory surface for copulation. The vaginal muscle and connective tissue layers are capable of great extension during the passage of the infant at birth, accommodating a head about 4 inches in diameter.

During sexual arousal the vagina secretes a clear, lubricating mucus from glands in its lining and, to a lesser degree, from the **vestibular glands**, or *Bartholin's glands*, near the vaginal opening. The latter correspond somewhat to the bulbourethral glands in the male but secrete far more mucus.

Other internal organs of the female reproductive system include the uterus, the uterine tubes, and the ovaries. As you can see in Figure 41.23, the **uterus** is a hollow, pear-shaped organ whose broader, dome-shaped end tilts forward in a slightly folded manner. Its thickened ring-like base, the **cervix**, opens into the vagina. The opening, the **cervical canal**, provides an entry for sperm and is also part of the birth canal. The cervix secretes an alkaline mucus that protects sperm and provides some of the nutrients they require. Between fertile periods, the mucus forms a plug in the cervical canal.

The walls of the uterus consist of three specialized layers. The inner layer is the versatile **mucosa**. It produces the soft, highly vascularized **endometrium** that will receive and support an embryo if fertilization occurs. (We will look at the cyclic production of the endometrium a little later in this chapter.) Below the mucosa lies the middle layer, a region of smooth muscle whose fibers run in several directions. The outermost layer is connective tissue. The uterus has an enormous capability to expand, as it must when accommodating a growing fetus.

Most mammals have paired uteri; the single uterus of humans and other primates (and bats as well) is an evolutionary specialization for species that customarily give birth to only one offspring at a time.

The **uterine tubes** (also called *Fallopian tubes* and *oviducts*) (see Figure 41.23b) are tubes that emerge from each side of the upper end of the uterus and extend for a few inches outward and then downward. They terminate in a cluster of finger-like processes known as the **fimbriae**. Curiously, there is no direct connection between the uterine tubes and the ovary; eggs are simply released into the moist surroundings and are drawn into the uterine tubes by active cilia on the fimbriae. It is in the uterine tubes that egg and sperm meet and fertilization occurs.

Ovary and Oocyte Development. The **ovaries** (see Figure 41.23b) are oval bodies, about 2.5 cm (1 in.) in length, that produce eggs and reproductive hormones. The eggs arise through meiosis (Figure 41.24) in the outermost tissue of the ovary. There, selected diploid cells known as **oogonia** enter meiosis, becoming **primary oocytes**. As we found earlier (see Chapter 9), meiosis in females begins during fetal life but stops during prophase I, the stage in which those cells remain until sexual maturity is reached. When meiosis resumes, the primary oocytes complete meiosis I and divide, becoming **secondary oocytes**. When they complete meiosis II, they become haploid eggs. Some prefer the term **ova** (singular, *ovum*).

Gamete production represents an important difference between males and females. Females are born with all the gametes they are ever going to have. The male, however, continues to produce new sperm cells throughout his reproductive lifetime—literally trillions of them. Nevertheless, the number of oocytes in the ovaries at birth—some 2 million—is much more than adequate, especially considering how few can ever mature. By puberty this number will have shrunk considerably, to approximately 200,000 oocytes in each ovary. Of course, nowhere near that many will mature. In her reproductive lifetime, each human female produces about 455 mature eggs, about one every four weeks for about some 35 years.

Hormonal Control of Human Reproduction

The reproductive activities of humans, as with all higher animals, are under the influence of hormones. The primary sites of reproductive hormone production are the anterior lobe of the pituitary gland, the gonads, and the placenta, all of which are under the influence of the hypothalamus. Essentially, the hypothalamus does two things: it initiates hormone release and monitors hormonal levels in the blood (see Chapter 36). The blood hormone levels form a negative feedback loop to the hypothalamus, enabling it to regulate its own secretions.

Gonadotropin-releasing hormones from the hypothalamus (see Chapter 36) prompt the anterior pituitary lobe to secrete **gonadotropins**. As their name implies, these hormones stimulate activity in the gonads (both the ovaries and the testes). Specifically, the gonadotropins stimulate the pro-

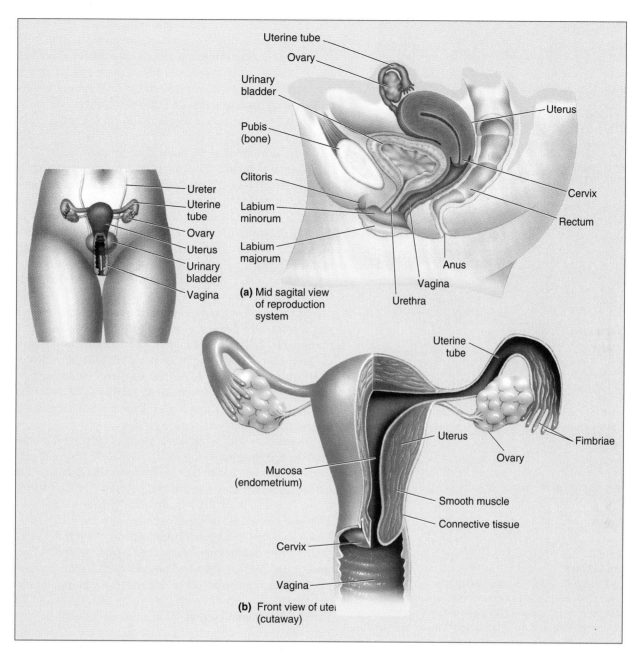

FIGURE 41.23
Internal Female Anatomy
(a) Internally, the vagina terminates at the cervix, which opens, via the cervical canal, into the uterus.
(b) The three tissue layers of the uterus are seen in this partially dissected view. The uterine tubes (Fallopian tubes) branch from the uterus and terminate at finger-like fimbriae, seemingly poised over the ovaries. Although the uterine tubes receive ova from the ovaries, note that there is no direct connection.

duction of the sex hormones and initiate the development of sperm and egg cells.

Two principal gonadotropins of the pituitary are **FSH (follicle-stimulating hormone)** and **LH (luteinizing hormone)**, both of which are polypeptides. Their targets are the ovaries in women and the testes in men. LH in men is also called *ICSH*, for *interstitial cell-stimulating hormone*, more appropriately named for its targets, the interstitial cells of the testes. We will turn first to hormonal action in males, where the processes are considerably simpler.

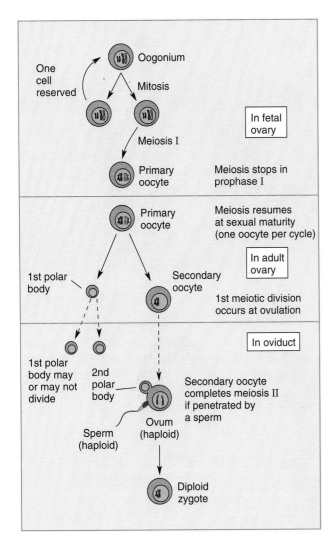

FIGURE 41.24
Meiosis in the Ovary
In humans, and many other animals, meiosis leads to just one functional gamete, the egg, and two or three nonfunctional polar bodies. (Sometimes the first polar body fails to divide so only two form.) The polar bodies are simply absorbed by the body and their constituents recycled. Note that meiosis I is completed before ovulation, but meiosis II is not generally completed unless fertilization has occurred. This means that the true haploid state is extremely brief, from about the time of sperm penetration up to the fusion of the two nuclei.

Male Hormonal Action. FSH is continuously secreted in males. Its specific targets are the seminiferous tubules of the testes, where it stimulates sperm production. FSH also teams up with LH to prompt the interstitial cells to produce **testosterone**, the primary male sex hormone.

Testosterone is vital to the development of the male characteristics, beginning in the embryo where its presence brings on **sexual differentiation** (the anatomical distinction between male and female: see Figure 42.18). These are called **primary**

sex characteristics. Its presence in sufficient amounts during the eighth or ninth month of development prompts the descent of the testes into the scrotum. Later on, between the ages of 10 and 17, testosterone is essential for those perturbing events associated with puberty. We refer to the development of **secondary sex characteristics**, such as voice changes, growth of body hair, bone and muscle development, and the enlargement of the testes and penis. Finally, the sex drive itself appears to be greatly influenced by the presence of testosterone.

Hormonal control in males occurs through negative feedback. High levels of testosterone in the blood form a negative feedback loop to the hypothalamus, where they inhibit the release of LHRH (luteinizing hormone releasing hormone). In this way the hypothalamus slows its stimulation of the anterior pituitary. The pituitary, in response, slows its release of LH, thus lowering the production of testosterone by the interstitial cells. So, as you can see, a feedback mechanism regulates testosterone levels.

Sperm production is controlled more directly. When sperm production peaks, the hormone **inhibin** is secreted into the blood by cells in the seminiferous tubules. When the hormone reaches the anterior pituitary, it inhibits the secretion of FSH, which results in a lower rate of sperm production. As sperm production falls, the release of inhibin slows, and FSH release increases once again. Hormonal control in males is reviewed in Figure 41.25.

In summary then, under the influence of releasing hormone from the hypothalamus, the pituitary secretes the gonadotropins FSH and LH. These hormones stimulate sperm production and initiate the release of testosterone by the testes. Levels of testosterone are sensed by the hypothalamus, which responds by adjusting the release of gonadotropins. The result is a fairly steady level of hormone production. Sperm production is controlled by the hormone inhibin, which acts directly on the anterior pituitary. As you will now learn, all of this is quite straightforward when compared to what happens in the human female.

Female Hormonal Action. The onset of puberty in girls most often occurs between 9 and 12 years of age. The beginning of fertility follows in two to three years. Both events are initiated by a rise in the pituitary gonadotropin FSH, which acts on the ovaries. The ovaries respond to FSH by producing estrogens. **Estrogens** are a family of steroids, the most important being **estradiol**.

During puberty, estrogens influence the development of secondary sex characteristics. Included are the growth of the breasts and nipples, broadening of the hips, and, less conspicuously, the growth of the uterus, the vaginal lining, the labia, and the clitoris. The appearance of pubic and axillary (underarm) hair, another signal of puberty, is initiated by the combined effects of estrogen and a pituitary hormone on the adrenal glands (see Chapter 36). In response to these hormones, the adrenals produce androgens (essentially, male hormones), which stimulate the growth of the coarser hair of the genitals and underarms.

Fertility is marked by the start of the **menstrual cycle**, in which the gonadotropins and ovarian hormones rise and fall with some regularity and the oocytes mature. Other female vertebrates may experience somewhat analogous cycles, but they occur only once or twice each year. The remainder of the time, most other female vertebrates are sexually unresponsive and infertile. The human female is considered to be unusual among mammals in always being potentially responsive.

The Menstrual Cycle. Ovulation, the release of an ovum from the ovary, averages once every 28 days. Intense preparatory activities in the uterus accompany this event. The **endometrium**, the temporary uterine lining, must reach a peak of readiness to support the young embryo, should fertilization follow. The two events—ovulation and endometrial growth (Figure 41.26)—are closely correlated by the cyclic action of pituitary and ovarian hormones.

In describing the menstrual cycle, it is common to place ovulation at midcycle, or the 14th day, and begin counting on the first day of menstrual flow. We will be referring to Figure 41.27, a calendar of events over the 28 days, as we proceed. Keep in mind that 28 days is an average, meaning that there is considerable variation.

Days 1 to 14: Proliferation. The first half of a menstrual cycle is known as the **proliferative phase** (see Figure 41.27). During the first three or four days (sometimes longer), blood and tissues of the endometrium are shed in what is called **menstruation** (the "period"). Even before this ends, hormonal activity prompting the next cycle will have begun.

During the proliferative phase, releasing hormones from the hypothalamus stimulate the pituitary to increase its output of FSH and LH. Their target is the **primordial follicle** (Figure 41.28), a structure that includes the primary oocytes and the cells clustered about them. It is at this time that the primary oocyte becomes active, resuming meiosis and becoming a **primary follicle**. As a primary follicle grows, it forms a fluid-filled cavity around the oocyte and becomes a **vesicular follicle**. With continued prompting by FSH and LH, the secondary follicle matures to form a short-lived **Graafian follicle**. By midcycle, or day 14, the vesicular follicle will protrude like a blister from the ovary's surface.

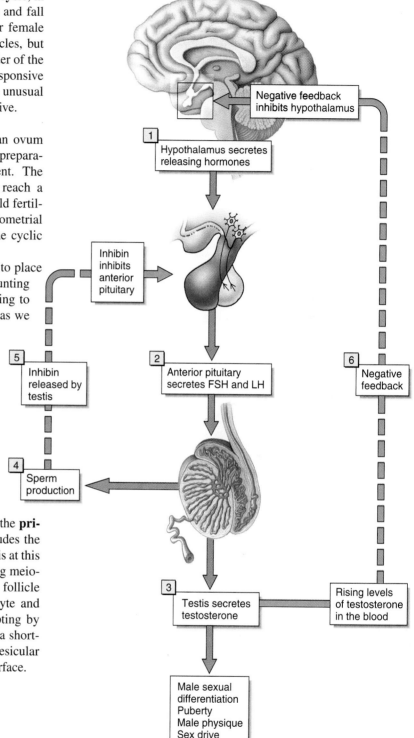

FIGURE 41.25
Hormonal Control in the Human Male
In males, the chain of events in hormonal control begins in the hypothalamus [1] as releasing hormones stimulate the anterior pituitary into [2] secreting FSH and LH. Their target is the testis, which responds by [3] secreting the sex hormone testosterone and [4] by accelerating sperm formation. Sperm production is regulated (left) by another hormone, inhibin [5], which is released when sperm formation reaches a set level. Inhibin has an inhibiting effect on the anterior pituitary, slowing its release of FSH. Rising testosterone (right) levels form a negative feedback loop [6] to the hypothalamus, slowing its secretion of releasing hormones.

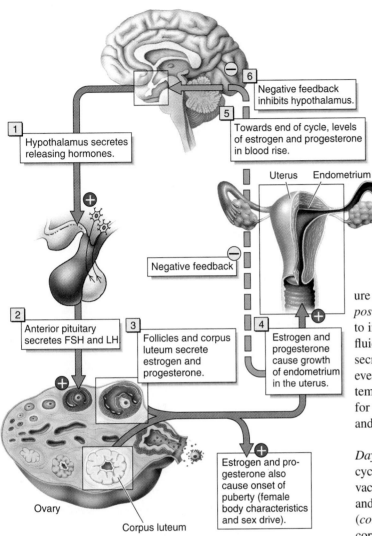

Uterus Endometrium

6 Negative feedback inhibits hypothalamus.

5 Towards end of cycle, levels of estrogen and progesterone in blood rise.

1 Hypothalamus secretes releasing hormones.

Negative feedback

2 Anterior pituitary secretes FSH and LH

3 Follicles and corpus luteum secrete estrogen and progesterone.

4 Estrogen and progesterone cause growth of endometrium in the uterus.

Ovary

Corpus luteum

Estrogen and progesterone also cause onset of puberty (female body characteristics and sex drive).

FIGURE 41.26
Hormonal Control in the Human Female

In females, the hormonal sequence begins when releasing hormones from the hypothalamus [1] stimulate the anterior pituitary [2] to secrete FSH and LH. Their specific target is an ovarian follicle, where a primary oocyte resides. The oocyte responds by growing and resuming meiosis. The surrounding cells respond [3] by secreting estrogens and progesterone. They support the cyclic growth of the endometrial lining of the uterus and maintenance of the lining during pregnancy [4]. The two hormones also support sexual development and sex drive. The entire sequence is subject to negative feedback by rising hormonal levels [5], [6].

ure 41.27). Rising levels of estrogens are believed to form a *positive* feedback loop, one that prompts the anterior pituitary to increase its output of LH. The target of LH is the enlarged fluid-filled **vesicular follicle** (or *Graffian follicle*). As LH secretion peaks, the follicle bursts, and **ovulation** occurs. The event is heralded by a slight, temporary elevation in body temperature (about 1°F), which is another important indicator for women trying to establish their fertile periods. The oocyte and its surrounding cells begin the journey to the uterine tube.

Days 15 through 28: Secretion. The **secretory phase** of the cycle follows ovulation (see Figures 41.27 and 41.28). The vacated vesicular follicle, under the continued influence of LH and FSH, forms a new structure called the **corpus luteum** (*corpus*, "body"; *luteum*, "yellow") (see Figure 41.28). The corpus luteum secretes two hormones, estrogen and **progesterone**. The target of both hormones is the endometrium, where progesterone prompts blood vessel growth and glycogen accumulation. The loosely packed cell layers become fluid-filled sinuses, penetrated with glandular tissue that produces other secretions. If fertilization has occurred, the tiny embryo will sink deep into this receptive tissue, where development will continue. Should this happen, progesterone will help maintain the endometrium throughout pregnancy.

The uterus remains receptive for several days following ovulation. Unless an embryo has snugly implanted in that time period, rising levels of estrogen and progesterone form a negative feedback loop that slows the FSH and LH secretion. The corpus luteum then slows its production of estrogen and progesterone, and the endometrium begins to degenerate. In a few days it will slough away, signaling the beginning of menstruation and the next 28-day cycle.

When pregnancy occurs, the fetus sends its own signals to the corpus luteum (Figure 41.29). The signals occur in the form of **human chorionic gonadotropin**, or **HCG**, a hormone that replaces FSH and LH from the hypothalamus. The corpus luteum is rejuvenated and continues to secrete estrogens and progesterone. (The presence of HCG in the urine,

As it grows, the follicle secretes estrogens. Their principal target, the endometrium, responds with rapid cell division, forming dense layers over the inner uterine surface (see Figure 41.27). The cell layers are soon penetrated by a profusion of blood vessels from the uterus. The result is a soft, highly vascularized lining, exactly what is required to nourish the early human embryo. The cervix responds to estrogen by secreting a thin alkaline mucus that helps neutralize the acidic vaginal environment. As the time of ovulation approaches, the amount of mucus increases, and it takes on a stretchy consistency, both of which indicate an impending fertile period. This is important information to women using the rhythm method of birth control, which we will discuss shortly.

As you can see, the reproductive hormones tightly coordinate the events in the ovary with those in the uterus. One result is that at the time of ovulation the uterus is nearing a fully receptive state. Should fertilization occur, the young embryo will find itself in a highly receptive environment.

Day 14: Ovulation. On the 14th day of the cycle, there is a rapid rise in LH, often referred to as the "LH surge" (see Fig-

incidentally, is the basis for many pregnancy tests.) HCG will diminish by the third month of pregnancy as the placenta begins secreting its own maintenance hormones.

Let us now summarize the important events of the menstrual cycle. In the cycle, the production of a mature ovum is correlated with the preparation of the uterus to receive it. The first half of the cycle is under the influence of FSH, LH, and estrogens, which prompt follicle and uterine growth. Ovulation is brought about by a surge in LH at midcycle. In the second half of the cycle, FSH and LH continue to support the secretion of progesterone and estrogen, which prompt final preparations in the uterus. In the absence of pregnancy, rising levels of estrogen and progesterone toward the end of the cycle produce a negative feedback loop, bringing an end to the cycle. In pregnancy, HCG from the embryo prompts the corpus luteum to go on secreting its uterus-supporting hormones.

Conception Control

A considerable variety of hormonal, physical, and chemical barriers to conception are available today (Figure 41.30). There are also remedies, should unwanted conception occur. The failure rates stated for these methods vary from zero to 25%. They are based on the first year of use.

Hormonal Control

Oral doses of synthetic hormones ("the pill") are used to inhibit both follicle development and ovulation. Should the two events occur anyway, and should fertilization follow, the synthetic hormones also alter cervical mucus secretion, making the vagina inhospitable to sperm, and they may impair implantation. The birth control pill has proved to be one of the most effective means of reproductive control. *The failure rate is 2%.* No one should use birth control pills without first consulting a physician. There are side effects to know about and there are people who should not use them at all. Smoking significantly increases the chances of undesirable side effects.

Recently, **Norplant**, a new hormonal method of contraception, was made available. Norplant is introduced under the skin as six slender capsules containing progestin. Progestin is a progesterone-like substance that is gradually released and, like hormones in the pill, inhibits ovulation. The capsules last five years and *the failure rate is only 1%.*

Implantation Barriers

Intrauterine devices (IUDs) physically prevent implantation of the fertilized egg. An IUD is a small plastic or metal object, often impregnated with birth control hormone, that is inserted into the uterus by a physician. It may be left in place for years. They were once quite popular until some brands caused serious infections. Improved brands are on the comeback today. *The failure rate is about 2%.* (Failure rate of the IUD without hormone is 5%.)

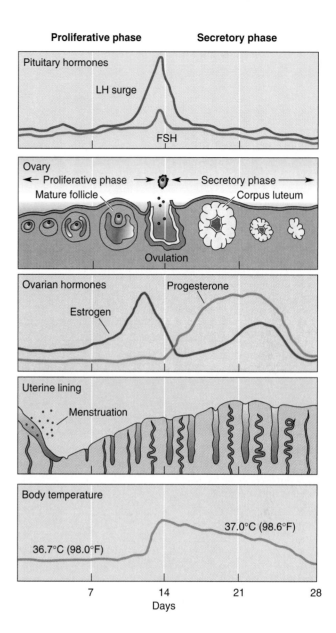

FIGURE 41.27
The Menstrual Cycle
Pituitary hormones: LH and FSH from the pituitary rise during the first two weeks (proliferative phase), suddenly peaking at midcycle. Both decrease, remaining constant for a time and dwindling during the last two weeks (secretory phase). *Ovary:* pituitary hormones stimulate follicle development which also peaks at midcycle, with ovulation following the LH surge (above). The empty follicle becomes a corpus luteum, which secretes the ovarian hormones shown next. *Ovarian hormones:* Estrogen, secreted by the maturing follicle, peaks shortly before midcycle, its level falling for a time as progesterone begins its increase. Both remain high but dwindle toward the cycle's end. *Uterine lining:* Following menstruation, the endometrium is reconstructed, reaching a peak condition by mid-cycle and remaining so through much of the second half. *Body temperature:* Typically, the body temperature is slightly subnormal (below 36.7° C) until ovulation, at which time it rises to 37° C, dwindling back down as the cycle continues.

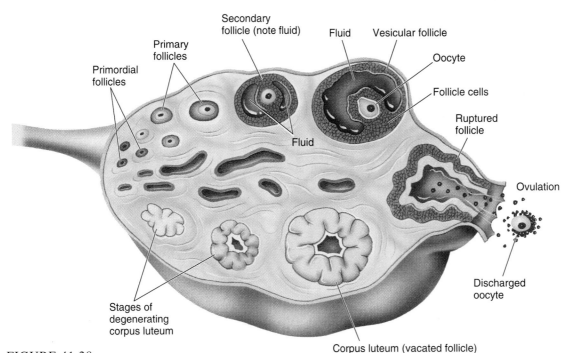

FIGURE 41.28
Developments in the Ovary
Starting with the upper left region, follow the sequence of activity in the ovary during the 28-day ovarian cycle. (Actually, the follicle is stationary and only one usually forms at a time. The artist is just showing transitions in the same diagram.) The ovary contains all the primary oocytes that remain after puberty. When one is targeted by FSH and LH, it begins its development. In the advanced vesicular follicle, the oocyte is a very large cell, surrounded by many smaller ones. When ovulation occurs, the oocyte will escape, still surrounded by many of the follicular cells. Note the large early corpus luteum, and how it diminishes toward the end of a cycle.

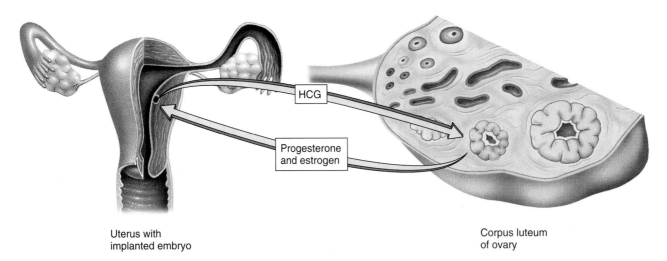

FIGURE 41.29
Signals from the Embryo
Once an embryo has implanted, an event that usually occurs toward the last week of the cycle, it begins sending its own hormonal signals to the corpus luteum. The hormone, HCG, is supplied by tissues that the embryo itself produces. The corpus luteum responds by continuing its secretion of progesterone—the pregnancy hormone.

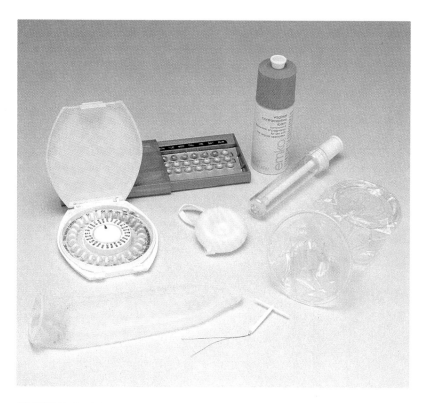

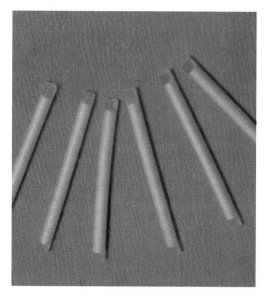

FIGURE 41.30
Contraceptive Devices
Seen here are hormonal, physical, and chemical barriers to conception; most are available at nominal cost.
Clockwise from the left are two brands of birth control pills, contraceptive foam and applicator, female
condom, an IUD, and a male condom. Norplant implants are seen in the smaller photo.

Sperm Barriers

Sperm barriers block sperm from entering the vagina or the uterus. The most common is the **condom**, a latex sheath worn over the penis. Its *failure rate, 10%,* can be cut to 5% *if* a quality condom with spermicide (sperm-killing agent) is used, the condom is always in place *before* intercourse begins, and it is removed promptly after ejaculation. (Sperm may be present in male lubricating fluids, released before ejaculation. Resuming intercourse with the condom in place can dislodge it and spill semen into the vagina.)

The **diaphragm** is a dome-shaped latex device that fits over the cervix. It must be fitted by a trained clinician and should be used with a spermicide. Much of its high *failure rate, about 20%,* can be attributed to improper use.

Chemical Agents

Chemical agents are spermicides that are introduced into the vagina just prior to intercourse. They are sometimes contained in foams and sponges. *The failure rate is 20%.*

Unaugmented Methods

Two birth control methods require no special agents. We mention these only because some people think they work. They don't. The first, called "withdrawal" or "premature withdrawal," involves withdrawing the penis just before ejaculation. Obviously this practice relies heavily on a man's total awareness and determination and a woman's confidence in the man. Even then, there is often a leakage of semen prior to ejaculation. Not surprisingly, the *failure rate is about 25%.*

Douching, as a contraceptive method, is using water or some other fluid to flush semen from the vagina. Traditionalists add a bit of vinegar to the water. A *failure rate of 60%* simply proves that sperm move faster than people (sperm enter the uterus within 90 seconds of ejaculation—too late for douching).

Rhythm

The **rhythm method**, or *natural method,* of birth control requires that copulation be avoided during the woman's fertile period. The fertile period begins 3–4 days before ovulation and ends 3–4 days afterward; but since this isn't rigidly controlled, it is prudent to allow even more time before and after ovulation (see Figure 41.27). The day of ovulation is determined by observing a slight temperature elevation (from 98.0 to 98.6°F), by an elastic consistency in the cervical mucus, and, in some women, a varying degree of ovarian pain. To be effective, the rhythm method requires faithful record-keeping and careful observations so that the time of ovulation is known with certainty. *The failure rate is 20%.* (An old joke

Essay 41.1 ● HUMAN SEXUALITY: THE SEXUAL RESPONSE

Much of what we know about the human sexual response today has grown out of the pioneering efforts of William H. Masters and Virginia E. Johnson, at the Reproductive Biology Research Foundation in St. Louis. Using human volunteers, Masters and Johnson developed observing and measuring techniques for studying many of the physiological aspects of the sex act. In their analysis, Masters and Johnson divided the complete sex act into four phases: *arousal, plateau, orgasm* and *resolution*. What follows is an idealized view of the sexual response; there are, of course, many variations.

Female Arousal. In women, the preliminary acts of sexual arousal, such as kissing and caressing (variations here are endless), bring on increased heart rate, faster breathing, and a rise in blood pressure. This is accompanied by firming of the nipples, spontaneous muscle contractions, sometimes a reddening of the skin, particularly about the genitals, face, breasts, and abdomen, and a rapid moistening of the vagina. The accumulation of blood in the genitals firms the clitoris and elevates and parts the labia majora and minora.

Male Arousal. In the male, arousal is characterized by somewhat similar events in circulation and respiration. Genital changes include erection, accompanied by a general contraction and elevation of the scrotum, and a moistening of the glans penis. Such events can occur notoriously fast in a male, often preceding correlating events in the female.

Female Plateau. As coitus proceeds, women enter the phase known as plateau,

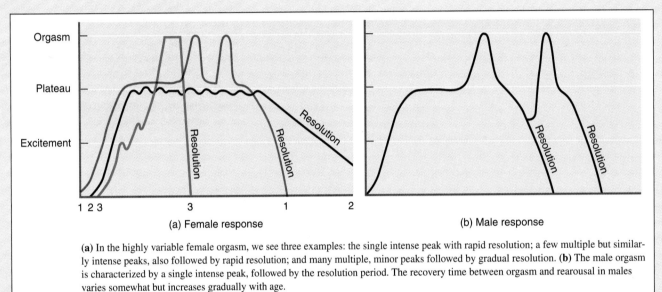

(a) In the highly variable female orgasm, we see three examples: the single intense peak with rapid resolution; a few multiple but similarly intense peaks, also followed by rapid resolution; and many multiple, minor peaks followed by gradual resolution. **(b)** The male orgasm is characterized by a single intense peak, followed by the resolution period. The recovery time between orgasm and rearousal in males varies somewhat but increases gradually with age.

goes, "What do you call people who practice the rhythm method? *Parents!*" The joke is funnier to some than others.)

Surgical Intervention

Sterilization. Conception can be permanently avoided by two common surgical procedures, **tubal ligation** in females and **vasectomy** in males. The procedure is not to be confused with contraception; the result here is *sterilization*. The first involves cutting and closing the uterine tubes, which prevents sperm and egg from meeting. The second is cutting and closing the ductus deferens, which prevents sperm from leaving the testes. Whereas vasectomy decreases semen volume somewhat, it has no other physiological effect on male orgasm. *Failure rate is below 1%.* (allowing for a few botched procedures). **Hysterectomy** in women (removal of the uterus) and **castration** in males (removal of the testes) are obviously sterilizing procedures as well but are carried out for different reasons from the tubal ligation and vasectomy.

Abortion. Pregnancy can be terminated through abortion—that is, through surgical procedures. The most common, the **D&C** (dilation and curettage—the "curette" is a spoon-shaped scraping instrument), involves dilating the cervix and scraping and suctioning fetal tissue from the uterine wall, a relatively simple procedure that does not usually require hospitalization. *Failure rate is below 1%.*

RU486 a new and controversial labor-inducing drug developed in France, is structurally similar to progesterone, enough so that it temporarily binds to and blocks the progesterone receptors in the uterus. With progesterone effectively eliminated from the uterus, the endometrium breaks down and the implanted embryo is dislodged and expelled. Study trials involving 2115 women *reveal a 4% failure rate.* The only undesirable side effects reported so far have been attributed to the addition of prostaglandins to the treatment. Prostaglandins stimulate uterine contractions, thus aiding the expulsion process, but such contractions can be quite painful.

sensation and movement heighten and vigorous pelvic thrusting is common. The swollen clitoris may withdraw into its hood at this time and is now exquisitely sensitive to touch. The labial swelling increases as does swelling in the lower portion of the vagina, especially the opening. The uterus may elevate, tilting backward somewhat. Physical activity may peak now, with intense thrusting of the pelvis, although there is a great deal of variation here. Characteristically, the facial muscles relax, producing a slack appearance and almost an absence of expression.

Male Plateau. As men approach plateau, the glans reaches its greatest size and the penis achieves its greatest curvature and rigidity. The testes increase in size and further elevate in the scrotum. Pelvic thrusting increases and secretions from the bulbourethral glands increase. As the intensity of the sex act peaks, orgasm begins. In both men and women, orgasm is a matter of complex muscular contraction, usually accompanied by intensely pleasurable sensations. These contractions are reflexive and involuntary, occurring in a steady series.

Female Orgasm. The orgasmic contractions in women emanate from the muscles and thickened tissue about the vaginal opening—joined to a varying extent by activity in the vagina and the uterus. During these contractions, *tenting* may occur: the cervix is drawn upward, increasing the diameter of the surrounding inner vagina. Some authorities claim that tenting creates a suction that causes semen to pool around the cervix, and subsequent contractions in the uterus are thought somehow to draw the semen in. Uterine and cervical contractions are caused at least in part by the hormone *oxytocin,* which is reportedly released from the pituitary at this time.

The involuntary contractions of orgasm occur in 0.8-sec intervals, spreading through the lower pelvis. In an intensely pleasurable orgasm, many of the other muscles of the body may begin to contract spasmodically. Most researchers in this field agree that clitoral stimulation, at first direct, but later indirect, triggers orgasm. Indirect stimulation involves the clitoral hood, which is tugged against the clitoris by the rhythmic thrusting of the penis against the labia minora. It is difficult to separate the physiological and anatomical factors involved in female orgasm from the psychological. Apparently all must be synchrony for the event to occur.

The duration of the orgasm in women is highly variable. Orgasms may occur with one intense peak or several, or the peaks may be far more numerous but less intense (see the graph). The ability of women to experience several orgasmic peaks is sharply contrasted to the usual single orgasmic peak to which men are limited.

Male Orgasm. In men, orgasm and ejaculation are one, but we can divide them into two phases: emission and ejaculation. In *emission,* the rhythmic, peristaltic contractions in the vas deferens, prostate, and seminal vesicles bring the semen to the base of the urethra. In *ejaculation,* the presence of semen in the base of the urethra triggers spinal reflexes that cause powerful striated muscles surrounding the urethra to contract. Thus the semen is forcefully expelled through the urethra and out of the penis. This occurs in several spurts, the first usually bearing more semen than the rest. The contractions in men also occur in intervals of about 0.8 sec but end in a few seconds. Intense pelvic thrusting and muscular contraction involving the whole body commonly accompany male orgasm.

Resolution. Following orgasm, both males and females retreat from the state of sexual excitement into what is known as *resolution,* a phase that has also been referred to as "afterglow." Most men lose erection rapidly and enter a refractory period in which rearousal can be quite difficult for a time (Some men measure the period with a stopwatch, others with a calendar). The refractory period varies, typically increasing with age. Resolution in women is somewhat different, since immediate rearousal is often possible. In any case, the quiet, relaxed, and often tender resolution phase may have an intense pleasure of its own.

Key Ideas

1. Sexual reproduction promotes genetic diversity in offspring that is not available through asexual reproduction.

REPRODUCTION AND THE SURVIVAL PRINCIPLE

1. Over the long run, individuals must replace themselves if the species is to survive. There are a great number of variations in the way animals reproduce, but all serve to promote this goal.

ASEXUAL REPRODUCTION

1. Asexual reproduction involves one individual without the input of genes from a second. Examples include **fission** (dividing in two), **budding** (new individuals branching from the adult), **fragmentation** (body breaking into parts and regenerating), **gemmule** formation (clusters of cells that can regenerate the adult), and **parthenogenesis** (development of an unfertilized egg).

SEXUAL REPRODUCTION

1. In **external fertilization** gametes are released into the surroundings, where fertilization and development occur. In **internal fertilization** the egg is fertilized within the female body.

SEXUAL REPRODUCTION IN INVERTEBRATES

1. External fertilization is common in aquatic invertebrates. Sponges release sperm into the watery surroundings, to be drawn into neighboring individuals by currents created by collar cells. Snails and barnacles are **monoecious** (one sex) and have well-developed copulatory organs, including the **penis** and **vagina**. Insects are usually **dioecious** (two sexes). The complex and species-specific copulatory organs help prevent interspecific mating.

SEXUAL REPRODUCTION IN VERTEBRATES

1. External fertilization in fishes may involve the mass shedding of great numbers of gametes or complex and intimate mating behavior, which may include elaborate courting and mating behavior and parental care.

2. In male reptiles, the **penis** is hidden in the cloaca, protruding out during mating. Some have **hemipenes**, paired copulatory organs.

3. In most bird species sperm transfer occurs as the cloacas are pressed together. Newly hatched ground-foraging birds move about and feed themselves. Songbirds hatch in a blind, featherless, and virtually helpless state.

4. The young of placental mammals develop in the mother's **uterus**, supported there by exchanges provided by the **placenta**. Monotremes (spiny anteaters and duck-billed platypuses) are reproductively primitive egg-layers. Marsupials are born in a much less developed state, completing development in the pouch.

HUMAN REPRODUCTION AND SEXUALITY

1. Male genitalia include the **penis** and **testes**, the latter enclosed in the sac-like **scrotum**. The scrotum responds to temperature changes by raising and lowering the testes. During **erection** the spongy spaces of the penis become engorged with blood.

2. Most of the testis is made up of highly coiled **seminiferous tubules**, in which a **germinal epithelium** produces sperm, and **interstitial cells** produce **testosterone.** Maturing sperm are swept by cilia to the **epididymis** for storage.

3. Sperm cells originate from undifferentiated cells, first becoming **primary spermatocytes**, which then enter meiosis I to form **secondary spermatocytes** and, after meiosis II, **spermatids**. The latter differentiate into **spermatozoa**.

4. The path of sperm during ejaculation is from the epididymis to the **ductus deferens**. Alkaline fluids containing sugars and prostaglandins are received from the **seminal vesicles**, and the **prostate** its secretions give the semen its characteristic color and odor.

5. The external female genitalia, or **vulva**, includes the **mons veneris**, a fatty mound overlying two folds or lips, the larger **labia majora** and the smaller **labia minora.** The minor lips join to produce a hood overlying the **clitoris**, a sensitive, erectile organ.

6. The internal female reproductive anatomy includes the **vagina, uterus, uterine tubes,** and **ovaries**. The lower uterine opening is a muscular ring called the **cervix**. The uterine wall is three-layered, with an innermost **mucosa**, which produces the vascular **endometrium**. The uterine tubes emerge from the uterus to terminate in finger-like, movable **fimbriae**. Their ciliated surfaces sweep the **oocytes** into the uterine tube.

7. Each ovary has an outer region that contains the oocytes, all of which form before birth. The transition is from oogonium to primary oocyte and then secondary oocyte. Meiosis stops in the fetus and resumes at sexual maturity.

8. The hormonal control of reproduction involves an interaction between the hypothalamus, pituitary, and gonads. Releasing hormone from the hypothalamus prompts the anterior pituitary to release two **gonadotropins**: **follicle-stimulating hormone (FSH)** and **luteinizing hormone (LH)**, which act on the ovaries or testes. (In males, LH is also called **interstitial cell-stimulating hormone** or **ICSH**.)

9. In males, FSH stimulates sperm production and works with LH to initiate **testosterone** production. Testosterone levels in the blood create a negative feedback loop back to the hypothalamus. Sperm production is governed by the hormone **inhibin**, which forms a negative feedback loop to the anterior pituitary.

10. Testosterone influences the development of the primary sex characteristics in the embryo, secondary sex characteristics at the onset of puberty, and sex drive.

11. Puberty in females is associated with FSH secretion, which prompts estrogen production and release by the ovaries. Estro-

gens influence development in the breasts, reproductive organs, and uterus, and general body growth. The onset of fertility begins with the monthly **menstrual cycle**.

12. The menstrual cycle in the human includes the release of an ovum and the growth of the uterine endometrium every 28 days. The **proliferative phase** begins with **menstruation** and ends with **ovulation**. FSH prompts **follicle** development and estrogen secretion. Estrogen prompts growth of the endometrium. LH levels rise at midcycle, bringing about ovulation. In the **secretory phase**, LH and FSH maintain the **corpus luteum**, which secretes both estrogens and **progesterone**, further influencing endometrial growth. At cycle's end negative feedback suppresses LH and FSH secretion, the corpus luteum fails, and the endometrium breaks down.

13. If fertilization and implantation of an embryo have occurred, cells associated with the embryo secrete **human chorionic gonadotropin**, which supports the corpus luteum for the first two months.

CONCEPTION CONTROL

1. **Hormonal birth control** involves synthetic estrogens and/or progesterone, which are believed to suppress ovulation. The **IUD** prevents implantation. **Sperm barriers** include **condoms** and **diaphragms**. **Chemical agents** include spermicides introduced directly into the vagina.

2. In the **rhythm method**, intercourse is avoided during fertile periods.

3. **Surgical intervention** includes the **vasectomy** (cutting and tying the ductus deferens) and **tubal ligation** (cutting and tying—or just tying—the uterine tubes). **Abortion** is the removal of an embryo or fetus or induction of its expulsion.

Application of Ideas

1. In evolutionary terms, hatching in the precocial (well-feathered; ready to move about and feed on their own) state is considered by ornithologists to be primitive, while the altricial state (generally blind, almost featherless, and helpless) is thought to be advanced. Offer reasons why such a hypothesis might be logical, taking into consideration the evolutionary history of birds and that precocial species are usually ground foragers.

2. Male and female pairing in humans is often described as a matter of "bonding"—that is, the pleasure bond, the pair bond, and so on. Both cultural and biological explanations of bonding have been proposed, and these are often in conflict. In the first, bonding is often considered to be a learned human behavior, approved and reinforced by society. In the second, bonding is considered to be genetic, a product of evolution, and fairly commonplace among higher animal species. Select one or the other of these points of view and offer supporting or opposing arguments. Can either assertion be tested? Can both be correct?

Review Questions

1. What appears to be the main adaptive advantage of sexual reproduction? (page 884)

2. Distinguish between monoecious and dioecious animals and list examples. (pages 887–888)

3. How are insects protected from accidental hybridization? Why is it especially important in this group? (page 889)

4. Contrast the two developmental conditions in bird hatchlings. How do these relate to food requirements? (pages 891–892)

5. Briefly describe how the monotremes and marsupials differ reproductively from the placental mammals. (page 892)

6. List the structures through which sperm move during ejaculation. (page 896)

7. List the five structures that make up the external human vulva—the external anatomy. (pages 896, 898)

8. List three structures that play a part in the hormonal control of reproduction, starting with one located in the brain. (pages 898–899)

9. Explain how negative feedback works in controlling testosterone and sperm production in males. (page 900)

10. In what two major aspects of life is testosterone important to males? (page 900)

11. Describe events in the ovary, hypothalamus, uterus, and anterior pituitary through the first days of the human menstrual cycle. Include any negative feedback loops. (pages 901–902)

12. What brings on the process of ovulation? (page 902)

13. Describe the output of hormones, and their roles, in the postovulatory phase. What events end the cycle? (page 902)

14. List several reasons why the rhythm system of birth control is subject to failure. (page 905)

15. List five birth control practices that are unacceptable from the standpoint of failure. (pages 903, 905–906)

Animal Development

C H A P T E R   O U T L I N E

GAMETES

The Sperm

The Egg

FERTILIZATION

Fertilization in Echinoderms

Fertilization in Mammals

EARLY DEVELOPMENT EVENTS

Becoming Multicellular: The First Cleavages

The Blastula

Gastrulation: Organizing the Germ Tissues

THE EMBRYO TAKES FORM: ORGANOGENESIS

Neurulation

Further Vertebrate Development

VERTEBRATE SUPPORTING STRUCTURES

The Support Systems of Reptiles and Birds

The Support System of Placental Mammals

DEVELOPMENT IN THE HUMAN

The First Trimester

Essay 42.1 The Emergence of Human Life

The Second and Third Trimesters

Birth

AN ANALYSIS OF DEVELOPMENT

Determination in the Egg and Zygote

Caenorhabditis elegans: A Model of Early Determination

Tissue Interaction in Vertebrates

The Role of Cell Migration

A Developmental Program in the Insect

A Developmental Program in the Vertebrate Wing

Summing Up

KEY IDEAS

REVIEW QUESTIONS

One cell divides into two, and two into four. The four again divide, and the process continues until eventually there are great numbers of cells. All the cells of the embryo stem from that one primordial cell, but yet, many divisions later, far down the line, descendant cells may come to be very different from one another. One may become an elongated nerve cell, while another becomes a pigmented cell within the eye's retina, and yet another an ordinary, unimpressive-looking liver cell—but a true virtuoso in biochemical versatility. Despite the variation, all the individual's cells are descendants of a single fertilized egg; they all carry the same genetic information in their chromosomes. How does this seeming miracle of development and differentiation occur? This is not a rhetorical question; we don't quite know. It seems that there should be some ready answer, some satisfying explanation.

To elaborate, let's note that the study of development has two aspects: the descriptive and the experimental—the familiar "what" and "how." The descriptive work—the story of "what" actually can be seen in a developing embryo—is already rather complete. However, there are serious, unanswered questions in the "how" department.

So we will take things one step at a time, beginning with the descriptive aspects of **embryogenesis**, the formation of the embryo. We will then move on to some fascinating studies of development, tracing the developmental path of tissues as they follow their individual fates. An underlying theme will be **determination**, the ongoing genetic commitment of cells and tissues to specific developmental pathways. Such commitment may begin very early, continuing until the last touches are made. We will see that the steps followed by cells as they become committed to developmental pathways become increasingly specific as time passes, until they are com-

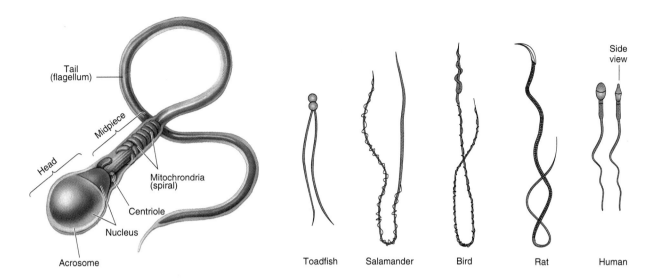

(a) Human sperm

(b) Vertebrate sperm

FIGURE 42.1
Sperm
(a) The human sperm, typical of animal sperm, contains an enlarged head that includes a highly condensed nucleus and an acrosome. The midpiece, just below the head, includes a centriole and a surrounding layer of mitochondria. The tail is a lengthy flagellum. **(b)** Animal sperm cells vary greatly in shape and length, yet each has similar structures.

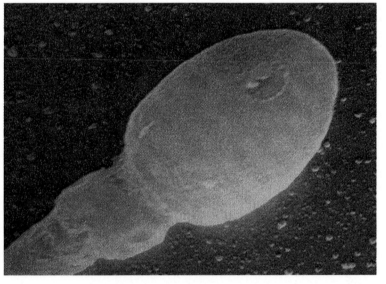

pletely differentiated (developmentally specialized for a specific role). A good beginning for us is a return to the gametes, the sperm and egg.

Gametes

The Sperm

The **sperm** is one of the most highly specialized of cells, but its role is straightforward: it must penetrate an egg and deliver its haploid chromosome complement. In so doing it will activate the egg, which is certainly not passive in the fertilization process. Once the sperm enters the egg, its chromosomes will fuse with the egg chromosomes, restoring the diploid number required by the new individual. The sperm's entire structure is devoted to this purpose.

The three major parts of the sperm are the head, midpiece, and tail (Figure 42.1a). The sperm head contains the nucleus,

which consists primarily of highly condensed chromosomes. At the tip of the sperm head lies the **acrosome**, actually a lysosome, that forms a cap-like structure over the sperm head. When released, the lysosomal enzymes aid the sperm in penetrating the egg. The sperm midpiece includes a sheath of mitochondria, arranged in a spiral. As you would expect, the mitochondria provide ATP energy to the flagellum that propels the sperm along. The midpiece also contains a single centriole and other cytoplasmic organelles typical of animal cells. Upon fertilization, the sperm centriole will join up with the egg centriole, producing the usual paired condition. The tail is a typical eukaryotic flagellum with the usual 9 + 2 microtubular arrangement. Animal sperm cells vary considerably in shape (Figure 42.1b), but most contain the elements listed.

The Egg

The **egg**, with its haploid nucleus, contains the other half of the offspring's required chromosome complement. Animal eggs vary considerably in content, organization, and size. For example, whereas a human egg (Figure 42.2a) is about the size of the dot over the letter "i," you would find it difficult to hide an ostrich egg behind this entire book. Such size differences, we will see, are largely due to differences in the amount of the yolk (stored foods). But size here is deceptive. In the yolky bird and reptile eggs, the cytoplasmic region destined to form the embryo, called the **blastodisc**, is just a small region of active

Eggs at life size

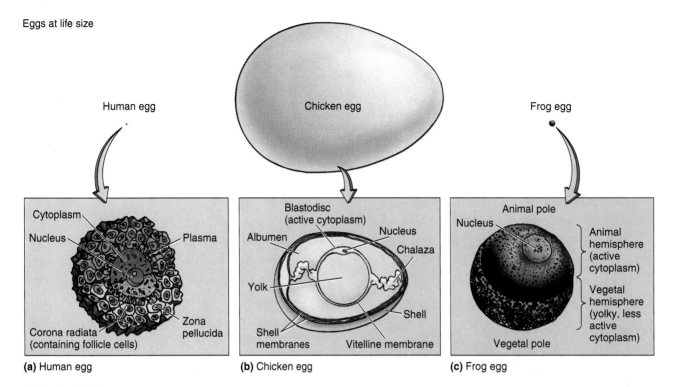

(a) Human egg   **(b)** Chicken egg   **(c)** Frog egg

FIGURE 42.2
Eggs
Animal eggs all contain a haploid nucleus but otherwise vary greatly. **(a)** The mammalian egg has two coverings. A dense zona pellucida directly surrounds the plasma membrane. Beyond that is the corona radiata, a layer of follicular cells held in place by a gluey material known as hyaluronic acid. **(b)** The chicken egg, typical of birds and reptiles, includes a large, dense yolky region. The light colored spot on its surface is the blastodisc—the active cytoplasm of the cell. The calcareous shell surrounds the shell membrane. The yolk is suspended by two dense, twisted proteinacious strands called chalazes. **(c)** The frog egg is polarized. It contains a moderate amount of yolk that is concentrated in the vegetal hemisphere. The animal hemisphere contains the nucleus and most of the metabolically active cell organelles.

cytoplasm and a haploid nucleus on the yolk surface (Figure 42.2b).

The varied yolk content of eggs in the different animal groups relates closely to the mode of development. The large yolk reserves in the eggs of birds and reptiles will see the embryo through its entire embryological development, which, as you know, occurs independently of the mother. In contrast, mammalian eggs have meager quantities of yolk, enough to sustain them for just a few days. But this is sufficient since the mammalian embryo soon implants in its mother's uterus, where its needs are met for the remainder of development.

The egg cell has the usual cellular organelles but is quite unlike other cells in many respects. In nearly all cases, animal eggs show **polarity**. This means that the cytoplasmic constituents are not equally distributed but occur in a gradient (Figure 42.2c). The unequal distribution results in essentially two hemispheric divisions. In most kinds of eggs, the metabolic "machinery" (mitochondria, Golgi bodies, endoplasmic reticulum, and so on) is concentrated in the more metabolically active region called the **animal hemisphere** and the yolky

food reserves in a less active region, the **vegetal hemisphere**. The apexes of the two hemispheres are the **animal pole** and **vegetal pole**, respectively.

Insect eggs, which differ in many ways from those of other animals, carry organization to the extreme. They have a well established two-directional polarity—both anterior–posterior and dorsal–ventral. In fact, the two regions in the egg and early embryo have their future already determined. They will contribute to specific regions in the adult (Figure 42.3).

Fertilization

Fertilization is the activation of the egg and the union of a haploid sperm nucleus with a haploid egg nucleus. Much of our knowledge about fertilization comes from studies of echinoderms (sea urchins, sea stars, and sand dollars), whose hardy gametes are readily available and easily manipulated. For this reason biologists have used the sea urchin to establish a fertilization model.

Fertilization in Echinoderms

Fertilization in the sea urchin, a representative echinoderm, occurs externally as individuals release clouds of sperm and egg, more or less simultaneously (see Chapter 41). The sea urchin egg (Figure 42.4) has two barriers that sperm must penetrate. The outermost is a thick region of mucopolysaccharide, known as the **jelly coat**. The jelly coat is thought to attract sperm and to stimulate the **acrosomal reaction** (Figure 42.4, left). In this reaction the lysosomal enzymes are released by the sperm and each sperm head forms a lengthy extension called the **acrosomal process**. The lysosomal enzymes digest a path through the jelly coat, thereby permitting the sperm to reach the second barrier, the vitelline membrane. The **vitelline layer**, a glycoprotein layer affixed to the plasma membrane, contains highly specific surface receptors that bind tightly to matching molecules on the acrosomal process of the sperm. The specific nature of this binding between sperm and egg is adaptive for an interesting reason. Since the receptors tend to be species specific—that is, different in each species—it helps prevent interbreeding. Considering that most interspecific animal hybrids are sterile, and in other ways unfit, this is important, particularly so in the aquatic environment where the gametes may simply be released into the watery surroundings.

Although many sperm attach to the vitelline layer, just one penetrates, its plasma membrane fusing with the egg plasma membrane. The fusion of the two releases the sperm nucleus and cytoplasm into the egg cytoplasm. Fertilization will be completed when the sperm and egg nuclei fuse. The binding of sperm to the egg surface brings on **activation**, a sudden surge in activity in the egg, which, up to now, existed in a metabolically sluggish state. One of the first things the activated egg does is form barriers against all other sperm cells.

Sperm Barriers. We've seen that just one sperm fuses with the egg's plasma membrane. This is critical, since an egg penetrated by more than one sperm, referred to as *polyspermy*, usually fails to develop. In the activated echinoderm egg, an elaborate pair of mechanisms, a *fast block* and a *slow block*, stop additional sperm from penetrating the egg.

In the fast block, which takes about 3 seconds, fusion of the sperm and egg membranes triggers an instant reaction in which calcium ions (Ca^{2+}) rapidly enter the egg, followed by sodium ions (Na^+). The inrush of positively charged ions produces a sudden shift in polarity, from -70 mV to $+10$ mV, that prevents sperm penetration for a short time. Researchers suggest that sperm receptors on the egg membrane can only fuse with the sperm when the membrane potential is negative. As soon as the polarity shifts to a positive condition the receptors become inoperative. The shift is temporary and the normal polarity soon returns. In the meanwhile, the egg makes other efforts to avoid polyspermy.

The slow block, called the **cortical reaction** (Figure 42.4, right), is chemical rather than electrical, and it is permanent. It culminates with the formation of an impenetrable **fertilization envelope** (also *fertilization membrane*). The cortical reaction

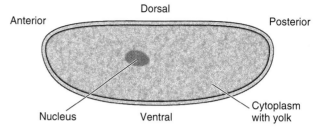

(a) Two dimensional polarization

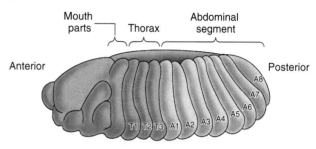

(b) Insect embryo (blastoderm stage)

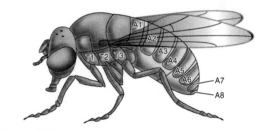

(c) Insect adult

FIGURE 42.3
The Insect Egg
(a) Even in its unfertilized state, the insect egg is highly polarized. (b) Regions of the young larva (see numbers) contain cells that are already committed to the formation of specific regions (c) in the adult (see corresponding numbers).

begins within 30 seconds of fusion as **cortical vesicles** from the underlying egg cytoplasm fuse with the plasma membrane and release their enzymes. The vitelline membrane rises up off the egg surface, carrying many bound sperm away from the plasma membrane, an act that ends any future chance of polyspermy. The fertilization envelope first appears as a small mound over the site of sperm entry, but as the reaction spreads the membrane soon rises up in a wave over the entire egg (Figure 42.5). This takes about three minutes.

The activated sea urchin egg is highly involved in sperm penetration and, in fact, may provide most of the effort. A **fertilization cone**, composed of microvilli extending from the egg cytoplasm, rises up and surrounds the sperm head, drawing it into the cytoplasm (see Figure 42.4, left).

Following sperm entry, the egg enters a period of intense metabolic activity. During this period, cell respiration in the egg increases dramatically and the egg cytoplasm undergoes a rise in pH. The latter appears to be critical to the initiation of develop-

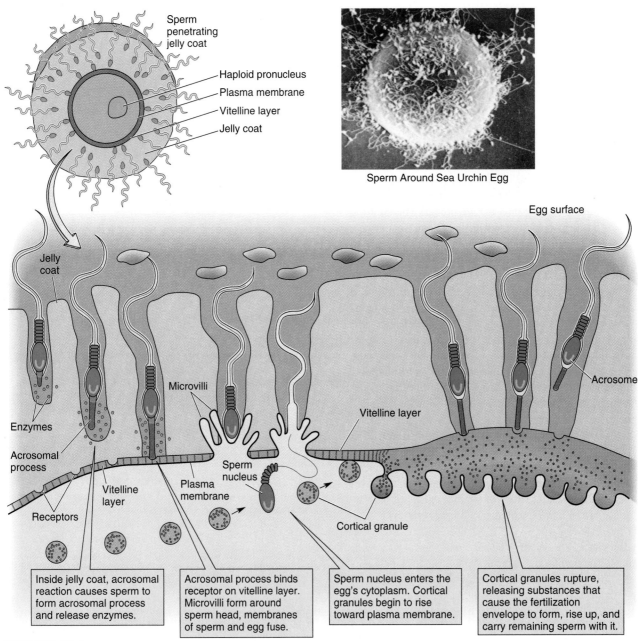

Sperm penetrating jelly coat

Haploid pronucleus
Plasma membrane
Vitelline layer
Jelly coat

Sperm Around Sea Urchin Egg

Egg surface

Jelly coat

Microvilli

Vitelline layer

Acrosome

Enzymes

Acrosomal process

Sperm nucleus

Vitelline layer

Plasma membrane

Receptors

Cortical granule

| Inside jelly coat, acrosomal reaction causes sperm to form acrosomal process and release enzymes. | Acrosomal process binds receptor on vitelline layer. Microvilli form around sperm head, membranes of sperm and egg fuse. | Sperm nucleus enters the egg's cytoplasm. Cortical granules begin to rise toward plasma membrane. | Cortical granules rupture, releasing substances that cause the fertilization envelope to form, rise up, and carry remaining sperm with it. |

FIGURE 42.4
Fertilization in the Echinoderm
During fertilization in the sea urchin (*left*), numerous sperm penetrate the jelly coat and bind to receptors on the egg surface. Fusion of sperm and egg membranes result in the release of the naked sperm head into the egg cytoplasm. The cortical reaction (*right*) leads to formation of the fertilization envelope, a permanent sperm barrier.

ment itself. This is soon followed by a period of protein synthesis and DNA replication, a prelude to the first of many cell divisions.

Fertilization is completed with the fusion of sperm and egg nuclei, or **pronuclei**, as they are called at this time. Fusion is preceded by a general swelling of the sperm nucleus and decondensation of sperm chromatin. The sperm pronucleus is then drawn to the egg pronucleus by the action of cytoplasmic microtubules. Figure 42.6 is a timetable of events in fertilization and activation.

Fertilization in Mammals

Fertilization in humans and other mammals is similar in basic ways to that of sea urchins, but there are structural differences to account for. The mammalian egg is surrounded by the **corona radiata**, a dense layer of follicle cells (there is no jelly coat) that accompanied the egg from the ovary during ovulation. Below the follicular cells is a glistening inner layer, the **zona pellucida**, which, like the vitelline membrane, is

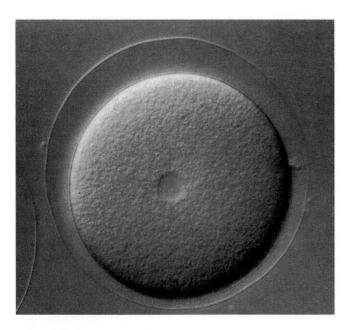

FIGURE 42.5
The Fertilization Envelope
The photograph shows a sea urchin zygote with a complete fertilization envelope around it. As it rises up from the egg surface, it carries numerous unsuccessful sperm along.

composed chiefly of glycoprotein and acts as a sperm binding surface (see Figure 42.2).

The acrosomal reaction in humans includes the release of lysosomal enzymes that digest a path around the cells of the corona radiata, enabling the sperm to reach and bind to the zona pellucida. Interestingly, mammalian sperm do not meet the egg head-on; instead, fusion occurs at the side of the sperm head.

Of the hundreds of millions of sperm released in a single ejaculation by a human male, only a few thousand usually reach the egg. Normally, just one sperm will penetrate the egg plasma membrane. However, digesting the path through the corona is actually a group effort, with many sperm pooling their enzymes to break through the barrier. For this reason, a large number of sperm are needed to accomplish fertilization, even though only one sperm normally joins the egg. This is why men with low sperm counts, while producing normal sperm, may be functionally infertile.

In contrast to the echinoderm, there is no electrical, fast block to polyspermy and no shift in polarity. In fact, most mammals do not form fertilization envelopes at all. Nevertheless, changes in the egg surface, in this case the zona pellucida, cause the egg to lose its binding properties and the excess sperm are released.

One sign of fertilization in humans is the appearance of a second polar body (Figure 42.7), the result of a second meiotic division. (Recall that the second meiotic division is not completed in human oogenesis unless fertilization has occurred. See Chapters 9 and 41.)

Early Development Events

The first task in development, it appears, is to provide the raw material from which the embryo can be molded. This means producing many cells through mitotic cell division. Early cell division in the animal embryo, referred to as **cleavage**, takes

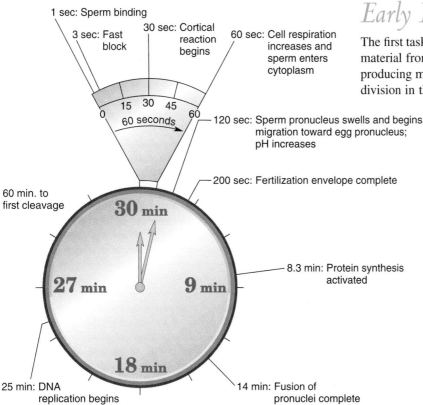

FIGURE 42.6
Timetable of Fertilization
The 30-minute clock reveals the fertilization and cleavage program of the echinoderm. Many of the activities, including sperm binding, formation of the fast block to polyspermy, and cytoplasmic penetration of the sperm, occur within the first 30 seconds. Within 120 seconds, the cortical reaction and fertilization envelope formation are well underway. Actual fusion of the sperm and egg pronuclei requires somewhat longer, completing in about 14 minutes. This is followed by DNA replication and other preparations for the first cleavage.

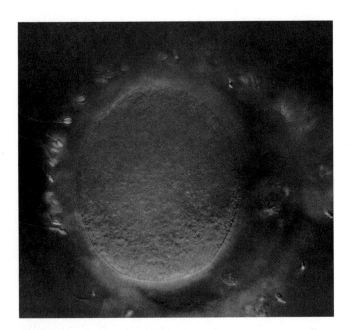

FIGURE 42.7
Human Fertilization
Most aspects of human fertilization are similar to that of the echinoderm, although no fertilization envelope rises up and no acrosomal process protrudes. Here many sperm cluster about the egg. A definite sign of fertilization is the appearance of a second polar body. Human sperm, like those of most mammals, do not meet the egg head-on. They bind to the zona pellucida through sites along the edge of the sperm head.

place through the constriction of microfilaments. Interestingly, the embryo doesn't grow during this period, so the ensuing cleavages result in smaller and smaller cells. We'll look into some of the contrasting cleavage patterns in echinoderms, humans, amphibians, birds, and reptiles.

Becoming Multicellular: The First Cleavages

The precise pattern taken by cleavage planes is largely dependent on the amount of yolk. Echinoderm eggs have little yolk, amphibians substantially more, and birds an enormous amount.

The cleavage pattern in echinoderms, with their scant yolk supply, is one of the simplest (Figure 42.8a). The first cleavage divides the zygote into cells of equal size and appearance. The second and third cleavages also produce daughter cells of similar size, yielding an embryo of eight cells that are similar in size. In subsequent cleavages, however, differences in cell size begin to appear. At the 16-cell stage a gradient of cells appears, which is maintained for a time. The embryo eventually becomes a simple ball of tiny cells just about the size of the original zygote. The simple ball of cells is called the **morula** (from "mulberry").

The first two cleavages in the frog zygote, with its moderate quantity of yolk, are vertical, bisecting the animal and vegetal hemispheres (Figure 42.8b). The daughter cells thus

contain equal amounts of yolk. The third cleavage forms at right angles to the first two, but high up in the animal hemisphere, resulting in four small cells and four larger cells. The small cells end up with the more active cytoplasm of the animal hemisphere, while the four large cells contain most of the less active yolky cytoplasm of the vegetal hemisphere. Cells of the more metabolically active animal pole divide considerably faster than those of the sluggish vegetal pole, thus becoming substantially smaller.

Bird and reptile eggs are large, their contents including the clear, watery "egg white" protein and the large, nutrient-laden yolk. The zygote is a small region, the blastodisc, on the yolk surface. Cleavages cannot cut through the entire yolk, so at first only partial cleavages occur (Figure 42.8c). Cell division is completed a short time later, as the **blastoderm**, a flattened two-layer island of cells, arises.

The Insect: A Very Different Pattern Early development in the fruit fly zygote differs considerably from that of the other examples (Figure 42.9). Rapid mitotic activity follows fertilization, but oddly, no cytoplasmic division occurs. In fact, no plasma membranes appear until much later, so the early embryo becomes one large multinucleate cell. After hundreds of nuclei have emerged, a few move to the posterior periphery of the egg, where they form plasma membranes.

The first cells, called **pole cells**, will give rise to germinal (gamete-forming) tissue in the adult. (It seems that reproduction has a high priority in insect life.) As we have emphasized, embryonic determination begins early in the insect. Nuclear division and migration in the fruit fly embryo continue until about 6000 nuclei are present in a monolayer just under the egg surface. Plasma membranes then form, and the new cell layer becomes the insect blastoderm. With this pattern established, the segmented form of the insect larva begins to emerge.

The Blastula

As the cells of amphibian and echinoderm embryos continue to divide, they arrange themselves into a hollow sphere known as a **blastula** (Figure 42.10a,b). The fluid-filled cavity within is called the **blastocoel**. The frog blastocoel forms off center, near the animal pole, with the pole cells forming a thin roof-like cover over the cavity.

Because of the great amount of yolk, the formation of a simple blastula is not possible in bird and reptile embryos. Its equivalent here is a flat, plate-like, two-layer blastoderm, below which lies a narrow cavity that forms as the blastoderm separates from the yolk (Figure 42.10c).

The blastula stage is represented in mammals by the **blastocyst**, which in its earliest period consists of just 32 cells (Figure 42.10d). The fluid-filled inner space is simply called the *cavity of the blastocyst*. A thin layer of cells, the **trophoblast**, forms the outer wall of the blastocyst, and an aggregation of cells to one side is the **inner cell mass**. The trophoblast plays a supporting role, whereas the inner cell mass forms most of the embryo.

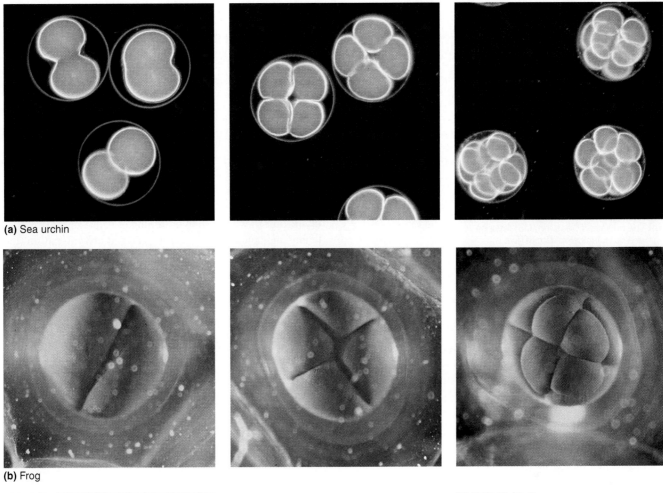

(a) Sea urchin

(b) Frog

(c) Chick

FIGURE 42.8
First Cleavages
Cleavage patterns differ among various animals. **(a)** Early divisions in the sea urchin are equal, but differences begin to be seen after the third cleavage. **(b)** In the frog, the first two divisions are parallel, dividing the zygote into four equal segments. The next cleavage is perpendicular to the first two and occurs nearer the animal pole. Future cleavages occur much more rapidly in this region and a large number of smaller cells soon appear. Cells in the vegetal hemisphere remain large. **(c)** Cleavage in the bird occurs in the thin blastodisc only. The early cleavages are incomplete, as shown in the side view.

The blastula and blastocyst represent the end of a preliminary period in development, one in which the events are relatively straightforward. All of this is about to change, as we reach a complex, critical event called gastrulation.

Gastrulation: Organizing the Germ Tissues

"It is not birth, marriage, or death, but gastrulation, which is truly the most important event of your life," quipped embryologist Lewis Wolpert in 1983. The reason is because nothing can proceed without a very special rearrangement of cells in the embryo, a process called **gastrulation**. At its completion, the embryo becomes a **gastrula**.

The cell rearrangements of gastrulation lead to the formation of three specific tissue layers, or **germ layers** as they are known: **ectoderm**, **endoderm**, and **mesoderm**. The word *germ* in developmental biology refers to beginnings, for it is from these layers that the formation of tissues, organs, and

Zygote

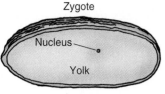

(a) Diploid zygote

Nuclei

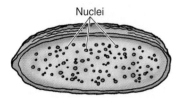

(b) The nucleus divides for nine divisions in a common cytoplasm to give a multinucleate cell.

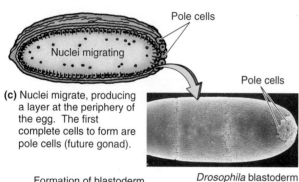

(c) Nuclei migrate, producing a layer at the periphery of the egg. The first complete cells to form are pole cells (future gonad).

Formation of blastoderm

Drosophila blastoderm

(d) Peripheral nuclei divide four more times; membranes form and they become somatic cells.

FIGURE 42.9
Early Insect Development

(a) Insect eggs have a moderate amount of yolk that is centrally located. **(b)** The early developmental events are unusual in that numerous mitoses occur without cytokinesis. **(c)** After about ten rounds of mitosis, nuclei begin migrating to the egg surface and the first plasma membranes appear. Pole cells, the first to appear, form at the posterior end. The pole cells are clearly seen in the SEM. **(d)** Eventually a layer of completed cells forms around the periphery, producing the insect blastoderm.

organ systems will begin. You may recall that the evolution of a third germ layer, the mesoderm, paved the way for the formation of true organ systems in animals (see Chapter 26).

Gastrulation involves several patterns of cell movement: **invagination** (an infolding or inpocketing of the surface), **involution** (inward rolling or turning of cells), **epiboly** (a flattening and spreading of a cell layer), and **cell migration** (the movement of cells to their functional positions).

The Sea Urchin. One of the simplest examples of gastrulation is provided by the sea urchin. As we see in Figure 42.11a, the process begins with invagination, an inpocketing of the blastula surface. (The result is like pushing your fist into a soft beach ball.) Newly formed cells at the site, called **mesenchyme**, migrate into the blastocoel and form attachments on its upper wall. The attachments then begin tugging on the invagination, helping it deepen. In this way a new cavity, the **archenteron** is created. Its external opening is called the **blastopore**. The archenteron, or "primitive gut" as it is also known, later forms the actual gut of the animal. Since the echinoderm is a deuterostome, the blastopore region represents the future anus, with a mouth later breaking through at the opposite end (see Chapter 26). Yet the ingrowth has a more immediate result: the outer cell layer is now the ectoderm, the lining of the archenteron is the endoderm, and cells in between are the mesoderm. From the gastrula, the sea

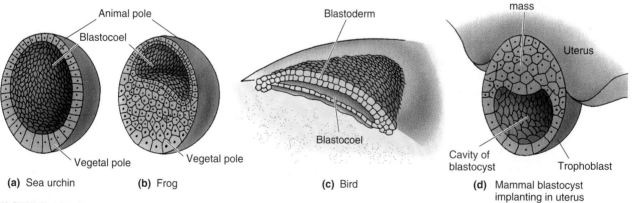

(a) Sea urchin   **(b)** Frog   **(c)** Bird   **(d)** Mammal blastocyst implanting in uterus

FIGURE 42.10
The Blastula Stage

Sea urchins **(a)** and amphibians **(b)** move from the morula state to a fluid-filled sphere known as a blastula. The central cavity is the blastocoel. Note the asymmetrical shape of the frog blastocoel, a result of larger yolky cells at the vegetal pole. **(c)** The equivalent of the blastula in the bird and reptile is a slender cavity below the blastoderm. **(d))** The equivalent in the human embryo is the blastocyst. Its outer cell layer, the trophoblast, surrounds the blastocyst cavity. A dense region of cells to one side, the inner cell mass, represents the embryo proper.

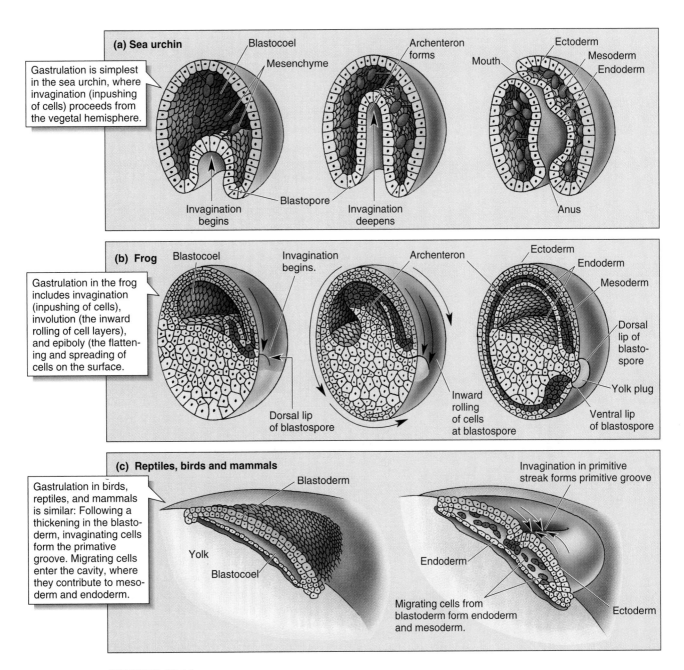

(a) Sea urchin

Gastrulation is simplest in the sea urchin, where invagination (inpushing of cells) proceeds from the vegetal hemisphere.

Blastocoel
Mesenchyme
Archenteron forms
Ectoderm
Mouth
Mesoderm
Endoderm
Invagination begins
Blastopore
Invagination deepens
Anus

(b) Frog

Gastrulation in the frog includes invagination (inpushing of cells), involution (the inward rolling of cell layers), and epiboly (the flattening and spreading of cells on the surface.

Blastocoel
Invagination begins.
Archenteron
Ectoderm
Endoderm
Mesoderm
Dorsal lip of blastospore
Yolk plug
Dorsal lip of blastospore
Inward rolling of cells at blastospore
Ventral lip of blastospore

(c) Reptiles, birds and mammals

Gastrulation in birds, reptiles, and mammals is similar: Following a thickening in the blastoderm, invaginating cells form the primative groove. Migrating cells enter the cavity, where they contribute to mesoderm and endoderm.

Blastoderm
Invagination in primitive streak forms primitive groove
Yolk
Blastocoel
Endoderm
Migrating cells from blastoderm form endoderm and mesoderm.
Ectoderm

FIGURE 42.11
Gastrulation

(a) Gastrulation in the sea urchin begins with an invaginating of the blastula surface, forming a blastopore. Newly formed mesenchyme cells, migrating into the blastocoel, will become mesoderm. The invagination deepens, forming a new cavity, the archenteron (future gut). It is composed chiefly of endodermal cells. As gastrulation continues, mesenchyme cells form attachments that literally pull the lengthening archenteron into the blastocoel. In its completed state the gastrula contains an outer layer of ectoderm, a middle layer of mesoderm, and an inner layer of endoderm. The archenteron eventually breaks through to the outside, establishing the mouth region and thereby completing the primitive gut. **(b)** Gastrulation in the frog also begins with invagination, but off to one side of the vegetal pole. More cells join the process, rolling into the newly forming blastopore and forming a new cavity, the archenteron. Epiboly, a spreading and flattening of cells at the animal pole, replaces those cells entering the blastopore. As gastrulation continues, the archenteron begins to crowd into the old blastocoel. Note that the archenteron is lined with endoderm and a region of newly formed mesoderm builds between it and the ectoderm. The archenteron will not break through to form the mouth region for some time. **(c)** Gastrulation in the bird embryo begins as cells of the newly thickened primitive streak involute, turning inward and entering the cavity below. They will organize into a mesoderm and endoderm layer. The archenteron will form as endoderm from below rises up to produce the tube-like outlines. Gastrulation in mammals is very similar.

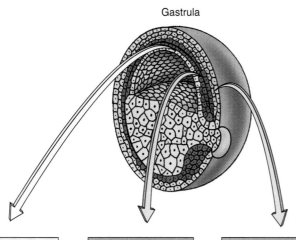

Gastrula

Endoderm	Mesoderm	Ectoderm

Primitive gut
Linings for:
Urinary bladder
Pharynx
Respiratory
  system
Pancreas
Liver
Intestine
	Skeleton
	Muscles
	Dermis
	Kidneys
	Gonads and
	  reproductive
	  structures
	Blood
	Heart
		Spinal nerves
		Adrenal medulla
		Spinal cord
		Brain
		Sympathetic nerves
		External coverings
		Inner ear
		Epidermis
		Milk-secreting glands
		Lens
		Mouth lining
		Oil glands
		Hair
		Sweat glands

FIGURE 42.12
The Germ Layers
The germ layers organize specific tissues as development
proceeds. Although each makes the primary contribution,
the resulting structures are invaded by the others. Thus
the organs contain derivatives of all three germ layers.

urchin develops into a ciliated larval feeding stage, from
which the adult will later emerge.

The Amphibian. The first sign of gastrulation in the frog
embryo (Figure 42.11b) is the appearance of a curved crease
on the surface. As in the sea urchin, it forms through invagination. The crease forms the **dorsal lip of the blastopore.** In
time, the blastopore will become a complete circle. Meanwhile, cells of the dorsal lip begin rolling inward. As the
inrolling, or involution, continues, the invagination grows
deeper, crowding into the old blastocoel and forming a new
cavity, the archenteron. Involution requires an ongoing supply
of cells, and this is provided by epiboly, the third process.
Here active surface cells of the animal pole undergo a flattening and spreading process that advances in a wave-like front
over the embryo.

By late gastrulation the three germ cell layers have formed.
A thin outer layer of cells makes up the ectoderm, a new middle
layer comprises the mesoderm, and an innermost layer forms
the endoderm. The archenteron itself is endodermally lined.

Reptiles, Birds, and Mammals. Gastrulation in reptiles and
birds (Figure 42.11c) is necessarily different. The blastoderm,

formed earlier, is a flat, two-layer structure, its layers separated
by a narrow, flattened cavity. The upper layer will contribute to
ectoderm, mesoderm, and endoderm, as its cells are rearranged.

Preparations for gastrulation in the blastoderm begin with
the migration of cells toward a thickening centerline known as
the **primitive streak.** The streak then lengthens and narrows.
Gastrulation begins with the formation of a crease, the **primitive groove,** along the midline of the streak. The groove deepens as involution occurs along its length, the inrolling cells
entering the cavity below. Some of the displaced cells will
become organized into mesoderm, whereas others will form
the endoderm. So gastrulation in the bird, as in the echinoderm
and amphibian, produces a three-layer embryo. An archenteron also forms in the reptile and bird gastrula,
but in these vertebrates the gut must arise through
additional cell rearrangements in the endoderm.

What about mammals? You will probably be
relieved to learn that gastrulation in mammals is
almost identical to that of birds and reptiles. Mammals also form a two-layer blastoderm, a primitive
streak, and primitive groove, and during gastrulation migrating cells from the surface will contribute to mesoderm and endoderm.

The developmental similarities in reptiles,
birds, and mammals may be viewed as compelling evidence of the close evolutionary relationship of the three vertebrate classes. The basic
developmental plan was presumably laid down by
the early reptiles, those in whom the first land
eggs evolved. The formation of a large, very yolky
egg and its developmental consequences led to
many developmental changes over the earlier
amphibian mode of development. Birds also made use of the
land egg; thus their development is very similar to the reptile's. Mammals (except for monotremes) abandoned the egg-
laying mode in favor of internal development. Yet the basic
ancestral pattern of gastrulation has been retained, although
modified to some extent. We'll be looking at other mammalian
modifications on the reptilian developmental plan shortly.

The Embryo Takes Form: Organogenesis

It is during **organogenesis,** the formation of organs and organ
systems, that the role of the three germ layers becomes clear
(Figure 42.12). The ectoderm makes contributions to body
coverings and neural structures, the latter including the brain
and spinal cord. Mesoderm, the middle germ layer, provides
cells for the skeleton, muscles, blood vessels, and gonads.
Endodermal tissue forms body linings, including those of the
digestive, respiratory, and excretory systems. Importantly, all
the organs eventually contain derivatives from all three germ
layers. For example, although the brain is principally ectoder-

mal in origin, it contains blood vessels that are both mesodermal (muscular wall) and endodermal (inner linings) in origin.

Neurulation

In **neurulation**, a striking phase seen only in vertebrate development, the crude outlines of the brain and spinal cord form, and the body axis, with distinct head and tail ends, is established. During the process, the gastrula is transformed into a **neurula**. The amphibian will serve as our representative example.

Neurulation in the frog begins with the formation of the **neural plate**, a flattened strip of ectoderm along the dorsal region of the gastrula (Figure 42.13a). The strip extends part way around the embryo, beginning at the dorsal lip of the blastopore (the future tail region).

As neurulation proceeds, the two edges of the neural plate thicken, forming what are called **neural folds**. The **neural groove**, a depression between the folds, also takes form (Figure 42.13b). Next, the neural folds rise up along their length, curve inward above the neural groove, and close, thus forming the **neural tube** (Figure 42.13c). In the closing, clusters of ectodermal cells known as **neural crest cells** become isolated between the neural tube and the overlying ectoderm. We mention these cells because they will later migrate far and wide from this location, contributing to a variety of structures (pigment cells, neural ganglia, connective tissue, Schwann cells, and elements of the autonomic nervous system).

Just before the start of neurulation, a critical structure, the rod-shaped **notochord**, takes form just below the region that will become the neural plate (see Figure 42.13a). Recall that the notochord, is present in all chordates at one time or another. Numerous experiments show that the notochord has a profound effect on the overlying ectoderm, prompting it to form the neural plate, which, in turn, forms the neural tube. In development terminology, the notochord is said to **induce** the ectoderm to form a neural tube. We will look further into the inductive process later in the chapter.

Further Vertebrate Development

As the frog neurula elongates and takes form, its head and tail end become evident, and it seems to lie perched over the spherical yolky region below. But as the yolk is absorbed and metabolized, the sphere shrinks, and the familiar body form of the tadpole arises. Within the body, the archenteron becomes more tube-like and will soon form the anal and mouth openings. Paired blocks of tissue called **somites** soon appear on either side of the notochord. The somites are forerunners of trunk muscles and the axial skeleton. Their serial organization reminds us again of the segmental plan of vertebrates.

Figure 42.14 illustrates these features and compares further development in representative vertebrates. Of particular interest is the formation of the **gill apparatus**, formed from **gill arches** and **gill slits**. In fishes, the embryonic gill apparatus does just what its name suggests—it forms gill structures. Terrestrial vertebrates, those that do not form gills, have retained

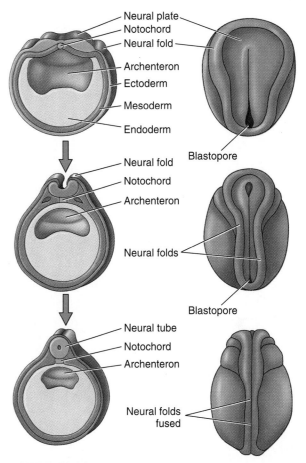

FIGURE 42.13
Neurulation
Neurulation in the amphibian begins with the formation of a flattened neural plate (*top*). Note the appearance of the notochord below. The edges thicken and rise up as neural folds (*middle*). The folds soon come together (*bottom*), forming the neural tube. The neural groove widens out at its anterior end, outlining the future brain region. With neurulation, the embryonic axis is established, with a widened head region at one end and a slender tail region at the other.

the gill apparatus, but it has been drastically modified for other purposes. Part of the first pair of arches in mammals, for example, contributes to the inner ear, while the next three pairs make contributions to the parathyroid glands, tonsils, and thymus. Not surprisingly, the presence of the gill apparatus in all terrestrial vertebrate embryos is often cited as strong evidence of evolutionary descent from the primal fishes (see Chapter 27).

Vertebrate Supporting Structures

All vertebrate embryos produce special structures known as **extraembryonic membranes**, namely the *amnion, yolk sac, chorion,* and *allantois.* The extraembryonic membranes are temporary structures that support and protect the embryo during development. As we found in Chapter 27, the amniotic egg

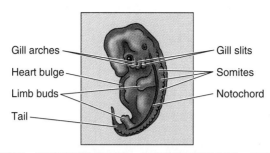

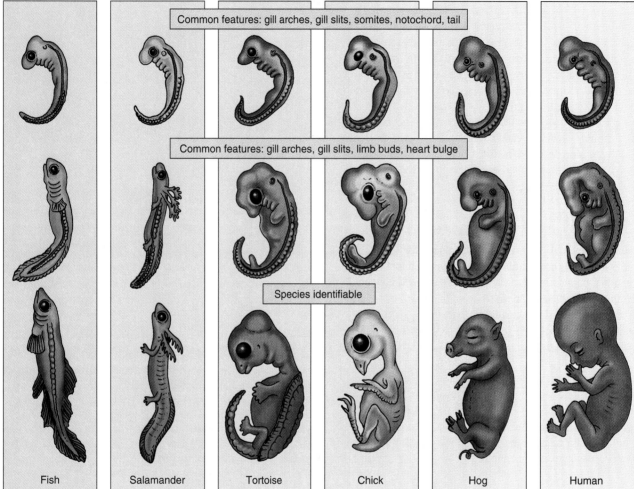

FIGURE 42.14
Further Vertebrate Development
The general vertebrate developmental plan is visible in representatives of the five vertebrate classes: fish, amphibian, reptile, bird, and mammal. The class representatives are arranged in vertical columns and each row shows comparable stages in the developmental events. Thus you can make your comparisons by viewing across the page. The specific features are pointed out in each row. The illustration is taken from the work of Karl Ernst von Baer, a very prominent 19th century Estonian embryologist.

originated in the early reptiles, and its general features have been retained in that group and in the birds and mammals.

The Support Systems of Reptiles and Birds

In reptiles and birds (Figure 42.15), the **yolk sac** is an extension of the primitive gut that expands over the surface of the yolk, carrying with it a profusion of blood vessels. Its role is to absorb food substances and transport them to the embryo.

The **amnion** grows up over the simple embryo, its folds meeting to enclose the **amniotic cavity**. The fluid-filled cavity acts as a shock absorber and also keeps the body lubricated so that the growing parts do not fuse together. The amnion also gives rise to the **chorion**, which grows up over the entire egg contents, forming a continuous membrane just under the shell. It will later fuse with the allantois to form the **chorioallantois.**

FIGURE 42.15
Extraembryonic Membranes in the Chick

(a) The four to five-day chick embryo lies within its sheltering, fluid-filled amniotic cavity. The blood vessels spreading out over the yolk are part of the yolk sac. (b) The first membrane to form is the amnion, which grows out over the newly emerging neurula. As the amnion is completed, its membranes continue their spreading growth, forming the chorion. The yolk sac emerges below the embryo. (c) At four days, the allantois has emerged and is joining the chorion. The vessels of the yolk sac are clearly visible. (d) At the ninth day, nearly halfway through development, the embryo has enlarged considerably. The chorioallantois surrounds the embryo and the yolk is diminishing.

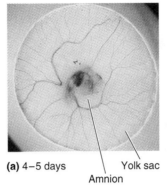

(a) 4–5 days
Yolk sac
Amnion

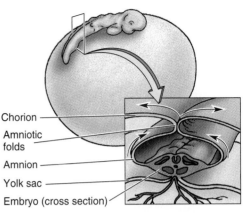

(b) Amnion forms, fuses over amniotic cavity and continues to form chorion. Yolk sac forms below.

Chorion
Amniotic folds
Amnion
Yolk sac
Embryo (cross section)

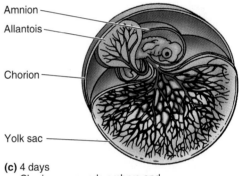

Amnion
Allantois
Chorion
Yolk sac

(c) 4 days
Chorion surrounds embryo and yolk. Allantois appears as a small vascular sac.

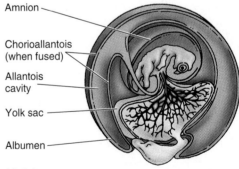

Amnion
Chorioallantois (when fused)
Allantois cavity
Yolk sac
Albumen

(d) 9 days
Chorioallantois has formed. Yolk diminishing.

The **allantois** is essentially endodermal in origin, beginning as a small pouch but later spreading into a full-fledged membrane as it fuses with the chorion. The network of blood vessels within the chorioallantois exchanges gases with the air outside the porous shell, while a cavity formed by the allantois functions as a nitrogen waste storage receptacle. The extraembryonic membranes remain behind when the bird or reptile hatches from the egg.

The Support System of Placental Mammals

The human embryo reaches the uterus in the trophoblast stage, about six or seven days after fertilization. There it contacts the soft endometrium, and implantation begins (Figure 42.16a,b). The trophoblast secretes enzymes that digest some of the endometrial tissue, and the embryo sinks into the minute, blood-filled cavity (Figure 42.16c). The blood provides the necessary food and oxygen and removes the embryo's meager metabolic wastes. But as the embryo grows, a more protected environment and a more efficient means of exchanging materials with the mother's body are needed. These needs are met by the extraembryonic membranes.

The Extraembryonic Membranes. About one week after implantation, cells of the trophoblast form the thin amniotic membrane, or amnion, that rises as a dome over the fluid-filled **amniotic cavity** (see Figure 42.16c). The floor of the cavity is the embryo itself. As in the bird, the amnion cushions the embryo, prevents adhesions from forming, and helps maintain a constant temperature. Finally, at the end of development, the bursting of the amnion (also called the "bag of water") will clearly signal the impending birth.

The yolk sac emerges as a simple, cell-lined cavity below the embryo (see Figure 42.16d). In mammals, where there are no yolk reserves to absorb, the yolk sac becomes the site of the first blood cell formation and houses certain lymphoid cells that later migrate into red bone marrow, where they give rise to stem cells (see Chapter 38). Other cells from the yolk sac migrate to the gonadal area, where they take up residence, later forming the germinal epithelium that gives rise to sperm and egg.

The chorion is formed by cells of the trophoblast. In time, the chorion produces finger-like cellular growths called **primary chorionic villi** (see Figure 42.16d). These growths break through older cells and deeply penetrate the uterine lining. When joined by blood vessels from the embryo, they become the highly branched **secondary chorionic villi**.

The allantois forms during the third week of life, when neurulation and the elongation of the embryo have begun (see Figure 42.16e). While it is reminiscent of its counterpart in

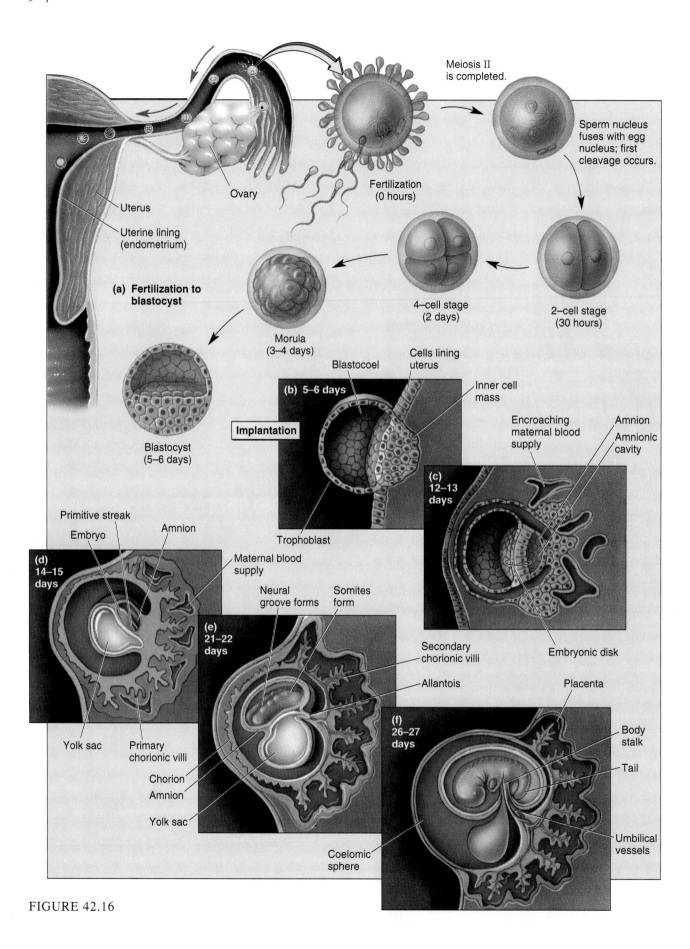

Meiosis II is completed.

Sperm nucleus fuses with egg nucleus; first cleavage occurs.

Ovary

Uterus

Uterine lining (endometrium)

Fertilization (0 hours)

2–cell stage (30 hours)

4–cell stage (2 days)

(a) Fertilization to blastocyst

Morula (3–4 days)

Blastocyst (5–6 days)

Blastocoel

Cells lining uterus

(b) 5–6 days

Implantation

Inner cell mass

Trophoblast

Encroaching maternal blood supply

Amnion

Amnionic cavity

(c) 12–13 days

Embryonic disk

Primitive streak

Embryo

Amnion

Maternal blood supply

(d) 14–15 days

Yolk sac

Primary chorionic villi

Chorion

Amnion

Yolk sac

Neural groove forms

Somites form

(e) 21–22 days

Secondary chorionic villi

Allantois

Placenta

(f) 26–27 days

Coelomic sphere

Body stalk

Tail

Umbilical vessels

FIGURE 42.16

FIGURE 42.16

Supporting Membranes and Early Growth in the Human
(a) During the five- or six-day journey through the uterine tube, the embryo reaches the blastocyst stage. (b) The trophoblast helps implant the embryo. (c) Cells of the trophoblast begin dividing at once, producing extensions that help carry out essential exchanges with the maternal blood. The amnion has formed above the simple plate-like embryo. (d) Here the yolk sac has formed and the primary chorionic villi deeply penetrate the surrounding blood sinuses. The embryonic axis is also well established. (e) The newest extraembryonic membrane, the allantois, has appeared and is producing blood vessels that will go on to form the umbilical cord. The chorionic villi are well organized and soon the placenta will take form. (f) The embryo is more recognizable now and the embryo lies within the surrounding coelomic sphere. Blood vessels have also formed, leading into the growing placenta.

the reptile and bird, the allantois does not store nitrogen wastes—they are constantly carried away in the mother's blood. In humans and other mammals, the allantois forms the **umbilical cord**, the structure containing blood vessels that link the placental circulation to that of the embryo.

The Placenta. The secondary chorionic villi, their blood vessels, and blood vessels from the endometrium of the uterus form the placenta (Figures 42.16f and 42.17). The chorionic villi are immersed directly in the mother's blood, where her blood passes through open sinuses. This facilitates exchanges between mother and embryo. Normally the blood of the mother and fetus do not mix, although we have noted situations in which this happens, some of them bringing on Rh incompatibility problems (see Chapter 11). The placenta eventually takes on a disk-like shape and reaches a diameter of about 20 cm and a thickness of 6 cm (8 in. by $2\frac{1}{2}$ in.—see Essay 42.1).

Development in the Human

Human development is traditionally divided into three-month sequences called **trimesters**. We can't get into all the detail here, but we will review some of the highlights. You will want to refer again to Figure 42.16 and Essay 42.1 as you proceed.

The First Trimester

The first trimester is marked by a number of far-reaching developmental events. As the embryo completes gastrulation and neurulation, the period of organogenesis begins (see Figure 42.16d,e). By the eighth week of development its human form is recognizable and the embryo is called a fetus. Yet, even then, it weighs only a gram and is about 30 mm in length (about 1.25 in.).

Nervous System. Neurulation begins at about day 18 or 19, closely following gastrulation. The neural folds rise and close over the embryonic disk, forming the dorsal hollow nerve cord. The anterior end enlarges to form the vesicles of the brain. By the fourth week, the major regions of the brain and spinal cord are recognizable. By week 12, the smoothly surfaced cerebrum will extend over much of the embryonic brain, and the cerebellum and medulla will have become distinct. While the nervous system is the earliest to emerge, many functional refinements will not be completed until long after birth.

Circulatory System. By the fourth week of life, the heart and circulatory system are clearly visible. The heart is at first a single tube-like structure, yet it efficiently pumps blood. As the blood is pushed along, it enters sinuses, forming channels that, when lined with endodermal cells, become blood vessels. By the sixth week, the tubular heart will have looped back on itself, taking on the four-chambered form. An opening between the atrial chambers, the **foramen ovale**, permits

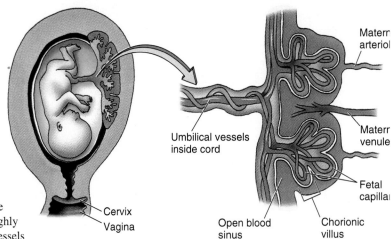

FIGURE 42.17

The Placenta
In the older fetus, the placenta takes up much of the uterine space. Its connections to the uterus occur in the form of highly vascular chorionic villi. The villi end in a maze of blood vessels that are surrounded by open sinuses containing slowly moving maternal blood. Numerous chorionic villi provide the surface area needed to carry on the essential exchanges between fetus and mother.

Essay 42.1 ● THE EMERGENCE OF HUMAN LIFE

(a) By the fifth week the human embryo is just 2 cm (0.7 in.) long. The hands have taken on a paddle shape, and darkened lines mark future centers of bone formation. (b) In another few days, the limbs have become better defined and blood vessels in the umbilical cord have progressed, as have surface vessels over the head and body. (c) At seven weeks, the embryo is 2.6 cm (1 in.) in length (crown to rump), weighing in at 2 grams (0.07 oz). The fleecy outlines of a well-formed placenta are also seen. (d) The limbs have lengthened in the nine-week fetus (it officially became a fetus at 8 weeks) and its fingers and toes are clearly defined. The eyes and ears are also taking form. (e) At the first trimester's end, all systems in the 6.4-cm long fetus ($2\frac{1}{2}$-in.) have neared a functional state. (f) Although it looks as if it could survive on its own, the 16-week fetus must still undergo a lot of growth and refinement.

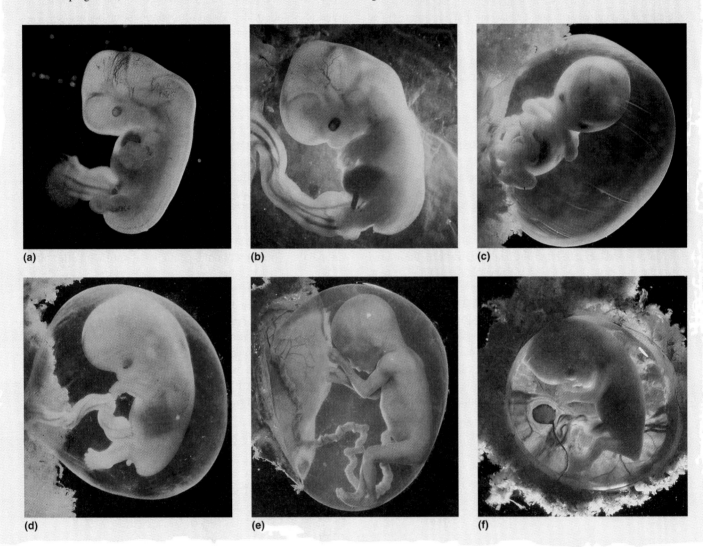

(a)   (b)   (c)

(d)   (e)   (f)

blood to bypass the budding, but nonfunctional lungs. Flaps forming the opening will close at birth.

Respiratory and Digestive Systems. By the fifth and sixth weeks, the crude outlines of the respiratory and digestive systems will have taken form. The simple tube-like gut gives rise to blind pouches and outpocketings that form the liver and pancreas. Similar branching in the pharynx gives rise to the trachea, bronchi, and lungs. Branches forming the lung buds soon merge to form the lungs.

Limbs. The limbs of humans appear as rounded buds during the fourth week. The arms and legs are distinguishable at six weeks, but fingers and toes require an additional week. The rudimentary hands and feet actually begin as simple webbed paddles, but fingers and toes take form through the breakdown and absorption of the webbing between.

Reproductive System. The reproductive system begins to develop during the first few weeks, but until the eighth week even a trained observer can't determine sex without a chromo-

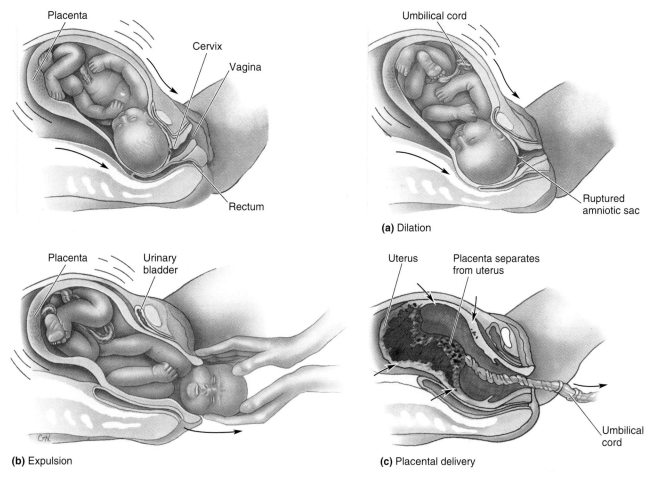

Placenta

Cervix

Vagina

Rectum

Umbilical cord

Ruptured amniotic sac

(a) Dilation

Placenta

Urinary bladder

(b) Expulsion

Uterus

Placenta separates from uterus

Umbilical cord

(c) Placental delivery

FIGURE 42.19
Birth
The birth process begins with **(a)** dilation, during which the cervix expands, labor contractions occur, and the baby's head crowns. **(b)** In expulsion, the second stage, the baby emerges through the birth canal. **(c)** The final stage involves the delivery of the placenta, which must complete its separation from the uterine wall. The uterus will immediately contract, a process that helps curb bleeding from the ruptured uterine vessels.

Lactation. Human mothers, like other female mammals, are prepared to nurse their newborn young immediately. The breasts, under the influence of estrogens and progesterone, have enlarged during pregnancy, and, with delivery, **colostrum** secretion begins. Colostrum, a clear yellowish fluid, differs from milk in that it contains more protein, vitamins, and minerals and less sugar and fat. It also contains maternal antibodies that can be important to the infant in its first days.

In a few days, the colostrum is replaced by whiter, thicker milk. Human milk is sufficient in all required vitamins except vitamin D and contains about 700 calories per liter. Interestingly, it is bacteria-free (which is not true of cow's milk, either in grocery stores or at the source). There are additional benefits of breastfeeding. Some studies reveal that breastfed infants are less susceptible to disease, anemia, and vitamin deficiencies. Then there is a sizable group of psychologists and physicians who believe that nursing is an important part of bonding between mother and infant.

An Analysis of Development

It is indeed fascinating to watch the changes in a developing embryo. Tissues appear, grow, move, reorganize, and disappear, all in a genetically choreographed ballet that easily arouses wonder in the most crusty and jaded of biologists. That wonder increases, though, when one moves from the *when* and *what* of development to the *why* and *how*. At this point, we come to the questions that many biologists view as the most challenging and exciting in biology.

For example, how does differentiation occur? That is, how do cells progress from a generalized, uncommitted state to a highly specialized state? How are the developmental fates determined, and when does determination begin? Finally, what is the role of the genes in development? How do genes produce the final phenotype? Until recently, we were woefully short on answers to such questions, but there have been

Fetal Circulation

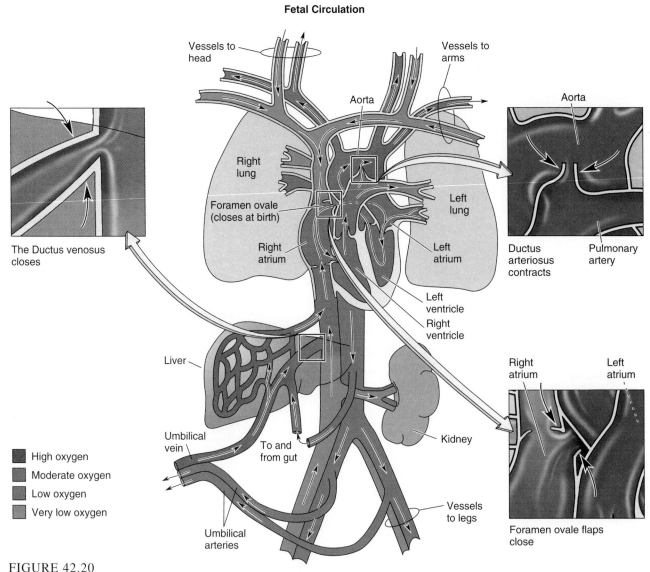

FIGURE 42.20
Circulatory Changes at Birth
Two pulmonary bypasses, the foramen ovale (between the atria) and the ductus arteriosis (between the pulmonary artery and aorta) must close at birth in order for pulmonary circulation to begin. In addition, the ductus venosus, a vessel that carries blood from the placenta to the liver, and arteries carrying blood from the fetus to the placenta must close off.

some amazing revelations, and we may be on the threshold of further breakthroughs in solving the mysteries of development. We will look first at some of the historical milestones and then tell you about some of the newer gains.

We will begin our look into the analytical side of development with a brief return to the polarized egg and zygote, where determination often begins. Determination, you will recall, is the ongoing commitment of cells to specific developmental pathways.

Determination in the Egg and Zygote

We've seen that animal egg cells typically reveal a degree of polarity. The dark pigmented vegetal hemisphere of the amph-

ibian egg can easily be distinguished from the lighter animal hemisphere (see Figure 42.2c). Furthermore, most of the organelles are clustered toward the animal pole and most of the stored yolky material toward the vegetal pole. Upon fertilization, and before even the first cleavages occur, the zygote becomes even more organized.

One of the earliest events is the formation of the **gray crescent**, a gray colored, crescent-shaped region that forms opposite the point of sperm penetration (Figure 42.21a). What is its significance? Careful observations show that the first division plane always bisects the gray crescent and we know that it is in this region that the blastopore appears at the start of gastrulation.

Clues to the developmental significance of the gray crescent were found about the start of the 20th century by Nobel

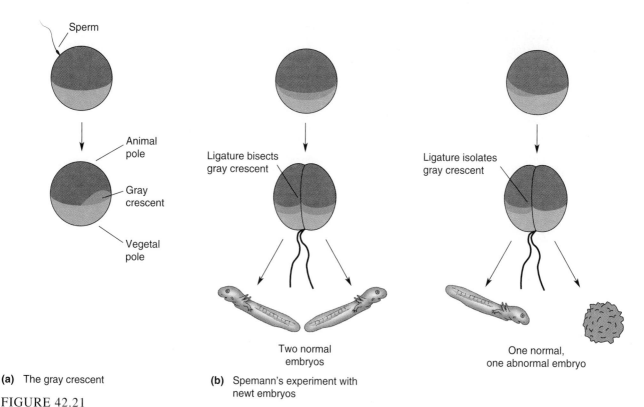

(a) The gray crescent

(b) Spemann's experiment with newt embryos

FIGURE 42.21
Early Determination in the Amphibian Embryo
(a) When fertilized, the amphibian egg forms the gray crescent, an area that helps organize the embryo.
(b) Its significance is seen in one of Spemann's experiments, where ligatures were tied *through* the gray crescent of one embryo and *alongside* the crescent in another. They show that elements of the gray crescent must be present in each cell for normal development to proceed.

laureate Hans Spemann. He succeeded in tying tiny ligatures around newt zygotes, partially dividing them in different planes (Figure 42.21b). Spemann found that at least part of the gray crescent had to be present if a cell was to go through development. Although the precise role of the gray crescent is still unknown, the experiments clearly show that a high degree of organization already exists in the amphibian zygote.

Caenorhabditis elegans:
A Model of Early Determination

A roundworm with the jaw-breaker name *Caenorhabditis elegans* (kano-RAB-dit-us) offers a special opportunity to trace the development of an embryo into the finished adult form. Why is this? First, *C. elegans* is easy to grow and maintain—its 1 mm-length makes housing easy and its $3\frac{1}{2}$-day life cycle enables the experimenter to produce many generations in a short time. Furthermore, *C. elegans* is monoecious—it produces both sperm and egg; but more than that, it is self-fertilizing—a rarity in animals. This helps keep the genetics simple to follow. In addition, the worm's transparent body makes its cells quite visible and readily identifiable.

Finally, and most significantly, *C. elegans* has just *959 somatic (nonreproductive) cells,* whose individual fates can all be traced back to just *six cells in the embryo!* Imagine the opportunities for studying differentiation here.

The six embryonic cells of *C. elegans,* or **founder cells** as they are known, have their developmental fates fully determined. They have been carefully tracked through development and growth, so a complete developmental map is available (Figure 42.22). This means that the original cells do not interact as developmental events progress, but each of the six goes its own way. In effect, the entire organism is assembled from pieces that are not interchangeable. This is referred to as **mosaic development**. As we will see, insects follow mosaic development as well.

Tissue Interaction in Vertebrates

The Primary Organizer. In 1921 Spemann and his associate, Hilde Mangold, devised some of the most remarkable experiments in the history of biology (Figure 42.23). They transplanted small pieces of presumptive (future) mesoderm from within the dorsal lip region of amphibian donor embryos

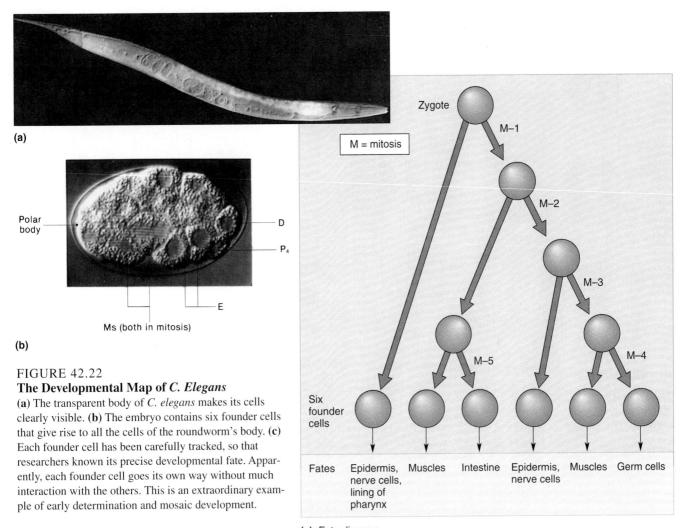

(a)

(b)

FIGURE 42.22
The Developmental Map of *C. Elegans*
(a) The transparent body of *C. elegans* makes its cells clearly visible. **(b)** The embryo contains six founder cells that give rise to all the cells of the roundworm's body. **(c)** Each founder cell has been carefully tracked, so that researchers known its precise developmental fate. Apparently, each founder cell goes its own way without much interaction with the others. This is an extraordinary example of early determination and mosaic development.

(c) Fate diagram

(newts) into recipient embryos of similar age. The transplant was inserted below the ectoderm of recipient embryos—in a region that would normally form *belly skin* in later stages. But rather than simple belly skin, the recipient embryo developed a second body axis (a notochord, neural tube, somites, and so on) in that location. In some cases, the embryo went on to form a second brain and spinal cord, and in others a second, fully formed embryo emerged.

Spemann and Mangold concluded that something in the transplanted material was organizing the tissue of the recipient, directing it along a different developmental pathway. Spemann labeled the donor dorsal lip mesoderm a **primary organizer**, and called the interaction between the organizer and the tissue it affected, **embryonic induction**, the ability of one tissue to redirect the developmental fate of another. Subsequent research has shown that such tissue interaction is a routine part of embryonic development.

Later transplant experiments have substantiated the controlling role of mesoderm over ectoderm. Birds, as you know, have scales on their feet and feathers on their legs. When mesoderm from the feet of chick embryos is introduced under

ectoderm in the upper leg, a patch of scales appears there rather than the usual feathers. When the reciprocal transplant is made, feathers appear on the feet! Again, the mesoderm is determining the fate of ectoderm.

Triple Induction in the Vertebrate Eye. Among the most complex of tissue interactions is the reciprocal influences that go on as the vertebrate eye takes form. Here tissues not only respond to each other, but their roles are interchangeable.

The developing vertebrate brain forms two bulb-like extensions that grow outward and contact the overlying covering of ectoderm. As you see in Figure 42.24, the first inducer is the bulging **optic vesicle**, which prompts the overlying ectoderm to form the **lens placode**. The lens placode returns the favor by then inducing the optic vesicle to form an **optic cup**. (So ectoderm can also induce!) The optic cup, in a final inducing step, stimulates the lens placode to form a lens. What we see is an embryonic "dialog" going on among the developing tissues of the eye.

Such interactions can be described in coldly analytical ways, but they can't help but show the fascinating nature of life. Those tissue interactions after all involve genes, some shutting down, others turning on, in a wonderful interplay that, in our example, will cause some cells to focus light and others to flash impulses to the brain as they respond to that light.

Search for the Organizer. The obvious question from the tissue interaction research is: What precisely is an organizer? Surely, it is something that moves from the organizing tissue to the target tissue; but what is it? After a search that lasted nearly 60 years, developmental biologists believe they have identified an important **morphogen**, or organizing substance. It is **retinoic acid**. A growing body of experimental evidence points to retinoic acid as a key substance in limb development in the chick embryo. We will come back to this important breakthrough later in the chapter.

The Role of Cell Migration

Cell migration is another important part of development. Earlier, in our discussion of neurulation, we noted the importance of neural crest cells in the continuing events of development. The neural crest cells, you may recall, form as isolated clusters between the neural tube and the overlying ectoderm (see Figure 42.13). Neural crest cells migrate far and wide in the young embryo, taking on many roles. Some move into the head region, forming the cartilages that are the forerunners of skull bones; others contribute to the adrenal medulla; and still others add to the eye's developing cornea.

Mesenchyme, another wandering young tissue, here originating from mesoderm in the block-like somites of the young chick embryo, migrates into the developing limb buds. There it gives rise to the muscles and skeleton of that limb.

Perhaps the most restless cells of all are the young neurons, which are of ectodermal origin. Their processes extend into every part of the body. Motor neurons, for example, arise from specific sources in the central nervous system and send their axons out toward target organs such as muscles. According to one theory, those that make their connections survive and become functional. Those that do not make such connections die off.

Studies of the regeneration of severed salamander limbs indicate how the selection is made. Cells in the regenerating limb stump secrete a substance called **nerve growth factor**, which stimulates newly arrived immigrant neurons to produce branches. In the absence of this stimulus, such as in neurons that reach the wrong target, the nerve cell simply dies. The central nervous system, "covering all bets," forms about twice as many motor neurons as are needed. Here is an example of how cell death is a normal part of development.

The developmental future of migrating cells often depends on cues from the new surroundings in which they find themselves, cues that arise either from resident cells or from the extracellular environment. One influential agent, hyaluronic acid, is a common constituent of connective tissue. In the formation of the cornea (the tough, transparent capsule over the pupil of the eye), neural crest cells migrating into the extracel-

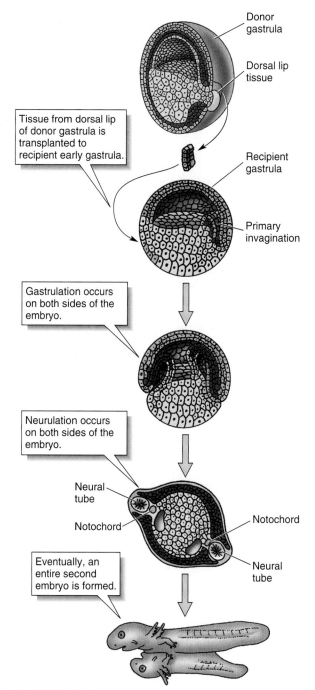

FIGURE 42.23
Embryonic Induction
In a classic experiment, Spemann and Mangold transplanted dorsal lip mesoderm from one early gastrula to another. Dorsal lip mesoderm of the donor was inserted under the ectoderm of the recipient, but on the side opposite the dorsal lip. The ectoderm overlying the transplant formed a second neural tube. Some embryos went on to produce fairly complete, "Siamese twin," embryos.

lular matrix differentiate into connective tissue. However, for the invading process to succeed, the cells already surrounding the matrix must first secrete hyaluronic acid. Otherwise, the migrating neural crest cells stop growing short of their goal.

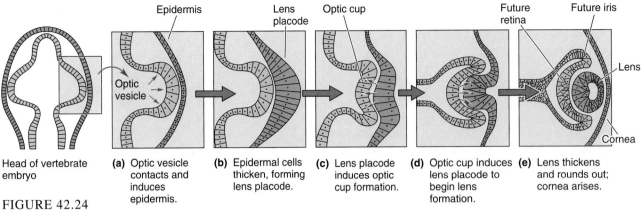

Head of vertebrate embryo

(a) Optic vesicle contacts and induces epidermis.

(b) Epidermal cells thicken, forming lens placode.

(c) Lens placode induces optic cup formation.

(d) Optic cup induces lens placode to begin lens formation.

(e) Lens thickens and rounds out; cornea arises.

FIGURE 42.24
Triple Induction
Proof that ectoderm, as well as mesoderm, can also become an inducing tissue is seen in the triple induction example. **(a, b)** Here optic vesicle induces epidermis to form lens placode. Lens placode, in turn, **(c)** induces optic vesicle to form optic cup. **(d, e)** Finally, the optic cup induces the lens placode to form a lens.

A Developmental Program in the Insect

Development in insects is similar, in principle, to that of the roundworm *C. elegans* but quite unlike that of vertebrates. Insect embryos undergo mosaic development (each structure derived from cells whose fate was set or determined very early in development and that have little influence on each other), and any experimental surgery on the early embryo, like that described for vertebrate embryos, will produce an abnormal or dead insect.

As we discussed earlier, the first actual cells to form in the embryo are the pole cells, and these are already committed to form germinal cells (see Figure 42.9). They will reside in the gonad, a structure that will not appear for several days. Such an early commitment of cells in the insect embryo makes it a very useful subject in the ongoing search for clues on how the developmental program works.

One of the clearest models of developmental determinism is seen in the *Drosophila* larva, where clusters of cells in the larva have their adult fate fully determined. These clusters, the **imaginal disks** (from *imago*, which means "adult"), enter an inactive state, remaining there until pupation. At that time, each imaginal disk gives rise to a specific adult structure. Other cells in the larva make little or no contribution to adult structure. In a sense, the larva itself does not change into an adult. Instead, it is a "temporary organism," one dedicated to the nourishment and support of dormant cells set aside for the task of producing the adult. The activation of these cells will occur when a shift in developmental hormones ushers in the events of pupation (see Chapter 36).

Determination and Imaginal Disks. The cells making up the imaginal disks appear to be very simple; there is no visible evidence that they are in a fully determined state. As we see in Figure 42.25, there are several paired imaginal disks. The pairs will give rise to antennae, legs, wings, genitals, and other organs. Each disk is, in turn, subdivided into compart-

ments, and each compartment is made up of cells that contribute to a specific part of the structure to which the disk is committed. In the wing disk, for example, one compartment will contribute cells to the anterior, or leading edge, of the wing, and the other to the posterior, or trailing part.

Homeotic Mutations. A particularly useful approach to the questioned relationship of genes to development in fruit flies has been the study of imaginal disk cell mutations. These are

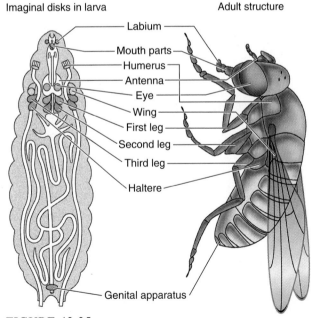

Imaginal disks in larva   Adult structure

- Labium
- Mouth parts
- Humerus
- Antenna
- Eye
- Wing
- First leg
- Second leg
- Third leg
- Haltere
- Genital apparatus

FIGURE 42.25
Imaginal Disks
Imaginal disks are cell clusters that are programmed to produce specific adult structures. They are set aside in the larva, where they remain inactive until pupation. Then control genes activate the clusters and adult structures emerge.

called **homeotic mutations**, and the genes affected are **homeotic genes**, or control genes, as we called them in an earlier chapter (see Chapter 14). As you may recall, homeotic genes regulate whole blocks of subservient genes, each of which, in the fruit fly, is responsible for some subpart of the structure assigned to cells of a given imaginal disk.

Because they disturb the regulation of such gene complexes, mutations in homeotic genes profoundly affect the usual developmental sequences. One of the most dramatic examples involves mutations in a large gene complex called *antennapedia* ("antenna-foot"). When active in cells near the eye, this complex normally dictates the formation of an antenna, but when the homeotic gene controlling the complex has mutated, the disk cells may give rise to a completely normal, but misplaced leg (Figure 42.26a)! Likewise, homeotic mutations in the so-called *bithorax* gene complex ("double chest"), a block of genes responsible for organizing the adult thorax (including the wings and halteres, or balancing organs), may bring about the production of extra thoracic regions, some bearing a second set of wings (Figure 42.26b)!

From such errors, developmental biologists determined that cells in a given disk have the potential to produce other structures, but as long as the homeotic genes are functioning normally, they are *inhibited* from doing so. At the moment it is not known just how the homeotic genes inhibit other genes, but molecular biologists suspect that they code for proteins that inhibit gene action by binding certain control regions of DNA (not unlike the *lac* operon, discussed in Chapter 14). Such a negative control mechanism would itself be affected by cues from the chromosome's surroundings. For example, it is known that when nuclei migrate to the outer region of the fruit fly embryo (see Figure 42.3), cues from the new surroundings bring on responses in certain control genes, which, in turn, initiate segmentation in the larva. The powerful analytical tools of molecular biology are now being brought to bear on the problem, and what biologists are finding is already casting light on some of the central puzzles in animal development.

The Homeobox: Molecular Biology of Development. As was mentioned in Chapter 14, molecular biologists have recently discovered that homeotic genes contain similar DNA segments, consisting of about 180 nucleotides and located in a region of DNA they named the **homeobox.** In *Drosophila*, the homeoboxes are believed to be the key to homeotic gene function, which means that they indirectly determine what the other genes in each cell of the imaginal disk are to do. It follows that a *single cue* from the surroundings may have far-reaching effects (Figure 42.27). We are confident that, in the fruit fly, such cues do affect the homeobox, thus activating particular imaginal disks. Such may be the case for control genes in other organisms. Should these hopes materialize, a major inroad into the understanding of development will have been made.

Even greater implications arise from our new knowledge of the homeobox. Molecular biologists, making use of search techniques from genetic engineering, have found strikingly similar stretches of nucleotides in other insects. This discovery prompted a broadened search, and, as a result, the homeobox was found in the genomes of annelid worms (a group closely related to arthropods). Then, to the great surprise of some, the homeobox was found in vertebrates, including humans. Some theorists are now suggesting that the homeobox occurs in all segmented animals.

How similar are homeoboxes? The protein product of the homeobox of the antennapedia gene complex of *Drosophila* is identical in 59 of its 60 amino acids to that of the homeobox of the frog *Xenopus*. Since the evolutionary lines of fruit flies and frogs must have diverged well over half a billion years ago (if, in fact, a single ancestral line ever existed—see Chapter 26), it looks as though natural selection has conserved the homeobox almost intact. To biologists, this suggests that its function is precise and vital. Biologists now speculate that the polypeptide product(s) of the homeobox may stimulate a battery of genes to work simultaneously, bringing on a major phase of development. We will watch these developments with avid interest.

A Developmental Program in the Vertebrate Wing

It's intriguing to ponder over the transition of the human hand from its crude paddle-like stage to the complex and versatile structure it becomes. On viewing a newborn baby, we invariably touch a tiny hand, test its tentative grasp, and marvel over the delicate, perfectly formed fingers. How does such a seeming developmental miracle take place? What assures the proper and timely assembly and arrangement of its parts?

For obvious reasons, we have little experimental data on morphogenesis in the human hand, but we are learning a great deal about limb development in other vertebrates, particularly about the formation of the wing. A favorite subject is the chick embryo with its very manageable three-week developmental period, its resilience, and, perhaps most importantly, its easy accessibility. Chick embryos will develop nicely out of the eggshell, in artificial environments where they are more available to the researcher.

The chick wing begins as a simple mound of tissue, the *wing bud*, appearing in the $3\frac{1}{2}$-day-old embryo. The wing bud consists only of a mass of mesenchyme cells surrounded by simple ectoderm that terminates in a thickening called the **apical ectodermal ridge** (Figure 42.28). But in just three more days, the skeletal elements begin to appear, and the wing takes on a paddle-like shape. After $9\frac{1}{2}$ days, each wing bone is nearly complete and clearly visible. The sequence in which the wing elements appear is highly ordered. Those elements closest to the body (the shoulder joint and upper arm) form first, and those furthest away (the digits) form last.

The wing's muscle and its connective tissues, including cartilage, bone, tendons, and dermis, are derived from mesenchyme tissue in the wing bud. The ectoderm, with its prominent apical ridge, is vital to the performance of the mes-

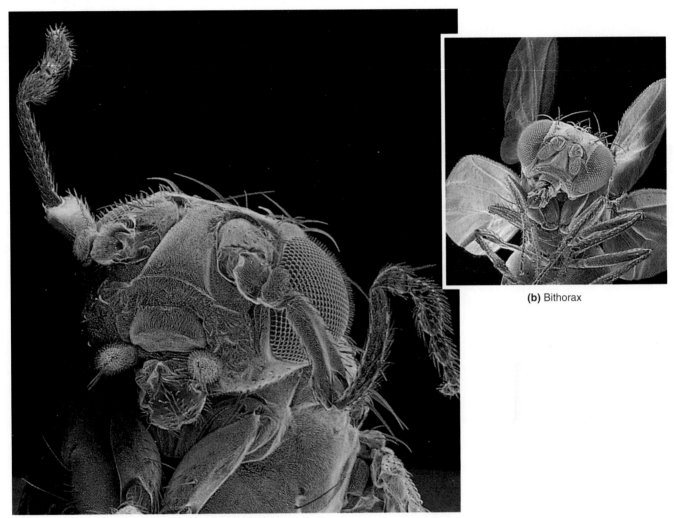

(a) Antennapedia

(b) Bithorax

FIGURE 42.26
Homeotic Mutants
Geneticists often make use of mutants to study the behavior of specific genes and track their hereditary patterns. Here they have been used to better understand homeotic control genes. **(a)** In antennapedia, a homeotic mutation produces legs instead of antennas. **(b)** In bithorax, a second thorax appears, complete with wings and bristles.

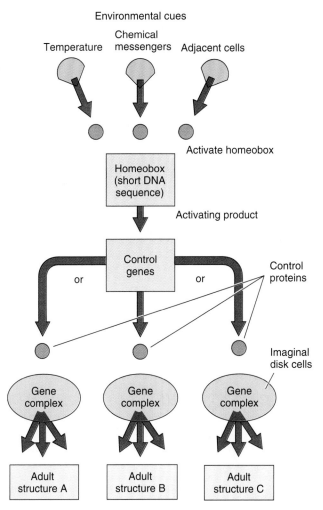

Environmental cues

Chemical
Temperature   messengers   Adjacent cells

Activate homeobox

Homeobox
(short DNA
sequence)

Activating product

Control
genes

or   or

Control
proteins

Imaginal
disk cells

Gene
complex   Gene
complex   Gene
complex

Adult
structure A   Adult
structure B   Adult
structure C

FIGURE 42.27
Homeobox Control
The homeobox, a nucleotide sequence in one of the chromosomes,
is activated by environmental cues. Its hypothetical role in the insect
is to code for control proteins that selectively activate gene complexes in imaginal disks. In this way, the adult structures are produced.

enchyme. As long as the ectodermal ridge is intact, formation
of the wing elements proceeds on schedule. Should the ridge
be surgically removed, development will be affected, the
degree depending on the age of the bud. From this fact, it
would seem that the apical ectodermal ridge might be in
charge, acting in effect as a primary organizer by inducing the
mesenchyme to differentiate. This idea has been tested.

As it turns out, the ectoderm isn't in charge at all. Replacing the apical ridge of an early bud with a late apical ridge
transplant or replacing the apical ridge of a late bud with an
early ridge has no effect at all. Development proceeds as
usual. So, while the presence of an apical ridge is essential, it
does not appear to control what happens in differentiating
mesenchyme. Determining what controls the mesenchyme
required still more experimentation.

The Zone of Polarizing Activity. A small cluster of mesenchyme cell, forming the **zone of polarizing activity**, or
ZPA, located along the posterior margin of the bud just within the ectoderm, appears to be in charge of limb development
in the chick. Again, transplant experimentation provides the
clues. When a portion of the ZPA and its overlying apical
ridge are removed from a donor and transferred to a location
on the *opposite side* of a recipient's limb bud, a second wing
with two sets of well-formed digits grows (Figure 42.29a).
But curiously, the digits formed by the transplant grow in a
reversed order. Thus the extra wing tip becomes the mirror
image of the original.

Researchers are now trying to understand just how the
ZPA affects the mesenchyme, and, since it must be present,
how the apical ridge fits into the picture. We'll concentrate on
the ZPA. Observations strongly suggest that the wing bud
mesenchyme cells differentiate according to their specific distance from the ZPA. This suggests that the ZPA may release
an active substance, a *morphogen,* that occurs in a concentration gradient across the limb bud.

The relationship between digit formation and distance
from the ZPA turns out to be very precise. The three digits of
birds are homologous to our own second, third, and fourth fingers. (Birds don't have the equivalents of thumbs and pinkies;
they have been lost in evolution.) Mesenchyme closest to the
ZPA forms the fourth digit, that somewhat further away forms
the third digit, and mesenchyme cells furthest from the ZPA
form the second digit. This is not a guess; the distance factor has
been determined through skillful transplant experimentation.

The pattern of digit formation in transplant experiments
depends on where the graft is placed. When ZPA mesenchyme from a donor is transplanted as far as possible from
the host's own ZPA, both the host and supernumerary wing
tips have the normal number of digits, but, as we've said, they
occur in reverse order (4-3-2:2'-3'-4') (Figure 42.29b). But
when the transplant is placed nearer the host ZPA, the resulting decrease in distance (morphogen concentration) inhibits
the formation of both second digits, so only the fourth and
third digits emerge (4-3:3'-4'). When the transplant is still
closer to the host's ZPA, the second digit from the donor ZPA
develops, but the host's second digit fails. Further, and
stranger still, is the rearrangement in digit order in both wing
tips: 4-3-3-4:4'-3'-2' (Figure 42.29c). This odd result, extra
digits in the host limb, is attributed to the high concentration
of morphogens from both host and donor ZPA.

Clues to the identity of the morphogen began to emerge
in the early 1980s when it was discovered that bathing the
amputated limb of a salamander with retinoic acid, a vitamin
A derivative, brought about the formation of two sets of limb
parts, just as occurred with the transplant of a ZPA in chick
limbs. Shortly afterward, the same results were produced in
chick embryos when tiny pieces of filter paper impregnated
with retinoic acid replaced ZPA at the transplant site. Retinoic
acid has since been extracted from the limb buds of chick

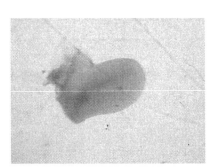

(a) Wing bud

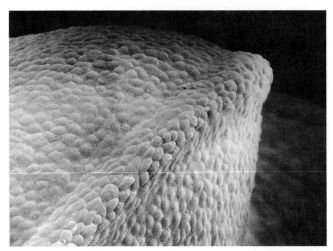

(b) Apical ridge

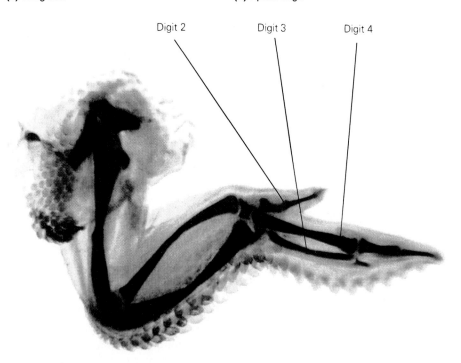

Digit 2   Digit 3   Digit 4

(c) Wing digits

FIGURE 42.28
Wing Development
(a) The wing begins development as a wing bud. **(b)** At the edge of the bud lies a thickening called the apical ridge. It contains undifferentiated mesenchyme with a covering of ectoderm. **(c)** The wing digits are numbered according to the vertebrate plan, with digits 1 and 5 having been lost through avian evolution.

embryos. It occurs in a gradient across the bud. This significant discovery may explain the peculiar effects of location on ZPA transplants. So it looks as though the breakthrough has finally come and the first inducing substance has been identified. There are some problems with this conclusion, so developmental biologists, who are used to such complications, remain guardedly optimistic.

Summing Up

What do we know about development? We know that cells in certain embryos pass through increasingly determined states, eventually to be committed to a specific fate in the embryo. We also know that tissues in many animals interact, some inducing others to follow specific developmental pathways. We've also seen where determination has a strong genetic basis, involving the selective activation and inactivation of genes, and how specific DNA sequences called homeoboxes may play key roles in activating gene complexes that contribute to the formation of tissues and organs. In addition, we have found that homeoboxes probably react to external cues from their environment and that the cellular environment in the embryo is a constantly shifting, changing entity. Then, in the complex interplay of cells that form the limbs, we realize that some cells determine the specific structures to be produced, whereas other cells determine the organization or pattern such parts will take. Furthermore, the events are carefully orchestrated in a strict but constantly shifting time frame.

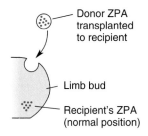

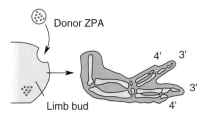

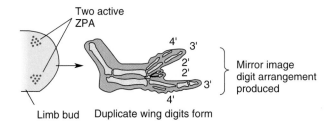

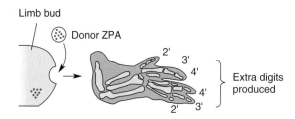

(a) Donor ZPA to opposite side of limb bud

(b) Donor ZPA nearer to recipient's ZPA

(c) Donor ZPA (zone of polarizing activity) close to recipient's ZPA

FIGURE 42.29
ZPA Transplants
(a) When the ZPA (zone of polarizing activity) of a donor is implanted opposite the host's ZPA on the limb bud, a mirror image pattern of digits is established. **(b)** A transplant placed nearer the host ZPA produces a different pattern of digit formation, whereas **(c)** one placed very close to the host ZPA results in the different pattern of donor-induced digits, plus extra host digits. The varying results are attributed to different concentrations of morphogen, released by the new ZPA.

Key Ideas

GAMETES

1. Animal **sperm** cells are made up of a head (condensed chromatin and an **acrosome**), a midpiece (mitochondria and centriole), and tail (9 + 2 flagellum).

2. **Egg** size varies according to yolk quantity, and they tend to be polarized. The more metabolically active region is the **animal hemisphere**; the less active region is the **vegetal hemisphere**.

Fertilization

1. The sea urchins **acrosomal reaction** includes enzyme release and the formation of an **acrosomal process**, the latter of which binds to the egg surface. Upon binding, their plasma membranes fuse, the egg forms a **fertilization cone**, and the sperm is drawn in.

2. Sperm entry brings on egg **activation**, in which metabolic activity is suddenly increased, an electrical fast block temporarily prevents **polyspermy**, and a **cortical reaction** produces a **fertilization envelope**.

3. Large numbers of human sperm release the acrosomal enzymes needed to penetrate the **corona radiata** and **zona pellucida**. No fast block or fertilization envelope forms.

EARLY DEVELOPMENT EVENTS

1. **Cleavage** patterns in mammals and echinoderms are similar, with equal-sized **daughter cells** appearing in the first three divisions. In amphibians, the third cleavage is unequal—rapid cell division produces small cells in the animal hemisphere, and slower division produces larger, yolk-filled cells in the vegetal hemisphere. Because of the large yolk, cleavage is incomplete at first in reptile and bird embryos.

2. Insect eggs are highly polarized. Mitosis without cytokinesis results in a multinucleate embryo. The earliest cells, **pole cells**, are committed to forming the adult gonad.

3. The **blastula** is a hollow ball of cells. The cavity within is called a **blastocoel**. The mammalian version of the blastula is the **blastocyst**. It is made up of a thin-walled **trophoblast** and **inner cell mass**.

4. In **gastrulation**, the inward movement of cells produces a **gastrula**, a three-layer stage containing an outer **ectoderm**, middle **mesoderm**, and inner **endoderm**. The inpouching forms a new cavity, the **archenteron**, which opens to the outside via the **blastopore**.

5. In reptiles, birds, and mammals, gastrulation is preceded by the formation of the **primitive streak**, a thickening in the epiblast layer. A **primitive groove** forms as cells begin to invade the

flattened cavity below, where they give rise to mesoderm and endoderm. The gut arises through infoldings of endoderm.

THE EMBRYO TAKES FORM: ORGANOGENESIS

1. The germ layers play specific roles in **organogenesis**: ectoderm lays down the framework of coverings and neural structures; mesoderm forms the skeleton, muscle, blood vessels, and gonads; and endoderm forms the linings of the gut and other cavities.

2. The formation of a **notochord** and **neural tube** establishes the body axis—the anterior–posterior organization. The notochord induces **neurulation**, which begins with the formation of the **neural plate**, whose edges, the neural ridges, rise up and come together, forming the neural tube. **Neural crest cells** gather between the tube and overlying ectoderm.

3. All vertebrate embryos produce a **gill apparatus**: the **gill arches** and **gill slits**. They give rise to gills in fishes, but in terrestrial animals they form other structures. Such similarity in vertebrate embryogenesis is strong evidence of evolution from a common ancestor.

VERTEBRATE SUPPORTING STRUCTURES

1. Special egg features include the protective leathery or hardened shell, a water and food supply, and **extraembryonic membranes**. The latter includes the **amnion** (shock absorber and lubricant), **yolk sac** (for absorbing food), **chorion** (for gas exchange), and the **allantois** (for gas exchange and nitrogen waste storage). The latter two fuse as the **chorioallantois.**

2. In humans, the blastocyst implants in the uterine wall, whereas for cells of the trophoblast provide the necessary exchanges until finger-like, cellular extensions form greater exchange surfaces. The first extraembryonic membrane to arise is the amnion, followed by the yolk sac. The yolk sac has no nutritional function but produces blood cells, lymphoid cells, and cells that will contribute to the gonads. The chorion, arising from the trophoblast, forms finger-like **primary chorionic villi**. When invaded by blood vessels, the fingers become secondary chorionic villi and later the **placenta**. In humans, the allantois forms the **umbilical cord**.

DEVELOPMENT IN THE HUMAN

1. Gastrulation and neurulation occur by the third week, followed by organogenesis. By the eighth week, most systems have begun formation. The embryo then becomes a fetus.

2. With neurulation, the crude outlines of the brain arise, and by the fourth week the major regions and the spinal cord are recognizable. The heart begins as a simple tube, which by seven weeks has changed to a four-chambered pump. The digestive and respiratory systems emerge from outpocketings and refinements in the primitive gut. Limb buds appear in the fourth week, with the limbs becoming discernible by six weeks. Fingers and toes go through a paddle-like stage.

3. The reproductive system begins in an indifferent state, with identical genital structures in both sexes. In the presence of testosterone, male external genitalia emerge. In the hormone's absence, female external genitalia emerge.

4. By the third trimester, all organ systems are fully formed, and most function. The central nervous system takes on its final form, and ossification of the cartilaginous skeleton has occurred.

5. The birth process begins with softening and dilation of the cervix, rupturing of the amnion, and uterine contractions. Crowning and fetal expulsion occur next, followed by placental expulsion.

6. For the respiratory system to function, the **foramen ovale**, between the atria, and the **ductus arteriosus**, between the pulmonary artery and aorta, must close.

THE ANALYSIS OF DEVELOPMENT

1. Some animal eggs are highly polarized, their organization already being critical to future events. Cleavages must divide the cytoplasmic elements in highly specific ways for development to succeed.

2. The **gray crescent**, which forms at fertilization in amphibians, is significant to establishing right and left halves, the future blastopore, and the body axis (dorsal region) of the amphibian embryo. Spemann determined that some gray crescent must be present for a cell to complete development.

3. The roundworm *C. elegans* is used in developmental studies because its relatively small number of somatic cells can be traced back to just six cells in the embryo. This is an extreme example of **mosaic development**.

4. In their study of tissue interaction, Spemann and Mangold determined by transplant that future mesoderm in the dorsal lip region acts as a **primary organizer**, capable of orchestrating the formation of the body axis and inducing ectoderm to form a neural tube.

5. Parts of the vertebrate eye form through a process of triple induction: (a) brain ectoderm induces overlying ectoderm to thicken, producing a **lens placode**; (b) the placode, in turn, induces the brain ectoderm to form an **optic cup**, which (c) influences the placode to invaginate and form the lens.

6. Migrating cells take chemical cues from their new surroundings. Hyaluronic acid cues mesenchyme migrating to the eye cells to form connective tissues in the cornea. Chemical agents that prompt developmental events are called **morphagens**.

7. Strong commitment in the insect occurs in the earliest developmental stages. The first complete cells, pole cells, are already committed to form germinal cells in the adult.

8. **Imaginal disks**, cell clusters in the *Drosophila* larva, are inactive but are destined to form specific adult structures such as antennae, wings, and legs. Mutations in the **homeotic genes** can result in misplaced or duplicated organs in the adult. Homeotic genes are believed to be control genes, those that code for inhibitory proteins that affect entire gene complexes. Homeotic genes have common gene sequences called **homeoboxes**, through which homeotic gene control may be exercised. Thus any change in a homeobox drastically affects development. Similar sequences are seen in a great many segmented animals, including humans.

9. The vertebrate wing (or any vertebrate forelimb) follows a strictly timed and organized developmental program. The early limb bud consists of mesenchyme and overlying ectoderm, the latter culminating in the **apical ectodermal ridge** at the leading edge. Within the posterior margin of the ridge lies the ZPA, which seems to determine the kinds and numbers of digits that will form. Transplants of the ZPA indicate that it contains a morphogen that determines the order and arrangement in which limb elements develop. The morphogen is believed to be retinoic acid, the first of its kind ever identified.

Review Questions

1. Using the terms vegetal and animal hemispheres, explain egg polarity. (page 912)

2. Describe the formation of two kinds of sea urchin sperm barriers. Why are such barriers essential? (page 913)

3. List four things that go on in the sea urchin egg to suggest that it is activated. (page 913)

4. Using simple sketches, make comparisons of blastulas, blastocysts, and blastoderms. In what animals does each occur? (pages 916–917)

5. Using simple sketches if you wish, briefly describe the events of gastrulation in the echinoderm. (page 918)

6. Name the primary germ layers formed during gastrulation and list several roles of each. (pages 917–918)

7. Summarize the events of neurulation. What is the apparent role of the notochord in this process? (page 921)

8. What roles are ascribed to the somites and gill apparatus in the vertebrate embryo? (page 921)

9. Outline the events from fertilization through implantation in humans. What is the role of the trophoblast in implantation and the early developmental days to follow. (pages 923–924)

10. Describe the formation of primary and secondary chorionic villi. What important structure forms from the chorion and endometrium? (page 923)

11. At what time in the life of a human embryo is its designation changed to "fetus"? What is the status of the nervous, circulatory, digestive, and reproductive systems at that time? (pages 925–927)

12. What is the role of the mesonephric and paramesonephric ducts in producing male and female sexual structures? Name the factor that determines sex at this time and explain what it does. (page 927)

13. How does the presence of the foramen ovale and ductus arteriosus affect circulation in the fetus? What would happen if they remained intact in the newborn baby? (page 928)

14. Why has the roundworm *C. elegans* become of such importance to developmental biologists? (page 931)

15. What is a primary organizer? Briefly explain how Spemann arrived at the notion of a primary organizer. (page 931–933)

16. Where do imaginal disks occur? What is their role in the development of adult fruit flies? (page 934)

17. Where do homeotic genes occur, and what is their role in fruit fly development? (page 935)

18. Cite two effects of homeotic mutations and explain why the effects are so dramatic. (page 935)

19. What is the apparent role of the homeobox? Why is this of such great importance to development? (page 935)

20. Briefly summarize the evidence suggesting that the ZPA of the chick wing bud, and not the apical ridge, directs wing formation. (pages 935–936)

21. Review the results of transplant experiments in which an older progress region is transferred to a young wing bud and a younger progress region to an older wing bud. What does this indicate about "time in service"? (pages 937–938)

CHAPTER 43
The Development and Structure of Animal Behavior

CHAPTER 44
Adaptiveness of Behavior

CHAPTER 45
The Ecology of Individuals and Populations

CHAPTER 46
Communities, Ecosystems, and Landscapes

CHAPTER 47
Global Ecology and Biomes

CHAPTER 48
The Human Impact (Applied Ecology)

PART SEVEN

Animal Behavior and Ecology

The Development and Structure of Animal Behavior

CHAPTER OUTLINE

THE DEVELOPMENT OF BEHAVIOR

Genes and Behavior

Hormones and Behavior

The Three Ways Hormones Can Influence Behavior

THE ETHOLOGIST'S CONCEPT OF INSTINCT

Sign Stimuli

Innate Release Mechanisms

Fixed Action Patterns

LEARNING

Types of learning

HOW INSTINCT AND LEARNING CAN INTERACT

LEARNING AND MEMORY

Essay 43.1 Lest Ye Forget

Theories on Information Storage

KEY IDEAS

APPLICATION OF IDEAS

REVIEW QUESTIONS

For some reason we humans have a great deal of interest in other animals. People in small towns all over the country will sit and watch film after TV film of cheetahs running down their prey. Farmers in Arkansas might come in from a day's work and watch something about penguins. (*Anything* they learn about penguins is probably more than they will ever need to know.) Why do we have such an intense interest in the behavior of other species?

There are probably a number of reasons for this interest. The first might be historical. Throughout our tenure on the planet, our lives have been inextricably tied to other species in ways that made our success partly dependent on what we knew about them. We hunted some of them, we avoided some of them, and in many cases their well-being signaled our well-being (Figure 43.1).

Today, we need to know something about animal behavior for other reasons. Our control of domestic animals, for example, depends on what we know of their behavior. It wouldn't do to put two stallions in the same pen with a group of mares—both the stallions and the mares would be likely to get hurt. Certain kinds of dogs will protect herds of sheep; others will run off with the coyotes. And cats won't guard the house.

Perhaps the most controversial reason for studying animal behavior is to learn more about ourselves. The argument against such an assumption is that if you can't learn about lions by studying gorillas, you surely can't learn much about humans by studying rats. It may be true that some researchers in the past have considered laboratory animals to simply be smaller, cheaper,

FIGURE 43.1
Cave Painting
Humans have been keen observers of other animals, as indicated by
these ancient cave paintings discovered in 1995.

and expendable humans, but we know too much now to
extrapolate very heavily from one species to another.

Others might say that the purest reason for studying the
behavior of other animals is simply to learn more about them
and thus to expand our knowledge of our world. In any case,
we will assume that the study of animal behavior is worth-
while and plunge ahead.

In this chapter we will first look at how behavior develops,
in particular, at the molding influences of genes and hormones.
Then we will consider the concepts of instinct and learning as
determinants of behavior, focusing on how the two interact to
produce an adaptive pattern. Finally, we will look at the role of
memory in determining behavior. In the next chapter we will
see how behavior can help animals fit more precisely into their
world, concentrating on ecological and social factors.

The Development of Behavior

Why does an animal do what it does? Why do particular patterns develop? A fledgling blue-footed booby (Figure 43.2) on
a nest cannot leave the nest yet, but from where it sits the
fuzzy youngster picks up nearby twigs and repeatedly tosses
them into the air, catching them by one end. This is precisely
the motion it will use, as an adult, to handle the fish that it will
catch. It will toss the fish into the air and catch them by their
heads so that the spines will be safely flattened for swallowing. One might wonder then if the behavior is genetically
based. Was the bird born with the ability to toss and catch? Is
it practicing? If practice results in learning, can one say the
behavior is genetically based? Or can it be an inherited pattern
that is improved upon by learning through practice? These are
the types of questions that behavioral biologists have been
working on for years. At one time, these researchers rather
clearly (too clearly) took one of two positions (particularly
regarding birds and mammals). Some said that behavior is, in
general, genetically based, while others believed that almost
all behavior is learned (the so-called "nature–nurture controversy"). Almost no one takes such rigid positions these days.
Instead, researchers ask how behavior develops, fully aware
that the development can involve both genetic and learned
components, as we will see later in this chapter. But first, let's
look at the influences of genes and hormones on behavior.

Genes and Behavior

It is important to understand that genes do not directly code
for behavior. That is, there is no gene for twig-tossing in blue-
footed boobies. Instead, genes code for proteins. These pro-

FIGURE 43.2
Blue-Footed Booby
Blue-footed booby chicks will toss sticks in the same manner that
they will use to position fish for swallowing.

teins then operate at the cellular level to influence both structure and function within an animal. With this distinction in mind, we can proceed to investigate some lines of evidence that show the effects of genes on behavior.

Inbreeding Experiments. One of the best lines of evidence for genetic influences on behavior comes from inbreeding experiments. Inbred lines, strains created by mating close family members with one another, provide a kind of control in that they hold relatively constant the genetic background of the animals being examined, because each inbred line is homozygous for almost every gene. In this way, the relative effects of genes and environment on behavior can be determined. The behavior of two inbred lines is compared while they interact with the same environment; any behavioral differences are likely to be due to genetic influences.

One of the best-analyzed strain comparisons has been in avoidance learning. (Avoidance learning involves learning to avoid an unpleasantry, such as a mild electric shock.) Avoidance learning in mice can be studied in a shuttle box, a two-compartment apparatus with an electrified floor (Figure 43.3). Mice from two different inbred strains are placed in one compartment, and a light flashes just before the mouse receives a light electrical shock. Some mice soon learn to run through a door into the other compartment to avoid a shock. Some mice never learn to run to safety. Different strains of mice show quite different abilities in avoiding the shocking experience, and the difference is believed to be genetically based.

Artificial Selection Experiments. The effects of genetics on behavior have also been shown by artificial selection. For example, pit bulls and fighting bulls have been bred for bravery and aggressiveness. Neither the dog nor the bull will retreat from pain. Sled dogs and certain kinds of draft horses have been bred for their willingness to push against a harness when asked. And very specific behaviors have been developed that seem to have no useful or adaptive qualities at all, such as in waltzing mice, a line of mice with equilibrium problems that cause them to spin and flounder about.

Hybridization Experiments. A third way to demonstrate genetic influences on behavior involves hybridization. The usual procedure is to mate two individuals of strains that exhibit distinctively different behaviors. Depending on the number of genes involved, the offspring may show one parental type or the other (say, one or two genes) or a blend of the two types (multiple genes).

Behavioral Effects of a Few Genes. Honeybees express the hygienic behavior of just one of the parent types. Personal hygiene is usually an important attribute, especially among certain groups of our own species, but for a social insect such as the honeybee (*Apis mellifera*) hive hygiene can be a matter of life and death (Figure 43.4). This is particularly true if the hive has been infected with *Bacillus larvae*, a bacterium that

FIGURE 43.3
Mice in Shuttle Box
The shuttle box is used to test for avoidance learning in mice. Upon a signal they must run through the door to avoid a shock.

kills larvae and pupae developing within the waxen cells of the comb. If the corpses are not removed, they accumulate and serve as a source of infection that will destroy the colony. Some strains of bees, such as the Brown and the Squires Resistant, are resistant to infection mainly because the workers practice good hive hygiene; they open the cells containing dead individuals and remove them. The Van Scoy and Squires Susceptible strains are vulnerable to the disease because their unhygienic workers allow the deceased brood to accumulate and decompose within the hive.

When a hygienic strain is crossed with an unhygienic strain, none of the hybrids remove corpses from the hive. Unhygienic behavior is said to be dominant over the hygienic behavior of removing the bodies of progeny that die from this bacterial infection. In an experiment, unhygienic hybrid offspring were mated with homozygous recessive individuals (a backcross—see Chapter 10) and 4 classes of offspring were obtained in 29 colonies.

1. The workers in 9 of the 29 colonies would uncap the cells containing dead bodies but failed to remove the corpses.

2. The workers of 6 of the colonies would remove the dead bodies from cells that were uncapped, but they would not uncap the cells themselves.

3. The workers of 8 colonies were unhygienic; they did not uncap cells or remove corpses.

4. The workers of the remaining 6 colonies were hygienic and would uncap cells and remove the dead bodies.

These results support the conclusion that hygienic behavior is controlled by two genes, one for uncapping (the *U gene*) and one for removing dead bodies (the *R gene*). If the hygienic behavior was controlled by a single gene, you would expect a cross of a hybrid (and therefore unhygienic) individual with

FIGURE 43.4
Honeybees
Some honeybees are genetically programmed to remove dead larvae as we see at lower right; other lines do not and the resulting risk of infection can wipe out the hive.

FIGURE 43.5
Peach-Faced Lovebird
Peach-faced lovebirds transport nest material tucked in their feathers.

a homozygous recessive (and therefore hygienic) individual to result in only two classes of offspring, hygienic and unhygienic, and each class would be expected to have equal numbers of colonies. Four equally sized classes of offspring is what is predicted if two genes are involved in the expression of the trait.

Behavioral Effects of Many Genes. Sometimes so many genes are involved that the behavior of hybrids does not resemble that of either parent but is instead intermediate between the parental forms. Consider, for example, the nesting behavior of lovebirds. As their name implies, these members of the parrot family pair while they are still young and remain true to their mates for life. The first part of nesting behavior, obtaining materials, is the same in the Peach-faced lovebird (*Agapornis roseicollis*) and Fisher's lovebird (*Agapornis personata fisheri*). Members of both species use their bills to clip paper, bark, or leaves into neat little ribbons of nesting material. The Peach-faced lovebird transports these strips by tucking them into the feathers of the lower back (Figure 43.5). The feathers are raised while the strips are tucked in and then lowered to hold them in place. Fisher's lovebird has a more direct method of carrying material to its nest—it holds the strips in its bill. These two species are closely enough related that they will mate in captivity.

The hybrid offspring of crosses between Peach-faced and Fisher's lovebirds display a compromise between the two parental forms of nesting behavior. Since both parental species cut strips of nesting material the same way, hybrids had no trouble with this part, but they were undecided about how to carry the strips. They attempted to tuck the material into the rump feathers, but the strips fell out. About 6% of the time they successfully carried the strips to the nest in their bill. Two months later, they learned not to waste as much time in

abortive tucking efforts and carried 41% of their strips to the nest in their bills. This kind of intermediate behavior among hybrids and the difficulty they had learning to use one behavior pattern at the expense of another are evidence that the behavior has a genetic basis and that a number of genes are involved.

Hormones and Behavior

A cat will avoid a German shepherd at all costs—unless, that is, the cat has kittens. A mother cat is likely to attack any large animal that comes near her offspring. The reason is, in part, because certain hormones compel her to behave protectively. By the same token, a mother seal will retrieve her pup if it should wander too near the surging sea. The retrieval behavior has been shown to be related to the presence of certain hormones. In fact, if the hormones are altered, the retrieval behavior changes. The effects of hormones are basically studied in two ways: (1) by removing the gland and administering hormones and noting their effect and (2) by correlating hormones with specific behavior.

In some mammalian species, castration (removal of the gonads) results in a decreased frequency of copulation. To find out if the change in behavior is due to the surgery or to the removal of the gonads (the source of the male hormone testosterone), the hormone is injected into some of the animals after the surgery. In those individuals, the copulatory tendencies reappear, indicating that this complex behavior is indeed related to the presence of the hormone.

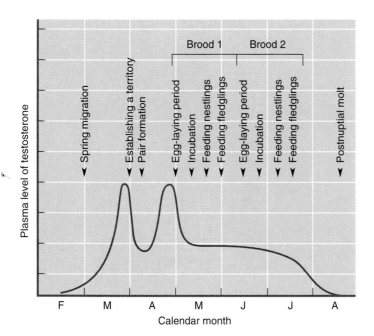

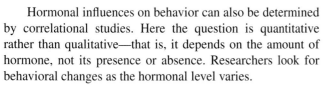

FIGURE 43.6
Testosterone and Male Territorial Behavior
Changes in circulating levels of testosterone occur during the breeding cycle of free-living male song sparrows (*Melospiza melodia*).

FIGURE 43.7
Zebra Finch
The song centers in this bird's brain are established at birth.

Hormonal influences on behavior can also be determined by correlational studies. Here the question is quantitative rather than qualitative—that is, it depends on the amount of hormone, not its presence or absence. Researchers look for behavioral changes as the hormonal level varies.

In one such study, blood samples taken from male song sparrows (*Melospiza melodia*) were analyzed over the course of a breeding season. The researcher, John Wingfield, found that testosterone levels were markedly correlated with certain behavioral patterns. Specifically, he found a close correlation between maximum levels of testosterone and the establishment of territorial behavior (Chapter 44), when a male defends a specific area (Figure 43.6).

The Three Ways Hormones Can Influence Behavior

Hormones can operate on behavior at three different levels: by influencing perception, by influencing the development of the central nervous system, and by influencing effectors.

Effects of Hormones on Perception. Hormones can influence the ability to detect certain stimuli, sometimes in unexpected ways. For example, the visual sensitivity (the ability to detect visual stimuli) of human females varies with the stages of the menstrual cycle. Visual sensitivity is greatest about the time of ovulation and declines abruptly with menstruation. After menstruation, visual sensitivity begins to increase again, peaking at ovulation.

Effects of Hormones on the Development of the Central Nervous System. Hormones can influence behavior by altering the morphology, the physiological activity, and the role of neurotransmitters in the central nervous system (the brain and spinal cord). For example, hormones can influence the size of the brain, cell number, cell branching, and the percentage of cells sensitive to particular hormones.

Certain brain changes, and the associated physiological activity, caused by hormones are reversible. For example, Fernando Nottebohm found that parts of the brain that control singing behavior in birds change in size in response to the waxing and waning of male hormones. Other changes in the brain due to hormones, though, may be more permanent. In the zebra finch, *Taeniopygia guttata* (Figure 43.7), sex differences in the brain's song-control centers are established around the time of hatching. At that time the song-control centers in males become permanently sensitive to certain hormones. Later, when the male hormones surge in their tiny bodies, the birds begin to sing.

Effects of Hormones on Effector Mechanisms. Hormones can influence behavior by altering effectors, such as muscle and neurons. An example involves the calling behavior of the clawed frog (*Xenopus laevis*) (Figure 43.8). Males of this species attract females by alternating fast and slow metallic "trills." Females, when declining males or when terminating breeding, produce slow, monotonous "clicks." (Sexually receptive females are silent.)

FIGURE 43.8
***Xenopus laevis*, the Clawed Frog**
The males and females produce different mating sounds as a result of hormonally controlled development.

The differences in the male and female sounds are due to the characteristics of the muscles and neuromuscular junctions of the larynx. Adult males have eight times as many muscles in their larynx. This difference arises under hormonal control during metamorphosis, as the frogs leave the tadpole stage. Before that time, the neuromuscular mechanisms of the sexes are identical. Apparently, male hormones (particularly androgen) stimulate the development of additional fibers that, with further specialization, lead to sexual differences in sound production.

The Ethologist's Concept of Instinct

The formal concept of instinct was developed in the 1940s by the Austrian Konrad Lorenz and refined by the Dutchman Niko Tinbergen in the 1950s. They called themselves *ethologists*. **Ethology** is concerned with four areas of inquiry: causation, development, evolution, and function (which, when preceded by *animal behavior,* can be remembered by a, b, c, d, e, and f—assuming one can remember the alphabet). Lorenz and Tinbergen, with Karl von Frisch (who worked on bee communication), were awarded the Nobel Prize in 1973—the first field biologists to be honored in this way. These pioneers have given rise to new generations of researchers who study animal behavior in the wild as they try to learn just how an animal's behavior helps it adapt to its environment.

Instinct may be defined as an inborn behavior with three components: the sign stimulus, innate releasing mechanism, and fixed action pattern.

Sign Stimuli

According to ethologists, instinctive behavior is triggered by certain very specific signals from the environment. Environmental factors that evoke, or release, instinctive patterns are called **sign stimuli**. The sign stimulus itself may be only a small part of any appropriate situation. For example, fighting behavior may be released in territorial male European robins, not only by the sight of another male, but even by the sight of a tuft of red feathers at a certain distance from the ground within their territories (Figure 43.9). Of course, such a response is usually adaptive because tufts of red feathers at that height are normally on the breast of a competitor. The point is that the instinctive act may be triggered by only *certain parts* of the environment.

FIGURE 43.9
Instinctive Attack
A male European robin in breeding condition will attack a tuft of red feathers placed in his territory. Since red feathers are usually on the breast of a competitor, it is to his reproductive advantage to behave aggressively at the very sight of them. The phenomenon illustrates that releasers of instinctive behavior need to meet only certain criteria—in other words, need to represent only one specific part of the total situation.

Innate Release Mechanisms

The exact mechanism by which sign stimuli work isn't known, but one theory is that there are certain neural centers called **innate releasing mechanisms** (**IRMs**), which, when stimulated by impulses set up by the perception of a sign stimulus, trigger a chain of neuromuscular events. It is these events that comprise instinctive behavior. The essence of the IRM is that it is genetically encoded, and once it is triggered by a sign stimulus, it results in the performance of a behavior called the *fixed action pattern.*

Fixed Action Patterns

It can't be denied that animals are born with certain behaviors that are indelibly stamped into their behavioral repertoire. Birds build nests by using peculiar sideways swipes of their heads, with which they jam twigs into the nest mass. And all dogs scratch their ears the same way, by moving their rear leg outside the foreleg. Such precise and identifiable patterns, which are innate, independent of the environment, and characteristic of a given species, are called **fixed action patterns**.

Of course, any behavioral pattern, in order to be effective, must be properly coordinated with the environment. The coordination of fixed action patterns with spatial variables in the environment is called **orientation**.

To illustrate the relationship between a fixed action pattern and its orientation, consider the fly-catching movements of a frog (Figure 43.10). For the fixed action pattern (the tongue flick) to be effective, the frog must carefully orient itself with respect to the fly's position. If the fixed action pattern is performed properly, the frog gains a fly. It is important to realize, however, that, in most cases, once the fixed action pattern is initiated it cannot be altered. If the fly should move after the tongue begins its motion, the frog will miss, since the fixed sequence of movements will be completed whether the fly is there or not. Thus the frog first *orients* and then *performs* the fixed action pattern.

To sum up the relationship between these three parts of the instinct idea, a *sign stimulus* in the environment activates the *innate releasing mechanism* in the nervous system, which then sends a signal to the muscles that causes a *fixed action pattern.*

Now we will consider the effects of learning on the development of behavior. Although we discuss innate and learned patterns separately, keep in mind that probably, in many cases, the two types of patterns operate together to produce an adaptive behavioral pattern.

Learning

Learning is the process of developing a behavioral response based on experience. As we consider learning in a wide range of animals, we should keep in mind that the importance of learning varies widely from one species to the next. For example, there are probably very few clever tapeworms. But then,

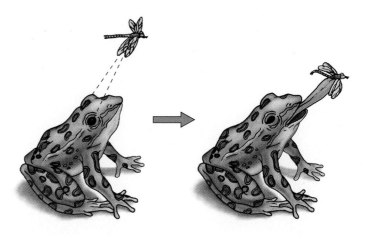

FIGURE 43.10
A Leopard Frog Orienting
The tongue flick, a fixed action pattern, is not released until the central nervous system is stimulated by the sight of an insect in the proper position (close and in the midline), after the frog orients toward it.

why should they be clever? They live in an environment that is soft, warm, moist, and filled with food. The matter of leaving offspring is also simplified; they merely lay thousands and thousands of eggs and leave the rest to chance—to sheer blind luck. In contrast, chimpanzees live in changeable and often dangerous environments, and they must learn to cope with a variety of complex conditions. Unlike tapeworms, they are long-lived and highly social, and the young mature slowly enough to give them time to accumulate information. The point is, each kind of animal is able to learn those things that are important to its survival and reproduction.

Types of Learning

Interestingly, different species may learn in different ways. However, certain kinds of learning seem to be common to a number of species. Here we will consider six kinds of learning: *habituation, classical conditioning, operant conditioning, latent learning, insight learning,* and *imprinting.*

Habituation. Learning *not* to respond to a stimulus is called **habituation**. In some cases, the first time a stimulus is presented, the response is immediate and vigorous. But if the stimulus is presented over and over again, the response to it gradually lessens and may disappear altogether. Habituation is not necessarily permanent, however. If the stimulus is withheld for a time after the animal has become habituated to it, the response may reappear when the stimulus is later presented again (a process called *sensitization*).

Habituation is important to animals in several ways. For example, a bird must learn not to waste energy by taking flight at the sight of every skittering leaf. A reef fish holding a territory may come to accept and pretty much ignore its neighbors but will immediately drive away a strange fish wandering through the area. The wandering fish is likely to be searching for a territory and therefore to be more of a threat.

It may also help explain why animals continue to avoid predators (which they are likely to see only rarely), while ignoring more common, harmless species. Habituation is often ignored in discussions of learning, perhaps because it seems so simple, but it may well be one of the more important forms of learning in nature.

Classical Conditioning. Classical conditioning was first described through the well-known experiments of the Russian biologist Ivan Pavlov. In **classical conditioning**, a behavior that is normally released by a certain stimulus comes to be elicited by a substitute stimulus. For example, Pavlov found that a dog would normally salivate at the sight of food. He then experimentally presented a dog with a signal light five seconds before food was dropped into a feeding tray. After every few trials, he presented the light without the food. He found that the dogs would still salivate and, furthermore, that the number of drops of saliva elicited by the light alone was in direct proportion to the number of previous trials in which the light had been followed by food.

Operant Conditioning. Learning to perform an act in order to receive a reward is called **operant conditioning**. Operant conditioning differs from classical conditioning in several important ways. Whereas in classical conditioning the reward (such as food) follows the stimulus (such as a light), in operant conditioning the reward follows the behavior (Figure 43.11). Also, in classical conditioning the experimental animal has no control over the situation. In Pavlov's experiment, all the dog could do was wait for lights to go on and food to appear. There was nothing the dog could do one way or the other to make it happen. In operant conditioning, the animal's own behavior determines whether or not the reward appears.

In the 1930s, B. F. Skinner developed an apparatus that made it possible to demonstrate operant conditioning. This device, now called a *Skinner box,* differed from earlier arrangements involving mazes and boxes from which the animal had to escape in order to reach the food. Once inside the Skinner box, an animal has to press a small bar in order to receive a pellet of food from an automatic dispenser (Figure 43.11). When the experimental animal (usually a rat or a hamster) is first placed in the box, it ordinarily responds to hunger with random investigation of its surroundings. When it accidently presses the bar, lo! a food pellet is delivered. The animal doesn't immediately show any signs of associating the two events, bar-pressing and appearance of food, but in time its exploring behavior becomes less random. It begins to press the bar more frequently. Eventually, it spends most of its time just sitting and pressing the bar. This sort of learning is based on the principle that if a behavior is rewarded, then the probability of that pattern reappearing is increased.

FIGURE 43.11
Rat in a Skinner Box
Skinner boxes are designed to promote operant conditioning. For example, a hungry rat may move randomly, searching for food, until it accidently presses a bar that delivers a food pellet. Each delivery means a greater probability that the rat will press the bar again, until finally the rat learns simply to press the bar each time it wants a pellet.

Latent Learning. Usually, learning is thought to be associated with some kind of reward. (In a sense, you learn in order to be rewarded.) However, **latent learning** occurs in the absence of an immediate reward. If rats are allowed to repeatedly run a maze without any sort of reward at the end, they will learn the maze anyway, but they seem to learn it very slowly. However, if they have first been allowed to wander through the maze (apparently without learning much about it) and food is then placed at the end, these same rats will learn the maze with remarkable swiftness. Apparently, learning had been occurring all along, but it was not evident until it was activated by a reward.

Latent learning may be important in the wild, for example, as an animal explores its range, looking into every nook and cranny. The learning that occurs under such circumstances is apparent when a predator appears on the scene and the animal immediately runs to a place of refuge.

Insight Learning. Solving problems, without experience, is called **insight learning**. Trial and error, of course, are important parts of certain other kinds of learning (such as operant conditioning), as an animal learns what works and what doesn't through experience. In insight learning, the trials apparently take place mentally. Such an ability is, at present, believed to be a higher order mental function found only in primates.

In an interesting experiment involving insight learning, researchers hung a stalk of bananas from the ceiling, out of reach of a chimpanzee on the floor. They placed a stick in the

FIGURE 43.12
Imprinting
In this famous photograph, we see a group of goslings that had been imprinted on Konrad Lorenz follow him down to the water.

FIGURE 43.13
A Crane Imprinted on Humans
Tex, the only female whooping crane at the International Crane Foundation breeding area in 1982, has been hand-reared and therefore hand imprinted on humans. She rejected the mate provided for her but could be enticed to lay eggs (artificially fertilized) by "dancing" with humans. She preferred Caucasian men of average size with dark hair.

room that would reach the bananas, believing that through insight the chimp would use the stick to knock them down. The chimp, however, carefully balanced the stick on end under the bananas, then quickly climbed the stick to the bananas, falling gently to the floor with the entire bunch, and all of them intact and unbruised (a better technique than the researchers had in mind).

Imprinting. Some years ago, Konrad Lorenz discovered that newly hatched goslings would follow whatever moving object they saw and that they would continue to identify with this object throughout the rest of their lives. Of course, under most circumstances any such object was likely to be a parent, and in this way they learned to identify their own species. Later, as they approached their first breeding season, they would seek out an individual with the traits of the individual they had followed soon after hatching. If they somehow were exposed to something else at that time, they would focus on whatever resembled that thing when it came time to breed. In an experiment, a group of goslings hatched in an incubator saw Lorenz first, and as they grew up, they would often dutifully fall into single file, following after him as he walked around the farm (Figure 43.12).

The sort of learning that takes place in such a brief period of a young animal's life has been called **imprinting**. It was once believed to take place through special learning processes, but most researchers now believe that it occurs through the same neural mechanisms that are involved in other types of learning.

Many animals also learn sexual identification during this critical period. Lorenz once had a tame jackdaw that he had hand-reared, and it would try to "courtship feed" him during mating season. On occasion when Lorenz turned his mouth away, he would receive an earful of worm pulp! The story of Tex, the dancing whooping crane, provides another example of this type of learning (Figure 43.13).

How Instinct and Learning Can Interact

We must keep in mind that a variety of factors are involved in an animal forming an adaptive behavioral pattern. Thus, in most cases, both instinct (genetics) and learning (experience) are involved. As an example of such interaction, consider the development of flight in birds. Flight is a largely innate pattern. A bird must be able to fly pretty well on the first attempt, or it will crash to the ground as surely as would a launched mouse. It was once believed that the little fluttering hops of

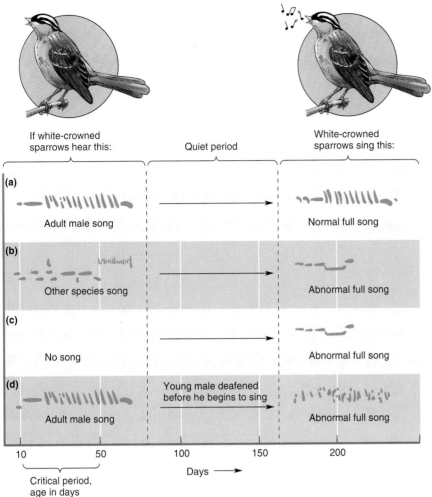

If white-crowned
sparrows hear this: Quiet period   White-crowned
                                                    sparrows sing this:

(a)

Adult male song   Normal full song

(b)

Other species song   Abnormal full song

(c)

No song   Abnormal full song

(d)

                        Young male deafened
                        before he begins to sing

Adult male song   Abnormal full song

10   50   100   150   200

Days ⟶

Critical period,
age in days

FIGURE 43.14
**Sonograms of Song-Learning in the
White-Crowned Sparrow**
A sonogram is a visual representation of a sound:
(**a**) shows the normal song of a wild male,
(**b**) shows that birds reared in isolation or hearing
the song of a different species produce a different
song, (**c**) shows that birds reared in silence sing
abnormal songs, and (**d**) indicates that deafened
birds sing much more complex songs, but quite
unlike the normal song.

nestling songbirds were beginning flight movements and that the birds were, in effect, learning to fly before they left the nest. But in a series of experiments, one group of nestlings was allowed to flutter and hop up and down, while another group was reared in boxes that prevented any such movement. At the time when the young birds would normally have begun to fly, both groups were released. Surprise! The restricted birds flew just as well as the ones that had "practiced."

One of the most interesting examples of the interaction of instinct and learning involves song learning in white-crowned sparrows. In a series of experiments by Peter Marlen and his colleagues related to those we just discussed regarding imprinting, white-crowned sparrows were reared in sound isolation and then watched to see what kind of song they sang. It turned out that they developed abnormal songs, quite different from the wild stock. Male white-crowned sparrows do not begin to sing until they are several months old, and the researchers found that if they were allowed to hear the normal white-crowned sparrow song, even briefly, between the ages of 10 and 50 days, when it came time for them to sing they would sing the correct song. They obviously are innately sensitive to the song of their species. Interestingly, if they were allowed to hear the song of other species during their develop-

ment, they would learn nothing about it. It was as if they were reared in isolation. Also, birds that were deafened after hearing the correct song could never sing it—they had to use that song as a model against which to compare their own sounds (Figure 43.14). Finally, even if a male heard the correct song after the age of 50 days, he would never be able to sing it. We see then that white-crowned sparrows can and do learn the correct song of their species, but that the learning is strongly influenced by genetics.

Learning and Memory

Many years had elapsed during which nothing of Cambray had any existence for me, when one day in winter my mother offered me some tea. . . . I raised to my lips a spoonful of the tea in which I had soaked a morsel of cake. No sooner had the warm liquid, and the crumbs with it, touched my palate than a shudder ran through my whole body. . . . And suddenly the memory returns. The taste was that of the little crumb of madeleine which on Sunday mornings at Cambray my Aunt Leonie used to give me, dipping it first in her own cup of real or of lime-flower tea. . . . Once I had recognized the

Essay 43.1 ● LEST YE FORGET

It is suspected that virtually everything we encounter is learned—stored away in the brain. Wilder Penfield and his group, working in Montreal, used electrodes to probe the brains (which have no pain receptors) of conscious patients and asked them to describe their sensations as various parts of the brain were stimulated. The results were startling. Some patients "heard" conversations that had taken place years before, some heard music or seemed to find themselves with old friends, long deceased. They seemed almost to "relive" the experiences, rather than simply remember them. The powers of recall

under such circumstances are phenomenal. A hypnotized bricklayer described markings on every brick he had laid in building a wall many years before. His recollections were carefully recorded, and then the wall was found and examined. The markings were there! (Our memory is not infallible, however. We also have a tendency to embellish partially recalled events.) Actually, there may be advantages to forgetting.

Although we may have entered every jot and tittle of our lives into our mind's ledger, we are not consciously able to recall very much of it. Sometimes we can't even recall

what we have tried to memorize—as you well know. But apparently the information is there, just the same. The fault lies with our recall techniques. The implications of this finding are enormous. If our research ever enables us to recall or relive earlier events at will, we might "reread" novels on long train trips, and all our exams would be virtually open-book tests. Perhaps terminally ill and pain-racked patients could be stimulated to relive happier, youthful days with loved ones as life dwindles in a changed body.

---

taste, all the flowers in our garden and in M. Swann's park, and all the water lilies on the Vivonne and the good folk of the village and their little dwellings and the parish church and the whole of Cambray sprang into being.

Marcel Proust, *Remembrance of Things Past*

There are indeed not only bits of information tucked away in our memories, but overall settings, moods, and impressions that can be recalled if properly summoned. In some cases, we can recall things at will, at times we simply cannot, and at other times, remembrances flock to our conscious thought that we would rather have suppressed (Essay 43.1). Obviously, we have not learned to use our memories as well as we might like. But our chances are better if we know more about how they work.

Memory is the storage and retrieval of information, and its characteristics are important in any consideration of learning. Memories are often triggered by unexpected stimuli. Just as the tea-soaked cake brought back a rich array of remembrances of the town of Cambray for Proust, so can our tucked-away memories be jolted to our consciousness by some small thing that, without the weight of the memory it stirs, would be insignificant. Proust's musing reminds us that memory is more than simple recollection. Each detail we have so neatly stored in our memories is inextricably bound to others in an interwoven fabric that may literally mold our identities, our sense of who we are. We can't pick at a thread without moving the whole cloth.

Theories on Information Storage

It now seems that there are two separate mechanisms by which information is stored, as evidenced in the differences between **long-term memory** and **short-term memory**. The existence of these two types of memory was first noticed in humans with brain concussions who were unable to recall what happened just before an accident but could remember what happened much earlier. On the basis of such findings, researchers postulated two memory systems that could process information independently. Information stored in the long-term memory system is relatively permanent (subject to very slow decay). Once information is processed into the long-term system, it is not easily disrupted and therefore may be recalled with minimum confusion after long periods of time. Short-term memory, on the other hand, is relatively easily disrupted and subject to rather rapid decay. (This is why cramming for exams is not a good idea. If the exam is postponed, or if something startles you on the way to class, all may be lost.) Also, there is evidence that the short-term memory can be overloaded, whereas the long-term memory can't be.

Apparently, short-term storage is necessary before long-term storage can occur. In a series of experiments, it was found that if an inedible object is presented to one arm of an octopus, over and over in rapid succession, that arm will come to reject it after learning that it is inedible. If the object is then immediately presented to other arms they will accept it. But if some time is allowed to elapse before the second

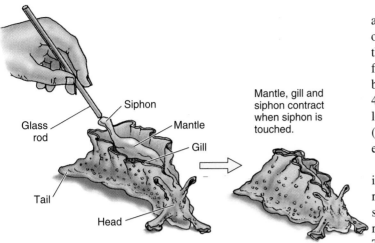

Glass rod

Siphon

Mantle

Gill

Tail

Head

Mantle, gill and siphon contract when siphon is touched.

FIGURE 43.15
The Siphon-Withdrawal Response of *Aplysia*
Habituation can reduce the response after repeated stimuli. The response can be restored by sensitization of the animal through mild electrical shocks just before lightly touching the siphon. The entire phenomenon can be understood in terms of events occurring at the synapse between sensory and motor cells.

presentation, all the arms will reject it. This finding suggests that the information acquired by one arm is only locally adaptive. It must filter from a hypothesized short-term center to a long-term system before a general adaptive pattern (total rejection) can appear. The experiment actually only suggests that something filters from neural centers associated with one arm to general centers.

In another experiment, a mouse was put into an arena (enclosed area) containing only a box with a hole. If the mouse ran into the hole, its feet received an electric shock, and it never went near that hole again. But if the mouse was given electroconvulsive or chemical shock treatment within about an hour after the unpleasant incident, it would forget the whole experience and helplessly run into the same hole the next day and every day as long as it survived the experiment. On the other hand, if the electroconvulsive treatment was delayed for two hours, it was not effective, and the mouse did not forget its experience. Mouse memory, it seems then, is labile for up to about an hour and then becomes fixed. We infer that transfer is made from short-term "holding" memory to permanent memory in that time.

In trying to understand how learning and memory are associated, scientists have turned to an animal called *Aplysia,* or the sea slug. The animal draws water in over its gills through a siphon that extends from its body. The researchers found that the siphon is quickly withdrawn if touched lightly, but that the animal quickly habituates to the touch (Figure 43.15). The habituation is now known to be due to reduced levels of neurotransmitter at the sensory and motor synapse. (There is only one such synapse between sensory and motor elements in this creature.)

The animal's response to touch could be reestablished if a light electrical shock were administered to its tail. The restored sensitization was due to a third neuron that synapsed with the axon terminals of the sensory cell. It released the neurotransmitter serotonin into the synapse. The serotonin triggered changes within the sensory cell that caused calcium channels to stay open longer, causing more neurotransmitter to be released into the synaptic cleft and an increased level of response in the motor neuron. Habituation (and its opposite effect, sensitization) then can be thought of as learning that can occur at the cellular level.

Calcium can also be important in learning and memory among vertebrates. According to one scenario, repetition causes a certain neural tract (pathway) to be repeatedly stimulated. In time, the continual stimulation over that tract causes the postsynaptic neurons to be flooded with calcium. The calcium then activates a dormant enzyme called *calpain,* which now has the ability to break down certain proteins. One such protein, called *fodrin,* lends structural form to the dendrites of neurons. As the enzyme calpain breaks down the protein fodrin (sounds like an episode from the epic fantasy *Lord of the Rings*), the reaction stimulates the formation of new receptors on the postsynaptic neuron, which can then be stimulated by neurotransmitters from the presynaptic cell. With these new receptors exposed, the postsynaptic neuron is much more sensitive to the transmitters (Figure 43.16), and impulses travel that neural pathway with increasing ease.

Repeated stimulation of this tract causes the continued breakdown of fodrin by calpain until the protein structure of the postsynaptic neuron actually breaks down, leaving the neuron free to change its shape or to make new connections, thus facilitating the passage of new impulses along the pathway. This new "wiring" of the brain then may be the physical basis of the memory that is so intimately associated with learning.

Now that we have some idea about the mechanisms of behavior, let's see how animals use these behaviors to fit better into their kind of world—the topic of our next chapter. Then we will take a closer look at that world in the following chapters, paying particular attention to how it is changing.

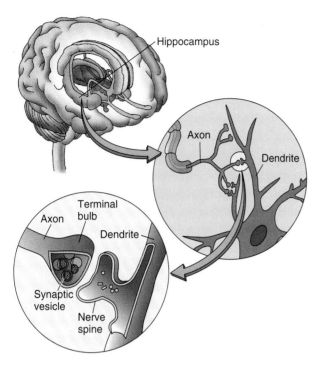

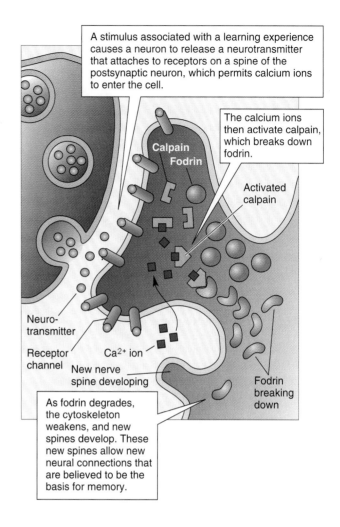

FIGURE 43.16
How Memories are Formed
A learning experience causes changes in the nervous system. At left, a stimulus associated with the experience causes a neuron to release a neurotransmitter that attaches to receptors of the next neuron. This, in turn, opens calcium gates, permitting the ion to enter the cell. The calcium then activates calpain, which begins to break down fodrin. As the fodrin degrades, more receptors begin to appear, allowing more calcium into the cell. The cell membrane weakens, and the cell begins to change shape as new spines develop. These new spines allow new neural connections that are believed to be the basis for memory.

Key Ideas

THE DEVELOPMENT OF BEHAVIOR

1. At one time researchers believed that each behavioral act was either genetically based or learned. Today researchers ask how behavior develops, admitting both genetic and learned influences. A number of lines of evidence show the clear effect of genes on behavior.

2. Inbreeding experiments involve mating members of closely related (inbred) lines. As they interact with the same environment, any behavioral differences are considered to be genetically based.

3. Genetic influences can be demonstrated by artificial selection—that is, managed breeding for specific behavioral traits.

4. In hybridization experiments, individuals of closely related strains, but with differing specific behaviors, are bred. If the resulting behaviors are of one parental type or the other, one or two genes are believed to be involved. If the resulting behavior is a blend of parental types, many genes are believed responsible.

5. Hormones can influence behavior, and the effect can be studied in two ways: by gland removal and hormone replacement, and by correlational analysis. In correlational studies, the behavior changes in proportion to the level of hormone administered.

6. Hormones can influence behavior by altering perception, development of the nervous system, and effectors.

THE ETHOLOGIST'S CONCEPT OF INSTINCT

1. The concept of **instinct** was developed by ethologists Lorenz, Tinbergen, and von Frisch. **Ethology** deals with causation,

development, evolution, and function. Three concepts are central to the ethologist's concept of instinct.

2. **Sign stimuli** are specific environmental signals that release instinctive behaviors.

3. **Innate releasing mechanisms** are neural centers that are activated by the perception of sign stimuli.

4. **Fixed action patterns** are genetically based behaviors that are performed at a signal from the innate releasing mechanism. They are performed independently of environmental influences.

LEARNING

1. **Learning** is a process of developing a behavioral response based on experience. Reinforcement is the result of an action that increases the probability of the action's being repeated.

2. **Habituation** is learning not to respond to a stimulus.

3. **Classical conditioning** involves a behavior that is normally released by one stimulus now being released by another stimulus.

4. **Operant conditioning** involves learning to perform an act in order to receive a reward.

5. **Latent learning** is learning that occurs in the absence of an immediate reward.

6. **Insight learning** involves solving problems without resorting to trial and error. The trials take place mentally.

7. **Imprinting** is a kind of learning that takes place only in a brief critical period of an animal's early life.

HOW INSTINCT AND LEARNING CAN INTERACT

1. Most behaviors are based on interactions between instinct (genetically based) and learning (experience-based). The degree to which experience can alter innate patterns depends largely on the species and the specific behavior in question. Most innate patterns can be improved upon by experience.

LEARNING AND MEMORY

1. **Memory** is the storage and retrieval of information. Two separate mechanisms are **long-term memory** and **short-term memory**. Information in long-term memory is slower to decay (forget and harder to disrupt. Short-term memory must precede long-term memory.

2. Studies on *Aplysia* show that some learning can be explained by changes at the cellular level. Both invertebrate and vertebrate learning can be related to the effect of calcium in neurons.

Application of Ideas

1. It was once thought that much behavior is either instinctive or learned. We now believe that instinct and learning interact to produce an adaptive behavioral response. What is the evidence?

2. The ethologist's concept of instinct has been said to apply, in its strictest sense, primarily to invertebrates. Why do you suppose this is? Consider the relative influence of cerebral processes in vertebrates and invertebrates.

Review Questions

1. Do genes code for behavior? State some evidence. (pages 945–947)

2. What is avoidance learning? (page 950)

3. What are hygienic bees, and how can they be used to determine how genes affect behavior? (page 946)

4. How do hormone replacement studies differ from correlation studies? (pages 947–948)

5. When in a human female's monthly cycle is visual acuity greatest? What does this show? (page 948)

6. With what four areas of inquiry is ethology concerned? (page 949)

7. What are the environmental signals that trigger instinctive actions called? What kinds of neural centers do they stimulate? (page 949)

8. Define fixed action patterns and provide two examples. (page 950)

9. Define learning. (page 950)

10. Define and give an example of habituation. (pages 950–951)

11. What kind of learning is shown by a bird perched on a familiar scarecrow's shoulder? How is such learning adaptive? (page 950)

12. Name and give an example of the kind of learning to which substitute stimuli are integral. (page 951)

13. Name and provide an example of the kind of learning where reward follows a behavior. (page 951)

14. What kinds of behavioral sequences are absent in insight learning? (pages 951–952)

15. Name and give an example of the kind of learning that involves a brief receptive period, usually very early in life. (page 952)

16. What happens to neurons when calpain breaks down fodrin? (page 955)

CHAPTER 44

Adaptiveness of Behavior

CHAPTER OUTLINE

PROXIMATE AND ULTIMATE CAUSATION

BEHAVIORAL ECOLOGY

Habitat Selection

Foraging Behavior

BIOLOGICAL CLOCKS

The Adaptiveness of Rhythms

The Range of Rhythms

ORIENTATION AND NAVIGATION

Homing Pigeons

Migration

THE SOCIAL BASIS OF COMMUNICATION

Visual Communication

Sound Communication

Chemical Communication

Tactile Communication

SOCIAL BEHAVIOR

Agonistic Behavior

Fighting, a Form of Aggression

Cooperation

Symbiosis

ALTRUISM

Humans and Reciprocal Altruism

SOCIOBIOLOGY

KEY IDEAS

APPLICATION OF IDEAS

REVIEW QUESTIONS

One of the most fundamental questions in the field of animal behavior is: Why do animals do what they do? Sometimes the answer is obvious. That mouse in your kitchen scurries away when you walk in because it is in its best interest to avoid such large creatures. Tarantulas living in holes in the desert floor avoid drowning by leaving their burrows and climbing into bushes at the first sign of rain. Chickens spend the night on low-lying limbs where foxes can't reach them. We can indeed see the adaptiveness of animals behaving in certain ways.

In other cases the explanations are far less apparent. Why do hammerhead sharks assemble in great numbers and then begin some long migration? How do some woodpecker pairs come to forage close together and others far apart? Why do some species seek out each other's company and others fight when they encounter each other? Questions of causation, the causes of behavior, can be answered at two levels, *proximate* and *ultimate*.

In this chapter we will first look at the differences in proximate and ultimate causation; then we will examine some ecological effects on behavior, particularly regarding choices of habitat and food. After exploring the role of communication in the lives of animals, we move on to focus on social behavior, altruism, and sociobiology.

Proximate and Ultimate Causation

In the Caribbean, several islands each harbor their own species of woodpecker. The Guadaloupe woodpecker, logically enough, lives on the French resort island of Guadaloupe, the Hispaniolan woodpecker inhabits the island of Hispaniola (comprised of Haiti and the Dominican Republic), and the Puerto Rican woodpecker, of course, lives in Puerto Rico. In each of these species the male is larger than the female, so the bill of the male is larger than that of the female. As a result of differences in bill size, the members of a pair are able to forage close together, because they take food of different sizes: they do not compete.

Another species of woodpecker inhabits Jamaica. The Jamaican woodpecker is slightly larger than its Caribbean colleagues, with little sexual difference in body size or bill size (Figure 44.1). Because males and females take the same kind of food, they must forage farther apart, with a reduced level of social interaction. The Jamaican woodpecker vocalizes less than the other species, but when it calls, it seems to call more loudly to its distant mate.

The behavior of the sexually monomorphic ("one form") Jamaican woodpecker and that of the sexually dimorphic ("two forms") species are clearly different. The reason for such differences may be explained in two ways.

Proximate causation involves immediate bases for a behavior—those usually involving psychological or physiological mechanisms. Proximate questions usually involve *how* the mechanisms of an animal operate to cause it to behave in a certain way. So we might answer the question of how sexually monomorphic species come to forage farther apart by considering whether the sexes have some aversive "feeling" for each other, or if some immediate physiological requirements are fulfilled by the sexes keeping a certain distance. Here we're trying to find out just what processes going on within the animal result in the performance of a certain behavior.

Ultimate causation involves the evolutionary and adaptive bases for a behavior. Ultimate questions involve *why* an animal behaves in a certain way. In other words, ultimate causation asks why the proximate mechanisms evolved in the first place. Thus if we ask why sexually monomorphic and sexually dimorphic species forage differently, we can answer in ultimate terms. Perhaps predatory pressures are different on the two types of animals. Maybe sexually monomorphic birds evolved a weaker social tendency because if they were drawn to each other and foraged closer together, they would be more likely to attract the attention of hawks or other predators. Or perhaps by both sexes evolving larger bills, they could each take a larger range of food sizes and thereby effectively increase their kinds of food sources. The tendency to look for food in different places then could have been the result of either predation or food pressures—both ultimate factors.

Behavioral Ecology

Behavioral ecology is the study of how the environment affects behavior. Remember, the environment can be viewed in the historical sense (ultimate causation) or the immediate sense (proximate causation). Behavioral ecology can involve studies of the physical habitat, interactions with other individ-

FIGURE 44.1
An Anatomical Correlation with Behavior
The sexually monomorphic Jamaican woodpecker has different foraging and social patterns than the sexually dimorphic Puerto Rican woodpecker.

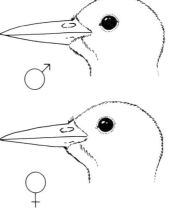

Jamaican woodpecker

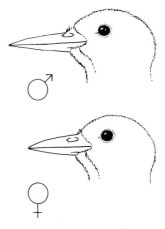

Puerto Rican woodpecker

Woodland habitat   Field habitat

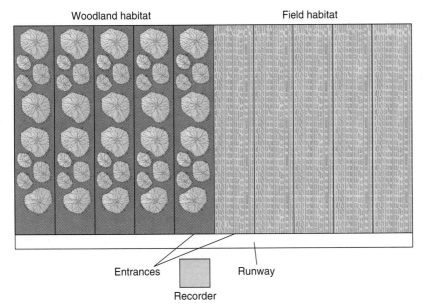

Entrances   Runway

Recorder

FIGURE 44.2
Wecker's Experimental Setup
Mice with different genetic makeups and different experiences were tested to see which kind of habitat they preferred. They were allowed to enter either simulated grassland or forest conditions from a runway. The results are described in the text.

uals, or less conspicuous influences on behavior, such as weather, oxygen levels, and humidity—any environmental variable that can influence behavior. As examples, we will consider the behavioral ecology of how animals choose a habitat (a place to live) and how they forage (get food).

Habitat Selection

Why do animals live where they do? In some cases it's because they simply landed there. Perhaps a larval form was carried by an ocean current to some place and simply deposited there. If it landed in a suitable place, it survived. If it didn't, it died. Simple as that. Other animals, though, are able to move around, assess an environment, and make a decision about whether to stay or move on. These animals are faced with two choices: (1) whether to settle to begin with and (2) once there, whether to stay or look for a better place.

An animal looking for a place to settle has several questions to ask. (Keep in mind that this is a shorthand way of saying it behaves *as if* it were asking these questions.) How good is a certain area compared to the best area? How much work would be involved in finding a better area? How dangerous would the search be? How great is the competition (should the animal give up a good area to search for a better area if the good area will be occupied as soon as the animal leaves, while other animals are also searching for the better areas and they may be all gone?) Animals that expend a great deal of energy in searching, at great risk, with high competition should decide quickly to take what may be an inferior area.

In some cases animals may be thrust into inferior areas by other, more dominant, animals. For example, in some species of gulls, pairs rest in close proximity, after the males fight for centrally located nests. Those pairs nesting at the periphery are subjected to greater predation and leave fewer offspring, so the competition for interior nests can be intense.

One of the questions regarding habitat selection is: Does an animal choose its habitat based on genetic propensities or through learning the best place to live? Stanley Wecker conducted a series of fascinating experiments on prairie deer mice to see to what degree their habitat preferences were genetically influenced. In his experiments, he used two kinds of mice—one of which lives in woodland, the other in grassland. Wecker had found that, in the laboratory, the grassland mice *could* do quite well under simulated forest conditions. He wondered then why they prefer grasslands in the wild. In one of his experiments he tested two groups of grassland mice, one reared in the laboratory and the other wild. The laboratory group was further subdivided into two groups—one of which had been reared under simulated field conditions, the other under simulated forest conditions.

Not surprisingly, Wecker found that the wild-caught field mice preferred the field-end of the enclosure (Figure 44.2). The "field"-reared laboratory mice made the same choice. However, the "forest"-reared laboratory mice showed no preference whatsoever. Subjecting them to the forest had obliterated any tendency to move toward the field but had not caused them to prefer the forest. Wecker's experiment suggested that habitat preference in prairie deer mice, at least, has a strong genetic component that can be influenced by experience (Table 44.1).

We can also imagine that the field mice would develop certain preferences, while in the field, based primarily on experience. For example, it wouldn't take long to learn not to climb around in thorns; certain kinds of grasses taste bad, while others are quite palatable; and sleeping with ants doesn't work. By experience then the mouse would fine-tune its behavior through proximate factors. The principle here might be phrased as: ultimate factors help to set limits on behavior, and proximate factors influence the behavior of the animal within those limits.

TABLE 44.1
The Results of Wecker's Experiment

Number of Mice Tested	Hereditary Background	Early Experience	Habitat Preference
12	Grassland	Grassland	Grassland
13	Laboratory	Grassland	Grassland
12	Grassland	Laboratory	Grassland
7	Grassland	Forest	Grassland
13	Laboratory	Laboratory	None
9	Laboratory	Forest	None

Source: Data from S.C. Wecker, "The Role of Early Experience in Habitat Selection by the Prairie Deer Mouse, *Perimaniculatus bairdi," Ecological Monographs* 33:307-325.

Foraging Behavior

When asked why he robbed banks, the infamous Willie Sutton is reputed to have replied, "Because that's where the money is." For years, biologists seemed to have tacitly assumed that the same reasoning applied to where animals forage, or search for food: they forage where they do because that's where the food is. In recent years, though, that answer has been deemed inadequate. Behavioral ecologists have, in fact, asked very specifically why animals forage where they do. Furthermore, they have asked how animals make food choices once there.

Of course, different species forage in different ways. The primary division of foraging styles exists between generalists and specialists. **Generalists** are those species with a broad range of acceptable food items. They are often opportunists and will take advantage of whatever is available, with certain preferences depending on the situation. Crows are an example of feeding generalists (Figure 44.3); they will eat anything from corn to carrion. **Specialists** are those with narrow ranges of acceptable food items. Some species are extremely specialized, such as the Everglade kite (Figure 44.4), which feeds almost exclusively on freshwater snails. There is a wide range of intermediate types between the two extremes, and in some species an animal will switch from being one type to being another depending on conditions, such as food availability or the demands of offspring.

Costs and Benefits of Foraging Behavior. Whatever the strategy, the question is: How does an animal forage most efficiently? More precisely, how does it maximize its gain relative to its expenditure? The gain is measured quite simply as the food value of the item. Expenditure is measured in two ways—by the energy the animal spends in searching for food and by the energy it expends in handling the food once it has been found.

Obviously, the animal will maximize its foraging success by spending the most time where the most food is. But is this what they do? J. N. M. Smith and H. P. A. Sweatman examined the feeding behavior of the great tit (Figure 44.5) by setting up a series of grids in an aviary. Each grid contained different densities of mealworms, a favorite food. The food density in each grid could be altered by experimenters. When a certain grid continued to hold the greatest food density for a time, the researchers found that the tits were soon spending

FIGURE 44.3
The Crow
The crow is a feeding generalist that boasts a large menu.

FIGURE 44.4
The Everglade Kite
A feeding specialist, this kite eats only snails.

FIGURE 44.5
***Parus major*, the Great Tit**
The tit has been the subject of intensive feeding studies by biologists.

almost all their time on that grid. However, the birds continued to hop over and sample from the other grids. The researchers wondered why the birds would waste time in these suboptimal sites, but the answers soon became clear. When the food density was suddenly reduced on the best grid, the birds immediately switched to the second-best grid. They spent little time in making a decision when their primary food source failed. Sampling all the grids, even after the best site had been determined, was adaptive after all.

Another question is: How do animals maximize their foraging efficiency when they are given a choice of food items with different food values? For example, all things being equal, larger food items are more profitable than smaller ones.

Thus the animal will be more successful if it takes larger items. Bluegill sunfish, for example, feed largely on *Daphnia,* the small crustaceans known as water fleas. Researchers found that sunfish will bypass very small water fleas even if they are nearby, in favor of large water fleas further away. In other words, they behave as if the extra travel were worth the extra gain (Figure 44.6).

The sunfish, of course, don't have to wrestle with the *Daphnia* once they find it, but striped bass may have to consider ease of handling. The bass feed on a variety of foods, including small fish called "shiners" and crayfish. The shiners are taken with a gulp, usually after a short chase, so after the bass catch the shiner, the handling required is minimal. The crayfish, on the other hand, is easier to catch but harder to handle. Not only do the pugnacious creatures fight back, requiring careful manipulation by the fish, but much of its body is indigestible chitin, which further reduces its value. So all things being equal, the bass is better off chasing the shiner.

Why then are crayfish taken at all? A predator's decision is often based on food availability. It will tend to take the items with the greatest net gain until their numbers are depleted to the degree that less desirable, but more numerous, food items become more attractive. Then it will generally begin to switch to the less desirable items—in which the food value may not be as high, but less time and energy are spent in searching.

Foraging, for most species, is thus a cost–benefit proposition. They tend to maximize the benefit while minimizing the cost to themselves. Since such studies have been undertaken, researchers have often been amazed at what appear to be analytical abilities of foragers. Of course, any such ability is largely programmed genetically, but learning may also play an important role. For example, young fish are often fairly good optimal foragers, but with age and experience their efficiency improves.

We will now turn our attention to two phenomena that are critical to the adaptation of life on earth. They have to do with how animals adjust with respect to time and place.

FIGURE 44.6
Optimal Foraging in the Bluegill
The bluegill sunfish will pass up smaller, nearby food items for larger ones, farther away, even if it means a greater energy expenditure.

Biological Clocks

You may have noticed that the small feline carnivore sharing your home becomes restless each evening. The eyes of the cat are dark-adapted, its claws are sheathed in silent pads, its hearing is remarkable, and it has a natural tendency to sneak around. Indeed, if you weren't feeding it, it would have its greatest success in stalking its prey at night. But with the lights on and the TV blaring in your apartment, how does your cat know when it is night outside? After all, you're probably a student (since few people would be reading this for pleasure), and, like students everywhere, you probably keep odd hours. You may burn the midnight oil, eat at 3:00 a.m., and turn off the lights and leave at dawn. You might think that such an odd schedule would throw your cat's schedule off too; but it doesn't. The cat knows when it's night—when to get up and stir around in preparation for the hunt that never happens.

How does the cat "know" when night falls outside its brightly lit home? Or better yet, how do plants "know" when it is nighttime, and time to fold their leaves? Although such rhythms were observed some 300 years before Aristotle, we still don't have the answers to some of the most fundamental questions about the timing of living things.

The Adaptiveness of Rhythms

In a cyclic and rhythmic world, life adapts by taking on its own cyclicity to help avail itself of the offerings of each of the earth's phases. One of the longest and most pronounced cycles is the earth's annual rhythm, brought about by the earth's tilt on its axis as it circles the sun in an elliptical orbit. This annual rhythm is expressed as the seasons. In many parts of the world, the seasons are marked by drastic differences. Thus an animal should begin to lay on fat for the lean, wintry season even while days are long and food is plentiful. The animal's body begins to "tell" the animal to accelerate its feeding schedule, one can say, because that body is the result of generations of ancestors who were subjected to seasonal cyclicity. In time, natural selection would have produced generations whose physiology anticipated the coming winter and who therefore "laid by in store."

Another cycle to which we earthlings are subjected is that of the moon, the **lunar cycle**. There has been a lot of anecdotal evidence that the moon has some influence on human behavior ("uh-oh, a full moon tonight"), but none is well-substantiated. Other species, however, may be very attuned to lunar cycles and, in fact, their lives may depend on it. Intertidal crabs, for example, must be most active, feeding on the flats when the tide is out, roughly twice a day (about every 12 hours, 24 minutes). Tidal activity is due to the gravitational pull of the moon as it circles the earth (every 24 hours, 50 minutes). The crabs not only time their periods of activity according to the lunar cycle, but they also change color according to the daily cycle of the sun (circadian cycle—see below) so that they are darker at night, making it more difficult for predators to see them.

The most obvious of the earth's rhythms is the daily or **circadian cycle** (*circa,* "about"; *diem,* "day"). The cat starts to prowl at night, about the time that most of us, for whatever reason (the evolutionary basis of sleep is unknown), must lapse into unconsciousness. The most obvious intuitive explanation is that we are visual animals and so we get our rest when we can't see anyway, a time when nocturnal predators may be up and about.

That circadian rhythms arise internally has been demonstrated by placing organisms under constant conditions of light, temperature, humidity, barometric pressure, and so on. Under such conditions, the organisms will show rhythmic behavior on a roughly 24-hour basis. However, the cycles may drift to 23 or 25 hours. Such rhythmic activity operates as if there were some sort of internal clock. The clock can be reset by exposure to some environmental cue, usually light.

Many living things can be "clock-shifted." Clock shifting involves allowing the animal to continue its circadian rhythm, but setting the rhythm ahead or behind of where it would normally be. Most organisms can be clock-shifted somewhat, but beyond a certain point some will not respond and will return to their normal cycling. Clock shifting is done by altering the onset of environmental cues normally associated with some phase of the day. For example, if a light is turned on a little earlier and turned off a little earlier each day in a cage in which an animal is kept, the animal will set its circadian clock according to the onset of the light.

Clock shifting is responsible for that traveler's bane called jet lag. As you cross time zones, say, flying across the country or to Europe, your body stays on its old schedule of sleep, hunger, digestion, and so on. Your body can only adjust up to an hour a day, so you can expect to feel a little logy for a time. (A good way to reset your clock before you leave is to be exposed to a bright light at the time that will be mid-day at your destination.)

The Range of Rhythms

Not all biological rhythms are so blatant as the circadian cycles of animals. There are far more subtle clocks measuring time. For example, cell division seems to occur in rather rhythmic spurts (peaking at noon), even in artificial laboratory cultures. If the heart of a hamster is removed and placed in a nutrient medium, it will continue to beat for days, its pace picking up at night, the hamster's normal running period. Even single, excised neurons will exhibit a daily rhythm of firing. Remarkably, sensitivity to drugs varies on a daily cycle. Physicians are beginning to learn about this cyclicity, perhaps spurred on by the finding that certain drugs can be beneficial at one time of day and kill the patient if administered a few hours later. Interestingly, most of us were born in the wee hours of the morning, and most of us will breathe our last in these same early hours.

The physiological basis of the clocks remains largely a mystery. Part of the uncertainty arises from contradictory findings. For example, the nucleus of unicellular organisms, such as the green alga *Acetabularia,* apparently controls the cell's timing, but if the nucleus is removed, the cytoplasm can maintain the rhythm on its own. Further confusing the issue, it appears that clocks function through neural mechanisms in some organisms and hormonal mechanisms in others, and *both* neural and hormonal mechanisms may operate within the same individual.

Scientists have been looking for the clock in mammals for a long time and have found two small groups of cells directly over the optic chiasma, where the two optic nerves cross. They are called the suprachiasmatic nuclei. If these nuclei are destroyed, the mammal loses all circadian rhythms. It will sleep, eat, or run around completely randomly—activities that normally are performed at certain times of the day or night.

But what about other kinds of animals? Birds have suprachiasmatic nuclei, but their circadian clock is found in the pineal gland, a clump of tissue between cerebral hemispheres. If the pineal gland of birds is removed, they lose all sense of circadian timing.

Other animals, such as the invertebrates, lack both suprachiasmatic nuclei and pineal glands, yet many of them show very precise diurnal timing. In these species, the clocks may be generalized mechanisms distributed among the cells or may be a property of the cells themselves. The fact that timing mechanisms vary among the species, though, is evidence for the powerful benefits associated with properly timing daily events.

Orientation and Navigation

On a variable and changing planet such as ours, proper positioning can be critical. Animals do, in fact, show a variety of abilities to position themselves and to find their way from one place to another. For example, it is best for certain aquatic insects to move to the protective, darkened bottom as larvae but, as they change to adults, to seek the brightly lit surface that signals a greater abundance of oxygen. Not only do many shorebirds stand so as to face cold winds that would otherwise ruffle their insulating feathers, but some of them also leave those cold winds to travel to warmer climes each fall.

There are undoubtedly people in South America who wonder where their robins go every spring. Those robins, of course, are "our" birds; they merely vacation in the south each winter. If you've been keeping a careful eye on the robins in your neighborhood, you may have noticed that a specific tree is often inhabited by the same bird or pair of birds each year. You may have wondered, since these birds winter thousands of miles away, how they find their way back to that very tree each spring. Such questions have perplexed researchers for years. And the more we learn, it seems, the more strange the story becomes.

Most researchers now believe that there are two primary mechanisms by which birds find their way around. One is a **compass sense**, which enables the animal to know direction.

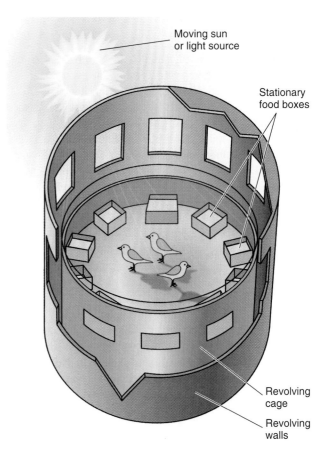

FIGURE 44.7
Kramer's Orientation Cage
The birds can see only the sky through the roof. The apparent direction of the sun can be shifted with mirrors.

Some birds couple this with an internal clock to form a time-direction mechanism with which the bird flies for a certain time in a certain direction. Certain European garden warblers, for example, are born with the ability to reach their wintering grounds in West Africa by flying southwest for a certain time (which takes them to the area of Gibraltar) and then turning and flying southeast for a certain period. Such a mechanism, of course, would be effective for getting the bird into a general area, but not for finding more precise locations.

The second mechanism birds apparently use to find their way around is a **map sense**. The map sense is one of the most fascinating and mysterious abilities of animals. An animal with a map sense seems to know its precise longitude and latitude no matter where it is. We will now take a look at two aspects of spatial organization, orientation and navigation, both of which are important to animals on the move.

Orientation is the ability to face in the right direction. **Navigation**, on the other hand, involves starting at point A and finding the way to point B. Much of the work with orientation of birds was triggered in the 1950s when a young German sci-

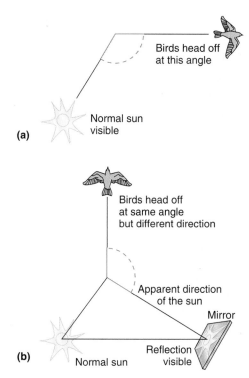

FIGURE 44.8
Kramer's Sun–Compass Orientation Experiment
(a) As the sun moves, caged starlings alter the direction of their attempted migratory movement with respect to it, thereby maintaining a constant heading. **(b)** Using mirrors, Kramer altered the apparent position of the sun. The birds shifted their migratory direction by the same angle.

entist, Gustav Kramer, devised some intriguing experiments on birds and came up with some startling conclusions.

Kramer found that caged migratory birds became very restless at about the time they would normally have begun migration in the wild. Furthermore, he noticed that as they fluttered around in the cage, they usually launched themselves in the direction of their normal migratory route (Figure 44.7). Kramer devised experiments with caged starlings and found that their orientation was, in fact, in the proper migratory direction—except when the sky was overcast. When they couldn't see the sun, there was no clear direction to their restless movements. Kramer therefore surmised that they were orienting by the sun. To test this idea, he blocked their view of the sun and used mirrors to change its apparent position. He found that under these circumstances the birds oriented with respect to the position of the "new" sun (Figure 44.8).

This sort of sun–compass orientation requires that starlings know both the time of day and the normal course of the sun. In other words, incredible as it seems, they apparently know where the sun is supposed to be in the sky at any given time. It seems that this ability is largely innate. In one experiment, a starling reared without ever having seen the sun was able to orient itself fairly well, although not as well as birds that had been exposed to the sun earlier in their lives. (This would appear to be another example of the interaction of

innate behavior, maturation, and learning.) Evidence for sun–compass orientation has also been found in a variety of other animals, including insects, fish, and reptiles.

The sun compass can't be the whole answer, of course. Possession of a compass may be necessary for navigation, but it isn't sufficient. Finding north would be easy; finding Bakersfield wouldn't, even if you had a compass and a map. As you sat on a rock, distraught and staring hopelessly at your devices, a bird might be passing overhead on its way to a certain tree in Argentina.

As night drew on and you hadn't budged from your rock, other bird species might pass overhead. As you strained to see their dim shapes through teary eyes, you might wonder: How do *they* navigate? Since it is night, they would obviously not be using the sun. Apparently, some birds navigate the way sailors once did—by the stars. In experiments somewhat analogous to the sun–compass work, night-migrating birds were brought in cages into a planetarium (Figure 44.9). A planetarium is essentially a theater with a dome-like ceiling onto which simulated night skies can be projected. Sure enough, the fly-by-night birds oriented in their cages according to the sky that was projected on the ceiling, even if it was different from the sky outside. In one experiment, birds that normally migrate north and south between western Europe and Africa were shown a simulated sky as it would appear that moment over Siberia. The birds, behaving as though they were thousands of miles off course, oriented toward the "west."

Homing Pigeons

The abilities of homing pigeons have fascinated people for years. (*Note:* All pigeons, even those that do no more than stand around decorating statues in the park, are technically homing pigeons.) A homing pigeon can be taken from its

FIGURE 44.9
Planetarium
The "stars" on a planetarium ceiling can be rotated to mimic the sky at various seasons.

home loft, put into a dark box, and transported to a distant location by a circuitous route. When it is released, there's a good chance it will soon show up in its home loft. Some make it quickly, others never turn up—there is a great deal of variation in abilities.

Studies with homing pigeons have proved inconclusive but instructive. Researchers have asked a wide range of questions. Do pigeons use a sun compass? If so, how do they home successfully on cloudy days? Do they use landmarks? Pigeons fitted with frosted contact lenses so that they cannot make out landmarks find their way back to the vicinity of their home loft, where they may flutter about directly overhead, or sit on the ground near the loft, unable to see it. Do pigeons use a magnetic compass? Small electromagnets have been attached to the heads of homing pigeons (Figure 44.10). Birds carrying magnets with reversed polarity tended to fly in opposite directions.

Evidently, then, pigeons use a variety of clues, including the sun compass, magnetic force fields, and visual details. Researchers have been foiled in trying to find a single mechanism. The pigeon evidently has backup systems.

Pigeons apparently have both the map sense and the compass sense required for navigation. Homing pigeons have further confused researchers by the fact that when they are navigating by the sun, their compass is time-dependent (that is, their directional sense can be changed by artificially altering their clocks). If they are navigating on cloudy days, however, clock-changes do not upset their directional sense.

It should be pointed out that the study of animal navigation and orientation is an extremely vigorous field at present, and new data are appearing almost daily. Any synthesis at this point would be extremely tenuous. However, these examples provide some idea of the value of innate characteristics in orientation, as well as the surprising sensitivity of some species to environmental cues.

Migration

Migration is the regular movement of animals between breeding and nonbreeding areas. It involves such costs and risks that one may well wonder how it arose at all. Why would an animal endure such hardships just to be somewhere else? That "somewhere else" must hold great advantages. In some cases the advantages are clear. The great caribou herds feed in summer on the plentiful grasses and lichens of the northern Arctic. In late summer, though, they begin a relentless trek hundreds of miles southward to the relative haven of the timbered areas. In this way, they extend their food reserves while wintering in a sheltered area.

Migration in Monarch Butterflies. In other cases, the adaptiveness and evolutionary history of a migratory pattern may not be so apparent. Lincoln Brower has, over years of intensive work, unraveled the story of migration in monarch butterflies. He found that they overwinter in enormous gatherings in a few places, particularly in the highlands of central Mexico, where masses hang like great living tapestries from only a few

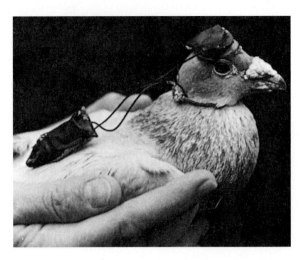

FIGURE 44.10
Homing Experiment
Pigeons carrying coils with reversed polarity on the head tended to home in opposite directions, indicating that pigeons can detect slight magnetic fields and that magnetism can influence their orientation.

specific trees in about ten separate sheltered valleys (Figure 44.11). The butterflies remain only to overwinter; they make little demand on the scarce resources of the area. The following spring the great masses break up and the butterflies begin to move northward to take advantage of the abundant, if temporary, milkweed food supply along the Gulf coast, from Texas to Florida. By May they will have laid all their eggs and played out their energy reserves. The next generation, appearing a short time later, continues on to the northern United States and Canada. These individuals and their offspring, in turn, lay eggs there and the late August–early September generation begins its long trek back to the sheltered valleys high in the Transvolcanic mountains of central Mexico.

Brower has determined that this multigenerational migration is a response to the short life span of the monarchs and a spatially variable food supply that proceeds northward from the Gulf coast as the seasons progress. When food is no longer available, the butterflies retreat to their valley, where they live off their reserves until the following spring. The perplexing problem of the adaptiveness and the evolution of monarch migration has now largely been solved. (Just how they navigate, though, remains a mystery.)

The Evolution of Migration. There has been a great deal of speculation on the evolution of migration. Some of the adaptive bases are apparent; some are not. The most obvious advantage, for example, in a robin migrating southward in the fall, is to escape the severe winter in the north. Not only is it cold up there, but the food supply dwindles in the winter, throwing overwintering species into stronger competition with their neighbors. Some species of food specialists may find that their entire food supply has vanished. So many species escape, not only to warmer climes, but to a better food supply.

FIGURE 44.11
Monarch Butterflies
Migrating monarchs gather for the winter in very special places, such as this tree in central Mexico.

If there is so much food in the warmer winter habitats, why don't migratory species just stay there? Why do they return to their summer homes at all? There are certain important advantages to rearing offspring in temperate or even polar areas in the spring. For example, days of the northern summer are long, and so the birds' working days can be extended; they can bring more food to their offspring in a given season. Food productivity is high in polar regions in summer because of increased sunlight for photosynthesis, optimal temperatures for photosynthesis, and nutrient-rich upswellings in polar oceans. And since, effectively, more food is available, more young can be raised. It is known that, generally, the farther north from the tropics a species breeds, the larger is its brood. And with more food, any brood can reach maturity faster.

Another advantage in returning to the temperate zone to breed is in escaping the high level of competition that exists in the species-packed tropics. The annual flush of life in the tem-

perate zones provides a predictable and otherwise underutilized supply of food that can be exploited readily by mobile species such as birds.

In the polar region, the breeding period is very short, but the brevity of the season may actually work to the advantage of nesting birds that are in danger of falling prey to predators. The short season means that a great number of birds must nest simultaneously; thus the likelihood of any single individual being taken by a predator is reduced. Also, since the birds come and go from both the summer and winter areas, their predators are denied a stable food supply. By leaving certain geographical areas each year, migratory species deprive many parasites and microorganisms of permanent hosts to which they can closely adapt. In addition, the long harsh winters in the frozen north reduce the numbers of parasites in that area. By the same token, predators that are unable to escape the rigorous northern climes might also be expected to be fewer in number the following spring.

There are three major theories regarding how bird migration may have evolved. According to one theory, birds evolved in the northern latitudes when the weather was gentler, more benign. Later, as the ice ages advanced, winters would have become increasingly severe, forcing the birds farther south. Those evolutionary lines that changed physiologically at a certain time of the year and then were prompted to move south before being caught in the deadly winter weather would have tended to survive. Their inclination to fly northward in the spring is considered an attempt to return to the ancestral breeding ground.

The second theory holds that birds evolved closer to the equator, and as their numbers grew, the increased competition there forced them northward, only to be driven to their ancestral homeland by the severe northern winters.

The third theory is based on continental drift (see Essay 20.1). According to this idea, birds, after originating in the balmy climes of Gondwanaland, followed the separating land masses, exploiting their seasonal offerings, only to return to the southerly areas as each winter drew on.

Let's now turn to another vigorous area in the study of animal behavior, one that also has stimulated a great deal of first-class detective work, but one that is more generally adaptive across a wide range of animals. This area is loosely referred to as communication.

The Social Basis of Communication

Communication, in its broadest sense, is an action by one animal that influences another animal. Whereas communication was once regarded as a means by which one animal let another animal know what it was going to do ("I'm baring my teeth; I'm going to bite!"), we now see that communication is a "reproductive enabling device" by which one animal manipulates those around it, thereby enhancing its own survival and reproductive output.

FIGURE 44.12
Graded Display
At right, the chaffinch *Fringilla coelebs* is showing a high-intensity threat. The posture at center is a medium-intensity threat, and that at the left is a low-intensity threat.

Low-intensity threat   Medium-intensity threat   High-intensity threat

Communication involves signaling, or displays, in which one animal performs an action that is perceived by another animal. The signals may be *directly* involved with reproductive success as a component of mating behavior or precopulatory displays. It may also be *indirectly* involved with reproductive success by helping the offspring avoid danger when parents give warning cries, or by simply helping the reproducing animal live better or longer so that it may successfully mate again. Remember, the charge to all living things is "reproduce, or your genes will be lost." Communication helps animals to carry out that reproductive imperative.

Let's consider a few general methods of communication, and in so doing perhaps we can learn something about the ways in which animals influence the behavior (or the probability of behavior) of other animals. That, after all, is what communication is all about.

Visual Communication

Visual communication is particularly important among certain fish, lizards, birds, and insects—and among some primates as well. Visual messages may be communicated by a variety of means, such as color, posture or shape, or movement and its timing. As was mentioned in Chapter 43, the color red can release territorial behavior in European robins. The female quickly solicits the attention of the male bird by assuming a head-up posture while fluttering her drooping wings. Some female butterflies attract male butterflies by the way they fly. As an example of communication by timing, fireflies are attracted to each other on the basis of their flash intervals, each species having its own frequency. (One predatory species "taps" communication lines by flashing at another species' frequency and then eating whoever comes to call.)

Because visual signals carry high information content, subtle variations in the message can be conveyed by gradations in the intensity of the display (Figure 44.12). (The same is true for some other means of communication, as we will see.) Of course, **graded displays** are useful only to the species that are sensitive and intelligent enough to be able to recognize such subtleties. Another advantage of visual signals is that the same message may be conveyed by more than one means, such as when an aggressive chimpanzee both stares and bares its teeth. Such redundancy may be used either to modify the message or to emphasize it in order to reduce the chance of error in interpretation.

A visual signal may be a permanent part of the animal, as in the elaborate coloration of the male pheasant and the striking facial markings of the male mandrill (Figure 44.13). These animals advertise their maleness at all times and are continually responded to as males by members of their own species. On appropriate occasions they may emphasize their "machismo" through behavior such as the strutting of the pheasant and the toothy glare of the mandrill. Such emphasis is of a temporary nature, so visual signals can be long-term or more temporary.

Short-term visual signals have the advantage that they can be started or stopped immediately. If a displaying animal suddenly spots a predator, it can freeze or dart for cover, and its former position won't be given away by any lingering images. Also, the recipient of a visual message is usually notified of the exact location of the sender. The recipient can then respond in terms of the sender's precise location, as well as its general presence and behavioral state (aggressive, amorous, or whatever).

Visual signs also have certain disadvantages. For example, the sender must be seen, and all sorts of things can block vision, from mountains and trees to fog and big hats. Visual signals are generally useless at night or in dark places (except for light-producing species). Since visibility weakens with distance, such signals are useless at long distances. Furthermore, as distance increases, the signal must become bolder and simpler; hence it carries less information. Other means of communication, as we will see, have their own advantages and disadvantages.

Sound Communication

Sound plays such an important part in communication in our own species that it may surprise you to learn that it is limited, for the most part, to arthropods and vertebrates. If you are familiar with the songs of the cricket and cicada, you are probably aware that insect sounds are usually produced by some sort of friction, such as rubbing the wings together or rubbing the legs against the wings. The key aspect of these sounds is cadence (timing), whereas with most birds and mammals, pitch or tone is more important.

A Male Mandrill

FIGURE 44.13
Permanent Versus Temporary Display
A male mandrill has permanent markings that advertise his sex, but his signals of anger are temporary and depend on his mood. It is probably often to his advantage to be perceived as a male, but less frequently advantageous to be perceived as an angry male.

Vertebrates use a number of different forms of sound communication. Fish may produce sound by means of frictional devices in the head area or by manipulation of the air bladder. Land vertebrates, on the other hand, usually produce sounds by forcing air through vibrating membranes in the respiratory tract. They use sound communication in other ways as well: rabbits thump the ground, gorillas pound their chests, and woodpeckers hammer hollow trees and drainpipes early on Sunday mornings.

Most of the lower vertebrates rely on signals other than sound, but some species of salamanders can squeak and whistle, and there is even one that barks. Frogs and toads advertise territoriality (willingness to defend their area) and choose mates at least partly by sound. Few reptiles communicate by sound, but large territorial bull alligators can be heard roaring in the swamps of the southern United States. Darwin described the roaring and bellowing of mating tortoises when he visited the Galápagos Islands.

Sound may vary in pitch (low and high), in volume, and in tonal quality. The last is apparent as two people hum the same note, yet their voices remain distinguishable. It is possible to show the characteristics of sound graphically by use of a sound spectrogram. In effect, this is a translation of sound into markings, which makes possible a more precise analysis of the sound.

The function of the message may dictate the characteristics of the sound. For example, Figure 44.14a shows a sound spectrogram of mobbing cries (calls to encourage other birds to help attack a predator) of several bird species. Such calls include brief, low-pitched *chuk* sounds, and their source is easy to pinpoint. When a bird hears the repetitious mobbing call of a member of its own species, it is able to locate the caller quickly and join it in driving away the object of concern—usually a hawk, crow, or other marauder. If an aerial predator such as a hawk is spotted flying overhead, the warning cry of songbirds is usually a high-pitched, extended *tseeeeee,* a sound that is difficult to locate (Figure 44.14b). The response of a bird hearing this call is quite different from its response to the mobbing call. The warning cry sends the listener heading for cover, often diving into deep, protective foliage, from which it may also take up the plaintive, hard-to-locate cry.

Sound signals have the advantage of being able to carry a variety of information through subtle variations in frequency, volume, timing, and tonal quality. Furthermore, they can be modulated (turned up or down). Obviously, louder sounds, with their higher energy levels, can carry over greater distances. In addition, sounds are transitory; they don't linger in the environment after they have been emitted. Thus an animal can cut off a sound signal should its situation suddenly change—for example, with the appearance of a predator. A further advantage is that many animals don't have to stop what they are doing to produce a sound. Unlike visual images, sounds can go around or through many kinds of environmental objects.

One disadvantage of sound communication is that it is rather useless in noisy environments. Thus some sea birds that live on pounding, wave-beaten shorelines rely primarily on visual signaling. Sound also weakens with distance, and the source of a sound is sometimes difficult to locate, especially under water.

Chemical Communication

You have probably seen ants rushing along single file as they sack your cupboard. You may also have taken perturbingly slow walks with dogs that stop to urinate on every bush. The behavior in both instances is based on chemical communication. The ants have laid down chemical trails that the others can follow, and the dogs are gloriously advertising their presence. In both cases, the animals are communicating by **pheromones** (from the Greek *pherein,* "to carry"; *horman,*

(a) Alarm calls

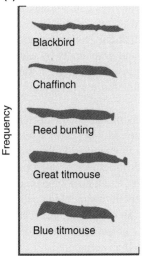

Blackbird

Chaffinch

Reed bunting

Great titmouse

Blue titmouse

Frequency

Time

(b) Mobbing calls

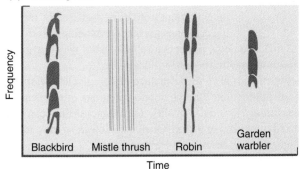

Blackbird   Mistle thrush   Robin   Garden warbler

Frequency

Time

Wren   Stonechat   Chaffinch

Frequency

Time

FIGURE 44.14
Calls

(a) Shown here are sound spectrograms of the mobbing calls of several species of British birds while attacking an owl. The sounds have qualities that make them easy to locate. Such calls are low-pitched *chuk* sounds. Different species of birds have developed similar calls through convergent evolution. That is, their calls serve much the same purpose and were developed under relatively similar conditions; thus the qualities of the sounds came to be somewhat alike. **(b)** The sound spectrograms shown here represent alarm calls of five species of British birds when a hawk flies over. Such calls are high-pitched, drawn-out, and difficult to locate. They too have achieved their similarity through convergent evolution.

"to excite"), chemicals that are produced by one animal and released into the environment and that influence the behavior of another animal, generally of the same species. Pheromones have been found in a number of species but have been most intensively studied in arthropods and mammals.

In insects, pheromones may incite very stereotyped behavior. For example, disturbed ants may produce an alarm chemical that causes other ants to drop whatever they are doing and to rush around in an agitated manner, ready to attack an intruder or to help the group in some other way. Insect pheromones can be very powerful; one molecule of bombykol is enough to excite a male *Bombyx* moth.

Pheromones can elicit quite complex behavior in mammals. For example, females of most mammal species elicit sexual responses in males (sometimes coupled with aggression toward other males) by pheromones. In other cases, pheromones can trigger hormonal responses. For example, if pregnant rats of certain species smell the urine of a strange male, some component of that urine will cause them to abort their fetuses and become sexually receptive again.

Chemical signals have the advantage of being extremely potent in very small amounts. Also, because of their persistence in the environment, the sender and receiver do not have to be precisely situated in order to communicate. In addition, chemicals can move around many sorts of environmental obstacles.

The specificity of chemicals, however, limits their information load. Moreover, because chemicals do linger in the environment, they may advertise the signaler to arriving predators as well as to the intended recipient of the message.

Tactile Communication

Tactile communication, or communicating by touch, is common among many forms of animals. Dogs, wolves, and other canids may communicate dominance by laying a paw on the subordinate's back. Chimpanzees, baboons, and other primates may build bonds by grooming each other. (Humans may hug.) Garden snails may sexually stimulate each other by, at first, intertwining each other's slippery bodies, then suddenly thrusting a chalky dart into each other. If they survive, they are ready to mate.

Perhaps the best-studied tactile communication is found in honeybees, in which a forager returns to the hive and communicates the direction and distance of a food source through dancing as the others gather around to touch the dancer (Figure 44.15).

Touch is a very direct means of communicating and has the advantage of precisely fixing signaler and recipient in time and space. It can also be changed rapidly (say, from a pat to a hit). Disadvantages include the necessity for signaler and recipient to be near each other.

Now let's see how communication is important in social behavior, focusing first on agonistic behavior, then cooperation, and closing with a look at the field called sociobiology.

Social Behavior

The famous student of animal behavior, Jane Goodall, who has spent much of her life among the chimpanzees of East Africa's Gombe Stream Preserve, once said, "One chimpanzee is no chimpanzee at all." Her point was that researchers should not attempt to study chimpanzee behavior by observing a single chimpanzee in a cage, because an isolated chimpanzee will behave quite abnormally. Chimpanzees, she noted, are highly social creatures that interact in extremely intensive and complex ways. If you want to know what chimpanzees are like, according to Goodall, you must watch them when they are with other chimpanzees.

The same statement might be made of any of a number of other creatures, including us. What would a termite be like without other termites? Or a human without other humans? Many of the earth's animals are indeed highly social species, and they interact with each other in subtle and complex ways. Yet there are some underlying themes. We will look at some of these principles here. In particular, we will consider how animals behave agonistically or cooperatively and on what adaptive bases populations may stay together.

Agonistic Behavior

Agonistic behavior is any behavior that helps to resolve conflict. It usually involves members of the same species. Agonistic behavior includes a wide range of behavioral patterns, such as aggression, threats, and submission. It does *not*, however, include predation. A lion is about as aggressive toward a wild pig as you are toward a hamburger. However, if the fleeing pig should turn around and charge a lion, the lion may, for an instant, show a phase of agonistic behavior—fear. At the other end of the agonistic spectrum is aggression, a behavior that eventually reduces the level of competition between two individuals. Aggression, as we will see, may take the form of fighting.

Fighting, a Form of Aggression

Let's consider the most obvious form of aggression—fighting. You can discount the old films you have seen of leopards and pythons battling to the death. Such fights simply aren't likely to happen. What is a python likely to have that a leopard needs badly enough to risk its life for, and vice versa? Although such fighting might occur in the unlikely event that one should try to eat the other, fighting is much more likely to occur between two animals that are competing for the same resources. The closest competitor is one that uses the same habitat in the same way, and the animal most likely to do this would be a member of the same species. Moreover, if there is a strong competition for mates, fighting is even more likely between members of the same sex within that species. And this is, in fact, where most fighting occurs—between members of the same sex of a given species.

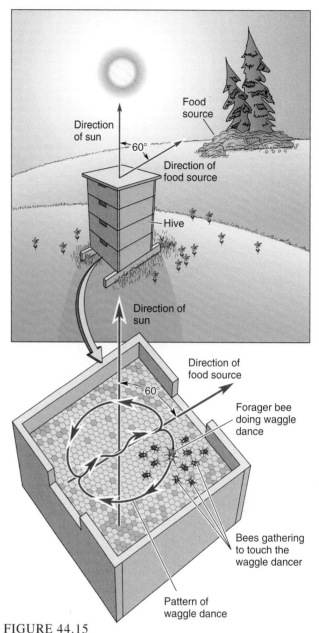

FIGURE 44.15
Honeybee Communication
Honeybees communicate very complex signals as they touch each other. Here a food source is communicated as the angle of the food from due north to the angle of the dance from a vertical line.

Of course, fighting between species sometimes occurs. The golden-fronted and red-bellied woodpeckers exclude each other from their respective territories. Lions may attack and kill African cape dogs at the site of a kill. The lions don't eat the dogs; they just exclude them.

Animals may fight in a number of ways, but combatants of the same species usually manage to avoid injuring each other (Figure 44.16). There are several apparent benefits to such a system. First, no one is likely to get hurt. The competitor is permitted to continue its existence, it is true, but the possibility

FIGURE 44.16
Fighting Male Rattlesnakes
Each male (*Crotalus horridus*) tries to push the other to the ground, but the deadly poisonous snakes never bite each other.

FIGURE 44.17
Male Pronghorn Antelopes Fighting
Although the horns of these medium-sized antelopes are formidable weapons, neither animal will attack the vulnerable flank of the other. Instead, a harmless pushing contest ensues as the tips of the ridged horns are engaged. These animals effectively employ horns and hoofs against species other than their own, but when confronted with a member of their own species, they are genetically constrained to behave in very circumscribed ways.

of having to compete again entails less risk than does serious fighting. Also, since animals are most likely to breed with the individuals around them, the opponent has a reproductive advantage. In fact, if the population is confined to a small area, the competitor might even be one's own mature offspring; it could also be a prospective mate. So even though the "motive" is selfish, not benevolent, it's best not to hurt each other.

When fighting occurs between potentially dangerous combatants, the fights are usually stylized and relatively harmless. For example, a horned antelope may gore an attacking lion, but when antelopes fight each other, the horns are almost never directed toward the exposed flank of the opponent, but toward the protected head (Figure 44.17). Such stylized fighting does, however, enable the combatants to establish which is the stronger animal. Once dominance is established, the loser is usually permitted to retreat.

On the other hand, all-out fighting may occur between animals that are unequipped to injure each other seriously, such as hornless female antelope (Figure 44.18), or between animals that are so fast that the loser can usually escape before serious injury, such as house cats.

We might wonder why male antelopes don't gore each other. An incurable romantic might assume that they simply don't want to hurt one another. In all likelihood, what the two antelopes "want" has very little to do with it. The fact is, they can't hurt each other. When the system works, antelope could no more gore an opponent than fly! In terms of the actual mechanism, it may be that the sight of an opponent's exposed flank acts as an inhibitor of butting behavior. Conversely, a facing view, under certain conditions, might serve as a release of very stereotyped fighting behavior.

Regardless of what we may or may not surmise about animal motivation, such standardized behavior must have evolved because it was beneficial to the ancestors of the present combatants. Thus we return to the original question: How does the animal benefit by refusing to do serious injury to the opponent?

One benefit is really not that subtle. The antelope might not gore its opponent because if it did so, it might get gored back. The situation is similar to that of two toughs in a barroom brawl. They slug and punch, and each is really trying to win the fight. On the other hand, each has a jackknife in his pocket—and both of them know it. Neither is willing to pull his knife, because then the other would be obliged to retaliate, and someone could get hurt that way. Either brawler would rather risk a loss in a fistfight than a draw in a knife fight.

Game Theory. Animal evolution has proceeded according to the same kind of logical accounting that may be described in a series of options called **game theory**, in which evolutionarily derived strategies dictate the behavior of animals in different

FIGURE 44.18
Hornless Female Antelopes Fighting
Unlike their male counterparts, hornless female Nilgai antelope
have no inhibitions against attacking the flank of the competitor, but
their butts are quite harmless, at least in the immediate sense.
Though the butt itself may be dangerous, it establishes dominance.
A loss, however, usually just means a temporary setback, so it
behooves the loser to accept it gracefully and attempt to breed
another time. Interestingly, horned males of the same species almost
never attack in this way, nor do horned females of other species.

situations. John Maynard Smith of the University of Sussex
has analyzed fighting strategies with mathematical models
and computer simulations to produce his game theory. What
he looked for was an **evolutionarily stable strategy**, which
he defined as an innate behavioral pattern that would outcompete all other behavioral patterns and would be stable against
the invasion of a new mutant pattern of behavior. He formed
his questions in terms of "hawks" and "doves." Is it best never
to fight and always retreat (dove)? Then a born fighter will
always win. Is it better always to fight (hawk)? No, because
there is too much risk of being beaten. Is it best to bluff consistently? No, the bluff will be called. (It's better to bluff
inconsistently and unpredictably.) Maynard Smith found two
behavioral "strategies" that proved to be the most stable in
populations, depending on the circumstances. The first strategy was called the **retaliator strategy**; a retaliator engages
only in ritual display and mock battle unless it is seriously
attacked, whereupon it will retaliate with just as much seri-
ousness. In a mathematical model, this behavior was always
the most successful in the long run, so it should not be surprising to see ritual display and mock battle so common in
nature. There's always the threat of real injury to any animal
that breaks the rules.

The second strategy, the **bourgeois strategy**, employs
quite a different set of rules. The bourgeois approach was
even more effective than retaliation, but it required following
a specific rule. In each encounter between adversaries, one
retreats immediately, so that neither wastes time and energy
fighting. We find the bourgeois strategy common in territorial
encounters. The owner of the territory will attack an intruder,
and the intruder almost always quickly retreats. However, the
strategy will prevail in a population even if the determination
of who concedes is completely arbitrary. Some of our traffic
rules work this way. The first car to the intersection has the
right of way; if two cars arrive simultaneously, the car on the
right has the right of way and the car on the left concedes (but
don't count on it). As long as everyone knows the rules, it is
advantageous to everyone to follow them.

Can we assume then that animals do not fight to the death
with their own kind? As a general rule they don't, but there
are many exceptions. Accidents may occur in normally harmless fighting, and this can lead to retaliation and escalation.
There are also some species that normally engage in dangerous fighting. If a strange rat is placed in a cage with a group of
established rats, the group may sniff at the newcomer carefully for a long time, but eventually they will begin to attack it
and do so repeatedly until they kill it. If escape is impossible,
male guinea pigs and mice often fight to the death. The males
of a pride of lions may kill any strange male they find within
their hunting area, and a pack of hyenas may kill any of
another pack that they can catch. Even gangs of male chimpanzees have been seen to ambush and kill isolated males
from other troops. Even game theory cannot predict every
case, but it gives us a framework with which to examine certain general principles. Let's now go to what may seem like
the opposite end of the social spectrum—cooperation. Keep
in mind, though, that even this seemingly "nicer" behavior is
still likely to be just another way for an animal to get its genes
into the next generation.

Cooperation

Cooperative behavior occurs both within species and between
species. As an example of *interspecific* (between species)
cooperation, consider the relationship of the rhinoceros and
the tickbird. The little birds ride around on the rhino's back. In
this way, the birds get free food, while the rhinoceros rids
itself of ticks and harbors a wary little lookout. Such relationships are well known in nature because of their inherent interest, but the highest levels of cooperation are most likely to
exist between members of the same species.

Let's consider a few examples of *intraspecific* (within
species) cooperative behavior. Porpoises are air-breathing
mammals, much vaunted in the popular press for their intelligence. In fact, certain of their actions support the claim.
Groups of porpoises will swim around a female in the throes
of birth and will drive away any predatory sharks that might
be attracted by the blood. They will also carry a wounded
comrade to the surface so that it can breathe. Their behavior in
such cases is highly flexible, rather than stereotyped. Such

(a) Arctic Musk Oxen

(b) Honeybees

FIGURE 44.19
Cooperation in Vastly Different Species
(a) Musk oxen form a defensive circle to protect the females and young. **(b)** These social insects show extremely high levels of cooperation, coordination, and self-sacrifice. The result is a hive in which the watchword is cold, sometimes brutal, efficiency.

flexibility indicates that their behavior is not solely a blind response to innate genetic influences.

Group cooperation among mammals is probably most common in defensive and hunting behavior. For example, musk oxen of the Himalayas form a defensive circle around the young at the approach of danger, standing shoulder to shoulder with their massive horns directed outward (Figure 44.19a). This defense is effective against all predators except humans, since it provides no defense against high-powered rifles. Other species, such as birds, wolves, African cape dogs, jackals, and hyenas, often hunt in packs and sometimes cooperate in bringing down their prey. In addition, they may bring food to members of the group that were unable to participate in the hunt.

We might expect mammals, with their high intelligence, to cooperate closely. But social behavior and cooperation are most highly developed in certain of the insects (Figure 44.19b). The complex and highly coordinated behavior patterns of insects are usually considered to be genetically programmed, and generally not influenced by learning. Some of the best examples of insect cooperation are found among the honeybees.

In honeybee colonies, the queen lays the eggs, and all other duties are performed by the workers, which are sterile females. Each worker has a specific job, but that job may change with time. For example, newly emerged workers prepare cells in the hive to receive eggs and food. After a day, or so, their brood glands develop, and so they begin to feed larvae. Later, they begin to accept nectar from field workers and pack pollen loads into cells. At about this time their wax glands develop, and they begin to build combs. Some of these "house bees" may become guards that patrol the area around the hive. Eventually, each bee becomes a field worker, or forager. She flies afield to collect nectar, pollen, or water, accord-

ing to the needs of the hive. Apparently, these needs are indicated by the "eagerness" with which the field bees' different loads are accepted by the house bees.

If a large number of bees with a particular duty are removed from the hive, the normal sequence of duties can be altered. Young bees may shorten or omit certain duties and begin to fill in where they are needed. Other bees may revert to a previous job where they are now needed again.

The watchword in a beehive is *efficiency*. In some species, the drones (males) exist only as objects of reproduction. Once the queen has been inseminated, the rest of the drones are quickly killed off by the workers; they are of no further use. The females themselves live only to work. They tend the queen, rear the young, and maintain and defend the hive. When their wings are so torn and battered that they can no longer fly, they either die or are killed by their sisters. But the hive goes on.

Symbiosis

In some cases, members of different species live together in close association, a relationship known as **symbiosis**. As a result of symbiotic relationships, one or both individuals can be benefitted, as with mutualism or commensalism, or in some cases one can be harmed, as in parasitism, where one organism derives some benefit at the expense of the other.

Mutualism is an interaction in which all involved species benefit. For example, cattle harbor cellulase-producing bacteria in their complex digestive system. The cattle derive energy-rich breakdown products of cellulose, vitamins, and amino acids from the mutualistic bacteria. The bacteria benefit from the steady food supply ingested by the cattle.

Commensalism is a species interaction in which one species benefits while the other is neither benefitted nor harmed. An example of commensalism is the association between cattle egrets and cattle (Figure 44.20). The birds, which feed on insects and other small organisms disturbed by moving cattle, benefit because their foraging success is greater near moving cattle than away from them. The cattle appear to be neither helped nor harmed, although observers report an occasional egret picking a parasite from the skin of a bovine associate. Another example is the interaction between oceanic birds, like gulls and terns, and predaceous fish. These birds are attracted to schools of feeding fish that injure more small fish than they can consume during feeding frenzies near the water surface. The birds quickly pick up these easy prey. The predaceous fish do not benefit and are probably not harmed by the opportunistic feeding of the birds.

Parasitism is a symbiotic relationship in which one organism (the parasite) lives on or in another organism (the host). The parasite gains sustenance from the host and, as a result, the host is harmed. As an example, consider human tapeworms (Chapter 26).

Next, we will look at relationships where one animal may be benefitted by another animal performing a behavior, but at a cost to the performer.

Altruism

You probably won't be shattered to learn that most of the "Lassie stories" aren't true. Consider what would happen to the genes of any dog that was given to rushing in front of speeding trains to save baby chickens. The reproductive advantages would be considerable to chickens, but dogs with those tendencies might be selected out of the population by the action of fast trains. In contrast, the genes of a "chicken" dog that spent his energy, not in chivalrous deeds, but in seeking out estrous females, would be expected to increase in the population.

Altruism may be defined as an act by one individual that benefits another, but at the first individual's expense. There are many apparent cases of altruism in the animal world, but our job here will be to ask if these are really altruistic acts, after all, and if so, how the behavior might have evolved.

It is easy to see how certain forms of altruism are maintained in a population. For example, pregnancy, in a sense, is altruistic. The prospective mother is swollen and slowed. Much of her energy goes to the maintenance of the developing fetus. While giving birth she is not only almost completely incapacitated but is in marked danger. Pregnancy is clearly detrimental to her. So why do females so willingly take the risk? It may help to understand the enigma if we remember that the population at any time is composed entirely of the offspring of individuals who have made such a sacrifice. Thus the females in the population are the descendants of generations who did reproduce, so they too are likely to be predisposed to make such a sacrifice. It makes little difference whether the tendency is genetic or learned. The tendency to

FIGURE 44.20
A Commensalism
Cattle egrets and cattle have a commensal relationship.

have offspring or *to do the things that result in having offspring* can be innate (genetically based) or transmitted culturally, as is so often the case in humans.

However, altruism on this basis doesn't explain why a bird may feed the young of another pair, or why an African hunting dog will regurgitate food to almost any puppy in the group. Why, also, would a bird that may have no offspring of its own give a warning cry at the approach of a hawk, alerting other birds at the risk of attracting the hawk's attention to itself? To answer such questions we must look past the answers that first come to mind. It may seem cynical, but we must start with the premise that birds don't give a hoot about each other. A bird that issues a warning call isn't thinking, "I must save the others." There is a simpler explanation of its behavior.

We should note that the biologically "successful" individual is the one that maximizes its reproductive output. One way of accomplishing this is for the organism to leave its own offspring, but another way is by helping relatives (who carry many of the same kinds of genes) to reproduce. We need only keep in mind that an individual also shares genes with a cousin, albeit fewer than with a son or a daughter. So it is reproductively beneficial for an individual not only to bear offspring but to assist relatives, as long as the cost is not too high. In fact, there is theoretically a point at which an individual could increase the success of its genes by saving its nieces and nephews (provided there were enough of them) rather than its own offspring. From the standpoint of effective reproductive output, the organism would be better off leaving 100 nieces than one daughter. The tendency to help a relative because of shared genes is called **kin selection**.

Hypothetical Mechanism for Maintenance of Altruism in Populations

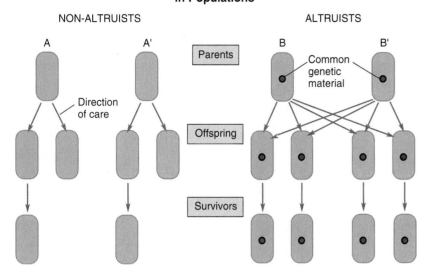

NON-ALTRUISTS

ALTRUISTS

A   A'   Parents   B   B'

Direction of care   Common genetic material

Offspring

Survivors

Populations A and A' do not have genes in common. Since they do not show altruistic behavior to each other's offspring, fewer offspring from each population survive.

Populations B and B' have genetic material in common. Since they show altruistic behavior to each other's offspring, more offspring from each population survive.

FIGURE 44.21
A Genetic Mechanism of Altruism
In this population, A and A' are nonaltruists. They behave in such a way as to maximize their own reproductive success but do nothing to benefit the offspring of other individuals. In another segment of the population (B and B', a gene for altruism has appeared that results in individuals benefitting the offspring of others in some way. It can be seen that, assuming the altruistic behavior is only minimally disadvantageous to the altruist, generations springing from B and B' are likely to increase in the population over those from A and A'. The altruistic behavior is likely to be greatest where B and B' are most strongly related, so that B shares the maximum number of genes in common with the offspring of B' and vice versa. The idea is that B, for example, can increase its own reproductive success by caring for the offspring of a relative with whom it has some genes in common. After all, reproduction is simply a way of continuing your own types of genes.

To illustrate, suppose a gene for altruism appears in a population. (Notice that this sets up a mechanism for the continuance of the behavior.) As you can see from Figure 44.21, altruistic behavior would most likely be maintained only in groups in which the individuals are related (that is, have some kinds of genes in common). Altruism might be expected then when there is a high probability that proximity indicates kinship, as we see in relatively stationary populations.

Keep in mind that no conscious decision on the part of the altruist is necessary. It simply works out that, when conditions are right, those individuals that behave altruistically increase their types of genes in the population, including the "altruism gene." Nonrelatives would benefit from the behavior of altruists, of course, but there is an increased likelihood that individuals near an altruist are related to it.

It has been determined mathematically that the probability of altruism increasing in a population depends on how closely the altruist and beneficiary are related (as well as on the risk to the individual, of course). Furthermore, the advantages to the beneficiary must increase as the kinship becomes more remote. For instance, an altruistic act that results in a risk of death of the altruist will be selected for if the net genetic gain to brothers and sisters who share one-half the altruist's genes is more than twice the loss to the altruist; to half-brothers who share one-quarter of the altruist's genes, four times the loss; and so on. To put it another way, an altruistic animal would gain reproductively if it sacrificed its life for more than two brothers, but not for fewer, and so on. Therefore we can deduce that,

in highly related groups, such as a small troop of baboons, a male might fight a leopard to the death in defense of the troop.

This model, developed by J. B. S. Haldane in 1932 and expanded by W. D. Hamilton in 1963, helps explain the extreme altruism shown by **eusocial insects** (*eu,* "true"), that is, those with castes—queens, workers, drones—such as honeybees and paper wasps. Since workers are sterile, their only hope of propagating their own genotype is to maximize the egg-laying output of the queen. In some species, the queen is inseminated only once (by a drone, a haploid male), so all the workers in a hive are sisters and have an average of three-fourths of their genes in common (Figure 44.22). In such a system, almost any sacrifice is worth any net gain to the hive and to the queen.

Humans and Reciprocal Altruism

In a brilliant essay, Robert Trivers expanded our understanding of the evolution of altruism by developing the theory of **reciprocal altruism**, which is any altruistic act that depends on the expectation of reciprocation. ("I'll scratch your back if you'll scratch mine.") Reciprocal altruism is an evolutionarily stable strategy with some complex rules. Help (altruistic acts) is given to others—even offered to strangers—when the cost or risk is not too great to the giver. The expectation is that some kind of help will be reciprocated at another time. If the expected reciprocation is not forthcoming, an evolutionarily derived emotion is experienced: moral indignation. The indi-

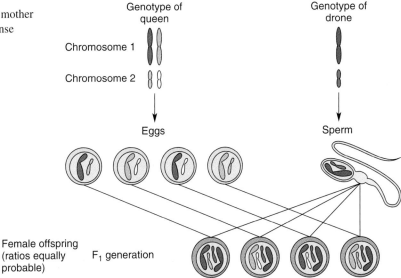

FIGURE 44.22
Relatedness in Social Insects
This chart shows the genetic relatedness between mother and offspring in eusocial insects that leads to intense altruistic behavior.

Relatedness between mother-offspring in eusocial insects

Each daughter shares ½ her chromosomes with the queen. To see the relatedness of the daughters, compare any genotype with these four possibilities – such as

100% 50% 75% 75%

Average: 75% – so daughters are likely to be more strongly related to each other than to their mother. Thus they are likely to sacrifice for each other, based on kin selection.

vidual that fails to reciprocate is scorned, turned out of the social group, no longer aided. To get ahead, everyone has to play by the rules—or appear to.

Some evidence of intraspecific reciprocal altruism has been reported in troops of social mammals, such as hunting dogs and baboons, but the evidence is not very strong for such behavior in any animals but humans. Trivers implies, in fact, that reciprocal altruism is the key to human evolution. The complexity of such behavior, entailing as it does memory of past actions, the calculation of risk, the foreseeing of the probable consequences of present actions, the possibility of advantageous cheating, and the need to be able to detect such cheating—all require a level of intelligence that is beyond most species. In the opinion of some anthropologists, it is exactly for the management of these elaborate social interactions that the human brain—and the conscious mind—evolved.

Sociobiology

Sociobiology applies principles of evolution to studies of social behavior in animals. Like other areas of ethology, especially behavioral ecology, it emphasizes the influence of ultimate factors on behavior (including that of humans).

Animals may live together (be social) on a number of adapted bases. We will mention six.

1. **Group foraging.** A group of animals may be more efficient at finding scattered food than a single individual. Of course, once food is found, it must be shared, but overall each individual gets more by searching with others.

2. **Group protection.** A group may offer protection by confusing a predator with sheer numbers. Birds about to be attacked by a falcon will draw closer together. The result is that so many potential targets confound the falcon's attempt to pick out one.

3. **"Selfish herd" effect.** An individual reduces its chance of being caught by a predator by using others as a form of cover. If a predator takes the nearest prey, then it is advantageous to be near other individuals that can act as a shield.

4. **Increased vigilance.** With more eyes and ears around, it is harder for a predator to sneak up.

5. **Reproductive coordination.** In a group, reproductive efforts can be coordinated so that offspring are produced together. The advantage for the animals may be to avail themselves of some fleeting commodity, such as a temporary food source, or to control predators on the offspring by overwhelming them. The predators can't take full

advantage of a single massive surge in food but would easily wipe the offspring out if they appeared one at a time.

6. **Mutual advantage to parent and offspring.** Offspring may survive and reproduce better eventually if they continue to remain with the parents for at least a time. Wolf packs, for example, are often composed of a single family group in which one male and one female do all the breeding. Younger animals stay around and help with hunting and rearing the pups until their time comes (if it ever does). Nonetheless, their odds are better this way than if they ventured out alone. In some birds, the previous year's offspring may stay around as "helpers." They help defend the nest area and bring food to their younger siblings. The parents gain assistance, and the offspring stand to inherit the territory upon the death of the parents.

We should keep in mind that these advantages are not mutually exclusive. Sociality probably rarely exists on a single adaptive basis but through a variety of adaptive effects.

Sociobiologists, we should add, also study the other end of the social spectrum—those animals that avoid each other. The most prominent examples of such antisocial behavior are found among predators. The adaptive basis is generally believed to be due to the difficulty many carnivores have in finding food. There just isn't enough to share. With predatory animals—such as leopards, jaguars, bears, and weasels—the limits of social behavior seem to be associating during periods of mating and, in some cases, rearing young.

Evolution and ecology have obviously played a great role in shaping the social behavior of animals. Just as such forces have molded the shape of an animal's head or the length of its leg, they have also strongly influenced how that animal behaves toward others.

At one time, not long ago, the topic of sociobiology was rife with controversy. The problem was that certain groups of people, some of whom were good scientists, strongly resisted placing humans under the sociobiological umbrella. It was considered politically and ethically inappropriate to suggest that human behavior might, in any way, be influenced by genes or evolution. Those arguments have largely died down, or at least been modulated, because of a continuing cascade of new information on the basis of human behavior and the knowledge that sociobiologists in no way seek to imply that biological influences on behavior do not leave room for change.

Key Ideas

PROXIMATE AND ULTIMATE CAUSATION

1. **Proximate causation** involves the more immediate bases for behavior and describes *how* the animal comes to act in a certain way.

2. **Ultimate causation** involves the evolutionary and adaptive bases for behavior and declares *why* the animal behaves in a certain way.

BEHAVIORAL ECOLOGY

1. **Behavioral ecology** is the study of how the environment affects behavior.

2. Wecker's experiments showed both genetic (ultimate) and learned (proximate) elements in habitat selection by mice.

3. The efficiency of foraging behavior is determined by how the animal maximizes gain relative to expenditure.

BIOLOGICAL CLOCKS

1. Most plants and animals respond to day–night, annual, lunar, and daily (circadian) cycles.

ORIENTATION AND NAVIGATION

1. Navigation involves the directional sense that enables an animal to get from one place to another. Compass sense is the ability to know direction. Map sense is the ability to know longitude and latitude. Time–direction mechanism is a compass sense coupled with a sense of time. Daytime migrators use the sun as a directional cue.

2. Orientation is the ability to face in the right direction.

3. The advantages of migration are not always clear. There are three major theories regarding how bird migration may have evolved: (a) the ice age caused winters in the north to be increasingly severe, (b) competition near the equator forced birds northward to breed, and (c) birds returned each year to the southern land masses where they originated.

THE SOCIAL BASIS OF COMMUNICATION

1. **Communication** is an action by one animal that influences another animal.

2. Visual messages may be sent by color, posture, shape, movement, or timing. Sound messages can be sent by cadence, pitch, or tone. Chemical signals involve molecules called **pheromones** that are produced by one animal and that influence another.

SOCIAL BEHAVIOR

1. **Agonistic behavior** helps to resolve conflicts among members of the same species.

2. Fighting, the most obvious form of aggression, usually occurs between competitors, does not involve injury, and is stylized. Fighting strategies have been analyzed as **evolutionarily stable strategies**, patterns that are selected over mutant variants.

3. Cooperation involves animals working together. Intraspecific cooperation occurs between members of the same species.

ALTRUISM

1. **Altruism** is an act that benefits another, but at the altruist's expense.

2. Altruism may continue in a population through **kin selection,** where relatives are cared for because they bear genes in common with the altruist.

3. The probability of altruism increasing in a population depends on how closely the altruist and beneficiary are related.

4. **Reciprocal altruism** involves performing an altruistic act with some expectation of the favor being returned.

SOCIOBIOLOGY

1. **Sociobiology** is the study of the effects of genetics and evolution on social behavior.

Application of Ideas

1. Make an observation of a behavior (either one you've read about or seen personally). Then describe the behavior in terms of probable ultimate and proximate causation.

2. Show how altruism is intimately related to recognition. To which levels of recognition (species, individual, or kin) would it appear more strongly?

Review Questions

1. A mother raccoon lies down on a cold day, and her litter of young, following close behind, immediately cuddle up to her belly. Explain the behavior in both ultimate and proximate terms. (page 959)

2. Distinguish between proximate and ultimate causation. (page 959)

3. How is clock shifting done? (page 963)

4. Distinguish between orientation and navigation. (pages 964–965)

5. Define communication in its broadest sense. (page 967)

6. How can visual communication carry such high information content? (page 968)

7. What kinds of animals are more likely to use graded displays? (page 968)

8. How does sound communication carry such high information content? (pages 968–969)

9. Do you think that agonistic behavior can include retreat? Why? (page 971)

10. Define an "evolutionarily stable strategy." (page 973)

11. Under what conditions are we most likely to see group cooperation among mammals? (page 974)

12. Distinguish between mutualism and commensalism. (pages 974–975)

13. Is any harm done in an altruistic act? Explain. (pages 975–976)

Individuals and Populations

CHAPTER OUTLINE

THE ECOLOGICAL HIERARCHY

Basic Versus Applied Ecology

THE ECOLOGY OF INDIVIDUALS

Environmental Factors

Resources and Regulators

Limiting Resources

Tolerance Ranges and Optima

Environment and Scale

Habitat and the Ecological Niche

The Principle of Competitive Exclusion

Indicator Species

Habitat Types

IMPORTANT ENVIRONMENTAL FACTORS

Temperature

Water

Light

Soil

THE ECOLOGY OF POPULATIONS

Dispersion Patterns

Population Changes in Time

The Difference Equation

Intrinsic Rate of Population Growth

Population Growth Models

Exponential Growth

Environmental Resistance and Logistic Growth

Demography

THE EVOLUTION OF REPRODUCTIVE STRATEGIES

The Theory of r and K Selection

POPULATION-REGULATING MECHANISMS

Density-Dependent Factors

Density-Independent Factors

KEY IDEAS

APPLICATION OF IDEAS

REVIEW QUESTIONS

The word "ecology" has made its way into the public vocabulary and has proved to be remarkably resilient there. Often, once a word enters that "great consciousness" it often becomes overused, misinterpreted, restructured, and finally battered into uselessness. ("Instinct" is an example.) But *ecology* has weathered the attacks in a remarkable fashion. The reasons are twofold. First, it has been freely interchanged with the word *environment* ("What are we doing to the ecology?"), a word with which people seem to be somewhat comfortable, so it has escaped intellectual massage by apparently being too easy a target. Second, people have found that it just isn't possible to deal with the true idea of ecology in a simplistic fashion. In its true sense, it is a complex and cumbersome concept, covering too much and touching too much to be handled tidily. We feel comfortable with it only if we don't know much about it.

In this chapter, we certainly want to convey an appreciation of the immensity of the problems associated with ecology. But more than that, we hope to show the great challenge of it all. (Ecologists may one day be recognized as the most important scientists on earth.) What, then, do ecologists study? In the original Greek the root word, *oikos*, means "house." Thus ecology is "the study of the house"—the place where we live, or the environment. The **environment**, technically, includes all those factors, both nonliving and living, that affect an organism. **Ecology**, then, is the study of the interaction between organisms and their environment. The key word is *interaction*.

We saved the study of interaction until late in our discussions of life because a certain amount of basic information about organisms and their physical world is necessary in order to understand their interactions. Certainly, ecology is where the sciences come together. Let's begin the discussion now by looking at ways in which ecologists have managed to organize and categorize the earth's environment.

The Ecological Hierarchy

From its beginnings in the late 1800s, ecology developed rapidly as a science. However, as it matured in the 20th century, it grew and became fragmented into a number of different subdisciplines. These disparate approaches were based mainly on whether you were looking at one species or many of them, and on whether your view was geographically restricted (a hillside) or more broad-based (the state of Montana). These approaches became organized into an *ecological hierarchy.*

The lowest level of the hierarchy, **physiological ecology**, focuses on the response of organisms or species to environmental factors. For example, a physiological ecologist might seek to understand how rising water temperatures affect the growth and reproduction of fish.

The next level is the ecology of the **population**, which is defined as an assemblage of individuals of a single species that live in the same place at the same time. Later in this chapter, you will learn that population ecologists are concerned with numbers of organisms and how environmental factors cause those numbers to increase or decrease with time.

Above the population is the ecology of the **community**, defined as an assemblage of interacting populations, forming an identifiable group. Community ecologists might be interested in understanding the environmental factors that cause one site to contain more species than a second site.

The next level of the hierarchy is the ecology of the **ecosystem**, defined as the sum total of all the species on a site, along with the physical factors, including the water, air, and soil. Ecologists commonly regard the ecosystem as being the basic unit of ecology, just as taxonomists regard the species as their basic unit. Ecosystem ecologists often study the passage of energy and nutrients among the various organisms that feed on each other. Often, ecosystems are considered to be rather restricted in geographical extent, such as a field or hillside.

Above the level of the ecosystem is the ecology of the **landscape**, which is composed of an assemblage of adjacent ecosystems covering perhaps dozens of square miles. Of all the levels of the hierarchy, the landscape level is the most recently recognized. Landscape ecologists are often called on to determine the size or shape of nature preserves for conserving rare species.

The level above the landscape is the ecology of the **biome**, which is defined as a large (normally covering hundreds or thousands of square miles) ecological unit. Ecologists recognize about a dozen major biomes; each one forms under a certain prevailing climate and has a characteristic type of plant and animal life. As we will see in Chapter 47, some examples of biomes include grasslands, deserts, and deciduous forests.

The highest level of the ecological hierarchy is the ecology of **biosphere**, which is defined as the sum total of all life on earth and the surrounding physical environment (all the soil, oceans, and atmosphere). As you might guess, biosphere-level scientists face a particularly daunting challenge because the object that they study (the earth) is so large. Biome-level analyses often require international cooperation and complex tools like satellites. A good example of a biosphere-level problem is that of global warming, the question of whether the earth really is growing warmer and, if so, what the causes and ramifications will be (see Essay 47.1).

Many ecologists like to organize the various levels of the ecological hierarchy into two broad domains: autecology and synecology. **Autecology** focuses on the ecological interactions experienced by a single individual or members of a single species. It is comprised of physiological ecology and population ecology. In contrast, **synecology** addresses the interactions of different species that live together. Synecologists do their work on communities, ecosystems, landscapes, biomes, and the biosphere. Thus the approach of synecology is usually more large-scale than that of autecology.

Basic Versus Applied Ecology

Many early ecologists conducted their work in natural communities that were largely unaffected by human disturbance. Investigations in such systems have led to many important basic ecological concepts like predator–prey relations, community organization, and energy flow in ecosystems. However, we know that most ecosystems throughout the world have been affected at least somewhat by human actions.

In 1962, Rachel Carson published her landmark book *Silent Spring,* which detailed the deleterious effects of pesticides on natural bird populations. Since then, many people have become very concerned about the actions of humans on our marine, freshwater, and terrestrial communities. During the late 1960s, a sociopolitical movement developed in which a large segment of the population, particularly those of college-age, sought to reverse the impacts that agricultural and industrial practices were having on natural populations. For want of a better term, these activists used the term "ecology" as a label for their cause, often much to the chagrin of professional ecologists. To the latter group, the individuals who protested nuclear energy plants or spent their time picking up litter were more accurately called **environmentalists**. The term **environmentalism** was used to define the sociopolitical movement concerned with the effects of humans on nature. Soon, **environmental science** became used to describe the application of scientific principles to understanding and solving human perturbations of nature. The term "ecology," or more precisely *basic ecology,* was reserved for the science that focused on understanding interactions of organisms under natural conditions, unimpacted by humans. In the 1970s, many professional ecologists were particularly sensitive to that distinction and purposely avoided studying ecosystems impacted by humans.

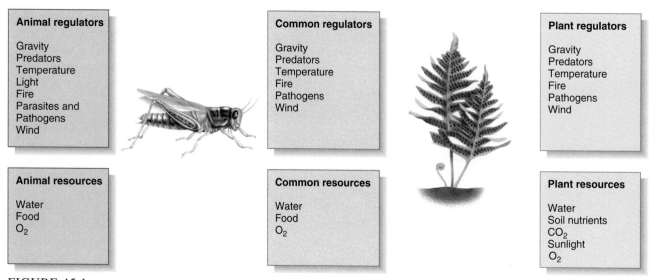

FIGURE 45.1
Resources and Regulators
Resources and regulators for a representative animal (a grasshopper) and a representative plant (a fern). Note that some resources and regulators are shared between the two types of organisms (for example, water, fire, wind), while some are different (for example, O_2 is a resource for the grasshopper while CO_2 is resource for the fern).

In the early 1980s, however, ecologists realized that the environmental problems were severe and deserved attention. The taboo against studying the impacts of humans on ecosystems had subsided, and professional ecologists increasingly studied the ecological effects of problems like acid precipitation, ozone depletion, global warming, water pollution, and human-induced loss of biological diversity. Thus was born the subdiscipline of **applied ecology**, solving specific problems of the environment through scientific means, with its specialties that included *restoration ecology* (devoted to reestablishing productive ecosystems on disturbed sites), *conservation biology* (devoted to preventing species from becoming extinct), and *sustainable agriculture* (devoted to growing crops using less-intensive cultivation methods), among others.

The Ecology of Individuals

Let's now consider some ecological principles at two levels. First we will look at the ecology of individuals, then at the ecology of populations, or groups of individuals. Essentially, then, we will be going from small scale to large scale. We begin with the effects of environmental factors on individuals.

Environmental Factors

Imagine what it would be like to spend all your life outdoors, exposed to blistering sunshine in the summer and freezing cold temperatures in the winter, drenched by soaking rains, subjected to attack by animals large and small. Thankfully, most of us are insulated from such extremes and hazards. Organisms living in the wild, like an octopus swimming at the bottom of the ocean or an oak tree growing in the middle of a forest, don't have the luxury of living in such a protected environment. Instead, they must constantly cope with the natural conditions within which they grow.

Looking at it a different way, ecologists have tried to figure out why some species, like sparrows, are so widespread, while others, like the endangered whooping crane, are rare. What prevents whooping cranes from being found on everybody's lawn? The answer lies, at least in part, in the fact that the natural environment limits the growth of every species. Some species are limited more than others. Over the years, ecologists have sought to identify the components of the natural environment that affect the growth, distribution, and reproduction of all species. Thanks to thousands of studies, we have come to appreciate that many factors are important, including temperature, moisture, fire, wind, light, food parasites, predators, competitors, and disease. Thus organisms deal with a complex environment in which numerous factors work independently and in concert to affect survival of individuals and species. While we know a lot about the manner in which environmental factors are important, scientists still have much to learn about this topic, and it is a fertile area of ongoing research.

Resources and Regulators

If you were to closely follow an organism in nature, you would soon see that some factors of the environment, like

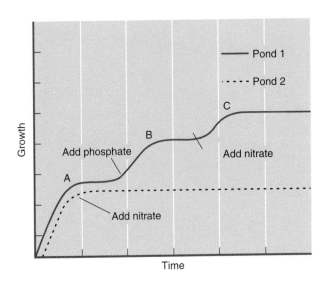

FIGURE 45.2
The Law of Limiting Resources
This graph shows the growth of two hypothetical populations of algae growing in separate ponds having identical water quality. Note how, in both ponds, the populations grow a little at first, but growth soon stops at level A. If phosphate is added to pond 1, that population quickly grows again, indicating that low phosphate was limiting the growth of the population. Note that the population soon stops growing (level B) because a second nutrient, nitrate, becomes limiting. When nitrate is added, the population grows again but stops at level C, because a third element is limiting. Note that in pond 2, nitrate is the first nutrient added, resulting in no additional growth. The difference in the algal growth response between pond 1 (phosphate added) and pond 2 (nitrate added) demonstrates that phosphate was the most critical limiting nutrient.

FIGURE 45.3
Critical Limiting Resources
Many organisms grow poorly when deprived of a critical, limiting resource. Plants growing in a nitrate-deficient soil (foreground) are yellow and stunted. When nitrate is present (background), the plant shows normal growth.

water and nutrients, are taken in by that organism and directly incorporated into its body. These are called **resources**. Other factors, such as temperature and gravity, are not consumed directly but instead influence the survival of the organism. These are called **regulators**.

Whether a particular factor is a resource or regulator depends greatly on the type of organism in question (Figure 45.1). For a plant, the primary resources include moisture, nutrients, carbon dioxide, and light. Other factors like temperature, fire, grazing animals, and gravity act as regulators because they affect the plant's rate of growth and ability to reproduce without being actively consumed. In contrast, the primary resources for an animal include water, food, and oxygen. Factors that are resources to plants, like soil nutrients, light intensity, and carbon dioxide, act as regulators to animals.

Limiting Resources

Every organism requires several resources. For example, you know that your own survival depends on your ability to acquire sufficient food, water, and oxygen. If any one of those resources is unavailable, you would die—even if the other two are present in abundance.

Early ecologists developed the **law of limiting resources**, which states that the survival and growth of an individual or a population may be limited by the resource that is in shortest supply. That limiting resource, however, can change. If that resource is somehow increased, the individual or population would quickly make use of it and would grow until a different resource became limiting.

An excellent example of the law of limiting factors can be found in ponds and lakes. There you can see populations of microscopic cyanobacteria and algae that photosynthesize and provide food for heterotrophic organisms. Populations of the cyanobacteria and algae are normally limited by the lack of critical nutrients, particularly phosphate (Figure 45.2). When phosphate is added, algal populations increase dramatically until a second resource—often nitrate—becomes limiting. If nitrate is added, the population grows again until a new resource (for example, calcium) becomes limiting.

The law of limiting factors can be put to practical use in a few ways. First, it can be used to increase the yield of crop plants. In many agricultural situations, nitrate availability is often limiting. Thus fertilizing with nitrate can increase plant productivity up to the point that a second nutrient becomes limiting—which, of course, can then be added as well (Figure 45.3, and see Figure 30.16). Second, many of the environmental pollutants that people introduce into the environment are actually important resources to certain species. For example, phosphate and nitrate often seep into lakes from septic systems and over fertilized farms and gardens (see Chapter 46). While those extra nutrients might seem to cause a beneficial increase in algal populations, the effect is often really negative because dense populations of algae tend to form. When that happens, the algae in lower parts of the mat die, fall to the bottom, and decay. As a result, oxygen is consumed and fish die due to asphyxiation. Thus maintaining low levels of phosphate in a lake is an important way to keep populations of algae in check.

Tolerance Ranges and Optima

Often, having too much of a good thing can be just as harmful as not having enough. In other words, most organisms are as sensitive to excesses in most resources as they are to deficiencies. Thus the growth of most species occurs within a certain **tolerance range** for each environmental factor. When conditions lie outside that range, a species cannot survive or grows poorly.

The performance of an organism in relation to its tolerance range often produces an **optimal curve** (Figure 45.4). In this relationship, the organism performs poorly at the extreme conditions (too high or too low) and it does well under conditions that are intermediate. The *optimum* is the part of the resource axis at which the best performance occurs. The *tolerance range* is the area in which a species can survive even under less than ideal conditions.

Patterns of Responses to Environmental Variables. Ecologists know that each organism or species responds to different environmental factors in different ways. Some species are able to survive under a broad set of environmental characteristics, while others require those that are more specific and special conditions. **Generalists** are species that do well under a broad range of environmental conditions, while **specialists** have a more specific set of needs. Generalists tend to be widespread and are usually found in a number of different habitats, while specialists are more restricted.

You may ask: Is it better to be a generalist or a specialist? Actually, either can be a successful strategy. Generalists have the advantage of being able to live in a variety of different habitats, so if any one habitat becomes unsuitable, the species can persist somewhere else. Conversely, specialists tend to do better than generalists in the one part of their range in which they are most adapted (Figure 45.5). They usually outcompete most generalists that occur there and can often be locally dominant. While it is true that specialists are locally successful, they are in peril if their limited range of sites becomes unsuitable. Most of the endangered species that we hear about so often are specialists whose existence is threatened by changes in their habitat.

Environment and Scale

We can easily see that many features of the environment change from place to place and from time to time. For example, water is more abundant in the eastern United States than in the desert southwest. Likewise, a given area can be very wet or bone dry depending on whether the previous month was rainy or not. Organisms must cope with this variability in both space

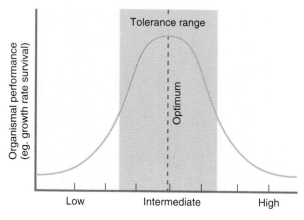

FIGURE 45.4
An Optimal Curve
Note how the performance of a species is best at an intermediate level of the environmental factor considered. Conditions at either extreme lead to poor performance. Those conditions are said to lie outside the tolerance range of the species.

and time, and each one has strategies that enable it to persist through periodic unfavorable conditions. Mobile animals have the ability to move to a different site that is not so stressful. Conversely, plants and other sessile organisms typically handle short-term stress by changing their physiology. For example, plants respond to short-term drought by closing their stomata, thus conserving water. However, such physiological acclimation is often not possible when the stress is long-term, for weeks or months. In these cases, certain organisms enter a dormancy period; others may die.

Since organisms differ in their size, it is important that ecologists accurately assess the environment in a way that is truly meaningful to the species in question. For example, a heavy rainstorm might prove to be a minor inconvenience to an elephant. However, to an earthworm that same rainstorm might pose a severe problem by flooding its small burrow and causing it to move up to the soil surface. Thus the **scale**, or relative size, of environmental variability is important and must be evaluated in terms of the organism being studied.

When evaluating the response of a species to a given set of environmental conditions, it is important to take *temporal* (time-based) fluctuations into account, especially periodic extremes. For example, we know that certain subtropical trees common to the Caribbean and southern Florida reach their northerly limit in central Florida. The most common explanation is that the trees do not survive hard winter freezes. However, people visiting northern Florida might wonder why those trees are not found there, especially if they visit during a winter in which a hard freeze does not occur. The explanation is that the trees are limited not by the *average* winter temperature but by hard freezes that occur perhaps only once every five to ten years.

(a) White-tailed deer

(b) Polar bear

FIGURE 45.5
Generalist and Specialist Species
(a) White-tailed deer is a generalist species that is equally at home in cold coniferous forests of Canada
and the warm Everglades of southern Florida. **(b)** The polar bear is a good example of a specialist,
because it is normally found only in frigid waters near the Arctic, and rarely ventures into warmer areas.

Habitat and the Ecological Niche

Habitat is easy to define (watch this). The **habitat** is where
the organism lives. The niche, on the other hand, is a bit more
conceptually complex. In a way, it's what the organism *does*
in its environment. Specifically, the **niche** (sometimes called
ecological niche) is the sum of all an organism's utilization of
its environment. For a songbird then, its habitat can be a
mixed forest in northern Florida, but its niche might involve
what it eats, where it perches, what it chases, what chases it,
and what its daily schedule is.

However, sometimes a species has great freedom in its
interactions and sometimes it does what it is forced to do. So we
have the **fundamental niche**, the species' interaction with its
environment under ideal circumstances. The fundamental niche
refers to what is possible. However, circumstances may force
the organism into certain interactions with its environment. So
we have the **realized niche**, those interactions between an
organism and the environment that actually occur. Competition,
predation, and habitat destruction may force an organism into
only certain interactions with its environment (Figure 45.6).

The Principle of Competitive Exclusion

In the first decades of this century, two biological mathemati-
cians, A. J. Lotka and V. Volterra, independently concluded
that two organisms occupying the same habitat cannot interact
with their environment in the same way indefinitely. They
concluded one would win out and the other would be driven
to local extinction.

FIGURE 45.6
Fundamental and Realized Niches
In **(a)**, we see the fundamental niche of a species with
respect to two environmental variables. Without
competitors, the species can live in a wide variety of
conditions (its fundamental niche). With competitors
(b), the range of conditions within which a species can
live is reduced, creating a realized niche. The realized
niche is normally what we see in nature.

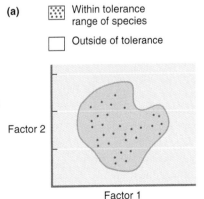

(a)

Within tolerance range of species

Outside of tolerance

Factor 2

Factor 1

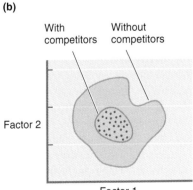

(b)

With competitors

Without competitors

Factor 2

Factor 1

(a) Caddis fly

(b) Bloodworm

FIGURE 45.7
Indicator Species
Invertebrates serve as excellent indicators of water quality in streams and rivers. **(a)** Caddis flies and mayflies in a stream are indicative of clean water. **(b)** The presence of bloodworms indicates a polluted stream.

Their idea was tested in 1934 by the Russian ecologist G. F. Gause. His experiment was exceedingly simple, as many of the best experiments are. He grew two cultures of *Paramecium, P. aurelia* and *P. caudatum,* in separate dishes in the laboratory. He fed them bacteria every day and noted that each population grew to its limits. Then he placed them together in the same dish, continued to give them plenty to eat, and watched as they attempted to utilize the same resources. The *P. aurelia* thrived as *P. caudatum* died out. Gause concluded that two species requiring the same resources cannot coexist in the same place, a concept that came to be known as the **principle of competitive exclusion**.

Indicator Species

Species that are specialists can give us important information about the prevailing ecological conditions at a site in which they occur. Thus scientists can tell a lot about a site merely by finding certain species there. **Indicator species** are those species that are used to identify environmental conditions at a specific place.

There are many examples of indicator species, some of which are painfully obvious (for example, the presence of living fish generally indicates that an area is inundated with water). The quality of fresh water is often determined by the types of invertebrates present (Figure 45.7). Many plants, like cattails, water lilies, and cranberries indicate that an area is a wetland. The sight of nesting redwinged blackbirds indicates wetland also. Oil prospectors look for the remains of certain kinds of unicellular algae to determine whether oil deposits are present.

Specialist species can also be valuable because they are sensitive to changes in environmental conditions; their inabil-

ity to survive can be a useful indicator. For years, people have used sensitive organisms to determine whether conditions are potentially unsuitable for human health. Perhaps the best known example is the use of canaries in mines to assess the quality of the air. If the canary died, it was taken as a suggestion to get out. Over the past few decades, scientists have developed and widely implemented the concept of a **bioassay**, which involves exposing organisms to a suspected pollutant and determining their ability to survive.

Habitat Types

Ecologists commonly classify the wide diversity of conditions in which organisms live into three main habitat types: terrestrial, freshwater, and saltwater. **Terrestrial habitats** are those on land. They include such places as forests, meadows, shrublands, deserts, and rock outcrops. Organisms in these habitats are adapted for living on the soil surface, within the soil, or among the plants that are rooted in the soil. We've seen again and again that a major challenge that many terrestrial organisms must meet is the retention of precious water. Each organism must prevent surface evaporation and excretory loss as much as possible. Without the buffering effects of water, terrestrial habitats can also experience tremendous variability in temperature over time, and from place to place. Despite those apparent drawbacks, terrestrial sites typically teem with life because they can be rich in nutrients, have high oxygen concentrations, and are often bathed in light—which provides the energy for most biological systems.

Freshwater habitats, watery environments with little dissolved salts, are found interspersed among terrestrial ones and are represented by lakes, ponds, creeks, and rivers. Organisms in these habitats don't have to worry about getting enough water; actually, they must cope with an excess. Moreover, water temperature changes slowly, so the environmental temperature is usually more constant. Instead, freshwater organisms have to confront a different set of environmental challenges, including the lack of critical nutrients, low oxygen concentrations, and light that is often limiting.

Saltwater habitats are those with high levels of dissolved salts and include the oceans and scattered salt lakes. They are the most common of all three habitat types, largely because about 75% of the earth's surface is covered by salt water. Inhabitants of saltwater habitats have the luxury of living in conditions that have relatively constant temperature and chemical composition. Those organisms do not have severe problems with water availability, as long as they are able to avoid losing water osmotically to their salty surroundings. Instead, ocean-dwelling creatures must often live in an environment in which dissolved inorganic nutrients are scarce, and light can be virtually absent.

Important Environmental Factors

For years, ecologists have focused on individual environmental factors. For example, we might be interested in understanding how the fish living in a given lake would respond to an increase or decrease in the temperature of the water. However, we have come to appreciate that several environmental factors often interact, and that focusing on a single factor might give only a partial picture (for example, raising the temperature of water causes it to hold less oxygen). Still, ecologists commonly accept that a few environmental factors are of particular importance and deserve special examination. Thus we will examine four of those factors: temperature, water, light, and soil.

Temperature

Organisms can often survive under a range of temperatures, but they do best under a certain part of that range, the **optimal temperature range** (Figure 45.8).

In view of this principle, all parts of the earth do not offer an equally accommodating thermal environment for any given organism or species. As we will see in Chapter 47, some parts of the earth—especially those near the poles and at the tops of tall mountains—are too cold for most species. Conversely, areas near the tropics and in our large deserts can become exceedingly hot. Temperatures also change over time—from season to season, and even between day and night. Thus organisms find their optimal temperatures only at certain places and at certain times. Such fluctuations are more pronounced on land than in open water habitats because, as mentioned, water tends to buffer temperature changes.

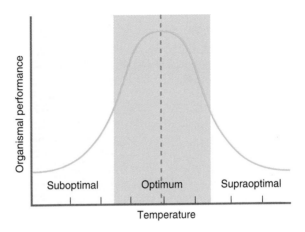

FIGURE 45.8
Optimal Response to Temperature
Organisms exhibit an optimal response to temperature, showing better performance at an intermediate temperature than either extreme. Two interacting factors are responsible for this optimal response to temperature. First, organisms may suffer at low temperatures because the rates of all chemical reactions, including those involved in vital metabolism, slow down as temperatures decline. Thus important processes like protein synthesis, diffusion of molecules through the cell membrane, and electron transport occur at a lower rate in low temperatures. Second, high temperatures cause other stresses, particularly the denaturation of proteins. One important result is that many enzymes lose their ability to catalyze reactions, and thus the organism suffers from poor metabolism.

Water

The fact that water is an important ecological factor should come as no surprise to anyone, considering the fact that our bodies are mostly composed of water and we certainly can't go long without it. We all know that the excessive loss of water leads to dehydration, which in most organisms causes death. Indeed plant ecologists now generally accept that most newly germinated seedlings die due to their inability to absorb enough water. However, most organisms begin their lives surrounded by water, either that found naturally or that carried within the mother.

As you might imagine, the effect of water on the growth and survival of organisms depends largely on the type of habitat in which they are found. The biggest difference is found in aquatic versus terrestrial habitats. As will be discussed in Chapter 47, organisms living in aquatic habitats must be able to move around and feed within a medium that has greater viscosity than air. On one hand, some water-dwelling creatures can swim and thus move about their habitat in a three-dimensional manner. However, water's high viscosity can also impede movement, so many mobile sea creatures have evolved a sleek, torpedo-shaped body form, and many produce a slimy, friction-reducing mucous covering.

FIGURE 45.9
Soils

Soil, the highly weathered, outer surface of the earth's crust, is of utmost importance to the establishment of a healthy ecosystem. Soils typically have *horizons* or horizontal layers. The uppermost region, the *litter layer,* is mainly dead material such as leaves, twigs, grasses, and the remains of other organisms. It is inhabited by soil microorganisms that begin the decomposition of the organic matter. Below this is the *humus,* which is partly decomposed organic matter and partly mineral soil. Below the humus lies a *subsoil* region that consists chiefly of minerals, similar in content to the rock below. There is little organic material, but the region is often penetrated by tree roots that absorb the minerals with soil water. The two final levels include *rock* in some state of weathering, brought about by water percolating from above, and the unweathered *bedrock* itself.

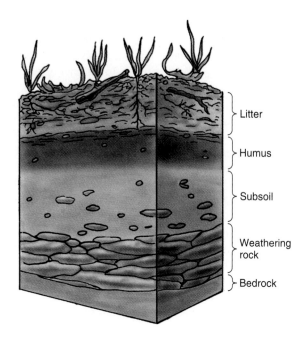

- Litter
- Humus
- Subsoil
- Weathering rock
- Bedrock

In terrestrial habitats, organisms must support themselves in a less buoyant medium. Thus terrestrial animals often have strong skeletons while many terrestrial plants have woody stems or lie low to the ground. The lower buoyancy of air also restricts most terrestrial organisms to the earth's surface, where they move around in a world that is essentially two-dimensional. Of course, birds, bats, and many insects are capable of flight. However, the great flexibility of movement comes at the cost of anatomical restrictions that force flying creatures to remain light. On the other hand, air offers much less resistance to movement and, as a result, the fastest organisms on earth are land dwellers.

Light

Light is defined as electromagnetic radiation between the wavelengths of 400 and 700 nm. Most light of biological significance is derived from the sun. Light is an extremely important environmental factor, because it acts as a regulator to almost all life and as a resource to photoautotrophic organisms like plants, algae, and cyanobacteria.

Light has four components of importance to organisms: intensity, directionality, spectral quality, and duration. **Intensity** describes brightness, and the level of intensity varies from place to place. Light intensity profoundly affects almost all organisms in nature. Many kinds of plants, including crops and weeds, require high intensities to survive, while other plants like the redwood sorel require less light and actually suffer when exposed to bright light. Animals also respond to light intensities. Squirrels do well in bright light, while bats thrive in dark or near-dark conditions.

Directionality describes the location of the light source. We know that the position of the sun changes continuously during the course of the day and throughout the year. Many plants, like sunflowers, are able to orient their leaves to track those changes. Plants in locations that are partially shaded bend toward the light thanks to a process called phototropism (see Chapter 31). Many soil invertebrates seek to move away from light and thus respond to light directionality behaviorally.

Spectral quality is a measure of the color of light. The shortest wavelengths of visible light, about 400 nm, are blue. The longest, about 670 nm, are red. Some organisms can see wavelengths that are beyond our perception. For example, insects and fish can see ultraviolet light, which has wavelengths of about 350 nm. Sunlight is a mixture of all the visible wavelengths, and we perceive that mix as white light. Light that passes through a canopy of leaves is depleted in the red and blue bands because chlorophyll preferentially absorbs those colors. As light passes into water, it becomes depleted in red and blue light, and therefore deep water is richest in green light.

The final biologically significant attribute of light is its duration (how long light lasts). Biologists use the term *photoperiod* to refer to the length of time during the day a given spot receives light. Organisms use photoperiod as a cue to control the timing of certain key life cycle events during the year. Animals link a variety of responses (at least in part) to photoperiod, including metamorphosis in insects, spawning in certain fish such as brook trout, and seasonal migratory behavior of some birds. Plants often start to flower when the photoperiod becomes greater than, or shorter than, a certain threshold (see Chapter 31).

Soil

Soil is usually defined as a porous mixture of finely pulverized rock, mixed with organic matter, that occupies the uppermost surface of the earth's crust (Figure 45.9). Even though most people call it "dirt" and don't think about it much, biologists recognize that soil is crucial for terrestrial plants because it provides support, water, and nutrients. Thus terrestrial areas lacking a good soil are typically barren and foster conditions that are hostile to plants and animals alike.

Most places have *mineral soils,* which are defined as soils whose mass is composed mainly of tiny rock fragments. Those fragments are generally classified as being coarse versus fine, depending on whether they are larger or smaller than 2 mm in diameter. Soils that have many coarse fragments are termed gravelly, stony, or rocky, depending on the size of the fragments. Such fragments can influence the growth of plants by impeding root penetration. Soil-dwelling animals are also often impaired by coarse fragments, though very large fragments can provide habitat for woodchucks and snakes. Fine fragments, in turn, are classified as being *sand* (0.02–2 mm wide), *silt* (0.002–0.02 mm wide), and *clay* (0.002 mm wide) depending on their size. *Soil texture* is the relative proportion of sand, silt, and clay in a soil; soils vary considerably in their textural properties. Some soils have a sandy texture; these drain very quickly and are often infertile. Soils that have a clay texture are often poorly drained and create wet zones, which inhibit the growth of oxygen-requiring creatures. The most ideal soil has a *loamy* texture, meaning that it has a good mix of all three particle size classes and has appropriate fertility and water-holding characteristics.

A second important component of most soils is the organic matter—the living or dead bodies of plants, animals, and microbes. Organic matter is important because it enhances both the fertility and water-holding capacity of the soil. Most mineral soils have a litter layer in which we find a surface organic layer that may reach a few inches thick covering a decomposing *humus* layer (see Figure 45.9). The uppermost part of the mineral soil itself also has considerable organic matter mixed among the inorganic rock fragments. Some of the organic matter is alive, consisting of plant roots, tiny arthropods, earthworms, roundworms, fungi, protozoans, and bacteria. These often overlooked organisms carry out the task of decomposing dead tissue and recycling the nutrients that it contains. Interestingly, the decomposition of organic matter proceeds slowly in wet areas like marshes and ponds. As a result, **peat** (poorly drained organic matter) may accumulate in such places. Peat has commercial value as fuel and potting soil.

Soil is a valuable resource that often takes centuries to form. You wouldn't know this from a look at the history of its use by humans. Earlier this century, soils were grossly misused through poor agricultural and construction practices. Thanks to our improved knowledge, coupled with federal, state, and local laws, those seeking to alter the soil must follow guidelines to reduce depletion and erosion.

The Ecology of Populations

Not too long ago, wolves roamed abundantly throughout the United States. Their mournful howls, their tracks left in the snow, and the remains of their kills were an integral part of the North American wilderness experience. But now, wolves no longer exist in many places where they were once common. In places where wolves do survive, they are rare at best. Today, only a few well-publicized bands precariously persist in scattered areas. What has happened to them? We hear reports that other seemingly abundant organisms like frogs, bluebirds, codfish, and the medicinally prized ginseng plant are also becoming scarce. At the same time, we can look around and see that many other kinds of organisms seem to be doing well. There's certainly no shortage of dandelions, houseflies, and pigeons. What is the key to their apparent success?

On one hand, we could look at the physical conditions that are available to each species. Perhaps some component of the environment is changing to the detriment of some of the species and to the benefit of others. For example, many birds that inhabit mature forests might be suffering because of the widespread loss of forested habitat, resulting in an increase in light intensity. At the same time, species preferring high-light conditions move in and thrive.

In many cases, we can make general observations of a species' commonness and rarity and correctly explain changes in terms of change in the physical features of the environment. In other cases, we need to focus on another aspect of the organism's biology, namely, its population biology.

The concept of a population is important to biologists, especially ecologists and geneticists. Simply put, a **population** is a group of individuals of the same species that share the same habitat. Populations are often easy to identify (Figure 45.10).

While that definition fits well with our everyday observations of organisms in nature, not all populations are so easy to identify. Imagine, for example, that you are driving along an interstate highway in September, seeing mile after mile of goldenrods along the side of the road. Would they all comprise the same population? Or what about all the swordfish scattered throughout the North Atlantic? In most instances, we would not consider such widely scattered individuals to be part of a single population. Instead, biologists add an additional condition: the individuals in a population *must interact with each other to the point of being able to interbreed.* You might guess from that definition that the size and degree of motility of a species affect the nature of its populations. For example, all the robins in a particular city park would probably represent a population, while all the carpenter ants in that same park would not belong to the same population because those at one end of the park would hardly have the opportunity to interact with those at the other end. (To be sure, however, the term "population" is often used to describe the number of individuals over a wide area; for example, "the *population* of northern right whales in the Atlantic Ocean stands at about 300 individuals.")

(a) Mackerel

(b) Geese

FIGURE 45.10
Examples of Populations
A school of mackerel (**a**) and a flock of geese (**b**), comprise two populations.

Thus the techniques that are used to study populations differ from those used to study individuals, and such techniques often emphasize statistics and mathematical modeling. Population ecologists are interested in determining the number of individuals in a population, understanding whether that population is increasing or decreasing, identifying the factors responsible for any observed population change, and characterizing the dispersal, longevity, and reproductive aspects of a given population.

Population ecology is crucial to understanding many important ecological and evolutionary phenomena. Ecologists can use information from population ecology to predict the success of a given species or assemblage of species. It is especially indispensable for wildlife managers and for those seeking to understand why species numbers are dwindling (like the frogs, bluebirds, codfish, and ginseng plants mentioned earlier). Concepts of population ecology are also often applied to our own species, such that epidemiologists seek to understand how disease affects human populations and insurance professionals regulate the premiums that they charge based on the expected longevity of the people that they serve. Indeed many people care deeply about the size of the human population, and the question of whether our population can continue to grow is hotly debated by politicians, religious leaders, economists, and environmentalists.

To geneticists and evolutionary biologists, populations are also of central importance. As you recall, both Gregor Mendel and Thomas Hunt Morgan examined the passage of traits in populations of peas and fruitflies, respectively. Evolution is often seen as operating at the level of the population, with changes in allelic frequencies being calculated on a population basis.

Though populations have been well studied over the past century, population biology remains an active area of research, and many additional questions remain to be answered. Population statistics integrate a variety of environmental, evolutionary, and physiological effects, most of which are not very well understood. Therefore many questions about populations are difficult to answer. Many researchers, at this very moment, are watching populations of tiny algae in little glass tanks, attempting to simplify and control conditions. Others are roaming around in fields or spending long hours watching holes in the ground, waiting for some small face to appear and be counted. A great many people, in fact, are asking very basic questions about how populations change. So our goals in the remainder of this chapter must be modest: to describe the basic concepts of populations and to identify a few of the ecological and evolutionary factors that can cause populations to change.

Dispersion Patterns

One attribute of populations that you have no doubt observed in nature is their **dispersion**, or the way in which individuals are distributed in a given area. Often, individuals will be clustered together, while in other cases they will be more spread out. Typically, biologists refer to three types of dispersion: *clustered* (aggregated), *regular* (evenly spaced), and *random* (irregularly spaced) (Figure 45.11).

Populations showing a clustered pattern (Figure 45.11a) are common in nature and are found among many different types of organisms. Clustered dispersion patterns are often due to environmental patchiness (heterogeneity). For example, cattails that prefer wet areas are often clustered in low places in the environment. Clustered patterns can also arise

(a) Clustered organisms **(b)** Regularly distributed organisms **(c)** Randomly distributed organisms

FIGURE 45.11
Dispersion Patterns
Populations show three different dispersion patterns in an area. **(a)** A clustered pattern occurs when individuals in a population are grouped together into dense bunches or aggregations, and there are sparse unpopulated areas in between. **(b)** A regular pattern occurs when individuals are evenly spaced in an area. **(c)** A random dispersion occurs whenever individuals are arrayed without any regard to the presence of an adjacent individual. Thus we find small dense clusters and sparse pockets of isolated individuals by chance.

due to limited dispersal from a parent. Strawberries and aspen, plants that reproduce vegetatively, are often found in clusters because the underground stems that produce the offspring can only be a limited length (a strawberry can't produce a 50-m long runner!). Being in a cluster also has its advantages (Chapter 44). Specifically, an individual animal in a flock, school, or herd can find a mate easier than if it's isolated. Also, an individual may run a lower risk of predation if it has others around to confuse a predator. Finally, animals in a group can regulate the environment better than if they are isolated (for example, small mammals huddled together can create a pocket of warmth in an otherwise cold area).

Regular dispersion patterns (Figure 45.11b) are relatively rare in nature and occur when a resource is scarce, and each individual is attempting to sequester that resource for itself. Thus individuals are rarely in close proximity to each other but will have some space in between. A good example of regular spacing occurs in animals that exhibit *territoriality,* a phenomenon in which animals establish an area for themselves and fight off all other individuals seeking to invade that area. Regular dispersion patterns can also be observed in plants, especially shrubs growing in the desert where water is scarce. Of course, thanks to planting practices, crop plants and trees in an orchard are almost always found in a regular pattern.

Random patterns can be found in a variety of organisms; think of trout in a lake, earthworms in a grassland soil, or maple trees in a forest (Figure 45.11c).

Population Changes in Time

You have probably walked through a grassy field in late summer and noticed grasshoppers scattering as you made your way. Then, as you walked through that same field again in late fall, the grasshoppers were no longer there. If so, you witnessed one of the most important attributes of any population: that its numbers typically change over time. Such changes are common and predictable in seasonal climates and can be observed in most populations.

Other sorts of population changes are less predictable and more puzzling. As some species dwindle, we hear of others exploding in the form of "plagues" or "invasions." In order to understand why a given species increases or decreases its numbers, we must first understand something about how populations generally behave. What determines the numbers of a population? What evolutionary and ecological factors are at work? What are the characteristics of population growth? Finally, what causes populations to fluctuate, and what sort of stabilizing influences might be at work?

The Difference Equation

For plants, mushrooms, algae, barnacles, and other sessile organisms, changes in numbers are due primarily to two important life cycle events: birth and death. Birth and death also bring about population change in birds, fish, mammals, insects, and other motile animals, but here immigration and emigration are also contributing factors. We can express that relationship in more mathematical terms according to the **difference equation**, which is given as

$$\frac{\Delta N}{\Delta t} = B + I - D - E$$

where $\Delta N/\Delta t$ = the change in numbers of individuals in a population over time, B = the birth rate (number of new births in that time interval), I = the immigration rate (the number of new individuals dispersing onto the site during that time interval), D = the death rate, and E = the emigration rate.

For example, a population of lobsters might experience 1000 births, 200 new immigrants, 700 deaths, and 100 emigrants in the course of a year. If so, the change in population size ($\Delta N/\Delta t$) would be +400 individuals per year (1000 + 200 − 700 − 100 = 400).

You should notice that $\Delta N/\Delta t$ can be a positive number, meaning the population is growing. That occurs when the birth and immigration rates exceed death and emigration. In declining populations, $\Delta N/\Delta t$ is a negative number, and that occurs when births and immigration are collectively less than deaths and emigration. Finally, when gains balance losses, $\Delta N/\Delta t$ can be zero, which means that the population is steady.

Intrinsic Rate of Population Growth

Regardless of which organisms they are studying, biologists invariably notice that every species has one property common to all other species: the ability to produce more than two offspring in its life. Thus if we look at the life span of a given individual (or pair of parents), the number of births always has the potential to be greater than the number of deaths. In other words, populations of all species have the capacity to grow. That property is of crucial importance to the success of any species, as it provides the impetus for it to flourish and potentially invade new habitats. Notice, however, that we are not saying that all species *will* increase their numbers under all circumstances, but instead they *can,* given appropriate conditions.

With this in mind, we can restate the difference equation presented above, but this time ignore immigration and emigration:

$$\frac{\Delta N}{\Delta t} = bN - dN$$

where b = the birth rate expressed in terms of the *average* number of new offspring produced *by each individual in the population* during a given time interval, and d = the average death rate during that time interval.

For example, over the course of a year, in a population of 1000 buffalo, we might see 50 new calves being born and 30 individuals dying. In that case, the birth rate (b) would be 50/1000 = 0.050 and the death rate (d) would be 30/1000 = 0.030. Thus

$$\frac{\Delta N}{\Delta t} = (0.050 \times 1000) - (0.030 \times 1000) =$$
$$50 - 30 = +20 \text{ individuals/year}$$

Often, b and d are considered to be constants and are the properties of a population of a given species, regardless of the number of individuals in that population (we will see later cases in which that assumption does not hold true, however).

Assuming for the moment that the difference $b - d$ is a constant, we can restate the previous equation as

$$\frac{\Delta N}{\Delta t} = (b - d)N$$

Thus for our buffalo population, where b = 0.05 births of individuals/year and d = 0.03 deaths of individuals/year, $\Delta N/\Delta t$ will be +2 individuals/year when there are 100 individuals in the population, +200 when N = 10,000, and +20,000 when N = 1,000,000. (What would be the value of $\Delta N/\Delta t$ for a population of 5000 buffalo?)

We can take the previous equation one step further by replacing $(b - d)$ with r, which is defined as the **intrinsic rate of growth** (the rate of population growth *per individual*):

$$\frac{\Delta N}{\Delta t} = rN$$

Note that r is a constant, and it represents growth rate of the population, for a specified period of time, on an individual basis. Since $\Delta N/\Delta t$ has the units individuals/time, r is expressed in units of time^{-1}.

Many ecologists like to distinguish between two forms of r. The first, r_{max}, is the maximum intrinsic rate of increase for a population that is under ideal conditions (lots of food and space, no predators or disease). The second, r_{obs}, is the observed intrinsic rate of increase for a population under a given set of actual conditions. Unfortunately, the two forms of r are often confused, leading to misinterpretations of ecological data.

Values of r_{max} vary widely from species to species, depending heavily on the reproductive abilities of the species in question. Species that reproduce early in their life, reproduce many times, and produce many offspring each time have higher r_{max} values than those that delay reproduction, reproduce only once, or produce only a few offspring. Thus bacteria, tapeworms, starfish, and maple trees have a high r_{max}, while elephants, hippopotamuses, and coconut trees have much lower r_{max} values. In general, larger species with long life spans have lower r_{max} values than smaller species with short life spans (Figure 45.12).

Population Growth Models

With this in mind, we are now ready to discuss models of population growth. Models are simplified or abstract versions of some larger, more complex entity and are used to make it easier to evaluate complex systems. Engineers make models of bridges and cars before constructing the full-blown versions, so that their behavior under various conditions can be understood. Biomedical researchers use animal models to test new medicines before they are administered to humans. The ideal gas laws normally covered in introductory chemistry courses are models. In fact, the difference equation discussed earlier is a simple model of population behavior.

Population biologists have created many kinds of mathematical models to explain and predict the behavior of populations under a variety of conditions. Some of the models are very simple, while others are very complex, involving many factors, differential calculus, and symbols taken from languages that few people speak (you might want to peruse some articles from the journal *American Naturalist* for examples of the latter). In general, simple models are easy to use and can be applied to many different types of organisms. Their predictive power is low, however. Conversely, complex models are harder to use, but they often have a high predictive power.

In this chapter, we will review two models that are commonly used in population biology: the exponential model and the logistic model. Both describe population growth, but as you will see one is more realistic than the other. You should also realize that they have broad implications that equally apply to human populations. Understanding them requires no more than a basic knowledge of mathematics.

Exponential Growth

One of the most basic models of population biology is the exponential growth equation, which is merely a slightly modified version of the last equation given above:

$$\frac{\Delta N}{\Delta t} = r_{max} N$$

This equation states that, in a growing population, the rate of change in population size is determined by the maximal intrinsic rate of increase (r_{max}) multiplied by the number of individuals in that population (N). Because r_{max} is always a positive number, new individuals will invariably be added to the population and N will grow with each new generation of offspring. More importantly, the population growth rate ($\Delta N/\Delta t$) will increase in each generation, leading to an ever-increasing rate of population growth.

Imagine that we are studying a bacterium that divides every half hour. If we begin with one individual, we get two after 30 minutes and four at the end of the hour. Thus population growth at the beginning is slow. By the end of the second hour we have 16 individuals, and 64 after three hours; thus things are starting to pick up. However, by ten hours we get one million cells, by 15 hours it's up to one billion, and by the end of one day we have 281 trillion!

If we plot such an increase on a graph where *N* is a function of *t*, we would see at first a gently increasing slope. But soon, as the numbers continue to double, the curve arches sharply upward until it seems to approach (but never reaches) the vertical (Figure 45.13). This type of population rise is called an **exponential increase**, and its growth curve has a J-shaped appearance, logically enough, called the **J-shaped curve.** Exponential increases are unlike linear or arithmetic increases, that progress in a straight line from 1 to 2 to 3, and so on.

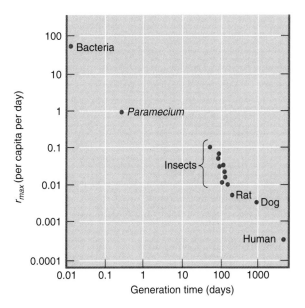

FIGURE 45.12
**Generation Time and r_{max}
(Maximum Population Growth Rate)**
Small organisms that mature quickly tend to have short generation times and high r_{max}. Larger organisms, which take longer to reach sexual maturity, usually have long generation times and low r_{max}. Note that here r_{max} is presented per capita per day, even for large animals.

Of course, the exact shape of the curve depends on r_{max}. Species with a low r_{max} would have a wider curve than those with a high r_{max}. Regardless of the species' intrinsic rate of increase, all species have the capacity to grow exponentially under ideal conditions—at least in theory—and overrun the earth.

Environmental Resistance and Logistic Growth

You may have noticed that the earth's surface is not covered by any single species. This is partly due to the fact that exponential growth never really continues for a prolonged period of time. Instead, something must act to limit further population increases when numbers get too high. The result is that population growth slows down to the point where the number of individuals stabilizes at some value or may even decrease over time.

In nature, population increases are inhibited by **environmental resistance**, which is that set of conditions in the environment that slow population growth. Components of environmental resistance often include the lack of food or space, as well as the accumulation of toxins and prevalence of disease often found in high-density conditions.

If populations cannot continue to grow indefinitely, you are probably wondering about the usefulness of the exponential growth model. As it turns out, early biologists recognized

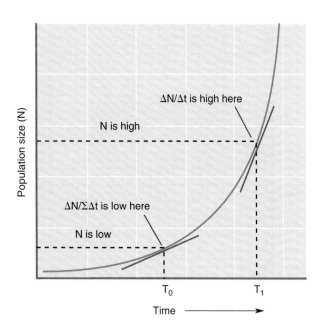

FIGURE 45.13
Exponential Population Growth
The exponential growth model predicts unlimited population increase under ideal conditions of unlimited resources. We can relate this curve to the exponential model equation, because the slope of the line is equal to $\Delta N/\Delta t$. Thus when there are few individuals in the population (N is low), $\Delta N/\Delta t$ is low, and the slope is essentially horizontal. When the number of individuals increases, $\Delta N/\Delta t$ increases, and the slope pitches upward to a 45° angle. And when the number of individuals is high, $\Delta N/\Delta t$ is high, and the slope is nearly vertical. At this point, the population is exploding.

the limitations of that equation and altered it to create something more realistic. The main feature of the new and improved equation is that it contains an additional term called the carrying capacity. **Carrying capacity** (symbolized as K) can simply be defined as the number of individuals of a species an environment can support, based on the available resources (particularly food and space). Carrying capacity is one of the most basic and widely discussed ecological concepts by scientists and nonscientists alike. Ecologists agree that the carrying capacity is not fixed; it may change drastically from one time to another. For example, a meadow's carrying capacity for grasshoppers will decline as trees invade over time and replace the grasses. Furthermore, carrying capacity is constantly affected by many factors, both *biotic* (living) and *abiotic* (non-living), on which we will elaborate later.

If the resources used by a given species are renewable, populations of that species will—in theory—stabilize at the carrying capacity. How then would we modify our exponential equation to show populations stabilizing at the carrying capacity? Biologists have done this by creating the **logistic growth equation** which is given as

$$\frac{\Delta N}{\Delta t} = r_{max} N \frac{(K - N)}{K}$$

This equation is the same as the exponential except for the inclusion of the part $(K - N)/K$, which serves as a modifier. Note that when the value of N is small compared to K (population densities are low), the numerator and denominator are essentially the same and the value of $(K - N)/K$ approximates 1. Thus the equation reverts to the exponential. As a result, $\Delta N/\Delta t$ is low because N is low. Conversely, when population densities are at the carrying capacity (when $N = K$), the expression $(K - N/K)$ becomes zero. As a result $\Delta N/\Delta t$ becomes zero because the term $r_{max} N$ is now multiplied by zero. Thus population growth ceases. The only time at which any appreciable population growth occurs is when N is about half of K. At that point, the expression $(K - N/K)$ is about 0.5. However, because N is now a respectable number, the $r_{max} N$ part of the equation is fairly high. Population growth rate is, in fact, often highest when N is at one-half K. That fact is important to people like fishermen and ranchers who harvest organisms and try to maximize population growth rates.

Just as exponential growth has a characteristic J-shaped curve, logistic growth has a characteristic **sigmoidal**, or **S-shaped**, curve (Figure 45.14). You will notice that, at low population densities, the logistic curve and the exponential curve are nearly identical. As time passes and N increases, the two diverge such that the sigmoidal lags behind the exponential. Ultimately, the sigmoidal curve stabilizes at the carrying capacity.

Biologically, logistic population growth should be viewed in terms of available resources. During the lag and log phases (see Figure 45.14) populations enjoy abundant resources. The birth rate is high and the death rate is low. During the saturation phase, however, resources become scarce. The birth rate generally begins to decrease while the death rate increases. Finally, the two reach equality and $r = 0$, completing the S-shaped curve. The population may then hover about this level.

You might wonder whether it is possible for a population to exceed its carrying capacity. The answer is yes, both in theory and in reality. According to the theory of the logistic model, whenever $N > K$, the expression $(K - N)/K$ becomes a negative number and thus $\Delta N/\Delta t$ becomes negative, meaning that the population should shrink to the carrying capacity. In reality, populations exceed their carrying capacity, especially when they overbreed at times when their N is just below K. For example, deer populations often exceed the carrying capacity in the northeastern United States, resulting in widespread starvation (unless numbers are artificially reduced by hunting).

The logistic equation predicts that species should remain at equilibrium. That is an idealized case that is hardly realized in nature. Instead, the number of individuals in a population fluctuates over time (Figure 45.15). Fluctuations may be caused by delays in the ability of a population to adjust its numbers relative to the carrying capacity. However, fluctuations can also be caused by a change in the carrying capacity, either through disturbance or by cyclic phenomena like the changing of seasons. As you will see, carrying capacity is also affected

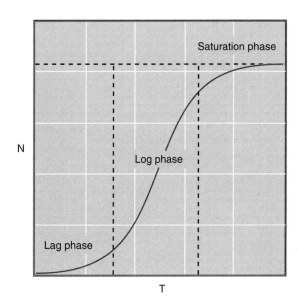

FIGURE 45.14
Logistic Population Growth
The logistic growth model predicts that populations will grow rapidly at first. However, as the number of individuals in the population (N) approaches the carrying capacity (K), the population growth rate eventually slows to zero, and the population stabilizes at K. The result is a *sigmoidal* or S-shaped curve (contrast to the J-shaped curve of exponential growth). Sigmoidal curves are often divided into three parts or phases. The first is called the *lag phase*—the period of slow growth that occurs when population numbers are low. The second is the *log phase,* which occurs when growth rate accelerates and becomes relatively rapid. The third is the *saturation phase,* during which population growth decelerates as N approaches K.

by other species in the ecosystem, so population changes in one species can alter the carrying capacity of a second.

You should realize that some species experience conditions in which resources are not renewable. In these circumstances, populations grow rapidly when conditions are abundant. However, when they deplete their resources they cannot sustain themselves, and populations crash (Figure 45.16). Population crashes are especially common to species like gypsy moths, cicadas, and microscopic algae growing in lakes. Population crashes also happen to species found in rapidly changing habitats like abandoned farm fields and fire-prone sites. Species found in such habitats must either have an excellent ability to disperse to other areas or must form dormant stages like spores, seeds, or cysts, enabling offspring to persist quietly in a habitat until favorable conditions return.

Finally, biologists now recognize that species found in low densities are often more prone to low birth rates and high death rates than would be predicted by the logistic model. In essence, low numbers pose an inherent risk to a species' continued survival in a habitat—a phenomenon called the **Allee effect**, after the scientist who first proposed it. Hazards associated with being rare include the inability to find a mate,

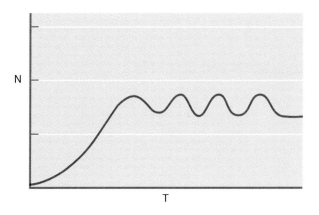

FIGURE 45.15
Carrying Capacity
Populations typically do not remain steady at the carrying capacity, but instead fluctuate.

excessive inbreeding and loss of genetic variability, effects of environmental stresses on isolated individuals (for example, from freezing and wind), and random catastrophes. All of these things may now be happening to endangered blue whales that were almost hunted to extinction. Such species are given legal protection and conservation biologists work hard to find ways to increase their numbers.

Demography

All species have a fairly well-defined **life history** that involves a beginning of life, a juvenile phase, a reproductive phase, and death. Of course, the exact nature and timing of these events vary enormously from species to species.

Biologists often want to know how many newborns in a population survive their juvenile phase. How many reproduce? How many offspring will a parent produce in its lifetime? In

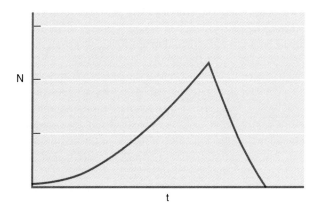

FIGURE 45.16
Population Crash
Some populations experience population crashes, especially when they deplete a critical resource.

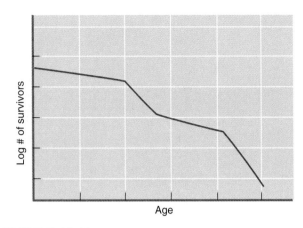

FIGURE 45.17
Survivorship Curve

A survivorship curve is created when the number of individuals from a given cohort that survive to progressively older ages is plotted. Note that the y axis is expressed on a logarithmic scale. A survivorship curve can either remain steady or decrease as you proceed from left to right on the graph. It can never go up, because new individuals cannot be added to a cohort once it is first established.

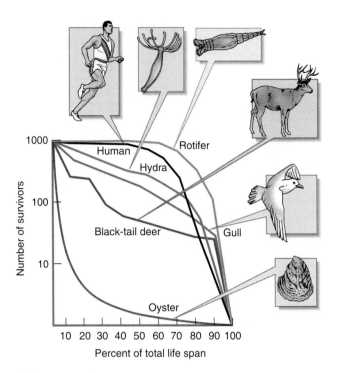

FIGURE 45.18
Species Survivorship Curves

Survivorship curves for six species, each beginning with a population of 1000. In black-tail deer, a high death rate is experienced among the young: the numbers drop off rapidly with the average individual completing only 6% of its average life span. Loss during the early period is even more dramatically seen with oysters, where great numbers of offspring are devoured by other marine animals. Those oysters that survive the juvenile period then experience low mortality. The average human and rotifer survive the rigors of early life to complete about two-thirds of their potential life spans while the average *Hydra* and gull complete about 40% and 25%, respectively.

addition to being of interest to biologists, these questions are also of enormous practical interest. For example, if you own a trout hatchery and want to stock a lake with 2000 fish, how many fingerlings (juvenile trout) should you rear to achieve the target number? How many female fish would you need to achieve that number? Conversely, if you own a one-acre meadow and want to grow a forest of 100 pine trees, how many seedlings should you plant? Could you obtain that number of seedlings from one parent tree?

To answer these questions, biologists examine two important parameters of a population: *survivorship* (how long one lives) and *fecundity* (how many offspring one leaves). (*Demographers* are biologists who study those two parameters.)

To understand survivorship, we must first consider the concept of a *cohort,* which is simply a group of individuals born at about the same time. For example, all people born in 1978 represent a single cohort. You probably progressed through elementary and secondary school with members of your cohort. Once a cohort is identified, you would then count the number of organisms in that cohort at birth or germination (of course, you could go back even further and count zygotes if you wanted to include embryological stages in your assessment). After a period of time (a week, month, or year), you would recount the number of survivors in your cohort. The difference would represent mortality. You would then repeat the process, making periodic counts, until the last individual died.

Having done that analysis, you would then have information on the number of survivors your cohort had at various ages. You would then plot that information on a graph where the x axis is age and the y axis is the number of survivors

(though for practical reasons, the latter axis is usually presented as a logarithmic scale). The points that you plot would then form a *survivorship curve,* which is a graph of the number of survivors at progressively older ages (Figure 45.17).

Survivorship analyses have been done for many decades and are common procedures in many types of biological studies. When we examine survivorship of different species, an almost infinite number of patterns can be found (Figure 45.18). However, in 1947, a scientist named Deevey defined three general categories that can be expected (Figure 45.19a). While Deevey's patterns are well recognized among most demographers, they may be modified versions of a single, generalized, survivorship pattern (Figure 45.19b).

As you might guess, survivorship curves do have considerable practical applications. For example, in the case of the trout hatchery or the property owner seeking to stock pine trees on his or her property, the survivorship pattern could guide decisions on how many fingerlings to rear or seedlings to plant.

FIGURE 45.19
Categories of Survivorship
(a) Deevey's three categories of survivorship.
Type I survivorship shows low mortality among
juveniles and young adults, but then heavy mortality among old adults. See humans and rotifers
in Figure 45.18. **Type II** survivorship shows a
constant rate of mortality through all stages of the
life span, giving a straight line that passes from
the upper left to the lower right of the graph. The
gull and black-tail deer in Figure 45.18 approximate a Type II pattern. **Type III** survivorship has
heavy mortality among juveniles and low mortality among those individuals lucky enough to survive to adulthood. Oysters have Type III survivor-

(a)
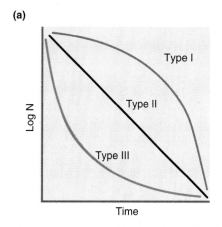

(b)
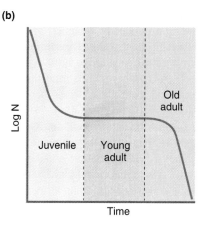

ship. **(b)** A generalized survivorship pattern consists of three stages: a juvenile phase that suffers heavy
mortality, a young-adult phase that experiences less mortality, and an old-adult phase that again suffers
heavy mortality. This general pattern makes sense biologically, because in most species (humans included)
the very young and very old are the least robust individuals and thus most prone to die off. In contrast,
young adults are stronger and have the higher survival rates.

Also, in the insurance industry, human-survivorship patterns
determine the premiums that companies charge their clients.

A second component of demography, besides survivorship, is *fecundity* or reproductive output. To determine fecundity, we count the number of offspring produced by females
of each age class and calculate an average for all members of
that age class. For example, in a hypothetical species with a
life span of four years, females aged less than two years old
might produce no offspring, two-year-olds might produce an
average of two offspring, three-year-olds an average of five
offspring, and four-year-olds an average of three offspring.
Thus you might expect a female living her full life span to
produce ten offspring. If all females survived to the full life
span, we would witness a population with a rapid growth rate.
However, survivorship analyses tell us that, in nature, not all
female newborns will survive to reproduce. Indeed many of
the two-year-old females producing two offspring might not
live to become three years old. Therefore, by examining both
survivorship and fecundity, we might determine that while a
female might produce 100 offspring in the course of her life,
only 1% of all newborns survive to reproduce. A population
having those demographic characteristics would be stationary—neither growing nor shrinking over time.

To allow us to simultaneously compare survivorship and
fecundity, demographers have developed a *life table,* which is
a type of chart that follows both parameters (Table 45.1). Each
life table consists of at least three columns of information.
The first lists the age, the second gives the survivorship, and
the third is the fecundity. You should note that, in life tables,
survivorship is expressed not as the number of survivors of the
original cohort (which can be any number) but instead as
some fraction of a constant original number—either 1.0 or
1000 (just as in baseball the success of a batter is often

expressed as a batting average, not as the total number of
hits). From survivorship we can calculate several additional
parameters such as the *mortality rate* (the proportion of individuals that die during a particular time interval), the survival
success of each age group from the previous year, and the
expected *longevity* (the amount of time individuals can expect
to remain alive, on average) for individuals in that age group.

TABLE 45.1
Life Table

A typical Life Table shows the survival and reproductive characteristics of a hypothetical population. In the example shown
here, for every 1000 newborns, 320 would live to age two and
90 would live to age four. None live beyond age five. Note also
that individuals would not start reproducing until they reach
age 3. Three-year-olds would produce three offspring each,
while four-year-olds would produce seven offspring. [As an
aside, a population having the characteristics shown here would
decline slowly—can you figure out how the life-table gives you
that kind of information?—Hint: calculate the replacement value
as described in the text.]

Age	Survivorship	Fecundity
0	1.000	0
1	0.850	0
2	0.320	0
3	0.110	3
4	0.090	7
5	0.000	0

In life tables, we commonly multiply the number of offspring produced by members of a certain age group with survivorship of that age. By summing all those products for all the ages, we then arrive at the *replacement value,* which is the average number of offspring that a newborn female can be expected to produce during her lifetime. If the replacement value is 1.0, the population is stationary. Values greater than 1.0 mean that the population is increasing, while those less than 1.0 mean that the population is declining (a replacement value of 0 means that the population is on the verge of extinction). Additional calculations enable us to arrive at the generation time, as well as annual rate of population change (r_{obs}). Thus life tables are powerful analytical tools that give us great insight into the behavior of a population and allow us to predict future trends.

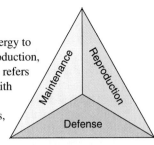

FIGURE 45.20
Energy

Each organism must allocate its energy to three functions: maintenance, reproduction, and predator defense. Maintenance refers to physiological tasks associated with remaining alive, like metabolism, homeostasis, muscular contractions, and neural impulses, as well as to general bodily growth. Reproduction refers both to direct investment in reproductive organs and nourishing offspring (costs are often larger for females than males) and behavioral activities like courtship displays and copulation itself. Predator defense includes protective structures, chemical defenses, and behavioral activities associated with escaping from predators.

The Evolution of Reproductive Strategies

The success of any species ultimately depends on one factor: the ability of individuals to produce sufficient numbers of offspring that can, in turn, also reproduce. To be successful, members of that species must battle the physical environment as well as members of other species that may be competing for the same limited resources. Biologists typically argue that, on an evolutionary scale, competing species that leave more successful offspring are "winners," while those that leave fewer offspring or reproductively unfit offspring are "losers." Thus species are in a race against each other to maximize their observed intrinsic rates of increase (r_{obs}).

People in the military are aware that warfare requires strategy, and that a variety of strategies exist that could prove successful or unsuccessful depending on the terrain, the weather, and the strategies employed by the opposition. (The same can be said for sports, dating, and earning a living.) In that context, a variety of life-history patterns have evolved as strategies for producing successful offspring. Some of the life-history events that are especially variable among species include the length of life, the length of the prereproductive stage, the number of times that an organism reproduces in its life, the *clutch size* (the number of offspring that it produces each time), and the relative size of those offspring. It is important to note, however, that even though species differ greatly in their life-history attributes, there are common life-history strategies that often tend to be successful in a given habitat. Often, organisms of very different evolutionary lineage (plants, animals, fungi) exhibit similar patterns, while species in the same genus may show sharply contrasting patterns.

The analysis of life-history strategies has been an active part of biology for the past 40 years. One reason for its popularity is that life-history components are relatively easy to study; it's not that difficult to count and weigh eggs, seeds, or newborns, or to look at survival and longevity patterns (obviously demographic analyses are important here). A second reason is that a rich body of mathematical models were developed, especially in the 1960s and 1970s, that provided many predictions as to which life-history strategies should be favored under a given set of ecological conditions. With those models in hand, field biologists took to the meadows, forests, streams, and oceans to verify or refute predictions made by "armchair ecologists."

Two important themes that recur throughout discussions of life-history attributes involve *energy* and *risk.* In terms of energy, we know that all activities of an individual organism depend on the energy that it's able to gather for itself, either autotrophically or heterotrophically. Since no organism has access to an unlimited amount of energy, an individual can't do everything. Instead, it must allocate (partition) its energy to particular structures or tasks at the expense of others. Biologists recognize three main energetic demands to which an organism can allocate its energy: maintenance, reproduction, and defense (Figure 45.20). Each individual organism is forced into an evolutionary "choice" as to whether it wants to allocate an ATP molecule to its own metabolism, to producing a new offspring, or to defense, as in avoiding some predator. It should be obvious to you that an organism must strike some kind of balance among the three. It would not be adaptive for an individual to devote all its effort to reproducing if it will be devoured by some predator beforehand. Conversely, an individual should not completely forego reproduction to survive and avoid predators, because it will then have won the battle, but lost the war (from an evolutionary perspective). Thus organisms must trade off one function for another.

How much energy should go toward enhancing reproduction versus survival versus defense? Simply, energy should be allocated to maximize fitness, that is, to maximize r_{obs}. If, in a particular environment, energy used in reproduction increases fitness more than does the same amount of energy used in maintenance or predator avoidance, then natural selection will prevail, and genotypes establishing the former pattern of allocation will replace others.

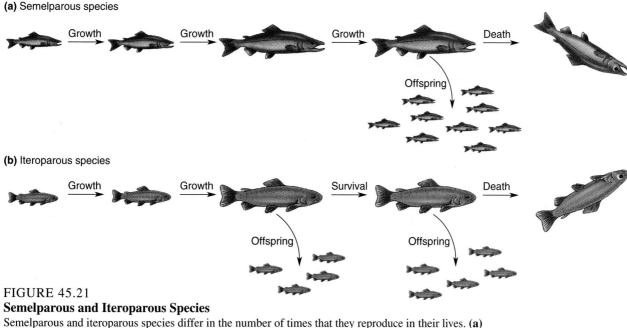

(a) Semelparous species

(b) Iteroparous species

FIGURE 45.21
Semelparous and Iteroparous Species
Semelparous and iteroparous species differ in the number of times that they reproduce in their lives. **(a)**
Semelparous species use all of their reproductive energy for a single event, after which they die. **(b)**
Iteroparous species reproduce several or many times in their lives and do not exhaust themselves in any
one reproductive event.

That answer, however, leads us to a second recurring theme in life-history studies: *risk*—the chance that conditions adverse to survival will occur. Simply put, each environment has a set of conditions that pose a threat to the continued well-being of an individual and its offspring. If an environment has abundant resources and few competitors, then an individual does not run much of a risk of losing some resource to a competing species. In theory, it would be adaptive for that species to devote its resources to reproduction. If predators are a problem, then the risk of being somebody's meal is high, and the well-adapted species is the one who devotes much of its energy to antipredator strategies.

Risk is often intermittent. Organisms that live in the northern United States and Canada are exposed to a periodic risk of freezing during the winter. That risk is fairly certain and organisms must have some way of dealing with periodic cold. Conversely, many hazards are infrequent and thus the risk is moderate. For example, animals living in the Everglades of South Florida generally enjoy freedom from severe storms, but occasionally hurricanes do pass through (as survivors of Hurricane Andrew well know). Thus, while an organism in the Everglades should not devote all its energy to counter the effects of a hurricane, natural selection would favor those who devote some energy to structures or behaviors that ensure survival through devastating storms.

Finally, you should realize that parents and their offspring often perceive the same environmental conditions differently. Specifically, tender offspring are often more prone to environmental hazards than their more robust parents. For that reason, variant individuals who do not produce offspring during stressful times would be more likely to succeed. For example, birds tend to avoid laying their eggs in the middle of the winter because there is a high risk that the newly hatched chicks will soon freeze. Conversely, if very severe conditions are imminent, through which even adults cannot survive, then those adults who produce dormant offspring are favored. Annual plants, fungi, and many invertebrates all employ that strategy. For an adult in an environment subject to unpredictable hazards, the benefit of delaying reproduction to gather more resources to ultimately produce more offspring must be balanced against the risk of being hit with that hazard before the onset of reproduction.

Organisms are faced with three primary evolutionary choices relating to their life history: the number of reproductive events, the timing of reproduction during the life span, and the number and size of each offspring during each reproductive event.

In terms of the number of reproductive events, we know that salmon, octopi, and ragweed produce only once before they die, whereas trout, lions, and goldenrod reproduce several or many times. *Semelparous* organisms reproduce only once (biologists informally call them "big bang" species). Conversely, *iteroparous* organisms reproduce recurrently (Figure 45.21).

In terms of the timing of reproduction, we know that some species start reproducing when they are rather young, whereas others delay reproduction. If a species is to produce as many offspring as possible, it might seem that those that start earliest would have an advantage over those that wait (Figure 45.22). Of course, species that delay reproduction

(a) A species that produces two offspring at the end of the first year.

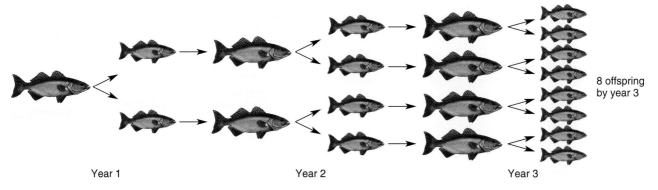

Year 1   Year 2   Year 3

8 offspring by year 3

(b) A species that waits until the end of the third year to reproduce.

Year 1   Year 2   Year 3

2 offspring by year 3

FIGURE 45.22
Species Prereproductive Period
Species that have a short prereproductive period can produce more offspring than those with a long prereproductive period. In this example, we have two semelparous species in which each individual produces two offspring before dying. **(a)** A species that matures quickly and reproduces after the end of its first year would produce eight offspring by the end of the third year. **(b)** A species that matures slowly and reproduces only after the third year would produce only two offspring.

often use that time to accumulate more resources to produce more offspring. Thus if the second species in Figure 45.22 produced ten offspring in the third year, it would have the advantage. Certainly, however, species that wait run the risk of dying before they can produce any progeny at all. Therefore, in considering which is more adaptive, the benefits of increased output must be weighed against the possibility of no output at all!

A similar argument can be made regarding the frequency of reproduction. Many iteroparous organisms, once they are mature, reproduce at regular, frequent intervals. Others delay subsequent reproductive events. Reproductive delays benefit those species that need the time to gather sufficient energy to reproduce again. Reproductive delays are also important to species that devote parental care to their newborns.

In terms of the number and size of offspring produced, we know that some species produce large clutches of many small offspring whereas others produce small clutches of a few large ones (Figure 45.23). While species that opt for large clutches (of small offspring) may seem to have a numerical advantage, larger offspring typically have more resources and better survival. A side benefit to producing small offspring though is that they are more easily dispersed. If the chance of colonizing a suitable site elsewhere is high, then an individual should produce small progeny.

Clutch size has been studied extensively in many animals, particularly birds. Birds often seem to produce one or two more eggs than can be supported by the environment. During years of normal resources, some of the chicks get enough food, while others starve. However, if resources prove especially abundant, then all the chicks survive. Using that strategy, birds can regulate their reproductive output to match an unpredictable resource base. The number of offspring attempted by some bird species may also reflect the vulnerability of the young to predators (Figure 45.24).

The Theory of r and K Selection

Beginning in the 1960s, biologists sought to develop a general theory that would predict the conditions that would favor one set of life-history traits over another. Before long, attention was focused on the effects of disturbance (disruption of habitat), resource level, and competition. Specifically, species found on uncrowded, resource-rich sites created by disturbance (like abandoned farm fields and new volcanic islands) should have different life-history attributes than species normally found in crowded, undisturbed sites. In essence, species found in uncrowded, rich sites are generally at the lag phase of logistic growth. Species in those sites should attempt to fill it up with their own offspring before a competing species does the same. Thus natural selection should favor attributes that maximize r_{max}, and for that reason, such species have been called ***r*-selected** species (some people call them *opportunistic*) (Figure 45.25a). Typical attributes associated with *r*-

(a) A species that produces a large clutch of small offspring

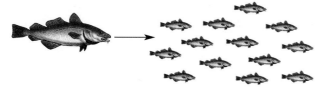

(b) A species that produces a small clutch of large offspring

FIGURE 45.23
Balancing Reproduction
Keeping total reproductive effort constant, species can trade off the clutch size with the size of the offspring: **(a)** a species that produces large clutches of small offspring and **(b)** a species that produces a small clutch of large offspring.

(a)   **(b)**

FIGURE 45.24
Clutch Size and Predation
Clutch size can be influenced by predation. **(a)** The cliff-nesting kittiwake lays two eggs each time it reproduces, which is approximately the number that the parents can feed. **(b)** The herring gull, like other ground-nesting species, commonly lay three eggs, though only two young can be fed successfully. The greater clutch size of the herring gull seems to be a compensation for the higher predation that they suffer because their nests are readily accessible to prowling foxes and other marauders. In contrast, kittiwake nests suffer little predation because they are inaccessible to mammalian predators. Thus the parents need only produce the number of young that they can feed.

selected species include semelparity (usually with a short life span), short prereproductive period, a large proportion of energy allocated to reproduction, and the production of many small offspring with good dispersal characteristics.

Conversely, **K-selected** (sometimes called *equilibrial*) species are found in crowded habitats and are normally found at their carrying capacity (Figure 45.25b). K-selected species tend to be large, to defend against predators, to be iteroparous, to be long-lived, to have a long prereproductive period, and to produce relatively few, large progeny on which they may expend considerable care (Table 45.2).

This **theory of *r* and *K* selection**, as it came to be known, generated an enormous amount of interest among ecologists in

the past 20 years. Numerous studies were undertaken; many supported the theory, but others did not. One of the problems with interpreting the theory was that *r*- and *K*-selected species really do not represent two categories, but instead represent two endpoints in a continuum. Most species occur in sites that have intermediate levels of density and disturbance. Therefore they show life-history attributes that are mixtures of *r* and *K* traits. Biologists concluded that the safest approach would be to compare closely related species that occur in different habitats.

TABLE 45.2
Some Characteristics of r And K Selection

	r-Selection	K-Selection
Climate	Variable and/or unpredictable	Fairly constant and/or predictable
Survivorship	High mortality when young; high survivorship afterwards	Either little mortality until a certain age, or constant death rates over a period of time
Intraspecific and interspecific competition	Variable, lax	Usually keen
Selection favors	Rapid development; high rate of population increase; early reproduction; small body size; single reproductive episode	Slower development; greater competitive ability; delayed reproduction; larger body size; repeated reproductive episodes
Length of life	Usually less than one year	Usually more than one year
Leads to	Productivity	Efficiency

SOURCE: Adapted from Pianka, E.R. 1970. On r- and K-selection. *American Naturalist* 104:592-97 (1970).

(a) Biting fly

FIGURE 45.25
r- and *K-*Selected Species

(a) The biting fly *Eumachro nythia* colonizes rich, uncrowded, temporary habitats and has many *r*-selected traits like short life span, high fecundity, and good dispersal. **(b)** The chimpanzee lives in an environment that is more crowded and stable. It has *K*-selected traits like a long life span, low fecundity (coupled with heavy parental investment in its young), and relatively poor dispersal.

(b) Chimp mother and baby

Population-Regulating Mechanisms

Throughout this section on population biology, we have emphasized that populations cannot increase forever. We noted that, in logistic growth, population size is controlled by the carrying capacity of the environment. What does that really mean in terms of the mechanisms for population regulation that actually operate in nature? Let's take a look.

First, though, it is important to note that the term "population control" is interpreted differently by different ecologists. To some, the term refers only to any factor in the environment that stops a population from growing indefinitely and overrunning the earth. As you will see, many factors in the environment accomplish that. To other ecologists, the term implies a more precise feedback system that keeps the density rather constant in an area.

Biologists typically categorize factors that limit population size into two groups: *density-dependent factors* and *density-independent factors*. These are both important in nature, though, as you will see, some populations are particularly prone to one of the two types.

Density-Dependent Factors

Density dependent factors are those events that bring about death and that have a greater impact as population density increases. As populations increase in density, conditions can develop that often make life more stressful for individuals in that population. Perhaps the most obvious scenario would be one in which less food is available to support the individuals in a crowded population. The result is that individuals would starve, which would increase the death rate. In addition, low food availability makes it harder for individuals to reproduce, and therefore birth rates would also decline. The effects of food shortages on animals like deer and birds are particularly well known to biologists. In plants, moderately high densities produce stunting, while very high densities lead to a phenomenon called *self-thinning,* whereby the smallest individuals die. (Actually, self-thinning occurs in different species at different population densities.)

Some density-dependent effects have little to do with food. For example, diseases can be communicated more easily from one individual to another when densities are high than when they are low. Similarly, individuals often produce toxic waste products, and these accumulate to the point of being fatal as densities increase. Animal ecologists know that high densities also produce stress that increases the likelihood of aggressive interactions (fighting) between individuals. An individual spending his/her time fighting has less time to feed and mate and may also die as a result of the encounter. Finally, individuals in high densities also attract predators more often than those in low densities.

FIGURE 45.26
Density-Independent Factor
Populations are often limited by density-independent factors. Here we see a mangrove population in the southern Everglades wiped out by Hurricane Andrew, which passed through the region in 1992.

You might notice that the agents of density-dependent population regulation are primarily **biotic** in nature, meaning that they are other living creatures, be they food source competitors, pathogens, or predators.

Density-Independent Factors

Density-independent factors are those events that bring about death but that are not related to the density of the population. These density-independent factors strike indiscriminately, killing individuals regardless of the size of the population and often decimating entire populations.

Density-independent factors are typically **abiotic** features of the environment such as floods, droughts, freezes, fires, and windstorms. In some cases, the effects are predictable and even seasonal, like the killing winter frosts familiar to those in the northern United States and Canada. In other cases, the factors are less predictable, such as forest fires and hurricanes (Figure 45.26).

Several decades ago, biologists found themselves engaged in a vigorous debate as to whether density-dependent or density-independent factors were most important in limiting natural populations. As it turned out, the answer depended on the type of organism being considered. Generally (though many exceptions exist), invertebrates like insects are under density-independent control, while vertebrates and plants seemed to be primarily under density-dependent control. Perhaps the best resolution of this debate was accomplished when the theory of r and K selection was developed. Species that are r-selected tend to be controlled by density-independent factors, while those that are K-selected are controlled by density-dependent ones.

As mentioned, many ecologists view population regulation more narrowly, as a process by which numbers are maintained within a certain range. In that way populations are like any homeostatic or negative feedback system. Just as a thermostat operates to keep temperatures within a certain range, population-regulating mechanisms operate to maintain numbers at a certain level. Typically, regulated populations are maintained by the density-dependent effects mentioned earlier. As populations rise beyond a certain level, food becomes limiting, fights break out, and disease becomes more prevalent. As densities fall, these pressures lessen.

If you see a population whose density remains constant from year to year, you might assume that the population is regulated. Unfortunately, constant numbers alone do not prove population regulation. Instead, you would have to prove regulation by an experimental manipulation of the system.

Finally, population biologists like to talk about "control from above" versus "control from below." To them, a population that is controlled from above is one whose numbers are limited by predators. A good example would be the limitation of rodent populations by weasels, raccoons, and predatory birds. Conversely, control from below refers to population control due to lack of resources. An excellent example would be the control of deer by lack of wintertime browse. Philosophers might debate whether it is better to be in a population that is controlled from below or that from above. In other words, is it better to worry about starving to death, or being somebody's bedtime snack?

Key Ideas

THE ECOLOGICAL HIERARCHY

1. **Ecology** is the study of the interrelationship of the organism and the environment.

2. The ecological hierarchy involves, from smallest to largest, **physiological ecology**, the **population**, the **community**, the **ecosystem**, the **landscape**, and the **biosphere**.

3. **Environmentalism** is the sociopolitical movement concerned with the effects of humans on nature.

4. **Applied ecology** is the solving of environmental problems through scientific means.

THE ECOLOGY OF INDIVIDUALS

1. **Resources** are environmental factors that are directly incorporated into the body. **Regulators** are factors that are not incorporated into the body.

2. **The law of limiting resources** states that the survival and growth of an individual or a population are limited by the resource that is in shortest supply.

3. **Generalists** are species that do well under a broad range of conditions. **Specialists** survive under a very narrow range.

4. **Scale**, or relative size, is important in determining how an organism will interact with its environment.

5. The **niche** is the sum of all an organism's interactions with its environment. The **fundamental niche** describes the possible interactions, while the **realized niche** describes the actual interactions.

6. **Indicator species** are those used to identify environmental conditions at a specific place.

IMPORTANT ENVIRONMENTAL FACTORS

1. Several important environmental factors on communities are temperature, water, light (including directionality and spectral quality), and soil.

2. Most mineral soils have an **organic horizon**, which describes the vertical layering of the soil.

THE ECOLOGY OF POPULATIONS

1. A population is a group of individuals of the same species that share the same habitat.

2. **Dispersion** describes the way individuals are distributed in a given area, generally referred to as clustered (aggregated), regular (evenly spaced), or random.

3. The difference in a population size during a particular interval can be described by the **difference equation**, which can yield a positive number if the population is growing and a negative number if it is decreasing.

4. The maximum intrinsic rate of increase can take two forms, r_{max}, which is the increase under ideal conditions, and r_{obs}, which is the increase under actual conditions.

5. With **exponential increase**, a population curve rises slowly at first, then rapidly. With **arithmetic increase**, it rises at a constant rate.

6. **Environmental resistance** can slow the rate of exponential increase.

7. **Carrying capacity** is the number of individuals an environment can support. A population's exponential increase will lead toward it, but the capacity will be dampened by environmental resistance.

8. Demographers study the relationship of survivorship and fecundity.

THE EVOLUTION OF REPRODUCTIVE STRATEGIES

1. Reproductive strategy is the means an individual uses to reproduce maximally and involves the allocation of time and resources.

2. The *r*-selected species are opportunistic species usually with a short life span, short prereproductive period, a large proportion of energy allocated to reproduction, and the production of many small offspring with good abilities to disperse.

3. The *K*-selected species are large species, often found in crowded habitats with high competition, a long prereproductive period, and relatively few, large offspring on which they may expend considerable care.

POPULATION-REGULATING MECHANISMS

1. **Density-dependent** factors tend to lower population numbers through mechanisms that have a greater effect as population density increases. These tend to be **biotic**.

2. **Density-independent factors** lower population numbers through mechanisms that are not affected by population density. These tend to be **abiotic**.

Application of Ideas

1. Review the definitions of ecology and niche. What do the similarities in wording suggest regarding the importance of mutual effects between life and its environment?

2. What sorts of environmental changes would tend to cause an *r*-selected species to become more *K*-selected? Make up a hypothetical case.

3. Can you think of an example where basic ecology has led to applied ecology?

Review Questions

1. Define ecology. Define it again in your own words. (page 980)

2. Define landscape. (page 981)

3. Would fighting within the same troop of baboons be an example of synecology? (page 981)

4. When an astronaut takes a space walk, is he or she living outside their tolerance range? (page 984)

5. If an organism is doing everything it can in its environment, has its fundamental niche reached its realized niche? (page 985)

6. Can you think of an example where the optimal conditions might differ from the preferred? (page 987)

7. What color is represented by the longest wavelengths of visible light? (page 988)

8. Define population. How does scale influence the definition, considering that the individuals must interact? Would clams some distance apart be more likely to be of the same population than fish half their size? (page 990)

9. Write and define the difference equation. (page 992)

10. What is the effect of environmental resistance on r_{max}? (pages 994–995)

11. What is one outcome when a population drastically overshoots the environment's carrying capacity? (page 995)

12. How can hunting influence the Allee effect? (pages 995–996)

13. Give three traits of *K*-selected species. (page 1001)

14. Name three density-dependent factors on population size. (page 1002)

CHAPTER 46

Communities, Ecosystems, and Landscapes

CHAPTER OUTLINE

ATTRIBUTES OF COMMUNITIES

Species Strata Composition

Frequency and Distribution

Diversity

Stability

**IMPORTANT COMMUNITY-LEVEL
PROCESSES**

The Roles of Competition and Niche in
Community Structure

The Role of Territoriality in Community Structure

The Roles of Disease and Parasitism in
Community Structure

The Role of Predation in Community Structure

The Role of Disturbance in Community Structure

**COMMUNITY DEVELOPMENT OVER TIME:
ECOLOGICAL SUCCESSION**

Primary Succession

Secondary Succession

Succession in Aquatic Communities

Three Alternative Models of Succession

Mature Communities

COMMUNITIES TO ECOSYSTEMS

Components of Ecosystem: Trophic Levels

Food Chains and Food Webs

Ecological Pyramids

Humans and Trophic Levels

Energy and Productivity

NUTRIENT CYCLING IN ECOSYSTEMS

The Nitrogen Cycle

From Producer to Consumer to Decomposer

The Nitrogen-Fixers

The Phosphorus and Calcium Cycles

The Carbon Cycle

KEY IDEAS

APPLICATIONS OF IDEAS

REVIEW QUESTIONS

*D*rive down any road connecting two cities, and chances are that you will see patches or vast expanses of "wild" plants off to either side. If you stop at a rest area and wander into the adjacent forest or meadow, you would undoubtedly notice that the plants there represent a mix of different species. If you spend a little more time, you would notice a sparrow here, a crow there, and perhaps a hawk soaring overhead. Poke around among the uppermost few inches of the soil and you might find several different kinds of ants, some earthworms, and a cricket or two. A very close look with a hand-lens would reveal mites, fungi, and roundworms. (If anyone asks, tell them you lost your keys.)

The point is that, in nature, individual species do not normally live by themselves as isolated populations. Instead, different species typically occur together, and the effects that they have on each other often greatly influence their ability to survive and reproduce. As you learned in the previous chapter, ecologists are fascinated by the interrelationships of coexisting species and have defined such assemblages as being small-scale communities and ecosystems, medium-scale landscapes, and broad-scale biomes. The branch of ecology dealing with such interspecific interactions is called **synecology**, and that topic will be the focus of the next two chapters. In this chapter we will look at smaller-scale assemblages (communities and ecosystems), while in Chapter 47 we will examine large-scale assemblages: biomes and the biosphere.

(a) Open Community

FIGURE 46.1
Open and Closed Communities
(a) An open community. Note the herb-dominated meadow of the lower slope (representing one community) that grades into a forest of conifers (representing a second community) near the summit. Various animal species including deer and rabbits freely migrate between the two communities. **(b)** A closed community bordering Puget Sound. Note that the spruce forest ends abruptly where the salt water begins. Very few species pass from one community to the other.

(b) Closed Community

Attributes of Communities

Because they are assemblages of different species, communities have properties that make them unique from individual organisms and populations. These properties are important to ecologists because they provide some insight into the health of a community and allow different communities to be compared on an objective basis. These different attributes are described below.

You should realize though that communities, like populations, can be difficult to assess and their boundaries hard to define. Some communities simply blend gradually into others and for this reason are called **open communities** (Figure 46.1a). Forest communities are like that, as different vegetation types blend together in mixed associations at borders. Open communities often show a free movement of organisms

like birds and large mammals that carry energy and nutrients from one community to another. Conversely, **closed communities** (Figure 46.1b) have more definite borders; few organisms pass from one community to another. Ponds and caves tend to contain distinct communities, since they are isolated from their surroundings, and the inhabitants have had to adapt to an entirely different range of conditions than those on the outside. In these closed communities, fewer organisms move in and out and thus they are more isolated in terms of energy and nutrients. You should realize though that no community is absolutely closed in the truest sense of the word, because some bird, insect, seed, or spore can always be blown in by chance.

Despite the fact that communities can sometimes be difficult to define, ecologists have been able to identify a rich array of attributes by which communities can be described and analyzed. These include *species composition, stratification, frequency, distribution, diversity,* and *stability.*

(a) Florida Pine Forest

(b) Maritime Hammock in Florida

FIGURE 46.2
Communities Differ in Species Composition
(a) This dense stand of Australian pine in South Florida is a monoculture because the tree produces such dense shade that no other plants can survive underneath. This species, which is not really a pine, was introduced to South Florida from Australia during the beginning of the 20th century for ornamental purposes. Since that time, it escaped and now grows throughout most of southern Florida, where it has become a serious problem in many spots. (b) A maritime hammock showing a mixture of trees along the Mantanzas River in Florida.

Species Composition

The most fundamental attribute of a community is its **species composition**, which is simply a list of species of which the community is comprised. Communities vary tremendously in their composition: at one extreme we have communities composed of only a single species (Figure 46.2a). They are rare in nature, but they can occur when conditions are extreme or when one species is so dominant that it chokes out everything else. At the other extreme are communities composed of dozens, even hundreds, of species that can be found in areas such as maritime hammocks (Figure 46.2b).

Strata

Like soils, ecological communities are often arranged into **strata**, horizontal layers occupied by different species. Much of the stratification, especially in terrestrial habitats, is caused by the vegetation, which creates structure, especially in forests (Figure 46.2). All terrestrial communities have a **ground layer** in which we find a diversity of herbaceous and low, woody plants as well as animals adapted to crawling, walking, or running on the surface (for example, antelopes, rabbits, grasshoppers, and frogs). The ground layer can be shaded, as in a forest, or brightly lit, as in a meadow. Below the ground layer we have a **soil layer** composed of plant roots and burrowing animals like earthworms, moles, and ants. The soil layer is typically dark and moist and exhibits a rather constant temperature. If we are in a forest, the highest layer is the **canopy**, composed of the foliage of the mature trees. Animals of the canopy, include birds, insects, and tree-dwelling mammals, like sloths, monkeys, and bats. Most forests also have an **understory** layer, which is between the canopy and ground layer (see Chapter 47). The understory is less brightly lit than the canopy and is home to the leaves of small trees and large shrubs, as well as many insects and birds. Because they are rather inaccessible, the canopy and understory layers of many forests are poorly understood. In some forests, *emergent trees* rise above the canopy here and there. Ecologists have recently devised elaborate methods to examine community interactions in canopies, especially in the tropics. One of the major findings of those studies is that the species diversity, particularly of insects, is much greater than previously estimated. Each year about 1000 new species of insects are identified in tropical forests alone.

Stratification also occurs in open-water communities, where the light-extinguishing property of water is the primary factor responsible for the sorting of organisms. Typically the surface layer of a large aquatic community (for example, an ocean or large lake) is rich in free-floating photosynthetic organisms like diatoms, dinoflagellates, or cyanobacteria. The organisms are accompanied by small heterotrophic species, like tiny crustaceans and the larvae of other invertebrates, that eat the photosynthesizers. Below the well-lit surface layer is a darker zone inhabited by medium sized and larger fish. Finally, the bottom is occupied by sessile forms and animals that scavenge the bottom. Of course, for shallow open-water habitats, the bottom might receive abundant sunlight. There we find beds of aquatic plants like turtlegrass, pondweed, or elodea as well as large red, green, or brown algae attached to the bottom.

FIGURE 46.3
Species Richness Versus Area
(a) Relationship between number of bird species and island area in the vicinity of New Guinea. (From P. R. Ehrlich, and J. R. Roughgarden, *The Science of Ecology.* Macmillan, New York, 1987.) (b) Relationship between number of flowering plant species and sampling area. (From M. Begon, J. L. Harper, and C. R. Townsend, *Ecology: Individuals, Populations and Communities.* Sinaver, Sunderland, MA, 1986.)

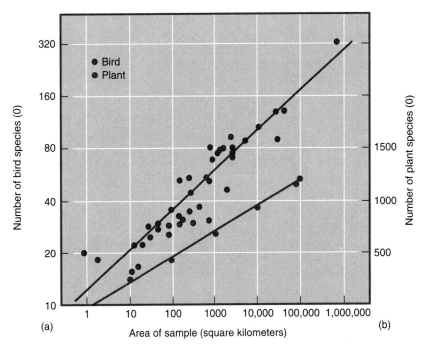

(a)   (b)

Frequency and Distribution

Another way to describe communities is in terms of how frequently species appear. **Frequency**, in ecology is a measure of how often we find a species in a community. Species that are commonly encountered are *frequent,* while those that are rarely encountered are, not surprisingly, *rare.* **Distribution**, or how species are arranged in a community, is also important as a function of frequency. Sparsely distributed species can sometimes be very important in the functioning of a community. For example, a grassland might consist of 90% coverage by grasses and 10% coverage by plants belonging to the pea (legume) family. Legumes are very important because they harbor populations of nitrogen-fixing bacteria, those that convert atmospheric nitrogen (N_2) into a form that plants can use. Thus legumes are rich in biological nitrogen, and they are good at "spreading their wealth" to other members of the community. Without legumes, the rest of the grassland would suffer from nitrogen deficiency and growth would be poor.

Diversity

Ecologists have long recognized that in some communities you can see many different kinds of species, while in others you see the same ones over and over again. To measure that aspect of a community, ecologists have developed the concept of diversity. **Diversity**, which is a measure of the variation in a community, has two components. The first is *richness,* which is the number of species in the community. Thus a sandy beach is a low-diversity community while the tropical rainforest is a high-diversity community. The second component is called *evenness,* which is the degree to which the different species are represented in a community. It is possible for a community of 50 individuals to be made up of five

species, but arranged so that 46 of them belong to species A and with one individual each from the other four species. That community would have a much lower diversity than one with ten individuals from each of the five species.

Diversity can also be influenced by the size of the area. **Alpha diversity**, describes the diversity within a given site, such as a community. Thus a cornfield with its one species would have lower diversity than a forest with its many species (compare Figures 46.2a and b). **Beta diversity** describes the diversity of a larger area that may include different communities developing on different kinds of sites. Thus an area that includes a forest, a meadow, and a pond would have greater beta diversity than an area of forest (Figure 46.3).

Ecologists often see a link between diversity and the "health" of a community. To that end, they have devoted a great deal of effort toward identifying which environmental factors promote high diversity and which cause a community to have low density. The answer, as you might expect, is complex, though a few factors seem to be most important.

One factor promoting high diversity, especially beta diversity, is the complexity of the habitat. An area that has the same physical conditions over a wide expanse will tend to have lower species diversity than an area with different conditions over the same size area. The reason is that some species tend to specialize on different kinds of sites; therefore we find more species when there are more nooks and crannies (opportunities to develop niches) to support them.

In addition, species diversity in a community is greater when that community is close to a larger community that contains a pool of species available for colonization. For example, imagine two new volcanic islands of equal size, one 20 miles from a mainland and the other 200 miles from that mainland. The first island will have greater diversity than the

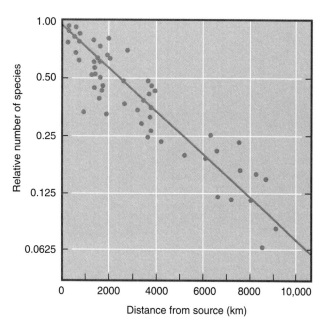

FIGURE 46.4
Species Richness as a Function of Distance from a Source
The relative number of bird species on an island decreases as the distance between the island and the mainland increases. The relative number of species is the actual number of species divided by the number of species the island would have if it were very close to the source area. (From P. R. Ehrlich and J. R. Roughgarden, *The Science of Ecology*. Macmillan, New York, 1987.)

more distant one, because it is easier for colonists to arrive at the closer island. This relationship between distance and diversity has been supported by observation (Figure 46.4).

Species diversity tends to be higher near the equator than at higher latitudes. One explanation for this trend is that equatorial weather is more benign, and organisms there do not suffer the food shortage and exposure to cold that polar species face.

Stability

Over time, some communities tend to remain the same while others change. Change can often be caused by some outside agent like fire, wind, or flood. Yet we can see that some communities are more sensitive and easily disrupted than others. To that end, ecologists have used the concept of stability to refer to the ability of a community to handle disturbance. Stability can refer to the ability of a community to *resist* being disturbed. For example, a thicket of prickly rose and blackberry shrubs is more resistant to predation by deer than a layer of birch and aspen saplings. Also, stability can refer to the *resilience* of a community (that is, its ability to recover quickly from a disturbance). For example, a community of oak trees defoliated by gypsy moth will recover quickly and may even produce a second set of leaves during a given growing season. In contrast, a community of pines defoliated by gypsy moths will likely die because pines lack the ability to quickly produce a second set of leaves.

We often hear of certain communities being called "fragile." That term is typically used to refer to communities that have low stability when faced with human disturbance. Again, their stability is low because they can easily be disturbed or because they take a long time to recover from disturbance (Figure 46.5).

For the past 30 years, ecologists have debated whether there is a link between diversity and stability. To some ecologists, diverse communities are more stable because several species can perform the same function (for example, pollination, photosynthesis, decomposition) and if something happens to one species, another can take over. To other ecologists, diverse communities are *less* stable because their greater complexity creates a greater chance that something will happen to at least one of the component species, thereby changing the community.

Important Community-Level Processes

Now that we know something about the structural aspects of communities, we can turn our attention to certain dynamic processes that occur within communities. These processes are important because they explain why certain communities have a certain type of species composition or distribution. They also provide clues as to how communities might behave in the future; such predictive aspects are important if we wish to manage or restore any community.

The Roles of Competition and Niche in Community Structure

As we saw in Chapter 45, competition involves a struggle for limited resources. Typically, competition involves two organisms battling over a limited resource. **Exploitative competition** is the use of the same resources in which one competitor has greater access than the other to the resources. **Interference competition** is actual fighting over resources.

Because the niche of an organism—the way in which it interacts with its environment—is often dependent on how it fares in competition with its neighbors, both kinds of competition are important in the structure of the community.

Perhaps the most famous field example of the role of competition and niche in community structure is the study of barnacles performed by Joseph Connell along the rocky coastline of Scotland. There you would note two common species coating the rocks. Adults of *Chthamalus stellatus* normally occur fairly high on the rocks, above the typical water line. In contrast, adults of *Balanus balanoides* occur on lower portions of the rocks, below the normal water level. That pattern might suggest that *Chthamalus* prefers the more exposed upper part of the rocks while *Balanus* prefers the less exposed lower portion. However, Connell noted that larvae of *Chthamalus* actu-

FIGURE 46.5
Example of a Fragile Ecosystem
The shrubby communities of the arctic tundra in Alaska are considered to be fragile because they take a long time to recover from disturbance. In this photograph we see tire tracks that remain from a truck that passed over the site about ten years earlier. Those tracks will likely take centuries to disappear.

Tracks Across the Tundra

ally settle on all portions of the rock, but those lower on the rock seem to be pushed away by the heavier-shelled *Balanus*. To test the system further, Connell removed *Balanus* and noted that the *Chthamalus* grew equally well on all parts of the rock. With *Balanus* present, the *Chthamalus* was *competitively displaced* on the more stressful upper part of the rock.

In the above example, *Balanus* acted as a bully, outcompeting the weaker *Chthamalus* for the most favorable site. However, like many bullies, *Balanus* could not handle the more stressful environment and the *Chthamalus* did better there. The ecological literature is filled with similar studies in which a good competitor does extremely well on low-stress sites, whereas the poorer competitor succeeds on the more stressful site.

Many studies show that coexisting species can have broad, overlapping niches when the level of available resources is high, but their niches shrink and separate when resources become more scarce. For example, in the tundra, two predatory birds, snowy owls and hawk-like jaegers, both feed on lemmings, but only when the lemmings are in abundance. When lemmings are rare, owls and jaegers switch to alternative prey that are different from each other to avoid competition. A study of 11 species of Panamanian stream fishes revealed that each species had a very specialized diet during the dry season when food was scarce. But in the wet season, the same abundant food sources were used by all the fishes.

From many such observations grows the hypothesis that competition in the wild rarely results in extermination, but rather encourages a subdivision of the habitat, each species coming to live where it does best. Ecologists use the term *niche separation* to describe the divergence of behavior or appearance that occurs when two species compete. A classical observation of niche separation was reported by the noted ecologist R. H. MacArthur. As Figure 46.6 reveals, five species of warblers that utilize the resources and shelter of spruce trees do not actually occupy the same niche. Although some overlap is inevitable, their feeding and nesting habits differ enough so that they are not in serious competition.

Niche separation in response to competition can be very subtle. For example, plants that grow together often position their roots at different depths within the soil to tap different layers of water and nutrients. Also, plants flower at different seasons, avoiding competition with pollinators. In fact, many properties found in organisms living today for which the adaptiveness is not immediately apparent might be reminders of competitive struggles that occurred in their ancestors generations earlier. For that reason, those feature have been referred to as "ghosts of competition past."

The Role of Territoriality in Community Structure

Generally we don't see interspecific competition becoming so strong that it threatens the existence of both species. Instead, potentially destructive competition between species is held in check by special forms of community organization.

One such organization is the formation of **territories**, which, as we saw in Chapter 45, is defined as any defended area. Territories generally contain some resource (such as food, nest sites, or mates) that must be defended against competitors. Typically, some territories are better than others because the environment is variable. In times of food shortage, the individuals holding the best territories will be more likely to survive. In some species of birds, the male must have a good territory in order to attract females. We see an example of this

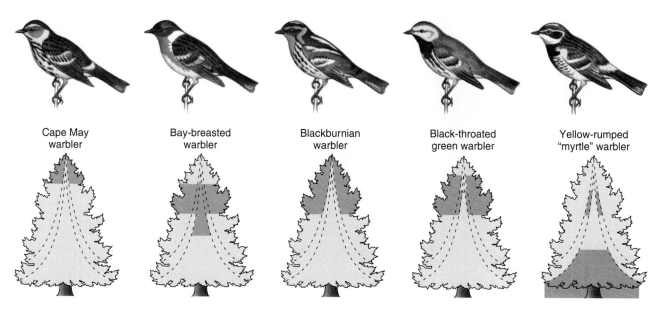

FIGURE 46.6
Subdividing the Habitat
Five species of North American warblers use spruce trees for feeding and nesting, but each prefers its own portion of the tree. The darkened areas indicate where each species spends at least half of its feeding time. By exploiting different parts of the tree, the species avoid direct competition; thus they can occupy the same general habitat.

sort of influence on community structure in certain blackbirds. Those males with low-quality territories may not attract mates.

Territory size is critical to the reproductive success of the territory holder. Territories that are too large will drain valuable energy and time required in defense; territories that are too small will not provide enough food for rearing offspring. Accordingly, studies show that birds appear to respond to both food supply and number of potential trespassers in setting their territory boundaries.

The Roles of Disease and Parasitism in Community Structure

Both disease and parasitism are common in nature and can certainly affect community structure. For example, before the 1820s, the Hawaiian Islands were home to several species of birds called honeycreepers that ranged throughout all elevations of the mountainous islands. In the mid-1820s, however, populations of tropical mosquitoes were introduced to the island of Maui by the careless actions of a visiting ship's crew. Those mosquitoes served as *vectors* (agents of dispersal) for several bird diseases. Soon the honeycreeper community below the elevation of 600 meters was decimated. In contrast, bird populations on the higher slopes remained intact because the mosquitoes could not survive the cooler temperatures there. Thus the presence of mosquito-borne bird disease played a great role in the structuring of bird communities on the Hawaiian Islands.

In some cases, disease may interact with predation to depress certain populations and thereby have a profound effect on community structure. For example, a two-week old caribou fawn can already outsprint a full-grown timber wolf, and healthy caribou seldom fall prey to wolves. However, caribou are subject to a hoof disease that lames them before it affects other parts of the body, and it is these lamed animals that a wolf is likely to cull out of a herd. In areas where the wolves were poisoned in order to protect the migrating caribou, this hoof disease spread unchecked and in a few seasons decimated entire caribou herds. With the caribou herd suffering in this way, opportunities arise for other herbivores that may have competed with the vast herds for grazing opportunities. Do you suppose that a weakened caribou herd would provide more or less food for wolves? Does your answer relate to the short-run or the long-run?

The effect of parasites on community structure is largely dependent on the length of the parasite–host association. Some parasites live in the host for long periods and also reproduce in the host. A long host-life is essential for the maximum fitness of these parasites, and, as might be expected, they have evolved a reduced virulence, minimizing their impact on host survival. Parasites with weaker effects live to reproduce and pass on their genes, but those with stronger effects die with their hosts before reproducing.

In contrast, other parasites live in or on the host for only brief periods and do not reproduce in the host. A long host-life is not critical to the survival and reproduction of such parasites.

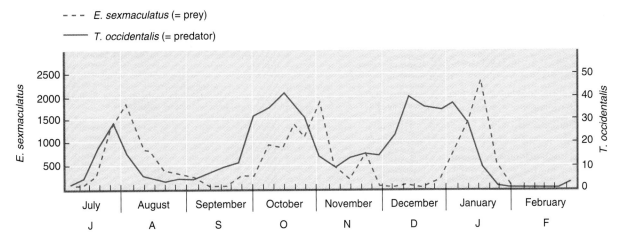

FIGURE 46.7
Predator–Prey Interactions
Populations of two mite species: *Typhlodromus occidentalis* (a predatory species) and the six-spotted *Eotetranychus sexmaculatus* (its prey) fluctuate over time and in synch with each other. Prey numbers increase first, followed by an increase in the density of the predator species. As the predator density increases, the prey diminishes in a population crash. The predator population soon declines as well, presumably because not enough food is available to support it.

In fact, increased virulence may evolve if it is tied to the extraction of large amounts of host resources that speed parasite development. Parasites developing before host death would not be harmed. Appropriate examples are nematodes of the family *Mermithidae.* Juvenile nematodes develop in the body cavity of their insect hosts, but before maturing they bore their way out of the host. This dramatic exit is often fatal to the host, especially if there are a large number of escaping worms. Mermithid nematodes are effective biological weapons in the nonchemical war that humans wage against insects that destroy crops.

Ecologists realize that the roles of parasites and disease on structuring communities can be very subtle and are not well understood. It is an active area of research today.

The Role of Predation in Community Structure

Predation, the relationship in which one organism (the predator) consumes the body of another organism (the prey), is an important ecological interaction that can be a powerful determinant of community structure. Predation has a dynamic influence on the number and quality of both predator and prey, because it acts as an important agent of natural selection on both groups.

Predators influence the numbers of prey by removing individuals from the prey population, yet they do not normally kill off the prey population. The reason can be explained by examining a theory developed in the early 1930s. In essence, the theory states that whenever the number of prey individuals rise, more food is available to predators, who respond with an increased birth rate. The number of predators does not rise immediately, however, since it takes time for the energy from food to be converted into new offspring. Because of this time lag, the prey enjoys a period in which the degree of predation is low, thus allowing prey populations to increase even more. Eventually the predator numbers rise to the point that they begin having a noticeable influence on the prey, particularly through reducing prey numbers. Then, as the prey begin to be killed off, the predators find themselves with less food, and so their own numbers soon fall off due to starvation or simply a failure to reproduce. At this point, prey populations have been depleted to the point that their densities are low relative to the amount of available resources. Such low densities enable prey populations to rise again, and the cycle begins anew.

As an example of predator–prey population dynamics, consider the curves produced by two mites (microscopic arthropods), one of which feeds on the other (Figure 46.7). Note that the two populations continually cycle, with the predator population lagging after the prey. We should add, however, that predator–prey relationships are not often as straightforward as seen in the mite example. For example, if the predator is able to hunt more than one prey type, the story can be complicated by prey-switching, in which the predators seek other, more available, food.

By regulating a prey population, predators can affect the level of competition experienced by the prey. Surviving prey often have access to more food and consequently have higher reproductive and growth rates. Also, other species that may have been held in check or even competitively excluded by the prey species may benefit indirectly by the predator's presence. As you remember from Connell's study of barnacles in Scotland, *Balanus* normally excluded *Chthamalus* from deeper portions of the intertidal zone. However, in the presence of a predatory snail that preyed selectively on the faster-growing and larger *Balanus,* both barnacle species coexisted in the deeper waters. Similarly, in other rocky intertidal areas inhabited by a

predatory starfish species, many species of barnacles and mussels coexisted and community diversity was high. When the
starfish was experimentally removed, populations of a few
species shot up, leading to an increased level of interspecific
competition. The result was that populations of some of the less
competitive barnacles and mussels were driven to extinction on
that site. Diversity of the prey community was thus lowered
when the predatory starfish was reduced. A species, such as a
predator, whose removal would profoundly affect the composition and dynamics of the surrounding community is called a
keystone species. The activities of a keystone species can indirectly increase the number of species in a community.

The Role of Disturbance in Community Structure

Most communities are subject to **disturbance**, a change in
environmental conditions that affect populations of the component species. Understanding the role of disturbance on
community structure has become an important area of study
for modern ecologists.

In essence, we know that disturbance comes in a great
many forms. Some disturbances, such as volcanic eruptions,
glaciers, and hurricanes, are large-scale and can devastate a
community. Other disturbances, such as the construction of a
beaver dam, the burrowing activities of a woodchuck, or the
ramming of a tree by an elephant, have a lesser effect. Yet
other disturbances, such as the onset of winter, are seasonal
and predictable, whereas some, such as a tornado, are more
erratic. Some disturbances are natural, while many others are
human-induced.

As you saw in the last chapter, communities that are frequently disturbed tend to harbor populations of opportunistic,
r-selected species. Communities relatively immune to disturbance are populated by equilibrial, *K*-selected species. Interestingly, terrestrial communities that are occasionally affected
by disturbance tend to have a reservoir of opportunistic
species like blackberry. They may just lie there for extended
periods, but following disturbance, these species quickly
sprout and flourish. In contrast, communities that are almost
never disturbed contain essentially no populations of opportunistic species. Thus, when they are disturbed, they take a
long time to recover. An excellent example is that many tropical rain forests, when cut down, require decades to regenerate,
if they recover at all.

One might wonder about the relationship between disturbance and species diversity within a community. In one way,
severe disturbance does tend to drive species locally extinct and
thus causes a reduction in diversity. However, communities not
disturbed at all often have one or two species "taking over" and
eliminating the others from the site. In another way, moderate
disturbance actually increases species diversity. To resolve this
apparent paradox, ecologists have developed the **intermediate
disturbance hypothesis** (Figure 46.8). According to this hypothesis, areas with intermediate levels of disturbance have more

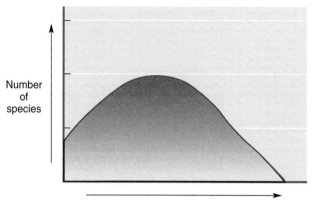

FIGURE 46.8
The Intermediate Disturbance Hypothesis
According to this hypothesis, communities that have intermediate
levels of disturbance have greater species diversity than those
subject to either little disturbance or severe disturbance.

species than do areas of lower or higher levels of disturbance.
At lower levels, competition is intense, and the resulting exclusion yields only a few surviving species. At higher levels, the
disturbance itself wipes out all but a few stress-tolerant species.
At intermediate levels—not strong enough to kill most species
but still strong enough to reduce the competitive impact of
dominant species—the number of species is highest because
competitively inferior and superior species as well as stress-
intolerant and stress-tolerant species survive.

Community Development Over Time: Ecological Succession

Our view of biotic communities has for the most part been
static—suspended in time—as a convenient way to describe
their attributes and key processes. However, communities
tend to change over time. They change not only in response to
climatic and geological forces but also in response to the
activities of their inhabitants. In some cases, the inhabitants
will alter the environment, which then influences the community in new ways.

The dynamic nature of communities became an important concept in ecological thought toward the end of the
1890s. Early in the 1900s ecologists adopted a view that communities respond to disturbance by undergoing a sequence of
events called **ecological succession** in which each species or
group of species gives way to the next. **Seres**, or individual
stages composed of a fairly well-defined cast of organisms,
are progressively replaced by assemblages belonging to later
seres. Ultimately, a mature "climax" community would come
to occupy the site indefinitely.

Two types of succession have been recognized. Many successional sequences involve **autogenic succession**, in which the change that occurs can be traced to the behavior of the species involved. For example, we know that as forests develop from old farmland in the eastern United States, certain trees, such as trembling aspen and pin cherry, dominate the site first, only to be replaced by other species, such as sugar maple and white oak. As it turns out, seedlings of aspen and cherry perform well in open sunlight but very poorly in shade. Maple and oak, in contrast, are more shade-tolerant. Thus, by their very growth, aspen and cherry create conditions that are unfavorable to themselves, but more favorable to later successional species. In contrast to autogenic succession is **allogenic succession**, which occurs when external forces effect change. A good example of allogenic succession occurs in ponds. Young ponds contain low-nutrient water and are inhabited by certain species of microorganisms and fish. As the pond ages, it receives runoff from the surrounding watershed, and that water is often nutrient rich. As a result of the progressively enriched water, the community in the pond changes to a different assemblage of organisms. In most instances, succession is a result of both autogenic and allogenic factors.

Allogenic succession is less predictable than autogenic succession. For example, an orderly progression of species during succession is often interrupted by the sudden bloom of unexpected opportunistic species, such as weeds. In addition, one would not like to leave the impression that one population gracefully gives up its place for the next. On the contrary, species are often quite persistent, seemingly resisting their own displacement. Ecologists have also categorized succession as either primary or secondary.

FIGURE 46.9
Pioneer Species
Lichens and mosses are able to exist under conditions that are inhospitable to other species. These hardy species slowly break down rock to produce soil and thus improve the site for other species. Since lichens and mosses are so successful in colonizing bare surfaces, ecologists refer to them as "pioneer" species.

Primary Succession

Primary succession is the establishment of a community where no community previously existed, such as on rocky outcroppings, newly formed deltas, sand dunes, emerging volcanic islands, and lava flows.

In primary succession, such as occurs on a rocky outcropping (Figure 46.9), the successional sequence begins with pioneer organisms, hardy drought-resistant species that can successfully establish themselves and reproduce in places previously not inhabited by their kind. They are specialized for the initial invasion of disturbed or uninhabited areas. Lichens (mutualistic assemblages of algae and fungi) are often the first pioneers to invade rocky outcroppings, where they are held fast by their tenacious, water-seeking, fungal component. Lichens are soil builders, producing weak acids that very gradually erode the rock surface. As organic products and sand particles accumulate in tiny cracks in the rock surface, opportunities arise for plants such as grasses and mosses to establish themselves and begin a new successional stage.

Plant roots penetrate the rocky crevices, exerting a remarkable turgor pressure, prying at the rocks and gradually widening the fissures. By then, certain insect and decomposer populations will also have established themselves. In time, the lichens that made the penetration of plant roots possible are no longer able to compete for light, water, and minerals, and they give way to the plants. Similarly, these plants will contribute to the soil-building for a time, and then they too will be replaced by fast-growing shrubs, and new populations of animals will invade.

The succession on bare rock outcroppings is, at first, an extremely slow process, with an individual stage often lasting hundreds of years or more. But once soil formation has begun, the process can accelerate. Primary succession in other places can also be slow. Ecologists estimate that succession from sand dune to mature forest community on the shores of Lake Michigan took about a thousand years.

Secondary Succession

Secondary succession involves the establishment of a new community where a previous community has been disrupted. Examples include neglected farms reverting to the wild and forest communities that have been subjected to "clear-cutting," the controversial lumbering practice in which all trees are removed from a stand.

In secondary succession, the principles are similar to those of primary succession, but the stages come and go at a more rapid pace. This rapid turnover is possible because the soil is already in place. On deserted farmland, weeds, grasses, shrubs, and saplings are often the first to appear. As secondary succession progresses, the initial invaders are eventually replaced by plants from the surrounding community. Weeds and small shrubs provide more organic matter to the soil, enabling fast-growing trees such as aspen, cherry, and pine to grow. As mentioned, these block the sun at the ground level and allow shade-tolerant species to develop. Eventually, there is a general blending with the surrounding community. Such a simple transition may require well over 100 years, depending on the community and the disturbance. For example, a microsuccessional process occurs in northern forests, in which maples and beeches are felled by periodic storms, creating gaps in the canopy. These gaps are soon populated by birch trees that grow quickly, create shade, and allow the beeches and maples to return.

Secondary succession in grassland communities, as you might expect, is much faster, taking perhaps 20–40 years. At the other extreme, fragile, disturbed tundra may require many hundreds or even thousands of years to recover (see Figure 46.5).

The successional sequences described here occur on land (terrestrial habitats). Succession that begins on exposed sites like rock outcrops and old strip mines is often called *xerarch succession,* because the original sites are rather dry (xeric). In contrast, the term *mesarch succession* is applied to sequences that begin on moist sites like old farmland and cleared forest.

Succession in Aquatic Communities

Aquatic communities also undergo community development or succession. Succession in lakes and ponds occurs as a product of *natural eutrophication,* changes brought about by the natural increase in nutrients carried in by streams and runoff from the land. Lakes and ponds that are rich in nutrients and high in productivity are called *eutrophic* ("true foods") lakes, while those that have limited nutrient supply and little productivity are called *oligotrophic* ("few foods") lakes. The general trend in freshwater bodies is toward increased eutrophication and thus increased community growth, but the loss of any essential nutrients can reverse the latter trend.

As community growth in a lake progresses, the sediments increase and the depth decreases. Plants crowd the shores, extending further and further into the lake, followed by increasing numbers of water-tolerant shore plants (Figure 46.10). In small, shallow lakes, we might find the process continuing to the development of a marsh and perhaps a tree-dominated swamp.

Three Alternative Models of Succession

We mentioned that autogenic succession occurs when organisms change the environment in such a way that it is improved for later successional species. That principle has been referred to as the **facilitation model** of succession (where each sere facilitates the arrival of the next) and seems especially valid for primary succession.

However, that facilitation doesn't always occur. A number of studies have revealed that nonpioneer plants grow better on plots in which pioneer plants had been experimentally removed. These results suggest that established plants may actually inhibit, not facilitate, the invasion or growth of other plants. Thus a new model, the **inhibition model**, was developed. According to the inhibition model, new kinds of plants appear only after established plants have died or been damaged. The established plants, if healthy, inhibit the development of new kinds of plants. Succession occurs despite inhibition because pioneer plants generally have shorter life spans than do later successional species, so they are replaced more often. As they die, they leave openings for other species to move in while the young pioneer plants are small and vulnerable. The result is that later successional species will eventually predominate.

The inhibition model of succession has some practical applications. For example, there are times in which we do not want a mature community to develop, such as immediately adjacent to a highway or under a power line. In those cases, we can sow the area with early successional species like crown vetch or any of several shrub species that will prevent later successional species from developing.

Studies of secondary succession in yet other communities suggest that pioneer plants in some cases neither help nor hinder later successional plants. Based on those studies, the **tolerance model** was developed, in which succession occurs because of differences in developmental rates and competitive abilities between pioneer and nonpioneer plants. Pioneers develop more quickly and so dominate the early stages of succession. But because pioneers are also likely to be less efficient exploiters of resources, they are eventually replaced by the slower developing but more efficient nonpioneers.

Mature Communities

Ecologists studying succession have noted that *developing* communities tend to produce more organic material than they use. In contrast, **climax** communities have been defined as showing an equilibrium between the production and utilization of organic matter. In the early stages of succession, the exchange rate between organisms and the environment is slow because mineral nutrients are largely stored in environmental reservoirs like the rocks. But as the climax state is reached, more of the nutrients cycle directly, through exchange pools, between the organisms and the decomposing material. The organisms themselves tend to become more diverse as the community enters the climax state. Ecologists also found that feeding relationships between organisms in a community become more complex as succession proceeds.

Within the past 20 years, ecologists increasingly abandoned the notion of climax, especially as originally stated. The

(a) Open water

(b) Sediment accumulation

(c) Invasion by marsh plants

FIGURE 46.10
Succession in a Pond
Ponds are created naturally by retreating glaciers, changes in the courses of rivers, and subsidence of land.
(a) Early in pond succession the water is open, and the edge is vegetated by aquatic plants like cattail and
bulrush. **(b)** As time proceeds, the bottom of the pond becomes shallower due to the accumulation of sediment. The water is colonized by submerged plants like pondweed and floating-leaved species like water
lily. **(c)** With the passage of time, more sediment accumulates to the point where water depth may be less
than 1 foot. In these sites, the community may consist of a diverse assemblage of marsh plants like rice-
cut grass and sedge.

reason for this shift is the widespread realization that even
"mature" communities are dynamic, though the change is very
slow from a human perspective. One reason for the dynamic
nature of mature communities can be traced to long-term
changes in weather. We know that temperature and rainfall patterns change over the course of centuries, if not millennia.
These changes in climate are correlated with the expansion
and contraction of glaciers, which periodically plow through
an area. As a result of climatic changes, and other large-scale
natural disturbances like fires, floods, and pest outbreaks, communities do not remain static. The scope of long-term community changes is made especially evident by studies of pollen
remains on lake bottoms. Above all, you should remember that
species are constantly evolving, and that fact alone should
argue against any notion that communities tend to become static entities over time.

Communities to Ecosystems

While the community has been studied extensively over the
past century, it is safe to say that the ecosystem holds a special
place in the hearts and minds of most ecologists. As defined in
the beginning of Chapter 45, an **ecosystem** consists of the
sum total of all species on a site, along with all the components of the physical environment on that site. In contrast, the
community is considered to be the assemblage of populations,

without the specific inclusion of the abiotic environment. In
that context, an ecosystem can carry out certain functions that
a community cannot—namely, the passage of energy and
nutrients among different groups of organisms.

By considering these processes, many ecologists regard
the ecosystem as being the fundamental identifiable unit of
ecology, just like the species is the fundamental unit of taxonomy. An important implication of the ecosystem concept is
that an ecosystem has certain *inputs, internal transformations, outputs,* and *feedback loops.* We saw in previous chapters that an individual organism processes food (an input) by
digestion and metabolism (transformations) to yield energy,
nutrients, and waste products (outputs) and, the whole process
is regulated by hormones and the nervous system (feedback
loops). Similarly, an ecosystem takes energy and nutrients
(inputs) and, through the component organisms (transformations), converts them into biomass, activity, and waste (outputs) using a system where organisms respond to each other's
behavior (feedback loops).

As we saw with the community, ecosystems are not isolated entities, and, in fact, it can be difficult to assess where one
ecosystem ends and another begins. The biggest difficulty arises due to the fact that component organisms and even many abiotic features of the environment can move in and out of an
ecosystem. For example, a migrating herd of zebras or a flock
of geese can easily carry energy and nutrients hundreds and
even thousands of miles. Similarly, a sudden windstorm can
pick up an inch of topsoil in Kansas and deposit it over an eight-

state area east of the Mississippi. Still, ecosystem ecologists have conducted numerous successful studies on ecosystem-level phenomena during the past 50 years, and their contributions are recognized as essential to the science of ecology.

Components of Ecosystem: Trophic Levels

As we have seen, organisms in an ecosystem all play important roles in converting nutrients and energy within a given area by certain groups feeding on other groups. The impetus for studying feeding interactions began in earnest in the 1920s with the work of the English ecologist Charles Elton. Elton studied the feeding relationships of animals on an island in the North Atlantic, and his work encouraged other ecologists to perform similar studies on other sites. Thanks to them we have some understanding of the complex feeding relationships across a broad spectrum of ecosystems. However, gaps in our knowledge still occur and research is ongoing. In their research, ecologists tend to categorize organisms into three groups based on their general metabolism and role within the community: producers, consumers, and decomposers.

Producers include autotrophic organisms such as green plants, algae, and many types of bacteria. They are crucial to the functioning of every ecosystem because they convert energy from sunlight or certain inorganic molecules into chemical-bond energy of organic molecules like carbohydrates and proteins. Most producers, particularly the plants and algae, also have a second function in that they absorb inorganic nutrients like nitrate, phosphate, sulfate, calcium, and potassium and convert them into organic forms like proteins, nucleic acids, and other molecules.

Consumers include heterotrophic organisms such as protozoans and animals. These organisms ingest the bodies of other organisms and, in doing so, create new organic matter from the organic matter in those bodies. As a result, consumers avail themselves of energy residing in the chemical bonds of the bodies of other organisms for their own use. In addition, consumers use the organic nutrients stored in the bodies of other organisms to make their own molecules. Ecologists often subdivide consumers into *primary consumers (herbivores)*, which feed off producers, and *secondary consumers (primary carnivores)*, which feed off herbivores. A representative primary consumer would be a rabbit (which eats plants), while a secondary consumer might be a snake (which eats rabbits). There are also *tertiary consumers (secondary carnivores)* that eat secondary consumers (such as a hawk that might eat a snake), and even *quaternary consumers* that eat tertiary consumers. Some consumers—primarily carnivores—are likely to cross trophic levels as they feed. Consider humans as an example. How many trophic levels do *you* occupy? Such variability in feeding behavior can complicate the way that we view ecosystems, as you will see.

Decomposers (also called *scavengers* or *detritivores*) are organisms that feed off decaying organisms. Generally, decom-posers are fungi or bacteria that invade the body of a dead organism, slowly break down the tissues, and give off pungent odors in the process. In so doing, decomposers absorb the organic molecules of the rotting carcass into their own bodies. Simultaneously, they convert the nutrients back into an inorganic form that can be returned to the soil or water, or transferred directly to another organism. For example, many soil-dwelling bacteria break down (or reduce) the remains and wastes of other organisms into simple products such as ammonia, sulfates, nitrates, phosphates, and the usual carbon dioxide and water. Thus the decomposers perform a vital ecological function of recycling the raw material of ecosystems, but they are often invisible to the naked eye and unappreciated.

Food Chains and Food Webs

We can see then that the components of an ecosystem interact to extract and convert energy and nutrients from level to level. It should be obvious that we can take the producers, consumers, and decomposers and assemble them into a **food chain**, which is a sequential, functional system based on feeding strategies (Figure 46.11). Let's take an example: the base of a representative food chain would be a community of green plants such as grasses that photosynthetically convert solar energy into chemical-bond energy and convert absorbed inorganic nutrients into an organic form (producers). The second component of the food chain would be an herbivore such as a mouse that nibbles the leaves and eats the seeds of our grass plants (primary consumers). In so doing, the mouse transfers energy and nutrients from the grass to itself. The third component of the food chain would be a primary carnivore like a weasel, which might kill and devour our hapless mouse, thereby transferring energy and nutrients to its own body (secondary consumers). The fourth component would be a secondary carnivore like a coyote, which includes weasels as part of its diet (tertiary consumers). Finally, the coyote, plus any unconsumed portions of the grass, mouse, or weasel (including feces), is ultimately subject to terminal breakdown by fungi and bacteria in the soil (decomposers).

Such a food chain does occasionally occur in certain circumstances and is called a *grazing food chain* because most of the organic matter entering the herbivores is derived from living plants that are grazed. However, another common type of food chain, called a *detritus food chain* (Figure 46.12), involves a different pathway. Here primary consumers are not vertebrates that feed off living plant tissue but instead are typically invertebrates that feed off decaying plant tissue. Detritus food chains are common in the upper layers in the soil and in creeks and rivers, where energy enters the system as fallen leaves, twigs, and reproductive structures. A representative detritus-based food chain would start with a fallen leaf that enters the top part of the soil (producer). After being softened by rainwater and trampled by passing deer, the leaf fragment enters the body of a burrowing earthworm (primary consumer). From there, the energy and nutrients are passed to a young robin just about ready to leave the nest (secondary consumer). The next recipient is a hawk that captures the robin (tertiary consumer).

We know, however, that the two food chains described above are painfully simplified versions of what really happens in nature. Perhaps the biggest oversimplification is the notion that grass is always eaten by mice, or that hawks always eat robins. Instead, we know that consumers can be quite variable in their diets and, by extension, individual organisms can be eaten by a variety of predators. Thus our food chain might show five different producers, seven herbivores, four primary carnivores, two secondary carnivores, and ten decomposers. Moreover, one adventurous consumer might include in its diet some plants, an herbivore, and a carnivore. The result is a flow diagram that is anything but straightforward and is instead a tangled mass of interconnections. Ecologists use the term **food web** to describe the more realistic system of interlaced trophic interactions that involve the diversity of producers, herbivores, carnivores, and decomposers that is normally found on a site (Figure 46.13).

Ecological Pyramids

While food webs are satisfying because they can accurately show the trophic interactions that can occur in an ecosystem, their complexity makes them unwieldy and often impractical to use. As an alternative, ecologists have resorted to the use of **ecological pyramids**, which are layered diagrams that express many important structural and functional aspects of ecosystems.

In essence, an ecological pyramid is a chart that consists of several layers (Figure 46.14). The bottom layer represents the producer trophic level. The next layer up represents the herbivores, the next layer the primary carnivores, and the next layer the secondary carnivores. The width of each layer is some measure of each trophic level (for example, numbers of individuals, biomass, or energy content). By looking at the shape of the pyramid, the different trophic levels for that parameter can be compared.

Pyramids of Numbers A pyramid of numbers is obtained when we count the total number of individuals in each trophic level. Note that in a grassland (see Figure 46.14a), the pyramid of numbers shows a broad base and it narrows as we proceed from herbivores to carnivores. Such a pattern makes sense for a grassland community in which the producers are represented by large numbers of individual grass plants. The herbivores include insects (grasshoppers) and mammals (mice and jackrabbits), which although numerous are far fewer than the plants. The carnivores are even fewer and include larger animals like snakes, carnivorous birds, wolves, and coyotes.

Not all communities show the pattern exhibited by the grassland. In a forest, note that the number of herbivores far exceeds the number of producers, thus producing an inverted pyramid (see Figure 46.14b). That makes sense when you

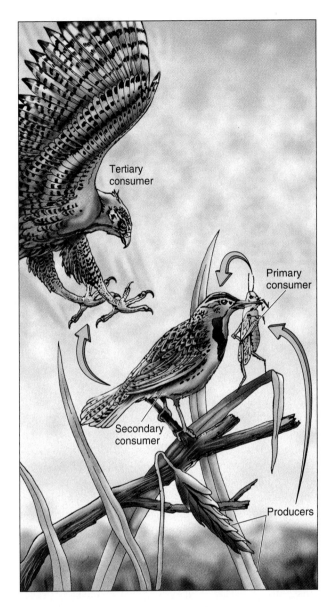

FIGURE 46.11
Simple Food Chain
This simple food chain consists of a producer (plant), a primary consumer (herbivore), a secondary consumer (carnivore), and a tertiary consumer (carnivore).

consider that the producers are mainly trees, and the herbivores include insects, rodents, and other small animals. Still you should note that, once again, the number of carnivores (especially secondary carnivores) is less. In fact, because of energetic reasons that will be explained below, virtually all communities have few secondary carnivores.

Pyramids of Biomass. Graphic depictions of biomass (the dry weight of organic matter) in an ecosystem can be quite

FIGURE 46.12
A Detritus-Based Food Chain
These food chains differ from the grazing food chain shown
in the previous diagram because their herbivores feed
mainly on litter. Often carnivores can participate in both
grazing and detritus-based food chains.

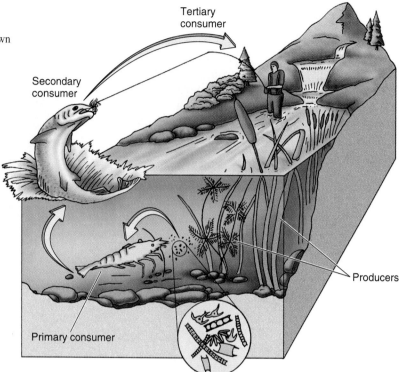

Ditritivores (decomposers)
feed off decaying producers,
primary, secondary, and
tertiary consumers.

revealing. Ecologists often estimate biomass by "harvesting"
several randomly selected square meters of a much larger area,
or perhaps harvest in narrow swaths across the designated area.
The ecologist is able to weigh all the individuals in these smaller areas and to extrapolate those measurements to the larger
area. If the organisms are then sorted according to known
trophic levels, dried, and weighed, the data can be plotted.

Usually pyramids of biomass show a progressive decrease
from producers to herbivores to carnivores (see Figure
46.14c). In fact, 99% of the earth's biomass is to be found in
the primary producer level. Much of that biomass is locked up
in dead wood found in tree trunks and branches, however. In
contrast, herbivores and carnivores are not able to store organic matter to the extent that plants can. Also, much of the plant
matter eaten by herbivores disappears (converted to carbon
dioxide, water, nitrogen waste, and so on) during metabolism.
In addition, some is not actually absorbed by the herbivore's
body, appearing in the feces. For the same reasons, the carnivore level generally supports less biomass than does the herbivore level. Thus we see the stepwise narrowing of the pyramid.

Pyramids of Energy. Energy pyramids represent the flow of
energy from one trophic level to the next and show the loss
that occurs in such transfers (see Figure 46.14d). The stepped
pyramid would be predicted by the second law of thermodynamics, which, as you may recall (Chapter 6), states that energy transfers are never perfect—there is a loss of useful energy
(free energy) with each transfer. This is true in living as well
as nonliving systems. In fact, loss can be considerable; the
energy transfers from one trophic level to the next average
about 10% in efficiency. Thus not much energy in a trophic
level is available at all to the next level. It turns out that 90%

of the energy in the food eaten by consumers is not stored as
body structures of one kind or another but is used in maintaining the animal (or is lost as waste). Consumers do more than
eat, and every act requires energy. Eventually, all the free
energy entering the earth's ecosystems is released as heat.
Heat, of course, is a low-grade form of energy, and there is no
way for organisms to gather heat energy for useful work. This
tremendous reduction in free energy is what ultimately limits
the number of trophic levels in nature (to four or so).

Humans and Trophic Levels

Energy pyramids may seem abstract and academic, but they
apply dramatically to human populations and suggest some
fundamental lessons in economics. Economics and ecology
share many things other than the prefix, including certain
premises and theories. Humans are basically **omnivores**, that
is, capable of feeding at several trophic levels. In large part,
availability of resources and economic state determine the
trophic level most used. Under poor economic conditions,
people generally feed at the primary consumer (herbivore)
level. The reason is clear: most of the earth's available energy
is stored at the primary level. Feeding at the secondary level
or higher involves costly (or wasteful) energy transfers, as
energy from the primary level is converted, for example, to
the bodies of cattle, sheep, hogs, and other animal stocks (Figure 46.15). But in most instances, while affluent people may

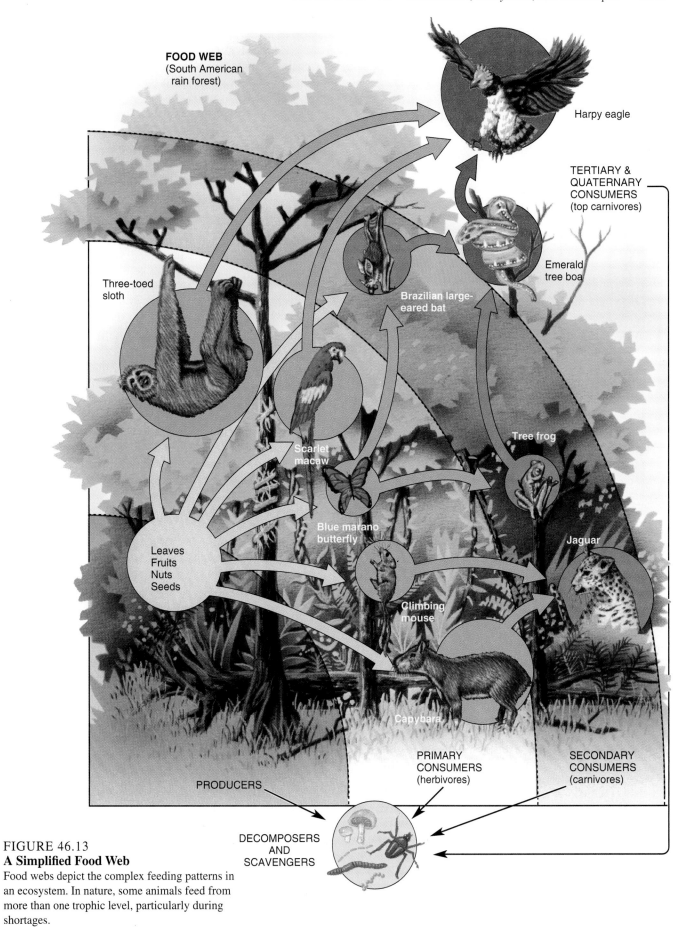

FIGURE 46.13
A Simplified Food Web
Food webs depict the complex feeding patterns in an ecosystem. In nature, some animals feed from more than one trophic level, particularly during shortages.

C₃ — Tertiary consumer
C₂ — Secondary consumer
C₁ — Primary consumer
P — Producer
R — Reducer

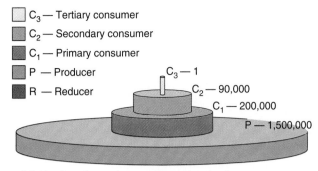

C_3 — 1
C_2 — 90,000
C_1 — 200,000
P — 1,500,000

(a) Number of organisms: summer grassland

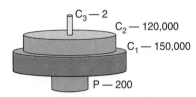

C_3 — 2
C_2 — 120,000
C_1 — 150,000
P — 200

(b) Number of organisms: summer temperature forest

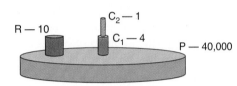

R — 10
C_2 — 1
C_1 — 4
P — 40,000

(c) Biomass: Panama tropical forest (g/m²)

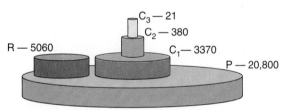

R — 5060
C_3 — 21
C_2 — 380
C_1 — 3370
P — 20,800

(d) Energy: fresh water organisms (kcal m²/yr)

FIGURE 46.14
Ecological Pyramids
Pyramids of numbers **(a,b)** depict the number of individuals in each trophic level by the width of the cylinder. **(a)** A pyramid of numbers for a grassland community. Values reflect number per 0.1 hectare (quarter acre). Note the progressive decrease from producers to herbivores to carnivores because producers in grasslands (grasses and herbs) are small and numerous. **(b)** A pyramid of numbers for a temperate forest is partly inverted because producers in forests (trees and shrubs) are large and fewer than the herbivores, which include insects and other small invertebrates. **(c)** A pyramid of biomass taken from a tropical forest shows a sharp decrease from producer to herbivore to carnivore. In that forest, 40,000 g/m² of producer biomass supports a total consumer biomass of 5 g/m². The large decomposer biomass shown to the left should be expected in a community where nutrient cycling is very rapid. **(d)** An energy pyramid from a freshwater community at Silver Springs, Florida. Energy flow values are expressed as kcal/m²/year. Energy pyramids should show a decrease from producers through the various levels of consumers because of energy losses at each trophic level (see text).

choose to eat steak and lamb, impoverished people tend to eat cereals and beans and peas. Of course, animals do make available the energy stored in plants that are unpalatable to humans like grasses, herbs, and legumes such as clover and alfalfa.

It is efficient to utilize fishes and other seafood (as do many Asians). Although there are costs in harvesting the high-protein seafood, humans cannot readily eat organisms belonging to the lower trophic levels in the ocean. Thus it is best to "wait" until the energy reaches the trophic levels occupied by fin and shellfish.

Based on the above relationships, many people argue that the earth could support a much larger population (or that the current human population would command less natural energy) if all humans ate only plant food. However, we must realize that food provides nutrients as well as energy. Humans feeding at the herbivore level may face problems with nutritional quality. For example, many common plant foods cannot provide certain essential amino acids (such as tryptophan, methionine, and lysine; see Chapter 38) that are critical to our own synthesis of proteins. Those amino acids are all found in meat. The absence of such essential amino acids from the daily diet can lead to such infamous protein deficiency disease as kwashiorkor, a chronic problem in some parts of Africa (Figure 46.16). To be sure, the critical amino acids can be provided if legumes such as soybeans and peas are included in the diet. However, such legumes must be eaten on a daily basis, because humans do not store amino acids very well, and excesses are rapidly converted to fats or carbohydrates.

Energy and Productivity

When ecologists focus on energy transformations within an ecosystem, they often discuss the terms *productivity* and *production*. **Production** is the amount of energy stored by a trophic level in a given area. The units of measurement are typically kilocalories/square meter. For example, a community of grasshoppers eating grass take in energy. The amount of energy that they contain at any one time is the production. In

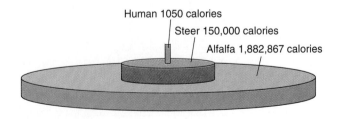

Human 1050 calories
Steer 150,000 calories
Alfalfa 1,882,867 calories

FIGURE 46.15
Humans as Second-Level Consumers
When humans behave as carnivores, energy must first flow from a producer to an herbivore, and that change entails a considerable loss of energy. Each 1000 calories of energy stored by a beef-eating human requires nearly 2 million calories of energy stored in the producers.

contrast, **productivity** is a rate, expressed as production per unit time. Thus if we wanted to know how much energy our grasshoppers took in during a 24-hour span, we would be considering their productivity.

Ecologists also distinguish between gross and net productivity. **Gross productivity** is the amount of energy taken in by photosynthesis or by consuming the bodies of other organisms. In contrast, **net productivity** is the amount of energy that remains after losses due to respiration are subtracted. In more concrete terms, your gross productivity is the total amount of energy that you took in from your food during the past day (or other time period). In contrast, your net productivity is the amount of energy contained within the new protein, fat, and carbohydrate that your body gained today. If you lost weight, your net productivity is a negative number.

Of course, ecologists are more interested in large-scale phenomena than in individual organisms. Therefore they are concerned with gross and net productivity of various levels. One particularly valuable measure is primary productivity, which measures the amount of sunlight taken in by an assemblage of producers and converted at least temporarily into organic matter. Primary productivity is such an important measure because it tells us how much usable energy an ecosystem has to work with. As shown in Table 46.1, annual productivity varies from place to place. In marine ecosystems, estuaries (places where fresh water meets the oceans) and coral reefs are particularly productive, while the open ocean has less productivity. The difference between these two regions is undoubtedly due to the fact that estuaries and reefs are nutrient-rich systems teeming with photosynthetic organisms. The open ocean is nutrient-poor and photosynthetic organisms are scarce.

On land, the tropical rain forest is the most productive ecosystem type while deserts and tundras (see Chapter 47) have low productivity. The disparity here is no doubt due to the fact that the tropical rain forests are warm and moist, which promotes high rates of photosynthesis. In contrast, desert plants are frequently parched, while tundra plants are exposed to prolonged subfreezing temperatures that tend to inhibit photosynthesis.

Ecosystem ecologists are not content to look at the productivity of producers alone. Instead, they have examined the energy flow for all components of an ecosystem: producers, herbivores, carnivores, and decomposers. Sunlight energy is first captured by producers during photosynthesis. Energy in each trophic level can have one of three fates. First, most (averaging 90%) of the energy is respired by the organism during metabolism and is given off as heat back to the environment. Second, if the organism is consumed, the energy in its body passes as chemical-bond energy to organisms belonging to the next higher trophic level. Third, if the organism dies before being eaten, its energy passes as food to scavengers, detritivores and decomposers. One important relationship to note is that energy is lost from each trophic level.

FIGURE 46.16
Protein-Deficient Diet
People living in poor countries often suffer from a diet that, while adequate in terms of calories, is deficient in nutrients—particularly amino acids supplied by protein. Victims of a protein-deficient diet, such as the child in the photograph, often develop a disorder called kwashiorkor. Symptoms include hair that is reddish, thin, and brittle, muscles that are poorly developed, skin that flakes, and an abdomen that is distended due to poor circulation. Often, simple dietary supplements such as a daily ration of peas or beans will provide the needed amino acids and alleviate the symptoms.

Such examinations of energy flow allow us to estimate **net community productivity**, which is net primary productivity minus heterotroph respiration in a community. Communities with a net community productivity greater than zero are growing communities that are accumulating biomass. They are frequently found in sites recovering from disturbance. In contrast, communities with a net community productivity less than zero are no longer adding new biomass, but are losing it. Such communities are said to be mature, having reached an equilibrium between the rates of energy assimilation and energy use.

TABLE 46.1
Estimated Gross Primary Production (Annual Basis) of the Biosphere and Its Distribution Among Major Ecosystems.

Ecosystem	Area, Millions of km	Gross Primary Productivity KCal/m²/yr	Total Gross Production 10¹⁶ KCal/yr
Marine			
Open ocean	326.0	1,000	32.6
Coastal zones	34.0	2,000	6.8
Upwelling zones	0.4	6,000	0.2
Estuaries and reefs	2.0	20,000	4.0
Subtotal	362.4	—	43.6
Terrestrial			
Deserts and tundras	40.0	200	0.8
Grasslands and pastures	42.0	2,500	10.5
Dry forests	9.4	2,500	2.4
Northern coniferous forests	10.0	3,000	3.0
Cultivated lands with little or no energy subsidy	10.0	3,000	3.0
Moist temperate forests	4.9	8,000	3.9
Fuel-subsidized (mechanized) agriculture	4.0	12,000	4.8
Wet tropical and subtropical (broad-leaved evergreen) forests	14.7	20,000	29.0
Subtotal	135.0	—	57.4
Total for biosphere (round figures, not including ice caps)	500.0	2,000	100.0

SOURCE: From *Fundamentals of Ecology*, 3rd Edition, by Eugene P. Odum, Copyright © 1971 by W. B. Saunders Company. Reprinted by permission of Holt, Rinehart and Winston, CBS College Publishing.

Nutrient Cycling in Ecosystems

While energy flows *through* an ecosystem, emerging eventually as heat and never to be reused, the nutrients essential to life tend to repeatedly *cycle* between organisms and the abiotic component of an ecosystem. Nutrients are the chemical building blocks of organisms and include elements of the familiar SPONCH series (Chapter 2)—sulfur, phosphorus, oxygen, nitrogen, carbon, and hydrogen—along with a host of others such as iron, cobalt, sodium, and chlorine. These elements are taken up by organisms as molecules or ions, depending on the species. For example, plants take up nitrogen in the form of nitrate, while animals incorporate nitrogen from the bodies of their prey. While some nutrients remain unchanged, others are incorporated into new molecules and structures. Some, particularly carbon, hydrogen, and oxygen, are metabolized for their chemical-bond energy. But eventually the elements reappear in metabolic wastes or, when death occurs, in the products of decay. Most of the elements cycle back to the producers as mineral ions, or mineral nutrients as we have called them before (Chapter 28). The pathways of elements as they are taken up from the physical environment and then released back to it are called **biogeochemical cycles**. As mentioned earlier, decomposers play a key role in such recycling.

The biogeochemical cycles have three major places where elements are accumulated. First, through feeding, metabolism, and growth, the elements are integrated into the bodies of living organisms. Second, they may be found in **exchange pools**, the readily available, water-soluble reserves of a mineral nutrient (such as nitrates in soil water that so easily enter plants). Third, mineral nutrients may be locked away in **reservoirs**, which are less available reserves (such as the elements in animal bones or shells, calcium in limestone, or atmospheric nitrogen).

Some of the biogeochemical cycles are rather short, involving only a few steps. For example, some of the water taken in from the environment by the plant simply passes through its vascular system to be released back into the environment through transpiration (see Chapter 30).

In describing biogeochemical cycles, we can again begin with the producers. Once they incorporate inorganic ions into their bodies, the substances are then available to be passed to a higher trophic level as food, eventually reaching the ever-waiting decomposers. Of course, uneaten dead plants pass their ions and molecules *directly* to the decomposer.

Keep in mind that elements are cycled through ecosystems. Each one has its own pathway. Instead of detailing each one, we will instead consider a few of the major elements that might provide a representation of the various forms that biogeochemical cycles can take.

The Nitrogen Cycle

We'll use the **nitrogen cycle** as our primary example of biogeochemical cycling (Figure 46.17). Nitrogen is a principal constituent of proteins, nucleic acids, chlorophyll, coenzymes, and many other biomolecules. Like other essential elements, nitrogen is found in both readily available exchange pools and in the less available reservoirs. The largest nitrogen reservoir is the atmosphere, where molecular nitrogen (N_2) accounts for about 78% of the dry atmospheric gases. Atmospheric nitrogen is not directly available to most of the earth's organisms. Plants and nearly all other producers must obtain their nitrogen primarily in the form of nitrate ions in order to incorporate it into amino acids. Nitrate ions (NO_3) are produced by soil and water bacteria in two complex biochemical pathways: *nitrogen fixation* and *nitrification*. **Nitrogen fixation** is the process by which atmospheric nitrogen is converted into an ionic form—typically ammonium (NH_4^+). Nitrogen fixation is accomplished mainly by bacteria. In contrast, **nitrification** involves the oxidative conversion of ammonium to nitrate. The process occurs in steps, each handled by a separate prokaryotic species. Thanks to nitrogen fixation and nitrification, soils have at least some available nitrogen in a readily usable form for plants.

From Producer to Consumer to Decomposer

Plants absorb nitrate ions and incorporate the nitrogen into nitrogen-containing molecules, particularly plant proteins. When plants are eaten, the amino acids pass into the consumer levels, where some are used to produce animal protein (and, of course, other nitrogen-containing molecules). The rest may be metabolized and the nitrogen waste product excreted in various forms (urea, uric acid, or ammonia—see Chapter 36). Eventually, all producers and consumers (and the nitrogen wastes) become food for the decomposers.

In decomposition, populations of microorganisms (bacteria and fungi), each with its role in a multistepped process, recycle the protein and nitrogen wastes. The decomposers break down organic wastes into simpler compounds such as ammonia (NH_3), carbon dioxide, and water. The ammonia readily ionizes in water, forming ammonium ions. In the next step, nitrification, the ammonium ions are acted on by bacteria such as the autotroph *Nitrosomonas,* which converts ammonium ions to nitrites (NO_2^-). A second group of nitrifiers, represented by *Nitrobacter,* then converts the nitrites to nitrates (NO_3^-). The nitrates join the exchange pools and cycle back to the plant. Nitrification is vital because ammonium ions are far less available to plants than are nitrates. Although ammonium ions are chemically easier to incorporate into amino acids than nitrate, in the soil these positively charged ions tend to interact with and cling tenaciously to negatively charged clay particles. On the other hand, the negatively charged nitrate ions remain mobile in the soil and are more readily available for uptake by the plant root.

So far, the nitrogen cycle may seem smooth-running and efficient, but there are complications that lead to losses from the exchange pools. For example, the soluble nitrogen-containing ions can be carried out of the producer's reach by leaching—removal by the downward percolation of water. Furthermore, should anaerobic conditions prevail, organisms such as *Pseudomonas denitrificans*—anaerobic soil bacteria that act as *denitrifiers*—can convert the nitrites and nitrates to nitrous oxide (N_2O, "laughing gas") and nitrogen gas (N_2). The two gases escape to enter the atmospheric reservoir. Obviously, denitrifiers can drastically deplete soil fertility. It is for this reason that anaerobic swamps and bogs are notoriously nitrate-poor.

The Nitrogen-Fixers

In a balanced ecosystem, the losses through denitrification can be recovered by the gain from nitrogen fixation (see Chapter 22). Nitrogen-fixing bacteria (and many cyanobacteria) take in atmospheric nitrogen, which they send through a complex biochemical pathway to combine with hydrogen, producing ammonia (see Essay 22.1). Their excesses are released into the soil or water, where nitrifying bacteria convert the ammonia to nitrite and nitrate. As you can see in Figure 46.17, denitrification and nitrogen fixation are not part of the main cycle of nitrification.

Nitrogen-fixing soil bacteria of the genus *Rhizobium* invade the roots of alfalfa, peas, and other leguminous plants, which respond by forming cyst-like nodules around the bacterial colony. In a mutualistic relationship, the bacteria absorb organic nutrients produced by the plant, and the plant gains usable nitrogen fixed by its guest bacteria. In rice paddies and other aquatic ecosystems, much of the nitrogen fixation is carried out by aquatic cyanobacteria like *Anabaena*. In natural terrestrial ecosystems, nitrogen-fixing bacteria live in mutualistic associations with the roots of wild legumes, alders, buckthorns, and locust. But as far as agriculture is concerned, it is the leguminous plant, chiefly alfalfa, that is most vital in harboring nitrogen-fixing bacteria.

Nitrogen may also be fixed the hard way, by lightning, and more gently by the photochemical action of the sun on

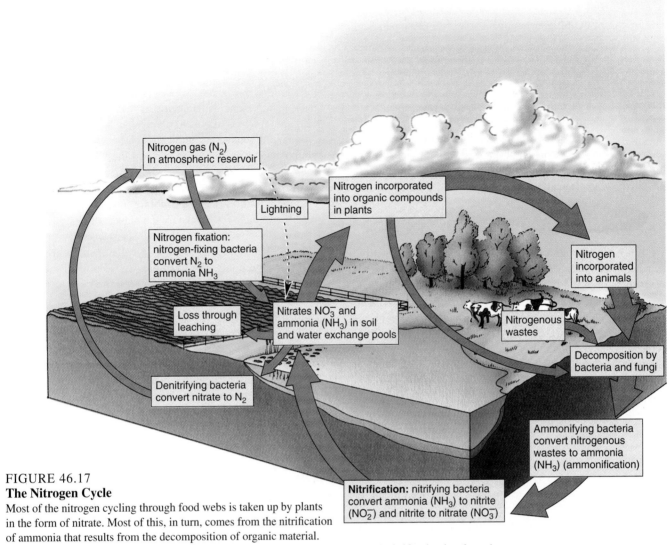

FIGURE 46.17
The Nitrogen Cycle
Most of the nitrogen cycling through food webs is taken up by plants
in the form of nitrate. Most of this, in turn, comes from the nitrification
of ammonia that results from the decomposition of organic material.
The addition of nitrogen from the atmosphere through fixation and its return via denitrification involve relatively small amounts compared to the local recycling that occurs in the soil or water.

certain pollutants such as oxides of nitrogen. However, nitrogen fixed through such atmospheric action amounts to less
than 10% of that fixed by organisms.

A problem has rather recently arisen regarding our
manipulation of nitrogen. It turns out that we are not using up
all our nitrate as one might have expected. In fact, we are
overloading the environment with nitrogenous products. The
problem has only become apparent since we began synthesizing ammonia through industrial means.

Synthetic fertilizers are applied liberally to the soil and
thus enter the natural nitrogen cycle. When we add the nitrogen from synthetic fertilizers to the nitrogen compounds produced by natural nitrogen fixation and by automobile exhaust
(another major new source), the total amount of available
nitrogen is astounding. C. C. Delwiche, at the University of
California at Davis, has calculated that there is a net gain of 9
million metric tons of fixed nitrogen to the biosphere each

year. Where is the surplus going? In California's central valley,
which receives more synthetic fertilizer than any place on
earth, the answer is: right into the water table and river systems
and from there into San Francisco Bay and on to the Pacific.

How does excess nitrogen affect the water supplied in the
soil? What is its effect on river systems and their estuarine
life? One visible effect is the sporadic choking of waterways
by uncontrolled, runaway algal growth. Such rapid growth
always accompanies what is called *cultural eutrophication,*
the sudden nutrient enrichment of lakes previously discussed.
We have yet to measure the effect of the nitrogen load on the
marine environment, but there even more nitrogen is added
from sewage dumped into the world's oceans.

Are we altering the nitrogen cycle in other ways? It's difficult to know just what's happening out there, but obviously
we need to find out. It is conceivable that we can feed the
world's burgeoning populations without calling the environ-

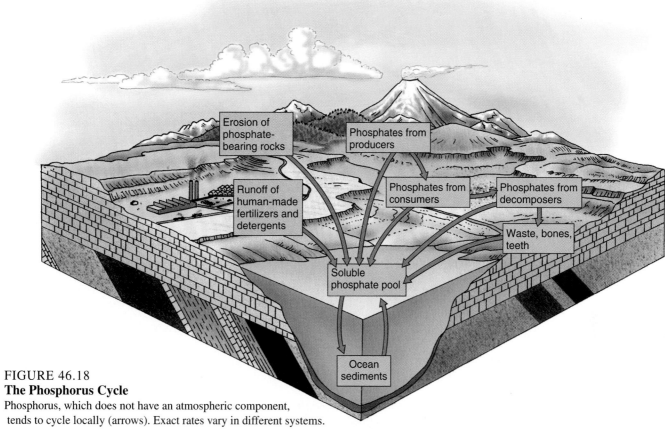

FIGURE 46.18
The Phosphorus Cycle
Phosphorus, which does not have an atmospheric component,
tends to cycle locally (arrows). Exact rates vary in different systems.
Generally, small losses from terrestrial systems caused by leaching are balanced by gains from the
weathering of rocks. In aquatic systems, as in terrestrial systems, phosphorus is cycled through food
webs. Typically, however, some phosphorus is lost from the ecosystem because of chemical processes
that cause precipitation or through settling of detritus to the bottom, where sedimentation may lock away
some of the nutrient before biotic processes can reclaim it. On a much longer time scale, this phosphorus
may become available to ecosystems again through geological processes such as uplifting and erosion.
This general pattern applies to many other nutrients, including trace elements.

ment down around us. By coming to understand the nitrogen cycle better, perhaps we will find it vital and even economically feasible to keep our agricultural practices within the limits of natural systems.

The Phosphorus and Calcium Cycles

Now let's briefly consider two more cycles, those of phosphorus (Figure 46.18) and calcium. Their cycles do not involve the atmosphere, but they do involve water. Phosphorus enters the roots of plants as the soluble phosphate ion HPO_4^{2-}. Phosphates are required for the production of many familiar molecules, such as ADP and ATP, phospholipids, nucleic acids, and the coenzymes of photosynthesis and respiration. Since ATP appears in the list, a shortage of phosphates will dramatically affect the energy-transforming processes of all organisms, including plants.

Calcium, another mineral nutrient, is critical to the proper functioning of cell transport, structure such as teeth, shells, and endoskeletons, and many enzymatic reactions. It enters living systems through the roots of plants as the cation Ca^{2+}. A shortage of calcium can disrupt transport processes in the plant cells, causing the plant to die. Although it functions in transport systems, the ion itself is rather immobile once in place. When local shortages of other ions occur, they can simply be shifted from other parts of the plant. But calcium must be taken in constantly by the roots.

The cycles of phosphorus and calcium are rather straightforward. Cellular phosphates, for example, are released into the phosphate pool by the action of decomposers. Since they are water-soluble, they may be recycled at once. However, phosphate incorporated into skeletal, tooth, or shell material is released very slowly by weathering. Calcium may also cycle very slowly, since it is commonly bound up in skeletons and

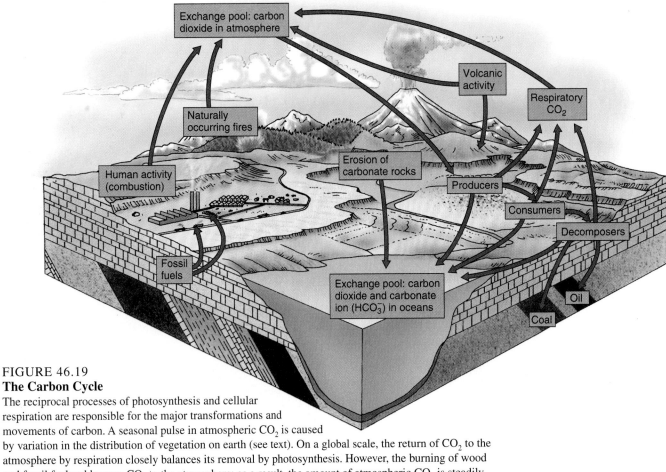

FIGURE 46.19
The Carbon Cycle
The reciprocal processes of photosynthesis and cellular
respiration are responsible for the major transformations and
movements of carbon. A seasonal pulse in atmospheric CO_2 is caused
by variation in the distribution of vegetation on earth (see text). On a global scale, the return of CO_2 to the
atmosphere by respiration closely balances its removal by photosynthesis. However, the burning of wood
and fossil fuels adds more CO_2 to the atmosphere; as a result, the amount of atmospheric CO_2 is steadily
increasing. Atmospheric CO_2 also moves into or out of aquatic systems, where it is involved in a dynamic
equilibrium with other inorganic forms, including bicarbonates.

shells. Calcium from such dense structures is very slowly
leached into the soil. In freshwater biomes, the reservoirs of
phosphorus and calcium may lie bound in the bottom sediments for long periods of time until currents agitate those
murky depths. In marine biomes (see Chapter 47), occasional
upwellings bring the reservoir sediments and dissolved ions to
the surface, where they reenter the cycle via the phytoplankton.

The Carbon Cycle

We are aware that life is based on carbon and, as with any key
element, its very availability may determine the sizes of populations. Like other elements, carbon cycles through the
ecosystem in a well-recognized pathway (Figure 46.19). In
terrestrial ecosystems, the main pool of carbon for photosynthesis is atmospheric carbon dioxide (CO_2). In the oceans and
other waters, plants and algae use the bicarbonate ion (HCO_3^-)

from dissolved carbon dioxide (and carbonate rock) as their
principal carbon source.

Paradoxically, as important as atmospheric carbon dioxide is to life, it is present in such small proportions (slightly
less than 0.04%) that it can almost be called a "rare gas." The
quantity is admittedly small when compared to other gases,
but it still represents an enormous amount in absolute terms.

In addition to the carbon found in the atmosphere, the
earth has a sizable reservoir in the form of carbonate rock
(such as limestone) and fossil fuels (natural gas, oil, coal,
peat). However, this reservoir is being steadily altered by
human and geological processes. Since the industrial revolution began, CO_2 from fossil fuels has been released into the
atmosphere in steadily increasing amounts. Today, atmospheric scientists estimate that 5–6 billion metric tons of CO_2
from fossil fuels are released into the air annually. Fortunately, as we shall see, about two-thirds of this amount is quickly
removed by the oceans and by photosynthesizers.

Key Ideas

SPECIES INTERACTIONS

1. In nature, populations do not occur as isolated entities but are integrated into larger assemblages called communities.

2. In any community, species interact—a relationship called **symbiosis**.

ATTRIBUTES OF COMMUNITIES

1. Some communities are **open**, meaning that their edges blend into adjacent communities and organisms move freely into and out of them. Other communities are **closed**, meaning that their borders are sharper and there is less exchange of component organisms.

2. The most basic community property is **species composition**, which is the list of species found in that community.

3. Communities are organized into vertical **strata** or layers, including the canopy, the understory, the ground layer, and the soil layer. Different species are arranged in the different layers.

4. Species within communities have different **frequencies**, which are a measure of how often a species can be found distributed throughout a community. Common species have a high frequency, while rare species have a low frequency.

5. Communities also have a certain **diversity**, which is a measure of the species richness (the number of species present) and the evenness (which is the degree to which individuals are allocated among the different species). **Alpha diversity** is the degree of diversity within a specific site, while **beta diversity** is the diversity across different sites. Several factors promote high diversity, including habitat complexity, area, proximity to a source of new individuals, and proximity to the equator.

6. Community stability has two components: resistance (ability to remain unchanged in the face of a change in the environment) and resilience (the ability to recover from change). Fragile communities have low stability with respect to human intervention. Many ecologists argue that complex communities are more stable than simple ones.

IMPORTANT COMMUNITY-LEVEL PROCESSES

1. **Competition** is a symbiosis in which both component species are hurt by the interaction, though one species usually wins the struggle. **Exploitative competition** is the direct battle for limited resources by two species. **Interference competition** is the activity by which one species prevents a second from gaining access to a resource. Competition occurs when niches of two species overlap. In the field, species often undergo niche separation, in which the competing species specialize on different resources, thus reducing competition.

2. **Territoriality** is behavior in which an area is defended.

3. Both disease and parasitism affect community dynamics by causing some individuals to become weakened, thereby influencing competition and predator–prey interactions.

4. **Predation** is a form of symbiosis in which one individual (the prey) is eaten by a second individual (the predator). In the process, the predator gains energy and nutrients from the prey. Predator populations often cycle with those of prey populations, though the fluctuation of the predator's population lags behind that of the prey's. Predation may affect competitive interaction, especially if the favored prey is a strong competitor. A keystone species is a species (often a predator) whose removal would profoundly affect the composition and dynamics of the surrounding community.

5. Virtually all communities are subject to **disturbance**, which is a change in the environmental conditions that cause a change in the component species. Disturbances can be small or large, predictable or erratic, natural or human-induced. Many species are adapted to disturbed conditions. The **intermediate disturbance hypothesis** states that communities receiving moderate levels of disturbance will have greater diversity than those receiving heavy disturbance or no disturbance at all.

COMMUNITY DEVELOPMENT OVER TIME: ECOLOGICAL SUCCESSION

1. **Ecological succession** is the orderly change in community composition following disturbance.

2. **Autogenic succession** describes that change that occurs thanks to the behavior of the component species. **Allogenic succession** is a change in community composition due to changes by some outside force.

3. **Primary succession** is the establishment of a community where no community had previously existed. **Secondary succession** is the reestablishment of a community where a preexisting community was disturbed. Primary succession occurs more slowly than secondary succession in terrestrial communities because soil must be built up.

4. Succession occurs in aquatic communities through eutrophication, the process by which water becomes enriched with nutrients.

5. According to the **facilitation model**, succession proceeds because early successional species improve the site for later successional species. The **inhibition model** states that early successional species prevent later species from entering. The **tolerance model** states that early successional species are faster growing than later successional species. All three models apply to different cases.

6. Historically, succession was thought to end in the development of a **climax** community, which was a mature community that did not change over time and whose composition depended on the prevailing climate. Ecologists now concede that even mature communities change due to long-term climate change, other natural disturbances, and evolution.

COMMUNITIES TO ECOSYSTEMS

1. **Ecosystems** are ecological entities that consist of the sum total of all species on a site, along with the physical conditions on that site. Ecologists regard the ecosystem as the basic unit of ecology.

2. The most important ecosystem-level processes include energy flow and nutrient cycling. Ecologists study inputs, transformations, and outputs of energy and materials in ecosystems.

3. **Producers** are autotrophs and include photosynthetic organisms like plants, algae, and certain monerans, as well as certain chemosynthetic bacteria. The photosynthesizers convert sunlight energy into chemical-bond energy using photosynthesis. They also incorporate inorganic nutrients and convert them into an organic form.

4. **Consumers** are heterotrophic organisms like animals and protozoans that eat autotrophs and other heterotrophic organisms. Consumers include herbivores that eat autotrophs, primary carnivores that eat herbivores, and secondary carnivores that eat primary carnivores. These trophic levels pass organic molecules, containing energy and nutrients, to each other.

5. **Decomposers** are chiefly fungi and bacteria that feed on decaying organisms. They are essential to ecosystem function because they convert nutrients from an organic to an inorganic form that is usable by producers.

6. **Food chains** consist of a simplified sequence of producers, consumers, and decomposers. In grazing food chains, herbivores eat living plant material. In detritus food chains, plant material is passed to the consumers only after it begins to undergo decomposition.

7. A **food web** is a more realistic system of interlaced trophic connections that show multiple predators and prey in an ecosystem.

8. **Ecological pyramids** are layered diagrams that express many important structural and functional aspects of ecosystems. Ecological pyramids can be used to represent numbers, biomass, or energy flow within communities. While producers may be numerically more or less plentiful than consumers, producers contain more biomass and energy than consumers. However, among consumers, we normally see a decrease in numbers, biomass, and energy as we go from herbivores, to primary carnivores, to secondary carnivores. That decrease is caused by the fact that, on average, only 10% of available energy is transferred to successive trophic levels.

9. As **omnivores**, humans feed at all trophic levels, but the impoverished tend to live at the primary consumer level. Feeding at higher levels is prohibitive because of the great losses involved in transfers.

10. **Production** is the amount of energy handled by a trophic level in a given area, while **productivity** is the rate of production per unit time.

11. **Gross productivity** is the amount of energy taken in, while **net productivity** is the amount of energy that remains after metabolic losses are subtracted.

12. Coral reefs, estuaries, and tropical rain forests are highly productive ecosystems. Open ocean, deserts, and arctic tundra are very unproductive.

13. **Net community productivity** (NCP) is the net primary productivity minus heterotroph respiration. NCP > 0 in growing ecosystems, while NCP = 0 in mature ecosystems.

NUTRIENT CYCLING IN ECOSYSTEMS

1. The movement of mineral ions and molecules in and out of ecosystems occurs through **biogeochemical cycles**. Most ions enter the living realm at the producer level.

2. Outside of life, nitrogen occurs in **exchange pools** and **reservoirs**. The largest reservoir is N_2 in the atmosphere, but it is only available to nitrogen fixers. Nitrate ions are made available in soil-exchange pools.

3. Plants incorporate nitrate or ammonium ions into protein. When plants die or are consumed or eaten, the protein goes to the consumer or decomposer level, but eventually all the incorporated nitrogen goes to the decomposers. During ammonification, decomposer bacteria reduce the nitrogen to ammonia, which forms ammonium ions. **Nitrification** is the two-step conversion of ammonium ions to nitrite and then to nitrate, which enters the exchange pools.

4. Denitrification is the conversion of fixed nitrogen back to nitrogen gas and is performed by certain bacteria under anaerobic conditions.

5. During **nitrogen fixation**, bacteria (including cyanobacteria) convert atmospheric nitrogen to ammonia, which then undergoes nitrification, forming nitrite and nitrate. Nitrogen-fixers include mutualists that live in the roots of leguminous plants.

6. The liberal use of synthetic nitrogen fertilizers in support of crops has produced worldwide soil and water excesses, and the balance between nitrogen fixation and denitrification has been lost. Two products of the nitrogen load are eutrophication—nutrient enrichment of waters—and the pollution of soil water supplies.

7. Cycles of phosphorus and calcium occur between living organisms and water. The two elements are taken up as soluble phosphate and calcium ions. Phosphates are used in producing ATP, nucleic acids, phospholipids, and tooth and shell materials, while calcium is essential to bone and shell development and in membrane activity.

8. Cellular phosphates cycle directly from living organisms to water and return, while phosphorus in skeletal material and teeth is freed very slowly. Calcium is also freed very slowly from skeletons and shells. Both accumulate in deep ocean and lake bottom sediments until upwellings and overturns redistribute them.

9. Carbon is an essential part of nearly all the molecules of life. The principal exchange pool on land consists of carbon dioxide gas, while the source in the waters is dissolved carbon dioxide gas and the carbonate ion. A large reservoir occurs in the form of limestone and fossil fuels.

10. Carbon enters the producer level during photosynthesis, where it is used initially to form carbohydrate. Producers, consumers, and decomposers all release carbon during cell respiration.

Application of Ideas

1. Can you think of any practical ways to reduce cultural eutrophication?

2. Why are closed communities generally not truly closed? Can you think of any that are? Which do you think is more inherently stable—a closed community or an open community?

Review Questions

1. Describe frequency in terms of distribution in a community. (page 1009)

2. Define community richness. (page 1010)

3. Distinguish between exploitative and interference competition. (page 1010)

4. Distinguish between predation and parasitism. (pages 1012–1013)

5. Why do areas of intermediate disturbance tend to have more species? (page 1014)

6. What is the main difference between primary and secondary succession in terms of rate? (pages 1015–1016)

7. Distinguish between xerarch and mesarch succession. (page 1016)

8. Describe the facilitation model of succession. (page 1016)

9. Contrast the above to the tolerance model. (page 1016)

10. Give four traits that an ecosystem shares with an individual. (page 1017)

11. Can you think of a true carnivore? (page 1018)

12. Define ecosystems. (page 1019)

13. Is there a risk from humans eating entirely from the producer level? (page 1022)

14. Distinguish between production and productivity. (pages 1022–1023)

15. Distinguish between pools and reservoirs. (page 1024)

16. Name three products of decomposer action. (page 1025)

CHAPTER 47

Biosphere and Biomes

CHAPTER OUTLINE

THE BIOSPHERE

Physical Characteristics of the Biosphere

Essay 47.1 The Changing Carbon Cycle and the Greenhouse Effect: A Destabilized Equilibrium

THE DISTRIBUTION OF LIFE: TERRESTRIAL ENVIRONMENT

THE BIOMES

Deserts

Grasslands

Tropical Savannas

Tundra

Tropical Rain Forests

Chaparral

Essay 47.2 The Destruction of Tropical Rain Forests

Temperate Deciduous Forests

Taiga

WATER COMMUNITIES

The Freshwater Province

Wetlands

The Marine Province

Essay 47.3 An Unusual Community: The Galápagos Rift

KEY IDEAS

APPLICATION OF IDEAS

REVIEW QUESTIONS

*P*erhaps the easiest way to see some of the greatest problems confronting ecologists is to look out the window. There's a world out there. (We hope that's true where you are.) The job of the ecologist is to make some sense of it, to look for themes, for commonalities, for differences, and for the influences that created them. If that window happens to be on a car, the problem seems to become magnified. A drive through northern Colorado reveals a landscape that may change drastically every few miles, so drastically that any theme or commonality is indeed difficult to detect. But! If that window is on an airplane, the job begins to seem easier. Now sweeping vistas may reveal themselves as far more homogeneous than a closer view would have suggested. Now patterns emerge and generalizations can be made. An analogy can be drawn with viewing a Jackson Pollock painting from close up: from there you can see the swipes, the drips, and the blobs in a confusion of color and form. But standing back, you may begin to see the form, the balance, and the movement.

Here then we will take two views of life on earth, one at the broadest level, as we consider the biosphere. Then we will take a step closer to the canvas of life and look at how that life is divided into biomes, keeping in mind that these must, to some degree, be considered artificial constructs based on certain sweeping generalities.

The Biosphere

The **biosphere** is where life exists; it is that thin layer of the earth wherein the wondrous properties of light and water interact to permit life, not only in the terrestrial surface but in the subterrestrial realm and the fresh- and marine waters as well. When we consider the biosphere, we find that we're dealing with great expanses but little depth. Things can't live very far above or below the earth's surface. More precisely, the habitable regions of the earth lie within an amazingly thin layer of approximately 14 miles. This includes the highest mountains and the deepest ocean trenches. If the earth were the size of a basketball, the biosphere would be about the thickness of one coat of paint.

Physical Characteristics of the Biosphere

Conditions within the biosphere are quite special. Our space probes have found no other place in the solar system where these conditions exist. The conditions here result from our distance from the sun, the presence of water, the makeup of our atmosphere, and the earth's solid crust. (Jupiter and Saturn, in fact, are gaseous balls.) Let's look at some of these factors a little more closely and then consider how various life forms are distributed over the earth.

Solar Energy. Only about half of the incoming solar radiation ever reaches the earth's surface. About 30% is reflected back into space by the earth and its atmosphere. Another 20% is absorbed by gases in the atmosphere. The remaining 50% reaches the earth's surface as light energy, where it is absorbed by the land and waters and is then radiated back into space as heat (Figure 47.1). But a great deal of work is accomplished by the 50% reaching the biosphere. (Fig. 47.1)

Surprisingly enough, of this light, considerably less than 1% will enter photosynthetic processes. Most of the energy is used in shuffling water around. After all, the heat from solar

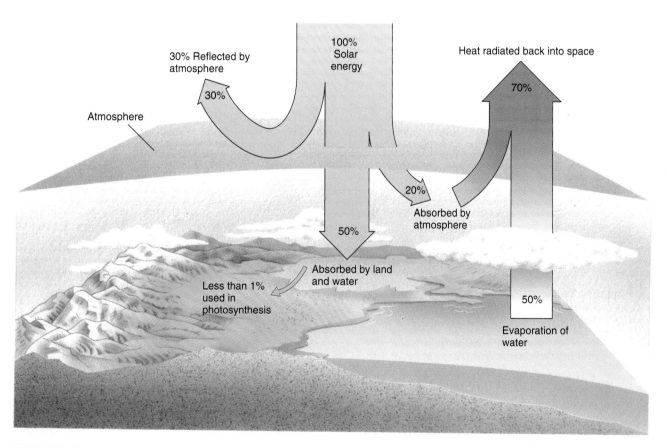

FIGURE 47.1
The Earth's Energy Budget
The relative constancy of conditions in the biosphere depends ultimately on an equilibrium between energy entering and energy leaving. Of the solar energy reaching the upper atmosphere, 30% is immediately reflected back into space. Another 20% is absorbed as heat by water vapor in the atmosphere. The remaining 50% reaches the earth's surface. Most of this energy reenters the atmosphere through evaporation from the earth's waters. The thin arrow represents energy used by organisms.

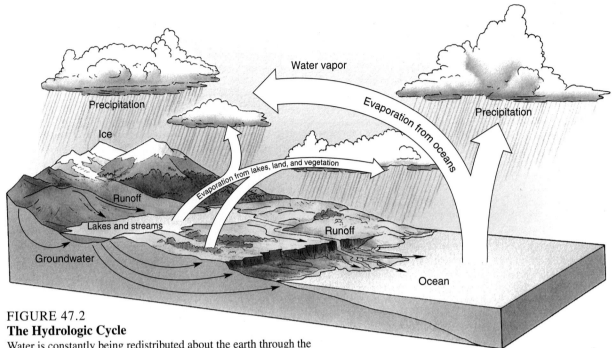

FIGURE 47.2
The Hydrologic Cycle
Water is constantly being redistributed about the earth through the
ongoing hydrologic cycle. The cycle, driven by solar energy, includes evaporation and
precipitation. The distribution of water on the earth is quite uneven, with the greatest percentage—almost
98%—occurring in the oceans. Visible fresh water—that of rivers, lakes, streams, and so on—makes up
far less than 1% of the earth's H_2O, an amount that is well exceeded by water locked into the polar ice
caps, snow-capped mountains, and glaciers—some 1.8%.

energy is responsible for most of the evaporation from the oceans, lakes, and rivers and, not insignificantly, from the leaves of plants through transpiration. As water absorbs this energy, its molecules gain heat—they move more rapidly—and are finally lifted into the atmosphere as water vapor. There the heat is lost as the water condenses into its liquid phase, falling as precipitation (such as rain or snow). So the absorption of solar energy is vital in the distribution of water over the earth, which, as we have seen, is also a way of redistributing heat, a factor that moderates our climate. The constant shifting of water between its liquid and gaseous phases over the earth is called the **hydrologic cycle** (Figure 47.2).

Water. You'll recall that much of the biochemistry of life is centered on the peculiar chemical traits of water (see Chapter 2). In addition, you may remember that water has a high specific heat—that is, it requires a relatively great input of energy to raise its temperature. Likewise, water loses energy very slowly (due to the hydrogen bonds between water molecules). Water then acts as a great stabilizer of temperature. It absorbs heat slowly and retains it well. Water vapor in the atmosphere is one of the reasons for the earth's comparatively moderate climates. This moisture helps hold heat and slows the radiation of heat from the earth's surface.

The Atmosphere. The earth's atmosphere may indeed be wispy and ethereal, but all life depends on this fragile veil.

Chemically, it is a protective envelope of gases—78% nitrogen, 21% oxygen, and less than 1% other gases, including small amounts of carbon dioxide and water vapor. In addition, there are small quantities of ions, such as nitrates, phosphates, and sulfates, that may mix with atmospheric water to create weakly acidic rain. (See Essay 48.1.)

Most of the earth's atmosphere clings close to the planet, not extending more than 5–7 miles above the surface. In addition to its reservoir of useful gases, the atmosphere is also vital to life in that it screens out much of the dangerous ultraviolet radiation that would otherwise make the earth's surface inhospitable.

As we also have noted, the atmosphere absorbs heat and, in so doing, acts as a great "heat sink," temporarily holding heat close to the earth's surface. In its role as a heat sink, the atmosphere can be compared to a florist's greenhouse. In a greenhouse, light energy is readily admitted through the windows, but as it strikes the surfaces in the interior, that energy is radiated back as heat energy, which cannot easily escape back through the windows. Thus greenhouses remain warm—even in the winter—but only as long as ample sunlight is available. The heat striking the earth is trapped in this manner by atmospheric carbon dioxide and water vapor. (People sometimes learn about the "greenhouse effect" the hard way—when they return from shopping after having left their pets in a closed car, finding the unfortunate animal overheated, dehydrated, or dead.)

Essay 47.1 ● **THE CHANGING CARBON CYCLE AND THE GREENHOUSE EFFECT: A DESTABILIZED EQUILIBRIUM**

In recent years, both physical and biological scientists have become increasingly concerned with the status of the carbon cycle. In the past 80 years alone, atmospheric carbon dioxide has increased by about 15%, and at the present rate of increase, human activities could easily double the present level over the next 40 years. Recently, a group of aroused scientists from the prestigious National Academy of Sciences informed Congress of the matter and initiated, of course, hearings. The hearings go on and the CO_2 continues to rise, but at least the problem is becoming recognized. The increasingly politically conservative Environmental Protection Agency concurs with the findings of the NAS, and they too are beginning to address the problem. However, there will be no quick fixes to the global problems of increased atmospheric carbon dioxide.

Some of the more hopeful solutions once suggested have not materialized. For example, not long ago some scientists said that the oceans would act as a great buffer by absorbing any overburden of atmospheric CO_2. Unfortunately, the oceans are not very good at absorbing the gas—the mixing of water and gas occurs only in the top 80 meters of the sea.

So what does an increasing level of CO_2 mean to us? It means trouble. The trouble arises because carbon dioxide does not absorb or reflect short (ultraviolet) light waves, but it does absorb and reflect the longer (infrared) light waves. Thus it freely admits solar energy into the biosphere but it slows the escape of heat that radiates from the surface of the earth. Carbon dioxide is thus a factor in the balance between energy entering and leaving the biosphere. So an increase in atmospheric carbon dioxide results in an increase in heat in the biosphere. The principle is similar to that in a greenhouse, which works by letting in the short light waves and retaining the longer waves of heat energy. Thus the carbon dioxide effect has become known as the *greenhouse effect*.

According to the experts, when this phase of the temperature cycle eventually passes, we may see some severe greenhouse effects. Warming of the earth would result in a warming of the ocean. If the ocean water gets warmer, the solubility of carbon dioxide will decrease. There are much greater reserves of dissolved carbon dioxide in the

Distribution of carbon in the biosphere

Plants (550)
Atmosphere (700)
Decomposers and humus (80)
Soil exchange pool (1,100)
Oceans (40,000)

Numbers represent billions of tons

ocean water than there are in the atmosphere, and an ocean temperature rise of only 1 or 2°C would unload even more carbon dioxide into the atmosphere. This, in turn, would increase the greenhouse effect and further raise the world's temperature, which would cause the release of still more carbon dioxide. To make matters worse, the accelerated melting and subsequent decreases in the size of the polar ice caps would further decrease the amount of solar energy reflected back into space. Again, we see what is known variously as a positive feedback loop, destabilized equilibrium, or vicious circle.

Of particular concern is the Antarctic ice pack, much of which is actually above sea level, either in floating masses or over the Antarctic continent itself. Obviously, the melting of such ice will raise the present sea level. By some estimates, this could be as

great as 6 meters (almost 20 ft), but more conservative estimates suggest that the mean sea level will increase by about 3 meters (almost 10 ft) by the year 2100. The most immediate effects will, of course, be felt in coastal regions and low-lying inland areas that have access to the sea.

But, quite possibly, flooding along the continents is not the most serious consequence to consider. A far greater problem may be the effect of slight increases in the earth's mean temperature on climatic patterns. Studies of model systems help in our predicting patterns of weather change. For example, in the western United States, a change of just a few degrees will alter precipitation enough to reduce the flow of the Colorado River by half. This could be disastrous to the great population centers of the Southwest, which are completely dependent on the Colorado River water for consumption and irrigation of crops. In other parts of the globe, there would be drastically increased river flow—for example, in the Niger and Nile in Africa; the Mekong, Volta, and Tigris-Euphrates in Asia; and the Sao Francisco in Brazil. As far as agriculture *per se* is concerned, Americans have long enjoyed the favorable climatic conditions that make us the world's leading producer of grains. Consider the political and economic ramifications of any significant climatic changes affecting this capability. For example, according to one of the predictions, the rain belt supporting American grain production could well move north, establishing itself in Canada and leaving the American grain belt in a semiarid condition.

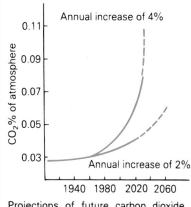

Projections of future carbon dioxide increases

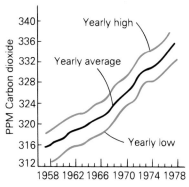

Recorded increases in atmospheric carbon dioxide

It is important to note that while the total energy the earth receives from the sun has historically been equaled by the escape of radiant energy, this equilibrium is being seriously disrupted as humans continue to pollute the atmosphere, especially by adding carbon dioxide. Every form of combustion—from breathing to burning fossil fuels to clearing forests and fields—releases carbon dioxide. The added carbon dioxide in the atmosphere contributes to the earth's greenhouse effect as it traps radiant heat, producing what is believed to be a gradual but inexorable increase in atmospheric temperature. (The greenhouse effect is examined in further detail in Essay 47.1.)

Climate in the Biosphere. One of the most striking effects of solar energy reaching the earth is seen in the great annual seasons of the planet. The cyclic seasonal changes occur as the earth follows its orbit, exposing different parts of its surface to the direct rays of the sun as it moves. This changing exposure occurs because the earth somehow ended up with a rotational axis that is not perpendicular to the sun's rays, but is 23.5° from vertical (Figure 47.3). Thus most of the earth's creatures are blessed with fluctuating but moderate surface conditions, and not the alternatives: great heat at the equator and huge expanses of perpetually frozen belts where the temperate zones are now located. This tilted axis also produces less dramatic phenomena, such as the tradewinds, patterns of rainfall and, of course, seasonal change. How do such changes arise?

The sun's rays fall more directly on the equator than any other part of the earth, so the equatorial regions are hot. This heat causes great warm air masses to rise, carrying with them large amounts of water vapor. Warmer air holds more moisture than does cooler air. The rising of warm air masses creates a void below, and lower, colder, and drier air masses rush in from the north and south. These masses, in turn, heat up, become moisture-laden, and rise. Thus we have air cells both north and south of the equator circulating in opposite directions (Figure 47.4). Because of the earth's rotational force, the cells are thrown off at an angle. This is called the **Coriolis effect**. These moving air cells produce what are called the tradewinds, the consistent winds so long used by sailing vessels. You can see in Figure 47.4 that there are other air cells in more northerly and southerly latitudes.

The movement of moisture-laden air from the equator creates the equatorial rainfall patterns. The rising air cells cool rapidly and lose most of their water as rainfall near the equator. As the air in the cells moves northward and southward, it becomes increasingly drained of moisture and thus yields less precipitation. Finally, at about 30° latitude, north and south, the lack of rain produces deserts.

Other factors may also contribute to the formation of deserts. One is called a **rain shadow**, a region of low rainfall on the lee (sheltered) side of a mountain range. Wherever moisture-laden prevailing winds encounter a high mountain range, most of the precipitation falls on the windward slopes of the mountains. As the air masses move up the mountain slopes, they cool and lose most of their moisture as rain or snow. The drier air mass moves down the leeward slopes (becoming warmer and therefore able to hold more moisture), then scurries over the desert and gathers up whatever moisture there is, intensifying the situation. As an example, the Great American Desert is partially a product of the rain shadow produced by the Sierra Nevada Mountains. Moisture-laden air, moving eastward from the Pacific, is confronted by the mountains. As the air masses rise, they cool, and most of the water is dumped on the western slopes, with very little to the leeward. In South America, where prevailing winds blow to the west, the rain shadow is produced by the Andes range, and the deserts that form west of the Andes along the coasts of Peru and Chile are some of the driest known.

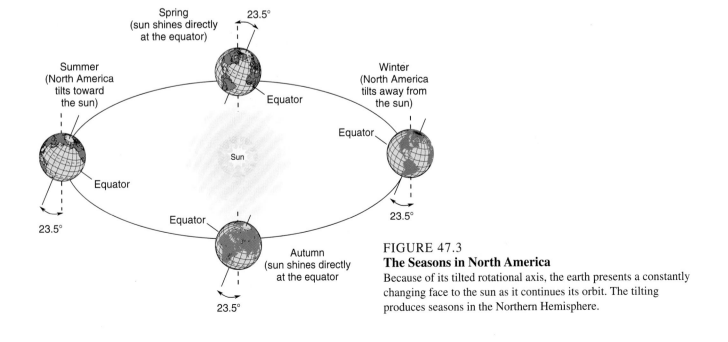

FIGURE 47.3
The Seasons in North America
Because of its tilted rotational axis, the earth presents a constantly changing face to the sun as it continues its orbit. The tilting produces seasons in the Northern Hemisphere.

FIGURE 47.4
Moving Air Masses and Rainfall Distribution

The distribution of precipitation on the earth is determined to a large measure by the formation of several groups of air cells (*wide arrows*). Air nearest the equator rises, carrying abundant moisture with it. As it rises, cooler, drier air rushes in underneath, and the cell rotates. The rising air mass cools and dumps most of its water in belts north and south of the equator. Rainfall is far more limited just above and below these belts, and here are found some of the earth's great deserts. Note that the great rotating equatorial cellsare thrown off center by the spinning of the earth—the Coriolis effect—so the winds do not simply blow north and south, but occur in easterly and westerly directions (*narrow arrows*).

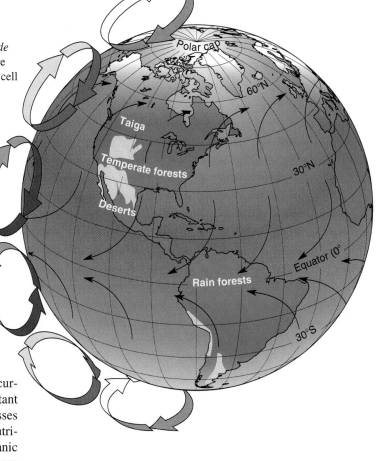

The moving air masses also help create the ocean currents (Figure 47.5). Ocean currents are, in turn, important because of their effects on the climate of nearby land masses and because they help mix the waters and distribute the nutrients and gases required by aquatic organisms. Major oceanic currents called **gyres** are found in the Northern and Southern Hemispheres. They are chiefly surface phenomena, with the flow extending down just 100–200 meters. The deepest flow of a major current is found in the Gulf Stream, extending as far down as 1000 meters. The movement of the major gyres transfers equatorial heat northward and southward, a feat that profoundly affects the climate of coastal regions.

The Distribution of Life: Terrestrial Environment

The earth's surface varies markedly from place to place, as does the life it supports. Vast, distinct, and recognizable associations of life are called **biomes**. More precisely, a biome is a particular array of plants and animals within a geographic area brought about by distinctive climatic conditions. Biomes are usually identified more by their plant associations than those of animals, not only because the first is far more obvious, but also because it determines the second. The specific plant associations are, as you would expect, a product of adaptation to several climatic factors, including precipitation, temperature, and light.

Biomes, in turn, may be subdivided into communities. As we saw in Chapter 45, a community is an assemblage of inter-

acting populations of organisms forming an identifiable group. A number of communities may exist within any biome. If you were to begin a walk across a biome, you might notice that although the general climate may remain the same, specific groups of plants and animals change somewhat. Thus in a desert biome, you might find both a mesquite community and a sage community. If you were somehow able to return to that biome hundreds of years later, you would find that it has changed; so change is another feature of biomes. Let's make a few more general observations about biomes and then describe several of the major ones.

The first point we might stress is that biomes are by no means homogeneous throughout their range. Within a given biome, it is possible to find different community types, thanks to variation in local environmental conditions, particularly topography, soils, climate, and disturbance (both natural and human-induced). Adding to this is the fact that the geographic ranges of many of the component species do not span the entire biome. For example, the eastern deciduous forest near the Great Lakes is dominated by beech and sugar maple, while that in Missouri is dominated by oaks and hickories. Ecologists have studied biomes extensively over the past century, but opportunities for additional research are still great, especially in the tropics. A map of the earth biomes is seen in Figure 47.6.

Biomes meet at ecotones (abrupt borders) and *transition zones* (gradual and perhaps containing species from both bio-

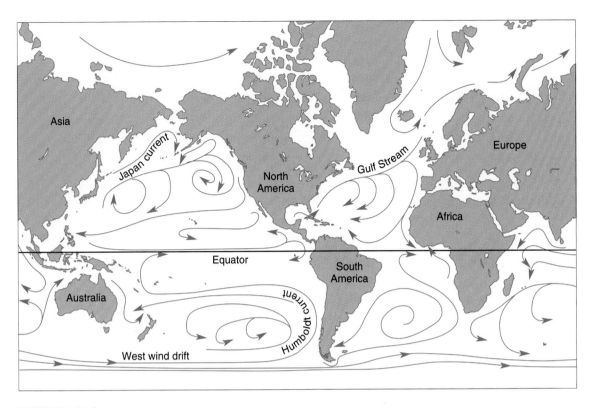

FIGURE 47.5
The Ocean Currents
The major ocean currents are produced by the earth's winds and modified by its rotational forces. The two great Pacific currents, the Japan current in the north and the Humboldt current in the south, carry cold water south and north along the west coasts of North America and South America, respectively. Note that the Atlantic Gulf Stream originates in a more tropical region and carries warm water north along the east coast of North America before heading out across the Atlantic, giving the east coast of North America its moderate climate.

mes). For example, we find a gradual transition across the United States from the east coast and Appalachian Mountains to the western coastline (Figure 47.7). The moist forests of the Appalachians slowly give way to drier oak–hickory forests and then to forests consisting almost entirely of oak. The forests become less luxuriant as they fade into the great American grasslands: that is, they did before the grasses yielded to the intensive agriculture of "America's breadbasket." The prairies were once seas of tall grasses, which gave way (where there was less precipitation) to shorter and shorter grasses. These yield to the Great American Desert, which is followed by the Sierra Nevada and coastal mountain ranges and finally the Pacific shore. Coniferous forests are common on the mountain slopes, and minor grasslands and deserts are found between some ranges. The borders of biomes are usually indistinct, with the mixture of plants seemingly engaged in an endless tug of war over boundaries. The principal determining factors in transitions are precipitation, temperature, and topography (see the biome map in Figure 47.6).

The distribution of biomes on the earth generally follows latitude, but this pattern is more obvious in the Northern Hemisphere than in the Southern Hemisphere. Going from the equator northward, we tend to move from tropical rain forest through desert, grassland, temperate deciduous forest, the taiga (coniferous forest), and finally tundra and ice cap. This latitudinal arrangement of biomes is not entirely orderly for several reasons, one of which is the terrain, or topography. Mountain ranges interrupt the orderly distribution of biomes and, as we have seen, are often responsible for the presence of deserts because they can block the movement of moisture.

Mountains also influence biomes in another interesting way: biome distribution is a product of altitude as well as latitude. This is because temperature decreases as altitude increases. Thus high mountains with permanent snow or ice are found in the tropics, and the usual biome transitions may be represented as well. Furthermore, the *altitudinal* transition may resemble the latitudinal one (grassland into deciduous forest, and so on; see Figure 47.8).

One of the best examples of altitudinal transition is the Ruwenzori mountain range in east central Africa. You ascend from tropical rain forest through broad-leaved evergreens, to deciduous and coniferous forests, then to alpine meadows, and finally to the barren, snowswept peaks. Once again we see that the factors determining the distribution of life are complex.

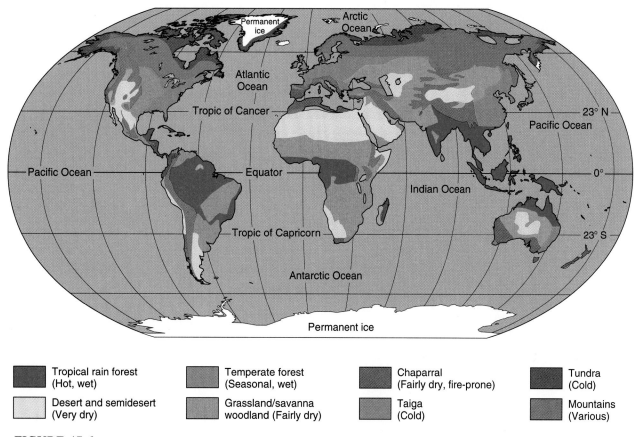

Tropical rain forest (Hot, wet)

Desert and semidesert (Very dry)

Temperate forest (Seasonal, wet)

Grassland/savanna woodland (Fairly dry)

Chaparral (Fairly dry, fire-prone)

Taiga (Cold)

Tundra (Cold)

Mountains (Various)

FIGURE 47.6
The Biomes
Each of the biomes can be identified by its dominant form of plant life. These forms are adapted to the climatic conditions, including precipitation, the availability of light, and, of course, temperature. Both latitude and altitude affect all these conditions.

The Biomes

Keeping in mind that the dividing lines are often indistinct and arbitrary and that the complexities are greater than any brief discussion can convey, we will now consider the nature of the earth's great biomes, beginning with the driest and perhaps most fabled.

Deserts

Deserts are areas that receive less than 25 cm (10 in) of rain each year and where evaporation very likely exceeds rainfall. Such places generally occur at about 30° latitude, north and south. You may be surprised to learn that deserts are not necessarily hot (even in the day) and not necessarily tropical. The largest desert is the Sahara, which, as you can see in Figure 47.6, covers nearly half of the African continent (and is getting bigger). Other larger expanses of desert are found in Australia, Asia, western North America, and South America (in fact, in every continent but Europe). Temperatures in these regions undergo dramatic day–night fluctuations and may vary as much as 30°C (54°F) in a 24-hour period. The reason for such extremes is the lack of buffering moisture, which resists temperature changes. In much of the year, the surface heats up rapidly in the daytime but cools down by evening. In spite of the long periods of drought, much of the actual topography of the desert floor is determined by water. Very seasonal but torrential rains cause flash flooding that continually remakes the face of the desert.

If you have never seen a desert, you might have an image of lifeless regions of drifting sand studded with a few palm-covered oases. Actually, there are such places, but most deserts are alive with plants and animals. Since dry areas are called *xeric* (Greek for "dry"), plants adapted to deserts are called *xerophytes*. Their adaptations can take a number of routes, but they have one common imperative: save water.

Native American perennials, such as cactus, ocotillo, Joshua, creosote, and palo verde (Figure 47.9), are adapted to living long periods on what little water they contain, while waiting for rain. Through evolution, the leaves of the cacti have been reduced to spines, thereby conserving water, and photosynthesis occurs mainly in the green stems. The stems are covered with a thick, waxy cuticle, and water is stored in the oversized cells of deeper tissues. The spines of cacti and the thorns of many other desert plants discourage animal browsing.

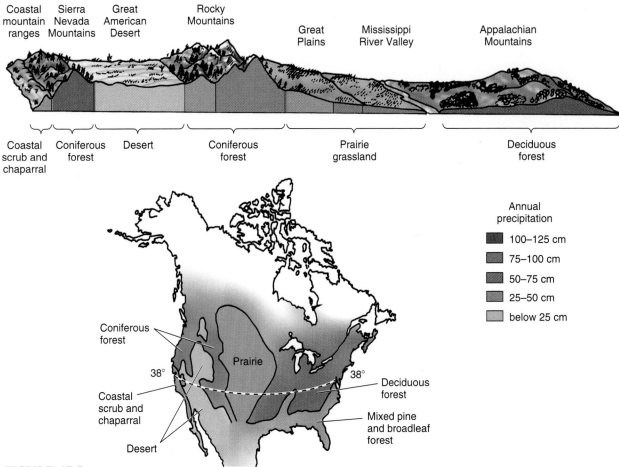

FIGURE 47.7

Profile Along the 38th Parallel Crossing the United States

The line transects several of the major biomes and helps indicate the conditions by which they were produced. For example, the American deserts lie in the rain shadow of the western mountain ranges. Rain and snow fall mostly on the western slopes. Moisture-laden air from the Atlantic, at the other end of the profile, drops its burden as it moves west, the total rainfall diminishing as it moves inland. Forests occur west to the Mississippi, but from there, prairie extends westward, dwindling into short grasses toward the Rocky Mountains. The Great American Desert extends between the Rocky and Sierra Nevada Mountains.

Compared to other plants, desert perennials tend to have fewer and more widely scattered stomata. This helps in conserving water, but the price is a considerably slower growth rate, since CO_2 uptake and photosynthesis are retarded. Other desert perennials have deep root systems able to tap whatever water seeps into the porous soil, very small leathery leaves, and the ability to lie metabolically dormant, growing very little, for long periods. Some desert cacti and other succulents have reversed their gas exchange cycles, taking advantage of the cooler and moister nights. This latter group is composed of the CAM plants discussed in Chapter 30.

The desert annuals have a different strategy for survival. After a spring rain (or *the* spring rain—there may only be one), countless dormant seeds germinate, seedlings quickly sprout and mature, and the desert becomes transformed as tiny flowers of all descriptions erupt into full bloom, a trembling and riotous offering of vibrant color and delicate forms.

Their life cycle is short, and they soon die. But their seeds are highly resistant, lying in the sand until subsequent rains propel them into their surge of growth and reproduction.

Just as plants have had to adapt to the rigors of the desert, so have animals. Animals have the advantage of being able to adapt not only anatomically and physiologically but also behaviorally. The desert is rich in animal life, including a variety of arthropods, resident birds, and often many seasonal migratory bird species. There are also reptiles and mammals. In the hot deserts, the primary problems of all these species include escaping daytime heat and avoiding water loss. Many desert animals beat the heat by simply staying out of the sun and becoming nocturnal. In fact, if you should visit the desert you may see only a few lizards, a bird or two, and a few insects—unless you want to traipse around at night with the snakes.

Desert mammals are largely represented by rodents, animals for which nocturnality has very clear advantages. They

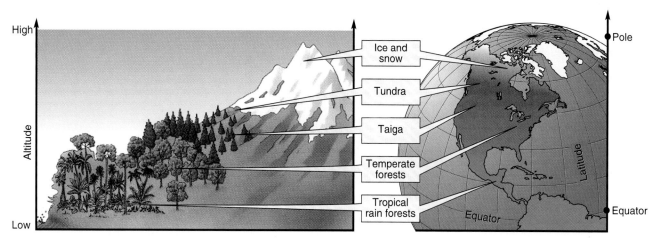

FIGURE 47.8
An Altitudinal Distribution of Biomes
The altitudinal variation provided by mountains often mimics latitudinal distribution. In this hypothetical model, a rain forest at sea level gives way to a deciduous forest at a higher elevation. Conifers replace the deciduous trees farther up, which yield to alpine meadows, and finally a rather typical tundra situation near the glacier.

tend to lose water quickly because they breathe rapidly and have a large surface area relative to their volume. Thus most rodents spend the desert days deep in insulating burrows, venturing out only at night to forage. Rodent-hunting predators such as owls and rattlesnakes follow suit, confining most of their activity to the cool of evening or night.

The few daytime animals, such as long-legged and fast-moving lizards, are preyed on by hawks and roadrunners, which are also adapted to daylight conditions. However, even the daytime animals restrict most of their activity to the morning and evening hours (Figure 47.10).

Perhaps we can best gain some insight into the adaptive strategies of desert animals by considering a mammal and an amphibian. ("Adaptive strategy" refers to the development of traits in response to an environmental condition and does not imply a purpose on the part of the organism.) These are the versatile kangaroo rat and the spadefoot toad. The kangaroo rat (*Dipodomys deserti*) of the southern California desert is particularly interesting because it doesn't drink. It survives on the water content in its food and supplements this with metabolic water (produced as a waste product in cell respiration; see Chapter 8). The kangaroo rat is also a great water miser. Its remarkably efficient kidney produces only highly concentrated urine (23% urea and 7% salt, compared to 6% urea and 2.2% salt in humans), and very little at that. The feces are also dry and crumbly. Most water loss, in fact, occurs through simple breathing. Even these special physiological features, however, wouldn't permit desert survival were they not coupled with nocturnality. The rat spends its day in a humid, hair-lined burrow, venturing forth in the cooler evening.

While amphibians are notably scarce in the desert, the fascinating spadefoot toad is a year-round resident of the American deserts. It also escapes heat and drying by burrowing, and while entombed it is also capable of **estivation**, a time of dormancy,

Joshua tree

FIGURE 47.9
Desert Plants
The North American desert plants have adapted to limited water; these are called xerophytes. Their thorny epidermis discourages browsing and helps shield the green surfaces from the harsh, direct sunlight. Many xerophytes have pulpy water-storing tissues within. Shown is the Joshua tree.

(a) Roadrunner

(b) Fennec fox

FIGURE 47.10
Day and Night Desert Foragers
Animals in the desert forage in two shifts. The day feeders are fewer in number but much more active.
Included in their ranks are the African ground squirrel, the roadrunner shown in **(a)**, and many insects. But
even these hardy foragers often remain in the shadows at midday. At sundown the second shift begins as
carnivores like the scorpion and the Fennec fox shown in **(b)** go quietly about their deadly business.

during long, hot, dry intervals. When sufficient rain falls, it springs (hops?) into action. For this creature, time is short. Like other toads, the spadefoot must reproduce in water. Finding a temporary pond, the male begins at once to call prospective mates in the briefest of courtships. Eggs, fertilized in the shallow ponds, hatch in a day or two. The young tadpoles complete their metamorphosis and emerge as adults in a few short weeks.

Although deserts seem tough and unyielding, they actually represent one of the more fragile biomes. Simple systems are always the most vulnerable, and deserts are essentially simple places, harboring few species compared to many other places. Because the desert is so vulnerable, the impact of humans can be great.

The recent experience of the region to the south of the Sahara is sobering. People of the Sahel region long supported themselves and their small herds of cattle by hand-drawn water from a few scattered deep wells. In recent decades, gasoline pumps were introduced, making water suddenly abundant. The result was an explosive increase in the cattle population, followed by a substantial increase in the human population. The cattle placed great pressure on the vegetation of the area until a recent drought—not unlike the many droughts the region has previously survived—killed most of the cattle and many of the people. As a result of such destructive patterns, the Sahara itself has expanded by thousands of square miles, probably permanently, demonstrating that well-intentioned but ill-conceived aid programs can backfire with tragic results.

Grasslands

Simply put, **grasslands** are areas dominated by grasses. In the Northern Hemisphere, they exist as huge inland plains and include such areas as the Asian *steppes* and (in times past) the *prairies* of North America (Figure 47.11). The Southern Hemisphere also has well-developed grasslands including the *pampas* of South America, the *veldt* of Africa, and the *outback* grassland in Australia that occupies half the continent.

Grasslands receive less rain—roughly 25–75 cm (10–30 in.)—than forest or savanna ecosystems, and much of that rain is seasonal. Also, grasslands are subject to periodic burning (often started by lightning) and to grazing pressure from a variety of large and small animals.

Yet grasses are ideally adapted to these factors. Unlike many other plants, they have narrow leaves that stand almost vertically, a trait that allows reduced evaporation. In addition, the active growing region of the plant is at the base of the leaf, near the ground level. Most other plants have their actively growing points at the tips of the stems (see Chapter 29). Being near the soil, the vital growing points are protected from grazing and most fires.

Grasslands are productive ecosystems and often support larger populations of animals than any other biome on earth. Thus you can find huge herds of grazing animals, particularly large-hoofed herbivores (see Figure 47.11). The original herbivores of the American prairie—the bison and the pronghorns—have been displaced by cattle and sheep, but in the plains of Africa there are still vast herds of migratory wildebeest, zebra, and other natural grazers.

Still, some grasslands can be harmed when they are overgrazed by cattle or cropped too short as sheep may do. The shrub- and cactus-covered wastelands area in Texas (55 million acres) was a stable, productive grassland before the cattle barons subjected it to heavy grazing. Much of the Sahara and the deserts of the Middle East, in fact, have been created by domestic grazing in past centuries. The barren Middle East, remember, was once called "the land of milk and honey."

While grasslands are home to a number of conspicuous mammals such as bison and pronghorn antelope, there are also many unobtrusive grass eaters, such as jackrabbits, rodents, and prairie dogs, as well as insects and seed-eating birds. These

FIGURE 47.11
Grasslands
Natural grasslands such as the South American pampas are slowly changing with the encroachment of civilization. The highly efficient producers of the grassland support an extensive food web, often including incredible numbers of large herbivores. These mammals, in turn, support a sizable predator population.

herbivores, in turn, provide food for a large number of carnivores, such as coyotes, badgers, rattlesnakes, hawks, and owls.

To the casual observer, grasslands often appear uniform, if not monotonous. However, grasslands vary markedly from place to place. For example, the grasslands in the western parts of the American prairie (eastern Wyoming and Colorado) are dominated by species that normally do not grow more than half a meter (1.5 feet) tall. In contrast, those to the east (Illinois and Indiana) are dominated by plants that exceed 3 meters. The difference in height is traced to the fact that the eastern tall-grass prairie receives more rainfall than the short-grass prairie to the west.

During the past century, the prairies were largely destroyed by farming and ranching. The western part of the prairie is now home to wheat and cattle production. The eastern part is now a great corn belt. Much of that conversion was hastened by the invention of a sturdy plow capable of breaking up dense sod (indeed, the first grassland farmers were termed "sodbusters"). Not only is the native vegetation now largely gone, but so too is much of the animal life. In particular, the normal grassland carnivores like wolves, cougars, coyotes, and foxes have virtually been eliminated in places by hunting and habitat destruction. Within the past fifteen years, however, there has been an effort to restore some of the native grasslands so that future generations can appreciate this fascinating biome.

Tropical Savannas

Tropical savannas are a special kind of grassland that forms at the borders of tropical rain forests. Unlike other grasslands, the savanna is frequently interrupted by scattered trees or groves (Figure 47.12). And unlike the tropical forests, the savannas have a prolonged dry season with an annual rainfall of 100–150 cm (40–60 in.). It is during the dry season that the savannas are subject to sweeping fires. The great annual droughts are also responsible for the movements of huge migratory herds of grazers in search of food.

The largest savannas occur in Africa, but savannahs are also found in South America and Australia. While grasses are the dominant form of plant life in the African savanna, the drab landscape is brought to life by palms, colorful acacias, and the strange, misshapen baobab tree, which appears to be growing upside down. The size of hooved animal populations exceed that of all other biomes and include the familiar zebras, giraffes, wildebeests, and numerous antelopes of the African plains. The African savanna is also the domain of familiar predators, such as the lion and cheetah. Just as in the grassland, the natural fauna of the savanna is being replaced by domestic grazers, especially cattle.

Tundra

The **tundra** is the northernmost land biome, characterized by Arctic plains that support a dense growth of mosses, lichens,

FIGURE 47.12
The Savanna
Large-hooved mammals are prominent among the savanna's herbivore population. On the African plains, predators include the large cats.

FIGURE 47.13
The Tundra
In summer, the treeless tundra becomes a veritable marsh as the snow melts. With little runoff, the landscape becomes dotted with innumerable small ponds. Below the water the soils of the tundra are perpetually frozen. The plants dotting the landscape include a number of dwarfed trees, grasses, and abundant reindeer moss.

and dwarf herbs and shrubs. Except for certain alpine meadows, it has no equivalent in the Southern Hemisphere. Tundra is located in a narrow band between northern taiga and the Arctic ice extending from the tip of the Alaskan peninsula around the earth and back to the Bering Sea (see Figure 47.6).

Travelers in the tundra may be struck by the absence of tall trees and shrubs. The annual precipitation is often less than 15 cm (5 in.), and much of this occurs as snow. So it's a dry place— at least during the long winter when everything is frozen. However, you wouldn't think it dry if you visited the tundra during its brief, damp summer. When spring and summer finally arrive, the upper few feet of soil thaw, leaving the *permafrost,* perpetually frozen soil, below. The surface thaw produces unusual conditions for a "desert." Ponds begin to form everywhere. Since water cannot percolate down past the icy permafrost, the plains of the tundra become a veritable bog (Figure 47.13).

The tundra receives less energy from the sun than does any other biome. Because it is near the pole, any sunlight there strikes at an acute angle, losing much of its energy after having traversed diagonally through the energy-absorbing atmosphere. The lack of sunlight drastically affects the growing season. During the brief six- to eight-week summer, plants must photosynthesize and store enough food to last through the rest of the year. And they must compress their reproductive period into this brief season, all of which means they can't put much energy into growth. So the tundra is carpeted by low-lying plants. Another factor that dictates that plants hug the ground is the constant high wind, particularly in regions of higher elevation. A tall plant would simply be buffeted to death or even ripped away.

Because of the permafrost, the plants of the tundra can't form deep anchoring root systems; so they form shallow, diffuse roots that often become entangled with the roots of their neighbors to form a continuous mat. In the wet season, any disturbance on hilly slopes can break the entire root mat loose from the wet soil underneath and send a great mass of plant conglomerate sliding down, exposing the barren soil below. Because of the limited growth of its plants, tundra recovery is a very long, very slow process.

Lichens and mosses are common in the tundra, but so are willows and birches. These latter, however, are almost unrecognizable, being only dwarfed symbols of their more southerly relatives. They, with the lichens and mosses, are joined by grasses, rushes, sedges, and other annuals to complete the summer ground cover (see Figure 47.13).

Animal life isn't as rare as one might expect in this most northerly biome. Where autotrophs are at work capturing and storing energy, there will always be opportunistic heterotrophs ready to harvest it. In fact, year-round residents of the tundra (Figure 47.14) include some rather large herbivores such as, in North America, the caribou and musk oxen, and, in Europe and Asia, reindeer. Other browsing animals found here are the ptarmigan, the snowshoe hare, and the ever-present and legendary lemmings. Lemming populations are clear indicators of how good the season is for producers. In good years, lemming populations soar.

Lemmings, in turn, determine the success of a number of predators, including the arctic fox, lynx, snowy owl, weasel, and arctic wolf. Also, the jaeger, a migratory bird, travels great distances to feed on the tiny rodents. The duration of the predators' stay, as well as their reproductive success, will depend on the number of lemmings. In years when lemmings are few, many of the predatory birds will migrate early, while the permanent residents will survive by switching to other prey.

A number of waterfowl and shore birds also migrate to the tundra. When they arrive in the spring they must mate and rear their broods quickly, before the brief summer is over. Why would some species travel long distances to breed in such a dismal and risky place? Perhaps because the long days permit extended feeding periods and less competition for food than might be present in the winter feeding grounds. If winters are harsh, resident competitors are few.

While invertebrates are certainly more limited in the frigid tundra, remarkable swarms of mosquitoes and tiny flies abound. They enter a frantic race to complete their reproductive activities before summer ends, since the adults only live for one season. Such insects survive by producing highly resistant immature stages that remain dormant through the long winter.

Winter comes early in the tundra, and with the rapidly shortening days the migratory animals disappear. Some caribou, for example, leave for the forested taiga, where winter food is more plentiful. Those that remain prepare for survival by whatever strategy they have developed. Lemmings retreat to food-laden burrows, while ptarmigans tunnel into snow banks to emerge periodically on foraging expeditions. Since the larger herbivores

Caribou

Lemming

FIGURE 47.14
Animals of the Tundra
Because of the paucity of species in the tundra, the food webs may be comparatively simple (and therefore easy to disrupt). Common herbivores in the tundra include large animals such as the caribou and small ones such as the lemming.

and predators don't hibernate, they must rove the barren, windswept landscape to feed on mosses, lichens, and each other.

Our own species, with the exception of a few Laplanders in northern Scandinavia and Finland, avoids the tundra. Even the hardy Eskimos prefer the Arctic coastline. For this reason, humans have traditionally had little effect on this most fragile ecosystem. This is fortunate, because the links in the food chains here are few and important. The sudden loss of a single predator or producer through human intervention could be disastrous. We know that recovery from damage is agonizingly slow and costly to the tundra communities. Our impact, however, is increasingly disruptive as we search out more oil and other minerals in this delicate biome.

North of the tundra (or in other places at higher altitudes), the vegetation gives way to barren and rocky soil similar to that of the Antarctic. Plant life is sparse and patchy. These dry windswept plains extend to the coastal ice floes and glaciers. Here the marine environment, in sharp contrast, is amazingly rich in life, even in its colder waters. Of course, the perpetual search for more petroleum threatens these rich waters as well.

Before leaving this biome, we should note that a similar vegetation type exists along the upper reaches of many tall mountains. Such communities are called *alpine tundra,* and they likewise have to cope with short growing seasons and long, snowy winters. Since they are at high elevations, plants and animals of alpine tundra are exposed to a thinner atmosphere, higher levels of ultraviolet radiation, and stronger winds than the arctic tundra. However, unlike the arctic tundra, the alpine tundra does not have permafrost, and therefore the soil does not remain waterlogged.

Now we move to the woods of the world, the great forest biomes.

Tropical Rain Forests

Forests are scattered over much of the world, but only in those places where the water supply is adequate. Rarely is water more available than in the **tropical rain forest**. These are

tropical woodlands with an annual rainfall of at least 250 cm (100 in.) and up to 450 cm (180 in.), typically found in lowlands. Rain falls throughout the year but is somewhat heavier during the "rainy season." The largest tropical rain forest is in the Amazon River basin in South America (Figure 47.15). The second largest is in the wilds of the Indonesian archipelago. And then there are those of the Congo basin in Africa, parts of India, Burma, Central America, and the Philippines.

Tropical rain forests are best described as lush, with a very large number of tree species growing to great heights. The floor is dark and wet, and the air is often cool and laden with rich smells. Unlike what we find in the other forest biomes, no single kind of plant dominates the forest. Any tree is likely to be a different species from its neighbors, and nearest trees of the same species are often miles apart. The exceedingly tall trees, ranging in height from 30 to 45 meters (about 100 to 150 ft) form a dense, continuous canopy overhead. Here and there the canopy is broken by even taller trees, ranging to 60 m, called *emergents,* whose top branches may have a different microclimate than that of the canopy because breezes are not blocked from reaching them. The crowns of shorter trees below the canopy form a layer called the *subcanopy.* The trees are invaded by large numbers of vines, and both trees and vines may be festooned with *epiphytes,* plants that live on the stems and branches of tall trees in the canopy, with no contact with the soil. They absorb water directly from the surrounding humid air. (One species surrounds its roots with a bucket-like base that collects water and drowned insects, the decay of the latter assuring the epiphyte of a continuing supply of nitrogen compounds.) The forest floor may have little to moderate foliage, but it is teeming with fungal and bacterial decomposers and insect scavengers. The darkness, warmth, and blanketing humidity there are ideal for rapid decomposition.

If the rain forest floors often don't have much foliage, then what about those reports of the "impenetrable jungle"? The answer is that jungles are special places in tropical forests. Essentially, they contain a low-growing tangle of plants, which is, in fact, almost impenetrable. Jungles arise where light reach-

FIGURE 47.15
The Tropical Rain Forest
Tropical rain forests receive enormous amounts of rain throughout the year. There is little seasonal change. No single plant species dominates the terrain, but a number of kinds form a dense canopy over the sodden earth. The humidity on the forest floor can be stifling.

es the forest floor. Jungle areas may be scattered through rain forests, but they are particularly common along river banks and steep slopes and in disturbed areas such as clearings and deserted farms. Along river banks the forest is called *wet jungle,* and it abounds with insects and reptiles. The idea that jungles represent tropical rain forests grew from descriptions by river travelers who weren't about to get out of the boat, so they missed seeing the relatively clear forest floor, often just a few hundred yards from the river banks.

One remarkable feature of the tropical rain forest is that it has exceedingly high species diversity. A 1 hectare ($2\frac{1}{2}$ acres) patch of rain forest is likely to contain 40–100 species of trees. In contrast, a similar area of temperate forest will contain only 5–15 species. The tropical rain forest also harbors a great many animal species, more than any other biome. Insects and

birds are particularly abundant, and reptiles, small mammals, and amphibians are common. Many of the animals are arboreal (tree dwellers), and many species are stratified according to the layers established by the plants, becoming specialists at occupying certain levels of the canopy and subcanopy (Figure 47.16). In one study of the Costa Rican rain forest, ecologists found 14 species of ground-foraging birds, 59 species occupying the subcanopy, and 69 in the upper canopy. They further found that about two-thirds of the mammals there were arboreal, as were a number of frogs, lizards, and snakes.

Tropical seasonal forests occupy a considerable portion of the tropical biomes. These differ from true tropical rain forests in that rains are, as you might expect, seasonal. A familiar example is the monsoon forest of Southeast Asia, although such forests exist in other tropical regions. In many instances, the trees there are deciduous, losing their leaves during the dry season. When the monsoons arrive with their torrential rains, the forest takes on some of the characteristics of the tropical rain forest. Some of the more highly valued hardwoods, such as Burmese teak, are found in the tropical seasonal forests.

The tropical rain forests of the world are rapidly disappearing. The Amazon basin rain forest of Brazil, the last really large area of undisturbed forest, is now in the process of being rapidly cleared for timber and farming and being exploited for minerals. The results may prove to be catastrophic for several reasons (see Essay 47.2).

Chaparral

Mediterranean scrub forest, or **chaparral** as it is known in California, is characterized by a dense growth of low evergreen shrubs and trees. As Figure 47.6 shows, it is rather insignificant among the forests of the world. It does, however, have unique characteristics and its own peculiar plant associations. Note, for example, that chaparral is exclusively coastal, found mainly along the Pacific coast of North America and the coastal hills of Chile, the Mediterranean, the coast of Africa, and southern Australia. This forest is unique in that it consists of broad-leaved evergreens, growing in subtropical regions marked by a marine air flow, low rainfall, and a long summer drought. Depending on the altitude, California's coastal chaparral receives between 25 and 75 cm (about 10–30 in.) of rainfall, almost all of it in the short winter rainy season. The seasonality of the rain means that the plants experience drought through most of the year.

Plants and animals adapted to chaparral have developed strategies similar to those of desert dwellers. In the chaparral, plants are chiefly represented by shrubs—mostly with dwarfed, gnarled, and scrubby stems—and scattered succulents (Figure 47.17). Leaves are generally small, with very waxy and tough cuticles. Many plants become dormant after the seasonal rains, remaining that way through the dry summers. Insects abound in the chaparral, including armies of beetles that live in the leaf litter covering the forest floor. Other inhabitants include vertebrates such as mule deer, rabbits, bobcats, rodents, lizards,

Essay 47.2   ●   **THE DESTRUCTION OF TROPICAL RAIN FORESTS**

Fact number one: more than half the world's people live in tropical and subtropical areas. Fact number two: about one-third of these people are extremely poor. These two statements have set the stage for a potential worldwide disaster. The problem is that the poor are most likely to have direct access to the world's tropical rain forests. The forests can meet some of their immediate needs, and so they are utilized on a pell-mell, devil-take-the-hindmost basis. Locally, the great forests are yielding to the unending search for firewood and new planting and grazing areas by people who have it tough and who are forced to take what they can when they can. But the rain forests are being destroyed by other groups too, often by the rich. Much of the Central and South American rain forest has been cleared for grazing areas in order to supply North American fast-food places with beef, which reduced the price of U.S. fast-food hamburgers by about a nickel. The beef is of relatively poor quality and, in fact, now comprises only about 6% of our fast-food needs. (Some chains, such as Burger King, no longer import beef from these areas.) Rain forests are also being cut to export hardwood logs. Unfortunately, such logging operations are often highly inefficient. (In some Malaysian forests, over 50% of the trees in an area are cut to obtain 3% that can be sold.) By 1980, almost half the world's tropical forests had been destroyed or severely disturbed. By the year 2000, almost no tropical rain forests will have escaped such devastation.

The greatest impact occurs through simply clearing the land, and the clearing is extensive. About 35 acres of tropical rain forest are cleared each minute. Much of this forest will never be recovered, especially that which is bulldozed, because the thin layer of topsoil with its nutrients is removed. Also, germinating seeds are removed, and watersheds are disrupted. Forests that are simply cut will recover fastest, since seeds, seedlings, and a protective undergrowth are left. Land that has been cleared by burning recovers at an intermediate rate; burned-over land could take about 80 years to recover unless rainfall patterns are disrupted—as explained below.

Land is burned in order to clear the trees and also to release the nutrients they have stored back to the earth through what is sometimes called "slash and burn agriculture." Here the trees are cut and burned, thereby releasing their stored nutrients back to the soil. Because of heavy rains that wash the nutrients away, the land is only fertile for a few years, and then the forest and the crops are both gone.

The extremely heavy rainfall that gave rise to this rain forest in the first place is almost entirely composed of moisture sent aloft from the steaming forest itself. Little additional moisture moves in from the ocean. Cutting down the trees greatly reduces the amount of moisture returned to the atmosphere within the Amazon basin, and this will ultimately reduce the rainfall there by a substantial amount. The rain that does fall into the cleared land will tend to run off into the river systems and on to the ocean.

In addition, the tropical rain forests, which are being cleared at an average rate of 1% per year, represent an enormous carbon dioxide "sink." This means that the lush growth absorbs a great deal of carbon dioxide from the atmosphere, locking it up in the car-bon compounds of the plants. While death and decay cause a steady turnover, the sudden removal of plants interrupts the normal cycling. Furthermore, since the plants are burned, the carbon stored in their molecules is suddenly released as carbon dioxide. Atmospheric scientists are particularly alarmed by this steadily increasing carbon dioxide burden.

Finally, you may be surprised to learn that tropical soils are very poor and infertile. The nutrients formed by decomposition are immediately recycled back into plant growth, so no reservoir of humus remains. In addition, the rains continually leach precious nutrients from the porous soil. This fact has become painfully clear to proponents of jungle agriculture. Once the native plants have been removed, the soil may rapidly change into a hard, water-resistant crust known as *laterite*. The term, which means "brick," rather aptly describes the reddened crust.

One of the greatest problems resulting from the destruction of the rain forest is the destruction of a promise we may not even know we had. We know of only about one-sixth of the three million species believed to exist there. Some of these species may have medicinal properties that could remedy specific problems of humankind in ways we can now only imagine. There may be resistant genes in plants related to our vulnerable crops (which could lend their resistance to our crops through genetic engineering). There may be new kinds of food, antibiotics, and therapies awaiting our discovery. But some scientists believe that, at current rates, one million tropical species may be gone by the year 2010. And with them, sadly, go promises untold.

---

snakes, and birds. Wrentits and towhees are typical residents, but a large number of birds are migratory.

Chaparral has a peculiar problem related to drought. It is often seared by fast-moving brush fires. Controlling the fires along the California coast is complicated by the rough, hilly terrain. Little of the chaparral escapes the periodic fires, but the plants have adapted to the stress. Some sprout quickly from burned stumps, and others scatter fire-resistant seeds over the burned ground. Ecologists have described the chaparral as a *fire subclimax community*. This means that the chaparral virtually never reaches maturity, and any given stand is always in some stage of recovery. The floor of chaparral is often heavily layered with leaf litter. Decomposition is slow because of the dryness. The accumulation of those leaves (many plants shed leaves year-round) magnifies the fire problem.

Temperate Deciduous Forests

If you live east of the Mississippi and north of Florida, the **temperate deciduous forests** may dominate your surroundings. Temperate deciduous forests are marked by trees that lose their leaves in winter. Winter is quite dramatic here as the

FIGURE 47.16
Canopy Dwellers
Each level of vegetation in a rain forest supports specific kinds of animal life. As examples, the dense canopy (the upper layer, broken by occasional taller "emergent" trees) supports monkeys and birds; the subcanopy, composed largely of smaller trees, supports many kinds of insectivorous birds; the smallest trees and bushes are home to many frogs and snakes; and the forest floor supports deer, capybara, and pigs. Ants are found in abundance at all levels.

trees rake the gnarly fingers of their bare branches at the winter sky. Each spring they blush anew with fresh faint buds that will change to the deep greens of summer. Temperate deciduous forests then are characterized by both leaf fall and changing seasons (Figure 47.18).

Temperate deciduous forests extend over much of the eastern United States, northward into southeast Canada. They are also found in the central and northern parts of Europe, including Great Britain, and reach into southern Norway and Sweden. A long finger of the forest pushes into the center of Russia. In eastern Asia, deciduous forests are found in China, Korea, and Japan. Although much less conspicuous in the Southern Hemisphere, they do exist in coastal Brazil, east Africa, the eastern coast of Australia, and across most of New Zealand.

Precipitation in the deciduous forest is rather evenly distributed throughout the year, with rainfall often averaging over 100 cm (39 in.)—enough to support a variety of trees. Typically, the larger trees such as oak, maple, beech, birch, and hickory form the canopy. In northerly deciduous forests, communities of beech and maple dominate, while in the south, the oak is common. In times past the oak was joined by the chestnut, but this majestic tree has been all but obliterated by a bark fungus (*Endothia parasitica*), accidentally introduced from China at the turn of the century. Interestingly,

FIGURE 47.17
The Chaparral
The chaparral may lack the lushness of many other forests, but its plants are tenacious and hardy. These scrubby plants resist an annual drought that would discourage most other plants.

whereas the adult chestnut trees are gone, shoots still emerge from the persistent roots, and some seeds are produced, but further growth is stopped by the fungus.

In temperate deciduous forests, moderate levels of light reach the forest floor and encourage the growth of younger trees, shrubs, ground-hugging plants, and a variety of annuals. The annuals begin to grow in early spring, or even late winter. They rather quickly produce their seeds and die off with the autumn frosts, contributing to the rich leaf litter. The litter and humus harbor an abundance of scavengers and decomposers, especially in the warm summer months. The forest floor is like a giant soft sponge, soaking up rain and contributing to the luxuriant forest growth.

A variety of animal life abounds in deciduous forests. Although we have displaced or killed off most of the larger mammals, in some areas they have been protected as game for hunters. The forests were once the home of deer, wolves, bears, foxes, and mountain lions. Mammals in these forests today are largely represented by rabbits, squirrels, raccoons, opossums, and rodents.

The largest predators are likely to be humans, owls, hawks, a few black bears, and occasional bobcats and badgers. In the northern deciduous forests, a few wolves have escaped having their skins trim cheap coats. Finally, as any camper knows, arthropods inhabit the forest in great numbers and can be found rummaging through the litter and foliage and crawling into tents.

Winter in the deciduous forest is heralded by one of the most beautiful and moving events in nature: autumn. As the abscission layers in the leaf petioles prepare to separate, and the green chlorophyll wanes, other colors come to dominate the woodland scene.

The leaves almost seem to celebrate the cool days, as browns, reds, and yellows mix with the greens of conifers. It is too soon over. And the fading light of the shorter days, which changed the leaves, also warns the animals of the approach of winter. Some birds leave; others change color and give up territories. Some mammals, grown heavy with fat, find holes in which to sleep or to hibernate. Others, unable to escape either behaviorally or physiologically, simply face the coming cold; some will not survive. Soon, the stark, cold days of winter settle in.

The deciduous forest biome produces trees with strong and flexible cell walls, and much of it has been cleared in order to satisfy our need for wood. The forests have also fallen simply because we had something else to put in that space. In the American Southeast, native hardwoods have been replaced by faster-growing, more marketable pines, planted in orderly rows. Forests have been hit hard since the 1800s. In any case, we have removed so much forest that today almost all American hardwood forests are, at best, second-growth areas. As such, they are in transition, with many regions dominated by scrub oak and other uninspiring invaders.

Taiga

The **taiga**, or boreal (northern) forest, is almost exclusively confined to the Northern Hemisphere. This is a moist, subarctic forest biome of Europe and North America, dominated by evergreens, particularly spruces, pines, and firs. The taiga forms immense forests across many northern climes, but it is also found at higher elevations in many more southerly mountain ranges (Figure 47.19).

If you have ever been on the eastern coastal plain of the United States, you probably passed through a well-developed coniferous forest dominated by various pine species. In the northern part of this zone, in areas such as Cape Cod or the Pinelands of New Jersey, pitch pine dominates. Further to the south in the Carolinas, eastern Georgia, and Florida, the forests are dominated by longleaf, loblolly, and slash pine. These forests are maintained by fire, and if fire is excluded, oaks and hickories invade, displacing the pines.

The taiga is unmistakable; there is nothing else like it. It is subject to long, cold winters and short summer growing seasons. It is difficult to generalize about rainfall because the taiga is so extensive; some places get a great deal of rain, some don't. But in many parts, there is much less rainfall than in deciduous forests. In the Canadian interior, for example, rainfall is between 50 and 100 cm (20–40 in.). The taiga is interrupted in places by extensive bogs, or **muskegs**, the remnants of large ponds. These wet regions are commonly dominated by spruce, but low-lying shrubs, mosses, and grasses form the spongy ground cover (see Figure 47.19). The northernmost taiga is touched by sunlight for only six to eight hours each day through much of the winter and bathed in sunlight for 19 hours in the summer.

The dominant plants of the taiga are conifers, but occasional communities of hardy, broad-leaved poplar, alder, wil-

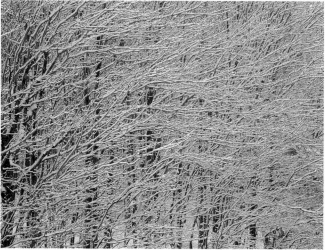

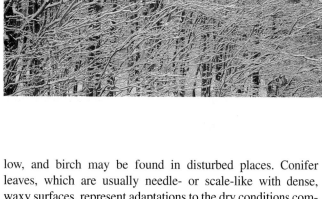

FIGURE 47.18
The Changing Deciduous Forest
The deciduous forest changes its appearance at different times of the year. The lovely green hillside of summer will explode in a riot of colors when autumn arrives. Both are in sharp contrast to the stark beauty of winter.

low, and birch may be found in disturbed places. Conifer leaves, which are usually needle- or scale-like with dense, waxy surfaces, represent adaptations to the dry conditions common in the taiga (see Figure 25.30). Such adaptations enable conifers to survive the winter with many of their leaves intact; thus they are sometimes referred to as evergreens. The leaves are adapted for water conservation, since much of the water reaching the taiga may be bound up (and made unavailable) in ice. (Recall the spines and waxy cuticles of desert plants.)

Leisurely strolls can be quite pleasant in a coniferous forest because there is very little underbrush. Those soft needles that feel so good under your boots are hard on other plants. They make the soil notoriously acid (and thus the conifers reduce their level of competition with other species). The straight trunks and dark, unobstructed forest floor lend a cathedral-like atmosphere to taiga forests, an atmosphere that some find compelling and others forbidding.

The taiga harbors large herbivores such as moose, elk, and deer (Figure 47.20). It is the last refuge of the grizzly bear and black bear, and wolves still roam here, as do lynx and wolver-

ines. While rabbits, porcupines, hares, and rodents abound, insect populations aren't as large as in deciduous forests, although there is an abundance of mosquitoes and flies in boggy regions during the summer.

There are other large coniferous forests that are not considered taiga. A number of temperate coniferous forests are to be found throughout the Northern Hemisphere. For example, the Olympic forest in western Washington is, on its western slopes, a bona fide temperate rain forest with 500 cm (200 in.) of rain per year near the glacial peaks. Here the Sitka spruce reaches its greatest size, approaching that of the famed California redwood. Fingers of the western forests follow the Cascade Mountains into Oregon and California.

Perhaps the most fascinating coniferous forests are the California coastal redwoods. Their gigantic demands for water are not met by rain alone, but also by coastal fogs. On the western slopes of the Sierra Nevada Mountains, we find the magnificent forests of giant sequoias. These awesome and splendid trees are among the world's oldest living things. Finally, not to be ignored, are the expanses of pines that dominate the coastal plain forest in the southeastern United States. These vast forests of longleaf, loblolly, and slash pine may be a temporary invasion of the usual oak–hickory communities that have been disturbed by human intervention and disease.

The coniferous forest is a continuing target of the lumber industry, but its very vastness has protected its more distant regions. The most controversial of foresting methods is called "clear-cutting." Unlike traditional selective foresting, clear-cutting involves clearing *every* tree from the land (Figure 47.21). As this practice continues, the timbering industry is

The Taiga
The taiga is an extensive biome, found almost exclusively in the Northern Hemisphere. Its dominant plant is the conifer, although communities of birch, willow, poplar, and alder are not unusual. Since the conditions of the taiga are duplicated in high mountains, similar communities are found there. Where bogs and marshes form, the taiga is interrupted. These areas are often referred to as muskegs. They represent a perpetual tug of war between aquatic and terrestrial environments, as plants continually invade the marshes, some failing, others becoming established.

careful to advertise that it is replanting the areas with foreign seedlings, and at great expense.

There is a problem with such reforestation, however. Mixed forests of genetically diverse and resilient plants are replaced by artificially developed and genetically homogeneous trees that are fast-growing and can be reharvested quickly. Because some kinds of conifers are now being routinely cloned through developments in tissue culture techniques, absolute genetic uniformity in some forests looms just over the horizon. Biologists are quite concerned with the dangers of genetic uniformity in any system, since without variation, destruction by an unforeseen invasion of a mutated fungus, bacterium, or virus could virtually destroy an extensive forest. All these possibilities aside, where will the jay, bear, and raccoon live while the clear-cut forest recovers?

A number of studies have shown that other kinds of damage result from clear-cutting forests. Following clear-cutting, runoff from rainfall increases substantially, as you might expect, but this is accompanied by what ecologists call a "nutrient flush." There is a threefold to 20-fold increase in the loss of mineral nutrients from the soil over what is normally recorded in forests. Most dramatic is a substantial loss of critical nitrogen.

Water Communities

The earth has been called "the watery planet," and not without good reason. A view of our planet from afar reveals that, in fact, most of the planet is covered not by land but by salt water. Any such view would also reveal great pockets and

(a) Elk male and female

(b) Lynx

FIGURE 47.20
Animals of the Taiga
Herbivores of the taiga include large mammals such as the elk shown in **(a)** and caribou. Porcupines are very common. The lynx shown in **(b)** and grizzly bear are still found in the taiga, protected somewhat from human intervention by the vastness of the area.

FIGURE 47.21
Clear-cutting
Clear-cutting—a controversial, potentially disastrous foresting practice—involves the widespread clearing of trees in large stretches of land, as can be seen in this view of an Oregon forest.

veins of fresh water and clouds, formed largely of water, moving silently across the earth's surface.

There are marked differences between water and land environments, differences that are critical to life. First, water is a more viscous and buoyant medium than air. As a result, many water dwellers have streamlined torpedo-shaped bodies and paddle-shaped appendages that enable efficient swimming. A host of microscopic water dwellers merely drift using flagella or oily inclusions.

Second, available food resources are often found dissolved or floating in the water. For water-dwelling autotrophs like algae that means that inorganic nutrients can be absorbed directly from the surrounding water. Extensive root systems are not needed. Water-dwelling consumers have the option of pursuing prey by active predation (like sharks) or by remaining essentially stationary and filtering water that flows by (like clams and sponges).

Third, water-dwelling organisms must obtain their oxygen from that dissolved in the water. Since water holds less oxygen than air, water-dwellers must cope with an environment that is comparatively deficient in oxygen.

Fourth, water absorbs light, especially when it contains lots of suspended particulate matter. Thus light intensities progressively decrease with depth to the point that deep oceans are virtually pitch black.

Finally, water-dwelling organisms face different water-balance challenges than those that live on land. Terrestrial organisms like plants, insects, reptiles, birds, and mammals are faced with a desiccating environment and thus must conserve water through structural or physiological adaptations (a waxy cuticle or water-impervious skin). Freshwater dwellers, especially those in streams and lakes, do not have that problem and actually must cope with too much water. It should be noted that saltwater organisms are faced with a "drying" environment because of osmotic factors, as the salt tends to draw water out of hypotonic fluids in living organisms.

Scientists classify open water habitats into two general **provinces** based on the concentration of dissolved ions: *freshwater* and *saltwater*. Freshwater occurs inland and consists of streams, rivers, ponds, lakes, and wetlands. In contrast, saltwater is mainly restricted to the oceans, though some saltwater lakes do occur inland (such as the Great Salt Lake of Utah and the Dead Sea in the Middle East). Many scientists also distinguish a third type called **brackish** water, which is intermediate in salinity and can be found in **estuaries** where rivers meet the ocean.

The Freshwater Province

The waters of the earth are similar to terrestrial realms in that they mold their inhabitants along certain general lines by presenting them with common ecological opportunities and risks. Life had been subjected to the molding effects of water for almost three billion years before it occupied dry land. Here we will discuss the water communities, beginning with fresh water, which has a relatively low level of ion concentration, particularly salts.

Rivers and Streams. Estuaries, where fresh water meets salt water, are fed by rivers and, on a smaller scale, streams. Rivers and streams are bodies of water that move in one direction. As we see in Figure 47.22, their headwaters (where they begin) are usually narrow and steep. Since headwaters are often spring-fed and at higher elevations, they may be fast and intensely cold. The falling, turbulent water tends to be mixed and thoroughly aerated. The communities in these areas are composed of relatively few species, such as trout (species that have adapted to the cold temperatures and oxygenated waters). A few invertebrates are found among the algae and mosses that cling tenaciously to the rocky banks. Further along, the riverbeds tend to become flatter and the rivers wider and slower. Here the waters carry more suspended material and may become more murky, relegating the penetration of light to the upper levels and shallow areas where algae

FIGURE 47.22
The Headwaters
In the headwaters where river systems begin, fast-moving rocky streams are typical.

and cyanobacteria may thrive. The fish species here need less oxygen, such as the slower, less active, bottom-feeding catfish, carp, and—lurking above them—bass. As the river widens and becomes even slower, suspended materials collected upriver may settle out, forming a nutrient-rich sediment that can support relatively large and complex communities.

Lakes and Ponds. As water moves along the channels cut by rivers and streams, it may pause in its inexorable cycle to rest for a time in still bodies called lakes (and, on a smaller scale, ponds).

Lakes and ponds differ not only in size but usually also in depth. Ponds are often shallow and therefore can be risky places to live. The shallowness means that light is likely to penetrate the entire pond, encouraging the growth of algae

and cyanobacteria, as well as bottom-dwelling plants. Thus the entire pond may rapidly become clogged with life that can eventually fill in the pond entirely and allow invasion by terrestrial plant life (see Figure 46.10). Another problem with shallowness is much more direct: in dry periods, the pond may simply evaporate entirely. The smallest ponds are called *vernal pools* and may contain water only for a few weeks. While they may seem transitory, seasonal ponds and vernal pools are extremely important to the maintenance of biological diversity—especially for amphibians who spawn there.

Lakes are grander places, and, compared to other major features of the earth, they may be of relatively recent geological origin. Most occur in northerly or alpine regions as products of the last glacial retreats (10,000–12,000 years ago). They are also produced through volcanic activity, as was the famed Crater Lake in Oregon, and through gradual uplifting of land, as were many of the shallow, acid lakes of Florida. The deepest lake in the world is much older than most other lakes: Lake Baikal in the Soviet Union, with a depth of 1750 m (5742 ft, or well over a mile), was formed during the Mesozoic era.

Lakes are characterized by pronounced zones (Figure 47.23). The area around the edge of the lake that is shallow enough for rooted water plants to grow is called the **littoral zone.** The deeper central portion of the lake is composed of a superficial **limnetic zone,** where light is present, and a deeper **profundal zone** that is below the limit of light penetration. Underlying the entire lake is the **benthic zone,** which comprises the basal layer of sediment. These zones are biologically significant because the littoral and limnetic zones support photosynthetic organisms: rooted plants like water lily and pondweed in the littoral, and suspended microscopic producers like diatoms, desmids, and cyanobacteria in the limnetic. The dark profundal zone can only support heterotrophs like fish and cladocerans. The benthic zone is generally a mucky ooze that is essentially anaerobic. As such, it supports those organisms with a very low need for oxygen, mainly bacteria.

Lakes also form layers of different temperatures. In the summer, the top portion of the lake becomes warm, while the

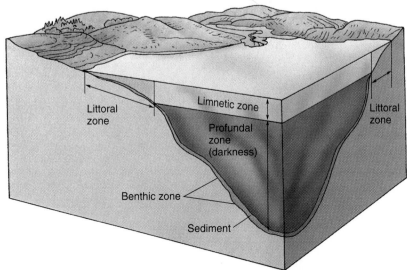

FIGURE 47.23
Lake Zonation
Lakes are organized into specific zones, as shown here.

lower waters remain cool. The transition between the two zones is often abrupt. If you have ever dived into the center of a deep lake, you have probably immediately regretted it as you hit the cold water several feet beneath the warmer surface. Mixing of water between the two layers is minimal, so the temperature difference is maintained. However, in the fall the surface temperature cools to the point where the waters reach maximum density and they begin to sink (see Chapter 2). At that point, the water mixes throughout the entire lake, a process called **fall overturn**. A new stratification occurs in the winter with the surface being frozen, and the lower layer being cold, liquid water. That stratification is disrupted in the spring with a **spring overturn**. Overturns are critical since they bring oxygen to the lake depths and nutrients from the sediments to the surface waters.

At those depths penetrated by light, photosynthesis stimulates the growth of algae and cyanobacteria and, with the suspended particles washed in by rivers and streams, can give a murkiness to the waters. If the suspended materials settle rapidly, though, the lake may maintain its clarity to some degree. The murkiness that signals productivity can be increased by the use of fertilizers, which, through runoff, find their way into the waters that feed the lake. (Even the private use of lawn fertilizers around Lake Tahoe threatens to turn that wonderfully clear body of water into a flat opaqueness.)

Lakes can produce complex communities as small fish and aquatic insects feed on the zooplankton (tiny protist and animal life that feeds on the producers near the surface). Below, larger fish feed on the insects and smaller fish. Around the shore snakes, salamanders, frogs, toads, and birds can be found. Raccoons, herons, and opossums may prowl the lake's damp edges, amidst a variety of predatory insects, such as dragonflies, damselflies, and caddis flies.

Wetlands

If you look closely in small depressions and at the edges of lakes and rivers, you would find communities of plants that are different from those on drier land or open water. These transitional areas between terrestrial and aquatic systems, called **wetlands**, develop in soils that are waterlogged for at least part of the year. For centuries, people considered wetlands to be waste areas and filled or drained them to make the land more "productive." In the 1970s and 1980s, however, the federal government, as well as those of many states, enacted laws that protected wetlands against further destruction.

Because of these laws, wetlands are defined in strict legal terms (unlike most other ecosystems). According to that legal definition, wetlands are those areas that have characteristic vegetation and soil that develop in response to periodic flooding or soil saturation. Thus wetlands are defined according to three criteria: vegetation, soils, and hydrology (the annual pattern of water flow).

Waterlogged soils cannot hold as much oxygen as typical upland soils, and therefore many plants cannot function normally. Instead, specialized plants called *hydrophytes* dominate the plant community. These plants either pump oxygen down into their roots or physiologically tolerate low oxygen conditions. Common hydrophytes include cattails, reeds, and many grass-like plants called sedges.

Wetlands are classified according to the dominant vegetation. Those dominated by grasses, sedges, and herbs are called *marshes*. Those dominated by shrubs are called *scrub-shrub wetlands*. Those dominated by trees are *swamps*. A *bog* is a special type of scrub-shrub wetland whose substrate is composed mainly of peat moss, of the genus *Sphagnum*. Bogs typically form in former glacial lakes and have water that is extremely acidic and low in nitrogen.

Wetlands are legally protected because they are valuable in a number of ways. First, their soils act as sponges and store water after a heavy rainstorm. Thus wetlands reduce downstream flooding. Second, wetland soils are excellent at absorbing particulate matter and dissolved substances and therefore serve as natural water purification systems. Third, wetlands are not common in the landscape and thus provide habitat for species that are relatively rare.

TABLE 47.1
Ocean Salinity

The ocean averages about 3.5% salts, usually expressed as 35 parts per thousand. Six elements, occurring as ions, comprise 99% of the minerals in sea water. Salinity varies, particularly in surface waters, but also in the depths; the 35 parts per thousand applies to measurements taken at 300 m (about 985 ft).

Element	% of Total Salts	Parts Per 1000 of Sea Water
Chlorine	55.0	18.98
Sodium	30.6	10.56
Magnesium	3.7	1.27
Sulfur (sulfate)	7.7	0.88
Calcium	1.2	0.40
Potassium	1.1	0.38
Less than 1% of salts:		
Bromine		0.065
Carbon		0.028
Strontium		0.013
Boron		0.005
Silicon		0.003
Fluorine		0.001
Nitrogen		0.0007
Aluminum		0.0005
Rubidium		0.0002
Lithium		0.0001
Phosphorus		0.0001
Barium		0.00005
Iodine		0.00005

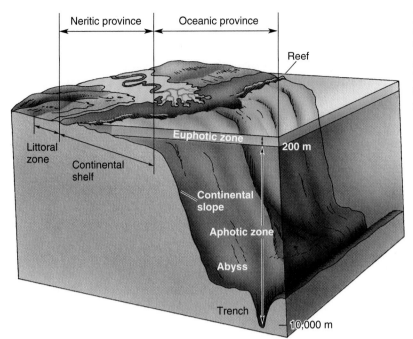

FIGURE 47.24
Organization of the Marine Environment
The marine environment is subdivided into the oceanic and neritic provinces, which, in turn, are divided into lighted (euphotic) and unlighted (aphotic) zones. Most marine life is concentrated in the neritic province, where more abundant mineral nutrients are found.

The Marine Province

About three-quarters of the earth's surface is covered with water, and most of it is salty (Table 47.1). The salinity of open ocean is a fairly constant 35 parts per thousand (3.5%). Chloride and sodium comprise about 86% of the ions. When we include magnesium, sulfate, calcium, and potassium (all important nutrients), the total reaches 99%. Other ions, including important nutrient elements like nitrogen and phosphorus, are found in only minute quantities. Seawater is slightly alkaline and strongly buffered at a pH around 8.1.

At the surface, the temperature of seawater can vary depending on latitude. In the tropics it can attain 27°C, while at the poles it can be as low as −3°C. However, as mentioned earlier in this chapter, the ocean has vast currents that cause equatorial warm water to flow poleward, and cold polar water to flow back to the equator. In contrast, deep ocean water is a rather constant and cold 3°C.

The weight of the water also creates pressure differences. At the surface, water has a pressure of 1 atmosphere. In stark contrast, the deepest parts of the ocean can exceed a bone-crushing 1000 atmospheres. Thus most organisms found at the surface cannot withstand the intense pressure below, though some like the sperm whale can tolerate dives to great depths.

Just as lakes can be divided into littoral, limnetic, and profundal zones, oceans are divided into several different provinces and zones. The part of the ocean in the relatively shallow areas along continents is termed the **neritic province**, while that in the deepwater open sea is the **oceanic province**. Along the coastline, from the high-tide "spray zone" out to a depth of 200m is the **littoral zone** (Figure 47.24). The marine environment is also subdivided into the **euphotic zone**, through which light penetrates, and the deeper **aphotic zone**, which is in perpetual darkness.

The Oceanic Province. Much of the open sea is devoid of visible life forms. Nonetheless, the oceanic province is the home of the **pelagic** organisms, those that drift in the open sea. Most of these are restricted to the euphotic zone—regions where light can penetrate. Almost all wavelengths of light, depending on the turbidity, are absorbed by the upper 100 m of water, although some shorter wavelengths (blue) may be detected at 300 m. Oceanic waters that are nearly devoid of particles (and thus of nutrients) are quite blue, while ocean waters with suspended particles (nutrient-rich) may be green, brownish, and sometimes even reddish. In such turbid waters, light may penetrate only 10 m or so.

Perhaps the greatest mysteries of the sea lie in the **abyssal region**, the deeper waters where depths can vary from 300 to nearly 11,000 m. The famed Marianas Trench is 10,680 m deep (over 6 miles, and deeper than Mt. Everest is high). Little was known about the ocean abyss until technology permitted its direct exploration. In 1960, the floor of the Marianas Trench was reached by Don Walsh and Jacques Piccard in the bathyscaph, *Trieste.* It was a momentous event, but scientists have visited these depths many times since then, and we are beginning to understand more about what is going on down there.

We know, for instance, that the ocean depths are places of darkness, tremendous pressure, and numbing cold. Nevertheless, these formidable, deep abyssal regions support a surprisingly large number of peculiar **benthic** (bottom-dwelling) scavengers. Tethered cameras focused on bait, miles deep, have photographed primitive hagfish, many species of bony fish, crustaceans, mollusks, echinoderms, and even an occasional shark. For years oceanographers and marine biologists assumed that since producer populations could not exist at such depths, benthic creatures of the aphotic zone had to rely for their energy on the continual "rain" of the remains of pelagic (surface-swimming) organisms as they settled to the ocean floor. But more recently, biologists have discovered that there are unusual communities of organisms living along rifts and vents in the seabed, and that these have producer populations. Obviously, the producers aren't photoautotrophs but are, instead, chemoautotrophic organisms. To learn more about the chemoautotrophs and the ocean rift community, see Essay 47.3.

The water below the photosynthesizers, but above the ocean floor, harbors some of the most peculiar creatures on earth: the deep-sea fishes (Figure 47.25). They are usually small and dark and keen of sight, with big toothy mouths. (They have to be able to handle just about any kind of food they come across.) These predators often produce their own light, with which they signal each other or attract prey. Some species cultivate luminescent patches of bacteria beneath their eyes, which become visible when the fish rolls down a specialized eye covering.

The Neritic Province. In the neritic province, the land masses touch the edge of the highly variable **continental shelf**, the shallow waters adjacent to the shores of a continent. The shelf is considered to be a submerged part of the continent. The neritic province ends at the continental slope, where the shelf drops off, the bottom receding to greater depths, often abruptly.

In many shallower areas of the shelf, such as the littoral zone, light penetrates to the ocean bottom. Such regions are continually stirred by waves, winds, and tides, keeping nutrients suspended and supporting many forms of swimming and bottom-dwelling marine life. Giant kelps and other seaweeds form extensive beds, offering hiding places for many fish species (see Figure 23.25). Let's look briefly at the organization of life and the flow of energy in this province.

As in lakes, the seas are home to many tiny species called **plankton**, which drift near the surface of the open ocean. The photosynthetic species are called the **phytoplankton**. Included are algal protists—mainly diatoms and dinoflagellates, all of which are microscopic. Also included are the more recently discovered minute, flagellated **nanoplankton**, which are clearly important ecologically but poorly understood. Phytoplankton, along with the marine algae—seaweeds—are the primary producers of the sea. Thus energy enters the marine ecosystem via these species. It has been estimated that 80–90% of the earth's photosynthetic activity is carried on by marine organisms. Thus the chain of life in the sea begins with tiny phytoplankton capturing the energy of the sun to build energy-containing compounds within their fragile bodies. But where there is energy, there is likely to be something trying to utilize it, and the tiny algae are fed upon by tiny planktonic animals only a little larger than themselves—the **zooplankton**. A variety of animals, from tiny fish to the great baleen whales, feed upon plankton of all sorts (Figure 47.26).

Among the more productive regions of the marine environment are the colder offshore waters, where **upwellings** occur. Upwellings are the movement of deep, nutrient-laden colder waters to the surface. While little is known about the vertical movement of water in the open sea, coastal upwellings are better understood. They are generally a seasonal phenomenon, where coastal winds blow either seaward or parallel to the coast, moving the surface layers, which are then replaced by deeper layers. Like thermal overturn in lakes, this stirring brings up nutrients that would otherwise be forever locked in the bottom sediments. The nutrients support photosynthetic organisms, which provide the base for marine food chains. The vast anchovy fisheries off the coast of Peru are dependent on such upwellings.

Another kind of shallow area that is among the richest of marine habitats is the coral reef, which occurs only in tropical regions where the ocean temperatures remain above 20°C. Corals, of course, are cnidarians that live in huge colonies, building heavy exoskeletons of calcium carbonate. As years

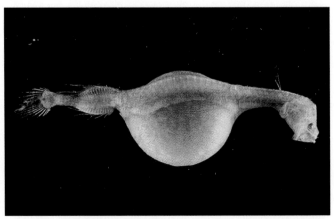

FIGURE 47.25
Fishes and the Abyss
The deep-sea fishes are among the most bizarre of all creatures. They live where there is no sunlight and very few other animals. Thus they must be equipped to capitalize on just about any living thing they might come across.

Essay 47.3 ● AN UNUSUAL COMMUNITY: THE GALÁPAGOS RIFT

The Galápagos rift community, a bizarre assemblage of animals and bacteria, centers around the vents of sulfide hot springs in an area of active sea floor spreading some 612 km (380 mi) from Darwin's islands. It was discovered in 1977 by geologists aboard the submarine *Alvin,* which was cruising 2500 m deep (8202 ft) at the time. The geothermal hot springs, spewing boiling-hot solutions of hydrogen sulfide and carbon dioxide, support an entire ecosystem based on the autotrophy called chemosynthesis. It is one of the most dense and productive communities on earth—near the vents, the mass of living tissue approaches 50 to 100 kg per square meter.

Prominent among these denizens are enormous, blood-red tube worms of a previously unknown group, apparently belonging to the pogonophores (an obscure phylum now thought to be related to annelids). Like all pogonophores, they lack all trace of a mouth, anus, or gut. They are chemosynthetic autotrophs, deriving all their energy and carbon needs directly from the oxidation of the hydrogen sulfide and the reduction of the carbon dioxide, probably with the help of intracellular bacterial symbionts. The redness of their flesh is due to heavy concentrations of oxygen-binding hemoglobin: oxygen from the surrounding cold seawater is needed both for the oxidation of H_2S and for the worm's own oxidative metabolism. The tube worms range up to nearly 3 m in length (nearly 10 ft) and are about the girth of a man's wrist.

The entire ecosystem, in fact, is based on hydrogen sulfide chemosynthesis and is entirely cut off from other photosynthesis-based ecosystems. Apart from the tube worms, the primary producers are chemosynthetic bacteria, which swarm in enormous numbers where the often superheated, carbon dioxide- and hydrogen sulfide-laden vent water mixes with the near freezing, oxygenated abyssal seawater. At least 200 bacterial species proliferate near the vents and were visible to observers in *Alvin* as milky clouds. Filter-feeding crabs, clams, mussels, smaller worms (dubbed "spaghetti" by their geologist discoverers), and barnacles live off the bacteria. A little higher on the food chain, larger crabs and a variety of fish scavenge on clumps of bacteria as well as on animal remains. Whelks, leeches, limpets, and miscellaneous worms complete the community. Some species were previously known, but others, such as the tube worms, are new discoveries with uncertain affinities. Similar communities have now been found on the East Pacific Rise Rift, and other deep-sea geothermal communities probably occur wherever there are appropriate sulfide springs. In fact, the rift ecosystems may be so widespread as to constitute a major earth community.

The Galápagos rift contrasts markedly with the rest of the deep-sea benthic communities, which are characterized by constant cold and a severely limited energy input. The usual benthic communities survive on what little organic material drifts down from the surface waters and are thus based ultimately on energy from distant photosynthesis. The animals of the cold water deep-sea bottom are generally slow-moving and slow-growing, playing a variety of refrigerated waiting games: scavengers patrol listlessly for the chance of a dead fish or a fecal pellet, and predators lie motionless in ambush for wandering scavengers. In the hot springs communities, however, food is virtually unlimited, the temperature is not so uniformly cold, and both growth and metabolism are relatively rapid. The Galápagos rift clams, for instance, grow up to a third of a meter (13 inches) long at a rate of 4 cm per year, some 500 times faster than their smaller cold-water relatives. The flesh of these clams, like that of the giant tube worms, is bright red with hemoglobin. This factor, along with their phenomenal growth rate, indicates a high metabolic rate.

Rich and active as they are, the deep-sea rift communities are ephemeral. The hot springs eventually die down, like volcanoes, leaving behind ghostly communities of empty clam shells. As the earth's crust shifts and new hot springs form, immigrants from established or dying geothermal communities arrive to begin the unusual chemosynthetic ecosystem anew. Rapid ecological succession is one more unusual characteristic of the rift community.

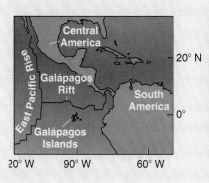

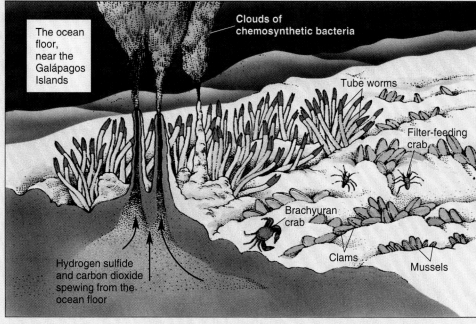

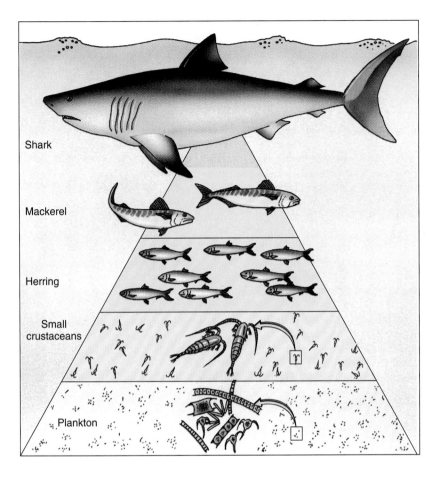

Shark

Mackerel

Herring

Small crustaceans

Plankton

FIGURE 47.26
The Food Pyramid of the Ocean
The tiny phytoplankton at the base (not drawn to scale) capture the energy of the sun. These are eaten by animals larger than themselves, which are eaten in turn by larger animals. At the top are the largest carnivores of the sea.

pass, the mass of exoskeletons grows, with the newer members of the colonies growing on top of the old. Coral atolls, common in the South Pacific, were shown by Darwin to represent coral growth at the top of submerged volcanoes. Barrier reefs, on the other hand, are coral deposits that form along coastlines, usually a short distance from the beach. The largest and most spectacular of these is the Great Barrier Reef, which extends some 1200 miles along the east coast of Queensland, Australia.

Reefs, by their irregular growth, form natural refuges for marine animals and thus set the stage for complex food chains. The complexity is due to the diversity of food types and a multitude of hiding places that forces predators to be able to exploit a number of different kinds of food, specializing in none. Coral reefs usually shelter sponges, encrusting algae, and bryozoans, as well as mollusks such as the octopus. Fishes abound in the reefs, and sharks patrol the deeper waters alongside.

Thus the ocean communities that are richest in life exist within the neritic province. Unfortunately, the world's neritic fishing grounds are being exploited at a rate that makes the likelihood of their recovery questionable. When increased fishing effort is not accompanied by increased catches, then it is fairly certain that the resource is in danger of depletion. Today's fishing technology is so sophisticated that even in

areas where catches have diminished, the last remnants of once-abundant species are being captured.

Coastal Communities. Where the oceans meet the land are found some of the most fascinating, diverse, yet physically stressful ecosystems of all—the coastal communities (though most people just call them "seashores"). This is the littoral zone, an area that includes the coast and the offshore waters. Part of the littoral zone is the **intertidal zone**—the area between high and low tide. Coastal communities can include sandy beaches, rocky coasts, bays, estuaries, and tidal mud flats. Such places have one thing in common—they abound with a rich diversity of life from every kingdom. The richness of life in the littoral zone is made possible by the availability of light, shelter (seaweeds, rocks, kelp beds), and nutrient runoff from rivers entering the sea. However, these organisms must cope with frequent periods of inundation by seawater interspersed with periods of exposure to air, thanks especially to the tides. That inundation–exposure cycle often leads to extreme fluctuations in temperatures, salinity, and levels of water availability. Physical pounding from wave action, especially during storms, is also often a problem.

Rocky shores, sandy beaches, and mud flats each have their own characteristic assemblage of species. The rocky tidal communities, for example, consist of a large variety of organisms, all adapted to avoid being swept away by the tremendous force of the waves. Starfish and periwinkles hold tightly to rocky substrate, using suction, while barnacles and rockweed use glues. Animals of the rocky coast are particularly prone to being left "high-and-dry" by low tides, so each must be able to conserve water during such intermittent periods. Thus barnacles and periwinkles have tough shells, while rockweeds and other coastal algae have a slimy outer layer that prevents desiccation.

Sandy beaches offer a distinct set of challenges for organisms that grow there. Such beaches may have a barren, desert-like appearance. However, hidden beneath the moist surface sand is a diverse mixture of crabs, sandworms, and bivalves. These creatures cannot use the loose, shifting substrate as a site for attachment. Instead, they are adapted for burrowing into the sand, where temperatures and levels of salinity are often fairly constant.

The third type of coastal community, the mud flat (Figure 47.27), is characterized by level areas dominated by cordgrass growing in a rich, black, mucky sediment. Thus the community is a type of marine wetland, called a **salt marsh**. These communities are particularly well established along **estuaries**, places where rivers run into oceans. Many people avoid mud flats because certain bacteria thrive in their anaerobic sediments producing hydrogen sulfide gas (which smells like rotten eggs). However, scientists have learned that salt marshes are extraordinarily productive ecosystems and make a large contribution to sustaining life in the oceans. Their productivity is due to the fact that rivers bring in nutrients washed from the land, and the tides continually churn up its nutritive riches. Many marine organisms spawn in such places, particularly various species of mollusks, arthropods, and fishes. Because of the great density of life, estuaries are important feeding grounds for amphibians, reptiles, birds, and mammals. Before the mid-1970s, estuarine salt marshes were under tremendous pressure from developers, who sought to turn them into

FIGURE 47.27
Mud Flats
Mud flats are saltwater areas in bays and estuaries that appear during intermittent low tides. At such times, their inhabitants are exposed to air, sunlight, and changing salinity and must adapt to such harsh conditions. Burrowing mollusks and worms are common, and predators from the shore often stalk these areas in search of such prey. The estuarine mud flat may harbor a great variety of organisms if the river has not been heavily polluted.

waterfront lots, complete with private boating docks. The wetland laws discussed earlier in this chapter have largely brought a halt to that activity. Still, that doesn't mean that estuaries are free from degradation by humans. As long as rivers are used as dumping places for manufacturing wastes, mine seepage, and sewage, inhabitants of salt marshes and adjacent estuarine water will suffer. In turn, that will affect life in the ocean and our relationship with that life.

Key Ideas

THE BIOSPHERE

1. The **biosphere** is that portion of the earth that supports life.

2. Water has very significant chemical and physical characteristics that include high specific heat, the resistance to temperature change. Moisture in the atmosphere provides for a moderate climate.

3. The atmosphere contains variable amounts of water and is 78% N_2, 21% O_2, and about 1% rarer gases, including CO_2. The atmosphere screens out harmful ultraviolet radiation. It also acts as a heat sink, producing conditions similar to those in a greenhouse.

4. About 30% of incoming solar energy is reflected from the atmosphere, while 20% is absorbed by the atmosphere, and the rest radiates back into space. Less than 1.0% is captured in photosynthesis, and most becomes involved in the evaporation of water. Evaporation and condensation constitute the **hydrologic cycle**.

5. Seasonal change is attributed to the earth's tilted rotational axis (23.5°).

6. Equatorial heat causes moisture-laden air to rise, forming giant rotating air cells that carry moisture and heat northward and southward from the equator. Rotational forces produce the

tradewinds. Most of the earth's large deserts occur north of the precipitation zone, although **deserts** also form as a result of **rain shadow**.

7. The air masses also produce the great ocean currents.

THE DISTRIBUTION OF LIFE: TERRESTRIAL ENVIRONMENT

1. **Biomes** are major, recognizable associations of plants and animals and can be subdivided into ecological communities.

2. Biome distribution often follows latitude, with transition zones from one biome to the next. Similar distributions are seen in altitudinal transitions, where increasing altitude emulates northerly changes.

THE BIOMES

1. **Deserts** receive less than 25 cm of highly seasonal precipitation per year. Without heat-retaining water, night–day temperature shifts are drastic. Most deserts occur at about 30° north or south latitude, or to the lee of high mountains.

2. Specific adaptations of xerophytes include water storage, fewer stomata, small leathery leaves, spines, long dormant periods, and specialized CAM cycles—all of which result in slower growth rates. Many annuals have rapid flowering cycles.

3. Most desert animal species are active at night, evening, and early morning. The desert kangaroo rat, survives by remaining in humid burrows and making use of highly efficient kidneys to conserve body water.

4. **Grasslands**, have more rainfall (25–75 cm/year) than deserts, but the rainfall is seasonal. Fires are a significant factor, but grasses recover because of their underground rhizomes and extensive diffuse root systems. Rhizomes and stolons (surface runners) form sods that prevent the invasion of other plants.

5. Grasslands survive grazing by basal regrowth, although most grasslands today have been converted to agriculture.

6. The **tropical savanna** is characterized by intermittent groves of trees in otherwise typical grassland. Rainfall averages between 100 and 150 cm/year. Savannas occur in Africa, South America, and Australia. Huge migratory populations of grazing mammals and large carnivores are common.

7. The northernmost biome is the **tundra**, an assemblage of dwarfed trees, grasses, and lichens. Precipitation is less than 15 cm, and winters are dry, but in the short summer, surface water collects in countless ponds. Below is the permafrost, permanently frozen soil. High winds and permafrost preclude tree growth. Lichens abound, and plant life includes mosses, grasses, sedges, and dwarf willows and birches.

8. Tundra mammals include caribou, musk oxen, reindeer, hares, and lemmings. Lemmings make up an important link in the food chain. Waterfowl and shore birds migrate in to nest in summer.

9. **Tropical rain forests** receive from 250 to 450 cm of evenly distributed rainfall per year. The highest rainfall is in the Amazon, Indonesia, and the Congo.

10. The foliage of trees in the tropical rain forest forms strata, with an overlying canopy 30–45 m high, and a subcanopy of smaller trees and vines. Epiphytes live entirely within the trees, receiving water mainly from the humid air.

11. Each animal species in the rain forest occupies a specific region in the canopy and subcanopy.

12. Clearing of the world's tropical rain forests may produce dramatic changes in climate and soil.

13. The **chaparral** is restricted to warm coastal regions, where rainfall is meager and seasonal (25–75 cm); the drought-like conditions are modified by a moist marine air flow. Chaparral plants are typically xerophytic; decomposition is slow, and ground litter is extensive. Rodents abound, as do migratory birds, but there are few large herbivores or carnivores. The chaparral is always in a state of recovery from fires that periodically sweep through.

14. **Deciduous forests** occur mainly in the north temperate zones and are characterized by warm summers and freezing winters. The principal plant adaptation is leaf-shedding and dormancy. Precipitation averages about 100 cm/year and is evenly distributed.

15. Deciduous forests include oaks, maples, beeches, birches, and hickory. A rich humus and deep litter support large soil populations. Larger animals adapt to the seasons by hibernation, migration, or simply facing the cold.

16. Because of human encroachment into the deciduous forest, large mammalian herbivores and predators have largely been displaced, and only smaller species persist.

17. The **taiga** extending in a northerly belt around the world, has long winters and short summers. Deciduous communities interrupt the conifers, around boggy **muskegs**. Light and low precipitation are growth-limiting factors.

18. The needle or scaly conifer leaf with its waxy surface and recessed stomata is well adapted to dryness.

19. Taiga herbivores include moose, elk, deer, rabbits, hares, porcupines, and numerous rodents. Carnivores include bears, wolves, lynx, and wolverines.

20. Much of the temperate coniferous forests in North America are heavily lumbered.

WATER COMMUNITIES

1. Rivers and streams originate at spring- and snow-fed headwaters and are clear and oxygen-rich at higher elevations. At lower elevations, river flow slows, and the waters become murky and contain less oxygen. As sediments accumulate further downriver, communities become more complex.

2. Ponds are generally shallow and tend to become choked with plant and algal growth.

3. Northern Hemisphere lakes tend to be permanent and of glacial or volcanic origin.

4. The marine environment is subdivided into the shallower, coastal **neritic province** and the deeper, open-sea **oceanic**

province. Each includes a light-penetrable **euphotic zone** and a dark **aphotic zone**.

5. Life is sparse in the nutrient-poor oceanic province and includes floating and drifting **pelagic** organisms, most of which are restricted to the upper 100–300 m, the euphotic zone.

6. The perpetually dark **abyssal region** (300–11,000 m) supports a few forms of **benthic** (bottom-dwelling) life that are adapted to the cold and immense pressure.

7. The neritic province includes the **continental shelf**. Nutrients from upwellings support the **phytoplankton**—the primary producers of the sea. Minute protists and animals make up the **zoo-**

plankton, on which many other marine animals feed. **Upwellings** are brought about by seasonal winds blowing seaward or parallel to the coast.

8. The **littoral zone** of the neritic province contains the coastal communities: sandy beaches, rocky coasts, bays, estuaries, and reefs. Along rocky coasts, organisms must adapt to wave surge and to exposure at low tide. Estuaries produce their own problems, including shifts in salinity that require osmoregulatory adaptations. Estuarine life, which often includes immature forms of deeper-water life, is constantly threatened by human encroachment and pollution.

Application of Ideas

1. Contrast the distribution of plant species in the tropical rain forest with that in the taiga and temperate deciduous forest. Suggest reasons for differences.

2. List the ways various animals have adapted to the desert biome. What general rules can be made for desert dwellers? Why are environmentalists more concerned with the human impact on this biome than they might be with most others?

3. Among the greatest environmental concerns today are those associated with the human encroachment on the tropical forests in South America and Asia. What is going on in these places, and how can the effects of this activity possibly be of significant global importance?

Review Questions

1. Describe the physical extent of the biosphere. (page 1033)

2. Explain the relationship between water in the biosphere and temperature stability. (page 1034)

3. Explain what happens to the total amount of solar energy impinging on the earth's atmosphere. What keeps the earth from continually heating up because of solar energy? (pages 1033–1034)

4. Describe the formation and movement of the equatorial air cells. What do they have to do with the formation and location of the major deserts? (page 1036)

5. Explain what causes rotating oceanic currents. How do they contribute to the distribution of heat over the globe? (page 1037)

6. Describe the physical conditions of the desert biome, including precipitation and its distribution, day–night temperature changes, and the effects of rainstorms. (page 1039)

7. List three significant ways in which plants and animals are adapted to desert life. (pages 1040–1042)

8. What characteristics of grasses enable them to recover from both grazing and fire? (page 1042)

9. Characterize by kind and number the major animal life in grasslands. What does this tell you about the grasses' efficiency of energy capture? (pages 1042–1043)

10. How does the tropical savanna biome differ from the grassland? (page 1042)

11. Characterize the physical conditions of the tundra, including precipitation, temperatures, seasons, and special soil conditions. (page 1044)

12. List five prominent plants of the tundra. What two conditions prevent the growth of tall trees? (page 1044)

13. Describe the organization of the tropical rain forest, including numbers of species, vertical stratification, and nature of epiphytes. (pages 1045–1046)

14. What are the unique physical conditions of the chaparral? What keeps it from becoming a desert? (page 1046)

15. What is the major adaptation of broadleafed plants to seasonal changes in the deciduous forest? What three responses do animals make to seasonal change? (pages 1048–1049)

16. What are three growth-limiting conditions in the taiga? (page 1049)

17. Describe the special adaptations of the conifer leaf. (page 1050)

18. Name and describe the two major divisions of the marine environment. In what way is each divided vertically? (page 1055)

19. What are three prominent physical conditions of the abyss? List five phyla of animals represented in the abyssal region. What is their primary food source? (page 1055)

20. What is an upwelling? How does this phenomenon affect the distribution of marine life? (page 1056)

CHAPTER 48

The Human Impact

CHAPTER OUTLINE

THE HISTORY OF THE HUMAN POPULATION

The First Population Surge

The Second Population Surge

The Third Population Surge

THE HUMAN POPULATION TODAY

Growth in the Developing Regions

Demographic Transition

THE FUTURE OF THE HUMAN POPULATION

Population Structure

Growth Predictions and the Earth's Carrying Capacity

Essay 48.1 What Have They Done to the Rain?

Essay 48.2 Holes in the Sky

KEY IDEAS

APPLICATION OF IDEAS

REVIEW QUESTIONS

*W*e turn our attention to a set of numbers that stimulated a great deal of concern and controversy in recent years: 5.6, 25, 9, 1.6, 43, and 33. You may not at first glance be particularly impressed; they may just look like numbers. However, a demographer will recognize them as the startling figures that represent, respectively, our human population (in billions), our crude birth rate, crude death rate, percentage annual growth, population doubling time (in years), and percentage of people below the age of 15 (see Tables 48.1 and 48.2).

It is just such numbers that have begun to reveal to us how pressing our problems are. For example, we can see that the population of the world will have doubled between the time most of today's college students were born and about the year 2015 (about the time their children are thinking about college).

The gloom-and-doom statistics are abundant, and anyone who is at all interested has probably heard enough of them by now. But just in case, here is one more: there are about 30 more humans living now than there were 10 seconds ago when you began reading this paragraph. By tomorrow at this time, about 245,000 more will have been added, and by next year, 90 million. The last is approximately the population of Mexico—and most of the newcomers can expect to live in a desperate style similar to that of the average citizen of that struggling nation (Figure 48.1).

We can generate such statements all day, but once we understand the problem, do we wallow in depression? Do we simply look away? Or do we join the ranks of hopeful and determined people who intend to learn as much as they can about the problem and then try to find ways to help? Obviously, we will assume you are in this last group, and so we'll begin by delving into the history of our numbers.

TABLE 48.1
World Population Data: 1970 And 1994

Region	Year	Total (millions)	Crude Birth Rate	Crude Death Rate	Natural Increase (Annual %)	Doubling Time (years)	% Below 15 Years of Age	Estimated Population in 2025 (millions)
World	1970	3632	34	14	2.0	35	37	8378
	1994	5607	25	9	1.6	43	33	
Africa	1970	344	47	20	2.6	27	44	1538
	1994	700	42	13	2.9	24	45	
Asia	1970	2045	38	15	2.3	31	40	5017
	1994	3392	25	8	1.7	41	33	
North America	1970	228	18	9	1.1	63	20	375
	1994	290	16	9	0.7	98	22	
Latin America	1970	283	38	9	2.9	24	42	679
	1994	470	27	7	2.0	35	36	
Europe	1970	462	18	10	0.8	88	25	731
	1994	728	12	11	0.1	1025	20	
Nations of Special Interest								
United States	1970	205	17.5	9.6	1.0	70	30	338
	1994	261	16.0	9.0	0.7	98	22	
People's Republic of China	1970	760	34.0	15.0	1.8	39	?	1504
	1994	1192	18.0	7.0	1.1	61	28	
India	1970	554	42.0	17.0	2.6	27	41	1376
	1994	912	29.0	10.0	1.9	36	36	
Mexico	1970	50.7	44.0	10.0	3.4	21	46	137
	1994	92.0	28.0	6.0	2.2	31	38	

Source of data: Population Reference Bureau.

TABLE 48.2
Basic Population Arithmetic

Crude birth rate = number of births per year per 1000 population

$$\left(\text{determined by: } \frac{\text{total births}}{\text{midyear population}} \times 1000\right)$$

Crude death rate = number of deaths per year per 1000 population

$$\left(\text{determined by: } \frac{\text{total deaths}}{\text{midyear population}} \times 1000\right)$$

Rate of natural increase (or decrease) = crude BR − crude DR

$$\textbf{Percent annual growth} = \frac{\text{rate of natural increase}}{10}$$

$$\textbf{Doubling time}_{\text{(approximately)}} = \frac{70}{\text{percent annual growth}}$$

$$\left(\text{Example: \% annual growth in the world in 1986 was 1.7: } \frac{70}{1.7} = 41.18 \text{ years}\right)$$

$$\textbf{General fertility rate} = \frac{\text{total births}}{\text{total women in reproductive years}} \times 1000$$

$$\left(\text{Example: In 1983 there were 3,164,000 births per 55,260,000 American women aged 15–44: } \frac{3,614,000}{55,260,000} \times 1000 = 65.40\right)$$

FIGURE 48.1
Exponential Growth—The Human Aftermath
A scene of squalor in Mexico City, the world's fastest growing metropolis. As the population exceeds 20 million, we see drastic unemployment and the continued shortage of dwellings, adequate water, and sewage facilities, all pointing to impending disaster unless heroic efforts to find a solution begin soon.

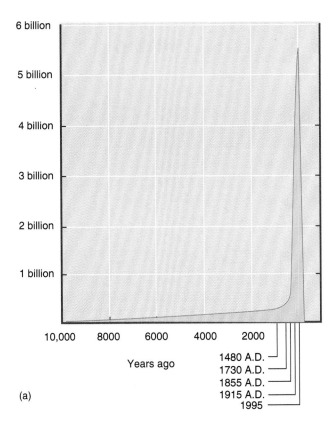

The History of the Human Population

The first thing we should know is that throughout most of our three- to four-million-year history, populations have remained fairly stable, but in the past million years that stability has been interrupted by three significant growth surges (Figure 48.2). But before that, there were only about 125,000 wandering hominids on the grassy East African plains, and they were not having a great deal of impact on their environment. These early hunters and gatherers were strongly subjected to the same factors that influence most populations of herbivores and carnivores today. Infants and children probably suffered high mortality rates but were quickly replaced in a species where, we believe, fertility was not a seasonal event. The average life span is estimated to have been 30 years, but the high infant mortality hides the possibility that some individuals lived much longer. Then, however, the population underwent a sudden, rapid increase.

The First Population Surge

The first growth surge in the human population, having begun about one million years ago, was probably brought about by the development of increasingly efficient tools, leading to more efficient ways of killing game and, in general, easier utilization of whatever was available in the surroundings. In addition, humans—inveterate wanderers—had by then pene-

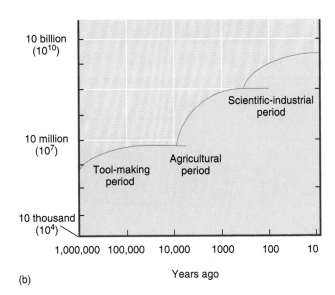

FIGURE 48.2
Two Ways of Viewing Human Population History
(a) The arithmetic plot of the past 10,000 years clearly reveals the soaring near-vertical rise of the J-shaped curve, a danger signal to most species. Hidden in the gradually rising line preceding this is the earlier population surge brought on by the switch from hunting and gathering to agriculture. **(b)** The logarithmic plot, where time is seemingly compressed, covers human history over the past million years. Here three growth surges are seen, the products, respectively, of advances in hunting and gathering technology, the rise of agriculture, and the recent era of modern health, industrial, and agricultural practices.

trated and established themselves on all the continents. Thus, by about 10,000 years ago, the earth probably supported about five million hunting and gathering humans (about the number in Croatia or El Salvador), who still had relatively little impact on their surrounding environment.

The Second Population Surge

About 10,000 years ago, the human population began its second growth surge, this time with more authority. With the advent of agriculture and the domestication of animals came increasing densities of local populations. There was less need to roam the countryside in search of food; in fact, there was a great need to stay put and tend the fields and livestock. With surplus food to store, winter and drought no longer exacted such a great toll on human life.

With the increased quantity and dependability of the food supply, humans probably experienced a lower death rate, particularly among the young. Furthermore, there was quite possibly an increase in the birth rate because of better nutrition. Large family size may have been encouraged because it meant more hands to till the fields. But life was by no means simple since the crops were subjected to the inconsistencies of weather and to infestation by insects and other herbivores whose own numbers responded to the novel food supply.

The unprecedented population growth of the early days of agriculture did not continue at its initial soaring rate but settled into a steadier, more gradual climb. Yet between the advent of agriculture and the time of Christ, the human population rose from five million to about 133 million. By 1650 A.D., it had reached an estimated 500 million.

History reveals that on a regional level this growth was interrupted many, many times by the decimating effects of disease, famine, and war. These are largely density-dependent and closely interrelated factors. A drastic example of the population-depressing effects of disease occurred in the 14th century, when one-fourth of Europe's population was killed by the "Black Death"—the bubonic plague. Such other diseases as typhus, influenza, and syphilis also took their toll on the crowded and incredibly filthy towns of the medieval period.

Interestingly, the loss of such numbers is insignificant in view of population growth today. At today's population growth rate, for example, the numbers lost to the 14th-century bubonic plague could be recouped in one year, while the number killed in all the wars of the last 500 years could be replaced in about six months.

The Third Population Surge

The third population growth surge began in Europe in the mid-17th century, after the unexplained decline of the plague. (Perhaps only those who were naturally immune were left alive.) A number of explanations have been advanced to account for this third surge. For one thing, the crowded populations of Europe expanded into the New World, with its array of opportunities and unexploited resources. More significant-

ly, the "germ theory" of disease, postulated later by Louis Pasteur and others, quickly led to ways of preventing and combating many dread diseases. By the late 19th century, public sanitation programs had begun, vaccines were developed, and rapid advances in food storage and transportation technologies led to a marked increase in food supplies. Between 1750 and 1850 the population of Europe doubled; that of the New World increased fivefold.

Populations were surging in other parts of the world as well. In China, the most heavily populated nation at that time, agriculture had made great gains, and a long period of comparative political stability followed the overthrow of the Ming Dynasty in 1644. India, however, had known little rest from turmoil and periodic famines. In 1770, the worst famine of all reportedly killed three million people in India. African populations are believed to have remained stable until about 1850, when the impact of imported European medical advances began to depress the death rate. However, recent experiences in Africa, namely, the Ethiopian drought and famine, remind us that we are not yet free of ancient hazards (Figure 48.3). Also, we are awaiting the full impact of the more recent threat, AIDS.

The third surge has continued into modern times, again supported by a host of innovations in industry, agriculture, and public health. Many of these innovations arose in industrial-

FIGURE 48.3
Famine in Ethiopia
In spite of modern achievements, undeveloped and developing nations are still subject to disruptive episodes of famine.

TABLE 48.3.
Doubling Times in Selected Nations

Doubling Time	Region			
	Africa	**Asia**	**Latin America**	**North America, Europe**
25 years or less	Algeria, Congo, Egypt, Ghana, Kenya, Libya, Nigeria, Niger	Iran, Iraq, Syria, Philippines, Pakistan	Ecuador, Guatemala, Honduras, Costa Rica, Nicaragua, El Salvador, Paraguay	None
26–35 years	Chad, Ethiopia, Angola, Morocco, South Africa	Afghanistan, Burma, India, Lebanon, Thailand, North Korea, Vietnam	Bolivia, Brazil, Mexico, Panama, Peru	Albania
36–69 years	Mozambique, Gabon	China, Indonesia, Israel, South Korea, Taiwan	Argentina, Chile, Cuba, Dominica, Jamaica	None
70+ years	None	Singapore, Japan	Barbados, Uruguay	All except Albania (Belgium: 1034 yr) (Italy: 3465 yr)
Negative growth	None	None	None	Denmark, Austria, Germany, Hungary

ized, developed countries and were exported to heavily populated developing regions[1]. In the developed nations, famine was all but eradicated with the advent of pesticides, chemical fertilizers, and high-yield crops. Potential disease epidemics were routinely controlled by vaccines, antibiotics, and insecticides.

In review, we can note that the human population grew from about five million at the dawn of agriculture to 500 million by 1650. By 1850, that population had doubled, reaching one billion. This amazing 200-year **doubling time** (the time it takes for the population to double) was only a hint of things to come. In the 80 years between 1850 and 1930, the numbers doubled again, to two billion. The next doubling took only 45 years, so by 1975 world population stood at four billion! By 1970, thoroughly alarmed population experts were predicting another doubling, to eight billion, by the year 2000—a span of only 35 years! (Doubling time is further explored in Table 48.3.)

The Human Population Today

To the surprise of nearly everyone, the rate of increase in world population growth slowed toward the end of the 1970s, and that slowing trend continues today (see Table 48.1). (Note the implications of the word *rate*. The occupants of a car approaching a cliff at 30 miles per hour might take little comfort in knowing that its rate is slowing and it will be traveling at only 15 miles per hour by the time it goes over the edge.)

At the end of 1994, demographers estimated the world's population to be over 5.6 billion. The annual growth rate had fallen slightly, and the doubling time had increased to 43 years. These data suggest some headway in our attempts to control our population, which can be at least partially attributed to changing attitudes of women toward their role in society, changes in preferences in family size, advances in methods of birth control, and the liberalization of abortion laws in developed nations.

Growth in the Developing Regions

The new data generated a measure of relief and even a growing optimism. But it turns out that the depressed growth rates occurred primarily in the most developed nations, those that could best support increased populations, such as the United States, Japan, Western Europe, and the former Soviet Union. Most poorer, developing nations of the world show few signs of controlling their populations. The annual rate of growth in those regions (excluding China) is ominous (averaging 2.4%), and the doubling time is 30 years (Table 48.3). While the crude birth and death rates (see Table 48.2) in the United States in 1994 were 16 and 9 per thousand, respectively, those rates in Africa were 42 and 13. Furthermore, the doubling time in Africa is an alarmingly brief 24 years. How do you suppose these shifting population trends will affect political, social, and economic stability in the near future? What effects are they having now? Figure 48.4 illustrates regional population forecasts.

---

[1]*Developed* nations are those that have a slow rate of population growth, a stable industrialized economy, a low percentage of workers employed in the agricultural sector, a high per capita income, and a high degree of literacy. *Developing* nations have the opposite traits. Of course, many countries have intermediate conditions, and some have a combination—a privileged upper class and a poor lower class.

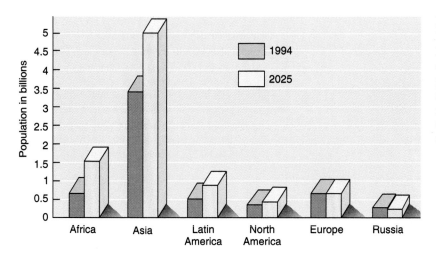

FIGURE 48.4
World Population by Region
In 1994 the earth's human population was unevenly distributed over the planet, bringing far greater pressures to bear in some areas than others. Most of the earth's people live in Asia and Africa, and this is not expected to change by 2020. The human population, at 5.6 billion in 1994, is expected to rise to 8.4 billion by 2025.

At first glance, the news from Latin America seemed hopeful. In the past few years there has been a decrease in the crude birth rate. But again, the numbers are misleading. In this region, the birth rate is over three times the death rate. Furthermore, many Latin American nations, already troubled by political unrest and dismal economic conditions, still face a doubling time of 35 years.

Asia has traditionally troubled demographers because so little is known about it and yet it has enormous reproductive potential. (It is instructive to keep in mind that some three-fifths of the people walking the earth today are Asians, and one in every five humans alive today is Chinese.) However, since 1970 the birth rate in China has been cut nearly 40%, with an equivalent fall in the death rate. Partly because of this shift in China, the doubling time for Asian populations has increased from 31 to 41 years. Massive birth-control programs in China and several other Asian nations have been instrumental in declining birth rates. Similar efforts in India have been somewhat successful, but India's annual increase is still 1.9%, and the doubling time of this nation of 912 million people is an alarming 36 years.

The successes Asia is currently enjoying, however, may be swamped by another problem, one of momentum. Specifically, 33% of these people are under the age of 15 (38% excluding China) and have yet to enter the breeding population. Asia could well be in for another population explosion.

It is difficult, if not impossible, to predict what the future holds, but there are ways of making educated guesses. Let's gaze into the crystal ball of demography and see how populations are forecast.

Demographic Transition

There is an interesting relationship between a country's developmental progress and its population structure: as nations undergo economic and technological development, their population growth tends to decrease. According to the **theory of demographic transition**, nations go through several developmental phases, the earliest of which is characterized by high birth and death rates and slow growth. As they begin to develop, the birth rate remains high, but the death rate falls. The result is that the population enters a rapid growth phase. Then, as industrialization peaks, the birth rate falls and begins to approximate the death rate. Population growth slows drastically, reaching a phase of very modest growth such as is seen in many European nations today.

One major prediction based on the theory of demographic transition is that population growth in developing parts of Asia, Africa, and Latin America will slow as they become further industrialized. But can the answer to the enigma of world population growth be that simple? Not quite. First, it is the developed nations that must pay the enormous costs of such a massive industrialization program, and this could place a great burden on their own economies. (Also, who needs more competition in the marketplace?) Second, there are severe risks in encouraging developing countries to go through such transitions. For example, if massive development programs stall in the second and most difficult phase, soaring populations will preclude the third stage and send the hopeful developing nation into a deadly population surge and devastating poverty cycle. Thus any such developmental program must be approached with extreme caution.

The Future of the Human Population

Our track record in predicting population growth is not very impressive; demographers in the United States were taken by surprise by the baby boom of the 1950s and they are still trying to explain the latest downward trend. However, increasingly sophisticated and precise calculations show some promise of improving the accuracy of our projections.

One of the more useful statistics in discussing human populations is the **general fertility rate** (**GFR**). Unlike the

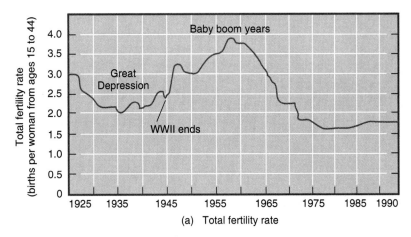

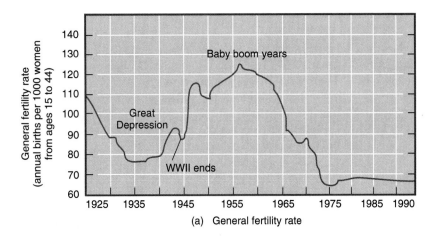

FIGURE 48.5
Fertility in the United States
The two plots include the total fertility rate (**a**) and general fertility rate (**b**). The depression years are seen as a valley, followed by a reproductive peak—the "baby-boom" years—that followed World War II. The more recent valley is good news to those alarmed by population growth, but keep in mind that attitudes toward family size change, and women born in the baby-boom years are still in the reproductive period of life.

birth rate, which reflects the number of babies born per total population in a given year, the GFR reflects the number of births per number of women of *reproductive age* per given year. Since the number of women of reproductive age varies from region to region and from nation to nation, comparisons based on the general fertility rate can be considerably more accurate. Reproductive age is somewhat arbitrarily defined as age 15–44, a figure that is most accurate for American women. The calculations for the GFR, together with other population statistics, can be seen in Table 48.2.

One of the highest general fertility rates ever recorded was in Iran: an incredible 200 live births per 1000 women of reproductive age. In that same year, the fertility rate in The Netherlands was 48.

A second useful indicator is the **total fertility rate (TFR)**, which is the number of children born to any woman who had conformed to the general fertility rate throughout her reproductive years. More simply, the TFR is really a prediction of how many children any woman is likely to have if the known general fertility rate remains constant.

The general fertility rate in the United States and much of Western Europe dipped sharply in the Great Depression years, falling close to the actual replacement level by 1936. The total

fertility rate in 1936 was about 2.2. (Replacement level is considered to be 2.1.) This trend was short-lived, however, for by the end of the 1930s the fertility rate had begun to climb, with the GFR peaking in the post-World War II years (1957) at 125 births per 1000 women, with the TFR reaching 3.7 (Figure 48.5). In the United States, this era became known as the "baby-boom" years. This was a time when the "good life" meant a new car, a home in the suburbs, and three or four kids playing in the yard.

Then came the awakening. Environmentally concerned scientists began to spread the alarm about the unprecedented population growth throughout the world. Soon after, reproductive rates in developed countries began to slow. The reasons may never be known, but perhaps people were taking the warnings seriously. By 1975, the total fertility rate in the United States, Japan, and most European nations had fallen to below 2.0. As of the time of this writing, the fertility rates in most of these nations show no signs of beginning a new upward trend.

Demographers have tried to understand what caused such a drastic change in reproductive behavior in the developed world. We can begin to guess at what happened, at least in the United States. It was a matter of attitude. The attitude of American women in their reproductive period toward domestic life

FIGURE 48.6
Age Structure Histograms

Age structure histograms break the population down into five-year age groups, revealing much about past history and permitting predictions of future trends to be made: (*left*) expanding nation (for example, Mexico), (*middle*) moderately stable nation (for example, United States), and (*right*) long-term stable nation (for example, Sweden).

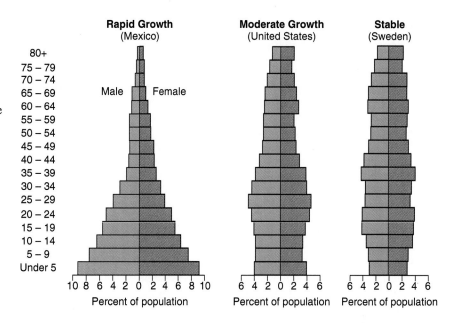

and family size changed radically in the 1960s and 1970s (as did that of women in many other developed nations). Women began to assume new roles in our industrialized society. They entered new areas of the work force, challenged male enclaves, demanded rights and privileges previously reserved for men, and placed less emphasis on raising children. Their new attitudes were duly noted by the demographers, who then predicted a reduced rate of population growth. But demographers are only too aware that such trends can change quickly, and if attitude is an important variable in forecasting population sizes, their data must remain current. In any case, measuring attitudes is a very risky business.

Whether the developed nations heeded the demographers' warnings or not, their rate of population growth has slowed. But what about the developing countries, those with poorer, often illiterate populations? After all, these are where most of the earth's teeming billions reside. It turns out, somewhat unexpectedly, that in the developing countries, the rates of increase are also declining, but the effect there is much more difficult to see. The problem is one of population momentum. With a declining death rate (particularly among the young), an increasing life span, and with great numbers of children who have yet to enter their reproductive years, downward trends in fertility will not have an appreciable effect for many years.

Using Mexico as an example once more, let's note that even at the peak of the baby boom in the United States, Mexican women experienced twice as many births per woman. Yet the subsequent dramatic decline seen in the United States also occurred in Mexico, and by the mid-1980s the total fertility rate in Mexico had gone from nearly 7 to below 5, and to 3.2 by 1994! Demographers estimate that it will continue to decline, reaching about 2.3 in 2025. Again, the problem is that, because of the large proportion of younger Mexicans in the population, the fertility *decline* will be accompanied by a whopping popu-

lation *increase* as great numbers of youths enter the breeding population. The Mexican population, by some estimates, will go from 91.8 million in 1994 to 138 million in 2025.

Population Structure

Knowing the age structure of any population is critical to understanding growth patterns and making predictions. One way to portray such data is by **age structure histograms**. In Figure 48.6, you can see how such diagrams are formed. Note the marked differences in the shapes of such histograms between developed and developing nations. In developed areas, recent population increases have been comparatively slow, so the base is not very wide. Also, people tend to live longer and thus to occupy the upper levels in greater proportions. In developing nations, on the other hand, the rate of increase is still expanding, swelling the lower levels. Obviously, these lower levels are important to population forecasting, since they represent future reproducers.

Growth Predictions and the Earth's Carrying Capacity

The fundamental question of how large the human population can become is irrevocably tied to what the earth can support—its human carrying capacity. If we have learned anything from population studies of other species, it is that this capacity cannot be exceeded for long without severe risk—especially the risk to the environment (see Essays 48.1 and 48.2). Any such damage would lower the environment's carrying capacity and set the stage for a devastating population crash.

The range of estimates of the earth's human carrying capacity is enormous. In other words, the experts cannot agree. Some population biologists believe that we have already

Essay 48.1 ● WHAT HAVE THEY DONE TO THE RAIN?

In the 1970s, it became clear that the rain was changing. In fact, in some areas the gentle raindrops were downright dangerous. The rain was becoming a dilute mixture of acids. It was first noticed in Scandinavia, then in the northeast United States and southeast Canada, then in northern Europe and Japan.

Rainwater, of course, had always been slightly acidic because the water dissolved atmospheric carbon dioxide, forming carbonic acids. But now the rain was showing alarming concentrations of the more dangerous sulfuric acid and nitric acid. Where were they coming from? They were the result of accumulations of nitrous oxides and sulfuric oxides in the atmosphere. The nitrous oxides, it turned out, were from power plant and automobile emissions; the sulfuric oxides were mainly from power plants and smelters. Dissolved in the water of cloud formations, they formed nitric acid and sulfuric acid, then

fell to earth to bathe our forests and cities and to fill our lakes with the corrosive mix.

The relative proportions of the two acids in rain depend on where you live. In the northeastern United States the acidity is primarily due to sulfuric acid; in California, to nitric acid. So we do have a choice.

The rain has caused the reduction and even the elimination of fish in many of our lakes. The rain apparently doesn't kill the fish; it just keeps them from reproducing. So no young fish are found as the old ones gradually go the way of all flesh. In fact, about 700 lakes in southern Norway are now *entirely devoid* of fish, and our own northeastern lakes are following one by one. As our Adirondack lakes reach pH levels of 5 (not uncommon), 90% have no fish whatever. They are also devoid of frogs and salamanders.

Entire patches of forests worldwide are sickening and dying as ecologists busily try

to find out just what effects the rain is having. In fact, such studies have masked inaction by the polluting countries. The Reagan administration (undoubtedly under heavy attack by industrial lobbyists) refused steadfastly for years to take action. Instead, it initiated one "study" after another, finally admitting in 1985 that there was a problem and that it had to do with industrial pollution.

Interestingly, the solution is clear to everyone. We simply need to reduce the levels of our effluent from power plants, smelters, and automobiles. Most of the technology exists, but its implementation would be too expensive for the polluters to willingly bear. Are we willing to pay higher prices for manufactured goods to save our lakes and rivers? The question is a fundamental one and is asked over and over in today's technological world.

---

exceeded our limits and that our present population represents a drastic overshoot. At the other end of the spectrum are the optimists who believe the human population can increase to 50 billion and still survive easily. (Biologists don't take this latter estimate very seriously.)

Recent estimates by more moderate population experts suggest that the human population could be sustained *temporarily* at 8–15 billion. From there, they suggest, our numbers could gradually decrease to new, more stable levels. From what

we know about population dynamics in other species, and our own history, we can predict the following: if we fail to restrain ourselves when we reach the higher numbers, we can expect not a gradual decrease in numbers but a massive increase in our death rate—a dieback or crash. It has been calculated that such a crash might kill 50–80% of the human population. Some population biologists believe that this is probably the way our population will stabilize. They suggest that the dieback will likely be due to a combination of famine, war, disease, and ecological

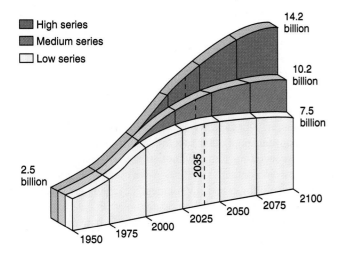

■ High series
■ Medium series
□ Low series

14.2 billion

10.2 billion

7.5 billion

2.5 billion

2035

1950 1975 2000 2025 2050 2075 2100

FIGURE 48.7
World Population Growth, 1950–2100: Three Scenarios
Three projections of world population based on different assumptions. The middle series assumes that, by 2035, each woman will have about two children. Even then, the population will double today's level. If "two-child families" are achieved by 2010, the lower curve results; the higher figure results if the "two-child family" is delayed until 2065.

Essay 48.2 ● HOLES IN THE SKY

Now let's explore the relationship between underarm deodorants and the death of the oceans. The propellant in underarm sprays is, in many cases, a class of molecules called chlorofluorocarbons (CFCs). These are essentially carbon molecules to which are attached chlorine and/or fluorine atoms. Chlorofluorocarbons are used in a variety of manufactured products, such as air conditioning, refrigeration, insulating foams (the type that keep our hamburger warm in fast-food places), plastics, and industrial solvents.

The problem is that these molecules are very stable. So after you spray under your arms, or after the insulated fastfood box begins to disintegrate, these long-lived little molecules are released into the air. Because they're light, they eventually, perhaps a few years later, end up in the upper atmosphere.

Paradoxically, these molecules would be safer for life if they stayed closer to earth, mingling with living things. The truth is, though, that they threaten life precisely because they drift upward and away from it. The reason is because at an altitude of about 15 miles, the CFCs break down the ozone layer. Ozone is O_3, formed by the sun breaking down atmospheric O_2 molecules, allowing them to rejoin as ozone. The chlorine in the CFCs attacks the ozone, breaking it back down into its components. There isn't much ozone up there to begin with. At sea level, all of it together would form a layer over the earth about as deep as a pencil lead is thick.

The ozone, however, is critical to life on earth. Primarily, it functions by blocking the sun's destructive ultraviolet light from the earth. Those rays are destructive on three primary bases. First, they increase the risk of skin cancer, particularly among light-skinned people, such as those who invented chlorofluorocarbons. Second, they depress the immune systems of humans, setting the stage for a host of illnesses. Third, they destroy the algae that form the first step in the ocean's food chains.

Ozone depletion was first discovered over the Antarctic. In fact, two-thirds of the springtime ozone over the Antarctic is now missing since the British began the measurements some years ago. In the northern latitudes where most people live, the ozone levels have declined by several percent since 1969. Now, an ozone hole is reported developing over the Arctic.

The manufacturers of CFCs have been reluctant to take action to reduce the levels of these chemicals over the earth. In fact, the Du Pont Company, which makes about $600 million a year on CFCs, took out ads in newspapers saying that the danger to the ozone layer was improved. The company has now agreed to phase out the manufacture of CFCs, but the phasing out will not be complete until the year 2000. Since the United States makes only 30% of the CFCs, the effort will have to be global, demanding more cooperation than one usually finds among industrial nations.

Part of the problem is that CFCs take 15–30 years or more to reach the altitude where they can do damage. Thus if we stop producing all CFCs now, the damage will continue for at least another 30 years.

---

disruption. With the exception of war (which is rare among other species), these are common density-dependent controls.

To leave the subject on a note of cautious optimism, we reiterate that world population growth is slowing down somewhat and that some of the more heavily populated nations have recently joined this trend. Furthermore, it is within the power of the world's family of nations to see that the decreasing trend is continued and accentuated. Our goal, if we are to avoid chaos, is to reach a *replacement level*. Essentially, it means that each couple simply replaces itself, by having only two children. (The popular belief that the United States has achieved this goal is inaccurate.) Because of the burgeoning numbers of young people, the future size of the world's population depends largely on when that replacement level is reached.

Scenarios portraying a stable world population are admittedly utopian, but they must not be regarded as impossible. Some population experts maintain that many slow-growing developed nations are now reaching the desired level and that others can do so within a few decades. The trend toward family planning and use of birth control is increasing, and one hopes that efforts to stimulate such interest in developing nations will continue. Under the best of circumstances, the world could reach some state of population stability in about 50 years (Figure 48.7).

The bottom line is that we must assume our species is not special; it is not exempt from the natural laws that govern population control in other species. Our only special feature is our mental capacity. We have the ability to analyze, predict, imagine, and finally to *choose*. Of course, we can choose by deciding not to choose, not to take a stand, not to be involved. But the time for that luxury is past. We must now learn as much as possible about the nature of overpopulation, apply ourselves to solving the problem, and be ready to stand accountable for our actions when we are judged by future generations.

Key Ideas

THE HISTORY OF THE HUMAN POPULATION

1. Human population numbers have remained relatively stable throughout much of our evolutionary history. Early hominids lived by hunting and gathering, their numbers regulated by the usual biotic and abiotic factors. Human population has experienced three significant growth surges. The first probably occurred with increasingly efficient tool-making. The second growth surge coincided with the development of agriculture and the domestication of animals.

2. In the mid-17th century, the third population surge began, possibly because of industrialization, expansion into the New World, the introduction of sanitation programs and sophistication of medical practices, advances in the preservation of food, and political stability in some areas. Between 1650 and 1850, the world population doubled (from 500 million to 1 billion). There were two billion people by 1930, and four billion by 1975; in 1970, some predicted that there would be eight billion people on earth by the year 2000.

THE HUMAN POPULATION TODAY

1. The rate of natural increase of the world's population slowed in the late 1970s and is continuing to decelerate.

2. The doubling time of the human population has increased in developed nations, but developing nations show few signs of controlling their populations. China has instituted major birth-control propaganda programs to stem its high rates of natural population increase.

3. According to the **theory of demographic transition**, as development increases, population growth tends to decrease. The earliest period is characterized by high birth and death rates, the next by high birth rates and lower death rates, and the next by low birth and death rates. This theory is the basis for the prediction that population growth in developing parts of Asia, Africa, and Latin America will slow as those areas become further industrialized. Programs based on the theory must be approached cautiously, however.

THE FUTURE OF THE HUMAN POPULATION

1. An important population growth indicator is the fertility rate. The **general fertility rate** (**GFR**) is the number of live births per 1000 women in their reproductive years per year, while the **total fertility rate** (**TFR**) is a prediction of the average number of children women will bear.

2. Fertility rates in the United States and Europe dipped during the early depression years but later rose, peaking in the post-WWII "baby-boom" years. The increase slowed again by the early 1970s.

3. The recent decline in fertility rates is attributed partly to changes in the attitude of women toward family size, improvements in birth-control, and increases in abortion.

4. Although declining growth rates have also occurred in developing nations, their effects will not be felt for years because of the large numbers of individuals yet to begin reproducing.

5. **Age structure histograms** are useful devices for portraying the age structure of a population. Such information is important in understanding growth patterns and in making predictions.

6. The basic issue in human population studies is the carrying capacity of the earth. Estimates vary greatly, and some population biologists believe that humans have already exceeded the carrying capacity and are in a phase of overshoot. A population crash could result in stabilization—or the population could gradually decrease, possibly through zero population growth programs, through which we could reach our replacement level.

Application of Ideas

1. Consider what we know about population control in other species and apply these factors to the future of the human species if voluntary population control measures fail. Considering such control, in what forms might we expect population reduction?

2. Why are the three historical human population growth surges usually understood in terms of decreased death rates?

Review Questions

1. Describe the innovations that brought about the first two great population surges in human history. (pages 1064–1065)

2. List the three major causes of death during the Middle Ages. Were these density-dependent or density-independent? Explain. (page 1065)

3. List three innovations that may explain the most recent human population growth surge, and explain how each may have affected the birth rate and the death rate. (page 1065)

4. What is the status of human population growth today? Specifically, where is growth most rapid? What are some of the possible political and social ramifications of this? (pages 1066–1067)

5. With simple drawings, depict the population age structure for Mexico, the United States, and Sweden. What future conditions can we predict according to the base of the diagram? According to the top? (page 1069)

6. What are experts telling us about the earth's carrying capacity? Should human numbers be maintained at our current level? Explain. (pages 1070–1071)

Appendix A

CLASSIFICATION OF ORGANISMS

The classification scheme was drawn from various commonly used sources in microbiology, botany, and zoology, but has been modified to accommodate the most recent ideas. Most taxa mentioned are also described in Chapters 22–27. Notably absent from this scheme are the viruses. Their taxonomic relationships have never been clarified, although some authors include them with the Monera while others have invented "Kingdom Virus."

THE PROKARYOTES

Kingdom Monera (also Prokaryota): Includes the prokaryotes or bacteria, single-celled and colonial organisms that generally lack membrane-bounded organelles, including an organized nucleus; and whose DNA is organized into a single, circular, main chromosome, sometimes supplemented by minute circular plasmids. Cell division is through fission, and sexual recombination is limited. Where present, the flagellum is tubular and rotating. Spore formation is common, and absorption is the usual mode of feeding by heterotrophs, including the parasites. Autotrophic bacteria include both chemoautrophs and photoautotrophs. Classification in this kingdom is unsettled and undergoing intensive revision. The newer taxonomies replace Kingdom Monera with two new kingdoms, Archaebacteria and Eubacteria, basing the new distinction in part on ribosomal RNA and cell wall and plasma membrane chemistry.

THE EUKARYOTES

All other kingdoms are eukaryotic. All have cells containing membrane-bounded organelles, their DNA heavily complexed with histone and other proteins, forming chromatin. Mitotic cell division is common, and the eukaryotic flagellum and cilium, where occurring, follows the 9 + 2 microtubular organization. Many undergo meiosis and exchange gametes in sexual reproduction.

Kingdom Protista: A polyphyletic kingdom with many unresolved taxonomic problems. Includes the protozoans, the fungus-like protists, and the algae. Protists include unicellular, colonial, and multicellular levels of organization. Nutrition is both heterotrophic and photoautotrophic. Sexual reproduction commonly occurs through meiosis and conjugation or the union of gametes, although it has not been observed in many species.

Protozoan Protists

Phylum Sarcomastigophora: Subphylum Mastigophora: Flagellated, unicellular heterotrophs that feed by phagocytosis and absorption. Includes parasites such as *Trypanosoma gambiense*, the agent of African sleeping sickness. Subphylum Sarcodina: Unicellular heterotrophs with ameboid movement that feed by phagocytosis and absorption; includes marine radiolarians and foraminiferans and *Entamoeba histolytica*, the parasite of amebic dysentery.

Phylum Apicomplexa: Unicellular, nonmotile, spore-forming heterotrophs that feed by absorption. Mainly parasitic, including *Plasmodium vivax*, an agent of human malaria.

Phylum Ciliophora: Ciliates. Unicellular, ciliated heterotrophs that feed by phagocytosis and absorption. Often large, with complex organelles, and sexual reproduction by meiosis and conjugation. Includes *Paramecium*.

Fungus-like Protists

Phylum Acrasiomycota: Cellular slime molds. Individual myxameba that fuse to form a pseudoplasmodium and later compound sporangia. Feed by phagocytosis. *Dictyostelium discoideum*.

Phylum Myxomycota: Plasmodial slime molds. Unicellular individuals fuse to form a true plasmodium and sporangia. Feed by phagocytosis. Some reproduce sexually. Includes *Physarum polycephalum*.

Phylum Oomycota: Water molds. Filamentous with cellulose cell walls. Flagellated stages. Some have large eggs and small nonmotile sperm. Many parasites (*Phytophthora infestans*).

Algal Protists

Phylum Pyrrhophyta: Dinoflagellates. Flagellated photoautotrophs with chitinous cell walls and a primitive fission and one-step meiosis. Chlorophylls *a* and *c* and carotenoids, starch storage. Red tide organisms include *Gonyaulax* and *Gymnodinium*.

Phylum Euglenophyta: Euglenoids. Unicellular, flagellated photoautotrophs and heterotrophs with red "eye spot." Chlorophylls *a* and *b* and carotenoids, paramylon starch storage. Includes *Euglena gracilis* and *Phacus*.

Phylum Chrysophyta: Yellow-green and golden-brown algae and diatoms. Unicellular and colonial photoautotrophs with pectin and glassy cell walls. Chlorophylls *a* and *c* and carotenoids and leucosin storage.

Phylum Rhodophyta: Red algae, seaweeds, Multicellular photoautotrophs with phycocyanin and phycoerythrin, chlorophyll *a*, and carotenoid pigments; floridean or carageenan storage. Sporic cycle. Separate sporophyte and gametophyte with varying dominance. *Polysiphonia, Porphyra*.

Phylum Phaeophyta: Brown algae: seaweeds, and kelps. Multicellular photoautotrophs with fucoxanthin, chlorophyll *a*, and carotenoid pigments; laminarin and mannitol storage. Many with sporic cycle. Separate sporophytes and gametophytes in some. Considerable tissue specialization. *Macrocystis, Necrocystis,* and *Sargassum.*

Phylum Chlorophyta: Green algae. Unicellular, colonial, and multicellular photoautotrophs, with chlorophylls *a* and *b* and carotenoids, starch storage. *Chlamydomonas, Ulva, Volvox.*

KINGDOM FUNGI

Multicellular heterotrophs with saprobic feeding. Many are parasites. Fungi usually have simple septate or nonseptate vegetative mycella, but many form complex, specialized, multicellular reproductive structures. The mycelia have chitinous walls. Fungi are primarily zygotic, that is, haploid with brief diploid interludes. Asexual reproduction is by mitotic spore formation, and sexual reproduction is often by conjugation, followed by a lengthy dikaryotic state.

Phylum Zygomycota: Bread molds. Extensive, simple, nonseptate mycelium, sexual reproduction through conjugation, and zygospore formation. *Rhizopus, Neurospora.*

Phylum Ascomycota: Sac fungi. Extensive, septate mycelium, asexual reproduction by conidiophore, sexual reproduction through conjugation, long dikaryotic state, and fertilization followed by meiosis and the development of ascospores in saclike asci. *Saccharomyces cerevisae* (baker's yeast). *Claviceps purpurea, Piziza.*

Phylum Basidiomycota: Club fungi. Extensive septate mycelium, often with a large, raised, fruiting body, the basidiocarp. In sexual reproduction conjugation leads to a long dikaryotic state. Fertilization occurs in the clublike basidia, followed by meiosis and the production of basidiospores. Mushrooms, shelf and bracket fungi. *Puccinia graminis* (wheat rust).

Deuteromycota. In older schemes, a fungal phylum. Most of its members are now assigned to the phylum Ascomycota; those remaining are awaiting assignment to one of the fungal phyla.

KINGDOM PLANTAE

A monophyletic kingdom containing multicellular photoautotrophs, most with protected embryos and great tissue specialization. Pigments include chlorophylls *a* and *b* and carotenoids, and the chief storage polysaccharide is starch. Sporic life cycle with the sporophyte clearly dominant in nearly all divisions. (Note: In plant taxonomy, division replaces phylum.)

Non-vascular Plants

Formerly the Division Bryophyta. Plants with limited supporting and vascular tissues, thus no true roots, stems, and leaves. Highly dominating gametophyte; brief dependent sporophyte stages; swimming sperm.

Division Hepatophyta: Liverworts. Often flattened ribbonlike growth, asexual reproduction by gemmae cups, simple rhizoids anchor the plant. *Marchantia.*

Division Bryophyta: Mosses. Upright or hanging body form, numerous leaflike scales, simple rhizoids. Sphagnum, *Polytrichum.*

Division Antherocerophyta: Hornworts. Flattened, spreading gametophyte; slender upright sporophyte grows continuously. *Anthoceros.*

Vascular Plants (Seedless)
(Vascular tissue, highly dominant sporophyte, embryo develops within separate gametophyte)

Division Psilophyta: Whisk ferns. Primitive group of four species, vascular stem but nonvascular scalelike leaves and rhizoids. Motile sperm; fertilization requires water. *Psilotum nudum.*

Division Lycophyta: Club mosses. Widely distributed; vascular roots, stems, leaves. Motile sperm; fertilization requires water. Homosporous and heterosporous species. *Lycopodium, Selaginella.*

Division Sphenophyta: Horsetails. Limited distribution; leaves highly reduced, slender stems bearing sporangia at tips. Motile sperm; fertilization requires water. *Equisetum.*

Division Pterophyta: Ferns. Widely distributed; foliage often treelike with complex leaves emerging from rhizomes. Motile sperm; fertilization requires water. Tree ferns, bracken ferns, sword ferns, and staghorn ferns.

Vascular Plants with Seeds: Gymnosperms
(Cones rather than flowers, seeds naked—not enclosed in fleshy fruits.)

Division Ginkgophyta: Maidenhair tree. One species, a large tree with bilobed, fan-shaped leaves. Motile sperm; plant provide fluids for sperm to swim to egg. *Ginkgo biloba.*

Division Cycadophyta: Cycads. Widespread, tropical; palmlike leaves, very prominent cones. Swimming sperm; plant provides fluids for sperm to swim to egg. *Zamia, Cycas.*

Division Gnetophyta: Widespread in deserts and warm temperature regions. Advanced, angiosperm characteristics in some: xylem with vessels, phloem with sieve tube members, bladelike leaf, and flowerlike pollen cones. *Gnetum, Welwitschia, Ephedra.*

Division Coniferophyta: Conifers. Widespread in temperate, subarctic regions and high altitudes; well adapted to dryness. Commonly with needlelike leaves. Nonmotile sperm; no fluids required for fertilization. Pines, spruce hemlock, cedar, larch, fir, juniper, redwood.

Vascular Plants with Seeds: Angiosperms
(Seed plants with flowers, fleshy fruits, and nonmotile sperm)

Division Anthophyta: Flowering plants. Seeds surrounded by fruit.

Class Dicotyledonae: Dicots. Worldwide distribution; floral parts in fours and fives or their multiples, two cotyledons in embryo, netveined leaves. Perennials with true secondary growth. Magnolia, oak, willow, maple, apple, rose, cucumber, bean, and pea.

Class Monocotyledonae: Monocots. Worldwide distribution; floral parts in threes or its multiples, one cotyledon in embryo, parallel-veined leaves. Many soft-bodied, herbaceous. No true secondary growth. Grasses, lily, palm, orchids.

KINGDOM ANIMALIA

A polyphyletic kingdom made up of motile, multicellular heterotrophs, generally with tissue specialization and organ and organ-system organization. Feeding commonly by engulfing, extracellular digestion, but phagocytosis and intracellular digestion are also common. Life cycle is gametic: sexual reproduction by the fusion of haploid gametes that are produced directly through meiosis.

Subkingdom Parazoa
(Primarily cell-level organization)

Phylum Porifera: Sponges. Nonmotile, filter-feeding, solitary or colonial adults with cellular level of organization and few cell types. Asexual reproduction by budding and sexual reproduction by internal fertilization. Swimming planula larva.

Class Calcaria: Calcium carbonate skeleton. *Leucosolenia.*

Class Hexactinellida: Silicon dioxide skeleton. *Euplectella* (Venus flower basket).

Class Demospongia: Skeleton of silicon dioxide or the protein spongin. *Cliona*, bath sponges.

Subkingdom Metazoa
(Tissue, organ, and organ-system levels of organization)

Radiate Phyla
(Diploblastic animals with radial symmetry)

Phylum Cnidaria: Radial animals with thin-walled bodies, tentacles armed with stinging cells. Saclike gastrovascular cavity, extracellular and intracellular digestion. Individuals exist as polyps, medusae, or alternating cycles of the two.

Class Hydrozoa: Solitary hydroids or those with colonial feeding polyps that bud reproductive medusae in which fertilization occurs. *Obelia, Hydra.*

Class Scyphozoa: Includes the jellyfish, a medusa in which feeding and sexual reproduction occur. Swimming larva forms a polyp stage that buds off young jellyfish. *Aurelia.*

Class Anthozoa: Anemones and corals, polyps that feed and reproduce. Development is usually direct from zygote to polyp. *Metridium*, sea fan, coral, brain coral.

Phylum Ctenophora: Radial, marine animals with thin-walled bodies, tentacles armed with glue cells. Saclike bodies like cnidarians, but with eight rows of combs consisting of fused cilia. Comb jelly. Venus girdle.

Bilateral, Protostome, Acoelomate Phyla
(Bilateral symmetry, triploblastic, spiral cleavage, blastopore area gives rise to mouth, solid body lacking a coelom)

Phylum Platyhelminthes: Flatworms. Dorsoventrally flattened, solid-bodied; marine, freshwater, and terrestrial worms, most with a highly branched gastrovascular cavity and some organ-system development.

Class Turbellaria: Free-living planarians, with extensive organ-system development, hermaphroditic. *Dugesia.*

Class Trematoda: Parasitic flukes with oral suckers, often with two or more hosts. *Clonorchis sinesis* (Asian liver fluke), *Schistosoma* (blood flukes).

Class Cestoda: Parasitic tapeworms with small scolex, numerous proglottids in the mature adults, often with two hosts. *Taenia solium* (pork tapeworm), *Taeniarhynchus* (beef tapeworm).

Phylum Rhynchocoela: Proboscis worms. Thin, ribbonlike worms with a complete, one-way digestive tract. Protrusible pharynx with piercing stylet. Marine.

Bilateral, Protostome, Pseudocoelomate Phyla
(Body cavity a false coelom, incompletely lined by mesodermal tissue)

Phylum Nematoda: Roundworms. Slender, threadlike worms with a fluid-filled body cavity and a complete digestive tract. Terrestrial, free-living predators and internal parasites. Widespread in soil and water—outnumbered only by arthropods. *Rhabditis, Ascaris lumbricoides* (giant roundworm), *Necator americanus* (hookworm), *Enterobias vermicularis* (pinworm).

Phylum Rotifera: Rotifers or "wheel animals." Complex, free-living, minute, ciliated animals with a complete digestive tract and well-developed organ systems. Sexes separate with parthenogenesis common. *Philodina.*

Bilateral, Protostome, Coelomate Phyla
(Triploblastic animals with a true—mesodermally lined—body cavity)

Phylum Ectoprocta, Phoronida, and Brachiopoda: Lophophorate phyla. Ciliated tentacles attached to a ridgelike lophophore. Moss animals, lampshells, and wormlike *Phoronis.*

Phylum Mollusca: Mollusks. Highly reduced coelom and segmentation, diversified through adaptive variations in foot, shell, and mantle. Open circulatory system. Chiefly marine and freshwater, some terrestrial.

Class Polyplacophora: Marine, shell in eight parts, lengthy foot surrounded by simple gills and mantle. *Mopalia* (a chiton).

Class Gastropoda: Aquatic and terrestrial, large muscular foot, often retractable foot and spiral shell, feed by radula. *Helix* (land snail), *Haliotis* (an abalone), whelks, limpets, slugs.

Class Bivalvia: Aquatic, hinged shells, digging retractable foot, filter feeders. *Mytilus* (a mussel), *Tagelus* (a clam), oyster, scallop.

Class Cephalopoda: Marine, enlarged head and brain, image-forming eyes, tentacles, internalized shell in most. *Octopus, Loligo* (a squid), and *Nautilus* (chambered nautilus).

Phylum Annelida: Segmented worms. Most with body divided into numerous metameres (segments), well-developed coelom, complete and specialized digestive tract, and closed circulatory system.

Class Oligochaeta: Terrestrial and aquatic, free-living and burrowing. Hermaphroditic. *Lumbricus terrestris* (earthworm).

Class Polychaeta: Marine, free-living burrowing, tube-dwelling and swimming forms. *Neries virens* (clam worm), fanworm, peacock worm.

Class Hirudinea: Mostly freshwater and terrestrial, parasitic, sucker mouth. *Hirudo* (medicinal leech).

Phylum Arthropoda: Largest animal phylum. Jointed-legged animals with chitinous exoskeleton, modified segmentation.

Subphylum Chelicerata (no jaws).

Classes Merostomata, Arachnida, and Pycnogonida: Terrestrial and aquatic, four pairs of legs, fangs rather than mandibles, two body regions and book lungs common. Horseshoe crab, spiders, mites, ticks, scorpions, and sea spiders.

Subphylum Crustacea (jaws).

Class Malacostraca: Mostly aquatic, hardened exoskeleton, head and thorax usually fused, many with stalked compound eyes: Crayfish, crabs, lobster, pill bug.

Class Maxillopoda: Enlarged maxilla for feeding, trunk reduced, many fixed in place on rocks, pilings and hulls: barnacles, tongue worms.

Class Branchiopoda: Mostly aquatic, leaf-shaped appendages for gas exchange: water fleas, brine, and fairy shrimp.

Class Ostracoda: Aquatic, body covered by hinged, bivalve carapace: ostracods.

Subphylum Uniramia (jaws).

Class Chilopoda: Terrestrial, highly segmented, flattened bodies, most segments with one pair of legs, stinging organ: centipedes.

Class Diplopoda: Terrestrial, highly segmented, cylindrical bodies, most segments with two pairs of legs: millipedes.

Class Insecta: Three-part body, two pairs of antennae, two pairs of wings in most, compound eye: beetles, moths, bees, ants, flies, and many others. The vast majority of animals are insects.

Phylum Onychophora: Intermediate between annelids and arthropods, terrestrial with jointed appendages, claws, wormlike body. *Peripatus.*

Bilateral, Deuterostome Phyla
(Radial cleavage, blastopore area gives rise to anus)

Phylum Echinodermata: Spiny-skinned animals. Marine, pentaradial as adults, endoskeleton of calcareous plates, commonly with spines and pedicellaria. All with a water vascular system.

Class Crinoidea: Cup-shaped body with branched arms, no suckers on the tube feet. Sea lily.

Class Holothuroidea: Cylindrical body, spines reduced, no pedicellaria. *Cucumaria* (a sea cucumber).

Class Echinoidea: Spherical or round, flattened body without arms, prominent movable spines, tube feet with suckers. *Lytechinus* (a sea urchin), *Echinarachnius* (a sand dollar).

Class Asteroidea: Prominent arms, short, blunt spines, numerous pedicellaria, tube feet with suckers. *Asterias* (a sea star).

Class Ophiuroidea: Prominent arms, movable spines, tube feet without suckers, no pedicellaria. *Ophiura* (a brittle star), *Gorgonocephalus* (a basket star).

Phylum Hemichordata: Hemichordates. Marine, burrowing, acorn-shaped proboscis, gill slits. Larvae similar to echinoderm's. *Saccoglossus* (an acorn worm).

Phylum Chordata: Chordates. Aquatic and terrestrial, gill slits, notochord, dorsal hollow nerve cord, postanal tail.

Subphylum Urochordata: Marine filter feeders, saclike body with gill slits, bilateral larva with all chordate characteristics. *Ciona* (a tunicate), *Thetys* (a salp).

Subphylum Cephalochordata: Marine filter feeder, fishlike body form, notochord in adults, prominent gill slits and segmented trunk muscles, sensory cirri. *Branchiostoma* (lancelet).

Subphylum Vertebrata: Vertebral column of bone or cartilage, increasing cephalization, ventral heart and dorsal aorta, no more than two pairs of limbs, separate sexes.

Class Agnatha: Fishes with jawless mouth, cartilaginous skeleton. Lamprey and hagfish.

Class Chondrichthyes: Marine fishes with cartilaginous skeleton and jaws, placoid scales and rows of replacement teeth, dorsoventrally flattened body. Sharks, rays, chimera.

Class Osteichthyes: Marine and freshwater fishes with bony skeleton and jaws. Includes Subclass Crossopterygii (lobe-finned fishes), Subclass Dipneusti (lungfishes), and Subclass Actinopterygii (ray-finned fishes), which include Superorder Chondrostei (sturgeons), Superorder Holostei (gar pike and bowfin), and Superorder Teleostei (perch, bass, tuna, eel, sunfish).

Class Amphibia: Freshwater and terrestrial amphibians, chiefly air breathers, moist vascularized skin, three-chambered heart, most requiring water for reproduction. Includes Order Anura (frogs and toads), Order Urodela (salamanders), and Order Apoda (caecilians).

Class Reptilia: Aquatic and terrestrial air breathers with dry, scaly skin and three-chambered hearts in most. Fertilization is internal, and development occurs in the independent, amniotic land egg. Includes Order Chelonia (turtles), Order Squamata (snakes and lizards), Order Crocodilia (crocodiles and alligators), and Order Rhychocephalia (tuatara).

Class Aves: Birds, principally terrestrial and flying; body modification adapting for flight include lightweight skeleton with fusion in the vertebrae, no teeth, covering of feathers, highly modified forelimbs, four-chambered heart, endothermy, and crosscurrent air-blood flow in lung. Hawks, penguins, peafowl, jays, bluebirds, ducks, turkey.

Class Mammalia: Mammals, terrestrial and aquatic air-breathers, body covering of hair, glandular skin, milk produced in mammaries; highly specialized, socketed teeth; skull bones highly fused, enlarged cerebrum, and increased learning capacity.

Subclass Prototheria: Monotremes, egg-laying mammals. Spiny anteater and duckbilled platypus.

Subclass Metatheria: Marsupial mammals, pseudoplacenta, immature young housed and nourished in marsupium (pouch), nearly all restricted to Australia and New Zealand. Opossums, kangaroos, wallabies, koalas, wombats.

Subclass Eutheria: Placental mammals, true placenta supports embryo development. Shrew, bat, anteater, rabbit, mouse, whale, aardvark, elephant, manatee, rhinoceros, giraffe, chimpanzee, and don't forget *us*.

Appendix B

ANSWERS TO SELECTED GENETICS PROBLEMS

The following are answers to the genetics problems that appear in the Application of Ideas sections in Chapters 10 and 11.

CHAPTER 10

1. a. 1/4
   b. $1/2 \cdot 1/2 = 1/4$
   c. $1/4 \cdot 1/4 = 1/16$
   d. $(1/4 \cdot 1/4) + (1/4 \cdot 1/4) = 1/8$
   e. In part C, the order of the children is given—so the multiplication law is all that is needed. In part D, no order is specified, so the addition law must be used because here are two possible orders in which the children may occur.

2. a. 1
   b. 0
   c. $0 \cdot 0 = 0$

3. a. 0
   b. 1

4. a. Ann—**aa**   Mike—**Aa** (how do you know this?)
      Sara—**Aa**   Martha—**aa**
   b. 1/2
   c. $1/2 \cdot 1/2 = 1/4$
   d. $(1/2 \cdot 1/2) + (1/2 \cdot 1/2) = 1/2$

5. a. Ralph–**AA**
   b. $1/2 \cdot 1/2 \cdot 1/2 \cdot 1/2 \cdot 1/2 \cdot 1/2 = 1/64$

6. a. $1/2 \cdot 1/2 = 1/4$
   b. $1/2 \cdot 1/2 = 1/4$
   c. $1 \cdot 1/2 = 1/2$
   d. $1/4 \cdot 1/4 = 1/16$
   e. Genotype would be **A_B_**; $3/4 \cdot 3/4 = 9/16$

CHAPTER 11

1. 9:3:3:1; 1:1:1:1

2. a. Parental strains are **AAbb** and **aaBB**; F_1 hybrid is **AaBb**.
      First backcross progeny: **AaBb**
                                  AAbb
                                  AaBb
                                  Aabb
      Second backcross progeny: **AaBB**
                                  AaBb

aaBB
aaBb
   b. Blue: **aaBB, aaBb, AAbb, Aabb**
      Purple: **AABB, AaBB, AABb, AaBb**
      Scarlet: **aabb**

3. a. X^B = brown spotted teeth
      X^a = hemophilia A
      X^c = colorblindness

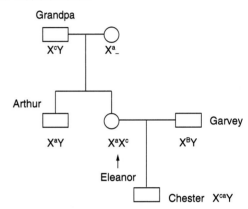

Grandpa

Arthur

Garvey

Eleanor

Chester $X^{ca}Y$

   b. She carries X^a; later she is shown to be X^c as well.
   c. Chester is also colorblind. Therefore she passed to him an X with both hemophilia and colorblindness. This must have resulted from a crossover between an X^a and an X^c, since she can pass only one X to a son.
   d. hemophilic—0
      colorblind—0
      brown teeth—1
   e. hemophilic—.5
      colorblind—.5
      brown teeth—0
   f. A son cannot be normal, as he would get either an X^c or an X^a from Eleanor, *unless* a crossover occurs to reconstitute a normal X.
      10% crossing over = 5% normal
                          5% colorblind and hemophilic
      90% non crossing over = 45% hemophilic
                              45% colorblind

4. a. broad-leaved is dominant
   b. sex linked
   c. F_1: males all narrow-leaved
             females all broad-leaved
      F_2 = males 50:50
             females all broad-leaved

Suggested Readings

PART ONE: MOLECULES TO CELLS

Albert, B., et al. 1995. *Molecular Biology of the Cell*. 3d ed. Garland Publishing, New York.

Darnell, James, et al. 1995. *Molecular Cell Biology*. W. H. Freeman, New York.

Darwin, C. 1859. *On the Origin of Species through Natural Selection*. A facsimile of the first edition. Harvard University Press, Cambridge, Mass.

de Beer, G. 1965. *Charles Darwin: A Scientific Biography*. Doubleday, New York.

de Duve, Christian. 1984. *A Guided Tour of the Living Cell*. W. H. Freeman, *Scientific American* Library, New York.

Dickinson, Robert E., and Ralph J. Cicerone. 1986. "Future Global Warming from Atmospheric Trace Gases." *Nature*, 319:109–115.

Folsome, C. E., ed. 1979. *Life: Origin and Evolution*. W. H. Freeman, *Scientific American* Library, New York.

Gould, S. J. 1977. *Ever Since Darwin: Reflections in Natural History*. W. W. Norton, New York.

Kerr, Richard A. 1988. "Is the Greenhouse Here?" *Science* 239: 559–561.

King-Hele, D. 1974. "Erasmus Darwin, Master of Many Crafts." *Nature* 247:87.

Kleinsmith, L. J., and V. M. Kish. 1995. *Principles of Cell and Molecular Biology*. 2d ed. HarperCollins, New York.

Lehinger. A. I., et al. 1993. *Principles of Biochemistry*, Worth, New York.

Margulis, L. L. To, and D. Chase. 1978. "Microtubules in Prokaryotes." *Science* 200:1118.

Moorehead, Alan. 1969. *Darwin and the Beagle*. Harper & Row, New York.

Oparin, A. I. 1938. *The Origin of Life*. Dover, New York.

Porter, E. 1971. *Galápagos*. Ballantine, New York.

Scientific American. 1986. *The Molecules of Life*. W. H. Freeman, New York.

Singer, S. J., and G. Nicolson. 1972. "The Fluid Mosaic Model of the Structure of Cell Membranes." *Science* 175:720.

Stryer, Lubert. 1988. *Biochemistry*. 3d ed. W. H. Freeman, New York.

Scientific American articles. New York: W. H. Freeman Co.

Allen, Robert Day. 1987. "The Microtubule as an Intracellular Engine." February.

Dautry-Varsat, Alice, and Harvey F. Lodish. 1984. "How Receptors Bring Proteins and Particles into Cells." May.

Govindjee and William J. Coleman. 1990. "Houseplants Make Oxygen." February.

Hayflick, L. 1980. "The Cell Biology of Aging." January.

Hinkle, P. C., and R. E. McCarty. 1978. "How Cells Make ATP." March.

Jacobs, William P. 1994. "Caulerpa." December.

Karplus, Martin, and J. Andrew McCammon. 1986. "The Dynamics of Proteins." April.

Linder, Maurine E. and Alfred G. Gilman. 1992. "G Proteins." July.

Margulis, L. 1971. "Symbiosis and Evolution." August.

Olson, Arthur J. and David S. Goodsell. 1992. "Visualizing Biological Molecules." November.

Pääbo, Svante. 1993. "Ancient DNA." November.

Rothman, James E. 1985. "The Compartmental Organization of the Golgi Apparatus." September.

Schopf, J. W. 1978. "The Evolution of the Earliest Cells." September.

Shulman, R. G. 1983. "NMR (Nuclear-Magnetic-Resonance) Spectroscopy of Living Cells." January.

Staehelin, L. Andrew, and Barbara E. Hull. 1978. "Junctions between Living Cells." May.

Stossel, Thomas P. 1994. "The Machinery of Cell Crawling." September.

Weber, Klaus, and Mary Osborn. 1985. "The Molecules of the Cell Matrix." October.

Youvan, Douglas C., and Barry L. Marrs. 1987. "Molecular Mechanisms of Photosynthesis." June.

PART TWO: MOLECULAR BIOLOGY AND HEREDITY

Avery, O. T., C. M. MacLeod, and M. McCarty. 1944. "Studies on the Chemical Nature of the Substance Inducing Transformation of Pneumococcal Types." *Journal of Experimental Medicine* 79:137.

Chedd, G. 1981. "Genetic Gibberish in the Code of Life." *Science 81*, November.

Crick, F. H. C. 1979. "Split Genes and RNA Splicing." *Science* 204:264.

DuPraw, E. J. 1970. *DNA and Chromosomes*. Holt, Rinehart and Winston, New York.

Jacob, F., and J. Monod. 1961. "Genetic Regulatory Mechanisms in the Synthesis of Proteins." *Journal of Molecular Biology* 33:318.

Kettlewell, H. B. D. 1956. "Further Selection Experiments on Industrial Melanisms in the Lepidoptera." *Heredity* 10:287.

Kleinsmith, L. J. and Kish, V. M. 1995. *Principles of Cell and Molecular Biology*. 2d ed. HarperCollins, New York.

Lewin, R. 1983. "A Naturalist of the Genome." *Science* 222:402.

Maynard Smith, J. 1971. "What Use is Sex?" *Journal of Theoretical Biology* 30:319.

Mendel, G. 1965. "Experiments in Plant Hybridization (1865)." Translated by Eva Sherwood in *The Origin of Genetics*, eds. C. Stern and E. Sherwood. W. H. Freeman, New York.

Meselson, M., and F. W. Stahl. 1958. "The Replication of DNA in *E. coli*." *Proceedings of the National Academy of Sciences* (U.S.) 44:671.

Okazaki, R. T., et al. 1968. "Mechanisms of DNA Chain Growth: Possible Discontinuity and Unusual Secondary Structure of Newly Synthesized Chains." *Proceedings of the National Academy of Sciences* (U.S.).

Russell, P. J. 1992. *Genetics*. 3d ed. HarperCollins, New York.

Sanger, F., et al. 1977. "Nucleotide Sequence of Bacteriophage fx174 DNA." *Nature* 265:687.

Scientific American. 1985. *The Molecules of Life.* W. H. Freeman, New York.

Shine, I., and S. Wrobel. 1976. *Thomas Hunt Morgan: Pioneer of Genetics.* University of Kentucky Press, Lexington, Ky.

Watson, J. D. 1968. *The Double Helix.* Atheneum, New York.

Watson, J. D. et al. 1987. *Molecular Biology of the Gene.* 4th ed. Benjamin/Cumming, Menlo Park, CA.

Watson, J. D., and F. H. C. Crick. 1953. "Molecular Structure of Nucleic Acids. A Structure of Deoxyribose Nucleic Acid." *Nature* 171:737.

***Scientific American* articles. New York: W. H. Freeman Co.**

Aharonowitz, Y., and G. Cohen. 1981. "The Microbiological Production of Pharmaceuticals." September.

Allison, A. C. 1956. "Sickle Cells and Evolution." August.

Anderson, W. F., and E. G. Diacumakos. 1981. "Genetic Engineering in Mammalian Cells." July.

Beardsley, T. 1992. "Diagnosis by DNA." October.

Bishop, J. M. 1982. "Oncogenes." March.

Brill, W. J. 1981. "Agricultural Microbiology." September.

Campbell, A. M. 1976. "How Viruses Insert Their DNA into the DNA of the Host Cell." December.

Chambon, P. 1981. "Split Genes." May.

Chilton, M. 1983. "A Vector for Introducing New Genes into Plants." June.

Crick, F. H. C. 1962. "The Genetic Code." October.

———. 1966. "The Genetic Code III." October.

Dickerson, R. E. 1972. "The Structure and History of an Ancient Protein." April.

Grivell, L. A. 1983. "Mitochondrial DNA." March.

Hopwood, A. 1981. "The Genetic Programming of Industrial Microorganisms." September.

Howard-Flanders, P. 1981. "Inducible Repair of DNA." November.

Hunter, T. 1984. "The Proteins of Oncogenes." August.

Kornberg, R. D., and A. Klug. 1981. "The Nucleosome." February.

McKusick, V. A. 1965. "The Royal Hemophilia." February.

Miller, O. L. 1973. "The Visualization of Genes in Action." March.

Novick, R. P. 1980. "Plasmids." December.

Pestka, S. 1983. "The Purification and Manufacture of Human Interferons." August.

Ptashne, M. 1989. "How Gene Activators Work." January.

PART THREE: EVOLUTION

Banks, H. P. 1975. "Early Vascular Plants: Proof and Conjecture." *Bioscience* 25:730.

Darwin, C. 1859, 1966. *On the Origin of Species.* Harvard University Press, Cambridge, Mass.

Dawkins, R. 1976. *The Selfish Gene.* Oxford University Press, New York and Oxford.

Dobzhansky, T. 1963. "Evolutionary and Population Genetics." *Science* 142:3596.

———. 1970. *Genetics of the Evolutionary Process.* Columbia University Press, New York.

Dodson, E. O., and P. Dodson. 1985. *Evolution: Process and Product.* 3d ed. Prindle, Weber and Schmidt, Boston.

Eldredge, N., and S. J. Gould. 1972. "Punctuated Equilibria: An Alternative to Phyletic Gradualism." In *Models in Paleobiology.* Edited by T. J. M. Schopf. W. H. Freeman, New York..

Gilbert, L. E., and P. H. Raven. 1975. *Coevolution of Plants and Animals.* University of Texas Press, Austin, Tex.

***Scientific American* articles. New York: W. H. Freeman Co.**

Bishop, J. A., and Laurence M. Cook. 1975. "Moths, Mechanism and Clean Air." January.

Eigen, M., et al. 1981. "The Orgin of Genetic Information." April.

Gilbert, L. E. 1982. "The Coevolution of a Butterfly and a Vine." August.

Gould, Stephen Jay. 1994. "The Evolution of Life." October.

Grant, P. R. 1991. "Natural Selection and Darwin's Finches." October.

Margulis, L. 1971. "Symbiosis and Evolution." August.

Orzel, Leslie E. 1994. "The Origin of Life on the Earth." October.

Rennie, J. 1991. "Are Species Specious?" November.

Sibley, C. G., and J. F. Ahlquist. 1986. "Reconstructing Bird Phylogeny by Comparing DNAs." February.

Stanley, S. M. 1984. "Mass Extinction in the Ocean." June.

PART FOUR: DIVERSITY

Ahmadjian, Vernon. 1982. "The Nature of Lichens." *Natural History,* March, pages 30–37.

Alexander, T. 1975. "A Revolution Called Plate Tectonics Has Given Us a Whole New Earth." *Smithsonian* 5:30.

Barth, Friedrich, G. 1985. *Insects and Flowers: The Biology of a Partnership.* Princeton University Press, Princeton, N.J.

Buschsbaum, Ralph, et al. 1987. *Animals without Backbones.* 3d ed. University of Chicago Press, Chicago.

Fox, G. E., et al. 1980. "The Phylogeny of Prokaryotes." *Science* 204:457.

Johanson, D. C., and M. A. Edey. 1981. "Lucy: The Inside Story." *Science 81.* March.

Johanson, D. C., and T. D. White. 1979. "A Systematic Assessment of Early African Hominids." *Science* 203:321.

Large, E. C. 1962. *The Advance of the Fungi.* Dover, New York.

Lee, J. J., S. H. Hutner, and E. C. Bovee, eds., 1985. *An Illustrated Guide to the Protozoa.* Society of Protozoologists, Lawrence, Kan.

Margulis, Lynn. 1981. *Symbiosis in Cell Evolution: Life and Its Environment on the Early Earth.* W. H. Freeman, New York.

McMenamin, M. A. S. 1982. "A Case for Two Late Proterozoic-Earliest Cambrian Faunal Province Loci." *Geology,* June.

Miller, S. L. 1935. "Production of Some Organic Compounds under Possible Primitive Earth Conditions." *Journal of the American Chemical Society* 77:2351.

Mulcahy, David L. 1981. "Rise of the Angiosperms." *Natural History* September, pages 30–35.

Nester, Eugene W., et al. *Microbiology.* 3d ed. Saunders, Philadelphia.

Raven, P. H. 1970. "A Multiple Origin for Plastids and Mitochondria." *Science* 169:641.

Rensberger, B. 1981. "Facing the Past." *Science 81,* October. (Neanderthal Man)

Schopf. J. W., and D. Z. Oehler. 1971. "How Old Are the Eukaryotes?" *Science* 193:47.

cerebellum a portion of the brain serving to coordinate voluntary movement, posture, and balance; it is located behind the cerebrum.

cerebral cortex the outermost region of the cerebrum, the "gray matter," consisting of several dense layers of neural cell bodies and including numerous conscious centers, as well as regions specialized in voluntary movement and sensory reception.

cerebral hemisphere either the right or left half of the cerebrum.

cerebrum the anterior, dorsal portion of the vertebrate brain, the largest portion in humans, consisting of two cerebral hemispheres and controlling many localized functions, among them voluntary movement, perception, speech, memory, and thought.

cervix the neck-like base and opening of the uterus.

C4 pathway a CO_2-concentrating adaptation of certain tropical and desert (C4) plants. The C4 pathway originates in leaf mesophyll cells, where highly efficient enzymes incorporate CO_2 into four carbon acids that then enter bundle sheath cells, release the CO_2 for use in the Calvin cycle, and recycle to the mesophyll cells.

chaparral a vegetation type common in California, characterized by a dense growth of low evergreen shrubs and trees.

character displacement where similar species share a niche, the tendency for physical differences to become emphasized through natural selection.

Chargaff's rule in DNA structure the observation that the quantity of adenine in DNA is equal to the quantity of thymine, while that of guanine is equal to that of cytosine.

charging enzyme any of a group of specific enzymes that covalently attach amino acids to their appropriate tRNAs.

chemical reaction the reciprocal action of chemical agents on one another; chemical change.

chemical synapse snyapses between neurons involving a space, the synaptic cleft, across which neurotransmitters must pass for a neural impulse to begin in the second neuron.

chemiosmosis the process in mitochondria, chloroplasts, and aerobic bacteria in which an electron transport system utilizes the energy of photosynthesis or oxidation to pump hydrogen ions across a membrane, resulting in a proton concentration gradient that can be utilized to produce ATP.

chemiosmotic phosphorylation the production of ATP using the energy of protons passing across a membrane and through ATP synthases (F_1 and CF_1 particles).

chemoautotroph an organism capable of utilizing simple inorganic substances as a source of energy and as a source of raw materials for its metabolic activities.

chemoreceptor a neural receptor sensitive to a specific chemical or class of chemicals.

chiasma (pl. *chiasmata*), the cross- or X-shaped configurations observed in homologous chromatids that have undergone recombination.

chitin a structural carbohydrate that is the principal organic component of arthropod exoskeletons.

chlorophyll *a* green photosynthetic pigment found in chloroplasts, cyanobacteria, and chloroxybacteria. It occurs in several forms, chlorophyll *a, b,* and *c*.

cholesterol a common sterol occurring in all animal fats, a vital component of plasma membranes, an important constituent of bile for fat absorption, a precursor of vitamin D, and too much of which is not good for you.

chorioallantois a highly vascular extraembryonic membrane of birds, reptiles, and some mammals, formed by the fusion of the chorion and the allantois.

chorion see *extraembryonic membrane.*

chorionic villi small finger-like processes of the chorion of the early mammalian embryo, especially before the formation of the placenta.

chromatid one of the two identical strands of the chromosome following replication and prior to cell division.

chromatin the substance of chromosomes, a molecular complex consisting of DNA, histones, nonhistone chromosomal proteins, and usually some RNA of unknown function.

chromosomal mutation a massive change in DNA, generally breakage involving a whole chromosome that has not been repaired or has been repaired improperly.

chromosome 1. in eukaryotes, an independent nuclear body carrying genetic information in a specific linear order, and consisting of one linear DNA molecule (in G_1) or two DNA molecules (in G_2), one centromere, and associated proteins. 2. in prokaryotes, an analogous circular DNA molecule. 3. the analogous DNA or genetic RNA molecule of a virus.

cilia (sing. *cilium*), fine, hair-like, motile organelles found in groups on the surface of some cells; shorter and more numerous than flagella, but similar in structure, they exhibit coordinated oar-like movement.

circadian rhythm any recurrent sequence of physiological or behavioral activities repeated on a daily basis.

cistron 1. also *structural gene,* a sequence of DNA specifying the sequence of a polypeptide chain. 2. any continuous genetic unit in which different recessive mutant lesions fail to complement one another in a double heterozygote.

citric acid cycle also *Krebs cycle,* tricarboxylic acid cycle, a cyclic series of chemical transformations in the mitochondrion by which pyruvate is degraded to carbon dioxide; NAD and FAD are reduced to NADH and $FADH_2$; and ATP is generated.

cladistics an approach to classification of organisms according to the order in time at which branches occur in the evolutionary tree.

clitoris in female mammals, an erectile, erotically sensitive organ of the vulva, homologous embryologically to most of the penis.

cloaca the common cavity into which the intestinal, urinary, and reproductive canals open in vertebrates other than placental and marsupial mammals.

clonal selection in immunology, the selection of specific populations of lymphocytes to begin the specific immune reaction.

clone 1. a group of genetically identical organisms derived from a single individual by asexual reproduction. 2 a group of identically differentiated cells derived mitotically from a single differentiated cell.

closed circulatory system a system in which blood is enclosed within arteries, veins, and capillaries throughout and is not in direct contact with cells other than those lining these vessels.

coacervates droplets with membrane-like surface layers that form spontaneously when certain substances are attracted together in water, a hypothetical step in the earliest formation of cells.

coccus (pl. *cocci*), any spherical bacterium (principally eubacteria), a condition found in many distantly related groups.

codominance the individual expression of both alleles in a heterozygote. See *dominance.*

codon 1. a series of three nucleotides in mRNA specifying a specific amino acid (or chain termination) in protein synthesis. 2. the colinear, complementary series of three nucleotides or nucleotide pairs in the DNA from which mRNA codon is transcribed.

coelom or *true coelom,* a principal body cavity, or one of several such cavities between the body wall and gut, entirely lined with mesodermal epithelium. Compare *pseudocoelom.*

coenzyme a small organic molecule required for an enzymatic reaction.

coevolution evolution of two closely interacting species, such as predator and prey, where changes in the first determine those of the second.

cohesion the attraction between the molecules of a single substance.

coleorhiza a protective sheath over the radicle in the embryo of grasses and grains.

coleoptile in the embryos and early growth of grasses, a specialized tubular structure, completely enclosing and protecting the plumule during emergence.

colinearity principle of colinearity, the finding that corresponding parts of a structural gene, mRNA, and polypeptide occur in the same linear order.

collagen in animals, a widely distributed fibrous protein of connective tissue that forms much of the structure of tendons and ligaments.

collenchyma in plants, a strengthening tissue, a modified parenchyma consisting of elongated cells with greatly thickened cellulose walls.

colon in mammals, the large intestine from the cecum to the rectum; including the ascending, transverse, descending, and sigmoid regions and the rectum.

colony 1. a group of animals or plants of the same kind living in a close semidependent association. 2. an aggregation of bacteria growing together as the descendants of a single individual, usually on a culture plate.

columnar epithelium epithelium consisting of one (simple columnar) or more (stratified columnar) layers of elongated, cylindrical cells.

common descent descent of two or more species (or individuals) from a common ancestor; for example, the similarity in blood chemistry of apes and humans is due to common descent.

community an assemblage of interacting plants and animals forming an identifiable group within a biome, as in salt.

community development also *ecological succession* and *succession.* 1. the process of change in the populations of an area as competing organisms alter their environment. 2. the sequence of identifiable ecological stages or communities occurring over time in the progress of bare rock to a climax community.

compact bone dense, hard bone with spaces of microscopic size.

companion cell in plants, a nucleated cell adjacent to a sieve tube member, believed to assist it in its functions.

competition seeking to gain what another is seeking to gain at the same time; a common struggle for the same object.

competitive inhibition enzyme inhibition involving molecules similar to the substrate that compete for the active site.

complement 1. (n.) a group of blood proteins that interact with antibody–antigen complexes to destroy foreign cells.

complete digestive tract a tubular digestive tract with an anal as well as an oral opening.

complete metamorphosis of insects, development includes distinct larval, pupal, and adult stages.

compound in chemistry, a pure substance consisting of two or more elements in a fixed ratio; consisting of a single molecular type.

compound eye an arthropod eye consisting of many simple eyes closely crowded together, each with an individual lens and a restricted field of vision, so that a mosaic image is formed.

compound microscope an optical instrument for forming magnified images of small objects, consisting of an objective lens that can be brought close to the object being examined, aligned with an ocular lens mounted at the other end of a body tube of fixed length.

concentration gradient a gradual, consistent decrease in the concentration of a substance along a line in space.

conditioned response an involuntary response that becomes associated with an arbitrary, previously unrelated stimulus.

conducting tissue see *vascular tissue.*

conifer an evergreen gymnosperm of the order Coniferales, bearing ovules and pollen in cones; included are spruce, fir, pine, cedar, and juniper.

conjugated protein a compound of one or more polypeptides with one or more nonprotein substance; for example, hemoglobin, which consists of four polypeptide chains and four heme groups.

conjugation in ciliates, a temporary cytoplasmic union in pairs, accompanied by meiosis and the exchange of haploid nuclei, in bacteria, the exchange of DNA through the pilus.

connective tissue a principal type of vertebrate supporting tissue, often with an extracellular matrix of collagen. Included are bone, cartilage, ligaments, and blood.

consolidation hypothesis the presumption that memory is first stored in short-term centers, where it is rapidly lost unless it is transferred to long-term centers, where its loss is more gradual.

consumer in ecology, an organism, especially an animal, that feeds on plants (primary consumer) or other animals (secondary consumer).

consummatory behavior the satisfying, stereotyped behavior that completes an instinctive act. Usually preceded by appetitive behavior.

continental drift a theory proposing that today's continents or land masses were once parts of a supercontinent called Pangea, which started to divide and drift apart some 200 million years ago. Also *plate tectonics.*

contraception also *birth control,* any process or method intended to prevent the sperm from reaching and fertilizing the egg, or preventing ovulation or implantation.

contractile vacuole an osmoregulatory, water-containing vacuole in protists, capable of filling and emptying through the contraction of microfilaments.

control a standard of comparison in a scientific experiment; a replicate of the experiment in which a possibly crucial factor being studied is omitted.

convergent evolution the independent evolution of similar structures in distantly related organisms; often found in organisms that have adopted similar ecological niches, as marsupial moles and placental moles.

coport membrane transporter that transports two different substances simultaneously.

Coriolis effect the deflection of a moving body of water or air caused by the rotation of the earth.

cork in plants, secondary tissue, produced by the cork cambium, consisting of cells that become heavily suberized and die at maturity, resistant to the passage of moisture and gases; the outer layer of bark.

corolla in flowers, the whorl of petals surrounding the carpels or carpel.

corona radiata ("radiating crown"), an aggregation of follicle cells surrounding the mammalian egg at ovulation.

corpus allatum in insects, one of the paired endocrine bodies that secretes juvenile hormone, which prompts retention of juvenile stages.

corpus callosum a broad, white neural tract in the mammalian brain that connects the cerebral hemispheres and correlates their activities.

corpus luteum (pl. *corpus lutea*), ("yellow body"), a temporary yellow endocrine body on the surface of the ovary, consisting of secretory cells filling a follicle after ovulation; regressing quickly if the ovum is not fertilized or persisting throughout pregnancy.

cortex 1. (zoology) the outer layer or rind of an organ, as adrenal cortex, kidney cortex. 2. (botany) the portion of stem between the epidermis and the vascular tissue.

cortical reaction during fertilization in echinoderms, the rupture of cortical granules, releasing chemicals that contribute to the formation of a fertilization envelope.

cotransport carrier a membranal carrier that, in each operating cycle, transports two different substances in the same or in opposite directions.

cotyledon also *seed leaf*, a food-storing structure in dicot seeds, sometimes emerging as first leaves; food-digesting organ in most monocot seeds; first leaves in a gymnosperm embryo.

countercurrent heat exchanger a heat-retaining mechanism in animals in which warm blood leaving the body core passes cooler blood returning from the extremities in closely paralleled vessels, thus providing the opportunity for heat to pass into the cooler blood along the full extent of the vessels.

covalent bond a relatively strong chemical bond in which an electron pair is shared by two atoms, alternately filling the outer electron shells of both.

crassulacean acid metabolism (CAM) in desert plants, a means of obtaining carbon dioxide without the risk of opening the stomata during the day. CAM plants utilize a biochemical system in which carbon dioxide is admitted into the plant at night and fixed into organic acids that can be broken back down during daylight, releasing carbon dioxide for the Calvin cycle.

crista (pl. *cristae*), a shelf-like fold of the inner mitochondrial membrane into the central cavity of the mitochondrion.

crop also *craw.* 1. an enlargement of the gullet of many birds, which serves as a temporary storage organ. 2. an analogous organ in certain insects and earthworms.

crosscurrent exchange in the bird lung, the efficient gas exchange produced by the flow of blood at right angles to the flow of air.

crossing over 1. the exchange of chromatid segments by enzymatic breakage and reunion during meiotic prophase. 2. a specific instance of such an exchange; a crossover.

cryptic coloration coloration that serves to conceal or to render inconspicuous.

cuticle a tough, often waterproof, nonliving covering, usually secreted by epidermal cells.

cutin a waxy, water-resistant substance covering the epidermis of leaves and stems.

cyanobacterium formerly blue-green alga, any of a large group of blue-green photosynthetic prokaryotes having as photopigments chlorophyll *a*, phycocyanin, and phycoerythrin, and producing oxygen as a photosynthetic waste product.

cycad a plant of an order of palm-like gymnosperms, known from the Triassic to the present.

cyclic AMP (cAMP) adenosine monophosphate in which the phosphate is linked between the 3' and 5' carbons of the ribose group; serves as an intracellular gene regulator under a variety of circumstances.

cyclic phosphorylation light reactions employing only photosystem I, during which electrons from P700 reaction center pass through the associated electron transport system and cycle back to P700, thus serving chemiosmosis only.

cytokinesis cytoplasmic cell division; actual division of the cell into two daughter cells.

cytokinin a plant cell hormone, mitogen, and plant tissue culture growth factor, which interacts with other plant hormones in the control of cell differentiation.

cytoplasm the semisolid, protein-rich matrix of a cell exclusive of the plasma membrane, the nucleus, or other large inclusions.

cytoplasmic streaming also *cyclosis,* the movement of cytoplasm within the cell, often in more or less regular, circular pathways.

cytoskeleton the internal structure of animal cells, composed of microtubules, intermediate filaments, and actin filaments; it controls the size, shape, and movement of the cell.

death rate in human populations, the number of deaths per 1000 population per year.

decarboxylation the enzymatic removal of carbon to a substrate. Compare *carboxylation.*

deciduous plant a perennial plant that seasonally drops its leaves.

decomposer also *reducer,* an organism that breaks down organic wastes and the remains of dead organisms into simpler compounds, such as carbon dioxide, ammonia, and water.

deductive reasoning logical progression proceeding from the general to the specific; making specific deductions based on a larger generalization or premise.

dehydration reaction an enzymatic reaction during which water is lost and a covalent bond forms between the reactants.

deletion 1. the removal of any segment of a chromosome or gene. 2. the site of such a removal after chromosome healing, considered as a mutation. See also *base deletion.* 3. the deleted chromosome.

demographic transition the proposal that in the development of nations, their population growth has distinct stages: (a) high birth and death rates and slow growth; (b) high birth rate, low death rate, and very rapid growth; and (c) low birth and death rates, and very slow growth.

denaturation the alteration of a protein so as to destroy its properties, through heating or chemical treatment (protein denaturation may be reversible or irreversible).

dendrite an extension of a neuron that conducts impulses toward the cell body and axon.

denitrification the conversion of nitrates and nitrites to nitrogen gas; *denitrifying bacteria,* common soil and manure bacteria that are responsible for denitrification.

density-dependent effects factors affecting population parameters, the degree of effect directly proportional to population density.

density-independent effects factors affecting population parameters independently of population density.

derived a character that has undergone an evolutionary change in the group being considered; the opposite of *primitive.* See also *advanced.*

dermis in animals, the inner mesodermally derived layer of the skin, beneath the ectodermally derived epidermis.

determinate cleavage early embryonic cleavages in which daughter cells begin the very early commitment to specific development fates; typical of protostomes and especially of insects.

determination the ongoing commitment of embryonic cells and tissues to specific developmental directions.

deuterostome a bilateral animal with radial cleavage and a mouth that does not arise, developmentally or phylogenetically, from the blastopore. Compare *protostome.*

development 1. the whole process of growth and differentiation by which a zygote, spore, or embryo is transformed into a functioning organism. 2. any part of this process, as the development of the kidney. 3. evolution.

diabetes mellitus a genetic disease of carbohydrate metabolism characterized by abnormally high levels of glucose in the blood and urine, and the inadequate secretion or utilization of insulin.

diaphragm 1. in mammals, a dome-shaped, muscularized body partition separating the chest and abdominal cavities, involved in breathing movement. 2. a dome-shaped, rubber, contraceptive device fitted over the cervix, often used with a spermicide.

diastole the period of expansion and dilation of the heart during which it fills with blood; the period between forceful contractions (systole) of the heart. See *systole.*

dicot a flowering plant of the angiosperm class Dicotyledonae, characterized by producing seeds with two cotyledons. Compare *monocot.*

differentiation in development, the process whereby a cell or cell line becomes morphogically, developmentally, or physiologically specialized. See *dedifferentiation.*

diffuse root system also *fibrous root system,* a root system of a plant in which there are many roots all about the same size.

diffusion the random movement of molecules of a gas or solute under thermal agitation, resulting in a net movement from regions of higher initial concentration to regions of lower initial concentration.

dihybrid cross a cross between two genotypically identical dihybrids.

dikaryotic in many fungi, a condition arising after conjugation, wherein parental plus and minus nuclei remain separated for a time before fusion occurs.

dilation and curettage (D & C) a surgical operation in which the cervix is forcefully dilated and the uterine mucosa scraped with a curette; a common method of induced abortion as well as a procedure for removing cysts and polyps.

dinoflagellate a flagellated, photosynthetic, marine protist of phylum Pyrrhophyta.

dioecious having separate sexes.

diploblastic a condition in animals in which the adult tissues are derived from just two embryonic germ tissue layers, ectoderm and endoderm. See also *triploblastic.*

diploid having a double set of genes and chromosomes—one set from each parent.

directional selection selection favoring one extreme of a continuous phenotypic distribution.

disaccharide a carbohydrate consisting of two simple sugar subunits.

disruptive selection natural selection during which extreme phenotypes receive favorable selection, whereas average phenotypes are selected against.

distal away from the center of the body, heart, or other reference point.

divergent evolution following speciation, the continued accumulation of differences between or among species, attributable to adaptive radiation.

diversity 1. variety; variability. 2. the range of types in a major taxon: plant diversity. 3. in ecology, a measure of the number of species coexisting in a community.

division a major primary category or taxon of the plant kingdoms, equivalent to phylum.

DNA, deoxyribonucleic acid, the genetic material of all organisms (except RNA viruses); in eukaryotes, DNA is confined to the nucleus, mitochondria, and plastids.

DNA hybridization the formation of a double helix, usually done experimentally to find DNA complementary to a radioactive piece of DNA called a probe.

DNA replication also *replication* or *DNA synthesis,* the semiconservative synthesis of DNA in which the double helix opens, the two strands separate, and each is used as a template for producing a new opposing strand.

dominance 1. the phenotypic expression of only one of the two alleles in a heterozygote. 2. a behavioral relationship between two animals where the subordinate individual withdraws or behaves submissively toward the dominant individual in any conflict, potential conflict, or interaction.

dominance hierarchy 1. also *pecking order,* behavioral interactions established in a group, in which every individual is dominant to those lower on the order and submissive to those above.

dorsal toward the back or (usually) upper surface of a bilateral animal.

dorsal lip in the animal gastrula, tissue associated with the early blastopore, the site of involution by presumptive mesoderm and endoderm.

dorsoventrally from top to bottom, or back to front, in humans.

double fertilization in flowering plants, fertilization of egg cell and central cell by two sperm entering from a pollen tube.

double helix the configuration of the native DNA molecule, which consists of two antiparallel strands wound helically around each other.

doubling time the time it takes a growing population to double in numbers.

drift *random drift,* also *genetic drift;* 1. the chance fluctuation of allele frequencies from generation to generation in a finite population. 2. the long-term consequences of such fluctuations, such as the loss or fixation of selectively neutral alleles.

ductus arteriosus in the mammalian fetus, a short broad vessel conducting blood from the pulmonary artery to the aorta, thus bypassing the fetal lungs.

ductus deferens also *vas deferens,* a tube within the spermatic cord that directs sperm from the epididymis to the urethra.

duodenum the first region of small intestine, just posterior to the pylorus.

ecdysis also *molting;* 1. in arthropods, the shedding of the outer, noncellular cuticle. 2. this shedding together with associated changes in size, shape, or function.

ecosystem the biotic and abiotic factors of an ecological community considered together.

ectoderm in animal development, the outermost of the three primary germ layers of an embryo: the source of all neural tissue, sense organs, the outer cellular layer of the skin and associated organs of the skin; also any of the tissues derived from embryonic ectoderm; also the outer cellular layer of a cnidarian.

ectoparasite a parasite that feeds on or attaches to the surface tissue of the host; fleas, lice, ticks, and athlete's foot fungus are human ectoparasites.

ectotherm an organism that makes use of external heat sources to increase its body temperature.

effector also effector organ, a bodily organ actively used in behavior, especially in communicationùdistinguished from receptor; for example, the human eyebrows and associated muscles constitute an effector organ, and a rather effective one at that.

egg activation following the binding of a sperm cell to its surface, the responses of an egg cell that lead to fertilization and the start of development.

ejaculation the ejection of semen from the penis during orgasm.

electrical synapse contact between neurons formed by gap junctions, in which an action potential passes directly from one neuron to the next.

electron acceptor a molecule that accepts one or more electrons in an oxidation–reduction reaction (and thus becomes reduced).

electron carrier a molecule that behaves cyclically as an electron acceptor and an electron donor.

electron donor a molecule that loses one or more electrons to an electron acceptor in an oxidation–reduction reaction (and thereby becomes oxidized).

electronegativity the relative ability of any element to attract electrons from other elements in a covalently bonded compound.

electron microscope a device for creating magnified images of small specimens through bombarding it with an electron beam and by subsequent magnetic focusing.

electron orbit also *electron orbital;* 1. the state of an electron as determined by its energy as it moves within an atom. 2. the space within which an electron pair moves within an atom.

electron shell an energy level of a group of electrons in an atom.

electron transport system also *electron transport chain,* a series of cytochromes and other proteins, bound within a membrane of a thylakoid, mitochondrion, or prokaryotic cell, that passes electrons and/or hydrogen atoms in a series of oxidation–reduction reactions that result in the net movement of hydrogen ions across the membrane.

element a substance that cannot be separated into simpler substances by purely chemical means.

Eltonian pyramid a concept in ecology that, because of thermodynamic inefficiency, organisms forming the base of a food chain are numerically abundant and comprise a large total biomass, while organisms of each succeeding level of the chain are successively less abundant and of smaller total biomass.

embryology the scientific study of early development in plants and animals.

embryo sac in flowering plants, the mature megagametophyte after division into six haploid cells and one binucleate cell, enclosed in a common cell wall.

endergonic (adj.), in a biochemical reaction, the expenditure of energy; moving from a state of lower potential energy to one of higher potential energy. Compare *exergonic.*

endocrine gland also *ductless gland,* a discrete gland that secretes hormones into the blood system.

endocrine hormone a hormone that acts at some distance from its source (usually transported in the blood), as opposed to *paracrine hormones,* which act on neighboring cells, and *autocrine hormones,* which remain within the producing cell. (For individual hormones see Table 36.1.)

endocrine system the endocrine glands taken together, and their hormonal actions and interactions.

endocytosis the process of taking food or solutes into the cell by engulfment (a form of active transport). See also *phagocytosis, pinocytosis;* compare *exocytosis.*

endoderm also *entoderm,* the innermost of the three primary germ layers of a metazoan embryo.

endodermis a single layer of cells around the stele of vascular plant roots that forms a moisture barrier in that the lateral cell walls are closely oppressed and waterproofed in bands, forming the Casparian strip.

endomembranal system within and surrounding the cytoplasm of eukaryotic cells, a number of dynamic, interconvertible membranes, including those surrounding the cell and the nucleus, and those forming the endoplasmic reticulum, Golgi complex, and various vacuoles.

endometrium in mammals, the mucous membrane tissue lining the cavity of the uterus; responds cyclically to ovarian hormones by thickening as preparation for implantation of the blastocyst.

endoplasmic reticulum (ER) internal membranes of the cell, usually a site of synthesis; *rough endoplasmic reticulum,* ER without bound ribosomes; the site of synthesis of noncytoplasmic proteins; *smooth endoplasmic reticulum,* ER without bound ribosomes, usually the site of synthesis of nonprotein materials.

endoskeleton a mesodermally derived supporting skeleton inside the organism, surrounded by living tissue, as in vertebrates and echinoderms.

endosperm a nutritive tissue of seeds, formed around the embryo in the embryo sac by the proliferation of the (usually) triploid endosperm nucleus to form a starch-rich mass; the endosperm may persist until germination or be resorbed by the cotyledon(s) during seed maturation.

endosperm mother cell a large, central, usually diploid or binucleate cell formed in the megagametophyte by fusion of haploid nuclei; when fertilized in double fertilization, forms the endosperm nucleus.

endospore an asexual resistant spore formed within a bacterial cell. See *spore.*

endosymbiont a mutualistic symbiont that resides within the cells of its symbiotic partner.

endothelium an epithelial tissue that forms the inner lining of blood and lymph vessels.

endotherm also *homeotherm* and *homoiotherm,* an organism with the ability to metabolically thermoregulate, also called "warm-blooded." See also *ectotherm.*

energy of activation the energy input needed to initiate a chemical reaction.

energy pyramid the Eltonian pyramid with regard to chemical energy rather than numbers or biomass. See *Eltonian pyramid.*

entropy in thermodynamics, the amount of energy in a closed system that is not available for doing work; also defined as a measure of the randomness or disorder of such a system. Negative entropy, free energy in the form of organization. See *energy, free energy.*

environment the surrounding conditions, influences, or forces that influence or modify an organism, population, or community.

environmental resistance the sum of environmental factors (for example, limited resources, drought, disease, predation) that restrict the growth of a population below its biotic potential (maximum possible population size).

enzyme a biologically active protein that catalyzes chemical reactions.

enzyme–substrate complex the unit formed by an enzyme bound by non covalent bonds to its substrate.

epicotyl in the embryo of a seed plant, the part of the stem above the attachment of the cotyledon(s); forms the epicotyl hook in the growth of some seeds.

epidermis 1. in plants, the outer protective cell layer in leaves and in root and stem primary growth, one cell thick and made waterproof with an outer layer of cutin; replaced in secondary growth by periderm. 2. in animals, the outer epithelial layer, derived from ectoderm.

epididymis in mammals, an elongated soft mass lying alongside each testis, consisting of convoluted tubules; the site of sperm maturation.

epiphyte a plant that grows nonparasitically on another plant or sometimes an object.

epistasis the masking of a trait ordinarily determined by one gene locus by the action of a gene or genes at another locus.

epithelium a tissue consisting of tightly adjoining cells that cover a surface or line a canal or cavity, and that serves to enclose and protect.

ergot a toxic fungal infective state of rye, which, when eaten, is capable of causing ergotism.

erythrocyte also *red blood cell,* a hemoglobin-filled, oxygen-carrying, circulating blood cell; enucleate in mammals, but having a physiologically inactive, condensed nucleus in other vertebrates.

esophagus the anterior part of the digestive tract; in mammals, it is muscularized and leads from the pharynx to the stomach.

essential amino acid one of the amino acids that the body cannot synthesize and thus must be provided by the diet if dietary diseases are to be avoided.

essential fatty acid one of the fatty acids that the body cannot synthesize and thus must be provided by the diet if dietary diseases are to be avoided.

estrus also *heat,* in the females of most mammalian species, the regularly recurring state of sexual excitement around the time of ovulation; the only time during which the female will accept the male and is capable of conceiving; the term is not applicable to people.

ethology the scientific study of animal behavior as it occurs in the organism's natural environment, or among free-ranging animals in the ethologist's environment.

euchromatin chromatin other than heterochromatin.

euphotic zone in the aquatic environment, those depths penetrated by light.

eutrophication any process in which nutrients and populations increase in a body of fresh water.

evolution 1. any process of formation, growth, or change. 2. descent with modification. 3. long-term change and speciation (division into discrete species) of biological entities. 4. the continuous genetic adaptation of organisms or populations through mutation, hybridization, random drift, and natural selection.

evolutionary stable strategy (ESS) an innate behavioral strategy (such as conflict) that confers greater fitness on individuals than can any alternative behavior that might arise by mutation or recombination.

exchange pool in biogeochemical cyles, the readily available reserves of a mineral nutrient, such as a soluble phosphate or nitrate pool in the soil water. Compare *reservoir.*

excitatory synapse a synapse in which the secretion of neurotransmitter stimulates neural impulses in the receiving neuron.

excretion the removal of metabolic wastes, particularly nitrogenous wastes, from the body.

exergonic (adj.), of a biochemical reaction or half-reaction, producing work, heat, or other energy; moving from a higher energy state to a lower energy state; as opposed to *endergonic.*

exocytosis the process of expelling material from the cell through the plasma membrane by the fusion of vacuoles, secretion granules, and so on with the plasma membrane, and their subsequent eversion.

exoskeleton an external skeleton or supportive covering, as in arthropods and armadillos; arthropod exoskeletons are formed by secretions of epidermal cells, largely chitin, proteins, and calcium carbonate.

exponential growth also *geometric growth,* population growth in which the population size increases by a fixed proportion in each time period, successive values forming an exponential series.

exponential growth curve population growth that when plotted takes on a J-shaped curve; growth that approaches the biotic potential of a species and is not immediately responsive to environmental resistance. Growth increments may be 2, 4, 8, 16, 32, and so on. Compare *arithmetic increase.*

expressed sequence also *exon,* the portion of mRNA remaining after the introns or unexpressed sequences have been removed. See also *intron.*

external fertilization fertilization outside the body, as in echinoderms and most bony fishes.

extinction a coming to an end or dying out of a species or other taxon (group of related organisms).

extracellular outside, between, or among cells.

extracellular digestion digestion outside cells (usually in the gut).

extraembryonic membrane any of several supporting membranes (amnion, chorion, allantois, yolk sac) produced by the embryo but not part of the embryo proper.

facilitated diffusion diffusion of specific molecules across a plasma membrane that is facilitated by reversible association with carrier molecules. No cellular energy is expended and net movement follows the concentration gradient.

facultative anaerobe an organism such as yeast that can use oxygen when it is available but is also able to live anaerobically.

FAD, FADH$_2$ flavin adenine dinucleotide, a coenzyme and hydrogen carrier in metabolism. FAD is the oxidized form and FADH$_2$ is the reduced form.

fascia a heavy sheet of connective tissue covering or binding together muscles or other internal structures of the body, often connecting with ligaments or tendons.

fascicle a distinct bundle of muscle fibers, surrounded by connective tissue and carrying nerves and blood vessels.

fatty acid an organic acid consisting of a linear hydrocarbon "tail" and one terminal carboxyl group.

fermentation anaerobic breakdown of glucose to form alcohol and carbon dioxide or lactate.

fertility rate the number of births per 1000 women from 15 to 44 years of age; a clearer indicator of reproductive activity in a population than the birth rate.

fertilization the union of gametes or gamete nuclei.

fertilization cone a cone-shaped mound of egg cytoplasm that rises to engulf the sperm.

fetus in vertebrates, an unborn or unhatched individual past the embryo state; in humans, a developing, unborn individual past the first 8 weeks of pregnancy.

fibrin an insoluble fibrous protein forming blood clots and also contributing to the viscosity of blood. See *fibrinogen.*

fibrinogen a globular blood protein (globulin) that is converted into fibrin by the action of thrombin as part of the normal blood clotting process.

fibrous protein a protein in which the secondary structure comprises zigzag beta sheets, in contrast to the alpha helix.

fibrous root system also *diffuse root system,* the roots collectively of a plant in which there are many equivalent roots rather than a prominent central root.

filter-feeder an animal that obtains its food by filtering minute organisms from a current of water.

fire-disclimax community an ecological community in continual transition because of recurring fires (for example., chaparral).

first law of thermodynamics the physical law that states that energy cannot be created or destroyed; later amended to allow for the interconversion of matter and energy.

fission the division of an organism into two (binary fission) or more organisms, as a process of sexual reproduction.

fitness 1. the state of being adapted or suited (for example, to the environment). 2. also relative fitness, Darwinian fitness, the relative expectation of surviving and reproducing of an individual or of a specific genotype, compared with that of the general population or of a standard genotype. 3. one-half the expected number of ffspring of a diploid genotype.

fixed action pattern a precise and identifiable set of movements, innate and characteristic of a given species. See also *instinctive pattern.*

flagellum (pl. *flagella*), 1. a long, whip-like, motile eukaryotic cell organelle, projecting from the cell surface; longer and fewer in number than cilia. 2. an analogous organelle of bacteria, consisting of a solid, rotating helical protein.

flame cell system in several invertebrate phyla, an osmoregulatory/excretory system consisting of ciliated flame bulbs composed of flame cells, along with associated conducting tubules in which fluids and nitrogen wastes are swept to external pores.

flower the part of a seed plant that bears the reproductive organs, including pistils, stamens, petals, and sepals.

fluid mosaic model also *Singer model,* a description of the plasma membrane as a phospholipid bilayer stabilized by specifically oriented proteins, with some proteins extending through to both surfaces and other proteins specific for the inner or outer surfaces.

fluke a parasitic trematode flatworm.

food chain a sequence of organisms in an ecological community, each of which is food for the next higher organism, from the primary producer to the top predator.

food web a group of interacting food chains; all of the feeding relations of a community taken together; the flow of chemical energy among organisms.

foramen ovale in the fetal mammalian heart, an opening in the septum between the atria; it normally closes at the time of birth.

foraminiferan a shelled marine sarcodine protist with a carbonate shell having numerous openings through which slender branching pseudopods are extended.

force filtration in the nephron, the movement of smaller molecules out of the blood and into Bowman's capsule by hydrostatic pressure in the glomerulus.

forebrain 1. the anterior of the three primary divisions of the vertebrate brain. 2. the parts of the brain developed from the embryonic forebrain.

fossil any remains, impression, or trace of an animal or plant of a former geological age, such as mineralized skeleton, a footprint, or a frozen mammoth.

fossil fuel coal, petroleum, or natural gas.

founder effect the population genetic effect of the chance assortment of genes carried by the successful founder or by a few founders of a subsequently large population.

F⁺ (fertility-positive) bacterium containing the plasmid responsible for producing a sex pilus through which a replica of the plasmid can pass to an F⁻ bacterium.

frame shift mutation a small insertion or deletion in a structural gene such that all mRNA codons downstream are misread in translation.

free energy energy available for work.

free radical a highly reactive atom or molecular unit with an unpaired electron.

frequency-dependent selection natural selection in which a phenotype is more fit than average when it is numerically rare, and is less fit than average when its relative number surpasses some equilibrium frequency value.

frontal lobe an anterior division of the cerebral hemisphere, believed to be a site of higher cognition.

fruit the seed-bearing ovary of a flowering plant.

fruiting body in fungi, slime molds, and myxobacteria, a structure specialized for the production of spores.

functional group in biochemistry, a side group with a characteristic chemical behavior or function.

fundamental niche the potential niche of an organism. Compare *realized niche.*

gamete haploid egg or sperm cells that unite during fertilization to form a zygote.

gametic cycle a life cycle with a dominating diploid state, the haploid state limited to gametes only, and fertilization restoring diploidy (some protists and all animals).

gametogenesis the process through which haploid gametes (sex cells: eggs and sperm) are produced, commonly including meiosis.

gametophyte in plants with alteration of generations, the haploid form, in which gametes are produced.

ganglion a mass of nerve tissue containing the cell bodies of neurons.

gap junction dense structures that physically connect plasma membranes of adjacent cells along with channels for cell-to-cell transport.

gastrointestinal tract the entire digestive tube from the mouth to and including the anus.

gastrovascular cavity also *coelenteron,* the cavity of cnidarians, ctenophores, and flatworms, which opens to the outside only via the mouth and serves the functions of a digestive cavity, crude circulatory system, and coelom.

gastrula an early metazoan embryo consisting of a hollow, two-layered cup with an inner cavity (archenteron) opening out through a blastopore.

gastrulation the process usually involving invagination, involution, and epiboly, whereby a blastula becomes a gastrula.

Gause's law the ecological principle that no two genetically isolated kinds of organisms can coexist (exist at the same time and place) while occupying the same ecological niche.

gene (variously defined), 1. the unit of heredity. 2. the unit of heredity transmitted in the chromosome and that, through the interaction with other genes and gene products, controls the development of hereditary character. 3. a continuous length of DNA with a single genetic function. 4. the unit of transcription of DNA. 5. a cistron or structural gene, that is, the sequence of DNA coding for a single polypeptide sequence.

gene flow the exchange of genes between populations through migration, pollen dispersal, chance encounters, and the like.

gene pool all the genetic information of a population considered collectively.

gene replacement therapy in medicine, the use of recombinant DNA to substitute for a gene that causes a condition needing remedy.

gene sequencing determining the specific sequence of nucleotides in a gene.

gene splicing the use of recombinant DNA techniques to form covalent bonds between DNA of different sources.

generative cell a cell in the microgametophyte (pollen) of a seed plant, capable of dividing to form two sperm cells.

genetic code the relationship by which the 64 possible codons (sequences of three nucleotides in DNA or RNA) each specify an amino acid or chain termination in protein synthesis.

genetic engineering the manipulation of genes through recombinant DNA techniques.

genetic equilibrium the state of a population wherein the frequency of certain alleles remains constant generation after generation.

genetic recombination any process, such as crossing over, that leads to new gene combinations on the chromosomes.

genetic variability a broad term indicating the presence of different genetic constitutions in a population or populations.

genetics 1. the science of heredity, dealing with the resemblances and differences of related organisms resulting from the interaction of their genes and the environment. 2. the study of the structure, function, and transmission of genes.

genome the full haploid complement of genetic information of a diploid organism, usually viewed as a property of the species.

genotype 1. the genetic constitution of an organism with respect to the gene locus or loci under consideration. 2. also *total genotype,* the sum total of genetic information of an organism.

genotype frequency the proportion of individuals in a population having a specified genotype.

genus (pl. *genera*), a group of similar and related species; a taxon smaller than family and larger than species.

geological era one of five divisions of the earth's history (Cenozoic, Mesozoic, Paleozoic, Proterozoic, and Archeozoic); see the geological timetable inside the front cover.

germ layer any of the three layers of cells formed at gastrulation, the partially differentiated precursors of specific body tissues and organs. See *ectoderm, endoderm, mesoderm.*

germinal epithelium 1. the epithelium of the gonads, which, through mitosis, gives rise to gametocytes. 2. also *germinal layer,* the innermost layer of an epithelium, containing mitotic cells.

germination the process whereby a seed and embryo ends dormancy, the stored materials of the seed being digested, the seed coat ruptured, and the plumule and hypocotyl begin to grow; *spore germination,* the end of dormancy in a spore.

gill arches also *pharyngeal arches;* in fish, a row of bony or cartilaginous curved bars extending vertically between the gill slits on either side of the pharynx, supporting the gills. 2. in land vertebrate embryos, a row of corresponding homologous rudimentary ridges that give rise to jaw, tongue, and ear bones.

gill clefts also *gill slits, pharyngeal clefts;* 1. in fish, the openings between the gills. 2. in vertebrate embryos, the corresponding and homologous grooves in the neck region, between the branchial arches.

gizzard 1. in birds, a thick-walled muscular enlargement of the alimentary canal, filled with small stones and used to grind seeds or other food. 2. any analogous structure in invertebrates, as in earthworms, rotifers, and gastrotrichs.

glomerulus (pl. *glomeruli*), the mass or tuft of capillaries within a Bowman's capsule.

glucagon a polypeptide hormone secreted by the pancreatic islets of Langerhans, whose action increases the blood glucose level by stimulating the breakdown of glycogen in the liver.

glucose also dextrose, blood sugar, corn sugar, grape sugar, a 6-carbon sugar, occurring in an open chain form or either of two ring forms; the subunit of which the polysaccharides starch, glycogen, and cellulose are composed, and a constituent of most other polysaccharides and disaccharides.

glycerol a triple alcohol component of neutral fats and of phospholipids.

glycogen a highly branched polysaccharide consisting of alpha glucose subunits; a carbohydrate storage material in the liver, muscle, and other animal tissues.

glycolipid a compound with lipid and carbohydrate subunits; for example, cerebrosides, gangliosides.

glycolysis a biochemical pathway including the enzymatic, anaerobic breakdown of glucose in cells, yielding ATP pyruvate and NADH.

glycoprotein a compound containing polypeptide and carbohydrate subunits.

goblet cell a goblet-shaped, mucus-secreting epithelial cell, found in the lining of the nasal cavity, bronchi, and elsewhere.

Golgi complex a stack or array of membranous vesicles, formed from endoplasmic reticulum and engaged in the modification and packaging of various protein substances.

gonad in animals, the primary sex gland, with endocrine functions, and the site of meiosis; an ovary or testis.

gonadotropin a hormone that stimulates growth or activity in the gonads; in vertebrates, a specific peptide hormone of the anterior pituitary.

gonorrhea a sexually transmitted bacterial disease caused by the diplococcus *Neisseria gonorrhoeae.*

graded display a communication pattern in which there exist intermediate states such as a behavioral continuum from low to high levels of motivation.

gradualism the Darwinian proposal that the pace of evolution is slow but steady with an ongoing accumulation of minor changes leading eventually to the formation of new species. Compare *punctuated equilibrium.*

Graafian follicle also *vesicular follicle,* a fluid-filled body in the ovary during the pre-ovulatory period, containing the maturing oocyte within a cluster of follicular cells.

granum (pl. *grana*), a stack of thylakoid disks in a chloroplast; seen as a series of minute, multiple green bodies containing most of the chlorophyll of the plant.

gravitropism (formerly geotropism), a plant growth response attributed to gravity.

greenhouse effect the warming principle of greenhouses, in which high-energy solar rays enter easily, while less energetic heat waves are not radiated outward; now especially applied to the analogous effect of increasing atmospheric concentrations of carbon dioxide through the burning of fossil fuels and forest biomass.

gross productivity the amount of biochemical energy captured by photosynthesis in a particular area per unit time.

ground meristem primary plant tissue from which the ground tissues, collenchyma, parenchyma, and sclerenchyma, are derived.

growth curve a graph of population numbers per unit time, especially in a period of rapid initial growth leading into population size stabilization or to a population crash.

guard cell in leaf and stem epidermis, either of a pair of crescent-shaped cells that with the pore make up the stoma.

habituation learning not to respond to environmental stimuli that may have no relevance to the organism.

Haldane–Oparin hypothesis the proposal that life arose spontaneously in the sea following a period of organic synthesis and under conditions that are no longer present on the earth.

half-life 1. of a radioisotope, the time it takes for half of the atoms in a sample to undergo spontaneous decay. 2. the time it takes for half of the amount of an introduced substance (such as a drug) to be eliminated by a natural system, either by excretion or by metabolic breakdown.

haploid having a single set of genes and chromosomes. Compare *diploid, polyploid.*

Hardy–Weinberg law a statement of the mathematical expectations of genotype frequencies given allele frequencies in a population of random mating diploid individuals.

haustoria specialized, host-penetrating hyphae of parasitic fungi.

Haversian system a Haversian canal together with its surrounding, concentrically arranged layers of bone, canaliculi, lacuna, and osteocytes.

heme group an iron-containing, oxygen-binding porphyrin ring present in all hemoglobin chains and in myoglobin. See *porphyrin group.*

hemipenis either of the paired hemipenes or copulatory organs of a male lizard or snake.

hemizygous a term used to describe the condition of X-linked genes in males, which are neither homozygous nor heterozygous.

hemocoel a body cavity, especially in arthropods and mollusks, formed by expansion of parts of the blood vascular system.

hemocyanin a copper-containing respiratory pigment occurring in solution in the blood plasma of various arthropods and mollusks.

hemoglobin an iron-containing respiratory pigment consisting of one or more polypeptide chains, each associated with a heme group.

hemophilia a genetic tendency to uncontrolled bleeding in humans, due to the lack of a necessary constituent of the blood clotting process; caused by recessive alleles at either of two sex-linked loci, it affects males almost exclusively.

heterochromatin chromatin that is relatively heavily condensed at times other than mitosis or meiosis, or that condenses early in prophase, and is genetically largely inert.

heterogametes the evolutionarily advanced state in gametes in which a large nonmotile egg and a small motile sperm form. See *isogametes.*

heterosporous a state in plants in which morphologically distinct spores, microspores, and megaspores are produced.

heterotherm an endothermic animal that thermoregulates only part of the time.

heterotroph a microorganism that requires organic compounds as an energy or carbon source.

heterozygous having two different alleles at a specific gene locus on homologous chromosomes of a diploid organism.

hindbrain the parts of the vertebrate brain derived from the embryonic hindbrain, including the cerebellum, pons, and medulla oblongata.

histone one of a class of small, highly basic proteins that complex with the nuclear DNA of higher eukaryotes, forming nucleosomes that presumably protect the DNA from degradation; one histone (histone I) appears to have a role in chromatin condensation and thus in one level of gene control.

homeobox a region within homeotic or control genes, consisting of some 100 nucleotides, whose base sequence is very similar in a variety of organisms. Homeoboxes are thought to play a key role in the activation of control genes.

homeostasis the tendency toward maintaining a stable internal environment in the body of a higher animal through interacting physiological processes involving negative feedback control.

homeotic gene a gene that controls the expression of entire blocks of genes and thus plays an important role in the development of a major structure. In *Drosophila,* homeotic genes control the activity of the imaginal disks.

hominids a primate group composed of humans and related extinct forms.

hominoid a primate group composed of humans, apes, and related extinct forms.

homologous 1. similar because of a common evolutionary origin. 2. derived by independent evolutionary modification from a corresponding body part of a common ancestor. 3. homologous chromosomes, which, because of common descent, have the same kind of genes in the same order and which pair in meiosis.

homologue also homologous chromosome, either of the two members of each pair of chromosomes in a diploid cell. See *homologous.*

homology similarity due to common descent.

homozygous having the same allele at a given locus in both homologous chromosomes of a diploid organism.

hormone a chemical messenger transmitted in body fluids or sap from one part of the organism to another, which produces a specific effect on target cells often remote from its point of origin and functions to regulate physiology, growth, differentiation, or behavior.

human chorionic gonadoptropin (HCG) a hormone secreted by the human chorion, prompting the corpus luteum to continue manufacturing progesterone.

human immunodeficiency virus (HIV) the retrovirus responsible for AIDS (acquired immune deficiency syndrome).

humoral response the manufacture and secretion of specific antibodies by plasma cells (activated B-cells) in response to the presence of antigen or stimulation by helper T-cells.

hybrid the offspring of two animals or plants of different races, breeds, varieties, species, or genera.

hydrocarbon 1. a chemical compound consisting of hydrogen and carbon only, often in chains or rings; for example, gasoline, benzene, paraffin. 2. a portion of an organic molecule consisting of hydrogen and carbon only; for example, the hydrocarbon portion of a fatty acid.

hydrogen bond a weak electrostatic attraction between the hydrogen of a side group and the oxygen or nitrogen of another side group.

hydrogen carrier a certain membranal element within an electron transport system that accepts both electrons and protons, transporting hydrogen across the membrane, whereupon its proton is released and its electron continues in the system. See *chemiosmosis.*

hydrogen ion 1. a proton. 2. as a convenient fiction, a proton in solution and bound to a water molecule; actually a hydronium ion (H_3O^+).

hydrologic cycle also *water cycle,* the cycle of water between its liquid and gaseous phases brought about by heat in the atmosphere.

hydrolytic cleavage also *hydrolysis,* the reaction of a compound with water such that the compound is split into two parts by the breaking of a covalent bond, and water is added in the place of the bond, an —OH group going to one subunit and an —H group going to the other.

hydrophilic "water loving," a characteristic of charged molecules in which they readily interact with water molecules.

hydrophobic with regard to a molecule or side group, tending to dissolve readily in organic solvents but not in water; resisting wetting; not containing polar groups or subgroups.

hydrostatic skeleton a supporting and locomotory mechanism involving a fluid confined in a space within layers of muscle and connective tissue; movement occurring when muscle contractions increase or decrease the hydrostatic pressure of the fluids.

hydroxyl group —OH, consisting of an oxygen and hydrogen covalently bonded to the remainder of the molecule; a constituent of alcohols, sugars, glycols, phenols, and other compounds.

hydroxide ion the anion (OH^+) of the dissociation products of water in the reaction.

hyperosmotic pertaining to a solution having a greater concentration of solutes than some reference solution.

hyperpolarization in a neuron, an inhibitory state caused by an influx of negative ions, whereby the threshold of stimulation is greater than during the usual resting state.

hyperthyroidism a condition caused by an excess of circulating thyroid hormone; symptoms in humans include nervousness, sleeplessness, hyperactivity, weight loss, and, if prolonged, a bulging of the eyeballs.

hypertonic (adj.), having a higher osmotic potential (for example, higher solute concentration) than the cytoplasm of a living cell (or other reference solution).

hypha (pl. *hyphae*), one of the individual filaments that make up a fungal mycelium.

hypocotyl the part of a plant embryo below the point of attachment of the cotyledon.

hypoglycemia an abnormally low level of blood glucose.

hypothalamus a portion of the floor of the midbrain, containing vital autonomic regulatory centers and closely associated functionally with the pituitary gland.

hypothesis a proposition set forth as an explanation for a specified group of phenomena, either asserted merely as a provisional conjecture to guide investigation (for example, working hypothesis) or accepted as highly probable in the light of established facts. See also *theory.*

hypothyroidism a condition caused by abnormally low levels of circulating thyroid hormone; symptoms include physical and mental sluggishness and weight gain.

hypotonic (adj.), having a lower osmotic potential (for example, a lower concentration of solutes) than the cytoplasm of a living cell (or other reference solution).

imbibition the taking up (absorption) of a fluid (often water) by a hygroscopic gel, colloid, or fibrous matrix; such as the ultramicroscopic spaces in a plant cell wall.

immune response the entire array of physiological and developmental responses involving specific protective actions aginst a foreign substance; including phagocytosis, the production of antibodies, complement fixation, lysis, agglutination, and inflammation.

immune system in vertebrates, widely dispersed tissues that respond to the presence of the antigens of invading microorganisms or foreign chemical substances.

implantation the act of attachment of the mammalian embryo (blastocyst) to the uterine endometrium.

impulse neural impulse; a wave of excitement (transitory membrane depolarization) transmitted through a neuron.

inborn error of metabolism a genetic defect in which an individual lacks one of the enzymes of a biochemical pathway.

incomplete dominance the situation when the F1 hybrid has a phenotype intermediate between the parents.

incomplete penetrance the condition when a trait, though encoded in the genotype, sometimes fails to be expressed in the phenotype.

indeterminate cleavage a lack of developmental commitment in newly cleaved cells in the early embryo, resulting for a time in daughter cells remaining developmentally flexible.

inducer 1. in molecular genetics, a small molecule that triggers the activity of an inducible enzyme. 2. in embryology, a substance that stimulates the differentiation of cells or the development of a particular structure.

induction the process by which the fate of embryonic cells of tissue is determined, especially as due to tissue interactions.

inductive reasoning logical progression proceeding from the specific to the general; reaching a conclusion based on a number of observations.

inflammatory response a nonspecific immune reaction brought on by the release of kinins, histamine, and other agents, that increases permeability in nearby capillaries and causes redness and swelling of tissue.

inhibitory block according to the classical definition of instinct, the neurological inhibitors of behavior that are selectively removed by the perception of the appropriate releaser.

inhibiting hormone any of several hypothalamic neurosecretions, targeted for the adenohypophysis (anterior pituitary), which responds by slowing the release of one of its hormones.

inhibitory synapse a synapse in which the secretion of neuroransmitter increases the threshold voltage requirement of the receiving neuron, thereby inhibiting it.

initiation the start of translation or polypeptide synthesis. See also *initiation complex.*

initiation complex the elements needed to begin initiation or polypeptide synthesis, a process requiring the presence of mRNA, methionine-charged tRNA, and the two ribosomal subunits.

initiator tRNA transfer RNA charged with methionine.

innate inborn, not due to learning.

innate releasing mechanism (IRM) a hypothetical center in the central nervous system that, when stimulated by the proper environmental cue, activates neural pathways that result in an instinctive behavior pattern.

inner cell mass in a mammalian blastocyst, the portion that is destined to become the embryo proper.

inner ear the portion of the ear enclosed within the temporal bone, consisting of a complex fluid-filled labyrinth and the associated auditory nerve; included are the three semicircular canals, the cochlea, and the round and oval windows. Compare *external ear, middle ear.*

inorganic molecule one not containing carbon, generally found occurring outside living organisms.

insertion 1. in genetics, the addition of extra genetic material into the middle of a chromosome, gene, or other DNA sequence; *base insertion,* the insertion of a single base pair into a DNA sequence; also *muscle insertion.* 2. in anatomy, the distal attachment of a tendon or muscle; compare *origin*.

instinct innate behavior involving appetitive and consummatory phases.

interbreeding 1. commonly breeding together. 2. hybridizing.

interferon cellular substance that interferes with replication by viruses, generally produced by virus-infected cells.

interleukin-1 or -2 chemical messengers released by antigen-presenting cells and helper T-cells that stimulate cell division in aroused T and B lymphocytes.

intermediary metabolism metabolic activity used in the biosynthesis of new substances and in the interconversion of existing substances.

intermediate host in parasitic relationships, a host in which a parasite may undergo asexual reproduction or at least some degree of development.

internal fertilization in animals, fertilization involving copulation, in which sperm release and fertilization occur within the female.

interneuron a neuron typically found in the spinal cord that synapses with sensory and motor neurons and other interneurons.

interstitial cells cells of the testis that have an endocrine function.

intracellular within cells.

intracellular digestion digestion within cells, in digestive vacuoles following phagocytosis.

intraspecific within a species; between members of the same species.

intrauterine device metallic contraceptive device that inhibits implantation when in place in the uterus.

intrinsic rate of increase the rate of population increase determined by subtracting the average death rate from the average birth rate.

intron also *intervening sequence*, a region of DNA separating two parts of a structural gene; transcribed but later removed from mRNA during post-transcriptional modification.

invagination the folding inward of a surface or tissue to make a cavity; specifically, the formation of a gastrula by the inward folding of part of the wall of the blastula.

inversion a chromosomal mutation in which part of the chromosome has been reversed in its relative orientation to the rest of the chromosome.

in vitro ("in glass"), in a test tube or other artificial environment. Compare *in vivo*.

in vivo in the living body of a plant or animal. Compare *in vitro*.

ion any electrostatically charged atom or molecule.

ion channel in the neural membrane, sodium and potassium (and chloride) channels that, through the opening and closing of gates, selectively admit or reject ions.

ionic bond a chemical attraction between ions of opposite charge.

ionization 1. the dissociation of a molecule into oppositely charged ions in solution. 2. the creation of ions by the energy of ionizing radiation.

islets of Langerhans also *islets, beta, alpha, and delta endocrine cells* within the pancreas that secrete the hormones insulin, glucagon, and somatostatin, respectively.

isogametes gametes that are identical in size and appearance, such as the (+) and (−) isogametes of Chlamydomonas.

isoosmotic a solution having the same solute concentration as some reference solution.

isotonic having the same osmotic potential (for example, the same concentration of solutes) as the cytoplasm of a living cell, or of some other reference fluid.

isotope a particular form of an element in terms of the number of neutrons in the nucleus; *radioactive* isotope, also radioisotope, an unstable isotope that spontaneously breaks down with the release of ionizing radiation.

juvenile hormone (JH) an insect hormone that prevents the differentiation of adult characters; secreted by the corpora allata of larval and subadult insects.

karyotype 1. a display of all chromosomes of an individual, arranged in order of decreasing size. 2. the total chromosome constitution of an individual. 3. the chromosomal makeup of a species.

keratin a cysteine-rich, fibrous, insoluble, intracellular structural protein making up most of the substance of the dead cells of hair, horn, nails, claws, feathers, and the outer epidermis.

kidney 1. in vertebrates, one of a pair of ducted excretory organs situated in the body cavity beneath the dorsal peritoneum, serving to excrete nitrogenous wastes and to regulate the balance of body ions and fluids. 2. any analogous organ in invertebrate metazoans.

kinetic energy the energy of motion.

kingdom one of the primary divisions of life forms; to the traditional plant and animal kingdoms Whittaker has added fungal, protist, and moneran (prokaryote) kingdoms.

K-selection population growth characterized as logistic in form, responding to environmental resistance and carrying capacity. Compare *r*-selection.

labia majora (sing. *labium majorum*), in human females, the outer, fatty, often hairy pair of folds bounding the vulva.

labia minora (sing. *labium minorum*), in the human female, the inner, thinner, highly vascular pair of folds surrounding the introitus, enfolding and extending downward from the clitoris.

lac operon (an inducible operon) in the chromosome of E. coli, composed of genes that code for three lactose-metabolizing enzymes and the region that controls their transcription. See also operon.

lacteal a lymphatic vessel of the intestinal villi.

lagging strand in DNA replication, the 5' end of a DNA strand, where nucleotides are first assembled into Okazaki fragments and then added in using the enzyme ligase.

lamella (pl. *lamellae*), 1. one of the bony concentric layers (ring-like in cross section) that surround a Haversian canal in bone. 2. one of the thin plates composing the gills of a bivalve mollusk. 3. any thin, plate-like structure.

larva in animals, an early, active, feeding stage of development during which the offspring may be quite unlike the adult.

larynx in terrestrial vertebrates, the expanded part of the respiratory passage just below the glottis and at the top of the trachea; in mammals, it contains the vocal cords and constitutes a resonating voice box or Adam's apple.

lateral line organ in nearly all fish, a sensory organ considered to be responsive to water currents and vibrations; thought to be distantly homologous with the inner ear.

lateral root also *branch root*, roots that arise from the pericycle of the primary root; the branch roots of taproot systems.

law a virtually irrefutable conclusion or explanation of a phenomenon, such as "the law of gravity" or "Mendel's first law."

law of alternate segregation also *Mendel's first law*, the observation, based on the regularity of meiosis, that heterozygous alleles separate in gamete formation, each allele going to approximately half the gametes produced.

law of independent assortment also *Mendel's second law,* the observation, based on the regularity of meiosis, that genetic elements on different chromosomes behave independently in the production of gametes.

laws of thermodynamics in physics, laws governing the interconversions of energy. See also *thermodynamics.*

leading strand in DNA replication, the 3' end of a DNA strand, where nucleotides are added one at a time to the growing strand.

leaf scar the mark left on a stem by a fallen leaf.

lenticel a pore in the stem of a plant for the passage of gases between the atmosphere and stem tissues.

leukocyte also *leucocyte,* a vertebrate white blood cell, including eosinophils, neutrophils, basophils, monocytes, and lymphocytes.

lichen a combination of a fungus and and an alga growing in a symbiotic relationship.

ligament a tough, flexible, but inelastic band of connective tissue that connects bones or that supports an organ in place. Compare *tendon.*

ligase an enzyme that heals nicks (single-strand breaks) in DNA.

light-harvesting complex a cluster of photosynthetic pigments (chlorophyll and accessory pigments), which receives energy from photons and transfers that energy to a single reaction center.

light-independent reaction that part of photosynthesis not immediately involved in chemiosmosis, specifically the fixation of CO_2 into carbohydrate using the NADPH and ATP produced by the light reaction (Calvin cycle).

light reaction also *light dependent reaction,* that part of photosynthesis directly dependent on the capture of photons; specifically, the photolysis of water, the thylakoid electron transport system, and the chemiosmotic synthesis of ATP and NADPH.

lignin an amorphous substance that gives wood its rigidity.

limbic system a region of the brain concerned primarily with emotions.

limnetic zone in freshwater bodies, the open water deep enough for a photic and an aphotic zone to occur and where food chains begin with phytoplankton.

linkage group a group of genes that tend to be inherited together, because they lie on the same chromosome.

linked (adj.), of two gene loci, not segregating independently.

lipase any fat-digesting enzyme.

lipid an organic molecule that tends to be more soluble in nonpolar solvents (such as gasoline) than in polar solvents (such as water).

liposome a spherical bilayer of phospholipids that forms spontaneously in water.

littoral zone the waters along the ocean or lake shore, limited to that region in which light penetrates to the bottom, supporting photosynthetic life.

liver in vertebrates, a large, glandular, highly vascular organ that serves many metabolic functions including detoxification, the production of blood proteins, food storage, the biochemical alteration of food molecules, and the production of bile.

locus (pl. *loci*), also *gene locus,* specific place on a chromosome where a gene is located.

logistic growth curve population growth plot that takes the form of a sigmoid curve; growth that is eventually subject to environmental resistance and hovers about the environment's carrying capacity.

long-term memory 1. learning that persists more than a few hours, the memory trace of which is physically located in a different part of the brain than short-term memory. 2. the part of the brain and the general neural function with which such persistent memory traces are associated.

Lotka–Volterra theory the proposal that predator population size rises and falls according to the population size of the prey.

lumen the cavity or channel of a hollow tubular organ or organelle.

lymph node a rounded, encapsulated mass of lymphoid tissue through which lymph ducts drain, consisting of a fibrous mesh containing numerous lymphocytes and phagocytes.

lymphatic system the system of lymphatic vessels, lymph nodes, lymphocytes, the thoracic duct, and the thymus, which together serve to drain body tissues of excess fluids and to combat infections.

lymphoid tissue tissue in which lymphocytes are activated and aggregate.

lymphocyte any of several varieties of similar-appearing leukocytes involved in the production of antibodies and in other aspects of the immune response. See *B-cell, T-cell.*

Lyon effect in human females, the genetic mosaic created by the random manner in which either the paternal or the maternal X chromosome is selected for permanent inactivation.

lysis (n.), the destruction or lysing of a cell by rupture of the plasma membrane.

lysogenic cycle a temperate bacteriophage life cycle in which its DNA is incorporated into a host chromosome.

lysogeny the incorporation of a bacteriophage chromosome into the bacterial host chromosome, together with mechanisms preventing further infection and lysis; in later cell generations the incorporated DNA may excise, replicate, and eventually cause cell lysis.

lysosome a small membrane-bounded cytoplasmic organelle, generally containing strong digestive enzymes or other cytotoxic materials.

lytic cycle the short cycle following bacteriophage invasion during which viral replication, capsid synthesis, viral assembly, and cell lysis occur, the latter releasing new infective phage particles.

macrophage a large phagocyte that forms from a monocyte.

major histocompatibility complex (MHC) genes that code for cell surface proteins and glycoproteins and that make individuals biochemically unique.

Malpighian tubule any of numerous blind, hollow, tubular structures that empty into the insect midgut and function as a nitrogenous excretory system.

mandible 1. the vertebrate lower jaw or jaw bone. 2. either of two laterally paired anterior mouth appendages of a mandibulate arthropod, which form strong biting jaws.

marsupial a mammal of the subclass Methatheria, usually with a pouch (marsupium) in the female; included are the kangaroo, wombat, koala, Tasmanian devil, opossum, and wallaby.

masturbation manipulation of one's sexual organs for erotic gratification.

mating type in fungi and various protists, a grouping of organisms incapable of sexual reproduction with one another but capable of such reproduction with members of one or more other groups or mating types.

matrix 1. also *inner compartment,* in the mitochondrion, the enzyme-laden region within the highly convoluted inner membrane, the site of the citric acid or Krebs cycle. 2. a secreted material surrounding cells of connective tissue.

medulla 1. the inner portion of a gland or organ; compare *cortex.* 2. also *medulla oblongata,* a part of the brain stem developed from the posterior portion of the hindbrain and tapering into the spinal cord.

megagametophyte the female gametophyte produced from a megaspore in flowering plants; the embryo sac consisting of eight haploid nuclei or cells.

megaspore a single large cell with one haploid nucleus formed after meiosis and a megasporocyte mother cell in plants, and from which the megagametophyte develops by mitosis.

megasporocyte also *megaspore mother cell,* in the ovule, a large diploid cell that will give rise to the megaspore by meiosis and the degeneration of three of the four haploid nuclei.

meiosis in all sexually reproducing eukaryotes, the process in which a diploid cell or cell nucleus becomes four haploid cells or nuclei through one round of chromosome replication and two rounds of nuclear division, the first of which involves the unique pairing of chromosome homologues.

melanin the characteristic animal surface pigmentation; also found in plants.

melanism an unusual development of a black or nearly black color in a species not generally so pigmented.

membranal pump any mechanism within the plasma membrane that actively transports molecules or ions into or out of the cell.

memory cell a mature, long-lived, B- or T-cell lymphocyte, specialized in retaining specific antigen information.

meninges tough, protective, connective tissues covering the brain and spinal cord.

menopause in human females, 1. the cessation of menstruation, usually occurring between the ages of 45 and 50. 2. the whole group of physical, physiological, and behavioral occurrences and changes associated with the cessation of menstruation.

menstrual cycle the entire cycle of hormonal and physiological events and changes involving the growth of the uterine mucosa, ovulation, and the subsequent breakdown and discharge of the uterine mucosa in menses; normally occurring at three- to five-week periods in nonpregnant women.

menstruation also *menses,* in nonpregnant females of the human species only, the periodic discharge of blood, secretions, and tissue debris resulting from the normal, temporary breakdown of the uterine mucosa in the absence of implantation following ovulation.

meristem also *meristematic tissue,* a plant tissue consisting of small undifferentiated cells that give rise by cell division both to additional meristematic cells and to cells that undergo terminal differentiation; all postembryonic plant growth depends on the meristem.

merozoite a stage of *Plasmodium* that reproduces asexually in a mammalian red blood cell, releasing the poisons that cause malarial symptoms.

mesenchyme patches of mesodermal tissue that has migrated from its original position in the gastrula, such as that giving rise to muscle, dermis, gonad, and skeletal tissue.

mesentery membranous folds of peritoneum to which the intestines are attached, which, while holding them in their coiled position, permit necessary movement and other changes.

mesoderm the middle of the three primary germ layers of the gastrula, giving rise in development to the skeletal, muscular, vascular, renal, and connective tissues, and to the inner layer of the skin and to the epithelium of the coelom (peritoneum).

messenger RNA (mRNA) 1. in prokaryotes, RNA directly transcribed from an operon or structural gene, containing one or more contiguous regions (cistrons) specifying a polypeptide sequence. 2. in eukaryotes, RNA transcribed from a structural gene, tailored and usually capped and polyadenylated in the nucleus and transported to the cytoplasm; containing a single contiguous region specifying a polypeptide sequence as well as leader and follower sequences.

metabolic pathway an orderly series or progression of enzyme-mediated chemical reactions leading to a final product, each step catalyzed by its own specific enzyme.

metabolism the total chemical changes and processes of living cells.

metabolite 1. a metabolic waste, especially one that is toxic. 2. an intermediate in a biochemical pathway.

Metazoa all animals other than sponges.

MHC restriction a condition of T-cells in which their ability to recognize other cells is restricted to those with specific class I or class II MHC proteins.

microevolution evolutionary change below the species level, including changes in gene frequencies brought about by natural selection and random drift.

microgametophyte a sperm-producing body developed from a microspore.

micrometer symbol μm, one-millionth of a meter.

microorganism also *microbe,* any organism too small to be seen readily without the aid of a microscope; such as bacterium, protist, or yeast.

microspore in seed plants, one of the four haploid cells formed from meiosis of the microspore mother cell, which undergoes mitosis and differentiation to form a pollen grain.

microsporocyte also *microspore mother cell,* in the anther of a flowering plant, the diploid cells that will undergo meiosis to form the haploid microspores.

microsporogenesis in flowering plants, the process of producing microspores, which develop into the male gametophyte.

microtubule a cytoplasmic hollow tubule composed of spherical molecules of tubulin, found in the cytoskeleton, the spindle, centrioles, basal bodies, cilia, and flagella.

microvilli (sing. *microvillus*), also *brush border,* tiny finger-like outpocketings of the plasma membrane of various epithelial secretory or absorbing cells, such as those of kidney tubule epithelium and the intestinal epithelium.

midbrain the middle of the three divisions of the vertebrate embryonic brain; the adult structures derived from the embryonic midbrain.

middle lamella a layer of cementing material between adjacent plant cell walls.

mineral nutrient an inorganic compound, element, or ion needed for normal growth of all organisms.

minimum mutation tree in systematics, a hypothetical phylogenetic tree selected because it represents the smallest number of evolutionary changes needed to account for known relationships.

missense mutation a base change in a structural gene that alters the coding property of the codon in which it appears, producing an abnormal polypeptide with one amino acid difference.

mitochondrion (pl. *mitochondria*), thread-like, self-replicating, membrane-bounded organelle found in every eukaryotic cell, functioning in oxidative chemiosmotic phosphorylation, apparently derived in evolution from a bacterial endosymbiont.

mitosis cell division or nuclear division in eukaryotes, involving chromosome condensation, spindle formation, precise alignment of centromeres, and the regular segregation of daughter chromosomes to daughter nuclei.

model a contrived biological mechanism, based on a minimal number of assumptions that, when applied, is expected to yield data consistent with past observations; a biological hypothesis with mathematical preditions.

mole 1. gram molecule, the quantity of a chemical substance that has a mass in grams numerically equal to its molecular mass in daltons; 6.023×10^{23} molecules of a substance.

molecular biology a branch of biology concerned with the ultimate physicochemical organization of living matter; the study of biological systems using biochemical methods.

molecule a unit of chemical substance, consisting of atoms bound to one another by covalent bonds.

Monera the prokaryote kingdom; bacteria.

monoclonal antibody a specific immunoglobulin produced by hybrid cells artificially cloned in the laboratory.

monocot a flowering plant of the angiosperm class Monocotyledonae, characterized by producing seeds with one cotyledon; included are palms, grasses, orchids, lilies, and irises.

monoecious producing both egg and sperm in the same individual. Compare *dioecious.*

mono hybrid cross a cross involving parents that differ only by a single trait, or a cross in which only a single trait is followed.

monophyletic (adj.), of a taxonomic group, deriving entirely from a single ancestral species that possessed the defining characters of the group, while not having given rise to organisms outside the group.

monosaccharide a sugar not composed of smaller sugar subunits (for example, glucose, fructose).

monotreme a platypus, echidna, or extinct egg-laying mammal of the order Monotremata, subclass Protheria.

mons veneris in women, a rounded, usually hairy bulge of fatty tissue over the pubic symphysis and above the vulva.

morph one of the particular forms of an organism that exists in two or more distinct forms in a single population.

mortality rate also *death rate,* the number of deaths per unit time occurring among a specified number of individuals (usually per 1000) in a given area or population.

morula in the early embryo, a simple ball of cells.

mosaic development the early determination and setting aside of cells for specific developmental roles whose fulfillment may be some time in the future.

motor end plate the terminal branching of the axon of a motor neuron as it forms multiple synapses with a muscle fiber.

motor neuron a neuron that innervates muscle fibers, and impulses from which cause the muscle fibers to contract.

mucosa the highly glandular mucous membrane lining of an organ.

mucous membrane lining tissue that specializes in secreting mucus.

mucus a viscid, slippery secretion rich in mucins, which is secreted by mucous membranes and which serves to moisten and protect such membranes.

Müllerian ducts also *paramesonephric ducts,* the embryological origin of oviducts in some female vertebrates.

Müllerian mimicry in a number of dangerous species, a common coloration or configuration serving as a common warning.

multicellular (adj.), consisting of a number of specialized cells that cooperatively carry out the functions of life.

multiple alleles the alleles of a gene locus when there are more than two alternatives in a population.

multiplicative law in mathematical probability theory, the statement that the probability that all of a group of independent events will occur is equal to the product of their individual probabilities.

mutagen a chemical or physical agent that causes mutations.

mutant 1. a new or abnormal type of organism produced by a mutation. 2. a mutated gene. 3. an individual that bears and is affected by a mutated gene. 4. (adj.), mutated.

mutate 1. (v.t.) to alter, cause a change, or cause a mutation or DNA change to occur in; to mutagenize. 2. (v.i.) to change in state or genetic condition, to become altered, to undergo a mutation.

mutation 1. any change in DNA. 2. a change in DNA that is not immediately and properly repaired. 3. any abnormal, heritable change in genetic material. 4. the act or process of mutating.

mutation rate the rate at which new mutations occur, generally in terms of mutations per locus per gamete per generation.

mutualism a mutually beneficial association between different kinds of organisms; symbiosis in which both partners gain fitness. Compare *symbiosis.*

mycelium (pl. *mycelia*); 1. the mass of interwoven hyphae that forms the vegetative body of a fungus. 2. the analogous filaments formed by certain filamentous bacteria.

mycorrhizae a mutualistic fungus–root association with the fungal mycelium either surrounding or penetrating the roots of a plant.

myelin sheath fatty sheath surrounding the axons of many vertebrate neurons. See also *Schwann cell.*

myosin a protein involved in cell movement and structure; especially muscle cells.

myosin bridge the more or less globular head of the generally fibrous protein myosin, which forms a movable bridge between a myosin myofilament and an actin filament.

N terminal also *amino terminal,* the beginning of a protein, characterized by the free amino group present on the first amino acid.

NAD (nicotine adenine dinucleotide) a coenzyme and hydrogen acceptor and carrier in cell metabolism (the oxidized form is NAD^+ and the reduced form, is NADH).

NADP (nicotine adenine dinucleotide phosphate) a coenzyme and hydrogen acceptor and carrier in photosynthesis (the oxidized form is $NADP^+$ and the reduced form is NADPH).

nastic response any movement in plants, particularly changes in leaf position, attributed to sudden turgor pressure changes in cells of the petiole.

natural killer (NK) cell a free-roving lymphocyte that identifies, binds to, and lyses cancerous and virus-infected cells as part of the nonspecific immune response.

natural selection the differential survival and reproduction in nature of organisms having different heritable characteristics, resulting in the perpetuation of those charactertistics and/or organisms that are best adapted to a specific environment. Also *survival of the fittest.*

navigation the ability to locate or maintain reference to a particular place without the use of landmarks.

negative feedback an automated control mechanism in which an action, brought about by a chemical or physical stimulus, directly or indirectly reduces the stimulus. Such an inhibiting effect constitutes a negative feedback loop.

nephridium an excretory organ primitive to many coelomate phyla and still found in annelids, brachiopods, mollusks, and some arthropods, occurring paired in each body segment in segmented animals, and typically consisting of a ciliated funnel (nephrostome) draining the coelom through a convoluted, glandular duct, or nephridiopore, to the exterior.

nephron a single excretory unit of a kidney, consisting of a glomerulus, Bowman's capsule, proximal convoluted tubule, loop of Henle, distal convoluted tubule, and collecting duct discharging into the renal pelvis.

neritic province the coastal sea from the low-tide line to a depth of about 183 m (600 ft), generally waters of the continental shelf.

nerve 1. (n.) a filamentous band of nerve cell axons and dendrites and protective and supporting tissue, that connect parts of the nervous system with other parts of the body. 2. (adj.) pertaining to the nerve or nervous system; for example, nerve cell, nerve net, nerve fiber.

net community productivity (NCP) the rate of community productivity by autotrophs after their energy utilization through respiration has been subtracted.

net productivity in ecology, the rate of production of energy or biomass by plants, being the gross productivity less the energy used by the plants in their own activities.

neural canal 1. the canal formed by the series of vertebral neural arches, through which the spinal cord passes. 2. the lumen (cavity) of the spinal cord.

neural folds in early vertebrate embryology, a pair of longitudinal ridges that arise from the neural plate on either side of the neural groove, which fold over and fuse to give rise to the neural tube.

neural groove in early vertebrate embryology, the linear depression in the neural plate between the neural folds that will invaginate to form the neural tube.

neural impulse a transient membrane depolarization, followed by immediate repolarization, traveling in a wave-like manner along a neuron.

neural plate in the vertebrate neurula, a thickened plate of ectoderm along the dorsal midline that gives rise to the neural tube, neural crests, and ultimately the nervous system.

neural tube 1. in early vertebrate embryology, the hollow dorsal tube formed by the fusion of the neural folds over the neural groove. 2. the spinal cord.

neuroglia also *glia,* supporting cells of the central nervous system.

neuromuscular junction the synapse between a neural motor end plate and a muscle fiber.

neuron also *nerve cell,* a cell specialized for the transmission of nerve impulses, consisting of one or more branched dendrites, a nerve cell body in which the nucleus resides, and a terminally branched axon.

neuropeptide a class of neurotransmitters produced and secreted in the central nervous system and consisting of short chains of amino acids.

neurotransmitter a short-lived, hormone-like chemical (for example, acetylcholine, norepinephrine) released from the terminus of an axon into a synaptic cleft, where it stimulates a dendrite in the transmission of a nerve impulse from one cell to another. See also *synapse, synaptic cleft.*

neutral mutation a change in DNA that has no measurable effect on the fitness of an organism.

neutralist a population geneticist or evolutionary theorist who argues that a substantial proportion of protein and DNA changes in evolution have been due to selectively neutral mutations fixed by random drift.

neutrophil the most common mammalian phagocytic leukocyte.

neutron one of the two common constituents of an atomic nucleus, having no charge and no effect on chemical reactions.

niche also *ecological niche,* the position or function of an organism in a community; the totality of the adaptations, specializations, tolerance limits, functions, biological interactions, and behavior of a species.

nitrification the chemical oxidation of ammonium salts into nitrites and nitrates by the action of soil bacteria.

nitrogen cycle a biogeochemical cycle in which nitrogen compounds pass from autotrophs (plants and algae) to heterotrophs (animals, fungi, protists, and bacteria) and back to autotrophs. New nitrogen enters the cycle through nitrogen fixation by bacteria, and losses of nitrogen occur through dentrification.

nitrogen fixation the chemical change of nitrogen from the stable, generally unavailable form of atmospheric nitrogen (N_2) to

soluble and more readily utilized forms such as ammonia, nitrate, or nitrite; usually by the action of cyanophytes or nitrogen-fixing bacteria, but sometimes by lightning, automobile engines, or synthetic fertilizer factories.

nitrogenous base 1. a purine or pyrimidine used in the synthesis of nucleic acids; specifically, adenine, cytosine, guanine, thymine, or uracil. 2. any related heterocyclic compound, such as xanthine or uric acid.

noncylic phosphorylation light reactions employing both photosystem II and photosystem I, during which electrons from water pass through both photosystems, reducing NADP to NADPH + H^+, thus serving both chemiosmosis and carbohydrate synthesis.

nondisjunction 1. the failure of a pair of homologous chromosomes to segregate to opposite poles in anaphase I of meiosis. 2. the failure of a pair of daughter chromosomes or daughter centromeres to segregate to opposite poles in mitotic anaphase or in anaphase II of meiosis.

nonhistone chromosomal protein a large class of proteins other than histones associated with the chromosomes, constituting about one-third of the substance of chromatin in animals and plants, highly variable from tissue to tissue and known in at least some cases to be involved in gene control.

nonspecific defense any of several fast-acting cellular and chemical responses made by the body against foreign substances, cancerous cells, or invading organisms, but not involving B-cell and T-cell lymphocytes.

nonvascular plants plants lacking tissues specialized for the transport of water and food and providing physical support (mosses, liverworts, and hornworts).

normal distribution also *normal curve, Gaussian distribution, bell-shaped curve,* the idealized, symmetrical distribution taken by a population of values centering around a mean, when departures from the mean are due to the chance occurrence of a large number of individually small independent effects; often approached in real populations.

notochord in all chordates at some point in development, a turgid, flexible rod running along the back beneath the nerve cord and serving as a skeletal support; replaced during development by the vertebral column in most vertebrates but persistent in adult coelocanths, cyclostomes, lancelets, and larvacean urochordates.

nucleic acid either DNA or RNA, DNA being a double polymer of deoxynucleotides and RNA being a polymer of nucleotides.

nucleolus (pl. *nucleoli*), a dark-staining body of RNA and protein found within the interphase nucleus of a cell, the site of synthesis and storage of ribosomes and ribosomal materials; disperses during mitosis and is reconstituted following nuclear envelope reorganization.

nucleotide 1. a compound consisting of a nitrogenous base and a phosphate group linked to the 1' and 5' carbons of ribose, respectively; the repeating subunit of DNA and RNA.

nucleus in all eukaryotic cells, a prominent, usually spherical or ellipsoidal membrane-bounded sac containing the chromosomes and providing physical separation between transcription and translation.

nutrition the process of being nourished, particularly the steps through which an organism obtains food and uses it for bodily processes.

obligate anaerobe an organism, most commonly a bacterium, that cannot utilize oxygen and often is killed by its presence.

oceanic province (n.), the open sea as distinguished from the neritic province; *oceanic* (adj.), pertaining to the open sea.

oceanic rift community a community along ocean floor rifts, where a variety of animal life is supported by chemosynthetic bacteria, producers that thrive near vents releasing heated water and hydrogen sulfide gas.

Okazaki fragment in DNA synthesis, the original form of a newly synthesized single strand, being a polynucleotide of some 200–300 bases produced at the lagging ends of each DNA strand.

olfaction 1. the sense of smell. 2. the process of smelling.

oligotrophic of a lake, rich in dissolved oxygen and poor in plant nutrients and algal growth, with clear water and no marked stratification.

omnivorous (adj.), feeding on both animal and plant material; literally, eating everything; *omnivore* (n.), an omnivorous animal, for example, pigs, people.

oncogene a cancer-causing gene.

oocyte an egg cell before maturation; *primary oocyte,* a diploid cell precursor of an egg, before meiosis; *secondary oocyte,* an egg cell after the formation of the first polar body.

oogonium (pl. *oogonia*), a cell that gives rise to oocytes.

open circulatory system a circulatory system in which the arterioles end openly into intercellular space, allowing blood to percolate directly through nonvascular tissues.

open communities biotic communities that blend into each other in a gradual transition.

open growth also *indeterminant growth,* in perennial plants the capability of continuous growth.

operant conditioning an animal learning to perform an act in order to receive a reward.

operator in an operon, the binding site of an inhibitor protein or inhibitor complex.

operon in prokaryotes, a region of DNA that includes structural genes and the genes controlling them; transcription may be inducible, remaining shut down until activated by an inducer substance, or repressible, remaining active until shut down by a repressor substance. Control regions generally consist of a promotor region (p), an operator region (o), and a regulator gene (*i*), which produces a repressor protein. See also *lac operon, tryptophan operon.*

opsonization the coating of an invading cell by antibodies whose exposed constant regions are keyed to matching receptors on phagocytes, which facilitates phagocytosis.

organ an organized assembly of various tissues performing some major body function; for example, the heart, brain, liver.

organelle a functionally and morphologically specialized part of a cell.

organic molecule a molecule containing carbon and generally produced by living organisms.

orgasm in humans, the climax of sexual excitement, usually accompanied in men by ejaculation and in women by rhythmic contractions of the cervix.

orientation the directing of bodily position according to the location of a particular stimulus; may be part of an instinctive action.

origin 1. evolutionary ancestry. 2. the fixed skeletal attachment of a muscle or tendon; compare *insertion.*

osmoconformer an aquatic organism that does not regulate the osmotic potential of its body tissues, but allows it to fluctuate with that of the environment.

osmoregulator an aquatic organism that maintains a constant internal osmotic potential in spite of fluctuations in the salinity of its environment.

osmosis the tendency of water to diffuse through a semipermeable membrane in the net direction from its higher to its lower concentration.

osmotic gradient any difference in the concentration of water molecules across a membrane, the difference between two fluids in terms of osmotic potential.

osmotic potential the tendency or capacity for water to move across a selectively permeable membrane into a second solution, such movement occurring because of the presence of a relatively high solute concentration (or less water) on the other side.

osmotic pressure the actual hydrostatic pressure that builds up in a confined fluid because of osmosis.

osteocyte a bone cell isolated in a lacuna of bone tissue.

outer compartment in the mitochondrion, the region between the outer and inner membranes into which protons are transported.

ovary 1. in animals, the (usually paired) organ in which oogenesis occurs and in which eggs mature. 2. in flowering plants, the enlarged, rounded base of a pistil, consisting of a carpel or several united carpels, in which ovules mature and megasporogenesis occurs.

overdominance if the phenotype of the heterozygote is more fit than those of both homozygotes, both alleles are said to be overdominant.

oviduct a tube, usually paired, for the passage of eggs from the ovary toward the exterior or to a uterus, often modified for the secretion of a shell or protective membrane; in humans, it is known as a uterine or fallopian tube.

oviparous (adj.), producing eggs that develop and hatch outside the mother's body. Compare *ovoviviparous, viviparous.*

ovoviviparous producing eggs that are fertilized internally, develop within the mother's body but without any direct connection with the maternal circulation; the young being released shortly before or after hatching. Compare *viviparous, oviparous.*

ovulation the release of one or more eggs from an ovary.

ovule 1. in animals, a small egg; an egg in the process of growth and maturation. 2. in seed plants, a rounded outgrowth of the ovary, consisting of the embryo sac surrounded by maternal tissue including a stalk, the nucleus, and one or more integuments.

oxidation 1. the loss of electrons from an element or compound. 2. also *dehydrogenation,* the loss of hydrogens from a compound.

oxidation–reduction reaction a chemical reaction in which one reactant is oxidized and another is reduced.

oxidative phosphorylation the production of ATP from ADP and phosphate in a process consuming oxygen, as by mitochondria or aerobic bacteria. See also *chemiosmosis, chemiosmotic phosphorylation, oxidation.*

oxidative respiration the breakdown of biochemicals to produce cellular energy, utilizing oxygen as the final electron acceptor. See *oxidative phosphorylation, respiration.*

oxygen debt a state of oxygen depletion after extreme physical exertion; measured by the amount of oxygen required to restore the system to its original state.

oxyhemoglobin hemoglobin carrying four oxygen molecules; the bright red arterial form of hemoglobin.

pancreas a large digestive and endocrine gland of vertebrates, which secretes various digestive enzymes into the duodenum by way of the pancreatic duct, and which also contains endocrine tissues in the form of interspersed islets of Langerhans, responsible for the production of the hormones insulin and glucagon.

parapatric living in separate but adjacent geographic regions not separated by geographical barriers. Compare *allopatric speciation, sympatric speciation.*

parasite an organism living in or on another living organism from which it obtains its organic sustenance to the detriment of its host. See also *ectoparasite.* Compare *symbiont, mutualism.*

parasympathetic division one of the two divisions of the vertebrate autonomic nervous system, the one that utilizes acetylcholine as a neurotransmitter, that increases the activity of smooth muscle and digestive glands, slows the heart, and dilates blood vessels. Compare *sympathetic division.*

parathyroid glands four small endocrine glands embedded in or adjacent to the thyroid gland and involved in the regulation of calcium ion levels in the blood.

Parazoa a subkingdom of the animal kingdom, consisting of sponges.

parenchyma in higher plants, a tissue consisting of thin-walled living cells that remain capable of cell division even when mature, and function in photosynthesis or food storage.

parthenogenesis asexual reproduction in which gametes develop without fertilization, either with or without having undergone meiosis; for example, male hymenopterans are produced parthenogenetically from haploid eggs, and *Daphnia* of either sex may be produced parthenogenetically from unreduced diploid eggs.

partial dominance a phenotype of a heterozygote in which the expressed alleles are halfway between the homozygous phenotypes.

partial pressure the independent pressure exerted by each gas in a mixture of gases.

passive immunization the conferring of temporary immunity against a specific disease by the injection of antibodies that act against the disease organism.

passive transport movement of fluids, solutes, or other materials without the expenditure of energy, for example, by diffusion, especially across a membrane.

pasteurization heating of wine or milk to a certain temperature for a given amount of time in order to kill pathogenic or otherwise undesirable bacteria without destroying the quality of the wine or milk.

pathogen an organism that is capable of causing disease in another organism; generally refers to viruses and parasitic bacteria and fungi.

pelagic (adj.), living in the open sea; *pelagic zone,* the region of the open sea beyond the littoral and neritic zones and extending from the surface to the depth of light penetration. Compare *abyssal region, neritic province, littoral zone.*

pelvic girdle the bones or cartilage supporting and articulating with the vertebrate hindlimbs; in humans, consisting of the sacrum, coccyx, and the paired and fused ischium, ilium, and pubis.

pelvic inflammatory disease, (PID) a sexually transmitted infection that has spread into the uterus, uterine tubes, and abdominal cavity.

pelvis 1. see *pelvic girdle;* 2. *renal pelvis,* the main cavity of the kidney, into which the nephrons discharge urine.

penicillin a group of closely related powerful antibiotics produced by fungi of the genus *Penicillium,* which function by disrupting the synthesis of bacterial cell walls.

penis 1. the copulatory organ of the male in any species in which internal fertilization is achieved by the insertion of a male body part into the female genital tract. 2. the erectile intromittant organ of male mammals, which also serves as a channel for the discharge of urine.

PEP cycle also known as *Hatch–Slack pathway* in C4 plants, a biochemical cycle during which carbon dioxide is first incorporated into 4-carbon compounds in the leaf mesophyll and later released into the Calvin cycle in the bundle sheath cells.

peptide a chain of two or more amino acids linked by peptide bonds, too short to be coagulated by heat or precipitated by saturated ammonium sulfate; most often seen as a partial digestion product or a protein or polypeptide.

peptide bond also peptide linkage, the dehydration linkage formed between the carboxyl group of one amino acid and the amino group of another.

peptide hormone any hormone consisting of one or more amino acids.

perennial 1. (adj.) continuing or lasting for several years. 2. (n.) a plant that lives for an indefinite number of years, as compared with annual or biennial.

pericycle a layer of parenchyma or sclerenchyma that sheaths the stele of the root and is associated with the formation of vascular cambium and lateral roots.

periderm in plants, a protective layer of secondary tissue derived from epidermal cells and consisting of cork, cork cambium, and underlying parenchyma.

periosteum tough connective tissue covering of bone.

peripheral nervous system nerves and receptors outside the central nervous system, including sensory and motor nerves of the somatic system and autonomic nervous system.

peristalsis successive waves of involuntary contractions passing along the walls of the esophagus, intestine, or other hollow muscularized tube, forcing the contents onward.

peritoneum the smooth, transparent membrane lining the abdominal cavity of a mammal; *peritoneal cavity,* the principal body cavity (abdominal coelom) of a mammal.

pH the negative logarithm of the hydronium ion concentration of a solution and a common measure of the acidity or alkalinity of a liquid; pH values of less than 7 indicating acidity, and values greater than 7 indicating alkalinity.

phage also *bacteriophage,* a virus that infects and lyses bacteria.

phagocyte 1. any leukocyte that engulfs particulate matter. 2. any cell that characteristically engulfs foreign matter, for example, cells of the reticuloendothelial system.

phagocytosis taking solid materials into the cell by engulfment and the subsequent pinching off of the plasma membrane to form a digestive vacuole.

phagosome a vacuole containing materials taken in by the cell through phagocytosis, also called a food vacuole in some instances.

phenetics an approach to evolutionary classification based on similarities and differences in observable characters.

phenotype the visible or otherwise detectable physical and chemical traits of an organism, as influenced by heredity and by the environment. Compare *genotype*.

phloem a complex vascular tissue of higher plants consisting of sieve tubes, companion cells, and phloem fibers, and functioning in transport of sugars and nutrients.

phloem fiber a fiber cell associated with phloem, with great strength and pliability, for example, the linen fibers of flax.

phloem ray the part of a vascular ray located in the phloem. See *vascular ray.*

phosphorylation the addition of a phosphate group to a compound, such as, the addition of phosphate to ADP to produce ATP and water.

photoautotroph a photosynthetic organism; one capable of utilizing light energy and simple inorganic substances in its metabolism.

photon a quantum of electromagnetic radiant energy.

photoperiodism the response of an organism to photoperiods, involving sensitivity to the onset of light or darkness and a capacity to measure time.

photorespiration a puzzling phenomenon in which abundant light energy is captured by photosynthesis but little or no net carbon dioxide fixation occurs; common in C_3 plants in bright sunlight on hot days.

photosynthesis the organized capture of light energy and its transformation into usable chemical energy in the synthesis of organic compounds.

photosystem I also *photosystem 700, P700,* the second of the two photosystems in the electron pathway of photosynthesis in cyanobacteria and chloroplasts, and the one involving the reduction of NADP to $NADPH_2$; it may be evolutionarily more ancient than photosystem II.

photosystem II also *photosystem 680, P680,* the first of the two photosystems in the electron pathway of photosynthesis in cyanobacteria and all photosynthetic eukaryotes, and the one involving the photolysis of water; it probably occurred later in evolution than photosystem I.

phototropism 1. the turning toward light of a growing plant stem. 2. phototaxis. 3. *negative phototropism,* the turning away from light, as of a root tip.

phylogenetic tree a graphical representation of the interrelations and evolutionary history of a group of organisms, indicating the relative order of successive divisions of the line of descent, coincident with past speciation events.

phylogeny 1. the evolutionary history of an organism or group of organisms. 2. a phylogenetic tree.

phylum 1. a major taxonomic unit of related, similar classes of animals, for example, Phylum Annelida. 2. a division of the plant kingdom.

physiology 1. a branch of biology dealing with the processes, activities, and phenomena of individual living organisms, organs, tissues, and cells. 2. the normal functioning of an organism.

phytoplankton photosynthesizing planktonic organisms.

pigment any chemical substance that absorbs light, whether or not its normal function involves light absorption: for example, chlorophyll, cytochrome *c,* hemoglobin, melanin.

pineal body also *pineal organ, pineal gland,* a small body arising from the roof of the third ventricle in all vertebrates, forming an eye-like photoreceptive organ in larval lampreys, tuatara, and some lizards; center of photoperiod responses in all vertebrates.

pinocytosis taking dissolved molecular food materials, such as proteins, into the cell by adhering them to the plasma membrane and invaginating portions of the plasma membrane to form digestive vacuoles; a form of active transport.

pioneer organism 1. an organism that successfully establishes residence and produces offspring in an area not previously inhabited by its kind. 2. a type of organism specialized for the initial invasion of a disturbed area, such as a landslide or burned-out region.

pistil in flowering plants, a unit comprised of one or more ovaries, a style, and a stigma.

pith thin-walled parenchymous tissue in the central strand of the primary growth of a stem; the dead remains of such tissue at the center of a woody stem.

pituitary gland also *hypophysis,* a small double gland lying just below the brain and intimately associated with the hypothalamus in all vertebrates, consisting of an anterior lobe and a functionally distinct posterior lobe.

placebo a substance having no pharmacological effect but administered as a control in testing experimentally or clinically the efficacy of a biologically active preparation.

placenta 1. in mammals other than monotremes and marsupials, the organ formed by the union of the uterine mucosa with the extraembryonic membranes of the fetus, which provides for the nourishment of the fetus, the elimination of waste products, and the exchange of dissolved gases. 2. in flowering plants, the part of the ovary to which the ovule and seeds attach.

plankton minute, drifting plant, algal, and animal life in marine and fresh waters.

plasma the fluid matrix of blood tissue, 90% water and 10% various other substances; distinguished from blood serum by the presence of fibrinogen and the absence of certain platelet-derived hormones.

plasma cell an activated B-cell lymphocyte, specializing in the production and secretion of antibodies.

plasma membrane the external semipermeable limiting layer of the cytoplasm. See also *membrane, fluid mosaic model.*

plasmid in bacteria, a small ring of DNA that occurs in addition to the main bacterial chromosome.

plastid any of several forms of a self-replicating, semiautonomous plant cell organelle, primarily as a chloroplast specialized for photosynthesis, a chromoplast specialized for pigmentation, or a leukoplast specialized for starch storage.

platelets minute, cell-like but enucleate, fragile, membrane-bounded cytoplasmic disks present in the vertebrate blood; rupture of platelets releases factors that initiate blood clotting.

plate tectonics the movement of great land and ocean floor masses (plates) on the surface of the earth relative to one another occurring largely in the Cenozoic era. See also *continental drift.*

pleiotropy the situation where one gene has many effects.

pleural cavity in mammals, either of two divisions of the coelom constituting the thoracic cavities, harboring the lungs and lined with the pleurae.

poikilotherm an animal that does not thermoregulate and whose body temperature hovers about that of its surroundings.

point mutation a small spontaneous change in DNA involving individual nucleotides, such as a single base change, a single base addition or deletion.

polar body either of two small cells produced by meiosis of an egg cell: *first polar body,* the recipient of the chromosomes going to one pole in anaphase I; *second polar body,* the recipient of the chromosomes going to one pole in anaphase II of the egg. Neither polar body divides again.

polarity in animal egg cells, the establishment of metabolic gradients that will later reveal themselves as the active animal and sluggish vegetal poles.

pollination the transfer of pollen from a stamen to a stigma, preceding fertilization of a flowering plant.

poly-A tail polyadenylic acid tail, a string of about 200 adenosine nucleotides added enzymatically to the 3' end of transcripts in the nucleus during the maturation of mRNA; thought to protect mRNA from degradation by cytoplasmic exonucleases.

polygenic inheritance inheritance involving many interacting variable genes, each having a small effect on a specific trait; also *quantitative inheritance.*

polygenic trait also *continuously varying trait, quantitative trait,* metric trait, a trait in which variation in a population is expressed in continuous increments about a mean and is attributable to the action and interaction of many variable gene loci and usually also to multiple variable effects of the environment.

polymer a generally large molecule, consisting of chemically bonded subunits, as in a polypeptide or polysaccharide.

polymerase an enzyme that causes polymerization; for example, RNA polymerase, DNA polymerase.

polymorphism 1. the existence within a population of two or more discrete, genetically determined forms other than variations in sex or maturity and apart from rare mutant forms; for example, black and spotted leopards. 2. the existence within a population of two or more alleles at a locus, at allele frequencies greater than some arbitrary value such as 1% or 5%; for example, ABO blood group polymorphisms.

polypeptide a continuous string of amino acids in peptide linkage, longer than a peptide. Compare *protein.*

polyphyletic of a taxonomic group, having two or more ancestral lines of origin.

polyploid having more than two complete sets of chromosomes; *autopolyploid,* having three or more complete sets of a specific complement of chromosomes; *allopolyploid,* having two or more full diploid sets of chromosomes derived from different ancestral diploid species.

polyribosome also *polysome,* a number of ribosomes attached to one messenger RNA molecule, each forming the same polypeptide.

polysaccharide a polymer of sugar subunits.

polyspermy an abnormal condition in which two or more sperm successfully penetrate the egg cytoplasm.

polyunsaturated of a fatty acid, having more than one carbon–carbon double bond.

pons a broad mass of nerve fibers running across the ventral surface of the mammalian brain.

population 1. (demography) the total number of persons inhabiting a given geographical or political area. 2. (genetics) an aggregate of individuals of one species, interbreeding or closely related through interbreeding and recent common descent, and evolving as a unit. 3. (ecology) the assemblage of plants and/or animals living in a given area; or all the individuals of one species in a given area. 4. (statistics) a finite or infinite aggregation of individuals under study.

population genetics the scientific study of genetic variation within populations, of the genetic correlation between related individuals in a population, and of the genetic basis of evolutionary change.

portal circuit an arrangement of blood vessels wherein blood passes through two successive capillary beds (connected by a vein) before its return to the heart.

posterior 1. in most bilateral organisms, away from the head end of an organism, the opposite of anterior.

posterior pituitary in humans, the posterior lobe of the pituitary, a stalked extension of neural tissue extending from the hypothalamus.

potential energy energy stored in chemical bonds, in nonrandom organization, in elastic bodies, in elevated weight, or any other static form in which it can theoretically be transformed into another form or into work.

precapillary sphincter in arterioles, rings of smooth muscle capable of regulating blood flow into capillaries; controlled by the autonomic nervous system.

precocial capable of walking and a high degree of independent activity from hatching or birth.

predation 1. the act of catching and eating. 2. being caught and eaten; for example, subject to predation. 3. a mode of life in which food is primarily obtained by killing and eating other animals.

predator an animal that habitually preys on other animals, a carnivorous animal.

pressure flow hypothesis a favored explanation of phloem sap movement: the active transport of sugars into the phloem leads to the osmotic uptake of water therein; the resulting increase in hydrostatic pressure provides the force needed to push the sap through the sieve elements.

presynaptic membrane the membrane of an axon or of the synaptic knob of an axon in the region of a synaptic cleft, into which it secretes neurotransmitters in the course of the transmission of a nerve impulse. See *synapse, synaptic cleft.*

prey switching behavior of a predator in response to the relative densities of two prey species, in which the predator ceases hunting one prey and begins to hunt the other.

primary growth of a plant, the initial growth or elongation of a stem or root, resulting primarily in an increase in length and the addition of leaves, buds, and branches. *Primary growth pattern,* the distinctive pattern of xylem, phloem, and other tissues in primary growth. Compare *secondary growth.*

primary host also *definitive host,* in parasitic relationships, a host in which a parasite may undergo sexual reproduction or at least reach a mature state.

primary immune response the slower, initial response against invasion of the body by organisms or foreign molecules, during which immature, inactive lymphocytes are activated into specialized B- and T-cell lymphocytes. See *secondary immune response.*

primary phloem phloem developed from apical meristem, that is, the phloem of primary growth.

primary producer in ecology, plants, algae, or other photosynthetic (or chemosynthetic) organisms that produce the food of a food chain.

primary succession the succession of vegetational states that occurs as an area changes from bare earth to a climax community.

primary xylem xylem in primary growth, which is produced by apical meristem rather than vascular cambium.

primitive a character or character state, characteristic of the original condition of the group under consideration; ancestral; not derived; of or like the earliest state within the group considered; old. Compare *advanced, derived.*

primitive groove in bird, reptile, and mammalian embryos, a lengthy indentation formed as gastrulation begins.

primitive streak in bird, reptile, and mammalian embryos, a thickening in the blastoderm formed by convergence of cells in preparation for gastrulation.

prion little known disease agent; protein shells capable of penetrating and killing cells.

probability 1. the relative likelihood of the occurrence of a specific outcome or event based on the proportion of its occurrence among the number of trials or opportunities for its occurrence, viewed as indefinitely extended. 2. the field of mathematics dealing with probability relationships.

procambium the part of a meristem that gives rise to cambium.

product enzyme product, one of the resulting compounds of an enzymatic reaction.

profundal zone in freshwater bodies, the dark waters below the photic zone, often having a low oxygen content and limited aerobic life that consists primarily of reducers and scavengers.

progeny testing determination of an organism's genotype by crossing it with a known recessive homozygous individual and observing the resulting offspring.

prokaryote also *procaryote,* any organism of the kingdom Monera, having no nucleus and a single circular chromosome of nearly naked DNA; a eubacterium or archaebacterium.

promoter a DNA sequence to which RNA polymerase must bind in order for transcription to begin. See *operon.*

prophage a name given to the temperate phage (a virus) when incorporated into a host chromosome.

prophase the first phase of mitosis and meiosis, the events of which include chromosomal condensation, dismantling of the nuclear envelope, organization of the spindle apparatus, and, in meiosis, synapsing and crossing over. See also *mitosis, meiosis.*

proplastid an undifferentiated organelle of plant egg cells and embryonic cells. It is the precursor of chloroplast and other plastids.

proprioception the sum of neural mechanisms involved in sensing, integrating, and responding to stimuli from within the body, including the integration of the movement of body parts, balance, and stance.

proprioceptor a sensory receptor that is located in internal tissues (as in skeletal, smooth, and heart muscle, tendons, carotid body) and responds to conditions of body positions, muscle tension, or internal chemistry.

prop root a type of adventitious root that emerges from the stem, grows down to the soil, and thus "props up" (supports) the plant (typical of corn and certain tropical fig trees).

prostaglandin any of a group of hormone-like substances derived from long-chain fatty acids and produced in most animal tissues.

prostate gland a pale, firm, partly muscular and partly glandular organ that surrounds and connects with the base of the urethra in male mammals; its viscid, opalescent secretion is a major component of semen.

protein a naturally occurring functional macromolecule consisting of one or more polypeptides held together by hydrogen bonds and van der Waals forces and often sulfhydryl linkages, frequently including one or more prosthetic groups.

proteinoid microspheres spherical aggregations of amino acid polymers with a membrane-like surface formation, known to grow and divide in a manner reminiscent of cells.

protocell the earliest form to exhibit characteristics associated with life. A coacervate-like body containing autocatalytic properties that assure faithful replication of specific chemical properties.

protoeukaryote a hypothetical predatory microorganism, capable of the engulfment of prey, that by the successive engulfment of a protomitochondrion, a protochloroplast, and perhaps a protocilium evolved into the ancestral eukaryotic cell, perhaps a billion years ago.

Proto-Laurasia according to the continental drift theory, one of two land masses (the other, Proto-Gondwana) preceding and giving rise to the supercontinent Pangea. The two earlier masses, according to one hypothesis, are the origin of the deuterostomes and proterostomes.

proton 1. one of the two particles composing the atomic nucleus in ordinary matter, having an electrostatic charge of +1 and a mass 1837 times that of an electron, and by

its numbers determining the chemical properties of the atom. 2. a hydrogen ion.

proton pump an active transport system using energy to move hydrogen ions from one side of a membrane to the other against a concentration gradient, as in chemiosmosis.

pseudocoelom also *pseudocoel,* "false coelom"; the body cavity in nematodes, rotifers, and certain other invertebrates, between the body wall and the intestine, which is not entirely lined with mesodermal epithelium; in which the gut is entirely endodermal and not muscularized.

psilopsid a simple plant of the tracheophyte subdivision Psilopsida, vascular plants lacking roots, cambium, leaves, and leaf traces and which have a reduced, subterranean, nonphotosynthetic gametophyte.

pulmonary circuit the passage of venous blood from the right side of the heart through the pulmonary arteries to the capillaries of the lung, where it is oxygenated and from which it returns by way of the pulmonary veins to the left atrium of the heart.

punctuated equilibrium a theory that evolution does not occur gradually but that life exists over long periods of time with little evolutionary change, interrupted periodically by great changes. Compare *gradualism.*

pyloric sphincter also *pyloric valve* and *pylorus,* a ring of muscle capable of closing off the opening between the stomach and the small intestine.

pyrimidine a nitrogenous base consisting of a six-membered heterocyclic ring; notably cytosine, uracil, and thymine, constituents of nucleic acids.

pyrogen any fever-producing substance or hormone.

radial symmetry the condition of having similar parts regularly arranged as radii from a central axis, as in a starfish.

radiation 1. the transfer of heat or other energy as particles or waves. 2. heat or other energy transmitted as particles or waves; see also *ionizing radiation.* 3. in evolution, the ongoing rise of new species and establishing of new niches.

radicle the lower portion of the axis of a plant embryo, especially the part that will become the root.

radioactive in elements, a state of instability in which radiation is emitted by the atomic nucleus.

rain shadow a region of low rainfall on the lee side of a mountain or mountain range.

range the geographical area occupied by a species.

reactant any element, ion, or molecule participating in a chemical reaction.

reaction center the part of a photosystem in which light-activated chlorophyll *a* transfers an electron to the electron transport system.

realized niche the niche of an organism as it actually is, generally only a portion of the fundamental niche.

receptor see *sensory receptor.*

recessive (adj.), an allele not expressed in a heterozygote. See *dominance.*

recessive lethal an allele that, when homozygous, results in death of the individual.

reciprocal altruism behavior functions to increase the fitness of the individual insofar as it increases the likelihood that the individual will be the recipient of beneficial behavior at another time.

reciprocal cross a cross in which pollen is exchanged between the flowers of two individual plants.

recombinant DNA a general term for the laboratory manipulation of DNA in which DNA molecules or fragments from various sources are severed and combined enzymatically and reinserted into living organisms.

recombination the process whereby the linkage relationship among genes on a chromosome are changed, because of the crossing over of DNA.

red alga any alga of the division Rhodophyta.

red marrow regions within the ribs, sterum, vertebrae, and hip bones where red blood cells are produced.

red tide seawater discolored by a dinoflagellate bloom in a density fatal to many forms of life.

reducing power the relative amount of energy released by a substance for each electron or hydrogen transferred from it in an oxidation–reduction reaction; the relative ability of one substance to transfer electrons or hydrogen atoms to another.

reduction of a substance, the addition of electrons or hydrogen atoms.

reduction division the first division of meiosis.

reflex arc in the vertebrate nervous system, a simple neural pathway involving as few as two or three neurons that sense and react to a stimulus.

refractory state a brief period when a neuron cannot generate a second impulse.

regeneration following the loss of a limb or other body part, regrowth of that part through dedifferentiation and repetition of the original developmental events.

regulatory region that part of a gene that controls the expression of the protein(s) encoded by the coding region.

releaser in animal behavior, any stimulus that evokes the release of instinctive behavioral patterns.

releasing hormone also *releasing factor,* a chemical messenger released by the hypothalamus that stimulates hormonal release by the pituitary.

REM sleep also *paradoxical sleep,* a normal period of sleep during which the muscles are very relaxed but the eyes move rapidly under closed lids; accompanied by high brain electrical activity.

renal circuit in mammals, the circuit of the branching renal arteries arising from the abdominal aorta and continuing through the glomerulus, the capillary beds of the nephron, and the renal veins to return to the vena cava.

renal pelvis the cavity of the kidney into which the collecting ducts deposit urine.

repair enzyme any of several different complexes of enzymes that recognize improper base pairing in DNA, excise a region of one of the strands, and rebuild the DNA according to the rules of Watson–Crick pairing; or that otherwise repair mutational damage, including double-strand chromosome breaks.

replica plating a screening technique in which an array of colonies on a nutrient-supplemented agar plate are transferred as a group to one or more additional plates of nutrient-deficient agar to test for nutritional mutants.

replication complex during replication, a grouping of the essential enzymes of that process including the unwinding enzyme, helicase, and DNA polymerase.

replication fork the point at which unwinding proteins separate the two DNA strands in the course of DNA replication.

repressible operon an operon governing a synthetic pathway and which is generally active but which can be inactivated by the presence of its normal metabolic product in the medium.

repressor protein in bacterial operons, a protein that binds the operator and prevents transcription either when bound to an inducing molecule (inducible operon) or when not bound (repressible operon).

reproductive isolation 1. the state of a population in which there is no mating between members of the group and members of other groups, and no immigration or emigration of individuals. 2. the state of a population or species in which successful matings outside the group are biologically impossible because of physical mating barriers, hybrid inviability, or hybrid sterility.

reservoir in biogeochemical cycles, the less readily available reserves of mineral nutrients, such as atmospheric nitrogen, which is only useful to nitrogen-fixing organisms. Compare *exchange pool.*

respiration 1. the physical and chemical process by which an organism supplies oxygen to its tissues and removes carbon dioxide. 2. the energy-requiring metabolic transformation of food or food storage molecules yielding energy.

respiratory center a region in the medulla oblongata that regulates breathing movements.

resting potential the charge difference across the membrane of a neuron or muscle fiber while not transmitting an impulse.

resting state a state of seeming inactivity in a neuron, but one in which the membrane activity maintains a polarized state in preparation for conduction.

restriction enzyme in bacteria, an enzyme that recognizes and serves a specific, short DNA sequence, thus protecting the cell from all but a few highly adapted, host-specific viruses; such enzymes have proved useful for experimental DNA manipulation.

reticular system also *reticular formation,* a major neural tract in the brain stem containing neural pathways to other parts of the brain and to the reticular activating system (RAS), an arousal center.

retinoic acid the first known morphogen, an active agent present in vertebrate limb buds that influences the development of the limb. See *tissue interaction.*

retrovirus a single-stranded RNA virus that, after undergoing reverse transcription and producing double-stranded DNA from its RNA coding, inserts into the host chromosome where it is replicated through many host cell generations. It may later escape from the chromosome to enter a period of viral reproduction and cell lysis. See *HIV, AIDS.*

reverse transcriptase an enzyme of certain RNA viruses that copies RNA sequences into single-stranded and double-stranded DNA sequences.

R-group a chemistry shorthand where R stands for the variable part of an amino acid.

Rh blood group system also *rhesus factor, Rh factor,* a polymorphic blood cell antigen system consisting of three variable antigenic sites controlled by eight alleles at a single locus.

rhodopsin a pigment in the rod cells of the retina that, in the presence of light, breaks down into opsin and retinal. The reaction stimulates action potentials in nearby neurons.

rhizoid 1. a root-like structure that serves to anchor the gametophyte of a fern or bryophyte to the soil and to absorb water and mineral nutrients. 2. a portion of a fungal mycelium that penetrates its food medium.

rhizome an underground horizontal shoot specialized for food storage.

ribosomal RNA also *rRNA,* the RNA that forms the matrix of ribosome structure, consisting of a large, intermediate, and two relatively small sequences.

ribosome an organelle consisting of a matrix of RNA and numerous specific proteins, binding a messenger RNA molecule, a charged tRNA, and a growing polypeptide chain during protein synthesis.

RNA also *ribonucleic acid,* a single-stranded nucleic acid macromolecule consisting of adenine, guanine, cytosine, and uracil on a backbone of repeating ribose and phosphate units; divided functionally into rRNA (ribosomal RNA), mRNA (messenger RNA), tRNA (transfer RNA), and viral RNA.

RNAase an enzyme capable of hydrolyzing RNA sugar–phosphate bonds.

RNA polymerase the enzyme or enzyme complex catalyzing transcription.

RNA replicase the enzyme capable of copying RNA into RNA.

rod one of the numerous long, rod-shaped sensory bodies in the vertebrate retina; responsive to faint light but not to detail or to variations in color. Compare *cone.*

root the portion of a seed plant, originating from the radicle, that functions as an organ of absorption, anchorage, and sometimes food storage, and differs from the stem in lacking nodes, buds, and leaves.

root pressure one force by which water rises into the stems from the roots.

r-selection a form of natural selection characterized by exponential population growth. Usually characterized by high reproductivity with little parental care. See *K-selection.*

saccule 1. a small sac. 2. the smaller of two chambers of the membranous labyrinth of the ear; compare *utricle.*

saltatory propagation the skipping movement of an impulse from one node of a myelinated neuron to another.

sap the watery fluid transported by phloem that circulates dissolved gases, sugars, other organic compounds, and mineral nutrients from one part of the plant to another.

saprobe an organism that reduces dead plant and animal matter.

sarcomere the contractile unit of striated muscle bounded by Z-disk partitions; consisting of regular hexagonal arrays of actin filaments bound to the Z-disk partitions and parallel myosin filaments regularly interspersed between them.

sarcoplasmic reticulum membranous, hollow tubules in the cytoplasm of a muscle fiber, similar to the endoplasmic reticulum of other cells.

savanna a tropical or subtropical grassland with scattered trees or shrubs, usually maintained by such human activities as burning and by foraging for firewood.

scanning electron microscope (SEM) a device for visualizing microscopic objects

Schwann cell one of the cells that constitute the myelin sheath, each cell being greatly flattened so as to consist almost entirely of cell membrane, and being repeatedly wrapped around an axon so as to form a sheath many layers thick.

seaweed a multicellular marine alga, of the red, green, or brown algal groups.

secondary growth growth in dicot plants that results from the activity of vascular cambium, producing primarily an increase in diameter of stem or root. Compare *primary growth.*

secondary immune response a rapid response to a second or subsequent invasion of the body by organisms or foreign molecules, during which memory cells quickly produce large numbers of active, specialized B- and T-cell lymphocytes. See also *primary immune response.*

secondary phloem in plants, phloem originating from secondary or vascular cambium.

secondary sex characteristics physical and physiological changes that occur at puberty, or seasonally in seasonal breeders.

secondary succession succession occurring in disturbed areas such as abandoned croplands.

secondary xylem in plants, xylem originating from the secondary or vascular cambium.

second law of thermodynamics the statement in physics that all processes occurring spontaneously within a closed system result in an increase of total entropy.

second messenger an intracellular chemical compound transferring a hormonal message from the cell membrane to the nucleus or cytoplasm.

seed the fertilized and ripened ovule of a seed plant, comprised of an embryo (miniature plant), including one, two, or more cotyledons, and usually a supply of food in a protective seed coat; capable of germinating under proper conditions and developing into a plant.

seed plant plant capable of forming seeds; gymnosperms and angiosperms.

segmentation also *metamerism,* the condition of being divided into segments, originally repetitions of nearly identical parts (as still seen in some annelids, chilopods, diplopods, and peripatus), but frequently followed in evolution by the specialization of different segments, as seen in most arthropods, some annelids, and vertebrates.

selectively permeable in cellular membranes, the characteristic of permitting selected substances to pass through while rejecting others.

semen a fluid containing sperm and released by male animals, usually involving ejaculation.

semicircular canals in vertebrates, a group of three loop-shaped tubular portions of the membranous labyrinth of the inner ear. Serves to sense balance, orientation, and movement.

seminal vesicle 1. in various invertebrates, a pouch in the male reproductive tract serving for the temporary storage of sperm. 2. in male mammals, paired outpocketings of the vas deferens producing much of the fluid substance of semen.

seminiferous tubule any of the coiled, thread-like tubules that make up the bulk of a testis and are lined with germinal epithelium from which sperm are produced.

semipermeable membrane a membrane that allows some substances to pass freely through it but that restricts or prevents the passage of other substances.

sensory neuron also *afferent neuron,* a neuron that conducts impulses carrying sensory information from a receptor to the brain or spinal cord.

sere a recognizable sequence of changes, occurring during community development or succession.

serial endosymbiosis hypothesis proposed evolution of the eukaryotic cell through the fusion of several prokaryotic lines, the sequence eventually giving rise to cells with combined phagocytic capabilities, chloroplasts, mitochondria, and flagella.

sex chromosome an X or Y chromosome, or any chromosome involved in the determination of sex.

sex-influenced trait a trait that can occur in either sex but is more common in one than the other; for example, breast cancer, ulcers, pyloric stenosis.

sex-limited trait a trait that affects members of one sex only; for example, prostate cancer, pattern baldness, endometriosis.

sexual intercourse also *coitus, copulation,* the entry of the erect penis into the vagina.

sexual recombination genetic recombination as the result of meiosis and biparental reproduction.

sickle-cell an abnormally crescent-shaped erythrocyte seen in the disease sickle-cell anemia.

sieve cell the more primitive food-conducting cells in vascular plants other than angiosperms. See also *sieve tube member.*

sieve element includes sieve tube members and sieve cells, the primary food-conducting structures of phloem.

sieve tube member in angiosperms, a thin-walled phloem cell having no nucleus at maturity and forming a continuous cytoplasm with other such cells to form sieve tubes, functioning in the transport of organic solutes, hormones, and mineral nutrients.

silent mutation a change in a DNA codon in which a synonymous codon forms (third letter change), having no effect on the amino acid sequence of the polypeptide.

sink as in "source" and "sink." In plants, a region where a manufactured product such as sucrose or starch is concentrated.

skeletal muscle also *voluntary muscle, striated muscle,* muscle attached to the skeleton or in some cases to the dermis, under direct control of the central nervous system, characterized by distinct striations at the subcellular level, with multinucleate unbranched muscle fibers. Compare *smooth muscle, cardiac muscle.*

slime mold a fungus-like protist possibly related to fungi, consisting of an ameboid, saprophytic plasmodium that on maturity coalesces into structured fruiting bodies. See *plasmodial slime molds.*

smooth muscle also *involuntary muscle,* the muscle tissue of the glands, viscera, iris, piloerection and other involuntary functions, consisting of masses of uninucleate, unstriated, spindle-shaped cells occurring usually in thin sheets.

sodium ion gate either of the two gates controlling sodium ion passage through an ion channel, including an activation gate and inactivation gate.

sodium/potassium ion exchange pump also *sodium pump,* a poorly understood molecular entity in the plasma membrane, capable of actively transporting sodium out of the cell and potassium in, at a cost of ATP energy.

solute something dissolved in solution.

somatic cells in the body, any cells other than germinal cells, especially of terminally differentiated tissues.

somatic mutation mutations that occur in somatic cells, cells that will not be passed on to the next generation.

somatic nervous system the voluntary nervous system, as distinguished from the visceral (autonomic) nervous system.

speciation 1. the division of a species into two or more species. 2. the process or processes whereby new species are formed.

species (pl., species) (variously defined), 1. a class of things or individuals having some common characteristics; a distinct type. 2. the major subdivision of a genus or subgenus, regarded as the basic category of biological classification and composed of related individuals that resemble one another through recent common ancestry and that share a single ecological niche. 3. in sexual organisms, a group whose members are the least potentially able to breed with one another but are unable to breed with members of any other group; a reproductively isolated group.

specific defense slow but highly specific cellular and chemical responses made by the B-cell and T-cell lymphocytes against foreign substances, cancerous cells, or invading organisms.

sperm 1. a male gamete. 2. a spermatozoan. 3. a spermatozoid. 4. (adj.) pertaining to a male gamete or male gamete function, as in sperm head, sperm nucleus, sperm cell.

spermatid one of the four haploid cells formed by meiosis of a spermatocyte, prior to maturation into a spermatozoan.

spermatocyte a cell giving rise to spermatozoa. See also *primary spermatocyte, secondary spermatocyte.*

spermatophore a packet containing sperm, produced by spiders and certain other invertebrates, and by certain salamanders.

spermatozoan (pl. *spermatozoa*), also *spermatozoon, sperm, sperm cell,* a motile male gamete of an animal, produced in great numbers in the male gonad, consisting of a sperm head containing the nucleus and an acrosome, a midpiece, and a single flagellum (sperm tail).

S phase (synthesis phase) one of the phases of the cell cycle, occurring in interphase between G_1 and G_2, the period of chromosome replication and DNA synthesis.

sphincter a ring of muscle surrounding a bodily opening or channel and able to close it off; for example, oral sphincter, anal sphincter.

spindle a cytoplasmic body present in mitosis and meiosis, in shape resembling the spindle of a primitive loom, and composed of tubulin microtubules and actin microfilaments; serving to segregate daughter chromosomes in anaphase.

spindle apparatus a functional unit consisting of the continuous and centromeric fibers of the spindle, the centromeres, the centrioles (or the analogous polar ground substance), and, when present, the asters.

spindle fiber one of the microtubule filaments constituting the mitotic spindle. *Centromeric spindle fiber,* one of the several microtubules attaching firmly to a centromere and extending to a spindle pole. *Polar fiber,* a microtubule extending from one pole, interlacing with polar spindle fibers of the other pole, and ending blindly somewhat short of the other spindle pole.

spirochaete spiral-shaped bacterium with a flexible cell wall and locomotion through the movement of axial filaments within the outer membrane.

spleen an abdominal organ consisting of lymphoid, reticular, and endothelial tissues with blood supply and red blood cells circulating freely in intercellular spaces; functions include the scavenging of debris and the maintenance of blood volume.

spliceosome the enzyme complex responsible for the removal of introns from mRNAs.

splicing the process of removing introns from mRNA.

spongy bone bone with a network of thin, hard walls surrounding open, nonbony pockets.

spore a minute unicellular reproductive or resistant body, specialized for dispersal, for surviving unfavorable environmental conditions, and for germinating to produce a new vegetative individual when conditions improve.

sporic cycle a life cycle in which generations alternate between a multicellular, diploid sporophyte (spore producer) and a multicellular haploid gametophyte (gametophyte producer), the phases sometimes occurring in separate individuals.

sporophyte in algae and plants having an alternation of generations, a diploid individual capable of producing haploid spores by meiosis; the prominent form of ferns and seed plants. Compare *gametophyte*.

stabilizing selection natural selection that tends to maintain the status quo, selection for the average and against extremes.

stamen the male reproductive structure of the flower, consisting of a pollen-bearing anther and the filament on which it is borne.

stele the cylindrical central portion of the axis of a vascular plant, including pith, xylem, and phloem, surrounded by a pericycle.

stem cell also hemocytoblast, generalized cell of red bone marrow from which all blood cells form.

steroid any of a class of lipid-soluble compounds on four interlocking saturated hydrocarbon rings; included are all sterols (alcoholic steroids), for example, cholesterol, estradiol, testosterone, cortisol.

steroid hormone a class of hormones consisting of the steroid molecule with various side group substitutions, believed to freely pass across the cell membrane and, once bound by a specific carrier protein, interact directly with the chromatin in gene control; included are the vertebrate sex hormones.

stigma the top, slightly enlarged, and often sticky end of the style, on which pollen adhere and germinate.

stimulus an aspect of the environment that influences the activity of a living organism or part of an organism, especially through a sense organ.

stoma (pl. *stomata*), pore in the epidermis of leaves, stems, and other plant organs, formed by guard cells, allowing the diffusion of gases into and out of intercellular spaces.

stomach a muscular dilation of the alimentary canal in vertebrates, between the esophagus and the duodenum, that functions in temporary storage, preliminary digestion, sterilization, and physical breakdown of ingested food.

stop codon also *termination codon,* in mRNA, one or more of the codons UAA, UAG, or UGA, signaling the end of polypeptide translation.

stroma the watery matrix surrounding the thylakoids of a chloroplast.

stromatolite a macroscopic geological structure of layered domes of deposited material attributed to the presence of shallow water photosynthetic prokaryotes.

structural protein protein that is incorporated into cellular or extracellular structures.

subspecies a more or less clearly defined, morphologically distinct, named geographic variety of a species; when named, the name forms a third part of the scientific name (genus, species, subspecies).

substitution 1. see *base substitution.* 2. *amino acid substitution,* a base substitution that changes the identity of one of the amino acids of a protein.

substrate 1. the base on which an organism lives. 2. a substance acted on by an enzyme. 3. a nutrient source or medium.

substrate-level phosphorylation the production in one or more steps of ATP from ADP, P_i, and an appropriate organic substrate; the capture of high-energy phosphate bonds directly from metabolic transformations. Compare *chemiosmotic phosphorylation.*

suppressor T-cell also T_s, specific subpopulation of T-cells whose role is to moderate, slow, and stop the specific immune responses.

surface–volume hypothesis the proposal that cells are restricted to a size that assures a surface–volume ratio that provides a sufficient membrane area to support the transport needed to maintain metabolic activity.

survivorship curve a graph with age on the X axis and zero to one on the Y-axis, the monotonic curve presenting the probability of surviving to age X for each X.

symbiosis 1. the living together in intimate association of two species. See *parasitism, commensalism, mutualism.*

sympatric speciation speciation within a population occupying a single habitat, without geographic isolation.

symplastic route in plants, the movement of water or other materials through the cytoplasm of adjoining cells (via plasmodesmata) as it makes its way to the stele. See also *apoplastic route.*

symport a membrane transporter that transports two or more substances simultaneously in the same direction. See also *antiport.*

synapse 1. (n.) the place at which nerve impulses pass from the axon or the synaptic knobs of an axon of one neuron to the dendrite or cell body of another. 2. (v.i.) of homologous chromosomes, to pair together, chromomere by chromomere, in zygotene of meiosis.

synapsis the pairing up and fusing of homologous chromosomes during zygotene of the first meiotic prophase, whereby preparations are made for crossing over.

synaptic cleft the minute space between the synaptic knob of one neuron and the dendrite or cell body of another, into which neurotransmitters are released in the transmission of nerve impulses between cells.

synaptonemal complex a complex structure composed of protein and RNA, the material of which is first formed between sister strands in leptotene of meiotic prophase and combines with a like structure to form the complex and accomplish the specific zipper-like pairing of homologous chromosomes in zygotene.

synecology ecological interactions of different species.

synonymous codon in the genetic code, two or more codons that specify the same amino acid.

synovial joint a freely movable joint surrounded by a fibrous capsule lined by a synovial membrane that secretes lubricating synovial fluid.

synthesist in science, one who attempts to deduce broad general principles from widely disparate observations, or who draws upon such observations to substantiate or refute such principles; one who reinterprets established relationships in the support of a new general idea or synthesis.

systole the period of heart contraction, particularly of the ventricles. Compare *diastole.*

systematics 1. the science of classifying organisms or groups of organisms on the basis of their evolutionary relatedness. 2. the science of determining the evolutionary relatedness of groups of organisms, particularly at higher taxonomic levels.

T$_4$ cell also *CD4 cell, helper T-cell,* a class of lymphocytes responsible for arousing cytotoxic T-cells and B-cells and prompting them to divide.

T$_8$ cell also *CD8 cell, cytotoxic T-cell,* a class of lymphocytes responsible for identifying, binding to, and destroying cancerous and virus-infected cells.

taiga moist, subarctic forest biome of Europe and North America dominated by spruces and firs.

tap root also *taproot;* 1. a root having a prominent central portion giving off smaller lateral roots in succession. 2. the central root of such a system, especially when it grows vertically and deep.

tap root system a root system consisting of a large primary root and its secondary and lateral branches. Compare *fibrous root system.*

target cell a cell acted on by a specific hormone, generally containing or bearing specific hormone receptor proteins not found in other cells.

taxon 1. any taxonomic category, such as species, genus, subspecies, or phylum. 2. any named group of related organisms.

taxonomist an individual skilled in identifying, describing, and classifying organisms, usually specializing in a particular group.

taxonomy the science dealing with the identification, naming, and classification of organisms.

T-cell a lymphocyte of a variety that matures in the thymus and interacts with invading cells and other cells of the immune system.

tendon a tough dense cord of fibrous connective tissue that is attached at one end to a muscle and at the other end to the skeleton, and transmits the force exerted by the muscle in moving the skeleton.

terminal bud the dormant bud at the stem tip, representing the next season's potential growth.

termination in transcription, a point when the ribosome reaches a stop codon, whereupon the polypeptide is released and the ribosomal subunits separate.

termination signal a DNA sequence that signals the end of a transcription sequence, dislodging RNA polymerase.

terminator codon also *termination codon,* stop codon, *nonsense codon,* one of the codons UAA, UAG, or UGA in mRNA, or the corresponding codons in DNA, signaling the end of polypeptide transcription.

territorial behavior also *territoriality,* defense of any area by animals.

territory the area defended by a territorial animal.

tertiary consumer a consumer at the third level. See *consumer.*

testcross 1. a cross of an individual of unknown genotype with a homozygous recessive to determine heterozygosity by progeny testing. 2. the cross of a known double heterozygote with a double homozygote to determine linkage relationships.

tetrapod also land vertebrate, any vertebrate of the classes Amphibia, Reptilia, Mammalia, and Aves, including some that don't have four feet (for example, birds, people, whales, dugongs).

T-even phage one general type of *Escherichia coli* bacteriophage, with a double-stranded DNA genome and a complex body consisting of a head, tailpiece, end plate, and attachment fibers; arbitrarily designated T_2, T_4, T_6, and so on.

thalamus a large subdivision of the diencephalon, consisting of a mass of nuclei in each lateral wall of the centrally located third ventricle of the brain.

thallus a plant body of a multicellular alga that does not grow from an apical meristem, shows no differentiation into distinct tissues, and lacks stems, leaves, and roots.

theory a proposed explanation whose status is still conjectural, in contrast to well-established propositions that are often regarded as facts. Theory and hypothesis are terms often used colloquially to mean an untested idea or opinion. A theory properly is a more or less verified explanation accounting for a body of known facts or phenomena, whereas a hypothesis is a conjecture put forth as a possible explanation of a specific phenomenon or relationship that serves as a basis for argument or experimentation.

thermal overturn in temperate zone lakes, the displacement of bottom waters by sinking colder, denser surface waters; generally occurring in spring and autumn.

thermiogenesis in endothermic animals, the production of body heat through an increase in the metabolic rate, that is, the release of energy from fuels.

thermodynamics 1. the branch of physics that deals with the interconversions of energy as heat, potential energy, kinetic energy, radiant energy, entropy, and work. 2. the processes and phenomena of energy interconversions.

thermoregulation. 1. an animal's control over its internal temperature. 2. the physiological mechanisms that maintain a body at a particular temperature in an environment with a fluctuating temperature. *Behavioral*

thermoregulation, maintaining a body temperature within acceptable limits by behavioral means, as by basking and by seeking out appropriate microenvironments.

threshold voltage the minimum voltage change needed to start an action potential in an axon.

thrombin in the blood clotting reactions, a proteolytic enzyme that catalyzes the conversion of fibrinogen to fibrin by the removal of two short peptide segments and, in turn, is produced from prothrombin by the action of thromboplastin.

thromboplastin protein released from damaged platelets, which catalyzes the conversion of inactive prothrombin to thrombin in the initiation of blood clotting.

thylakoid a membranous structure of chloroplasts and cyanophytes consisting of a thylakoid membrane containing chlorophyll, accessory pigments, photocenters, and the photosynthetic electron transport chain; an inner lumen that becomes acidic during active photosynthesis; CF_1 particles attached to the outer sides of the thylakoid membrane, which are the sites of chemiosmotic phosphorylation.

thymus a glandular body above the lungs, involved in T-cell lymphocyte development.

thyroid also thyroid gland, a large endocrine gland in the lower neck region of all vertebrates, arising as a ventral median outpocketing of the pharynx and believed to be homologous with the endostyle of cephalochordates, lamprey larvae, and urochordates; secretes thyroxine and calcitonin.

tissue a group of associated cells, identical in structure and function.

total fertility rate the predicted general fertility rate, determined through attitudinal studies and other forecasting indicators.

totipotency during development, or throughout life, the retention of complete genetic and developmental potential in the nucleus of a cell.

toxin a poisonous substance produced by a biological organism.

trachea (pl. *tracheae*); 1. in land vertebrates, the main trunk of the air passage between the lungs and the larynx, usually stiffened with rings of cartilage. 2. one of the air-conveying chitinous tubules comprising the respiratory system of an insect.

tracheid a long tubular xylem cell or its lignified, empty cell wall, which functions in support and the conduction of water, distinguished from xylem vessels by having tapered, closed ends communicating with other tracheids through pits.

transcribed strand the DNA strand of a structural gene that is physically involved in transcription.

transcription the process of RNA synthesis as the RNA nucleotide sequence is directed by specific base pairing with the nucleotide sequence of the transcribed strand of a DNA cistron.

transduction the transfer of a host DNA fragment from one bacterium to another by a viral particle.

transfer RNA (tRNA) any of a class of relatively short RNAs consisting of three or four loops and a double-stranded stem. Specific varieties become covalently linked to specific amino acids, then function in transcription to recognize appropriate mRNA codons and to transfer the amino acid to the growing polypeptide chain.

transformation in a bacterium, the direct incorporation of a DNA fragment from its medium into its own chromosome.

translation polypeptide synthesis as it is directed by an mRNA on a ribosome; the transfer of linear information from a nucleotide sequence to an amino acid sequence according to the structures of the genetic code.

translocation 1. the step in protein synthesis in which a transfer RNA molecule that is covalently attached to the growing polypeptide chain is moved (translocated) from one ribosomal tRNA attachment site to the other. 2. a chromosome rearrangement in which a terminal segment of one chromosome replaces a terminal segment of another, nonhomologous chromosome. 3. the directed movement of materials from one part of a plant to another part, especially the directed movement of sugars through phloem.

transpiration 1. the evaporation of water vapor from leaves, especially through the stomata. 2. the physical effects of such evaporation taken together. *Transpiration pull,* the pulling of water up through the xylem of a plant utilizing the energy of evaporation and the tensile strength of water.

transposition in molecular genetics, the movement of DNA from one place in the genome to another.

transposons segments of DNA capable of autonomous movement through a genome.

triploblastic in most animals, the existence in the embryo of three primary germ layers—the ectoderm, mesoderm, and endoderm—from which all other tissues are formed.

trophic level a level in a food pyramid.

tropism a turning toward or away from a stimulus, usually accomplished by differential growth. See *gravitropism, phototropism.*

tropomyosin a low-molecular-weight filamentous protein that accompanies the globular protein actin in making up actin filaments.

troponin a protein of low molecular weight that binds calcium in muscle contraction, a specific variant of the ubiquitous calcium-binding molecule calmodulin.

true-breeding (adj.), of an organism or strain, when mated with individuals like itself producing offspring like itself; specifically, homozygous for all relevant loci.

tubal ligation sterilization of a female mammal by cutting and tying the oviducts.

tundra a biome, characterized by level or gently undulating treeless plains of the arctic and subarctic, supporting a dense growth of mosses and lichens and dwarf herbs and shrubs, seasonally covered with snow and underlain with permafrost.

turgor the normal state of turgidity and tension in living plant cells, created by osmosis. *Turgor pressure,* the actual hydrostatic pressure developed by the fluid in a turgid plant cell.

ultraviolet light also *ultraviolet radiation,* UV, electromagnetic radiation having a shorter wavelength than visible light and longer than x-rays.

uniport membrane transporter that transports a single substance. See *symport, coport, antiport.*

unsaturated capable of accepting hydrogens; *unsaturated fat,* a triglyceride with one or more carbon–carbon double bonds.

upwelling the coming to the surface of water from the depths of the ocean, which is associated with the introduction of mineral nutrients and a consequent richness in productivity and biomass.

ureter either of the pair of ducts that carry urine from the kidney to the bladder in mammals, or from the kidney to the cloaca in other vertebrates. Compare *urethra.*

urethra the canal that carries urine from the mammalian bladder to the exterior, and in males serves also for the transmission of semen.

uterine tube also *oviduct, Fallopian tube,* in mammals, a tube opening near the ovary, which receives released ova and in which fertilization occurs.

uterus 1. in female mammals, a muscular, vascularized, mucous-membrane-lined organ for containing and nourishing the developing young prior to birth and for expelling them at the time of birth. 2. an enlarged section of the oviduct of various vertebrates and invertebrates modified to serve as a place of development of the young or of eggs.

vagina 1. an expandable canal that leads from the cervix of the uterus of a female mammal to the external orifice (introitus), serving to receive the penis in copulation and as a birth canal in parturition. 2. any canal of similar function in other animals.

variable age of onset of a trait, not being observed at birth but becoming manifest at some variable time in later development.

variable expressivity of a trait, having different degrees in different individuals with the same or similar genotype.

variable region the portion of an antibody concerned with the binding of an antigen, which varies from antigen to antigen.

vascular bundle a unit of the vascular system of plants, a strand of xylem associated with a strand of phloem and usually a sheath of fibrous sclerenchyma, in the primary growth pattern and in leaves and petioles.

vascular cambium the cylinder of lateral meristem that produces xylem on its inner side and phloem on its outer side in secondary growth, contributing thus to growth in circumference.

vascular plant a plant with xylem and phloem; a tracheophyte.

vascular ray spoke-like layers of parenchyma tissue that extend through the wood, vascular cambium, and phloem.

vascular system 1. the circulatory system of an animal. 2. the xylem and phloem of a vascular plant.

vascular tissue plant tissues specialized for conducting water and foods; See also *xylem, phloem.*

vector 1. an organism that transmits a parasite, virus, bacterium, or other pathogen from one host to another. 2. in molecular genetics, a DNA molecule capable of accepting additional DNA to be carried and replicated.

vegetal hemisphere in eggs with polarity established, the less metabolically active, yolky region.

ventral nerve cord common feature of many invertebrate phyla, the main longitudinal nerve cord of the body, being solid, paired, and ventral, with a series of ganglionic masses.

ventricle a cavity or a body part or organ: 1. one of the large muscular chambers of the four-chambered heart. 2. one of the systems of communicating cavities of the brain, consisting of two lateral ventricles and a median third ventricle.

venule a small vein.

vertebral column also *spinal column,* the articulated series of vertebrae connected by ligaments and separated by intervertebral disks that in vertebrates forms the supporting axis of the body and of the tail in most forms.

villus (pl. *villi*), a small, slender, vascular, finger-like process: 1. one of the minute processes that cover and give a velvety appearance to the inner surface of the small intestine, that contain blood vessels and a lacteal, are covered with microvilli, and serve in the absorption of nutrients. 2. one of the branching processes of the surface of the implanted blastodermic vesicle of most mammals.

virion, also *virus particle,* an individual virus.

viroid infectous agents of plants consisting of naked RNA segments.

virulence 1. disease-producing capability. 2. the capacity of an infective organism to overcome host defenses.

virus a noncellular organism consisting of DNA or RNA enclosed in a protein coat, often together with a few enzymes; replicating only within a host cell, utilizing host ribosomes, enzymes, and energy.

vitamin any organic substance that is essential to the nutrition of an organism, usually by supplying part of a coenzyme.

viviparity the condition of producing live young.

viviparous (adj.), producing live young from within the body of the female, following development of the young within a uterus in intimate association with the tissues of the mother. Compare *ovoviviparous, oviparous.*

water mold an aquatic protist of the oomycetes.

water potential the potential energy of water to move, as a result of concentration, gravity, pressure, or solute content.

water vascular system a system of vessels in echinoderms, containing seawater, used in the movement of tentacles and tube feet and possibly in excretion and respiration.

Watson strand either of the two strands of DNA, the other being designated the Crick strand.

Wernicke's area a region of the cerebrum concerned with deciphering speech.

wild type 1. of a phenotype, normal or typical in appearance; not mutant, rare, or abnormal. 2. of an allele, normal and common in the population; not mutant; giving rise to a typical phenotype when homozygous.

xerophyte a plant adapted to dry areas.

X-linked also *sex-linked,* the condition, common to many genes of functions unrelated to sex, of being present on the X chromosome; thus males have only one copy rather than the normal diploid two, and recessive X-linked alleles are always expressed in males, thus radically affecting patterns of inheritance.

xylem one of the two complex tissues in the vascular system of vascular plants, consisting of the dead cell walls of vessels, tracheids, or both, often together with sclerenchyma and parenchyma cells, functioning chiefly in the conduction of water and in support.

xylem vessel water-conducting tissue in plants, consisting of vessel elements arranged end to end.

Y chromosome one of the two heteromorphic sex chromosomes of mammals, flies, and certain other insects such that XY individuals are male and XX individuals are female; in mammals carrying the primary male sex determinant for the H-Y antigen, a determinant for increased height, and a few other genes; in general, nearly devoid of genes not concerned with maleness. Compare *X chromosome.*

yeast a unicellular sac fungus, especially *Saccharomyces cerevisiae,* which is used in the production of bread, beer, wine, and raised doughnuts.

yolk sac 1. one of the extraembryonic membranes of a bird, reptile, or mammal, that in birds and reptiles grows over and encloses the yolk mass and in placental mammals encloses a fluid-filled space; the first site of blood cell and circulatory system formation. 2. an analogous structure in elasmobranchs and cephalochordates.

zero population growth (ZPG) a point in population dynamics where the birth rate equals the death rate.

zooplankton animal life drifting at or near the surface of the open sea.

Z scheme a graphic presentation of the oxidation–reduction reactions occurring in the light reaction and electron transport chain in photosynthesis.

zygospore a diploid fungal or algal spore formed by the union of two similar sexual cells, having a thickened and usually ornamented wall that serves as a resistant resting spore.

zygote a cell formed by the union of two gametes; a fertilized egg.

zygotic cycle a primitive life cycle with a dominating haploid state and one in which gametes form through mitosis. Upon fertilization, a brief diploid state is followed by a meiosis and a return to the haploid state (some protists, all fungi).

Acknowledgments

Chapter 1

1.1/George Richmond/By permission of the Darwin Museum Down House

1.2/© The Bettman Archive

1.3/© Jen and Des Bartlett/Bruce Coleman Inc.

Page 8 /(top)© Wayne Lynch/DRK Photo (center left) © Ed Robinson/Tom Stack & Associates (center right) © Joe McDonald/Visuals Unlimited (bottom left) © Ed Reschke (bottom right) © Joe McDonald/Tom Stack & Associates

1.6, A & B/© Jim Strawser/Grant Heilman Photography

1.7 (left)/© Zig Leszczynski/Animals Animals (right) A & M Shah/Animals Animals

1.8 (left)/© R. Fleming/Bruce Coleman Inc.
1.8 (right)/ © Brian Seed/TSW

1.9/© Fritz Polking/Peter Arnold, Inc.

1.10/© Bruce Watkins/Animals Animals

1.11/© Stephen J. Krasemann/DRK Photo

1.12/© Stpehen J. Kraseman/Drk Photo

1.13/© Dr. Frank M. Carpenter

Chapter 2

2.1/© Ph. Avouris, IBM

2.10/© Dr. Dennis Kunkel/Phototake

2.14 (left)/© William E. Ferguson Photography
2.14 (right) © Photo Researchers

2.16 (left)/© Ken W. Davis/Tom Stack & Associates
2.16 (right)/© Rod Planck/Tom Stack & Associates

2.17/©Dwight R.Kuhn

2.18/©Jeanette Thomas/Visuals Unlimited

2.22 (left)/©Patrice Ceise/Visuals Unlimited
2.22 (right)/ ©William Amos/Bruce Coleman Inc.

Chapter3

Page 48 /(left)© Don W. Fawcett/Visuals Unlimited (center) © Dr. Jeremy Burgess/Science Photo Library/Photo Researchers (right) © Biophoto Associates/Photo Researchers

3.5/Brown, R.M. and J.H.M. Wilson, 1977. In *International Cell Biology* 1976–1977, ed., B.R. Brinkley and K.R. Porter, pp. 267–283. ©1977 by the Rockefeller University Press

3.6/© Jane Burton/Bruce Coleman Inc.

3.12/© Cabisco/Visuals Unlimited

Chapter 4

4.1 (top left, top center, top center right, bottom left, bottom center)/© Ed Reschke
4.1 (top right)/© Clay Adams
4.1 (bottom center right)/Carolina Biological Supply Company/Phototake
4.1 (bottom right)/© Manfred Kage/Peter Arnold

Page 72 /(top)© Triarch/Visuals Unlimited (top center) © John Gerlach/Visuals Unlimited

Page 73 /(top) from *Living Images* by Gene Shih and Richard Kessel/Reprinted courtesy of Jones & Bartlett Publishers, Inc., Boston/Scott Foresman (center left) © Jeremy Burgess/Science Photo Library/Photo Researchers, Inc. (center right) University of California (bottom) © KP/Beck

4.5/From *Behold Man* © Lennart Nilsson, National Geographic Society, Inc.

4.6/© Stanley C. Holt/Biological Photo Service

4.7 (top left)/© Dr. Don W. Fawcett/Roy Jones/Photo Researchers
4.7 (top right)/© John Taylor/University of California, Berkeley
4.7 (center)/© Dwight R. Kuhn
4.7 (bottom right)/© K.R. Porter/Photo Researchers

4.8 (top left, top right)/© Biophoto Associates/Photo Researchers
4.8 (bottom)/© Dr. Eva Frei and Professor R.D. Preston

4.10 (top)/© M. Schliwa/Visuals Unlimited
4.10 (bottom)/© John J. Woloscwick/Department of Anatomy, University of Illinois at Chicago/Biological Photo Service

4.11/© P.R. Burton, University of Kansas/Biological Photo Service

4.12/Courtesy of W.W. Franke

4.13 (top)/© P.R. Burton, University of Kansas/Biological Photo Service
4.13 (bottom)/© L.E. Roth/Biological Photo Service

4.14 (left)/Dr. B.R. Brinkley and Donna Turner, Department of Cell Biology, Baylor College of Medicine
4.14 (right)/© Stanley Holt/Biological Photo Service

4.15 (top)/© K.G. Murti/Visuals Unlimited
4.15 (center)/© Dwight R. Kuhn
4.15 (bottom)/© William L. Dentler/Biological Photo Service

4.17 (bottom)/© Don Fawcett/Photo Researchers

4.18 (top left)/Courtesy D.W. Fawcett
4.18 (top right, bottom right)/© K. Tanaka
4.18 (center right)/© M. Bielinska

4.19/© Dr. John Taylor

4.20 (top)/© John Taylor/University of California, Berkeley
4.20 (bottom left)/© Birgit Satir, Department of Anatomy, Albert Einstein College of Medicine
4.20 (bottom right)/© Dr. Don W. Fawcett, Roy Jones/Photo Researchers

4.22/© Frederick, S.E., and E.H. Newcomb, 1969. *Cell Biology* 43: 343

4.23/© G.F. Leedale/Science Source/Photo Researchers

4.24 (top left)/Runk/Schoenberger/Grant Heilman Photography
4.24 (top right)/Biophoto Associates/Science Source/Photo Researchers

4.25/© K.R. Porter/Photo Researchers

Chapter 5

5.1 (left)/© J.F. Hoffman, Yale University

5.1 (right)/Photo Researchers

5.3/© R.B. Park

5.4/© R.B. Park

5.5/© Susumu Ito, Harvard Medical School

5.6/© Biology Media/Science Source/Photo Researchers

5.14/(bottom right) © Kevin Collins/Visuals Unlimited

5.14/ (right center) © Runk/Schoenberger/ Grant Heilman Photography

5.20 (left)/© D.W. Fawcett/Science Source/ Photo Researchers

5.20 (center)/© D.W. Fawcett/Science Source/Photo Resources

5.20 (right)/Dr. Birgit H. Satir, Albert Einstein College of Medicine, NY

5.21/M.M. Perry and A.B. Gilbert, Agricultural Research Council's Poultry Research Center, Edinburgh

5.23 (center)/Courtesy of R.S. Decker

5.24 (left)/Reproduced from D.S. Friend and N.B Gilula (1972)/*J. Cell Biology* 53:758–776 by copyright permission of the Rockefeller University Press, New York

5.24 (right)/Courtesy of R.S. Decker

Chapter 6

6.1/© Michael Fogden/DRK Photo

6.2/© NASA

6.3/© Adrian Davies/Bruce Coleman Inc.

6.5/Archive Photos

6.8/Regents of University of California

6.12/© Stephen Trimble/DRK Photo

Chapter 7

7.1/© Adrian P. Davies/Bruce Coleman Inc.

7.6/Courtesy of William E. Barstow

7.17 (left)/© Wayne A. Bladholm

7.17 (right)/© Alan Pitcairn/Grant Heilman Photography

7.18/Courtesy of Raymond Chollet/ University of Nebraska

Chapter 8

8.10/© R. Bhatnagar/Visuals Unlimited

Chapter 9

9.1/© Lester Bergman & Associates

9.2 (left)/© Howard Sochurek/Medical Images Inc.

9.2 (right)/© Barabara Hamkalo

9.4/© Carolina Biological Supply/Phototake

9.5 (top)/© Conley Reider/Biological Photo Service

9.5 (bottom left & right)/© J. Pickett Heaps/ Science Source/Photo Researchers

9.6/Kessal, R.G., and C.Y. Shih. 1974. *Scanning Electron Microscopy in Biology: A Student's Atlas on Biological Organization.* New York, Heidelberg, Berlin: Springer Verlag © 1974

Chapter 10

10.1/© Barry Runk/Grant Heilman Photography

10.3/© The Bettman Archive

Chapter 11

11.1/© Carl Purcell/Photo Researchers

11.3/© Dr. E.R. Degginger

11.4/© John D. Cunningham/Visuals Unlimited

11.5 (left)/© Dr. E.R. Degginger

11.5 (center, right)/© Runk Schoenberger from Grant Heilman Photography

11.6/© Dr. E.R. Degginger

11.7/© Runk/Schoenberger from Grant Heilman Photography

11.9/© Walter Chandoa

11.11/© Yoav Levy/Phototake

11.12/© Runk/Schoenberger from Grant Heilman Photography

11.16 (top)/Columbia University, Columbiana Collection

11.16 (bottom)/© G.I Bernard/Animals Animals

11.21/© Dr. Murray L. Barr

11.22/Courtesy Macmillan Science Co., Inc.

Chapter 12

12.1/© Bruce Iverson

12.2/© Alex Rakosy/Custom Medical Stock Photo

12.6/Watson and Crick in front of the DNA model. Photographer: A.C. Barrington Brown. In Watson, J.D. 1968 *The Double Helix*. New York: Atheneum, p. 215 © 1968 by J.D. Watson

12.13/National Institute of Health

Chapter 13

13.4/© Miller, O.L., Jr. and B.R. Beatty, 1969. *Science* 164: 955–957

13.5/Science Source/Photo Researchers

13.17/© Miller, O.L, Jr., B.A. Hamkalo, and C.A. Thomas, Jr. 1970. *Science* 169: 392–395

Chapter 14

14.10/© Eastrom J., and W. Beerman. 1962 *J. Cell Biology* 14: 374. Reprinted by copyright permission of the Rockefeller University Press

Chapter 15

15.1/© Tom Broker/Rainbow

15.3/© Bruce Iverson

Chapter 16

16.7/© Dr. Rob Ferl

16.8/© Dan McCoy/Rainbow

16.12/© John Gordon/Phototake

Chapter 17

17.10 (left)/UPI/Bettman Archive

17.10 (right)/© G.R. Roberts

Chapter 18

18.1/© The Bettman Archive

18.2/© Renee Lynn/Photo Researchers

18.4/© Hans Reinhard/Bruce Coleman Inc.

18.7 (left)/© Art Wolfe/Tony Stone Images
18.7 (right)/Bob Daemmrich/The Image Works

18.8/© James Carmichael Jr./NHPA

18.10/© Breck P. Kent/Animals Animals

18.11/© David C. Houston/Bruce Coleman Inc.

18.13/© Charlie Heidecker/Visuals Unlimited

Chapter 19

19.7/© John Gerlach/Visuals Unlimited

Chapter 20

20.1 (left)/© Kjell B. Sandved/Bruce Coleman Inc.
20.1 (top right)/© Robert P. Carr
20.1 (bottom right)/© Stephen J. Kraseman/DRK Photo

20.2 (top left)/© J. Van Wommer/Bruce Coleman Inc.
20.2 (top right)/© Bob and Clara Calhoun/Bruce Coleman Inc.
20.2 (center left)/© Russ Kinne/Photo Researchers
20.2 (center)/© Mark Carlson/Tom Stack & Associates
20.2 (center right)/© Leonard Lee Rue III/Bruce Coleman Inc.
20.2 (bottom left)/© Jen and Des Bartlett/Photo Researchers
20.2 (bottom center)/© Clem Haagner/Bruce Coleman Inc.
20.2 (bottom right)/© Warren Garst/Tom Stack & Associates

20.3 (left)/© Erwin and Peggy Bauer/Bruce Coleman Inc.
20.3 (right)/© Jen and Des Bartlett/Bruce Coleman Inc.

20.11/© Pat & Tom Leeson/Photo Researchers

20.13/© Grant Heilman Photography

20.14/© Jungle Larry's Safari Land, Inc.

20.15 (top)/© C.C. Lockwood/Bruce Coleman Inc.
20.15 (bottom)/© Robert P. Carr/Bruce Coleman Inc.

20.16/© Bruce S. Cushing/Visuals Unlimited

20.17/© Michael Habicht/Animals Animals

20.18 (bottom left)/© Tom McHugh/Photo Researchers
20.18 (bottom right)/© Peter Knowles/The Photo Library, Sydney

20.19/© William E. Ferguson

Chapter 21

21.4/© Sidney Fox/Visuals Unlimited

21.6/© Sidney Fox/Visuals Unlimited

Chapter 22

22.2/© Pasteur Institute, Paris

22.3 (left)/© Gene M. Milbrath, PhD
22.3 (right)/© Norm Thomas/Photo Researchers

22.4/© Harold W. Fisher

22.5/© Lee D. Simon/Photo Researchers

22.8/© John Colwell/Grant Heilman Photography

22.9/© Dr. Erskine Palmer/Alayne Harrison

22.11/© Robert Noonan

22.12/© Jerome Wexler/Photo Researchers

22.15 (left)/Centers for Disease Control, Atlanta
22.15 (center)/© Runk/Schoenberger/Grant Heilman Photography
22.15 (right)/© Ed Reschke

22.18/© Manfred Kage/Peter Arnold Inc.

22.19/Dr. G. deHaller, Courtesy of Dr. C.F. Robinow

22.20 (left)/© John J. Cardamone Jr., University of Pittsburgh School of Medicine, Microbiology Dept.
22.20 (right)/Courtesy Dr. R. Wyckoff/National Institutes of Health

22.22/Micrograph courtesy of M.L. Higgins and L. Daneo Moore

22.24/© Lester V. Bergman & Associates

22.25 (left)/© Cabisco/Visuals Unlimited
22.25 (center)/© Runk/Schoenberger/Grant Heilman Photography
22.25 (right)/© David M. Phillips/Visuals Unlimited

22.26/© Biological Photo Service

22.27 (left)/© Arthur M. Siegelman/Visuals Unlimited
22.27 (right)/© Leonard Lessin/Peter Arnold

22.28 (top left)/© Sinclair Stammers/Science Source/Photo Researchers
22.28 (top right)/© J.R. Waaland, University of Washington/Biological Photo Service
22.28 (center)/© Sherman Thomson/Visuals Unlimited

22.29/© Norma J. Lang

Chapter 23

23.1 (top left)/Granger Collection
23.1 (top right)/Granger Collection

23.2/© Runk/Schoenberger/Grant Heilman Photography

23.3 (left)/Courtesy of G.A. Antipa
23.3 (right)/© Runk/Schoenberger/Grant Heilman Photography

23.6/© Stephen T. Brentano, University of Iowa

23.7 (top left)/© M. Abbey/Visuals Unlimited
23.7 (top right)/© Manfred Kage/Peter Arnold
23.7 (bottom)/© Dr. Rudolf Rottger

23.9 (top left)/© M. Abbey/Photo Researchers
23.9 (center)/© Manfred Kage/Peter Arnold
23.9 (top right)/© Dr. E.R. Degginger

23.10/© Philip Sze/Visuals Unlimited

23.11 (left)/© Runk/Schoenberger/Grant Heilman Photography
23.11 (right)/© Ed Reschke

23.12 (top)/© Dr. E.R. Degginger
23.12 (bottom)/© John Shaw

23.13/© V. Duran/Visuals Unlimited

23.14/© R. Calentine/Visuals Unlimited

23.17 (left)/© Gordon Leedale/Biophoto Associates/Photo Researchers
23.17 (right)/© Eric V. Grave

23.18/© Robert H. Pelham/Bruce Coleman Inc.

23.19/© T.E. Adams/Visuals Unlimited

23.20/© Jan Hinsch/Science Photo Researchers

23.22/© John D. Cunningham/Visuals Unlimited

23.24/© Runk/Schoenberger/Grant Heilman Photography

23.25 (left)/© Bob Evans/Peter Arnold
23.25 (right)/© Marty Snyderman

23.27/© Doug Wechsler

23.28/© Biophoto Associates/Photo Researchers

23.29/© Kim Taylor/Bruce Coleman Ltd.

Chapter 24

24.1/© Grant Heilman Photography

24.2/© James W. Richardson/Visuals Unlimited

24.7 (top left)/© Charles Marden Fitch
24.7 (top right)/© Runk/Schoenberger/Grant Heilman Photography
24.7 (bottom left)/© Bruce Iverson
24.7 (bottom right)/© G. Shih–R. Kessel/Visuals Unlimited

24.9/ © Stanley Fleger/ Visuals Unlimited

24.10/© Larry West/Valenti Photo

24.11/© Eric V. Grave/Phototake

24.12/© N. Allin and G.L. Barron/Courtesy NSERC Cananda, Can. J. Bot. 57:187–193

24.13/© A Ahmadjian/Visuals Unlimited

24.14 (top)/© Dr. James W. Richardson
24.14 (bottom left, right)/© James L. Castner

24.15 (top left)/S. Rannels/Grant Heilman Photography
24.15 (top right)/© John Eastcott/DRK Photo
24.15 (bottom)/© Kerry T. Givens/Tom Stack & Associates

24.16/From C.Y. Shih and R.G Kessel, *Living Images*, Science Books International, Boston,1982

Page 487/ From *Discover the Invisible*, Eric V. Grave, 1984, Prentice-Hall, Inc. Used with permission

24.17/© Stanley Flegler/ Visuals Unlimited

Chapter 25

25.1/© Larry Lefever/Grant Heilman Photography

25.3/© Dr. Linda E. Graham, University of Wisconsin

25.6 (left)/© Ed Reschke/Peter Arnold
25.6 (right)/© John Cunningham/Visuals Unlimited

25.8/© Runk/Schoenberger/Grant Heilman Photography

25.10/© William E. Ferguson

25.11 (top left)/© R.F. Evert
25.11 (bottom)/© William E. Ferguson

25.12/© William E. Ferguson

25.18/© The Peabody Museum of Archaeology and Ethnology, Yale University

25.19/© Dough Wechsler

25.20/© Runk/Schoenberger/Grant Heilman Photography

25.22/© William E. Ferguson

25.23 (top left)/© Ed Cooper
25.23 (top right)/© Rod Planck/Tom Stack & Associates
25.23 (bottom right)/© John D. Cunningham/Visuals Unlimited

25.25 (left)/© Runk/Schoenberger/Grant Heilman Photography
25.25 (right)/From: C.Y. Shih and R.G. Kessel, *Living Images*, Science Books International, Boston, 1982

25.26/© Dr. E.R. Degginger

25.27 (top left,bottom)/From: C.Y. Shih and R.G. Kessel, *Living Images*, Science Books International, Boston, 1982
25.27 (top right)/© Ray F. Evert/University of Wisconsin

25.28 (top left, bottom)/© Brian Parker/Tom Stack & Associates
25.28 (top right)/© William E. Ferguson

25.29/© William E. Ferguson

25.30 (left)/© Walter Chandoha
25.30 (right)/© Gustav Verderber/Visuals Unlimited

25.32 (top left)/© William H. Allen
25.32 (top right)/© Steve Salum/Bruce Coleman Inc.
25.32 (bottom left)/© Charlton Photos

Chapter 26

26.1/© Christopher Newbert/Bruce Coleman Inc.

26.9/© Kim Taylor/Bruce Coleman Ltd.

26.10/© Kim Taylor/Bruce Coleman Ltd.

26.12 (right)/© Neil McDaniell/Photo Researchers

26.13/© Runk/Schoenberger/Grant Heilman Photography

26.18/© Larry Jensen/Visuals Unlimited

26.19/© R. Knauft/Biology Media/Photo Researchers

26.22/© Frieder Sauer/Bruce Coleman Lt.

26.23 (top)/©James H. Carmichael/Photo Researchers
26.23 (bottom)/© Tom Mchugh/Steinhart Aquarium/Photo Researchers

26.26/© E.S. Ross

26.28/© Zig Leszczynski/Animals Animals

26.31/© Scott Camazine/Photo Researchers

26.33/© John Cancalosi/Peter Arnold

26.35 (left)/© Alex Kerstiten/Sea of Cortez Enterprises
26.35 (center, right)/© James L. Castner

26.36 (left)/© Robert Calentine/Visuals Unlimited
26.36 (right)/© Manfred Kage/Peter Arnold, Inc.

26.40/© Dwight R. Kuhn

26.41 (top)/© Fred Bavendon/Peter Arnold, Inc.
26.41 (center)/© William E. Ferguson
26.41 (bottom)/Richard K. LaVal/Animals Animals

26.42 (top)/© Robert Dunne/Photo Researchers
26.42 (bottom)/© Al Grotell

26.43/© Al Grotell

Chapter 27

27.3/© Mike Newman/Photo Researchers

27.4/© Heather Angel/Biophotos

27.7 (top left)/© Dr. Giuseppe Mazza
27.7 (top right)/© Heather Angel/Biophotos
27.7 (bottom)/© Steinhart Aquarium/Photo Researchers

27.9 (left)/© Tom McHugh/Photo Researchers
27.9 (right)/© Wolf H. Fahrenbach

27.10 (left)/© Jeffrey L. Rotman
27.10 (right)/© Peter David/Planet Earth Pictures/Seaphot

27.15/© Joe McDonald/Bruce Coleman Inc.

27.17/© Palaontaologisches Museum

27.18/Courtesy Dr. Carl Welty

27.19 (left)/© William E. Ferguson
27.19 (right)/© Dwight R. Kuhn

27.21 (top left)/© The Photographic Library of Australia Pty.Ltd.
27.21 (top right)/© Stephen J. K. Kraseman/Peter Arnold, Inc.
27.21 (bottom)/© Andrew Rakoczy/Bruce Coleman Inc.

27.24/© Bruce Coleman Ltd.

27.25/© Jen & Des Bartlett/Bruce Coleman Inc.

27.26/© Carolina Biological supply/Phototake

Page 594 /(top left) © Barry Runk/Grant Heilman Photography
(top right) © John Gerlach/Visuals Unlimited
(center left) © Derek Fell
(center) © Bruce Coleman Inc.
(center right) © Runk/Schoenberger/Grant Heilman Photography
(bottom left) © Alfred B. Thomas/Earth Scenes
(bottom right) © W.H. Hodge/Peter Arnold, Inc.

27.29/© FPG International

Chapter 28

28.2/© Doug Wechsler

28.3/© Dr. Thomas Eisner

28.4 (left)/© Eric Chrichton/Bruce Coleman Ltd.

28.4 (right)/© Lester Bergman

28.5 (left)/© William Ferguson
28.5 (right)/© Rod Planck

Page 602 /(left) © G.R. Roberts
(right) © G.R. Roberts

Page 603 /(top left) © Robert P. Carr/Bruce Coleman Inc.
(left center) © G.R. Roberts
(center) © 1974 John Ebeling
(right center) © G.R. Roberts
(bottom left) © John Shaw
(bottom right) © John N.A. Lott/Biological Photo Service

28.11 (top left)/© Lester Bergman & Associates
28.11 (top right, bottom)/© Walter Chandoha

Chapter 29

29.2/© Francois Gohier/Photo Researchers

29.4/© Ray F. Evert

29.5 (top)/From: C.Y. Shih and R.G. Kessel, *Living Images*, Science Books International, Boston, 1982
29.5 (bottom left)/© SUNY College of Environmental Science & Forestry
29.5 (bottom right)/© Dr. Jeremy Burgess/ Science Photo Library/Photo Resources

29.7/© Robert and Linda Mitchell

29.9/© Ed Rescke

29.10/© Omikron/Science Source/Photo Resources

29.11/© John D. Cunningham/Visuals Unlimited

29.12/© J. Robert Waaland/Biological Photo Service

29.13/© Bruce Iverson

29.15 (top)/© Rod Planck/Photo Researchers

29.16/© Ray F. Evert

29.17 (left)/© Dwight R. Kuhn
29.17 (center)/© Lott/Biological Photo Service

29.20 (top)/© Runk/Schoenberger/Ray F. Evert

Chapter 30

30.8 (top)/© Runk/Schoenberger/Grant Heilman Photography

30.8 (bottom)/© E. Zeiger and N. Burstein, Stanford University/Biological Photo Service

30.13/© Robert and Linda Mitchell

30.15/© S.A. Wilde

30.16/© Gerald Sanders

30.17 (center)/© Dwight R. Kuhn
30.17 (right)/© Runk/Schoenberger/Grant Heilman Photography

Chapter 31

31.1/© E.S. Ross

31.4/© Breck P. Kent/Earth Scenes

31.5/© Runk/Schoenberger/Grant Heilman Photography

31.9/© Biophoto Associates/Photo Researchers

31.10/© Lynn M. Stone

31.11 (bottom)/© Rod Planck

31.12 (top left)/© M. Evans, Ohio State University
31.12 (top right, bottom)/© R. Moore

31.13/© R. Calentine

31.14/© Lillian N. Bolstad/Peter Arnold, Inc.

31.15/© William Ferguson

31.16/© Runk/Schoenberger/Grant Heilman Photography

31.17 (left, center)/© Eric Crichton/Bruce Coleman Ltd.
31.17 (right)/© Wayne A Bladholm

Chapter 32

32.2/© Dr. Giuseppe Mazza

32.7/© Ed Reschke

32.8 (left)/From: *Tissues and Organs: A Text-Atlas of Scanning Electron Microscopy*, by Richard G. Kessel and Randy H. Kardon, © 1979, W.H. Freeman and Company
32.8 (right)/© Hans Pfletschinger/Peter Arnold, Inc.

32.9 (top)/© Dr. Don W. Fawcett/Visuals Unlimited
32.9 (bottom)/© Science Photo Library/ Custom Medical Stock Photo

32.13/© Carolina Biological Supply/Phototake

32.15 (top)/© Biophoto Associates/Photo Researchers
32.15 (center)/From: *Tissues and Organs: A Text-Atlas of Scanning Electron Microscopy*, by Richard G. Kessel and Randy H. Kardon, © 1979, W.H. Freeman and Company
32.15 (bottom)/© Franzini Armstrong/Photo Researchers

32.16/© Biophoto Associates

32.17/© Omikron/Science Source/Photo Researchers

Chapter 33

33.1/© Irwin and Peggy Bauer/Bruce Coleman Inc.

33.3/© Prof. Dr. S. Cedric Ruine

33.4/From: *Tissues and Organs: A Text-Atlas of Scanning Electron Microscopy*, by Richard G. Kessel and Randy H. Kardon, © 1979, W.H. Freeman and Company

33.11 (left)/© J. Heuser/D.W. Fawcett/Visuals Unlimited
33.11 (right)/© E.R. Lewis, T.E. Everhart and Y.Y. Zeeri/Visuals Unlimited

Chapter 34

34.2/© Zig Leszczynski/Animals Animals
37.14/From: *Tissues and Organs: A Text-Atlas of Scanning Electron Microscopy*, by Richard G. Kessel and Randy H. Kardon, © 1979, W.H. Freeman and Company

34.6/© Martin M. Rotker

34.15/Scientific Catalog, © J. Allan/ Courtesy DREAMSTAGE Hobson and Hoffman-LaRoche Inc.

34.17/© Tom McHugh/Photo Researchers

34.20/© Larry West/Bruce Coleman Inc.

34.21/© Michael P. Gadomski/Bruce Coleman Inc.

34.23/© STU/Visuals Unlimited

34.25/© 1992 Science Photo Library/Custom Medical Stock Photo

34.26/© 1992 Scienc Photo Library/Custom Medical Stock Photo

Chapter 35

35.1/© Bruce Coleman Inc.

35.3/© Eric & David Hosking/Photo Researchers

35.8/© Barry Runk/Grant Heilman Photography

35.9/© Dan Guravion/Photo Researchers

35.11/© Doug Allan/Oxford Scientific Films/Animals Animals

35.12/© Souricat/Animals Animals

Chapter 36

36.1/© Ronald Thomason/Bruce Coleman Inc.

36.7/© Bob Cranston

36.12 (left)/© The Bettman Archive
36.12 (right)/© Wide World Photos

36.13/© Mac E. Hadley, University of Arizona

36.15 (left)/N. Tully/Sygma

36.15 (right)/© AFIP/Science Source/Photo Researchers

36.18/© Ed Reschke

36.21/© Robert and Linda Mitchell

Chapter 37

37.1/© Leonard Lee Rue III/Photo Researchers

37.5/© Jeff Rotman/Peter Arnold, Inc.

37.6 (left)/© Fritz Polking GDT/Peter Arnold, Inc.
37.6 (right)/© Nathan W. Cohen

37.7 (top left)/© Leonard Lee Rue III/Photo Researchers
37.7 (top right)/© Bruce Coleman Inc.

37.7 (bottom left)/© J. Dunning/Academy of Natural Sciences, Philadelphia/VIREO

Chapter 38

38.1/© Carolina Biological Supply/Phototake

38.3/© CNRI/Science Photo Library/Photo Researchers

38.4/© Lennart Nilsson

38.6 (right, center)/© Carolina Biological Supply/Phototake
38.6 (left)/© Biophoto Associates/Photo Researchers

38.8/From:*Behold Man.* © Lennart Nilsson, National Geographic Society

38.10/© Blair Seitz/Photo Researchers

38.13/© Don W. Fawcett/Photo Researchers

Chapter 39

39.2 (top)/© Al Grotell

39.2 (center left)/© Nancy Sefton
39.2 (center right)/© Ed Reschke
39.2 (bottom)/© Al Grotell

39.4/© Bruce Iverson

39.6 (left)/© G.I. Bernard/Animals Animals
39.6 (right)/© William H. Beatty/Visuals Unlimited

39.8/© Prof. Dr. rer. nat. Dr. med. Han-Rainer Dincker

39.10/© B.F. King, University of California/Biological Photo Service

39.16/© Centre National de Recherches Iconographiques

Chapter 40

40.4/© Dr. Don W. Fawcett/Hekko Chemes/Photo Researchers

40.5/© Lennart Nilsson, *The Body Victorious*

40.6/© John D. Cunningham/Visuals Unlimited

40.8/Zucker-Franklin, D., M.F. Greaves, C.E. Grossi and A.M. Marmont.1981. *Atlas of Blood Cells:Function and Pathology*, Vol. 2. Philadelphia: Lea & Febiger.© 1981 Edi. Ermes s.r.l.—Milan, Italy

40.18/© Morton H. Nielsen and Ole Werdelin/University of Copenhagen

40.19/© Dr. Gilla Kaplan

40.26/© Dr. Warner C. Greene, University of California, San Francisco

Chapter 41

41.1 (left)/© Dwight R. Kuhn
41.1 (right)/© Leonard Lee Rue III/Photo Resources

41.2 (left)/© Biophoto Associates/Photo Researchers
41.2 (right)/© C. Allan Morgan/Peter Arnold, Inc.

41.3/© Dwight R. Kuhn

41.4/© Susan Ernst, Department of Biology, Tufts University

41.5 (left)/© Heather Angel/Biofotos
41.5 (right)/© Nancy Sefton/Photo Researchers

41.7/Tom McHugh/Photo Researchers

41.9/© Dwight R. Kuhn

41.10/© Hilgert Ikan/Peter Arnold, Inc.

41.11 (left)/© Dwight R. Kuhn
41.11 (right)/© Visuals Unlimited

41.12/© Nathan W. Cohen

41.13/© Jow McDonald/Visuals Unlimited

41.14 (left)/© VIREO
41.14 (right)/© Dwight R. Kuhn

41.15/After Mangold and Tiedemann, from Balinsky, 1975 Sum. Fig. p. 980. Nilsson, L 1977. *A Child is Born*. New York: Dell Publishing Co., Inc.

41.16/© Jacana/Photo Researchers

41.17/© Ken M Higheill/Photo Researchers

41.21/© Dr. G. Schatten/Photo Researchers

Chapter 42

42.1/From: *Tissues and Organs: A Text-Atlas of Scanning Electron Microscopy*, by Richard G. Kessel and Randy H. Kardon, W.H. Freeman and Company, copyright © 1979

42.4/© Dr. G. Schatten/Science Photo Library/Photo Researchers

42.5/© Biology Photo Service

42.7/© Lennart Nilsson, *Being Born*

42.8 (top)/© Cabisco/Visuals Unlimited
42.8 (center)/© D.P. Wilson/FLPA

42.10/© Dr.Anthony P. Machowald, Case Western Reserve University

42.15/© Oxford Scientific Films/Animals Animals

42.22/ Courtesy of Einhard Schierenberg, Zoologisches Institut der Universität Köln

Page 926/After Mangold and Tiedemann, from Balinsky, 1975 Sum. Fig. p.980 Nilsson L 1977. *A Child is Born.* New York: Dell Publishing Co., Inc.

42.26/David Scharf/ Peter Arnold, Inc.

42.28 (top left, bottom)/Courtesy Professor Dennis Summerbell, National Institute for Medical Research, London, UK.
42.28 (top right)/© Katherine Tosney, University of Michigan

Chapter 43

43.1/© Sygma

43.2/© Robert Wallace

43.3/© Dr, Philip G. Zimbardo

43.4/© National Audubon Society/Photo Researchers

43.7/© Hans Reinhard/Bruce Coleman Inc.

43.8/© Zig Leszczynski/Animals Animals

43.9/© BBC Natural History Unit. From *The Discovery of Animal Behavior* by John Sparks, 1981. HarperCollins Publishers/ BBC Co-production

43.11/© Richard Wood/The Picture Cube

43.12/Thomas McAvoy, Life Magazine, © Time Warner Inc.

Chapter 44

44.3/© John Shaw/Tom Stack & Associates

44.4/© Neil Bromhall/Genesis Films Ltd./ Oxford Scientific Films/Animals Animals

44.5/© Patricia Caulfield/Animals Animals

44.9/© David Parker/Science Photo Library/ Photo Researchers

44.10/© Charles Walcott

44.11/© Lincoln Brower

44.13/© George Harrison

44.16/© Gordon Wiltsie/Bruce Coleman Inc.

44.17/© Charles G. Summers, Jr./Charles G. and Rita Summers

44.19 (left)/© Stephen J. Kraseman/DRK Photo

44.19 (right)/© Stephen Dalton/NHPA

44.20/© Grant Heilman Photography

Chapter 45

45.3/© Runk/Schoenberger/Grant Heilman Photography

45.5 (left)/© Len. Rue, Jr.
45.5 (right)/© Konrad Wothe/Oxford Scientific Films/Animals Animals

45.7 (left)/© William H. Mullins/Photo Researchers
45.7 (right)/© E.R. Degginger/Bruce Coleman Inc.

45.10 (left)/© Larry Lipsky/DRK Photo
45.10 (right)/© Noah Satat/Animals Animals

45.24/© Dr. E.R. Degginger

45.25 (left)/© E.S. Ross
45.25 (right)/© Tom McHugh/Photo Researchers

45.26/© Ted Levin/Earth Scenes

Chapter 46

46.1 (top)/© Arthur C. Smith III/Grant Heilman Photography
46.1 (bottom)/© Breck P. Kent/Earth Scenes

46.2 (left)/© Zig Leszczynski/Earth Scenes
46.2 (right)/© Jeff Greenberg/Photo Researchers

46.5/© John Shaw/Bruce Coleman Inc.

46.9/© Robert Wallace

46.10/© John Ebeling

46.16/© U.S. News & World Report

Chapter 47

47.9/© Rod Planck

47.10 (left)/© Sullivan and Rogers/Bruce Coleman Inc.
47.10 (right)/© Nancy Adams

47.11/© David C. Fritts

47.12/© David C. Fritts

47.13/© David C. Fritts

47.14 (left)/© David C. Fritts
47.14 (right)/© Steven McCutcheon

47.15/© David C. Fritts/Animals Animals

47.16 (top left, top center)/© Loren McIntyre/Woodfin Camp & Associates
47.16 (top right)/© Jany Sauvanet/Photo Resources
47.16 (center)/©Michael DeMacker/Visuals Unlimited
47.16 (bottom)/© Ken Schafer and Martha Hill/Ton Stack & Associates

47.17/© Tom McHugh/Photo Researchers

47.18 (top left)/© William Ferguson
47.18 (top right, bottom)/© Rod Planck

47.19/© G.R. Roberts

47.20 (left)/© Charles G. and Rita Summers
47.20 (right)/© Dr. E.R. Degginger

47.21/© David R. Frazier

47.22/© Doug Sokell/Tom Stack & Associates

47.25 (left)/© J.M. Bassot/H.Chaumeton/ Nature
47.25 (right)/© Peter David/Planet Earth Pictures-Seaphot

47.27/© William E. Ferguson

Chapter 48

48.1/© Sergio Dorantes/Gamma Liaison

48.3/© Thomas S. England/Photo Researchers

definition of, G–6
   oxidized, 142*f*
   reduced, 142*f*
coenzyme Q, 183
coevolution, 395, 397*f*
   definition of, G–6
cofactors, 135
cohesion, 36, 637
   definition of, G–6
cohort, 996
colchicine, 392*f*
Coleochaete, 492, 492*f*
coleoptile, 604, 651
   definition of, G–6
coleorhiza, 604
   definition of, G–6
colinearity, 281, 281*f*
   definition of, G–6
collagen, 57, 61, 63*f*, 675, 678*f*
   definition of, G–6
collar cells, 789, 789*f*
collecting ducts, reabsorption in, 759
collenchyma, 615
   definition of, G–6
colon, 801
   definition of, G–6
colony, 445
   definition of, G–6
colorblindness
   inheritance of, 251–252, 252*f*
   test for, 252, 252*f*
colostrum, 929
columnar epithelium, 675
   definition of, G–6
commensalism, 975, 975*f*
common descent, definition of, G–6
communication
   chemical, 969–970
   social basis of, 967–970
   sound, 968–969, 970*f*
   tactile, 970, 971*f*
   visual, 968
community(ies), 981
   attributes of, 1007–1010
   closed, 1007, 1007*f*
   definition of, G–6
   ecological succession in, 1014–1017
   mature, 1016–1017
   open, 1007, 1007*f*
   structure of
      competition and niche in, 1010–1011
      disease and parasitism in, 1012–1013
      disturbance in, 1014
      predation in, 1013–1014

territoriality in, 1011–1012
community development, definition of, G–6
community-level processes, 1010–1014
compact bone, 677
   definition of, G–6
companion cell(s), 616
   definition of, G–6
comparative anatomy, evolutionary proof
      through, 361–362
compass sense, 964
competition, 1010–1011
   definition of, G–6
   exploitative, 1010
   interference, 1010
competitive exclusion, principle of, 985–986
competitive inhibition, 136
   definition of, G–6
complement, 860
   action of, 860*f*
   activation of, 867
   definition of, G–6
complementary, definition of, 264
complementary DNA, 322, 322*f*
complete metamorphosis, definition of, G–6
compound, 22
   definition of, G–6
compound eye(s), 553, 555*f*, 732
   definition of, G–6
compound microscope, definition of, G–6
concentration gradient, 110
   definition of, G–6
conceptacles, 465
conception, control of, 903–907
conclusion, 12
conditioned response, definition of, G–6
conditioning
   classical, 951
   operant, 951, 951*f*
condoms, 905
   latex, in HIV prevention, 880
conduction, 746–747
cones, 734
conformation, 59
conidia, 479
conidiophores, 480, 480*f*
conifer, 513, 515
   definition of, G–6
Coniferophyta, 502, 504*f*
   adaptations for dryness, 516*f*
   reproduction in, 515–516
conjugated protein, definition of, G–6
conjugating molds. *See* Zygomycota

conjugation, 312–313, 314*f*, 433
   definition of, G–6
conjugation tube, 312
connective tissue, 675, 676*f*–677*f*
   definition of, G–6
connective tissue matrix, 675
Connell, Joseph, 1010–1011
conservation biology, 982
consolidation hypothesis, definition of, G–6
constant region, 866
constitutive, 293
constitutive genes, 293
constitutive heterochromatin, 301
consumer(s), 1018, 1025
   definition of, G–6
consummatory behavior, definition of, G–6
contact receptors, 726
continental drift, 388
   definition of, G–6
continental shelf, 1056
continuous variation, 244, 244*f*
contraception, 904–907, 907*f*
   barrier methods of, 904–905
   definition of, G–6
   hormonal methods of, 904
   surgical methods of, 907
   unaugmented methods of, 905
contractile vacuole(s), 93, 451
   definition of, G–6
contraction, 387–389, 684–686
   fast, 689
   graded, 688–689
   molecular interactions during, 686
   slow, 689
control, 11
   definition of, G–6
control enzyme, 137
controlled experiments, 11, 11*f*
convection, 746–747
convergent evolution, 395, 396*f*
   definition of, G–6
Cooksonia, 503*f*
cooperation, 973–974, 974*f*
coport(s), 115
   definition of, G–6
coport transport, 116*f*
copper, 22*t*
copulation, 887
Coriolis effect, 1036
   definition of, G–6
cork, 626
   definition of, G–6

cork cambium, 626

corn
development of, 604
seedling growth in, 611, 612f
variegated coloration in, 340–341, 343f

cornea, 734

corn plant, as experimental organism, 306

corolla, 593, 593f
definition of, G–6

corona radiata, 914
definition of, G–7

coronary arteries, 825

coronary thrombosis, 825

coronary veins, 825

corpora allata, 771

corpus allatum, definition of, G–7

corpus callosum, 718
definition of, G–7

corpus luteum, 902
definition of, G–7

cortex, 755
definition of, G–7
in roots, 617

cortical reaction, 913
definition of, G–7

cortical vesicles, 913

corticosteroids, 779

cortisol, 779
actions of, 784t

cotransport carrier, definition of, G–7

cotyledon(s)
definition of, G–7
seed development in, 601–602

countercurrent exchange, 757

countercurrent gas exchanger, 837, 839f

countercurrent heat exchanger(s), 743–744
definition of, G–7
in humans, 746f

covalent bond(s), 26–28, 27f
definition of, G–7
nonpolar, 27–28
polar, 27

Cowper's glands, 897

C3 pathway, 168f

C4 pathway
definition of, G–5
leaf cells and, 168f

C3 plants, 164
anatomy of, 165f
definition of, G–4

C4 plants, 164, 165f
anatomy of, 165f
and evolution, 168–169

cranium, 678

crassulacean acid metabolism (CAM), 642
definition of, G–7

crayfish, circulatory system in, 814f

creatine phosphate, 192

cretinism, 776

Creutzfeldt-Jakob disease, 425

Crick, Francis, 258, 263–266, 264f

cri-du-chat syndrome, deletion causing, 340, 340f

Crinoidea, 559

crista(e), 94, 183
definition of, G–7

crocodiles, heart of, 817–818

Crocodilia, 573

crop
in birds, 793
definition of, G–7
in earthworm, 790

crosscurrent exchange, 842
definition of, G–7

crossing over, 212–213f
definition of, G–7
frequency of, 246f
and genetic recombination, 245–247

crown gall, 326, 438, 438f

crude birth rate, 1064t
calculation of, 1064t

crude death rate, 1064t
calculation of, 1064t

Crustacea, 552

crustaceans, 553, 554f

crustose lichens, 483, 485f

cryogenic surgery, 748

cryptic coloration, definition of, G–7

Ctenophora, 537, 540f

C-terminal, 59, 285

cuboidal epithelium, 675

cuticle, 493
definition of, G–7

cutin, 80, 614
definition of, G–7

cuttings, 606

cyanobacteria, 438–439, 439f
definition of, G–7
structure of, 439, 440f

cycad, definition of, G–7

Cycadophyta, 502, 503f, 513–514

cyclic AMP, 296, 767f, 767–769
actions of, 767f
definition of, G–7

cyclic phosphorylation, definition of, G–7

cyclic photophosphorylation, 156, 161, 163f

cyclohexane, 31f

cyclosporine, 475

Cyclostomata, 568, 569f

cysteine, 58f

cytochromes, 143, 156, 183

cytokinesis, 208–209
in animal cells, 208, 208f
definition of, G–7
in plant cells, 208, 209f

cytokinin(s), 654–656
action of, 653t
definition of, G–7
and differentiation, 655f

cytoplasm, 69
definition of, G–7
in prokaryotic cell, 75

cytoplasmic streaming, 74, 81, 83f
definition of, G–7

cytosine, 65, 262

cytoskeleton, 81–83, 82f
definition of, G–7
function of, 97t

cytostome, 451

D

daltons, 23

Daphnia, 215–217

dark clock, and flowering, 663–664, 665f

Darnell, James E., Jr., 408

Dart, Raymond, 584

Darwin, Charles, 4–9, 359, 378
The Descent of Man, 15
The Expression of the Emotions in Man and Animals, 15
The Formation of Vegetable Mould Through the Action of Worms, 15
genetics of, 220–221
on gradualism, 396
historical context of, 350, 350f
impact of, 14–15
Insectivorous Plants, 651
on origin of life, 403
On the Origin of Species, 13–14, 352, 353
The Power of Movement in Plants, 651
snapdragon experiment, 221, 221f
theory of evolution, 350–353
The Voyage of the Beagle, 12

Darwin, Francis, 651–652

daughter cells, 88

day-neutral plants, 663

deamination, 194, 751, 751f

death, 129f

death rate, definition of, G–7

decarboxylating enzyme, 168

decarboxylation, definition of, G–7

deciduous plant, definition of, G–7

decomposer(s), 76, 433, 1018, 1025
  in cheese, 435, 435*f*
  definition of, G–7
  economically useful, 435
  inhibition of, 434–435

decomposition, by fungi, 474

dedifferentiation, 656

deductive reasoning, 10, 10*t*
  definition of, G–7

de Duve, Christian, 90

dehiscent, 602

dehydration reaction(s), 45–47
  definition of, G–7

deletion(s), 339–340
  definition of, G–7

Delwiche, C.C., 1027

demographic transition
  definition of, G–7
  theory of, 1067

demography, 995–998

denaturation, 61
  definition of, G–7

dendrite(s), 693
  definition of, G–7

dendrometer, 638

denitrification, definition of, G–7

density-dependent factors, 1002–1003
  definition of, G–7

density-independent factors, 1003, 1003*f*
  definition of, G–7

dentition, in mammals, 794, 794*f*

deoxyribonucleic acid (DNA), 62–65, 64*f*
  chemical modification of, 303
  chemical properties of, 263–264
  conversion to proteins, 277*f*
  damage to, at molecular level, 335–336
  definition of, G–8
  diagrammatic representation of, 267*f*
  in early cells, 409
  of eukaryotes, 273
  Feulgen staining of, 259, 259*f*
  5′ end of, 265
  and functions of nucleus, 88
  and genetic information, 272–274
  as genetic material, 258–275
    discovery of, 259–261
  major groove in, 265
  minor groove in, 265
  packaging of, 301–303
  primary lesions in, 336, 336*f*

repair of, 336–337
replication of, 265–272, 269*f*
  in circular DNA, 267, 272*f*
  in eukaryotes and prokaryotes, 268–272
  experiments on, 267
in reproduction, 201
versus RNA, 277, 278*f*
stability of, 335
structure of, 64–65, 261–265
synthesis of, 265–267, 268*f*
technology of, 319–325
three-dimensional structure of, 265, 266*f*
3′ end of, 265
Watson and Crick model of, 264–265, 265*f*
x-ray diffraction of, 263, 264*f*

deoxyribose, 64

derived, definition of, G–7

derived traits, 385

dermal brancheae, 833

dermal tissues, 614

dermis, definition of, G–7

descending colon, 801

The Descent of Man (Darwin), 15

desert(s), 1039–1042
  foragers in, day and night, 1042*f*
  plants of, 1039, 1041*f*

desmosomes, 122*f*, 123

desmotubules, 123

detection, by nervous system, 692

detergents, 36

determinate cleavage, definition of, G–7

determination, 910
  definition of, G–8
  and imaginal disks, 934

Deuteromycota, 476, 481–482

deuterostome(s), 558–559
  definition of, G–8
  development of, 531, 531*f*
  evolution of, 566*f*

developed nations, 1066

developing nations, 1066
  population growth in, 1066–1067

development, 910–941
  analysis of, 929–938
  definition of, G–8
  early events in, 915–920
  influence on phenotypes, 241–242
  molecular biology of, 935

dextrose, 45

diabetes mellitus, 781
  definition of, G–8

diapause, 770

diaphragm, 905
  definition of, G–8

diastole, 823
  definition of, G–8

diatom(s), 461*f*, 461–462
  reproduction in, 461–462, 462*f*

diatomaceous earth, 462

dicot, definition of, G–8

Dicotyledonae, 519, 520*f*

dictyosomes, 90

difference equation, 991

differentiation, definition of, G–8

diffuse roots, 620

diffuse root system, definition of, G–8

diffusion, 109–111, 110*f*–111*f*
  definition of, G–8
  facilitated, 113–114, 115*f*
  and free energy levels, 110, 111*f*
  of ions, 111
  in living cells, 111

digestion, 47, 788–806
  chemistry of, 801–805
  definition of, 788
  extracellular, 789
    definition of, G–10
  hormones of, 806*t*
  in hydra, 789*f*
  intracellular, 789
    definition of, G–15
  regulation of, 803–804
  in sponges, 789*f*

digestive enzymes, 801*t*

digestive system(s), 788–795
  of humans, 795–801, 796*f*
  of rotifer, 543, 544*f*
  sac-like, 789–790
  tube-within-a-tube, 790–791
  in vertebrates, 791–795

digestive tract, 542*f*
  complete, definition of, G–6

digestive vacuoles, 91

dihybrid cross, 228, 229*f*
  definition of, G–8

dihybrid testcross, 231*t*

dikaryotic, definition of, G–8

dikaryotic state, in fungi, 475

dilation and curettage (D & C), 906
  definition of, G–8

dinoflagellate(s), 458
  definition of, G–8

dinosaurs, 573–575
  extinction of, 577

dioecious, 495, 888
  definition of, G–8

Dionaea muscipula. See Venus flytrap

dipeptidase, 805

dipeptide, 57

diploblastic, definition of, G–8

diplococcus, 428

diploid, definition of, G–8, 210

direct exchange, 813, 813f

directionality, 988

directional selection, 371–372, 372f
  definition of, G–8

disaccharide(s), 45–47
  definition of, G–8

disease, 1012–1013

dispersion, 990

dispersion patterns, 990–991, 991f

display, 968–970
  graded, 968, 968f
  permanent versus temporary, 968, 969f

disruptive selection, 372, 372f
  definition of, G–8

dissociation, 28, 29f

distal, definition of, G–8

distal convoluted tubule, 755
  reabsorption in, 759

distance receptors, 726

distribution, 1009

disturbance, role in community structure, 1014, 1014f

disulfide linkage, 33, 58

divergent evolution, 394–395
  definition of, G–8

diversity, 1009–1010
  definition of, G–8

division, definition of, G–8

DNA. See deoxyribonucleic acid

DNA fingerprinting, 327, 329

DNA hybridization, definition of, G–8

DNA hybridization probe, 323

DNA–modifying agents, 336

DNA polymerase, 270–271

DNA replication, definition of, G–8

DNA sequencing, 324

Dolichotis patagonum, 5–6, 7f, 395

dome, 473

dominance, 222–227
  definition of, G–8
  incomplete (partial), 237–238, 238f

dominance hierarchy, definition of, G–8

dominance relationships, 237–241

dorsal, definition of, G–8

dorsal lip, 920
  definition of, G–8

dorsal root, 720

dorsal root ganglia, 720

dorsoventrally, definition of, G–8

double covalent bond, 26

double fertilization, 514, 520, 598
  definition of, G–8

double helix, 65
  definition of, G–8

doubling time, 1066t, 1066
  calculation of, 1064t
  definition of, G–8
  in selected nations, 1066t

douching, 905

Down's syndrome, 217

downy mildew, 457, 457f

drift, definition of, G–8

Drosera, 647, 648f

Drosophila melanogaster, 247, 247f
  antennapedia gene in, 301
  chromosome map for, 248f
  as experimental organism, 306
  eye color in, 248–249, 249f
  replication in, 271
  salivary chromosomes, decondensation of chromatin in puffs of, 302, 302f
  sex chromosomes in, 249–250, 250f
  sex linkage in, 249f, 250

drupe, 602

duck-billed platypus, 579, 580f.
    See also monotremes
  eggs of, 893f

ductus arteriosus, 928
  definition of, G–8

ductus deferens, 896
  definition of, G–8

Dugesia, 540, 540f

duodenum, 797
  definition of, G–8

Dutch elm disease, 478

dynein arms, 86

dynein walking, 86

E

ear, 730, 731f

earth
  early, 403, 405f
  timetable of, 404f

earthworm
  body plan of, 550f
  circulatory system in, 814f
  digestive system, 790, 790f

ecdysis, 770
  definition of, G–8

ecdysone, 770

Echinodermata, 527, 558–559, 559f
  fertilization in, 912–914, 914f
  gastrulation in, 917–920, 919f
  nervous system of, 709, 709f

Echinoidea, 559

ecological hierarchy, 981–982

ecological pyramid(s), 1019–1020, 1022f
  of biomass, 1019–1020
  of energy, 1020–1022
  of numbers, 1019

ecological succession, 1014–1017
  in aquatic communities, 1016, 1017f
  facilitation model of, 1016
  inhibition model of, 1016
  models of, 1016
  primary, 1015
  secondary, 1015–1016
  tolerance model of, 1016

ecology
  applied, 982
  basic, 981
  definition of, 980
  of individuals, 982–987
  of populations, 989–998

ecosystem(s), 981, 1017–1028
  components of, 1018
  definition of, G–8
  nutrient cycling in, 1024–1028

ectoderm, 917
  in cnidarians and ctenophorans, 538
  definition of, G–8
  derivatives of, 920f

ectoparasite(s), 551
  definition of, G–8

Ectoprocta, 545f, 545–546

ectotherm(s)
  behavioral adaptations in, 750
  definition of, G–9

ectothermic regulation, 748

ectothermy, 743

Ediacara fauna, 527, 528f

effector, 693
  definition of, G–9

efferent vessel, 755

egg, 556, 911–912, 912f
  amniotic, 573, 575f
  determination in, 930–931
  of insect, 913f

egg activation, definition of, G–9

egg cell, 520, 521f
  in flower, 598

Einstein, Albert, 155

ejaculation, 896
  definition of, G–9

elastin, 61, 675

elaters, 495

Eldridge, Niles, 396

electrical synapse(s), 701–702
    definition of, G–9

electrochemical gradient, 144

electroencephalogram, 720–721, 722f
    in sleep, 722f, 722–723

electromagnetic spectrum, 150, 151f

electron(s), 22, 23f
    and chemical properties of elements, 24–25
    ejection of, 153, 159f
    energy shells, 24, 24f, 24t
    orbitals, 24–25
    valence, 25

electron acceptor, definition of, G–9

electron carrier(s), 141
    definition of, G–9

electron donor, definition of, G–9

electronegativity, 27
    definition of, G–9

electron flow, 185f

electron microscope
    definition of, G–9
    high-voltage, 71
    scanning, 71, 73f
    specimen preparation and staining for, 71
    transmission, 70–71, 72f

electron orbit, definition of, G–9

electron shell, definition of, G–9

electron transport, 174

electron transport system, 143, 143f, 148,
        182–183, 184f
    definition of, G–9
    energetics of, 186f
    in photosynthesis, 155–156

electrophoresis, 328–329

element(s), 21
    chemical properties of, 24–25
    definition of, G–9
    essential to life, 22t
    table of first, 20, 26f

elongation, 285–287
    formation of peptide bond in, 286, 286f
    translocation in, 286, 286f

Elton, Charles, 1018

Eltonian pyramid, definition of, G–9

embryo
    and corpus luteum, 904f
    of plant, 600
        development of, 600, 601f, 602–604

embryogenesis, 910

embryology, definition of, G–9

embryonic induction, 932, 933f

embryo sac, 520, 521f
    definition of, G–9
    in flower, 598

emergent trees, 1008, 1046

endergonic, definition of, G–9

endergonic reactions, 30, 129, 130f

endocrine gland(s), 765
    definition of, G–9

endocrine hormone(s), 765
    definition of, G–9

endocrine system
    definition of, G–9
    human, 771–785, 772f

endocytosis, 90, 118f, 118–120
    definition of, G–9
    receptor-mediated, 119f, 119–120

endoderm, 917
    in cnidarians and ctenophorans, 538
    definition of, G–9
    derivatives of, 920f

endodermis
    definition of, G–9
    in roots, 617

endomembranal system, 93–94
    definition of, G–9

endometrium, 898, 901
    definition of, G–9

endonucleases, 805

endoplasmic reticulum (ER), 88, 89f. See
        also rough endoplasmic reticulum;
        smooth endoplasmic reticulum
    definition of, G–9
    function of, 97t

endoskeleton(s), 532, 671, 674
    definition of, G–9

endosperm, definition of, G–9

endosperm mother cell, definition of, G–9

endospore(s), 430, 431f
    definition of, G–9

endosymbiont(s), 96
    definition of, G–9

endothelium, 814
    definition of, G–9

endotherm(s)
    behavioral adaptations in, 747–748
    definition of, G–9

endothermic regulation, 743–744

endothermy, 742

Endothia parasitica, 478, 1049

energy, 126–128. See also free energy
    of activation, 130, 131f
    allocation of, 998, 998f
    definition of, 127
    earth's budget of, 1033f

and electrons, 156f
    forms of, 127
    kinetic, 38, 127
        definition of, G–15
    as life characteristic, 16, 16f
    for muscular activity, 191–192, 192f
    potential, 127
        definition of, G–23
    and productivity, 1022–1023
    solar, and biosphere, 1033
    transfers of, 127f
    transformations of, 128
        and human activity, 127f

energy of activation, definition of, G–9

energy pyramid, definition of, G–9

Englemann, Theodore, 166

Ensatina escholtzi, species definition in,
        380–381, 381f

Entamoeba histolytica, 450

enterocoel, 532

enterogastrone, 806t

enterokinase, 805

entropy, 128, 129f
    definition of, G–9

envelope, 414

environment, 980
    definition of, G–9
    influence on phenotypes, 241, 242f
    and scale, 984

environmentalism, 981

environmentalists, 981

environmental resistance, 993
    definition of, G–9

environmental science, 981

enzymatic reaction, 132–133

enzyme(s), 46, 132–138
    and activation energy, 132f
    activation of, mechanisms of, 136
    characteristics of, 138
    control of, mechanisms of, 135–137
    definition of, G–9
    digestive, 801t
    functions of, 63t, 138
    inhibition of, mechanisms of, 136
    rate of action, factors affecting, 134–135
    specificity of, 132
    teams of, 135

enzyme saturation, 134–135

enzyme-substrate complex, 133–134,
        134f
    definition of, G–9

eosinophils, 862

Ephedra, 514, 520

epiboly, 918

epicotyl, 604
    definition of, G–10
epidermis
    definition of, G–10
    in plants, 493, 614
epididymis, 894
    definition of, G–10
epiglottis, 795
    in humans, 843
epinephrine, 767–769, 780
    actions of, 784t
    structure of, 766f
epiphyte, 504
    definition of, G–10
epistasis, 241, 242f
    definition of, G–10
epithelial tissue, 675
epithelium, 675, 676f
    definition of, G–10
Epstein–Barr virus, 415
Equus, evolution of, 360
ER. See endoplasmic reticulum
erectile tissue, 894, 895f
ergot, 478
    definition of, G–10
ergotism, 478
erythrocyte(s), 815, 815f
    definition of, G–10
erythropoietin, 851
Escherichia coli, 426
    in defense, 858
    as experimental organisms, 306
    functions of, 435
    growth rate of, 433
    in human digestive tract, 801
    mapping, 315f, 315–316
    phages of, DNA experiments with, 261
    polyribosomal translation in, 288, 288f
    replication in, 271
    tryptophan in, 296
esophagus
    definition of, G–10
    in earthworm, 790
    in humans, 795–796
essential amino acid(s), 807
    definition of, G–10
essential fatty acid(s), 53
    definition of, G–10
ester linkage, 52
estivation, 1041
estradiol, 782, 900
    structure of, 766f
estriol, 782
estrogen(s), 782, 900

actions of, 784t
estrone, 782
estrus, definition of, G–10
estuaries, 1059
ethology, 949–950
    definition of, G–10
ethylene, 31f, 656–657
    action of, 653t
    and flowering, 666
    and hypocotyl hook, experiment on, 656f
    and senescence, 657
Eubacteria, 425, 428–432
    cell wall of, 429–430, 430f
    characteristics of, 426t
    form and arrangement in, 428–429, 428f
    origins of, 409
    outer membrane of, 430
    phylogeny of, 427f
    structure of, 429f, 429–430
euchromatin, 301
    definition of, G–10
Euglena gracilis, 448, 460, 461f
Euglenophyta, 458, 460
    characteristics of, 459t
eukaryotes, 74
    DNA of, 273
    life cycles in, 447
    origins of, 409
euphotic zone, 1055
    definition of, G–10
Euplotes, 453f
eusocial insects, 976
eutrophication
    definition of, G–10
    natural, 1027
evolution
    of Amphibia, 572f
    of animals, 527–528, 531–532
    biogeography and, 361f
    of bryophytes, 500–501
    convergent, 395, 396f
    C4 plants and, 168
    Darwinian, 12–15
    definition of, G–10
    of deuterostome, 566f
    divergent, 394–395
    of eukaryotic cell, 96–97
    evidence of, 358–363
    and extinction, 397–399
    of fungi, 487–488
    of gene structure, mutation and, 344
    and genetics, 353
    of hominids, 586f
    of humans, 584–588
    of jaws, 568–569, 570f
    landmarks in, 532

as life characteristic, 17, 17f
    of limbs, 572–573
    of mammals, 576–584
    of migration, 966–967
    Onychophora in, 558
    patterns of, 394–395
    of plants, 490–493
    of Primates, 582f
    primitive and advanced features in, 518
    rate of, 396
    of reproductive strategies, 998–1001
    sex and meiosis and, 214–217
    testing hypotheses of, 14
    theories of, Darwin's, 350–353
    timeframe of, 529f
    of vascular plants, 501–502, 502f
    of vertebrate lung, 837–839
evolutionarily stable strategy (ESS),
    definition of, G–10
evolutionary tree, based on mutations in
    DNA, 363f
exchange pool(s), 1024
    definition of, G–10
excision repair, 336, 337f
excitatory synapse(s), 702
    definition of, G–10
excretion, 751–754
    definition of, G–10
excretory pore, 754
excretory system, human, 755–760
    anatomy of, 755, 756f–757f
excurrent siphon, in mollusks, 548
exergonic, definition of, G–10
exergonic reactions, 30, 129, 130f
exocytosis, 118, 118f
    definition of, G–10
exon(s), 282
    definition of, G–10
exonucleases, 805
exoskeleton(s), 551, 671–674
    definition of, G–10
exotoxin, 429
experiment(s), 11
    controlled, 11, 11f
    field, 12
    mark-and-recapture, 12
experimental organisms, 306
    garden pea as, 222
    humans as, 272
expiratory centers, 853
exploitative competition, 1010
exponential growth, 993, 994f
    consequences of, 1063f
    definition of, G–10

exponential growth curve, definition of, G–10

expressed sequence, definition of, G–10

The Expression of the Emotions in Man and Animals (Darwin), 15

expressivity, variable, 242, 243*f*

external exchange, 833

external fertilization, 887, 887*f*
  definition of, G–10
  in invertebrates, 887
  in vertebrates, 889–890

external genitalia, female, 896–898, 897*f*

extinction, 397–399
  definition of, G–10
  human activity and, 397–399

extracellular, definition of, G–10

extraembryonic membrane(s), 921–925
  in chick, 923*f*
  definition of, G–10
  in placental mammals, 923–925
  of reptile egg, 573

eye(s)
  compound, 553, 554*f*, 732
  development of, 932–933, 934*f*
  human, 733–734, 735*f*
  path of light in, 734
  simple, 732

F

F^+, definition of, G–11

facilitated diffusion, 113–114, 115*f*
  definition of, G–10

facilitation model, of succession, 1016

facultative anaerobe(s), 190, 434
  definition of, G–10

facultative heterochromatin, 301

FAD, 140
  definition of, G–10

$FADH_2$, definition of, G–10

fall overturn, 1054

false plasmodium, 456

families, 382

famine, 1065*f*

fangs, 793

fascia, 683
  definition of, G–10

fascicle(s), 683
  definition of, G–10

fat(s), 51–52
  body, brown, 187–189
  polyunsaturated, 52
  saturated, 52, 53*f*
  unsaturated, 52–53, 53*f*

fatigue, 191–192

fats
  in diet, 806
  digestion of, 803, 803*f*

fatty acid(s), 51, 51*f*
  definition of, G–10
  essential, 53
  metabolism of, 192–193

Fc regions, 866

feather fluffing, 746, 747*f*

fecundity, 996–997

feedback, 741

Felis, 385

fermentation, 174, 188–192
  alcoholic, 190–191, 191*f*
  definition of, G–10
  lactate, 191–192, 191*f*
    in muscle tissue, 191
  pathways of, 190–192

ferns, 509–510, 510*f*
  gametophyte of, 512*f*
  life cycle of, 509–510, 511*f*
  sporangia of, 512*f*

ferredoxin, 156

fertility, in United States, 1068*f*

fertility rate. *See also* general fertility rate; total fertility rate
  definition of, G–10

fertilization, 600*f*, 912–915
  definition of, G–10
  double, 515, 520, 598
  in echinoderms, 912–914, 914*f*
  external. *See* external fertilization
  human, 916*f*

internal. *See* internal fertilization
  in mammals, 914–915

fertilization cone, 913
  definition of, G–10

fertilization envelope, 913, 915*f*

fetus, definition of, G–10–11

Feulgen, Robert, 259

F_2 generation
  genotype of, Mendel's predictions and results for, 229*t*
  in Mendel's experiments, 223*t*
  phenotype of, Mendel's predictions and results for, 228*t*

F_3 generation, in Mendel's experiments, 225*f*

fiber(s), 49
  of sclerenchyma, 615

fibrils, of cellulose, 48*f*, 49

fibrin, 816
  definition of, G–11

fibrinogen, 815–816

definition of, G–11

fibroblasts, 675

fibroin, 553

fibrous protein(s), 57
  definition of, G–11

fibrous root system, definition of, G–11

fiddleheads, 509

field experiments, 12

fighting, 971–973, 972*f*

filament, 520, 521*f*, 593, 593*f*

filamentous forms, 429

filterable viruses, 417

filter-feeder, definition of, G–11

fimbriae, 898

Fire algae. *See* Pyrrhophyta

fire-disclimax community, definition of, G–11

first law of thermodynamics, 128
  definition of, G–11

fish
  bony, 570–572, 571*f*
    digestive system of, 791
  external fertilization in, 889, 889*f*
  gills of, 836–837, 839*f*
  heart of, 817–818
  internal fertilization in, 890–891, 891*f*
  jawless, 567–568
    extinct, 568–569
  parental care in, 889, 890*f*
  thermoregulation in, 744*f*

fission, 75, 432, 433*f*, 885
  definition of, G–11

fitness, definition of, G–11, 357–358

Fitzroy, James, 5–9

fixed action pattern(s), 950
  definition of, G–11
  and orientation, 950, 950*f*

flagellates, 448–449, 449*f*

flagellin, 76, 431

flagellum(/a), 76, 83–86
  of bacteria, 431, 432*f*
    basal body, 431
    filament of, 431
    hook of, 431
  definition of, G–11
  function of, 97*t*
  movement in, 85–86
  structure of, 85–86

flame bulb(s), 540, 753

flame cells, 540, 753, 753*f*

flame cell system, definition of, G–11

flatworms, 538–541
  free-living, 539–541

flavin mononucleotide, 183, 184

flavoproteins, 183

flightless grasshoppers, sympatric speciation in, 392

floats, 465

floral parts, accessory, 593

floridean starch, 463

florigen, 666

flower(s), 593, 593f
  complete versus incomplete, 595
  composite, 595
  definition of, G–11
  disk, 595
  diversity in, 594
  perfect versus imperfect, 595
  pistillate, 595
  pollination of. See pollination
  ray, 595
  sexual activity in, 597–600
  staminate, 595
  structure of, 595–597
  variation in, 593–595

flowering
  dark clock and, 663–664, 665f
  initiation of, phytochrome and, 664–666
  light and, 662–666

flowering plants. See angiosperm(s)
  reproduction in, 592–609

fluid mosaic model
  definition of, G–11
  of plasma membrane, 107–108, 108f–109f

fluke(s), 541
  definition of, G–11

fluorine, 22t

FMN. See flavin mononucleotide

foliose lichens, 483, 485f

follicle-stimulating hormone, 773, 899
  actions of, 784t

food chain(s), 1018–1019, 1019f
  definition of, G–11
  detritus, 1018–1019, 1020f
  grazing, 1018

food vacuoles, 93

food web(s), 1019, 1021f
  definition of, G–11

foot, of mollusks, 546

foraging
  group, 977
  optimal, 962, 962f

foraging behavior, 961–962
  costs and benefits of, 961–962

foramen magnum, 720

foramen ovale
  at birth, 928

definition of, G–11
  during development, 925–926

foraminiferan(s), 449, 450f, 451
  definition of, G–11

force filtration, 755
  definition of, G–11

forebrain, 711, 713–719
  definition of, G–11

foregut, in insects, 790–791

foreskin, 894

The Formation of Vegetable Mould Through the Action of Worms (Darwin), 15

fossil(s), 358, 358f
  definition of, G–11

fossil fuel, definition of, G–11

fossil record, 358f, 358–359
  on animal evolution, 527–528

founder cells, 931

founder effect, 374, 375f
  definition of, G–11

fovea, 734

Fox, Sidney, 406

F plasmids, as agents of gene transfer, 313–314, 314f

fragmentation, 447, 494, 885

frameshift mutation(s), 339, 339f
  definition of, G–11

Franklin, Rosalind, 263

free energy, 129
  of chemiosmotic gradient, 144
    tapping, 144
  definition of, G–11

free radical(s), 336
  definition of, G–11

freeze fracture technique, 73f, 105f

freezing, physiologic adaptations to, 748–750, 750f

frequency, 1009

frequency-dependent selection, 355, 356f
  definition of, G–11

freshwater environment, 752–753

freshwater habitats, 986–987

freshwater province, 1052–1054

frogs
  external fertilization in, 890, 890f
  lungs in, 840, 840f

frontal lobe, 715, 715f
  definition of, G–11

fructose, 45, 45f

fructose-1, 6-biphosphate, 162, 175

fruit(s), 600
  aggregate, 603
  definition of, G–11

development of, 602–603
  multiple, 603

fruit fly, 246f, 247

fruticose lichens, 483, 483f

fruiting body, definition of, G–11

fucoxanthin, 460

Fucus, life cycle of, 465, 466f

functional group, 33, 34t
  definition of, G–11

fundamental niche, 985, 985f
  definition of, G–11

fungal body, 472–473, 473f

fungi imperfecta, 481–482

fungus(/i), 384, 472–489, 473f
  club. See Basidiomycota
  in decomposition, 474
  dikaryotic state in, 475
  ecological importance of, 474–475
  evolution of, 487–488
  feeding in, 474, 474f
  in lichens, 482–483, 484f
  medical and commercial uses of, 475
  mutualism in, 475
  parasitism in, 474–475
  and plant roots, 645–646, 646f
  relationships of, 476
  reproduction in, 475
  zygotic life cycle in, 476, 477f

funiculus, in flower, 597

G

galactose, 47

Galápagos finches, 9, 9f, 350, 390

Galápagos islands, 7–9, 8f, 388

Galápagos rift, 1057

Galápagos tortoise, 8f

Galileo, 68

gallbladder, 800

gametangia, 477

gamete(s), 210, 911–912
  definition of, G–11

game theory, 972–973
  strategies in, 973

gametic cycle, 447, 448f
  definition of, G–11

gametogenesis, definition of, G–11

gametophyte, definition of, G–11

gametophyte generation, 447

gamma-interferon, 872

gamonts, 451

ganglia
  in annelids and mollusks, 709

in planarians, 540

ganglion, definition of, G–11

ganglion cells, 734

gap junction(s), 121f, 121–122
  definition of, G–11

garden pea
  as experimental organism, 222, 306
  flower and pollination in, 224f
  Mendel's experiments with, 222–227

Garrod, A.E., 272–274
  Inborn Errors of Metabolism, 272

gases, exchange of, 832–855
  in alveolus, 848, 848f
  external, 833
  external phase of, 836f
  in human respiration, 847–848
  internal, 833
  internal phase of, 836f
  surfaces for, 833–843

gasohol, 190

gastric ceca, 791

gastric lipase, 797

gastrin, 805, 806t

gastrointestinal tract, definition of, G–11

Gastropoda, 547, 547f–548f

gastrovascular cavity, 537, 790
  definition of, G–11

gastrula, 917
  definition of, G–11

gastrulation, 917–920, 919f
  definition of, G–11

Gause, G.F., 986

Gause's law, definition of, G–11

Gazzaniga, Michael, 717

GDP. *See* guanosine diphosphate

gemmae, 494

gemmae cups, 494, 495f

gemmules, 535, 885

gene(s), 62, 225, 276–291
  and behavior, 945–947
  on chromosomes, 244–254
  control of, steroid hormones in, 769, 769f
  definition of, G–11–12, 277
  inducible, 298–301
  interactions of, 241
  linked, 230
  mapping, 247f, 247–248
  regulation of, 292–305, 293f
    eukaryotic, 298–304
    opportunities for, 298, 299f
    post-transcriptional, 303–304
    prokaryotic, 293–298
  steroid-hormone-induced, 300–301, 302f
  structure of, 293f

gene flow, definition of, G–12

gene machine, constructing probe with, 323, 325f

gene pool, 366
  definition of, G–12

general fertility rate, 1067
  calculation of, 1064t

generalists, 596
  with respect to environmental conditions, 984, 985f
  with respect to foraging behavior, 961, 961f

generation time, and maximum population growth rate, 993, 993f

generative cell, 598
  definition of, G–12
  in pollen, 512

gene replacement therapy, 328–330
  definition of, G–12
  microinjection in, 330f

gene sequencing, definition of, G–12

gene splicing, definition of, G–12

Genest, Jacques, 781

genetic code, 59
  definition of, G–12
  mRNA and, 280–281

genetic drift, 373
  definition of, G–8

genetic engineering, 318–333
  in agriculture, 326–327
  applications of, 325–330
  definition of, G–12, 318
  in medicine, 327–330
  in plants, 326–327, 326f–327f
  social issues in, 330–332

genetic equilibrium, 368
  definition of, G–12

genetic recombination, 245f, 245–247
  definition of, G–12

genetics
  definition of, G–12
  evolution and, 353

genetic system, early, 408f

genetic variability, definition of, G–12

genome, definition of, G–12

genotype, 222
  definition of, G–12

genotype frequency, definition of, G–12

genus, 382
  definition of, G–12

geological era, definition of, G–12

germinal epithelium, 214, 895
  definition of, G–12

germination, 522, 611, 612f
  in bacteria, 430

conditions for, 611
  definition of, G–12

germ layer(s), 920f
  in cnidarians and ctenophorans, 538
  definition of, G–12

germ theory, 416

giant armadillo, 359, 359f

giant kelp, 464f

gibberellins, 653–654
  action of, 653t
  and seed germination, 655f
  and stem elongation, 655f

gill(s), 484, 836
  aquatic, 836–838
  of arthropods, 836–837, 838f
  of fishes, 837, 839f
  of mollusks, 546
  tracheal, 836

gill apparatus, development of, 921, 922f

gill arches, 837
  definition of, G–12
  development of, 921, 922f

gill clefts, definition of, G–12

gill filaments, 837

gill slits, development of, 921, 922f

Ginkgo biloba, 513, 514f

Ginkgophyta, 502, 503f, 513

giraffe(s)
  evolution of, 398f–399f
  inheritance in, 349, 349f
  natural selection in, 373f

gizzard, 543
  in birds, 793
  definition of, G–12
  in earthworm, 790

glans, 891
  human, 894

glial cells, 693

gliding bacteria, 431–432

gliding joint, 678

global warming, 982, 1034

globular proteins, 57

globulins, 815

Gloeocapsa, 440f

glomerulus, 755
  definition of, G–12

glottis, 795
  in frogs, 840

glucagon, 778
  actions of, 773t
  definition of, G–12

glucocorticoids, 779

gluconeogenesis, 192

glucose, 45, 45f, 50t
    definition of, G–12
    metabolism of, 188f
        ATP yields from, 187t
        parts of, 175f
    structural formula of, 46
    utilization of, 174
glucose-1-phosphate, 162
glutamic acid, 58f
glutamine, 58f
glyceraldehyde-3-phosphate, 149, 162, 175
glycerol, 51, 51f
    definition of, G–12
glycine, 58f
glycocalyx, 105, 106f, 798
glycogen, 47, 48f, 50t
    definition of, G–12
glycolipid(s), 53, 105–106
    definition of, G–12
glycolysis, 175–177, 176f, 179f
    control of, 176–177, 180f
    definition of, G–12
    energy yield of, 175–176, 177t
    highlights of, 175–176, 176f
glycoprotein(s), 53, 61–62, 105–106
    definition of, G–12
glycosidic linkage, 46
glyoxysomes, 93
Gnetophyta, 502, 503f, 514
goblet cell, definition of, G–12
goiter, 776, 777f
golden-brown algae. See Chrysophyta
Golgi, Camillo, 90
Golgi complex, 90, 91f
    definition of, G–12
    function of, 97t
gonad(s). See also ovary(ies); testes
    definition of, G–12
    hormones of, 782–783
gonadotropin(s), 773, 898
    definition of, G–12
gonadotropin-releasing hormone, 328
Gondwanaland, 389
gonorrhea, 436f, 437
    definition of, G–12
Gonyaulax catanella, 459f
Goodall, Jane, 971
Gould, Stephen Jay, 396
G protein, 768
Graafian follicle, 901
    definition of, G–12
grackles, 379
graded display, 968, 968f

definition of, G–12
gradualism, 396, 398f–399f
    definition of, G–12
Gram-negative bacteria, 430, 431f
Gram-positive bacteria, 430, 431f
Grand Canyon squirrels, 388f
granum(/a), 94, 151
    definition of, G–12
grasshopper, 673, 673f
    circulatory system in, 814f
grasslands, 1042–1043, 1043f
Graves' disease, 776, 777f
gravitropism, 658
    definition of, G–12
    in roots, 660f
gravity receptors, 732
gray crescent, 930
great tits, foraging behavior in, 961–962, 962f
green algae. See Chlorophyta
greenhouse effect, 1034
    definition of, G–12–13
greenhouse gases, 427
Griffith, Fred, 259–261
gross production, 1024f
gross productivity, 1023
    definition of, G–13
ground layer, 1008
ground meristem, 604, 613
    definition of, G–13
    derivatives of, 615–617, 615f–617f
group foraging, 977
group protection, 977
growth curve, definition of, G–13
growth hormone, 772, 774
    actions of, 784t
growth hormone release-inhibiting hormone, 772
growth hormone releasing hormone, 772
growth response
    differential, 658–661
    light and, 651–652, 652f
GTP. See guanosine triphosphate
guanine, 65, 262
guanosine diphosphate (GDP), 180
guanosine triphosphate (GTP), 180
guard cell(s), 615
    definition of, G–13
    and water transport, 640–643
guard hair, 745
gustatory receptors, 728
gut, organization of, 797f
guttation, 636

gymnosperms, 502, 503f, 504, 513–516
    decline of, 516–517
gynoecium, 593
gyres, 1037
gyri, 715–716

H

habitat(s), 985
    selection of, 960, 960f, 961t
    types of, 986–987
habitation, definition of, G–13
hagfish, 569f
habituation, 950–951
hair, 726
hair cells, 732
Haldane, J.B.S., on origin of life, 403
Haldane–Oparin hypothesis, definition of, G–13
half-life, 23
    definition of, G–13
halophiles, 427
haploid, definition of, G–13, 211
Hardy, G.H., 367
Hardy–Weinberg equilibrium, 368–370, 369f
Hardy–Weinberg principle, 367
    definition of, G–13
    implications of, 367–368
    restrictions of, 369–370
Harvey, William, 781
haustoria, 474, 474f
    definition of, G–13
Haversian system(s), 678
    definition of, G–13
hearing. See auditory reception
heart
    branchial, 548
    circulation through, 819–821
        left side, 821
        right side, 819–821
    as endocrine gland, 781
    four-chamber, 817–818
    function of, 822–823
    hormones of, 784t
    in humans, valves of, 821, 822f
    regulation of, 821–822
        extrinsic versus intrinsic, 821–822
    three-chamber, 817
    two-chamber, 817
heartbeat, control of, 821, 823f
heartwood, 627
heat
    and body size, 745, 745f
    generation of, 744–745

loss of
  increasing, 747
  slowing, 745–747
heat shock element, 299
heat shock factor (HSF), 299
heat shock protein *(Hsp)* genes, 298–300
  structure of, 300*f*
heavy chains, 865
Heimlich maneuver, 800
helical bacteria, 429
helicase, 267
heliozoans, 449
helper T-cell lymphocytes, 422
heme group(s), 61, 156
  definition of, G–13
hemicelluloses, 49
Hemichordata, 565, 565*f*
hemipenes, 891, 892*f*
  definition of, G–13
hemizygous, definition of, G–13, 250
hemocoel(s), 814
  definition of, G–13
hemocyanin, 837
  definition of, G–13
hemoglobin, 849, 850*f*
  definition of, G–13
  structure of, 60*f*, 61
hemophilia
  definition of, G–13
  in European royal families, 253, 253*f*
  inheritance of, 252–254
hemoproteins, 61
Henslow, John, 5
hepatic portal circuit, 824–825
hepatic portal vein, 817, 825
Hepatophyta, 494–495
herbivores, 1018
hereditary traits, resistance to blending, 221, 221*f*
hermaphroditic, 535
Hershey, Alfred, 261
heterochromatin, 301
  constitutive, 301
  definition of, G–13
  facultative, 301
heterocysts, 439
heterogametes, 447, 447*f*
  definition of, G–13
heteromorphic, 463
heterosporous, 495, 507, 597
  definition of, G–13
heterotherm(s), 742

definition of, G–13
heterotroph(s), 76
  definition of, G–13
heterotrophic, definition of, 409
heterozygote advantage, 354
heterozygous, definition of, G–13, 225
Hfr bacteria, 314
hibernation, 743, 743*f*
Hill, Robert, 167
hindbrain, 719–720
  definition of, G–13
Hindenburg explosion, 129–130, 130*f*
hindgut, in insects, 791
hinge joint, 678
hippocampus, 719
hirudin, 551
Hirudinea, 550
histamine, 859
histidine, 58*f*
histone, definition of, G–13
historicity, 14
HIV. *See* human immunodeficiency virus
holdfasts, 463, 465
Holmes, Sherlock, 260
Holothuroidea, 559
homeobox, 301, 935
  actions of, 937*f*
  definition of, G–13
homeostasis, 741–742
  definition of, G–13
homeotherms, 742
homeotic gene(s), 301, 934
  definition of, G–13
homeotic mutations, 934–935, 936*f*
homing pigeons, 965–966
  experiments with, 966, 966*f*
Hominidae, 580
hominids, 580
  definition of, G–13
  distribution of, 587*f*
  early, 585*f*
  evolution of, 586*f*
hominoid, definition of, G–13
Hominoidea, 580
Homo erectus, 585–587
Homo habilis, 585
homologous, definition of, G–13
homologous chromosomes, 211–212
homologous structures, 361, 362*f*
homologue(s), 212
  definition of, G–13

homology, definition of, G–13
Homo sapiens neanderthalensis, 587*f*, 587–588
Homo sapiens sapiens, 588
homosporous, 495, 507
homosporous plants, 496
homozygous, definition of, G–13, 225
honeybees
  behavior experimentation with, 946, 947*f*
  tactile communication in, 970, 971*f*
Hooke, Robert, 68–69, 643
Hooker, Joseph, 13–14, 351–352
hormonal control, 764–785
  characteristics of, 766
  versus neural, 765
hormone(s)
  in arthropods, 770–771
  and behavior, 948, 948*f*
  and central nervous system development, 948
  control of osmoregulation, 759*f*
  definition of, G–13
  of digestion, 806*t*
  and effector mechanisms, 948–949
  effects of, 765*f*
  functions of, 63*t*
  and human reproduction, 898–903
    in females, 900–901, 903*f*
    in males, 900, 901*f*
  in invertebrates, 769–771
  molecular structure of, 765, 766*f*
  and perception, 948
  in plants, 652–658, 653*t*
    applications of, 658
  and target cell, 766–769
  types of, 765
hornworts, 494, 495–497, 497*f*
  life cycle of, 496–497
horse, 580*f*
  evolution of, 360
horsetails, 507–509, 509*f*
host(s)
  intermediate, 541
    definition of, G–15
  primary, 541
    definition of, G–23
human(s)
  circulatory system, 819–829, 820*f*–821*f*
    circuits in, 824–825, 825*f*
    development of, 925–926
  development of, 925–929, 926*f*
    early, 924*f*–925*f*
    first trimester, 925–927
    second and third trimesters, 928
    supporting structures in, 924*f*–925*f*
  evolution of, 584–588

human(s) *continued*
  excretory system, 754–760
    anatomy of, 754–755, 756f–757f
  as experimental organisms, 272
  and extinction of other species, 398
  fertilization in, 916f
  genetic future of, 375–376
  immune system. *See* immune system
  impact on ecosystem, 1062–1071
  meiosis in, 214–217, 216f
  nervous system, 712–723
    development of, 925
    organization of, 712f
  pedigrees for traits of, 232
  and reciprocal altruism, 976–977
  reproduction and sexuality in, 894–903
  reproductive system. *See* reproductive
      system
  respiratory system, 842f, 843–854
    air flow in, 843–846
    development of, 926
  as second-level consumers, 1022f
  sexual response in, 906–907
  specialization in, 583f, 583–584
  and trophic levels, 1020–1022
human chorionic gonadotropin (HCG),
      902–903
  definition of, G–13
human immunodeficiency virus (HIV),
      422–425, 423f, 879, 880f. *See also*
      AIDS
  definition of, G–13
  life cycle of, 424f–425f
  progression of, 881, 881f
  targets of, 880
  transmission of, 880–881
human population
  current state of, 1066–1067
  future of, 1067–1071
  growth predictions, 1069–1071, 1070f
  history of, 1063f, 1063–1066
  percentage of, below 15 years of age, 1064t
  second, 1065
  surges in, 1063f
    first, 1063–1065
  third, 1065–1066
  world, 1066t
    by region, 1067f
humans, digestive system(s), 795–801, 796f
humoral response, 864–868
  definition of, G–13
Huntington's disease, 242
Huxley, Thomas, 14, 359
hybrid(s)
  breakdown of, 393
  definition of, G–13
  inviability of, 393

polyploidy in, 389f
  sterility of, 389f, 394
hybridization, 329
  sympatric speciation through, 388
hybridization experiments, 946
hydra, 536f
hydration shells, 36, 36f
hydrocarbon(s), 31
  definition of, G–13
hydrogen, 21, 22t
  electron shells in, 24t
hydrogen bond(s), 29–30, 30f
  definition of, G–13
hydrogen carrier, definition of, G–13–14
hydrogen ion, 38
  definition of, G–14
hydroids, of moss, 497
hydrologic cycle, 34, 35f, 1034, 1034f
  definition of, G–14
hydrolysis, 47, 801–802
hydromedusae, 537
hydronium ion, 39
hydrophilic, definition of, G–14
hydrophobic, definition of, G–14
hydrophobic interaction, 54, 55f
hydrophobic molecules, 36
hydrophytes, 1054
hydrostatic pressure, 113
hydrostatic skeleton, 671f, 671–674
  definition of, G–14
hydroxide ion, 39
  definition of, G–14
hydroxyl group, 34t
  definition of, G–14
hydrozoans, 537
  life cycle of, 538f
hymen, 898
hyperallergic reaction, 879f
hypercholesterolemia, 120
hyperosmotic, definition of, G–14
hyperpolarization, definition of, G–14
hyperthyroidism, 776
  definition of, G–14
hypertonic, definition of, G–14
hypertonic medium, 113, 113f
  cell responses to, 114f
hypha(e), 472–473, 473f
  definition of, G–14
  feeding, 474f
  metabolic activity in, 473, 474f
hypocotyl, 604
  definition of, G–14

hypocotyl hook, 611
  ethylene and, experiment on, 656f
hypoglycemia, 780
  definition of, G–14
hypothalamus, 714, 719
  control of pituitary gland by, 772, 773f
  definition of, G–14
  hormones of, 784t
  in thermoregulation, 748, 749f
hypothesis, 10–12
  definition of, G–14
hypothyroidism, 776
  definition of, G–14
hypotonic, definition of, G–14
hypotonic medium, 113, 113f
  cell responses to, 114f
hysterectomy, 906
H zone, 684

I

I band, 684
i gene, 294
ileocecal valve, 801
ileum, 797
ilium, 678
imaginal disks, 934, 934f
imbibition, 36
  definition of, G–14
immediate disturbance hypothesis, 1014,
      1014f
immune response, definition of, G–14
immune system, 106, 856–883
  defenses of, 857t, 857f
  definition of, G–14
  problems with, 878–882
immunity
  active, 876
  passive, 876
immunoglobulin, definition of, G–2
immutability, 349
implantation, definition of, G–14
imprinting, 952, 952f
  on humans, 952, 952f
impulse, definition of, G–14
inborn errors of metabolism
  definition of, G–14
  genetic engineering in diagnosis of,
      327–328
Inborn Errors of Metabolism (Garrod), 272
inbreeding experiments, 946
incisors, 794
incomplete dominance, 237–238, 238f

definition of, G–14

incomplete metamorphosis, 557

incomplete penetrance, 241–242, 243f
definition of, G–14

incurrent siphon, in mollusks, 547

incus, 730

indehiscent, 602–603

independent assortment, 229f
chromosomal basis of, 230f
law of, 228–230

indicator species, 986, 986f

individuals
ecology of, 982–987
environmental factors affecting, 982,
987–989
patterns of response to, 984

indoleacetic acid (IAA), 651

induced fit, 133

induced fit hypothesis, definition of, G–14

inducer, 294
definition of, G–14

induction, definition of, G–14

inductive reasoning, 9–10, 10t
definition of, G–14

indusium, 509

inert, definition of, 25

inferior vena cava, 819

inflammatory response, 859, 859f
definition of, G–14

inflorescences, 595

influenza virus, 418f

information storage, theories of, 954–955

Ingenhousz, Jan, 166

Ingles, Lloyd, 373

inheritance, of acquired characteristics, 349,
349f

inhibin, 900

inhibiting hormone(s), 772
definition of, G–14

inhibition model, of succession, 1016

inhibitor, 136

inhibitory block, definition of, G–14

inhibitory synapse(s), 702–704, 704f
definition of, G–14

initiation, 285, 285f
definition of, G–14

initiation complex, 285
definition of, G–14

initiation signal, 282

initiatory tRNA, definition of, G–14

innate, definition of, G–14

innate release mechanisms, 950

definition of, G–14

inner cell mass, 916
definition of, G–14

inner ear, 730, 731f
definition of, G–14–15

inner sheath, 86

inorganic molecule, definition of, G–15

inorganic phosphate, 32, 139

Insectivorous Plants (Darwin), 651

insects, 554–557. *See also* Arthropoda
abdomen of, 555
body plan, 555f
brain of, 710, 711f
development of
early events in, 916, 918f
hormonal control of, 770–771
program for, 934–935
digestive system, 790–791, 791f
eggs of, 913f
eusocial, 976
fertilization in, 888f, 888–889
head of, 555
mouthparts of, 556f, 557
movement in, 672–673, 673f
relatedness in, 977f
respiration in, 836, 837f
thorax of, 555

insertion, 673
definition of, G–15

insight learning, 951–952

inspiratory centers, 853

instinct
definition of, G–15
ethologist's concept of, 949–950
and learning, 952–953
learning and, in white-crowned sparrows,
953, 953f

instinctive attack, 949, 949f

insulin, 778–779
actions of, 784t
production of, 325–326

integral proteins, in plasma membrane, 103

integration, by nervous system, 692

integuments, in flower, 597

intensity, 988

interbreeding, definition of, G–15

intercalated disks, 682

intercalating agents, 336, 336f

interference competition, 1010

interferon(s), 860–861
definition of, G–15
gamma, 872

interleukin(s)
IL–1, 871

definition of, G–15
IL–2, 873
definition of, G–15

intermediary metabolism, 162, 193
definition of, G–15

intermediate filaments, 82, 83f

internal exchange, 833

internal fertilization, 887
definition of, G–15
in invertebrates, 887–889
in vertebrates, 890–894

internal nares, 839

interneuron(s), 693, 694f
definition of, G–15

internode, 622, 694

interphase, 202
G_1 phase of, 202
G_2 phase of, 202
of meiosis, 211–212
S phase of, 202

interstitial cells, 894–895
definition of, G–15

intertidal zone, 1058

intervening sequence(s). *See also* intron(s)
definition of, G–15

intervertebral disks, 678

intestine, in earthworm, 790

intracellular, definition of, G–15

intrauterine device, 903
definition of, G–15

intrinsic rate of growth, 992
definition of, G–15

intron(s), 282
definition of, G–15
splicing out, 282f

invagination, 918
definition of, G–15

inversion(s), 340, 341f
definition of, G–15

invertebrates, 526–563
endoskeletons in, 674, 675f
exoskeletons in, 671–674
nervous systems of, 709–710
support and movement in, 671–674

inverted repeats, 342

in vitro, definition of, G–15

in vivo, definition of, G–15

involution, 918

iodine, 22t

ion(s), 28
complex, 29
definition of, G–15
diffusion of, 111

ion channel(s), 698–700
    definition of, G–15
ion gates, 698–700
ionic bond(s), 28–29, 29f
    definition of, G–15
ionization, 39
    definition of, G–15
ionizing radiation, damage from, 336
iris, 734
iron, 22t
    deficiency of, in plants, 647f
iron-sulfur proteins, 183
islets of Langerhans, 778, 779f
    definition of, G–15
    hormones of, 784t
isobutane, 31f
isogametes, 447, 447f
    definition of, G–15
isoleucine, 58f
isomorphic, 463
isosmotic, definition of, G–15
isotonic, definition of, G–15
isotonic medium, 112, 113f
    cell responses to, 114f
isotope(s), 23
    definition of, G–15
iteroparous species, 999, 999f
Iwanowski, Dimitri, 416–417

J

Jacob, Francois, 293–294
    model of operon, 295f
Jacobson's organ, 728, 793
jawless fishes, 567–568
    extinct, 568–569
jaws, evolution of, 577–578
jejunum, 797
jelly coat, 913
Johanson, Donald, 584
Johnson, Virginia E., 906
joints, 678, 681f
J-shaped curve, 993
jungle, 1045
juvenile hormone (JH), 771
    definition of, G–15
juxtaglomerular complex, 755

K

K. See carrying capacity
Kamen, Martin, 166

kangaroo, 580f
karyotype, definition of, G–15
karyotyping, 212
Kennedy, J.S., 643
keratin, 61, 63f
    definition of, G–15
ketone group, 34t, 45
ketose family, 45
Kettlewell, H.B.D., 357
kidney(s), 755
    definition of, G–15
kinase, 323
kinetic energy, 38, 127
    definition of, G–15
kinetin, 655
kinetochore, 203
kingdom(s), 382–384, 385f
    definition of, G–15
kinins, 859
kin selection, 975–976
Klinefelter's syndrome, 251
knee-jerk reflex, 705
Koch, Robert, 416
Kramer, Gustav, 965
Krause's corpuscles, 727f, 728
Krebs, Hans A., 177
Krebs cycle. See citric acid cycle
K selection, 1000–1001
    characteristics of, 1001t
    definition of, G–15
    species, 1002f
kuru, 425

L

labial palps, in mollusks, 548
labia majora, 896
    definition of, G–15
labia minora, 896
        definition of, G–15
labium, 557
Labrador retrievers, nonblending inheritance
        in, 221, 221f
labrum, 557
Labyrinthodonts, 572, 572f
lac operon, 294–296, 295f
lac operon, definition of, G–15
lactase, 802
lactation, 929
lacteal, 798
    definition of, G–15

lactose, 47, 47f
    digestion of, by inducible operon,
        294–296
lactose intolerance, 47
lacunae, 678
ladder, 709, 709f
lagging strand, definition of, G–15
lakes, 1053–1054
    zonation of, 1053, 1053f
Lamarck, Jean Baptiste de, 349
lamella(e), 94, 151, 678, 837
    definition of, G–15
Laminaria, life cycle of, 465, 465f
laminarin, 464
lampreys, 566, 569f
lancelets, 566, 568f
landscape, 981
land snails, variation in, 354, 355f
language processing, in brain, 718, 718f
large intestine, in humans, 801
larva, 556
    definition of, G–15
laryngopharynx, 795
larynx, 843, 846f
    definition of, G–15
late blight, 457, 457f, 615
latent learning, 951
lateral bud, 623
lateral canal, 559
lateral fissure, 715, 715f
lateral line organ, 570, 726, 726f
    definition of, G–15–16
lateral root, definition of, G–16
Laurasia, 391
law, 10
    definition of, G–16
law of alternate segregation, definition of,
        G–16
law of conservation of energy, 128
law of entropy, 128
law of independent assortment, 228–230
    definition of, G–16
law of segregation, 227–228
laws of thermodynamics, 127–128
    definition of, G–16
    life and, 128–129
LDL receptor, 120, 120f
5′ leader, 282
leading strand, definition of, G–16
leaf(/leaves), 623–625
    anatomy of, 623–625, 625f
    dicot, 624f

anatomy of, 624*f*
evolution of, 503*f*
monocot, venation in, 625*f*
photoreceptor in, 665, 665*f*
water movement in, 636–637
water potential in, 637*f*
leaf abscission, 657
leaf abscission layer, 657*f*
leaf blade, 623
leaf cells, and C4 pathway, 168*f*
leaf fall, definition of, G–1
leaf generation, 606, 607*f*
leaf hairs, 624
leaf mesophyll, 164, 623
leaf primordia, 622
leaf scales, 494
leaf scar, 623
    definition of, G–16
leaf sheath, 625
Leakey, Louis, 585
Leakey, Mary, 585
Leakey, Richard, 585
learning, 950–952
    insight, 951–952
    instinct and, 952–953
    and instinct, in white-crowned sparrows, 953, 953*f*
    latent, 951
    and memory, 953–956
    types of, 950–952
Lederberg, Esther, 310
Lederberg, Joshua, 310, 313
leech, 550, 551*f*
left atrium, 817
left brain, abilities of, 717
left ventricle, 821
leguminous plants, 32
lens, 734
lens placode, 932
lenticel(s), 627
    definition of, G–16
leopards, development of spots of, 357, 357*f*
leptoids, of moss, 497
lethal alleles, 240–241, 241*f*
    in selection, 371
leucine, 58*f*
leucoplasts, 94
leukocidins, 862
leukocyte(s), 815
    definition of, G–16
    nonphagocytic, 862
    phagocytic, 861*f*, 861–862

action of, 861*f*
levels of organization, 532, 534*f*
Levene, P.A., 259
library(ies), 321–325
    screening, 323*f*, 323–325
lichen(s), 482–484, 484*f*
    definition of, G–16
life
    characteristics of, 15–17
    distribution of, 1037–1038
    and laws of thermodynamics, 128–129
    origin of, Haldane–Oparin hypothesis of, 403–404
        experiments on, 404, 406*f*
    origins of, 402–411
life history, 995
life table, 997, 997*t*
ligament(s), 678
    definition of, G–16
ligand, 119
ligase, 267
    definition of, G–16
light
    effects of, on organisms, 988–989
    as energy source, 150–151
    and flowering, 662–666
    and growth response, 651–652, 652*f*
light chains, 865
light-dependent reactions, 150, 156–160
light-harvesting complex, 151–152, 153*f*
    definition of, G–16
light-independent reaction(s), 150, 160–162
    definition of, G–16
light microscope, 71, 72*f*
    specimen preparation and staining for, 70
light reaction(s), 149–150
    cyclic, 159, 161, 161*f*
    definition of, G–16
    noncyclic, 157–159, 159*f*
light waves, 149
lignin, 49
    definition of, G–16
limbic system, 714, 719, 719*f*
    definition of, G–16
limbs
    evolution of, 577
    human, development of, 926
limiting resources, 983–984
    critical, 983*f*
    law of, 983*f*
limnetic zone, 1053
    definition of, G–16
limpet *(Acmaea limatula)*, 751–752, 752*f*
linkage, 244–245

linkage group, 244, 245*f*
    definition of, G–16
linked, definition of, G–16
linked genes, 230
Linnaeus, Carolus, 349, 382
lipase, 803
    definition of, G–16
lip cells, 509
lipid(s), 49–56
    definition of, G–16
    functions of, 56*t*
lipid-anchored proteins, in plasma membrane, 104
lipid bilayer(s), in plasma membrane, 103–104, 104*f*
lipoproteins, 61
liposome(s), 103
    definition of, G–16
littoral zone, 1053, 1055, 1056
    definition of, G–16
liver, 800*f*
    definition of, G–16
    in digestion, 800
liver cell, 767–769
liverworts, 494, 496
    life cycle of, 496, 498*f*
locus, 248
    definition of, G–16
logistic growth, 993–995, 995*f*
logistic growth curve, definition of, G–16
logistic growth equation, 994
long-day plants, 662
long-term memory, 954
    definition of, G–16
loop of Henle, 755
    function of, 757–759
lophophorates, 545–546
lophophore, 545
Lorenz, Konrad, 949, 952, 952*f*
Lotka, A.J., 985
Lotka–Volterra theory, definition of, G–16
low-density lipoprotein(s), 120*f*
lower epidermis, in plants, 624
luciferase, 458
luciferin, 458
Lucy *(Australopithecine afarensis)*, 584
lumen, 88, 151
    definition of, G–16
lunar cycle, 963
lung(s), in vertebrates, 838–843
    evolution of, 838–840

lung cancer, 844–845
luteinizing hormone, 773, 899
    actions of, 784t
Lycophyta, 502, 503f, 506–507, 507f
Lycopodium, 506, 507f, 507t
Lyell, Charles, 7, 13–14
    Principles of Geology, 7
lymph, 828
lymphatic collecting ducts, 828
lymphatics, 828
lymphatic system, 828, 829f
    definition of, G–16
lymph capillaries, 828
lymph node(s), 828, 863–864, 864f
    definition of, G–16
lymphocyte(s), 106, 863f
    activity of, 865t
    definition of, G–16
    diversity of, 876–877
        origin of, 877f
lymphoid tissue(s), 862f, 863–864
    definition of, G–16
    primary, 863
    secondary, 863
lymphotoxin, 872
lymph vessels, 828
Lyon hypothesis, 250–251
    definition of, G–16
Lyon, Mary, 251
Lysichiton americanum, 596
lysine, 58f
lysis, definition of, G–16
lysogenic cycle, 307–308
    definition of, G–16
lysogenic state, 418
lysogeny, 307, 309f
    definition of, G–16
lysosome(s), 79f, 90–92, 92f
    definition of, G–16
    function of, 97t
lysozymes, 859
lytic cycle, 307, 418
    definition of, G–16

M

MacArthur, R.H., 1011
MacDougal, D.T., 638
macrocyst, 456
Macrocystis, 464f
macroevolution, 378
macromolecules, 32, 44

macronucleus, of paramecium, 451
macronutrients, 646–647
macrophage(s), 422, 861–862
    definition of, G–16
madreporite, 559
maggot fly, sympatric speciation in, 392
magnesium, 22t
major histocompatibility complex (MHC),
        868
    definition of, G–16
    and self-recognition, 868
major histocompatibility complex proteins,
        868, 868f
malaria, and sickle cell anemia, 354–355
malate, 168
malleus, 730
Malpighian tubule(s), 754, 754f
    definition of, G–16
maltase, 802
Malthus, Thomas, 13, 351
Mammalia, 575–584
mammals. *See also* marsupial(s);
        monotremes; placental mammals
    cold-adapted, 746, 747f
    digestive system of, 793–794
    divergence of, 581f
    evolution of, 578–579
    fertilization in, 892–894, 914–915
    foraging in, 793–794
    gastrulation in, 920
    heart of, 817–818
mammary gland(s), 575, 892
mandible(s), 553
    definition of, G–16
    in insects, 557, 790
manganese, 22t
Mangold, Hilde, 931
mangos, flowering in, 666
mangrove, gas exchange in, 645, 645f
mannitol, 464
mantle, of mollusks, 546
mantle cavity, of mollusks, 546
Manx cats, 371
mapping, 247f, 247–248
    of *Escherichia coli,* 315f, 315–316
map sense, 964
map units, 247
mara, 5–6, 7f, 359, 394–395
Marchantia
    gametophyte, 495f
    life cycle of, 494–495, 496f
    sexual structures of, 497f
marine environment, 751–752

organization of, 1055f
marine flatworm, 528f
marine iguana, 8f
marine province, 1055–1059
mark-and-recapture experiment, 12
Marlen, Peter, 953
Marsh, O.C., 359
marshes, 1054
marsupial(s), 359–361, 578–579, 580f
    Australian, 395, 396f
    definition of, G–16
    fertilization in, 892, 894f
masked booby, 8f
mass, 22
    of atoms, 22–23
    of molecule, 23
Masters, William H., 896
masturbation, definition of, G–17
mating type, definition of, G–17
matrix, 94, 182
    definition of, G–17
matter, 21
maxillae, in insects, 557, 790
maximum population growth rate,
        generation time and, 992, 993f
Mayr, Ernst, 353, 379
McCarty, M., 260
McClintock, Barbara, 340–343, 343f
McLeod, C., 260
meadowlarks, interbreeding in, 393, 394f
medicine, genetic engineering in, 327–330
medulla, 755
    definition of, G–17
medulla oblongata, 719
medusa, 537
megagametophyte(s), 507
    definition of, G–17
    in flower, 598
megaphylls, 506
megasporangium(/a), 507
    in flower, 597
megaspore(s), 507
    definition of, G–17
megasporocyte, 515
    definition of, G–17
    in flower, 597
megasporogenesis, 598f–599f
    in flower, 597
meiosis, 210–214, 210f
    complications of, 217
    definition of, G–17
    and evolution, 215–218

in females, 214–215, 216*f*
in humans, 215–218
I, 210*f*, 212–213
II, 211*f*, 213–214
versus mitosis, 215*f*
in ovary, 900*f*
Meissner's corpuscles, 726, 727*f*
melanin, definition of, G–17
melanism, definition of, G–17
melanocytes, 774
melanocyte-stimulating hormone, 774
actions of, 784*t*
and skin pigmentation, 774, 775*f*
melatonin, 782
actions of, 784*t*
membranal pump, definition of, G–17
membrane potential, 115
membrane proteins, 104–105, 105*f*
memory
experiments on, 954–955
formation of, 956*f*
learning and, 953–955
long-term, 954
short-term, 954
theories of, 954–955
memory cell(s), 864
definition of, G–17
Mendel, Gregor Johann, 221, 222*f*, 353
experiments of, 222–227, 223*t*
testcrosses by, 230
Mendelian genetics, 220–235
decline and rise of, 232–233
Mendel's laws
chromosomal basis for, 231
first, 227–228
second, 228–230
meninges, 712
definition of, G–17
menopause, definition of, G–17
menstrual cycle, 901–903
definition of, G–17
menstruation, 901
definition of, G–17
meristem, definition of, G–17
meristematic tissue, 604
merozoite(s), 451
definition of, G–17
Meselson, Matthew, 270–271
mesenchyme, 918
definition of, G–17
mesentery(ies)
definition of, G–17
in Rotifera, 544, 545*f*
mesoderm, 538–541, 917

definition of, G–17
derivatives of, 920*f*
mesoglea, 537
mesonephric ducts, 927
mesosome, 75, 429
messenger RNA, 88, 279–283
definition of, G–17
eukaryotic, 282*f*
and genetic code, 280–281
storage of, regulation through, 303–304
structure of, 281–283
variations in, regulation through, 304
metabolic pathway(s), 135, 137*f*
allosteric control of, 138*f*
definition of, G–17
metabolic water, 184
metabolism, definition of, G–17
metabolite, definition of, G–17
metamerism, 549
metamorphosis
complete, 556, 556*f*
hormonal control of, 770–771, 771*f*
incomplete, 556
metaphase
meiosis I, 212–213
mitosis, 204
metaphase plate, 204
metastasis, 845
Metazoa, definition of, G–17
methane, 31
geometry of, 31, 31*f*
methanogens, 426
methionine, 58*f*
methylation, 303
5-methyl cytosine, 303*f*
methyl group, 34*t*
Meyers, Robert E., 717
MHC restriction, 869
definition of, G–17
mice, behavior experimentation with, 946, 946*f*
micelles, 54, 55*f*, 103
microevolution, 366–377
definition of, G–17
microfibrils, of cellulose, 49, 50*f*
microgametophyte(s), 507
definition of, G–17
micrometer, definition of, G–17
micronucleus, of paramecium, 451
micronutrients, 647
microorganism(s), 414
definition of, G–17
friendly, in human defenses, 858

microphylls, 506
micropyle, 516, 597
microscope
high-voltage electron, 71
invention of, 68
light, 71
scanning electron, 71
transmission electron, 70–71
microsporangia, 507
microspore(s), 507, 593
definition of, G–17
microsporocyte(s), 515
definition of, G–17
in flower, 597
microsporogenesis, 598, 598*f*–599*f*
definition of, G–17
microtubular organizing centers, 205
microtubule(s), 82–84, 83*f*
structure of, 84*f*
microvilli, 74, 79*f*, 81, 83*f*
definition of, G–17
of small intestine, 797
midbrain, 719–720
definition of, G–17
middle ear, 730, 731*f*
middle lamella, definition of, G–17
midgut, in insects, 791
midrib, 623
Miescher, Friedrich, 259
migration, 966–967
evolution of, 966–967
in monarch butterflies, 966, 967*f*
starling route for, 964*f*
Miller, Stanley, 404
Miller, T.E., 643
Miller–Urey experiment, 404, 406*f*
millipedes, 553
mimicry, in plants, 596, 597*f*
Mimosa pudica, *662*
mineral nutrient, definition of, G–17
mineralocorticoids, 779
minerals, 135
in diet, 807–809, 809*t*
essential to plants, 647*f*
minimum mutation tree, definition of, G–17
minus strand RNA viruses, 420
missense mutation, definition of, G–18
Mitchell, Peter, 143–144, 167
mitochondria, 94–97, 99*f*, 144*f*, 183*f*
in aerobic cell respiration, 177–181
definition of, G–18
development of, 96
function of, 97*t*

mitochondria *continued*
  heat production by, 187–189
  inner compartment of, 94, 183
  inner membrane of, 94, 183
  outer compartment of, 94, 183
  outer membrane of, 94, 183
  stationary electron carriers of, 143
mitosis, 203–210, 205*f*, 208*t*–209*t*
  advantages of, 210
  definition of, G–18
  versus meiosis, 215*f*
mitral valve, 821
model, definition of, G–18
modern synthesis, 353
molars, 794
mole, 38
  definition of, G–18
molecular biology
  definition of, G–18
  evolutionary evidence in, 362–363
  materials for, 319, 319*f*
molecular mass, 23
molecular orbitals, 27, 28*f*
molecule, 22
  definition of, G–18
Mollusca, 527, 546*f*, 546–548
  body plan of, 547*f*
mollusks
  cephalopod, brain in, 709, 710*f*
  ganglia of, 709
  movement in, 671
  shells of, growth of, 673–674, 674*f*
molting
  in arthropods, 551
  definition of, G–8
  hormones affecting, 770, 770*f*
molting hormone (MH), 770
molt-inhibiting hormone (MIH), 770
molybdenum, 22*t*
monarch butterflies, migration of, 966, 967*f*
Monera, 382–384, 425–440. *See also* bacteria
  definition of, G–18
monoamines, 720
monoclonal antibody(ies), 877–878
  definition of, G–18
  production of, 878*f*
monococcus, 428
monocot, definition of, G–18
Monocotyledonae, 519, 520*f*
monocytes, 861
Monod, Jacques, 293–294
  model of operon, 295*f*
monoecious, 495, 887
  definition of, G–18

monohybrid cross, 222
monohybrid, definition of, G–18
monomers, 44
  of life, 405
monophyletic, definition of, G–18
monosaccharide(s), 45
  definition of, G–18
monotremes, 578–579, 580*f*
  definition of, G–18
  fertilization in, 892–893
mons pubis, 896
mons veneris, definition of, G–18
Morgan, Thomas H., 246*f*, 247–249
morph, 354
  definition of, G–18
morphogen, 933
mortality rate, 997
  definition of, G–18
morula, 916
  definition of, G–18
mosaic development, 931
  definition of, G–18
moss(es), 494, 497–500
  gametophyte of, 497, 498*f*
  life cycle of, 498–500, 500*f*
  sexual structures of, 499*f*
  sporophyte of, 499*f*
motor cells, 661
motor end plate, definition of, G–18
motor impulses, 693
motor neuron(s), 693, 694*f*
  definition of, G–18
movement, 670–689
movement reception, 732
mRNA. *See* messenger RNA
mucin, 797
mucosa, 898
  definition of, G–18
  of esophagus, 796–797
mucous membrane(s)
  in defense, 858
  definition of, G–18
mucus, definition of, G–18
mud flats, 1059, 1059*f*
mules, sterility of, 388, 394, 395*f*
Müllerian ducts, definition of, G–18
Müllerian mimicry, definition of, G–18
multicellular, definition of, G–18
multicellular organisms, 445–446
multiple alleles, 238–240, 239*f*
  definition of, G–18
multiplicative law, 227

definition of, G–18
Munch, Ernst, 644
muscle(s), 675, 676*f*
  contracted, 686*f*
  human, 683*f*
  insertion of, 673
  origin of, 673
  relaxed, 686*f*
  in vertebrates, 681–689
muscular dystrophy, 242–243
muscularis, of esophagus, 796
mushroom(s)
  cap, 484
  cautions with, 484–485
  growth and sexual reproduction in, 484
  life cycle of, 486*f*
  stalk, 484
muskegs, 1049
mutagen, 335
  definition of, G–18
mutualism, 435, 974
  definition of, G–18
  in fungi, 475
mutant, definition of, G–18
mutate, definition of, G–18
mutation(s), 273
  base substitution, 337, 338*f*
  and cancer, 342
  chain-termination, 337–339
  chromosomal, 335, 340
  definition of, G–18
  evolutionary tree based on, 363*f*
  and evolution of gene structure, 344
  frameshift, 339
    definition of, G–11
  homeotic, 934–935, 936*f*
  missense, definition of, G–18
  neutral, 337, 338*f*
    definition of, G–19
  and phenotype, 335*f*
  point, 335, 336–339
    definition of, G–23
    in noncoding regions of DNA, 339
  and selection, 370
  sickle cell, 337, 338*f*
  silent, 337, 338*f*
    definition of, G–27
  somatic, 334, 335*f*
    definition of, G–27
  Toguchi, 337, 338*f*
mutation rate, definition of, G–18
mycelium, 472–473
  definition of, G–18
mycoplasmas, 437, 437*f*
mycorrhizae, 475, 645–646
  definition of, G–18

myelin, and impulse velocity, 700–701

myelin sheath, 693
  definition of, G–18

myofibrils, 684

myofilaments, 684

myosin, 684, 687f
  definition of, G–18

myosin bridge(s), 686, 687f
  definition of, G–18

myosin heads, 686

myxameba, 456

myxedema, 776

myxobacteria, 432

Myxomycota, 454, 455t

myxospores, 432

N

N-acetylglucosamine, 49

NAD (nicotinamide adenine dinucleotide), 140
  definition of, G–18

NADP⁺, reduction of, 157

NADP (nicotinamide adenine dinucleotide
  phosphate), 140
  definition of, G–18

NADP reductase, 156

Na⁺/K⁺. *See* sodium/potassium ion exchange
  pump

nanoplankton, 1056

nasopharynx, 795

nastic movements, 661–662

nastic response, definition of, G–18

natural increase, 1063t
  rate of, calculation of, 1063t

natural killer (NK) cell(s), 862
  definition of, G–18

natural selection, 351, 352f, 355–358
  and apomixis, 607–608
  definition of, G–19
  in giraffes, 373f
  outcomes of, 370–372
  theory of, 12–13

navigation, 964–967
  definition of, G–19

Neanderthals, 587f, 587–588

nectar guides, 596

negative feedback, definition of, G–19

negative feedback loop, 137, 741–742, 742f

Neisseria gonorrhoeae, 436f, 437

nematocysts, 537

Nematoda, 527, 541–543
  movement in, 671

in soil, 543f

neon, orbitals of, 25f

nephridia, 546, 550, 753f, 753–754
  definition of, G–19

nephron(s), 755
  definition of, G–19
  function of, 755–760, 757t
    control of, 759–760
  microanatomy of, 755
  transport processes in, 758f

neritic province, 1055–1056, 1058
  definition of, G–19

nerve(s), 675, 676f, 695, 696f
  definition of, G–19

nerve cord, in chordates, 565

nerve growth factor, 933

nerve net, 537, 709, 709f

nerve rings, 709, 709f

nervous system(s)
  cells of, 693–695
  functions of, 692–693
  human, 712–723
    development of, 925
    organization of, 712f
  in invertebrate, 709–710
  of vertebrates, 710–712

net community productivity (NCP), 1023
  definition of, G–19

net productivity, 1023
  definition of, G–19

neural canal, definition of, G–19

neural codes, 725

neural control, versus hormonal, 765

neural crest cells, 921

neural folds, 921
  definition of, G–19

neural groove, 921
  definition of, G–19

neural impulse(s), 693
  definition of, G–19
  origin and transmission of, 695–701

neural plate, 921
  definition of, G–19

neural tube, 921
  definition of, G–19

neuroglia, definition of, G–19

neuromuscular junction(s), 684, 693
  definition of, G–19

neuron(s), 692–707
  definition of, G–19
  structure of, 693–694, 695f
  types of, 694f

neuropeptide(s), 720, 775
  definition of, G–19

neurosecretions, 765

Neurospora crassa, as experimental
  organism, 306

neurotransmitter(s), 693
  definition of, G–19

neurula, 921

neurulation, 921, 921f

neutralist(s)
  definition of, G–19
  versus selectionists, 374–375

neutral mutation, 337, 338f
  definition of, G–19

neutron(s), 22, 23f
  definition of, G–19

neutrophil(s), 861
  definition of, G–19

newborn, physiologic changes in, 928
  circulatory, 930f

nexin, 86

niche, 985, 1010–1011
  definition of, G–19
  fundamental, 985, 985f
  realized, 985, 985f

niche separation, 1011, 1012f

nicotinamide, 140

Nikolas II (Czar of Russia), 253

nitrification, 1025
  definition of, G–19

nitrogen, 21, 22t, 32

nitrogen base, 64
  definition of, G–19

nitrogen cycle, 1025, 1026f
  definition of, G–19

nitrogen fixation, 440, 1025
  definition of, G–19

nitrogen-fixing bacteria, 32, 32f, 1025–1027

nitrogen wastes, production of, 751, 751f

nodes, 622, 694

noncompetitive inhibition, 136

noncyclic phosphorylation, definition of, G–19

noncyclic photophosphorylation, 155–160,
  158f

nondisjunction, definition of, G–19

nonhistone chromosomal protein, definition
  of, G–19

nonoxynol-9, in HIV prevention, 880

nonpolar covalent bonds, 27–28

nonpolar molecules, water and, 36

nonspecific defense(s), 857–862
  cellular, 861–862, 862t
  chemical, 858–861, 860t
  definition of, G–19
  primary, 857–858

nonvascular plants, definition of, G–19

norepinephrine, 780
    actions of, 784t

normal distribution, 370
    definition of, G–19–20

Norplant, 903

Nostoc, 440f

notochord, 921
    in chordates, 565
    definition of, G–20

N-terminal, 59, 285
    definition of, G–18

nucellus, in flower, 597

nuclear envelope, 87, 201

nuclear pores, 87, 201

nucleases, 805

nucleic acid(s), 62–65
    definition of, G–20
    digestion of, 805
    structure of, 64–65

nucleoid, 75

nucleolus, 87
    definition of, G–20
    function of, 97t

nucleosomes, 201

nucleotide(s), 64, 64f
    definition of, G–20

nucleus (atomic), 22, 23f

nucleus (cellular), 77, 86–88, 87f, 201f
    definition of, G–20
    function of, 97t
    in reproduction, 201–203

nutrition, 806–809
    definition of, G–20

nyctinastic movements, 661, 662f

nymph, 557

O

Obelia, life cycle of, 538f

obligate anaerobe(s), 426, 434
    definition of, G–20

occipital lobe, 714, 715f

ocean
    food pyramid of, 1058f
    salinity of, 1054t

ocean currents, 1037, 1038f

oceanic province, 1055–1056
    definition of, G–20

oceanic rift community, definition of, G–20

ocelli, 557

octopus, 549

oils, 51–52

okapi, evolution of, 398f–399f

Okazaki fragment(s), 268
    definition of, G–20

Oken, Lorenz, 69

olfaction, definition of, G–20

olfactory bulb, 729

olfactory epithelium, 729

olfactory receptors, 728
    in humans, 729

Oligochaeta, 550

oligotrophic, definition of, G–20

omasum, 794

ommatidia, 554, 555f, 732–733, 735f

omnivores, 1020

omnivorous, definition of, G–20

Oncidium, 597f

oncogene(s), 342
    definition of, G–20

oncogenic viruses, 415

onion, seedling growth in, 611

On the Origin of Species (Darwin), 13–14, 352

Onychophora, 558, 558f

oocyte(s)
    definition of, G–20
    development of, 898
    primary, 214, 898
    secondary, 214, 898

oogenesis, 214–215, 216f

oogonia, 457, 898
    definition of, G–20

Oomycota, 454, 455t, 457–458

Oparin, A.P., on origin of life, 403

open communities, 1007, 1007f
    definition of, G–20

open growth, 611–612
    definition of, G–20

operant conditioning, 951, 951f
    definition of, G–20

operator, 294
    definition of, G–20

operculum, 498, 837
    in mollusks, 547

operon, 293–298
    definition of, G–20
    inducible
        characteristics of, 298f
        for digesting lactose, 294–296, 295f
    repressible
        characteristics of, 298f
        for synthesis of tryptophan, 296–298, 297f
    structure of, 294f

Ophiuroidea, 559

Ophys speculum, 597f

opioid neurotransmitters, 720

opossum, 361

opsin, 734

opsonization, 868
    definition of, G–20

optic cup, 932

optic nerve, 734

optic nerve tract, 734

optic vesicle, 932

optimal curve, 984, 984f

optimal temperature range, 987, 987f

Opuntia cactus, 8f

oral cavity, in humans, 795

oral contraceptives, 903

oral groove, 451

orbitals, 24–25
    molecular, 27, 28f
    of neon, 25f
    2p, 24
    3p, 25
    1s, 24
    2s, 24
    3s, 25

organ(s), 532
    definition of, G–20

organelle(s), 77
    definition of, G–20
    functions of, 98t

organic horizon, 989

organic molecule, definition of, G–20

organization, levels of, 532, 534f

organ of Corti, 731f, 732

organogenesis, 920–921
    in vertebrates, 921, 922f

organ system, 532

orgasm
    definition of, G–20
    in females, 907
    in males, 907

Orgel, Leslie E., 408

orientation, 950, 964–967
    definition of, G–20
    experiments on, 964–965, 965f
    fixed action patterns and, 950, 950f

origin, 673
    definition of, G–20

origins of replication, 267–270, 271f

Oscillatoria, 439, 439f

osculum, 535

osmoconformer(s), 751–752

definition of, G–20

osmoregulation, 751–754
 hormonal control of, 760f

osmoregulator(s), 752
 definition of, G–20

osmosis, 111–113, 112f
 in cell, 112–113
 definition of, G–20
 encouragement of, 117, 117f

osmotic environment, 751

osmotic gradient, definition of, G–20

osmotic potential, definition of, G–20

osmotic pressure, 112
 definition of, G–20

Osteichthyes, 570–572, 571f

osteoblasts, 678, 778

osteoclasts, 678, 777

osteocyte(s), 678
 definition of, G–20

ostia, 814

ostracoderms, 567

Ostriches, 13, 14f

outer compartment, definition of, G–20

outer ear, 730, 731f

ova, 898

oval window, 730

ovarian cycle, 903f
 follicular development during, 903f

ovary(ies), 593, 593f
 definition of, G–20
 of flower, reproductive events in, 597–598
 hormones of, 784, 782
 in humans, 898, 902f
 in insects, 889
 in plants, 520, 521f

overdominance, definition of, G–20

oviduct(s), 889
 definition of, G–20
 in insects, 889

oviparous, definition of, G–20

ovipositor(s), 556, 889

ovoviviparous, definition of, G–20–21

ovulation, 901–903
 definition of, G–21

ovule(s)
 definition of, G–21
 in flower, 597

oxaloacetate, 168, 177

oxidation, 140–141
 in citric acid cycle, 180
 definition of, G–21

oxidation-reduction reaction, definition of,
 G–21

oxidative phosphorylation, definition of,
 G–21

oxidative respiration. *See* cell respiration

oxidizing agent, 141

oxygen, 21, 22t
 as electron acceptor, 184
 partial pressure in body, 849f
 from plants, 166
 transport of, 849–851

oxygen debt, 192
 definition of, G–21

oxyhemoglobin, 849
 definition of, G–21

oxytocin, 775–776
 actions of, 784t
 in female orgasm, 897
 structure of, 766f

ozone layer, holes in, 1071

P

P680, 155
 restoration of, 157

P700, 155

Paal, Arpad, 651–652

pacemaker, 821

Pacinian corpuscles, 726, 727f

pairing, 213f

palate, in vertebrates, 839

palisade parenchyma, 623

palmitic acid, 193

palomino horses, incomplete dominance in,
 238, 239f

palps, 790
 sensory, 557

pancreas, 800, 800f
 definition of, G–21
 in digestion, 800

pancreatic amylase, 802

pancreatic duct, 800

Pangea, 391

Panthera, 385

papillary muscles, 821

parabronchia, 841

paracrine hormones, 765

paramecium, 451–454
 reproduction in, 451–454, 454f

Paramecium caudatum, 453f

paramesonephric ducts, 927

paramylon, 460

parapatric, definition of, G–21

parapodia, 551

parasite(s), 76, 433, 474
 definition of, G–21

parasitism, 975, 1012–1013
 in fungi, 475

parasympathetic division, definition of,
 G–21

parasympathetic nerves, responses to, 723t

parathyroid glands, 776f, 777–778
 definition of, G–21
 hormones of, 784t

parathyroid hormone, 777
 actions of, 784t
 in regulation of body calcium, 778f,
 777–778

Parazoa, definition of, G–21

parenchyma, 615
 definition of, G–21

parietal cells, 797

parietal lobe, 715, 715f

parthenogenesis, 445, 544, 886, 886f
 definition of, G–21

partial dominance, 237–238, 238f
 definition of, G–21

partial pressure, 847–848
 definition of, G–21

passive immunity/immunization, 876
 definition of, G–21

passive transport, 109–114
 definition of, G–21

Pasteur, Louis, 416

pasteurization, 435
 definition of, G–21

Patagonian hare, 5–6, 7f, 359, 394–395

patellar response, 705

pathogen(s), 76, 434
 definition of, G–21

peach-faced lovebirds, behavior
 experimentation with, 947, 947f

peat, 989

pectins, 49

pedigree, 232, 232f

pelagic, definition of, G–21

pellicle, 451

pelvic girdle, 678
 definition of, G–21

pelvic inflammatory disease (PID), 436
 definition of, G–21

pelvis, definition of, G–21

Pelycypoda, 547

penetrance, incomplete, 241–242, 243f

Penfield, Wilder, 954

penicillin, 309, 481
 definition of, G–21

Penicillium, 481
Penicillium camembertii, 481
Penicillium roquefortii, 481
penis
    definition of, G–21
    human, 894
        erectile tissue of, 895, 895f
    in invertebrates, 887
PEP carboxylase, 168
PEP cycle, definition of, G–21
peppered moth
    mark-and-recapture experiment on, 357,
        357t
    natural selection in, 355–357, 356f
pepsin, 797
pepsinogen, 797
peptide, 57
    definition of, G–21
peptide bond, 56, 59
    definition of, G–21
peptide hormone(s), 767–769
    definition of, G–21
peptidoglycan(s), 75, 429
perennial(s), 611
    definition of, G–21
perforin, 872
pericarp, 604
pericycle, 618
    definition of, G–21
periderm, 615
    definition of, G–21
    formation of, 628f
    growth of, 626–627
periosteum, 677
    definition of, G–21
peripheral nervous system, 712, 723
    definition of, G–21
peripheral proteins, in plasma membrane,
    104
peristalsis, 796, 797f
    definition of, G–21
peritoneum
    definition of, G–22
    in Rotifera, 544, 545f
peritubular capillaries, 755
permafrost, 1044
permeability, of plasma membrane,
    109
peroxisomes, 92–93, 93f
petals, 519, 521f, 593, 593f
Peters, John, 781
petiole, 623
Peziza, reproduction in, 480,
    481f

pH
    definition of, G–22
    and enzyme activity, 135, 135f
Phaeophyta, 458, 464f, 464–465
    characteristics of, 459t
phage
    definition of, G–22
    DNA experiments with, 260–261
phagocyte, definition of, G–22
phagocytic vesicle, 118
phagocytosis, 118, 118f
    definition of, G–22
phagosome(s), 93, 118
    definition of, G–22
pharyngeal gill slits, in chordates, 565
pharynx
    in earthworm, 790
    in fish, 838
    in humans, 843
    in planarians, 540
phenetics, 386, 387f
    definition of, G–22
    species designation through, 386f
phenotype(s), 222
    conditional, 241–244
    definition of, G–22
phenylalanine, 58f
    metabolic pathway of, 273f
phenylketonuria, 243, 272–273
pheophytin, 155
pheromones, 969–970
Philadelphia chromosome, 340, 342f
phloem, 501, 616–617
    definition of, G–22
    pressure flow in, 644f
    primary, definition of, G–24
    role of, 633
    secondary, 625
        definition of, G–26
    transport by, 643–644
    transport in, mechanism of, 644
phloem fiber(s), 616
    definition of, G–22
phloem parenchyma, 616
phloem ray, definition of, G–22
phloem sap, 643
    sampling, aphids in, 643f, 643–644
phloem tissue, 617f
Phoronida, 545–546
phosphate, 32
phosphate group, 32, 34t
phosphocreatine, 192
phosphoenolpyruvate, 168
phosphofructokinase, 177
3-phosphoglycerate, 162, 176

phospholipid bilayer, 55
phospholipids, 53–54, 54f
    function of, 56t
    in water, 54, 55f, 104f
phosphoproteins, 62
phosphoric acid, 33f
phosphorus, 21, 22t, 32
    electron shells in, 24t
phosphorus cycle, 1027f, 1027–1028
phosphorylation, 139, 140f
    in Calvin cycle, 164
    chemiosmotic. *See* chemiosmotic
        phosphorylation
    definition of, G–22
    substrate-level. *See* substrate-level
        phosphorylation
photoautotroph(s), 434, 438
    definition of, G–22
photoautotrophic, 77
photon(s), 148–149
    definition of, G–22
photoperiodism, 662–663, 664f, 988
    definition of, G–22
photophosphorylation
    cyclic, 155, 161, 163f
    noncyclic, 155–159, 160f
photoreactivation repair, 336
photoreceptor, 460
photorespiration, 162–169
    definition of, G–22
photosynthesis, 77, 94, 148–171, 152f
    versus carbohydrate metabolism,
        173f
    cyclic events of, 161, 163f
    definition of, G–22, 149
    investigations in, 150–151
    light in, 150
    light-independent reactions in, 160–162
    light reactions of, 156–160
    participants in, free energy levels of, 150f
    parts of, 149, 151
    water in, 150
photosynthetic pigments, and absorption of
    light, 158f
photosystem I, 151, 153f
    definition of, G–22
    in photosynthesis, 157
photosystem II, 151, 153f
    definition of, G–22
    in photosynthesis, 157
phototropism, 658–659
    definition of, G–22
pH scale, 39–40, 40f
phylogenetic tree, definition of, G–22
phylogeny, 382
    definition of, G–22

phylum, definition of, G–22

physiological ecology, 981

physiology, definition of, G–22

phytochrome, and flower initiation, 664–666

Phytophthora infestans, 457, 457*f*

phytoplankton, 458, 1056
  definition of, G–22

pigment, definition of, G–22

pili, 75, 312, 429

Pilobolus, 478

pine, life cycle of, 515, 517*f*

pineal body, 782–783, 783*f*
  definition of, G–22
  hormones of, 782–783, 784*t*

pineapples, flowering in, 666

pinna, 730

pinocytosis, 118*f*, 119
  definition of, G–22

pioneer organisms, 497, 1015, 1015*f*
  definition of, G–22

pistil, 519, 521*f*, 593, 593*f*
  definition of, G–22

pitcher plant, 647, 648*f*

pith, 622
  definition of, G–22

pituitary dwarfs, 774, 775*f*

pituitary giants, 774, 775*f*

pituitary gland
  control by hypothalamus, 772, 773*f*
  definition of, G–22
  function of, 771–776

pivotal joint, 678

placebo, 11
  definition of, G–22

placenta, 892, 893*f*, 925, 925*f*
  definition of, G–22
  in flower, 597

placental mammals, 579, 580*f*
  development of, support system in,
    923–925
  fertilization in, 892

Placodermi, 568–569

planarians, 540*f*, 540–541

planetarium, 965, 965*f*

plankton, 1056
  definition of, G–22

plantae, 384

plant cell, osmotic responses of, 114*f*

plants, 490–525, 491*f. See also* flowering
    plants
  alternation of generations in, 491, 492*f*
  cells, elongation of, 620*f*
  climbing, 650, 651*f*
  definition of, 491
  in deserts, 1040, 1041*f*

differentiation in, 614–615
  cytokinins and, 655*f*
evolution of, 490–493
  adaptations to land, 493, 493*f*
flowering. *See* angiosperm(s)
flowers of. *See* flower(s)
genetic engineering in, 326–327,
    326*f*–327*f*
growth of, 610–630
growth response in, 651–652, 652*f*
hormones in, 652–658, 653*t*
  applications of, 658
insectivorous, 647, 648*f*
minerals essentials to, 647*f*
nonvascular. *See* bryophytes
nutrition in, 645–647
regulation in, 651–667
response in, 651–667
response to touch, 662, 663*f*
seed. *See* seed plants
seedless. *See* seedless plants
structural support in, water and, 113,
    115*f*
tendrils, 659
tissue organization in, 613–617
transport in, 632–649
  of food, 643–644
  of gas, 644–645, 645*f*
  of minerals, 633–643, 645–647
  of water, 633–643, 634*f*, 640*f*
vascular. *See* vascular plants

plant starches, 47

planula, 537

plaques, 307, 308*f*, 826

plasma, 815
  definition of, G–22

plasma cell(s), 873
  definition of, G–22

plasma membrane, 53, 79*f*, 80–81, 102–108,
    103*f*
  definition of, G–22
  fluidity of, 108, 110*f*
  function of, 74, 97*t*
  functions of, 80
  permeability of, 109
  recycling, 119
  structure of, 80–81, 81*f*

plasmid(s), 75, 312, 314*f*, 429
  definition of, G–22

plasmid genes, 316

plasmodesmata, 121*f*, 122–123, 637

plasmodial slime molds, 455*f*, 455–456

plasmodial stage, 455

Plasmodium polycephalum, 455

Plasmodium vivax, 451, 452*f*

plasmolysis, 113

plastid(s), 94
  definition of, G–22

plastiquinone, 155

plastocyanin, 156, 157

plastoquinone, 157

plateau phase
  in females, 906–907
  in males, 907

platelets, 815–816
  definition of, G–22

plates, 389

plate tectonics, definition of, G–23

Plato, 348–349

Platyhelminthes, 527, 538–541

pleiotropy, 243
  definition of, G–23

Plesiadapis, 582*f*

pleurae, 843

pleural cavity, 843
  definition of, G–23

plumule, 604

plus strand RNA viruses, 420

Pneumococcus, 259*f*
  transformation in, 259–261, 260*f*

poikilotherm(s), 742
  definition of, G–23

point mutation(s), 335, 336–339
  definition of, G–23
  in noncoding regions of DNA, 339

polar body, 215
  definition of, G–23

polar covalent bonds, 27

polarity, 912
  definition of, G–23

polar nuclei, in flower, 598

polar spindle fibers, 204

pole cells, 916

pollen, 512–513, 513*f*

pollen grains, 593

pollen sacs, 593

pollen tube(s), 512, 598

pollination, 512, 595–598, 600*f*
  definition of, G–23
  by insects, plant devices for, 596, 596*f*
  specialized, 393, 394*f*
  by wind, 596–597, 597*f*

pollinators, animal, floral adaptations to,
    595*t*, 595–596

poly-A tail, 282
  definition of, G–23

Polychaeta, 550

polydactyly, inheritance of, 242, 243*f*

polygenic inheritance, 244, 244*f*
  definition of, G–23

polygenic trait, definition of, G–23

Polyisophonia, 463*f*

polymer(s), 44
    definition of, G–23
    formation of, 405–407
polymerase, definition of, G–23
polymerase chain reaction, 330
polymorphism, definition of, G–23
polyp, 537
polypeptide(s), 56, 59–61
    definition of, G–23
    synthesis of, 88–90
polyphyletic, definition of, G–23
Polyplacophora, 547
polyploid, definition of, G–23
polyploidy
    in hybrids, 389f
    sympatric speciation through, 388–389
polyribosome(s), 89, 287, 288f
    definition of, G–23
polysaccharide(s), 47–49, 48f
    definition of, G–23
polysomes. See polyribosome(s)
polyspermy, definition of, G–23
polytene, 301
Polytrichum, 500f
polyunsaturated, definition of, G–23
polyunsaturated fat, 52
pome, 602
ponds, 1053–1054
Pongidae, 580
pons, 720
    definition of, G–23
poppy, 603f
population(s), 981, 989, 990f. See also
        human population
    changes in, in time, 991
    crash of, 995, 995f
    definition of, 366, G–23
    ecology of, 989–998
    growth of
        exponential. See exponential growth
        intrinsic rate of, 992
        logistic. See logistic growth
        models of, 992
    regulation of, mechanisms of, 1002–1003
    structure of, 1069
population bottleneck, 373–374, 374f
population genetics, 366
    definition of, G–23
    of G.H. Hardy, 367
Porifera, 527, 532–535
porocytes, 535
portal circuit, definition of, G–23
Porter, Keith, 120

positive feedback loop, 742, 742f
postanal tail, in chordates, 565
posterior, definition of, G–23
posterior chamber, 734
posterior pituitary, 771
    definition of, G–23
    hormones of, 784t, 774–776
    hypothalamus and, 772, 773f
postsynaptic membrane, 702
postsynaptic neuron, 702
potassium ion gates, 698, 699f, 700
potassium ions, and stomata, 641–642, 642f
potential energy, 127
    definition of, G–23
potometer, 638
The Power of Movement in Plants (Darwin),
        651
praying mantis, 557f
preadaptation, 839
precapillary sphincter(s), 827
    definition of, G–23
precocial, definition of, G–23
predation, 1013–1014
    definition of, G–23
predator(s), 474, 1013–1014
    definition of, G–23
    interactions with prey, 1013f
predictions, 10–11
prefrontal area, 715, 715f
premolars, 794
preproductive period, 1000, 1000f
pressure flow hypothesis, 644
    definition of, G–23
presynaptic membrane, 702
    definition of, G–23
presynaptic neuron, 702
prey, 1013–1014
    interactions with predators, 1013f
prey switching behavior, definition of, G–23
Priestley, Joseph, 150
primary bronchi, 843
primary chorionic villi, 923
primary endosperm nucleus, 598
primary follicle, 901
primary growth, 611
    definition of, G–23
primary immune response, 863, 870–874,
        875f
    definition of, G–23–24
    versus secondary, 876f
primary lesions, 336, 336f
primary oocytes, 214, 898

primary organizer, 931–932
    search for, 933
primary producer, definition of, G–24
primary sex characteristics, 900
primary spermatocyte, 214, 895
primary succession, 1015
    definition of, G–24
primary transcript(s), 278
    nuclear storage of, regulation through, 303
    processing of, alternative, 303, 304f
primase, 268
Primates, 579–583
    brachiation in, 583f
    evolution of, 582f
primitive, definition of, G–24
primitive groove, 920
    definition of, G–24
primitive streak, 920
    definition of, G–24
primitive traits, 385
primordial follicle, 902
principle of competitive exclusion, 985–986
Principles of Geology (Lyell), 7
prion(s), 423, 425
    definition of, G–24
probability
    definition of, G–24
    and genetics, 227f, 227–228
procambium, 604, 613
    definition of, G–24
    derivatives of, 616f, 615–617
producers, 1018, 1025
product(s), 24
    definition of, G–24
production, 1022
    gross, 1024f
productivity, 1023
    gross, 1023
    net, 1023
profundal zone, 1053
    definition of, G–24
progenote, 409, 409f
progeny testing, definition of, G–24, 225
progesterone, 782, 902
    actions of, 784t
proglottids, 541
prokaryote(s), 74, 425
    definition of, G–24
prolactin, 774
    actions of, 784t
proliferative phase, 901, 903f
proline, 58f
promoter, 277, 293

definition of, G–24

pronuclei, 914

prophage, 418
definition of, G–24

prophase
definition of, G–24
meiosis I, 212–213
mitosis, 203–204

proplastid(s), 94
definition of, G–24

proprioception, 729
definition of, G–24
in invertebrates, 729
in vertebrates, 729

proprioceptor, definition of, G–24

prop root(s), 620
definition of, G–24

Prosomii, 579

prostacyclin, 783–785

prostaglandin(s), 765, 783–785
actions of, 784*t*
definition of, G–24
structure of, 766*f*

prostate gland, 896
definition of, G–24

proteases, 803
activation of, 805

protection, group, 977

protein(s), 56–62
assembly of, 284–288, 288*f*
conjugated, 61–62
functions of, 63*t*
deficiency of, 1022, 1023*f*
definition of, G–24, 56–57
in diet, 807
digestion of, 803–805, 804*f*
fibrous, functions of, 63*t*
functions of, 63*t*
metabolism of, preparation for, 193
structural, 61, 63*f*
structure of, 59–61, 60*f*
primary, 59
quaternary, 61, 62*f*
secondary, 59
tertiary, 59, 62*f*

protein kinase, 768

proteinoid microspheres, 406, 407*f*
definition of, G–24

protein recognition sites, 868

protein Z, 156

prothallus, 509, 512*f*

prothoracic gland, 770

prothrombin, 816

Protista, 384, 444–471, 446*f*

animal-like, 448–454, 449*t*
characteristics of, 445–448
evolution of, 447–448
fungus-like, 454–458, 455*t*
life cycles in, 447
movement in, 446
nutrition in, 446, 446*f*
organization of, 445–446
phylogeny of, 448
plant-like, 458–468
reproduction in, 447

Protoavis, 575

protocell(s), 408
definition of, G–24

protoderm, 604, 613
derivatives of, 614, 614*f*

protoeukaryote, definition of, G–24

Proto-Laurasia, definition of, G–24

proton, definition of, G–24

protonema, 498

protonephridia, 540

proton gradient, 144, 184–186
uses of, 187–189

proton pump, 117, 185*f*
definition of, G–24

protons, 22, 23*f*

proto-oncogenes, 342

protostomes, development of, 531, 531*f*

protozoa, 384, 448–454

proventriculus, 793

provinces, 1052

proximal convoluted tubule, 755
reabsorption in, 755–757

proximate causation, 959

pseudocoelom, 541
definition of, G–24

pseudopodia, 82

psilophyta, 504
definition of, G–24

psilopsid, definition of, G–24

psilotophytes, 502, 503*f*, 504, 506*f*

Psilotum, 513

P site, 285

Pterophyta, 502, 503*f*, 509–510, 510*f*

Ptychodiscus brevis, 459*f*

Puccinia graminis, 485

pUC19 plasmid, 321, 321*f*

pulmonary arteries, 819

pulmonary circuit, 817
definition of, G–24

pulmonary semilunar valve, 821

pulvinus, 661

punctuated equilibrium, 396, 398*f*–399*f*
definition of, G–24

Punnett, Reginald Crandall, 226, 244, 367

Punnett square, 226, 226*f*

pupa, 556

pupil, 734

Purkinje fibers, 822

pyloric sphincter, 797
definition of, G–24

pyloric stenosis, 376

pyrimidine, definition of, G–24

pyrimidine dimers, 336, 336*f*

pyrogen(s), 859
definition of, G–24

Pyrrhophyta, 458–460, 459*f*
characteristics of, 459*t*

pyruvate, 168
in cell respiration, 179

Q

quiescent center, 617

R

rabbits
absent from Patagonia, 5–9, 359
coat colors in, multiple alleles and, 239, 239*f*
evolution of, 394

radial canal, 559

radial cleavage, 531, 531*f*

radial spokes, 86

radial symmetry, definition of, G–25

radiation, 746–747
definition of, G–25

radicle, 604
definition of, G–25

radioactive, definition of, G–25

radioactivity, 23

radioisotopes, 23

radiolarians, 449, 450*f*, 451

radula, 546, 548*f*

rainfall, distribution of, 1036, 1037*f*

rain shadow, 1036
definition of, G–25

range, definition of, G–25

ras genes, 342

rays, 569–570, 570*f*
cartilaginous skeleton in, 569

reactant(s), 24
definition of, G–25

reaction center, 151–152
  definition of, G–25
realized niche, 985, 985f
  definition of, G–25
reasoning
  deductive, 10, 10t
  inductive, 9–10, 10t
receptacle, 593, 593f
receptor-mediated endocytosis, 119f,
    119–120
receptor potentials, 725
receptor sites, 766
recessive, definition of, G–25
recessive lethal alleles, 240–241, 241f
  definition of, G–25
recessive trait, 222
reciprocal altruism, 976–977
  definition of, G–25
reciprocal cross, definition of, G–25
recombinant(s), 245
recombinant DNA, 319, 320f
  definition of, G–25
recombinant plasmids, identification of
    colonies with, 322f
recombination
  in bacteria, 310–316
  definition of, G–25, 310
recombination frequency, 246
rectal valves, 801
rectum, 801
red algae. See also Rhodophyta
  definition of, G–25
red blood cells. See erythrocyte(s)
red eyespot, in Chlamydomonas, 466
red marrow, 677
  definition of, G–25
redox reactions, 140–141
red tide, 460, 460f
  definition of, G–25
reducing agent, 141
reducing power, 143
  definition of, G–25
reduction, 140–141
  in Calvin cycle, 162
  definition of, G–25
reduction division, 213
  definition of, G–25
reductionism, 15, 21
reflex arc, 705, 705f
  definition of, G–25
refractory period, 698
refractory state, definition of, G–25

regeneration, 885
  in Calvin cycle, 162
  definition of, G–25
regeneration pathway, 162
  rationale for, 162
region of cell division, 617
region of elongation, 617
region of maturation, 617
regulators, 982f, 982–983
regulatory region(s), 277, 292, 293f
  alternative, 304f
  definition of, G–25
reindeer moss, 483
relative refractory period, 698
releaser, definition of, G–25
releasing hormone(s), 772
  definition of, G–25
REM sleep, 722
  definition of, G–25
renal arteries, 755, 825
renal circuit, 755, 825
  definition of, G–25
renal pelvis, 755
  definition of, G–25
renal veins, 754, 825
renin, 759
repair enzyme, definition of, G–25
replacement value, 998
replica plating, 310, 311f
  definition of, G–25
replication
  definition of, 267
  of DNA. See deoxyribonucleic acid,
    replication of
  origins of, 267–270, 271f
replication complex, 267
  definition of, G–25
replication fork, definition of, G–25
replisome, 267
repressible operon, definition of, G–25
repressor, 293
repressor protein, definition of, G–25
reproduction, 884–903. See also asexual
    reproduction; sexual reproduction
  in bacteria, 433–434
  balancing, 1000, 1001f
  in Coniferophyta, 515–516
  delay in, 1000
  in diatoms, 461–462, 462f
  of eukaryotic cells, 200–219
  in flowering plants, 592–609
  in fungi, 475–476
  in humans, 894–903
    hormonal control of, 898–903

as life characteristic, 16, 17f
  in paramecium, 451–454, 454f
  in protists, 447
  reasons for, 885, 885f
  and survival principle, 884–885
reproductive coordination, 977–978
reproductive isolation
  definition of, G–25
  mechanisms of, 392–394
    behavioral barriers, 393
    ecological barriers, 393
    gametic barriers, 393
    mechanical barriers, 393, 394f
    postzygotic barriers, 393–394
    prezygotic barriers, 393
reproductive output, natural selection and,
    352, 353f
reproductive strategies, evolution of,
    998–1002
reproductive system
  development of, 926–927
  female, 896–898
    external anatomy of, 896, 897f
    internal anatomy of, 899
  male, 894–896
    anatomy of, 894–895, 895f
Reptile Age, 573
  survivors of, 574f
Reptilia, 573–575
  decline of, 577
  development of, support systems in,
    922–923
  egg characteristics in, 573
  gastrulation in, 920
  heart of, 817
  internal fertilization in, 891
  lungs of, 840
  ruling, 576f
reservoir(s), 1024
  definition of, G–25
residual air, 847
resistance, in pathogens, 862
resolution phase, 906
resolving power, 70
resources, 982f, 982–983
respiration, 832–855
  breathing movements, 846–847, 847f
  cell (oxidative). See cell respiration
  control of, 851–854, 853f
  definition of, G–25
respiratory center, definition of, G–25
respiratory interface(s), 833
  in animals, 834f
  complex, 836–843
  internalized, 836
  simple body, 833–836, 835f

respiratory lining, 843*f*

respiratory system(s), human, 842*f*, 843–854
    air flow in, 843–846
    development of, 926

respiratory tree, 843

response
    in animals, 693*f*
    as life characteristic, 16, 16*f*
    by nervous system, 693

resting potential, 696–697
    definition of, G–25
    reestablishment of, 698

resting state, definition of, G–25

restoration ecology, 982

restriction enzyme(s), 319–320, 320*t*
    definition of, G–25
    sites of action of, 320*t*

restriction fragment, 319

restriction fragment polymorphism, 327

restriction site, 319

retaliator strategy, 973

rete mirabile, 744

retia, 734

reticular system, 714, 719
    definition of, G–25–26

reticulum, 794

retina, 734

retinal, 734

retinoic acid, 933
    definition of, G–26

retrovirus, definition of, G–26

reverse transcriptase, 308
    definition of, G–26

R group(s), 57–58, 58*f*

R–group, definition of, G–26

Rh blood group system, definition of, G–26

Rhea, 13, 14*f*

rhinoceros, African versus Indian, 14, 15*f*

rhizoid(s), 477, 494
    definition of, G–26

rhizome(s), 501, 606
    definition of, G–26

Rhizopus, 477, 478*f*
    reproduction in, 477, 479*f*

Rhodophyta, 458, 462*f*, 462–463
    characteristics of, 459*t*
    life history of, 463*f*

rhodopsin, 734
    definition of, G–26

Rhynia, 501, 502*f*

rhyniophytes, 501

rhythm method, 905–906

rhythms
    adaptiveness of, 963
    range of, 963–964

rib cage, 678

ribonucleic acid (RNA), 62–65
    definition of, G–26
    versus DNA, 277, 278*f*
    in early cells, 408–409
    as enzyme, 408–409
    structure of, 277, 278*f*
    synthesis of, 277–278
    types of, 278–284

ribose, 139, 277

ribosomal RNA, 87–88, 279–280
    definition of, G–26
    5S, 280
    5.8S, 280
    18S, 280
    28S, 280
    40S subunit of, 280
    60S subunit of, 280

ribosome(s), 88–90, 90*f*
    bound, 89, 287–288, 289*f*
    definition of, G–26
    free, 287–288
    function of, 97*t*
    rRNA and, 279–280, 280*f*

ribulose-1, 5-biphosphate, 163–164

ribulose-1, 5-biphosphate carboxylase, 163–164

right atrium, 817

right brain, abilities of, 717

right ventricle, 819

ring canal, 559

ring chromosome, 340, 341*f*

risk, 999

rivers, 1052
    headwaters of, 1052–1053, 1053*f*

RNA. *See* ribonucleic acid

RNAase, definition of, G–26

RNA polymerase(s), 277
    definition of, G–26

RNA replicase, 309
    definition of, G–26

RNA viruses, 308, 310*f*
    minus strand, 420
    plus strand, 420

rod(s), 734
    definition of, G–26

root(s)
    adventitious, 620
    aerial, 620
    definition of, G–26
    differentiation in, 617

    ion passage through, 646*f*
    lateral, growth of, 622*f*
    mature, 621*f*
    mycorrhiza and, 645–646, 646*f*
    primary growth in, 617–620
    prop, 620
    tap or diffuse, 620
    water movement in, 639*f*, 639–640

root apical meristem, 604, 617

root cap, 617

root hairs, 615

root pressure, 636
    definition of, G–26
    detection of, 636*f*

root systems, 620
    types of, 622*f*

root tip, 617–620, 618*f*

Roth, Thomas, 120

Rotifera, 543–545, 544*f*

rough endoplasmic reticulum, 78*f*, 88, 89*f*

round window, 730

R6 plasmid, 316, 316*f*

rRNA. *See* ribosomal RNA

r selection, 1000–1002
    characteristics of, 1001*t*
    definition of, G–26
    species, 1002*f*

RU486, 906

Ruben, Samuel, 166

Ruffini's corpuscles, 727

rumen, 794

ruminants
    digestion in, 794–795
    stomach in, 795*f*

runners, 606

S

saccule, 732
    definition of, G–26

sac fungi. *See* Ascomycota

sacroiliac joint, 678

sacroplasmic reticulum, 684

sacrum, 678

salamanders, species definition in, 380–381, 381*f*

saliva, 795

salivary glands, 795

salivary amylase, 795

Salkowski, E. and H., 652

salmonellosis, 436

saltatory propagation, 701, 701*f*
    definition of, G–26

salt marsh, 1059

saltwater habitats, 987

sampling error, 232

Sanger, F., 324

sap, definition of, G–26

saprobe(s), 434, 474
    definition of, G–26

Saprolegnia, 458f

sapwood, 627

sarcina, 428

sarcodines, 449–450

sarcolemma, 684

Sarcomastigophora, 448

sarcomere, 684
    definition of, G–26

sarcoplasmic reticulum, 684
    definition of, G–26

Sargassum, 464, 464f

Sarracenia, 647, 648f

saturated fat, 52, 53f

savanna
    definition of, G–26
    tropical, 1043f, 1043

scale, environment and, 984

scanning electron microscope (SEM),
    definition of, G–26

scapulae, 678

schizocoel, 532

Schleiden, Matthias Jakob, 69

Schwann cell, definition of, G–26

Schwann, Theodor, 69

scientific method, 9–12

sclera, 734

sclereids, 615

sclerenchyma, 615

scolex, 541

scrotum, 894

scrub-shrub wetlands, 1054

scutellum, 604

scyphomedusa, 537, 539f

scyphozoans, 537
    life cycle of, 539f

seasons, in North America, 1036, 1036f

sea urchins. *See* Echinodermata

seaweeds, 458

secondary bronchi, 843

secondary chorionic villi, 923

secondary growth, 611
    definition of, G–26

secondary immune response, 874–876
    definition of, G–26

versus primary, 876f

secondary oocyte(s), 215, 898

secondary phloem, 625

secondary sex characteristics
    definition of, G–26
    in females, 900
    in males, 900

secondary spermatocyte(s), 214, 895

secondary succession, 1015–1016
    definition of, G–26

secondary xylem, 625

second law of thermodynamics, 128
    definition of, G–26

second messenger(s), 767–769
    definition of, G–26
    multiple, 768f, 769

secretin, 806t

secretory phase, 902–903, 903f

Sedgwick, Adam, 12

seed(s), 513, 513f
    definition of, G–26
    development of, 600–605, 601f
        without sex, 606–608
    dispersal of, 604–605, 606f
    dormancy of, 604
    germination of, gibberellins and,
        655f
    monocot versus dicot, 605f

seed coats, 600

seedless plants, 502

seedling, 611
    germination of, 611
    growth of, 611–613, 612f–613f
        primary versus secondary, 611–613

seed plants, 502, 510–516
    definition of, G–26
    reproductive adaptation in, 512–513

segmentation, definition of, G–26

segregation
    law of, 227–228
    and probability, 227–228

Selaginella, 507, 507t
    life cycle of, 508f

selection. *See also* artificial selection; *K*
        selection; natural selection; *r* selection
    balancing, 354
    clonal, 863, 870–871
        definition of, G–5
    directional, 371–372, 372f
        definition of, G–8
    disruptive, 372, 372f
        definition of, G–8
    frequency-dependent, 355, 356f
        definition of, G–11
    kin, 975–976

mutation and, 370
    species, 359, 359f
    stabilizing, 370, 372f
        definition of, G–28

selectionists, versus neutralists, 374–375

selectively permeable, 109
    definition of, G–26

selenium, 22t

selfish herd effect, 977

self-replicating mechanisms, 407, 407f

self-thinning, 1002

self-tolerance, 869, 870f

semelparous species, 999, 999f

semen, 896
    definition of, G–26

semicircular canals, 732
    definition of, G–26

seminal vesicle(s)
    definition of, G–27
    in insects, 889

seminiferous tubule(s), 894
    definition of, G–27

semipermeable membrane, definition of,
    G–27

senescence, 656
    ethylene and, 657

senses, 723–736

sensilla, 728, 728f

sensory and motor cortex, 715–716, 716f

sensory neuron(s), 693, 694f
    definition of, G–27

sensory organs, 725

sensory palps, 557

sensory receptors, 693, 725

sepals, 519, 521f, 593, 593f

septa, 477

septate, 477

septum, 817

sere(s), 1014
    definition of, G–27

serial endosymbiosis hypothesis
    definition of, G–27
    in eukaryotic cell evolution, 96, 98–99

serine, 58f

serosa, of esophagus, 796

Sertoli cells, 895

Servetus, Miguel, 781

setae, 551

sex
    in bacteria, 310–311, 311f
    chromosomes and, 248–254
    influence on phenotype, 242

rationale for, 214–217
sex chromosomes
    abnormalities in, 251
    definition of, G–27
sex determination, 250
sex-influenced trait, 242
    definition of, G–27
sex-limited trait, 242
    definition of, G–27
sex pili, 312, 314f, 429
sex steroids, 779–780
sexual differentiation, 900
    in human development, 927, 927f
sexual intercourse, definition of, G–27
sexuality, in humans, 894–903
sexually transmitted diseases, 436–437
    agents of, 436f
sexual recombination, definition of, G–27
sexual reproduction
    in animals, 886–887
    in ascomycota, 479–481, 482f
    in flowering plants, 593–605
    in invertebrates, 887–889
    in vertebrates, 889–894
sharks, 569–570, 570f
    cartilaginous skeleton in, 569
    characteristics of, 571f
    digestive system of, 791–792, 792f
    internal fertilization in, 890–891, 891f
shell, of mollusks, 546
shoot, anatomy of, 623f
shoot apical meristem, 604
short-day plants, 662
short-term memory, 954
Siamese cats, 241, 242f
sickle-cell anemia, and malaria, 354–355
sickle-cell, definition of, G–27
sickle-cell mutation, 337, 338f
sieve cell(s), 616
    definition of, G–27
sieve element(s), 616
    definition of, G–27
sieve plates, 616
sieve tube(s), 616
sieve tube member(s), 616
    definition of, G–27
sight. *See* visual reception
sigmoidal curve, 994
sigmoid colon, 801
signal sequence, 287–288
sign stimuli, 949, 949f
silent mutation, 337, 338f

definition of, G–27
Silent Spring (Carson), 981
silicon, 22t
simple epithelium, 675
simple eyes, 732
Singer model, definition of, G–11
single-factor cross, 224f
sink, 643
    definition of, G–27
sinoatrial node, 821
sinusoids, 825
skeletal muscle, 682f, 682–687
    anatomy of, gross, 683–684
    contraction of. *See* contraction
    definition of, G–27
    organization of, 684f–685f
    ultrastructure of, 684
skeleton(s)
    human, 680f
        versus gorilla, 681f
    of vertebrates, 675–678
        organization of, 678–681
skin breathers, 833–836
skin, in defense, 858, 858f
Skinner, B.F., 951
Skinner box, 951, 951f
skin receptors, 727f
sleep, electrical activity during, 722f, 722–723
sleep movements, 661, 662f
slime mold(s), 448, 454–456, 455t
    definition of, G–27
slug, 456
small intestine, in humans, 797–799, 796f
    lining of, 719f
Smith, J.N.M., 961
Smith, John Maynard, 973
smoking, lung cancer and, 844–845
smooth endoplasmic reticulum, 88, 89f
smooth muscle, 682, 682f
    definition of, G–27
snails, fertilization in, 887–888, 888f
snake(s)
    digestion in, 792f, 792–793
    reproductive structures in, 891, 892f
social behavior, 971–977
    advantages of, 977–978
        to parents and offspring, 978
sociobiology, 977–978
sodium, 22t
sodium bicarbonate, 800
sodium chloride, 28, 29f
    regulation of, 760

sodium ion, 28
sodium ion gate(s), 698–700, 699f
    definition of, G–27
sodium/potassium ion exchange pump
        (Na^+/K^+), 115–117, 116f, 696, 697f
    definition of, G–27
soft palate, 795
soil(s), 988f, 989
    particles of, 989, 988f
    types of, 989
soil layer, 1008
solar energy, and biosphere, 1033
solar system, young, 403f
solar tracking, 660–661, 661f
solute(s), 34
    definition of, G–27
solvent, 34
somatic cells, definition of, G–27
somatic mutation(s), 334, 335f
    definition of, G–27
somatic nervous system, 723
    definition of, G–27
somatic tissue, 214
somites, 921
soredia, 482
sori, 509
source, 643
specialists
    as to environmental conditions, 984, 985f
    as to foraging behavior, 961, 961f
speciation
    definition of, G–27, 387
    mechanisms of, 387–392
species, 379, 382
    definition of, G–27, 379–381
        problems in, 379–381
    naming of, 382
    origin of, 378–400
    variation within, 379
species composition, 1008, 1008f
species richness, 1009
    versus area, 1009f
    as function of distance from source, 1010f
species selection, 359, 359f
specific defenses, 857, 863–870
    definition of, G–27
specific heat, 38
spectral quality, 988
Spemann, Hans, 931–932
sperm, 911, 911f
    definition of, G–27
    route of, 896
spermatangia, 463

spermatheca, in insects, 889

spermatia, 463

spermatic cord, 896

spermatid(s), 895
    definition of, G–27

spermatocyte(s)
    definition of, G–27
    primary, 214, 895
    secondary, 214, 895

spermatogenesis, 214, 217f, 895, 897f

spermatogonia, 895

spermatophore(s)
    definition of, G–27
    in insects, 889

spermatozoa, 895
    definition of, G–27

sperm barriers, 905

sperm ducts, 888

sperm receptacles, 550

Sperry, Roger W., 717

Sphagnum, 497, 1054

S phase, definition of, G–27

Sphenodonia, 573

Sphenophyta, 502, 503f, 507–509, 509f

spherosomes, 90

sphincter(s), 684
    definition of, G–27

spicules, 535

spider, 552–553
    silk glands of, 553f

spinal canal, 678, 720

spinal cord, 720, 721f

spindle, 206f
    definition of, G–27
    in mitosis, 203–204

spindle apparatus, definition of, G–27

spindle fiber(s), 83
    definition of, G–27–28

spindle fiber attachment, 206f

spinnerets, 553, 554f

spiny anteater, 579. See also monotremes

spiracles, 836
    of insects, 557

spiral cleavage, 531, 531f

spiral valve, 791

spirilla, 429

spirochaete, definition of, G–28

spirochetes, 429

spleen, definition of, G–28

spliceosome(s), 282
    definition of, G–28

splicing, 282

alternative, 304f
    definition of, G–28
    of DNA, 320–325

split-brain experiments, 717

SPONCH, 21

sponges, 532–535
    anatomy of, 534f
    biology of, 535
    calcareous, 535
    fertilization in, 887
    glassy, 535
    proteinaceous, 535

spongin, 535

spongocoel, 535

spongy bone, 677
    definition of, G–28

spongy parenchyma, 624

sporangiophores, 456, 477, 479f–480f

sporangium(/a), 456, 477, 479f–480f, 495

spores
    definition of, G–28
    in fungal reproduction, 475, 475f

sporic cycle, 447, 448f
    definition of, G–28

sporophylls, 506

sporophyte
    definition of, G–28
    of moss, 501f

sporophyte generation, 447

sporopollenin, 493

sporozoans, 451

sporozoites, 451

spring overturn, 1054

Squamata, 573

squamous epithelium, 675

stability, 1010

stabilizing selection, 370, 372f
    definition of, G–28

Stahl, Franklin, 270–271

stamen(s), 520, 521f, 593, 593f
    definition of, G–28

standard conditions, 847

Stanley, Wendell, 417

stapes, 730

staphylococcus, 428

starch, digestion of, 802

start signal, 282

stationary electron carriers, 143

statocysts, 732

STD. See sexually transmitted diseases

stele, 618
    definition of, G–28

stem
    elongation of, gibberellins and, 655f
    primary growth in, 620–623, 623f
    secondary growth in, 625–630, 626f
    transition to secondary growth, 627f,
        627–628
    water movement in, 637–639
    woody, older, 628f, 627–630

stem cell(s), 816
    definition of, G–28

stems, monocot versus dicot, 624f

sterilization, 906

sternum, 678

steroid(s), 55
    definition of, G–28
    function of, 56t

steroid hormone(s)
    definition of, G–28
    and gene control, 769, 769f
    in regulation of gene expression,
        300–301, 302f

Steward, F.C., 629

sticky ends, 319–320

stigma, 520, 521f, 593, 593f
    definition of, G–28

stimulus, 693
    definition of, G–28

stipes, 465

stolons, 477, 606

stoma(ta), 493, 497, 615, 641f
    definition of, G–28
    mechanism of, 641f
    potassium ions and, 641–642, 642f
    turgor and, 640–641

stomach
    in birds, 793
    definition of, G–28
    in humans, 797, 798f
    in ruminants, 795f
    secretions of, regulation of, 806f

stomatal apparatus, 624, 640

stone canal, 559

stone cells, 615

stop codon(s), 281
    definition of, G–28

storage diseases, 91

strata, 1008

stratified epithelium, 675

streams, 1052

streptococcus, 428

stroboli, 506

stroke volume, 823

stroma, 94, 151
    definition of, G–28

stromatolite(s), 425, 426*f*
   definition of, G–28
structural formulas, 26
   reading, 46
structural protein, definition of, G–28
Sturtevant, Alfred, 247
style, 520, 521*f*, 593, 593*f*
subatomic particles, 22
suberin, 80, 615
submucosa, of esophagus, 796
subspecies, 381
   definition of, G–28
subsporangial swelling, 478
substitution(s), 337
   definition of, G–28
substrate, 59, 132
   concentration of, and enzyme activity, 134*f*, 134–135
substrate-level phosphorylation, 143, 174
   in citric acid cycle, 180
   definition of, G–28
succession. *See* ecological succession
sucrase, 802
sucrose, 45, 46*f*, 50*t*
sulfhydryl group, 32, 34*t*
sulfur, 21, 22*t*, 32–33
sundew, 647, 648*f*
sunflower, 603*f*
superior vena cava, 819
support, 670–689
surface tension, of water, 37, 37*f*
surface-volume relationship, 76*f*
surface-volume theory
   of cell size, 74
   definition of, G–28
survivorship, 996
   categories of, 997*f*
survivorship curve(s), 996, 996*f*
   definition of, G–28
   of species, 996*f*
suspensor, 603
sustainable agriculture, 982
Sutton, Walter, 231
sutures, 678
swallowing, 795, 796*f*
swamps, 1054
swarming, 456
Sweatman, H.P.A., 961
swim bladder, 840
   in bony fish, 571–572
symbiosis, 974–975
   definition of, G–28

symmetry
   bilateral, 535*f*, 537–541
      in flowers, 593
   pentaradial, 709
   radial, 535*f*, 558
      in flowers, 593
sympathetic ganglia, 723
sympathetic nerves, responses to, 723*t*
sympatric speciation, 388–392
   in animals, 389–392
   definition of, G–28
   in plants, 388–389
symplastic route, 639
   definition of, G–28
symport(s), 115
   definition of, G–28
synapses, 701, 702*f*–703*f*
   actions at, 702
   chemical, 702–705
   definition of, G–28
   electrical, 701–702
   excitatory, 702
   inhibitory, 702–704, 704*f*
   recovery at, 704–705
synapsis, 212
   definition of, G–28
synaptic cleft, 702
   definition of, G–28
synaptic end bulbs, 693
synaptonemal complex, 212
   definition of, G–28
synecology, 981, 1006
   definition of, G–28
synergids, 598
synonymous codon(s), 281
   definition of, G–28
synovial joint, 678
   definition of, G–28
synthesis
   definition of, 265
   of DNA, 265–267, 268*f*
synthesists, 15
   definition of, G–28–29
syphilis, 436*f*, 437
system, 532
systematics, 382–384
   definition of, G–29, 382
systemic circuit, 817, 825
systole, 823
   definition of, G–29
Szent-György, Albert, 21

T

tactile reception, 725–726
   in invertebrates, 725*f*, 725–726
   in vertebrates, 726
taiga, 1049–1051, 1051*f*
   animals of, 1050, 1051*f*
   definition of, G–29
tapeworms, 541, 542*f*
tap root(s), 620
   definition of, G–29
tap root system, definition of, G–29
target cell(s), 300, 766
   definition of, G–29
   hormones and, 766–769
taste buds, 729, 730*f*
   in humans, 795
TATA binding (TAB) proteins, 299
TATA box, 299
Tatum, Edward, 313
taxon(/a), 382
   definition of, G–29
   hierarchical organization of, 383*f*
      images of, 384*f*
taxonomist, definition of, G–29
taxonomy, 382–384
   definition of, G–29, 382
Tay-Sachs disease, 92
T–cell(s), 863
   arousal of, 871–873, 872*f*
   cytotoxic, 869, 872–873, 873*f*
   definition of, G–29
   dual recognition site, 869, 869*f*
   function of, 864–865, 868–869
   helper, 422, 869
      activation of B-cells, 873, 874*f*
   memory, 871
   selection of, 872*f*
   suppressor, 874
      definition of, G–28
   virgin, 869
T_4 cell, definition of, G–29
T_8 cell, definition of, G–29
tectorial membrane, 732
teichoic acid, 429
teliospore, 485
telophase
   meiosis I, 212–213
   mitosis, 206–208
temperate bacteriophages, 418
temperate deciduous forests, 1047–1049, 1050*f*
temperature
   effects of, 987

temperature *continued*
  and enzyme activity, 135, 136*f*
    optimal range, 987, 987*f*
temporal lobes, 715, 715*f*
tendon(s), 683
  definition of, G–29
tendrils, 659
tension, 637
tenting, 907
terminal bud, 623
  definition of, G–29
terminal bud scale scars, 623
termination, 287, 287*f*
  definition of, G–29
termination signal, definition of, G–29
terminator codon, definition of, G–29
terrestrial environment, 753–754, 1037–1038
  invertebrates of, 753–754
  vertebrates of, 754
terrestrial habitats, 986
territorial behavior, definition of, G–29
territoriality, 991, 1011–1012
territory, definition of, G–29
tertiary bronchi, 843
tertiary consumer, definition of, G–29
test, 450
testcross, 230, 231*f*
  definition of, G–29
  dihybrid, 231*t*
testes, 894, 896*f*
  hormones of, 782, 784*t*
  in insects, 888
testosterone, 782, 895, 900
  actions of, 784*t*
  and male territorial behavior, 948*f*
  structure of, 766*f*
tetanus, 436
tetrad, 212
Tetrahymena, 453*f*
tetraploidy, synthetic, 391*f*, 392
tetrapod, definition of, G–29
T-even phage, definition of, G–29
thalamus, 714, 718–719
  definition of, G–29
thallus, definition of, G–29
theory, 10
  definition of, G–29
theory of demographic transition, 1067
therapsids, 578, 579*f*
thermal overturn, 38
  definition of, G–29
thermal proteinoids, 406

thermiogenesis, definition of, G–29
thermoacidophiles, 427
thermodynamics
  definition of, G–29
  laws of, 127–128
thermophiles, 427
thermoreception, 727–728
  in invertebrates, 727
  in vertebrates, 727*f*, 727–728
thermoregulation, 742–750
  in birds and mammals, 744–748
  categories of, 742–743
  definition of, G–29
  in fish, and circulation, 744*f*
  in humans, mechanisms of, 748
  hypothalamus in, 748, 749*f*
  reasons for, 742
  through basking, 750, 750*f*
thigmonastic response, 662
thigmotropism, 658, 661*f*
threonine, 58*f*
threshold voltage, 698
  definition of, G–29
thrombin, 816
  definition of, G–29
thromboplastin(s), 816
  definition of, G–29
thromboxane, 783
thylakoid(s), 94, 151–156, 153*f*
  definition of, G–29
thymine, 65, 262
thymosin, 782
  actions of, 784*t*
thymus
  definition of, G–29
  hormones of, 782, 784*t*
thyroid, 773, 777*f*, 777–778
  definition of, G–29
  hormones of, 784*t*
  regulation of metabolic rate, 778*f*
thyroid-stimulating hormone, 773
  actions of, 784*t*
thyroxine, 773, 776, 777*f*
  actions of, 784*t*
  structure of, 766*f*
tight junctions, 122*f*, 123
tiglons, 393, 393*f*
Tinbergen, Niko, 949
tissue(s), 532, 535–537
  definition of, G–29
  in vertebrates, 675, 676*f*
tobacco mosaic virus, 417*f*–418*f*
  discovery of, 416–417
Toguchi mutation, 337, 338*f*

tolerance model, of succession, 1016
tolerance range, 984
tomato, 602*f*
tongue, 795
tongue rolling, inheritance of, 242, 243*f*
total fertility rate, 1068
  definition of, G–30
totipotency, 656
  definition of, G–30
toxin, definition of, G–30
trace elements, in diet, 807–809
tracers, 24
trachea
  definition of, G–30
  in humans, 843
tracheae, 836, 837*f*
  of insects, 557
tracheal gills, 836
tracheid(s), 501, 514, 616
  definition of, G–30
tracheoles, 836
  of insects, 557
tracheophyte, 501
3′, trailer, 282
transcribed strand, definition of, G–30
transcription, 277–278, 279*f*
  control of, 298–301
  definition of, G–30, 277
  simultaneous, 278, 280*f*
transcription termination signal, 278
transduction, 312, 313*f*
  definition of, G–30
trans face, 90
transfer RNA, 279, 283–284
  definition of, G–30
  structure of, 283*f*
transformation, 259–261, 260*f*
  in bacteria, 312, 312*f*
  definition of, G–30
transition state, 131
translation, 284–288, 288*f*
  definition of, G–30
translocation(s), 286, 286*f*, 340, 342*f*, 644
  definition of, G–30
transpiration, 633
  definition of, G–30
  in leaf, 636–637
  rates of, 633*t*
transport, mechanisms of, 108–120
transport proteins, 104
transposable elements, 343*f*
  mutation by, 343*f*
transposase, 342

transposition(s), 335, 340–343
definition of, G–30
transposons, 340
definition of, G–30
as mutagenic agents, 343
structure of, 342, 343f
transverse colon, 801
trees, annual rings in, 627
trematodes, 541
Treponema pallidum, 436f, 437
Trichinella spiralis, 543, 544f
trichinosis, 543
trichocysts, 451
Trichomonas vaginalis, 449f
Trichonympha campanula, 449f
Trichophyton mentagrophytes, 481
tricuspid valve, 821
trigger, 238, 239f
triglycerides, 51–52
formation of, 52, 52f
function of, 56t
triiodothyronine, 773, 776
actions of, 784t
trilobita, 552
trilobites, 552, 553f
triple covalent bond, 26
triple induction, in vertebrate eye, 932, 934f
triploblastic, definition of, G–30
trisomy 21, 217
Triticale grain, 389, 392f
Trivers, Robert, 976
tRNA. *See* transfer RNA
trophic level(s), 1020
definition of, G–30
humans and, 1020–1022
trophoblast, 916
tropical rain forests, 1045–1046, 1046f
canopy and sub-canopy of, 1045
animals of, 1044, 1048f
destruction of, 1047
tropisms, 658–661
tropomyosin, 686
definition of, G–30
troponin, 686
definition of, G–30
trout, interbreeding in, 393
trp operon, 296–298, 297f
true-breeding, 222
definition of, G–30
Trypanosoma gambiense, 449, 449f
trypsin, 804
tryptophan, 58f

synthesis of, repressible operon for, 296–298, 297f
T-tubules, 684
tubal ligation, 906
definition of, G–30
tube cell, 598
in pollen, 512
tube feet, 559
tubers, 606
tubular secretion, 759
tubulin, 82
tule elk, 373, 374f
tundra, 1044f, 1043–1045
animals of, 1044, 1045f
definition of, G–30
tunicates, 565, 567f
turbellarians, 539–541
turgor, 113
definition of, G–30
and stomatal opening, 640–641
turgor pressure, 113
Turner's syndrome, 251
tympanal organs, 730
tympanum, 730
tyrosine, 58f
metabolic pathway of, 273f

U

UAA, 281
UAG, 281
UGA, 281
Ulothrix, 468
ultimate causation, 959
ultraviolet light, definition of, G–30
Ulva, 468, 468f
umbilical cord, 925
underhair, 745
understory, 1008
undulating membrane, 449
unicellular organisms, 445
uniport(s), 115
definition of, G–30
uniport transport, 116f
Uniramia, 552, 555
uniramians, 554f
unsaturated, definition of, G–30
unsaturated fat, 52–53, 53f
upper epidermis, in plants, 623
upwelling(s), 1056
definition of, G–30

uracil, 277
uranium-238, half-life of, 23
uredinia, 485
uredospore, 485
ureter(s), 755
definition of, G–30
urethra, 755, 896
definition of, G–30
Urey, Harold, 404
urinary bladder, 755
Urochordata, 565
Urodela, 573
uterine tube(s), 898
definition of, G–30
uterotonics, 775
uterus, 898
definition of, G–30
utricle, 732
Utricularia, 647, 648f

V

vaccination, 416f
vacuole(s), 74, 93, 93f
contractile, 93
digestive, 91
food, 93
function of, 97t
vagina
definition of, G–30
in invertebrates, 887
vagus nerve, in regulation of digestion, 805
valence electrons, 25
valence shell, 25
valine, 58f
valves, in mollusks, 547
vanadium, 22t
van der Waals forces, 54–55
van Helmont, Jan Baptiste, 166
van Leeuwenhoek, Antony, 445, 445f
van Niel, C.B., 166
variable, 11
variable age of onset, definition of, G–30
variable expressivity, 242, 243f
definition of, G–30
variable region(s), 865
definition of, G–30
variation, 353–355, 354f
continuous, 244, 244f
genetic, maintenance of, 354–355
sources of, 354
variegated coloration, 341, 343f

vascular bundle, definition of, G–30

vascular cambium, 615, 625
definition of, G–30

vascular plants, 501–510
definition of, G–30
evolution of, 501–502, 504, 504f–505f
first, 503f
seeded. *See* seed plants
seedless. *See* seedless plants
taxonomic organization of, 504
vascular tissues in, 503f

vascular rays, 626
definition of, G–31

vascular system, definition of, G–31

vascular tissue, definition of, G–31

vas deferens. *See* ductus deferens

vasectomy, 906

vasoconstriction, 824

vasodilation, 824

vasopressin, 776

vector(s), 319, 1012
definition of, G–31

vegetal hemisphere, 912
definition of, G–31

vegetal pole, 912

vegetative propagation, 605–606, 607f

vein(s), 819, 827–828
structure of, 819f

ventral nerve cord(s), 550, 709, 709f
definition of, G–31

ventral root, 720

ventricle, 817
definition of, G–31
left, 821
right, 819

venule(s), 819
definition of, G–31

Venus flytrap *(Dionaea muscipula)*, 647
response to touch, 662

vertebral column, 567, 677
definition of, G–31

Vertebrata, 567–584
characteristics of, 567
classes of, 567, 568f
development of
supporting structures in, 921–925
tissue interaction in, 931–933
limbs and posture in, 573, 573f
muscle in, 681–689
nervous system of, 710–712
relationships in, 568f
skeletons of, 675–678
tissues in, 675, 676f

vesicles, 90, 91f

vesicular follicle, 901–902

vessel elements, 616

vessels, 616

vestibular apparatus, 730, 732, 733f

vestibular glands, 898

vestigial organs, 361–362, 362f

vibrio, 429

Victoria (Queen of England), 253

vigilance, in groups, 977

villi
definition of, G–31
of small intestine, 797–798

Virchow, Rudolf, 69

virion, 417
definition of, G–31

viroid(s), 423, 425
definition of, G–31

virulence, definition of, G–31

virus(es), 414–425. *See also*
bacteriophage(s)
of animals, 418–421
attack on animals, 420f–421f
assembly and release in, 421
binding in, 419
penetration in, 420
transcription and replication in, 420
uncoating in, 420
bacterial. *See also* bacteriophages
binding by antibody, 867
characteristics of, 417
classification of, 419t
definition of, G–31
dimensions of, 415, 415f
discovery of, 416–417
enveloped, attack by, 421f
filterable, 417
genetic differences in, 417
genetics in, 306–309
naked, attack by, 420f
oncogenic, 415
origins of, 415–416
of plants, 421–422, 422f
shapes of, 417, 418f

visible light, 150

visual reception, 732–736
in invertebrates, 732–733
signal transduction in, 734–736
in vertebrates, 733–736

vital capacity, 847

vitamin(s), 135
definition of, G–31
in diet, 808t, 807

vitelline layer, 913

vitreous humor, 734

viviparity, definition of, G–31

viviparous, definition of, G–31

vocal cords, 843

voice box, 843

Volterra, V., 985

Volvox, 466–468, 467f

von Baer, Karl Ernst, 922f

von Frisch, Karl, 949

von Linne, Carl. *See* Linnaeus, Carolus

Vorticella, 453f

The Voyage of the Beagle (Darwin), 12

vulva, 897

W

wall, 193

Wallace, Alfred Russel, 13, 352

wall pressure, 113

water
in biosphere, 1034
content of organisms, 29, 34
cycle of, 34, 35f
in desert, 33, 33f
effects of, on organisms, 987–988
and heat, 37–38
hydrogen bonding and, 33–40, 39f
hydrogen bonding in, 30
metabolic, 184
molecular interactions, 36f
molecule formation, 28f
movement of, testing forces in,
638
and nonpolar molecules, 36
in photosynthesis, 156
properties of, 34–38
surface tension of, 37, 38f
transport in nephron, control of,
760

water communities, 1051–1059

water mold(s), 448, 457, 458f
definition of, G–31

water movement
in leaf, 636–637
in root, 639f, 639–640
in stem, 637–639

water potential, 633–635
in cells, 635f
definition of, G–31
in leaf, 637f
in total plant, 635f

water vascular system, 559, 560f
definition of, G–31

Watson, James, 258, 263–266, 264f

Watson strand, definition of, G–31

waxes, 55
functions of, 56t

web, 553

Wecker, Stanley, 960
experiment on habitat selection, 960, 960f, 961t

Wegener, Alfred, 391

weight, 22

Welwitschia, 514f, 514–515

Went, Fritz, 651–652, 653f

Wernicke's area, 718, 718f
definition of, G–31

wetlands, 1054

wheat rust, 485–486

whisk ferns, 504, 506f

white blood cells. *See* leukocyte(s)

whooping crane, imprinted on humans, 952f

whorls, 593

Wilcox, R.S. Stimson, 725–726

wild type, definition of, G–31

Wilkins, Maurice, 263

wings, development of, 937f
program for, 936–938

withdrawal, 905

Woese, Carl R., 409, 426

Wolpert, Lewis, 917

X

X chromosomes, 250–251, 251f

Xenopus laevis (clawed frog), hormonal effects on, 948, 949f

xeroderma pigmentosum, 336

xerophyte, definition of, G–31

X-linked, definition of, G–31

xylem, 501, 616
definition of, G–31
primary, definition of, G–24
role of, 633
secondary, 625
definition of, G–26

xylem tissue, 618

xylem vessel(s), 514
definition of, G–31

Y

Y chromosome, definition of, G–31

yeast(s), 480
definition of, G–31
life cycle of, 480, 482f

yellow-green algae. *See* Chrysophyta

yellow marrow, 677

yield, in Calvin cycle, 162

yield pathway, 162

yolk, of reptile egg, 573

yolk sac, 922
definition of, G–31
of reptile egg, 573

Y-organ, 770

Z

Z disks, 684

Zea mays, as experimental organism, 306

zebra finch, hormonal effects on, 948, 948f

zero population growth (ZPG), definition of, G–31

zinc, 22t

zona pellucida, 914

zone of polarizing activity (ZPA), 937–938
transplants of, 939f

zooplankton, 1056
definition of, G–31

zoospores, 457

Z–scheme, 157, 160f
definition of, G–31

Zygomycota, 476–477

zygospore, 466, 476, 477
definition of, G–31

zygote
definition of, G–31
determination in, 930–931

zygotic cycle, 447, 448f
definition of, G–31

THE METRIC SYSTEM

METRIC PREFIXES

(Units: gram, meter, and liter are common suffixes)

Prefix	Multiple	Symbol
(greater than one)		
deka	10	da
hecto	10^2	h
kilo	10^3	k
mega	10^6	M
(less than one)		
deci	10^{-1}	d
centi	10^{-2}	c
milli	10^{-3}	m
micro	10^{-6}	μ
nano	10^{-9}	n
pico	10^{-12}	p

METRIC LENGTH

1 meter	×	10	= dekameter (10 m)
(the unit)		100	= hectometer (10^2 m)
		1,000	= kilometer (10^3 m)
		1,000,000	= megameter (10^6 m)

1 meter	÷	10	= decimeter (10^{-1} m)
		100	= centimeter (10^{-2} m)
		1,000	= millimeter (10^{-3} m)
		1,000,000	= micrometer (10^{-6} m)
		1,000,000,000	= nanometer (10^{-9} m)
		1,000,000,000,000	= picometer (10^{-12} m)
		10,000,000,000	= Angstrom (Å) (10^{-10} m)
			(an older unit of measurement)

METRIC-ENGLISH CONVERSIONS

Length

English (USA)	= Metric
inch	= 2.54 cm, 25.4 mm
foot	= 0.30 m, 30.48 cm
yard	= 0.91 m, 91.4 cm
mile (statute) (5,280 ft)	= 1.61 km, 1609 m
mile (nautical) (6077 ft, 1.15 statute mi)	= 1.85 km, 1850 m

Metric	= English (USA)
millimeter	= 0.039 in
centimeter	= 0.39 in
meter	= 3.28 ft, 39.37 in
kilometer	= 0.62 mi, 1,091 yd, 3273 ft

To Convert	To	Multiply
inches	centimeters	in × 2.54
feet	centimeters	ft × 30
centimeters	inches	cm × 0.39
millimeters	inches	mm × 0.039

METRIC WEIGHTS OR MASSES

1 gram (the unit)	×	1,000 = kilogram

1 gram	÷	1,000 = milligram (mg) (10^{-3} g)
		1,000,000 = microgram (μg) (10^{-6} g)
		1,000,000,000 = nanogram (ng) (10^{-9} g)
		1,000,000,000,000 = picogram (pg) (10^{-12} g)

METRIC-ENGLISH CONVERSIONS

Weight

English (USA)	= Metric
grain	= 64.80 mg
ounce	= 28.35 g
pound	= 453.60 g, 0.45 kg
ton (short—2000 lb)	= 0.91 metric tons (907 kg)

Metric	= English (USA)
milligram	= 0.02 grains (0.000035 oz)
gram	= 0.04 oz
kilogram	= 35.27 oz, 2.20 lb
metric ton (1000 kg)	= 1.10 tons

To Convert	To	Multiply
ounces	grams	oz × 28.3
pounds	grams	lbs × 453.6
pounds	kilograms	lbs × 0.45
grams	ounces	g × 0.035
kilograms	pounds	kg × 2.2

METRIC VOLUME

METRIC-ENGLISH CONVERSIONS

English (USA)	= Metric
cubic inch	= 16.39 cc
cubic foot	= 0.03 m^3
cubic yard	= 0.765 m^3
ounce	= 0.03 l (30 ml or cc)*
pint	= 0.47 l
quart	= 0.95 l
gallon	= 3.79 l

Metric	= English (USA)
milliliter	= 0.03 oz
liter	= 2.12 pt
liter	= 1.06 qt
liter	= 0.27 gal

1 liter ÷ 1,000 = milliliter or cubic centimeter (10^{-3} l)
1 liter ÷ 1,000,000 = microliter (10^{-6} l)

* *Note*: 1 ml = 1 cc

To Convert	To	Multiply
fluid ounces	milliliters	oz × 30
quarts	liters	qt × 0.95
milliliters	fluid ounces	ml × 0.03
liters	quarts	l × 1.06